2011 IEEE International Solid-State Circuits Conference Digest of Technical Papers

(ISSCC 2011)

AA002377

San Francisco, California, USA
20 - 24 February 2011

Editors:

Laura C. Fujino

IEEE Catalog Number: CFP11ISS-PRT
ISBN: 978-1-61284-303-2

Copyright © 2011 by the Institute of Electrical and Electronic Engineers, Inc
All Rights Reserved

Copyright and Reprint Permissions: Abstracting is permitted with credit to the source. Libraries are permitted to photocopy beyond the limit of U.S. copyright law for private use of patrons those articles in this volume that carry a code at the bottom of the first page, provided the per-copy fee indicated in the code is paid through Copyright Clearance Center, 222 Rosewood Drive, Danvers, MA 01923.

For other copying, reprint or republication permission, write to IEEE Copyrights Manager, IEEE Service Center, 445 Hoes Lane, Piscataway, NJ 08854. All rights reserved.

***This publication is a representation of what appears in the IEEE Digital Libraries. Some format issues inherent in the e-media version may also appear in this print version.**

IEEE Catalog Number:	CFP11ISS-PRT
ISBN 13:	978-1-61284-303-2
ISSN:	0193-6530

Additional Copies of This Publication Are Available From:

Curran Associates, Inc
57 Morehouse Lane
Red Hook, NY 12571 USA
Phone: (845) 758-0400
Fax: (845) 758-2633
E-mail: curran@proceedings.com
Web: www.proceedings.com

TABLE OF CONTENTS

Electronics for Healthy Living 3

PLENARY SESSIONS:

Session 1 Overview: Plenary Session 6

New Interfaces to the Body through Implantable-System Integration 9
Stephen Oesterle, Paul Gerrish, Peng Cong
Game-Changing Opportunities for Wireless Personal Healthcare and Lifestyle 15
Jo De Boeck
Eco-Friendly Semiconductor Technologies for Healthy Living 22
Oh-Hyun Kwon
Beyond the Horizon: The Next 10x Reduction in Power - Challenges and Solutions 31
Jan Rabaey

TECHNOLOGIES FOR HEALTH:

Session 2 Overview: Technologies for Health /Technology Directions 32

A 0.24nJ/b Wireless Body-Area-Network Transceiver with Scalable Double-FSK Modulation 34
Joonsung Bae, Kiseok Song, Hyungwoo Lee, Hyunwoo Cho, Long Yan, Hoi-Jun Yoo
A 75µW Real-Time Scalable Network Controller and a 25µW ExG Sensor IC for Compact Sleep-
Monitoring Applications 36
Seulki Lee, Long Yan, Taehwan Roh, Sunjoo Hong, Hoi-Jun Yoo
A 3µW Wirelessly Powered CMOS Glucose Sensor for an Active Contact Lens 38
Yu-Te Liao, Huanfen Yao, Babak Parviz, Brian Otis
A 90nm CMOS SoC UWB Pulse Radar for Respiratory Rate Monitoring 40
Domenico Zito, Domenico Pepe, Martina Mincica, Fabio Zito
A Broadband THz Imager in a Low-Cost CMOS Technology 42
*Franz Schuster, Hadley Videlier, Antoine Dupret, Dominique Coquillat, Maciej Sakowicz, Jean-Pierre Rostaing,
Michaël Tchagaspanian, Benoît Giffard, Wojciech Knap*
A Programmable Implantable Micro-Stimulator SoC with Wireless Telemetry: Application in
Closed-Loop Endocardial Stimulation for Cardiac Pacemaker 44
Shuenn-Yuh Lee, Y-C. Su, M-C. Liang, J-H. Hong, C-H. Hsieh, C-M. Yang, Y-Y. Chen, H-Y. Lai, J-W. Lin, Q. Fang
A 660pW Multi-Stage Temperature-Compensated Timer for Ultra-Low-Power Wireless Sensor Node
Synchronization 46
Yoonmyung Lee, Bharan Giridhar, Zhiyoong Foo, Dennis Sylvester, David Blaauw
A Low-Power Fully Integrated RF Locked Loop for Miniature Atomic Clock 48
David Ruffieux, Matteo Contaldo, Jacques Haesler, Steve Lecomte

RF TECHNIQUES:

Session 3 Overview / RF: RF Techniques 50

Spur-Free All-Digital PLL in 65nm for Mobile Phones 52
*Robert Bogdan Staszewski, Khurram Waheed, Sudheer Vemulapalli, Fikret Dulger, John Wallberg, Chih-Ming
Hung, Oren Eliezer*
A 5.3GHz Digital-to-Time-Converter-Based Fractional-N All-Digital PLL 54
Nenad Pavlovic, Jos Bergervoet

A 2.5GHz 32nm 0.35mm2 3.5dB NF -5dBm P_{1dB} Fully Differential CMOS Push-Pull LNA with Integrated 34dBm T/R Switch and ESD Protection..56

Chang-Tsung Fu, Hasnain Lakdawala, Stewart S. Taylor, Krishnamurthy Soumyanath

A 65nm CMOS Pulse-Width-Controlled Driver with 8V_{pp} Output Voltage for Switch-Mode RF PAs up to 3.6GHz 58

David A. Calvillo-Cortes, Mustafa Acar, Mark P. van der Heijden, Melina Apostolidou, Leo C.N. de Vreede, Domine Leenaerts, Jan Sonsky

A Low-Power Process-Scalable Superheterodyne Receiver with Integrated High-Q Filters................60

Ahmad Mirzaei, Hooman Darabi, David Murphy

A 40nm CMOS Highly Linear 0.4-to-6GHz Receiver Resilient to 0dBm Out-of-Band Blockers 62

Jonathan Borremans, Gunjan Mandal, Vito Giannini, Tomohiro Sano, Mark Ingels, Bob Verbruggen, Jan Craninckx

A 1.0-to-4.0GHz 65nm CMOS Four-Element Beamforming Receiver Using a Switched-Capacitor Vector Modulator with Approximate Sine Weighting via Charge Redistribution................................64

Michiel C. M. Soer, Eric A. M. Klumperink, Bram Nauta, Frank E. van Vliet

A Harmonic Rejection Mixer Robust to RF Device Mismatches..66

Aslam A. Rafi, Alessandro Piovaccari, Peter Vancorenland, Tyson Tuttle

ENTERPRISE PROCESSORS & COMPONENTS:

Session 4 Overview / High-Performance Digital: Enterprise Processors & Components............................68

A 5.2GHz Microprocessor Chip for the IBM Enterprise™ System..70

M. J. Saccamango, F. Malgioglio, P. Meaney, D. Plass, Y.-H. Chan, M. Mayo, G. Mayer, L. Sigal, R. Dude, R. Averill, M. Wood, T. Strach, H. Smith, B. Curran, E. Schwarz, L. Eisen, D. Malone, S. Weitzel, P.-K. Mak, T. McPherson, C. Webb

Dynamic Hit Logic with Embedded 8Kb SRAM in 45nm SOI for the zEnterprise™ Processor................72

Antonio R. Pelella, Yuen H. Chan, Bargav Balakrishnan, Pradip Patel, Daniel Rodko, Richard E. Serton

A 32nm Westmere-EX Xeon® Enterprise Processor..74

Shankar Sawant, Utpal Desai, Gururaj Shamanna, Lokesh Sharma, Mandar Ranade, Anil Agarwal, Sampath Dakshinamurthy, Rajagopal Narayanan

Godson-3B: A 1GHz 40W 8-Core 128GFLOPS Processor in 65nm CMOS..76

Weiwu Hu, Ru Wang, Yunji Chen, Baoxia Fan, Shiqiang Zhong, Xiang Gao, Zichu Qi, Xu Yang

Design Solutions for the Bulldozer 32nm SOI 2-Core Processor Module in an 8-Core CPU......................78

Tim Fischer, Srikanth Arekapudi, Eric Busta, Carl Dietz, Michael Golden, Scott Hilker, Aaron Horiuchi, Kevin A. Hurd, Dave Johnson, Hugh McIntyre, Samuel Naffziger, James Vinh, Jonathan White, Kathryn Wilcox

40-Entry Unified Out-of-Order Scheduler and Integer Execution Unit for the AMD Bulldozer x86-64 Core..80

Michael Golden, Srikanth Arekapudi, James Vinh

Clock Generation for a 32nm Server Processor with Scalable Cores..82

Shenggao Li, Ashwin Krishnakumar, Edward Helder, Roan Nicholson, Vivian Jia

A 32nm 3.1 Billion Transistor 12-Wide-Issue Itanium® Processor for Mission-Critical Servers...............84

Reid J. Riedlinger, Rohit Bhatia, Larry Biro, Bill Bowhill, Eric Fetzer, Paul Gronowski, Tom Grutkowski

PLLS:

Session 5 Overview / Analog: PLLs..86

A 2.9-to-4.0GHz Fractional-N Digital PLL with Bang-Bang Phase Detector and 560fs$_{rms}$ Integrated Jitter at 4.5mW Power 88

Davide Tasca, Marco Zanuso, Giovanni Marzin, Salvatore Levantino, Carlo Samori, Andrea L. Lacaita

An Injection-Locked Ring PLL with Self-Aligned Injection Window 90

Che-Fu Liang, Keng-Jan Hsiao

A 0.4-to-3GHz Digital PLL with Supply-Noise Cancellation Using Deterministic Background Calibration 92

Amr Elshazly, Rajesh Inti, Wenjing Yin, Brian Young, Pavan Kumar Hanumolu

A 0.1-f_{ref} BW 1GHz Fractional-N PLL with FIREmbedded Phase-Interpolator-Based Noise Filtering...........94

Dong-Woo Jee, Yunjae Suh, Hong-June Park, Jae-Yoon Sim

A Scalable Sub-1.2mW 300MHz-to-1.5GHz Host-Clock PLL for System-on-Chip in 32nm CMOS 96

Hyung-Jin Lee, Alexandra M. Kern, Sami Hyvonen, Ian A. Young

A 570f$_{rms}$ Integrated-Jitter Ring-VCO-Based 1.21GHz PLL with Hybrid Loop........98
Akihide Sai, Takafumi Yamaji, Tetsuro Itakura

A Rotary-Traveling-Wave-Oscillator-Based All-Digital PLL with a 32-Phase Embedded Phase-to-Digital Converter in 65nm CMOS........100
Koji Takinami, Richard Strandberg, Paul C. P. Liang, Gregoire Le Grand de Mercey, Tony Wong, Mahnaz Hassibi

SENSORS & ENERGY HARVESTING:

Session 6 Overview / IMMD: Sensors & Energy Harvesting........102

A Low-Power 3-Axis Digital-Output MEMS Gyroscope with Single Drive and Multiplexed Angular Rate Readout........104
Luciano Prandi, Carlo Caminada, Luca Coronato, Gabriele Cazzaniga, Fabio Biganzoli, Riccardo Antonello, Roberto Oboe

A 50mW CMOS Wind Sensor with ±4% Speed and ±2° Direction Error........106
Jianfeng Wu, Youngcheol Chae, Caspar P. L. van Vroonhoven, Kofi A. A. Makinwa

A Telemetric Stress-Mapping CMOS Chip with 24 FET-Based Stress Sensors for Smart Orthodontic Brackets........108
Matthias Kuhl, Pascal Gieschke, Daniel Rossbach, Sascha Alexander Hilzensauer, Patrick Ruther, Oliver Paul, Yiannos Manoli

A 21b ±40mV Range Read-Out IC for Bridge Transducers........110
Rong Wu, Johan H. Huijsing, Kofi A. A. Makinwa

A ±1.5% Nonlinearity 0.1-to-100A Shunt Current Sensor Based on a 6kV Isolated Micro-Transformer for Electrical Vehicles and Home Automation........112
Frederic Rothan, Helene Lhermet, Brice Zongo, Cyril Condemine, Henri Sibuet, Patrick Mas, Miguel Debarnot

Indirect X-ray Photon-Counting Image Sensor with 27T Pixel and 15e$_{rms}$ Accurate Threshold........114
Bart Dierickx, Benoit Dupont, Arnaud Defernez, Nayera Ahmed

A 1.32pW/frame•pixel 1.2V CMOS Energy-Harvesting and Imaging (EHI) APS Imager........116
Suat U. Ay

5µW-to-10mW Input Power Range Inductive Boost Converter for Indoor Photovoltaic Energy Harvesting with Integrated Maximum Power Point Tracking Algorithm........118
Yifeng Qiu, Chris Van Liempd, Bert Op het Veld, Peter G. Blanken, Chris Van Hoof

A Self-Supplied Inertial Piezoelectric Energy Harvester with Power-Management IC........120
Ethem Erkan Aktakka, Rebecca L. Peterson, Khalil Najafi

MULTIMEDIA & MOBILE:

Session 7 Overview / Energy-Efficient Digital: Multimedia & Mobile........122

A 216fps 4096×2160p 3DTV Set-Tps Box SoC for Free-Viewpoint 3DTV Applications........124
Pei-Kuei Tsung, Ping-Chih Lin, Kuan-Yu Chen, Tzu-Der Chuang, Hsin-Jung Yang, Shao-Yi Chien, Li-Fu Ding, Wei-Yin Chen, Chih-Chi Cheng, Tung-Chien Chen, Liang-Gee Chen

A Highly Parallel and Scalable CABAC Decoder for Next Generation Video Coding........126
Vivienne Sze, Anantha P. Chandrakasan

A 275mW Heterogeneous Multimedia Processor for IC-Stacking on Si-Interposer........128
Hyo-Eun Kim, Jae-Sung Yoon, Kyu-Dong Hwang, Young-Jun Kim, Jun-Seok Park, Lee-Sup Kim

A 57mW Embedded Mixed-Mode Neuro-Fuzzy Accelerator for Intelligent Multi-core Processor........130
Jinwook Oh, Junyoung Park, Gyeonghoon Kim, Seungjin Lee, Hoi-Jun Yoo

A 28nm 0.6V Low-Power DSP for Mobile Applications........132
Gordon Gammie, Nathan Ickes, Mahmut E. Sinangil, Rahul Rithe, J. Gu, Alice Wang, Hugh Mair, Satyendra Datla, Bing Rong, Sushma Honnavara-Prasad, Lam Ho, Greg Baldwin, Dennis Buss, Anantha P. Chandrakasan, Uming Ko

A MIMO WiMAX SoC in 90nm CMOS for 300km/h Mobility........134
Gene C. H. Chuang, Pang-An Ting, Jen-Yuan Hsu, Jiun-You Lai, Shun-Chang Lo, Ying-Chuan Hsiao, Tzi-Dar Chiueh

A 70Mb/s -100.5dBm Sensitivity 65nm LP MIMO Chipset for WiMAX Portable Router 136
Jyh-Shin Pan, Ming-Yang Chao, Eric Yeh, Wen-Wei Yang, Ching-Wen Hsueh, Shyuan Liao, Jian-Bang Lin, Shun-An Yang, Chin-Tai Liu, Tsai-Pao Lee, Jin-Ru Chen, Chia-Hua Chou, Min Chen, Den-Kai Juang, Jen-Hao Yeh, Chieh-Wei Liao, Po-Hung Chen, Kaipon Kao, Chia-Hsin Wu, Wen-Tso Huang, Shih-Hsien Liao, Chih-Heng Shih, Chien-Hsun Tung, Yen-Po Lee

A Direct Digital Frequency Synthesizer with Minimized Tuning Latency of 12ns 138
Alan Willson, Mukund Ojha, Shilpa Agarwal, Thriven Lai, Tzu-chieh Kuo

ARCHITECTURES & CIRCUITS FOR NEXT-GENERATION WIRELINE TRANSCEIVERS:

Session 8 Overview / Wireline: Architectures & Circuits for Next Generation Wireline Transceivers 140

11.3Gb/s CMOS SONET-Compliant Transceiver for Both RZ and NRZ Applications 142
Namik Kocaman, Adesh Garg, Bharath Raghavan, Delong Cui, Anand Vasani, Keith Tang, Deyi Pi, Haitao Tong, Siavash Fallahi, Wei Zhang, Ullas Singh, Jun Cao, Bo Zhang, Afshin Momtaz

A Full-Duplex 10GBase-T Transmitter Hybrid with SFDR >65dBc Over 1 to 400MHz in 40nm CMOS 144
Gaurav Chandra, Moshe Malkin

A 40Gb/s TX and RX Chip Set in 65nm CMOS 146
Ming-Shuan Chen, Yu-Nan Shih, Chen-Lun Lin, Hao-Wei Hung, Jri Lee

10:4 MUX and 4:10 DEMUX Gearbox LSI for 100-Gigabit Ethernet Link 148
Goichi Ono, Keiki Watanabe, Takashi Muto, Hiroki Yamashita, Koji Fukuda, Noboru Masuda, Ryo Nemoto, Eiichi Suzuki, Takashi Takemoto, Fumio Yuki, Masayoshi Yagyu, Hidehiro Toyoda, Akihiro Kambe, Tatsuya Saito, Shinji Nishimura

A 12.5+12.5Gb/s Full-Duplex Plastic Waveguide Interconnect 150
Satoshi Fukuda, Yasufumi Hino, Sho Ohashi, Takahiro Takeda, Satoru Shinke, Masahiro Uno, Kenji Komori, Yoshiyuki Akiyama, Kenichi Kawasaki, Ali Hajimiri

A Highly Digital 0.5-to-4Gb/s 1.9mW/Gb/s Serial-Link Transceiver Using Current-Recycling in 90nm CMOS 152
Rajesh Inti, Amr Elshazly, Brian Young, Wenjing Yin, Marcel Kossel, Thomas Toifl, Pavan Kumar Hanumolu

A 1-to-6Gb/s Phase-Interpolator-Based Burst-Mode CDR in 65nm CMOS 154
Behrooz Abiri, Ravi Shivnaraine, Ali Sheikholeslami, Hirotaka Tamura, Masaya Kibune

A 14Gb/s High-Swing Thin-Oxide Device SST TX in 45nm CMOS SOI 156
Christian Menolfi, Thomas Toifl, Michael Rueegg, Matthias Braendli, Peter Buchmann, Marcel Kossel, Thomas Morf

WIRELESS & MM-WAVE CONNECTIVITY:

Session 9 Overview / Wireless: Wireless & mm-Wave Connectivity 158

A 60GHz 16QAM/8PSK/QPSK/BPSK Direct-Conversion Transceiver for IEEE 802.15.3c 160
Kenichi Okada, Kota Matsushita, Keigo Bunsen, Rui Murakami, Ahmed Musa, Takahiro Sato, Hiroki Asada, Naoki Takayama, Ning Li, Shogo Ito, Win Chaivipas, Ryo Minami, Akira Matsuzawa

A 65nm CMOS Fully Integrated Transceiver Module for 60GHz Wireless HD Applications 162
Alexandre Siligaris, Olivier Richard, Baudouin Martineau, Christopher Mounet, Fabrice Chaix, Romain Ferragut, Cedric Dehos, Jerome Lanteri, Laurent Dussopt, Silas D. Yamamoto, Romain Pilard, Pierre Busson, Andreia Cathelin, Didier Belot, Pierre Vincent

A 60GHz CMOS Phased-Array Transceiver Pair for Multi-Gb/s Wireless Communications 164
Sohrab Emami, Robert F. Wiser, Ershad Ali, Mark G. Forbes, Michael Q. Gordon, Xiang Guan, Steve Lo, Patrick T. McElwee, James Parker, Jon R. Tani, Jeffery M. Gilbert, Chinh H. Doan

A 65nm CMOS 4-Element Sub-34mW/Element 60GHz Phased-Array Transceiver 166
Maryam Tabesh, Jiashu Chen, Cristian Marcu, Lingkai Kong, Shinwon Kang, Elad Alon, Ali Niknejad

An 87GHz QPSK Transceiver with Costas-Loop Carrier Recovery in 65nm CMOS 168
Shih-Jou Huang, Yu-Ching Yeh, Huaide Wang, Pang-Ning Chen, Jri Lee

A 65nm Dual-Band 3-Stream 802.11n MIMO WLAN SoC 170
Shahram Abdollahi-Alibeik, David Weber, Hakan Dogan, William W. Si, Burcin Baytekin, Abbas Komijani, Richard Chang, Babak Vakili-Amini, MeeLan Lee, Haitao Gan, Yashar Rajavi, Hirad Samavati, Brian Kaczynski, Sang-Min Lee, Sotirios Limotyrakis, Hyunsik Park, Phoebe Chen, Paul Park, Mike Shuo-Wei Chen, Andrew Chang, Yangjin Oh, Jerry Jian-Ming Yang, Eric Chien-Chih Lin, Lalitkumar Nathawad, Keith Onodera, Manolis Terrovitis, Sunetra Mendis, Kai Shi, Srenik Mehta, Masoud Zargari, David Su

A 0.46mm^2 4dB-NF Unified Receiver Front-End for Full-Band Mobile TV in 65nm CMOS 172
Pui-In Mak, Rui Martins

An All-Digital 8-DPSK Polar Transmitter with Second-Order Approximation Scheme and Phase Rotation-Constant Digital PA for Bluetooth EDR in 65nm CMOS 174
Hiroyuki Kobayashi, Shouhei Kousai, Yoshiaki Yoshihara, Mototsugu Hamada

A Digital-Intensive Receiver Front-End Using VCO-Based ADC with an Embedded 2nd-Order Anti-Aliasing Sinc Filter in 90nm CMOS 176
Jaewook Kim, Wonsik Yu, Hyun-Kyu Yu, SeongHwan Cho

NYQUIST-RATE CONVERTERS:

Session 10 Overview / Data Converters: Nyquist-Rate Converters 178

A 480mW 2.6GS/s 10b 65nm CMOS Time-Interleaved ADC with 48.5dB SNDR up to Nyquist 180
Kostas Doris, Erwin Janssen, Claudio Nani, Athon Zanikopoulos, Gerard Van der Weide

A 12b 1GS/s SiGe BiCMOS Two-Way Time-Interleaved Pipeline ADC 182
Robert Payne, Charles Sestok, William Bright, Manar El-Chammas, Marco Corsi, David Smith, Noam Tal

An 800MS/s Dual-Residue Pipeline ADC in 40nm CMOS 184
Jan Mulder, Frank M. L. van der Goes, Davide Vecchi, Jan R. Westra, Emre Ayranci, Christopher M. Ward, Jiansong Wan, Klaas Bult

A 16b 80MS/s 100mW 77.6dB SNR CMOS Pipeline ADC 186
Janet Brunsilius, Eric Siragusa, Steve Kosic, Frank Murden, Ege Yetis, Binh Luu, Jeff Bray, Phil Brown, Allen Barlow

A 0.024mm^2 8b 400MS/s SAR ADC with 2b/Cycle and Resistive DAC in 65nm CMOS 188
Hegong Wei, Chi-Hang Chan, U-Fat Chio, Sai-Weng Sin, Rui Martins, Franco Maloberti

A Resolution-Reconfigurable 5-to-10b 0.4-to-1V Power Scalable SAR ADC 190
Marcus Yip, Anantha P. Chandrakasan

A 12b 1.25GS/s DAC in 90nm CMOS with >70dB SFDR up to 500MHz 192
Wei-Hsin Tseng, Chi-Wei Fan, Jieh-Tsorng Wu

A 56GS/s 6b DAC in 65nm CMOS with 256×6b Memory 194
Yuriy M. Greshishchev, Daniel Pollex, Shing-Chi Wang, Marinette Besson, Philip Flemeke, Stefan Szilagyi, Jorge Aguirre, Chris Falt, Naim Ben-Hamida, Robert Gibbins, Peter Schvan

NON-VOLATILE MEMORY SOLUTIONS:

Session 11 Overiview / Memory: Non-Volatile Memory Solutions 196

A 151mm^2 64Gb MLC NAND Flash Memory in 24nm CMOS Technology 198
Koichi Fukuda, Yoshihisa Watanabe, Eiichi Makino, Koichi Kawakami, Junpei Sato, Teruo Takagiwa, Naoaki Kanagawa, Hitoshi Shiga, Naoya Tokiwa, Yoshihiko Shindo, Toshiaki Edahiro, Takeshi Ogawa, Makoto Iwai, Osamu Nagao, Junji Musha, Takatoshi Minamoto, Kosuke Yanagidaira, Yuya Suzuki, Dai Nakamura, Yoshikazu Hosomura, Hiromitsu Komai, Yuka Furuta, Mai Muramoto, Rieko Tanaka, Go Shikata, Ayako Yuminaka, Kiyofumi Sakurai, Manabu Sakai, Hong Ding, Mitsuyuki Watanabe, Yosuke Kato, Toru Miwa, Alex Mak, Masaru Nakamichi, Gertjan Hemink, Dana Lee, Masaaki Higashitani, Brian Murphy, Bo Lei, Yasuhiko Matsunaga, Kiyomi Naruke, Takahiko Hara

A 4Mb Embedded SLC Resistive-RAM Macro with 7.2ns Read-Write Random-Access Time and 160ns MLC-Access Capability 200
Shyh-Shyuan Sheu, Meng-Fan Chang, Ku-Feng Lin, Che-Wei Wu, Yu-Sheng Chen, Pi-Feng Chiu, Chia-Chen Kuo, Yih-Shan Yang, Pei-Chia Chiang, Wen-Pin Lin, Che-He Lin, Heng-Yuan Lee, Pei-Yi Gu, Sum-Min Wang, Frederick T. Chen, Keng-Li Su, Chen-Hsin Lien, Kuo-Hsing Cheng, Hsin-Tun Wu, Tzu-Kun Ku, Ming-Jer Kao, Ming-Jinn Tsai

A 32Gb MLC NAND Flash Memory with V_{th} Margin-Expanding Schemes in 26nm CMOS 202
Tae-yun Kim, Sang-Don Lee, Jin-su Park, Ho-youb Cho, Byoung-sung You, Kwang-ho Baek, Jae-ho Lee, Chang-won Yang, Misun Yun, Min-su Kim, Jong-woo Kim, Eun-seong Jang, Hyun Chung, Sang-o Lim, Bong-Seok Han, Yo-Hwan Koh

95%-Lower-BER 43%-Lower-Power Intelligent Solid-State Drive (SSD) with Asymmetric Coding and Stripe Pattern Elimination Algorithm 204
Shuhei Tanakamaru, Chinglin Hung, Atsushi Esumi, Mitsuyoshi Ito, Kai Li, Ken Takeuchi

An Offset-Tolerant Current-Sampling-Based Sense Amplifier for Sub-100nA-Cell-Current Nonvolatile Memory 206

Meng-Fan Chang, Shin-Jang Shen, Chia-Chi Liu, Che-Wei Wu, Yu-Fan Lin, Shang-Chi Wu, Chia-En Huang, Han-Chao Lai, Ya-Chin King, Chorng-Jung Lin, Hung-Jen Liao, Yu-Der Chih, Hiroyuki Yamauchi

A Low-Voltage 1Mb FeRAM in 0.13µm CMOS Featuring Time-to-Digital Sensing for Expanded Operating Margin in Scaled CMOS 208

Masood Qazi, Michael Clinton, Steven Bartling, Anantha P. Chandrakasan

A 4Mb Conductive-Bridge Resistive Memory with 2.3GB/s Read-Throughput and 216MB/s Program-Throughput 210

Wataru Otsuka, Koji Miyata, Makoto Kitagawa, Keiichi Tsutsui, Tomohito Tsushima, Hiroshi Yoshihara, Tomohiro Namise, Yasuhiro Terao, Kentaro Ogata

A 7MB/s 64Gb 3-Bit/Cell DDR NAND Flash Memory in 20nm-Node Technology 212

Ki-Tae Park, Ohsuk Kwon, Sangyong Yoon, Myung-Hoon Choi, In-Mo Kim, Bo-Geun Kim, Min-Seok Kim, Yoon-Hee Choi, Seung-Hwan Shin, Youngson Song, Joo-Yong Park, Jae-Eun Lee, Chang-Gyu Eun, Ho-Chul Lee, Hyeong-Jun Kim, Jun-Hee Lee, Jong-Young Kim, Tae-Min Kweon, Hyun-Jun Yoon, Taehyun Kim, Dong-Kyo Shim, Jongsun Sel, Ji-Yeon Shin, Pansuk Kwak, Jin-Man Han, Keon-Soo Kim, Sungsoo Lee, Young-Ho Lim, Tae-Sung Jung

DESIGN IN EMERGING TECHNOLOGIES:

Session 12 Overview / Technology Directions: Design in Emerging Technologies 214

A 95mV-Startup Step-Up Converter with V_{TH}-Tuned Oscillator by Fixed-Charge Programming and Capacitor Pass-On Scheme 216

Po-Hung Chen, Koichi Ishida, Katsuyuki Ikeuchi, Xin Zhang, Kentaro Honda, Yasuyuki Okuma, Yoshikatsu Ryu, Makoto Takamiya, Takayasu Sakurai

100V AC Power Meter System-on-a-Film (SoF) Integrating 20V Organic CMOS Digital and Analog Circuits with Floating Gate for Process-Variation Compensation and 100V Organic PMOS Rectifier 218

Koichi Ishida, Tsung-Ching Huang, Kentaro Honda, Tsuyoshi Sekitani, Hiroyoshi Nakajima, Hiroki Maeda, Makoto Takamiya, Takao Someya, Takayasu Sakurai

Real-Time Current-Waveform Sensor with Plugless Energy Harvesting from AC Power Lines for Home/Building Energy-Management Systems 220

Shingo Takahashi, Nobuhide Yoshida, Kenichi Maruhashi, Muneo Fukaishi

A 3.9ns 8.9mW 4×4 Silicon Photonic Switch Hybrid Integrated with CMOS Driver 222

Alexander Rylyakov, Clint Schow, Benjamin Lee, William Green, Joris Van Campenhout, Min Yang, Fuad Doany, Solomon Assefa, Christopher Jahnes, Jeffrey Kash, Yurii Vlasov

A 820GHz SiGe Chipset for Terahertz Active Imaging Applications 224

Erik Öjefors, Janus Grzyb, Yan Zhao, Bernd Heinemann, Bernd Tillack, Ullrich R. Pfeiffer

A 130µA Wake-Up Receiver SoC in 0.13µm CMOS for Reducing Standby Power of An Electric Appliance Controlled by An Infrared Remote Controller 226

Hiroaki Ishihara, Toshiyuki Umeda, Katsuya Ohno, Shigeyasu Iwata, Fumi Moritsuka, Tetsuro Itakura, Manabu Ishibe, Keijiro Hijikata, Yasunori Maki

Programmable Cell Array Using Rewritable Solid-Electrolyte Switch Integrated in 90nm CMOS 228

Makoto Miyamura, Shogo Nakaya, Munehiro Tada, Toshitsugu Sakamoto, Koichiro Okamoto, Naoki Banno, Shinji Ishida, Kimihiko Ito, Hiromitsu Hada, Noboru Sakimura, Tadahiko Sugibayashi, Masato Motomura

6W/25mm² Inductive Power Transfer for Non-Contact Wafer-Level Testing 230

Andrzej Radecki, Hayun Chung, Yoichi Yoshida, Noriyuki Miura, Tsunaaki Shidei, Hiroki Ishikuro, Tadahiro Kuroda

GHz-Range Continuous-Time Programmable Digital FIR with Power Dissipation that Automatically Adapts to Signal Activity 232

Mariya Kurchuk, Colin Weltin-Wu, Dominique Morche, Yannis Tsividis

ANALOG TECHNIQUES:

Session 13 Overview / Analog: Analog Techniques 234

A Simple LED Lamp Driver IC with Intelligent Power-Factor Correction 236

Jong Tae Hwang, Kunhee Cho, Donghwan Kim, Minho Jung, Gyehyun Cho, Seunguk Yang

A 1.2A Buck-Boost LED Driver with 13% Efficiency Improvement Using Error-Averaged SenseFET-Based Current Sensing 238

Sachin Rao, Qadeer Khan, Sarvesh Bang, Damian Swank, Arun Rao, William McIntyre, Pavan Kumar Hanumolu

Filterless Integrated Class-D Audio Amplifier Achieving 0.0012% THD+N and 96dB PSRR When Supplying 1.2W 240

Mykhaylo Teplechuk, Tony Gribben, Christophe Amadi

A 5.9nV/√Hz Chopper Operational Amplifier with 0.78μV Maximum Offset and 28.3nV/°C Offset Drift 242

Yoshinori Kusuda

A Current-Feedback Instrumentation Amplifier with a Gain Error Reduction Loop and 0.06% Untrimmed Gain Error 244

Rong Wu, Johan H. Huijsing, Kofi A. A. Makinwa

A 6.7nV/√Hz Sub-mHz-1/f-Corner 14b Analog-to-Digital Interface for Rail-to-Rail Precision Voltage Sensing 246

Chinwuba D. Ezekwe, Johan P. Vanderhaegen, Xinyu Xing, Ganesh K. Balachandran

A 36V JFET-Input Bipolar Operational Amplifier with 1μV/ C Maximum Offset Drift and −126dB Total Harmonic Distortion 248

Martijn F. Snoeij, Mikhail V. Ivanov

A 3.3V-Supply 120mW Differential ADC Driver Amplifier in 0.18μm SiGe BiCMOS with 108dBc IM3 at 100MHz 250

Gwilym F. Luff

HIGH-PERFORMANCE EMBEDDED MEMORY:

Session 14 Overview / Memory: High-Performance Embedded Memory 252

A 64Mb SRAM in 32nm High-k Metal-Gate SOI Technology with 0.7V Operation Enabled by Stability, Write-Ability and Read-Ability Enhancements 254

Harold Pilo, Igor Arsovski, Kevin Batson, Geordie Braceras, John Gabric, Robert Houle, Steve Lamphier, Frank Pavlik, Adnan Seferagic, Liang-Yu Chen, Shang-Bin Ko, Carl Radens

A 4R2W Register File for a 2.3GHz Wire-Speed POWER™ Processor with Double-Pumped Write Operation 256

Gary S. Ditlow, Robert K. Montoye, Salvatore N. Storino, Sherman M. Dance, Sebastian Ehrenreich, Bruce M. Fleischer, Thomas W. Fox, Kyle M. Holmes, Junichi Mihara, Yutaka Nakamura, Shohji Onishi, Robert Shearer, Dieter Wendel, Leland Chang

An 8MB Level-3 Cache in 32nm SOI with Column-Select Aliasing 258

Don Weiss, Michael Dreesen, Michael Ciraula, Carson Henrion, Chris Helt, Ryan Freese, Tommy Miles, Anita Karegar, Russell Schreiber, Bryan Schneller, John Wuu

A 28nm High-Density 6T SRAM with Optimized Peripheral-Assist Circuits for Operation Down to 0.6V 260

Mahmut E. Sinangil, Hugh Mair, Anantha P. Chandrakasan

HIGH –PERFORMANCE EMBEDDED SOCS & COMPONENTS:

Session 15 Overview / High-Performance Digital: High-Performance SoCs & Components 262

A Fully Integrated Multi-CPU, GPU and Memory Controller 32nm Processor 264

Marcelo Yuffe, Ernest Knoll, Moty Mehalel, Joseph Shor, Tsvika Kurts

An 80Gb/s Dependable Communication SoC with PCI Express I/F and 8 CPUs 266

Sugako Otani, Hiroyuki Kondo, Itaru Nonomura, Atsuyuki Ikeya, Minoru Uemura, Yasushi Hayakawa, Takeshi Oshita, Satoshi Kaneko, Katsushi Asahina, Kazutami Arimoto, Shin'ichi Miura, Toshihiro Hanawa, Taisuke Boku, Mitsuhisa Sato

A Fully-Integrated 3-Level DC/DC Converter for Nanosecond-Scale DVS with Fast Shunt Regulation 268

Wonyoung Kim, David M. Brooks, Gu-Yeon Wei

A Low-Power Integrated x86-64 and Graphics Processor for Mobile Computing Devices 270

Srinivasa Rao Gutta, Denis Foley, Ajay Naini, Robert Wasmuth, Don Cherepacha

A Programmable Adaptive Phase-Shifting PLL for Clock Data Compensation Under Resonant Supply Noise 272

Dong Jiao, Chris H. Kim

A Side-Channel and Fault-Attack Resistant AES Circuit Working on Duplicated Complemented Values 274

Marion Doulcier-Verdier, Jean-Max Dutertre, Jacques Fournier

MM-WAVE DESIGN TECHNIQUES:

Session 16 Overview / RF: mm-Wave Design Techniques 276

A 21.7-to-27.8GHz 2.6-Degrees-rms 40mW Frequency Synthesizer in 45nm CMOS for mm-Wave Communication Applications 278

Juan F. Osorio, Cicero S. Vaucher, Bill Huff, Edwin V. D. Heijden, Anton de Graauw

A mm-Wave Quadrature VCO Based on Magnetically Coupled Resonators 280

Ugo Decanis, Andrea Ghilioni, Enrico Monaco, Andrea Mazzanti, Francesco Svelto

A 6.5mW Inductorless CMOS Frequency Divider-by-4 Operating up to 70GHz 282

Andrea Ghilioni, Ugo Decanis, Enrico Monaco, Andrea Mazzanti, Francesco Svelto

A 60GHz Antenna-Referenced Frequency-Locked Loop in 0.13µm CMOS for Wireless Sensor Networks 284

Kuo-Ken Huang, David D. Wentzloff

A 220-to-275GHz Traveling-Wave Frequency Doubler with -6.6dBm Power at 244GHz in 65nm CMOS 286

Omeed Momeni, Ehsan Afshari

Distributed Active Radiation for THz Signal Generation 288

Kaushik Sengupta, Ali Hajimiri

A 120GHz 10Gb/s Phase-Modulating Transmitter in 65nm LP CMOS 290

Noël Deferm, Patrick Reynaert

A 1.5GHz-Modulation-Range 10ms-Modulation-Period 180kHz$_{rms}$-Frequency-Error 26MHz-Reference Mixed-Mode FMCW Synthesizer for mm-Wave Radar Application 292

Hiroki Sakurai, Yuka Kobayashi, Toshiya Mitomo, Osamu Watanabe, Shoji Otaka

A Short-Range UWB Impulse-Radio CMOS Sensor for Human Feature Detection 294

Ta-Shun Chu, Jonathan Roderick, SangHyun Chang, Timothy Mercer, Chenliang Du, Hossein Hashemi

183GHz 13.5mW/Pixel CMOS Regenerative Receiver for mm-Wave Imaging Applications 296

Adrian Tang, Mau-Chung Frank Chang

BIOMEDICAL & DISPLAYS:

Session 17 Overview / IMMD: Biomedical & Displays 298

A 160µW 8-Channel Active Electrode System for EEG Monitoring 300

Jiawei Xu, Refet Firat Yazicioglu, Pieter Harpe, Kofi A. A. Makinwa, Chris Van Hoof

A 0.013mm^2 5µW DC-Coupled Neural Signal Acquisition IC with 0.5V Supply 302

Rikky Muller, Simone Gambini, Jan M. Rabaey

An AC-Powered Optical Receiver Consuming 270µW for Transcutaneous 2Mb/s Data Transfer 304

Steffen Lange, Hongcheng Xu, Christian Lang, Holger Pless, Joachim Becker, Hans-Jürgen Tiedtke, Eckhard Hennig, Maurits Ortmanns

A Neural Stimulator Front-End with Arbitrary Pulse Shape, HV Compliance and Adaptive Supply Requiring 0.05mm^2 in 0.35µm HVCMOS 306

Kriangkrai Sooksood, Emilia Noorsal, Joachim Becker, Maurits Ortmanns

A Low Noise Current Readout Architecture for Fluorescence Detection in Living Subjects 308

Roxana T. Heitz, David B. Barkin, Thomas D. O'Sullivan, Natesh Parashurama, Sanjiv S. Gambhir, Bruce A. Wooley

A Cubic-Millimeter Energy-Autonomous Wireless Intraocular Pressure Monitor 310

Gregory Chen, Hassan Ghaed, Razi-ul Haque, Michael Wieckowski, Yejoong Kim, Gyouho Kim, David Fick, Daeyeon Kim, Mingoo Seok, Kensall Wise, David Blaauw, Dennis Sylvester

A 160×128 Single-Photon Image Sensor with On-Pixel 55ps 10b Time-to-Digital Converter 312

Chockalingam Veerappan, Justin Richardson, Richard Walker, Day-Uey Li, Matthew W Fishburn, Yuki Maruyama, David Stoppa, Fausto Borghetti, Marek Gersbach, Robert K. Henderson, Edoardo Charbon

Bidirectional OLED Microdisplay: Combining Display and Image Sensor Functionality into a Monolithic CMOS Chip 314

Bernd Richter, Uwe Vogel, Rigo Herold, Karsten Fehse, Stephan Brenner, Lars Kroker, Judith Baumgarten

A 0.014mm^2 9b Switched-Current DAC for AMOLED Mobile Display Drivers 316
Hyun-Sik Kim, Jin-Yong Jeon, Sung-Woo Lee, Jun-Hyeok Yang, Seung-Tak Ryu, Gyu-Hyeong Cho

A 10b Resistor-Resistor-String DAC with Current Compensation for Compact LCD Driver ICs 318
Chih-Wen Lu, Ping-Yeh Yin, Ching-Min Hsiao, Mau-Chung Frank Chang

ORGANIC INNOVATIONS:

Session 18 Overview / Technology Directions: Organic Innovations 320

An 8b Organic Microprocessor on Plastic Foil 322
Kris Myny, Erik van Veenendaal, Gerwin H. Gelinck, Jan Genoe, Wim Dehaene, Paul Heremans

A 3.3V 6b 100kS/s Current-Steering D/A Converter Using Organic Thin-Film Transistors on Glass 324
Tarek Zaki, Frederik Ante, Ute Zschieschang, Joerg Butschke, Florian Letzkus, Harald Richter, Hagen Klauk,
Joachim N. Burghartz

**A 1V Printed Organic DRAM Cell Based on Ion-Gel Gated Transistors with a Sub-10nW-per-Cell
Refresh Power** 326
Wei Zhang, Mingjing Ha, Daniele Braga, Michael J. Renn, C. Daniel Frisbie, Chris H. Kim

Fully Printed Organic CMOS Technology on Plastic Substrates for Digital and Analog Applications 328
Anis Daami, Cécile Bory, Mohamed Benwadih, Stéphanie Jacob, Romain Gwoziecki, Isabelle Chartier, Romain
Coppard, Christophe Serbutoviez, Lidia Maddiona, Enzo Fontana, Antonino Scuderi

LOW-POWER DIGITAL TECHNIQUES:

Session 19 Overview / Energy-Efficient Digital: Low-Power Digital Techniques 330

A Voltage-Scalable Biomedical Signal Processor Running ECG Using 13pJ/cycle at 1MHz and 0.4V 332
Maryam Ashouei, Jos Hulzink, Mario Konijnenburg, Jun Zhou, Filipa Duarte, Arjan Breeschoten, Jos Huisken,
Jan Stuyt, Harmke de Groot, Francisco Barat, Johan David, Johan Van Ginderdeuren

An 82µA/MHz Microcontroller with Embedded FeRAM for Energy-Harvesting Applications 334
Michael Zwerg, Adolf Baumann, Rüdiger Kuhn, Matthias Arnold, Ronald Nerlich, Marcus Herzog, Ralph Ledwa,
Christian Sichert, Volker Rzehak, Priya Thanigai, Bjoern Oliver Eversmann

**Comparison of 65nm LP Bulk and LP PD-SOI with Adaptive Power Gate Body Bias for an LDPC
Codec** 336
Julien Le Coz, Philippe Flatresse, Sylvain Engels, Alexandre Valentian, Marc Belleville, Christine Raynaud,
Damien Croain, Pascal Urard

**A 77% Energy-Saving 22-Transistor Single-Phase-Clocking D-Flip-Flop with Adaptive-Coupling
Configuration in 40nm CMOS** 338
Chen Kong Teh, Tetsuya Fujita, Hiroyuki Hara, Mototsugu Hamada

A 62mV 0.13µm CMOS Standard-Cell-Based Design Technique Using Schmitt-Trigger Logic 340
Niklas Lotze, Yiannos Manoli

A 0.27V 30MHz 17.7nJ/transform 1024-pt Complex FFT Core with Super-Pipelining 342
Mingoo Seok, Dongsuk Jeon, Chaitali Chakrabarti, David Blaauw, Dennis Sylvester

HIGH-SPEED TRANSCEIVERS & BUILDING BLOCKS:

Session 20 Overview / Wireline: High-Speed Transceivers & Building Blocks 344

**A 0.076mm^2 3.5GHz Spread-Spectrum Clock Generator with Memoryless Newton-Raphson
Modulation Profile in 0.13µm CMOS** 346
Sewook Hwang, Minyoung Song, Young-Ho Kwak, Inhwa Jung, Chulwoo Kim

A 4-Channel 10.3Gb/s Transceiver with Adaptive Phase Equalizer for 4-to-41dB Loss PCB Channel 348
Yasuo Hidaka, Takeshi Horie, Yoichi Koyanagi, Takashi Miyoshi, Hideki Osone, Samir Parikh, Subodh Reddy,
Toshiyuki Shibuya, Yasushi Umezawa, William W. Walker

**A 1.0625-to-14.025Gb/s Multimedia Transceiver with Full-rate Source-Series-Terminated Transmit
Driver and Floating-Tap Decision-Feedback Equalizer in 40nm CMOS** 350
Shaolei Quan, Freeman Zhong, Wing Liu, Pervez Aziz, Tai Jing, Jen Dong, Chintan Desai, Hairong Gao, Monica
Garcia, Gary Hom, Tony Huynh, Hiroshi Kimura, Ruchi Kothari, Lijun Li, Cathy Liu, Scott Lowrie, Kathy Ling,
Amaresh Malipatil, Ram Narayan, Tom Prokop, Chaitanya Palusa, Anil Rajashekara, Ashutosh Sinha, Charlie
Zhong, Eric Zhang

Analog-DFE-Based 16Gb/s SerDes in 40nm CMOS That Operates Across 34dB Loss Channels at Nyquist with a Baud Rate CDR and 1.2V$_{pp}$ Voltage-Mode Driver 352

Andrew K. Joy, Hugh Mair, Hae-Chang Lee, Arnold Feldman, Clemenz Portmann, Neil Bulman, Eugenia Cordero Crespo, Peter Hearne, Patty Huang, Ben Kerr, Pulkit Khandelwal, Franz Kuhlmann, Shaun Lytollis, Joaquim Machado, Casey Morrison, Scott Morrison, Shahriar Rabii, Dushmantha Rajapaksha, Vishnu Ravinuthula, Giuseppe Surace

An 8.4mW/Gb/s 4-Lane 48Gb/s Multi-Standard-Compliant Transceiver in 40nm Digital CMOS Technology 354

Mehrdad Ramezani, Mohamed Abdalla, Ayal Shoval, Marcus Van Ierssel, Afshin Rezayee, Angus McLaren, Chris Holdenried, Jennifer Pham, Eric So, David Cassan, Saman Sadr

A Pattern-Guided Adaptive Equalizer in 65nm CMOS 356

Shayan Shahramian, Clifford Ting, Ali Sheikholeslami, Hirotaka Tamura, Masaya Kibune

A 6Gb/s Receiver with 32.7dB Adaptive DFE-IIR Equalization 358

Yi-Chieh Huang, Shen-Iuan Liu

A 5.4Gb/s Adaptive Equalizer Using Asynchronous-Sampling Histograms 360

Wang-Soo Kim, Chang-Kyung Seong, Woo-Young Choi

CELLULAR:

Session 21 Overview / Wireless: Cellular 362

A SAW-less GSM/GPRS/EDGE Receiver Embedded in a 65nm CMOS SoC 364

Ivan Siu-Chuang Lu, Chi-yao Yu, Yen-horng Chen, Lan-chou Cho, Chih-hao Eric Sun, Chih-Chun Tang, George Chien

A 9-Band WCDMA/EDGE Transceiver Supporting HSPA Evolution 366

Magnus Nilsson, Sven Mattisson, Nikolaus Klemmer, Martin Anderson, Torkel Arnborg, Peter Caputa, Staffan Ek, Lin Fan, Henrik Fredriksson, Fabien Garrigues, Henrik Geis, Hans Hagberg, Joel Hedestig, Hu Huang, Yevgeniy Kagan, Niklas Karlsson, Henrik Kinzel, Thomas Mattsson, Thomas Mills, Fenghao Mu, Andreas Mårtensson, Lars Nicklasson, Filip Oredsson, Ufuk Ozdemir, Fitzgerald Park, Tony Pettersson, Tony Påhlsson, Markus Pålsson, Stephane Ramon, Magnus Sandgren, Per Sandrup, Anna-Karin Stenman, Roland Strandberg, Lars Sundström, Fredrik Tillman, Tobias Tired, Satish Uppathil, Joel Walukas, Eric Westesson, Xuhao Zhang, Pietro Andreani

A 65nm CMOS SoC with Embedded HSDPA/EDGE Transceiver, Digital Baseband and Multimedia Processor 368

Alberto Cicalini, Sankaran Aniruddhan, Rahul Apte, Frederic Bossu, Ojas Choksi, Dan Filipovic, Kunal Godbole, Tsai-Pi Hung, Christos Komninakis, David Maldonado, Chiewcharn Narathong, Babak Nejati, Deirdre O'Shea, Xiaohong Quan, Raj Rangarajan, Janakiram Sankaranarayanan, Andrew See, Ravi Sridhara, Bo Sun, Wenjun Su, Klaas van Zalinge, Gang Zhang, Kamal Sahota

A Receiver for WCDMA/EDGE Mobile Phones with Inductorless Front-End in 65nm CMOS 370

Federico Beffa, Tze Yee Sin, Alexander Tanzil, David Ivory, Bernard Tenbroek, Jon Strange, Walid Ali-Ahmad

A Compact SAW-less Multiband WCDMA/GPS Receiver Front-End with Translational Loop for Input Matching 372

Xin He, Harish Kundur

A Multiband LTE SAW-less Modulator with -160dBc/Hz RX-Band Noise in 40nm LP CMOS 374

Vito Giannini, Mark Ingels, Tomohiro Sano, Bjorn Debaillie, Jonathan Borremans, Jan Craninckx

A Fully Digital Multimode Polar Transmitter Employing 17b RF DAC in 3G Mode 376

Zdravko Boos, Andreas Menkhoff, Franz Kuttner, Markus Schimper, Jose Moreira, Hans Geltinger, Timo Gossmann, Peter Pfann, Alexander Belitzer, Thomas Bauernfeind

A Low-Power Wideband Polar Transmitter for 3G Applications 378

Michael Youssef, Alireza Zolfaghari, Hooman Darabi, Asad Abidi

DC/DC CONVERTERS:

Session 22 Overview / Analog: DC/DC Converters 380

A Fully Integrated Power-Management Solution for a 65nm CMOS Cellular Handset Chip 382

Arnold James D'Souza, Ravpreet Singh, Raja J. Prabhu, Gajendranath Chowdary, Ankit Seedher, Shyam Somayajula, Nageswara Rao Nalam, Lionel Cimaz, Stephane Le Coq, Praveen Kallam, Siddharth Sundar, Shanfeng Cheng, Sanjay Tumati, Wenchang Huang

A Digitally Controlled DC-DC Converter for SoC in 28nm CMOS 384

Franz Kuttner, Harun Habibovic, Thomas Hartig, Michael Fulde, Gernot Babin, Andreas Santner, Peter Bogner, Claus Kropf, Harald Riesslegger, Uwe Hode

20μA to 100mA DC-DC Converter with 2.8 to 4.2V Battery Supply for Portable Applications in 45nm CMOS 386
Saurav Bandyopadhyay, Yogesh K. Ramadass

A Digitally Controlled DC-DC Buck Converter with Lossless Load-Current Sensing and BIST Functionality 388
Tao Liu, Hyunsoo Yeom, Bert Vermeire, Philippe Adell, Bertan Bakkaloglu

Zero-Order Control of Boost DC-DC Converter with Transient Enhancement Using Residual Current 390
Tae-Hwang Kong, Young-Jin Woo, Se-Won Wang, Sung-Wan Hong, Gyu-Hyeong Cho

Robust and Efficient Synchronous Buck Converter with Near-Optimal Dead-Time Control...................392
Sungwoo Lee, Seungchul Jung, Jin Huh, Changbyung Park, Chun-Taek Rim, Gyu-Hyeong Cho

A 90% Peak Efficiency Single-Inductor Dual-Output Buck-Boost Converter with Extended-PWM Control 394
Weiwei Xu, Ye Li, Zhiliang Hong, Dirk Killat

Spurious-Noise-Free Buck Regulator for Direct Powering of Analog/RF Loads Using PWM Control with Random Frequency Hopping and Random Phase Chopping 396
Chengwu Tao, Ayman A. Fayed

IMAGE SENSORS:

Session 23 Overview / IMMD: Image Sensors...398

An 80μV$_{rms}$-Temporal-Noise 82dB-Dynamic-Range CMOS Image Sensor with a 13-to-19b Variable-Resolution Column-Parallel Folding-Integration/Cyclic ADC 400
Min-Woong Seo, Sungho Suh, Tetsuya Iida, Hiroshi Watanabe, Taishi Takasawa, Tomoyuki Akahori, Keigo Isobe, Takashi Watanabe, Shinya Itoh, Shoji Kawahito

A Sub-Electron Readout Noise CMOS Image Sensor with Pixel-Level Open-Loop Voltage Amplification..402
Christian Lotto, Peter Seitz, Thomas Baechler

A 16 Mfps 165kpixel Backside-Illuminated CCD 406
Takeharu G. Etoh, Dung H. Nguyen, Son V. T. Dao, Cuong L. Vo, Masatoshi Tanaka, Kohsei Takehara, Tomoo Okinaka, Harry van Kuijk, Wilco Klaassens, Jan Bosiers, Michael Lesser, David Ouellette, Hirotaka Maruyama, Tetsuya Hayashida, Toshiki Arai

A 300mm Wafer-Size CMOS Image Sensor with In-Pixel Voltage-Gain Amplifier and Column-Level Differential Readout Circuitry 408
Yuichiro Yamashita, Hidekazu Takahashi, Shin Kikuchi, Keisuke Ota, Masato Fujita, Satoshi Hirayama, Taikan Kanou, Sakae Hashimoto, Genzo Momma, Shunsuke Inoue

A 128×96 Pixel Event-Driven Phase-Domain ΔΣ Based Fully Digital 3D Camera in 0.13μm CMOS Imaging Technology 410
Richard J. Walker, Justin A. Richardson, Robert K. Henderson

An Angle-Sensitive CMOS Imager for Single-Sensor 3D Photography 412
Albert Wang, Patrick R. Gill, Alyosha Molnar

A 1/13-inch 30fps VGA SoC CMOS Image Sensor with Shared Reset and Transfer-Gate Pixel Control 414
Robert Johansson, A. Storm, C. Stephansen, S. Eikedal, T. Willassen, S. Skaug, T. Martinussen, D. Whittlesea, G. Ali, J. Ladd, X. Li, S. Johnson, V. Rajasekaran, Y. Lee, J. Bai, M. Flores, G. Davies, H. Samiy, A. Hanvey, D. Perks

A 1/2.33-inch 14.6M 1.4μm-Pixel Backside-Illuminated CMOS Image Sensor with Floating Diffusion Boosting 416
Sangjoo Lee, Kyungho Lee, Jongeun Park, Hyungjun Han, Younghwan Park, Taesub Jung, Youngheup Jang, Bumsuk Kim, Yitae Kim, Shay Hamami, Uzi Hizi, Mickey Bahar, Changrok Moon, JungChak Ahn, Duckhyung Lee, Hiroshige Goto, Yun-Tae Lee

An APS-C Format 14b Digital CMOS Image Sensor with a Dynamic Response Pixel...........................418
Dan Pates, Jeong-Ho Lyu, Shinji Osawa, Isao Takayanagi, Toshiaki Sato, Tim Bales, Katsuyuki Kawamura, Eduard Pages, Shinichiro Matsuo, Tetsuji Kawaguchi, Tadashi Sugiki, Norio Yoshimura, Junichi Nakamura, John Ladd, Zhiping Yin, Russell Iimura, Xiaofeng Fan, Scott Johnson, Aditya Rayankula, Rick Mauritzson, Gennadiy Agranov

A 17.7Mpixel 120fps CMOS Image Sensor with 34.8Gb/s Readout 420
Takayuki Toyama, Koji Mishina, Hiroyuki Tsuchiya, Tatsuya Ichikawa, Hiroyuki Iwaki, Yuji Gendai, Hirotaka Murakami, Kenichi Takamiya, Hiroshi Shiroshita, Yoshinori Muramatsu, Toshihiro Furusawa

TRANSMITTER BLOCKS:

Session 24 Overview / RF: Transmitter Blocks 422

A 40nm Wideband Direct-Conversion Transmitter with Sub-Sampling-Based Output Power, LO Feedthrough and I/Q Imbalance Calibration 424
Emanuele Lopelli, Silvian Spiridon, Johan van der Tang

A Flip-Chip-Packaged 1.8V 28dBm Class-AB Power Amplifier with Shielded Concentric Transformers in 32nm SoC CMOS 426
Yulin Tan, Hongtao Xu, Mohammed A El-tanani, Stewart Taylor, Hasnain Lakdawala

A Switched-Capacitor Power Amplifier for EER/Polar Transmitters 428
Sang-Min Yoo, Jeffrey S. Walling, Eum Chan Woo, David J. Allstot

An EDGE/GSM Quad-Band CMOS Power Amplifier 430
Woonyun Kim, Ki Seok Yang, Jeonghu Han, Jaejoon Chang, Chang-Ho Lee

A Compact 1V 18.6dBm 60GHz Power Amplifier in 65nm CMOS 432
Jiashu Chen, Ali M. Niknejad

CDRS & EQUALIZATION TECHNIQUES:

Session 25 Overview / Wireline: CDRs & Equalization Techniques 434

A 5Gb/s Adaptive DFE for 2x Blind ADC-Based CDR in 65nm CMOS 436
Behrooz Abiri, Ali Sheikholeslami, Hirotaka Tamura, Masaya Kibune

A 0.5-to-2.5Gb/s Reference-less Half-Rate Digital CDR with Unlimited Frequency Acquisition Range and Improved Input Duty-Cycle Error Tolerance 438
Rajesh Inti, Wenjing Yin, Amr Elshazly, Naga Sasidhar, Pavan Kumar Hanumolu

A TDC-less 7mW 2.5Gb/s Digital CDR with Linear Loop Dynamics and Offset-Free Data Recovery 440
Wenjing Yin, Rajesh Inti, Amr Elshazly, Pavan Kumar Hanumolu

A Digital Wideband CDR with ±15.6kppm Frequency Tracking at 8Gb/s in 40nm CMOS 442
Hui Pan, Magesh Valliappan, Wei Zhang, Kambiz Vakilian, Seong-Ho Lee, Hamid Hatamkhani, Mario Caresosa, Karo Khanoyan, Haitao Tong, Duke Tran, Anthony Brewster, Ichiro Fujimori

A 20Gb/s Digitally Adaptive Equalizer/DFE with Blind Sampling 444
Yu-Ming Ying, Shen-Iuan Liu

A 15Gb/s 0.5mW/Gb/s 2-Tap DFE Receiver with Far-End Crosstalk Cancellation 446
Meisam Honarvar Nazari, Azita Emami-Neyestanak

A 10Gb/s Half-UI IIR-Tap Transmitter in 40nm CMOS 448
Halil Cirit, Marc J. Loinaz

A 13.8mW 3.0Gb/s Clock-Embedded Video Interface with DLL-Based Data-Recovery Circuit 450
Sungchun Jang, Heesoo Song, Seokmin Ye, Deog-Kyoon Jeong

LOW-POWER WIRELESS:

Session 26 Overview / Wireless: Low-Power Wireless 452

A 7.9μW Remotely Powered Addressed Sensor Node Using EPC HF and UHF RFID Technology with -10.3dBm Sensitivity 454
Hannes Reinisch, Martin Wiessflecker, Stefan Gruber, Hartwig Unterassinger, Günter Hofer, Michael Klamminger, Wolfgang Pribyl, Gerald Holweg

An Isolator-less CMOS RF Front-End for UHF Mobile RFID Reader 456
Eun-Hee Kim, Kwyro Lee, Jinho Ko

A 2.4GHz ULP OOK Single-Chip Transceiver for Healthcare Applications 458
Maja Vidojkovic, Xiongchuan Huang, Pieter Harpe, Simonetta Rampu, Cui Zhou, Li Huang, Koji Imamura, Ben Busze, Frank Bouwens, Mario Konijnenburg, Juan Santana, Arjan Breeschoten, Jos Huisken, Guido Dolmans, Harmke de Groot

A 120μW MICS/ISM-Band FSK Receiver with a 44μW Low-Power Mode Based on Injection-Locking and 9x Frequency Multiplication 460
Jagdish Pandey, Jianlei Shi, Brian Otis

A GPS/Galileo SoC with Adaptive In-Band Blocker Cancellation in 65nm CMOS 462
Chia-Hsin Wu, Wen-Chieh Tsai, Chun-Geik Tan, Chun-Nan Chen, Kuan-I Li, Jui-Lin Hsu, Chi-Lun Lo, Hsin-Hua Chen, Sheng-Yuan Su, Kun-Tso Chen, Min Chen, Osama Shana'a, Shu-Hung Chou, George Chien

A 0.05-to-10GHz 19-to-22GHz and 38-to-44GHz SDR Frequency Synthesizer in 0.13μm CMOS 464
Sujiang Rong, Howard C. Luong

A 4.6GHz MDLL with -46dBc Reference Spur and Aperture Position Tuning 466
Tamer A. Ali, Amr A. Hafez, Robert Drost, Ronald Ho, Chih-Kong Ken Yang

OVERSAMPLING CONVERTERS:

Session 27 Overview / Data Converters: Oversampling Converters 468

A 4GHz CT ΔΣ ADC with 70dB DR and −74dBFS THD in 125MHz BW 470
Muhammed Bolatkale, Lucien J. Breems, Robert Rutten, Kofi A. A. Makinwa

An 8mW 50MS/s CT ΔΣ Modulator with 81dB SFDR and Digital Background DAC Linearization 472
John G. Kauffman, Pascal Witte, Joachim Becker, Maurits Ortmanns

A Third-Order DT ΔΣ Modulator Using Noise-Shaped Bidirectional Single-Slope Quantizer 474
Nima Maghari, Un-Ku Moon

A 250mV 7.5μW 61dB SNDR CMOS SC ΔΣ Modulator Using a Near-Threshold-Voltage-Biased CMOS Inverter Technique 476
Fridolin Michel, Michiel Steyaert

A 84dB SNDR 100kHz Bandwidth Low-Power Single Op-Amp Third-Order ΔΣ Modulator Consuming 140μW 478
Aldo Pena Perez, Edoardo Bonizzoni, Franco Maloberti

A 1.7mW 11b 1-1-1 MASH ΔΣ Time-to-Digital Converter 480
Ying Cao, Paul Leroux, Wouter De Cock, Michiel Steyaert

A 120dB-SNR 100dB-THD+N 21.5mW/Channel Multibit CT ΔΣ DAC 482
Abhishek Bandyopadhyay, Michael Determan, Sejun Kim, Khiem Nguyen

A 108dB-DR 120dB-THD and 0.5V$_{rms}$ Output Audio DAC with Inter-Symbol-Interference-Shaping Algorithm in 45nm CMOS 484
Lars Risbo, Rahmi Hezar, Burak Kelleci, Halil Kiper, Mounir Fares

DRAM & HIGH-SPEED I/O:

Session 28 Overview / Memory: DRAM & High-Speed I/O 486

An 8.4Gb/s 2.5pJ/b Mobile Memory I/O Interface Using Simultaneous Bidirectional Dual (Base+RF) Band Signaling 488
Gyung-Su Byun, Yanghyo Kim, Jongsun Kim, Sai-Wang Tam, H-H Hsieh, P-Y Wu, C Jou, Jason Cong, Glenn Reinman, Mau-Chung Frank Chang

A 2.7Gb/s/mm^2 0.9pJ/b/Chip 1Coil/Channel ThruChip Interface with Coupled-Resonator-Based CDR for NAND Flash Memory Stacking 490
Noriyuki Miura, Yasuhiro Take, Mitsuko Saito, Yoichi Yoshida, Tadahiro Kuroda

A 12Gb/s Non-Contact Interface with Coupled Transmission Lines 492
Tsutomu Takeya, Lan Nan, Shinya Nakano, Noriyuki Miura, Hiroki Ishikuro, Tadahiro Kuroda

A 4.8Gb/s Impedance-Matched Bidirectional Multi-Drop Transceiver for High-Capacity Memory Interface 494
Woo-Yeol Shin, Gi-Moon Hong, Hyongmin Lee, Jae-Duk Han, Sunkwon Kim, Kyu-Sang Park, Dong-Hyuk Lim, Jung-Hoon Chun, Deog-Kyoon Jeong, Suhwan Kim

A 1.2V 12.8GB/s 2Gb Mobile Wide-I/O DRAM with 4×128 I/Os Using TSV-Based Stacking 496
Jung-Sik Kim, Chi Sung Oh, Hocheol Lee, Donghyuk Lee, Hyong-Ryol Hwang, Sooman Hwang, Byongwook Na, Joungwook Moon, Jin-Guk Kim, Hanna Park, Jang-Woo Ryu, Kiwon Park, Sang-Kyu Kang, So-Young Kim, Hoyoung Kim, Jong-Min Bang, Hyunyoon Cho, Minsoo Jang, Cheolmin Han, Jung-Bae Lee, Kyehyun Kyung, Joo-Sun Choi, Young-Hyun Jun

A 40nm 2Gb 7Gb/s/pin GDDR5 SDRAM with a Programmable DQ Ordering Crosstalk Equalizer and Adjustable Clock-Tracking BW 498
Seung-Jun Bae, Young-Soo Sohn, Tae-Young Oh, Si-Hong Kim, Yun-Seok Yang, Dae-Hyun Kim, Sang-Hyup Kwak, Ho-Seok Seol, Chang-Ho Shin, Min-Sang Park, Gong-Heom Han, Byeong-Cheol Kim, Yong-Ki Cho, Hye-Ran Kim, Su-Yeon Doo, Young-Sik Kim, Dong-Seok Kang, Young-Ryeol Choi, Sam-Young Bang, Sun-Young Park, Yong-Jae Shin, Gil-Shin Moon, Cheol-Goo Park, Woo-Seop Kim, Hyang-Ja Yang, Jeong-Don Lim, Kwang-Il Park, Joo Sun Choi, Young-Hyun Jun

A 58nm 1.8V 1Gb PRAM with 6.4MB/s Program BW 500
  Hoeju Chung, Byung Hoon Jeong, ByungJun Min, Youngdon Choi, Beak-Hyung Cho, Junho Shin, Jinyoung Kim, Jung Sunwoo,Joon-min Park, Qi Wang, Yong-jun Lee, Sooho Cha, Dukmin Kwon, Sangtae Kim, Sunghoon Kim, Yoohwan Rho, Mu-Hui Park, Jaewhan Kim, Ickhyun Song, Sunghyun Jun, Jaewook Lee, KiSeung Kim, Ki-won Lim, Won-ryul Chung, ChangHan Choi, HoGeun Cho, Inchul Shin, Woochul Jun, Seokwon Hwang, Ki-Whan Song, KwangJin Lee, Sang-whan Chang, Woo-Yeong Cho, Jei-Hwan Yoo, Young-Hyun Jun

A 1.6V 1.4Gb/s/pin Consumer DRAM with Self-Dynamic Voltage-Scaling Technique in 44nm CMOS Technology 502
  Hyun-Woo Lee, Ki-Han Kim, Young-Kyoung Choi, Ju-Hwan Shon, Nak-Kyu Park, Kwan-Weon Kim, Chulwoo Kim, Young-Jung Choi, Byong-Tae Chung

An Embedded DRAM Technology for High-Performance NAND Flash Memories 504
  Daisaburo Takashima, Mitsuhiro Noguchi, Noboru Shibata, Kazushige Kanda, Hiroshi Sukegawa, Shuso Fujii

A 700MHz 2T1C Embedded DRAM Macro in a Generic Logic Process with No Boosted Supplies 506
  Ki Chul Chun, Wei Zhang, Pulkit Jain, Chris H. Kim

Tutorials 510

FORUMS:

F1: Advanced Transmitters for Wireless Infrastructure 512
F2: Ultra-Low Voltage VLSIs for Energy Efficient Systems 514
F3: Towards Personalized Medicine and Monitoring for Healthy Living 516
F4: Design of "Green" High-Performance Processor Circuits 518
F5: Image Sensors for 3D Capture 520
F6: High-Speed Transceivers: Standards, Challenges, and Future 522

SHORT COURSE:

Cellular and Wireless LAN Transceivers: From Systems to Circuit Design 524

EVENING SESSIONS:

EP1: Good, Bad, Ugly - 20 Years of Broadband Evolution: What's Next? 525
EP2: 20-22nm Technology Options and Design Implications 526
ES0: Student Research Preview 527
ES1: Data Converter Breakthroughs in Retrospect 528
ES2: Wireless Sensor Systems: Solution & Technology 529
ES3: Future System and Memory Architectures: Transformations by Technology and Applications 530
ES4: Body Area Network: Technology, Solutions, and Standardization 531
ES5: Gb/s+ Portable Wireless Communications 532
ES6: Technologies for Smart Grid and Smart Meter 533
Author Index 534

PROGRAM COMMITTEE

EXECUTIVE COMMITTEE

CONFERENCE CHAIR

Anantha Chandrakasan
*Massachusetts Institute of
Technology,
Cambridge, MA*

WEB MANAGER

Bill Bowhill
*Intel,
Hudson, MA*

EDUCATIONAL EVENTS CHAIR

Willy Sansen
*K.U. Leuven-ESAT MICAS,
Leuven, Belgium*

EXECUTIVE DIRECTOR

Dave Pricer
Charlotte, VT

**ITPC FAR EAST REGIONAL
CHAIR**

Hoi-Jun Yoo
*KAIST,
Daejeon, Korea*

**SSCS ADCOM
REPRESENTATIVE**

Bryan Ackland
*NoblePeak Vision,
Wakefield, MA*

**SECRETARY, FORUM CHAIR
AND DATA TEAM CHAIR**

Trudy Stetzler
*Texas Instruments,
Stafford, TX*

**ITPC FAR EAST REGIONAL VICE
CHAIR**

Makoto Ikeda
*University of Tokyo,
Tokyo, Japan*

**DIRECTOR OF PUBLICATIONS
AND PRESENTATIONS**

Laura Fujino
*University of Toronto,
Toronto, Canada*

DIRECTOR OF FINANCE

Bryant Griffin
Penfield, NY

**ITPC FAR EAST REGIONAL
SECRETARY**

Mototsugu Hamada
*Toshiba,
Kawasaki, Japan*

**DIRECTOR OF AUDIOVISUAL
SERVICES**

John Trnka
Rochester, MN

PROGRAM CHAIR

Wanda Gass
*Texas Instruments,
Dallas, TX*

ITPC EUROPEAN REGIONAL CHAIR

Bram Nauta
*University of Twente,
Enschede,
The Netherlands*

**PRESS AND PUBLICITY CHAIR
AND AWARDS CHAIR**

Kenneth C. Smith
*University of Toronto,
Toronto, Canada*

PROGRAM VICE CHAIR

Hideto Hidaka
*Renesas Electronics,
Itami, Japan*

**ITPC EUROPEAN REGIONAL
VICE CHAIR**

Aarno Pärssinen
*Nokia,
Helsinki, Finland*

**PRESS AND PUBLICITY
VICE-CHAIR**

James Goodman
*Kluless Technologies,
Ottawa, Canada*

STUDENT FORUM CHAIR

Jan van der Spiegel
*University of Pennsylvania,
Philadelphia, PA*

**ITPC EUROPEAN REGIONAL
SECRETARY**

Eugenio Cantatore
*Eindhoven University of Technology,
Eindhoven, The Netherlands*

DIRECTOR OF OPERATIONS

Melissa Widerkehr
*Widerkehr and Associates,
Montgomery Village, MD*

TECHNICAL EDITORS

Vincent Gaudet, *University of Waterloo, Waterloo, Canada*
Glenn Gulak (Editor-at-Large), *University of Toronto, Toronto, Canada*
James W. Haslett, *University of Calgary, Calgary, Canada*
Shahriar Mirabbasi, *University of British Columbia, Vancouver, Canada*
Kostas Pagiamtzis, *Gennum, Burlington, Canada*
Kenneth C. Smith (Editor-at-Large), *University of Toronto, Toronto, Canada*

PROGRAM COMMITTEE

INTERNATIONAL TECHNICAL PROGRAM COMMITTEE

PROGRAM CHAIR: Wanda Gass, *Texas Instruments, Dallas, TX*

PROGRAM VICE CHAIR: Hideto Hidaka, *Renesas Electronics, Itami, Japan*

ANALOG SUBCOMMITTEE

Chair: Bill Redman-White, *NXP Semiconductors, Southampton, United Kingdom*
Ivan Bietti, *STMicroelectronics, Grenoble, France*
Gyu-Hyoeong Cho, *KAIST, Daejeon, Korea*
Yoshihisa Fujimoto, *Sharp, Osaka, Japan*
Ian Galton, *University of California, San Diego, La Jolla, CA*
Baher Haroun, *Texas Instruments, Dallas, TX,*
Jed Hurwitz, *Gigle Networks, Edinburgh, Scotland*
Minkyu Je, *Institute of Microelectronics, A*STAR, Singapore, Singapore*
Wing Hung Ki, *HKUST, Hong Kong, China*
Peter Kinget, *Columbia University, New York, NY*
Kimmo Koli, *ST-Ericsson, Turku, Finland*
Tsung-Hsien Lin, *National Taiwan University, Taipei, Taiwan*
Chris Mangelsdorf, *Analog Devices K.K, Tokyo, Japan*
Bram Nauta, *University of Twente, Enschede, The Netherlands*
Francesco Rezzi, *Marvell Semiconductor, Pavia, Italy*
Jafar Savoj, *Xilinx, San Jose, CA*

DATA CONVERTERS SUBCOMMITTEE

Chair: Venu Gopinathan, *Texas Instruments, Bangalore, India*
Andrea Baschirotto, *University of Milan-Bicocca, Milano, Italy*
Lucien Breems, *NXP, Eindhoven, The Netherlands*
Aaron Buchwald, *Mobius Semiconductor, Irvine, CA*
Klaas Bult, *Broadcom, Bunnik, The Netherlands*
Marco Corsi, *Texas Instruments, Dallas, TX*
Dieter Draxelmayr, *Infineon Techologies, Villach, Austria*
Michael Flynn, *University of Michigan, Ann Arbor, MI*
Gabriele Manganaro, *Analog Devices, Wilmington, MA*
Yiannos Manoli, *University of Freiburg, Freiburg, Germany*
Tatsuji Matsuura, *Renesas Electronics, Gunma, Japan*
Un-Ku Moon, *Oregon State University, Corvallis, OR*
Boris Murmann, *Stanford University, Stanford, CA*
Katsu Nakamura, *Analog Devices, Wilmington, MA*
Shanthi Pavan, *Indian Institute Of Technology,Chennai, India*
Michael Perrot, *SiTime, Sunnyvale, CA*
Kong-pang Pun, *Chinese University of Hong Kong, Hong Kong, China*

ENERGY-EFFICIENT DIGITAL SUBCOMMITTEE

Chair: Tzi-Dar Chiueh, *National Taiwan University, Taipei, Taiwan*
Kazutami Arimoto, *Renesas Electronics, Hyogo, Japan*
Ming-Yang Chao, *MediaTek, Hsinchu, Taiwan*
Wim Dehaene, *KULeuven, Leuven, Belgium*
Vasantha Erraguntla, *Intel Technology, Bangalore, India*
Jos Huisken, *Holst Centre, Eindhoven, The Netherlands*
Stephen Kosonocky, *Advanced Micro Devices, Fort Collins, CO*
Byeong-Gyu Nam, *Samsung Electronics, Yongin, Korea*
Michael Phan, *Qualcomm, Raleigh, NC*
Raney Southerland, *ARM, Austin, TX*
Masaya Sumita, *Panasonic, Moriguchi, Japan*
Pascal Urard, *STMicroelectronics, Crolles, France*
Alice Wang, *Texas Instruments, Dallas, TX*

HIGH-PERFORMANCE DIGITAL SUBCOMMITTEE

Chair: Stefan Rusu, *Intel, Santa Clara, CA*
Lew Chua-Eoan, *Qualcomm, San Diego, CA*
Don Draper, *True Circuits, Los Altos, CA*
Tim Fischer, *AMD, Fort Collins, CO*
Joshua Friedrich, *IBM, Austin, TX*
Anthony Hill, *Texas Instruments, Dallas, TX*
Tanay Karnik, *Intel, Hillsboro, OR*
Sonia Leon, *Intel, Santa Clara, CA*
Takashi Miyamori, *Toshiba, Kawasaki, Japan*
Shannon Morton, *Icera, Bristol, United Kingdom*
Tobias Noll, *RWTH Aachen University, Aachen, Germany*
Vladimir Stojanovic, *MIT, Cambridge, MA*
Se Hyun Yang, *Samsung, Yongin, Korea*

IMAGERS, MEMS, MEDICAL AND DISPLAYS SUBCOMMITTEE

Chair: Roland Thewes, *TU Berlin, Berlin, Germany*
JungChak Ahn, *Samsung Electronics, Yongin, Korea*
Jan Bosiers, *DALSA Professional Imaging, Eindhoven, The Netherlands*
Iliana Chen, *Analog Devices, Somerset, NJ*
Timothy Denison, *Medtronic, Minneapolis, MN*
Maysam Ghovanloo, *Georgia Institure of Technology, Atlanta, GA*
Christoph Hagleitner, *IBM Research, Ruschlikon, Switzerland*
Makoto Ikeda, *University of Tokyo, Tokyo, Japan*
Sam Kavusi, *Bosch, Palo Alto, CA*
Shoji Kawahito, *Shizuoka University, Hamamatsu, Japan*
Kofi Makinwa, *Technical University of Delft, Delft, The Netherlands*
Young-Sun Na, *LG Electronics, Seoul, Korea*
Tetsuo Nomoto, *Sony, Kanagawa, Japan*
Jun Ohta, *Nara Institute of Science & Technology, Nara, Japan*
Aaron Partridge, *SiTime, Sunnyvale, CA*
Albrecht Rothermel, *University of Ulm, Ulm, Germany*
Johannes Solhusvik, *Aptina, Oslo, Norway*

MEMORY SUBCOMMITTEE

Chair: Kevin Zhang, *Intel, Hillsboro, OR*
Leland Chang, *IBM T. J. Watson Research Center, Yorktown Heights, New York*
Joo Sun Choi, *Samsung, Hwasung, Korea*
Sungdae Choi, *Hynix Semiconductor, Icheon, Korea*
Roberto Gastaldi, *Numonyx, Agrate Brianza, Italy*
Satoru Hanzawa, *Hitachi Central Research Laboratory, Tokyo, Japan*
Heinz Hoenigschmid, *Elpida Memory, Munich, Germany*
Nicky C.C. Lu, *Etron Technology, Hsinchu, Taiwan*
Cormac O'Connell, *TSMC, Ottawa, Canada*
Harold Pilo, *IBM, Essex Junction, VT*
Peter Rickert, *Texas Instruments, Richardson, TX*
Frankie Roohparvar, *Micron Technology, San Jose, CA*
Yasuhiro Takai, *Elpida Memory, Sagamihara, Japan*
Daisaburo Takashima, *Toshiba, Yokohama, Japan*
Ken Takeuchi, *University of Tokyo, Tokyo, Japan*
Tadaaki Yamauchi, *Renesas Electronics, Itami, Japan*

RF SUBCOMMITTEE

Chair: Nikolaus Klemmer, *Texas Instruments, Dallas, TX*
Pietro Andreani, *Lund University, Lund, Sweden*
Andreia Cathelin, *STMicroelectronics, Crolles Cedex, France*
Jan Craninckx, *imec, Leuven, Belgium*
Hooman Darabi, *Broadcom, Irvine, CA*
Brian Floyd, *North Carolina State University, Raleigh, NC*
Joseph Golat, *Motorola, Algonquin, IL*
Kari Halonen, *Aalto University, Espoo, Finland*
Songcheol Hong, *KAIST, Daejeon, Korea*
Mike Keaveney, *Analog Devices, Limerick, Ireland*
Shoji Otaka, *Toshiba, Kawasaki, Japan*
Harald Pretl, *Infineon / DICE, Linz, Austria*
Chris Rudell, *University of Washington, Seattle, WA*
Bogdan Staszewski, *Technical University of Delft, Delft, The Netherlands*
Francesco Svelto, *Università degli Studi di Pavia, Pavia, Italy*
Masoud Zargari, *Irvine, CA*
Jing-Hong Conan Zhan, *MediaTek, HsinChu, Taiwan*
Michael Zybura, *RF Micro Devices, Scotts Valley, CA*

TECHNICAL DIRECTIONS SUBCOMMITTEE

Chair: Siva Narendra, *Tyfone, Portland, OR*
Pascal Ancey, *STMicroelectronics, Crolles, France*
Ahmad Bahai, *National Semiconductor, Santa Clara, CA*
Azeez Bhavnagarwala, *Advanced Micro Devices, Boxborough, MA*
Shekhar Borkar, *Intel, Hillsboro, OR*
Alison Burdett, *Toumaz Technology, Abingdon, United Kingdom*
Eugenio Cantatore, *Eindhoven University of Technology, Eindhoven, The Netherlands*
Eric Colinet, *CEA-LETI, Grenoble, France*
Christian Enz, *CSEM, Neuchatel, Switzerland*
Donhee Ham, *Harvard University, Cambridge, MA*
Uming Ko, *Texas Instruments, Houston, TX*
Tadahiro Kuroda, *Keio University, Yokohama, Japan*
Masaitsu Nakajima, *Panasonic, Osaka, Japan*
David Scott, *TSMC, Plano, TX*
Satoshi Shigematsu, *NTT Electronics, Yokohama, Japan*
Chris Van Hoof, *imec, Leuven, Belgium*
Hoi-Jun Yoo, *KAIST, Daejeon, Korea*

WIRELESS SUBCOMMITTEE

Chair: David Su, *Atheros Communications, San Jose, CA*
Arya Behzad, *Broadcom, San Diego, CA*
Didier Belot, *STMicroelectronics, Crolles, France*
Gangadhar Burra, *Texas Instruments, Dallas, TX*
George Chien, *MediaTek, San Jose, CA*
Jan Crols, *AnSem, Heverlee, Belgium*
Ranjit Gharpurey, *University of Texas at Austin, Austin, TX*
Mototsugu Hamada, *Toshiba, Kawasaki, Japan*
Stefan Heinen, *RWTH Aachen University, Aachen, Germany*
Myung-Woon Hwang, *FCI, Sungnam, Kyunggi, Korea*
Domine Leenaerts, *NXP Semiconductors, Eindhoven, The Netherlands*
Sven Mattisson, *Ericsson, Lund, Sweden*
Ali Niknejad, *University of California at Berkeley, Berkeley, CA*
Yorgos Palaskas, *Intel, Hillsboro, OR*
Aarno Parssinen, *Nokia, Helsinki, Finland*
Zhihua Wang, *Tsinghua University, Beijing, China*
Taizo Yamawaki, *Renesas Electronics, Gunma, Japan*

WIRELINE SUBCOMMITTEE

Chair: Franz Dielacher, *Infineon Technologies, Villach, Austria*
Ken Chang, *Xilinx, San Jose, CA*
SeongHwan Cho, *KAIST, Daejon, Korea*
Daniel Friedman, *IBM Thomas J. Watson Research Center, Yorktown Heights, NY*
Jack Kenney, *Analog Devices, Somerset, NY*
Jerry (Heng-Chih) Lin, *Ralink Technology, Hsinchu, Taiwan*
Miki Moyal, *Intel, Haifa, Israel*
Masafumi Nogawa, *NTT Microsystem Integration Laboratories, Atsugi, Japan*
Bob Payne, *Texas Instruments, Dallas, TX*
Tatsuya Saito, *Hitachi, Tokyo, Japan*
Naresh Shanbhag, *University of Illinois at Urbana-Champaign, Urbana, IL*
Ali Sheikholeslami, *University of Toronto, Toronto, Canada*
Jae-Yoon Sim, *Pohang University of Science and Technology, Pohang, Korea*
John T. Stonick, *Synopsys, Hillsboro, OR*
Koichi Yamaguchi, *NEC, Sagamihara, Japan*
Takuji Yamamoto, *Fujitsu Laboratories, Kawasaki, Japan*

PROGRAM COMMITTEE

EUROPEAN REGIONAL COMMITTEE

ITPC EUROPEAN REGIONAL CHAIR
Bram Nauta, *University of Twente, Enschede, The Netherlands*

ITPC EUROPEAN REGIONAL VICE CHAIR
Aarno Pärssinen, *Nokia, Helsinki, Finland*

ITPC EUROPEAN REGIONAL SECRETARY
Eugenio Cantatore, *Eindhoven University of Technology, Eindhoven, The Netherlands*

Members: Pascal Ancey, *STMicroelectronics, Crolles, France*
Pietro Andreani, *Lund University, Lund, Sweden*
Andrea Baschirotto, *University of Milan-Bicocca, Milano, Italy*
Didier Belot, *STMicroelectronics, Crolles, France*
Ivan Bietti, *STMicroelectronics, Grenoble, France*
Jan Bosiers, *DALSA Professional Imaging, Eindhoven, The Netherlands*
Lucien Breems, *NXP, Eindhoven, The Netherlands*
Klaas Bult, *Broadcom, Bunnik, The Netherlands*
Alison Burdett, *Toumaz Technology, Abingdon, United Kingdom*
Andreia Cathelin, *STMicroelectronics, Crolles Cedex, France*
Eric Colinet, *CEA-LETI, Grenoble, France*
Jan Craninckx, *imec, Leuven, Belgium*
Jan Crols, *AnSem, Heverlee, Belgium*
Wim Dehaene, *KULeuven, Leuven, Belgium*
Franz Dielacher, *Infineon Technologies, Villach, Austria*
Dieter Draxelmayr, *Infineon Techologies, Villach, Austria*
Christian Enz, *CSEM, Neuchatel, Switzerland*
Roberto Gastaldi, *Numonyx, Agrate Brianza, Italy*
Christoph Hagleitner, *IBM Research, Ruschlikon, Switzerland*
Kari Halonen, *Aalto University, Espoo, Finland*
Stefan Heinen, *RWTH Aachen University, Aachen, Germany*
Heinz Hoenigschmid, *Elpida Memory, Munich, Germany*
Jos Huisken, *imec, Eindhoven, The Netherlands*
Jed Hurwitz, *Gigle Networks, Edinburgh, Scotland*
Mike Keaveney, *Analog Devices, Limerick, Ireland*
Kimmo Koli, *ST-Ericsson, Turku, Finland*
Domine Leenaerts, *NXP Semiconductors, Eindhoven, The Netherlands*
Kofi Makinwa, *Technical University of Delft, Delft, The Netherlands*
Yiannos Manoli, *University of Freiburg, Freiburg, Germany*
Sven Mattisson, *Ericsson, Lund, Sweden*
Shannon Morton, *Icera, Bristol, United Kingdom*
Miki Moyal, *Intel, Haifa, Israel*
Tobias Noll, *RWTH Aachen University, Aachen, Germany*
Harald Pretl, *Infineon / DICE, Linz, Austria*
Bill Redman-White, *NXP Semiconductors, Southampton, United Kingdom*
Francesco Rezzi, *Marvell Semiconductor, Pavia, Italy*
Albrecht Rothermel, *University of Ulm, Ulm, Germany*
Johannes Solhusvik, *Aptina, Oslo, Norway*
Bogdan Staszewski, *Technical University of Delft, Delft, The Netherlands*
Francesco Svelto, *Università degli Studi di Pavia, Pavia, Italy*
Roland Thewes, *TU Berlin, Berlin, Germany*
Pascal Urard, *STMicroelectronics, Crolles Cedex, France*
Chris Van Hoof, *imec, Leuven, Belgium*

FAR EAST REGIONAL COMMITTEE

ITPC FAR EAST REGIONAL CHAIR
Hoi-Jun Yoo, *KAIST, Daejeon, Korea*

ITPC FAR EAST REGIONAL VICE CHAIR
Makoto Ikeda, *University of Tokyo, Tokyo, Japan*

ITPC FAR EAST REGIONAL SECRETARY
Mototsugu Hamada, *Toshiba, Kawasaki, Japan*

Members: JungChak Ahn, *Samsung Electronics, Yongin, Korea*
Kazutami Arimoto, *Renesas Electronics, Hyogo, Japan*
Ming-Yang Chao, *MediaTek, Hsinchu, Taiwan*
Tzi-Dar Chiueh, *National Taiwan University, Taipei, Taiwan*
Gyu-Hyoeong Cho, *KAIST, Daejeon, Korea*
SeongHwan Cho, *KAIST, Daejon, Korea*
Joo Sun Choi, *Samsung, Hwasung, Korea*
Sungdae Choi, *Hynix Semiconductor, Icheon, Korea*
Vasantha Erraguntla, *Intel Technology, Bangalore, India*
Yoshihisa Fujimoto, *Sharp, Osaka, Japan*
Venu Gopinathan, *Texas Instruments, Bangalore, India*
Satoru Hanzawa, *Hitachi Central Research Laboratory, Tokyo, Japan*
Songcheol Hong, *KAIST, Daejeon, Korea*
Myung-Woon Hwang, *FCI, Sungnam, Kyunggi, Korea*
Minkyu Je, *Institute of Microelectronics, A*STAR, Singapore, Singapore*
Shoji Kawahito, *Shizuoka University, Hamamatsu, Japan*
Wing Hung Ki, *HKUST, Hong Kong, China*
Tadahiro Kuroda, *Keio University, Yokohama, Kanagawa, Japan*
Jerry (Heng-Chih) Lin, *Ralink Technology, Hsinchu, Taiwan*
Tsung-Hsien Lin , *National Taiwan University, Taipei, Taiwan*
Nicky C.C. Lu, *Etron Technology, Hsinchu, Taiwan*
Chris Mangelsdorf, *Analog Devices K.K, Tokyo, Japan*
Tatsuji Matsuura, *Renesas Electronics, Gunma, Japan*
Takashi Miyamori, *Toshiba Center for Semiconductor Research and Development, Kawasaki, Japan*
Young-Sun Na, *LG Electronics, Seoul, Korea*
Masaitsu Nakajima, *Panasonic, Osaka, Japan*
Byeong-Gyu Nam, *Samsung Electronics, Yongin, Korea*
Masafumi Nogawa, *NTT Microsystem Integration Laboratories, Atsugi, Japan*
Tetsuo Nomoto, *Sony, Kanagawa, Japan*
Jun Ohta, *Nara Institute of Science & Technology, Nara, Japan*
Shoji Otaka, *Toshiba, Kawasaki, Japan*
Shanthi Pavan, *Indian Institute Of Technology, Chennai, India*
Kong-pang Pun, *Chinese University of Hong Kong, Hong Kong, China*
Tatsuya Saito, *Hitachi, Tokyo, Japan*
David Scott, *TSMC, Plano, TX*
Satoshi Shigematsu, *NTT Electronics, Yokohama, Japan*
Jae-Yoon Sim , *Pohang University of Science and Technology, Pohang, Korea*
Masaya Sumita, *Panasonic, Moriguchi, Japan*
Yasuhiro Takai, *Elpida Memory, Sagamihara, Japan*
Daisaburo Takashima, *Toshiba, Yokohama, Japan*
Ken Takeuchi, *University of Tokyo, Tokyo, Japan*
Zhihua Wang, *Tsinghua University, Beijing, China*
Koichi Yamaguchi, *NEC, Sagamihara, Japan*
Takuji Yamamoto, *Fujitsu Laboratories, Kawasaki, Japan*
Tadaaki Yamauchi, *Renesas Electronics, Itami, Japan*
Taizo Yamawaki, *Renesas Electronics, Gunma, Japan*
Se Hyun Yang, *Samsung, Yongin, Korea*
Jing-Hong Conan Zhan, *MediaTek, HsinChu, Taiwan*

ISSCC 2011 TIMETABLE

Sunday, February 20th	ISSCC 2011 TUTORIALS		
8:00AM	**T1**: Integrated LC Oscillators	**T2**: Embedded Memories for SoC: Overview of Design, Test and Applications and Challenges in the Nano-Scale CMOS	
10:00AM	**T3**: Ultra Low-Power and Low-Voltage Digital-Circuit Design Techniques	**T4**: Layout – The Other Half of Nanometer Analog Design	**T5**: DPLL-Based Clock and Data Recovery
12:30PM	**T6**: Practical Power-Delay Design Trade-offs	**T7**: Distortion in Cellular Receivers	
2:30PM	**T8**: Noise Analysis in Switched-Capacitor Circuits	**T9**: Interfacing Silicon with the Human Body: A Primer on Applications, Interface Circuits and Technologies for the Medical Market	

	ISSCC 2011 FORUMS	
8:00AM	**F1**: Advanced Transmitters for Wireless Infrastructure	**F2**: Ultra-Low Voltage VLSIs for Energy-Efficient Systems

	ISSCC 2011 EVENING SESSIONS		
	7:30 PM ES0: Student Research Review	**8:00 PM ES1:** Data-Converter Breakthroughs in Retrospect	**8:00 PM ES2:** Wireless Sensor Systems: Solutions & Technology

Monday, February 21st	ISSCC 2011 PAPER SESSIONS				
8:15AM	**Session 1:** Plenary Session				
1:30PM	**Session 2:** Technologies For Health	**Session 3:** RF Techniques	**Session 4:** Enterprise Processors & Components	**Session 5:** PLL	**Session 6:** Sensors & Energy Harvesting
5:15PM	Social Hour: Poster Session - DAC/ISSCC Student-Design-Contest Winners; Author Interviews, Womens Reception				

	ISSCC 2011 SESSIONS		
8:00PM	**ES3**: Future System and Memory Architectures: Transformations by Technology and Applications	**ES4**: Body Area Networks (BAN): Technology, Solutions, and Standardization	**EP1**: Good, Bad, Ugly - 20 Years of Broadband Evolution: What's Next?

Tuesday, February 22nd	ISSCC 2011 PAPER SESSIONS				
8:30AM	**Session 7:** Multimedia & Mobile	**Session 8:** Architectures & Circuits for Next Gen Wireline Xceivers	**Session 9:** Wireless & mm-Wave Connectivity	**Session 10:** Nyquist Rate Converters	**Session 11:** Non-Volatile Memory Solutions
1:30PM	**Session 12:** Design in Emerging Technologies	**Session 13:** Analog Techniques	**Session 14:** High-Performance Embedded Memory / **Session 15:** High-Performance SoCs & Components	**Session 16:** mm-Wave Design Techniques	**Session 17:** Biomedical & Displays
5:15PM	Industrial Demo Session (4-7), Author Interviews, Social Hour				

	ISSCC 2011 EVENING SESSIONS		
8:00PM	**ES5**: Gb/s+ Portable Wireless Communications	**ES6**: Technologies for Smart Grid and Smart Meter	**EP2**: 20-22nm Technology Options and Design Implications

Wednesday, February 23rd	ISSCC 2011 PAPER SESSIONS				
8:30AM	**Session 18:** Organic Innovations / **Session 19:** Low-Power Digital Techniques	**Session 20:** High-Speed Transceivers & Building Blocks	**Session 21:** Cellular	**Session 22:** DC/DC Converters	**Session 23:** Image Sensors
1:30PM	**Session 24:** Transmitter Blocks	**Session 25:** CDRs & Equalization Techniques	**Session 26:** Low-Power Wireless	**Session 27:** Oversampling Converters	**Session 28:** DRAM & High-Speed I/O
5:15 PM	Author Interviews				

Thursday, February 24th	ISSCC 2011 SHORT COURSE		
8:00 AM	Cellular and Wireless LAN Transceivers: From Systems to Circuit Design		

	ISSCC 2011 FORUMS		
8:00AM	**F3**: Towards Personalized Medicine and Monitoring for Healthy Living	**F4**: Design of "Green" High-Performance Processor Circuits	**F5**: Image Sensors for 3D Capture	**F6**: High-Speed Transceivers: Standards, Challenges, and Future

ISSCC 2011 / SESSION 1 / PLENARY / AWARDS

February 21, 2011 / 10:30 AM

ISSCC AWARDS
2011 ISSCC Distinguished Technical Paper

*"Filterless Integrated Class-D Audio Amplifier Achieving 0.0012%
THD+N and 96dB PSRR When Supplying 1.2W"*

Mykhaylo Teplechuk, Anthony Gribben, Christophe Amadi

Dialog Semiconductor, Edinburgh, Scotland

2010 Lewis Winner Award for Outstanding Paper

*"A 45nm SOI Embedded DRAM Macro for POWER7 32MB
On-Chip L3 Cache"*

John Barth[1] Don Plass[2], Erik Nelson[1], Charlie Hwang[2],
Gregory Fredeman[2], Michael Sperling[2], Abraham Mathews[2],
William Reohr[4], Kavita Nair[2], Nianzheng Cao[2]

[1]IBM, Essex Junction, VT; [2]IBM, Poughkeepsie, NY; [3]IBM, Austin TX
[4]IBM T.J. Watson, Yorktown Heights, NY

2010 Jack Kilby Award for Outstanding Student Paper

*"A 90GHz-Carrier 30GHz-Bandwidth Hybrid Switching Transmitter
with Integrated Antenna"*

Amin Arbabian[1], Bagher Afshar[1], Jun-Chau Chien[1], Shinwon Kang[1],
Steven Callendar[1], Ehsan Adabi[1], Stephano Dal Toso[2], Romain Pilard[3],
Daniel Gloria[3], Ali Niknejad[1]

[1]University of California, Berkeley CA; [2]University of Padova, Padova, Italy
[3]ST Microelectronics, Crolles, France

2010 Jack Raper Award for
Outstanding Technology-Directions Paper

*"Demonstration of Integrated Micro-Electro-Mechanical Switch
Circuits for VLSI Applications"*

Fred Chen[1], Matthew Spencer[2], Rhesa Nathanael[2], Chengcheng Wang[3],
Hossein Fariborzi[1], Abhinav Gupta[2], Hei Kam[2], Vincent Pott[2], Jaeseok Jeon[2],
Tsu-Jae King Liu[2], Dejan Markovic[3], Vladimir Stojanovic[1], Elad Alon[2]

[1]Massachusetts Institute of Technology, Cambridge, MA
[2]University of California, Berkeley, CA
[3]University of California, Los Angeles, CA

2010 Jan Van Vessem Award for
Outstanding European Paper

"Design Issues and Considerations for Low-Cost 3D TSV IC Technology"

Geert Van der Plas[1], Paresh Limaye[1], Abdelkarim Mercha[1], Herman Oprins[1],
Cristina Torregiani[1], Steven Thijs[1], Dimitri Linten[1], Michele Stucchi[1],
Katti Guruprasad[1], Dimitrios Velenis[1], Domae Shinichi[1], Vladimir Cherman[1],
Bart Vandevelde[1], Veerle Simons[1], Ingrid De Wolf[1], Riet Labie[1], Dan Perry[2],
Stephane Bronckers[1], Nikolas Minas[1], Miro Cupac[1], Wouter Ruythooren[1],
Jan Van Olmen[1], Alain Phommahaxay[1], Muriel de Potter de ten Broeck[1],
Ann Opdebeeck[1], Michal Rakowski[1], Bart De Wachter[1], Morin Dehan[1],
Marc Nelis[1], Rahul Agarwal[1], Wim Dehaene[4], Youssef Travaly[1], Pol Marchal[1],
Eric Beyne[1]

[1]IMEC, Leuven, Belgium; [2]Panasonic, Leuven, Belgium,
[3]Qualcomm, Leuven, Belgium; [4]K.U. Leuven, Leuven, Belgium

2010 ISSCC Award for Outstanding Forum Presenter

"Substrate Noise: A Few Observations"

Marcel J.M. Pelgrom, with Jacob Bakker, Sergei Kapora

NXP Semiconductors, Eindhoven, The Netherlands

2011 ISSCC Silkroad Award

*"A 0.024mm² 8-bit 400MS/s SAR ADC with 2-bit per Cycle and
Resistive DAC in 65nm CMOS"*

Hegong Wei

University of Macau, Macau, China

2011 DAC/ISSCC Student-Design-Contest Winners

*"A 90nm CMOS Data Flow Processor using Fine Grained DVS for
Energy Efficient Operation from 0.3V to 1.2V"*

S. Arrabi, Y. Shakhsheer, K. Craig, S. Khanna, J. Lach, B. H. Calhoun

University of Virginia, Charlottesville, Virginia

*"A 1900MHz-Band GSM-Based Clock-Harvesting Receiver
with -87dBm Sensitivity"*

Jonathan K. Brown, David D. Wentzloff
University of Michigan, Ann Arbor, MI

*"SRAM Dynamic Stability Characterization Using Pulsed
Word-lines in 45nm CMOS"*

Seng Oon Toh, Borivoje Nikoli

University of California, Berkeley, CA

*"Design and Implementation of Centip3De, a 7-layer
Many-Core System"*

David Fick, Ronald G. Dreslinski, Bharan Giridhar, Gyouho Kim,
Sangwon Seo, Matthew Fojtik, Sudhir Satpathy, Yoonmyung Lee,
Daeyeon Kim, Nurrachman Liu, Michael Wiekowski, Gregory Chen,
Trevor Mudge, Dennis Sylvester, David Blaauw

University of Michigan, Ann Arbor, MI

"A Flexible Wireless Receiver system with a 7b 21 MS/s SAR DAC"

David T. Lin, Li Li, John Bell, Ming-Hao Wang, Michael P. Flynn,

University of Michigan, Ann Arbor, MI

*"Augmented Reality Headset based on a Heterogeneous Multi-core
Object Recognition Chip"*

Seungjin Lee, Jinwook Oh, Junyoung Park, Joonsoo Kwon, Hoi-Jun Yoo
KAIST Daejeon, Korea

*"A 0.9-V 11-bit 25-MS/s 0.58-mW Binary-Search SAR ADC
in 90-nm CMOS"*

Ying-Zu Lin, Ya-Ting Shyu, Guan-Ying Huang,
Chun-Cheng Liu Soon-Jyh Chang

National Cheng Kung University, Tainan, Taiwan

*"A Video Stabilization System with Background Motion Estimation
and Smoothing for Digital Camera"*

Chih-Lun Fang Hui-Min Chuang Tsung-Han Tsai

National Central University, Jhongli City, Taiwan

ISSCC 2011 / SESSION 1 / PLENARY / AWARDS

IEEE SOLID-STATE CIRCUITS SOCIETY AWARDS

2011 IEEE Leon K. Kirchmayer Graduate Teaching Award

*"For inspirational teaching and student mentoring in the field
of advanced mircroelectronic devices and circuits"*

John D. Cressler

Georgia Institute of Technology

2011 IEEE Donald O. Pederson Award in Solid-State Circuits

"For leadership in analog integrated design"

Willy Sansen

Katholieke Universiteit Leuven, Leuven, Belgium

2011 IEEE Gustav Robert Kirchhoff Award

*"For crucial conceptual research contributions to the behavior
and the use of electrical circuits an systems"*

Charles A. Desoer (posthumous)

University of California, Berkeley

2009 Journal of Solid-State Circuits Best Paper Award

*"A 10-Gb/s Compact Low-Power Serial I/O With DFE-IIR Equalization
in 65-nm CMOS"*

Byungsub Kim[1], Yong Liu[2], Timothy O. Dickson[2], John F. Bulzacchelli[2], Daniel J. Friedman[2]

[1]Massachusetts Institute of Technology, Cambridge, MA
[2]IBM T. J. Watson Research Center, Yorktown Heights, NY

2011 IEEE Fellows

Paul Davis
Reading, PA

"For development of bipolar integrated circuits"

Changhyun Kim
Samsung Electronics, Hwasung, Korea

*"For contributions to low voltage, high performance, high density
memory design"*

Peter Rene Kinget
Columbia University, New York, NY

"For contributions to analog and radio frequency integrated circuits"

Uming Ko
Texas Instruments, Dallas, TX

"For leadership in ultra-low power circuit techniques"

2011 IEEE Fellows *(continued)*

Ram Krishnamurthy
Intel, Hillsboro, OR

*"For contributions to high performance and low power digital circuits
for microprocessors"*

Kofi Makinwa
Delft University of Technology, Delft,The Netherlands

"For the development of precision analog circuits"

Yoshinobu Nakagome
Renesas Technology,Takasaki, Japan

*"For pioneering development of low-voltage dynamic random access
memory circuits and low-leakage complementary metla-oxide
mechanical system circuits"*

Ken O
University of Texas, Dallas, Richardson, TX

*"For contributions to ultra-high frequency complementary
metal-oxide semiconductor circuits"*

Dennis Sylvester
University of Michigan, Ann Arbor, MI

"For contributions to energy-efficient for integrated circuits"

Kunio Uchiyama
Hitachi, Tokyo, Japan

"For contributions to power-efficient microprocessors"

Chih-Kong Yang
University of California, Los Angeles, CA

*"For leadership in enhancement of input-output efficiency
in integrated circuits"*

Kevin Zhang
Intel, Hillsboro, OR

*"For contributions to static random access memory for
high-performance microprocessors"*

2011 IEEE INTERNATIONAL SOLID-STATE CIRCUITS CONFERENCE

DIGEST OF TECHNICAL PAPERS

First Edition

February 2011

IEEE Catalog Number CFP11ISS-PRT

978-1-61284-303-2/11 $26.00 © 2011 IEEE

Foreword

Electronics for Healthy Living

It is my pleasure to welcome you to the 58th International Solid-State Circuits Conference. The Conference continues its outstanding tradition of presenting the most-advanced and innovative work, both from industry and academe, worldwide, in the area of integrated circuits and systems. This year, the geographical distribution of the accepted technical papers illustrates the truly international character of the Conference: 38% of the accepted papers are from North America, 33% from the Far East, and 29% from Europe. Of all of these, 47% are from academe, 42% are from industry, and 11% from institutions/labs.

The Conference Theme for 2011 is "Electronics for Healthy Living." Electronics play a significant role in enabling a healthier lifestyle: Technology in the hospital enables doctors to diagnose and treat illnesses that might have gone undetected just a few years ago; External monitors provide us with a good assessment of our health risk and vital-sign status; Those with chronic diseases can live a more normal life with implanted devices that sense, process, actuate, and communicate; Body Area Networks can be connected to a monitoring program running on our mobile phone; Those with disabilities also benefit from electronics that improve their lifestyle. The three Plenary speakers will share their vision of achieving healthier lives. In support of the conference theme, innovative solutions to get another 10 times reduction in power will be needed for implantable devices, wearable devices, and data centers which support cloud computing. New this year to the Conference, there will be a Plenary Roundtable to discuss next-generation challenges and solutions. The Roundtable panelists will share their ideas for delivering the next breakthroughs required for deployment of power-efficient products in the future.

The ISSCC Technical Program consists of 211 outstanding papers, distributed over 27 thematic sessions. Besides the regular paper presentations, the Conference continues to offer a wide variety of high-quality educational events, adding to the already significant value of attending ISSCC. This year, there are 9 tutorials covering a wide range of topics: integrated LC oscillators, embedded memories, ultra-low-power digital circuit design, nanometer analog layout, digital PLL design, digital power-delay tradeoffs, distortion in cellular receivers, noise analysis of switch-capacitor circuits, and interfacing silicon with the human body. These tutorials are given by top experts in the field and offer an opportunity for participants to experience an introduction and overview of important developments in each of these topics. The Conference also offers 6 Forums whose intent is to cover more-advanced topics for experts in the field. This year, Forums deal with transmitters for wireless infrastructure, ultra-low-voltage circuit design, monitoring for healthy living, sensors for 3D-image capture, and high-speed transceivers. As well, a full-day Short Course is available. It deals with the popular topic of wireless transceivers. In addition, there are two types of evening sessions available to all attendees: Panel discussions and Special-Topic Sessions, which provide participants with the opportunity to learn about a timely topic in a more-relaxed setting. One of the two panels will look at 20 years of broadband evolution, and another will discuss designs at the 20/22nm node. Each panel will provide a stimulating evening of discussion and interaction with some of the most knowledgeable visionaries in their field. Six Special-Topic Sessions (spread over three evenings) will offer wide opportunities for attendees to be informed about topics of increasing importance in the field of solid-state circuits. These topics have been carefully selected by the Program Committee for their timeliness and relevance: data-converter breakthroughs, wireless sensor systems, future memory architectures, Body-Area-Network standards, high-speed portable wireless circuits, and smart-grid technologies. New this year, we are introducing an Industrial Demo Session (IDS) during the Social Hour on Tuesday evening.

Preparing the way for our future presenters and participants, two student initiatives have been organized: the Student Research Preview and the ISSCC/DAC Student Design Contest. Both events allow graduate students from around the world, who are beginning to work in the field, to interact with each other and experts in the field. The students are given the opportunity to present their work and benefit from immediate feedback.

The quality and high standards that we associate with ISSCC is by-and-large due to the diligent work of the International Technical-Program Committee (ITPC). This year, ITPC has 160 members distributed over ten technical subcommittees. The members represent academia and industry on 4 continents. Each member has made a tremendous contribution to the ITPC by reading and reviewing large numbers of submitted papers, planning and organizing evening sessions and educational events, preparing materials (Advance Program, Press-Kit, and Digest), in a timely manner, as well as performing session chair/organizer duties. I would, in particular, like to acknowledge the leadership and guidance of the Technical-Subcommittee Chairs: Bill Redman-White (Analog), Venu Gopinathan (Data Converters), Tzi-Dar Chiueh (Energy-Efficient Digital), Stefan Rusu (High-Performance Digital), Roland Thewes (Imagers, MEMS, Medical, and Displays), Kevin Zhang (Memory), Nikolaus Klemmer (RF), Siva Narendra (Technology Directions), David Su (Wireless), and Franz Dielacher (Wireline). Also, I would like to thank the members of the Regional Committees: Bram Nauta, Aarno Parssinen, and Eugenio Cantatore from the European Region, and Hoi-Jun Yoo, Makoto Ikeda, and Mototsugu Hamada from the Far-East Region. Their help and support was essential for the smooth preparation of the Conference.

Many other individuals have played an essential role in making the Conference possible. Specifically, I would like to thank Albert Theuwissen, Program Chair of ISSCC 2011 and Hideto Hidaka, Program Vice-Chair of ISSCC 2011 for their help; Melissa Widerkehr and Diane Melton of Widerkehr Associates for their invaluable help with Conference operations and arrangements. Thanks are also due to representatives of MIRA Digital Publishing and S3 Publishing for their assistance with the electronic manuscript submission, formatting and prevoting, in putting the Advance Program together, as well as the Digest. I must also thank the Technical Editors: Vincent Gaudet, Glenn Gulak, James Haslett, Shahriar Mirabbasi, and Kostas Pagiamtzis. Also special thanks to Laura Fujino and Kenneth Smith for their unrelenting help with many aspects of the Conference including the paper submission process, the Advance Program, the Press Kit, the Digest and associated editing, and all CDs and DVDs, and to Jim Goodman for his persistent pursuit of the Press-preparation process. Also thanks to Willy Sansen for his coordination of the Tutorials; to Trudy Stetzler for her guidance in putting the Forums together; to John Long for organizing the Short Course; to Jan Van der Spiegel and Kerry Bernstein for their organization of the student activities; to Hoi-Jun in his long hours organizing the first year of the Industrial Demo Session; to Bill Bowhill for coordinating the launch of the new Conference website; to John Trnka for his excellent work with planning and coordinating the audio-visual services for the Conference; and to David Pricer for dealing efficiently with Corporate Relations, and providing wise advice on Conference operations.

We are indebted as well to an unusual group of volunteer-graduate students from the University of Toronto, who through their individual technical expertise ensure the orderly conduct of the presentations of each session, as well as countless other behind the scenes activities.

Finally, an individual who deserves special recognition is Anantha Chandrakasan, Conference Chair, for the enthusiastic support he has given me and his visionary leadership that will maintain ISSCC as the premier Conference in solid-state circuits.

Enjoy ISSCC 2011!

Wanda Gass

Wanda Gass, ISSCC 2011 International Technical-Program Chair.

Reflections

What you see before you this year, is the result of many years of continuous iterative refinement of submission process and information processing. This year, however, as the economic situation continues, we continue to provide a reduced-featured Digest, in which the continuation pages (typically including a micrograph and occasionally summary data) have been eliminted from only the print version, but are available in the Digest CD, in the Conference DVD, and in IEEE xPLORE. As well, some other items, such as, the Glossary and hotel maps have been eliminated from the print version but included in the CD/DVD. Also, in the print version various sections have been compressed to save space: Tutorials, Forums, etc. Finally, both to reduce cost and to be more green, we continue to use partially recycled paper along with a new just-in time printing process.

Again, this year, the technical editorial staff included six Editors (V. Gaudet, G. Gulak, J. Haslett, S. Mirabbasi, K. Pagiamtzis, and K.C. Smith), under the direction of a managing editor (Laura Fujino). However, to reduce costs, the amount of technical and language editing has been dramatically reduced. Abstracts have been eliminated from the Digest, with slightly edited versions available in the Advance Program posted at www.isscc.org.

In recognition of the large amount of work leading to the Digest open before you, I wish to acknowledge a great many individuals: Wanda Gass, Hideto Hidaka, Anantha Chandrakasan, members of the ITPC, and all of the authors, for their individual contributions; Brad Philips, Chris Gooch, and Mira Digital Publishing, for Web-based and other prepatory support, including improvement and facilitation of the paper-review and pre-voting process, as well as the Digest printing and Digest CD; Steve Bonney, and S³ Digital Publishing, for author and Session-Chair interaction, for figure layout, for paper formatting, for pre-press preparation, and for general assistance; Melissa Widerkehr, and Widerkehr and Associates for general interfacing, problem-solving, and coordination; Vincent Gaudet (University of Waterloo), James Haslett (University of Calgary), Shahriar Mirabbasi (University of British Columbia), Kostas Pagiamtzis (Gennum), our editors, with Glenn Gulak (University of Toronto) and K.C. Smith (University of Toronto), our editors-in-reserve, for heroic effort on a tight schedule.

My thanks to you all!

Laura Chizuko Fujino
ISSCC Director of Publications & G. Gulak, Presentations

February 2011

978-1-61284-303-2/11 $26.00 © 2011 IEEE

Due to formatting issues there is a gap in pagination.

Page 5

ISSCC 2011 / SESSION 1 / PLENARY / OVERVIEW

Session 1 Overview

Plenary Session

Chair: **Anantha Chandrakasan,** *Massachusetts Institute of Technology, Cambridge, MA, ISSCC Conference Chair*

Co-Chair: **Wanda Gass,** *Texas Instruments, Dallas, TX ISSCC Technical Program Chair*

The Plenary Session begins with opening remarks and a welcome to attendees from the Conference Chair, Anantha Chandrakasan. Then the three Plenary speakers will introduced by the Program Chair, Wanda Gass and their visionary presentations given. After a short break, the Awards Presentation will be given, moderated by the Conference Chair. The Session will conclude with the inaugural Technology Roundtable which will be moderated by Jan Rabaey, University of California, Berkeley.

The three Plenary presentations and the Technology Roundtable support the Conference theme: "Electronics for Healthy Living". Electronics play a significant role in enabling a healthier lifestyle. Wearable technology which communicates via Body Area Networks enables monitoring of our vital-sign status, providing continual assessment of our health risk. Those with chronic diseases can live a more normal life with implanted devices that sense, process, actuate, and communicate. Our lifestyles can also be improved as we strive to make our environment better by consuming less energy in both the operation of electronic systems and manufacturing of semiconductor devices themselves.

The first Plenary presentation by Dr. Stephen Oesterle of Medtronics will provide the clinical perspective on how implantable electronics has changed healthcare. From the first implantable cardiac pacemaker, semiconductor electronics have been applied both to save lives and improve quality-of-life. To continue to improve treatment, however, a systems approach must be taken when designing implantable devices. This requires attention to three key areas: the interfaces to the device, how information flows and is processed, and energy management. The primary goal is to make the physical interface between the device and the body as small as possible to enable less-invasive procedures with less morbidity. Systems which incorporate sensors, algorithms and closed-loop feedback, can create seamless information exchange and offer opportunities for better therapies and management of chronic disease. New methods for storing and managing power translate to the use of a smaller battery for the device. Future circuit advancements, when integrated as part of a well-defined system, will make significant contributions to improving patient outcomes, and easing clinical burden.

In the second Plenary presentation, Professor Jo De Boeck will give us insight into new advancements at imec that will make wearable health-monitoring devices available for use by everyone. Silicon is playing a key enabling role in the emerging healthcare paradigm where healthcare is personalized, predictive, preventive, and participatory, all at the same time. In multiple studies, even relatively simple wireless ECG (Electro-Cardiogram) patches are starting to impact the efficiency of care and the reduction of hospitalization time. The next-generation multisensory smart patch will be a game changer in healthcare, and a prime driver for a technology roadmap with radical steps in efficiency and effectiveness of system energy consumption, signal acquisition, signal conditioning, on-board signal processing and decision making, and wireless data transmission. We will see patches and other body-worn devices connect to the healthcare infrastructure with further extended power autonomy and with the ability to smell, to listen, and even to feel. The well-targeted use of these future-generation devices for monitoring stress, emotions, and so on, will revolutionize how we live, play, and work.

Dr. Oh-Hyun Kwon from Samsung will deliver the third Plenary presentation where he will discuss ways the semiconductor industry can improve our environment. Industrial advances have made human life more productive and convenient. However, it has also created unprecedented levels of environmental destruction which, ironically, has become a serious threat to our health. This crisis has been caused mainly by an extensive amount of energy consumption and green-house gas creation, and, therefore, the key solution is to reduce energy requirements of all industry. As semiconductor products are pervasive, reduction of their power consumption is one effective way to contribute. Energy efficiency in semiconductor products is achievable through the creation of low-power multi-core processors, green memory solutions, high-bandwidth memory interfaces, 3D-packaging technologies, low leakage/ low-supply process technology, low-power design methodologies, and smart-grid power systems. To address the inefficiency of fabrication equipment, semiconductor manufacturers must lead a collaboration with equipment suppliers which will result in reduced energy consumption of individual pieces of equipment, as well as, of entire manufacturing facilities. In this continuing process, energy-efficient semiconductor products and ecofriendly manufacturing processes are the answers to new environmental challenges, likely to enable an improving environment, ultimately saving us and our planet!

In keeping with the Conference Theme, innovative solutions leading to another 10 times reduction in power will be needed for implantable devices, wearable devices, and even data centers. New this year to the Conference, will be a Technology Roundtable for the discussion of next-generation challenges and solutions for reducing energy consumption of semiconductor devices. Six renowned Roundtable panelists will share their insights into the future challenges for energy reduction in digital design, memory design, RF design, communication devices, manufacturing, and design tools. Three domain experts will question the panelists on the nature of the future low-power roadmap.

Overall, the Plenary Session will provide an excellent introduction to ways that electronics will enable us to live healthier lives. We thank our Plenary Speakers and Roundtable participants in advance for their eager willingness to lead off this year's exciting program!

978-1-61284-303-2/11 $26.00 © 2011 IEEE

February 21, 2011 / 8:15 AM

1.1   New Interfaces to the Body Through Implantable System Integration   8:30AM

Stephen Oesterle, *Senior VP, Medtronic, Minneapolis, MN*

The pace of technological change continues to enable new possibilities of how we interface with our world. An engine of that change is highlighted by Moore's law, which has driven the semiconductor market for the last 40 years. This engine enabled a diverse set of capabilities for sensing, processing, actuating, and communicating with the environment; it is also being applied to revolutionize health care.

From the first implantable cardiac pacemaker, semiconductor electronics were applied to both save lives and improve quality of life. The initial success from pacing transferred into new applications, including modulation of cellular networks for treating neurological disorders and systems that grow ever closer to realizing an artificial pancreas. Emerging technological possibilities promise to continue this trend of expanding applications with greater capabilities. To optimize treatment, however, a systems approach must be taken when designing both the implanted device and the entire care pathway.

The system to support an implantable device requires attention to three key design abstractions: the interfaces to the device, how information flows and is processed, and energy management. As Moore's law slows for simple planar scaling of integrated circuits, electronics packaging technologies and other complimentary work termed "More than Moore" are enabling new system strategies that address these abstracted constraints in a symbiotic manner:

Interfaces: The primary physical interface is the interaction of the device and the human body. The goal is to make the device as small as possible while not trading off other performance metrics. Smaller devices enable less-invasive procedures with less morbidity, yielding simpler and lower-cost procedures to an expanded group of implanters. Advanced miniaturization also enables new form factors for shorter leads, potentially leadless systems, and independent devices that communicate with each other as a body area network. Another key interface is the transfer of data to and from the implant, highlighting the need for information management.

Information Flow and Processing: Better connected devices represent the increased expectations patients and physicians have for seamless information exchange. Increased amounts of data, however, must also be processed into high-value information and communicated securely to physicians and electronic medical records. Systems that incorporate sensors, algorithms and closed-loop feedback offer these opportunities for better therapies and management of chronic disease. The challenges to realize the full promise of "smart" systems include: sensor biostability, power-efficient signal processing, acceptable algorithm sensitivity, and specificity in real-world environments.

Energy Management: In order to fully reap the benefit of a smaller smarter implant, an acceptable energy strategy must also be identified. Advancements in integration are also introducing new methods for storing and managing power. Lower power and energy requirements for circuit functions directly translate to being able to use a smaller battery for the device.

Despite recent technological advancements, significant unmet needs remain, and integrated circuit technologies have the opportunity to be further optimized for application in the medical-device space. This requires careful attention to system-level objectives concerning what problems are being solved. While none of these new technologies will transform care in isolation, if integrated as part of a well-defined system, they can make significant contributions to improving patient outcomes and easing clinical burden.

1.2   Game-Changing Opportunities for Wireless Personal Healthcare and Lifestyle   9:05AM

Jo De Boeck, *Senior VP, imec / Holst Centre, Leuven, Belgium*

Silicon is playing a key enabling role in the emerging healthcare paradigm: disease-centric medical care, patient-centric decision making and therapy, proactive personalized and ubiquitous diagnosis and treatment. The necessity of healthcare which is personalized, predictive, preventive, and participatory, all at the same time, implies massive amounts of measurement, data, and associated ICT infrastructure.

In this vision, the data will be taken mostly "on the move" in residential or desolate settings, with minimal intervention from trained professionals. Patient-centric data, currently mostly self-reported, are key in performing clinical trials, reaching a diagnosis, checking on treatment evolution, evaluating post-treatment health-related quality of life, and, increasingly also in the creation of patient social networks. The quality of self-reported data is often questioned. Connected devices that provide such data in an acceptable unobtrusive way, with guaranteed quality, privacy, and identity, will make all the difference. Such infrastructure will further enable daily monitoring of so-called high- yielding micro-events that announce a problematic situation well before symptoms arise.

In multiple studies, even relatively simple wireless ECG (Electro-Cardiogram) patches are starting to impact the efficiency of care and the reduction of hospitalization time. The next-generation multi-sensory smart patch will be a game changer in health care and a prime driver for a technology roadmap with radical steps in efficiency and effectiveness of system energy consumption, signal acquisition, signal conditioning, on-board signal processing and decision making, and wireless data transmission.

We will see patches and other body-worn devices connect to the healthcare infrastructure with further extended power autonomy and with the ability to smell, to listen, and even to feel. We will need technologies to deliver these in flexible stretchable formats. To achieve this, cost-effective and ecological manufacturing challenges will also have to be tackled. For each application, hard constraints will have to be met on system reliability (QoS), cost, and energy budget.

Game changers at the level of the technology are only meaningful and efficient if they are driven by application needs. There is rapid progress in-close and in-depth interactions between the medical and the electronics community illustrated by the world-wide healthcare-related trials with wireless sensors in body-area-network configurations. This technology validation for personal diagnostic and theranostic products clearly drives game-changing circuit-, system-, and business-model innovation.

Many visionary applications such as brain-computer interfaces sound like magic. However, with every new generation of technology and application algorithms, wearable wireless systems become less obtrusive and more autonomous. The well-targeted use of these future generations for monitoring stress, emotions, etc, will revolutionize how we live, play, and work. None of these developments heralds a "Brave New World", but, instead, fosters and strengthens the individual in their quest for a healthier, happier, and longer life.

ISSCC 2011 / SESSION 1 / PLENARY / OVERVIEW

1.3   Eco-Friendly Semiconductor Technologies for Healthy Living   9:40AM

Oh-Hyun Kwon, *President, Samsung Electronics, Giheung, Korea*

In the history of human kind, we have witnessed remarkable development and growth of industry. Everywhere, from home appliances to space shuttles, such industrial growth has made human life more productive and convenient. However, it has also created unprecedented levels of pollution, environmental destruction and climate change, and, ironically, has become a serious threat to healthy human life.

To rescue mankind from the global environmental crisis, all industry is now expected to convert to eco-friendly technologies in a significantly greater degree than before. Recent studies have shown that this crisis is caused mainly by an extensive amount of energy consumption and green-house gas creation, and, therefore, the key effort is to reduce energy requirements of all industry.

Thus, as an important segment of world industry, the semiconductor manufacturers must participate in this effort to slow down the environmental crisis and enable healthy human life. As semiconductor products are pervasive, constituting a large portion of global energy consumption, reduction of their power consumption is one effective way to contribute. As well, the semiconductor manufacturing process consumes large amounts of energy; thus, the industry itself must reduce the energy used in its processing and make it more efficient.

From the semiconductor product perspective, it is useful to expand the usage of semiconductor products and in replacing conventional sizable mechanical parts, which are very energy hungry with relatively small and energy-efficient semiconductor products. But, the semiconductor industry must make an effort to improve energy-efficiency of each such product.

Energy efficiency in semiconductor products is achievable through coordination of the entire cycle of product development: design (including architecture), design methodology, process technology, and packaging. Moreover, co-operation with other industries can be very useful in reducing world-wide energy consumption: early interaction with software-industry providers and users will increase product efficiency; interaction with the electrical- power equipment-manufacturing, generation, and distribution industries, will increase the efficiency of the world's electric-power infrastructure and likely impact the design and application of products. Recent progress in the creation of low-power multi-core processors, green memory solutions, high-bandwidth memory interfaces, 3D-packaging technologies, low-leakage/low-supply process technology, low-power design methodologies, and smart-grid power systems are already indicative of such achievements.

Concerning the use of energy in semiconductor manufacturing itself, the industry is reviewing the main causes of energy inefficiency in manufacturing equipment and operation, with a view to reducing the large standby-power loss within and across manufacturing equipment. To address the inefficiency of fabrication equipment, semiconductor manufacturers must lead a collaboration with equipment suppliers. Such collaboration is likely to lead to reduced energy consumption of individual pieces of equipment, as well as, of entire manufacturing facilities.

The semiconductor industry continues to respond to society's ongoing challenges for products supporting a better life. In this continuing process, energy-efficient semiconductor products and eco-friendly manufacturing processes are the answers to new environmental challenges, likely to enable an improving environment, ultimately saving us and our planet!

978-1-61284-303-2/11 $26.00 © 2011 IEEE

ISSCC 2011 / SESSION 1 / PLENARY / 1.1

February 21, 2011 / 8:30 AM

1.1 New Interfaces to the Body through Implantable-System Integration

Stephen Oesterle, Paul Gerrish, Peng Cong

Medtronic, Minneapolis, MN

1. Introduction

The pace of technological change continues to enable new methods for interfacing with our world. An engine of that change has been the increasing capability in the electronics industry, as highlighted by Moore's law, driving the semiconductor market for the last 40 years. This engine enabled a diverse set of capabilities for sensing, signal processing, actuating, and communicating with the environment. These technologies are also being applied to revolutionize general healthcare, in particular through the implantable-device industry.

2. The Proliferation of Technologies for Therapeutic Applications

From the first battery-powered cardiac pacemaker, semiconductor electronics has enabled implantable medical devices to both save lives and improve patients' quality of life. This progression is illustrated by the pipeline of implantable defibrillators shown in Fig. 1.1.1; while, over the past few decades, the volume of the implant decreased by an order-of-magnitude, its sensing, diagnostic, telemetry, and therapeutic capabilities have significantly increased. The initial success of cardiac pacing also transferred semiconductor technology into new therapy areas, including the measurement of glucose levels linked with insulin pumps that grow ever closer to an artificial pancreas. Semiconductor technology has also expanded to directly interface with the body's electrical network; the breadth of neuromodulation technologies now impacting diseases such as Parkinson's, and chronic pain, is shown in Fig. 1.1.2. Emerging technologies with greater capabilities promise to continue this trend of expanding applications, through increasingly smaller smarter and better-connected devices. Yet to optimize treatment, a systems approach must be taken to designing not only the implanted device, but to the entire care pathway.

3. Designing an Implantable Device Requires System-Level Thinking

The creation of an implantable device requires, at a minimum, attention to three key design abstractions that must be balanced as system constraints: the physical interfaces with the device; information flow and processing; and the energy source with its requisite power management. As illustrated in Fig. 1.1.3, these three core design abstractions must be further decomposed into a diverse set of technologies that are impacted by the electrical design. As Moore's law slows for simple planar scaling of integrated circuits, electronics packaging technologies, sensor technology, and other complimentary work termed "More than Moore", are enabling new strategies that address these three design abstractions in a symbiotic manner (Fig. 1.1.4) [1][2].

4. Smarter Interfaces to the Body and the World through Novel Packaging Technology

The primary interface in an implantable medical system is where the device and the human body interact. The design intent is to make the device as small as possible, while not trading-off other performance metrics. Smaller devices enable less-invasive procedures with less-morbidity and improved cosmesis, enabling simpler and lower cost procedures to an expanded group of potential implanters. In addition, patient acceptance increases for procedures targeting prevention rather than therapy. Advanced miniaturization enables new form factors that allow for shorter leads, completely leadless devices, and independent devices that communicate with each other as part of a body-area network. Novel packaging also provides for the elimination of external packaging, and new types of downsized components and building blocks, both electrical and mechanical. One example of the significant volume decrease achieved by packaging is shown in Fig. 1.1.5, where the hybrid design for a bradycardia

pacemaker, shown in the original design scale, can achieve a five times decrease in form factor using advanced component packaging. Being able to take advantage of these new techniques and components enables the entire electronic system to be packaged in the smallest-possible volume. This downsizing enables simplified delivery of the implanted device, with less chance of complications.

Advanced packaging also moves the electronics assembly from the traditional two-dimensional assembly to a three-dimensional one. Being able to take advantage of all three-dimensions allows optimization of form factor for implantable devices. Technologies employed to achieve this range from today's flexible interconnect and die stacking to tomorrow's reconstructed wafer and through-silicon-via 3D assembly constructs. We are exploring these technologies for all aspects of electronic systems including MEMS sensors (Fig. 1.1.6), RF telemetry, passives integration, and ultra-low-power circuitry.

Technologies that have been traditionally applied to semiconductor processing are also increasingly being applied to other components of electronics systems. One critical example is that of high-value discrete-capacitor integration. Capacitors are a key component in an implantable system, as they provide energy management through charge pumps and storage, as well as isolation safety for electrodes. These capacitors have traditionally been made with ceramics or tantalum as discrete capacitors, but now 3D integrated capacitors using silicon substrates have made inroads into this space. If the capacitance density continues to increase on its current trajectory, per Fig. 1.1.7, many of the older forms of capacitor may be displaced, as integrated capacitors have the advantage that they can be packaged densely like an integrated circuits [3].

5. Smarter Information Management throughout the Medical-Implant System

Another key interface is where the data is transferred to and from the implant. Better-connected devices fulfil the increased expectations patients and physicians have relative to seamless information sharing across different platforms. Wireless technologies continue to expand in different formats in different geographies. Consumers of these technologies have expectations that communication between devices will 'just work'. Medtronic was successful in developing Medical-Implant Communication Service (MICS) band distance telemetry worldwide for medical-device communication and the Carelink network for managing this data (Fig. 1.1.8). These developments provided a streamlined method for remotely transfering implantable diagnostics and monitoring to remote clinician sites. However, this approach requires dedicated custom hardware for communicating with an implanted device. Networks beyond the immediate environment of the implanted medical device are fueling an expanded infrastructure that the medical-device industry can coordinate with and leverage for the future. Increased amounts of data processed into high value information, and communicated seamlessly and securely to physicians and electronic medical records, will be realized by smarter implanted systems and communications with external networks as they continue to evolve.

Information flow and processing extend beyond telemetry, and include new algorithms and processing capabilities embedded within the device itself. Smarter devices, with systems that incorporate sensors and closed-loop feedback, offer opportunities for better therapies and management of chronic disease. Closed-loop systems can be fully automatic, contained within the implanted device, or the "system" can be defined to include physician or patient response to sensor-output information provided by the implanted device. The challenge is to architect these systems within the constraints of power usage, sensitivity, and specificity.

An example of a recent closed-loop innovation enabled by emerging semiconductor technology is illustrated in Fig. 1.1.9. To help our chronic pain patients, Medtronic has architected a new "reflex" for a spinal-chord stimulator. Historically, spinal-chord-stimulator patients were required to titrate their therapy with an external controller as they changed their posture and activity levels, as a result of the dynamic motion of the electrodes with respect to the

978-1-61284-303-2/11 $26.00 © 2011 IEEE

spinal chord. The new reflex creates a feedback loop using a micropower three-axis accelerometer [8], and a custom algorithm to automatically titrate therapy based on a patient's posture and activity levels. The device is now CEmarked and undergoing clinical trials in the United States (investigational device use only).

In addition to improving the quality of therapy, closing the loop with embedded sensors and electronics should also enable other advantages. These include more quantitative clinical follow up, which should ultimately result in less-expensive procedures, greater ease of use, and improved therapy outcomes. Smarter systems might also enable improved performance of devices under different use conditions. Increasingly, there is an expectation that medical devices should not impose additional constraints on compatibility with a more sophisticated ecosystem within the hospital, clinic, or other environments. MRI compatibility is one such example of what has become the standard of care and exclusion of this diagnostic due to device compatibility is not well tolerated. A device that is smart enough to recognize the type of environment it is in, and respond accordingly, simplifies care.

Challenges to realizing the full promise of closed-loop systems include long-term sensor biostability, power-efficient signal processing, algorithm sensitivity and specificity, as well as the time to prove clinical efficacy. In some cases, new business models for physician reimbursement, and care to provide information and not just more data, must be considered to gain full acceptance of these systems.

6. Energy Supply and Power Management

In order to fully reap the benefit of a smaller smarter implant, an acceptable energy strategy must also be identified. Advancements in integration are also introducing new methods for storing and managing power. Increasing density in Li ion primary and rechargeable batteries, and other power sources, with volume-efficient packaging approaches enable more energy at smaller scales, such as that shown in Fig. 1.1.10, translating to smaller devices. As transistors become smaller, and applications for portable devices expand, there is increasing capability available for electronics to use less power as well, and use more dynamic control of supplies as needed by the instantaneous therapy requirements. Lower power and energy requirements for circuit functions directly translate to being able to use a smaller battery for the device.

7. Integrating the System for Breakthrough Implantable Designs

In summary, taking a system-level approach to design, the implantable-device designer can architect small and volumetric-efficient form factors bearing little resemblance to traditional implantable medical devices. The toolkit includes integrating discrete elements and sensors, planarizing components including power sources, so that they can be stacked in three dimensions, and removing legacy-packaging approaches that limit form factors. Other advancements include exploiting increased circuit density to run increasingly-sophisticated algorithms for information management.

As shown in Fig. 1.1.11, the prototypes for the next generation of medical devices promises to offer an almost order-of-magnitude decrease in volume, while maintaining or increasing therapeutic and diagnostic capability. This jump in performance enables disruptive opportunities for the use of devices in medical practice.

8. Enabling the Future: Using Technology to Drive the Scientific Discovery Required for Future Devices

A common dilemma when designing new medical devices is the need to first obtain a better scientific understanding of how the disease affects the body's natural dynamics. The need to better understand physiology motivates the use of advanced semiconductor technology to create clinical-research tools capable of answering key questions. One area where this need is particularly keen is in the area of treating diseases of the nervous system.

Since the Food and Drug Administration (FDA) approved Deep Brain Stimulation (DBS) as a treatment for essential tremor in 1997, Parkinson's disease in 2002, Dystonia in 2003, and Obsessive Compulsive Disorder in 2009, more than 75,000 people have had these devices implanted. Although the infrastructure for these systems, shown in Fig. 1.1.12, is known to be safe and effective for helping treat those diseases, scientists are highly motivated to better understand how DBS fundamentally works. A challenge for further progress is the lack of knowledge of human brain physiology and how disease impacts normal function. This lack of knowledge obstructs the development of what might be more-effective implantable devices.

To help catalyze the next generation of innovation in DBS, we at Medtronic are architecting a chronic implantable research tools to identify and apply the bio-markers relevant to a broad class of potential therapies. Such research devices would leverage technology and clinical infrastructure of existing systems. As illustrated in the functional-block diagram of Fig. 1.1.13, the core infrastructure of the device is the existing stimulator and telemetry system found in the released neurostimulators depicted in Figure 1.1.12. This design methodology allows us to leverage an implant base of approximately 75,000 devices already implanted in patients, for faster and safer clinical translation.

The prototype builds on top of this architecture with the addition of three major hardware subsystems, enabled by semiconductor breakthroughs. These are a biopotential (ECoG/LFP (Electrocorticogram/Local Field Potential)) brain-activity sensing module, that amplifies and processes field potentials measured from the existing electrodes; a three-axis accelerometer for measuring activity and posture; and an algorithmic microprocessor for signal classification, telemetry streaming, and system control. A supplemental recording memory is included within the algorithmic-processor block for chronically logging event-based waveforms for later upload. The hardware system is controlled by embedded firmware that can be downloaded via telemetry, which facilitates research into classification algorithms. Great care is taken to ensure the reliability of the prototype device, with focus on ESD performance, MRI coupling, and cross-channel stimulation isolation through the electrodes.

In order to provide the necessary sensor-signal fidelity with acceptable power dissipation, we have custom integrated our sensing channels. For the brain, we focus of the measurement of field potentials, as we believe this currently represents, for neurological disease studies, the best trade-off of chronic operation and desired information. Within this design space, a key challenge of architecting a brain interface circuit is the low signal level that must be detected with micropower circuitry [4][5]. Figure 1.1.14a illustrates the Brain-Activity-Sensing Integrated Circuit (BASIC). As described in [6], the BASIC implements a Short-Time Fourier Transform (STFT) estimation by using a modified chopper-amplification scheme to extract power at a given frequency from the brain activity. The custom design is able to achieve ~$1\mu V_{rms}$ signal resolution using 5μW/channel. This sensor has proven to be equivalent to off-the-shelf equipment for decoding brain states in motor-prosthesis applications. While the direct measurement of brain activity provides an exciting opportunity for better understanding disease states, we are also exploring increasing specificity using sensor-fusion methods.

Multi-sensor capability is a critical feature of the research tool and many future devices. While the direct measurement of neuronal activity is desirable for understanding and treating brain disease, it might need to be supplemented with additional information to improve the specificity of algorithms. For example, in the case of Parkinson's disease, akinesia and bradykinesia are highly correlated with biomarkers in the sub-thalamic nucleus of the basal ganglia, while resting tremor is not. The chronic measurement of activity, tremor, and motion, might therefore provide useful complimentary measures of patient state for the understanding of epilepsy, movement disorders, and psychological conditions [7]. To that end, we also designed and fabricated a custom-integrated micropower three-axis accelerometer. As described in [8] and illustrated in Fig. 1.1.14b, the activity sensor consists of a MicroElectroMechanical (System) (MEMS) polysilicon accelerometer that transduces acceleration to a differential capacitance measurement on the order of 1fF/g and a micropower custom-design integrated circuit to convert the capacitance change to a voltage signal that is digitized by the algorithm processor. The sensor can survive shocks in excess of 10kg, which is possible in a 1m drop of an IPG (Implantable Pulse Generator) onto an operating-room floor. A custom micropower ASIC was designed to convert the capacitance change to a voltage signal that is digitized by the algorithm processor. The sensor can measure ±5g, and through extended-life testing has demonstrated stabilities of ±5mg (one sigma) with a noise floor of 1mg/√Hz, while drawing only 300nA/axis. This sensor is from the same family as that in the spinal-chord stimulator described earlier.

By using advanced circuit-design technology, the prototype research device adds a multi-channel sensing capability to a chronic implantable system. The research device is currently being used for chronic preclinical animal studies to better understand the fundamental nature of disease in a natural environment, and potential optimization of treatment with adaptive neuromodulation. The understanding of basic physiology in a real-world model of disease, with the flexibility to explore new therapy algorithms linking sensed signals to adaptation of stimulation, is critical for implementing a practical next generation of closed-loop neurological devices.

9. Challenges Remain, but Opportunities Merit the Effort!

This paper has described the opportunities that semiconductors offer to medical implementation and the treatment of human disease. However, care must be taken using these deep-submicron IC technologies to achieve the desired ultra-low-power and high-reliability outcome. Current leakage and changes in transistor performance over the life of the part must be fully understood and managed effectively, in order to meet system lifecycle goals for both manufacturability and reliability [4]. Integrated-circuit processes and circuit building blocks, as well as new design-flow techniques and tools must be optimized in order to meet these goals. Similarly, chronic performance of sensors and novel sensors remain to be addressed as well, although industrial and academic labs are laying the foundation with aggressive prototyping efforts.

From a broader perspective, none of these new technologies will transform care by themselves, but if integrated as part of a well-defined system to meet customer needs, they can make significant contributions to improving patient outcomes. But, this requires careful alignment of system-level objectives to what problems are being solved. Planning for the technology life cycle with attention to reliability, manufacturability, and supply-chain logistics, are just as important as the performance and cost measures the technologies are judged against. However, when new technologies are combined to make new systems that address chronic-disease management needs, there are exciting possibilities for the future to restore health, alleviate pain, and extend life.

References:
[1] R. Tummala, "SOP: what is it and why? A new microsystem-integration technology paradigm- Moore's law for system integration of miniaturized convergent systems of the next decade," *Advanced Packaging, IEEE Transactions on*, vol. 27, issue 2, pp. 241-249, September 2004.
[2] F. Roozeboom, A. Kemmeren, "More than Moore: Towards Passive and Si-Based System in Package Integration", *Invited paper at the 20th SB Micro Conference, Electrochemical Society Proceedings*, vol. 2005-08, pp. 16-31, June, 2005.
[3] J. Klootwijk, K. Jinesh, W. Dekkers, J. Verhoeven, F. van den Heuvel, H. Kim, D. Blin, M. Verheijen, R. Weemaes, M. Kaiser, J. Ruigrok, F. Roozeboom, "Ultrahigh Capacitance Density for Multiple ALD-Grown MIM Capacitor Stacks in 3D Silicon," *IEEE Electron Device Letters*, vol. 29, issue 7, pp. 740-742, July 2008.
[4] P. Gerrish, E. Herrmann, L. Tyler, K. Walsh, "Challenges and Constraints in Designing Implantable Medical ICs", *IEEE Transactions on Device and Materials Reliability*, vol. 5, No. 3, September 2005.
[5] M. Porter, P. Gerrish, L. Tyler, S. Murray, R.Mauriello, F. Soto, G. Phetteplace, S. Hareland, "Reliability Considerations for Implantable Medical IC's," *IEEE International Reliability Physics Symposium Proceedings 2008*, pp.516-523, April 2008.

[6] A. Avestruz et al., "A 5 μW/Channel Spectral Analysis IC for Chronic Bidirectional Brain Machine Interfaces," *IEEE Journal of Solid-State Circuits*, vol. 43, no. 12, pp. 3006-3024, 2008.
[7] M. Weinberger et al., "Beta Oscillatory Activity in the Subthalamic Nucleus and Its Relation to Dopaminergic Response in Parkinson's Disease," *J Neurophysiol*, vol. 96, no. 6, pp. 3248-3256, Dec. 2006.
[8] T. Denison, K. Consoer, W. Santa, M. Hutt, and K. Mieser, "A 2μW Three-Axis MEMS-based Accelerometer," in *Instrumentation and Measurement Technology Conference Proceedings, 2007. IMTC 2007. IEEE*, pp. 1-6, 2007.

ISSCC 2011 / SESSION 1 / PLENARY / 1.1

Figure: 1.1.1: Historical downsizing of implantable defibrillators.

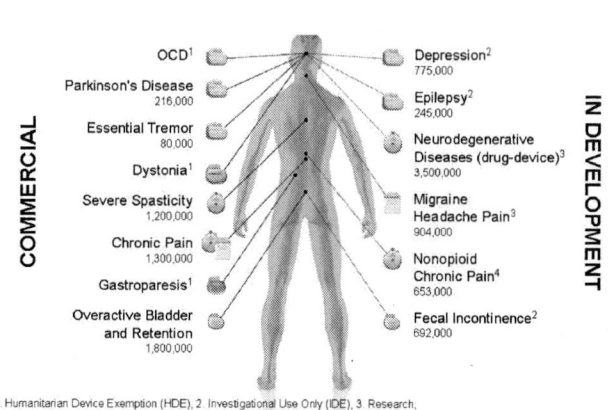

Figure: 1.1.2: Proliferation of electrical technology in the treatment of diseases of the nervous system.

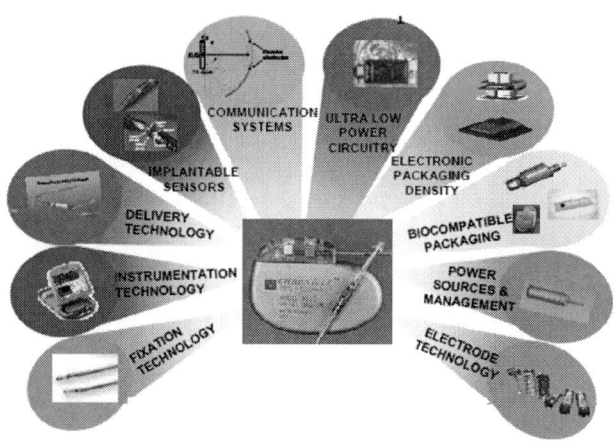

Figure: 1.1.3: Technologies for implanted devices.

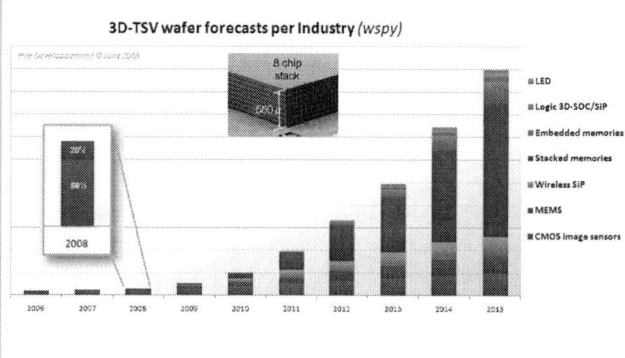

Figure: 1.1.4: Growth of "More-than-Moore" technologies in the electronic marketplace.

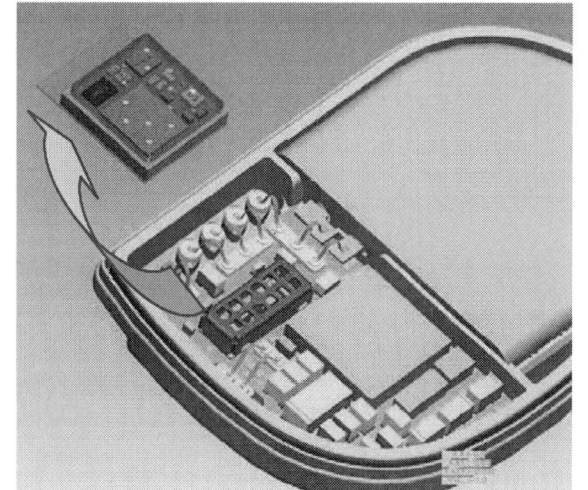

Figure: 1.1.5: Downsizing of pacemaker electronics using new packaging technologies.

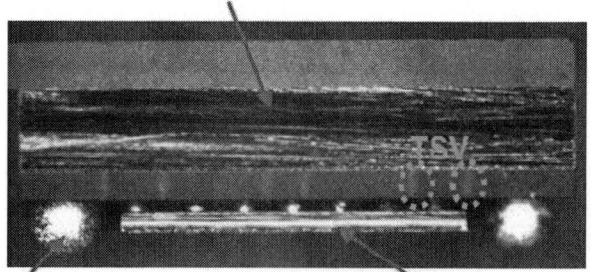

Figure: 1.1.6: 3D-subsystem electronic-component assembly.

February 21, 2011 / 8:30 AM

Figure: 1.1.7: Density increase of integrated 3D capacitors on a silicon substrate.

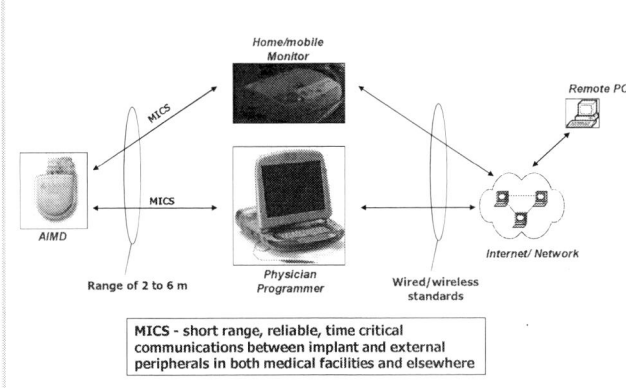

Figure: 1.1.8: Medtronic Carelink System using the MICS-band telemetry system.

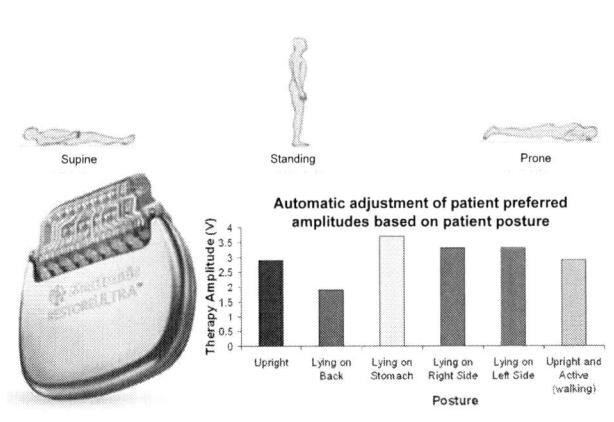

Figure: 1.1.9: Exploring a new spinal-chord "reflex" using MEMS sensors and implantable algorithms (investigational device use only).

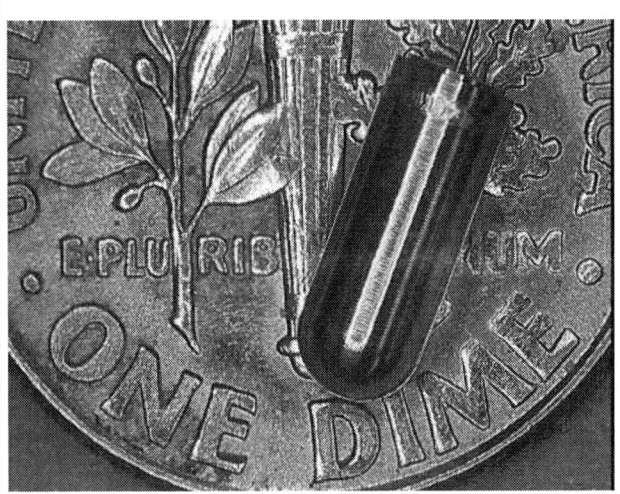

Figure: 1.1.10: Lower-power electronics leads to smaller batteries.

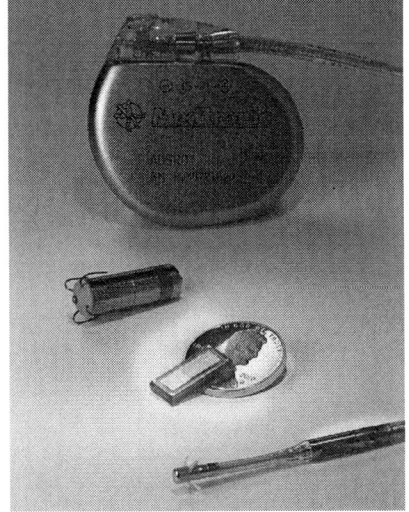

Figure: 1.1.11: Dramatically downsized implantable electrical systems.

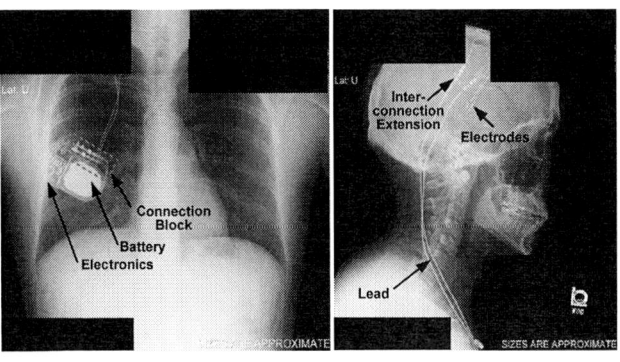

Figure: 1.1.12: General device architecture for an implantable system.

978-1-61284-303-2/11 $26.00 © 2011 IEEE

ISSCC 2011 / SESSION 1 / PLENARY / 1.1

February 21, 2011 / 8:30 AM

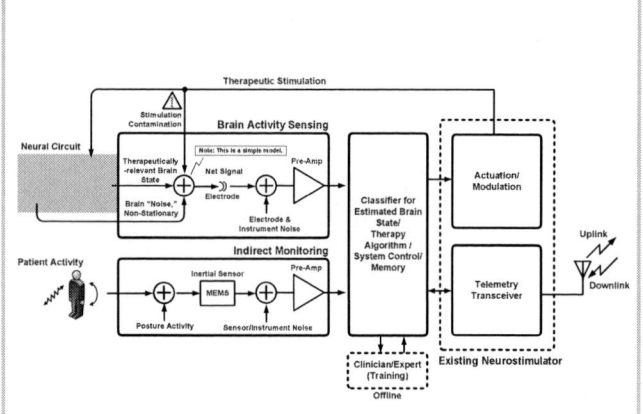

Figure: 1.1.13: Functional block diagram for bi-directional BMI leveraging existing implantable system platform.

Figure: 1.1.14:(a) Brain-Activity Sensing-Interface IC (BASIC).

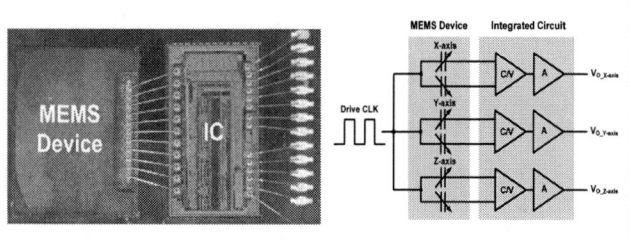

Figure: 1.1.14:(b) MEMS acceleration sensor.

ISSCC 2011 / SESSION 1 / PLENARY / 1.2

February 21, 2011 / 9:05 AM

1.2 Game-Changing Opportunities for Wireless Personal Healthcare and Lifestyle

Jo De Boeck

imec, Leuven, Belgium & Holst Centre, Eindhoven, The Netherlands

1.0 Introduction

In recent years, Personalized, Predictive, Preventive, and Participatory healthcare have become more than just buzzwords. Silicon is playing an important enabling role in this gradual, but certain revolution of our healthcare system: Silicon will become more essential, in view of the many challenges in realizing ubiquitous monitoring, real-time diagnostics, and patient-centric therapies. By reviewing world-wide technology breakthroughs, as well as healthcare-related trials with wireless sensors in Body-Area-Network (BAN) configurations, we will demonstrate that application validation for personal diagnostics and theranostic products is driving game-changing circuit and system innovation. Visionary applications such as brain-computer interfaces sound like magic! However, with every new generation of technology and application algorithms, wearable wireless systems become less obtrusive, higher performing, and more autonomous. The envisaged large-scale deployment of wearable healthcare and lifestyle add-ons that can monitor systemic factors such as stress and emotions, will revolutionize how we live, play, and work. But, none of these developments is heralding a "Brave New World"; instead, they will foster and strengthen the role and impact of each individual along the path to a longer, healthier and happier life.

2.0 "How are you?"

2.1 Self-Reported Information

Christine, one of the 21,500 Multiple Sclerosis (MS) patients that share self-reported health data on the web (www.patientslikeme.com, Fig. 1.2.1), evaluates her condition this week with a score of 71 on this site's MS rating scale. She shares the type and frequency of the prescription medication she took over the past years and the evolution of the primary symptoms like anxiety, bladder and bowel problems, brain fog, depression, emotional lability, excessive daytime sleepiness, general fatigue, mood swings, pain, sexual dysfunction, and stiffness or spasticity. No fewer than 40 secondary symptoms are scored further down the list. Except for evaluation of heart rate, blood pressure, temperature and weight, very few of these assessments are by direct measurement. Also, the data are not taken by trained medical staff and the time and duration of observation is highly variable. From a strict medical viewpoint, questions can be posed concerning the validity of this type of Internet "service" but patients clearly see the benefit, as do the pharmaceutical companies, seeking direct feedback and subjects for clinical trials. Without doubt, this information is more complete and precisely logged than that provided in the private setting of a doctor's practice, concerning recently developed symptoms or feedback on health-related quality of life during recovery follow-up.

2.2 Brush Your Teeth, Wear Your Sensors ... "Just Checking"

The Continua Health Alliance [1] describes the use of recording the presence (or absence) of high-yield micro-events as standard activities, and daily-life parameters that are related to the behavior and well-being of healthy or non-diagnosed people. These high-yield micro-events, when monitored frequently and regularly, can support preventive healthcare and can be indicative of a developing situation such as the onset of dementia or diabetes that would otherwise go unnoticed. Also, this type of monitoring helps gathering important information (for example, during the yearly flu season), and helps to understand the pre-hospital patterns that develop in the community. In this scenario, we envision an environment in which we are surrounded by a network of evaluators: weight scales, work-out tools, interactive gaming coaches, tooth brushes that report activity, toilets with analytical capabilities, pillows measuring sleep stages, and so on. Sensors in the environment will detect social contact, weather conditions, GPS coordinates, calendar items, and events such as falling. Industry consortia will certify the interoperability, accessibility and self-management of these networked devices. In some cases, such as detected predisposition for certain conditions, the individual will be asked to show some discipline and maybe provide measurements on relevant parameters

such as weight, temperature, heart rate and heart rate variability (HRV), blood pressure, glucose level, galvanic skin response (GSR), and so on, using certified easy-to-operate networked devices. In many cases, these companion devices would be just checking in to see how you are doing, and their activity would mostly go unnoticed. In general, unobtrusive ambulatory monitoring will become highly relevant. The sensors will disappear in shirts and shoes, or will be disguised in trendy non-stigmatizing accessories. Early products using BAN technology for health and wellness applications are listed in Table I in Fig. 1.2.2.

In this scheme, data mining and handling must be trustworthy and secure. Web-based services (such as Microsoft Health Vault, Google Health, and so on) will store and make available such data to whomever is authorized. When the situation warrants, each individual will be asked to change behavior or see a medical professional.

2.3 Remote and Virtual Doctors

Increasingly, we grow accustomed to employing massive amounts of medical-measurement data from the moment we enter the doctor's office and the healthcare system. Diagnosing, treatment, and hospital care are supported by technology-based techniques such as precise analytical testing, continuous monitoring of a myriad of parameters, and information presentation using "gaming-quality" 3D imaging. Increasingly, overnight diagnosing services, such as radiology, can be delivered from remote locations. With the goal of guiding us toward a healthier lifestyle, to spot early patterns that enable faster diagnosis, to enhance the options of residential care, and to improve timeliness of self-medication, ambulatory monitoring will provide remote doctors with objective and higher-quality data (Fig. 1.2.1). The impact will be most dramatic in places where the healthcare infrastructure is underdeveloped or hard to set up. There is an urgent need for high-quality robust networked ambulatory-monitoring, backed by a virtual health-service infrastructure in many places in the world where the nearest medical centre is hundreds of kilometers away.

3.0 Game-Changing Opportunities

3.1 Listen to Your Heart: The First-Generation Health Patch

Cardiac monitoring is one of the earliest adopters of wearable healthcare technology. Already, this technology has impacted the efficiency of care and reduction of hospitalization, as has been shown in various studies [2]. The monitoring applications are many; the contact to the body is relatively easy; the monitoring can be done in an unobtrusive and non-stigmatizing manner. Figure 1.2.3 illustrates some of the patch-format digital cardiac monitoring devices currently in practice, typically aimed for use in a hospital setting, or for recovering patients. The look and feel of these emerging devices makes them much more acceptable than Holter-type monitors. In a recent study [3], patient satisfaction, arrhythmia detection, and costs were compared in 20 patients ranging in age from 36 to 95, undergoing 7-day conventional Holter monitoring, and monitoring with a commercial "Nuvant" ECG patch [4]. Overall, patient satisfaction for a variety of daily activities on a 1 to 10 scale was as follows: the Nuvant device scored 8.3, the holter device 4.3. According to the study, for an analysis of a 7-day interval by a cardiac physiologist, the ECG patch required 15 minutes for data interpretation, while the Holter monitor required 3 hours. Moreover, the average delay between recording and analysis of the Holter monitor was about 4 days, while the Nuvant device having a centralized database which is monitored around the clock, allows instant analysis and urgent attention as necessary. Significant arrhythmias were detected in 3 patients (3 via Nuvant, 1 via Holter).

Heart-rate variations are the response of the autonomous nervous system to epileptic seizures. In a pre-clinical study on patients with tonic-clonic, generalized tonic and hypermotor seizures, epilepsy was monitored through ECG analysis, and seizures were detected with satisfactory sensitivity. The beat-detection algorithm embedded in the ECG patch was tailored to deal with the artifacts inherent to ambulatory monitoring systems, and achieved accurate computation of the instantaneous heart rate even under high levels of noise. The ambulatory sensing platform consisted of a full-custom ultra-low power ECG front end [5], a commercial ultra-low power microcontroller and a 2.4GHz radio for short-distance communication. The sensor node enables full ECG

ISSCC 2011 / SESSION 1 / PLENARY / 1.2

streaming with 24/7 autonomy for 1 week. The ECG-based seizure-detection system allows beat-to-beat analysis and detection of epileptic seizures in real-time, sending an alarm and storing all data locally for further inspection [6]. In this mode, the platform consumed an average 8.9mW of power, giving 30 hours of autonomy. Figure 1.2.4 shows the power breakdown over the different components and tasks. Data transmission accounted for 6% of the power and 23% was consumed by the optional SD card for on-board data logging. The activity of the MCU is divided into background (20%), epilepsy "signaling" (5%), and beat detection (75%).

3.2 Analog-Assisted Digital

Feature extraction from ECG in the analog domain is a radically new approach which can lead to dramatic power savings and increased functionality. In ECG, precise definition of the R peak is essential. In the traditional approach, the ECG signal is amplified and digitized and an on-board DSP performs the detection of the QRS complex in the digital domain, after which the R peak of the QRS complex is localized. As illustrated by the example in Fig 1.2.4, the power dissipated by digital signal processing accounts for about 50% of the power usage of the wireless sensor node. Of this, 95% goes to the feature extraction and only 5% to the detection and classification of R within the QRS. Hence, feature extraction through analog-signal preprocessing may significantly improve the power dissipation of the beat detection algorithms, if it can be achieved with minimal power dissipation. An original circuit was designed [7] that performs such analog preprocessing. The power dissipation for this analog preprocessor is only 2µW, with the result that the DSP needs only to perform the rest of the beat detection algorithm, but no longer the feature-extraction process. This results in an 800µW power saving in the DSP, an 11 times reduction in power.

An additional benefit of analog signal preprocessing can be the compression of the data, which can be used to further reduce the power dissipation of the algorithms. This can be done by implementing a variable sampling interval, where the sampling rate of the ADC is controlled by an activity monitor in the analog domain that consumes only 400nA. Using the activity-based sampling, the output of the ADC can be compressed as much as 6 times, which leads the power savings both in the digital and wireless transmission domains. The system that incorporates the analog-signal preprocessor ASIC is shown in Fig. 1.2.5. The combination of feature extraction in the analog domain, and adaptive sampling can reduce the power dissipation of wireless ECG-monitoring system as much as 7 to 1 as shown in Fig. 1.2.6.

An important challenge in ambulatory biomedical-signal monitoring systems is the false positives that occur during physical activities and indicate the need for improved signal integrity and motion-artifact reduction. The indirect approach would be to measure activity with accelerometers and gyroscopes, but this is a power-hungry solution and overall body movement is not necessarily causing artifacts that can be easily associated with the ECG signal. A direct approach to dealing with motion artifacts is to make use of the cause of the artifact, such as, the electrode-tissue interface. By measuring the electrode-tissue contact impedance in real-time the artifact can be detected, as illustrated by Fig. 1.2.7.

3.3 Ultra-Low-Energy Computing

Recent work on designing ultra-low power systems has focused on the sub-threshold regime [8-10], with an energy efficiency of a few pJ/cycle reported, using a clock in the KHz range. The low frequency combined with limited processing capacity, small on-chip memory, and low computation precision prevents the use of these systems for complex ambulatory monitoring beyond a simple ECG algorithm. For more-complex biomedical processing for example in the case of real-time Electro-Encephalogram (EEG) analysis, a biomedical processor with a very high energy efficiency and medium compute power is needed. Low-voltage complex systems with more computational power are demonstrated in [11-13]. An event-driven system with comparable energy per cycle, more computation power, and larger available frequencies was realized [14], through the implementation of effective system partitioning. This approach allows duty cycling, SIMD instructions, power gating, voltage scaling, multi-clock domains, multi-voltage domains, and extensive clock gating. The system has sufficient computational power to run multi-channel EEG processing or a complex ECG algorithm with feature extraction and motion-artifact cancelation. It consumes 13pJ/cycle at 0.4V running an ECG algorithm. With a wide voltage range of [0.4V to 1.2V] and frequency range of [1MHz to 100MHz], the proces-

sor is capable of adapting to a wide variety of application power and performance needs. For other interesting advancements see [15].

Also higher energy efficiencies for a given task can be envisioned in future platform generations where novel ultra-low power Application-Specific-Instruction-Set-Processor (ASIP) concepts are introduced. Potentially more than 2000 MOPS/mW (less than 0.5pJ per operation) can then become feasible in 40nm CMOS technology [16].

3.4 Ultra-Low-Energy BAN Radios

Bluetooth-Low Energy has been adopted by the Continua Health Alliance for wireless monitoring devices, while Zigbee Health Care (started in IEEE.802.15.4) is selected for wireless communication with such assisting devices as motion detectors and sleep monitors. These state-of-the-art, short range, standardized radios for WPAN consume up to 100 times more than what is available for advanced healthcare applications. Their power consumption is at the level of 50 to 100nJ/bit (Fig. 1.2.8). IEEE is drafting new standards, proposing different physical layers tackling specific application needs: in-and-around-body communication, medical BAN, and consumer BAN (IEEE 802.15.6 and IEEE 802.15.4 MBAN SG).

Standard radio solutions will co-exist in the market with proprietary solutions optimized for required performance in healthcare networks at the lowest possible power. Examples are Sensium's wireless sensor kits which do not specifically target the healthcare market; Nordic's nRF24AP2-USB single chip 8-channel ANT solution (Nordic 2.4GHz nRF24L01+ transceiver core and 8-channel ANT protocol stack) for use in product applications such as wireless sport, fitness, and health monitoring. To motivate the goals of future systems, let us assume we allow a (mean) radio power consumption of 20µW. Depending on the duty cycle of the application, the peak power consumption of the radio can be ~2mW (with a duty cycle of 1%). Unfortunately, the peak power consumption of the best solutions above, as shown in Fig. 1.2.9, is still ~10 times above this budget.

With this level of power consumption, a patch cannot be "on the air" frequently. Although local processing of data with help of a biomedical processor, assisted by analog preprocessing, could alleviate this power-consumption problem, for many applications this is not the solution, and a more energy efficient radio is needed. In the quest for radios with an energy efficiency of 1nJ per bit, early solutions are appearing in the literature: human body communication radios [17], ultra-low power ISM band SoC [18], and a BAN radio [19] with optimized sensitivity for high data rates. The latter 1nJ/bit solution is a single chip OOK transceiver fully optimized for low data rate on- and off-body communication. It operates in the 2.36 to 2.4GHz medical BAN and 2.4 to 2.48GHz ISM bands, and is specifically targeted at WBAN applications, such as for the ECG patch and motion monitoring. The transmitter delivers pulse-shaped OOK with 0dBm peak power, and the receiver front-end supports up to 5Mbps with sensitivity of -74 to -84dBm. Including the digital baseband, the transmitter and receiver consume 2.59mW and 715µW, respectively, while continuously operating. Power consumption is reduced to a fraction of these values for lower data rates. The transceiver is implemented in 90nm CMOS and occupies 2.4×1.85mm^2.

Impulse Radio UWB (IR UWB) is adopted in the IEEE802.15.4a standard for the realization of low-cost small and power-efficient wireless-sensor nodes and WPAN devices, and is currently proposed as one of 3 PHYs in the draft of the IEEE802.15.6 WBAN standard [20]. This air interface offers a unique combination of communication and precision localization. The communication technique is typically used for BAN communication in consumer applications, such as audio streaming, video or data-file transfer with rates between a few hundreds of kb/s and a few tens of Mb/s. The air interface is based on transmission of pulses with long intervals in between, thus allowing wide-range duty cycling of the RF circuitry. Overall, this technique reduces the average power consumption and enables streaming communication at a few nJ/b.

3.5 Robust Quality of Service (QoS) Within Strict Constraints

The wireless-sensor technologies and network implementations in e-health must meet a very challenging combination of constraints and objective trade-offs.

In our design, we face hard constraints on system reliability (expressed as a minimum QoS level that should always be guaranteed for a certain use scenario), cost, and energy budget. At the same time, concern for system bulk, chip area, and design time, involve very difficult trade-offs. This combination is quite different from the requirements for typical microprocessor related domains where best effort realizations are more acceptable. To reconcile the three major constraints is a very tough problem to solve [21].

Especially due to adverse impact on reliability, the use of advanced deep submicron technologies has until now been very much delayed or even avoided in such safety-critical applications. But, with the increasing need for complex algorithms and data handling, and strict cost and energy targets, the use of such advanced technologies is practically unavoidable in the future.

Currently, a popular solution to address some of these issues is to use subthreshold design for energy reduction, but regrettably, this comes with the price of unacceptable noise margins and reliability compromise in deep submicron technologies. Another solution is to add redundancy at the platform level, but also that quickly leads to unacceptable increases in system cost and power overhead (and thus energy budget) demands.

So instead, a more disruptive combination of approaches and directions must be investigated fundamentally, to solve this challenging combination of requirements. We believe it will require a multi-disciplinary approach involving all abstraction levels of the design. But, in addition, due to the need for employing heavily scaled technologies each of these approaches will have to be fully "technology-aware".

This approach will lead to changes in memory organization and processor components of computing platforms to make them truly ultra-low-power in terms of achievable task functionality per available mJ. The conventionally-used MOPS/mW is not an appropriate measure for this evaluation. In addition, at the system and middleware levels, we will need to exploit the available dynamism that is present in all modern applications and systems, to reduce the averaged-out energy demand.

Finally, to mitigate the impact of unreliable devices and wires in the scaled technologies, we will need circuit, architecture, and middleware support to obviate the degradation and wear-out effects while still providing full functional and parametric operation guarantees. Going for worst-case design approaches is simply leading to too much overhead in energy and system cost. In the end the mitigation approach will lead to the introduction of truly "self-adaptive smart" systems.

3.6 Always Ready to Go with Energy Harvesting

The use of wireless sensor nodes would not be feasible if battery replacement or cumbersome charging procedures are required. Smart-wireless-sensor technology for healthcare will at first (and even in the long run for some applications), rely on primary batteries that provide enough energy for the time of application. However, for long-term re-use, there be will applications that will request rechargeable batteries (in remote locations, implantable, and so on). A game-changing technology is autonomous energy harvesting on the sensor node. The absence of the need for intervention by the user or care giver will make this approach attractive, provided the cost is acceptable. Since energy buffering will remain important, a (smaller) secondary battery or super-capacitor and more complex power management will add to the bill of materials. The list of available sources of energy around the body is not that long: photovoltaic sources (optimized for indoors), heat flow through thermo-electric elements, and electromagnetic radiation are obviously available. Radio-active sources have also been suggested [22]. For implantable devices future scenarios for micro-bio fuel cells have been proposed [23]. All have drawbacks and limitations. Thermo-electric harvesting devices can only make use of a very small temperature difference between the skin and the surroundings, and careful thermal design is required. Most thermal harvesting devices with a BAN-acceptable form factor currently generate power of only a few mW (Fig. 1.2.10) [24 to 26]. Thermolife proposed 5200 BiTe thermocouples on a kapton tape to provide 123µW at a temperature difference of 4°C (42.5mA at 2.9V) [27]. Our team has demonstrated a set of 14 BiTe TEGs integrated in a shirt, sufficient to continuously power an ECG system (Fig. 1.2.10(e)). Intelligent power-management circuits and appropriate on-sensor energy storage systems are essential ingredients of energy-autonomous sensor implementations.

With components currently available, we feel it is possible to design a fully-autonomous ECG patch performing only beat detection by 2012. The main challenge remaining is integration and power management. Individual components are optimized for varying supply voltages and clock frequencies. Efficient power-management circuits, providing the plurality of voltages needed for digital (0.2 to 1.2V), analog (0.8 to 2.4V) are needed. Simple beat detection is only a start: applications and algorithms will grow in complexity, improving and adding functionality.

3.7 Multi-Sensor Health Patch

Multi-parameter sensor integration using advanced CMOS or hybrid integration technologies can add functionality to a patch or other ambulatory platforms. Such functionalities could be movement, acceleration (needing MEMS components), direction (as compass and gyro), sound (as microphone), gas (as e-nose), fluid analysis (as Lab-on-Chip), image and light. Table II (Fig. 1.2.11) lists some of the frequently-occurring diseases and their potential for monitoring. Such physical, chemical, and mechanical sensors will require extremely-low power read-out and signal conditioning. For example, emotion monitoring would require a BAN with measurement of respiration, galvanic skin conductance, skin temperature and ECG. We recently compared such a BAN system with the standard Cortisol test in a pre-clinical study involving 15 patients diagnosed with psychiatric stress disorder, and 15 healthy controls. Average error rates in separating between patients and healthy controls showed that with skin conductance alone the error was 35%, with ECG alone 30%, and with the combination of both 25%. Increasing the number of parameters indicative of the Autonomous Nervous System (ANS), and EEG is expected to decrease the error to well below 20%. Back-muscle EMG and voice are important additional stress indicators as well. When the signals from the ANS can be included, levels of physiological arousal can be separated. The direct chemical measurement, for example of Cortisol, can of course further improve the correct interpretation of a patient's situation.

As can be seen in Table II (Fig. 1.2.11), gas detectors could become very important for monitoring context awareness (such as air quality for patients that suffer from respiratory problems) and body gases (such as breath analysis), or body fluids for analysis such as saliva and blood. Current commercial e-nose concepts are bulky and costly (Fig. 1.2.12(a,b)) [28]. Besides the need for ultra-low power read-out and the implementation of light-weight algorithms, the demanding request for sensitivity, selectivity, and calibration are present. MOS-based and MEMS-based approaches have been demonstrated in the lab with sub-µW power consumption. The most promising implementations for BAN application will allow room temperature use without the need for (excessive) heating to calibrate and regenerate. It is to be expected that such sensors will be ready for inclusion on a mobile platform that enables a number of new applications (Fig. 1.2.12), besides the medical applications that benefit from gas-sensing referred to in Table II (Fig. 1.2.11).

3.8 Flex, Stretch, and Green

In spite of the very remarkable progress in wireless cardiac monitoring, the current devices lack the convenience of sports accessories. One technological enabler is smart packaging using ultrathin flexible and even stretchable electronics (Fig. 1.2.13) [29] and textile integration of electronics [30, 31]. As this form-factor requirement puts the packaging technology very much on the forefront, system-package co-design is essential, especially in view of electromagnetic compatibility (such as battery-antenna-body coupling). Because flexible packaging needs ultra-thin circuits, the effect of thinning on digital and RF circuits is of critical concern; the current indication is that thinning down to 10µm is possible without jeopardizing performance, for the flexibility required in BANs. This flexibile technology will certainly lead to re-use in other application areas like food and drug smart-packaging.

While thinned silicon circuits will remain essential for the intelligence in the patches, there is a rapid evolution in the performance of organic and oxide circuits on foil. When some of the components in the patch could be realized in cheap roll-to-roll manufacturing, this could be the start of a whole new set of implementations. It is likely that one will see OLED devices and organic detec-

978-1-61284-303-2/11 $26.00 © 2011 IEEE

ISSCC 2011 / SESSION 1 / PLENARY / 1.2

tor devices integrated in smart bandages (Fig. 1.2.13). One of the remaining issues is the recycling and re-use of all such patches and medical BAN devices, which is an area of research in its own right.

Acknowledgements:
The author would like to acknowledge colleagues that have contributed to this paper: C. Van Hoof, H. De Groot, G. Dolmans, K. Philips, M. Ashouei, J. Huisken, R. van Schaijk, R. Vullers, M. Crego-Calama, S. Brongersma, V. Pop, J. Penders, F. Yazicioglu, R. Lauwereins, F. Catthoor, B. Gyselinckx, H. De Man. Further the author acknowledges the industrial partners in imec / Holst Centre's Human++ program, and AgentschapNL (Dutch Ministry of Economic Affairs) for the financial support of Holst Centre (www.holstcentre.com, Eindhoven).

References:
[1] B. Piniewski, et al., "Empowering Healthcare Patients with Smart Technology", *Computer* ,Vol. 43, 7 pp. 27-34, 2010; see www.continuaalliance.org
[2] www.connected-health.org
[3] Private Communication: TM Yung, et al., "A comparison of a mobile cardiac telemetry system with conventional cardiac monitoring for the investigation suspected cardiac arrhythmias." Wiltshire Cardiac Centre, The Great Western NHS Foundation Trust, Swindon, United Kingdom.
[4] Nuvant (using PiiX device in Fig 1.2.3(b)), Corventis, www.corventis.com
[5] R. F. Yazicioglu, et al., "A 60 μW 60nV/√Hz Readout Front-End for Portable Biopotential Acquisition Systems", *ISSCC Dig. Tech. Papers*, Feburary 2006; R.F. Yazicioglu, et al., "Ultra-Low-Power Wearable Biopotential Sensor Nodes", *Proc IEEE Eng Med Biol Soc.*, pp. 3205-8, 2009.
[6] F. Massé, et al., "Miniaturized Wireless ECG-Monitor for Real-Time Detection of Epileptic Seizures", *Wireless Health Conference*, San Diego, Oct 5-7, 2010
[7] R. F. Yazicioglu, et al., "A 30μW Analog Signal Processor ASIC for Biomedical Signal Monitoring," *ISSCC Dig. Tech. Papers*, February 2010.
[8] J. Kwong, et al., "A 65nm Sub-Vt Microcontroller with Integrated SRAM and Switched-Capacitor DC-DC Converter", *ISSCC Dig. Tech. Papers*, February 2008.
[9] M. Seok, et al., "Phoenix Processor: A 30pW Platform for Sensor Applications", *Dig. Symp. VLSI Circuits*, June 2008.
[10] S. C. Jocke, et al., "A 2.6-μW Sub-Theshold Mixed-Signal ECG SoC", *Dig. Symp. VLSI Circuits*, June 2009.
[11] S.R. Sridhara, et al., "Microwatt Embedded Processor Platform for Medical System-on-Chip Applications", *Dig. Symp. VLSI, June* 2010.
[12] G. Chen, et al., "Millimeter-Scale Nearly Perpetual Sensor System with Stacked Battery and Solar Cells", *ISSCC Dig. Tech. Papers*, February 2010.
[13] J.Kwong, et al., "An Energy-Efficient Biomedical Signal Processing Platform", ESSCIRC, September 2010.
[14] M. Ashouei, et al., "Voltage-Scalable Biomedical Signal Processor Running ECG at 13pJ/cycle 1MHz 0.4V", *ISSCC Dig. Tech. Papers*, February 2011.
[15] M.Zwerg, et.al. "A 82μA/MHz Microcontroller with Embedded FeRAM for Energy Harvesting Applications"; N.Lotze, et.al., "A Standard Cell Based Design Technique for Digital Circuits Fully Operational Down to 62mV in 0.13μm CMOS"; M.Seok, et.al., "A 0.27V, 30MHz, 17.7nJ/Transform 1024-pt Complex FFT Core with Super-Pipelining", *ISSCC Dig. Tech. Papers*, February 2011.
[16] F. Catthoor, et al., ``Ultra-Low Power Domain-Specific Instruction-Set Processors'', ISBN 978-90-481-9527-5, Springer, Heidelberg, Germany, June 2010.
[17] P.P. Mercier et al., 'A 110uW 10Mb/s eTextiles Transceiver for Body Area Networks with Remote Battery Power', *ISSCC Dig. Tech. Papers*, February 2010.
[18] E. Le Roux et al., 'A 1V RF SoC with an 863-to-928MHz 400kb/s Radio and a 32 Dual-MAC DSP Core for Wireless Sensor and Body Networks', *ISSCC Dig. Tech. Papers*, February 2010.
[19] M. Vidojkovic et al., 'A 2.4GHz ULP OOK Single-Chip Transceiver for Healthcare Applications', *ISSCC Dig. Tech. Papers*, February 2011.
[20] http://www.ieee802.org/15/pub/TG6.html
[21] Tutorial on System Reliability Issues at [ref HOT TOPIC session - Cross-Layer Optimization to Address the Dual Challenges of Energy and Reliability, *DATE*, Dresden, March 2010.
[22] Rajesh Duggirala,et al., "Radioisotope Thin-Film Fueled Microfabricated reciprocating Electromechanical Power Generator", *J. of Electromechanical Systems* Vol. 17, No 4, pp. 837-849, 2008.
[23] D. Danilov, et al., "Modeling and Management of An Integrated Bio-Fuel Cell/Micro-Battery Energy Storage System", Keynote Power MEMS, Leuven, December, 2010.

[24] H. Böttner, et al., "New Thermoelectric Components Using Microsystem Technologies", *J. of Microelectromechanical Systems*, Vol. 13, No. 3, pp. 414-420 (2004)
[25] Z. Wang, et al., "Characterization and Optimization of Polycrystalline Si70%Ge30% for Surface Micromachined Thermopiles in Human Body Applications", *J. Micromech. Microeng.* 19, 2009.

[26] H. Lhermet, et al., "Efficient Power Management Circuit: Thermal Energy Harvesting to Above IC microbattery Energy Storage", *ISSCC Dig Tech Papers*, February 2007.
[27] I. Stark, "Thermal Energy Harvesting with Thermo Life", *Proc. of the International Workshop on Wearable and Implantable Body Sensor Networks*, pp. 19-22, BSN 2006.
[28] Electronic Sensor Technologies (www.estcal.com)
[29] D. Brosteaux, et al., "Design and Fabrication of Elastic Interconnections for Stretchable Electronic Circuits", *IEEE Electron Dev. Lett.* 28(7), pp. 552-554, 2007
[30] T. Loher, et al., "Stretchable Electronic Systems for Wearable and Textile Applications", *Proc. of 9ᵗʰ IEEE VLSI Packaging Workshop* (VPWJ2008), pp. 9-12, 2008.
[31] J. Yoo, et al., "A wearable ECG acquisition system with compact planar-fashionable circuit board based shirt," *IEEE Trans. on Information Technology in Bio-Medicine*, Vol. 13(6), 2009.

February 21, 2011 / 9:05 AM

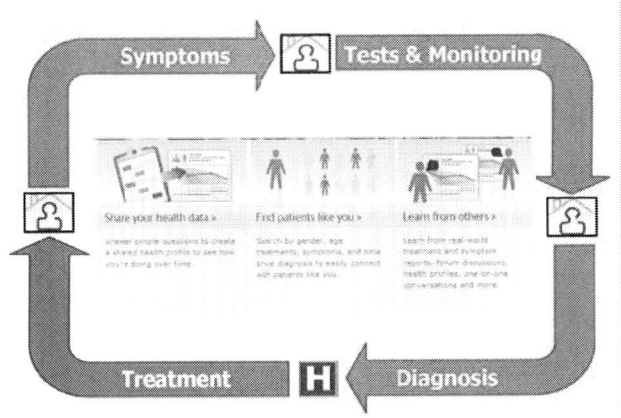

Figure 1.2.1: Web-based information platforms are set up which serve the need of patients that are looking to compare their (early-ustage) symptoms, the way they feel and the effectiveness of their eventual treatment.
Example from www.patientslikeme.com.

Application domain	Product	Company	# wireless nodes	Signals recorded
Health	VitalSense	Respironics (Philips)	2	Core body temperature and skin temperature; additional wired sensors available
	Bioharness™	Zephyr	1	ECG, respiration, skin temperature, 3D-acceleration
Fitness	Nike+	Nike & Apple	1	Activity
	Polar	Polar	2	Heart rate, Activity
	Bodymedia FIT	Bodymedia	1	Motion, Heat flux and skin temperature
	ActiPed	FitSense Technology	1	Activity
Wellness	DirectLife	Philips	1	Activity
	FitBit	FitBit	1	Activity
Sleep	AXbo	AXbo	1	Activity
	SleepTracker	SleepTracker	1	Activity
	Personal sleep coach	Zeo	1	EEG/EOG

Note that 'Activity' may refer to several combined parameters like distance travelled, speed, calories burned, etc

Figure 1.2.2: Table I.

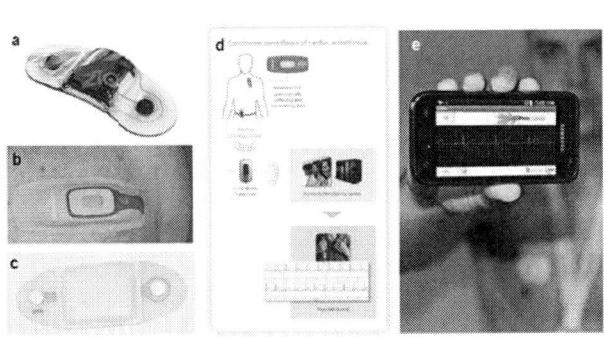

Figure 1.2.3: Some of the patch-format digital cardiac monitoring devices currently availble. (a) Zio by iRhythm, (b) Piix by Corventis, (c) Digital Plaster by Toumaz. (d) Wireless communication of data to a server and helpdesk, closing the care-giver loop. (e) ULP ECG sensor platform, linked to android mobile (Holst Centre/imec).

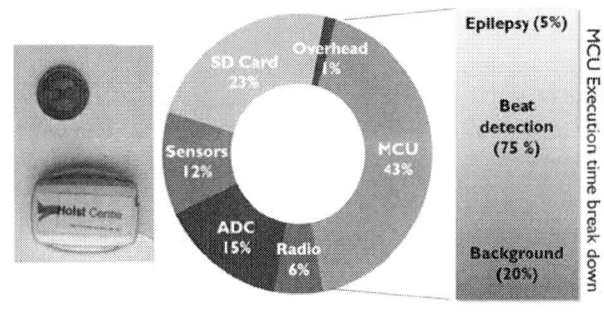

Figure 1.2.4: Power breakdown of a Wireless Sensor node for cardiac monitoring.

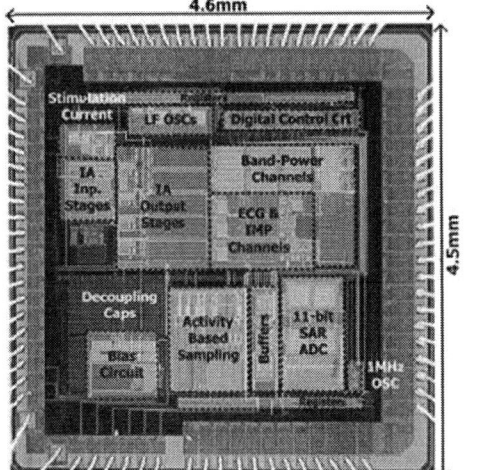

Figure 1.2.5: Analog-assisted digital processing for radical reduction of power consumption in an ECG application.

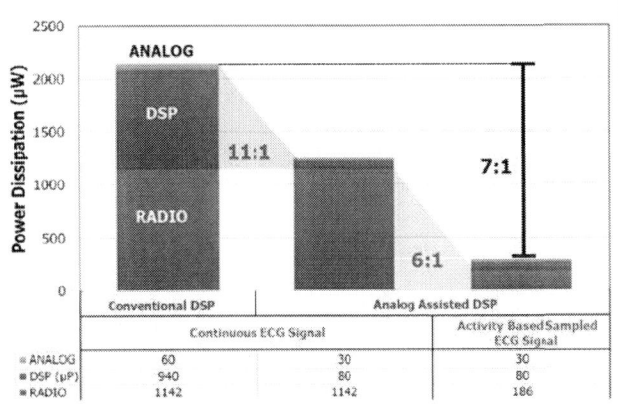

Figure 1.2.6: Results of power savings when applying analog-assisted digital signal processing and activity based sampling.

978-1-61284-303-2/11 $26.00 © 2011 IEEE

ISSCC 2011 / SESSION 1 / PLENARY / 1.2

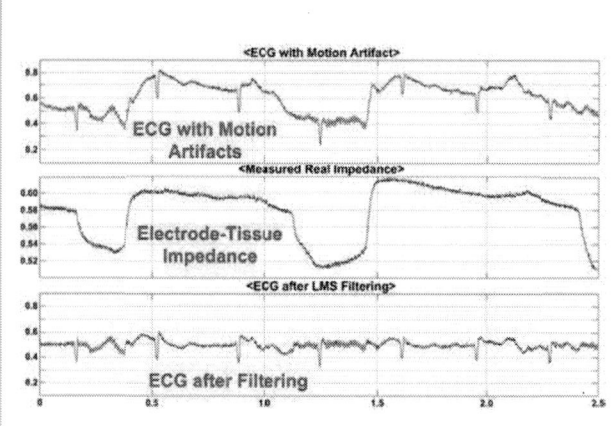

Figure 1.2.7: Real-time measurement of the electrode-tissue impedance can help to eliminate the motion artifacts in an ambulatory ECG implementation.

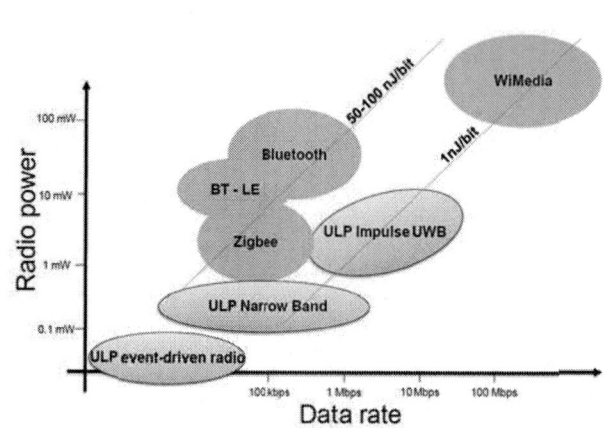

Figure 1.2.8: Ultra-Low-Power radios for wireless healthcare-assisting devices.

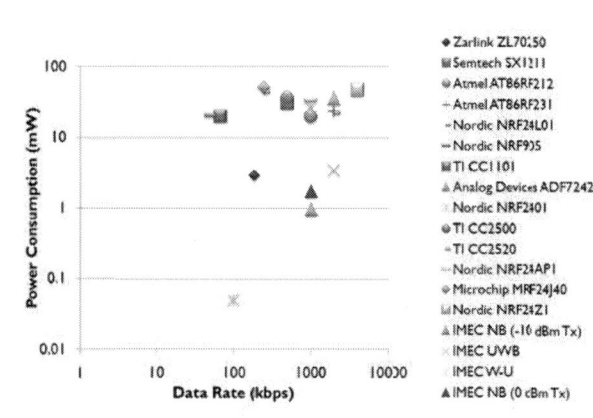

Figure 1.2.9: Power consumption of recent Ultra-Low-Power radio implementations.

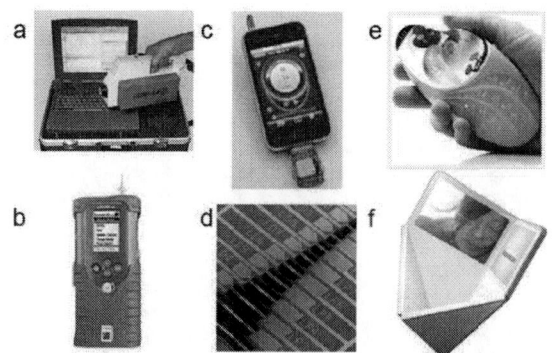

Figure 1.2.10: (a) Micropelt TEG (shown without cooler) Courtesy H. Böttner, FhG IPM; (b) imec TEG SiGe DT=1oC, 2 V; (c) LETI, 4mW/cm2 at DT=1oC / 1V at DT=60oC; (d) ThermoLife; 5200 BiTe thermocouples on kapton tape ; 123mW at DT = 5K (42.5µA at 2.9V); (e) thermo-electric generator and ECG-shirt with 14 3x4 cm² thermal harvesting devices (shown in insert). These are complemented with photovoltaic energy harvesters. The power harvested is 0.9mW standing still indoors, 1.5 – 2.7mW walking indoors, 3mW outdoors. Radio transmits a 2 channel ECG signal (2kb/s) over a range of 10m.

Disease	# affected	Wireless solutions
Alzheimer	5 M	Vital signs, location, activity, balance
Asthma	23 M	RR, FEV1, air quality, oximetry, pollen count
COPD	10 M	RR, FEV1, air quality, oximetry
Depression	21 M	Med compliance, activity, communication
Diabetes	24 M	Glucose, hemoglobin A1C
Heart Failure	5 M	Cardiac pressures, weight, blood pressure, fluid status, RR
Hypertension	74 M	Continuous blood pressure, Med compliance
Obesity	80 M	Smart scales, glucose, caloric in/out, activity
Sleep Disorder	40 M	Sleep phases, quality, oximetry, apnea, vital signs

Courtesy of Don Jones (Qualcomm)

Figure 1.2.11: Table II.

Figure 1.2.12: E-nose early examples, current technology and visions. (a,b) 7100 zNose, Electronic Sensor Technology (California), Cyranose 320, Cyrano Science (Pasadena); (c) i-phone with sensor module (NASA, nov 2009); (d) MEMS based e-nose sensor working at 1µW level; (e) health conscious phone that smells food properties (http://www.yankodesign.com   Designer: Kristina Lee (f) concept Nokia Scentsory Concept where an e-nose samples the odor of caller environment and transmit to recipient electronically.   http://www.tuvie.com/scentsory-nokia-future-of-mobile-phone-concept/ Designer Kimberli Hu.

978-1-61284-303-2/11 $26.00 © 2011 IEEE

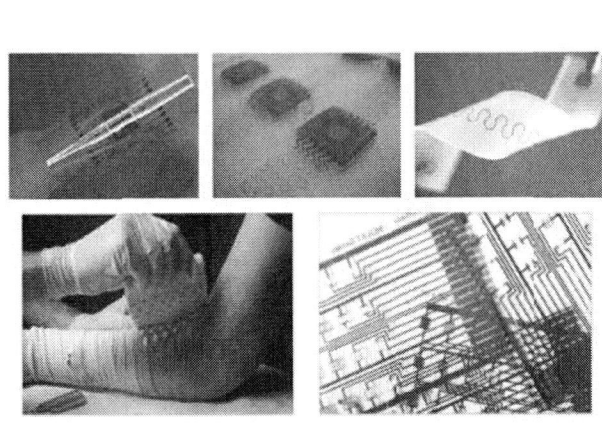
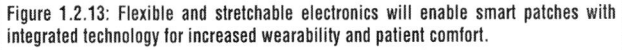

Figure 1.2.13: Flexible and stretchable electronics will enable smart patches with integrated technology for increased wearability and patient comfort.

ISSCC 2011 / SESSION 1 / PLENARY / 1.3

1.3 Eco-Friendly Semiconductor Technologies for Healthy Living

Oh-Hyun Kwon

President, Samsung Electronics, Giheung, Korea

1.0 Introduction

In the history of human kind, we have witnessed remarkable development and growth of industry. Everywhere, from home appliances to space shuttles, such industrial growth has made our lives more productive and comfortable. However, this comes with an alarming increase in energy consumption, primarily involving fossil fuels, and increasing greenhouse-gas emissions. Such activity has led to unprecedented levels of pollution, environmental destruction and climate change, and, ironically, has become a serious threat to human health [1].

To save mankind from the resulting life-threatening environmental crisis, all industry is expected now to convert to eco-friendly technologies to a significantly greater degree than before. In particular, immediate action is needed to reduce energy demands of all industry. For example, saving 200TWh of energy would eliminate 144M tons of carbon dioxide emission, and moreover, would result in a cost saving of $13B [2]. This is equivalent to the carbon dioxide emission from 27.5M passenger vehicles for a year, and, interestingly, is the carbon sequestered by 3.7B tree seedlings grown for 10 years [3] (Fig. 1.3.1).

As an important segment of industry, semiconductor manufacturers must participate in this effort to ameliorate the environmental crisis, and reduce its impact on human health. In view of the pervasive nature of semiconductor products, and the fact that they currently consume a large portion of global energy output, improving their energy efficiency is one effective way to make a difference. As well, semiconductor manufacturers, themselves, consume large amounts of energy; thus, the semiconductor industry must reduce the energy used in its manufacturing processes.

This paper will discuss how we, the semiconductor industry, can contribute to reducing global energy consumption and greenhouse-gas emission, eventually making human life healthier (of course, without sacrificing convenience and productivity). In the following sections, we will present two different perspectives: (1) semiconductor products, and (2) semiconductor manufacturing processes.

2.0 Low-Power Semiconductor Products

For information and communications technologies and residential consumer electronics, the world-wide electrical-energy consumption in 2010 is estimated to be as large as 800TWh [4]. Most such devices contain a large number of semiconductor products, making semiconductor components one of the largest sources of energy consumption in our everyday lives, even though they are usually less power consuming than similar products based on other technologies.

Therefore, along with the efforts to replace conventional mechanical and other energy intensive parts by relatively energy-efficient semiconductor products, we also need to make efforts to improve energy efficiency of the semiconductor devices themselves. Both approaches can make significant contributions in global energy saving, and will be the main topic of this section.

Even though semiconductor products span a wide spectrum of applications, we group them into two major categories, whose power consumptions are particularly significant, and they cover most key product segments: (1) processors representing logic devices, and (2) memory representing storage devices. We will review the past and the present of key component technologies in each product category, and investigate how they will evolve in the future to become more eco-friendly and therefore healthier.

2.1 Processors

Processors, often equated with human brains, are amongst the most sophisticated and complicated semiconductor logic devices available. Designing such processors is a very complicated task, which requires careful combination of five key technologies: (1) semiconductor process, (2) architecture, (3) circuit, (4) design methodology, and (5) package.

During the previous few decades, processors and related technologies have been developed for performance to achieve high-bandwidth and low-latency data processing. However, minimizing the power consumption of the processor without compromising its performance creates a whole new set of challenges in the aforementioned key technologies, and requires a new way of thinking. In this section, we will discuss the evolution and the co-development of the key technologies required to achieve low power consumption in high-performance processors and their contribution to reducing global energy consumption.

2.1.1 Process

The power consumption in processors is composed of dynamic and static components, each of which critically depend on factors such as the operating voltage, the leakage current, and the geometry of process devices such as, capacitance. Among the key factors, low-voltage operation is the main contributor to reducing the dynamic power. When we migrated two nodes from 180nm to 90nm, the dynamic power decreased by ~30% per node through scaling-down of the operating voltage. However, beyond 90nm, we experience the well-known short-channel effects, which make it hard to lower the operating voltage. In case of static power consumption, two main contributors are gate leakage and sub-threshold leakage. Unfortunately, they have increased in the recent process technology, due to process shrink and the short-channel effect, respectively.

Now, High-κ/Metal-Gate (HK/MG) field-effect transistors (FETs) overcome such challenges in both dynamic and static power consumptions and have been considered as a breakthrough technology. HK/MG FETs enable us to reduce not only the gate leakage, but also the threshold voltage of transistors, which in turn allows for a lower operating voltage without sacrificing sub-threshold leakage and performance. For example, a 32nm HK/MG process provides 20% less dynamic power consumption and 20~30% better performance than the preceding 45nm p-SiON process. Moreover, the HK/MG FETs enable more-stable low-voltage operation of Static Random Access Memory (SRAM) cells than that of conventional p-SiON gate FETs, due to an improvement in mismatch.

A low-capacitance technology for Back-End-of-Line (BEOL) Ultra-Low- (ULK) dielectric of 2.4 or less can also reduce the total capacitance, which can lower dynamic power by about 10% over the 45nm technology node having a BEOL low- dielectric of 2.7. However, reliability risks such as Time-Dependant Dielectric Breakdown (TDDB) need to be resolved for volume production.

We expect that HK/MG with ULK dielectric will be the mainstream technology to achieve low power and high performance processes down to 20nm. For the nodes beyond 20nm, however, HK/MG is expected to be combined with a FinFET structure (Fig. 1.3.2), providing up to a 20% performance improvement and up to 10% power reduction, compared to the planar device. This will be a strong candidate for the mainstream technology beyond the 20nm process node, due to its superior short-channel characteristics, with reduced gate leakage, although there are still technical hurdles in designing 3D devices and maturing the process.

2.1.2 Architecture

Over time, there has been a continuous performance demand placed on processors, especially for general-purpose CPUs and 3D-graphics processors. In the past, the processor industry has focused on improving performance by increasing clock frequency, Instruction-Level Parallelism (ILP) and cache size. However, this approach requires long and fine-grain pipelines with sophisticat-

978-1-61284-303-2/11 $26.00 © 2011 IEEE

ed control logics, and resulted in diminishing returns in performance and a super-linear increase in power consumption.

The industry has responded to these issues by designing lower-frequency multi-core processors instead of super-fast single-core processors. The multi-core approach allows for continuous performance scaling (through increased chip core count) by exploiting the ever-increasing transistor density. Such designs inherently consume less power as the cores are implemented based on low-power architectures with various per-core power management methods. Recent studies show that the transition from a single- to dual-core processor can save 20 to 30% of processor-power consumption for the same performance [5]. We expect the trend toward integration of more cores in a single processor will continue.

To meet a wide range of application needs with high energy efficiency, the processor architecture can be further optimized to utilize heterogeneous processing engines: some of them serve to provide Thread-Level-Parallel (TLP) engines for multi-tasking/multi-threading applications, while others are designed as hard-wired engines customized for specific and frequently-used functions. For certain application areas, we have seen a dramatic energy saving by employing hard-wired engines. For example, we observed that H.264 video encoding and 3D-graphics rendering with hard-wired engines consume only 1/136 and 1/80, respectively, of the power of software-only solutions [5].

Combining these two approaches provides better energy efficiency; we can already find this trend in a number of current-generation processors. Figure 1.3.3 shows a die photo of one of the latest mobile-application processors, which consists of a dual-core CPU, a quad-core graphics processor, and a few hard-wired engines, including a video codec, an image signal processor, and a video post-processor. We expect that this trend will continue in order to keep up with future power and performance requirements, and respond to the increasing demand of applications.

2.1.3 Circuit

Commonly-used low-power circuit-design techniques include clock gating, clock-tree gating, power gating, multi-threshold/multi-channel libraries, and voltage islands. Along with these techniques, significant efforts have been devoted to lowering the operating voltage of circuit structures, since operating voltage has the biggest impact on power consumption. However, circuit structures sensitive to process variation tend to be unstable at lower voltages, limiting the extent of voltage down-scaling.

Embedded SRAMs are typically one of the largest structures in processors and are designed at the minimum geometry. Consequently, SRAMs are particularly sensitive to process variation and have a higher failure rate at lower voltages than other logic structures. To overcome this problem, SRAM designers have employed various techniques: one of them is the read-assist and write-assist circuit technique, which improves SRAM cell stability. Also, replacing 6-Transistor (6T) memory cells with 8-Transistor (8T) cells is an attractive option because the write-only and read-only ports of an 8T cell can be independently optimized for low voltage operation, without conflict [6]. In conventional 6T cells, improving one (such as, write stability) typically degrades the other (such as, read stability).

The use of transistors which are insensitive to process variation would allow for lower operating voltage, and thus higher energy efficiency. For example, FinFET technology provides transistors immune to statistical dopant fluctuation effects [7], and thus less sensitive to process variation, which in turn, enables energy-efficient circuit design for future diminishing technologies. However, currently it may limit flexibility in logic gate sizing and analog circuits, which may need further study of FinFET-based circuit optimization.

2.1.4 Design Methodology

Low-power design methodology refers to CAD tools and methods that optimize power while maintaining target performance. For this, architecture, logic, circuits, transistors, as well as process should be considered in a holistic way.

Figure 1.3.4 describes a typical example of low-power-design methods which have been used in developing mobile application processors. With a combination of these methods, we have seen reductions as much as 50% in active current, 55% in leakage current, and 99.5% in total current in stop mode, while maintaining the required parametric yield for volume production [8].

So far, the low-power-design methods listed in Fig. 1.3.4 have provided sizable energy savings in processors. However, simply applying these methods may no longer be enough. With ever-increasing functionality demands and processor complexity, it will further require careful architecture-level exploration, based on various usage scenarios for an optimal combination of architectural parameters and implementation options required to satisfy speed, power, thermal, and area constraints.

Furthermore, we must address the variation, reliability, and parametric yield associated with continuously-shrinking processes, since otherwise, conventional approaches would result in prohibitive design margins. For example, in a 32nm process, statistical timing analysis for process variation improves timing margins by 3 to 5%, leading to about 5% reduction in total power consumption. With ongoing improvement in traditional statistical methods for practical use, the design window reduction by scaling of supply and body biasing voltages and better-than-worst-case design- by error-correction will play key roles in eliminating such penalties [9].

In addition, applying asynchronous-design techniques can further reduce power consumption. However, these techniques can be used only in some limited blocks due to practical issues, such as lack of CAD tools. Experimental results show that asynchronous-design techniques can save around 50% of the power, for some logic functions. Given the benefits such as low power consumption and resilience to voltage drops and parameter variability, we expect that the aforementioned techniques will appear, as the technology matures, for designs that require aggressive power reduction.

2.1.5 Package

3D package technology including Package on Package (PoP) (Fig. 1.3.5) has been used for its smaller form-factor and short interconnects, giving low-power benefits due to lower capacitive loads, for example in high-bandwidth memory interfaces. Furthermore, the bulk of the materials used in packages and the system board, can be reduced by 3D packaging. This, combined with Pb-free and halogen-free solutions, will contribute to environmental conservation.

Lowering the power-supply voltage for the processor core requires a low-inductance power-delivery network, which can be implemented by finer- pitch flip-chip bump and embedded-capacitor technologies. As presented in [10], the power-supply voltage for high-clock-frequency processor cores can be lowered, using enhanced power-delivery networks. While lowering IO-power-supply voltage saves power, the smaller noise margins require more advanced Signal Integrity (SI) in the design of memory interfaces. SI has been achieved through the co-design of package-level signal delivery-networks and IO circuits. While a low-capacitance technology using BEOL ULK dielectric of 2.4 or less can reduce dynamic-power consumption, the porous dielectric of ULK silicon devices requires enhanced mechanical-stress management of chip-package interactions. To protect the ULK dielectric, there are two solutions to be combined: one is the use of the robust BEOL structure of silicon devices to buffer the stress, and the other is the use of stress-free package technology employing advanced materials matched in terms of their Coefficient of Thermal Expansion (CTE).

Wide IO memory interfaces can be a key technology to provide required performance for future tablet PCs and server processors. One of the most promising solutions for implementing wide IO interfaces is the use of more-advanced 3D-package solutions, such as Through-Silicon Via - System-in-Package (TSV-SiP) shown in Fig. 1.3.6. As the table in Fig. 1.3.7 shows, the wide IO memory interface enabled by TSV technology offers considerable power reduction (up to 75%) by reducing the load capacitance of interconnect and IO circuits.

ISSCC 2011 / SESSION 1 / PLENARY / 1.3

Assume that there exist one billion smart phones in the world and that they are used for watching videos or playing 3D games for 10 hours everyday. If we implement future processors for smart phones with a suitably-small feature-size technology, and use all the advanced low-power technologies for process, architecture, circuit, design methodology, and package, we can save as much as 2 to 2.5TWh of energy consumption, solely through the smart phone industry [5]. Though this amount alone does not seem big enough, extrapolation to all electronic devices would show that the impact of using low-power processors adds up enormously. Therefore, all of our effort to improve energy efficiency of processors is a big step towards resolving our present energy challenges.

2.2 Memory

Efforts to achieve low power in memory can be largely divided into those associated with system memory and storage devices. System memory devices have evolved to process large amounts of data very quickly, in order to keep up with enhancement in processor performance in server and PC platforms. These days, low power consumption has become very critical as the power required by memory is a relatively large portion of that used in server and mobile applications. Especially as memory-intensive applications, such as cloud computing and virtualization, increase, energy efficiency is becoming more critical. Meanwhile, storage devices such as Hard-Disk Drive (HDD) have evolved with a focus on very-fast storage of large amounts of data in a small physical area. NAND flash memory has proliferated as the storage medium in music players since 2004. Now, it is being applied for storage in PCs and servers, due to its favorable characteristics of low power, high performance, high reliability, and small form factor. Solid-State Drives (SSDs), consisting of a controller and NAND flash memory, have been developed, and are beginning to replace conventional HDDs. Such a trend is rapidly growing in server applications requiring low power and high reliability. As the capacity requirement of system memory and storage devices increases, their portion of power consumption is growing, and the lifetime energy cost is rising significantly. Consequently, the Total Cost of Ownership (TCO), rather than the initial system build-cost is becoming a more important factor, along with the need for low-power devices, in such systems.

So far, the low power consumption of system memory, specifically Dynamic Random Access Memory (DRAM), has been driven mainly through process-technology shrinkage and low supply voltage. Process technology has been shrinking every year, and the world's first 30nm-class product has been launched in 2010, following 40nm-class product in 2009, and 50nm-class product in 2008. The smaller geometry available in advanced process technology required the application of lower supply voltages in DRAM devices: 3.3V for Single Data Rate (SDR) Synchronous DRAM (SDRAM), 2.5V for Double Data Rate (DDR) SDRAM, 1.8V for DDR2 SDRAM, and 1.5V for DDR3 SDRAM. Currently, the introduction of the 30nm process has required the use of 1.35V, providing reductions of 45% in operating power and 55% in standby power over 50nm-class product with the same capacity, while improving performance by at least 20%. In addition, new design architectures and design technologies for low power, such as power gating and the use of dual driver voltage per operation, have contributed significantly to power optimization. With these enhancements, 30nm-class devices can support even 1.25V operation, providing an additional 30% operating-power reduction at a given speed. Thus, the dramatic enhancement of transistor performance in advanced process technologies has resulted in high performance under low operating voltages. The main challenges to shrinking process technology annually have been in reducing the leakage current of cell-array transistors, and maintaining the cell capacitance at a specific level independent of technology scaling. Currently, the issue of high leakage current has been solved through new channel doping methods such as Localized ASymmetric Channel doping (LASC), and novel cell-array-transistor structures including Recessed-Channel Array Transistors (RCAT) and FinFETs. The issue of cell capacitance has been overcome by changing the cell capacitor structure and cell capacitor dielectric material [11].

So far, as explained above, the effort to implement low-power memory has been focused on process-technology shrink and low supply voltage. Of course, development of advanced process technologies will continue. But, due to the increase in technology complexity and the difficulty of equipment devel-opment, the inter transition interval for new process technology will increase, and new approaches for overcoming such challenges will be necessary. In 2011, 20nm process technology will be introduced, with 10nm and sub-10nm class technologies to be developed and launched. The supply voltage in these technologies will decrease to 1.2V, 1.0V, and sub-1.0V, accordingly. However, additional power reduction is unlikely without new technologies and new materials, because the definition of cell capacitors and cell-array transistors becomes more difficult as process technology enters into the sub-10nm range. In system memory, new technologies such as Wide IO, TSV, and optical IO along with new non-volatile memory devices such as Phase Change RAM (PRAM), Spin-Transfer Torque Magnetoresistive RAM (STTMRAM), and Resistive RAM (RRAM), will be strong candidates in this process. Such new approaches will further drive low power implementation in system memory. Around 90% additional savings are targeted by 2020 [12] (Fig. 1.3.8).

In storage devices, the focus so far has been on the reduction of the power consumption by mechanical parts such as platters, spindles, and arms for HDDs, because mechanical action takes a significant portion of HDD power consumption. However, since SSDs do not have mechanical parts, there exists only electrical power consumption by semiconductor components such as the controller and the NAND-flash memory itself [13]. Therefore, power consumption can be reduced to 70% or less of that for HDDs (Fig. 1.3.9). Using SSDs in real system environments, a minimum of 3.5 times energy efficiency could be achieved. The low power nature of current SSD system designs has been sufficient to lower system power until now, but the need to reduce power consumption of the basic SSD system itself, is increasing. Specifically, much effort is being directed at technology development to reduce the power consumption of the controller and the NAND-flash memory.

Power consumption in SSD depends primarily on the controller. In particular, the key factors for reducing controller power consumption are minimizing standby power and optimizing firmware for operational algorithms and power management. Similar technologies to those described in Fig. 1.3.4 are applied to reduce the power consumption of the controller. In addition to the effort to reduce the controller power, increasing focus is being placed on the reduction of the power consumed by the storage-media part of the NAND flash itself. To implement large-capacity storage devices, terabyte-scale NAND flash memories need to be mounted on a board, and are activated simultaneously for programming and processing of large amounts of data. Thus, methods to reduce the power of the NAND flash and to manage the number of NAND flash devices operating at one time, need further study. New NAND flash memories such as 3D-cell (vertically-stacked cell) NAND and 3-bit or 4-bit per cell NAND are being developed to reduce the power consumption per bit, and research to apply new material based memory such as PRAM and RRAM into storage devices is underway.

The effect of power savings by greener memory is expected to be very significant when considering total power consumption of electronic applications that include memory devices. For example in a particular server application, using conventional 48GB memory based on 60nm class 1Gb DDR2 at 1.8V which is replaced by a corresponding amount of 30nm class 4Gb DDR3 at 1.35V, and 2TB of 15K RPM HDD is replaced with a storage pair consisting of 146GB 15K RPM HDD and 1.44TB SSD in a server system, about 5600KWh of electricity can be saved annually [14]. Assuming that all the servers used in the world in 2010 [15] have the same configuration as above for system memory and storage device, 178.3TWh of electricity could be saved, annually. This corresponds to an 11.6B-dollarsaving, and 128M tons of carbon dioxide emission [2, 3]. Thus, green memory devices can contribute to building eco-friendly data centers not only by saving power, but also the cost of cooling.

3.0 Green Manufacturing

Energy reduction in semiconductor manufacturing itself is another crucial way in which the semiconductor industry can contribute to healthier human life. Recent studies show that in 2008 the semiconductor manufacturing facilities around the world consumed more than 40TWh [16], an amount which is larger than that of the annual residential energy consumption in the state of Michigan with a population of 10 million [2]. Moreover, the amount of energy consumed by semiconductor manufacturing continues to increase; the consumption in 2008 was 47% larger than that of 2001, and demonstrates a 7%

978-1-61284-303-2/11 $26.00 © 2011 IEEE

annual growth rate for the last two years [16] (Fig. 1.3.10). Therefore, reducing energy consumption by the semiconductor manufacturing process will play a significant role in reducing global energy consumption.

The largest part of the energy consumption in semiconductor manufacturing originates in the manufacturing equipment and the process itself. This Section will examine various ways to reduce energy consumed in semiconductor manufacturing, particularly from the perspective of equipment and process.

3.1 Semiconductor-Manufacturing Equipment

In semiconductor manufacturing, there exists a clear trade-off between manufacturing quality and its energy consumption. In order to ensure high quality results during manufacturing, manufacturers typically maintain a certain amount of operating margin to the equipment, consequently consuming more than the minimally-required energy. In particular, equipment employed to control plasma temperature and vacuum levels in the processing chambers tends to use enormous amounts of energy.

Figure 1.3.11 compares the best and the worst energy-efficient equipment for four manufacturing process steps; it clearly illustrates that there exists great potential for improving energy efficiency (by as much as 58%). It is mainly the role of the equipment suppliers to improve the energy efficiency of each type of equipment; but, we can further exploit this energy efficiency opportunity by establishing collaboration between equipment suppliers and semiconductor manufacturers. Industry level consortia for manufacturers such as the World Semiconductor Council (WSC) and International SEMATECH Manufacturing Initiative (ISMI) already provide the opportunity for expanding the scope of this collaboration to the equipment suppliers [17]. Thanks to these efforts, per-wafer energy efficiency of equipment has improved by 12.2% over the last four years. This has resulted in a saving of approximately 4.9TWh, which is as large as the total amount of electrical energy consumed in San Francisco in 2002 [18].

Another manufacturing area where energy efficiency can be improved is that of equipment standby. In the case of home appliances, average standby power consumption may be less than 1W, while their active power consumption is several hundreds of watts. Unfortunately, the same is not true for semiconductor manufacturing equipment. As depicted in Fig. 1.3.12, approximately 72.9% of the power needed for full-blown operation is still required for standby. This is because most equipment needs to stay at close-to-operation status during standby, and restart quickly. If we can reduce the average standby-to-operation ratio of all equipment to less than 50%, we can save 2.32TWh annually. However, considering the need for rapid process restart, there are many technical challenges in improving equipment standby-energy efficiency. One way to meet such challenge is to enhance the communication and control mechanisms between Automated Material-Handling Systems (AMHS) and lot-scheduling systems, thereby predicting process restart in advance.

Lastly, another source of power reduction in semiconductor manufacturing is in the area of utility systems. Utility systems supply gases, Process Cooling Water (PCW), Ultra Pure Water (UPW), and exhaust, according to the manufacturing equipment specifications. Utilities are one of the largest energy consumers in the manufacturing process, as they consume more than 30% of total semiconductor-manufacturing energy.

There is a large variation among the utility specifications provided by all equipment makers. In order to satisfy all possible specifications, the centralized utility systems must use the worst-case specification, resulting in excessive energy consumption. For example, the range of PCW pressure specifications can be 3 to 6kgf/m² to cover the most equipment requirements. Thus, a centralized PCW supply system must sustain a pressure of at least 6kgf/m² for all equipment. After careful study by equipment suppliers and semiconductor manufacturers, excessive energy can be avoided by optimizing and standardizing the utility specifications and the supply systems. For the PCW pressure example discussed above, we estimate a possible energy reduction of more than 20%.

3.2 Semiconductor Manufacturing Process

As we have seen in Section 2, migrating to the next generation of semiconductor technology helps improve energy efficiency of semiconductor products. In addition, we must reduce the energy requirement of manufacturing processes: for example, development of future semiconductor devices must lower the thermal budget of processes for forming shallow junctions, and must reduce the Radio-Frequency (RF) energy levels to minimize the process-induced damage in devices. A study estimates that the electrical energy consumption for heating elements can be reduced by 80% if the process temperature of SiO_2 deposition is reduced from 780°C to 300°C. Ultimately, the electrical energy consumption to drive heating elements can be minimized if the process temperature is reduced to room temperature. Moreover, it is advantageous to keep the temperature of the equipment as steady as possible, since energy dissipation is proportional to temperature fluctuation, in accordance with the equation of $\Delta Q = MC\Delta T$, in which Q stands for energy, M for mass, C for specific-heat capacity, and T for temperature. We can expect that a process with low thermal budget is more likely to have less temperature fluctuations, and therefore will consume less energy during the manufacturing process.

Despite the merits listed above, low thermal budget and low power processes tend to increase impurity concentration in a thin film. This in turn reduces the endurance of the thin film during subsequent manufacturing process steps. In order to overcome such problems, new material should be developed to form a robust and impurity-free thin film at low temperature with low RF power. Also, in order to increase dopant-activation efficiencies with a low thermal budget, advanced equipment should be developed. For example, more-advanced annealing processes can be considered such as millisecond anneal including Laser Spike Annealing (LSA) and Flash-Lamp Annealing (FLA).

4.0 Conclusion

The semiconductor industry has responded to ongoing desire for comfortable and productive life. Now, the industry faces a new need for healthier and eco-friendly living. Toward a long term answer to healthier living, the semiconductor industry has investigated technologies for alternative energy solutions such as solar cells, and for healthcare applications such as biochips. As the first step, industry needs to find and capitalize on the opportunity for improving energy efficiency of its products and their manufacturing processes, which can provide an immediate yet effective answer for healthier living.

Energy efficiency in semiconductor products is achievable through a holistic solution encompassing the entire product-development cycle: process, architecture, circuit, design methodology, and packaging. Recent progress in the creation of low-power multi-core processors, green memory solutions, 3D-packaging technologies, low-leakage/low-voltage process, and low-power design methodologies are already indicative of such achievements.

As for the use of energy in semiconductor manufacturing itself, the semiconductor industry is reviewing the main causes of energy inefficiency in manufacturing equipment and processes. Controlling excessive operating margins in equipment and utility systems, along with the development of the next-generation process can lead to a more energy-efficient manufacturing flow.

Furthermore, we propose "Interdisciplinary Collaboration" to capture all the possible energy saving opportunities in the areas relevant to the semiconductor industry. Interdisciplinary Collaboration in this context refers to the broad cooperation, not only within the semiconductor industry, but also with the other relevant industries including software, system, and infrastructure. We can further expand the scope of collaboration to academia and industry level consortiums such as WSC and ISMI. We believe all these eco-friendly Interdisciplinary Collaboration efforts will initiate and accelerate industry-wide energy saving activities, and can make a tremendous contribution towards the relief of the current environmental crisis.

978-1-61284-303-2/11 $26.00 © 2011 IEEE

ISSCC 2011 / SESSION 1 / PLENARY / 1.3

References:

[1] International Energy Agency (IEA), http://www.iea.org

[2] U.S. Energy Information Administration (EIA), http://www.eia.doe.gov

[3] U.S. Environmental Protection Agency, http://www.epa.gov, July 2010

[4] IEA, Gadgets and Gigawatts: Policies for Energy Efficient Electronics, 2009

[5] Samsung Electronics, "Multi-core Designs for Mobile Application Processors", July 2010

[6] L. Chang, et al, "Stable SRAM Cell Design for the 32nm Node and Beyond", Dig. Symp VLSI Technology, pp.128-129, June 2005

[7] T. King, "FinFETs for Nanoscale CMOS Digital Integrated Circuits", ICCAD-2005, Nov. 7, 2005

[8] JY Choi, et al, "Improving Parametric Yield in DSM ASIC/SOC Design", DAC UT, 2009

[9] S.Bhunia, S. Mukhopadhyay, K. Roy, "Process Variations and Process-Tolerant Design", Int'l Conf. on VLSI Design, pp. 699-704, 2007

[10] H.Lee, et al, "Power Delivery Network Design for 3D SIP Integrated over Silicon Interposer Platform", ECTC, 2007

[11] Kinam Kim, et al, "Memory Technologies for sub-40nm Node", Technical Digest of 2007 IEDM, pp. 27-30, 2007

[12] Samsung Electronics, "Green Memory Projection", Sep. 2010

[13] Samsung Electronics, "Samsung Green SSD Promotion", Sep. 2010

[14] Standard Performance Evaluation Corporation (SPEC), http://www.spec.org, Oct. 2009

[15] International Data Corporation (IDC), 32M units@2010, http://www.idc.com, 2009

[16] World Semiconductor Council 14th meeting, May 2010, Seoul

[17] International SEMATECH Manufacturing Initiative (ISMI), http://ismi.sematech.org

[18] San Francisco Energy Watch, "San Francisco Energy Information", http://www.sfenergywatch.org/energy.html

February 21, 2011 / 9:40 AM

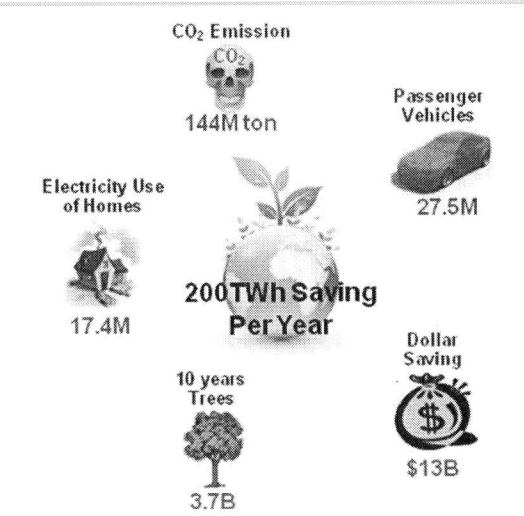

Figure 1.3.1: Environmental impact from energy reduction of 200TWh.

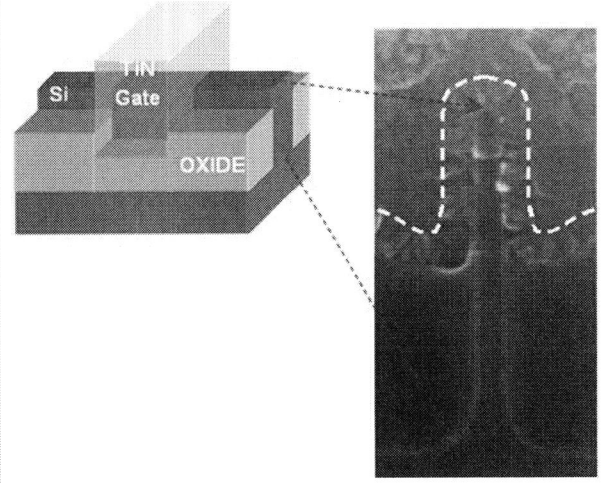

Figure 1.3.2: Schematic and cross-sectional images of a FinFET.

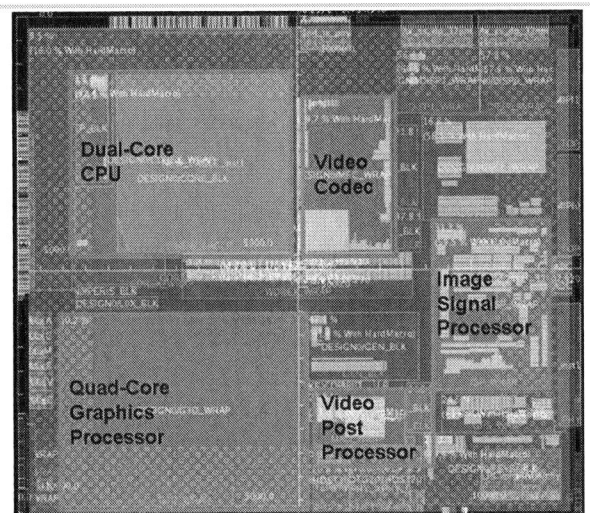

Figure 1.3.3: An example of current-generation application processor.

Applied Low Power Methods	Power Benefits
Clock gating	~50% active current reduction
Power gating	~55% leakage current reduction
MVTH (dopant dose & channel lengths)	~99.5% stop mode current reduction
DVFS	
Adaptive body biasing	~4% parametric yield increase
Statistical timing/leakage	

MVTH: Multiple Threshold Voltage
DVFS: Dynamic Voltage and Frequency Scaling

Figure 1.3.4: Applied low-power methods for mobile-application processors.

Figure 1.3.5: An example of current-generation 3D packaging, Flip Chip Package on Package (FCPoP).

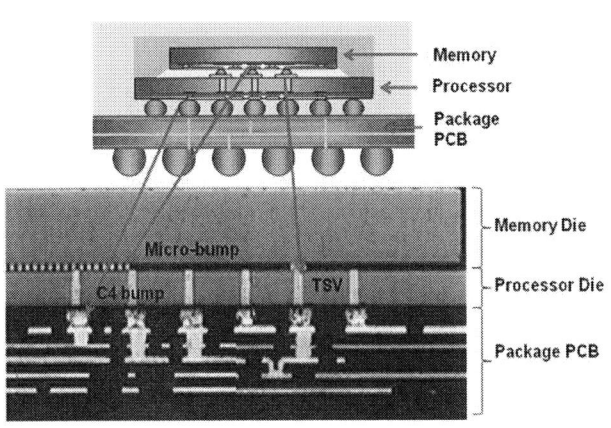

Figure 1.3.6: TSV-SiP for Wide IO: Top - Schematic; Bottom - Cross-sectional photo of actual TSV-SiP.

978-1-61284-303-2/11 $26.00 © 2011 IEEE

	Conventional 3D Package (FC-PoP) with LPDDR2	TSV-SIP with Wide IO memory
Memory I/O Power Consumption	176 mW	44 mW

Figure 1.3.7: Comparison of memory IO power consumption with LPDDR2 and Wide IO, which is implemented in FC-PoP and TSV-SiP.

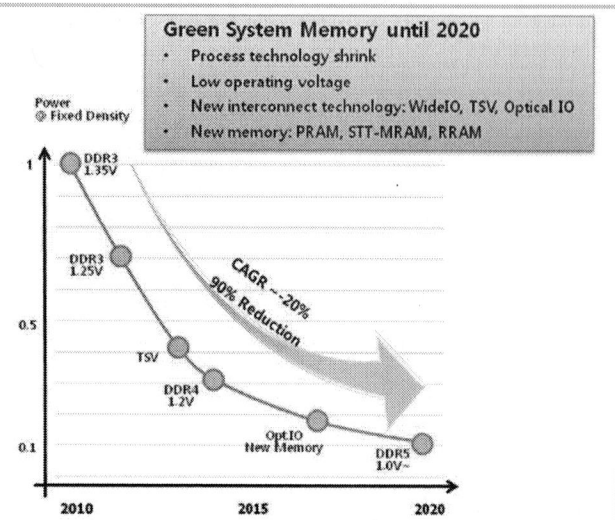

Figure 1.3.8: Power projection of green system memory to 2020.

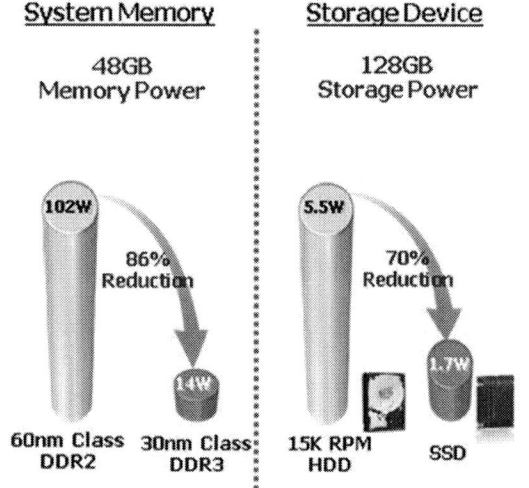

Figure 1.3.9: Power savings in system memory and storage devices.

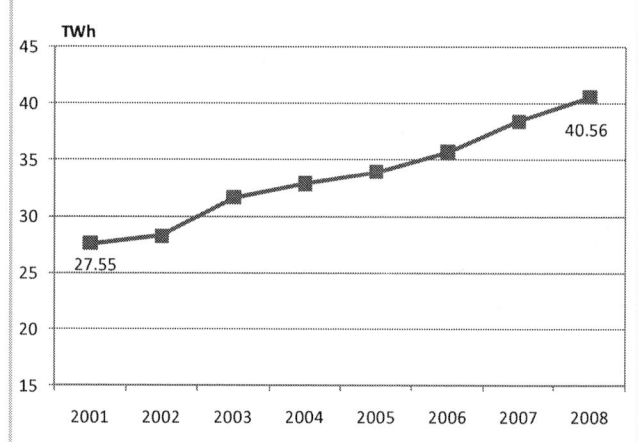

Figure 1.3.10: Electric power consumption by the semiconductor industry (WSC).

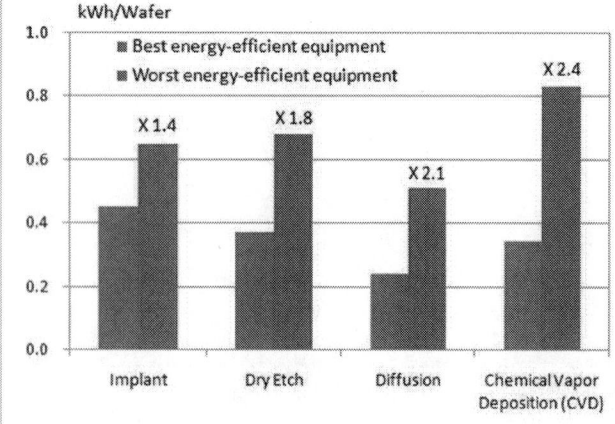

Figure 1.3.11: Comparison of per wafer energy consumption for various manufacturing operations.

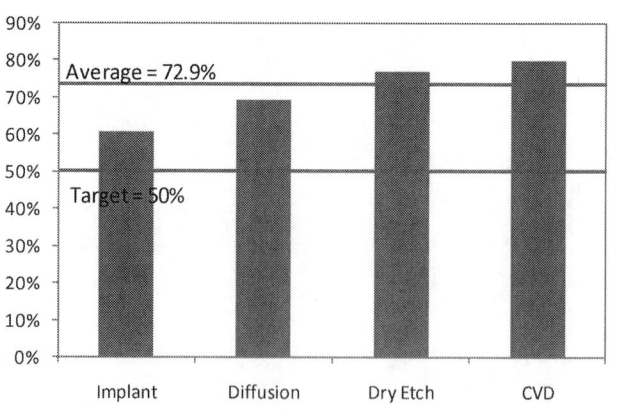

Figure 1.3.12: Ratio of standby power to active operating power consumption for various manufacturing operations.

Due to formatting issues there is a gap in pagination.

Pages 29-30

ISSCC 2011 / SESSION 1 / PLENARY / 1.4
February 21, 2011 / 10:50 AM

1.4 Beyond the Horizon: The Next 10x Reduction in Power - Challenges and Solutions 10:50AM

The energy efficiency of electronic circuits has dramatically improved over the past two decades. At the same time, computation, storage, and communication demands continue to grow with emerging wireless multimedia devices. In this inaugural Plenary Technology-Roundtable event, experts will discuss the opportunities to achieve the next order-of-magnitude reduction in energy consumption across various domains, including analog, digital, RF, and memory. The line between analog and digital continues to blur, as analog circuits are enhanced by applying digital corrections to compensate for increased analog component variability with process scaling. As well, digital will incorporate more analog to become more adaptive; for example, to optimize operating voltages at a fine-grain to match workloads and process variations. Memory circuits will need to use a system-level approach which requires bit-cell optimization, low-voltage operation with integrated regulators, 3D Through-Silicon Vias (TSV), and process optimization. RF transceivers will continue to trend toward highly-digital architectures.

The role of process-technology innovation and CAD tools will also be discussed. Future process technology will deliver new transistor structures and higher-mobility channel materials for low-voltage digital circuits. TSVs will be important in reducing I/O power and the length of on-chip interconnects. For RF, integrated inductors and transformers with significantly lower resistance will be the challenge. Future CAD tools optimizing energy will focus on co-design of packaging, architecture, power sources, and antenna to provide the best system solution. Domain experts will challenge the distinguished panelists to suggest directions and help create a roadmap for next-generation energy-efficient electronics.

Moderator:

Jan Rabaey
*University of California,
Berkeley, Berkeley, CA*

Domain Experts:

Hugo DeMan
*imec,
Leuven, Belgium*

Mark Horowitz
*Stanford University,
Palo Alto, CA*

Takayasu Sakurai
*University of Tokyo,
Tokyo, Japan*

Panelists:

Jack Sun
*TSMC,
Hsin-Chu, Taiwan*

Dan Dobberpuhl
*Consultant,
Monterey, CA*

Kiyoo Itoh
*Hitachi,
Tokyo, Japan*

Philippe Magarshack
*STMicroelectronics,
Crolles, France*

Asad Abidi
*University of California,
Los Angeles,
Los Angeles, CA*

Hermann Eul
*Infineon Technologies,
Neubiberg, Germany*

978-1-61284-303-2/11 $26.00 © 2011 IEEE

ISSCC 2011 / SESSION 2 / TECHNOLOGIES FOR HEALTH / OVERVIEW

Session 2 Overview

Technologies for Health / *Technology Directions*

Session Chair: *Uming Ko, Texas Instruments, Dallas, TX*

Session Co-Chair: *Eric Colinet, CEA-LETI, Grenoble, France*

Technology advancement not only enables smaller biosensors, health-care monitoring and communication systems, but also reduces their power dissipation. Ease-of-use and a reduction in cost-of-ownership fuel portability and the increasing success of next-generation medical electronics. In this session, the latest advancements in technologies and healthcare applications are presented, encompassing Wireless Body Area Networks (WBAN), body sensor networks, wireless sensors, Ultra Wide Band (UWB) and Terahertz (THz) imaging. In addition, the session features two accurate timers for ultra-low-power wireless sensor nodes and high-speed communication applications.

WBAN is an emerging technology that combines health care and consumer electronics applications around the human body. Paper 2.1 [KAIST] presents the very first WBAN that fulfils the IEEE 802.15.6 Task Group specifications. Its extremely low energy consumption of 0.24nJ/b makes it very attractive for future body-area network applications. This is achieved by resonance matching, contact impedance monitoring, and a low-power scalable double-FSK modulation scheme. To enable excellent low-power performance, circuit techniques like a reconfigurable LNA/driver, current-reuse wide-band demodulation, and a divider-based LO generator with duty-cycle correction are exploited. The 2.5×5mm² transceiver is fabricated in 0.18μm CMOS.

Paper 2.2 [KAIST] presents a wearable body-sensor network that realizes continuous sleep monitoring by ExG (EEG, EMG, EOG, and ECG) extraction from a sleeper's face, using 20mm fabric patch sensors. The average power consumption is 75μW for the real-time scalable network controller and 25μW for the ExG sensors. The form factor of this work is 10× smaller and 70× more energy-efficient than prior art, achieving a communication energy of 0.33pJ/b.

The increased incidence of diabetes makes non-invasive and continuous glucose monitoring a pressing demand. A wireless glucose sensor fully integrated into a contact lens is demonstrated in Paper 2.3 [U Washington]. The system has a measured sensitivity of 1.67μAmm⁻²mM⁻¹. It is wirelessly powered, communicates by load-shift keying (LSK) at 2.4GHz and achieves a measurement range of 0-to-2 mM while consuming 3μW from a regulated 1V supply. The whole chip area is 0.5mm² and requires no external components.

Paper 2.4 [UC Cork; Tyndall National Inst; U Pisa; U Reggio Calabria] reports what we believe to be the first implementation, including experimental evidence, of an SoC UWB (3.1 to 10.6GHz) pulse radar front-end for respiratory rate monitoring. The radar sensor, based on a correlation receiver implemented in 90nm CMOS, consumes 73mW from a 1.2V supply. It can detect target movements up to 2cm at 70cm distance and the respiratory rate of a person under test at distance of 25cm.

Paper 2.5 [MINATEC; U Montpellier 2] presents a low-cost CMOS imager for Terahertz (THz) imaging based on a 0.13μm CMOS technology. A MOSFET is coupled to a bow-tie antenna. Self-mixing allows direct conversion to the low-frequency band used for light modulation. The imager includes an in-pixel low-noise amplifier and multiplexing circuitry for single video output. Measured electrical performance and images demonstrate a maximum responsivity of 90kV/W (12% higher than state-of-the-art), with a pixel power dissipation of only 97μW. This is a 57× improvement compared to prior art (5.5mW).

Paper 2.6 [National Chung Cheng U; National Yang-Ming U; National Chiao Tung U; National Taiwan U; RMIT U] highlights a 48μW, 0.81cm³ implantable micro-stimulator SoC (IMSoC) with smart power management, sensors for closed loop stimulation, and a wireless telemetry/recharging system. The wireless telemetry, dissipating 10× less power than prior art in a 20× smaller footprint, handles the adjustable parameters for *in vivo* stimulation. Rat intracardiac electrograms are employed in an animal study.

978-1-61284-303-2/11 $26.00 © 2011 IEEE

Paper 2.7 [U Michigan] emphasizes that accurate synchronization cycle time measurement is required for ultra-low power wireless sensor nodes with a stringent power budget. A multi-stage temperature-compensated gate-leakage-based timer reduces rms jitter by 8.1× and synchronization uncertainty by 4.1× with a power consumption of just 0.66nW. Effective temperature sensitivity is reduced to 31ppm/°C, a 15× improvement from the prior art. The sub-nW timer makes synchronization of ultra-low-power sensor nodes a reality.

Very accurate local clocks play a fundamental role in modern communications, where high-precision references enable fast communication data rates. Paper 2.8 [CSEM] presents the first low-power CMOS RF lock loop highly integrated on-chip, for a Miniature Atomic Clock (MAC). An Allan deviation 1-second intercept point of $\sigma_y = 4 \times 10^{-10}$ is measured on a 10MHz clock using an RF loop locked on a ^{87}Rb cell for CPT interrogation.

Presenters:

1:30 PM

2.1 A 0.24nJ/b Wireless Body-Area-Network Transceiver with Scalable Double-FSK Modulation

J. Bae, KAIST, Daejeon, Korea

2:00 PM

2.2 A 75µW Real-Time Scalable Network Controller and a 25µW ExG Sensor IC for Compact Sleep-Monitoring Applications

S. Lee, KAIST, Daejeon, Korea

2:30 PM

2.3 A 3µW Wirelessly Powered CMOS Glucose Sensor for an Active Contact Lens

Y-T. Liao, University of Washington, Seattle, WA

3:15 PM

2.4 A 90nm CMOS SoC UWB Pulse Radar for Respiratory Rate Monitoring

D. Zito, University College Cork, Cork, Ireland;
Tyndall National Institute, Cork, Ireland

3:45 PM

2.5 A Broadband THz Imager in a Low-Cost CMOS Technology

F. Schuster, CEA-LETI-MINATEC, Grenoble, France;
Université Montpellier 2 -CNRS
UMR, Montpellier, France

4:15 PM

2.6 A Programmable Implantable Micro-Stimulator SoC with Wireless Telemetry: Application in Closed-Loop Endocardial Stimulation for Cardiac Pacemaker

S-Y. Lee, National Chung Cheng University,
Chia-Yi, Taiwan

4:30 PM

2.7 A 660pW Multi-Stage Temperature Compensated Timer for Ultra-Low-Power Wireless Sensor Node Synchronization

Y. Lee, University of Michigan, Ann Arbor, MI

4:45 PM

2.8 A Low-Power Fully Integrated RF Locked Loop for Miniature Atomic Clock

D. Ruffieux, CSEM, Neuchatel, Switzerland

978-1-61284-303-2/11 $26.00 © 2011 IEEE

ISSCC 2011 / SESSION 2 / TECHNOLOGIES FOR HEALTH / 2.1

2.1 A 0.24nJ/b Wireless Body-Area-Network Transceiver with Scalable Double-FSK Modulation

Joonsung Bae, Kiseok Song, Hyungwoo Lee, Hyunwoo Cho,
Long Yan, Hoi-Jun Yoo

KAIST, Daejeon, Korea

Wireless Body Area Network (WBAN) is an emerging technology that combines health care and consumer electronic applications around the human body. There are 3 PHY schemes discussed in the IEEE 802.15.6 Task Group for WBAN standardization [1]: ultra-wide-band (UWB) PHY, narrow-band (NB) PHY, and body channel communication (BCC) PHY. The BCC, which uses the human body as a communication channel based on the near-field coupling mechanism, has advantages over UWB and NB in energy efficiency because it provides low path-loss without the body shadowing effect in low-frequency bands below 150MHz [2-4]. However, the previous body channel transceivers (BCTs) were not optimized for WBAN because only phenomenological circuit models were used for the body channel analysis [5] and were unable to satisfy requirements such as energy efficiency, scalability of QoS, interference mitigation, and coexistence at once.

In this paper, we report a BCT that not only consumes the lowest energy with very high sensitivity but also is fully WBAN compatible. It is possible because: (1) the signal propagation principle is more thoroughly investigated, (2) resonance matching (RM) and context-aware sensing (CAS) are adopted, and (3) low-power double-FSK modulation is exploited for the full satisfaction of WBAN requirements.

According to the experimental data, a BCC signal path can be divided into 2 parts: forward path, and return path as shown in Fig. 2.1.1. Electrodes in contact or in close proximity to the human body constitute the forward path while the floated ground electrodes of TX and RX form the closed-loop return path by capacitive-coupling to the earth ground. The signal path loss has band-pass characteristics, mainly determined by the return path loss because the small capacitance of C_R in the return path has the highest impedance value compared with the contact impedance (electrode-to-body) and body impedance. RM cancels the C_R effect by inserting a resonating series inductor in the return path. The contact impedance in the forward path varies its value dynamically and even 30dB overall signal path loss variation is observed when the electrode is in contact with or apart from the body. To compensate for channel quality degradation due to contact impedance variation, the CAS observes the contact impedance by recognizing if the electrode is capacitively coupled or resistively coupled to the human body, and then automatically determines the BCT operation mode for the better power efficiency.

Figure 2.1.2 shows the overall architecture of the BCT using scalable double-FSK, which is based on UWB-FM [6]. The BCC uses a 40-to-120MHz frequency band while the CAS utilizes a chopper-stabilized AC current-injection source of 1MHz to monitor the differential contact impedance between 2 electrodes. On the TX side, from the frequency synthesizer and divider chain, a low-modulation-index sub-band FSK signal S(t) is transformed into a high-modulation-index wide-band FSK signal W(t) that drives the electrodes. On the RX side, the delay-line-based wide-band demodulator converts the RF carrier signal into a sub-band FSK signal that is demodulated by low-frequency direct-conversion RX circuits. Combined with RM and CAS, the double-FSK BCT has 3 low-power features: reconfigurable differential LNA/driver, current-reuse wide-band demodulator, and divider-based LO generation with duty-cycle corrector (DCC), which together reduce the power consumption by half.

Figure 2.1.3 shows the fully differential reconfigurable LNA of RX and TX driver with RM scheme. The dual-resonance networks at GND electrodes are matched over a wide frequency range between RX and TX. As a result, RM in both TX and RX enhances the return signal path by 4dB as shown in Fig. 2.1.3. The capacitively cross-coupled common-gate LNA in RX provides low impedance to the series RM network and balances the RF signal gain between differential inputs. Moreover, the noise figure (3 to 16dB), input referred 1dB compression point (-18 to -6dBm), and gain (13 to 22dB) of the LNA can be adjusted by controlling the size of M_1 and M_2 with constant bias current. In contrast, the inverter-based TX driver provides a high impedance to effectively drive the parallel RM network. The voltage swing of TX output, from 1 to 6V, controlled by a 5b digital code, can be also modified at 50Ω load to increase its power efficiency.

The wide-band FSK demodulator simultaneously demodulates multiple signals even though the received signal has a negative SNR. A fixed time delay with Gilbert-multiplier is adopted for wide-band FSK demodulation as shown in Fig. 2.1.4. To reduce power, the bias current of I_B is shared between the Gilbert multiplier and current-driven lattice all-pass filter. The delayed signal from M_1 and lattice LC tank is fed to sources of M_2 and M_3, where it is multiplied with the non-delayed signal, yielding the low-frequency sub-band signal at the output. To avoid frequency offset in demodulation, the frequency at the phase shift of $\pi/2$ is modified by controlling C_{tune} with a varactor. Similarly, to ensure a linear group delay in the operating frequency range, a variable MOS resistor, R_{tune}, is added. The output of the demodulator is obtained by a 1Mb/s FM signal with -40dBm RF input power.

Since the sub-band signal is located in a low frequency band below 20MHz, the reference clock is divided to obtain a sub-band LO signal without using a high-power frequency synthesizer. The output of the divider chain must have an exact 50% duty cycle to generate an accurate I/Q signal. However, the LO signal generated by a pulse swallow-based programmable divider cannot guarantee 50% duty cycle due to the prescaler operation. To ensure 50% duty cycle, the DCC circuit is implemented as in Fig. 2.1.5(a). The complementary charge-pump circuits detect the imbalance in the duty cycle of CLK_{OUT} and convert it into the mismatch between I_{C1} and I_{C2}. This mismatch is regulated to be equal by DCC loop gain for the duty-cycle correction. When 11% duty-cycle signal, which is generated by /23 using a pulse swallow divider, is applied to DCC, CLK_{OUT} is improved up to 50% duty cycle while consuming 80μW.

Figure 2.1.5(b) shows the micrograph of the WBAN transceiver. It is fabricated in 0.18μm CMOS and its area is 2.5×5mm² including pads. The scalable double-FSK signals are modulated from 4 sub-bands with variable data rate of 10kb/s to 10Mb/s and its TX output spectra are shown in Fig. 2.1.6. Each sub-band is distinguished by a wide-band FSK demodulator with corresponding sub-band carrier frequencies. To demonstrate WBAN coexistence in a low-SNR condition, 2 users that occupy the same sub-band (2MHz) with different sub-band carrier frequencies of 0.5MHz and 1MHz are applied with -40dBm RF input. The time-domain waveforms show the output of the wide-band demodulator using the RF inputs of 2 users. The performance and comparisons of Fig. 2.1.7 shows that the BCT supports all requirements for WBAN in terms of sensitivity, data rate, scalable BER, interference rejection, and co-existence thanks to scalable double-FSK modulation. The RM and CAS reduce the power consumption of the LNA and TX driver. In addition, the bias-reuse wide-band demodulator, and divider-based LO generation with DCC make the power-hungry frequency synthesizer unnecessary. As a result, the RX consumes 2.4mW with a data rate of 10Mb/s. Its minimum detectable signal is 250μV, which is 80× better than [2]. The energy consumption is 0.24nJ/b, which is the most energy efficient among the reported BCTs in our table.

References:

[1] IEEE 802.15 WPAN™ Task Group 6: Body Area Networks (BAN), Nov. 2007 [Online]. Available: http://www.ieee802.org/15/pub/TG6.html

[2] S. Song, et al., "A 0.9V 2.6mW Body-Coupled Scalable PHY Transceiver for Body Sensor Applications," *ISSCC Dig. Tech. Papers*, pp. 366-367, Feb. 2007.

[3] N. Cho, et al., "A 60kb/s-to-10Mb/s 0.37nJ/b Adaptive-Frequency-Hopping Transceiver for Body-Area Network," *ISSCC Dig. Tech. Papers*, pp. 132-133, Feb. 2008.

[4] A. Fazzi, et al., "A 2.75mW Wideband Correlation-Based Transceiver for Body-Coupled Communication," *ISSCC Dig. Tech. Papers*, pp. 204-205, Feb. 2009

[5] N. Cho, et al., "The Human Body Characteristics as a Signal Transmission Medium for Intrabody Communication," *IEEE Trans. Microwave Theory and Techniques*, vol. 55, pp. 1080-1086, May. 2007.

[6] J. Gerrits, et al., "Principles and Limitations of Ultra-Wideband FM Communications Systems," *EURASIP J. Applied Signal Processing*, vol. 2005, no. 3, pp. 382-396, Mar. 2005.

ISSCC 2011 / February 21, 2011 / 1:30 PM

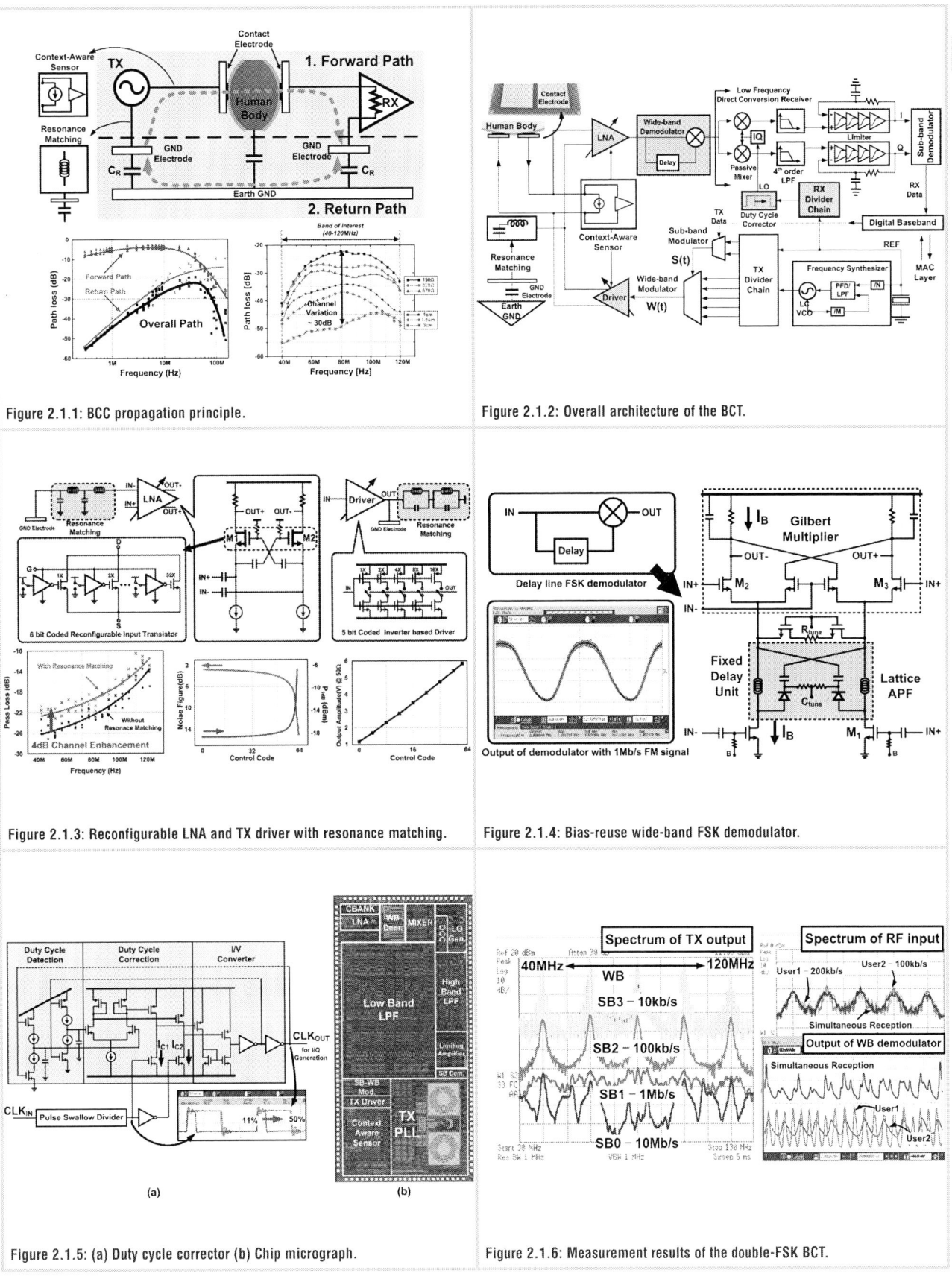

Figure 2.1.1: BCC propagation principle.

Figure 2.1.2: Overall architecture of the BCT.

Figure 2.1.3: Reconfigurable LNA and TX driver with resonance matching.

Figure 2.1.4: Bias-reuse wide-band FSK demodulator.

Figure 2.1.5: (a) Duty cycle corrector (b) Chip micrograph.

Figure 2.1.6: Measurement results of the double-FSK BCT.

978-1-61284-303-2/11 $26.00 © 2011 IEEE   35

Technology	0.18 μm RF CMOS
Die Area	2.5mm X 5mm
Frequency Band	40 MHz - 120 MHz
Modulation	FSK - FSK (Double FSK)
LNA Gain	13 to 22 dB
Post LNA Amp. Gain	0 to 30 dB
Sensitivity	-66 to -40 dBm
Data Rate	10 Mb/s to 1kb/s
BER @ -62dBm	10^{-5} @ 10Mb/s & 10^{-12} @ 10kb/s
Interference Rejection	6 to 36 dB
# of Co-existence	1 to 15
Supply Voltage	1.0 V

		Transmitter
Power Break-down	TX PLL	1.0 mW
	Divider/Other	0.8 mW
	Driver	0.2 to 2.0 mW
	Total	2.0 to 3.8 mW
		Receiver
	LNA	0.6 mW
	Post LNA Amp.	0 to 0.8 mW
	WB Demdulator	0.4 mW
	LP Filter X 2	0.4 X 2 mW
	Limitter X 2	0.2 X 2 mW
	DCC & CAS	0.1 mW
	Divider/Other	0.1 mW
	Total	2.4 to 3.2 mW
Energy/bit		0.24nJ/b

Parameters	ISSCC 2007 [2]	ISSCC 2008 [3]	ISSCC 2009 [4]	This Work
Technology	0.18μm CMOS	0.18μm CMOS	0.13μm CMOS	0.18μm CMOS
Supply Voltage	0.9V	1V	1.2V	1V
Modulation	3-Level PPM	AFH FSK	Correlation Direct Digital	Double FSK
Frequency Band	10-70MHz	30-120MHz	1-30MHz	40-120MHz
Data Rate	10kb/s - 10Mb/s	60kb/s - 10Mb/s	8.5Mb/s	1kb/s - 10Mb/s
Sensitivity	-30dBm	-65dBm	-60dBm	-66dBm
# of Coexistence	128	1	1	15
Power Consumption	2.6mW	3.7mW	2.75mW	2.4mW
Energy/bit	0.26nJ/b	0.37nJ/b	0.32nJ/b	0.24nJ/b

Figure 2.1.7: Performance summary and comparison.

ISSCC 2011 / SESSION 2 / TECHNOLOGIES FOR HEALTH / 2.2

2.2 A 75μW Real-Time Scalable Network Controller and a 25μW ExG Sensor IC for Compact Sleep-Monitoring Applications

Seulki Lee, Long Yan, Taehwan Roh, Sunjoo Hong, Hoi-Jun Yoo

KAIST, Daejeon, Korea

Recently, a wearable body-sensor network realized continuous sleep monitoring by ExG (EEG, EMG, EOG, and ECG) extraction from a sleeper's face [1]. At least 14 sensors were placed on the face, and were managed by a network controller for sleep monitoring. However, the system in [2] was too bulky and heavy (127×63.5×28mm³, 210g) to wear during sleep, and consumed high power (>100mW). Another system used a CMOS IC sensor for smaller power consumption (5 to 10mW/sensor), but it was still inconvenient to wear due to its cumbersome size (49×19×6mm³) [3]. This paper presents a 5g sub-500μW compact sleep-monitoring system based on a low-power network controller and ExG sensors, and Planar Fashionable Circuit Board (P-FCB) technology [4]. The entire sleep monitor operates on a 1.5V 10mAh single tiny coin battery (φ=5.8mm, T=2.1mm, 200mg) stacked on the controller chip. The power dissipation of 14 sensors and a controller is below 425μW, which is <1.5% of previous works [2-3]. Its low power consumption is possible by the Linked-List Manager (LLM)-based protocol such as Adaptive Dual-Mode Control (ADMC) and Continuous Data Transmission (CDT), and low-swing clock and data transceivers.

Figure 2.2.1 shows the real-time scalable sleep monitoring system. It consists of a network controller (NC) patch, 14 ExG sensor node (SN) patches (φ=20mm, T=2mm), and the conductive yarn wearable band (W-Band), which has 5 sublines (P, D, C, G, and R-line). Each patch has 3 P-FCB layers (Layer-1, an electrode; Layer-2, a power plane; and Layer-3, electronics) with <5mm thickness [4]. Since the system uses the power through the P-line from single tiny coin battery stacked on the NC patch, the power consumption of the overall system must be <1.5mW to ensure 10h lifetime. Moreover, the NC dynamically controls the sensors such as deactivating the detached sensor and also activating the secondary sensor automatically in real-time.

Figure 2.2.2 shows the overall architecture of NC, SN chips, and its network topology. The NC chip is composed of: (1) LLM for low-power real-time scalable network management, (2) ADMC for low-power dual-mode operation, (3) a 16b RISC-based bio-signal processor as a host for local signal filtering, compression, and storage, (4) a network interface (NI) for interfacing with W-Band, and (5) an external interface for storage extension. The ADMC has 2 operation modes: the SO mode for ExG sensing in all SNs at 20kHz clock, and the SC mode for both sensing and communication between SNs and the NC at 20MHz clock. To reduce power consumption, the SC mode duration, t_{SC}, should be much less than the SO mode duration, t_{SO}. An SN chip is composed of: (1) NI, (2) an SN controller for MAC and local storage, and (3) an analog front-end including 2 amplifiers and a 10b low-power SAR-ADC. To deal with 4 different ExG signals, the first amplifier maximizes the power noise efficiency and the second amplifier provides programmable gain and bandwidth from 48.5 to 60dB and from 0.4-to-40Hz to 0.4-to-300Hz, respectively. The SAR-ADC is implemented in a fully differential structure to optimize the trade-off between higher SNR, lower power consumption, and smaller chip area.

Figure 2.2.3 shows the LLM-based leave detector and join controller for real-time scalable network. SN information is automatically recorded in the linked list of LLM when the SN is connected with the W-Band. When one SN is supposed to leave the network during network operation (②), the leave detector can recognize that event by checking if the data is in the out-of-allowed signal dynamic range (±750mV differential). If the data is continuously out of range over 1Kb, the leave controller in LLM removes the SN information from the linked list. (0.5s @ 2kHz) If an SN wants to join the network (③), it sends a join request packet right after the last SN's data transmission. The join controller in LLM adds the new node to the linked list in 8μs. Compared to the previous static network controller in [3,5], the LLM-based network controller can achieve real-time dynamic network management, eliminating unnecessary power consumption.

Figure 2.2.4 shows the command-based ADMC scheme. When the total data length from all SNs becomes short due to leaving SNs, the ADMC adaptively reduces the duty cycle. Also if a new SN has joined, the ADMC scheme increases the SC mode duty cycle to accommodate the new SN. The measurement result in Fig. 2.2.4 shows that the data transaction time is reduced by 40% when one SN is left from the network resulting in the same amount of power reduction.

The CDT protocol for low-power data collection from multiple sensors is shown in Fig. 2.2.5. In this protocol, a 13b postamble packet with the 4b post ID of the linked list from the previous SN automatically starts the next SN rather than a data-request command from NC as the first SN. Average power consumption of clock and data transceivers is investigated as a function of the local memory depth, and it is found that a depth of 20 can achieve the lowest power consumption. Therefore, each SN transmits 20×10b per transaction so that a network with 14 SNs operates in SC mode with 1.7% duty cycle. The CDT can save the command and ACK bit for each SN reducing its communication time by 23.5% of [3].

Figure 2.2.6 shows the low-swing clock and data-transmitter circuits in NC and SN with their measured waveforms. In NC, the low-swing clock transmitter converts the 1.5V full-swing clock to a 1V-swing clock, which is clamped to 1V swing by the diode-connected load (MP_1, MN_1) in the output stage. In SN, the clock recovered by a cross-coupled latch-type receiver is used as an SN system clock and switching signal for $\frac{1}{2} \cdot V_{DD}$ generation. Under the loading effect of the data transmitter, the output voltage becomes 0.75V and 0.3V with 20MHz and 20kHz clocks, respectively. The ExG signal measurement result through SPI interface is shown in Fig. 2.2.6. Using the low-power techniques of ADMC and CDT, power is reduced by up to 58%. The low-swing clock and data transceivers reduce system power by an additional 17.4%.

The lower part of Fig. 2.2.5 shows the chip micrograph and the system performance summary. NC and SN chips are fabricated in 0.18μm 1P6M standard CMOS and occupy 6.4mm² and 4.2mm², respectively, including pads. Average power consumption is 75μW for the NC chip and 25μW for the SN chip. Its transaction energy is only 0.33pJ/b, which is the lowest energy consumption among wireline-based body sensor network transceiver reported in [6-7]. Also, it can dynamically manage the leaving and joining SNs within 500ms in the worst case. As shown in Fig. 2.2.7, the total system has only 1/9th the size and <1.5% power consumption, which arguably makes it more comfortable to wear than the previous system [3].

References:

[1] B. V. Vaughn, et al., "Technical Review of Polysomnography," *Chest*, vol. 134, no. 6, pp.1310-1319, Dec. 2008.

[2] H. A. Kayyali, et al., "Remotely attended home monitoring of sleep disorders," *Telemed. J. E. Health*, vol. 14, no. 4, pp.371-374, May, 2008.

[3] N. Vicq, et al., "Wireless Body Area Network for Sleep Staging," *Proc. IEEE BIOCAS*, pp.163-166, Nov. 2007.

[4] J. Yoo, et al., "A 5.2mW Self-Configured Wearable Body Sensor Network Controller and a 12μW 54.9% Efficiency Wirelessly Powered Sensor for Continuous Health Monitoring System," *ISSCC Dig. Tech. Papers*, pp.290-291, Feb. 2009.

[5] A. C. W. Wong, et al., "A 1V, Micropower System-on-Chip for Vital-Sign Monitoring in Wireless Body Sensor Networks," *ISSCC Dig. Tech. Papers*, pp.138-139, Feb. 2008.

[6] P. P. Mercier, et al., "A 110μW 10Mb/s eTextile Transceiver for Body Area Networks with Remote Battery Power," *ISSCC Dig. Tech. Papers*, pp.496-497, Feb. 2010.

[7] J. Yoo, et al., "A 1.12pJ/b Inductive Transceiver With a Fault-Tolerant Network Switch for Multi-Layer Wearable Body Area Network Applications," *IEEE J. Solid-State Circuits*, vol. 44, no. 11, pp.2999-3010, Nov. 2009.

978-1-61284-303-2/11 $26.00 © 2011 IEEE

ISSCC 2011 / February 21, 2011 / 2:00 PM

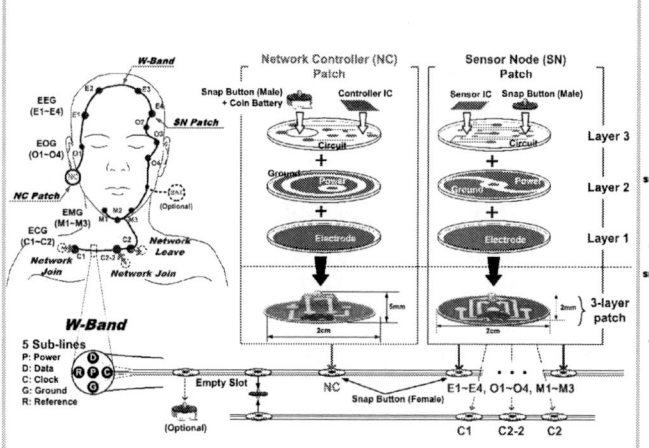

Figure 2.2.1: Compact sleep-monitoring system with a real-time scalable network.

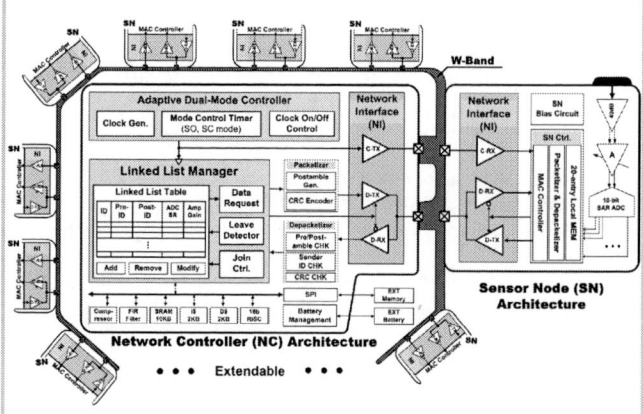

Figure 2.2.2: Overall architecture of a network controller (NC) and sensor nodes (SN) with their network topology.

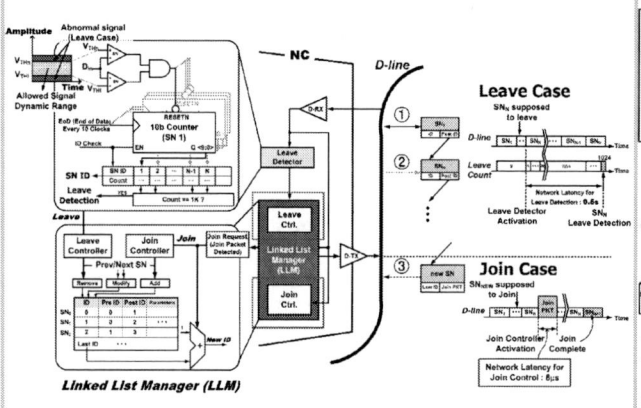

Figure 2.2.3: Linked-List Manager (LLM)-based leave detector and join controller for real-time scalable network.

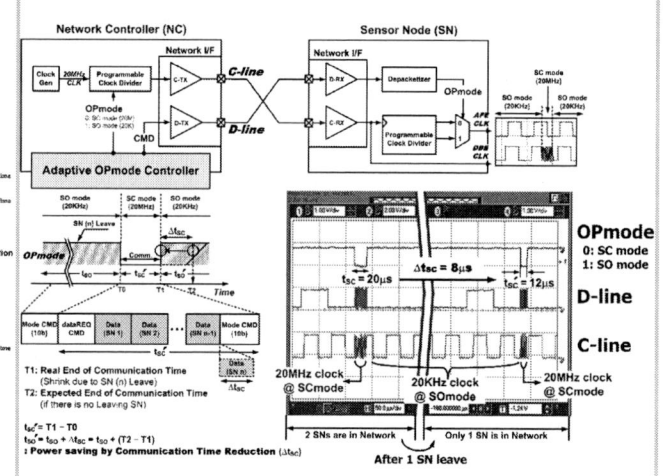

Figure 2.2.4: Command-based Adaptive Dual-Mode Control (ADMC) scheme for low power consumption.

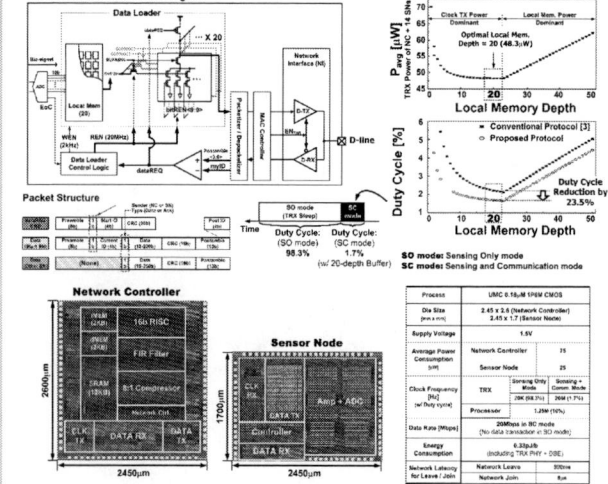

Figure 2.2.5: Continuous Data Transmission (CDT) of multiple sensors (Upper part). Chip micrograph and performance summary (Lower part).

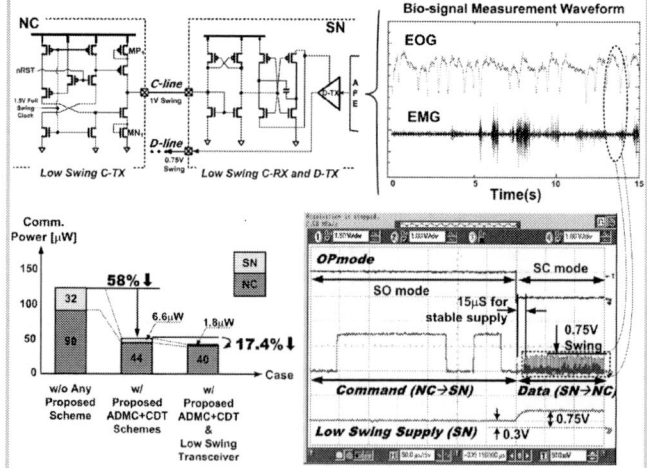

Figure 2.2.6: Low swing clock and data transceiver circuits with measurement results.

978-1-61284-303-2/11 $26.00 © 2011 IEEE

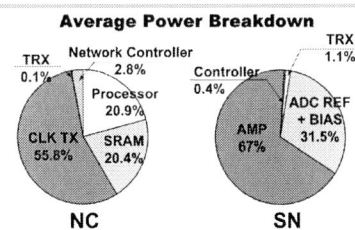

Average Power Breakdown

NC

SN

Performance Comparison

	[3] BioCAS 2007	This Work
Size / 1 SN	49 x 19 x 6 mm^3	π x (20/2)^2 x 2 mm^3
Lifetime (System)	10-h with 3 prismatic lithium battery	35-h with 1 coin battery (10mAh, 1.5V)
ExG Channel	5	14
MAC Protocol	Duty-cycled TDMA w/ Preallocated Time Slot	Duty-cycled TDMA w/ Adaptive Time Slot (ADMC)
Network Reconfiguration	NO	YES (real-time scalable network)
SN-NC Comm.	RF	Wireline (W-Band)

Figure 2.2.7: Chip power breakdown, and performance comparison.

ISSCC 2011 / SESSION 2 / TECHNOLOGIES FOR HEALTH / 2.3

2.3 A 3μW Wirelessly Powered CMOS Glucose Sensor for an Active Contact Lens

Yu-Te Liao, Huanfen Yao, Babak Parviz, Brian Otis

University of Washington, Seattle, WA

The increase in the diabetes population makes glucose monitoring a pressing demand for clinical and continuous use. Non-invasive sensing would allow a painless, convenient solution compared to traditional skin-piercing glucose meters. Among various body fluids, tear fluid, which is correlated to the glucose concentration in blood [1], is directly accessible on the eye and can provide a unique opportunity to develop an interface between a sensor and the human body. The current technique is to collect tear fluid samples in capillary tubes and assay the samples for glucose *ex situ* using standard laboratory instrumentation. Integrating sensors into a contact lens would provide a way to continuously and reliably sense metabolites in tear fluids.

We present an integrated wirelessly powered glucose sensor allowing a functional active system on a contact lens. This sensor IC consists of power management, readout circuitry, wireless communication interface, and energy-storage capacitors in a standard CMOS process with no external components. Fig. 2.3.1 shows the electrochemical glucose sensor generating a current proportional to the glucose concentration. The main principles are as follows:

$$D-Glucose+O_2 \xrightarrow{GOD} H_2O_2 + D-Gluconolactone$$
$$H_2O_2 \rightarrow 2H^+ + O_2 + 2e^-$$

Amperometric results of a continuous flow test for low concentrations of glucose in tear film are shown. The sensor is composed of the working electrode (WE), counter electrode (CE) and reference electrode (RE). Using thin-film microfabrication techniques, the sensor is fabricated with 3 metal layers (Ti(10nm)/Pd(10nm)/Pt(100nm)) on a transparent polyethylene terephthalate (PET) polymer, and then molded to a lens shape with a diameter of 1cm, as shown in Fig. 2.3.1. The sensing area is 0.22mm^2. Glucose oxidase (GOD) and titanium isopropoxide are used to create a GOD/titania sol-gel membrane on the sensor area, which creates a measured sensitivity of 1.67μAmm^{-2}mM^{-1}.

Our previous work demonstrated the potential value of integrating silicon chips, antenna, and an LED on a contact lens [2]. The basic functional elements required are energy delivery, sensing signal readout circuitry, signal processing, and communication subsystems. Each of these blocks must be designed for minimum energy consumption. An on-eye sensor would clearly not allow an on-board power source or battery, so power must be delivered wirelessly using far-field electromagnetic radiation or near-field inductive coupling. Near-field coupling requires a large inductance/transformer, making integration difficult in size-constrained applications. By considering the path loss, antenna efficiency and regulations on maximum RF power exposure to the human eye, we adopt a 2.4GHz carrier frequency for power delivery and data communication.

Another design challenge is the size of chip, which is limited to approximately 0.5×0.5mm^2 due to the radius of curvature of the human eye. All active and passive components must be integrated into this area. Figure 2.3.2 shows the sensor readout architecture. A 3-stage full-wave rectifier is used to convert the incoming 2.4GHz RF power to a DC voltage supply, which is subsequently filtered by an on-chip 800pF capacitor. Since fluctuations in incident RF power cause supply variations, on-chip regulation is necessary. A sub-μW low-power regulator and bandgap reference [3] provide a stable supply for the sensing electronics. We use a separate digital and analog supply-regulation technique [4] to reduce the noise coupling into the sensing element from the oscillator and logic switching noise. To provide isolation from digital and analog supply without adding an extra regulator, the regulator pass transistor is separated. This topology achieves 30dB isolation between digital and analog supplies. Figure 2.3.2 shows the low-power regulator schematic and the measured on-chip regulation with different input power levels. The minimum input power of the chip is 0dBm since the chip impedance is directly matched to the antenna without an external matching network.

Figure 2.3.3 shows the current readout circuitry. This readout circuitry includes a potentiostat, providing a stable 400mV voltage difference between the working and reference electrodes. The sensing current is injected into an oscillator-based current-to-frequency converter. The ring oscillator nominally operates at 250kHz, consumes 300nA, and is sensitive to additional injected current. The linear readout current range is 750nA. The bandgap voltage reference and a sub-threshold voltage divider [5] provide excellent stability to environmental variations. The measured frequency stability over 6 hours is shown in Fig. 2.3.3, showing a maximum frequency deviation less than 1kHz. The temperature variation is compensated to first-order through differential measurement of a reference oscillator and the sensing oscillator. Frequency-modulated load shifting key (FM-LSK) is used to transmit the sensing data to the reader. The backscattered signal encodes the glucose concentration as an offset-modulated frequency.

To measure the performance of the glucose sensor, we connect the molded sensor with the readout IC and then use a syringe pump to control a volume of test solution flowing through the sensor. A buffer solution (pH 7.4), consistent with physiological conditions, is added after each concentration to flush the remaining ions of the previous solution. The glucose level in tear-film is reported in the range of 0.1 to 0.6mM. Figure 2.3.4 shows the measured oscillator frequency change with different glucose concentrations from 0 to 2mM. The linear correlation coefficient is 0.9968 from 25 tested samples. Figure 2.3.5 shows the measured backscattered signal from the chip. The glucose concentration of 0.25mM results in a 450kHz frequency deviation of the backscattering carrier. The glucose sensor system consumes 3μW and can be powered over 15cm from an effective isotropically radiated power (EIRP) of 0.1W power source at 2.4GHz. The efficiency of the glucose sensor system is limited to the size of antenna, storage capacitance, and matching network. The frequency deviation of the backscattered carrier can be demodulated by a conventional FM receiver. Figure 2.3.6 is a performance summary of our CMOS glucose sensor system and Fig. 2.3.7 shows the micrograph of the chip. The small chip area, high level of integration, and low power of our system allows application in multiple bio-sensing tasks, including smart contact lenses.

Acknowledgments:

We would like to thank Dr. Ilkka Lahdesmaki, Jagdish Pandey, and Andy Lingley for helpful discussion in sensor design as well as the National Science Foundation for support.

References:

[1] J.T. Baca, C.R. Taormina, E. Feingold, D.N. Finegold, J.J. Grabowski, S.A. Asher, "Mass spectral determination of fasting tear glucose concentrations in nondiabetic volunteers," *Clin. Chem.* vol. 53, pp. 1370-1372, 2007.

[2] J. Pandey, Y.-T. Liao, A. Lingley, Babak Parviz, B. Otis, "Toward an active contact lens: integration of a wireless power harvesting IC," *IEEE Biomedical Circuits and Systems Conf.*, pp. 125-128, Nov. 2009.

[3] D. Yeager, F. Zhang, A. Zarrasvand, and B. Otis, "A 9.2μA gen 2 compatible UHF RFID sensing tag with -12dBm sensitivity and 1.25μVrms input-referred noise floor," *IEEE ISSCC Dig. Tech. Papers*, Feb. 2010.

[4] M.M. Ahmadi, G. Jullien, "A wireless-implantable microsystem for continuous blood glucose monitoring," *IEEE Trans. on Biomedical Circuits and Systems*, vol. 3, pp. 169-179, June 2009.

[5] K. N. Leung, P.K.T. Mok, "A sub-1-V 15ppm/°C CMOS bandgap voltage reference without requiring low threshold voltage device," *IEEE J. Solid-State Circuits*, vol. 37, pp. 526-530, Apr. 2002.

ISSCC 2011 / February 21, 2011 / 2:30 PM

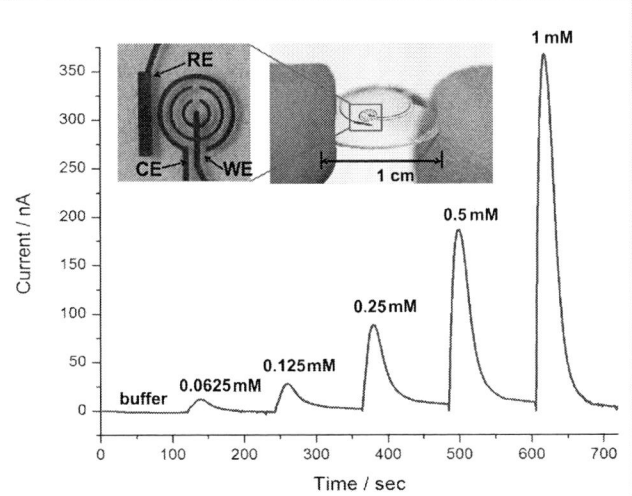

Figure 2.3.1: Photograph of assembled glucose sensor on a plastic substrate. The plot shows continuous glucose response measurements.

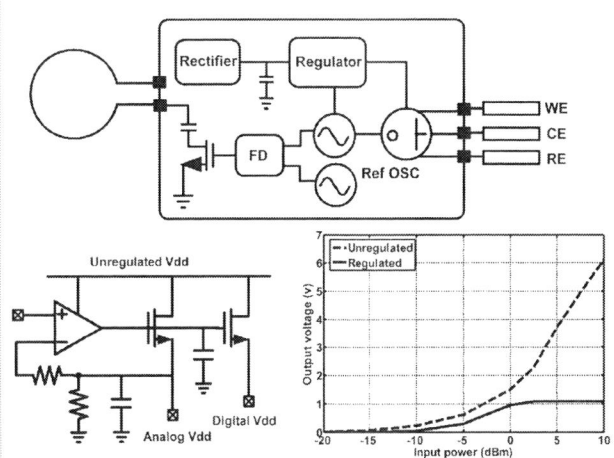

Figure 2.3.2: Block diagram of the wireless glucose sensor readout architecture and measured on-chip regulation.

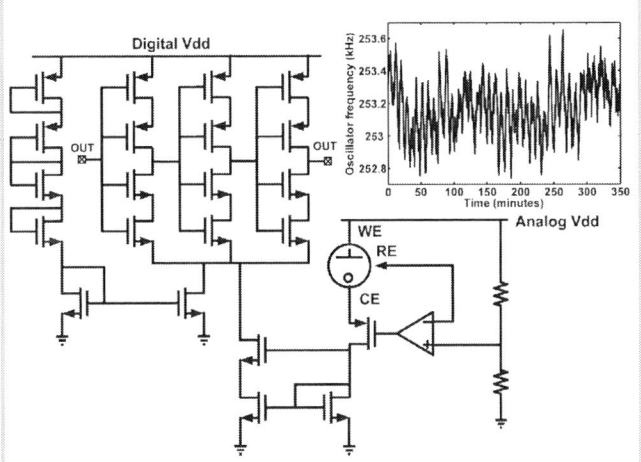

Figure 2.3.3: Current-sensing readout circuitry and measured frequency drifts.

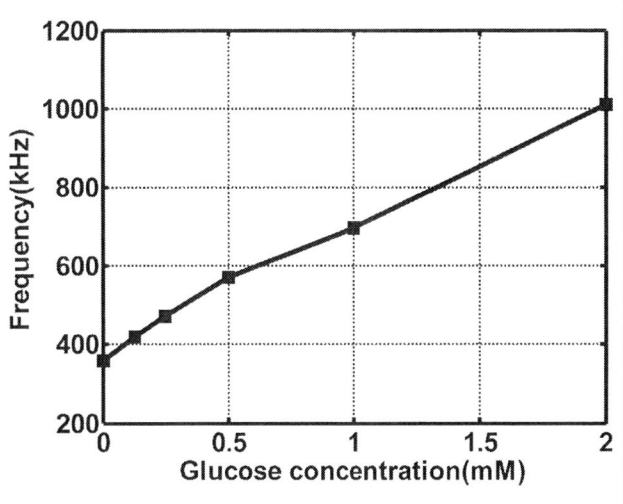

Figure 2.3.4: Measured oscillator output frequency vs. glucose concentration.

Figure 2.3.5: Measured wireless transmission results (backscattering).

	This work	[4]
Modulation scheme	LSK	LSK
Carrier frequency	2.4 GHz	13.56 MHz
Power consumption	3 µW	110 µW
Full scale measured current	750 nA	1 µA
External component	No	Yes
Read range	15 cm (EIRP= 0.1W)	4 cm
Tempco (ppm) (35-45°C)	8300	N/A
Glucose detect range	0-2 mM (tear)	0-40 mM (blood)
Glucose sensitivity (µAmm^{-2} mM^{-1})	1.67	N/A

Figure 2.3.6: Performance summary and comparison.

978-1-61284-303-2/11 $26.00 © 2011 IEEE

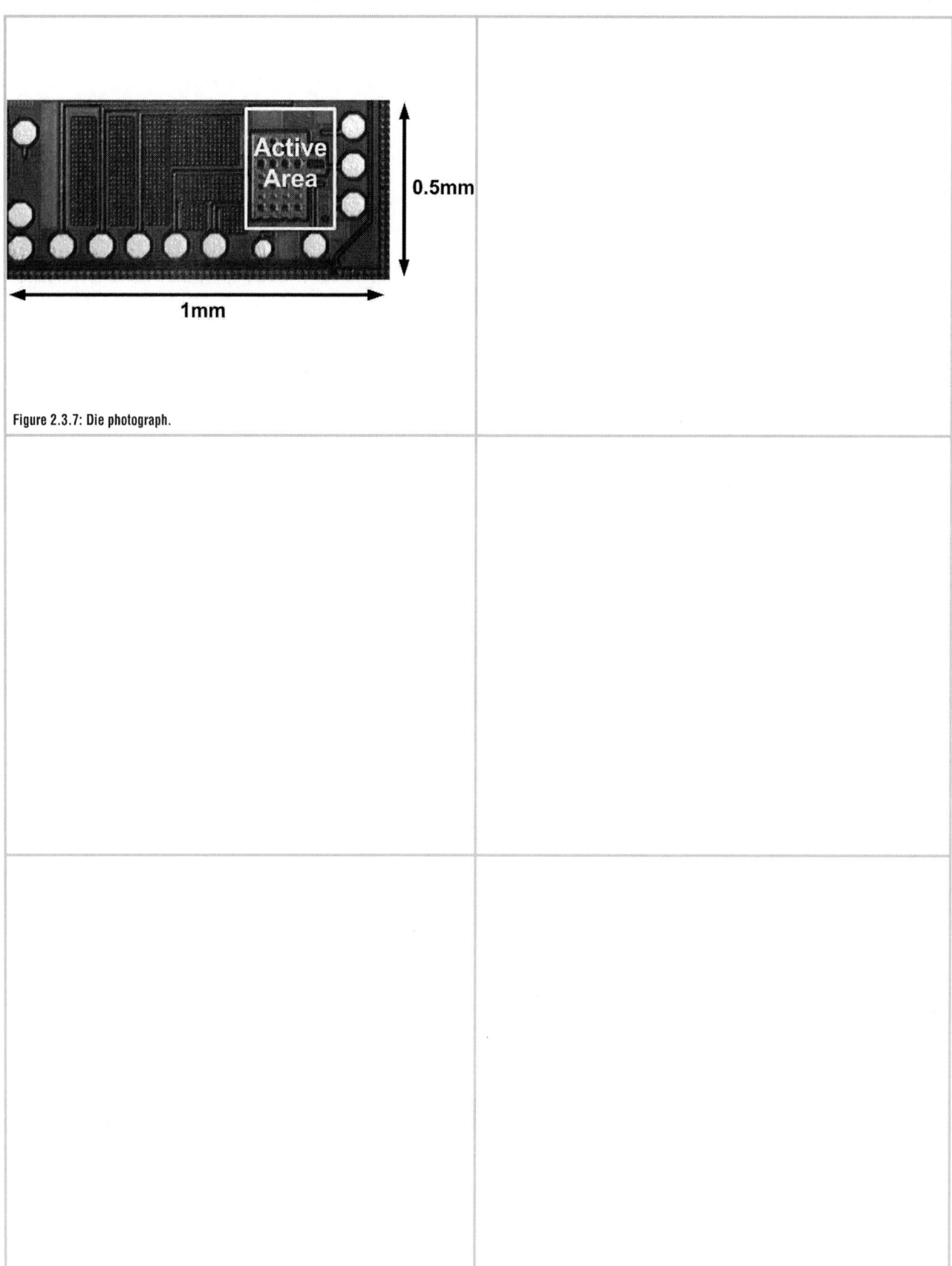

Figure 2.3.7: Die photograph.

ISSCC 2011 / SESSION 2 / TECHNOLOGIES FOR HEALTH / 2.4

2.4 A 90nm CMOS SoC UWB Pulse Radar for Respiratory Rate Monitoring

Domenico Zito[1,2], Domenico Pepe[2], Martina Mincica[2,3], Fabio Zito[4]

[1]University College Cork, Cork, Ireland,
[2]Tyndall National Institute, Cork, Ireland,
[3]University of Pisa, Pisa, Italy,
[4]University of Reggio Calabria, Reggio Calabria, Italy

UWB technology (3.1 to 10.6GHz) allows new applications for both data communication and sensing (FCC, reference in [1]). Due to their potential in terms of resolution and extremely low level of EIRP spectral density (< -41.3dBm/MHz, see mask in Fig. 2.4.1) [1], UWB radars are very attractive for a large set of civil and military sensing applications, as ground penetrating, surveillance, localization, intra-wall and through-wall detections and biomedical imaging. Moreover, with respect to continuous-wave (CW) radars [2], UWB radar transceivers present a lower circuit complexity since no frequency conversions are required, leading to lower power consumption (P_C) for longer battery autonomy.

In spite of the implementations of efficient building blocks [3] and especially pulse generators [1], UWB radar front-ends fully integrated in a single silicon die, tested and working, have not been reported yet in the literature. This paper reports what we believe to be the first implementation, including experimental evidence of a SoC UWB pulse radar based on a correlation receiver, in 90nm CMOS technology by STMicroelectronics.

Figure 2.4.1 reports the UWB pulse radar sensor, which is implemented to be used mainly (but not only) for contactless detection of vital signs (e.g. respiration) [4]. The pulse generator (PG) transmits short pulses towards the target (e.g. human body) with a pulse repetition frequency (f_{PR}). The signals reflected by the target are captured by the RX antenna, amplified by the LNA and multiplied with a delayed replica of the transmitted pulses generated on-chip by the Shaper. The delay can be varied in order to span the range of interest and identify the target (*ranging mode*, RM). When the target is detected, the delay can be fixed in order to monitor a fixed range gate (*tracking mode*, TM). Since vital signs vary within a few Hertz, an integrator 3-dB band (B_{3dB}) of 100Hz allows an accurate detection. Averaging a large number of pulses (in the order f_{PR}/B_{3dB}) allows us to increase the SNR_{out} (e.g. 10^5 pulses for f_{PR}=10MHz). Therefore, the output voltage is directly sensitive to the target movement, e.g. the chest movement due to the pulmonary activity in case of respiratory rate monitoring.

All the circuits are designed according to the specifications derived by the feasibility study [4]. They have fully differential topologies for better immunity to EMI and noise, and linearity.

Figure 2.4.2 shows the PG designed by following the principle in [1]. It provides a monocycle pulse on the 100Ω-diff antenna when activated by the negative edge of the digital signal *command* occurring at f_{PR}. The PG, tested individually on the same die of the SoC radar, exhibits pulses with a duration time (t_D) of 350ps and amplitude 650mV$_{pp-diff}$ (including the losses of microprobes and cables) corresponding to ~900mV$_{pp-diff}$ on-chip. The energy consumption (E_C) per transmitted pulse is 19.8mW×350ps=6.9pJ. The PG exhibits the highest pulse amplitude, voltage (55%) and energy (3%) efficiencies, and robustness to PVT variations among those reported in the reference list of [1]. The Shaper adopts the same structure in [1], but it requires a lower bias current (2mA) for its shaping network since it has a different load impedance (i.e. the input impedance of the multiplier).

Figure 2.4.2 shows the UWB LNA implemented according to the design in [5]. It consists of a common-gate stage, which realizes a wideband input impedance matching to the 100Ω-diff antenna, and two subsequent common-source stages which increase the overall gain. Figure 2.4.2 reports the main performance of the LNA tested as stand-alone. Note that $|S_{21}|$=22.7dB at 5GHz, $|S_{11}|$<-10.5dB and NF=6.5dB (including the noise contributions of UWB baluns, probes and cables) over the all UWB band. The input P_{1dB} is -19.7dBm. This circuit solution exhibits one of the highest transducer gains (references in [5]).

Figure 2.4.3 shows the UWB multiplier, which exploits a p-MOSFET (M_P) common-gate pair input stage in order to provide a low-complexity input matching

to the LNA output impedance (~200Ω-diff) over a wide band. The multiplier stage consists of a p-MOSFET (M_Q) Gilbert's quad. The common-mode feedback (CMFB) regulates the common-mode output voltage through V_{CM} and V_C. V_{REF} (0.6V) is an external dc voltage reference. The multiplier has been tested as stand-alone by applying two sequences of pseudo-Gaussian monocycle pulses with amplitude 170mV$_{pp-diff}$ (V_P) and 650mV$_{pp-diff}$ (V_Q), and t_D=700ps (V_P is expected to be the echo amplified by the LNA, whereas V_Q the delayed replica generated on-chip by the Shaper). Figure 2.4.3 reports the output voltage (V_O) for 3 different delays between the input pulses V_P and V_Q. The NF is 14.4dB at 4GHz. The P_{1dB} is -3.5dBm and -2dBm for P and Q inputs, respectively. This circuit solution allows us to efficiently implement the multiplication between short pulses, with improved performance with respect to the state of the art (SoA) in terms of P_C, NF and P_{1dB}.

Figure 2.4.4 shows the schematic of the integrator. It consists of a 3-stage amplifier with RC feedback and output buffer. The stages use identical differential pair amplifiers. The voltage gain is 58dB and B_{3dB}=147Hz.

Figure 2.4.4 also shows the 5b programmable monotonic delay generator (DG) designed following the principle in [6]. The 5 bits allow us to select the output current of the inverter (M_8 and M_9) in order to vary the slope of the voltage ramp (high to low) across its load capacitance. By varying the bias current, the DG can provide delays in the range 1 to 30ns.

The UWB radar sensor has been tested in 3 different experimental test setups (TS1, TS2 and TS3). In TS1, we tested the radar testchip only (no antennas) as in Fig. 2.4.5. The PG output is connected to the LNA input by means of microprobes, cables and attenuators. The propagation delay of cables is fixed (~8ns), which corresponds to a static target. Figure 2.4.5 reports the output voltage for RM and TM. In RM, DG is driven by a 5b ramp in order to span the range of interest (around 8ns). In TM, we emulated the target movement by driving the DG with 2 digital words, corresponding to full and no correlations between echo and on-chip replica, as in case (a) and case (c) of Fig. 2.4.3. These results demonstrate that the testchip works properly. Note that the dc offset (~50mV), which can be eliminated by means of digital processing, is not an issue for this application aimed at detecting the movement rate.

Figure 2.4.6 shows TS2 and TS3, in which the radar testchip (packaged in QFN32) is mounted on a test-board including the antennas (2.3dBi at 3.5GHz, $|S_{11}|$<-10dB in the band from 2.8 to 5.4GHz, which covers the range of interest 3 to 5GHz). In TS2, the radar detected 3 targets with different areas (26×26, 13×26 and 13×13cm^2) for movements (front-back) up to d=2cm around a distance (D) up to 70cm. Figure 2.4.6 reports the output voltage for the target of area 26×26cm^2 (d=2cm, D=70cm). E_C per received pulse is 59.4mW×0.6ns=36pJ (t_D of the echo captured by the antenna is ~0.6ns). For TS3, the measurements show that the radar detects the respiratory rate (~0.5Hz) of the person under test placed at D=25cm. All these results demonstrate that the UWB radar sensor proposed detects properly the targets in the operating scenarios.

Acknowledgements:
This work has been funded by the European Commission (FP6-2004-IST-4-026987). We thank SFI and IRCSET for the financial support.

References:
[1] F. Zito, D. Pepe, D. Zito, "UWB CMOS Monocycle Pulse Generator", *IEEE Trans. Circuits and Systems I*, vol. 57, pp. 2654–2664, Oct. 2010.
[2] A.D. Droitcour, O. Boric-Lubecke, V.M. Lubecke, J. Lin, "0.25μm CMOS and BiCMOS single-chip direct-conversion Doppler radars for remote sensing of vital signs", *IEEE ISSCC Dig. Tech. Papers*, pp. 348–349, Feb. 2002.
[3] D.C. Daly et al., "A pulsed UWB receiver SoC for insect motion control", *IEEE ISSCC Dig. Tech. Papers*, pp. 200–201, Feb. 2009.
[4] D. Zito, et al., "Wearable System-on-a-Chip UWB Radar for Health Care and its Application to the Safety Improvement of Emergency Operators", *IEEE Dig. Int. Conf. Engineering in Medicine and Biology Society*, pp. 2651–2654, Aug. 2007.
[5] D. Pepe, D. Zito, "22.7dB Gain –19.7dBm ICP$_{1dB}$ UWB CMOS LNA", *IEEE Trans. Circuits and Systems II*, vol. 56, pp. 689–693, July 2009.
[6] M. Maymandi-Nejad, M. Sachdev, "A monotonic digitally controlled delay element", *IEEE J. Solid-State Circuits*, vol. 40, pp. 2212–2219, Nov. 2005.

978-1-61284-303-2/11 $26.00 © 2011 IEEE

ISSCC 2011 / February 21, 2011 / 3:15 PM

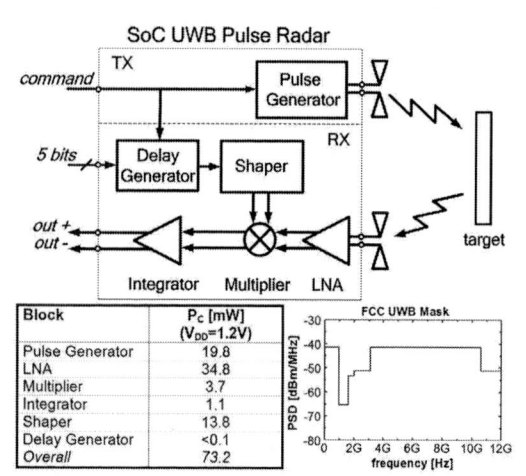

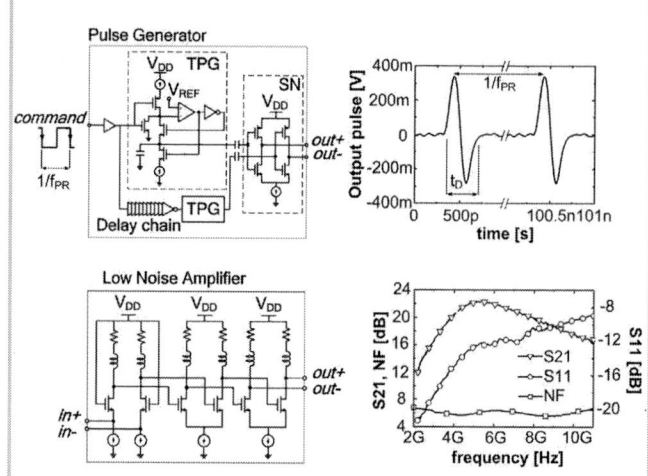

Figure 2.4.1: Diagram of the SoC UWB radar front-end. The overall P_C (bias inclusive) is lower than CW radars for similar applications (e.g. >80mW [2]).

Figure 2.4.2: Simplified schematic and measurements of PG and LNA. In PG, TPG is for 'triangular pulse generator', SN is for 'shaping network'.

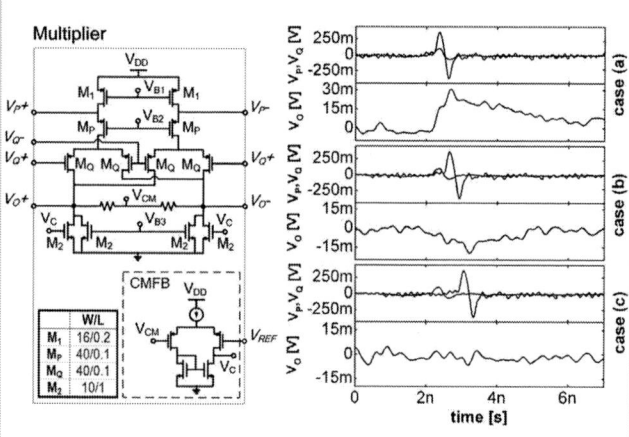

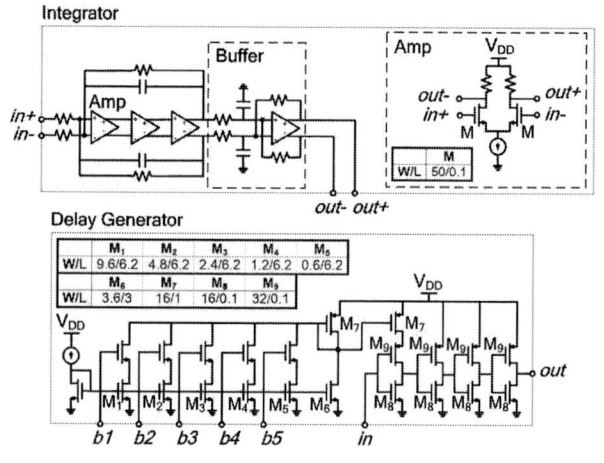

Figure 2.4.3: Schematic and measurements of the multiplier. V_O is captured by the probe 1134A (impedance 50kΩ||0.27pF) and oscilloscope DSO 54855A.

Figure 2.4.4: Schematic of the integrator and DG.

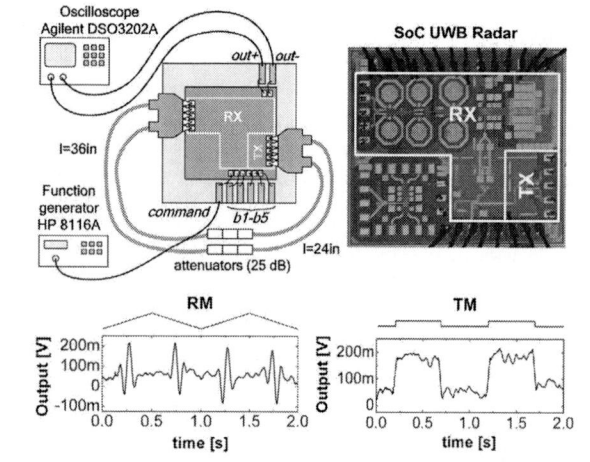

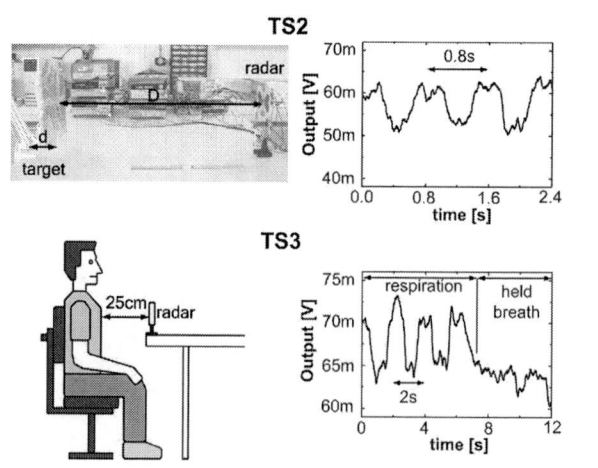

Figure 2.4.5: TS1 and output voltages for RM and TM (f_{PR}=40MHz). The die size is 1.5×1.3mm² (including the multiplier as stand-alone device).

Figure 2.4.6: TS2 and TS3, and output voltages in TM (f_{PR}=40MHz). For TS2, the target is made of a 0.5cm-thick board covered by an aluminium foil.

978-1-61284-303-2/11 $26.00 © 2011 IEEE

ISSCC 2011 / SESSION 2 / TECHNOLOGIES FOR HEALTH / 2.5

2.5 A Broadband THz Imager in a Low-Cost CMOS Technology

Franz Schuster[1,2], Hadley Videlier[2], Antoine Dupret[1],
Dominique Coquillat[2], Maciej Sakowicz[2], Jean-Pierre Rostaing[1],
Michaël Tchagaspanian[1], Benoît Giffard[1], Wojciech Knap[2]

[1]CEA-LETI-MINATEC, Grenoble, France,
[2]Université Montpellier 2 -CNRS UMR, Montpellier, France

Terahertz (THz) technology has become of large interest over the last 10 years. THz rays are an alternative to X-rays for imaging through thin materials and their non-ionizing character makes them inherently health-safe. The THz domain is also suitable for heterodyne detection and the use of radar techniques to perform 3D imaging. Commercial applications range from non-destructive testing, security screening of objects or persons, and medical imaging to secure communications.

Among the multitude of existing THz detectors, silicon field-effect transistors have shown to be suitable for cost-effective video-rate imaging, offering the advantages of room temperature operation, integration of read-out electronics on the same chip, and straightforward array fabrication. The first demonstration of sub-THz and THz detection by CMOS field-effect transistors in silicon was made in 2004 [1] and it was shown shortly later that these devices can reach a noise equivalent power competitive to the best conventional room-temperature THz detectors [2]. The first CMOS focal-plane arrays (FPAs) for imaging at 600 and 645GHz were demonstrated in 2008 and 2009 [3, 4]. Further reduction of the system costs can be achieved by designing a versatile image sensor that operates at a wide range of THz frequencies. This paper presents a prototype THz imager in 0.13µm CMOS with imaging capability from 300GHz to 1THz, low-noise in-pixel amplifiers and multiplexing circuitry for single video output. The 0.13µm CMOS technology enables short enough gate lengths for optimum detection performance up to a few THz, while being significantly cheaper than 90nm or silicon on insulator (SOI) technologies. Furthermore, it is dense enough to be compatible with the THz pixel size.

Figure 2.5.1 presents the architecture of the 3×4 pixel imager. Each pixel consists of a differential bow-tie antenna, a single nMOSFET as detecting element, and a single-ended base band amplifier with capacitive feedback. The pixel size is 190×190µm². The pixel outputs are multiplexed to the array output in a standard way via line and column switches controlled by two shift registers. The antenna couples the free-space THz radiation to the detector nMOSFET. The bow-tie type has been chosen in order to achieve a wide detection bandwidth and allow different possible working frequencies from 300GHz to 1THz. The bow-tie shapes are drawn in each of the metal layers M1 to M6 and interconnected by via arrays to reduce conduction losses. In order to decrease substrate losses in the standard resistivity silicon substrate (ρ=10Ωcm), we grind down the dies to a total thickness of 130µm. The detector nMOSFET has a gate length of 130nm and a gate width of 250nm. Even though being far above its cut-off frequency, it effectively rectifies the received THz radiation, leading to a dc detection voltage ΔU between source and drain. The rectification phenomenon is explained by the non-resonant case of the Dyakonov-Shur plasma wave theory [5] or alternatively by distributed resistive self-mixing [4, 6]. In order to avoid $1/f$ noise in the amplifier and external read-out electronics we modulate either the THz radiation itself (mechanical chopper) or the detector's gate bias. Therefore the source-drain detection signal ΔU becomes a low frequency signal at the modulation frequency (400Hz–30kHz). The FET is placed between the bow-tie tips, thus no transmission lines are required for the antenna connection.

Figure 2.5.2 shows the circuit schematic of the amplifier. It is designed to achieve a good noise performance, while consuming less than 100µW to be compatible with the use in bigger arrays (e.g. 160×120 pixels). The amplifier consists of a pMOSFET cascoded common source stage and a source follower. pMOS devices have been chosen over nMOS one because of less $1/f$ noise at the low frequencies used for modulation. The large input pMOSFET $P0$ (W = 480µm, L = 300nm) is drawn in a custom "waffle" ("Manhattan") layout to minimize the gate, source and drain resistances for better noise performance. There is one gate contact for every 1.14µm-wide gate segment. The amplifier is used in feed-back mode and the closed-loop gain is fixed to 31dB by two metal-insulator-metal (MIM) capacitors with C'=2fF/µm² in a dedicated layer stack above the $M6$ layer. The amplifier has a simulated input referred noise of 16nV/Hz$^{0.5}$ at 30kHz, a 2MHz bandwidth, and a consumption of 97µW under the 1.2V supply. This corresponds to the pixel consumption as the detector nMOSFET is not current biased.

Figure 2.5.3 presents the measured detection voltage V_{out} after the amplifier (peak-to-peak value) and the voltage responsivity R_v versus the gate bias at 300GHz illumination with an all-electronic source based on frequency multipliers. The inset shows a raster scan image of the focused 300GHz beam. The responsivity of the detector is determined by normalizing the surface integral of the beam image to the physical pixel size A_{pix} = (190µm)² and the total beam power in the detector plane P_{beam} that is previously measured with a free-space absolute power meter (from Thomas Keating). The maximum responsivity value is 90kV/W which is still 12% higher than the one reported in [4] with 43dB on-chip amplification. We have furthermore characterized the detectors with a backward wave oscillator (BWO) source at 1.05THz and the responsivity is still ~1.8kV/W even at that high frequency (see also [7]). All measurements were done at room temperature.

In order to validate the pixels' functionality through an imaging application, the experimental set-up in Fig. 2.5.4 is used. To increase the image acquisition time compared to mechanical modulation with a chopper we modulate the detector's gate bias with a 30kHz square wave between 0.2 and 1.0V, switching the detector from a high-sensitivity state to a low-sensitivity state as can be seen in Fig. 2.5.3. Figure 2.5.5 shows a raster scan image at 300GHz of different tree leaves. The image consists of 600×225 scanned points and the spatial resolution is close to the wavelength of 1mm. The inner structure of the leaves is very well revealed as zones with higher thickness and/or higher moisture content become visible. Figure 2.5.6 shows a photograph of the test chip containing the THz imager as well as other THz pixels and test structures. The table gives an overview of the imager's characteristics and compares them to the CMOS THz focal plane array in [4].

The paper demonstrates a complete acquisition chain of a 300GHz to 1THz image sensor with on-chip multiplexing in a 0.13µm bulk silicon CMOS technology. The pixel consumes less than 100µW and has a responsivity of 90kV/W at 300GHz and of 1.8kV/W at 1.05THz respectively. High resolution and contrast THz images revealing the inner structure of tree leaves are presented. These results show that multi-frequency and room temperature imaging systems are possible in cost-effective CMOS technologies.

References:
[1] W. Knap, F. Meziani, N. Dyakonova, N. Lusakowski, *et al.*, "Plasma wave detection of sub-terahertz and terahertz radiation by silicon field-effect transistors," *Applied Physics Letters*, vol. 85, no. 4, p. 675, 2004.
[2] R. Tauk, F. Teppe, S. Boubanga, W. Knap, *et al.*, "Plasma wave detection of terahertz radiation by silicon field effect transistors: Responsivity and noise equivalent power," *Applied Physics Letters*, vol. 89, p. 253511, 2006.
[3] U. Pfeiffer and E. Ojefors, "A 600-GHz CMOS focal-plane array for terahertz imaging applications," *European Solid-State Circuits Conf.*, pp. 110–113, 2008.
[4] E. Ojefors, U. Pfeiffer, A. Lisauskas, and H. Roskos, "A 0.65 THz Focal-Plane Array in a Quarter-Micron CMOS Process Technology," *IEEE J. Solid-State Circuits*, vol. 44, no. 7, pp. 1968–1976, July 2009.
[5] W. Knap, M. Dyakonov, D. Coquillat, F. Teppe, *et al.*, "Field Effect Transistors for Terahertz Detection: Physics and First Imaging Applications," *Journal of Infrared, Millimeter and Terahertz Waves*, vol. 30, no. 12, pp. 1319–1337, 2009.
[6] A. Lisauskas, U. Pfeiffer, E. Öjefors, P. H. Bolivar, *et al.*, "Rational design of high-responsivity detectors of terahertz radiation based on distributed self-mixing in silicon field-effect transistors," *Journal of Applied Physics*, vol. 105, no. 11, p. 114511, 2009.
[7] F. Schuster, H. Videlier, M. Sakowicz, F. Teppe, *et al.*, "Imaging above 1 THz Limit with Si-MOSFET Detectors," *IEEE Int. Conf. Infrared, Millimeter, and Terahertz Waves*, 2010.

978-1-61284-303-2/11 $26.00 © 2011 IEEE

ISSCC 2011 / February 21, 2011 / 3:45 PM

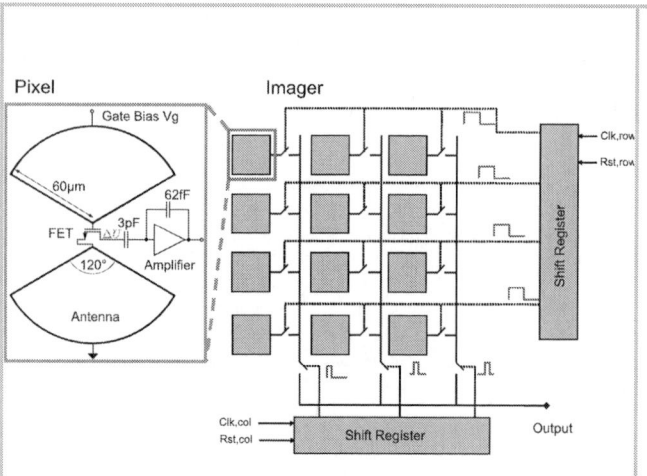

Figure 2.5.1: Architecture of the 3×4 pixel imager prototype.

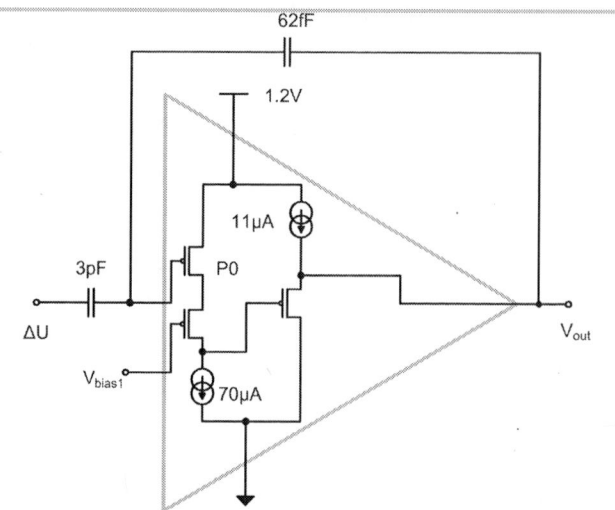

Figure 2.5.2: Circuit schematic of the in-pixel base band amplifier with 31dB closed-loop gain.

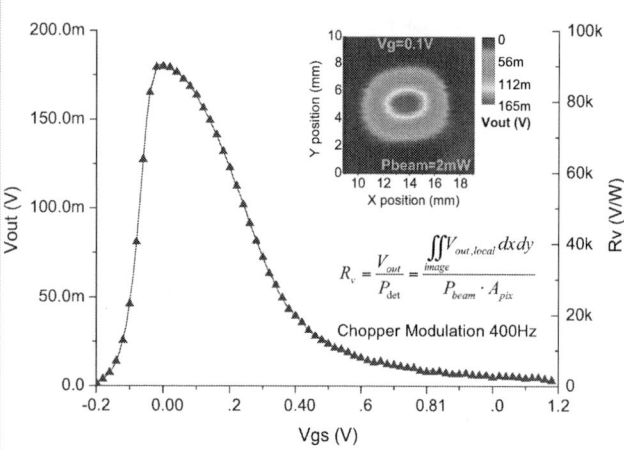

Figure 2.5.3: Measured detection signal and responsivity over gate bias at 300 GHz. Inset: Raster scan image of the source beam.

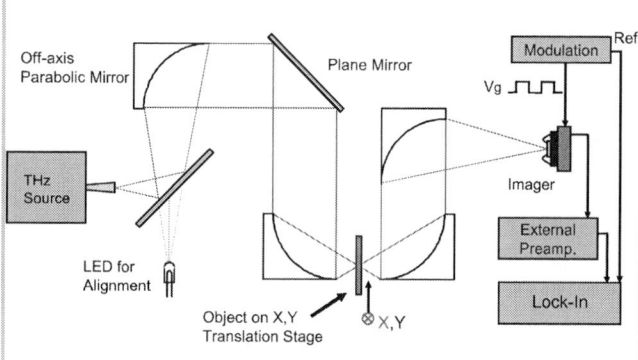

Figure 2.5.4: Measurement set-up for imaging. The detector's gate bias is modulated with a 30kHz square wave.

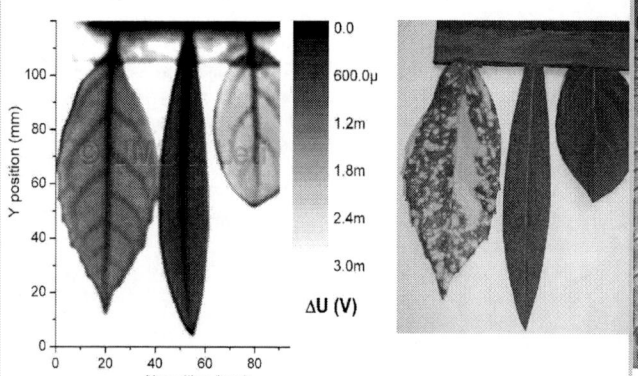

Figure 2.5.5: Transmission mode raster scan image at 300 GHz and its visible counterpart.

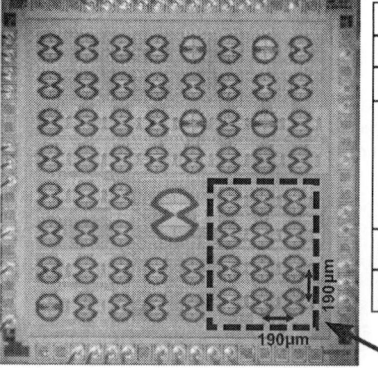

3 x 4 Pixel THz Imager

	This Work	Ref. [4]
Array Size	3 x 4	3 x 5
Pixel Size	190µm x 190µm	150µm x 200µm
In-Pixel Amplifier		
Consumption	97µW	5.5mW
Gain	31dB	43dB
Input Noise	16nV/Hz$^{0.5}$ @f$_{mod}$=30kHz	n. c.
Responsivity	90kV/W @300GHz 1.8kV/W @1.05THz	80kV/W @645GHz
Pixel Multiplexing	Yes	No

Figure 2.5.6: Chip photograph and key characteristics of the imager in this work compared to the circuit in [4].

978-1-61284-303-2/11 $26.00 © 2011 IEEE

ISSCC 2011 / SESSION 2 / TECHNOLOGIES FOR HEALTH / 2.6

2.6 A Programmable Implantable Micro-Stimulator SoC with Wireless Telemetry: Application in Closed-Loop Endocardial Stimulation for Cardiac Pacemaker

Shuenn-Yuh Lee[1], Y-C. Su[1], M-C. Liang[1], J-H. Hong[1], C-H. Hsieh[1], C-M. Yang[1], Y-Y. Chen[2], H-Y. Lai[3], J-W. Lin[4], Q. Fang[5]

[1]National Chung Cheng University, Chia-Yi, Taiwan,
[2]National Yang-Ming University, Taipei, Taiwan,
[3]National Chiao Tung University, Hsinchu, Taiwan,
[4]National Taiwan Univeristy, Hospital Yun-Lin Branch, Taiwan,
[5]RMIT University, Melbourne, Australia

The use of closed-loop implantable telemetry devices [1,2] is increasing as recent clinical studies have shown their efficiency and usefulness in detecting and treating various cardiac arrhythmias. In this paper, we present a single-chip closed-loop system providing closed-loop feedback of sensed cardiac patterns. This device promises more optimized stimulation parameters to cure rapid arrhythmias compared with open-loop devices. Our system incorporates more functionality comparing with the conventional micro-stimulators, yet consumes less power. It is powered by implanted rechargeable batteries periodically charged by radio-frequency (RF) coupling. This improvement of the powering system can prevent improper device operation or shutdown due to RF coupling and loading variations. Furthermore, this system eliminates the need for periodic surgery for battery replacement and improves quality of life of the patient.

In this work, an implantable micro-stimulator SoC (IMSoC) is designed as cardiac defibrillator for remote delivery of power by means of coupled coils supplying power to a bio-implanted rechargeable battery. The IMSoC consists of a smart control circuit for power management, an RF-coupling power system for battery charging [3], and 2 ultra-small rechargeable batteries as power supplies. The micro-stimulator circuit embedded in the IMSoC is externally powered and controlled by a single encoded RF carrier. Stimulus parameters, such as pulse width, amplitude, and frequency, can be programmed [4] from the remote controller.

Using an implantable SoC with closed-loop functionality, stimulation can be dynamically adjusted according to the real-time heart rate (HR). To continuously acquire intracardiac electrograms (EGMs), a monitoring analog front-end (MAFE) is implemented to sense and detect the peak value of EGMs, and to achieve the closed-loop control of HR. To verify a prototype of the IMSoC in vivo, this study presents an example of an implantable RF power converter used to regulate endocardial pacing with closed-loop physiological feedback from a conscious rodent model.

The block diagram of the closed-loop stimulation system with external device is presented in Fig. 2.6.1. The external device, including a battery-based digital signal processor (DSP) in pocket, transmits encoded data and charging energy to an in-body circuit via a set of coils. The modulated digital data acquired from the sensing channel is transmitted out-of-body through the same coils to save on coil area. A MAFE and pacing channel detect and stimulate the paced neuron, respectively, allowing the system controller to immediately program the stimulus parameters. The powering interface smartly manages batteries, providing low- and high-voltage power supplies for system operation and micro-stimulation, respectively.

As shown in Fig. 2.6.2, the PSK demodulator based on a phase-locked loop (PLL) technique is proposed to recover the clock and data from received 256kHz carriers. The PLL synchronizes certain signals, such as PSK data, oscillating signal, and recovered data, by detecting the phase difference of received signals. To avoid errors resulting from phase transitions, the signal "CP reset" is employed to ensure the pumped voltage is stable. The signal "D_{ini}" is helpful to fetch the correct data from MX3 while the PLL is locked. In the telemetry system, initial identification bits, parity bits, and end codes are used together to sustain the system controller with an on-line accuracy protection function. An acknowledgement signal (Ack) is generated to notify complete reception.

Figure 2.6.3 demonstrates the circuit-level diagram of MAFE. Potentials of the stimulated objective are detected and amplified by the preamplifier with a differential structure to achieve a programmable gain under low supply voltage. The capacitance ratio between C_{d1}/C_{d2} and C_f determines the amplification of 21/31dB, respectively. Blanking switches are implemented in the amplifier to prevent the saturation of MAFE from the large stimulation pulse. Following the signal-processing stage, which is composed of biquad low-pass (LP)/band-pass (BP) filters and an analog-to-digital converter (ADC) with real-time threshold detection, the cardiac waveform of the stimulated objective and its peak value can be obtained. A programmable-gain 2rd-order switched-capacitor amplifier (SCAMP) acting as a sample-and-hold circuit prior to the comparator is developed with a gain of 2×/5×, respectively. This real-time detection ADC consumes 41nW by reusing most components for the functions of threshold detection and A/D conversion. Based on the timing diagram shown in Fig. 2.6.3, the real-time detection ADC implements 3 functions during each sampling period: amplification while resetting, threshold detection, and A/D conversion.

As shown in Fig. 2.6.4, a 2-battery-based power-supply network in the powering system provides a nonstop energy source. Moreover, the use of charge and supply detectors recognizes a working battery that is charged at least to 72% of its capacity. The idle battery is connected with the charging current path of the charger to wait for the RF-coupling energy received from rectifier via coil L_2. To optimize the efficiency of the high-potential supply, the battery supplying voltage is allotted to a charge-pumping circuit with 4 stages for stimulation. The algorithm of the telemetry system in Fig. 2.6.2, and the pacing channel shown in Fig. 2.6.4, which includes a D/A converter and a 3.2V pulse generator, are combined. A capacitor-less charge-pump circuit is used to provide 3.2V for the pulse generator. This enables the stimulation function, including programmable arguments regarding the specification such as triggered pulse duration/amplitude and detection frequency, to be implemented according to the detection result of MAFE. The frequency of threshold detection ($f_{detection}$) is much higher than that of stimulation ($f_{stimulation}$), which is lower than the physiological signal frequency. The pulse of triggered stimuli occurs if the stimulation indicators miss the peak value of the detected signal. The stimulation pulse is inhibited if the peak value is detected.

Figure 2.6.5 demonstrates in vivo experiments conducted by implanting the IMSoC into rats and one measured closed-loop behavior of triggering and inhibition. The IMSoC receives input RF power and the configuration of stimulus parameters from the battery-based DSP kept by the rat's side. The maximum operation distance of the IMSoC is approximately 25–45mm in the animal experiment. The MAFE amplifies EGMs measured from the implanted catheter and subsequently recognizes the R-beat at a detection rate of 1kHz. When one R-beat is present within a stimulation period of 400ms, the stimulus is inhibited by the controller. Conversely, intracardiac stimulation (-3.2V; pulse width: 5ms) is performed at the detected R-R beat interval longer than 400ms in the conscious rodent experiment. Simultaneously, the stimulation is visually indicated by illuminating the light emitting diode (LED). One stimulus artifact appears after 20ms of intracardiac stimulation, as shown in Fig. 2.6.5. The specification summary of the entire IMSoC is presented in Fig. 2.6.6, including digital blocks, powering interface, pacing channel, and MAFE.

Acknowledgements
The authors would like to thank the CIC and the NSC of Taiwan, R.O.C. for their support of this work (under grant no. NSC-99-2628-E-194-032), and the help of M-Y. Huang, C-P. Wang, and C-J. Cheng.

References
[1] G. Wang, W. Liu, M. Sivaprakasam, and G. A. Kendir, "Design and Analysis of an Adaptive Transcutaneous Power Telemetry for Biomedical Implants," *IEEE Trans. Circuits and Systems-I: Regular Papers*, vol. 52, no. 10, pp. 2109–2117, Oct. 2005.
[2] P. Cong, N. Chaimanonart, W. H. Ko, and D. J. Young, "A Wireless and Batteryless 130mg 300µW 10b Implantable Blood-Pressure-Sensing Microsystem for Real-Time Genetically Engineered Mice Monitoring," *ISSCC Dig. Tech. Papers.*, Feb. 2009, pp. 428–430.
[3] P. Li and R. Bashirullah, "A wireless power interface for rechargeable battery operated medical implants," *IEEE Trans. Circuits Syst. II, Exp. Briefs*, vol. 54, no. 10, pp. 912–916, Oct. 2007.
[4] M. Ghovanloo and K. Najafi, "A wireless implantable multichannel microstimulating system-on-a-chip with modular architecture," *IEEE Trans. Neural Syst. Rehabil. Eng.*, vol. 15, no. 3, pp. 449–457, Sept. 2007.

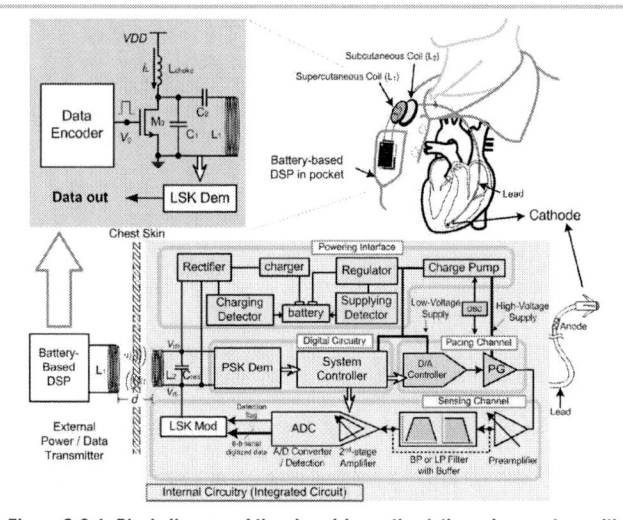

Figure 2.6.1: Block diagram of the closed-loop stimulation micro-system with external device.

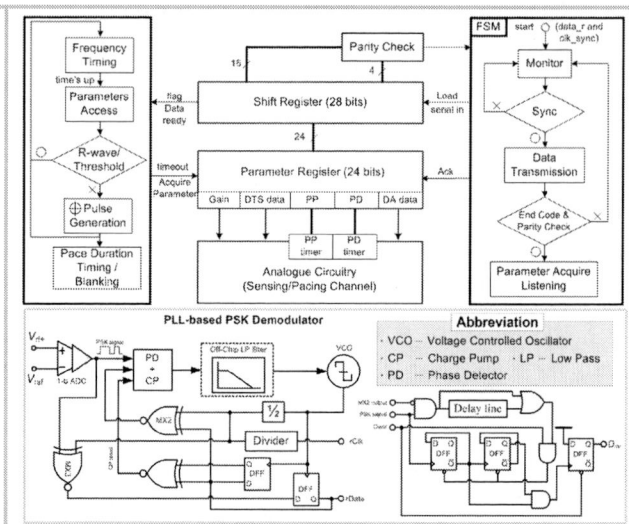

Figure 2.6.2: System controller and PSK demodulator.

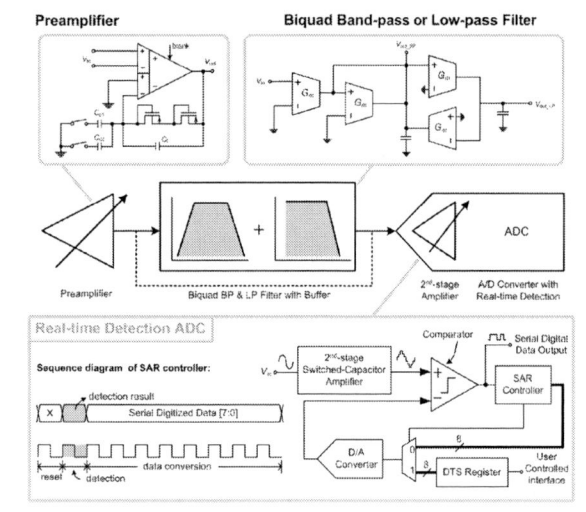

Figure 2.6.3: Monitoring analog front end.

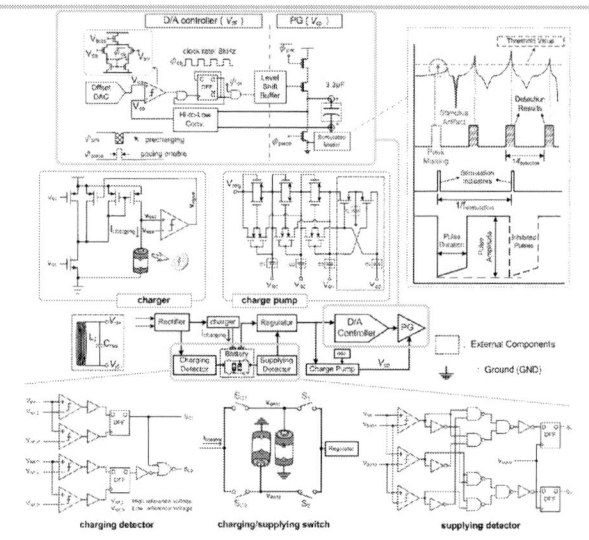

Figure 2.6.4: Power management and stimulus generator circuitry.

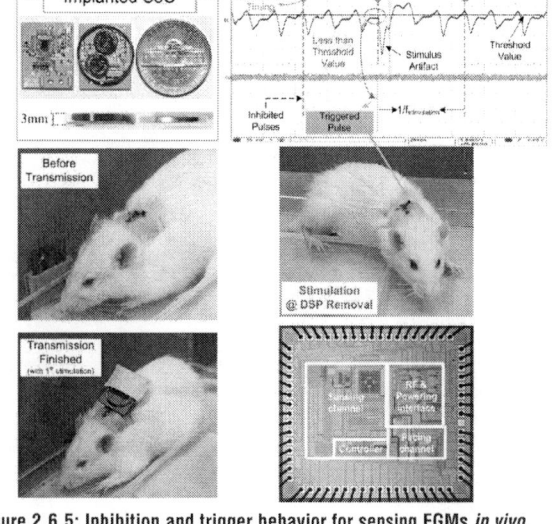

Figure 2.6.5: Inhibition and trigger behavior for sensing EGMs *in vivo*.

General		Digital Blocks	
Technology	TSMC 0.35 μm 2P4M	PSK Demodulator	power = 1.76 μW @ 1 V
Chip Area	1.25 × 1.85 mm²	System Controller	power = 260 nW @ 1 V operation frequency = 32 kHz
Carrier Frequency	256 kHz		
Power Interface and Pacing Channel		Programmable ranges	Stimulation frequency : 0.5–31.25 Hz Stimulus amplitude : 0–3.2 V Stimulus duration : 62.5 μs–1.94 ms
Rectifier	2 V output with induced voltage of 2.9 V$_{rms}$		
		Monitoring Analog Front End @ 1.4 V	
Regulator	power = 7 μW @ 1 V		
Charge Pump	power = 10 μW output voltage = 3.2 V @ 1 V input voltage pumping clock frequency = 18 kHz	Preamplifier	power = 40 nW Input referred noise = 1.1 μV$_{rms}$ 3 dB bandwidth = 0–140 Hz DC gain = 21/31 dB
Power Management (Supply Detector, Charge Detector, and Charger)	power = 20 μW charging current = 2 mA battery : 2 × V6HR, 6 mAh @ 1.2 V supply voltage 1000 times recharging	Bi-quad LP/BP Filter	power = 126 nW with buffer 3 dB bandwidth (LP) : 0–15 Hz 3 dB bandwidth (BP) : 15–60 Hz
D/A Controller	power = 42 nW @ 1 V operation frequency = 8 kHz	Real-time Detection ADC	power = 71 nW with SCAMP resolution = 8 bit sampling frequency = 800 Hz sample-and-hold gain = 2/5 times
Pulse Generator	power = 15.1 μW @ 3.2 V (stimulation period = 490 ms) (stimulus duration = 125 μs)		

Figure 2.6.6: Features and measured performance summary.

ISSCC 2011 / SESSION 2 / TECHNOLOGIES FOR HEALTH / 2.7

2.7 A 660pW Multi-Stage Temperature-Compensated Timer for Ultra-Low-Power Wireless Sensor Node Synchronization

Yoonmyung Lee, Bharan Giridhar, Zhiyoong Foo, Dennis Sylvester, David Blaauw

University of Michigan, Ann Arbor, MI

Recent work in ultra-low-power sensor platforms has enabled a number of new applications in medical, infrastructure, and environmental monitoring. Due to their limited energy storage volume, these sensors operate with long idle times and ultra-low standby power ranging from 10s of nW down to 100s of pW [1-2]. Since radio transmission is relatively expensive, even at the lowest reported power of 0.2mW [3], wireless communication between sensor nodes must be performed infrequently. Accurate measurement of the time interval between communication events (i.e. the synchronization cycle) is of great importance. Inaccuracy in the synchronization cycle time results in a longer period of uncertainty where sensor nodes are required to enable their radios to establish communication (Fig. 2.7.1), quickly making radios dominate the energy budget. Quartz crystal oscillators and CMOS harmonic oscillators exhibit very small sensitivity to supply voltage and temperature [4] but cannot be used in the target application space since they operate at very high frequencies and exhibit power consumption that is several orders of magnitude larger (>300nW) than the needed idle power. A gate-leakage-based timer was proposed [5] that leveraged small gate leakage currents to achieve power consumption within the required budget (< 1nW). However, this timer incurs high RMS jitter (1400ppm) and temperature sensitivity (0.16%/°C). A 150pW program-and-hold timer was proposed [6] to reduce temperature sensitivity but its drifting clock frequency limits its use for synchronization. The quality of a timer is not captured well by RMS jitter since it ignores the averaging of jitter over multiple timer clock periods in a single synchronization cycle. Instead, we propose the uncertainty in a single synchronization cycle of length T as new metric and use this synchronization uncertainty (SU) to evaluate different timer approaches. The timer period is a random variable X(n), with mean and sigma, μ and σ. Given a synchronization cycle time T, consisting of N timer periods, we define SU as the standard deviation of T as given by $\sqrt{T/\mu} \times \sigma$, assuming X(n) is Gaussian. Note that a smaller clock period increases N and results in more averaging and a lower SU with fixed jitter (σ/μ).

The timer in [5] has a high SU since it is triggered with a low-gain Schmitt trigger and it has a long period (~10s). To combat this, we introduce: (1) a multi-stage structure with a high-gain triggering buffer, (2) boosted capacitance charging, (3) the use of a zero-threshold-voltage transistor (ZVT) for faster gate leakage discharge, and (4) closed-loop temperature compensation to reduce temperature sensitivity. The structure of the proposed multi-stage gate-leakage based timer and its waveforms are shown in Fig. 2.7.1. In a stage, a load capacitor (C_L) is charged with the combined gate leakage current of a ZVT and a PMOS transistor. As C_L is charged, the output driving the next stage is triggered by a buffer stage, which shows higher gain than a traditional Schmitt trigger previously used [5]. This places the next stage in a charging state while the current stage discharges. At any given time, only one stage is in a charging state while all others discharge. This allows n-1 more discharging time than charging time in an n stage timer and increases the voltage swing on C_L (Q[n]). Longer discharge time lowers the slope at node Q[n] at the end of discharging state (from -238mV/s to 20mV/s for n from 3 to 10), which makes the initial capacitor node voltage for next following charging stage less sensitive to uncertainty. Each stage has low and high supply voltage domains and the use of two voltage domains allows us to boost the gate-leakage current with a higher supply voltage, which steepens the charging transition on Q[n] by 5× and reduces uncertainty at the triggering point (Fig. 2.7.1).

We achieve temperature sensitivity compensation by exploiting the opposite temperature dependencies of gate leakage in ZVT and PMOS during the charging state as shown in Fig. 2.7.2. By using arrays of ZVT and PMOS transistors and selecting an appropriate combination of transistor sizes for charging, linear temperature dependency can be eliminated. This compensation scheme results in a residual second-order dependency. To minimize the impact of this second-order dependency, we propose an adaptive scheme in which, for each temperature range, a controller automatically selects a pre-stored transistor size configuration, which minimizes the second-order dependency (top left of Fig. 2.7.2 and

Fig. 2.7.5, left). The optimal configurations are determined and stored during post-silicon testing. Each time when the sensor node processor wakes up, it computes time by calculating the elapsed time using the stored period for proceeding configuration and the number of cycles during the last standby state. The transition between configurations occurs synchronously when the first stage starts a new charging state; this allows an exact period calculation and prevents noise injection during capacitor charging. Un-selected ZVTMOS transistors are driven to 400mV to minimize leakage by placing them in accumulation mode.

A test chip was designed and fabricated in 0.13μm CMOS with the proposed multi-stage gate-leakage timer (MGT). Measured results in Fig. 2.7.3 show that as the number of stages increases, duty cycle (the ratio of charging time to timer period) decreases, increasing voltage swing and reducing jitter by up to 8.1×. Boosting of the charging gate-leakage current (Fig. 2.7.2) leads to higher jitter, particularly for low stage counts. However, the key SU metric for an interval of 1 hr is reduced by 3× due to the shorter clock period, which enhances statistical averaging. Together with multi-staging and boosted charging, SU is reduced by 3.6×. With small stage counts (<5) power consumption increases. This is due to the higher average node voltage of Q[n] resulting in higher leakage current for the triggering buffer (Fig. 2.7.4). With high stage counts (>7), power increases due to static leakage of added stages. A proposed MGT with 9 stages was tested for 24 hours allowing us to compute the SU for a large number of synchronization intervals. We also tested a baseline 3-stage MGT without boosted charging or ZVT transistors. The SU distribution had expected value of 196ms for 1-hour synchronization intervals. It also shows that the proposed timer reduced the expected SU by 4.1× compared to the baseline. Since the period of the timer is not truly Gaussian, the measured SU was larger than the theoretical calculation based on jitter. Power supply sensitivity was 0.42%/mV from 650 to 750mV for low supply and was 0.49%/mV from 1.15 to 1.25V for high supply. This necessitates voltage regulation using an ultra-low power voltage reference such as the one proposed in [7].

The period of the temperature-compensated MGT for -20 to 60°C with selected configurations is shown in Fig. 2.7.5. A five-configuration scheme and its temperature range is shown as an example (Fig. 2.7.5, left). For each configuration, period deviation as a function of temperature is shown and worst period deviation was 0.28% (Fig. 2.7.5, top right). With a single configuration, the maximum deviation in period over -20 to 60°C was 3% while the use of 10 configurations reduced this to 0.25%, giving an effective temperature sensitivity of 31ppm/°C. Measured results from a closed loop, temperature compensated MGT is shown in Fig. 2.7.6 when the temperature oscillates between 20 and 30 °C. The closed-loop temperature compensation reduces SU by 4.8×. A second test chip where the proposed closed-loop temperature compensation was implemented on-die was also tested and Fig 2.7.6, top left, shows how the configurations track with temperature.

Acknowledgement:
The authors thank MuSyC for partial funding support and Samsung Scholarship Foundation.

References:
[1] B. Warneke, et al., "An Ultra-Low Energy Microcontroller for Smart Dust Wireless Sensor Networks," *ISSCC Dig. Tech. Papers*, pp. 316-317, Feb. 2004.
[2] G. Chen, et al., "Millimeter-Scale Nearly Perpetual Sensor System with Stacked Battery and Solar Cells," *ISSCC Dig. Tech. Papers*, pp. 288-289, Feb. 2010.
[3] M. Crepaldi, et al., "An Ultra-Low-Power Interference-Robust IR-UWB Transceiver Chipset Using Self-Synchronizing OOK Modulation," *ISSCC Dig. Tech. Papers*, pp. 226-227, Feb. 2010.
[4] M. McCorquodale, et al., "A 0.5-to-480MHz Self-Referenced CMOS Clock Generator with 90ppm Total Frequency Error and Spread-Spectrum Capability," *ISSCC Dig. Tech. Papers*, pp. 350-351, Feb. 2008.
[5] Y. Lin, et al., "A sub-pW timer using gate leakage for ultra low-power sub-Hz monitoring systems," *IEEE Custom Integrated Circuits Conf.*, pp. 397-400, Sep. 2007.
[6] Y. Lin, et al., "A 150pW Program-and-Hold Timer for Ultra-Low-Power Sensor Platforms," *ISSCC Dig. Tech. Papers*, pp. 326-327, Feb. 2009.
[7] M. Seok, et al., "A 0.5V 2.2pW 2-Transistor Voltage Reference," *IEEE Custom Integrated Circuits Conf.*, pp. 577-580, Sep. 2009.

978-1-61284-303-2/11 $26.00 © 2011 IEEE

ISSCC 2011 / February 21, 2011 / 4:30 PM

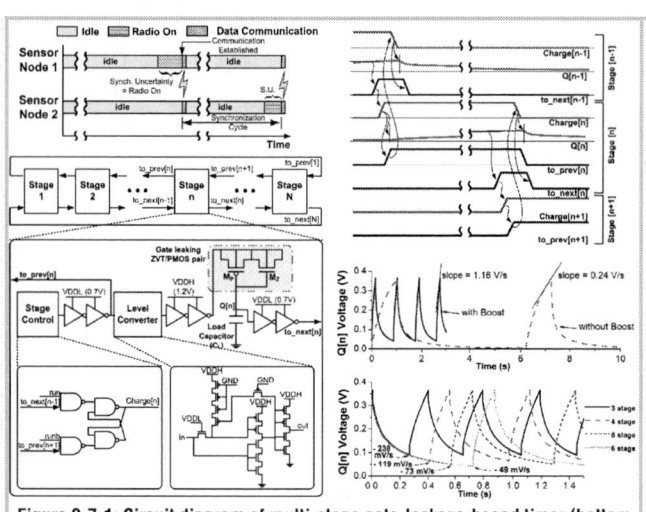

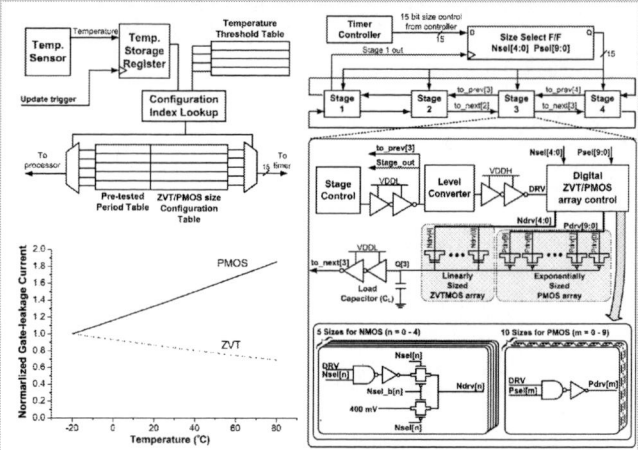

Figure 2.7.1: Circuit diagram of multi-stage gate-leakage-based timer (bottom left) and simulated waveform showing effect of multi-stage and boosted charging (bottom right).

Figure 2.7.2: Circuit diagram of temperature-compensated timer (right) and its controller (top left). Compensation is realized by exploiting opposite temperature dependency of ZVTMOS and PMOS gate leakage current as shown in simulated data (bottom left).

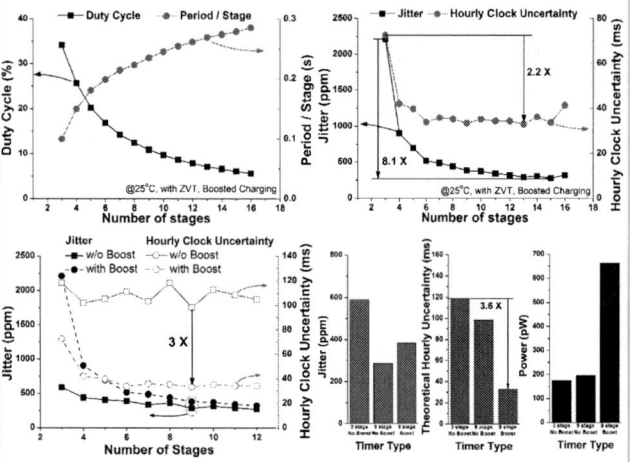

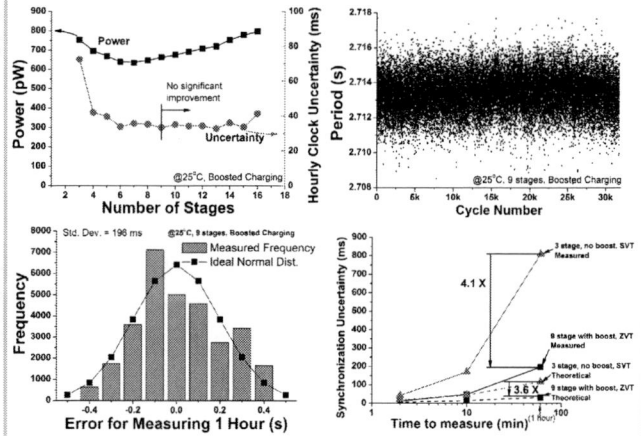

Figure 2.7.3: Measured results shows that, with larger number of stages, duty cycle decreases (top left) and jitter is reduced (top right). With boosted charging, uncertainty is reduced by 3× (bottom left).

Figure 2.7.4: Measured power and uncertainty trade-off (top left) and a continuous 24-hour measurement (top right). Synchronization uncertainty (SU) distribution for 1-hour measurement (bottom left) and uncertainty reduction was 4.1× (bottom right).

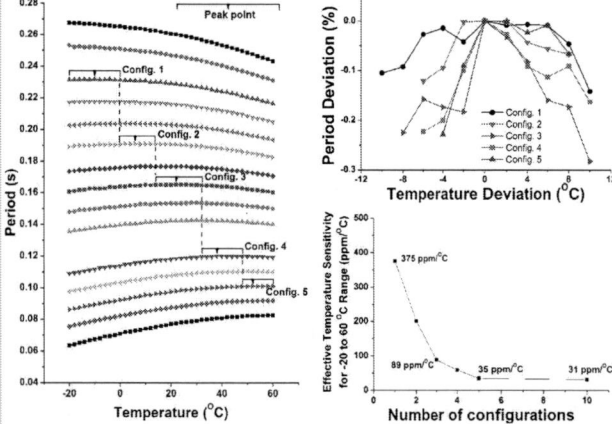

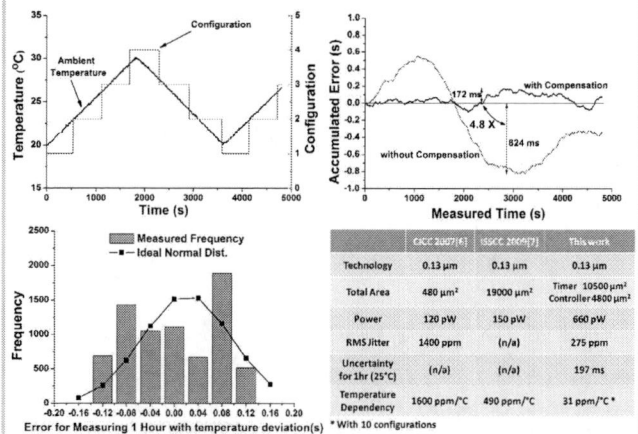

Figure 2.7.5: Period of temperature-compensated timer with selected configurations (left) and period vs. temperature deviation for selected configurations (top right). Maximum period variation for -20 to 60°C range decreases with use of more configurations (bottom right).

Figure 2.7.6: Closed-loop control of temperature-compensated timer (top left) and accumulated time measurement error with given temperature profile (top right). Synchronization uncertainty distribution (bottom left) and comparison with other works (bottom right).

978-1-61284-303-2/11 $26.00 © 2011 IEEE

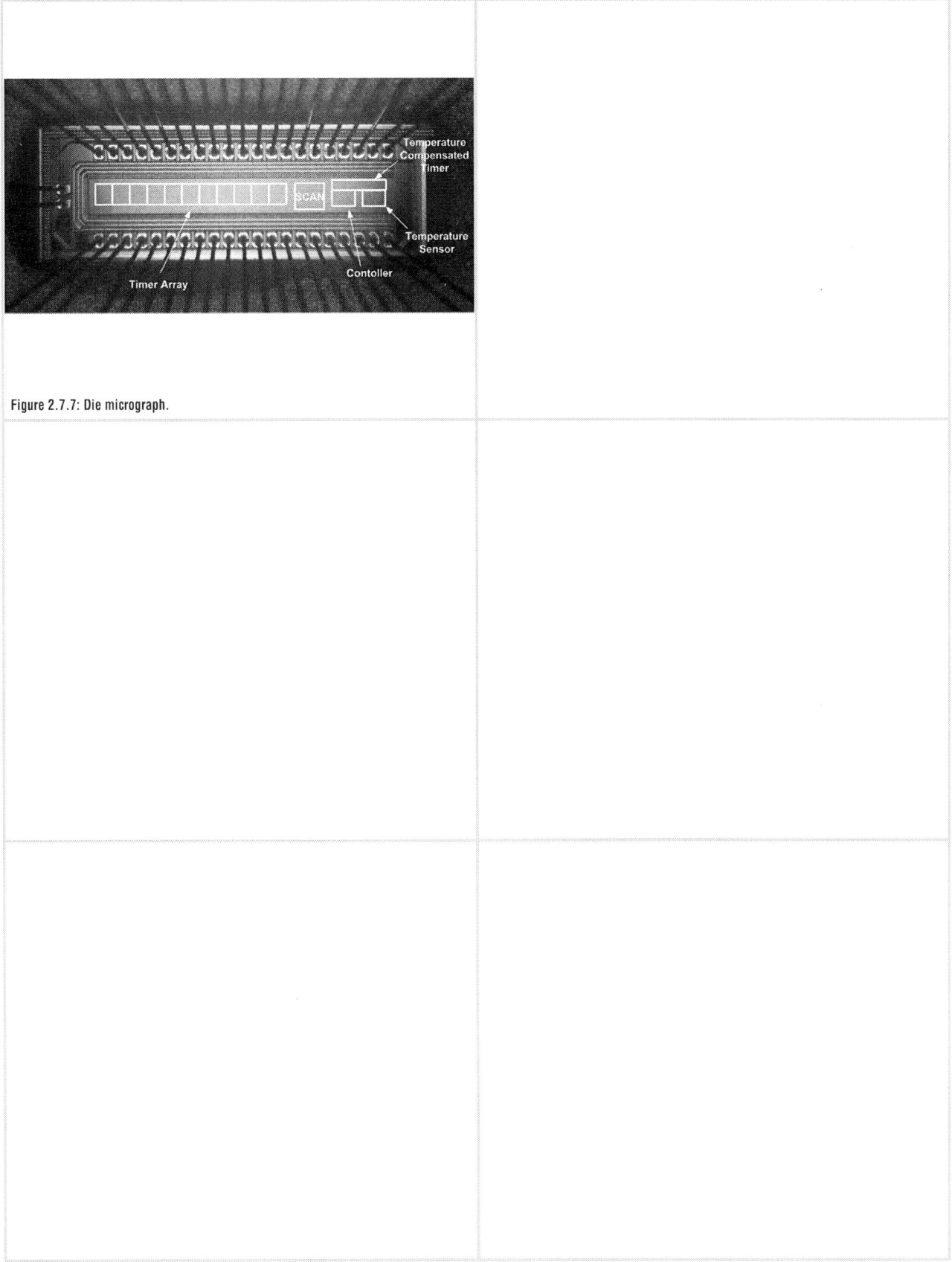

Figure 2.7.7: Die micrograph.

ISSCC 2011 / SESSION 2 / TECHNOLOGIES FOR HEALTH / 2.8

2.8 A Low-Power Fully Integrated RF Locked Loop for Miniature Atomic Clock

David Ruffieux, Matteo Contaldo, Jacques Haesler, Steve Lecomte

CSEM, Neuchatel, Switzerland

Very accurate local clocks play a fundamental role in modern communication and navigation applications. High-precision references enable fast communication data rates, while in navigation they allow longer holdover operation times in the absence of a synchronization signal, for example from the Global Positioning System (GPS). Even if this high accuracy can be achieved with atomic clocks, only the recent developments in photonics and MEMS processes allowed reaching the low power consumptions and small sizes needed for hand-held devices [1], paving the way to the realization of Miniature Atomic Clocks (MAC). However, reaching the target of overall device volumes <1cm³ and power consumptions <30mW [2] requires further miniaturization and improved design of all the system aspects, including the RF control electronics, which has to consume low power without affecting the clock performances. State-of-the-art implementations as [3] and [4] reached interesting low-power consumption performances, but using discrete electronic components that limit system miniaturization. This work presents a fully integrated frequency locked loop designed for MAC applications, tested in combination with a buffered ⁸⁷Rb cell system based on coherent population trapping (CPT) interrogation.

The CPT interrogation [2] consists in modulating the bias current of a wavelength-tuned VCSEL at an RF frequency (ν_{RF}) that is half the ground state hyperfine splitting of ⁸⁷Rb (ν_{HF} /2 = 3.417GHz). Figure 2.8.1 illustrates the principle. Let's consider that a ramp on the laser diode (LD) bias current sweeps the pumping photon energy (E=hν), producing two peaks of absorption corresponding to transitions between each of the two ground states toward the excited state. When RF amplitude modulation of the LD is added, the absorption spectrum is composed of each individual laser wavelength (carrier plus sidebands). For $\nu_{RF} = \nu_{HF}$ /2, a transmission peak hundreds of Hertz wide (CPT signal) is observed and this resonance provides the high-Q frequency reference (>10⁶) of the MAC. To maintain the system locked around that peak and provide a highly stable clock, the frequencies of both the RF modulating signal and VCSEL must be controlled within feedback loops. In particular, locking of the RF frequency onto the atoms is obtained by introducing frequency modulation of the RF carrier, which allows deriving a signed error signal used to form the RF frequency-locked loop.

Figure 2.8.2 shows the block diagram of the integrated frequency-locked loop. A transimpedance amplifier (TIA) converts the photocurrent of the PD into a voltage. The signal is then chopped coherently with the FM modulation of the RF carrier before being integrated and low-pass filtered. The resulting voltage drives the frequency tuning port of a 40MHz voltage-controlled crystal oscillator (VCXO). The remainder of the system is equivalent to a direct-modulation transmitter featuring a high-resolution fractional-N PLL, whose loop bandwidth is chosen to minimize the RF close-in noise that will interact with the atoms. For increased flexibility, the low pass filter was implemented with external components. The standard 40MHz crystal was chosen for small size (available down to 1.6×1.2 mm²), possibility to derive the traditional 10MHz reference after integer division and low in-band phase noise plateau thanks to a reduced PLL ratio. An external bias-T is used to couple the RF power and the LD biasing current before driving the VCSEL. The typical RF drive level is from -2 to 2dBm. Frequency modulation of the RF carrier is performed at a programmable rate and depth with a 2-FSK unfiltered square wave signal. Both are typically in the kHz range with a lock-in amplifier loop bandwidth of 1Hz. A resolution of the RF carrier frequency of 10⁻¹² (ppt) is obtained with a 40b 2ⁿᵈ-order ΔΣ modulator. The N divider, whose fractional value should be close to 85.425, is implemented with a cascade of dynamic dividers [5] to achieve low power dissipation.

The photocurrent generated by the PD contains a strong DC component onto which a weak CPT signal, superimposed at the FM rate, has to be amplified. A low noise TIA (Fig. 2.8.3) with a band-pass filtering characteristic is implemented with 2 transistors, an external capacitor and a current source. The 2 transistors form a current conveyor that reverse-bias the PD at a constant voltage independent of the photocurrent. At a higher frequency however, the capacitor freezes the gate of the top-most transistor that presents at its source a large transimpedance. The latter is maximized with a large overdrive voltage and using a native transistor, resulting in high gain and much reduced noise compared to the PD shot noise. After amplification of the AC square-wave signal by the transconductance and current mirror ratios, coherent chopping is performed so that the difference of its two components is integrated on a large external capacitor until the error signal vanishes. The resulting saw-tooth voltage is further low-pass filtered with an external RC network before it drives the VCXO differential accumulation varactor. The differential VCXO based on [6], was designed for maximum tuning range (~100ppm). A 7b switched-capacitor bank with a large on/off ratio is used for coarse frequency tuning down to ppm level so that ppt stability could be reached on the analog varactor.

The RF locked loop is fabricated in a standard digital 0.18μm CMOS technology. The chip (Fig. 2.8.7) has overall dimensions of 2.1×2.3mm² comprised of pads and several test structures, which reduce the active area to about one half. Figure 2.8.4 shows the measured phase noise of the RF signal (black), corresponding to −85.6dBc/Hz at 1kHz offsets from the 3.417GHz carrier. At the PLL cutoff frequency of 250kHz, the phase noise is −98dBc/Hz. The phase noise of the RF signal is compared with what is achievable with typical laboratory equipment. The dark gray curve depicts the noise measured using a bulky oven controlled crystal oscillator (OCXO) in combination with a N5181A RF analog signal generator. The light gray curve shows the phase noise resulting from the use of the N5181A synthesis together with the integrated 40MHz VCXO output after division by 4.

Figure 2.8.5 depicts the MAC Allan deviation measured on the 10MHz clock when locked through the IC RF loop to the physics package (square). The clock exhibits a 1-second intercept point of $\sigma_y(\tau = 1) = 4 \times 10^{-10}$, improving as $\sigma_y(\tau) \propto \tau^{-1/2}$ up to some tens of seconds. Work is ongoing to explain the abnormal drift observed for integration time τ > 20s also measured when using the IC VCXO in combination with the N5181A synthesizer (triangle pointng down). The free VCXO, whose stability is also plotted (triangle pointing up), seems to be the problem. The last curve (circle) represents the Allan deviation measured when locking the laboratory equipment to the ⁸⁷Rb cell.

The power consumption of the presented circuit is summarized in Fig. 2.8.6, together with a comparison with state-of-the-art RF electronics for MAC. The TIA and VCXO contribute 0.31mW to the overall consumption, while the PLL comprised of the LC VCO consumes 7.89mW. The overall consumption of 26.35mW at an output power of 0dBm suffers from the low efficiency of the power stage, which has been initially designed to reach output power up to 10dBm. Redesign towards 0dBm would necessitate the elimination of the power stage, resulting in an overall estimated power consumption of about 12mW.

References:
[1] J. Kitching, et al., "Microfabricated atomic clocks," *IEEE International Conference on Micro Electro Mechanical Systems*, pp. 1-7, Feb. 2005.
[2] R. Lutwak, et al., "The chip-scale atomic clock – coherent population trapping vs. conventional interrogation," *34ᵗʰ Annual Precise Time and Time Interval (PPTI) Meeting*, pp. 539-550, Dec. 2002.
[3] R. Lutwak, et al., "The chip-scale atomic clock – recent developments," *IEEE International Frequency Control Symposium*, pp. 573-577, Apr. 2009.
[4] J.F. DeNatale, et al., "Compact, low-power chip-scale atomic clock," *IEEE Position, Location and Navigation Symposium*, pp. 67-70, May 2008.
[5] D. Ruffieux, et al., "A narrowband multi-channel 2.4 GHz mems-based transceiver," *IEEE J. Solid-State Circuits*, vol 44, no. 1, pp. 228-239, Jan. 2009.
[6] D. Ruffieux, "A high-stability, ultra-low-power quartz differential oscillator circuit for demanding radio applications," *European Solid-State Circuits Conf.*, pp. 85-88, Sep. 2002.

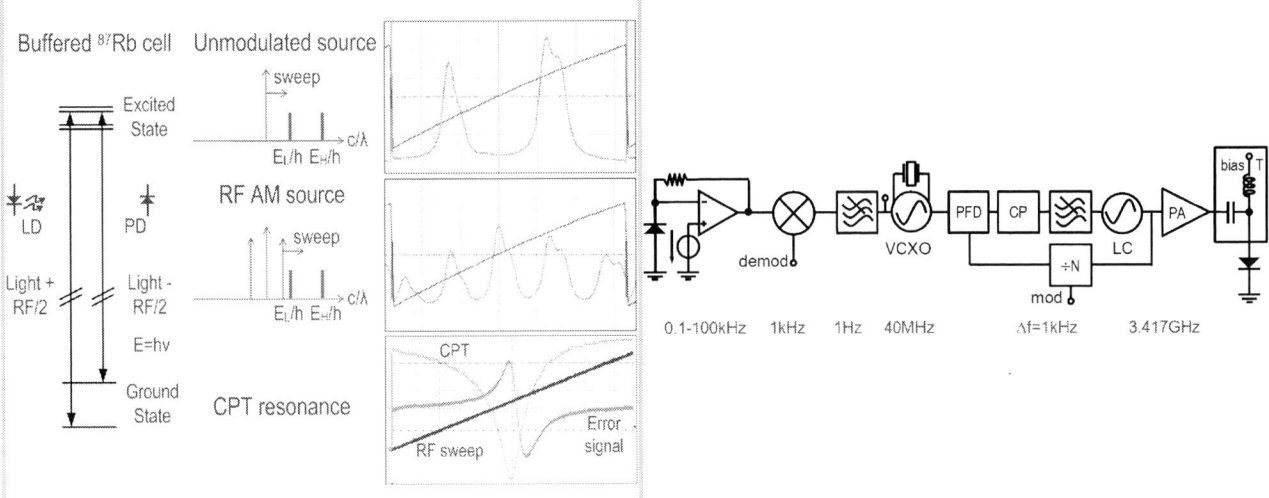

Figure 2.8.1: Physical principle of ^{87}Rb CPT interrogation.

Figure 2.8.2: Architecture of the atomic clock RF frequency-locked loop.

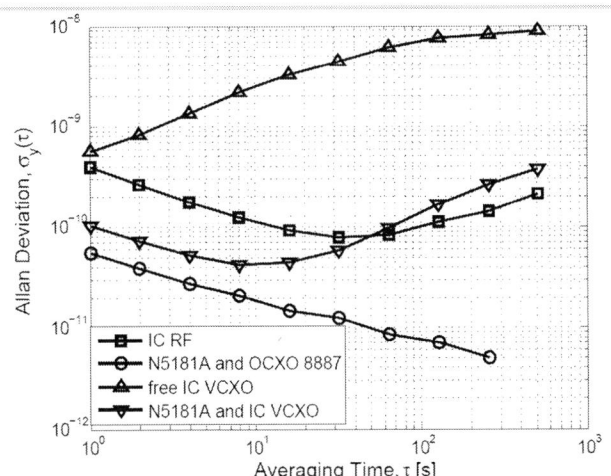

Figure 2.8.3: TIA, demodulator, loop filter and VCXO transistor-level schematic.

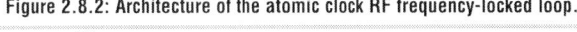

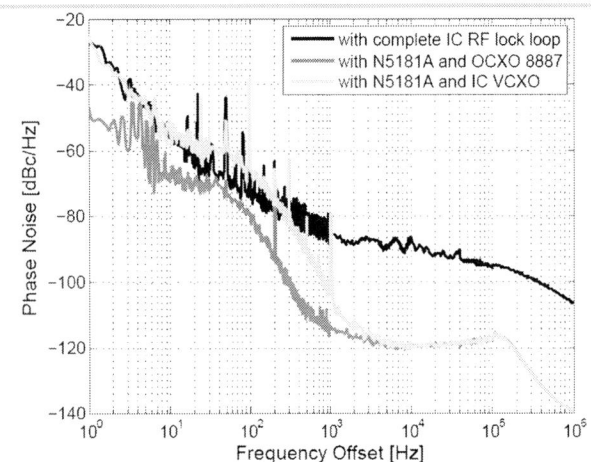

Figure 2.8.4: Measured RF lock loop integrated circuit phase noise (black curve) compared with an Agilent N5181A synthesizer referenced to a commercial OCXO (Oscilloquartz 8887, dark gray curve) or to the divided by four small 40MHz IC VCXO (light gray curve).

Figure 2.8.5: Allan deviations of the miniature atomic clock measured at 10MHz with different configurations for the RF generation and frequency lock loops.

IC consumption breakdown

Block	Pow. Cons. [mW]	Supply [V]
TIA	0.17	1.8
VCXO	0.14	1.8
VCO	5.71	1.5
DIVIDERS	1.83	1.8
PFD+CP	0.3	1.8
MOD/DEMOD	0.05	1.8
PREPA	3.49	1.8
PA	14.66	1.5
TOT	26.35	-

Comparison with published work

	RF pow. cons. [mW]	$\sigma_y(\tau=1)$	Single chip
This work	26	4.0 E-10	yes
[3]	51	5.1 E-11	no
[4]	15	1.0 E-10	no

Figure 2.8.6: Chip consumption breakdown and comparison with published work.

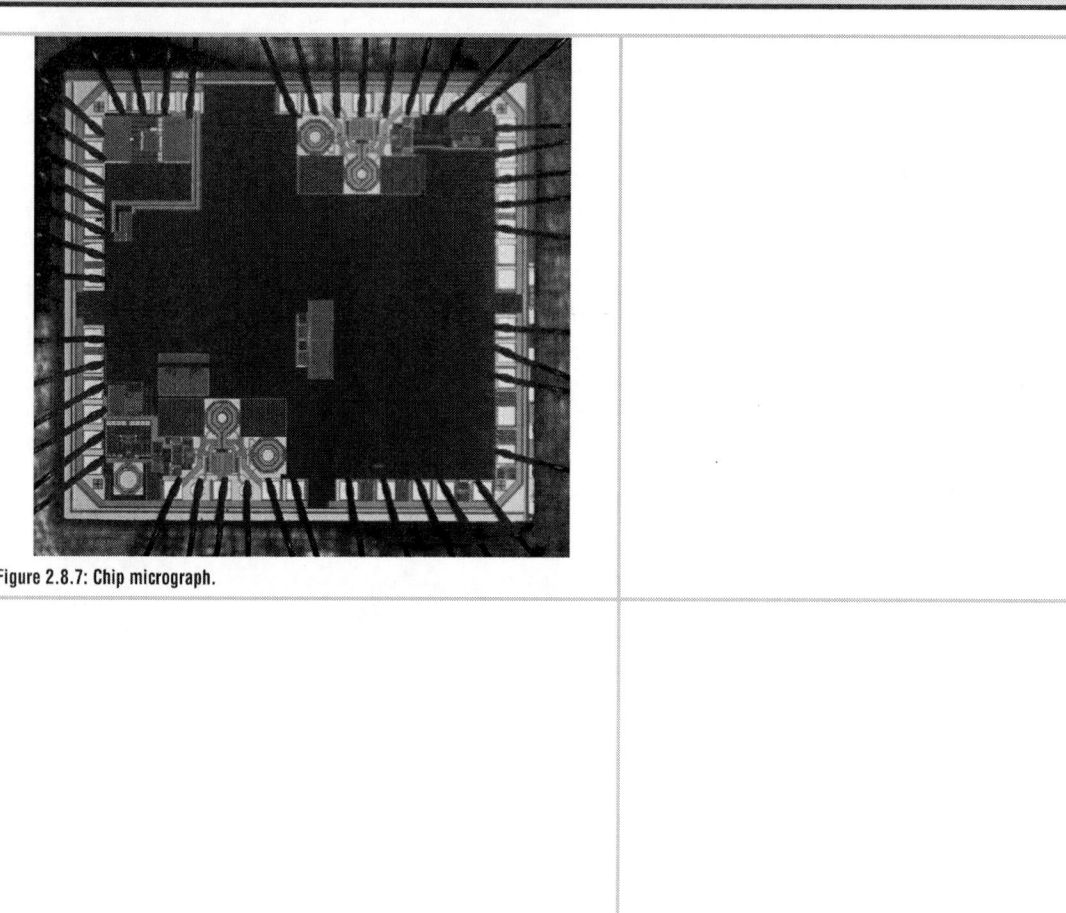

Figure 2.8.7: Chip micrograph.

ISSCC 2011 / SESSION 3 / RF TECHNIQUES / OVERVIEW

Session 3 Overview / RF

RF Techniques

Session Chair: *Jan Craninckx, imec, Leuven, Belgium*

Session Co-Chair: *Jing-Hong Conan Zhan, Mediatek, HsinChu, Taiwan*

Wireless circuits have become the foundation of the way we live our lives, and this trend will only continue further in the future. Even in this mature field, new techniques are developed every year, and these inventions will allow the further integration of new technologies into every consumer product that will in no time become indispensible for the whole population, without people realizing the complex systems they have in their pockets.

The key enabler for this evolution is cost reduction and feature enrichment through implementation in advanced digital-based CMOS technologies, as will be evident from the papers presented in this session.
Digital PLLs use less area and moreover incorporate such a rich variety of digital processing techniques that they actually outperform their analog counterparts. RF receivers as well use all the capabilities of deeply scaled CMOS to obtain state-of-the-art performance, and on top of that architectures and techniques that can handle large blockers are conceived. That avoids the need for expensive external RF blocker filters, which was the only remaining roadblock for multiband worldwide mobile terminals.

The first two papers of this session describe advances in digital phase-locked loops. Paper 3.1 [Texas Instruments] shows how the power of high-speed digital processing and dithering is used to overcome all spurious issues in an ADPLL. Implemented in 65nm CMOS within an area of $0.35mm^2$, it dithers the crystal oscillator slicer to avoid ill-shaped quantization noise at near-integer fractional-N operation. It is also capable of phase modulation up to a bandwidth higher than the reference frequency, and succeeds in complete TX spectral mask requirements without any exceptions.

Reversing the obvious TDC into a Digital-to-Time converter (DTC), Paper 3.2 [NXP Semiconductors] enables reduction in the phase detector quantization noise, which allows for easier and lower power implementation. The 5.3GHz fractional-N ADPLL implemented in a 65nm CMOS technology achieves an in-band phase noise floor of -96 dBc/Hz and in-band fractional spur power of -45 dBc after calibration.

The proof that CMOS scaling will continue also for RF circuits is given in Paper 3.3 [Intel]. In this 32nm Hi-k MG completely digital technology, an ESD-protected 2.5GHz T/R switch and LNA are implemented, making clever use of the possibilities and the restrictions of the technology. The LNA uses nested coupled inductors, compatible with flip-chip bumps, and achieves 3.5dB NF, 11dB Gain, -5dBm P_{1dB} while drawing 11mA from a 1.8V supply. The TX thin-oxide switch has 1.1dB insertion loss and provides +300/-200V CDM ESD protection.

Paper 3.4 [Delft University of Technology] shows that high voltages needed in a base station PA driver are compatible with baseline 65nm CMOS, by using thin-oxide extended-drain high-voltage MOS devices at no extra mask cost. The pulse-width-controlled RF power driver delivers a maximum output swing of $8.04V_{pp}$ to a 50Ω load with 9V supply, from 0.9 to 3.6GHz.

The next two papers of this session present blocker-resilient receiver architectures paving the way for possible SAW filter removal. In Paper 3.5 [Broadcom], a superheterodyne 65nm CMOS receiver utilizes impedance transformation to create a high-Q filter at RF and IF frequencies. The receiver consumes 12mA and occupies $0.67mm^2$ while achieving 2.8dB NF and IIP3 of -8.5dBm.

A second 40nm CMOS receiver operating for both LNA-first and Mixer-first modes is presented in Paper 3.6 [imec]. The complete RX chain achieves 3dB NF, +10dBm out-of-band IIP3 and +80dBm IIP2, while tolerating 0dBm blockers at 20MHz offset using $2mm^2$.

978-1-61284-303-2/11 $26.00 © 2011 IEEE

ISSCC 2011 / February 21, 2011 / 1:30 PM

The last two papers of this session address problems faced by broadband receiver front-ends. Paper 3.7 [University of Twente] presents a frequency insensitive approximation for sine weighting through discrete-time switched-capacitor vector modulators for accurate beamsteering. Implemented in 65nm CMOS, the 4-element phased array demonstrates one-to-one mapping between control settings and phase shift. The array occupies 0.44mm^2, draws 308mW from 1.2V, and results in a 1.4° phase and 0.4dB gain error (RMS).

Paper 3.8 [Silicon Laboratories] demonstrates a 110nm CMOS harmonic rejection mixer for tuner applications. Critical matching is only required for IF amplifiers instead of for RF switches, providing a rejection ratio in excess of 52dB for the 3rd-, 5th- and 7th- order harmonics without calibration, while also achieving an IIP2 of 75dBm.

As is apparent from the papers presented in this session, continuing circuit and architecture innovations in advanced digital CMOS technologies pave the way to further integration and cost reduction!

Presenters:

1:30 PM

3.1 Spur-Free All-Digital PLL in 65nm for Mobile Phones

R. B. Staszewski, Delft University of Technology, Delft, The Netherlands

2:00 PM

3.2 A 5.3GHz Digital-to-Time-Converter-Based Fractional-N All-Digital PLL

N. Pavlovic, NXP Semiconductors, Eindhoven, The Netherlands

2:30 PM

3.3 A 2.5GHz 32nm 0.35mm^2 3.5dB NF -5dBm P$_{1dB}$ Fully Differential CMOS Push-Pull LNA with Integrated 34dBm T/R Switch and ESD Protection

C-T. Fu, Intel, Hillsboro, OR

2:45 PM

3.4 A 65nm CMOS Pulse-Width-Controlled Driver with 8V$_{pp}$ Output Voltage for Switch-Mode RF PAs up to 3.6GHz

D. A. Calvillo-Cortes, Delft University of Technology, Delft, The Netherlands

3:15 PM

3.5 A Low-Power Process-Scalable Superheterodyne Receiver with Integrated High-Q Filters

A. Mirzaei, Broadcom, Irvine, CA

3:45 PM

3.6 A 40nm CMOS Highly Linear 0.4-to-6GHz Receiver Resilient to 0dBm Out-of-Band Blockers

J. Borremans, imec, Leuven, Belgium

4:15 PM

3.7 A 1.0-to-4.0GHz 65nm CMOS Four-Element Beamforming Receiver Using a Switched-Capacitor Vector Modulator with Approximate Sine Weighting via Charge Redistribution

M. C. Soer, University of Twente, Enschede, The Netherlands

4:45 PM

3.8 A Harmonic Rejection Mixer Robust to RF Device Mismatches

A. A. Rafi, Silicon Laboratories, Austin, TX

978-1-61284-303-2/11 $26.00 © 2011 IEEE

ISSCC 2011 / SESSION 3 / RF TECHNIQUES / 3.1

3.1 Spur-Free All-Digital PLL in 65nm for Mobile Phones

Robert Bogdan Staszewski[1], Khurram Waheed[2], Sudheer Vemulapalli[2],
Fikret Dulger[2], John Wallberg[2], Chih-Ming Hung[2], Oren Eliezer[2]

[1]Delft University of Technology, Delft, The Netherlands,
[2]Texas Instruments, Dallas, TX

After the first-ever all-digital PLL (ADPLL) [1] for Bluetooth radios has proven benefits of CMOS scaling and integration, demonstrators for more challenging wireless standards have emerged [2-6]. In the ADPLL, however, the digitally-controlled oscillator (DCO) and time-to-digital converter (TDC) quantize the time and frequency tuning functions, respectively, which can lead to spurious tones and phase noise increase. As such, finite TDC resolution can distort data modulation and spectral mask at near integer-N channels, while finite DCO step size can add far-out spurs and phase noise. Also, a major underreported issue is an injection pulling of the DCO due to harmonics of the digital activity at closely-spaced frequencies, which can also create spurs. This work addresses all these problems and demonstrates RF performance matching that of the best-in-class traditional approaches.

The finite TDC resolution t_{inv} of 10 to 20ps produces low-enough flat quantization noise for satisfactory RF operation with ADPLL bandwidths of up to 150-300kHz. However, at integer-N channels, and especially when the TDC resolution is an integer multiple of the DCO period, the quantization noise is ill-shaped and can concentrate within the loop bandwidth. To solve the problem, FREF dithering of up to several t_{inv} is used by delaying the FREF clock by means of slowing down the edges [6,7] through adjustment of an inverter driving strength to its load capacitance ratio. Unfortunately, degrading FREF clock edges not only adds significant noise but also makes it more sensitive to various aggressors. Figure 3.1.1 shows the proposed *noise-free* method, in which the crystal oscillator (XO) slicer combines the programmable edge delay and performs time shift Δt_{cd} by *dynamically* adding intentional mismatch transistors $M_{cd1,2}$ to the differential input pair $M_{p1,2}$. This way, the programmable voltage offset $\Delta V = \sim 30mV$ gets converted to coarse dither time offset $t_{cd} = \sim 100ps$ through the sinusoidal waveform slope at the origin. The coarse dither, when engaged at near integer-N channels by synchronously toggling at 2.4MHz rate, uses only two levels, $-\Delta t_{cd}$ and $+\Delta t_{cd}$, so its transfer function is perfectly linear. The exact Δt_{cd} value is not critical as long as it spans several t_{inv}. The high toggling rate places the resulting mixing products outside of the higher-order ADPLL loop filter.

A second supplementary method, fine dithering, is added by connecting 16 unit-weighted transistors M_{fd1-16} in parallel with M_{bp2} to change the mirroring ratio, thus affecting the bias current of the differential pair, and ultimately the delay of $\sim 9ps/LSB$. They are digitally-controlled by 3rd-order $\Sigma\Delta$ MASH sequence.

Figure 3.1.2 shows the top-level diagram of the multirate ADPLL, which features support of modulating samples of much higher rate than the reference clock. In fact, FREF does not play any role in the data modulation. Consequently, XO could be free-running and the reference frequency adjustment performed through the frequency command word (FCW). The phase error ϕ_E samples at FREF rate get converted to channel-dependent DCO/16 rate by the sample-rate converter (SRC) and merged with the modulating samples of the same rate. The fractional bits get further dithered by the $\Sigma\Delta$ modulator operating at DCO/8 rate. This way, the injection-pulling spurs of the prior implementations [1,2], with the input at FREF rate, are avoided.

The single DCO gain-normalization multiplier of the prior implementations gets split into two parts: a fine-precision multiplier in the data modulation path and a coarse multiplier (right bit shift) of the filtered ϕ_E. This allows making hitless periodic normalization adjustment which could be problematic if ϕ_E had a large dc component.

The ADPLL operates in the phase domain by counting the number of the DCO clock edges (variable phase – $R_V[i]$), sampling it on FREF ($R_V[k]$) and comparing it to the accumulated value of FCW (reference phase – $R_R[k]$). Fine resolution of R_V is obtained through the TDC-based interpolator, whose normalized output $\varepsilon[k] = [0,1)$ signifies the position of the FREF edge with respect to the two neighboring DCO edges.

The retimed FREF clock, CKR, which is used for the ϕ_E generation and filtering, is obtained in a *metastability-free* manner by *speculative* resampling of FREF by the rising and falling DCO clock edges, and using the path (CKR_P or CKR_N), which is farther away from metastability, based on the arbitration signal SEL_EDGE from the TDC. The arbitration signal is simply a tapped delay of a quarter of the DCO period. A similar speculative mechanism is used for 3 LSB bits of the variable phase R_V that are counted asynchronously in a carry-ripple manner. Finally, the LSB bits of R_V get merged with 5 MSB bits of R_V, which is based on a synchronous counter.

Figure 3.1.3 reveals a technique to lower spurs due to injection pulling. They are likely to happen when the higher harmonic of the digital baseband (DBB) clock falls into the vicinity of the DCO LC-tank resonant frequency. The coupling mechanism could be magnetic (DCO inductor, bondwires), capacitive (long parallel wires), through the substrate, through ground/supply common IR drop, etc. The injection-pulling force gets reduced when the DCO itself is used to clock the DBB, rather than the accompanying PLL. In addition, applying clock dithering further reduces the spurious tones by at least 5dB.

Figure 3.1.4 demonstrates effectiveness of the coarse dithering in both CW and GSM modulated modes. The carrier is 200kHz and 400kHz away from the 46th harmonic of the 38.4MHz FREF. Since the DCO LC tank resonates at 2× of the high-band frequency, the injection-pulling spurs will be 400kHz and 800kHz away from the carrier, respectively. The quantization noise, however, has a complex pattern, which is analyzed in [8]. Engagement of the coarse dithering eliminates the ill-shaped quantization noise (mainly causing modulation distortion) and injection pulling (mainly causing spurs).

Figure 3.1.5 shows a typical measured far-out phase noise in the receive bands of GSM-850 and GSM-900 during the 2-point GFSK modulation of the DCO, as part of a design-of-experiments (DoE) with 7 IC's to fully cover the manufacturing process corners. The higher bands (DCS-1800 and PCS-1900) also show similar behavior. It proves the virtually spurious-free ADPLL operation that guarantees GSM-compliant SAW-less transmit operation. Note that the GSM spec of -112dBc/100kHz corresponds to -162dBc/Hz with RBW=100kHz.

The table in Fig. 3.1.6 summarizes the ADPLL performance, when used as a GSM transmitter, which is equal to or better than the best-in-class conventional designs. The transmitter data is also given since the ADPLL is intimately tied within the 2.5G transmitter. Estimated area of the ADPLL is 0.35mm², which is mostly occupied by the DCO. The current consumption is 32mA and 38mA in low-band and high-band, respectively. Figure 3.1.7 shows the chip micrograph of the transceiver implemented in TI's 65nm digital CMOS.

References:
[1] B. Staszewski et al., "All-digital phase-domain TX frequency synthesizer for Bluetooth radios in 0.13μm CMOS," *ISSCC Dig. Tech. Papers*, pp. 272–273, Feb. 2004.
[2] R. B. Staszewski et al., "All-digital PLL and GSM/EDGE transmitter in 90nm CMOS," *ISSCC Dig. Tech. Papers*, pp. 316–317, 600, Feb. 2005.
[3] C.-M. Hsu et al., "A low-noise, wide-BW 3.6GHz digital $\Delta\Sigma$ fract.-N frequency synthesizer with a noise-shaping TDC and quantization noise cancellation," *ISSCC Dig. Tech. Papers*, pp. 340-341, Feb. 2008.
[4] H.-H. Chang et al., "A fractional spur-free ADPLL with loop-gain calibration and phase-noise cancellation for GSM/GPRS/EDGE," *ISSCC Dig. Tech. Papers*, pp. 200-201, Feb. 2008.
[5] C. Weltin-Wu et al., "A 3GHz fractional-N all-digital PLL with precise time-to-digital converter calibration and mismatch correction," *ISSCC Dig. Tech. Papers*, pp. 344-345, Feb. 2008.
[6] C. Weltin-Wu et al., "A 3.5GHz wideband ADPLL with fractional spur suppression through TDC dithering and feedforward compensation," *ISSCC Dig. Tech. Papers*, pp. 468-469, Feb. 2010.
[7] R. Staszewski et al., "Elimination of spurious noise due to time-to-digital converter," *IEEE Dallas Circuits and Systems Workshop*, pp. 67-70, Oct. 2009.
[8] S. D. Vamvakos et al., "Noise analysis of time-to-digital converter in all-digital PLLs," *IEEE Dallas Circ. and Sys. Workshop*, pp. 87-90, Oct. 2009.

978-1-61284-303-2/11 $26.00 © 2011 IEEE

ISSCC 2011 / February 21, 2011 / 1:30 PM

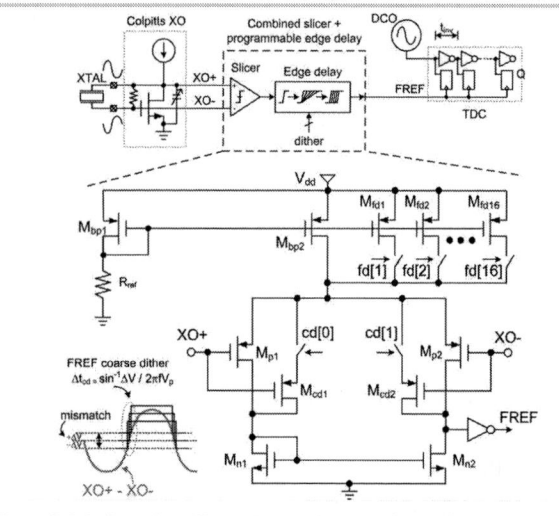

Figure 3.1.1: Crystal oscillator slicer with coarse (3-level) and fine (16-level) delay control.

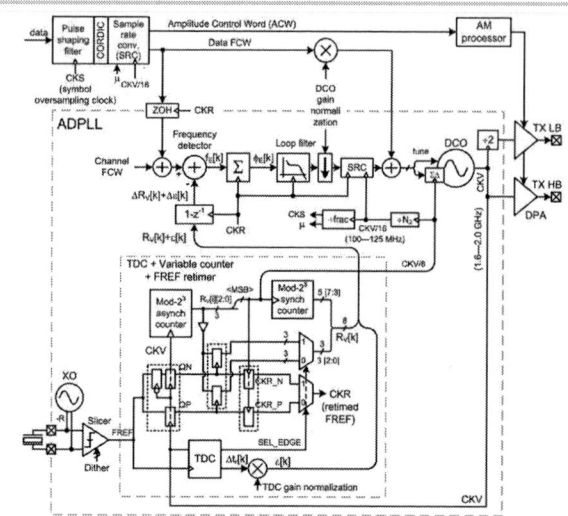

Figure 3.1.2: Top-level diagram of multirate wideband all-digital PLL.

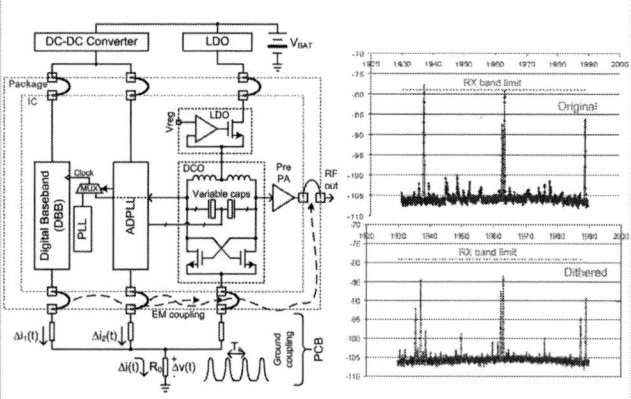

Figure 3.1.3: Digital baseband (DBB) clock is synchronous to the DCO clock. Dithering the DBB clock lowers the DBB induced spurs.

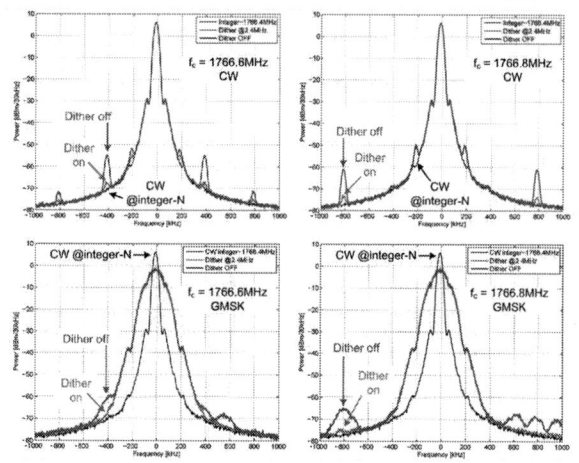

Figure 3.1.4: Measured effect of coarse dither at 2.4MHz on CW-(top) and GMSK-modulated (bottom) spectra at 200kHz (left) and 400kHz (right) away from 1766.4MHz integer-N channel.

Figure 3.1.5: Measured TX WBN in the RX bands in 7 IC's covering all process corners: (top) GSM-850: 20-to-45MHz offsets from 848.8MHz; (bottom) GSM-950: 10-to-45MHz offsets from 914.8MHz.

TX GMSK Parameter	Measured						3GPP Spec		Units
	LB			HB			LB	HB	
	min	mean	max	min	mean	max			
RMS phase error	0.5	0.7	1.1	1.1	1.5	2.7	5		deg
Peak phase error	1.8	2.4	3.7	4.2	6	9.1	20		deg
400kHz mask	-73	-71	-68	-68	-67	-66	-60	-58	dBc
20MHz phase noise	-169	-167	-166	-160	-160	-160	-162	-151	dBc/Hz
SoC output power	7			6			N/A		dBm
TX current consumption	58			62			N/A		mA
ADPLL current consumption	32			38			N/A		mA
Supply voltage	1.2						N/A		V
TX area	0.8						N/A		mm^2
ADPLL area (w/o XO)	0.35						N/A		mm^2
Technology	65nm digital CMOS						N/A		N/A

Figure 3.1.6: ADPLL key performance when transmitting in GSM GMSK mode and implementation table.

978-1-61284-303-2/11 $26.00 © 2011 IEEE

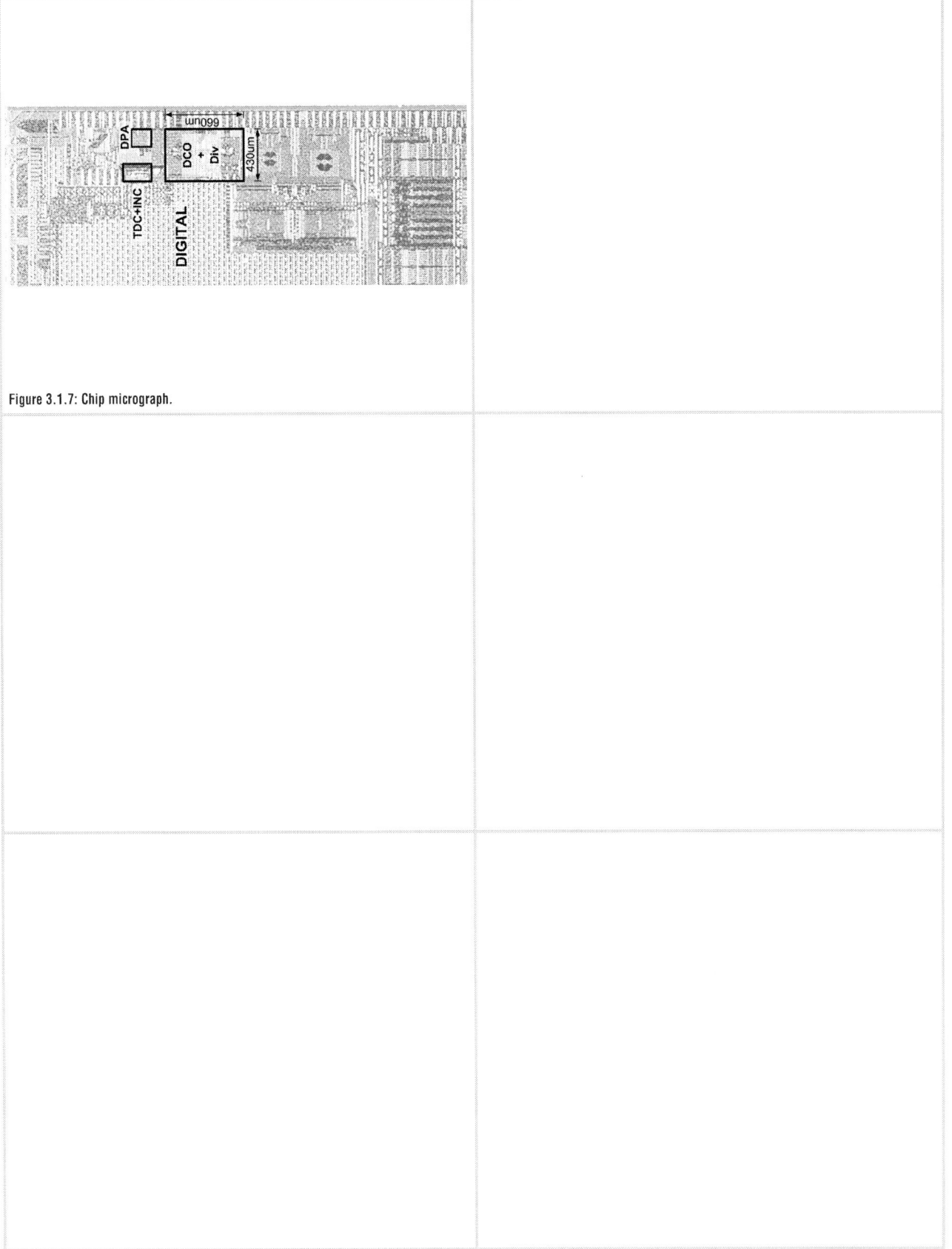

Figure 3.1.7: Chip micrograph.

ISSCC 2011 / SESSION 3 / RF TECHNIQUES / 3.2

3.2 A 5.3GHz Digital-to-Time-Converter-Based Fractional-N All-Digital PLL

Nenad Pavlovic, Jos Bergervoet

NXP Semiconductors, Eindhoven, The Netherlands

Advanced deep-submicron CMOS processes are well-suited for a digital implementation of phase-locked loop-(PLL) based frequency synthesizers. Recently, several RF all-digital phase-locked loops (ADPLL) have been reported [2,4,5]. While ADPLLs come close to achieving the phase-noise performance of analog PLLs, the in-band spur level requirement is still challenging. In this paper we present a new digital-to-time-converter-(DTC) based ADPLL architecture where the time resolution can be easily scaled.

The block diagram of the DTC-based ADPLL is shown in Fig. 3.2.1. The ADPLL is operating in the true phase domain [1]. The ADPLL control loop updates the DCO frequency so that the DCO phase R_{DCO}, measured by counting the DCO clock edge transitions, follows the reference phase R_{REF} (R_{REF_i} is integer and R_{REF_f} is fractional part), produced by accumulating the frequency control word (FCW) on each reference clock cycle. The DTC is added to lower the quantization noise in the fractional-N mode.

The DCO phase measurement and synchronization between asynchronous reference and DCO clock domain is achieved by using a sampler-based counter (SBC). The counter implementation is shown in Fig. 3.2.2. The reference signal f_{ref} is sampled with the sampler tree clocked by clock signals generated from an asynchronous divide-by-2 chain driven by the DCO clock f_{DCO}. The sampling element "L" is realized with a gated inverter for all stages except the last stage, where a latch cell from a standard static CMOS digital library is used. The odd and even samples, represented as charge on internal capacitors, are sent in parallel to the next stage working at half of the speed. Only the first sampler and divider are working at the DCO frequency. The sampler tree output is a thermometer representation of the f_{ref} edge transition inside the frame (in the example shown in Fig. 3.2.2 the frame has 4 DCO periods, in the implementation we use 16 periods) sampled on a half-DCO-period grid. Only one sampler output (bin) could be in a metastable state. The digital decision is done at low frequency, clk_frame, to have enough time to resolve metastability (clk_frame = f_{DCO}/16 for a 4b counter used in the test chip). The extra low-speed counter is counting frames, R_{DCO_frame}. The phase of the DCO, R_{DCO}, is obtained by combining the output of the frame counter with information from the edge detector R_{DCO_edge}. The DCO phase is expressed in the normalized DCO domain with half of integer resolution. The edge detector also produces the digital clock, ckr, for the digital loop.

The phase detector is implemented using the jitter model of the synchronizer [3]. For the integer-N mode, where in the locking state the f_{ref} edge coincides with the DCO edge, the expectation value of counter output $R_S = 0.5$ is added to the phase difference R_{REF_i}- R_{DCO}. In this way a bang-bang operation is achieved with a single-path architecture while still having faster locking time compared with other solutions because a counter instead of a divider is used [2]. In the integer-N mode the measured absolute jitter, integrated from 1kHz to 10 MHz, is 300fs rms, as shown in Fig. 3.2.5. The phase noise is dominated by the noise of the input reference buffer.

In the fractional-N mode, on each reference clock cycle the reference signal is a-priori delayed to create f_{ref}_DTC signal edge close to the next DCO edge, based on the fractional part of the reference accumulator R_{REF_f} as shown in Fig. 3.2.3. Unlike in integer mode the expectation is not $R_S = 0.5$ but becomes a function of the DTC quantization error, which can be a-priori computed. The phase detector output is now R_{REF_i}- R_{DCO}+ R_S(DTC_error). The counter output, when subtracted from the reference phase, behaves as a one-bit phase quantizer. Averaged values of three consecutive digital outputs of the sampler, for a frequency channel with a small fractional part, are shown in Fig. 3.2.4 (this shows the on-average occurrence of "1" and "0"). The moment when the f_{ref} edge passes across the DCO edge boundary is clearly visible in the activities of the two rising-edge bins. Also the average output of the counter shows the jitter model behaviour of the synchronizer.

The calibration of the average DTC unit element delay, DTC_delay, is based on a closed-loop measurement of the fractional spur in the digital domain, Fig. 3.2.1. The phase detector output R_E is processed with a second-order IIR filter that implements an efficient Goertzel DTFT algorithm. The estimation of the average delay of the DTC is updated so as to lower the power of the fractional spur. The sign-error LMS algorithm is used to adaptively adjust the DTC_jitter parameter of the phase detector. The filtered output of R_E*sgn(DTC_error) will increase/decrease when the DTC_jitter parameter is underestimated/overestimated, and it is used in a negative-feedback configuration to find the optimum value of the DTC_jitter parameter.

The DTC is realized by a digitally controlled Vernier delay line. The delay element is constructed as a cascade of two CMOS inverters each with tuneable capacitive load (digitally controlled MOS capacitors), Fig. 3.2.3. Extra buffers are added to isolate the next element from the tuned delay of the previous step. The fixed part of the delay due to the Vernier principle does not influence the loop operation. An advantage of the DTC is that its resolution could be improved by cascading a fine DTC and a coarse DTC. In contrast, the TDC approach would require a finer spacing in the whole delay line, resulting in many more elements. The DCO with tuning range 4.9 to 6.9GHz is an LC oscillator. An 8b binary-weighted PVT bank compensates for process variation. The frequency acquisition is done with a 7b thermometer-coded coarse bank The 9b thermometer-coded fine-tuning bank has a minimuml step of 26kHz. The capacitors are implemented as switched-metal fringe capacitors. The DCO is updated through a $\Delta\Sigma$ quantizer.

The ADPLL is fabricated in 65nm CMOS and has an active area of 1.3mm^2 (30% is used by the LDO). A die micrograph of the IC is shown in Fig. 3.2.7. The reference frequency is 48MHz. The chip consumes 20mA from a regulated supply. The sampler-based counter consumes 2.7mA for 6GHz sampling frequency. The phase noise of the DCO is -124dBc/Hz at 1MHz offset from the carrier for 6mA of current consumption. The measured DTC step size is 4.7ps for 1.1 V supply (supply voltage is programmable by the LDO). The average current consumption of the DTC is only 200uA as it is clocked with the reference frequency. The measured in-band phase noise for the fractional-N mode is -96dBc/Hz, and the fractional in-band spur level is -45dBc. The worst-case channel is the one with in-band fractional spurs at 20kHz offset, as shown in Fig. 3.2.6. The fractional spurs are results of the DTC nonlinearity and can be improved with DTC mismatch calibration. The reference spurs were measured to be below -67dBc. The described DTC-based ADPLL has a phase noise performance [4] or comparable to the TDC-based ADPLL [5]. The measured in-band phase noise is 24dB better compared to the 1b phase detector fractional-N PLL architecture, adjusted for the higher reference frequency and the lower VCO output frequency reported in [6].

Acknowledgements:
The authors thank the support from Jan-Peter Frambach, Paul Mateman, Ulrich Mohlmann and BL Car Hamburg colleagues.

References:
[1] A. Kajiwara and M. Nakagawa, "A new PLL frequency synthesizer with high switching speed," *IEEE Trans. Veh. Technol.*, vol. 41, pp. 407-413, Nov. 1992.
[2] Nicola Da Dalt, Edwin Thaller, et al., "A Compact Triple-Band Low Jitter Digital LC PLL With Programmable Coil in 130-nm CMOS," *IEEE J. Solid-State Circuits*, Vol. 40, No. 7, pp. 1482- 1490, July 2005.
[3] Linsay Kleeman, "The jitter model for metastability and its application to redundant synchronizers," *IEEE Transaction on Computers*, 39(7), pp. 930-942, July 1990.
[4] R. B. Staszewski, J. L. Wallberg, S. Rezeq, et al., "All-digital PLL and transmitter for mobile phones," *IEEE J. Solid-State Circuits*, vol. 40, no. 12, pp. 2469-2480, 2005.
[5] Enrico Temporiti, Colin Weltin-Wu, et al., "A 3 GHz Fractional All-Digital PLL With a 1.8 MHz Bandwidth Implementing Spur Reduction Techniques," *IEEE J. Solid-State Circuits*, vol. 44, no. 3, pp.824-834, March 2009.
[6] Mark A. Ferriss, Michael P. Flynn, "A 14mW Fractional-N PLL Modulator With a Digital Phase Detector and Frequency Switching Scheme," *IEEE J. Solid-State Circuits*, vol. 43, no. 11, pp. 2464-2471, Nov. 2008.

978-1-61284-303-2/11 $26.00 © 2011 IEEE

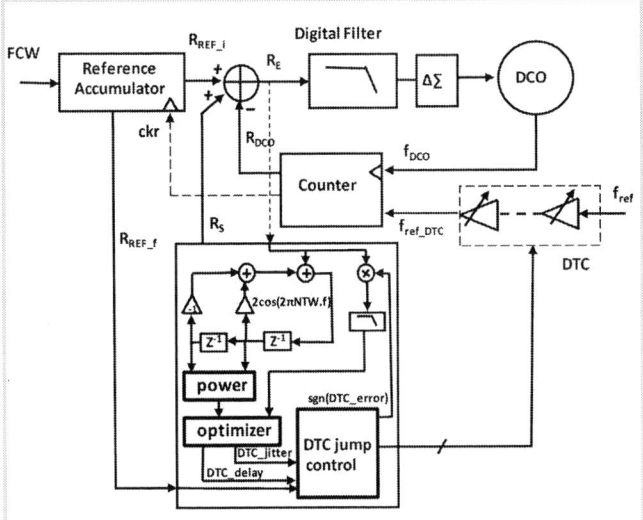

Figure 3.2.1: Block diagram of DTC-based ADPLL architecture.

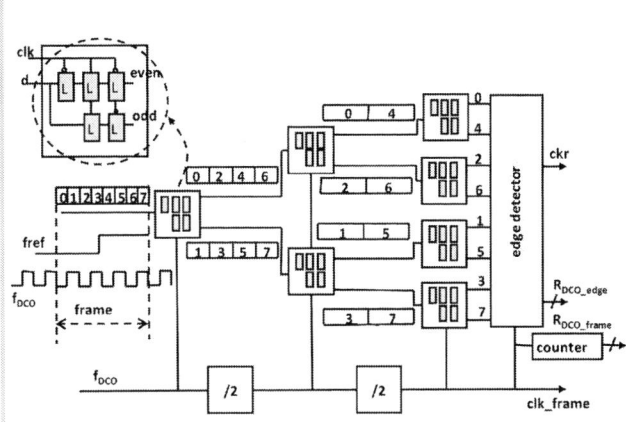

Figure 3.2.2: Example of 2b sampler-based counter (one extra fractional bit is available for a half of DCO period resolution).

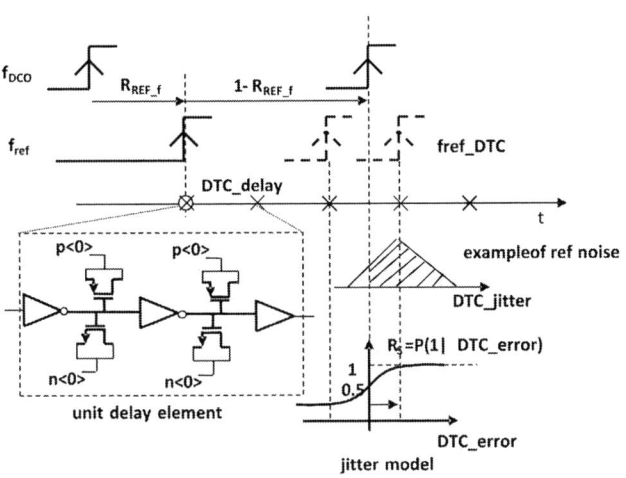

Figure 3.2.3: DTC control in a fractional-N mode with a jitter model of the synchronizer.

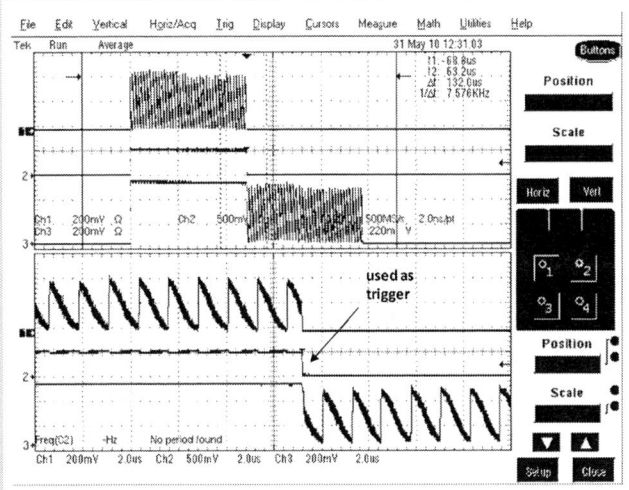

Figure 3.2.4: Averaged values of three consecutive digital outputs of sampler for a frequency channel with a small fractional part.

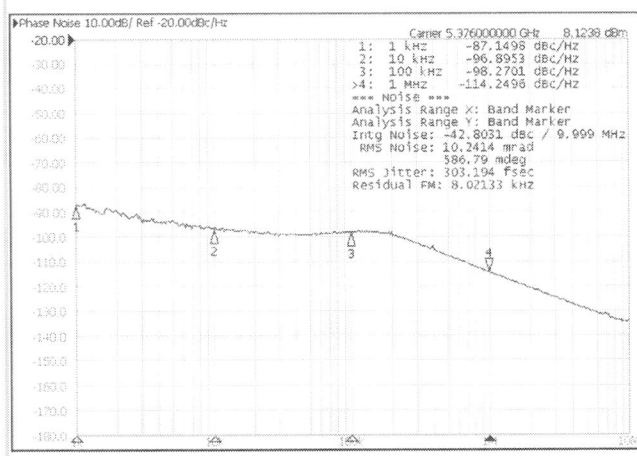

Figure 3.2.5: Phase noise in integer-N mode.

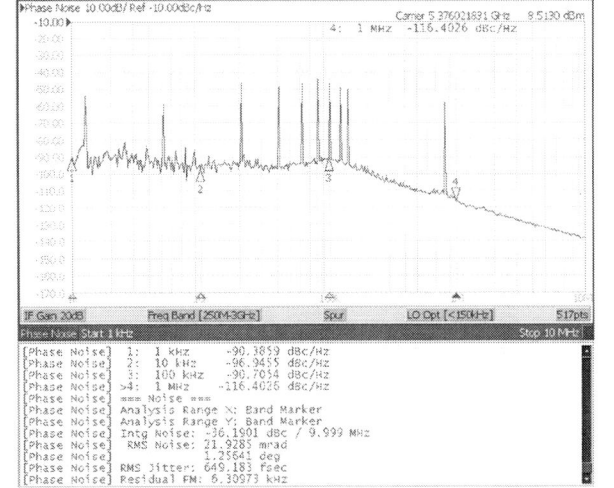

Figure 3.2.6: Phase noise in the fractional-N mode. The worst-case channel is selected that creates in-band fractional spurs at 20kHz frequency offset.

ISSCC 2011 PAPER CONTINUATIONS

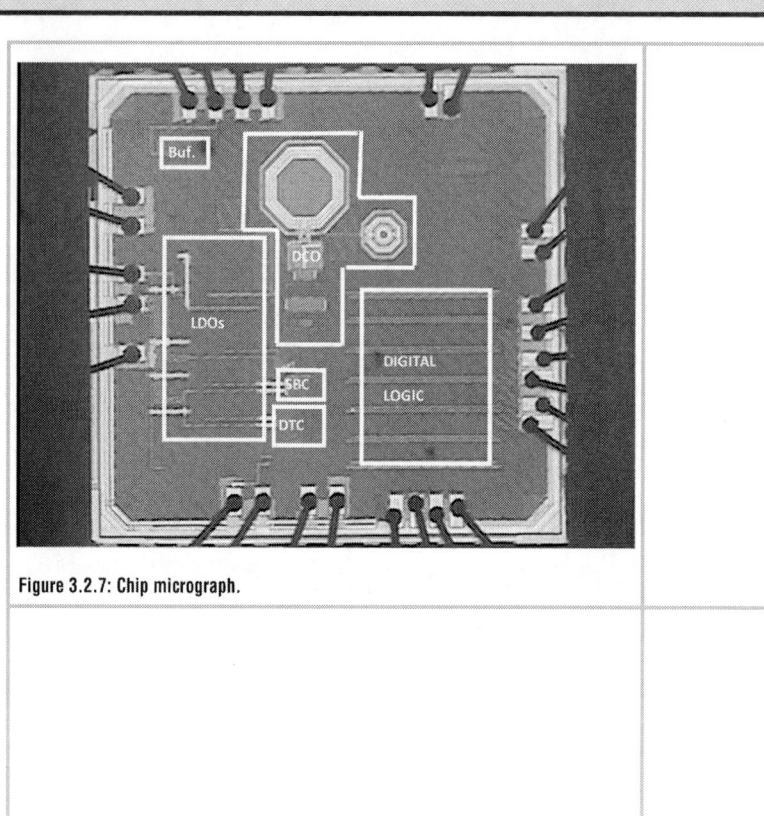

Figure 3.2.7: Chip micrograph.

ISSCC 2011 / SESSION 3 / RF TECHNIQUES / 3.3

3.3 A 2.5GHz 32nm 0.35mm² 3.5dB NF -5dBm P_{1dB} Fully Differential CMOS Push-Pull LNA with Integrated 34dBm T/R Switch and ESD Protection

Chang-Tsung Fu, Hasnain Lakdawala, Stewart S. Taylor, Krishnamurthy Soumyanath

Intel, Hillsboro, OR

Process scaling enables SoC integration of radio and large digital systems at reduced cost and area. Furthermore, the crowded spectrum requires high linearity receivers to enable co-existence in the 2.4GHz ISM band. Inductorless LNA designs have been studied to make use of the excess f_T in advanced CMOS technologies to lower cost [1]. However, an inductively tuned LNA has the advantage of improved out-of-band rejection, and can be readily integrated with a transmit/receive (T/R) switch to provide isolation in the TX mode without the need for additional inductors [2]. In this paper, a 2.5GHz fully differential tuned LNA with integrated T/R switch is designed in a High-K metal gate 32nm digital CMOS process, and packaged in an SoC-compatible flip-chip package. Reliability constraints of the package severely limit the ability to depopulate soldering bumps, and RF components must be designed taking the bump location into account. The LNA achieves a 3.5dB NF, -5dBm P_{1dB} at 2.5GHz while drawing 11mA from a 1.8V supply. LNA performance is enabled by (1) use of a push-pull topology that exploits the equal strength of p and n transistors to improve linearity, (2) use of nested coupled inductors (NCI) for low-noise input matching and to reduce area. The T/R switch handles 34dBm of power with an insertion loss of 1.1dB at 2.5GHz. T/R switch performance is enabled by (1) reuse of LNA gate inductor to enable low RX mode loss [2], (2) use of remote body-contacted TX switch with high power handling and ESD protection for a transformer-coupled PA.

Figure 3.3.1 shows the block diagram of the LNA and integrated T/R switch. The TX switch consists of a minimum gate length NMOS transistor with a remote body contact. The RX switch is combined with the input matching network of the LNA and contributes negligible loss to the RX NF. The RF I/O has 50ohm differential impedance, which is transformed to single-ended 50ohm using an off-chip balun. Figure 3.3.2 illustrates the combination of PMOS and NMOS source-inductive-degenerated LNA stacked to create the push-pull LNA. The similar performance of the p and n transistors in the 32nm hi-K metal gate process [3] improves the push-pull LNA to achieve a good NF, and allows for similar-valued source-degeneration inductors, which is important for area reduction. The push-pull topology improves the linearity by combining the complementary response of the gm non-linearity for the p and n transistors to obtain a constant transconductance with changes in input voltage [4]. Additionally, the effective AC gain to the p and n gates from the input, and the design among cascode transistors and load resistance affect the linearity. The schematic of the fully differential LNA with the Rx switch is illustrated in Fig. 3.3.3. Bias of the p and n transistors is generated by a replica bias circuit and is optimized for maximum linearity. The gate voltage is chosen to maximize output swing while maintaining sufficient f_T. Thin-gate devices with the minimum channel length are chosen to maximum f_T. The P_{1dB} is limited by the supply voltage and is maximized in this design to 1.8V by stacked PMOS and NMOS transistors with thin-gate devices while meeting process reliability. An output CMFB circuit maintains the output CM voltage at 0.9V. Nonlinearity cancellation is robust: simulations indicate P_{1dB} variation <1dB across PVT and bias current changes. The input impedance of the push-pull LNA is 25ohm single-ended, with a 50ohm input impedance for both p and n portions.

One of the main disadvantages of the push pull topology is the need for two more inductors over a regular p or n design. The solution presented in this paper is to nest each of the inductors inside one another by making the observation that the gate and the source inductors carry either in-phase or 180° out-phase currents. The NCI as illustrated in Fig. 3.3.4 exploits mutual inductance between inductors. This reduces the physical inductance and hence the overall area. The

effective source and gate inductance (L_g' and L_s') are made larger by M, the mutual inductance between them. The two multi-turn gate inductors (L_{gp} and L_{gn}) are coupled to the four source inductors (L_{s2n}, L_{s1n}, L_{s1p} and L_{s2p}). By coupling, the self-resonance frequencies of the effective L_s's are close to those of L_g's, around 5GHz. Dimensions of the NCI are constrained by the location of the bumps that cannot be moved or depopulated for high packaging reliability. The NCI design includes a single grounded bump at the center. The performance degradation due to this is negligible as long as the internal diameter of the NCI is larger than that of the bump.

The RX switch is incorporated in the LNA design along with the gate inductor and can tolerate >34dBm of TX power without damage to the LNA. The inclusion of the RX switch increases parasitic capacitance at the LNA input. Simulation indicates a >32dB isolation from the input pad to the LNA gate in the TX mode. The TX switch is implemented using an n transistor with remote body contact so that the drain and the bulk voltages track to prevent breakdown [2]. The minimum-channel-length device without deep NWell is used to minimize insertion loss and maximize linearity. The measured insertion loss in the TX mode between 2.3 and 2.7GHz is less than 1.3dB. This includes all the losses seen in a packaged transceiver, including interconnect, bumps, package traces, and balls. The TX switch at the transceiver input also provides ESD protection for a transformer-coupled PA with a grounded center tap, a topology commonly used in CMOS designs. During a positive or negative ESD event the MOS parasitic BJT is triggered and the pad discharges through the PA ground. The resistance on the TX path is kept low due to the low insertion loss requirement in the TX mode and metal electromigration limits which helps to maintain good RF performance.

The LNA with T/R switch were tested using a low-cost flip-chip matrix molded BGA package and FR4 board. The LNA survives a charge device model (CDM) ESD test of +300V/-200V on packaged parts. Figure 3.3.6 shows the measured S_{11}, NF, Gain, P_{1dB}, and IIP3 of the LNA in RX mode with balun and PCB traces de-embedded. S_{11}<-13dB, gain>11dB and NF<3.5dB are obtained from 2.3 to 2.7GHz. The reverse isolation of the LNA is >38dB. The measured P_{1dB} at 2.5GHz is -5dBm, and IIP3 is +5dBm. The LNA draws 11mA from a 1.8V supply. Power-handling capability of the TX switch, shown in Fig. 3.3.6, is measured by sweeping the power at the PA port in the TX mode. The breakdown at TX power level higher than 34dBm at 2.5GHz is non-destructive and performance is recoverable if without further damage by heating. The micrograph of the LNA with integrated T/R switch and ESD, occupying a total area of 0.35mm², is shown in Fig. 3.3.7.

Acknowledgments:
The authors would like to thank: J. Duster, H. Xu, Y. Tan, R. Bishop, W. Kwong, B. Carlton, H. Alavi, S. McKenney, Q. Fan, M. El-Tanani, J. Rizk, T. Kamgaing, J. Lin, and P. Vandervoorn in Intel for their constructive discussion and help.

References:
[1] J.-H. C. Zhan, S. S. Taylor, "A 5 GHz resistive-feedback CMOS LNA for low-cost multi-standard application," *ISSCC Dig. Tech. Papers*, pp.721-722, Feb. 2006.
[2] A. Kidwai et al, "A Fully Integrated Ultra-Low Insertion Loss T/R Switch for 802.11b/g/n Application in 90 nm CMOS Process", *IEEE JSSC*, vol. 44, May 2009, pp 1352-1360.
[3] C.-H. Jan, et al., "A 32nm SoC platform technology with 2nd generation high-k/metal gate transistors optimized for ultra low power, high performance, and high density product applications ", *IEDM Tech. Dig.*, pp. 1-4, Dec. 2009.
[4] D. Im et al, "A wideband CMOS low noise amplifier employing noise and IM2 distortion cancellation for a digital TV tuner," *IEEE JSSC*, vol. 44, pp. 686–698, Mar. 2009.

978-1-61284-303-2/11 $26.00 © 2011 IEEE

ISSCC 2011 / February 21, 2011 / 2:30 PM

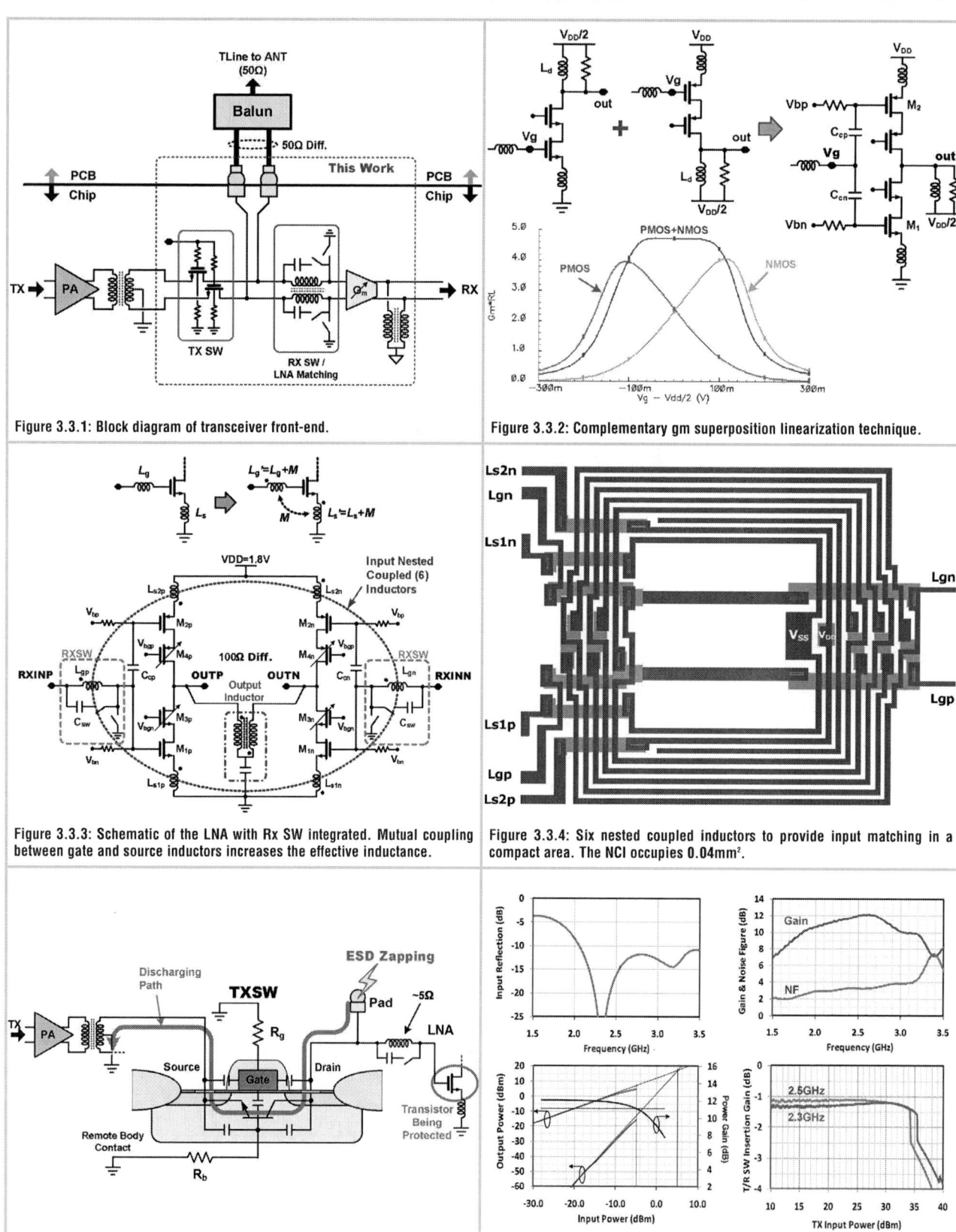

Figure 3.3.1: Block diagram of transceiver front-end.

Figure 3.3.2: Complementary gm superposition linearization technique.

Figure 3.3.3: Schematic of the LNA with Rx SW integrated. Mutual coupling between gate and source inductors increases the effective inductance.

Figure 3.3.4: Six nested coupled inductors to provide input matching in a compact area. The NCI occupies 0.04mm².

Figure 3.3.5: ESD protection of TX Switch for output transformer-coupling PA.

Figure 3.3.6: LNA & T/R Switch measured results.

978-1-61284-303-2/11 $26.00 © 2011 IEEE

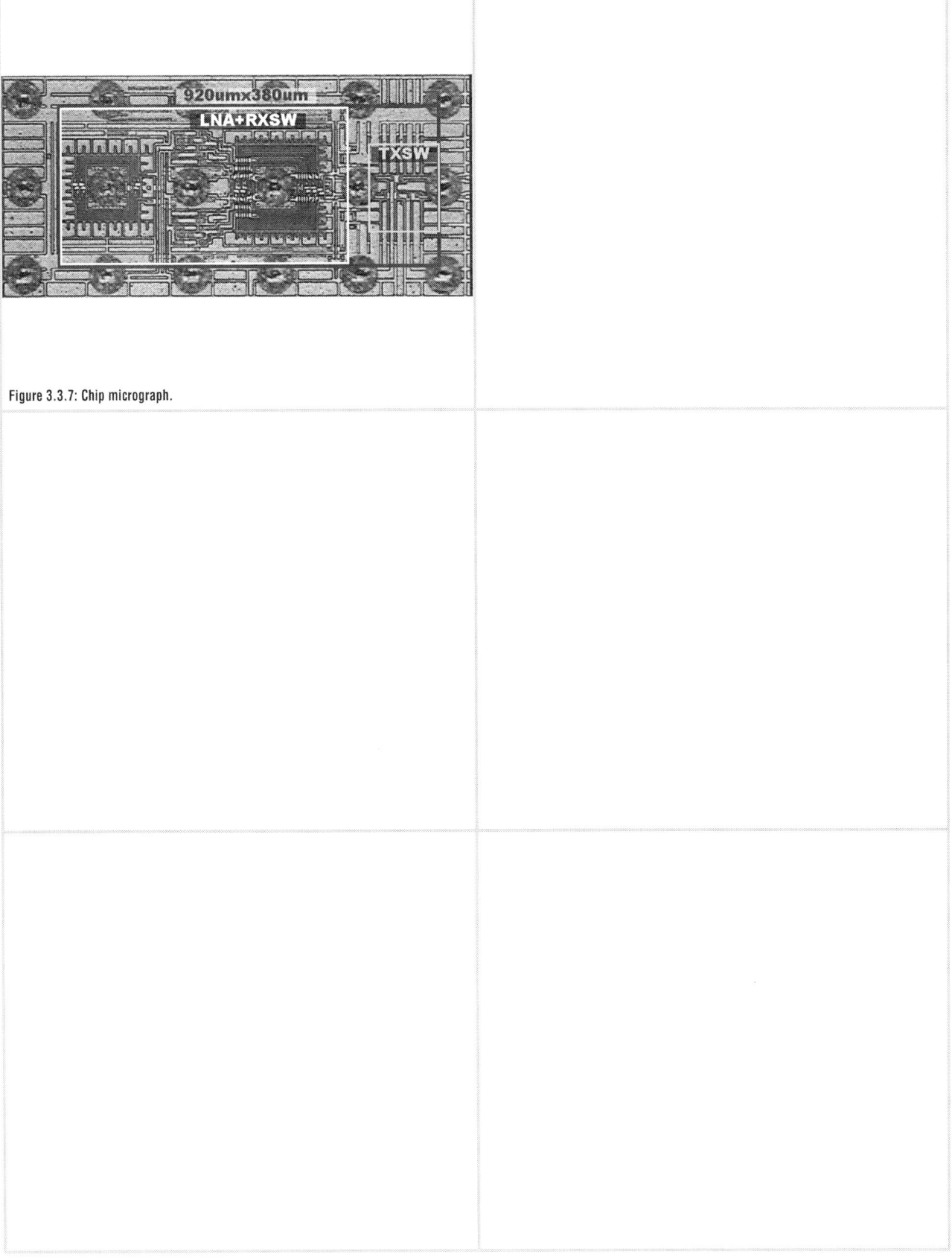

Figure 3.3.7: Chip micrograph.

ISSCC 2011 / SESSION 3 / RF TECHNIQUES / 3.4

3.4 A 65nm CMOS Pulse-Width-Controlled Driver with 8V_{pp} Output Voltage for Switch-Mode RF PAs up to 3.6GHz

David A. Calvillo-Cortes[1], Mustafa Acar[2], Mark P. van der Heijden[2], Melina Apostolidou[2], Leo C.N. de Vreede[1], Domine Leenaerts[2], Jan Sonsky[3]

[1]Delft University of Technology, Delft, The Netherlands,
[2]NXP Semiconductors, Eindhoven, The Netherlands,
[3]NXP-TSMC Research Center, Leuven, Belgium

State-of-the-art wireless communication radios are implemented in deep-submicron CMOS, including the RF power amplifiers (PAs). However, in wireless infrastructure systems, the RF PA is often realized in an LDMOS or a compound technology to obtain the required large output powers. For next-generation reconfigurable infrastructure systems, the switch-mode PAs (SMPA) seem to offer the required flexibility for multiband multimode transmitters. In order to interface the high-power devices of the SMPA with the digital CMOS blocks of the transmitter, a wideband RF CMOS driver capable to generate high voltage (HV) swings is required. In this way, digital signal processing can be directly applied to control the required input pulse shapes of the SMPA.

An LDMOS or GaN SMPA typically has an input capacitance of several picofarads and needs to be driven by pulsed signals with signal amplitudes exceeding 5V_{pp}. Therefore, the SMPA CMOS driver must provide both HV and Watt-level powers at RF. Unfortunately, deep -submicron CMOS is severely limited in its maximum supply voltage due to reliability issues. Although device stacking is the most common HV enabling technique [1,2], it has complexity and performance limitations at RF [2]. In this work, an HV CMOS driver topology is presented in which Extended-Drain MOS (EDMOS) transistors [3] are used to avoid device stacking. The proposed RF driver is fabricated in a baseline 65nm low-standby-power CMOS process using thin-oxide EDMOS devices without any extra mask or process step. The measured f_T of these devices exceed 30GHz and 50GHz for PMOS and NMOS, respectively, and their breakdown voltage limit is 12V. The proposed high-speed driver delivers 8V_{pp} output swing up to 3.6GHz, which allows driving wide-bandgap-based SMPAs, such as GaN.

Figure 3.4.1 depicts the schematic diagram of the proposed driver. The output stage consists of an EDMOS-based inverter. The EDMOS devices can be driven directly by low-voltage high-speed standard transistors, simplifying the integration of the output stage with other digital and analogue CMOS circuits on a single die. Each EDMOS transistor is driven by a tapered buffer (buffer A and B in Fig. 3.4.1), implemented as 3 CMOS inverter stages. The schematic of the tapered buffers is shown in Fig. 3.4.2. The two buffers have different DC levels to ensure that each CMOS inverter operates reliably within 1.2V, i.e. V_{DD1}-V_{SS1}=V_{DD0}-V_{SS0}=1.2V. In order to use different sets of supply voltages and to allow identical AC operation, the two buffers were built identically and placed inside individual Deep N-Wells (DNW). The output swing of the driver is determined by V_{DD1}-V_{SS0} and can be chosen freely up to the breakdown limit of the EDMOS devices, while the internal driver operation remains unchanged. A DC level shifter splits the input signal for each of the buffers.

The presented CMOS driver has pulse-width control of its output square wave, which is accomplished by using the pulse-width modulation (PWM) by variable-gate-bias technique [4]. PWM control provides a means to perform fine adjustment/tuning functionality to enhance the performance in an advanced SMPA. The bias level at the first inverter (M3) of buffer A and B shifts up/down the RF sinusoidal input signal with respect to the inverter's own switching threshold. A change on this bias voltage will vary the pulse width at the output of inverter M3. Then, this PWM signal will propagate through the remaining inverters M2 and M1, and will be combined at the output stage (EDMOS) of the RF driver.

All the inverters (M0 to M3) employ a unitary PMOS-to-NMOS transistor size ratio in order to ensure layout symmetry between the two RF paths prior to the output stage. The total transistor widths for the M0, M1, M2 and M3 stages are 4032μm, 1440μm, 480μm and 240μm, respectively, and have minimum gate length. The layout of each transistor was split up into many unit transistor layout-parameterized cells (P-cell) and was optimized to obtain maximum frequen-cy performance. Each P-cell contains an asymmetric-layout multi-fingered transistor with finger width of 3μm, with its own guard ring and all its interconnections up to metal 7, and it is layout-scalable [5].

The driver includes large on-chip AC-coupling and –decoupling parallel-plate interdigitated metal-fringe capacitors. Capacitors C_{in} (16pF) implement the DC level shifter along with two DC input biasing lines ($BIAS_{a,b}$). The output can be taken DC-coupled or AC-coupled using an on-chip capacitor C_{out} (49pF). AC-coupling allows driving power transistors that require a negative gate biasing (e.g. GaN). Four thick and wide power lines were routed inside the chip, two on metal 6 (V_{SS0} and V_{SS1}) and two on metal 7 (V_{DD0} and V_{DD1}). The internal supply lines are decoupled with the capacitors C_0, C_1, C_2 and C_3 (97pF, 96pF, 437pF and 104pF, respectively). Dedicated ESD protection circuitry was added to protect every single pin of the CMOS chip.

The fabricated prototype die was mounted on a PCB for test purposes, and was measured in a 50Ω load environment. Time-domain signals were captured using a high-speed digital sampling oscilloscope. Figure 3.4.3 shows the time-domain waveforms of the DC-coupled output of the RF driver at 3V, 5V, 7V and 9V supply voltage with an input sine wave at 2.1GHz. The maximum swing measured was 8.04V_{pp} for a 50Ω load and 9V supply. The on-resistance of the driver was measured as 4.6Ω. Figure 3.4.4 gives the measured pulse width (expressed as duty-cycle) control range as a function of the normalized DC bias levels, i.e. $BIAS_{a,b}$-$V_{SS0,1}$. In the same figure, two time-domain waveforms are shown for different duty-cycle conditions. At 2.4GHz and 5V supply, a duty-cycle control range of 30.7 to 71.5% was observed. In Fig. 3.4.5, the 10-to-90% and 20-to-80% rise and fall times are shown for the operating frequency range. The RF driver maintains its pulse shape behavior with 8V_{pp} up to 3.6GHz. Measurements performed at 2.4GHz showed no performance degradation with 5V and 9V supply after 24 hours of continuous operation.

Figure 3.4.6 summarizes the measured performance of the driver and provides a comparison with other works. The proposed driver reaches a much larger output voltage swing and higher operating frequency as compared to the previous state of the art in CMOS. This CMOS driver has similar performance as a SiGe-BiCMOS equivalent circuit [6], and additionally it has pulse-width control functionality. The photograph of the fabricated die is given in Fig. 3.4.7. The total chip area is 1.99mm², while the active area (EDMOS and buffers) is 0.16mm².

This work presented the first broadband PWM-controlled RF SMPA driver reaching 8.04V_{pp} up to 3.6GHz using a 1.2V baseline 65nm CMOS technology. The CMOS driver can serve as a key building block for next-generation reconfigurable multiband multimode transmitters for wireless infrastructure systems, interfacing digital CMOS circuitry with high-power transistors.

Acknowledgments:

This work was supported by the Programme Alβan (scholarship E07M401622MX) and the Mexican National Council for Science and Technology (scholarship 230803). The assistance of E. van der Heijden, M. de Langen and K. Buisman is gratefully acknowledged.

References:

[1] B. Serneels, et al., "A High-Voltage Output Driver in a Standard 2.5V 0.25μm CMOS Technology," *ISSCC Dig. Tech. Papers*, pp. 146–147, Feb. 2004.

[2] A. Mazzanti, et al., "Analysis of Reliability and Power Efficiency in Cascode Class-E PAs," *IEEE J. Solid-State Circuits*, Vol. 41, pp. 1222-1229, May 2006.

[3] J. Sonsky, et al., "Innovative High Voltage Transistors for Complex HV/RF SOCs in Baseline CMOS," *Int. Symp. on VLSI Tech., Syst. and App.*, pp. 115-116, Apr. 2008.

[4] E. Cijvat and H. Sjoland, "Two 130nm CMOS Class-D RF Power Amplifiers Suitable for Polar Transmitter Architectures," in *9th Int. Conf. on Solid-State and IC Tech.*, pp. 1380-1383, Oct. 2008.

[5] M. Acar, et al., "Scalable CMOS Power Devices with 70% PAE and 1, 2 and 3.4 Watt Output Power at 2GHz," *IEEE RFIC Symp. Dig.*, pp. 233-236, Jun. 2009.

[6] S. Heck, et al., "A Switching-Mode Amplifier for Class-S Transmitters for Clock Frequencies up to 7.5GHz in 0.25μm SiGe-BiCMOS," *IEEE RFIC Symp. Dig.*, pp. 565–568, May 2010.

978-1-61284-303-2/11 $26.00 © 2011 IEEE

ISSCC 2011 / February 21, 2011 / 2:45 PM

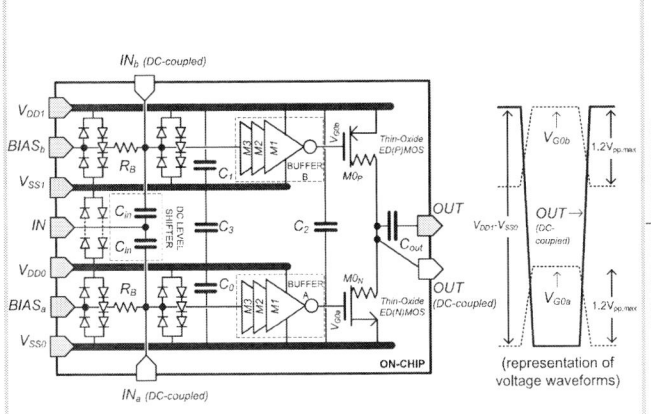

Figure 3.4.1: Schematic of the RF driver with relevant internal voltage waveforms.

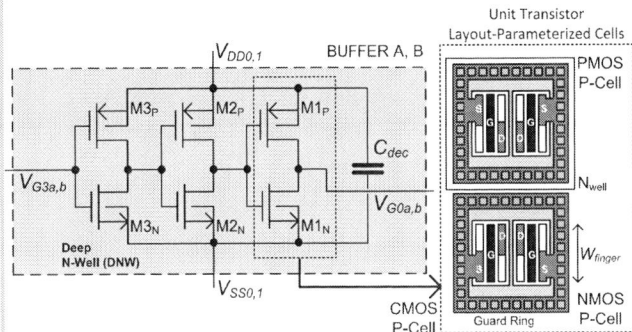

Figure 3.4.2: Schematic of the tapered buffers with top view representation of transistor P-cells.

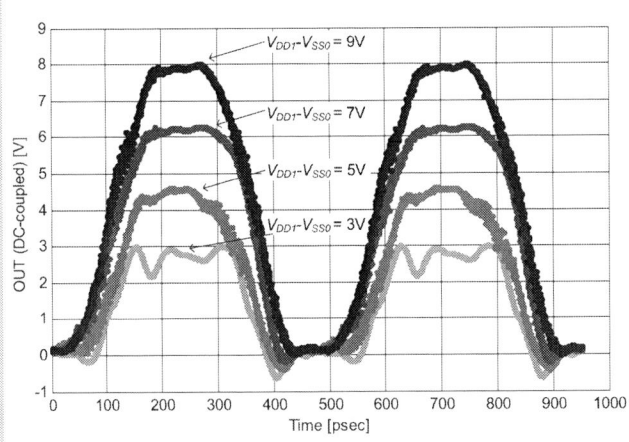

Figure 3.4.3: Measured time-domain waveforms with $V_{DD1} - V_{SS0}$ = 3V, 5V, 7V, 9V at 2.1GHz.

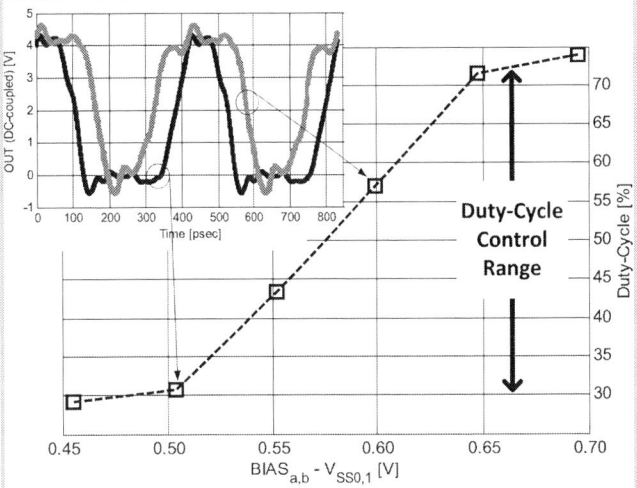

Figure 3.4.4: Measured duty-cycle performance versus normalized biasing levels at 2.4GHz and $V_{DD1} - V_{SS0}$ = 5V.

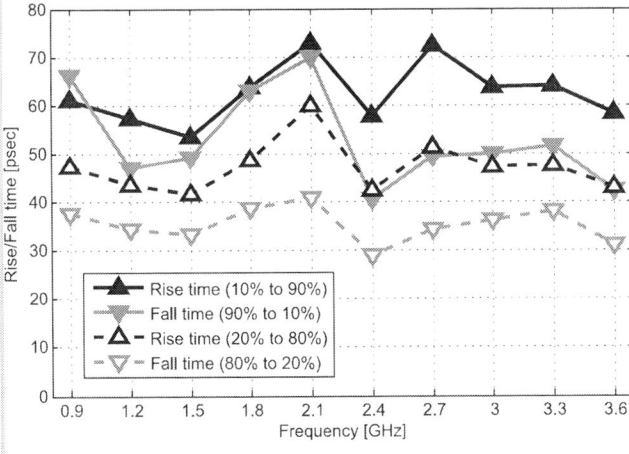

Figure 3.4.5: Measured rise and fall times versus frequency for a continuous-wave signal with 50% duty-cycle at $V_{DD1} - V_{SS0}$ = 6V.

Parameter	[1]	[6]	This work
Technology	0.25um	0.25um	65nm
	CMOS	SiGe-BiCMOS	CMOS
Nominal Supply Voltage	2.5V	4.3V	1.2V
Operating Supply Voltage	7.5V	4.3V	Up to 9V
Load	50Ω	50Ω	50Ω
Maximum Output Swing	6.46V_{pp}	4V_{pp}	8.04V_{pp}
ON-Resistance	5.9Ω	Not specified	4.6Ω
Operating Frequency	10MHz	0.5 to 3.5GHz	0.9 to 3.6GHz
Duty-Cycle Range	50% fixed	50% fixed	30.7 to 71.5% @ 2.4GHz
Square Wave Performance			
Pout @ Frequency, Supply Voltage	208mW @ 10MHz, 7.5V	145mW @ 0.9GHz, 4.3V 130mW @ 2.1GHz, 4.3V 116mW @ 3.5GHz, 4.3V	216mW @ 0.9GHz, 9V 221mW @ 2.1GHz, 9V 179mW @ 3.5GHz, 9V

Figure 3.4.6: Measured performance and process specifications of the driver in this work and the previous state of the art.

ISSCC 2011 PAPER CONTINUATIONS

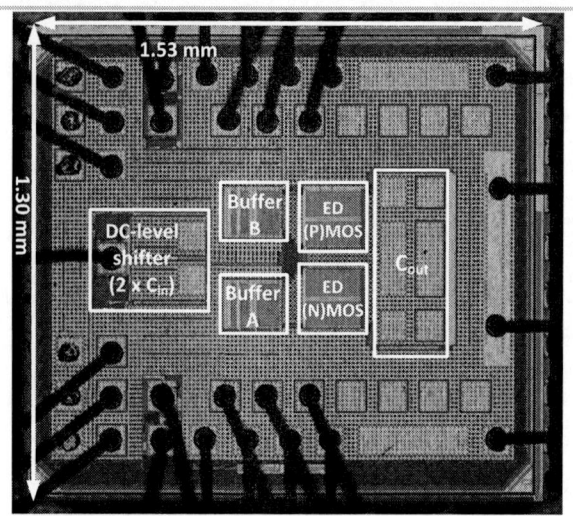

Figure 3.4.7: RF driver micrograph. The total chip area is 1.99mm², while the active area is 0.16mm².

ISSCC 2011 / SESSION 3 / RF TECHNIQUES / 3.5

3.5 A Low-Power Process-Scalable Superheterodyne Receiver with Integrated High-Q Filters

Ahmad Mirzaei[1], Hooman Darabi[1], David Murphy[2]

[1]Broadcom, Irvine, CA, [2]University of California, Los Angeles, CA

So far most integrated receivers are *zero or low-IF*, benefitting from a simple structure, and a high level of integration as *image rejection* is not a major concern, and *channel selection* is performed by low-frequency lowpass filters [Fig. 3.5.1(a)]. Dominated by linear capacitors and resistors, the lowpass filters which are typically active-RC occupy a large die area that *does not scale* with the technology. In addition, constraints such as 1/f noise, 2nd-order nonlinearity, and DC offset demand large device sizes for the IF blocks, further exacerbating the scalability issue. On the other hand, a superheterodyne architecture [Fig. 3.5.1(b)] does not experience the above shortcomings due to a considerably higher IF, but needs *external filters* for image rejection and channel selection.

Recently it has been shown that *frequency-translated BPFs* offer high-Q filtering with their center frequency precisely controlled by a clock [1,2]. To address the scalability concerns and to achieve the same level of integration as the zero or low-IF receivers, we propose the architecture shown in Fig. 3.5.1(c). Similar to the superheterodyne architecture the IF is high. However, the image and other in- and out-of-band blockers are progressively filtered throughout the RX chain both at the RF and baseband by utilizing different forms of integrated *frequency-translated BPFs* that are evolved from the original N-path filtering concept [3].

Figure 3.5. 2 depicts the details of the receiver. The incoming RF signal at the input is converted to differential using an on-chip balun followed by a self-biased inverter-based LNA. To prevent compression of the LNA by out-of-band blockers, a high-Q BPF, enabled by the LO and its divided-down clocks, and centered at f_{RF} is placed at the secondary of the transformer (FTBPF1 in Fig. 3.5.2). Following the LNA is a quadrature current-driven passive mixer, loaded by a common-gate transimpedance amplifier (TIA), that has a bandpass transfer function centered at f_{IF} both at its input and output. The I/Q TIAs drive inverter-based amplifiers, which are loaded by another set of frequency-translated filters. The IF filter however, is *complex* and acts as a high-Q BPF at f_{IF} for one quadrature sequence that contains the desired channel, while attenuating image components at the other. The filtered I/Q signals go through another set of inverter-based amplification followed by a similar complex filter. The resulting IF signal has now experienced adequate filtering, and blockers are weak enough, that it can be digitized by a bandpass ADC with modest dynamic range requirements.

The proposed architecture offers several advantages to reduce the *area* and more importantly allow *technology scaling* that most of the RF receivers lack today. First, since the BPFs are controlled by the clock, the proposed structure eliminates the need for frequency calibration common in the conventional receivers. Secondly, all the filters consist of minimum-channel-length *switches* and *MOS capacitors* that occupy small area, consume no power, and scale well with the technology. Moreover, since 1/f noise and 2nd-order nonlinearity are no longer an issue, the RF and IF amplifiers are minimum-channel devices, making the structure very friendly to the technology scaling. Finally, the DC offset and the area overhead caused by the DC correcting circuitry are not a concern anymore.

The circuit of the on-chip RF filter is depicted in Fig. 3.5.3. It consists of a 4-phase high-Q filter clocked at f_{LO}. Since the receiver IF is high, unlike the conventional design where the switches are connected to four *lowpass* impedances [1], they are connected to a differential *complex bandpass* impedance that is centered at f_{IF}. The complex filter is designed to exhibit an input that is bandpass for one quadrature sequence, but wideband low-impedance for the opposite one. Similar to the conventional 4-phase filter in which the baseband impedance is translated to f_{LO}, the input impedance seen from the complex filter is also frequency translated. In contrast, with the proposed topology only the existing LO is utilized to synthesize a high-Q filter at $f_{LO}+f_{IF}$ instead. The complex impedance

is realized by 16 switches connected to one input, and 16 baseband capacitors with another set of switches added to appropriately sample the other input to the same 16 capacitors using the same clock phases. The switches are toggled by 1/16th duty-cycle clock phases at f_{IF} created by dividing the LO signal by 16. Any two switches that sample the two quadrature inputs to the same capacitor are toggled by two clock phases that are 90° shifted with respect to each other. The resulting filter at $f_{LO}+f_{IF}$ offers more than 10dB attenuation to the image and other out-of-band blockers, limited by the switch ON resistance. One drawback of this BPF is that it folds blockers located at $f_{LO}+17\times f_{IF}$ or $f_{LO}-15\times f_{IF}$.... The receiver rejection is more than 39dBc, and since the IF is high, they are subject to further attenuation by the preselect filter. This reveals why 16 phases are utilized to realize the complex filter.

The LNA is composed of two self-biased inverters that are capacitively coupled to the I/Q downconversion mixers clocked by the same 25% duty-cycle signals at f_{LO} (Fig. 3.5.4). The mixer reflects the input impedance of the TIA, which is designed to be bandpass to the RF side through a simple frequency shifting. This prevents the blockers from causeing large voltage swings at internal nodes, improving linearity. Moreover, the 2nd-order bandpass response of the TIA along with the RF filter provide *anti-aliasing* for the following IF filter whose closest image replica is at a conveniently far offset of $7\times f_{IF}$.

Illustrated in Fig. 3.5.5, the IF filtering is performed with a pair of *cross-coupled* 8-phase frequency-translated filters. The switches are clocked by 1/8th duty-cycle signals at f_{IF} and each pair needs four baseband capacitors. The blocker located at the image is further attenuated by another similar complex filter at IF. The total attenuation is ultimately limited by the quadrature accuracy of the LO clocks driving the downconversion mixer. The image rejection is on average 51dB with a standard deviation of 2dB over 5 samples.

For the proof of the concept, a test chip was fabricated in 65nm CMOS. The incoming signal is assumed to be in the 1.8-to-2GHz range, with the *sliding IF* to be around 110MHz. The clock signals for the 16-phase and 8-phase filters are derived from the main LO at 3.6GHz, and through a chain of divide-by-two's. The receiver provides 55dB of gain when all the filters are ON, and 56dB when only the IF filters are ON. At the maximum gain when IF filters are activated the measured noise figure is 2.8dB, and the in-band IIP3 is -8.5dBm. The measured transfer functions for two cases when all RF/IF filters are activated and when only the IF filters are enabled are shown in Fig. 3.5.6. The receiver consumes 13.5mA, and the LO chain including the dividers draws 15/7.5mA with the RF filter ON/OFF. The corresponding battery current is 11.6mA in nominal mode with the RF filter OFF. The entire receiver including the LO path occupies an active area of 0.76mm², shown in Fig. 3.5.7. The receiver presented here exceeds the requirements of most applications today, achieves a low power consumption at a small die area, yet addresses the scalability issue of the conventional radios as is mainly built of basic CMOS building blocks, namely, inverters, switches, and MOS capacitors.

References:

[1] A. Mirzaei, et. al., "A 65nm CMOS Quad-Band SAW-less Receiver for GSM/GPRS/EDGE", *IEEE VLSI Symposium*, pp. 179-180, 2010.

[2] A. Ghaffari, et. al., "A Differential 4-Path Highly Linear Widely Tunable On-Chip Bandpass Filter", *IEEE Radio Frequency Integrated Circuits Symposium (RFIC)*, pp. 299-302, 2010.

[3] D. V. Grunigen, and et. al., "An Integrated CMOS Switched-Capacitor Bandpass Filter Based on N-path and Frequency-Sampling Principles," *IEEE Journal of Solid-State Circuits*, vol. SC-18, no. 6, pp. 753–761, 1983.

ISSCC 2011 / February 21, 2011 / 3:15 PM

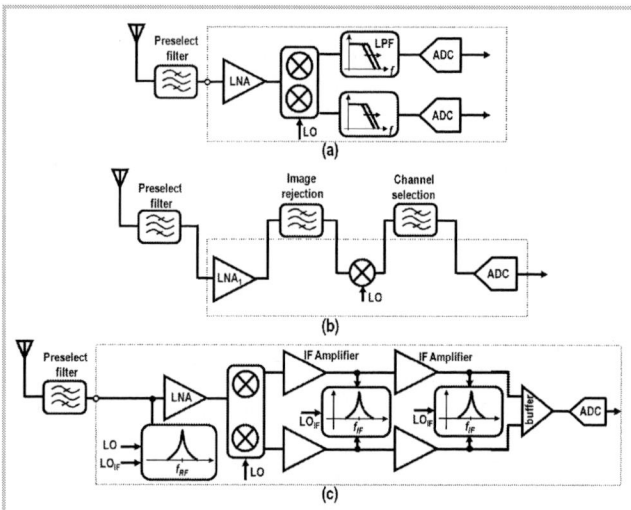

Figure 3.5.1: (a) Zero-IF receiver architecture. (b) Conventional heterodyne receiver. (c) Proposed heterodyne receiver architecture.

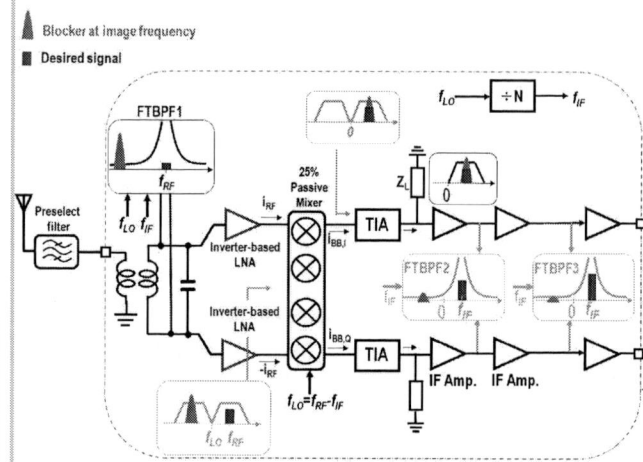

Figure 3.5.2: Superheterodyne receiver architecture composed of frequency-translated bandpass filters.

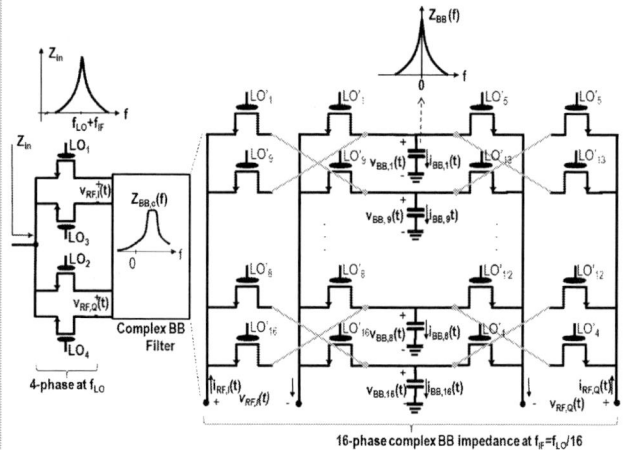

Figure 3.5.3: High-Q frequency-translated bandpass filter centered at $f_{RF}=f_{LO}+f_{IF}$.

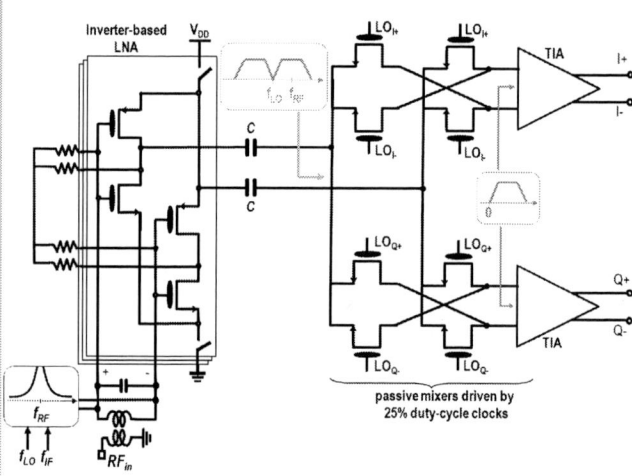

Figure 3.5.4: Front-end circuit optimized for minimum IQ cross-talk and maximum conversion gain.

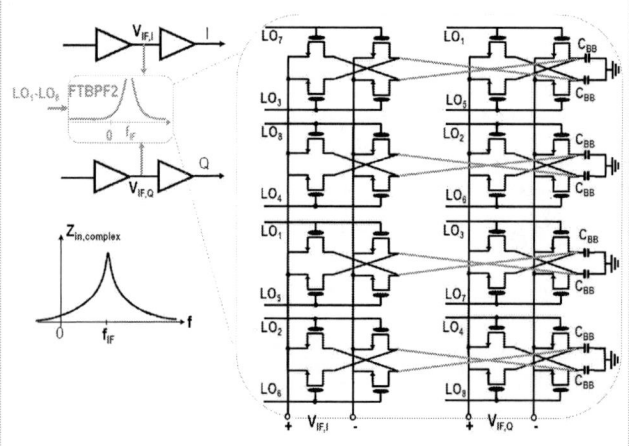

Figure 3.5.5: Circuit of frequency-translated bandpass filters with complex inputs.

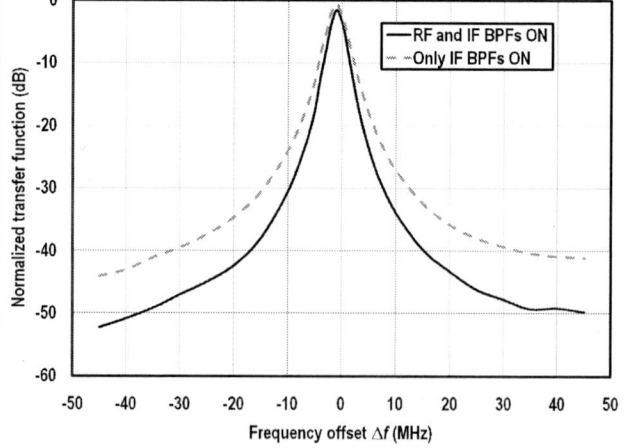

Figure 3.5.6: Measured transfer functions at around f_{RF} (normalized to the case of no high-Q bandpass filtering).

978-1-61284-303-2/11 $26.00 © 2011 IEEE

ISSCC 2011 PAPER CONTINUATIONS

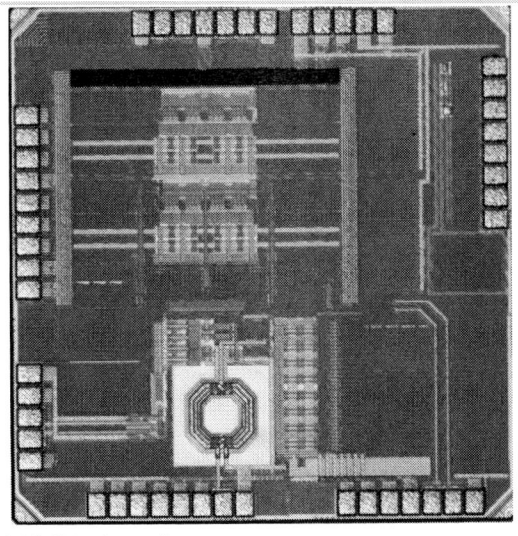

Figure 3.5.7: Chip micrograph.

ISSCC 2011 / SESSION 3 / RF TECHNIQUES / 3.6

3.6 A 40nm CMOS Highly Linear 0.4-to-6GHz Receiver Resilient to 0dBm Out-of-Band Blockers

Jonathan Borremans[1], Gunjan Mandal[1], Vito Giannini[1], Tomohiro Sano[2], Mark Ingels[1], Bob Verbruggen[1], Jan Craninckx[1]

[1]imec, Leuven, Belgium, [2]Renesas Electronics, Itami, Japan

SDRs come of age ([1,2]) and transcend beyond just acquiring the reconfigurability to replace any standard radio: they develop toward systems where a simplified antenna interface can be used, with most dedicated filtering removed. This requires a receiver accommodating much higher linearity and resilience against out-of-band interference than a standard radio, still achieving competitive sensitivity (especially in the absence of interference). Mixer-first front-ends with excellent linearity have been reported [3]. However, their NF (including 1/f in absence of the LNA gain) is not competitive, and they may suffer from large LO feedthrough to the antenna (LOFT). Moreover they lack receiver functionality such as gain and filtering, which cannot be simply added without compromising linearity. A receiver with mixer-at-the-antenna-based bandpass filter [4] similarly may suffer from LOFT and increased NF. This work presents a full software-defined receiver with 3dB NF that tolerates 0dBm blockers with acceptable blocker NF at maximum gain. It achieves +10dBm out-of-band (OB) IIP3 and >+70dBm IIP2. Such a receiver is to operate using no other than harmonic-rejection filtering.

The blocker-resilient receiver architecture is depicted in Fig. 3.6.1. The main-path RF front-end embodies a linear LNA and *voltage-sampling* mixer. An RF blocker filter *at the LNA output* is formed by low-pass reconfigurable differential sampling capacitor C_{BB} which turns into a bandpass filter at RF centered around the LO frequency. Indeed, a passive mixer upconverts the impedance at the baseband terminals to the RF side [4-6]. Hence, a baseband cut-off at 2MHz provides low impedance for out-of-band blockers at RF beyond 2MHz from the carrier. In this work, the cut-off tunes from 1.5 to over 20MHz, and filters a blocker at 20MHz offset by about 15dB. The voltage sampling offers a *high* impedance in-band, unlike the low-impedance case in [1,2,4], which has a benign effect on NF, while out-of-band linearity benefits from the RF bandpass filtering.

Since in this architecture blocker filtering occurs at the LNA output, the latter should carry the input blocker to the output without compressing. To handle blockers up to 0dBm *at the input and internal nodes of the LNA* with such high linearity, the headroom of a 2.5V supply is leveraged. Dedicated startup circuitry in Fig. 3.6.2 ensures that no terminal-to-terminal voltages over the triple-well devices exceed the technology reliability limits: neither during start-up, nor during standard operation. Consequently, reliability is not jeopardized. At startup the LNA bias is enabled first, next the startup circuitry provides the supply causing the internal LNA nodes to ramp up to their operating point. The startup circuitry never enables the internal supply when the LNA is not biased properly (e.g. in absence of the 1.1V supply).

A capacitive cross-coupled common-gate LNA provides moderately low noise (2.5dB) for high linearity (~+10dBm IIP3) since no passive voltage amplification occurs at the input [7]. A small low-Q inductor provides shunt peaking for relatively wideband operation. The LNA is centered at 0.7 to 3GHz: out of this band NF is less imperative. The front-end provides ample gain (~15dB) to suppress the noise of the receiver's baseband, and 2b gain selection through switched R_{LG}.

When lower RF gain is needed at low NF, a mixer-first mode [3] can be activated by enabling LO_2 in Fig. 3.6.1. In this mode LO_1 is disabled, and so is the standard mixer by opening its switches. The bypass mixer reuses the relatively big filtering capacitor C_{BB}. Input matching is now provided by the parallel impedance of both mixer and LNA. The latter is reconfigured to draw less current (less g_m) and to provide the now higher required input impedance (see Fig. 3.6.2). At this lower RF gain the G_m of the first baseband stage is increased to suppress baseband noise (especially 1/f).

In mixer-first mode, the translational effect of the mixer and capacitor (Fig. 3.6.1) filters out-of-band blockers directly at the antenna. Nevertheless, since the mixer targets operation up to 6GHz, the tolerable switch size is limited. The switch resistance, which limits the lower bound of the out-of-band impedance provided by the mixer-based filter, therefore hardly provides an impedance lower

than 50Ω. Consequently the filtering profile at the antenna is limited to just a few dB. A NF down to 6.5dB is attained (better than when lowering the LNA gain further), with proper linearity. The advantage of the configuration in this work, is that the mixer-first operation can be disabled when not required, while the additional area of the system is a mere eight switches and a set of LO drivers. Elegantly, the LNA bypass switch *is* the auxiliary mixer, which avoids linearity issues associated to other LNA bypass solutions.

IIP2 of both mixers, important for modulated blockers and cross-modulation in an SDR, can be calibrated through a voltage DAC adjusting the gate bias voltage of the switches (Fig. 3.6.4). Calibration (which can be done by TX loop-back) with I- and Q-DAC boosts IIP2 independently on both channels to levels in excess of +70dBm.

The voltage sampling front-end sets severe challenges on OB linearity (+20dBm) and noise (1nV/$\sqrt{Hz}$) of the first baseband stage, not to degrade the achieved front-end's performance in the full receiver chain. A variable G_m stage followed by a 2nd-order filtering TIA biquad addresses these requirements. It provides a linear high-input-impedance interface as well as initial conditioning of the signal for the rest of the chain. A passive pole, G_m-C filter and VGA finalize the receive chain. Together with the mixer pole formed by C_{BB}, 6th-order filtering is provided. The baseband tunes from 0.4 to 30MHz and <0-to-55dB gain. Similar to [1], a 6-to-12GHz fractional-N PLL with divider chain offers any possible 100MHz-to-6GHz quadrature LO. A 25% duty-cycle generator delivers either LO_1 or LO_2.

The receiver has been implemented in 40nm LP digital CMOS (Fig. 3.6.7). It consumes 30 to 55mW and 30 to 40mA in the PLL. Figure 3.6.3 reports conversion gain (70/60dB), S_{11} and NF (3/6.5dB) in LNA-/mixer-first mode. The excellent IB/OB-IIP3 are +6/+10dBm respectively, and the calibrated IIP2 exceeds +70dBm. The EVM is 3%. The 1dB blocker compression point $B_{1dB,CP}$ [2] vs. baseband frequency in Fig. 3.6.5 reaches -8dBm at 20MHz offset (similar for mixer-1st) at highest gain, limited by baseband. It improves to -5.5dBm with 6dB baseband gain back-off with no impact on NF. Naturally $B_{1dB,CP}$ tracks the RF and baseband filtering profile. LO leakage to antenna in mixer-1st operation is lower than -65dBm, below the 3GGP specification.

Finally, Fig. 3.6.5 shows blocker NF, limited by – amongst others – reciprocal mixing of blocker and LO phase noise. With a 0dBm blocker at 20MHz offset, the blocker NF is below 15dB (e.g. 3GPP requirement). In [4], the NF is 15dB at 20MHz for a -5dBm blocker. This work therefore reports the receiver with highest blocker resilience for low NF, highest linearity and frequency range among the works in Fig. 3.6.6. Unlike other solutions, the receiver handles blockers well in any mode. It therefore needs not to be configured to a dedicated (noisy) blocker-tolerant mode.

Acknowledgements:
The authors acknowledge H. Suys, M. Libois and B. Debaillie.

References:
[1] M. Ingels, V. Giannini, J. Borremans, G. Mandal, B. Debaillie, P. Van Wesemael, T. Sano, T. Yamamoto, D. Hauspie, J. Van Driessche, J. Craninckx, "A 5mm² 40nm LP CMOS 0.1-to-3GHz multistandard transceiver", *ISSCC Dig. Tech. Papers*, pp. 458-459, Feb. 2010.
[2] Z. Ru, N. A. Moseley, E. Klumperink, B. Nauta, "Digitally Enhanced Software-Defined Radio Receiver Robust to Out-of-Band Interference", *J. Solid-State Circuits*, Vol. 44, No. 12, pp. 3359-3375, Dec. 2009.
[3] M. Soer, E. Klumperink, Z. Ru, F. E. van Vliet, B. Nauta, "A 0.2-to-2.0GHz 65nm CMOS receiver without LNA achieving >11dBm IIP3 and <6.5 dB NF", *ISSCC Dig. Tech. Papers*, pp.222-223, Feb. 2009.
[4] A. Mirzaie, A. Yazdi, Z. Zhou, E. Chang, P. Suri, H. Darabi, "A 65nm CMOS Quad-Band SAW-Less Receiver for GSM/GPRS/EDGE", *VLSI Dig. of Tech. Papers*, pp.179-180, June 2010.
[5] B. W. Cook, A. Berny, A. Molnar, S. Lanzisera, K. S. J. Pister, "Low-Power 2.4-GHz Transceiver With Passive RX Font-End and 400-mV Supply", *J. Solid-State Circuits*, Vol. 41, No. 12, pp. 2757-2766, Dec. 2006.
[6] J. Borremans, G. Mandal, B. Debaillie, V. Giannini, J. Craninckx, "A sub-3dB NF Voltage-Sampling Front-End with +18dBm IIP3 and +2dBm Blocker Compression Point", *Proc. of ESSCIRC*, Sept. 2010.
[7] H. Darabi, "A Blocker Filtering Technique for SAW-Less Wireless Receivers", *IEEE J. Solid-State Circuits*, Vol. 42, No. 12, Dec. 2007.

978-1-61284-303-2/11 $26.00 © 2011 IEEE

ISSCC 2011 / February 21, 2011 / 3:45 PM

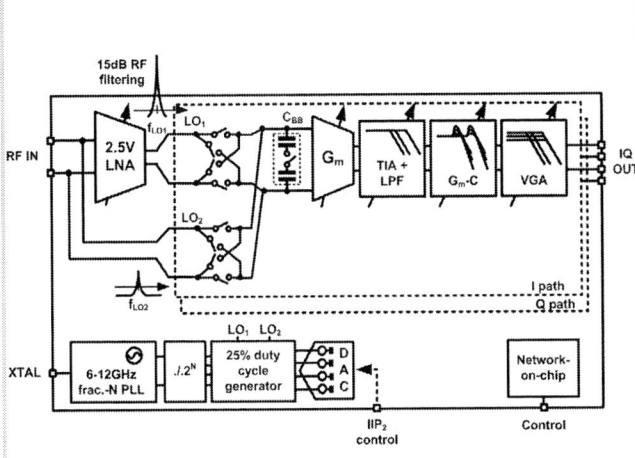

Figure 3.6.1: Highly linear software-defined receiver.

Figure 3.6.2: a) LNA schematic, (b) LNA start-up circuitry, (c) Input matching in LNA-first and mixer-first operation.

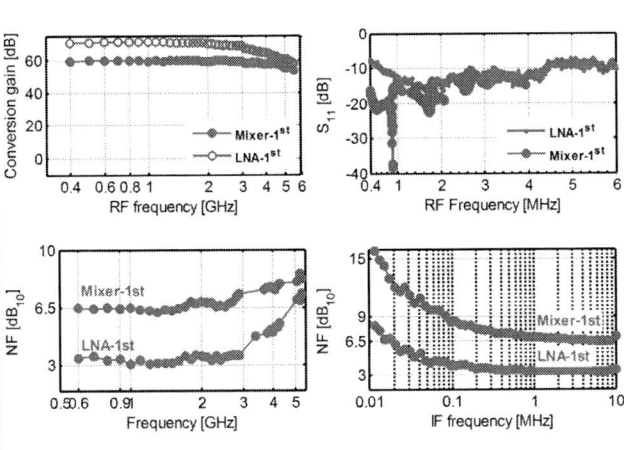

Figure 3.6.3: Measured conversion gain, S_{11}, NF of the receiver.

Figure 3.6.4: Measured conversion gain for baseband bandwidth set to 20MHz, EVM (64 subcarrier OFDM, 16QAM @2GHz), OB-IIP2/IIP3, and example IIP2 vs. calibration.

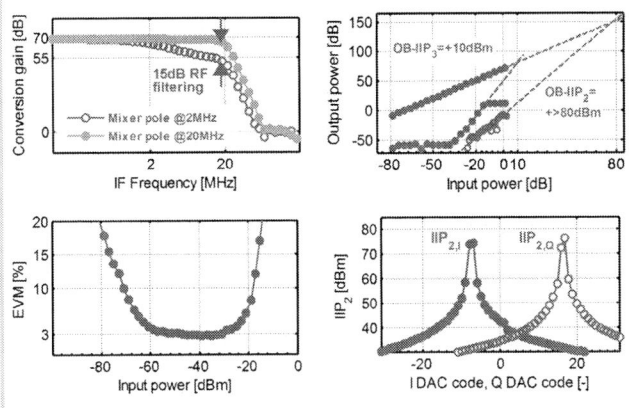

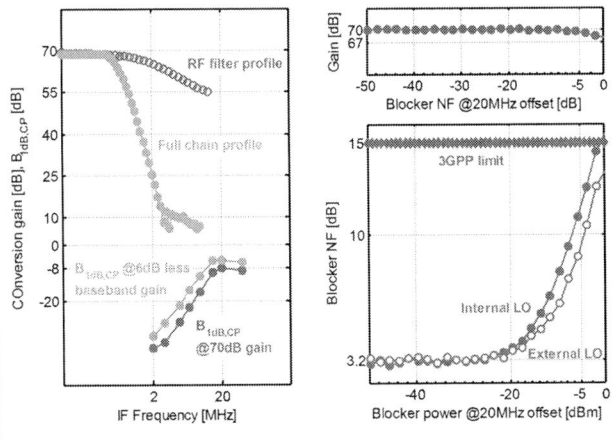

Figure 3.6.5: Blocker compression point at 70dB gain, filtering profile and measured compression and blocker NF vs. blocker power.

Figure 3.6.6: State-of-the-art overview of SDR receivers.

	This	Ingels [ISSCC'10]	Mirzaie [VLSI'10]	Bagheri [ISSCC'06]	Ru [JSSC'10]
Technology	40nm	40nm	65nm	65nm	65nm
System	Full RX	Full RX	Full RX	Full RX	Front-end
Frequency [GHz]	0.4-6	0.1-6	0.8-1.99	0.8-5	0.4-0.9
Gain [dB]	70	77	78	36	34
NF [dB]	3	2.6	3.1	5	4
NF @-4dBm blocker at 20MHz offset [dB]	10/12	>20	15	-	-
IB-IIP₃ [dBm]	+6	-7.5..0	-12.4	-3.5	+3.5
OB-IIP₃ @20MHz [dBm]	+10	-17..-4.6	-	-	-
IIP₂ [dBm]	+70	+53..+59	+45..+50	+45..+65	+51
B₁dB,CP @20MHz [dBm]	>-8	-30	~-9	-	<-20
Power cons [mW]	30-55* {64-100}	66-143	55mA from battery	68-98*	60*
Area [mm²]	2	2	2.44	6.96*	1*

* does not include synthesizer, IB-IIP₃ @LNA-1ˢᵗ, min gain - OB-IIP₃ @max gain.

978-1-61284-303-2/11 $26.00 © 2011 IEEE

ISSCC 2011 PAPER CONTINUATIONS

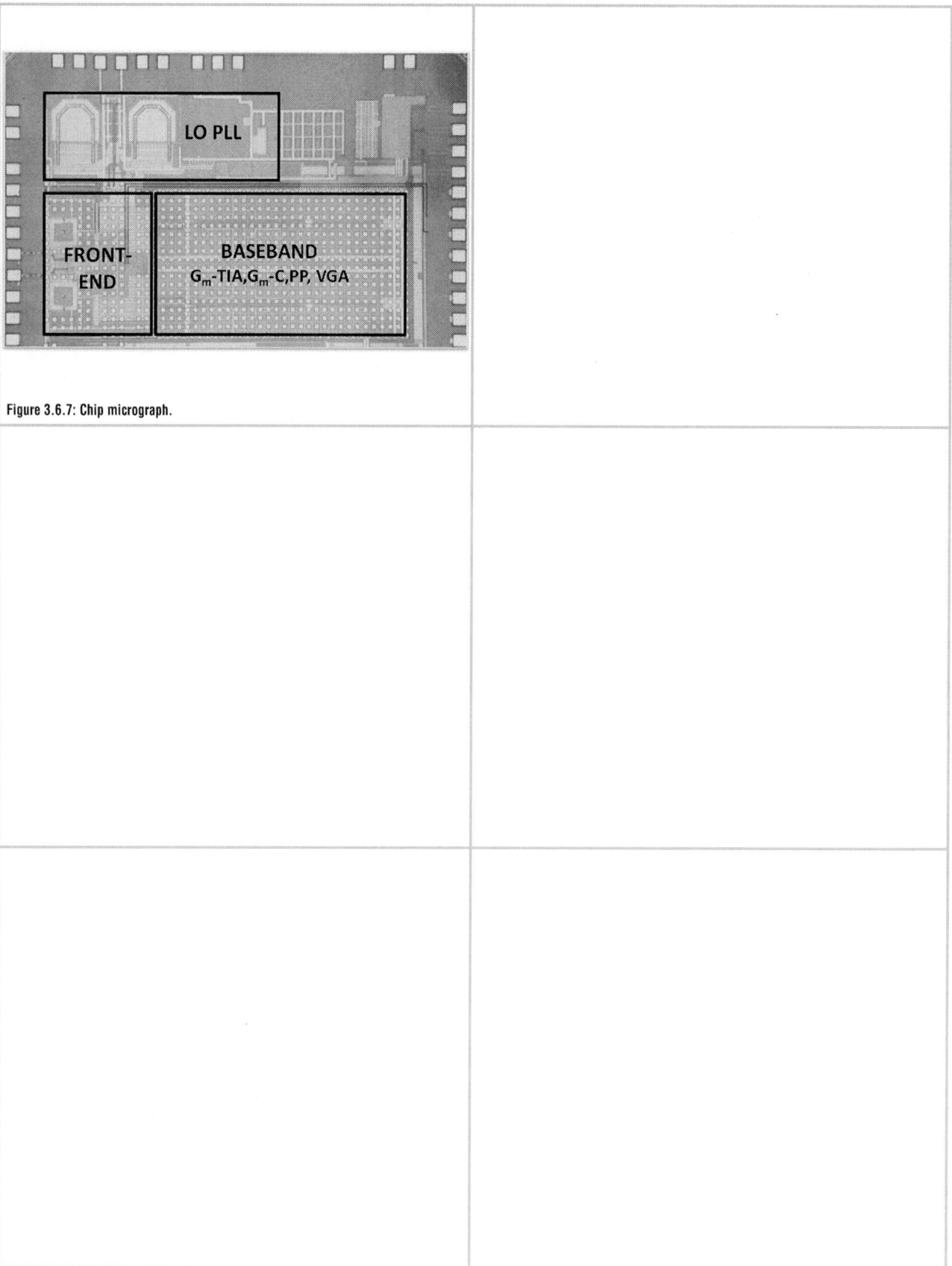

Figure 3.6.7: Chip micrograph.

ISSCC 2011 / SESSION 3 / RF TECHNIQUES / 3.7

3.7 A 1.0-to-4.0GHz 65nm CMOS Four-Element Beamforming Receiver Using a Switched-Capacitor Vector Modulator with Approximate Sine Weighting via Charge Redistribution

Michiel C.M. Soer[1], Eric A.M. Klumperink[1], Bram Nauta[1], Frank E. van Vliet[1,2]

[1]University of Twente, Enschede, The Netherlands,
[2]TNO Science and Industry, The Hague, The Netherlands

Phased-array receivers provide two major benefits over single-antenna receivers [1]. Their signal-to-noise ratio (SNR) doubles for each doubling in the number of elements, resulting in extended range. Secondly, interferers can be rejected in the spatial domain for increased link robustness. These arrays can be implemented by phase shifting and summing the signals from antenna elements with uniform spacing. For accurate interference rejection, a phase shifter with uniform phase steps and constant amplitude is desired. Several types of continuous-time phase shifters have been published, e.g. using injection locking [2], phase selection [3] and vector modulation [1,4,5,6]. This paper proposes a phased-array receiver architecture with a *discrete-time* vector modulator that takes advantage of the high linearity and good matching of switched-capacitor circuits, which are highly compatible with advanced CMOS. A simple charge-redistribution circuit is presented that performs a rational approximation of the sine and cosine needed for the vector modulator weights.

The principle of a vector modulator phase shifter is plotted in Fig. 3.7.1 (left) in the form of a phasor diagram. The desired phasor Z with arbitrary phase can be formed by the combination of orthogonal phasors X and Y. These phasors themselves are weighted versions of unity-amplitude phasors I and Q. A positive I and Q correspond to phasors in the first quadrant with a phase shift between 0° and 90°, while the other three quadrants are reached by using negative I and/or Q signals. Because the weights relate to the sine and cosine of the desired phase, a *uniform* step in phase will require a *non-uniform* step in the weights. In a straightforward implementation with uniform weight quantization, this will result in a large number of quantization steps for a much smaller number of phase steps. We propose to approximate the sine and cosine with rational functions $\alpha / (\alpha+\frac{3}{4})$ and $(1-\alpha) / (1-\alpha+\frac{3}{4})$ respectively (where α ranges between 0 and 1), which are plotted in Fig. 3.7.1 (right). With this rational mapping, a *near-uniform* step in phase corresponds to a *uniform* step in α. Moreover, the rational functions are equivalent to the transfer function of a two-element voltage divider, which is easily implemented in analog circuits.

To prove this concept, a 4-element phased-array receiver front-end is implemented in a zero-IF downconverter architecture (Fig. 3.7.2). For each element, a common-gate transistor realizes impedance matching and pre-mixer gain. A 4-phase clock is generated from an off-chip differential clock by division-by-two with dynamic transmission gate flipflops and used to drive a combined mixer/phase shifter implemented with switched capacitors. Summing of the element outputs is performed in the current domain through transconductances loaded by a common resistor. The stages after the common gate are copied four times and clocked with different LO phases to generate a differential in-phase and quadrature output for easy demodulation. The vector modulator phase shifter starts with a 4-phase 25% duty cycle passive mixer [7] that provides downconversion, generates the I and Q vectors (positive and negative) for the vector modulator, and converts the continuous-time signal to discrete time [8]. The discrete-time mixer outputs are buffered, weighted with switched-capacitor circuits, and then summed to form the phase-shifted signal. Because the switched-capacitor circuits run at the same frequency as the mixer, a post-mixer decimation filter is not needed.

One of the four mixer/phase-shifter slices is worked out in detail in Fig. 3.7.3. Not shown are the static switches that select the I and Q polarity. Summing into the final phasor Z is achieved by clocking the I and Q paths with opposite clock phases and connecting the path outputs. The required voltage divider for the sine/cosine approximation is formed by a charge-redistribution switched-capacitor circuit. During the first clock phase, the *variable* capacitor is charged to the voltage of the mixer output and the charge on the *fixed* capacitor is removed. During the second clock phase, the two capacitors are connected together and charge redistributes until the capacitor voltages are equal. Conservation of charge dictates that the resulting voltage is equal to the initial voltage times the ratio of the variable capacitance to the variable and fixed capacitance together. The correct transfer function is achieved if the variable capacitance ranges between 0 and C, and the fixed capacitor is $\frac{3}{4}C$. As the sum of the two variable capacitors is constant, capacitance is effectively switched *between* the I and Q path and a *single* switchable capacitance bank can be used for each slice, where special care is taken to scale the switch parasitics with the capacitors. The 3b capacitor control enables 8 phases per quadrant, for a total of 32 possible phases (5b). Since the phase shift only depends on the capacitor ratios, the behavior of the proposed vector modulator is insensitive to bias, temperature and frequency.

The chip is implemented in 65nm CMOS (Fig. 3.7.7), with an active area of 0.44 mm². Figure 3.7.4 (left) plots the expected and measured vector modulator phasors. When compared to ideal, constant amplitude, uniform phase steps, the phase and amplitude errors of these phasors are plotted in Fig. 3.7.4 (right). The deviation from the expected rational functions of Fig. 3.7.1 is mainly caused by the parasitic capacitance on node Z of the vector modulator (Fig. 3.7.3), which creates coupling between the X and Y vectors. This decreases the phase error, but increases the gain error, resulting in a balanced RMS phase and gain error of 1.4° and 0.4dB respectively. Figure 3.7.5 (top) plots the gain, double sideband noise figure (DSB NF) and 1dB compression point (P_{1dB}) of a single element for different LO frequencies. The -3dB bandwidth after downconversion is 65MHz. Because the total 4-element array has an SNR improvement of 6dB, the sensitivity of the array is equivalent to that of a 4-to-5dB NF single antenna receiver. With a 5b vector modulator, it is possible to make 32 different beams in a phased array configuration. Figure 3.7.5 (bottom) plots eight array beam patterns, measured on packaged samples with four phase-controlled signal generators at the inputs. The beam shapes fit to theory and the depths of the nulls are more than 25dB below the main beam peak. This enables the suppressing of interferers by placing nulls on their angular location, which relaxes the linearity requirement of the baseband circuitry considerably.

The performance is summarized in Fig. 3.7.6. A low RMS phase mismatch of 0.2° and gain mismatch of 0.04dB between elements is achieved due to the good matching of the switched-capacitor implementation. Applications in crowded frequency bands, like the 2.4GHz ISM band, are enabled by the high -1dBm third-order intercept point (IP3). The proposed approximate-sine weighting is elegantly mapped on a charge-redistribution circuit and results in *all* vector modulator settings being used for effective phase shifts.

Acknowledgements:
We thank D. Leenaerts, G. Wienk, H. de Vries and NXP for chip fabrication.

References:
[1] J. Paramesh, R. Bishop, K. Soumyanath, et al., "A 1.4V 5GHz Four-Antenna Cartesian-Combining Receiver in 90nm CMOS for Beamforming and Spatial Diversity Applications", *ISSCC Dig. Tech. Papers*, pp. 210-211, Feb. 2005.
[2] S. Patnaik, N. Lanka, R. Harjani, "A Dual-Mode Architecture for a Phased-Array Receiver Based on Injection Locking in 0.13µm CMOS", *ISSCC Dig. Tech. Papers*, pp. 490-491, Feb. 2009.
[3] K. Scheir, S. Bronckers, J. Borremans, et al., "A 52GHz Phased-Array Receiver Front-End in 90nm Digital CMOS", *ISSCC Dig. Tech. Papers*, pp. 184-185, Feb. 2008.
[4] S. Jeon, Y. Wang, H. Wang, et al., "A Scalable 6-to-18GHz Concurrent Dual-Band Quad-Beam Phased-Array Receiver in CMOS", *IEEE J. Solid-State Circuits*, vol. 43, no. 12, pp. 2660-2673, Dec. 2008.
[5] K. Raczkowski, W. De Raedt, B. Nauwelaers, et al., "A Wideband Beamformer for a Phased-Array 60GHz Receiver in 40nm Digital CMOS", *ISSCC Dig. Tech. Papers*, pp. 40-41, Feb. 2010.
[6] K. Koh, G.M. Rebeiz, "An X- and Ku-band 8-Element Phased-Array Receiver in 0.18-µm SiGe BiCMOS Technology", *IEEE J. Solid-State Circuits*, vol. 43, no. 6, pp. 1360-1371, Jun. 2008.
[7] B.W. Cook, A.D. Berny, A. Molnar, et al., "An Ultra-Low Power 2.4GHz RF Transceiver for Wireless Sensor Networks in 0.13µm CMOS with 400mV Supply and an Integrated Passive RX Front-End", *ISSCC Dig. Tech. Papers*, pp. 1460-1469, Feb. 2006.
[8] M.C.M. Soer, E.A.M. Klumperink, P.T. de Boer, et al., "Unified Frequency Domain Analysis of Switched-Series-RC Passive Mixers and Samplers", *IEEE Trans. Circuits and Systems-I*, vol. 57, no. 10, pp. 2618-2631, Oct. 2010.

978-1-61284-303-2/11 $26.00 © 2011 IEEE

ISSCC 2011 / February 21, 2011 / 4:15 PM

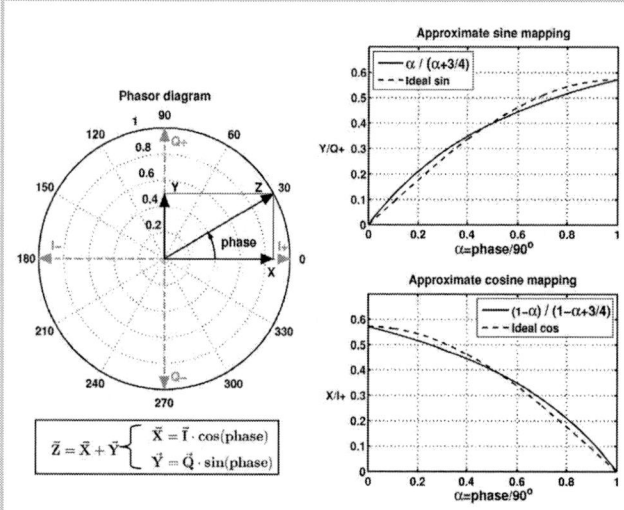

Figure 3.7.1: Vector modulator principle and sine/cosine approximation.

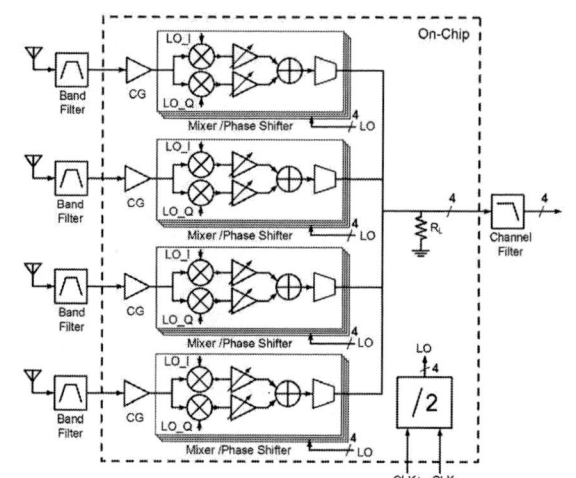

Figure 3.7.2: Phased-array zero-IF receiver front-end architecture.

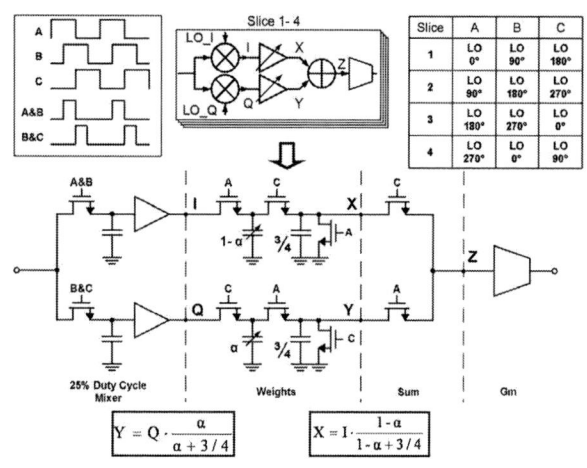

Figure 3.7.3: Switched-capacitor phase shifter (α=phase/90°).

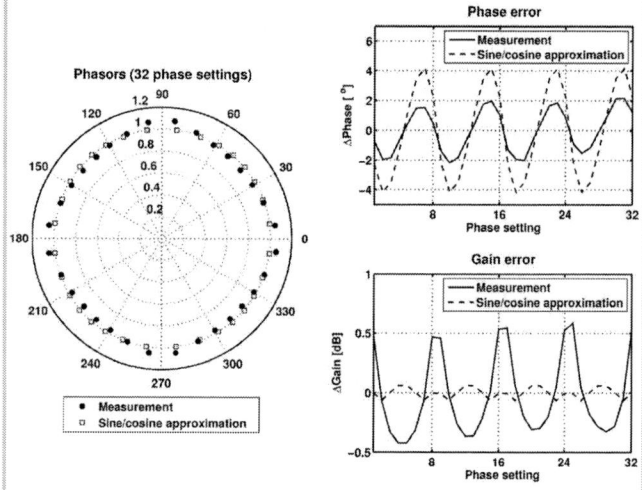

Figure 3.7.4: Measured 5b phase-shifter performance.

Figure 3.7.5: Measured gain, DSB NF, P_{1dB} and array beam patterns.

Technology	65nm CMOS
Die area	0.9x1.2mm^2
RF frequency band	1 - 4GHz
Power consumption	308mW @ 1.2V [1]
Single Element Performance:	@ 2.5 GHz RF
Gain	16dB
Noise figure	10dB DSB
Input-referred 1-dB compression point	-14dBm
Input-referred in-band IP3	-1dBm
Input-referred in-band IP2	40dBm
Vector modulator control	5 bits
Phase shift step	11¼°(5 bits)
Phase error (RMS)	1.4°
Amplitude error (RMS)	0.4dB
Phased Array Performance:	
Array directivity	6dB (4 elements)
Element phase mismatch (RMS)	0.2°
Element amplitude mismatch (RMS)	0.04dB
Peak-to-null ratio	> 25dB

[1] 120mW static, 188mW dynamic @ 2.5GHz RF.

Figure 3.7.6: Performance summary.

978-1-61284-303-2/11 $26.00 © 2011 IEEE

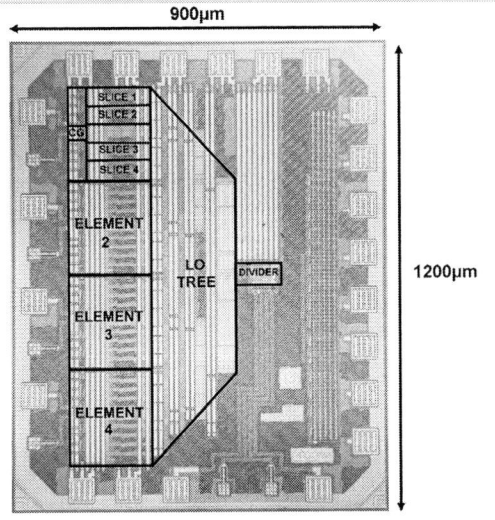

Figure 3.7.7: Chip micrograph.

ISSCC 2011 / SESSION 3 / RF TECHNIQUES / 3.8

3.8 A Harmonic Rejection Mixer Robust to RF Device Mismatches

Aslam A Rafi, Alessandro Piovaccari, Peter Vancorenland, Tyson Tuttle

Silicon Laboratories, Austin, TX

Harmonic rejection (HR) mixers are key building blocks in wideband receivers such as TV tuners and software-defined radio receivers employing low-IF or direct downconversion. In TV tuners, in-band HR in excess of 72dB is desired. A mixer capable of high harmonic-rejection ratio (HRR) is needed to relax the requirements on pre-mixer RF filters, which are expensive and bulky. HRR for conventional HR mixers [1] is limited by mismatches amongst devices operating at RF frequencies (shown in Fig. 3.8.1). Reducing mismatches by increasing device area to achieve high HRR comes at a severe cost of power, area and RF bandwidth. This matching requirement at RF frequencies limits the achievable HRR to be only between 30 and 40dB [2,3]. Recently, two-stage HR mixers have been proposed [4]. While [4] has reduced gain errors, phase errors that limit the achievable HRR are still determined by device matching at RF frequencies.

Here we present a new HR mixer that has significantly reduced sensitivity to mismatches in devices operating at high frequencies. HRR for this mixer is primarily determined by resistor and capacitor matching in the low frequency IF section. Significantly improved HRR is achieved as using large resistor areas for better matching does not cause a power or bandwidth penalty in the IF section. Additionally, this active mixer rejects flicker noise and improves second-order intermodulation (IIP2) and image rejection (IR) performance.

The concept of the proposed rotational HR Mixer is shown in Fig. 3.8.2. Multiphase IF outputs are generated by multiplying the RF signal with a clock, called master LO, having N times the desired LO frequency and rotating the product to obtain N IF outputs through the rotational switch S. The switch S transitions from one IF output to the next only when there is no signal through it. This reduces the sensitivity to the noise introduced by the switch and the phase errors caused by multiplication. These are explained in detail later. The low-frequency IF signals after downconversion at IF_0-IF_{N-1}, are then scaled by weights proportional to sine wave samples and summed to produce IF_{out}, where all LO harmonics except $mN \pm 1$ are rejected (m being an integer).

In the multiphase mixer shown in Fig. 3.8.3, RF transconductor current is cyclically rotated to the N IF ports, through master LO transistors, MP-MN, and rotational switches, RP_0-RP_{N-1} and RN_0-RN_{N-1}. Only a single-balanced mixer is shown here for clarity. The current flows into any one of the N IF ports for $1/N^{th}$ of the time period of LO. During the positive and negative phases of MLO, RF transconductor current flows through the paths via MP/RP_k and MN/RN_k to an IF_k, respectively. Rotation of RF current to IF_{k+1}, continues in the next cycle of MLO through RP_{k+1} and RN_{k+1}. The transitions at the gates of rotational switches happen only when they carry no instantaneous current. An N-stage shift register having master-slave flip-flops is used to generate the rotational pulses. Gating signals for RP_0-RP_{N-1} and RN_0-RN_{N-1} are triggered off of the falling edges of MLO_p and MLO_n respectively.

Mismatches in the master LO transistors modulate the pulse widths of instantaneous currents through MP and MN, shown as i_p and i_n, in Fig. 3.8.3. If in the presence of mismatches, the current pulse i_p widens, then i_n narrows by an equal amount. The sum of i_p and i_n has the same pulse width as the case without mismatches. The period of conduction of RF signal current to any IF port and thereby conversion gain, is unaffected by these transistor mismatches. Since the delay caused by mismatches in the time of conduction is the same for every IF output, there is no relative phase shift between signals at any two IF ports. Thus, mismatches do not cause any phase or gain errors. Gates of rotational transistors are set up to the right voltage prior to conduction and at the time of conduction merely serve as cascodes. This reduces the phase and gain errors caused by mismatches in these transistors.

Multi-phase mixer outputs are buffered through super source-followers as shown in Fig. 3.8.4. Loop Gain of this buffer at the low IF frequency helps to significantly reduce its gain and phase errors. Gains proportional to sine-wave coefficients are set by conductances proportional to sine-wave samples. For N=8, 12 and 16, the following half-sine-wave coefficients are used: [0 17 24 17], [$6^1/_5$ 17 $23^1/_5$ $23^1/_5$ 17 $6^1/_5$], [0 $9^1/_5$ 17 $22^1/_5$ 24 $22^1/_5$ 17 $9^1/_5$], respectively. The integer part is realized using unit resistors in parallel, whereas the fractional part is realized using unit resistors in series. This results in a systematic HRR in excess of 60dB for all rejected harmonics. Quadrature phase output is obtained by using the above sine wave coefficients, shifted by N/4, as shown in Fig. 3.8.5. For N=6, half sine wave coefficients [17 $23^1/_5$ $6^1/_5$] and [17 $-6^1/_5$ $-23^1/_5$] are used to generate quadrature outputs.

Compared to conventional commutating mixers, performance parameters that are affected by RF device matching are improved for this mixer and the actual HRR is primarily determined by matching amongst the multiphase mixer's RC loads and sine/cosine weighted resistors, each operating at the low IF frequency. For the same reason, IIP2 and IR performance are also enhanced. Flicker noise of master LO transistors modulates the pulse widths of i_p and i_n [5], as shown in Fig. 3.8.3. Since the pulse width of $i_p + i_n$ is unaffected, flicker noise is also rejected at any single-ended output of this multiphase mixer.

This mixer is fabricated in 110nm CMOS process as part of a receiver. As shown in Fig. 3.8.4, multiphase mixer cores with different N, share the same RC loads and IF section. Mixers with different N are designed to operate over different RF frequency ranges. This minimizes the frequency range of MLO required for the desired RF band while pushing the un-rejected harmonic beyond the band edge. Multiphase mixer and unity gain buffers draw 10mA and 12mA respectively from 2.7V. Multiphase rotational pulse generation and master clock buffering consumes 8mA from 1.3V with a master clock of 1.6GHz in 16-phase mode. Less digital current is consumed for lower N. The multiphase mixer load resistors occupy 0.008mm² and the sine/cosine-weighted resistors with switches, 0.026mm² (Fig. 3.8.7).

Figure 3.8.5 shows the histograms of measured third HRR for N=6,8,12 and 16 at LO frequencies of 300MHz, 200MHz, 150MHz and 100MHz respectively. A randomly selected sample of 100 parts is used for this test. HRR from both 3LO + IF and 3LO - IF is measured as part of an image-reject receiver and worst case is reported in Fig. 3.8.6. For single-stage HR mixers, HRR reported in literature is only between 30dB and 40dB [2,3,4]. Measured 3rd, 5th and 7th HRRs for this mixer are in excess of 52dB, 54dB and 55dB respectively. Even HRR is in excess of 60dB. The single-stage HR mixer realizes HRR in excess of 52dB, without any trimming or calibration. Given the reduced sensitivity to phase errors in LO generation, this mixer can be used in a two-stage configuration to further enhance HR. A high IIP2, in excess of 75dBm, is measured over 100 parts. Figure 3.8.6 summarizes other mixer parameters extrapolated from receiver measurements.

Acknowledgements:
We thank Dr. T. R. Viswanathan of Silicon Laboratories for reviewing this work.

References:
[1] J. Weldon, J. Rudell, L. Lin et al, "A 1.75GHz Highly Integrated Narrowband CMOS Transmitter with Harmonic-Rejection Mixers", *ISSCC Dig. Tech. Papers*, pp. 160-161, Feb. 2001.
[2] R. Bagheri, A. Mirzaei, S. Chehrazi et al, "An 800MHz-to-5GHz Software-Defined Radio Receiver in 90nm CMOS", *ISSCC Dig. Tech. Papers*, pp. 480-481, Feb. 2006.
[3] Z. Ru, E. Klumperink and B. Nauta, "A Discrete-Time Mixing Receiver Architecture with Wideband Harmonic Rejection", *ISSCC Dig. Tech. Papers*, pp. 322-323, Feb. 2008.
[4] Z. Ru, E. Klumperink, G. Wienk, B. Nauta, "A Software-Defined Radio Receiver Architecture Robust to Out-of-Band Interference", *ISSCC Dig. Tech. Papers*, pp. 230-232, Feb. 2009.
[5] H. Darabi, A. Abidi, "Noise in RF-CMOS mixers: a simple physical model", *IEEE J. Solid-State Circuits*, pp. 15-25, vol. 35, no. 1, Jan. 2000.

978-1-61284-303-2/11 $26.00 © 2011 IEEE

ISSCC 2011 / February 21, 2011 / 4:45 PM

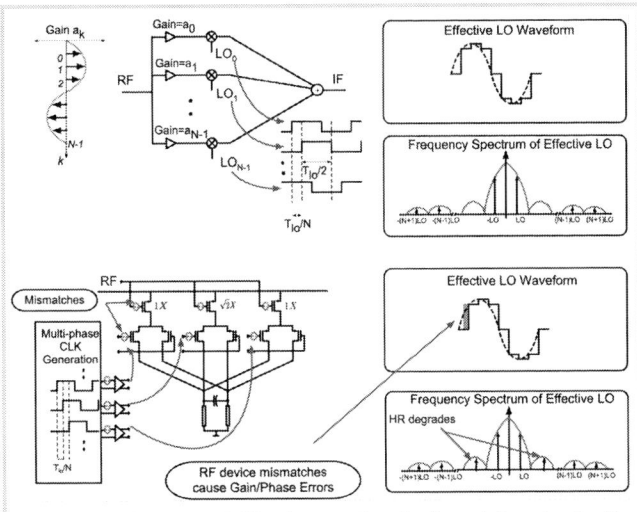

Figure 3.8.1: Conventional HR scheme and an implementation showing the effect of RF device mismatches.

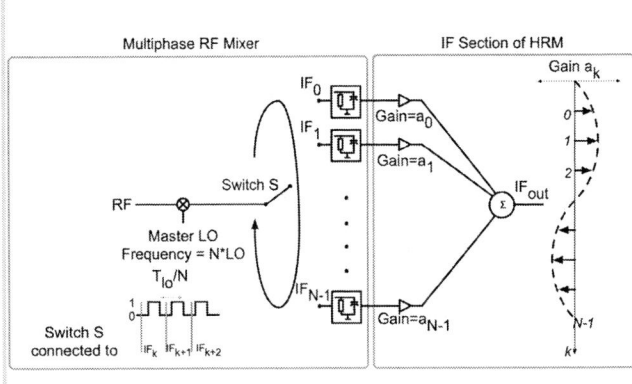

Figure 3.8.2: Concept of rotational HR Mixer.

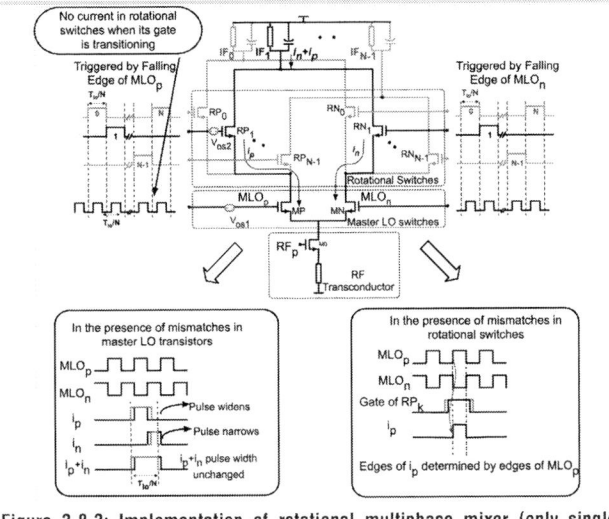

Figure 3.8.3: Implementation of rotational multiphase mixer (only single balanced mixer shown for clarity).

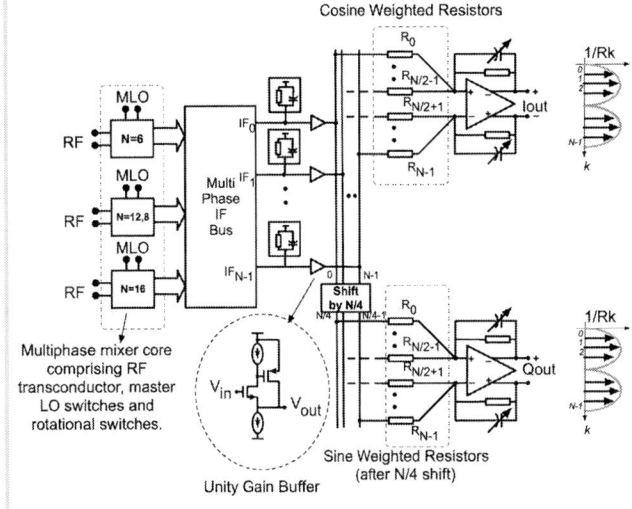

Figure 3.8.4: Architecture of rotational HR Mixer.

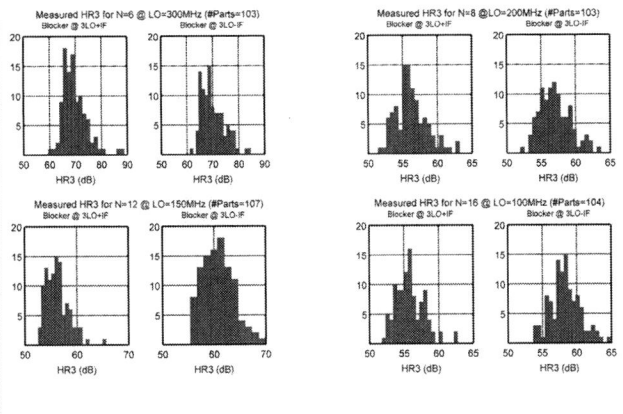

Figure 3.8.5: Histograms of measured third harmonic rejection for N=6,8,12 and 16.

Technology	110nm CMOS
Vdd	2.7V (Multiphase mixer, Unity Gain Buffers and Opamps) 1.3V (Clock)
Current Consumption	10mA (Multiphase Mixer) 12mA (Unity Gain Buffers) 8mA (clock generation in 16 phase mode)
Complex Gain of Multiphase mixer[a]	15dB
DSB NF	11dB
IIP3	12dBm
IIP2[b]	>75dBm
3rd order HR[b]	>52dB
5th order HR[b,c]	>54dB
7th order HR[b,c]	>55dB

3rd Harmonic Rejection Ratio (>100 samples)

Number of Mixer Phases (N)	LO Frequency	First Un-rejected Harmonic Frequency	HR3
6	300MHz	1.5GHz	>59dB
8	200MHz	1.4GHz	>52dB
12	150MHz	1.65GHz	>52dB
16	100MHz	1.5GHz	>52dB

[a] Complex Gain is defined as $Gain = \sqrt{2} \cdot \frac{v(IF_0) - v(IF_{N/2})}{v(RF_p) - v(RF_n)}$
This does not include gain in the stage following unity gain buffers.

[b] Number of parts tested >100.

[c] 5th order HR is measured for 8 phase mixer and 7th order HR is measured for 12 phase mixer.

Figure 3.8.6: Measured performance summary.

978-1-61284-303-2/11 $26.00 © 2011 IEEE

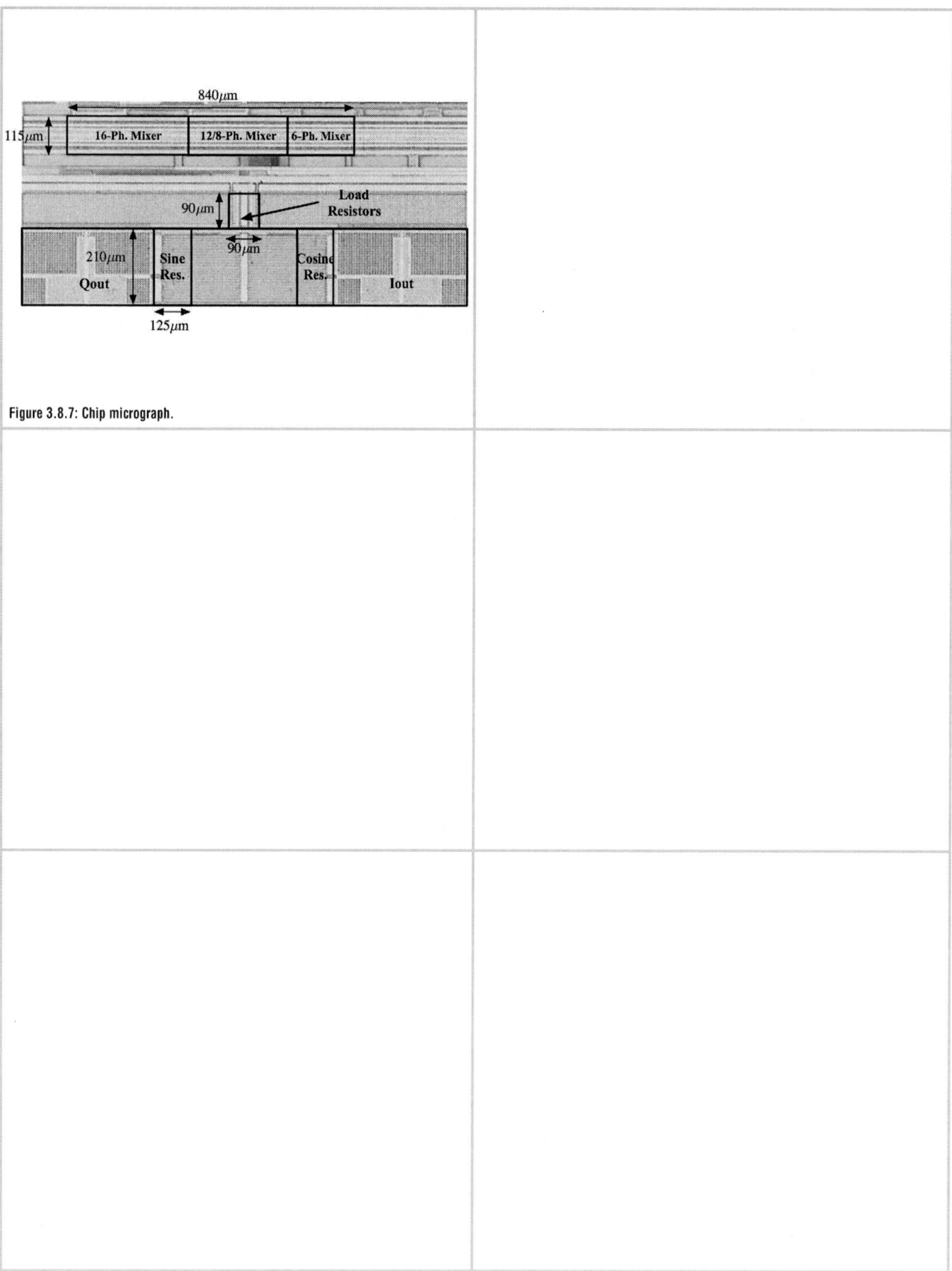

Figure 3.8.7: Chip micrograph.

ISSCC 2011 / SESSION 4 / ENTERPRISE PROCESSORS & COMPONENTS / OVERVIEW

Session 4 Overview / *High-Performance Digital*

Enterprise Processors & Components

Session Chair: *Joshua Friedrich, IBM Systems and Technology Group, Austin, TX*

Session Co-Chair: *Takashi Miyamori, Toshiba, Kawasaki, Japan*

Session 4 focuses on significant new innovations in the development of enterprise class microprocessors. This year's enterprise designs feature the fastest clock frequency, the highest x86 core count, and the highest energy efficiency, and the highest transistor count ever achieved in commercial microprocessors. The trend of dramatically increasing the number of threads, special function units, and off-die interconnect contained within a single die also continues. These designs from IBM, Intel, AMD, and the Chinese Academy of Sciences also employ a variety of power-efficient circuit and micro-architectural techniques to continue Moore's Law in the face of the dramatic power challenge plaguing deep submicron technology.

The first paper describes the microprocessor chip for the IBM zEnterprise 196 system, which is the first commercial processor chip to ship at a frequency in excess of 5 GHz. The 512mm2, 5.2GHz chip was built in IBM's 45nm Silicon-on-Insulator CMOS process with 13 levels of copper interconnect. It contains 4 processor cores, 1.5MB of private SRAM L2 cache per core, and an on-chip, high-speed 24MB shared L3 cache built from IBM's unique embedded DRAM technology that combines DRAM density with a high-speed digital logic process. To meet the incredibly high frequency objective, the design team made major cycle time enhancements, reduced core power consumption at constant PVT conditions by over 25% compared to the 65nm Z core, and improved the on-die power distribution. At the same time, the design added out-of-order instruction processing to further improve performance at constant frequency.

Paper 4.2 highlights a particularly crucial high speed component of IBM's zProcessor core – the dynamic hit logic and embedded 8Kbit SRAM used for L1 cache hit detection. The 14 bit hit logic described in this paper uses programmable launch and reset clocks along with a mix of dynamic and highly skewed static CMOS to achieve record operating frequencies (above 6.5GHz at lab conditions). In addition, its innovative use of a "search-for-a-hit" scheme minimizes power consumption without degrading performance. Array BIST provides both the hit logic and SRAM with full "at-speed" test coverage for the first time.

Paper 4.3 from Intel's lab in Bangalore, India, describes the next generation enterprise Xeon® processor consisting of 10 Westmere 32nm cores and a shared, inclusive L3 cache integrated on a monolithic die with link based IOs. The die contains the highest number of x86 cores ever integrated onto a single chip and contains significant innovations and circuit optimizations over its predecessor, Nehalem, targeting idle power reduction, robust high speed IO links at the next generation process node, and performance per watt improvements. The processor is implemented in 32nm CMOS using high-κ metal gate transistors and nine copper interconnect layers. It supports individually controlled power gating of each of the ten cores and implements macro clock gating of uncore and cache functions to drive idle power to new lows.

Energy efficient processing is also a key area of innovation, and the Godson-3B processor from the Chinese Academy of Sciences, described in Paper 4.4, reports the highest energy efficiency (3.2GFlops/Watt) among all state-of-art high-performance processors. Godson-3B is an 8-core design implemented on 65nm CMOS LP/GP mixed process with 7 layers of Cu metallization. It contains 582.6M transistors within 299.8mm2 area and operates at frequencies of up to 1.05GHz. Its peak performance is 128/256GFlops for double/single-precision with 40W power consumption.

Paper 4.5 from AMD describes the new x86-64 based 2-core CPU module (Bulldozer) that contains 213M transistors in 11-metal layer 32nm high-k metal-gates SOI CMOS process. In addition to the micro-architecture improvements, the components, such as the L1 and L2 caches, the integer unit and the Floating Point unit, are designed to achieve higher frequency, lower power consumption, and lower gate counts per cycle than the 45nm AMD core while maintaining IPC (Instructions per Cycles). It operates over 3.5GHz in an area (including 2MB L2 cache) of 30.9mm2.

The next paper 4.6 from AMD presents details of the 40-entry unified, out-of-order scheduler and integer unit of the Bulldozer. The scheduler issues four operations per cycle and the integer execution unit supports single-cycle bypass between four functional units. Instead of dynamic logic, skewed gates with static standard cell are extensively used. Furthermore, to minimize power consumption, the design reduces switching activity by manipulating pointers rather than moving large amounts of data within queues.

978-1-61284-303-2/11 $26.00 © 2011 IEEE

ISSCC 2011 / February 21, 2011 / 1:30 PM

Paper 4.7 from Intel describes the clock generation system of a multi-core processor for server applications, fabricated on a 9-metal layer 32nm CMOS process. A growing number of CPU cores and integration of high speed I/O ports, such as QuickPath Interconnect, PCI express and DDR3, presents significant challenges to the clock generation and distribution systems. The proposed clock system architecture delivers low skew, low power and low latency with modularity and scalability for a ring-like architecture in spite of these challenges.

Finally, Paper 4.8 from Intel is the Itanium processor implemented in a 32nm CMOS process with 9 layers of copper interconnection. It contains 3.1 billion transistors within a 18.2 mm by 29.9 mm die. This transistor count represents a 50% increase over the previous Itanium processor and is the highest transistor count ever reported for a single die. The processor includes eight multi-threaded cores, a ring based system interface, and more than 54MB of cache memory. A three level cache hierarchy consists of a first level single cycle 16K Instruction (I) and Data (D) cache that is backed by two second level caches (a nine cycle 512K I cache and an eight cycle 256K D cache) and a 32MB the last level cache. The design implements twice as many as cores as the previous Itanium while lowering the thermal design power (TDP) by 15W to 170W. High speed links allow for peak processor-to-processor bandwidth of up to 128 GB/s and memory bandwidth of up to 45GB/s.

Together, these innovative enterprise processors and design components represent a new height of performance, integration, and energy efficiency. By continuing to deliver exponentially increasing performance and capability, these efficient engines will enable computers to analyze our world's most challenging scientific problems and manage the needs of increasingly complex global enterprise.

Presenters:

1:30 PM

4.1 A 5.2GHz Microprocessor Chip for the IBM zEnterprise™ System

J. Warnock, IBM Systems and Technology Group, Yorktown Heights, NY

2:00 PM

4.2 Dynamic Hit Logic with Embedded 8Kb SRAM in 45nm SOI for the zEnterprise™ Processor

A. R. Pelella, IBM Systems and Technology Group, Poughkeepsie, NY

2:15 PM

4.3 A 32nm Westmere-EX Xeon® Enterprise Processor

S. Sawant, Intel, Bangalore, India

2:30 PM

4.4 Godson-3B: A 1GHz 40W 8-Core 128GFLOPS Processor in 65nm CMOS

W. Hu, Chinese Academy of Sciences, Beijing, China; Loongson Technologies, Beijing, China

3:15 PM

4.5 Design Solutions for the Bulldozer 32nm SOI 2-Core Processor Module in an 8-Core CPU

H. McIntyre, AMD, Sunnyvale, CA

3:45 PM

4.6 40-Entry Unified Out-of-Order Scheduler and Integer Execution Unit for the AMD Bulldozer x86-64 Core

M. Golden, AMD, Sunnyvale, CA

4:15 PM

4.7 Clock Generation for a 32nm Server Processor with Scalable Cores

S. Li, Intel, Santa Clara, CA

4:45 PM

4.8 A 32nm 3.1 Billion Transistor 12-Wide-Issue Itanium® Processor for Mission-Critical Servers

R. J. Riedlinger, Intel, Fort Collins, CO

978-1-61284-303-2/11 $26.00 © 2011 IEEE

ISSCC 2011 / SESSION 4 / ENTERPRISE PROCESSORS & COMPONENTS / 4.1

4.1 A 5.2GHz Microprocessor Chip for the IBM zEnterprise™ System

J. Warnock[1], Y. Chan[2], W. Huott[2], S. Carey[2], M. Fee[2], H. Wen[3],
M.J. Saccamango[2], F. Malgioglio[2], P. Meaney[2], D. Plass[2], Y.-H. Chan[2],
M. Mayo[2], G. Mayer[4], L. Sigal[5], D. Rude[2], R. Averill[2], M. Wood[2],
T. Strach[4], H. Smith[2], B. Curran[2], E. Schwarz[3], L. Eisen[3], D. Malone[2],
S. Weitzel[3], P.-K. Mak[2], T. McPherson[2], C. Webb[2]

[1]IBM Systems and Technology Group, Yorktown Heights, NY,
[2]IBM Systems and Technology Group, Poughkeepsie, NY,
[3]IBM Systems and Technology Group, Austin, TX,
[4]IBM Systems and Technology Group, Boeblingen, Germany,
[5]IBM Research, Yorktown Heights, NY

The microprocessor chip for the IBM zEnterprise 196 (z196) system is a high-frequency, high-performance design that adds support for out-of-order instruction execution and increases operating frequency by almost 20% compared to the previous 65nm design, while still fitting within the same power envelope. Despite the many difficult engineering hurdles to be overcome, the design team was able to achieve a product frequency of 5.2GHz, providing a significant performance boost for the new system.

The chip (refer to die photo) has 4 processor cores, with 1.5MB of private SRAM L2 cache per core and 24MB of shared DRAM L3 cache. In addition, there are 2 co-processors for data encryption and compression, a DDR3 RAIM memory controller and an I/O bus (GX) controller. A high-frequency clock grid covers each processor core, with a half-frequency grid for the L2 and the rest of the chip ("nest"), supporting synchronous communication from the cores to the L2 and L3. Much of the logic in the nest was geared down to operate at ½ the nest clock frequency (1/4 of the core clock rate). The chip has 1.4B transistors in 512mm², and uses 45nm SOI CMOS technology with 2 additions compared to other IBM 45nm servers[1]. Two high-performance wiring planes are added (13 levels in total) for solving critical RC issues and improving the L3 access latency. Also, low-threshold voltage devices are added for use on the most critical timing paths.

The core microarchitecture [2] leverages the 65nm design, but added the ability to execute instructions out of order. Instructions are fetched from a 64KB I-cache, then decoded and expanded into multiple simpler instructions. Up to 6 simple instructions/cycle are sent to the sequencing unit, from which up to 5 instructions/cycle are sent for execution. The core contains 2 load-store units (LSU) with a 128KB D-cache, 2 fixed-point units (FXU) and binary and decimal floating-point units. The logic is hardened against soft errors at the circuit level [3], and by extensive parity and residue error checking. This checking circuitry is estimated to add roughly 20 to 25% to the digital-logic area. On detection of an error, a recovery unit stops the processor and restarts from a previously checkpointed state.

Starting from the previous design in 65nm technology (at 4.4GHz), the plan was to keep about the same cycle time, measured in technology-independent FO4 inverter delays, in order to fully leverage the technology device speedup to maximize single-thread performance. The circuit design styles and methodology are similar to the 65nm design, although with a different clocked-storage-element strategy [3]. There are many tight timing paths, given the wire RC non-scaling from the 65nm design, logic added for out-of-order operation, and strict test- and error–checking-coverage requirements.

The L1 D-cache access, in a 4-cycle loop, is particularly critical (Fig. 4.1.1). Address generation in cycle A0, uses highly tuned static circuits, low V_T devices, and cycle stealing into A1. The D-cache is partitioned into 8 macros with a 2-stage access pipeline supporting 2 reads or 1 write per cycle. Each read port is 512 entries deep, 8-way set associative, and provides 4 bytes of data after a late-select mux. A 6T SRAM cell (0.462µm²) with 16 cells per local bitline and a single-phase read head are optimized for power and performance. The 8:1 set late-select hit signals to the D-cache are generated by the Set-Predict (SETP) macro. The SETP array uses short local bitlines (8 cells) and a dual-phase read head for faster access, with dynamic comparators and hit drivers. Both SETP/D-cache are accessed in parallel in cycle A1 with outputs in A2, and then captured by the formatter before A3. A key feature of the z196 arrays is the extensive programmable timing control for all critical paths. This flexibility enables hardware-based array tuning for maximum frequency, as well as ensuring robust timing and functional margins.

Another fundamental critical timing path was the single-cycle execution loop through the FXU. In previous designs, the FXU operands were multiplexed and stored with a pulsed-clock dynamic MUX operand latch [4]. Process variability

and a lower n:p strength ratio drove the implementation to a pulsed latch solution (Fig. 4.1.2) with static inputs, which was found to require a shorter pulse width, giving a more manageable hold time.

The 24MB on-chip L3 cache uses high-speed DRAM modules [5] to give a large performance boost relative to the 65nm design. The L3 fetch return path, starting from the eDRAM macros (only 2 are shown for simplicity), is shown in Fig. 4.1.3, along with a timing diagram. The L3 is fully shared with equal access times from any core. Each core has 2 non-blocking interfaces to a low-order-address bit slice. Fetch bandwidth from each of the 2 slices is 32B per nest cycle (1 nest cycle = 4 core cycles) yielding an L3-to-cores bandwidth of >300GB/s. Buses and logic paths are engineered for minimum access time, providing an L3 hit of 45 core clock cycles. The lower level caches are store-through, so the L3 is required to process all stores generated from the 4 cores. This drives the cache to be highly interleaved, allowing for more parallel store processing, but making it more sensitive to cache busy times. To help mitigate the busy time impact and improve latency, the EDRAM macros are half-sized compared to [5] and further split into two banks, so that half can be accessed while the other half is busy.

A focused effort on power dissipation reduction is a key contributor to the project's success, allowing power-constrained frequency improvement beyond the gains offered by technology scaling alone. Power is analyzed and budgeted during concept planning and tracked through the entire design process. A comprehensive methodology calculates leakage and dynamic power with various workloads. The maximum sustained thermal power is worked at the microarchitectural level and throughout the detailed physical implementation. Overall core power efficiency was improved by about 25% compared to the previous design, which is essential to achieving operating frequency above 5GHz. The chip power breakdown while running a typical workload is shown in Fig. 4.1.4a, while Fig. 4.1.4b shows the DC leakage components (DC leakage is about 30% of the total power).

Instantaneous peak power due to switching activity was also a concern. More than 10µF on-chip decoupling capacitance is added using deep trench capacitors to maintain the stability and integrity of power rails. With multiple supplies on the chip for logic, SRAM and DRAM, it is important to get enough capacitance on each supply. Early hardware showed significant noise on the DRAM supply voltage rail (Fig. 4.1.5a), caused by an almost-simultaneous access/refresh of large numbers of DRAM banks. The initial on-chip decoupling capacitance for this rail, of about 265nF, was insufficient to meet the desired noise specifications. Later hardware added 1.5µF more capacitance on the array supplies, reducing the noise by over 2× (Fig. 4.1.5b).

Finally, hardware-based design tuning provides the last performance increment, leveraging clock-edge controls and other fine-grained clock control overrides for post-silicon frequency optimization. A plot of minimum operating voltage at 5.4GHz versus process speed (Fig. 4.1.6) shows good tracking with raw device speed. In addition, the dashed line shows the mean of the results with local clock control bits at their default settings, indicating about a 1.5% frequency improvement from the programmed local control settings.

In conclusion, the work described above culminated in a successful product release at 5.2GHz, pushing a commercial microprocessor frequency above the 5GHz mark and providing up to a 40% performance improvement for System z legacy workloads compared to the previous design. This result relied on specific high-performance technology features in combination with a design process and methodology that kept a tight focus on timing, power, and noise throughout the entire chip design and bringup process.

Acknowledgements:
The authors wish to thank the whole System z team and the IBM technology development and manufacturing teams for their contributions to the success of this project.

References:
[1] D. Wendel et al, "The Implementation of POWER7™: A Highly Parallel and Scalable Multi-Core High-End Server Processor", *ISSCC Dig. Tech. Papers*, pp. 102-103, Feb. 2010.
[2] B. Curran, "The Next-generation System z Micro-Processor, Hot Chips 2010
[3] J. Warnock et al, "POWER7™ Local Clocking and Clocked Storage Elements", *ISSCC Dig. Tech. Papers*, p. 178-179, Feb. 2010.
[4] B. Curran, et al., "4GHz+ Low-Latency Fixed Point and Binary Floating-Point Execution Units for the POWER6 Processor", *ISSCC Dig. Tech. Papers*, p. 1728, Feb. 2006.
[5] J. Barth et al, "A 45nm SOI Embedded DRAM Macro for POWER7™ 32MB On-Chip L3 Cache", *ISSCC Dig. Tech. Papers*, pp. 342-343, Feb. 2010.

978-1-61284-303-2/11 $26.00 © 2011 IEEE

ISSCC 2011 / February 21, 2011 / 1:30 PM

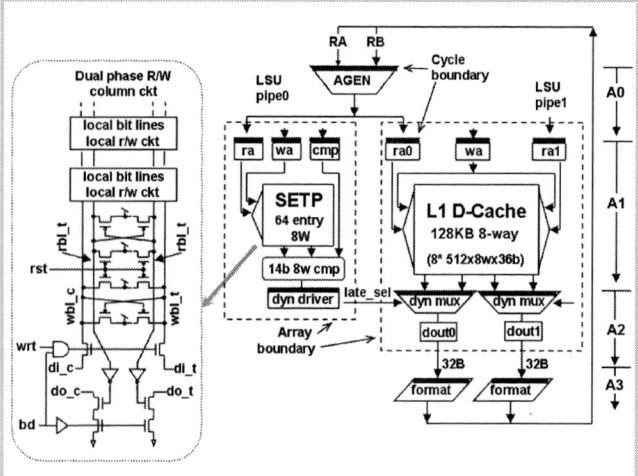

Figure 4.1.1: L1 D-cache critical 4-cycle access loop.

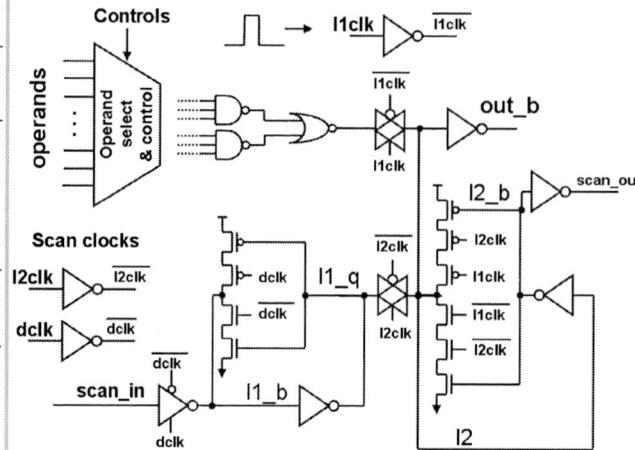

Figure 4.1.2: Operand mux and pulsed-clock latch (also provides "propagate" for downstream adder).

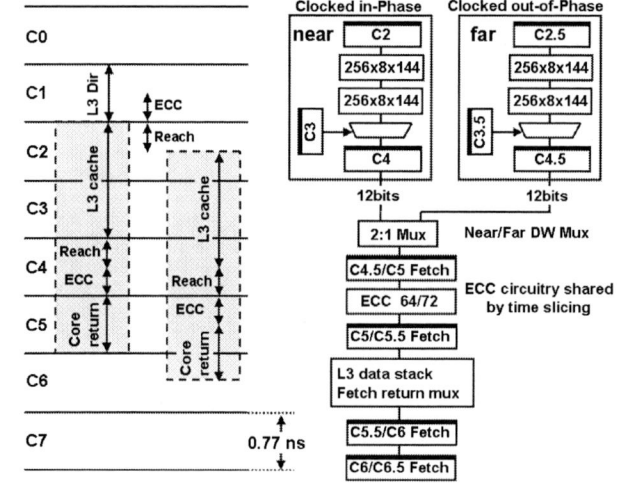

Figure 4.1.3: L3 Cache access for near/far L3 blocks. L3 DRAM access is 2 nest cycles (1.54ns)

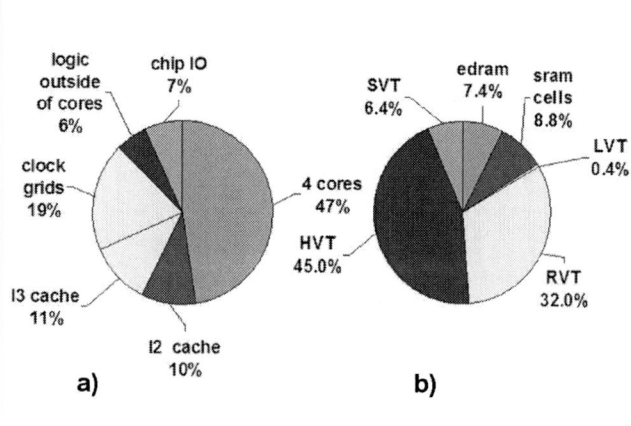

Figure 4.1.4: Chip power analysis: a) Total chip power breakdown by component b) DC leakage breakdown by device type.

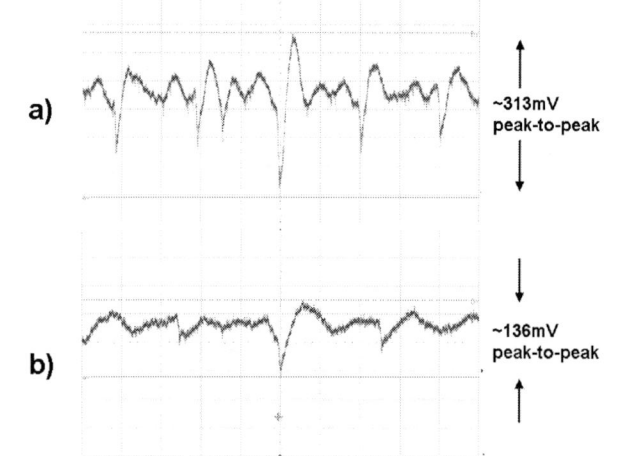

Figure 4.1.5: DRAM supply voltage noise with functional exerciser: a) Early hardware, 265nF decap on array supplies b) Later hardware, 1740nF decap.

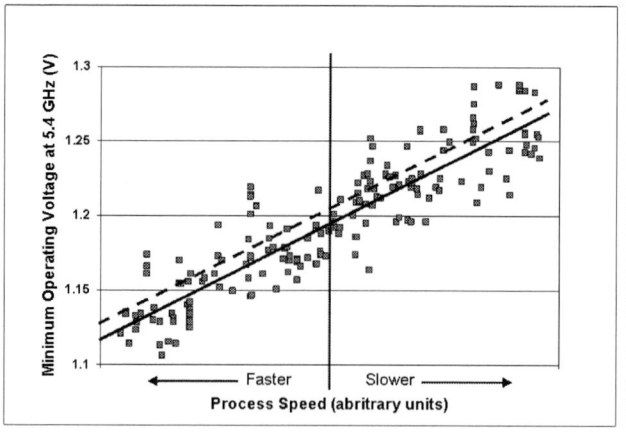

Figure 4.1.6: Min operating voltage at 5.4 GHz vs silicon process speed. Dashed line shows mean with default settings for local clock control bits.

978-1-61284-303-2/11 $26.00 © 2011 IEEE

ISSCC 2011 PAPER CONTINUATIONS

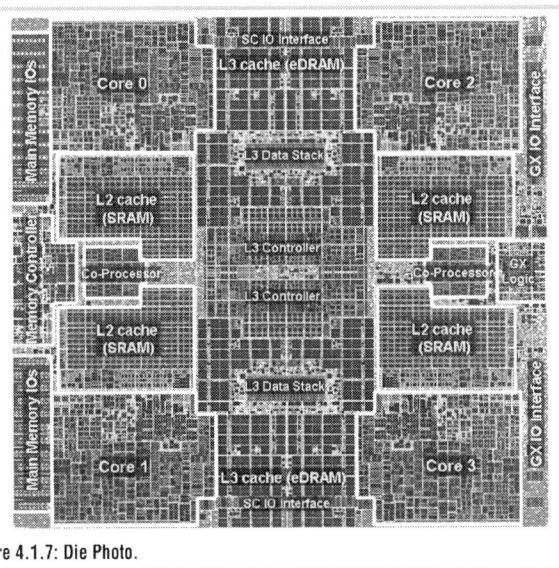

Figure 4.1.7: Die Photo.

ISSCC 2011 / SESSION 4 / ENTERPRISE PROCESSORS & COMPONENTS / 4.2

4.2 Dynamic Hit Logic with Embedded 8Kb SRAM in 45nm SOI for the zEnterprise™ Processor

Antonio R Pelella, Yuen H Chan, Bargav Balakrishnan, Pradip Patel, Daniel Rodko, Richard E Serton

IBM Systems and Technology Group, Poughkeepsie, NY

With the push to ever higher core frequencies, more logic functions are making their way onto critical path SRAMs in the L1 cache look up structure. Described in this paper is a 14 bit dynamic hit logic scheme with an embedded 8K bit SRAM in IBM's 45nm SOI [3]. The hit logic uses a "search-for-a-hit" scheme (i.e., XOR's followed by AND functions, pre-charged to a *miss*) to provide optimal performance, timing, and power. A custom microcode programmable Array-Built-In-Self-Test (ABIST) engine tests both the SRAM and hit logic function jointly, resulting in comprehensive "at-speed" test coverage to guarantee circuit functionality and timing margins. The SRAM is organized as a 64×15b×8W (way, or set) array and uses a 6T SRAM cell (1Read/1Write, 0.462µm²) in a "domino" hierarchal dual read bitline approach [1]

In the Load/Store Unit (LSU) of the zEnterprise 196 microprocessor core, this hit logic/SRAM combination is implemented by the Set Predict (SETP) macro to provide "late-select" control (LS<0:7>) to the 8 way set associative L1 data cache [2]. Figure 4.2.1 shows the block diagram of the SETP macro. It has 8 sets where each set has a 14-bit dynamic hit logic comparator and control logic that is tightly coupled to a 64-entry, 15 bit SRAM. A key feature of the SETP macro is extensive programmable timing control for all internal and external critical paths to allow hardware tuning for maximum frequency operation. It has flexible cycle-steal adjustments for the address, data_in and write control. In addition, the dynamic hit logic has a separate cycle steal for the compare-in address (cmp_in) and programmable hit launch and reset clocks. Referring to Fig. 4.2.7, the SETP macro is physically partitioned into 4 quadrants with 2 sets per quadrant. The right and left quadrants share 32 wordline drivers located in the center. The upper 4 sets are mirrored around a central region to create the bottom 4 sets, with the 1st level decode, clocks, input latches, write and timing control logic in between. Referring to Fig. 4.2.2, to deliver maximum read performance, the 32 cell bitline stack is divided into four segments of eight cells per local bitline (LBL) with dual-phase read head. The global bit select circuit (gbitsel) is placed in the center with two 8-cell subarrays above and two 8-cell subarrays below to reduce RC effects on the global read bitlines. The gbitsel has Fast-Read-Before-Write (FRBW) blocking circuits to prevent false reads during write operations [1]. The gbitsel also provides dual-phase dynamic outputs to directly drive the 14-bit dynamic hit logic comparator.

Figure 4.2.3 shows a comparison between two hit logic approaches: a Search-for-a-Hit (SFH) with a compare followed by an AND function, and a Search-for-a-Miss (SFM) with a compare followed by an OR function. SFH is suitable for bit compare widths ranging from 2 bits to 16 bits, and SFM is suitable for compare widths of 8 bits and higher. However, compare circuits in the 14 bit range are in a "gray zone" where either approaches can be justified depending on the technology and design goals. The two comparators illustrated in Fig. 4.2.3 assume a 2-input AOI for the XOR function and 2-input NANDs & NORs gates for the AND & OR functions. With this assumption, a 14-bit SFH will require 6 gate delays; whereas a SFM requires only 5 gate delays. Also, OR functions have a natural tendency to be faster than AND functions. However, dynamic wired-OR functions tend to have slow rise and fall times and need extra timing margin to cover for leakage, noise, and process variability, especially in skewed process corners (i.e. strong PFETs and weak NFETs). Also the timing control for a SFM need to be more conservative because any noise at the pull-down NFET inputs will be multiplied by the number of NFETs wired-ORed together. For these reasons, the performance of a 14-bit SFH is equivalent to a SFM. For SETP with 8 sets, a SFH has less power with 1 output to reset per cycle vs 7 for a SFM. To fulfill z processor's design goals for robust noise margin, circuit simplicity, performance and less power, a 14-bit Search-for-a-Hit scheme is used here.

A detail schematic of the SETP hit logic is shown in Fig. 4.2.4. The SRAM's dual-phase dynamic outputs allows the use of a single delay AOI as the XOR verses a 3-gate-delay static XOR. To prevent timing problems, a true-complement generator (TCG) converts the "cmp_in" bits from static to dynamic dual phase data before entering the compare logic. The TCG_CLOCKs are programmable with variable delay and pulse width. The hit logic gates are highly slewed towards the evaluate edge, whereas the pre-charge edge arrives early and is less critical. With this scheme, the-last-to-rise of either the dual phase SRAM data or the TCG data can trigger the compare. The duration of these two signals' pulse overlap is critical for circuit functionality. They can be adjusted with either the programmable SRAM Word decode clock or TCG_CLOCK. The hit driver is the last AND gate in the path and performs a dynamic-to-static conversion before sending out the late select (LS) signal. Once the hit driver has captured the data, the upstream logic gates will pre-charge with the adjustable falling edge of either the SRAM or TCG data. In addition to the traditional ABIST functions, the hit driver also includes an ABIST OVERRIDE input to allow for a unique test where the SETP is exercised by the full ABIST patterns while still sending valid hit signals over to the L1 cache. In this test mode, the SETP array and L1 cache can be tested in parallel with valid LS signals launched from the SETP. Due to the timing difficulty of mixing static and dynamic, the ABIST OVERRIDE signal is staged through a Timing Control Latch which is only allowed to update when the Hit Driver is in a pre-charge state. Also shown in Fig. 4.2.4 is the ABIST Capture latch feeding an XOR with the ABIST HIT LOGIC EXPECT BIT. That result is ORed with the ABIST Compare latch output to form a sticky latch.

The at-speed test scheme is shown in Fig. 4.2.5. To ensure internal setup and hold time specifications at the SRAM-hit logic interface, at-speed testing is employed. The traditional method for testing SRAM's with logic is to have high speed ABIST testing for the SRAM and relatively limited pattern coverage of Logic BIST (LBIST) to test the remaining logic downstream from the array. For example, during LBIST, the SRAM portion of the macro is typically in a write-through mode. As a result, read patterns are not included thus limiting test coverage. Also, LBIST test mode includes scan operations to load functional latches with data. This is followed by a single system clock, then more scan operations to unload the resulting test data. Therefore, the at-speed ABIST for both SRAM + hit logic delivers system-like high speed multi-cycle testing to catch fails that may be missed by other methods. This scheme also includes an ABIST Mux to uniquely test each memory cell; therefore, isolating the fail to the SRAM or to the hit logic.

Figure 4.2.6 shows a typical SETP hardware shmoo of Vdd (0.8 to 1.3V) vs. frequency (3.9 to 6.5 GHz). The "." indicates a Pass, an "h" indicates "hit logic only" fails, and an "s" indicates that both SRAM and the resulting compare failed. Several "h" locations (including "holes" in the shmoo) would have been passing without the at-speed "SRAM+hit logic" ABIST scheme described above.

Acknowledgements:
Thanks to Uma Srinivasan and Bill Huott for the hardware test data.

References:
[1] A. Pelella et al, "A 8K Domino Read SRAM with Hit Logic and Parity Checker", *ESSCIRC Dig. Tech. Papers*, pp 359-362, 2005.
[2] D. W. Plass and Y. H. Chan, "IBM POWER6 SRAM arrays", IBM J. RES. & DEV. VOL. 51, NO. 6, November 2007, pp 747-756.
[3] S. Narasimha et al, "High Performance 45-nm SOI Technology with Enhanced Strain, Porous Low-k BEOL, and Immersion Lithography", *IEDM* pp. 1-4, 2006.

ISSCC 2011 / February 21, 2011 / 2:00 PM

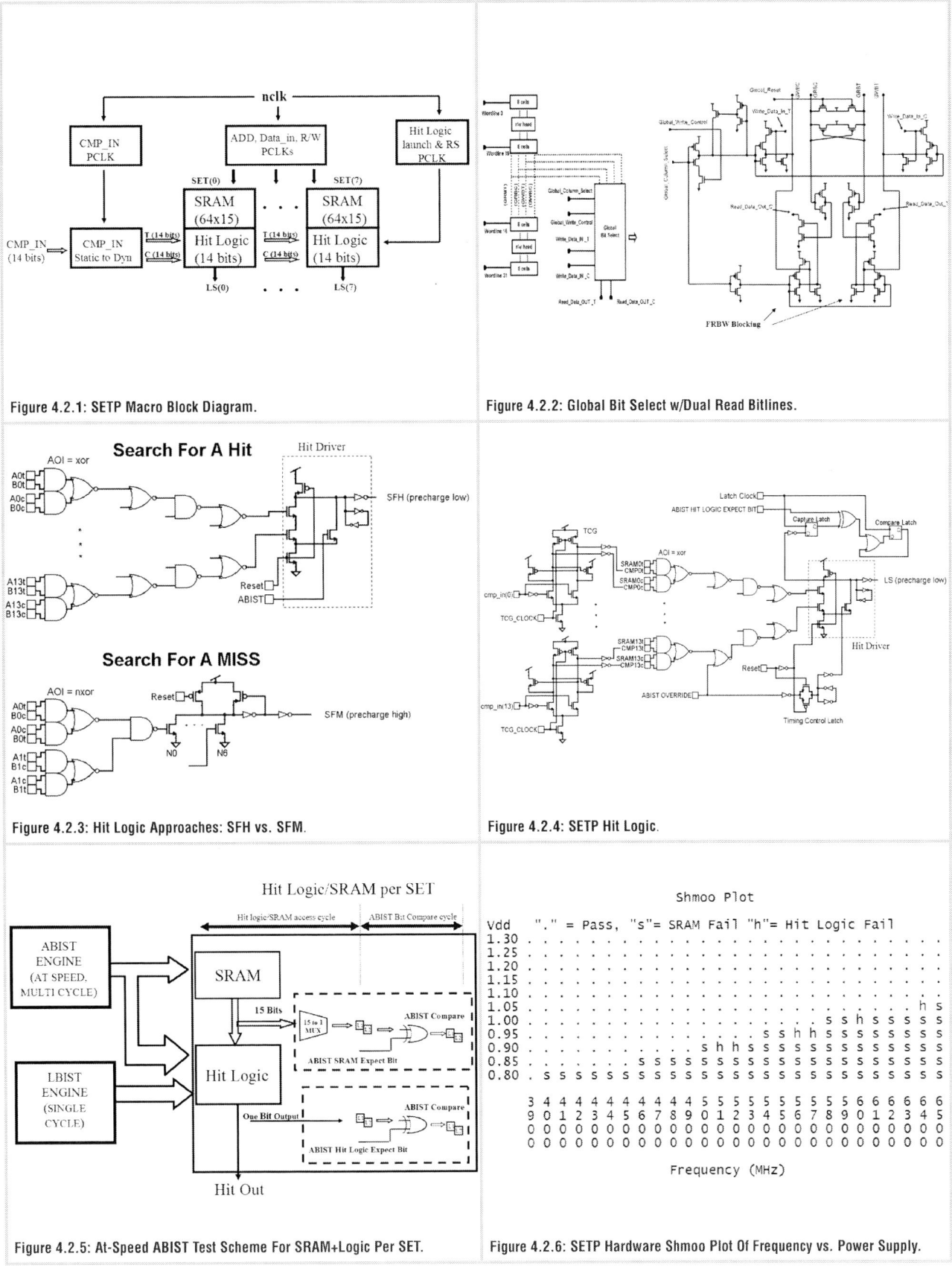

Figure 4.2.1: SETP Macro Block Diagram.

Figure 4.2.2: Global Bit Select w/Dual Read Bitlines.

Figure 4.2.3: Hit Logic Approaches: SFH vs. SFM.

Figure 4.2.4: SETP Hit Logic.

Figure 4.2.5: At-Speed ABIST Test Scheme For SRAM+Logic Per SET.

Figure 4.2.6: SETP Hardware Shmoo Plot Of Frequency vs. Power Supply.

ISSCC 2011 PAPER CONTINUATIONS

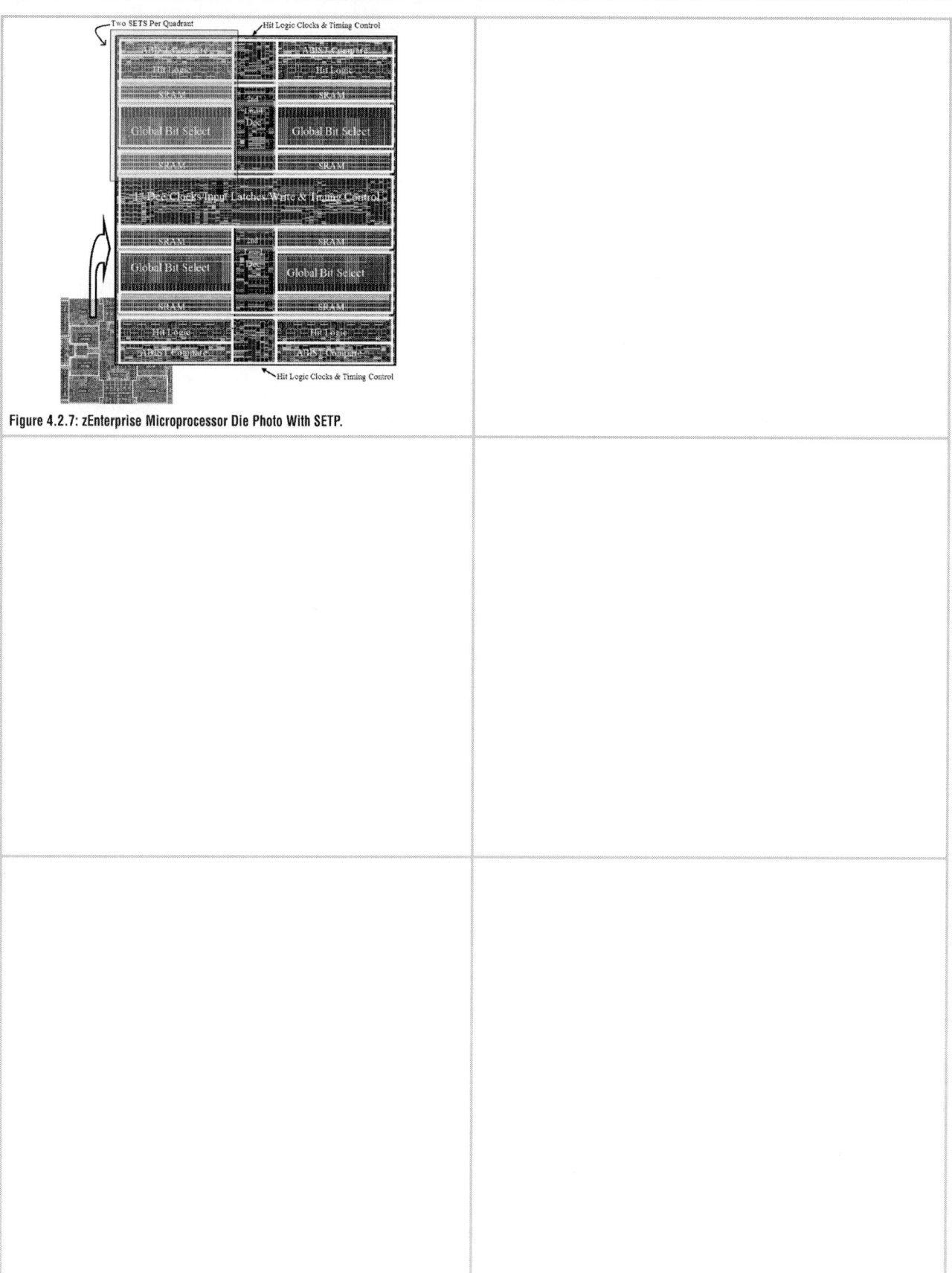

Figure 4.2.7: zEnterprise Microprocessor Die Photo With SETP.

ISSCC 2011 / SESSION 4 / ENTERPRISE PROCESSORS & COMPONENTS / 4.3

4.3 A 32nm Westmere-EX Xeon® Enterprise Processor

Shankar Sawant, Utpal Desai, Gururaj Shamanna, Lokesh Sharma,
Mandar Ranade, Anil Agarwal, Sampath Dakshinamurthy,
Rajagopal Narayanan

Intel, Bangalore, India

The next-generation enterprise Xeon® processor [1] consists of 10 Westmere 32nm cores [2] and a shared inclusive L3 cache (LLC) integrated on a monolithic die, with link-based I/Os. This paper focuses on the innovations and circuit optimizations over the predecessor [3] targeting idle power reduction, robust high-speed I/O links, and performance per watt improvements. The processor is implemented in 32nm CMOS using high-κ metal gate transistors and nine copper interconnect layers [4].

Core C6 power management is supported with individually controlled power gates for each of the 10 cores (Fig. 4.3.1). Core state is retained in a dedicated SRAM block, while the power gates are switched-off. Macro clock gating (MCG) is implemented in the uncore to reduce switching power from local clock network and associated logic. Power consumption in this state is limited to critical uncore circuits that monitor incoming snoops, interrupts, and machine check events. The on-die power controller (PCU) controls entry/exit from this mode. The MCG mode signal is overloaded on existing fine-grained DFX infrastructure (Fig 4.3.2) to maximize circuit shut-down granularity right down to the regional clock buffers, synchronized to a common clock edge. An additional level of idle power reduction is provided by extending the package C6 states to disable memory I/O links (SMI-Disable) by shutting off the bias currents to transmit-driver, receiver-VOC, phase-interpolators and DLL. The PC6 state (the deepest processor sleep state) entry is coordinated by CPU firmware to negotiate a guaranteed platform idle. SMI-disable is a sub-state of PC6 leveraging memory link error, fast reset init sequence to exit from this state. To minimize exit latency, several components of the full reset sequence like the VOC calibrate, and voltage offset tuning are bypassed in this exit mode. Also the IO-PLL and compensation loops are kept alive to shorten link bring-up, by saving on PLL lock times. Green VID (GVID) flow is implemented in High Volume Manufacturing (HVM) to minimize uncore operating voltage to further optimize idle power. Unlike the conventional approach to bin parts at highest voltage meeting TDP power, the GVID flow bins parts with highest voltage meeting idle power and TDP power.

The uncore is designed to reduce dynamic power usage, to maximize available TDP budget for ten cores. Fine grain clock gating is done in datapath structures and automated power aware algorithms are used as part of the control block synthesis implementation. Contrary to conventional wisdom a low-leakage process is NOT selected for this multi-core processor. Power distribution numbers (Fig. 4.3.3) show that the dynamic power component of the total TDP is significant, indicating the need to operate uncore at a lower voltage. A low leakage process would reduce intrinsic transistor performance, which is compensated by increasing the voltage to maintain frequency parity, and hence increase overall processor power.

To enable robust I/O data rate over 20" FR4 channel, 2nd order CTLE with temperature compensation and 3-stage clock amplifier with inbuilt duty-cycle corrector (DCC) circuits have been implemented in link's receiver with 1100mV$_{pp}$ differential TX driver swing. The 2nd order CTLE is implemented as two-stage amplifier (Fig. 4.3.4). A 1st stage amplifier is implemented with controllable source-degeneration resistance (R1) and capacitor(C1). The 1st stage provides required peaking, or delta between low frequency and high frequency gain, by reducing DC/low frequency gain. A 2nd stage amplifier is implemented with controllable resistance (R2) across the diode-connected load. The 2nd stage provides higher gain and bandwidth at the operating frequencies of interest, via its peaking response, while maintaining the DC gain similar to a wide-band amplifier. The combination of 1st and 2nd stage DC gain and peaking controllability achieve desired performance for given interconnect and to solve saturation issues for shorter channels. Several per-lane features are implemented in the I/O ports to optimize performance/power and to improve yield, like; per-lane termination resistance control, TX/RX EQ coefficient, bias current controls and DCC. The I/O links speed is decoupled from the uncore voltage by retaining only 20% of the I/O digital control circuits on the high-speed I/O clock. A timeslot-valid generated based on the Bresenham Algorithm is used to qualify the nominal uncore

clock to derive a gated uncore clock (GUCLK), that runs at half the I/O-link frequency, for the bulk of the I/O port (80%). Iso-performance is achieved by doubling the datapath width to compensate for the slower uncore clock. The decoupling enables scaling the uncore voltage to achieve low power SKUs while maintaining the full I/O link rate. Statistical DOE models indicate link performance is impacted due to jitter amplification in the off-chip link. The I/O clock design focused on reducing the forwarded (Tx) clock jitter. A per-lane DCC feature is added to recover duty-cycle distortion in the Tx clock (Fig 4.3.5a). Post-silicon results show an improvement of ~2% in the duty cycle of the I/O clocks with this feature. The Tx clock is implemented as pseudo-differential while the Rx clock is low-swing fully differential. Other jitter reduction techniques include jitter-attenuating DLLs on Rx clock path, and improving the bias voltages for Tx buffers. Figure 4.3.5b shows the forwarded I/O clock, captured by a BERT, on an electrical validation system that has a 7" trace. The clock eye is 106ps/420mV at 10^{-6} BER and 87.5ps/406mV at 10^{-15} BER.

The clock generation leverages the implementation from the 8-core predecessor and extends it for the two additional cores. Once powered on, the frequency of the various clock domains is fixed, except for the cores, enabling on-demand performance increase through the turbo-boost™ feature. The output clock from the uncore PLL is first distributed vertically using a binary tree distribution. Matched distributions, embedded in horizontal spines, with their point of divergence at the vertical spine, further distribute the global UCLK. This topology reduces the POD between clock end points and eliminates the need to use de-skew compensators used in prior designs, providing an iso-skew profile at 10% lower global clock distribution power.

The processor memory system supports 3 levels of on-die cache hierarchy. L3 cache is organized as 10 independent SRAM modules, each 24 way set-associative, connected via ring bus [2] architecture. 0.212μm^2 6T bit-cell is used in L3 cache data array to achieve maximum array efficiency and to minimize SRAM leakage power. 0.272μm^2 6T bit-cell is used in L3 cache tag/status arrays to support back-back read/write operation. L3 cache achieves a standby VMIN of 700mV and an active VMIN of 800mV. Inline DECTED in L3 data and SECDED in L3 tag, error correction schemes are incorporated in arrays to prevent data corruption due to soft error and erratic bit failures. Segmented sleep techniques [6] implemented in L3 cache provide leakage power savings of 1.7W without degrading processor frequency. Core and cache recovery schemes are supported to realize lower configuration parts.

Power/iiming convergence of ring bus paths [5] was a prime concern due to ring bus width (~1000 signals) and 2× higher interconnect delay in 32nm over 45nm node. To achieve a balance between timing/power vectors, a dual Vt de-coupled master-slave flop is designed. This dual Vt MS flip-flop achieves a sub-10ps Tsetup time for both rise/fall transitions by controlled skewing of clock inverters I3A and I3B (Fig. 4.3.6) and also achieves power savings due to 60% reduction in data pin input capacitance. Inverter I5 in the de-coupled dual Vt MS flip flop helps eliminate data write-back problem existing in conventional MS flops when master latch opens on rising edge of clock. This 10 core processor with its targeted idle power reduction techniques, and emphasis on robust I/O links and performance per watt, provides a refresh to its predecessor 8 core design.

Acknowledgements:
The authors gratefully acknowledge the work of the talented and dedicated Intel team that implemented this processor.

References:
[1] D Nagaraj, et al., "Westmere-Ex: A 20 thread server CPU", Hot Chips 2010.
[2] N Kurd, et al, "Westmere: A family of 32nm IA processors", *ISSCC Dig. Tech. Papers*, Feb. 2010.
[3] S Rusu, et al., "A 45nm 8-Core Enterprise Xeon Processor", *ISSCC Dig. Tech. Papers*, Feb. 2009.
[4] S. Natarajan, et al., "A 32nm Logic Technology Featuring 2nd-Generation High-k + Metal-Gate Transistors, Enhanced Channel Strain and 0.171um^2 SRAM Cell Size in a 291Mb Array", *IEDM Tech. Digest*, Dec. 2008.
[5] C Park, et al., "A 1.2TB/s On-Chip Ring Interconnect for 45nm 8-Core Processor", *ISSCC Dig. Tech. Papers*, Feb. 2010.
[6] Y. Wang, et al., "A 4.0 GHz 291 Mb Voltage-Scalable SRAM Design in a 32 nm High-k + Metal-Gate CMOS Technology", *IEEE J. Solid-State Circuits*, vol 45, Issue1, Jan. 2010.

978-1-61284-303-2/11 $26.00 © 2011 IEEE

ISSCC 2011 / February 21, 2011 / 2:15 PM

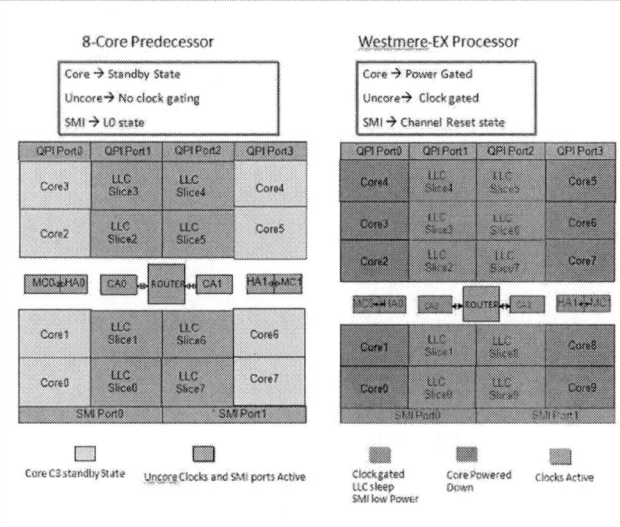

Figure 4.3.1: Idle power reduction features in block diagram.

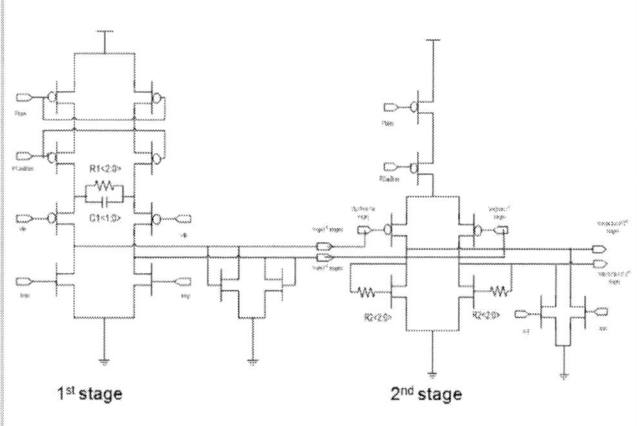

Figure 4.3.2: Idle power reduction techniques, and their contribution to overall idle power savings.

Figure 4.3.3: Process choice for this multi-core processor.

Figure 4.3.4: 2nd order CTLE circuit, with follow-on 2nd stage to increase gain of the amplifier.

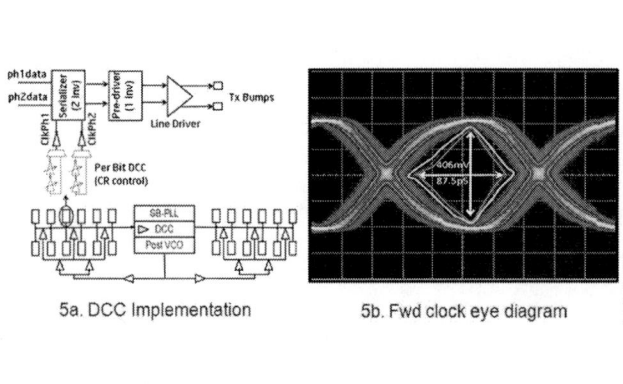
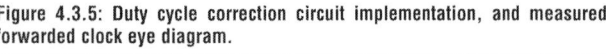

Figure 4.3.5: Duty cycle correction circuit implementation, and measured forwarded clock eye diagram.

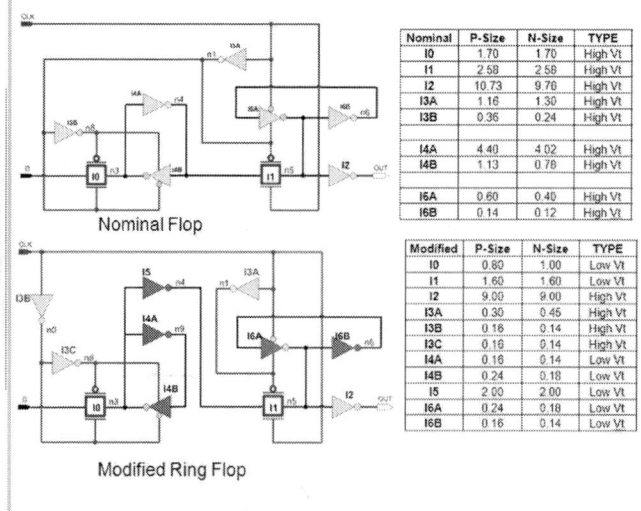

Figure 4.3.6: Modified flop used in the ring bus design.

978-1-61284-303-2/11 $26.00 © 2011 IEEE

ISSCC 2011 / SESSION 4 / ENTERPRISE PROCESSORS & COMPONENTS / 4.4

4.4 Godson-3B: A 1GHz 40W 8-Core 128GFLOPS Processor in 65nm CMOS

Weiwu Hu[1,2], Ru Wang[1,2], Yunji Chen[1,2], Baoxia Fan[1,2], Shiqiang Zhong[1,2], Xiang Gao[2], Zichu Qi[1,2], Xu Yang[2]

[1]Chinese Academy of Sciences, Beijing, China,
[2]Loongson Technologies, Beijing, China

As the latest product of Godson processor series, the Godson-3B processor is an 8-core high-performance general-purpose processor implemented in 65nm CMOS low-power general-purpose mixed process with 7 layers of Cu metallization. Godson-3B contains 582.6M transistors (including 4MB L2-cache) within 299.8mm^2 area. The number of signal pins in Godson-3B is 654. The highest frequency of Godson-3B is 1.05GHz, and the peak performance is 128GFLOPS (double-precision) or 256GFLOPS (single-precision) at 1GHz frequency with 40W power consumption. Godson-3B has an energy efficiency of 3.2GFLOPS/W. Other state-of-art high-performance processors have energy efficency of ~1.3GFOPS/W (Westmere [2]) and ~1.5GFLOPS/W (Power7 [3,4]). As shown in Fig. 4.4.1, Godson-3B contains 2 nodes while being able to scale to 16 nodes through inter-chip connection. Each node contains four cores, four L2-cache banks, one HyperTransport (HT) controller, one DDR2/3 controller, and the interconnection network connecting these components together. The interconnection network takes the 128-bit AXI standard interface with cache coherence extension.

The core in Godson-3B is GS464V eXtra Processing Unit (XPU). It is MIPS64-compatible while providing additional 200 instructions for x86 emulation. As shown in Fig. 4.4.2, GS464V adopts a nine-stage dynamic pipeline. In each cycle, at most 5 instructions can be issued to the 5 functional units (2 fixed-point functional units ALU1 and ALU2, 2 floating-point/vector functional units VPU1 and VPU2, and 1 memory functional unit).

An important feature of GS464V is the two-issue 256-bit vector computation ability, which is illustrated by the black modules in Fig. 4.4.2. In each cycle, a 28-entry vector queue can dispatch out-of-order two vector instructions to the two 256-bit vector functional units. Each VPU can perform 4 double-precision or 8 single-precision floating-point multiply and add (MADD) operations, or at most 32 fixed-point operations per cycle. To keep compatibility with MIPS64, each VPU can also perform one MIPS64 floating-point instruction per cycle. GS464V defines a vector instruction set to meet the requirements of vector compiler as well as important applications (e.g., scientific computation, signal processing, multimedia). The vector instruction set of GS464V includes a series of shuffle-computation mixed instructions that eliminate the necessity for a dedicated shuffle unit (although the VPUs can still perform shuffle operation). To reduce the required memory bandwidth of the vector unit, GS464V adopts a 4w8r 128-entry 256-bit vector register file, and a programmable memory coprocessor (vecdma in Fig. 4.4.2), which can directly transfer data between vector register file and cache/memory [1]. Hence, a 1GHz GS464V needs 2.95μs to solve the 1024-point single-precision complex FFT, which is less than 3.67μs needed by 2.66H Core i7.

To reduce the dynamic power consumption of Godson-3B, three methods are employed. First, Godson-3B benefits from a mix of static and dynamic design. Godson-3B is mainly based on static CMOS. Only its timing-critical components (such as register file, RAM, CAM) employ dynamic circuits. Second, clock-gating technique is extensively used in Godson-3B. There are coarse-grain gating cells visible to operating system for each main component of Godson-3B. There are also fine-grain gating cells covering 85% FFs in Godson-3B. Meanwhile, there are mode-gating cells to disable unused circuits in the current mode. For example, BIST logic can be gated in functional mode. Third, Godson-3B employs dynamic frequency–scaling (DFS) scheme to reduce dynamic power. To reduce the static power, Godson-3B employs three kinds of cells with different threshold voltage. Less than 0.1% cells are the low-threshold voltage cells, which have high speed as well as high static power.

The power proportions of core and uncore parts in Godson-3B are shown in Fig. 4.4.3. Eight cores contribute about 72.25% of the total power (28.9W) for high switching activities, while the power proportion of uncore is about 27.75% (11.1W) due to their relatively low frequencies. The macro cells, which are the

dynamic circuit parts of Godson-3B, contribute more than 40% of the overall power. The power dissipation of clock network is less than 24% of the overall power, which demonstrates the effectiveness of our clock tree design.

Figure 4.4.4 shows the propagation path of core clock domain, the main clock domain of Godson-3B. The clock is distributed to all IP modules with balanced latency in each node. There are 5 custom-designed skew-measure modules in the chip. They can perform on-line clock skew measurement in inter-IP level, intra-IP level, and intra-node level. The circuit structure of the skew-measure module is shown in Fig. 4.4.4. Two clock signals are coarsely phase shifted by DCDLs at first. Then, a group of parallel SDCs perform fine calibration. Skew between clocks can be obtained by checking the control signals of DCDL and the structures of SDC after PD. The measuring range is ±241.34ps with 9.47ps accuracy. Furthermore, 10 on-chip custom-designed DCDL modules can provide latency adjustments for timing critical modules after fabrication. The adjusting range is ±105.74ps with 29.66ps accuracy. DCDL is designed with strict symmetrical structures, which is based on complementary symmetry NAND-gate. DC is designed with NMOS loading connected INV. In skew-measure, SDC is composed of DC and selected loading capacitance. In general, the skew measure and adjustment scheme can provide ~7% frequency improvement for Godson-3B.

In Godson-3B, there are three identical PLLs for core, HT and DDR respectively. The output frequency of the PLL is in the range between 700MHz and 3.2GHz. The PLL, whose block diagram is illustrated in Fig. 4.4.5, employs a pseudo low dropout regulator (PLDO) to avoid 1.8V power supply noise for analog blocks such as bandgap, current array and voltage-current converter. The PLL also adopts a current controlled oscillator (ICO) to overcome the sensitivity to the power supply noise and temperature. To remove the impact of process variations on the oscillator, an open-loop calibration is performed during the starting up of the PLL. As a consequence, the VCO gain can be reduced from 3GHz/V to 600MHz/V.

The DDR2/3 interface of Godson-3 can achieve a maximum bandwidth of 12.8GB/s for DDR2 mode at 1.8V, and 25.6GB/s for DDR3 mode at 1.5V. With the on-die termination (ODT) circuits, impedance matching can be realized for high-speed data transmission requirements. To reduce the standby power, power management circuit is employed to automatically shut down the non-operating parts of the DDR interface.

The I/O interface of Godson-3B complies with HT 1.0 specification, which supports a maximum bandwidth of 12.8GB/s with the HBM 2000V ESD protection. Current-mode LVDS driver with common-mode feedback maintains the amplitude of the output differential signals. It can tolerate variations of power supply of 0.9 to 1.3V, process corners and temperatures from -40 to 125C. An I/Q clock is generated by the PLL for HT, where the I-clock distribution of transmitters adopts a low-skew differential clock tree. The HT interface transmits both Q-clock and data simultaneously to decrease BER in the receivers. Each channel of the transceiver can operate at 1.6Gb/s with a BER of better than 10^{-15}. Figure 4.4.6 shows the eye diagram when the transmitters work at 1.6Gb/s on a 20cm channel.

Acknowledgements:
This work is partially supported by National S&T Major Project (No. 2009ZX01028-002-003, 2009ZX01029-001-003 and 2010ZX01036-001-002), National 863 Program of China (No. 2008AA010901 and 2009AA01Z125), and National Natural Science Foundation of China (No. 60736012 60921002, 60803029 and 61003064).

References:
[1] W. Hu and Y. Chen, "GS464V: A High-Performance Low-Power XPU with 512-Bit Vector Extension", in *Proceedings of Hot Chips*, 2010.
[2] M. Chowdhury et al, "Architectural Innovations in WSM-EP", in *Proceedings of Hot Chips*, 2010.
[3] R. Kalla et al, "Power7: IBM's Next-Generation Server Processor", *IEEE Micro*, 30(2), 2010.
[4] D. Wendel et al, "The implementation of POWER7™: A Highly Parallel and Scalable Multi-core High-end Server Processor", in *Proceedings of ISSCC*, 2010.

ISSCC 2011 / February 21, 2011 / 2:30 PM

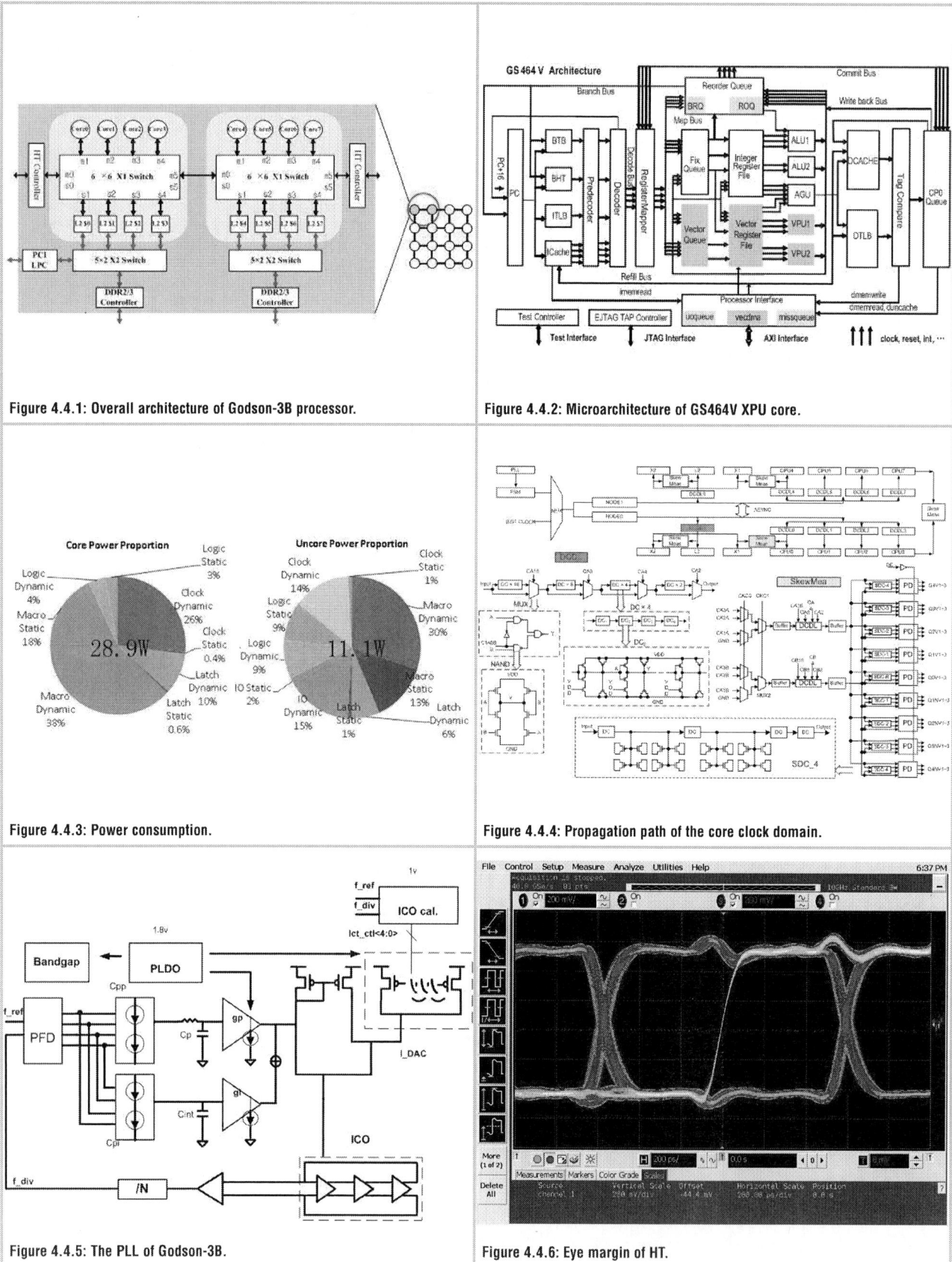

Figure 4.4.1: Overall architecture of Godson-3B processor.

Figure 4.4.2: Microarchitecture of GS464V XPU core.

Figure 4.4.3: Power consumption.

Figure 4.4.4: Propagation path of the core clock domain.

Figure 4.4.5: The PLL of Godson-3B.

Figure 4.4.6: Eye margin of HT.

ISSCC 2011 PAPER CONTINUATIONS

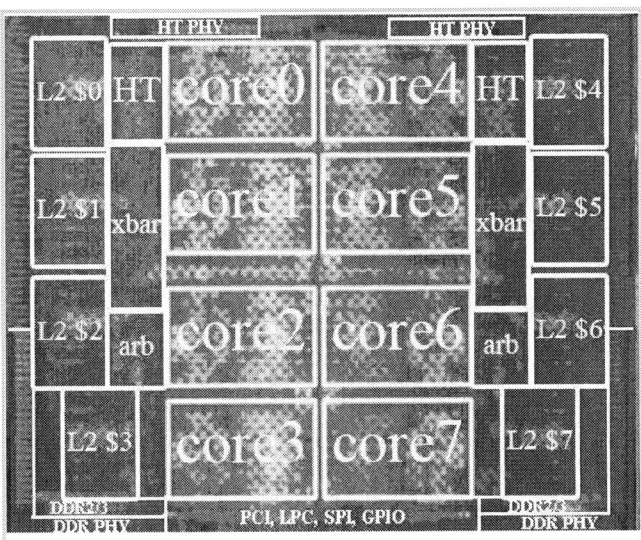

Figure 4.4.7: Godson-3B die photo.

ISSCC 2011 / SESSION 4 / ENTERPRISE PROCESSORS & COMPONENTS / 4.5

4.5 Design Solutions for the Bulldozer 32nm SOI 2-Core Processor Module in an 8-Core CPU

Tim Fischer[1], Srikanth Arekapudi[2], Eric Busta[1], Carl Dietz[3],
Michael Golden[2], Scott Hilker[2], Aaron Horiuchi[1], Kevin A. Hurd[1],
Dave Johnson[1], Hugh McIntyre[2], Samuel Naffziger[1], James Vinh[2],
Jonathan White[4], Kathryn Wilcox[4]

[1]AMD, Fort Collins, CO,
[2]AMD, Sunnyvale, CA,
[3]AMD, Austin, TX,
[4]AMD, Boxborough, MA

AMD's 2-core "Bulldozer" module contains 213 million transistors in an 11-metal layer 32nm HKMG SOI CMOS process and is designed to operate from 0.8 to 1.3V. This new micro-architecture [1] improves performance and frequency while reducing area and power compared to a previous AMD x86-64 CPU in the same process [2]. To achieve these goals, the design reduced the number of FO4 inverter delays/cycle by more than 20%, achieving higher frequencies in the same power envelope even with increased core counts. The 2-core CPU module area (including 2MB L2 cache) is 30.9mm^2 (Fig. 4.5.7).

The module design contains 84 unique custom macros and 317,000 scannable flops. Module-level VSS power gating (CC6) reduces leakage power by 95% when both cores are idle [2]. Transistor Vts across the design are mostly regular (47%) and long-channel regular (46%).

The Bulldozer micro-architecture is cycle-based, using soft-edge flip-flops (SEF) to provide high-frequency performance, process variation tolerance, and low power consumption (Fig. 4.5.1). Performance and process tolerance are provided by a 2-clock design: early and late clocks (ECLK, LCLK) create a soft timing edge, allowing limited cycle stealing. Power is reduced in low-power SEFs by internally gated slave latch clocks. The majority of flops (78%) are low-power, using high-performance flops only on timing-critical paths.

In contrast to leveraged power-optimized CPU designs [2,4], Bulldozer's ground-up design requires co-development of power efficiency, timing, and functionality. Initially, micro-architectural power is optimized using a power-aware high-level performance model. Next, before schematic completion, the team tracks and analyzes RTL-based clock and flip-flop activity (a proxy for switching power) to meet clock gating goals. Finally, a new power model enables early mixed schematic/layout analysis of transistor-level power. This enables aggressive power optimizations while the implementation is still malleable. The result is a design with low power consumption for typical applications, making it well-suited to active power management and boost (Fig. 4.5.2).

The L1 caches are split, with I-cache residing in the instruction unit and a D-cache located in each load/store unit of the 2-cores. The 2-way, 64KB I-cache consists of an 8×2 array of 4KB bank macros, with 2 more arrays for pre-decode bits. Load/store area in the 2 cores is at a premium, so the D-cache uses a 4-way 16KB array with performance features described later in the paper. Both L1 caches use an 8T storage cell. The change from a 6T cell in 45nm to 8T in 32nm was required to improve low-voltage margin and read timing and to reduce power. Use of the 8T cell also eliminated a difficult D-cache read-modify-write timing path. Reads use a 2-level pre-charged local/super bitline structure with delayed-onset keeper, single-rail, full-swing signals, and glitch latches.

Several D-cache performance features reduce conflicts and power. First, micro-banking reduces read conflicts to the same rate as a previous 16-bank 64KB design by retaining 16 logical banks. Next, when both read ports access the same 128b word (a common occurrence), port-A data feeds both ports, saving power and avoiding conflicts (Fig. 4.5.3). Load/store unit power features include a static TLB with power filter to avoid TLB reads if the page has not changed, and extensive use of static wakeup CAMs to avoid retries.

Each core integer unit [3] contains a 40-entry out-of-order scheduler supporting single-cycle wake and pick of up to four instructions. To save wake array power, source/destination compares use a pre-decoded dynamic AND "match CAM"

instead of a traditional mismatch CAM. The integer datapath supports 0-cycle result bypass to dependent instructions. To remove critical-path wire delay, the physical register file arrays and address generator (AGEN) incrementor are replicated. Most register file read bits are single-ended, but dual-rail reads for a few critical bits supply clocked data to the dynamic shifter.

The floating-point (FP) unit adds many features within the same basic pipeline as previous AMD x86-64 processors:

- a) new multiply-accumulate functionality and instructions, including AVX, AES, SSSE3, SSE4.1, SSE4.2, XOP, and PCLMULQDQ;
- b) dual-threaded support;
- c) 4-wide instruction issue; and,
- d) increased size of performance-critical structures [2, 4].

To minimize bus lengths, the datapath is split into high and low halves (Fig. 4.5.4), each acting on 64 bits for AVX/SSE/MMX or 80 bits for X87 operations. Four fully pipelined FMACs each support one EP/DP or two SP ops per cycle with 5-cycle latency to dependent ops. Optimized algorithms and dense, hand-placed datapaths minimize area, reduce cap and RC delay for timing, and reduce power. Extensive fine-grain clock gating allows FP to achieve an idle active power of less than 2% of peak active power.

A 160-entry 10R/6W FP register file (FPRF) reads 2 or 3 sources for 4 instructions, and writes up to 6 results, per cycle. The FPRF is split horizontally into two 10-bank arrays of 91 and 73 bits respectively. Each array aligns with a split FP datapath, and differs significantly from the replicated integer unit RF array design [3]. A flexible 2 and 3-level read bitline structure is used to reduce power and latency across the large data arrays (Fig. 4.5.5).

The cache unit (CU) contains L2 data TLB, and L2 cache interface datapath and controls. The CU and other units have large, complex random-logic control blocks that required the use of design automation techniques for efficiency. The complex logic and aggressive 12 to 14 gates useable between flops requires a variety of approaches to meet timing. The use of "bounds" statements guides structured logic placement without forcing explicit locations. Route estimation during cell placement models a metal stackup in which dense lower layers are up to 20× more resistive than higher, fast layers. Many arrays also rely on design automation or reuse. Across the core, many queues and buffers are built from compiled array structures and the larger CU memory arrays for the TLB and pre-fetch history tables are copies of I-cache and I-TLB arrays.

The 2MB L2 cache array has a 64B line size and 6-cycle internal pipeline. The 0.258μm^2 6T bitcell comprises 128KB slices of eight 16KB data array macros each. Figure 4.5.6 shows each data array macro uses a traditional sense amp with 128 rows per bitline and supports both row and column repair. The L2 data array macro contains four banks of sub-arrays with two-cycle throughput, allowing a full cycle for read or write and a full cycle for pre-charge. As shown in Fig. 4.5.6, wordlines are active during cycle 1. Early in cycle 2, the isolation PFETs turn on, passing differential from the main to the sense bitlines. After a programmable self-timed delay, the sense enable fires, turning off isolation PFETs and word lines. The sense enable is active for the rest of cycle 2, allowing the sense amp to latch the output data. A column circuit interlock prevents isolation PFETs from turning off before the sense enable fires. Interlock logic also disables isolation PFETs during the first cycle of the access while the sense amp bitlines are in pre-charge.

References:
[1] Butler, M. "Bulldozer: A new approach to multithreaded compute performance." Hot Chips 22, August 24, 2010.
[2] Jotwani, R. et al. "An x86-64 Core Implemented in 32nm SOI CMOS," *ISSCC Dig. Tech. Papers,* pp. 106-107, Feb. 2010.
[3] Golden, M. et al. "40-entry unified, out-of-order scheduler and integer execution unit for the AMD Bulldozer x86-64 two-core CPU module", *Accepted to ISSCC 2011.*
[4] Golden, M. et al., "A 2.6GHz Dual-Core 64bx86 Microprocessor with DDR2 Memory Support," *ISSCC Dig. Tech. Papers,* vol. pp. 325-332, Feb. 2006.

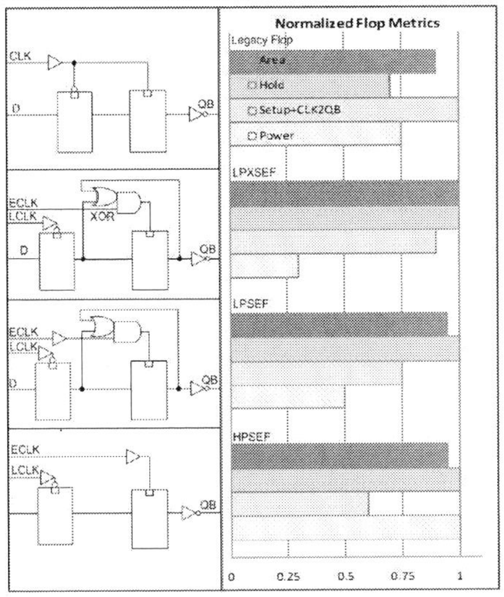

ISSCC 2011 / February 21, 2011 / 3:15 PM

Figure 4.5.1: Normalized Soft-Edge Flop (SEF) metrics (lower value is better).

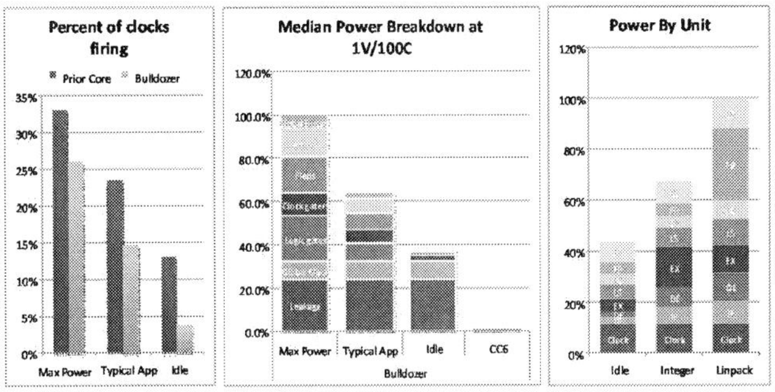

Figure 4.5.2: Bulldozer Power Consumption.

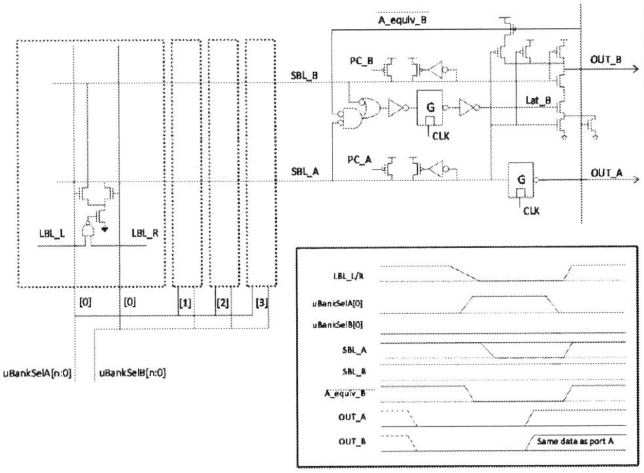

Figure 4.5.3: L1 D-cache Half-bank Output Latches.

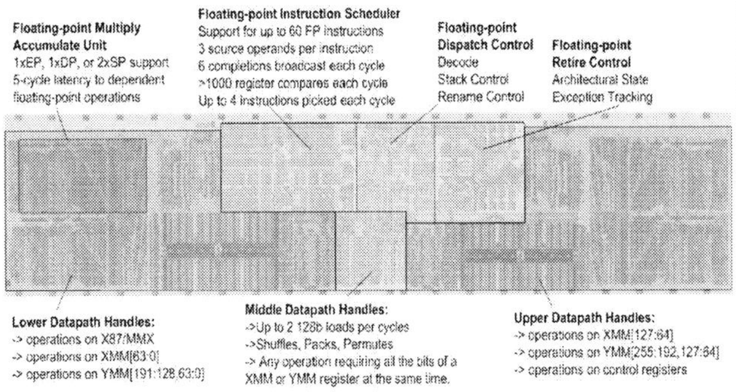

Figure 4.5.4: Floating-point Unit Floorplan.

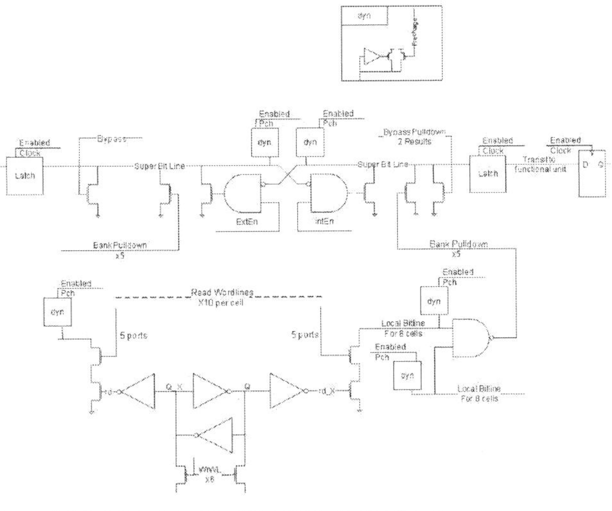

Figure 4.5.5: FP Register File Read Structure and Bypass.

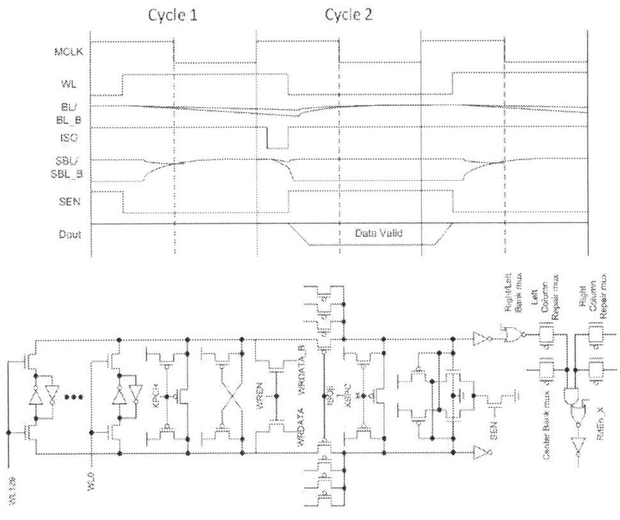

Figure 4.5.6: Cache Macro Read Datapath and Timing Diagram.

978-1-61284-303-2/11 $26.00 © 2011 IEEE

ISSCC 2011 PAPER CONTINUATIONS

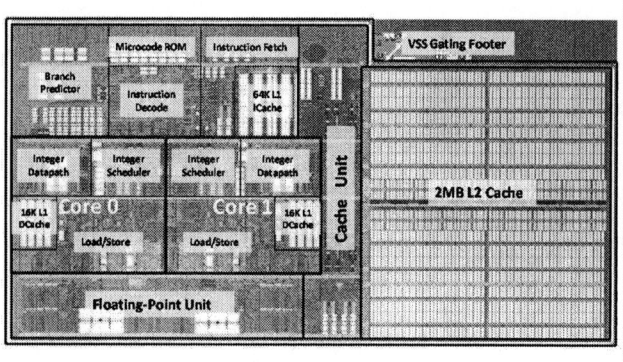

Figure 4.5.7: Bulldozer 2-core Processor Module Die Photo.

ISSCC 2011 / SESSION 4 / ENTERPRISE PROCESSORS & COMPONENTS / 4.6

4.6 40-Entry Unified Out-of-Order Scheduler and Integer Execution Unit for the AMD Bulldozer x86-64 Core

Michael Golden, Srikanth Arekapudi, James Vinh

AMD, Sunnyvale, CA

AMD's two-core Bulldozer module [1,2] implements the AMD x86-64 microarchitecture in an 11-layer 32-nm SOI HKMG technology. The 40-instruction out-of-order unified integer scheduler issues up to four operations per cycle and supports single-cycle wake-up of dependent operations. The 2.37mm^2 integer execution unit supports single-cycle data bypass among four independent functional units. Compared to previous AMD x86-64 cores [3-6], project goals reduce the number of FO4 inverter delays per cycle by more than 20%, while maintaining constant IPC, to achieve higher frequency and performance in the same power envelope, even with increased core counts.

Critical paths (Fig 4.6.1) are implemented without exotic circuit techniques or heavy reliance on full-custom design. Dynamic logic appears only when required for density. Inputs and outputs of dynamic gates tolerate duty-cycle variation by flowing through timing test points if clock arrives early. In lieu of dynamic logic, extensive use of skewed static CMOS standard-cell gates speeds transmission of the evaluate edge versus the pre-charge edge. To minimize power consumption, the architecture favors tag movement relative to data movement and pointer management relative to shifting, collapsing data structures. Duplicated register files and functional units ameliorate wire-limited critical paths.

The integer scheduler issues up to four out-of-order operations per cycle. Unlike a reservation station, which stores data or tags, the scheduler stores only tags: the physical register numbers (PRN) of destination and source operands. To save power, all structures except the physical register file (PRF) and functional units manipulate 8-bit PRNs instead of 64-bit instruction data. Instead of a power-hungry shifting, collapsing structure to preserve age information, an ancestry table keeps track of the oldest instruction in the scheduler. The oldest instruction will be picked, if ready. If not, a priority encoder scans ready instructions in their physical order, which does not correspond to their age order.

When an operation is picked for execution, the destination PRN is read from a register file and broadcast to a fully associative CAM, which stores up to four source operands for every instruction in the scheduler (Fig. 4.6.2). To save power, pairs of destination bits are decoded before broadcast, reducing the number of switching bitlines. An extra pull-down drives PRNs of results from the load/store unit into the source CAM. Rather than pulling down on a "mismatch" signal if a single bit mismatches in the CAM, a logical AND structure is used. This has two benefits. Fewer CAM structures switch – match is less likely than mismatch – and, because it is known *a priori* whether a source will match a destination from an arithmetic logic unit (ALU) or an address generation unit (AGU), a single CAM wire can share pulldowns between AGU and ALU destinations. One hundred sixty "current match" signals indicating which source operands match the destinations picked in the current cycle flow through glitch latches into post-wake and pick logic (Fig. 4.6.3). Post-wake and pick logic are implemented with static CMOS logic, with beta ratios skewed higher or lower than the natural process beta ratio (Fig. 4.6.4). Skewed gates favor the evaluate edge of the glitch latch output over the reset edge. Although wide OR functions in the post-wake and pick logic are amenable to domino logic, re-clocking the glitch latch outputs would make the design susceptible to duty-cycle variation. Source readiness information flowing in from register rename logic depends on pick results from the previous cycle, and is also critical. Though standard cell flops launch this path, the data is gated with the clock and fed into a glitch latch local to the picker to allow it to flow quickly through the skewed gates.

The pick result must be clocked and sent back to the wake-up array for the next cycle. A low-latency edge-triggered clock gater reduces clocking overhead (Fig. 4.6.5). The device has short setup time, determined only by having the output of the data inverter switch past the input. The cross-coupled PMOS keeper devices allow the dynamic node to recover from any false evaluation before the input switches full-rail, and do not fight the NMOS evaluation devices. A self-resetting NMOS keeper holds data for a full clock phase. The fast combination of a two-high NMOS stack followed by an inverter determines the clock-to-output latency.

The pick signal is driven directly into the instruction payload register file (PLDRF) as a read word line. Operand source PRNs in PLDRF flow out of glitch latches into dynamic bypass compare logic. Although most bit reads in the payload are single-ended, source PRNs are read dual-rail and full-swing. This (1) allows the use of cross-coupled keepers, against which evaluate pull-downs do not have to fight; (2) eliminates inverters in data paths; (3) permits the use of skewed logic gates for both true and complementary data; and, (4) increases tolerance of duty-cycle variation by allowing late-arriving bits to flow through clock gaters.

The execution unit supports single-cycle operand bypass from an instruction to a dependent instruction. Two ALU ops and two AGU ops can be executed in a cycle. AGU ops include increment/decrement (INC), address generate, and x86-64 LEA instructions. The ALUs and the INC units participate in single-cycle operand bypass. Four result buses, one from each of the units, feed four writes ports into the PRF. Each execution unit can receive data from any of the four write ports or from a PRF read. Each execution unit has two sources, requiring eight read ports in the PRF. To remove wire delay (and congestion) from the critical bypass path, the PRF is duplicated. The register file is divided into halves; the critical half, further from the output driver, has fewer pull-downs on the local bitline, reducing slew rates on critical words with minimal area impact (Fig. 4.6.6). Duplicating the INC removes the vertical wire delay from the worst-case bypass path, which is from any INC to the far ALU.

The ALU itself consists of multiple blocks, each of which executes a family of ALU ops. The results of these blocks are multiplexed hierarchically onto the ALU result bus, implemented in footless domino logic. Results from the adder and shifter are timing-critical and pull down directly on the results bus [1]. Like data results, flag results also have a single-cycle loop. Flag logic generates carry, overflow, sign, and auxiliary flags in parallel to the data result, and zero and parity flags from the final data result. Flag bypass is simpler because only ALUs produce and consume flags.

The shift unit supports all shift and rotate operations with a footless domino barrel-shifting core that performs a rotate operation through the carry bit. Set-up of data to the clock gaters driving into the shifter, which launch data on the falling edge of the clock, is susceptible to duty-cycle variation. To ameliorate this problem, the critical six bits that drive into the shift count decoder that controls the barrel shifter are driven dual-rail throughout the entire EX datapath, including the PRF read. All other functional units in the AGU and the ALU are built from standard cell logic. This includes the way predictor array (Fig. 4.6.4). The critical INC and adder units drive data onto the results bus using the same edge-triggered clock gater used in the picker.

Acknowledgements:
This design is the result of years of effort by AMD's entire Bulldozer team.

References:
[1] Fischer, T.; Arekapudi, S.; Dietz, C.; Golden, M.; Hilker, S.; Hurd, K.; Johnson, D.; McIntyre, H.; Vinh, J.; White, J.; Wilcox, K.; "Design Solutions for the Bulldozer 32nm SOI 2-core processor module in an 8-core Orochi CPU," *submitted to ISSCC 2011*.
[2] Butler, M.; "AMD 'Bulldozer' Core - a new approach to multithreaded compute performance for maximum efficiency and throughput," Hot Chips 22, Aug. 2010.
[3] Jotwani, R.; Sundaram, S.; Kosonocky, S.; Schaefer, A.; Andrade, V.; Constant, G.; Novak, A.; Naffziger, S.; "An x86-64 core implemented in 32nm SOI CMOS," *ISSCC Dig. Tech. Papers*, pp. 106-107, Feb. 2010.
[4] Dorsey, J.; Searles, S.; Ciraula, M.; Johnson, S.; Bujanos, N.; Wu, D.; Braganza, M.; Meyers, S.; Fang, E.; Kumar, R.; , "An Integrated Quad-Core Opteron Processor," *ISSCC Dig. Tech. Papers*, pp. 102-103, Feb. 2007.
[5] Golden, M.; Arekapudi, S.; Dabney, G.; Haertel, M.; Hale, S.; Herlinger, L.; Kim, Y.; McGrath, K.; Palisetti, V.; Singh, M.; "A 2.6GHz Dual-Core 64bx86 Microprocessor with DDR2 Memory Support," *ISSCC Dig. Tech. Papers*, pp. 325-332, Feb. 2006.
[6] Keltcher, C.N.; McGrath, K.J.; Ahmed, A.; Conway, P.; , "The AMD Opteron processor for multiprocessor servers," *Micro, IEEE* , vol. 23, no. 2, pp. 66- 76, March-April 2003.

ISSCC 2011 / February 21, 2011 / 3:45 PM

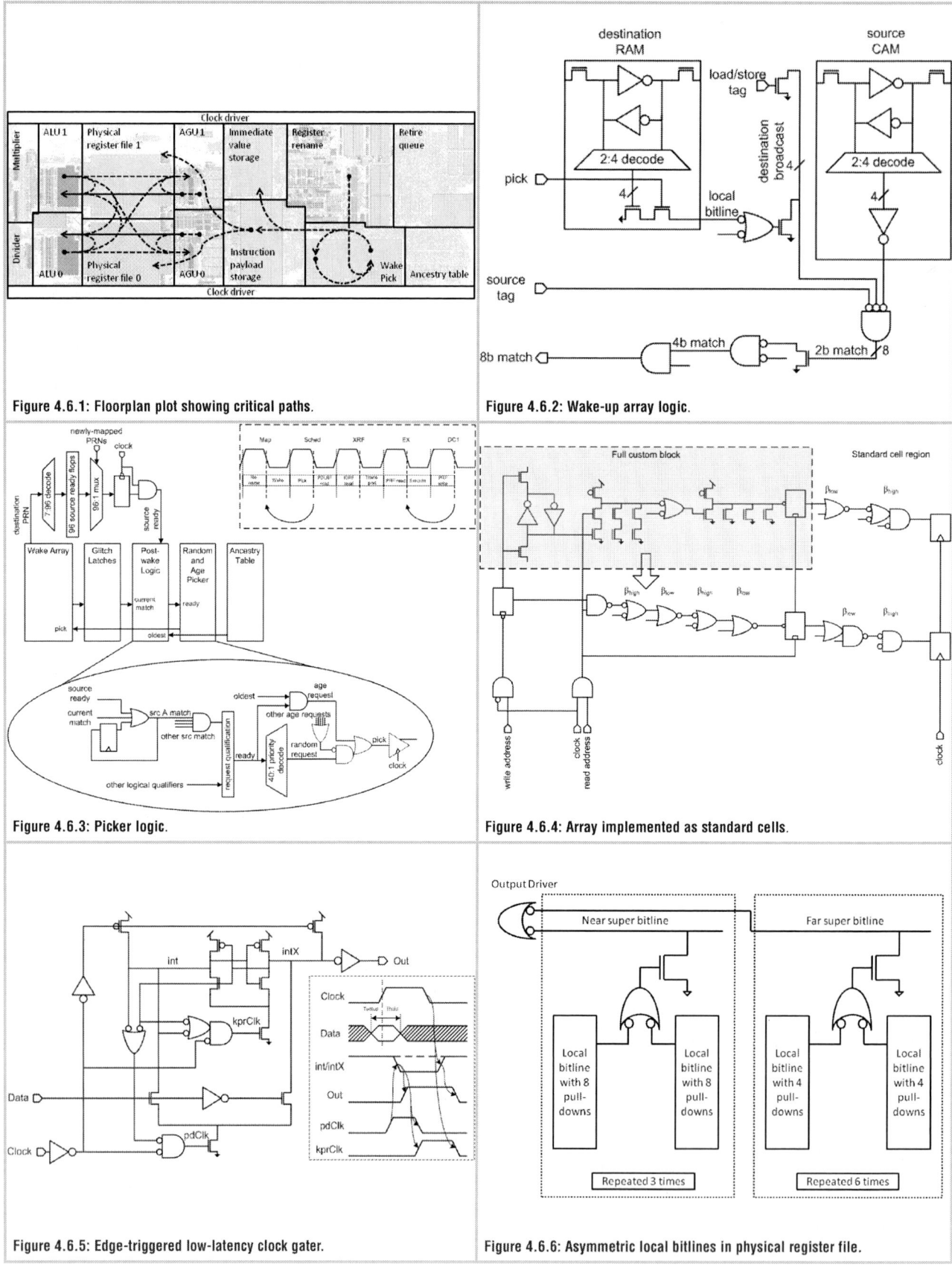

Figure 4.6.1: Floorplan plot showing critical paths.

Figure 4.6.2: Wake-up array logic.

Figure 4.6.3: Picker logic.

Figure 4.6.4: Array implemented as standard cells.

Figure 4.6.5: Edge-triggered low-latency clock gater.

Figure 4.6.6: Asymmetric local bitlines in physical register file.

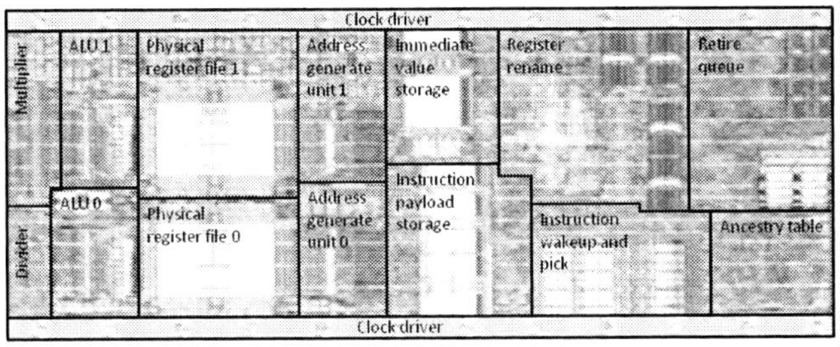

ISSCC 2011 PAPER CONTINUATIONS

Figure 4.6.7: Die photograph showing floorplan.

ISSCC 2011 / SESSION 4 / ENTERPRISE PROCESSORS & COMPONENTS / 4.7

4.7 Clock Generation for a 32nm Server Processor with Scalable Cores

Shenggao Li, Ashwin Krishnakumar, Edward Helder, Roan Nicholson, Vivian Jia

Intel, Santa Clara, CA

Within a given power envelope, the performance of a multi-core enterprise processor is greatly affected by inter-core (including I/O) data throughput and data transport latency. This paper presents the implementation of a clock system targeting low-power low-skew high-data throughput and low latency for a next-generation Xeon® server processor [1,2] with scalable cores in a 32nm 9-metal digital CMOS process.

Figure 4.7.1 is the die diagram of the processor featuring 8 SandyBridge cores [3], a shared last-level cache (LLC), a dedicated power control unit (PCU), and multiple high-speed I/O ports including Intel® QuickPath Interconnect (QPI) at 6.4GT/s or greater, PCI Express (PCIe) at 2.5/5GT/s, and DDR3 memory interface at 1600MT/s. The cores, cache, and I/O ports are attached through traffic-stops to an on-die ring-like bidirectional interconnect fabric running at the core clock rate. This highly modular ring architecture allows the number of cores and cache size to be scalable for various applications. The QPI link further offers the flexibility to form a multi-socket system [2]. The processor is powered by four major power domains, with VCCU serving the cores and caches, VCCP serving PCIe/QPI I/Os and PCU, VCCD serving the DDR3 I/Os, and VCCSA serving the system agents (i.e., PCIe/QPI/Memory logics). All power domain voltages (0.8 to 1.05V except VCCD, which is 1.35 to 1.5V) are controllable by PCU to allow voltage and clock frequency reduction or boost depending on thermal dynamics on the die. The clock architecture in Fig. 4.7.2 is designed to accommodate the design scalability, and power/clock domain crossing needs. The root of the clock system is an external 100MHz base clock (BCLK) providing reference clocks to the PCU, QPI, PCIe, and RCLK PLLs. The remaining PLLs are synchronized through the jitter filtering RCLK PLL, with a 100MHz reference clock (RxCLK) to the CPLLs/UPLL in the core and cache area, and a 133MHz reference clock (RyCLK) to the MPLLs in the memory system. Ideally, all PLLs and all clocks in the processor are synchronized to the same BCLK edge (defined as "Time0"). Clock jitter and path mismatches cause the end points in each clock tree to deviate from Time0. By taking into account the deviation, clock domains of same frequency can be crossed over synchronously using simple "de-skew" devices such as latch or D-flipflop. Clock domains of different frequencies are crossed over asynchronously using FIFO.

Data transport latency along the high-speed interconnect ring is a dominating factor on clock skew budgeting. In a 2-socket configuration with N cores per socket, every extra de-skew flop at each traffic-stop crossing causes a round-trip latency penalty of ~2(N+4) core clock cycles for an off-die data fetch. Assuming a non-scalable clock skew, the latency penalty gets worse when the clock cycle shrinks. Minimizing the inter-traffic-stop clock skew is the main motivation for adopting a single UPLL (see Fig. 4.7.1) for all the traffic stops. This is, however, at the expense of up to one cycle latency penalty at the core-cache crossings, due to the random UPLL-to-CPLL jitter (budgeted at 100ps). Skew reduction at the memory-agent and traffic-stop crossing is another design focus, by noting that RxCLK and RyCLK are distributed across a span of over 15×10mm². With matched clock trees between RxCLK and RyCLK, the estimated residual skew due to device mismatch and spatial voltage/temperature difference is still significant, and needs to be compensated using programmable delay lines. To ease the skew compensator design, the point of divergence (POD) on Rx|yCLK is pushed away from the RCLK PLL to the center of the die. Rx|yCLK is then retimed by a 400MHz clock at the POD. This reduces the skew range to be compensated by more than 50%.

Within each clock domain, one or more spine structure is used to deliver clocks across a large area to a sea of "clock islands", as shown in Fig. 4.7.3. Each clock island has a clock driver unit with a 1× and/or a 2× clock output (e.g., 800MHz and 1600MHz in the memory system) to serve a sub-region of about 200×200µm². The 1× clock is re-aligned to the 2× clock by synchronously resetting hundreds of distributed by-2 dividers. Static skew between the 1x and 2×

clocks is negligible, in comparison with that of an alternative solution where the 1× and 2× clocks are distributed through two separate central spines. The islands are attached to the spine through balanced clock trees. The spine, the balanced trees, and the islands are individually tuned to minimize the global skew. Spines in the same clock domains are calibrated against each other using delay compensators with skew resolution of 5ps. An island distribution topology enables both skew and power reduction. It allows the fine tuning of pre-silicon clock skews to 10ps or below in each domain. Unused clock islands are disabled. A sparse clock grid and less metal utilization near each island helps in saving power.

To reduce clock jitter, a fifth power supply (1.8V) dedicated to PLLs and thermal sensors is routed through metal layers inside the package and down to the PLL power bumps directly. On-die voltage regulator is used to provide >20dB PSRR. All PLLs except the PCIe/QPI PLLs are based on a 3-stage ring-oscillator optimized for jitter. The oscillator is tuned by a capacitor array coarsely and a MOS varactor finely at each delay stage. To meet the tight jitter budget of the PCIe/QPI IO, LC-tank based VCO is chosen for the PCIe/QPI PLLs. The inductor is a 3-turn symmetrical structure using top level metals (M7-8, Q = 5, L = 0.8nH). It is the first time on-chip planar inductors are being used in an Intel® server processor. Being an indispensible device for RF and high-speed ICs, the inductors are yet incompatible with low cost digital CMOS process. Trade-offs have to be made with regard to fitting an inductor in a sea of package bumps with degraded performance, versus depopulating bumps around the inductor that may cause non-uniformity on packaging stress. Power grid discontinuity, routing congestion, and metal density non-uniformity are of other concerns since metal is blocked below and near inductors. Figure 4.7.4 illustrates the PCIe/QPI IO clock system. The LC-VCO is AC-coupled to two pseudo-differential CMOS buffers to drive a CMOS by-2 (or by-4) divider with 4-phase (or IQ phase) outputs needed for data recovery at ½ or ¼ of the VCO rate. The IQ clocks are distributed differentially using 2 stages of AC-coupled current-mode-logic (CML) buffers to drive 5mm on-die passive transmission lines [4]. All transceiver lanes are tapped to the T-lines. It is noted that precision resistors are not available in this process. NWELL based resistive loads on CML logic circuits have to be calibrated against an external 50Ω resistor. The bandwidth of the NWELL resistors is limited by parasitics. It is a challenge to design the post VCO buffer and IQ generators in CML logic to operate near and beyond 10GHz. The circular ring structures in Fig. 4.7.4 and Fig. 4.7.5 offer a solution to 4-phase (by-2 divider) and 8-phase (by-4 divider) generation. The back-to-back connected inverters are used both for duty-cycle correction, but also prevent the ring from entering a "dead-lock" state. Figure 4.7.5 presents the state transition table of the ring emerging from any initial state and resolving to a desired phase relationship. The simplicity of this circuit allows it to operate at speed higher than any other non-regenerative divider structure, in addition to providing IQ phase and 50% duty cycle.

Figure 4.7.6 shows the measured phase noise of the LC PLL, at the aforementioned by-2 divider and by-4 divider outputs, with a jitter of 1ps_{rms} integrated from 100Hz to 1GHz. Long term rms jitter for ring-oscillator based PLLs is less than 2ps, measured through a shared debug port. Time-resolved emission (TRE) is used for clock skew probing. A maximum skew of 14ps is obtained for a group of representative clock nodes in the cache area. The maximum compensator offset from its nominal setting is 2-LSB (1 LSB = 5ps) collected from delay compensators in the cache area.

Acknowledgment:
The authors thank Ernest Knoll, Yair Talker, Stephane Smith, Rohit Mittal, Luke Tong, and ST Chen for their contributions, Simon Tam, Stefan Rusu, and Fulvio Spagna for valuable inputs, and Abhi Kolla for managerial support.

References:
[1] Stefan Rusu, Simon Tam, et al, "A 45nm 8-core Enterprise Xeon Processor", *ISSCC Dig. Tech. Papers*, pp. 56-57, Feb. 2009.
[2] Nasser Kurd, et al, "Westmere: A family of 32nm IA Processors", *ISSCC Dig. Tech. Papers*, pp. 96-97, Feb 2010.
[3] Marcelo Yuffe, Ernest Knoll, et al, "A Fully Integrated Multi-CPU, GPU and Memory Controller 32nm Processor", *ISSCC Dig. Tech. Papers*, Feb. 2011.
[4] G. Balamurugan, et al, "A Scalable 5–15 Gbps, 14–75 mW Low-Power I/O Transceiver in 65 nm CMOS", *IEEE J. Solid State Circuits*, Vol. 43, No. 4, pp 1010-1018, Apr. 2008.

978-1-61284-303-2/11 $26.00 © 2011 IEEE

ISSCC 2011 / February 21, 2011 / 4:15 PM

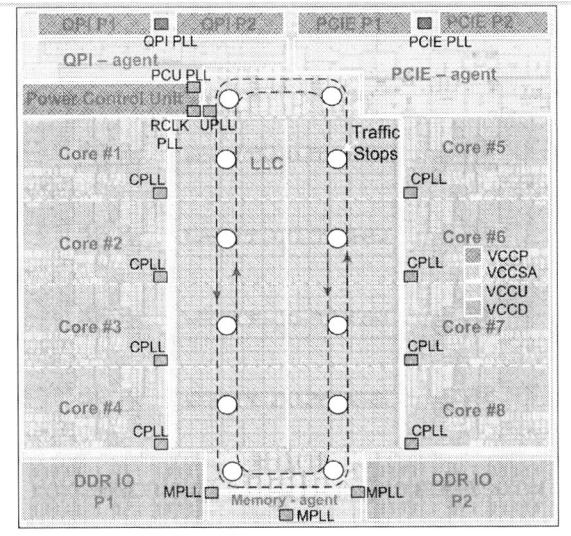

Figure 4.7.1: Die micrograph of the processor.

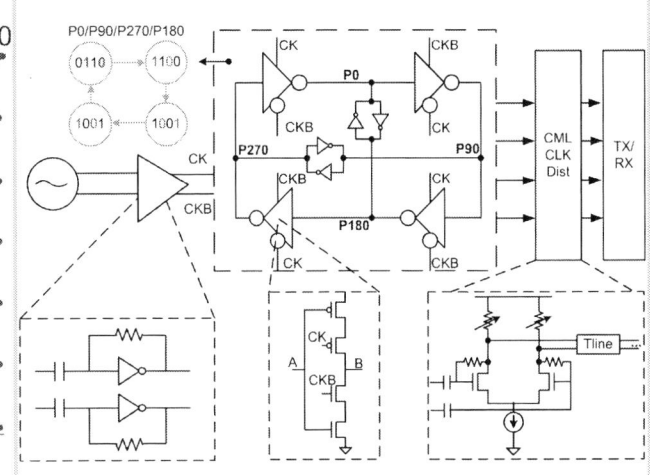

Figure 4.7.2: PLL and clock distribution architecture.

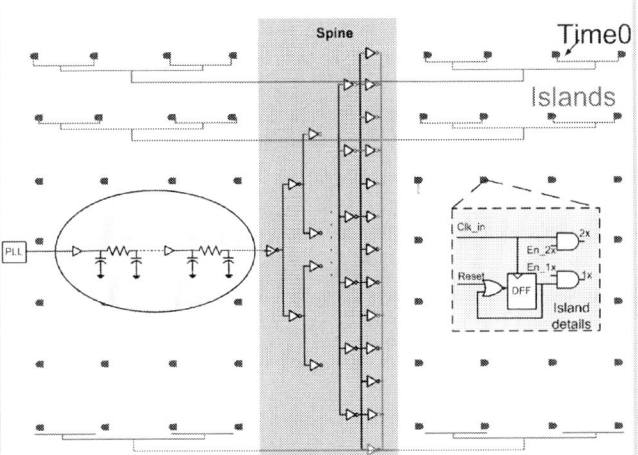

Figure 4.7.3: Islands based clock distribution.

Figure 4.7.4: PCIe/QPI IO clock generation and distribution.

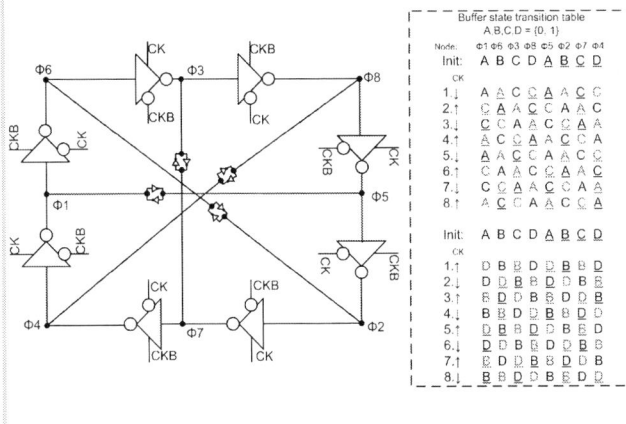

Figure 4.7.5: A by-4 divider with 8 phase generation.

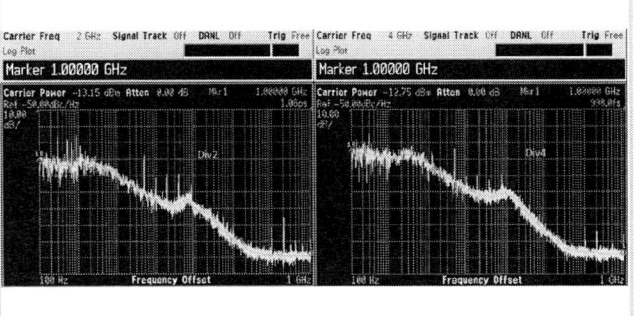

Figure 4.7.6: LC PLL Phase noise measurement at by-2, by-4 divider outputs.

978-1-61284-303-2/11 $26.00 © 2011 IEEE

Figure 4.7.7: Die Photo.

ISSCC 2011 / SESSION 4 / ENTERPRISE PROCESSORS & COMPONENTS / 4.8

4.8 A 32nm 3.1 Billion Transistor 12-Wide-Issue Itanium® Processor for Mission-Critical Servers

Reid J. Riedlinger[1], Rohit Bhatia[1], Larry Biro[2], Bill Bowhill[2], Eric Fetzer[1], Paul Gronowski[2], Tom Grutkowski[1]

[1]Intel, Fort Collins, CO,
[2]Intel, Hudson, MA

The next generation in the Intel® Itanium® processor family, code named Poulson, has eight multi-threaded 64 bit cores. Poulson is socket compatible with the current Intel® Itanium® Processor 9300 series (Tukwila) [1]. The new design integrates a ring-based system interface derived from portions of previous Xeon® and Itanium® processors, and includes 32MB of Last Level Cache (LLC). The processor is designed in Intel®'s 32nm CMOS technology utilizing high-K dielectric metal gate transistors [2] combined with nine layers of copper interconnect. The 18.2×29.9mm² die contains 3.1 billion transistors, with 720 million allocated to the eight cores (Fig. 4.8.1). A total of 54MB of on die cache is distributed throughout the core and system interface. Poulson implements twice as many cores as Tukwila while lowering the thermal design power (TDP) by 15W to 170W and increases the top frequency of the I/O and memory interfaces by 50% to 6.4GT/s.

The design introduces a new core micro-architecture and floor plan that significantly improves frequency, and power efficiency. The core implements an 11-stage in-order, decoupled frontend and backend pipeline which employs replay and flush mechanisms versus the previous global stall micro-architecture. The decoupled pipelines enable an increase in resource utilization and throughput. The frontend pipeline fetches six instructions per cycle while the backend executes and retires up to twelve instructions per cycle. The backend execution resources include six ALUs, two integer units, two floating-point units, two memory units and three branch units distributed across twelve ports. A 96-entry distributed instruction buffer, replicated for each thread, decouples the frontend and backend pipelines while storing replay information. To further reduce data access penalties the core implements new features including: a hardware data prefetcher, data access hints and improved concurrent TLB accesses. The core improves performance through fine grain multi-threading, replicated instruction buffers and D-side TLBs, a 4 cycle integer multiplier, and an additional 32 entries to the integer register file. A three-level cache hierarchy supports these parallel execution resources with the first level single cycle 16K Instruction (I) and Data (D) cache that is backed by two second level caches a nine cycle 512K I cache and an eight cycle 256K D cache.

The core floor plan is optimized around the performance-sensitive single-cycle integer execution and first-level data cache (Fig. 4.8.2). Timely delivery of virtual addresses to the first level TLB require placing it within the IEU data path. Way muxing and data rotation circuitry is positioned directly below the integer vertical centerline to speed data return from the FLD to the IEU. The floor plan optimizations helped enable a 25% reduction in cycle time for the pre-silicon design target over the previous generation core.

To further improve Vmin operation the pre-silicon frequency analysis is primarily done at low voltage with heuristic runs at high voltage. Vmin sensitive circuits, including pulse latches, dynamic logic and NFET-only latches are avoided to enable robust low-voltage operation and lower power. Register files (RF) use fully interrupted feedback cells for low-voltage writes and static global bitlines where possible. (Fig. 4.8.3) A full-scan methodology supports an ATPG scan-based testing flow that enables excellent manufacturing test coverage. All RFs contain a local direct-access testing port. A fine-grain clock vernier system facilitates speed path debug through the insertion of 45000 clock skew adjustment points. The clock vernier [3] circuit has separate edge controls enabling both duty cycle adjustment and insertion delay modification to improve frequency or robustness without doing a silicon stepping. (Fig. 4.8.4)

Techniques such as eliminating domino logic in large data-paths, replacing architectural stalls with replays, eliminating glitches, and increasing dynamic clock gating efficiency to over 85% effectively reduce dynamic power in the core

by an additional 60% beyond the technology scaling (Fig. 4.8.5). Post-timing analysis circuit downsizing maximizes the use of lower-leakage devices in the core (>81%) and uncore (76%), reducing overall leakage by 30%.

An improved digital activity sensor includes the ability to monitor the >30% dynamic power consumed in data patterns and reacts to power events in under 1µs. This system monitors 1834 architectural and data events to predict core power consumption. To monitor the power supply in real time, the design features an on-die droop measurement [4] with the ability to introduce controlled droop events to improve debug. The power supplies are divided into four individually regulated core pairs with additional supplies for the system interface/large caches and the I/O subsystem. The power on Poulson is distributed as follows: ~55% across the 8 cores, ~35% in the uncore and ~10% consumed in the I/O. Core pairs can be turned off for core defeatured configurations and for power savings. Individual regulation of the core pairs allows for optimization of frequency by compensating for within-die technology variations with voltage adjustments unique to each core pair. Core power supplies are bypassed by an embedded array capacitance in the central layers of the package to improve di/dt-induced noise.

The system interconnect is provided by an integrated 10-port router that connects to external IO and processors via four full-width and two half-wide point-to-point 6.4GT/s Quickpath™ (QPI) link interfaces. The ring-based system interface provides a theoretical peak bandwidth of 700GB/s (Fig. 4.8.6). The chip includes two integrated memory controllers each supporting two scalable memory interconnect (SMI) links operating in lockstep. These four SMI ports provide a 6.4GT/s connection to up to 512GB memory per socket. The SMI and QPI links provide reliability, availability, and scalability (RAS) support for features such as clock and lane failover.

In order to enable mission critical servers and the associated high RAS requirements several improvements were made to the design. These improvements include:
- Large cache arrays covered by ECC including the large L3 utilizing DECTED and protecting the MLI/MLD with inline SECDED.
- Extensive parity protection and parity interleaving on nearly all RFs.
- End-to-end parity protection with recovery-support on all critical internal buses and data paths including the ring.
- Residue protection on key logic and execution data paths.
- The adoption of radiation-hardened (RAD) sequential latching elements for vulnerable architectural and state.

The adoption of these techniques requires design methodology and automation enhancements to ensure the RAS goals are achieved without dramatically increasing die area and power. The random logic synthesis (RLS) flows optimally select hardened sequential usage based upon attack vulnerability and timing criticality. Additionally, a new error micro-architecture combines uniform logging and configuration with key hardware/firmware hooks enabling increased availability and serviceability. These techniques enabled 2× the cores and 1.5× the transistors with a lower overall susceptibility to radiation induced error events as compared to the previous generations.

Acknowledgements:
The authors thank the entire design teams from Fort Collins, CO and Hudson, MA.

References:
[1] Stackhouse, B.; et. al.; , "A 65 nm 2-Billion Transistor Quad-Core Itanium® Processor," *IEEE J. Solid-State Circuits*, 2009.
[2] Packan, P; et. al.; , "High Performance 32nm Logic Technology Featuring 2ʳᵈ Generation High-k + Metal Gate Transistors," *IEDM 2009*,
[3] Mahoney, P.; Fetzer, E.; et al., "Clock distribution on a dual-core, multi-threaded Itanium®-family processor," *ISSCC Dig. Tech. Papers*, Feb. 2005.
[4] Petersen, R.; et. al.; , "Voltage transient detection and induction for debug and test," *Test Conference*, pp.1-10, Nov. 2009.

978-1-61284-303-2/11 $26.00 © 2011 IEEE

ISSCC 2011 / February 21, 2011 / 4:45 PM

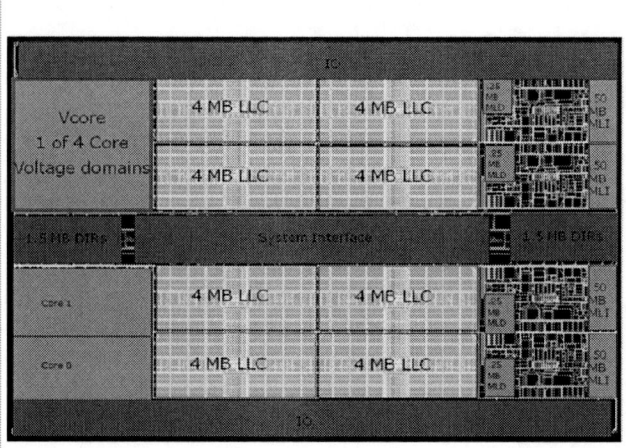

Figure 4.8.1: Poulson Processor Block Diagram.

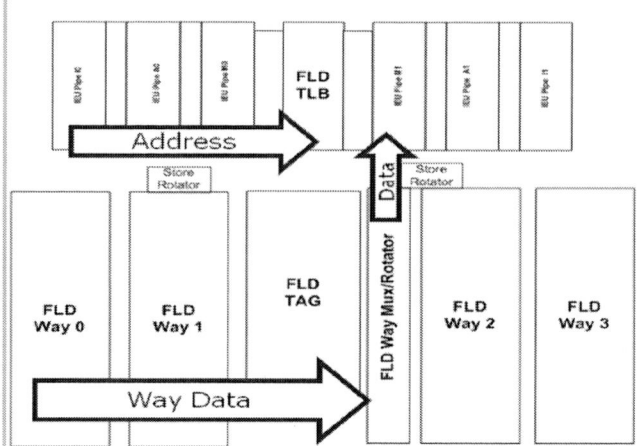

Figure 4.8.2: EXE/FLD datapath integration.

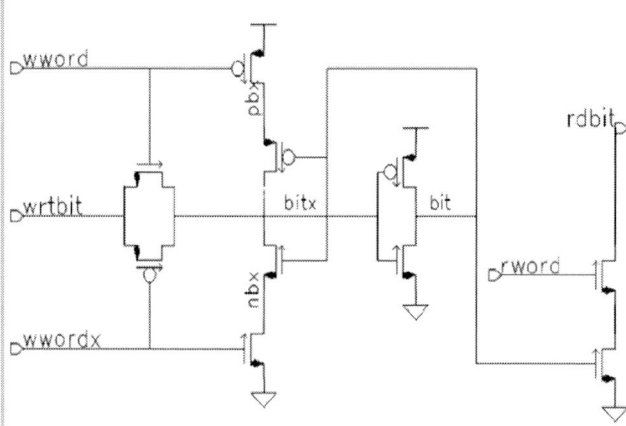

Figure 4.8.3: Fully interrupted RF bit cell.

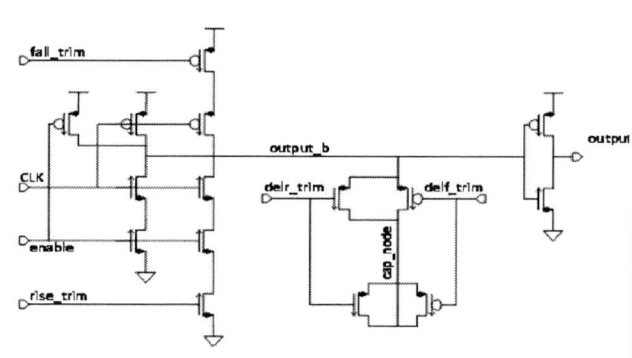

Figure 4.8.4: Clock Vernier Circuit.

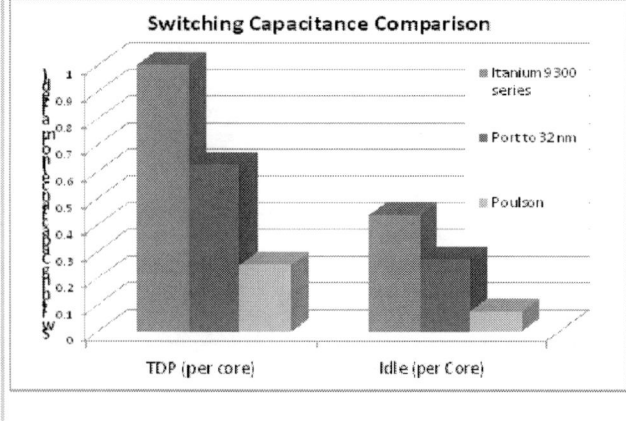

Figure 4.8.5: Power Reduction in Poulson Core.

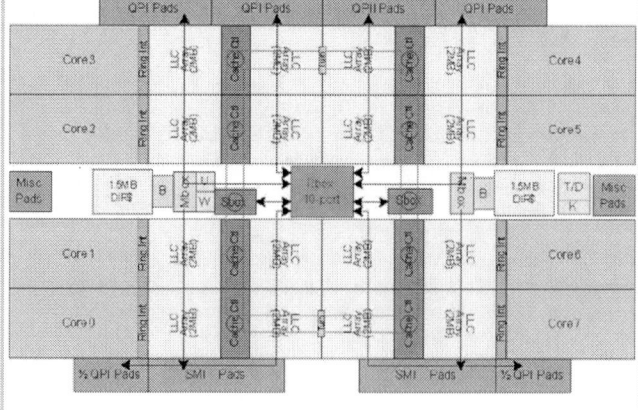

Figure 4.8.6: Poulson Block Diagram.

978-1-61284-303-2/11 $26.00 © 2011 IEEE

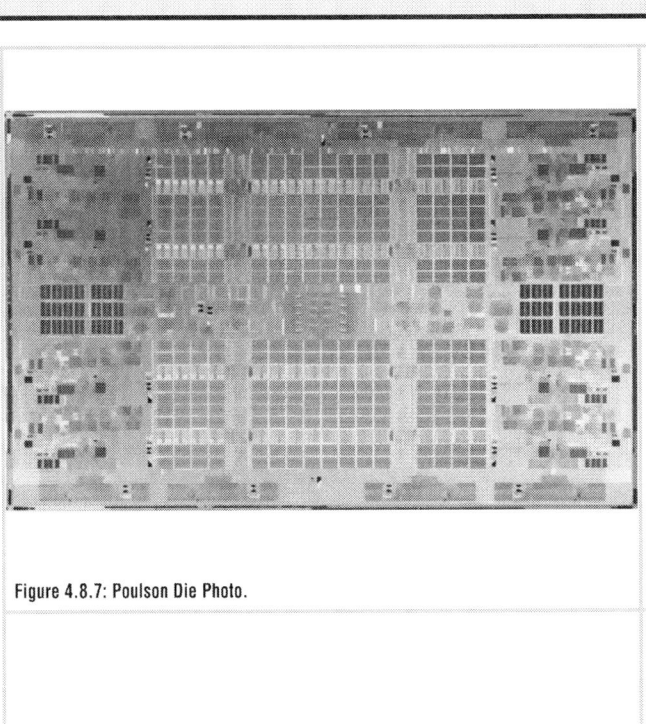

Figure 4.8.7: Poulson Die Photo.

ISSCC 2011 / SESSION 5 / PLLs / OVERVIEW

Session 5 Overview / Analog

PLLs

Session Chair: *Ivan Bietti, STMicroelectronics, Grenoble, France*

Session Co-Chair: *Tsung-Hsien Lin, National Taiwan University, Taiwan*

Frequency Synthesizers and clock generators are essential building blocks in almost all modern electronic systems. The Phase-locked loop (PLL) is the most suitable circuit architecture to perform the extremely diverse tasks required by the very different applications. In wireless transceivers, PLLs are used to generate high-frequency local oscillator signals with extremely low phase noise for up-conversion and down-conversion of the transmitted and received signals. For high-speed data communications very low-jitter clock signals are required. When used as clock generators for large processors, wide frequency range, fast locking time and very-low-power quiescent dissipation are absolutely mandatory. These tough specifications need to be achieved at the lowest possible power, both for the stringent requirements of portable systems, but also to reduce the issues associated with heat dissipation.

The relentless scaling of CMOS technology and the associated increase in computational power have allowed the introduction of sophisticated digital techniques to overcome the analog limitations of the scaled devices. This trend continues and has made the performance of the partly and all-digital fractional PLLs comparable to those achieved by the integer analog ones in terms of phase-noise and jitter. On the other hand, the usage of digital techniques gives superior flexibility and portability, lower sensitivity to the analog parameters and allows easier supply scaling. Digital techniques are also effective for adaptive supply noise cancellation, which is vital to maintain the PLLs' high performance in a real, noisy environment.

Paper 5.1 (Politecnico of Milan) presents a new architecture for an ADPLL which represents a step forward in the design of these kind of synthesizers. Thanks to a 10b digitally compensated and finely programmable delay line, the time-to-digital converter (TDC) required by a classical ADPLL for accurate phase error detection is eliminated and replaced by a simple bang-bang phase detector. The combination of the achieved jitter and power results is the best ever presented for an ADPLL: jitter is lower than $0.56ps_{rms}$ and the circuit burns less than 5mW bringing to FoM very close to -240dB.

Paper 5.2 (MediaTek) uses injection locking on a ring-oscillator-based PLL. The problem of aligning the phase of the injected reference clock to that resulting from the PLL loop is elegantly solved by driving both the VCO S/H used for phase comparison and the injecting circuit with the same exact signal. The PLL, integrated in 65nm CMOS, achieves an overall jitter integrated from 1kHz to 40MHz of 2.4ps while the PLL runs at 432MHz and consumes 6.9mW.

Paper 5.3 (Oregon State University) introduces an adaptive and quite efficient way to cancel the effect of the supply noise on the PLL jitter. This is particularly important for ring-oscillator-based PLLs since the supply sensitivity of this VCO is quite high. It is based on a background digital calibration using a test signal injected on the VCO supply. The effectiveness of the technique works over a wide range of frequencies (0.4 to 3GHz) and PVT variations. The achieved results are impressive: the effect of $30mV_{pp}$ supply noise is completely removed and the PLL performances are equivalent to those with a clean supply.

Paper 5.4 (POSTECH) implements a jitter reduction technique for the quantization noise of a fractional-N PLL, effective both in and out of band. Multiple phases of a ring oscillator are used in combination with a phase interpolator to reduce the quantization noise while an embedded 8-tap FIR is used to digitally filter the $\Delta\Sigma$ noise in front of the Phase Frequency detector. The phase noise reduction is 34dB, resulting in an in-band noise of -106dBc/Hz at 100kHz and an out of band noise of -107.5dBc/Hz at 6MHz.

978-1-61284-303-2/11 $26.00 © 2011 IEEE

ISSCC 2011 / February 21, 2011 / 1:30 PM

Paper 5.5 (Intel) presents a host-clock generator in 32nm CMOS. Several techniques are used to keep the design flexible, portable and process insensitive. Furthermore the main PLL parameters are automatically programmed according to the input/output frequency change to maintain unconditional loop stability. This is vital for Host-clock generators since the frequency working range is wide: 300MHz to 1.5GHz in this paper. The prototype is realized in 32nm CMOS. The current draw varies from 0.4 to 1.15mA, properly tracking the output frequency, from a 1V supply.

Paper 5.6 (Toshiba) instead addresses the noise reduction problem by maximizing the loop bandwidth while keeping control of the reference spur. This is done using a high CMRR differential amplifier placed in front of a differentially controlled oscillator. The resulting jitter for the 65nm PLL prototype is 570fs$_{rms}$ while the reference spur is better than -40dBc even for a reference to bandwidth ratio of 10.

Finally Paper 5.7 (Panasonic Wireless Research Laboratories) builds an ADPLL around a rotary-travelling-wave oscillator using its 32 phases, accurately spaced by 3.9ps. This technique does not require the period normalization circuits as in LC-tank based PLLs while keeping very good phase noise performance. The prototype, realized in 65nm CMOS, achieves -108dBc/Hz in-band phase noise for a 1MHz loop BW.

Presenters:

1:30 PM

5.1 A 2.9-to-4.0GHz Fractional-N Digital PLL with Bang-Bang Phase Detector and 560fs$_{rms}$ Integrated Jitter at 4.5mW Power

D. Tasca, Politecnico di Milano, Milan, Italy

2:00 PM

5.2 An Injection-Locked Ring PLL with Self-Aligned Injection Window

C-F. Liang, MediaTek, Hsinchu, Taiwan

2:30 PM

5.3 A 0.4-to-3GHz Digital PLL with Supply-Noise Cancellation Using Deterministic Background Calibration

A. Elshazly, Oregon State University, Corvallis, OR

3:15 PM

5.4 A 0.1-f$_{ref}$ BW 1GHz Fractional-N PLL with FIR-Embedded Phase-Interpolator-Based Noise Filtering

D-W. Jee, Pohang University of Science and Technology, Pohang, Korea

3:45 PM

5.5 A Scalable sub-1.2mW 300MHz-to-1.5GHz Host-Clock PLL for System-on-Chip in 32nm CMOS

H-J. Lee, Intel, Hillsboro, OR

4:15 PM

5.6 A 570fs$_{rms}$ Integrated-Jitter Ring-VCO-Based 1.21GHz PLL with Hybrid Loop

A. Sai, Toshiba, Kawasaki, Japan

4:45 PM

5.7 A Rotary-Traveling-Wave-Oscillator-Based All-Digital PLL with a 32-Phase Embedded Phase-to-Digital Converter in 65nm CMOS

K. Takinami, Panasonic, Cupertino, CA

978-1-61284-303-2/11 $26.00 © 2011 IEEE

ISSCC 2011 / SESSION 5 / PLLs / 5.1

5.1 A 2.9-to-4.0GHz Fractional-N Digital PLL with Bang-Bang Phase Detector and 560fs$_{rms}$ Integrated Jitter at 4.5mW Power

Davide Tasca, Marco Zanuso, Giovanni Marzin, Salvatore Levantino, Carlo Samori, Andrea L. Lacaita

Politecnico di Milano, Milan, Italy

State-of-the-art digital fractional-N PLLs intended for modern wireless systems make use of high-resolution and high-linearity time-to-digital converters (TDCs) in order to meet the stringent integral phase noise requirements [1, 2]. Those high-performance TDCs complicate the synthesizer design and dissipate large part of the power budget, leading to poor jitter-power compromise. This paper introduces a fractional-N PLL based on a 1b TDC, achieving jitter of 560fs$_{rms}$ (from 3kHz to 30MHz) at 4.5mW power consumption, even in the worst-case of fractional spur falling within the PLL bandwidth. The circuit synthesizes frequencies between 2.92 and 4.05GHz with 70Hz resolution.

Figure 5.1.1 shows the PLL block diagram, where the standard-cell-based digital blocks are filled in gray. The TDC is a single D-flip-flop, which in the following will be referred to as the bang-bang phase detector (PD). Bang-bang PDs are never used in $\Delta\Sigma$ fractional-N PLLs for wireless systems, because they act as hard limiters on the timing error between reference and divider output and their nonlinearity would be responsible for the huge undesirable spurs and noise at the output. In an attempt to resolve this issue, the synthesizer in [3] includes the bang-bang PD in a loop with the $\Delta\Sigma$ modulator, but this results into a type-I PLL with poor phase noise. Moreover, the coarse quantization of the time error provided by a bang-bang PD does not allow to cancel the $\Delta\Sigma$ noise at the TDC output as in [2] or [4]. In order to design a low-noise bang-bang fractional-N PLL, we have implemented the divider as the cascade of an integer-N divider and a 10b controllable delay stage. In this way, the deterministic quantization error of the $\Delta\Sigma$ modulator is reduced below random noise. Theoretical analysis shows that if random noise at bang-bang PD input dominates, the average of the PD output is linearly related to the timing error [5]. This property both avoids noise folding and in-band spectral regrowth, which are typical properties of nonlinear detectors, and allows applying the adaptive cancellation techniques of quantization noise, similar to the ones presented in [4]. The undesirable dependence of the bang-bang PD gain on input jitter is eliminated by an automatic bandwidth regulation system, which sets the loop-bandwidth-to-reference-frequency ratio independent of practical PVT variations of jitter and other analog parameters in the loop.

Figure 5.1.2 shows a time model of the fractional-N divider, illustrating its operating principle. Assuming a frequency control word FCW=N+2$^{-n_f}$ (with integer numbers N and n_f), the output of a 1st order $\Delta\Sigma$ modulator c[k] switches between N and N+1. The integer divider, driven by c[k], is modeled as an integrator with gain equal to T_{dco} (i.e., the DCO period). In the proposed fractional divider, the $\Delta\Sigma$ quantization error q[k] is converted into delay by the 10b controllable delay stage, which is modeled as a block with gain t_g. The correct unwrapped ramp at the output is obtained if t_g is equal to T_{dco}. The unavoidable analog inaccuracies of t_g are corrected by multiplying q[k] by $\hat{g} = T_{dco}/t_g$. The digital correlator in Fig. 5.1.1 estimates the correct gain $\hat{g}$ in the background. The proposed delay stage allows reducing the peak-to-peak time quantization error from T_{dco} to $2^{-10}t_g$. Taking into account the longest T_{dco} of 345ps at 2.9GHz, a 10b controllable delay is sufficient to get a resolution of about 340fs. Compared to conventional digital PLLs, a high-resolution controllable delay, i.e., a digital/time converter (DTC), is used in place of a high-resolution TDC. This choice leads to easier design and better jitter-power compromise.

A low-power 10b controllable delay with sufficient linearity is realized by segmenting its conversion characteristic, as shown in Fig. 5.1.3. The MSB a_0[k] of the delay control selects either phase P0 or P1, which are time shifted by T_{dco}/2. P0 and P1 are obtained by consecutively sampling the integer divider output with the two 0°/180° outputs of the differential DCO. Before sampling, in order to prevent metastability, an automatic control loop adds a regulated delay to the divider output. The remaining bits of the delay control, divided into two words a_1[k] and a_2[k], switch two scaled banks of nMOS load capacitors of the MUX stage. The additional 3 bits allow overlapping the sub-characteristics even in the

presence of PVT variations. Automatic adaptation of the sub-characteristic gain is performed by the correlation algorithm shown in Fig. 5.1.1. The source-coupled logic circuit used in the delay stage reduces supply sensitivity, while the dynamic logic circuit in the integer divider saves power.

The mismatch between P0 and P1, i.e., the delay error Δt with respect to T_{dco}/2, is the main limit of quantization noise cancellation, in this scheme. The digital correlation algorithm in Fig. 5.1.3 corrects this impairment operating in the background. When a_0[k] is 1 and Mux selects P1, the error Δt is added into the loop and detected by the PD. Correlating the PD output e[k] with a_0[k], a signal is obtained which is proportional to Δt. By multiplying by a_0[k] and subtracting the result from the delay control, Δt is cancelled out. Since a_0[k] is a single bit, the multipliers included in the corrector are implemented by digital multiplexers. A similar correlation algorithm (not shown in Fig. 5.1.3) corrects for mismatches in the coarse capacitor bank of the delay stage.

The 14b segmented DCO has 4 tuning sub-characteristics. Three scaled banks of capacitors, which are switched by a background frequency control loop, realize the coarsest characteristics. A 5b DAC controlling a MOS varactor implements the finest one. The synthesizer has been fully integrated in 65nm CMOS and it occupies a core area of about 0.22mm^2 (a die micrograph is shown in Fig. 5.1.7). The measured phase noise spectrum is shown in Fig. 5.1.4, in the worst-case of fractional spur falling at 100kHz offset (i.e., channel at 100kHz offset from the integer one). The rms jitter (from 3kHz to 30MHz), including random noise and in-band spurs, is about 560fs$_{rms}$. The 312kHz (i.e., f_{ref}/128) bandwidth optimizes the random noise component of jitter. Measurements have been repeated, with no off-chip regulation of loop parameters, over the fractional synthesized channels from 2962 to 4042MHz in steps of 40MHz and from 3520MHz (integer channel) to 3540MHz with logarithmic frequency spacing (see Fig. 5.1.5). For fractional channels causing only out-of-band spurs, RMS jitter reduces to 420fs$_{rms}$. Reference spur is below −72dBc in all cases. The core power consumption (excluding pad driver and crystal oscillator) is 4.5mW, and the majority of the power is used in the controllable delay (49%) and the DCO (38%). This leads to a figure of merit (FoM), defined as the product of the jitter variance by the dissipated power in mW [6], of −238.3dB, in the worst case. Figure 5.1.6 compares the measured performance of the proposed circuit with other low-jitter digital and analog fractional-N PLLs. Thanks to the elimination of the high-resolution TDC, this PLL architecture achieves good jitter-power compromise over wide tuning range. The integer-N implementation with the best FoM [6] is reported on the same plot as benchmark.

Acknowledgment:
The authors wish to thank S. Pellerano and the Radio Integration Research Lab of Intel Corp. for sponsoring this research, M. Zuffada and E. Temporiti of STMicroelectronics for supporting chip fabrication, and D. Canavesi, A. Fenaroli, F. Pepe and S. Vilardi for their help in the design.

References:
[1] C. Weltin-Wu, E. Temporiti, D. Baldi, M. Cusmai, and F. Svelto, "A 3.5GHz Wideband ADPLL with Fractional Spur Suppression through TDC dithering and Feedforward Compensation," *ISSCC Dig. Tech. Papers*, pp. 468-469, Feb., 2010.
[2] J. Borremans, K. Vengattaramane, V. Giannini, and J. Craninckx, "A 86MHz-to-12GHz Digital-Intensive Phase-Modulated Fractional-N PLL Using a 15pJ/Shot 5ps TDC in 40nm digital CMOS," *ISSCC Dig. Tech. Papers*, pp. 480-481, Feb., 2010.
[3] M. Ferris and M. Flynn, "A 14mW Fractional-N PLL Modulator with an Enhanced Digital Phase Detector and Frequency Switching Scheme," *ISSCC Dig. Tech. Papers*, pp. 352-353, Feb., 2007.
[4] M. Zanuso, S. Levantino, C. Samori, A. L. Lacaita, "A 3MHz-BW 3.6GHz Digital Fractional-N PLL with Sub-gate-delay TDC, Phase-interpolation Divider, and Digital Mismatch Cancellation," *ISSCC Dig. Tech. Papers*, pp. 476-477, Feb., 2010.
[5] N. Da Dalt, "Linearized Analysis of a Digital Bang-Bang PLL and Its Validity Limits Applied to Jitter Transfer and Jitter Generation," *IEEE Trans. on Circuits and Systems–I*, vol. 55, no. 11, pp. 3663-3675, Dec., 2008.
[6] X. Gao, E.A.M. Klumperink, G. Socci, M. Bohsali, and B. Nauta, "A 2.2GHz Sub-Sampling PLL with 0.16ps$_{rms}$ Jitter and −125dBc/Hz In-Band Phase Noise at 700µW Loop-Components Power," *IEEE Symp. on VLSI Circuits*, pp. 139-140, June, 2010.

978-1-61284-303-2/11 $26.00 © 2011 IEEE

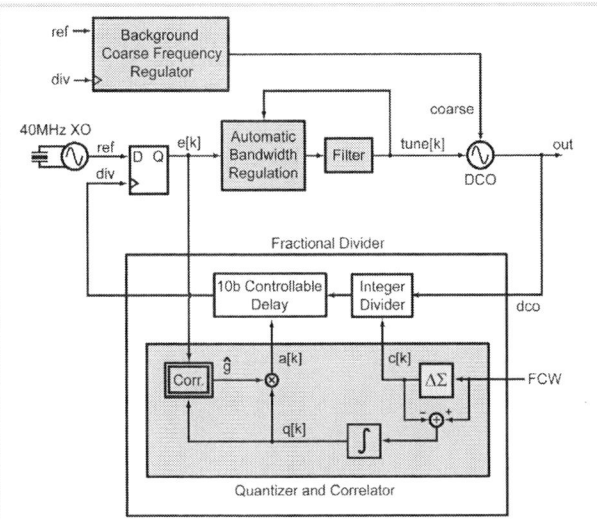

Figure 5.1.1: Digital fractional-N bang-bang PLL block diagram.

Figure 5.1.2: Fractional divider: equivalent time model and principle of operation.

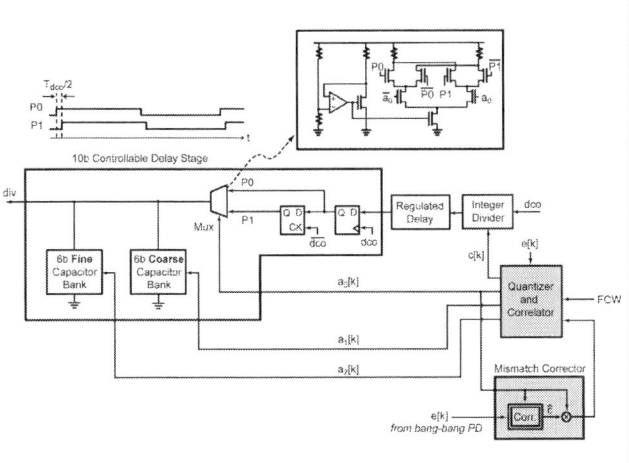

Figure 5.1.3: Fractional divider implementation.

Figure 5.1.4: Measured phase noise (worst case over tuning range).

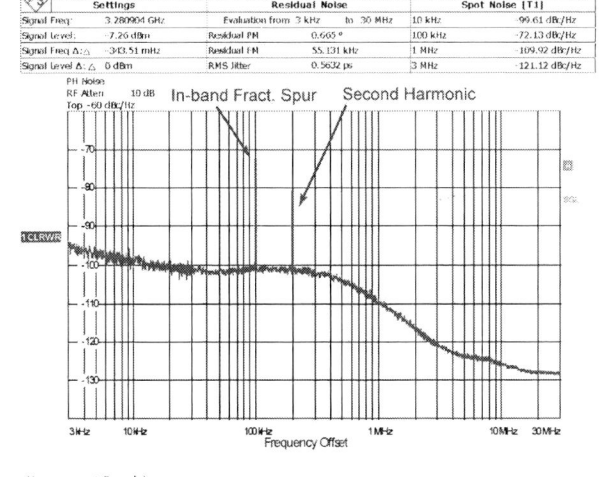

Figure 5.1.5: Measured jitter over the fractional synthesized channels.

Figure 5.1.6: Performance summary and comparison with other low-jitter PLLs.

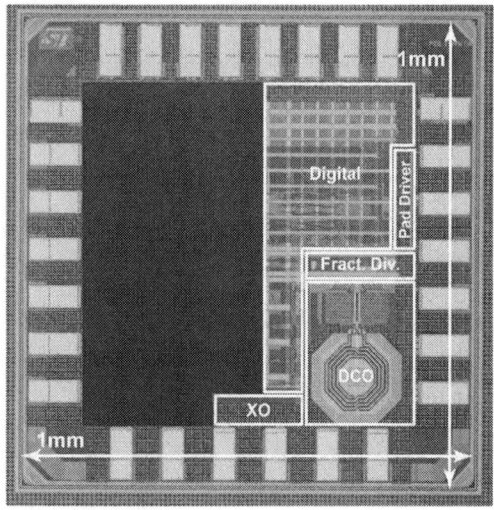

Figure 5.1.7: Die micrograph.

ISSCC 2011 / SESSION 5 / PLLs / 5.2

5.2 An Injection-Locked Ring PLL with Self-Aligned Injection Window

Che-Fu Liang, Keng-Jan Hsiao

MediaTek, Hsinchu, Taiwan

In modern analog front-ends, there is an increasing demand on high- performance analog-to-digital converters (ADCs), which require high sampling frequency and low-jitter sampling clock. This makes low-jitter phase-locked loops (PLLs) with jitter on the order of few picoseconds desirable. Unfortunately, due to stringent limit on die area, sometimes a PLL with a ring oscillator is the only choice. To get better phase noise, a wider loop bandwidth is needed to suppress the noise of the voltage-controlled oscillator (VCO). However, due to the discrete-time nature of the operations, the loop bandwidth is limited to one-tenth of the crystal oscillator (XTAL) frequency. One way to solve this problem is to use the injection-locking technique. This method exploits the clean reference clock but has several production problems. One is the frequency offset between injection signal and VCO, and this can be solved by using the injection-locked PLL architecture [1, 2]. However, in [1, 2] extra loops are still needed to adjust the injection window due to on-chip variations. In this work, an injection-locked ring PLL (ILRPLL) architecture is proposed to solve this problem. Using the concept of sub-sampling PLLs [3], the injection window is aligned automatically without feedback adjustment. A 432MHz ILRPLL is realized in ATV/DTV system to justify this technique.

The proposed ILRPLL is shown in Fig. 5.2.1. This technique is based on a traditional type-2 ring PLL. A 4-stage ring oscillator is used as the VCO core, and 3 additional transistors with 2 capacitors are added to realize the injection-locked VCO (ILVCO) with sample-and-hold (S/H) circuits, which will be discussed later. In the steady state, the ILVCO itself generates 2 differential phase error signals, SHP and SHN, to voltage-to-current convertor (VI) so as to adjust the voltage on loop filter and to make the loop locked. A pulse generator [3] is also needed to generate a 27MHz pulse train with pulse width of 50ps, which will be used in the ILVCO. Note that the traditional phase-frequency detector (PFD), charge pump (CP), and divider are still added in the auxiliary loop to set up the initial VCO frequency. If necessary, this ILRPLL can be configured as traditional charge-pump PLL for normal use.

To suppress the VCO noise, an injection-locked methodology in [4] is adopted. Compared to other injection schemes, this method minimizes the disturbance on the VCO and keeps the circuits simple because only one NMOS switch is needed. The circuit schematic and timing diagram are shown in Fig. 5.2.2. Designed with a traditional 4-inverter VCO delay cell, the injection pulse may correct the phase of VCO no matter whether the VCO is phase leading or lagging. Assuming the phase errors of VCO before and after injection are ϕ and ϕ', respectively, we define the injection strength β:

$$\beta = (\phi - \phi') / \phi \le 1 \qquad (1)$$

With infinite injection energy, ϕ' is eliminated totally and maximal β is achieved. Though this injection scheme is useful, it is still problematic when it is implemented in a PLL rather than a free-running VCO. In an ILRPLL, both phase-locking and injection-locking mechanisms are introduced to suppress the noise without suffering from the frequency offset between the ILVCO and the harmonics of the injection signal. However, the injection timing is important to make it a constructive disturbance rather than a destructive one. To solve this problem, two S/H circuits are added in the ILVCO to sample the phase of the VCO and serve as a phase detector (PD). As illustrated in Fig. 5.2.3, besides the injection transistor M0, two transistors M1 and M2 with two capacitors C1 and C2 form S/H circuits [3]. Because now both the injection-locked and phase-locked mechanisms are controlled by the same injection pulse, the injection timing is aligned automatically and no additional loop is required.

The S/H circuits in ILVCO generate two signals, SHP and SHN, to represent the phase difference. Then one VI is implemented to convert the voltage-domain phase error signals to current domain ones. The structure of the VI circuit is shown in Fig. 5.2.4. It takes SHP and SHN as inputs and generates a single-ended current output. A pulser is needed to reduce gain appropriately and maintain the loop stability, which will be discussed further. In order to reduce the effect of on-chip variations, the bias currents in pulser and VI are mirrored so as to track each other. According to simulation results, the combinational gain of S/H and VI remains within ±20% around corners. Note that the additional noise on VI will be suppressed significantly due to the high gain [3], which makes the in-band phase noise better.

With the aid of the feed-forward injection path, the phase noise of ILVCO below 5MHz is substantially suppressed. However, the injection-locked technique suppresses the VCO noise with 1-order high-pass transfer function [1] and this is not good enough for VCOs in deep-submicron processes, which have strong flicker noise. Hence, the phase-locked mechanism may provide extra suppression for VCO in-band noise. This will be done by the loop formed by S/H, VI, and loop filter in the proposed circuit. In this work, the loop bandwidth is less than 2MHz, which is a little smaller than the traditional one. Note that the loop is still a discrete time system and the maximum bandwidth is limited. In injection-locked PLLs, the feed-forward injection path and the feedback loop respond to the phase error simultaneously so the maximum stable bandwidth is actually reduced. Modifying the analysis done in [5] and assuming a first-order LPF for simplicity, the maximal natural frequency of the loop (F_{nmax}) can be calculated as:

$$F_{nmax} \cdot \pi < [(1+\xi^2 - 0.5 \cdot \beta)^{\frac{1}{2}} - \xi] F_{ref} \qquad (2)$$

where F_{ref} means the XTAL frequency and ξ is the damping factor of the loop. From eqs. (1) and (2), we know that a large β results in a lower F_{nmax}. In this work, the F_{nmax} is designed to be 2.1MHz so the loop natural frequency never exceeds that on all corners.

This ILRPLL is realized in a pure digital 55nm CMOS technology with complete ESD protection circuits and a standard LQFP256 package. The measurements are done in a system-on-chip (SoC) environment. The measured spectrums before/after injection are shown in Fig. 5.2.5. Configured as a traditional ring PLL, the integrated jitter from 1kHz to 40MHz is measured to be 4.8ps. After applying the injection, the integrated jitter is reduced to 2.4ps, including the reference spur of −70.7dBc. The current draw is 3mA from a 1.2V supply and 1mA from a 3.3V supply, including all the currents needed to drive other blocks. A 10b 216MS/s ADC is measured with this ILRPLL and passes the ATV/DTV system requirements. The measured XTAL phase noise is provided in Fig. 5.2.6 for reference. The die micrograph and performance summary are both shown in Fig. 5.2.7. The area of this ILRPLL is 195×153µm^2 with buffers and level shifters.

References:
[1] S. Ye, L. Jansson, and I. Galton, "A Multiple-Crystal Interface PLL With VCO Realignment to Reduce Phase Noise," *IEEE J. Solid-State Circuits*, vol. 37, no. 12, pp. 1795-1803, Dec., 2002.
[2] J. Lee, H. Wang, W.-T. Chen, et al. "Subharmonically Injection-Locked PLLs for Ultra-Low-Noise Clock Generation," *ISSCC Dig. Tech. Papers*, pp. 92-93, Feb., 2009.
[3] X. Gao, E.A.M. Klumperink, M. Bohsali, and B. Nauta, "A 2.2GHz 7.6mW Sub-Sampling PLL with −126dBc/Hz In-Band Phase Noise and 0.15ps$_{rms}$ Jitter in 0.18µm CMOS," *ISSCC Dig. Tech. Papers*, pp. 392-393, Feb., 2009.
[4] B.M. Helal, C.-M. Hsu, K. Johnson, M.H. Perrott, "A Low Noise Programmable Clock Multiplier based on a Pulse Injection-Locked Oscillator with a Highly-Digital Tuning Loop," *IEEE RFIC Symposium*, pp. 423-426, June, 2008.
[5] M. Van Paemel, "Analysis of a Charge-Pump PLL: A New Model," *IEEE Trans. Communications*, vol. 42, no. 7, pp. 2490-2498, July, 1994.

978-1-61284-303-2/11 $26.00 © 2011 IEEE

ISSCC 2011 / February 21, 2011 / 2:00 PM

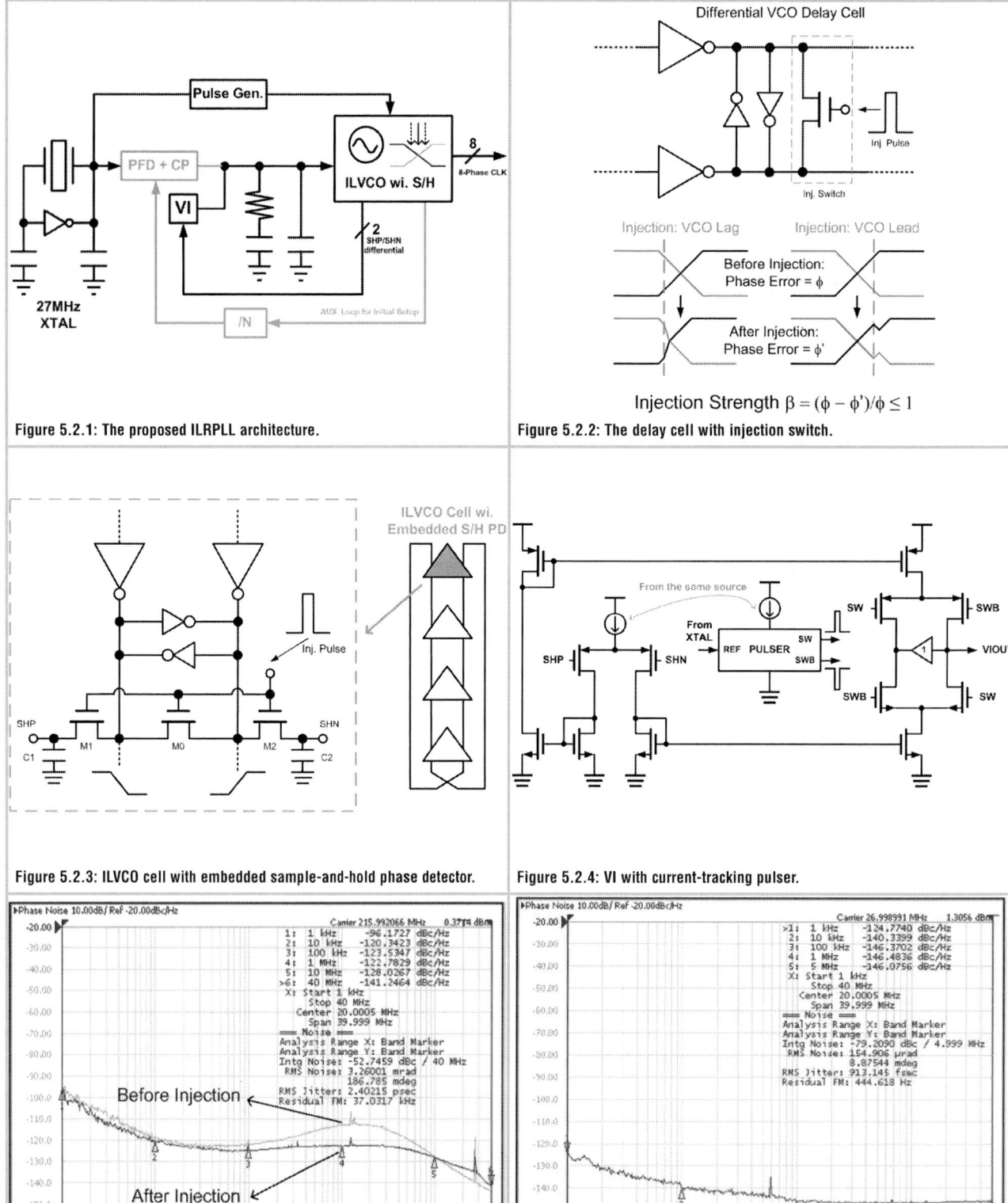

Figure 5.2.1: The proposed ILRPLL architecture.

Figure 5.2.2: The delay cell with injection switch.

Injection Strength $\beta = (\phi - \phi')/\phi \le 1$

Figure 5.2.3: ILVCO cell with embedded sample-and-hold phase detector.

Figure 5.2.4: VI with current-tracking pulser.

Figure 5.2.5: Measured ILRPLL phase noise before and after injection.

Figure 5.2.6: Measured XTAL phase noise.

ISSCC 2011 PAPER CONTINUATIONS

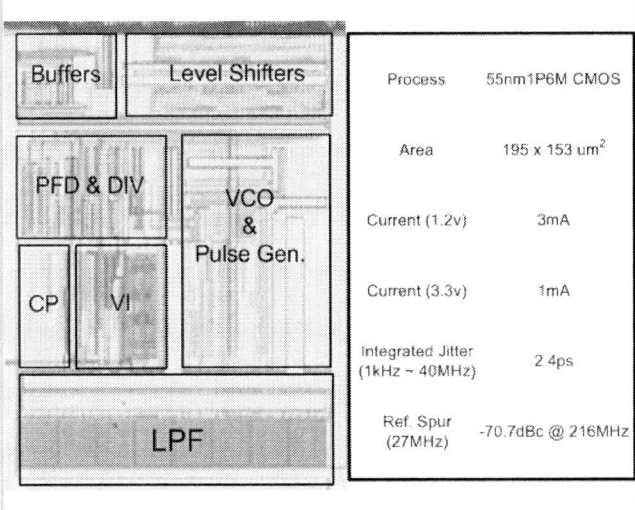

Process	55nm1P6M CMOS
Area	$195 \times 153 \ um^2$
Current (1.2v)	3mA
Current (3.3v)	1mA
Integrated Jitter (1kHz ~ 40MHz)	2 4ps
Ref. Spur (27MHz)	-70.7dBc @ 216MHz

Figure 5.2.7: Die micrograph and performance summary.

ISSCC 2011 / SESSION 5 / PLLs / 5.3

5.3 A 0.4-to-3GHz Digital PLL with Supply-Noise Cancellation Using Deterministic Background Calibration

Amr Elshazly, Rajesh Inti, Wenjing Yin, Brian Young, Pavan Kumar Hanumolu

Oregon State University, Corvallis, OR

Digital phase-locked loops (DPLLs) have recently emerged as a viable alternative to classical charge-pump analog PLLs [1-4]. By obviating the need for a large loop filter capacitor and a high-performance charge pump, DPLLs offer area savings and easier scalability to newer processes. The ability to reconfigure the digital loop filter dynamically offers flexibility in setting the loop response and helps to optimize the locking behavior of the DPLL [1]. However, conflicting bandwidth requirements to simultaneously suppress TDC quantization and oscillator phase noise mandate either a high-resolution TDC or a low-noise oscillator to minimize jitter [2]. Further, much like in an analog PLL, the ring oscillator is susceptible to supply noise, which especially limits the jitter performance of a DPLL integrated into a large digital system. A low-dropout regulator is commonly used to shield the DCO from supply noise at the expense of additional area, power, and voltage headroom [3]. Alternatively, an open-loop supply-noise cancellation scheme can operate at a lower supply voltage, but its accuracy is highly sensitive to process variations [5]. Analog foreground calibration compensates for process variation but is susceptible to voltage, temperature, and frequency variations [6]. In this paper, we present a deterministic test-signal-based background calibration scheme that leverages the highly digital nature of the DPLL to adaptively cancel the supply noise in the DCO. The prototype DPLL achieves nearly perfect supply-noise cancellation over an output frequency range of 0.4 to 3GHz while consuming 2.65mW at 1.5GHz.

Figure 5.3.1 shows the block diagram of the Type-2 DPLL with the proposed supply-noise cancellation scheme. It consists of separate proportional and digital integral paths, feedback divider, supply noise gain-calibration logic, and supply-noise-insensitive DCO. The proportional control is implemented using a 3-state PFD that directly drives the oscillator through a 3-level current-mode DAC, thus eliminating the TDC quantization error in the proportional path. A flip-flop (FF) acts as an early/late detector on PFD outputs and drives the digital accumulator with the sign of the phase error. The low-bandwidth digital integral path suppresses the phase quantization error of the FF. A 1-to-4 demux is used to ease the speed requirements of fully synthesized digital control logic. The digital $\Delta\Sigma$ modulator truncates the 14b accumulator output, D_I, to 15-levels and drives a current-mode DAC. A 2nd-order passive low-pass filter suppresses the out-of-band quantization error and drives the integral control voltage input of the oscillator.

The DCO supply-noise immunity is greatly improved by deliberately injecting an appropriate magnitude cancellation current, I_C, in proportion to the supply noise. Because the DCO supply sensitivity depends on PVT variations and oscillation frequency, a test-signal-based digital background calibration scheme is used to determine the compensation gain accurately and achieve excellent broadband supply-noise immunity under all operating conditions.

A low-frequency deterministic digital test signal, D_{TEST}, is converted to an analog voltage and added to the DCO supply voltage. Because the supply-noise compensation circuit is insensitive to the *source* of the noise, adjusting the compensation gain to cancel D also suppresses the supply noise. In the absence of supply-noise cancellation (I_C=0), owing to the low-pass characteristic of $D_I(s)/V_{DD}(s)$ transfer function shown in Fig. 5.3.2, a scaled version of the test signal appears at the output of the accumulator. In other words, the loop rejects the low-frequency supply noise due to the well-known band-pass characteristic of the DCO supply-noise transfer function, $\Phi_{OUT}(s)/V_{DD}(s)$ (Fig. 5.3.2). Note that in an over-damped DPLL, the bandwidth and the center frequency of the transfer functions $D_I(s)/V_{DD}(s)$ and $\Phi_{OUT}(s)/V_{DD}(s)$ are equal to the lower (F_{P1}) and the higher (F_{P2}) of the two closed-loop poles, respectively. Consequently, a test signal whose frequency is lower than F_{P1} appears at the accumulator output with little attenuation while it is heavily suppressed at the DPLL output.

With supply-noise cancellation ($I_C \neq 0$), ideally, D_{TEST} would be cancelled in the DCO. However, DCO supply-noise compensation gain error causes a portion of D_{TEST} to leak to the accumulator output. As a result, correlating the accumulator output with D_{TEST} yields an estimate of the gain error. Adjusting the gain of the cancellation path until D_{TEST} completely disappears at the accumulator output results in perfect calibration of the DCO supply-noise compensation gain. Under this condition, as indicated by the dotted curves in Fig. 5.3.2, excellent broadband supply-noise immunity is achieved independent of the DPLL bandwidth.

The schematic of the ring oscillator with supply-noise cancellation is shown in Fig. 5.3.3. It is composed of 3-stages of pseudo-differential delay cells with noise-canceling transistors, M_{n1}-M_{n6}, at their outputs and a digitally controlled resistor to inject the test signal into the VCO supply. Transistor M_0 couples supply noise to the gates of transistors M_{n1}-M_{n6} with a digitally calibrated gain. To minimize headroom penalty, voltage drop across the variable resistor is designed to be less than 20mV. The integral and proportional control voltages tune the delay by varying the strength of the latch load and the output time constant, respectively. In simulation, transistors M_{n1}-M_{n6} degrade the DCO phase noise by 3dB at 1MHz offset from the carrier. Because the phase noise contribution from M_{n1}-M_{n6} scales inversely with the calibration code, varying the proportional path current (I_{BW} in Fig. 5.3.1) accordingly implements bandwidth tracking and mitigates jitter degradation.

The DPLL is implemented in 0.13μm CMOS and its active area occupies 0.08mm². An integrated supply-noise monitor is used to observe on-chip supply voltages and guarantee the fidelity of the injected supply noise. The DPLL achieves an operating range of 0.4 to 3GHz and consumes 2.65mW at 1.5GHz while operating from a 1V supply. The area and power penalty due to calibration/cancellation circuitry is less than 12.5% and 9.5% at 1.5GHz, respectively. As shown in Fig. 5.3.4, the measured long-term absolute jitter (50k hits) at 1.5GHz is 5ps$_{rms}$ (47ps$_{pp}$) and is unchanged even in the presence of 10mV test signal. The jitter with and without cancellation is 22ps (170ps$_{pp}$) and 5.9ps (50ps$_{pp}$), respectively, in the presence of 20mV$_{rms}$ white Gaussian supply noise. Figure 5.3.5 quantifies the measured dynamic sensitivity by plotting peak-to-peak jitter versus the frequency of a 30mV supply-noise tone. Without cancellation, sensitivity is highest in the vicinity of the DPLL bandwidth. With cancellation, minimal jitter degradation is observed over a broad noise frequency range (0.1MHz to 2GHz). Figure 5.3.5 also illustrates that cancellation is most effective over 0.8-to-2.5GHz output frequency range. The compensating current saturates outside this range resulting in only partial cancellation of the supply noise. The performance summary of the DPLL and comparison to state-of-the-art is shown in Fig. 5.3.6. The normalized power is 4 times lower than DPLLs and analog PLLs with supply-noise cancellation. Die micrograph is shown in Fig. 5.3.7.

Acknowledgments:
This work was supported by CDADIC and the National Science Foundation under CAREER EECS-0954969. Dongbu HiTek provided IC fabrication.

References:
[1] S.Y. Yang and W.Z. Chen, "A 7.1mW 10GHz All-Digital Frequency Synthesizer with Dynamically Reconfigurable Digital Loop Filter in 90nm CMOS," *ISSCC*, pp. 90-91, Feb., 2009.
[2] C.-M. Hsu, M.Z. Straayer, and M.H. Perrott, "A Low-Noise, Wide-BW 3.6GHz Digital $\Delta\Sigma$ Fractional-N Frequency Synthesizer with a Noise-Shaping Time-to-Digital Converter and Quantization Noise Cancellation," *ISSCC*, pp. 340-341, Feb., 2008.
[3] W. Grollitsch, R. Nonis, and N. Da Dalt, "A 1.4ps$_{rms}$-Period-Jitter TDC-Less Fractional-N Digital PLL with Digitally Controlled Ring Oscillator in 65nm CMOS," *ISSCC*, pp. 478-479, Feb., 2010.
[4] M. Chen, D. Su, S. Mehta, "A calibration-free 800MHz fractional-N digital PLL with embedded TDC," *ISSCC*, pp. 472-473, Feb. 2010.
[5] M. Mansuri and C.-K.K. Yang, "A Low-Power Low-Jitter Adaptive-Bandwidth PLL and Clock Buffer," *ISSCC*, pp. 430-440, Feb., 2003.
[6] T. Wu, K. Mayaram, and U. Moon, "An On-chip Calibration Technique for Reducing Supply Voltage Sensitivity in Ring Oscillators," *IEEE Symp. VLSI Circuits*, pp. 128-129, Jun., 2006.

ISSCC 2011 / February 21, 2011 / 2:30 PM

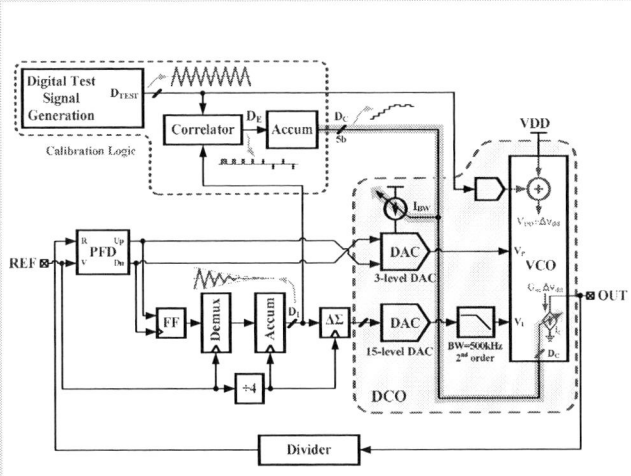

Figure 5.3.1: Block diagram of the proposed digital PLL.

Figure 5.3.2: DCO supply noise to DPLL and accumulator output transfer functions. The evolution of D_I with and without cancellation is shown at the bottom.

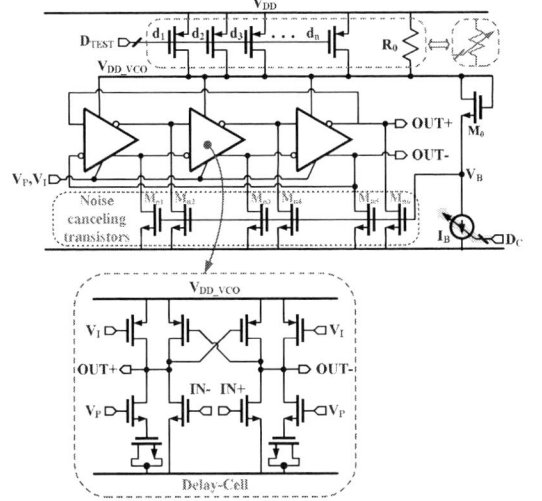

Figure 5.3.3: Schematic of the proposed ring oscillator with test-signal injection and supply-noise cancellation.

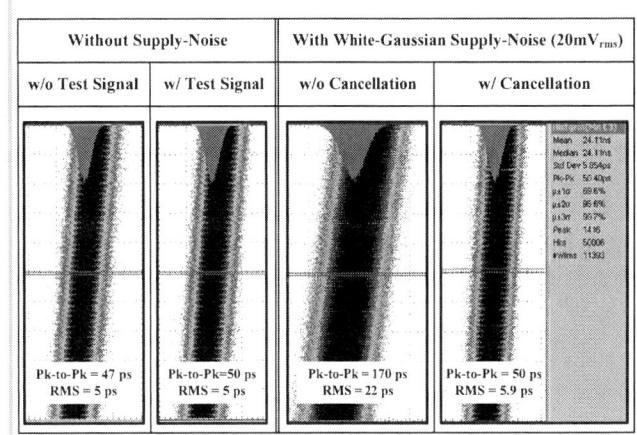

Figure 5.3.4: Measured jitter histograms at 1.5GHz output frequency.

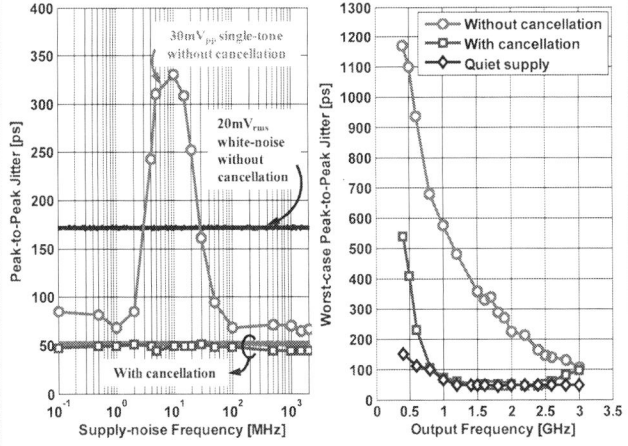

Figure 5.3.5: Measured peak-to-peak jitter as a function of supply-noise frequency at 1.5GHz output frequency, and worst-case peak-to-peak jitter versus DPLL output frequency for a 30mV$_{pp}$ single-tone supply noise.

Figure 5.3.6: Performance summary and comparison with state-of-the-art.

978-1-61284-303-2/11 $26.00 © 2011 IEEE

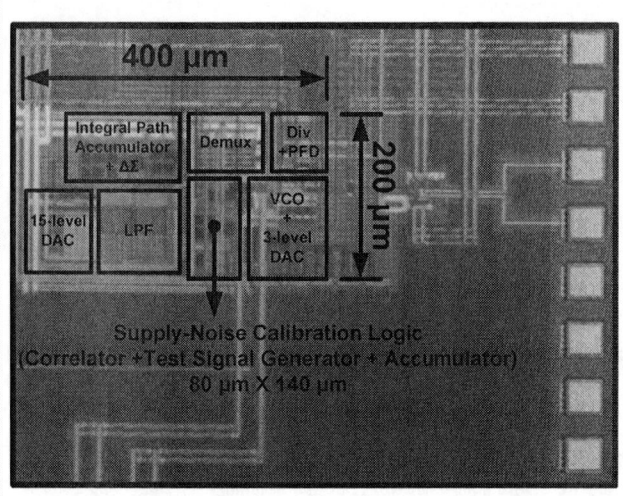

Figure 5.3.7: DPLL die micrograph.

ISSCC 2011 / SESSION 5 / PLLs / 5.4

5.4 A 0.1-f_{ref} BW 1GHz Fractional-N PLL with FIR-Embedded Phase-Interpolator-Based Noise Filtering

Dong-Woo Jee, Yunjae Suh, Hong-June Park, Jae-Yoon Sim

Pohang University of Science and Technology, Pohang, Korea

In the design of a fractional-N PLL, the trade-off between in-band VCO noise and $\Delta\Sigma$ quantization noise constrains the choice of loop bandwidth. Various circuit schemes have been proposed to relax such constrains with noise canceling methods [1, 2] at the cost of significant extra power and chip area, and with FIR filtering techniques [3,4] utilizing multiple charge pumps (CPs), PFDs and dividers. To reduce the $\Delta\Sigma$ quantization noise, fractional phase rotation [4] has been also a popular approach as an alternative to the dual modulus divider. However, high-resolution phase interpolators (PIs) suffer from nonlinearities due to random mismatches among phase steps and systematic imperfections in circuit operation when the interpolated vector approaches quadrant boundaries, and such nonlinearities eventually limit the amount of noise reduction in PI-based PLL. This work presents a 1GHz $\Delta\Sigma$ fractional-N PLL based on the noise filtering by FIR-embedded PI. The proposed PI scheme greatly improves phase linearity by a dual-referenced interpolation and realizes FIR filtering without using multiple CPs, PFDs, and dividers. The designed fractional-N PLL shows a comparable phase-noise performance to that of an integer-N PLL even with loop bandwidth of $0.1 \times f_{ref}$.

Figure 5.4.1 shows the block diagram of the proposed PLL. The 4-stage ring VCO provides 8 seed phases for the phase interpolation. With the divide-by-N divider only, the proposed FIR-embedded PI operates as a fractional dual-modulus frequency divider. The PI consists of eight dual-referenced interpolators (DIs) and a phase blender. The DI has 5b resolution with one step of 31.25ps at 1GHz and continuously rotates the output phase to perform the fractional frequency division. The output of the $\Delta\Sigma$ modulator is applied to a D-FF chain to generate delayed (z^{-1}, z^{-2}, ... z^{-7}) outputs. Each output is used to control a DI. If the control output is 1, the output phase of the DI leads by one step while it stays the same if the control output is 0, resulting in a dual-modulus frequency division with resolution improved by 1/32. The phase blender averages the eight outputs of DIs and functions as an FIR filter.

Figure 5.4.2 illustrates the concept of noise filtering performed by the proposed PI. $\Delta\Sigma$ quantization noise, $x[i-k]_{k=0\sim7}$, is theoretically reduced to 1/32 at $p[i-k]_{k=0\sim7}$, since the quantization step is reduced by 5b DI. The $p[i-k]_{k=0\sim7}$ are averaged by the phase blender. The resultant noise, $y[i]$, is derived as $1/8 \cdot \sum_{k=0}^{7} x[i-k]/32$, functioning as an 8-tap FIR filter on the quantization noise. At $0.1 \times f_{ref}$, the simulated noise power is less than -100dBc/Hz, which is close to the practical limit of the designed ring VCO phase noise. The division factor of 8 by the filtering reduces the maximum noise power by 18dB at the high frequency, resulting in the effective improvement of resolution in the phase interpolation. Since the phase blender averages the outputs of the eight DIs, random phase error of each DI has identical contribution in generation of the output phase. Therefore, the phase blender additionally reduces the effect of random phase mismatch of DI by a factor of $1/\sqrt{8}$, assuming the random errors are not correlated.

The nonlinearity of conventional PI operation is the most critical bottleneck determining the noise performance. In digitally controlled conventional PI, the linear increase or decrease in weighting code causes a systematic nonlinear phase interpolation since the phase step decreases as the interpolated vector approaches to the quadrant boundaries. The failure to have uniform phase steps degrades the interpolator performance and eventually limits the amount of noise suppression. Figure 5.4.3 shows the concept of the proposed DI. In addition to the traditional PI control with quadrant boundaries of 0°, 90°, 180°, and 270°, another PI with quadrant boundaries of 45°, 135°, 225°, and 315° is also utilized for the dual-referenced interpolation. With the weight control by a 45°-outphased counter, the two interpolators generate the same vectors. Then, the two outputs of PIs are simply tied together for averaging. Without complicated weighting control, the opposite nonlinearites of 0° and 45°-outphased PIs are compensated by each other. Simulation shows that the proposed DI scheme

reduces the noise due to PI nonlinearity by 15dB, and suppresses the noise, $y[i]$, to lower than -100dBc/Hz at $0.1 \times f_{ref}$.

Figure 5.4.4 (a) shows the schematic of the unit PI circuit of DI. For 5b resolution, 2 MSBs are applied to Sel$\varphi\pm$ and Sel$\psi\pm$ for quadrant selection, and 3 LSBs are encoded to 8b thermometer code (CODE<0:7>) to generate eight-phase steps in each quadrant. Figure 5.4.4 (b) shows the schematic of the phase blender. Outputs of 8 differential pairs are simply tied together for FIR filtering. The current-source ratio for each differential pair can be considered as FIR tap coefficients. In this work, all coefficients are set to 1 ($I_{ref0-7} = I_{ref}$), resulting in a simple averaging.

The PLL is implemented in a 0.13μm CMOS technology (Fig. 5.4.7). Several options are included for the performance comparison. 1) integer-N PLL with a fixed division ratio of 32 to obtain the reference level of the minimum noise without $\Delta\Sigma$ modulator, 2) conventional fractional-N PLL with a dual-modulus (divide-by-32 and divide-by-33) divider only, 3) the proposed PI with FIR embedded, and 4) the proposed PI with FIR disabled (all DIs are driven by the same control code). Figure 5.4.5 shows the measured phase noise at 1.0478GHz (1.024GHz for option 1) with the reference frequency of 32MHz and the loop bandwidth of $0.1 \times f_{ref}$. The proposed FIR-embedded PI scheme (option 3) reduces the phase noise by 34 dB, compared to the conventional dual-modulus divider scheme (option 2). To compare the phase noise for option 1, 3 and 4, smoothed plots of the phase noise are taken. The proposed scheme (option 3) shows an in-band phase noise of -106dBc at 100kHz offset and out-of-band noise of -107.5dBc at 6MHz offset. Both results are almost equal to the minimum achievable level indicated by integer-N option, -107dBc for 100kHz offset and -109dBc for 6MHz offset. Without FIR function (option 4), the out-of-band noise is -101dBc for 6MHz offset. The FIR filtering becomes effective when the offset frequency is above 1.5MHz. The filtering (option 3) suppresses the noise level down to the minimum level indicated by the integer-N option around 4MHz, which is the first zero position of the FIR filter. The other zeros at higher frequencies are not clearly seen since the noise level already reaches the VCO noise limit. The measured reference spur is -66dBc, which is also almost equal to that of the integer-N option (-67dBc). Since the input phase difference of PFD is greatly reduced by the pre-filtering with FIR-embedded PI, the proposed scheme results in little periodic disturbance at loop filter output and achieves a low reference spur close to that of the integer-N option. The core area is 0.31mm². The FIR-embedded PI occupies 0.04mm² and consumes 9.6mW. Total power consumption is 16.8mW. Figure 5.4.6 shows the performance comparison table with previously reported fractional-N PLLs [1-4].

Acknowledgement:
This work was supported by IDEC and Dongbu HiTek for chip fabrication, and Basic Science Research Program through NRF grant funded by Korea Ministry of Education, Science and Technology (2010-0007955).

References:
[1] A. Swaminathan, K.J. Wang, and I. Galton, "A Wide-Bandwidth 2.4 GHz ISM Band Fractional-N PLL With Adaptive Phase Noise Cancelation," *IEEE J. Solid-State Circuits*, vol. 42, no. 12, pp. 2639-2650, Dec., 2007.
[2] H. Hedayati, W. Khalil, and B. Bakkaloglu, "A 1 MHz Bandwidth, 6 GHz 0.18 μm CMOS Type-I $\Delta\Sigma$ Fractional-N Synthesizer for WiMAX Applications," *IEEE J. Solid-State Circuits*, vol. 44, no. 12, pp. 3244-3252, Dec., 2009.
[3] X. Yu, Y. Sun, L. Zhang, et al., "A 1 GHz Fractional-N PLL Clock Generator with Low-OSR $\Delta\Sigma$ Modulation and FIR-Embedded Noise Filtering," *ISSCC Dig. Tech. Papers*, pp. 346-347, Feb., 2008.
[4] X. Yu, Y. Sun, W. Rhee, et al., "A $\Delta\Sigma$ Fractional-N Synthesizer With Customized Noise Shaping for WCDMA/HSDPA Applications," *IEEE J. Solid-State Circuits*, vol. 44, no. 8, pp. 2193-2200, Aug., 2009.

ISSCC 2011 / February 21, 2011 / 3:15 PM

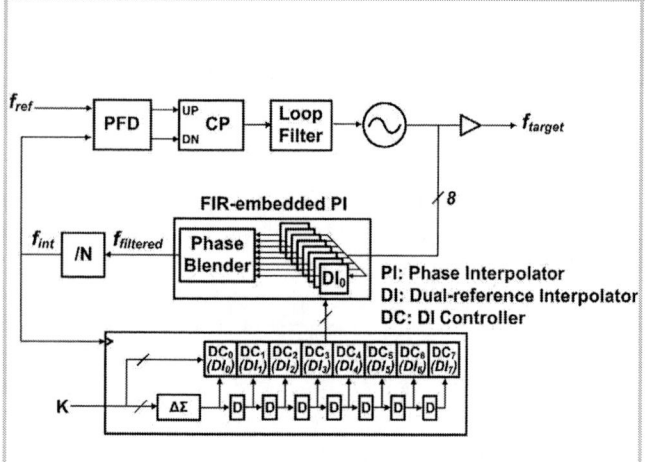

Figure 5.4.1: Top block diagram of the proposed fractional-N PLL.

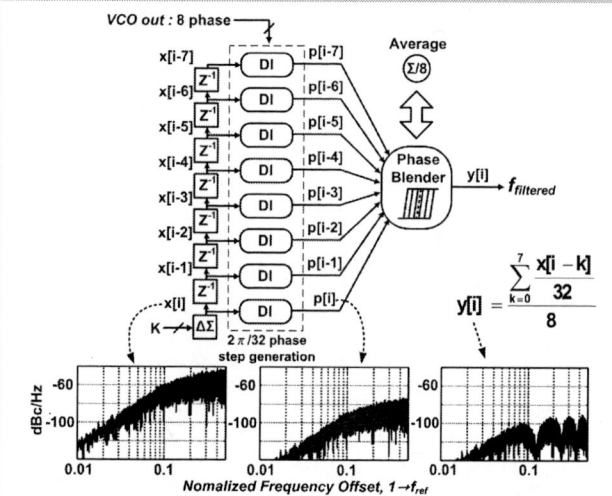

Figure 5.4.2: Noise-filtering concept of the proposed FIR-embedded phase interpolator.

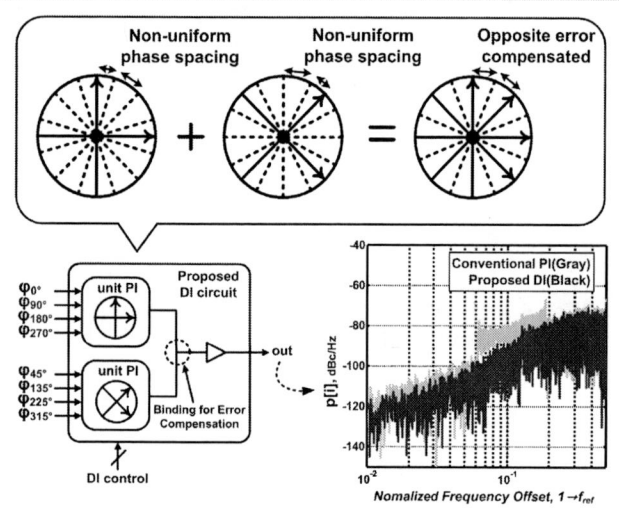

Figure 5.4.3: The proposed dual-referenced interpolation.

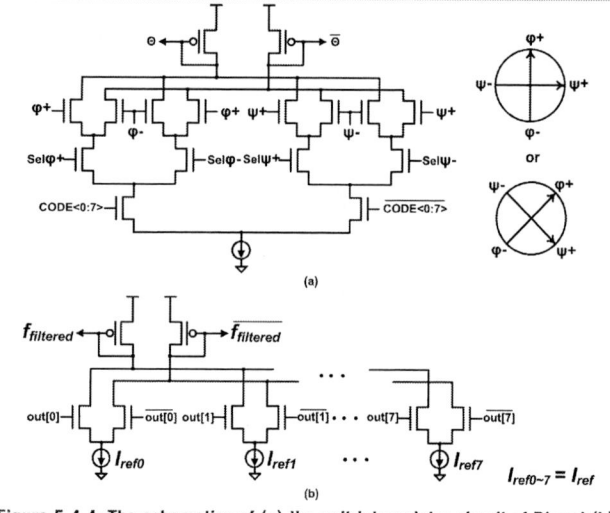

Figure 5.4.4: The schematics of (a) the unit interpolator circuit of DI and (b) the phase blender (b).

Figure 5.4.5: The measured phase noise with different options.

Figure 5.4.6: Performance comparison with other fractional-N PLLs.

	This Work	[1]	[2]	[3]	[4]
Output Frequency	1 GHz	2.4 GHz	6.12 GHz	1 GHz	2 GHz
f_{BW}/f_{REF}	0.1	0.061	0.026	0.037	0.0076
Inband Phase Noise (@100kHz)	-106 dBc/Hz	-101 dBc/Hz	-102 dBc/Hz	-85 dBc/Hz @10kHz	-92 dBc/Hz
Noise Canceling Method	FIR embedded PI	DAC	DAC	FIR	FIR
Calibration	None	Adaptive LMS DAC Calibration	None	None	None
Phase Noise Cancelation	34 dB	33 dB	28 dB	15 dB	7 dB@3MHz Offset
Reference Spur	-66 dBc	-53 dBc	-78 dBc	N.A.	-62.84 dBc
Core Power	16.8 mW	38 mW	47 mW	6.1 mW	17.2 mW
Core Area	0.31 mm²	4.8 mm²	3.24 mm²	0.5 mm²	1.5 mm²
Technology (CMOS)	0.13 μm	0.18 μm	0.18 μm	0.18 μm	0.18 μm
VCO Type	Ring	LC	LC	Ring	LC

978-1-61284-303-2/11 $26.00 © 2011 IEEE

ISSCC 2011 PAPER CONTINUATIONS

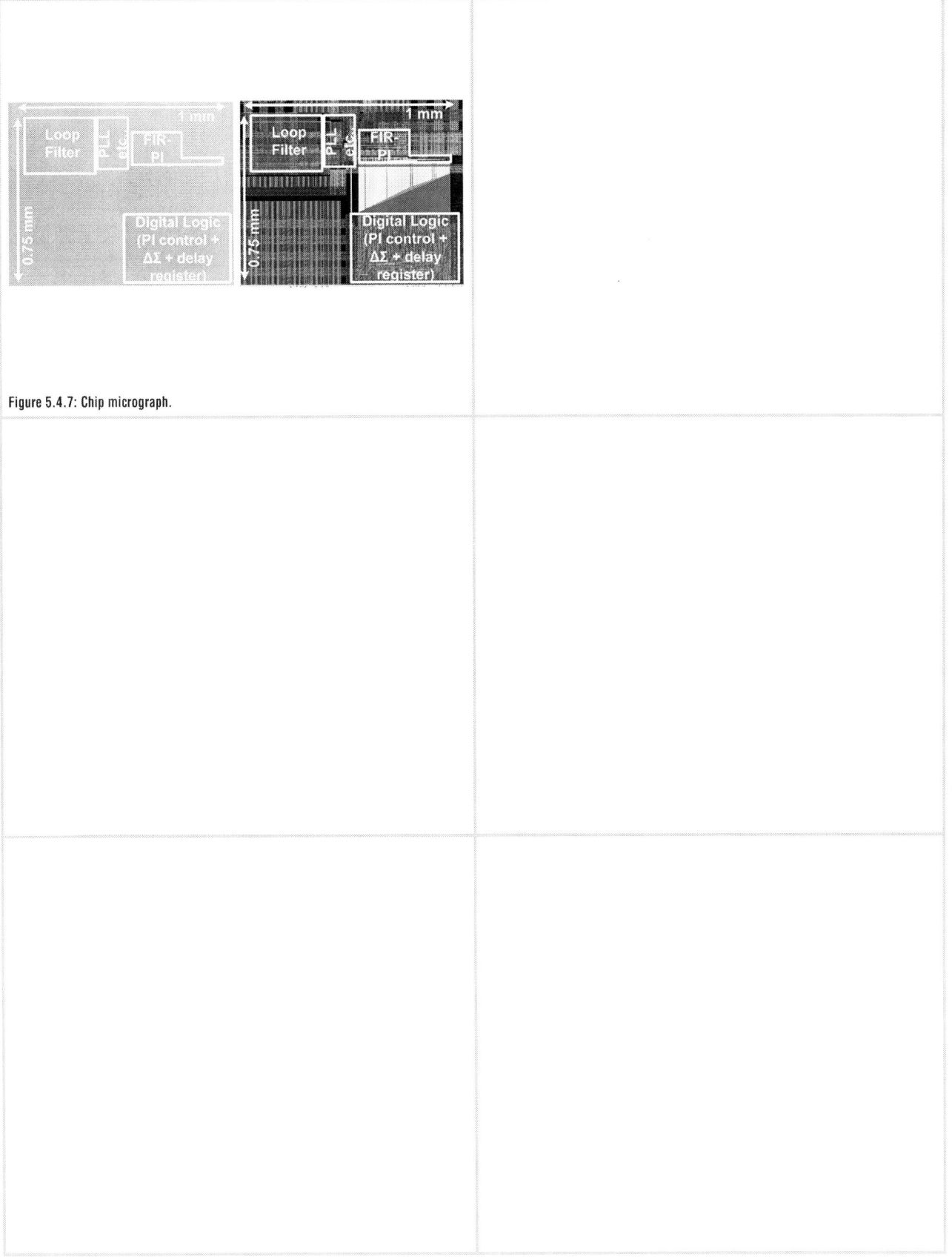

Figure 5.4.7: Chip micrograph.

ISSCC 2011 / SESSION 5 / PLLs / 5.5

5.5 A Scalable Sub-1.2mW 300MHz-to-1.5GHz Host-Clock PLL for System-on-Chip in 32nm CMOS

Hyung-Jin Lee, Alexandra M. Kern, Sami Hyvonen, Ian A. Young

Intel, Hillsboro, OR

System-on-chips (SoCs) are being widely adopted in mobile applications, and are driven by the need for longer battery life, their power budget continues to decrease. In addition, the phase-locked loop (PLL) for the SoC host clock has to be a very low power circuit to support the always-on always-connected (AOAC) feature for SoCs integrated into hand-held devices. The proposed PLL, implemented in a high-k metal-gate 32nm logic CMOS technology, provides process scalability to the next process technology node, uncompromised system response, and loop stability under process variation and minimum power envelop constraints. As the jitter requirements for the host clocking PLL are not stringent, the proposed architecture emphasizes the power efficiency over the jitter performance.

Figure 5.5.1 shows the architecture of the proposed PLL which includes a first-order resistor-less loop filter (LF). The PLL consists of a current-controlled saw-tooth-wave oscillator (CCO), a non-current-steering charge pump (CP), a feedback divider with a spread spectrum generator (SSG) based on a digital delay-locked loop (DLL), a digital automatic frequency controller (AFC), and an integrated reference current generator driving the CP.

The resistor-less first-order LF creates a proportional path gain with a pair of identically sized capacitors C_{p1} and C_{p2} and commutator switch pairs (SW$_1$ and SW$_2$). The switch pairs of SW$_1$ and SW$_2$ reverse the charge polarity of C_{p2} in every reference clock cycle erasing previous-cycle charge in C_{p1} and C_{p2}, while simultaneously taking present-cycle charge from the CP. The effective resistance in the LF can therefore be expressed as T_{ref}/C_p, where T_{ref} is the reference clock period and C_p is the total capacitance of C_{p1} and C_{p2}. As a result, the zero frequency (f_{zero}) of the loop filter is proportional to the reference clock frequency (f_{fer}). The static capacitor C_i provides an integral path gain, and the ratio of the static to the dynamic capacitor (C_i/C_p) along with the f_{ref} determines the loop response.

A major advantage of this design is that the PLL loop response because it is determined by capacitance ratio is independent of temperature, supply voltage, and process skew variation. From a design perspective, the loop-filter architecture allows process shifting into the next generation process technology with minimum circuit optimization. Since C_i/C_p determines the loop response, the capacitor sizes can be made small, unless parasitic capacitance becomes significant or switched-capacitor thermal noise becomes dominant. Note that this passive-only loop-filter architecture avoids any noisy active component and amplifier offset voltage sensitivity unlike the previous work [1] that uses an Opamp.

The proposed CCO shown in Fig. 5.5.2 is composed of a voltage-to-current converter (V2C), a sawtooth-wave generator (SWG), a divide-by-2 block, and a differential rail-to-rail clock generator (CG). The divide-by-2 block is for creating a 50% clock duty cycle. The voltage to current gain of the V2C, which is a resistive source-degenerated amplifier, is inversely proportional to the adjustable source resistance (R_{src}).

The CG consists of a non-overlapping clock generator (NOCG) and a comparator. The comparator is triggered when the sawtooth wave exceeds the reference voltage, which is a PMOS threshold voltage (V_{thp}) for design simplicity. The M3 and M7 transistors are for preventing unnecessary current leakage during a signal transition. The NOCG generates a differential clock interleaving the switches S1 and S2 in the SWG. The non-overlapping clocking allows time for C_{osc} to be fully discharged between the switching. The discharging of C_{osc} improves phase noise performance at low carrier offset frequencies by suppressing long-term thermal noise accumulation. The frequency range of the CCO is 600 to 3000MHz, and the overall PLL frequency range after the divide-by-2 is 300 to 1500MHz.

The digital AFC mitigates process skew variation by means of trimming the source resistor ladder in the V2C for coarse frequency tuning. The AFC senses the CCO clock frequency (f_{cco}), and increases or decreases the number of series resistors only in one direction until the f_{cco} reaches a given target frequency within a reasonable error range. The AFC provides accurate frequency tuning capability in the CCO over large process skew variation, allowing the design to be implemented in a standard CMOS process without costly precision resistor options. The AFC runs once at the power-up, so it does not contribute to active power dissipation.

The input voltage (V_{ctl}) to the CCO finely tunes f_{cco} by modulating the CCO current (I_c). The size of C_{osc} and I_c determines the ramp speed of the sawtooth wave, which corresponds to f_{cco}. For simplicity, f_{cco} can be expressed by $f_{cco}=(V_{ctl}+V_{static})/(2C_{osc}\cdot R_{src}\cdot V_{thp})$, where V_{static} is a static bias voltage and its corresponding current combines with the current of the V2C branch to reduce the frequency control sensitivity (i.e., K_{cco}). Since $K_{cco}=df_{cco}/dV_{ctl}= 1/(2C_{osc}\cdot R_{src}\cdot V_{thp})$, the K_{cco} is independent of V_{ctl}, and hence it is constant over a wide V_{ctl} range for a given R_{src} trim as shown in Fig. 5.5.3. The ratio of K_{cco} and f_{cco} remains constant as the frequency range is adjusted by digitally tuning R_{src}. The simpler CCO architecture compared to the previous work [2] contributes to its lower power consumption. The phase noise performance can be improved at the cost of higher current dissipation by increasing C_{osc} and I_c.

The non-current steering CP is a low power design with negligible static current and virtually no dynamic current flow in the steady state. The reference current generator is another copy of the V2C and the reference current drives the N_{bias}. Since the reference current is also synchronous to f_{cco}, just like I_c, the CP gain (I_{cp}) is proportional to f_{cco} and also to the reference clock frequency (f_{ref}) for a given PLL feedback division ratio (N_{div}).

As discussed earlier, since the system parameters, such as f_{zero}, K_{cco}, and I_{cp} are related to f_{ref}, and to f_{cco} for a given N_{div}, they enable the PLL to automatically adjust its loop bandwidth (BW) according to f_{ref} without affecting loop stability. Unconditional loop stability at various f_{ref} requires two conditions [3]; (a) constant f_n/f_{ref} ratio, where f_n is the PLL natural frequency, and (b) constant damping ratio independent of f_{ref}. It can be proven that both (a) and (b) are satisfied if the closed loop gain and f_{zero} are linearly proportional to f_{ref}, which is true of the proposed architecture.

The overall size of the PLL in 32nm logic CMOS [4] is 0.046mm². The lock time is less than 3.5µs including the start-up AFC operation, which completes the initial coarse frequency acquisition within 1µs as shown in Fig. 5.5.4. Figure 5.5.5 shows the phase noise performance of the free-running CCO, the reference clock, and the PLL output. The phase noise plot shows the PLL loop BW is scaled by the reference clock frequency. Note that for clarity the phase-noise performance shown in Fig. 5.5.5 has been re-scaled for 1GHz clock frequency. The overall PLL current draw from a 1V supply is 0.4 to 1.15mA for the PLL output clock frequency range of 300 to 1500MHz. Single-cycle rms jitter at 1GHz PLL output clock frequency is 2.36 to 3ps and the long-term rms jitter settles around the 40 to 60ps range according to the N-cycle rms jitter plot shown in Fig. 5.5.6.

References:

[1] P. J. Lim, "An Area-Efficient PLL Architecture in 90-nm CMOS," *Symposium on VLSI Circuit*, pp. 48-49, June, 2005.

[2] S.L.J. Gierkink and E. van Tuijl, "A Coupled Sawtooth Oscillator Combining Low Jitter With High Control Linearity," *IEEE J. Solid-State Circuits*, vol. 37, no. 6, pp.702-710, June, 2002.

[3] J. G. Maneatis, "Low-jitter Process-Independent DLL and PLL Based on Self-Biased Techniques," *IEEE J. Solid-State Circuits*, vol. 31, no. 11, pp. 1723-1732, November, 1996.

[4] C.-H. Jan, M. Agostinelli, M. Buehler, et al. "A 32nm SoC Platform Technology with 2nd Generation High-k/Metal Gate Transistors Optimized for Ultra Low Power, High Performance and High Density Product Applications," *IEDM Dig. Tech. Papers*, pp. 647-650, December, 2009.

ISSCC 2011 / February 21, 2011 / 3:45 PM

5

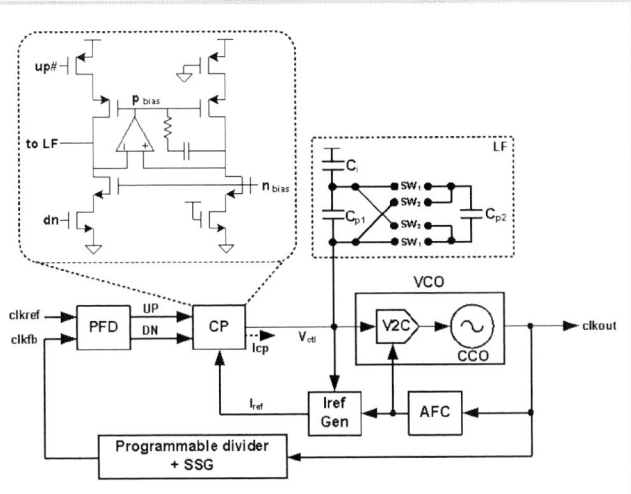

Figure 5.5.1: Block diagram of the scalable low-power phase-locked loop and non-current-steering charge pump.

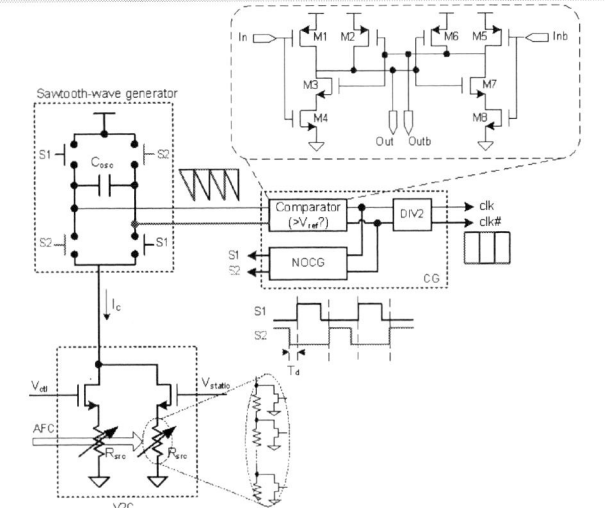

Figure 5.5.2: Low-power current-controlled sawtooth wave oscillator.

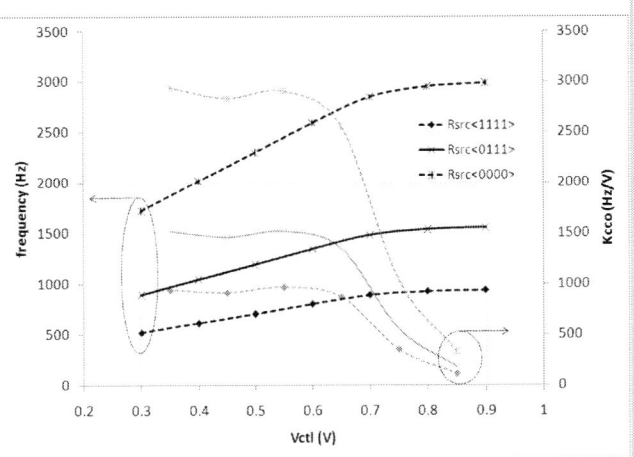

Figure 5.5.3: Measured oscillator frequency versus control voltage (V_{ctl}) curves and frequency gain (K_{cco}) curves for different resistor settings.

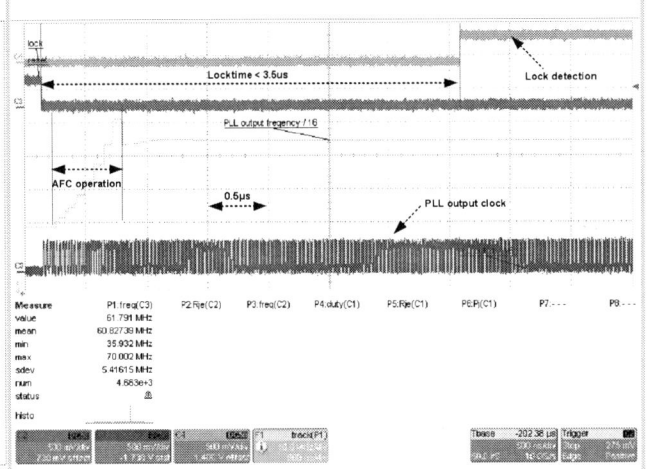

Figure 5.5.4: Settling behavior of the PLL showing the lock-indicator output signal and the track of the PLL output clock frequency.

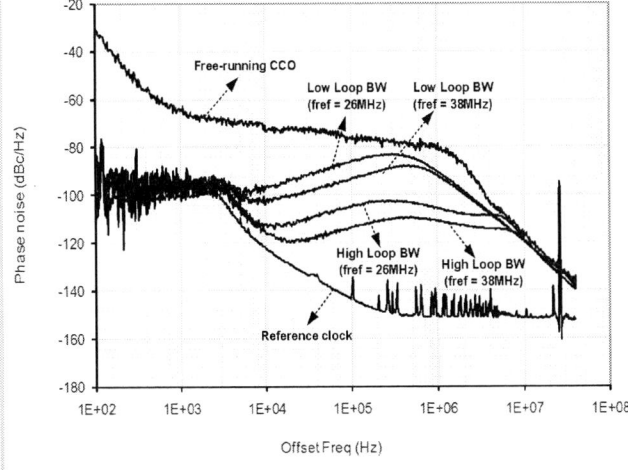

Figure 5.5.5: Measured phase-noise performance of the free-running oscillator, the PLL output in low and high loop bandwidth configurations, and the reference clock at 1GHz PLL output clock frequency.

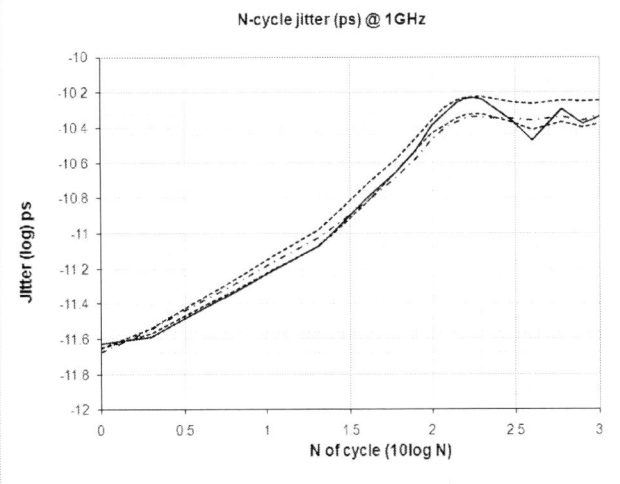

Figure 5.5.6: Measured rms N-cycle jitter performance from four parts.

978-1-61284-303-2/11 $26.00 © 2011 IEEE

ISSCC 2011 / SESSION 5 / PLLs / 5.6

5.6 A 570fs$_{rms}$ Integrated-Jitter Ring-VCO-Based 1.21GHz PLL with Hybrid Loop

Akihide Sai, Takafumi Yamaji, Tetsuro Itakura

Toshiba, Kawasaki, Japan

Sampling clock jitter significantly degrades the circuit performance and dynamic range of an ADC [1]. This paper presents a 570fs$_{rms}$ integrated-jitter 1.21GHz PLL with a hybrid loop. A ring VCO has a much inferior phase noise characteristic as compared to an LC VCO, but its area efficiency is attractive. To suppress the phase noise of a ring VCO, a wide-loop-bandwidth PLL with sufficiently low in-band noise is indispensable. An all-digital PLL (ADPLL) [2] has insufficiently low in-band phase noise because of the quantization error of the time-to-digital converter (TDC) without employing additional techniques such as power-hungry time amplification [3]. On the other hand, a conventional analog PLL has a superior in-band phase noise but needs a large loop-filter capacitor to maintain a wide tuning range due to a high control sensitivity of the ring VCO. The dual-tuning topology is useful for minimizing the size of the loop filter while maintaining low in-band phase noise [4, 5]. However, it also suffers from strong reference spurs, as it is the case in the conventional analog PLL having a wide loop bandwidth. The proposed PLL employs a hybrid loop consisting of a type-II ADPLL and a type-I analog PLL. The type-II ADPLL enables the wide tuning range without a large loop-filter capacitor. The loop-filter capacitor in the analog PLL is also minimized since it does not need to cover a wide tuning range. The analog PLL eliminates the residual quantization error of the TDC in the ADPLL and achieves a sufficiently low in-band phase noise. Overall, the proposed PLL suppresses the phase noise contribution from the ring digital/voltage-controlled oscillator (DVCO).

Figure 5.6.1 shows the proposed integer-N PLL with a hybrid loop. The TDC in the type-II ADPLL utilizes the multi-phase output signals $CKV+k\pi/32$ (k=-32, -31, ...0, ...30, 31) from the ring DVCO and does not require an additional delay line. The operation of the proposed PLL is as follows. First, the frequency and the phase of the ring DVCO output, CKV, are coarsely locked to the integer multiple of the reference signal REF by the ADPLL. Note that CKV and REF still have a difference that reflects the TDC quantization error at this point. Then, the proposed type-I analog PLL takes over the locking operation. The 3-input phase/frequency detector (PFD) utilizes REF and the 2 clock signals from the ring DVCO, CKV and CKV-$\pi/2$. The PFD outputs the phase error between the rising edges of REF and CKV-$\pi/2$ as UP and the phase error between the rising edges of CKV and CKV-$\pi/2$ as DN. The difference between UP and DN represents TDC quantization error as shown in the bottom left portion of Fig. 5.6.1, which represents the case that CKV lags REF. The difference signal, $UP - DN$, is sent to the ring DVCO as a negative-feedback signal through the loop filter and the amplifier. The amplifier output voltage, $V_{ctlp} - V_{ctln}$, increases until the TDC quantization error is corrected. Then, the pulse widths of UP and DN become equal at every reference edge. Eventually, the analog PLL reaches the lock. The proposed 3-input PFD does not require a divider since CKV is coarsely locked to REF by the ADPLL. Since the sensitivity of the ring DVCO differential control voltage is lower than 100MHz/V, the analog PLL, including the PFD, the loop filter and the amplifier, do not degrade the in-band phase noise. For power saving, the clocking of all the digital blocks in the ADPLL is stopped during the analog PLL operation. As a result, the proposed PLL achieves the in-band phase noise below −120dBc/Hz with a low power consumption while keeping the wide tuning range from 1.045 to 1.485GHz.

Figure 5.6.2 shows the details of the proposed 3-input PFD in the analog PLL. The 3 signals, REF, CKV and CKV-$\pi/2$ are inputs of the PFD. The PFD detects the phase error between the rising edges of REF and CKV-$\pi/2$ as UP in the same manner as the conventional PFD. The PFD also has to detect the phase error between the rising edge of CKV and CKV-$\pi/2$ as DN. If only the phase error between CKV and CKV-$\pi/2$ is detected (in the same manner as in the case of the conventional PFD), DNi signal is generated and it repeats at every CKV rising edge as shown in the bottom right portion of Fig. 5.6.2. Therefore, the window signal, DNw is generated at every reference edge by detecting the phase error between the rising edge of REF and the falling edge of CKV. Finally, DN signal is generated by multiplying DNw and DNi. Although the locking range of the proposed 3-input PFD is quite narrow, 55MHz, the proposed PLL covers the wide tuning range, from 1.045 to 1.485GHz, by the coarse tuning of the ADPLL.

Figure 5.6.3 shows the concept of the reference spurs reduction technique in the proposed analog PLL. In an ideal case, the reference spurs are small since the sensitivities of the ring DVCO, K_{vp} and K_{vn}, are matched. In a practical case, however, there is a mismatch between K_{vp} and K_{vn} and the common-mode signal in the control voltages, V_{ctlp} and V_{ctln}, is converted to the differential-mode signal, which dominates the reference spurs. Two cases are shown in the bottom right part of Fig. 5.6.3, with or without the amplifier. Suppose the amplifier has 20dB differential gain A_{dm} and 40dB CMRR. Without amplifier, the analog PLL needs to set the unity-gain frequency and the loop-filter cut-off frequency, f_{LPF}, at $f_{REF}/100$ and $f_{REF}/10$, respectively, to suppress the reference spurs by 60dB. With the amplifier, on the other hand, the PLL can set them at $f_{REF}/10$ and f_{REF} or higher with keeping both the same phase margin and the 60dB reference spurs suppression thanks to the CMRR. In this case, the phase noise of the ring DVCO is sufficiently suppressed below the stability limit of the PLL, $f_{REF}/10$, and the loop-filter capacitor size is also minimized by a tenth or less as compared to the case without the amplifier. The amplifier is utilized without feedback configuration to prevent the degradation of the CMRR due to the feed-through. In addition, a cascode structure is applied to the tail current source of the first input stage to boost the CMRR as shown in the bottom left part of Fig. 5.6.3. Although the open-loop gain of an amplifier generally changes with process variations, the differential gain A_{dm} of the amplifier, which is determined by $g_{mp}R_{load}$, is accurately implemented since a g_m constant biasing circuit makes g_{mp} proportional to $1/R$. The loop bandwidth of the analog PLL is set at the optimal point, 4MHz, which is close to $f_{REF}/10$.

Figure 5.6.4 shows the schematic of the ring DVCO. The ring DVCO consists of the 32 delay stages to output the 64-phase signals, $CKV+k\pi/32$ (k=-32, -31, ... 0, ...30, 31). The individual stages utilize the differential inverter with cross-coupled pairs. Through the PMOS and NMOS variable resistor blocks, the delay of the stage is digitally controlled by the output of the ADPLL in 3 modes, PVT, ACQ and TR, and controlled by the output of the analog PLL, V_{ctlp} and V_{ctln}.

The PLL chip is fabricated in a standard 65nm CMOS process and occupies an active area of 0.4×0.3mm^2 (see Fig. 5.6.7). The left part of Fig. 5.6.5 shows the measured PLL spectrum at 1.21GHz output and the reference spur is −42.2dBc. The right side of Fig. 5.6.5 shows the measured PLL phase noise at 1.21GHz output. The in-band phase noise at 1MHz offset is −119.6dBc/Hz and integrated rms jitter is 570fs. Figure 5.6.6 summarizes the PLL performance. For fair comparison, the power consumption of each ring-VCO-based PLL [6, 7] is normalized to the regulated supply voltage since the proposed PLL does not use the on-chip LDO. Compared with those PLLs, this design achieves the lowest jitter and the best FoM in [8].

References:

[1] R.H. Walden, "Analog-to-Digital Converter Survey and Analysis," *IEEE J. Selected Areas in Communications*, vol.17, No.4, pp. 539-550, Apr., 1999.

[2] R.B. Staszewski, J.L. Wallberg, S. Rezeq, et al., "All-Digital PLL and Transmitter for Mobile Phones," *IEEE J. Solid State Circuits*, vol. 40, no. 12, pp. 2469-2482, Dec., 2005.

[3] M. Lee, M.E. Heidari, and A.A. Abidi, "A Low Noise, Wideband Digital Phase-locked Loop based on a New Time-to-Digital Converter with Subpicosecond Resolution," *IEEE Symp. VLSI Circuits*, pp. 112-113, Jun., 2008.

[4] R. Nonis, N. Da Dalt, P. Palestri, and L. Selmi, "Modeling, Design and Characterization of a New Low-Jitter Analog Dual Tuning LC-VCO PLL Architecture," *IEEE J. Solid-State Circuits*, vol. 40, no. 6, pp. 1303-1309, Jun., 2005.

[5] S. Williams, H. Thompson, M. Hufford, and E. Naviasky, "An Improved CMOS Ring Oscillator PLL with Less Than 4ps Accumulated Jitter," *IEEE CICC*, pp. 151-154, Sept., 2004.

[6] W. Grollitsch, R. Nonis, and N. D. Dalt, "A 1.4ps-Period-Jitter TDC less Fractional-N Digital PLL with Digitally Controlled Ring Oscillator in 65nm CNMOS," *ISSCC Dig. Tech. papers*, pp. 478-479, Feb., 2010.

[7] D.M. Fischette, A.L.S. Loke, M.M. Oshima, et al., "A 45nm SOI-CMOS Dual-PLL Processor Clock System for Multi-Protocol I/O," *ISSCC Dig. Tech. papers*, pp. 246-247, Feb., 2010.

[8] X. Gao, E.A.M, Klumperink, M. Bohsali, and B. Nauta, "A Low Noise Sub-Sampling PLL in Which Divider Noise is Eliminated and PD/CP Noise is Not Multiplied by N," *IEEE J. Solid State Circuits*, vol. 44, no. 12, pp. 3253-3263, Dec., 2009.

ISSCC 2011 / February 21, 2011 / 4:15 PM

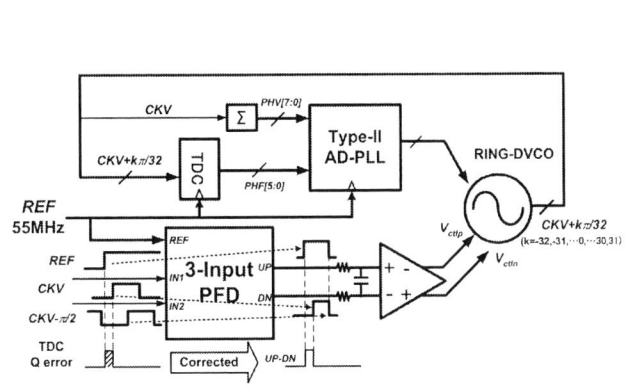

Figure 5.6.1: Block diagram of the proposed integer-N PLL with hybrid loop.

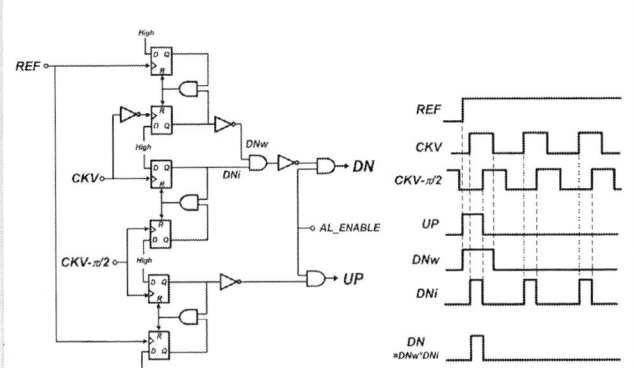

Figure 5.6.2: Detail of the proposed 3-input PFD in the analog PLL.

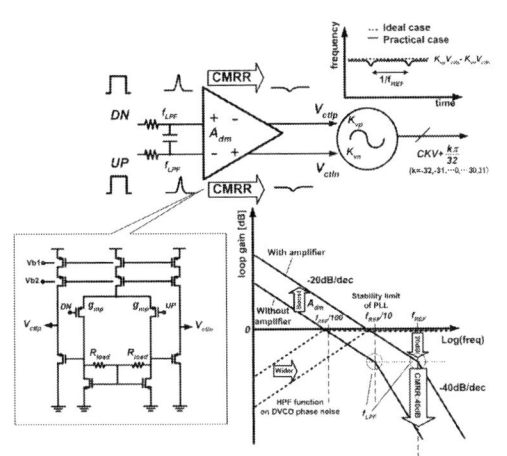

Figure 5.6.3: Concept of the proposed reference spurs reduction technique using the CMRR of the amplifier in the analog PLL.

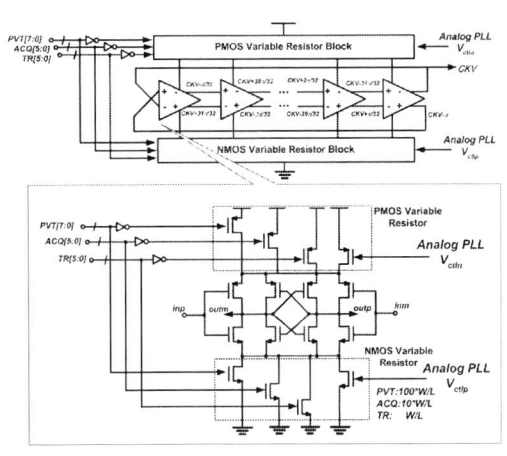

Figure 5.6.4: Circuit schematic of the ring digital/voltage-controlled oscillator (DVCO).

Figure 5.6.5: Measured PLL output spectrum and phase noise at 1.21GHz output.

PLL Type	[6] Digital	[7] Analog	This work Hybrid
Output Frequency	3 GHz	2.5 GHz	1.21 GHz
REF Frequency	25 MHz	100 MHz	55 MHz
Integrated Jitter	1.42 ps	0.99 ps	0.57 ps
Power wo/w LDO	9.6/11.6 mW	33.6/70 mW	51.6/- mW
FoM wo/w LDO	-227.2/-226.3 dB	-224.8/-221.6 dB	-227.7/- dB
Supply Voltage	1.3 V	2.5 V	1.2 V
Reglated Supply Voltage	1.0 V	1.2 V	-
Technology	65 nm CMOS	45 nm SOI	65 nm CMOS

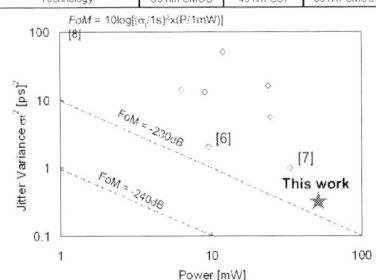

Figure 5.6.6: Performance comparison with recently reported ring-VCO-based PLLs.

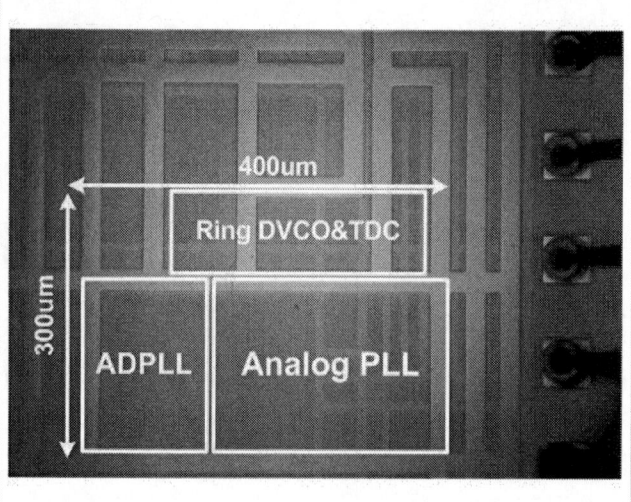

Figure 5.6.7: Chip Micrograph of the test chip in 65nm CMOS.

ISSCC 2011 / SESSION 5 / PLLs / 5.7

5.7 A Rotary-Traveling-Wave-Oscillator-Based All-Digital PLL with a 32-Phase Embedded Phase-to-Digital Converter in 65nm CMOS

Koji Takinami, Richard Strandberg, Paul C. P. Liang,
Gregoire Le Grand de Mercey, Tony Wong, Mahnaz Hassibi

Panasonic, Cupertino, CA

All-digital phase-locked loops (ADPLLs) have recently become more popular as possible alternatives to conventional analog charge-pump-based PLLs [1]. Currently, most of the ADPLLs are based on a time-to-digital converter (TDC) utilizing inverter delay chains. There have been tremendous efforts to improve TDC performance, i.e., maximizing resolution and reducing power consumption, but they normally require additional complex circuits.

Recently, the ring-oscillator-based ADPLLs have been presented in [2, 3], which substantially simplify the architecture of ADPLL by eliminating inverter delay chains as well as calibration routines for period normalization that are required for the conventional TDC-based approach. However, due to the poor phase noise of the ring oscillators and the limited time resolution available from inverters, those techniques are not applicable to modern wireless systems with stringent phase noise requirements.

In this work, we demonstrate a low-phase-noise 4GHz ADPLL with an embedded phase-to-digital converter (PDC), where the rotary traveling-wave oscillator (RTWO) [4] is used as a digitally controlled oscillator (DCO). By using 32 multiphase signals available from the RTWO, the analog phase information is naturally converted into the digital domain, which simplifies the ADPLL architecture while maintaining excellent phase noise both close-in and far-out.

The typical circuit topology of the RTWO is illustrated in Fig. 5.7.1. A cross-connected differential transmission line forms a closed-loop to establish the required feedback. In the steady-state, only one direction of the signal is sustained and propagates along the transmission line. By tapping off the signals from different positions on the resonator, multiphase signals are easily obtained. Since the RTWO relies on the resonance of the distributed LC-tank, unlike ring oscillators, it can achieve excellent phase noise.

Figure 5.7.2 shows the architecture of the proposed ADPLL. Multiphase signals from the RTWO are latched by the reference clock (REF), which provides a pseudo-thermometer code representing the instantaneous phase of oscillation. The bubble correction block eliminates the unexpected bubbles in the outputs. By barrel shifting the outputs with the rotator block, the delay between the fractional phase and the integer phase (counter outputs) are adjusted within fractional phase resolution. This is achieved by minimizing the spurs at the ADPLL output. The remaining skew potentially causes an instantaneous 2π error, but this is corrected by the glitch correction algorithm [5]. The direction of the oscillation is determined by comparing the phase between non-adjacent taps. Based on the direction, either a clockwise look-up table (LUT) or a counter-clockwise LUT is selected to decode the pseudo-thermometer code into binary code. The coefficients of the LUT used to suppress the fractional spurs due to nonlinearity of the PDC are estimated by measuring the statistical distribution of the fractional phase outputs [6]. The integer phase is obtained by counting the number of oscillation period by using a counter. The high-speed $\Delta\Sigma$ modulator is used to enhance frequency resolution.

In the RTWO, since one lap exactly corresponds to half the oscillation period, the PDC gain does not vary over PVT. This removes the need for real-time calibration for period normalization. Besides eliminating power hungry inverter chains, the phase resolution is solely determined by the physical distance between the tap positions. Therefore, by increasing the number of taps, finer phase resolution is easily obtained.

The architecture of the PDC is shown in Fig. 5.7.3. Each of the latches is implemented by a sense-amplifier-based flip-flop. To minimize frequency pulling, a dummy cell, which is clocked by the opposite polarity, is connected. Since the latches are clocked by a low-frequency reference clock and no static current is drawn, current draw is relatively small. To minimize skew, the reference clock is distributed from the center of the resonator to each latch by a clock tree. To achieve a more uniform distribution of the voltage waveform along the resonator, five cross-over sections are introduced. The 4GHz high-speed counter consists of a 2b counter followed by a 6b synchronous counter.

The unit cell of the DCO is shown in Fig. 5.7.4. The capacitor bank consists of coarse 5b MIM varactors for PVT variation (Δf=120MHz) and fine nMOS varactors for acquisition (Δf=1MHz) and tracking (Δf=30kHz). 32 unit cells are distributed along the ring, resulting in a 255-bit thermometer code for both acquisition and tracking. They are programmable from ×0.5 to ×2 to have better control of the frequency step for the entire frequency band. The layout of these blocks is carefully done to maintain good uniformity in the distributed resonator.

The prototype is fabricated in a 65nm CMOS process. The DCO tunes from 2.6 to 4.5GHz, drawing 22mA for the DCO core from a 1.5V supply. The phase noise of the free running DCO is −148dBc/Hz at 20MHz offset. Other analog circuits draw 5mA from a 1.2V supply, excluding the output buffer which uses 18mA.

Figure 5.7.5 shows the measured phase noise. The 1st-order $\Delta\Sigma$ modulator is running at 516.75MHz (the divide-by-8 version of the DCO output frequency). When the ADPLL is operated as an integer synthesizer (f_0=53×78MHz=4.134GHz), the in-band phase noise of −110dBc/Hz at 200kHz offset is achieved. The reference spur is measured as −56dBc at 78MHz offset, which is due to insufficient cancellation of the frequency pulling by the dummy cells. When the ADPLL is operated as a fractional synthesizer, fractional spurs are observed. One of the worst-case conditions of f_0=(53+1/128)×78MHz is also shown in Fig. 5.7.5, where the fractional spur falls inside the loop bandwidth. A fractional spur appears at 609kHz (=78MHz/128) due to the PDC nonlinearity. By updating the LUT based on the statistical distribution of the fractional phase, this spur is suppressed from −30dBc to −40dBc. The spur-reduction techniques presented in the recent literature can also be used to mitigate those spurs, which is a subject for future study.

Figure 5.7.6 gives a performance summary. The theoretical phase resolution is given by π/N_{tap} where N_{tap} represents the number of taps. With 32-taps, the achievable phase resolution is calculated as 5.6°, which corresponds to 3.9ps (1.1ps$_{rms}$) time resolution at 4GHz. With conservative design of this prototype, the total power consumption is relatively large, but the fractional phase detection draws only 350μA (0.42mW), which is orders of magnitude smaller than the state-of-the-art TDC-based ADPLL [5]. Even though we use 32 taps in the prototype, we can further improve the PDC resolution by increasing the number of taps or increasing the DCO frequency. The drawback will be additional layout complexity and a slight increase in the current draw. Figure 5.7.7 shows a die micrograph. The overall active area is 780×780μm². The digital block placed inside of the resonator occupies 500×500μm² where only 10% of the area is used in the prototype.

Acknowledgements:
The authors thank Rick Booth for helpful discussions and advice; Wayne Lee and Dan Garland for technical support; Keyu Chen for the layout work.

References:
[1] R.B. Staszewski, J.L. Wallberg, S. Rezeq, et al., "All-Digital PLL and Transmitter for Mobile Phones," *IEEE J. Solid-State Circuits*, vol. 40, no. 12 ,pp. 2469-2482, Dec., 2005.
[2] M.S.-W. Chen, D. Su, and S. Mehta, "A Calibration-Free 800MHz Fractional-N Digital PLL with Embedded TDC," *ISSCC Dig. Tech. Papers*, pp. 472-473, Feb., 2010.
[3] W. Grollitsch, R. Nonis, and N. Da Dalt, "A 1.4ps$_{rms}$-Period-Jitter TDC-Less Fractional-N Digital PLL with Digitally Controlled Ring Oscillator in 65nm CMOS," *ISSCC Dig. Tech. Papers*, pp. 478-479, Feb., 2010.
[4] J. Wood, T.C. Edwards, and S. Lipa, "Rotary Traveling-Wave Oscillator Arrays: A New Clock Technology," *IEEE J. Solid-State Circuits*, vol. 36, no. 11 ,pp. 1654-1665, Nov., 2001.
[5] M. Lee, M.E. Heidari, and A.A. Abidi, "A Low-Noise Wideband Digital Phase-Locked Loop Based on a Coarse-Fine Time-to-Digital Converter With Subpicosecond Resolution," *IEEE J. Solid-State Circuits*, vol. 44, no. 10, pp. 2808-2816, Oct., 2009.
[6] C. Weltin-Wu, E. Temporiti, D. Baldi, and F. Svelto, "A 3GHz Fractional-N All-Digital PLL With Precise Time-to-Digital Converter Calibration and Mismatch Correction," *ISSCC Dig. Tech. Papers*, pp. 344-345, Feb., 2008.

978-1-61284-303-2/11 $26.00 © 2011 IEEE

ISSCC 2011 / February 21, 2011 / 4:45 PM

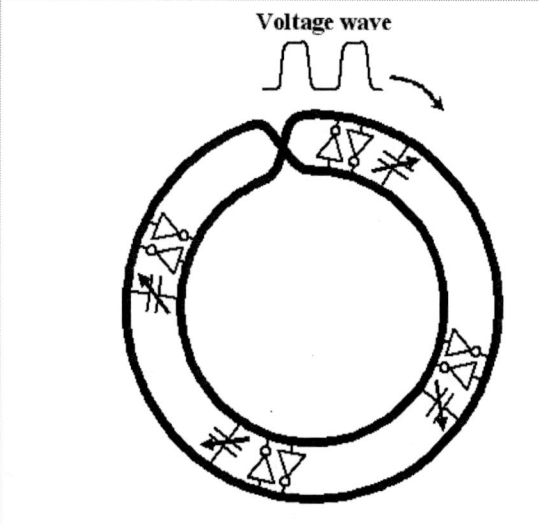

Figure 5.7.1: Typical RTWO circuit topology.

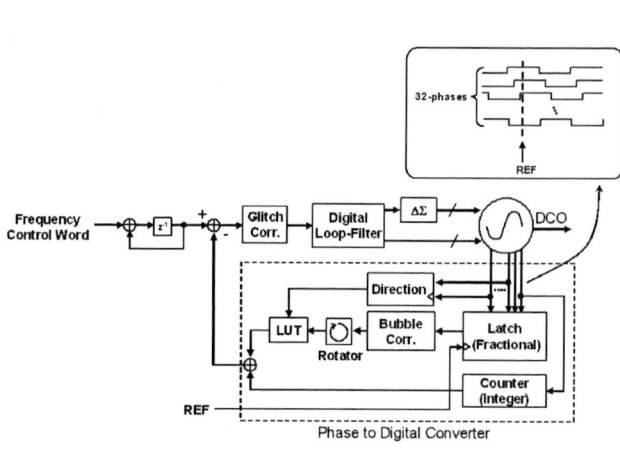

Figure 5.7.2: Block diagram of the proposed ADPLL.

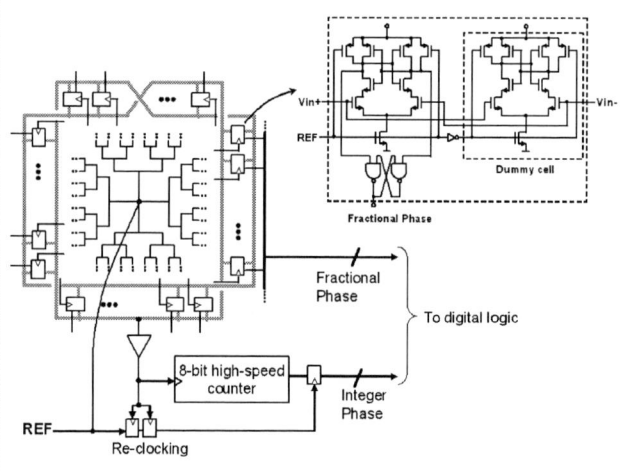

Figure 5.7.3: Phase-to-digital converter implementation.

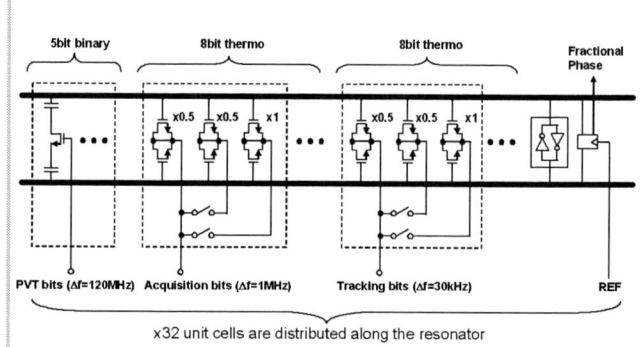

Figure 5.7.4: Schematic of the DCO unit cell.

Figure 5.7.5: Measured phase noise at integer ratio (left) and fractional ratio (right).

Technology	65nm CMOS
Supply voltage	1.5V (DCO core), 1.2V (others)
Active area	780x780μm²
DCO core current	22mA
Other analog current exclude output buffer	5mA
Digital core current	5mA
Reference frequency	78MHz
Output frequency	2.6 to 4.5GHz
Output power	+3dBm
In-band phase noise	-108 to -110dBc/Hz
DCO phase noise	-148dBc/Hz at 4 GHz output, 20 MHz offset
PDC resolution	5.6deg (3.9ps at 4GHz)

Figure 5.7.6: Performance summary.

ISSCC 2011 PAPER CONTINUATIONS

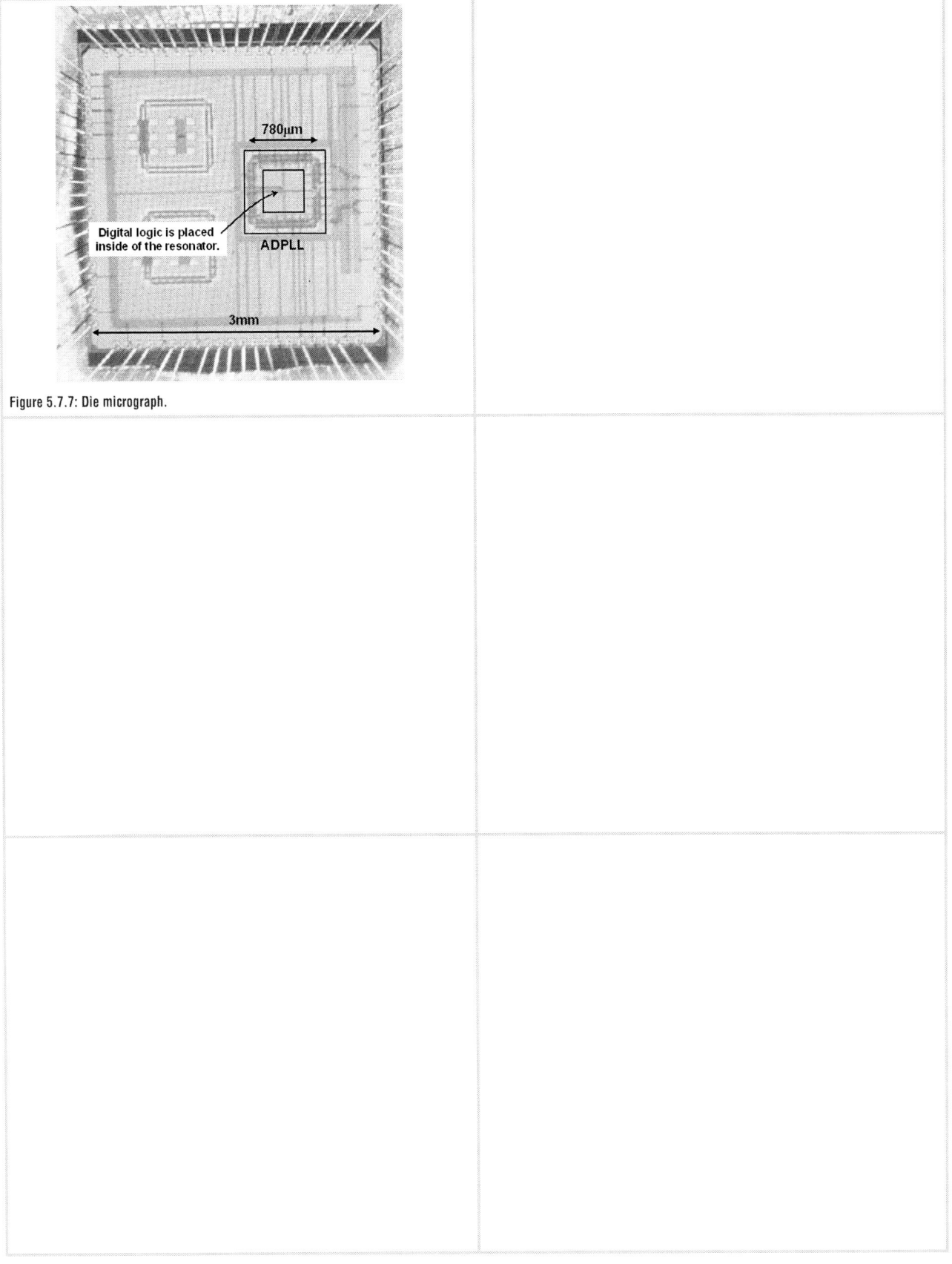

Figure 5.7.7: Die micrograph.

ISSCC 2011 / SESSION 6 / SENSORS & ENERGY HARVESTING / OVERVIEW

Session 6 Overview / IMMD

Sensors & Energy Harvesting

Session Chair: *Aaron Partridge, SiTime, Sunnyvale CA*

Session Co-Chair: *Christoph Hagleitner, IBM Research, Ruschlikon, Switzerland*

Sensors and energy harvesting are growing in importance, as many applications require environmental inputs and often must be placed in remote or inaccessible locations. Advanced MEMS sensors require highly precise interface circuitry that must usually transduce small or delicate signals and mitigate sensor limits, while in some cases the signals are large and must be isolated. When sensor systems function in inaccessible environments it can be difficult to power them. In these cases it is often advantageous to scavenge or harvest power from the system's immediate environments.

Paper 6.1 [STMicroelectronics; U Padova] presents a 3-axis gyroscope drive and transduction circuit. A 3.2×3.2mm² 3-axis 24μm-thick polysilicon surface micromachined gyro combining 3 tuning forks in a single vibrating element is sensed by a 0.13μm HCMOS multiplexed digital readout, achieving a 0.03dps/√Hz rate noise density, a ±0.04dps/°C ZRO and ±2% cross-axis sensitivities.

Paper 6.2 [Delft U] presents a low-power thermal wind speed and direction sensor. A 2-D thermal wind sensor is realized in a standard CMOS process. Two 2nd-order thermal $\Delta\Sigma$ modulators control and digitize the flow-dependent heat distribution in the sensor. The sensor measures wind speed and direction with errors of less than ±4% and ±2°, respectively. It dissipates only 50mW, 9× less than a previous CMOS design and less than that of MEMS-based wind sensors.

Paper 6.3 [U Freiburg; IMTEK] presents a telemetric stress sensor for orthodontic brackets. A CMOS stress-mapping system with 24 sensors is sensitive to in-plane shear stress or differences of normal stresses with a resolution better than 25kPa. The telemetrically powered system consisting of a chip and microcoil communicates at 13.56MHz. Its dimensions of 2×2.5mm² permit the development of smart brackets for a direct force feedback in orthodontic treatments.

Paper 6.4 [Delft U] presents a precision bridge transducer and ADC. The paper presents a 21b read-out IC (ROIC) with ±40mV full scale for precision bridge transducers and thermocouples. The ROIC employs dynamic element matching to achieve an INL of 5ppm and a gain drift of 1.2ppm/°C, multistage chopping to achieve a 1mHz 1/f noise corner at 16.2nV/√Hz, and nested chopping to achieve 200nV offset, while drawing only 270μA from a 5V supply.

Paper 6.5 [CEA-LETI – MINATEC; Schneider Electric] presents a shunt current sensor with integrated transformer signal coupling. An integrated current sensor including a shunt, 2 micro-transformers for 6 kV isolation, a chopper IC and a readout IC is presented. Current measurements are performed with a ±1.5 % nonlinearity over a 0.1-to-100A range. The signal BW ranges from DC to 20kHz, and the overall power consumption is 16mW. The microsystem fits into a 13×7.6mm² SO20 package.

Paper 6.6 [Caeleste] presents a small and sensitive imaging X-ray sensor containing A 16×16 pixel X-ray photon-counting array for indirect detection, i.e. in combination with a scintillator. To count charge packets smaller than 100 electrons it has a noise floor and a comparator threshold of about 15 e⁻$_{rms}$. Counting happens in a nonlinear fashion in the analog domain, yielding pixels with 27-to-40 transistors.

Paper 6.7 [U Idaho; National U Singapore] presents an implantable retinal image sensor that is powered by the light falling on the imaging array. A CMOS energy-harvesting and imaging (EHI) APS imager capable of 7.4fps video capture and 3.5μW power generation is designed, fabricated, and tested in 0.5μm CMOS. It has a 54×50 array of 21μm² EHI pixels, a 10b supply-boosted SAR-ADC and charge-pump circuits consuming only 14.25μW from 1.2V resulting in the lowest power imager with 1.32pW/frame·pixel.

978-1-61284-303-2/11 $26.00 © 2011 IEEE

ISSCC 2011 / February 21, 2011 / 1:30 PM

Paper 6.8 [imec - Holst Centre; Philips Research Laboratories] presents a circuit to convert optical power over a wide brightness range to readily usable voltage. A fully autonomous inductive boost converter for indoor photovoltaic harvesting with maximum power point tracking circuit is implemented in a commercial 0.25µm CMOS process. The converter can handle input power from 5µW up to 10mW and charge a battery or a super-capacitor up to 5V. Its control circuit consumes between 0.8 and 2.1µA depending on the input power level, resulting in a peak end-to-end efficiency of 70% when tracking a maximum input power of 17µW.

Paper 6.9 [U Michigan] presents a transduction circuit for a MEMS energy harvester that converts vibration to usable circuit power. A self-supplied energy harvester platform is developed including a MEMS harvester integrated with a power management IC, for autonomous charging of an energy reservoir. The volume of the system is <0.3cm³, and it can charge a reservoir from 0 to 1.25V in <15min under 0.3g vibration input at 69Hz.

6

Presenters:

1:30 PM

6.1 A Low-Power 3-Axis Digital-Output MEMS Gyroscope with Single Drive and Multiplexed Angular Rate Readout

L. Prandi, STMicroelectronics, Cornaredo, Italy

2:00 PM

6.2 A 50mW CMOS Wind Sensor with ±4% Speed and ±2° Direction Error

J. Wu, Tsinghua University, Beijing, China;
Delft University of Technology, Delft,
The Netherlands

2:30 PM

6.3 A Telemetric Stress-Mapping CMOS Chip with 24 FET-Based Stress Sensors for Smart Orthodontic Brackets

M. Kuhl, University of Freiburg - IMTEK, Freiburg,
Germany

3:15 PM

6.4 A 21b ±40mV Range Read-Out IC for Bridge Transducers

R. Wu, Delft University of Technology, Delft,
The Netherlands

3:45 PM

6.5 A ±1.5% Nonlinearity 0.1-to-100A Shunt Current Sensor Based on a 6kV Isolated Micro-Transformer for Electrical Vehicles and Home Automation

F. Rothan, CEA-LETI-MINATEC, Grenoble, France

4:00 PM

6.6 Indirect X-ray Photon-Counting Image Sensor with 27T Pixel and 15e⁻$_{rms}$ Accurate Threshold

B. Dierickx, Caeleste, Antwerp, Belgium;
Vrije Universiteit Brussel,
Brussels, Belgium

4:15 PM

6.7 A 1.32pW/frame·pixel 1.2V CMOS Energy-Harvesting and Imaging (EHI) APS Imager

S. U. Ay, University of Idaho, Moscow, ID

4:45 PM

6.8 5µW-to-10mW Input Power Range Inductive Boost Converter for Indoor Photovoltaic Energy Harvesting with Integrated Maximum Power Point Tracking Algorithm

Y. Qiu, imec - Holst Centre, Eindhoven,
The Netherlands

5:00 PM

6.9 A Self-Supplied Inertial Piezoelectric Energy Harvester with Power-Management IC

E. Aktakka, University of Michigan, Ann Arbor, MI

978-1-61284-303-2/11 $26.00 © 2011 IEEE

ISSCC 2011 / SESSION 6 / SENSORS & ENERGY HARVESTING / 6.1

6.1 A Low-Power 3-Axis Digital-Output MEMS Gyroscope with Single Drive and Multiplexed Angular Rate Readout

Luciano Prandi[1], Carlo Caminada[1], Luca Coronato[1], Gabriele Cazzaniga[1], Fabio Biganzoli[1], Riccardo Antonello[2], Roberto Oboe[2]

[1]STMicroelectronics, Cornaredo, Italy,
[2]University of Padova, Vicenza, Italy

Motivated by the increasing demand of integrated inertial-sensing solutions for motion processing and dead-reckoning navigation in handheld devices and low-cost GPS navigators, this paper reports the details of a 3-axis silicon MEMS vibratory gyroscope that fulfills the pressing market requirements for low power consumption, small size and low cost. Thanks to a compact mechanical design that combines a triple tuning-fork structure within a single vibrating element, our solution achieves satisfactory performance in terms of thermal stability, cross-axis error, and acoustic noise immunity by using a small die size. Furthermore, the presence of a single primary vibration mode for the excitation of the 3 tuning-forks, together with the possibility of sensing the pickoff modes in a multiplexing fashion, allows to design a small-area, low-power ASIC.

The mechanical structure design is explained with the aid of Fig. 6.1.1. The structure comprises 4 suspended plates coupled to each other by means of 4 folded springs connected to their outer corners, and elastically connected to a central cross-shaped hinge by an additional set of central coupling springs. The primary mode of vibration (driving mode) consists of an in-plane inward/outward radial motion of the plates: on the whole, the structure cyclically expands and contracts, similarly to a "beating heart" (hence, the name). Primary actuation is provided with a set of comb-finger electrodes placed on a pair of opposite plates; the mechanical motion is then propagated to the second pair by means of the coupling folded springs at the corners. The secondary modes of vibration (sensing modes) consist of an in-plane, opposite-phase motion of the second pair of plates (yaw mode), and two out-of-plane, opposite-phase motions of both pairs (pitch and roll modes). The yaw mode is sensed by a set of parallel-plate electrodes located on the second pair of plates, whereas the pitch and roll motions are detected by sensing the capacitive variations between each plate and an electrode placed underneath; additional comb-finger electrodes are reserved for sensing the vibrating motion of the driving mode. The mechanical coupling between the 2 proof masses of each sensing pair allows to read the secondary vibrating motions in differential mode, thus improving rejection of external linear accelerations and vibrations; moreover, each secondary mode has a single resonant sensing frequency, instead of 2 independent frequencies that require accurate matching to avoid performance degradations, such as in designs with uncoupled proof masses. The overall mechanical structure has frequency-unmatched primary and secondary modes, with a nominal primary resonant frequency of 20kHz. This design choice, combined with a high Q-factor, guarantees a satisfactory level of acoustic noise isolation.

The Coriolis force exciting a secondary mode is proportional to the velocity of the driving mode and the input angular rate, and directed orthogonally to both the driving axis and sensor rotation axis. The angular rate measurement is obtained from the sensed Coriolis acceleration by demodulation, once the driving mode is oscillated at constant amplitude [1].

The primary mode is excited to oscillate at resonance by closing a feedback loop around the micro-resonator made up of the resonating masses and the drive-readout/drive-forcing comb-finger electrodes. In the feedback path, the capacitive unbalancing generated by the oscillating motion of the primary mode is transduced into a voltage signal by a differential charge amplifier (CA); then, a band-pass (BP) switched-capacitor (SC) amplifier removes the residual offset and provides the necessary phase adjustment to have a total loop phase shift of 360° at the resonant frequency, which is required for enforcing a sustained oscillation in the electromechanical loop. The BP amplifier output is interpolated by a

2nd-order continuous-time (CT) low-pass (LP) Chebyshev filter and amplified by a variable-gain amplifier (VGA). The VGA gain is automatically tuned by an outer automatic gain control (AGC) loop to regulate and verify the amplitude of the sustained oscillation at the CT-LP filter output to a constant set-point value. Finally, the VGA output, boosted by a charge-pump, is fed back to the comb-drive actuating electrodes. All internal timings are generated by a PLL synchronized with the CT-LP filter output.

A single, time-division-multiplexed open-loop readout interface is used to retrieve the angular rate measurements out of the Coriolis accelerations along the 3 sensing axes. A differential CA front-end converts the capacitive unbalancing induced by the Coriolis movement into a voltage signal, which is then synchronously AM-demodulated using a carrier in-phase with the velocity of the primary mode motion. A 12b SAR ADC performs internal A/D conversion at a rate of 6.06kHz/axis; a 100/200/400/800Hz output data rate (ODR) is selected by changing the decimation factor of the output sinc-decimators. The final output wordlength is equal to 16b. The compensation of the quadrature error is performed at the CA input with a purely passive structure consiting of a dynamically reconfigurable bank of calibrated capacitors [2].

The micro-mechanical element is fabricated with the STMicroelectronics (STM) proprietary thick-film epitaxial polysilicon surface-micromachining process (ThELMA) on a 3.2×3.2mm² die; the ASIC is implemented on a 2.5×2.5mm² die with a 0.13μm min channel length CMOS process. The MEMS and ASIC dies are stack-assembled in a single 4×4×1.1mm³ plastic LGA package. Thanks to the STM mechanical structure design, involving only attachment points to the substrate (instead of multiple attachments to the cap and the substrate), the sensor output exhibits excellent immunity to external mechanical stresses applied to the package.

The characterization results of 33 different samples are reported in Figs. 6.1.3–6.1.5. The overall performances are excellent: the average noise density level is <0.03dps/√Hz (with BW=40Hz and ODR=200Hz – note: dps = degree per second), and the zero-rate output (ZRO) and sensor scale factor (So) are very stable over temperature – with FS=2000dps, the ZRO temperature sensitivity is less than ±0.04dps/°C, while the scale-factor change over the temperature range of -40 to +85°C is within ±2% of the factory-trimmed value (So=70mdps/LSB with FS=2000dps). The design robustness is certified by the tight statistical distributions of the ZRO and So temperature sensitivities. Thanks to the highly symmetric mechanical design, the cross-axis sensitivities, measured as a percentage of the nominal selected full scale, are always below ±2% and mainly due to mounting tolerances during packaging. High immunity to acoustic noise is evident from the sensor output response to an acoustical stimulus (i.e. sine noise at 90 dBSPL with frequency sweeping in the range 500Hz to 25kHz with steps of 5Hz) reported in Fig. 6.1.6. The plots show the average values of the pitch/roll/yaw angular rate outputs for each frequency of the sinusoidal acoustical stimulus (tests are performed with FS=2000dps, ODR=200Hz and output BW=50Hz).

Regarding power consumption, with a supply voltage in the range of 2.16 to 3.6V, the current absorbtion is 6.1mA during normal operation, 1.5mA in sleep mode (sensing electronics switched off, but with the driving microresonator still operative to reduce the power-on time), and 5μA in power-down mode.

References:
[1] J. Geen, et al., "Single-Chip Surface Micromachined Integrated Gyroscope with 50°/h Allan Deviation," *IEEE J. Solid-State Circuits*, pp. 1860-1866, Dec. 2002.
[2] R. Antonello, R. Oboe, L. Prandi, F. Biganzoli, and C. Caminada, "Open loop Compensation of the Quadrature Error in MEMS Vibrating Gyroscopes," *Annual Conference of IEEE Industrial Electronics Society (IECON)*, Nov. 2009.

978-1-61284-303-2/11 $26.00 © 2011 IEEE

ISSCC 2011 / February 21, 2011 / 1:30 PM

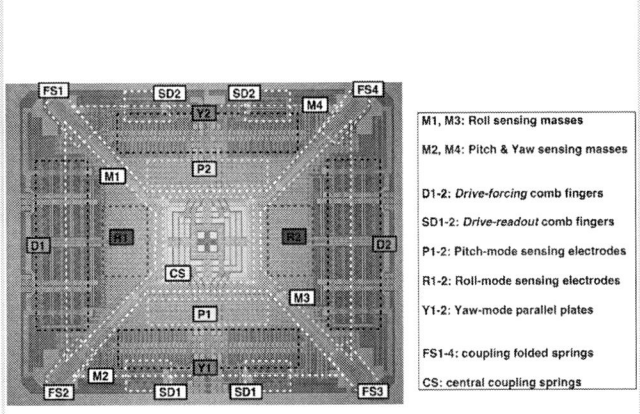

Figure 6.1.1: Micrograph of the MEMS gyroscope (actual die size = 3.2×3.2mm²).

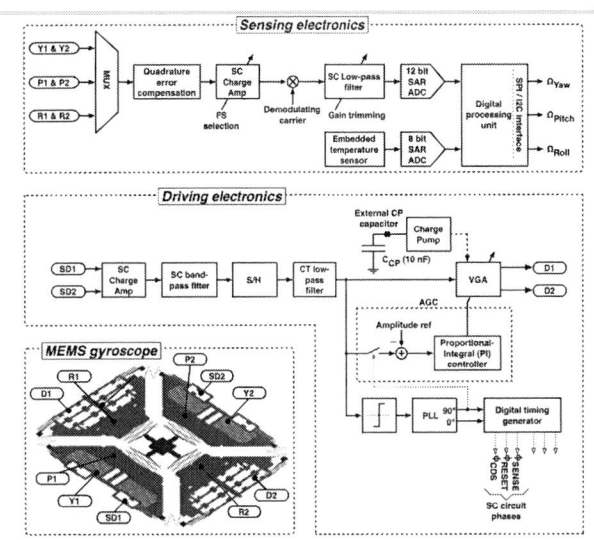

Figure 6.1.2: System architecture block diagram.

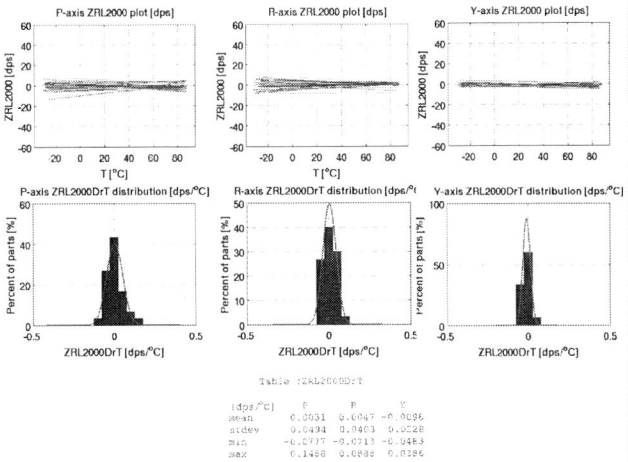

Figure 6.1.3: ZRO stability over temperature (top row); statistical distribution of the ZRO temperature sensitivity over a set of 33 samples (bottom row). Tests are performed with FS=2000dps.

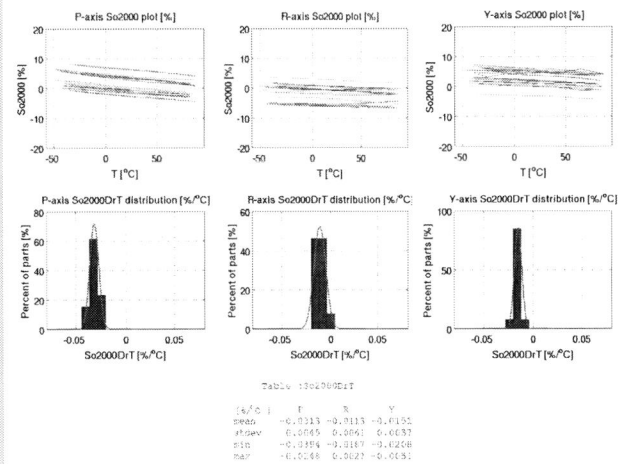

Figure 6.1.4: Scale factor (So) variation over temperature (top row); statistical distribution of the scale factor temperature sensitivity over a set of 33 samples (bottom row). Tests are performed with FS=2000dps.

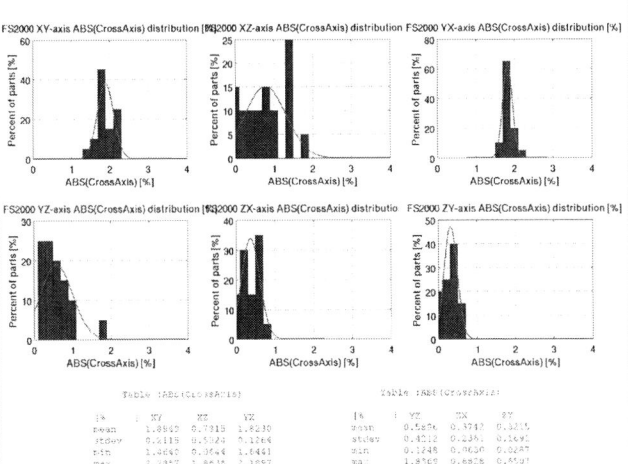

Figure 6.1.5: Statistical distributions of the cross-axis sensitivities over a set of 33 samples. Tests are performed with FS=2000dps.

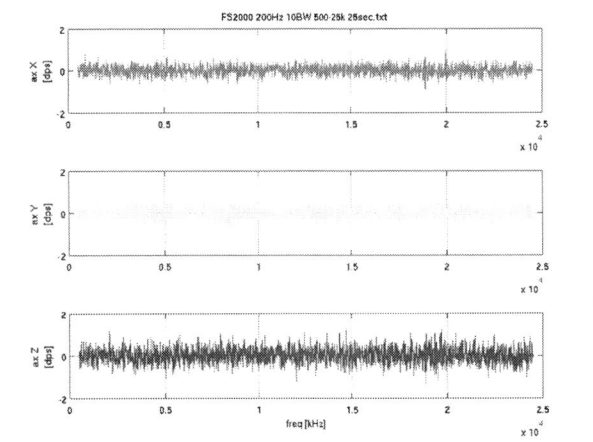

Figure 6.1.6: Sensor output response to an acoustical stimulus (sine noise at 90 dBSPL with frequency sweeping in the range 500Hz to 25kHz with steps of 5Hz).

978-1-61284-303-2/11 $26.00 © 2011 IEEE

ISSCC 2011 PAPER CONTINUATIONS

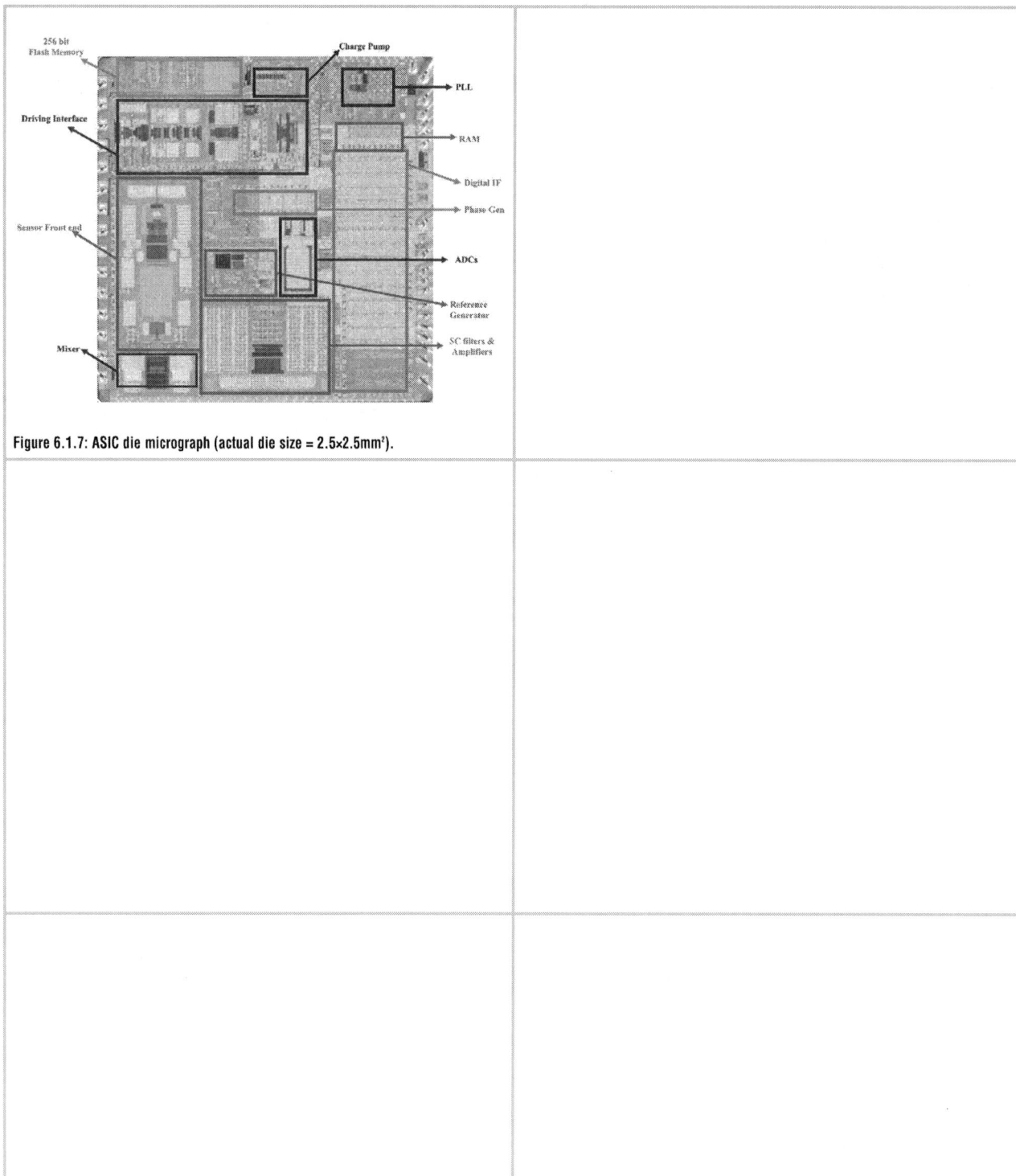

Figure 6.1.7: ASIC die micrograph (actual die size = 2.5×2.5mm²).

ISSCC 2011 / SESSION 6 / SENSORS & ENERGY HARVESTING / 6.2

6.2 A 50mW CMOS Wind Sensor with ±4% Speed and ±2° Direction Error

Jianfeng Wu[1,2], Youngcheol Chae[2], Caspar P.L. van Vroonhoven[2], Kofi A.A. Makinwa[2]

[1]Tsinghua University, Beijing, China,
[2]Delft University of Technology, Delft, The Netherlands

This paper describes a smart CMOS wind sensor that measures wind speed and direction without moving parts. It combines a 2-D thermal flow sensor and the required readout circuitry on a standard CMOS chip. The flow-dependent heat distribution in the chip is controlled and digitized by two thermal delta-sigma (T$\Delta\Sigma$) modulators. Both wind speed and direction can then be determined from their bitstream outputs. For wind speeds ranging from 1 to 25m/s, the sensor's error is less than ±4% (speed) and ±2° (direction), while consuming 9× less power than a previous design [1].

The wind sensor consists of a heated silicon chip glued to one side of a thin ceramic disc (Fig. 6.2.1). The wind is passed over the other side, and so the chip makes good thermal contact with the flow while still being protected from direct mechanical contact. The wind cools the sensor asymmetrically, thus inducing a temperature gradient δT in the chip. Wind speed and direction can then be determined by measuring orthogonal components δT_{ns} and δT_{ew} of this gradient with on-chip thermopiles [2].

Alternatively, the sensor can be read out by *canceling* δT by dynamically adjusting the power dissipated in its 4 heaters (Fig. 6.2.1). In [1], this was implemented by two T$\Delta\Sigma$ modulators, which canceled δT_{ns} and δT_{ew} by driving opposing pairs of heaters with pulses of constant power P_{ref}. Wind speed and direction could then be determined by measuring orthogonal components δP_{ns} and δP_{ew} of the resulting power gradient δP. In [1], an extra control loop maintained the sensor at a constant temperature above ambient temperature. This constant-temperature difference (CTD) mode ensured that $|\delta P|/P_{ref}$ had a simple square-root law relationship with wind-speed (Fig. 6.2.2). To cope with the inaccuracy of the off-chip ambient temperature sensor, however, a temperature difference of about 10°C was used, which required up to 450mW of heater power. In this work, the sensor is operated in constant power (CP) mode, i.e. without the extra control loop. This simplifies the system, enables a 9× reduction in heater power, and increases the sensor's sensitivity at low wind speeds (Fig. 6.2.2) [3].

For a desired sensor resolution, the minimum heater power is, in principle, only limited by the thermal noise of the thermopiles and the readout circuitry. In practice, however, the offset of the readout circuitry is the main limiting factor, as this must be less than the thermopiles' output amplitude, which decreases with heater power. As in [1], the sensor can be read out by embedding it in the feedback loops of 2 auto-zeroed T$\Delta\Sigma$ modulators. The polarity of the thermopiles' output is then determined by a latched comparator preceded by an auto-zeroed preamp (Fig. 6.2.3a). Pairs of opposing heaters serve as a single-bit thermal DAC, while the sensor's thermal inertia serves as the loop filter. The modulator's feedback loop maintains $\delta T = 0$ and simultaneously digitizes δP. However, its resolution will be limited by the poor noise shaping provided by the thermal loop filter [1].

The resolution of each T$\Delta\Sigma$ modulator can be improved by augmenting its thermal filter with an electrical integrator (Fig. 6.2.3b). The integrator provides additional noise shaping and so the resulting "second" order modulator will have significantly more resolution than the "first" order T$\Delta\Sigma$ modulator used in [1]. To achieve lower offset and 1/f noise, the integrator is chopped rather than auto-zeroed. Chopping at the modulators' sampling frequency f_s effectively cancels the resulting chopper ripple by ensuring that it is only sampled at zero-crossings. Since the thermal filter behaves like a rather lossy integrator, the resulting modulator is stable, i.e. the integrator's output is always bounded, despite the absence of an explicit feed-forward or feedback path.

Each 2nd-order T$\Delta\Sigma$ modulator is built around a g_m-C integrator (Fig. 6.2.3b). To ensure that the 100kΩ thermopiles are the dominant source of noise, its transconductance is set to 20μS. A 120pF integration capacitor then ensures that the amplitude of the chopper ripple (f_s = 10kHz) is below 100mV, and that the integrator's pole is low enough (<1Hz) to obtain good noise shaping in the sensor's bandwidth (a few Hz). The integrator's output is then digitized by a latched comparator. To compare the performance of 1st and 2nd-order T$\Delta\Sigma$ modulators, an auto-zeroed amplifier was also realized (Fig. 6.2.3a). This consists of a folded-cascode OTA, an auxiliary g_m-stage and an 80pF offset-storage capacitor. Via a pair of multiplexers, the sensor can then be switched between 1st and 2nd-order modes.

The 4×4mm^2 chip is realized in a 0.7μm CMOS process (Fig. 6.2.7). Four poly-silicon resistors act as heaters, while four p$^+$/Al thermopiles sense on-chip temperature differences. Opposing thermopiles sense the same temperature difference and so are connected in series. The readout circuits are located in the middle of the chip.

The power spectrum of the sensor's bitstream outputs was measured at wind speeds of 0m/s and 25m/s (Fig. 6.2.4). In 1st-order mode, with P_{ref} = 25mW and f_s = 10kHz, the thermal filter results in 12b resolution into a 1Hz bandwidth. Above 100Hz, the noise shaping ceases, because the thermal filter's output then drops below the input-referred noise of the auto-zeroed OTA, thus effectively breaking the feedback loop [4]. In 2nd-order mode, the extra integrator increases the resolution to 15b. The measurements show that the modulator's low-frequency noise is now significantly less than the noise caused by wind turbulence (at 25m/s) or random air currents (at 0m/s). In fact, in order to measure the modulator's true noise floor at 0m/s, the sensor had to be shielded by a small box. Unlike a conventional $\Delta\Sigma$ modulator, however, the 2nd T$\Delta\Sigma$ modulator outputs a square-wave with a flow-dependent duty-cycle. Its output spectrum thus exhibits strong tones at a fundamental frequency (about 40Hz) and its harmonics. The fundamental frequency is determined by the sensor's thermal inertia and so varies by only a few percent from sample to sample. A complex fixed-length sinc^4 decimation filter would normally be required to suppress the tones and achieve 15b resolution into a 1Hz bandwidth. In this work, a simpler tracking sinc^2 filter is used instead. The filter's length is dynamically set to 20 periods of the output square-wave, which ensures that the tones are precisely notched out. For flexibility, the filter was implemented off-chip.

The decimated outputs of the differential modulators are sinusoidal functions of wind direction whose amplitude is a monotonic function of wind speed (Fig. 6.2.5). However, their offset (due to packaging asymmetry) is also a weak function of wind speed. By using a physics-based calibration model [3], both wind speed and direction can be determined within 2 iterations of a Newton-Raphson-based recursive algorithm implemented on an off-chip microprocessor. Measurements on 3 samples show that in 1st-order mode the sensor achieves a resolution of 0.08m/s, which improves to 0.01m/s in 2nd-order mode. The sensor's measured error is less than ±4% (speed) and ±2° (direction) respectively for wind speeds between 1 and 25m/s (Fig. 6.2.6). The interface circuitry draws 25μA, and so the sensor's total power dissipation is mainly due to the heaters. At 50mW, this is 9× lower than [1], and is less than that of similarly robust MEMS-based sensors [5,6].

References:

[1] K.A.A. Makinwa and J.H. Huijsing, "A smart CMOS wind sensor," *ISSCC Dig. Tech. Papers*, pp. 432–479, Feb. 2002.

[2] B.W. van Oudheusden, "Silicon thermal flow sensors," *Sensors and Actuators A*, vol. 30, pp. 5–26, 1992.

[3] K.A.A. Makinwa and J.H. Huijsing, "Constant Power Operation of a 2-D Flow Sensor using Thermal Sigma-Delta Modulation Techniques," *Trans. on Instrumentation and Measurement*, vol. 51, no. 4, pp. 840–844, Aug. 2002.

[4] O. Leman, F. Mailly, L. Latorre, and P. Nouet, "Noise Analysis of a First-order Thermal $\Sigma\Delta$ Architecture for Convective Accelerometers," *Analog Integrated Circuits and Signal Processing*, vol. 63, no. 3, pp. 415–423, 2010.

[5] S. Kim, T. Nam, and S. Park, "Measurement of flow direction and velocity using a micromachined flow sensor," *Sensors and Actuators A: Physical*, vol. 114, pp. 312-318, 2004.

[6] G. Shen, M. Qin, Q. Huang, H. Zhang, and J. Wu, "A FCOB packaged thermal wind sensor with compensation," *Microsystems Technology*, vol. 16, pp. 511–518, 2010.

978-1-61284-303-2/11 $26.00 © 2011 IEEE

ISSCC 2011 / February 21, 2011 / 2:00 PM

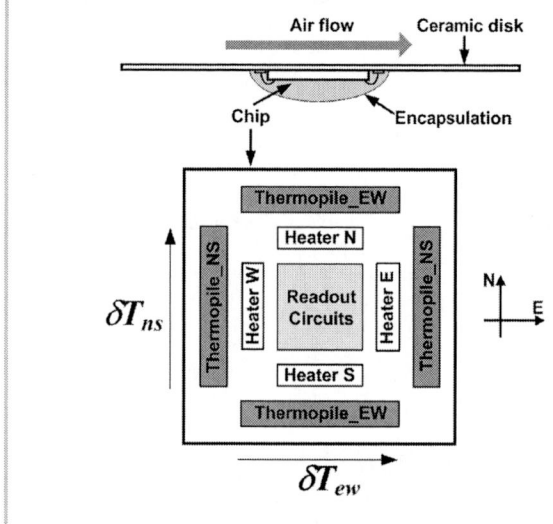

Figure 6.2.1: Block diagram of the smart wind sensor.

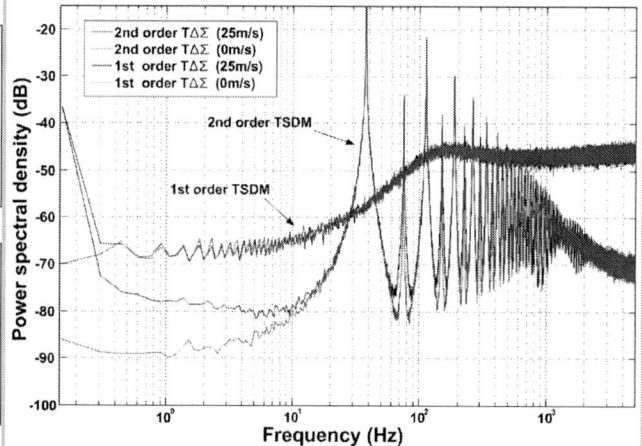

Figure 6.2.2: Wind sensor characteristic in CP and CTD modes.

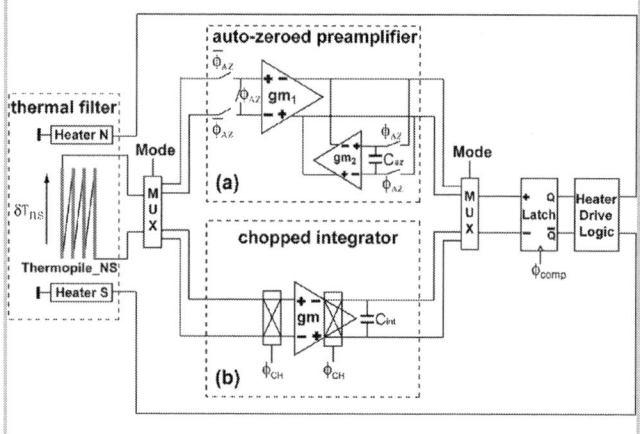

Figure 6.2.3: Block diagram of (a) auto-zeroed 1st order thermal ΔΣ modulator (b) chopped 2nd order thermal ΔΣ modulator.

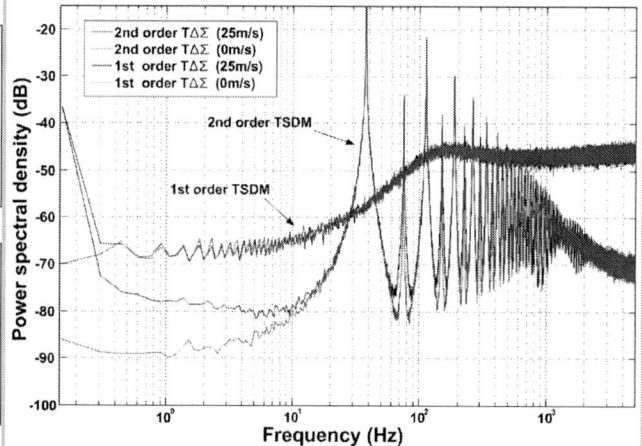

Figure 6.2.4: Measured output spectrum of the 1ˢᵗ and 2ⁿᵈ-order thermal ΔΣ modulators at wind speeds of 0m/s and 25m/s.

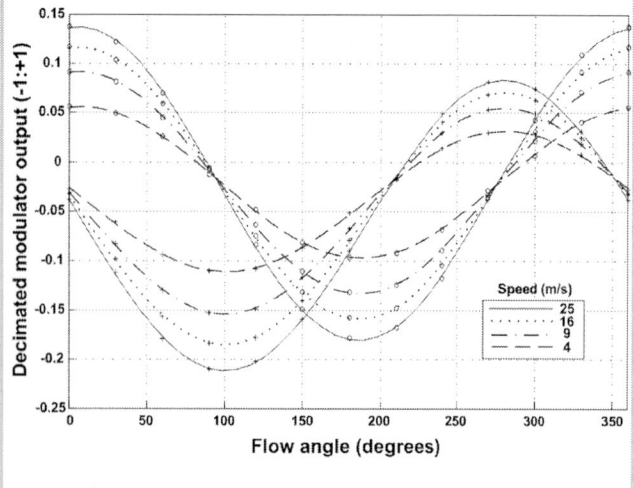

Figure 6.2.5: Measured output of the north-south (o) and east west (+) modulators.

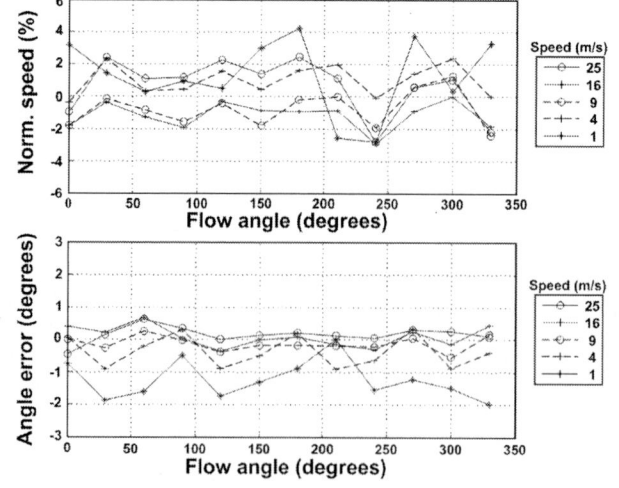

Figure 6.2.6: Measured wind speed and direction error.

978-1-61284-303-2/11 $26.00 © 2011 IEEE

ISSCC 2011 PAPER CONTINUATIONS

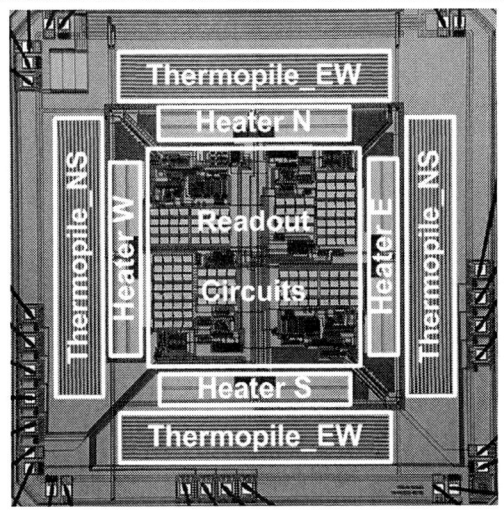

Figure 6.2.7: Wind sensor micrograph.

ISSCC 2011 / SESSION 6 / SENSORS & ENERGY HARVESTING / 6.3

6.3 A Telemetric Stress-Mapping CMOS Chip with 24 FET-Based Stress Sensors for Smart Orthodontic Brackets

Matthias Kuhl[1], Pascal Gieschke[1], Daniel Rossbach[1],
Sascha Alexander Hilzensauer[1], Patrick Ruther[1], Oliver Paul[1],
Yiannos Manoli[1,2]

[1]University of Freiburg - IMTEK, Freiburg, Germany,
[2]HSG-IMIT, Villingen-Schwenningen, Germany

With ongoing miniaturization in technology and increasing complexity in assembly and packaging, stress-sensing microsystems gain importance for the evaluation of IC packages [1]. Additionally, integrated stress-sensor systems are attractive for miniaturized force and torque sensing in various fields. The chip presented here is designed for the next generation of smart brackets, i.e. intelligent orthodontic brackets serving the orthodontist with measured loads applied to each treated tooth during therapy [2]. Furthermore, wireless communication is beneficial for achieving access to harsh environments and to reduce maintenance [3], and is mandatory for clinical smart brackets.

The system presented in this paper as shown in Fig. 6.3.1 generates a map of 2 in-plane stress components on the chip surface by integrating 24 stress sensors. For offset compensation each sensor can be biased in 4 directions by a MUX. A variable-gain differential difference amplifier (DDA) together with a 10b SAR ADC performs the sensor readout. For power and data transmission a telemetric interface circuit working at 13.56MHz with ASK modulation is integrated. For telemetric coupling, a 35-winding microcoil with gold leads is developed. Chip and coil occupy an area of 2×2.5mm² each and can be connected via flip-chip bonding. The circuit is implemented in a 0.35μm CMOS process, the coil in an electrodeposited gold process.

The application flow can be described as follows: the power required to wake-up the measurement system is 138mW from an electromagnetic field transmitted at 13.56MHz. After start-up and as long as a minimum of 29mW are available the system remains operating and alternates between 3 states: (1) charging the buffer capacitor, (2) reading out one sensor direction, and (3) sending the sampled data. With each cycle another sensor element is selected.

24 stress sensors of 2 kinds are placed at strategic positions on the chip. Each of them forms a square of active elements and can be described by 4 transistors in a Wheatstone bridge arrangement (Fig. 6.3.2 top). Since the sensors are symmetrical, the current and sensing direction is rotated by 90° to eliminate several non-mechanical signal contributions [4]. The sensitivity to normal or shear stress depends on the orientation within the crystal structure and polarity of doping, leading to 45° NMOS sensors for shear stress and 0° PMOS sensors for the difference of normal stresses (Fig. 6.3.2 bottom). Both types of sensors are characterized with 3.3V as drain-source and gate-source voltage. Sensitivities of -990±2.89μV/MPa and 910±4.68μV/MPa are determined under well-defined mechanical stresses with an average current consumption of 209μA for NMOS and 47μA for PMOS sensors.

The optimal sensor positions are verified by finite element analysis, modeling the chip within an orthodontic bracket [2]. Furthermore, low-stress areas are identified for the circuitry, especially for the analog elements of the sensor readout.

The sensor readout is performed by a DDA with 16 digitally selectable amplifications ranging from 1 to 271. The attached 10b SAR ADC forwards the digitized sensor value to a gain controller (GC), which determines the ideal gain value with respect to range on the basis of a binary tree search. If the digitized sensor value is out of the DDA's linear range the gain is reduced; otherwise it is increased (Fig. 6.3.3). The highest measured value in the linear region is stored. After the gain is set successfully, the calibrated readout is performed a second time and the average value together with the sensor number, bias direction and determined gain is stored in a 22b serial shift register for data transmission.

With the ADC clocked at 847.5kHz and a settling time of 80μs after each gain alteration, the readout time is 600μs for each sensor direction. The sensor readout consisting of DDA, ADC, and GC consumes 290μA at 3.19V.

A telemetric interface performs the transmission of power into and data out of the system and is designed for 13.56MHz. A high-voltage full-wave rectifier allows input voltages up to 18V$_{pp}$. A combination of a coarse and a fine voltage limiter together with a bandgap reference and a power-on-reset (POR) circuit enables a stabilized output voltage with defined startup behavior. In measured chip-to-chip variations, the coarse limiter ties the rectified output voltage during the system startup to 3.15±0.05V. After 250μs the fine limiter controlled by the bandgap reference clamps the output voltage to 3.19±0.01V (Fig. 6.3.4 top). To avoid malfunctions caused by memory errors, a hysteretic comparator activates the system above 3.1V and resets it below 940mV. With the clamps being the only voltage control, on-chip buffer capacitors with a total of 1.18nF are integrated for voltage stabilization.

For on-chip timing a 13.56MHz power signal is extracted and 2 counters generate an 847.5kHz signal for the digital control logic and the ADC and a 13.56kHz signal for the speed of data transmission. The data transmission itself is realized by ASK modulation of the received power signal. 22b per sensor direction are transmitted after each readout at 13.56kHz, modulated with 847.5kHz.

A microcoil for power and data transmission is fabricated using an electrodeposited Au process in thick photoresist on Pyrex wafers followed by SU-8 encapsulation. With a wire track width of 5μm, height of 20μm and pitch of 15μm, 35 windings fit in an area of 2×2.5mm². The resulting resistance, inductance and capacitance are 30Ω, 2μH and 1pF, respectively. This results in a self-resonant frequency of 113MHz (Fig. 6.3.4 bottom).

Measurement results show that the readout and data transmission of all 24 sensors take a total of 530ms. Without applied stresses the readout standard deviation per sensor is 1.9LSB, resulting in a theoretical maximum resolution of 22.1kPa for the NMOS and 24.0kPa for the PMOS sensors.

Defined mechanical stresses applied to the chip using a bending and a torsion bridge reveal a sensitivity of -926μV/MPa for the NMOS and 880μV/MPa for the PMOS sensors (Fig. 6.3.5). The decreased sensitivity of the NMOS sensors and their nonlinearity below 5mV are caused by a decreased common-mode range of the DDA compared to simulation results. A resolution of 22μV corresponding to 23.6kPa for the NMOS and 24.8kPa for the PMOS sensors is measured at maximum gain.

The system with microcoil in combination with an external MuRata series LQW SMD coil is fully operational up to a distance of 2mm, clearly separating power and data into 2 channels by the modulation frequency. The telemetrically powered system receives up to 250mW, of which 1.75mW are consumed by the sensor biasing and signal processing.

The complete system size is 2×2.5mm² (Fig. 6.3.6), enabling the integration into the basis of orthodontic brackets, with chip and coil being connected via flip chip. Figure 6.3.7 summarizes the measured specifications in comparison to [3] and [4]. Compared to the only published wireless stress system [3], the resolution is increased by a factor of 5 with a reduction of 75% in area and 70% in power. The system thus paves the way towards the implementation of smart brackets for orthodontic therapy while introducing direct sensory feedback.

Acknowledgements:
The authors would like to thank the German Research Foundation (DFG) for financial support through projects MA 2193/7-1 & PA 792/5-1.

References:
[1] J. Roberts, et al., "Characterization of Microprocessor Chip Stress Distributions During Component Packaging and Thermal Cycling," *Electronic Components and Technology Conference (ECTC)*, pp. 1281-1295, June 2010.
[2] B.G. Lapatki, et al., "Smart Bracket for Multi-dimensional Force and Moment Measurement," *Journal of Dental Research 86*, pp. 73-78, Jan. 2007.
[3] M. Suster, et al., "A Wireless Strain Sensing Microsystem with External RF Power Source and Two-Channel Data Telemetry Capability," *ISSCC Dig. Tech. Papers*, pp. 380-381, Feb. 2007.
[4] P. Gieschke, et al., "CMOS Integrated Stress Mapping Chips with 32 N-Type or P-Type Piezoresistive Field Effect Transistors," *Int. Conf. Micro Electro Mechanical Systems (MEMS)*, pp. 769-772, Jan. 2009.

ISSCC 2011 / February 21, 2011 / 2:30 PM

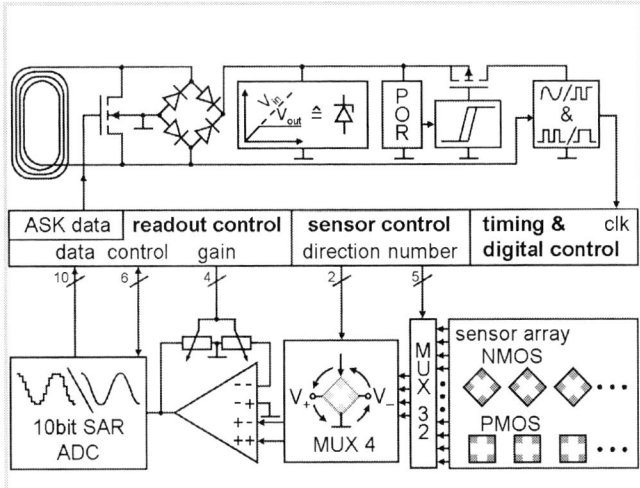

Figure 6.3.1: Schematic of the telemetric stress mapping system.

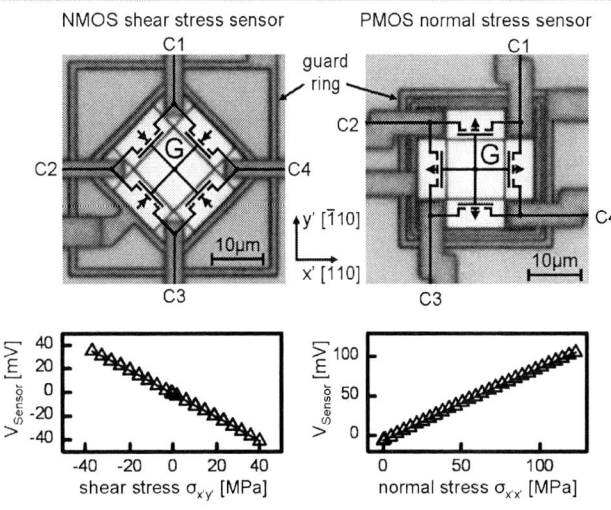

Figure 6.3.2: Sensor micrographs with overlaid schematics and characterization results.

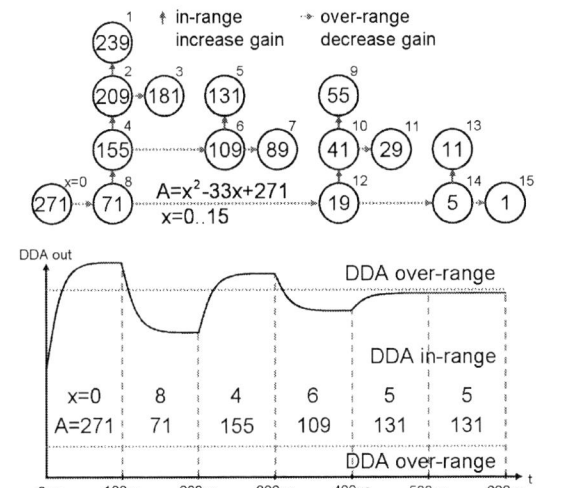

Figure 6.3.3: DDA behavior: binary search tree for gain evaluation and example of one sensor readout.

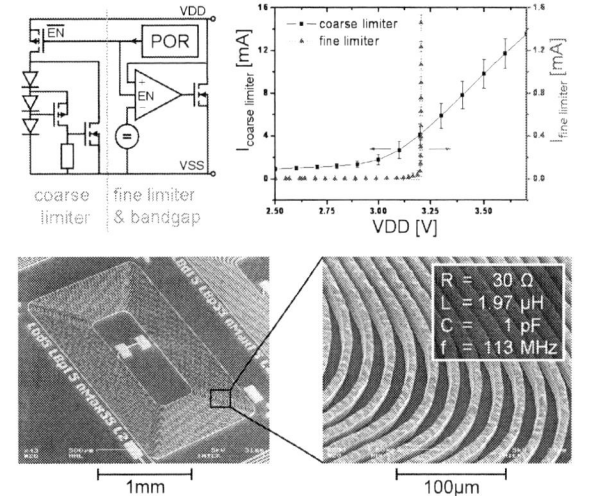

Figure 6.3.4: Telemetric elements: schematics and measurements of the voltage limiters and SEM pictures of the utilized coil.

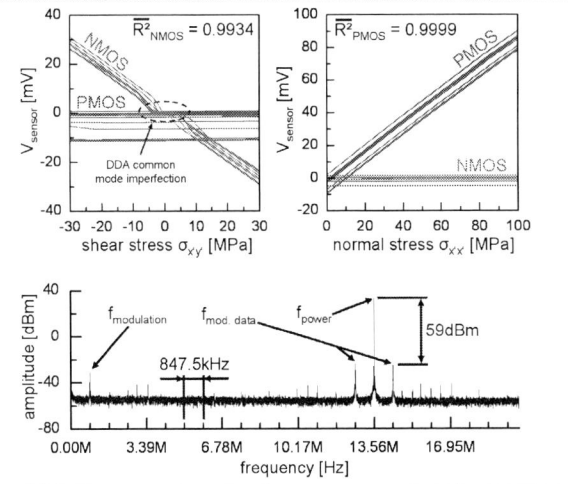

Figure 6.3.5: Measurement results: stress measurements on the bending and torsion bridge showing in each case the sensitive and non-sensitive sensors as well as the received PSD at the external coil.

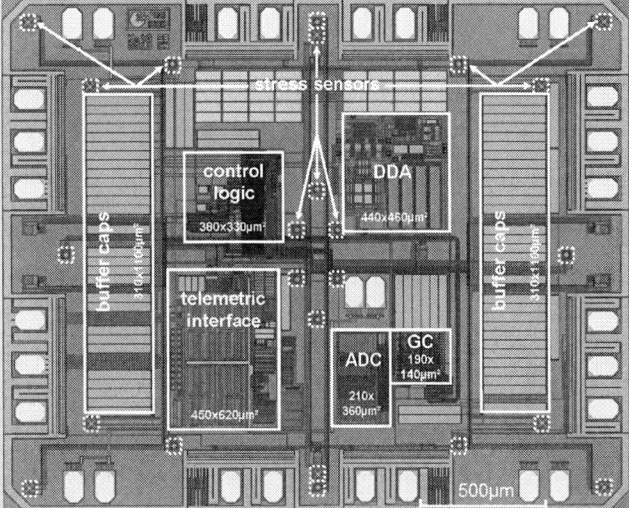

Figure 6.3.6: Optical chip micrograph.

978-1-61284-303-2/11 $26.00 © 2011 IEEE

	this work	[4], 2009	[3], 2007
technology	0.35µ CMOS	0.6µ CMOS	1.5µ CMOS
frequency	13.56MHz	DC	50MHz
VDD	3.19V	5V	3V
system size	2x2.5mm²	2x2.5mm²	3x7mm²
stress sensors	24	32	1
sensor principle	piezoresistive FET	piezoresistive FET	capacitive MEMS
sensitivity	-926µV/MPa 880µV/MPa	-448µV/MPa 477µV/MPa	816µV/µε (5443µV/MPa for Si)
resolution	23.6kPa / 24.8kPa	62.3kPa / 58.5kPa	0.87µε (130kPa for Si)
power consumption	1.75mW (w/o RF interface)	18.17mW	6mW (w/o RF interface)

Figure 6.3.7: Measured system performance in comparison to [3] and [4].

ISSCC 2011 / SESSION 6 / SENSORS & ENERGY HARVESTING / 6.4

6.4 A 21b ±40mV Range Read-Out IC for Bridge Transducers

Rong Wu, Johan H. Huijsing, Kofi A.A. Makinwa

Delft University of Technology, Delft, The Netherlands

Precision thermocouples and bridge transducers such as strain gauges and thermistors require read-out ICs with low noise, high accuracy and low drift. In such applications, the sensor and the read-out IC (ROIC) are usually calibrated as a single system, and so in addition to low thermal and 1/f noise, the ROIC should exhibit very low offset and gain drift (a few ppm/°C) to maintain system accuracy over temperature. This paper describes a 21b ROIC that meets this challenge.

The ROIC consists of a 3-stage current-feedback instrumentation amplifier (CFIA) and an incremental $\Delta\Sigma$ modulator (Fig. 6.4.1). The CFIA provides the high input impedance required for accurate bridge readout and relaxes the requirements on the ADC's noise and offset. Its closed-loop gain is given by $(G_{m3}/G_{m4})(R_1+R_2)/R_1$, which, if G_{m3} and G_{m4} are well-matched, can be accurately set by the external resistors R_1 and R_2.

The 1st and 2nd stages of the CFIA (Fig. 6.4.2) are chopped to achieve low offset and low noise [4]. The resulting output ripple must then be suppressed to prevent aliasing errors due to the ADC's sampling action. A continuous-time ripple-reduction loop (RRL) reduces the output ripple caused by the 1st stage's up-modulated offset to about $100\mu V_{pp}$ [4]. Furthermore, by setting $f_s = f_{ch1}$ and introducing a 90° phase shift to compensate for the phase-shift of the CFIA's Miller compensation network, any residual ripple will be sampled at zero-crossings (Fig. 6.4.1). However, the amplitude and phase shift of the 2nd stage's ripple is not so well defined, and is a function of the CFIA's closed loop gain. To avoid aliasing errors, the chopper clock of the 2nd stage is randomized in a bitstream-controlled manner [5]. The resulting ripple is then a zero-mean signal that is orthogonal to the ADC's bitstream and therefore does not cause in-band intermodulation products.

The ROIC's gain error and drift are mainly caused by the mismatch of the CFIA's input and feedback transconductances, G_{m3} and G_{m4} (Fig. 6.4.1). This is reduced by a dynamic element matching (DEM) scheme in which a multiplexer (MUX) periodically swaps their inputs. The required DEM switches were merged with the input choppers and so contribute no extra thermal noise. In practice, G_{m3} and G_{m4} will be at different common-mode levels. Swapping their inputs then results in large spikes at the CFIA's output. To avoid digitizing these spikes, the MUX state is altered during the reset period (1ms) at the start of every ADC conversion. The same goes for the low-frequency (LF) choppers CH_{LP1} and CH_{LP2}, which chop the entire analog signal path [6] at half the DEM frequency to further reduce the offset. The resulting low-frequency ripple and modulated G_m mismatch are then averaged in the decimation filter. Each decimated output is thus the result of four ADC conversions.

To further reduce the gain error and drift, a back-end gain error correction (GEC) scheme is used to cancel the static mismatch Δ of G_{m3} and G_{m4} (Fig. 6.4.1). It consists of an off-chip 6b DAC, which trims this mismatch by fine-tuning the tail currents of G_{m3} and G_{m4} via transconductor G_{m6}. By applying a fixed DC signal to the CFIA, the appropriate value of V_{cal} (defined by the DAC) can be determined within two DEM periods. First, the maximum calibration voltage $V_{cal,max}$ is applied to G_{m6}, and the output-referred mismatch error $(\Delta+\Delta_{cal,max})$ is determined from the difference in the results of the two conversions of one DEM period. Next, the minimum calibration signal $-V_{cal,max}$ is applied to G_{m6}, so that $(\Delta-\Delta_{cal,max})$ can also be determined. Since the DAC's range is designed to be larger than the worst-case static mismatch, the value of V_{cal} that minimizes the mismatch error can then be found by linear interpolation.

The output of the CFIA is digitized by a 2nd-order single-bit $\Delta\Sigma$ modulator operated in incremental mode (Fig. 6.4.3). It employs an input-feedforward topology, which relaxes the linearity and slewing requirements on the integrators' opamps [7]. The same sampling capacitor C_{s1} is used to sample both the input and the DAC feedback signals. As a result, the ADC's gain accuracy is not limited by component matching. The ADC's input-referred noise density was designed to be 6dB less than that of the CFIA (at a gain of 100), and also to have a 1mHz 1/f noise corner. The thermal noise requirement was met by setting $C_{s1} = 3pF$, while the 1/f noise of the 1st integrator's opamp A_1 was suppressed by correlated double sampling (CDS). To obtain a 1mHz 1/f noise corner, the opamp A_1 must have a gain in excess of 160dB. This was achieved by using a 2-stage Miller-compensated topology, with a gain-boosted 1st stage and a class-A 2nd stage.

The chip was fabricated in a standard 0.7µm CMOS process, and has an active area of 6mm² (Fig. 6.4.7). The measured output noise spectrum of the ROIC is shown in Fig. 6.4.4-left. To eliminate the low-frequency lobe that would occur due to the interaction between the ROIC's residual offset and the Hann window used, the ROIC's offset was subtracted before the FFT was computed (from 2^{24} samples). It can be seen that the ROIC's noise floor is flat from 1mHz to 10Hz, which corresponds to a resolution of 21b with respect to a full-scale range of ±40mV.

At a gain of 100, measurements on 10 samples show that the ROIC achieves low offset (<200nV) and low drift (<7.6nV/°C). The use of DEM and GEC reduces its gain error from 0.6% to ±0.00165% when G_{m3} and G_{m4} are at the same CM voltage (2.5V). This drops to 0.12% in the worst-case when one of the Gm's is at 0V. The use of DEM also improves the gain drift from 6.1ppm/°C to 4.3ppm/°C, and the INL from 35ppm to 5ppm (Fig. 6.4.4-right). The use of DEM and GEC improves the gain drift to 1.2ppm/°C, but only slightly improves the INL. The CFIA is the main source of INL, since the ADC's measured INL is less than 1ppm. At low gain settings, saturation of the CFIA's input stages eventually limits its linear input range (INL <10ppm) to ±120mV.

To test the ROIC's performance, it was used to read out a dual thermistor bridge (Fig. 6.4.1), which measures the temperature drift in a large (96cm³) oven-stabilized aluminum block. The bridge was also read out by a Keithley 2002 7-½ digit multimeter, which also reads out a Pt-100 reference sensor. The three measurement systems were set-up for a conversion time of 0.25s, which for the ROIC meant that 5000 samples (@ f_s=20kHz) were decimated by a sinc³ filter. The measurement results are shown in Fig. 6.4.5. Due to its lower sensitivity, the resolution of the Pt-100 is much less than that of the thermistor bridge. The 0.7µK (rms) temperature-sensing resolution achieved by the thermistors and the ROIC is roughly 2× better than that achieved by the thermistors and the Keithley, despite the fact that the ROIC only draws 270µA and is much more compact.

The performance of the ROIC is summarized in Fig. 6.4.6 and compared with the state-of-the-art. The ROIC achieves sub-200nV offset and a 16.2nV/√Hz thermal noise floor with a 1mHz 1/f noise corner, while only consuming 270µA current (CFIA 220µA, ADC 50µA) from a 5V supply. Its offset and drift performance exceed the state-of-the-art. These qualities make this ROIC suitable for demanding bridge transducer applications, which require low thermal and 1/f noise, low drift, and low power consumption.

Acknowledgement:
The authors would like to thank Y.Chae, M. Pertijs, Z. Chang and P. Haak, for their suggestions and support.

References:
[1] A. Thomsen, et al., "A DC Measurement IC with 130nV$_{pp}$ Noise in 10Hz", *ISSCC Dig. Tech. Papers*, pp. 334-335, Feb. 2000.
[2] ADS 1282 datasheet,
http://focus.ti.com/docs/prod/folders/print/ads1282.html
[3] CS5530 datasheet,
http://www.cirrus.com/en/products/pro/detail/P1108.html
[4] R. Wu, et al., "A Chopper Current-Feedback Instrumentation Amplifier with a 1mHz 1/f noise corner and an AC-Coupled Ripple Reduction Loop", *J. Solid-State Circuits*, vol. 44, no. 12, pp. 3232–3243, Dec. 2009.
[5] M. A.P. Pertijs, et al., "A Sigma-Delta Modulator With Bitstream-controlled Dynamic Element Matching", *Proc. ESSCIRC*, pp.187-190, Sep, 2004.
[6] D. McCartney et al., "A Low-Noise Low-Drift Transducer ADC", *J. Solid-State Circuits*, vol.32, no.7, pp. 959–967, July 1997.
[7] V. Quiquempoix, et al., "A Low-Power 22-bit Incremental ADC", *J. Solid-State Circuits*, vol.41, no.7, pp. 1562-1571, July 2006.

978-1-61284-303-2/11 $26.00 © 2011 IEEE

ISSCC 2011 / February 21, 2011 / 3:15 PM

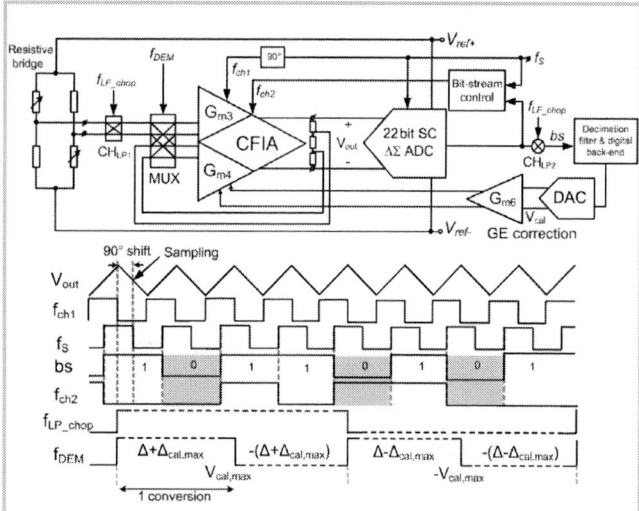

Figure 6.4.1: Block & timing diagrams of a bridge-readout system.

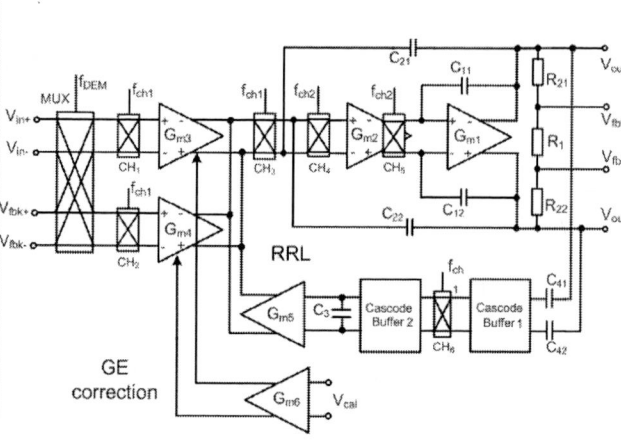

Figure 6.4.2: Block diagram of the CFIA with ripple reduction loop (RRL) and DEM.

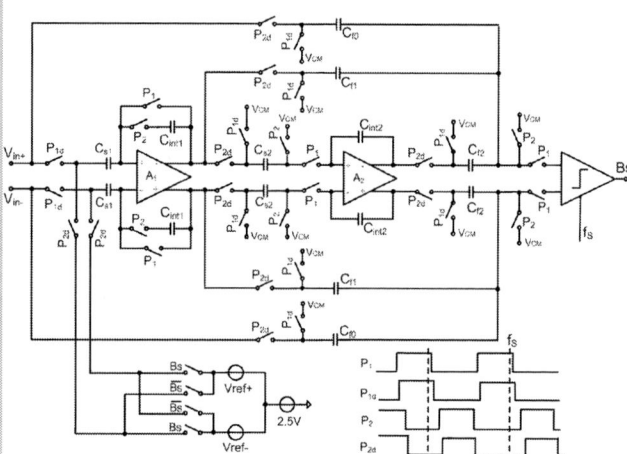

Figure 6.4.3: Simplified schematic diagram of the 2nd-order $\Delta\Sigma$ modulator.

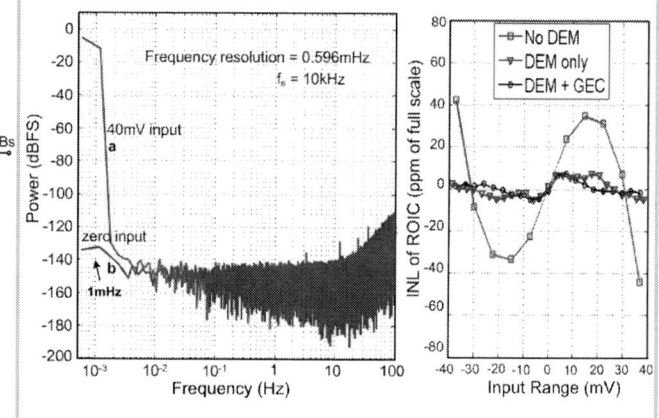

Figure 6.4.4: (left) Measured output spectrum of the ROIC with DEM and LF-chopping off. (right) Measured INL of ROIC.

Figure 6.4.5: Temperature drift as measured by a 1) thermistor bridge + ROIC; 2) thermistor bridge + Keithley; 3) Pt-100 + Keithley.

	This work	CS5530 [3]	ADS1282 [2]	Thomsen [1]	McCartney [6]
Year	2011	2009	2007	2000	1997
Supply current	270µA	7mA	4.5mA	8.2mA*	18mA*
Supply voltage	5V	5V	5V	±2.5V and 5V	5V
1/f noise corner	1mHz	25mHz	25mHz	--	10mHz
Input range	± 40mV	± 78mV	± 263mV	± 28mV	± 10mV
CMRR	130dB (typ)	120dB(typ)	110dB (typ)	--	120dB(typ)
Differential input impedance	80MΩ	--	1GΩ	--	--
Input offset current	3.4pA	1.2nA	1nA	1nA	--
Input referred noise density	16.2nV/√Hz	12nV/√Hz	5nV/√Hz	6.2nV/√Hz	22nV/√Hz
Gain drift	<1.2 ppm/° C	2ppm/° C (typ)	9ppm/° C (typ)	15ppm/° C (typ)	2ppm/° C (typ)
Gain error (uncalibrated)	<0.12%	1%	1%	--	--
Offset drift	<7.6 nV/° C	10nV/° C (typ)	20nV/° C (typ)	<70nV/° C	<10nV/° C (typ)
Offset (uncalibrated)	< 200 nV	< 9.5 µV	< 200 µV	--	1 µV
Nonlinearity	< 5 ppm	< 30 ppm	< 4ppm	--	<15ppm
Conversion time	0.1s	0.13s	0.004s	0.0083s	0.5s

* Total supply current (analog + digital)

Figure 6.4.6: Performance of the ROIC versus the state-of-the-art.

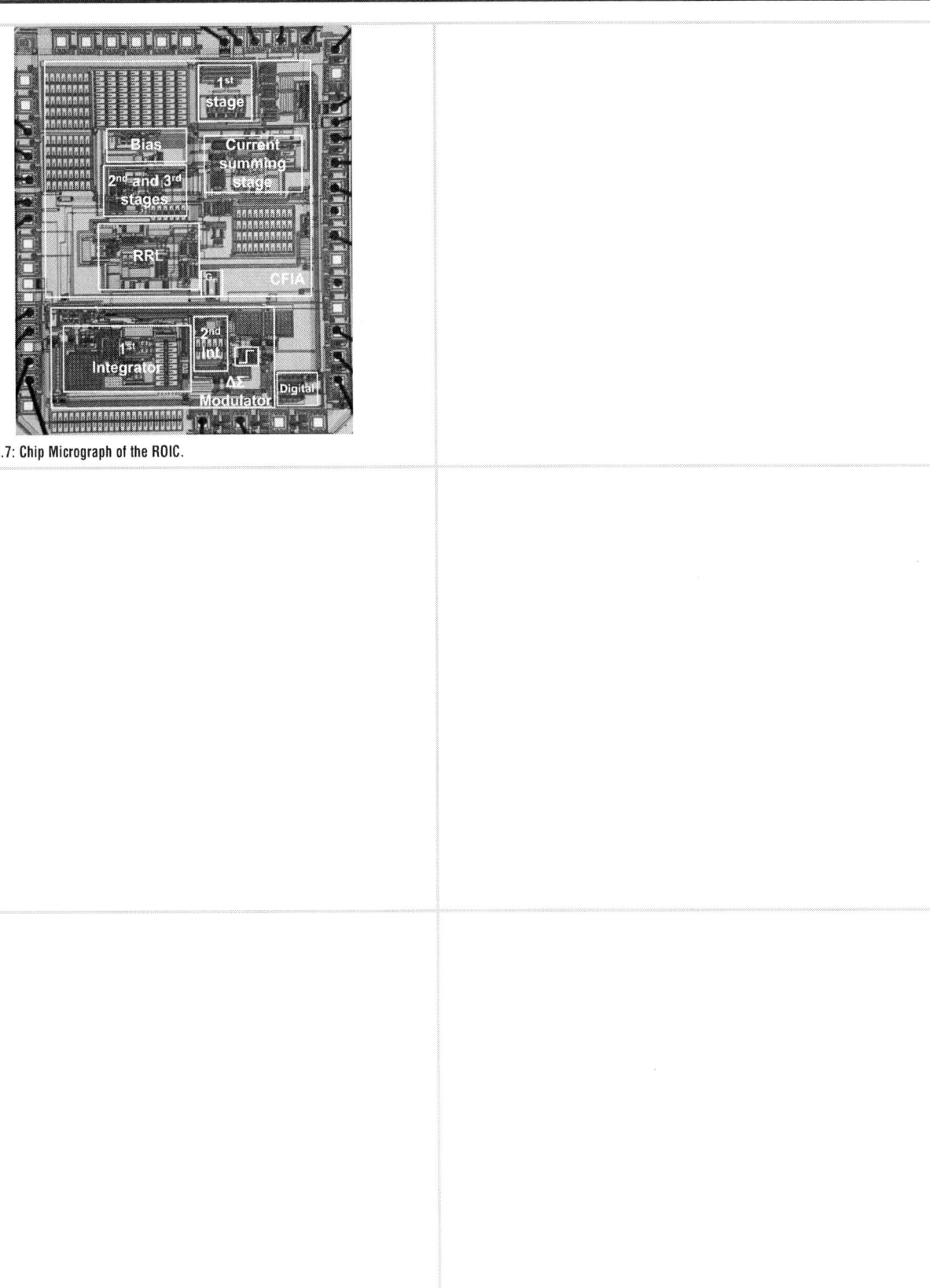

Figure 6.4.7: Chip Micrograph of the ROIC.

ISSCC 2011 / SESSION 6 / SENSORS & ENERGY HARVESTING / 6.5

6.5 A ±1.5% Nonlinearity 0.1-to-100A Shunt Current Sensor Based on a 6kV Isolated Micro-Transformer for Electrical Vehicles and Home Automation

Frederic Rothan[1], Helene Lhermet[1], Brice Zongo[1], Cyril Condemine[1], Henri Sibuet[1], Patrick Mas[2], Miguel Debarnot[2]

[1]CEA-LETI-MINATEC, Grenoble, France,
[2]Schneider Electric, Grenoble, France

A wide range of current-sensing techniques has been developed to satisfy various electrical and electronics applications requirements [1,2]. In high-voltage applications, the main issue is the electrical isolation with accurate measurement at low signal levels. Isolated shunt technology is an attractive and versatile solution for current sensing. Electrical isolated products usually need a power supply at both low and high voltage sides [3,4]. An original topology that needs only one power supply at the secondary (low voltage) side is reported in [5]. In this paper, we introduce an implementation of an integrated shunt current-measurement microsystem based on a 6kV isolated micro-transformer. This work finds its applications in many domains such as hybrid or electric vehicle battery monitoring or motor control, system building automation or smart grids, where small size and low cost are required.

The microsystem architecture is illustrated in Fig. 6.5.1 in the case of a motor (M) power-tracking application, based on the motor's shunt-current measurement. It is comprised of a shunt, 2 micro-transformers for 6kV isolation, a chopper IC for current switching and a readout IC. The 2 electromagnetic micro-transformers isolate the high-voltage system part, including the motor to be monitored, the shunt and the chopper IC, from the low-voltage chopper external clock and the current readout IC (ROIC). As a consequence, 2 ASICs with different electric grounds are necessary. The current I_{shunt} to be measured flows through a 1mΩ shunt resistance. The resulting ($R_{shunt} \times I_{shunt}$) voltage drop is chopped at high frequency by 2 transistors (chopper ASIC) driven through the isolation command transformer. Consequently, the resulting AC high-frequency current I_{sense} can be read through electromagnetic coupling by the readout ASIC that amplifies, demodulates and filters the signal I_{read}. The peak amplitude of I_{read} is proportional to the motor current I_{shunt}.

Micro-transformers have been manufactured with standard microelectronic technologies (Fig. 6.5.3(a)). Two stacked coaxial electrochemical deposited copper coils isolated by a thick (10µm) SiO_2 layer provide a 6kV electrical isolation. To achieve micronic inter-conductor spacers and thick conductive lines, a damascene technique is employed. Special attention was granted to the final passivation layers to prevent copper from corrosion and to take into account the surface arcing issue due to possible high voltage presence. The core transformer die area is 600×600µm², with 28-turn windings for the primary and secondary coils leading to a unity transformer ratio. The output impedance is set to 40Ω. For design purposes, a micro-transformer equivalent electrical model taking into account magnetic and capacitive effects was derived from micro-transformer characterization results. The behavior is modeled from DC to 2GHz.

The chopper ASIC is composed of 2 NMOS power switches to chop the current. One switch is placed on each of the readout transformer inputs. This symmetrical structure balances and cancels out the common parasitic signals between the 2 differential readout IC inputs. The switches' W/L ratios are 300µm/0.13µm to minimize the signal attenuation, while offering a low gate capacitance to reduce the current spikes on the command transformer. The NMOS gate driving signal is generated by the command micro-transformer from an external clock at 20MHz typical frequency, with 1.6V amplitude and 2ns rise and fall times. However, this square waveform is distorted and its amplitude is attenuated by the micro-transformer. This is modeled and taken into account for the readout IC design. The chopper ASIC operates without any external supply voltage, making it inherently compatible with high voltage isolation.

The readout ASIC (Fig. 6.5.2) is dedicated to the current measurement through a micro-transformer. The ASIC input signal is a current spike, with about 1ns rise and fall times, a 5ns typical width and a 700nA to 700µA peak amplitude range,

corresponding to the 0.1-to-100A shunt current range to be measured. The first stage is a closed-loop high-bandwidth resistive transimpedance amplifier (TIA). A S/H envelope detector followed by a low-pass filter allows recovery of the input signal. To amplify the input signal with low distortion, the TIA bandwidth is set to 80MHz. The amplifier input impedance must be lower than 4Ω to avoid transformer overloading. Consequently, the closed-loop gain is limited to 700V/A. To obtain the specified 2% resolution over the whole range, one voltage amplifier is inserted between the TIA and the envelope detector for the 1-to-10A range and two are inserted for the 0.1-to-1A range. These 2 voltage amplifiers have a gain of 10 and 80MHz BW, and are AC coupled to remove any residual offset. All 3 amplifiers are based on a 1GHz GBW Miller operational amplifier.

The resulting voltage is then sampled and held to detect its maximum. A 2nd-order low-pass filter with a 25kHz cutoff frequency allows recovering the original waveform. The sampling signal is generated from the external clock, allowing a synchronous demodulation. An on-chip adjustable delay line enables adjusting the delay between the sample clock and the chopper by 250ps steps to ensure proper signal sampling. In order to reduce any disturbance impact, the analog chain is fully differential.

The micro-transformer standards require 6kV impulse voltage for a voltage withstand test (1.2/50µs) and 3.2kV 50Hz voltage for a dielectric test. First, a test bench with current limitation is designed to allow determination of threshold voltage for arcing and visual inspection of chips. Then standardized tests are performed (Fig. 6.5.3(b)): more than 80% of the micro-transformers passed the impulse test up to 7.4kV.

The chopper and readout ASICs are implemented in a CMOS ST 0.13µm technology. They are pad-limited for test purposes and occupy a core area of 0.0015mm² and 0.39mm², respectively. The chopper IC experimental tests show that the ASIC is functional with no external supply. The switches' ON resistance is 3Ω, as expected from electrical simulations. The Bode magnitude plot of the whole ROIC amplification chain is shown in Fig. 6.5.4. It exhibits a maximum gain of 97.8dB at 18MHz, with a BW of 2.5MHz to 55MHz, and a power consumption of 15.6mW for a 1.2V nominal supply voltage.

The system is characterized with bare dies mounted on a PCB. The measured system response (output voltage vs. shunt current) shows a nonlinearity lower than ±1.5% in the 3 current ranges (Figs. 6.5.4 and 6.5.5). The whole microsystem fits into a 13×7.6mm² SO20 package. The micro-transformer and ASICs dimensions can be further shrinked to fit in a smaller package.

In order to achieve shunt current measurement of high-voltage systems through a 6kV isolated micro-transformer, a chopper and a readout chipset was designed and tested. A shunt current is measured with this microsystem with less than ±1.5% nonlinearity over the whole 0.1-to-100A range.

Acknowledgements:
The authors would like to thank R. Cuchet, M. Audoin, S. Borel, A. Walter from CEA-Léti and Y. Lembeye, J.-C. Crebier and O. Deleage from G2ELab for their contribution to the micro-transformer design, technology modeling and characterization, S. Bacquet and M. Ranieri for their contribution to the system test.

References:
[1] S. Ziegler, R. Woodward, H.-C. Lu, and L. Borle, "Current sensing techniques: a review", *IEEE Sensors Journal*, vol. 9, pp. 354-376, Apr. 2009.
[2] C. Xiao, L. Zhao, T. Asada, W. Odendaal, and J. van Wyk, "An overview of integratable current sensor technologies", *Industry Applications Conference*, vol. 2, pp. 1251-1258, 2003.
[3] M. Münzer, W. Ademmer, B. Strzalkowski, and K. Kaschani, "Insulated signal transfer in a half bridge driver IC based on a coreless transformer technology", *Int. Conf. Power Electronics and Drive Systems*, vol. 1, pp. 93-96, 2003.
[4] J. Xu, L. Zhao, Z. Liang, and J. van Wyk, "Design of an embedded bi-planar coil-based integrated current sensor for power module integration", *Applied Power Electronics Conference and Exposition*, vol. 1, pp. 369-374, 2005.
[5] Y. Cadoux, D. Leonard, and B. Reymond, "Electrically insulated current measuring device", *Schneider Electric patent n° WO2008087275*, 2008.

978-1-61284-303-2/11 $26.00 © 2011 IEEE

ISSCC 2011 / February 21, 2011 / 3:45 PM

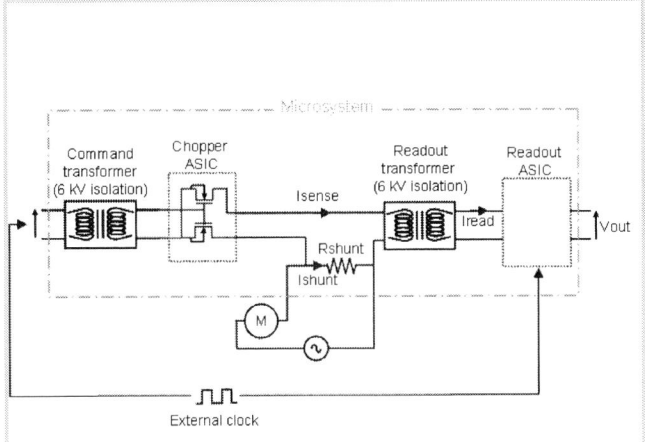

Figure 6.5.1: System architecture.

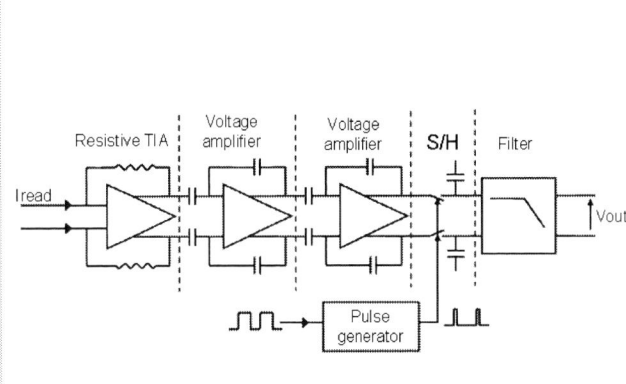

Figure 6.5.2: Readout ASIC architecture.

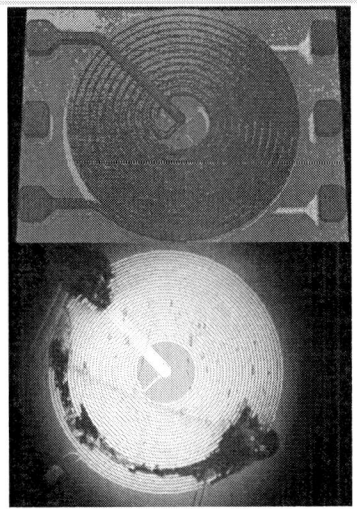

Figure 6.5.3: Micro-transformer photographs: (a) optical interferometry and (b) arc traces after destructive tests.

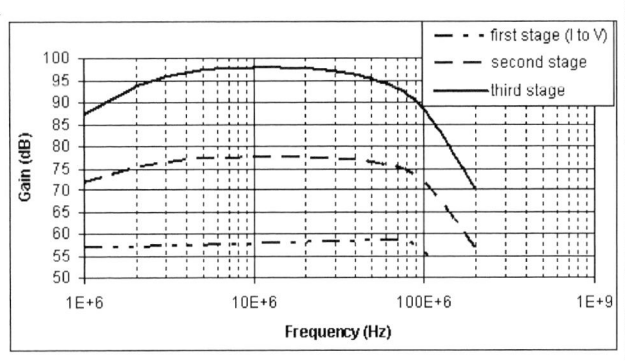

Figure 6.5.4: Analog amplification chain transfer functions.

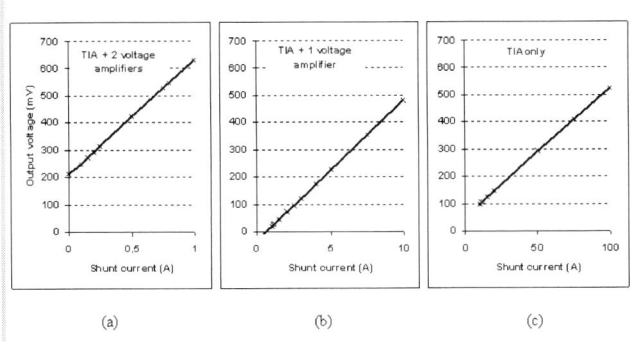

Figure 6.5.5: Output voltage versus shunt current: (a) 0.1 to 1A, (b) 1 to 10A, and (c) 10 to 100A.

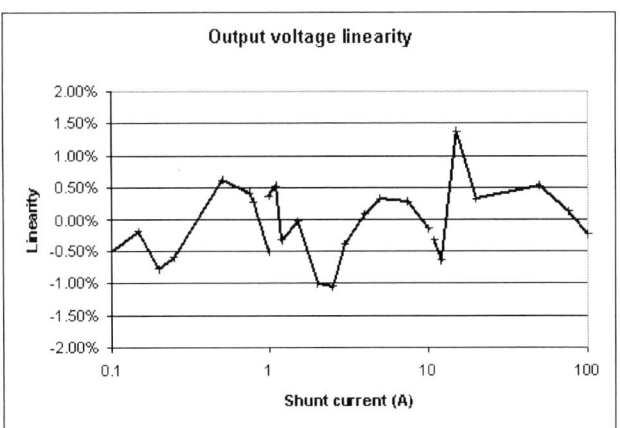

Figure 6.5.6: Output voltage linearity.

978-1-61284-303-2/11 $26.00 © 2011 IEEE

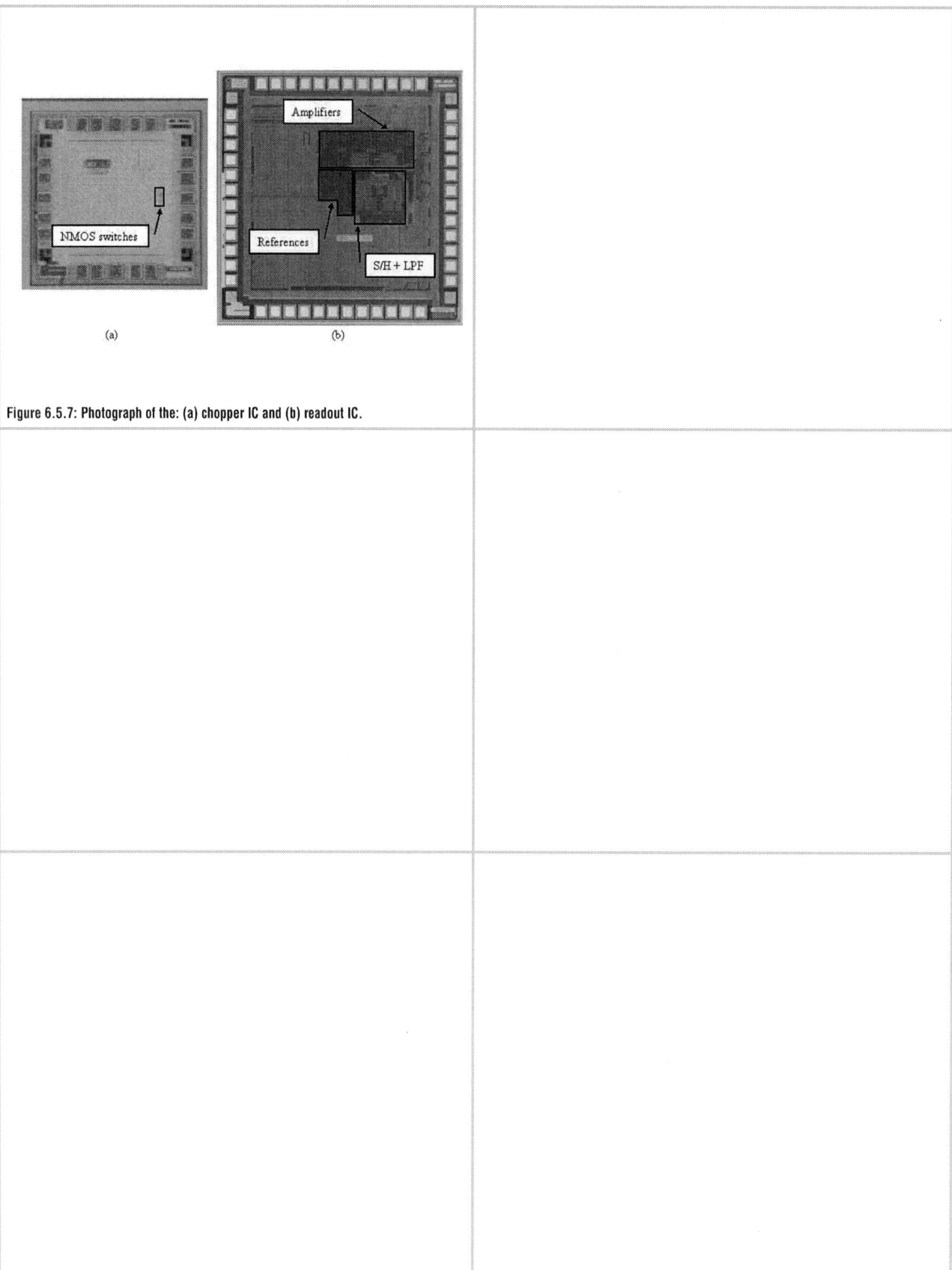

Figure 6.5.7: Photograph of the: (a) chopper IC and (b) readout IC.

ISSCC 2011 / SESSION 6 / SENSORS & ENERGY HARVESTING / 6.6

6.6 Indirect X-ray Photon-Counting Image Sensor with 27T Pixel and 15e⁻rms Accurate Threshold

Bart Dierickx[1,2], Benoit Dupont[1], Arnaud Defernez[1], Nayera Ahmed[1]

[1]Caeleste, Antwerp, Belgium,
[2]Vrije Universiteit Brussel, Brussels, Belgium

In X-ray imaging, as in other imaging domains, the ultimate sensitivity and signal-to-noise ratio are obtained when each incoming photon is counted - the so-called quantum limit. Present state-of-the-art digital radiography is largely "charge integration" based, which results in a read noise that is composed of the quantum-limited photon shot noise, but also of electronic read noise and excess noise due to the non-reproducible charge packet sizes per absorbed X-ray photon.

State-of-the-art photon-counting imagers [1,2,3] use direct detection, i.e. the X-ray-to-charge conversion happens by the photo-electric effect in a high-Z semiconductor photoconductor or photodiode with about 3 to 5eV/e⁻. For energies used in medical imaging, each X-ray photon generates a charge packet of 5,000 to 40,000 electrons, which are processed by hybrid-connected electronic readout circuits. In *indirect* detectors, or scintillators, the X-photon is absorbed in the scintillator volume, where it creates a secondary visible light flash, which is detected by a Silicon photodiode in an underlying readout IC (Fig. 6.6.1). Scintillators are far more easy and economical to manufacture than high-Z detector hybrids. Yet the efficiency of the overall indirect detection process is poor, so that the size of the charge packet for an X-photon is in the range 100 to at most 1,000 electrons, depending on scintillator characteristics [4,5]. Photon counting with scintillators thus requires a very low-noise and accurate threshold detection concept, with an accuracy and noise floor far below 100 electrons [6].

In this paper we solve the challenges of high charge sensitivity and comparator accuracy by circuit design and realize a high optical coupling between the scintillator and the Si photodiode and pixel.

Another barrier to implementation of large arrays is the photon-counting pixel's complexity and yield. Digital domain counting pixels contain several 100's of transistors [1]. Our pixel design makes use of an analog representation [7] of the digital count. When properly done, it bears no count accuracy degradation compared to the digital representation. The analog counter consists of a low number of transistors and capacitors. The total number of transistors in our pixels, for the various variants, is between 27 and 49.

In this paper, we present the pixel design, simulation and measurement results under visible light and X-ray illumination.

Figure 6.6.2 shows the general topology used in all our pixel variants. Key circuits in our implementation are:

(a) *The pulse shaper* detects and amplifies with sufficient reproducibility a charge packet of 100 electrons or less. The pulse shaper (Fig. 6.6.3) is a CTIA with a slow RC feedback that returns it to an equilibrium point with a time constant of a few µs. In order to offer a signal with sufficient amplitude to the comparator, we choose CTIA (charge to voltage) gains between 0.1 and 0.4fF. This low capacitance is the Miller capacitance of the single-ended amplifier. As a resistive feedback that realizes a time constant of a few µs is in the order of $R \approx 10µs/1fF \approx 10G\Omega$, a physical resistor is out of the question. The resistive feedback is realized by an OTA in a similar way as in [1].

(b) *The analog counter.* Instead of counting photons in a digital fashion, we create an analog signal that is an accurate measure for the count, which can be read out and converted to a digital number outside of the IC. The actual circuit in Fig. 6.6.4 is compact and yields an analog signal that is an exponential decaying function of the count. This nonlinear analog counter has large, accurate, steps at low counts, and gradually smaller steps towards higher counts, where the required accuracy relaxes as the X-photon shot noise becomes more important.

In Fig. 6.6.4 the linear law applies to a voltage step that is 1/20th of the range; the exponential law shown applies to a ratio C2/C1=20. In the real circuit this ratio is 100. One can show that an overall accuracy better than the RMS on the count (which is identical to the X-Photon Shot Noise) can be obtained for counts between 0 and a several 1,000's.

Figure 6.6.7 is a picture of the demonstrator IC. The 16×16 pixel array contains 8 pixel circuit variants and 5 photodiode variants. The array is read via an X-Y addressing scheme. The in-pixel circuitry is small so that the fill factor is about 75 to 80%, ensuring a high overall quantum efficiency, with good optical coupling of the light flashes in the GdOS scintillator sheet into the photodiodes.

Devices are evaluated with electrical inputs signals, visible light pulses, and X-ray illumination. Using the electrical input shown in Fig. 6.6.2 allows calibrating the charge-to-voltage conversion factor. Visible light pulses are used in the optical lab for characterization of speed, power, and the ratio of false and missed counts as a function of the charge packets size and circuit topology. Maximum count speed is between 100kHz and 1MHz, i.e. the circuit can discriminate pulses that are one to a few µs apart. There is a clear tradeoff of count speed versus power. The power dissipation is in the order of 2 to 20µW per pixel, not including the circuitry used for the real-time observation, as this would also not be present in a final product.

X-ray measurements are done in conditions representative for typical medical applications. Figure 6.6.5 shows a real-time trace of the comparator signal, with and without X-radiation. One recognizes clearly the X-ray photon pulses. Calibration versus an electrical input shows that these charge packets are in the range of 100-to-300e⁻ per packet. The lower trace of Fig. 6.6.5 is recorded without radiation and can thus be considered to bear only noise. The noise is estimated at 10 to 15e⁻RMS. The shape suggests that it is largely 1/f noise dominated, which we have not yet verified. Figure 6.6.6 shows the real-time trace of the analog count output of the same pixel.

Acknowledgements:
We thank Nico Buls, Claire Bourgain, Gert Van Gompel and Peter Covens of the Universitair Ziekenhuis Brussel for the assistance with the X-ray measurements.

References:
[1] R. Ballabriga, et al., "The Medipix3 Prototype, a Pixel Readout Chip Working in Single Photon Counting Mode With Improved Spectrometric Performance", *IEEE Trans. Nuclear Science*, vol. 54, no. 5, 2007, and references therein.
[2] K. Spartiotis, et al.,"A photon counting CdTe gamma- and X-ray camera", *Nuclear Instruments and Methods in Physics Research*, vol. 550, p.267-277, 2005.
[3] W. Barber, et al., "Large Area Photon Counting X-Ray Imaging Arrays for Clinical Dual-Energy Applications", *NSS-MIC*, Oct 30, 2009.
[4] E. Miyata, et al., "High resolution X-ray photon-counting detector with scintillator-deposited charge-coupled device", *IEEE Trans. Nuclear Science*, vol. 52, no. 2, p. 576, 2006.
[5] B. Dierickx, et al., "Towards photon counting X-ray image sensors", *OSA Symposium*, June 2010.
[6] C. Lotto and P. Seitz, "Charge Pulse Detection with Minimum Noise for Energy-Sensitive Single-Photon X-Ray Sensing", *European Optical Society Symposium*, June 2009, and references therein.
[7] M. Perenzoni, et al., "A Multi-Spectral Analog Photon Counting Readout Circuit for X-Ray Hybrid Pixel Detectors", *IMTC*, Ap. 2006.

978-1-61284-303-2/11 $26.00 © 2011 IEEE

ISSCC 2011 / February 21, 2011 / 4:00 PM

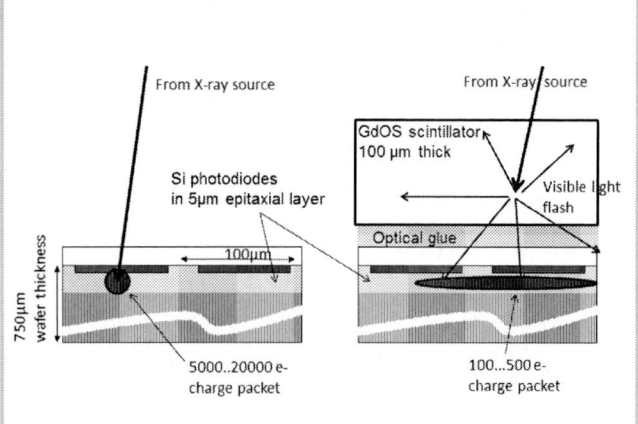

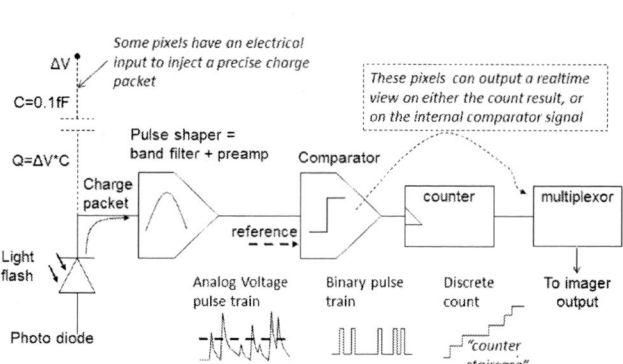

Figure 6.6.1: Direct (left) vs. indirect (right) X-photon detection in the present device. Direct detection is about 600 times less probable in the present configuration.

Figure 6.6.2: X-photon counting pixel general circuit topology.

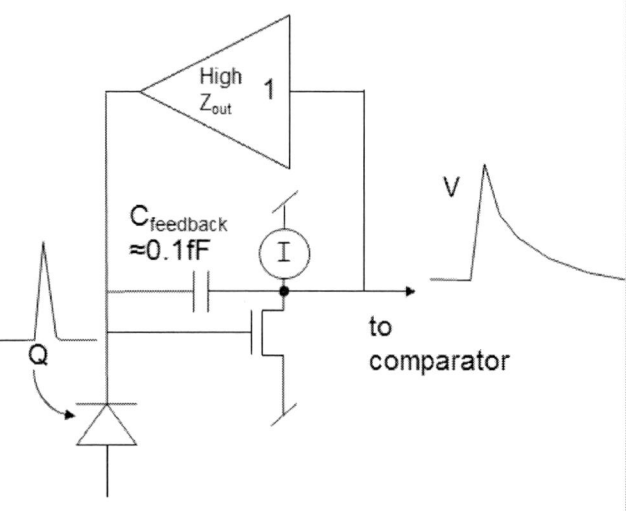

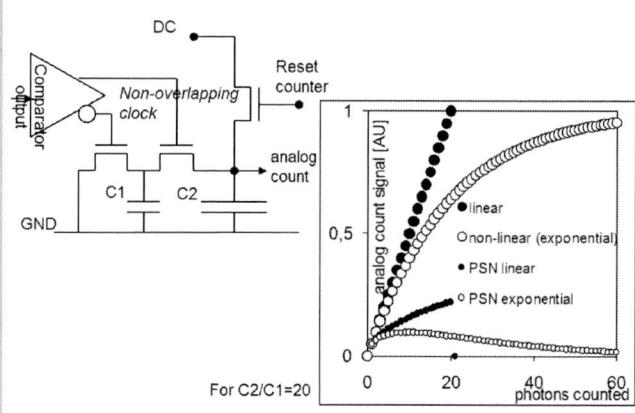

Figure 6.6.3: Pulse-shaper part circuit topology.

Figure 6.6.4: Analog counter and output multiplexor circuit. Insert: linear (as reference) and exponential counts, and their corresponding Photon Shot Noise.

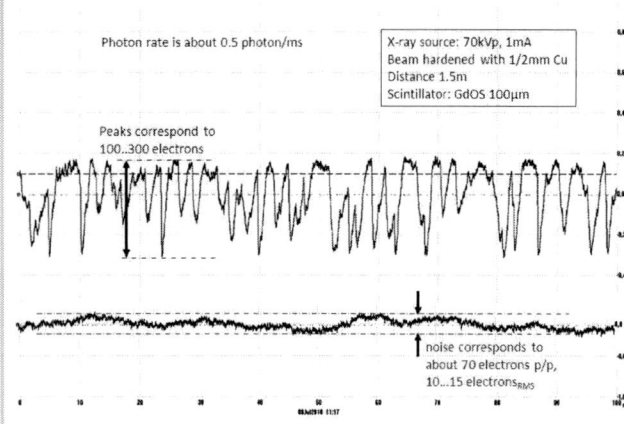

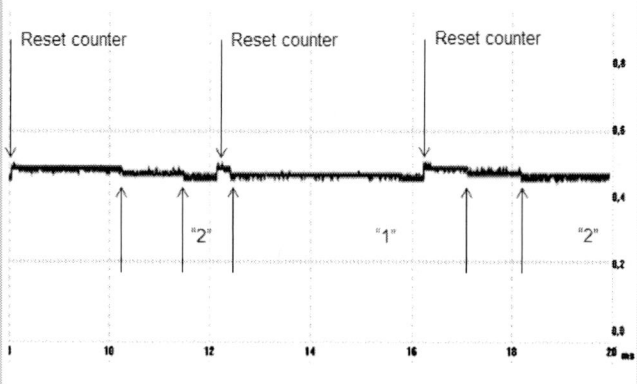

Figure 6.6.5: Real-time observation of comparator output for indirectly detected X-rays. Upper trace: under X-radiation; lower trace: radiation off.

Figure 6.6.6: Real-time observation of the analog counter signal, same setting as Fig. 6.6.5. In these 3 consecutive periods, the counter counted 2, 1 and 2.

Figure 6.6.7: Micrograph of the 16×16 pixel on 100 µm pitch demonstrator imager. Size 2.5×2.5mm. 0.18µm CMOS on high resistivity epi.

ISSCC 2011 / SESSION 6 / SENSORS & ENERGY HARVESTING / 6.7

6.7 A 1.32pW/frame·pixel 1.2V CMOS Energy-Harvesting and Imaging (EHI) APS Imager

Suat U. Ay

University of Idaho, Moscow, ID

Recent advances in video sensor networks and implantable biomedical devices –e.g. retinal prostheses [1]– necessitate very low-voltage, low-leakage, and energy-efficient image sensors that preferably produce their own power from ambient sources. A natural energy source for an image sensor that produces video images from impinging "sufficient" amount of light energy is the light itself. This in mind, a CMOS image sensor that can both produce power from light and capture video images on same focal plane is developed. The CMOS energy harvesting and imaging (EHI) active pixel sensor (APS) is incorporated in a 54×50 array along with low-power supporting electronics. It is designed in a mature 0.5μm 2P3M CMOS process that has only high-V_t transistors.

A mixed-signal circuit design technique called supply boosting (SBT) [2,3] is utilized to overcome dynamic range and switch overdrive issues emanating from the low supply voltage of 1.2V and high-V_t. In SBT, the supply voltage is boosted locally once without compromising reliability margins of the process in use, while analog signal processing such as level shifting, amplification, or comparison is performed. SBT is used in designing both a pixel source follower (PSF) and an improved version of a supply-boosted successive approximation register type (SB-SAR) ADC from [3] for the EHI imager.

A circuit diagram of the EHI imager is shown in Fig. 6.7.1. The EHI pixel is composed of 2 photodiodes. One works as the integrating-type photodiode (D2) for imaging, and the other works as the micro solar cell (D1) for energy harvesting. The pixel contains 4 NMOS transistors. M1 is for resetting the FD node. M2 and M3 are the source follower and select transistors. M4 is the enable transistor for energy harvesting. Anodes of D1s are connected to an energy-harvesting bus (EHB). Depending on the operation mode, EHB is either connected to ground (imaging mode, IM) or to an on-chip charge pump (CP) or directly to a load (energy harvesting mode-EHM). In IM, D1 and D2 work in parallel to discharge the FD node. The FD node voltage is buffered to a column analog signal processor (ASP) like a regular 3T APS pixel, and processed for imaging. During EHM, M4 is turned on shorting D2, and forcing D1 to work as a regular solar cell.

Timing diagrams for IM and EHM are shown in Fig. 6.7.2. A 1.2V supply makes buffering FD node voltages impossible because of the high threshold voltage of the NMOS transistor (+0.8V), and its backgate bias. Thus, pixel reset and select signals are both boosted while SBT is utilized for PSF by boosting the pixel supply voltage (V_{AAB}) during row read. The pixel supply (V_{AAB}) is driven by the supply booster circuit (Fig. 6.7.1). When IN is asserted high (V_{AA}), V_{AAB} is boosted close to 2V_{AA}, and drops to V_{AA} when IN is low. An on-chip 170pF booster capacitor is used due to the very low PSF current consumption. Charge-based column ASPs, coupled to a single global charge amplifier [5], and a 10b SB-SAR ADC are used in the design to achieve low power. Reference currents are generated on chip. 6b programmable current DACs are used to generate bias currents with 60nA steps at 1.2V supply.

The EHI pixel layout is shown in Fig. 6.7.3. Die area for the imager is 2×2mm² while the pixel size is 21μm². An Nwell/P+/P-sub structure is used to form D1 (P+/N-well) and D2 (N-well/P-sub) [4]. P+ is directly connected to EHB, while the N-well is connected to the drain of reset (M1) and gate of PSF (M2) transistors. A pixel fill factor of 32% was achieved due to the N-well design rules, the extra transistor (M4) area, and the extra routings of EHB and ground signals.

Two operation modes are available during EHM: direct drive (DD) and charge pump (CP). In both cases, all D1s are connected between EHB and ground through M4. During CP-EHM, all control signals except CP clock are stopped. The reference generator and iDACs are also turned off right after the CP comparator bias is sampled. Thus, only the CP comparator consumes static power.

External capacitors C_{P1}=1mF and C_{P2}=10nF are used to measure CP efficiency and rise time. The measured rise time is 270s achieving an efficiency level between 70% and 89%, depending on the light level. Rise time depends on the C_{P1} value and can be reduced further by choosing it close to C_{P2}. During CP-EHM, total power consumption of the chip is <1μW. CP control logic is designed such that the charge from C_{P2} is transferred to C_{P1} completely in 8 clock cycles while sequentially triggering clock phases as shown in Fig. 6.7.2.

Energy-harvesting efficiency and current-voltage/power-voltage (I-V/P-V) characteristics of the EHI pixels are measured for different light levels between 1,000lux (overcast daylight) and 60,000lux (sunny daylight) at DD-EHM by using an external load resistor (R_L), as shown on Fig. 6.7.4a. At the maximum power point (MPP), the 54×50 EHI pixel array produces 380mV and 8.75μA, resulting in 3.35μW of power for 60,000lux illumination. Power reduces to 2.1μW and 1.0μW for 20,000lux and 1,000lux (normal and overcast daylight), respectively. Short-circuit voltages of the cells are between 400 and 450mV. Harvesting efficiency is also measured to be around 9% as shown on Fig. 6.7.4b at the MPP of the EHI pixels.

Imaging mode measurements are performed at various supply voltages (1.2 to 1.6V), 7.4fps frame rate, ¼ pixel saturation, and optimum bias settings with which the EHI imager produces low FPN images, as shown on Fig. 6.7.5. Full chip power at 1.5V is 27.4μW. Detailed block powers are measured at 1.2V. Pixel array power consumption is directly measured to be 22nA resulting in 26.4nW power consumption. This is achieved by using SBT while biasing each PSF to 78nA, and fast row sampling resulting in 0.52% per frame PSF activity. The SB-SAR ADC consumes 3.13μW while running at 20kS/s, less than half of its maximum speed. The global amplifier consumes 2.17μW, while iDAC and reference generator consume 3.21μW. 16 timing and control signals are generated off chip. Thus, the largest power is consumed on pads, which is about 5.64μW. Total leakage current when the chip is powered down is 80nA. As a result, total power consumption is 14.25μW for 1.2V supply and 7.4fps.

An imager figure of merit (iFOM) is defined to quantify the total energy consumption per pixel for one code quantization of effective pixel signals and for driving them off chip. Thus, total power consumption of the chip is used for calculating iFOM in J/pixel-code. Both FOM (from [6]) and iFOM are calculated and shown on Fig. 6.7.6. The EHI imager has 1.32pW/frame-pixel FOM and 696fJ/pixel-code iFOM achieving lowest power consumption. These do not include the EHI imager's ability to harvest energy from light and could be further reduced if a low-power MPP tracer and regulator is incorporated on-chip to drive chip power from a larger C_{P1}.

Acknowledgments:
This work was made possible by support from Micron Technology Foundation.

References:
[1] L. Theogarajan, et al., "Minimally Invasive Retinal Prosthesis," *ISSCC Dig. Tech. Papers*, pp.99-108, Feb. 2006
[2] A. Mesgarani, et al., "Supply Boosting Technique for Designing Very Low-Voltage Mixed-Signal Circuits in Standard CMOS," *MWSCAS Dig. Tech. Papers*, pp.893-896, Aug. 2010.
[3] S. U. Ay, "A sub-1Volt 10-bit Supply Boosted SAR ADC Design in Standard CMOS," *Int. Journal on Analog Integrated Circuits and Signal Processing*, Aug. 2010.
[4] M. Ferri, et al., "Integrated Micro-Solar Cell Structures for Harvesting Supplied Microsystems in 0.35-μm CMOS Technology," *IEEE Sensors Conference*, pp.542-545, Oct. 2009.
[5] K.-B. Cho, et al., "A 1.5-V 550-μW 176x144 Autonomous CMOS Active Pixel Image Sensor," *IEEE Tran. Elec. Dec.*, vol. 50/1, pp. 96-105, Jan 2003.
[6] K. Kagawa, et al., "A 3.6pW/frame-pixel 1.35V PWM CMOS Imager with Dynamic Pixel Readout and no Static Bias Current," *ISSCC Dig. of Tech. Papers*. pp. 54-55, Feb. 2008.

978-1-61284-303-2/11 $26.00 © 2011 IEEE

ISSCC 2011 / February 21, 2011 / 4:15 PM

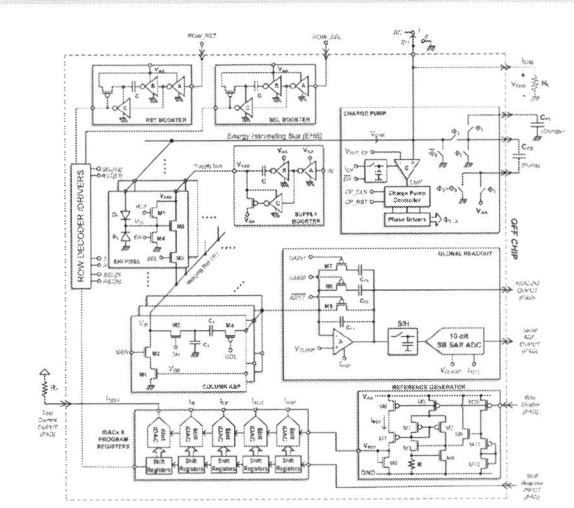

Figure 6.7.1: Circuit diagram of the EHI CMOS APS imager.

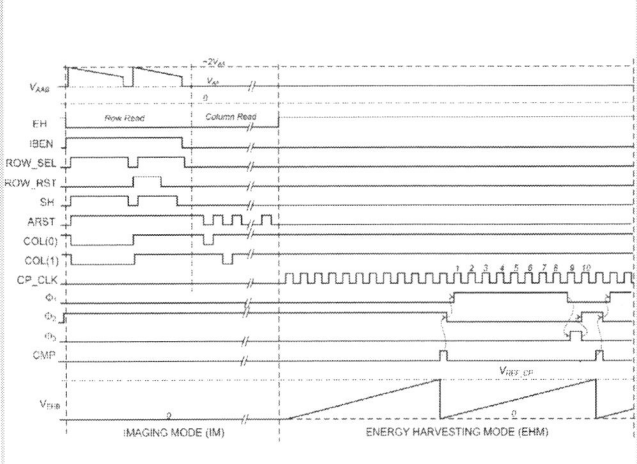

Figure 6.7.2: Timing diagram of the EHI CMOS APS imager.

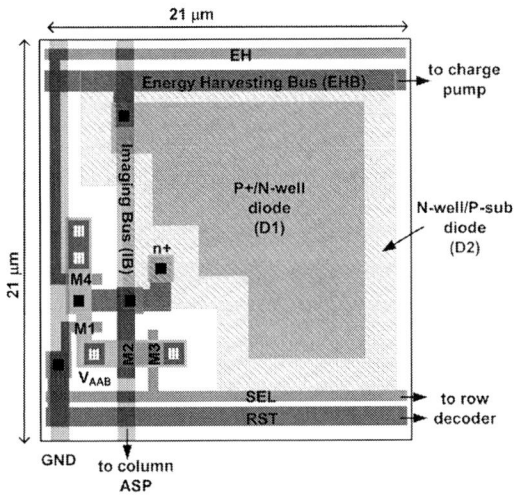

Figure 6.7.3: Energy harvesting and imaging (EHI) pixel layout.

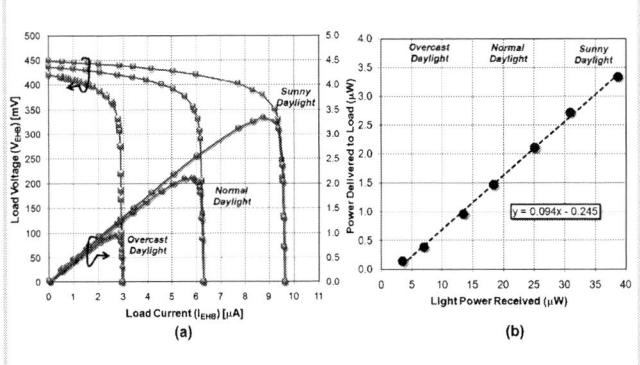

Figure 6.7.4: Measured energy-harvesting characteristics of the EHI imager: (a) voltage-current/voltage-power (I-V/V-P) curves, (b) input-ouput power characteristics at maximum power point (MPP) for harvesting efficiency.

Figure 6.7.5: Full chip power consumption and captured images at different supply voltages.

	Cho (JSSC-03) [5]	Kagawa (ISSCC-08) [6]	This Work
CMOS Technology	0.35-μm (2P3M)	0.35-μm (2P3M)	0.5-μm/5V (2P3M)
Threshold Voltages	+0.65V/ -0.85V	+0.6V/ NA	+0.80V / -0.90V
Pixel Size	5 μm²	10 μm²	21 μm²
Pixel Count	176 x 144	128 x 96	54 x 50
Power Supply voltage	1.5V	1.35V	1.2V
Fill Factor	30%	18.5%	32%
Frame Rate	30fps	9.6fps	7.4 fps
ADC	8bit	8bit	10bit
ADC Type	SAR	Ramp	SB-SAR
Random Noise (dark)	0.85 LSB	0.95 LSBrms	0.98 LSBrms
Power Consumption (dark) (whole chip)	560μW	55.2μW	14.25μW
Power Consumption (dark) (pixel array)	4.4μW	0.42μW	0.0264μW
FOM (pixel array) (pW/frame.pixel)	5.79	3.56	1.32
IFOM (whole chip) (fJ/pixel.code)	2826	914	696

Figure 6.7.6: Performance comparison of the EHI imager and similar low-power imagers from [5] and [6].

978-1-61284-303-2/11 $26.00 © 2011 IEEE

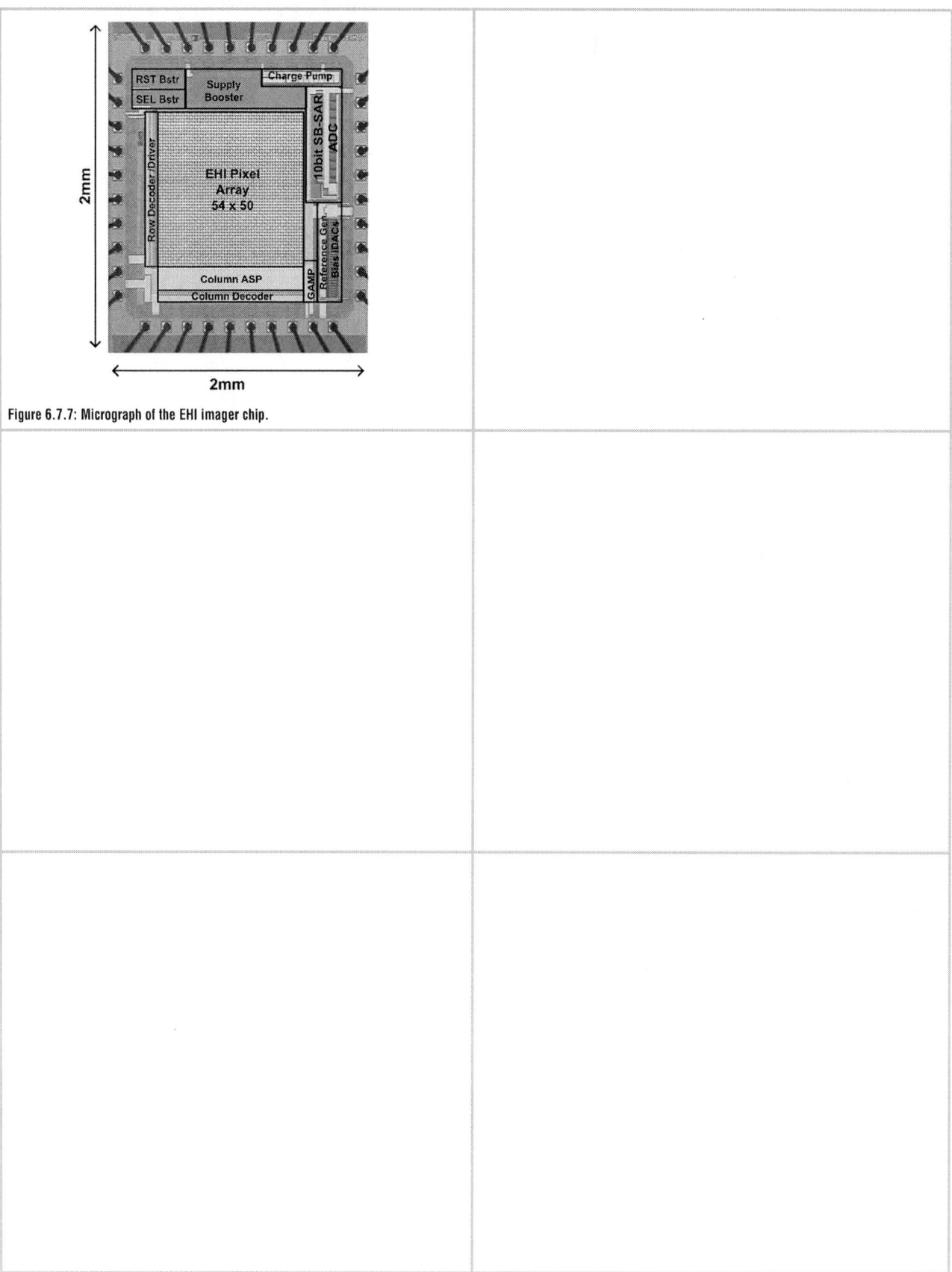

Figure 6.7.7: Micrograph of the EHI imager chip.

ISSCC 2011 / SESSION 6 / SENSORS & ENERGY HARVESTING / 6.8

6.8 5µW-to-10mW Input Power Range Inductive Boost Converter for Indoor Photovoltaic Energy Harvesting with Integrated Maximum Power Point Tracking Algorithm

Yifeng Qiu[1], Chris Van Liempd[1], Bert Op het Veld[2], Peter G Blanken[2], Chris Van Hoof[3,4]

[1]imec - Holst Centre, Eindhoven, The Netherlands,
[2]Philips Research Laboratories, Eindhoven, The Netherlands,
[3]imec, Leuven, Belgium,
[4]KU Leuven, Leuven, Belgium

Energy harvesting provides a means to supply wireless sensor networks in building environments with autonomous and sustainable power [1,2]. Indoor light can be converted into electricity by a solar cell and stored in a rechargeable battery or super-capacitor [3]. Depending on the indoor illumination level and the installed location as well as the orientation of the solar cell, the output power of an amorphous silicon solar cell can vary from less than a few µW/cm² up to hundreds of µW/cm² [3]. To manage the variation in both the illumination and panel size, a power-management circuit that is able to efficiently handle a wide range of input power is necessary. A solar cell can be modeled as a light-controlled current source in parallel with a diode. Its output current is determined by the output voltage in an exponential relation. At one point, the solar cell reaches its maximum power point (MPP). This MPP can be tracked by tuning either the output voltage or the load impedance [4].

We report an inductive boost converter targeting solar cells at indoor conditions. The converter bridges a solar cell array and an energy storage system (ESS), which can be either a Li-ion battery or a super-capacitor. As shown in Fig. 6.8.1, the converter consists of two major functional blocks: the power train and the control circuit. The power train consists of an inductor and two power switches (NMOS and PMOS). An on-chip linear regulator directly supplied by V_{BAT}, provides a voltage domain of 1V for the control circuits. The linear regulator takes its reference voltage from a nano-power reference circuit operating under this 1V domain. A relaxation oscillator based on charging and discharging a capacitor, generates a 100kHz clock for the whole system.

The converter operates in discontinuous conduction mode (DCM) and uses pulse-skipping modulation (PSM), as illustrated in Fig. 6.8.2. When operating in DCM, the inductor current I_L always decreases to 0 at the end of T_{PMOS}. PSM means the converter skips an N_{skip} number of clock cycles before the next T_{NMOS} is generated. By controlling N_{skip} the average input current of the converter is changed and this influences the operation point of the solar cell. The N_{skip} is tunable as integer numbers from 1 to 511, which determines the input power range that the converter can handle. It is possible to reach the maximum power point of the source by controlling N_{skip}, which is handled by a maximum power point tracking (MPPT) algorithm, shown in Fig. 6.8.1.

The MPPT algorithm is based on the classic Hill-climbing method [5]. A simplified flow chart of the algorithm is given in Fig. 6.8.3. This method uses an iterative feedback process. While it does not require the characteristics of the solar cell to be known in advance, it needs to monitor a feedback parameter. We chose to monitor the output current of the converter, which is the current through the PMOS. The reason is V_{BAT} hardly changes during one group measurement. Therefore the output current represents the power converted. Maximizing this current is equivalent to maximizing the input power. The algorithm has two sessions, *Search* and *Long-wait*. In the *Search* session, the algorithm carries out a series of group measurements. Each group measurement consists of 3 measurements of the converter output current using three N_{skip} values (N_0, N_{Less} and N_{More}). The 3 values correspond to 3 different operation points on the power curve, as illustrated in Fig. 6.8.3 (upper-right). N_{Less} and N_{More} are calculated from N_0 based on the selected step size and there are 3 step sizes. During each measurement, a high-side current sensor outputs a scaled copy of the current through the PMOS switch and charges one of the 3 algorithm capacitors in Fig. 6.8.1, which are associated with the 3 N_{skip} values.

At the end of each group measurement, the voltages accumulated on the algorithm capacitors are compared with each other by the evaluation comparators in

Fig. 6.8.1. The one with the highest voltage wins and the N_{skip} value associated with it becomes the N_0 of the next group measurement, from which a new N_{Less} and a new N_{More} are calculated. Each time the first value of a group measurement (N_0) in 2 consecutive group measurements wins, the algorithm switches to a finer step size. If this happens when the current step size is the finest, the algorithm declares that the MPP state is reached and this N_0 is the optimal N_{skip} value at this moment. The *Search* session then ends and the algorithm enters the *Long-wait* session, during which the converter operates with the optimal N_{skip} value. The *Search* session is started again after the *Long-wait* session. A wide input power range results in a huge variation of the algorithm capacitor voltages, as illustrated in Fig. 6.8.3 (lower-right). To address this issue, the scaling factor of the current sensor is adapted during the Hill-climbing procedure by monitoring the voltages accumulated on the capacitors in order to expand the dynamic range of the algorithm.

The efficiency of the converter without MPPT was measured under various conditions, as shown in Fig. 6.8.4. The converter is able to convert input power from 5µW up to 10mW. The measured peak efficiency is around 87%. The whole control circuit consumes a static current of 0.65µA from the battery when not converting power. While converting minimal power (5µW) and maximum power (10mW), the current consumption is 0.8µA and 2.1µA, respectively. The measured performance of the converter with MPPT is shown in Fig. 6.8.5. For the convenience of measurement without losing generality, we simulated a solar cell array with a current source in parallel with three diodes in series, which gives MPP voltages (around 70% of open-circuit voltage) well within the input voltage range when the current is adjusted from 10µA to 1.2mA. The algorithm succeeded in tracking MPP with a relatively flat end-to-end efficiency curve, peaking at 70%. Since the algorithm was implemented as a demonstration of feasibility, Fig. 6.8.5 also shows some mismatch loss due to the algorithm timing, which was not yet optimized.

When V_{BAT} is below 1V, the converter first charges the ESS through the body diode of the PMOS switch while a startup circuit (Fig. 6.8.1) ensures that the NMOS switch remains turned off and the PMOS switch subsequently is turned on so that the solar cell can continue to charge the ESS until V_{BAT} reaches 1V, after which the startup circuit is disabled. This feature is useful when a super-capacitor is left in a dark area for an extended period.

In conclusion, this self-contained and fully autonomous converter features an input power range spanning from 5µW up to 10mW while consuming very little power. A comparison with state-of-the-art publications is shown in Fig. 6.8.6. The method used for MPPT is generic for energy source with one global maximum on their power curves. It is thus possible to use it with TEG and RF harvesters provided their output voltages fall within the input range. The presented circuit is fabricated in a commercial TSMC 0.25µm CMOS process with 5 metal layers and uses one compact SMD inductor of 1mH. A die photo is shown in Fig. 6.8.7.

References:
[1] H. Shao, C.-Y. Tsui, and W.-H. Ki, "The Design of a Micro Power Management System for Applications Using Photovoltaic Cells With the Maximum Output Power Control," *IEEE Trans. VLSI Systems*, vol. 17, no. 8, pp. 1138-1142, Aug. 2009.
[2] J.E. Carlson, K. Strunz, and B.P. Otis, "A 20 mV Input Boost Converter With Efficient Digital Control for Thermoelectric Energy Harvesting," *IEEE J. Solid-State Circuits*, vol. 45, no. 4, pp. 741-750, Apr. 2010.
[3] W.S. Wang, T. O'Donnell, and L. Ribetto, et al., "Energy harvesting embedded wireless sensor system for building environment applications," *Int. Conf. Wireless VITAE*, pp. 36-41, May 2009.
[4] D. Shmilovitz, "On the control of photovoltaic maximum power point tracker via output parameters," *IEE Proc. Electric Power Applications*, vol. 152, no. 2, pp. 239-248, Mar. 2005.
[5] T. Esram and P.L. Chapman, "Comparison of Photovoltaic Array Maximum Power Point Tracking Techniques," *IEEE Trans. Energy Conversion*, vol. 22, no. 2, pp. 439-449, June 2007.
[6] I. Doms, P. Merken, R. Mertens, and C. Van Hoof, "Integrated capacitive power-management circuit for thermal harvesters with output power 10 to 1000µW," *ISSCC Dig. Tech. Papers*, pp. 300-301, Feb. 2009.

978-1-61284-303-2/11 $26.00 © 2011 IEEE

ISSCC 2011 / February 21, 2011 / 4:45 PM

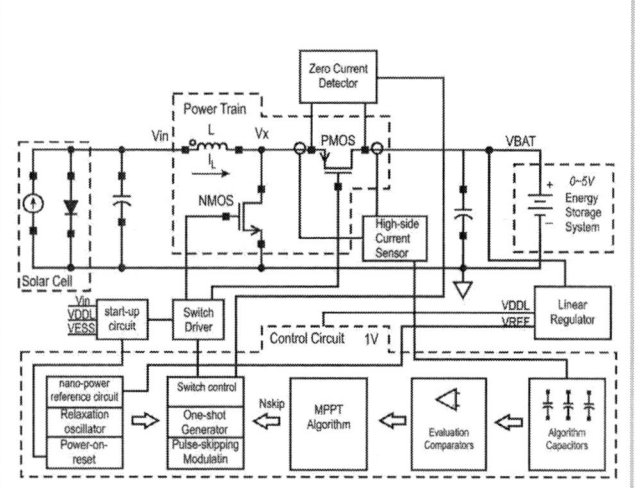

Figure 6.8.1: Block diagram of the system.

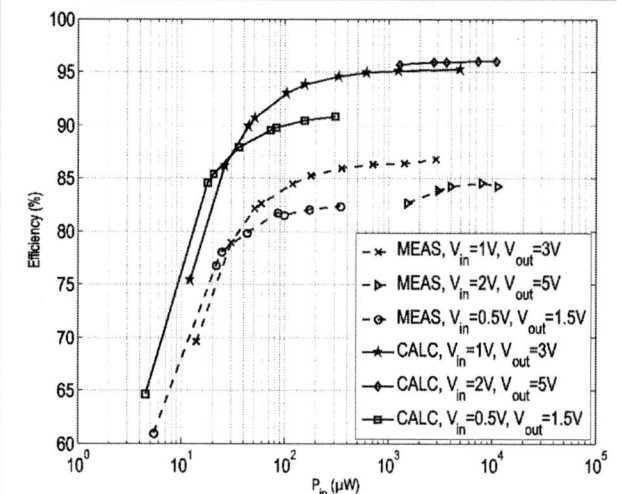

$$I_{in,avg} = \frac{V_{in} D_1^2 T_s}{2L} \cdot \frac{V_{out}}{V_{out} - V_{in}}$$

$$D_1 = \frac{D_1 T_S}{(1 + N_{skip}) T_{CLK}}$$

Figure 6.8.2: An illustration of DCM and PSM influencing the operation point of the solar cell. The dashed curves are sweep of N_{skip} (using a fixed $D_1 T_S$) and the solid curves are sweep of solar cell power.

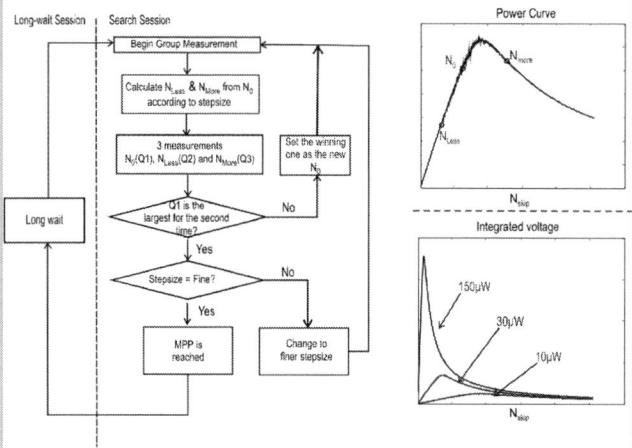

Figure 6.8.3: Simplified diagram of the maximum power point tracking algorithm.

Figure 6.8.4: Measured (dashed) and calculated (solid) converter efficiency with respect to input power.

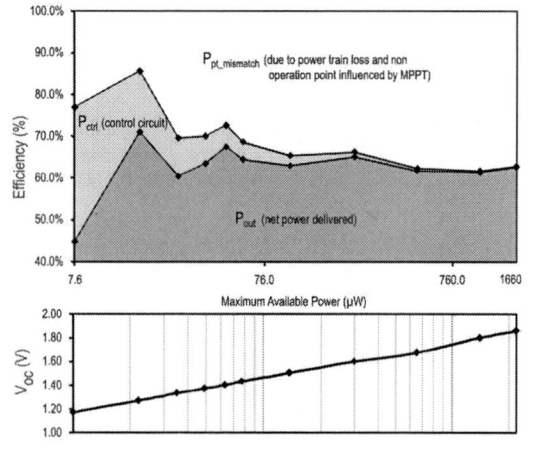

Figure 6.8.5: Experimental end-to-end efficiency using a discrete equivalent circuit as a solar cell simulator (V_{BAT}=3V). Also shown is the corresponding open-circuit voltage.

	[6] (ISSCC,09)	[2] (JSSC,10)	This work
Input voltage range	>0.6V	20mV to 250mV	0.5~2V
Output voltage	0~3.3V	1V	0~5V
Power throughput	Pin=10µW ~ 1mW	Pout=1µW ~ 100µW	Pin=5µW ~ 10mW
Power consumption of control circuit	2µW @ Pin=100µW 7µW @ Pin=1mW	1.6µW @ Pin=54µW 1.1µW @ Pin=233µW	2.4µW @ Pin=8µW 3.5µW @ Pin=1.66mW
	(V_{BAT} = 2V)	(V_{BAT}=1V)	(V_{BAT} = 3V)
Function	Harvester interface +MPPT	Harvester interface + voltage regulator	Harvester interface + MPPT
Architecture	Reconfigurable capacitive charge-pump	Inductive boost converter	Inductive boost converter
Maximum efficiency converter: η=P_out/P_in	82%	78%	87%
Maximum efficiency end to end: η=P_out/P_mp	70%	N/A	70%
Startup voltage	Vin>0.6V and VBAT>2V	VBAT > 0.6V	Vin or VBAT > 1V

Figure 6.8.6: Comparison with state-of-the-art publications.

978-1-61284-303-2/11 $26.00 © 2011 IEEE

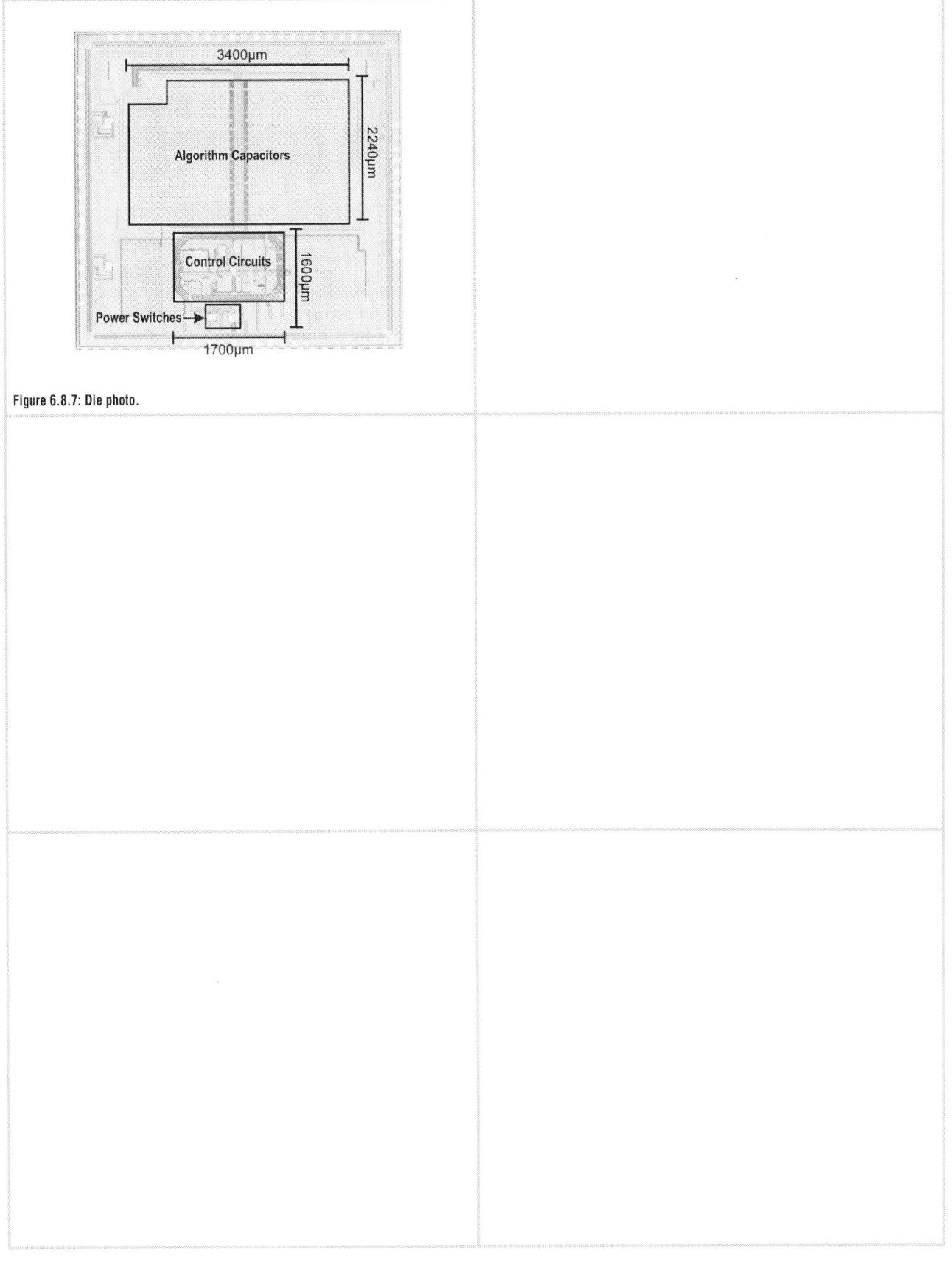

Figure 6.8.7: Die photo.

ISSCC 2011 / SESSION 6 / SENSORS & ENERGY HARVESTING / 6.9

6.9 A Self-Supplied Inertial Piezoelectric Energy Harvester with Power-Management IC

Ethem Erkan Aktakka, Rebecca L. Peterson, Khalil Najafi

University of Michigan, Ann Arbor, MI

Harvesting energy from ambient vibrations is a promising technology for fully autonomous wireless sensor nodes, which can give birth to new applications in biomedical, industrial, and environmental monitoring. There have been independent solutions in increasing the harvesting efficiency either on the mechanical harvester [1] or on its power management circuitry [2,3]. Recently, a piezoelectric MEMS harvester using AlN was demonstrated to generate enough energy to autonomously power a wireless temperature sensor with a full-bridge rectifier built with off-the-shelf components [1]. Meanwhile, AC-DC converters for piezoelectric harvesters have been designed to enable efficient power extraction [2], or efficient rectification of low-voltage outputs [3], and have been tested with commercial meso-scale piezoelectric beams. However, to realize an efficient stand-alone energy generator platform, it is necessary to integrate these efforts into a single low-volume system. This paper presents a self-supplied energy generator, which includes a MEMS harvester hybridly integrated with its power management circuitry for autonomous charging of an energy reservoir (Fig. 6.9.1). The proposed packaging of the generator is <0.3cm³. Initial testing results are obtained with an unpackaged MEMS harvester.

The MEMS harvester is fabricated with a recently developed process technology [4], which involves aligned solder bonding and thinning of bulk PZT ceramics on silicon. This technology offers advantages in fabrication flexibility and device performance over existing piezoelectric thin film deposition methods. Although the introduced low-temperature PZT-on-Si process allows post-CMOS integration of the piezoelectric MEMS and its circuitry on the same silicon die, hybrid integration is preferred due to the large difference in the harvester and circuit die sizes. The harvester is packaged with a Si bottom cap and Si-Glass top cap, both incorporating a recess for free movement of the harvester proof mass. The leads from the harvester are fed to the top glass substrate through vertical low-resistivity Si vias. The CMOS chip is wire-bonded on this top cap, and the connections between surface-mount device (SMD) components and the chip is routed through 1μm thick aluminum interconnects on the glass.

The power-management IC utilizes the TSMC 0.18μm technology. The system is completely self-supplied by the harvester vibration energy, and has no dependence on a previously charged energy reservoir. Power management is achieved with three sub-circuits, MOS switches and an active diode for low-dropout rectification, a shunt pass system to increase harvesting efficiency (similar aim with the "bias flip rectifier" described in [2]), and a trickle battery charger (Fig. 6.9.2). A supply-independent bias [5] is used to act as current mirror to the comparators in the system, and limit the overall power consumption. This V_{BIAS} is also used as the reference voltage (V_{REF}) in the trickle charger to define the voltage regulation levels.

The rectification of the piezoelectric output is achieved by incorporating two parts (Fig. 6.9.3). First, four CMOS switches output the modulus of V_{PIEZO} by converting the negative half cycles into positive ones ($V_{MODULUS}$) [6]. Second, an active diode rectifies this voltage to store the charge on a temporary energy reservoir ($V_{STORAGE}$), which is also used as the power supply of the active circuitry (V_{DD}). The PMOS gate used in the active diode is bulk-regulated in order to connect its n-well to the highest potential available for minimum leakage [7]. This two-stage active rectification scheme minimizes the voltage drop on the path between V_{PIEZO} and $V_{STORAGE}$, although there is a minimum input threshold requirement to drive the initial CMOS switches. Also, between these two stages a voltage limiter, implemented with a series of on-chip diodes, is used to protect the circuit against possible peaks from the harvester.

A parallel shunt system is placed across the inputs from the piezoelectric harvester, in order to increase the harvesting efficiency ($P_{IN-LOADED}/P_{IN-UNLOADED}$). When the harvester's output current, I_{PIEZO}, changes polarity, the harvester power starts to be wasted to discharge and recharge its self-capacitance (C_{PIEZO}) in reverse polarization. In order to partially avoid this situation, the charge on C_{PIEZO} is shunted to zero whenever I_{PIEZO} crosses zero in either direction. This

condition is verified by checking when $V_{MODULUS}$ drops down $V_{STORAGE}$. Because of the Schottky diodes (D1, D2) used in the shunt path, no tuning for precise timing control is necessary to drive M1 and M2, as in [2]. Instead, one of these gates remains on during the whole recharging period of C_{PIEZO}.

The energy stored in the temporary reservoir ($C_{STORAGE}$) is dumped into a final reservoir by a trickle charger (Fig. 6.9.4). Just to illustrate the operation principle, assume that the system starts with zero voltage in both reservoirs. With a vibration input, the MEMS harvester starts to supply power, and there is a start-up period where $C_{STORAGE}$ is charged passively. When there is enough vibration energy to charge this capacitor up to a minimum level (V0), the active circuit becomes operational, and a temporary dropout occurs in $V_{STORAGE}$. Now, $C_{STORAGE}$ continues to be charged, but more efficiently with the active diode and shunt system fully operational. When $V_{STORAGE}$ reaches a certain value (V2), the bulk-regulated PMOS gates between the temporary and final energy reservoirs are on, and the scavenged energy is transferred to the battery through a current pulse, until $V_{STORAGE}$ drops down to the level of either V1 or $V_{BATTERY}$. Then, the gates are turned off, and $C_{STORAGE}$ is charged again back to the level of V2. This charging scheme continues till $V_{BATTERY}$ is fully charged to V3 value. At this point, the gates turn off not to overcharge the battery above its rated voltage. In order to avoid the system to be locked at this stable operation point, $V_{STORAGE}$ is put into scanning between V1 and V4 levels, so the system can check whether the battery needs to be recharged again. When the input vibration ends, $V_{STORAGE}$ decays down to zero, and the back-flow of battery charge is prevented with a NAND gate powered by the battery. The power consumption of this NAND gate added with the leakage back to the $C_{STORAGE}$ results in a near-zero (<5pA) standby current draw. For testing flexibility, the battery's final charged voltage level can be adjusted by the option to use an external signal instead of internally generated V_{BIAS} for the comparison signal V_{REF}. Alternatively, a different set of resistors, which are used to obtain fractions of $V_{STORAGE}$, can be chosen to define the new voltage-regulation levels.

The overall system is tested by charging a final energy reservoir at different vibration levels (Fig. 6.9.5). When the MEMS harvester is connected to an optimum resistive load without any power-management circuitry, it can supply 24.1μW and 63.9μW under 0.5g and 1.0g vibration levels, respectively. The Normalized Power Density (Power / Volume / Acceleration²), and bandwidth of the generator is compared with the current state-of-the-art micro-fabricated inertial harvesters (Fig. 6.9.6). When the harvester is connected with its power management IC, it can charge a 20mF ultra-capacitor up to 1.31V in 20min under 0.5g vibration, and in ~8min under 1.0g vibration input.

Acknowledgements:
This work is supported by the DARPA Hybrid Insect MEMS program under Grant # N66001-07-1-2006.

References:
[1] R. Elfrink, et al., "First Autonomous Wireless Sensor Node Powered by a Vacuum-Packaged Piezoelectric MEMS Energy Harvester," *IEEE International Electron Devices Meeting*, pp. 543-546, Dec. 2009.
[2] Y.K. Ramadass and A.P. Chandrakasan, "An Efficient Piezoelectric Energy Harvesting Interface Circuit Using a Bias-Flip Rectifier and Shared Inductor," *ISSCC Dig. Tech. Papers*, pp. 296-297, Feb. 2009.
[3] D. Kwon and G.A. Rincon-Mora, "A Single-Inductor AC-DC Piezoelectric Energy-Harvester/Battery-Charger IC Converting ±(0.35 to 1.2V) to (2.7 to 4.5V)," *ISSCC Dig. Tech. Papers*, pp. 494-496, Feb. 2010.
[4] E.E. Aktakka, R.L. Peterson, and K. Najafi, "A CMOS Compatible Piezoelectric Vibration Energy Scavenger Based on the Integration of Bulk PZT Films on Silicon," *IEEE International Electron Devices Meeting*, Dec. 2010.
[5] E. Dallago, et al., "Active Self Supplied AC-DC Converter for Piezoelectric Energy Scavenging Systems with Supply Independent Bias," *IEEE Int. Symp. Circuits and Systems*, pp. 1448-1451, May 2008.
[6] C. Peters, et al., "A CMOS Integrated Voltage and Power Efficient AC/DC Converter for Energy Harvesting Applications," *Journal of Micromechanics and Microengineering*, vol. 18, no. 10, pp. 104005+, Sept. 2008.
[7] M. Ghovanloo and K. Najafi, "Fully Integrated Wideband High-Current Rectifiers for Inductively Powered Devices," *IEEE J. Solid-State Circuits*, vol.39, pp. 1976-84, Nov. 2004.

978-1-61284-303-2/11 $26.00 © 2011 IEEE

ISSCC 2011 / February 21, 2011 / 5:00 PM

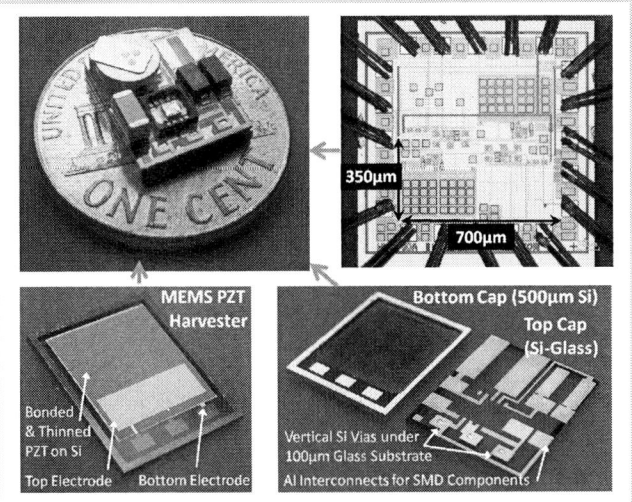

Figure 6.9.1: Photo of the generator, with MEMS harvester, its packaging, placement of the chip and SMD components, and chip micrograph.

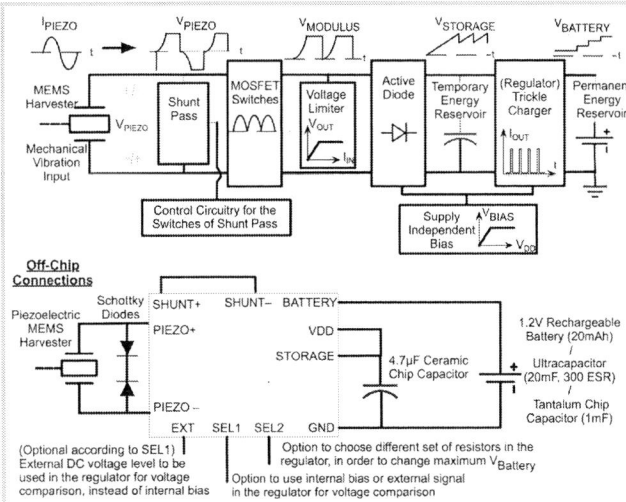

Figure 6.9.2: Overall view of the power-management circuitry, showing the connections between sub-circuits and off-chip components.

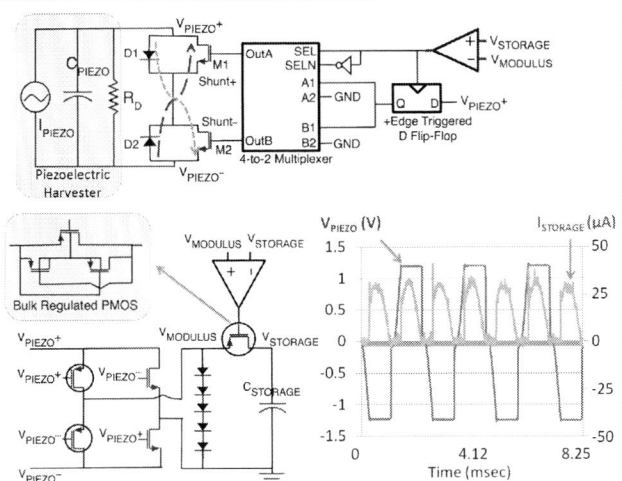

Figure 6.9.3: Active rectification is used to allow low drop-out, and the shunt-pass to increase the power extraction from the harvester.

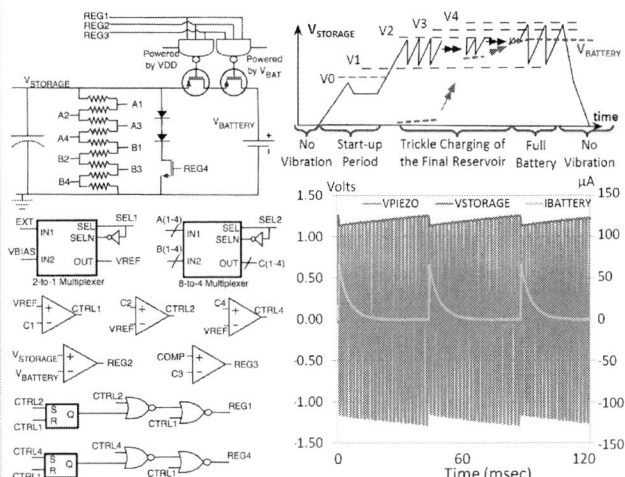

Figure 6.9.4: The trickle charger allows the system to charge a large energy reservoir, while it is self-supplied from the temporary reservoir.

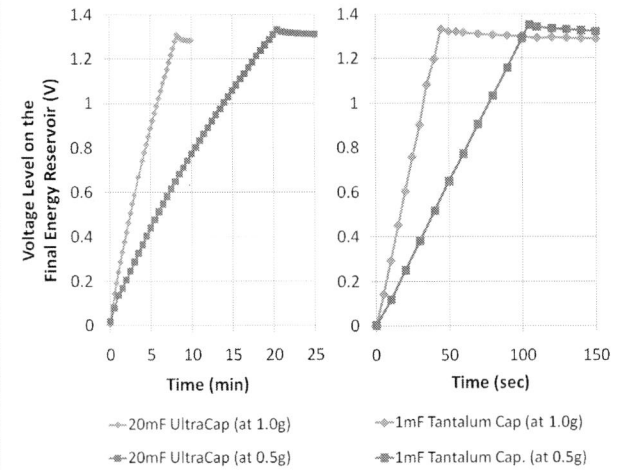

Figure 6.9.5: Charging of a final energy reservoir under different vibration levels.

Ref.	Vibration Harvester	Harvester Size (mm³)	Res. Freq.	P_IN-UNLOADED (Output on R_OPT)	N.P.D. (µW /mm³/g²)	Band-width	Sys. Volume (w/ Circuitry)
This Work	MEMS Harvester with thinned PZT unimorph	27.3	429 Hz	1.8µW at 0.1g	6.45	16.0 Hz	< 0.3 cm³
			419 Hz	67.9µW at 1.0g	2.49	26.3 Hz	
[4]		12.1	80 Hz	3.75µW at 0.3g	3.44	2.5 Hz	N/A
[1] IMEC	MEMS Harvester with sputtered AlN unimorph	28.7	353 Hz	17µW at 0.64g	1.45	-	1 cm³
			325 Hz	85µW at 1.75g	0.97	3 Hz	
[2] MIT	Meso-scale Commercial PZT bimorph	130.3	225 Hz	at 3.35g	-	-	N/A
[3] Gatech	Meso-scale Commercial PZT bimorph	228.8	100 Hz	40 µW	-	-	N/A

Ref.	Circuitry	Process Tech. & Active Area	Power Consumption	P_OUT-BATTERY/ P_IN-UNLOADED	Additional Requirements
This Work	Rectifier w/ Active Diode & Shunt- Pass, Trickle Charger	0.18µm CMOS 0.25 mm²	< 1 µW	50-60%	N/A
[1] IMEC	Schottky Diode Bridge, µC, Temp. Sensor, Radio	Off-the-shelf N/A	10 µW whole system	60%	–
[2] MIT	Bias Flip Rectifier, Inductor Arbiter, DC-DC Buck-Converter	0.35µm CMOS 4.25 mm²	< 2 µW	–	Initially charged battery (1.8V)
[3] Gatech	Rectifier-Free AC-DC Charger w/ LC Transfer Network	2µm BiCMOS 0.90 mm²	0.92µW (4.6nJ/cycle)	75%	Initially charged battery (2.5V)

Figure 6.9.6: Performance metrics, and comparison of the results with the state-of-the-art.

978-1-61284-303-2/11 $26.00 © 2011 IEEE

ISSCC 2011 / SESSION 7 / MULTIMEDIA & MOBILE / OVERVIEW

Session 7 Overview / *Energy-Efficient Digital*

Multimedia & Mobile

Session Chair: *Pascal Urard, STMicroelectronics, Crolles, France*

Session Co-Chair: *Michael Phan, Qualcomm, Raleigh, NC*

Since the release of the Avatar movie, 3DTV has become a primary focus for the multimedia industry, in order to deliver the most realistic 3D experience. This technology will soon reach your home, and then your mobile devices. Additionally, augmented reality blurs the line between what is real and what is computer-generated by enhancing what we see, creating a plethora of new applications. More embedded intelligence, more processing capability and always-improving energy efficiency are the main driving forces for these innovative products.

Paper 7.1 [NTU] demonstrates a state-of-the-art design for 3DTV decoding solutions, in H.264 format. It targets applications such as Quad Full HD, stereoscopic 3DTV, Multi-viewer stereoscopic 3DTV and Virtual Reality. This design is composed of an MVC decoder and a free-viewpoint engine able to synthesize multiple points of view. It is able to perform Quad Full HD (QFHD) / 2160p at 216fps consuming only 69.5mW at 240MHz under 0.9V supply and is realized in 40nm CMOS in 5.76mm².

Next we explore what could be the near-future H.265/HEVC video standard with Paper 7.2 [MIT]. This paper presents Context-Based Adaptive Binary Arithmetic Coding (CABAC), which addresses a key bottleneck in video-coding standards. The proposed coding scheme enables parallelization at the algorithmic level. In this 65nm CMOS chip, the decoder takes full advantage of this innovation and implements an easily scalable 5×16 parallel architecture that achieves QFHD/2160p decoding at 186fps while consuming only 77mW with a clock frequency of 125.5MHz under a 1.0V supply.

A heterogeneous multimedia processor for 3D graphics, image processing, and augmented reality integrated with a reconfigurable transceiver pool is discussed in Paper 7.3 [KAIST]. This 0.13μm CMOS multimedia solution is implemented in IC-stacking on SI-interposer. It embeds a configurable vector-processing unit for frame-level parallelism, a unified filtering unit with a memory-access-efficient texturing algorithm and a programmable shader integrating multiple cores, which achieves a throughput of 1.6Gpixels/s and 25Mpolygons/s, and consuming 275mW at 1.2V.

Paper 7.4 [KAIST] presents the first portable embedded neuro-fuzzy accelerator for object recognition and an intelligent reconfigurable integrated system, which realizes low power consumption and high-speed recognition, prediction and optimization for artificial intelligence applications. This analog/digital mixed-mode processor is realized in 0.13μm CMOS and achieves 57mW power consumption at 1.0V for the analog part and 1.2V for the digital part, leading to a power efficiency of 655GOPS/W in 13.5mm².

Paper 7.5 [TI] is a 28nm low-power CMOS processor based on a 4-issue, 32-register version of the TMS320C64x+ VLIW DSP. By utilizing ultra-low-voltage standard-cell libraries, SRAM with 6T bit-cells using hierarchical sensing, wordline boosting, pre-read during write techniques, and new statistical time analysis methodology. This DSP demonstrates working prototypes from 331MHz under 1.0V supply consuming 145mW, down to 14.4MHz under 0.6V consuming 5.9mW.

Paper 7.6 [ITRI] is a 49mm² WiMAX IEEE802.16e SoC consuming 600mW at 1.2V in 90nm CMOS. It implements a 2×2 MIMO digital baseband transmitter and receiver, and associated MAC and integrates an ARM-926, Flash memory, SDRAM controller, AES engine, and USB2.0. It also allows possible future extension to 802.16m/LTE. This WiMAX system is integrated in a dongle with relevant RF circuit and provides 5Mb/s downlink at 300km/h for mobile users in high-speed trains, and up to 30Mb/s reception in low-mobility-use cases.

978-1-61284-303-2/11 $26.00 © 2011 IEEE

ISSCC 2011 / February 22, 2011 / 8:30 AM

The chipset solution presented in Paper 7.7 [MediaTek] is fully compliant with the IEEE 802.16e Mobile WiMAX corrigendum 1 and 2 and WiMAX forum Wave2 requirements. This chipset consists of a dual-band 2×2 MIMO RF transceiver chip and a fully integrated WiMAX modem/router chip. The RF transceiver chip is implemented in 11.05mm^2 in 65nm CMOS and has a power consumption of 364mW at 2.8V. Implemented in the same technology in 24.99mm^2, the modem/router chip targeting portable routers has a power consumption of 632.7mW at 1.3V. The total chipset reports a margin of up to 7dB versus WiMAX RCT requirements. This ensures that users in the field will get high link quality service.

Paper 7.8 [UCLA] presents the first Direct Digital Synthesizer (DDS) to eliminate Phase-Accumulator (PA) pipelining, thus opening the door to dynamic frequency hopping. It uses PA rounding and gets 2's complement conditional negations via 1's complement negation with no carry ripple. The chip operates at 260MHz and reports a power consumption of 16.5mW at 1.8V as an average over typical frequency control words. This gives an FoM of 0.0635mW/MHz which is the lowest reported to date for any 16b DDS. Its outputs have a 113dB SFDR and a 98dB SNR, virtually the best possible for a 16b DDS. This chip is fabricated in 0.18μm CMOS and has an area of 0.16mm^2.

Presenters:

8:30 AM

7.1 A 216fps 4096×2160p 3DTV Set-Top Box SoC for Free-Viewpoint 3DTV Applications

P-K. Tsung, National Taiwan University, Taipei, Taiwan

9:00 AM

7.2 A Highly Parallel and Scalable CABAC Decoder for Next-Generation Video Coding

V. Sze, Massachusetts Institute of Technology, Cambridge, MA

9:30 AM

7.3 A 275mW Heterogeneous Multimedia Processor for IC-Stacking on Si-Interposer

H-E. Kim, KAIST, Daejeon, Korea

10:15 AM

7.4 A 57mW Embedded Mixed-Mode Neuro-Fuzzy Accelerator for Intelligent Multi-core Processor

J. Oh, KAIST, Daejeon, Korea

10:45 AM

7.5 A 28nm 0.6V Low-Power DSP for Mobile Applications

G. Gammie, Texas Instruments, Dallas, TX

11:15 AM

7.6 A MIMO WiMAX SoC in 90nm CMOS for 300km/h Mobility

G. C. Chuang, ITRI, Hsinchu, Taiwan

11:45 AM

7.7 A 70Mb/s -100.5dBm Sensitivity 65nm LP MIMO Chipset for WiMAX Portable Router

J-S. Pan, MediaTek, Hsinchu, Taiwan

12:15 PM

7.8 A Direct Digital Frequency Synthesizer with Minimized Tuning Latency of 12ns

A. Willson, University of California, Los Angeles, CA

978-1-61284-303-2/11 $26.00 © 2011 IEEE

ISSCC 2011 / SESSION 7 / MULTIMEDIA & MOBILE / 7.1

7.1 A 216fps 4096×2160p 3DTV Set-Top Box SoC for Free-Viewpoint 3DTV Applications

Pei-Kuei Tsung, Ping-Chih Lin, Kuan-Yu Chen, Tzu-Der Chuang, Hsin-Jung Yang, Shao-Yi Chien, Li-Fu Ding, Wei-Yin Chen, Chih-Chi Cheng, Tung-Chien Chen, Liang-Gee Chen

National Taiwan University, Taipei, Taiwan

3DTV promises to become the mainstream of next-generation TV systems. High-resolution 3DTV provides users with a vivid watching experience. Moreover, free-viewpoint view synthesis (FVVS) extends the common two-view stereo 3D vision into virtual reality by generating unlimited views from any desired viewpoint. In the next-generation 3DTV systems, the set-top box (STB) SoC requires both a high-definition (HD) multiview video-coding (MVC) decoder to reconstruct the real camera-captured scenes and a free-viewpoint view synthesizer to generate the virtual scenes [1-2].

There are three main challenges to design an efficient high-resolution 3DTV STB SoC: (1) High processing capability of both the real-view decoder and virtual-view synthesizer is required to support various 2D/3D applications in HDTV. For example, 66.2TOPS of computation is consumed to synthesize one virtual view in quad full-HD (QFHD) at 30fps. (2) In order to support FVVS, including 3D translation and 3D rotation (6D), matrix-based warping is needed for every pixel. In this case, two adjacent pixels in the reference view may be warped to the arbitrary positions in the virtual view according to their depth and epipolar geometry. Horizontal raster-scan-based scheduling [4-5] cannot deal with these irregular pixel relationships and can only support 1D horizontal shifts in the view synthesis. (3) The block-based memory accessing of reference pixels is not suitable for FVVS because of the processing nature of the irregular pixel access. 31.5GB/s system memory bandwidth is thus required for each virtual view in QFHD. State-of-the-art 3DTV chips cannot solve the issues above [4-6].

Our 3DTV STB SoC is summarized as follows. First, a hardware-oriented 6D FVVS flow is introduced along with the corresponding architecture to solve the first two design challenges. A maximum 1911MPixel/s throughput is achieved, and is 9-to-40.5× higher than the previous works [4-6]. Second, the cache-based texture reorder architecture with the dynamic warping reference frame selection (DWRFS) scheme reduces the external memory bandwidth by 95.7% in view synthesis. Finally, the precision-optimized Homographic Transform (HT) and the single-iteration inpainting save 68% area in the warping engine and 93.3% of computing cycles, respectively.

Figure 7.1.1 shows the 6D FVVS flow and the target applications. The MVC decoder reconstructs the real-view videos, the corresponding camera matrix, and the depth values from the bitstream. The view synthesizer then generates the virtual views from the real views. As shown in the left of Fig. 7.1.1, the pixels on one specific epipolar line of the virtual-view frame can only be found along the corresponding epipolar line in the reference frame. Therefore, the processing schedule along the epipolar lines avoids the conflict of memory accessing compared with the traditional horizontal raster-scan schedule. To support the various epipolar geometries, 7 kinds of block patterns are used to access reference pixels with slopes between ±45°. The accessed pixels are then reordered into an 8×8 block in the texture-reorder stage. For the slopes larger than ±45°, rotating the scan order of blocks converts the effective slope within ±45° and thus extends the supporting rotation angle to ±180°. The warping stage is capable of performing the accurate geometry transformation to further support the continuous epipolar geometry from the 7 discrete slopes. After the warping, the occlusion regions on the virtual-view blocks are then filled in the inpainting stage. The results are outputted after inverse reordering in the final stage.

Figure 7.1.2 shows the system architecture. The first 3 stages are the MVC decoder while the remaining 4 stages constitute the 6D FVVS flow. The parallel mode is shown at the bottom of Fig. 7.1.2. Since the textures in one virtual view are the subset of the neighboring virtual view in the 3D-translation-only cases, one reference block loaded from the system bus can be reused to generate multiple virtual views in parallel. In this way, the throughput is boosted, and the data access bandwidth is reduced. Furthermore, a full-utilization mode is designed by

further increasing the parallelism of views and reusing the decoder bus to increase the output bandwidth. The throughput of 216fps corresponding to 9 views @ 24fps is achieved for QFHD, and is 12.5-to-40.5× higher than the previous works [4-5].

Figure 7.1.3 shows the bandwidth-reduction schemes. In the software-based algorithm [3], multiple reference views are jointly utilized in the warping to generate one virtual view. The textures provided by these reference views have large overlapped regions. In order to avoid the bandwidth to access the redundant information, the DWRFS sets one reference view as the main reference, and others are loaded only for the occlusion regions. Furthermore, the block-based pixel accesses on the system bus conflict with the epipolar-based processing pattern in the texture reordering and inverse reordering stages. A texture-reordering cache and a 16×24 inverse reordering buffer are designed to efficiently load the reference pixels and write back the results in the bus burst mode as shown on the left and right sides of Fig. 7.1.3. The system memory bandwidth for view synthesis is reduced by 95.7%.

Figure 7.1.4 shows the warping and inpainting engines. The warping engine requires 256 matrices corresponding to 256 depth values in HT. To save the large memory requirements from these 256 matrices, a linear interpolation (LI) scheme is used. Only 3 warping matrices are saved in LI. For other depth values, matrices are linearly interpolated from these 3 matrices. As a result, area is reduced by 68%. The inpainting engine is shown on the bottom-right of Fig. 7.1.4. In the software-based inpainting algorithms [3], the occluded pixels are padded from the neighboring pixels by iteratively calculating and updating the gradient of the textures. In the proposed inpainting algorithm, the pixels are filled in a single iteration by analyzing the cause of occlusion of each pixel. The inpainting process can thus be parallelized, and the cycle count is reduced by 93.3%.

Figure 7.1.5 shows the detailed chip specifications and the example FVVS results. The core size is 5.76mm² including 1416K logic gates and 19.9KB on-chip SRAM in TSMC 40nm CMOS. 6D FVVS functionality, H.264/AVC High-Profile decoding, and MVC High-Profile decoding are supported in a single chip. The maximum FVVS capability is 4096×2160p 216fps (24fps @ 9 views), and the maximum decoding capability is 4096×2160p 30fps.

The performance evaluation is shown in Fig. 7.1.6. Compared with the previous works, the chip supports 12.5-to-40.5× view synthesis, 1.25-to-26.7× decoding, and 9-to-40.5× overall system capabilities. Furthermore, 6D FVVS is supported in addition to the conventional 1D horizontal shift. The total power efficiency of the chip is 27.5MPixel/mW, which is 6.6-to-229× higher than those of the previous works. The chip micrograph is shown in Fig. 7.1.7.

Acknowledgement:
The authors thank TSMC University Shuttle Program and Morly Hsieh for process support and National Chip Implementation Center for chip testing. This work is funded by National Science Council and TSMC.

References:
[1] Joint Video Team of ISO/IEC MPEG & ITU-T VCEG, "Joint Draft 8.0 on Multiview Video Coding," ISO/IEC JTC1/SC29/WG11 and ITU-T SG16 Q.6, July, 2008.
[2] MPEG-FTV Group, "Draft Report on Experimental Framework for 3D Video Coding" ISO/IEC JTC1/SC29/WG11 MPEG2010/N11273, April. 2010.
[3] MPEG-FTV Group, "Reference Softwares for Depth Estimation and View Synthesis" ISO/IEC JTC1/SC29/WG11 M15377, April. 2008.
[4] S. H. Kim, et al., "A 36fps SXGA 3D Display Processor with a Programmable 3D Graphics Rendering Engine," *ISSCC Dig. Tech. Papers*, pp. 276–277, Feb. 2007.
[5] S. H. Kim, et al., "A 116fps 74mW Mobile Heterogeneous 3D-Media Processor for 3D Display Contents," *Symposium on VLSI Circuits*, pp. 258-259, June 2009.
[6] T. D. Chuang, et al., "A 59.5mW scalable/multi-view video decoder chip for Quad/3D Full HDTV and video streaming applications," *ISSCC Dig. Tech. Papers*, pp. 330–331, Feb. 2010.

ISSCC 2011 / February 22, 2011 / 8:30 AM

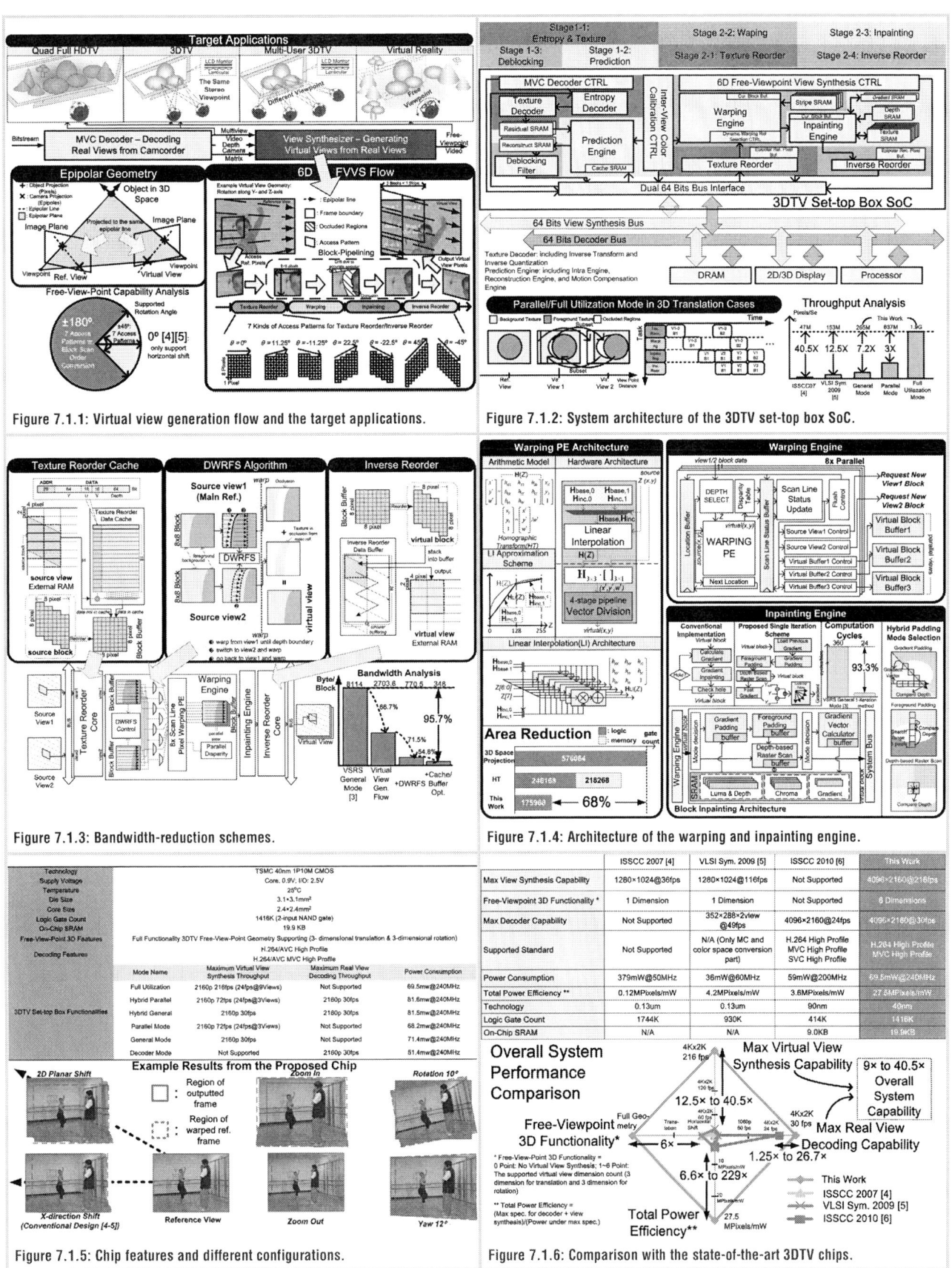

Figure 7.1.1: Virtual view generation flow and the target applications.

Figure 7.1.2: System architecture of the 3DTV set-top box SoC.

Figure 7.1.3: Bandwidth-reduction schemes.

Figure 7.1.4: Architecture of the warping and inpainting engine.

Figure 7.1.5: Chip features and different configurations.

Figure 7.1.6: Comparison with the state-of-the-art 3DTV chips.

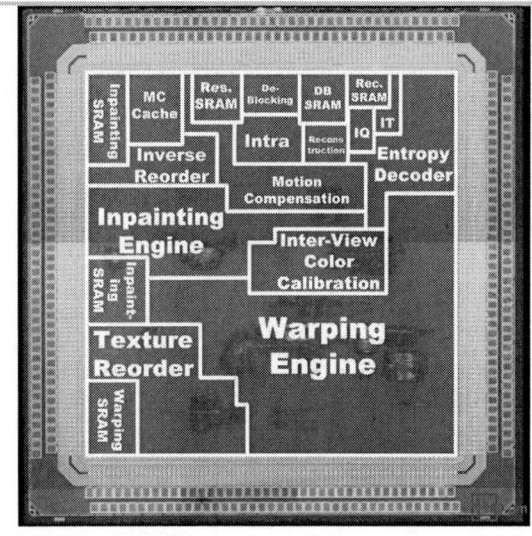

Figure 7.1.7: Chip micrograph.

ISSCC 2011 / SESSION 7 / MULTIMEDIA & MOBILE / 7.2

7.2 A Highly Parallel and Scalable CABAC Decoder for Next Generation Video Coding

Vivienne Sze, Anantha P. Chandrakasan

Massachusetts Institute of Technology, Cambridge, MA

Future video decoders will need to support high resolutions such as Quad Full HD (QFHD, 4096×2160) and fast frame rates (e.g. 120fps). Many of these decoders will also reside in portable devices. The next-generation standard called High-Efficiency Video Coding (HEVC), which is being developed as a successor to H.264/AVC, not only seeks to improve the coding efficiency but also to account for implementation complexity and leverage parallelism to meet future power and performance demands [1]. Parallel processing increases the throughput for higher performance, which can be traded-off for lower power with voltage scaling. This paper presents a silicon prototype for a pre-standard algorithm developed for HEVC ("H.265") called Massively Parallel CABAC (MP-CABAC) that addresses a key video decoder bottleneck. The test chip has over an order-of-magnitude higher throughput than state-of-the-art H.264/AVC CABAC engines [2-4], while maintaining high coding efficiency. Architecture and joint algorithm-architecture optimizations, which modify the MP-CABAC algorithm, are used to reduce critical path delay and memory size.

Context-based Adaptive Binary Arithmetic Coding (CABAC) is a well known bottleneck in existing H.264/AVC decoders. Although CABAC provides high coding efficiency, its tight feedback loops (Fig. 7.2.1) make it difficult to parallelize and limit the overall decoder throughput. The feedback loops are tied to the binary symbols (bins); thus, the throughput and performance of the CABAC engine are measured in bins/cycle and bins/second, respectively. Speculative computation is often used to increase the throughput at the cost of increased power consumption [2-4]. Unlike the rest of the video decoder which can use macroblock-line (wavefront) parallelism, CABAC can only be parallelized across frames [5]; consequently, buffering is required between CABAC and the rest of the decoder which increases external memory bandwidth.

Massively Parallel CABAC (MP-CABAC), previously developed by the authors [6], is currently under consideration for HEVC, and has been adopted into the standard body's JM-KTA working software [7]. It enables parallel processing, while maintaining the high coding efficiency of CABAC, by using a combination of two forms of parallelism shown in Fig. 7.2.1: interleaved entropy slices (IES) and syntax element partitions (SEP). IES enables several slices to be processed in parallel, allowing the entire decoder to achieve wavefront parallel processing without increasing external memory bandwidth [8]. SEP enables different syntax elements (e.g. motion vectors, coefficients, etc.) to be processed in parallel with low area cost [9]. Figure 7.2.1 shows the MP-CABAC data structure, where each frame is composed of several IES, and each IES is composed of five SEP. The MP-CABAC test chip presented in this paper supports up to 16 IES per frame with 80 arithmetic decoders running in parallel.

IES are processed in parallel by several slice engines as shown in Fig. 7.2.2. IES FIFOs are used between slice engines to synchronize IES, which is required due to top block dependencies. The properties of the neighboring blocks (A and B) are used for context selection and are stored in the IES FIFOs and last line buffer. Figure 7.2.2 shows a joint algorithm-architecture optimization in the context selection logic that reduces the last line buffer size by 50%. To enable scalability, the number of slice engines is configurable; a multiplexer connects the output of the last enabled slice engine to the last line buffer. To reduce power, the clocks to the disabled slice engines are turned off using hierarchical clock gating. Over a 9× increase in throughput is achieved with 16 IES per frame using the architecture in Fig. 7.2.2.

SEP are processed in parallel by several arithmetic decoders (AD) within the slice engine as shown in Fig. 7.2.2. Syntax elements are assigned to 5 different partitions based on their workload (i.e. number of bins). The FSM of the context modeler (CM) and de-binarizer (DB) is divided into smaller FSMs for each SEP. The register-based context memory is also divided into smaller memories for each SEP. Thus, the context memory and the FSM are not replicated, which keeps the area cost low. The slice engine contains 5 different partition engines, each with a small FSM, context memory and AD. Dependencies between SEP are managed using SEP FIFOs, allowing SEP of different macroblocks to be

processed concurrently. During the stall cycles, the partition engine clock is disabled with hierarchical clock gating to reduce power. Using this slice engine architecture, up to five bins can be decoded in parallel, with an average throughput increase of 2.4×.

Figure 7.2.3 shows the architecture of the partition engine. CM selects the context (i.e. state and most probable symbol (MPS)) based on the syntax element being processed. AD uses this context and encoded bits from the bitstream controller to decode a bin. The bin is fed back to CM to update the context memory and to DB to compute the syntax element. Several techniques are used to reduce critical path delay. First, the engine is pipelined by inserting a register between CM and AD for a 40% reduction. Next, the critical path in AD is reduced using 3 optimizations: (1) Leading-Zero (LZ) detection is done using a look-up table (LUT) in parallel with least-probable symbol interval (rLPS) LUT to speed up renormalization. (2) Early range shifting enables renormalization of rLPS to occur in parallel with the range and offset subtractions. (3) Offset renormalization is moved to the beginning of the next cycle so that it occurs in parallel with the rLPS look up. These architectural optimizations reduce the critical path delay of AD by 11%.

Finally, a joint algorithm-architecture optimization, highlighted as (4) in Fig. 7.2.3 and shown in detail in Fig. 7.2.4, further speeds up AD. Subinterval reordering changes the order of the least and most probable symbol subintervals (rLPS and rMPS). Placing rLPS at the bottom of the range enables the offset comparison to occur in parallel with the subtraction for rMPS, which reduces the critical path by an additional 11% without affecting coding efficiency.

The MP-CABAC test chip, shown in Fig. 7.2.7, was implemented in 65nm CMOS. Figure 7.2.5 shows a summary of the chip features and an example trace and distribution of the number of decoded bins-per-cycle for a video sequence. Figure 7.2.6 shows the trade-off between measured power, performance and coding efficiency across a wide operating range. Scaling the number of IES per frame from 1 to 16 increases the performance range by an order of magnitude, and reduces the minimum energy per bin by 3× to 10.5pJ/bin with less than 5% coding penalty. An average throughput of 24.11bins/cycle is achieved across several HD video sequences, which is 10.6× higher than the state-of-the-art H.264/AVC CABAC implementations [2-4]. The MP-CABAC approach can be combined with techniques used in [2-4] for an additional throughput increase of 1.3 to 2.3×. The MP-CABAC test chip decodes the max H.264/AVC bit-rate (300Mb/s) with an 18MHz clock at 0.7V, consuming 12.3pJ/bin. At 1.0V, it has a performance of 3026Mbins/s for a bit-rate of 2.3Gb/s, enough for real-time QFHD at 186fps, or equivalently 7.8 views of QFHD at 24fps for multiview video coding (MVC). MP-CABAC helps enable a fully parallel video decoder.

Acknowledgements:
Thanks to Texas Instruments for algorithm support, chip fabrication and funding.

References:
[1] "Joint Call for Proposals on Video Compression Technology", ITU-T SG16/Q6, 39th VCEG Meeting: Kyoto, 17-22 Jan. 2010, Doc. VCEG-AM91
[2] T.-D. Chuang, et al., "A 59.5mW Scalable/Multi-View Video Decoder Chip for Quad/3D Full HDTV and Video Streaming Applications," *ISSCC Dig. Tech. Papers*, Feb. 2010.
[3] P. Zhang, et al., "Variable-bin-rate CABAC engine for H.264/AVC high definition real-time decoding," *IEEE Trans. on VLSI Systems*, March 2009.
[4] Y.-C. Yang, et al., "High-Throughput H.264/AVC High-Profile CABAC Decoder for HDTV Applications," *IEEE Trans. CSVT*, Sept. 2009.
[5] S. Nomura et al., "A 9.7mW AAC-Decoding, 620mW H.264 720p 60fps Decoding, 8-Core Media Processor with Embedded Forward-Body-Biasing and Power Gating Circuit in 65nm CMOS Technology," *ISSCC Dig. Tech. Papers*, Feb. 2008.
[6] V. Sze et al., "Massively Parallel CABAC", ITU-T SG16/Q6, 38th VCEG Meeting: London / Geneva, 1-8 July 2009, Doc. VCEG-AL21
[7] KTA Reference Software, 2.7 [Online]. Available: http://iphome.hhi.de/suehring/tml/download/KTA/
[8] D. F. Finchelstein, et al., "Multi-Core Processing and Efficient On-Chip Caching for H.264 and Future Video Decoders," *IEEE Trans. CSVT*, Nov. 2009.
[9] V. Sze, A.P. Chandrakasan, "A High Throughput CABAC Algorithm Using Syntax Element Partitioning," *IEEE Int. Conf. on Image Processing*, Nov. 2009.

978-1-61284-303-2/11 $26.00 © 2011 IEEE

ISSCC 2011 / February 22, 2011 / 9:00 AM

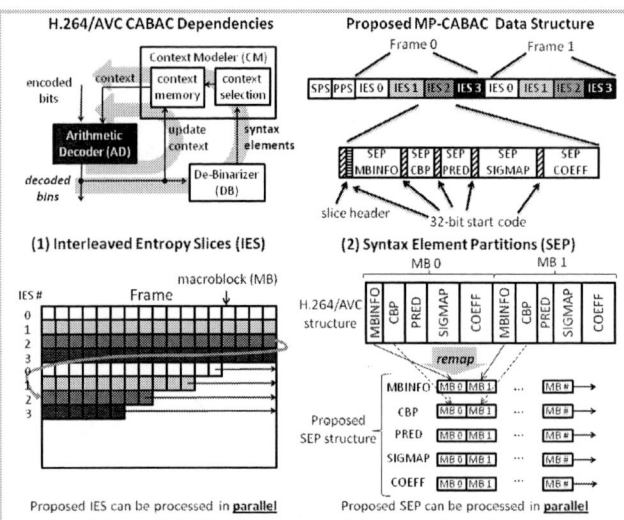

Figure 7.2.1: MP-CABAC data structure: Two forms of parallelism for highly parallel decoding.

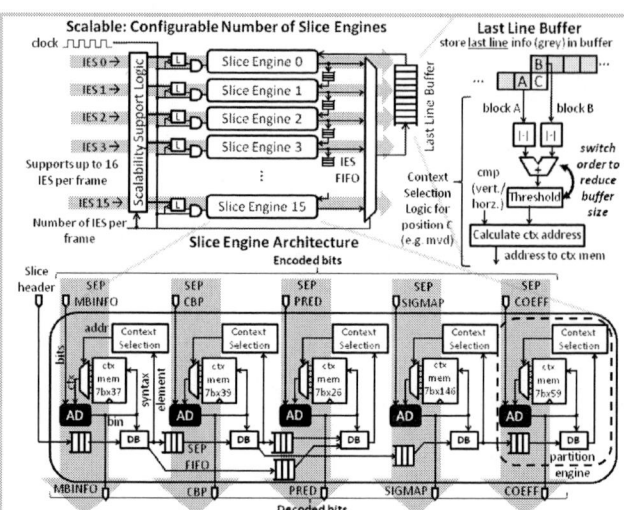

Figure 7.2.2: Scalable architecture of MP-CABAC for parallel processing of IES and SEP. Last line buffer supports up to 4k×2k (QFHD) resolutions.

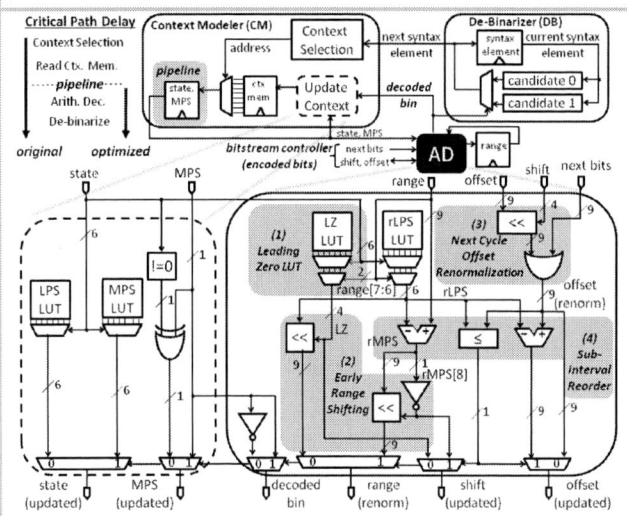

Figure 7.2.3: Partition engine speed up: insert pipeline register after context memory and four AD optimizations (highlighted).

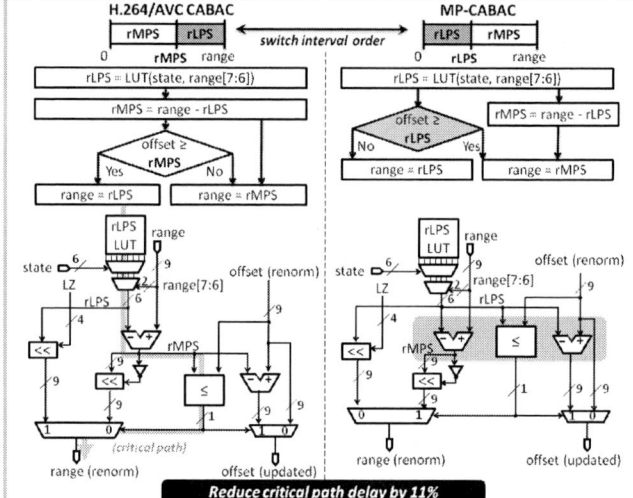

Figure 7.2.4: Subinterval Reordering: Joint algorithm and architecture optimization to reduce critical-path delay in the arithmetic decoder.

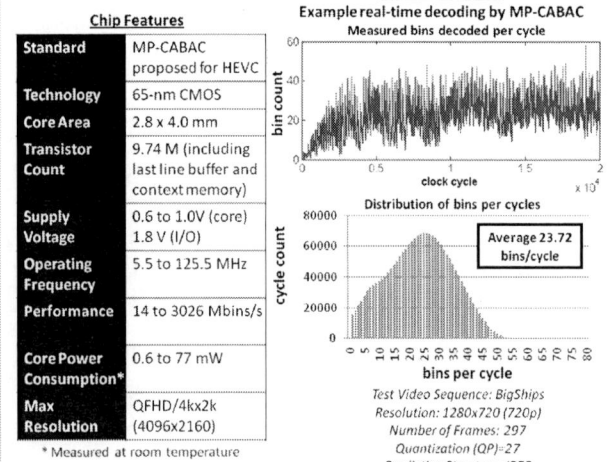

Figure 7.2.5: Features of the MP-CABAC test chip and plot showing the real-time performance for the BigShips video sequence.

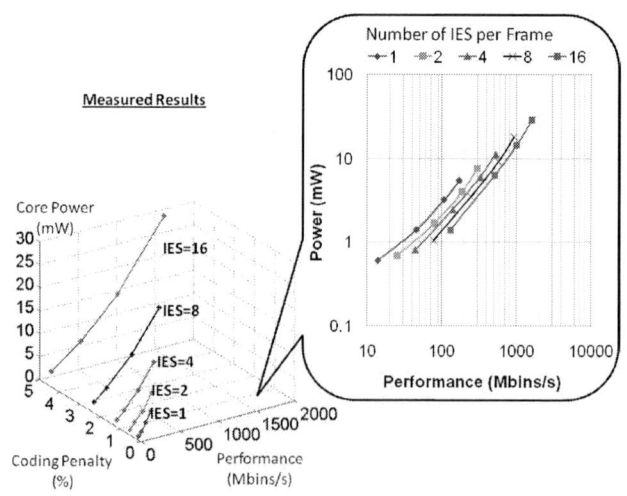

Figure 7.2.6: Measured results showing trade-off between power-performance-coding efficiency of the MP-CABAC test chip.

978-1-61284-303-2/11 $26.00 © 2011 IEEE

ISSCC 2011 PAPER CONTINUATIONS

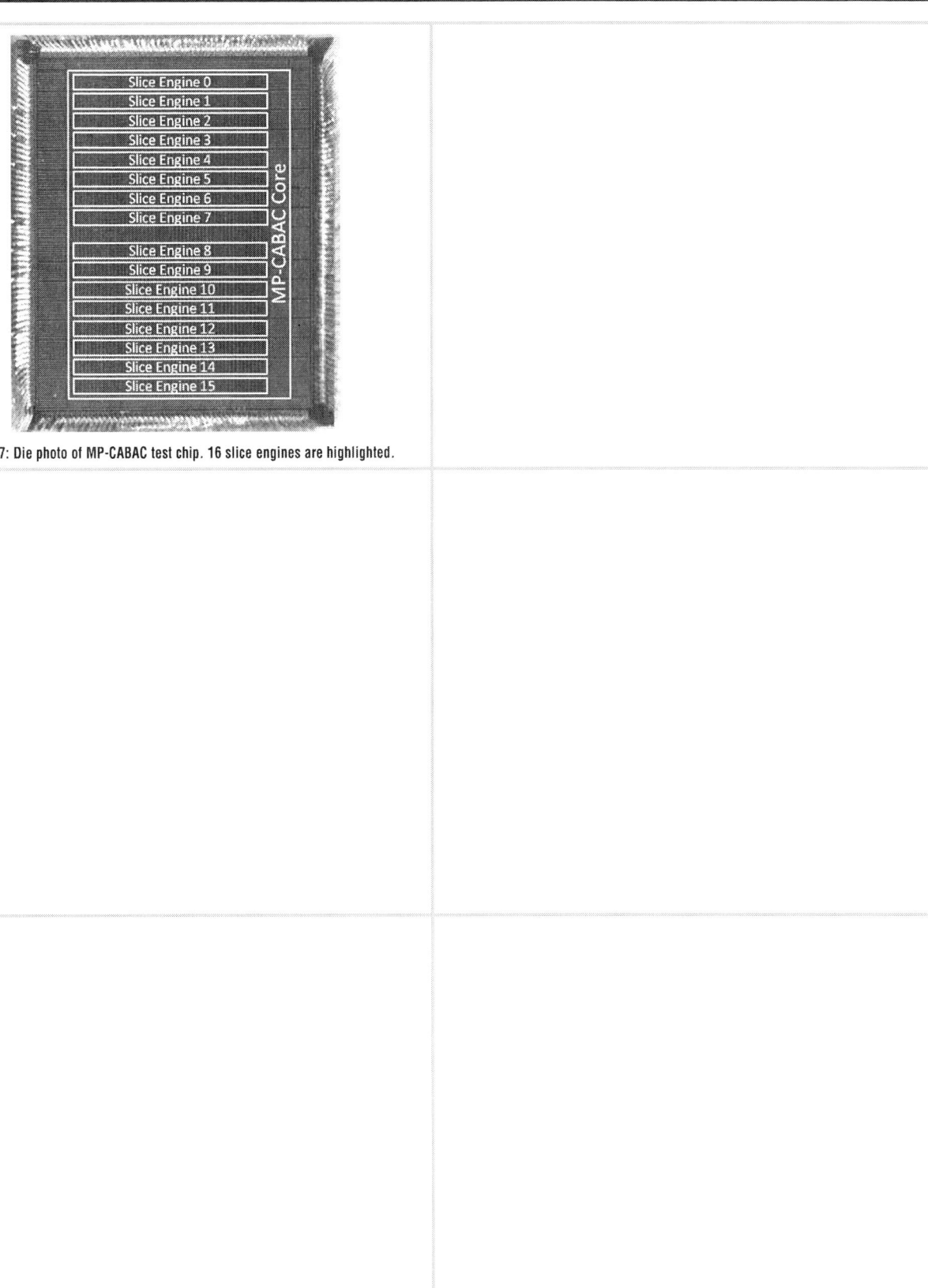

Figure 7.2.7: Die photo of MP-CABAC test chip. 16 slice engines are highlighted.

ISSCC 2011 / SESSION 7 / MULTIMEDIA & MOBILE / 7.3

7.3 A 275mW Heterogeneous Multimedia Processor for IC-Stacking on Si-Interposer

Hyo-Eun Kim, Jae-Sung Yoon, Kyu-Dong Hwang, Young-Jun Kim, Jun-Seok Park, Lee-Sup Kim

KAIST, Daejeon, Korea

Most data-intensive operations for multimedia applications such as image processing, vision, and 3D graphics require high external memory bandwidth. In augmented-reality (AR) processors [1], both 3D graphics and vision operations are required, so memory bandwidth becomes even more critical. In [1], however, memory bandwidth is not considered, floating-point processing is not supported, and there is no cache memory for texturing, which is a performance bottleneck of common graphics pipelines. In this work, a heterogeneous multimedia processor is presented to process various mobile multimedia applications in a single chip on Si-interposer for high memory bandwidth. The implemented processor has 4 key features: (1) A transceiver pool (TRx) that reconfigures strength of output drivers according to the channel loss for IC-stacking on Si-interposer, (2) A mode-configurable vector processing unit (MCVPU) for frame-level parallelism, (3) An energy-efficient unified filtering unit (UFU) with adaptive block selection (ABS) algorithm for memory-access-efficient texturing, and (4) a unified shader (US) with floating-point scalar processing elements (SPE) and partial special function units (PSFU) to enhance graphics processing performance and quality. With these techniques, we achieve 1.7× frame rate and 8× memory bandwidth improvement in full AR operation.

Figure 7.3.1 shows a target prototype of the entire system. The processor and DRAM silicon dies are stacked on a Si-interposer redistribution-layer. A TRx based on channel characteristics of the Si-interposer enables data communication between two dies while minimizing channel loss. Through-silicon-vias (TSV) constitute the interface for external data communication and power supply. The processor is composed of 4 programmable IPs (MCVPUs, a UFU, a US, and a RISC), 3 application-specific IPs (vision/graphics specific processing units (VSPU/GSPU), and a pose estimator), and a custom-designed analog IP, TRx. MCVPUs perform both data-intensive low-level image processing and per-pixel operations for vision and 3D graphics, respectively. Sixteen types of filtering operations for vision and 3D graphics are supported in UFU with a level-0 data buffer for energy-efficiency. The US performs floating-point geometry processing. The VSPU receives parallel data from MCVPUs and generates descriptors for pose estimation. The pose estimator makes a transformation matrix using descriptor vectors, and stores it into the constant-memory of the US for matrix multiplication. The GSPU interpolates vertex attributes for fragment rasterization and generates each fragment's level-of-detail (LOD) of texture maps for the ABS algorithm. The ABS algorithm includes tile-based rasterization (TBR), which exploits spatial locality of pixel fragments [3]. Each IP is independently controlled by IP-level clock gating, so different operation modes can be organized with different combination of various IPs.

In general, a processor communicates with an external memory through metal wires on a PCB. The length of metal wire is normally centimeter-order, so the channel loss is considerable when using high-speed data communication. Stacking the processor and DRAM dies on Si-interposer overcomes this drawback, since the channel length of Si-interposer is much shorter than a PCB's. It also realizes system miniaturization [2]. TRx is designed for channel (1.6GHz) loss compensation of Si-interposer as shown in Fig. 7.3.2. Eight select-signals are generated from a linear-feedback-shift-register, and serialize multiple data bits into a single output stream. Strength of output drivers is configured as 8 steps according to the channel loss. A serialized input stream is demuxed into parallel streams in a receiver. TRx results in 8× memory bandwidth improvement compared to previous work [1].

Descriptor generation with pose estimation and 3D object augmentation are completely independent operations, so these 2 operations can be processed in parallel. In this case, performance is substantially improved, but required memory bandwidth also increases linearly. So, our system is appropriate for frame-level parallelism. Four MCVPUs and a UFU can be configured as independent clusters for frame-level parallelism of AR. The MCVPU consists of 8 PEs, each of which has 6-way VLIW architecture for 16b arithmetic, logic operations, and data

communication. Eight PEs are organized as an 8-way SIMD unit for low-level image processing (IMG-mode), and two 4-way SIMD units for 3D graphics (3D-mode). An image-processing operation configures 4 MCVPUs as IMG-mode, and uses UFU as a filtering unit for low-level image processing. A 3D graphics operation configures 4 MCVPUs as 3D-mode for pixel shading, and uses UFU as a texture unit. Three MCVPUs in IMG-mode and MCVPU3 with UFU in 3D-mode are organized as independent processing clusters for frame-level pipelining as shown in Fig. 7.3.1. Frame-level pipelining requires a large set of registers and memory spaces. This increases energy and area cost. In our processor, however, CMEM of US is used for frame-level pipelining. The proposed AR configuration achieves 1.7× higher frame rate compared to previous work [1].

Most low-level vision operations consist of various filtering operations. For a single filtering operation, multiple instructions are required in a general processing core. In UFU, however, a single instruction is enough for each filtering operation. Figure 7.3.3 shows the UFU. The UFU supports 16 filtering operations for graphics and vision. The previously proposed CFU supports 7 filtering operations [4]. UFU includes a 512B L0 buffer and an 8KB L1 cache. The L0 buffer improves energy efficiency by the addition of small hardware area by limiting direct references to L1 SRAM cache. In a texturing operation, energy consumed in the UFU is reduced by 80% compared to the conventional texture unit, and 15% compared to the TFM [5]. The ABS algorithm dynamically controls the size of a texture block to be fetched from an external memory. It reduces stall cycles caused by external memory accesses, and then enhances texturing performance. The center point of each tile, area and shape of the current triangle determine the size of a texture block to be fetched as described in Fig. 7.3.4. Texturing performance is improved by 17.9% compared to the A-index [6], and 9.1% compared to the TBR without ABS [3].

In a general graphics pipeline, a floating-point datapath is used in geometry processing for its precision. The US consists of 4 homogeneous single-precision floating-point SPEs, and each SPE has a 4-way VLIW architecture. Each SPE has its own PSFU, but the look-up table (LUT) is shared by 4 SPEs with table loader (TBLD) as shown in Fig. 7.3.5. TBLD exploits access patterns to the shared LUT by using multi-fetch and bank-partition schemes [7]. The US reduces total latency of full shading operation by 53%, 30% compared to 1-issue SIMD and 2-issue SIMD architecture, respectively.

Figure 7.3.6 summarizes chip features. The UFU with the ABS algorithm improves energy efficiency and performance, so the energy-delay product is reduced by 70%. The US reduces the area-delay product by 38.5% compared to the 2-issue VLIW SIMD. This chip is fabricated in 0.13μm 1P6M CMOS. It integrates 1.46M logic gates and an analog IP, TRx, within a 4×4mm² die. MCVPUs operate at maximum 200MHz, and the other digital IPs operate at 100MHz.

Acknowledgements:
Chip fabrication was supported by IDEC at Korea Advanced Institute of Science and Technology (KAIST), Korea. This work was supported by the IT R&D program of MKE/KEIT. [KI002134, Wafer Level 3D IC Design and Integration]

References:
[1] Jae-Sung Yoon, et al., "A Graphics and Vision Unified Processor with 0.89uW/fps Pose Estimation Engine for Augmented Reality," *ISSCC Dig. Tech. Papers*, pp. 336-337, Feb. 2010.
[2] J.U. Knickerbocker, et al., "3D Silicon Integration," *Proc. of Electronic Components and Technology Conf.*, pp. 538-543, May 2008.
[3] J. McCormack, et al., "Tiled polygon traversal using half-plane edge functions," *Proc. of SIGGRAPH Conf. on Graphics Hardware*, pp. 15-21, 2000.
[4] Chih-Hao Sun, et al., "CFU: Multi-Purpose Configurable Filtering Unit for Mobile Multimedia Applications on Graphics Hardware," *Proc. of Conf. on High Performance Graphics*, pp.29-36, 2009.
[5] B. V. N. Silpa, et al., "Texture Filter Memory – A power-efficient and scalable texture memory architecture for mobile graphics processors," *Proc. of Int. Conf. on Computer-Aided Design*, pp. 559-564, Nov. 2008.
[6] C. H. Kim, et al., "Adaptive selection of an index in a texture cache, " *Proc. of Int. Conf. on Computer Design*, pp. 295-300, Oct. 2004.
[7] Young-Jun Kim, et al., "Bank-Partition and Multi-Fetch Scheme for Floating-Point Special Function Units in Multi-Core Systems," *Proc. of Int. Symp. on Circuits and Systems*, pp. 1803-1806, May 2009.

ISSCC 2011 / February 22, 2011 / 9:30 AM

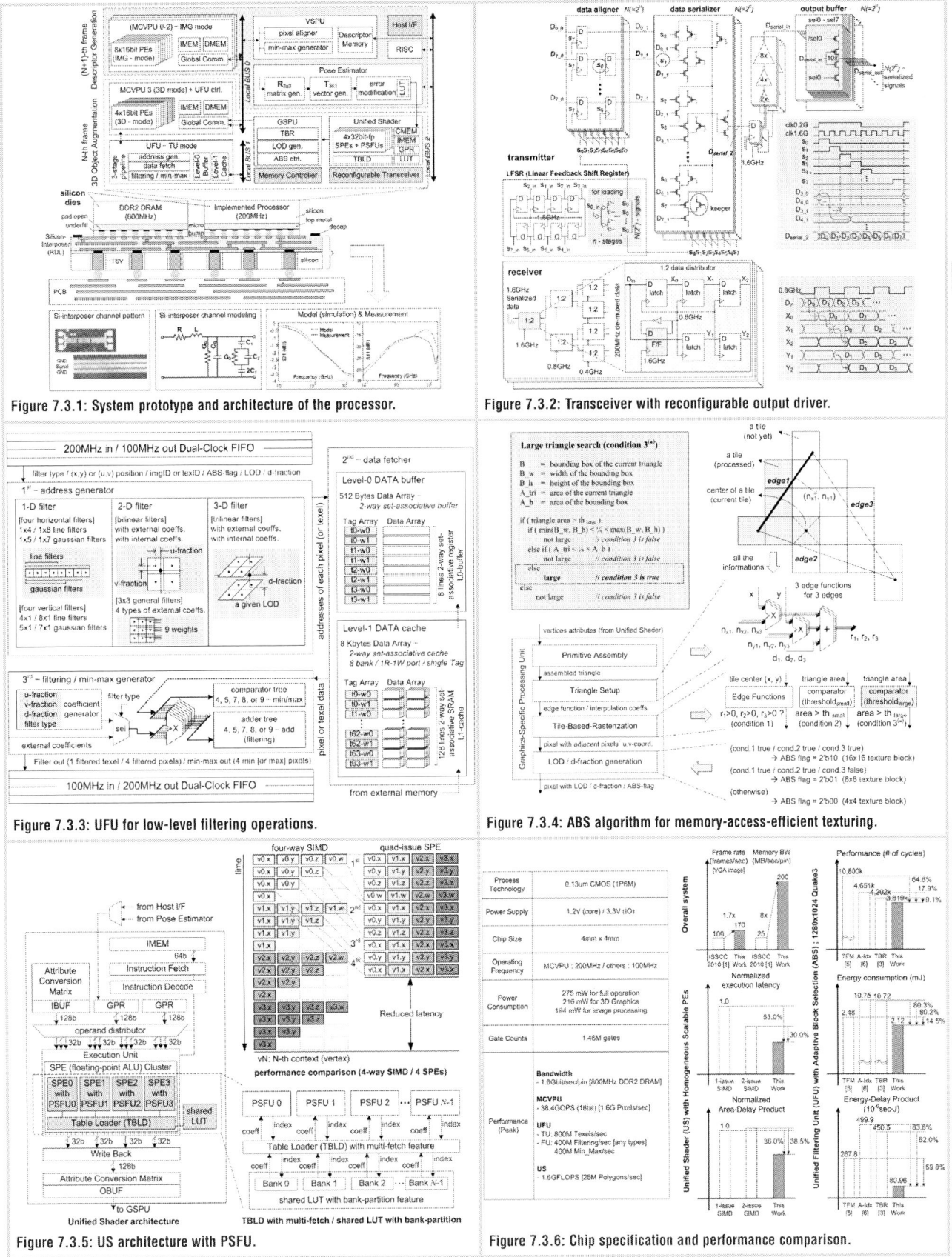

Figure 7.3.1: System prototype and architecture of the processor.

Figure 7.3.2: Transceiver with reconfigurable output driver.

Figure 7.3.3: UFU for low-level filtering operations.

Figure 7.3.4: ABS algorithm for memory-access-efficient texturing.

Figure 7.3.5: US architecture with PSFU.

Figure 7.3.6: Chip specification and performance comparison.

7

978-1-61284-303-2/11 $26.00 © 2011 IEEE   129

ISSCC 2011 PAPER CONTINUATIONS

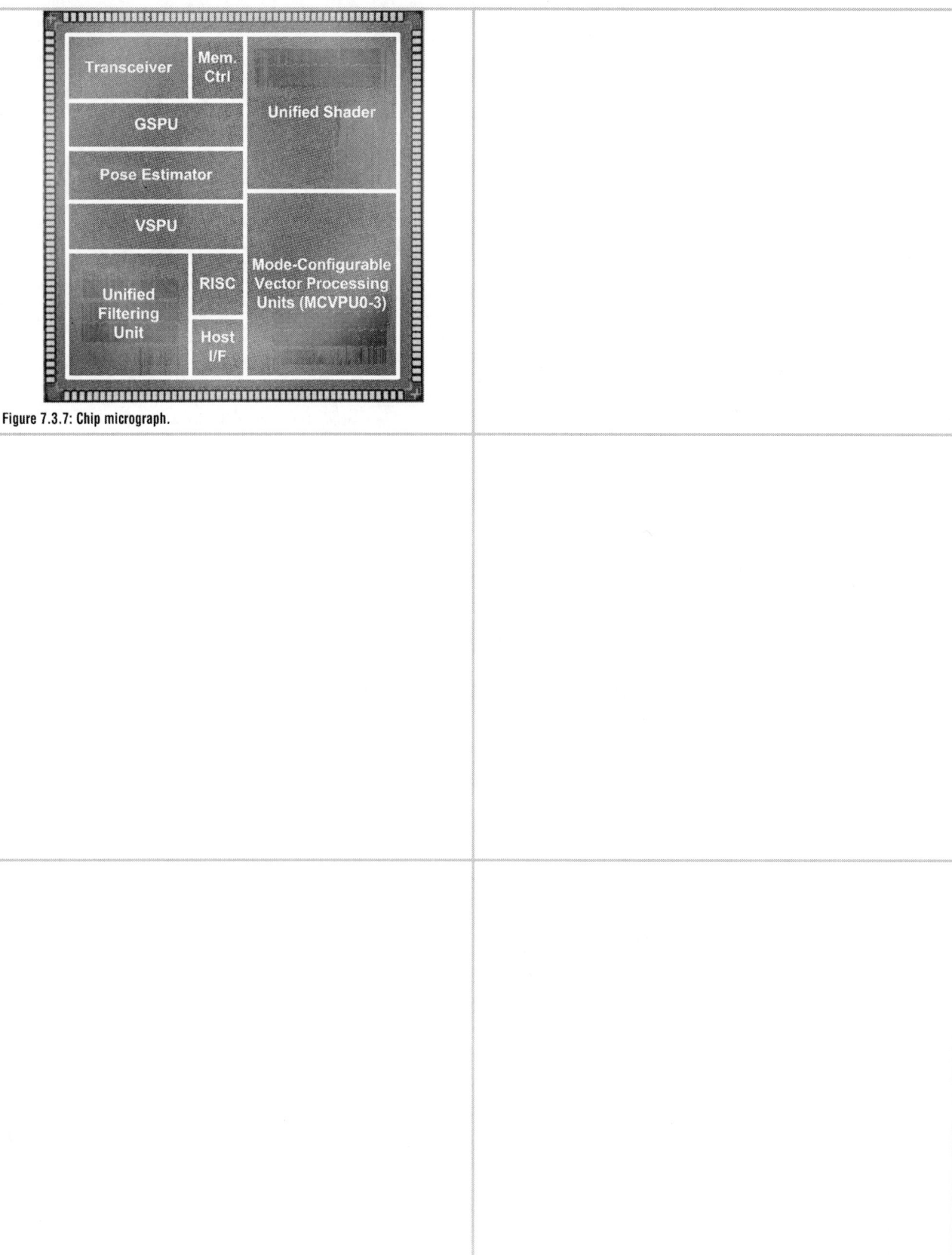

Figure 7.3.7: Chip micrograph.

ISSCC 2011 / SESSION 7 / MULTIMEDIA & MOBILE / 7.4

7.4 A 57mW Embedded Mixed-Mode Neuro-Fuzzy Accelerator for Intelligent Multi-core Processor

Jinwook Oh, Junyoung Park, Gyeonghoon Kim, Seungjin Lee, Hoi-Jun Yoo

KAIST, Daejeon, Korea

Artificial intelligence (AI) functions are becoming important in smartphones, portable game consoles, and robots for such intelligent applications as object detection, recognition, and human-computer interfaces (HCI). Most of these functions are realized in software with neural networks (NN) and fuzzy systems (FS), but due to power and speed limitations, a hardware solution is needed. For example, software implementations of object-recognition algorithms like SIFT consume ~10W and ~1s delay even on a 2.4GHz PC CPU. Previously, GPGPUs or ASICs were used to realize AI functions [1-2]. But GPGPUs just emulate NN/FS with many processing elements to speed up the software, while still consuming a large amount of power. On the other hand, low-power ASICs have been mostly dedicated stand-alone processors, not suitable to be ported into many different systems [2].

This paper presents a portable embedded neuro-fuzzy accelerator: the intelligent reconfigurable integrated system (IRIS), which realizes low power consumption and high-speed recognition, prediction and optimization for AI applications, as shown in Fig. 7.4.1. The key design features of the IRIS are: (1) a reconfigurable analog processing-element cluster (APEC) that computes various NN/FS topologies, (2) a global/local learning accelerator (GLA) that speeds up learning while reducing power, and (3) fast analog/digital mixed-mode operation with delay prediction based adaptive settling time. As a result, the IRIS can reduce power consumption by 54% and processing delay of NN/FS applications by 83% compared to a multicore processor implementation to perform various NN/FS operations with 665GOPS/W power efficiency and ~100µs processing delay.

Figure 7.4.2 shows the IRIS block diagram. It consists of the APEC, a versatile adaptive neuro-fuzzy inference system (VANFIS) [2], the GLA, a neuro-fuzzy RISC (NFRISC) controller, and a network-on-chip (NoC) network interface (NI). The APEC contains 32×32 processing elements (PE) capable of 1024 simultaneous NN MAC operations, and 32 normalization and nonlinear function circuits for neural-network activation functions. Each PE contains a 6b multiplying DAC (MDAC) and 6b SRAM cells. The weight memory (WMEM) loads the multiplying weights to the PE's SRAM buffer, and the data vector memory (DVMEM), with the DACs, provides analog multiplicands to the PEs. The GLA performs quick NN learning algorithms in collaboration with APEC. The NFRISC controls the overall system with the configuration special function register (CSFR), which stores NN/FS information, such as topologies, dimension and learning algorithms. It also compensates for the process variations in analog circuits. The NI with DMA enables its seamless integration within an NoC-based multicore processor.

Figure 7.4.3 shows the APEC architecture, which supports various NN topologies by reconfiguring connections of its analog processing elements. The normalized nonlinear function circuit (NNLF), which combines normalization and nonlinear functions, outputs the normalized voltage ratio of V_{SS} and V_{tot} to the output current terminal. I_{SS} limits the output voltage swing to reduce the analog current settling time. The 2 output voltages, V_{norm} and V_{NL}, are determined by the total input currents of V_{tot}, and the nonlinear function relationship, respectively. As a result, the NNLF increases the APEC throughput while reducing power consumption. It takes 6µs to multiply a 32×32 matrix by a 32-dimension vector and 50ns for one inner-product operation. Several NN topologies, including the multilayer perceptron (MLP), the radial basis function network (RBFN), and the recurrent neural network (RNN), can be generated by changing the signal paths of APEC. For example, the RBFN operation can be achieved by connecting the nonlinear function layer before normalization and DMEM access; otherwise the MLP operation uses in-order connection of each layer. With the help of APEC, the IRIS is 70% faster than a previous multicore processor [3] for NN operations. The compact and low-power features of analog NN circuits provide 71.2% power and 43% area reduction with consideration of 7.7% standby leakage power and 4% area overhead of counter-based current source calibration, while achieving 10.3× higher GOPS than an equivalent digital implementation.

The GLA architecture in Fig. 7.4.4 is composed of the programmable SIMD-PU, a dedicated local parameter calculator (LPC), and a global parameter calculator (GPC). The SIMD-PU executes 16-way parallel data processing and adopts the memory-mapped PE configuration structure for learning operations. The LPC, which performs local spatial parameter learning algorithms, increases local parameter-update speed by 22% compared to SIMD-PU. On the other hand, the GPC evaluates cross-validated training results, and learns more robustly based on temporal global information. A cross-validation buffer (CVB) shuffles weight and data vector values for the 128-entry FIFOs to enable more robust validation of learning data sets. As a result, the GLA reduces learning time by 71.4% compared to a previous multi-core processor [3] while the average power consumption is decreased by 53.8% for machine learning algorithms, namely perturbation, back-propagation and reinforcement.

Using the delay predictor of NFRISC, which can predict the data-transfer delays between the GLA and the APEC, the IRIS enables mixed-mode operation with fast learning speed and reduced power consumption, as shown in Fig. 7.4.5. Since the analog signal settling time is comparatively large and is dependent on the amount of data transfer and process variations in analog circuits, the NFRISC predicts the data-transfer delay between IF and EX stages of its pipeline, and accordingly configures the APEC settling time through the CSFR and APEC process variation calibration. The mixed-mode operation can achieve 35.7% processing time reduction with the help of NFRISC delay prediction. When the GLA sends the last parameter's load signal *wl_32*, by the predicted delay information, *APEC_EN* is set high for the duration of predicted delay. The APEC operation is disabled by the GLA's done signal, *APEC_DONE*. Figure 7.4.5 shows the APEC learning sequence when the target output is 0. The evaluation (EV) and learning (LN) stages of IRIS are pipelined to increase the learning throughput.

With the help of the GLA, the IRIS performs object recognition for VGA image sequences based on temporal/spatial learning algorithms, as shown in Fig. 7.4.6. It learns pixel-level object features first and then, the temporal contextual information such as previous frames' object information. In 200MHz/1.2V digital and 1.0V analog voltages, the IRIS achieves 1mJ/frame energy efficiency compared to 5.3mJ/frame for local parameter learning. Its average power consumption is measured to be 53.6% of the previous multicore processor [3] with 655GOPS/W power efficiency, which is 3.3× higher than the equivalent digital vector processing unit and 1.2× higher than related works [3-6].

Figure 7.4.7 shows the chip micrograph and its performance summary. The 5×2.7mm² portable neuro-fuzzy accelerator is fabricated in 0.13µm 1P8M CMOS and dissipates 57mW for object recognition, including Gaussian filtering, feature clustering, object detection and DB matching. As a result, it achieves 8.5× higher energy efficiency than the previous multi-core processor [3].

References:
[1] H. Graf, et al., "A Massively Parallel Digital Learning Processor," *Adv. Neural Information Processing Systems 21*, pp. 529-536, 2009.
[2] J. Oh, et al., "1.2mW On-Line Learning Mixed Mode Intelligent Inference Engine for Robust Object Recognition," *Dig. Symp. VLSI Circuits*, pp. 17-18, June 2010.
[3] S. Lee, et al., "A 345mW Heterogeneous Many-Core Processor with an Intelligent Inference Engine for Robust Object Recognition," *ISSCC Dig. Tech. Papers*, pp. 332-333, Feb. 2010.
[4] T. Kurafuji, et al., "A Scalable Massively Parallel Processor for Real-Time Image Processing," *ISSCC Dig. Tech. Papers*, pp. 334-335, Feb. 2010.
[5] T. Chen, et al., "A Multimedia Semantic Analysis SoC (SASoC) with Machine-Learning Engine," *ISSCC Dig. Tech. Papers*, pp. 338-339, Feb. 2010.
[6] S. Arakawa, et al., "A 512GOPS Fully-Programmable Digital Image Processor with full HD 1080p Processing Capabilities," *ISSCC Dig. Tech. Papers*, pp. 312-313, Feb. 2008.

978-1-61284-303-2/11 $26.00 © 2011 IEEE

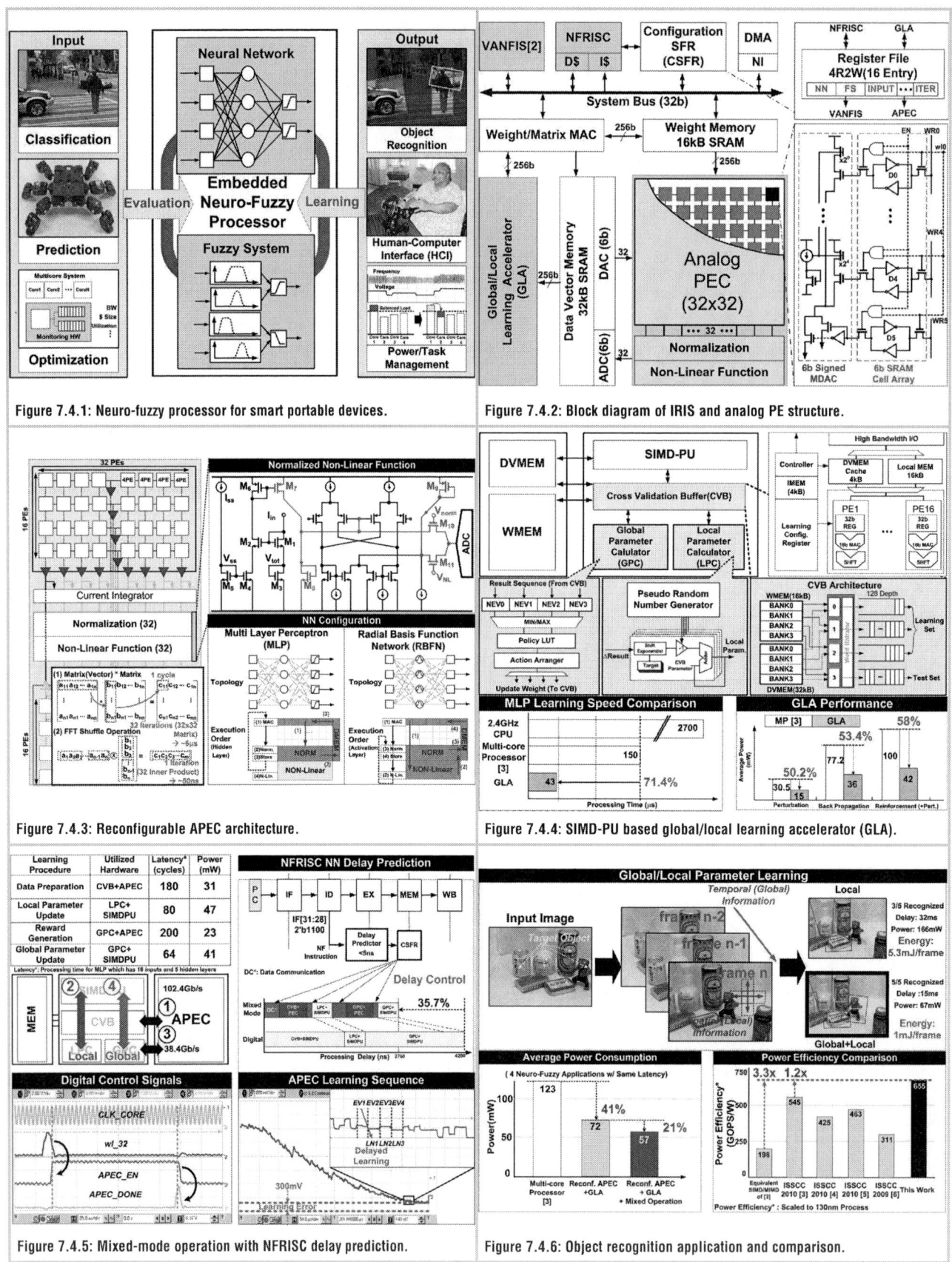

Figure 7.4.1: Neuro-fuzzy processor for smart portable devices.

Figure 7.4.2: Block diagram of IRIS and analog PE structure.

Figure 7.4.3: Reconfigurable APEC architecture.

Figure 7.4.4: SIMD-PU based global/local learning accelerator (GLA).

Figure 7.4.5: Mixed-mode operation with NFRISC delay prediction.

Figure 7.4.6: Object recognition application and comparison.

ISSCC 2011 PAPER CONTINUATIONS

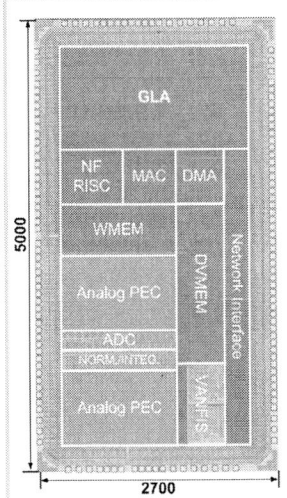

Technology	130nm 1P8M Logic CMOS	
Die Size	2.7mm x 5.0mm	
Operating Frequency	Digital	200MHz @ 1.2 V
	Analog	10MHz @ 1.0 V (32 Matrix MAC)
Peak Performance	Digital	14 GOPS
	Analog	35.14 GOPS
	Total	49.14 GOPS
Power Consumption	Digital	61 mW
	Analog	14 mW
	Total	75mW
Power Efficiency	655 GOPS/W	
Energy Efficiency	1mJ/Frame Object Recognition @ VGA Image	

Figure 7.4.7: Die micrograph. and feature summary.

ISSCC 2011 / SESSION 7 / MULTIMEDIA & MOBILE / 7.5

7.5 A 28nm 0.6V Low-Power DSP for Mobile Applications

Gordon Gammie[1], Nathan Ickes[2], Mahmut E Sinangil[2], Rahul Rithe[2], J. Gu[3], Alice Wang[1], Hugh Mair[1], Satyendra Datla[1], Bing Rong[1], Sushma Honnavara-Prasad[1], Lam Ho[1], Greg Baldwin[1], Dennis Buss[1], Anantha P Chandrakasan[2], Uming Ko[1]

[1]Texas Instruments, Dallas, TX,
[2]Massachusetts Institute of Technology, Cambridge, MA,
[3]Texas Instruments (now with MaxLinear), Dallas, TX

A multimedia applications processor is fabricated using a 28nm low-power process technology for ultra-low-power applications. Based on a 4-issue, 32-register version of the TMS320C64x+ VLIW DSP, this System on Chip (SoC) includes 32kB L1 and 128kB L2 caches, and I2S, SPI, UART, MultiMediaCard, and external memory interfaces (Fig. 7.5.1). The design incorporates over 600k instances of custom low-voltage logic cells and 43 instances (1.6 Mb) of 6T SRAM. Utilizing ultra-low-voltage (ULV) optimized standard-cell libraries and 6T SRAM macros, and demonstrating a new statistical static timing analysis (SSTA) methodology, the SoC scales as designed from high performance at 1.0V down to ultra-low power at 0.6V.

The 28nm low-power (LP) technology (Fig. 7.5.2) used for this DSP SoC design is a custom process with a dual-gate poly/SiON gate stack, double patterning at gate, high-NA 193i lithography and epitaxial S/D SiGe for pMOS performance enhancement. Typical strain techniques are also used for nMOS performance enhancement. The ultra-low-K dielectric dual damascene metal stack includes a thick top Cu level and an Al level that can be used for power and signal routing. The integration techniques described in [1] are utilized to support SoC components (e.g. multi-V_t, multi-channel length, analog and I/O transistors, capacitors and diodes) along with a custom 0.12µm^2 6T SRAM bitcell with minimal added process cost.

Performance of logic circuits operating in or near sub-threshold is highly sensitive to any variation of the threshold voltage (V_t), and some circuits can cease to function at the extremes of V_t variation. To quantify the functionality of each combinational cell, the static noise margin is determined using the procedure reported in [2]. To maintain sufficient reliability and performance at ULV, a custom digital cell library is developed, using a variation-driven design flow. Figure 7.5.3(a) shows typical functional failure modes for logic cells. Through selective adjustment of transistor sizes, a beta ratio of 1 is found to provide optimal performance and ensure functionality over a wide voltage range. To maintain proper clock duty cycle, clock-tree cells require a higher beta ratio (stronger pMOS), with the optimal ratio ranging from 1.5 at high voltage to 2.0 at ULV. The increase in beta ratio is driven by the increased difference in drive current between pMOS and nMOS transistors at ultra-low supply voltages. A beta ratio of 1.5 is chosen for clock buffers to ensure performance goals at high voltage are met. To alleviate flip-flop failures from data slip through or reverse conduction, an inverter is inserted between the master latch and pass-gate in order to delay the turn-on of the master stages and avoid reverse current flow. In the SRAM, a hierarchical bitline is used to improve read stability, with wordline boosting to ensure write-ability, and a pre-read during write to avoid half-select-related disturbances (Fig. 7.5.3(b)) [3].

Timing closure is a challenging problem for ULV designs. As shown in Fig. 7.5.4(a) for a representative library cell, the local 3σ delay variation is larger than the 3σ global corner delay by 1.5×. As the supply voltage decreases from 1.0 to 0.5V, the 3σ global corner delay increases by 15× and the standard deviation of the local delay increases by 100×. The relative impact of local variation varies with the drive strength: as drive strength increases, the standard deviation of local variation decreases and the PDF becomes more Gaussian (Figure 7.5.4(b)). During synthesis, the use of low-drive-strength cells on critical paths was reduced by applying delay-derating factors. The derating factors were calculated for each library cell, and are proportional to the magnitude of the local variation impact at ULV. Additionally, all clock cells are restricted to drive strengths of 8× or higher. The area impact of these drive-strength increases was less than 5%.

Predicting path-level delay distributions for timing closure is another challenge of ULV design. Modeling delay with Gaussian PDFs, as traditional statistical static timing analysis (SSTA) tools do, leads to a consistent underestimate of the actual delay distribution by 10 to 70%. For this design, setup and hold time margins at 0.5V are verified using a ULV SSTA design methodology based on Nonlinear Operating Point Analysis for Local Variations (NLOPALV). NLOPALV models cell delay distributions to within 5% [4], and can be used in conjunction with existing static timing analysis (STA) tools to predict path delays to within 8% [5]. To manage run-time, the analysis is performed in four passes of successively increasing accuracy. In the first pass, non-critical paths are identified and discarded using standard STA with 3σ cell delays. Summing 3σ cell delays gives highly pessimistic results compared to the actual 3σ variation of the overall paths. 92% of setup paths and 95% of hold paths are eliminated in this pass. Three subsequent passes reduce the pessimism by running NLOPALV on the capture clock tree only, capture and launch clock trees, and finally the entire timing path including the datapath. Figure 7.5.5(a) shows the number of paths analyzed decreases dramatically with each pass. The plots in Figure 7.5.5(b) show the distribution of analyzed path delays at the end of each pass. This 4-pass process integrates with and makes use of conventional timing closure tools. This analysis found 87 hold violations at 0.5V before hold-fixing, and verified that no hold violations remained after hold-fixing. The setup analysis shows that the design achieves 14MHz at 0.5V.

The DSP SoC design is fabricated and demonstrated (Fig. 7.5.6) to be operational from 587MHz at 1.0V (113mW) down to 3.6MHz at 0.34V (720µW) when operating from external memory (caches disabled). At the ULV target voltage of 0.5V, the maximum frequency is 43.4MHz, as compared to 14MHz from SSTA at worst-case conditions, end of life, and with margins. The on-chip caches are functional for supply voltages above 0.6V. When executing from cache, the chip scales from 145mW at 331MHz (1.0V) down to 5.9mW at 14.4MHz (0.6V). For lower voltage and reliable ULV cache operation in production, redundancy and repair should be implemented. Active and leakage power scale by 60× and 8.5×, respectively, when executing from cache, and by 1240× and 39× when executing from external memory. The measured leakages are representative of early development silicon with transistors not yet at final leakage targets for the technology. The minimum energy-per-cycle occurs at 0.75V (cache on), or 0.5V (external memory) and is expected to reduce slightly as leakage is reduced.

References:
[1] K. Benaissa, G. Baldwin, S. Liu, P. Srinivasan, F. Hou, B. Obradovic, S., Yu, H. Yang, R. McMullan, V. Reddy, C. Chancellor, S. Venkataraman, H. Lu, S. Dey, and C. Cirba,, "New Cost-Effective Integration Schemes Enabling Analog and High-Voltage Design in Advanced CMOS SOC Technologies", *VLSI Technology Symposium,* pp. 221-222, June 2010.
[2] J. Kwong and A. Chandrakasan, "Variation-Driven Device Sizing for Minimum Energy Sub-threshold Circuits", *International Symposium on Low Power Electronics and Design*, pp. 8-13, Oct. 2006.
[3] M. Sinangil, H. Mair, and A. Chandrakasan "A 28nm High-Density 6T SRAM with Optimized Peripheral Assist Circuits for Operation down to 0.6V", *ISSCC Dig. Tech. Papers,* in press, Feb. 2011.
[4] R. Rithe, S. Chou, J. Gu, A. Wang, S. Datla, G. Gammie, D. Buss, A. Chandrakasan, "Cell Library Characterization at Low Voltage using Non-linear Operating Point Analysis of Local Variations", *International Conference on VLSI Design,* in press, 2011.
[5] R. Rithe, J. Gu, A. Wang, S. Datla, G. Gammie, D. Buss, and A. Chandrakasan, "Non-Linear Operating Point Statistical Analysis for Local Variations in Logic Timing at Low Voltage", *Design Automation and Test in Europe Conference (DATE),* pp. 965-968, March 2010.

978-1-61284-303-2/11 $26.00 © 2011 IEEE

ISSCC 2011 / February 22, 2011 / 10:45 AM

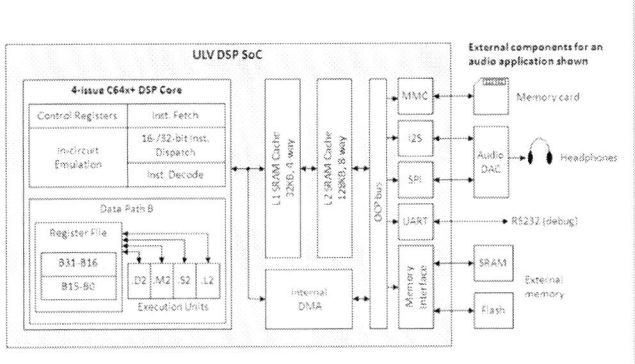

Figure 7.5.1: Series (C64x+) based DSP system with peripherals for low-power applications.

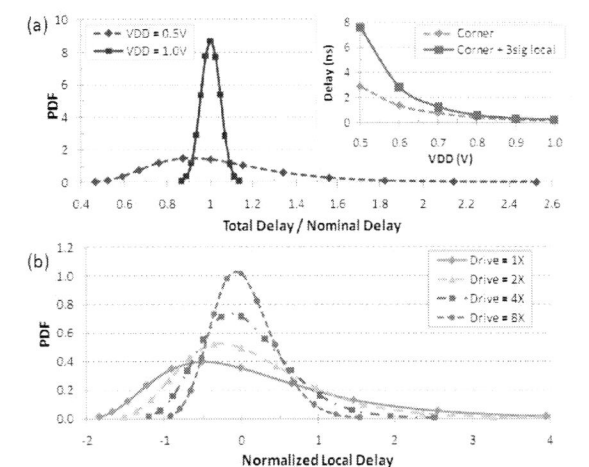

Process	28nm Low Power CMOS
Lithography	high-NA 193i
Performance	pMOS: Epitaxial S/D SiGe
	nMOS: Strain techniques
Gate stack	dual-gate poly/SiON
	double patterning
Metal stack	ULK dual damascene
	Thick top Cu level and Al level
Logic	SVT, LVT, multi-channel length
Bit	$0.12\mu m^2$ 6-T SRAM bitcell

Figure 7.5.2: 28nm low-power process.

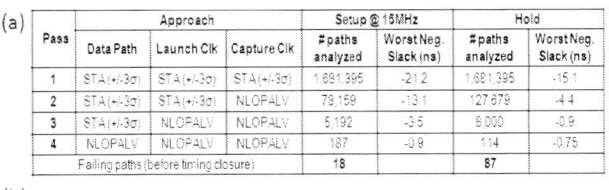

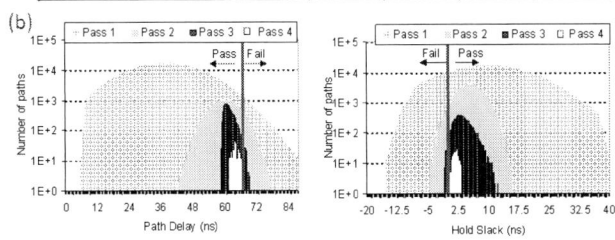

Figure 7.5.3: (a) Typical functional failure modes for logic cells. (b) ULV SRAM with 6T bit-cell, hierarchical sensing, wordline boosting, and pre-read during write.

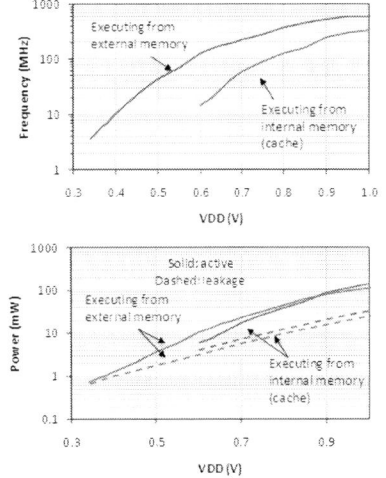

Figure 7.5.4: (a) Delay PDF of a representative library cell at $V_{DD} = 1.0V$ and $V_{DD} = 0.5V$. The horizontal axis is normalized the respective 3σ global corner delays. (b) Impact of drive strength on PDF.

(a)

Pass	Approach			Setup @ 15MHz		Hold	
	Data Path	Launch Clk	Capture Clk	#paths analyzed	Worst Neg. Slack (ns)	#paths analyzed	Worst Neg. Slack (ns)
1	STA(+/-3σ)	STA(+/-3σ)	STA(+/-3σ)	1,691,395	-21.2	1,691,395	-15.1
2	STA(+/-3σ)	STA(+/-3σ)	NLOPALV	79,159	-13.1	127,679	-4.4
3	STA(+/-3σ)	NLOPALV	NLOPALV	5,192	-3.5	8,000	-0.9
4	NLOPALV	NLOPALV	NLOPALV	187	-0.9	114	-0.75
Failing paths (before timing closure)				18		87	

(b)

Figure 7.5.5: (a) SSTA 4-pass progressive NLOPALV methodology (b) Distribution of paths for hold and setup analysis using the progressive application of NLOPALV.

Figure 7.5.6: Measured power and performance, comparing operation from internal and external memories.

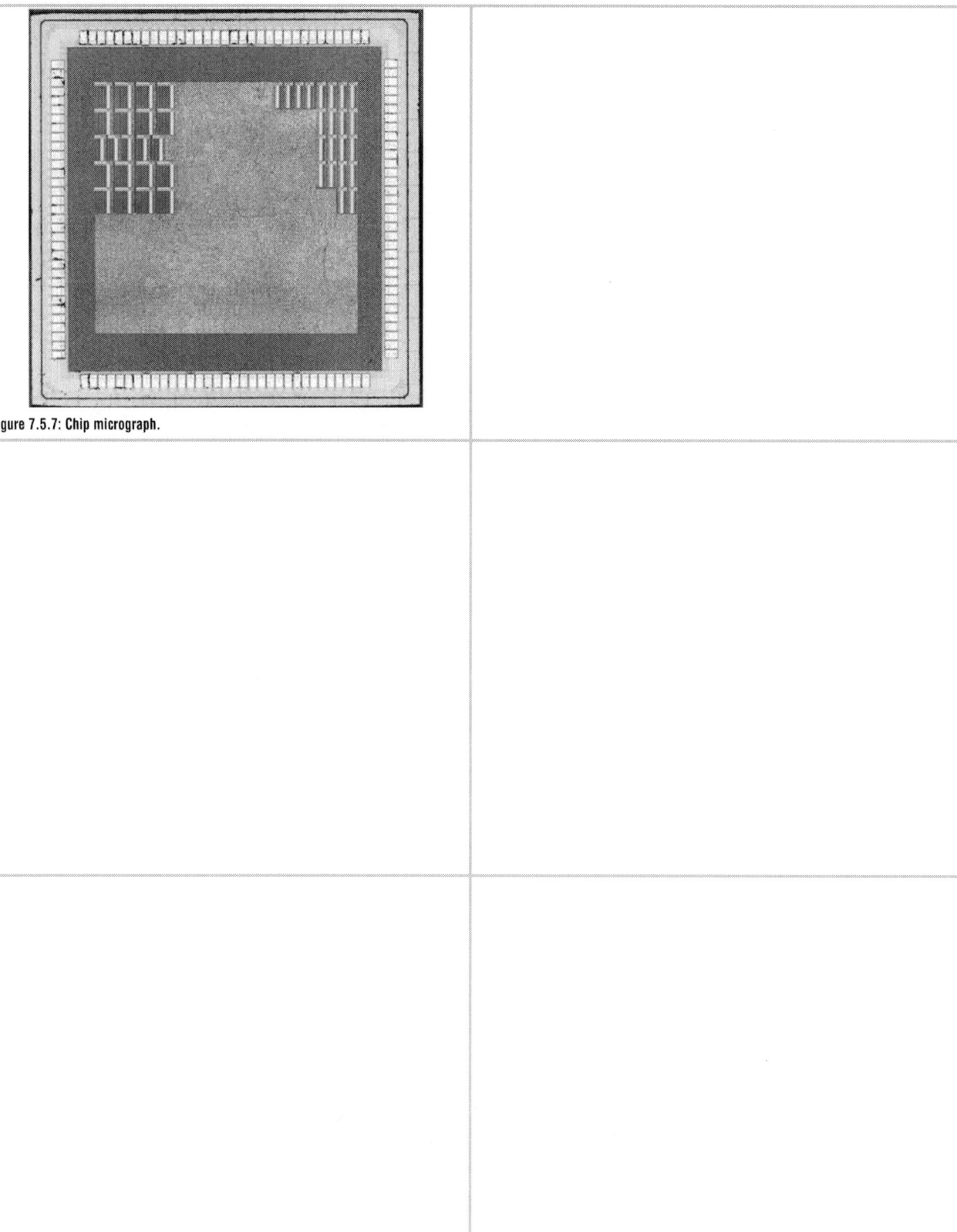

Figure 7.5.7: Chip micrograph.

ISSCC 2011 / SESSION 7 / MULTIMEDIA & MOBILE / 7.6

7.6 A MIMO WiMAX SoC in 90nm CMOS for 300km/h Mobility

Gene C.H. Chuang[1], Pang-An Ting[1], Jen-Yuan Hsu[1], Jiun-You Lai[1], Shun-Chang Lo[1], Ying-Chuan Hsiao[1], Tzi-Dar Chiueh[2]

[1]ITRI, Hsinchu, Taiwan,
[2]National Taiwan University, Taipei, Taiwan

The IEEE 802.16 standard for broadband wireless access, known as Worldwide Interoperability for Microwave Access (WiMAX), provides high throughput over long-range transmission. The key developments in the physical (PHY) layer include Orthogonal Frequency-Division Multiple Access (OFDMA) and the Multiple-Input-Multiple-Output (MIMO) antenna technology, which enhance transmission quality and throughput.

In this paper, we present a WiMAX baseband SoC implemented in 90nm and capable of high-speed transmission at 300km/h. This SoC is integrated in a USB device (Fig. 7.6.7) that also includes an RF transceiver for MIMO WiMAX transmission. The baseband transmitter (Fig. 7.6.1) generates a WiMAX waveform according to the IEEE 802.16e Rev2 specification [1]. In this time-division-duplex WiMAX system, the uplink subframe is composed of OFDM symbols of 10MHz bandwidth, which are aggregated by 173 guard, 83 pilot, and 768 data subcarriers. The MAC, using an ARM926 processor, arranges the data and control packets in a Protocol Data Unit (PDU) format, and feeds these packets to the baseband transmitter. Finally, the 50MHz clocked 10b DACs convey baseband I and Q signals to the RF transceiver chip.

On the receiving side, RF signals from 2 antennae are down-converted from 2.5GHz to baseband. The WiMAX waveforms are digitized by 4 on-chip 10b 50MS/s ADCs, thus representing one ADC per I/Q channel per antenna. The block diagram of the PHY receiver is shown in Fig. 7.6.2. Due to 300km/h mobility, the power of the received preamble signal can fluctuate by more than 12dB in 20ms. So we use a preamble window-search mechanism for gating the AGC reference power calculation, as in Fig. 7.6.3, and adjust the input signal power to the targeted level. Note that the entire gain adjustment should be finished within the CP (Cyclic Prefix) period, including the time for signal-strength measuring, the AGC decision making, the gain-change command passing to the RF chip via SPI serial bus, and the gain-adjusting by the analog circuits in the RF chip, all within 11.2μs. The gain-controlled 50MHz digitized samples are then decimated and interpolated into 11.2MHz samples. The baseband receiver uses an 802.16e defined preamble to perform estimations on symbol timing offset and carrier frequency offset at the initial synchronization stage, where delay-correlations and time-domain matched filters are used. Moreover, both time and frequency offsets are tracked and smoothed each 5ms radio frame. After the compensation of time and frequency offsets, the time-domain samples are transformed to frequency-domain subcarriers via FFT. Afterward, the Channel Estimation is conducted for each spatial stream using corresponding pilot subcarriers. The preamble-based initial estimation is followed by sparse-pilot channel tracking in the data region using MMSE interpolation filters. Both initial estimation and channel tracking use the same hardware but with different filter coefficients. Since the Doppler spread is around 700Hz at the mobility of 300km/h, the coherence time, inversely related to the Doppler spread, is around 1.4ms. As this coherence time is about the period of a dozen OFDM symbols in an 802.16e WiMAX system, as in Fig. 7.6.3, we adopt a simplified linear interpolation for time-domain and a cluster-based 4-tap Wiener filter for frequency-domain channel estimation. For single-antenna mode, the estimated channels and received subcarriers are used to calculate soft values by the Channel Equalization function. For MIMO mode, the soft values are generated by an MRC-based STBC decoder or a fixed-complexity Sphere Decoder.

In addition to the Channel Estimator, the MIMO detector consists mainly of 3 functions: STBC decoder for matrix A processing, and the QR Decomposition followed by the Sphere Decoder for matrix B processing. Using CORDICs, QR Decomposition disintegrates the channel matrix H into the multiplication of a unitary matrix Q and a triangular matrix R. The QR Decomposition reorders the column entries of the channel matrix according to their 2-norm values. Two columns of the smallest 2-norm values are placed in the last 2 columns of the R

matrix; then the remaining columns are sorted by increasing 2-norm values. At the searching phase of the Sphere Decoder [2], a 2-layer full search is executed first for the last 2 symbols and Successive Interference Cancellations (SIC) are applied for each symbol layer. As shown in Fig. 7.6.4, the parallel mechanism is implemented by 16 SUs (Search Unit) with 4-layer pipelined Processing Elements (PE) in each SU. One PE handles one layer SIC and up to 16 layers SIC run in parallel. The hardware controller schedules the duty cycle of each SU according to modulation types, namely QPSK, QAM16 or QAM64, while the minimum distance value is updated after each PE calculation. Finally, the maximum likelihood symbol is picked from all searching nodes with the smallest distance. The generated soft values can be used to perform HARQ chase combing if they belong to retransmitted bursts. All soft values are grouped into one or more FEC blocks followed by block-based de-interleaving and de-puncturing. To improve the bit error rate, either a Viterbi decoder for tail-biting Convolutional Codes or a Convolutional Turbo Code (CTC) decoder is employed. A CTC decoder runs up to 250MHz clock rate so that a Max-log-MAP decoder operates up to 5 iterations. CTC can be clocked by a lower rate; with fewer MAP iterations are performed thus saving power if the frame error rate is lower than required. Finally, the decoded information bits are de-randomized and sent to the MAC functions.

The MAC architecture is designed to be flexible and compatible with future 4G solutions. Executed on an ARM microprocessor, the UMAC layer handles functions on protocol behaviors, while the hardwired LMAC layer deals with time-critical functions, such as DL-MAP Information Elements (MAP-IE) interpretation. To accommodate 4G communication in a unified framework, we have designed the MAC architecture with a maximum degree of portability by following the LTE layer structure conceptually. The MAC architecture has 6 sub-layers, namely, Medium Access Control (MAC), Radio Link Control (RLC), Radio Resource Control and Management (RRCM), PHY abstraction, RTOS abstraction, and API abstraction. This MAC solution enables the QoS (Quality of Service) data delivery service and provides sleep and idle modes that facilitate system-level power optimization schemes.

To briefly summarize the measured performance of this MIMO 802.16e USB dongle system [3], Fig. 7.6.5 shows the CINR and downlink data rate versus the SNR for ITU Pedestrian B (PB) 3km/h and Vehicular A (VA) channel models of 3km/h, 120km/h, and 300km/h. As the SNR changes from 0 to 21dB, the measured CINR increases from 4.2 to 17dB and the downlink data rate reaches 5Mb/s at 300km/h mobility, where the channel mobility is emulated in real time by an Elektrobit C8 channel emulator. The measurements confirm that the AGC, synchronization and channel estimation modules indeed achieve robust tracking at 300km/h.

A summary of communication features and the electrical measurements are tabulated in Fig. 7.6.6, and a chip micrograph is shown in Fig. 7.6.7. The CTC runs by a variable clock from 50MHz up to 250 MHz, the AMBA clock is 200MHz, while all other modules run at 50MHz. Embedding printed PIFA antennas and a 2.5GHz CMOS RF transceiver chip, the WiMAX modem dongle is compliant with the IEEE 802.16e standard.

Acknowledgements:
Authors acknowledge the contributions of the Analog/CAD/Broadband teams, ChiTien Sun, JianYu Chen, and Ming-Chien Tseng.

References:
[1] IEEE 802.16e-2005, "Local and Metropolitan Networks — Part 16: Air Interface for Fixed and Mobile Broadband Wireless Access Systems, Amendment 2: Physical and Medium Access Control Layers for Combined Fixed and Mobile Operation in Licensed Bands and Corrigendum1," 2006.
[2] L. G. Barbero and J. S. Thompson, "A Fixed-Complexity MIMO Detector Based on the Complex Sphere Decoder", *IEEE International Workshop on Signal Processing Advances in Wireless Communications*, July 2006
[3] WiMAX Forum™ Mobile Radio Conformance Tests MRCT Release 1.0 Approved Specification Revision 2.2.1.

978-1-61284-303-2/11 $26.00 © 2011 IEEE

ISSCC 2011 / February 22, 2011 / 11:15 AM

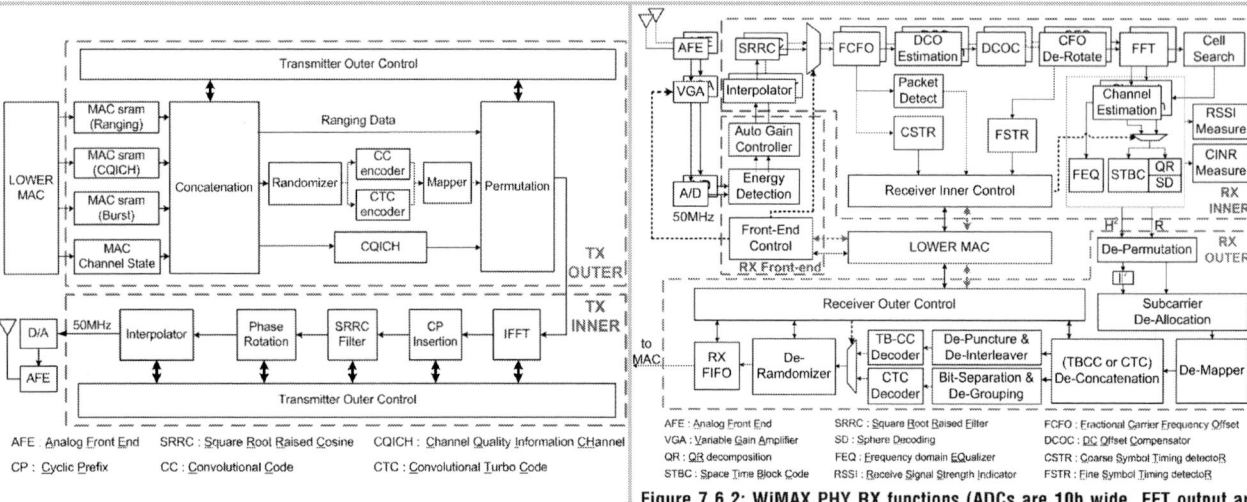

Figure 7.6.1: WiMAX PHY TX functions (CC/CTC output are 8b, IFFT output is 16b, and DAC outputs are 10b wide).

Figure 7.6.2: WiMAX PHY RX functions (ADCs are 10b wide, FFT output are 16b, SD and STBC in are 14b and out are 16b, CTC input are 24 parallel data of 9b wide, CC in are 9b and output are 8b, outputs of RSSI and CINR are 14b).

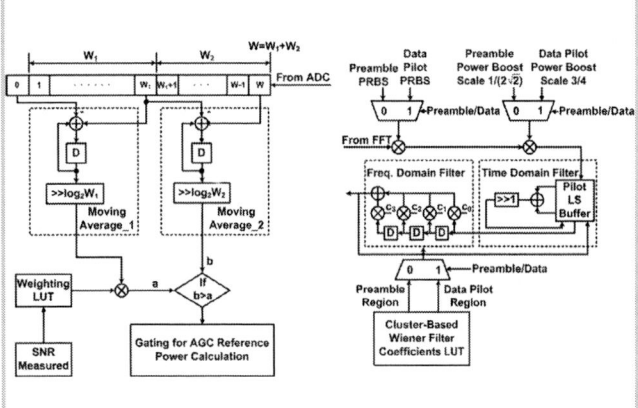

Figure 7.6.3: Block diagrams of Energy Detector for AGC and the Wiener filter for Channel Estimation.

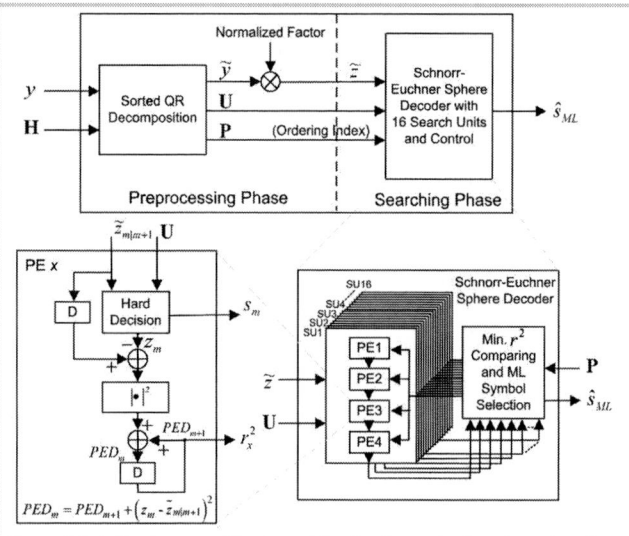

Figure 7.6.4: Hardware architecture of the fixed complexity sphere decoder.

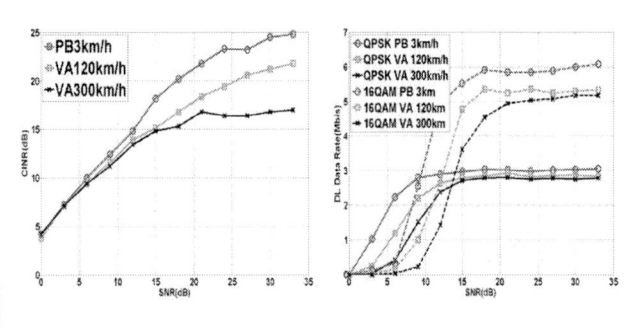

Figure 7.6.5: CINR measurements and data rates for moving speeds up to 300km/h (based on the WiMAX dongle PCB with RF chip and mobility emulation by an Elektrobit Propsim C8 radio channel emulator).

Figure 7.6.6: Main features and performance summary.

ISSCC 2011 PAPER CONTINUATIONS

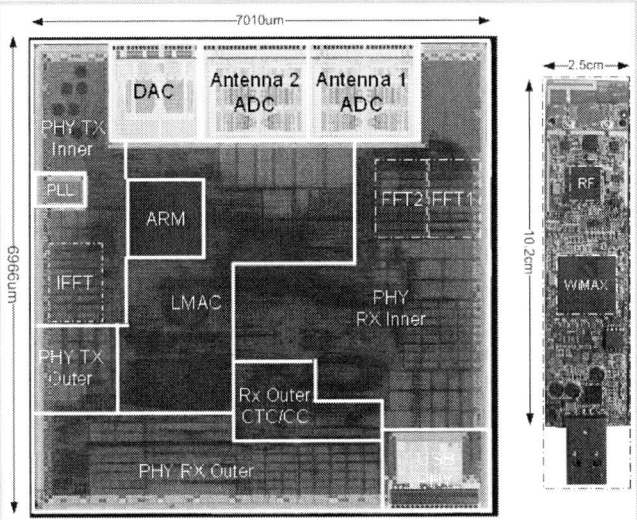

Figure 7.6.7: Chip micrograph.

ISSCC 2011 / SESSION 7 / MULTIMEDIA & MOBILE / 7.7

7.7 A 70Mb/s -100.5dBm Sensitivity 65nm LP MIMO Chipset for WiMAX Portable Router

Jyh-Shin Pan, Ming-Yang Chao, Eric Yeh, Wen-Wei Yang,
Ching-Wen Hsueh, Shyuan Liao, Jian-Bang Lin, Shun-An Yang,
Chin-Tai Liu, Tsai-Pao Lee, Jin-Ru Chen, Chia-Hua Chou,
Min Chen, Den-Kai Juang, Jen-Hao Yeh, Chieh-Wei Liao,
Po-Hung Chen, Kaipon Kao, Chia-Hsin Wu, Wen-Tso Huang,
Shih-Hsien Liao, Chih-Heng Shih, Chien-Hsun Tung, Yen-Po Lee

MediaTek, Hsinchu, Taiwan

In this paper, we present a low-power high-performance WiMAX chipset fully compliant with IEEE 802.16e specification corrigendum 1, 2 for mobile broadband access and WiMAX Forum System Profile Wave2. The chipset is comprised of a 632.7mW modem/router chip and a 364mW RF transceiver chip, both developed in 65nm CMOS. In [1], a fully programmable SIMD processor is used for WiMAX baseband and only parts of physical-layer functions are implemented; the FEC processor, media access control (MAC) processor, application processor, peripherals, etc. are excluded. Compared to hard-wired ASIC design, the use of a SIMD processor may cause higher power consumption and larger area if exactly the same functionality and algorithm are implemented. The baseband of our chipset is realized by HW accelerators and controlled by a simple DSP to minimize area and power. The modem/router chip also includes an analog frontend (AFE), MAC, memory controller, application processor, and interfaces such as PCM, PCI, USB 2.0 and fast Ethernet to provide rich and highly integrated CPE applications for portable router, VoIP phone, printer server, storage server, NAS, etc.

Figure 7.7.1 shows the block diagram of the WiMAX chipset. It is capable of handling a maximum WiMAX peak throughput of 70Mb/s for MIMO STC Matrix-B zone with 64QAM 5/6 modulation and coding scheme (MCS). The baseband part of the modem/router chip consists of a DSP, RX digital frontend (DFE), synchronization, FFT/IFFT, inner receiver (channel/noise variance estimation, MIMO detection, etc.), outer receiver (Turbo/Viterbi decoder, etc.), and TX. The baseband DSP is in charge of activating/deactivating HW modules, RF control, RX sequence control, power-saving control, timing control, handover control, idle mode control, etc. Three approaches in baseband are utilized to achieve low power: (1) DSP dynamically turns off AEF and unused blocks in the RF IC according to the DL and UL traffic. (2) In the WiMAX idle state, all clocks are turned off. Only the real-time clock (RTC) is alive to maintain the internal timer and to sense a wake-up event. (3) HW modules request clocks from the baseband clock controller dynamically according to data streams. When the data streams are finished, their clocks turn off by themselves.

Figure 7.7.2 shows the energy-efficient SW architecture. Two power-saving modules (PWS) are built-in with the OS scheduler. Each has 3 running levels: normal, standby, and hibernate. In normal mode, all threads are in the active state and control HW components. The PWS dynamically adjusts the system internal clock rate according to CPU utilization monitored by the OS. Fig. 7.7.3 shows the throughput versus power consumption of the whole WiMAX router system (including WiFi transceiver, PA, and DRAM; shown in inset). When the idle ratio of the system goes higher than a threshold, the PWS puts the system into standby mode, waking up within the smallest period at the lowest speed only to satisfy the minimum wireless protocol requirements while keeping the connection alive. However, in standby mode, where the WiMAX connection is usually at the idle state, the system can jump to normal mode seamlessly, within several clocks (about 50ns), as needed by disabling the succeeding clock pausing after waking up for paging. All major HW components have a corresponding control thread as an agent. An idle thread means the component can be shut down by its agent. When all agents shut down their clients, the system enters hibernate mode and cuts down all wireless connections. The PWRSV module co-operates with baseband DSP to enable HW-oriented deep power-saving mode. The system pauses the clocks with pre-defined wakeup events or timers when both DSP/PHY and MAC/SOC are ready to be paused. Furthermore, after DSP/PHY enters sleep mode the PWRSV module also handles power-on/off of the AFE.

The transceiver chip adopts a direct-conversion architecture associated with system optimization, circuit design techniques and built-in self-calibrations to achieve high performance [2] and to cope with PVT variations for mass-production. To benefit from MIMO technology, the radio is required to have higher dynamic range. The RF chip employs dual-band (2.3 to 2.7GHz and 3.3 to 3.8GHz) 2×2 MIMO architecture, as illustrated in Fig. 7.7.4. In the RX chain, the RF signal is received by a differential LNA, then directly down-converted by quadrature mixers. The down-converted signals are then filtered by LPFs and further amplified by programmable-gain amplifiers (PGAs), each with their own DC offset-cancellation loops. The LPF's cut-off frequency is calibrated to the desired bandwidth (3, 5, 7, 8.75, or 10MHz) to block adjacent-channel interference (ACI). In the TX chain, the quadrature signals from DACs are applied to LPFs, then directly up-converted by quadrature mixers to the RF band. The RF signal is then amplified through a PGA, capable of driving loads through an integrated transformer without the need for an external balun. The transceiver uses many auto and digital-assisted calibration circuits, including TX LO and IQ, RX DC and IQ, resistor for bias current, LPF, and VCO calibration.

The RF transceiver performance summary is given in Fig. 7.7.5. The low-noise high-linearity RX can achieve 2.5dB and 2.8dB noise figure (NF) at low-band (2.3 to 2.7GHz) and high-band (3.3 to 3.8GHz), respectively, as well as error vector magnitude (EVM) better than -35dB within a -10 to -60dBm input signal range while supporting MIMO 64-QAM OFDM. The RX sensitivity is better than WiMAX Forum Mobile Radio Conformance Test (MRCT) [4] with a margin of 7.5dB and 7.2dB for low-band and high-band, respectively. The TX EVM is better than -37dB and -36dB at -2dBm TX power for low-band and high-band, respectively, within a 0 to -60dBm output power range. A single fractional-N frequency synthesizer supporting multiple reference clock frequencies is implemented to provide low-noise LO signal [3]. The integrated phase noise of the synthesizer for both bands is about 0.4 degree.

Figure 7.7.6 shows the measured performance summary of the WiMAX modem/router chipset and power consumption of the whole WiMAX router system (including WiFi transceiver, PA, and DRAM) at 12Mb/s and 8Mb/s WiMAX RX and TX throughput, respectively. The sensitivity of the chipset outperforms MRCT [4] not only in AWGN but also in multipath fading channel. The margin of the chipset is more than 9dB for most cases. For the lowest sensitivity, the chipset could reach up to -100.5dBm on an AWGN channel, compared to -98.9dBm in [5], and -96.1dBm in the VA60 channel.

References:
[1] A. Nilsson, et al., "An 11mm2 70mW Fully-Programmable Baseband Processor for Mobile WiMAX and DVB-T/H in 0.12μm CMOS," *ISSCC Dig. Tech. Papers*, 2008.
[2] D. Chowdhury, et al., "A Fully Integrated Dual-Mode Highly Linear 2.4 GHz CMOS Power Amplifier for 4G WiMax Applications," *IEEE J. Solid-State Circuits*, vol. 44, pp. 3393-3402, Dec. 2009.
[3] M. Locher, et al., "A Versatile, Low Power, High Performance BiCMOS MIMO/Diversity Direct Conversion Transceiver IC for WiBro/WiMAX (802.16e)," *IEEE J. Solid-State Circuits*, vol. 43, pp. 1731-1740, Aug. 2008.
[4] "WiMAX Forum Mobile Radio Conformance Tests Release 1.0 Approved Specification (Revision 2.2.0)," WiMAX Forum, 2008.
[5] J.-J. Cho, et al., "The Highly Integrated Mobile WiMAX Module Using Embedded PCB & SIP Technology," *IEEE CICC*, 2009.

ISSCC 2011 / February 22, 2011 / 11:45 AM

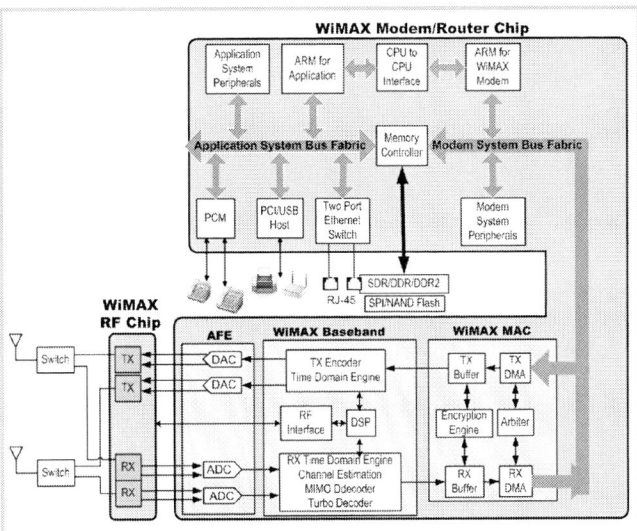

Figure 7.7.1: WiMAX chipset block diagram.

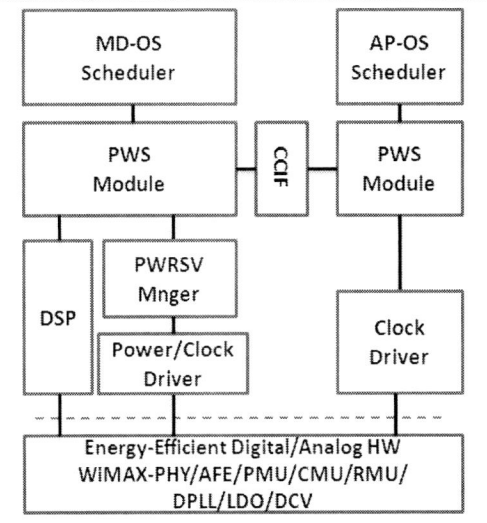

Figure 7.7.2: Energy-efficient SW architecture.

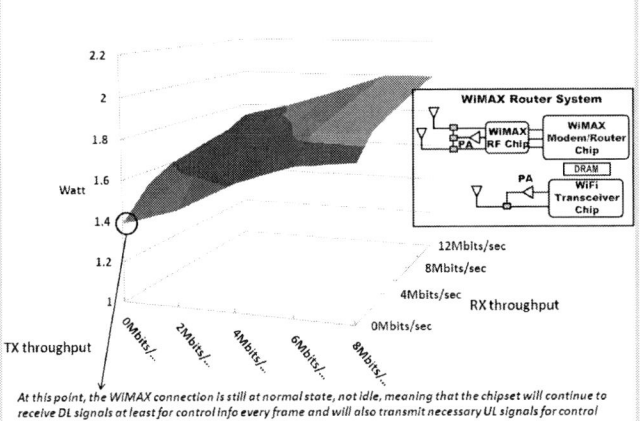

At this point, the WiMAX connection is still at normal state, not idle, meaning that the chipset will continue to receive DL signals at least for control info every frame and will also transmit necessary UL signals for control and maintenance, even though there's no user data throughput at both TX and RX. The WiFi side is likewise.

Figure 7.7.3: TX/RX throughput versus power consumption in normal mode.

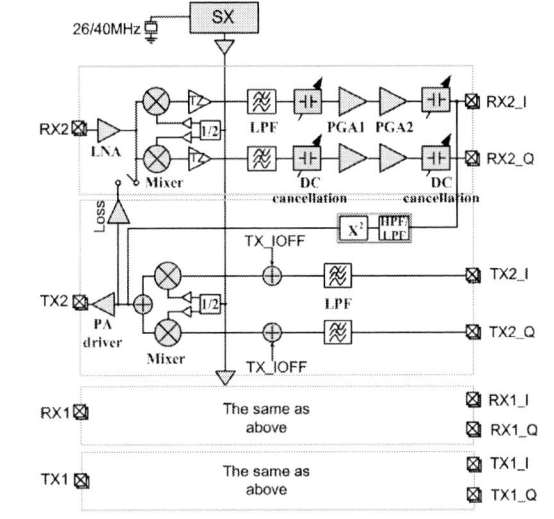

Figure 7.7.4: Dual-band 2×2 MIMO WiMAX transceiver architecture.

	Measurement		WiMAX Forum spec
Frequency band	2.3~2.7G	3.3~3.8G	
Power supply	2.8V	2.8V	
2RX mode current	130mA	143mA	
2RX mode power consumption	364mW	400.4mW	
2TX mode current	190mA	207mA	
2TX mode power consumption	532mW	579.6mW	
RX NF @ max RF gain	2.5dB	2.8dB	
RX IIP3 @ max RF gain	-13dBm	-13dBm	
RX 10MHz 64QAM5/6 sensitivity*	-79dBm	-78.7dBm	-71.5dBm
RX 10MHz QPSK1/2 sensitivity*	-96dBm	-95.7dBm	-88.5dBm
RX 10MHz 64QAM3/4 ACI (I-S)	32dB	32dB	4dB
RX 10MHz 16QAM3/4 ACI (I-S)	37dB	37dB	11dB
RX 10MHz 64QAM3/4 AACI (I-S)	40dB	40dB	23dB
RX 10MHz 16QAM3/4 AACI (I-S)	45dB	45dB	30dB
TX EVM @-2dBm (16QAM3/4)	-37dB	-36dB	-24dB
Integrated phase noise (1K~BW/2)	0.4degree	0.4degree	
Chip size	11.05mm² in 65nm CMOS		

* The sensitivity is measured at one of the two RX antenna in Preamble region

Figure 7.7.5: RF IC performance summary.

WiMAX modem/router chip			
Technology	65nm CMOS		
Chip size	24.99mm²		
Input clock (from RF chip)	20/40MHz		
Power supply	3.3V(IO), 2.8V(RFIO), 1.8V (DRAM), 1.3V (Core)		
Max RX peak throughput	70Mbps		
Power consumption	Normal mode	Standby mode	Hibernate mode
	632.7mW(*)	60mW	60mW
Measured sensitivity of the chipset(***)			
SIMO, AWGN			
MCS	This work (dBm)	MRCT (dBm)	Margin (dB)
QPSK1/2	-100.5	-91	9.5
16QAM1/2	-95	-85.3	9.7
64QAM5/6	-83.4	-74	9.4
MIMO STC Matrix-A, Ped-B@3km/h, High correlation			
MCS	This work (dBm)	MRCT (dBm)	Margin (dB)
QPSK1/2	-96.4	-87.2	9.2
16QAM1/2	-90.9	-81.7	9.2
64QAM5/6	-79.7	-70.4	9.3
MIMO STC Matrix-A, Veh-A@60km/h, High correlation			
MCS	This work (dBm)	MRCT (dBm)	Margin (dB)
QPSK1/2	-96.1	-87	9.1
16QAM1/2	-90.6	-81.3	9.3
64QAM5/6	-79	-69.6	9.4
Whole router system power consumption			
Power consumption(**)	Normal mode	Standby mode	Hibernate mode
	2160mW(*)	329.92mW	64.73mW

(*) At WiMAX RX and TX throughput of 12Mbps and 8Mbps, respectively
(**) Including WiFi transceiver, PA, and DRAM
(***) Measured at one of the two RX antenna in data zone, which is about 4dB lower than Preamble for SIMO case

Figure 7.7.6: Measured performance and whole router system power.

978-1-61284-303-2/11 $26.00 © 2011 IEEE

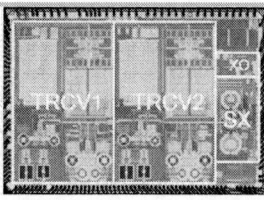

Figure 7.7.7: Die Micrograph.

ISSCC 2011 / SESSION 7 / MULTIMEDIA & MOBILE / 7.8

7.8 A Direct Digital Frequency Synthesizer with Minimized Tuning Latency of 12ns

Alan Willson, Mukund Ojha, Shilpa Agarwal, Thriven Lai, Tzu-chieh Kuo

University of California, Los Angeles, CA

A downside for all direct digital synthesizer (DDS) architectures is that every DDS has a phase accumulator (PA) whose normalized phase value ϕ must be updated for each (sin $2\pi\phi$, cos $2\pi\phi$) output-pair produced, and such updating introduces a rather long carry-ripple. PA lengths of 32-b are commonplace and 48-b or longer PA can be found in commercial DDS products. When a DDS with high data-rate is needed, the long PA carry-ripple can present a serious bottleneck—one usually overcome by some form of PA pipelining—but then, only at the cost of a significant increase in "tuning latency" as well as added power consumption and chip-area. Ref. [1] explains how (for a mere 24-b PA) such pipelining introduces a 55-cycle latency, setting the system's frequency-hopping limit at 700/55 = 12.7MHz, for a DDS generating outputs at 700MHz. All such PA pipelining difficulties are completely eliminated by the architecture reported here.

Our new DDS architecture, beyond speeding up PA updating, provides other major benefits, including reduced computation for each DDS output value, hence lower power consumption, and effortlessly providing an increase in DDS output SNR by facilitating the use of carry-ripple-free phase-*rounding* rather than conventional phase truncation. (Note: phase-rounding is not found in any previous DDS architecture.)

The Fig. 7.8.1 DDS example has a 32-b PA, split into two 16-b parts. These PA parts get repeatedly updated by simultaneous additions of FCW_H and FCW_L, the two parts into which the frequency control word (FCW) is split, thus reducing the PA-update carry-ripple bottleneck by half. By using a "single-bit register" to hold the C_out bit for the FCW_L addition—with this stored bit becoming a C_in bit for the *next* FCW_H addition—the contents of the upper and lower PA halves (ϕ_H and ϕ_L) are always correct, and any desired FCW input change can take effect at an arbitrary instant (a much desired attribute for any DDS) by simultaneously changing FCW_H and FCW_L and continuing with the PA updates. The issue we must address is to (efficiently—no long carry ripple can be tolerated!) incorporate C_out into the DDS phase-to-amplitude-mapping computations. This is done within the Fig. 7.8.1 "Excess Fours Fine Rotation" block.

In our Fig. 7.8.1 design prototype we must use 19 MSBs of each phase value ϕ, from which we get the sin $2\pi\phi$ and cos $2\pi\phi$ outputs. Thus, we need the three MSBs of ϕ_L as well as all 16-b of ϕ_H (and C_out). We also take ϕ_{20}, which gets rounded into the 19-b phase value. We do this, where a conventional DDS would simply use 19-b phase truncation, since we can do so (as will be explained) without the 20-b carry ripple that would normally be incurred. Thus, not only is PA pipelining avoided or, equivalently, we avoid the slow-down resulting from a 32-b ripple for PA updating, we also avoid a conventional system's 20-b carry-ripple delay—the cost of which has heretofore dismissed phase-rounding as a practical DDS consideration.

Another new-system advantage is the avoidance of yet another carry-ripple computation: the one normally encountered in getting the two's complement conditional negation (cn) of truncated phase angles whenever such angles lie within odd octants. Fig. 7.8.2 shows a state-of-the-art DDS example [2] where this carry ripple occurs within the "two's complement cn-mapping" block. By contrast, the new system employs a faster *ones' complement* negation (part of which can be seen in Fig. 7.8.1 as the parallel XOR operations using ϕ_3, the even/odd-octant designating bit). Notice that the π/4-multiplication block of Fig. 7.8.2 has been eliminated in the Fig. 7.8.1 DDS (at the cost of a nontrivial, but reasonable increase in ROM size). Both systems employ an Output Stage that maps computed sin/cos values for Octant-0 angles—within [0, π/4]—into the correct octant (given by MSBs $\phi_1\phi_2\phi_3$ within [0, 2π]).

The Fig. 7.8.1 fine-stage computations are organized into 3-bit-rotation groups (Fig. 7.8.3) providing an approximately 2/3 reduction in the number of fine angle sub-rotations (one rotation for each 3-bit group), where the state-of-the-art DDS [2] provides a reduction factor of only 1/2. Each coarse-ROM rotation includes the offset angle "100100100" comprising multiple 3-b *fine-stage* rotations, each

being $\phi_a\phi_b\phi_c$ = "100," i.e., the value 4 for each 3-bit group—hence the "excess fours" name. With this offset, the appropriate rotations for each fine sub-stage are as shown in the BIAS = 0 column of Fig. 7.8.5. Such rotations are accomplished by the computationally efficient circuit of Fig. 7.8.4, where A = ϕ_a, B = ϕ_b, and C = ϕ_c and where fixed, hard-wired shifted ROM data are routed by MUXes to a carry-save adder.

We can increase by one bit the amount of a fine sub-stage rotation angle—in place, on the fly—by simply using the BIAS = 1 column of Fig. 7.8.5 rather than the BIAS = 0 column, while still using the Fig. 7.8.4 circuit. [BIAS = 1 rotations are just BIAS = 0 rotations, moved up one line in Fig. 7.8.5—with a "100 (4)" entry added in the last line.] This capability is important! It facilitates the advantages mentioned previously: getting two's complement negation via ones' complement negation, employing phase rounding at no cost over that for phase truncation, and avoiding PA pipelining. Operationally, it is evident that BIAS = 1 outputs can be obtained in Fig. 7.8.5 by simply "flipping the whole BIAS = 0 column top/bottom" and negating all entries. Hence, in Fig. 7.8.4 we change to column 2 by just inverting all control bits (A = ~ϕ_a, B = ~ϕ_b, and C = ~ϕ_c) and negating the output's sign.

Since a two's complement negation can be obtained by adding "1" to a ones' complement negation, the ones' complement negation is converted into a two's complement negation by simply modifying the control bits of just the last 3-bit excess fours sub-group (here, the $\phi_{17}\phi_{18}\phi_{19}$ group in Fig. 7.8.3) to use BIAS = 1 while doing a conditional ones-complement negation for *all* fine-stage groups. Similarly, the use of ϕ_{20} to modify (conditionally invert) the $\phi_{17}\phi_{18}\phi_{19}$-group control bits accomplishes the desired "phase rounding." Moreover, a similar "rounding" of C_out into the 16 bits of ϕ_H (by modifying the $\phi_{14}\phi_{15}\phi_{16}$ control bits of Fig. 7.8.3) causes it to be included in the fine-stage rotation, thereby overcoming the need for PA pipelining. It can be shown that *all* these control-bit modifications can be employed simultaneously, yielding a DDS that efficiently (with no costly carry-propagations!) performs all desired functions. Moreover, it can be shown that the same type of solution can apply for the numerous and varied other ways a PA can be broken into smaller segments.

The Fig. 7.8.7 TSMC-prototype includes BIST for testing. The die has two separate cores, one is a DDS with 16-b quadrature outputs. With a 1.8V supply, this 16-b DDS was tested at a variety of FCW settings: its average power dissipation was 16.5mW at 260MHz. Its SFDR is 113dB and its SNR is 98dB (virtually the best possible SNR for 16-b outputs). From Fig. 7.8.6 comparisons its power dissipation is seen well below all other 16-b DDS; it is comparable to that of the 13-b DDS of [3] (after adjusting for its 0.25-μm CMOS). Our chip's time-domain error is always less than one LSB (unlike that of [3]). Notably, our DDS uses only standard Artisan TSMC 0.18-μm CMOS library cells—an advantage for commercial applications.

Acknowledgment:
This work was supported by grants from Analog Devices, Inc. and Broadcom Corp., whom we thank, and we thank TSMC for chip fabrication services.

References:
[1] F. Lu, H. Samueli, J. Yuan, and C. Svensson, "A 700-MHz 24-b pipelined accumulator in 1.2-μm CMOS for application as a numerically controlled oscillator," *IEEE J. Solid-State Circuits*, vol. 28, pp. 878-886, Aug. 1993.
[2] A. Torosyan, "Method and apparatus for improved direct digital frequency synthesizer," U.S. Pat. 7 539 716 B2, May 26, 2009.
[3] A. G. M. Strollo, D. De Caro, and N. Petra, "A 630MHz. 76mW direct digital frequency synthesizer using enhanced ROM compression technique," *IEEE J. Solid-State Circuits*, vol. 42, pp. 350-360, Feb. 2007.
[4] F. Curticapean, K. I. Palomaki, and J. Niittylahti, "Quadrature direct digital frequency synthesizer using an angle rotation algorithm," *Proc. IEEE ISCAS*, May 2003, pp. 81-84.
[5] Y. Wu, D. Fu, and A. Willson, "A 415MHz direct digital quadrature modulator in 0.25-μm CMOS," *Proc. IEEE CICC*, Sept. 2003, pp. 287-290.
[6] A. Madisetti, A. Y. Kwentus, and A. N. Willson, Jr., "A 100-MHz, 16-b, direct digital frequency synthesizer with a 100-dBc spurious-free dynamic range," *IEEE J. Solid-State Circuits*, vol. 34, pp. 1034-1043, Aug. 1999.
[7] Y. Song and B. Kim, "Quadrature direct digital synthesizers using interpolation-based angle rotation," *IEEE Trans. VLSI*, vol. 12, pp. 701-710, July 2004.

978-1-61284-303-2/11 $26.00 © 2011 IEEE

ISSCC 2011 / February 22, 2011 / 12:15 PM

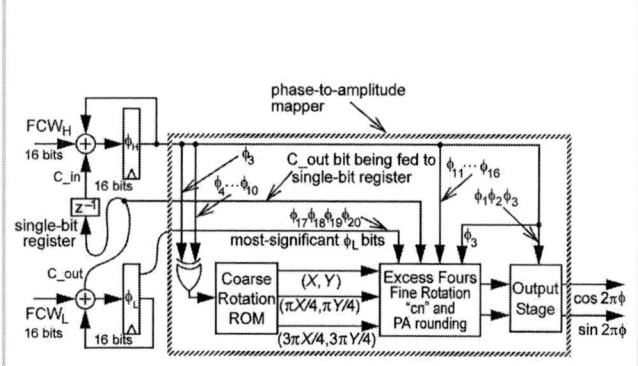

Figure 7.8.1: Prototype quadrature DDS requiring no phase-accumulator pipelining.

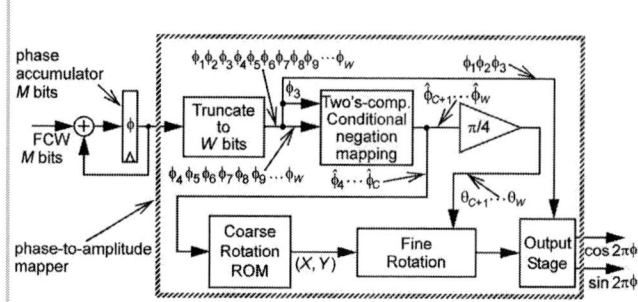

Figure 7.8.2: Conventional state-of-the-art DDS, e.g., Ref [2].

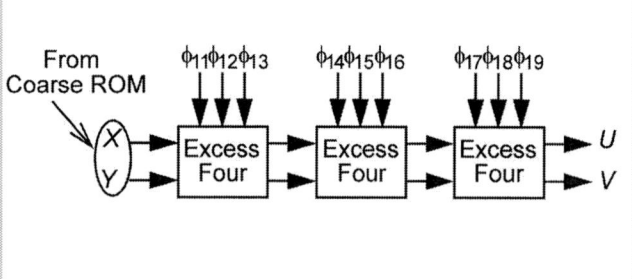

Figure 7.8.3: Example of Fine Stage sub-rotation blocks, where Coarse Rotation uses $\phi_4...\phi_{10}$.

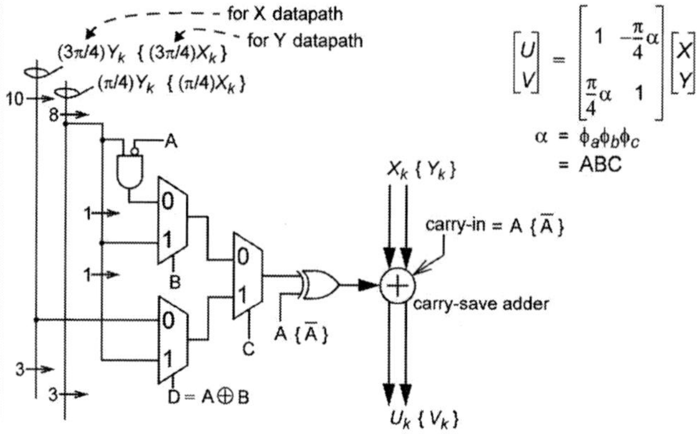

Figure 7.8.4: Hardwired-shift and MUX fine-stage rotation circuit.

$$\begin{bmatrix} U \\ V \end{bmatrix} = \begin{bmatrix} 1 & -\frac{\pi}{4}\alpha \\ \frac{\pi}{4}\alpha & 1 \end{bmatrix} \begin{bmatrix} X \\ Y \end{bmatrix}$$

$$\alpha = \phi_a\phi_b\phi_c = ABC$$

bit pattern $\phi_a\phi_b\phi_c$	if BIAS = 0 treated as	if BIAS = 1 treated as
000 (0)	–100 (–4)	–011 (–3)
001 (1)	–011 (–3)	–010 (–2)
010 (2)	–010 (–2)	–001 (–1)
011 (3)	–001 (–1)	000 (0)
100 (4)	000 (0)	001 (1)
101 (5)	001 (1)	010 (2)
110 (6)	010 (2)	011 (3)
111 (7)	011 (3)	100 (4)

Figure 7.8.5: Excess Fours System.

Circuit	Technique	PA (bits)	Output (bits)	SFDR (dBc)	Process (µm)	Area (mm²)	Data Rate (MHz)	Power (mW/MHz)
This work	excess fours	32	16	113	0.18	0.16	260	0.0635
Wu [5]	angle rotation (modulator)	32	16	100	0.25	0.43	415	0.602
Curticapean [4]	angle rotation	32	16	114	0.35	0.46	100	1.15
Song [7]	interpolation-based angle rotation	32	16	100	0.35	1.4	150	2.33
Madasetti [6]	angle rotation	36	16	100	1.0	12	100	14.0
Strollo [3]	multipartite table lookup	32	13	90	0.25	0.063	630	0.121

Figure 7.8.6: Comparison with recently reported quadrature-output DDS designs.

978-1-61284-303-2/11 $26.00 © 2011 IEEE

Figure 7.8.7: Die Photo: HP DDS is the 16-b quadrature DDS example.

ISSCC 2011 / SESSION 8 / ARCHITECTURES & CIRCUITS FOR NEXT GENERATION WIRELINE TRANSCEIVERS / OVERVIEW

Session 8 Overview / Wireline

Architectures & Circuits for Next Generation Wireline Transceivers

Session Chair: *Daniel Friedman, IBM T.J. Watson Research Center, Yorktown Heights, NY*

Session Co-Chair: *Koichi Yamaguchi, Renesas Electronics, Kawasaki, Japan*

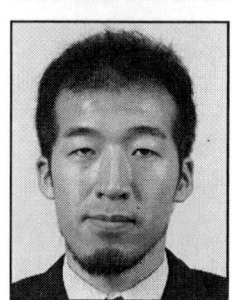

Meeting the challenges for next generation communication systems demands a combination of circuit and architecture innovation. Circuit innovation, by extending the performance achievable using CMOS technologies, yields increasingly flexible standard-compliant designs, allows data rates to be pushed to new levels, and enables the creation of new ways to mitigate deep submicron technology constraints. Architecture innovation will have longer-term impact in areas ranging from addressing serial link power challenges to creating entirely new serial link application spaces.

This session includes 8 papers that extend serial link state-of-the-art through both circuit and architecture avenues. It starts with the presentation of new approaches for increasing flexibility and then for improving echo-cancellation performance in the context of standard-compliant I/O. Various approaches to enable increases in data rates follow, driving not only toward 100GbE and 40Gb/s serial link solutions, but even to 12.5Gb/s full duplex data transmission over a plastic waveguide. The final papers in the session describe new architectures for power efficient link design, for achieving ultra-fast burst-mode CDR lock, and new ways to achieve target transmitter output swing using advanced-node thin-oxide devices.

Paper 8.1 (Broadcom) describes a SONET transceiver capable of supporting the NRZ and RZ data formats deployed in optical networks. To achieve this goal, a new transmitter architecture and a new receiver threshold adjustment circuit are introduced and integrated into a 65nm CMOS transceiver demonstrated at 11.3Gb/s, achieving 17ps TX rise/fall times and 5mV$_{pp\text{-diff}}$ RX sensitivity with 0.54UI jitter tolerance. The transceiver consumes 214mW from a 1.0V supply.

An approach to achieving echo cancellation in a 10GBase-T transmitter that relaxes DAC linearity requirements is introduced in Paper 8.2 (Teranetics). Through the use of echo-cancelling DAC elements matched to transmit DAC elements, along with DSP-based control, the transmitter achieves a residual linear echo <-30dBc and residual transmitter distortion <-65dBc. These results are demonstrated over a bandwidth of 1 to 400MHz with the transmitter driving a 50Ω load with 2V$_{pp}$ output swing while consuming 250mW.

The state-of-the-art for high data-rate CMOS equalizing serial link designs is extended in Paper 8.3 (National Taiwan University), presenting a transmitter plus receiver chip set operating at 40Gb/s. In this work, a multi-tap FIR filter realized using an LC-based delay line is presented as an alternative to latch-based equalizer implementations. This approach is applied to both the transmitter and the receiver, and the resulting chip set is demonstrated with BER<10^{-12} over a channel with 19dB loss at 20GHz while consuming 457mW (135mW in the TX from a 1.2V supply and 322mW in the RX from a 1.6V supply).

Paper 8.4 (Hitachi) describes a gearbox LSI, a critical element of future 100Gb/s Ethernet systems. This 65nm CMOS design achieves its high performance while only consuming 2W from 1.0V and 1.8V supplies. A key power-reduction strategy is circuit architecture changes that enable the replacement of current-mode circuits with static CMOS circuits. The 25Gb/s interface in the LSI achieves a BER<10^{-12} with 34.3mV input sensitivity.

In Paper 8.5 (Sony; Caltech), millimeter-wave wireless techniques are used to enable bidirectional 12.5Gb/s+12.5Gb/s communication over a plastic waveguide. This 40nm CMOS design, which suggests a path to achieve some of the benefits of optical links without requiring electrical-optical or optical-electrical conversion, uses 57GHz and 80GHz carrier frequencies and is demonstrated to operate over a 120mm waveguide at a BER of 10^{-12} while consuming 143mW from a 1.1V supply.

978-1-61284-303-2/11 $26.00 © 2011 IEEE

ISSCC 2011 / February 22, 2011 / 8:30 AM

A charge-recycling approach to serial link transceiver design is presented in Paper 8.6 (Oregon State University; IBM Zurich Research Labratory). This 90nm CMOS design operates from 0.5 to 4Gb/s and uses stacked digital transmitter and receiver clock generators as a key technique to reduce power consumption; at 3.2Gb/s its power efficiency is 1.9mW/Gb/s with a BER<10^{-12} .

Paper 8.7 (University of Toronto; Fujitsu Laboratories) introduces a new phase interpolator-based approach for implementing a burst-mode CDR. The technique is demonstrated in a 65nm CMOS test chip, achieving lock in less than 1 UI for data rates between 1 and 6Gb/s while consuming 22mW from a 1.2V supply.

Finally, Paper 8.8 (IBM Zurich Research Labratory; Miromico) describes techniques enabling the realization of a high-swing high-data-rate source series terminated transmitter using thin-oxide devices in a 45nm CMOS SOI technology. The 4-tap FFE transmitter achieves >1.28V$_{pp}$ output swing at 14Gb/s while consuming 85.5mW from a dual supply (a main supply of 1V and an output driver supply of 2V).

This session presents work ranging from extending current state-of-the-art serial link designs to demonstrating possibilities for future link data-rate increases to proposing entirely new ways to consider serial-link implementation. The advances reported here will be crucial to increasing the capabilities of next-generation high-performance systems and to enabling serial link introduction into new application spaces.

8

Presenters:

8:30 AM

8.1 11.3Gb/s CMOS SONET-Compliant Transceiver for Both RZ and NRZ Applications

N. Kocaman, Broadcom, Irvine, CA

9:00 AM

8.2 A Full-Duplex 10GBase-T Transmitter Hybrid with SFDR >65dBc Over 1 to 400MHz in 40nm CMOS

G. Chandra, Teranetics, San Jose, CA

9:30 AM

8.3 A 40Gb/s TX and RX Chip Set in 65nm CMOS

M-S. Chen, National Taiwan University, Taipei, Taiwan

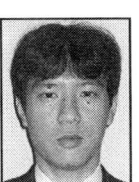

10:15 AM

8.4 10:4 MUX and 4:10 DEMUX Gearbox LSI for 100-Gigabit Ethernet Link

G. Ono, Hitachi, Tokyo, Japan

10:45 AM

8.5 A 12.5+12.5Gb/s Full-Duplex Plastic Waveguide Interconnect

S. Fukuda, Sony, Tokyo, Japan

11:15 AM

8.6 A Highly Digital 0.5-to-4Gb/s 1.9mW/Gb/s Serial-Link Transceiver Using Current-Recycling in 90nm CMOS

R. Inti, Oregon State University, Corvallis, OR

11:45 AM

8.7 A 1-to-6Gb/s Phase-Interpolator-Based Burst-Mode CDR in 65nm CMOS

B. Abiri, University of Toronto, Toronto, Canada

12:00 PM

8.8 A 14Gb/s High-Swing Thin-Oxide Device SST TX in 45nm CMOS SOI

C. Menolfi, IBM Zurich Research Laboratory, Rueschlikon, Switzerland

978-1-61284-303-2/11 $26.00 © 2011 IEEE

ISSCC 2011 / SESSION 8 / ARCHITECTURES & CIRCUITS FOR NEXT GENERATION WIRELINE TRANSCEIVERS / 8.1

8.1   11.3Gb/s CMOS SONET-Compliant Transceiver for Both RZ and NRZ Applications

Namik Kocaman, Adesh Garg, Bharath Raghavan, Delong Cui, Anand Vasani, Keith Tang, Deyi Pi, Haitao Tong, Siavash Fallahi, Wei Zhang, Ullas Singh, Jun Cao, Bo Zhang, Afshin Momtaz

Broadcom, Irvine, CA

An 11.3Gb/s CMOS SONET-compliant transceiver is designed to work in both RZ and NRZ data formats. The TX driver exhibits 17ps rise/fall times, $0.25ps_{rms}$ RJ, and $2ps_{pp}$ DJ. The RX has a multi-stage vertical threshold adjustment circuit. It achieves $5mV_{pp-diff}$ RX input sensitivity with 0.54UI jitter tolerance. The transceiver core area occupies 1.36mm² in 65nm CMOS and consumes 214mW.

Two main data formats are widely deployed in optical networks: return to zero (RZ) and non-return to zero (NRZ). The RZ data format sends full-rate clock during "1", while stays low during "0" (Fig. 8.1.1). This is equal to transmitting pulses at half baud period, i.e., 50ps for 10Gb/s operation. Transmitting narrower pulses in RZ signaling compensates pulse widening caused by dispersion in fiber transmission. After the fiber link, received RZ data will have more horizontal eye opening providing longer reach compared to NRZ. On the other hand, RZ transmitter requires twice the bandwidth compared to NRZ due to narrower data pulses and sharper rise/fall times. In addition, received RZ data exhibit severe amplitude attenuation and vertical eye asymmetry. Thus, RZ receiver requires better input sensitivity with vertical threshold adjustment. Up to now, above implementation challenges have prevented the use of low-cost CMOS technology in multi-GHz RZ transceivers. Current semiconductor solutions are limited to 2-chip designs; an NRZ transmitter followed by an NRZ to RZ converter fabricated in non-CMOS processes [1, 2]. This paper discusses the design and measured results of the first single-chip, multi-GHz dual mode transceiver capable of operating in both RZ and NRZ formats in standard 65nm CMOS technology. The TX serializes 16-bit incoming data into 8.5 to 11.3Gb/s bit stream [3]. The RX de-serializes 8.5 to 11.3Gb/s bit stream into 16-bit data output [4]. The RZ operation is enabled via an NRZ-to-RZ converter embedded in the TX driver. On the RX side, a vertical threshold adjustment (TA) circuit distributed over cascaded, high-gain limiting amplifier (LA) stages adjusts the slicer vertical sampling point of the received RZ data to improve input sensitivity and slicer margins.

Figure 8.1.2 shows the high speed TX path including retimer, FIR, data/clock drivers, phase interpolator (PI), duty-cycle adjuster (DCA), and AND gate. The retimer followed by a FF generate main-tap and post-tap inputs for the FIR which enables de-emphasis in NRZ mode. The FIR coefficients are set via adjustable tail currents of the buffers which share the same output load in data driver. To ensure sharp rise/fall times and delay matching, shunt-peaked CML structures are employed in both clock and data paths. The resistor loads of CML circuits are calibrated to reduce process variation. By comparing an on-chip resistor with an external 1% precision resistor, the load resistance variation is reduced by more than 75% [5].

The PI has different roles in different modes. In NRZ mode, the PI adjusts the timing between TSDP/N and TSCKP/N. In RZ mode, the PI is used to shift clock phase with respect to data to control stringent timing at the input of AND gate. Then DATAP/N is ANDed with the full speed clock CK2P/N to generate RZ data. Unused post-tap section as well as clock driver is shut off in RZ mode to save power.

Figure 8.1.3 shows the circuit details of the NRZ-to-RZ converter, consisting of the DCA, AND gate, and the dummy load. The DCA is a single-pole high pass filter realized via series capacitors with separate DC biasing on the P and N sides followed by a CML buffer. In NRZ mode, the DCA is disabled by turning off the tail transistor M5. CK2P is pulled up to VDD, and CK2N is pulled down to VSS via S1 switches. As a result, AND gate is configured as a CML buffer to pass through NRZ data. The gates of cascode transistors M3 and M4 are also connected to VSS which significantly reduces CK1P/N coupling on to CK2P/N through Cgd of M1 and M2, and prevents RZ path circuits inducing periodic jitter on to NRZ data. In RZ mode, the DCA adjusts the data pulse width from 40 to 60 percent by changing the duty cycle of CK2P/N at the input of AND gate.

The shunt-peaked CML-based AND gate has a dummy load (DL) (Fig. 8.1.3). Since DATAP/N are exposed to changing capacitive loading via tail transistor M8 controlled by CK2N, RZ output at the AND gate exhibits data dependent jitter. In the DL, M9/M10 act exactly opposite to M6/M7 since tail transistor M11 is controlled by complementary phase CK2P. Therefore, the DL balances the capacitive loading at DATAP/N inputs during different phases of CK2P/N and reduces data dependent jitter in RZ mode. During NRZ operation, the DL is turned off. Overall, the combination of PI, DCA, and AND gate with dummy load enables low jitter operation in both RZ and NRZ modes without significant power penalty. After the AND gate, data reaches the shunt-peaked CML-based TX driver through cascaded CML buffers. Since RZ data has a non-zero DC level, it creates different output DC voltages at the P and N sides. At the TX driver output, this causes differences in amplitudes, rise/fall times, and return loss at the P and N sides. This is corrected by sourcing extra current into the N side and adding a matched capacitive load to P side.

In the RX, threshold adjustment (TA) is divided into 3 parts and distributed over 8 stages of the high-gain (>60dB) limiting amplifier (LA) (Fig. 8.1.4). The 1st TA is located at the RX input and provides the required range with coarse step size (20mV). The 2nd and 3rd TAs are located at the output of 3rd and 5th LA stages. Referred back to RX input, the 2nd and 3rd TAs have smaller step sizes (1.25mV and 0.15mV), respectively. Compared to single-stage TA implementations, this distributed scheme provides not only fine granularity and wide range (±340mV at RX input), but also less capacitive loading for the LA stages. Each TA stage consists of 16 identical differential pairs connected in parallel and sourcing current into resistive loads of the LA stage output. Digitally controlled input transistors steer current to create a DC voltage difference between D3P and D3N. This adjustment counteracts the vertical asymmetry present in the incoming data and optimizes the slicer threshold.

The 11.3Gb/s RZ/NRZ transceiver is integrated together with a 16-bit wide LVDS transceiver as well as loopback and diagnostic functions such as BER checker and PRBS generator. Duo-binary encoding and decoding are also implemented in TX and RX paths, respectively. The chip supports data rates from 8.5 to 11.3Gb/s. The measured TX driver eye diagrams for both NRZ and RZ formats with different duty cycles are depicted in Fig. 8.1.5. The RZ output exhibits $0.25ps_{rms}$ RJ, $2ps_{pp}$ DJ, and 17ps rise/fall times (Fig. 8.1.6). The measured RX input sensitivity is $5mV_{pp-diff}$ at 11.3Gb/s. RX jitter tolerance exceeds SONET requirements, achieving 0.54UI (pp) at high frequency. Fabricated in a 65nm CMOS process, TX core area is 0.8×0.8mm² while RX core occupies 0.8×0.9mm² (Fig. 8.1.7). From a 1.0V supply, RX and TX consume 99mW and 115mW, respectively. To our knowledge, this is the first reported multi-gigahertz SONET compliant CMOS transceiver with capability to transmit and receive in both RZ and NRZ format. It achieves more than three-fold power reduction compared to present two-chip RZ solutions [1, 6].

References:
[1] Inphi Corporation, "13707RZ 12.5Gb/s NRZ-to-RZ Converter"
[2] T. Suzuki, Y. Kawano, Y. Nakasha, et al., "A Novel 50-Gbit/s NRZ-RZ Converter with Retiming Function Using InP HEMT Technology," *IEEE Comp. Semiconductor Integrated Circuit Symposium*, pp. 41-44, Oct., 2006
[3] M. Green, A. Momtaz, K. Vakilian, et al., "OC-192 Transmitter in Standard 0.18mm CMOS," *ISSCC Dig. Tech. Papers*, pp. 248-249, Feb., 2002.
[4] J. Cao, A. Momtaz, K. Vakilian, et al., "OC-192 Receiver in standard 0.18Ém CMOS" *ISSCC Dig. Tech. Papers*, pp. 200-201, Feb., 2002.
[5] A. Momtaz, D. Chung, N. Kocaman, et al., "A Fully Integrated 10-Gb/s Receiver with Adaptive Optical Dispersion Equalizer in 0.13-μm CMOS,", *IEEE J. Solid-State Circuits*, vol. 42, no. 4, pp. 872-880, Apr., 2007.
[6] T. Masuda, H. Suzuki, H. Iizuka, et al., "A 250mW Full-Rate 10Gb/s Transceiver Core in 90nm CMOS using a Tri-State Binary PD with 100ps Gated Digital Output," *ISSCC Dig. Tech. Papers*, pp. 438-439, Feb., 2007.

978-1-61284-303-2/11 $26.00 © 2011 IEEE

ISSCC 2011 / February 22, 2011 / 8:30 AM

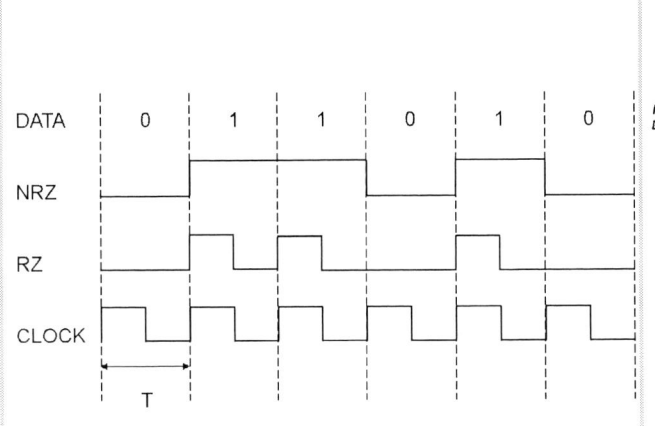

Figure 8.1.1: Illustration of return-to-zero (RZ) and non-return-to-zero (NRZ) Data Formats (T=symbol period).

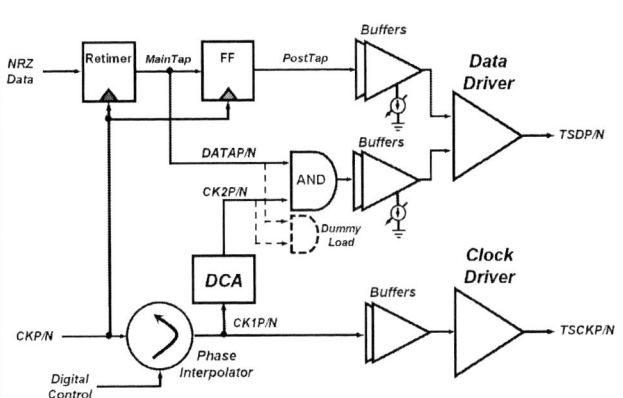

Figure 8.1.2: High-Speed TX Path.

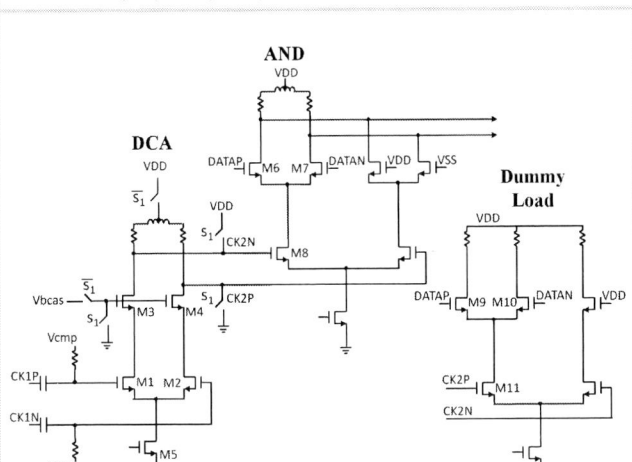

Figure 8.1.3: Circuit Details of duty-cycle adjustment (DCA), AND gate, and dummy Load.

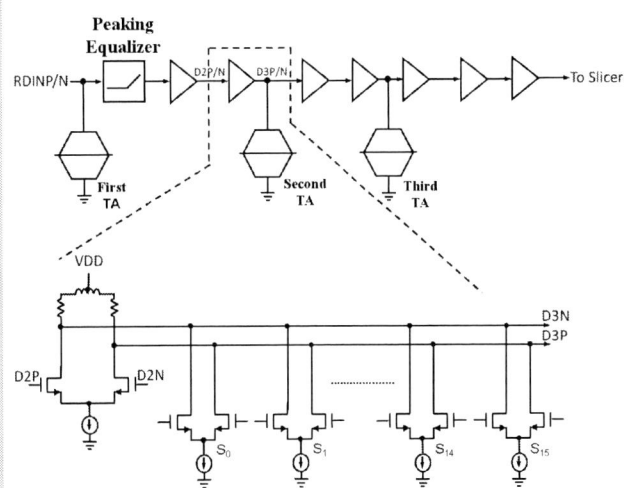

Figure 8.1.4: Implementation Details of limiting amplifier and threshold adjustment.

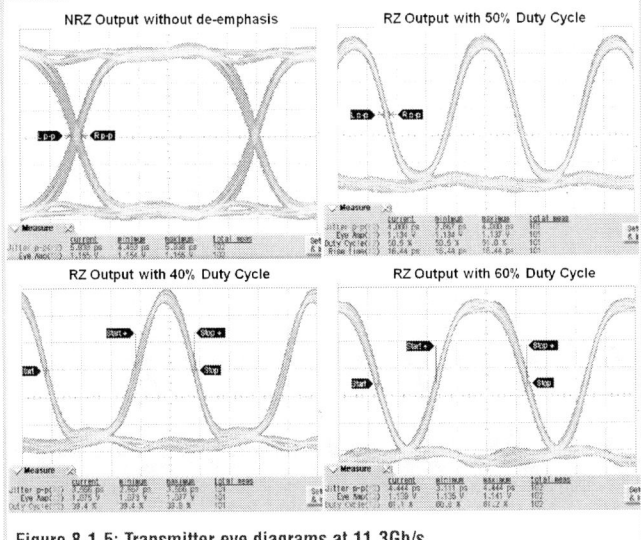

Figure 8.1.5: Transmitter eye diagrams at 11.3Gb/s.

Process Technology	7-metal 65nm CMOS
Power Supply Voltage	1.0V
Transmitter @11.3Gbps	
Output Amplitude	1Vpp-diff
Random Jitter	0.25ps (rms) (50KHz-80MHz)
ISI	2ps(pp)
Transition Time (trise/fall : 20-80%)	17ps
Duty Cycle Adjustment Range	40-60% with 0.67% step size
Receiver @11.3Gbps	
Jitter Tolerance	0.54UI @ high-frequency (>10MHz)
Min. Input Sensitivity	5mVpp-diff @11.3Gbps
	(TX/RX running asynchronous)
Core Area	
Receiver	0.8x0.9mm²
Transmitter	0.8x0.8mm²
Power Consumption	
Receiver	99mW
Transmitter	115mW

Figure 8.1.6: Performance Summary.

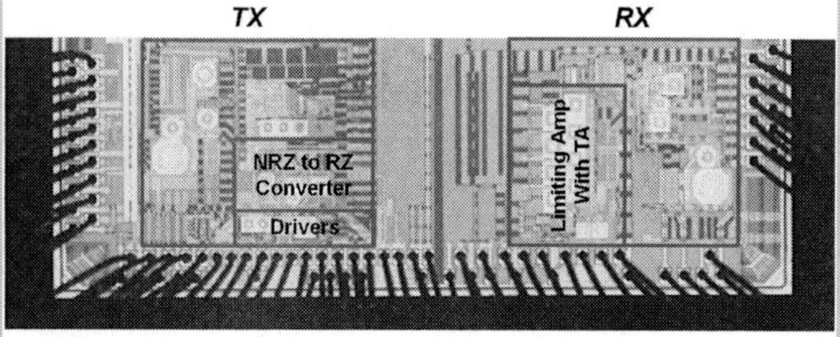

Figure 8.1.7: RZ transceiver IC die micrograph.

ISSCC 2011 / SESSION 8 / ARCHITECTURES & CIRCUITS FOR NEXT GENERATION WIRELINE TRANSCEIVERS / 8.2

8.2 A Full-Duplex 10GBase-T Transmitter Hybrid with SFDR >65dBc Over 1 to 400MHz in 40nm CMOS

Gaurav Chandra, Moshe Malkin

Teranetics, San Jose, CA

A transmitter and echo cancellation hybrid for IEEE 802.3an 10GBase-T Ethernet standard is presented, that utilizes DSP techniques to enhance the linear cancellation, and analog non-linearity cancellation to eliminate the need of a high linearity transmitter. Implemented in 40nm CMOS, the transmitter has a residual distortion < −65dBc and residual linear echo < −30dBc over a bandwidth of 1 to 400MHz. The transmitter, including the echo-cancellation circuitry and on-chip DSP engine, consumes 250mW power and occupies 0.9mm².

The IEEE 802.3an 10GBase-T Ethernet standard requires full-duplex transmission and reception over each of the 4 twisted pairs in a 100m UTP cable. Both the transmitter and receiver operate in the same frequency band of 1 to 400MHz. This necessitates very good linear echo cancellation to ease the dynamic range requirements of receive circuitry, as well as very good transmitter distortion, which due to its non-linear nature can not be easily cancelled by traditional DSP techniques. Historically, for Gigabit Ethernet and DSL, a line driver followed by passive hybrid has been the architecture of choice for full duplex transmission [1]. However, the large swing, low distortion and wide bandwidth requirements of 10GBase-T make it very difficult to implement such architecture. Instead, for high bandwidths the transmit DAC (TXDAC) driving the line directly has proven to be higher performance and more energy efficient as it approaches the theoretical class-A efficiency [2]. Echo cancellation is performed by having an additional DAC, which also needs to be very high linearity while minimizing additional power consumption [3].

In a previous implementation [3], echo-cancellation DAC (ECDAC) was driven by an FIR filter to adaptively eliminate the linear echo, and its non-linearity was made negligible compared to the TXDAC. The proposed architecture focuses instead on making the distortion of TXDAC and ECDAC highly correlated, as shown in Fig. 8.2.1. A 10b 800MS/s TXDAC directly drives the line impedance and also the receiver. A 10 800MS/s ECDAC, scaled to be 20% of TXDAC, operates on the same data as TXDAC. The ECDAC is terminated with a digitally adjustable trimmed resistor and a programmable capacitor array, providing a first order match to the line impedance. The residual linear echo term is further suppressed by two 6b 800Ms/s auxiliary echo-cancellation DACs (AECDACs), running on opposite phases of the transmit clock, and each scaled to be 5% of TXDAC. These are summed with the ECDAC in current mode. The two phase operation provides the ability to correct for return loss impairments beyond the 400MHz range, which can often limit performance. The AECDACs see only a very small swing associated with the residual echo, and hence their linearity requirements are significantly relaxed.

Figure 8.2.2 illustrates the DSP engine associated with the proposed transmit architecture. The inputs to the AECDACs are generated by adaptive FIR filters, whose coefficients are generated by an LMS engine seeking to minimize the overall receiver power over a large bandwidth. Unlike the TX and ECDAC quantization, which sees a first order linear cancellation, the AEC quantization noise may limit the receiver noise floor. Hence, the AEC datapath is chosen to be 12b. However, the significantly smaller size of the AECDAC makes it very difficult to implement a truly 5% scaled version of TXDAC in the analog datapath. Hence, the analog data to AECDACs is rounded off to 6b precision, and the residual 6b quantization noise is passed through filters Qfilter 0 and Qfilter 180, that model the linear analog signal path from the AECDACs to the ADC. This noise is then digitally subtracted from the received signal path. A similar approach, operating on the TXDAC quantization, was proposed in [4], but only simulation results were reported. Also, unlike [4], the channel information of the two-phase AEC-DAC path is unknown, and is adapted by means of introducing additional dither signals and LMS engines that perform the cross-correlation with received signal to estimate the filter coefficients.

Figure 8.2.3 illustrates the circuit design and layout of the composite TX and ECDAC. The DAC is implemented as a current-steering architecture with 6-4 segmentation. The layout of each ECDAC unit element is interleaved with a corresponding TX DAC unit element, and is driven by a shared latch and clock tree. This ensures that any distortion arising due to the skew in clock tree and latch rise/fall time is common to both the DACs. This also relaxes the jitter requirements on the PLL by providing first order cancellation. For the current source array, a common centroid layout is not followed as it increases the routing parasitics, potentially causing significant performance degradation at higher frequencies. Since the layout is interleaved, any non-linearity due to mismatch gradient in the horizontal direction is expected to be cancelled significantly at the receiver input, including any IR drop introduced systematic gradient. As a further precaution, the decoding is scrambled to suppress linear and quadratic gradient. To minimize impairments due to random mismatch, the current source devices are sized large enough to achieve the INL/DNL at 12b level.

Although the devices in ECDAC are scaled by an exact ratio of 5:1, the interconnect parasitics do not scale by the same ratio. As a result, the rise/fall times of internal nodes of the EC DAC can witness significant degradation, causing higher non-linearity. To overcome this problem, all the internal nodes of the ECDAC are bootstrapped with the corresponding nodes of the TXDAC, using small coupling capacitors as shown in Fig. 8.2.3. This matches the glitch energy of both the DACs for all the transitions. The dominant source of the non-linearity difference then remains to be the output nodes actually being different. This non-linearity difference is mitigated by matching the output swings to a first order, and making the capacitive output impedance of both of the DACs very good at high frequencies by careful circuit design and layout.

Figure 8.2.4 shows the 2-tone SFDR plot over 1 to 400MHz for the TXDAC, and the 2-tone SFDR plot of the equivalent SFDR of the cancelled TX and ECDAC, as seen by the receiver. This is obtained by measuring the 2-tone SFDR of the residual transmitter signal, and adjusting for the linear cancellation. The TX DAC stand-alone achieves > 55dBc SFDR with a $2V_{pp}$ swing, which easily meets the IEEE 802.3an specification for the stand-alone transmitter. But the equivalent SFDR measured at the receiver input is > 65dBc, illustrating up to 10dB of non-linearity cancellation.

Figure 8.2.5 shows the frequency spectrum of transmit DAC as seen on the line for a 2-tone 390MHz input, and the corresponding spectrum of the cancelled TX signal as measured by the receiver. The received spectrum has a −51dBc spurious with respect to the residual TX echo, which translates to -65dBc with respect to the actual transmit power.

Figure 8.2.6 shows the transmitter performance with the broadband input signal. With the AEC path turned on, 10 dB improvement in linear echo is observed, resulting in an overall linear echo of −30dBc averaged over 1 to 400MHz. Also, the resultant average non-linearity plus quantization noise level is at −148dBm/Hz (−66dBc) with a transmit power of −82dBm/Hz. Figure 8.2.7 shows a die micrograph.

References:
[1] P. Roo, S. Sutardja, S. Wei, et al., "A CMOS Transceiver Analog Front-End for Gigabit Ethernet over CAT-5 Cables," *ISSCC Dig. Tech. Papers*, pp. 310-311, Feb., 2001.
[2] C.-H. Lin, F. van der Goes, J. Westra, et al., "A 12b 2.9GS/s DAC with IM3 < -60dBc beyond 1GHz in 65nm CMOS," *ISSCC Dig. Tech. Papers*, pp. 74-75, Feb., 2009.
[3] S. Gupta, J. Tellado, S. Begur, et al., "A 10Gb/s IEEE 802.3an Compliant Ethernet Transceiver for 100m UTP Cable in 0.13μm CMOS," *ISSCC Dig. Tech. Papers*, pp. 106-107, Feb., 2008.
[4] P. J. Hurst and A. Norell, "DAC Quantization-Noise Cancellation in an Echo-Canceling Transceiver," *IEEE TCAS-II*, vol. 55, no. 2, pp. 111 -115, Feb., 2008.

ISSCC 2011 / February 22, 2011 / 9:00 AM

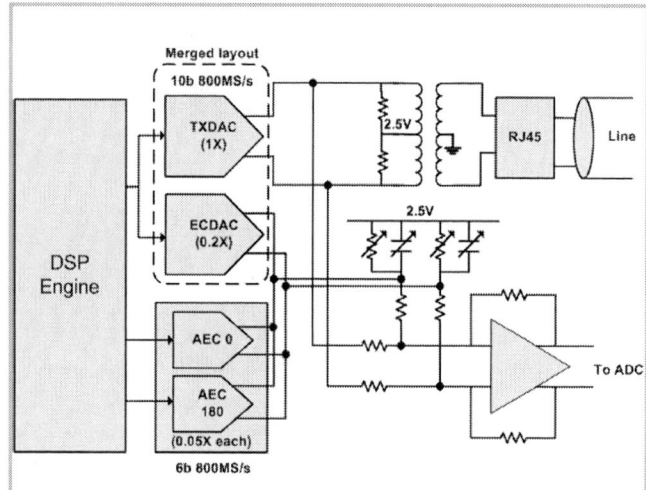

Figure 8.2.1: Overall architecture of proposed transmit and echo-cancellation circuit.

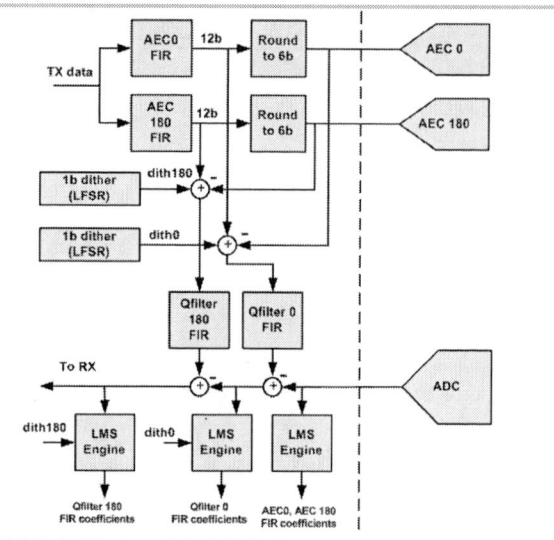

Figure 8.2.2: Architecture of the DSP Engine.

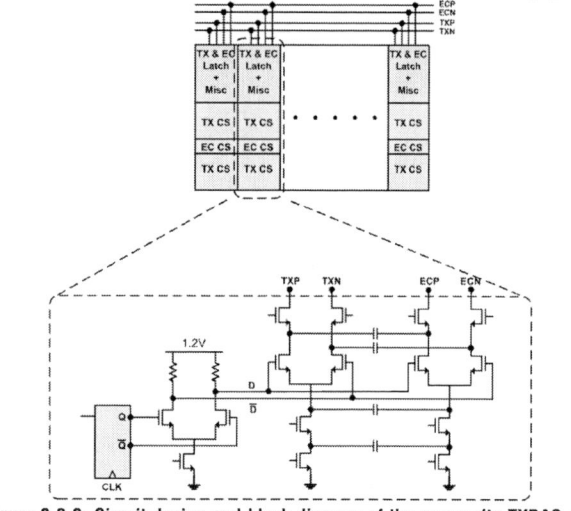

Figure 8.2.3: Circuit design and block diagram of the composite TXDAC and ECDAC.

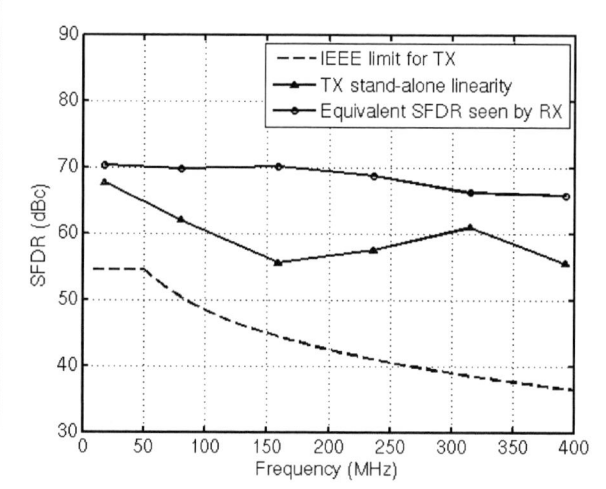

Figure 8.2.4: Measured performance of TX stand-alone, and the cancelled TX signal measured through receive ADC.

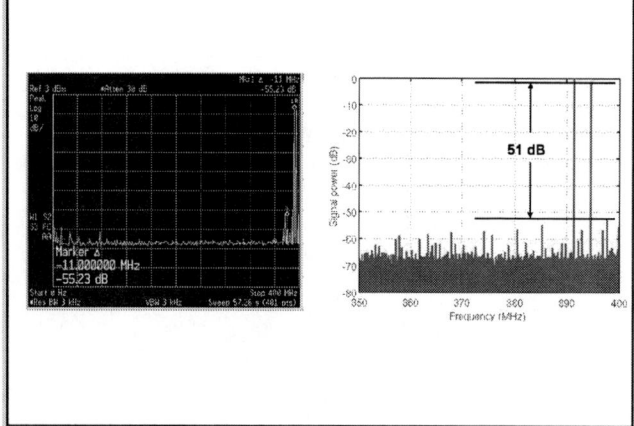

Figure 8.2.5: Spectral plot for 2-tone high-frequency input, for TX (left) and the cancelled receive signal (right).

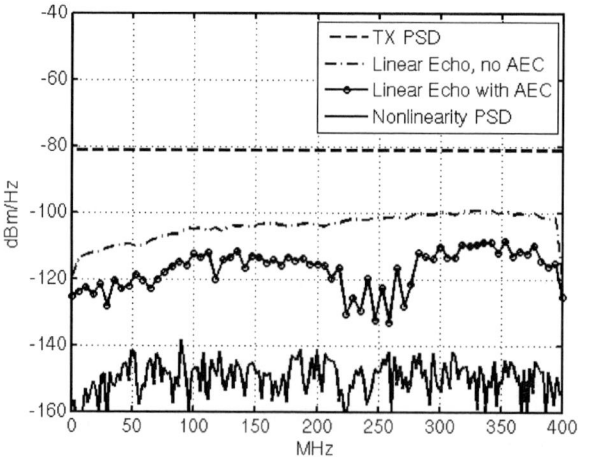

Figure 8.2.6: Transmit, echo and non-linearity PSD with broadband spectrum.

8

978-1-61284-303-2/11 $26.00 © 2011 IEEE

ISSCC 2011 PAPER CONTINUATIONS

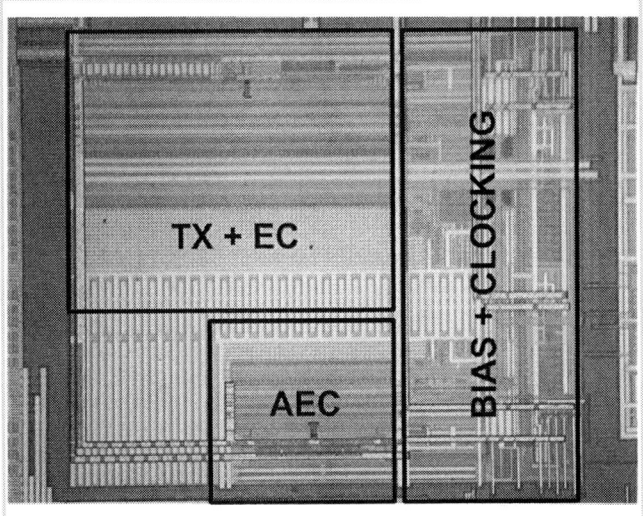

Figure 8.2.7: Zoomed in die micrograph of the transmitter.

ISSCC 2011 / SESSION 8 / ARCHITECTURES & CIRCUITS FOR NEXT GENERATION WIRELINE TRANSCEIVERS / 8.3

8.3   A 40Gb/s TX and RX Chip Set in 65nm CMOS

Ming-Shuan Chen, Yu-Nan Shih, Chen-Lun Lin, Hao-Wei Hung, Jri Lee

National Taiwan University, Taipei, Taiwan

Next generation optical and electrical communications such as chip-to-chip serial links or 100GbE require very-high-speed transceivers. At tens of Gb/s, both transmitters and receivers suffer from inadequate bandwidth and high power consumption. One major difficulty arises from the performance degradation of FIR-based FFEs as the FF's CK-Q delay becomes significant to one bit period. Using passive components as delay elements [1][2] can relax this issue to some extent, but the untunable delay is quite vulnerable to PVT variations. Traditional DFEs also suffer from speed limitation in its feedback loop, and parallelization schemes usually introduce complex circuits and high power consumption. This paper presents a full-rate 40Gb/s transceiver prototype significantly alleviating the above issues.

Figure 8.3.1 shows the transmitter, which consists of a 20GHz PLL, a 5-tap FFE with tunable delay, and a 1UI delay generator. To create exactly 1-bit delay, conventional FFE approach needs flip-flops, which would either suffer from limited bandwidth or large CK-Q delay. Both issues would devastatingly destroy the filter's performance. Even in 65nm CMOS, a flip-flop-based FFE can only operate at 20Gb/s. To increase speed, we propose a 5-tap FFE incorporating delay elements $[T_b$ (=25ps) + remainder $\Delta t]$, which is made of LC networks for each tap, and compensation elements between the nodes of the combiner. In other words, the 40Gb/s input data through the FIR are combined by exactly 25ps separation with different weighting coefficients α_{-2}, α_{-1}, ...α_2. The 1UI delay generator is responsible for creating one T_b delay. Each T_b delay element is actually a voltage-controlled delay line (VCDL) composed of inductors and varactors. By tuning the control voltage (V_{ctrl}), the 1UI delay generator locks the 10GHz input (from the 20GHz PLL and ÷2 circuit) and the delayed version to form a 90° phase difference, forcing T_b delay to be exactly 25ps. As demonstrated in Fig. 8.3.2(b), this "artificial" transmission line provides good matching for broadband and narrowband signals. Distributing V_{ctrl} through the FIR filter, we reproduce the 5-tap 40Gb/s delay line with high accuracy. Note that T_b is immune from PVT variations because of the closed-loop control.

The T_b-delay cell is illustrated in Fig. 8.3.2(a). The differential pair M_{1-2} propagates input data into an emulated transmission line made of 7 segments of LC network. By tuning V_{ctrl} from 0 to 1.2V, the delay from D_{in} to D_{out} varies from 29.7ps to 22.9ps (27.2%). In addition to accuracy, another important advantage of this approach over conventional flip-flop-based delay cell is that most parasitic capacitances are absorbed into the transmission line. By the same token, the 40Gb/s FIR combiner must have resonating elements between taps, or the routing parasitics would degrade the output data considerably. Fig. 8.3.2(b) shows the proposed combiner. Following our layout arrangement, taps are separated by one LC element (=Δt). Note that although the 25ps delay is created from the 10GHz clock, the artificial transmission line still provides great consistency for broadband data. The maximum deviation between group delays of 40Gb/s data and 10GHz clock across the whole tuning range is less than 1ps [Fig. 8.3.2(b)].

The receiver is depicted in Fig. 8.3.3. It employs a 3-tap adaptive FIR filter as a front-end equalizer and demultiplexing circuits [3]. A typical DFE suffers from stringent speed requirement in its feedback path, and even though quite a few methods such as sub-rate or loop-unrolled structures have been proposed, they usually involve complicated design and significant power. Here, instead of using a DFE, we place a 3-tap FIR in the RX to overcome the difficulties. Following similar approach as we illustrated in the TX, this adaptive equalizer is capable of providing up to 14dB boosting. Again, driven by an external clock of 40GHz, a 1UI delay generator is responsible for a proper control voltage V_{ctrl}. The 40Gb/s data is therefore sampled by three flip-flops (FF$_1$, FF$_2$, and FF$_3$) and one latch (L$_1$) with half-rate clock (20GHz), giving rise to three consecutive bits (D_{n-1}, D_n, and D_{n+1}). Note that the operation here is very similar to an Alexander phase detection [4] but with half-rate clock (since only data eyes need to be sampled). Meanwhile, the reference swing generator and the adaptation logic detect the average data swing V_{SW}. For one data bit to be optimally equalized, the eye centers of D_{n-1}, D_n, and D_{n+1} should present a magnitude very close to V_{SW}, other-

wise the data is either under or over compensated. Based on this observation, the 3 coefficients α_{-1}, α_0, α_1 can be adjusted accordingly. Here two slicers determine the boosting condition for logic ZERO and ONE. Due to half-rate operation, a 2-to-1 selector picks the corresponding results (E_n) depending on the present bit. After phase alignment, the 4 outputs (D_{n-1}, D_n, D_{n+1}, and E_n) are XORed, 1:8 DEMUXed, and finally sent to the adaptation logic, which is implemented purely in digital domain to save power. Since the overall tail current of the FIR combiner is a constant, the data common-mode level (V_{CM}) must be provided so as to create V_{SW}. Two resistors R_b in Figure 8.3.2(b) (in gray) accomplish this requirement. The averaging length of producing V_{SW} is much longer than the adaptation bandwidth, making the comparing reference very stable.

The adaptation logic is shown in Fig. 8.3.4(a). Since the least-mean-square (LMS) algorithm for eye opening has been realized in the XORs, it takes the majority voting result from the 8×1.25b/s inputs to determine the polarity. A 12-bit counter serves as a digital loop filter, in which a 3-bit bandwidth control is included to adjust the convergence speed. The 6-bit MSBs are fed into the iDACs to change the FIR coefficients. As a result, α_{-1}, α_0, and α_1 are tuned in a way that the data ISI is minimized, while 3 iDACs provide current-mode outputs with constant total amount. The voting result for $E_n \oplus D_n$ is to produce present swing information for reference swing generator to create V_{SW} [Fig. 8.3.4(a)]. Here, M_1-M_2 pair is fully tilted, and the third iDAC provides the tail current directly to the reference swing generator. The opamp plus M_3-M_6 loop makes the common-mode level of V_{SW} exactly equal to V_{CM} of the FIR filter. The phase detector in the 1UI delay generator is illustrated in Fig. 8.3.4(b). To lock two 10GHz clocks in quadrature (90° phase difference), we utilize SSB mixer to distill the phase error. That is, for a phase difference θ between two inputs (with magnitudes A_1 and A_2), the output is given by $k_1A_1A_2\cos\theta$, where k_1 denotes the mixer gain. The sinusoidal characteristic forces the phase to lock at $\pi/2$ with an approximately liner behavior in the vicinity. Note that non-idealities (such as second-order harmonic caused by mismatch) are easily suppressed by the RC network on the output port.

The transceiver is fabricated in a 65nm CMOS technology. The transmitter consumes 135mW from a 1.2V supply, of which 29mW dissipates in 20GHz PLL (TX side), 77mW in 5-taps FIR, 29mW in 1UI delay generator. The receiver consumes 322mW from a 1.6V supply. To evaluate the performance, the TX and RX are mounted on a Rogers board (RO4003) with different channel lengths. The 20GHz PLL presents an operation range of 20GHz (19.4-21GHz), suggesting the TX output data rate can be tuned from 38.8-42Gb/s. The transmitter's outputs with 0dB and 9.5dB pre-emphasis are illustrated in Fig. 8.3.5(a) and (b). Shown in Fig. 8.3.5(c) is the spectrum of the 20GHz PLL in TX under locked condition. The phase noise measures –91.57dBc/Hz at 1MHz offset, and the integrated jitter (from 10Hz to 6.5GHz) is given by 0.5766ps,rms. For the RX, the equalized data (40Gb/s) for different channel lengths (5cm and 20cm) are depicted in Fig. 8.3.6(a). Figure 8.3.6(b) illustrates the demuxed data (10Gb/s) for a 10cm channel. The BER testing has been conducted as well. For 40Gb/s PRBS of 2^{31}–1, the transceiver achieves BER of less than 10^{-12} until channel length =17.5cm, which has 18dB loss at 20GHz. For 20cm channel (19dB loss at 20GHz), BER begins to increase as the data pattern becomes longer [Fig. 8.3.6(c)]. Figure 8.3.7 shows the die photographs, which occupy 0.9×0.7mm² for TX and 1.1×0.6mm² for RX, respectively. A table summarizing the performance of this work is also included in Fig. 8.3.7.

Acknowledgment:
The authors thank the TSMC University Shuttle Program for chip fabrication.

References:
[1] H. Wu, J. Tierno, P. Pepeljugoski, et al., "Differential 4-tap and 7-tap Transverse Filters in SiGe for 10Gb/s Multimode Fiber Optic Link Equalization," *ISSCC Dig. Tech. Papers*, pp. 180-181, Feb., 2003.
[2] A. Momtaz and M. M. Green, "An 80 mW 40 Gb/s 7-Tap T/2-Spaced Feed-Forward Equalizer in 65 nm CMOS," *IEEE J. Solid-State Circuits*, vol. 45, no. 3, pp. 629-639, Mar., 2010.
[3] H. Sugita, K. Sunaga, K. Yamaguchi, and M. Misuno, "A 16Gb/s 1st-Tap FFE and 3-Tap DFE in 90nm CMOS," *ISSCC Dig. Tech. Papers*, pp. 162-163, Feb., 2010.
[4] J.D.H. Alexander, "Clock Recovery from Random Binary Data," *Electronics Letters*, vol. 11, no. 22, pp. 541-542, Oct., 1975.

978-1-61284-303-2/11 $26.00 © 2011 IEEE

ISSCC 2011 / February 22, 2011 / 9:30 AM

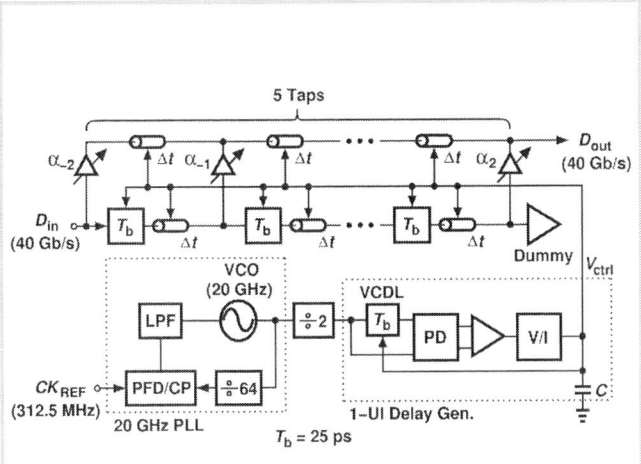

Figure 8.3.1: Transmitter architecture.

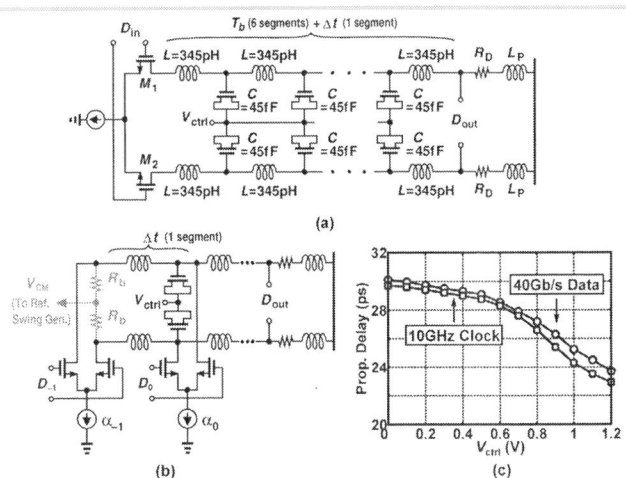

Figure 8.3.2: (a) Voltage-controlled delay line (T_b), (b) Simplified FIR combiner and delay (gray part for RX FIR only), (c) Propagation delay for 40Gb/s data and 10GHz clock.

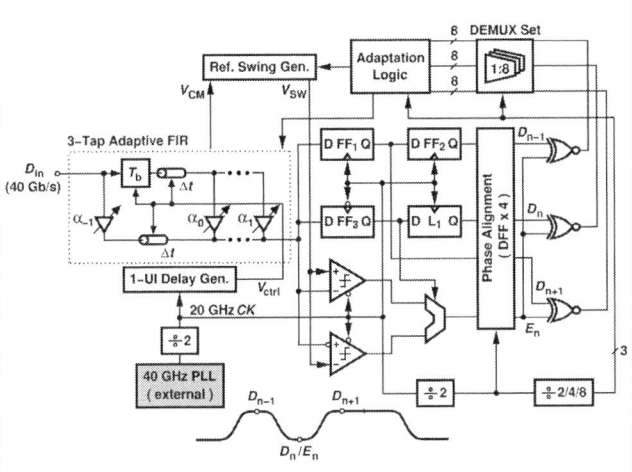

Figure 8.3.3: Receiver with adaptive FIR equalizer.

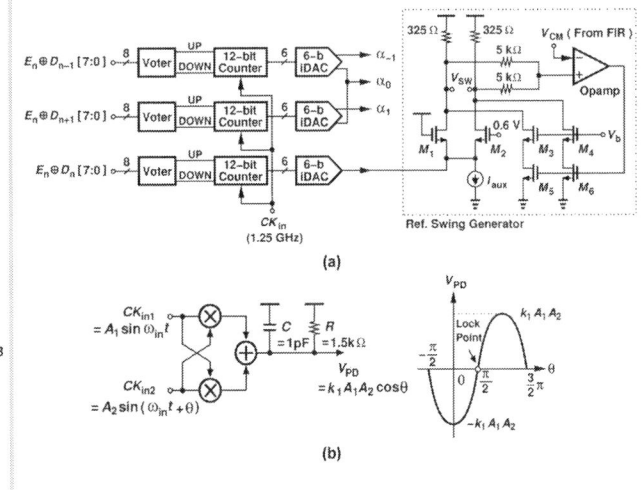

Figure 8.3.4: (a) FIR adaptation circuits, (b) Mixer-based PD.

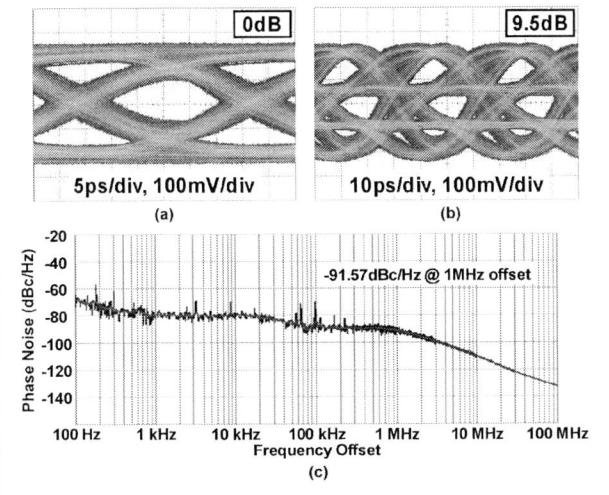

Figure 8.3.5: With (a) 0dB, (b) 9.5dB pre-emphasis, (c) Phase-noise plot of 20GHz PLL.

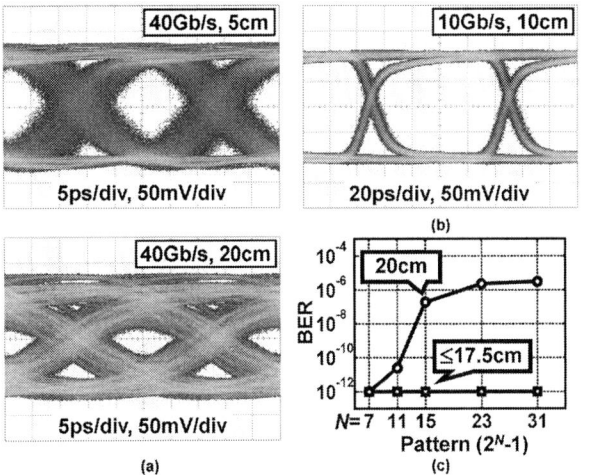

Figure 8.3.6: (a) Equalized 40Gb/s data in RX for 5cm and 20cm channel, (b) Demuxed 10Gb/s data, (c) BER for different channel lengths.

978-1-61284-303-2/11 $26.00 © 2011 IEEE

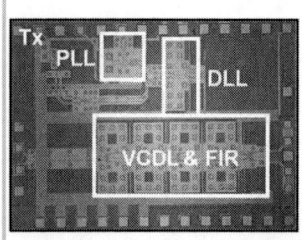
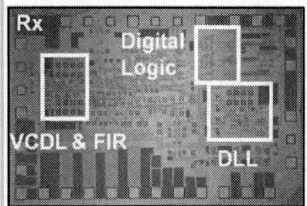

Data Rate	40Gb/s (38.8~42Gb/s)	
Architecture	5-Tap FIR (Tx) 3-Tap FIR (Rx)	
BER	$< 10^{-12}$, $2^{31}-1$ PRBS for ≤ 17.5cm channel (18-dB Loss @ 20GHz)	
	$< 10^{-12}$, $2^{7}-1$ PRBS for 20cm channel (19-dB Loss @ 20GHz)	
Output Data Jitter (10Gb/s)	10cm channel:	1.72ps,rms 11.56ps,pp
	20cm channel:	3.88ps,rms 20.44ps,pp
Power	Tx: 135mW, Rx: 322mW	
Supply	Tx: 1.2V, Rx: 1.6V*	
Chip Area	Tx: 0.9 x 0.7mm^2 Rx: 1.1 x 0.6mm^2	
Technology	65nm CMOS	

* 1.2V used in Digital Logic

Figure 8.3.7: Chip micrograph and performance summary.

ISSCC 2011 / SESSION 8 / ARCHITECTURES & CIRCUITS FOR NEXT GENERATION WIRELINE TRANSCEIVERS / 8.4

8.4 10:4 MUX and 4:10 DEMUX Gearbox LSI for 100-Gigabit Ethernet Link

Goichi Ono[1], Keiki Watanabe[1], Takashi Muto[1], Hiroki Yamashita[1], Koji Fukuda[1], Noboru Masuda[1], Ryo Nemoto[2], Eiichi Suzuki[1], Takashi Takemoto[1], Fumio Yuki[1], Masayoshi Yagyu[1], Hidehiro Toyoda[1], Akihiro Kambe[1], Tatsuya Saito[1], Shinji Nishimura[1]

[1]Hitachi, Tokyo, Japan
[2]Hitachi, Ibaraki, Japan

The 100-gigabit Ethernet (100GbE) was standardized as IEEE 802.3ba in 2010 [1]. The optics module must be equipped with a "gearbox" LSI—which switches between 10×10Gb/s data signals on the physical-coding-sublayer side and 4×25Gb/s data signals on the physical-media-dependent side. A gearbox LSI based on 0.13 μm SiGe BiCMOS consumes 8W of power [2], which is about half of the total power consumption of the optics module. Aiming to reduce the power consumption, a 50Gb/s 2:5 DEMUX based on CMOS technology is developed [3]. In addition, since the architecture in reference [2] consists of two LSIs (MUX and DEMUX), loop-back operation, which is required in the IEEE standard, is impossible. In response to these circumstances, we have developed a 100GbE gearbox LSI combining a 10:4 MUX and a 4:10 DEMUX. This gearbox LSI—implemented in 65nm CMOS—decreases power dissipation by 75% compared to that of a conventional LSI.

Figure 8.4.1 shows a block diagram of the 100GbE gearbox LSI—which consists of three blocks: a 10×10Gb/s interface, a 4×25Gb/s interface, and a logic area. As for the architecture of the gearbox LSI, the logic-circuit area contains the functions for 10:4 MUX, 4:10 DEMUX, loopback operation of the 10Gb/s and 25Gb/s interfaces. The 10Gb/s and 25Gb/s interfaces have both a transmitter (TX) and a receiver (RX), and it converts 10Gb/s and 25Gb/s serial data to 625Mb/s parallel data. The gearbox LSI can therefore operate in three data-transfer modes: upstream (10×10Gb/s to 4×25Gb/s), downstream (4×25Gb/s to 10×10Gb/s), and loopback operation.

Low-power technologies for the 10Gb/s and 25Gb/s interfaces are also developed. The developed 10Gb/s interface is based on our previous work [4]. Although CMOS circuit technology cannot provide operation speed of over 20Gb/s, it has an advantages compared with SiGe technology in terms of low power dissipation. Therefore, by decreasing the clock frequency and the effective circuit-operation speed to available CMOS operation speed, the gearbox LSI can replace current-mode logic (CML) circuits to the CMOS circuits. This is the first key technique for decreasing power dissipation. Figures 8.4.2 and 8.4.3, respectively, show a block diagram of the TX and the RX blocks. The TX consists of a 40:2 MUX, a 2:1 MUX, a data phase control, a 3-tap-feed-forward equalizer (FFE), and a driver. The RX is composed of a receiver with a linear equalizer and a 1-tap decision-feedback equalizer (DFE), a PLL with phase rotation, a phase detector (PD), a 4:40 DEMUX, and a CDR logic circuit. Half-rate operation of the TX and quarter-rate operation of the RX that lower circuit operation speed and clock frequency are used to apply a low-power CMOS circuit to under 12.5Gb/s operation circuit block. In particular, a dynamic-operation-type CMOS circuit (with lower power consumption than a static CMOS circuit) is used for 40.2 MUX and 4:40 DEMUX.

The second key technique for low power is reducing the frequency of clock distribution. In the architecture the RX, a PLL that generates the RX operating clock of 4-phase 6.25GHz from a 625MHz reference clock is placed in each of the four channels. As a result, a low-frequency clock of 625MHz can be distributed instead of a 6.25GHz clock, thereby reducing the power consumption of the clock buffers. Since a RX PLL is placed in the RX of each channel, the PLL with small area is required. However, as the phase noise of the voltage-controlled oscillator (VCO) in the PLL must be reduced to satisfy transmission performance, a large VCO (such as an LC-based VCO) is required. By using two control methods for frequency of the VCO, namely analog control through a charge pump and a low-pass filter (loop bandwidth of 1MHz) and digital control directly connected to the VCO (loop bandwidth: 100 MHz), the loop bandwidth is increased to 100MHz. These proposed methods reduce the influence of VCO phase noise on the RX PLL, so small size PLL with a ring-based VCO can be used.

Other techniques for lowering power consumption in the TX are to separate the driver and the FFE, 2:1 MUX with 1-V operation, and data-phase control between 40:2 and 2:1 MUX. The structure of a conventional driver combines the functions for generating waveforms for equalization (by the FFE) and driving the load connected output terminal (by the driver). The output impedance of the driver is set to 50Ω to achieve impedance matching with the transmission line. Although generating waveforms for the FFE does not require 50Ω impedance, the FFE with a combined structure consumes unnecessary power. Moreover, large transistors that increase output-terminal load and decrease bandwidth are used for the FFE owing to 50Ω impedance. Accordingly, a driver with a separated FFE and driving function was designed (Fig. 8.4.2). Since the driver enables the FFE with over 50Ω output impedance, it achieves both low power and high bandwidth.

A 2:1 MUX generally is configured with three transistors between power supply and ground. However, at 1V operation, sufficient source-drain voltage is not supplied to each transistor. An NMOS transistor for clock switching is thus inserted between the power supply and the CML common node.

For data transmission from the 40:2 MUX to 2:1 MUX, data-phase control is used. The last stage of the 40:2 MUX outputs two 12.5-Gb/s data signals by using two-phase 6.25GHz clocks with phase difference of 90°. On the other hand, in the 2:1 MUX, phase adjustment between the two input data signals from the 40:2 MUX and the 12.5GHz clock distributed by the TX PLL is required. Consequently, in the data-phase control block, the same MUX circuit as that of the last stage in the 40:2 MUX is used as a replica circuit, the phases of its output and the 12.5GHz clock are compared, and two interpolators that generate 6.25GHz 40:2 MUX operating clocks are adjusted.

Figures 8.4.4 and 8.4.5 show measured TX and RX performance of the 25Gb/s interface. The phase noise of the TX clock generated by 12.89GHz PLL with LC-VCO is measured. Integrated phase noise is 429fs at 12.89GHz. The 25.78Gb/s-operation eye diagram at the LSI output terminal (insertion loss of −2dB at 12.5GHz) with a $2^{31}-1$ PRBS pattern is confirmed. The TX achieves a low jitter of $3.3ps_{pp}$. It confirms that the linear equalizer of the RX can achieve about 10dB equalization (Fig. 8.4.5). Bit error rate versus input sensitivity during loopback operation is also measured. The minimum input power under error-free operation is −25.3dBm ($34.3mV_{pp}$), which represents sufficient input sensitivity for 100GbE applications.

Figure 8.4.6 summarizes the features of the gearbox LSI and power dissipation of the 25Gb/s interface. Total power dissipation of the gearbox LSI is 1.992W. The power consumption of the developed gearbox LSI is 75% lower than that of a conventional SiGe-based gearbox LSI. The power dissipations of the 10Gb/s and 25Gb/s interfaces and the logic area are 0.3, 1.398, and 0.294W, respectively. The power dissipation per channel of the 25Gb/s interface is 187 (TX), 105 (RX), 57.5 (TX clock distribution), and 349.5mW (14mW/Gb/s) in total. The power consumption of the presented 25Gb/s interface is decreased by more than 35% as compared to that of other 25Gb/s-class interfaces [5, 6]. Figure 8.4.7 shows a chip micrograph of the gearbox LSI.

Acknowledgments:
This work is partially supported by "Universal Link" project of the National Institute of Information and Communications Technology (NICT), which is administrated by the Ministry of Internal Affairs and Communications (MIC), Japan. The authors would like to thank S. Umai and M. Kuwata for evaluating the test chip. We would also like to thank T. Kiyuna and M. Kakimi for designing the test chip.

References:
[1] IEEE P802.3ba 40 Gb/s and 100 Gb/s Ethernet Task Force; http://grouper.ieee.org/groups/802/3/ba/index.html
[2] 100 Gb/s MUX; http://www.semtech.com/images/datasheet/smi10021.pdf 100 Gb/s DEMUX; http://www.semtech.com/images/datasheet/smi10031.pdf
[3] K.-C. Wu and J. Lee, "A 2×25Gb/s Deserializer with 2□5 DMUX for 100Gb/s Ethernet Applications," *ISSCC Dig. Tech. Papers*, pp. 374-375, Feb., 2010.
[4] K. Fukuda, H. Yamashita, G. Ono, et al., "A 12.3mW 12.5Gb/s Complete Transceiver in 65nm CMOS," *ISSCC Dig. Tech. Papers*, pp. 368-369, Feb., 2010.
[5] A. Amirkhany, A. Abbasfar, J. Savoj, et al., *IEEE Symp. VLSI Circuits*, pp. 38-39, Jun., 2007.
[6] N. Nedovic, S. Parikh, A. Kristensson, et al., "A 2×22Gb/s SFI5.2 CDR/Deserializer in 65nm CMOS Technology," *IEEE Symp. VLSI Circuits*, pp. 10-11, Jun., 2009.

978-1-61284-303-2/11 $26.00 © 2011 IEEE

ISSCC 2011 / February 22, 2011 / 10:15 AM

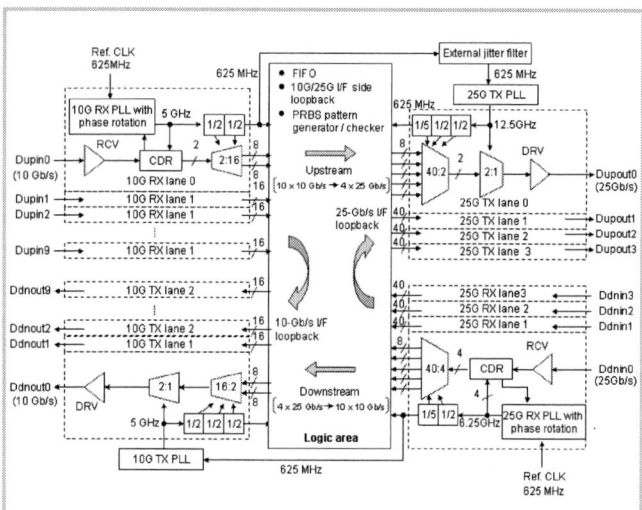

Figure 8.4.1: 100GbE gearbox LSI block diagram.

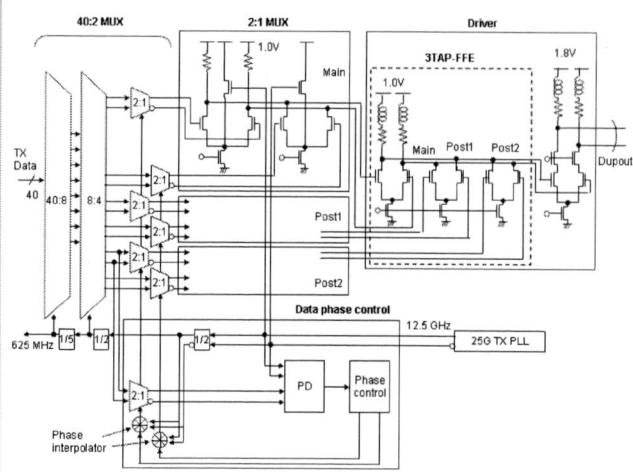

Figure 8.4.2: Block diagram of the presented 25Gb/s transmitter.

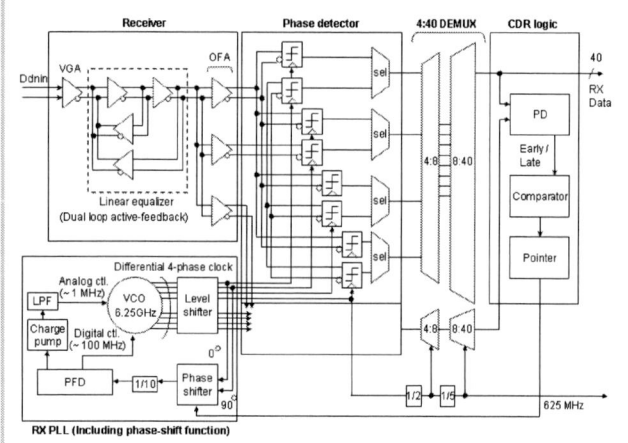

Figure 8.4.3: Block diagram of the presented 25Gb/s receiver.

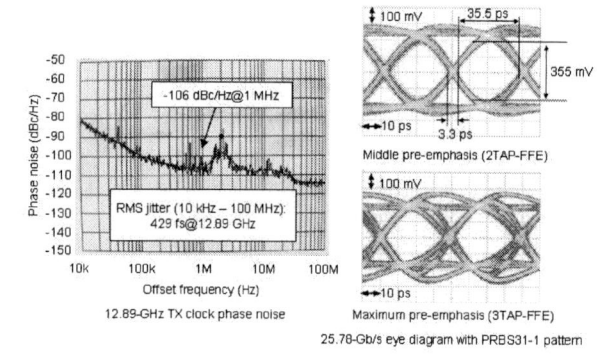

Figure 8.4.4: Measured transmitter performance.

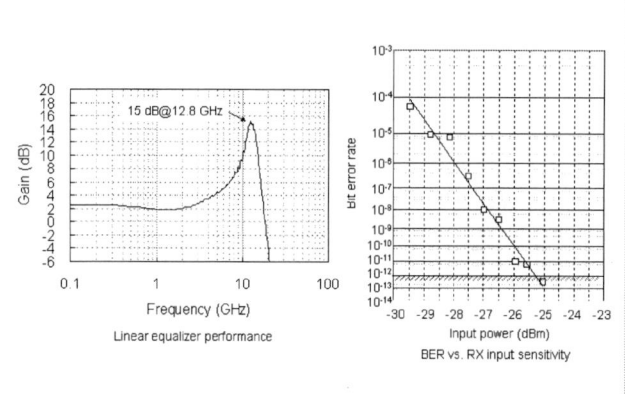

Figure 8.4.5: Measured receiver performance.

	This work	Ref. [2]
Year	ISSCC 2011	Press release 2009
Chip structure	100-Gb/s Gearbox	100-Gb/s 10:4 MUX 100-Gb/s 4:10 DEMUX
Data rate	10×10G → 4×25G 4×25G → 10×10G MUX/DEMUX combined chip	10×10G → 4×25G 4×25G → 10×10G MUX/DEMUX separated chip
Loopback	10×10G ↔ 10×10G 4×25G ↔ 4×25G	-
Technology	65-nm CMOS	130-nm SiGe BiCMOS
TX clock jitter	0.429 ps-rms@12.89 GHz (10 kHz – 100 MHz)	-
Input sensitivity	34.4 mVp-p	60 mVp-p
Power dissipation	Total 1.992 W 25-Gb/s interface 1.40 W 10-Gb/s interface 0.30 W Logic area 0.29 W	8 W (MUX + DEMUX)
Chip area	6.3×3.7 mm	-
Power supply	1.0/1.8 V	1.2/1.5/2.8 V

Measured 25-Gb/s interface power dissipation (mW)	
TX	187
RX	105
TX clock distribution (/ch)	57.5
Total (/ch)	349.5

Figure 8.4.6: Chip summary and comparison.

8

978-1-61284-303-2/11 $26.00 © 2011 IEEE   149

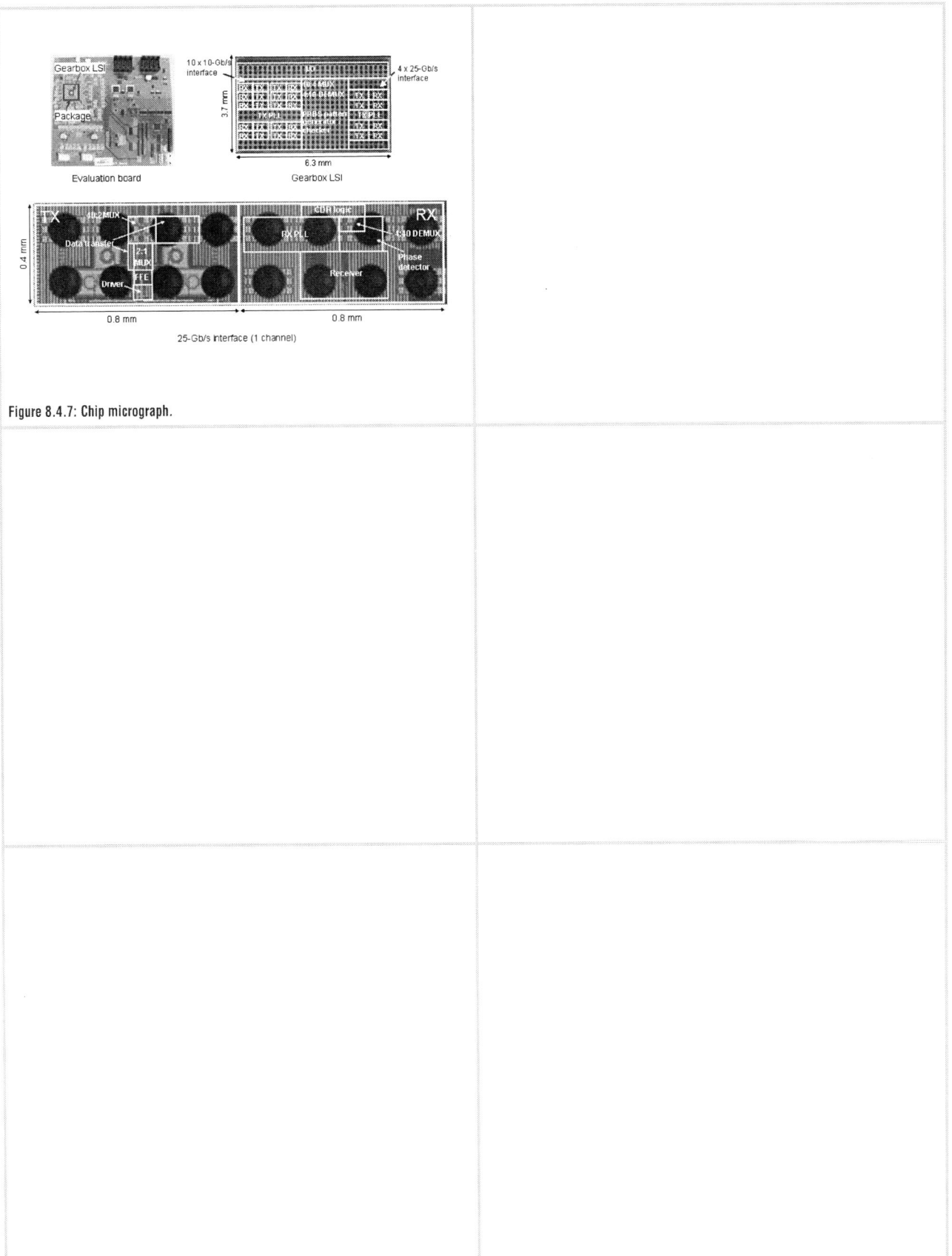

Figure 8.4.7: Chip micrograph.

ISSCC 2011 / SESSION 8 / ARCHITECTURES & CIRCUITS FOR NEXT GENERATION WIRELINE TRANSCEIVERS / 8.5

8.5 A 12.5+12.5Gb/s Full-Duplex Plastic Waveguide Interconnect

Satoshi Fukuda[1], Yasufumi Hino[1], Sho Ohashi[1], Takahiro Takeda[1],
Satoru Shinke[1], Masahiro Uno[1], Kenji Komori[1], Yoshiyuki Akiyama[1],
Kenichi Kawasaki[1], Ali Hajimiri[2]

[1]Sony, Tokyo, Japan
[2]California Institute of Technology, Pasadena, CA

The growing demand for high-speed, low-cost, and low-overhead I/Os in today's electronic systems, has been addressed by three general categories of interconnects: electrical, optical, and wireless. The electrical interconnects are the oldest and have improved the most, where bit rates in excess of 20Gb/s are achieved over a pair of conductors [1]. At such high bit rates, these serial links must handle transmission line loss, dispersion, impedance mismatches, and electromagnetic crosstalk among multiple lines requiring sophisticated designs, often needing equalization, with their own cost and overhead limitations [2]. Optical fibers as interconnects do not suffer from similar bandwidth limitations or cross-talk issues. However, they require additional electrical-to-optical (EO) and optical-to-electrical (OE) conversion devices for generation and detection of optical signals [3], which impose serious constraints on power consumption, cost, and footprint of optical interconnects. Wireless connection at millimeter-wave frequencies can also be used for short distance connections [4]. While they provide the most versatility and are a promising option, additional development is still necessary to scale them to a highly parallel system with multiple channels running concurrently.

This paper presents a 12.5+12.5Gb/s full-duplex plastic waveguide interconnect solution based on millimeter-wave signal transmission. The plastic waveguide is simply a long solid piece of plastic that provides a very simple, versatile, flexible, and low-cost transmission medium that has the main advantages of optical fiber in isolation and bandwidth, without the need for costly EO and OE. The dielectric waveguide does not need to be connected electrically like the wire or aligned to micron-level accuracy like optical fibers. It can be bent and twisted without significant impact on the signal. Compared to the wireless link discussed earlier, it offers additional signal isolation and confinement. Thus, it can be extended over much longer distances due to the low attenuation in the waveguide (as opposed to free space) and multiple independent lines can be run in parallel to increase the bandwidth.

In our proposed plastic waveguide link, the TXs and RXs are fully integrated in CMOS, and the waveguide couplers can be fabricated in a conventional resin package without additional cost. In our existing setting there are a transmitter and a receiver operating at different carrier frequencies on each side of the waveguide, making it possible to realize a full-duplex solution. Because of the smaller fractional bandwidth for the millimeter-wave transmission, no equalization circuit is required.

Figure 8.5.1 presents a diagram of our proposed solution. It consists of a pair of transceivers A and B, and a plastic waveguide. Transceiver A contains a 57GHz RX and an 80GHz TX, and Transceiver B contains an 80GHz RX and a 57GHz TX. This combination allows for a bi-directional full-duplex transmission. An alternative is to place both TX on one side and both RX on the opposite side, doubling the one-directional simplex data rates.

In Transceiver A, a differential baseband signal is fed to baseband inputs, upconverted with an on-chip 80GHz TX LO and an 80GHz upconversion mixer. The output is amplified with an 80GHz TX amplifier and coupled to the plastic waveguide at the 80GHz port. The signal travels over the waveguide, and is coupled to an 80GHz LNA in Transceiver B. The amplified output from the 80GHz LNA is delivered to an 80GHz downconversion mixer and to an 80GHz RX LO via the injection path circuit, where the signal is coherently downconverted to the baseband. Similarly, in Transceiver B, a baseband signal is upconverted to 57GHz, and coupled to the waveguide at the 57GHz port. The propagated 57GHz signal is downconverted to the baseband in Transceiver A.

The isolation between 57GHz and 80GHz channels is a challenge in implementing this 12.5+12.5Gb/s frequency division full-duplex system. To reduce the interference due to the leaked signal from the 80GHz TX to the 57GHz RX, a trap circuit is implemented in the 57GHz LNA. The trap is realized utilizing a series resonance comprising inductor L_1 and capacitor C_1 as shown in Fig. 8.5.2. In simulation, the trap reduces the gain at 80GHz by 15dB. In the 80GHz LNA, the top and the second-top metal layers in the inductors L_2, L_3, L_4 and L_5 are connected together with vias to reduce the series resistances. The simulated Q of the inductor is improved by 50% at 80GHz compared to the one with the top metal layer only. Even though the fractional bandwidth is reduced with higher inductor Q, enough bandwidth is still maintained at 80GHz, and comparable gain to the 57GHz LNA has been achieved.

The 57GHz TX has a saturated output power (P_{sat}) of 0dBm at 57GHz, a 3dB bandwidth of 8GHz, and a power consumption of 29mW at a supply voltage of 1.1V. The 57GHz RX achieves a peak gain of 31dB and a 3dB bandwidth of 7.5GHz with a power consumption of 42mW. The 80GHz TX has a P_{sat} of 0dBm at 80GHz, a 3dB bandwidth of 9.5GHz, and a power consumption of 27mW. The 80GHz RX achieves a peak gain of 21dB and a 3dB bandwidth of 10.5GHz with a power consumption of 45mW. The measurement results are summarized in Fig. 8.5.3. By improving the coupling efficiency, the output power can be reduced. Therefore further reduction in power consumption is possible.

To evaluate the performance of the system, each transceiver chip is mounted on a printed circuit board with a chip-to-waveguide coupler. A pair of quasi-Yagi antennas for 57GHz and for 80GHz is utilized as couplers, as shown in Fig. 8.5.4. The measured coupling efficiency to waveguide is −10dB at the 80GHz port for a 5cm-long waveguide. The measured isolation between the 57GHz and 80GHz ports is more than 20dB from 40GHz to 110GHz.

Baseband-to-baseband (BB-to-BB) transfer functions are measured with a spectrum analyzer and a signal generator, as shown in Fig. 8.5.4. Many different materials can be used to make the waveguide. As an example, a 120 mm polystyrene waveguide with a thickness of 1.1mm and a width of 8mm is employed. The exact thickness and width of the waveguide are not critical in the system, and reliable signal transmission with waveguide of different sizes and materials has been demonstrated. This plastic waveguide solution does not require a common DC grounding voltage between TX and RX for the operation and is completely isolated electrically.

Figure 8.5.5 shows the demonstrator setup and measured eye diagrams. The fabricated demonstrator achieves full-duplex transmission of 12.5Gb/s ASK modulated signal in each direction over the 120mm polystyrene waveguide. The observed BERs for both 57GHz and 80GHz channels are less than 10^{-12} for 2^7-1 PRBS. As another example, uncompressed video at 4.3Gb/s has been transmitted over a 1000mm plastic waveguide.

With advancement of CMOS process, it is possible to shift the operating frequency higher, transmitting more channels on the same waveguide. The results are summarized in Fig. 8.5.6. The demonstrated chips are fabricated in 40nm low-power logic CMOS. The chips have active area of 0.4mm^2 and their micrographs are shown in Fig. 8.5.7.

References:
[1] K. Sunaga, H. Sugita, K. Yamaguchi, and K. Suzuki , "An 18Gb/s Duobinary Receiver with a CDR-Assisted DFE," *ISSCC Dig. Tech. Papers*, pp. 274-275, Feb., 2009.
[2] S. Ibrahim and B. Razavi, "A 20gb/s 40mW Equalizer in 90nm CMOS Technology," *ISSCC Dig. Tech. Papers*, pp. 170-171, Feb., 2010.
[3] C. Schow, F. Doany, C. Chen, et al., "A <5mW/Gb/s/link, 16×10Gb/s Bi-Directional Single-Chip CMOS Optical Transceiver for Board-Level Optical Interconnects," *ISSCC Dig. Tech. Papers*, pp. 294-295, Feb., 2008.
[4] K. Kawasaki, Y. Akiyama, K. Komori, et al., "A Millimeter-Wave Intra-Connect Solution," *ISSCC Dig. Tech. Papers*, pp. 414-415, Feb., 2010.

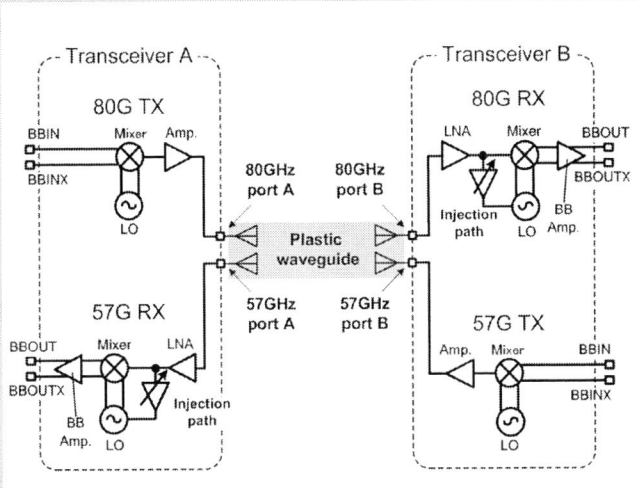

Figure 8.5.1: Block diagram of the presented full-duplex plastic waveguide interconnect.

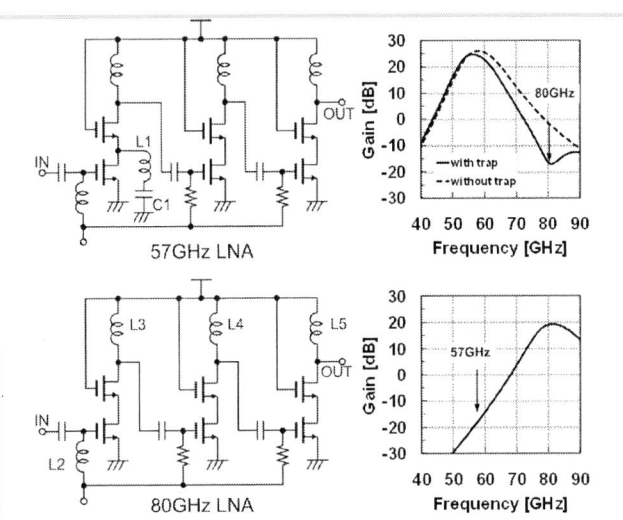

Figure 8.5.2: Schematics of the 57GHz and 80GHz LNA and their simulated gains.

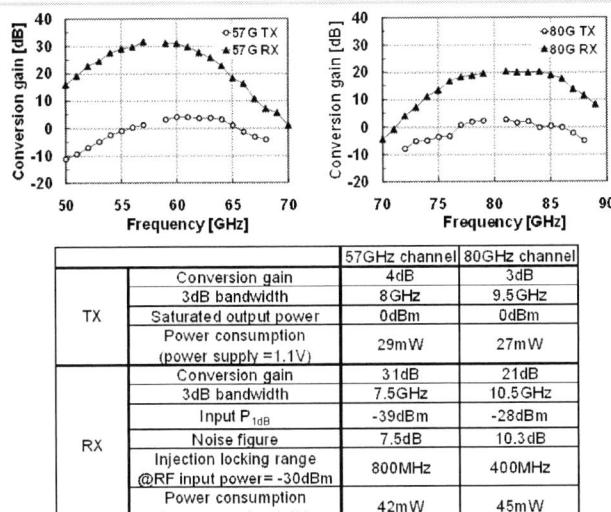

		57GHz channel	80GHz channel
TX	Conversion gain	4dB	3dB
	3dB bandwidth	8GHz	9.5GHz
	Saturated output power	0dBm	0dBm
	Power consumption (power supply =1.1V)	29mW	27mW
RX	Conversion gain	31dB	21dB
	3dB bandwidth	7.5GHz	10.5GHz
	Input P_{1dB}	-39dBm	-28dBm
	Noise figure	7.5dB	10.3dB
	Injection locking range @RF input power= -30dBm	800MHz	400MHz
	Power consumption (power supply =1.1V)	42mW	45mW

Figure 8.5.3: Measured performance of TXs and RXs.

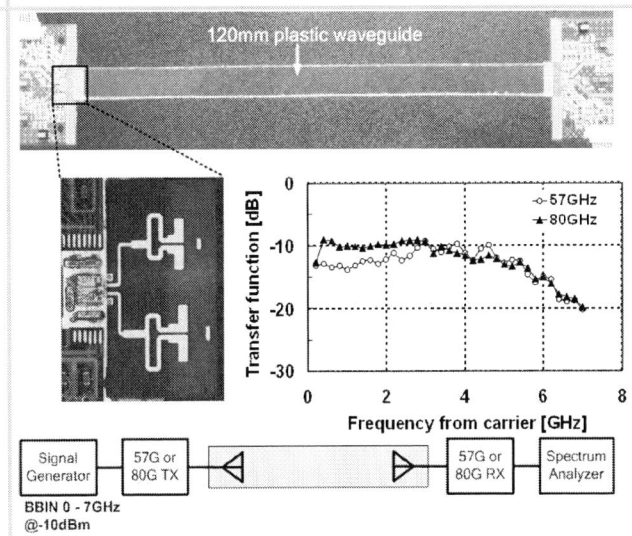

Figure 8.5.4: BB-to-BB transfer function and the measurement setup.

at 57GHz RXout
12.5Gb/s BER < 10^-12

at 80GHz RXout
12.5Gb/s BER < 10^-12

Figure 8.5.5: Demonstrator setup and the eye diagrams at RX outputs.

	[2]	[3]	[4]	This work
Category	Wireline	Optical	Wireless	mm-Waveguide
Duplex	One way	Full-duplex (multiple one way lanes)	One way	Full-duplex
Transmission lane	Trace on FR4 board	Polymer waveguide	Free-space	Plastic waveguide
Data rate per transmission lane [Gb/s/lane]	20	10	11	25
Transmission distance [mm]	450	1000	14	120
Additional conversion device	Not required	Required	Not required	Not required
Equalization	Required	Not required	Not required	Not required
Micron-level alignment accuracy	Not required	Required	Not required	Not required

Figure 8.5.6: Comparison with other work.

ISSCC 2011 PAPER CONTINUATIONS

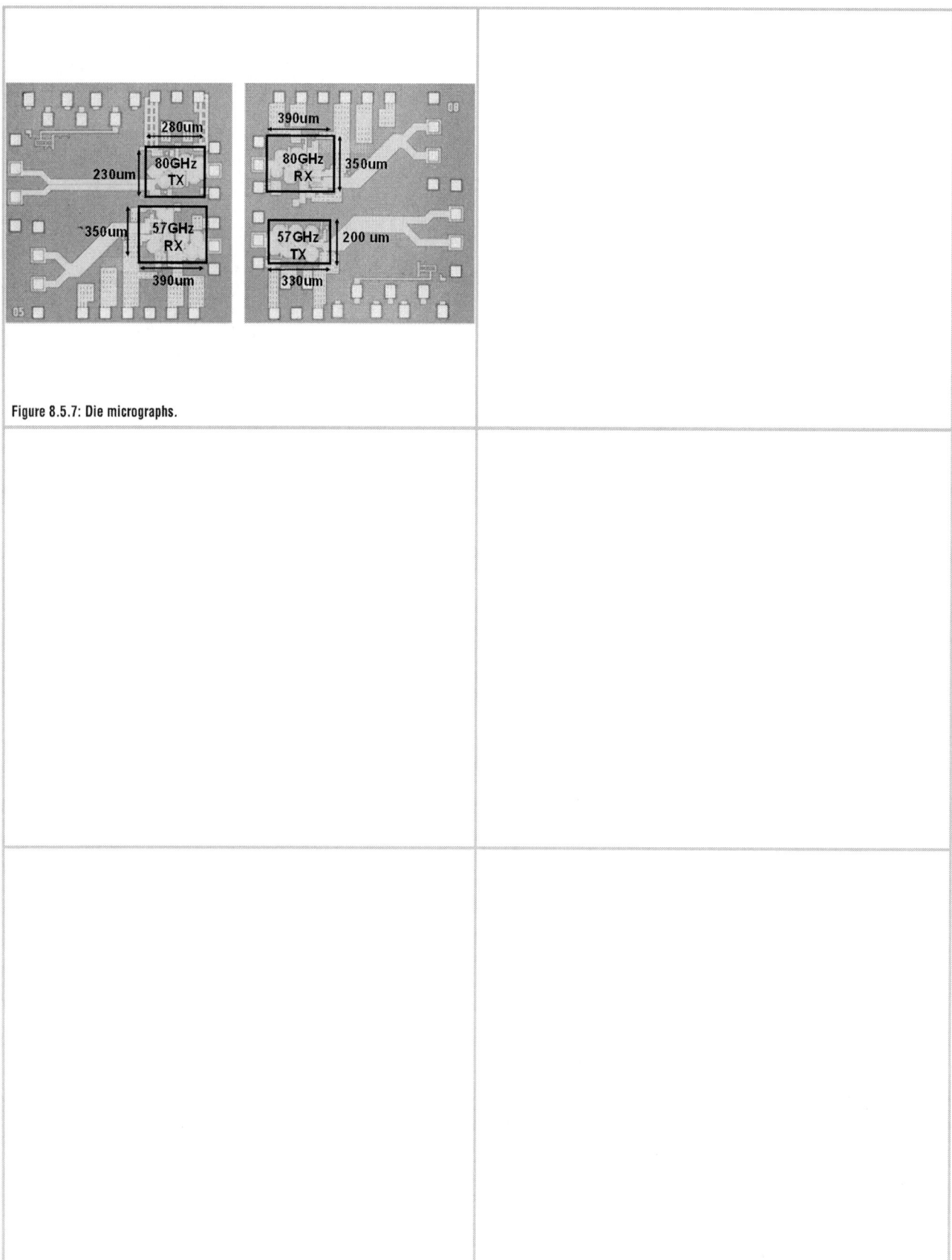

Figure 8.5.7: Die micrographs.

ISSCC 2011 / SESSION 8 / ARCHITECTURES & CIRCUITS FOR NEXT GENERATION WIRELINE TRANSCEIVERS / 8.6

8.6 A Highly Digital 0.5-to-4Gb/s 1.9mW/Gb/s Serial-Link Transceiver Using Current-Recycling in 90nm CMOS

Rajesh Inti[1], Amr Elshazly[1], Brian Young[1], Wenjing Yin[1], Marcel Kossel[2], Thomas Toifl[2], Pavan Kumar Hanumolu[1]

[1]Oregon State University, Corvallis, OR
[2]IBM Zurich Research Laboratory, Zurich, Switzerland

Ever-growing demand for higher communication bandwidth in high performance compute systems is driving the need for energy-efficient multi-Gb/s I/O serial links. Improved power efficiency was demonstrated using adaptive supply regulation [1, 2]. However, power losses in the DC-DC converter needed to generate the optimal supply voltage and the difficulty in operating analog circuits at low voltages limit the power savings. Instead of scaling the supply with the data rate, we seek to operate with two fixed voltages and eliminate the need for a high-efficiency DC-DC converter. To this end, this paper presents a serial link using a highly efficient current recycling-based *implicit DC-DC conversion* to generate 0.6V from a 1.2V supply. Highly digital clocking circuits capable of operating at 0.6V maximize power savings. A 0.5-to-4Gb/s serial-link transceiver is designed in a 1.2V LP 90nm CMOS process to operate with a short channel and plesiochronous timing. The transceiver dissipates 1.9mW/Gb/s at 3.2Gb/s.

Figure 8.6.1 shows the serial link transceiver. The transmit PLL (TxPLL) operates from an internally generated 0.6V supply and generates half-rate quadrature clocks. The clock phases are level shifted to 0 to 1.2V and are used to multiplex the pattern generator outputs to full rate. A voltage-mode driver launches data at 100-to-200mV swing onto the channel. On the receiver side, a continuous-time linear equalizer (CTLE) compensates for moderate channel loss and feeds the data samplers. A digital bang-bang CDR recovers the half-rate clock and data.

Internal mid-supply voltage, V_{MID}, is generated using *implicit DC-DC conversion*. Ideally, by stacking TxPLL and the receive-side frequency-locked loop (RxFLL) as shown in Fig. 8.6.1, all of the current from the TxPLL is recycled in the RxFLL, thereby self regulating V_{MID} to half the supply voltage, $V_{DD}/2$ [3]. However, in practice, because of inevitable mismatches in the supply currents of the TxPLL and the RxFLL, V_{MID} strays away from $V_{DD}/2$. A simple push-pull regulator provides the mismatch current and forces V_{MID} to $V_{DD}/2$.

A performance limitation of this *implicit DC-DC conversion* may arise from the ripple on the V_{MID} node, which could cause additional timing jitter. Because the link performance is more sensitive to Tx clock jitter, the TxPLL is connected in the bottom stack, and all of its sensitive voltages are referred to ground to minimize the impact of V_{MID} ripple voltage. Additional suppression of the high frequency ripple voltage is provided by the TxPLL feedback. As a result of connecting the RxFLL in the top stack, its NMOS transistors suffer from the body effect. Instead of a making these transistors in a separate PWELL at the cost of an additional mask, they are enlarged to mitigate the effects of increased threshold voltage.

The 0.25GHz-to-2GHz TxPLL capable of operating with a 0.6V supply is implemented using a highly digital architecture shown in Fig. 8.6.1. A conventional PFD drives the digitally controlled oscillator (DCO) directly and implements the proportional control. An all-digital integral path driven by the sign of the phase error introduces the pole at DC needed to realize the Type-2 behavior. Due to elimination of a charge-pump and the use of an inverter-based ring oscillator, the TxPLL is well suited for low voltage operation. The RxFLL is very similar to the TxPLL, except that a cycle-slip detector (CSD) is employed in place of the PFD.

Figure 8.6.2 shows the schematic of the DCO. It is composed of a 2-stage pseudo-differential ring oscillator with two separate tuning ports driven by the proportional and integral paths. Transistors M_{1-2} and M_3 controlled by the pulse-width modulated PFD outputs (P_{UP}, P_{DN}) and the integral control voltage (V_I), respectively, tune the DCO frequency by regulating the delay cell ground voltage. The integral control voltage is generated by truncating the 14-bit frequency control word with a digital delta-sigma modulator and converting the resulting 4-bit output to an analog voltage using an R-2R DAC. Leveraging the constant output impedance of the R-2R DAC, a 2nd order passive low-pass filter is used to sup-

press quantization noise. Transistor M_4 is used to extend the tuning range of the DCO. Because the proportional path solely determines the bandwidth of the over-damped TxPLL, transistors M_{1-2} are sized to achieve the desired bandwidth as illustrated in Fig. 8.6.2. A similar DCO is used in the RxFLL with transistors M_{1-2} driven by early/late outputs of the half-rate bang-bang phase detector (!!PD).

Figure 8.6.3 shows the block diagram of the transmitter. Half-rate PRBS data is converted to full-rate by a 2:1 CMOS multiplexer and transmitted using a low swing voltage-mode (VM) driver. A regulator buffers the output of the phase-domain impedance locked loop (ILL) and sets the voltage swing of the pre-driver to ensure 50Ω output impedance of the VM driver. ILL is implemented as a simple PLL that tunes M_1' gate voltage to match the phase shift in the two M_1'-C and R_TC paths. As a result, by matching M_1' to M_1, the output swing of the pre-driver settles to V_{PDRV} needed to achieve 50Ω termination impedance. Because there is no static current, the ILL-based tuning helps lower power compared to a conventional voltage-based approach [4].

Figure 8.6.4 shows the block diagram of the receiver. A conventional source-degenerated CTLE equalizes moderate channel loss and drives a bank of four data/edge samplers at half-rate. The !!PD generates 2 pairs of early/late (E/L) signals that control the DCO directly. In the digital integral path of the CDR, the E/L signals are de-serialized by a factor of 16 before feeding to the accumulator.

Prototyped in an LP 90nm CMOS process and TQFP64 package, the transceiver operates over 0.5-to-4Gb/s data rates with a BER<10^{-12}. Figure 8.6.5 shows the measured results of the transmitter at 3.2Gb/s. As illustrated by the transmitted eye, the horizontal and vertical eye opening at the board edge is 69mV$_{ppd}$ and 267ps, respectively. When transmitting a PRBS31 sequence, there is negligible reduction in the vertical opening while the horizontal eye opening decreases by 14ps. Long-term jitter of the half rate 1.6GHz clock generated by the TxPLL is 4.6ps$_{rms}$ and 40ps$_{pp}$ (> 100K hits). The measured TxPLL power is 640μW and its loop bandwidth is 12MHz. The measured transmit driver resistance is 55±3Ω, within the output swing range of 100 to 200mV$_{ppd}$. The BER bathtub curve at the input of the receiver shows an eye opening of 0.4UI for BER<10^{-12}. The recovered clock jitter is 11.7ps$_{rms}$ and 69ps$_{pp}$.

The power dissipation of each of the building blocks is summarized in Fig. 8.6.6. At 3.2Gb/s, transmitter, receiver, and clock generation circuits consume 1.9mW, 2.8mW, and 1.4mW, respectively. The power efficiency at 3.2Gb/s is 1.9mW/Gb/s and improves to 1.8mW/Gb/s at 4Gb/s. The performance is compared to state-of-the-art low-power transceivers in Fig. 8.6.6. The proposed solution leverages all-digital clock generation/recovery circuits operating at low supply voltages to reduce power. It achieves comparable power efficiency even with using a dedicated PLL and a wide tracking bandwidth CDR. A die micrograph is shown in Fig. 8.6.7.

Acknowledgments:
National Science Foundation (NSF) under CAREER EECS-0954969 and Semiconductor Research Corporation (SRC) under contract 2007-HJ-1597 provided partial financial support.

References:
[1] J. Kim and M.A. Horowitz, "Adaptive Supply Serial Links with Sub-1V Operation and Per-Pin Clock Recovery," *ISSCC Dig. Tech. Papers*, pp. 268-269, Feb., 2002.
[2] G. Balamurugan, J. Kennedy, G. Banerjee, J.E. Jaussi, M. Mansuri, F. O'Mahony, B. Casper, and R. Mooney, "A Scalable 5-15Gbps, 14-75mW Low Power I/O Transceiver in 65nm CMOS," *IEEE J. Solid-State Circuits*, vol. 43, no. 4, pp. 1010-1019, April, 2008.
[3] S. Rajapandian, Z. Xu, and K.L. Shepard, "Implicit DC-DC Down Conversion Through Charge-Recycling," *IEEE J. Solid-State Circuits*, vol. 40, no. 4, pp. 846-852, April, 2005.
[4] R. Palmer, J. Poulton, W. J. Dally, J. Eyles, A.M. Fuller, T. Greer, and M.A. Horowitz, "A 14mW 6.25 Gb/s Transceiver in 90nm CMOS for Serial Chip-to-Chip Communications," *ISSCC Dig. Tech. Papers*, pp. 440-441, Feb., 2007.
[5] K.L.J. Wong, M. Mansuri, H.Hatamkhani, and C.K.K. Yang, "A 27mW 3.6-Gb/s I/O Transceiver," *Symp. VLSI circuits*, pp. 99-102, Feb., 2003.

978-1-61284-303-2/11 $26.00 © 2011 IEEE

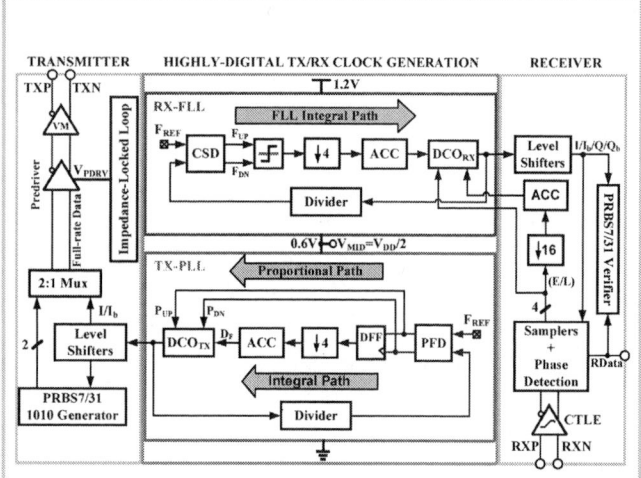

Figure 8.6.1: Proposed serial-link transceiver.

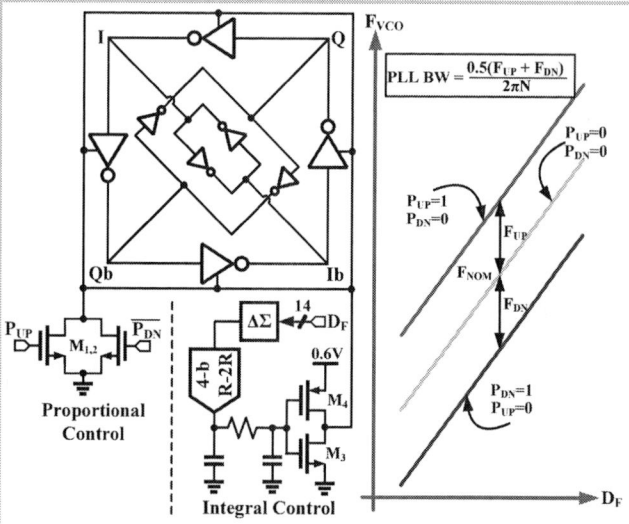

Figure 8.6.2: Schematic of the low-voltage DCO.

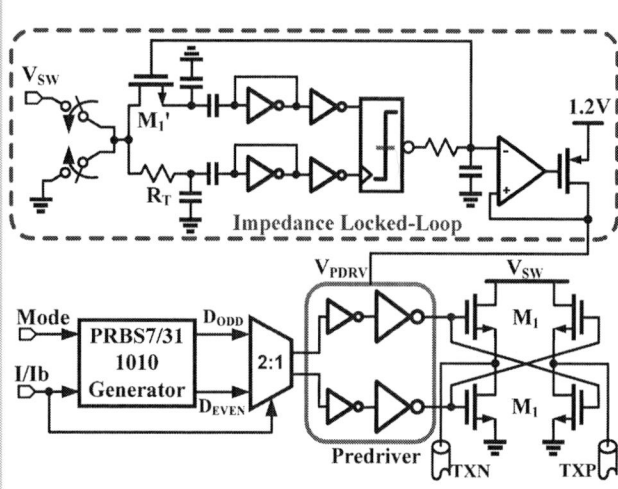

Figure 8.6.3: Block diagram of the transmitter.

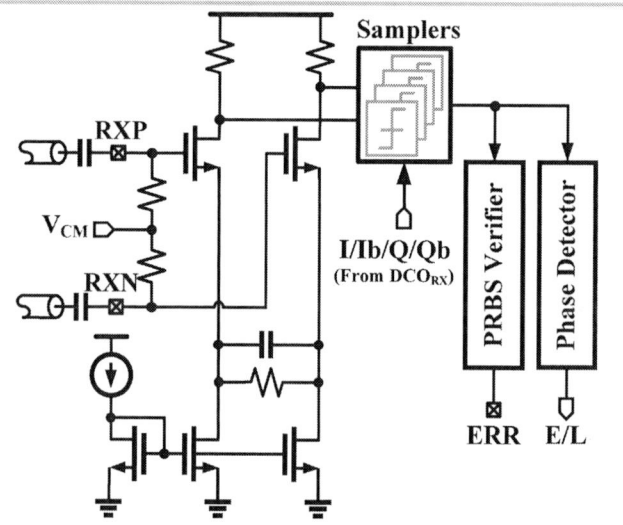

Figure 8.6.4: Block diagram of the receiver.

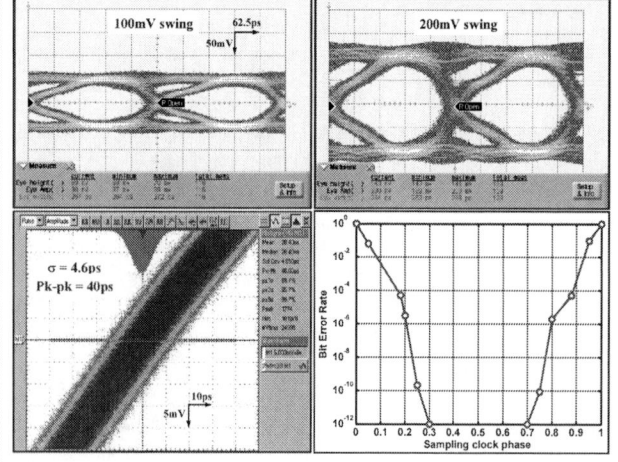

Figure 8.6.5: Measurement results of the transceiver: Transmitter eye at 100/200mV$_{ppd}$ swings, TxPLL clock jitter, and BER bathtub curve at the receive end.

Simulated power at 3.2Gbps in mW

Transmitter	
2:1 Mutiplexer	0.2
Predriver	0.3
ILL + Regulator	0.36
VM driver	1.1
Total	1.96
Receiver	
CTLE	1.2
4 Samplers	1.1
E/L mapping	0.1
CDR logic	0.2
Total	2.6
Clocking	
TxPLL	0.64
RxPLL	0.62
Level shifters	0.2
Total	1.46
Link Total	6

Measured power at different data rates

Data rate [Gbps]	4	3.2	1	0.5
Tx-Power [mW]	2.1	1.9	1.2	1
Rx-Power [mW]	3	2.8	1.8	1.5
Clocking [mW]	1.8	1.4	0.9	0.8
Total [mW]	6.9	6.1	3.9	3.3

	ISSCC'07 [1]	JSSC'08 [2]	This work
Techology	90nm	65nm	90nm
Supply voltage	1V (Fixed)	0.68-1.05V (Variable)	1.2V (Fixed)
Data rate	6.25Gbps	5-15Gbps	0.5-4Gbps
Implementation	Half-rate	Half-rate	Half-rate
Clocking	Mesochronous	Source synchronous	Plesiochronous
Link FOM	2.2 @ 6.25Gbps	2.7 @5Gbps	1.9 @3.2Gbps
Transmitter			
Driver	VM, Diff	CML, Diff	VM, Diff
Swing	200mVppd	100mVppd	100mVppd
FOM	0.8mW/Gbps	1.5mW/Gbps	0.6mW/Gbps
Receiver			
Input coupling	DC	AC	AC
Clock recovery	PR-PLL	None	Digital CDR
Tracking BW	128Hz	N/A	14MHz
Equalization	CTLE	CTLE	CTLE
FOM	1.4mW/Gbps	1.2mW/Gbps	0.9mW/Gbps
Clock Generation			
Oscillator	LC	N/A	Ring
Frequency	3.125GHz	2.5-7.5GHz	0.25-2GHz
Implementation	Analog	N/A	Digital
PLL BW	5MHz	N/A	12MHz
Phase noise @ 1MHz offset	-108dBc @ 3.125GHz	N/A	-106dBc @ 1.6GHz
Jitter	3.9mUI$_{rms}$	N/A	7.3mUI$_{rms}$
FOM	1.4mW/GHz	N/A	0.8mW/GHz

Figure 8.6.6: Detailed simulated/measured power distribution and comparison to the state-of-the art low-power links.

978-1-61284-303-2/11 $26.00 © 2011 IEEE

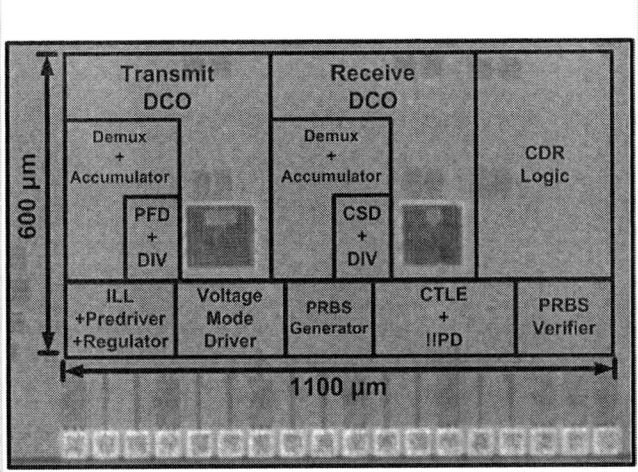

Figure 8.6.7: Chip micrograph.

ISSCC 2011 / SESSION 8 / ARCHITECTURES & CIRCUITS FOR NEXT GENERATION WIRELINE TRANSCEIVERS / 8.7

8.7 A 1-to-6Gb/s Phase-Interpolator-Based Burst-Mode CDR in 65nm CMOS

Behrooz Abiri[1], Ravi Shivnaraine[1], Ali Sheikholeslami[1],
Hirotaka Tamura[2], Masaya Kibune[2]

[1]University of Toronto, Toronto, Canada
[2]Fujitsu Laboratories, Kawasaki, Japan

Burst-mode clock and data recovery circuits (BMCDR) are widely used in passive optical networks (PON) [1] and as a replacement for conventional CDRs in clock-forwarding links to reduce power [2]. In PON, a single CDR performs the task of clock and data recovery for several burst sequences, each originating from a different source. As a result, the BMCDR is required to lock to an incoming data stream within tens of UIs (for example 40ns in GPON). Previous works use either injection locking [3, 4] or gated VCO [5, 6] to achieve this fast locking. In both cases, the control voltage of the CDR's VCO is generated by a reference PLL with a matching VCO to guarantee accurate frequency locking. However, any component mismatch between the two VCO's results in a frequency offset between the reference PLL frequency and the CDR's VCO frequency, and hence in a reduction of the CDR's tolerance for consecutive identical digits (CID). For example, [7] reports a frequency offset of over 20MHz (2000ppm) for 10Gb/s operation. We present a BMCDR that is based on phase interpolation (PI), eliminating the possibility of local frequency offset between the reference and recovered clock. We demonstrate 1 to 6Gb/s operation in 65nm CMOS with a locking time of less than 1UI.

The principle of operation for the proposed PI-based BMCDR is illustrated in Fig. 8.7.1. Since the main function of a BMCDR is to align the rising edge of the clock with the data transition, we use the data to sample-and-hold (S/H) two quadrature clocks (CK_I and CK_Q) with a pair of dual-edge-triggered S/H circuits. The samples of CK_I and CK_Q at a data transition, denoted by β and α respectively, are then used to interpolate between CK_I and CK_Q. This, as shown on Fig. 8.7.1, produces a recovered clock (CK_{REC}) with a rising edge that is aligned with the data transition. To demonstrate this analytically, assume $sin(2\pi ft)$ and $-cos(2\pi ft)$ represent CK_I and CK_Q respectively. A data transition at $t = t_0$ yields the PI coefficients $\alpha = CK_Q(t_0) = -cos(2\pi ft_0)$ and $\beta = CK_I(t_0) = sin(2\pi ft_0)$. As a result, $CK_{REC}(t) = \beta CK_Q(t) - \alpha CK_I(t) = sin(2\pi f(t-t_0))$ which is a clock whose zero crossing coincides with the data transition.

In our implementation, CK_I and CK_Q are provided externally; however, in an integrated system CK_I and CK_Q may be provided from a PLL or be generated from the forwarded clock. In the case of using a received clock, there is no frequency mismatch between CK_I/CK_Q and the embedded clock in the data. When using a PLL, a small frequency offset may be present between CK_I/CK_Q and the input due to a frequency error in the PLL reference. For sufficient transition density, the PI's output phase is updated at every data transition, hence any frequency offset between data and the reference clock is tracked.

Figure 8.7.2 shows the circuit implementation of the dual-edge triggered S/H. Dual-edge sampling is implemented by connecting the outputs of two S/Hs operating at opposite data edges. Each S/H is implemented as a master/slave configuration triggered by either the rising or the falling edge of data. The S/Hs are able to track up to 6GHz clock signals and as such no extra capacitor besides the existing parasitic capacitance is required. A buffer is inserted between the pass transistors and the transmission gates to prevent charge sharing and to avoid any kickback from the output transmission gates to the input pass gates. To provide rail-to-rail levels for switching, a CML-to-CMOS converter is placed before the S/Hs.

The circuit implementation of the analog PI is shown in Fig. 8.7.3. A differential trans-conductance stage is used to convert the coefficients (α and β) to current. The tails currents of CK_I and CK_Q are used as weighting factors in the PI. The operation of $-\alpha CK_I + \beta CK_Q$ is performed in current mode at the nodes of the resistors and then converted to an output voltage, CK_{REC}. Since α and β are differential signals, their signs can be changed. This allows the PI to create phases from 0° to 360°. Immunity to charge injection from the S/H is enhanced by providing inputs to the PI differentially.

Figure 8.7.4 shows the block diagram of the entire receiver along with the recovered data for a 6Gb/s PRBS10 input. The output of the clock recovery unit is buffered to drive the input of the clock divider and flip-flop. To compensate for the delay of the buffer in the clock path, a replica of the CK_{REC} buffer is placed in the data path. The recovered clock and data are buffered further (not shown) to be observed off-chip for direct probing. The measured peak-to-peak clock jitter is 24ps and the eye opening for recovered data is 500mV. The CDR recovers a 6Gb/s PRBS7 sequence with a BER of less than 10^{-12} in the presence of a 100MHz frequency offset between CK_I/CK_Q and the input.

Figure 8.7.5 shows the measured characteristics of the PI. The input data has been phase shifted and the corresponding incremental phase shift of phase interpolator output at 4GHz and 6GHz has been recorded. At 4GHz and 6GHz, the recovered clock is shown for 30° input phase steps. The maximum deviation from ideal interpolation is 6.5° at 4GHz and 2° at 6GHz. The PI's latency and unequal delay between the clock path and the data path may produce a phase offset in the CDR. At 6Gb/s this delay was simulated to be less than one tenth of a UI and as such was left uncompensated.

To measure the time it takes for the clock edge to align with the data edge (locking time), a deliberate frequency offset is introduced. This frequency offset combined with a long CID introduces a phase jump that allows the CDR locking behaviour to be observed (Fig. 8.7.6). In the 1Gb/s and 6Gb/s cases, the clock is delayed by the PI in order to align with the data edge. In the 2.5Gb/s case, the clock is reversed whereas in the 4Gb/s case the clock is advanced. A Centellax OTB3P1A PRBS generator is used to create PRBS10 patterns from 1 to 6Gb/s. An Agilent DCA-J 81600C digital communication analyzer (with 8611A 20GHz electrical module) is used to capture sharp phase transitions in the recovered clock.

The chip is fabricated in a 65nm CMOS process. The receiver area is 250×70μm², of which 50×70μm² is occupied by the clock recovery circuitry (S/H and PI). The chip operates from a 1.2V supply and consumes a total power of 22mW (excluding output drivers), of which 3.8mW is consumed by the clock recovery circuit.

Acknowledgement:
The authors acknowledge the use of the University of Toronto's Advanced Digital Systems Lab and thank Saeid Rezaei for his help with the measurements.

References:
[1] C. Liang and S. Liu, "A 20/10/5/2.5Gb/s Power-scaling Burst-Mode CDR Circuit Using GVCO/Div2/DFF Tri-mode Cells," *ISSCC Dig. Tech. Papers*, pp. 224-608, Feb., 2008.
[2] N. Miura, Y. Kohama, Y. Sugimori, H. Ishikuro, T. Sakurai, and T. Kuroda, "An 11Gb/s Inductive-Coupling Link with Burst Transmission," *ISSCC Dig. Tech. Papers*, pp. 298-614, Feb., 2008.
[3] J. Lee and M. Liu, "A 20Gb/s Burst-Mode CDR Circuit Using Injection-Locking Technique," *ISSCC Dig. Tech. Papers*, pp. 46-586, Feb., 2007.
[4] K. Maruko, T. Sugioka, H. Hayashi, Zhiwei Zhou, Y. Tsukuda, Y. Yagishita, H. Konishi, T. Ogata, H. Owa, T. Niki, K. Konda, M. Sato, H. Shiroshita, T. Ogura, T. Aoki, H. Kihara, and S. Tanaka, "Burst-Mode CDR Using Dual-Edge Injection-Locked Oscillator," *ISSCC Dig. Tech. Papers*, pp. 364-365, Feb., 2010.
[5] M. Nogawa, K. Nishimura, S. Kimura, T. Yoshida, T. Kawamura, M. Togashi, K. Kumozaki, and Y. Ohtomo, "A 10 Gb/s Burst-Mode CDR IC in 0.13μm CMOS," *ISSCC Dig. Tech. Papers*, pp. 228-229, Feb., 2005.
[6] L. Cho, C. Lee, and S. Liu, "A 33.6-to-33.8Gb/s Burst-Mode CDR in 90nm CMOS," *ISSCC Dig. Tech. Papers*, pp. 48-49, Feb., 2007.
[7] J. Terada, K. Nishimura, S. Kimura, H. Katsurai, N. Yoshimoto, and Y. Ohtomo, "A 10.3125Gb/s Burst-Mode CDR Circuit using a $\Delta\Sigma$ DAC," *ISSCC Dig. Tech. Papers*, pp. 226-227, Feb., 2008.

978-1-61284-303-2/11 $26.00 © 2011 IEEE

ISSCC 2011 / February 22, 2011 / 11:45 AM

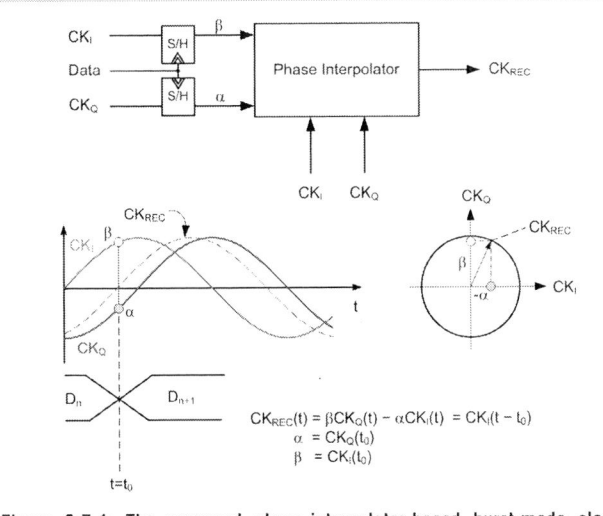

$$CK_{REC}(t) = \beta CK_Q(t) - \alpha CK_i(t) = CK_i(t - t_0)$$
$$\alpha = CK_Q(t_0)$$
$$\beta = CK_i(t_0)$$

Figure 8.7.1: The proposed phase-interpolator-based burst-mode clock recovery.

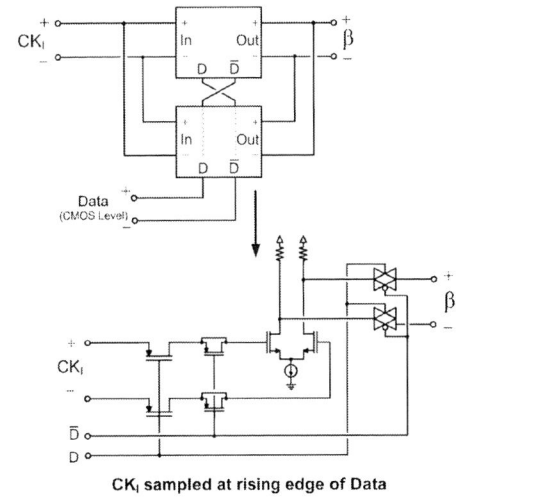

CK$_I$ sampled at rising edge of Data

Figure 8.7.2: Dual edge triggered sample and hold.

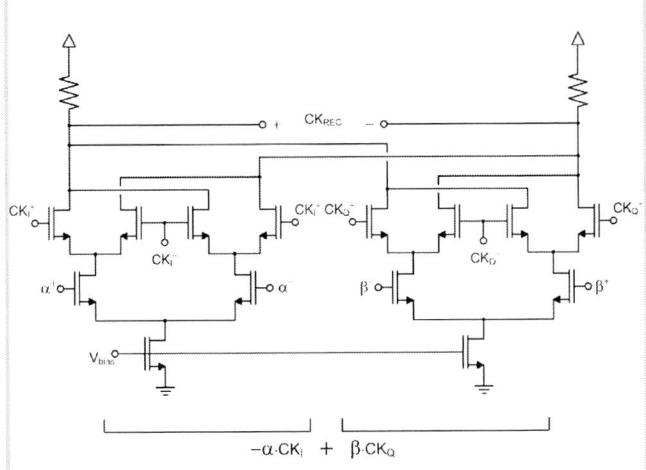

$$-\alpha \cdot CK_i + \beta \cdot CK_Q$$

Figure 8.7.3: Phase interpolator implementation.

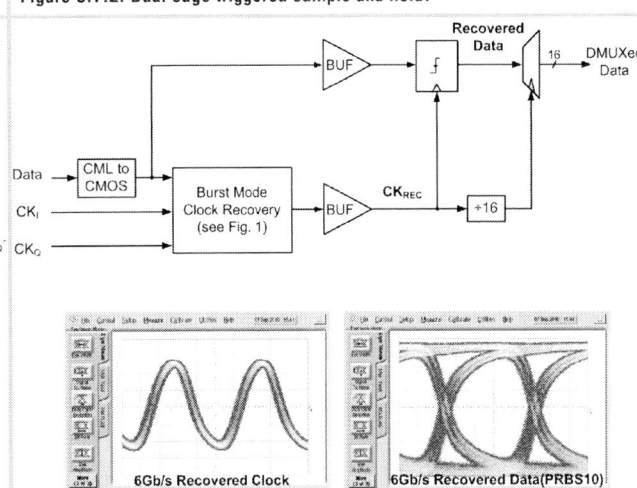

Figure 8.7.4: Receiver block diagram (top) and measured 6Gb/s recovered clock and data (bottom).

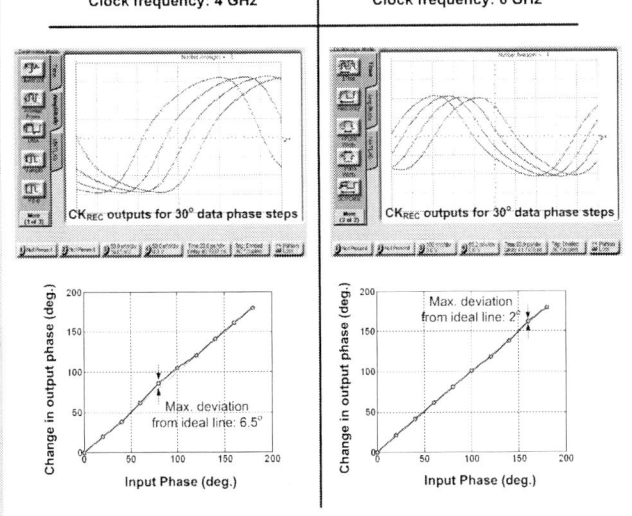
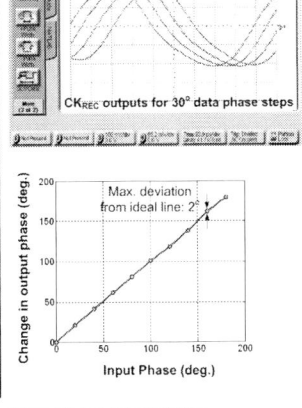

Figure 8.7.5: Measured characteristics of the phase interpolator.

Figure 8.7.6: Measured transient response of CDR to phase jump at 1 to 6Gb/s.

8

978-1-61284-303-2/11 $26.00 © 2011 IEEE

ISSCC 2011 PAPER CONTINUATIONS

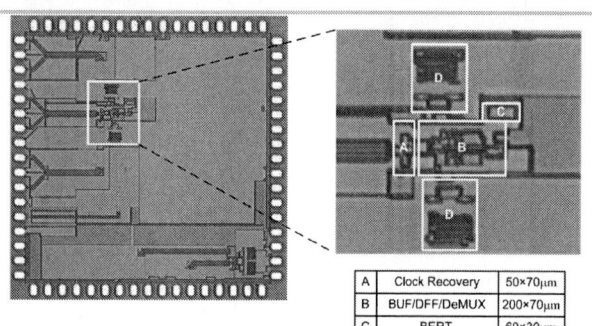

A	Clock Recovery	50×70μm
B	BUF/DFF/DeMUX	200×70μm
C	BERT	60×30μm
D	CML Output Drivers	90×140μm

Design	ISSCC 2005	ISSCC 2007	ISSCC 2010	This Work
Technology	0.13μm CMOS	90nm CMOS	40nm CMOS	65nm CMOS
Data rate	10Gb/s	20Gb/s	1.3-5.2Gb/s	1-6Gb/s
Locking Method	Gated VCO	Injection Lock	Injection Lock	Phase Interpolator
Locking time	5UI	1UI	< 20UI	< 1UI
Power	1.2W (entire RX)	102mW (CDR Core)	12.4mW (CDR Core)	22mW (CDR Core + DMUX)

Figure 8.7.7: Die micrograph and comparison table.

ISSCC 2011 / SESSION 8 / ARCHITECTURES & CIRCUITS FOR NEXT GENERATION WIRELINE TRANSCEIVERS / 8.8

8.8 A 14Gb/s High-Swing Thin-Oxide Device SST TX in 45nm CMOS SOI

Christian Menolfi[1], Thomas Toifl[1], Michael Rueegg[2], Matthias Braendli[1], Peter Buchmann[1], Marcel Kossel[1], Thomas Morf[1]

[1]IBM Zurich Research Laboratory, Rueschlikon, Switzerland
[2]Miromico, Zurich, Switzerland

The limited supply voltage of today's state-of-the-art CMOS technologies makes the design of high-speed transmitters at signaling swings above the typical 1V supply a challenging task. Higher-voltage TX amplitude is not only required in older I/O standards and legacy applications, but also in emerging electro-optical extensions where high voltage swing combined with high-speed operation is desired. Higher swing also helps meet certain I/O standards in applications where losses introduced by high-density package constraints can be compensated to some extent.

The source-series terminated (SST) driver is a versatile building block in a multi-standard I/O TX thanks to its potential for low-power operation, its low area consumption, high CMOS-style circuit content, and flexible termination capability [1]. Also, the SST driver supports single-ended output and differential operation.

It is well known that the maximum achievable peak-to-peak output swing of a differential voltage mode driver is limited to Vdd under impedance-matched conditions. To increase the available signal swing, a straightforward solution is to increase the output driver supply and to use thick-oxide devices [2]. However, these devices feature a larger minimum channel length and, as a result, tend to be slower and exhibit a larger associated switch capacitance than their thin-oxide counterparts. Furthermore, they are driven at the elevated output driver supply, making the pre-drive power — a substantial part of the power budget in many SST-based TXs — intolerably high. A solution is desired that maintains the logic swing of all internal nodes at the nominal Vdd supply, with thin-oxide devices employed in the high-voltage output driver for lower switch capacitance and higher speed. This contribution presents a high-swing SST TX based entirely on thin-oxide devices and using a split supply approach for driving the NMOS and the PMOS branches of the output driver, along with a static voltage protection scheme [3].

A schematic of the thin-oxide high-voltage driver is shown in Fig. 8.8.1. The output driver is operated from a high-voltage supply Vddh, which can be flexibly set between 1.2 to 2V and determines the available output swing. A low-side 1V Vdd supply is used to drive the pull-down NMOS devices, whereas a high-side 1V supply between Vddh-Vdd (high-side ground) and Vddh is used for the PMOS pull-up branch. The pull-up and pull-down branches are protected by a thin-oxide NMOS or PMOS cascode protection device biased at Vdd or Vddh-Vdd, respectively. The protection devices ensure that none of the stacked thin-oxide devices are exposed to voltages exceeding the nominal 1V Vdd supply. The cascode protection scheme used allows a maximum high-voltage supply of Vddh=2Vdd. The driver has been implemented for half-rate operation and includes a 2:1 multiplexer in the output stage. Multiplexing of the output data at the last stage ensures tight output timing control and avoids the introduction of data-dependent timing inaccuracies due to ISI and potential floating body history effects pertinent to SOI technologies. Given the different switching activities of the stacked data and clock transistors in the multiplexer, the ratio between data and clock switch size has been found to be 2:1 for minimized power consumption. Last, static NMOS footer devices and PMOS header devices are provided to compensate for PVT variations of the driver's output impedance.

The use of a split high/low side supply pre-drive scheme requires voltage level translators to interface with the logic and to provide the data, clock, and control signals in the appropriate supply domain. In this design, three types of thin-oxide voltage level translators provide simultaneous output in both supply domains. A dc voltage level translator depicted in Fig. 8.8.2a is used for low-speed control signals with the primary goal of saving area. The circuit consists of a conventional voltage level shifter with the addition of cascode protection devices in the NMOS and PMOS branches [3]. High-speed ac coupled level translators have been chosen for the timing-critical data and clock signals.

Fig. 8.8.2b shows the schematic of a data voltage level translator. The differential input data at the low-side supply is ac-coupled to the high-side supply domain. A cross-coupled inverter pair at the high side keeps the high-side data signal stable in the absence of toggling input data. Note that the voltage across the coupling capacitor remains constant under stationary conditions and that all voltage swings of the internal nodes are limited to the nominal Vdd supply. The ac coupling capacitors have been implemented with longer channel body-contacted thin-oxide devices that feature the largest capacitance per area in our technology. Additional enable logic is provided for appropriate capacitor bias at startup to prevent potential over-voltage damage of the coupling capacitors. The clock voltage level translator depicted in Fig 8.8.2c uses ac coupling along with a cross-coupled PMOS pair to translate the incoming differential clock to the high-side supply domain. Given the tight tracking requirements between the low-side and the high-side clock (ck2_lo, ck2_hi), the translator directly drives the output stage clock transistors. Simulations indicate that a tracking accuracy as low as 1ps can be achieved. Charge sharing slightly reduces the swing of the high-side clock node ck2_hi. This has been taken into account in the output driver PMOS clock transistor sizing.

To prove the principle of the split supply high-swing SST driver, a TX test-chip has been implemented. The architecture is similar to an earlier implementation [1] enhanced for high-swing operation. A block diagram of the architecture is shown in Fig. 8.8.3. The design features a quarter-rate data interface d0…d3 and a half-rate CML clock input ck2cml along with a half-rate 4-tap FIR shift register for FFE operation. The half-rate SST driver has been split into 32 equally sized, high-swing SST unit segments, each of which can be configured to operate as one of the four FIR taps and contains the individual data and clock level translators to cope with the split supply output driver stage. Fig. 8.8.4 shows the layout of the TX macro implemented in 45nm SOI technology. The core size is 250×110μm². An area overhead on the order of 20% has been observed, compared with a conventional nominal supply SST TX. A dual TX macro along with an on-chip data pattern generator has been integrated in a wafer-probable test chip. A chip micrograph is shown in Fig. 8.8.7. Figure 8.8.5 shows two eye diagrams measured at 12Gb/s and 14Gb/s at a driver supply Vdh=2V. The observed RJ is <200fs$_{rms}$ and DJ is <11ps with manually tuned FFE. Measured amplitudes are 1.875V$_{pp-diff}$ at 6Gb/s, 1.36V$_{pp-diff}$ at 12Gb/s and 1.28V$_{pp-diff}$ at 14Gb/s. Note that the measurement setup introduces attenuation on the order of −1.5dB at 7GHz while the driver's output pole located around 14GHz accounts for another −1dB at 7GHz. Figure 8.8.6 shows the plot of the total power consumption versus different Vdh and data rates. Power at 14Gb/s varies from 70.6mW at Vdh=1.2V to 85.5mW at Vdh=2V. A constant offset in power is observed between the different Vdh supply configurations regardless of the data rate. This indicates that the pre-drive power is independent of the Vdh supply, and confirms the proper operation of the split supply approach.

References:
[1] C. Menolfi, T. Toifl, P. Buchmann, et al., "A 16Gb/s Source-Series Terminated Transmitter in 65nm CMOS SOI," *ISSCC Dig. Tech. Papers*, pp. 446-447, Feb., 2007.
[2] M. Kossel, C. Menolfi, J. Weiss, et al., "A T-Coil-Enhanced 8.5 Gb/s High-Swing SST Transmitter in 65 nm Bulk CMOS with <-16 dB Return Loss over 10 GHz Bandwidth," *IEEE J. Solid-State Circuits*, vol. 43, no. 12, pp. 2905-2920, Dec., 2008.
[3] H. Sanchez et al., "A Versatile 3.3/2.5/1.8-V CMOS I/O Driver Built in a 0.2 μm, 3.5-nm Tox, 1.8-V CMOS Technology," *IEEE J. Solid-State Circuits*, vol. 34, no.11, pp. 1501-1511, Nov., 1999.

978-1-61284-303-2/11 $26.00 © 2011 IEEE

ISSCC 2011 / February 22, 2011 / 12:00 PM

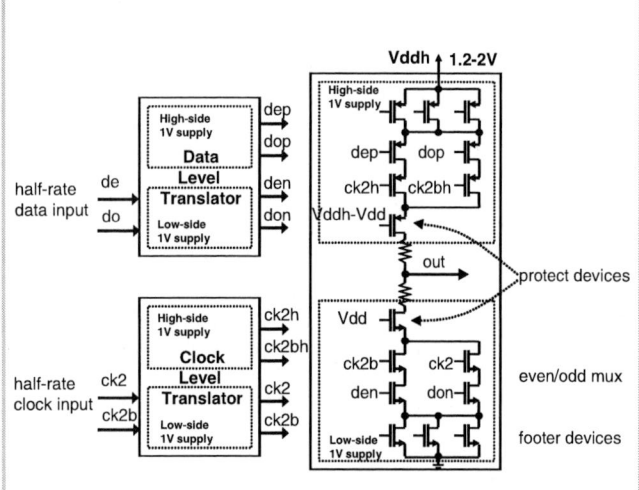

Figure 8.8.1: Schematic of the thin-oxide high-voltage SST driver.

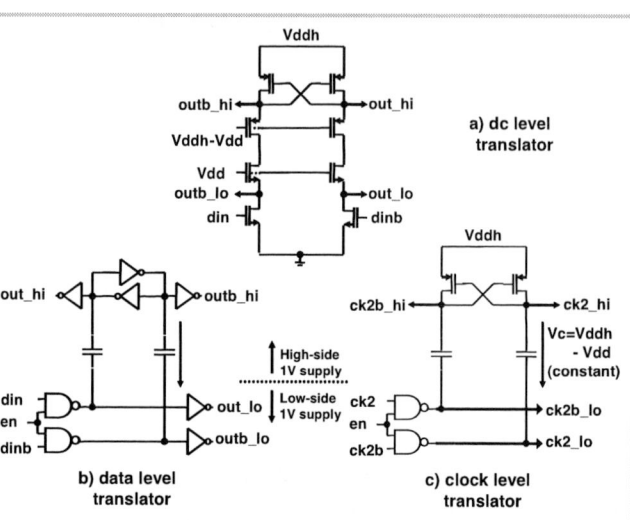

Figure 8.8.2: Schematic of the three types of thin-oxide voltage level translators.

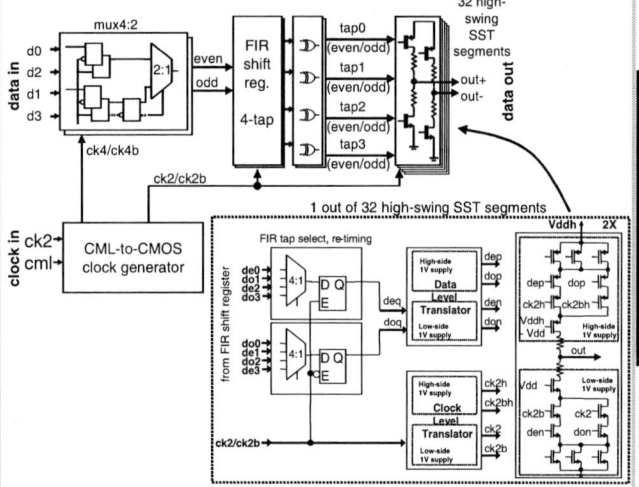

Figure 8.8.3: Block diagram of the high-swing SST TX implemented.

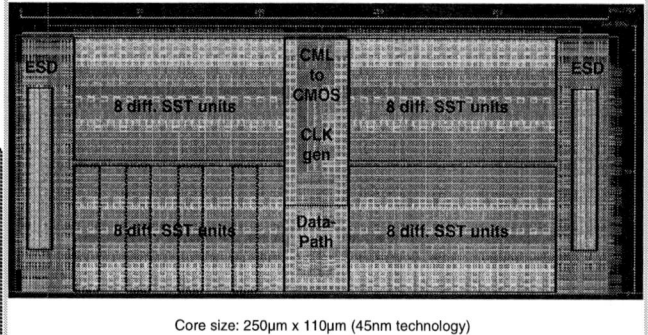

Core size: 250μm x 110μm (45nm technology)

Figure 8.8.4: Layout of the high-swing TX macro implemented.

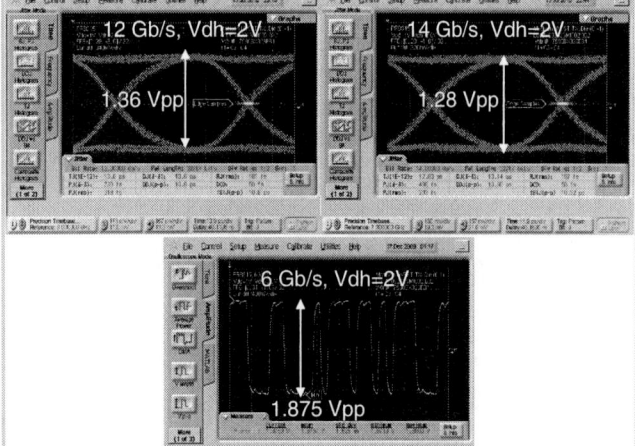

Figure 8.8.5: Eye diagrams measured at 12Gb/s, 14Gb/s, and 6Gb/s time response.

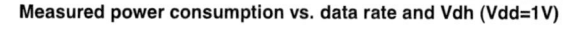

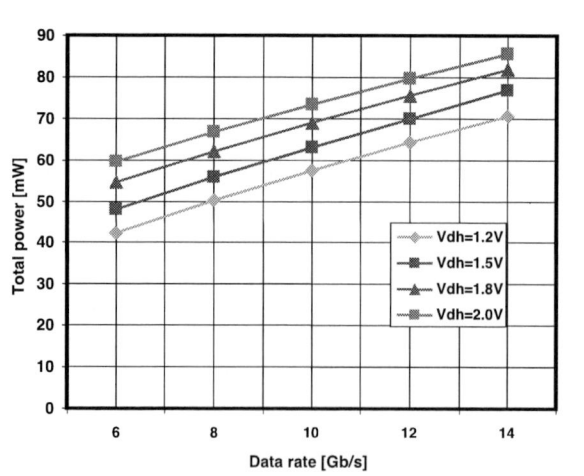

Figure 8.8.6: Total power consumption versus data rate measured at different driver supplies Vdh.

978-1-61284-303-2/11 $26.00 © 2011 IEEE   157

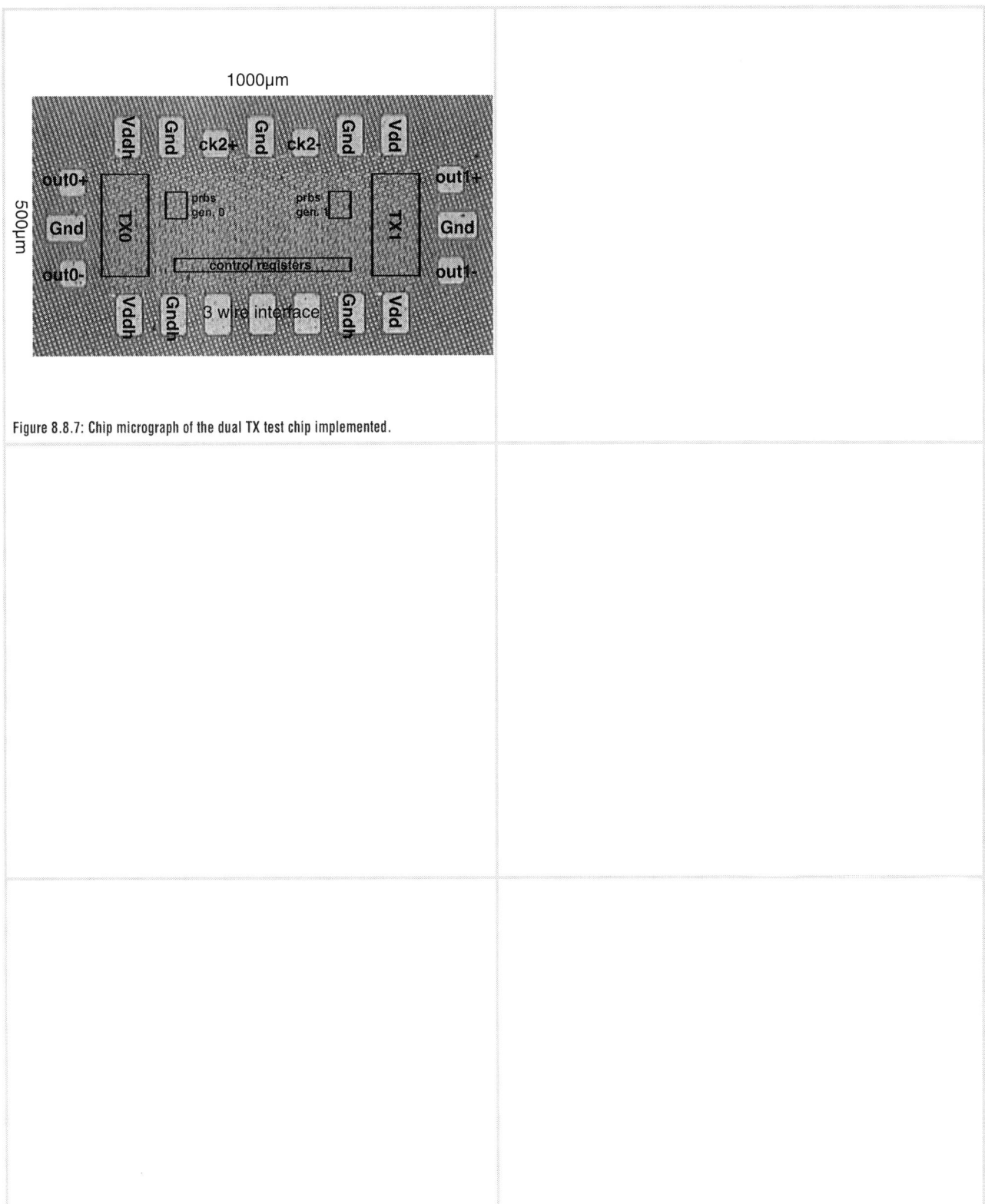

Figure 8.8.7: Chip micrograph of the dual TX test chip implemented.

ISSCC 2011 / SESSION 9 / WIRELESS & mm-WAVE CONNECTIVITY / OVERVIEW

Session 9 Overview / Wireless

Wireless & mm-Wave Connectivity

Session Chair: *Gangadhar Burra, Texas Instruments, Dallas, TX.*

Session Co-Chair: *Yorgos Palaskas, Intel, Hillsboro, OR*

High-throughput and low power continue to demand advances in CMOS wireless technologies for connectivity in various applications. The papers in this session show these developments in the area of mm-Wave 60GHz-and-up systems, as well as conventional connectivity platforms such as WLAN, Bluetooth and mobile TV. The over-arching trends at 60GHz are phased-arrays for wider range, integrated antennas and low power to target mobile applications. Traditional connectivity technologies show higher levels of integration, digital-centric architecture implementations and overall innovative ideas for system cost reduction.

Paper 9.1 from the Tokyo Institute of Technology discusses a 60GHz transceiver compliant with the IEEE802.15.3c standard. The receiver shows a measured NF of <6.8dB and the transmitter PA efficiency is 8.8%. The 65nm CMOS transceiver with in-package antenna achieves 3.5Gb/s up to a distance of 2.7m and consumes 186mW in transmit and 106mW in receive modes respectively.

Paper 9.2 from CEA-LETI and STMicroelectronics discusses a fully integrated 60GHz transceiver also in 65nm CMOS. The transmitter with an external in-module PA achieves a P_{sat} of greater than 16dBm. The transceiver has a throughput rate of 3.8Gb/s at up to 1 meter with WirelessHD-compliant EVM parameters. This is obtained with a single in-module 5dBi antenna built on an HTCC substrate manufactured in an industrial packaging line.

The first beam-forming 60GHz transceiver compliant with both WirelessHD and draft IEEE 802.11ad standards is described in Paper 9.3 from SiBeam. This implementation contains up to 32 elements with antennas integrated in a ceramic package and achieves 3.8Gb/s at a non-line of sight range of 10 meters. The 65nm digital CMOS transceiver is capable of dynamic beam-steering in the presence of line-of-sight obstructions. Tradeoffs between rate, range, and power consumption are described.

Integrating multiple elements of a 60GHz phased-array radio requires careful optimization of power dissipation per element. Paper 9.4 from UC Berkeley discusses techniques that address achieving a low per-element power consumption of 34mW, including LO generation and distribution. The paper describes a stacked mixer/phase rotator and zero-voltage switching PA that improve transmitter efficiency. The techniques are demonstrated on a 65nm CMOS, phased-array transceiver with integrated PLL and 4-element receiver and transmitter.

Due to the high data rates involved in multi-Gb/s mm-Wave systems, digital baseband processing can also consume significant power. To overcome this issue, Paper 9.5 from National Taiwan University revisits an older analog technique, the Costas loop, to perform synchronization and demodulation in an 87GHz QPSK system. The 65nm CMOS chipset is combined with a low-cost waveguide adaptor that can directly drive a horn antenna for a point-to-point 3.5Gb/s link.

Wireless LAN SoC innovation is addressed in this session with a presentation of a 3x3 MIMO SoC in Paper 9.6 by Atheros. Architecture techniques associated with enhanced I/Q mismatch correction as well as phase-noise impact reduction are discussed. Both high (5GHz) and low (2.4GHz) band results with an integrated power amplifier show an output power of 25dBm and over-the-air throughput of >300Mb/s. The 65nm CMOS chip occupies 22m² and achieves an EVM floor of -39dB/-36dB at 2.4/5GHz.

Paper 9.7 from the University of Macau and Instituto Superior Tecnico, Lisbon, describes a multiband TV tuner covering the VHF-III, UHF and L bands for mobile applications. The paper addresses the reduction of the overall system cost by creating a single

978-1-61284-303-2/11 $26.00 © 2011 IEEE   158

ISSCC 2011 / February 22, 2011 / 8:30 AM

front-end for all bands without an external balun and describes techniques that achieve a 4dB noise figure and a +32dBm IIP2 across all bands. Implemented in 65nm CMOS, the device consumes 55mW and occupies an area of 0.46mm^2.

The next two papers discuss digital-intensive techniques for implementing transmitters and receivers respectively in scaled CMOS processes. Paper 9.8 from Toshiba Semiconductor describes an 8-DPSK Polar transmitter for Bluetooth EDR. The phase path uses a digital PLL with correction of the inherent DCO nonlinearity. The amplitude is introduced by a digital PA with constant output impedance for good linearity. The 0.56mm^2, 65nm CMOS chip delivers 0dBm output power while dissipating 42mW.

Recently VCO-based ADCs have been proposed for easy scalability and porting to future CMOS nodes. Paper 9.9 from KAIST presents a 0.2-to-1.8GHz receiver using a VCO-based ADC. The design takes advantage of the inherent integration of a VCO to implement a 2nd order Sinc^2 anti-aliasing filter for better interference rejection. The 90nm CMOS IC achieves -94dBm sensitivity over a 1MHz bandwidth, with 50dB rejection of aliased signals due to the Sinc^2 filter.

60GHz systems are moving to productization of high definition video over the air and other throughput-intensive wireless applications. New low-power and phased-array technologies are making 60GHz more feasible for mobile applications. In the interim, traditional connectivity technologies continue to optimize system performance and cost via innovative digital and RF architectures.

9

Presenters:

8:30 AM

9.1 A 60GHz 16QAM/8PSK/QPSK/BPSK Direct-Conversion Transceiver for IEEE 802.15.3c

K. Okada, Tokyo Institute of Technology, Tokyo, Japan

9:00 AM

9.2 A 65nm CMOS Fully Integrated Transceiver Module for 60GHz Wireless HD Applications

A. Siligaris, CEA-LETI-MINATEC, Grenoble, France

9:30 AM

9.3 A 60GHz CMOS Phased-Array Transceiver Pair for Multi-Gb/s Wireless Communications

C. Doan, SiBEAM, Sunnyvale, CA

9:45 AM

9.4 A 65nm CMOS 4-Element Sub-34mW/Element 60GHz Phased-Array Transceiver

M. Tabesh, University of California, Berkeley, CA

10:15 AM

9.5 An 87GHz QPSK Transceiver with Costas-Loop Carrier Recovery in 65nm CMOS

S-J. Huang, National Taiwan University, Taipei, Taiwan

10:45 AM

9.6 A 65nm Dual-Band 3-Stream 802.11n MIMO WLAN SoC

S. Abdollahi-Alibeik, Atheros Communications, San Jose, CA

11:15 AM

9.7 A 0.46mm^2 4dB-NF Unified Receiver Front-End for Full-Band Mobile TV in 65nm CMOS

P-I. Mak, University of Macau, Macau, China

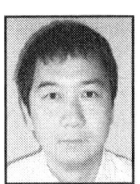

11:45 AM

9.8 An All-Digital 8-DPSK Polar Transmitter with Second-Order Approximation Scheme and Phase Rotation-Constant Digital PA for Bluetooth EDR in 65nm CMOS

H. Kobayashi, Toshiba Semiconductor, Kawasaki, Japan

12:00 PM

9.9 A Digital-Intensive Receiver Front-End Using VCO-Based ADC with an Embedded 2nd-Order Anti-Aliasing Sinc Filter in 90nm CMOS

J. Kim, KAIST, Daejeon, Korea

978-1-61284-303-2/11 $26.00 © 2011 IEEE

ISSCC 2011 / SESSION 9 / WIRELESS & mm-WAVE CONNECTIVITY / 9.1

9.1 A 60GHz 16QAM/8PSK/QPSK/BPSK Direct-Conversion Transceiver for IEEE 802.15.3c

Kenichi Okada, Kota Matsushita, Keigo Bunsen, Rui Murakami,
Ahmed Musa, Takahiro Sato, Hiroki Asada, Naoki Takayama,
Ning Li, Shogo Ito, Win Chaivipas, Ryo Minami, Akira Matsuzawa

Tokyo Institute of Technology, Tokyo, Japan

This paper presents a 60GHz direct-conversion transceiver using 60GHz quadrature oscillators as shown in Fig. 9.1.1. The 65nm CMOS transceiver realizes the IEEE802.15.3c full-rate wireless communication for every 16QAM/8PSK/QPSK/BPSK mode. The maximum data rates with an antenna built in the package are 8Gb/s in QPSK mode and 11Gb/s in 16QAM mode within a BER of < 10^{-3}, and the transmitter and the receiver consume 186mW and 106mW, respectively.

The transmitter design is shown in Fig. 9.1.2. The transmitter consists of a 4-stage PA, I/Q mixers and a quadrature oscillator. A direct-conversion architecture is employed because of energy efficiency [1]. The PA is implemented with a low-loss transmission line, which has a loss of 0.7dB/mm. A MIM transmission line (MIM TL) is also used for the de-coupling, which is characterized as a scalable transmission line. A MIM capacitor array is arranged along the MIM TL to lower the characteristic impedance. Transistors in the PA have a finger width of 2µm, and the total gate width of the final stage is 80µm. A double-balanced Gilbert mixer is used, and only one side is outputted in consideration of power consumption, LO leak, and layout area. The measured output power is shown in Fig. 9.1.2, and it is 9.5dBm at 1dB-compression. The conversion gain is 18.3dB, which is measured through the antenna. The large-signal measurement is calibrated with the saturated output power, which is measured by a probe station. The peak PAE is 8.8%. The PA consumes 114.6mW, and the two mixers consume 46.0mW from a 1.2V supply.

The receiver design is shown in Fig. 9.1.3. The receiver consists of a 4-stage LNA, I/Q passive mixers, and a quadrature oscillator. The LNA has a CS-CS topology to improve the noise figure [2], and is connected to the passive mixer through a parallel-line transformer and a 2-stage differential amplifier. Since a transformer balun generally causes an imbalance in differential signals, the differential amplifiers are used to compensate the imbalance with common-mode rejection in the matching blocks. Moreover, 2Ω resistors are inserted into the power line to avoid a parasitic oscillation. The measured conversion gain and noise figure are also shown in Fig. 9.1.3. The LNA realizes a gain control, and the conversion gain is 17.3dB in high-gain mode and 4.7dB in low-gain mode. The lower cut-off frequency of the IF amplifier is less than 4MHz. The entire noise figure is less than 6.8dB in the high-gain mode, and the measured IIP3 of the LNA is –5dBm in the low-gain mode. The LNA consumes 20.7mW and the two mixers with IF amplifiers consume 60.8mW from a 1.0V supply.

The LO consists of a quadrature injection-locked oscillator (QILO) [3] and a 20GHz PLL. The QILO design is shown in Fig. 9.1.4. The QILO works as a frequency tripler with a 20GHz injection-lock input, and it has a tail I/Q coupling. The I/Q coupling is carefully designed so that it can robustly keep the I/Q balance. The poly-phase filter for the 20GHz injection is not used to improve the I/Q mismatch over the entire frequency range. The measured free-running frequency is from 54 to 61GHz, and the phase noise is –85dBc/Hz at 1MHz offset in the free-running mode. Two quadrature oscillators are used, one for the TX and the other for the RX, to avoid insertion loss in the 60GHz LO distribution, which also contributes to maintain I/Q phase balance. The core area of the QILO is only 0.014mm², and it consumes 14.9mW from a 1.0V supply. The LO buffers consume 10.0mW and can be turned off in sleep mode. The 20GHz PLL in [4] is used for the injection-locking signal. The core area of the PLL is 1.2mm², and it consumes 66mW from a 1.2V supply. The PLL has a 2-stage divide-by-4 CML divider, a divide-by-5 static divider, and a programmable divider, /27, /28, /29 and /30, to generate 58.32GHz, 60.48GHz, 62.64GHz and 64.80GHz with a 36MHz reference, respectively. The measured frequency range of the PLL is from 17.9GHz to 21.2GHz. The overall phase noise is –94.2dBc/Hz@1MHz-offset at 60.48GHz, which is measured through the entire TX path including QILO, mixer and PA.

The antenna built in a package can radiate in the direction parallel to the printed circuit board [5]. The antenna is connected with the CMOS chip through a 270µm bonding wire. There are no 60GHz connections between the package and the board. This implementation is practically cost-effective since usual PCB materials can be applicable. The antenna in the package has a 2dBi gain, which is designed with the wire parasitics. The beam widths in the E- and H-planes are 120° and 72°, respectively.

The transceiver is fabricated in 65nm CMOS technology. The core areas of the transmitter and the receiver including all matching blocks are 3.5mm² and 3.8mm², respectively. Figure 9.1.5 shows the measured spectrum in QPSK mode with the IEEE802.15.3c spectrum mask. The input I/Q signal is generated by an arbitrary waveform generator (AWG) with a symbol rate of 1.76GS/s and a roll-off factor of 25%. The TX output signal is received by a horn antenna and is measured by a spectrum analyzer with a downconversion mixer. The measured spectrum meets the IEEE802.15.3c standard.

Figure 9.1.6 shows the measured constellation and performance summary, and Fig. 9.1.7 shows the die photo. Two test boards are used as TX and RX, and the 20GHz PLL on a probe station provides the injection signal for the boards. An AWG generates I/Q modulated signals for 16QAM/8PSK/QPSK/BPSK modes, and an oscilloscope is used to evaluate the constellation, EVM, and BER with a built-in software. Full-rate communication speed is possible for channel 1 (57.24 to 59.40GHz) and channel 2 (59.40 to 61.56GHz) of IEEE802.15.3c within a BER of < 10^{-3}. The measured EVM is from 12% (-18dB) to 14% (-17dB), and it can be improved up to 4% (-28dB) by using decision feedback equalization (DFE) realized by the software. The π/2 modulation series are also capable. Figure 9.1.6 also shows the communication distance range using the 2dBi antenna, and the low-gain mode of the LNA is used for short-distance receiving. The minimum BER is also confirmed up to <10^{-7} in QPSK mode (limited by measurement time). The symbol rate is 1.76GS/s with a roll-off factor of 25%, and the data rates with 2.16GHz BW are 1.76, 3.52, 5.28 and 7.04Gb/s for BPSK, QPSK, 8PSK and 16QAM, respectively. The maximum data rates using wider bandwidth in QPSK and 16QAM with a 25% roll-off are at least 8Gb/s and 11Gb/s within a BER of < 10^{-3}.

Acknowledgments:
This work was partially supported by MIC, MEXT, STARC, NEDO, Canon Foundation, and VDEC in collaboration with Cadence Design Systems, Inc., and Agilent Technologies Japan, Ltd. The authors thank Dr. Hirose, Dr. Suzuki, Dr. Sato, and Dr. Kawano of Fujitsu Laboratories, Ltd., Dr. Taniguchi of JRC, Dr. Hirachi of AMMSys Inc., Dr. Noda, Mr. Kondo, Mr. Yamagishi, and Dr. Fukuzawa of SONY, and Prof. Ando of Tokyo Institute of Technology for their valuable discussions and technical supports.

References:
[1] C. Marcu, D. Chowdhury, C. Thakkar, J.-D. Park, L.-K. Kong, M. Tabesh, W. Yanjie, B. Afshar, A. Gupta, A. Arbabian, S. Gambini, R. Zamani, E. Alon, and A. M. Niknejad, "A 90nm CMOS low-power 60GHz transceiver with integrated baseband circuitry," *ISSCC Dig. Tech. Papers*, pp. 314-315, Feb. 2009.
[2] N. Li, K. Bunsen, N. Takayama, Q. Bu, T. Suzuki, M. Sato, T. Hirose, K. Okada, and A. Matsuzawa, "A 24dB gain 51-68GHz CMOS low noise amplifier using asymmetric-layout transistors," *ESSCIRC Dig. Tech. Papers*, pp. 342-345, Sep. 2010.
[3] W. Chan, and J. Long, "A 56-65GHz injection-locked frequency tripler with quadrature outputs in 90-nm CMOS," *IEEE J. Solid-State Circuits*, vol. 43, no. 12, pp. 2739-2746, Dec. 2008.
[4] A. Musa, R. Murakami, T. Sato, W. Chaivipas, K. Okada, and A. Matsuzawa, "A 58-63.6GHz quadrature PLL frequency synthesizer in 65nm CMOS," *A-SSCC Dig. Tech. Papers*, pp.189-192, Nov. 2010.
[5] R. Suga, H. Nakano, Y. Hirachi, J. Hirokawa, and M. Ando, "Cost-effective 60-GHz antenna-package with end-fire radiation from open-ended post-wall waveguide for wireless file-transfer system," *IMS Dig. Tech. Papers*, pp. 449-452, May 2010.

978-1-61284-303-2/11 $26.00 © 2011 IEEE

ISSCC 2011 / February 22, 2011 / 8:30 AM

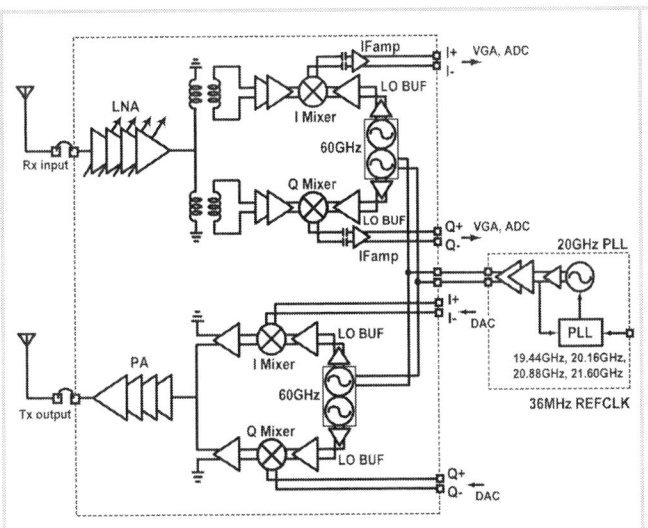

Figure 9.1.1: Block diagram of the 60GHz direct-conversion transceiver.

Figure 9.1.2: Schematics of the transmitter and measured output power.

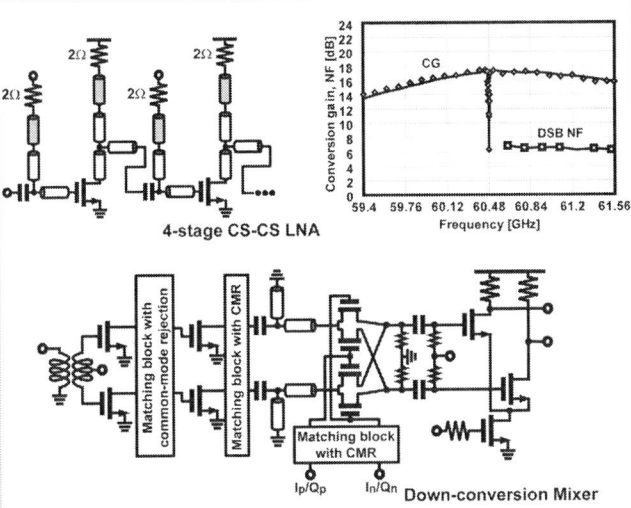

Figure 9.1.3: Schematics of the receiver, measured CG and NF.

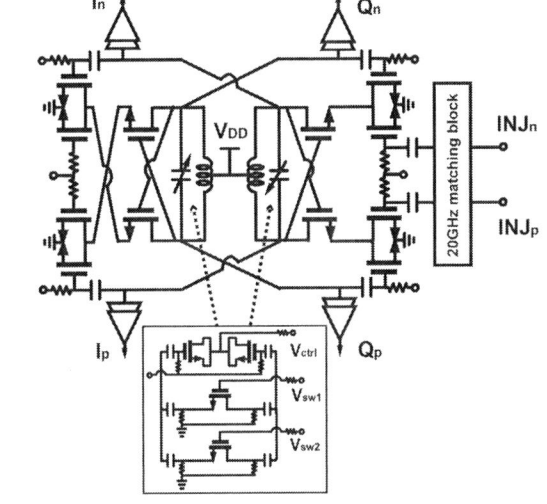

Figure 9.1.4: Schematics of the quadrature injection-locked oscillator.

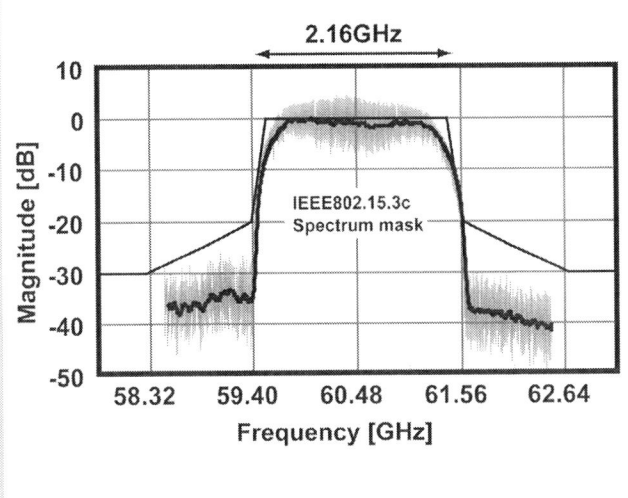

Figure 9.1.5: Measured spectrum for QPSK at TX output.

Constellation	1585 points	3170 points	4755 points	6340 points
Modulation	BPSK	QPSK	8PSK	16QAM
Data rate within 2.16GHz-BW	1.76Gb/s	3.52Gb/s	5.28Gb/s	7.04Gb/s
EVM	-18dB (-24dB with DFE)	-18dB (-28dB with DFE)	-17dB	-17dB
Distance (BER < 10^{-3})	0.5 – 274 cm	0.5 – 270 cm	0.5 – 20 cm	0.5 – 17 cm

Tx		Rx		PLL [4]	
CG	18.3dB	CG	17.3dB (high-gain mode) / 4.7dB (low-gain mode)	Frequency	17.9-21.2GHz
P_{1dB}	9.5dBm	NF	<6.8dB (high-gain mode)	Phase Noise through Tx @60.48GHz	-94.2dBc/Hz @1MHz-offset
P_{SAT}	10.9dBm	IIP3	-5dBm (only for LNA)	Ref. spur	<-58dBc
PAE	8.8% (only for PA)	P_{DC}	106mW	P_{out}	-2dBm
P_{DC}	186mW			P_{DC}	66mW

Figure 9.1.6: Measured constellation and performance summary.

ISSCC 2011 PAPER CONTINUATIONS

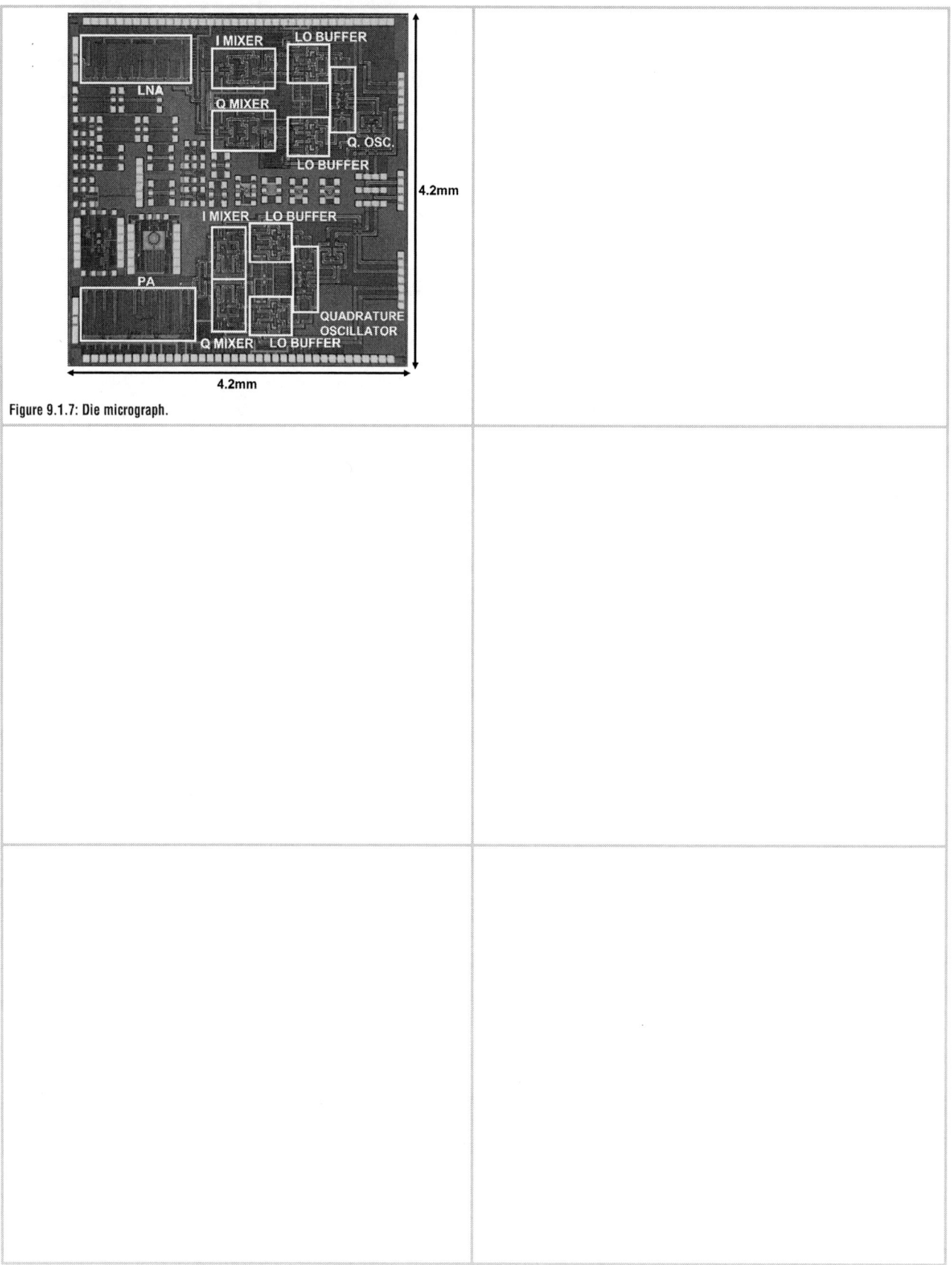

Figure 9.1.7: Die micrograph.

ISSCC 2011 / SESSION 9 / WIRELESS & mm-WAVE CONNECTIVITY / 9.2

9.2 A 65nm CMOS Fully Integrated Transceiver Module for 60GHz Wireless HD Applications

Alexandre Siligaris[1], Olivier Richard[2], Baudouin Martineau[2],
Christopher Mounet[1], Fabrice Chaix[1], Romain Ferragut[3],
Cedric Dehos[1], Jerome Lanteri[1], Laurent Dussopt[1],
Silas D. Yamamoto[2], Romain Pilard[2], Pierre Busson[2],
Andreia Cathelin[2], Didier Belot[2], Pierre Vincent[1]

[1]CEA-LETI-MINATEC, Grenoble, France,
[2]STMicroelectronics, Crolles, France,
[3]STMicroelectronics, Grenoble, France

This paper presents a fully integrated 60GHz transceiver module in a 65nm CMOS technology for wireless high-definition video streaming. The CMOS chip is compatible with the WirelessHD™ standard, covers the four channels and supports 16-QAM OFDM signals including the analog baseband. The ESD-protected die (9.3mm²) is flip-chipped atop a High Temperature Cofired Ceramic (HTCC) substrate, which receives also an external PA and the emission and reception glass-substrate antennas. The module occupies an area of only 13.5×8.5mm². It consumes 454mW in receiver mode and 1.357W in transmitter mode (357mW for the transmitter and 1W for the PA).

The transceiver architecture is shown in Fig. 9.2.1. It consists of a zero-sliding IF transceiver in which the LO frequencies are related to the RF frequency as follows: $F_{LO1}=F_{LO2}/2=F_{RF}/3$. The RX and the TX share a common frequency synthesis phase-locked to an external reference clock at 36MHz [1]. The chip includes a serial bus interface that encodes the channel selection, configures the analog parts and manages the power.

The LNA (Fig. 9.2.2) is a four-single-ended stages amplifier; the first one uses a common source topology improving NF and the last three stages use cascode topology to increase gain and stability. The last two cascode stages include an AGC that do not affect the NF and matching when switching from high-gain (HG) to low-gain (LG) mode. The measured gain of a stand-alone version is 10dB (LG) and 19dB (HG) with a 3dB bandwidth of 9GHz (57.5 to 66.5GHz). The measured NF is between 4.5dB and 5.5dB in the above mentioned band. The top of Fig. 9.2.3 shows the RX downconversion chain schematic. RF signals are downconverted to the 20GHz IF band with a differential mixer, while a tunable impedance load connected at the IF mixer output optimizes the gain for each channel. The mixer is followed by an IF amplifier (IFA) including channel gain tuning and AGC with 10dB voltage-gain difference. Then IF signals are demodulated and converted to I/Q base band (6MHz to 880MHz) by two double-balanced Gilbert cells. The RX analog baseband (ABB) comprises a 3rd-order Chebyshev Gm-C filter with 900MHz 3dB cut-off frequency, two gain stages (12dB and 10dB respectively), and three programmable CMOS attenuators for each I and Q channel. The ABB measured gain spans from -6dB to 16dB (0.88dB mean step) with a 3dB cut-off frequency of 850MHz (bottom-right of Fig. 9.2.2). The RX power consumption is 454mW; 74mW for the RF front-end, 80mW for the PLL and 300mW for the ABB.

The TX upconversion chain schematic is shown in the bottom of Fig. 9.2.3. The TX analog baseband includes two differential I and Q 5th-order Gm-C filters with 830MHz bandwidth. The modulator is composed of double-balanced Gilbert mixers with tunable loads. Each I and Q modulator is followed by a 20GHz IF amplifier with tunable load. The I and Q IF signals are combined into the upconverter (20 to 60GHz) by a current sum on the Gilbert cell transconductors. The Pre-PA (PPA) of Fig. 9.2.4 is constituted of two differential cascode stages with differential baluns optimizing the frequency band operation. The circuit reliability is ensured thanks to cascode topology [2]. The bottom of Fig. 9.2.4 shows the measured power gain (14 to 16.4dB) and output power (13dBm P_{sat}) of a stand-alone PPA circuit for the four channel frequencies of the WirelessHD™ standard. The output 1dB compression point is 7 to 8dBm and the power added efficiency is 7 to 8%. Measured S-parameters are shown in the last inset of Fig. 9.2.4. The TX power consumption is 357mW; 207mW for the RF front-end, 80mW for the PLL and 70mW for the ABB.

Figure 9.2.5 shows a micrograph of the transceiver (2.8×3.3mm²). For measurement purposes, the transceiver is wire-bonded on a PCB. The RF 60GHz input and output are measured on-chip with V-band RF probes while baseband signals are measured on the PCB. The receiver achieves a maximum conversion gain (in

HG mode) of 41dB at the 3rd and 4th channels, and 33dB and 36dB at the 1st and 2nd channels respectively (Fig. 9.2.6). The output 1dB compression point is -1dBm and -6dBm for the HG and the LG modes respectively. In order to validate the transceiver's functionality, data frames at 3.8Gb/s rate corresponding to the WirelessHD™ HRP2 mode (OFDM 16-QAM modulation) have been generated with an arbitrary waveform generator. Both TX and RX on-chip performance have been measured with Agilent's "Connected Solution". The top left of Fig. 9.2.6 shows an example of the TX constellation at the 4th communication channel (after demodulation). Similar results were obtained on the other channels. The TX exhibits an EVM of 9.2%, which is compliant with the WirelessHD™ requirements. The RX measurements exhibit also a clear constellation with an EVM around 11%, also compliant with the WirelessHD™ requirements.

A full-module solution for wireless communication is proposed in the bottom right of Fig. 9.2.5. The module consists of a 5-layers HTCC substrate [3] carrying the 65nm CMOS transceiver die in a flip-chip configuration, the TX and RX antennas and an external PA [2]. The two antennas are identical, and are implemented as Copper-track folded dipoles on a glass substrate and flip-chip mounted atop a cavity on the HTCC substrate. They present an omnidirectional gain of 5dBi, including the interconnect transmission lines losses, and are spaced at 500µm in order to ensure minimal interference. The module has been fully assembled on an industrial packaging line, thus providing a fully industrial packaging solution for mm-Wave transceivers. First experiments with this module demonstrated a wireless transmission range above 1 meter in an office-like space for the 16-QAM OFDM HRP2 (3.8Gb/s). In the bottom left of Fig. 9.2.6 the output spectrum of the TX module at the 4th channel is shown (wired without antenna). The bottom right of Fig. 9.2.6 shows the received spectrum of the ABB output after 1 meter wireless transmission. The mean value of the RX SNR is higher than 20dB in all four standard channels, which is compliant with 16-QAM OFDM demodulation. Finally, the table in Fig. 9.2.7 shows a summary of the progress on 60GHz transceivers in Silicon technologies.

To conclude, this work describes a 65nm CMOS 60GHz transceiver, assembled on a low-cost module with high aperture antennas in an industrial implementation. The module has achieved a 16-QAM OFDM modulation wireless link at the four WirelessHD™ standard channels over 1 meter distance.

Acknowledgements:
The authors would like to acknowledge Kyocera for the module manufacturing, Hilal Hezzedine in ST Tours for the antennas manufacturing, Caroline Arnaud and Daniel Gloria in ST Crolles and David Grieve in Agilent Technologies for the measurement support, and the European Community funded Qstream project. In memory to Willy Beulé.

References:
[1] O. Richard, A. Siligaris, F. Badets, C. Dehos, C. Dufis, P. Busson, P. Vincent, D. Belot, and P. Urard, "A 17.5-to-20.94GHz and 35-to-41.88GHz PLL in 65nm CMOS for Wireless HD Applications," *ISSCC Dig. Tech. Papers*, pp. 252-253, Feb. 2010.
[2] B. Martineau, V. Knopik, A. Siligaris, F. Gianesello, D. Belot, "A 53-68GHz 18dBm Power Amplifier with 8 combined ways in standard 65nm CMOS," *ISSCC Dig. Tech. Papers*, pp.428-429, Feb. 2010.
[3] J. Lanteri, L. Dussopt, R. Pilard, D. Gloria, S.D. Yamamoto, A. Cathelin, H. Hezzeddine "60 GHz Antennas in HTCC and Glass Technology," *4th European Conference on Antennas and Propagation (EuCAP-2010)*, Barcelona, Spain, 12-16 April 2010.
[4] S. Pinel, S. Sarkar, P. Sen, B. Perumana, D. Yeh, D. Dawn, J. Laskar, "A 90nm CMOS 60GHz Radio," *ISSCC Dig. Tech. Papers*, pp. 130-131, Feb. 2008.
[5] C. Marcu, D. Chowdhury, C. Thakkar, Ling-Kai Kong, M. Tabesh, Jung-Dong Park, Yanjie Wang, B. Afshar, A. Gupta, A. Arbabian, S. Gambini, R. Zamani, A.M. Niknejad, E. Alon, "A 90nm CMOS low-power 60GHz transceiver with integrated baseband circuitry," *ISSCC Dig. Tech. Papers*, pp. 314-315, Feb. 2009.
[6] S.K. Reynolds, A.S. Natarajan, T. Ming-Da, S. Nicolson, C.Z. Jing-Hong, L. Duixian, D.G. Kam, O. Huang, A. Valdes-Garcia, B.A. Floyd, "A 16-element phased-array receiver IC for 60-GHz communications in SiGe BiCMOS," *IEEE Radio Frequency Integrated Circuits Symposium*, pp. 461-464, 2010.
[7] E. Cohen, C. Jakobson, S. Ravid, D. Ritter, "A thirty two element phased-array transceiver at 60GHz with RF-IF conversion block in 90nm flip chip CMOS process," *IEEE Radio Frequency Integrated Circuits Symposium*, pp. 457-460, 2010.

ISSCC 2011 / February 22, 2011 / 9:00 AM

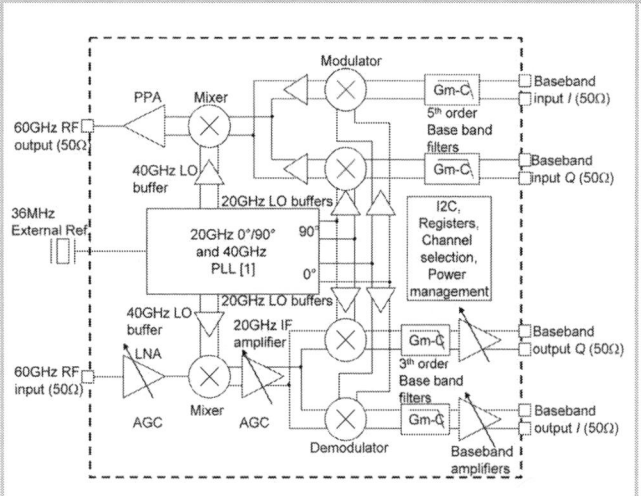

Figure 9.2.1: Architecture of the zero-sliding IF transceiver.

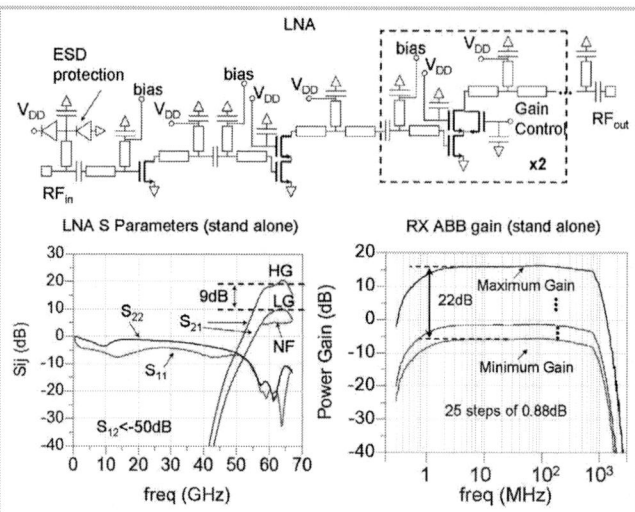

Figure 9.2.2: Schematic of the four stages LNA. Measured S parameters of the LNA and RX ABB gain.

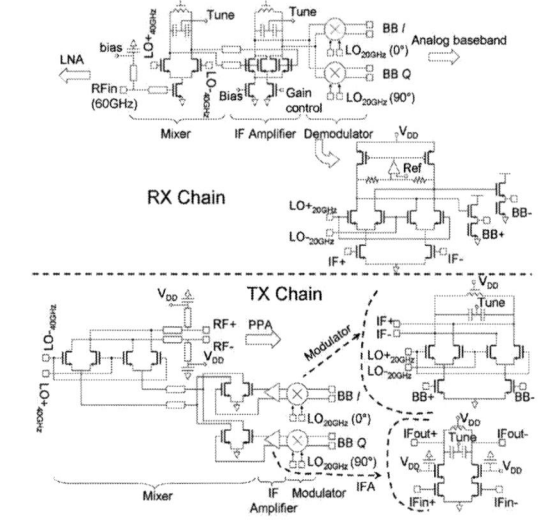

Figure 9.2.3: RX and TX chain schematics.

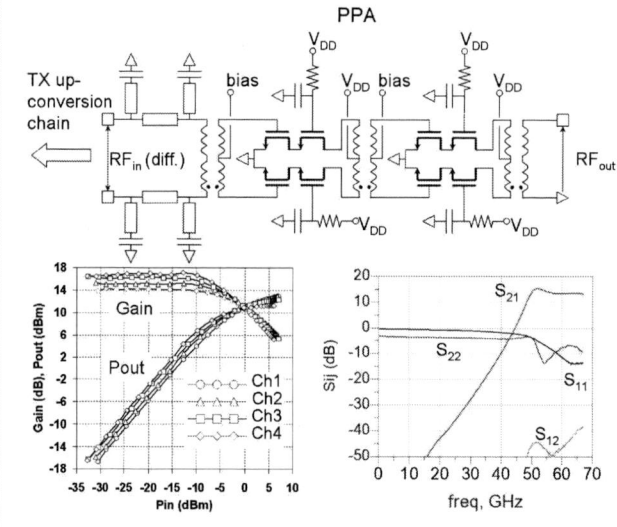

Figure 9.2.4: PPA schematic. Power and S-parameter measurements.

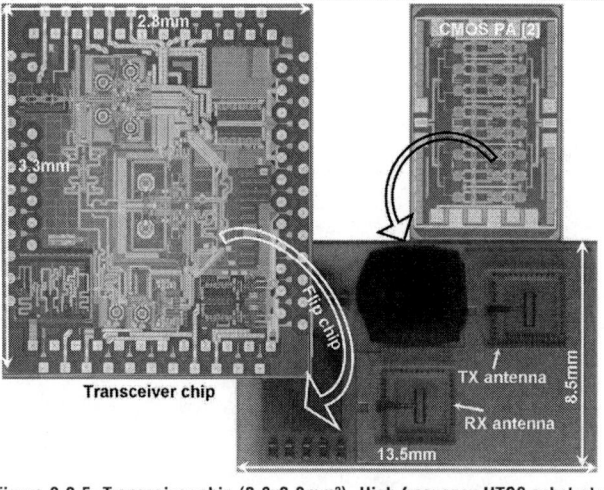

Figure 9.2.5: Transceiver chip (2.8x3.3mm²). High-frequency HTCC substrate carrying the CMOS die, the glass antennas and the external CMOS PA (13.5 x 8.5 mm²).

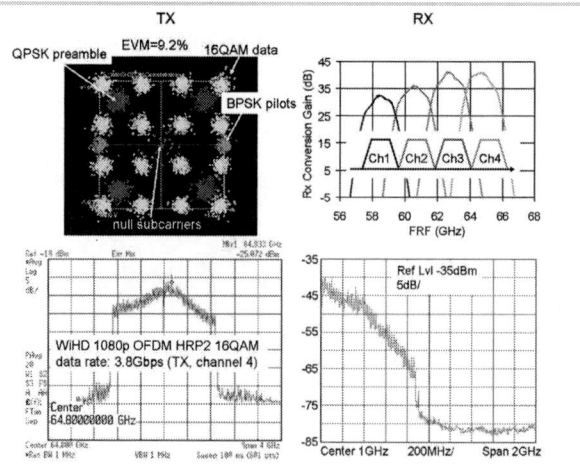

Figure 9.2.6: TX 16-QAM OFDM constellation with 9.2% EVM. Emitted spectrum at the output of the TX module. RX conversion gain. Received analog baseband signal at the 4th channel over 1m distance.

978-1-61284-303-2/11 $26.00 © 2011 IEEE

163

Reference	Tech.	P_{DC} (mW) RX / TX / PLL	Emitted Power	Channels	Integration	Modulation scheme	Package
S.Pinel, ISSCC 2008 [4]	90nm CMOS	129 / 113 / 80	Psat (dBm) 8.4	4	Full TRX, PLL and ABB	QPSK 16QAM	FR4
C.Marcu, ISSCC 2009 [5]	90nm CMOS	138 / 170 / 76	Psat (dBm) 11	1	Full TRX, PLL and ABB	QPSK	N/A
S.K.Reynolds, RFIC 2010 [6]	0.12µm SiGe BiCMOS	RX phased array: 1800	-	4	Phased array x16, RX, PLL	OFDM 16QAM	Organic BGA
E. Cohen, RFIC 2010 [7]	90nm CMOS	RX / TX phased array: 500 / 500	Psat (dBm) 8 (array)	N/A	Phased array x32, TX, RX, no PLL	N/A	Alumina
This work	65nm CMOS	74+300* / 207+70*+1000**/ 80	Psat (dBm) >16	4	Full TRX, PLL and ABB & filters	OFDM 16QAM	HTCC + glass Antennas

* Analog Base Band (ABB) **external CMOS PA

Figure 9.2.7: Summary of the presented work and comparison to state of the art fully integrated 60GHz transceivers.

ISSCC 2011 / SESSION 9 / WIRELESS & mm-WAVE CONNECTIVITY / 9.3

9.3 A 60GHz CMOS Phased-Array Transceiver Pair for Multi-Gb/s Wireless Communications

Sohrab Emami, Robert F Wiser, Ershad Ali, Mark G Forbes, Michael Q Gordon, Xiang Guan, Steve Lo, Patrick T McElwee, James Parker, Jon R Tani, Jeffery M Gilbert, Chinh H Doan

SiBEAM, Sunnyvale, CA

Recent advances in silicon technology, mm-Wave integrated circuit/antenna/package design, and beam-forming techniques at 60GHz, together with the emergence of suitable wireless standards, have enabled consumer electronics products to support wireless transmission of multi-Gb/s data such as high-definition (HD) audio/video content [1,2]. Further expansion into portable and mobile platforms will require lower power consumption, smaller form factor, and lower cost. This paper describes a fully integrated, low-cost 60GHz phased-array transceiver pair, implemented in 65nm standard digital CMOS and packaged with an embedded antenna array, capable of robust 10m non-line of sight (NLOS) communication. The array is configurable from 32 elements to 8 or fewer elements, making the transceiver pair suitable for both fixed, high-data-rate and portable, low-power applications. To enhance the robustness of the multi-element design, dynamic phase shifters allow the beam direction to be changed in real time to adapt to changing environments without interruption of the multi-Gb/s data stream. The transceiver pair supports the WirelessHD and draft 802.11ad (WiGig) standards at maximum data rates of 7.14Gb/s and 6.76Gb/s, respectively.

Figure 9.3.1 shows the block diagram of the Source and Sink transceivers along with the frequency plan. The baseband (BB) and IF circuit blocks are fully differential while the RF/mm-Wave blocks are single-ended. The design methodology for the mm-Wave circuit blocks relies on accurate high-frequency device models and the use of coplanar waveguide transmission lines for power matching [3]. The Source and Sink transceivers share common circuit blocks in the BB to RF signal paths, and in the LO generation. However, the 60GHz signal distribution and number of RF transmit (TX) and receive (RX) elements for the two chips are different to allow for an asymmetrical link. In the high-data-rate direction, the Source uses 32 TX elements to transmit to 32 RX elements on the Sink. For the reverse lower-data-rate direction, the Sink uses 8 TX elements to transmit to 4 RX elements on the Source. The number of active elements can be configured on both transceivers allowing a trade-off between power dissipation and performance.

The BB I/Q channels in the TX path include a VGA and a 1GHz low-pass filter. Upconversion from BB to IF is performed by a double-balanced I/Q mixer. The upconversion to RF is performed by a double-balanced mixer power matched to the first splitter stage. The RF signal is distributed to the M TX elements via a tree of active 1:2 3dB splitters. Each RF TX element consists of an active phase shifter (PS) and power amplifier (PA). The PA provides more than 22dB of gain over 8GHz of bandwidth to ease linearity requirements of the preceding stages. Programmable bias of the PA allows for a trade-off between power consumption and linearity. The OP_{1dB} can be varied from 0dBm to +9dBm with a peak power-added efficiency of 22%. A power detector (PD) at the PA output allows the transmitted power to be controlled.

The RX path consists of N elements, each containing a low-noise amplifier (LNA) and PS, followed by passive 2:1 power combiners. The 3-stage LNA provides more than 20dB of gain, with two coarse gain steps to ease linearity requirements of subsequent stages. After power combining, the downconversion to IF is performed by a single-gate mixer loaded by an IF VGA. A double-balanced I/Q mixer downconverts to BB and is followed by the BB VGA and filter. The BB and IF VGAs provide fine gain control in addition to the gain steps available at RF.

The transceivers use a dual-conversion sliding IF architecture with local oscillator (LO) frequencies at $f_{RF}/5$ and $4f_{RF}/5$. The synthesizer generates the $f_{RF}/5$ LO centered at one of two channels, 12.096GHz or 12.528GHz. Spot phase noise of the carrier at 75kHz and 1MHz offsets when centered at 60.48GHz was measured at the TX output to be -66dBc/Hz and -96dBc/Hz, respectively. The phase-noise

profile satisfies both WirelessHD and WiGig requirements for 64-QAM transmissions. The $4f_{RF}/5$ LO is generated by a frequency quadrupler. The LO path for the TX RF mixer and simplified circuit schematic of the quadrupler are shown in Fig. 9.3.2. The quadrupler consists of two stages of frequency doubling and a cascoded output buffer. Cascodes M_{1A} and M_{1B} are biased at subthreshold and form a push-push doubling stage that sums the drain currents to select even-order harmonics. M_2 is also biased at subthreshold and performs another frequency doubling. Explicit filtering of unwanted harmonics at the output of the second doubler is not required due to the selectivity provided by subsequent matching networks along the LO chain. The quadrupler output buffer, M_3, improves reverse isolation and provides a match to the directional coupler. The amplitude of the $4f_{RF}/5$ LO is monitored by the LO PD and controlled by a feedback loop. At the mixer LO port, a dual transformer architecture is used to convert the single-ended LO to a well-balanced differential signal. The 50Ω damping resistor at the virtual ground of the second transformer stabilizes the mixer. The mixer achieves better than 40dBc of LO suppression.

Bare die were tested at RF frequencies on a probe station, and packaged die were tested over-the-air paired with digital BB chips designed for the WirelessHD 1.0 standard. The CMOS ICs are flip-chip bonded to a multiple-metal-layer interconnected ceramic package with integrated antennas. Full ESD protection is provided on all I/Os. Fig. 9.3.3 shows the received constellation achieving an EVM of -19.2dB for a 16-QAM 512 carrier OFDM signal with channel bandwidth of 1.76GHz and PHY rate of 3.8Gbps. In this mode the system is transmitting 3.0Gb/s of uncompressed 1080p/60Hz video in a 10m NLOS scenario. The transmitted spectrum and WirelessHD spectral mask [4] shown in Fig. 9.3.3 exhibits minimal spectral regrowth at an EIRP of +28dBm.

Figure 9.3.4 shows a computed array factor for the Source and Sink arrays for a typical operating link. Paired with the WirelessHD BB chip, whenever an interferer blocks the current path the system performs a dynamic beam search to find the new optimal path using the electronically steerable phased array. Beam search takes less than 1ms. The system also allows for adaptation of the PHY data rate to accommodate for difficult wireless environments. Figure 9.3.5 shows the trade-off between the PHY rate and range for different Source and Sink configurations in both the LOS and typical NLOS cases. A further trade-off between power consumption and data rate/range can be made by adjusting the number of active elements. For example, dropping the number of active Source TX elements from 32 to 8 reduces the power consumption by more than half, at the cost of a 4x decrease in range. In this scenario, the link still supports multi-Gb/s data rate in both directions.

Performance of the transceivers is summarized in Fig. 9.3.6, and die micrographs are shown in Fig. 9.3.7. The areas of the Source and Sink die are 72.7mm² and 77.2mm² respectively including pads. Both transceivers meet worldwide in-band and spurious emission regulatory limits. The presented transceiver pair achieves the required low-cost, robust performance, and low-power configurations to extend multi-Gb/s data-rate capabilities to both fixed and portable applications.

Acknowledgements:
We thank C. Shen, P. Tran, J. Garcia, S. Bennett, and K.N. Toussi for help with layout, test, and verification.

References:
[1] J. M. Gilbert, C. H. Doan, S. Emami, and C. B. Shung, "A 4-Gbps Uncompressed Wireless HD A/V Transceiver Chipset", *IEEE Micro*, pp. 56-64, March-April 2008.
[2] A. Valdes-Garcia, et al., "A SiGe BiCMOS 16-Element Phased-Array Transmitter for 60GHz Communications", *IEEE Int. Solid-State Circuits Conf. Dig. Tech. Papers*, pp. 218-219, Feb. 2010.
[3] C. H. Doan, S. Emami, A. M. Niknejad, and R. W. Brodersen, "Millimeter-wave CMOS design," *IEEE J. Solid-State Circuits*, vol. 40, pp. 144-155, Jan. 2005.
[4] WirelessHD Specification, Revision 1.0, January 3, 2008 [*Online*]. Available: www.wirelessHD.org.

978-1-61284-303-2/11 $26.00 © 2011 IEEE

ISSCC 2011 / February 22, 2011 / 9:30 AM

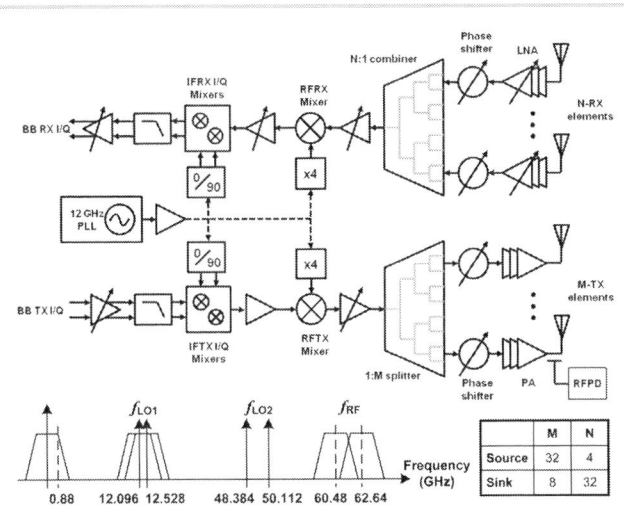

Figure 9.3.1: System block diagram and frequency plan.

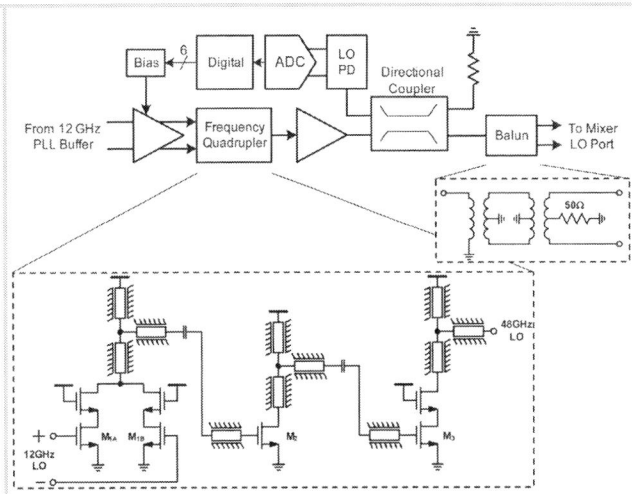

Figure 9.3.2: Block diagram of LO distribution and schematics of the frequency quadrupler and double-transformer balun.

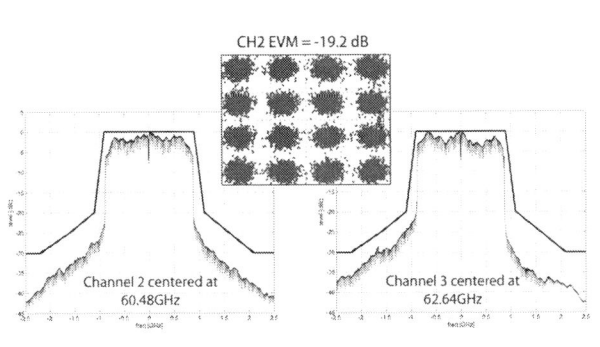

Figure 9.3.3: Transmit spectrum and WirelessHD spectral mask for Channels 2 and 3, and the received constellation for Channel 2.

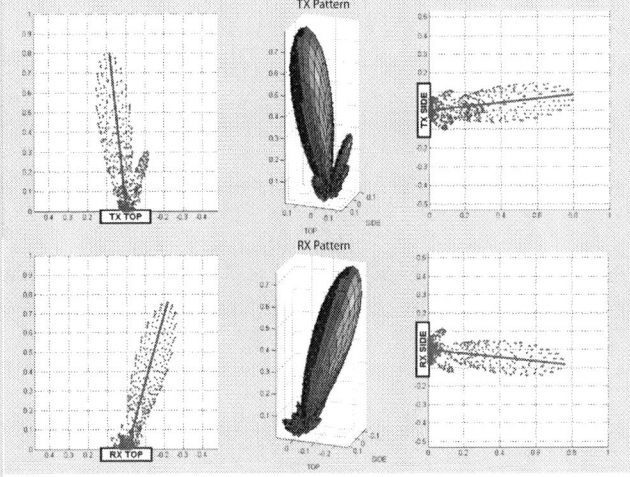

Figure 9.3.4: Computed amplitude patterns for the Source and Sink transceivers.

	Data Rate	Forward Channel Range† (m)	
		32 Element Source To 32 Element Sink	8 Element Source To 32 Element Sink
LOS	3.8 Gb/s	50	13
	1.9 Gb/s	184	47
	0.95 Gb/s	390	101
NLOS*	3.8 Gb/s	16	4
	1.9 Gb/s	58	15
	0.95 Gb/s	123	32

* NLOS distances are dependent on the quality of any reflection in the signal path.
NLOS range here assumes a reflection with 11 dB loss exists between Source and Sink.

† Range is the distance at which the packet error rate exceeds 10^{-3}

Figure 9.3.5: Range of the system for various PHY rates for both LOS and NLOS cases.

Technology	65nm Digital CMOS			
Supply Voltage	1.0V, 2.5V(PLL, I/O)			
Supported Channels	59.40GHz ~ 61.56GHz, 61.56~63.72GHz			
Data Transfer Rate	4 Gb/s			
Link Margin	14dB, operating at 10m, 1080p/60Hz 4:4:4 video with BER≤10^{-11}			
Source Chip Performance				
Die Size	8.95mm x 8.12mm			
Number of RF Elements	32 TX, 4 RX			
EIRP	+28 dBm			
EVM	<-19 dB @ +28 dBm EIRP			
Mode	32 TX elements on		8 TX elements on	
Temperature	25°C	75°C	25°C	75°C
Tx Power Consumption	1.82W	2.16W	895mW	1.03W
Sink Chip Performance				
Die Size	8.64mm x 8.93mm			
Number of RF Elements	32 RX, 8 TX			
RX Sensitivity	-72 dBm at each antenna element, HRP full rate			
RX Max Signal (all RX antenna on)	-32 dBm at each antenna element			
RX Noise Figure	< 10dB			
Mode	32 RX elements on		4 RX elements on	
Temperature	25°C	75°C	25°C	75°C
Rx Power Consumption	1.25W	1.54W	711mW	786mW

Figure 9.3.6: Summary table of the transceiver performances.

978-1-61284-303-2/11 $26.00 © 2011 IEEE

ISSCC 2011 PAPER CONTINUATIONS

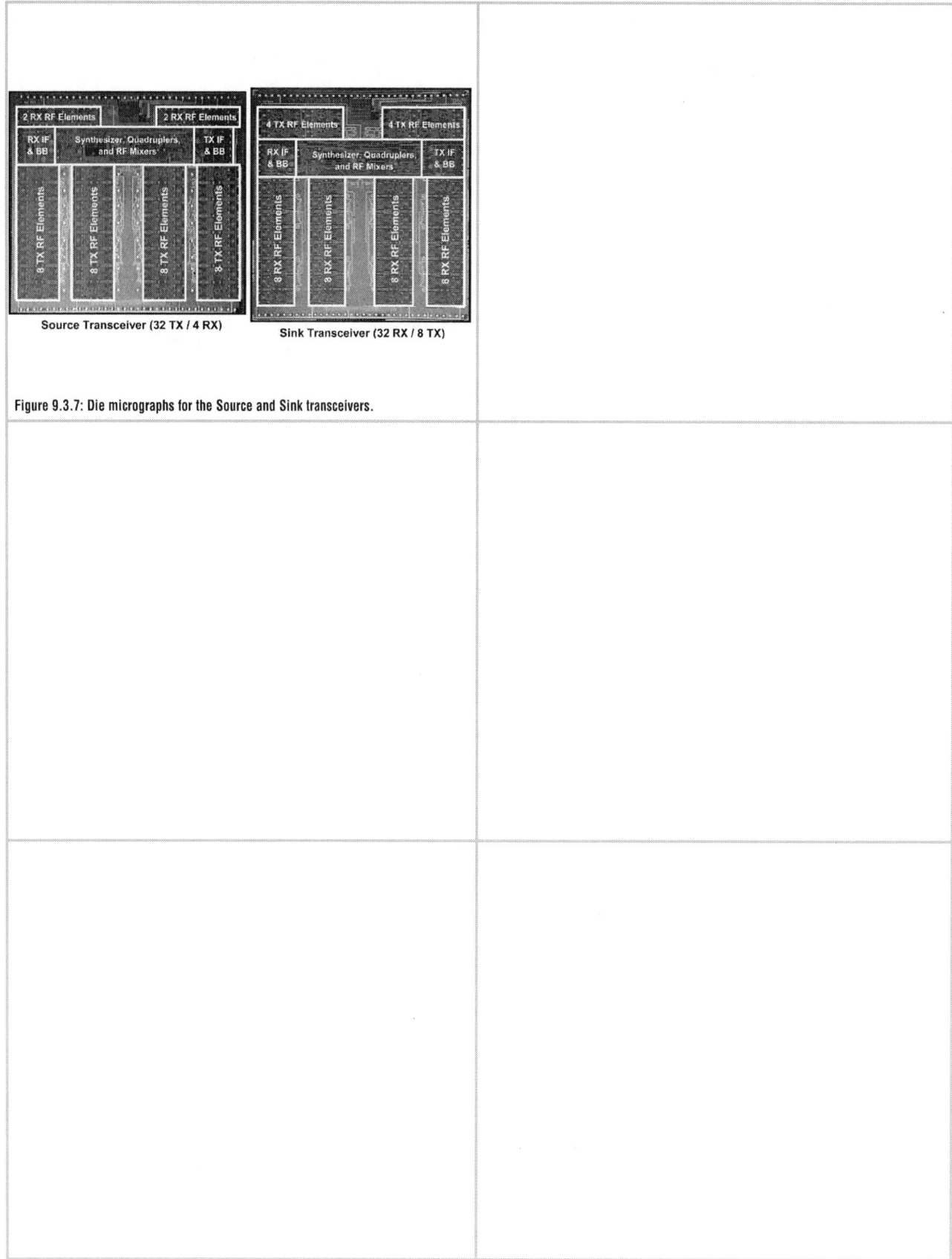

Figure 9.3.7: Die micrographs for the Source and Sink transceivers.

ISSCC 2011 / SESSION 9 / WIRELESS & mm-WAVE CONNECTIVITY / 9.4

9.4 A 65nm CMOS 4-Element Sub-34mW/Element 60GHz Phased-Array Transceiver

Maryam Tabesh, Jiashu Chen, Cristian Marcu, Lingkai Kong,
Shinwon Kang, Elad Alon, Ali Niknejad

University of California, Berkeley, CA

The 60GHz band has received significant attention as an enabler for multi-Gb/s wireless communication. Practical mm-Wave systems will require relatively large phased arrays in order to robustly overcome path-loss and fading issues. Despite significant progress [1,2], CMOS implementations of 60GHz phased arrays have so far been area and power hungry. This paper therefore presents a 60GHz 4-element 65nm CMOS phased-array transceiver consuming <34mW/element (including LO synthesis and distribution) that utilizes baseband (BB) phase shifting and holistic impedance optimization.

While all phased-array architectures require some form of mm-Wave combining or splitting, phase shifters at RF tend to be large and can introduce substantial losses on either the RF or LO signal paths. On the other hand, phase shifting at BB can achieve improved phase-shifting resolution as well as much lower loss [3], hence this architecture was adopted (Fig. 9.4.1). This approach requires I/Q mixers in each element, so optimizing the interface impedance and power consumption of these components is essential.

Specifically, in the transmitter (TX) (Fig. 9.4.2), maximizing the current efficiency of the mixers is critical not only in reducing mixer power, but also in maximizing the impedance at the LO ports. Combining the phase rotator/quadrature mixer into a single structure is attractive since their bias currents can be shared, but in a conventional BB phase rotator analog current-mode subtraction of the scaled I and Q signals limits the current efficiency to 50 to 71%. This design therefore utilizes steering of two current digital-to-analog converters: one carrying cosine weighting, the other sine weighting. As shown in Fig. 9.4.2, the current distribution switches (S1-S8) are controlled by the quadrant of the input data and the rotation angle in order to steer the cosine/sine currents to the appropriate I/Q channel, improving the current efficiency to 71 to 100% and increasing the mixer LO-port impedance by 40%.

Using a 2:1 transformer balun to current combine the outputs of the quadrature mixer and increase the voltage swing at the input of the power amplifier further optimizes TX efficiency. Since QPSK modulation has small peak-to-average power ratio, this enables the use of a zero-voltage-switching amplifier with harmonic tuning. The second harmonic reactance of the tuning network is chosen to maximize efficiency, resulting in 0dBm output power and 4.9dB gain, with 34% efficiency in simulation including the loss of the tuning network and RF pad. The measured output power (Fig. 9.4.5) and efficiency of a single element are -1.5dBm and 20% respectively with 8GHz 3dB output power bandwidth (BW). The TX consumes a total worst-case (45° phase shift) power of 27mW/element (including LO dist.) and achieves 5° of phase-shift resolution.

Impedance optimization is equally important in the receiver (RX) (Fig. 9.4.3). For a given gain and BW requirement, device and current scaling is used to reduce power consumption in the three-stage, inductively degenerated, low-noise amplifier (LNA). This design includes integrated ESD protection as part of the input matching network. Cascode devices are used due to their high maximum stable gain, unconditional stability at these frequencies, and high reverse isolation.

Transistor width and impedance scaling is also used to reduce the power consumption in the mixer. However, scaling the mixer devices is limited by passive losses, noise, and BW considerations. Progressively smaller switch sizes require larger inductors for matching at the LO port. Further reduction in switch size also results in a higher required overdrive and hence larger LO power for the same gain due to BW limitations. Holistic optimization of the RX and LO chain resulted in the choice of a single-balanced (instead of double-balanced) mixer to reduce the LO power. A series-resonant LO trap is used to reduce the LO leakage at the output of the mixer. A series-gain-enhancement tuning network, implemented with 80Ω CPW transmission line, reduces high-frequency loss and thus increases the mixer gain [4].

The BB phase shifter consists of a bank of current-summed g_m stages with digitally controlled polarities and input sources. The relative gains of I and Q channels are controlled with 3 bits of resolution, and a polarity bit controls the phase quadrant to achieve a measured worse-case phase resolution of 11°. To reduce power consumption, a partial I/Q sharing structure [5] is implemented here,

resulting in only 12 g_m cells in each phase shifter. The outputs from the 4 elements of the phased-array receiver are summed in current mode and fed into a BB buffer to drive off-chip loads.

A single-element RX path consumes only 27mW (including LO dist.) and has measured gain and input-referred 1dB compression point (IP_{1dB}) of 24dB and -29dBm respectively, with S_{11} better than -10dB from 56 to 65GHz. An average noise figure (NF) of 6.8dB was measured over 2GHz of IF BW with minimum NF as low as 6.3dB. The measured receiver RF 3dB BW was higher than 6.5GHz while the BB BW was 1.8GHz including board and bonding limitations. Figure 9.4.5 includes relevant receiver measurements.

The 60GHz LO for both the TX and RX is generated by a fully integrated, integer-N, charge-pump-based phase-locked loop (PLL) that has been optimized for minimum integrated output phase noise. The VCO has 3 bits of coarse tuning and an analog varactor equivalent to 2LSB of the coarse bank. The overall measured tuning range is from 57.9 to 65.56GHz (12.4%) with free-running phase noise of -110dBc/Hz at 10MHz offset. The differential VCO outputs are buffered to isolate the VCO and each drives a single-ended distribution network for the RX and TX, respectively. The divider chain consists of an injection-locked divide-by-2 prescaler followed by a CML divide-by-4, a TSPC divide-by-2, and finally a programmable CMOS divider.

The LO distribution network (Fig. 9.4.4) was designed to minimize power consumption while maintaining scalability for larger phased arrays. Maintaining a constant impedance with transmission lines and matched power splitters allowed arbitrary routing of the LO signal, which scales well for larger arrays. In this design, in-phase splitting is performed by Wilkinson dividers to each of the 4 elements, followed by local hybrids to generate the quadrature LO. The Wilkinson dividers utilize meandered CPW transmission lines to reduce the area required. The hybrid is a transformer-based lumped design that requires very little area by comparison (area is only $0.002\lambda^2$). Finally, a buffer is required to drive the mixer impedance and to provide gain. For a small number of elements, the gain of this buffer easily offsets the loss of the distribution network. However, for larger arrays, further buffering would be required within the distribution network by, for example, using active signal splitters [1]. Nevertheless, on a per-element basis, the power consumption is dominated by the mixer buffers, which are always required and are thus optimized to maximize drain efficiency while driving the mixer impedance with a sufficient voltage swing for high conversion gain.

The circuit is fabricated in a 65nm CMOS technology and measures 2.5x3.5mm² (Fig. 9.4.7). Measurements were performed in a chip-on-board configuration and by directly probing mm-Wave signals. The total measured power consumption is 137mW in both RX and TX mode, including 29mW for the synthesizer. The RX phase-rotation constellation (Fig. 9.4.6) was measured by applying a 60.5GHz carrier to the LNA and a 60GHz LO with the BB analog outputs observed on an oscilloscope. The TX phase-rotation constellation (Fig. 9.4.6) was calculated from S-parameters. Finally, the measured constellations from all channels were used to synthesize the resulting TX and RX array patterns (Fig. 9.4.6).

Acknowledgement:
The authors acknowledge BWRC sponsors, NSF Infrastructure Grant No. 0403427, chip fabrication donated by ST Microelectronics, C2S2, and Berkeley Design Automation.

References:
[1] A. Valdes-Garcia, *et. al.* "A SiGe BiCMOS 16-element phased-array transmitter for 60GHz communications," *ISSCC Dig. Tech. Papers*, vol. 53, pp.218-219, Feb. 2010.
[2] E. Cohen, *et. al.* "A thirty two element phased-array transceiver at 60GHz with RF-IF conversion block in 90nm flip chip CMOS process," *IEEE Radio Frequency Integrated Circuits Symp.*, pp.457-460, May 2010.
[3] K. Raczkowski, *et. al.* "A wideband beamformer for a phased-array 60GHz receiver in 40nm digital CMOS," *ISSCC Dig. Tech. Papers*, vol. 53, pp.40-41, Feb. 2010.
[4] B. Afshar, *et. al.* "A Robust 24mW 60GHz Receiver in 90nm Standard CMOS," *ISSCC Dig. Tech. Papers*, vol. 51, pp.182-605, Feb. 2008.
[5] C. Marcu, *et. al.* "A 90nm CMOS low-power 60GHz transceiver with integrated baseband circuitry," *ISSCC Dig. Tech. Papers*, vol. 52, pp.314-315,315a, Feb. 2009.
[6] S. K. Reynolds, *et. al.* "A 16-Element Phased-Array Receiver IC for 60-GHz Communications in SiGe BiCMOS," *IEEE Radio Frequency Integrated Circuits Symp.*, pp.461-464, May 2010.

978-1-61284-303-2/11 $26.00 © 2011 IEEE

ISSCC 2011 / February 22, 2011 / 9:45 AM

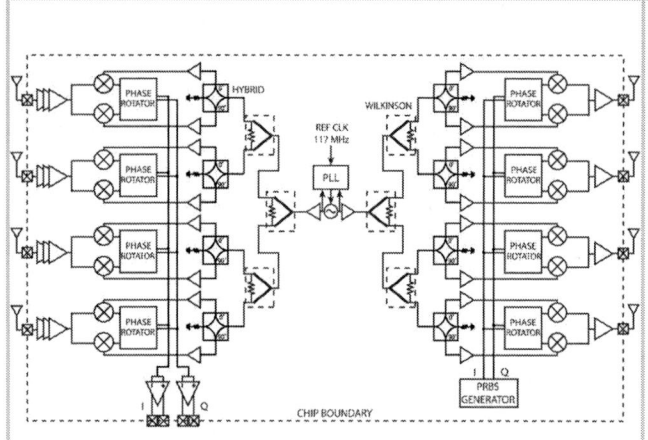

Figure 9.4.1: Block diagram of the phased-array transceiver, including 4 RX and 4 TX elements, integrated BB phase rotators, VCO/PLL and LO distribution networks.

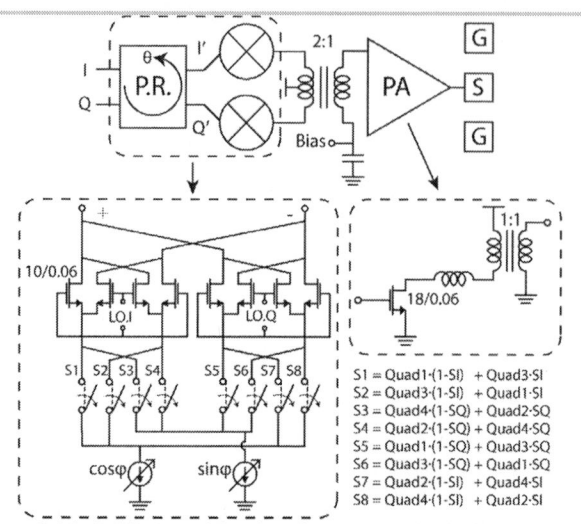

Figure 9.4.2: Architecture of the phased-array transmitter utilizing a stacked mixer/phase rotator architecture and ZVS mode switching PA.

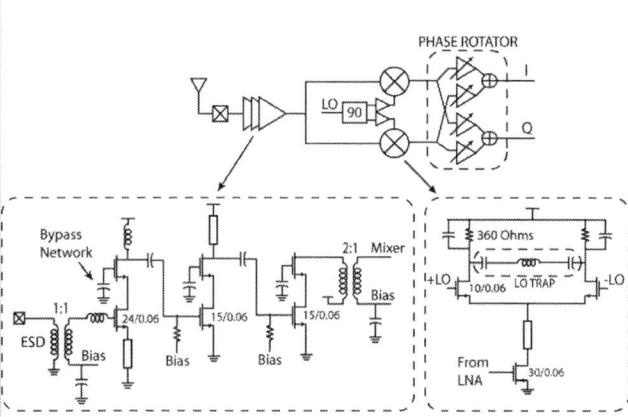

Figure 9.4.3: Schematic of the receiver chain including the transformer-coupled ESD input stage, three-stage LNA, single-balanced mixer, and baseband I/Q phase rotator.

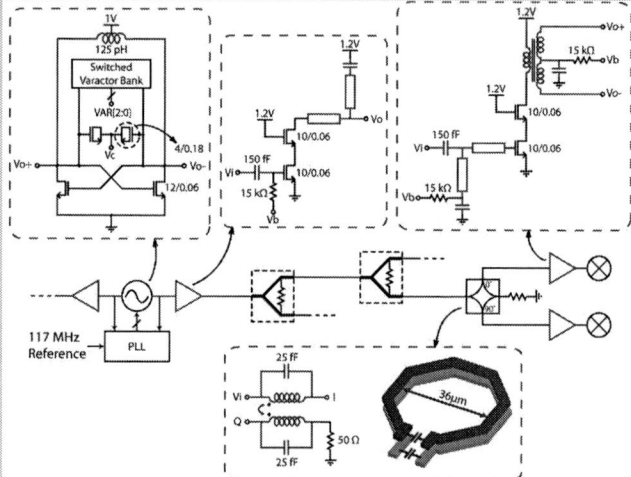

Figure 9.4.4: Schematic of VCO, LO buffers, and LO distribution chain including Wilkinson power splitters and transformer-coupled lumped quadrature hybrid.

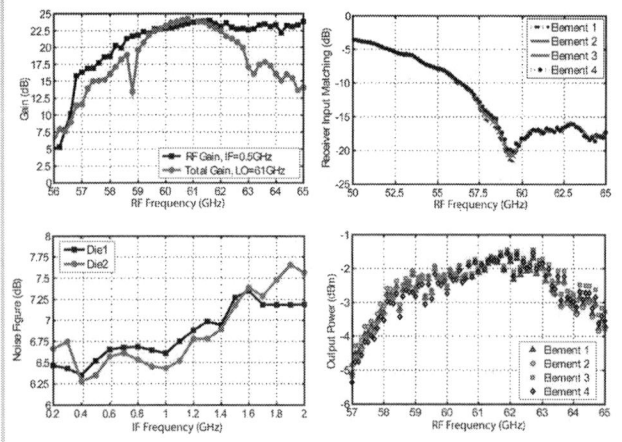

Figure 9.4.5: Measurements of RX gain, BW, S_{11}, and NF (LO frequency=61GHz); and TX output power. RF BW measured by sweeping LO and RF together maintaining IF=0.5GHz.

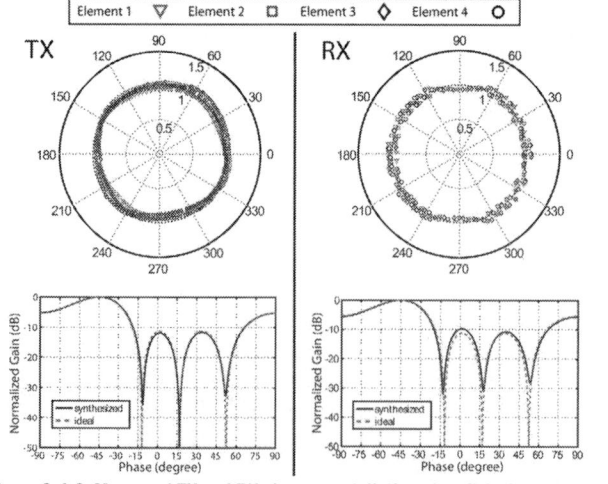

Figure 9.4.6: Measured TX and RX phase constellations for all 4-elements and synthesized array patterns with array steered to 45° angle. Phase rotator resolution: TX=5°, RX=11° (with +/-0.5dB gain variation).

9

ISSCC 2011 PAPER CONTINUATIONS

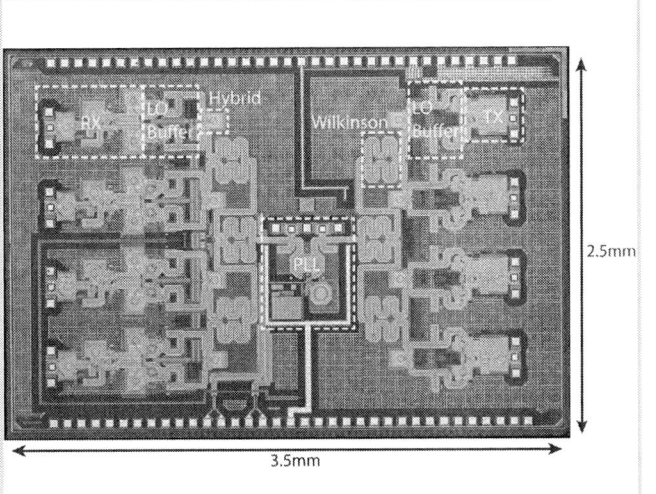

Figure 9.4.7: Die Photo (left side: RX, center: LO generation and distribution, right: TX).

ISSCC 2011 / SESSION 9 / WIRELESS & mm-WAVE CONNECTIVITY / 9.5

9.5 An 87GHz QPSK Transceiver with Costas-Loop Carrier Recovery in 65nm CMOS

Shih-Jou Huang, Yu-Ching Yeh, Huaide Wang, Pang-Ning Chen, Jri Lee

National Taiwan University, Taipei, Taiwan

Modern high-speed wireless data links such as 60GHz RF and point-to-point communications activate research on Gb/s transceivers for V-band (50 to 75GHz) and W-band (75 to 110GHz). Conventional approaches in SiGe or III-V compounds simply consume too much power and occupy too large area, in addition to the integration difficulty. In CMOS realizations, baseband processors and high-speed digitizers also increase design complexity and power consumption. This paper introduces a fully integrated CMOS QPSK transceiver with baseband-less carrier- and data-recovery circuitry, significantly reducing power consumption while achieving up to 3.5Gb/s data rate.

The QPSK transmitter is shown in Fig. 9.5.1. Providing a prototype for the 81-to-86GHz band and 94GHz band, we set up the carrier at 87GHz. An integer-N frequency synthesizer provides 77.3GHz LO and 9.7GHz IF signals, forming an 8/9-1/9 up-convert architecture. It contains a VCO running at 77.3GHz, ÷8 and ÷16 circuits, a type-IV PFD, a V-to-I converter, and an on-chip 2nd-order loop filter. The frequency arrangement is also applied to the RX, making no potential frequency offset between the two. A QPSK modulator generates 4 phases (0°, 90°, 180°, and 270°) of the 9.7GHz IF signal and sends them to the up-convert mixer, creating the 87GHz RF signal. A mm-Wave balun converts differential RF inputs into single-ended mode to drive the PA. With the input capacitance of M_1 and M_2 absorbed by the mixer's resonance network, the balun achieves conversion loss of 1.4dB if the transformer coupling factor is 0.6. The PA's output is fed into a matched microstrip line, which connects to the waveguide adapter.

To perform coherent demodulation, the LO clock in the RX must be synchronized in phase and frequency. Unlike typical baseband approaches that require high-speed ADCs (several GS/s) and signal processors (>1GHz), we realize the carrier recovery in the analog domain (i.e., Costas loop). As depicted in Fig. 9.5.2, the receiver is composed of an LNA, a down-convert mixer, an IF amplifier, a 77.3GHz clock source (VCO) and its ÷8 descendants (9.7GHz I//Q signals), and a phase detector (PD) and a frequency detector (FD) for carrier recovery. After LNA amplification, the incoming RF signal is downconverted into IF by the VCO's clock. Assuming the signal at node A is given by

$$V_A(t) = D_I(t)\cos(\omega_{IF}t + \Delta\phi) + D_Q(t)\sin(\omega_{IF}t + \Delta\phi),$$

where D_I, D_Q denote the embedded data, ω_{IF} is the LO_2 frequency (=2π×9.7GHz), and $\Delta\phi$ is the phase error between IF and LO_2. After mixing and limiting, the two outputs $V_D(t)$ and $V_E(t)$ become $D_I(t)$ and $D_Q(t)$, respectively, given that −45°<$\Delta\phi$<45°. The V_C (V_B) and V_D (V_E) are further mixed together, creating the final output V_F proportional to sin($\Delta\phi$). As a result, the (V/I)$_1$'s output I_{CP1} presents a sinusoidal characteristic as shown in the upper-left corner of Fig. 9.5.2. Denoting pumping current as I_P when $\Delta\phi$ = π/4, the PD [together with (V/I)$_1$] gain is given by √2I_P. The approximately linear behavior in the vicinity of the origin makes itself a linear PLL with two correlated phase-adjusting mechanisms, i.e., LO_1 loop and LO_2 loop. Upon lock, the two demodulated data are found in nodes D and E, as expected.

In addition to phase alignment, the VCO frequency has to be corrected at power on. It is possible to form another loop and push the VCO frequency into the lock range by means of an external crystal. A much better way to do so is to extend the Costas loop one step further. As shown in Fig. 9.5.2, examining the cross- and auto-correlation of signals V_B and V_C twice gives rise to an output containing the frequency error information. Denoting the frequency error between LO_2 and IF signals as $\Delta\omega$, we follow the same calculation procedure and obtain signals in nodes G and H:

$$V_G(t) = -A_V\cos(4\Delta\omega t)$$
$$V_H(t) = A_V\sin(4\Delta\omega t),$$

where A_V represents the swing magnitude designed in our CML blocks. These two signals are separated by 90°. Whether V_G is leading or lagging V_H depends on the polarity of $\Delta\omega$, which can be easily detected by sampling one signal with the other [1]. As a result, the two limiters sharpen the sinusoids as square

functions, and send them to an edge-triggered flip-flop. The error polarity is therefore applied to (V/I)$_2$, which provides pumping current I_{CP2} (=5I_{CP1}) to adjust the VCO control voltage accordingly. Note that to minimize undesired disturbance on the control line, the FD loop must be disabled when the loop is locked. Observing that V_G will stay low when $\Delta\omega \approx 0$, we apply V_K (V_G after limiting) to (V/I)$_2$ to automatically shut it down when the frequency locking is achieved. Similar to that in [1], this automatic shut-off mechanism saves significant power and area.

The QPSK modulator is illustrated in Fig. 9.5.3(a). Here, a DMUX parallelizes the input data and a mapping logic rearranges the sequences. After retiming, the two outputs are created to control the IF 2-to-1 selector and the clock synthesization mixer, where the former picks the proper phase (0°, 90°) and the latter the polarity (180°, 270°). To achieve such a high frequency, the VCO and dividers involve mm-Wave techniques [Fig. 9.5.3(b)]. It incorporates a cross-coupled LC oscillator with thick-oxide varactors, which is followed by two stages of tuned amplifiers (M_3-M_5) as buffers. A direct injection-locked frequency divider is employed here as the 1st divider stage, where the injection signal is ac-coupled to the gate of the switch M_7 [2]. Simulation shows that its lock range is approximately equal to 4GHz when V_b = 0.8V.

The interconnection between chip and antenna is of great importance. As shown in Fig. 9.5.4(a), we realize the coplanar strip to waveguide by a transition fabricated on a single-layer dielectric substrate [3]. With the chip flipped onto the microstrip line, the RF signal is coupled through the substrate to the matching element, which is connected to the waveguide (entrance of horn) tightly. The measured insertion loss of the transition is depicted in Fig. 9.5.4(b), where the maximum loss from 81 to 86GHz is less than 5.3dB. Figure 9.5.4(c) illustrates a photo of the assembly. The LNA, PA, and RF mixer designs are based on our previous work in 60 and 77GHz [2,4]. The 3-stage LNA achieves gain of 18.5dB, and the 5-stage PA 13dB gain and 6dBm P_{1dB}.

The transmitter and receiver are fabricated in 65nm CMOS technology. The TX and RX consume 212mW and 166mW, respectively, from a 1.2V supply. Figure 9.5.5(a) shows the TX's output spectrum under 650Mb/s QPSK modulation with bit length of 2^{31}−1, revealing a sinc function centered at the carrier frequency (87GHz). The TX presents a phase noise of −85.8dBc/Hz at 1MHz offset. The spectrum of the carrier recovered in the RX is also demonstrated in Fig. 9.5.5(b), which reveals a phase noise of −77.7dBc/Hz at 1MHz offset and −88dBc/Hz at 10MHz offset. Figure 9.5.6(a) shows the recovered (and demuxed) data in the RX. With D_{in} = 2.5Gb/s, the recovered data jitter measures 34ps,rms and 200ps,pp, respectively. In this testing, both the TX and the RX are connected to 24dBi horn antennas, separated by 1 meter of distance. The BER as a function of data rate is also investigated. Figure 9.5.6(b) depicts the results for different data patterns. BER < 10^{-11} can be obtained for input of 3.5Gb/s 2^7−1 PRBS. Figure 9.5.7 shows the die micrograph, which occupies 1.32×1mm². In estimation, the longest distance for D_{in} = 2.5Gb/s is approximately equal to 2km given that 48dBi dish antennas are available. Note that a conventional TRX without carrier recovery [5] suffers from synchronization difficulty and cannot operate independently. A table summarizing the performance of this work is also included in Fig. 9.5.7.

Acknowledgment:
The authors thank the TSMC University Shuttle Program for chip fabrication.

References:
[1] J. Lee et al., "A 75-GHz Phase-Locked Loop in 90-nm CMOS Technique," *IEEE J. Solid-State Circuits*, vol. 43, pp. 1414-1426, June 2008.
[2] Y. Li et al., "A Fully-Integrated 77GHz FMCW Radar System in 65nm CMOS," *ISSCC Dig. Tech. Papers*, pp. 216-217, Feb. 2010.
[3] H. Iizuka et al., "Millimeter-Wave Microstrip Line to Waveguide Transition Fabricated on a Single Layer Dielectric Substrate," *IEICE Tran. Commun.*, pp. 1169-1177, June 2002.
[4] J. Lee et al., "A Low-Power Fully Integrated 60GHz Transceiver System with OOK Modulation and On-Board Antenna Assembly," *ISSCC Dig. Tech. Papers*, pp. 316-317, Feb. 2009.
[5] C. Marcu et al., "A 90 nm CMOS Low-Power 60 GHz Transceiver with Integrated Baseband Circuitry," *IEEE J. Solid-State Circuits*, vol. 44, pp. 3434-3447, Dec. 2009.

978-1-61284-303-2/11 $26.00 © 2011 IEEE

ISSCC 2011 / February 22, 2011 / 10:15 AM

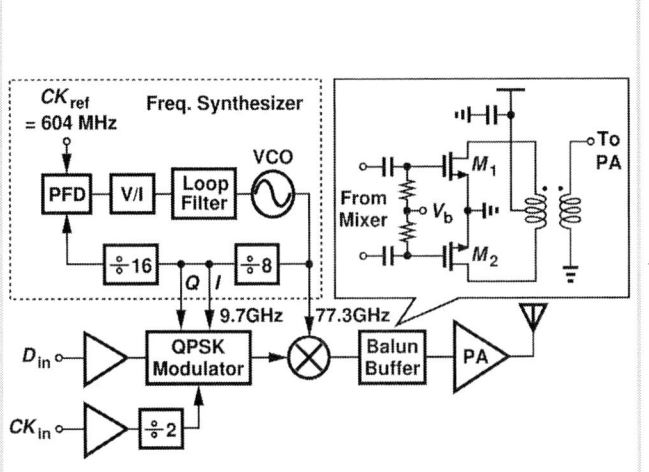

Figure 9.5.1: Transmitter architecture.

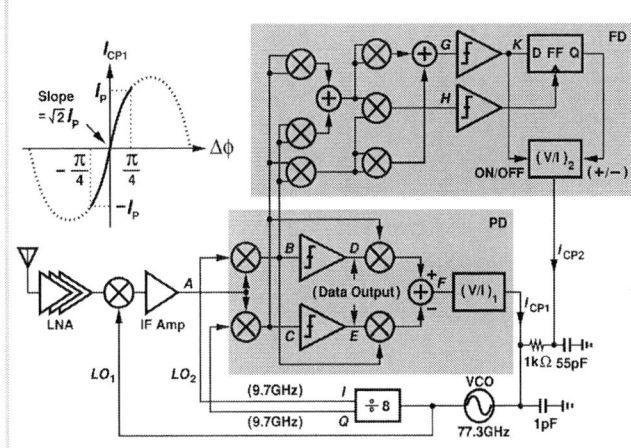

Figure 9.5.2: Simplified receiver architecture.

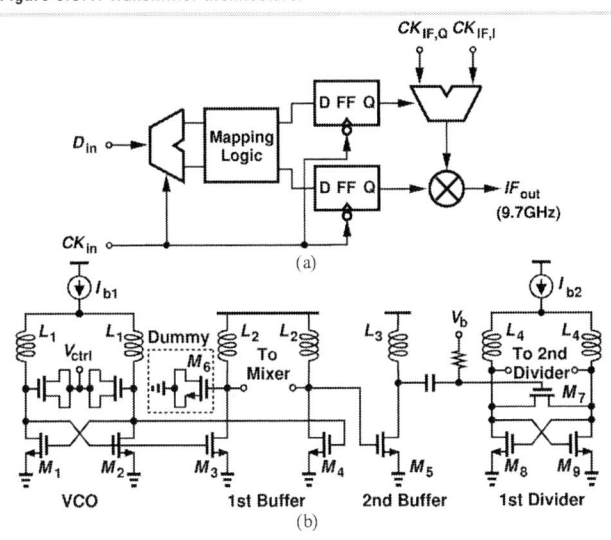

Figure 9.5.3: (a) QPSK modulator, (b) VCO and 1st divider stage.

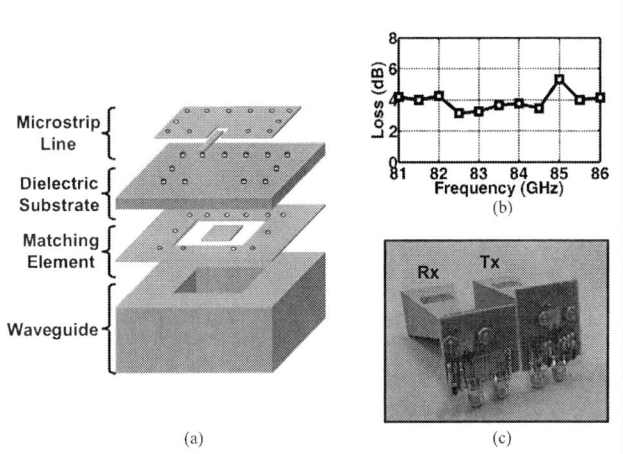

Figure 9.5.4: (a) Waveguide adapter, (b) its insertion loss, (c) assembly photo.

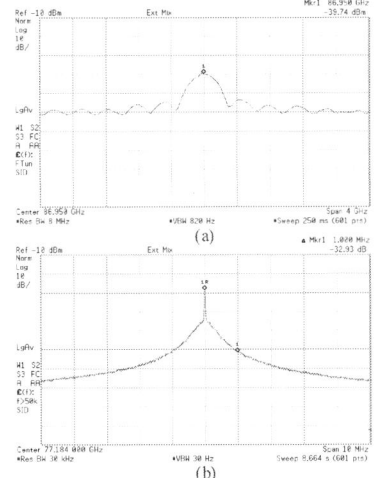

Figure 9.5.5: (a) TX's output spectrum with $2^{31}-1$ 650Mb/s input data, (b) recovered carrier spectrum (77.2GHz).

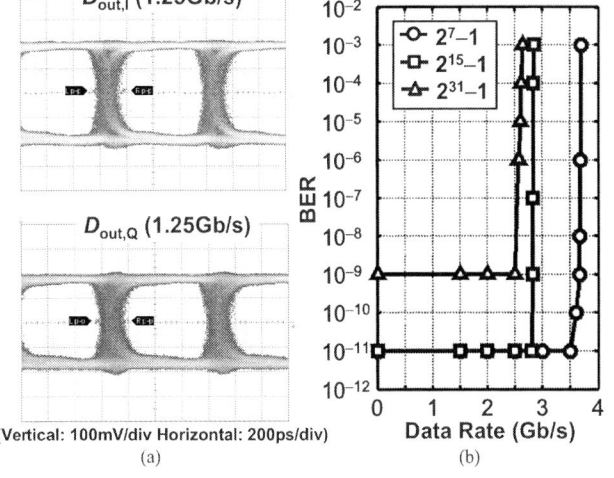

Figure 9.5.6: (a) Recovered data (2^7-1 PRBS, 2.5Gb/s), (b) BER as a function of data rate.

9

978-1-61284-303-2/11 $26.00 © 2011 IEEE

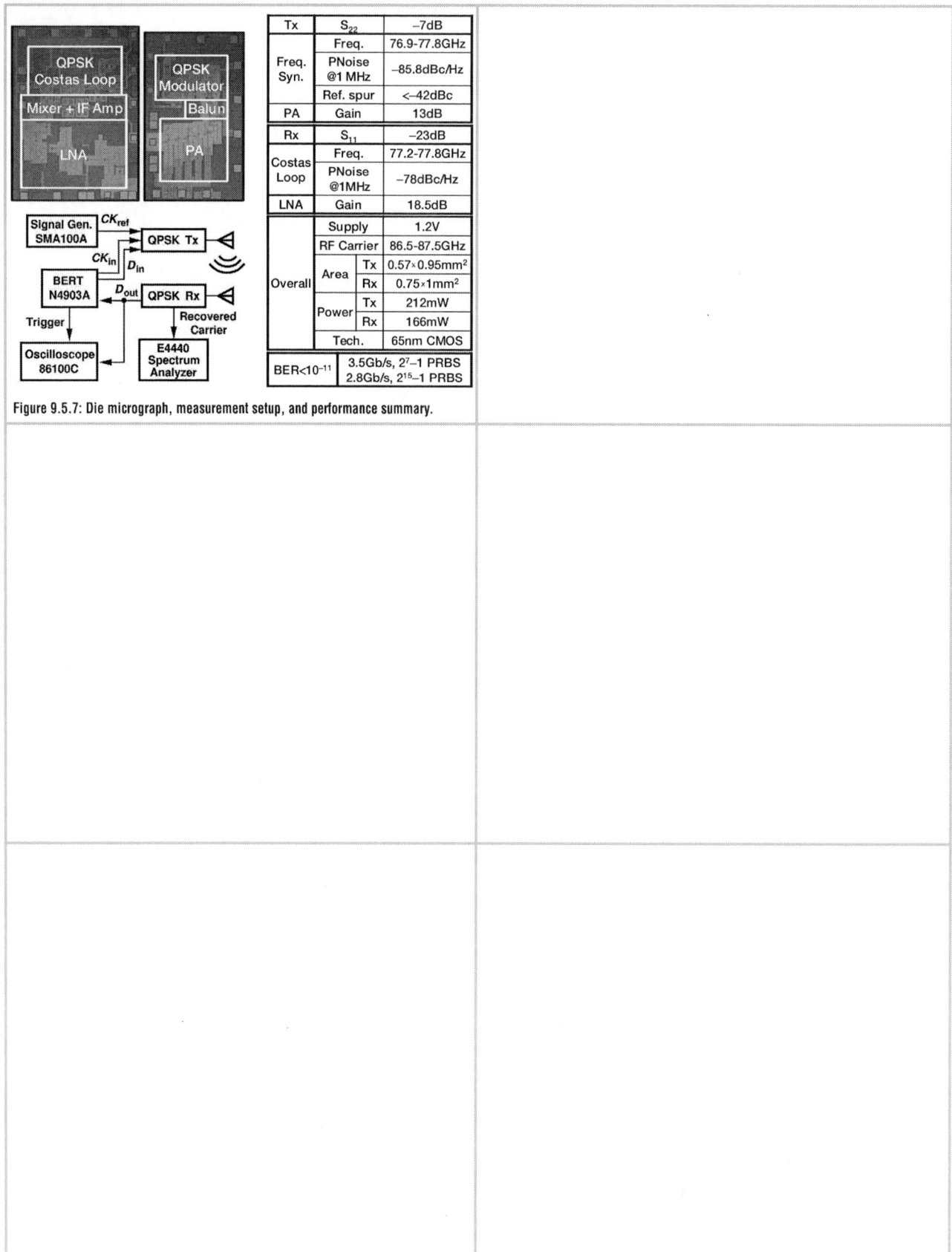

Figure 9.5.7: Die micrograph, measurement setup, and performance summary.

ISSCC 2011 / SESSION 9 / WIRELESS & mm-WAVE CONNECTIVITY / 9.6

9.6 A 65nm Dual-Band 3-Stream 802.11n MIMO WLAN SoC

Shahram Abdollahi-Alibeik[1], David Weber[1], Hakan Dogan[1],
William W. Si[1], Burcin Baytekin[1], Abbas Komijani[1], Richard Chang[1],
Babak Vakili-Amini[2], MeeLan Lee[1], Haitao Gan[1], Yashar Rajavi[1],
Hirad Samavati[1], Brian Kaczynski[1], Sang-Min Lee[1],
Sotirios Limotyrakis[1], Hyunsik Park[1], Phoebe Chen[2], Paul Park[2],
Mike Shuo-Wei Chen[1], Andrew Chang[1], Yangjin Oh[1],
Jerry Jian-Ming Yang[1], Eric Chien-Chih Lin[3], Lalitkumar Nathawad[1],
Keith Onodera[1], Manolis Terrovitis[1], Sunetra Mendis[1], Kai Shi[1],
Srenik Mehta[1], Masoud Zargari[2], David Su[1]

[1]Atheros Communications, San Jose, CA,
[2]Atheros Communications, Irvine, CA,
[3]Atheros Communications, Hsinchu, Taiwan

The rapid commercialization of the IEEE 802.11n WLAN standard has increased the demand for higher data-rate and longer-range fully integrated MIMO SoCs that are backward-compatible with legacy IEEE 802.11a/b/g networks. This paper introduces a 3-stream, 3×3 MIMO WLAN SoC that utilizes three antennas to improve throughput, range, and link robustness. This chip integrates three dual-band transceivers, digital physical layer, media access controller, and a PCI express interface in a 65nm CMOS process. Improved EVM is achieved by reducing transmit and receive I/Q mismatch with calibration, and reducing the integrated phase noise with a reference clock doubler.

The block diagram of one transceiver in the 3×3 MIMO radio SoC is shown in Fig. 9.6.1. The 2.4GHz radio employs a direct-conversion architecture with an LO at twice the channel frequency, while the 5GHz radio utilizes a sliding-IF architecture with an LO at 2/3 the channel frequency [1]. A fractional-N delta-sigma RF synthesizer that can operate from 3 to 4GHz generates the required LO signals. The RF section consists of two paths, one for each of the available spectrum at 2.4GHz and 5GHz. A single analog baseband block is shared between the 2.4GHz and 5GHz paths. To further reduce area, the same biquad filter is used for both receive and transmit operations. When configured for transmit, a second transresistance amplifier is used to loop the transmitted signal back to the receiver to enable digital calibration. The measured system noise figure is 4dB at both the 2.4GHz and 5GHz bands. The integrated power amplifiers for both bands have saturated output powers greater than 25dBm. When transmitting a 5.5GHz 64-QAM OFDM signal at 14dBm, the output stage of each integrated 5GHz PA consumes 140mA from a 3.3V supply.

For robust operation at the highest PHY rate of 450Mb/s, and to allow margin for PA nonlinearity, each transceiver of the 3×3 MIMO radio is targeted to achieve an EVM floor of at least -36dB. All factors contributing to EVM were carefully considered. Every stage of the design, from the DAC to the PA of the transmitter, and from the LNA to the ADC in the receiver, were designed to have an adequate dynamic range to support the large peak-to-average ratio of the OFDM signals. In addition, EVM deterioration due to I/Q mismatch and phase noise was analyzed and addressed in detail.

Systematic amplitude and phase mismatch of I/Q signals in baseband circuits and mixers can be minimized through careful design. However, random mismatches of otherwise identical circuits can result in an EVM as poor as -29dB. Many I/Q mismatch calibration methods have been reported based on envelope detection [2], self-mixing [3], and peak detection followed by a comparator [4]. These calibration methods require additional circuitry and die area. In this design, the existing transmit-to-receive loopback path used for carrier leak calibration [5] is leveraged for I/Q calibration. I/Q calibration is performed by sending a sinusoidal tone from the transmitter DAC, and looping back the upconverted RF signal to the receiver, as shown in Fig. 9.6.1. The combined I/Q mismatch of the transmitter and receiver is decoded from the loop-back signal. Separating I/Q mismatches of the transmitter and the receiver requires a second set of loopback data. This is accomplished by enabling a phase shifter added in the RF loopback path. With this new set of data, the I/Q mismatch of the transmitter can

be separated from that of the receiver without prior knowledge of the phase-shift amount. Once the I/Q mismatch is derived, the transmit data can be digitally pre-distorted before the DAC, while the receive data is digitally corrected after the ADC.

Figure 9.6.2 shows how a phase shift is introduced in the loopback path. A phase shift of larger than 30° is targeted, because smaller phase shifts can result in poor calculation accuracy. In the phase-shift mode, several changes are made in the loopback path: (1) Loopback dc current is reduced by turning off transistors Mbiasp and Mbiasn, (2) Capacitors Cp and Cn are added, (3) Transistors M3p and M3n are added in the path to generate another pole. Figure 9.6.3 shows the measured 64-QAM constellations before and after I/Q mismatch calibration. Before calibration, the EVM was limited by I/Q mismatch to -29.5 dB. The overall EVM is improved to -39.4dB, after I/Q mismatch contribution to the total EVM has been minimized.

Increasing the reference clock frequency is an effective way of reducing the integrated phase noise of a fractional-N RF synthesizer. It results in a lower feedback divider ratio, which reduces the noise multiplication factor of the charge pump and reference clock. In addition, a higher reference clock frequency provides a larger oversampling ratio that effectively reduces the in-band quantization noise of the delta-sigma modulator.

The easiest way to increase the effective reference clock frequency of a synthesizer without using a higher frequency crystal is to use a frequency doubler. However, a modest duty-cycle error in the input clock can create large reference clock spurs. Thus, the duty-cycle correction loop shown in Fig. 9.6.4 is utilized. The input clock is first low-pass filtered to generate an average voltage of $V_{dd}/2$ if the input duty cycle is precisely 50%. Two comparators with thresholds set at $V_{dd}/2 \pm V_{err}$ determine if the duty cycle of the clock is within the desired accuracy. If the average is larger than $V_{dd}/2 + V_{err}$ or smaller than $V_{dd}/2 - V_{err}$, the accumulator is incremented by 1 or -1, respectively. The output of the accumulator is fed to a digital-to-delay cell, whereby a larger accumulator output code provides more delay. To avoid dithering, V_{err} is chosen such that its equivalent delay is larger than that of a unit delay. This circuit can correct up to ±16% of duty cycle error.

Figure 9.6.5 shows the measured phase noise at the transmitter output with and without the clock doubler. The measured integrated phase noise at 2.462GHz of -46.0dBc shows a 2dB improvement with the frequency doubler. No increase in the reference spur level is observed, confirming the proper operation of the duty-cycle correction.

A summary of the measured performance of the WLAN SoC is shown in Fig. 9.6.6. This SoC meets all the IEEE 802.11n specifications. It achieves an over-the-air TCP throughput rate greater than 310Mb/s and PHY rates of up to 450Mb/s. The die area is 22mm^2, of which 47% is occupied by the analog and RF circuits. A chip micrograph is shown in Fig. 9.6.7.

Acknowledgements:
The authors gratefully acknowledge the work of the entire WLAN team at Atheros, especially the algorithm, CAD, digital design, physical design, characterization, and system integration groups.

References:
[1] L. Nathawad, et al., "A Dual-Band CMOS MIMO Radio SoC for IEEE 802.11n Wireless LAN," *ISSCC Dig. Tech. Papers*, pp 358-359, Feb. 2008.
[2] A. Behzad, et al., "A Fully Integrated MIMO Multiband Direct Conversion CMOS Transceiver for WLAN Applications (802.11n)," *ISSCC Dig. Tech. Papers*, pp 560-561, Feb. 2007.
[3] J. Craninckx, et al., "A WLAN direct up-conversion mixer with automatic image rejection calibration," *ISSCC Dig. Tech. Papers*, pp 546-547, Feb. 2005.
[4] Y.H. Hsieh, et al., "An auto-I/Q calibrated CMOS transceiver for 802.11 g," *ISSCC Dig. Tech. Papers*, pp 92-93, Feb. 2005.
[5] S. Mehta, et al., "An 802.11g WLAN SoC," *IEEE J. Solid-State Circuits*, vol. 40, pp. 2483-2491, Dec. 2005.

978-1-61284-303-2/11 $26.00 © 2011 IEEE

ISSCC 2011 / February 22, 2011 / 10:45 AM

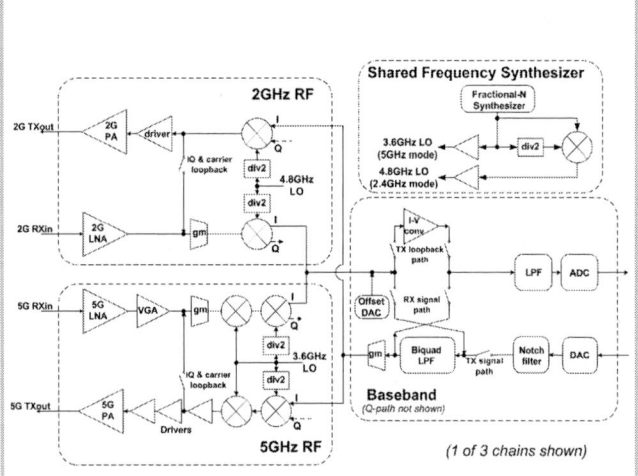

Figure 9.6.1: Block diagram of the dual-band 3×3 MIMO WLAN radio.

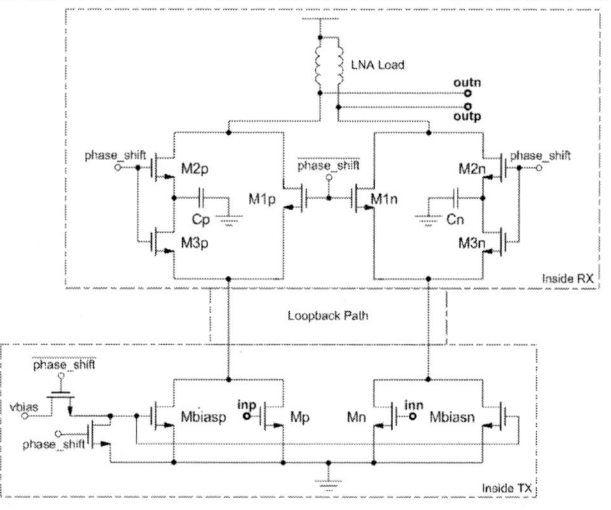

Figure 9.6.2: RF phase shift schematic in the TX-RX loopback path.

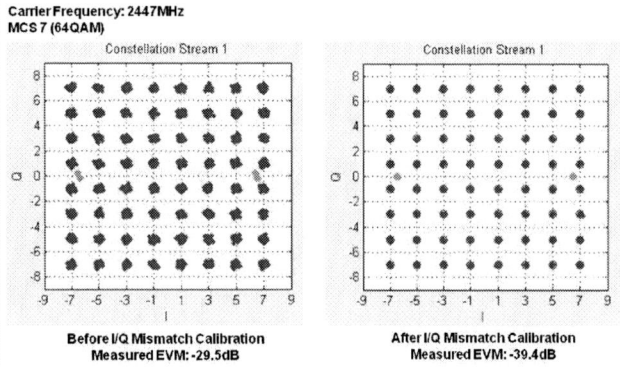

Figure 9.6.3: Measured MCS7 64-QAM constellations before and after I/Q mismatch calibration.

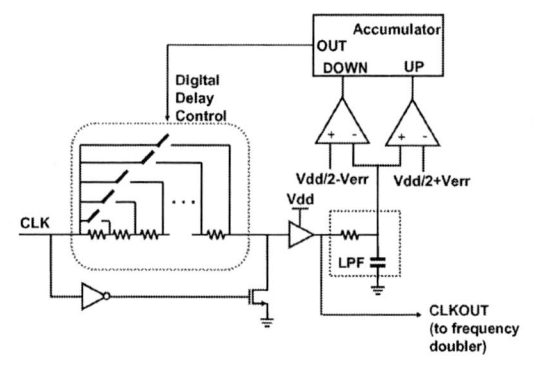

Figure 9.6.4: Synthesizer reference-clock duty-cycle correction loop used before frequency doubler.

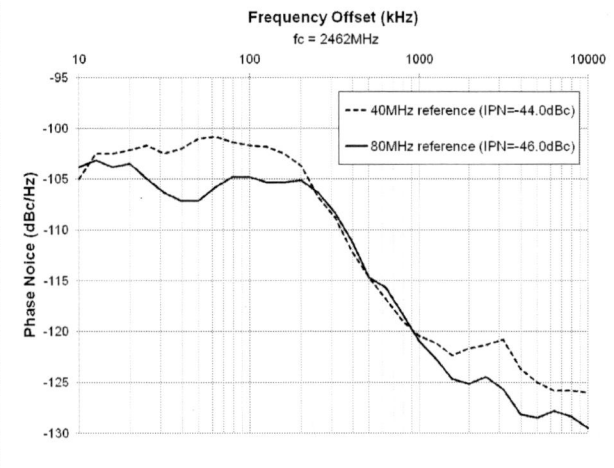

Figure 9.6.5: Phase-noise measurements at the 2.4GHz transmitter output without and with reference clock frequency doubler.

Technology	65nm CMOS	
Die Area	22mm²	
Radio+Analog area (including data converters)	10.4mm²	
TX EVM (for single chain)		
2.4GHz	-39dB @ Pout = -5dBm	
5 GHz	-36dB @ Pout = -5dBm	
Integrated Phase Noise (10kHz to 10MHz)		
2.4GHz	-46dBc	
5 GHz	-39dBc	
Receiver Noise Figure		
2.4GHz	4dB	
5 GHz	4dB	
Power Amplifier Psat		
2.4GHz	26dBm	
5 GHz	25.5dBm	
SoC Power consumption (3-stream, MCS23, HT40)	1.2V	3.3V
RX 2G	310mA	36mA
RX 5G	360mA	36mA
TX 2G (Pout = -5dBm)	198mA	177mA
TX 5G (Pout = -5dBm)	208mA	268mA
Package	QFN-108	
Max over-the-air TCP/IP throughput	311Mbps	

Figure 9.6.6: Performance summary.

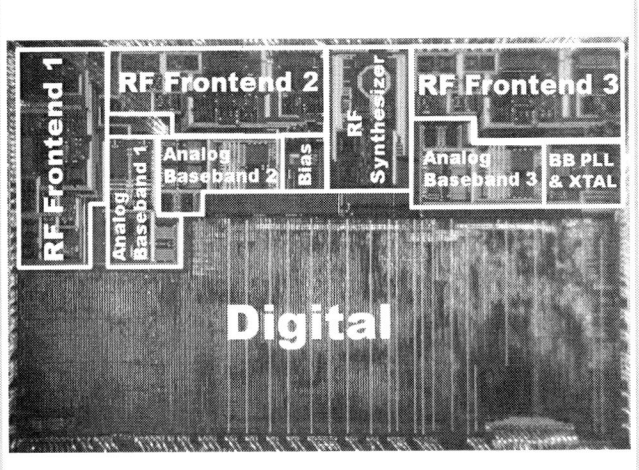

Figure 9.6.7: Die micrograph.

ISSCC 2011 / SESSION 9 / WIRELESS & mm-WAVE CONNECTIVITY / 9.7

9.7 A 0.46mm² 4dB-NF Unified Receiver Front-End for Full-Band Mobile TV in 65nm CMOS

Pui-In Mak[1], Rui Martins[1,2]

[1]University of Macau, Macau, China,
[2]Instituto Superior Tecnico, Lisbon, Portugal

This paper describes a unified receiver front-end (RFE) for mobile TV covering the VHF-III (174 to 248MHz), UHF (470 to 862MHz) and L (1.4 to 1.7GHz) bands. The RFE (Fig. 9.7.1) is tailored to avert any external balun, or dedicated narrow-band radios, commonly found in prior arts [1, 2]. The featured advances are threefold: 1) a *gain-boosting current-balancing balun-LNA* addresses the imbalanced bias current ($I_{CS}=4I_{CG}$) and load ($R_{CG}=4R_{CS}$) conditions of the noise-canceling balun-LNA [3], improving IIP2 and wideband output balancing while eliminating output buffers; 2) a *current-reuse mixer-LPF* realizes quadrature/harmonic-rejection (HR) mixing and 3rd-order current-mode post filtering in one block, enhancing linearity and noise just where both are demanding, while saving power and area; 3) a *direct injection-locked 4-/8-phase LO generator* (LOG) relaxes the master LO frequency (f_{LO}) by avoiding frequency division. A high f_{LO} has been the speed limit of wideband RFEs with HR, such as [4], where a 7.2GHz f_{LO} is employed for 0.9GHz RF reception after divide-by-8. Here, the LOG uses 2 chains of 8-phase corrector (8PC) and 1 chain of 4-phase corrector (4PC) to optimize the phase precision in each band. A master LO_{in} (e.g., from VCO) injection locks either one of the chains to synthesize a 4-/8-phase LO_{out} at f_{LO}. This LOG scheme particularly suits this application, as the problem of VCO pulling is irrelevant here in the absence of the PA.

Figure 9.7.2 shows the proposed balun-LNA. It is composed of a gain-control attenuator (ATT), a single-to-differential stage (S2D) and a differential current balancer (DCB). The S2D is described first: a common-gate amplifier (CGA) M1 generates an in-phase output. An external inductor L_{ext} assists M1 to achieve a wideband impedance match. A common-source amplifier (CSA) M2 generates an anti-phase output, and serves as a gain-booster creating a loop gain (1+A) around M1, to lower its noise and input resistance R_{in}. Unlike the noise-canceling balun-LNA [3], where M2 features a higher g_m (4×) than M1 to minimize NF, M1-M2 here are of equal bias and size. This aim ensures M1-M2 to feature the same bias current I_{DC} and load R_L, that otherwise degrade IIP2 and output balancing. A balanced R_L nullifies the need of output buffers. The NF is addressed via ac-coupled gain-boosting. An extra g_{mx} realized as a self-biased inverter amplifier (M3-M4) enhances the gain of M1-M2 (i.e., lower NF), and handily biases them.

The DCB acts as a current-control current source with unity gain, forcing $i_{sig+}=-i_{sig-}=i_{sig}$. In principle, a cascode amplifier (M5-M6) with cross-coupled capacitors (C3-C4) might realize the DCB. However, accounting for the finite output resistance of M5-M6, a double-cascode amplifier (M5-M8, C3-C6) is adopted to improve the balancing precision. The DCB also benefits the S2D by 1) mapping the loop gain around M1 into a definable ratio, i.e., $1+A=1+g_{mx}/g_{m1}$, and 2) lowering the swing at v_{o1p} and v_{o1n}, where distortion coming from the nonlinear output resistance of M1-M4 can be minimized.

Dual voltage supplies enlarge the design headroom in nm-length processes [1, 2]. In simulations, the 1.2/2.5V balun-LNA exhibits 25.6dB gain, 2.5dB NF and +44.8/ +5.6dBm IIP2/IIP3 while dissipating 11.6mW. The metrics were achieved with $R_{in}=33\Omega$ to improve gain and NF while keeping $S_{11}<-13$dB. The output gain (phase) imbalance is <0.037dB (<1.87°). Such a phase imbalance is limited by the value of C3-C6 (2pF) that was not oversized due to parasitics.

The ATT (not shown) is an nMOS-only resistive network. It offers an 18dB gain-control range with a 6dB step size, while ensuring $S_{11}<-13$dB at all gain levels. L_{ext} dc-shorts the ATT to ground and allows each nMOS, serving as switch+ resistor, to enjoy a large V_{GS} of 1.2V. A replica of the ATT is placed at the source of M2 to match its bias condition with M1 against gain change.

On-chip HR mixing can relax the rejection profile of the off-chip SAW filters [4]. A polyphase mixer with gain ratios of 1:√2:1 and an 8-phase LO [5] enhances the HRR_3 and HRR_5 in receiving the VHF-III or UHF band. For the L band, a 4-phase LO is sufficed for quadrature mixing as the LO harmonics are far out of

band. The mixer core is a switched-g_m cell (Fig. 9.7.3) for its low-voltage operation and the LO path will only induce a common-mode noise at the differential outputs. M_{m1}-M_{m4} utilize resistive source degeneration to improve linearity. The LO buffers use weak latches to reduce duty-cycle distortion.

The mixer is current-reused with a gyrator-C current-mode Biquad $H_{Biq}(s)$ [6] for *low-noise* and *linear* post filtering. The former is owing to the zero exhibited in the noise transfer functions of M_{f1}-M_{f4}, lowering the in-band noise [6]. C_{f1} and the current-mode operation contribute to the latter, suppressing the adjacent channels prior to I-V conversion. The Biquad and the load (R_{BB}-C_{BB}), in conjunction, realize a 3rd-order non-OpAmp LPF sufficient to move the rest of the BB functions to the digital domain via practical ADCs [2]. Joint-tuning of C_{f1}, C_{f2} and C_{BB} allows constant-Q BW adjustment. M_{f1}-M_{f4} are thick-oxide MOSFETs, with big W and L, to minimize NF.

A dividerless 4-phase LOG [7] has been reported for the 0.37-to-2.5GHz band, but the phase error is limited to 5°. Here, we propose a scheme combining 4- and 8-phase LOGs; each is dedicated to one TV band to achieve a phase error <1° (simulation). As shown in Fig. 9.7.4, LO_{in} drives a chain of six 8PCs to synthesize an 8-phase LO_{out}. Each 8PC involves 32 inverters classified into 2 types and 3 sets. L-type features a larger device size than S-type for optimizing the phase correctability of each 8PC. Set A is to interpolate the intermediate phases. Set B is to enlarge the locking range by suppressing self-oscillation. Set C is for injection lock. In tests, with a signal generator serving as the LO_{in}, the phase noise of the LOG measures a 10dB/dec roll-off, which is well below that achieved in [1] using a PLL+VCO as LO_{in}. Thus, the phase noise penalized by the LOG should be tolerable when the PLL+VCO is present.

The RFE fabricated in 65nm CMOS measures 35±1dB gain, 4±0.2dB NF and 43-to-55mW power consumption over the covered bands (Fig. 9.7.5a). The BB gain responses (Fig. 9.7.5b) at different gain steps show 60dB/dec rejection, and an untuned cutoff frequency of around 12MHz. The distortion is justified by the out-of-channel linearity metrics [1]; with two tones applied at [$f_{LO}+20$MHz, $f_{LO}+31$MHz], the RFE shows an IIP2/IIP3 of +32/-3.4dBm at the highest gain, and +35/+11dBm at the lowest gain (Fig. 9.7.5c). The on-chip HRR_3/HRR_5 is 35/39dB, confirming the high phase precision of the LOG. Note that to increase the data accuracy, the reported IIP2/HRR_3/HRR_5 is an average value of 12 samples with σ of 3.9/1.8/2.4dB. The in-band S_{11} measures <-10dB at all gain steps (Fig. 9.7.5d). The die occupies 0.46mm² (Fig. 9.7.7).

Benchmarking with the state-of-the-art wideband RFEs that attain 4dB NF (Fig. 9.7.6), this work succeeds in extending the operating BW and BB selectivity, while reducing external parts, chip area and f_{LO} that can ease the design of the PLL+VCO.

Acknowledgements:
The work is funded by Macao FDCT. We thank B. Razavi (UCLA) and F. Maloberti (U. Pavia) for suggestions and K.-F. Un (U. Macau) for assistance.

References:
[1] I. Vassilios, K. Vavelidis, N. Haralabidis et al., "A 65nm CMOS Multistandard, Multiband TV Tuner for Mobile and Multi-Media Applications," *IEEE J. Solid-State Circuits*, vol. 43, pp. 1522-1533, Jul. 2008.

[2] M. Jeong, B. Kim, Y. Cho et al., "A 65nm CMOS Low-Power Small-Size Multistandard, Multiband Mobile Broadcasting Receiver SoC," *ISSCC Dig. Tech. Papers*, pp. 460-461, Feb. 2010.

[3] S. Blaakmeer, E. Klumperink, D. Leenaerts et al., "Wideband Balun-LNA with Simultaneous Output Balancing, Noise-Canceling and Distortion-Canceling" *IEEE J. Solid-State Circuits*, vol. 43, pp. 1341-1350, Jun. 2008.

[4] Z. Ru, E. Klumperink, G. Wienk et al., "A Software-Defined Radio Receiver Architecture Robust to Out-of-Band Interference," *ISSCC Dig. Tech. Papers*, pp. 230-231, Feb. 2009.

[5] J. Weldon, J. Rudell, L. Lin et al., "A 1.75-GHz Highly Integrated Narrowband CMOS Transmitter with Harmonic-Rejection Mixers," *ISSCC Dig. Tech. Papers*, pp. 160-161, Feb. 2001.

[6] A. Liscidini, A. Pirola and R. Castello, "A 1.25mW 75dB-SFDR CT Filter with In-Band Noise Reduction," *ISSCC Dig. Tech. Papers*, pp. 336-337, Feb. 2009.

[7] K. H. Kim, P. W. Coteus, D. Dreps et al., "A 2.6mW 370MHz-to-2.5GHz Open-Loop Quadrature Clock Generator," *ISSCC Dig. Tech. Papers*, pp. 458-459, Feb. 2008.

978-1-61284-303-2/11 $26.00 © 2011 IEEE

ISSCC 2011 / February 22, 2011 / 11:15 AM

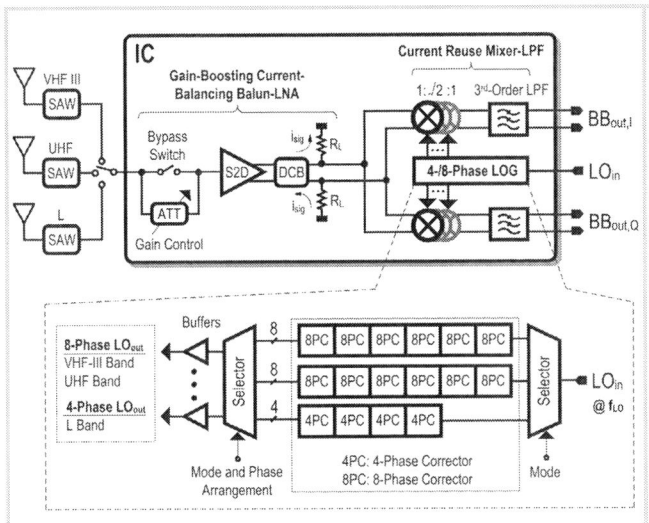

Figure 9.7.1: Unified RFE for full-band mobile TV.

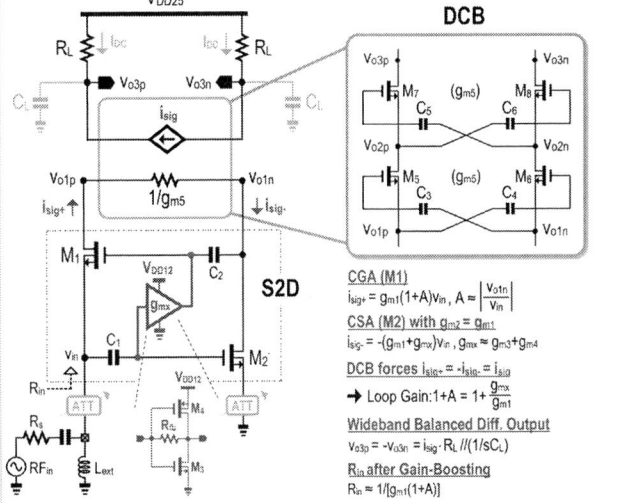

CGA (M1)
$i_{sig+} = g_{m1}(1+A)v_{in}$, $A = \left|\frac{v_{o1n}}{v_{in}}\right|$

CSA (M2) with $g_{m2} \cong g_{m1}$
$i_{sig-} \cong -(g_{m1}+g_{mx})v_{in}$, $g_{mx} \cong g_{m3}+g_{m4}$

DCB forces $i_{sig+} \cong -i_{sig-} \cong i_{sig}$

→ Loop Gain: $1+A = 1 + \frac{g_{mx}}{g_{m1}}$

Wideband Balanced Diff. Output
$v_{o3p} = -v_{o3n} = i_{sig} \cdot R_L // (1/sC_L)$

R_{in} after Gain-Boosting
$R_{in} \approx 1/[g_{m1}(1+A)]$

Figure 9.7.2: Balun-LNA. ATTs are bypassed at high-gain mode.

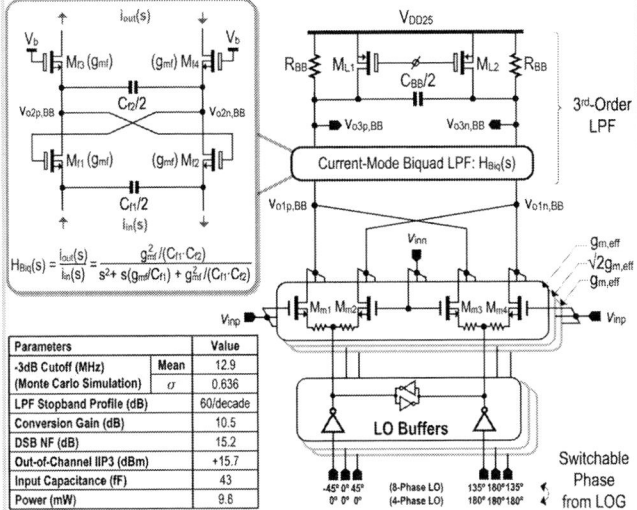
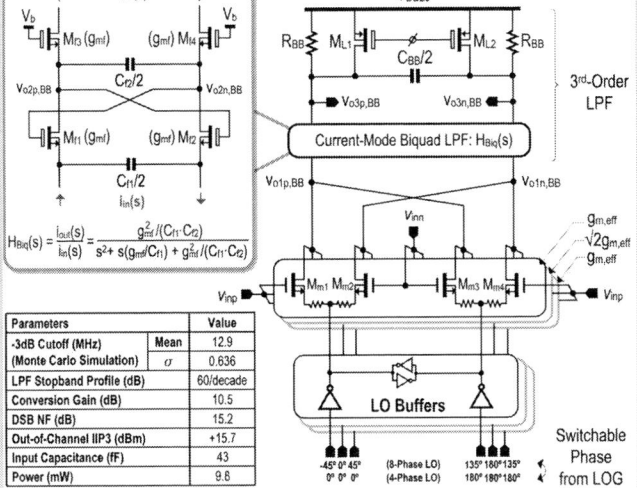

Parameters		Value
-3dB Cutoff (MHz)	Mean	12.9
(Monte Carlo Simulation)	σ	0.636
LPF Stopband Profile (dB)		60/decade
Conversion Gain (dB)		10.5
DSB NF (dB)		15.2
Out-of-Channel IIP3 (dBm)		+15.7
Input Capacitance (fF)		43
Power (mW)		9.8

Figure 9.7.3: Mixer-LPF (I channel) and its simulated performances.

Figure 9.7.4: 8-phase chain of the LOG and its measured phase noise.

Figure 9.7.5: Measured (a) NF, gain and power. (b) BB response. (c) IIP2 and IIP3. (d) S11.

Parameters	This work	ISSCC'09 [4]	[Z. Ru et al, JSSC May'10]
Operation Frequency f_{RF} (GHz)	0.17 to 1.7	0.4 to 0.9	0.3 to 0.8
Required Master LO Frequency f_{LO} (GHz)	$f_{LO} = f_{RF}$ (4 and 8 Phases)	$f_{LO} = 8 f_{RF}$ (8 Phases)	$f_{LO} = 4 f_{RF}$ (8 Phases)
Maximum Gain (dB)	35	34	22 to 28 #
RF Gain Control (dB)	17 to 35	No	No
External Components	1 Inductor	2 Inductors and 1 Balun	2 Inductors
Area (mm²)	0.46	1	0.5
BB Filter Order	3rd-Order LPF (1 biquad + 1 real)	2nd-order LPF (2 real poles)	1st-order IIR LPF (minor channel selectivity)
Power (mW) @ f_{RF} (GHz)	55 @ 1.7	60 @ 0.9	18 @ 0.8
Input Impedance Matching	Matched	Matched	Unmatched
DSB NF (dB)	4 [Spec: 4] *	4	0.8 to 4.3 #
IIP3 (dBm)	-3.4 [Spec: -5] *	3.5	-14 to -9 #
IIP2 (dBm)	32 [Spec: 27] * (Balun LNA)	46 (Differential LNA)	38 to 49 # (Balun LNA)
HRR₃ (dB)	35	60	60
HRR₅ (dB)	39	64	60
Supply Voltage (V)	1.2 and 2.5	1.2	1.2
Technology	65nm CMOS	65nm CMOS	65nm CMOS

(*) - from [1]   (#) - In-band variation

Figure 9.7.6: Summary of measured performances and comparison.

978-1-61284-303-2/11 $26.00 © 2011 IEEE

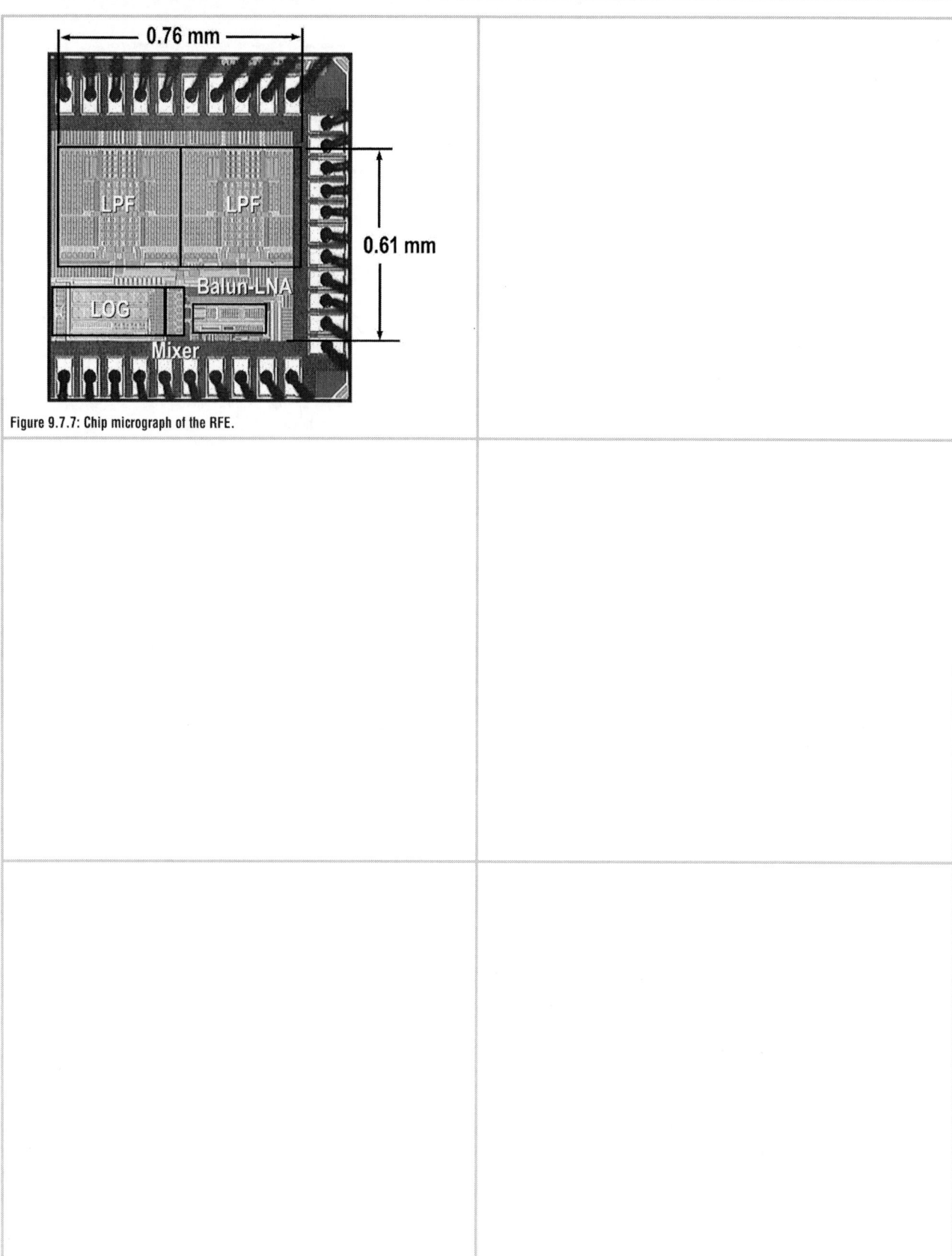

Figure 9.7.7: Chip micrograph of the RFE.

ISSCC 2011 / SESSION 9 / WIRELESS & mm-WAVE CONNECTIVITY / 9.8

9.8 An All-Digital 8-DPSK Polar Transmitter with Second-Order Approximation Scheme and Phase Rotation-Constant Digital PA for Bluetooth EDR in 65nm CMOS

Hiroyuki Kobayashi, Shouhei Kousai, Yoshiaki Yoshihara, Mototsugu Hamada

Toshiba Semiconductor, Kawasaki, Japan

As single-chip implementation of Bluetooth transceivers gains popularity [1,2], area reduction in the analog/RF part is inevitable. While all-digital polar transmitters achieving a small die size are attracting attention and their application to Bluetooth Basic Rate and GSM standards is now the reality [3,4], due to their shortcoming in the modulation accuracy above their loop bandwidth, it is very difficult to apply it to non-constant envelope modulations. This paper reports on a polar transmitter that is capable of 8-DPSK modulation for the Bluetooth EDR standard. It implements the second-order approximation scheme in the gain calculation of the DCO and a phase rotation-constant digital PA. It achieves DEVM of 6.1% at 0dBm output in sending 8-DPSK signal for Bluetooth EDR, drawing 35mA from a 1.2V supply and occupies 0.75×0.75mm².

Figure 9.8.1 shows the block diagram of the proposed all-digital polar transmitter. The differences from the conventional one are highlighted, which are the second-order approximation scheme in the gain calculation of the DCO and the phase rotation-constant digital PA (PRC-DPA). By virtue of the PRC-DPA, the phase information can be completely separated from the amplitude information, which eliminates a pre-distortion function in the phase modulation path and its calibration procedure such as the use of a table map of the phase rotation information against the signal amplitude.

Figure 9.8.2 shows how the second-order approximation scheme in the gain calculation of the DCO works. The DCO frequency is inversely proportional to the square root of the tank LC product. However, conventional polar transmitters have approximated it by a linear function. When a polar transmitter is applied to 8-DPSK, frequency information deviates from $f_C - f_{REF}/2$ to $f_C + f_{REF}/2$. As shown in the top-left, there is a significant error of the offset tuning word (OTW) in the conventional first-order approximation scheme as the DCO frequency (f_{DCO}) deviates from its center, f_C. We approximate the DCO frequency by a parabolic function and integrate the error compensation calculation in the phase modulation path. The approximation is validated by Taylor expansion of $(1+x)^{-1/2}$ where $x \ll 1$. Supposing that the DCO frequency is parabolic, the modulation coefficient, K, is a first-order function against the DCO frequency while K is a constant at K_C in the conventional first-order approximation scheme as shown in the bottom-left of the figure. In the proposed scheme, before starting the modulation, the PLL locks at three different frequencies, f_C, $f_C - f_{REF}/2$, and $f_C + f_{REF}/2$ to obtain respective offset tuning words, OTW_C, OTW_{LOW}, and OTW_{HIGH}. The angle of the line connecting ($f_C - f_{REF}/2$, OTW_{LOW}) and ($f_C + f_{REF}/2$, OTW_{HIGH}) is mathematically identical to K_C in the conventional first-order approximation as long as the DCO frequency is a parabolic function. The angle of K in this scheme is derived from the error of OTW, Δ, at $f_C - f_{REF}/2$ or $f_C + f_{REF}/2$ and is calculated as 4Δ. Once the function is obtained, it is implemented in the phase-modulation path as a block diagram shown in the right of the figure. Though it requires an additional multiplier and an adder, the area and power penalty is negligible.

Figure 9.8.3 depicts the schematic of the proposed phase rotation-constant digital PA (PRC-DPA). Conventional digital PA issues are the phase rotation and linearity variations at the output against output power. It is due to the output impedance variation when the number of 'ON' transistors is changed according to the output power. The proposed DPA has a differential topology and keeps the number of 'ON' transistors constant, irrespective of output power. The left side of the figure shows the whole system of the proposed DPA. It has 67 identical element amplifiers. The phase information comes from the DCO as $RFIN_P$ and $RFIN_N$. The amplitude information comes from the CORDIC as 14-bit digital information, out of which, 6 bits from the MSB is transformed as a thermometer code to control 64 element amplifiers, and the remaining 8 bits go to the $\Delta\Sigma$ modulator that is in charge of fine power control. Dynamic element matching is also implemented in the 64 element amplifiers. On the right side of the figure, the circuit diagram of the element amplifier is shown. It has 4 current paths each to differential outputs, $RFOUT_P$ and $RFOUT_N$. Supposing that $EN=1$, the leftmost and the second paths to $RFOUT_P$ and rightmost and the second paths to $RFOUT_N$. are activated and the element amplifier outputs the unitary power. Supposing that $EN=0$, the second and the fourth paths to $RFOUT_P$ and the second and the fourth paths to $RFOUT_N$ are activated and the element amplifier outputs no power. Note that in each case two 'ON' transistors are connected to the differential outputs so that the output impedance is kept constant.

Figure 9.8.4 shows the TX constellation and spectrum. The error vector magnitude is 6.1% at 0dBm output and meets the requirement of Bluetooth EDR specification. The output spectrum meets the spectral mask requirements.

Figure 9.8.5 shows the calibration sequence of the modulation coefficients in the proposed second-order approximation scheme. From time 0, the PLL is going to lock the frequency to f_C until 50µsec. After the lock-up to f_C, the PLL stores the offset tuning word at frequency f_C, OTW_C, next, it shifts its frequency setting to $f_C - f_{REF}/2$ and records OTW_{LOW}, and then it shifts its setting to $f_C + f_{REF}/2$ and records OTW_{HIGH}. It takes about 160µsec to obtain the necessary parameters to modulate the 8-DPSK signal and to lock the frequency to f_C again, which is less than the 200µsec generally assigned to the PLL lock-up process in Bluetooth transceivers. The right side of the figure shows the repeatability of the calibration results. After 100 trials, K_C and Δ are obtained within the variation of 0.099% and 12.2% in terms of the standard deviation to average.

Figure 9.8.6 shows the simulation results of the proposed phase rotation-constant digital PA (PRC-DPA). Since it is difficult to measure the phase relation of the input and output of the DPA, the effect of the phase rotation-constant digital PA is verified by simulation. The output voltage and phase are plotted against the amplitude control word, showing that the output voltage grows in proportion to the amplitude control word while the output phase is almost constant. The variation of the phase is less than 0.014 degrees from -32dBm output to +4dBm output. The characteristics enable us to separate the phase information completely from the amplitude information. As a result, the proposed transmitter does not require calibration and/or pre-distortion procedures for the phase rotation and non-linearity against the transmitted power.

Figure 9.8.7 is the die micrograph of the proposed all-digital polar transmitter and the feature comparison with recent Bluetooth transceivers [2]. It is fabricated in a 65nm standard CMOS technology and occupies about 0.75×0.75mm². The size of transmitter is significantly smaller than previous reports owing to its all-digital implementation. It draws 35mA from a 1.2V single supply when outputting 0dBm signal. The power consumption is also smaller than the previous report even considering 2dB difference of their output powers.

References:
[1] B. Marholev et al., "A Single-Chip Bluetooth EDR Device in 0.13µm CMOS," *ISSCC Dig. Tech. Papers*, pp. 558-559, Feb. 2007.
[2] D. Weber et al., "A Single-Chip CMOS Radio SoC for v2.1 Bluetooth Applications," *ISSCC Dig. Tech. Papers*, pp. 364-365, Feb. 2008.
[3] R. B. Staszewski et al., "All-digital phase-domain TX frequency synthesizer for Bluetooth radios in 0.13µm CMOS," *ISSCC Dig. Tech. Papers*, pp. 272- 273, Feb. 2004.
[4] R. B. Staszewski et al., "All-Digital PLL and GSM/EDGE Transmitter in 90nm CMOS," *ISSCC Dig. Tech. Papers*, pp. 316-317, Feb. 2005.

ISSCC 2011 / February 22, 2011 / 11:45 AM

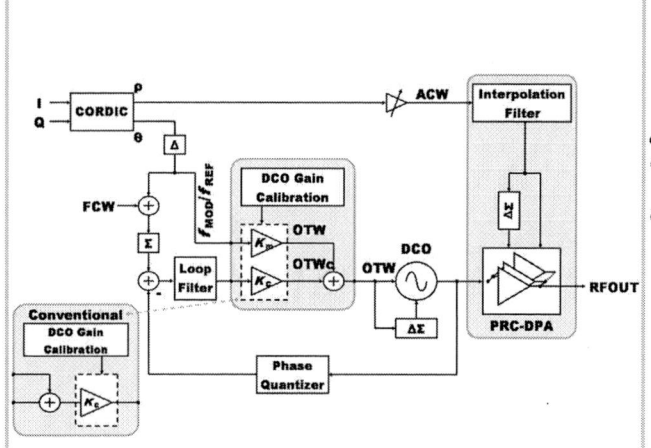

Figure 9.8.1: Block diagram of the proposed transmitter.

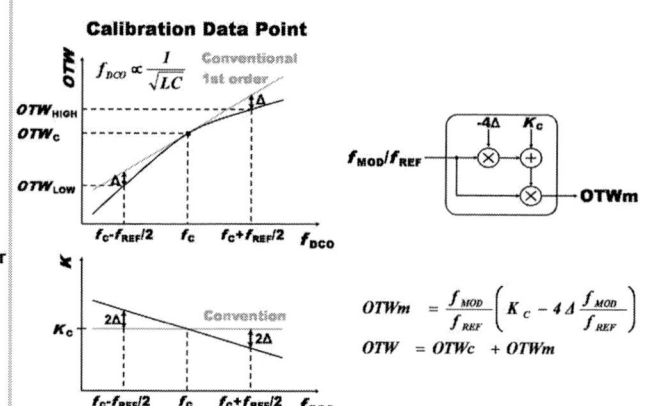

Figure 9.8.2: Second-order approximation scheme in the gain calculation of DCO.

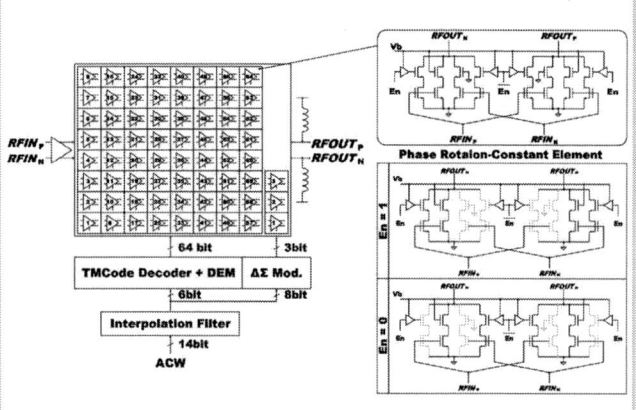

Figure 9.8.3: Schematic of the proposed phase rotation-constant digital PA.

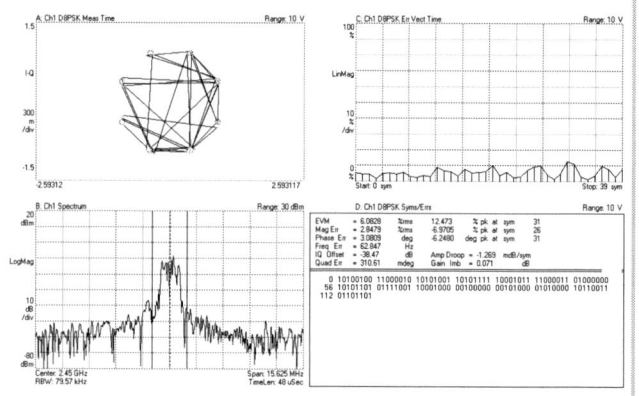

Figure 9.8.4: Measured constellation and spectrum.

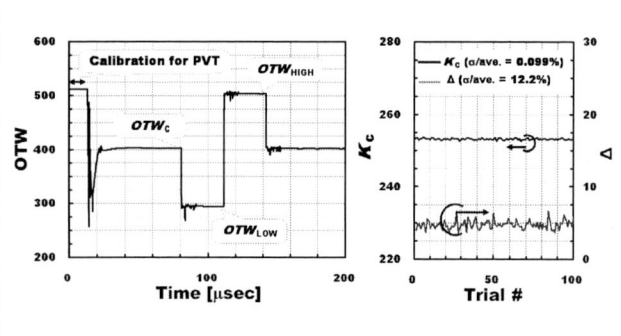

Figure 9.8.5: Calibration sequence of the modulation coefficients in the proposed second order approximation scheme.

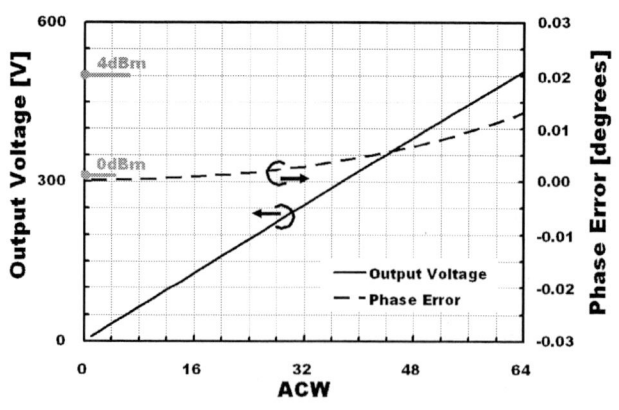

Figure 9.8.6: Output power and phase response of the proposed phase rotation-constant digital PA.

9

978-1-61284-303-2/11 $26.00 © 2011 IEEE

ISSCC 2011 PAPER CONTINUATIONS

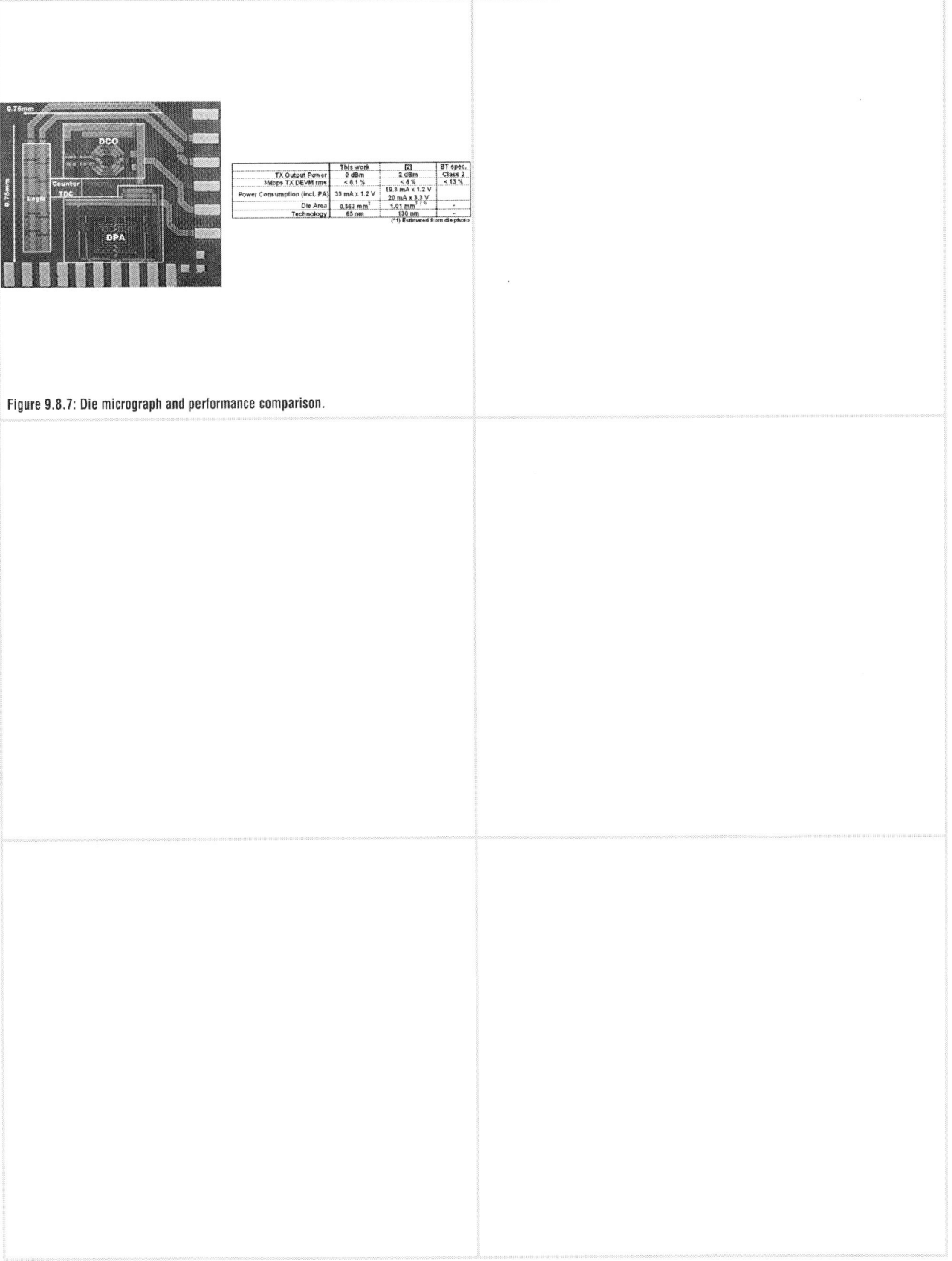

	This work	[2]	BT spec.
TX Output Power	0 dBm	2 dBm	Class 2
3Mbps TX DEVM rms	< 8.1 %	< 8 %	< 13 %
Power Consumption (incl. PA)	35 mA x 1.2 V	19.3 mA x 1.2 V 20 mA x 3.3 V	
Die Area	0.563 mm²	1.01 mm²⁽*¹⁾	-
Technology	65 nm	130 nm	-
		⁽*¹⁾ Estimated from die photo	

Figure 9.8.7: Die micrograph and performance comparison.

ISSCC 2011 / SESSION 9 / WIRELESS & mm-WAVE CONNECTIVITY / 9.9

9.9 A Digital-Intensive Receiver Front-End Using VCO-Based ADC with an Embedded 2nd-Order Anti-Aliasing Sinc Filter in 90nm CMOS

Jaewook Kim[1], Wonsik Yu[1], Hyun-Kyu Yu[2], SeongHwan Cho[1]

[1]KAIST, Daejeon, Korea,
[2]ETRI, Daejeon, Korea

One of the recent trends in multimode multiband (MMMB) receivers is to remove the analog filter or variable-gain amplifier (VGA) in the receiver chain and employ a wide-dynamic-range ADC directly after the mixer or include the mixer in the ADC [1,2]. While such architecture provides ease of programmability once the signals are digitized, it puts a large burden on the ADC and anti-alias filter. Hence, ADCs typically use high-performance analog circuits for wide dynamic range, even though it is difficult to implement these circuits using low-voltage nanoscale CMOS processes. A promising ADC architecture for an MMMB receiver is the VCO-based ADC, since it offers 1st-order noise-shaping from its open-loop digital-intensive nature, thus allowing high sampling rate and high SNR [3]. Furthermore, the VCO-based ADC provides an inherent anti-aliasing 1st-order Sinc filter due to the innate integrating ability of the VCO [4]. Unfortunately, for multiband receivers that do not have an RF pre-filter, a Sinc filter alone does not provide enough out-of-band rejection and hence higher-order anti-aliasing filters are required. In order to solve this problem, we propose a 2nd-order anti-aliasing Sinc filter (Sinc^2 filter) that provides double the rejection ratio of a Sinc filter. Furthermore, the proposed technique is highly digital and can be embedded in a VCO, resulting in little overhead.

The operation principle of the proposed Sinc^2 filter is shown in Fig. 9.9.1, where the key concept is that the 2nd-order Sinc function is the time-domain convolution of the signal and a triangular waveform. The time-domain convolution with sampling can be achieved by multiplying the input signal by a triangular waveform, integrating it, and repeating this process every sampling period (T_S). As the triangular waveform has a period of $2T_S$, two-way time-interleaving is required for seamless operation of the Sinc^2 filter. Note that a two-way time-interleaved VCO-based ADC has quantization noise nulls at DC and π [5]. While the operations required for the Sinc^2 filter can be difficult to implement in conventional designs, it can be easily implemented by using a VCO-based ADC that operates in the time domain. That is, the integrator can be implemented by using a VCO that converts the input voltage to frequency and integrates it into phase. The difference operator ($1-z^{-1}$) is implemented by using a reset counter that quantizes the VCO phase by counting the rising edges of the VCO output during a sampling period [4]. To further simplify the complexity, the triangular signal that is multiplied to the input is approximated as a three-level staircase waveform. While there is only 3dB theoretical difference in the rejection ratios between these two waveforms, there is negligible difference if other imperfections such as clock jitter and circuit mismatch are considered. Multiplication of a three-level staircase waveform is realized by adding the outputs of two multipliers that receive square waveforms with different duty cycles.

The schematic of the proposed Sinc^2-filter-embedded VCO-based ADC is shown in Fig. 9.9.2. The ADC employs a 16-phase sub-feedback differential ring VCO so as to increase the tuning range while maintaining the number of phases for enhanced time-resolution and high SNR. The Sinc^2 filter is realized in the delay cell of the ring VCO. As the frequency of the VCO is determined by the control transistors (M_C), M_1 and M_2 are added to M_C in order to multiply the desired signal by a square waveform. The outputs of the multipliers are added in the current-domain, which changes the VCO frequency accordingly. It can be seen that the overhead to achieve the Sinc^2 filter is small, as only a control transistor and passive mixers are required in the delay cell of the ring VCO. Depending on the target SNR, the sampling frequency can be programmable from 800MHz to 1.4GHz, where the lower frequency is limited due to the one-bit quantization of the counter [4]. Scaling sampling frequency also enables scaling of power consumption, since the VCO-based ADC is implemented in mostly digital circuits.

The proposed technique has been implemented in a 1P9M 90nm CMOS process. In order to verify its operation, other typical receiver circuits such as a wideband LNA and passive mixer have been added before the ADC for zero-IF operation. The block diagram of the implemented signal path is shown in Fig. 9.9.3. The output of the passive mixer is directly connected to the control transistors of the ring VCO. Note that an implicit 1st-order low-pass filter is formed by the on-resistance of the switch transistor and the input capacitance of the ring VCO. Also note that four ADCs are required for differential I/Q signals, as the ADC receives single-ended input. The wideband LNA is based on a noise-canceling architecture using common-gate and common-source amplifiers as shown in Fig. 9.9.4 [6]. In order to achieve wideband input impedance matching, an external LC ladder filter and a termination resistor (R_T) are adopted.

The die micrograph of the proposed receiver front-end is shown in Fig. 9.9.7. The core area is less than 0.4mm^2, where the LNA accounts for 0.06mm^2 and the differential I/Q ADCs occupy 0.19mm^2. The measured output spectra with the Sinc filter and the proposed Sinc^2 filter are shown in Fig. 9.9.5 (a) and (b), respectively. The conventional Sinc filter is obtained by bypassing the multiplier in the ring VCO. A -45dBm 1.001GHz wanted signal and -25dBm 1.810GHz out-of-band interferer, which is aliased to 10MHz, are applied with 1GHz LO and 800MHz ADC sampling clock. The proposed Sinc^2 filter achieves an out-of-band rejection ratio of 50.2dB, which is almost twice as large as the 25.5dB of the Sinc filter. The rejection ratio of the Sinc filter and Sinc^2 filter when the frequency of the out-of-band interferer is changed from 1.8001GHz to 1.81GHz is shown in Fig. 9.9.5 (c), where the interferer is aliased to 100kHz and 10MHz. It can be seen that the rejection ratios of the two filters are relatively constant even when the aliased frequency is close to DC. Simulation results reveal that this is due to the clock jitter and mismatch among the delay cells. The measured output spectrum with -28dBm two-tone test is shown in Fig. 9.9.5 (d), where it can be seen that in-band IIP3 is -6.78dBm. Note that the quantization noise nulls at DC and at π are seen due to the two-way time interleaving of the VCO-based ADC. The SNDR as a function of the RF input power is shown in Fig. 9.9.6 (a) when the LO frequency is 1GHz, sampling frequency is 1.4GHz and the input bandwidth is 1MHz and 10MHz. Assuming that the required system SNR is 10dB, the proposed receiver achieves sensitivity levels of -94dBm and -74dBm for 1MHz and 10MHz bandwidth, respectively. The SNR versus RF carrier frequency from 0.2 to 1.8GHz is shown in Fig. 9.9.5 (c) for the case when the RF input power is -35dBm and the sampling frequency is 1.4GHz. The performance is summarized in Table. 1. The measured IIP2 is 31.2dBm. As sampling frequency is changed from 800MHz to 1.4GHz, the current consumption of the ADC varies from 20.3mA to 27.4mA, where 9mA of constant current is consumed in the VCO. The LNA dissipates 12mA and the LO buffer dissipates 2.1mA from a 1.35V power supply.

Acknowledgements:
The authors would like to thank IC Design Education Center (IDEC) for their support.

References:
[1] R. Winoto, et.al., "A Highly Reconfigurable 400-1700MHz Receiver Using a Down-Converting Sigma-Delta A/D with 59-dB SNR and 57-dB SFDR over 4-MHz Bandwidth" *IEEE Symp. VLSI Circuits Dig. Tech. Papers*, pp. 142-143, 2009.

[2] J. Lee, et.al., "A 1.8-to-2.4-GHz 20mW Digital-Intensive RF Sampling Receiver with a Noise-Canceling Bandpass Low-Noise Amplifier in 90nm CMOS, *IEEE Symp. RFIC Dig. Tech. Papers*, pp. 293-296, 2010.

[3] F. Opteynde., "A Maximally Digital Radio Receiver Front-End," *IEEE ISSCC Dig. Tech. Papers*, pp. 450-451, 2010.

[4] J. Kim et al, "Analysis and Design of Voltage Controlled Oscillator-Based Analog-to-Digital Converter," *IEEE Trans. Circuits and Systems I*, Vol. 57, No.1, pp. 18-30, 2010.

[5] Y.G. Yoon, et.al, "A Time-Based Bandpass ADC Using Time-Interleaved Voltage-Controlled Oscillators," *IEEE Trans. Circuits and Systems I*, Vol. 55, No.11, pp. 3571-3581, 2008.

[6] W. H. Chen et al. "A Highly Linear Broadband CMOs LNA Employing Noise and Distortion Cancellation," *IEEE J. Solid-State Circuits.*, Vol. 43, pp. 1164-1176.

978-1-61284-303-2/11 $26.00 © 2011 IEEE

ISSCC 2011 / February 22, 2011 / 12:00 PM

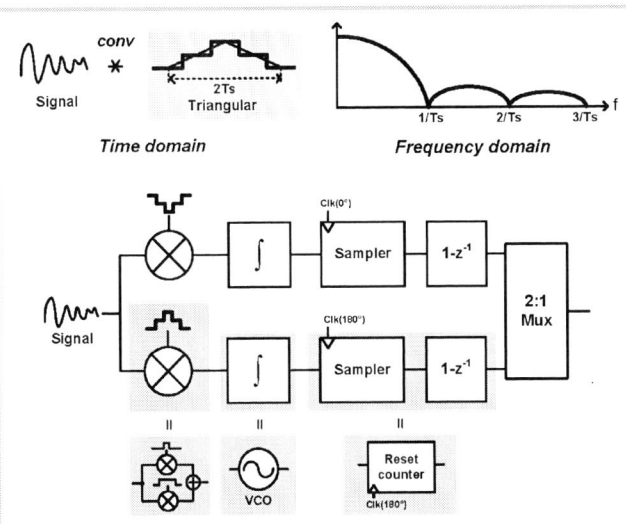

Figure 9.9.1: Operation principle of the Sinc^2 filter.

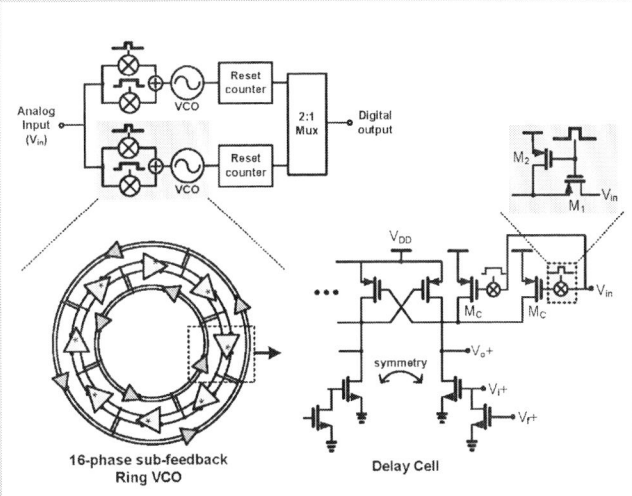

Figure 9.9.2: Proposed Sinc^2 filter-embedded VCO-based ADC.

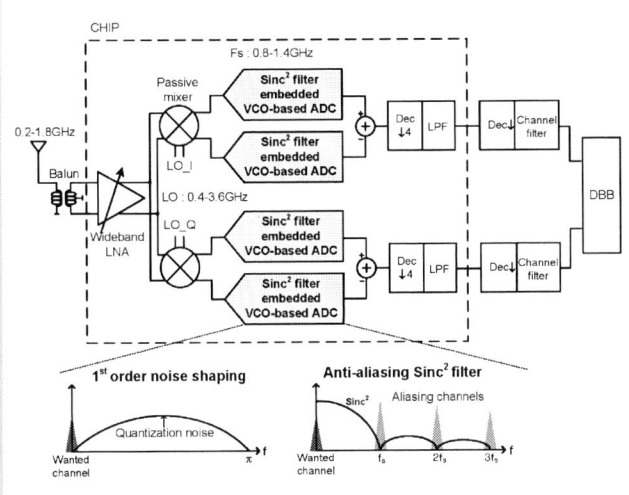

Figure 9.9.3: Digital-intensive MMMB receiver architecture.

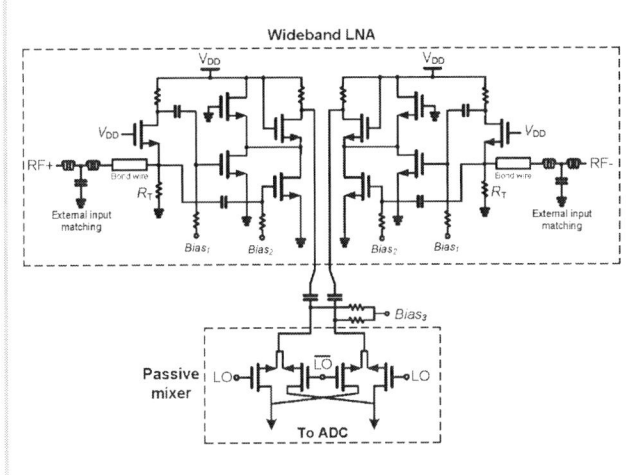

Figure 9.9.4: Wideband LNA and passive mixer.

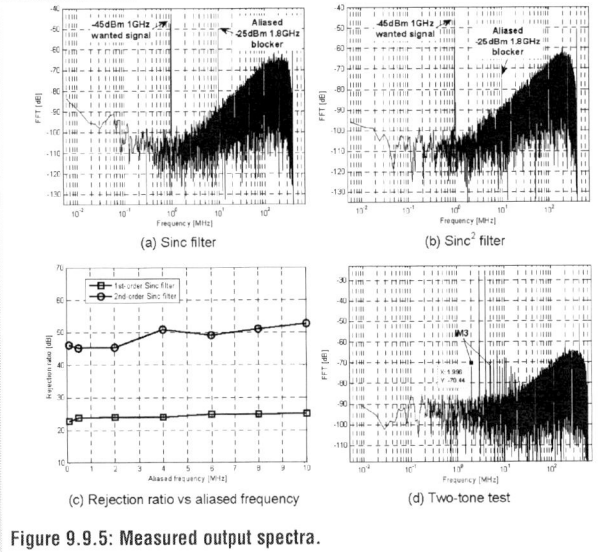

Figure 9.9.5: Measured output spectra.

Figure 9.9.6: Performance summary.

RF frequency	0.2-1.8GHz
Sampling frequency	0.8-1.4GHz
Sensitivity (SNRsys=10dB)	-94dBm @1MHz BW
IIP3	-6.78dBm
IIP2	> 31.2dBm
S11 (0.2-1.8GHz)	< -7.5dB
Out-of-band Rejection from Sinc^2 filter	50.2dB @10MHz BW
Technology	90nm CMOS
Supply voltage	1.35V
Current consumption (Fs : 0.8-1.4GHz)	31.1-36mA
Area	0.4mm^2

978-1-61284-303-2/11 $26.00 © 2011 IEEE

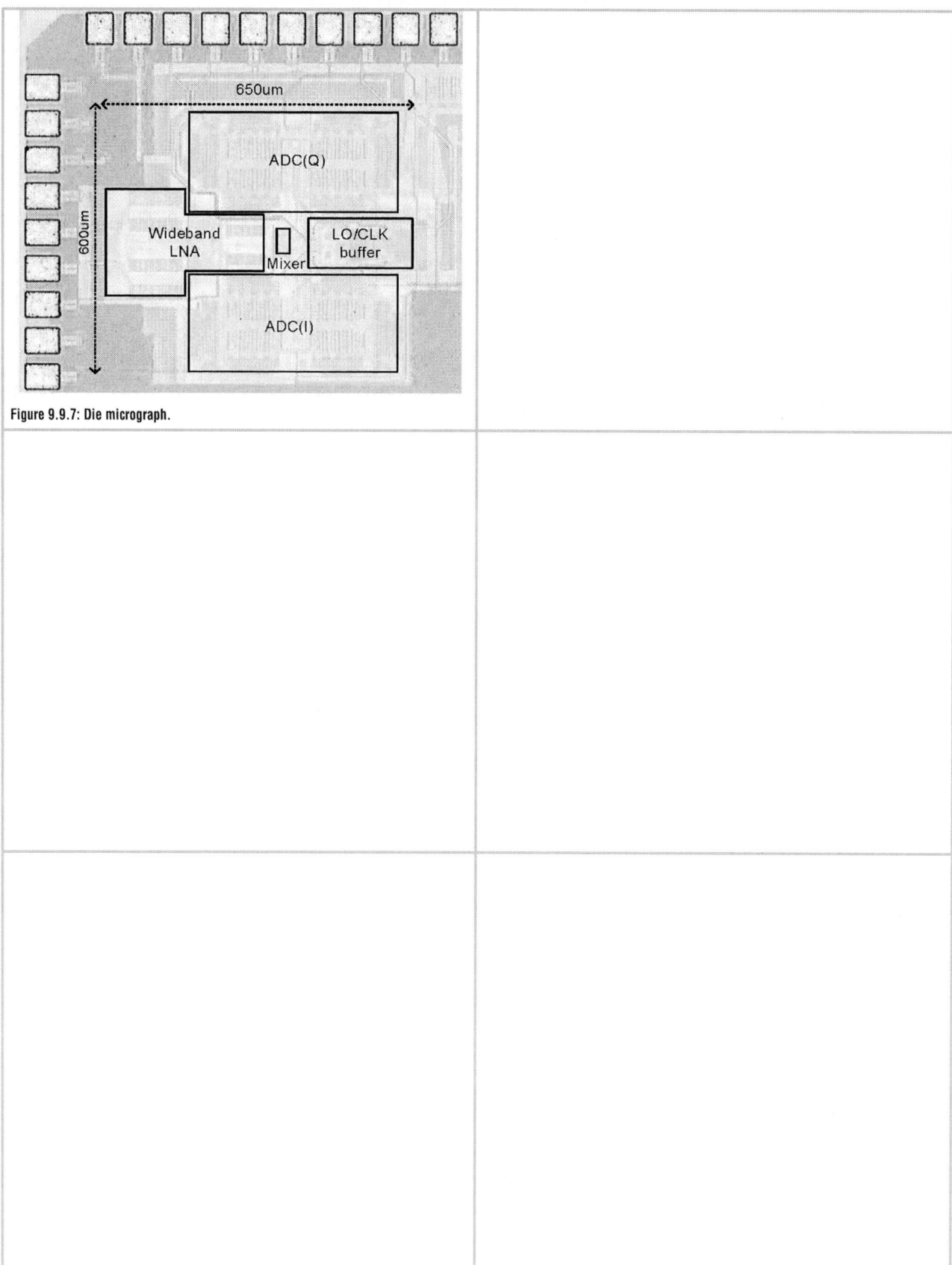

Figure 9.9.7: Die micrograph.

ISSCC 2011 / SESSION 10 / NYQUIST-RATE CONVERTERS / OVERVIEW

Session 10 Overview / Data Converters

Nyquist-Rate Converters

Session Chair: *Michael Flynn, University of Michigan, Ann Arbor, MI*

Session Co-Chair: *Michael Perrott, SiTime, Sunnyvale, CA*

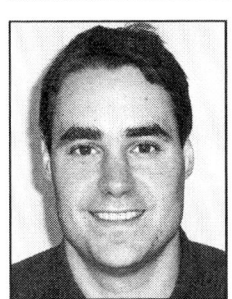

The demands placed on high-speed communication are steadily increasing as services such as cloud computing, video-on-demand, and ever increasing web access, call for higher date rates with dramatically improved energy efficiency. In particular, optical network data rates are now approaching 100Gb/s, wireless cellular base stations are processing wide-spectrum content for multi-standard support, and cable modems and medical-imaging systems are striving to achieve GHz bandwidths. In response to these challenges, new analog-to-digital converter (ADC) and digital-to-analog converter (DAC) architectures are being developed that dramatically increase conversion rates and conversion efficiency by leveraging advanced CMOS, SiGe technology, and clever circuit techniques.

Paper 10.1 from NXP Semiconductors presents a very high speed ADC that leverages 65nm CMOS and an interleaving technique to achieve 2.6GS/s operation. To achieve such a high speed, an on-chip clock generator feeds 4 interleaved track-and-holds operating at 650MHz, which in turn feed 16 SAR stages for a total of 64-times interleaving. By also utilizing on-chip gain, offset, and DAC linearity calibrations, an SNDR of 48.5dB is achieved at the Nyquist frequency of 1.3GHz while consuming only 0.48W of power. This combination of speed and precision enables simultaneous capture of many channels in the 48 to 1002MHz TV band while offering a cost and power effective implementation since only one ADC is required.

Paper 10.2 from Texas Instruments focuses on improving resolution while still achieving 1GS/s operation through the use of 2-way interleaving within a SiGe BiCMOS technology. To achieve a peak SNR of 62dB and SFDR>67dB, DACs are included to address timing skew and bandwidth mismatch between the interleaved stages. By sharing most of the circuits between the interleaved pipeline stages, matching is improved between the stages and a power consumption of 575mW is achieved. Potential applications include high performance cable modems and signal processing approaches to power amplifier linearization.

Paper 10.3 from Broadcom focuses on reduced power consumption at 800MS/s conversion rate while still maintaining SNDR levels of 59dB. To achieve this high speed sample rate, 4-way interleaving is utilized with pipeline stages that are implemented in 40nm CMOS. To counter the reduced intrinsic gain offered by this process, a dual-residue technique is employed which leverages matching between pairs of open loop gain stages to remove the need for accurate gain such that only offset is the key concern for each pipeline stage. A fast background offset calibration algorithm takes care of this issue, and the overall ADC consumes 105mW.

Paper 10.4 from Analog Devices introduces a pipeline architecture that achieves a high resolution of 77dB SNDR at 80MS/s sample rates and an input frequency of 10MHz while also consuming only 100 mW in 0.18µm CMOS. Key circuit techniques include a dynamically driven deep N-well input sampling switch, an offset-cancelled comparator, and a back-gate- voltage biased amplifier. The resolution and bandwidth offered by this ADC opens the door to multi-standard wireless receivers that can be easily accommodate a wide range of data rates and modulation standard.

Paper 10.5 from the University of Macau, Instituto Superior Tecnico, and University of Pavia focuses on the issue of power efficiency and low area by leveraging a resistor ladder to provide sub-ranging reference levels within a SAR architecture such that 2-bit conversion is achieved per conversion cycle. This approach allows 400MS/s conversion rates to be achieved without interleaving, and an SNDR>40dB with only 4mW of power consumption. Implemented in 65nm CMOS technology, the ADC occupies an area of 0.024mm^2. By achieving such low power and area with high sample rates, this ADC structure opens the door to new opportunities in mobile communication.

978-1-61284-303-2/11 $26.00 © 2011 IEEE

ISSCC 2011 / February 22, 2011 / 8:30 AM

Paper 10.6 from MIT presents a reconfigurable 5 to 10bit SAR ADC, implemented in 65nm CMOS, with power consumption that is linear related to sampling speed. A reconfigurable DAC architecture and voltage scaling down to 0.4V maintain energy efficiency at low resolutions. The DAC power scales exponentially with resolution and voltage scaling further reduces the energy-per-conversion. Applications, such as sensor networks and medical monitoring, benefit from ADCs that can digitize with varying bandwidths and resolutions.

Finally, a pair of presentations advances the state-of-the-art in high performance CMOS current-steering digital to analog converters (DACs). As the first of these papers, Paper 10.7 from National Chiao Tung University describes a 1.2GS/s 12bit DAC, implemented in 90nm CMOS, that achieves SFDR greater that 70dB up to 500MHz. A digital random return-to-zero technique enhances dynamic performance. Background calibration ensures matching of compact current cells. The device consumes 128mW and occupies an active area of 0.83mm^2.

Paper 10.8 from Ciena presents a 56GS/s 6bit DAC that achieves a new record for DAC performance. A compact design facilitates very high speed clocking and data distribution. At 56GS/s the DAC achieves SFDR greater than 30dBc and ENOB greater than 4.3b up to an output frequency of 26.9GHz. In test pattern generation mode and running at 56GS/s power is less than 750mW. This very high speed DAC, implemented in 65nm CMOS, helps enable 100Gbps optical data communication.

The advancements provided by these ADC and DAC techniques will provide further fuel to the steady trend of using sophisticated signal processing to make better of use bandwidth to enable increasing data rates with better energy efficiency in next generation communication systems.

Presenters:

8:30 AM

10.1 A 480mW 2.6GS/s 10b 65nm CMOS Time-Interleaved ADC with 48.5dB SNDR up to Nyquist

K. Doris, NXP Semiconductors, Eindhoven, The Netherlands

9:00 AM

10.2 A 12b 1GS/s SiGe BiCMOS Two-Way Time-Interleaved Pipeline ADC

R. Payne, Texas Instruments, Dallas, TX

9:30 AM

10.3 An 800MS/s Dual-Residue Pipeline ADC in 40nm CMOS

J. Mulder, Broadcom, Bunnik, The Netherlands

10:15 AM

10.4 A 16b 80MS/s 100mW 77.6dB SNR CMOS Pipeline ADC

J. Brunsilius, Analog Devices, San Diego, CA

10:45 AM

10.5 A 0.024mm^2 8b 400MS/s SAR ADC with 2b/Cycle and Resistive DAC in 65nm CMOS

H. Wei, University of Macau, Macau, China

11:15 AM

10.6 A Resolution-Reconfigurable 5-to-10b 0.4-to-1V Power Scalable SAR ADC

M. Yip, Massachusetts Institute of Technology, Cambridge, MA

11:45 AM

10.7 A 12b 1.25GS/s DAC in 90nm CMOS with >70dB SFDR up to 500MHz

W-H. Tseng, National Chiao Tung University, Hsinchu, Taiwan

12:15 PM

10.8 A 56GS/s 6b DAC in 65nm CMOS with 256×6b Memory

Y. M. Greshishchev, Ciena, Ottawa, Canada

10

978-1-61284-303-2/11 $26.00 © 2011 IEEE

ISSCC 2011 / SESSION 10 / NYQUIST-RATE CONVERTERS / 10.1

10.1 A 480mW 2.6GS/s 10b 65nm CMOS Time-Interleaved ADC with 48.5dB SNDR up to Nyquist

Kostas Doris, Erwin Janssen, Claudio Nani, Athon Zanikopoulos, Gerard Van der Weide

NXP Semiconductors, Eindhoven, The Netherlands

Trends in cable TV reception for data and video require simultaneous capture of many channels, e.g., 16, arbitrary located in the 48-to-1002MHz TV band. The challenges of integrating more than two zero-IF tuners on a single die [1] could be simplified with a low-power 10b ADC that can digitize the entire TV band and be suitable for integration with baseband DSP. This work presents a 64× interleaved 2.6GS/s 10b 65nm CMOS ADC with on-chip calibrations, combining interleaving hierarchy with an open-loop buffer array operated in feedforward-sampling and feedback-SAR mode. The ADC achieves an SNDR of 48.5dB at Nyquist and consumes only 0.48W.

Time-interleaving of multiple low-rate SAR ADCs, e.g., 64 or more, offers power efficiency but so far has been limited to low resolutions [4]. At 10b level, flat interleaving for that many units, i.e., one T/H per ADC, requires an input buffer with stringent bandwidth, linearity and noise requirements to drive the large T/H input load. Clocking with skew and jitter close to 0.1ps becomes challenging and claims substantial power. On-chip calibration for timing, bandwidth, offset, gain mismatches and DAC nonlinearity presents another major challenge. These issues have limited 10b time-interleaved SAR ADCs to 16 units and 1.8GS/s [5]. An interleaving hierarchy that is based on an input power splitter and a few T/Hs that drive many ADCs with unity gain buffers can relax some of these issues [4]. However, at 10b level the load seen by each T/H limits bandwidth and linearity whereas a power splitter introduces SNR loss.

The proposed ADC architecture is shown in Fig. 10.1.1. It combines interleaving hierarchy with an open-loop multiplexed buffer topology operating in feedforward-sampling and feedback-SAR mode. The sampling front-end (FE) consists of 4 interleaved T/Hs at 650MHz that are optimized for timing accuracy and sampling linearity. The back-end (BE) consists of four ADC arrays, each consisting of 12 54MHz 10b current-mode non-binary SAR ADCs, with an additional 4 units for redundancy. Calibration per ADC deals with offset, gain and DAC nonlinearity at start-up and makes non-binary-to-binary code conversion prior to data recombination. A CML clock generator generates all necessary clock signals from a 2.6GHz reference. The proposed architecture (a) decouples the FE from the BE, which allows for more ADCs to be used per T/H and (b) eliminates distortion stemming from open loop buffers interfacing the two.

Figure 10.1.2 shows the T/H that is based on a bootstrapped switch-capacitor topology with feedthrough compensation. To decouple the T/H from the ADC array the T/H output is passed through a 1 to 16 demux to a buffer driving the SAR ADC unit. During each track-and-hold cycle only one small buffer is visible to the main sampling capacitor. This reduces significantly the nonlinear load seen at the sampling node. The use of only 4 T/Hs receiving a common 2.6GHz clock in combination with a sampling scheme based on [5] with special attention paid to the H-tree interconnect geometry, steep slopes, supply tracks and mismatch allows for low skew without calibration and offers low jitter and large bandwidth without a need for an input buffer to interface with external 50W source. Thus, this approach combines the T/H-ADC interfacing simplicity of flat interleaving with the clocking simplicity and bandwidth advantages of hierarchical interleaving of a few T/Hs.

Figure 10.1.3 shows the SAR ADC and elaborates further the T/H-ADC interfacing. The SAR ADC uses a differential non-binary topology. Current-steering is used for main and calibration DACs for high speed, immunity to supply noise, gain and offset calibration simplicity and for temperature stability. In a charge-redistribution SAR ADC, sampling and DAC functions are usually merged. The buffer interfacing T/H and ADC unit then limits linearity and restricts signal swing at the cost of more noise, or it is replaced by closed loop topologies at the cost of speed. In this ADC, the buffer charges first the unit SAR ADC re-sampling capacitor to the input signal in the track phase. Then, in the SAR phase, the buffer is introduced in the loop passing the DAC signal through it with another

multiplexer input in the T/H. Thus, both sampling and DAC paths are distorted in the same way effectively cancelling the nonlinearity. This technique (a) allows more than an order of magnitude higher linearity than what is normally achieved with open-loop buffers and (b) enables maximum voltage swing which results in a higher SNR.

The SAR operation is completed in 12 cycles of 650MHz. In the 1st cycle, the input signal is tracked and held in the T/H while the i-th ADC is set in tracking phase. At the end of this cycle the signal held in the T/H is re-sampled at the i-th preamplifier input and the unit enters the SAR phase. At the same time the demux sets unit i+1 in feedforward mode to track the T/H output. In the next 11 cycles the i-th comparator steers the DAC current in the two resistor loads according to the non-binary SAR algorithm. In the last cycle, the data are sent to the calibration logic. Offset and gain mismatches are calibrated with 12b DACs. These DACs have less than 2 radix in order to ensure low DAC DNL requirements and therefore keep the area small. During start-up the ADC offset is calibrated first by applying a zero differential input and unbalancing the preamplifier's output until zero output is reached. Gain correction is applied next modifying the range of the main DAC to match the external voltage reference. Off-chip reference voltages are provided directly to the sampling nodes with on-chip low-noise buffers. The main DAC calibration is similar to [6] to reduce total area and obtain a compact layout.

A differential low-swing CML clock generator ensures sampling with high immunity to supply and substrate noise without generating noise itself. The fs/4 clocks for the ADC arrays (Fig. 10.1.1) are generated from the 2.6GHz reference by means of a divider and have adjustable delay to maximize T/H settling time.

The ADC is realized in baseline 7-metal 65nm CMOS process and is integrated as part of a DOCSIS 3.0 direct-sampling receiver, which is placed in a LGA132 package. The ADC consumes 480mW from 1.6/1.3/1.2V supplies at 2.6GS/s (excluding LVDS). The on-chip calibration logic occupies 50% of the total 5.1mm^2 and consumes 36mW from 1.2V. Figure 10.1.4 shows spectral plots with and without calibration. The SNDR without calibration is 32.4dB and improves to 52.8dB with calibration. The total offset and gain mismatch tone power is 10dB below the total thermal noise level. Figure 10.1.5 shows a spectral plot with a 1.25GHz input and the performance over frequency for a full scale 1.4V$_{pp-diff}$ input. The SNDR of 52.8dB at low frequencies is dominated by thermal noise, and stays above 48.5dB up to Nyquist thanks to the low clock skew (estimated <400fs$_{rms}$, dominated by mismatch) and high linearity. The more than 5GHz input bandwidth in combination with less than 110fs$_{rms}$ clock jitter (including the signal and clock sources) results in an SNR of 49dB up to 4GHz. A performance summary and a comparison to prior-art ADCs with more than 1GS/s operation are shown in Fig. 10.1.6. A die micrograph is shown in Fig. 10.1.7.

Acknowledgements:
The authors thank BL TV Front-End team for inputs, top-level and packaging, H.v.d. Ploeg and M. Vertregt for their contributions, Y. Tang for helping in layout, and L. Lo Coco and L. Warmerdam for support.

References:
[1] F. Gatta, R. Gomez, Y.J. Shin, et al., "An Embedded 65nm CMOS Baseband IQ 48MHz-1GHz Dual Tuner for DOCSIS 3.0," *IEEE J. Solid-State Circuits*, vol. 44, no. 12, pp. 3511-3525, Dec., 2009.
[2] E2V Semiconductors, "AT84AS008," http://www.e2v.com, Web-datasheet accessed on Sep. 12, 2010.
[3] Datasheet, ADC12D1600, http://www.national.com/ds/DC/ADC12D1600.pdf, accessed on Sep. 12, 2010.
[4] Y.M. Greshishchev, J. Aguirre, M. Besson, et al., "A 40GS/s 6b ADC in 65nm CMOS," *ISSCC Dig. Tech. Papers*, pp. 390-391, Feb., 2010.
[5] S. M. Louwsma, A.J.M. van Tuijl, M. Vertregt, and B. Nauta, "A 1.35 GS/s, 10b, 175 mW Time-Interleaved AD Converter in 0.13μm CMOS", *IEEE J. Solid-State Circuits*, vol. 43, no. 4, pp. 778-786, Apr., 2008.
[6] W. Liu, Y. Chang, S.-K. Hsien, et al., "A 600mW 30mW 0.13μm CMOS ADC Array Achieving Over 60dB SFDR with Adaptive Digital Equalization," *ISSCC Dig. Tech. Papers*, pp. 82-83, Feb., 2009.

978-1-61284-303-2/11 $26.00 © 2011 IEEE

ISSCC 2011 / February 22, 2011 / 8:30 AM

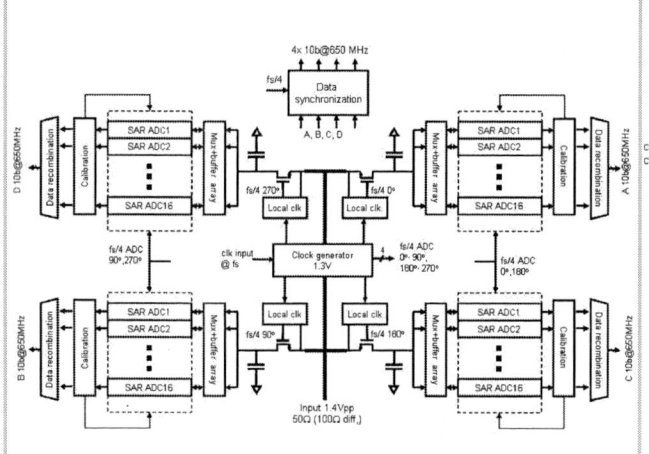

Figure 10.1.1: Hierarchical time-interleaved ADC architecture with multiplexed buffer array.

Figure 10.1.2: T/H implementation showing the multiplexed buffer array.

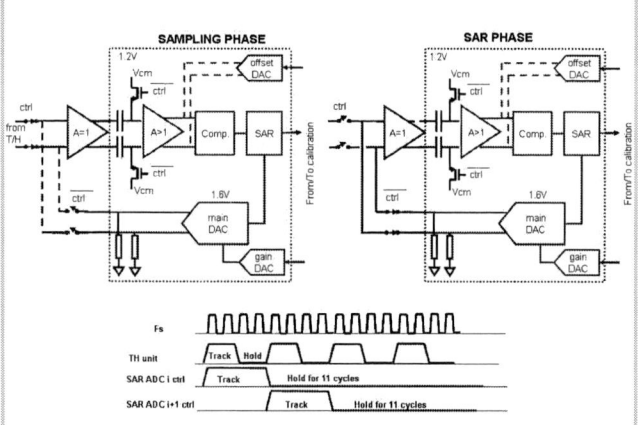

Figure 10.1.3: SAR ADC operation with T/H buffer in sampling and SAR operation.

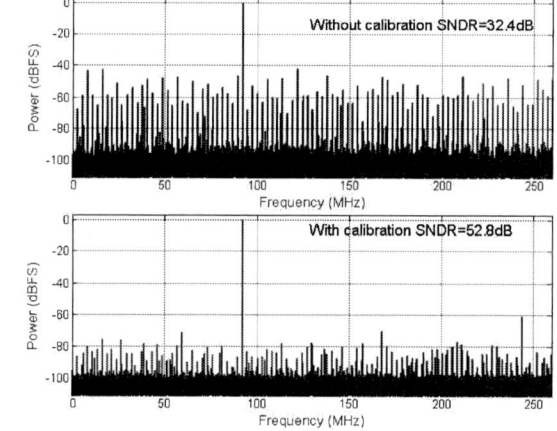

Figure 10.1.4: ADC output spectrum with a 92MHz input with and without calibrations. The output is 2.6GS/s decimated by 5.

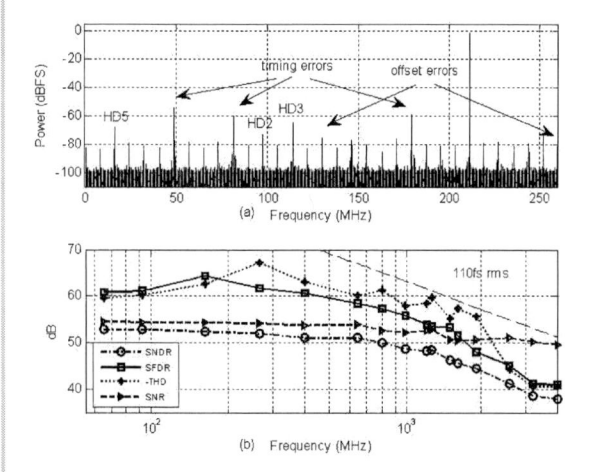

Figure 10.1.5: (a) ADC output spectrum with a 1.25GHz input at 2.6GS/s (decimated by 5), and (b) performance versus frequency at 2.6GS/s.

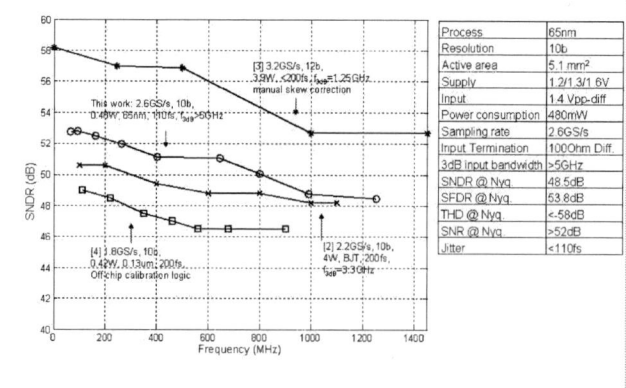

Figure 10.1.6: Comparison with prior art and performance summary.

Process	65nm
Resolution	10b
Active area	5.1 mm²
Supply	1.2/1.3/1.6V
Input	1.4 Vpp-diff
Power consumption	480mW
Sampling rate	2.6GS/s
Input Termination	100Ohm Diff.
3dB input bandwidth	>5GHz
SNDR @ Nyq	48.5dB
SFDR @ Nyq	53.8dB
THD @ Nyq	<-58dB
SNR @ Nyq	>52dB
Jitter	<110fs

978-1-61284-303-2/11 $26.00 © 2011 IEEE

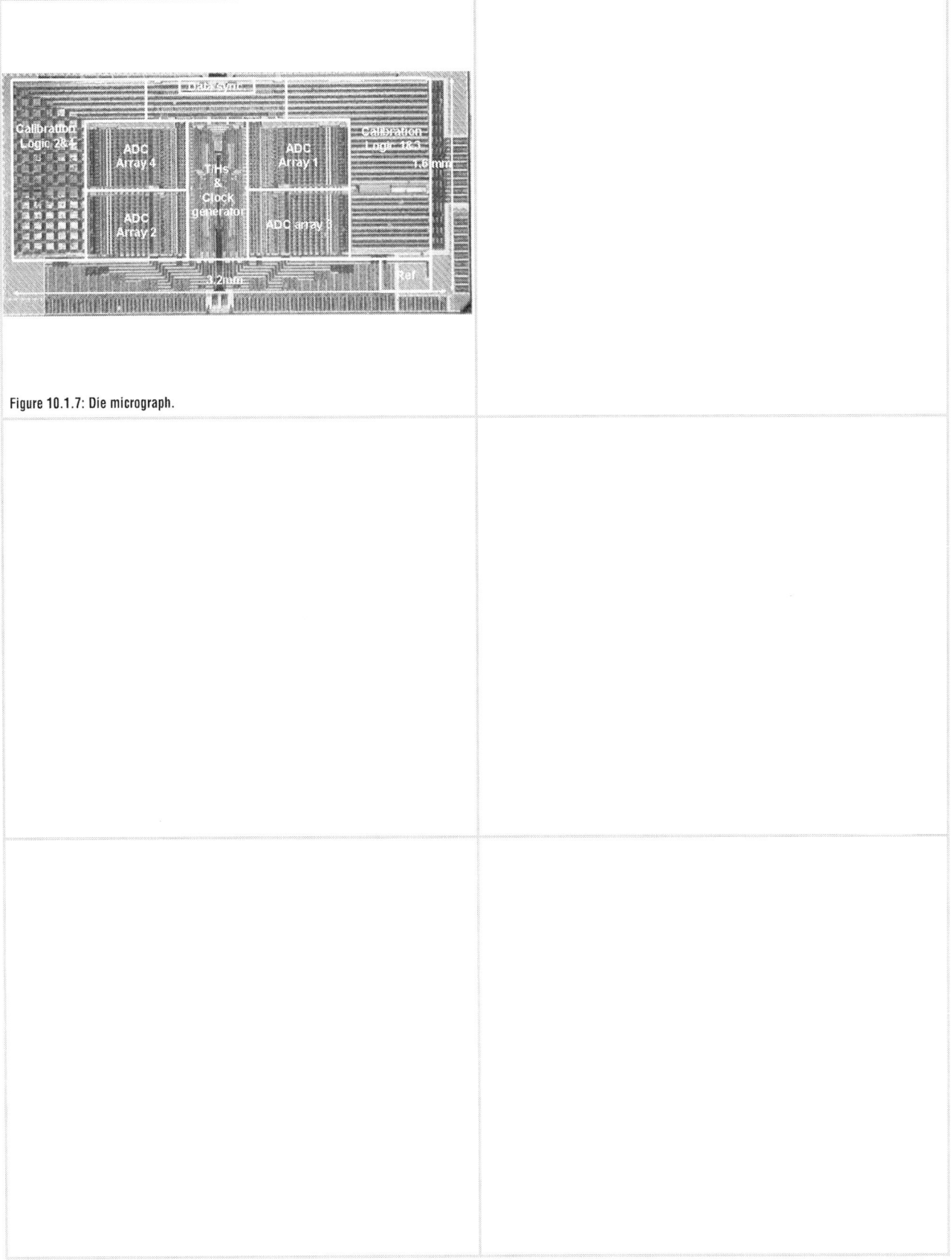

Figure 10.1.7: Die micrograph.

ISSCC 2011 / SESSION 10 / NYQUIST-RATE CONVERTERS / 10.2

10.2 A 12b 1GS/s SiGe BiCMOS Two-Way Time-Interleaved Pipeline ADC

Robert Payne[1], Charles Sestok[1], William Bright[1], Manar El-Chammas[1], Marco Corsi[1], David Smith[1], Noam Tal[2]

[1]Texas Instruments, Dallas, TX
[2]Texas Instruments, Raanana, Israel

Pipelined ADCs designed in analog BiCMOS technologies can offer good linearity and high SNR performance for input signals with reasonable voltage swings. Such ADCs, however, face two critical design challenges: the process limits the sampling rate, and the pipeline architecture limits power efficiency. This paper introduces a two-way time-interleaved (TI) switched-current 1Gs/s 12b pipelined ADC in SiGe BiCMOS that addresses these issues.

Time-interleaving multiple low-rate ADCs increases the effective sampling rate (f_s) [1]. Each ADC branch is designed to meet the desired dynamic range and overall TI bandwidth (BW). Contrary to the goal of increased f_s, the added ADC loading reduces system BW since n times more load reduces BW by n. In addition, interaction between the multiple-input samplers, package, and board parasitics must be prevented. Finally, mismatch between the ADC branches produces interleaving artifacts in the output signal spectrum, limiting the achievable SFDR. Important mismatches occur in offset, gain, sample timing [1] and track-and-hold (T/H) BW [2].

Pipelined ADCs convert one sample per clock cycle. Each clock cycle is divided into two modes: track, when the input is followed and sampled, and hold, when a sub-ADC makes a coarse decision, a DAC generates a precise reconstruction of the decision, and a residue amplifier (ResAmp) subtracts the DAC output from the held signal and amplifies the result for the next stage quantizer. This architecture has low power efficiency since the sub-ADC, DAC, and ResAmp have 50% duty cycle. Furthermore, these circuits consume most of the total power in a switched-current design.

In pipeline ADCs, amplifier sharing, where adjacent pipeline stages use the same amplifier on opposite clock phases, has been used to reduce power consumption [3]. While amplifier count is halved, power savings is not necessarily maximized since the relaxed noise requirements of later pipeline stages due to the inter-stage gain ordinarily leads to power scaling. For example, in the first two stages, the 2nd amplifier may consume less than ¼ the power of the 1st amplifier, so removing the 2nd has limited benefit. More power reduction is achieved by exploiting the timing of a two-way TI ADC. Since the T/Hs hold on opposite clock phases, the same quantizer (DAC and amplifier) can be reused, resulting in a 2× power and area reduction. Efficiency is better than that of amplifier sharing since the 1st-stage circuits are reused where highest performance is needed.

Figure 10.2.1 shows the block diagram of the pipeline ADC. Here, two T/Hs are driven by a shared input buffer and clocked on opposite phases of an $f_s/2$ frequency clock. Each T/H output is quantized by a 4b flash ADC. A current-mode DAC and residue amplifier with resistive feedback are shared on opposite hold phases using muxes in the analog and digital paths. These shared blocks consume more than 60% of the stage power and area; therefore their reuse is a substantial benefit. Sharing also minimizes any possible gain and offset mismatch. The 2nd- and 3rd-stage use the same architecture followed by a final 4b flash. Pipeline correction, DAC trimming, and offset correction is performed with 14b precision logic (13b from the pipeline itself) and reduced to a 12b output.

Circuit design techniques alleviate the input loading problem in TI ADCs. The input buffer and a T/H are shown in Fig. 10.2.2 (two are used pseudo-differentially in each branch). The input is terminated by a 50Ω resistor to a common-mode (CM) voltage and drives two emitter-follower (EF) buffers. The first EF buffer (Q1) drives the sampling switches of the two T/Hs and the second EF buffer (Q2) drives the bootstrap circuits to isolate them from the signal path. In addition to preventing interaction between the TI branches, the buffers also allow the design to scale to 8-way interleaving (4Gs/s), where simulations indicate an input BW greater than 2GHz is achievable. The low output impedance of the EF

drives the top-plate sampling switch (M1) and achieves high SFDR. A 20Ω resistor between M1 and the 600fF sampling capacitor improves the linearity of M1.

Analog calibration techniques address the mismatches between branches. Timing skew is adjusted using a 10b DAC to supply the clock buffer power supply which changes its propagation delay. BW is adjusted similarly by changing the power supply to the bootstrap circuit to vary the on-resistance of M1. Since all other circuits are shared, the only gain mismatch between branches is the output EF (Q3) which did not need calibration.

The residue amplifier is shown in Fig. 10.2.3. It consists of muxed input gm-stages, a folded-cascode gain stage, and output EFs. The amplifier is connected fully differentially, as shown in the inset. Feedback forces (INP-DACP)+(INM-DACM)≈0V, but does not enforce INP=DACP and INM=DACM. The CM gain of the amplifier is cancelled, which allows optimizing the differential settling of the amplifier independent of the CM. Separate CMFB forces the output CM to track the input CM. DAC current is balanced by a constant current source connected into the middle of the feedback network. The output EFs use base current cancellation provided by QA, QB, QA', and QB' to reduce the impact of load current on the open-loop gain. QA'/QB' sense the current in QA/QB and subtract the base current from the signal path, effectively increasing the impedance at the base of QA/QB and the open-loop gain from 71dB to 92dB.

Off-chip tuning of the built-in calibration DACs removes the TI spurs. Resolution of the calibration circuits is set by a 65dBc SFDR specification and determined by finding the mismatches that produce distortion 65dB below the input signal. DC offset (which is removed digitally on-chip) requires 11b matching. Gain mismatch should be <500ppm and is achieved without calibration. Timing skew (t_{skew}/T_S) is less than 200ppm. BW mismatch simulations use a two-tone input signal and assumed T/H BW is 5× the sampling frequency of a single ADC branch and yield less than 2000ppm required mismatch [4]. Figure 10.2.4 illustrates the range of the timing skew (measured) and BW (simulated) calibration circuits. Mismatch cancellation is adjusted using cost functions derived from signal statistics [5]. For wide-sense stationary inputs, matched TI ADCs produce wide-sense stationary outputs. If there are significant mismatches, the interleaved output is cyclo-stationary and the estimated signal statistics depend on the ADC branch. The calibration circuits are tuned to minimize cost functions that measure the amount of cyclo-stationarity in the ADC output.

The test chip is fabricated in a 0.18µm 1.8V CMOS 3.3V SiGe complementary BJT process and is tested in an 84-pin multi-row QFN package. Figure 10.2.5 shows the measured SNR and SFDR plots for a range of input frequencies and sampling rates. Figure 10.2.6 summarizes the design and shows the spectrum with and without the TI spurs compensated. The die micrograph is shown in Fig. 10.2.7. The ADC occupies 1.8mm² while the TI correction DACs and voltage regulators occupy 0.55mm².

Acknowledgements:
The authors thank Texas Instruments for enabling this project through the support of the Kilby Labs research center. In addition, this work also benefited greatly through interaction with the cable modem team in Israel, ADC teams in Dallas and India, and the Systems and Applications R&D organization.

References:
[1] W.C. Black Jr. and D.A. Hodges, "Time Interleaved Converter Arrays," *IEEE J. Solid-State Circuits*, vol. 15, no. 6, pp. 1022-1029, Dec., 1980.
[2] N. Kurosawa, H. Kobayashi, K. Maruyama, H. Sugawara, and K. Kobayashi, "Explicit Analysis of Channel Mismatch Effects in Time-Interleaved ADC Systems," *IEEE Trans. Circuits and Systems I*, vol. 48, no. 3, pp. 261-271, Mar., 2001.
[3] P. Yu and H.S. Lee, "A 2.5-V, 12-b, 5-MSample/s Pipelined CMOS ADC," *IEEE J. Solid-State Circuits*, vol. 31, no. 12, pp. 1854-1861, Dec., 1996.
[4] P. Satarzadeh, B.C. Levy, and P.J. Hurst, "Adaptive Semi-Blind Calibration of Bandwidth Mismatch for Two-Channel Time-Interleaved ADCs," *IEEE Trans. Circuits and Systems I*, vol, 56, no. 9, pp. 2075-2088, Sep., 2009.
[5] J. Elbornsson, F. Gustaffson, and J.E. Eklund, "Blind Equalization of Time Errors in a Time-Interleaved ADC System," *IEEE Trans. Signal Processing*, vol. 53, no. 4, pp. 1413-1424, Mar., 2005.

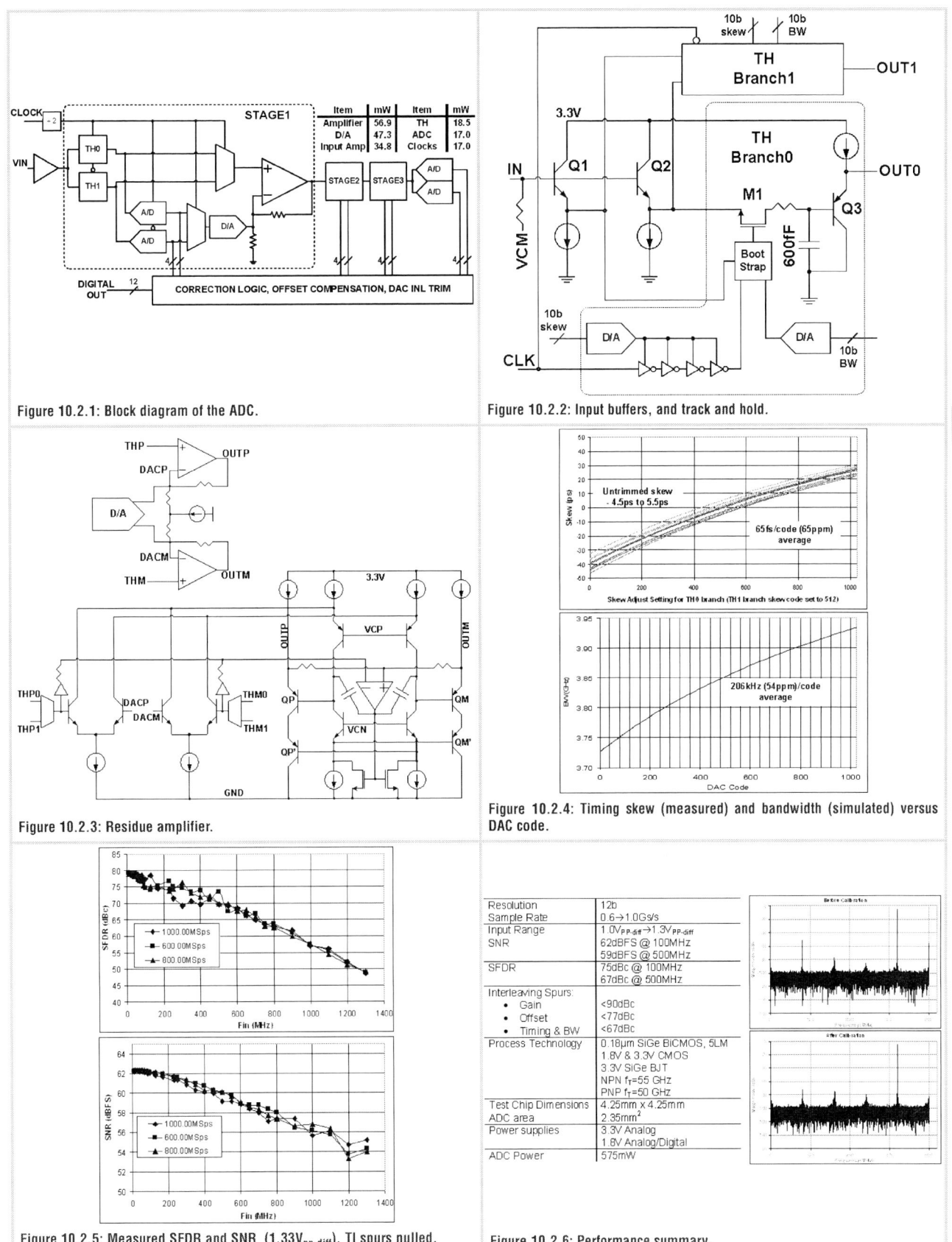

Figure 10.2.1: Block diagram of the ADC.

Figure 10.2.2: Input buffers, and track and hold.

Figure 10.2.3: Residue amplifier.

Figure 10.2.4: Timing skew (measured) and bandwidth (simulated) versus DAC code.

Figure 10.2.5: Measured SFDR and SNR (1.33V$_{pp-diff}$), TI spurs nulled.

Figure 10.2.6: Performance summary.

Resolution	12b
Sample Rate	0.6→1.0Gs/s
Input Range	1.0V$_{PP-diff}$→1.3V$_{PP-diff}$
SNR	62dBFS @ 100MHz
	59dBFS @ 500MHz
SFDR	75dBc @ 100MHz
	67dBc @ 500MHz
Interleaving Spurs:	
• Gain	<90dBc
• Offset	<77dBc
• Timing & BW	<67dBc
Process Technology	0.18µm SiGe BiCMOS, 5LM
	1.8V & 3.3V CMOS
	3.3V SiGe BJT
	NPN f$_T$=55 GHz
	PNP f$_T$=50 GHz
Test Chip Dimensions	4.25mm x 4.25mm
ADC area	2.35mm^2
Power supplies	3.3V Analog
	1.8V Analog/Digital
ADC Power	575mW

ISSCC 2011 PAPER CONTINUATIONS

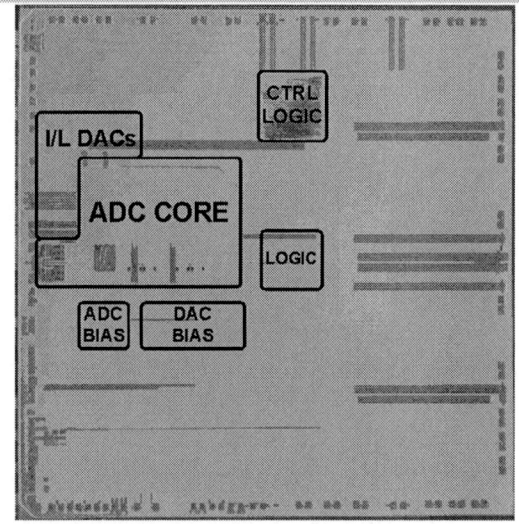

Figure 10.2.7: Die micrograph.

ISSCC 2011 / SESSION 10 / NYQUIST-RATE CONVERTERS / 10.3

10.3 An 800MS/s Dual-Residue Pipeline ADC in 40nm CMOS

Jan Mulder, Frank M.L. van der Goes, Davide Vecchi, Jan R. Westra, Emre Ayranci, Christopher M. Ward, Jiansong Wan, Klaas Bult

Broadcom, Bunnik, The Netherlands

The 800MS/s 12b pipeline ADC presented here achieves a 59dB peak SNDR while consuming 105mW, resulting in an FOM of 0.18pJ/conversion-step. With digital power dissipation decreasing with technology much faster than analog power consumption, power efficient ADC designs have to make use of calibration. A major advantage offered by the dual-residue ADC architecture [1] is that the only calibration required is a calibration of the offset voltages of the MDAC amplifiers; a simple algorithm has been implemented that reaches convergence very rapidly and tracks the offsets for temperature drift and aging. Furthermore, the relaxed open-loop gain and bandwidth requirements of the MDACs allowed for a low-power implementation. Low power consumption is essential especially in applications where multiple high-speed ADCs have to be implemented on a single chip, such as 10GBase-T Ethernet.

Instead of processing a single residue signal throughout the ADC pipeline, each MDAC stage processes two residue signals simultaneously. As illustrated in Fig. 10.3.1 for the first MDAC stage (MDAC1), the input signal V_{in} is sampled on two identical input capacitors. Based on the output of the first coarse ADC, two DC reference voltages $V_{ref,0}$ and $V_{ref,1}$ are selected from a reference ladder, which satisfy the inequality $V_{ref,0} < V_{in} < V_{ref,1}$. Two residue signals are obtained by subtracting $V_{ref,0}$ and $V_{ref,1}$ from V_{in}, respectively. After amplification by two residue amplifiers A_0 and A_1, output voltages $V_{out,0}$ and $V_{out,1}$ result, where $V_{out,0} > 0$ and $V_{out,1} < 0$. The position of the zero crossing, denoted by Z, that can be obtained when interpolating between $V_{out,0}$ and $V_{out,1}$, now represents the MDAC residue signal. It can easily be shown that Z, which equals $V_{out,0} / (V_{out,0} - V_{out,1})$, does not depend on the *absolute gain* of the residue amplifiers; only *gain matching* is required. This is a major advantage of this architecture as it relaxes the open-loop gain and settling requirements of the amplifiers, which allows for a low-power implementation. At the input of the second MDAC, the ADC reference voltages are not required anymore. Instead, interpolation between $V_{out,0}$ and $V_{out,1}$ is used to zoom in on the exact position of the zero crossing. The required interpolation factors are determined by the output code of the second coarse ADC [1].

Voltage-mode amplifiers are used in MDAC1 to amplify the residue voltages, as shown in Fig. 10.3.2. An advantage of this type of amplifiers is that the input capacitors are not discharged during the amplification phase. This relaxes the speed requirements for the on-chip reference buffers, which helps to limit their power consumption. A two-stage design is used to implement the opamps. The input stage uses a single transistor M_1, instead of a differential pair, to limit the amplifier noise. It is followed by a folded-cascode transistor M_2 and the output stage transistor M_3. A small Miller capacitance around M_3 and a feedback capacitor around R_{fb2} provide the amplifier with sufficient phase margin. Although the open-loop gain of the amplifier typically is only 40dB, the performance of this pipeline ADC is not compromised, owing to the dual-residue principle.

The INL of the ADC is sensitive to the *difference in offset voltage* of the two residue amplifiers comprising an MDAC stage [2]. This design uses a simple analog background calibration algorithm, illustrated in Fig. 10.3.3, which reaches convergence within only a few thousand clock cycles. During the reset phase, when the inputs of the amplifiers are shorted to a DC voltage, the outputs are equal to their respective offset voltages. A comparator is used to determine the sign of the offset of a particular residue amplifier. Depending on the result, the DAC output voltage is increased or decreased. By repeating this measurement, the DAC output voltage converges to the point where the amplifier offset is near zero. To limit the impact of the comparator noise, several subsequent comparator decisions are averaged before updating the DAC input code. A single comparator is used to subsequently measure the offset of both residue amplifiers comprising an MDAC stage. This way, the offset of the comparator itself is not important, since the *common offset voltage* of the amplifiers does not compromise the ADC performance. The offset DACs have a 9b resolution. To guarantee

monotonicity, their design is based on a resistor string. A differential pair is used to convert the DAC voltage into a current that is added to the output current of the amplifier input stage. Although in principle the DAC resolution can be scaled down in subsequent MDAC stages, the same DAC is used throughout the ADC to limit design time.

To achieve an overall sampling frequency F_s of 800MS/s, four ADC lanes are interleaved, each running at 200MS/s, as depicted in Fig. 10.3.4. Each two ADC lanes share one common reference ladder. The T/H in front of the ADC is 4× interleaved as well. To minimize the timing errors between the T/H sampling clocks, special care has been taken in the layout. Calibration of these errors is therefore not required. The offset and gain errors that exist among the interleaved ADC lanes have been removed by means of digital calibration, implemented on-chip. The offset errors are first calculated by averaging the output data of the ADC lanes, and then subtracted digitally from the output data. Subsequently, the gain errors are estimated by calculating the rms-value of the output data of the ADC lanes. Digital gain blocks are used next to correct for the differences in rms-value.

For ADCs used in SoCs, decreasing the quantization noise is often a cost- effective way of decreasing the total ADC noise contribution. In the presented ADC design, the quantization noise has been made negligible. This is accomplished by choosing an overall ADC resolution of 12 bits, implemented by cascading seven stages. The first MDAC stage resolves 5 bits. Choosing such a relatively large number of bits in the first stage has several advantages. First, it allows the two residue amplifiers to have a relatively large gain, approximately 7 in this design, which decreases the noise contribution of the subsequent stages. Secondly, it relaxes the requirements with respect to gain matching, bandwidth matching and distortion of the residue amplifiers in MDAC1. The second MDAC stage resolves 2.5 bits and has a gain of approximately 2.5. Owing to the large gain of MDAC1, aggressive power scaling could be used. The next 4 MDAC stages all resolve 1.5b per stage and have a gain of ~2. The designs of MDAC2 through MDAC6 are based on single-stage charge amplifiers, without gain-boosting or any other techniques to improve the open-loop gain. The final stage is a 2b flash ADC.

The ADC is integrated as part of a 10GBase-T Ethernet transceiver, fabricated in a 40nm CMOS process. Figure 10.3.7 shows a die micrograph of the ADC, which occupies an area of 0.88mm². The ADC is preceded by a programmable-gain amplifier (PGA), which is used to optimize the loading of the ADC. In the measurement setup, the on-chip transmitter is used as the input signal to the receiver. As the PGA could not be bypassed, note that the reported measurement results include the noise and distortion contributions of the PGA.

The ADC operates from a dual 1V/2.5V power supply. It consumes 105mW, which includes 15mW for the on-chip ADC reference buffers. Figure 10.3.5 shows the INL and DNL, measured at 800MS/s with a two-tone input signal around 37MHz. The INL equals +2.4/–2.5LSB and is dominated by the third-order distortion caused by the PGA. The DNL equals +0.3/–0.4LSB. Figure 10.3.6 shows the ADC output spectrum, measured at F_{sig}=360MHz, and the SNDR and SFDR as a function of the input frequency, F_{sig}. A peak SNDR of 59dB is achieved. At low frequencies, the SNDR drops due to the band-pass characteristic of the PGA. The ADCs figure of merit, defined by FOM=P/(min{2×ERBW,F_s}×2^{enob}), equals 0.18pJ/conversion-step.

References:
[1] C. Mangelsdorf, H. Malik, S.H. Lee, S. Hisano, and M. Martin, "A Two-Residue Architecture for Multistage ADCs," *ISSCC Dig. Tech. Papers*, pp 64-65, Feb., 1993.
[2] H. van der Ploeg, G. Hoogzaad, H.A.H. Termeer, M. Vertregt, and R.L.J Roovers, "A 2.5-V 12-b 54-Msample/s 0.25-μm CMOS ADC in 1-mm² with Mixed-Signal Chopping and Calibration," *IEEE J. Solid-State Circuits*, vol. 36, no. 12, pp 1859-1867, Dec., 2001.

978-1-61284-303-2/11 $26.00 © 2011 IEEE

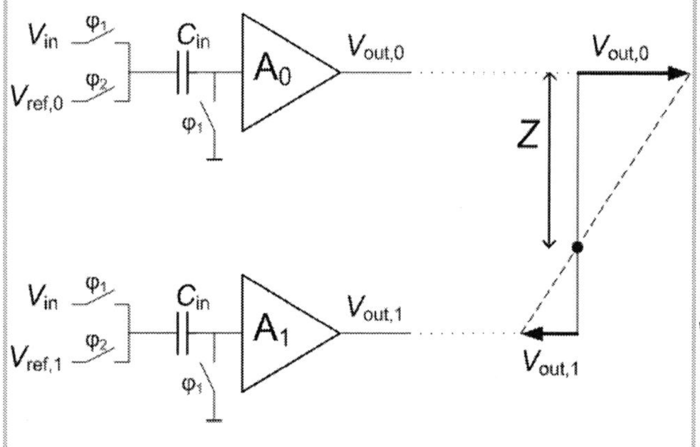

ISSCC 2011 / February 22, 2011 / 9:30 AM

Figure 10.3.1: First MDAC stage based on the dual-residue principle.

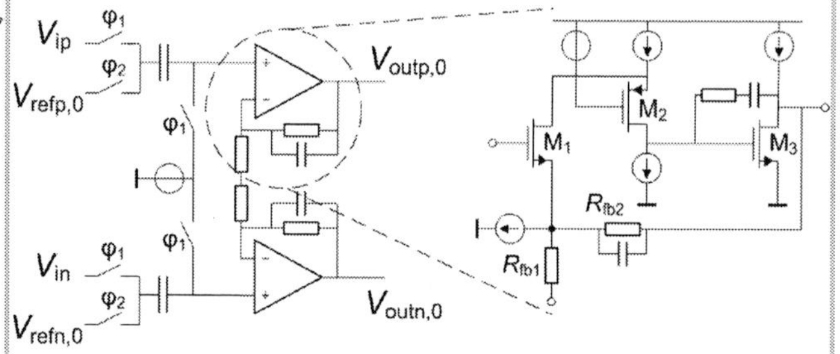

Figure 10.3.2: Implementation of one of the residue amplifiers in the first MDAC stage.

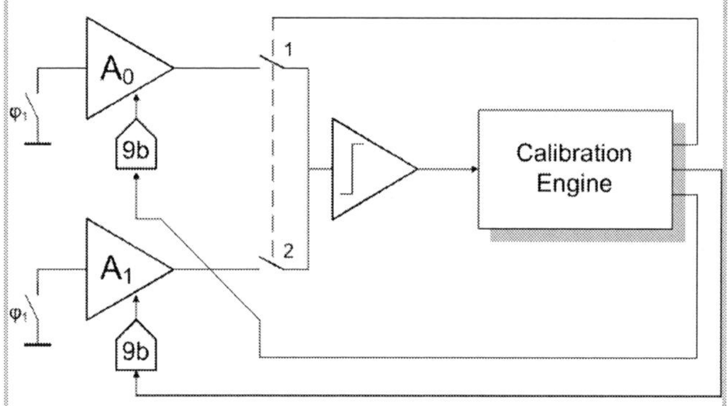

Figure 10.3.3: Background offset calibration of the MDAC residue amplifiers.

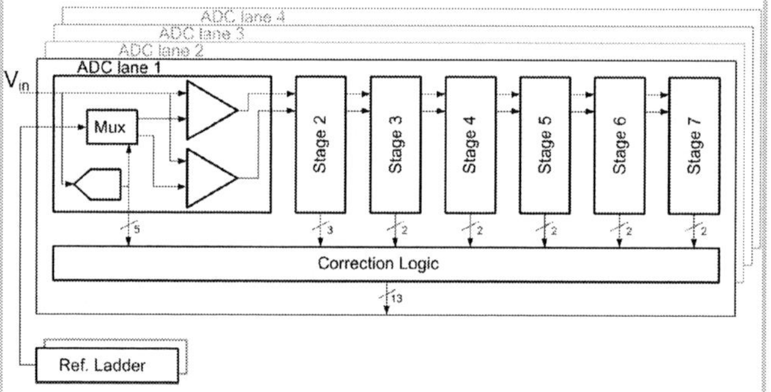

Figure 10.3.4: Block Diagram of the ADC.

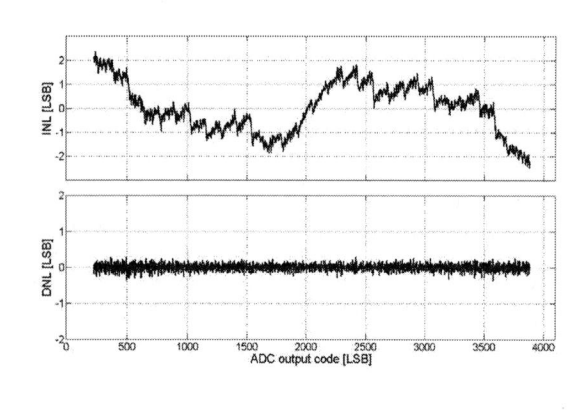

Figure 10.3.5: Measured INL and DNL.

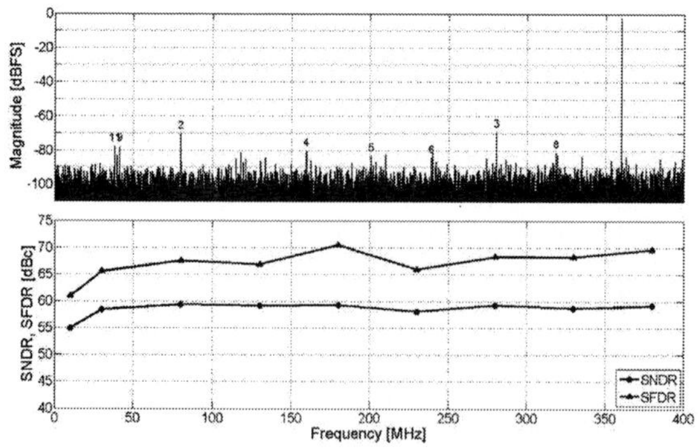

Figure 10.3.6: Measured output spectrum at F_{sig} = 360MHz, and SNDR and SFDR versus F_{sig}.

10

978-1-61284-303-2/11 $26.00 © 2011 IEEE

185

Figure 10.3.7: Die micrograph of the ADC.

ISSCC 2011 / SESSION 10 / NYQUIST-RATE CONVERTERS / 10.4

10.4 A 16b 80MS/s 100mW 77.6dB SNR CMOS Pipeline ADC

Janet Brunsilius[1], Eric Siragusa[1], Steve Kosic[1], Frank Murden[2],
Ege Yetis[1], Binh Luu[1], Jeff Bray[1], Phil Brown[1], Allen Barlow[1]

[1]Analog Devices, San Diego, CA
[2]Analog Devices, Greensboro, NC

The high channel count of many modern communication systems increasingly requires high-performance ADCs that consume very little power. The 16b pipeline ADC described here achieves 77.6dBFS SNR, 77.6dBFS SNDR and 95dBc SFDR at 80MS/s with a 10MHz input. With a 200MHz input, the ADC achieves 71.0dBFS SNR, 69.4dBFS SNDR and 81dBc SFDR. The complete ADC including reference, clock, and digital circuitry consumes 100mW from a 1.8V supply. This compares favorably with recently reported ADCs in this performance class [1-3]. In this paper, several architectural and circuit techniques used to achieve this performance are presented. The techniques include a dynamically driven deep N-well input sampling switch, an offset-cancelled comparator, and a back-gate voltage-biased MDAC amplifier. The ADC is fabricated in a 1P5M 0.18μm CMOS process with deep N-well (DNW) isolation.

For design efficiency, the pipeline ADC has four 4b stages followed by a 4b flash. A SHA-less architecture is used to achieve low noise and low power [4]. In a SHA-less architecture, the input sampling occurs on separate MDAC and sub-ADC (flash) capacitors (Fig. 10.4.1). Sampling errors due to a bandwidth mismatch between these two paths can be modeled as a flash offset and thus consume a portion of the redundancy range. Careful circuit matching is required as the mismatch errors increase with input frequency and can degrade IF sampling performance.

In the flash, separate capacitors sample the input and reference in the track phase (Fig. 10.4.1). The capacitors are reset for the next track phase after the comparators have latched. The advantage of this approach is reduced signal-dependent kickback to the input and reference networks. However, the additional switches increase area and clock loading. In addition, the increased capacitance at the comparator input attenuates the input signal and results in an effective increase in input-referred flash offsets. This is addressed by the offset-cancelled comparator described below.

The MDAC also uses separate input and reference sampling capacitors. These capacitors are not reset since the residue amplifier removes all of the charge during the hold phase. As with the flash, this approach prevents signal-dependent kick-back to the input and reference. It also decouples the input, reference and MDAC common-mode voltages to allow for a larger variation of the input common-mode. However, due to the increased capacitive load at the input of the amplifier, the feedback factor of the residue amplifier is reduced. This requires increased amplifier power to achieve the same noise and bandwidth. To address this, the DAC reference is doubled sampled onto half-sized capacitors [5]. This results in the same amount of charge being transferred but reduces the KT/C noise and improves the MDAC feedback factor. In addition, the reference load is reduced and equalized over the clock phases. Drawbacks of double sampling are increased area and clock loading and degraded matching of the smaller capacitors. The capacitors are sized for noise and the first 2 pipeline stages require calibration for matching. After factory calibration, the correction values are stored in on-chip memory. The capacitors in the backend stages match adequately without calibration.

The calibration also corrects for inter-stage gain errors, but it cannot correct for continuous-time non-linearity due to the input-sampling network. The input network to the ADC can limit the input bandwidth and the linearity performance of the ADC, especially affecting IF sampling applications. To improve the linearity performance, the input-dependent on-resistance of the input switch is reduced by bootstrapping the gate and driving the back gate during the track phase. Since the input drives the back gate, it is loaded by the parasitic non-linear capacitance due to the reverse-biased PN junction between the P-well and the buried DNW (C_{PW}) (Fig. 10.4.2). This capacitance can limit the linearity performance. To reduce this capacitance, a dynamically driven DNW technique is employed. In

the hold phase, the DNW is connected to VDD, while in the track phase it is not driven. The parasitic capacitance, C_{PW}, causes the DNW voltage to float to approximately $V_{DD} + \alpha \cdot V_{IN}$, where α is determined by the relative size of C_{PW} and the parasitic capacitance due to the reverse-biased PN junction between the DNW and the substrate (C_{DNW}). This bootstrapping increases the reverse-bias voltage of both junctions which reduces both C_{PW} and C_{DNW}. The input thus drives the series combination of two smaller non-linear capacitors. The total capacitance is reduced by 75% in this implementation. The reduction in non-linear capacitance along with the increased input bandwidth improves IF sampling performance. The input sample capacitance is 6pF and the measured input bandwidth is 700MHz.

In multi-bit per stage architectures, comparator power consumption is often dictated by offset requirements. In this design, the power is reduced by using a comparator offset-cancellation scheme (Fig. 10.4.3). The comparator's preamplifier is an NMOS differential pair with active PMOS loads that mirror the current to a switch-controlled cross-coupled NMOS latch. In the sample phase, the current in the latch drives diode-connected NMOS loads along with a pair of series connected offset-sampling capacitors that are connected between the differential outputs via a switch. In the next phase, the capacitors are reconfigured between the drains and gates of the NMOS devices to create a positive feedback cross-coupled latch, which reverses the polarity of the sampled offset and thus performs the cancellation as the circuit latches. The cancellation is limited by the non-zero Gds and parasitic capacitance at the gate of the NMOS devices (Fig. 10.4.4). In this implementation approximately 75% of the offset is cancelled.

A significant power and noise contributor is the residue amplifier. To achieve the required performance targets, a two-stage amplifier with complementary push-pull inputs in both stages is used (Fig. 10.4.5). The input stage is cascoded and the back gates of the NMOS input pair are biased to provide adequate headroom for the associated NMOS current source. A replica input stage mitigates the effects of input common-mode variation and generates the bias current for the input stage. The output stage is ground referenced, with the back-gates adjusted via voltage boosting circuits. This creates positive and negative bias voltages that maximize the output voltage swing while allowing the input stage to operate within an optimal range. A feedback circuit is used to sense the output common-mode and adjust the input-stage current. The amplifier is compensated using a combination of regular and cascode-node Miller capacitors. The open-loop gain of the amplifier is 92dB and the closed-loop bandwidth is 230MHz. In stage1 of the ADC, the total hold-phase noise is $107\mu V_{rms}$ and the power is 21mW. Subsequent stages are scaled for noise and power.

Figure 10.4.6 shows measured performance results and a comparison to prior work. The pipeline core consumes 70% of the total power. The ADC achieves 40% FOM improvement over prior work with comparable performance. The converter area is 9.9mm² in a 0.18μm CMOS process.

Acknowledgements:
The authors would like to thank Art Cepeda, Tom Pilling, Chris Carney, Kaushal Shrestha, Gina Kelso, Jon Hall, Tony Meyers, Cory Stroud and other contributing members from the High Speed Converters group.

References:
[1] S. Dervarajan, L. Singer, D. Kelly, et al., "A 16B 125MS/s 385mW 78.7dB SNR CMOS Pipeline ADC," *ISSCC Dig. Tech. Papers*, pp.86-87, Feb., 2009.
[2] A. Panigada and I. Galton, "A 130mW 100MS/s Pipelined ADC with 69dB SNDR Enabled by Digital Harmonic Distortion Correction," *ISSCC Dig. Tech. Papers*, pp. 162-163, Feb., 2009.
[3] M. Anthony, E. Kohler, J. Kurtze, et al., "A Process Scalable Low-Power Charge-Domain 13-bit Pipeline ADC," *IEEE Symp. VLSI Circuits*, pp. 222-223, June, 2008.
[4] I. Mehr and L. Singer, "A 55-mW, 10-bit, 40-Msample/s Nyquist-Rate CMOS ADC," *IEEE J. Solid-State Circuits*, vol. 35, no. 3, pp. 318-325, Mar., 2000.
[5] E. Siragusa and I. Galton, "A Digitally Enhanced 1.8V 15-bit 40-MSample/s CMOS Pipelined ADC," *IEEE J. Solid-State Circuits*, vol. 39, no. 12, pp. 2126-2138, Dec., 2004.

ISSCC 2011 / February 22, 2011 / 10:15 AM

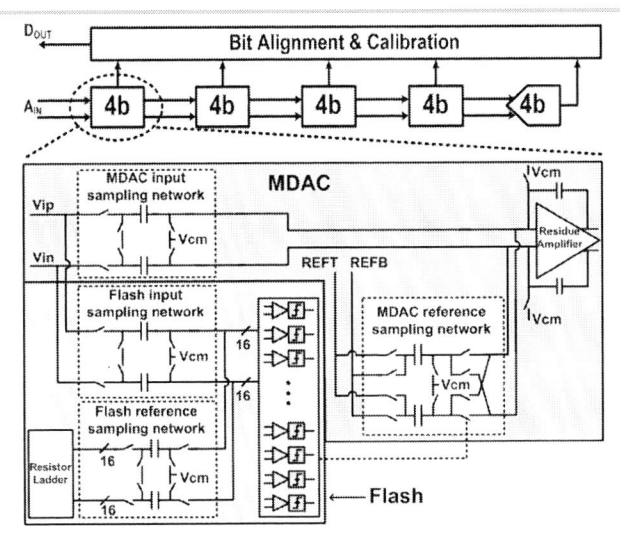

Figure 10.4.1: ADC block diagram and simplified first-stage schematic.

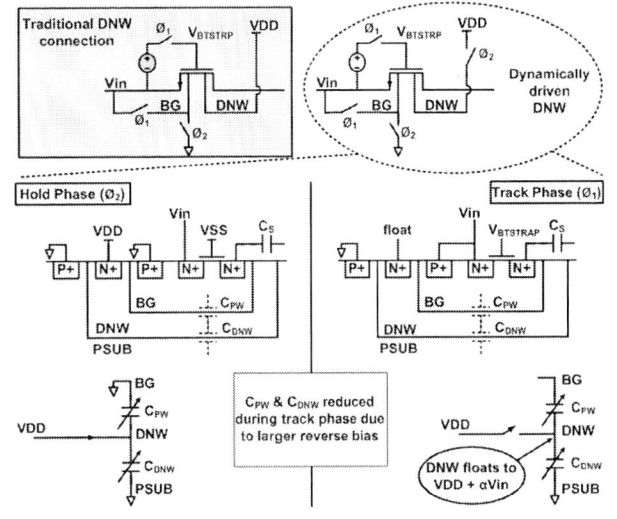

Figure 10.4.2: Input switch.

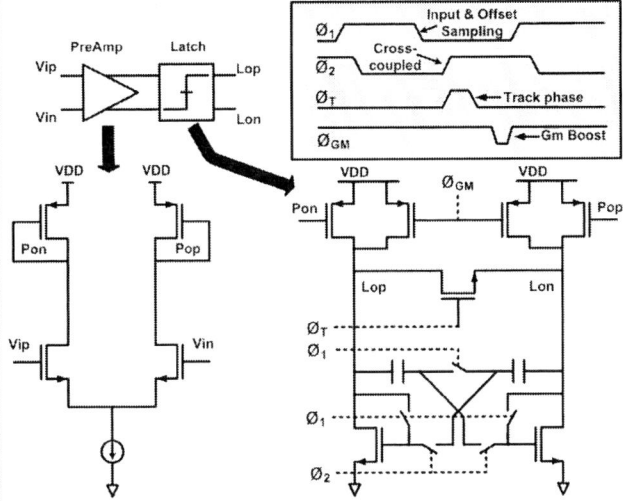

Figure 10.4.3: Simplified flash comparator schematic.

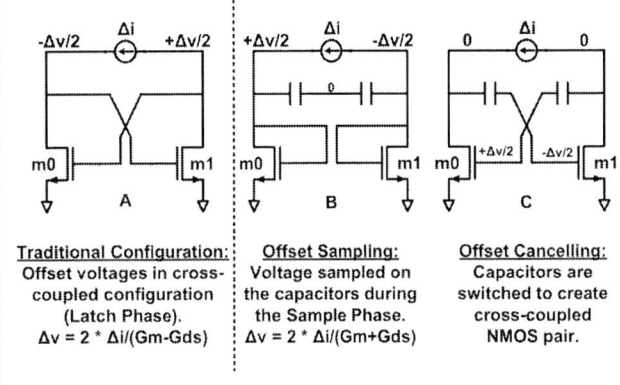

Figure 10.4.4: Simplified schematic demonstrating comparator offset cancellation.

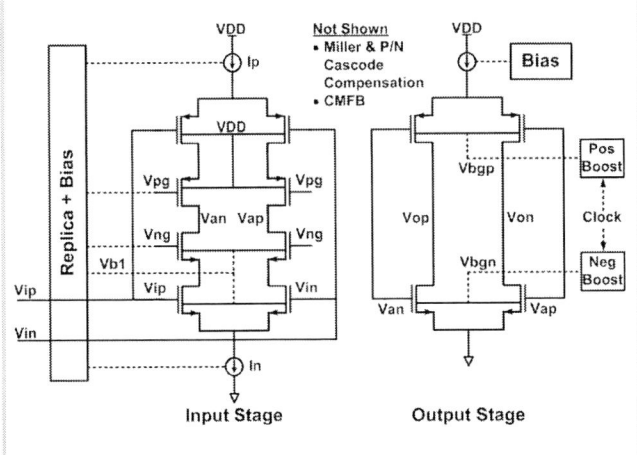

Figure 10.4.5: Simplified residue amplifier schematic.

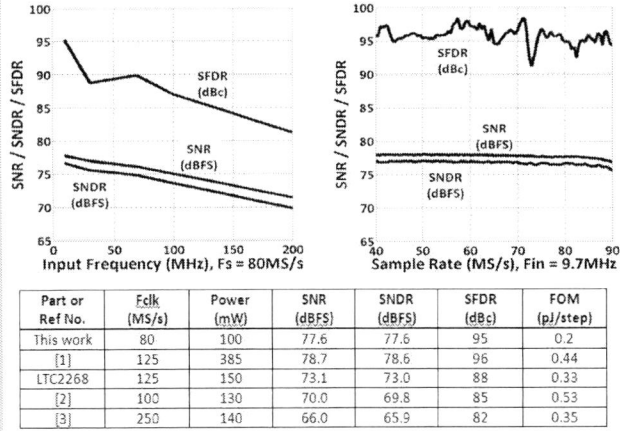

Part or Ref No.	Fclk (MS/s)	Power (mW)	SNR (dBFS)	SNDR (dBFS)	SFDR (dBc)	FOM (pJ/step)
This work	80	100	77.6	77.6	95	0.2
[1]	125	385	78.7	78.6	96	0.44
LTC2268	125	150	73.1	73.0	88	0.33
[2]	100	130	70.0	69.8	85	0.53
[3]	250	140	66.0	65.9	82	0.35

$$FOM = \frac{P_{tot}}{2^{ENOB} \cdot F_{CLK}} \quad \text{(1st Nyquist zone data)}$$

Figure 10.4.6: Measured performance and comparison to prior work.

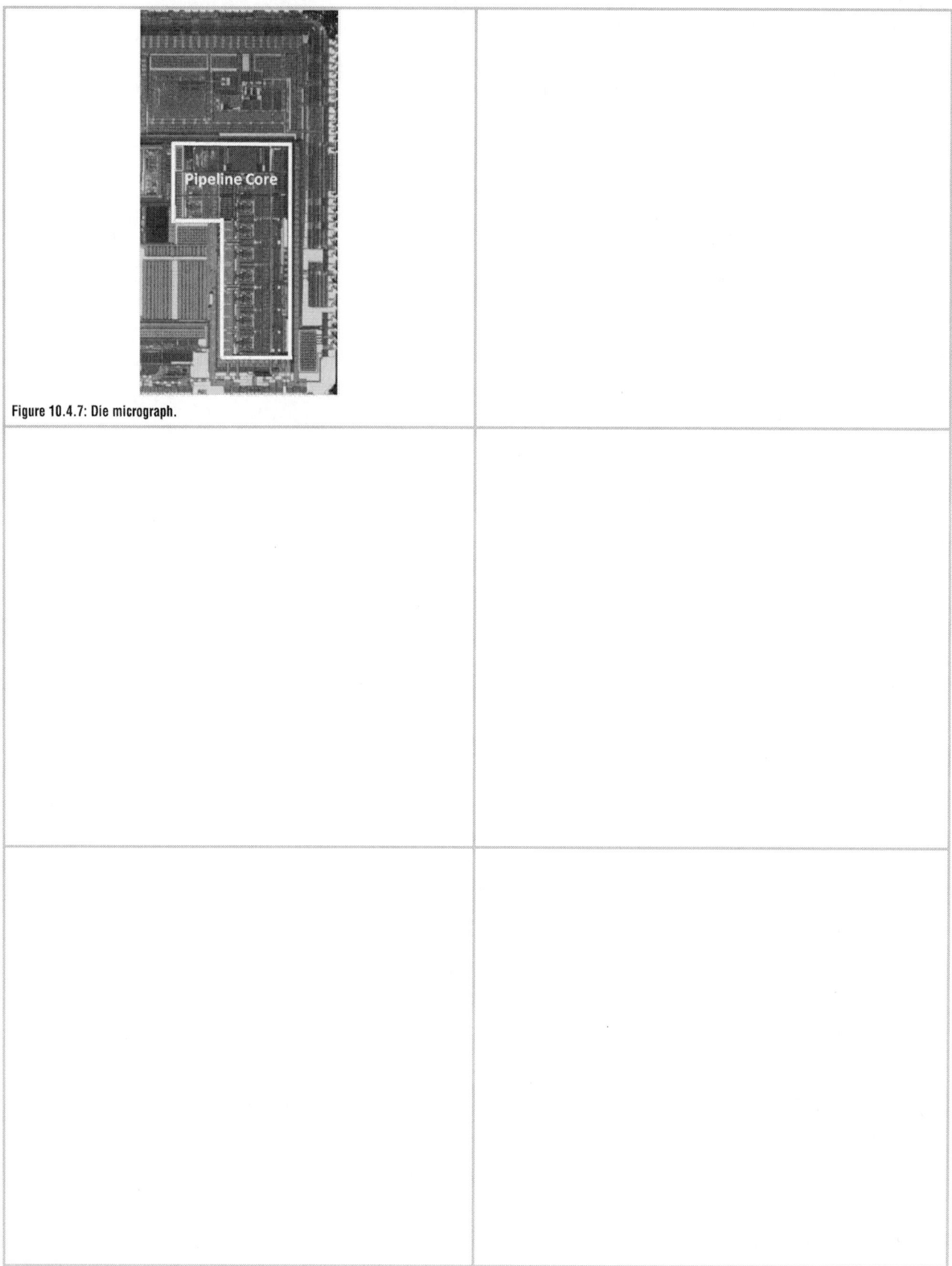

Figure 10.4.7: Die micrograph.

ISSCC 2011 / SESSION 10 / NYQUIST-RATE CONVERTERS / 10.5

10.5 A 0.024mm² 8b 400MS/s SAR ADC with 2b/Cycle and Resistive DAC in 65nm CMOS

Hegong Wei[1], Chi-Hang Chan[1], U-Fat Chio[1], Sai-Weng Sin[1], Seng-Pan U[1], Rui Martins[1,2], Franco Maloberti[3]

[1]University of Macau, Macau, China
[2]Instituto Superior Tecnico, Lisbon, Portugal
[3]University of Pavia, Pavia, Italy

The successive-approximation (SA) algorithm is traditionally used for low-bandwidth applications because it requires n clock cycles or more to obtain n-bit resolution. However, the use of modern nanometer CMOS technologies and special design solutions overcome the speed limit, enabling conversion rates in the hundreds of MHz with very low power consumptions [1]. This design uses the successive-approximation method to obtain 8b up to 400MS/s with very low power using a 1.2V supply. Key features of the architecture are a resistive DAC and a 2b-per-cycle conversion with interpolated sampling front-ends and shift registers. A cross-coupled bootstrapping network is also implemented to alleviate the signal-dependent clock feed-through. The very compact layout leads to a silicon area of 0.024 mm².

Converting more than one bit per cycle in SA schemes requires using multiple reference voltages that scale along the conversion cycle. Since this need leads to complex and multiple capacitor-based DACs [2], this design uses a Kelvin divider and an effective switch-selection network. Moreover, dynamic bit registers and synchronous successive approximation operation avoid the possible speed bottleneck established by a conventional SAR logic.

Figure 10.5.1 shows the block diagram of the proposed architecture. The shift registers control 170 switches to provide two differential reference voltages, V_{rH} and V_{rL}. Two sampling front-ends generate the difference between input and references. The differential signals serve a 3-level interpolation network with three fast comparators. An on-chip foreground offset calibration circuit minimizes the offset of the comparators. The scheme adjusts the comparator offset with digitally controlled MOS-capacitance located at the output of the comparator. The use of interpolation reduces the number of switches and shift registers, and results in diminishing consumed power and area.

Figure 10.5.2 shows the schematic of the sampling front end. Capacitors C_S sample the input signal during the sampling phase, Φ_S, and hold it for the entire conversion period. The reference voltages at the left terminal of C_S shift the differential voltages V_{in} at the comparator input. The use of a resistive DAC enables a very fast settling with a relatively small dynamic and reasonable power, since it is required to charge only the parasitic capacitances including the parasitics of switches, input capacitance of the comparator and parasitics of the C_S, but not the C_S itself due to the high-impedance node at the comparator input.

Clock feed-through occurring when sampling the input signal is a key limit to the overall accuracy. Bootstrapping the sampling switch, driven with an almost constant V_{GS}, minimizes the clock feed-through. However, the rising edge of the bootstrapped clock phase depends on V_{in}. This determines a second-order signal-dependent clock feed-through term that is alleviated by the cross-connected capacitance C_C. The value of C_C matches the parasitic C_{gd} of M_{SW}.

Figure 10.5.3 shows the resistive DAC consisting of 128 taps connected between V_{ref} and ground. It also includes 170 switches which has been reduced by 1/3 from the original 255 switches, due to the interpolation. The 127 shift register unit with the AND function is laid out with the corresponding switch locally. The inputs of the shift registers are the corresponding approximation phases and the determined digital bits. As shown in Fig. 10.5.3, the two active switches in the 4ᵗʰ cycle are quite close to each other to be able of sharing one shift register, thus achieving power and area optimization. V_{ref} is at mid-supply to allow the use of

single NMOS switches. This choice enables rapid settling having minimum parasitic capacitances. Figure 10.5.3 also highlights an example of switching operation for selecting the positive reference voltage from the resistive DAC. The selection for obtaining the complementary reference is similar. The ADC approximates the input with four steps. At the first step, all the digital bits given to the shift registers are reset to '0' and Φ_1 will enables two switches to provide $3V_{ref}/4$ and V_{ref} to $V_{rH,p}$ and $V_{rL,p}$, respectively. The following steps are similar but with the control of the determined digital bits. At the last cycle, since the difference of $V_{rH,p}$ and $V_{rL,p}$ is twice of V_{LSB} ($V_{ref}/2^8$), only 128 resistive taps of the DAC would be required. As the matching of the resistive DAC is not an issue in 8b, reducing the taps by half can save significant area and also alleviate the gradient effect in the DAC.

The already determined bits select the voltages needed for the next conversion step by closing the related switches. An alternative circuit, different from the AND gates structure shown in Fig. 10.5.4 (a), makes the selection while saving power and area. Figure 10.5.4 (b) presents that special configuration of inverters, using our method for the 1011 selection control, referring to the one used at the 3ʳᵈ-step in Fig. 10.5.3. The cascade uses the output of an inverter as ground connection of the subsequent one. The consumption is low and the speed is very high. The operation, similar to pass transistor logic, foresees at the beginning of the conversion cycle, all zeros at the input for setting the outputs V_X at V_{DD}. Only 1011 at the input brings the control of the last inverter, used to drive the N-channel switch, to zero. The operation is very fast because the speed only depends on the transition time of the last inverter, being the controls of others already set.

The prototype of the ADC is fabricated in 65nm CMOS with low-V_T option. Figure 10.5.7 shows the chip micrograph. The ADC core occupies 154×158µm². The on-chip foreground digital calibration is implemented to alleviate the comparator offset and occupies 35×117µm². The large INL and DNL of 5 LSB are optimized to 1 LSB by the calibration. Figure 10.5.5(a) shows that the SNDR is flat and above 46dB from 50MS/s to 350MS/s. At the maximum conversion rate of 400MS/s, the SNDR is 44.5dB. Figure 10.5.5(b) shows the measured SNDR versus the input frequency. The ADC consumes 4mW at 400MS/s from a 1.2V supply, resulting in a peak FOM of 73 fJ/conversion-step. The reference DAC consumes 500µW from the mid supply using about 12% of the total power. At 250MS/s, the ADC consumes 1.8mW from a 1V supply, resulting in a peak FOM of 42 fJ/conversion-step. Figure 10.5.5(c) shows the frequency spectrum for a low-frequency input and Fig. 10.5.5(d) shows the spectrum for a near-Nyquist tone. Figure 10.5.6 summarizes the performance of the proposed ADC. It achieves a smaller active area and a lower FOM as compared to the previously reported 7b+ 200M+S/s ADCs. It is also faster than the previous single-channel successive approximation ADC with SNDR>40dB.

Acknowledgment:
This work was financially supported by University of Macau and Macao Science & Technology Development Fund (FDCT).

References:
[1] W. Liu, Y. Chang, S.-K. Hsien, et al., "A 600MS/s 30mW 0.13µm CMOS ADC Array Achieving Over 60dB SFDR with Adaptive Digital Equalization," *ISSCC Dig. Tech. Papers*, pp. 82-83, Feb., 2009.
[2] Z. Cao, S. Yan, and Y. Li, "A 32mW 1.25GS/s 6b 2b/Step SAR ADC in 0.13µm CMOS," *ISSCC Dig. Tech. Papers*, pp. 542-543, Feb., 2008.
[3] Y. Shimizu, S. Murayama, K. Kudoh, and H. Yatsuda, "A Split-Load Interpolation-Amplifier-Array 300MS/s 8b Subranging ADC in 90nm CMOS," *ISSCC Dig. Tech. Papers*, pp. 552-553, Feb., 2008.
[4] L. Brooks and H.-S. Lee, "A Zero-Crossing-Based 8b 200MSs Pipelined ADC," *ISSCC Dig. Tech. Papers*, pp. 460-461, Feb., 2007.
[5] Y.-D. Jeon, Y.-K. Cho, J.-W. Nam, et al., "A 9.15mW 0.22mm2 10b 204MS/s Pipelined SAR ADC in 65nm CMOS," *IEEE CICC*, Sept., 2010.

978-1-61284-303-2/11 $26.00 © 2011 IEEE

ISSCC 2011 / February 22, 2011 / 10:45 AM

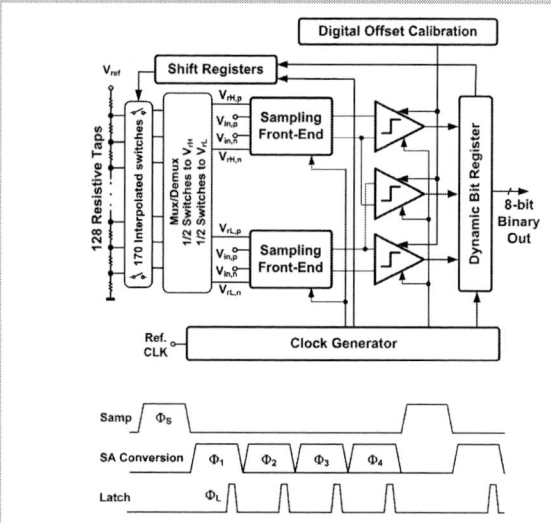

Figure 10.5.1: ADC architecture and timing diagram.

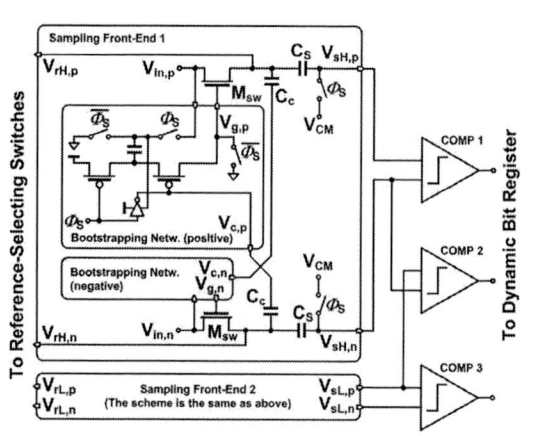

Figure 10.5.2: Interpolated sampling front-ends with cross-coupled bootstrapping network.

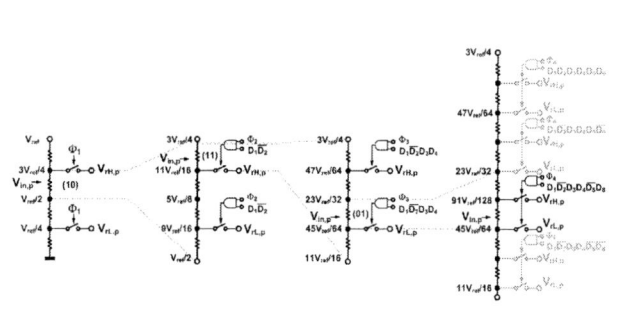

Figure 10.5.3: Switching of the reference voltage.

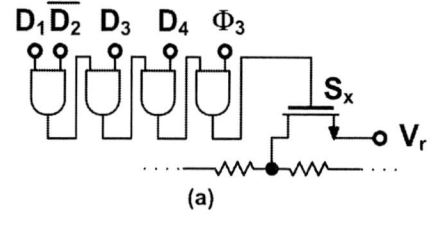

(a)

(b)

Figure 10.5.4: Implementation of the shift-register unit in 3rd conversion cycle with (a) AND gates and (b) inverters.

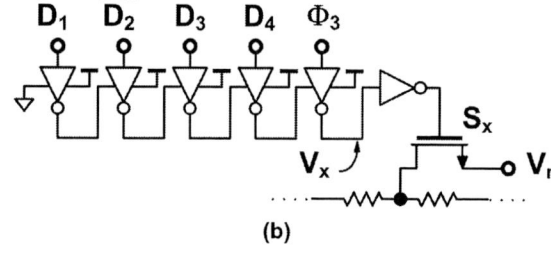

(a) SNDR vs. conversion rate

(b) SNDR vs. input frequency

(c) FFT spectrum at 2MHz input (Output Decimated by 25)

(d) FFT spectrum at Nyquist input (Output Decimated by 25)

Figure 10.5.5: Measured SNDR versus conversion rate or input frequency.

Specifications	ISSCC'09 [1]	ISSCC'08 [3]	ISSCC'07 [4]	CICC'10 [5]	This Work	
Architecture	TI-SAR	2-Step	Pipelined	TI-SAR	SAR	
Technology (nm)	130	90	180	65	65	
Resolution (bits)	8	8	8	10	8	
Sampling Rate (MS/s)	600	300	200	204	400	250
Supply Voltage (V)	1.2	1.2	1.8	1	1.2	1
SNDR (dB)	47	46.1	40.3	55.2	44.5	46.7
Power (mW)	30	34	8.5	9.15	4	1.8
FOM (fJ/Conv.-step)	208	680	510	95.4	73	42
Active Area (mm²)	1.1	0.29	0.05	0.22	0.024	

Figure 10.5.6: Performance summary and comparison with the state-of-the-art ADCs.

978-1-61284-303-2/11 $26.00 © 2011 IEEE

ISSCC 2011 PAPER CONTINUATIONS

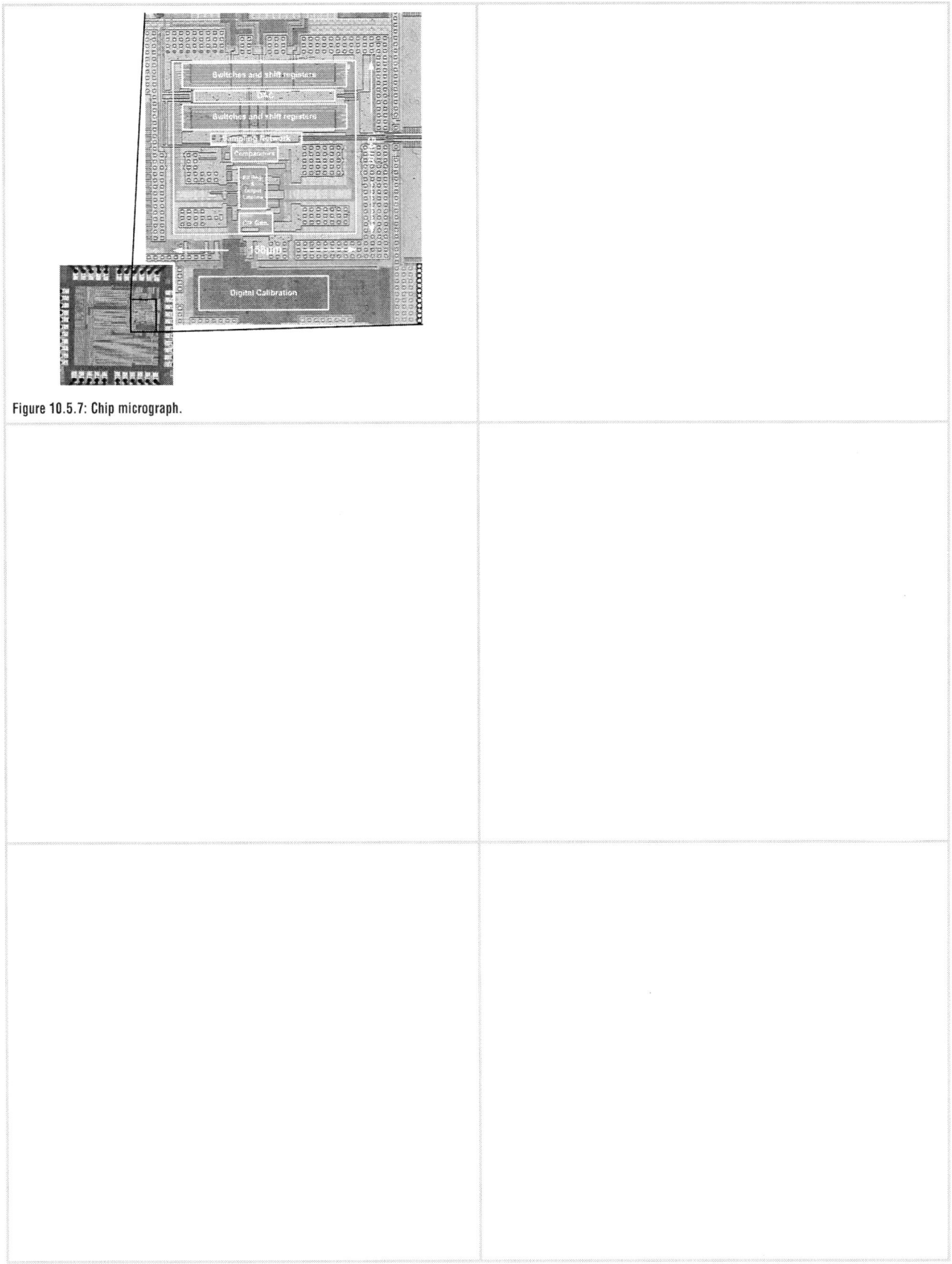

Figure 10.5.7: Chip micrograph.

ISSCC 2011 / SESSION 10 / NYQUIST-RATE CONVERTERS / 10.6

10.6 A Resolution-Reconfigurable 5-to-10b 0.4-to-1V Power Scalable SAR ADC

Marcus Yip, Anantha P. Chandrakasan

Massachusetts Institute of Technology, Cambridge, MA

Applications such as sensor networks and medical monitoring often require ADCs that can digitize signals with varying bandwidth and dynamic range requirements. In energy-constrained systems, it is beneficial to adapt the ADC performance to the signal to avoid consuming power on unnecessary bandwidth or accuracy. Therefore, this paper presents a single reconfigurable SAR ADC whose power scales with resolution and sample rate to improve energy efficiency, while reducing system complexity and cost.

Recent SAR ADCs have achieved ultra-low FOMs but most lack the ability to reconfigure their resolution [1, 2]. With fixed-resolution ADCs, a straightforward way to decrease resolution is to truncate bits, but this leads to a 2× degradation in energy efficiency (or alternatively, the FOM, calculated as $P/(2^{ENOB} \cdot 2f_{in})$) with each dropped bit. In [3], a 3-to-8b SAR ADC is presented, but it is not optimized for scalability over frequency. The ADC presented in this paper has 6 resolution modes from 5 to 10b and scalable sample rate up to 2MS/s at 1V and 5kS/s at 0.4V. Several techniques are used to enable power scaling across the entire performance range. First, a resolution-reconfigurable DAC improves power scaling with resolution. Second, a fully dynamic architecture with boosted sampling and no bias currents enables ultra-low-voltage operation which provides CV² power savings [4]. Lastly, the use of a low-leakage 65nm process together with leakage power-gating extends efficiency to low sample rates.

Figure 10.6.1 shows a block diagram of the fully differential ADC. Each conversion begins with the assertion of the CNV input pulse and consists of 4 phases as in [5]: DAC zeroing, input sampling, bit cycling and sleep. First, the DAC is zeroed to remove all residual charge so that the top plates of the DAC settle passively to the input common-mode during sampling, eliminating the need for a common-mode reference. Then, the SAR control logic starts bit cycling at the appropriate capacitor to achieve resolution scaling. The resolution mode is digitally controlled with 3 bits (RES[2:0]) and thermometer and one-hot decoders. After the bits are resolved, the ADC is put into sleep mode where the clock is gated (CLKG) and the digital circuits are power-gated with a low-leakage device. The duration of the sleep phase is variable to enable sample-rate scaling. Finally, to ensure adequate sampling linearity at low voltages, a charge pump circuit is used [4].

Resolution scaling in the DAC can be done in two ways. First, it is possible to bit cycle starting at the MSB capacitor and stop at the desired resolution [5]. However, this approach is avoided because most of the DAC power is consumed by cycling the large MSB capacitors and only linear power savings is achieved. The alternative is to start somewhere in the middle of the array and always bit cycle to the LSB capacitor. However, the MSB capacitors attenuate the DAC output which increases the resolution requirement of the comparator. To address this, switches are inserted between the top plates of the MSB capacitors so that the inactive MSB capacitors can be decoupled from the DAC as shown in Fig. 10.6.2. The DAC uses a reconfigurable 1 to 6b main split-capacitor array [6] to lower DAC switching energy and a 4b sub-DAC to reduce area. Resolution is not scaled in the sub-DAC because it cannot easily be done without adjusting the value of the attenuation capacitor C_C. The MSB capacitor of size 32C is split into its own MSB sub-array (5b main-DAC and 4b sub-DAC). In 8b mode for example, switches R[4:3] turn off, reconfiguring the DAC into a 4b main split-capacitor array with a 4b sub-DAC. A unit capacitance of 65fF using metal-metal finger capacitors is used for 10b matching. The effect of non-linearity in the top plate switches is minimized by using boosted minimum-sized switches. In addition to exponential power savings, the reconfigurable DAC also has the added advantage of exponentially scaling its input capacitance, thus saving power in the ADC driver.

The comparator is a regenerative Strongarm latch and a preamplifier is avoided to enable ultra-low-voltage operation (Fig. 10.6.1). The measured offset from 10 samples has a standard deviation of 11mV, which is compensated by 4-bit capacitor banks providing ±50mV of offset tuning in 3mV steps [2]. The additional capacitance increases the comparator power by only 2% at the largest setting.

Exponential power reduction with respect to resolution is achieved only in the DAC, which consumes 25% of the total power in the 10b mode. The power consumed by the rest of the ADC scales only linearly with resolution due to the binary nature of the SAR algorithm. In order to achieve further power reduction, aggressive voltage scaling is used. Figure 10.6.3 illustrates the measured effect of resolution and voltage scaling on the FOM at 200kS/s. When only DAC resolution scaling is enabled, it provides a 1.7× improvement in the FOM in the 5b mode when compared to truncating 5 bits from fixed 10b data. When both DAC resolution and voltage scaling are used, the FOM in the 5b mode is improved by 5×.

Although the maximum sample rate decreases with supply voltage, even at 0.4V, it is adequate for monitoring applications with modest bandwidth requirements (~5kS/s) (Fig. 10.6.4). The quadratic reduction of energy-per-conversion with voltage is also shown in Fig. 10.6.4. At high voltages, CV² losses degrade the FOM, whereas at low voltages, leakage and reduced ENOB (due to reduced sampling linearity and leakage through the sampling switch during bit cycling) degrade the FOM due to the low sample rates. The opposing trends lead to an optimum FOM point of 0.5V and 0.55V for 5b and 10b modes respectively (Fig. 10.6.4) [4].

The ADC power also scales linearly with sample rate down to leakage levels of 53nW at 1V and 4nW at 0.4V. Leakage becomes significant at sample rates below 2kS/s and limits the energy efficiency at low voltages. To address this, a high-V_T NMOS device (Fig. 10.6.1) gates the digital leakage power during sleep mode. At sample rates below 1kS/s, leakage power-gating of the digital circuits reduces total power by up to 14%. The effectiveness of this technique could be improved by extending the gating to the analog circuits as well. Lastly, leakage-power gating has the potential for greater impact as CMOS processes continue to scale.

The ADC core occupies an area of 0.212mm² and the die micrograph is shown in Fig. 10.6.7. At 0.6V, the average INL and DNL in 10b mode are +0.57/−0.56LSB and +0.58/−0.13LSB, respectively from 10 samples (Fig. 10.6.5). At 0.55V and 20kS/s, the 10b SNDR and SFDR with a 9.763kHz input tone are 55dB and 69dB, respectively (Fig. 10.6.5). The ENOB vs. input frequency for all resolution modes at 0.55V is also plotted in Fig. 10.6.5. Figure 10.6.6 summarizes the ADC performance and shows a comparison with the state-of-the-art. The ADC achieves an FOM of 22.4fJ/conversion-step at 0.55V in the 10b mode. The techniques of voltage and DAC resolution scaling maximize efficiency at low resolutions, resulting in an FOM that increases by only 7× over the 5b scaling range, improving upon a 32× degradation that would otherwise arise from truncation of bits from an ADC of fixed voltage and resolution. The overhead of reconfigurability is minimal with the exception of a few extra logic gates. The presented techniques enable the ADC to remain energy efficient over a wide range of resolutions and sample rates.

Acknowledgements:
This work was funded by DARPA and the NSERC fellowship. The authors would like to thank N. Verma and B. Ginsburg for technical discussion and Cambridge Analog Technologies for providing I/O pads.

References:
[1] M. van Elzakker, E. van Tuijl, P. Geraedts, et al., "A 1.9μW 4.4fJ/Conversion-step 10b 1MS/s Charge-Redistribution ADC," *ISSCC Dig. Tech. Papers*, pp. 244-245, Feb., 2008.
[2] G. Van der Plas and B. Verbruggen, "A 150MS/s 133μW 7b ADC in 90nm Digital CMOS Using a Comparator-Based Asynchronous Binary-Search Sub-ADC," *ISSCC Dig. Tech. Papers*, pp. 242-243, Feb., 2008.
[3] S. O'Driscoll and T.H. Meng, "Adaptive Resolution ADC Array for Neural Implant," *Proc. IEEE EMBC*, pp. 1053-1056, Sep., 2009.
[4] D.C. Daly and A.P. Chandrakasan, "A 6b 0.2-to-0.9V Highly Digital Flash ADC with Comparator Redundancy," *ISSCC Dig. Tech. Papers*, pp. 554-555, Feb., 2008.
[5] N. Verma and A.P. Chandrakasan, "A 25μW 100kS/s 12b ADC for Wireless Micro-Sensor Applications," *ISSCC Dig. Tech. Papers*, pp. 822-823, Feb., 2006.
[6] B.P. Ginsburg and A.P. Chandrakasan, "An Energy-Efficient Charge Recycling Approach for a SAR Converter with Capacitive DAC," *Proc. IEEE Int. Symp. Circuits and Systems*, pp. 184-187, May, 2005.

978-1-61284-303-2/11 $26.00 © 2011 IEEE

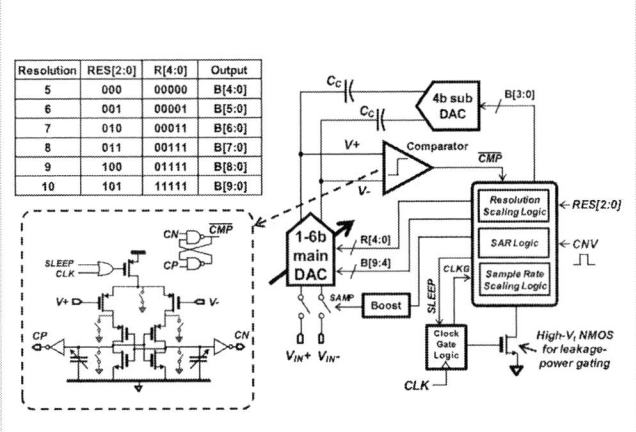

Resolution	RES[2:0]	R[4:0]	Output
5	000	00000	B[4:0]
6	001	00001	B[5:0]
7	010	00011	B[6:0]
8	011	00111	B[7:0]
9	100	01111	B[8:0]
10	101	11111	B[9:0]

Figure 10.6.1: Reconfigurable ADC block diagram with the comparator schematic and DAC reconfiguration logic table.

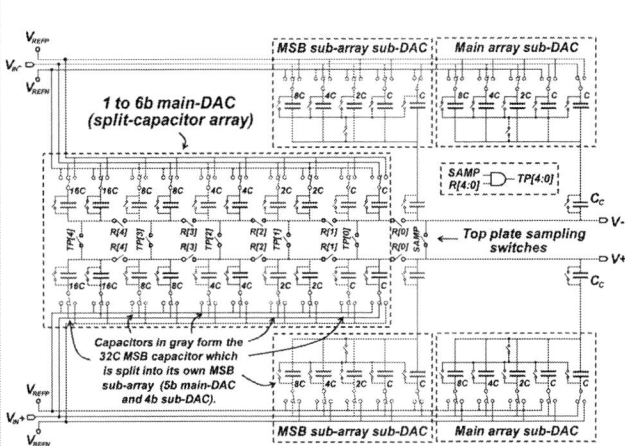

Figure 10.6.2: Reconfigurable DAC schematic showing the combination of a 1 to 6b main split-capacitor array and a 4b sub-DAC to the main array.

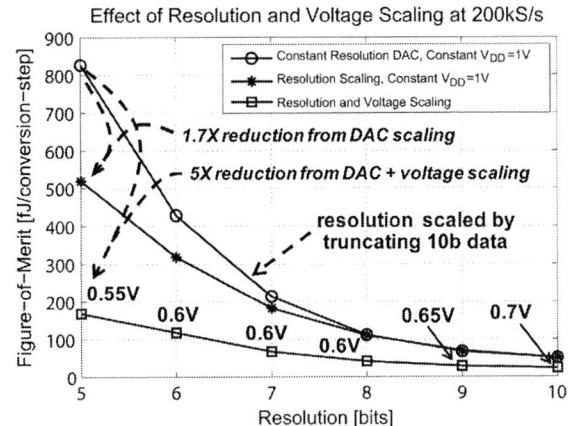

Figure 10.6.3: Measured effect of voltage scaling (with respect to 1V) and DAC resolution scaling on the ADC FOM at 200kS/s.

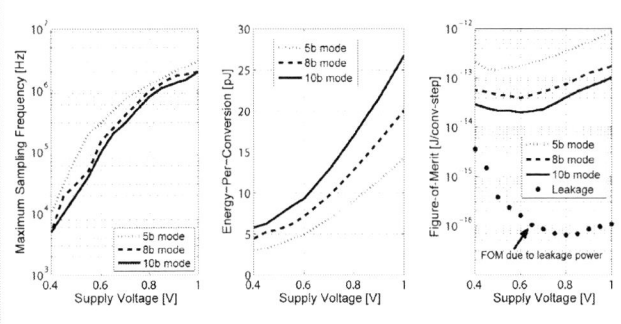

Figure 10.6.4: Maximum sampling frequency, energy-per-conversion (power/f_S) and FOM versus supply voltage, showing the optimum FOM at 0.55V in 10b mode.

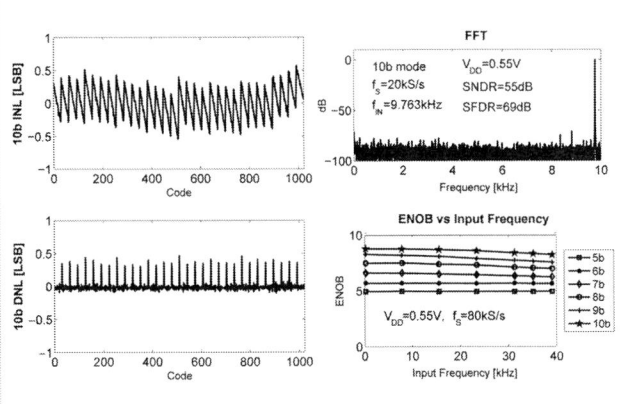

Figure 10.6.5: 10b INL and DNL, FFT and ENOB versus input frequency. The sawtooth pattern in the INL is due to parasitic capacitance on the top plate of the sub-DAC, resulting in a systematic mismatch with the main-DAC.

Active Die Area	0.212 mm² (65nm low-leakage CMOS)					
Supply Voltage (V$_{DD}$)	0.4V to 1V (differential input range is ±V$_{DD}$)					
Maximum Sampling Frequency (all resolutions)	5 kS/s @ 0.4V, 2 MS/s @ 1V					
Resolution Mode	5b	6b	7b	8b	9b	10b
Differential Input Capacitance [fF]	65	130	260	520	1040	2080
INL [LSB] @ 0.6V, 100kS/s	0.07	0.33	0.33	0.43	0.50	0.57
DNL [LSB] @ 0.6V, 100kS/s	0.11	0.35	0.40	0.51	0.55	0.58
Dynamic Performance @ 0.55V, f$_S$ = 20kS/s (*except for 5b data @ 0.5V, f$_S$ = 60kS/s)						
SFDR [dB] @ Nyquist	*44.0	48.5	54.6	61.2	63	68.8
SNDR [dB] @ Nyquist	*30.4	36.6	41.5	47.0	51.2	55.0
ENOB	*4.77	5.79	6.60	7.51	8.21	8.84
Power [nW]	*234	116	133	146	159	206
FOM [fJ/conversion-step]	143.4	105.1	68.5	40.0	26.8	22.4

Figure 10.6.6: ADC performance summary and comparison with the state-of-the-art* (*data adopted from B. Murmann, "ADC Performance Survey 1997-2010", http://www.stanford.edu/~murmann/adcsurvey.html).

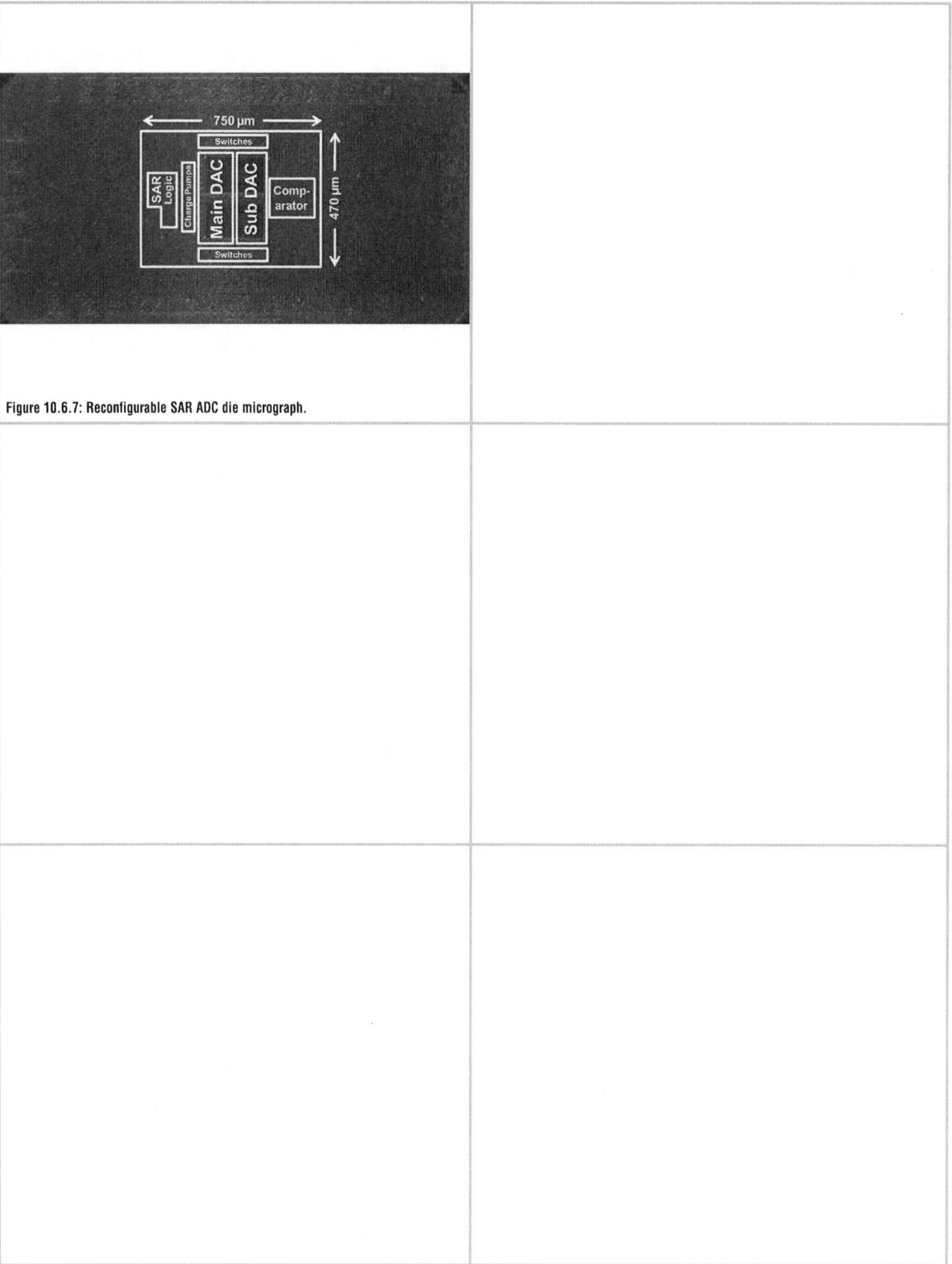

Figure 10.6.7: Reconfigurable SAR ADC die micrograph.

ISSCC 2011 / SESSION 10 / NYQUIST-RATE CONVERTERS / 10.7

10.7 A 12b 1.25GS/s DAC in 90nm CMOS with >70dB SFDR up to 500MHz

Wei-Hsin Tseng, Chi-Wei Fan, Jieh-Tsorng Wu

National Chiao Tung University, Hsinchu, Taiwan

The current-steering DACs are commonly used in generating high-frequency signals [1-4]. A current-steering DAC comprises current cells of various sizes. Each of them contains a current source and a current switch. The DAC static linearity, specified as differential nonlinearity (DNL) and integral nonlinearity (INL), is mainly determined by the mutual matching and the output resistance of the current sources. The DAC also exhibits dynamic distortion. It is manifested as spurious-free dynamic range (SFDR) degradation. The SFDR decreases rapidly with increasing input frequency. There are two major sources of dynamic distortion, code-dependent switching transients (CDST) and code-dependent output-loading variation (CDLV). Switching transients are temporal disturbances in DAC output when the current switches in current cells make transitions. The output loading of a DAC varies when the output impedances of current cells change due to the transposition of their current switches. This DAC applies a digital random return-to-zero (DRRZ) technique to mitigate the CDST effect. Compact current cells are designed to minimize the CDLV effect. The current mismatches of the current cells are corrected by background calibration.

Figure 10.7.1 shows the DAC block diagram. All functional blocks, except the output resistive loads R_{L1} and R_{L2}, are integrated on the same chip. The main DAC is segmented into a 6-bit equally weighted MSB DAC (M-DAC) and a 6-bit binary-weighted LSB DAC (L-DAC). The differential output currents from the M-DAC and L-DAC are tied together and connected to R_{L1} and R_{L2} to generate differential output voltage $V_0 = V_{o1} - V_{o2}$. The M-DAC comprises 63 identical current cells. Each current cell is designed to output a nominal current of 64I, where I is the DAC unit current. The L-DAC comprises 7 current cells which output a current of 1I, 1I, 2I, 4I, 8I, 16I, and 32I, respectively. There are two 1I current cells in the L-DAC so that a differential V_0 of zero can be realized. In our design, I=4µA and $R_{L1}=R_{L2}=25\Omega$ yield a V_0 signal range of $0.8V_{pp}$.

The DAC has a DRRZ setup. Figure 10.7.2 shows its operation. When CK is high, the DAC is in the $D_i[k]$ data phase. Its input encoder generates 63 $B_j[k]$ binary control signals to drive the M-DAC and 7 $L_j[k]$ binary control signals to drive the L-DAC. The DAC output V_0 reflects the value of input $D_i[k]$. When CK is low, the DAC is in the $Z[k]$ zero phase. The 64 $R_j[k]$ binary control signals drive the M-DAC and L-DAC respectively and make V_0=0. The entire L-DAC is treated as a single MSB current cell controlled by $R_0[k]$. The $R_j[k]$ control signals are generated from a pseudo-random number generator (PRNG). Their values change every clock cycle. The injection of $R_j[k]$ randomizes the switching transients so that they appear as noises in the V_0 output but not as distortions. Thus, the CDST effect is mitigated. The PRNG is a 32-bit linear feedback shift register. Its 32 outputs and their complements form the 64 $R_j[k]$ signals.

Figure 10.7.3 shows the circuit schematic of the j-th current cell of the M-DAC. MOSFETs M11-M18 and the four inverters form a level-sensitive MUX-latch. When CK is high, the $B_j[k]$ signal is loaded into the latch. When CK is low, the $R_j[k]$ signal is loaded into the latch. The MUX-latch is operated under a 1.2V supply. MOSFETs M5-M6 function as a current switch together. MOSFET M1 and M3 form a cascoded current source with a fixed current of 48I. MOSFET M2 and M4 form another cascoded current source whose output current is mirrored from the $I_{c,j}$ current. Both current sources are operated under a 2.5V supply. M1-M6 are MOSFETs with thick gate oxide. Disregarding the matching requirement, smaller devices are chosen for the current switches and current sources to minimize the CDLV effect. A calibration setup adjusts $I_{c,j}$ to ensure current matching among the current cells.

The calibration runs in the background and calibrates each M-DAC current cell sequentially. The output current of the j-th M-DAC current cell is denoted as $I_{M,j}$ and the total output current of the L-DAC as I_L. The calibration adjusts $I_{M,j}$ to be equal to I_L. Assume the j-th M-DAC current cell is under calibration. Its current mismatch against the L-DAC is $\Delta I = I_L - I_{M,j}$. During the $Z[k]$ zero phase, while the control signal for the L-DAC is $R_0[k]$, the control signal for the j-th current cell is set to the complement of $R_0[k]$. Thus, the mismatch term $R_0[k]$ 1 DI is embedded in V_0 during the $Z[k]$ zero phase. The calibration extracts ΔI from V_0 and then accordingly adjusts $I_{c,j}$ in order to reduce ΔI to zero. As illustrated in Fig. 10.7.1 and Fig. 10.7.2, V_0 is sampled during the zero phase, yielding $V_z = V_{z1} - V_{z2}$. When CK is high, V_z=0. The V_z is correlated with $R_0[k]$ by using a chopper. A low-pass filter (LPF) extracts the averaged value of V_p, yielding $V_m = \Delta I \times R_L$. A continuous-time delta-sigma modulator ($\Delta\Sigma M$) is used to digitize V_m. The $\Delta\Sigma M$ operates at 1/16 CK frequency. The decimation filter (DF) following the $\Delta\Sigma M$ is an accumulator that dumps its content every 2^{18} D_s samples. The digital code D_m is then quantized into D_a which is an integer in {0, ±1, ±2, ±4}. The D_a is added to the content of the j-th accumulator (ACC). There are 63 ACCs. Their outputs, $D_{c,j}$, control 63 calibration DACs (C-DACs) respectively. Each C-DAC is a 7-bit current-steering DAC with a I/8 resolution. The output of the j-th C-DAC, $I_{c,j}$, adjusts the output of the j-th M-DAC current cell, $I_{M,j}$.

Figure 10.7.4 shows the schematic of the calibration analog signal path. MOSFETs M1-M4 form the V_0 sampler. When CK is high, the V_{z1} and V_{z2} nodes are connected to a V_0 common-mode voltage, V_{CM}. MOSFETs M5-M8 form the V_z chopper. The RC pairs, R_{F1}-C_{F1} and R_{F2}-C_{F2}, are the LPFs with a 26.5kHz bandwidth. The LPF output, V_m, is converted into current by a transconductor G_m and then integrated by the following 1st-order $\Delta\Sigma M$. The internal feedback in the DSM comprises a current source I_s and a current switch. Voltage V_m is reset by switch S1 before every calibration measurement. One calibration cycle, during which all 63 M-DAC current cells are updated once, requires 214ms.

The DAC is fabricated in a 90nm CMOS technology. Figure 10.7.7 shows the chip micrograph. Active area is 1100×750µm². It consumes 56mW from a 1.2V supply and 72mW from a 2.5V supply. Figure 10.7.5 shows the measured DNL and INL. The DNL is +4.5/−0.55LSB before calibration and improved to +0.45/0.5LSB after calibration. The INL is +10.1/−6.0LSB before calibration and improved to +1.0/−1.2LSB after calibration. Figure 10.7.6 shows the measured SFDR at various input frequencies. The sampling rate is 1.25GS/s. The SFDR is below 57dB before calibration. When both DRRZ and calibration are active, the SFDR is better than 70dB up to 500MHz input frequency and is better than 66dB up to 625MHz. Also shown in Fig. 10.7.6 is the theoretical CDLV limit based on the output impedances of the current cells. This limit does not confine the measured SFDR up to 625MHz signal frequency. After calibration, the SFDR is measured again but with DRRZ disabled. The DAC output is non-return-to-zero (NRZ). In Fig. 10.7.6, the SFDR degradation of the NRZ DAC reveals the effect of CDST and attests the function of DRRZ.

Acknowledgements:
The authors thank Faraday Technology, Hsin-Chu, Taiwan for engineering support and United Microelectronics, Hsin-Chu, Taiwan for chip fabrication.

References:
[1] C-H. Lin, F. van der Goes, J. Westra, J. Mulder, Y. Lin, E. Arslan, E. Ayranci, X. Liu, and K. Bult, "A 12b 2.9GS/s DAC with IM3 <−60dBc Beyond 1GHz in 65nm CMOS," *ISSCC Dig. Tech. Papers*, pp. 74-75, pp. ,Feb., 2009.
[2] Bob Jewett, Jacky Liu, and Ken Poulton, "A 1.2GS/s 15b DAC for Precision Signal Generation," *ISSCC Dig. Tech. Papers*, pp. 110-11, , Feb., 2005.
[3] K. Doris, J. Briaire, D. Leenaerts, M. Vertregt, and A. van Roermund, "A 12b 500MS/s DAC with >70dB SFDR up to 120MHz in 0.18µm CMOS," *ISSCC Dig. Tech. Papers*, pp. 116-117, Feb., 2005.
[4] B. Schafferer and R. Adams, "A 3V CMOS 400mW 14b 1.4GS/s DAC for Multi-Carrier Applications," *ISSCC Dig. Tech. Papers*, pp. 360-361, Feb., 2004.

978-1-61284-303-2/11 $26.00 © 2011 IEEE

ISSCC 2011 / February 22, 2011 / 11:45 AM

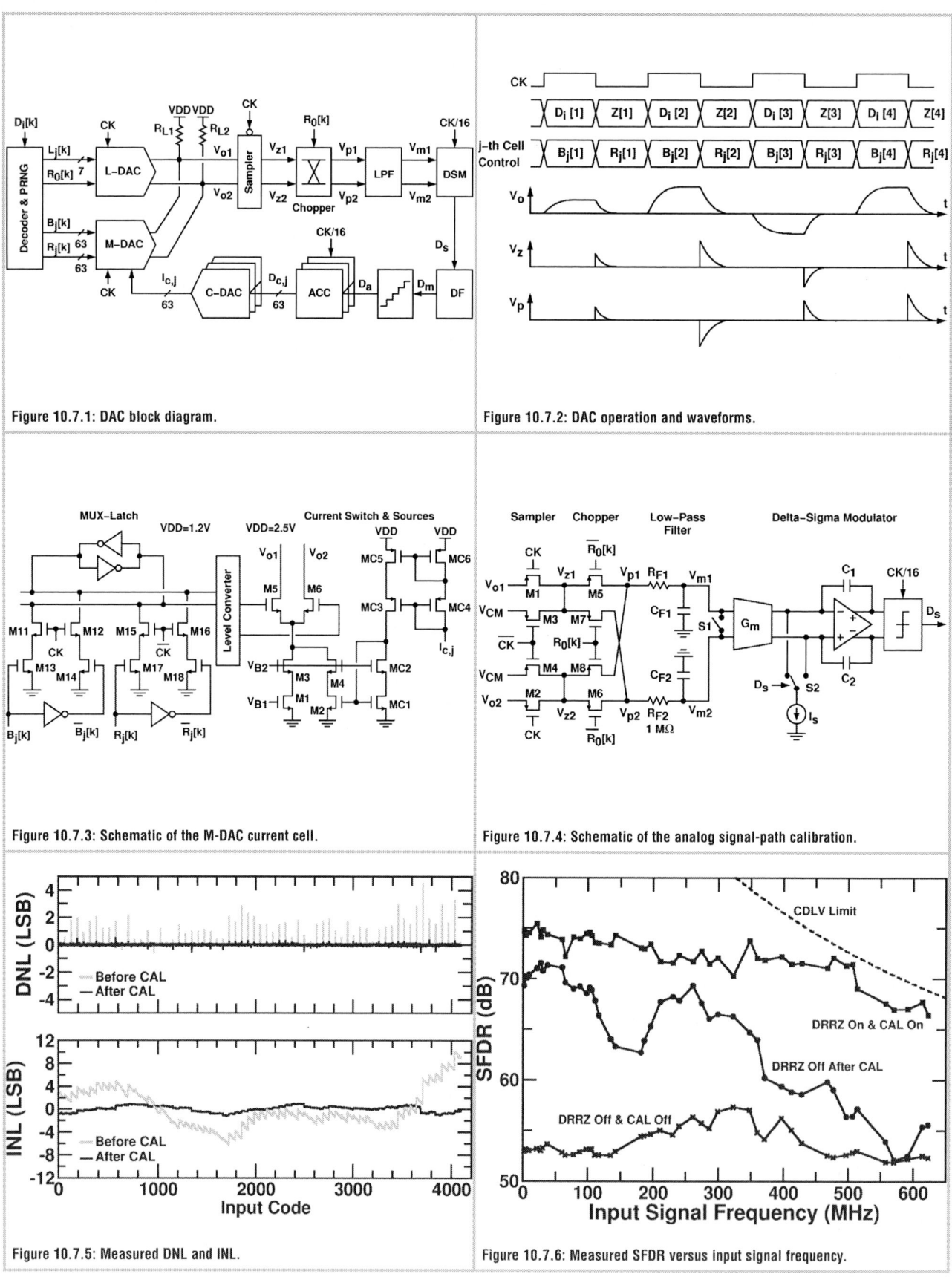

Figure 10.7.1: DAC block diagram.

Figure 10.7.2: DAC operation and waveforms.

Figure 10.7.3: Schematic of the M-DAC current cell.

Figure 10.7.4: Schematic of the analog signal-path calibration.

Figure 10.7.5: Measured DNL and INL.

Figure 10.7.6: Measured SFDR versus input signal frequency.

10

978-1-61284-303-2/11 $26.00 © 2011 IEEE

ISSCC 2011 PAPER CONTINUATIONS

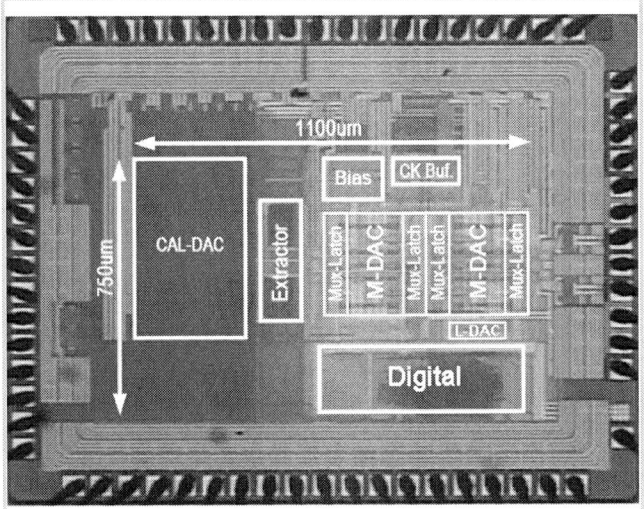

Figure 10.7.7: Chip micrograph.

ISSCC 2011 / SESSION 10 / NYQUIST-RATE CONVERTERS / 10.8

10.8 A 56GS/s 6b DAC in 65nm CMOS with 256×6b Memory

Yuriy M. Greshishchev, Daniel Pollex, Shing-Chi Wang,
Marinette Besson, Philip Flemeke, Stefan Szilagyi, Jorge Aguirre,
Chris Falt, Naim Ben-Hamida, Robert Gibbins, Peter Schvan

Ciena, Ottawa, Canada

Modern optical systems increasingly rely on DSP techniques for data transmission at 40Gbs and recently at 100Gbs and above. A significant challenge towards CMOS TX DSP SoC integration is due to requirements for four 6b DACs (Fig. 10.8.1) to operate at 56Gs/s with low power and small footprint. To date, the highest sampling rate of 43Gs/s 6b DAC is reported in SiGe BiCMOS process [1]. CMOS DAC implementations are constraint to 12Gs/s with the output signal frequency limited to 1.5GHz [2-4]. This paper demonstrates more than one order of magnitude improvement in 6b CMOS DAC design with a test circuit operating at 56Gs/s, achieving SFDR >30dBc and ENOB>4.3b up to the output frequency of 26.9GHz. Total power dissipation is less than 750mW and the core DAC die area is less than 0.6×0.4 mm².

A critical element in CMOS DAC design at sampling rates Fs = 56Gs/s is a small footprint, so that the clock distribution and data path delays are minimized. In addition, short interconnect guarantees low load capacitance for the driver circuitry and further facilitates die size reduction with speed performance improvement. There are two key obstacles in circuit size reduction: circuit topology with relatively large devices (for an example, poly resistors in CML –style logic) and interconnect metal width. Minimum width is limited by electro-migration (EM) reliability rules. A CMOS logic topology with a minimum DC current in interconnects helps to reduce the EM factor. As a result, 56Gs/s 16:1 MUX circuitry is implemented using a combination of CMOS, pseudo-differential CMOS and transmission gate style of logic. Compactness of the 16:1 MUX design then requires clock phase alignment solution to provide the MUX with precise timing, similar to the phase calibration described in [4].

DAC architecture (Fig. 10.8.2) contains a 256×6b data memory with control register, 16:1 MUX, DAC current-steering matrix, DAC current sources, and finally clock generation and phase alignment block. The memory size provides DAC output data pattern length programmability up to 256b at 56Gs/s; sufficient for time domain 256b PRBS, or frequency domain 256-points FFT testing, such as SFDR and ENOB. The DAC current-steering structure combines two segments: 15 thermometer- encoded MSBs and 2 binary LSBs; there are 17 current sources and current-steering switches in total. Thermometer encoding improves DAC linearity and minimizes output glitch energy [5]. The 4b to 2b split in segmentation provides a balance between circuit complexity and DAC overall performance. There are different techniques to generate the remaining binary-weighted LSBs currents across the on-chip 50Ω load [4,6,7]. Two last solutions prevent the use of series inductive peaking (L1 in Fig. 10.8.2) with 50Ω load at the output. This is why the DAC uses binary-weighted currents of $Io/2$ and $Io/4$, where Io is the unary current value. The DAC full scale single-ended output current, FS= $15.75 \cdot Io$. All 17 currents are generated in the DAC current sources block with 2.5V thick oxide devices. Currents are matched so they don't impair the 6b DAC DC-linearity performance. The current sources block is remotely located in the layout freeing-up space for the 56Gs/s current switches combined with the 2:1 MUX circuit (MUX-CS block in Fig. 10.8.2).

The DAC 16:1 MUX structure is similar to the one described in [4]. The DAC current-steering circuit (MUX-CS in Fig. 10.8.2) combines 2:1 MUX and a 56Gs/s current switch cell (Fig. 10.8.3). The 2:1 MUX is based on CMOS transmission gate logic. The current switch is a PMOS differential pair (M2, M3), isolated with a cascode transistor M1. That configuration makes switching performance insensitive to the interconnect length at the current input Io ($Io/2$ or $Io/4$). The M1 gate voltage, Vcs, is set to guarantee reliability with maximum gate oxide and drain-source voltages. Vcs is critical only with regards to the required output voltage swing that is typically 300mV per differential side. The MUX-CS timing (see Fig. 10.8.3) is precisely tuned with 14GHz phase rotator located in the clock

generation block. The DAC employs eight identical 4-15b binary to thermometer encoders, incorporated in a 6b to 17b encoder block (Fig. 10.8.4 left). The thermometer part is B2THERM block with 4MSBs input and 15 outputs. This block uses a similar approach as described in [8]. The two LSBs at the bottom are co-located in this block for exact delay matching at the output. The pseudo-differential 14Gs/s flip-flop in the 4:2 MUX is based on a latch (Fig. 10.8.4 right) that is a modified NMOS-only version of circuit [9]. A third logic level (SEL, SELB) is introduced to multiplex the inputs, while additional two inverters at the outputs (Q, QB) improve fan-out at required speed.

The DAC is characterized on-wafer with test platform interconnect bandwidth in excess of 40GHz. Time-domain and frequency-domain characteristics are both measured with Agilent digital scope equipped with 50GHz sampling heads and low-jitter synchronization option. Calibration of 28GHz reference clock source showed that its noise performance exceeds 7b. The 256b PRBS pattern is programmed via PC interface into the data memory, and then cyclically read to observe 56Gb/s eye (Fig. 10.8.5). The DAC offers an ability to pre-compensate the pattern for the frequency loss at its output before it reaches the sampling scope. The upper plot in Fig. 10.8.5 is obtained with no frequency compensation, while the bottom one with 10% single-pole pre-emphasis at 28GHz. In both cases the eye is wide open and symmetric despite no final re-timing operation at 56Gs/s. For the frequency domain measurements, the 256b memory is programmed to synthesize DAC output frequency, $fo=prime$ ($Fs/256$), where $prime$ is a prime number. The output is captured with 4× oversampling by digital scope and post-processed with 1K-FFT to obtain SFDR, ENOB (Fig. 10.8.6). The DAC ENOB is computed similarly to an ADC, where SNDR would represent all "noise and distortion" spectral components in the specified bandwidth. The ENOB and SFDR is > 5.9b and 42dBc respectively at frequencies below 1GHz, and then reduced to 4.3b and 30dBc at frequencies approaching 26.9GHz. In the frequency range close to 10GHz, ENOB >5b and SFDR > 38dBc.

The DAC is fabricated in a 65nm CMOS technology. Figure 10.8.7 shows a die micrograph of the DAC and summary of on-wafer measured performance. The DAC core occupies 600×400mm² area. The metal to metal and MOS decoupling capacitors are extensively used for power supply filtering, as well analog and digital parts of the circuit are powered separately. In 56Gbs/s pattern-generation mode and 400mVpp differential output eye amplitude, power dissipation including the memory is < 750mW from a dual 1.1V/2.5V supply.

Acknowledgements:
The authors thank ST Microelectronics for manufacturing support.

References:
[1] T. Sugihara, T. Kobayashi, T. Konishi, et al., "43 Gb/s DQPSK Pre-Equalization Employing 6-bit, 43GS/s DAC Integrated LSI for Cascaded ROADM Filtering," OFC/NFOEC , pp. 1-3, March, 2010.
[2] B.C. Kim, M.-H. Cho, Y.-G. Kim, and J.-K. Kwon, "A 1 V 6-bit 2.4 GS/s Nyquist CMOS DAC for UWB Systems," *IEEE Inernational Microwave Symp.*, pp. 912-915, May, 2010.
[3] X.Wu, P. Palmers, and M. Steyaert, "A 130 nm CMOS 6-bit Full Nyquist 3 GS/s DAC," *IEEE J. Solid-State Circuits*, vol.43, no.11, pp. 2396-2403, Nov., 2008.
[4] J. Savoj, A. Abbasfar, A. Amirkhany et al., "A 12-GS/s Phase-Calibrated CMOS Digital-to-Analog Converter," *IEEE Symp. VLSI Circuits*, pp. 68-69, June, 2007.
[5] C.-H. Lin and K. Bult, "A 10-b 500-MSample/s CMOS DAC in 0.6mm ," IEEE J. Solid-State Circuits, vol. 33, no. 12, pp. 1948-1958, Dec., 1998.
[6] P. Schvan, D. Pollex, T. Bellingrath, "A 22 GS/s 6b DAC with Integrated Digital Ramp Generator," *ISSCC Dig. Tech. Papers*, pp. 122-123, 588, Feb., 2005
[7] D. Baranauskas and D. Zelenin, "A 0.36W 6b up to 20GS/s DAC for UWB Wave Formation," *ISSCC Dig. Tech. Papers*, pp. 580- 581, 675, 2006.
[8] T. Miki , Y. Nakamura, M. Nakaya, et al., "An 80-MHz 8-bit CMOS D/A Converter," *IEEE J. Solid-State Circuits*, vol. 21, pp. 983–988, Dec., 1986.
[9] Y. Jiren and C. Svensson, "New Single-Clock CMOS Latches and Flip-Flops with Improved Speed and Power Savings," *IEEE J. Solid-State Circuits*, vol. 32, no.1, pp.62-69, Jan., 1997.

ISSCC 2011 / February 22, 2011 / 12:15 PM

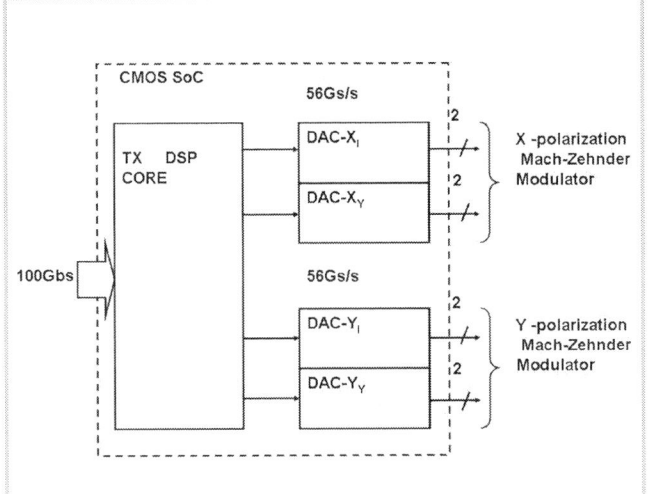

Figure 10.8.1: TX DSP SoC architecture for optical DP QPSK.

Figure 10.8.2: DAC architecture with 256×6b data memory.

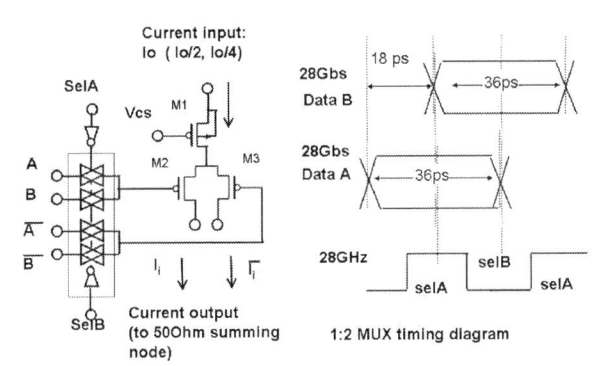

Figure 10.8.3: Unit current steering cell with 2:1 MUX.

Figure 10.8.4: Block diagram of 6b to 17b encoder (left) and 14GHz 2: 1 MUX latch (right).

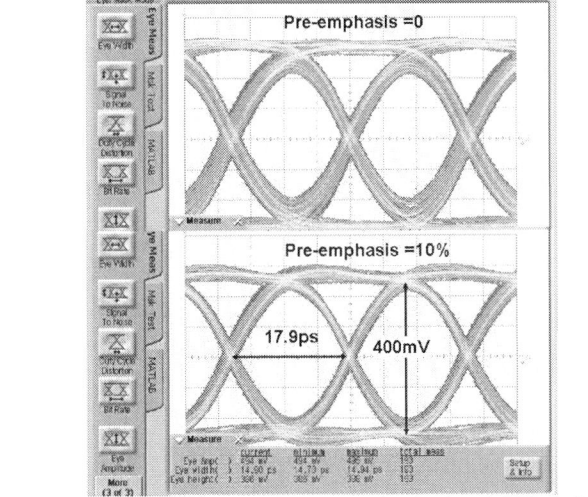

Figure 10.8.5: 56Gb/s 256b PRBS eye with 0% (top) and 10% (bottom) pre-emphasis @28GHz.

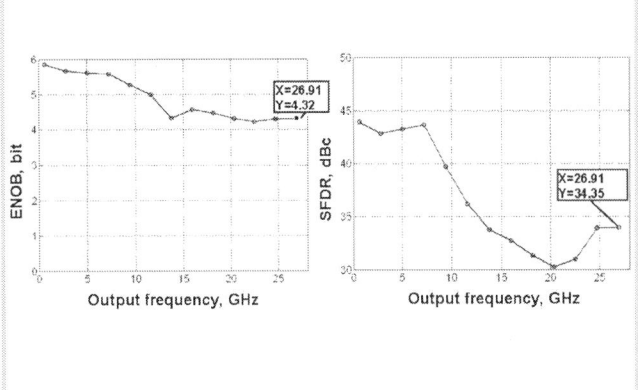

Figure 10.8.6: ENOB and SFDR versus output frequency at Fs=56GHz.

978-1-61284-303-2/11 $26.00 © 2011 IEEE

195

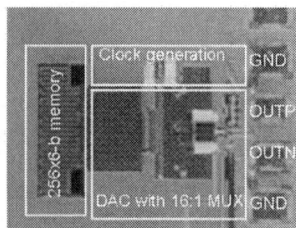

DAC on-wafer performance

Resolution	6b
Conv. Rate	56 GS/s
Output Freq	26.9 GHz
Full scale	Nom 0.6 V diff.
ENOB $F_{out} < 1GHz$ $F_{out} < 10$ GHz $F_{out} < 26.9GHz$	> 5.9b > 5b > 4.3b
SFDR $F_{out} < 10$ GHz $F_{out} < 26.9GHz$	38 dBc 30 dBc
Power	≤ 0.75 W
DAC core	0.6 x 0.4 mm²
Process	65nm CMOS

Figure 10.8.7: DAC die micograph with test circuitry and performance summary

ISSCC 2011 / SESSION 11 / NON-VOLATILE MEMORY SOLUTIONS / OVERVIEW

Session 11 Overview / Memory

Non-Volatile Memory Solutions

Session Chair: *Frankie Roohparvar, Micron Technology, San Jose, CA*

Session Co-Chair: *Sungdae Choi, Hynix Semiconductor, Icheon, Korea*

The question on everyone's mind is: "When is the NAND party over?" As other emerging technologies chase the leader with the advantages they bring to bear, beating the cost of NAND is very difficult. The hope for these technologies is that NAND would eventually run into a wall. However, NAND engineers manage to keep moving ahead and getting around barriers that laws of physics erect for them. There are always people proclaiming that the end of NAND is just around the corner, but that corner seems like a mirage. In the meanwhile, the emerging memories are not sitting idle either. They are making very good progress and are finding areas of opportunities where their technical advantages over NAND allow them be adopted in some applications. This year has a very exciting series of papers that showcase the progress that both NAND and emerging memories are making.

As NAND scaling issues are getting more and more difficult, the solution is moving towards a managed memory solution paradigm. One can think of a floating gate on NAND as a holding cell to capture and imprison electrons. Due to cost cutting measures everywhere, these prisons are also getting smaller and smaller; and these electrons are very good escape artists. We are getting to the point where we may have to give each individual electron a name tag and keep track of them individually since they hold our data. Use of more sophisticated controller technology as a part of non-volatile memory solutions is akin to making higher security prison systems. More and more sophisticated controllers are utilized to help with the issues that NAND is facing with process migration. These solutions are incorporated on the memory controllers that accompany the NAND devices and help the overall system provide acceptable storage solutions.

In this year's session, there are a series of exciting papers describing the progress being made in all these fronts. The session reports on eight papers ranging from the most advanced technology node for NAND to the smallest reported die size for NAND as well as the type of work being done on controllers to help with the NAND management. On the emerging memory side, there are papers on RRAM and conductive-bridge resistive memories as well as FeRAM. The session also has a short paper describing a very innovative low-current sense-amplifier circuit that is a key element in being able to detect the state of nonvolatile memories as their current is reduced.

Paper 11.1 reports on a 64Gb MLC NAND device with 14MB/s programming and 266MB/s data transfer fabricated in 24nm. This chip is the smallest die size reported to date for this density at 151mm^2 using a dual-plane architecture. A newly introduced two-program algorithm related to bitline precharging during a program operation is introduced that improves the program throughput of the device by 5% and reduces program operation current by 6%.

Paper 11.2 describes a 4Mb embedded SLC Resistive RAM that implements a 1T1R cell consisting of a NMOS selector and a HfO$_2$-based bipolar resistive memory element. This device features a 7.2ns read-write random-access time and 2bit-per-cell capability featuring 160ns write-verify operation. The high performance is achieved using a parallel-series reference-cell scheme and a process- and temperature-aware dynamic-bitline-bias circuit.

Paper 11.3 reports on a 32Gb MLC NAND Flash memory in 26nm technology. This device uses an innovative approach of applying negative voltage to the wordline of the NAND device as well as a new bit line compensation scheme to make the V_{th} distribution narrower. These allow for a larger V_{th} margin window that NAND cells can be programmed in, hence generating better reliability and performance.

Paper 11.4 describes techniques used by a solid-state drive (SSD) controller to improve the BER of Flash by 95% as well as reduce the power consumption by 43%. The controller algorithm uses an asymmetric coding scheme to reduce the population of high V_{th} states, resulting in an improvement of BER and a Stripe Pattern Elimination Algorithm avoiding the worst case writing pattern to achieve program current reduction

978-1-61284-303-2/11 $26.00 © 2011 IEEE

Paper 11.5 reports a sense amplifier that can detect a very low current of sub 100nA. The circuit described uses a very innovative technique of offset-tolerant current sampling. Compared to previous sense amplifiers, this circuit achieves seven times faster access time which is demonstrated on a 90nm OTP macro.

Paper 11.6 describes a 1Mb low-voltage FeRAM in 0.13μm with 1024 1T1C cells per bitline This device features a time-to-digital sensing scheme for expanded operating margin addressing the challenges of sensing small charge operating in low-voltage regime.

Paper 11.7 reports on a 4Mb conductive-bridge resistive memory in 0.18μm CMOS technology. A folded BL cell array architecture as well as a direct-sensing method in programming operation using a common sense amplifier enable 2.3GB/s read and 216MB/s program performances. Using a planned 64B program size and a 2 verify cycle operation, the device achieves a program throughput exceeding 1GB/s.

Paper 11.8 introduces a 64Gb TLC NAND Flash memory in 27nm technology providing both high density and high speed I/O. A proposed 2-step verify ISPP scheme and fail-page copy-back program function enhance the program performance. The data is transferred to the page buffer through a 200Mb/s DDR Interface. This device is the smallest geometry reported to date.

The lineup of papers this year is a testament to the ingenuity of the engineering community working on these technologies. At some point in the future of NAND, we have to face the fact that there is a wall out there; however, one can take comfort in the fact that these clever engineers keep finding ways not to crash into it.

Presenters:

8:30 AM

11.1 A 151mm² 64Gb MLC NAND Flash Memory in 24nm CMOS Technology

K. Fukuda,

Toshiba Semiconductor, Yokohama, Japan

9:00 AM

11.2 A 4Mb Embedded SLC Resistive-RAM Macro with 7.2ns Read-Write Random-Access Time and 160ns MLC-Access Capability

S-S. Sheu, ITRI, Hsinchu, Taiwan

9:30 AM

11.3 A 32Gb MLC NAND Flash Memory with V$_{th}$ Margin-Expanding Schemes in 26nm CMOS

S-D. Lee, Hynix Semiconductor, Icheon, Korea

10:15 AM

11.4 95%-Lower-BER 43%-Lower-Power Intelligent Solid-State Drive (SSD) with Asymmetric Coding and Stripe Pattern Elimination Algorithm

S. Tanakamaru, University of Tokyo, Tokyo, Japan

10:45 AM

11.5 An Offset-Tolerant Current-Sampling-Based Sense Amplifier for Sub-100nA-Cell-Current Nonvolatile Memory

M-F. Chang, National Tsing Hua University, Hsinchu, Taiwan

11:00 AM

11.6 A Low-Voltage 1Mb FeRAM in 0.13μm CMOS Featuring Time-to-Digital Sensing for Expanded Operating Margin in Scaled CMOS

M. Qazi, Massachusetts Institute of Technology, Cambridge, MA

11:15 AM

11.7 A 4Mb Conductive-Bridge Resistive Memory with 2.3GB/s Read-Throughput and 216MB/s Program-Throughput

W. Otsuka, Sony, Kanagawa, Japan

11:45 AM

11.8 A 7MB/s 64Gb 3-Bit/Cell DDR NAND Flash Memory in 27nm-Node Technology

K-T. Park, Samsung Electronics, Hwasung, Korea

978-1-61284-303-2/11 $26.00 © 2011 IEEE

ISSCC 2011 / SESSION 11 / NON-VOLATILE MEMORY SOLUTIONS / 11.1

11.1 A 151mm² 64Gb MLC NAND Flash Memory in 24nm CMOS Technology

Koichi Fukuda[1], Yoshihisa Watanabe[1], Eiichi Makino[1], Koichi Kawakami[1], Junpei Sato[1], Teruo Takagiwa[1], Naoaki Kanagawa[1], Hitoshi Shiga[1], Naoya Tokiwa[1], Yoshihiko Shindo[1], Toshiaki Edahiro[1], Takeshi Ogawa[1], Makoto Iwai[1], Osamu Nagao[1], Junji Musha[1], Takatoshi Minamoto[1], Kosuke Yanagidaira[1], Yuya Suzuki[1], Dai Nakamura[1], Yoshikazu Hosomura[1], Hiromitsu Komai[1], Yuka Furuta[1], Mai Muramoto[1], Rieko Tanaka[1], Go Shikata[1], Ayako Yuminaka[1], Kiyofumi Sakurai[2], Manabu Sakai[3], Hong Ding[3], Mitsuyuki Watanabe[3], Yosuke Kato[3], Toru Miwa[3], Alex Mak[4], Masaru Nakamichi[3], Gertjan Hemink[3], Dana Lee[4], Masaaki Higashitani[4], Brian Murphy[4], Bo Lei[4], Yasuhiko Matsunaga[1], Kiyomi Naruke[1], Takahiko Hara[1]

[1]Toshiba Semiconductor, Yokohama, Japan,
[2]Toshiba Memory Systems, Yokohama, Japan,
[3]Sandisk, Yokohama, Japan,
[4]Sandisk, Milpitas, CA

NAND flash memories are now indispensable for our modern lives. The application range of the storage memory devices began with digital still cameras and has been extended to USB memories, memory cards, MP3 players, cell phones including smart phones, netbooks, and so on. This is because higher storage capacity and lower cost are realized through means of technology scaling every year. Emerging markets, such as solid-state drives (SSDs) and data-storage servers, require lower bit cost, higher program and read throughputs, and lower power consumption,

To meet these market demands, we develop a 64Gb 2b/cell NAND flash memory in 24nm CMOS technology. Figure 11.1.1 shows a micrograph and key features of the chip. The small die size of 151mm² enables the 64Gb 2b/cell NAND flash memory chip to fit in a 12×18 BGA package. The chip contains two 32Gb memory planes with 8KB page size and 2K blocks. A block consists of 64 data wordlines (WLs) with 2 dummy WLs [1] and 256 pages. That is, a block size is 2MB. 14MB/s program throughput is achieved when 16KB or 2 planes are programmed at the same time, which is comparable or even higher performance than MLC NAND flash memories reported in the previous 3Xnm technology generation [2,3]. A high-speed DDR interface is incorporated and the data transfer performance is improved [2]. The measured output eye-diagram is shown in Fig. 11.1.2, demonstrating the capability of 266MB/s data-transfer rate.

To realize a high-capacity chip with small die size, reduction of peripheral areas including row-decoder and bitline control circuit is important as well as cell-size scaling. A low-resistivity material, which has a resistance 1/3 of the previous material, is adopted as a wordline material in 24nm technology to minimize wordline RC delay. The material enables 2-physical-plane configuration with 16KB-length wordlines, instead of 4-plane configuration with 8KB-length wordlines. Thus, the area of row-decoder is reduced by half, and the die size is reduced by 5%.

Figure 11.1.3 shows a new bitline hook-up architecture. In read and program operations, a selected bitline and an unselected bitline are connected to a bitline control circuit (BLCRL) and VPRE by hook-up transistors (BLSe, BLSo, BIASe, and BIASo), respectively. Since in an erase operation all bitlines are precharged up to an erase voltage applied to a cell well, where memory cells reside, the bitline hook-up is previously composed of HV transistors with thick gate-oxide and a long gate-length [4]. In the new architecture, the bitline hook-up is composed of LV transistors with normal gate-oxide and short gate-length, and placed in the cell well. One additional HV transistor BLS is located outside of the cell well, and is used to isolate the BLCRL circuit composed of LV transistors from bitlines. During an erase operation, gates of LV hook-up transistors are left floating and the potential of the gates is raised close to the erase voltage by capacitive coupling between the gates and the cell well. Thus, gate-oxide breakdown of the LV transistors is avoided. Since an LV transistor occupies smaller layout area, the total area of bitline hook-up is reduced by 33%, which results in 0.6% smaller die size.

Moreover, significant effort is spent minimizing peripheral circuit area. Since only 3 metal layers are used for lower fabrication cost, the top metal wiring is used for various purposes such as power-buses, high-speed signals, and miscellaneous signals. Especially when the peripheral circuit area is placed along the longer side of a cell array area and the aspect ratio of the area is large, like this chip, the congestion of the top layer wiring is severe and the peripheral circuit area tends to increase. To overcome the situation, all bonding pads of control and I/O signals are placed in the left half, so that a sequence control circuit can be located near the center of the peripheral circuit area. The top-layer signal wirings from the control circuit, which cause the congestion, can be distributed both in the left side and the right side of the control circuit. Thus, the congestion of top layer wiring is circumvented, and 1% die size reduction is obtained. With all the effort spent in die size reduction, 79% cell area efficiency is achieved.

Two program algorithms, *precharge detect* and *smart precharge*, are introduced to improve a program throughput and to reduce operating current consumption, respectively. Both are related to bitline precharging during program operation. In a NAND flash memory, when a memory cell selected by a WL is inhibited from being programmed, the bitline connected to the memory cell needs to be precharged to V_{DD} (internal power supply) before an incremental step-up program pulse (ISPP) is applied to the WL. All unselected bitlines are precharged to V_{DD} to inhibit programming of memory cells, while each selected bitline is forced to 0V (program) or precharged to V_{DD} (inhibit) depending on the program status of the memory cell. The bitlines that are program-inhibited are precharged with a constant current to avoid an undesirable current spike, which may cause a supply-voltage drop and a malfunction of a memory system. About 80% of the bitline capacitance is to adjacent bitlines in 24nm technology. As a result, the total amount of bitline precharge varies significantly depending on how many cells are programmed at a time.

Figure 11.1.4 shows how the precharge detect algorithm works. An unselected bitline is precharged to V_{DD} by a VPRE driver, while a selected bitline is precharged to V_{DD} through a BLC transistor in BLCRL as shown in Fig. 11.1.5. The gate voltage of the BLC transistor is controlled by a BLC driver, so that the gate voltage follows VPRE potential and the precharge speeds of both unselected and selected bitlines are comparable. The diode-connected transistor in the BLC driver works as a replica of a BLC transistor in BLCRL shown in Fig. 11.1.3. The number of inhibited bitlines increases and the total amount of bitline precharge decreases as ISSPs are repeatedly applied. Thus, the necessary time for bitline precharge gets shorter in later ISSPs. By detecting the completion of bitline precharge by a VPRE detector, a redundant wait time for the completion of bitline precharge can be removed and the program throughput is improved by 5%.

The smart precharge algorithm is shown in Fig. 11.1.6. Memory cells are programmed from "E" (erased) state to "A", "B", and "C" states in order with ISPPs. In the past, a program bitline voltage of 0V is kept applied to the selected bitlines that are connected to memory cells to be programmed to "B" or "C" states (B-bitline and C-bitline), before the voltage of ISPP reaches high enough for "B" or "C" states programming. Consequently, the amount of bitline precharge is fairly large in the early stage of ISSPs, especially at the first ISPP. In the new architecture, a program-inhibit bitline voltage V_{DD} is kept applied to B-bitlines and C-bitlines until the number of ISPP reaches to the pre-determined numbers, N_A and N_B, respectively. Thus, the total amount of bitline precharge is reduced and 6% smaller program operation current is achieved. The undesirable precharge current spike is also suppressed.

Acknowledgements:
The authors appreciate M. Momodomi, S. Mori, K. Kanazawa, H. Mukai, T. Yamamura, T. Watanabe, K. Shimizu, T. Ikeda, T. Okuno, Y. Fukuda, T. Shimizu, A. Imamoto, T. Hisada, S. Lee, M. Doki, H. Chien, C.-M Wang, Y. Fong, T. Kamei, K. Quader and the entire Toshiba and SanDisk development team in design, layout, CAD, verification, device, evaluation, test and process teams.

References:
[1] K. Kanda et al., "A 120mm² 16Gb 4 Multi Level NAND Flash Memory with 43nm CMOS Technology", *ISSCC Dig. Tech. Papers*, pp. 430-431, Feb. 2008.
[2] K. Hyunggon et al., "A 159mm² 32nm MLC NAND-Flash Memory with 200MB/s Asynchronous DDR Interface", *ISSCC Dig. Tech. Papers*, pp.442-443, Feb. 2010.
[3] G.G. Morotta et al., "A 3bit/Cell 32Gb NAND Flash Memory at 34nm with 6MB/s Program Throughput and with Dynamic 2b/Cell Blocks Configuration Mode for a Program Throughput Increase up to 13MB/s", *ISSCC Dig. Tech. Papers*, pp. 444-446, Feb. 2010.
[4] T. Hara et al., "A 146mm² 8Gb NAND Flash Memory with 70nm CMOS Technology", *ISSCC Dig. Tech. Papers*, pp. 44-45, Feb. 2005.

978-1-61284-303-2/11 $26.00 © 2011 IEEE

ISSCC 2011 / February 22, 2011 / 8:30 AM

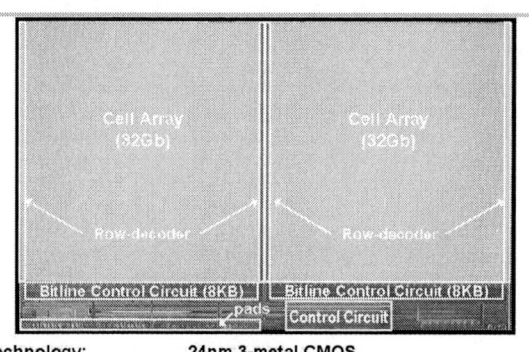

Technology:	24nm 3-metal CMOS
Capacity:	64Gb (2bit/cell)
Die Size:	151mm²
Organization:	8K bytes x 256 pages x 2K blocks x 2 planes x 8 I/O
Block Size:	2MB
Power Supply:	2.7V ~ 3.6V
Data Transfer Rate:	266MB/s
Program Throughput:	14MB/s (typical)
Iccop:	< 25mA@16KB operation

Figure 11.1.1: Die Micrograph and Key Features.

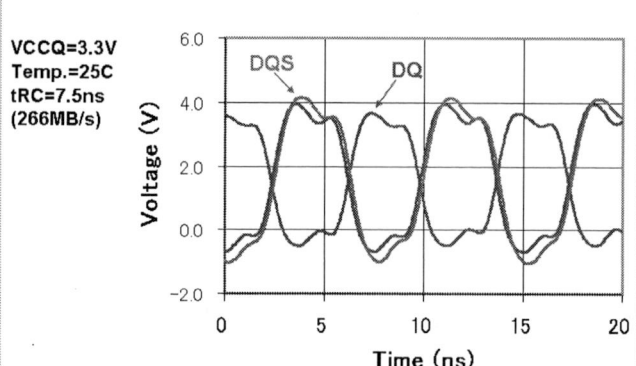

VCCQ=3.3V
Temp.=25C
tRC=7.5ns
(266MB/s)

Figure 11.1.2: Output Data Eye-Diagram.

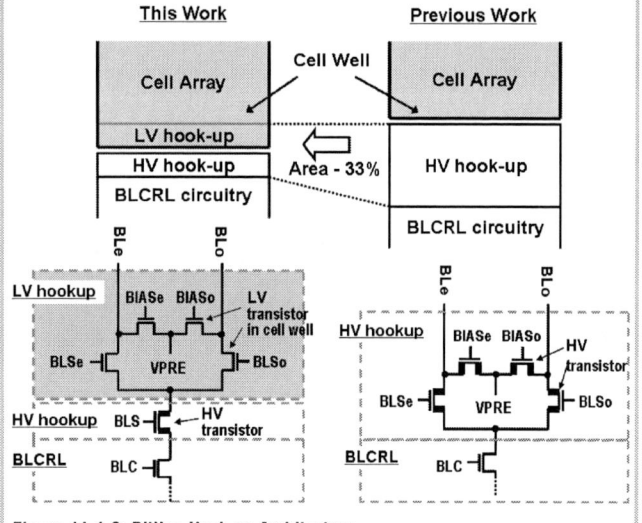

Figure 11.1.3: Bitline Hook-up Architecture.

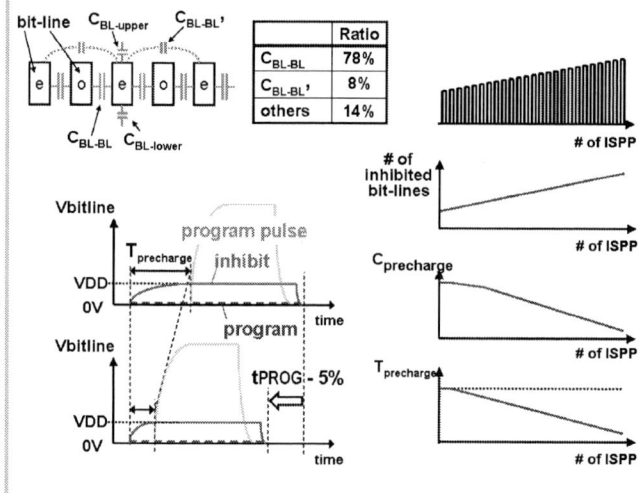

	Ratio
C_{BL-BL}	78%
C_{BL-BL}'	8%
others	14%

Figure 11.1.4: Precharge Detect Algorithm.

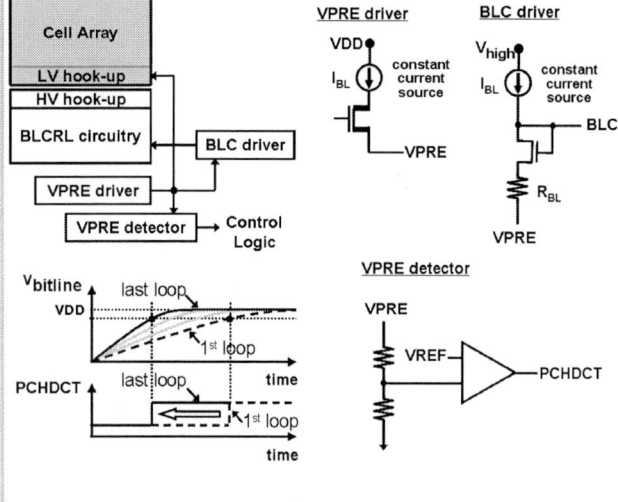

Figure 11.1.5: Driver and Detector Circuits for Bitline Precharge.

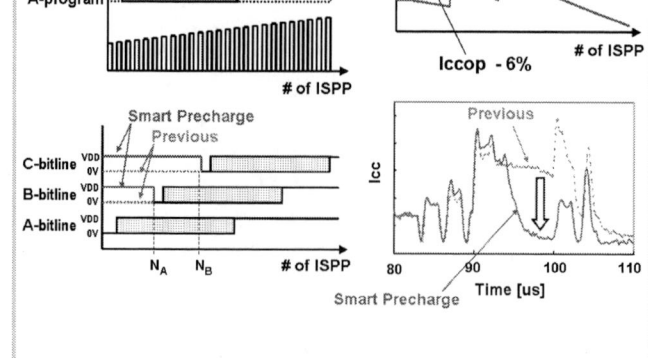

Figure 11.1.6: Smart Precharge Algorithm.

978-1-61284-303-2/11 $26.00 © 2011 IEEE

ISSCC 2011 / SESSION 11 / NON-VOLATILE MEMORY SOLUTIONS / 11.2

11.2 A 4Mb Embedded SLC Resistive-RAM Macro with 7.2ns Read-Write Random-Access Time and 160ns MLC-Access Capability

Shyh-Shyuan Sheu[1], Meng-Fan Chang[2], Ku-Feng Lin[2], Che-Wei Wu[2], Yu-Sheng Chen[1,2], Pi-Feng Chiu[1,2], Chia-Chen Kuo[2], Yih-Shan Yang[2], Pei-Chia Chiang[1], Wen-Pin Lin[1], Che-He Lin[1], Heng-Yuan Lee[1], Pei-Yi Gu[1], Sum-Min Wang[1], Frederick T. Chen[1], Keng-Li Su[1], Chen-Hsin Lien[2], Kuo-Hsing Cheng[3], Hsin-Tun Wu[1], Tzu-Kun Ku[1], Ming-Jer Kao[1], Ming-Jinn Tsai[1]

[1]ITRI, Hsinchu, Taiwan,
[2]National Tsing Hua University, Hsinchu, Taiwan,
[3]National Central University, Jhongli, Taiwan

Several emerging nonvolatile memories (NVMs) including phase-change RAM (PCRAM) [1-3], MRAM [4-5], and resistive RAM (RRAM) [6-8] have achieved faster operating speeds than embedded Flash. Among those emerging NVMs, RRAM has advantages in faster write time, a larger resistance-ratio (R-ratio), and smaller write power consumption. However, RRAM cells have large cross-die and within-die resistance variations (R-variations) and require low read-mode bitline (BL) bias voltage (V_{BL-R}) to prevent read disturbance. This work proposes process/resistance variation-insensitive read schemes for embedded RRAM to achieve fast read speeds with high yields. An embedded mega-bit scale (4Mb), single-level-cell (SLC) RRAM macro with sub-8ns read-write random-access time is presented. Multi-level-cell (MLC) operation with 160ns write-verify operation is demonstrated.

Figure 11.2.1 shows the characteristic of the 1T1R RRAM cell, which consists of an NMOS switch transistor and an HfO$_2$-based bipolar resistive memory device [8]. The RRAM is capable of two direct overwrite operations: SET (write-0) and RESET (write-1). The SET operation changes the RRAM from a high-resistive state (HRS) to a low-resistive state (LRS), by applying a SET-voltage (V_{SET}) to the BL and 0V to the source line (SL). The RESET operation changes the RRAM from having a low resistance (R_L, LRS) to having a high resistance (R_H, HRS) by applying a RESET voltage (V_{RESET}) to the SL, and 0V to the BL. The mean R-ratio (R_H/R_L) of our RRAM exceeds 100. The write operation consumes a small current (<25µA) and a fast 0.3ns switch time. However, the V_{BL-R} cannot exceed 0.3V for read disturbance consideration.

Current sensing (CS) [2-6] is commonly used in NVMs to achieve a faster read speed than conventional voltage sensing (VS) [7] for NVMs, especially for long BL and small read cell current (I_{CELL}) memories. The CS requires a reference-current (I_{REF}) to compare with the to-be-sensed I_{CELL}. To achieve good process-voltage-temperature (PVT) tracking of I_{CELL}, many NVMs use reference cells [4-6] to generate the required I_{REF}. However, RRAM devices have large cross-die and within-die R-variations. R-variations occurring on reference-cells [4] cause wide I_{REF} distribution and low sensing yield for conventional single-cell [3], parallel-cells or R_H-R_L-average reference-cell schemes [4-6], especially for the large R-ratio RRAM. A dynamic BL-bias scheme achieves better V_{BL-R} accuracy [6] with a faster settling time than a static BL-bias scheme. When V_{BL-R} has large fluctuations due to process and temperature variations, RRAM suffers read disturbance. Figure 11.2.2 shows our read schemes and RRAM R-variation effects. To achieve high read speeds, this macro uses differential-input-current sensing with a dynamic V_{BL-R} bias. A parallel-series reference-cell (PSRC) scheme narrows the I_{REF} distribution against R-variation. Also, this work develops a process-temperature-aware dynamic BL-bias circuit (PTADB) to achieve a fast BL precharge with small V_{BL-R} fluctuations for read disturbance consideration.

Figure 11.2.3 shows the PSRC scheme. Each PSRC unit consists of p serial-connected sub-reference-blocks (SRBs). Each SRB has q parallel-connected reference-cells. Unlike using both LRS and HRS cells in previous reference-cell schemes [4-6], this work uses LRS cells only as the reference cells for SLC operations. It is because the LRS cells dominated the I_{REF} generation when R-ratio is large. Moreover, the within-die R-variation range of R_L is 10× smaller than R_H for our RRAM. The PSRC scheme prevents the I_{REF} generation from

being dominated by one ultra-low resistance LRS cell (a tail-bit). Though the series-parallel [5] scheme can generate the mid-point resistance, its I_{REF} derived from the mid-point resistance does not yield the mid-point current of two neighboring MLC states. From the statistical analysis, the PSRC scheme narrows the SLC I_{REF} distribution and reduces the σ-I_{REF} by 35 to 80% compared to previous schemes. For MLC operation, the PSRC achieves a 99% smaller mean-I_{REF} deviation from the mid-point current compared to the series-parallel approach. Thus, the PSRC scheme is suitable for both SLC and MLC RRAMs against R-variation.

Figure 11.2.4 shows the PTADB scheme, which consists of a process-temperature detector (PTD) and a process- temperature compensated feedback amplifier (PTFA). The PTD circuit compares the PMOS-NMOS transistor strength across various process corners and operating temperatures, then generates a k-bit code. The PTFA, a common-source amplifier (P1) with a current bias (N1), receives the k-bit code from the PTD to adjust its output voltage (V_{CLP}) for controlling the BL-clamping NMOS transistor (N_{CLP}). Unlike the conventional scheme of using a fixed current bias for the V_{CLP} generation, the PTFA uses a dynamic current-bias for N1 to enlarge V_{CLP} swing ranges for accelerating the BL precharge and clamping operation. Compared to conventional dynamic BL-bias schemes [6] the PTFA reduces 50% V_{BL-R} variation across global process corners and achieves a 24% faster read access time.

An RRAM testchip, with a 4Mb RRAM macro and test-mode circuits, is fabricated using an HfO$_2$-based RRAM device and a 0.18µm CMOS logic process. The testchip access time ($T_{AC-CHIP}$) includes the access time of the embedded RRAM macro (T_{AC}) and the path delay time ($T_{P-DELAY}$). The $T_{P-DELAY}$ consists of the delay times of on-chip and load-board wirings. A dummy path is implemented to measure $T_{P-DELAY}$ for extracting the T_{AC}. Figure 11.2.5 shows the measured I_{REF} distribution and read access time. The PSRC scheme has a narrower I_{REF} distribution than the single-cell I_{REF} scheme, with a 44% reduction in σ-I_{REF}. Since the RRAM device has fast write speed, the read-write access time is dominated by the read operation. The measured random-access and burst-read T_{AC} of the 4Mb RRAM macro is 7.2ns and 3.6ns, respectively. Figure 11.2.6 shows the measured SLC and MLC results. The read circuits are functional from 1.8V to 700mV and detect a wide-range of I_{CELL} variation, across a wide range of WL voltage. These tests confirm that our read circuit shows good resistance-tracking capability and is scalable for a lower V_{DD}. The MLC write-verify procedure is presented. The Level-2~Level-4 writings are the SET operations but with different WL voltages. This RRAM achieves 2b/cell MLC operation with 160ns random access time. Figure 11.2.7 shows the die micrograph.

Acknowledgements:
The authors thank TSMC University Shuttle Program for the read-circuit testchip fabrication.

References:
[1] S. Hanzawa, et al., "A 512kB embedded Phase Change Memory with 416kB/s write through at 100µA cell write current," *ISSCC*, pp. 474-475, Feb. 2007
[2] G. D. Sandre, et al., "A 90nm 4Mb embedded Phase-Change memory with 1.2V 12ns read access time and 1MB/s write throughput", *ISSCC*, pp. 268-269, Feb. 2010
[3] Y. N. Hwang, et al., "MLC PRAM with SLC write-speed and robust read scheme," *Symp. VLSI Tech.*, pp. 201-202, June 2010
[4] K. Tsuchida, et al.,"A 64Mb MRAM with clamped-reference and adequate-reference schemes," *ISSCC*, pp. 258-259, Feb. 2010
[5] M. Durlam, et al., "A 1-Mbit MRAM based on 1T1MTJ bit cell integrated with copper interconnects," *IEEE J. Solid-State Circuits*, pp. 769-773, May 2007
[6] S. Dietrich, et al., "A nonvolatile 2-Mbit CBRAM memory core featuring advanced read and program control," *IEEE J. Solid-State Circuits*, pp. 839-845, April 2007
[7] P. Schrogmeier, et al., "Time discrete voltage sensing and iterative programming control for a 4F^2 multilevel CBRAM," *Symp. VLSI Circuits*, pp. 186-187, June 2007
[8] H. Lee, et al., "Evidence and solution of over-RESET problem for HfO$_x$ based resistive memory with sub-ns switching speed," *IEDM*, Dec. 2010 (accepted).

ISSCC 2011 / February 22, 2011 / 9:00 AM

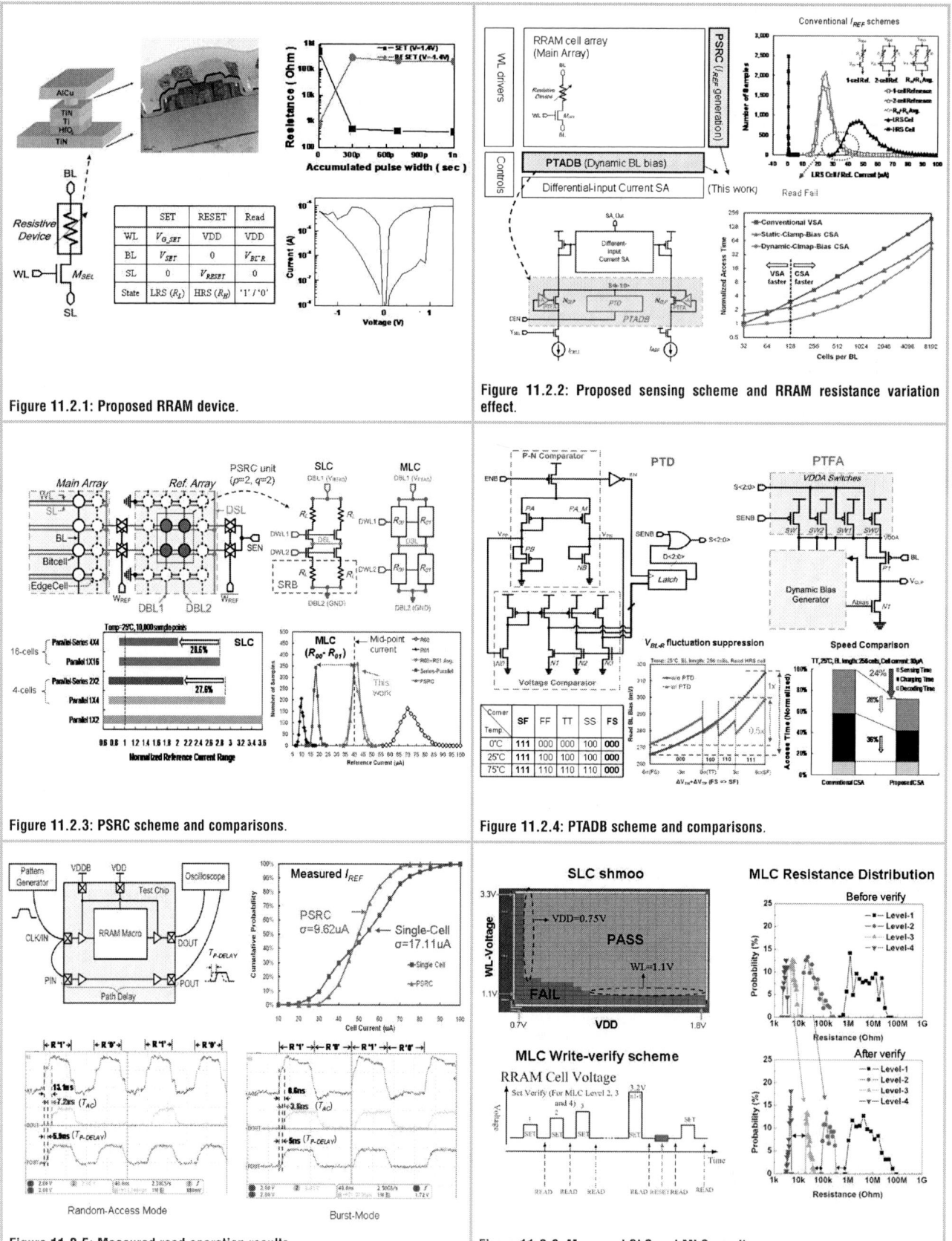

Figure 11.2.1: Proposed RRAM device.

Figure 11.2.2: Proposed sensing scheme and RRAM resistance variation effect.

Figure 11.2.3: PSRC scheme and comparisons.

Figure 11.2.4: PTADB scheme and comparisons.

Figure 11.2.5: Measured read operation results.

Figure 11.2.6: Measured SLC and MLC results.

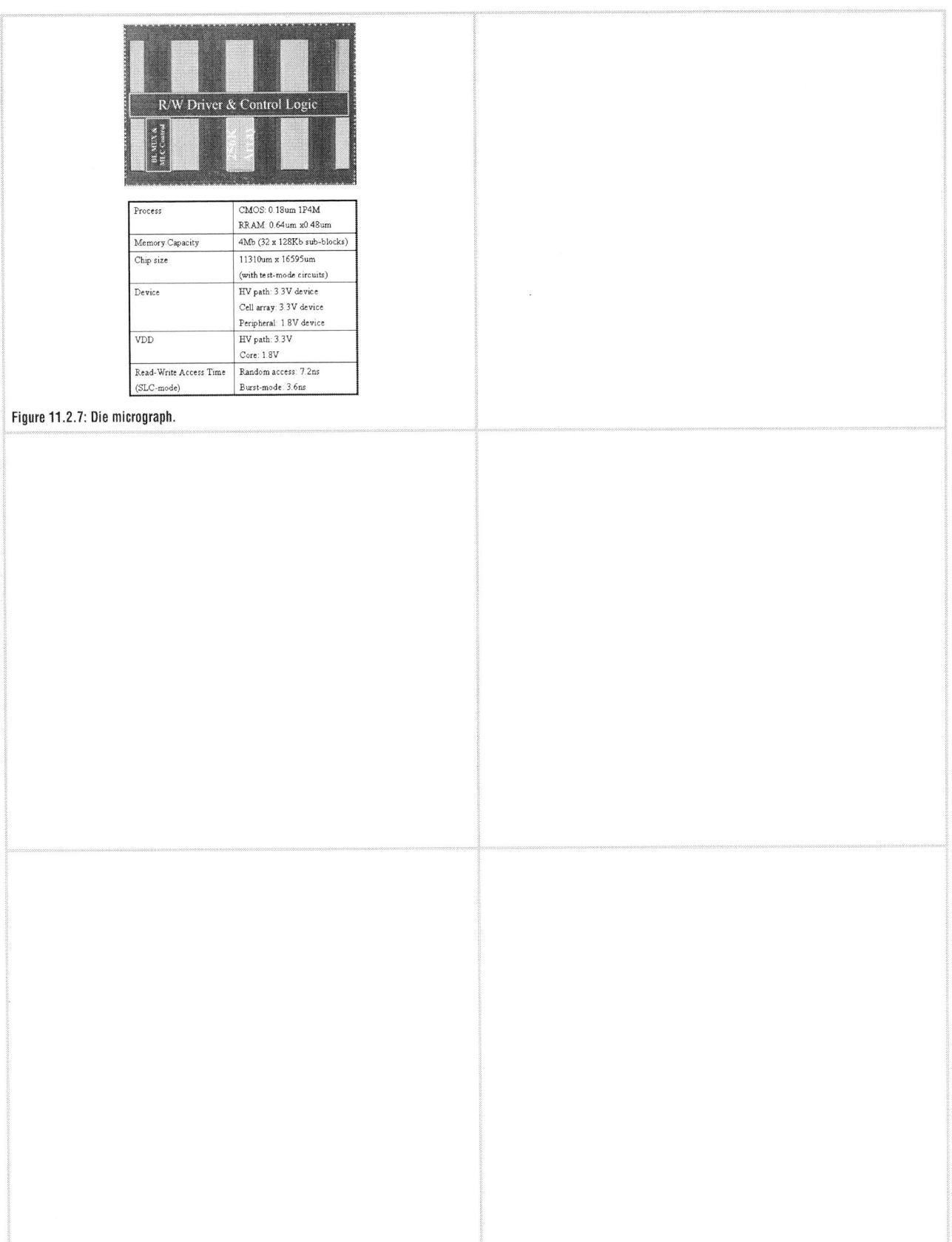

Process	CMOS: 0.18um 1P4M
	RRAM: 0.64um x0.48um
Memory Capacity	4Mb (32 x 128Kb sub-blocks)
Chip size	11310um x 16595um
	(with test-mode circuits)
Device	HV path: 3.3V device
	Cell array: 3.3V device
	Peripheral: 1.8V device
VDD	HV path: 3.3V
	Core: 1.8V
Read-Write Access Time	Random access: 7.2ns
(SLC-mode)	Burst-mode: 3.6ns

Figure 11.2.7: Die micrograph.

11.3 A 32Gb MLC NAND Flash Memory with V_{th} Margin-Expanding Schemes in 26nm CMOS

Tae-yun Kim, Sang-Don Lee, Jin-su Park, Ho-youb Cho,
Byoung-sung You, kwang-ho Baek, Jae-ho Lee,
Chang-won Yang, Misun Yun, Min-su Kim, Jong-woo Kim,
Eun-seong Jang, Hyun Chung, Sang-o Lim, Bong-Seok Han,
Yo-Hwan Koh

Hynix Semiconductor, Icheon, Korea

As the NAND flash memory market grows rapidly due to various applications, such as USB devices, MP3 players, SSDs, cellular phones, and cameras, there is a requirement for high-density and low-cost devices. Two different approaches to meet these requirements are increasing data per cell and area scaling. 3b/cell or 4b/cell NAND flash memories were introduced as an effective way to lower cost [1,2]. However, these devices suffer from program performance degradation since tighter V_{th} distribution is required. On the other hand, area scaling is a candidate to achieve low cost while maintaining high program performance even though there are several hurdles to overcome, such as FG coupling and charge retention [3]. As the cell size gets smaller, the V_{th} distribution widens and the erase-write cycling margin is decreased by the floating-gate coupling ratio [4].

Our 26nm 32Gb MLC flash memory device has an area of 181.5mm² using the following schemes: negative wordline, all-bitline (ABL) parallel program and ABL with BL compensation.

The method of negative wordline scheme is introduced to control the V_{th} distribution using the negative voltage region under 0V. For this method, negative bias is directly supplied to the control gate of a selected memory cell in cell strings. Figure 11.3.1(a) shows the conventional V_{th} distribution and Fig. 11.3.1(b) shows V_{th} distribution using a negative WL technique, which allows program status under 0V. MLC V_{th} distribution has 4 statuses. One status is erase status, the others are program statuses PV1, PV2 and PV3 that are classified according to the verify level. Negative WL technique is implemented to widen the V_{th} allocation window by setting PV1 status in negative region

The 26nm 32Gb MLC flash memory device has a negative charge pump that can generate a voltage level from -3.3 to 0V for applying negative bias to WL. A new high-voltage (HV) switch, which is a combination HV switch, can transfer both a high-voltage bias and a negative bias. By using triple-well high-voltage NMOS (TW-HVN) device instead of an existing HV NMOS device there is no problem to prevent reverse bias current while transferring negative bias. Figure 11.3.2 illustrates the integrated block decoder circuit, which can transfer both high voltage and negative voltage. One aspect of the combination HV bias switch for transferring only high voltage is that ngof supply 0V to the well of a triple HVN device. Another aspect of combination HV bias SW for transferring negative bias is to supply negative bias to two nodes. One node is a well of TW_HVN in pass transistor group. The other node is BLK_VPP, which acts like a switch gate bias in unselected blocks. These operations effectively prevent supplying negative bias to unselected blocks by supplying positive bias to BLK_VPP nodes in selected blocks.

The left fig of Fig. 11.3.3 illustrates the measured waveform of negative bias of the circuit. It shows the waveform of positive bias and negative bias without any transformations in an integrated circuit. The program pulse is applied at the first pulse and the following negative-bias pulse is applied for the verify in a selected word line. The right of Fig. 11.3.3 is the V_{th} distribution of a memory cell measured with negative WL bias. This scheme has enough margin for disturbance by setting the level between PV2 and PV3 to over 1V.

All-bitline parallel-program operation (ABL parallel program) is a method for improving interference in a memory cell, which operates in accordance with the status during a program operation. The scheme controls the method for applying program and verify pulse at the same time when PV state read for PV1 and PV3. PV3 status read has the largest interference since the PV3 state is farthest from erase state and is affected by FG coupling, even though unselected cells are inhibited at the program operation. PV1 status verify operates after a memory cell interfered by PV3 program pulse in accordance with this scheme as shown in Fig. 11.3.4. Thereby, PV1 status read compensates memory cells of the memory array from interferences. The ABL parallel scheme provides a way to reduce a fixed V_{th} distribution less than 100mV compared with previous ABL scheme and also creates tighter V_{th} distribution by about 150mV than previous EOBL structure. This is almost 10% reduction.

A bitline compensation scheme is used to get tighter V_{th} distribution during ABL Programming. Figure 11.3.5, shows that if adjacent cells pass verify-levels then its channel will be boosted from 0V to boosting level. Because of channel-to-gate coupling, the boosted channel bias makes adjacent cell's gate bias rise. This widens V_{th} distribution of selected cells because of step variation of program pulse. To compensate for the wide V_{th} distribution because of adjacent channel-to-gate coupling, when adjacent cells are in program inhibit mode, boost BLs of program inhibit mode cells from Vinhibit1 (VCC − ΔV/α) to Vinhibit (VCC) so that it can be compensate step variation of selected cell's gate by coupling effect, where α is the BL-to-BL coupling ratio). For example if α = 0.5 and ΔV=100mV then Vinhibit = VCC-100mV/0.5 ≈ VCC-200mV.

The 32Gb 26nm flash memory device has an improved V_{th} distribution margin with the negative wordline scheme. Moreover the ABL parallel program operation causes the Vt distribution width to get smaller by 10% than basic ABL program operation. Key device features are summarized in Fig. 11.3.6 and die micrograph is shown in Fig. 11.3.7.

Acknowledgements:
The authors would like to thank Layout Team, DV Team, Device Team, Product Team, and Process Team for great support and development

References:
[1] Seung-Ho Chang et al., "A 48nm 32Gb 8-Level NAND Flash Memory with 5.5MB/s Program Throughput", in *ISSCC Dig. Tech. Papers*, pp. 240-241, Feb. 2009.
[2] Changhyuk Lee et al., "A 32Gb MLC NAND Flash Memory with Vth Endurance Enhancing Schemes in 32nm CMOS" in *ISSCC Dig Tech Papers*. pp. 446-447 Feb. 2010.
[3] Ki-Tae Park et al., "A Zeroing Cell-to-Cell Interference Page Architecture With Temporary LSB Storing and Parallel MSB Program Scheme for MLC NAND Flash Memories", in *Symp. VLSI Circuits*, pp. 188-189, Jun. 2007.
[4] M. Bauer et al., "A multi-level-cell 32Mb flash memory," in *ISSCC Dig Tech. Papers*, pp132-133. Feb. 1995.
[5] K. D. Suh et al., "A 3.3V 32Mb NAND Flash memory with incremental step pulse programming scheme," in *ISSCC Dig Tech Papers*. pp. 128-129. Feb. 1995.

ISSCC 2011 / February 22, 2011 / 9:30 AM

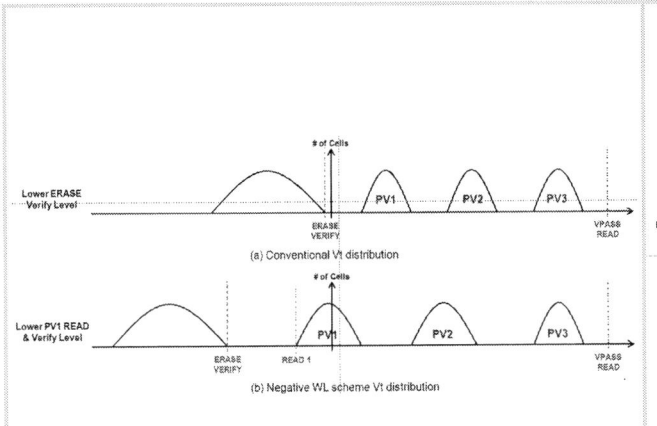

Figure 11.3.1: Negative WL Vt distribution.

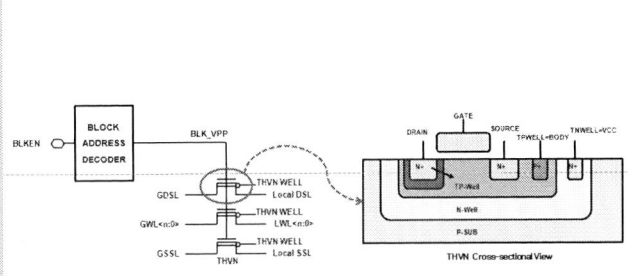

Figure 11.3.2: Negative-bias high-voltage switch with THVN device.

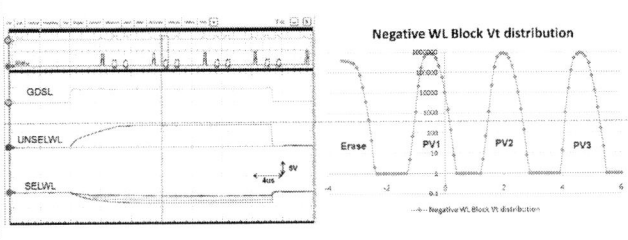

Figure 11.3.3: Word line Negative bias & Vt distribution measured data.

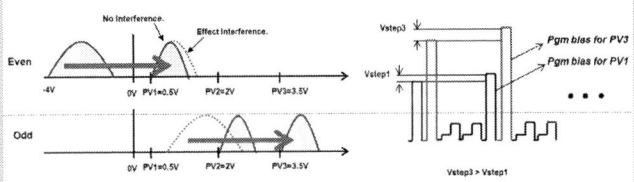

Figure 11.3.4: ABL parallel program method.

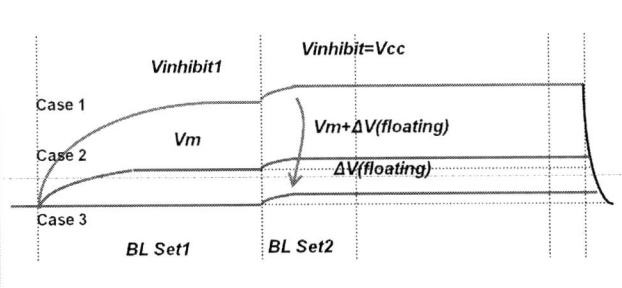

Figure 11.3.5: Bitline control for bitline compensation.

Technology	3metal 26nm CMOS
Density	32G
Organization	Page size : 8KB ABL 256pages / Block 2K block / Plane 2plane / chip
Power supply	2.7V~3.6V
Die size	181.5mm²

Figure 11.3.6: 26nm 32Gb MLC NAND Flash key features.

978-1-61284-303-2/11 $26.00 © 2011 IEEE

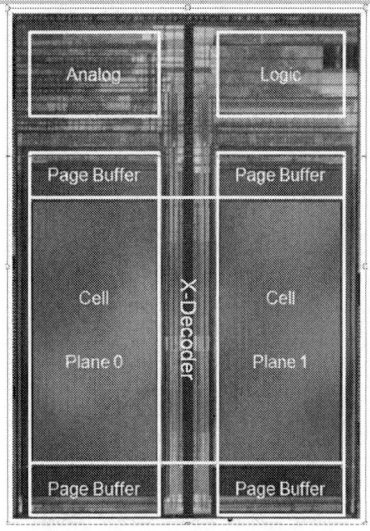

Figure 11.3.7: Die Micrograph of 26nm 32Gb MLC NAND Flash.

ISSCC 2011 / SESSION 11 / NON-VOLATILE MEMORY SOLUTIONS / 11.4

11.4 95%-Lower-BER 43%-Lower-Power Intelligent Solid-State Drive (SSD) with Asymmetric Coding and Stripe Pattern Elimination Algorithm

Shuhei Tanakamaru[1], Chinglin Hung[1], Atsushi Esumi[2], Mitsuyoshi Ito[2], Kai Li[2], Ken Takeuchi[1]

[1]University of Tokyo, Tokyo, Japan,
[2]SIGLEAD, Yokohama, Japan

This paper presents intelligent solid-state drives (SSDs), which decrease memory errors by 95% and reduce power consumption by 43%. Figure 11.4.1 shows the measured memory cell error in the data retention and program disturb of 4X, 3X and 2Xnm NAND flash memories. As the memory size decreases, both data retention and program disturb errors increase due to the interference, random telegraph noise and reduced electrons [1]. In the scaled NAND, the electric field in the channel increases [2] and the program disturb due to GIDL-induced hot electron injection becomes more significant (Fig. 11.4.1(c)). In conventional SSDs, 20 to 40b correction per 1KB codeword error-correcting code (ECC) is used to correct errors [3]. As stronger codes, such as LDPC, are developed [4], the capability of ECC is close to the Shannon limit of a few percent error correction. Thus, the additional high-reliability scheme is required. As the feature size decreases, the power consumption increases due to the increased bit-line capacitance of NAND [5]. As the space between bitlines decreases, the inter bitline capacitance increases. To overcome reliability and power problems in SSDs, this paper describes two technologies. Asymmetric coding improves memory-cell reliability by 95% without access-time penalty. Stripe pattern elimination algorithm eliminates the worst-program data pattern and decreases the power during the program by 43% without circuit area or access time overhead.

Figure 11.4.2 shows the measured error analysis. The key observation is that for both data retention and program disturb errors, the number of "0→1"-errors and "1→0"-errors are NOT equal. The origin of the data retention error is the electron ejection from the floating gate due to the stress-induced leakage current (SILC). During the data retention, the memory cell V_{TH} moves to V_{THi}, the thermally equilibrium V_{TH}, where no electron exists in the floating gate. V_{THi} is around 0V. Most errors correspond to the V_{TH} decrease of "10→00" or "00→01" due to the higher electric field across the tunnel oxide (Fig. 11.4.2(a)). Also, if the memory cell V_{TH} is higher, the program voltage is also higher. The enhanced voltage stress to memory cells increases the trap density in the tunnel oxide. As a result, SILC increases and the data retention errors also increase. The program disturb error is caused by the electron injection due to GIDL to the floating gate. Thus, the program disturb error corresponds to the V_{TH} increase of "01→00" or "00→10" (Fig. 11.4.2(b)). In the multi-level cell NAND flash memory, two bits stored in a memory cell are assigned to two different page (row) addresses, upper and lower pages [6]. For the data retention, the major error is "0→1" of the lower page and "1→0" of the upper page. In contrast, the major error of the program disturb is "1→0" of the lower page and "0→1" of the upper page.

Figure 11.4.3 shows the randomizing coding and asymmetric coding schemes. Considering data retention errors are over 100× more than program disturb errors (Fig. 11.4.1(c)), the coding improves data retention by increasing "1"- and "0"-data of the lower and upper pages, respectively. In randomizing coding, the population of "10" and "00" is about 25% of the total data. In the this SSD, asymmetric coding encoder modifies the data programmed to NAND so that at least 60% of the lower and upper pages are "1" and "0", respectively. As a result, "10" and "00" occupy about 16% and 24% of the total data. By decreasing the population of "10" and "00", the data retention error decreases because 1) the lower program voltage decreases the voltage stress to memory cells and 2) the lower memory cell V_{TH} decreases the electric field across the tunnel oxide and reduces SILC.

The code length, data unit where Asymmetric Coding is applied, is 16. If Data unit1 in Fig. 11.4.3 contains more than or equal to 8 bit of "0", Data unit1 is flipped to increase the number of "1" and the flag is set to "1". On the other hand, if Data unit2 contains less than 8 bit of "0", the data is not modified. As shown in Fig. 11.4.4(a), the smaller code length realizes a higher population of "1" with a drawback of a larger overhead due to the flag. If the code length is 16, the overhead is 6.3%. The number of "1" of the lower page and that of "0" of the upper page becomes 60% of the total data. Figure 11.4.4(b) shows the measured data retention error vs. the program voltage stress. In Asymmetric Coding, the population of high V_{TH} states, "10" and "00", is lower and the program voltage stress to memory cells is reduced. As a result, the data retention error decreases by 91%. Figure 11.4.4(c) shows the measured errors vs. the data pattern during the data retention. Again, as the proposed asymmetric data pattern decreases the population of the high V_{TH} states, the data retention improves by 40%. Due to two effects mentioned above, the total retention error decreases by 95% (Fig. 11.4.4(d)).

Figure 11.4.5 shows the stripe pattern elimination algorithm (SPEA). During the program of NAND, the selected and unselected bit-lines are biased to V_{CC} and V_{SS}, respectively. When bitlines are alternately biased to V_{CC} and V_{SS}, which corresponds to the column-stripe data pattern "1010…", all inter bitline capacitance is charged and a large current flows. On the other hand, when every bit-line is biased to V_{CC} with all "1" data pattern, inter bit-line capacitance diminishes and the program current decreases. SPEA Encoder in Fig. 11.4.3 modifies the original data from the host to avoid the worst-case column-stripe pattern and save the power. SPEA calculates the number of "1" in even and odd columns. Then, SPEA calculates the difference. If the difference is larger than a threshold value, N_{TH}, the data programmed to NAND is modified so that the odd column data are arranged first and the even column data are arranged next (Fig. 11.4.5). As a result, the column-stripe pattern which consumes the maximum power is changed to decrease the power. In the proposed SSD (Fig. 11.4.3), data from the host is first modified by SPEA Encoder to save the power, then modified again with the Asymmetric Coding Encoder to improve the reliability and finally programmed to NAND. Figure 11.4.6 shows the measured power consumption. With SPEA, the program current of 4Xnm and 3Xnm NAND decreases by 35% and 43%, respectively. SPEA is more effective in the scaled NAND because as the memory cell size decreases, the bitline capacitance as well as the power increase.

Figure 11.4.7 shows the micrograph of our SSD. Besides the on-board 16 NAND chips, another NAND is implemented in the daughter board for the reliability/power test. Asymmetric coding reduces memory cell errors by 95% and realizes the highest reliability by working with advanced ECC such as LDPC. SPEA decreases the program current by 43%. SPEA can be combined with other low-power memory technologies [5,7] and minimize the power of SSDs.

Acknowledgments:
This work is supported by CREST/JST. The authors appreciate Dr. K. Johguchi.

References:
[1] K. Takeuchi, "NAND successful as a media for SSD," *ISSCC*, Tutorial T-7, 2008.
[2] J. D. Lee *et al.*, "A New Programming Disturbance Phenomenon in NAND Flash Memory by Source/Drain Hot-Electrons Generated by GIDL Current," *Non-Volatile Semiconductor Memory Workshop (NVSMW)*, pp. 31-33, 2006.
[3] M. Abraham, "NAND Flash Trends for SSD/Enterprise," *Flash Memory Summit*, 2010.
[4] Y. Y. Tai, "Error Control Coding for MLC Flash Memories," *Flash Memory Summit*, 2010.
[5] K. Takeuchi, "Novel Co-design of NAND Flash Memory and NAND Flash Controller Circuits for sub-30nm Low-Power High-Speed Solid-State Drives (SSD)," *Symposium on VLSI Circuits Dig. Tech. Papers*, pp.124-125, 2008.
[6] K. Takeuchi *et al.*, "A Multipage Cell Architecture for High-Speed Programming Multilevel NAND Flash Memories," *Symposium on VLSI Circuits Dig. Tech. Papers*, pp.67-68, 1997.
[7] C. Lee *et al.*, "A 32Gb MLC NAND-Flash Memory with V_{TH}-Endurance-Enhancing Schemes in 32nm CMOS," *ISSCC Dig. Tech. Papers*, pp. 446-447, 2010.
[8] K. Takeuchi *et al.*, "A 56nm CMOS 99mm^2 8Gb Multi-level NAND Flash Memory with 10MB/s Program Throughput," *ISSCC Dig. Tech. Papers*, pp. 144-145, 2006.

ISSCC 2011 / February 22, 2011 / 10:15 AM

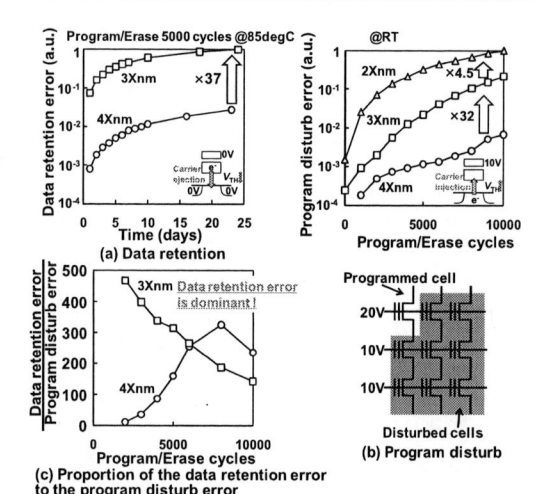

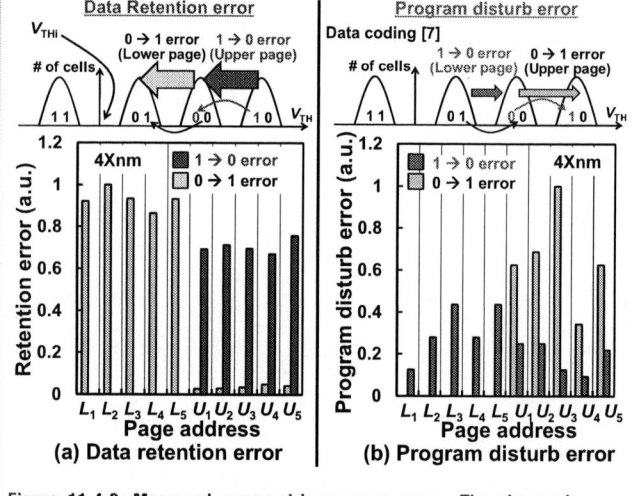

Figure 11.4.1: Measured memory cell error rate in SSDs.

Figure 11.4.2: Measured asymmetric memory errors. The observed errors strongly depend on the data pattern.

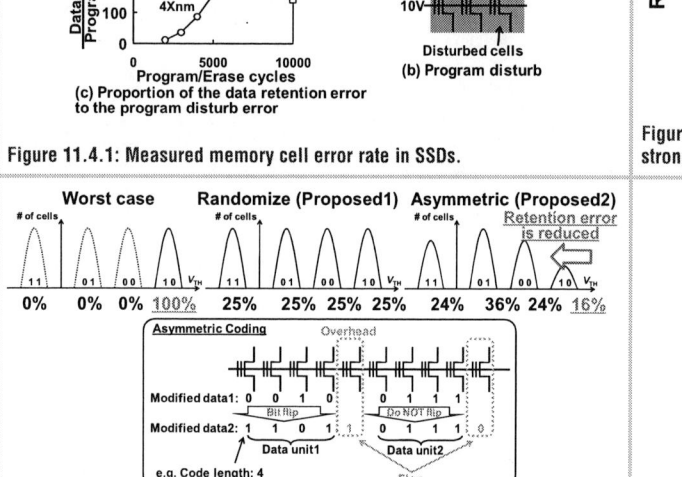

Figure 11.4.3: Proposed Asymmetric Coding. The program data of NAND is modified to decrease the population of "10" and "00".

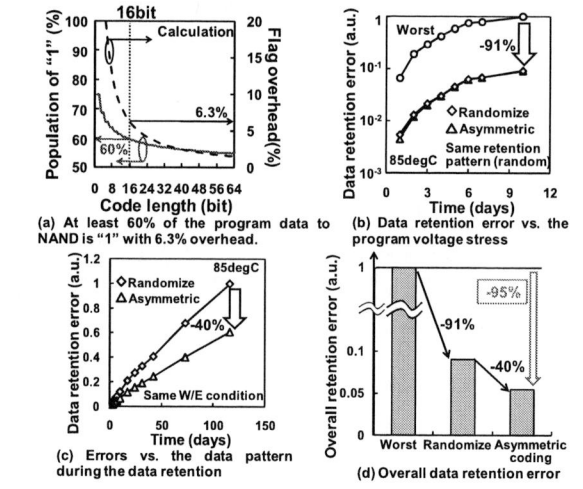

Figure 11.4.4: Measured reliability improvement of Asymmetric Coding.

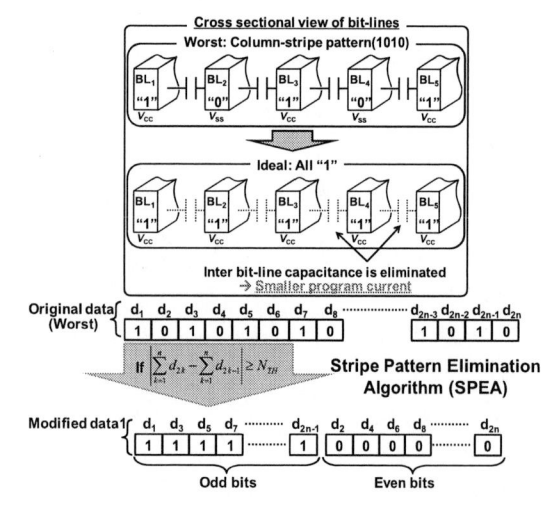

Figure 11.4.5: Proposed Stripe Pattern Elimination Algorithm (SPEA).

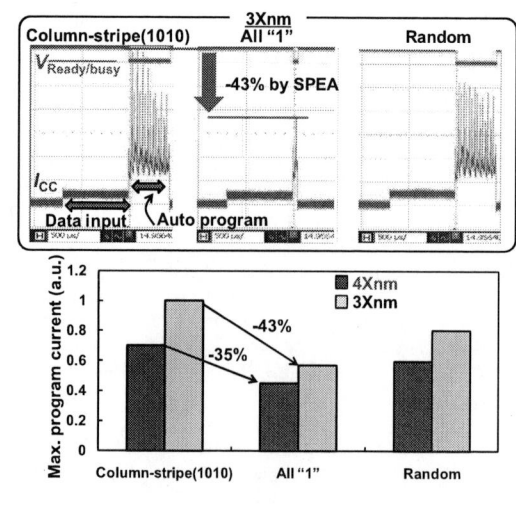

Figure 11.4.6: Measured power reduction with SPEA.

978-1-61284-303-2/11 $26.00 © 2011 IEEE

205

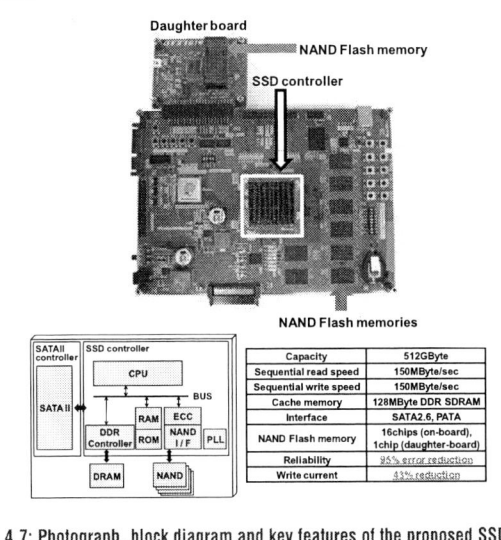

NAND Flash memories	
Capacity	512GByte
Sequential read speed	150MByte/sec
Sequential write speed	150MByte/sec
Cache memory	128MByte DDR SDRAM
Interface	SATA2.6, PATA
NAND Flash memory	16chips (on-board), 1chip (daughter-board)
Reliability	95% error reduction
Write current	43% reduction

Figure 11.4.7: Photograph, block diagram and key features of the proposed SSD.

ISSCC 2011 / SESSION 11 / NON-VOLATILE MEMORY SOLUTIONS / 11.5

11.5 An Offset-Tolerant Current-Sampling-Based Sense Amplifier for Sub-100nA-Cell-Current Nonvolatile Memory

Meng-Fan Chang[1], Shin-Jang Shen[1], Chia-Chi Liu[1], Che-Wei Wu[1],
Yu-Fan Lin[1], Shang-Chi Wu[1], Chia-En Huang[1,2], Han-Chao Lai[1,2],
Ya-Chin King[1], Chorng-Jung Lin[1], Hung-Jen Liao[2], Yu-Der Chih[2],
Hiroyuki Yamauchi[3]

[1]National Tsing Hua University, Hsinchu, Taiwan,
[2]TSMC, Hsinchu, Taiwan,
[3]Fukuoka Institute of Technology, Fukuoka, Japan

Decreasing read cell current (I_{CELL}) has become a key trend in nonvolatile memory (NVM). This is not only due to device size and V_{DD} scaling while keeping the same threshold voltage (V_{TH}), but also to the growing spread of the following applications: 1) multiple-level-cell (MLC) [1-2] to achieve smaller area-per-bit; 2) lower-V_{DD} [3] to save power consumption; 3) Logic-process-compatible one-time programming memories (OTP) for embedding into mobile chips. A smaller I_{CELL} leaves the sense amplifiers (SAs) operation vulnerable to 1) bitline (BL) level offset due to noise, bias and load (C_{BL}) mismatches and 2) V_{TH} variation. As device size and BL-pitch is continually scaled down, the above factors have become major showstopper for SAs. To tolerate these offsets, small-I_{CELL} NVMs suffer from slow read speed or high read fail probability. Thus, a more largely offset tolerant SA is a prerequisite to achieve faster read speeds. In this study, we propose a new offset tolerant current-sampling-based SA (CSB-SA) to achieve 7× faster read speed than previous SAs for sensing small I_{CELL}. A fabricated 90nm 512Kb OTP macro, using the CSB-SA and our CMOS-logic-compatible OTP cell [4], achieves 26ns macro random access time for reading sub-200nA I_{CELL}. Measurements also confirmed that this 90nm CSB-SA could achieve sub-100nA sensing.

Many small-I_{CELL} NVMs employ voltage-mode SA (VSA) [2] with a long BL developing time to tolerate SA offset, at the cost of a reduced read speed. Current-mode SA (CSA) achieves faster read speeds than VSA [1]. Cascode-current-load or resistive-divider-like CSAs (RD-CSAs) [1], [5], achieve sub-100nA sensing, but require long BL settling times to achieve high-accuracy 1st-stage voltage difference. The inverter-offset-compensated SA (IOC-SA) [6] reduces the SA offset. However, BL offset and BL settling time still limits its advantages with regard to VSA/CSA. In comparison with I_{CELL} and a reference-current (I_{REF}), current-mirror CSA (CM-CSA) [7], has fast read speeds but cannot sense small I_{CELL} due to its input-stage V_{TH} mismatch. Figure 11.5.1 compares the concepts of CSB-SA with previous SAs. CSB-SA uses the same MOS device for current sampling and current-ratio amplifying. This enables V_{TH}-independent current sampling schemes for its differential I_{CELL} and I_{REF} inputs. This is significantly different from CM-CSA, using different MOS devices for current-mirroring or I-V conveying, which results in increased vulnerability to V_{TH} mismatch. In addition, CSB-SA uses sampled current to generate fast 1st-stage voltage difference at its BL-decoupled small-load internal nodes. Unlike VSA or RD-CSA, which have to develop their 1st-stage voltage on the heavy-load BL using continuous I_{CELL} driving. IOC alleviates SA V_{TH}-mismatch but with a complex multi-step V_{TH}-nulling process and numerous switching devices. IOC also does not cancel the SA offset due to transistor width/length or T_{OX} variations, and is still vulnerable to BL noise/C_{BL} mismatch. In our CSB-SA, the sampled currents are insensitive to transistor and C_{BL} mismatch. Thus, CSB-SA is able to achieve faster read speed while tolerating V_{TH}/BL offset for sensing small I_{CELL}.

The CSB-SA operations are presented in Fig. 11.5.2. In phase-1 (current sampling), S1~S4 are turned on to connect the BL and dummy-BL (DBL, at I_{REF} side) to the diode-connected M1 and M2, respectively. The M1/M2 provide large precharge current (I_{PRE}) to BL/DBL at the beginning of phase-1. When the precharge time (T_{PRE}) is sufficient, the node voltage SA1 (V_{SA1}) and SA2 (V_{SA2}) are high, while the I_{PRE} is low. The drain current of the M1 and M2 is ($I_{M1}=I_{CELL}+I_{PRE1}$) and ($I_{M2}=I_{REF}+I_{PRE2}$), respectively. The gate voltages at the end of phase-1 for M1 (V_{G1}) and M2 (V_{G2}) are stored on C2 and C1, respectively. When the I_{PRE} is low enough to be disregarded, the I_{M1}/I_{M2} is equal to I_{CELL}/I_{REF}, despite various M1-M2 V_{TH} mismatch conditions. In phase-2 (current-ratio amplifying), the S3/S4 is switched off to disconnect the SA1/SA2 from BL/DBL.

At the beginning of phase-2, the M1/M2 charges the SA1/SA2 with the current sampled at phase-1. For a given T_{P2} period, M1/M2 raises the V_{SA1}/V_{SA2} by the amount of $\Delta V_{SA1}/\Delta V_{SA2}$. Due to AC-coupling behavior of C1/C2, the ΔV_{SA1} reduces the V_{G2} and ΔV_{SA2} reduces the V_{G1}. This results in an amplification of the current ratio ($CR=I_{M2}/I_{M1}$) or current difference ($\Delta I_{M1-M2}=I_{M1}-I_{M2}$) between M1 and M2. This CR-amplification speeds up the development of difference (ΔV_{SA}) between V_{SA1} and V_{SA2}. In phase-3 (2nd-stage amplifying), the EN turns on the NMOS-latch and pulls down the SA2 while M1 continually charges the SA1 to V_{DD}. Finally, the digital output is generated at node SA1 and SA2.

Although V_{TH} mismatch between M1 and M2 and different I_{CELL} cause various ΔV_{SA} at the end of phase-1 (ΔV_{SA-P1}), the CSB-SA does not suffer from sensing failure. This is because the ΔV_{SA-P1} can be overcome by ΔI_{M2-M1} if T_{P2} is sufficient, unlike un-recoverable sensing error due to V_{TH} mismatch in CM-CSA. As the Fig. 11.5.3 shows, the V_{SA1}-V_{SA2} crossover point of V_{TH}-mismatch case occurs later than that of the no-mismatch case, but both are able to sense the 100nA I_{CELL}. Although the ΔI_{M2-M1} is small when sensing small I_{CELL}, the required T_{P2} for tolerating large device mismatch is insignificant and remains independent of C_{BL}, thanks to the small parasitic load at node SA0 and SA1 (disconnection from BL) during phase-2. The T_{P2} penalty to compensate for the 150mV V_{TH}-mismatch is 0.4% for the access-time of a macro with 2048 cells per BL. Unlike the CM-CSA suffering from low yield due to V_{TH} variations, the CSB-SA can achieve 100% yield for sensing sub-100nA. This macro uses an inactive sub-array to provide the DBLs for I_{REF} without area penalty, as Fig. 11.5.4 shows. This enables the I_{CELL} and I_{REF} branches have improved common-mode BL precharge behavior. Reducing the T_{PRE}, the I_{PRE} increases but does not significantly influences the yield of CSB-SA. As switch-point analysis shows, CSB-SA has 100% yield if its ΔI_{M1-M2} exceeds 3.8nA~7.8nA when the I_{PRE2} is within 100nA. This analysis indicates that CSB-SA is not highly sensitive to I_{PRE}. When reading a 100nA I_{CELL} with a 50nA I_{REF}, the T_{PRE} can be reduced by 76% if I_{PRE2}=50nA is used. Thus, CSB-SA can employ a shorter T_{PRE} to achieve a faster speed than previous SAs.

Figure 11.5.5 shows our OTP cell, which consists of two NMOS transistors in series with a parasitic U-shaped nitride storage node formed by the two merged nitride spacers. This cell is logic-process compatible and does not need any additional process steps/masks. The 90nm cell size is only 0.26μm², which is smaller than other logic-process programmable NVMs (i.e. eFuse). A 90nm 512Kb asynchronous NOR-type OTP macro was fabricated. The measured I_{CELL} of this macro ranges between 160nA and 1μA. One SA in each testchip can use external I_{CELL}/I_{REF} for exploration of detectable I_{CELL}. From the I_{REF}=50nA test (Fig. 11.5.6), the detectable Read1 and Read0 I_{CELL} are 70nA~85nA and 25nA~5nA, respectively. This OTP macro is functional from V_{DD}=1.4V to 0.85V. The minimum access time of this OTP macro at V_{DD}=1.2V is 26ns. Compared to previous SAs, the CSB-SA achieved 7× faster access time for 100nA I_{CELL} on a 2048-cells BL. Figure 11.5.7 shows the die photo.

Acknowledgements:
The authors thank CIC and TSMC University Shuttle Program for chip and testkey manufacturing.

References:
[1] R.-A. Cernea, et al. "A 34MB/s MLC write throughput 16 Gb NAND with all bit line architecture on 56nm technology," *IEEE J. Solid-State Circuits*, vol. 44, no. 1, pp.186–194, Jan. 2009.
[2] D. Nobunaga, et al., "A 50nm 8Gb NAND Flash memory with 100MB/s program throughput and 200MB/s DDR interface," *ISSCC Dig. Tech. Papers*, pp. 426-427, Feb. 2008.
[3] M.-F. Chang, et al., "A 0.29V embedded NAND-ROM in 90nm CMOS for ultra-low-voltage applications," *ISSCC Dig. Tech. Papers*, pp. 266-267, Feb 2010.
[4] C. E. Huang, et al., "A new self-aligned nitride MTP cell with 45nm CMOS fully compatible process," *IEDM Dig. Tech. Papers*, pp.91-94, Dec. 2007.
[5] C. J. Chevallier, et al., "A 0.13μm 64Mb multi-layered conductive metal-oxide memory," *ISSCC Dig. Tech. Papers*, pp.260-261, Feb. 2010.
[6] J. Javanifard, et al., "A 45nm self-aligned-contact process 1Gb NOR Flash with 5MB/s program speed," *ISSCC Dig. Tech. Papers*, pp.424-425, Feb. 2008.
[7] M.-K. Seo, et al. "A 130-nm 0.9V 66-MHz 8-Mb (256Kx32) local SONOS embedded Flash EEPROM," *IEEE J. Solid-State Circuits*, vol. 40, no. 4, pp. 877-883, April 2005.

978-1-61284-303-2/11 $26.00 © 2011 IEEE

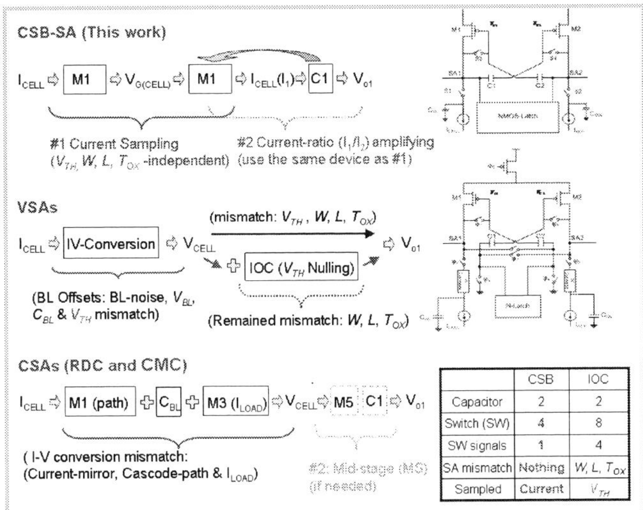

	CSB	IOC
Capacitor	2	2
Switch (SW)	4	8
SW signals	1	4
SA mismatch	Nothing	W, L, T_{OX}
Sampled	Current	V_{TH}

Figure 11.5.1: Concepts of proposed current-sampling-based sense amplifier (CSB-SA).

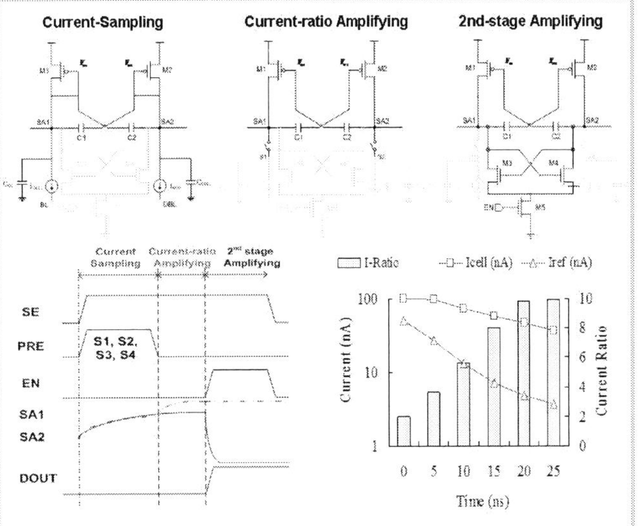

Figure 11.5.2: Phase breakdown of the proposed CSB-SA

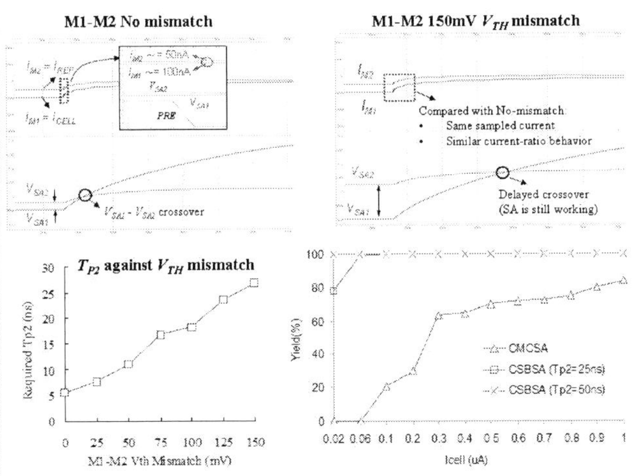

Figure 11.5.3: Analysis and comparison of CSB-SA against variations in V_{TH}.

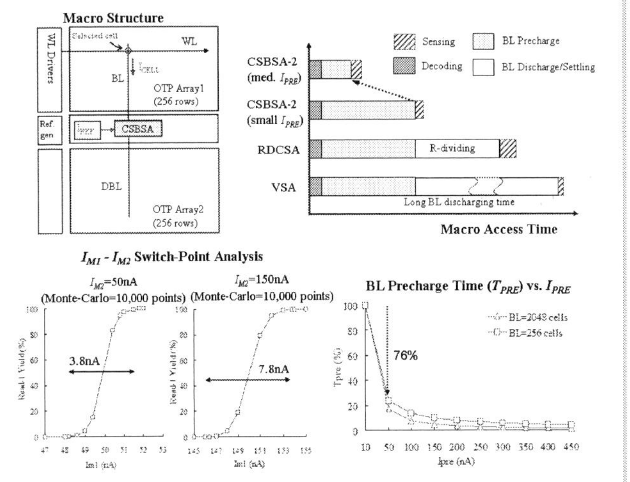

Figure 11.5.4: Macro structure and speed analysis.

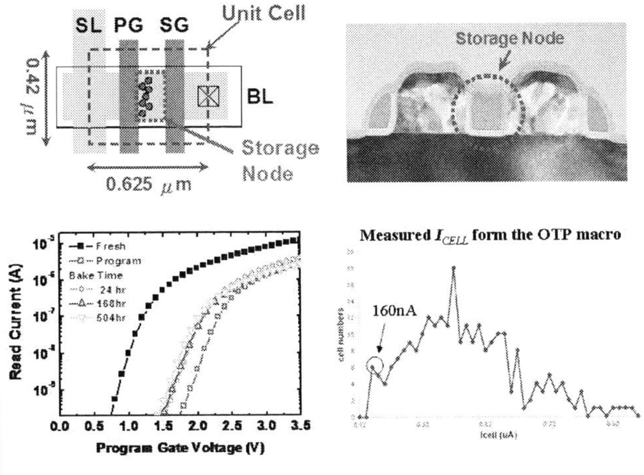

Figure 11.5.5: Structure and characteristics of proposed OTP cell.

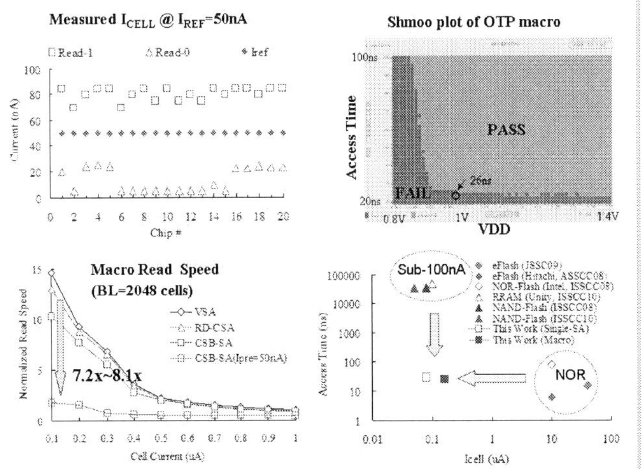

Figure 11.5.6: Measurement results.

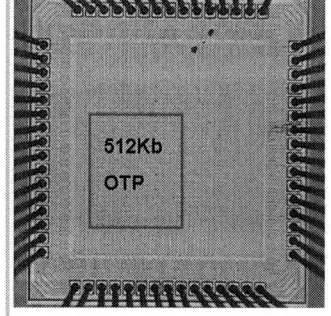

Technology	90nm LP CMOS Logic
Capacity	512Kb
Testchip Size	1.39mm x 1.32mm
Bitcell size	0.26um^2
Typical VDD	1.2V
Access time	26ns @ 1.2V

Figure 11.5.7: Die photo.

ISSCC 2011 / SESSION 11 / NON-VOLATILE MEMORY SOLUTIONS / 11.6

11.6 A Low-Voltage 1Mb FeRAM in 0.13μm CMOS Featuring Time-to-Digital Sensing for Expanded Operating Margin in Scaled CMOS

Masood Qazi[1], Michael Clinton[2], Steven Bartling[2], Anantha P. Chandrakasan[1]

[1]Massachusetts Institute of Technology, Cambridge, MA,
[2]Texas Instruments, Dallas, TX

Low-power portable electronics such as implantable medical devices require low-access-energy non-volatile memory to deliver longer battery lifetime and richer functionality. Ferroelectric random access memory (FeRAM) technology is a good candidate for both storage [1] and non-volatile RAM [2]. The power and supply voltage of FeRAM need further reduction, and this work presents a solution in anticipation of FeRAM scaling to advanced technology nodes for which the bitcell charge reduces and transistors operate at 1V and below. Specifically, a time-to-digital converter (TDC) sensing scheme is developed to capture the diminishing charge signal from the memory element at low supply voltage.

The 1T1C cell in Fig. 11.6.1 based on the technology in [3] contains a ferroelectric capacitor that exhibits a hysteresis in stored charge versus bias V_{FE}. During read, the plateline (PL) is pulsed across the cell in series with the floating bitline (BL) capacitance C_{BL} which presents a load to extract charge from the cell hysteresis and converts the stored charge into a voltage. After-pulse sensing—the PL returns to 0V—cancels much of the variable non-hysteretic cell capacitance, especially for high-capacitance, low-swing bitlines.

Figure 11.6.1 shows the simulated 5.6σ voltage signal between "1" and "0" versus BL length at a modest supply of 1.35V. An optimum C_{BL} balances the desirable extraction of charge against undesirable voltage attenuation, favoring a BL of 256 cells. Per contra, the charge separation continues growing with BL length because V_{FE} during read increases. A BL with 1024 cells halves the voltage separation but doubles the charge separation, motivating FeRAM with longer BLs and improved array efficiency—provided that charge can be directly sensed. Data in Fig. 11.6.1 for a BL of 128 8T8C cells also illustrate that simply up-sizing the FeRAM capacitor for a fixed BL length does not greatly improve the voltage signal—since relative loading from the BL degrades—although the separation in charge increases by 5.7×. Because of these charge considerations, a high-density 1Mb FeRAM with 1024 1T1C cells per BL has been designed to exploit the 2.2× charge signal from increased BL loading.

Figure 11.6.2 introduces the TDC-based read path as a means to directly sense charge separation and Fig. 11.6.3 shows waveforms of the associated signals. During read, a local plateline (LPL) pulses high to transfer the cell charge to the BL and master bitline (MBL). After the LPL returns to 0V, the ST signal initiates the first of two delay measurements. As the rising edge of "start" propagates through 32 TDC slices, a current source converts the BL charge to a delay defined by the crossing of V_{ref1} which is detected by a continuous-time comparator. The positive edge of CMP captures the state of the TDC slices and additional control logic discharges MBL to 0V in preparation for a second delay measurement. The ST signal also initiates the reference delay measurement in which the same BL is charged through the same current source to a different threshold V_{ref2} that is detected by the same comparator. The rise of CMP again captures the state of the TDC but this time also performs digital subtraction through the feedback NOR gate within each slice. Namely, if the signal delay is longer than the reference delay (Read 0), the TDC subtraction slices that observe only the first edge but not the second edge (TQ15-TQ23) will produce "1" on their outputs after the second delay measurement completes. A 32-input OR gate detects this condition. If signal delay is shorter than reference delay (Read 1), all TDC outputs either transition to "0" (TQ9-TQ31) or remain at "0" (TQ0-TQ8). Thus the subtraction between signal and reference is accomplished with logic instead of analog amplification by a differential transistor pair. The waveforms in Fig. 11.6.3 also illustrate how timing skews and comparator offsets are digitally cancelled such that the comparator offset distribution need not fit within the bitline voltage distributions.

The quantity $(V_{ref1}-V_{ref2})C_{BL}$ corresponds to a reference charge which must be greater than a worst-case "0" and less than a worst-case "1" charge after relaxation. For experimental flexibility the reference charge in this work is delivered

by off-chip voltages. This TDC sensing is compatible with techniques that generate a midpoint reference on-chip with dummy capacitors [4] although the inability to track relaxation in a reference cell remains a challenge.

The 32-input OR gate and TDC register are implemented with dynamic circuits to minimize area. The current-starved delay element contains a dynamic output inverter clocked by the "stop" signal in order to save cross-over current and terminate activity in the delay line immediately upon completion of a measurement. The TDC consumes an area of 110μm², and because charge is converted into delay, the continuous signal in time is multiplexed through a simple daisy chain of OR gates. Thus 2^{20} 1T1C cells multiplex to eight 5bit TDCs while preserving 64.4% array efficiency.

The principle of charge sensing permits widening of operating margin by slowing down: the current source charging rate can be reduced while maintaining the TDC delay resolution (set by VCTL). A fixed charge difference between "1" and "0" can be expanded to a longer delay separation between t_0 and t_1. The limit of this expansion depends on two main factors. First, parasitic current from unselected bitcells and columns may overpower too low of a BL charging current. Second, offsets in comparators or local fluctuation of the BL charging current across segments may dominate the signal budget of the TDC measurement. The TDC sensing scheme is compatible with low-offset comparator design techniques to attain further margin.

Figure 11.6.4 illustrates the chip architecture and the timing diagram for externally supplied signals. The high degree of column interleaving—1 out of 64—avoids neighbor BL coupling encountered in DRAM-style designs because 63 BLs are grounded between every active BL. Unselected cells on an active row have both the LPL and BL grounded to preserve data. The hierarchical partitioning of the 1024-cell column into eight vertically-routed LPL groups, each driving 128 1T1C cells, reduces the delay and switching energy of the LPL.

Measured performance and energy are in Fig. 11.6.5. From 1.5V to 1.0V core supply (with the off-chip WL supply set to 0.7V above the core supply), all bits in the 1Mb prototype pass a checkerboard and inverse checkerboard pattern. The chip was verified to retain the pattern for 30min at 27°C under power-down with the core and WL supplies shorted to ground. Power is measured also under a checkerboard pattern to average over both data states. Random access performance spans 5.03MHz (1.5V) to 1.37MHz (1.0V) and supply scaling enables a 1.96× decrease in energy per accessed bit from 19.2pJ to 9.77pJ per bit. Idle power is only 251nW to 95nW with no extra wake-up time required before an access. Further voltage scaling of the TDC-based sensing scheme has been measured down to 0.8V in a 128kb 8T8C version of the chip, in which the cell area is increased by shorting eight 1T1C cells in parallel through a decoder modification. Therefore, the TDC-based sensing can extend to future FeRAM process technologies that have the potential to reduce the voltage thresholds of the cell hysteresis. Figure 11.6.6 summarizes the key features of the high-density 1T1C design for low-power operation.

Acknowledgements:
This work was funded in part by the C2S2 Focus Center, one of six research centers funded under the Focus Center Research Program (FCRP), a Semiconductor Research Corporation entity. The authors thank Scott Summerfelt and Borna Obradovic for technical discussion, and Bharat Rajaram and Kelly Krantz for their support.

References:
[1] D. Takashima, et al., "A scalable shield-bitline-overdrive technique for 1.3V Chain FeRAM," *ISSCC Digest Tech. Papers*, pp. 262-263, 2010.
[2] B. Jeon, M. Choi, Y. Song and K. Kim, "A nonvolatile ferroelectric RAM with common plate folded bit-line cell and enhanced data sensing scheme," *ISSCC Digest Tech. Papers*, pp. 38-39, 426, 2001.
[3] T. S. Moise, et al., "Demonstration of a 4 Mb, high density ferroelectric memory embedded within a 130 nm, 5 LM Cu/FSG logic process," *IEDM Tech. Digest*, pp. 535-538, 2002.
[4] K. Yamaoka, et al., "A 0.9 V 1T1C SBT-based embedded non-volatile FeRAM with a reference voltage scheme and multi-layer shielded bit-line structure," *ISSCC Digest Tech. Papers*, pp. 50-512, 2004.

978-1-61284-303-2/11 $26.00 © 2011 IEEE

ISSCC 2011 / February 22, 2011 / 11:00 AM

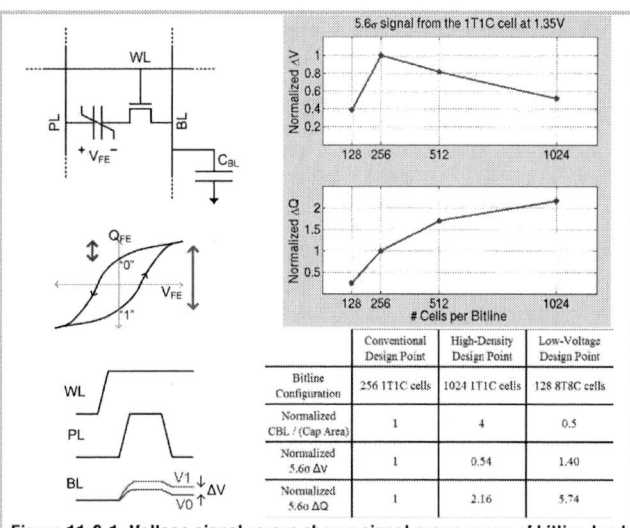

Figure 11.6.1: Voltage signal versus charge signal over a range of bitline load capacitance from Monte Carlo simulation of the bitcell.

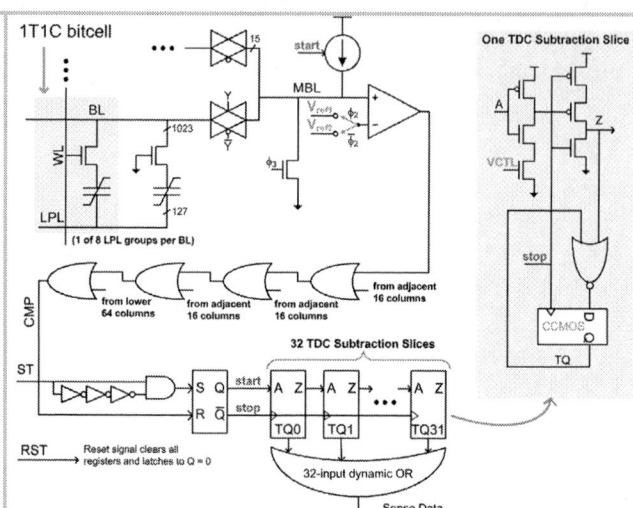

Figure 11.6.2: Simplified schematic of the TDC-based read path with external signals ST and RST from which other control signals derive.

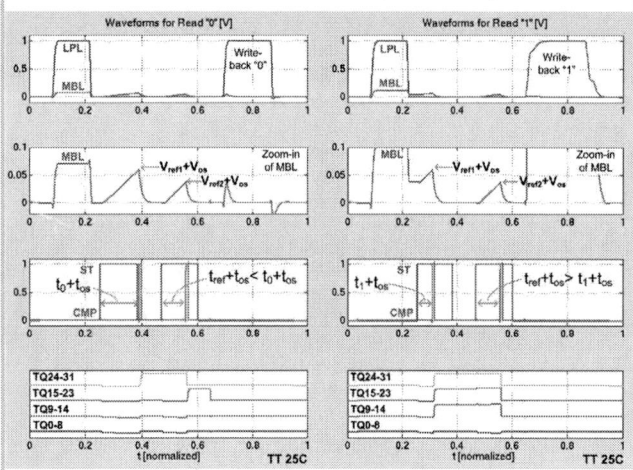

Figure 11.6.3: Simulation waveforms of the extracted 1T1C critical path for read "0" and read "1" at 1.0V, illustrating TDC delay subtraction.

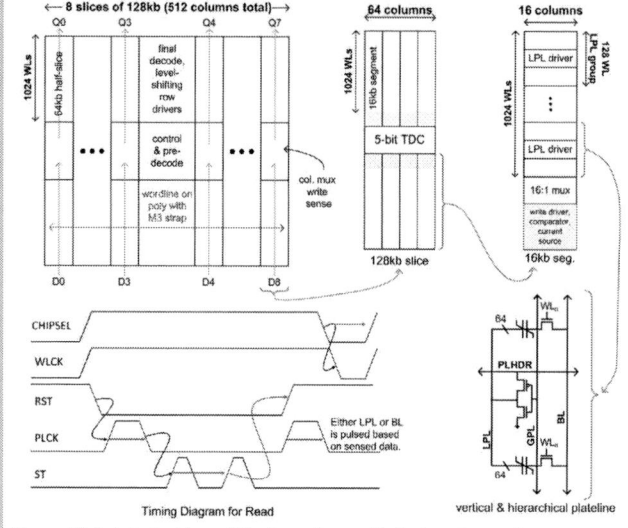

Figure 11.6.4: Full chip architecture along with timings for read.

Figure 11.6.5: Measured random access cycle energy and performance scaling from 1.5V to 1.0V.

	High-Density 1T1C Design
Organization	128k words of 8 bits
Technology	130nm LP CMOS
Array Efficiency	66.4%
Full Macro-level Density	0.936 Mb/mm²
Operating Voltage (CORE / WL)	1.5V / 2.2V to 1.0V / 1.7V
Read Cycle Time	200ns to 730ns
Read Power	772uW to 107uW
Idle Power	251nW to 95nW
Standby Power	0W

Figure 11.6.6: Summary of test chip characteristics.

978-1-61284-303-2/11 $26.00 © 2011 IEEE

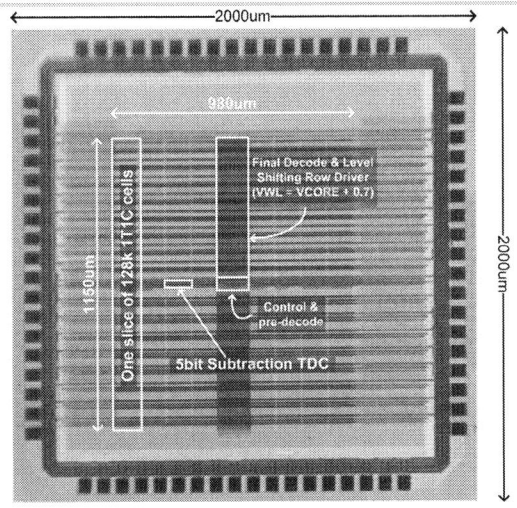

Figure 11.6.7: Die photo of 1Mb FeRAM prototype and identification of important structures.

ISSCC 2011 / SESSION 11 / NON-VOLATILE MEMORY SOLUTIONS / 11.7

11.7 A 4Mb Conductive-Bridge Resistive Memory with 2.3GB/s Read-Throughput and 216MB/s Program-Throughput

Wataru Otsuka[1], Koji Miyata[1], Makoto Kitagawa[1], Keiichi Tsutsui[1], Tomohito Tsushima[1], Hiroshi Yoshihara[2], Tomohiro Namise[2], Yasuhiro Terao[2], Kentaro Ogata[2]

[1]Sony, Kanagawa, Japan, [2]Sony LSI Design, Nagasaki, Japan

The growing demand for higher performance in the storage and access of data in various consumer electronic and computing devices has driven the development of nonvolatile memory (NVM) technologies. The promising candidates for future NVM such as FeRAM and PCM have demonstrated shorter access time, faster programming and wide read/write bandwidth in the chip and the memory macro [1-2]. Resistive memory (ReRAM) is also one of alternative NVMs, because of its low operating voltage, high speed and good scalability. Several types of ReRAM characteristics have been investigated on memory array [3-6]. However, most are limited in terms of memory array performance because of not having suitable read/write circuit for ReRAM. In this work, we present a 4Mb conductive bridge ReRAM test macro realizing 2.3GB/s read-throughput, 216MB/s program-throughput and robust reliability results by using read/write fully functional device technology with direct sense in programming (DSIP) method.

The micrograph of 4Mb test macro built in a 0.18μm CMOS technology and features are shown in Fig. 11.7.1. The test macro is organized into 16 tiles of 256Kb each. Dual-layered conductive bridge elements that employ CuTe-based conductive material and GdOx thin insulators are used in 1T-1R memory cells in the same manner as previously reported elsewhere [6]. The resistive element has top and bottom electrodes and is of the bipolar switching where high resistance state (HRS) turns into low resistance state (LRS) when the top electrode is positively biased and reversely when the bottom electrode is positively biased. The top electrodes form a plate covering a half of the tile. Bringing HRS into LRS is defined as "erase" assuming that the operation involving driving the large top electrode is done by blocks. The opposite is defined as "program" done by individual bits.

The memory cell and circuit diagram are shown in Fig. 11.7.2. Each tile has 64 write drivers and sense amplifiers in the middle of the folded arrangement of BLs. All of the 1024 (64 per tile) sense amplifiers are activated simultaneously in read operation and the 128 (8 per tile) write drivers are activated simultaneously in program operation. Figure 11.7.3 shows schematic of the sense amplifier, reference cell and write driver. In read operation, the Vo pair is compared while the BLs are maintained by VBIAS at low read voltages to prevent destruction of data. In program operation, the program pulse length and height are controlled by WRTEN and VGRST, respectively. For the verify functionality that is implemented in program operation, DSIPEN activates DSIP circuits, which is described in the latter part. BLEQ provides equalization control for fast and accurate sensing results.

One of the most common challenges in realizing ReRAM is to support a sufficient read speed. Though applications demand high performances in read-throughput, a lot of issues have to be addressed to sense a huge number of cell resistances within a very limited time in a robust manner. Our solution includes the folded BL cell arrays along with symmetric design of the sense amplifiers that operate in parallel. Figure 11.7.4 shows simulated internal signals and observed outputs in read operation. The resistance threshold and read voltage are set at 100kΩ and 0.3V, respectively. A read operation is executed in two steps. The first step starts when a read command is received (/RAS LOW at a clock edge with address provided). The address is decoded and transferred in 3 clocks and then the sense amplifier output is latched at 16ns (2 clocks) in the middle of Vo transition. After that, the BLs and WLs are put back on stand-by in 2 clocks. In total, the read cycle is 56ns (7 clocks). The second step starts when a data request command is received (/CAS LOW). The latched data is available in 3 cycles. This macro is able to handle read commands at an interval of 7 cycles; a read-throughput of 2.3GB/s is achieved as it is designed. With the critical read time as little as 2 cycles, 8.0GB/s read-throughput would be moved closer to if it had an adequate addressing and output interface.

Another challenge is about program speed that is affected by write-to-write verify. In general, programming verify takes place slowly due to low sense currents. Note that a resistance threshold for programming verify is always set higher than what for read operation to secure read margin. The DSIP program operation is our solution to the issue. Figure 11.7.5 shows simulated waveforms of the DSIP operation in comparison to the conventional verify scheme. It is assumed that programming is done after three consecutive fails in verify, so the sequence consists of 4 sets of generating program pulse, verify read and judgment. Given that most of bits are successfully done by an only single attempt, the number could be reduced further than 4. Program voltage increments as verify fails for robust programming. For the DSIP operations, the write drivers turn off when WRTEN drops at 5ns, but selected cells keep connected to sense amplifiers. Consequently, Vo voltages drop as the charge drains through the memory cells or reference cells. The time constant of the voltage drop is determined by the fixed capacitance on the sense node and cell resistance. At a right timing, the Vo voltages are compared to confirm if memory cells are successfully switched. The attempts run over with a cycle time of 16ns for a 0.3MΩ verify resistance. On the other hand, the conventional scheme that utilizes normal read circuit takes 110ns for a single attempt, resolved into 95ns to verify 1MΩ, 5ns to program, 5ns to feedback and 5ns to equalize BLs. DSIP takes advantage of higher sense currents because not only does it read at the higher voltages, but also has an ability to lower reference resistance; it owes the fact that the element resistance in HRS is negatively logarithmic against bias, at a rate where raising read voltage by 1.5V decreases the resistance by one order of magnitude. As a result, 216MB/s program-throughput is achieved given the cycle time of 74ns for 4 verifying cycles and the program size of 16B. On the planned 64B program size and 2 verifying cycles, program-throughput is over 1GB/s.

The yielded chip map taken from an 8-inch wafer and resistance distributions of LRS and HRS collected with DSIP over the 4Mb cells are shown in Fig. 11.7.6. A chip yield of 67% is obtained and the yielded chips are well distributed across the wafer. Yield loss contributors are write errors corresponding to any of 4Mb not passing verify during erase and/or program, read errors that mean any of read error bits after passing the write test, and incomplete circuitry. The resistance distributions are collected by scanning the read threshold and counting error bits. A sufficient HRS-to-LRS resistance separation of nearly one order of magnitude to ensure a robust read operation is obtained from a typical yielded chip.

In conclusion, the 4Mb test macro has been developed in 180nm technology, demonstrating feasibility of the conductive bridge ReRAM integration with competitive performance. Design issues on the read and program operations have been discussed and addressed by presenting our solutions that have been implemented in the macro and proved effective.

Acknowledgements:
We gratefully acknowledge the contributions of K. Aratani, T. Shiimoto and K. Yamamichi for their technical support, N. Nagashima and K. Hayashi for the managerial support.

References:
[1] H.Shiga, et al., "A 1.6GB/s DDR2 128Mb Chain FeRAM with Scalable Octal Bitline and Sensing Scheme," *ISSCC Dig. Tech. Papers*, pp.464-465, Feb. 2009.
[2] G.Sandre, et al., "A 90nm 4Mb Embedded Phase-Change Memory with 1.2V 12ns Read Access Time and 1MB/s Write Throughput," *ISSCC Dig. Tech. Papers*, pp.268-269, Feb. 2010.
[3] Y.Lin, et al., "A Novel TiTe Buffered Cu-GeSb Te/SiO$_2$ Electrochemical Resistive Memory (ReRAM)," *Dig. Symp. VLSI Technology*, pp.91-92, June 2010.
[4] M.Wang, et al., "A Novel Cu$_x$Si$_y$O Resistive Memory in Logic Technology with Excellent Data Retention Distribution for Embedded Applications," *Dig. Symp. VLSI Technology*, pp. 89-90, June 2010.
[5] S.Kawabata, et al., "CoOx-RRAM Memory Cell Technology using Recess Structure for 128Kbits Memory Array," *IEEE International Memory Workshop*, pp. 60-61, May 2010.
[6] K.Aratani, et al., "A Novel Resistance Memory with High Scalability and Nanosecond Switching," *IEDM Dig. Tech. Papers*, pp. 783-786, Dec. 2007.

978-1-61284-303-2/11 $26.00 © 2011 IEEE

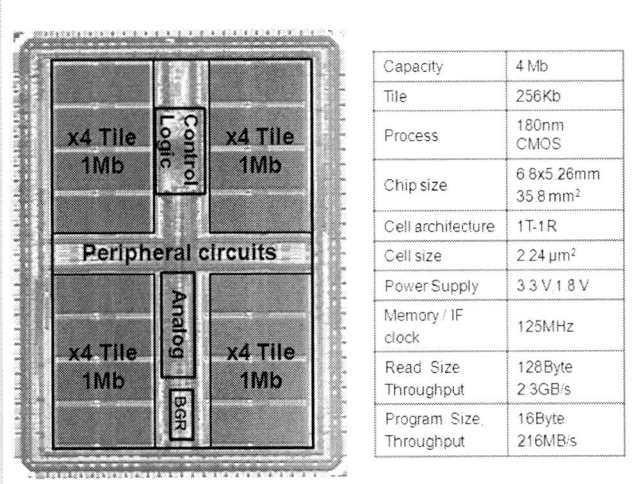

Capacity	4 Mb
Tile	256Kb
Process	180nm CMOS
Chip size	6.8x5.26mm 35.8 mm²
Cell architecture	1T-1R
Cell size	2.24 µm²
Power Supply	3.3 V,1.8 V
Memory / IF clock	125MHz
Read Size, Throughput	128Byte 2.3GB/s
Program Size, Throughput	16Byte 216MB/s

Figure 11.7.1: Die micrograph and features.

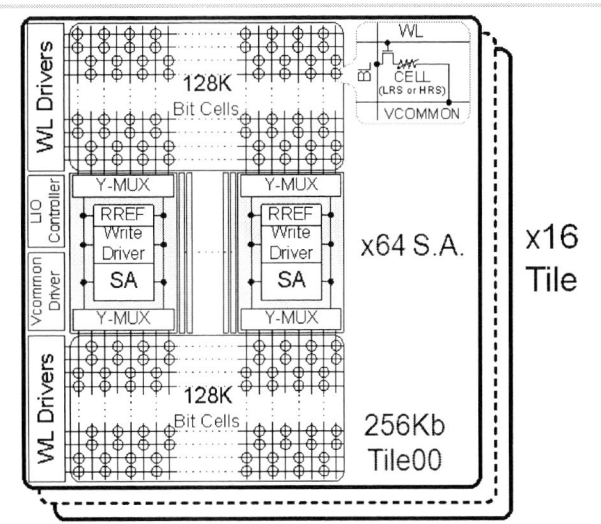

Figure 11.7.2: Array organization.

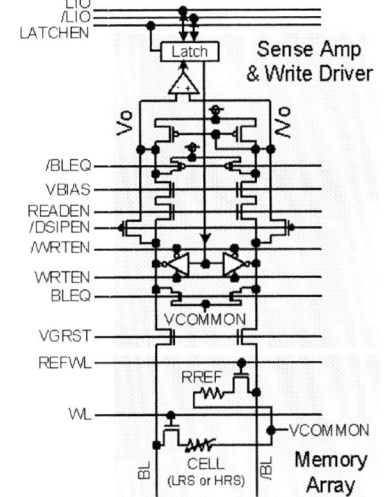

Figure 11.7.3: Sense amplifier, reference cell and write driver system.

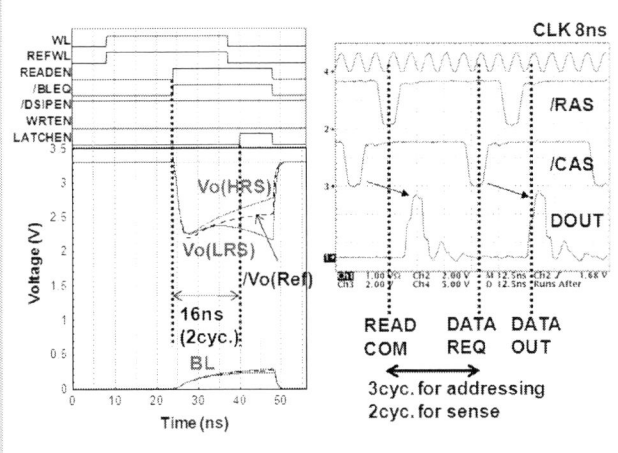

Figure 11.7.4: Simulated internal signals and observed outputs in read operation.

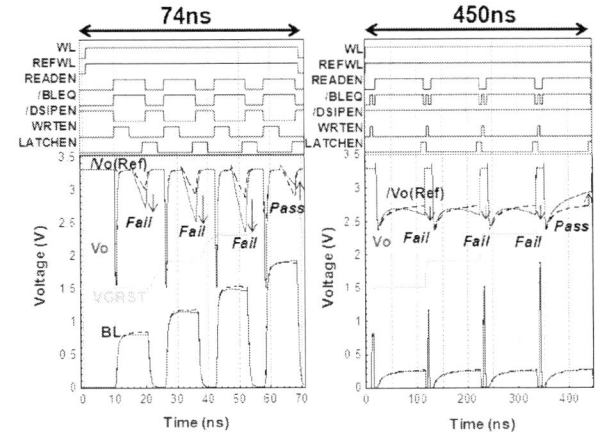

Figure 11.7.5: DSIP program operation waveforms in contrast to the conventional.

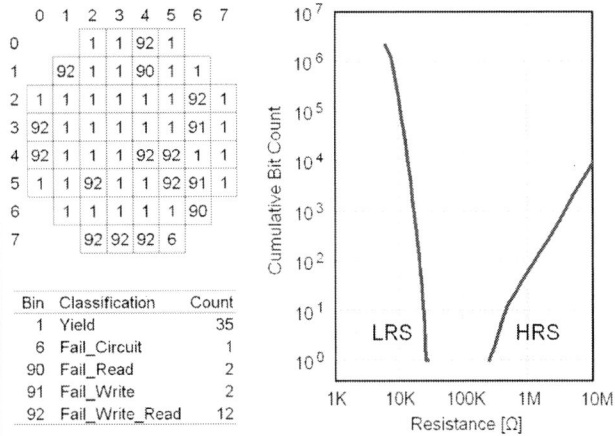

Bin	Classification	Count
1	Yield	35
6	Fail_Circuit	1
90	Fail_Read	2
91	Fail_Write	2
92	Fail_Write_Read	12

Figure 11.7.6: Final test result on an 8-inch wafer and resistance distribution over 4Mbit memory cells in a checkerboard pattern.

ISSCC 2011 / SESSION 11 / NON-VOLATILE MEMORY SOLUTIONS / 11.8

11.8 A 7MB/s 64Gb 3-Bit/Cell DDR NAND Flash Memory in 20nm-Node Technology

Ki-Tae Park, Ohsuk Kwon, Sangyong Yoon, Myung-Hoon Choi,
In-Mo Kim, Bo-Geun Kim, Min-Seok Kim, Yoon-Hee Choi,
Seung-Hwan Shin, Youngson Song, Joo-Yong Park, Jae-Eun Lee,
Chang-Gyu Eun, Ho-Chul Lee, Hyeong-Jun Kim, Jun-Hee Lee,
Jong-Young Kim, Tae-Min Kweon, Hyun-Jun Yoon, Taehyun Kim,
Dong-Kyo Shim, Jongsun Sel, Ji-Yeon Shin, Pansuk Kwak,
Jin-Man Han, Keon-Soo Kim, Sungsoo Lee, Young-Ho Lim,
Tae-Sung Jung

Samsung Electronics, Hwasung, Korea

Recently, the demand for 3b/cell NAND flash has been increasing due to a strong market shift from 2b/cell to 3b/cell in NAND flash applications, such as USB disk drives, memory cards, MP3 players and digital still cameras that require cost-effective flash memory. To further expand the 3b/cell market, high write and read performances are essential [1]. Moreover, the device reliability requirements for these applications is a challenge due to continuing NAND scaling to sub-30nm pitches that increases cell-to-cell interference and disturbance. We present a high reliability 64Gb 3b/cell NAND flash with 7MB/s write rate and 200Mb/s asynchronous DDR interface in a 20m-node technology that helps to meet the expanding market demand and application requirement.

Figure 11.8.1 shows the micrograph and device feature of the 64Gb 3b/cell DDR NAND flash memory chip, fabricated in 20nm-node CMOS technology. The chip has two 32Gb memory planes. Each plane consists of 2732 blocks with 8KB page size and 1.5MB block size. The block consists of 64-cell strings with 2-dummy WLs to reduce NAND string overhead and abnormal disturbance [2]. To realize a small die area with 65.3% cell efficiency and 200Mb/s high-speed DDR interface in the 3b/cell NAND chip, a one-sided page buffer structure, 2-way interleaving and 2-stage pipeline architecture [3] are used. To further reduce chip size, a shared block decoder scheme is used, as shown in Fig. 11.8.2. Compared to a conventional block decoder where every block's WLs transfer gates have their own block decoder circuit [4], two different block WLs transfer gates are shared with single block decoder in this chip. By selecting the drive lines of WLs transfer gate, only the WLs of selected memory array are driven. Using the shared block decoder scheme reduces the row-decoder area by 25%, which results in 4.2% chip size reduction. In addition to that, 30% pump area reduction is also achieved because that total output loading of pump circuits is reduced compared to the conventional scheme. The results in a further 0.3% chip-size reduction. With simple bad-block-remapping logic in peripheral circuits, conventional bad-block replacement can be fully supported.

Controlling V_{th} distribution without performance degradation in 3b/cell NAND chip is a critical challenge at the 20nm-node. A 2-step verify ISPP scheme of programming [5] is helps to achieve a tighter V_{th} distribution width than that of conventional ISPP programming [6] that is used widely in MLC NAND flash. As shown in Fig. 11.8.3, the BL voltage of a programmed cell is raised from 0V to a predetermined low voltage during 1st step verify, which is slighty lower than the target verify level. Because it suppresses FN tunneling current at programming, it slows down a rate of V_{th} shift, realizing a tighter V_{th} distribution compared to the conventional ISPP scheme. The 2-step verify ISPP scheme, however, requires two times more verify operations for each target V_{th} state, causing an increase in program time. This is especially true for 3b/cell NAND, where over 66% of total program time is spent in the verify operation. In this chip, in order to remove the extra verify overhead time, a verify-skip 2-step tunneling ISPP scheme is introduced, as shown in Fig. 11.8.3. The verify-skip 2-step tunneling ISPP scheme uses the 2nd verify level of the previous target state as the 1st step verify for next target state. To obtain 2-step tunneling phenomenon effectively, the time of forcing the BL voltage should be delayed to after a few program pulses are applied. This is performed by a counting data latch in the page buffer, when passing the verify level of previous target state, as shown in Fig. 11.8.4. Thus, a tight V_{th} distribution width using 2-step tunneling rate is realized without an extra verify operation. Compared to the conventional 2-step verify ISPP scheme, this scheme achieves 13% better program performance.

As the page size of NAND flash has increased up to 16KB in two-plane program mode, the SRAM buffer of the off-chip NAND controller has been also increased to store and manipulate the page data. Furthermore, additional SRAM buffer entries are typically required in the NAND controller to backup the page data that is already transferred to NAND flash chip. This is because when a page program is not performed successfully, the page data stored in the page buffer of the NAND chip is lost. To ensure correct program operation, the NAND controller retains the original page data when it re-sends the page data to NAND chip to complete the page program successfully. The increased page size becomes a critical issue for NAND controllers in terms of die size and power consumption. It results in lower performance and cost efficiency of the embedded flash system. In our 3b/cell NAND chip, a fail-page copy-back program function is supported in order to solve this issue. Figure 11.8.5 shows the fail-page copy-back program implemented in the 3b/cell NAND chip. In the NAND chip, the page data that is transferred from NAND controller is loaded into cache-latch (LAT<c>) in the page buffer. Then, the page data is transferred to a data latch (LAT<n>) that works for program verification and to another data latch (LAT) that works as a storage latch to maintain the original page data. If page program fails, the stored original page data is transferred from the LAT<n> to the LAT and then the page program operation is re-invoked with a different target address. Additional data latches for backing up the page data are not required because that the LAT is also used for intelligent page copy program function [7].

Figure 11.8.6 shows a measured 3b/cell V_{th} distribution of the fabricated chip. The verify-skip 2-step tunneling ISPP scheme and parallel program using dual-pulse ISPP for 2nd page program [3,8] result in a tightly controlled 8-level 3b/cell V_{th} distribution. Furthermore the 3b/cell device reliability can be extended with a read-retry scheme by using moving read level which is controlled by off-chip NAND controller. In addition to high reliability, 7MB/s write performance and 200Mb/s read performance enable the 3b/cell DDR NAND chip enable to replace a most of 2b/cell NAND applications mentioned above.

Acknowlegement:
The authors specially appreciate Keonho Lee, Youngwook Jeong, Dongkyoo Park, Jaesun Yun, Sejun Park, Suk-Kang Sung, Choong-Ho Lee, Jinhyun Shin, Hyunchul Back and also appreciate flash design team, flash product engineering team, CAE team, process and device engineering teams.

References:
[1] Y. Li, et al., "A 16Gb 3b/Cell NAND Flash Memory in 56nm with 8MB/s Write Rate", *ISSCC Dig. Tech. Papers*, pp. 506-507, Feb., 2008.
[2] K.-T. Park et al., "Scalable Wordline Shielding Scheme using Dummy Cell beyond 40nm NAND Flash Memory for Eliminating Abnormal Disturb of Edge Memory Cell", *Ext. Abst. Of SSDM*, pp. 298-299, Sep. 2006.
[3] H. Kim et al., "A 159mm² 32nm 32Gb MLC NAND-Flash Memory with 200MB/s Asynchronous DDR Interface", *ISSCC Dig. Tech. Papers* , pp. 442-443, Feb., 2010.
[4] T. Futatsuyama, et al., "A 113mm² 32Gb 3b/cell NAND Flash Memory", *ISSCC Dig. Tech*. Papers, pp. 242-243, Feb., 2009.
[5] T. Tanaka, et al., US Patent, US6643188, 2003
[6] K.-D. Suh et al., "A 3.3V 32Mb NAND Flash Memory with Incremental Step Pulse Programming Scheme", *ISSCC Dig. Tech. Papers*, pp. 128-129, Feb., 1995.
[7] K. Takeuchi et al., "A 56nm CMOS 99mm² 8Gb Multi-level NAND Flash Memory with 10MB/s Program Throughput", *ISSCC Dig. Tech. Papers*, pp. 144-145, Feb., 2006.
[8] K.-T. Park et al., "A Zeroing Cell-to-Cell Interference Page Architecture With Temporary LSB Storing and Parallel MSB Program Scheme for MLC NAND Flash Memories", *IEEE J. Solid-State Circuits*, vol. 43, no. 4, pp. 919-928, Nov. 2008.

978-1-61284-303-2/11 $26.00 © 2011 IEEE

ISSCC 2011 / February 22, 2011 / 11:45 AM

Bits per Cell	3
Density	64Gb
Technology	20nm-node Triple-well 3-Metal CMOS
Organization	8kB × 192 pages × 5464 blocks × 8
Program Performance	7MB/s
Read Performance	200Mbps (Asyn-DDR)
Erase Performance	1.5ms
Power Supply	2.7V ~ 3.6V / 1.7V ~ 1.9V

Figure 11.8.1: Die micrograph and key features.

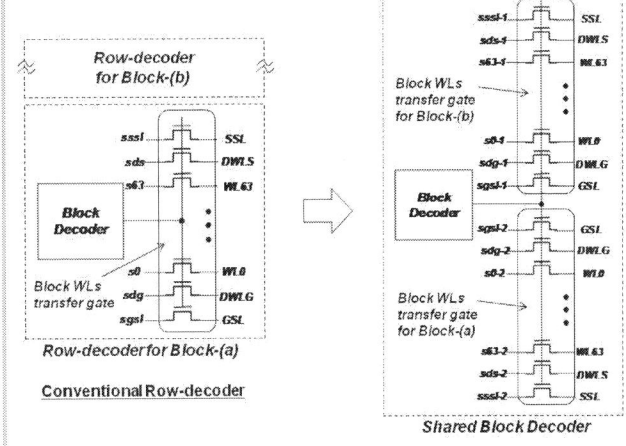

Figure 11.8.2: Shared block decoder scheme.

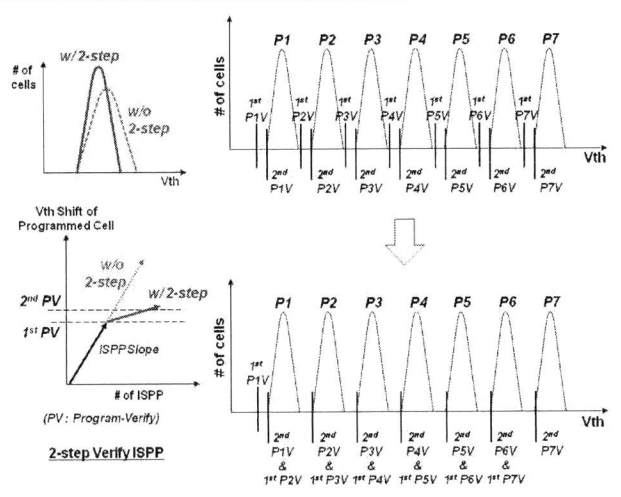

Figure 11.8.3: Verify-skip 2-step tunneling ISPP.

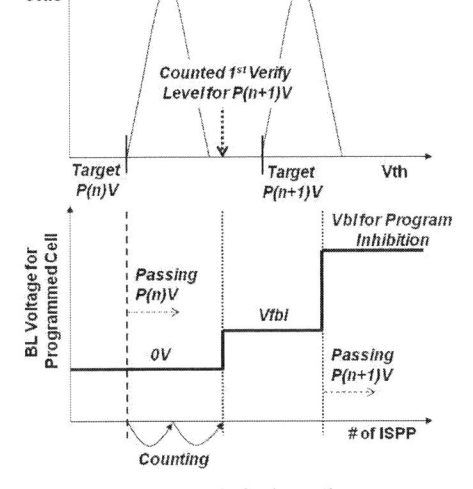

Figure 11.8.4: Forcing BL voltage by latch counting.

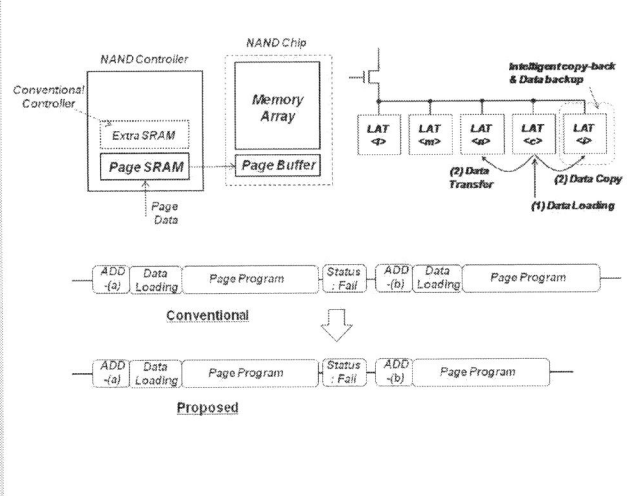

Figure 11.8.5: Fail-page copy-back program.

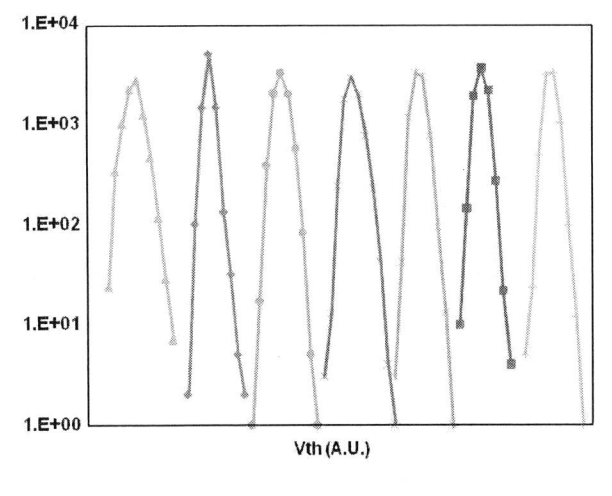

Figure 11.8.6: Measured 8-level Vth distribution.

978-1-61284-303-2/11 $26.00 © 2011 IEEE

ISSCC 2011 / SESSION 12 / DESIGN IN EMERGING TECHNOLOGIES / OVERVIEW

Session 12 Overview / *Technology Directions*

Design in Emerging Technologies

Session Chair: *Azeez Bhavnagarwala, Advanced Micro Devices, Boston MA*

Session Co-Chair: *Tadahiro Kuroda, Keio University, Yokohama, Japan*

New advances in designing in emerging technologies of energy harvesting, smart power grid, silicon photonics, terahertz imaging and power management are reported in this session.

Harvesting energy from the environment by using thermoelectric generators or photovoltaic cells provides a solution for battery-free sensor networks or electronic healthcare systems. The power-management circuit converting harvested energy to practical supply voltage is in focus and one of the main challenges is the extremely low voltage that the energy harvesting generators can provide. The session begins with paper 12.1 from the University of Tokyo on a sub-100mV startup voltage step-up converter to a 0.9V output with 72% conversion efficiency without any external clock or mechanical assist in 65nm CMOS.

A smart meter is an enabling technology for realizing the smart grid. In order to further reduce the energy loss in the power grid rather than simply replacing conventional power meters with smart meters, an extremely fine-grain power monitoring system is desirable. However, this requires an enormous number of low-cost power meters. Printable organic devices on flexible films have great potential to realize low-cost power meters. In Paper 12.2, also from the University of Tokyo, a low-cost 100V AC power meter based on system-on-a-film (SoF) concept is demonstrated.

While energy management systems in homes or buildings may offer monitoring, reporting, and controlling of energy usage, these functions simply track intermittently the energy consumption and operational states of appliances and equipment. Further, since their sensing devices are not small, they cannot be attached to every appliance or piece of equipment. Paper 12.3 from NEC reports a battery-free realtime current waveform sensor in 90nm CMOS. It is small enough to be attached to any piece of equipment for continuously power consumption monitoring. Its key features are plug-less energy-harvesting (EH) from AC power lines, real-time sensing and transmitting. A half cycle time-to-digital convertor (TDC) is used to detect detailed information on current waveforms with the precise time-resolution and asymmetrical power consumption to help achieve high power RF transmission with only a limited EH power supply. It successfully demonstrates transmission of sensed current waveform information from several appliances such as a hair dryer, a TV, and a notebook PC.

The emerging field of silicon photonics targets monolithic integration of optical components in the CMOS process, potentially enabling high bandwidth, high density interconnects with dramatically reduced cost and power dissipation. Paper 12.4 from the IBM TJ Watson Research Center reports a monolithic 4×4 silicon photonic router with a custom 90nm bulk CMOS driver, routing 3×40Gb/s WDM data with a BER <10^{-12}, less than -10dB cross-talk and 7dB loss.

The spatial resolution of micro-wave and millimeter-wave imaging systems can be improved by increasing their operating frequencies into the submillimeter-wave range (300GHz to 3THz). Electronic terahertz sources and receivers are presently dominated by III-V semiconductor and waveguide packaging technologies. In Paper 12.5, a 0.25μm BiCMOS process with an upgraded SiGe HBT device technology is used to demonstrate an 820GHz TX/RX chipset for active terahertz imaging applications.

Standby power from household electric appliances driven by AC power such as TVs is lost at an AC-DC converter even when the appliances are turned off. Although an infrared remote controller is usually used in these appliances, no wake-up circuit has been reported for the infrared remote controller. Paper 12.6 proposes a receiver SoC in 0.13μm CMOS with wake-up function for the infrared remote controller.

978-1-61284-303-2/11 $26.00 © 2011 IEEE

ISSCC 2011 / February 22, 2011 / 1:30 PM

Programmable devices such as SRAM-based FPGAs have the major challenges of power consumption and circuit area due to the excessive standby leakage current and a large footprint memory cell. In Paper 12.7 a programmable cell array and a 32×32 crossbar switch using a nonvolatile and rewritable solid-electrolyte switch (NanoBridge) on a 90nm CMOS platform is reported. A 72% reduction in chip-area compared with that of a standard-cell-based design is also achieved.

Wafer-level testing detects manufacturing problems and removes non-functional devices early in the fabrication process. It is performed by placing a probe card directly above a device under test (DUT) and establishing a mechanical contact between them with an array of probes. This is an invasive technique that may cause damage to fragile low-k dielectric layers and deformation of pads or bumps. One solution is wireless probing. With a number of proposed techniques for establishing high-speed inductive-coupling data links and measuring DC analog signal wirelessly, the primary obstacle to non-contact wafer-level testing is supplying power to the DUT. Paper 12.8 proposes an inductive power transfer system in 0.18μm CMOS, capable of delivering up to 6W of DC power to the on-chip load.

GHz-range applications that operate in a variety of signal situations and/or multiple standards require highly programmable responses that cannot be provided by analog circuits. Paper 12.9 enables a five-orders-of-magnitude improvement in frequency capability compared to earlier work, using 65nm CMOS, making continuous time DSP a candidate for wideband GHz low-dynamic range applications, such as those found in pulse radio, spectrum sensing, and channel equalization.

Presenters:

1:30 PM

12.1 A 95mV-Startup Step-Up Converter with V$_{TH}$-Tuned Oscillator by Fixed-Charge Programming and Capacitor Pass-On Scheme

P-H. Chen, University of Tokyo, Tokyo, Japan

2:00 PM

12.2 100V AC Power Meter System-on-a-Film (SoF) Integrating 20V Organic CMOS Digital and Analog Circuits with Floating Gate for Process-Variation Compensation and 100V Organic PMOS Rectifier

K. Ishida, University of Tokyo, Tokyo, Japan

2:30 PM

12.3 Real-Time Current-Waveform Sensor with Plugless Energy Harvesting from AC Power Lines for Home/Building Energy-Management Systems

S. Takahashi, NEC, Kawasaki, Japan

3:15 PM

12.4 A 3.9ns 8.9mW 4×4 Silicon Photonic Switch Hybrid Integrated with CMOS Driver

A. Rylyakov, IBM T. J. Watson Reseach Center, Yorktown Heights, NY

3:45 PM

12.5 A 820GHz SiGe Chipset for Terahertz Active Imaging Applications

E. Öjefors, University of Wuppertal, Wupppertal, Germany

4:15 PM

12.6 A 130μA Wake-Up Receiver SoC in 0.13μm CMOS for Reducing Standby Power of An Electric Appliance Controlled by An Infrared Remote Controller

H. Ishihara, Toshiba, Kawasaki, Japan

4:30 PM

12.7 Programmable Cell Array Using Rewritable Solid-Electrolyte Switch Integrated in 90nm CMOS

M. Miyamura, NEC, Sagamihara, Japan

4:45 PM

12.8 6W/25mm^2 Inductive Power Transfer for Non-Contact Wafer-Level Testing

A. Radecki, Keio University, Yokohama, Japan

5:00 PM

12.9 GHz-Range Continuous-Time Programmable Digital FIR with Power Dissipation that Automatically Adapts to Signal Activity

M. Kurchuk, Columbia University, New York, NY

978-1-61284-303-2/11 $26.00 © 2011 IEEE

ISSCC 2011 / SESSION 12 / DESIGN IN EMERGING TECHNOLOGIES / 12.1

12.1 A 95mV-Startup Step-Up Converter with V_{TH}-Tuned Oscillator by Fixed-Charge Programming and Capacitor Pass-On Scheme

Po-Hung Chen[1], Koichi Ishida[1], Katsuyuki Ikeuchi[1], Xin Zhang[1],
Kentaro Honda[1], Yasuyuki Okuma[2], Yoshikatsu Ryu[2],
Makoto Takamiya[1], Takayasu Sakurai[1]

[1]University of Tokyo, Tokyo, Japan,
[2]Semiconductor Technology Academic Research Center, Yokohama, Japan

Harvesting energy from the environment by using a thermoelectric generator (TEG) or photovoltaic cells provides a solution for battery-free sensor networks or electronic healthcare systems. In these systems, the harvested energy is supplied at a very low voltages, requiring a low-startup-voltage power circuit for kick-start from low voltage. A previous sub-100mV-startup-voltage boost converter [1] was implemented by using a mechanically assisted step-up process that needs vibration at startup and the application is rather limited. In this paper, a 95mV startup voltage step-up converter without any mechanical stimulus extends the applicability of energy harvesting. The circuit converts a 100mV input to a 0.9V output with 72% conversion efficiency without any external clocks or mechanical switches. A capacitor pass-on scheme eliminates an additional external output capacitor that functions only at the startup.

The minimum startup voltage of a step-up DC-DC converter in standard CMOS technology is limited by the oscillator since a clock signal is required for a charge pump (CP). Practically, the minimum supply voltage of an oscillator (V_{DDMIN}) is limited by PMOS-NMOS V_{TH} unbalance caused by within-die and die-to-die V_{TH} variations. The within-die V_{TH} variation issue can be solved by increasing the channel width and thus it is easily solved in this application. The problem is the die-to-die variation, which can usually be adjusted by controlling the body bias of the MOSFET. In the startup circuit, however, the body-biasing circuit is not functional since the body-biasing circuit needs another oscillator, which does not function below V_{DDMIN}. To solve this problem, post-fabrication V_{TH} programming is applied to the oscillator.

Figure 12.1.1 compares the conventional and our step-up converter architectures. In the conventional approach, two external capacitors are required. One (C_{OUT1}) is used in the charge-pumping startup circuit and the other (C_{OUT2}) is for the boost converter. C_{OUT1} is needed to separate the boost converter from the CP at the start-up. Otherwise the start-up CP needs to drive the boost converter control circuit, which is seen as leakage current (I_{Leak}) for the CP. In our scheme, C_{OUT1} is eliminated by using the output capacitor of the booster, C_{OUT2}, as the charge buffer of the CP as shown in Fig. 12.1.1(b). Once the C_{OUT} is charged without leakage, C_{OUT} is "passed on" with its charge inside to the boost converter. By doing so, the large external capacitor of C_{OUT1} is eliminated.

The detailed block diagram of the proposed step-up converter and the timing chart are shown in Fig. 12.1.2. The capacitor pass-on controller contains a CP for PMOS super cut-off (PSC) and two voltage detectors D1 and D2. The CP for PSC provides the overdrive voltage to the gate of PMOS switch and reduces the leakage current. At startup, C_{OUT} is charged by the CP and V_{START} rises. When V_{CP1} (=V_{START}) is pumped to the preset trip voltage of D1, the node Y changes from low to high to pass on the C_{OUT} to the boost converter and C_{OUT} becomes the output capacitor of the boost converter. The charge stored in C_{OUT} activates the clock generator for the boost converter and start driving the boost converter. If the load is connected at the same time when the capacitor is passed on, the charge stored in C_{OUT} drains out and boost converter fails to start up. To overcome this problem, a voltage detector D2 with 60mV higher trigger voltage comparing to D1 is added to delay the load connection timing a bit. This ensures the output switch M2 to turn on after the boost converter starts working and the output is boosted smoothly.

Since the detector D1 has to be connected to C_{OUT} at startup, it acts as the load of the CP. Thus the power consumption of the detectors should be minimized. The circuit schematic of the voltage detector and its simulated waveforms are shown in Fig. 12.1.3. It consumes less than 1.6nW when $V_{SENSE} < V_{Trigger}$. To

reduce the power consumption, the gate of M_{D3} is connected to its source and only off-current flows. V_{DETECT} changes sharply from low to high when drain current of M_{D1} and M_{D2} get larger than off-current of M_{D3} exponentially. The trigger voltage can be written as (1) in Fig. 12.1.3, which can be tuned by designing the size of M_{D2} and M_{D3}. M_{D1} is applied to almost double the trigger voltage without using unreasonable size ratio of M_{D2} and M_{D3}.

To make sure the circuit functions even under the die-to-die process variation, the V_{TH} of transistors in the ring oscillator, which generates the CP clocks, are trimmed by fixed-charge programming, as shown in Fig. 12.1.4. The upper part shows the detailed circuit schematic when post-fabrication V_{TH} programming is in progress. Since the V_{TH} of NMOS, V_{TN}, was higher than that of PMOS, V_{TP}, in the measured chips, V_{TP} is increased to reduce the V_{DDMIN} of the oscillator here. The V_{TP} is programmed by applying the high reverse body bias to all PMOSFETs (V_{NWELL}=8.5V, V_{PWELL}=0) when the ring oscillator is oscillating under 1V power supply. If V_{TN} programming is required, the high reverse voltage to V_{PWELL} and 1V supply to V_{NWEL} can be applied. In the PMOS programming, holes are injected into the gate oxide and the fixed charge changes V_{TP}. To verify V_{DDMIN} improvement by this post-fabrication V_{TH}-trimming, three test chips of 15-stage ring oscillators are measured. Figures 12.1.4(b) and 12.1.4(c) show V_{DDMIN} change under stress and after stress. V_{DDMIN} is improved (decreased) by 34% compared with the initial value. In Fig. 12.1.4(c), the measured dependence of V_{DDMIN} after stress is shown. The change is as small as 1mV after 3 days of stress and gets stable.

The step-up converter containing the startup circuit with V_{TH}-trimmed oscillator is realized in 65nm CMOS technology. The chip micrograph is shown in Fig. 12.1.7 with 0.17mm^2 active area. Fig. 12.1.5(a) shows measured waveforms of the step-up converter with the capacitor pass-on mechanism. The circuit is measured under 95mV input voltage without any external clocks or mechanical switches. The CP for startup pumps up the 10nF external capacitor C_{OUT} and passes on the capacitor to the boost converter after 262ms. The power efficiency versus the current delivered to the load under 100mV input is shown in Fig. 12.1.5(b). It provides maximum efficiency of 72% at 1.5mA output current with 0.9V output voltage.

The performance comparison between our circuit and the state-of-the-art step-up converters applying startup mechanism is shown in Fig. 12.1.6. The proposed circuit achieves the lowest startup voltage except for reference [1], which requires mechanical vibration to assist the startup. The proposed circuit also demonstrates the feasibility of countermeasures for the process variation that limits the minimum startup voltage of an on-chip ring oscillator.

Acknowledgement:
This work was carried out as a part of the Extremely Low Power (ELP) project supported by the Ministry of Economy, Trade and Industry (METI) and the New Energy and Industrial Technology Development Organization (NEDO).

References:
[1] Y. K. Ramadass and A. P. Chandrakasan, "A Batteryless Thermoelectric Energy-Harvesting Interface Circuit with 35mV Startup Voltage," *IEEE ISSCC Dig. Tech. Papers*, pp. 486-487, Feb 2010.
[2] E. Carlson et al., "20mV input Boost Converter for Thermoelectric Energy Harvesting," *IEEE J. Solid-State Circuits*, vol. 45, pp. 741-750, Apr., 2010.
[3] P. Chen et al., "0.18-V Input Charge Pump with Forward Body Biasing in Startup Circuit using 65nm CMOS," *IEEE Custom Integrated Circuit Conference*, pp.239-242, Sep. 2010.
[4] S. Matsumoto et al., "A Novel Strategy of a Control IC for Boost Converter with Ultra Low Voltage Input and Maximum Power Point Tracking for Single Solar Cell Application," *International Sym. on Power Semiconductor Devices & ICs*, pp. 180-183, June 2009.

978-1-61284-303-2/11 $26.00 © 2011 IEEE

ISSCC 2011 / February 22, 2011 / 1:30 PM

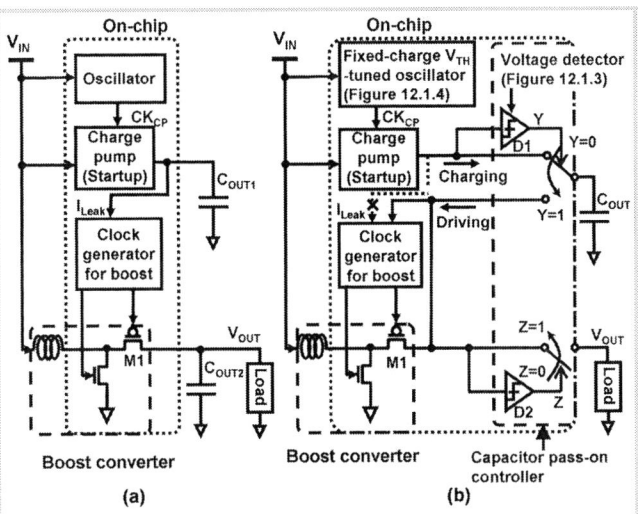

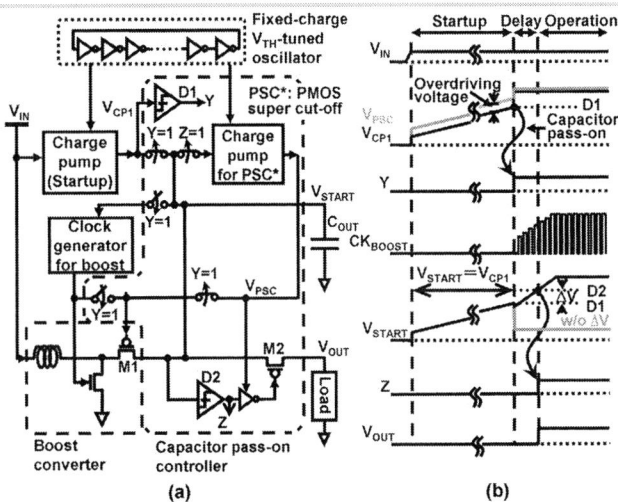

Figure 12.1.1: The system architecture of (a) the conventional and (b) the proposed startup mechanism in step-up DC-DC converter.

Figure 12.1.2: (a) The detailed block diagram of the proposed step-up converter. (b) Timing chart illustrating the circuit operation sequences.

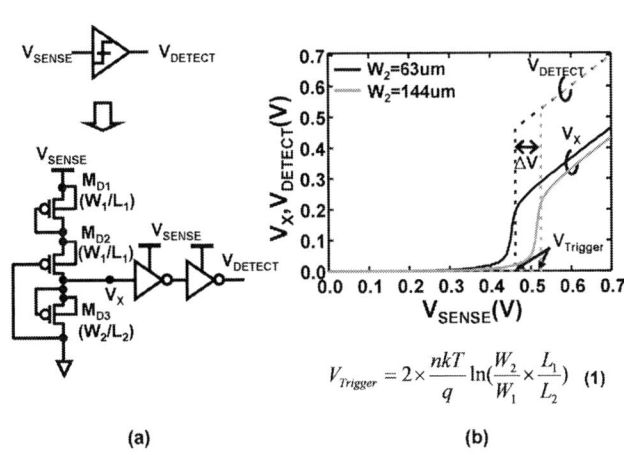

$$V_{Trigger} = 2 \times \frac{nkT}{q} \ln(\frac{W_2}{W_1} \times \frac{L_1}{L_2}) \quad (1)$$

Figure 12.1.3: (a) Circuit schematic of 1.6-nW voltage detector (b) Simulated waveforms of the proposed voltage detector.

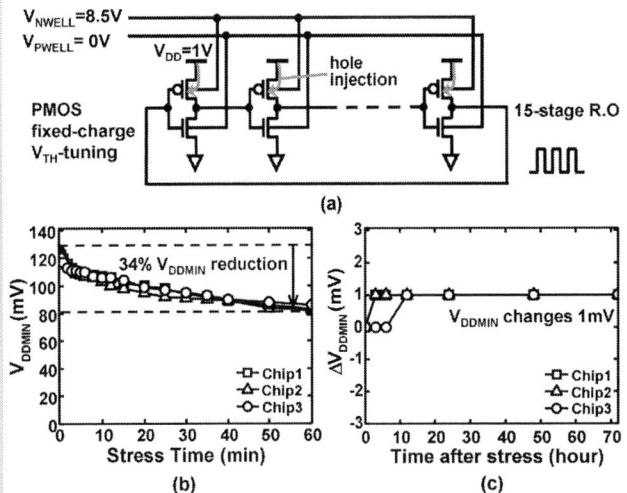

Figure 12.1.4: (a) Circuit schematic of V_{TH}-tuned oscillator. (b) Measured V_{DDMIN} change under stress. (c) Measured V_{DDMIN} change after stress.

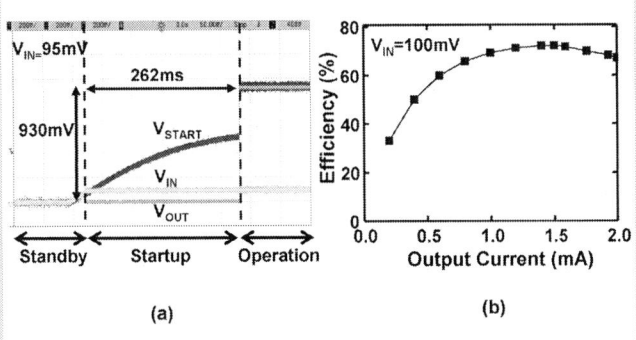

Figure 12.1.5: (a) Measured startup waveforms of the proposed step-up converter. (b) Measured conversion efficiency versus output current.

	Startup mechanism	Min. Startup Voltage	Typical Input Voltage (V_{IN})	Output Voltage @V_{IN}	Peak Efficiency @V_{IN}	Process
[1]	Mechanical switch	35mV	50mV	1.8V	58%	350-nm CMOS
[2]	External Voltage	650mV	100mV	1V	75%	130-nm CMOS
[3]	Charge pump	180mV	180mV	0.74V	N/A	65-nm CMOS
[4]	Charge pump	360mV	500mV	7V	82%	350-nm SOI-BCD
This Work	Charge pump with V_{TH}-tuned oscillator	95mV	100mV	0.9V	72%	65-nm CMOS

Figure 12.1.6: Comparison of recently published step-up converter applying startup mechanism.

12

978-1-61284-303-2/11 $26.00 © 2011 IEEE

ISSCC 2011 PAPER CONTINUATIONS

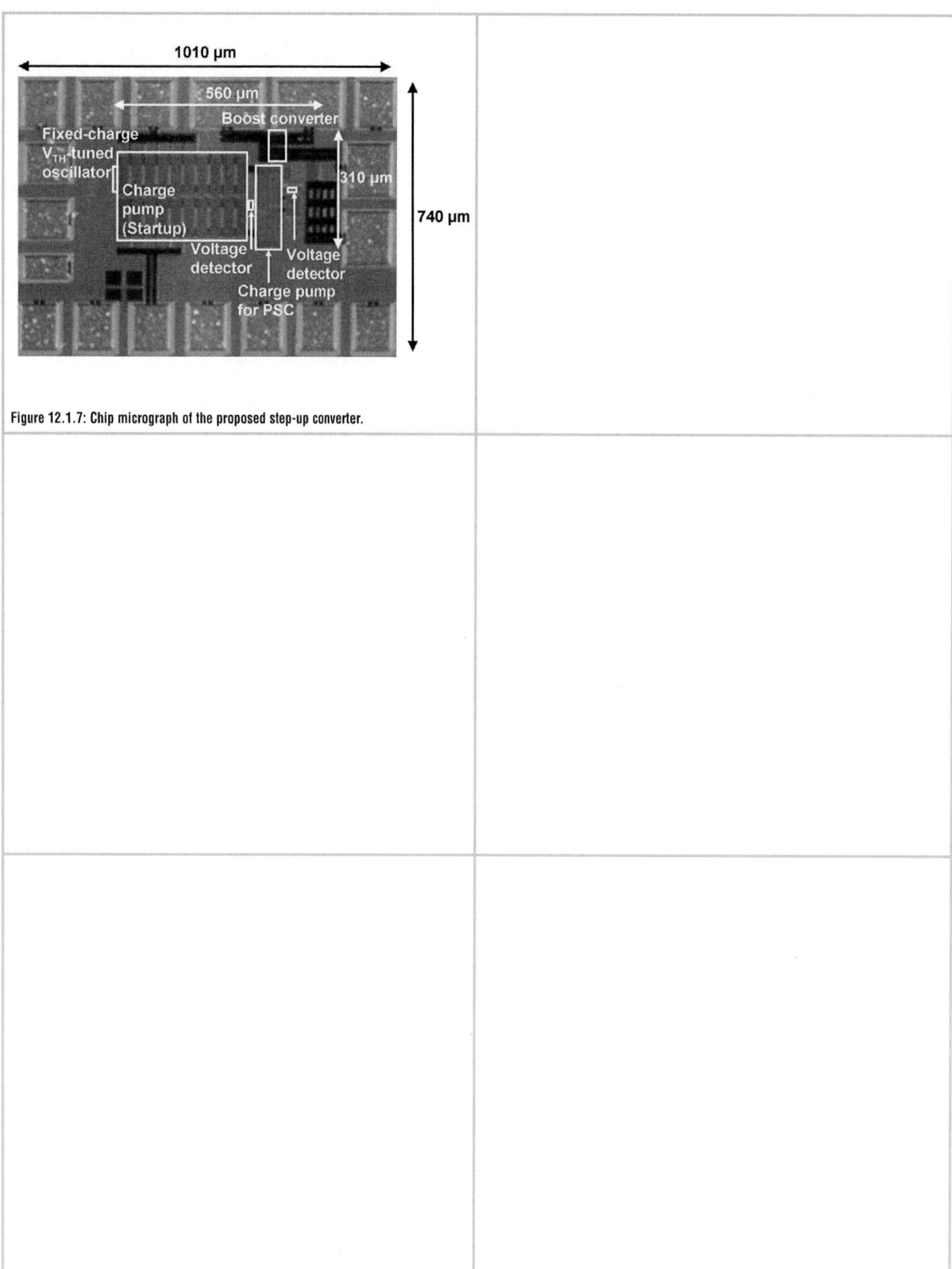

Figure 12.1.7: Chip micrograph of the proposed step-up converter.

ISSCC 2011 / SESSION 12 / DESIGN IN EMERGING TECHNOLOGIES / 12.2

12.2 100V AC Power Meter System-on-a-Film (SoF) Integrating 20V Organic CMOS Digital and Analog Circuits with Floating Gate for Process-Variation Compensation and 100V Organic PMOS Rectifier

Koichi Ishida[1], Tsung-Ching Huang[1], Kentaro Honda[1],
Tsuyoshi Sekitani[1], Hiroyoshi Nakajima[2], Hiroki Maeda[2],
Makoto Takamiya[1], Takao Someya[1], Takayasu Sakurai[1]

[1]University of Tokyo, Tokyo, Japan,
[2]Dai Nippon Printing, Chiba, Japan

A smart meter is essential for realizing the smart grid. In order to further reduce the energy loss in the power grid, an extremely fine-grain power monitoring system is desirable and it will require an enormous number of low-cost power meters. Existing power meters, however, do not meet the cost and size requirements. On the other hand, organic devices on flexible films have great potential to realize low-cost power meters. In this paper, a 100-V AC power meter based on System-on-a-Film (SoF) concept is demonstrated.

Figure 12.2.1 shows the photograph of the proposed 100-V AC power meter on a flexible film, including: (a) analog circuits composed of a 20-V organic CMOS opamp for AC current sensing, (2) logic circuits composed of a 20-V organic CMOS frequency divider for integrating the measured current, (3) AC-to-DC power converter composed of a 100-V organic PMOS rectifier to generate 20-V DC power for the power meter, (4) an OLED [1] bar indicator, and (5) an AC connector inserted between the power plug and the AC outlet are fully integrated on a 200×200mm^2 flexible film. The entire sheet can be folded and the total size of the proposed AC power meter can be shrunk to 70×70mm^2. In this work, the system dimensions are mainly determined by the design-rule of the organic transistors and can be further reduced by scaling of the design-rule. Figure 12.2.2 shows the block diagram of the proposed 100-V AC power meter. The measured 100V 50Hz AC load current i_L is first converted into the sense voltage (v) by means of the sense resistor (R). The converted sense voltage v is then amplified by the amplifier and rectified into V_{SENSE}, which is compared with the triangular waveform (V_{TRI}) by the comparator. The output of the comparator enables or disables the 10-bit counter. five most-significant bits in the counter are connected to the OLED bar indicator. To get the accumulated results, the maximum integration time of the power meter is designed to be 43min. The 0.05-Hz clock for the input of the triangular waveform generator and the 0.4-Hz clock for the counter are generated by a 10-bit frequency divider, for which the clock is generated by a half-wave rectifier from 100-V 50-Hz AC signal. The required DC power for the power meter is provided by converting the 100-V 50-Hz AC power into 20-V DC power by the full-wave rectifier.

We implement a full-wave rectifier using 100-V organic PMOS. The current consumption of the system, mainly consumed by the 5-digit OLED bar indicator, is around 2mA. Since the driving capability of organic NMOS is weaker than that of PMOS by an order, we choose an all-PMOS full-wave rectifier. In a typical rectifier as shown in Fig. 12.2.3(a), each PMOS operates at the pinch-off region. To increase the output current of the rectifier, two PMOS diodes are replaced with a pair of cross coupled PMOS operating in the linear region as shown in Fig. 12.2.3(b). Both two rectifiers are implemented by PMOSs with gate length and width of 20μm and 100mm, respectively, which can supply up to 2-W DC power, the highest power level ever reported. Figure 12.2.3 (c) shows the comparison of the rectifiers. While the PMOS diode rectifier supplies 2.1-mA output current at 20V, the cross-coupled PMOS rectifier can increase the output current by 24%. Figure 12.2.3(d) shows the measured waveform of the 21.9-V output voltage of the cross-coupled PMOS rectifier.

In our 20-V CMOS, DNTT-based PMOS has 8 times higher carrier mobility than NTCDI-based NMOS [2, 3]. In addition, our CMOS inverter gain was only 3.2 at 20V and this leads to functional errors in the large scale logic circuits. To solve the problem, we designed the frequency divider with high-gain Pseudo-CMOS inverters [4]. The Pseudo-CMOS inverter uses only PMOS. The gain of 148, static noise-margin of 6.7V, and 156-Hz oscillation frequency of a 3-stage ring-oscillator can be achieved at 20V supply voltage. Figure 12.2.4 shows the schemat-

ics and measured waveform of the Pseudo-CMOS inverter and the proposed frequency divider. In the divider, NMOSs are used only for transmission gates, in which high gain is not required. Thanks to high gain Pseudo-CMOS, the divider successfully operates at 50Hz and 20V. In the frequency divider, the dynamic slave latch, which consists of only an inverter and the parasitic capacitance as the charge keeper, is used to reduce the number of transistors.

A major challenge in organic analog circuit design is to compensate for large process variations. The offset voltage in the differential pair of the amplifier due to the device mismatch should be reduced to lower than the sense voltage generated by the sense resistor R in Fig. 12.2.2. Some variation compensation techniques using back gate biasing [5, 6] were presented. However, the back gate voltages of each organic device should be biased throughout the operation time. To tackle this problem, we employ floating gate (FG) PMOS technology [7] to compensate for the process variations in the input stage of the opamp. The cross section of the device with FG is shown in Fig. 12.2.5(a). By applying -60V 100ms-width pulses to the gate terminal as shown in Fig. 12.2.5(b), holes can be injected into FG, which increases V_{TH} of the FG-PMOS. The key advantages include: 1) the variation compensation can be carried out by single high voltage source with the fixed voltage, 2) the controllability is provided by varying the voltage pulse width and the number of pulses, and 3) once the variation compensation is completed, no external DC voltage sources are required throughout the operation time. As shown in Fig. 12.2.5(c), the input differential pair of the opamp is composed of two PMOS (M1, M2) with FG and five IO pads are added for the use of mismatch compensation.

Figure 12.2.6 shows measured results of the variation compensation. Figure 12.2.6(a) shows I_D-V_{GS} characteristics of M1 and M2 in Fig. 12.2.5(c). V_{TH} of M1 is monotonically shifted toward V_{TH} of M2 by increasing the compensation time. As the result, V_{OUT}-V_{IN} characteristics of the opamp can be modified as shown in Fig. 12.2.6(b). Figure 12.2.6(c) shows the dependence of differential voltage gain at V_{INP} = V_{INN} = 15V on the compensation time derived from Fig. 12.2.6(b). The initial gain of 2.7 can be raised to 4.9 by applying 2 pulses (=200ms). Figure 12.2.6(d) shows the dependence of V_{OUT} on the compensation time. V_{OUT} is shifted from 8.4V to 12.5V by applying 2 pulses. In this way, the performance of the opamp can be optimized. Figure 12.2.7 shows the photographs of organic circuits and summarizes key features. The AC power meter consists of 609 transistors and the total area excluding AC connector is 200 × 200mm^2 (unfolded form) or 70 × 70mm^2 (using form).

Acknowledgement:
This study was partially supported by JST/CREST and the Special Coordination Funds for Promoting and Technology. We also thank Prof. K. Takimiya, Hiroshima Univ. and Drs. H. Kuwabara, and M. Ikeda, Nippon Kayaku Co., Ltd. for high purity DNTT.

References:
[1] H. Nakajima et al., "Flexible OLEDs poster with gravure printing method," Society for Information Display 2005 Digest Vol.XXXVI, Book2, pp.1196-1199, May 2005.
[2] T. Yamamoto and K. Takimiya, "Facile Synthesis of Highly π-Extended Heteroarenes,
Dinaphtho[2,3-b:2',3'-f]chalcogenopheno[3,2-b]chalcogenophenes, and Their Application to Field-Effect Transistors," Journal of American Chemical Society,vol.129, no.8, pp. 2224-2225, Aug. 2007.
[3] H. E. Katz et al., "A Soluble and Air-stable Organic Semiconductor with High Electron Mobility," Nature, vol.404, pp478-481. Mar. 2000.
[4] T.-C. Huang et al., "Pseudo-CMOS: A Novel Design Style for Flexible Electronics," Design, Automation & Test in Europe Conference & Exhibition (DATE) 2010, pp. 154-159, Mar. 2010.
[5] M. Takamiya et al., "An Organic FET SRAM with Back Gate to Increase Static Noise Margin and its Application to Braille Sheet Display," IEEE Journal of Solid-State Circuits, Vol. 42, No. 1, pp. 93 - 100, Jan. 2007.
[6] H. Marien et al., "An Analog Organic First-Order CT ΔΣ ADC on a Flexible Plastic Substrate with 26.5dB Precision," ISSCC Dig. of Tech. Papers, pp.136-137, Feb. 2010.
[7] T. Sekitani et al., "Organic Nonvolatile Memory Transistors for Flexible Sensor Arrays," Science, vol.326, no.5959, pp.1516-1519, Dec. 2009.

978-1-61284-303-2/11 $26.00 © 2011 IEEE

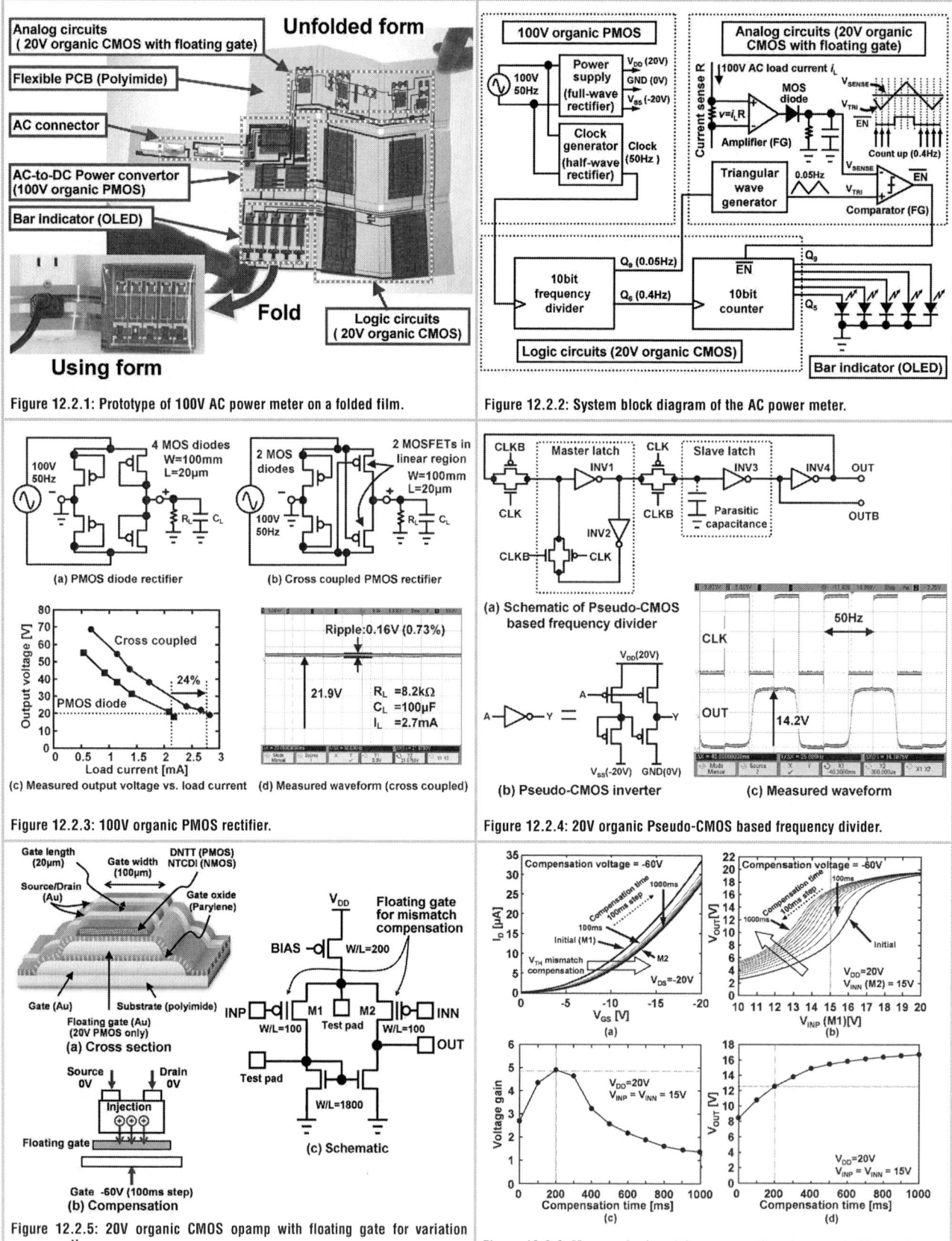

Figure 12.2.1: Prototype of 100V AC power meter on a folded film.

Figure 12.2.2: System block diagram of the AC power meter.

Figure 12.2.3: 100V organic PMOS rectifier.

Figure 12.2.4: 20V organic Pseudo-CMOS based frequency divider.

Figure 12.2.5: 20V organic CMOS opamp with floating gate for variation compensation.

Figure 12.2.6: Measured mismatch compensation of opamp in Fig. 12.2.5.

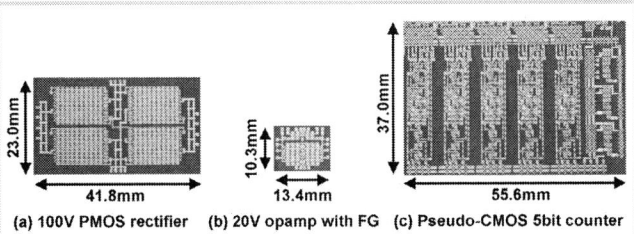

(a) 100V PMOS rectifier   (b) 20V opamp with FG   (c) Pseudo-CMOS 5bit counter

Organic transistors	
100V PMOS	Pentacene(0.5 cm²/Vs),
20V CMOS	PMOS:DNTT(0.7 cm²/Vs), NMOS:NTCDI (0.09 cm²/Vs)
Gate oxide material, thickness	Parelene, 560nm(100V), 360nm(20V), 240+240nm(20V+FG)
Compensation voltage	-60V(100ms step)
Organic LED	
Materials	TBADN, TCTA, TTPA
100V AC power monitor	
100V PMOS maximum power dissipation	2W (20mA@V_{GS}=-48V, V_{DS}=-100V)
Ring oscillator frequency of Pseudo-CMOS	156Hz@20V
Number of transistors	609Tr.
Total area excluding AC connector	200x200mm²(unfolded), 70x70mm²(folded)

Figure 12.2.7: Photographs and key features.

ISSCC 2011 / SESSION 12 / DESIGN IN EMERGING TECHNOLOGIES / 12.3

12.3 Real-Time Current-Waveform Sensor with Plugless Energy Harvesting from AC Power Lines for Home/Building Energy-Management Systems

Shingo Takahashi, Nobuhide Yoshida, Kenichi Maruhashi, Muneo Fukaishi

NEC, Kawasaki, Japan

Home/building energy-management systems (EMSs) driven by information technology are expected to be key to the achievement of an upgraded energy infrastructure, such as Smart Grid. While EMS may offer monitoring, reporting, and control of energy usage, these functions simply track intermittently the energy consumption and operational states of certain appliances and equipments. Further, since their sensing devices are not small, they cannot be attached to every appliance or piece of equipment. Some EMSs gather information, rather, at the power-distribution-board level or the multi-output-tap level. We report here a TX-integrated battery-free real-time current-waveform sensor. It is small enough to be attached to any piece of equipment for continuous power-consumption monitoring. Its key features are : 1) plugless energy harvesting (EH) from AC power lines, 2) real-time sensing and transmitting that use a half-cycle time-to-digital convertor (TDC) to detect detailed information on current-waveforms with the precise time-resolution, and 3) asymmetrical power consumption, which helps to achieve high-power RF transmission with a limited EH power supply. Our sensor provides the 1mW power supply from an EH unit, as well as the -5.5dBm RF output power at the 50kS/s sampling with a 1mW EH power supply. We also demonstrate transmission of sensed current-waveform information of several appliances such as a hair dryer, a TV, a notebook PC.

Figure 12.3.1 shows the TX-integrated current-waveform sensor and some application examples. The sensor is roughly the size of a 25-cent coin, and one in attached either to the AC power cable of an appliance to be monitored or inside the electrical outlet to which that appliance is connected. Its circuit utilizes the energy harvested from the AC power line to sample information on current-waveforms of individual appliances and to transmit sampling data to an EMS server. Each electrical appliance has a current-waveform signature that has a different frequency element, ranging from 50Hz to 2KHz under the IEC61000-3-2 standard, and the sampling data enables the EMS server to detect in real-time what appliances are being used.

Figure 12.3.2 shows an overall block diagram and a timing chart for the current-waveform sensor, which contains an energy-harvesting unit (EH unit), a sampling unit (ADC), a transmitter (TX), and a sampling control unit (SC unit). It samples the current-waveform of an AC power line and transmits the sampled data. The ADC unit employs a half-cycle TDC based on pulse-width modulation (PWM) [1,2] in a real-time sensing and transmitting. The time interval between the starting point of the ramp and the instant when the input signal crosses the ramp voltage is measured by the TDC, and the input signal amplitude is calculated from the measured time vector. On the other hand, at the intervals corresponding to those between the highest position and ending point of the ramp, the SC unit triggers the TX to send the ADC data. The EH unit obtains power via a current transformer (CT) connected to an AC power line and generates a supply voltage using a CMOS rectifier. In an EMS, this CT is also used to sense and transmit current-waveforms taken from individual appliances. Since a current-waveform is symmetrical with respect to the zero-cross position, the sampling period can be set to be within an upper half cycle from zero-cross position. The SC unit switches on-states between the ADC and the TX, and thus, the sampling of current-waveforms and the transmitting of sampling data are conducted alternately. The loss of data, such as that which an instantaneous power-outage might cause, can be prevented by making very short the period that the chip stores data before transmission.

Figure 12.3.3 illustrates the EH architecture and three conventional power supply schemes: AC-powered, battery-powered, and powered by energy harvested from radio waves. The architecture does not need additional components for sensing because the CT itself provides current-waveform information and power (see also Fig. 12.3.2). The AC-powered scheme requires a voltage transformer (VT), an AC-DC converter, and a large capacitor (around 1mF at 50/60Hz) to flatten and smooth the DC power sufficiently for use in other circuits. The battery-powered scheme has the disadvantage that battery size often restricts flexibility and makes the size of a sensor device too large. Moreover, the need for battery replacement is also a drawback. EH from radio waves, like RFID, has the disadvantage of needing an antenna and a storage capacitor of as much as 1μF [3]. In addition, a number of factors limit the amount of power harvested from radio waves, e.g., distance from radio wave source, source frequency, interference from physical of obstacles, etc. These may make it difficult to generate a sufficient power supply at any given time. Our power supply is synchronized to the period of current-waveform sampling, and power only needs to be supplied to the chip.

Figure 12.3.4 shows our asymmetry power management (APM) scheme for high-power RF transmission, employing a relaxation oscillator that switches between the ADC and the TX and generates triangle waveforms for PWM reference signal to the ADC. When the TX is transmitting sampling data from the ADC, power consumption in the SC unit should be as low as possible because the RF output power of the TX should be as high as possible. In conventional sampling-timing generation, which was composed of a clock generator and a frequency divider, power consumption remains constant. However, in the APM, the relaxation oscillator based on a Schmitt trigger circuit makes it possible to reduce power consumption when the TX is active over that when the ADC is active, which makes it possible to provide more RF power during TX periods.

Figure 12.3.5 shows measurement results. The left-hand side graph in Fig. 12.3.5 illustrates the AC characteristics of the EH unit. The lower-left-graph shows measured output voltage to the sensor circuit vs. the frequency of simulated current-waveform swept from 10Hz to 2MHz. This result shows that the harvested power is able to be successfully supplied to our sensor so that all electrical equipments connected to an AC power line are in conformity with the IEC61000-3-2 standard. With our design, energy can be harvested from AC power lines, and no batteries are needed for current sensing operation. The lower-right-hand graph in Fig. 12.3.5 shows the measured RF output power of the TX with APM. The RF power is sufficiently supplied to a chip to send current-waveform data to an EMS server. The upper-right-hand side of Fig. 12.3.5 lists the measured performance values for our proposed current-waveform sensor chip.

Figure 12.3.6 illustrates the demonstration of real-time current-waveform detection for four appliances. The left-hand side graph in Fig. 12.3.6 shows both actual and demodulated current-waveforms for a plasma TV. Our chip can successfully detect detailed information on current-waveforms. The right-hand side graph in Fig. 12.3.6 also shows the current-waveforms for a hair dryer, a note PC, and a DC power source. These results indicate that the proposed current-waveform sensor with plugless energy harvesting from AC lines will greatly contribute to detecting what appliances are being utilized in the EMS.

Figure 12.3.7 shows a die micrograph of our current-waveform sensor. This chip was fabricated using 90nm CMOS process. Total chip size is 560×300μm² excluding the bonding pad.

References:

[1] A. Naraghi et al., "A 9b 14uW 0.06 mm² PPM ADC in 90 nm Digital CMOS," *ISSCC Dig. Tech. papers, pp. 168-169*, Feb 2009.

[2] T. Shimamura et al., "Nano-Watt Power Management and Vibration Sensing on a Dust-Size Batteryless Sensor Node for Ambient Intelligence Applications," *ISSCC Dig. Tech. papers, pp. 52-53*, Feb 2010.

[3] D. Yeager et al., "A 9.2uA Gen 2 Compatible UHF RFID Sensing Tag with -12dBm Sensitivity and 1.25uV$_{rms}$ Input-Referred Noise Floor," *ISSCC Dig. Tech. papers, pp. 52-53*, Feb 2010.

ISSCC 2011 / February 22, 2011 / 2:30 PM

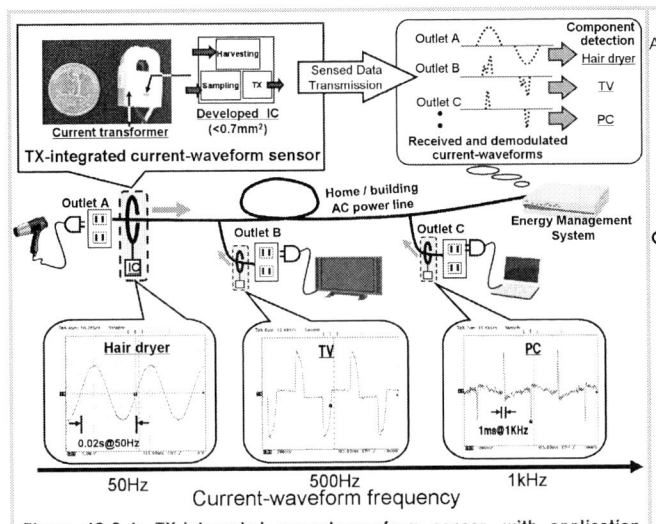

Figure 12.3.1: TX-integrated current-waveform sensor, with application examples.

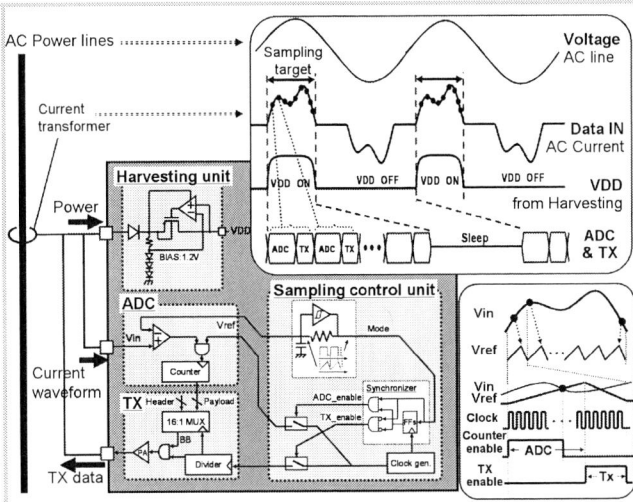

Figure 12.3.2: Overall block diagram and timing chart for the current-waveform sensor.

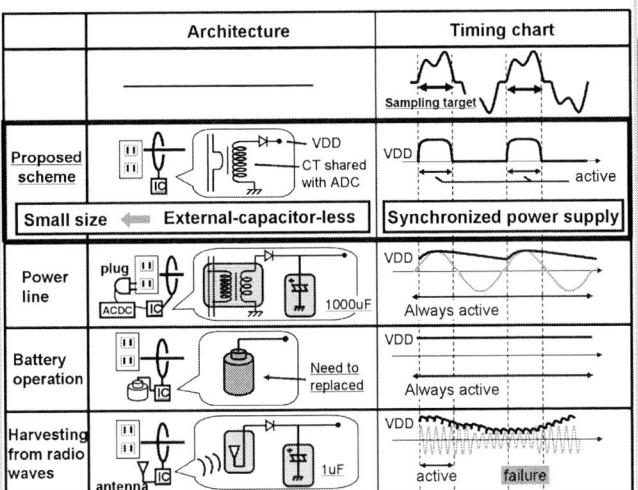

Figure 12.3.3: Proposed power supply scheme and conventional schemes.

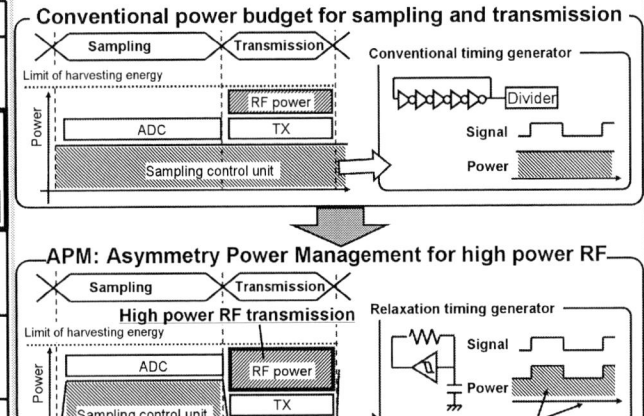

Figure 12.3.4: Asymmetry power management scheme for high power RF tansmission.

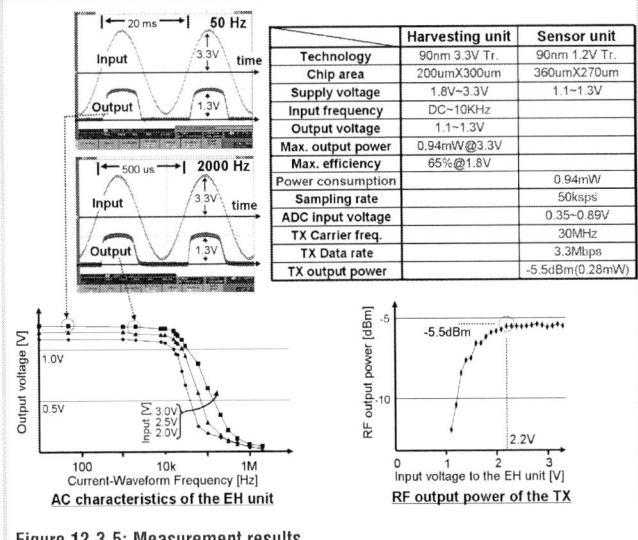

Figure 12.3.5: Measurement results.

	Harvesting unit	Sensor unit
Technology	90nm 3.3V Tr.	90nm 1.2V Tr.
Chip area	200umX300um	360umX270um
Supply voltage	1.8V~3.3V	1.1~1.3V
Input frequency	DC~10KHz	
Output voltage	1.1~1.3V	
Max. output power	0.94mW@3.3V	
Max. efficiency	65%@1.8V	
Power consumption		0.94mW
Sampling rate		50ksps
ADC input voltage		0.35~0.89V
TX Carrier freq.		30MHz
TX Data rate		3.3Mbps
TX output power		-5.5dBm(0.28mW)

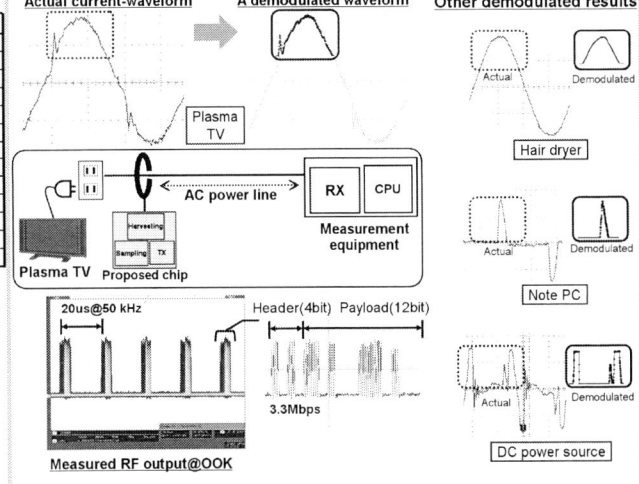

Figure 12.3.6: System demonstration of appliances detection with our current-waveform sensor.

978-1-61284-303-2/11 $26.00 © 2011 IEEE

ISSCC 2011 PAPER CONTINUATIONS

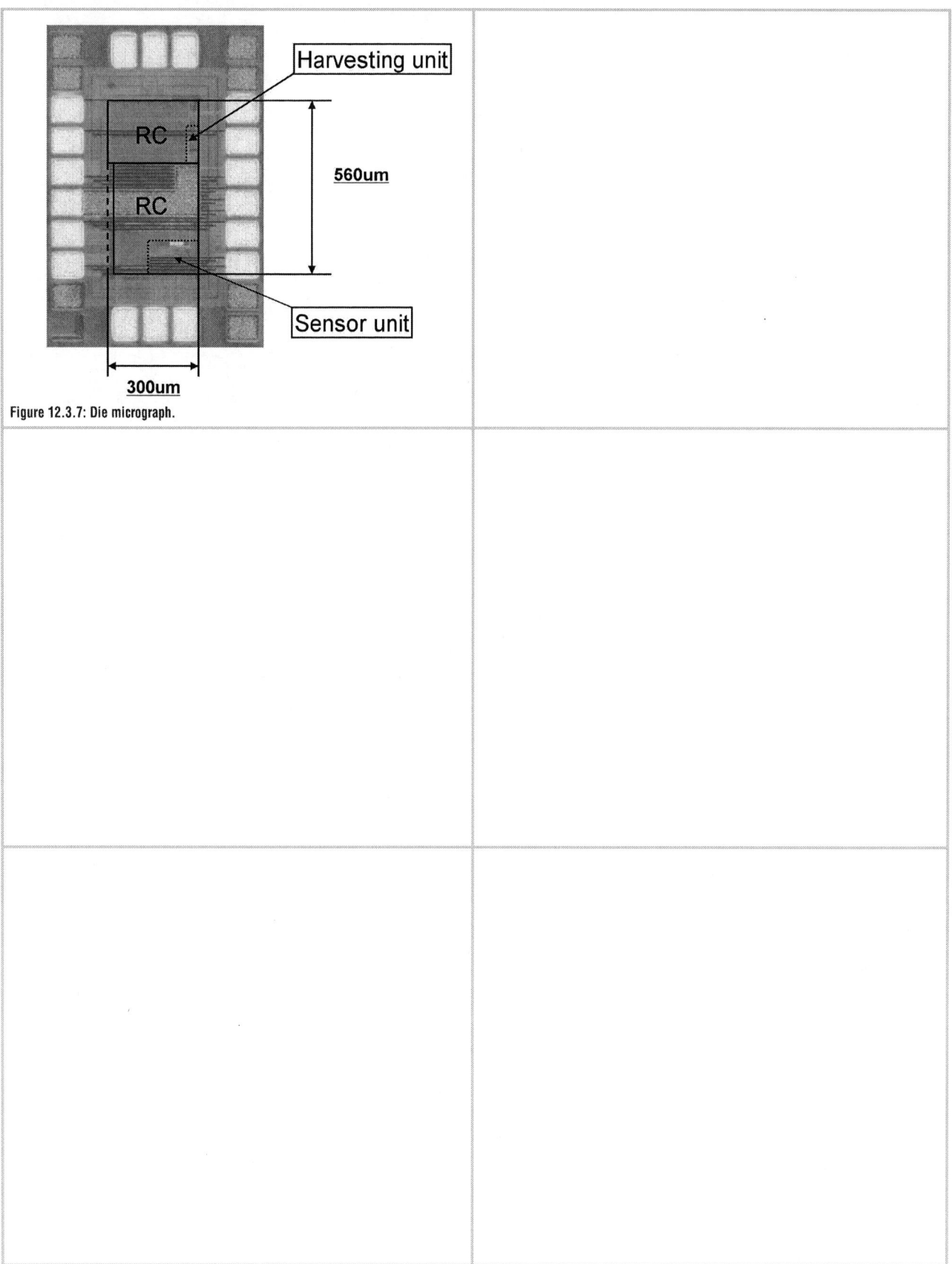

Figure 12.3.7: Die micrograph.

ISSCC 2011 / SESSION 12 / DESIGN IN EMERGING TECHNOLOGIES / 12.4

12.4 A 3.9ns 8.9mW 4×4 Silicon Photonic Switch Hybrid Integrated with CMOS Driver

Alexander Rylyakov, Clint Schow, Benjamin Lee, William Green, Joris Van Campenhout, Min Yang, Fuad Doany, Solomon Assefa, Christopher Jahnes, Jeffrey Kash, Yurii Vlasov

IBM T. J. Watson Reseach Center, Yorktown Heights, NY

The emerging field of silicon photonics [1-3] targets monolithic integration of optical components in the CMOS process, potentially enabling high bandwidth, high density interconnects with dramatically reduced cost and power dissipation. A broadband photonic switch is a key component of reconfigurable networks which retain data in the optical domain, thus bypassing the latency, bandwidth and power overheads of opto-electronic conversion. Additionally, with WDM channels, multiple data streams can be routed simultaneously using a single optical device. Although many types of discrete silicon photonic switches have been reported [4-6], very few of them have been shown to operate with CMOS drivers. Earlier, we have reported two different 2×2 optical switches wirebond packaged with 90nm CMOS drivers [7,8]. The 2×2 switch reported in [7] is based on a Mach-Zehnder interferometer (MZI), while the one reported in [8] is based on a two-ring resonator.

In this work we report a 4×4 silicon photonic switch composed of 6 MZI based switches actuated by integrated p-i-n diodes, 6 waveguide crossings and 8 I/O fiber couplers as shown in Fig. 12.4.1. The switch is flip-chip bonded with a CMOS driver also shown in Fig. 12.4.1. Hybrid integration of photonic devices and controlling electrical circuits is an important step towards a full monolithic integration. It also offers significant advantages on its own by decoupling the optimization, fabrication and test of the optical and electrical components, increasing the design exploration space and shortening turnaround times. The driver shown in Fig. 12.4.1 is an all-static CMOS circuit, fabricated in a standard digital IBM 90nm process. The driver features an array of 6 output buffers, with predrivers, controlled by a serial interface. Several versions of the driver chip have been designed and evaluated. All variants operate with ample speed (50ps to 100ps transition times) to drive ns-scale photonic switches. The photonic switch electrical load was modeled as a forward biased p-i-n diode. The diode model (Fig. 12.4.1) includes a series resistance, a charge-dependent capacitance and a current source of magnitude determined by carrier lifetime. The driver output inverter stages were sized to minimize the output impedance and to cover a wide range of capacitive loads and steady state currents in the p-i-n diode resulting from variations in carrier lifetime. The structure of the MZI based 2×2 switch is shown in Fig. 12.4.2. The optical signal, travelling in the on-chip waveguide, is split into two arms and then merged again. One arm contains a p-i-n diode that implements an optical phase change through carrier injection. In the ON state the input signal is routed to the T_{11} port, with T_{12} output attenuated by 21 dB. In the OFF state the input is routed to the T_{12} port, while the T_{11} output is attenuated by 18dB within the 30nm optical bandwidth of the device (Fig. 12.4.2). The 4×4 switch is composed of 6 2×2 switches as shown in Fig. 12.4.1, connecting each of the 4 inputs to any one of the other 3 outputs; note that in this specific topology U-turns are not allowed, e.g., the West input (W_{IN}) cannot be routed to the West output (W_{OUT}), etc. The 4×4 switch has a total of 9 different valid routing states, with each valid state requiring 0, 1, 2 or 4 2×2 switches turned on.

A 4×4 switch can be built out of many different types of 2×2 switches. An interesting alternative to the MZI is a ring resonator based switch (Fig. 12.4.3). The two-ring resonator switch is a smaller device, measuring only 25×50µm², compared to 300x50µm² of MZI. The device also offers significant savings in power dissipation at the expense of narrower optical bandwidth and increased sensitivity to temperature.

Measured optical transition times of the 2×2 MZI switch are shown in Fig. 12.4.4. Testing was done with a continuous-wave 1530nm optical signal injected into one input port at a time. A 100MHz PRBS7 switching signal was applied to a wirebonded single-channel CMOS driver circuit held at 1.0V supply. The extinction ratio and crosstalk were both measured to be better than 15dB. The measured OFF to ON and ON to OFF 10/90 transition times were 3.9ns and

1.4ns, respectively. The slower transient occurs during the carrier injection phase. Power dissipation was measured with 50% duty cycle data and was found to be 2mW from a 1.0V supply. At speeds below 10MHz, typical for the switch application, power dissipation is dominated by the steady-state contribution of the injected carrier recombination in the switch.

The 4×4 switch was tested with a flip-chip bonded CMOS driver. A custom hybrid integration process was developed that included the following steps: NiAu plating of both photonic and CMOS bond pads, injection-molded solder processing to transfer PbSn solder balls to the CMOS chip and a flux-free flip-chip solder reflow to complete the assembly. The overall size of the micro-assembly was 1×2×2mm³. Correct routing of each of the 4 inputs to any one of the other 3 was verified by routing 40Gb/s data (Fig. 12.4.5). A 1531nm wave was modulated with a LiNbO$_3$ modulator driven by a 40Gb/s PRBS7 data and edge-coupled to the photonic chip with a lensed fiber probe. The output, also edge-coupled, was amplified with an erbium-doped fiber amplifier (EDFA), wavelength filtered and received. A WDM transmission demonstration is shown in Fig. 12.4.6. In this experiment, 3 different wavelengths were passively combined and simultaneously modulated with 40Gb/s PRBS7 data. One kilometer of standard single-mode fiber was used to decorrelate the wavelength channels in time via fiber chromatic dispersion. Three different routing states of the 4×4 switch were tested, routing the 3×40 Gb/s signal from the East input to each of the 3 possible outputs, as shown in Fig. 12.4.6. Power sensitivity curves were taken for each state and each wavelength, demonstrating BER < 10^{-12} at -10dBm average optical power. All wavelength channels and all configurations were found to be within a 0.5 dB power spread at a BER of 10^{-12}. Since each of the 2×2 switches dissipates 4mW in the ON state, the total power of the 4×4 switch depends on a particular routing state and can be as high as 16 mW, when 4 of the 6 MZI switches are turned on. Assuming uniform utilization of all 9 valid states and low frequency of routing state transitions, the average power of the 4×4 switch would be 8.9mW. Insertion loss of the switch was in the range from 3 to 6dB, depending on the routing state. Off-chip coupler loss was around 1dB. In all 9 valid states, the measured worst case crosstalk between channels was less than -10dB. The micrographs of the CMOS and the photonic dies are shown in Fig. 12.4.7. CMOS drivers with predrivers occupied 144×520µm². The total area of the 4×4 photonic switch was 300×1600 µm².

Acknowledgements:
The authors gratefully acknowledge support from DARPA under contract HR0011-08-C-0102. The views, opinions, and/or findings contained in this article are those of the authors and should not be interpreted as representing the official views or policies, either expressed or implied, of DARPA or the Department of Defense. Approved for public release, distribution unlimited.

References:
[1] D. Kucharski, et al.," 10Gb/s 15mW Optical Receiver with Integrated Germanium Photodetector and Hybrid Inductor Peaking in 0.13µm SOI CMOS Technology", IEEE ISSCC Dig. Tech. Papers, pp. 360-361, Feb. 2010.
[2] I.Young, et al., "Optical I/O Technology for Tera-Scale Computing", IEEE ISSCC Dig. Tech. Papers, pp. 468-469, Feb. 2009.
[3] S. Assefa, et al., "Reinventing nanophotonic avalanche photodetector for on chip optical interconnects," Nature, pp.84-88, March 2010.
[4] Y. Vlasov, et al., "High-throughput silicon nanophotonic wavelength-insensitive switch for on-chip optical networks," Nature Photonics, vol. 2, pp. 242–246, 2008.
[5] J. Van Campenhout, et al., "Low-power, 2×2 silicon electro-optic switch with 110-nm bandwidth for broadband reconfigurable optical networks", Optics Express 17(26), pp. 24020-24029, 2009.
[6] B. G. Lee, et al., "High-Performance Modulators and Switches for Silicon Photonic Networks-on-Chip," IEEE JSTQE, vol. 16, no. 1, pp. 6-20, 2010.
[7] B. G. Lee, et al., "Broadband silicon photonic switch integrated with CMOS drive electronics," in Proc. Conf. Lasers Electro-Optics (CLEO) 2010, paper CThJ1, 2010.
[8] B. G. Lee, et al., "Comparison of Ring Resonator and Mach-Zehnder Photonic Switches Integrated with Digital CMOS Drivers", IEEE Photonics Society Annual Meeting, 2010.

978-1-61284-303-2/11 $26.00 © 2011 IEEE

ISSCC 2011 / February 22, 2011 / 3:15 PM

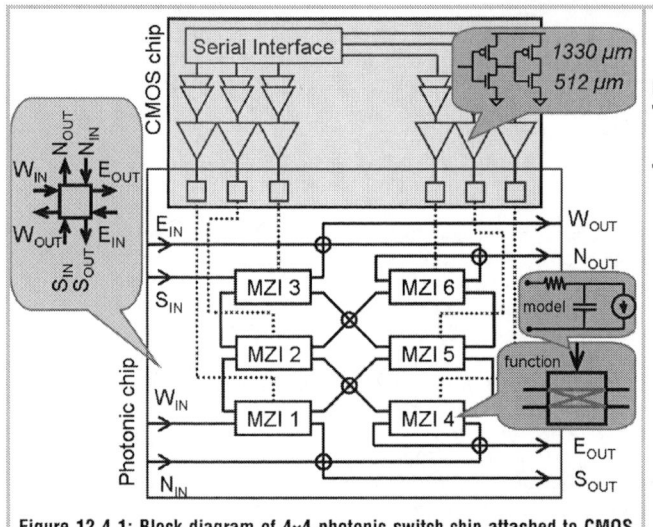

Figure 12.4.1: Block diagram of 4×4 photonic switch chip attached to CMOS driver chip.

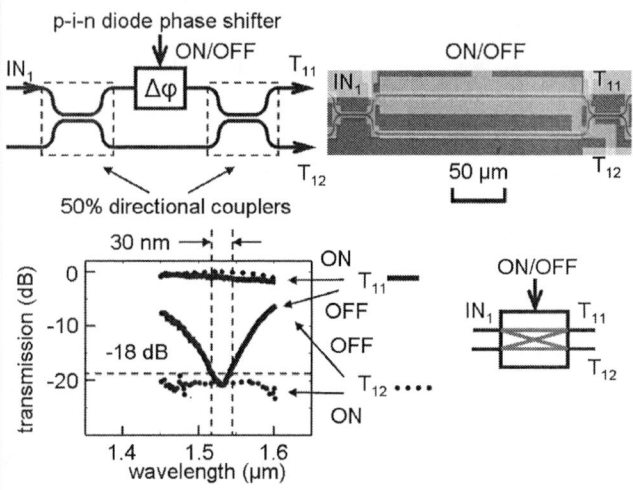

Figure 12.4.2: Mach-Zehnder interferometer (MZI) based 2×2 switch.

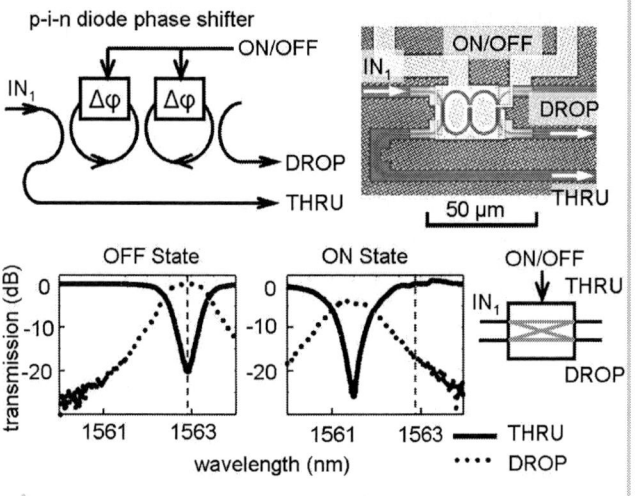

Figure 12.4.3: Two-ring resonator based 2×2 switch.

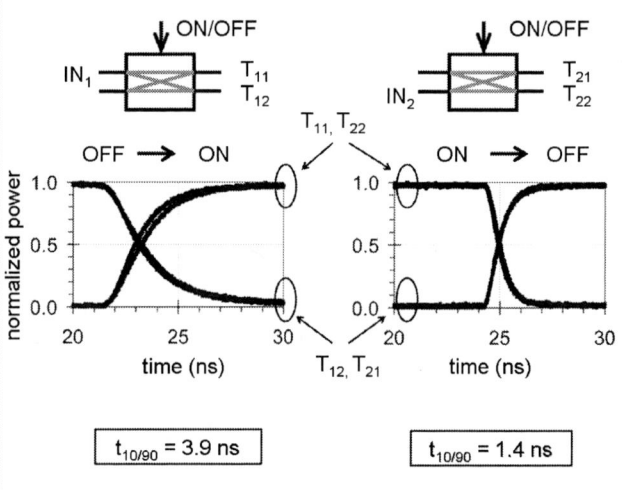

Figure 12.4.4: Measured optical switching times of a 2×2 MZI switch.

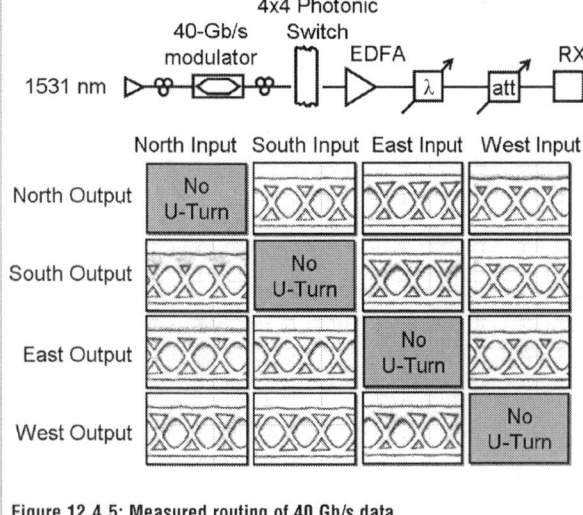

Figure 12.4.5: Measured routing of 40 Gb/s data.

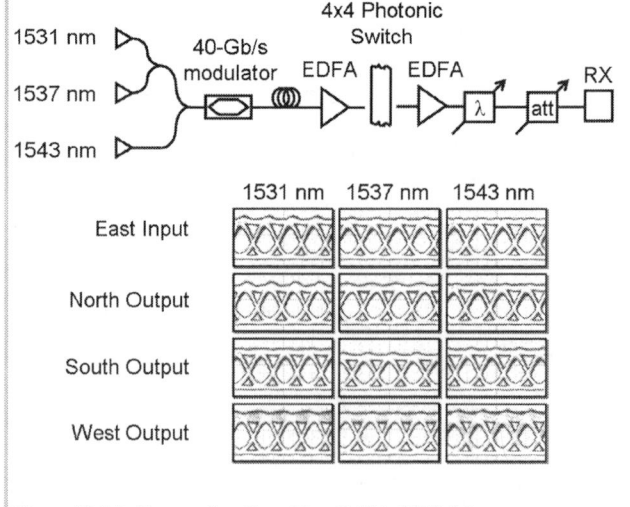

Figure 12.4.6: Measured routing of 3 × 40 Gb/s WDM data.

12

978-1-61284-303-2/11 $26.00 © 2011 IEEE

ISSCC 2011 PAPER CONTINUATIONS

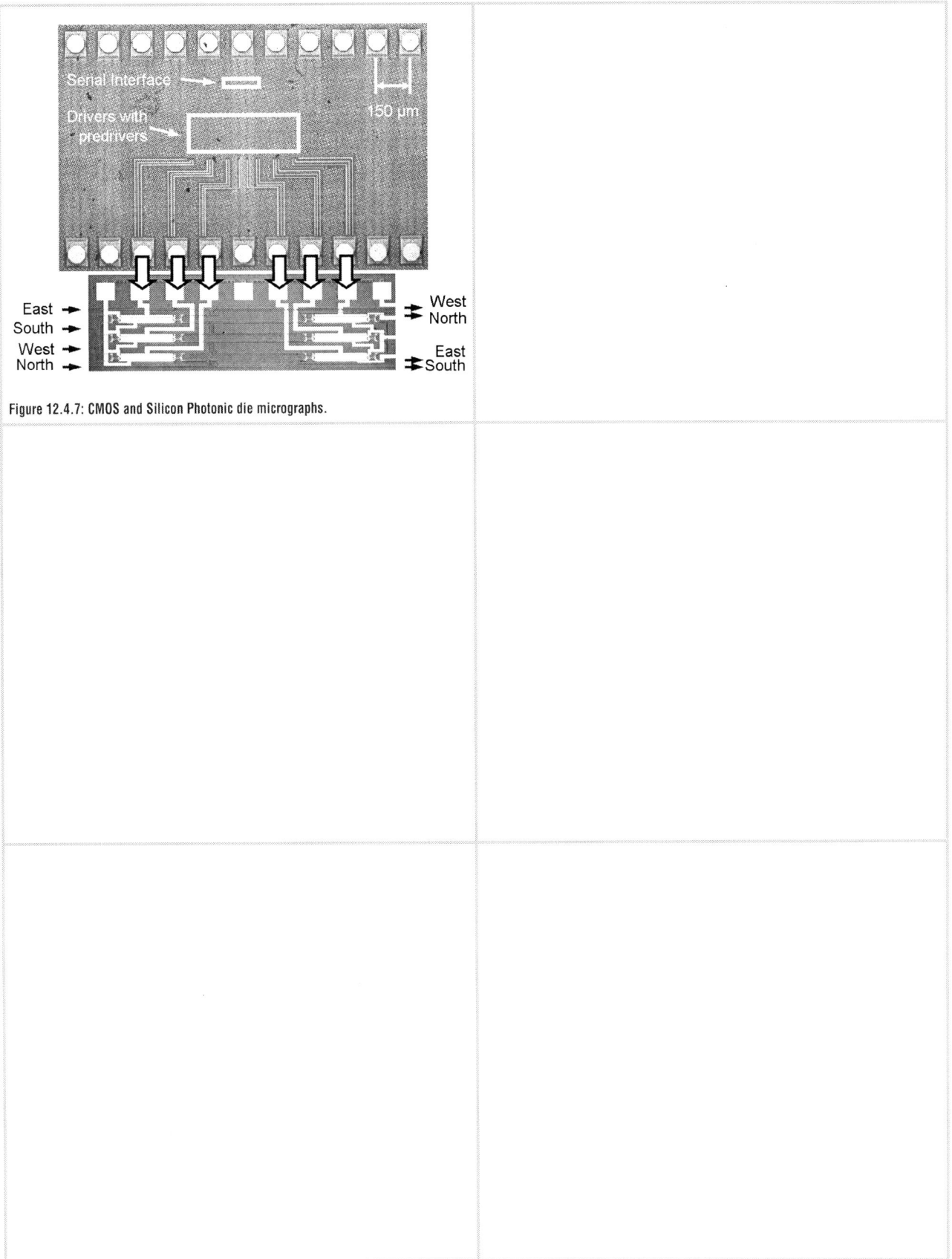

Figure 12.4.7: CMOS and Silicon Photonic die micrographs.

ISSCC 2011 / SESSION 12 / DESIGN IN EMERGING TECHNOLOGIES / 12.5

12.5 A 820GHz SiGe Chipset for Terahertz Active Imaging Applications

Erik Öjefors[1], Janus Grzyb[1], Yan Zhao[1], Bernd Heinemann[2],
Bernd Tillack[2], Ullrich R Pfeiffer[1]

[1]University of Wuppertal, Wuppertal, Germany,
[2]IHP, Frankfurt (Oder), Germany

The spatial resolution of microwave and mmWave imaging systems can be improved by an increase of their operating frequencies into the submillimeter-wave range (300GHz to 3THz). Electronic terahertz sources and receivers are presently dominated by III/V semiconductor and waveguide packaging technologies. Multipliers and mixers capable of an output power of 1.4mW at 875GHz [1], as well as a conversion loss of 9.25dB at 850GHz [2], have been demonstrated in remote sensing and radio astronomy applications. However, short-range active terahertz imaging systems exhibit less stringent link-budgets, thus opening up new opportunities for highly integrated low-cost silicon TX and RX front-ends. Previous circuit design examples include a 410GHz CMOS VCO in a 45nm technology capable of an output power of -47dBm [3], as well as a 650GHz subharmonic receiver with a 44dB noise figure [4]. In this paper, recent advances in SiGe HBT device technology [5] are used to demonstrate a 820GHz TX/RX chipset for active terahertz imaging applications.

The chipset consists of a spatial power combining transmitter chip and a receiver array as shown in the block diagram in Fig. 12.5.1. To overcome the limitation of power generation at 820GHz, the transmitter comprises four parallel antenna-integrated ×45 frequency multiplier chains fed from a common 18.2GHz input signal. This configuration improves the antenna gain and increases the total radiated power by 6dB, respectively. Each on-chip differential patch antenna is fed by a ×5 frequency multiplier, which is used to quintuple the output of a ×9-section, previously presented in [6], up to 820GHz. The receiver consists of four mixers sub-harmonically pumped by a shared ×9 multiplier chain. Each of the mixers is connected to one on-chip differential patch antenna of a 2×2 array. This multi-pixel receiver can be used as a focal-plane sub-array in order to reduce the acquisition time in a raster-scanned imaging setup, or, as part of a synthetic aperture imaging system.

Figure 12.5.2 shows the circuit schematic of one of the four final ×5 multiplier stages of the transmitter. The balanced design is based on a differential HBT stage (Q1/Q2), where each device is 4×(0.12×0.96)um² large. The base terminals of Q1/Q2 are driven with a 8dBm 160GHz differential signal from the ×9 multiplier, while open quarter-wave stubs (TLOC1/TLOC2) provide a ground return path for the generated 820GHz harmonic frequency component. Shunt stubs (TLSC1/TLSC2) at the base terminals improve the impedance match to the 100Ω differential 164GHz multiplier chain. The output LC-tank network is tuned to the 5th harmonic. It consists of a 35μm long U-shaped differential inductor L1/L2 at the top metal level (M5) and 15fF MIM capacitors C1/C2. Multiplier simulations based on foundry-level device models suggest a simulated output power of -25dBm delivered into a 100Ω differential load.

The output of each transmitter channel is fed through microstrip transmission lines (TL3/TL4) to a capacitively edge-coupled differential patch antenna. The transmitter patch antennas are implemented in the top metal of the backend and are arranged in a 2×2 matrix with a 190μm element spacing in order to provide spatial power combining with a total simulated (HFSS) antenna directivity of 12dBi, an efficiency of 66%, and a total gain of 10dBi. The spatial power combining of four -25dBm channels through this array leads to an effective isotropic radiated power of -9dBm.

Figure 12.5.3 shows the circuit schematic of the receiver harmonic mixer, which uses a balanced device configuration in order to suppress LO noise. Due to limitations in the available LO drive level, a single-balanced mixer is preferred instead of a double-balanced design. The 0.12×0.96μm² large devices Q1/Q2 are pumped in common-mode with a 0dBm 164GHz LO signal generated from the shared ×9 LO multiplier chain. Like the transmitter antennas, a single receiver patch antenna exhibits a 6dBi simulated directivity, and provides the differential RF signal to the emitters of Q1/Q2. Two 820GHz quarter-wave shorted stubs in

parallel with the RF input serve as the LO ground return path. Resistive collector loads and output emitter-follower buffers are used at the IF output in order to drive external measurement equipment. The simulated conversion gain of the harmonic mixer circuit is -15dB for a 0dBm LO signal.

The transmitter is characterized in free-space at a distance of 30cm. A calibrated WR-1.2 waveguide harmonic mixer and a 25dBi horn antenna are used to measure the TX EIRP from 820-850GHz with a maximum output power of -17dBm at 823GHz as shown in Fig. 12.5.4. The model-to-hardware correlation for the transmitter is within 8dB, which is reasonable considering that the HBT device models are operated far above their validated frequency range and EM-modeling of antenna and interconnect losses are limited. The bandwidth limitation of the transmitter is due to the narrow-band tuning of the integrated 164GHz ×9 multiplier chain.

Figure 12.5.5 shows the single-channel receiver conversion gain for a 190MHz IF frequency. A similar setup as for the free-space characterization of the transmitter is used, where the external mixer is replaced by a -19dBm 823GHz frequency multiplied source. The measured conversion gain includes the receiver antenna gain (simulated directivity 6dBi), and therefore, represents the gain relative to a receiver with an isotropic antenna. A maximum measured conversion gain of -22dBi is obtained at 823GHz, which is in reasonable agreement with the harmonic mixer simulation results. The measured IF noise floor of -149dBm/Hz together with the conversion gain yields a minimum receiver NF of 47dB (relative to a receiver with an isotropic antenna) in a room-temperature environment. Like in the case of the TX, the tuning range of the RX is limited by the bandwidth of the 164GHz LO multiplier chain.

The device technology is based on a 0.25μm BiCMOS process with an upgraded SiGe HBT module. Further lateral scaling (minimum emitter area 0.12×0.96μm²), as well as a modified collector and emitter formation, enables a performance enhancement from f_T/f_{max}=250GHz/380GHz (used in [6]) to 280GHz/435GHz used in this paper. The open-base emitter-collector breakdown voltage remains at 1.7V. A further optimized version of this technology with peak f_T/f_{max} values of 300GHz/500GHz is presented in [5].

The chipset results are summarized in Fig. 12.5.6. Die micrographs of the 4.3×0.75mm² and 2.3×0.57mm² large TX and RX chips are shown in Fig. 12.5.7. Unlike previous works, the inclusion of a transmitter with sufficient power generation capability provides an all silicon chipset for active terahertz imaging applications.

Acknowledgments:
This work was partially funded by the European Commission within the project DOTFIVE (no. 216110) and the European Science Foundation through an European Young Investigator Award. The authors would like to thank Hans Keller, University of Wuppertal, Germany, for his support.

References:
[1] A. Maestrini, et al., "A Frequency-Multiplied Source With More Than 1 mW of Power Across the 840–900-GHz Band", IEEE Trans. MTT, vol. 58, no. 7, July 2010.
[2] B. Thomas, et al., "A Broadband 835–900-GHz Fundamental Balanced Mixer Based on Monolithic GaAs Membrane Schottky Diodes", IEEE Trans. MTT, vol. 58, no. 7, July 2010.
[3] E. Seok, et al., "A 410GHz CMOS Push-Push Oscillator with an On-Chip Patch Antenna", ISSCC Dig. Tech. Papers, pp. 472-473, Feb. 2008.
[4] E. Öjefors, U.R. Pfeiffer, "A 650GHz SiGe Receiver Front-End for Terahertz Imaging Arrays", ISSCC Dig. Tech. Papers, pp. 430-431, Feb. 2010.
[5] B. Heinemann et al., "SiGe HBT Technology with f_T/f_{max} of 300GHz/500GHz and 2.0 ps CML Gate Delay," accepted for IEDM 2010.
[6] U.R. Pfeiffer, E. Öjefors, and Y. Zhao, "A SiGe Quadrature Transmitter and Receiver Chipset for Emerging High-Frequency Applications at 160GHz", ISSCC Dig. Tech. Papers, pp. 416-417, Feb. 2010.

978-1-61284-303-2/11 $26.00 © 2011 IEEE

ISSCC 2011 / February 22, 2011 / 3:45 PM

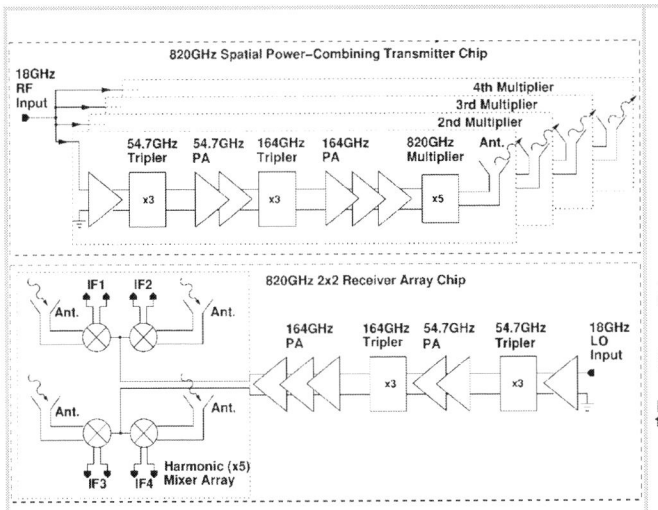

Figure 12.5.1: Block diagram of the 820GHz TX/RX chip-set.

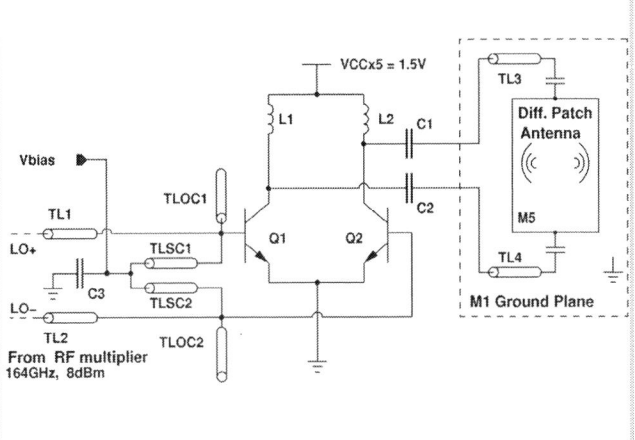

Figure 12.5.2: Simplified circuit schematic of one ×5 multiplier stage of the 820GHz source.

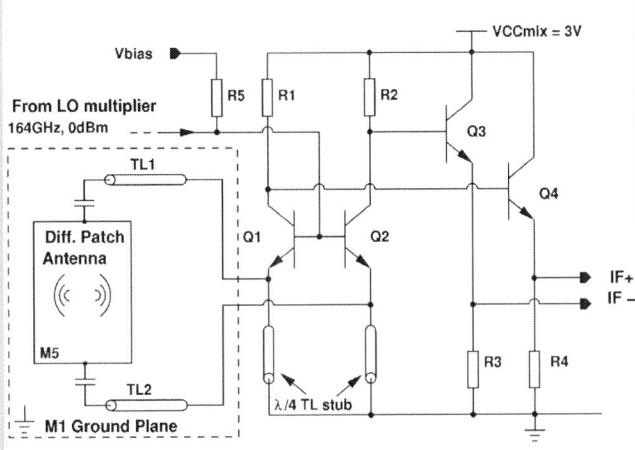

Figure 12.5.3: Schematic diagram of one ×5 harmonic mixer of the 820GHz receiver array.

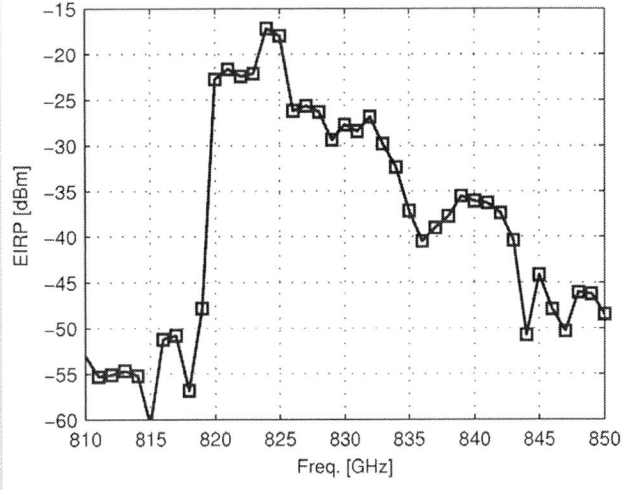

Figure 12.5.4: Measured transmitter effective isotropic radiated power over the 810 to 850GHz frequency range.

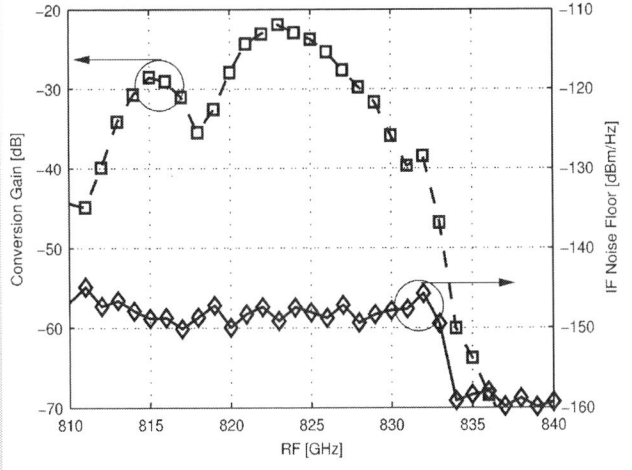

Figure 12.5.5: Measured conversion gain of a single channel of the 2×2 receiver array.

DC Specification ($T_A = +25$ °C)		
PARAMETER	**TRANSMITTER**	**RECEIVER**
CIRCUIT DESIGN		
Configuration	2×2 Spatial Power Combining	2×2 Array Receiver
Die size	4.3×0.75mm²	2.3×0.57mm²
Technology	0.25μm SiGe BiCMOS, $f_t/f_{max} = 280/435$GHz	
DC POWER REQUIREMENTS		
Voltage Supply	+4V, +1.5V	+4V, +3V
Power Dissipation	3.7W	1.2W
Quiescent Current, I_{DC}	900mA @4V, 20mA @1.5V	300mA @4V, 20mA @2.5V
AC TEST CONDITION		
Frequency	820 - 845GHz	810 - 830GHz
RF/LO drive (18.3GHz)	14dBm	4dBm
Effective isotropic radiated power[1]	-17dBm (823GHz) to -40dBm	
Conversion Gain / Channel[1]	-	-22dB (823GHz) to -45dB
Noise Figure[1]	-	47dB
ANTENNA		
Simulated Directivity	12 dBi	6 dBi / Channel

[1] Referenced to a transmitter/receiver with an isotropic (0 dBi) antenna.

Figure 12.5.6: Summary of the TX and RX measurement results.

Figure 12.5.7: Micrograph of the 820GHz chipset consisting of a ×4 spatial power combining transmitter chip and the 2×2-element receiver array.

ISSCC 2011 / SESSION 12 / DESIGN IN EMERGING TECHNOLOGIES / 12.6

12.6 A 130µA Wake-Up Receiver SoC in 0.13µm CMOS for Reducing Standby Power of An Electric Appliance Controlled by An Infrared Remote Controller

Hiroaki Ishihara[1], Toshiyuki Umeda[1], Katsuya Ohno[2], Shigeyasu Iwata[2], Fumi Moritsuka[1], Tetsuro Itakura[1], Manabu Ishibe[2], Keijiro Hijikata[2], Yasunori Maki[2]

[1]Toshiba, Kawasaki, Japan,
[2]Toshiba, Ome, Japan

In the context of efforts to achieve zero emissions of carbon dioxide, wake-up circuits have attracted attention as a promising approach for reducing standby power [1]. The dominant consumers of standby power are household electric appliances driven by AC power such as TVs. In these appliances, power is lost at an AC-DC converter even when the appliances are turned off. Although an infrared remote controller (IRC) is usually used in these appliances, an RF signal is utilized in the wake-up circuits reported so far and no wake-up circuit has been reported for the IRC. This paper describes a receiver (RX) SoC with wake-up function for the IRC.

Figure 12.6.1 shows the concept of the proposed low-standby-power IRC system. In a conventional system, the main AC-DC converter is turned off when an appliance is in OFF state, i.e., standby mode, to reduce the power consumption. But power must still be supplied to an infrared RX and a microcomputer (µCOM), because the RX waits for the IRC signal. The standby power of the conventional system is several hundred mW, and the loss of the sub power supply circuit is the dominant portion of the standby power. The energy-related products (ErP) directive requires the standby power should be less than 500mW for TVs in 2011 [2].

Our system consumes no power from the AC supply in the standby mode. The sub power supply circuit and µCOM are turned the power off by a relay connected to the AC supply, and only the low power RX with wake-up function is activated. The power of the RX is less than 0.5mW, 1/1000th of the ErP requirement. The power is supplied to the RX from a charged ultra capacitor (C_{sup}), which has very large capacitance of more than 1F. The RX can distinguish an ON command transmitted from the IRC without using a µCOM. When the ON command is distinguished, the RX turns on the relay to activate the appliance. When the voltage across the C_{sup} becomes less than a predetermined value in the standby mode, the relay is turned on to recharge the C_{sup}. The conventional RX consumes about 270 to 1000µA DC current. In the proposed system, our target current is lower than 140µA, assuming that the C_{sup} is 4.5F, the RX operates at 3.6V to 2.2V supply, and the C_{sup} is recharged every 12 hours. The total current of the RX including the wake-up function is 130µA, which is smaller than the target value, and less than 1/2 of the conventional RXs.

Figure 12.6.2 shows the block diagram of the RX system. The signal from the IRC is detected by the photodiode (PD). The PD output signal is amplified and filtered by the TIA, the VGA and the BPF. The BPF output is converted into digital signals by the ADC. The filter function is partly realized by the digital filter to relax the specification of the analog BPF, and reduce the power consumption. The AGC controller detects the signal amplitude and controls the gain of the analog circuits. The gain range of the VGA and TIA is 39dB and 35dB, respectively. The voltage provided by the C_{sup} is converted to 1.5V by the integrated series regulator. The supply level monitor detects the voltage of the C_{sup} to control the recharge timing. The 228kHz and 38kHz clock are generated from 114kHz reference.

When the appliance is in the standby mode, the RX operates as a wake-up RX as described before. In the standby mode, the RX distinguishes the ON command and controls the relay to activate the appliance without using the µCOM. When the appliance is in the operation, the function of the RX is the same as that of the conventional RX. The received signal is transferred to the µCOM, and the µCOM distinguishes the controller commands. The RX has a timer for scheduled operations such as scheduled recording. When the operation is scheduled, the timer turns on the relay at scheduled time even if the ON command is not distinguished to execute the operation.

The RX must work even in the presence of strong ambient light. The ambient light causes DC photocurrent (I_{DC}), and the reverse bias voltage of the PD (V_{PD}) is reduced as I_{DC} increases, while the V_{PD} should be larger than 0.4V to ensure the proper operation of the PD. Therefore, an I_{DC} rejection circuit is required to bypass the I_{DC} and provide a sufficiently large V_{PD} in case large ambient light exists. In [3], the I_{DC} rejection technique with active inductor realized by an operational transconductance amplifier (OTA) is presented. But the technique is unsuitable for low-power operation because the OTA consumes bias current even if the ambient light does not exist. Figure 12.6.3 shows the proposed I_{DC} rejection circuit. The supply voltage of the circuit is directly provided by the C_{sup} without using the regulator to maximize the available V_{PD}. The transistor M1 converts the input current into voltage signal, and the passive filter composed by the R1 and the C1 rejects the high-frequency component. Only the DC current of M1 is multiplied by 5 because M2 has 5 times larger size than M1, and provided to the PD. When the ambient light does not exist, the proposed circuit consumes no extra current, and is suitable for low-power operation. The measured result at 2.2V supply shows that the proposed circuit can provide more than 0.4V V_{PD} even with 50µA I_{DC}. The signal gain loss with 50µA I_{DC} is less than 1dB.

Figure 12.6.4 shows the baseband circuit, which has 3 stages of the VGA, 2 stages of the HPF and 2 stages of the LPF. Each stage of the VGA is composed by an operational amplifier with a resistive feedback. Each stage of the filters is designed with a multi-feedback active RC configuration, and has the 2nd-order characteristic. Therefore, the BPF is configured by a 4th-order HPF and a 4th-order LPF. When the VGA gain is set to a large value, each VGA stage may cause a large DC offset. But the offset cancel circuit is not needed because the output of the VGA stages are connected to the HPF or capacitor coupled and large DC offsets are not transferred to the next stage.

The BPF should pass the signal around 38kHz, which is the sub-carrier frequency of the IRC signal, and eliminate the undesired signals. The passband calibration of the BPF is needed because the process variations of the integrated passive components directly affect the filter characteristic. The passband is digitally calibrated by the calibration controller. In the calibration, 38kHz clock is used as a reference signal. The passband is controlled by the 4b digital code (F_{CNT}). The reference signal is input to the BPF at each F_{CNT} setting, and output amplitude is detected. After that, the F_{CNT}, which corresponds to the maximum detected amplitude, is selected. In calibration, the signal amplitude may exceed the input range of the ADC for more than one F_{CNT} setting. This deteriorates the calibration accuracy owing to the uncertainty in detecting the maximum amplitude. To improve the accuracy, various amplitude reference signals are used, and the amplitude is controlled by the variable attenuator with 3b digital control signal.

Figure 12.6.5 shows the relationship between the frame error rate and the input current. Our target receiving performance is more than 10m communication distance. With the PD HPI6FGR4, the required minimum sensitivity is 1nA peak-to-peak at the TIA input. The measured sensitivity is less than 600pA, and is sufficient for 10m communication. Figure 12.6.5 also shows the communication range with HPI6FGR4 in the line of sight condition. In this measurement, a test IRC whose output power is lower than typical controller is used for a margin. Figure 12.6.5 shows more than 10m communication distance can be achieved from -20 to 90 degrees.

Figure 12.6.6 shows the performance summary. The current of the digital and the analog part is 42µA and 85µA, respectively. The IC is fabricated in 0.13µm CMOS process, and the chip area is 2.8×2.8mm².

References:
[1] X. Haung et al., "A 2.4GHz/915MHz 51µW wake-up receiver with offset and noise suppression", *ISSCC Dig. Tech. Papers*, pp. 222-223, Feb., 2010.
[2] DIRECTIVE 2009/125/EC establishing a framework for the setting of ecodesign requirements for energy-related products.
[3] W. Yong-Shen, Xu li, L. Feng-chang, "DC photocurrent rejection of high transimpedance gain preamplifier in infrared wireless optical receiver", *Proc. of IEEE 8th international Conference on ASIC*, pp. 250-253, Oct, 2009.

978-1-61284-303-2/11 $26.00 © 2011 IEEE

ISSCC 2011 / February 22, 2011 / 4:15 PM

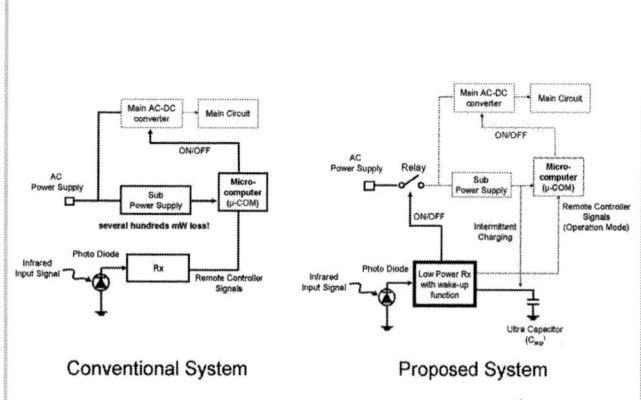

Figure 12.6.1: The concept of the low-standby-power system for an infrared remote controller.

Figure 12.6.2: Block diagram of the receiver SoC.

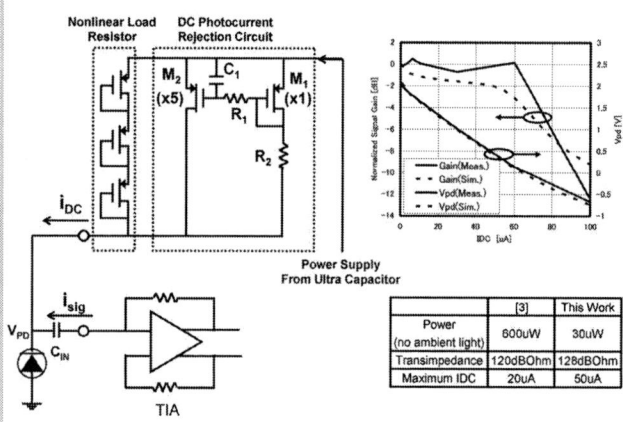

Power (no ambient light)	[3]	This Work
Power (no ambient light)	600uW	30uW
Transimpedance	120dBOhm	128dBOhm
Maximum IDC	20uA	50uA

Figure 12.6.3: The proposed DC photocurrent rejection circuit.

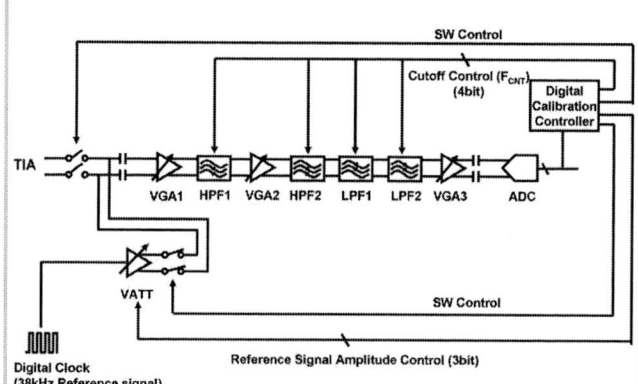

Figure 12.6.4: Block diagram of the baseband circuit.

Figure 12.6.5: Measured frame-error rate vs. input current, and communication range with photodiode HPI6FGR4.

Total		
Technology		0.13um CMOS
Chip Area		2.8mm x 2.8mm
Current		130µA
TIA		
Current		20µA
Maximum PD DC Photocurrent		50uA
Trans Impedance	High Gain	128 [dBOhm]
	Low Gain	93 [dBOhm]
Output Noise@38kHz		280nV/√Hz
VGA and BPF		
Current		22µA
Maximum Gain		49dB
Gain Control Range		39dB
Input Referred Noise@38kHz		120nV/√Hz
Regulator, BGR		
Current		0.73µA
ADC, Digital, Oscillator and Clock generator		
Current		85µA

Figure 12.6.6: Performance summary

978-1-61284-303-2/11 $26.00 © 2011 IEEE

ISSCC 2011 PAPER CONTINUATIONS

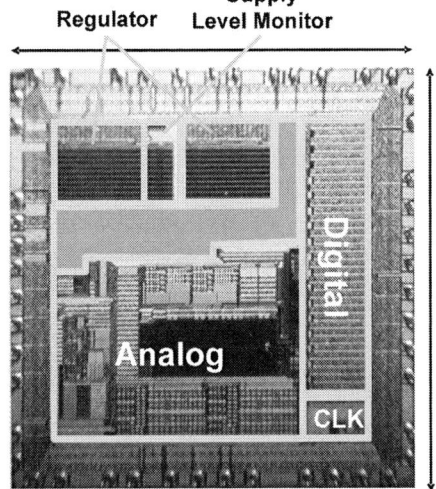

Figure 12.6.7: Chip Micrograph.

ISSCC 2011 / SESSION 12 / DESIGN IN EMERGING TECHNOLOGIES / 12.7

12.7 Programmable Cell Array Using Rewritable Solid-Electrolyte Switch Integrated in 90nm CMOS

Makoto Miyamura[1], Shogo Nakaya[2], Munehiro Tada[1],
Toshitsugu Sakamoto[1], Koichiro Okamoto[1], Naoki Banno[1],
Shinji Ishida[1], Kimihiko Ito[1], Hiromitsu Hada[1], Noboru Sakimura[1],
Tadahiko Sugibayashi[1], Masato Motomura[2]

[1]NEC, Sagamihara, Japan,
[2]NEC, Kawasaki, Japan

Programmable devices such as SRAM-based FPGAs have the major challenges of power consumption and circuit area due to the excessive standby leakage current and the threshold voltage variation in highly scaled SRAM. Back-end-of-line (BEOL) device, which is integrated in the interconnect layers, is attractive for reducing the performance gap between FPGA and cell-based ASIC [1-4]. In this paper, we demonstrate the fundamental operations of a programmable cell array and a 32×32 crossbar switch using a nonvolatile and rewritable solid-electrolyte switch (nanobridge or NB). A 72% reduction in chip-area compared with that of a standard-cell-based design is achieved on a 90nm CMOS platform.

The NB is composed of a solid-electrolyte sandwiched between the Cu and Ru electrodes. By applying a positive voltage to the Cu electrode, a Cu bridge is formed in the solid-electrolyte and the switch turns on. When a negative voltage is applied, the Cu atoms in the bridge are reverted to the Cu electrode and then the switch turns off. The transition between the low resistive (on) state and high resistive (off) state is repeatable as shown in the I-V curve (Fig. 12.7.1) and each state is nonvolatile. The on-resistance of the NBs can be tuned by a switching current to be around 1kΩ, which is suitable for the application of a signal line. Turn-on is achieved within less than 2ns [4], enabling short and parallel programming. In addition, the NBs are programmed at a low voltage of around 3.3V, which allows 2.5 or 3.3V multi-oxide (MOX) transistors to be used without an internal voltage boost. These characteristics, summarized in the table in Fig. 12.7.1, can mitigate the difficulty in incorporating the NBs into the state-of-the-art CMOS platform compared with the antifuse-based devices [5].

Figure 12.7.2 shows the concept of the NB-based programmable cell array. The NBs are placed at each cross-point between the two interconnect metals, realizing a footprint of $4F^2$, where the unit F represents a half pitch of the interconnect metal. For FPGA, the routing switches composed of SRAMs and pass transistors consume the major area of the chip. When the NBs replace them and the switch plane is placed on the logic plane, the chip area and the routing length are reduced. In addition, the low parasitic capacitance of the NBs (~0.1fF) is also beneficial for lowering the power consumption.

A schematic diagram of the NB-based programmable cell is depicted in Fig. 12.7.2. The following architectures are newly proposed, aiming at area benefit and high functionality. First, the NBs are used not only for the crossbars connecting the functional blocks and horizontal/vertical wires, but also for the level selectors and bus holders. In conventional FPGAs, the look-up tables (LUTs) and bus holders are composed of transistors, and these components require a substantial amount of chip area. In our architecture, the logic function of the programmable cell is configured with the level selectors. Second, the programming transistors are shared by many NBs so that a minimal number of programming transistors are involved in an individual programmable cell. For example, TR_I in Fig. 12.7.2 can address both of the crossbar switch and the level selector. It is essential for NB-based cell to minimize the number of MOX transistors, which occupy a large area in the cell. Third, the polarities of the NBs on the schematics are aligned so as to maximize the layout density of the NBs. Programming drivers are placed on the outer edge of the programmable cell array, employing a multi-level voltage output and a set/reset verification. The programming of a test vehicle involving a single programmable cell is demonstrated in Fig. 12.7.3. Seven NBs are set to be the on-state to connect the horizontal lines and function block (MUX2). The correct MUX2 output-pattern is confirmed by a tester measurement using the corresponding input patterns of IN0, IN1, and SEL.

A 32×32 crossbar circuit is investigated to verify the programming of the NBs on a practical scale. As shown in Fig. 12.7.4, 1kb NBs are integrated between the metal 4 and 5 layers, and aligned in the 32×32 matrix with a minimum line pitch of 2F. The row and column lines work for the programming lines as well as the signal routing paths. A half-selected programming scheme is used to set or reset the individual NBs. In the set operation, the selected row and column lines are set at Gnd and V_{set}, respectively, and all the other lines are set at $1/2V_{set}$. The set-voltage distribution needs to fall on a range between $1/2V_{set}$ and V_{set} for the correct programming. In Fig. 12.7.4, a tight set-voltage distribution of 0.8V and a wide write margin of 1V are confirmed, leading to a high immunity to a write disturb. A histogram of the path resistances is also characterized, where the distribution of the resistances is clearly separated as the on- and off-state, and its on/off ratio reaches around 10^5.

The high on/off ratio in the crossbar is imperative for signal integrity. In Fig. 12.7.5, the NBs diagonally placed are programmed to be the on-state and the others are to be the off-state. When a signal is transferred to a particular input, the signal via the programmed path is observed and there is no crosstalk between the other paths. Moreover, the crossbar shows an excellent re-programmability. After a half of the diagonal elements are set to the off-state, the opposite diagonal elements are then programmed to be the on-state, as shown in Fig. 12.7.5. The high on/off ratio after reprogramming proves that the NBs are applicable for the reprogrammable switches.

According to the above-mentioned results, a macro chip integrating a 1k programmable cell array can be designed and is currently being manufactured on a 90nm CMOS with five metal layers including the switch plane. The individual programmable cell has a functionality nearly equivalent to a conventional 4-input LUT. The switch plane is effectively overlaid on the logic plane. The switch plane involves 354 NBs and the underneath logic layer is composed of core transistors for the logic blocks and 3.3V MOX transistors for the logic buffers and programming transistors. For comparison, a standard-cell-based macro chip is also designed with the same architecture. Die micrographs of each chip are presented in Fig. 12.7.6. An 81% reduction in programmable cell area and a 72% reduction in the total chip area including the 1k programmable cells, programming drivers, and peripherals are achieved while using a practical design.

Acknowledgements:
A part of this work was supported by the Ministry of Economy, Trade and Industry, through New Energy and Industrial Technology Development Organization (NEDO), under the Project of Strategic Development for Energy Conservation Technology.

References:
[1] M. Lin, A. E. Gamal, Y. Lu, and S. Wong, "Performance benefits of monolithically stacked 3D-FPGA," *Int. Symposium on Field-Programmable Gate Arrays (FPGA'06) Dig. Tech. Paper*, pp. 113-122, Feb. 2006.
[2] T. Sakamoto, S. Kaeriyama, H. Sunamura, et al., "A Nonvolatile Programmable Solid Electrolyte Nanometer Switch," *ISSCC Diag. Tech. Papers*, pp. 290-291, Feb. 2004.
[3] K. Terabe, T. Hasegawa, T. Nakayama, and M. Aono, "Quantized conductance atomic switch," *Nature*, vol. 433, no. 7021, pp. 47–50, 2005.
[4] M. Tada, T. Sakamoto, K. Okamoto, et al., "Polymer Solid-Electrolyte (PSE) Switch Embedded in 90nm CMOS with forming-free and 10nsec Programming," *Int. Electron Device Meeting Tech. Dig.*, 16.5, Dec. 2010.
[5] S. Chiang, Rahim Forouhi, Wenn Chen, et al., "Antifuse Structure Comparison for Field Programmable Gate Arrays," *Int. Electron Device Meeting Tech. Dig.*, pp. 611-614, Dec. 1992.

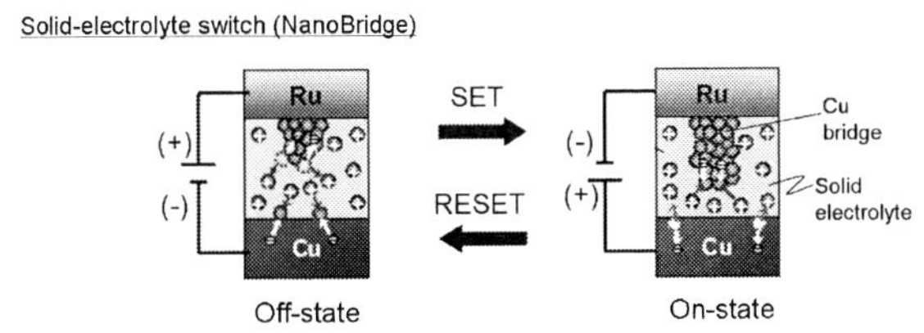

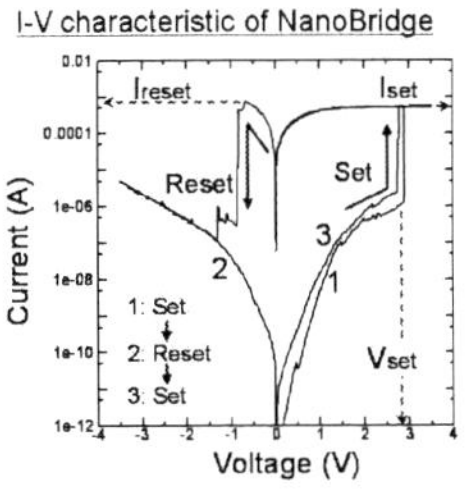

Comparison of typical characteristics

	NanoBridge	Antifuse (BEOL)	SRAM
Nonvolatility	Yes	Yes	No
Rewrite	> 1000	Write Once	–
Area	$4F^2$	$4F^2$	~$100F^2$
V_{set}	3.3V	> 8V	Vdd
τ_{set}	< 2ns	~1 msec	–
I_{set}	< 0.2mA	5mA	< 0.1 mA
I_{reset}	2mA (peak)	–	< 0.1 mA

Figure 12.7.1: NanoBridge characteristics.

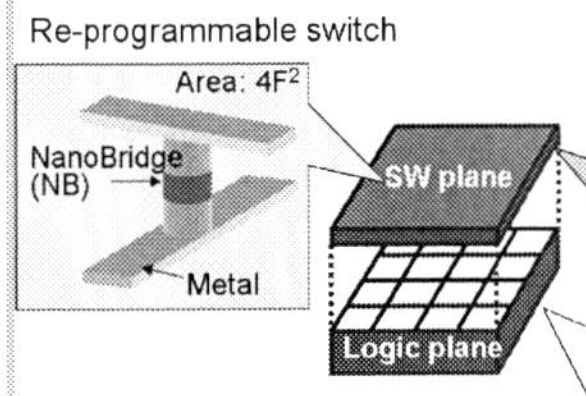

	NB-based programmable cell
Advantages	• Re-Programmable • Switch array ($4F^2$) over logics • Low C_{input} (~ 0.1fF) • Low power consumption
Design features	• Polarity-aware design • Programming Tr. sharing • NB for crossbar, bus holder and logic selectors • Multi-level programming driver for SET/RESET

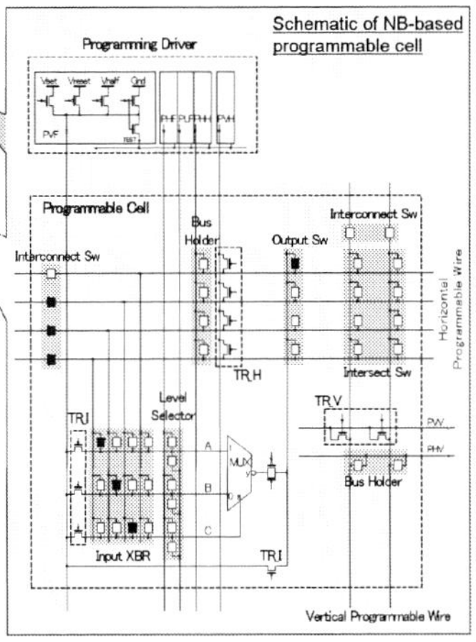

Figure 12.7.2: Concept of NB-based programmable cell array.

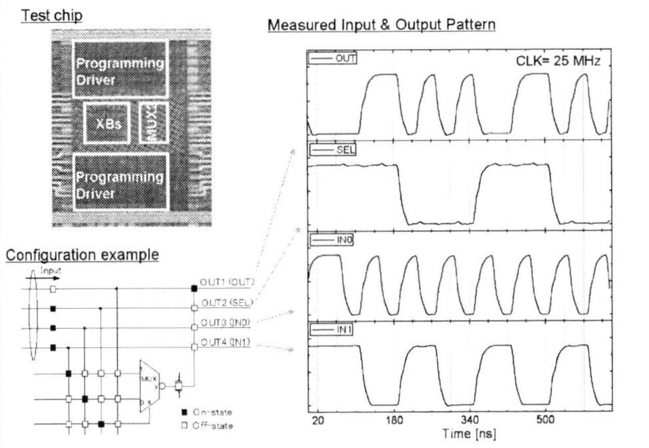

Figure 12.7.3: Measured pattern of a programmed cell in a test chip.

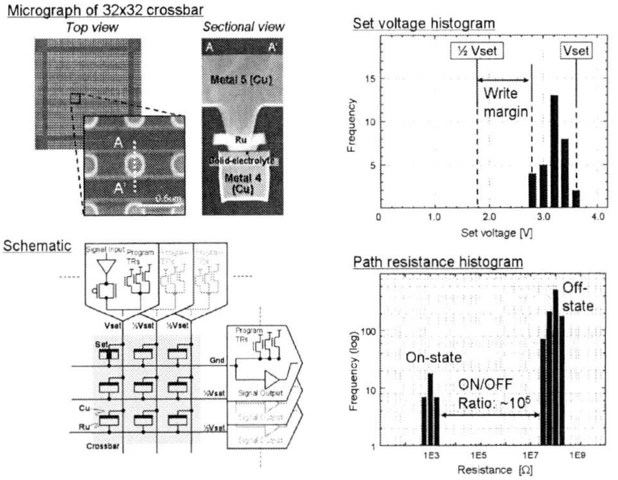

Figure 12.7.4: 32 x 32 crossbar switch.

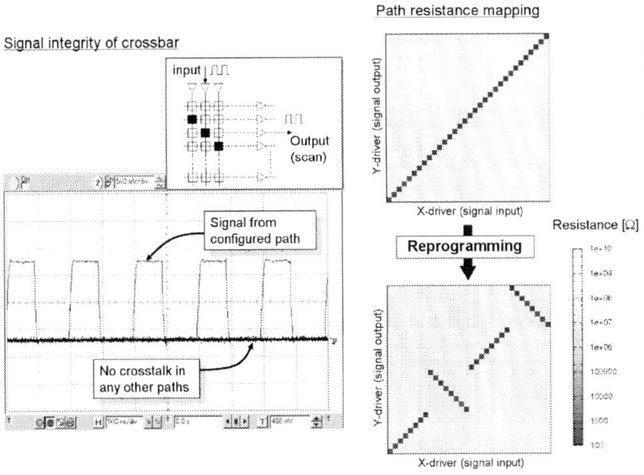

Figure 12.7.5: 32 x 32 crossbar switch operation.

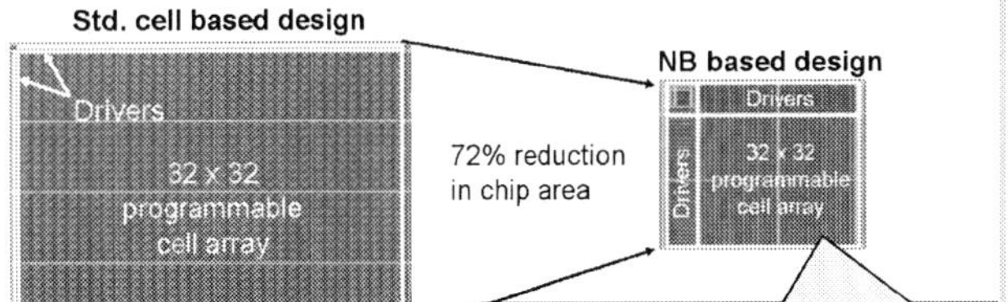

Implementation summary

	Std. cell based	NB based
Cell area	95 x 66 μm^2 (-81%)	38 x 31 μm^2
Chip area (Incl. 32x32 cells, Driver, I/O, etc.)	7.38 mm^2 (-72%)	2.09 mm^2
Transistor	Core	Core + 3.3 V
Metal layer #	5 Metal layers	
Path switch	Multiplexer	NanoBridge
Rout	~1 kΩ	~1 kΩ
C_{input}	~2 fF	~0.1 fF

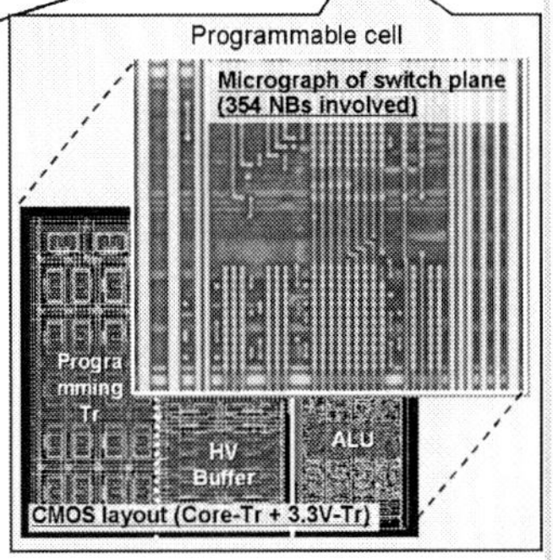

Figure 12.7.6: Physical implementation summary.

ISSCC 2011 / SESSION 12 / DESIGN IN EMERGING TECHNOLOGIES / 12.8

12.8 6W/25mm² Inductive Power Transfer for Non-Contact Wafer-Level Testing

Andrzej Radecki, Hayun Chung, Yoichi Yoshida, Noriyuki Miura, Tsunaaki Shidei, Hiroki Ishikuro, Tadahiro Kuroda

Keio University, Yokohama, Japan

Wafer-level testing allows detection of manufacturing errors and removes non-functional devices early in the fabrication process. It is commonly performed by placing a probe card directly above a device under test (DUT) and establishing a mechanical contact between them by means of an array of probes. This is an invasive technique that may damage fragile low-k dielectric layers and deform pads or bumps. More importantly, it is very difficult to flip thinned wafers face up for probing if they were earlier positioned face down for back grinding. Additional difficulty in handling of thinned wafers arises if dies have to be flipped again for bumping. One solution to above problems is wireless probing. With a number of proposed techniques for establishing high-speed inductive-coupling data links [3] and measuring DC analog signal wirelessly [4], the largest remaining obstacle to non-contact wafer-level testing is supplying power to the DUT. This is because wireless power transfer solutions reported earlier [1,5] do not provide an output power that is sufficient for testing modern high performance devices.

This paper presents an inductive power transfer system that is capable of delivering up to 6W of DC power to the on-chip load. The system, depicted in Fig. 12.8.1, is partitioned into 8 power transfer channels, each containing two circuits: a power transmitter implemented in the probe card, and an on-chip power receiver. This paper focuses on the design of the power receiver, as the receiver performance limits the output power of the whole system.

To maximize power output following steps are adopted. First, the inductor coils are sized to maximize their self and mutual inductances and to minimize their area. Then, the operating frequency is selected to maximize the quality factor of on-chip inductors and to minimize rectifier losses. Lower frequency reduces the quality factor of on-chip inductors, higher frequency results in lower rectifier efficiency. Second, a new coil arrangement is used to further increase the quality factor of inductors and to reduce their area. Third, a new synchronous rectifier capable of operating at chosen frequency is developed.

Figure 12.8.2 depicts the inductor coil arrangement. Coils are laid out in such a way that each of them couples with two others. By adjusting phases of coil currents it is possible to change the power transfer efficiency and the output voltage ripple. In the ideal case, if phases of inductor currents are shifted 180° with respect to each other, magnetic fluxes produced by neighboring segments of these inductors add. Conversely, if the currents are in phase, the magnetic fluxes cancel each other. This coupling phenomenon has a strong effect on the effective self inductance and the quality factor of inductors. The resulting power transfer efficiency at 180° phase difference is 50% higher than it is at 0°.

Although the best efficiency can be achieved at 180° phase difference, it is highly beneficial to reduce the phase offset to 135° in return for much smaller output voltage ripple. If 0° or 180° phase offset is used, the rectifier output current in non-continuous, leading to an excessive output voltage ripple, which is difficult to remove using a smoothing capacitor. 90° phase offset solves this problem but any mismatch in coil currents will affect the ripple. 45° and 135° phase offsets further decrease the voltage ripple and makes the circuit less sensitive to matching of inductor currents.

The empty space between power transmission coils provides a relatively interference-free environment that can be used for implementing data transmission channels using inductive coupling. Assuming symmetry of the inductor currents the calculated maximum magnetic flux through any sub-window of the 600×600µm area located in the center of the chip is smaller than 0.1% of the magnetic flux produced by a single power transmission coil. This space can fit 4 data transmission channels with a communication range of up to 100µm.

Figure 12.8.3 presents the design of the synchronous rectifier. At large output currents using a diode rectifier is not advisable, leading not only to increased power losses but also to the risk of latch-up. Synchronous rectifiers reduce both problems but their applications were conventionally limited to frequencies of up to tens of megahertz [6]. This limitation is caused by the delay in the feedback path formed by the comparator and buffers driving large transistor switches. The proposed circuit avoids this problem by employing a DLL circuit, which is synchronized to the input RF signal and produces all the control signals necessary for driving transistor switches. This concept relies on the fact that the RF signal envelope varies slowly and that small timing errors cause insignificant reduction of the rectifier efficiency. The circuit has to be assisted by a conventional diode rectifier during start-up, which can be obtained from the existing structure by resetting control signals *Ctrlp* and *Ctrln* using an input *En*.

The rectifier occupies 0.2mm² of the die area, which is determined mostly by $R_{DS,on}$ of MOS transistor switches. Because of this, the area is expected to scale quadratically with the process feature size. Resulting smaller gate area of transistor switches reduces power consumption of the rectifier, shifting the optimum carrier frequency higher, which consequently decreases inductor losses. If implemented in a 40nm CMOS process, the estimated area of the rectifier would be 0.01mm² and the optimum carrier frequency would approach 1GHz raising the power transfer efficiency and enabling proportional reduction of the area occupied by inductors.

The power receiver circuit is implemented using a 0.18µm, 1P6M, 2µm-thick top metal, p-substrate CMOS process. Test chips are thinned and mounted on top of transmitter coils to simulate a scenario in which power is transferred from the back side of the wafer. Figure 12.8.4 shows measurement results of the total power efficiency and supply voltage ripple at 135° and 180° phase offsets. The efficiency at 135° phase offset is about 6.9% lower than it is at 180° but this drop is compensated by an 85% reduction in the output voltage ripple. Figure 12.8.5 presents measured characteristics of a single power transfer channel. The peak power transfer efficiency with the diode rectifier reaches 12%. Synchronous rectification increases the efficiency to 17%.

The power transmitter is implemented as a discrete Class-E amplifier followed by an impedance-matching circuit. The matching circuit transforms impedance Z_T to resistive impedance Z_L that acts as a damping for the resonant circuit L_1, C_3 and thus enables the transistor M_1 to switch at low V_{DS} voltage.

Figure 12.8.6 presents a performance summary of the inductive power transfer system. The system provides up to 6W of output power to the load and occupies 25mm² of the die area. This compares favorably with other reported solutions [1,5]. As the width of inductor coils is optimized for maximum coupling at distances lower than 100µm the reduction of the figure of merit can be observed at larger distances. The power receiver circuit operates in synchronous mode at a carrier frequency of 150MHz and it is designed for a target supply voltage of 1.8V.

References:

[1] Y. Yuxiang, et al., "Simultaneous 6Gb/s Data and 10mW Power Transmission using Nested Clover Coils for Non-Contact Memory Card," Symp. on VLSI Cir, pp. 199-200, Jun. 2010.

[2] N. Miura, et al., "An 11Gb/s Inductive-Coupling Link with Burst Transmission," ISSCC Dig. Tech. Papers, pp. 298-299, Feb. 2008.

[3] Y. Yoshida, et al., "Wireless DC voltage transmission using inductive-coupling channel for highly-parallel wafer-level testing" ISSCC Dig. Tech. Papers, pp. 470-471, Feb. 2009.

[4] C. V. Sellathamby, et al., "Non-contact Wafer Probe Using Wireless Probe Cards," IEEE International Test Conference, 6 pp. - 452, Nov. 2005.

[5] S. Slupsky, et al., "A Non-Contact Interconnect Technology for Reducing Semiconductor Chip Cost," CMOS Emerging Technologies Workshop, 2006.

[6] S. Guo, et al., "An Efficiency-Enhanced CMOS Rectifier With Unbalanced-Biased Comparators for Transcutaneous-Powered High-Current Implants," IEEE J. Solid-State Circuits, vol. 44, no. 6, pp. 1796-1804, Jun. 2009.

978-1-61284-303-2/11 $26.00 © 2011 IEEE

ISSCC 2011 / February 22, 2011 / 4:45 PM

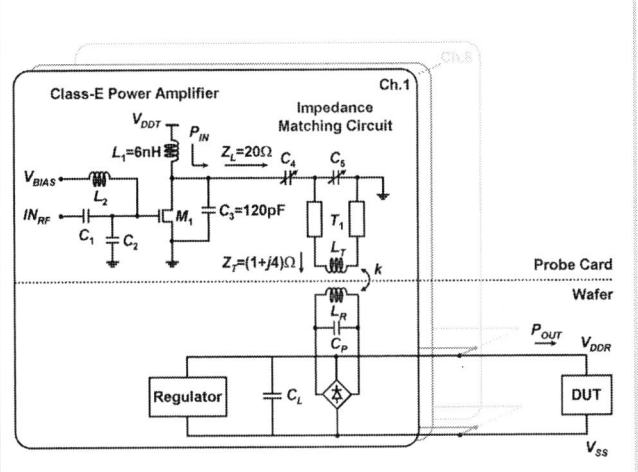

Figure 12.8.1: Power transceiver.

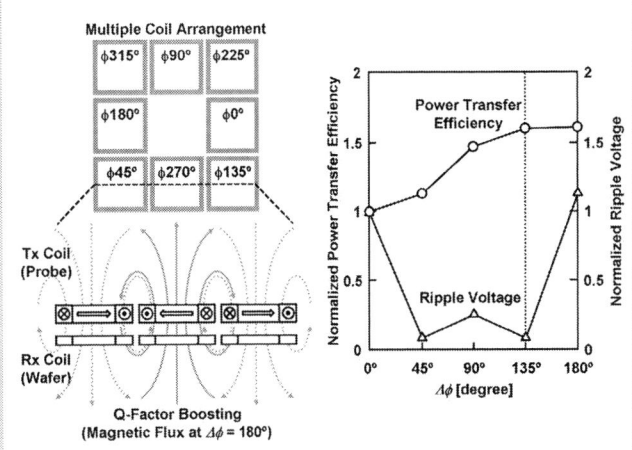

Figure 12.8.2: Multiple coil arrangement.

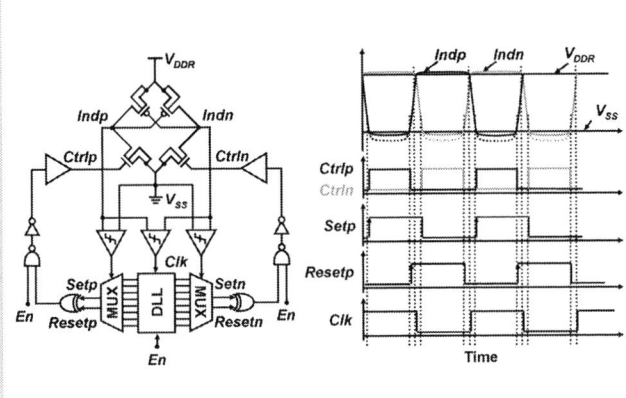

Figure 12.8.3: DLL-based synchronous rectifier.

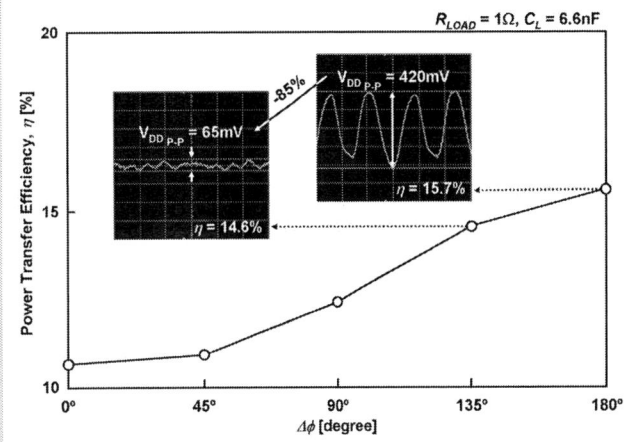

Figure 12.8.4: Measurement results.

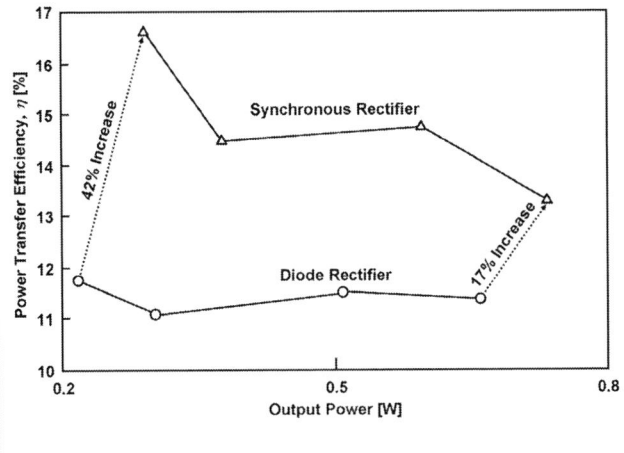

Figure 12.8.5: Measured efficiency of synchronous and diode rectifiers.

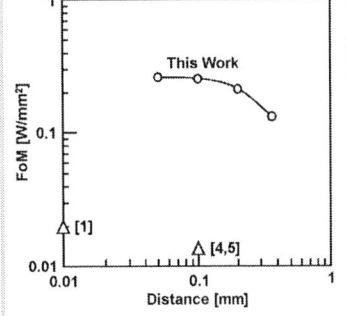

Received Power, P	6W
Die Area, A	5x5mm²
FoM = P / A	0.24W/mm²
Distance	0.05~0.32mm
Efficiency	17%
Carrier Frequency	150MHz
Ripple Voltage	65mV
Supply Voltage	1.8V
Process	0.18μm CMOS

Figure 12.8.6: Performance summary.

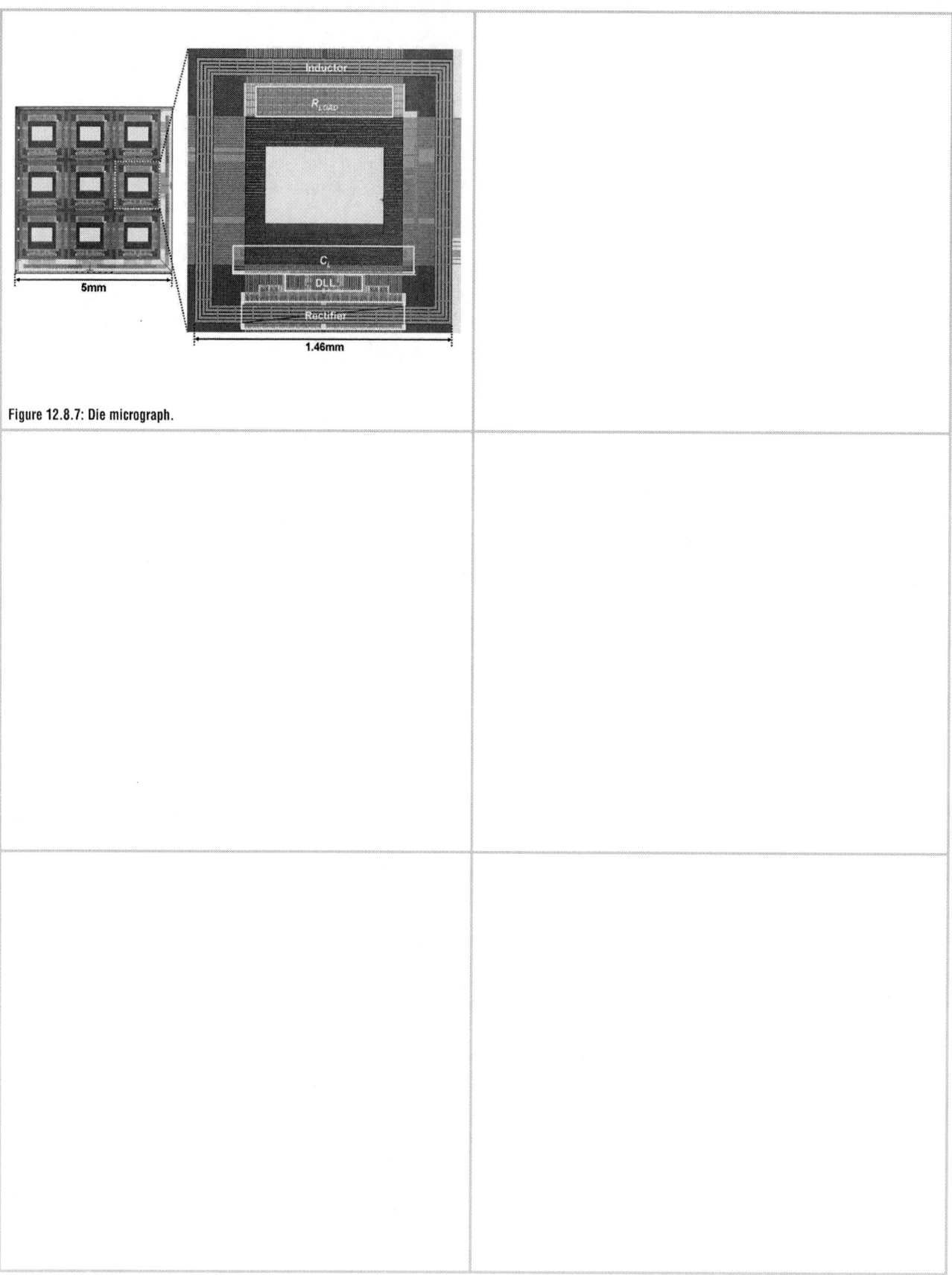

Figure 12.8.7: Die micrograph.

ISSCC 2011 / SESSION 12 / DESIGN IN EMERGING TECHNOLOGIES / 12.9

12.9 GHz-Range Continuous-Time Programmable Digital FIR with Power Dissipation that Automatically Adapts to Signal Activity

Mariya Kurchuk[1], Colin Weltin-Wu[1], Dominique Morche[2], Yannis Tsividis[1]

[1]Columbia University, New York, NY,
[2]CEA-LETI-MINATEC, Grenoble, France

GHz-range applications that operate in a variety of signal situations and/or multiple standards require highly programmable responses that cannot be provided by analog circuits. Conventional digital solutions suffer from aliasing, thus requiring a complicated antialiasing filter and/or extremely high clock speeds with high power dissipation. An alternative is continuous-time (CT) DSP [1], which uses level-crossing sampling [2] but without a clock. It offers activity-dependent power dissipation, is alias-free and has lower EMI emissions. This technique has so far been demonstrated in the voice band [3] but cannot be pushed beyond the MHz range because it involves extremely narrow pulse widths that cannot be handled by digital logic. This work bypasses this timing problem, enabling a five-orders-of-magnitude improvement in frequency capability compared to [3], thus making CT DSP a candidate for wideband GHz low-dynamic-range applications, such as those found in pulse radio, spectrum sensing, and channel equalization. Presented is a 3b 6-tap CT DSP system with wide programmability that is implemented in ST 65nm technology.

Figure 12.9.1 shows the system block diagram (top), one level of the ADC and encoder (middle), and representative waveforms (bottom). No clock is used. A CT flash ADC constantly tracks the input and toggles each thermometer-code output LEV_m whenever the mth quantization level is crossed, where $m=0...6$. The ADC does not convert the output to a binary word in order to avoid handling extremely small times, T_{LC}, between consecutive level crossings (which, for GHz signals, can be less than an inverter's propagation delay). The CT comparator is realized as an offset-compensated inverter with a capacitively-coupled input. During a reset phase, the capacitor at the mth-level is charged to the difference between the mth quantization level voltage $V_{q,m}$ and the inverter's threshold, so that during operation the inverter switches when the input crosses that level. Thus the input is represented by the sum of the LEV_m outputs. The capacitors are periodically refreshed, e.g. during input lulls.

Arbitrarily narrow pulses can still occur in LEV_m, which would get distorted or disappear in signal processing logic. To resolve this problem, alternative edges of LEV_m are encoded and then processed in parallel by the DSP as rising-edge R_m and falling-edge F_m signals. These signals toggle when the input crosses the mth threshold in a rising and falling direction, respectively. "Per-edge" encoding guarantees the minimum pulse width, shown in Fig. 12.9.1 as T_{EDGE}, to be extended to the minimum signal's period. While the edges of LEV_m are processed in parallel, the information in the timing of the edges is preserved. The resulting contribution to the output (Fig. 12.9.2, bottom) is the same as what would have been produced if the corresponding LEV_m signal had been processed directly. While the timing issue is resolved in the DSP, the ADC must still be able to generate narrow LEV_m signals, limiting the speed of the system. If the SNDR requirements are modest, a small number of bits can be used, which eases the ADC requirements. We have used 3 bits for a target SNDR of 20dB. An analog adder performs the functions of a decoder, adder and DAC simultaneously.

Figure 12.9.2 shows one of the rising-and-falling edge processors, comprised of tunable digital delay elements as tap delays and charge pumps as programmable filter coefficients. The nth-tap's charge pump, $n=0...5$, reconstructs the per-level signal resulting from processing the per-edge signals $R_{m,n}$ and $F_{m,n}$. The charges produced at each tap of every level are deposited onto a load capacitor serving as the adder, the DC value across which is maintained by a control circuit. When $R_{m,n}$ toggles, it triggers a pulse generator, which turns on a controlled current source for the duration of the pulse $T_{PW}\approx65$ ps, thus depositing a charge on the capacitor. When $F_{m,n}$ toggles, the charge is removed from the capacitor, returning the voltage to the original value. The filter coefficients are programmed via biases V_{POn}, V_{NOn} and can be designed using conventional dig-

ital filter design tools. The energy-efficient tunable delay element [4] is a buffered current-starved inverter with positive feedback, which triggers when an internal node voltage reaches a threshold. The tap delays are digitally programmable in the range of 95ps to 170ps via biases V_{PD}, V_{ND}.

The CT DSP system was measured though a variable-gain input buffer and an output buffer, both providing 50Ω load match. These buffers, as well as input AC coupling and ADC speed, result in a usable system -3dB bandwidth 0.8 to 3.2GHz. Measured results are presented directly without post processing. Figure 12.9.3 shows output power versus input frequency for several cases, indicating a good match to the ideal responses. Figure 12.9.4 shows the output spectra of the system in several DSP configurations for two-tone inputs. The spectra contain tones at the input frequency, as well as harmonics, intermodulation products and half-harmonics, the latter due to the per-edge encoding. These are the only components due to quantization; there is no quantization noise floor in this CT system [1,3].

The parameters and performance of the system, which occupies a core area of 0.08mm², are summarized in Fig. 12.9.5. SFDR (ratio of the power of the fundamental to that of the maximum spur) and SNDR are given for both a "1-tap DSP", basically a cascade of the ADC and the DAC (a worst-case situation, as it has the least signal gain and no filtering of the noise and harmonics), and for a representative filter. The results for the latter are better because, while noise increases when more taps are used, the signal gain is also higher, and there is some filtering of noise and distortion. Another interesting test involved an unwanted signal outside the system usable bandwidth, at 4GHz, show in top left of Fig. 12.9.4; this resulted in an in-band SNDR of over 19.5dB, in contrast to the case of an equivalent Nyquist-rate system, which would alias the unwanted tone into the baseband and degrade the SNDR below 0dB. The system can handle a 3.2GHz input, yielding an effective sampling rate of (3.2GHz)(14samples/cycle)=45GS/s. A state-of-the-art clocked DSP [5] offers higher resolution but operates with a throughput of 2.1GS/s, corresponding to a maximum input frequency of 1.05GHz.

The power dissipation of the system depends on input activity, and can be shown to be given by:

$$P = P_{\mathrm{STATIC}} + \left(E_{\mathrm{S,ADC}} + E_{\mathrm{S,DSP}} \times (\# \text{ of taps})\right) \times \left\lfloor \frac{A_{\mathrm{IN}}}{A_{\mathrm{MAX}}/M} \right\rfloor \times f_{\mathrm{IN}}$$

where E_S is the energy dissipated in processing one sample ($E_{\mathrm{S,ADC}}$=36fJ, $E_{\mathrm{S,DSP}}$=40fJ), P_{STATIC} is the static power (1.1mW), which includes the power of the ADC and the DC control circuit in the analog adder, M is the number of levels (7), A_{MAX} is the input range and A_{IN} is the amplitude of the signal with a frequency of f_{IN}. Test features and bias generation circuits, designed for flexibility in this exploratory version, consume an additional 3.3mW of power. Figure 12.9.5 shows that the power dissipation of the system, excluding all bias and test circuits, varies linearly with input frequency in the band of interest and increases with input amplitude. As the frequency goes beyond 3.2GHz, the outer-most ADC levels fail to trigger, and the power begins to decay.

The clockless system presented makes programmable CT digital filtering possible in the 0.8 to 3.2GHz range with graceful degradation at higher frequencies and no aliasing, offering activity-dependent power dissipation.

References:
[1] Y. Tsividis, "Continuous-time digital signal processing," *Electron. Letters*, vol. 39, no. 21, pp. 1551-1552, Oct. 16, 2003.
[2] N. Sayiner, H. N. Sorensen, and T. R. Viswanathan, "A level-crossing sampling scheme for A/D conversion", IEEE Trans. Circ. Syst. II, vol. 43, no. 4, pp. 335-339, April 1996.
[3] B. Schell and Y. Tsividis, "A continuous-time ADC/DSP/DAC system with no clock and with activity-dependent power dissipation," *IEEE J. Solid-State Circuits*, vol. 42, pp. 2472-2481, Nov. 2008.
[4] M. Kurchuk and Y. Tsividis, "Energy efficient asynchronous delay element with wide controllability," in *Proc. 16th IEEE ISCAS*, Paris, France, pp. 3837-3840, May 2010.
[5] A. Agrawal et al., "A 320mV to 1.2V on-die fine-grained reconfigurable fabric for DSP/Media accelerators in 32nm CMOS", *2010 IEEE ISSCC Digest*, vol.

ISSCC 2011 / February 22, 2011 / 5:00 PM

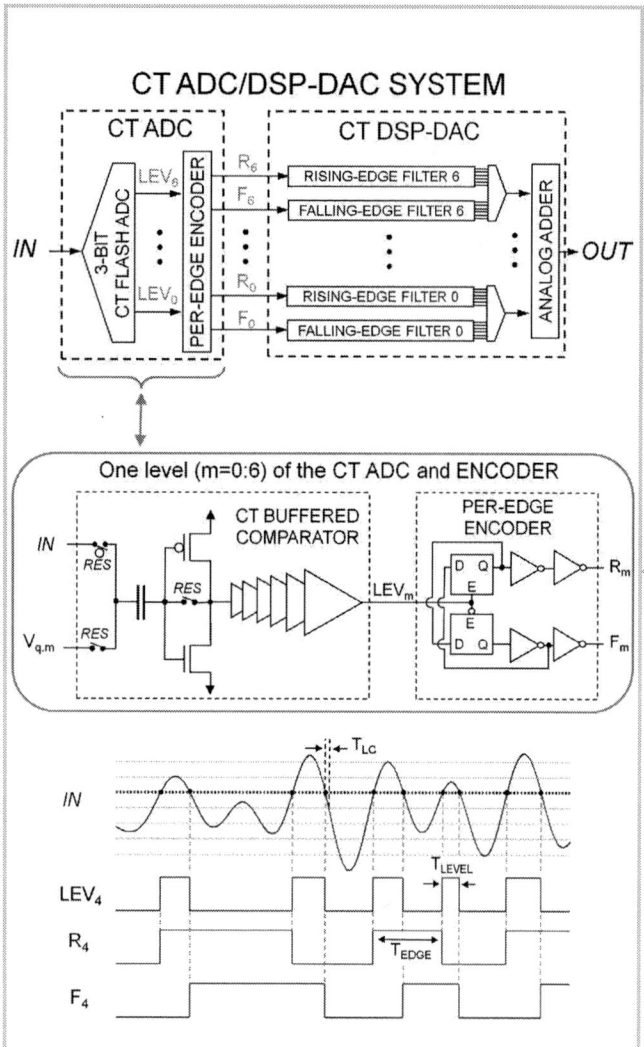

Figure 12.9.1: Block diagram of the continuous-time ADC/DSP-DAC and example signals.

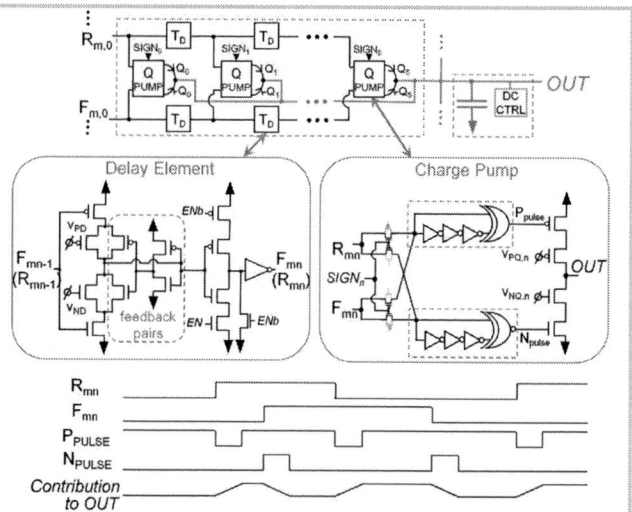

Figure 12.9.2: Rising-/falling-edge processors, circuit of a delay cell, simplified circuit of a charge pump and example signals.

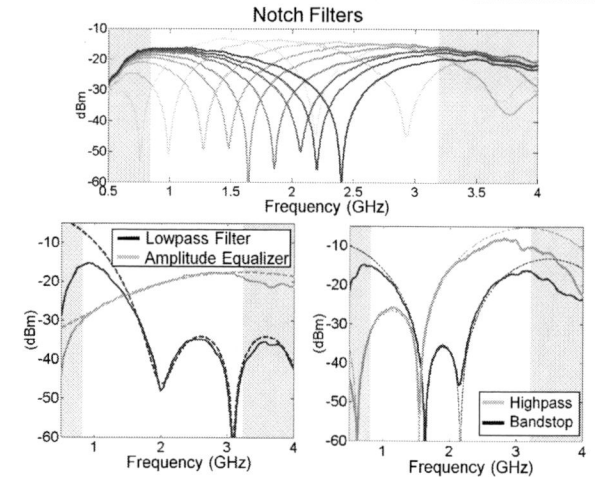

Figure 12.9.3: Output power vs. frequency for a full-scale input, including the effects of AC coupling and input/output buffers (ideal responses, dashed, are vertically aligned with measurements for an easy comparison).

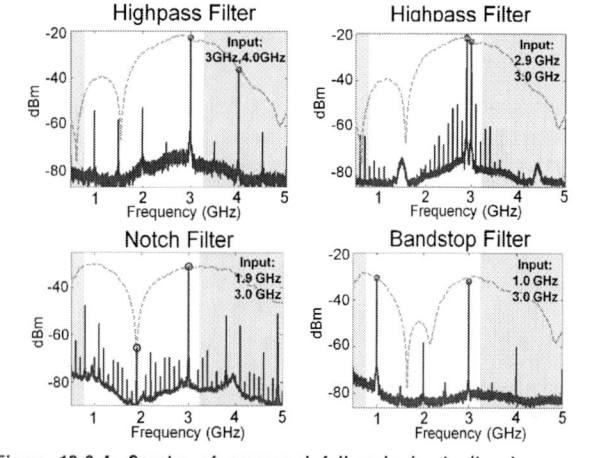

Figure 12.9.4: Spectra of processed full-scale inputs (two-tone, equal amplitude) for several filter configurations (measured frequency responses shown dashed). Measurements include the effects of roll-off and non-linearity of the input driver and output buffers.

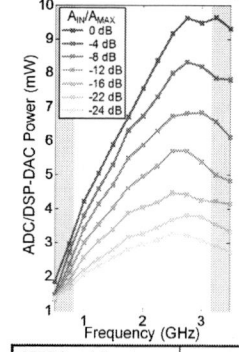

Process/supply voltage	65nm /1.2V
Input Range	0.3 to 0.9 V
ADC Resolution	3 bits
DSP programmability	1 to 6 taps
tap delay range	95 to 170 ps
System bandwidth	0.8 GHz 3.2 GHz
Maximum effective sampling rate	45 GHz

Core area	0.08 mm²
ADC area	0.0036 mm²
DSP-DAC area (without bias and test circuits)	0.029 mm²

SFDR for a full-scale 1GHz input, ADC-DAC*/6-tap Filter**	23.5dB/ 31.5dB
Minimum In-band SNDR for a full-scale 1GHz input ADC-DAC*/6-tap Filter**	20.3dB/ 25.3dB

* no filtering
** Lowpass filter with 2.1GHz bandwidth

ADC static power (in reset)	0.82 mW (1.45mW)
ADC energy per sample per tap	36 fJ
DSP-DAC static power	0.27 mW
DSP-DAC energy per sample per tap	40 fJ
Typical DSP-ADC bias and test circuit power	3.3 mW

Figure 12.9.5: System power dissipation and a summary of system parameters and measurement results.

978-1-61284-303-2/11 $26.00 © 2011 IEEE

ISSCC 2011 PAPER CONTINUATIONS

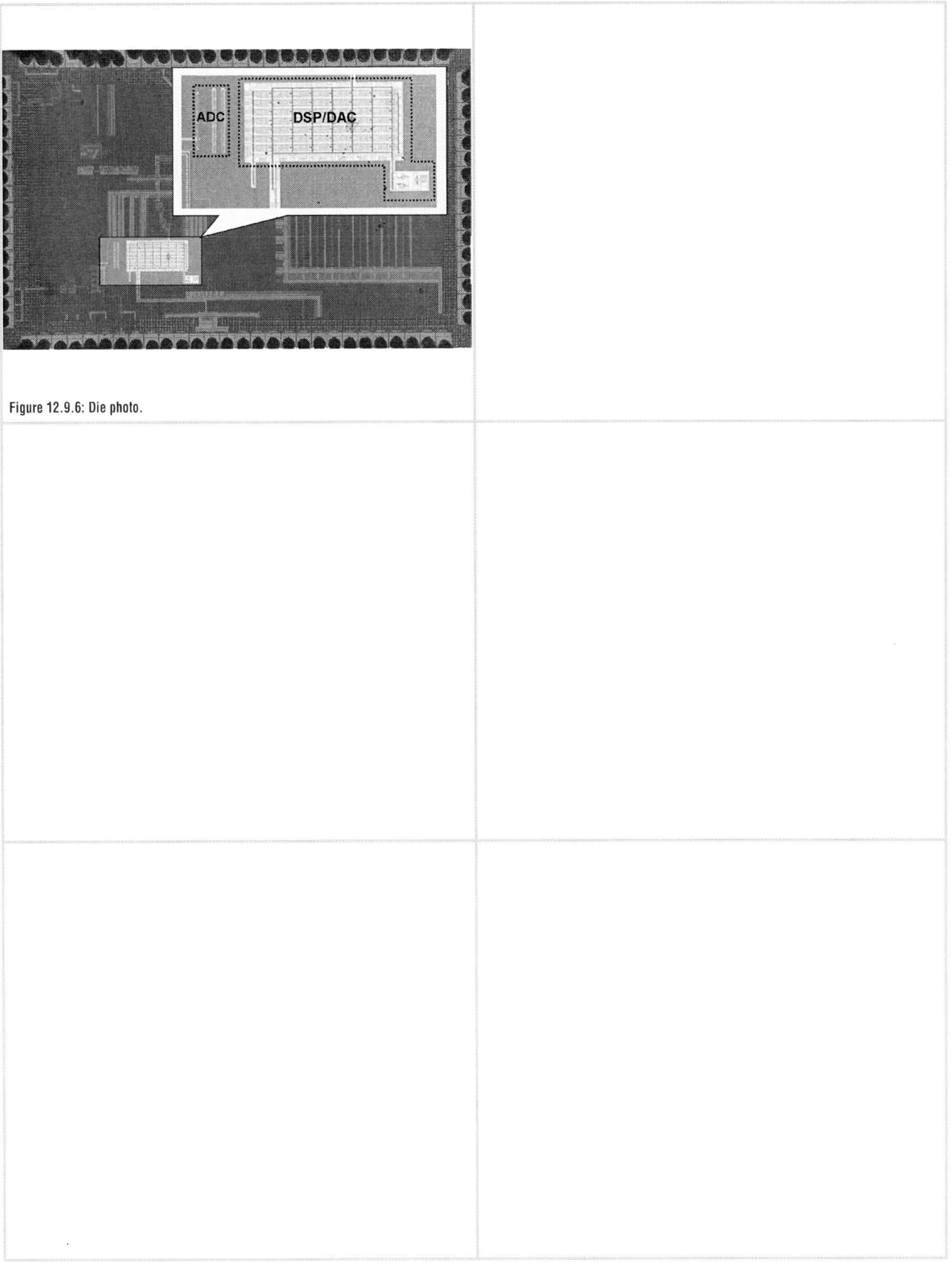

Figure 12.9.6: Die photo.

ISSCC 2011 / SESSION 13 / ANALOG TECHNIQUES / OVERVIEW

Session 13 Overview / Analog

Analog Techniques

Session Chair: *Jafar Savoj, Xilinx, San Jose, CA*

Session Co-Chair: *Kimmo Koli, ST-Ericsson, Turku, Finland*

Design of state-of-the-art analog circuits for energy-efficient power delivery and high-performance amplification pose significant challenges in modern electronic systems. Advances with LED drivers have enabled a new frontier for energy efficient lighting technology. Progress in audio amplifier design has resulted in longer battery life for mobile devices and has affected our daily life with prolonged delivery of entertainment and information. Operational and instrumentation amplifiers have made interfacing of our real analog world to digital processing systems possible.

This session starts with description of two LED drivers as high brightness and high efficiency of LED devices begin a new era for lighting and LCD backlight technology and demonstrate superior performance in comparison to incandescent lamps. With projected growth of LED technology, design of high efficient drivers is an essential ingredient of green systems.

Paper 13.1 (Fairchild Semiconductor) describes a simple 3-pins buck-type LED driver with intelligent power factor correction function using digital control algorithm. The LED lamp driver is based on a peak- current-controlled PWM method. The proposed controller utilizes a sine-wave reference signal, in phase with AC line voltage, and synchronizes the LED current phase accordingly. The circuit achieves power factors of 0.98 and 0.92 and the THD of 18% and 16% at 110V_{ac} and 220V_{ac} with 5W LED load, respectively.

Paper 13.2 (Oregon State University; National Semiconductor) describes a 1.2A Buck-Boost LED driver, fabricated in a 0.5µm CMOS process, which regulates the LED current without any series current source or series sense resistor. The circuit employs a highly accurate current sensing scheme to directly regulate LED current and achieve more than 13% efficiency improvement. The LED driver operates at switching frequencies up to 2MHz with an off-chip 2.2µH inductor and a 10µF capacitor. The converter can drive the full load at an LED forward voltage of 3.6V. The proposed error-averaged current sensing technique achieves better than 2.8% of sensing accuracy.

High power efficiency, small size and reduced heat dissipation are highly desirable in battery-powered mobile systems and switched-mode Class-D amplifiers can readily satisfy these requirements. A full adoption of class-D technology in mobile systems has been limited due to limitations on signal distortion and noise, poor power supply noise rejection, electromagnetic interference (EMI) and the requirements for external LC filters. Paper 13.3 (Dialog Semiconductor) demonstrates a filter-less class-D audio amplifier with uniform pulse-width modulation architecture that achieves 0.0012% THD plus noise, delivering 1.2W into an 8Ω load with 93% power efficiency and 96dB of PSRR. The amplifier is fabricated in standard CMOS process and packaged in WLCSP with total chip area of 1.44mm². It achieves 103dB SNR with quiescent current consumption of 4mA and maximum output power of 3.1W to loads as small as 4Ω. The circuit operates with a supply voltage ranging from 2.5V up to 5.5V.

Many operational amplifiers utilizing chopping techniques successfully reduce offset and low frequency drift, yet suffer from unwanted modulated output ripples. Analog techniques target unconditional stable chopper operation with the smallest input noise PSD.

Paper 13.4 (Analog Devices) reports a 5.9nV/√Hz unconditionally stable chopper operational amplifier with 1.47mA current at 2.5-5.5V supply voltages and 1.26mm² die area, achieved by phase compensation using current attenuation. In addition, adaptive clock level shift and backgate biasing for the input chopping allows optimization for noise and offset, realizing 0.78µV maximum offset with a worst-case 28.3nV/C drift. The circuit is fabricated in 0.35µm CMOS.

Introduction of a gain error reduction loop suppresses modulated output ripples of a chopping amplifier and improves gain accuracy, gain drift, and linearity of current-feedback instrumentation amplifiers without resorting to trimming techniques. As described in Paper 13.5 (TU Delft), the gain error reduction loop continuously cancels the gain error of the amplifier and achieves gain error as low as 0.06%. This result represents a 4× improvement in power efficiency compared to the prior art. The circuit consumes 290µA from a 5V supply.

978-1-61284-303-2/11 $26.00 © 2011 IEEE

ISSCC 2011 / February 22, 2011 / 1:30 PM

Many sensors demand energy-efficient precision voltage sensing interface with low noise performance at very low frequencies. Paper 13.6 (Robert Bosch) describes a digital sensor interface that minimizes noise folding with boxcar sampling and stabilizes the overall interface gain with global feedback. The combination of nested chopping and auto-zeroing allows the interface to thoroughly reject low-frequency noise. The circuit is fabricated in a 0.35μm CMOS process and achieves a noise floor of 6.7nV/√Hz down to 0.1mHz while dissipating 6.6mW. The circuit draws only 2mA from a 3.3V supply.

High-voltage operational amplifiers are also used in signal conditioning circuits for industrial applications to ensure compatibility with legacy equipment. In such applications, good DC precision, low offset, offset drift and noise are essential requirements. Paper 13.7 (Texas Instruments) describes a 36V JFET-input bipolar operational amplifier with a maximum offset drift of 1μV/°C over a temperature range of -40 to 125°C, which represents a 3x improvement on the state-of-the-art. This is achieved with a drift-compensating circuit incorporated in the input stage that relies on a wafer-level 2-temperature laser trimming method. The amplifier has a GBW of 11MHz, a flat-band noise of 5.1nV/√Hz, a slew-rate of 20V/μs, a -126dB (0.00005%) total harmonic distortion plus noise ratio, and a quiescent current of 1.8mA. This combination of high slew rate and good noise-to-power ratio is accomplished through the use of a linearized class-AB boosting circuit in the input stage. Maximum supply voltage is 40V.

Fully differential operational amplifiers are considered a superior method of driving 100 to 500 MS/s ADCs with 12 to 16 bits of resolution. Prior art has utilized dielectrically isolated (DI) Silicon Germanium complementary bipolar processes for implementation of such blocks. Paper 13.8 (Intersil) describes the implementation of an operational amplifier using a 0.18μm SiGe NPN-only RF BiCMOS process. The circuit achieves 108dBc of IM3 at 100MHz using a feedforward nested Miller architecture. Operating from a 3.3V supply voltage, the circuit delivers a $2V_{pp-diff}$ composite two tone output at 100MHz into 200 ohms. The circuit consumes only 120mW from a 3.3V supply and achieves an input noise of 0.85nV/√Hz and a 3dB bandwidth of 2.2GHz.

Presenters:

13

1:30 PM

13.1 A Simple LED Lamp Driver IC with Intelligent Power-Factor Correction

J. Hwang, Fairchild Semiconductor, Bucheon, Korea

3:45 PM

13.5 A Current-Feedback Instrumentation Amplifier with a Gain Error Reduction Loop and 0.06% Untrimmed Gain Error

R. Wu, Delft University of Technology, Delft, The Netherlands

2:00 PM

13.2 A 1.2A Buck-Boost LED Driver with 13% Efficiency Improvement Using Error-Averaged SenseFET-Based Current Sensing

S. Rao, Oregon State University, Corvallis, OR

4:15 PM

13.6 A 6.7nV/√Hz Sub-mHz-1/f-Corner 14b Analog-to-Digital Interface for Rail-to-Rail Precision Voltage Sensing

C. D. Ezekwe, Robert Bosch, Palo Alto, CA

2:30 PM

13.3 Filterless Integrated Class-D Audio Amplifier Achieving 0.0012% THD+N and 96dB PSRR When Supplying 1.2W

M. Teplechuk, Dialog Semiconductor, Edinburgh, United Kingdom

4:30 PM

13.7 A 36V JFET-Input Bipolar Operational Amplifier with 1μV/°C Maximum Offset Drift and -126dB Total Harmonic Distortion

M. F. Snoeij, Texas Instruments, Erlangen, Germany

3:15 PM

13.4 A 5.9nV/√Hz Chopper Operational Amplifier with 0.78μV Maximum Offset and 28.3nV/°C Offset Drift

Y. Kusuda, Analog Devices, Wilmington, MA

4:45 PM

13.8 A 3.3V-Supply 120mW Differential ADC Driver Amplifier in 0.18μm SiGe BiCMOS with 108dBc IM3 at 100MHz

G. F. Luff, Intersil, Harlow, United Kingdom

978-1-61284-303-2/11 $26.00 © 2011 IEEE

ISSCC 2011 / SESSION 13 / ANALOG TECHNIQUES / 13.1

13.1 A Simple LED Lamp Driver IC with Intelligent Power-Factor Correction

Jong Tae Hwang, Kunhee Cho, Donghwan Kim, Minho Jung, Gyehyun Cho, Seunguk Yang

Fairchild Semiconductor, Bucheon, Korea

High-brightness and high-efficiency LED technology has opened a new era in lighting devices [1, 2] and LCD backlight devices [3]. Prior to emergence of high-performance LED devices, the fluorescent lamps have been in a strong position in lighting systems as compared to the incandescent lamps since they have a better efficiency [4]. However, LED dramatically improves the performance in both life time and efficiency. In addition, LED is gaining popularity as a green solution. If the manufacturing cost issue is going to resolve in near future, widespread use of LED lighting is being expected. Therefore, LED related products will be placed in a very important position and the market requires effective driving solutions.

A step-down buck converter is one of the widely used topologies in off-the-shelf LED lamp drivers owing to its simplicity of use. However, the conventional buck driver which uses DC supply voltage from the rectified AC input has the apparent reactive power loss due to the different shape of supply voltage and current of the AC line and LED driver circuit. Since, lots of LED lamps are required in general to illuminate a large area with uniform brightness, the power factor (PF) and total harmonic distortion (THD) of AC line current are key characteristics of a driver IC. The PF can be improved by a controller-based active or valley-fill passive PF control methods, however, these approaches both increase manufacturing cost and power consumption. In addition, the conventional method requires a bulky capacitor to implement the rectifier. In general, the lifetime of LED reaches about 50,000 hours, however, a bulky capacitor such as electrolytic capacitor has a much shorter lifetime of about 3,000 hours. Thus, the capacitor determines the life time of the LED lamp system. This paper proposes the simple 3-pin buck-type LED lamp driver which has a power factor correction (PFC) function. This IC can directly drive the LEDs from AC outlet source, and does not require bulky capacitors and line voltage sense resistors for PFC.

Figure 13.1.1 shows the simplified schematic of the proposed LED lamp driver based on a peak current controlled PWM method. The conventional buck-type LED controller compares the current from switching MOSFET with DC level of reference signal, and outputs a constant duty. However, the proposed controller uses a sine wave reference signal which is in phase with AC line voltage and controls the LED current to be in phase with AC line voltage. As a result, the PF becomes very close to unity and the LED current shows low THD. Since, in the conventional continuous-conduction mode (CCM) PFC controller, the resistive divider is used to extract the sine reference from the AC line and thus this method increases power consumption, variable peak current according to the amplitude of AC line voltage, and is hard to integrate on an IC.

To sense the AC line voltage without using a resistor divider, a high-voltage JFET is used. When the JEFT drain voltage is very high, the JFET can be considered as a current source and its source voltage sustains a constant value (=9.7V). When the JFET drain voltage lowers and the JFET operates in the triode region, it can be considered as a resistor. Thus, the source voltage becomes lower than 9.7V. Consequently, it is possible to detect a zero-crossing time of the AC input by sampling the source voltage of JFET when the power MOSFET is off (SH_CLK). The digital sine-wave generator (DSG) uses this AC zero-crossing detection signal (AC_ZCD) to regenerate sine-wave information as shown in Fig. 13.1.2. The timing generator samples AC_ZCD and counts the period of AC_ZCD (T_{AC_LINE}) and AC_ZCD LOW time (T_{ZCD}) by using fixed 86kHz clock signal (CLK0). Since, the period of SinClk is determined by integer{(T_{AC_LINE}/T_{CLK0})/64}, the residue, mod{(T_{AC_LINE}/T_{CLK0})/64}, could distort the sine wave and decrease the overall THD performance of the converter. Therefore, the random voting is done by digital $\Delta\Sigma$ modulator. Finally, the timing generator makes 64 SinClk cycles to generate 6b digital sine-wave based on a sine pattern table without any residue.

Figure 13.1.3 shows the full schematic of the proposed LED driver. The driver contains a shunt regulator which generates 9.0V VDD from AC line voltage to supply whole power to the IC. To protect the IC in the unusual conditions, under voltage lockout (UVLO), over voltage protection (OVP) and thermal shutdown (TSD) block are implemented. As explained, the DAC generates the sine wave reference when AC_ZCD signal is detected. However, if AC_ZCD is not detected over consecutive 7 cycles, operation mode automatically changes from AC to DC mode. At DC mode, DAC generates constant DC level. The DC level becomes $1/\sqrt{2}$ times lower than peak of sine wave. Thus, the average rms current of LED sustains a constant level in both modes.

The chip is fabricated in a 0.5μm BCDMOS process which supports 500V LDMOS and 700V JFET. Figure 13.1.4 shows the drain voltage and LED current waveforms at AC mode and DC mode with soft start. After soft start, the profile of LED current quite well follows the sine wave form (in AC mode). In AC mode, the peak current level is 350mA. At DC mode, it can see that current peak reduces to 250mA. To verify the operation of timing generator, the frequency of AC line voltage is varied from 45Hz to 100Hz and vice versa. The timing generator successfully tracks the AC line frequency and the IC performs PFC function well as shown in Fig. 13.1.4. When AC frequency abruptly changes from 100Hz to 45Hz, during the transition period, the IC operates in DC mode as the AC_ZCD signal is not detected within 7 cycles of SinClk after DAC_OUT is zero. The driver returns to AC mode after sensing AC_ZCD signal over 7 times.

The proposed LED driver can operate at the universal input range from 50V_{AC} to 320V_{AC} at 60Hz AC mode and up to 450V_{DC} at DC mode. The test circuit uses 5.5mH inductor, variable number of LEDs from 5 to 16 (2.5 to 7W LED load) and 1μF MLCC capacitor. Since, the proposed LED driver follows the peak current control method, the peak current level is almost constant against the different AC voltage and the number of LEDs as shown in Fig. 13.1.5. However, the rms current level is reduced when the number of LEDs increases since the current ripple is proportional to the number of LEDs. The variation of LED peak current is under ±3% over the temperature range from −20°C to 160°C. The achievable PF, as shown in Fig. 13.1.6, is 0.98 and 0.92 at 110V_{AC} and 220V_{AC} with 5W LED load, respectively, and the THD is under 20%. The PF and THD are measured with AC input common-mode noise filter. Since the filter distorts current shape, PF and THD are slightly deteriorated. Nevertheless, the THD meets the standard IEC 1000-3-2 (EN61000-3-2) at 110V_{AC} and 220V_{AC} regardless of the number of LEDs. Figure 13.1.7 shows the micrograph of the proposed LED driver assembled in an SOT-223 package.

References:
[1] Steve Winder, *Power Supplies for LED Drivers*, Elsevier, 2008.
[2] R.P. de Vries, "Leaner and Greener: Adapting to a Changing Climate of Innovation," *ISSCC Dig. Tech. Papers*, pp. 8-13, Feb., 2009.
[3] S. Hong, J. Han, D. Kim, and O. Kwon, "A Double-Loop Control LED Backlight Driver IC for Medium-Sized LCDs," *ISSCC Dig. Tech. Papers*, pp.116-117, Feb., 2010.
[4] J. Hwang, D. Kim, M. Jung, and D. Kim, "550V SiP Compact Fluorescent Lamp Ballast IC," *ESSCIRC*, , pp.576-579, Sep., 2006.

ISSCC 2011 / February 22, 2011 / 1:30 PM

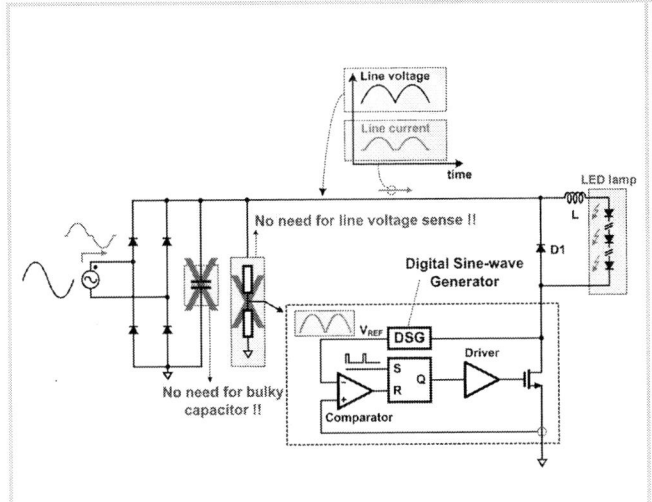

Figure 13.1.1: Simplified schematic of the proposed LED driver.

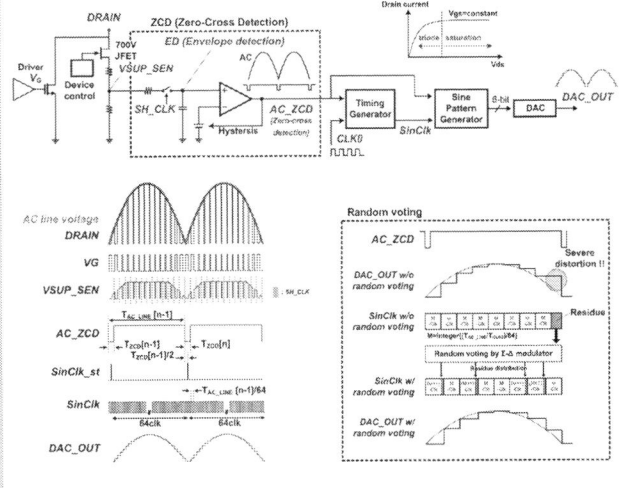

Figure 13.1.2: Principle of the digital sine-wave generator.

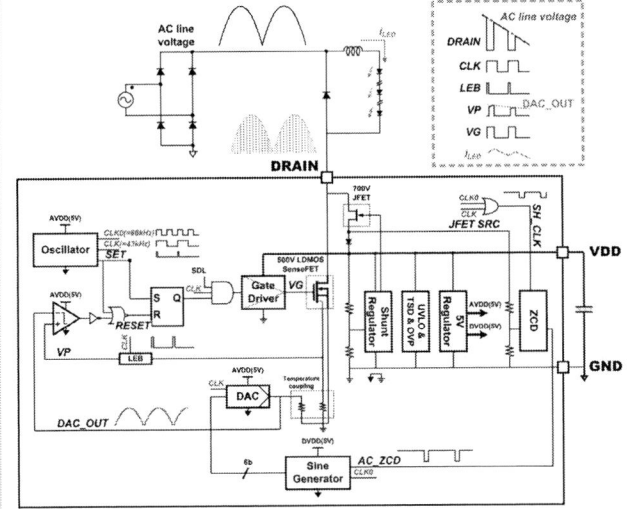

Figure 13.1.3: Full diagram of the proposed LED driver.

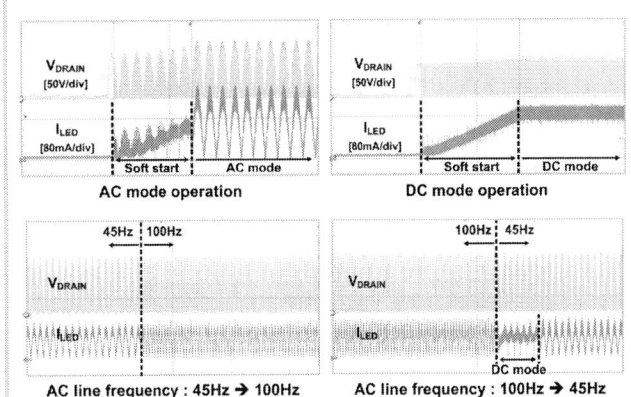

Figure 13.1.4: Measured waveform showing the AC and DC operation with soft start. The LED driver tracks the AC line frequency by changing the digital sine-wave frequency.

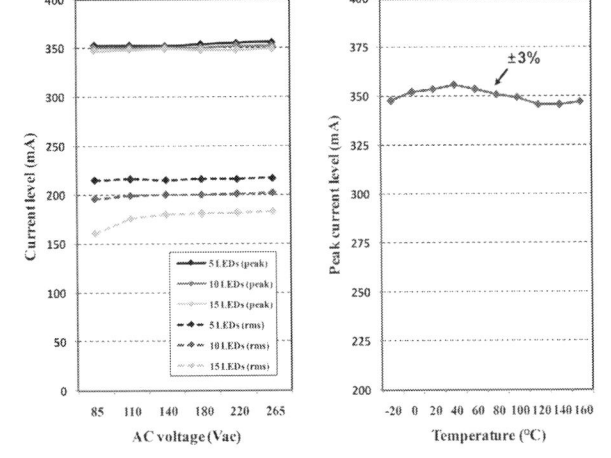

Figure 13.1.5: Measured peak current level and rms current level at AC voltage variation with different number of LEDs, and measured peak current level at temperature variation.

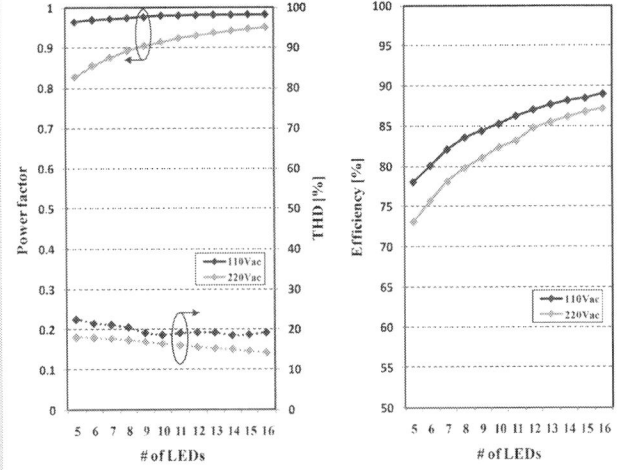

Figure 13.1.6: Measured power factor, THD, and efficiency at different number of LEDs and AC voltage.

13

978-1-61284-303-2/11 $26.00 © 2011 IEEE

ISSCC 2011 PAPER CONTINUATIONS

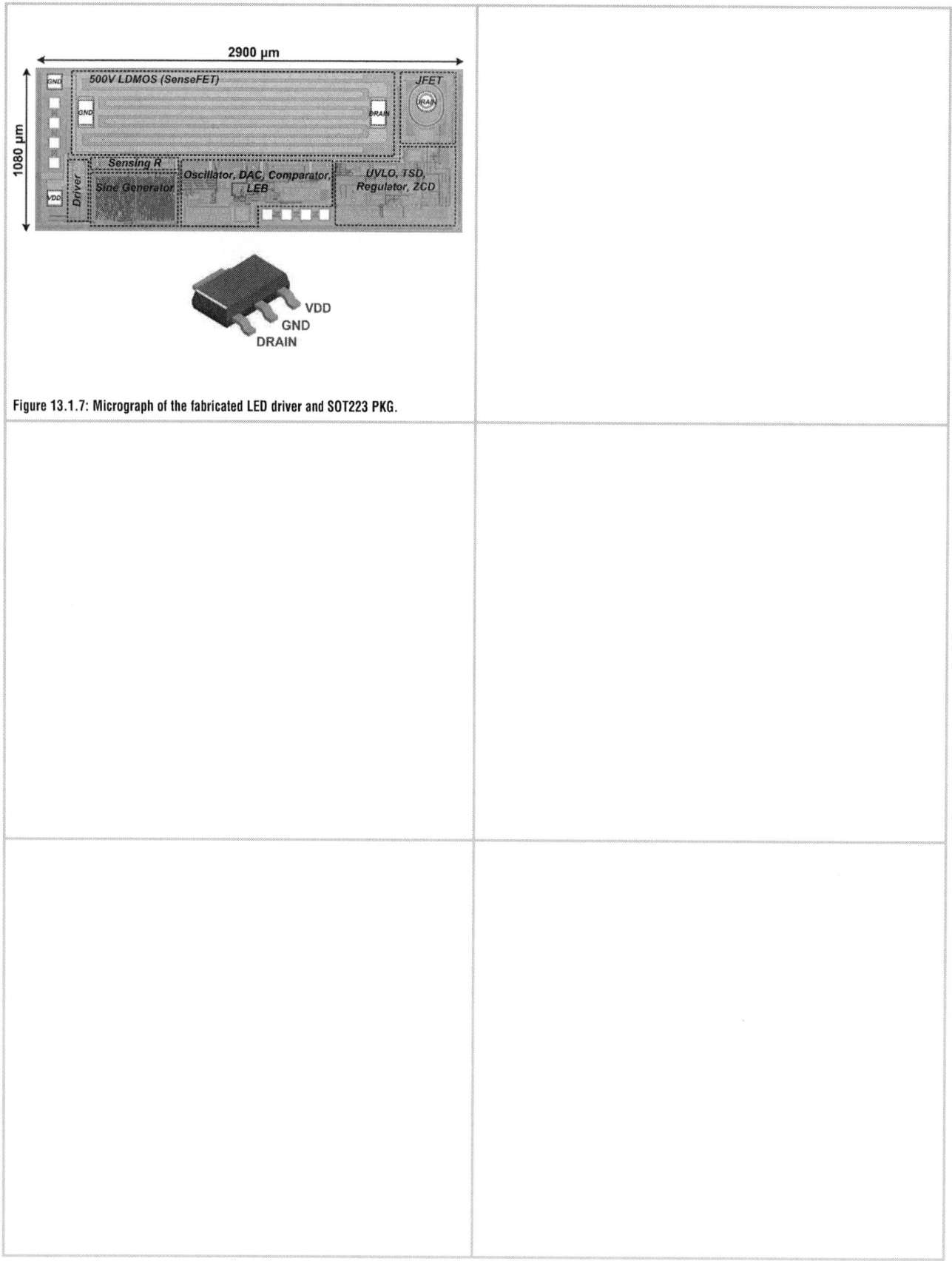

Figure 13.1.7: Micrograph of the fabricated LED driver and SOT223 PKG.

ISSCC 2011 / SESSION 13 / ANALOG TECHNIQUES / 13.2

13.2 A 1.2A Buck-Boost LED Driver with 13% Efficiency Improvement Using Error-Averaged SenseFET-Based Current Sensing

Sachin Rao[1], Qadeer Khan[1], Sarvesh Bang[2], Damian Swank[2], Arun Rao[2], William McIntyre[2], Pavan Kumar Hanumolu[1]

[1]Oregon State University, Corvallis, OR
[2]National Semiconductor, Grass Valley, CA

High-current LED drivers suffer from a significant efficiency loss due to the presence of a current regulation element (CRE) in series with the LED. In a conventional driver, either a series current source [1] or a sense resistor [2, 3] acts as a CRE to regulate the LED current (I_{LED}). In this paper, we seek to improve the efficiency by eliminating the series CRE. To this end, we employ a highly accurate current sensing scheme to directly regulate I_{LED} and achieve more than 13% efficiency improvement.

Figure 13.2.1 shows the proposed CRE-less LED driver architecture. The converter output directly drives the LED and the feedback circuitry regulates I_{LED} to the desired value. To achieve seamless operation over the entire Li-ion battery voltage range (3 to 5.5V), a buck-boost converter is employed. Efficiency is improved by controlling the buck-switches P_{BU}/N_{BU} and boost-switches P_{BO}/N_{BO} by two separate duty-cycles D_{BUCK} and D_{BOOST}, respectively. The converter operates in 3 different modes: (1) when $V_{IN} > V_{LED}$, P_{BO} is ON, N_{BO} is OFF and the converter operates in the buck-mode ($D_{BOOST}=0$), (2) when $V_{IN} < V_{LED}$, P_{BU} is ON, N_{BU} is OFF ($D_{BUCK}=1$) and the converter operates in the Boost-mode and, (3) when V_{IN} is close to V_{LED} the converter operates in the Buck-Boost mode. Smooth transition between the 3 modes is achieved by instantaneously changing D_{BUCK} and D_{BOOST} using appropriate feed-forward voltages V_{F1} and V_{F2}.

Current through P_{BU} and N_{BU} is sensed by IP- and IN-sensors, respectively, and fed to the LED current estimator block that determines average LED current, $\overline{I_{LED}}$, in the form of the voltage $\overline{V_{SENSE}}$. A G_M-C integrator accumulates the error between V_{REF}, which represents the desired LED current, and $\overline{V_{SENSE}}$ and drives the duty-cycle generator. The high DC gain of the G_M-C integrator forces $V_{REF} = \overline{V_{SENSE}}$ in steady-state, thus regulating $\overline{I_{LED}}$ to the desired value.

IP and IN current sensors are central to the performance of the proposed LED driver. Traditional current sensing schemes suffer from an accuracy and efficiency tradeoff. For instance, a filter-based current sensor is lossless but is susceptible to PVT variations of the filter components [4]. Ideally, tuning and calibration techniques can mitigate component variation but the high-precision analog circuitry needed to implement them complicates the design, incurs power penalty, and limits the sensor accuracy to about ±8% in practice [4]. On the other hand, a conventional sense-FET based current sensor simplifies the design but its accuracy is directly affected by the mismatch between the power-FET and the sense-FET [5]. Large mirror ratios of the order of 1000:1 needed to minimize power loss in the sense-FET combined with the offsets in the amplifiers used to equalize the terminal voltages of power-FET and sense-FET, further exacerbate the matching problem and limit the sensor accuracy to about ±20% [4]. We use error-current averaging to decouple the tradeoff between efficiency and accuracy and auto-zeroing to mitigate errors due to amplifier offsets.

Figure 13.2.2 shows the proposed IP-sensor circuit. It consists of an array of sense-FETs, P_{S1} to P_{SN}, that are nominally matched to P_{BU} individually. A one-hot encoded circular shift register selects a different sense-FET during each switching cycle. The moving average of the sensed current over N cycles yields an accurate estimate of P_{BU} current. Averaging improves matching between the power-FET and the sense-FET, ideally by a factor of √N. Because only one sense-FET is enabled at any time, this approach incurs no power penalty, thereby breaking the fundamental tradeoff between accuracy and efficiency in a conventional sense-FET based approach. In our implementation, N=32 is used to reduce mismatch-induced error by more than 5 times. Special attention is paid to layout techniques to reduce systematic errors due to routing metal parasitic. Similarly, IN-sensor has 32 sense-FETs N_{S1} to N_{S32} that are matched to N_{BU} individually. IN-sensor architecture is identical to that of the IP-sensor.

An auto-zeroed (AZ) amplifier forces V_{DS} of the power-FET and sense-FETs to be equal. To alleviate the settling requirements during the AZ phase, two amplifiers operate in a ping-pong fashion as shown in Fig. 13.2.2, wherein Stage-1 senses the current while Stage-2 is being auto-zeroed and vice-versa. The sense phase Φ_{PS} and the auto-zeroing phase Φ_{AZ} operate in a non-overlapping manner. An intermediate hold phase Φ_H is used to reduce sensing errors during the dead time of the converter (Fig. 13.2.2).

Figure 13.2.3 shows the simplified block diagram of the LED current estimator. Noting that in steady-state $I_{LED} = (1-D_{BOOST})\ I_L$ independent of both the converter mode of operation and losses; this circuit generates an output voltage $\overline{V_{SENSE}} = \alpha \overline{I_{LED}}$, where α is equal to $R_{SENSE}/1024$ in our implementation. Because IP- and IN-sensor outputs are valid only during the Φ_{PS} and Φ_{NS} phases, respectively, switches S_2 and S_3 selectively route the sensed currents to obtain a continuous current of $I_L/1024$ at node V_{SENSE}. This current is scaled by a factor of $(1-D_{BOOST})$ by enabling switch S_7 only when the Boost-side NFET N_{BO} is OFF and is passed through the sense resistor R_{SENSE}. The voltage drop across R_{SENSE} is averaged using a low-pass filter resulting in $\overline{V_{SENSE}} = R_{SENSE} \overline{I_{LED}}/1024$. Dummy resistors R_D and switches S_5/S_8 are added to match the impedances in all the branches.

The two other sources of error in estimating $\overline{I_{LED}}$ are variation in R_{SENSE} and the sensed current inaccuracy during the dead-time. Highly accurate thin-film resistors are used to minimize R_{SENSE} variation. Alternatively, an accurate external R_{SENSE} can be used with little cost penalty. Because the dead-time is only a small fraction of the switching cycle, errors in the sensed current during this time have negligible impact on the sensed $\overline{I_{LED}}$.

The LED driver is fabricated in a 0.5μm CMOS process and operates at switching frequencies up to 2MHz with an off-chip 2.2μH inductor and a 10μF capacitor. The converter can drive up to 1200mA at an LED forward voltage of about 3.6V. Figure 13.2.4 shows the converter start-up, I_L, and V_{SENSE} in all the 3 modes of operation. For better illustration, I_L is displayed with a 50mA offset on all the plots. Because a different sense-FET is selected in each switching cycle, V_{SENSE} differs from cycle to cycle by the amount of mismatch between the sense-FET and the power-FET. However, V_{SENSE} averaged over 32 switching cycles, obtained from the LED current estimator, yields an accurate measure of the LED current.

Figure 13.2.5 shows the measured $\overline{I_{LED}}$ for different output current settings. Better than ±2.8% accuracy is achieved over 3-5.5V V_{IN} range. This represents a factor of 3 improvement over the state-of-the-art [4]. The measured standard deviation of $\overline{I_{LED}}$ across 7 devices is less than 1.6%. Measured efficiency plots for the proposed and conventional drivers at 600mA and 1200mA LED currents shown in Fig. 13.2.6 illustrate 4.5% to 13% improvement. Because of the absence of CRE, the proposed converter output voltage is lower by about 250mV resulting in a higher efficiency improvement in the boost and buck-boost modes. The peak efficiencies are 90.7% and 86% at 600mA and 1200mA currents, respectively. Figure 13.2.7 shows a die micrograph. Active die area is 5mm².

References:
[1] National Semiconductor, part number LM3553, 1.2A Dual Flash LED Driver System with I²C Compatible Interface, 2008.
[2] Linear Technology, part number LTC3454, 1A Synchronous Buck-Boost High Current LED Driver, 2005.
[3] ST Microelectronics, part number STCF06, 1.5A White LED Driver with I²C Interface, 2010.
[4] H. P. Forghani-Zadeh and G. A. Rincon-Mora, "An Accurate, Continuous, and Lossless Self-Learning CMOS Current-Sensing Scheme for Inductor Based DC-DC Converters," *IEEE J. Solid-State Circuits*, vol. 42, no. 3, pp. 665-679, Mar., 2007.
[5] C. Lee and P. Mok, "A Monolithic Current-Mode CMOS DC-DC Converter with On-Chip Current-Sensing Technique," *IEEE J. Solid-State Circuits*, vol. 39, no. 1, pp. 3-14, Jan., 2004.

978-1-61284-303-2/11 $26.00 © 2011 IEEE

ISSCC 2011 / February 22, 2011 / 2:00 PM

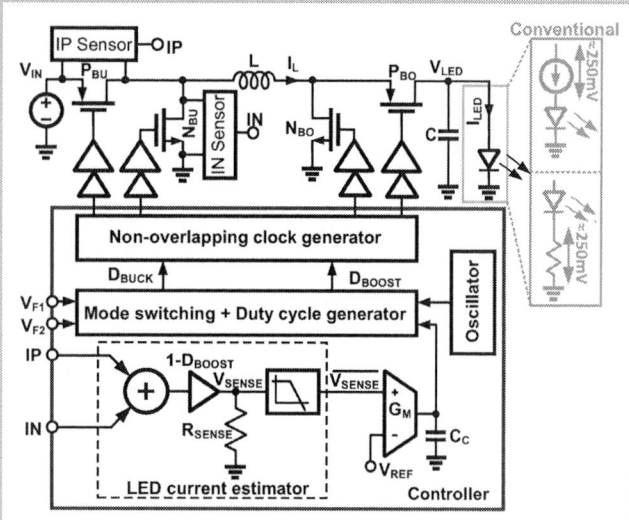

Figure 13.2.1: Block diagram of the proposed LED driver.

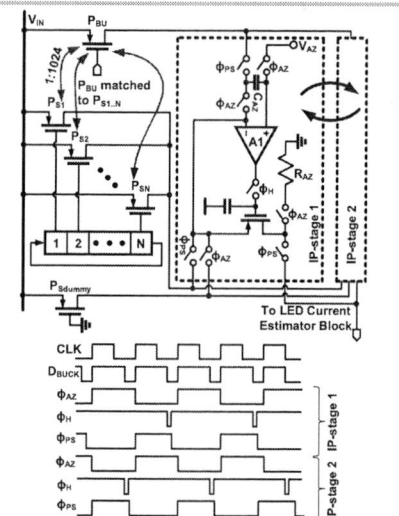

Figure 13.2.2: IP-Sensor block diagram and the associated timing diagram.

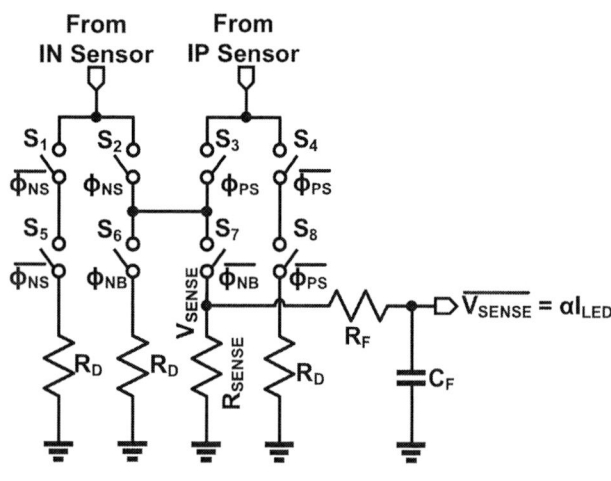

Figure 13.2.3: LED current estimator (LCE) block diagram.

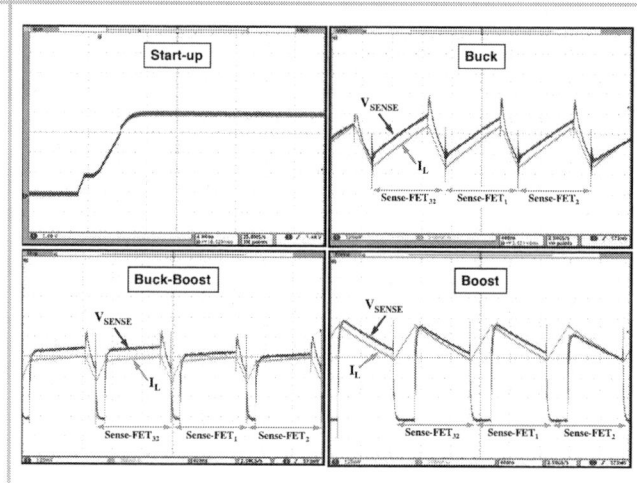

Figure 13.2.4: Measured results: start-up, inductor current (I_L), and sensed voltage (V_{SENSE}).

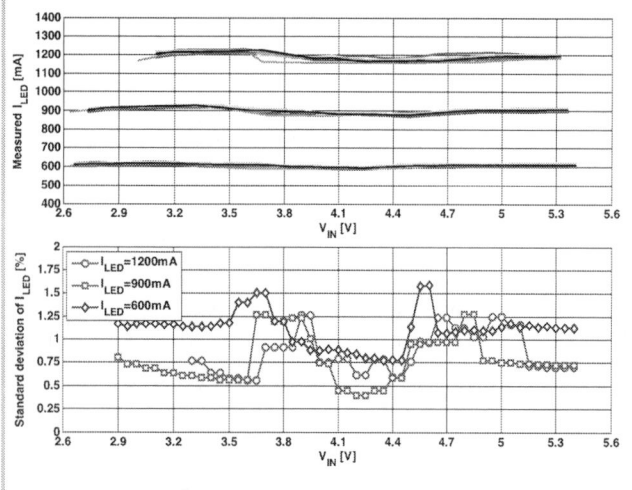

Figure 13.2.5: Measured I_{LED} and its standard deviation at 600mA, 900mA, and 1200mA for seven devices.

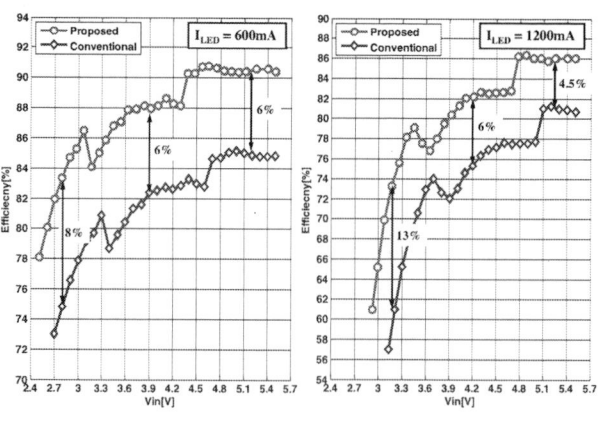

Figure 13.2.6: Measured efficiency versus V_{IN} for 600mA and 1200mA I_{LED} settings.

978-1-61284-303-2/11 $26.00 © 2011 IEEE

ISSCC 2011 PAPER CONTINUATIONS

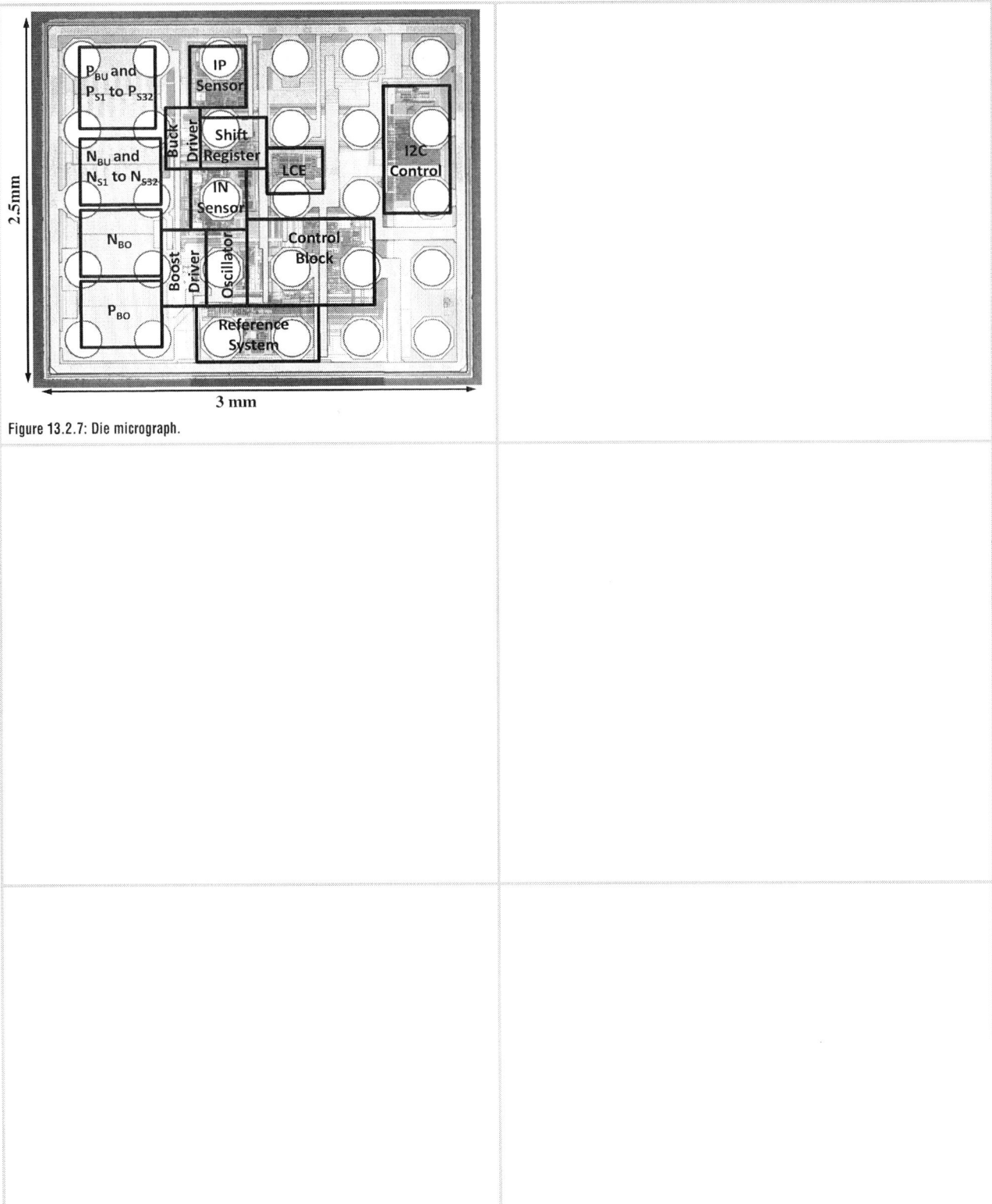

Figure 13.2.7: Die micrograph.

ISSCC 2011 / SESSION 13 / ANALOG TECHNIQUES / 13.3

13.3 Filterless Integrated Class-D Audio Amplifier Achieving 0.0012% THD+N and 96dB PSRR When Supplying 1.2W

Mykhaylo Teplechuk, Tony Gribben, Christophe Amadi

Dialog Semiconductor, Edinburgh, United Kingdom

High power efficiency, small size and reduced heat dissipation are highly desirable in battery-powered mobile systems and switched-mode class-D amplifiers can readily satisfy these requirements. However, full adoption of class-D technology in mobile systems has been somewhat limited due to concerns such as signal distortion and noise, poor power supply noise rejection, electromagnetic interference (EMI) and the requirement for external LC filters.

Recently, filterless class-D architectures have become popular [1-4]. These architectures can generally be divided into two categories: pulse-width modulation (PWM) type [1, 3] and $\Delta\Sigma$-modulation type [2]. In addition, sliding-mode architectures have also been researched [4]. The recently reported $\Delta\Sigma$ class-D amplifier [2] demonstrates that class-D amplifiers can achieve very low total harmonic distortion levels, THD+N of 0.001% @100Hz. By comparison, signal distortion in conventional PWM type amplifiers remain relatively high, typical THD+N@1kHz > 0.03% (see Fig. 13.3.6). This level of distortion in conventional filterless PWM-type architectures can be attributed to high frequency inter-modulation distortion. In conventional "filterless" architectures, the output of the amplifier is directly connected to a load and the load (speaker) itself provides filtering, thereby demodulating the audio signal. The feedback signal, which is high frequency, rail-to-rail modulated, is taken before any external filtering is applied and is fed back to the input of the loop filter block. In such an arrangement, the on-chip loop filter processes the in-band audio signal and the high frequency, rail-to-rail feedback signal. The amount by which the loop filter can attenuate the high frequency content of the feedback signal is restricted by the need to provide high, in-band, open-loop gain. Due to the finite attenuation of the amplifier at the clock (modulation) frequency, some residual, high frequency, signal-dependent ripple is present at the output of the loop filter block, i.e., the input of the modulation block. This residual signal (frequency higher or equal to the amplifier main clock frequency) inter-modulates with the main amplifier modulation frequency and with the input signal. Some inter-modulation products become signal-dependent and can fold (alias) back in-band to form harmonics of the input audio signal. These harmonics can effectively be viewed as distortion products of the amplifier. For that reason, the poor THD+N performance of conventional filterless PWM type amplifiers can be attributed to these inter-modulation (aliasing) products. Conventional PWM-type amplifiers [1, 3, 4 and 5] suffer from the above problem and exhibit high levels of distortion, Figure 13.3.6.

The proposed, improved ('true') filterless architecture, which uses uniform pulse-width modulation (UPWM), is illustrated in Fig. 13.3.1(a). It eliminates the problem of unwanted inter-modulation products by sampling and holding the signal at the output of the loop filter before the conventional PWM block. The block diagram in Fig. 13.3.1(a) illustrates the fully-differential architecture. It is comprised of a loop filter, a feedback/gain block, a pre-modulation block that performs sample and hold, a conventional PWM block, an oscillator, output drivers and an H-bridge power stage. In the proposed architecture, the oscillator generates a triangular wave and a rectangular wave with a duty cycle of 50% to drive the sample and hold. The sample and hold in conjunction with the conventional PWM block, achieves UPWM. Both differential outputs of the loop filter are sampled on two sampling capacitors (C3). In phase 1, SW1 is closed and SW2 is opened, the loop filter output is stored on to the first sampling capacitor while the charge that was stored on the second sampling capacitor is being modulated by conventional PWM block. In phase 2, SW1 is opened and SW2 closed, the output of the loop filter is sampled on the second sampling capacitor while the charge stored on the first capacitor is being modulated. Figure 13.3.1(b) illustrates one cycle of operation of this UPWM system. The sampling frequency is equal to the main PWM modulation frequency which is half the frequency of the residual high frequency ripple. The voltages at the input (V_i) and output (V_o) of the pre-modulation block as well as sampling (V_{unif}) signal that drives switches

SW1 and SW2 and modulation signals (V_{pwm}) are all shown. As can be seen, by uniformly sampling with a clock frequency half that of the residual ripple, the high frequency ripple is removed. In a frequency domain, this operation is equivalent to a transmission zero (very deep "notch") at twice the sampling frequency ($2 \times f_{clk}$). As a result the pre-modulation block enables true filterless operation of a switched-mode amplifier.

The chip is fabricated in a standard CMOS process and packaged in wafer-level chip-scale package (WLCSP). It can operate from 2.5 up to 5.5V supply. Measured THD+N for 4 and 8Ω loads versus output power (@V_{bat}=5.5V) is shown in Fig. 13.3.2. All measurements were performed with the 4 or 8Ω load in series with 33μH inductance. From Fig. 13.3.2, it can be seen that amplifier demonstrates 98.4dB THD+N (0.0012%) at 1.2W output signal (V_{bat}=5.5V into 8Ω). Figure 13.3.3 shows output spectrum of 2W signal (V_{bat}=5.5V into 4Ω) with THD+N = 96.5dB (0.00149%) and third harmonic distortion (HD$_3$) of 101dB. Figure 13.3.4 demonstrates additional advantage of the proposed architecture, in addition to the low THD+N performance simultaneously high power supply rejection ratio (PSRR) performance is achieved PSRR at 217Hz is –96dB (200mV$_{p-p}$ square wave ripple signal on the power supply). Figure 13.3.5 illustrates the efficiency of the amplifier plotted versus output power. Efficiency in excess of 90% at 1.3W output power is achieved (V_{bat}=5V, into 4Ω). Amplifier shows SNR of 103dB with quiescent current 4mA. Maximum achieved output power at THD+N=10% (Vbat=5.5V into 4Ω) is 3.6W. Figure 13.3.6 compares performance of the presented true filterless UPWM architecture to that of the state-of-the-art amplifiers. A chip micrograph is shown in Fig. 13.3.7. The chip area is 1.2×1.2mm^2.

Acknowledgement:

The authors would like to thank all those at Dialog Semiconductor, whose invaluable help made this design possible, in particular, Taner Dosluoglu, Jim Brown, Rufilynn Taruc, Martin Fiala, Stefan Hausser, Wolfgang Cramer, Florin Pop and Markku Vaatanen.

References:

[1] P. Muggler, W. Chen, C. Jones, et. Al., "A Filter Free Class D Audio Amplifier with 86% Power Efficiency", *IEEE ISCAS*, pp. 1036-1039, May, 2004.

[2] E. Gaalaas, B.Y. Liu, and N. Nishimura, "Integrated Stereo Delta-Sigma Class D Audio Amplifier," *ISSCC Dig. Tech. Papers*, pp. 120-121, February, 2005.

[3] B. Forejt, V. Rentala, J.D. Arteaga, and G. Burra, "A 700+-mW Class D Design with Direct Battery Hookup in a 90-nm Process," *IEEE J. Solid-State Circuits*, vol. 40, no. 9, pp. 1880-1887, September, 2005.

[4] M. A. Rojas-González and E. Sánchez-Sinencio "Two Class-D Audio Amplifiers with 89/90% Efficiency and 0.02/0.03% THD+N Consuming Less than 1mW of Quiescent Power," *ISSCC Dig. Tech. Papers*, pp. 450-452, February, 2009.

[5] S. Samala, V. Mishra, and K. C. Chakravarthi, "45nm CMOS 8Ω Class-D Audio Driver with 79% Efficiency and 100dB SNR," *ISSCC Dig. Tech. Papers*, pp. 86-88, February, 2010.

978-1-61284-303-2/11 $26.00 © 2011 IEEE

ISSCC 2011 / February 22, 2011 / 2:30 PM

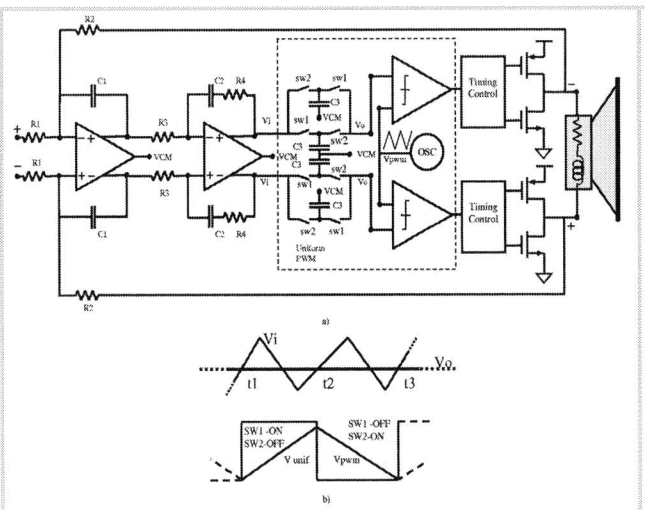

Figure 13.3.1: Filterless 2nd-order class-D amplifier a) block diagram, b) UPWM timing diagram.

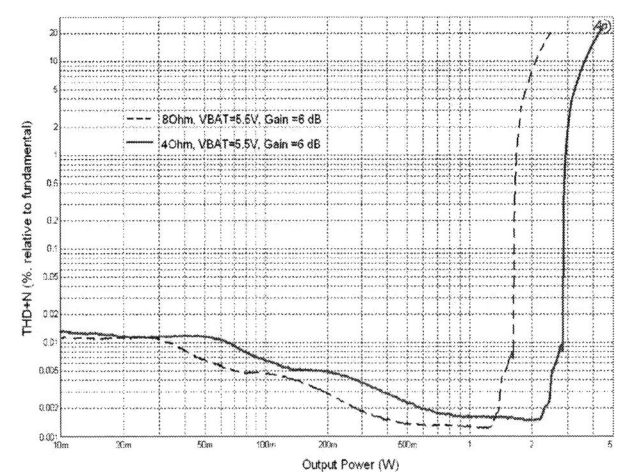

Figure 13.3.2: THD+N versus output power (1kHz sine) for 4 and 8Ω.

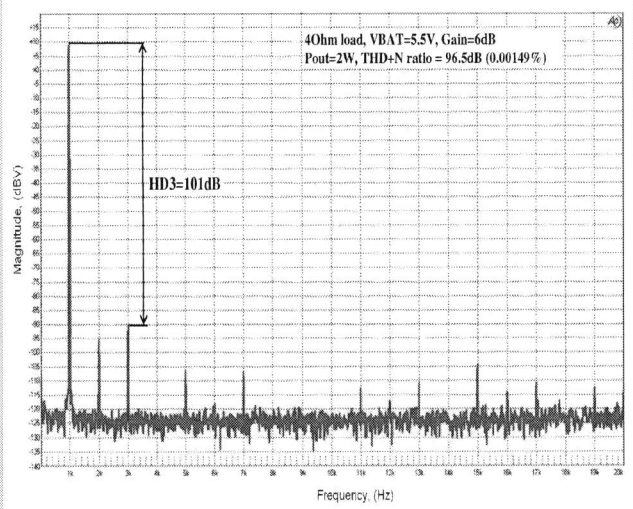

Figure 13.3.3: FFT of 2W 1kHz sine and 4Ω load (THD+N = 96.5dB).

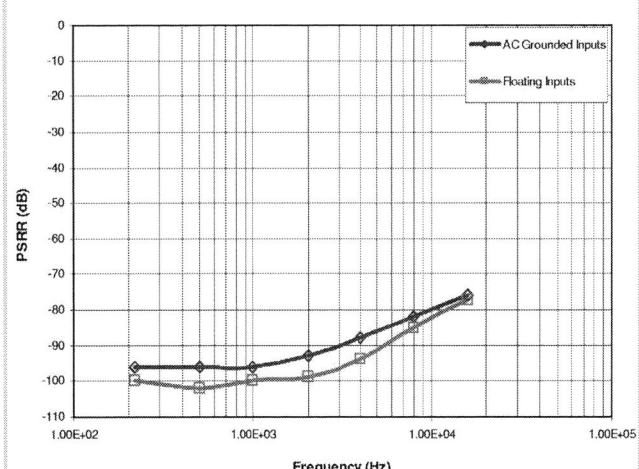

Figure 13.3.4: PSRR measurement (V_{ripple}=200mV$_{p-p}$).

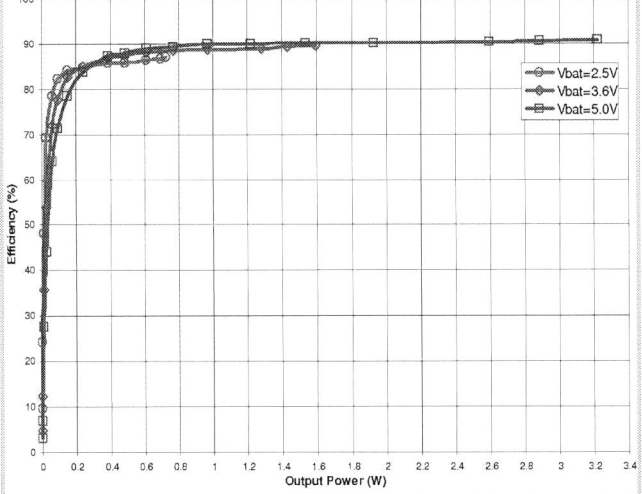

Figure 13.3.5: Efficiency versus output power for 4Ω load (maximum output power measured @ THD+N=10%).

Design	Presented	[1]	[2]	[3]	[4] BMA (TMA)	[5]
SNR (dB), A-weighted	103	-	103	98.5	94(92)	100
THD+N (%) @ 100 Hz	0.00093	-	0.001	-	-	-
THD+N (%) @ 1kHz	0.00122	0.04 - 0.4	0.00384	0.03	0.02(0.03)	0.1
PSRR (dB) @ 217Hz	96	84	65	70	77(81)	-
η (%)	93	79	88	75.5	89(90)	79
Supply (V)	2.5-5.5V	2.5-6.5	12-30V	2.7-5.4	2.7	2.3-4.8
Load (Ω)	4	8	6	8	8	8
I$_Q$(mA)	4	2.5	12	3	0.250(0.21)	3
Fs (kHz)	1000	250	450	410	450	-
Max. Pout (W) @1kHz THD+N <1%	3.1	3	22	0.7	0.25	1
Area (mm²)	1.44	2.25	10.15	0.44	1.49(1.39)	0.42
Process	CMOS	BiCMOS	BCDMOS	DCMOS	CMOS	CMOS
Architecture	UPWM	PWM	ΔΣ	PWM	SMC	PWM
FOM 1	250	8	88	8	178(143)	3
FOM 2	7906	125	156	27	1260(1603)	-

$$FOM_1 = \frac{\eta}{I_Q \cdot THD \cdot 10^5}; \quad FOM_2 = \frac{\eta}{I_Q \cdot THD \cdot PSRR \cdot 10^6};$$

Figure 13.3.6: Performance summary.

978-1-61284-303-2/11 $26.00 © 2011 IEEE

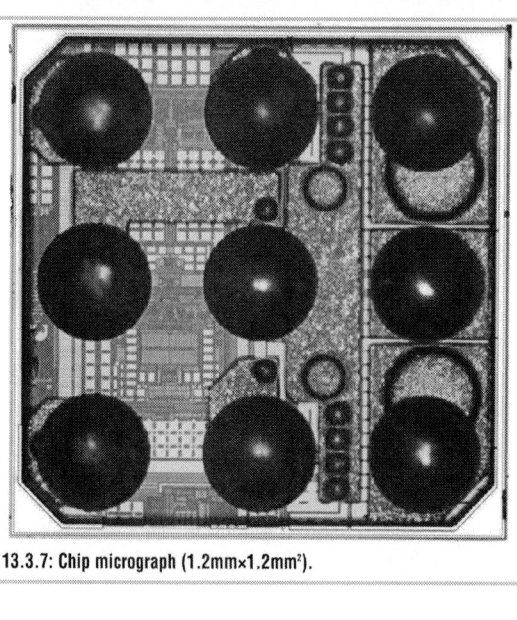

Figure 13.3.7: Chip micrograph (1.2mm×1.2mm²).

ISSCC 2011 / SESSION 13 / ANALOG TECHNIQUES / 13.4

13.4 A 5.9nV/√Hz Chopper Operational Amplifier with 0.78µV Maximum Offset and 28.3nV/°C Offset Drift

Yoshinori Kusuda

Analog Devices, Wilmington, MA

Many auto-zero or chopper operational amplifiers have been reported with low offset and low-offset drift. The resulting baseband noise can also be a significant error source, and thus reducing the total error down to sub µV levels at DC and low frequencies has been targeted. Chopping is more suitable to lower the baseband noise PSD, as it is primarily determined by the broad-band thermal noise floor and not aliased by the high frequency noise [1]. Reducing the thermal noise floor requires higher value of input transconductance [1]. Consequently, more capacitance or higher output transconductance is required to stabilize the overall feedback loop with regular Miller compensation. A 6.5nV/√Hz conditionally stable chopper operational amplifier has been reported, using multi-pole feedforward compensation techniques [2]. Unconditionally stable chopper operational amplifiers presented so far exhibit more than 10nV/√Hz noise PSD [1,3-5]. This paper reports a 5.9nV/√Hz unconditionally stable chopper operational amplifier with 1.47mA supply current and 1.26mm² die area, achieved by phase compensation using current attenuation. In addition, adaptive clock level shift and backgate biasing for the input chopping allows optimization for noise and offset, realizing 0.78µV maximum offset with a worst-case 28.3nV/°C drift.

Figure 13.4.1 shows the overall amplifier diagram including a high DC gain path and a high frequency path. To achieve low noise and low offset, the high-DC-gain path has a chopped G_{m1} with a 3.2mS transconductance, followed by G_{m2} and G_{m3}. It also contains auto-correction feedback (ACFB) to null out the initial offset of G_{m1} through G_{mnull}, which otherwise would become modulated output ripples [6]. The ACFB senses the modulated ripples at CHOP2's output by G_{m5}, and demodulates it by CHOP3 to form a DC feedback correction signal at G_{mnull}'s input. On the other hand, the baseband signal at CHOP2's output is modulated up to the chopping frequency by CHOP3, and filtered out by a switched capacitor notch filter (SC-NF). Therefore the ACFB doesn't disturb the desired baseband signal from the overall input. In contrast to [6], it has an active integrator stage with G_{m6}, C5, and C6, to enhance the loop gain of ACFB to further suppress the output ripple. The open-loop unity-gain frequency is set to $G_{m1}/\{2\pi \times 4(1+ C2)\}$, which is maintained below 4MHz, thanks to the factor of 1/4 by the current attenuator of F1. A fully differential output stage with G_{m3} and G_{m3B} drives both C1 and C2, to avoid impedance mismatches at CHOP2's output and avoid charge injection related offsets. As the G_{m3B} doesn't have to drive large external load capacitors, its transconductance is designed to be 1/4 of G_{m3}. The G_{m4} forms the high-frequency path to bypass a phase shift created by ACFB. It has the same unity gain frequency as the high DC gain path to avoid a pole-zero doublet.

The circuit diagram of G_{m1}, F1, and G_{mnull} is shown in Fig. 13.4.2. As a rail-to-rail input stage, G_{m1} has both P-channel and N-channel differential pairs. The tail currents depend on the input common-mode voltage (Vcm). Setting Itailp+Itailn=256µA is done by adjusting the bias voltage of Vbp1, Vbn1, Vbp2, and Vbn2. The F1 has 4:1 current mirrors in the negative signal path with Mn21 and Mn22, and in the positive signal path with Mn31 and Mn32. Degeneration resistors will reduce MOSFET current sources current noise contributions, especially for Rn22, Rn23, Rp22, and Rp23 which are larger resistors. Otherwise, their current noises can be critical contributors to the input referred noise when the signal is attenuated. The bias currents are set lower to keep the IR drop at the same level as the others, at the cost of slew rate performance for large signal response. The G_{mnull} generates a current based on the DC feedback correction signal to null out Gm1's initial offset. Its transconductance is reduced by Rp11 and Rp12, to have less sensitivity from the error voltage in SC-NF due to charge injection and kT/C noise.

Figure 13.4.3 shows the input chopping switches (CHOP1) and the clock generator. To reduce its noise contribution, the on-resistance is targeted below 125Ω with any input common mode voltage (Vcm). The chopping frequency is chosen to be 200kHz to mitigate flicker noise from G_{m1}. However, charge-injection related errors, such as offset and input bias current, tend to increase proportionally

with the size of switches, clock swing, and clock frequencies. To achieve the targeted on-resistance with smaller sizes, the switches are made up of NMOS devices with the backgate driven by Vcm through a common mode buffer. In addition, the gate is driven to Vcm+1.8V for its on state and Vcm for its off state. It maintains a constant and sufficient gate-source voltage to achieve the targeted on-resistance for the rail-to-rail Vcm range, as well as constant amount of channel charge injection and clock feedthrough without redundant clock swing. Two identical adaptive clock boosters provide such clock levels based on the potential of Vcm. Following to an 800kHz oscillator, a timing generator provides 200kHz chopping and SC-NF clocks with 90 degree phase shift, required by the filtering function [1, 6]. An on chip LDO will provide constant 0 to 1.8V clock levels regardless of external supply voltages. This level is enough to turn on the internal NMOS switches in CHOP2, CHOP3, and SC-NF, as their source potentials are biased at 0.625V. Delay elements are added to those internal switches to compensate the phase delay in the signal path occurred at G_{m1}, reducing chopping related offset and distortion.

The design has been fabricated in a 0.35µm CMOS process with 5V I/O devices in a 0.89×1.42mm² die area, drawing 1.47mA supply current at 2.5 to 5.5V supply voltages. Among 30 units tested, the input offset and offset drift are 0.34µV and 11.3nV/°C (mean), and 0.78µV and 28.3nV/°C (worst-case) respectively after adding 6σ distribution. Typical open-loop gain, CMRR, and PSRR are over 150dB at DC. The measured open-loop gain and phase with a 100pF load capacitance are plotted in Fig. 13.4.4, showing regular −20dB/dec roll off, 3.7MHz unity gain frequency, and 63° phase margin. A slight hump at the chopping frequency of 200kHz is due to the notch created by ACFB, which is suppressed by the high-frequency path. Figure 13.4.5 shows the input noise PSD measured with close-loop gains of 1, 10, and 100, which indicates flat and similar noise floors of 6.3nV/√Hz, 6.0nV√Hz, and 5.9nV/√Hz, respectively from 1Hz to 180kHz. A lower noise PSD of 5.3nV/√Hz has been measured, when the N-channel input pair is working with a gain of 100. The peak PSD at 200kHz is 49nV/√Hz and 24nV/ Hz with gains of 1 and 10, respectively, which includes both modulated kT/C noise from SC-NF and the residual ripple from the initial offset. Not shown here is a spike at twice the chopping frequency due to charge injection from the input chopping switches. With a gain of 1, it is 2.5µVrms mean and 8.9µVrms worst-case after adding 6σ distribution for 20 units. Fig. 13.4.6 compares the performance with other chopper operational amplifiers including state-of-the-art designs.

Acknowledgements:
The author would like to thank Meredith Dauphinee and Feliks Borzyszkowski for the mask layout design, and Yen Truong and Ken Low for the test.

References:
[1] R. Burt and J. Zhang, "A Micropower Chopper-Stabilized Operational Amplifier using a SC Notch Filter with Synchronous Integration inside the Continuous-Time Signal Path," *IEEE J. Solid-State Circuits*, vol. 41, no 12, pp. 2729-2736, Dec., 2006.
[2] M. Kejariwal, P. Ammisetti, and A. Thomsen, "A 250+dB Open Loop Gain Feedforward Compensated High Precision Operational Amplifier," *ESSCIRC*, pp. 187-190, Sept., 2002.
[3] A.T.K. Tang, "A 3µV-Offset Operational Amplifier with 20nV/√Hz Input Noise PSD at DC Employing Both Chopping and Autozeroing," *ISSCC Dig. Tech. Papers*, pp. 362-387, Feb., 2002.
[4] National Semiconductor, "LMP2021 Datasheet," 2009.
[5] Q. Fan, J.H. Huijsing, and K.A.A. Makinwa, "A 21nV/√Hz Chopper-Stabilized Multipath Current-Feedback Instrumentation Amplifier with 2µV Offset," *ISSCC Dig. Tech. Papers*, pp. 80-81, Feb., 2010.
[6] Y. Kusuda, "Auto Correction Feedback for Ripple Suppression in a Chopper Amplifier," *IEEE J. Solid-State Circuits*, vol. 45, no 8, pp. 1436-1445, Aug., 2010.

978-1-61284-303-2/11 $26.00 © 2011 IEEE

ISSCC 2011 / February 22, 2011 / 3:15 PM

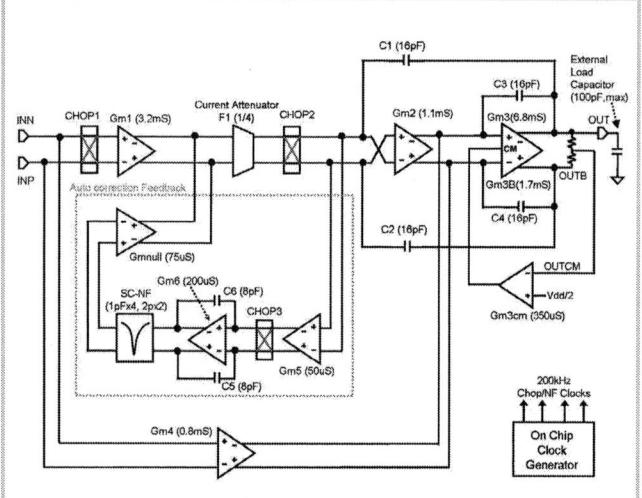

Figure 13.4.1: Block diagram of the proposed chopper amplifier.

Figure 13.4.2: Circuit diagram of G_{m1}, F_1, and G_{mnull}.

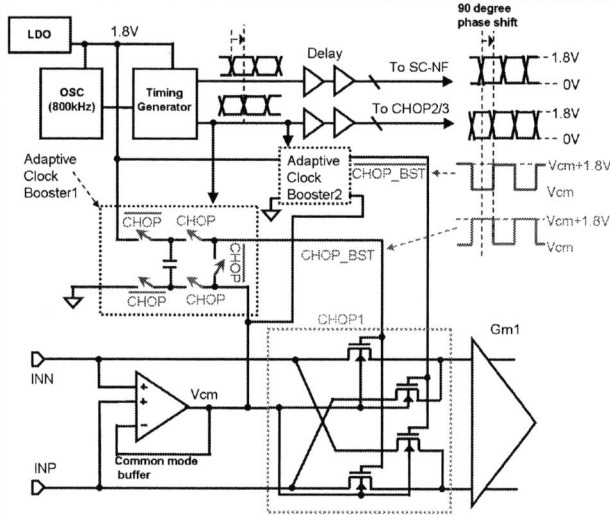

Figure 13.4.3: Input chopping switches and clock generator.

Figure 13.4.4: Open-loop gain and phase.

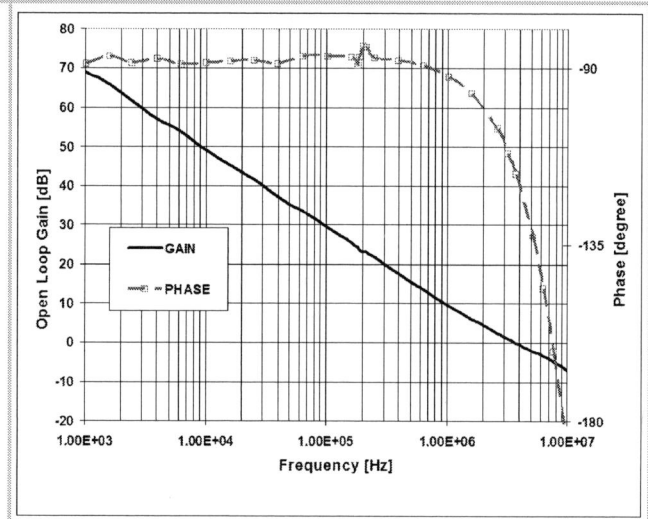

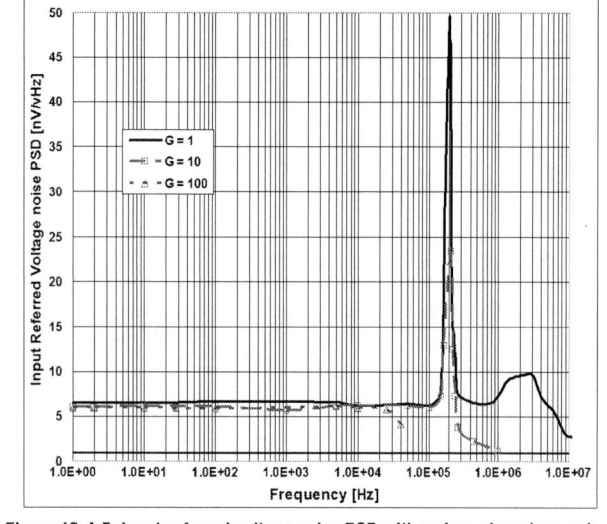

Figure 13.4.5: Input-referred voltage noise PSD with various close-loop gains.

Figure 13.4.6: Performance comparison with previous work.

	This work	[2]	[4]	[5]
Year published	2011	2002	2009	2010
Chopping frequency	200kHz	200kHz	Not shown	30kHz
Maximum Offset voltage	0.78 μ V	4 μ V	5 μ V	1 μ V
Maximum Offset drift	28.3nV/C	Not shown	20nV/C	Not shown
Input bias current	72pA	450pA	25pA	Not shown
Input voltage noise PSD (e_n)	5.9nV/√ Hz	6.5nV/√ Hz	11nV/√ Hz	10.5nV/√ Hz
0.1-10Hz peak-to-peak noise	108nV$_{p-p}$	* 123nV$_{p-p}$	260nV$_{p-p}$	* 198nV$_{p-p}$
Unity gain frequency	4.0MHz	** 5.0MHz	5MHz	1.8MHz
Current dissipation (I_q)	1.47mA	1.8mA	1.1mA	143μA
Die area	1.26mm²	Not shown	Not shown	1.8mm²
$e_n{}^2$ x I_q (nV/√ Hz●mA) Figure of merit	51.2	76.1	133	15.8

(* calculated as 6●√(10-0.1)● en)
(** Conditionally stable operational amplifier)

978-1-61284-303-2/11 $26.00 © 2011 IEEE

ISSCC 2011 PAPER CONTINUATIONS

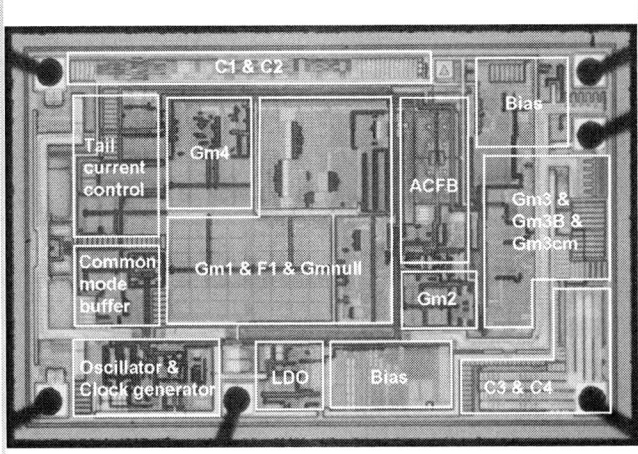

Figure 13.4.7: Die mocrograph of the proposed chopper amplifier (1420µm x 890µm).

ISSCC 2011 / SESSION 13 / ANALOG TECHNIQUES / 13.5

13.5 A Current-Feedback Instrumentation Amplifier with a Gain Error Reduction Loop and 0.06% Untrimmed Gain Error

Rong Wu, Johan H. Huijsing, Kofi A.A. Makinwa

Delft University of Technology, Delft, The Netherlands

Current-feedback instrumentation amplifiers (CFIAs) have significant advantages over the classic three-opamp topology: better power efficiency [1, 2], higher CMRR and rail-sensing capability [4]. Their main disadvantage, however, is their limited gain accuracy, which is determined by the mismatch of two transconductors. Using resistor-degenerated differential pairs, 0.1% gain error and improved linearity have been achieved [3, 4]. However, this incurs either increased noise or increased power. This paper describes a chopper CFIA that employs a continuous-time (CT) gain error reduction loop (GERL) without incurring a noise penalty. Without trimming, it achieves a gain error of less than 0.06% and a maximum gain drift of 6ppm/°C in a power efficient manner (NEF=11.2).

A block diagram of the CFIA is shown in Fig. 13.5.1. Input and feedback transconductors G_{m3} and G_{m4} convert the input and feedback voltages into currents, which are then applied to the virtual ground established by a 2-stage opamp (G_{m2} and G_{m1}). The overall feedback ensure that these currents cancel and thus the CFIA's gain is given by $V_{out}/V_{in} = \{(R_1+R_{21}+R_{22})/R_1\}(G_{m3}/G_{m4})$. To achieve low offset and $1/f$ noise, the 1st and 2nd stages are chopped at 32kHz and 512kHz respectively [1]. However, the up-modulated offset and $1/f$ noise then cause output ripple. A CT offset-reduction loop (ORL) reduces the ripple caused by the 1st stage's offset. The loop synchronously demodulates the output ripple and uses the resulting amplitude information to null the ripple, and hence the offset (Fig. 13.5.1). The 2nd stage's higher-frequency ripple is suppressed by the Miller-compensation network [1].

If precision gain-setting resistors R_1, R_{21} and R_{22} are used, the CFIA's gain accuracy will be limited by the mismatch of G_{m3} and G_{m4}. To reduce this, dynamic element matching (DEM) and a GERL are employed, in a manner similar to the combination of chopping and an ORL (Fig. 13.5.1). Applying DEM to G_{m3} and G_{m4}, i.e. swapping them between the input and feedback paths, will average out their mismatch [5], but also cause output ripple at the DEM frequency. The GERL suppresses this DEM ripple by dynamically adjusting the values of G_{m3} and G_{m4}.

The GERL senses the DEM ripple via capacitors C_{51} and C_{52}. The current through these capacitors is applied to a cascode buffer CB_3, demodulated via chopper CH_9, and integrated on C_6. The resulting voltage $V_{int,GE}$ is applied to transconductor G_{m6}, whose differential output current nulls the mismatch of G_{m3} and G_{m4} by adjusting their bias currents. The noise contribution of the GERL is negligible, since $G_{m6} = G_{m3}/400$ by design. The interaction between the GERL and the ORL is minimized by ensuring that one DEM cycle corresponds to four chopping periods, i.e. $f_{DEM} = f_{chop1}/4 = 8$kHz.

Unlike the ORL, which feeds back an additive offset-compensating signal, the GERL feeds back a *multiplicative* gain-compensating signal, which adjusts the ratio of G_{m3} and G_{m4}. The output DEM ripple is then the product of the residual mismatch and the output signal, and so the gain of the GERL will be signal dependent. To guarantee negative feedback, the feedback polarity must be controlled by the polarity of the input signal. This is done by a polarity-reversing switch (CH_{10}), driven by an auto-zeroed quantizer Q_1 (Fig. 13.5.1). Furthermore, the GERL's loop gain decreases with signal amplitude, and is zero for zero input. In this state, leakage causes $V_{int,GE}$ to drift with a time constant of several seconds, meaning that the GERL has to re-settle whenever a finite input signal reappears: within 10ms for a 20mV step input at a CFIA gain of 200.

To avoid the need for re-settling, a leakage-free digital integrator can be used. The result is the digitally assisted GERL shown in Fig. 13.5.2. Here, the analog integrator is replaced by a comparator (Q_3), an up-down counter, a 1b 1st-order $\Delta\Sigma$ DAC and an RC low-pass filter. To mitigate the effect of the comparator's offset, a chopped integrate-and-dump pre-amplifier is realized by integrating the demodulated DEM ripple on C_6 for seven DEM periods. The comparison is then made. During the next DEM period, SW_1 resets the voltage on C_6. The compara-

tor's output increments or decrements the 10-bit counter, whose output drives the DAC. A 3-level quantizer Q_2 controls the polarity of the loop via CH_{10} and also ensures that the integrator state is "frozen" for small (< 8mV) CFIA output signals. At steady state, the DAC's output will toggle between two LSBs. 10-bit resolution is enough to ensure that the resulting tone is well below the CFIA's noise level and that the DEM ripple can be reduced to the same level as the analog GERL. Since the counter is updated at a rate of $f_{DEM}/8 = 1$kHz, the digitally assisted GERL has a 1s (worst-case) start-up time.

The use of DEM and the GERL ensures good gain accuracy, which means that the input transconductors can be implemented as power-efficient PMOS differential pairs (Fig. 13.5.3). Since the input and feedback stages typically operate at different common-mode (CM), their CMRR is enhanced by cascoding the input transistors with low-threshold devices [1]. During the DEM transitions, the CM voltages of the G_m stages change abruptly. As a result the parasitic capacitances between the substrate and the n-wells of the input devices must be charged and discharged, causing large CM current spikes in the input stages. To divert these spikes, the n-wells of the input transistors M_1, M_2, and their cascodes M_3, M_4 are actively bootstrapped by class-AB buffers consisting of M_5-M_{10}. M_7-M_9 act as level shifters to accommodate the bias voltage of the class-AB stages. The source followers M_5 and M_6 provide a low-impedance path to ground, while M_{10} provides a low-impedance path to the supply. As a result, the bootstrap circuit effectively diverts the DEM spikes to the supply rails.

The 5mm^2 chip (Fig. 13.5.7) is realized in a 0.7µm CMOS process. Both analog and digitally-assisted GERLs are implemented. For flexibility, the 1st-order $\Delta\Sigma$ DAC and the counter are implemented in an FPGA. Measurements on 30 samples show that the CFIA achieves 3µV offset and 15nV/°C offset drift. With a 30mV input voltage and a gain of 100, the output DEM ripple is suppressed by 40dB, to below 47µV (Fig. 13.5.4), which varies maximum 0.52µV/°C over temperature. The use of DEM and the GERL reduces the CFIA's gain error from 0.6% to 0.01% when G_{m3} and G_{m4} are at the same CM voltage (2.5V). Under these conditions, the use of DEM reduces the gain drift from 300ppm/°C to 9ppm/°C (Fig. 13.5.5-left), and switching on the GERL reduces it even further, to 6ppm/°C (11 samples). Due to the input stages' limited CMRR, the gain error increases to 0.06% when G_{m3} is at 0V. The use of DEM improves the CFIA's linearity from 25ppm to 6ppm (at a gain of 100), which the GERL then reduces to 4ppm (Fig. 13.5.5-right).

The measured performance of the CFIA is summarized in Fig. 13.5.6 and compared with the state-of-the-art [2-4]. Without trimming, it achieves a gain error of less than 0.06%, a maximum gain drift of 6ppm/°C. Compared to a CFIA with similar gain accuracy [4], this represents a 4× improvement in power efficiency, while compared to a CFIA with similar power efficiency [2], this represents a 9× improvement in gain accuracy. These results confirm that the combination of DEM and a GERL is a power-efficient manner of improving the gain accuracy, gain drift and linearity of a CFIA.

Acknowledgement:
The author would like to thank Y. Chae for the suggestions and support.

References:
[1] R. Wu, K.A.A. Makinwa, J.H. Huijsing, "A Chopper Current-Feedback Instrumentation Amplifier with a 1mHz $1/f$ Noise Corner and an AC-Coupled Ripple Reduction Loop," *IEEE J. Solid-State Circuits*, vol. 44, no. 12, pp. 3232–3243, Dec., 2009.
[2] Q. Fan, J.H. Huijsing, and K.A.A. Makinwa, "A 21nV/√Hz Chopper-Stabilized Multipath Current-Feedback Instrumentation Amplifier with 2µV Offset," *ISSCC Dig. Tech. Papers*, pp. 80-81, Feb., 2010.
[3] J.F. Witte, J.H. Huijsing, K.A.A. Makinwa, "A Current-Feedback Instrumentation Amplifier with 5µV Offset for Bidirectional High-Side Current-Sensing," *ISSCC Dig. Tech. Papers*, pp. 74-75, Feb., 2008.
[4] M.A.P. Pertijs and W.J. Kindt, "A 140dB-CMRR Current-Feedback Instrumentation Amplifier Employing Ping-Pong Auto-Zeroing and Chopping," *IEEE J. Solid-State Circuits*, vol. 45, no. 10, pp. 2044–2056, Oct., 2010.
[5] P. G. Blanken and S.E.J. Menten, "A 10µV-Offset 8kHz Bandwidth 4th-Order Chopped $\Sigma\Delta$ A/D Converter for Battery Management," *ISSCC Dig. Tech. Papers*, pp. 388-389, Feb., 2002.

978-1-61284-303-2/11 $26.00 © 2011 IEEE

ISSCC 2011 / February 22, 2011 / 3:45 PM

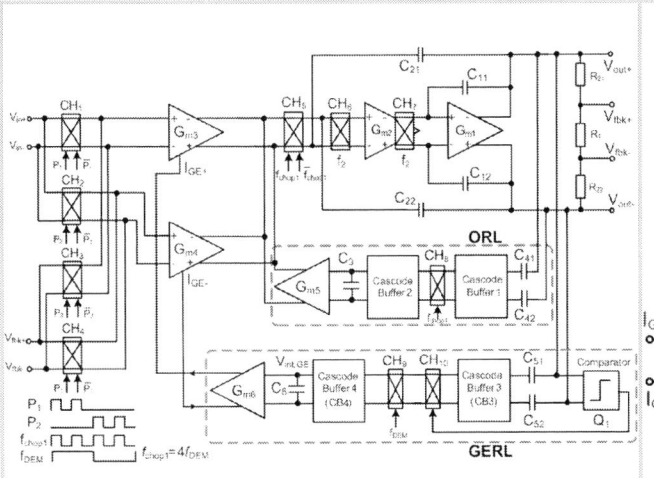

Figure 13.5.1: Block diagram of a CFIA with an offset-reduction loop (ORL) and a gain error reduction loop (GERL).

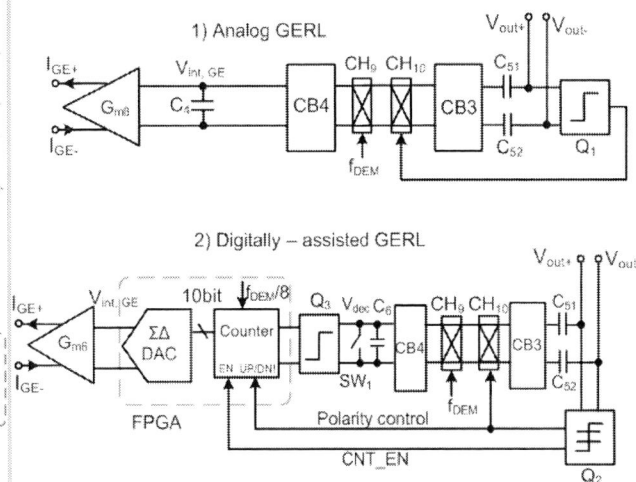

Figure 13.5.2: Block diagram of the analog GERL and digitally assisted GERL.

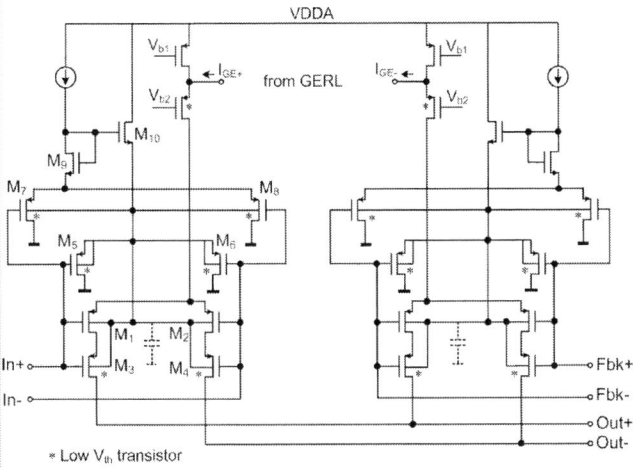

Figure 13.5.3: Schematic diagram of the input and feedback G_m stages G_{m3} and G_{m4} with class-AB bootstrapping of the back-gates.

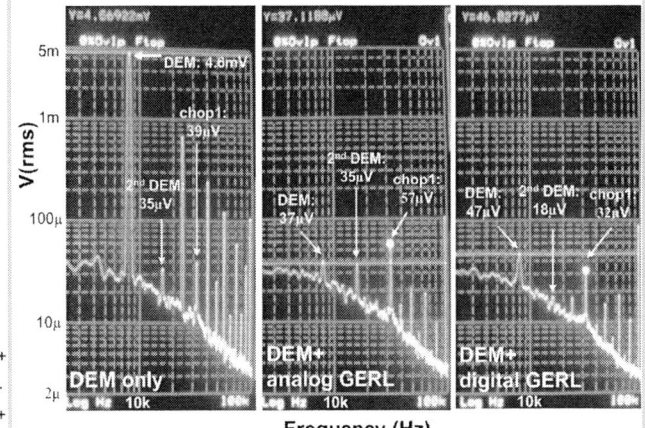

Figure 13.5.4: Output ripple measurement with GERL "on" and "off" (at a gain of 100, $f_{DEM} = f_{chop1}/4 = 8kHz$).

Figure 13.5.5: Measured gain drift histogram and INL of the CFIA (1. No DEM; 2. DEM only; 3. With DEM and the GERL).

	This work	Witte [3]	Pertijs [4]	Fan [2]
Year	2011	2008	2010	2010
Supply voltage	5V	2.8 to 5.5V	3.0 to 5.5V	5V
Supply current	290μA (analog GERL)	850μA	1.7mA	143μA
Input noise PSD	17nV/√Hz	136nV/√Hz	27nV/√Hz	21nV/√Hz
1/f noise corner	1mHz	---	---	1Hz
Input offset current	250pA	---	---	---
Input Impedance	180MΩ	---	---	---
CMRR	127dB	130dB	139dB	137dB
PSRR	130dB	114dB	138dB	---
Gain error (Absolute)	Same CM: 0.01% Dif CM:0.06% (Untrimmed)	0.1% (Trimmed)	0.1% (Untrimmed)	0.53% (Relative) (Untrimmed)
Gain drift	3ppm/° C (typ) 6ppm/° C (max)	---	---	---
Offset	< 3 μV	< 5 μV	< 3 μV	< 2 μV
Offset drift	15 nV/° C	---	50 nV/° C	--
GBW	800kHz (stable when gain> 20)	640kHz	800kHz	900kHz
INL	4ppm	---	---	---
NEF	11.2 (analog GERL)	152.9	43	9.6

Figure 13.5.6: Performance summary and comparison with the state-of-the-art instrumentation amplifiers.

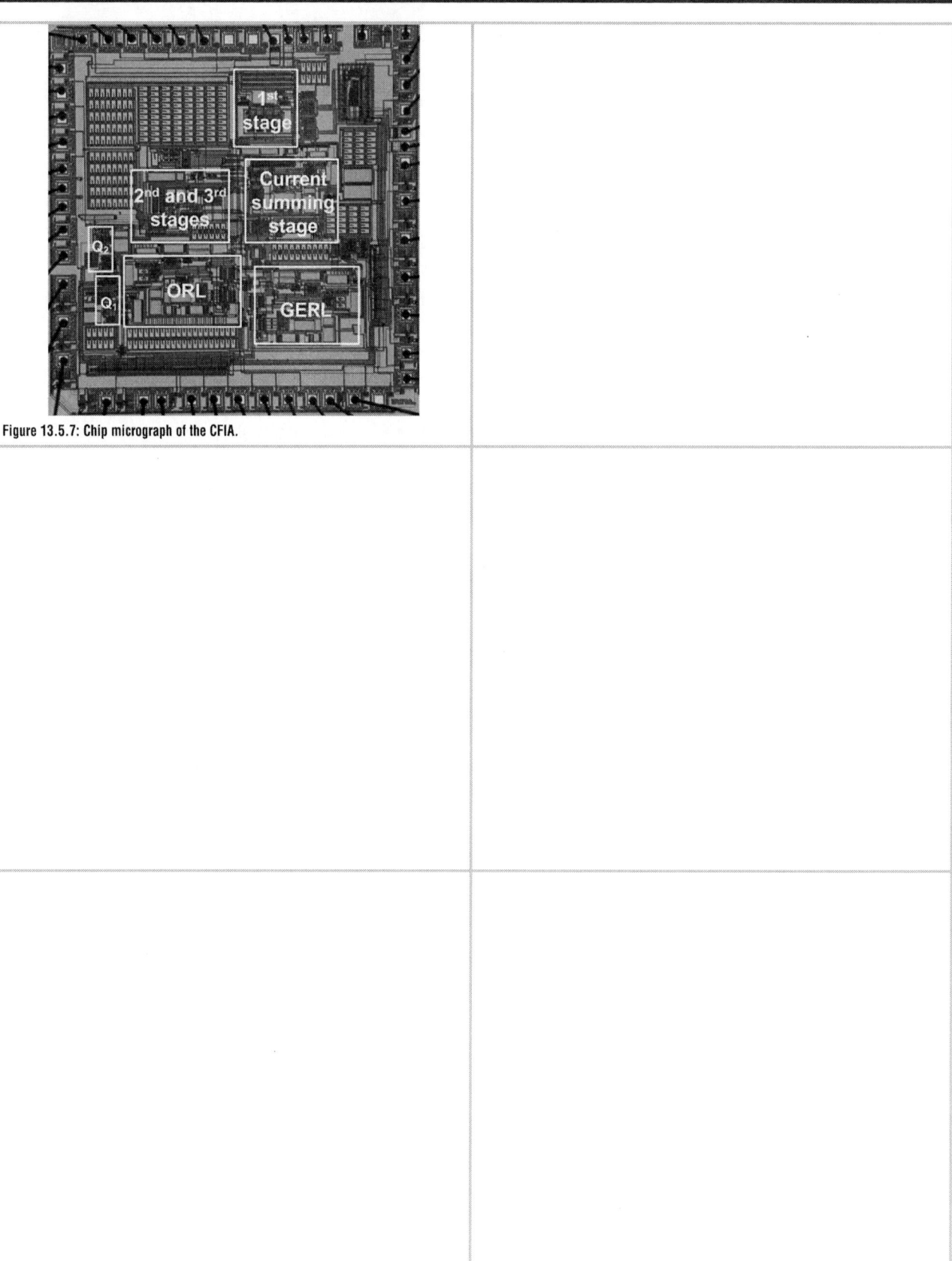

Figure 13.5.7: Chip micrograph of the CFIA.

ISSCC 2011 / SESSION 13 / ANALOG TECHNIQUES / 13.6

13.6 A 6.7nV/√Hz Sub-mHz-1/f-Corner 14b Analog-to-Digital Interface for Rail-to-Rail Precision Voltage Sensing

Chinwuba D. Ezekwe, Johan P. Vanderhaegen, Xinyu Xing, Ganesh K. Balachandran

Robert Bosch, Palo Alto, CA

Many sensors demand energy-efficient precision voltage sensing interfaces with low noise performance down to very low frequencies. Examples include micro Kelvin resolution temperature sensors for wafer stepper temperature stabilization [1] and others [2]. In many such applications, analog-to-digital conversion is inevitable, which mandates sampling. So far, sensor interfaces have demonstrated low offset drift using chopping/auto-zeroing while preserving the intrinsic SNR of the transducer by anti-aliasing filtering before sampling. Integration with a boxcar window, or boxcar sampling, is a particularly convenient way to realize anti-aliasing filtering in the sampling process [3]. However, implementations of the technique, due to their use of just a voltage-to-current converter whose output current is integrated over a time window, e.g. in a sigma-delta modulator [2], remain susceptible to gain drift resulting from either transconductance or time window drift. Meanwhile, related work suggests that embedding the boxcar sampler in a feedback loop effectively removes this limitation [4]. Whereas the loop in [4] is closed electromechanically, the design presented here overcomes the same limitations by using a purely electrical feedback loop to establish a well-defined gain that is insensitive to boxcar sampler parameter drift.

The interface, shown in Fig. 13.6.1, includes a switch network that selects which among several sources is active and at the same time modulates the selected input signal to separate it from the low-frequency errors lurking in the signal chain. The modulation is nested chopping [2] with fast chopping at 1MHz and slow chopping every 4096 fast-chopping periods (only fifteen fast-chopping periods shown for convenience). The modulated signal feeds into the closed-loop boxcar sampler where the 1MHz chopping is first demodulated, then integrated, then sampled at 1MHz, and then remodulated and fed back with a small feedback factor to null the input of the integrator so that the closed-loop output holds a demodulated and accurately amplified discrete-time representation of the input. The DC accuracy of the output is insensitive to integrator parameter variation since the integrator is in the forward path of the feedback loop. The oversampled nature of the output fits well with a sigma-delta type converter. The modulator, which operates as an incremental converter, is momentarily reset on every edge of the slow chopping clock. The incremental operation is ideal for multiplexed readout of a multi-axis sensor as is the case in this design. The slow chopping, which addresses the residual offset and flicker noise contributed mainly by the ADC, is demodulated digitally. The demodulated output is filtered then decimated by a factor of two to the final output rate of about 120Samples/sec.

The closed-loop boxcar sampler, shown in Fig. 13.6.2, consists of a G_m-Opamp-C integrator followed by a SHA, with both wrapped in a capacitive feedback loop. The two-stage Miller opamp maximizes current efficiency using high gm/Ids, which makes the right-half plane zero introduced by G_{m3} and C_m difficult to cancel precisely using the traditional triode MOSFET-capacitor series connection. This design uses the feed-forward path through G_{m4} to reliably cancel that right-half plane zero. The integrator is designed to have DC gain in excess of 140dB to achieve 56dB of overall closed-loop gain with 80dB of linearity. The modulated nature of the input signal enables the use of capacitors to couple the input and feedback signals to the V-to-I converter. Besides providing signal summation that is noiseless unlike the traditional resistive approach, capacitors offer the ability to still present a high-Z load to the transducer while using a single input stage for improved noise efficiency over the traditional dual-input current-feedback topology such as that used in [1]. They also enable rail-to-rail common-mode input capability and conveniently store V-to-I input offset for coarse offset cancellation to prevent the offset induced ripple from saturating the output after 56dB of gain. Choppers demodulate the AC signal and modulate away both the residual offset of the V-to-I and the offset of the opamp input stage. Those chop-

pers run on delayed clocks to allow the SHA to sample the integrator output free of any disturbance from the chopping and hold it for feedback and for further processing by the ADC. Feedback switches remodulate the SHA output back to the V-to-I input. Circuit operation starts with the autozeroing of the V-to-I and resetting of C_{int}, followed by N correlated double integration phases, where N (typically 4096 in this design) is the number of input samples the ADC needs to produce a decimated output. A guard phase in between integration phases (akin to the charge injection dead-band in [2]) rejects artifacts from the modulation transients for improved offset rejection. The SHA samples the integrator output at the end of every other integration phase.

The V-to-I converter sets the noise floor and accounts for half of the power dissipation of the interface. This design uses the telescopic cascode shown in Fig. 13.6.3 for its high noise efficiency. The restricted output swing is acceptable here since the outputs drive the virtual ground inputs of the opamp-C stage. The discrete-time feedback breaks the strict tradeoff between linearity and noise efficiency, enabling the use of high gm/Id input devices. Cascodes that track the common-mode input level keep the drain-to-source voltages of the input devices approximately constant for better common-mode rejection. Resistors degenerate the current sources to reduce their noise contribution.

The incremental analog-to-digital converter, shown in Fig. 13.6.4, consists of a single-loop, single-bit, second-order $\Delta\Sigma$ modulator followed by a decimation filter [5]. The input feed-in path to the quantizer ensures that the first comparator decision of each conversion cycle contains information pertinent to the input signal and also minimizes integrator swings. The decimation filter, which consists of a cascade of two integrators, facilitates the trade-off of noise and conversion speed simply via the over-sampling ratio. Both the modulator and decimation filter are reset during the auto zeroing of the boxcar sampler. Input signal sampling occurs on the fast chopping clock.

Fabricated in a 0.35μm CMOS process, the interface, whose ADC reference is supplied from off-chip, draws 2mA from 3.3V and occupies 1.2mm² of active area. Figure 13.6.5 shows the output spectrum measured with the interface driven by an on-chip sensor resistor bridge biased at zero current to isolate the flicker noise performance of the interface from external interference that can induce a drift in the sensor signal. Thanks to the combination of nested chopping and autozeroing, the noise power spectral density remains flat down to 0.1mHz at 8.3nV/√Hz of which the interface contributes 6.7nV/√Hz. Figure 13.6.6 shows the output spectrum measured with the interface driven by a 5Hz test signal with an amplitude of 820μV (interface full-scale = 1.65mV). The peak SFDR of 80dB validates the efficacy of the closed-loop approach presented here in achieving a fixed and highly linear gain while exploiting the noise/power-efficiency of boxcar sampling in a precision-voltage-sensing analog-to-digital application.

Acknowledgements:
The authors thank Jonas Handwerker, Thomas Rocznik, and Martin Krämer for their assistance during the design and testing of the experimental prototype.

References:
[1] R. Wu, K. A. A. Makinwa, and J. H. Huijsing, "A Chopper Current-Feedback Instrumentation Amplifier with a 1mHz 1/f Noise Corner and an AC Coupled Ripple-Reduction loop," *ISSCC Dig. Tech. Papers*, pp. 322-323, Feb., 2009.
[2] J.C. van der Meer, F.R. Riedijk, E. van Kampen, K.A.A. Makinwa, and J.H. Huijsing, "A Fully Integrated CMOS Hall Sensor with a 3.65 μT 3σ Offset for Compass Applications," *ISSCC Dig. Tech. Papers*, pp. 246-247, Feb., 2005.
[3] L.R. Carley and T. Mukherjee, "High-Speed Low-Power Integrating CMOS Sample-and-Hold Amplifier Architecture," *IEEE CICC*, pp. 543-546, May, 1995.
[4] C. D. Ezekwe and B. E. Boser, "A Mode-Matching ΣΔ Closed-Loop Vibratory Gyroscope Readout Interface with a 0.004°/s/√Hz Noise Floor over a 50Hz Band," *IEEE J. Solid-State Circuits*, vol. 43, no. 12, pp. 3039-3048, Dec., 2008.
[5] J. Markus, J. Silva, and G. C. Temes, "Theory and Applications of Incremental ΔΣ Converters," *IEEE TCAS-I*, vol.51, no.4, pp. 678-690, April, 2004.

978-1-61284-303-2/11 $26.00 © 2011 IEEE

ISSCC 2011 / February 22, 2011 / 4:15 PM

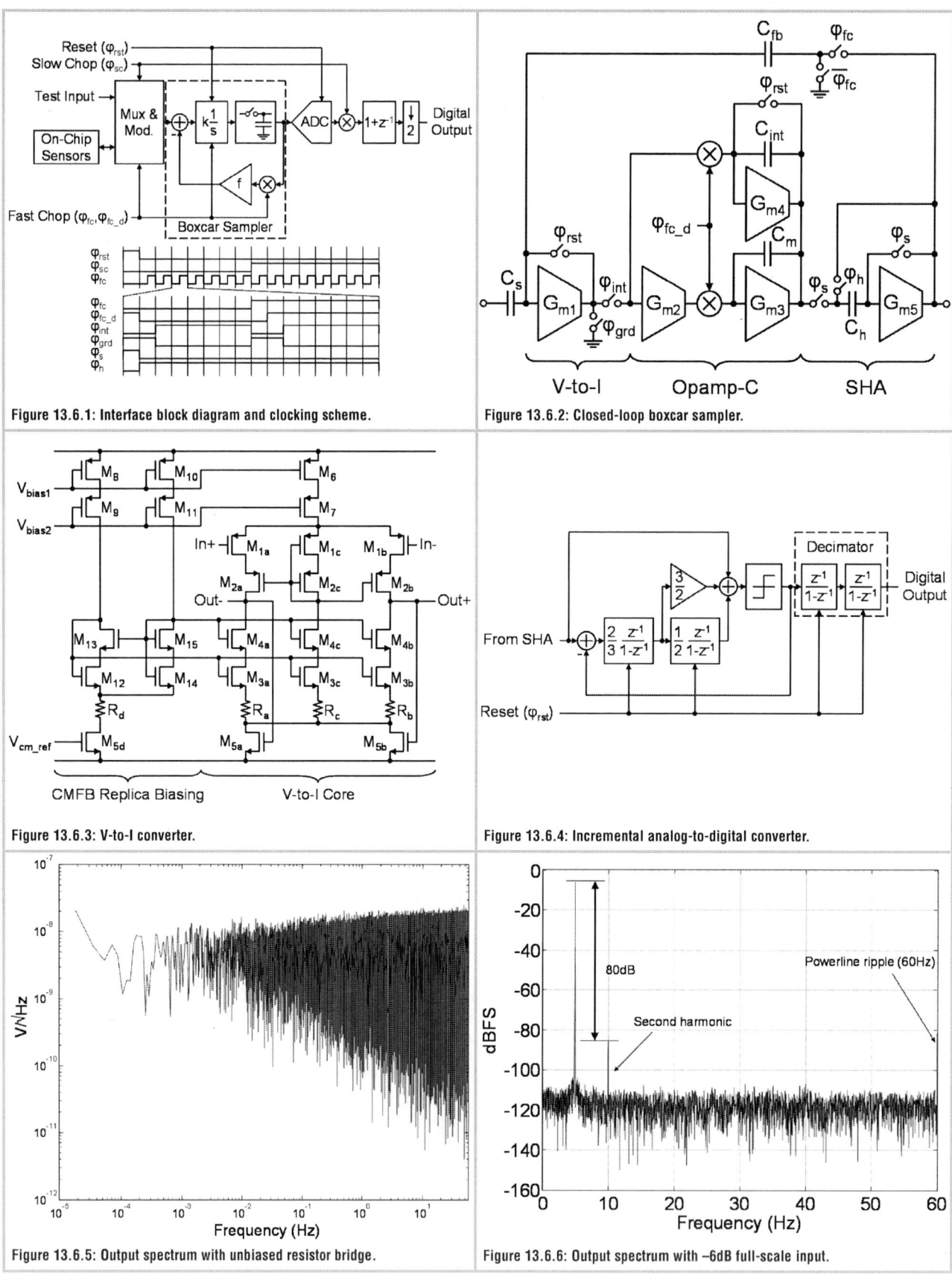

Figure 13.6.1: Interface block diagram and clocking scheme.

Figure 13.6.2: Closed-loop boxcar sampler.

Figure 13.6.3: V-to-I converter.

Figure 13.6.4: Incremental analog-to-digital converter.

Figure 13.6.5: Output spectrum with unbiased resistor bridge.

Figure 13.6.6: Output spectrum with –6dB full-scale input.

13

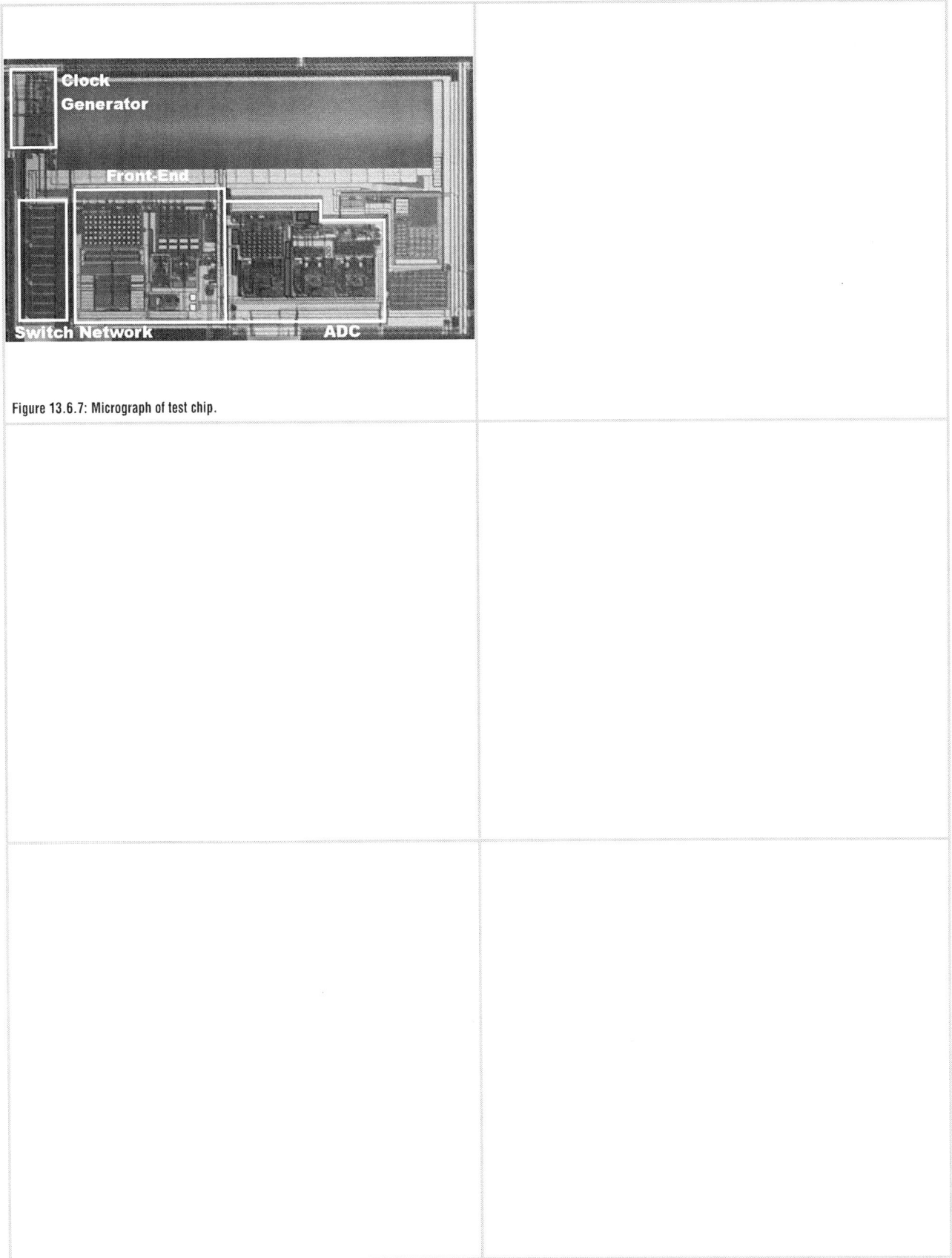

Figure 13.6.7: Micrograph of test chip.

ISSCC 2011 / SESSION 13 / ANALOG TECHNIQUES / 13.7

13.7 A 36V JFET-Input Bipolar Operational Amplifier with 1μV/°C Maximum Offset Drift and −126dB Total Harmonic Distortion

Martijn F. Snoeij, Mikhail V. Ivanov

Texas Instruments, Erlangen, Germany

A 36V JFET-input bipolar operational amplifier is presented with a maximum offset drift of 1μV/°C over a temperature range of −40 to 125°C, which represents a 3x improvement on the state-of-the-art. This is achieved with a drift-compensating circuit incorporated in the input stage that relies on a wafer-level 2-temperature laser-trimming method. The opamp has a GBW of 11MHz, a flat-band noise of 5.1nV/√Hz, a slew-rate of 20V/μs, a −126dB (0.00005%) total harmonic distortion plus noise (THD+N) ratio, and a quiescent current of 1.8mA. This combination of high slew rate and good noise-to-power ratio is accomplished through the use of a linearized class-AB boosting circuit in the input stage.

High-voltage (>30V) operational amplifiers (opamps) are often used in signal-conditioning circuits for industrial applications to ensure compatibility with legacy equipment. In such applications, good DC precision, i.e., low offset, offset drift and noise are essential requirements. A second application of such opamps is in high-end audio systems, where high voltages are needed to achieve a high SNR (>110dB), and good AC performance, i.e., a high THD+N ratio, high slew-rate and low-noise, is of key importance. In both these applications, a low quiescent current is essential, as the high supply voltage leads to significant power consumption and heat generation, especially in multi-channel systems. Although dynamic offset cancellation techniques can be used to decrease offset and offset drift, they may cause spurious output signals and require the use of CMOS technology, which typically yields a poorer noise-to-power ratio, especially at high supply voltages. Therefore, the superior noise-to-power ratio, 1/f noise and DC precision of bipolar technologies is often preferred. The low (<50pA) input bias current that is needed to process high-impedance input signals, mandates a JFET input stage.

The proposed opamp uses a two-stage topology, as shown in Fig. 13.7.1. The input stage consists of a differential JFET pair (JN_1, JN_2) loaded by a degenerated current mirror (Q_1, Q_2, R_1, R_2). As in [1], the rail-to-rail output stage (Q_3, Q_4) uses class-AB biasing with level-shifters Q_5 and Q_6, biased by transistors Q_7 to Q_{10} and current sources I_1 to I_4. Because a bipolar output stage needs a significant base current drive, a non-inverting current-gain driver is added between the input and output stage. In order to improve high-frequency behavior, feed-forward capacitors C_{FF} are added to bypass the driver. Frequency compensation is achieved using a normal Miller capacitor C_M and active Miller capacitor C_{MA}.

The offset and offset drift of the presented opamp is primarily determined by its input JFET devices. Unlike bipolar transistors, the offset and drift of JFETs are poorly correlated and thus both need to be trimmed. Although a room-temperature drift trimming method was shown in [2], it requires a BiCMOS process and is thus unsuitable for this design. Instead, a 2-temperature wafer-level laser-trimming method is used, along with a drift-trim circuit that adds a differential current to the output of the input stage (Fig. 13.7.1). First, the offset is trimmed to zero at 90°C by means of trim-resistors R_1 and R_2 (Fig. 13.7.1), resulting in an offset drift characteristic as shown in Fig. 13.7.2a. Second, the offset is trimmed to zero again at 27°C (room temperature) by use of the drift trim circuit, which is designed to output an adjustable temperature slope current I_{drift}, that is always zero at 90°C (Fig. 13.7.2b). By using this current, the offset drift will also be removed, since the temperature drift of I_{drift} will be exactly opposite to the opamp's drift. Since no information needs to be stored between the two trimming steps, this method is also suitable for packaged devices. The choice of trim temperatures sets the range with the lowest drift.

The drift trim circuit (Fig. 13.7.3) consists of a current source I_{ctat}, that is inversely proportional to temperature (set by V_{be} characteristics) and I_{ptat}, which is proportional to temperature (set by ΔV_{be} characteristics). Both sources are mirrored via transistors Q_1 to Q_4 and Q_5 to Q_8, respectively. This results in 3 currents $I_{trim1,2,3} = I_{ptat} − I_{ctat}$. Since I_{ctat} and I_{ptat} have opposite temperature dependen-

cies, I_{trim} will be zero for some temperature; to ensure that this occurs at 90°C, I_{trim3} is measured via probe-only pad P_{test} and is trimmed to zero by adjusting I_{ctat}. To prevent saturation of Q_4 or Q_8 when the probe-pad is not connected, clamps Q_9 and Q_{10} drain I_{trim3}. Currents I_{trim1} and I_{trim2} are dropped across trimmable resistors R_1 and R_2, resulting in a voltage V_{drift}, which has the desired adjustable temperature dependency, since it is zero for all temperatures when $R_1=R_2$ (during 90°C trimming step), has positive temperature slope for $R_1>R_2$, and negative slope for $R_2>R_1$. The transconductance stage gm_1 converts V_{drift} into a differential current I_{drift} that is summed with the input stage current (Fig. 13.7.1). The offset and gain error of gm_1 is trimmed out along with the offset and offset drift of the opamp.

In order to improve the opamp's slew-rate and high-frequency THD without degrading its noise/power ratio, input class-AB boosting [3] (also known as slew boosting) is used (Fig. 13.7.4). JFETs JN_1 and JN_2 copy the opamp's inputs onto V_A and V_B. Transistors Q_1 to Q_6 form a translinear loop, which can be explained as follows. Because of transistors Q_1 and Q_2, V_C follows the minimum of V_A and V_B; similarly, V_D and V_E follow the maximum of V_A and V_B via Q_3-Q_4 and Q_6. In quiescent condition, $V_A=V_B=V_E$. As a result, current I_1 is split proportional to the emitter area of Q_5, and Q_1 and Q_2 and the sizing of R_1 and R_2, resulting in a quiescent tail current of 200μA. During slewing, $V_A>V_B$ or $V_A<V_B$; thus, V_E will increase, while V_C will decrease, which reduces the V_{be} of Q_5. If the difference between V_A and V_B is large enough, Q_5 will turn off, and the full current I_1 will be mirrored, resulting in a well-defined increase of I_{tail} to 700μA, or 3.5x. When the opamp outputs a sine wave, its limited open-loop gain will cause an attenuated sine wave to appear across its inputs. The response of the slew-boost circuit to this sine wave needs to be linear to ensure a good THD+N ratio. Resistors R_1 and R_2 degenerate Q_5, and Q_1 and Q_2 and so linearize the translinear loop.

The opamp is realized in a fully complementary 36V SOI SiGe bipolar process (Fig. 13.7.7). Figure 13.7.5a depicts the offset versus temperature of 282 random samples that represent the entire production distribution without pre-selection. The drift trimming method essentially eliminates the linear drift component. The largest residual drift occurs at lower temperatures, as this is furthest away from the two trimming temperatures. Figure 13.7.5b depicts a histogram of the offset drift, which is defined as the maximum minus the minimum offset that occurs over the 6 measured temperatures, divided by the total temperature range. Fig. 13.7.6 summarizes the performance parameters and compares it to other commercially available high-voltage JFET-input opamps. It shows that the presented work exceeds prior art in all DC precision specifications, while it pairs a superior noise-efficiency factor (NEF) to a high slew-rate (20V/μs) and excellent THD+N (−126dB).

Acknowledgement:
The authors would like to acknowledge Sergey Alenin, Henry Surtihadi and Vadim Ivanov for their contributions to the design.

References:
[1] D. Monticelli, "A Quad CMOS Single-Supply Opamp with Rail-to-Rail Output Swing," *ISSCC Dig. Tech. Papers*, pp. 18-19, Feb., 1986.
[2] M. Bolatkale, M.A.P Pertijs, W.J. Kindt, et al., "A BiCMOS Operational Amplifier Achieving 0.33μV/°C Offset Drift using Room-Temperature Trimming," *ISSCC Dig. Tech. Papers*, pp.76-77, Feb., 2008
[3] J.H. Huijsing, *Operational Amplifiers – Theory and Design*, Kluwer Academic Publishers, 2001.
[4] ADA4627 Data Sheet, Analog Devices, www.analog.com, Aug., 2010.
[5] AD8610 Data Sheet, Analog Devices, www.analog.com, Aug., 2010.
[6] LT1057 Data Sheet, Linear Technology, www.linear.com, Aug., 2010.

ISSCC 2011 / February 22, 2011 / 4:30 PM

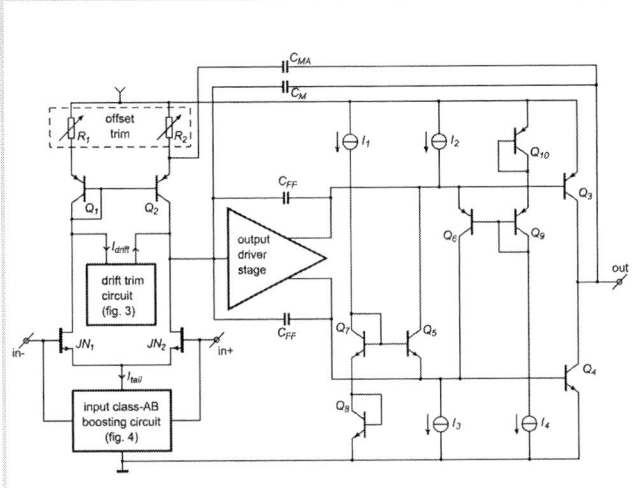

Figure 13.7.1: Simplified block diagram of the operational amplifier.

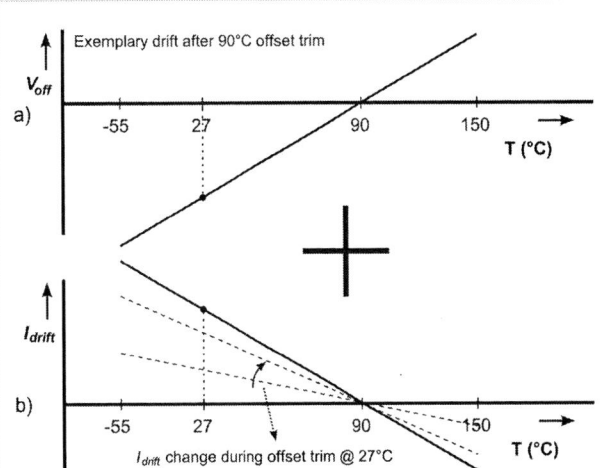

Figure 13.7.2: Drift trimming principle: a) Opamp offset drift after offset trim at 90°C b) Current that is added by the drift trim circuit.

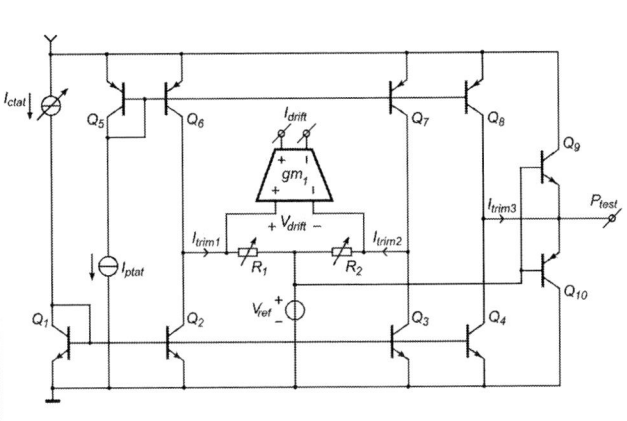

Figure 13.7.3: Simplified diagram of the drift-trim circuit.

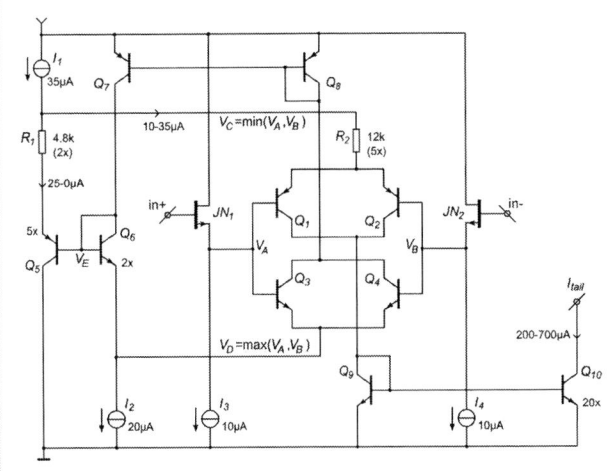

Figure 13.7.4: Input class-AB boosting circuit.

a)

b)

Figure 13.7.5: a) Measured offset versus temperature b) Measured offset drift histogram.

Parameter	This work	ADA4627 [4]	AD8610 [5]	LT1057 [6]
Max. supply voltage (V)	40	36	27.3	40
Max. drift (uV/C)	1	3*	3.5*	12*
Max. offset (uV)	120	300*	250*	800*
1/f noise 0.1-10Hz (uVpp)	0.25	0.7	1.8	2.1
Noise @ 1kHz (nV/√Hz)	5.1	6.1	6	14
GBW (MHz)	11	19	25	5
Slew rate (V/us)	20	56	60	13
THD+N ratio (dB)	-126	-127	-	-
CMRR/PSRR (dB)	140/140	110/108	110/110	98/102
Quiescent current (mA)	1.8	7.0	3.0	1.5
NEF	8.3	19.7	12.7	20.9

* Specification reflecting the entire production distribution ("Low-grade")

Figure 13.7.6: Performance summary and comparison with state-of-the-art JFET-input opamps.

13

Figure 13.7.7: Chip micrograph of the opamp (dual version). The die size is 1.4mm × 1.5mm.

ISSCC 2011 / SESSION 13 / ANALOG TECHNIQUES / 13.8

13.8 A 3.3V-Supply 120mW Differential ADC Driver Amplifier in 0.18μm SiGe BiCMOS with 108dBc IM3 at 100MHz

Gwilym F. Luff

Intersil, Harlow, United Kingdom

Fully differential (operational) amplifiers (FDA) have become the preferred method of driving 1st and 2nd Nyquist zone signals into 100 to 500 Ms/s 12 to 16 bit ADCs. Previously the best performance (95dBc IM3 at 100MHz [1]) came from designs in dielectrically isolated (DI) Silicon Germanium complementary bipolar (CB) processes [2]. These have not been developed at the 0.18μm geometry node, as the cost of shrinking these esoteric processes is not justified by the low volume markets they serve. Here 108dBc IM3 at 100MHz is obtained using a 0.18μm SiGe NPN-only RF BiCMOS. NPN-heavy circuits and a feedforward nested Miller architecture achieve this at 3.3V supply voltage and 36mA total supply current.

To achieve over 100dB of third-order linearity above 100 MHz requires both linear amplifying stages and copious loop gain. Without complementary PNPs, amplifying stages have 40dB or less DC open loop gain, limiting lower frequency (<100MHz) performance. Multipath nested Miller compensation (MNMC) [3] gives an overall 6dB/octave roll-off with high DC gain from 3 cascaded stages without bandwidth shrinkage. Figure 13.8.1 is the signal-path block diagram, including the common-mode control. This adds voltage buffers to a differential MNMC configuration to deliver current gain at the output. In this scheme the inner Miller feedback capacitors could be connected to the voltage buffer outputs or to their inputs. The optimum combination for stability of this inner loop resembles a nested Miller amplifier, with 1/3 of the capacitance connected around the 3rd gain stage and the remaining 2/3 of it from the voltage buffer outputs. In operation this amplifier has four nested feedback loops: an internal loop in the voltage buffer, inner and output Miller feedback loops and the external gain setting feedback loop. The frequency compensation is tuned for a closed loop gain of 4, but the part is stable for gains of two or more.

In the FDA application circuit (Fig. 13.8.2) the feedback resistors couple any common mode signal at the input directly to the output, however low the common mode gain of the amplifier. To absorb this current the output voltage buffers' low source impedance is augmented by a dual loop CMFB. The two capacitors Cb give a low gain 3GHz bandwidth CMFB path, while the two Gm stages (G_{mA} and G_{mB}) and Ca form a high-gain lower-bandwidth path to accurately set the DC common-mode level.

The input stage G_{m1} and G_{m2} are bipolar differential pairs for low noise. They need a high bandwidth common mode control to absorb any common mode feedback through the miller feedback capacitors. Figure 13.8.4 is the schematic of the input stage G_{m1}, showing the parallel connection of a cross-coupled PMOS pair and PMOS diode loads. To control this common-mode voltage without complicating the high-speed load, the input-stage positive supply is fed from a tracking LDO regulator. The input stage of this regulator is a replica PMOS stripe, with its gate fed from the common-mode reference voltage. A 6mA tail current gives sufficient slew rate for signals up to 1GHz applications, and lowers the input referred noise below 1nV/√(Hz).

The output buffer (Fig. 13.8.3) is a parallel connection of a flipped (or White) voltage follower and an emitter follower. The flipped follower Q1, Q0, and Q44 pulls down well, while the emitter follower Q31 pulls up to give a push-pull class AB action. The resistor R4 is set equal to the r_e of Q1 to combine the outputs of the two parallel followers. The DC feedback of the flipped follower is through level shifter Q25, M11, M6, and R1. The level shifter has high output impedance from the common gate stage M11, to stabilize the quiescent current over power supply variation. At high frequencies the level shifter is bypassed by C18, although this creates a pole-zero doublet. The zero is set by C18 and the composite G_m of Q25 and M11 and the pole is set by C18 and $\beta.r_e$ of Q0 and Q44. This can cause an undesirable long settling 'tail' for both differential and common mode signals. Resistor R1 between bases of Q0 and Q44 reduces the differential DC gain to unity, equal to the high frequency feed-forward gain through C18. This minimizes the effect of the differential pole-zero doublet created by the feed forward capacitors.

To give the $3V_{pp-diff}$ output required by 16 bit ADCs, the supply voltage can be raised above the 3.6V maximum rating of the MOSFETs. Voltage sharing cascodes (such as M6 and M13 in the output stage) are arranged so that no 3V device sees the full supply voltage in either the full power operating state or the low power 'shutdown' state. 5V NPNs are used in the bias circuits, and a self contained voltage clamp limits the drain and gate voltages of the LDO pass transistor. The bases of 3V NPNs are fed from defined low impedances, so that their voltage breakdown is in the V_{CER} region. In the shut-down state, the main CMFB circuit and the output buffers run at much reduced current (300uA total) to maintain the DC voltages at the output pins. This prevents the ADC driver delivering full power full bandwidth transients during power up and power down that could damage the costly ADC.

The design is implemented in a 0.18μm SiGe process, featuring 90GHz f_t 3V HBTs, 3.3V CMOS, MiM capacitors and 4 level metal including a low-resistance thick top metal. Low capacitance ESD diodes, an RC-triggered SiGe-based ESD clamp at each corner and top metal ESD buses complete a robust ESD design. A 16 pin TQFN package has low lead inductance and good heat transfer. The pinout in Fig. 13.8.2 minimizes PCB inductance in the supplies and off chip feedback network. The strict symmetry of the power pins improves 2nd-order distortion.

This design achieved its simulated distortion performance in 1st silicon. Figure 13.8.5 shows the measured IM3 results, for a $2V_{pp-diff}$ composite 2 tone output into a 200Ω load. The IM3 improves rapidly below 100MHz as the multiple nested feedback loops increase in gain. The measured DC open-loop gain is 97dB, corresponding to a 68kHz low-frequency pole. A closed-loop 3dB bandwidth of 2.2GHz maintains low-frequency gain flatness. The simple input stage gives 0.85nV/√Hz input referred noise. Supply current is 36mA at 3.3V (120mW), with correct operation from 3.0 to 4.5V. ESD robustness is 3kV HBM, 300V MM and 1kV CDM. Figure 13.8.6 shows the measured performance, compared with some recent $2V_{pp-diff}$ FDA ADC drivers in CB SiGe BiCMOS processes. The figure of merit (FOM) is the ratio in dB between the OIP3 and the quiescent power consumption P_Q adjusted for the feedback noise gain G_N (This is significant for [5]).

$$FOM = OIP3 + G_N/2 - P_Q$$

Where OIP3 and P_Q are in dBm, and G_N is the noise gain in dB. Our 7dB advantage over [6] is a 4.4× improvement in power efficiency normalized for linearity and gain.

Our results confirm that technology scaling in SiGe BiCMOS can overcome the advantages of CB DI processing [4]. This NPN-only 0.18μm SiGe BiCMOS design achieves 108 dBc IM3, delivering $2V_{pp-diff}$ composite two tone output at 100MHz. It has twice the linearity at less than half the power of designs in 0.35μm complementary SiGe BiCMOS, taking 120mW from a 3.3V supply.

Acknowledgements:
The author would like to thank Michael Steffes and all the other members of the ADC driver product team.

References:
[1] M. Steffes and X. Ramus, "Low-Power, High-Intercept Interface to the ADS5424 14-bit, 105-MSPS Converter for Undersampling Applications," *Texas Instruments Analog Applications Journal (SLYT223)*, pp. 10-18.
[2] B. El-Kareh, S. Balster, W. Leitz, et.al.," A 5V Complementary-SiGe BiCMOS Technology for High Speed Precision Analog Circuits", *IEEE BCTM*, pp. 211-221, Sept., 2003.
[3] R.G.H. Eschauzier, L.P.T. Kerklaan, and J.H.A. Huijsing, "A 100-MHz 100-dB Operational Amplifier with Multipath Nested Miller Compensation Structure," *IEEE J. Solid State Circuits*, vol. 27, no. 12, pp. 1709-1717, Dec., 1992.
[4] M.A.A. Ali, A. Morgan, C. Dillon, et al., "A 16b 250MS/s IF-Sampling Pipelined A/D Converter with Background Calibration", *ISSCC Dig. Tech. Papers*, pp. 292-293, Feb., 2010.
[5] LMH6554 datasheet, www.national.com/ds/LM/LMH6554.pdf, National Semiconductor, 2009.
[6] ADL5562 datasheet, Analog Devices, www.analog.com/static/imported-files/data_sheets/ADL5562.pdf, 2010.
[7] LTC6400-14 datasheet, http://cds.linear.com/docs/Datasheet/640014fb.pdf, Linear Technology, 2008.

978-1-61284-303-2/11 $26.00 © 2011 IEEE

ISSCC 2011 / February 22, 2011 / 4:45 PM

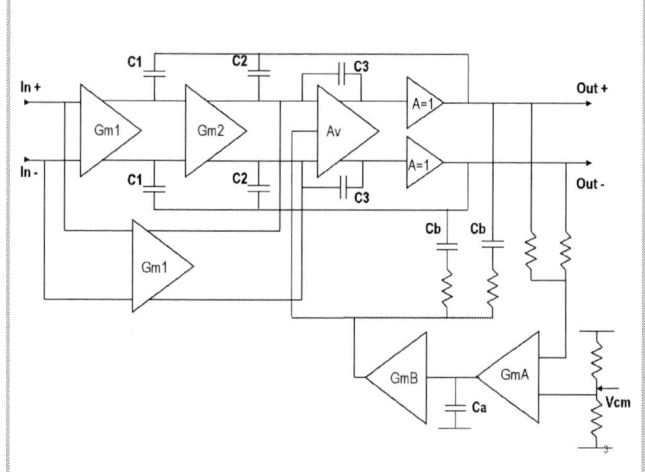

Figure 13.8.1: Signal path and common-mode control architecture.

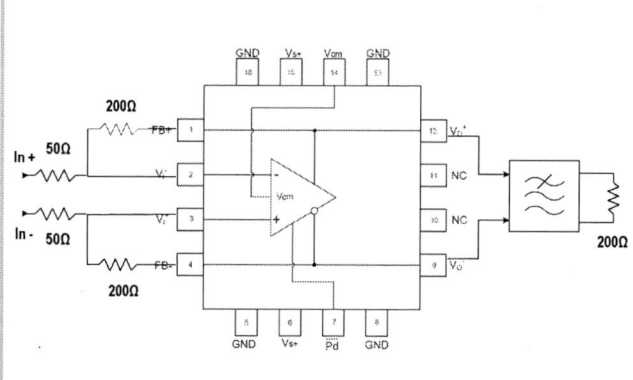

Figure 13.8.2: FDA ADC driver application and pinout.

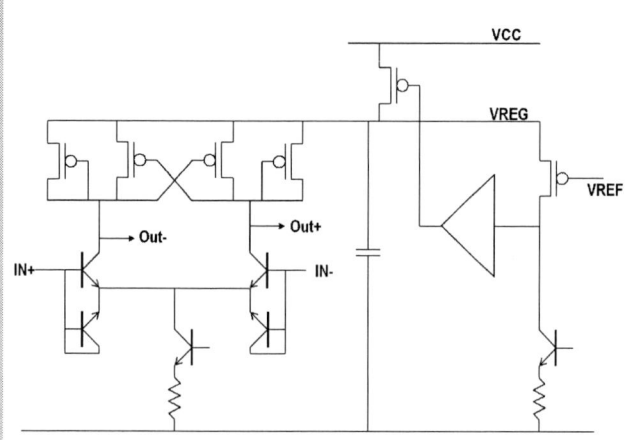

Figure 13.8.3: Schematic of the input stage.

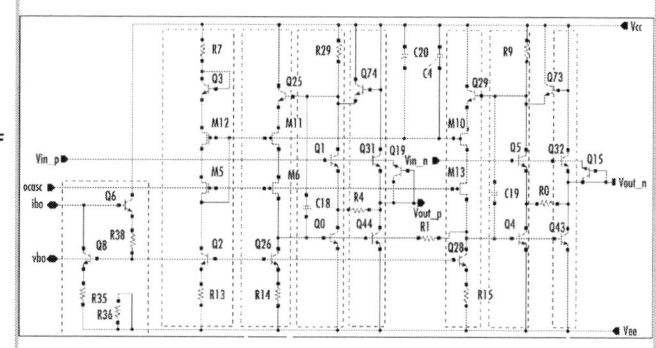

Figure 13.8.4: Schematic of the output stage.

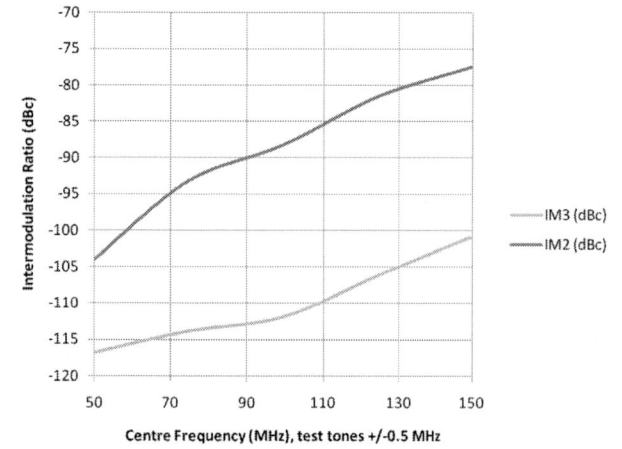

Figure 13.8.5: Inter-modulation results, 2V$_{pp\text{-diff}}$ composite into 200Ω.

PARAMETER	This Work	Ref [1]	Ref [5]	Ref [6]	Ref [7]	Units
IM3 at 140MHz	-101	-87	-99	-95	-80	dBc
OIP3 (into 200Ω)	48.5 (A$_V$=4)	41.5 (A$_V$=5)	47.5 (A$_V$=1)	45.5 (A$_V$=4)	38 (A$_V$=5)	dBm
HD3 at 140MHz	-93		-87	-81	-73	dBc
HD3 at 100MHz	-101	-81		-85		dBc
HD2 at 140MHz	-72			-82	-78	dBc
HD2 at 100MHz	-74	-68		-85		dBc
Input ref'd noise	0.85	1.9	0.9	2.1	1.1	nV/√(Hz)
3dB BW (A$_V$=4)	2.2	0.6	1.6	3.9	2.4	GHz
Supply Voltage	3.3	5	5	3.3	3.3	V
Power diss. (P$_Q$)	120	190	260	265	280	mW
FOM (OIP3+G$_N$/2-P$_Q$)	33	24	25	26	19	dB
Process Features	0.18µ	0.35µ DI CB	0.25µ DI CB	0.35µ DI CB	CB	
ESD (HBM)	3000	2000	2000			V

Figure 13.8.6: Performance summary and comparison.

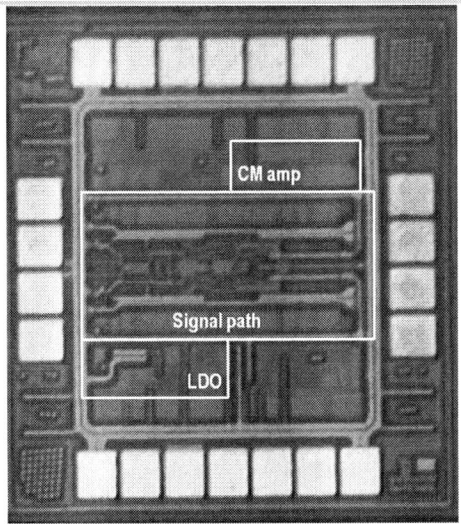

Figure 13.8.7: Die micrograph.

ISSCC 2011 / SESSION 14 / HIGH-PERFORMANCE EMBEDDED MEMORY / OVERVIEW

Session 14 Overview / Memory

High-Performance Embedded Memory

Session Chair: *Leland Chang, IBM T. J. Watson Research Center, Yorktown Heights, NY*

Session Co-Chair: *Peter Rickert, Texas Instruments, Richardson, TX*

Embedded memory plays a crucial role in today's VLSI applications – from high-performance computing to low-power consumer electronics. While scaling of technology feature size to the 32nm and 28nm nodes has enabled ever larger and higher performance on-die memories, it has also created growing challenges for the embedded memory designer. Growing device variability and power limitations are driving innovative solutions to maintain robustness and area efficiency in such aggressively scaled memories. In particular, peripheral circuit assist features have become the key to maintaining cell read and write margins to enable low voltage operation for dense SRAM caches. New strategies ranging from circuit-level techniques to fundamental changes in array architecture can also enable significant gains in area and power efficiency.

The first paper 14.1 from IBM describes peripheral circuit assist features implemented in 32nm high-k metal-gate SOI-CMOS to enable 0.7V operation of a $0.154\mu m^2$ bit cell. A 64Mb macro demonstrates stability margin enhancement by using a regulation scheme to reduce the bitline precharge voltage level, thus limiting charge injection into the cell. Improved write margin is achieved by using a negative bitline technique with increased boost voltage as compared with previous work. A bitcell-tracking delay monitor circuit is also used to improve process-limited performance and yield.

The next paper 14.2 from IBM introduces architectural techniques to significantly improve the area, power, and performance of multi-ported register file arrays. A 144×78b macro for a 45nm SOI-CMOS 2.3GHz POWER™ processor is presented with double-pumped write ports that are operated twice in a single cycle and replicated read ports that are combined from duplicate data copies. A compact 2R1W memory cell is thus leveraged to perform a 4R2W function with near 2× area and read power reduction, low 190ps read latency, and fast error correction. The macro operates at up to 2.76GHz at a supply voltage of 0.9V.

Paper 14.3 presents the design of the 8MB level-3 cache in 32nm SOI-CMOS for AMD's next-generation Bulldozer architecture that operates above 2.4GHz at 1.1V. Area efficiency is improved by the use of a column-select aliasing technique, in which column select wires are shared between odd and even pairs for reads and writes, while leakage power is minimized by supply gating and floating bitlines. An efficient redundancy scheme is also implemented using centralized redundancy blocks instead of storing all redundant data in the data macro itself.

Finally, Paper 14.4 from MIT describes a 128kb SRAM macro featuring a $0.12\mu m^2$ bit cell fabricated in a low-power 28nm CMOS process. A hierarchical bit-line architecture for low local bitline capacitance combined with delayed wordline boosting after a pre-read phase during write cycles improves stability and write margins to provide functionality down to 0.6V. The performance of this memory scales from 20MHz to 400MHz in the 0.6V to 1V operating voltage range while active power consumption scales from 2.8mW to 68.5mW.

The area, performance, and power of embedded memory will continue to improve well beyond the 32nm and 28nm nodes. The innovations presented in this session, including peripheral assist circuits, new architectural techniques, and low-power circuits, will drive such continued scaling into the nano-technology regime for high-performance embedded memories.

Presenters:

14.1 A 64Mb SRAM in 32nm High-k Metal-Gate SOI Technology with 0.7V Operation Enabled by Stability, Write-Ability and Read-Ability Enhancements

I. Arsovski,

IBM Systems and Technology Group, Essex Junction, VT

1:30 PM

14.2 A 4R2W Register File for a 2.3GHz Wire-Speed POWER™ Processor with Double-Pumped Write Operation

R. K. Montoye,

IBM T. J. Watson Reseach Center, Yorktown Heights, NY

2:00 PM

14.3 An 8MB Level-3 Cache in 32nm SOI with Column-Select Aliasing

D. Weiss,

AMD, Fort Collins, CO

2:30 PM

14.4 A 28nm High-Density 6T SRAM with Optimized Peripheral-Assist Circuits for Operation Down to 0.6V

M. E. Sinangil,

Massachusetts Institute of Technology, Cambridge, MA

2:45 PM

ISSCC 2011 / SESSION 14 / HIGH-PERFORMANCE EMBEDDED MEMORY / 14.1

14.1 A 64Mb SRAM in 32nm High-k Metal-Gate SOI Technology with 0.7V Operation Enabled by Stability, Write-Ability and Read-Ability Enhancements

Harold Pilo[1], Igor Arsovski[1], Kevin Batson[1], Geordie Braceras[1], John Gabric[1], Robert Houle[1], Steve Lamphier[1], Frank Pavlik[1], Adnan Seferagic[1], Liang-Yu Chen[2], Shang-Bin Ko[2], Carl Radens[2]

[1]IBM Systems and Technology Group, Essex Junction, VT,
[2]IBM Systems and Technology Group, Hopewell Junction, NY

A 64Mb SRAM macro is fabricated in a 32nm high-k metal-gate (HKMG) SOI technology [1]. Figure 14.1.1 shows the $0.154\mu m^2$ bitcell (BC). A 2× size reduction from the previous 45nm design [2] is enabled by an equal 2× reduction in BC area. No corner rounding of BC gates allows tighter overlay of gate electrode and active area. The introduction of HKMG provides a significant reduction in the equivalent oxide thickness, thereby reducing the Vt mismatch. This reduction allows aggressive scaling of device dimensions needed to achieve the small area footprint. A 0.7V VDD_{MIN} operation is enabled by three assist features. Stability is improved by a bitline (BL) regulation scheme. Enhancements to the write path include an increase of 40% of BL boost voltage. Finally, a BC-tracking delay circuit improves both performance and yield across the process space.

Stability assist by reducing the pass-gate (PG) strength is an effective method for decreasing failure rate. However, several of these methods [4,5] interfere with other operations and require external modulation to preserve the balance between stability and write-ability. Figure 14.1.2 shows a BL regulation system that does not degrade the write margin. To improve stability, BLs are precharged to a reduced level (VBLH) [3]. VBLH lowers the bump level of the internal "low" node to improve the noise margin of the cross-coupled inverter's trippoint. Figure 14.1.2 shows the improvement in failure rate as a function of VBLH level, plotted as a fraction of VDD supply. The regulator is designed to operate with a VBLH range of 68 to 78% of VDD. An increase in failure rate as VBLH is lowered beyond the operating range is caused by reverse stability fails; the PG connected to the "high" side of the internal BC node begins to conduct and is discharged from VCS (BC power supply) to VBLH. The voltage reference, Ref is PVT compensated by body-contacted device T0. Diode-connected PFET, T1 modulates the body of T0 to compensate for changes in V_T and maintain a near constant reference. The distributed regulator output device, Treg manages the high-current demand of the BL restore. Treg is physically embedded with the BL and Sense-Amplifier (SA) pre-charge FETs. BC leakage causes the VBLH supply to drift towards VDD, which decreases the stability advantage. To prevent VBLH from drifting, a single pull-down device, TL is driven by the push-pull regulator (Gnbias). Overlap current between Treg and TL is minimized with a 40mV dead zone built into the regulator. The voltage translation from the VBLH to the VDD domain occurs in the transition from the SA data lines DLT/DLC to the global read data lines GBLT/GBLC.

Negative BL boosting is an effective technique to improve write margin [4]. The BL boost increases the V_{GS} of the PG, facilitating the discharge of the internal BC node. Limitation in the boost voltage in previous work [4] is caused by partial capacitor discharge, which reduces the charge transfer into the BLs. Another limitation is the leakage from the unselected transistors in the write path that begin to conduct when source voltages are boosted below GND. Figure 14.1.3 shows a schematic of the write driver with boost control that features a 40% improvement in boost voltage compared to the previous design [2]. Boosted node Nboost connects to eight physical BL pairs, segmented into upper (Ntu/Ncu) and lower half (Ntl/Ncl) partitions. Nboost is pre-charged to GND by Nd at the end of the write cycle. Boost capacitor, Cboost is also charged during this time by the transition of WS1n to VDD. To write a "1" into the BC, BLT is discharged to GND through bit-switch device Nt0 and segment device Ntu. Shortly after BLT reaches GND, the gate of Nd is shut off and WS1n transitions to GND to boost BLT below GND. Ntu is selected by the combination of true write data, WDTn and upper write select, WSELn. Boost voltage is increased as the gates of the three unselected segment devices (Gcu/Gtl/Gcl) are also boosted below GND. A 0V V_{GS} across the unselected segment devices guarantees full isolation and no loss of charge.

To minimize the gate-dielectric stress of the write path, a boost control scheme reduces the boost voltage at higher VDD where full assist is not required. The amount of boost is determined by the separation of WS0n and WS1n that control Nd and Cboost, respectively. An array edge-cell BL is used as a write path load to accurately time the initiation of the boost after BL is discharged to GND. A write bias generator, Wbias sets the V_{GS} of T0 (Wbias − NS) to modulate the delay of WS0n with respect to WS1n. At high VDD, Wbias is lowered by the increased strength of the four-NFET stack and WS0n is delayed compared to WS1n. The transition of WS1n before WS0n depletes the charge on Cboost by on-device Nd and the boost is attenuated. At lower VDD, WS0n switches low prior to WS1n to prevent the charge of Cboost to drain across Nd and the boost is maximized. Figure 14.1.4 compares waveforms at 0.7V (-198mV boost) and 1.0V (-66mV boost). The timing relationship between WS0n/WS1n for the two VDD cases is shown. The maximum BL boost as a function of VDD is also plotted in Fig. 14.1.4. The dotted lines represent the BL boost level without attenuation. Voltage stress is reduced by 200mV at 1.2V/-40°C.

BC optimization requires unique V_T implants that are decoupled from logic transistors. For improved yield and performance a BC-aware delay controls critical SRAM timings (see Fig. 14.1.5). A small memory array is configured to discharge a capacitor load (node Nbitcell). The WL activation is driven by the macro clock. To reduce the effects from random variations, 16 BCs are activated in parallel. The BC terminals are biased at higher metal levels to guarantee the correct state of the internal nodes at power-up. An edge-cell BL from an adjacent array is connected to the output of the 16-bit array to capture BL capacitance characteristics. A logic delay is added in parallel to the 16-bit delay element; this prevents the overall delay from becoming too fast and guarantees worst-case signal margin at the slow corners. Figure 14.1.5 shows WL to SET time as a function of VDD. At 0.7V, the required SET delay times, t_{SET} are indicated for two BCs (typical and +50mV V_T on PG). Delay times for logic-only delay elements (dotted lines) are compared with BC-tracking elements. Considerable performance improvements are gained at higher VDD for the BC-aware delay element.

Figure 14.1.6 shows hardware results at 85°C. The fail-count is compared with assist features disabled (left) and enabled (right). The VCS supply (array and WL-driver) is plotted against the VDD supply (periphery). SRAM operation is shown down to 0.7V. An overall improvement of 400mV of VCS is observed when these features are enabled compared to the default state. The area overhead for the stability and write assist is 1.5% and 1.2%, respectively. An overall array efficiency of 71.6% is achieved with a 128-BL sub-array configuration. Figure 14.1.7 shows the micrograph of the test chip and design features. The 64Mb SRAM is built from 128 512Kb macros used as the principal building block for high-performance ASIC SoC.

References:

[1] Greene, B., et al., "High Performance 32nm SOI CMOS with High-k/Metal Gate and $0.149\mu m^2$ SRAM and Ultra Low-k Back End with Eleven Levels of Copper," *Symposium on VLSI Technology*, June 2009.

[2] Pilo, H., et al., "A 450ps Access-Time SRAM Macro in 45nm SOI Featuring a Two-Stage Sensing-Scheme and Dynamic Power Management," *ISSCC Digest of Technical Papers*, pp. 378-379, Feb. 2008.

[3] Bhavnagarwala, A., et al., "A Sub-600mV, Fluctuation Tolerant 65nm CMOS SRAM Array with Dynamic Cell Biasing," *Symposium on VLSI Circuits*, pp. 78-79, June 2007.

[4] Fujimura, Y., et al., "A Configurable SRAM with Constant-Negative-Level Write Buffer for Low-Voltage Operation with $0.149\mu m^2$ Cell in 32nm High-k Metal-Gate CMOS," *ISSCC Digest of Technical Papers*, pp. 348-349, Feb. 2010.

[5] Kolar, P., et al., "A 32nm High-k Metal Gate SRAM with Adaptive Dynamic Stability Enhancement for Low-Voltage Operation," *ISSCC Digest of Technical Papers*, pp. 346-347, Feb. 2010.

978-1-61284-303-2/11 $26.00 © 2011 IEEE

ISSCC 2011 / February 22, 2011 / 1:30 PM

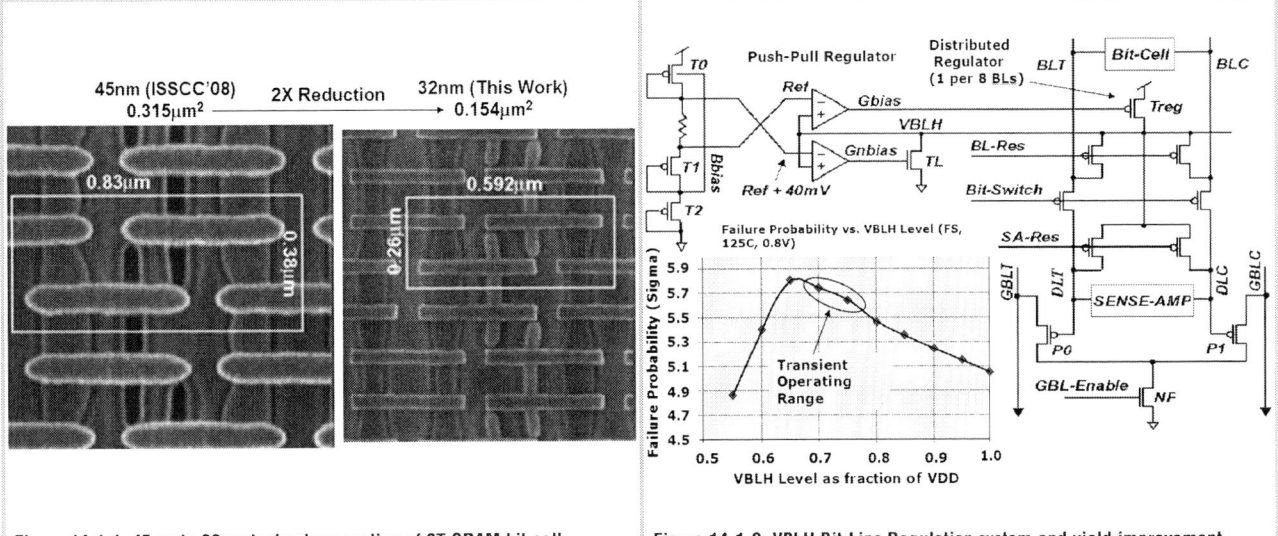

Figure 14.1.1: 45nm to 32nm technology scaling of 6T SRAM bit-cell.

Figure 14.1.2: VBLH Bit-Line Regulation system and yield improvement.

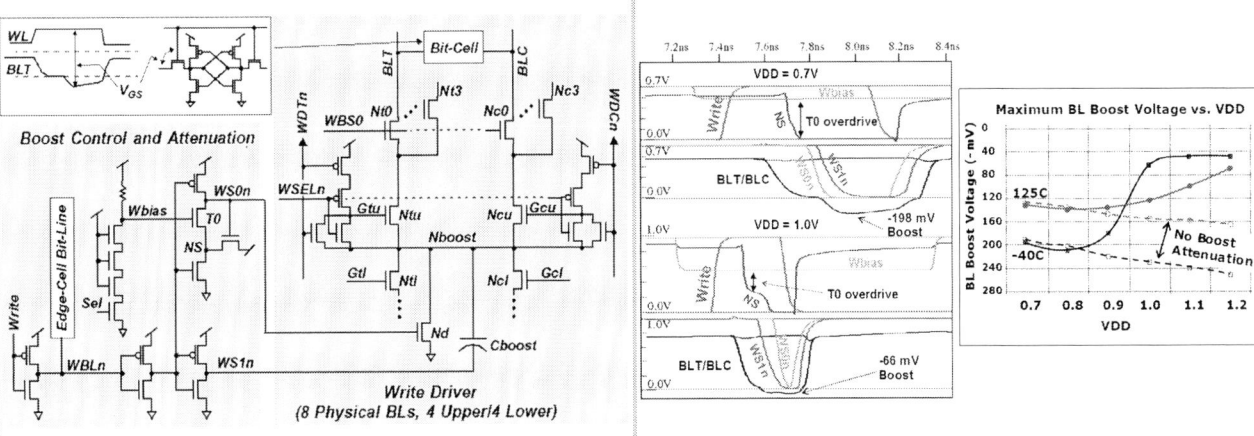

Figure 14.1.3: Write driver with Boost control and attenuation.

Figure 14.1.4: Write cycle simulation waveforms and attenuation results.

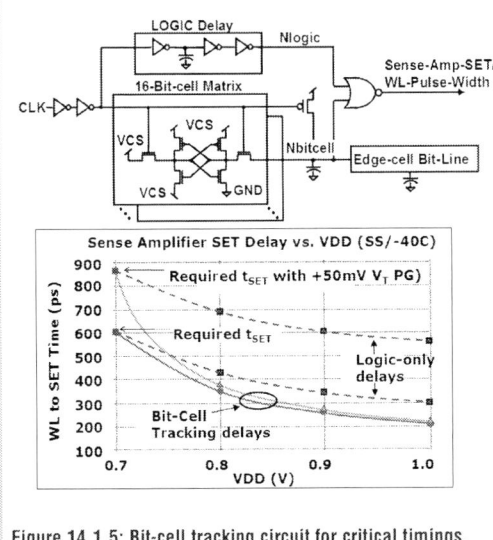

Figure 14.1.5: Bit-cell tracking circuit for critical timings.

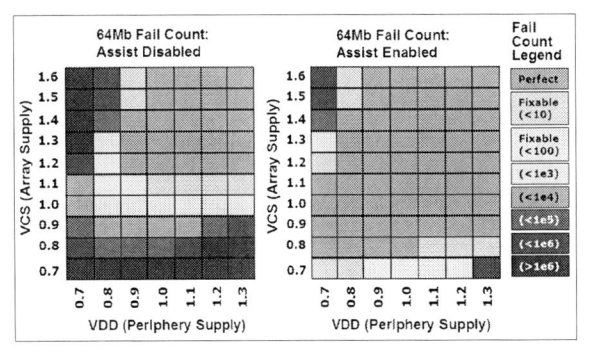

Figure 14.1.6: 64Mb Hardware results: 85°C VCS/VDD Voltage shmoo.

978-1-61284-303-2/11 $26.00 © 2011 IEEE

ISSCC 2011 PAPER CONTINUATIONS

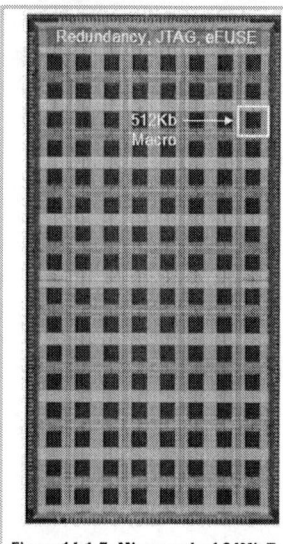

Technology	32nm PD SOI with High-k Metal Gate
Cell Size	0.154μm²
512Kb Macro Size	331μ x 339μ
Sub-Array Configuration	128 Word-Line x 128 Bit-Line
Operating Voltage	Core: 0.7V – 1.0V (0.9V typ.) SRAM: 0.7V – 1.0V (0.9V typ.)
Performance Target	1.4GHz (Slow process, 0.8V, 0C)

Figure 14.1.7: Micrograph of 64Mb Test Chip and Features.

ISSCC 2011 / SESSION 14 / HIGH-PERFORMANCE EMBEDDED MEMORY / 14.2

14.2 A 4R2W Register File for a 2.3GHz Wire-Speed POWER™ Processor with Double-Pumped Write Operation

Gary S. Ditlow[1], Robert K. Montoye[1], Salvatore N. Storino[2], Sherman M. Dance[2], Sebastian Ehrenreich[3], Bruce M. Fleischer[1], Thomas W. Fox[1], Kyle M. Holmes[4], Junichi Mihara[5], Yutaka Nakamura[5], Shohji Onishi[5], Robert Shearer[2], Dieter Wendel[6], Leland Chang[1]

[1]IBM T. J. Watson Reseach Center, Yorktown Heights, NY,
[2]IBM Systems and Technology Group, Rochester, MN,
[3]Hoerner & Sulger, Schwetzingen, Germany,
[4]IBM Systems and Technology Group, Essex Junction, VT,
[5]IBM Systems and Technology Group, Kyoto, Japan,
[6]IBM Systems and Technology Group, Boeblingen, Germany

In multi-ported register files, memory cell size grows quadratically with the total number of ports due to wordline and bitline wiring. Reducing the number of physical access ports in a memory cell can thus lead to significant area and power savings as well as latency improvement. Double-pumped register files operate access ports twice in a single clock period to reduce area by halving the number of physical ports in the memory cell—a technique often confined to low-frequency applications. Replication of a memory cell in separate arrays halves the number of physical read ports in each copy. In this work, double-pumped write ports and replicated read ports are applied to a 4R2W register file in a high-performance microprocessor product [1]. This paper describes detailed implementation and measured hardware characteristics of this array and demonstrates a fast error correction scheme. The techniques used balance high efficiency and low latency and thus differ from previous work, in which double-pumped ports perform a write followed by a read in a very large register file [2] or where write ports are double-pumped without cell-level read port reduction [3].

Counter to intuition, double-pumping and cell duplication both work to reduce overall power and area by reducing the number of cell write and read ports, respectively. Due to more efficient wiring and contact sharing, a 2R1W register file cell is ~3 to 4× smaller than a 4R2W cell (Fig. 14.2.1), which reduces cell dimensions and thus both wordline and bitline lengths by nearly a factor of two. The single physical write port is double-pumped (early and late write operations on the rising and falling edge of the clock) to yield two effective write ports. To achieve fast 190ps read latency, the read ports are not double-pumped since a late read operation would not meet latency targets; instead, the 2R1W cell subarray is replicated (with common write operations occurring on two duplicate copies of the data) so that four read ports are functionally achieved while still maintaining low word and bitline capacitances. The two techniques are complementary and mutually inclusive, as both are needed to enable the small 2R1W cell size, which, in turn, enables double-pumped write operation and improved read performance. Even with subarray duplication, the 3 to 4× smaller cell size achieves a near 2× macro-level area reduction over a traditional 4R2W design. This area reduction also results in a corresponding decrease in leakage power. Due to reduced read bitline capacitance and smaller drivers, read power and read bitline latency can both be improved by ~2×. Write power is not dramatically affected as reduced write bitline capacitance balances subarray duplication.

As system reliability challenges mount in future technologies, error checking algorithms can add significantly to system-level register file access latency. The availability of two identical copies of the stored data enables error correction in contiguous bits using a simple interleaved parity code (Fig. 14.2.2). Parity bits are generated for small data blocks, which minimizes system-level write latency as compared with traditional ECC generation. If an error is detected in any of the four read ports while reading one of the two copies of the data, a state machine recovers the correct data from the other copy. In such a situation, the Instruction Unit stalls and flushes the pipeline and then inserts a new "select" instruction. This instruction reads the erroneous register from both arrays and rewrites each array with the correct data as chosen by the two parity values (if both arrays have errors, an unrecoverable error is signaled).

In the double-pumped write path, write ports are operated twice per clock cycle using pulses (LCLK and DCLK) triggered off both edges of the global clock (Fig. 14.2.3). Through static CMOS multiplexors, these two pulses select and combine the early and late versions of the predecoded address. For both the MSB (most significant bits) and LSB (least significant bits) paths, the early and late addresses share predecode drivers and wires, which reduces area and capacitance and results in improved latency. Due to statistical process variation, merging pulses (a logical AND) tends to reduce worst-case pulse widths. To enable high frequency operation, predecoded versions of the MSB and LSB, which are both pulsed signals, are merged as late in the path as possible near the final WWL driver. All circuits in the write address path utilize skewed beta ratios in conjunction with wide WWL wires to minimize latency. To select and merge early and late data for the WBL, transmission-gate multiplexors, which minimize WBL switching caused by input activity, are utilized. A small feedback inverter is also used to latch the WBL state when neither the early or late signals are asserted.

To achieve a fast read path, the I/O block is placed between the top and bottom halves of each subarray. This enables short global RBLs with a maximum of five dots hierarchically coupled to local RBLs with eight memory cells (Fig. 14.2.4). For robustness, the global RBL uses a standard keeper while the local RBL uses a delayed keeper to speed evaluation. Further read latency improvement is achieved by the use of a fast dynamic latch with a built-in NAND function to combine the upper and lower global RBL signals as well as logical reoptimization of the bypass multiplexor to minimize further inversions from the inverted read port of the 2R1W cell. Predecoder latency is improved by skewing beta ratios in all circuits handling the MSB, which are clocked half-cycle pulses that set RWL timing. Predecoding of the LSB is performed with static gates, with the last stage of gain for each 16b LSB group placed close to the RWL driver to sharpen slew rates.

To achieve double-pumped write operation at multi-GHz-range frequencies, early and late pulse widths and separations were tuned to minimize latency while ensuring robustness. Critical timing margins (Fig. 14.2.5) are ensured between successive writes (early followed by late in the same cycle and late followed by next-cycle early) and in a read-after-write (late write followed by next-cycle read) through rigorous Monte Carlo-based statistical simulation. While the early and late WWL pulse widths must both be sufficient to complete a cell write operation, it must also be guaranteed that a WWL pulse falls sufficiently to avoid collision with an immediately succeeding WWL pulse. During this separation time, the WBL must transition while maintaining write data setup and hold times. A nominal WWL pulse width/separation target of 90/127ps was found to satisfy all critical timing requirements. Read-after-write critical timing must ensure that the late cell write has reached completion before assertion of the RWL. Process-induced variability and RC delays across the array were considered in this analysis as well as jitter on the back-edge of the clock (off which the late clock pulse is derived), which is especially important since such mid-cycle clock uncertainty is traditionally not well controlled in front-edge-triggered clock networks.

Measured results in 45nm SOI-CMOS demonstrate operation of the double-pumped write register file of up to 2.76GHz at a supply voltage of 0.9V (Fig. 14.2.6). This results in a read latency of 190ps, active power dissipation with all ports active of ~28mW, and leakage power of ~31mW. 1.6GHz operation is maintained down to a supply voltage of 0.7V.

Acknowledgement:
The authors thank R. Redmond and R. R. Robertazzi for characterization assistance.

References:
[1] C. Johnson, et al., "A Wire-Speed Power™ Processor: 2.3GHz 45nm SOI with 16 Cores and 64 Threads," *ISSCC Dig. Tech. Papers*, pp. 104-105, Feb. 2010.
[2] E. S. Fetzer, et al., "A Fully Bypassed Six-Issue Integer Datapath and Register File on the Itanium-2 Microprocessor," *IEEE J. Solid-State Circuits*, vol. 37, no. 11, Nov. 2002, pp. 1433-1440.
[3] D. Wendel, et al., "The Implementation of POWER7™: A Highly Parallel and Scalable Multi-Core High-End Server Processor," *ISSCC Dig. Tech. Papers*, pp. 102-103, Feb. 2010.

978-1-61284-303-2/11 $26.00 © 2011 IEEE

ISSCC 2011 / February 22, 2011 / 2:00 PM

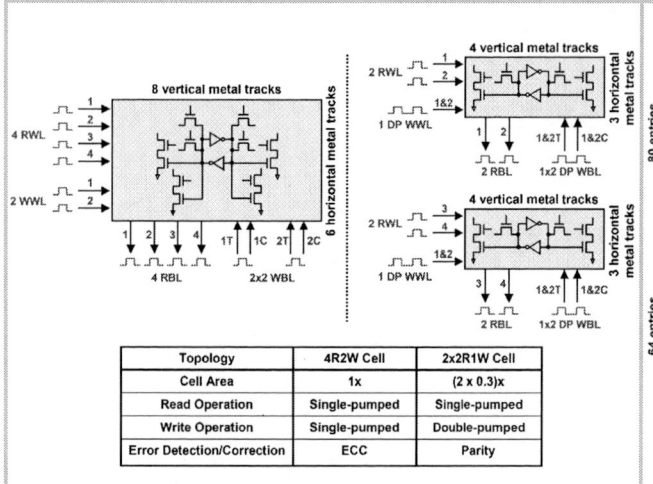

Figure 14.2.1: Implementation comparison: Standard 4R2W cell vs. 2 copies of a 2R1W cell.

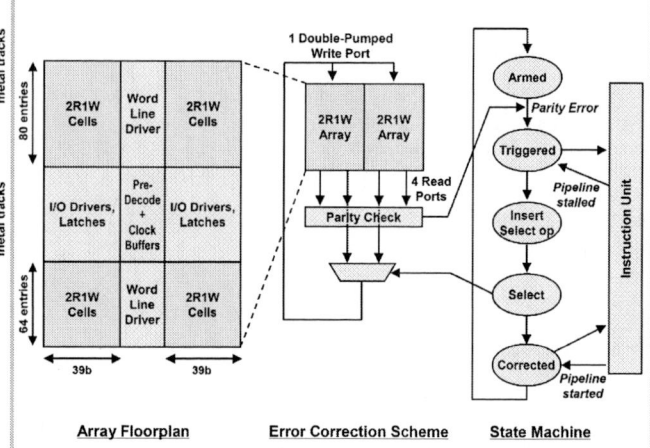

Figure 14.2.2: Floorplan depicting replicated 2R1W subarrays and parity-based error correction scheme.

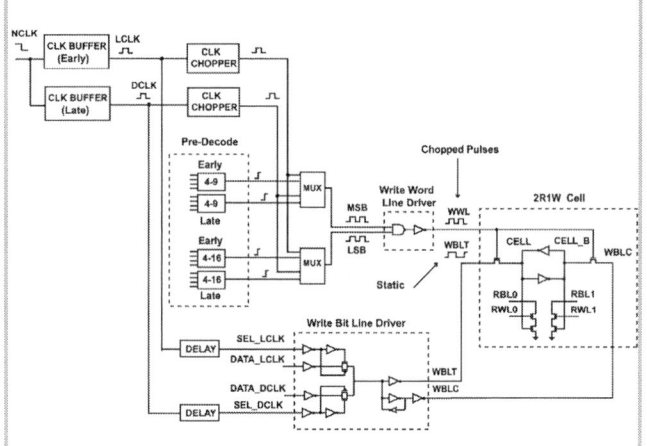

Figure 14.2.3: Circuit cross-section of double-pumped write path.

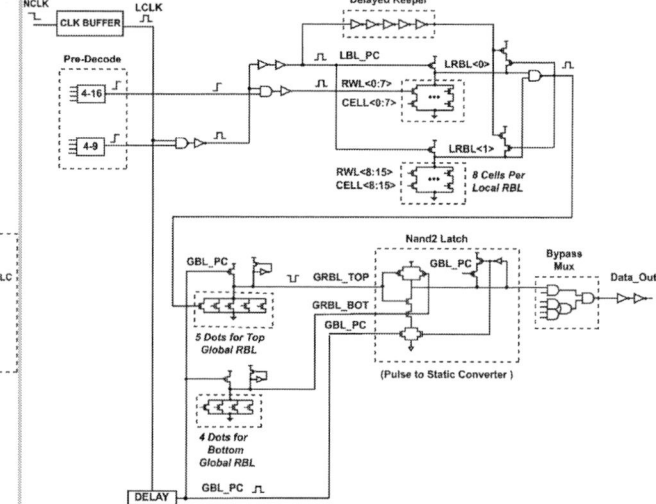

Figure 14.2.4: Circuit cross-section of fast read path.

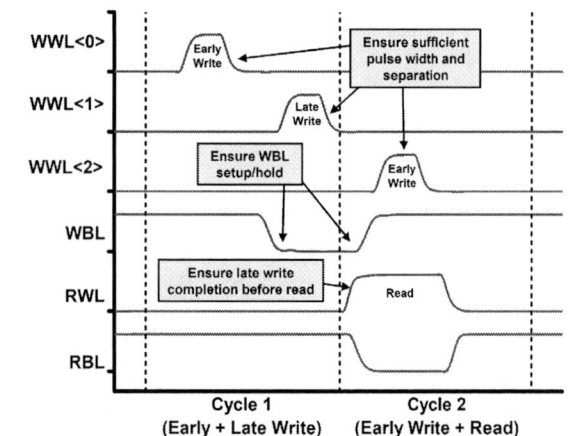

Figure 14.2.5: Simulated waveforms depicting critical timing margins.

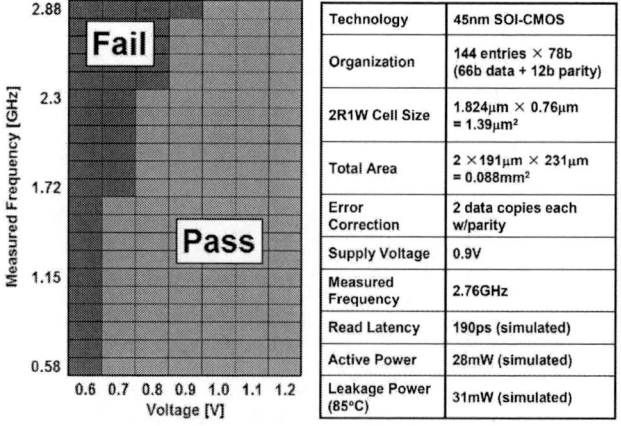

Figure 14.2.6: Measured characteristics of the fabricated array in 45nm SOI CMOS.

14

ISSCC 2011 PAPER CONTINUATIONS

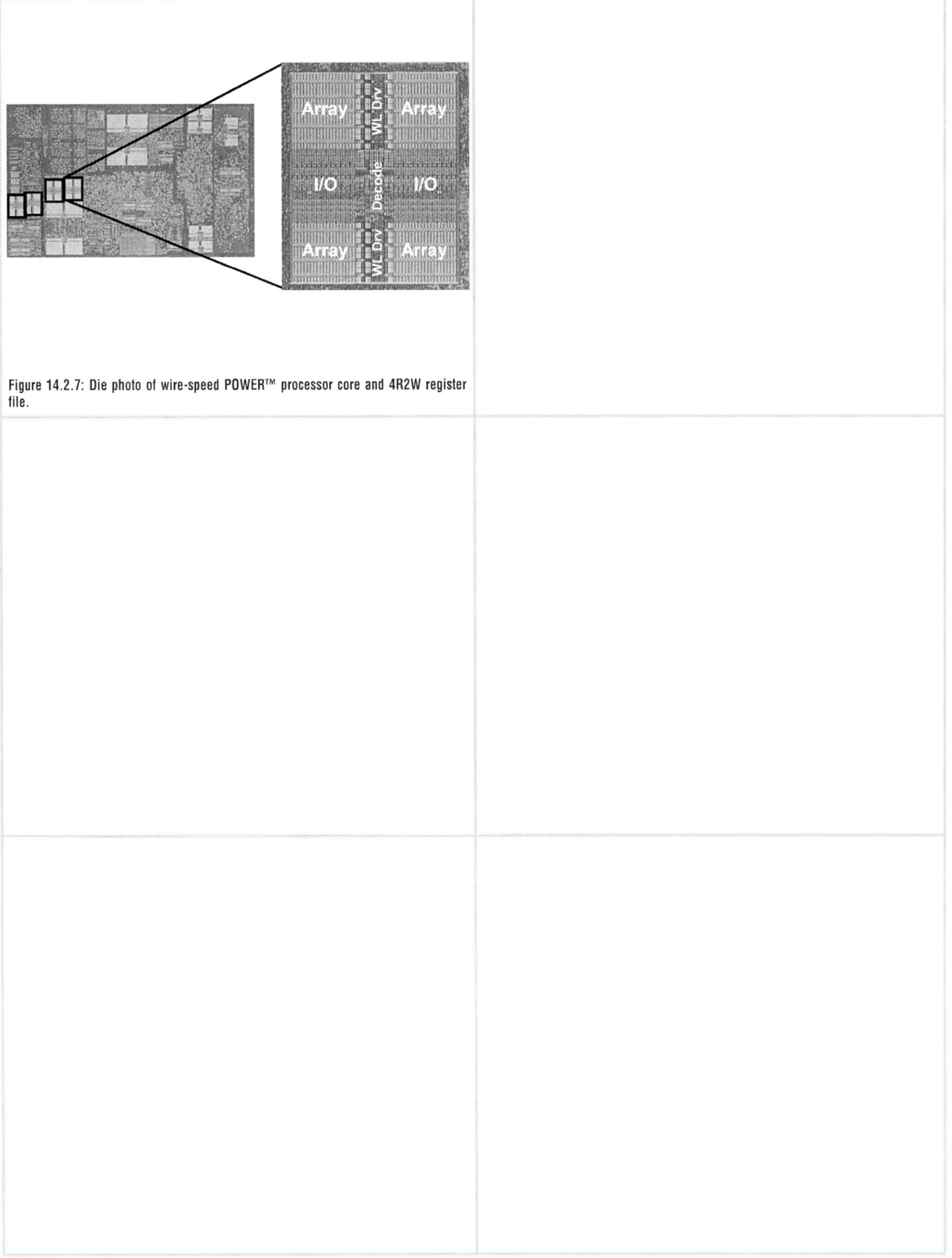

Figure 14.2.7: Die photo of wire-speed POWER™ processor core and 4R2W register file.

ISSCC 2011 / SESSION 14 / HIGH-PERFORMANCE EMBEDDED MEMORY / 14.3

14.3 An 8MB Level-3 Cache in 32nm SOI with Column-Select Aliasing

Don Weiss[1], Michael Dreesen[1], Michael Ciraula[1], Carson Henrion[1], Chris Helt[1], Ryan Freese[1], Tommy Miles[1], Anita Karegar[1], Russell Schreiber[2], Bryan Schneller[1], John Wuu[1]

[1]AMD, Fort Collins, CO,
[2]AMD, Austin, TX

High-performance multi-core processors require efficient multi-level cache hierarchies to meet high-bandwidth data requirements. Because level-3 (L3) cache is typically the largest cache on the die, the drive to lower cost places pressure on density, yields, and test time. Performance-per-watt goals and total power constraints also compel a variety of circuit techniques to reduce power. The next-generation server processor codenamed "Orochi", implemented on a 32nm high-k metal-gate SOI process with 11 metal layers, consists of four 2-core modules using AMD's next-generation architecture, code named "Bulldozer", with 2MB of dedicated L2 cache per module and an 8MB shared L3 cache [1].

The L3 cache is divided into 4 independent 2MB subcaches, shown in Fig. 14.3.1, with each subcache further divided into 4 banks. The interface to the subcache consists of the following ports: 2 tag read, 2 tag write, 2 data read and 1 data write. Concurrent operations on different ports are supported when accesses do not target the same bank. The data portion of the subcache is divided into 4 sequential regions, each running one phase behind the previous region and containing 1/4 of the 512b cache line. The combination of half-cycle delay for control signals reaching a region and half-cycle delay for read data crossing that region means a cache line is returned in a burst of 4 cycles during a read. Similarly, the subcache requires 4 cycles to write a full cache line. A sequential region contains 8 64KB macros, each containing 64 data, 6 ECC and 2 repair I/O's, with 2 macros in each sequential region accessed per operation. The macro operates in a flow-through manner, illustrated in Fig. 14.3.2, to enable high speed and area efficiency while reducing clock power [2,3].

The data array uses short bitlines containing 32 cells for robust read stability. To achieve high density, a compact single-ended read circuit is used for every 16 columns, with 8 columns on each of its 2 sides (Fig. 14.3.3). Pre-MUX read sensing achieves high-speed read operations and low-voltage performance. Writes are performed with true and complement data driven through an 8:1 NFET write-MUX with cross-coupled PFETs on the bitlines. Traditionally, separate read and write 8:1 MUXes would require 16 column-select signal tracks in layout, consuming additional area. Using the same column-select signals to drive the read and write MUXes could halve the number of required tracks, but would defeat the purpose of pre-MUX sensing because the load inside the write-MUX would be exposed to the read bitline, dramatically reducing speed.

A solution dubbed column-select aliasing (CSA) uses only 8 wires to drive both MUXes by aliasing odd and even pairs of column-selects for reads and writes. Figure 14.3.4 illustrates the concept with a pair of columns as an example. CSEL<0> is used as "column 0 select" for reads and "column 1 select" for writes; CSEL<1> is used as "column 1 select" for reads and "column 0 select" for writes. Therefore, the bitline selected for a read is isolated from the load inside the write-MUX to maintain high speed, while the same set of column select signals can be reused for write operations. The local write drivers are tri-stated during read operations to avoid contention with the SRAM cells on the aliased columns. Because of its significant area savings, CSA was also used in the L3 tag and least-recently used (LRU) arrays. Different column select enhancements were applied to other areas of the chip. For example, the L2 tag employs a shared, oversized write driver to simultaneously write all columns of a 4:1 MUX configuration in an invalidate operation which shortens the invalidate time.

Power reduction is another L3 cache design priority, and this design employs several techniques to reduce static power. For example, all of the large wordline and column select drivers in data macros are power-gated in groups of eight, accounting for 28% data macro static power reduction, or 20% reduction in overall subcache static power. Similarly, tag wordline drivers are power-gated to reduce the overall subcache static power by another 6% (Fig. 14.3.5). In traditional SRAM designs, bitlines remain precharged at V_{DD} when bitcells are not accessed, resulting in subthreshold leakage current into the logical 0 sides of the SRAM cells and junction leakage current from the N+ regions connected to the bitlines. By allowing the bitline voltages to float, these leakage components are decreased, reducing overall standby power. 18% power savings have been reported in 32nm bulk SRAM using floating-bitline mode (FBM) [4], while this design's 10% standby power reduction due to FBM represents a contrasting data point for 32nm SOI SRAM that has negligible junction leakage. V_{DD} on the read NAND gates is gated in FBM, which protects the NAND gate from crowbar current while the inputs are floating. Headers are not used for the cross-coupled bitline PFETs to keep area small, so as a compromise only half the bitlines receive the full benefits of FBM. In addition to power gating, high-V_T transistors are used throughout the design to reduce power further, and the SRAM bitcell is chosen for its use of low-leakage SRAM transistors at the expense of larger bitcell area. The combination of these techniques reduces the leakage power enough to eliminate the need for more complex and costly solutions, such as array sleep or regulated supplies.

Along with density and power, focus is also placed on yield and testability enhancements. Traditional row and column redundancies in each macro usually have several drawbacks, including requiring multiple BIST passes for identifying and programming replacement elements, inefficient use of area by including redundancy storage and control circuitry in all macros, and increased pressure on timing paths by placing the redundancy comparison and data insertion circuitry in critical paths. To address these drawbacks, this design uses centralized row and column redundancy blocks for its data macros. Redundant row data is stored in independent latch arrays separate from the data macros, and each row redundancy macro can repair any 12 rows of data across 16 data macros. The redundant column data is stored in each data macro; however, the repair address and CAM are stored in a separate column redundancy macro. Each column redundancy macro can repair a total of 24 columns across 16 data macros, replacing up to two columns per data macro. All redundancy MUXes are centrally located at the subcache interface. This centralized redundancy scheme offers lower area overhead, removes the speed penalty typically associated with row redundancy, and provides more flexibility in utilizing available repair elements. In addition, the centralized redundancy scheme reduces test time by requiring only 1 BIST pass for the data region, rather than needing additional passes after repairs are made.

This L3 cache is designed to operate above 2.3GHz at 1.1V, and its key technology, density, and bandwidth features are summarized in Fig. 14.3.6.

Acknowledgements:
The authors thank Mike Leary, Bill Hughes, Gregg Donley, Benjamin Tsien, Steve Foster, Sho-Chien Kang, Pouya Razavi, Nick Fournier, John Chan, the L2 team, the layout team, and GLOBALFOUNDRIES for their contributions and support.

References:
[1] T. Fischer, et al., "Design Solutions for a 32nm SOI 2-core processor module in an 8-core CPU," *ISSCC Dig. Tech. Papers*, Feb., 2011
[2] J. Dorsey, "An Integrated Quad-Core Opteron Processor," *ISSCC Dig. Tech. Papers*, pp. 102-103, Feb., 2007
[3] J. Wuu, et al., "The Asynchronous 24MB On-Chip Level-3 Cache for a Dual-Core Itanium-Family Processor," *ISSCC Dig. Tech. Papers*, pp. 488-489, Feb., 2005
[4] Y. Wang, et al., "A 4.0 GHz 291Mb Voltage-Scalable SRAM Design in 32nm High-K Metal-Gate CMOS with Integrated Power Management," *ISSCC Dig. Tech. Papers*, pp. 456-457, Feb., 2009

978-1-61284-303-2/11 $26.00 © 2011 IEEE

ISSCC 2011 / February 22, 2011 / 2:30 PM

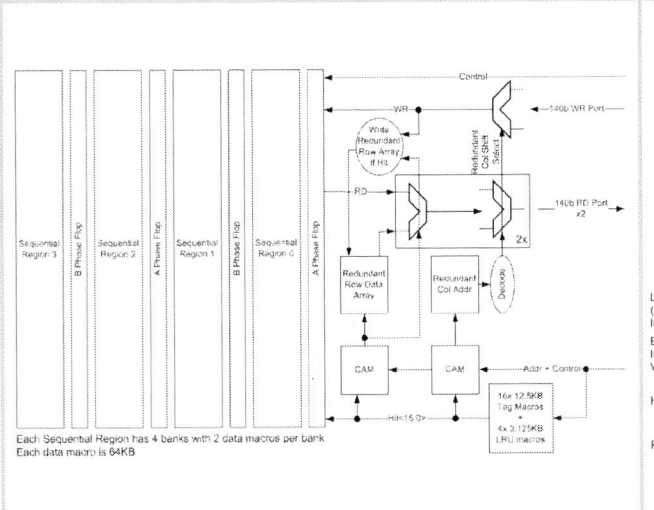

Each Sequential Region has 4 banks with 2 data macros per bank
Each data macro is 64KB

Figure 14.3.1: 2MB Subcache Organization.

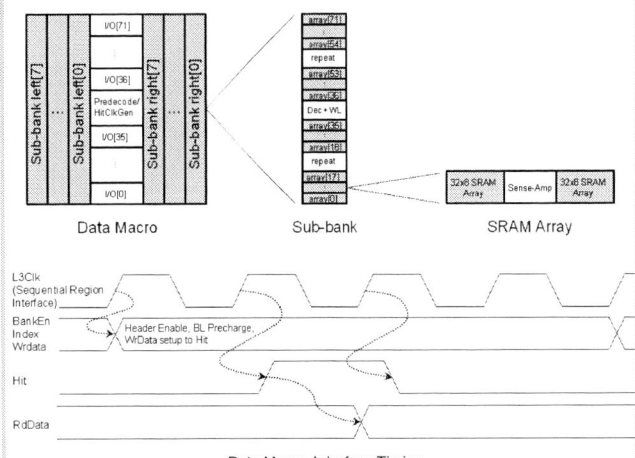

Figure 14.3.2: Data Macro Organization and Timing.

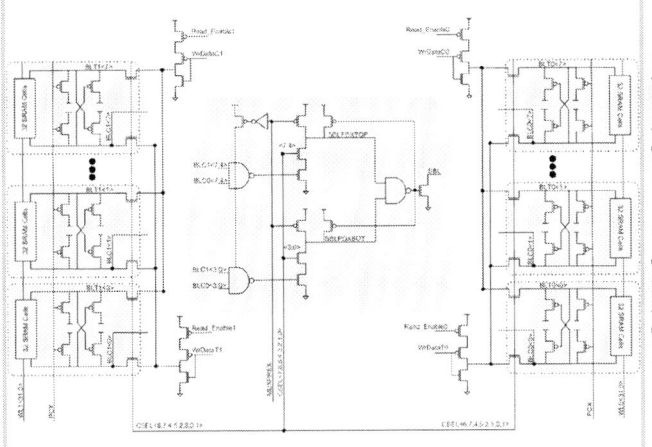

Figure 14.3.3: Sensing Circuit.

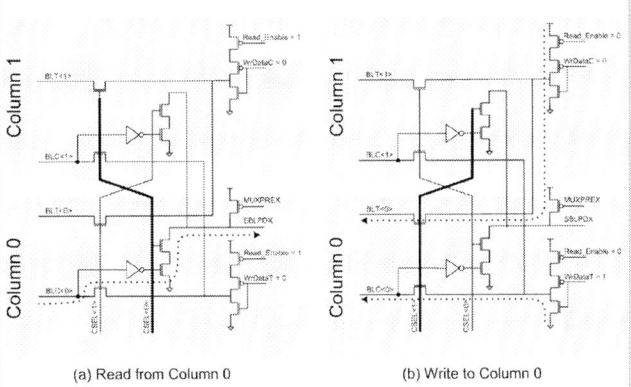

(a) Read from Column 0   (b) Write to Column 0

Figure 14.3.4: Column Select Aliasing Illustration.

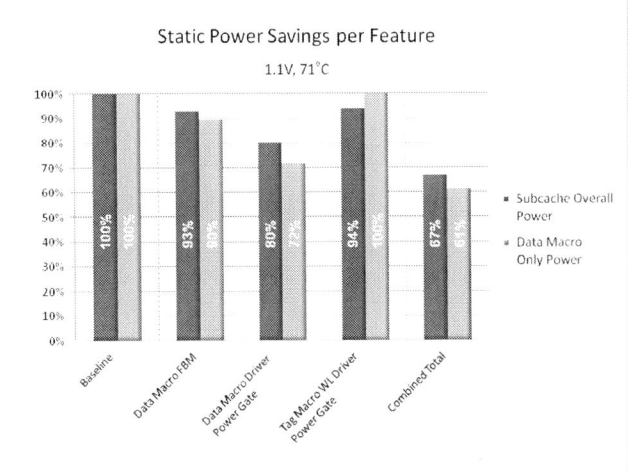

Figure 14.3.5: Static Power Reduction Features Summary.

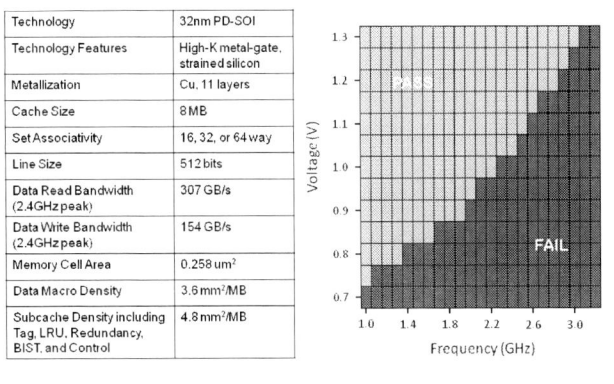

Technology	32nm PD-SOI
Technology Features	High-K metal-gate, strained silicon
Metallization	Cu, 11 layers
Cache Size	8MB
Set Associativity	16, 32, or 64 way
Line Size	512 bits
Data Read Bandwidth (2.4GHz peak)	307 GB/s
Data Write Bandwidth (2.4GHz peak)	154 GB/s
Memory Cell Area	0.258 um²
Data Macro Density	3.6 mm²/MB
Subcache Density including Tag, LRU, Redundancy, BIST, and Control	4.8 mm²/MB

Figure 14.3.6: Feature Summary and Shmoo Plot.

978-1-61284-303-2/11 $26.00 © 2011 IEEE

Figure 14.3.7: Die Photo.

ISSCC 2011 / SESSION 14 / HIGH-PERFORMANCE EMBEDDED MEMORY / 14.4

14.4 A 28nm High-Density 6T SRAM with Optimized Peripheral-Assist Circuits for Operation Down to 0.6V

Mahmut E. Sinangil[1], Hugh Mair[2], Anantha P. Chandrakasan[1]

[1]Massachusetts Institute of Technology, Cambridge, MA,
[2]Texas Instruments, Dallas, TX

An increasing amount of embedded memory is used in today's ICs and consequently the design of low-power, high-density SRAM is becoming critical. With technology scaling, it is becoming increasingly more challenging to achieve reliable low voltage operation for SRAMs due to process variation. Recent work proposed various design innovations to address this problem [1-2]. Conventional 6T SRAM bit-cell provides high area density but fails to operate as supply voltage scales down. Instead, the work in [3] utilizes an asymmetrical 6T bit-cell that can operate down to 0.7V in 45nm CMOS. An alternative 8T bit-cell topology can potentially operate at lower V_{min} [4-5], yet it leads to ~40% larger bit-cell area and suffers from half-select problem if used in a column-interleaved architecture. In this work, we present a 28nm high-density 6T SRAM operating down to low supply voltages. Hierarchical bit-line architecture, pre-read phase during write cycles and signal boosting for better write-ability provide functionality down to 0.6V. Figure 14.4.1 shows an SEM image of the 6T bit-cell used in this work. This bit-cell has an area of $0.12\mu m^2$ providing very high transistor density.

Static noise margin (SNM) has been the standard way of analyzing SRAM bit-cell stability using static DC analysis. However, recent work [3] demonstrates that read operation is a dynamic phenomenon and smaller bit-line capacitance results into significant read-stability improvement. In our work, transient simulations are performed to characterize bit-cell stability at various voltages as shown in Fig. 14.4.1. DC Voltage sources are placed in series between cross-coupled inverters and values of these sources are swept in successive transient analyses until bit-cell can no longer retain its state during a read operation. Read margin with 1mV accuracy is calculated using a three step search algorithm with 100, 10 and 1mV step sizes to reduce simulation time. Results confirm that reducing number of bit-cells sharing a bit-line substantially improves read stability at the expense of larger area due to replicated periphery. Choosing 32 bits-per-localBL results in 15% area overhead for our work while extending operating voltage range from 0.85 to 0.6V at a constant normalized read margin μ/σ of 0.5.

Figure 14.4.2 shows the array architecture of the memory. Local R/W circuit is shared by two memory cell sub-arrays of 32 rows and 256 columns. Long global bit-lines (globalBL/globalBLB) traverse the entire memory while bit-cells are connected to short local bit-lines (localBL/localBLB) of the sub-arrays. During a read operation, voltage differentials first appear on local bit-lines and then are transferred to corresponding global bit-lines through a sensing inverter and a pull-down device. A 4-to-1 column interleaving ratio is used to minimize local R/W circuit area. During an access, active column is selected by clmnSel[0:3]. Write drivers are implemented with pull-down NMOSs controlled by data and dataB signals together with cross-coupled PMOSs between local bit-line pairs. Finally, PMOS devices controlled by pchgB signal are used to pre-charge local and global bit-lines as well as internal nodes before the sensing inverters.

Conventionally, high-density SRAMs employ a small signal sensing scheme where signal development on bit-lines is passed on by PMOS column switches and subsequently amplified by a complex sensing circuitry [6]. For this work, placing a complex sense-amplifier in local R/W circuit is not practical for area considerations and a large-signal sensing scheme is more appropriate. Figure 14.4.3 shows one side of the differential read path evaluating the use of PMOS vs. NMOS column switches. At the beginning of a read cycle, localBL, input of the sensing inverter (int) and globalBL nodes are pre-charged to V_{DD}. A transition on localBL from V_{DD} to ground trips the inverter and causes globalBL to discharge and finally data is stored in a cross-coupled NAND latch. Voltage transfer characteristic of the sensing inverter is designed to favor a low-to-high transition at its output to provide shorter read access time but this can change significantly due to process variation. Since a conventional PMOS switch is not suited to drive low voltages, the resulting falling edge at int is slow. With an

inverter trip point (V_M) deviation of ±50mV around $V_{M,nominal}=250mV$, an NMOS switch improves the falling edge transition time from localBL to int by 18× from 90ns to 5ns at 0.5V. Monte Carlo analysis depicts overall read access time distribution. NMOS switch provides a narrow distribution with 2× shorter worst-case delays than PMOS column switch implementation at 0.5V.

Figure 14.4.4 shows the word-line (WL) boosting scheme that is used to improve write-ability of the bit-cells. At the beginning of a write cycle, WL is first pulled high to V_{DD} by asserting block-wise WLdrive signal. Half-selected bit-cells on an active row are pre-read during this period where voltage differential generated on bit-lines is preserved by the cross-coupled PMOSs. Negative edge of the clk triggers boost signal for a short pulse period during which charge-pump circuit shown in Fig. 14.4.4 is turned on. Boosted voltage node $V_{DD,BST}$ is used to power up WL drivers of a sub-array. 100mV of voltage boosting necessary for robust operation is provided by carefully selecting the value of boost capacitor, C_{BST}, with respect to the total capacitance connected to $V_{DD,BST}$. To reduce effect of parasitics, body terminal of PMOSs in boost circuit as well as WL drivers are connected to a static high voltage, V_H. To achieve high area density, C_{BST} as well as the boosting circuitry are shared across 32 rows of a sub-array and C_{BST} capacitors are embedded beneath the metal routing of the address decoder. Likewise, data/dataB and clmnSel drivers utilize similar shared boosting circuitry to elevate signal voltage levels for a short pulse period during write operation.

In this work, two cross-coupled PMOS (CC-PMOS) devices are placed between each local bit-line as shown in Fig. 14.4.5. Voltage differential that is created on local bit-lines during read and write operations are preserved by CC-PMOS structure. Simulated waveforms demonstrate a sample case where localBL discharges to 0V before WL boosting takes place causing functional failure. Addition of CC-PMOS devices fight bit-cell transistors and preserve the differential between local bit-lines. In layout, these devices are designed to fit into the bit-cell NWELL strip introducing less than 3% area overhead.

Figure 14.4.6 summarizes the features and performance characteristics of the test chip fabricated in a low power 28nm CMOS process. Four SRAM macros with local bit-line lengths ranging from 4 to 32 together with their test structures are placed on a test die (Fig. 14.4.7). 128kbit SRAM macro with local bit-line length of 32 bits works down to 0.6V. Operating at the nominal voltage (1V), one or two bit failures per chip observed and can be repaired assuming redundancy (not implemented in this test chip). Assuming two row and two column redundancy the chip is able to achieve 400MHz at 1V and down to 20MHz at 0.6V. Lower bit-error rate (BER) may be expected as the technology matures. Power consumption of the test chip scales by ~25× from 68.5 to 2.8mW.

Acknowledgements:
We would like to thank Texas Instruments for funding and chip fabrication.

References:
[1] H. Nho, et al., "A 32nm High-K Metal Gate SRAM with Adaptive Dynamic Stability Enhancement for Low-Voltage Operation," *ISSCC Dig. Tech. Papers*, pp. 346-347, Feb., 2010.
[2] Y. Fujimura, et al., "A Configurable SRAM with Constant-Negative-Level Write Buffer for Low-Voltage Operation with $0.149\mu m^2$ Cell in 32nm High-K Metal-Gate CMOS," *ISSCC Dig. Tech. Papers*, pp. 348-349, Feb., 2010.
[3] A. Kawasumi, et al., "A Single-Power-Supply 0.7V 1GHz 45nm SRAM with An Asymmetrical Unit-β-ratio Memory Cell," *ISSCC Dig. Tech. Papers*, pp. 382-383, Feb., 2008.
[4] L. Chang, et al., "Stable SRAM Cell Design for the 32nm Node and Beyond," *IEEE Symposium on VLSI Technology*, pp. 128-129, June, 2005.
[5] N. Verma, A. Chandrakasan, "A 65nm 8T Sub-V_t SRAM Employing Sense-Amplifier Redundancy," *ISSCC Dig. Tech. Papers*, pp. 328-329, Feb., 2008.
[6] K. Zhang, et al., "The Scaling of Data Sensing Schemes for High Speed Cache Design in Sub-0.18μm Technologies," *IEEE Symposium on VLSI Circuits*, pp. 226-227, June, 2000.

978-1-61284-303-2/11 $26.00 © 2011 IEEE

ISSCC 2011 / February 22, 2011 / 2:45 PM

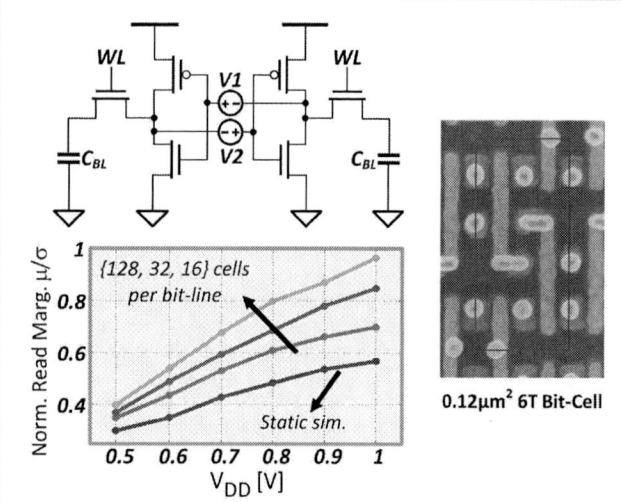

Figure 14.4.1: Static and dynamic read margin simulation results for the high-density 6T cell (SEM photo shown).

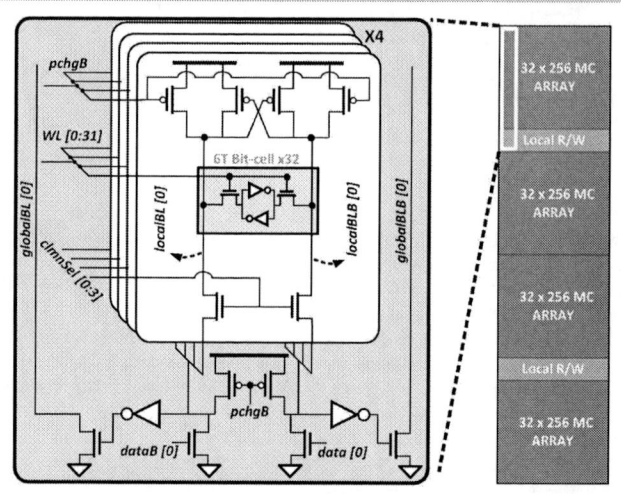

Figure 14.4.2: Array architecture with hierarchical bit-lines. Local R/W circuit decouples long globalBLs from the bit-cells.

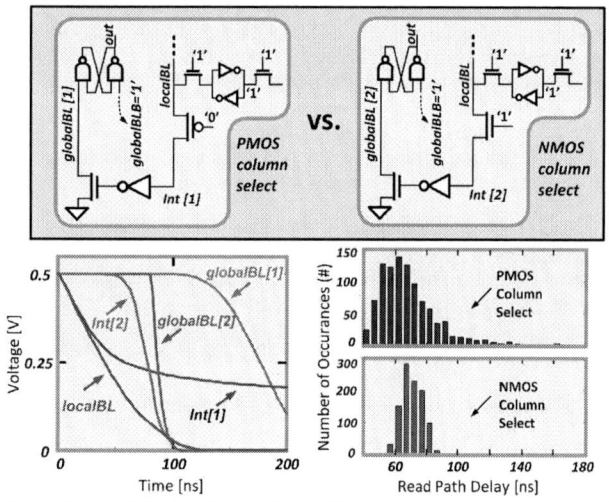

Figure 14.4.3: NMOS column select with skewed inverter improves read access time at low voltages.

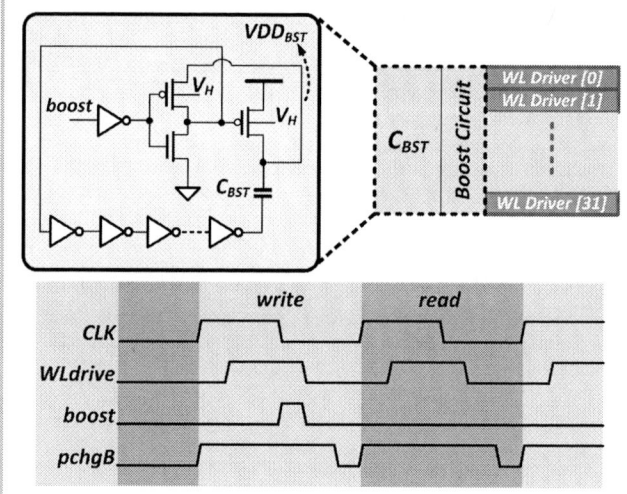

Figure 14.4.4: Timing diagram for critical signals during read and write cycles and voltage boosting circuit implementation.

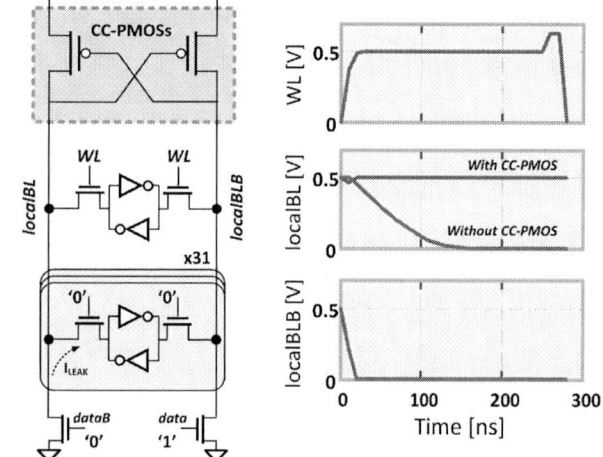

Figure 14.4.5: Cross-coupled PMOS (CC-PMOS) transistors ensure voltage differential on local bit-lines during read and write cycles.

Test-Chip Summary	
Process Technology	28nm CMOS
Features	
Bitcell	6T - 0.12μm²
Sub-Array Size	32 rows x 256 cols
SRAM Size	128kbit
SRAM Area	0.028mm²
Performance	

V_{DD}	Power	Frequency
0.6 V	2.8mW	20MHz
0.8 V	21.1mW	105MHz
1 V	68.5mW	400MHz

Figure 14.4.6: Performance summary of the test chip. 128kbit SRAM achieves operation down to 0.6V at 20MHz.

ISSCC 2011 PAPER CONTINUATIONS

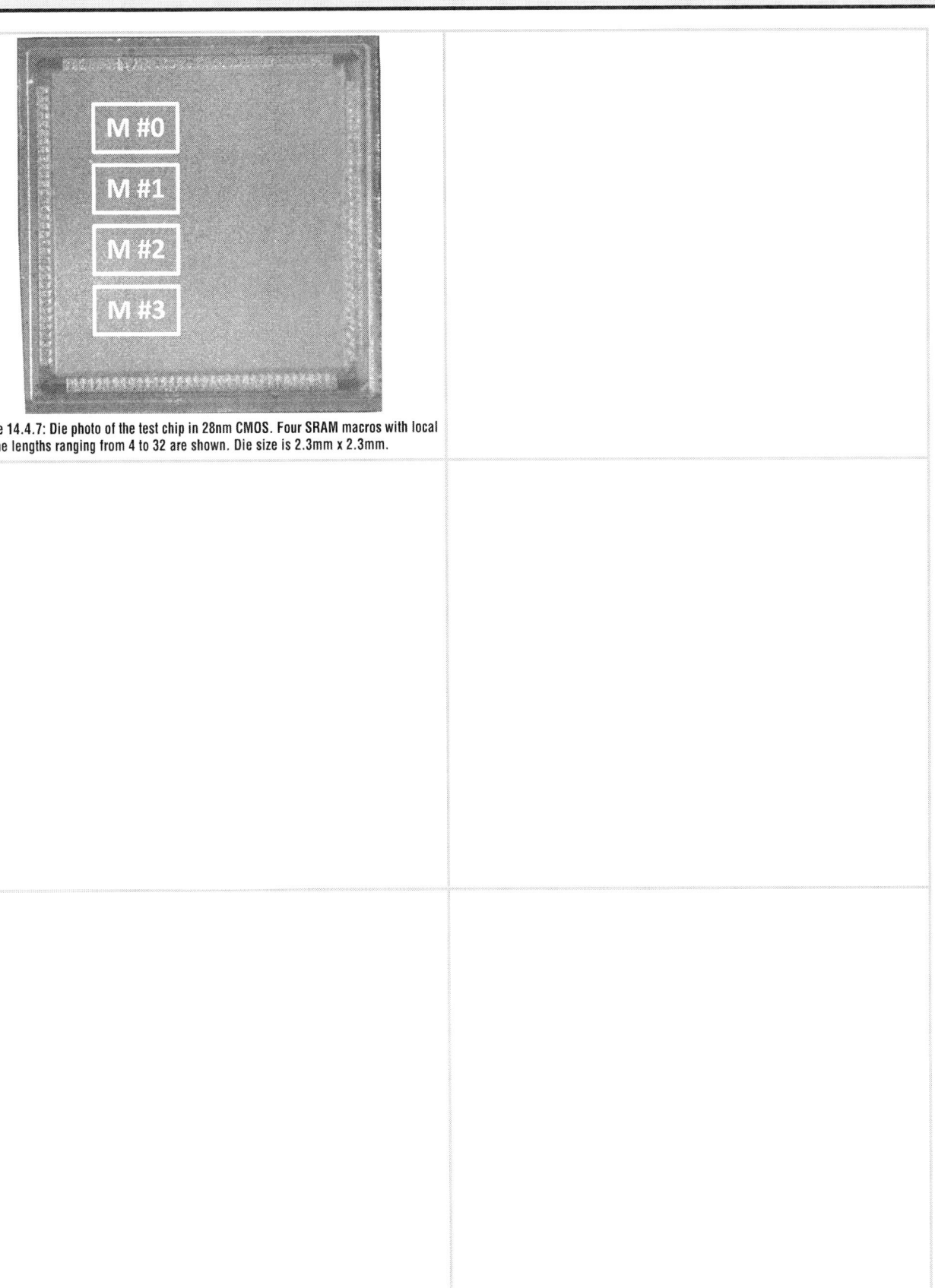

Figure 14.4.7: Die photo of the test chip in 28nm CMOS. Four SRAM macros with local bit-line lengths ranging from 4 to 32 are shown. Die size is 2.3mm x 2.3mm.

ISSCC 2011 / SESSION 15 / HIGH-PERFORMANCE SoCs & COMPONENTS / OVERVIEW

Session 15 Overview / *High-Performance Digital*

High-Performance SoCs & Components

Session Chair: *Shannon Morton, Icera, Bristol, United Kingdom*

Session Co-Chair: *Lew Chua-Eoan, Qualcomm, San Diego, CA*

Integration rules. Why?

Here's why. The number of transistors available on a die continues to grow in line with Moore's law. In the previous decade, designers hit a wall in terms of how to make a single core processor utilize the available transistors efficiently. So they added larger caches. When the performance advantages of these caches flat-lined, they added multiple processor cores. This trend has dominated the industry for many years. Until now. A revolutionary shift is underway.

Now, integration rules. Processors are integrating other chips and components from the system onto the processor die. This not only enables better use of the available silicon area, it also improves communication, reduces system complexity and cost, and lowers overall system power. This session presents state-of-the-art high-performance processors that integrate system-level components, and describes those components in detail.

For many years high-end microprocessors (CPU) and graphics-processor (GPU) chips have competed for bragging rights on highest performance, largest complexity, and most cutting-edge technology and design techniques. Now, finally, these two behemoths of the chip industry have come together, as disclosed in Papers 15.1 and 15.4 from Intel and AMD.

Paper 15.1 (Intel) describes a fully integrated multi-CPU, GPU, and memory controller in 40nm technology. Known as "Sandy Bridge", the chip integrates four high performance X86 cores, an optimized graphics-processing unit, DDR3 memory controllers, and a 20-lane PCIe interface. Power dissipation ranges from 95W for a high-end desktop to only 17W for an optimized mobile product, employing impressive design techniques to enable low-voltage operation. Additional design improvements include improved branch prediction, the addition of a micro-op cache, an advanced floating-point unit, and the addition of vector extension (AVX) operations. The out-of-order cores are incorporated with a shared 8MB L3 and Ring Bus core communication topology. The design includes 6 power domains and independent PLLs per scalable domain to minimize skew.

Paper 15.4 (AMD) integrates two X86-64 "Bobcat" cores (each with a 512KB L2 cache) with a dedicated Radeon HD5000 series graphics and multimedia engine, DDR3 memory controller, client north bridge, and a 4X PCIe link. This "Zacate" SOC has over 450 million transistors and is manufactured in 32nm technology with 10 metal layers. The design employs advanced power management techniques including separate supply rails for CPU and GPU domains, and rigorous power gating to minimize power.

Similarly, switch fabric chips are increasing in complexity, requiring more compute power and higher integration. Paper 15.2 (Renesas Electronics, University of Tsukuba) is an 80Gb/s communication SOC in 45nm CMOS technology with four 4X PCIe links (instead of Infiniband 4X) and eight CPU cores for flow control, packetization, and adaptive network routing, thereby enabling highly robust communication. A direct DDR3 interface is also provided, and an internal high speed system bus enables low-latency communication. The 80Gb/s communication SOC achieves 0.04W/Gb/s and can revert to a single lane with CDR disabled in low BW mode to save transmission power.

These chips integrate some of the biggest and most important components of the system, such as the CPU, GPU, memory controllers, and channel controllers, but there are other system components also on the leading edge of monolithic integration. Examples include on-die DC/DC converters and advanced encryption engines. Integrating these can save cost and enable greater security for end users.

978-1-61284-303-2/11 $26.00 © 2011 IEEE

ISSCC 2011 / February 22, 2011 / 3:15 PM

In Paper 15.3 (Harvard University), a novel approach to integrating a DC/DC converter on chip is presented. Unlike pure switched-capacitance converters or buck converters requiring large external inductors, this paper describes a hybrid approach merging the best of both. An on-die inductor and a 3-level switched capacitor are merged on a monolithic die in 0.130μm technology. The converter delivers up to 0.85A load current with reasonable efficiency, and can regulate output voltages from 0.4V to 1.4V given an input voltage of 2.4V. The design offers voltage scaling across 1V within 20ns setting time with a small footprint. Overall, integrating such a DC/DC converter should provide a net cost benefit to the system.

Clock generators form the backbone of high performance SOCs, and optimizing their performance is critical to the success of such products. In Paper 15.5 (University of Minnesota) a programmable adaptive phase-shifting PLL is presented that compensates for resonant supply noise by modulating the output frequency. Using such a technique, the clock phase stretches in line with the data path delay whenever a short-term supply droop event takes place. This ultimately enables a higher average case frequency or a lower operating voltage to be achieved. Improvements of up to 7.4% in maximum operating frequency were achieved for a standard pipeline on a 65nm chip.

Paper 15.6 (CEA-LETI-MINATEC) tackles an important issue that concerns us all: security. Every day you trust encryption engines to keep your personal details secure, but just how immune are they to focused attacks? This paper enlightens us on the unscrupulous methods available to would-be electronic thieves, and provides a novel mechanism for thwarting such attacks on die. Complementary data paths are implemented to maintain uniform power consumption, and the data paths are mixed to prevent targeted fault attacks from being detected. A test-chip occupying 1336×1411.8μm² in 0.130μm CMOS validates that secret keys are not disclosed even after more than one million acquisitions.

The integration provided by these high-speed SOCs is impressive and continues to progress following Moore's Law. The industry faces huge challenges associated with utilizing the available silicon technologies in the most efficient way possible—attaining the best energy per unit work performed. More importantly though, the higher levels of integration also enable lower overall system cost, while allowing for higher levels of functionality. The continued SOC integration trend will lead to greater overall cost savings and greater user satisfaction as we scale towards smaller geometries.

15

Presenters:

3:15 PM

15.1 A Fully Integrated Multi-CPU, GPU and Memory Controller 32nm Processor

M. Yuffe, Intel, Haifa, Israel

3:45 PM

15.2 An 80Gb/s Dependable Communication SoC with PCI Express I/F and 8 CPUs

S. Otani, Renesas Electronics, Itami, Japan

4:00 PM

15.3 A Fully-Integrated 3-Level DC/DC Converter for Nanosecond-Scale DVS with Fast Shunt Regulation

W. Kim, Harvard University, Cambridge, MA

4:15 PM

15.4 A Low-Power Integrated x86-64 and Graphics Processor for Mobile Computing Devices

D. Foley, AMD, Boxborough, MA

4:45 PM

15.5 A Programmable Adaptive Phase-Shifting PLL for Clock Data Compensation Under Resonant Supply Noise

D. Jiao, University of Minnesota, Minneapolis, MN

5:00 PM

15.6 A Side-Channel and Fault-Attack Resistant AES Circuit Working on Duplicated Complemented Values

A. Tria, CEA-LETI-MINATEC, Gardanne, France; Ecole Nationale Supérieure des Mines de Saint-Etienne, Gardanne, France

ISSCC 2011 / SESSION 15 / HIGH-PERFORMANCE SoCs & COMPONENTS / 15.1

15.1 A Fully Integrated Multi-CPU, GPU and Memory Controller 32nm Processor

Marcelo Yuffe, Ernest Knoll, Moty Mehalel, Joseph Shor, Tsvika Kurts

Intel, Haifa, Israel

This paper describes the 32nm Sandy Bridge processor that integrates up to 4 high performance Intel Architecture (IA) cores, a power/performance optimized graphic processing unit (GPU) and memory and PCIe controllers in the same die. The Sandy Bridge architecture block diagram is shown in Fig. 15.1.1 and the floorplan of a four IA-core version is shown in Fig. 15.1.2. The Sandy Bridge IA core implements an improved branch prediction algorithm, a micro-operation (Uop) cache, a floating point Advanced Vector Extension (AVX), a second load port in the L1 cache and bigger register files in the out-of-order part of the machine; all these architecture improvements boost the IA core performance without increasing the thermal power dissipation envelope or the average power consumption (to preserve battery life in mobile systems). The CPUs and GPU share the same 8MB level-3 cache memory. The data flow is optimized by a high performance on die interconnect fabric (called "ring") that connects between the CPUs, the GPU, the L3 cache and the system agent (SA) unit that houses a 1600MT/s, dual channel DDR3 memory controller, a 20-lane PCIe gen2 controller, a two parallel pipe display engine, the power management control unit and the testability logic. An on die EPROM is used for configurability and yield optimization.

The modular ring interconnect enables the 4-core die to be easily converted into a 2-core die by "chopping" out two cores and two L3 cache modules as described in Fig. 15.1.2. Additional optimizations can be done by reducing the number of execution units of the GPU or by reducing the L3 cache size. Sandy Bridge is offered in three different C4 package types, PGA for mobile computers, LGA for desktop systems and BGA for small form factor systems. For each die flavor an optimized package stack-up was developed with 10 layers for the highest power consumption chips and 8 or 6 layers for the smaller dies.

The die is powered by 6 different power planes. The IA cores and the L3 cache share the same power plane. Power gates were uniformly spread in the cores, enabling the power control unit (PCU) to shut off any core independently. The L3 cache also incorporates power gates; this allows keeping the majority of the cache in a low power data retention mode while only the section of the cache that is accessed is fully connected to the power supply. This embedded power gate approach simplifies the package design reducing the number of package layers. The GPU is connected to its own power plane allowing independent power voltage optimization. The other power planes are used for the SA, I/O and analog circuitry. The voltage of all these power planes is controlled by the PCU using serial VID (SVID), a dedicated low frequency serial bus that connects the PCU with the external voltage regulators in a daisy chain manner.

One of the challenges of sharing the same power plane between the cores and the L3 cache is that the minimum voltage needed to keep the L3 cache data may limit the minimum operating voltage of the cores, increasing the overall average power of the system. Previous processors solved this problem by connecting the L3 cache to a separated higher voltage power plane; however this approach considerably increases the power dissipated by the L3 cache itself and taking into account Sandy Bridge implements 3MB, 4MB or 8MB of L3 cache capacity (depending on the chip configuration) the power dissipated by the cache memory accounts for a big portion of the overall power consumption of the die. Several circuit and logic design techniques have been developed to minimize the Vccmin of the L3 cache and the register files of the chip to bring it to a lower level than the core logic. Figure 15.1.3 shows the Vccmin distribution of the cache base-line and its improvement in Sandy Bridge. Fig 15.1.4 illustrates one of these techniques in a register file. Fabrication variations may cause RF writeability degradation at low voltages (for example for wafers where T_P comes out stronger than T_N); this technique weakens the memory cell pull up device effective strength solving the low voltage write-ability issue caused by a too strong T_P device in the memory cells. The effective size of the shared PMOS is set during production testing by enabling any combination of the three parallel transistors T_1, T_2 and T_3.

The processor clocking architecture shown in Fig. 15.1.5 includes 13 PLLs driving independent clock domains for the core and cache slices, the GPU, SA as well as the four independent I/O regions [1,2]. Most of the PLLs use a 100MHz reference clock provided by the external system clock generator that is integrated into the Platform Control Hub (PCH) chip. The reference for the slices and GPU PLLs is generated by a dedicated RCLK PLL to minimize the clock skew over the different power plane domains. The SA PLL generates the clock needed by the SA logic as well as the clock for the display engine and the 133MHz reference clock for the main memory system. This clocking architecture ensures high synchronicity between the independent clock domains at the BCLK reference edges even when the domains are operating at different frequencies; to ensure data transfer between different clock domains with minimum latency low jitter PLLs (long term jitter $\sigma < 2ps$) are used. The clock distribution ensures low clock power and good clock skew performance by a combination of vertical clock spines and embedded clock compensators controlled by on die compensation state machines. Figure 15.1.6 shows the clock distribution in one of the slices. The measured clock skew is 16ps, with on-die clock compensators canceling the within die random variations.

Temperature control is extensively used in modern processors to maximize performance. When the die is hot the frequency is lowered, when the die is cold the PCU takes advantage of the thermal head room to increase the frequency of the CPU. Temperature information is also used to control the system fan and to shut the CPU down in case of a catastrophic thermal event. Sandy Bridge has taken this method one step further by using two different types of thermal sensors. The first is a diode-based thermal sensor described in [3]. This sensor compares the diode voltage, which has a negative temperature coefficient to a reference voltage to output the temperature. This sensor functions over the full temperature range of operation, providing information for throttling, catastrophic function and fan regulation. Due to die area considerations there is only one such sensor per core (sensors are also located in the GPU and in the SA). Sandy Bridge has also implemented a miniaturized CMOS-based thermal sensor [4] that has a substantially reduced area ($5100um^2$) compared to the diode sensor, but has a more limited temperature range, being accurate between 80-100°C; since this sensor is small it can be placed at several locations inside the core providing an accurate picture of the core hot spots.

Sandy Bridge introduces the Generic Debug eXternal Connection (GDXC), a debug bus that allows monitoring the traffic between the IA cores, GPU, caches and SA on the processor internal ring. GDXC allows chip, system or software debuggers to sample ring data traffic as well as ring protocol control signals and drive it to an external logic analyzer in a packet manner through a dedicated on-package probe array. A post processing software is used to recover and analyze the data.

The Sandy Bridge thermal dissipation power (TDP) ranges from 17W to 45W for a 2-core and a 4-core mobile part and all the way to 95W for a high-end desktop part. The IA cores and GPU are powered from independent 0.7V to 1.15V variable voltage power supply sources, all controlled by the SVID bus. The DDR3 interface uses a 1.5V power plane while the PCIe interface uses a 1.05V power plane. The die photo is shown in Fig. 15.1.7.

References:

[1] Rusu, Stefan et al.: A 45nm 8-Core Enterprise Xeon Processor, ISSCC-2009

[2] Fayneh, E, Knoll, E: Clock Generation and Distribution for Intel Banias Mobile Microprocessor, VLSI Circuits Symposium, 2003

[3] Duarte, D., Geannopoulos, G., Mughal, U., Wong, K.L. and Taylor, G., "Temperature Sensor Design in a High Volume Manufacturing 65nm CMOS Digital Process", Proceedings of the 2007 IEEE Custom Integrated Circuits Conference, (CICC '07), pp. 221-224, Sept. 2007.

[4] K. Luria and J. Shor, "Miniaturized CMOS Thermal Sensor Array for Temperature Gradient Measurement in Microprocessors", Proceedings of the IEEE 2010 International Symposium of Circuit and Systems, Paris France, May 30 2010.

978-1-61284-303-2/11 $26.00 © 2011 IEEE

ISSCC 2011 / February 22, 2011 / 3:15 PM

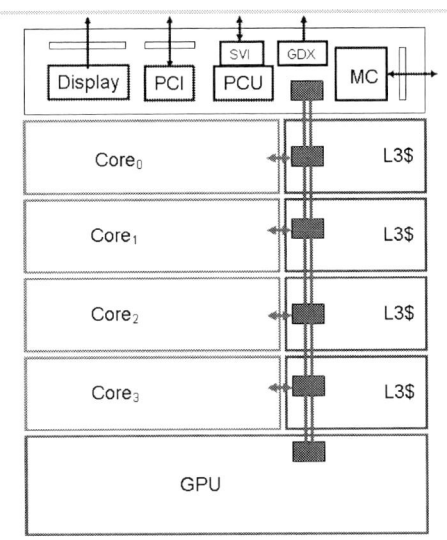

Figure 15.1.1: Sandy Bridge block diagram.

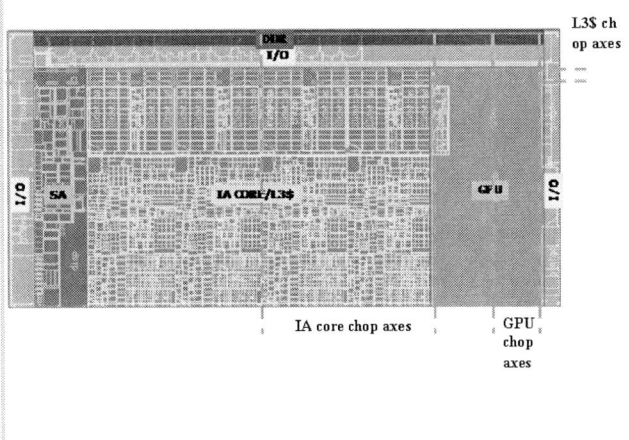

Figure 15.1.2: Sandy Bridge floorplan, power planes and choppability axes.

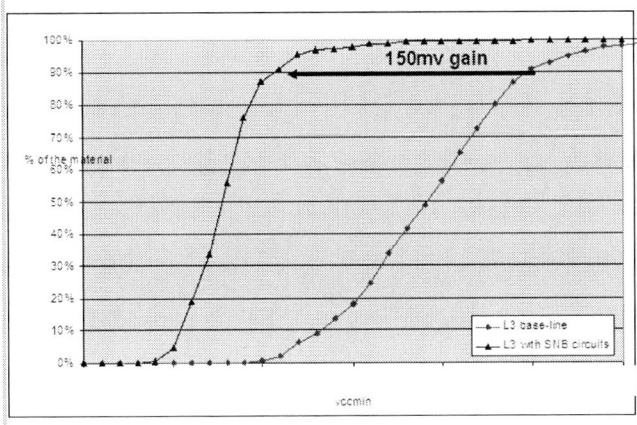

Figure 15.1.3: Vccmin improvement after applying SNB Vccmin reduction circuits.

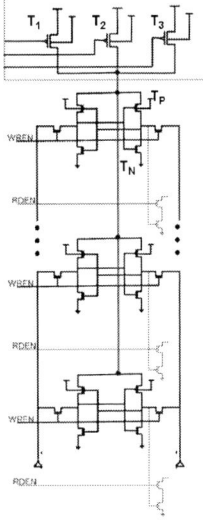

Figure 15.1.4: Register file shared strength control PMOS devices.

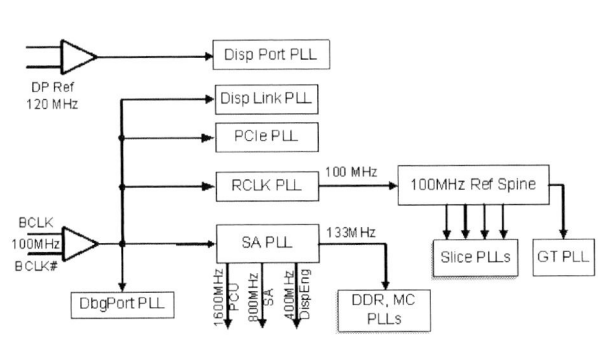

Figure 15.1.5: Sandy Bridge clock generation block diagram.

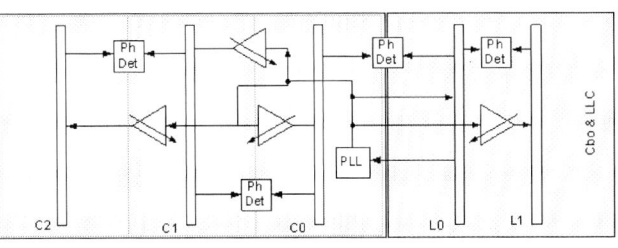

Figure 15.1.6: Slice clock distribution block diagram.

ISSCC 2011 PAPER CONTINUATIONS

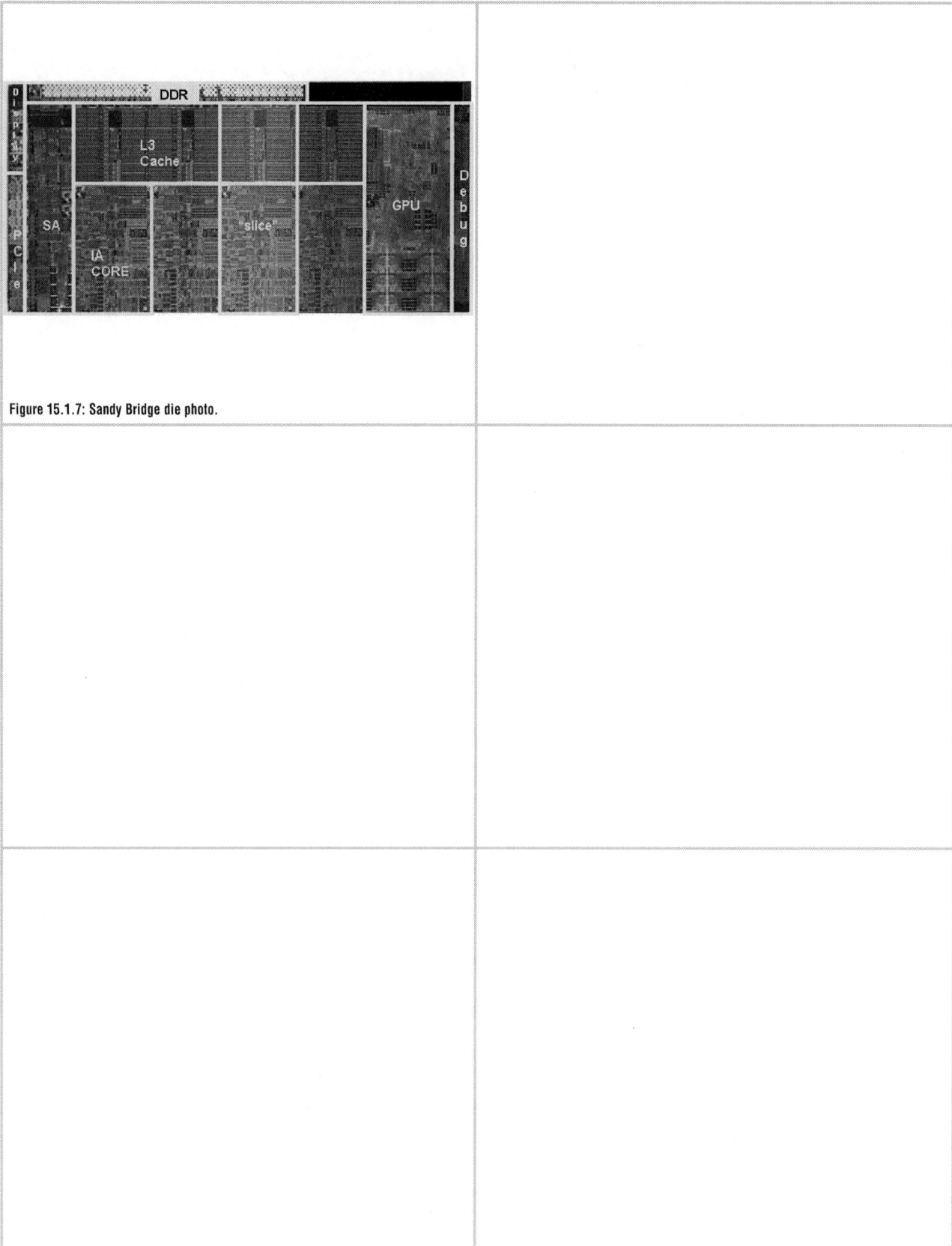

Figure 15.1.7: Sandy Bridge die photo.

ISSCC 2011 / SESSION 15 / HIGH-PERFORMANCE SoCs & COMPONENTS / 15.2

15.2 An 80Gb/s Dependable Communication SoC with PCI Express I/F and 8 CPUs

Sugako Otani[1], Hiroyuki Kondo[1], Itaru Nonomura[2], Atsuyuki Ikeya[2], Minoru Uemura[2], Yasushi Hayakawa[1], Takeshi Oshita[1], Satoshi Kaneko[1], Katsushi Asahina[2], Kazutami Arimoto[1], Shin'ichi Miura[3], Toshihiro Hanawa[3], Taisuke Boku[3], Mitsuhisa Sato[3]

[1]Renesas Electronics, Itami, Japan,
[2]Renesas Electronics, Kodaira, Japan,
[3]University of Tsukuba, Tsukuba, Japan

InfiniBand is widely used as a low-latency and high-bandwidth network for high-performance computing (HPC) clusters [1]. However, power consumption and system cost become large in exchange for high performance on small-scale clusters or embedded systems. To cope with these problems, we use PCI Express (PCIe) [2] technology as a direct communication link between computing nodes. Point-to-point bidirectional packet communication is implemented by using PCIe basic operations such as memory read/write between the host and the device. Therefore, PCIe technology can be applied to inter-node communication and thereby eliminate communication overhead and extra power consumption caused by the protocol conversion via the network device [3].

We develop a robust communication SoC with the PCIe I/F. The SoC has four PCIe Rev.2.0 ports with four lanes each, and employs a control processor with eight CPU cores. This communication SoC acts as a communication link in an HPC cluster. In embedded systems, the communication SoCs can extend their role to the computing nodes and give the system flexibility (Fig 15.2.1). When a request is received from a node processor via the PCIe port, the SoC generates a packet header, and then the packet is sent to the appropriate destination port. The SoC also analyzes a packet header from a node and forwards the packet to another node or passes it to the node processor. By using the PCIe I/F, the SoC behaves as a communication interface to other processing nodes as well as a communication switch or router to realize a multi-node parallel platform.

Figure 15.2.2 shows the block diagram of the communication SoC. This chip includes two blocks, the control processing block and the transfer-processing block. These two blocks are connected with a bus bridge. The control-processing block integrates a control processor with eight CPUs which are connected to a common pipelined bus [4,5]. The control-processing block also has an internal multi-layer bus to connect the controller to peripheral IOs and a DDR3-600 I/F. The 512KB L2 cache accesses both the DDR3 I/F and the internal bus in parallel. The control-processing block performs data processing and data flow control, which consists of adaptive network routing and packet header analysis. The transfer-processing block has four PCIe ports which can transfer packets using up to four lanes, 512KB SRAM which is allocated for temporary packet storage, and an interrupt generator (INTGEN), which are used to achieve high transfer performance and low power consumption. All the main modules in the transfer-processing block are connected by a high-speed internal system bus, which provides both high speed and flexible control. The features of the SoC are described in Fig. 15.2.3.

The INTGEN generates an interrupt request to the control processor to send information such as notification of the completion of a sequential transfer. In addition, the INTGEN supports automatic transfer mode and handles transfer processing without using the control-processing block. Automatic transfer mode can dramatically reduce processing time by 20% compared to the conventional mode, which uses the CPU interrupt handler. The 256MB DDR3-600 I/F is accessed via both the control-processing block and the transfer processing block in parallel. This high-speed and large memory contributes to improving processor performance. The DDR memory is also used as a large packet buffer if the packet size is larger than 512KB SRAM.

Each PCIe port has a link controller, PHY, local DMA controller (DMAC), and local RAM (the packet buffer). The maximum transfer speed of the latest Rev.2.0 standard is 5.0Gb/s, double that of the earlier Rev.1.1 (2.5Gb/s). Therefore, the total maximum transfer rate to each destination is 20Gb/s. The theoretical peak bandwidth is actually 2GB/s due to 8b/10b encoding for the embedded clock and error detection. As a network communication interface, this SoC with four PCIe

ports realizes a high performance communication link with a theoretical peak bandwidth of 4×20Gb/s.

Figure 15.2.4 shows the PCIe PHY analog block diagram. A TX equalizer is used to realize high-speed transmission of 5.0GT/s up to 40-inch FR4 trace. The power dissipation of PHY has been reduced by 22%, using the "Partially-regulated LC-VCO" while suppressing the PLL/CDR jitter down to $10ps_{pp}$. The maximum power consumption per port of this SoC is 0.8W at 20Gb/s. So, the data transfer performance is 0.04W/(Gb/s). The data transfer performance of 4X InfiniBand is 0.083W/(Gb/s) [6]. Thus, this SoC provides 51.5% less power consumption than 4X InfiniBand. Figure 15.2.5 shows the TX eye diagram, which is measured at 5.0GT/s, with a de-emphasis ratio of 3.2dB.

Each PCIe port has an optional upconfiguration function that allows the transfer rate to be switched dynamically by software. Figure 15.2.6 illustrates the function. When the required transfer volume is higher, each lane operates at the maximum transfer speed of 5.0Gb/s (a); when the transfer volume is lower, only one lane operates at a transfer speed of 2.5Gb/s for low power consumption (b). The PCIe port also supports the active-state power management (ASPM) protocol. When the PCIe link is idle, the link can transition to the low-power state by stopping the CDR. The combination of these power management functions contributes substantially in reducing power consumption of the overall system. Using the PCIe I/F with four lanes and the transfer speed of Rev.2.0 improves power efficiency 48% over the same data transfer performance of Rev.1.1 using one lane.

To achieve a highly dependable network, the communication SoC continuously monitors the network condition and performs adaptive routing dynamically. In the transfer-processing block, the functions for the error detection, flow control, and retransmission control in the specifications are processed by the hardware automatically. When a link error that cannot be corrected automatically occurs, the control-processing block reduces the number of lanes to remove the defective lane by re-initializing the link. The controller also monitors the communication between the adjacent nodes on each port by using daemon programs on Linux. The detour routing is applied when the controller finds a fault on a link or node. The heterogeneous multicore system in the control processing block operates the fault handling and error monitoring / logging efficiently by using IRQ affinity and multiple group mode for CPU snooping [5]. Furthermore, the cpu-hotplug mechanism on the Linux Kernel can dynamically suspend and resume a CPU responding to system load. These mechanisms are useful for power awareness and dependability.

The test-chip is fabricated in 45nm low-power CMOS (8 layers, triple-V_{th}). The die is packaged in a flip-chip BGA with 1008 pins. Three different power-supply voltages are used: 1.2V for the cores, PCIe-related modules, most of the logic circuits, and memories; 1.5V for the DDR3 interface; and 3.3V for the other peripherals. Figure 15.2.7 shows the chip micrograph.

Acknowledgements:
This work is supported by a JST/CREST program entitled "Computation Platform for Power-aware and Reliable Embedded Parallel Processing Systems" in the research area of "Dependable Embedded Operating Systems for Practical Use."

References:
[1] The InfiniBand architecture specification. InfiniBand Trade Association. http://www.infinibandta.org/specs/
[2] PCI Express Base Specification, Rev. 2.0, PCI-SIG, Dec. http://www.pcisig.com/
[3] T. Hanawa, et al, "Low-power and high-performance communication mechanism for dependable embedded systems", Proceedings of 2008 International Workshop on Innovative Architecture for Future Generation Processors and Systems, 2009.
[4] S. Kaneko et al., "A 600MHz single-chip multiprocessor with 4.8GB/s internal shared pipelined bus and 512kB internal memory," *IEEE J. of Solid-State Circuits*, Vol.39, No.1, pp.184-193, Jan. 2004
[5] H. Kondo et al., "Design and Implementation of a Configurable Heterogeneous Multicore SoC With 9 CPUs and 2 Matrix Processors", *IEEE J. of Solid-State Circuits*, Vol.43, No.4, pp.892-901, Apr. 2008
[6] "QLogic TrueScale™ InfiniBand, the Real Value", http://www.qlogic.com/

978-1-61284-303-2/11 $26.00 © 2011 IEEE

ISSCC 2011 / February 22, 2011 / 3:45 PM

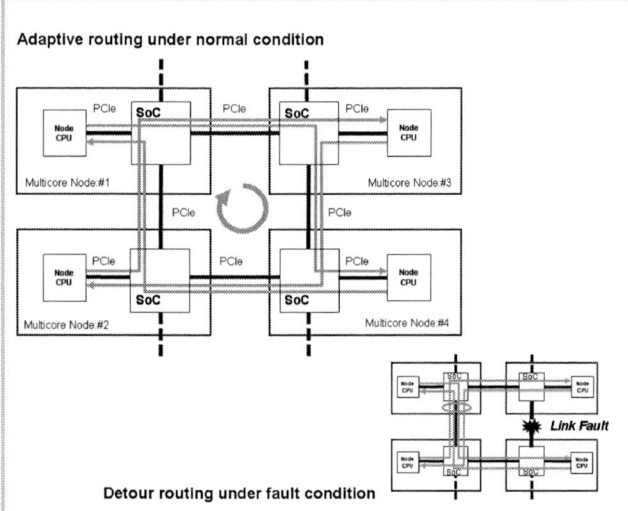

Figure 15.2.1: Dependable system in HPC cluster.

Figure 15.2.2: Block diagram of communication SoC.

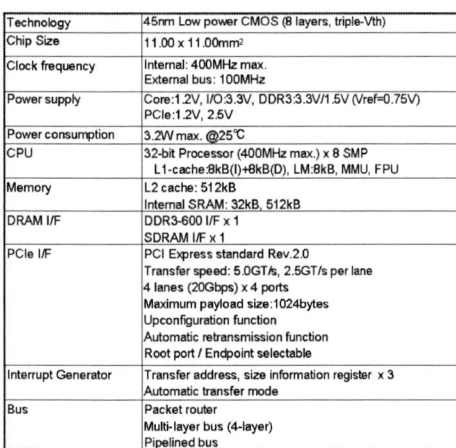

Technology	45nm Low power CMOS (8 layers, triple-Vth)
Chip Size	11.00 x 11.00mm²
Clock frequency	Internal: 400MHz max. External bus: 100MHz
Power supply	Core:1.2V, I/O:3.3V, DDR3:3.3V/1.5V (Vref=0.75V) PCIe:1.2V, 2.5V
Power consumption	3.2W max. @25°C
CPU	32-bit Processor (400MHz max.) x 8 SMP L1-cache:8kB(I)+8kB(D), LM:8kB, MMU, FPU
Memory	L2 cache: 512kB Internal SRAM: 32kB, 512kB
DRAM I/F	DDR3-600 I/F x 1 SDRAM I/F x 1
PCIe I/F	PCI Express standard Rev.2.0 Transfer speed: 5.0GT/s, 2.5GT/s per lane 4 lanes (20Gbps) x 4 ports Maximum payload size:1024bytes Upconfiguration function Automatic retransmission function Root port / Endpoint selectable
Interrupt Generator	Transfer address, size information register x 3 Automatic transfer mode
Bus	Packet router Multi-layer bus (4-layer) Pipelined bus

Figure 15.2.3: Features of SoC.

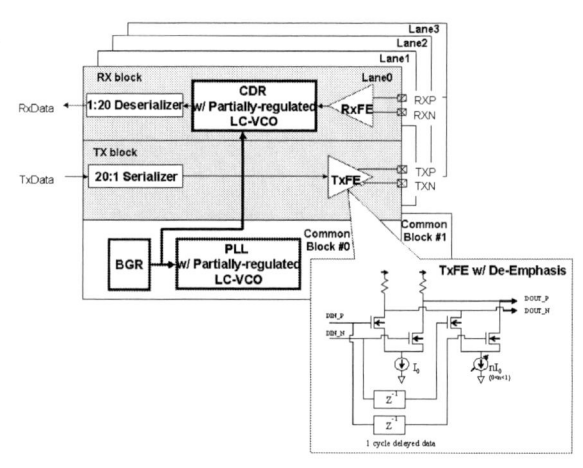

Figure 15.2.4: PCIe PHY analog block diagram.

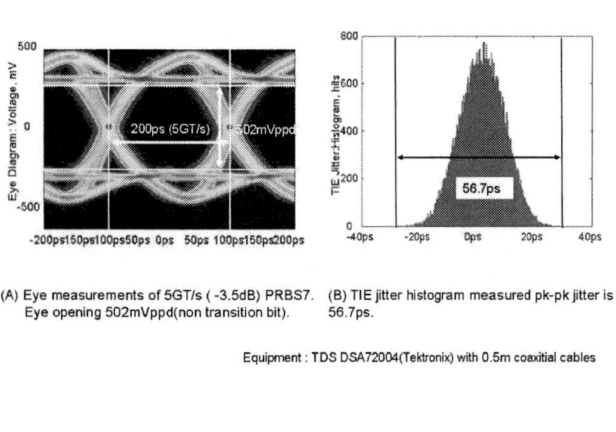

(A) Eye measurements of 5GT/s (-3.5dB) PRBS7.
Eye opening 502mVppd(non transition bit).

(B) TIE jitter histogram measured pk-pk jitter is
56.7ps.

Equipment : TDS DSA72004(Tektronix) with 0.5m coaxitial cables

Figure 15.2.5: Eye measurement (PRBS7) results.

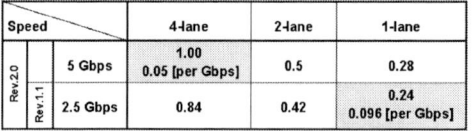

Speed		4-lane	2-lane	1-lane
Rev.2.0	5 Gbps	1.00 0.05 [per Gbps]	0.5	0.28
Rev.1.1	2.5 Gbps	0.84	0.42	0.24 0.096 [per Gbps]

0.05/0.096 = 0.52 (up to 48% reduction)

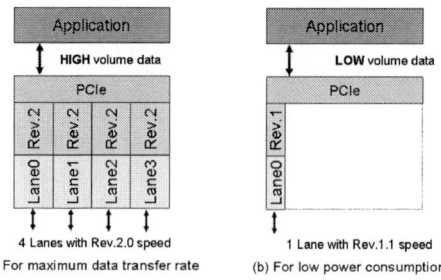

(a) For maximum data transfer rate

(b) For low power consumption

Figure 15.2.6: Upconfiguration function and power dissipation ratio.

978-1-61284-303-2/11 $26.00 © 2011 IEEE

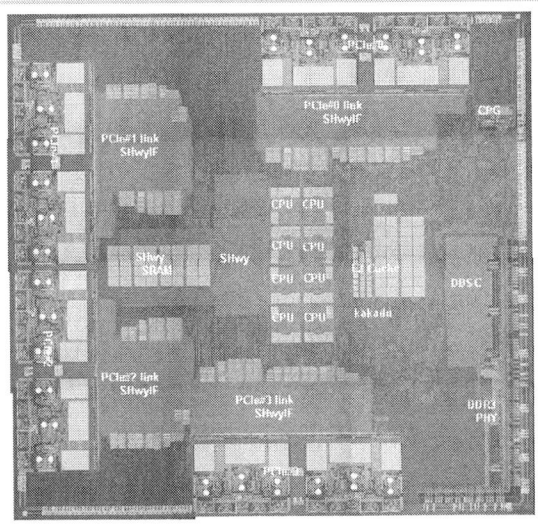

Figure 15.2.7: Micrograph of SoC.

ISSCC 2011 / SESSION 15 / HIGH-PERFORMANCE SoCs & COMPONENTS / 15.3

15.3 A Fully-Integrated 3-Level DC/DC Converter for Nanosecond-Scale DVS with Fast Shunt Regulation

Wonyoung Kim, David M Brooks, Gu-Yeon Wei

Harvard University, Cambridge, MA

In recent years, chip multiprocessor architectures have emerged to scale performance while staying within tight power constraints. This trend motivates per-core/block dynamic voltage and frequency scaling (DVFS) with fast voltage transition. Given the high cost and bulk of off-chip DC/DC converters to implement multiple on-chip power domains, there has been a surge of interest in on-chip converters. This paper presents the design and experimental results of a fully integrated 3-level DC/DC converter [1] that merges characteristics of both inductor-based buck [2-4] and switched-capacitor (SC) converters [5]. While off-chip buck converters show high conversion efficiency, their on-chip counterparts suffer from loss due to low quality inductors. With the help of flying capacitors, the 3-level converter requires smaller inductors than the buck converter, reducing loss and on-die area overhead. Compared to SC converters that need more complex structures to regulate higher than half the input voltage, 3-level converters can efficiently regulate the output voltage across a wide range of levels and load currents. Measured results from a 130nm CMOS test-chip prototype demonstrate nanosecond-scale voltage transition times and peak conversion efficiency of 77%.

Shown in Fig. 15.3.1, the 3-level converter comprises a set of power FETs (P_{top}, PN_{mid}, N_{bottom}), a flying capacitor (C_{fly}), an inductor, and an output capacitor. The power FETs use thin-oxide devices with small parasitic resistance in a stacked structure to support input voltages (V_{in}) up to twice the maximum gate-source voltage allowed by the process technology. Level shifters generate input signals for P_{top} and PN_{mid}. This converter regulates the output voltage (V_{out}) by iterating through four steps per switching period (T). In steps 1 and 3, the converter relies on C_{fly}, whereas in steps 2 and 4, it relies on the inductor to regulate V_{out}. Duty cycle (D) sets the fraction of a switching period the converter operates in each step. When D equals 0.5 (50%), steps 2 and 4 essentially disappear and the converter operates like a SC converter, regulating the output to ~V_{in}/2. As D deviates away from 0.5, V_L swings between V_{in}/2 and V_{in} ($D > 0.5$), or 0 and V_{in}/2 ($D < 0.5$). The square-wave signal on V_L causes inductor current (I_L) ripple to vary with D and is filtered by the output inductor and capacitor to generate V_{out}. The 3-level converter offers advantages over conventional inductor-based converters that use large inductors [2,4] or high switching frequencies [3] to reduce I_L ripple, which otherwise causes large V_{out} ripple and increases resistive loss. Since the frequency of voltage swing on V_L is twice the converter's switching frequency, the converter can use lower switching frequencies and/or smaller inductors. Moreover, the amplitude of voltage swing on V_L is V_{in}/2, which further reduces I_L ripple and parasitic switching losses.

Figure 15.3.2 presents a block diagram of the test-chip prototype that comprises a pair of 2-phase, 3-level converters arranged as two identical sectors. To reduce ripple on V_{out}, the power FETs for each phase operate off of clock signals offset by 180 degrees. Low-impedance switches can connect the two sectors together to create a single 4-phase converter with each phase offset by 90 degrees. Otherwise, the test chip implements two independent 2-phase converters. An ability to disable power FETs also enables multiple 3-level converter configurations consisting of one to four phases. A programmable load in each sector facilitates experimental measurements by sinking up to 0.5A in 25mA steps as steady or pseudorandom patterns of current. The converter relies on a slow digital feedback loop to regulate V_{out} to a desired level by adjusting D. A supplemental shunt regulator suppresses output voltage fluctuations by detecting when V_{out} crosses low or high thresholds and injecting or extracting current to compensate for sudden load current fluctuations [6].

Data captured from a real-time oscilloscope (plotted in Fig. 15.3.3) demonstrates the converter can regulate the output voltage across a wide range—from 0.4 to 1.4V while operating off of a 2.4V input voltage—and rapidly scale V_{out} by 1V within 15 to 20ns. Such high-speed voltage transitions at nanosecond timescales enable complex digital systems to leverage temporally fine-grained DVFS and improve energy efficiency [7].

Figure 15.3.4 summarizes conversion efficiency measurements. The converter operates in open loop with fixed duty cycles ranging from 40% to 65% in 5% steps to facilitate measurements across a wide range of conditions. Two converter sectors can also operate with duty cycles that differ by 5% to implement finer steps. Since duty cycle is fixed during open-loop measurements, IR drop due to parasitic resistance causes a spread in output voltages with respect to load currents for the same duty cycle. Figure 4(a) aggregates all of the measured efficiencies collected across a range of static load current conditions (0.3 to 0.8A), duty cycles (40 to 65%), switching frequencies (50 to 160MHz), and number of phases (1 to 4). Efficiency peaks at 77% for low load current conditions at 50% duty cycle. Figure 15.3.4(b) compares measured data for 50% duty cycle operation using 2 and 4 phases. IR losses degrade efficiency as load current increases, worse for the 2-phase configuration. Higher switching frequency can also degrade efficiency at low load currents due to higher switching losses. Figure 15.3.4(c) plots the upper range of efficiency measurements for the 4-phase configuration by picking the best efficiency data across different duty cycle settings. Trend line overlays again illustrate the spread in output voltages due to IR drop. Since the 3-level converter merges characteristics of both SC and buck converters, efficiency peaks for 50% duty cycle. As duty cycle deviates from 50%, inductor current ripple grows and the corresponding increase in resistive losses degrades conversion efficiency. Figure 15.3.4(d) adds results for the 2-phase configuration (symbols with outlines) to show that fewer phases can sometimes improve efficiency at low load currents.

Figure 15.3.5 presents histogram plots of measured voltage noise, due to pseudorandom current patterns generated by the programmable loads, with and without the supplemental shunt regulator turned on. These results verify the shunt regulator can appreciably squeeze the noise distribution together and reduce peak-to-peak voltage excursions. Moreover, connecting the power domains reduces voltage noise as a result of larger output capacitance and some canceling of the pseudorandom load currents.

While the shunt regulator—reacting to threshold crossings—reduces voltage fluctuations, it has two drawbacks. First, internal circuit delays limit how quickly this feedback loop can sense and react. Second, simply relying on thresholds provides limited information as to the magnitude of voltage noise and the appropriate response needed to suppress it. One solution is to use a prediction-based shunt regulator that leverages microarchitecture-level information to reliably predict upcoming voltage droops [8]. Figure 15.3.6 presents snapshots of measured voltage droops due to two consecutive 80ns wide current pulses of 100mA and 150mA. Predictive current shunting reduces the maximum voltage droop by over 40% compared to simply reacting to threshold crossings.

Figure 15.3.7 presents a die micrograph and a list of specifications for the test chip.

Acknowledgements:
This work was supported by NSF CNS-0720566 and CCF-0102344. We thank UMC for chip fabrication.

References:
[1] V. Yousefzadeh, et al., "Three-level buck converter for envelope tracking applications," *IEEE Trans. Power Electronics*, Mar. 2006.
[2] J. Wibben and R. Harjani, "A high-efficiency DC–DC converter using 2nH integrated inductors," *IEEE JSSC*, Apr. 2008.
[3] M. Alimadadi, et al., "A 3GHz switching DC-DC converter using clock-tree charge-recycling in 90nm CMOS with integrated output filter," *ISSCC*, 2007.
[4] S. Abedinpour, et al., "A multi-stage interleaved synchronous buck converter with integrated output filter in a 0.18μm SiGe process," *ISSCC*, 2006.
[5] H. Le, et al., "A 32nm fully integrated reconfigurable switched-capacitor DC-DC converter delivering 0.55W/mm² at 81% efficiency," *ISSCC*, 2010.
[6] E. Alon and M. Horowitz, "Integrated regulation for energy-efficient digital circuits," *IEEE JSSC*, Aug. 2008.
[7] W. Kim, et al., "System level analysis of fast, per-core DVFS using on-chip switching regulators," *IEEE HPCA*, 2008.
[8] V. J. Reddi, et al., "Voltage emergency prediction: Using signatures to reduce operating margins," *IEEE HPCA*, 2009.

978-1-61284-303-2/11 $26.00 © 2011 IEEE

ISSCC 2011 / February 22, 2011 / 4:00 PM

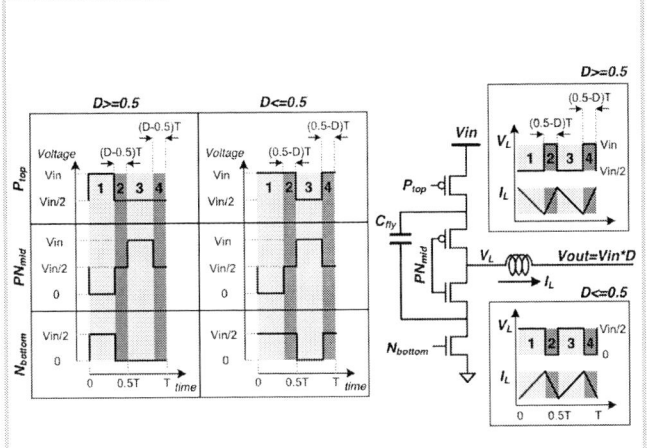

Figure 15.3.1: Schematic of the proposed 3-level power converter. Signal timing diagrams illustrate different operating modes.

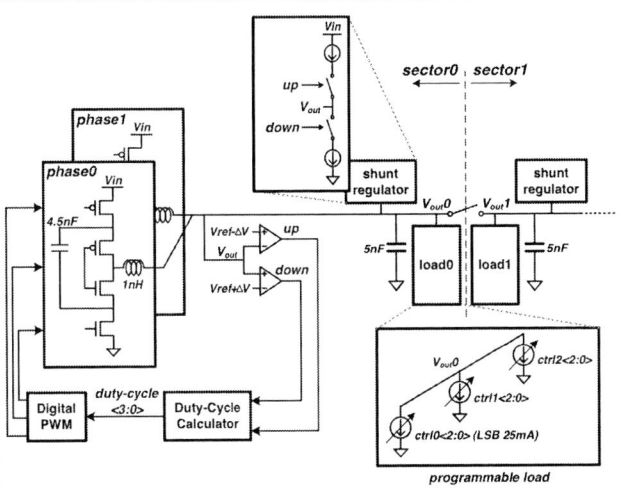

Figure 15.3.2: Block diagram of 3-level converter with slow digital feedback control and fast shunt regulation.

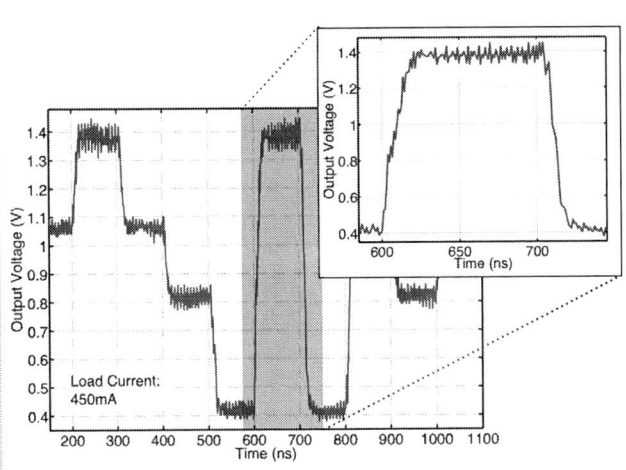

Figure 15.3.3: Measured snapshot of fast dynamic voltage scaling. Voltage scales from 1.4V to 0.4V and vice versa in 15-20ns.

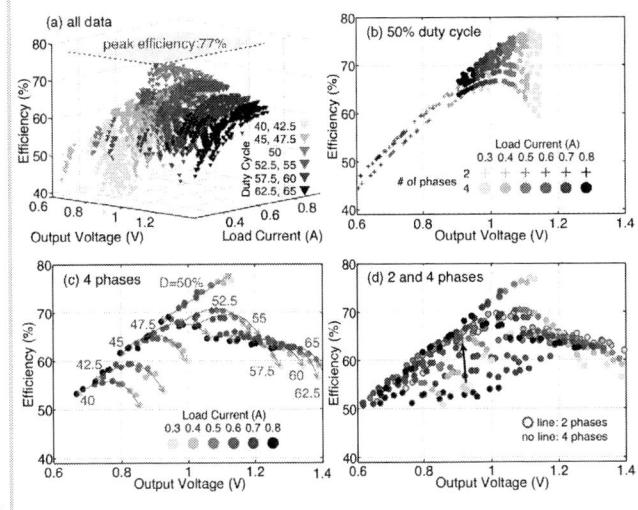

Figure 15.3.4: Measured conversion efficiency.

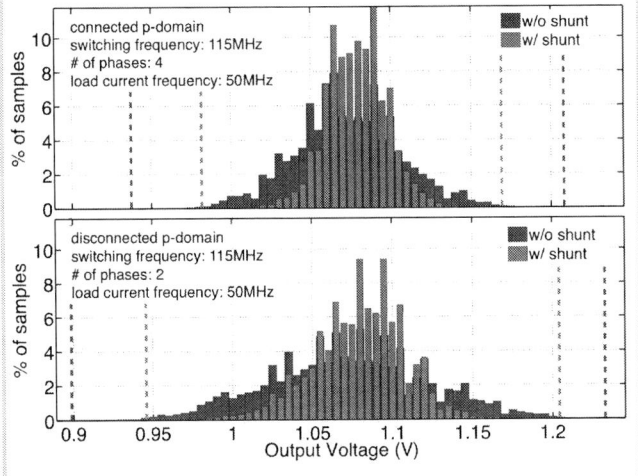

Figure 15.3.5: Histogram of measured voltage noise with and without shunt regulator for connected and disconnected power domains of two sectors.

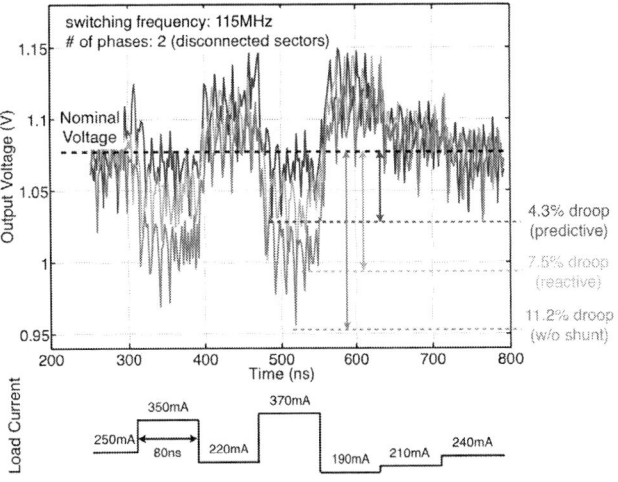

Figure 15.3.6: Comparison of measured on-die voltage noise without shunt regulator, with reactive shunt, and with predictive shunt.

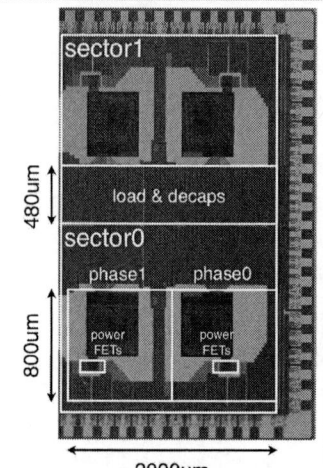

Technology	130nm CMOS
Load Power	0.2-1W
Input Voltage	2.4V
Output Voltage	0.4-1.4V
Inductor per phase	1nH
Total Flying Capacitance	18nF
Total Decoupling Capacitance	10nF
Switching Frequency	50-200MHz
Peak Efficiency	77%

Figure 15.3.7: Die micrograph of the converter with dimensions of main blocks. Flying capacitors with patterned ground planes sit under the inductors to save on-die area while maintaining inductor quality. The table shows converter specifications.

ISSCC 2011 / SESSION 15 / HIGH-PERFORMANCE SoCs & COMPONENTS / 15.4

15.4 A Low-Power Integrated x86-64 and Graphics Processor for Mobile Computing Devices

Srinivasa Rao Gutta[1], Denis Foley[2], Ajay Naini[1], Robert Wasmuth[3], Don Cherepacha[4]

[1]AMD, Hyderabad, India,
[2]AMD, Boxborough, MA,
[3]AMD, Austin, TX,
[4]AMD, Markham, Canada

AMD's first Fusion Accelerated Processor Unit (APU) codenamed "Zacate" (Fig. 15.4.1) combines a pair of x86 CPUs cores codenamed "Bobcat", 1MB L2 Cache, Client Northbridge (CNB), with a DirectX® 11 Radeon™ HD5000 graphics/multimedia controller on a single die. The CNB provides an interface to a single 64b DDR3 memory channel, which can operate at up to DDR3-1066. The Fusion architecture implements an efficient form of unified memory architecture (UMA) where a portion of system memory is reserved as graphics frame buffer memory. The graphics memory controller (GMC) arbitrates between graphics, video and display memory accesses and presents a well-ordered stream of system memory requests through the CNB over dedicated 256b wide read and write busses. These GMC requests bypass all of the CNB coherency mechanisms allowing for fast direct access to memory and exposing most of the available memory bandwidth (8.53GB/s). Compared to two chip solutions, use of the on-die integrated GPU significantly reduces memory latency, improves request ordering, and reduces power. The APU supports display formats including VGA, LVDS, Display Port, DVI or HDMI™. A ×4 Gen2 PCIe® Unified Media Interface (UMI) to an external Fusion Controller Hub (FCH) is supported for system I/O. An additional 4× PCIe Gen2 link supports I/O to external Discrete Graphics chip.

The SoC including the Bobcat cores is entirely synthesized from RTL and implemented using standard ASIC style SAPR flows. The die shown in Fig. 15.4.7 has over 450 million transistors and is manufactured using 10-layer metal 40nm bulk CMOS technology. As shown in Fig. 15.4.2, logic comprises 68% of die transistors, with approximately 56% HVT and 42% SVT. To limit leakage power, less than 2% LVT devices are used only in critical paths. The Bobcat core is implemented as a single physical tile consisting of 1.1 million instances and uses only 7 custom memory macros for branch predictor arrays, TLB, L1 tag and data arrays. The GPU core logical RTL is physically partitioned into multiple physical tiles with varying numbers of cell instances and memory macros generated using standard memory compilers. The GPU has zero grout space between the tiles and uses feed-through repeater bundles to connect the tiles.

The SAPR methodologies are carefully chosen at every step to have an optimal balance between performance, power and area. The CPU and GPU core operate on separate voltage supplies and support dynamic voltage and frequency scaling to optimize power consumption. The design is optimized for power and performance at discrete process and voltage points. The CPU is optimized for power and performance at 1.2V and 0.8V. The GPU is optimized at 3 voltage points 1.0V, 0.9V and 0.8V.

The design has 16 functional, 10 scan and 11 debug mode clocks and clock relationships which made the clock planning and implementation for the complete chip challenging. A Digital Frequency Synthesizer (DFS) is used to generate 9 functional clocks used by the CPUs, CNB, and the GPU as shown in Fig. 15.4.3. The system PLL provides 4 phase-offset references to the DFS which combines the phases to generate the required clock frequencies. The DFS allows the CPU to continue normal instruction execution during frequency transitions. The high speed CPU clock is distributed as a mesh network with 6-8 levels of local buffers and GPU clock (SCLK) is implemented with X-tree & H-tree topologies feeding into the Clock Tree Synthesis (CTS) branches inside a physical tile. All other clocks are implemented as balanced buffer trees from DFS to the physical tile. The CNB and GPU clocks are balanced globally with respect to each other within a ½ cycle of the fastest clock using programmable delay buffers pre-placed beside the global trees.

On-die power gating is implemented using a leakage optimized HVT PFET header switch to turn off the VDD rails. The header uses parallel stacked PFET transistors with separate enables for each PFET as shown in Fig. 15.4.4. The enables are daisy chained and have programmable delay to control the timing. This helps limit the inrush current and voltage transients. In contrast to other x86 designs

using power-gating ring [1,2], the global power grid in the GPU uses M7 & M8 with headers placed on a regular grid as shown in Fig. 15.4.4. The distance between the headers is calculated as a function of minimum horizontal M7 grid pitch Y. The Memory macro power gating is implemented using headers arranged as a ring in 1 or 2 rows around the macro. The grid is designed to drive a current density of $0.5A/mm^2$ and achieve a 2% static IR drop. The total gate width of GPU headers is 1.93m and resistance is 0.6mΩ. The CPU power grid is designed to drive higher current density and minimize frequency impact due to voltage drop associated with the header.

Bobcat and its L2 cache together share a power island as shown in Fig. 15.4.5. Both core power islands are supplied by a single variable VDD rail. Core clocks can be varied independently and VDD voltage is selected to be the lowest voltage required to support the highest of the selected core clocks. If either of the cores is idle its caches can be flushed, its state saved to memory and power to the core island gated off. If both cores are idle and power gated, VDD can be further reduced to eliminate leakage in even the power gating structures. GPU and CNB share a variable VDDNB rail. GPU and CNB clocks are varied independently based on activity. VDDNB is selected based on the highest voltage request from either GPU or CNB. The GPU supports several power islands allowing for driver controlled power gating of the video acceleration engine, and independent dynamic power gating of the GFX core and the GMC. The display controller supports a static screen refresh stutter mode where the controller requests data periodically from memory. Between requests the memory is kept in self-refresh and the CNB enters a low power state. By stuttering the display requests to memory, the time spent in self-refresh for the memory can be maximized, and the time that clocks are running in the CNB can be minimized. The combination of power savings techniques yields an APU with an average MobileMark® (MM07) power consumption of less than 1.8W.

As shown in Fig. 15.4.6, the Bobcat core provides nearly 90% of mainstream CPU performance [3] with approximately 50% reduction in area and dynamic power compared to the same reference core (excluding L2) if scaled to the same technology. The front-end contains a 32k 2-way L1 instruction cache with a return stack, indirect dynamic and advanced conditional branch predictors with a high-capacity ITLB. The pipeline supports out-of-order execution, retiring up to two complex uOPs per cycle – a significant differentiating factor compared with simpler in-order CPUs. The load-store unit enables out-of-order accesses to an 8-way 32KB data cache with a high-capacity two-level DTLB, and 8-stream data pre-fetcher. The bus unit allows for 8 outstanding data accesses and 2-outstanding fetch accesses with a 512KB 16-way L2 cache with optimized data buffering to the CNB for high memory bandwidth.

The GPU delivers a peak of 80GFLOPS, which provides sufficient compute capability for gaming or compute applications enabled by DirectCompute and OpenCL™. As shown in Fig. 15.4.6, the power efficient APU outperforms current mainstream UMA chipsets. The low-latency, high bandwidth GMC memory interface increases performance per Watt by allowing a higher percentage of the power to be used for computation rather than in moving large amounts of data across high power chip-to-chip interconnects. The advanced, multi-level memory arbitration units resolve performance issues with previous Integrated Graphic Processors (IGP) by enabling the GPU to achieve memory efficiency similar to a discrete GPU without compromising CPU performance. GPU memory accesses are sent in efficient streams of reads and writes to avoid DDR penalties while CPU accesses are optimized for low read latency. These improvements yield a 17% increase in Zacate UMA memory efficiency compared to its two-chip predecessor.

Acknowledgments:
The authors would like to acknowledge the talented AMD Engineering community for their contribution to the SoC.

References:
[1] R. Jotwani, et al. "An x86-64 Core Implemented in 32nm SOI CMOS", *ISSCC Dig. Tech. Papers*, pp.106-107, Feb. 2010
[2] T. Fischer, et al. "Design Solutions for the Bulldozer 32nm SOI 2-core processor module in an 8-core CPU". Accepted to ISSCC 2011.
[3] M. Golden, et al., "A 2.6GHz Dual-Core 64bx86 Microprocessor with DDR2 Memory Support," *ISSCC Dig. Tech. Papers*, pp.325-332, Feb. 2006.

ISSCC 2011 / February 22, 2011 / 4:15 PM

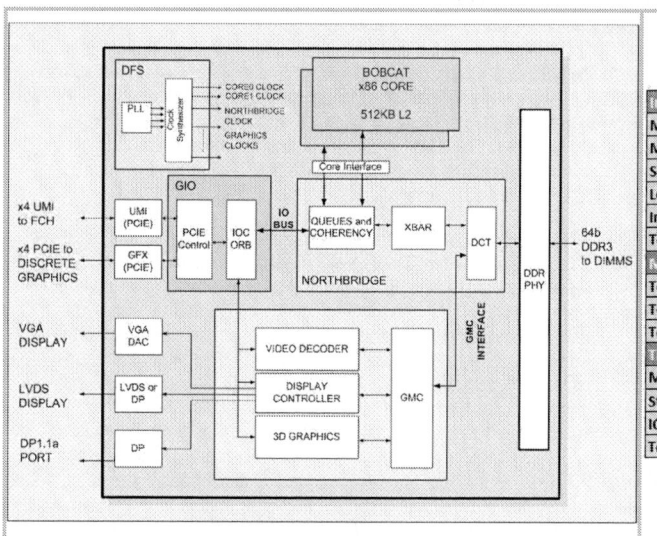

Figure 15.4.1: "Zacate" Architectural Block Diagram.

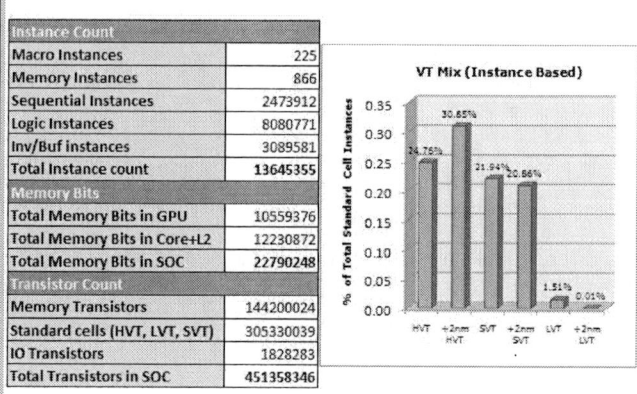

Figure 15.4.2: "Zacate" SoC Statistics.

Instance Count	
Macro Instances	225
Memory Instances	866
Sequential Instances	2473912
Logic Instances	8080771
Inv/Buf instances	3089581
Total Instance count	13645355
Memory Bits	
Total Memory Bits in GPU	10559376
Total Memory Bits in Core+L2	12230872
Total Memory Bits in SOC	22790248
Transistor Count	
Memory Transistors	144200024
Standard cells (HVT, LVT, SVT)	305330039
IO Transistors	1828283
Total Transistors in SOC	451358346

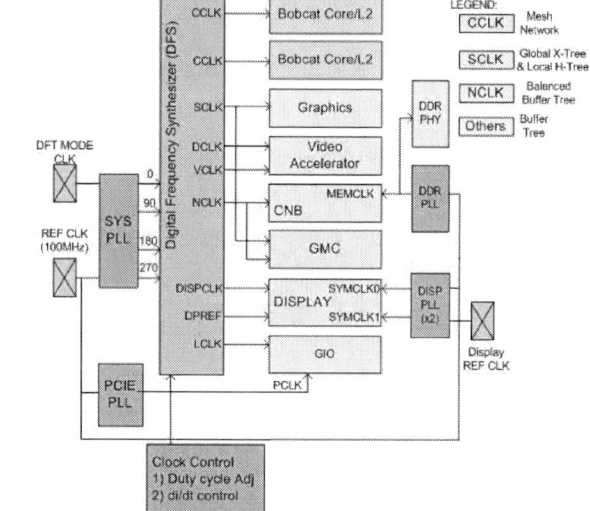

Figure 15.4.3: Clock Generation.

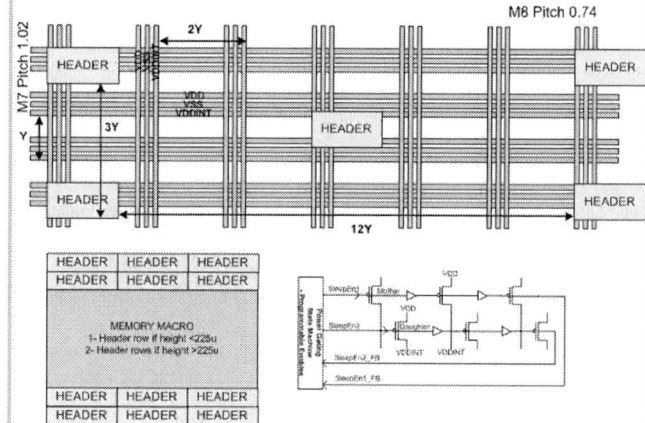

Figure 15.4.4: Power Gating Methodology.

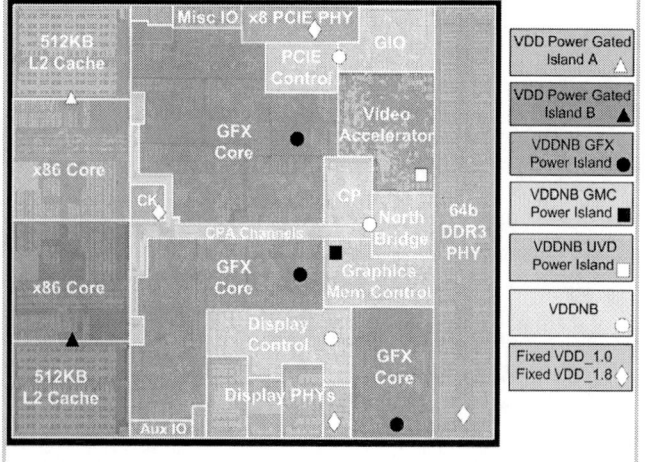

Figure 15.4.5: "Zacate" Die picture highlighting Power Islands and Logical Units.

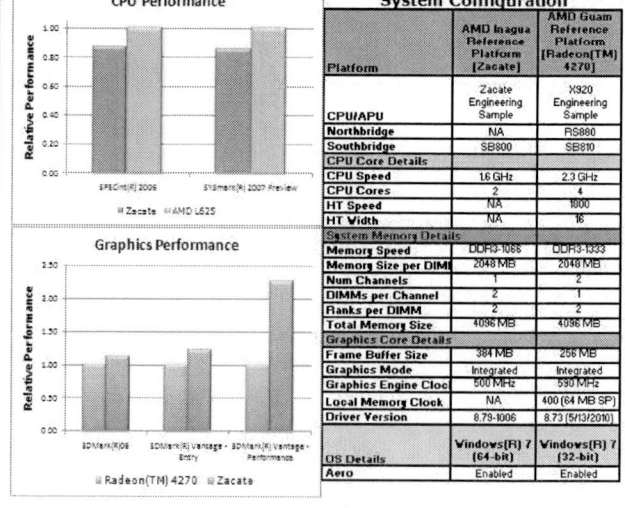

Figure 15.4.6: "Zacate" CPU & Graphics Relative Performance.

	AMD Inagua Reference Platform [Zacate]	AMD Guam Reference Platform [Radeon(TM) 4270]
Platform		
CPU/APU	Zacate Engineering Sample	X920 Engineering Sample
Northbridge	NA	RS880
Southbridge	SB800	SB810
CPU Core Details		
CPU Speed	1.6 GHz	2.3 GHz
CPU Cores	2	4
HT Speed	NA	1800
HT Width	NA	16
System Memory Details		
Memory Speed	DDR3-1066	DDR3-1333
Memory Size per DIMM	2048 MB	2048 MB
Num Channels	1	2
DIMMs per Channel	2	1
Ranks per DIMM	2	2
Total Memory Size	4096 MB	4096 MB
Graphics Core Details		
Frame Buffer Size	384 MB	256 MB
Graphics Mode	Integrated	Integrated
Graphics Engine Clock	500 MHz	590 MHz
Local Memory Clock	NA	400 (64 MB SP)
Driver Version	8.79-1006	8.73 (5/13/2010)
OS Details	Windows(R) 7 (64-bit)	Windows(R) 7 (32-bit)
Aero	Enabled	Enabled

ISSCC 2011 PAPER CONTINUATIONS

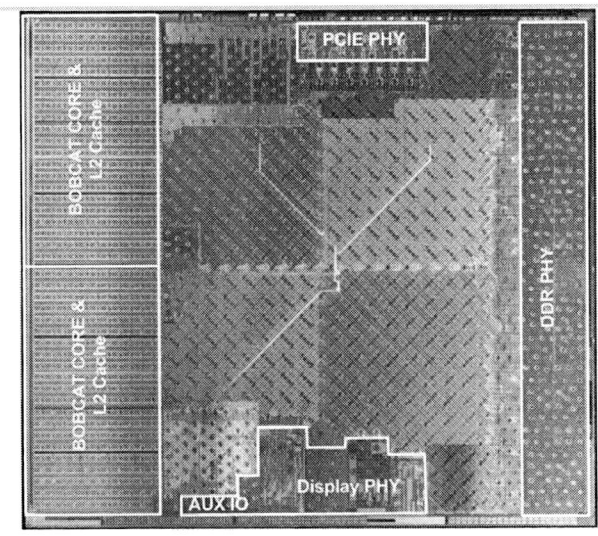

Figure 15.4.7: "Zacate" Die Photograph.

ISSCC 2011 / SESSION 15 / HIGH-PERFORMANCE SoCs & COMPONENTS / 15.5

15.5 A Programmable Adaptive Phase-Shifting PLL for Clock Data Compensation Under Resonant Supply Noise

Dong Jiao, Chris H. Kim

University of Minnesota, Minneapolis, MN

Power supply noise has become one of the main performance-limiting factors in sub-1V technologies. Resonant supply noise caused by the package/bonding inductance and on-die capacitance has been reported as the dominant supply noise component in high performance microprocessors [1,2]. Resonant noise frequency typically resides in the 40MHz to 300MHz frequency band but can be made as low as 7MHz with a dedicated metal-insulator-metal capacitor technology [3].

Recently, adaptive clocking schemes have been proposed to mitigate the impact of resonant noise on circuit performance. Here, the clock period is intentionally modulated by the resonant noise such that the increased clock period partially compensates for the increased datapath delay, which is also modulated by the resonant noise. Figure 15.5.1 (top) illustrates the concept of the adaptive clocking scheme. In a "constant-period clock" scenario, sampling failures can occur due to the increased datapath delay under resonant noise. In contrast, the adaptive clocking scheme stretches out the clock period to compensate for the increased datapath delay such that sampling failures are avoided. The clock period can either be modulated in the clock generation block, for example using a phase-locked loop (PLL) [4], or in the clock tree while the clock edge is propagating [5,6]. A brief analysis of the adaptive clocking scheme is shown in Fig. 15.5.1 (bottom left). The four waveforms represent the supply voltage with resonant noise, and the clock period modulation effect seen by the PLL, the clock distribution and the local registers, respectively. The minimum supply voltage occurs at point "A", which is also the point when the datapath delay is worst. Suppose the adaptive PLL produces the longest clock period at "B" [4] and the clock cycle is stretched to its maximum at "C" when the supply voltage has the sharpest negative slope. Since the clock cycle is modulated by both the PLL and the clock path, the net effect results in the maximum clock cycle occurring somewhere between "B" and "C", denoted as "D". Once we account for the clock path delay, local registers see the maximum clock cycle at time "E". To achieve optimal timing compensation between the clock cycle and the datapath delay, "E" needs to be aligned with the maximum datapath delay ("A") with the same phase and amplitude. Therefore, a certain amount of phase shift and proper adjustment of the clock period's sensitivity to supply noise are required for the best possible timing compensation, as shown as "B_{opt}". Previous designs, however, did not consider both effects and are not able to adapt to different design parameters. Motivated by these observations, we design an adaptive phase-shifting PLL design, in which both the phase shift and the supply noise sensitivity of the clock can be digitally programmed and adjusted. A comparison between our work and previous designs is given in Fig. 15.5.1 (bottom right).

Figure 15.5.2 shows the schematic of the phase-shifting PLL. The phase shifting and noise sensitivity adjustment are implemented with a supply-tracking modulator that consists of three binary-weighted capacitor banks and a bias-generation circuit. The capacitor arrays and transistors M1 and M2 form a high-pass filter to provide the desired programmable phase shift. The equivalent capacitance and the clock period's sensitivity to supply noise can be expressed as $C_{eq}=C_f||(C_u+C_d)$ and $S_V=C_u/C_d$, respectively, which are both digitally programmable. By choosing proper configurations of the three capacitor banks, the resonant noise can be coupled to the bias voltage of the voltage-controlled oscillator (VCO) to generate the desired adaptive clock signal.

A 1.2V, 65nm testchip is designed to verify the effectiveness of the phase-shifting PLL (Fig. 15.5.3). The adaptive clock signal is generated by the PLL and then propagates through the clock distribution networks. We implement eight clock trees with different buffer types (i.e., inverter, differential, and RC-filtered inverter [5]) and different interconnect lengths. A separate 40pF decoupling capacitor (decap) can be enabled to reduce the supply noise seen by the clock trees. The datapath under test consists of two D-flip-flops and both logic-dominated and

interconnect-dominated circuit paths. An XOR gate is used to compare the sampled results from the datapath with the reference data, and any sampling error will generate a pulse at the XOR output, which increments a 10-bit ripple counter. As a result, the transition in the ith bit of the counter output (i.e., BER<9:0>) indicates that $2i$ sampling errors have occurred. By measuring the average period of the counter output and the clock frequency, the bit-error rate (BER) can be conveniently calculated. The noise injection block has individual devices clocked by an on-chip VCO and a clock pattern synthesis circuit. The clock pattern can be selected from 1, 2, 8 or 32 pulses for every 32 clock cycles to emulate a first-droop or a sinusoidal noise waveform. The test chip also includes an array of linear feedback shift registers for injecting random supply noise and a local supply noise monitor [1].

Figure 15.5.4 (left) shows an example of the BER data measured at different clock frequencies. Without loss of generality, we define the maximum operating frequency as the point when the BER is 10^{-6}, and denote it as F_{max} in this paper. The noise waveforms measured from the supply noise monitor when injecting a first-droop noise and a sinusoidal supply noise are shown in Fig. 15.5.4 (right).

Figure 15.5.5 shows the measured F_{max} while sweeping the phase shift and supply noise sensitivity values. The chip is tested for a supply voltage of 1.2V and 1.0V using a sinusoidal noise waveform. As can be seen from the figure, F_{max} improves by more than 5% for both cases when an optimal configuration is chosen. We also see a large discrepancy in the optimal configurations between the two cases (i.e., 1.2V and 1.0V). This is because the timing compensation is affected by various design parameters such as clock frequency, clock path delay, noise frequency, and so on. The PLL is flexible and can adapt to different operating conditions and clock network designs by configuring the phase shift and supply noise sensitivity.

The PLL is tested under different supply noise frequencies. For this test, an inverter-based clock tree is chosen and the noise pattern is configured to emulate the first-droop noise. Measurement results in Fig. 15.5.6 (left) show a 4% F_{max} improvement for noise frequencies between 40 and 300MHz. As the noise frequency increases, the performance improvement becomes smaller. This is because the clock distribution delay makes it difficult, or even impossible, for the adaptive clock to compensate for the datapath delay variation if the noise period is too short. Different clock trees are also tested and the results are shown in Fig. 15.5.6 (right). Here, clock tree names with "_C" have a 40pF decap enabled in the clock tree supply and "short" or "long" refers to the interconnect length between the clock buffers. For a 74MHz sinusoidal noise, the F_{max} is consistently improved by 3.4% to 7.3% verifying the flexibility of the design. The chip micrograph is shown in Fig. 15.5.7.

Acknowledgements:
This work was supported by the Semiconductor Research Corporation under award 2008-HJ-1804.

References:
[1] J. Gu, H. Eom, and C.H. Kim, "On-chip Supply Noise Regulation Using a Low Power Digital Switched Decoupling Capacitor Circuit," *J. Solid-State Circuits*, vol. 44, no. 6, pp. 1765-1775, Jun. 2009.
[2] J. Xu, P. Hazucha, M. Huang, P. Aseron, et al., "On-Die Supply-Resonance Suppression Using Band-Limited Active Damping", *ISSCC Dig. Tech. Papers*, pp. 286-603, Feb. 2007.
[3] D. Wendel, R. Kalla, R. Cargoni, et al., "The Implementation of POWER7™: A Highly Parallel and Scalable Multi-Core High-End Server Processor," *ISSCC Dig. Tech. Papers*, pp. 102-103, Feb. 2010
[4] N. Kurd, P. Mosalikanti, M. Neidengard, J. Douglas and R. Kumar, "Next generation Intel® core™ micro-architecture (Nehalem) clocking," *J. Solid-State Circuits*, vol. 44, no. 4, pp. 1121-1129, Apr. 2009.
[5] K. L. Wong, T. Rahal-Arabi, M. Ma and G. Taylor, "Enhancing microprocessor immunity to power supply noise with clock-data compensation," *J. Solid-State Circuits*, vol. 41, no. 4, pp. 749-758, Apr. 2006.
[6] D. Jiao, J. Gu, and C. H. Kim, "Circuit Design and Modeling Techniques for Enhancing the Clock-Data Compensation Effect under Resonant Supply Noise," *J. Solid-State Circuits*, vol. 45, no. 10, pp. 2130-2141, Oct. 2010.

978-1-61284-303-2/11 $26.00 © 2011 IEEE

ISSCC 2011 / February 22, 2011 / 4:45 PM

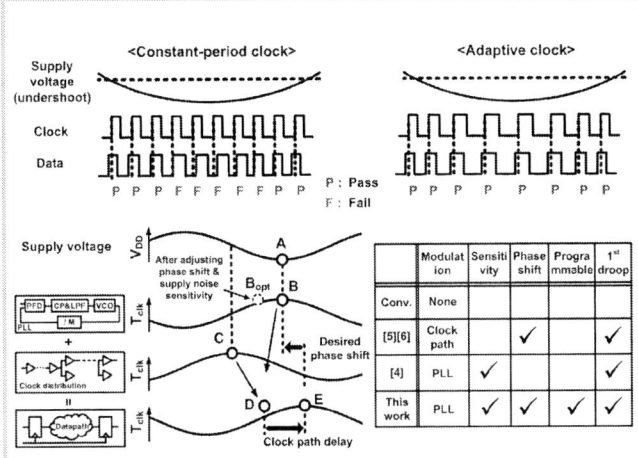

Figure 15.5.1: Illustration of adaptive clocking scheme for clock data timing compensation under resonant supply noise.

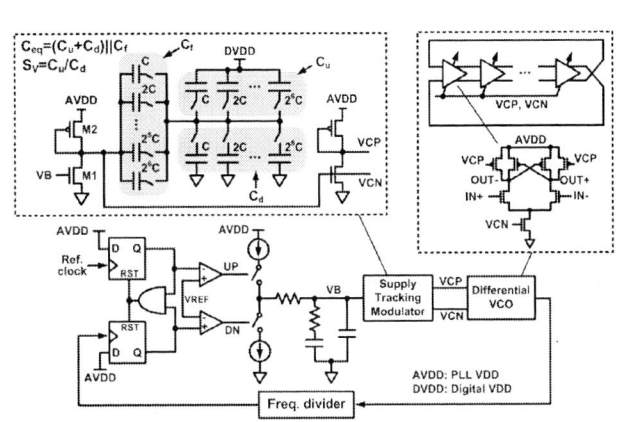

Figure 15.5.2: Schematic of the proposed phase-shifting PLL design.

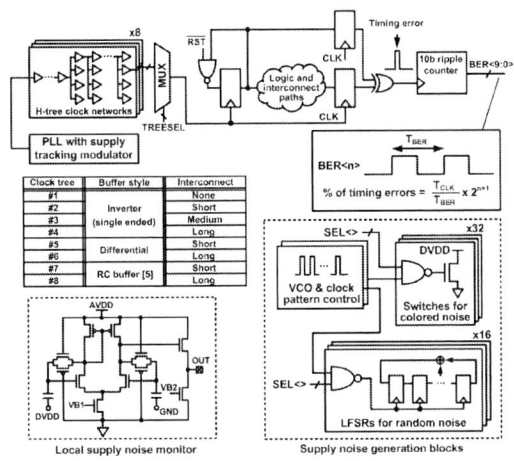

Figure 15.5.3: Block diagram of the 65nm test chip.

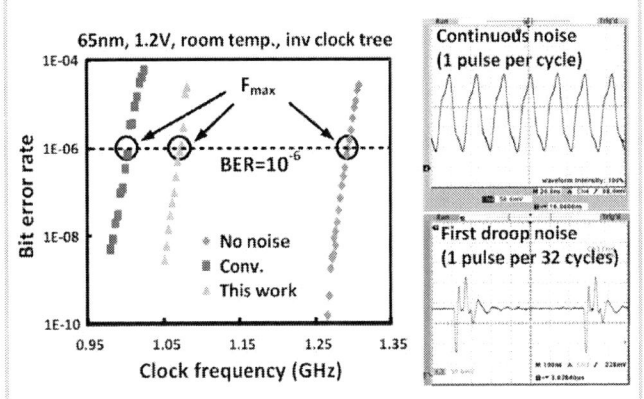

Figure 15.5.4: Measured BER versus clock frequency (left). Example supply noise waveforms generated by noise injection circuits (right).

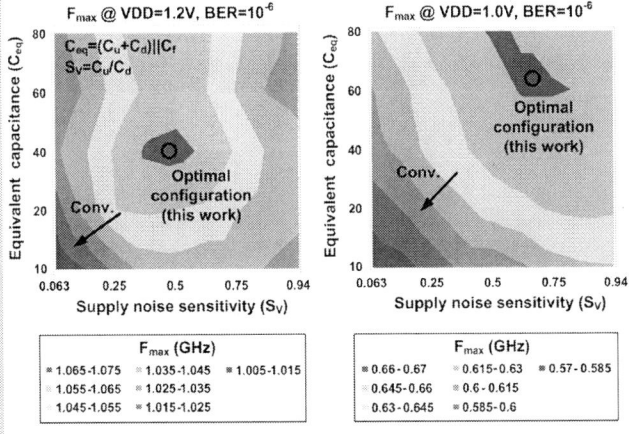

Figure 15.5.5: Measured results at 1.2V and 1.0V showing the F_{max} (@ BER=10^{-6}) dependency on phase shift and supply noise sensitivity.

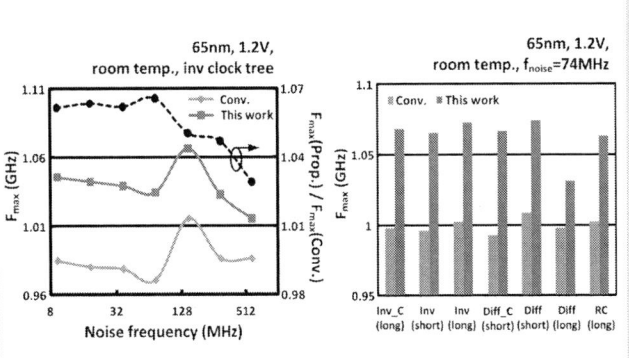

Figure 15.5.6: Measured F_{max} for different noise frequencies and clock trees.

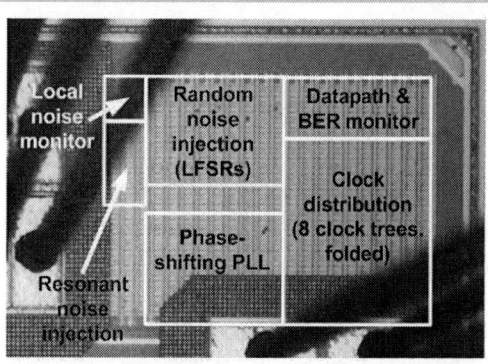

Technology	65nm LP CMOS	Supply voltage	1.2V
Total area	350 x 250 μm^2	PLL area	120 x 100 μm^2
Regulation frequency	40MHz-300MHz	F_{max} improvement	3.4%-7.3%

Figure 15.5.7: Chip micrograph and performance summary of the test chip.

ISSCC 2011 / SESSION 15 / HIGH-PERFORMANCE SoCs & COMPONENTS / 15.6

15.6 A Side-Channel and Fault-Attack Resistant AES Circuit Working on Duplicated Complemented Values

Marion Doulcier-Verdier[1,2], Jean-Max Dutertre[2], Jacques Fournier[1,2], Jean-Baptiste Rigaud[2], Bruno Robisson[1,2], Assia Tria[1,2]

[1]CEA-LETI-MINATEC, Gardanne, France,
[2]Ecole Nationale Supérieure des Mines de Saint-Etienne, Gardanne, France

Cryptographic circuits can be subjected to several kinds of side-channel and fault attacks in order to extract the secret key. Side-channel attacks can be carried by measuring either the power consumed or the EM waves emitted by the cryptographic module and trying to find a correlation between the given side-channel and the data manipulated [1]. Concerning fault attacks, in the case of differential fault attacks (DFA) [2], a cryptographic calculation is corrupted in such a way as to retrieve information about the secret key. Faults can be induced by different means such as lasers, voltage glitches, electromagnetic perturbations or clock skews. Several counter-measures, like in [3], have been separately proposed to tackle either kind of attack. In this paper, we describe the implementation of an AES chip where duplicated and complemented data paths provide resistance against both side-channel and fault attacks.

The AES [4] is an iterative algorithm, consisting of 'rounds', where the 128b input *plaintext* (and the intermediate values) is represented as a 4×4 matrix of bytes. Each *round* involves the use of a different *round-key* which is iteratively derived from the input *secret key*. Our architecture (depicted in Fig. 15.6.1), which we shall call a tamper-resistant (TR) AES, is a secure implementation of the 128b-key AES algorithm where each *round* is executed during one clock cycle. The initial aim of this design is to protect against DFA: the chosen security strategy consists in detecting induced errors and to spread them so that the resulted erroneous *cipher-text* can no longer be used for differential cryptanalysis [5]. Detection is achieved by duplicating the data paths of the RND_EXE and KEY_EXPANDER blocks and checking data consistency. In addition, this duplication is done in a dual manner, thus creating an original data path working on the input message and key and a dual data path working on complemented values of the input message and key. The logic gates used on the dual data path are also complemented, for example the XOR gates from the original data path are replaced by XNOR gates. This modification comes with no additional cost and brings another benefit: most leakage models used in side-channel attacks are based on the Hamming Weight or Hamming Distance models [6] and working on complemented data in parallel balances this Hamming Weight/Distance leakage in terms of power or EM emissions. The error detection is implemented in the SUB-BYTES module (the upper part of Fig. 15.6.2) whereby for the error matrix, the bits within each byte of the error matrix are stored in the reverse order (most significant bit swapped with least significant bit and so on). Further spreading of the errors is achieved by cross-changing wires between the original and the dual data paths at the level of the SHIFT_ROWS module. Figure 15.6.2 illustrates what happens when an error is generated on one of the data paths: the resulting cipher-text has too many faulty bytes to be exploited by differential cryptanalysis.

Several analyses have been performed at the register transfer level (RTL) and at the netlist level in order to detect and correct any potential security flaw (both based on power simulations and fault injection simulations) and hence have a design-time validation of the counter-measures implemented. This AES is designed in HCMOS9gp 0.13μm STM technology with a typical working condition of 1.20V, 50MHz maximum frequency at 27°C. The resulting die is 1336×1411.8μm^2 in size (Fig. 15.6.7). The TR-AES itself uses 27400 gates (including the communication interface), which represents an overhead of 67% when compared to a non-secure AES in the same technology (16500 gates).

We first test the TR-AES chip against side-channel attacks. In a first set of experiments, we observe the power consumed by the TR-AES during an encryption process and we try to carry correlation power analysis (CPA) [1]: for each byte of the intermediate data matrix of 4×4 bytes, we calculate correlation curves between the power traces and the Hamming Weight of the intermediate values (analyses are done both for intermediate values output of the first ADD-RND-KEY and of the SUB-BYTES of the first round) for each of the 256 possible val-

ues of the round-key byte. The correct round-key byte is thus obtained for the correlation curve, which has the highest peak compared to the other ones (Fig. 15.6.3). Despite using a large number of curves (~1,000,000 acquisitions), no significant data dependant leakage is found. Likewise, we try electromagnetic analysis (EMA) to find the same type of data dependant leakage but this time by measuring the EM radiations emitted by the chip. Compared to power measurements, EM measurements can be done in a more localized way and provide curves with better signal-to-noise ratios. We use a horizontal probe with a diameter of 150μm and having an amplifier offering a gain of 30dB. We first identify the regions of the chip where the most significant data-dependant variances are seen in the measured EM waves. For each of those regions, we carry out correlation EMA attacks [1]. Despite the small size and low power consumption of the chip, we can clearly identify the AES calculation (Fig. 15.6.4). Nevertheless, no significant data-dependant correlation is obtained (Fig. 15.6.5). Both results illustrate the efficiency of the dual data representation of the TR-AES against power and EM attacks, which have been proven to work on non-secure implementations of the AES.

We also test the TR-AES against fault attacks. The injection means used is a green laser source (532nm wavelength) with a spot size that is varied between 6 and 12μm, with the energy set to its minimum value (0.2 to 5 nJ) in order to avoid damaging the chip. In this case, we reproduce the effect illustrated in Fig. 15.6.2, i.e., fault detection and error spreading. In other words, the laser actually injects faults (even single-bit faults in some cases) during the AES calculation but the error is spread by our mechanism: faults induced into one byte of the intermediate values at the beginning of a given *round* are spread across at least seven other bytes at the end of the same *round*. The resulting erroneous cipher-text could not be used for differential fault cryptanalysis, which would have allowed us to find the secret key.

This paper describes an ASIC AES chip having a complemented duplicated implementation, which constitutes a counter-measure against both fault attacks (detection via duplicated data paths) and side-channel attacks (working on complemented data in parallel). Compared to other techniques that have been proposed and tested on ASIC AES chips, our scheme's originality is based on the fact that the errors induced during fault attacks are spread across the AES calculation rendering the erroneous cipher-text irrelevant for any differential cryptanalysis attack. At the same time, for no additional cost, by working on complemented values in the duplicated data path, we achieve resistance against side-channel attacks. The benefits of our chip compared to that of Tokunaga & Blaauw [3], which was implemented in the same technology, are summarized in the table of Figure 15.6.6.

Acknowledgements:
This work was funded by the SECRICOM project under the EC FP7-SEC-2007 grant 218123.

References:
[1] E. Brier, C. Clavier & F. Olivier, "Correlation Power Analysis with a leakage model", in the *Proceedings of CHES 2004*, vol. 3156 of Lecture Notes in Computer Science, pages 16-29, 2004.
[2] E. Biham & A. Shamir, "Differential fault analysis of secret key cryptosystems", in *Proceedings of CRYPTO'97*, Lecture Notes in Computer Science, pages 513–525, Springer, 1997.
[3] C. Tokunaga & D. Blaauw, "Secure AES engine with a local switched-capacitor current equalizer," *ISSCC Dig. Tech. Papers*, pp. 64-65,65a, 8-12 Feb. 2009.
[4] NIST, "Announcing the Advanced Encryption Standard (AES)", *Federal Information Processing Standards Publication, n° 197*, November 26, 2001.
[5] M. Joye, P. Manet & J-B. Rigaud, "Strengthening hardware AES implementations against fault attack", *IET Information Security*, vol. 1, issue 3, pages 106-110, September 2007.
[6] R. Mayer-Sommer, "Smartly analyzing the simplicity and the power of Simple Power Analysis on smartcards", in the *Proceedings of CHES 2000*, vol. 1965 of the Lecture Notes in Computer Science, pages 78-92, 2000.

978-1-61284-303-2/11 $26.00 © 2011 IEEE

ISSCC 2011 / February 22, 2011 / 5:00 PM

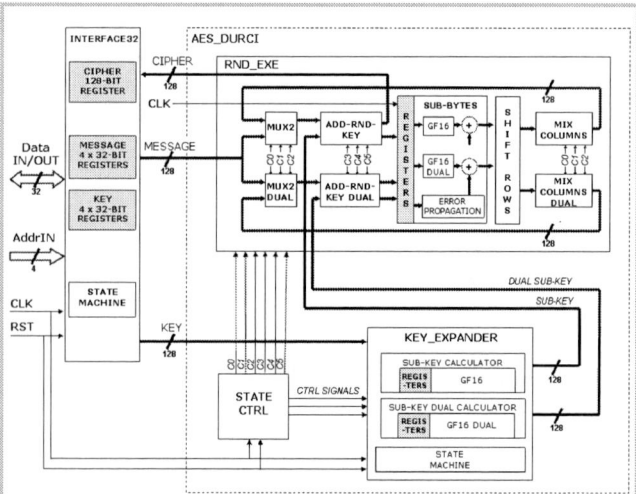

Figure 15.6.1: TR-AES block diagram illustrating the duplicated 'dual' data paths on the RND_EXE and on the KEY_EXPANDER.

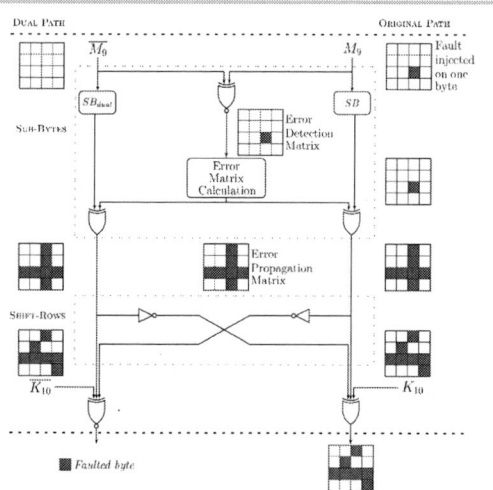

Figure 15.6.2: Faults generated on one 'data path' during the last AES round are propagated across the intermediate value matrix.

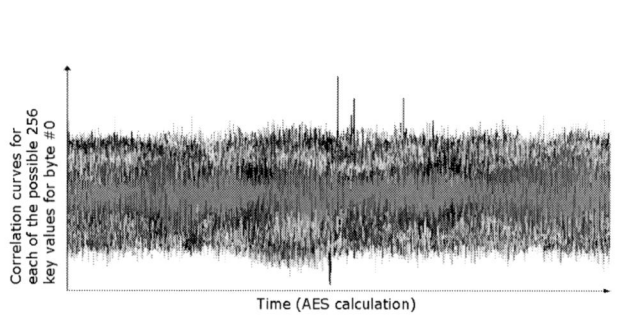

Figure 15.6.3: Correlation curves obtained from a CPA on a non-protected AES. The right key is revealed by the curve with the highest peak.

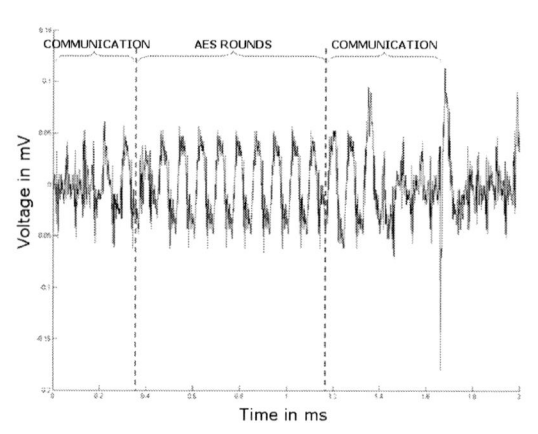

Figure 15.6.4: An EM measurement curve of the TR-AES where we can see the different stages of the AES calculation.

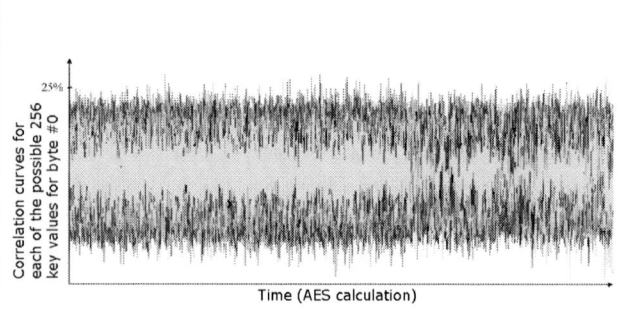

Figure 15.6.5: Correlation curves based on a CEMA on the TR-AES. No distinctive peak can be seen. The secret key byte was not revealed.

	AES with Switch-Capacitor Current Equalizer [3]	TR-AES (ours)
Area Overhead (Including communication)	25 % (7.2%)	104 % (67%)
Impact on max. frequency	Halved	None
Power Analysis resistant	Yes (10M[1])	Yes (1M[1])
Electro-Magnetic analysis resistant	Not tested	Yes (0.8M[1])
'Laser' Fault resistant	Not	Yes

[1] M = millions of curves

Figure 15.6.6: Comparison between the TR-AES (our chip) and that of Tokunaga & Blaauw [3].

978-1-61284-303-2/11 $26.00 © 2011 IEEE

ISSCC 2011 PAPER CONTINUATIONS

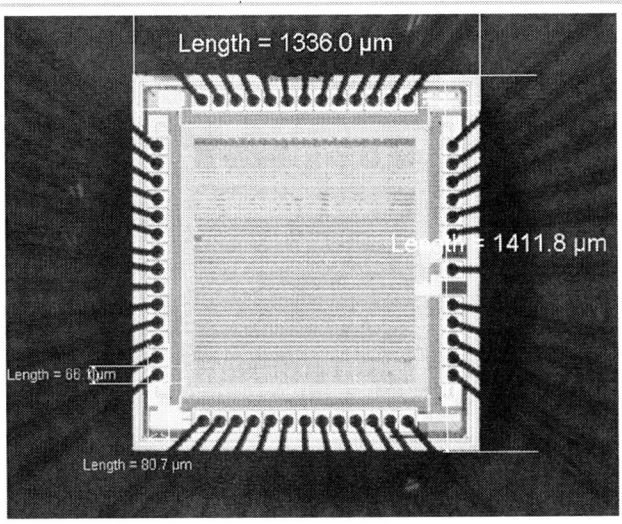

Figure 15.6.7: Picture of the TR-AES chip.

ISSCC 2011 / SESSION 16 / mm-WAVE DESIGN TECHNIQUES / OVERVIEW

Session 16 Overview / RF

mm-Wave Design Techniques

Session Chair: *Andreia Cathelin, ST Microelectronics, Crolles, France*

Session Co-Chair: *Brian Floyd, North Carolina State University, Raleigh, NC*

The increasing capabilities of advanced silicon technologies are enabling new market opportunities at 60GHz and above. For these markets to succeed, highly integrated, low-cost and low-power-consumption solutions are required. This session includes circuit demonstrations from 5 to 300GHz for high-speed communications, radar, and imaging applications, implemented in 130- to 45nm CMOS technology. The 60GHz band offers great opportunities for high-data-rate communications for consumer wireless high-definition video streaming and ultra-fast file transfer applications.

The first part of this session focuses on 60GHz circuits, with an emphasis on low-power and low-phase-noise local oscillator generation. The second part of the session focuses on circuits for radar, imaging, and data communications at frequencies up to 300GHz. These papers establish new benchmarks for high-frequency data communications, imaging, and radar systems. They demonstrate the high performance and low-power capabilities of advanced CMOS technologies for 3-to-300GHz operation.

One critical challenge for 60GHz radios is the power-efficient generation of low-phase-noise local oscillator signals. These LO signals can be generated either directly at the fundamental frequency or at a sub-harmonic and multiplied up to the mm-Wave band. Paper 16.1 (NXP Semiconductors) presents a sub-harmonic approach, where a 21.7-to-27.8GHz frequency synthesizer is realized in 45nm CMOS. Power consumption is 40mW and phase noise at 1MHz offset is -93.4 dBc/Hz, when used with a times-two frequency multiplier.

Papers 16.2 and 16.3 present building blocks for a fundamental frequency 60GHz synthesizer. In paper 16.2 (University of Pavia), a quadrature VCO is demonstrated in 65nm CMOS, leveraging weakly coupled transformers to reduce phase noise and power consumption. The VCO operates from 56 to 60.4GHz and achieves -95dBc/Hz phase noise at 1MHz offset.

In Paper 16.3 (University of Pavia), a low-power frequency divider is presented. This inductorless divider uses standard differential amplifiers clocked as dynamic latches, operates from 20 to 70GHz, and consumes 6.5 mW.

Finally, Paper 16.4 (University of Michigan) presents an approach to eliminate the reference oscillator for a 60GHz PLL, suitable for short-range wireless sensor networks. The 0.13µm CMOS circuit measures the standing wave on an on-chip antenna to frequency lock the oscillator to the resonant frequency of the antenna with a standard deviation of 195MHz around 59.3GHz.

Another important and growing research area is the development of sources and detectors at 70 to 300GHz for applications including data communications, radar, imaging, and spectroscopy. Paper 16.5 (Cornell University) presents a 220-to-275GHz frequency doubler implemented in 65nm CMOS that is capable of generating -6.6dBm output power while consuming 40mW of power.

Paper 16.6 (Caltech) presents a distributed approach to on-chip generation, multiplication, and radiation of a 300GHz signal. A 2x2 array of these distributed active radiators is implemented in 45nm CMOS, generates 80µW of output power at 291GHz, and achieves an EIRP of -1dBm.

Paper 16.7 (KU Leuven) presents a 120GHz transmitter capable of >10Gb/s data rates. The circuit integrates an LO buffer, a quadrature hybrid, a 4:1 multiplexer functioning as a phase modulator, and a six-stage driver amplifier, all in 65nm CMOS.

978-1-61284-303-2/11 $26.00 © 2011 IEEE

ISSCC 2011 / February 22, 2011 / 1:30 PM

Paper 16.8 (Toshiba) presents an 82.1-to-83.8GHz FMCW synthesizer designed for 77GHz automotive radar and implemented in 65nm CMOS. A combination of analog and digital PLL techniques is applied to achieve a 1.5GHz modulation range, 10ms modulation period, and 180kHz frequency error.

Paper 16.9 (University of Southern California) presents an impulse-based ultra-wideband wireless sensor for the monitoring of people's movement and vital signs. The 0.13μm CMOS radar circuit operates over 2 to 5GHz, consumes 695mW, and achieves a range and resolution of <15m and <1cm, respectively.

Finally, Paper 16.10 (UCLA) presents a 183GHz receiver for mm-Wave imaging. A regenerative architecture is employed to reduce power consumption to 13.5mW per pixel while providing time-encoded outputs of the detected radiation.

Presenters:

1:30 PM

16.1 A 21.7-to-27.8GHz 2.6-Degrees-rms 40mW Frequency Synthesizer in 45nm CMOS for mm-Wave Communication Applications

J. F. Osorio, NXP Semiconductors, Eindhoven, The Netherlands

2:00 PM

16.2 A mm-Wave Quadrature VCO Based on Magnetically Coupled Resonators

U. Decanis, University of Pavia, Pavia, Italy

2:15 PM

16.3 A 6.5mW Inductorless CMOS Frequency Divider by 4 Operating up to 70GHz

A. Ghilioni, University of Pavia, Pavia, Italy

2:30 PM

16.4 A 60GHz Antenna-Referenced Frequency-Locked Loop in 0.13μm CMOS for Wireless Sensor Networks

K-K. Huang, University of Michigan, Ann Arbor, MI

2:45 PM

16.5 A 220-to-275GHz Traveling-Wave Frequency Doubler with -6.6dBm Power at 244GHz in 65nm CMOS

O. Momeni, Cornell University, Ithaca, NY

3:15 PM

16.6 Distributed Active Radiation for THz Signal Generation

K. Sengupta, California Institute of Technology, Pasadena, CA

3:30 PM

16.7 A 120GHz 10Gb/s Phase-Modulating Transmitter in 65nm LP CMOS

N. Deferm, KU Leuven, Leuven, Belgium

3:45 PM

16.8 A 1.5GHz-Modulation-Range 10ms Modulation-Period 180kHz$_{rms}$-Frequency-Error 26MHz-Reference Mixed-Mode FMCW Synthesizer for mm-Wave Radar Application

H. Sakurai, Toshiba, Kawasaki, Japan

4:15 PM

16.9 A Short-Range UWB Impulse-Radio CMOS Sensor for Human Feature Detection

T-S. Chu, University of Southern California, Los Angeles, CA; National Tsing Hua University, Hsinchu, Taiwan

4:45 PM

16.10 183GHz 13.5mW/Pixel CMOS Regenerative Receiver for mm-Wave Imaging Applications

A. Tang, University of California, Los Angeles, CA

16

ISSCC 2011 / SESSION 16 / mm-WAVE DESIGN TECHNIQUES / 16.1

16.1 A 21.7-to-27.8GHz 2.6-Degrees-rms 40mW Frequency Synthesizer in 45nm CMOS for mm-Wave Communication Applications

Juan F. Osorio[1], Cicero S. Vaucher[1], Bill Huff[2], Edwin v.d. Heijden[1], Anton de Graauw[1]

[1]NXP Semiconductors, Eindhoven, The Netherlands,
[2]NXP Semiconductors, San Diego, CA

This work presents a 21.7-to-27.8GHz frequency synthesizer in a 45nm CMOS process that combines a tuning range of 24.8%, a residual phase modulation of 2.57°rms (with integrated phase noise from 100kHz to 100MHz), and a total power dissipation of 40mW. Combined with a frequency multiplier-by-two circuit and a divider-by-two circuit in a sliding-IF configuration, the PLL provides the four source frequencies required by the IEEE 802.15.3c 60GHz communication standard. In addition, the attained phase noise makes it suitable for microwave links with higher-order modulation schemes used as the back-bone for 3G/LTE base-station networks.

The PLL diagram is shown in Fig. 16.1.1. It consists of a PFD supplied by a 1.1V source, a programmable charge pump (CP) supplied with 1.8V, an external loop filter, a negative-g_m VCO, a fixed prescaler, a chain of 2/3-dividers and two buffers.

The VCO diagram is shown in Fig. 16.1.2; it is a cross-coupled pair oscillator with AC coupling in the feedback network. The variable capacitance is composed of a thin-oxide accumulation-mode varactor and three binary-weighted programmable capacitors. The thin-oxide varactor has a better Q and gives a larger tuning range than a thick-oxide varactor. To cope with the low breakdown the drain of the cross-coupled VCO is biased at 0.9V and a control signal from 0 to 1.8V is used. This provides large tuning range and optimal phase-noise performance simultaneously. The VCO draws a maximum current of 13.5mA from a 0.9V supply. Two buffers are used to drive a 100Ω differential output and the divider. They consume a total of 7mA from the 1.1V supply.

The inset in Fig. 16.1.2 demonstrates the locking range of the PLL for every band of the VCO. It also shows the four frequencies needed, with the sliding-IF scheme mentioned above, to fulfill the standard.

The divider chain is composed of one divider-by-two followed by a programmable multi-modulus divider based on a chain of 2/3-divider cells, see Fig. 16.1.1. The last 2/3-divider can be disabled to reduce the lowest division ratio [1]. The division range spans from 256 to 1022 in steps of two. The divider-by-two and the first two 2/3-dividers are implemented using CML, in contrast with the Injection-Locked Frequency Dividers (ILFD) used in some of the previous comparable CMOS PLL's [2,3]. The CML approach increases the robustness of the divider chain, as it is inherently broadband, and does not require calibration or tracking schemes to increase the operation range. The last 7 divider cells use Swing-Restored Pass-Transistor Logic (SRPT) and consume 400µW operating at 3.48GHz from the 1.1V supply.

In Fig. 16.1.3 the architecture of the first 2/3-divider cell is shown, together with one of the latches. The divider input signal is AC-coupled to obtain the 0.75mV optimum DC level that results in maximum speed for an implementation using low-V_t transistors. The first two 2/3-divider cells were measured stand-alone, with a 50-Ω input buffer, showing a maximum operating frequency of 18GHz and a self-oscillation frequency of 14GHz, see Fig. 16.1.3. Using a high frequency divide-by-2/3 circuit after a single divider-by-2 stage, in contrast with [4], where several divide-by-two stages are used, allows to have a higher reference frequency and as a consequence a lower divider ratio in the feedback path. Moreover, the four frequencies of the standard are synthesized without a delta-sigma modulator [2]. These two aspects translate into a better overall phase noise performance.

The divider output is taken from the modulus signal (modo) of the last SRPT 2/3-divider cell, as shown in Fig. 16.1.1. This signal has low phase noise, as it is re-timed by the high frequency clock signal (clko) coming from the second 2/3-divider cell. The combined input-referred phase noise of the frequency divider

and the PFD-CP is lower than -142dBc/Hz at 100kHz, as can be concluded from the close-in phase noise of Fig. 16.1.4 (which includes also the phase noise of the reference). The current consumption of the divider chain is 16mA from a 1.1V supply.

Figure 16.1.4 shows the measured phase noise at 23.32GHz. The phase noise of the VCO is -121dBc/Hz at 10MHz offset. The close-in phase noise is -88dBc/Hz at 100kHz offset. The integrated noise from 100kHz to 100MHz is 2.57°rms. The PLL bandwidth is set to minimize the integrated phase noise. The programmable CP compensates for the VCO gain variation in different channels. The inset shows a consistent phase noise performance in the four different channels; the integrated phase noise for the four channels ranges from 2.57°rms to 2.92°rms.

The block diagram and the measured results of the LO multiplier and divider chain that complements the presented frequency synthesizer are shown in Fig. 16.1.5. The circuit is driven by a differential pair with an LC load. The multiplier requires a large swing at the input, so two single-ended rail-to-rail buffers are used at its input. The multiplier-by-2 is composed of two NMOS transistors with their sources and drains connected together and of two resistors for biasing. The AC voltage at the resistor terminals contains the second harmonic of the differential input, which is amplified by a chain of buffers. The divider is a CML divider followed by two buffers for the I and the Q signals. The total power consumption is 105mW and the core area 0.049mm².

Figure 16.1.6 shows how this work compares with other frequency synthesizers for the same application. This PLL shows the largest tuning range of all the designs and the best integrated phase noise of all the CMOS implementations. The integrated phase noise (referred to 61.5GHz) of our PLL is almost three times better than what has been reported so far for a CMOS PLL intended to cover the 60GHz band [3]. In addition, similar performance as the SiGe implementation was achieved at a much lower power dissipation level. The third column shows the results for the PLL combined with the multiplier-by-two and the divider-by-two; in that case the total power consumption is 145mW. The different circuits are compared using the following figure-of-merit FOM = $f_c^2/(\phi_{rms} \times P_{diss})$. The FOM shows that, among the listed designs, our circuit achieves state-of-the-art performance at 24.8GHz and 61.5GHz. In addition, this PLL features a very large tuning range of 24.8%, the largest among the works listed in the comparison table. The higher FOM is achieved by combining different techniques. The use of high-frequency 2/3-divider cells enables an integer-N architecture with smaller N factor which results into a lower close-in phase noise. Moreover, thanks to an inherent re-timing of the divider chain the close-in phase noise is further reduced. Finally, the use of thin-oxide varactors under optimal biasing enables a low-phase-noise VCO without compromising the power consumption and achieving a large tuning range.

Figure 16.1.7 shows the chip micrograph. The active area is 0.14mm² including all the PLL except the filter. The chip was tested on a probe station with the external filter mounted on the DC biasing probe.

Acknowledgements:
The author's acknowledge Gerard van der Weide, Raf Roovers, Domine Leenaerts, Frank Leong, Robert Rutten and Remco van de Beek for their contributions to this work.

References:
[1] C. S. Vaucher, "Architectures for RF frequency synthesizers", *Kluwer Academics Publishers*, 2002.
[2] S. Pellerano, R. Mukhopadhyay, A. Ravi, J. Laskar, and Y. Palaskas, "A 39.1-to-41.6GHz ΣΔ fractional-n frequency synthesizer in 90nm CMOS", *ISSCC Dig. Tech. Papers*, pp. 484 –630, Feb. 2008.
[3] K. Scheir, G. Vandersteen, Y. Rolain, and P. Wambacq. "A 57-to-66GHz quadrature PLL in 45nm digital CMOS", *ISSCC Dig. Tech. Papers.*, pp. 494 –495, Feb. 2009.
[4] O. Richard, A. Siligaris, F. Badets, C. Dehos, C. Dufis, P. Busson, P. Vincent, D. Belot, and P. Urard, "A 17.5-to-20.94GHz and 35-to-41.88GHz PLL in 65nm CMOS for wireless HD applications", *ISSCC Dig. Tech. Papers*, pp. 252 –253, Feb. 2010.
[5] B.A. Floyd, "A 16-18.8-GHz sub-integer-N frequency synthesizer for 60-GHz transceivers", *IEEE JSSC*, vol. 43, no. 5, pp. 1076 –1086, May 2008.

978-1-61284-303-2/11 $26.00 © 2011 IEEE

ISSCC 2011 / February 22, 2011 / 1:30 PM

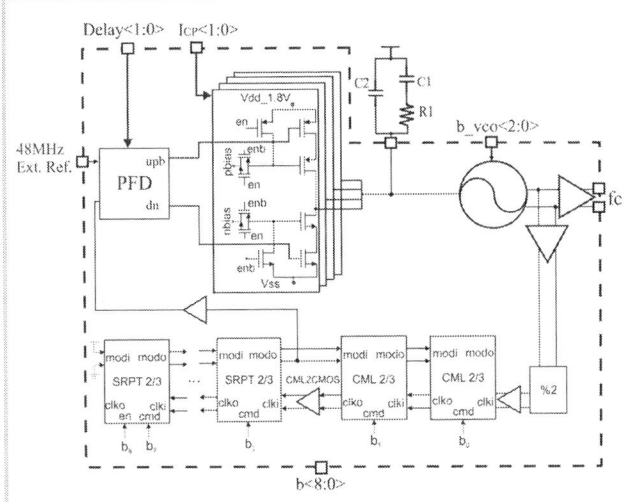

Figure 16.1.1: PLL Architecture.

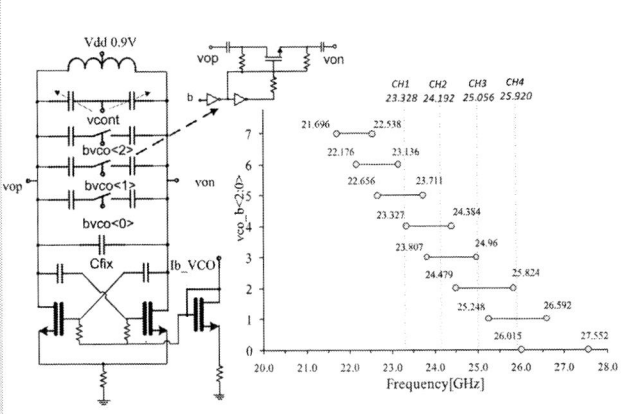

Figure 16.1.2: VCO and measured maximum and minimum locking frequencies for every VCO band.

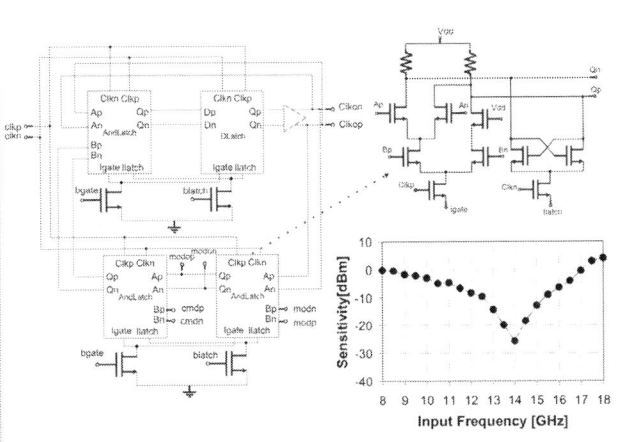

Figure 16.1.3: CML 2/3-Divider, showing one of the And-Latches and the measured sensitivity curve.

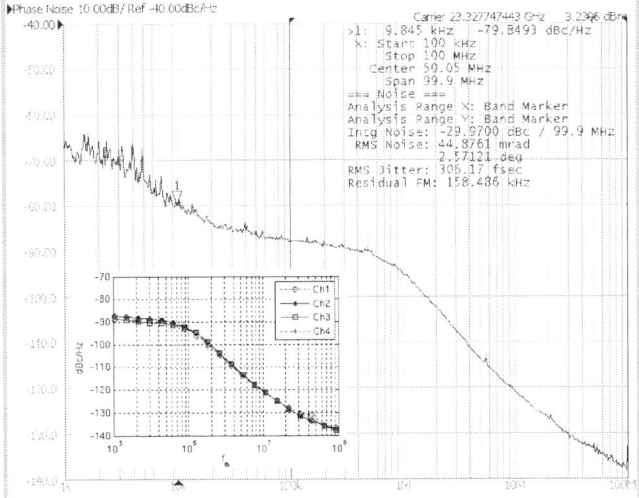

Figure 16.1.4: Measured phase noise at 23.328GHz. The insert shows a consistent phase noise performance for the four channels in the standard.

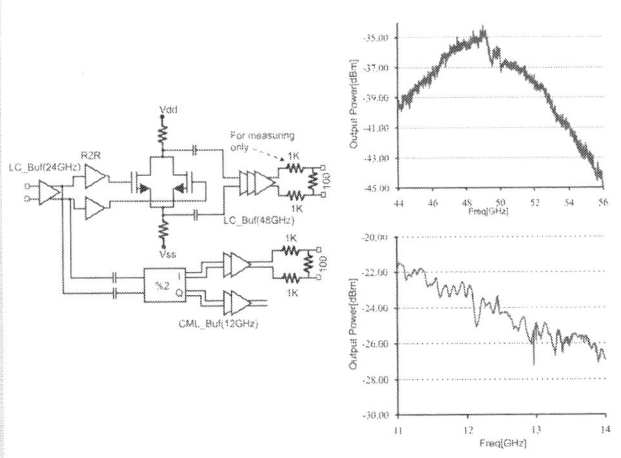

Figure 16.1.5: LO multiplier and divider chain that complements the frequency synthesizer for the 802.15.3c communication standard.

	This work	This work +LO Path	[2]	[3]	[4]	[5]
Tech. [nm]	45 CMOS	45 CMOS	90 CMOS	45 CMOS	65 CMOS	130 Si-Ge BiCMOS
Supply[V]	0.9, 1.1, 1.8	0.9, 1.1, 1.8	1.2	1.1	1.2, 1.8	1.2, 2.7
Frequency range[GHz]	21.69 to 27.85 (24.9%)	43.38 to 55.7 10.85 to 13.93 (24.9%)	39.1 to 41.6 (6.2%)	57 to 66 (14.63%)	17.5 to 20.94 35 to 41.88 (17.9%)	15.92-18.81 (16.5%)
Integrated Phase noise 100KHz-100MHz [deg. rms]	2.57 (24.77GHz)	6.42 (61.5GHz)	18.20* (40.5GHz)	18.34* (61.5GHz)	9.15* (20.88GHz) 26.95 (61.5GHz)	1.4 ** (17.37GHz)
In Band Phase Noise [dBc/Hz]	-88	-81	-60	-75	-55	-90
Extrapolated VCO Phase Noise at 1MHz offset [dBc/Hz]	-101 (24.77GHz)	-93 (61.5GHz)	-90 (40.5GHz)	-75 (61.5GHz)	-100 (20.88GHz) -86.6 (61.5GHz)	-106
Power[mW]	40*	145*	64	78*	80*	144*
FOM [GHz²/(dgrms*mW)]	5.96	4.06	1.40	2.64	1.75	1.49
Reference Freq [MHz]	48	48	50	100	36	285.714
Type	Integer	Integer	Integer	Integer	Integer	Sub-Integer
Division Rate	256 to 1022	256 to 1022	512 to 2032	512 to 8184	620 to 1240	56.5-64
Ref. Spur[dBc]	-50	-50	-54	-42	<-50	-60
Area[mm2]	0.48x0.29	0.19	1.77x0.87***	0.99x0.830***	1.1x1	0.65x1
Filter Type	External	External	External	Internal	External	Internal

* Estimated from the phase noise plot
** From 100kHz to 1GHz
*** Complete test structure including IO-pads
' Including the IO buffers
'' This circuit includes a VCO regulator, a VCO automatic amplitude control and a circuit to select the VCO band
† That includes the PLL and the LO multiplier and divider.

Figure 16.1.6: Summary and comparison.

16

978-1-61284-303-2/11 $26.00 © 2011 IEEE

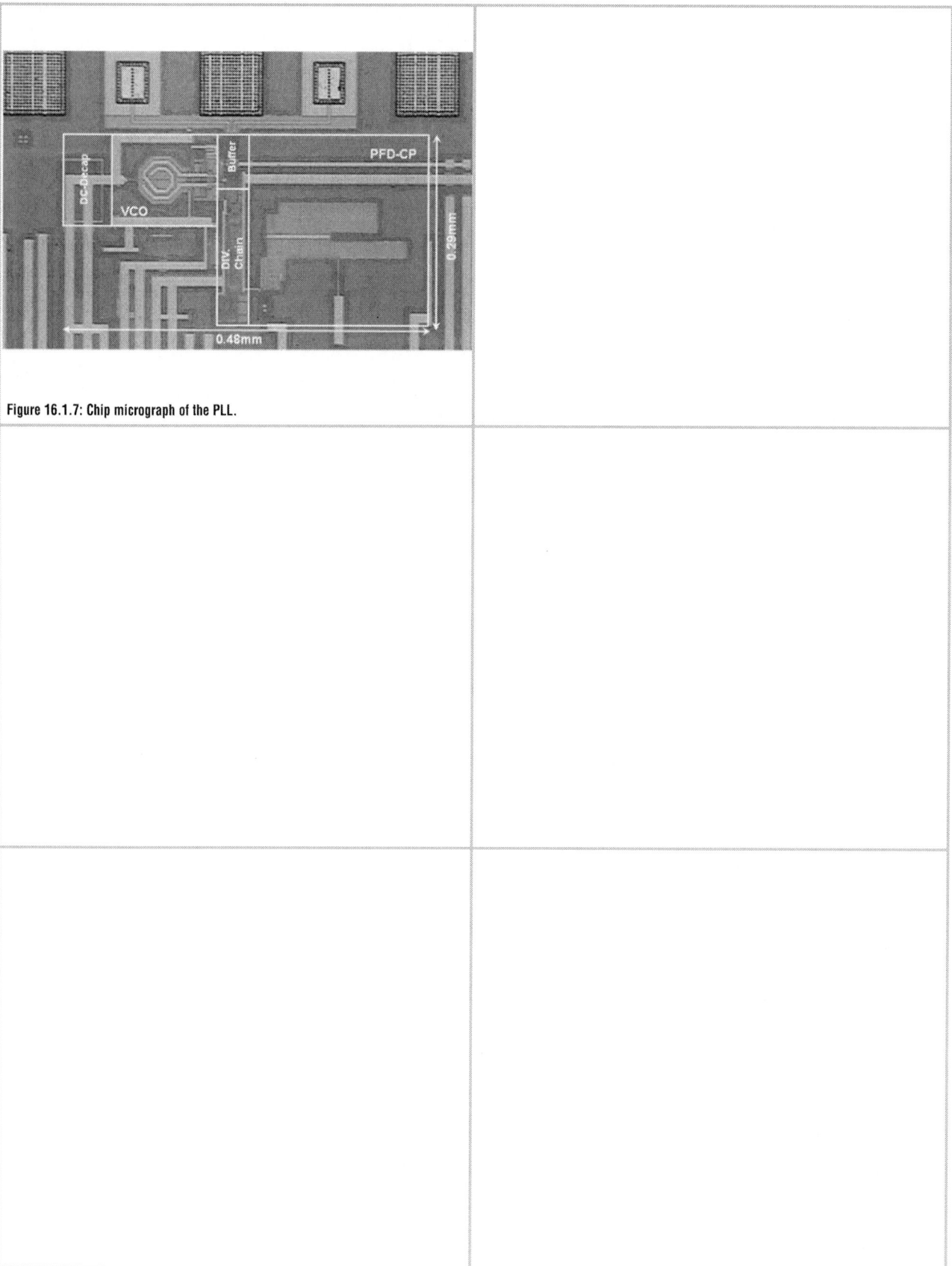

Figure 16.1.7: Chip micrograph of the PLL.

ISSCC 2011 / SESSION 16 / mm-WAVE DESIGN TECHNIQUES / 16.2

16.2 A mm-Wave Quadrature VCO Based on Magnetically Coupled Resonators

Ugo Decanis[1], Andrea Ghilioni[1], Enrico Monaco[2,3], Andrea Mazzanti[1], Francesco Svelto[1]

[1]University of Pavia, Pavia, Italy,
[2]University of Modena e Reggio Emilia, Modena, Italy,
[3]Istituto Universitario di Studi Superiori di Pavia, Pavia, Italy

Wireless signal processing at mm-Waves would benefit from the availability of a quadrature signal reference, enabling direct-conversion transceiver architectures and providing phase rotators drivers in phased arrays systems [1]. They are furthermore attractive for clock recovery in ICs for wireline applications. Search of a compact quadrature generator at around 60GHz with low phase noise at moderate power is the topic of this work. Discarding a double-frequency VCO followed by dividers-by-two given the high frequency range of operation, the most suitable topology borrowed by RF solutions is represented by cross-coupled LC voltage-controlled oscillators [2]. However, the oscillation frequency dependence on the biasing current makes it susceptible to phase noise, close-in in particular [3]. At mm-Waves, this is exacerbated by core devices of small dimensions to such an extent that $1/f^3$ noise remains dominant up to more than ~10MHz, making it unsuitable for stringent applications. On the contrary a ring of two VCOs magnetically coupled to each other, as shown in Fig. 16.2.1, has an oscillation frequency dependence on inter-stage passive components only, low $1/f^3$ noise together with good quadrature accuracy. The quadrature oscillator has been realized in a 65nm CMOS technology and prototypes show the following performances: 56-to-60.3GHz tunable oscillation frequency, phase noise better than -95dBc/Hz at 1MHz offset in the tuning range, 1.5° maximum phase error while consuming 22mA from a 1V supply.

By inspection of Fig. 16.2.1, the ring oscillates at $f_o = 1/2\pi\sqrt{LC(1-k^2)}$ where k is the magnetic coupling coefficient of the transformer, when the transimpedance ($Z=V_o/I_{in}$) of each pair of coupled resonators is 90°. Because the oscillation frequency is independent of active device parameters, no direct modulation of $1/f$ noise originates $1/f^3$ phase noise in the output spectrum. To gain insight, Z is plotted versus frequency for a 60GHz resonance frequency of the coupled resonators and for different values of the kQ product, with Q the quality factor of each LC. Coupled resonators realize a filter function of order higher than a classical LC resonator, typically used as a VCO load, with the effect of a steeper phase variation at resonance. This is desirable to reduce phase noise and opens up to a quadrature oscillator with phase-noise figure of merit competitive with single LC VCOs. The higher kQ the steeper the phase variation versus frequency at resonance [4]. But at the same time, the transimpedance magnitude decreases, suggesting that an optimum kQ value exists to optimize the phase noise performance of the oscillator.

Capacitive coupling of two parallel LC resonators also lends itself to realize the VCO ring, as demonstrated at RF frequency in [4]. But this approach requires four differential inductors, which are area consuming and, foremost at mm-Waves, significantly complicate signals routing. Magnetic coupling halves the area of the resonators by implementing inductive components in a transformer fashion. Figure 16.2.2 suggests that the transformer's spirals are weakly coupled for optimum performance. The low-k transformer is shown in Fig. 16.2.3, together with a simple lumped model. The spiral in closed loop L_{sh} serves the purpose of reducing the equivalent coupling coefficient k_{eq}, shielding the two transformer windings L_{int} and L_{ext}. Intuitively, the current in L_{ext} induces a current in L_{sh} with opposite sign. The magnetic fields induced on L_{int} by L_{ext} and L_{sh} tend to cancel each other. By inspection, the equivalent coupling k_{eq} between L_{int} and L_{ext} is derived as $k_{eq}=k_1-k_2k_3$. The shield is made of two turns shorted together, embedding the secondary winding, in order to maximize k_3. L_{sh} has also the effect of reducing inductance and Q of L_{ext} in particular. From electromagnetic simulations, L_{int} is 100pH with Q=19 and L_{ext} is 110pH with Q=15. The self-resonance frequencies are in excess of 120 GHz. The transformer k_{eq} is 0.15. The reduced Q of L_{ext} has only a minor effect, since the resonators Q-factor is primarily determined by the load capacitors C, made of three digitally controlled varactors plus an A-MOS for analog variation. LC resonators in each pair are designed to have nominally the same resonance. In order to hold true in the

entire tuning range, the fraction of variable capacitance over parasitic should be the same. Because parasitics of different kinds are involved a slight mismatch is expected. However, systematic differences up to 10% in the LC product have a negligible impact in the transimpedance response and oscillator performances.

Digitally-controlled variable resistors (R_{bias}) replace active current sources for oscillator biasing, due to $1/f$ noise reasons. The transconductors are pseudo-differential. R_{bias} are connected between the supply and the inductors' center-tap, biasing the output nodes around $V_{dd}/2$ in order to explore the tuning characteristic of the AMOS varactors in their region of steepest variation. The risk of common-mode oscillations is minimized by placing a high value resistor (R_{cm}) between transformer center taps, as shown in Fig. 16.2.1. R_{cm} drastically reduces the coupled resonators quality factor under common mode excitation while the loop gain is not affected when a differential signal propagates.

The quadrature VCO has been closed in a phase-locked loop for characterization. A quadrature direct-conversion mixer has also been integrated for testing purposes. Figure 16.2.7 shows the chip micrograph. The VCO draws 22mA from 1V and occupies an active area of 0.075mm². A relatively narrow loop bandwidth of 10kHz has been chosen so as to characterize the phase noise close to the carrier. The quadrature VCO is tunable from 56GHz to 60.35GHz, i.e. 7.4% around central frequency. Figure 16.2.4 shows the phase noise, measured by downconverting a 58.6GHz frequency tone from a signal source to 200MHz. Because the injected tone is more spectrally pure than the quadrature reference generated on chip, the phase noise of the down-converted signal reflects the phase noise of the quadrature oscillator. The $1/f^3$ corner is ~900kHz and phase noise at 1MHz and 10MHz offset are -95dBc/Hz and −117dBc/Hz respectively. Phase noise values at 1MHz offset within the tuning range are reported in the same figure, showing maximum variation of 2dB in the tuning range. The phase noise figure of merit calculated at 1MHz ranges from -177 to -179dBc/Hz, significantly better than previously published results as shown in Fig. 16.2.6. I-Q accuracy is measured at baseband by downconverting an input tone by means of the quadrature mixer driven by the quadrature VCO. Figure 16.2.5 shows the scope screen-shot with quadrature signals downconverted at 200MHz. The same measurement repeated at different VCO frequencies reveals a phase error <1.5° and an amplitude error <1dB. From simulations, assuming a relative mismatch between the coupled resonators of the two stages of 1%, phase error is ~2°. The table in Fig. 16.2.6 compares this work with state-of-the-art mm-Wave quadrature VCOs [2,5,6].

Acknowledgment:

This work has been carried out within the *Studio di Microelettronica*, a joint research laboratory between Università degli Studi di Pavia and STMicroelectronics, and partially supported by Italian national funding program FIRB, contract # RBA06L4S5.

References:

[1] A. Natarajan, A. Komijani, X. Guan, A. Babakhani, A. Hajimiri: "A 77GHz Phased–Array Transceiver With On–Chip Antennas in Silicon: Transmitter and Local LO-Path Phase Shifting", *IEEE J. Solid State Circuits,* vol. 41, no. 12, pp. 2807-2819, Dec. 2006.

[2] K. Scheir, S. Bronckers, J. Borremans, P. Wambacq, Y. Rolain: "A 52GHz Phased-Array Receiver Front-End in 90nm Digital CMOS", *IEEE J. Solid State Circuits,* vol. 43, no. 12, pp. 2651-2659, Dec. 2008.

[3] A. Mazzanti, P. Andreani: "A Time Variant Analysis of Fundamental $1/f^3$ Phase Noise in CMOS Parallel LC-Tank Quadrature Oscillators" *IEEE Trans. on Circuits and Systems I,* vol. 56, no.10, pp. 2173-2180, Oct. 2009.

[4] A. M. ElSayed, M. I. Elmasry: "Low-Phase-Noise LC Quadrature VCO Using Coupled Tank Resonators in a Ring Structure", *IEEE J. Solid State Circuits,* vol. 36, no. 4, pp. 701-705, Apr. 2001.

[5] E. Laskin, M. Khanpour, S.T. Nicolson, A. Tomkins, P. Garcia, A. Cathelin, D. Belot, S.P. Voinigescu: "Nanoscale CMOS Transceiver Design in the 90-170GHz Range", *IEEE Trans. on Microwave Theory and Techniques,* vol. 57, no. 12, pp. 3477-3490, Dec. 2009.

[6] K. Scheir, G. Vandersteen, Y. Rolain, P. Wambacq: "A 57-to-66GHz quadrature PLL in 45nm digital CMOS", *ISSCC Dig. Tech Papers,* pp.494-495, 495a, Feb. 2009.

978-1-61284-303-2/11 $26.00 © 2011 IEEE

ISSCC 2011 / February 22, 2011 / 2:00 PM

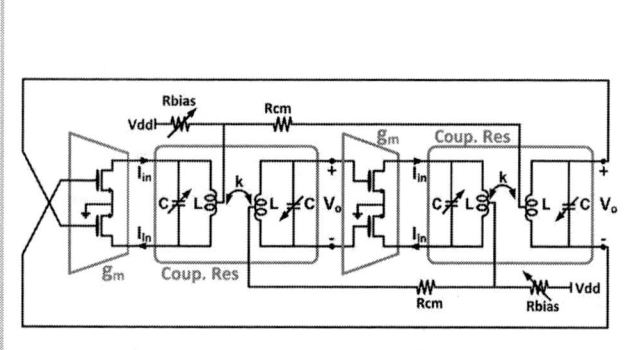

Figure 16.2.1: Quadrature VCO schematic.

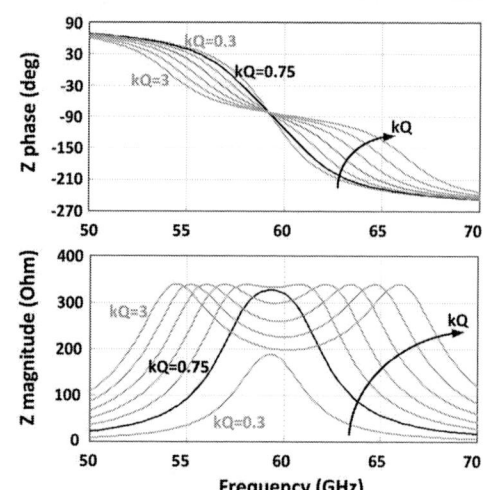

Figure 16.2.2: Phase and magnitude transimpedance of magnetically coupled resonator vs frequency for different kQ values.

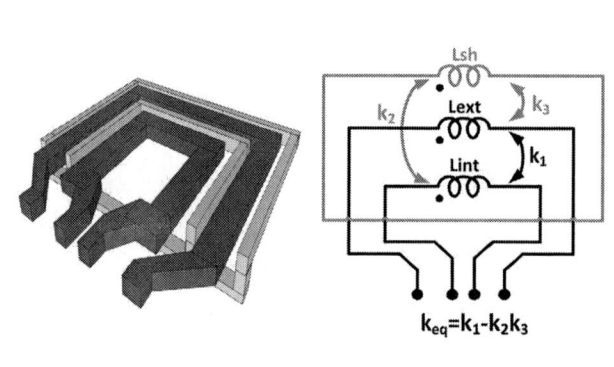

Figure 16.2.3: Low-k transformer. 3D view and simplified lumped model.

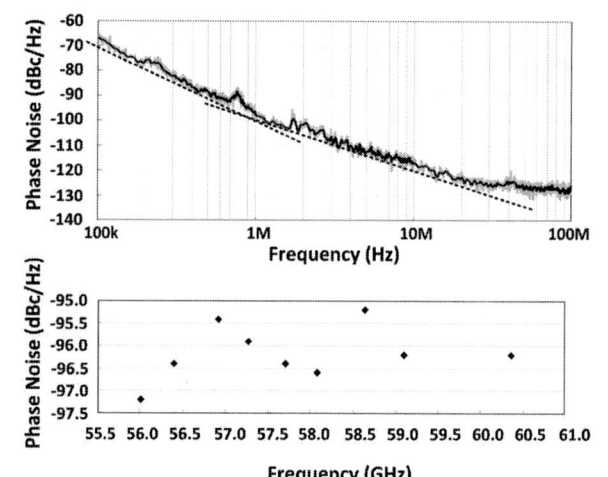

Figure 16.2.4: Phase noise measured from 58.6GHz (top plot). Phase noise at 1MHz offset within the tuning range (bottom plot).

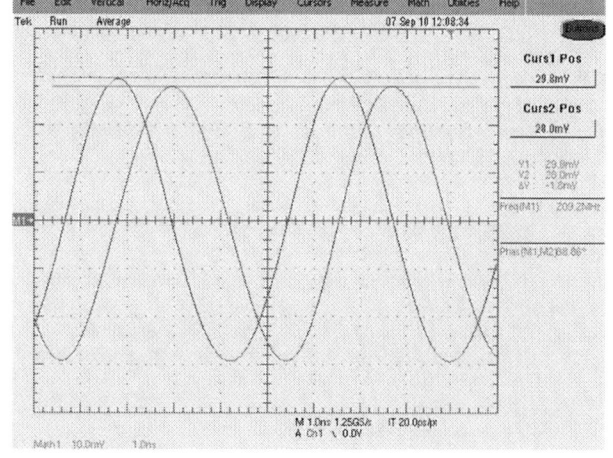

Figure 16.2.5: Quadrature signals downconverted to 200MHz.

Ref	[2]	[5]	[6]	This work
Tech	90nm CMOS	65nm CMOS	45nm CMOS	65nm CMOS
Frequency (GHz)	48	93.1	61.6	**58.2**
T.R. (GHz)	8	4	9	**4.35**
P.N. @ 1MHz (dBc/Hz)	-85	-90	-75	**-95 / -97**
FOM (dBc/Hz)	-165	-172.7	-156	**-177 /-179**
Phase Error	n.a.	n.a.	n.a.	**< 1.5°**
Area (mm²)	n.a.	n.a.	n.a.	**0.075**
Power (mW)	22.7	43.2	28	**22**

Figure 16.2.6: Overview of the state-of-the-art for mmW quadrature VCOs.

16

978-1-61284-303-2/11 $26.00 © 2011 IEEE

ISSCC 2011 PAPER CONTINUATIONS

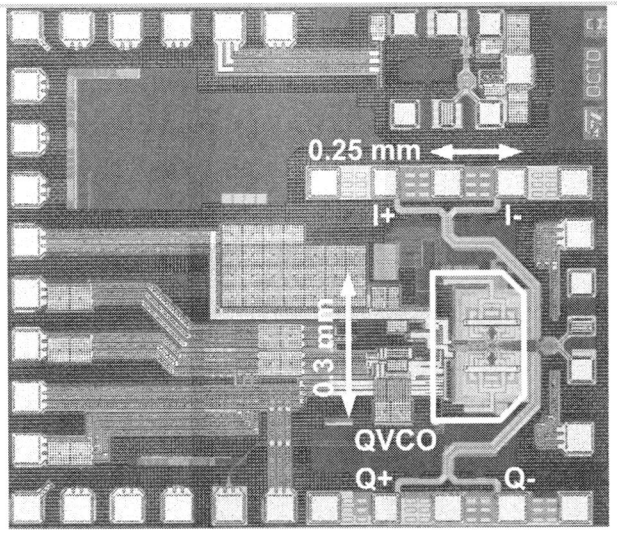

Figure 16.2.7: Chip micrograph.

ISSCC 2011 / SESSION 16 / mm-WAVE DESIGN TECHNIQUES / 16.3

16.3 A 6.5mW Inductorless CMOS Frequency Divider-by-4 Operating up to 70GHz

Andrea Ghilioni[1], Ugo Decanis[1], Enrico Monaco[2,3], Andrea Mazzanti[1], Francesco Svelto[1]

[1]University of Pavia, Pavia, Italy,
[2]University of Modena e Reggio Emilia, Modena, Italy,
[3]Istituto Universitario di Studi Superiori di Pavia, Pavia, Italy

With a cut-off frequency in excess of 250GHz, nanometer-scale CMOS technology is rapidly expanding from Radio Frequency to mm-Waves applications. Frequency dividers are key building blocks for LO generation in wireless transceivers and clock synchronization in front-ends for wire-line and optical communications. Dividers based on traditional static CML latches work over a wide band but power dissipation at mm-Waves is extremely large. To save power, recently reported mm-Wave PLLs propose tunable narrowband dividers, based on injection-locking techniques, together with digital calibration algorithms [1,2]. On the other hand, for division factors higher than 2, the frequency locking range of injection-locked oscillators is very limited, mandating fine and frequent calibrations. This paper introduces clocked differential amplifiers, working as dynamic CML latches, to realize high speed and low power mm-Wave dividers. The solution is very compact, which is particularly desirable at mm-Waves to ease chip layout and shorten IC interconnections, minimizing signal losses. A frequency divider-by-4 has been realized in a 65nm bulk CMOS technology and prototypes prove an operating frequency programmable from 20 to 70GHz. The frequency range in each sub-band spans from 10% to 17%, corresponding to a 2.5x to 4x improvement compared to injection-locked dividers-by-4. Maximum power dissipation is 6.5mW and occupied area is only 15μm x 30μm.

The divider, shown in Fig 16.3.1, is made up of four latches in a feedback loop, driven by complementary clock phases, leading to an output signal at ¼ the frequency of the input clock. A differential amplifier, shown in the inset of Fig.16.3.1, with the tail current source (M_b) enabled by the clocking signal, is proposed as high speed dynamic CML latch. When En is high, the differential amplifier charges the parasitic capacitors according to the input data. When En is low, the output is isolated from the input, and the capacitors hold the sampled data. But load resistors R_L, realized by PMOS devices (M_L) biased in triode, tend to discharge parasitics. The maximum hold time determines a lower bound for the divider operating frequency. Nonetheless, the frequency range can be easily extended over a broad range by changing the bias voltage of M_L. This changes the equivalent load resistors and the capacitors' charge/discharge time constant. To gain insight and derive guidelines to maximize bandwidth for given load resistors value, the plots in Fig. 16.3.2 show the simulated input and output signals for one of the divider's latches assuming a square-wave 60GHz clock. To simplify waveforms inspection, the square waves produced by an ideal static latch are also reported. The operation of the latch can be described dividing the period of the output signal in eight distinct time slots. Operations in T_{1-4}–$T_{1-4'}$ are the same but with complementary voltages. Before T_1 the output is low. During T_1, when the clock is high, the latch samples the input and the output differential signal tends asymptotically to $R_L I_B$. During T_2 Ck is low, the latch enters the hold mode and the output evolves toward zero. The logic state is maintained provided the signal does not fall below $V_{min} \propto V_{ov}$, the minimum voltage to switch the biasing current of the cascaded pair. V_{ov} is the overdrive voltage of M_{1-2} in Fig. 16.3.1. During T_3, where the clock is high and the latch samples the input, still in the high state, the output is refreshed. During T_4 the latch is in hold mode again.

A correct divider operation mandates the output signal crossing V_{min} during rise transient in T_1, while not falling below V_{min} in the hold mode. The following key dependences emerge: a large voltage swing and low V_{ov} are desirable to achieve both maximum frequency and wide bandwidth. On the contrary, the R_L-C_L time constant shifts minimum and maximum frequencies together and does not significantly affect the bandwidth. The details of the circuit design choices involve further trade-offs, e.g. device V_{ov} and parasitic capacitor versus transistor width, rather involved with analytical approach but easily found with a computer simulation.

A simulated sensitivity curve with the PMOS loads biased for the maximum operating frequency of 70GHz, with a sinusoidal clock signal, is reported in Fig. 16.3.3. Selected sizes of components are reported in Fig. 16.3.1. The average current consumption is 1.5mA per latch from 1V supply. The sensitivity curve of a divider based on static CML latches is also reported, for comparison, in Fig. 16.3.3. The schematic and components size of a single latch are shown in the inset. Width and bias of the PMOS loads are the same while width of all the NMOS devices is halved, leading to the same average current consumption and the same parasitic capacitance loading the outputs. The divider has a much wider fractional bandwidth but, despite the same R_L-C_L time constant, the maximum frequency is limited to only 38GHz. The higher operating frequency of the dynamic latch is due to shorter rise and fall times, during commutation of the outputs. The reason is twofold: (1) for the same average power dissipation the dynamic latch charges and discharges the load capacitors with current pulses of double amplitude and (2) the self discharge of the capacitors, just before a transition (T_4-T_4' in Fig. 16.3.2), reduces the required capacitors' voltage swing at each commutation.

Test chips have been realized by STMicroelectronics. For testing, the complementary clock phases are generated with a spiral transformer, shown in Fig. 16.3.1. The clock transistors (M_b) are biased through a center tap on the secondary winding. The voltages V_{bias} and V_{tune} are provided by 8b DACs programmed by a serial interface. The divider is buffered by a two-stage amplifier driving the 50Ω impedance of the instruments. When not driven by the external clock signal, the divider displays a self-oscillation frequency continuously tunable between 5 and 18GHz. Under signal injection, the divider locks in a sub-band, selected by changing the PMOS bias voltage, V_{tune}. Figure 16.3.4 shows the measured sensitivity curves. The operating input frequency covers the 20-to-70GHz range in nine bands. The locking range in each sub-band, with 0dBm input power, varies from 10% to 17%. Core power dissipation ranges from 1.5mW at minimum frequency to 6.5mW at maximum frequency. The supply is 1V. Figure 16.3.5 compares the phase noise measured at the input and at the output when the divider is driven by a 60GHz signal. The output phase noise is ~12dB lower, as theoretically expected from a divider-by -4, up to ~7MHz frequency offset demonstrating a negligible degradation introduced by the divider. Measurements at larger offsets are limited by the instrument noise floor.

Experimental results are summarized and compared against recently reported mm-Wave injection-locked dividers in Fig. 16.3.6. The proposed divider has the widest operating frequency range with a locking range in each sub-band which is 2.5 to 4 times larger than reported dividers-by-three and four. Core silicon area is 15μm x 30μm, the lowest reported to date. The die micrograph is shown in Fig. 16.3.7.

Acknowledgment:
This work has been carried out within the *Studio di Microelettronica*, a joint research laboratory between Università degli Studi di Pavia and STMicroelectronics, and partially supported by Italian national funding program FIRB, contract # RBA06L4S5.

References:
[1] K. Scheir et al., "A 57-to-66GHz quadrature PLL in 45nm digital CMOS", *ISSCC Dig. Tech Papers*, pp.494-495,495a, Feb. 2009.
[2] Stefano Pellerano et al., "A 39.1-to-41.6GHz ΔΣ Fractional-N Frequency Synthesizer in 90nm CMOS", *ISSCC Dig. Tech Papers*, pp.484-485,495a, Feb. 2008.
[3] Hsien-Ku Chen et al., "A mm-Wave CMOS Multimode Frequency Divider", *ISSCC Dig. Tech. Papers*, pp. 280-281, Feb. 2009.
[4] Pierre Mayr et al., "A 90GHz 65nm CMOS Injection-Locked Frequency Divider", *ISSCC Dig. Tech. Papers*, pp. 198-199, Feb. 2007.
[5] Ken Yamamoto and Minoru Fujishima, "70GHz CMOS Harmonic Injection-Locked Divider", *ISSCC Dig. Tech. Papers*, pp. 2472,2481, Feb. 2006.

978-1-61284-303-2/11 $26.00 © 2011 IEEE

ISSCC 2011 / February 22, 2011 / 2:15 PM

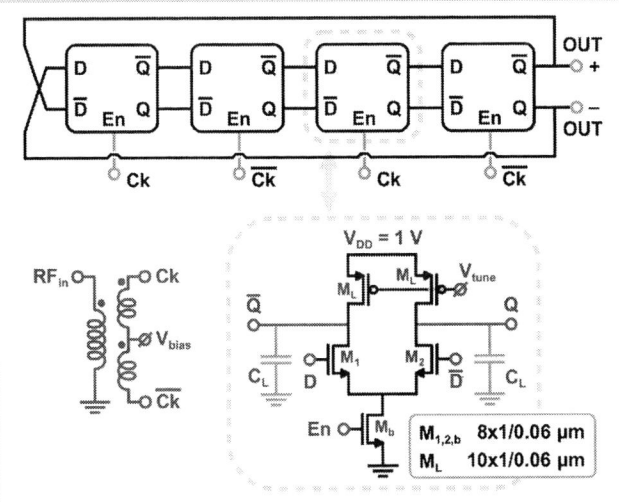

Figure 16.3.1: Block diagram of the divider-by-4 and schematic of the dynamic CML latch.

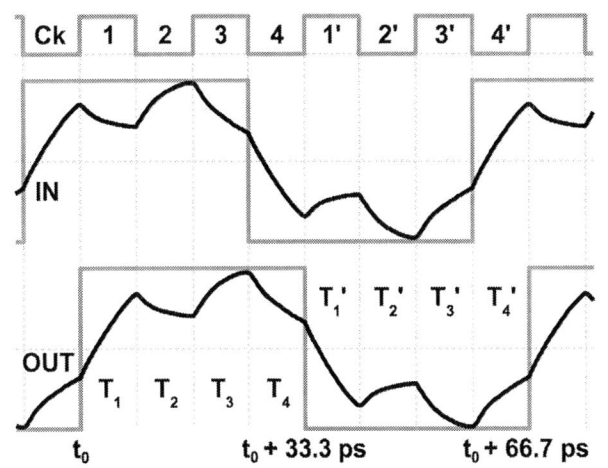

Figure 16.3.2: Simulated input and output voltage waveforms for one divider's latch. Square waves produced by an ideal static latch are reported for reference.

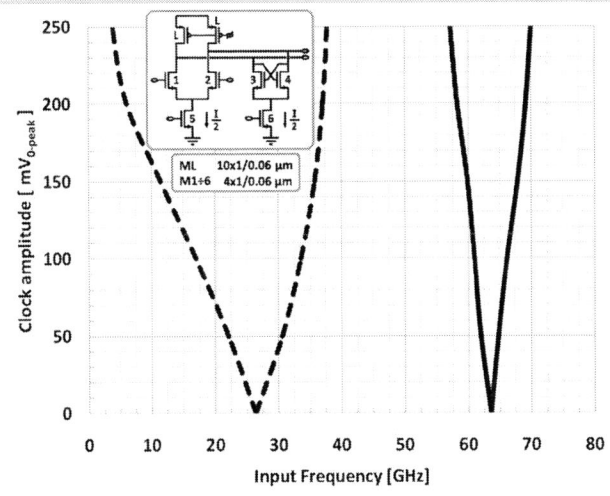

Figure 16.3.3: Simulated sensitivity curves for the divider based on dynamic (solid lines) and static (dotted lines) CML latches.

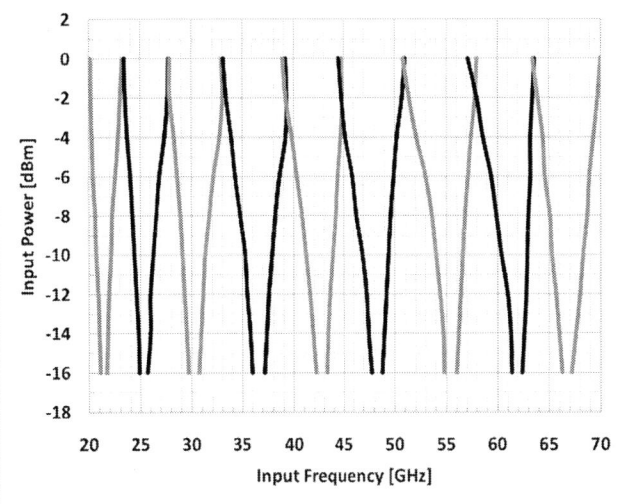

Figure 16.3.4: Measured sensitivity curves for different ML bias voltage.

Figure 16.3.5: Phase noise measured at divider's input (top curve) and output (bottom curve).

Ref	f_{in}/f_{out} (div.)	Frequency f_{min} / f_{max} [GHz]	Locking Range in sub-bands	Power [mW]	Area [µm x µm]	Technology
This work	4	20 / 70	10 - 17%	6.5	15 x 30	65nm CMOS
Injection Locked LC oscillators						
[3]	3	53.9 / 57.8	3.9%	3.12	130 x 180	130nm CMOS
[4]	4	79.7 / 81.6	2.35%	12	106 x 330	65nm CMOS
[5]	4	62.9 / 71.6	3.2%	2.75	110 x 130	90nm CMOS
Injection Locked RC oscillators						
[1]	4	56 / 67.5	3.9%	6.3	~ 50 x 80	45nm CMOS
[2]	4	> 38 / 44.5	2.4%	10	350 x 220	65nm CMOS

Figure 16.3.6: Measurements summary and comparison with state of the art.

ISSCC 2011 PAPER CONTINUATIONS

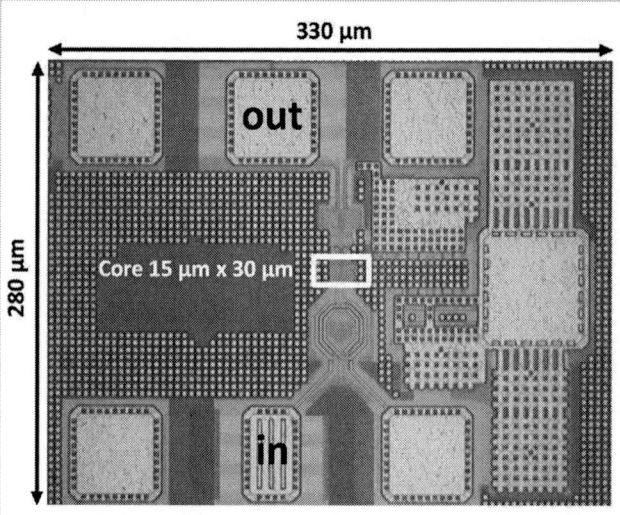

Figure 16.3.7: Chip micrograph.

ISSCC 2011 / SESSION 16 / mm-WAVE DESIGN TECHNIQUES / 16.4

16.4 A 60GHz Antenna-Referenced Frequency-Locked Loop in 0.13µm CMOS for Wireless Sensor Networks

Kuo-Ken Huang, David D. Wentzloff

University of Michigan, Ann Arbor, MI

Some applications of wireless sensor networks (WSN) require long-term deployments and vanishingly-small unit volumes to make them cost effective and unobtrusive. Energy- and area-efficient circuits, as well as eliminating off-chip components, are critical to meeting these objectives. A conventional WSN node can be fully integrated in CMOS except for the battery, crystal reference, and antenna. Typically, a PLL and a crystal reference are used for RF synchronization and to position the wireless signal in a desired band for FCC compliance. Crystals, however, require custom packaging that currently prohibits scaling to mm³ sizes. Moreover, an off-chip antenna increases the size of the node and raises the cost. In this paper, a fully integrated 60GHz frequency-locked loop (FLL) in 0.13µm CMOS is presented using an on-chip patch antenna as both the radiator and reference for frequency generation. The proposed technique efficiently integrates the antenna, eliminates the need for a crystal reference, is FCC compliant, and ensures the node transmits at the antenna's peak efficiency. The substrate beneath the antenna is shielded by an intermediate metal layer ground plane, freeing up space for active circuits and routing beneath the patch. By integrating circuits in CMOS underneath the patch, and stacking the die on e.g. a thin-film battery, a fully integrated WSN node in mm³ scale is feasible.

Figure 16.4.1 shows a block diagram of the proposed FLL and the concept of a fully integrated WSN node. An OOK signal may be modulated through the bias circuit of the VCO, amplified by the PA, and radiated through the antenna. The frequency of the VCO is regulated by measuring amplitudes of the standing wave at two points on the antenna. There are two taps located on the edge of the antenna feeding back signals to envelope detectors (EDs). The magnitude difference of the two EDs serves as a measure of the error between the VCO frequency and the peak-efficiency frequency of the antenna. After envelope detection, an error amplifier and integrator provide gain as the controller in the feedback loop, and ensure zero steady-state error of the tracking frequency. The LPF stabilizes the FLL, and is designed with a cutoff frequency of 100MHz. All blocks are integrated on chip, and the baseband circuits indicated in Fig. 16.4.1 are located beneath the patch antenna M4 ground plane for area efficiency. There remains 1.08mm² of area beneath the antenna available for additional circuits.

Figure 16.4.2 illustrates how the patch antenna is used to detect its own frequency of maximum radiation efficiency by utilizing the standing-wave pattern on it. A transmission line model is commonly used in patch antenna design and characterization, where the open load end and source end of a transmission line with large width represent the two radiation edges of the patch antenna. When a frequency tone is applied from the source, it will generate a standing wave pattern along the length of the antenna. The standing wave magnitude is always fixed at the open end due to the boundary condition there. Under this condition, the location of the first electric field null differs along the length dimension according to the input frequency. If the half-wavelength of the applied tone equals the length of the antenna, the electric field null will be located in the middle of the length. Higher frequency moves the null closer to the open end, and lower frequency moves it away. Taking the magnitude difference of two tap points that are equally spaced from the center of the antenna results in a monotonic curve versus frequency as plotted in Fig. 16.4.2. The theory line is derived by the standing wave equation of the transmission line model, and the simulation line takes matching between the PA output and antenna input into account. The measured data is acquired from the output of the EDs as frequency is swept across the range of the VCO. The zero-crossing point represents the target center frequency, and the frequency at which the antenna radiation efficiency peaks. Therefore, a feedback loop is applied to regulate the EDs output difference to zero and thus the VCO frequency to the antenna center frequency.

The center frequency of an integrated patch antenna varies with process variation. Figure 16.4.3 highlights the accuracy of using an on-chip patch antenna as a frequency reference. The physical dimensions of the antenna determine the target locking frequency, and the frequency variation mainly comes from variation on width, length and the metal thickness. These variations are relatively small compared with the dimensions of the antenna, which occupies an area of

1220 x 1580µm². The measured S11 plot shows that the center frequency is 59.8GHz and the bandwidth is 1.2GHz. 20 replica antennas with the same dimensions and process were measured, and the mean and standard deviation of the center frequency are 59.7GHz and 65.1MHz, respectively. The mean and standard deviation of the bandwidth are 1.03GHz and 67.9MHz, respectively. This results in a 3σ variation in center frequency of 3270ppm, ensuring it is FCC compliant in the 60GHz unlicensed band. In order for the transmitter to reliably communicate with e.g. an energy-detection receiver with an identical patch antenna (not included in this work), process variation should not cause the center frequency to fall outside the bandwidth of the receive antenna. Assuming worst-case 3σ variation on all parameters from process variation, two patch antennas would always overlap. Based on the measured S11 of 20 dies from a single wafer, no missed alignment was observed.

Practically, the properties of the antenna are affected by application-specific scenarios in which objects may be placed near the radiating element. In this case, the transmission line theory still applies, but the center frequency and the bandwidth will change. Full EM simulations show that when metal is placed directly in front of the antenna at a distance of 500µm, the center frequency shifts by 67.6MHz (0.1%). This value is equivalent to the 1σ variance on center frequency due to process variation alone, and is not enough to compromise FCC compliance in the 60GHz ISM band. The bandwidth is also impacted by metal in front of the antenna, and it reduces by 17%. The frequency shifts caused by nearby metal and by process variation are uncorrelated, so the center frequency of a blocked antenna has the same distribution as the unblocked one, but with a shifted mean. Therefore, including worst-case 3σ process variation and the effects of interfering metal 500µm away, two patch antennas can still communicate.

The schematic of the FLL is shown in Fig. 16.4.4. A differential VCO is designed so that one of the VCO outputs feeds the antenna and forms the feedback loop, while the other output is used for testing. The core resonator of the VCO is composed of a half-wavelength transmission line and varactors. The patch antenna is drawn in M8 with a global M4 ground plane for the antenna allowing room for circuits beneath. The inputs of the two EDs are high impedance loads so that the two taps on the edge of the antenna do not significantly affect the standing wave pattern. Envelope detection down-converts the signal from 60GHz to DC. The error amplifier and the integrator provide 20dB and 27dB DC gain, respectively. The LPF is realized using a distributed transmission line and metal comb capacitors designed with a self-resonance frequency above 60GHz.

The FLL is fabricated in 0.13µm CMOS and the power consumption (excluding the dummy PA to test pads) is 29.6mW. Figure 16.4.5 shows the output spectrum of the VCO under locked condition at 1GHz and 10MHz spans. The locked frequency is at 59.27GHz, which is within the variation of the replica antenna. Because no crystal reference is multiplied by a PLL, there are no reference spurs on the spectrum. 15 FLLs from a single wafer were tested, and Fig. 16.4.6 compares the frequency distribution of the free-running VCO when the feedback loop is off, and when operating in locked mode. The mean center frequency is 59.34GHz, with a standard deviation of 195MHz. Compared with the 503MHz standard deviation of the free-running VCO, the FLL provides an improvement in frequency variation, while eliminating the need for an external reference and tracking the peak-efficiency frequency of the integrated antenna. The measured results of the FLL are summarized in Fig. 16.4.6. A die micrograph is shown in Fig. 16.4.7. The FLL occupies 1.60x1.78mm² without pads.

Acknowledgements:
The authors would like to thank MOSIS for IC fabrication, and Mona Jarrahi, Anthony Grbic, and Jack East for measurement support. The work was partially supported by NSF under grant No. 9986866.

References:
[1] Jri Lee, "A 75-GHz PLL in 90-nm CMOS Technology," *ISSCC Dig. Tech. Papers,* pp. 432-433, Feb. 2007.
[2] Chihun Lee, and Shen-Iuan Liu, "A 58-to-60.4GHz Frequency Synthesizer in 90nm CMOS," *ISSCC Dig. Tech. Papers,* pp. 195-197, Feb. 2007.
[3] Jri Lee, Yenlin Huang, Yentso Chen, Hsinchia Lu, and Chiajung Chang, "A Low-Power Fully Integrated 60GHz Transceiver System with OOK Modulation and On-Board Antenna Assembly," *ISSCC Dig. Tech. Papers,* pp. 315-316, Feb. 2009.

ISSCC 2011 / February 22, 2011 / 2:30 PM

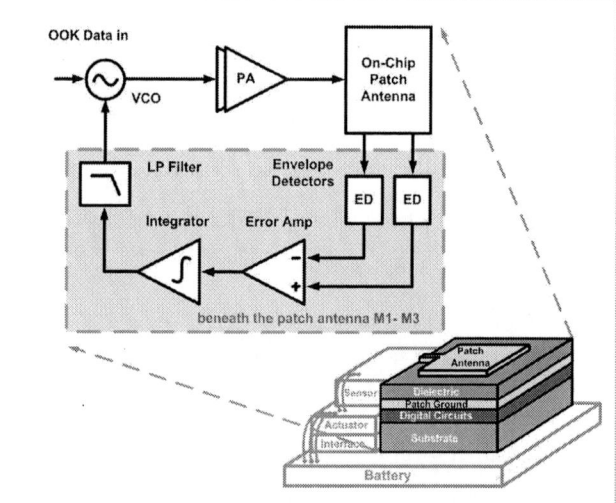

Figure 16.4.1: The proposed architecture of the frequency-locked loop.

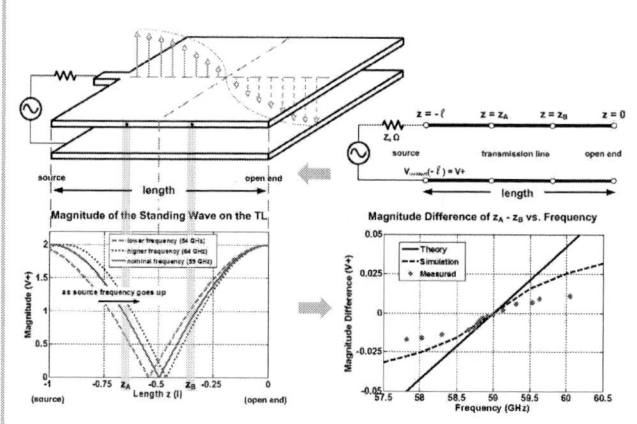

Figure 16.4.2: Standing-wave pattern on the on-chip patch antenna.

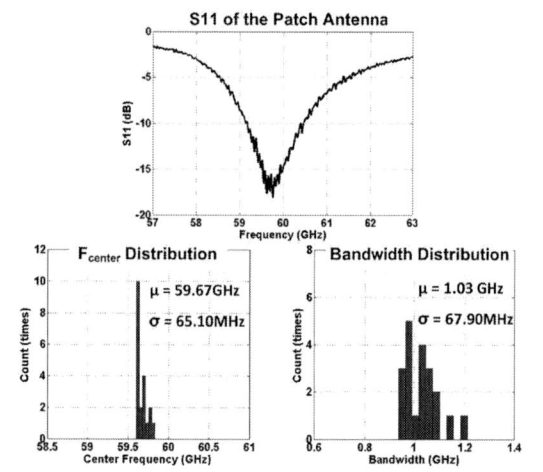

Figure 16.4.3: Measured performance of 20 replica patch antennas with identical process and geometry.

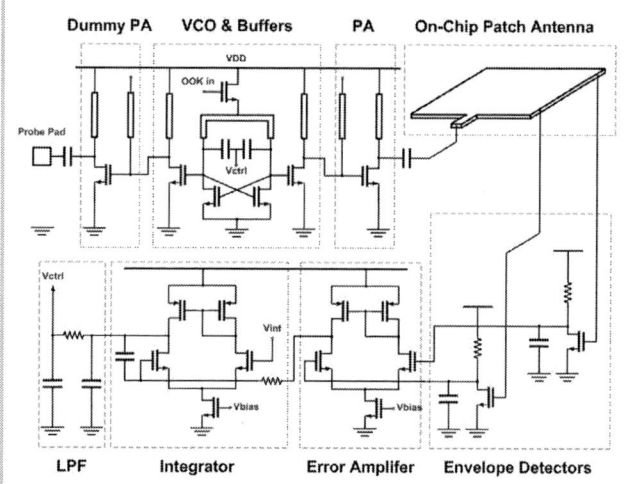

Figure 16.4.4: Schematic of the frequency-locked loop.

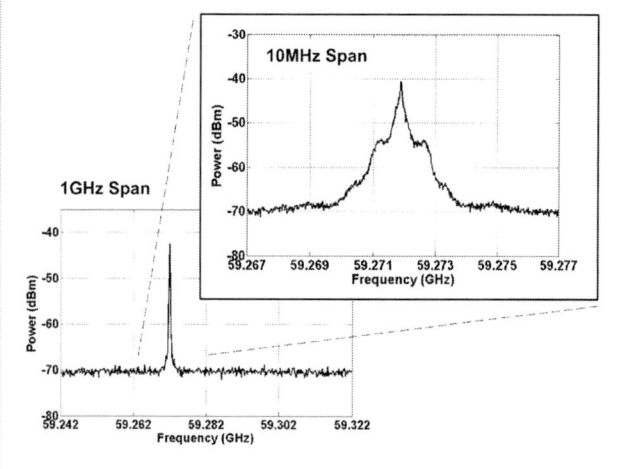

Figure 16.4.5: Output spectrum of the VCO.

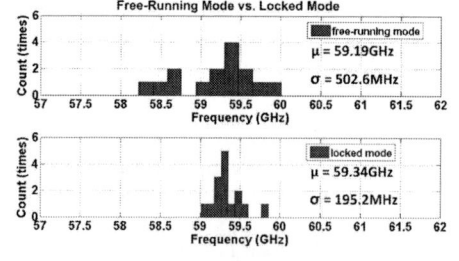

Figure 16.4.6: Frequency distribution across dies and the summary table.

Performance Summary				
	[1]	[2]	[3]	This Work
Type	PLL	Frequency Synthesizer	OOK TX	FLL
Technology	90nm CMOS	90nm CMOS	90nm CMOS	0.13μm CMOS
Power	88.0mW	80.0mW	183.0mW	29.6mW
Area	0.80mm²	0.95mm²	0.43mm²	2.85mm²
Frequency	75GHz	60GHz	60GHz	60GHz
On-Chip Antenna	N/A	N/A	No	Yes

16

978-1-61284-303-2/11 $26.00 © 2011 IEEE

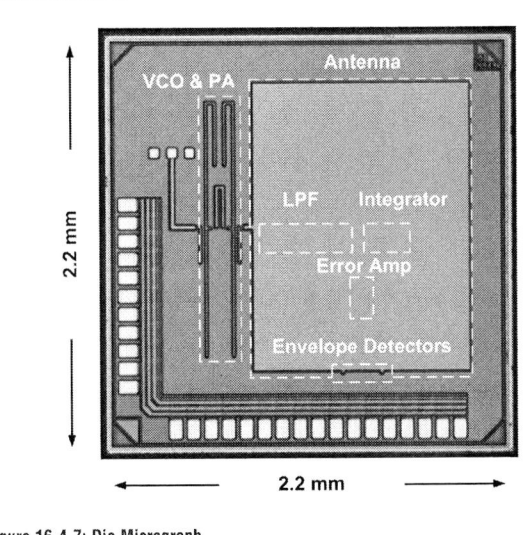

Figure 16.4.7: Die Micrograph.

ISSCC 2011 / SESSION 16 / mm-WAVE DESIGN TECHNIQUES / 16.5

16.5 A 220-to-275GHz Traveling-Wave Frequency Doubler with -6.6dBm Power at 244GHz in 65nm CMOS

Omeed Momeni, Ehsan Afshari

Cornell University, Ithaca, NY

There is a growing interest in using CMOS technology in the mm-Wave and terahertz frequency ranges for applications such as spectroscopy, imaging, compact range radars, and remote sensing [1]. Tunable signal sources are one of the main blocks in any such systems. Fundamental frequency voltage-controlled oscillators (VCOs) are widely employed for signal generation. However, CMOS VCOs suffer from low tuning range and low output power due to poor quality factor of varactors at high mm-Wave frequencies [2]. To alleviate these drawbacks CMOS frequency multipliers have been proposed for frequencies below 150GHz [2,3]. At higher frequencies, multipliers are implemented using compound semiconductors [4,5]. In this paper we propose a traveling-wave frequency doubler that operates from 220 to 275GHz in a 65nm CMOS process. Output power of -6.6dBm and conversion loss of 11.4dB are achieved at 244GHz.

Figure 16.5.1 shows the schematic of the proposed frequency doubler. The input signal at f_0 is applied to two identical parallel paths that are connected together at point "A". Each path can be viewed as a discrete transmission line that consists of microstrip inductors, L_g, and input capacitance of two transistors, C_g. Point "A" in Fig. 16.5.1 is the common-mode node for the incident signals and therefore the signals reflect back into the transmission lines from this point. The sum of incident and reflected signals is applied to the gate of each transistor. Due to nonlinearity of the transistors the 2nd harmonics at $2f_0$ are generated in the four branches. As discussed later, these $2f_0$ components are in-phase from all four branches and hence add constructively at node V_{out}. L_d and C_d are used to match the transistor output to R_L at $2f_0$. They also create an inductive load for the transistor at f_0 resulting in a negative real impedance at the input of the transistor at this frequency. This negative resistance partially cancels the loss of the input transmission lines and hence boosts the gate-source voltage amplitudes, resulting in stronger harmonic generation.

Figure 16.5.2 shows the half circuit of the input network. The incident wave travels from the signal source and reflects back at point "A". Assuming a good matching at the input, the gate voltage of each transistor can be written as the sum of incident and reflected signals as shown in Fig. 16.5.2. Here, φ is the phase shift of the signal when it travels through the last inductor, $L_g/2$, and reflects back. It can be seen from Fig. 16.5.2 that the sum of the phases of the incident and reflected signals at both nodes V_1 and V_2 are equal to $2k+\varphi$. As a result, the 2nd-harmonic components at the outputs are in phase from both transistors. The 1st- and 3rd-harmonic components are not in phase and partially cancel out at the output. Since the operation principle of this doubler is not a function of the input frequency, its bandwidth is high and only limited to the bandwidths of the input and output matching networks.

The input is matched to a 50Ω source at f_0 using C_{in}, L_{in}, L_g and C_g. In order to guarantee the wave propagation along the two parallel transmission lines, their cut-off frequency has to be higher than the highest input frequency, which is around 140GHz. Since for strong harmonic generation high input power is applied to the doubler, the input matching should be designed for large signals. Figure 16.5.3 shows the simulated input reflection coefficient for an input power of 3dBm. In this simulation L_g=50pH and the transistor size of W=28μm corresponding to C_g=42fF are used. These values result in a cut-off frequency of 220GHz for the input transmission lines, which is well above the highest input frequency. The values of L_g and C_g are also selected to ensure that the amplitude of the generated standing wave at each gate is large enough for strong harmonic generation.

The output impedance is matched to a 50Ω load at $2f_0$ using L_d and C_d. At the same time, L_d and C_d form an inductive load for the transistor at f_0 so that the negative input resistance of the transistors cancels out part of the loss of the input transmission lines resulting in higher voltage swing. Optimum values for maximum output power are found to be L_d = 42pH and C_d = 8fF. These values are also selected to prevent the input network from oscillation. Fig. 16.5.3 shows the simulated output reflection coefficient.

A 65nm CMOS process with a top metal thickness of 1.3μm was used to implement this circuit. All the inductors and capacitors are implemented using microstrip transmission lines and metal-to-metal capacitors of the pads, respectively. The Sonnet electromagnetic simulator was used to design all the passive components. Simulation shows a peak power of -2dBm and a conversion loss of 7dB at 270GHz. The simulated output power is more than -8dBm at all other frequencies from 250 to 290GHz for an input power of 5dBm. Other harmonics are at least 13dB lower than the 2nd harmonic at 270GHz.

A GGB 140-GSG probe and a Cascade i325-GSG probe with built-in bias tees were used to probe the input and output signals, respectively. Figure 16.5.4 shows the setup for the frequency measurement using an OML WR-3.0 harmonic mixer. With no input signal, no output signal was detected. As the input power reaches −7.4dBm the output power becomes detectable. By sweeping the LO frequency and observing the IF, the LO harmonic number and the signal frequency can be determined. The detectable output signals from 220 to 275GHz were measured to have twice the frequencies of the input. The measurement setup is limited to the range of 220 to 280GHz because of the lower and the higher cutoff frequency of the WR-3.0 and the WR-8.0 waveguides, respectively. A typical measured output spectrum with the 48th harmonic of the LO frequency is shown in Fig. 16.5.5. For this spectrum, the input frequency and power are 118.5GHz and 4.5dBm, respectively. The losses of the probe and all the other components are calibrated using network analyzers by Cascade, VDI, and OML. Using the conversion loss of the mixer and the loss of the waveguides, we measured the peak power at 237GHz to be 0dBm.

To measure the output power more accurately, we used an Erickson PM4 power meter. Figure 16.5.4 illustrates this power measurement setup. Figure 16.5.5 shows the output power and the conversion loss as a function of the output frequency using the power meter. In this plot the input power is kept constant at 3dBm for all frequencies. This input power level is the highest power that our setup can generate across the entire band. The peak measured power and conversion gain using this setup occurs at 244GHz. At higher and lower frequencies, input and output matching networks limit the output power. Figure 16.5.6 shows the output power and conversion loss as a function of input power at 244GHz. A peak output power of -6.6dBm with a conversion loss of 11.4dB is achieved at this frequency. It can be seen in Fig. 16.5.6 that the output power is not saturated and higher output power can be achieved by providing higher input power. The maximum input power that our setup can provide at 122GHz is 4.8dBm. The circuit consumes 40mW from a 1.2V supply. The comparison with the state of the art is provided in Fig. 16.5.6. This work has the highest bandwidth and output power at this frequency range in CMOS and is comparable with mHEMT GaAs doublers. The die micrograph is shown in Fig. 16.5.7.

Acknowledgement:
The authors would like to thank the TSMC University Shuttle Program, the C2S2 Focus Center, MOSIS and Virginia Diodes, Inc. for their support.

References:
[1] O. Momeni and E. Afshari, "High Power Terahertz and Millimeter-Wave Oscillator Design: A Systematic Approach," *IEEE J. Solid-State Circuits*, vol. 46, no. 3, Mar. 2011.
[2] E. Monaco, M. Pozzoni, F. Svelto, and A. Mazzanti, "Injection-Locked CMOS Frequency Doublers for μ-Wave and mm-Wave Applications," *IEEE J. Solid-State Circuits*, vol. 45, no. 8, pp. 1565-1574, Aug. 2010.
[3] C. Mao, C. S. Nallani, S. Sankaran, E. Seok, and K. K. O, "125-GHz diode frequency doubler in 0.13-μm CMOS," *IEEE J. Solid-State Circuits*, vol. 44, no. 5, pp. 1531-1538, May 2009.
[4] M. Abbasi, R. Kozhuharov, C. Kärnfelt, I. Angelov, I. Kallfass, A. Leuther and H. Zirath, "Single-chip frequency multiplier chains for millimeter-wave signal generation," *IEEE Trans. Microwave Theory and Techniques*, vol. 57, no. 12, pp. 3134 - 3142, Dec. 2009.
[5] I. Kallfass, A. Tessmann, H. Massler, D. Lopez-Diaz, A. Leuther, M. Schlechtweg and O. Ambacher, "A 300 GHz active frequency-doubler and integrated resistive mixer MMIC," *European Microwave Integrated Circuits Conference*, pp. 200 - 203, Sep. 2009.
[6] D. Huang, T. R. LaRocca, L. Samoska, A. Fung, M. C. F. Chang, "324GHz CMOS frequency generator using linear superposition technique," *ISSCC Dig. Tech. Papers*, pp. 476 - 629, Feb. 2008.

978-1-61284-303-2/11 $26.00 © 2011 IEEE

ISSCC 2011 / February 22, 2011 / 3:15 PM

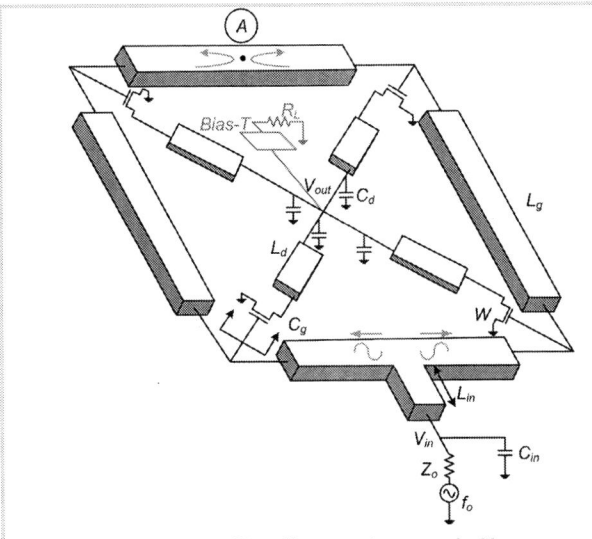

Figure 16.5.1: The proposed traveling-wave frequency doubler.

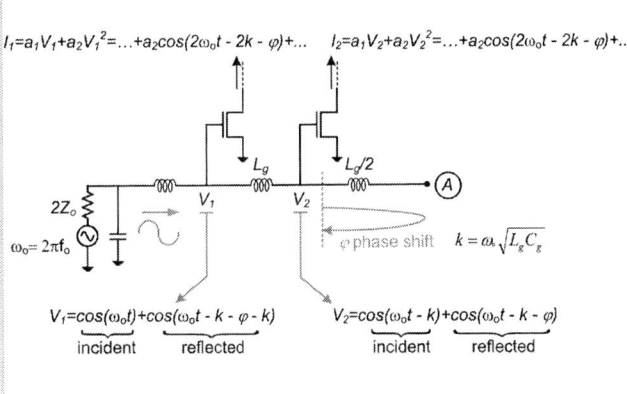

$$I_1 = a_1V_1 + a_2V_1^2 = \ldots + a_2\cos(2\omega_o t - 2k - \varphi) + \ldots$$
$$I_2 = a_1V_2 + a_2V_2^2 = \ldots + a_2\cos(2\omega_o t - 2k - \varphi) + \ldots$$

$$\omega_o = 2\pi f_o$$

$$\varphi \text{ phase shift} \quad k = \omega_o\sqrt{L_g C_g}$$

$$V_1 = \cos(\omega_o t) + \cos(\omega_o t - k - \varphi - k)$$
incident — reflected

$$V_2 = \cos(\omega_o t - k) + \cos(\omega_o t - k - \varphi)$$
incident — reflected

Figure 16.5.2: Half circuit of the input network of the frequency doubler.

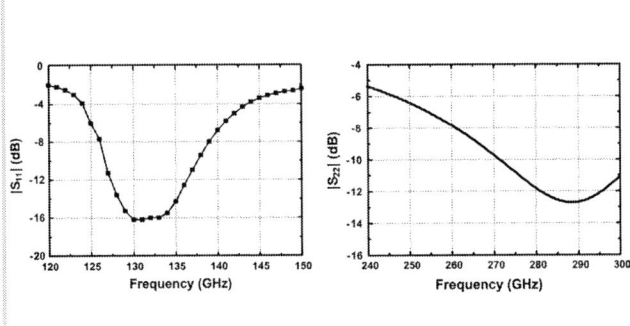

Figure 16.5.3: Simulated large-signal input reflection coefficient (S_{11}) and output reflection coefficient (S_{22}) of the frequency doubler.

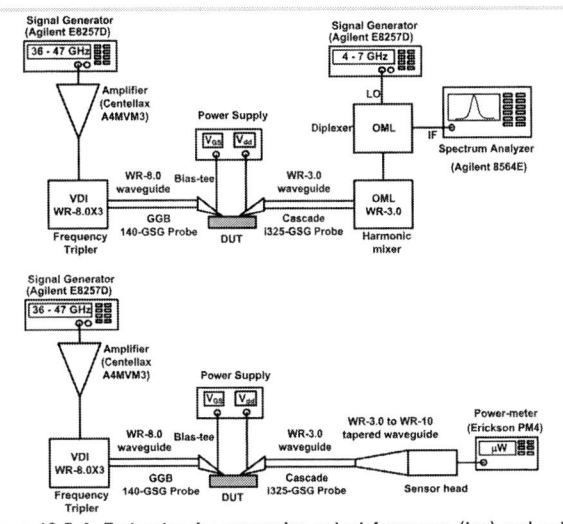

Figure 16.5.4: Test setup for measuring output frequency (top) and output power (bottom).

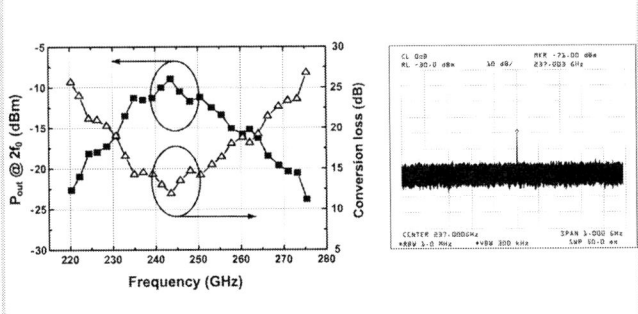

Figure 16.5.5: Measured output power and conversion loss as a function of the output frequency using the power meter for an input power of -3dBm (left) and the measured output spectrum with the 48th harmonic of the LO frequency and the input power of 4.5dBm at 118.5GHz (right).

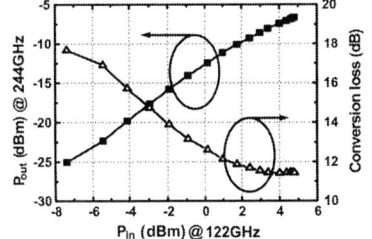

Ref.	This work	[2]	[3]	[4]	[5]	[1]*	[6]*
Frequency (GHz)	244	115	125	180	300	482	324
Tuning Range	22.2%	13.1%	12.7%	23.5%	21.4%	<1%	1.2%
Power (dBm)	-6.6 / 0	-2.6	-1.5	0	-6.4	-7.9	-46
Power Measurement	Power meter/ mixer	Mixer	Mixer	Power meter	Power meter	Power meter	Mixer
Con. Loss (dB)	11.4	NA	10	6.5	7.4	NA	NA
DC Power (mW)	40	12	0	92.5	NA	61	12
Type	Traveling wave	Injection locking	Schottky diode	x6 multiplier chain	Single transistor	Triple-push	Superposition oscillator
Technology	65nm CMOS	65nm CMOS	0.13μm CMOS	100nm GaAs mHEMT	50nm GaAs mHEMT	65nm CMOS	90nm CMOS

* These are CMOS oscillators, not frequency multipliers. They are cited to show the maximum output powers on CMOS.

Figure 16.5.6: Measured output power and conversion loss as a function of input power at 244GHz (top) and a comparison table (bottom).

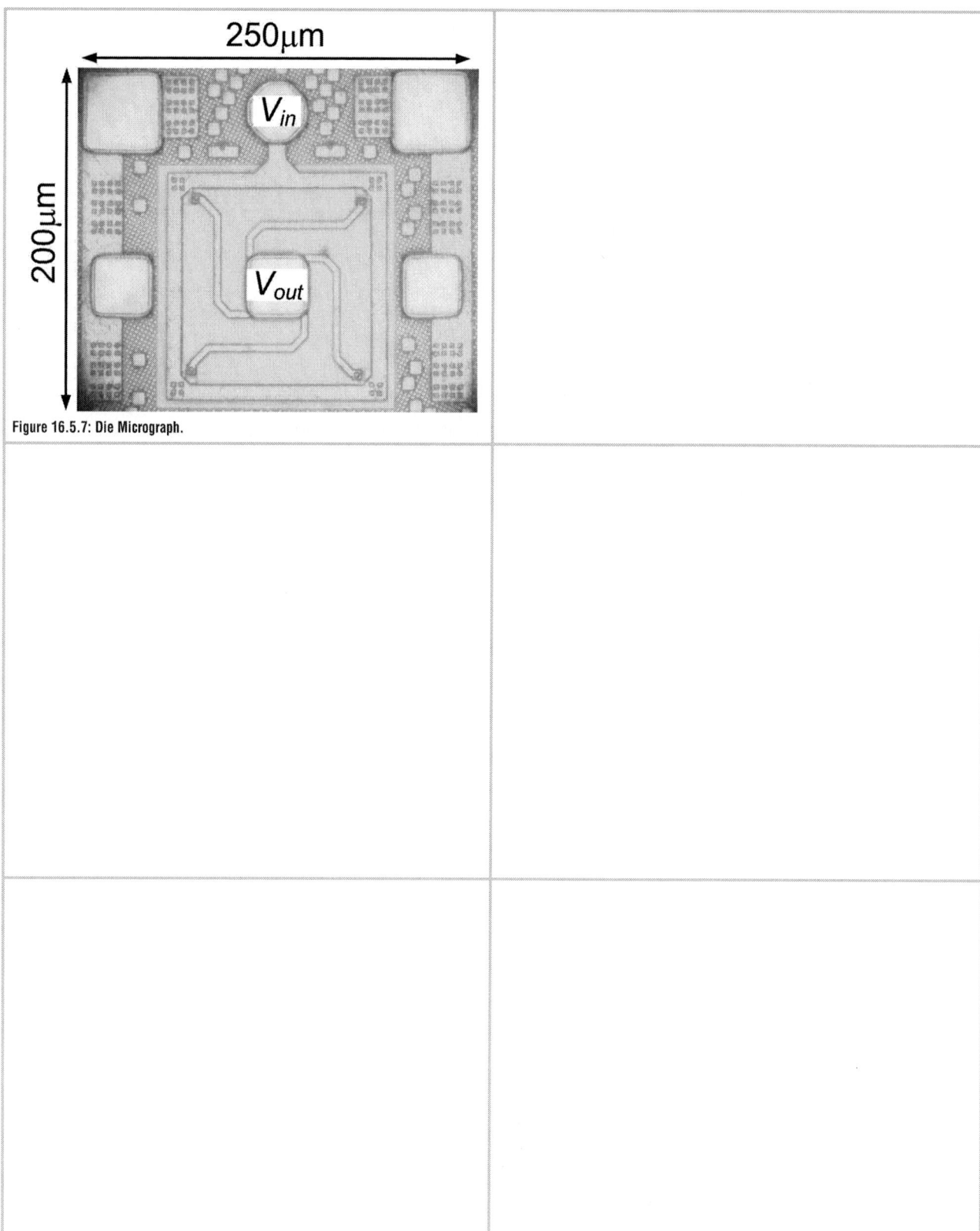

Figure 16.5.7: Die Micrograph.

ISSCC 2011 / SESSION 16 / mm-WAVE DESIGN TECHNIQUES / 16.6

16.6 Distributed Active Radiation for THz Signal Generation

Kaushik Sengupta, Ali Hajimiri

California Institute of Technology, Pasadena, CA

Despite the aggressive scaling of silicon-based IC's over the past few decades, the transistor characteristics have yet to improve so that the 'THz'-range (~300GHz-to-3THz) circuits can be effectively designed using the conventional techniques. This has led the few attempts at signal generation at these frequencies in CMOS to produce very small power levels (e.g., tens of nano-watts)[1,2]. The broad range of applications that could benefit from efficient power generation justifies novel approaches that allow high power generation and efficient radiation in CMOS. This can be achieved by removing the artificial boundaries between levels of abstraction such as electromagnetics, antenna, propagation and circuits; when we can truly leverage the advantages of the new design space that lies in the confluence of these separate treatments leading to more optimal design [3].

Two major challenges to a fully integrated 'THz' signal source with high enough power for practical applications in CMOS are: effective signal generation above transistor cut-off frequencies and efficient electromagnetic radiation out of silicon. Traditional ways to generate high-frequency signals above f_{max} of devices such as using varactors, nonlinear transmission lines or push-push oscillators [1] and radiating through conventional tuned antennas suffer from lack of power scalability due to parasitic scaling, modeling inaccuracies leading to poor efficiency and low power. Also radiation through traditional antennas (e.g., dipole) in silicon leads to leaky substrate modes that are often remedied with off-chip structures such as dielectric lenses.

In this paper, we introduce the distributed-active-radiator (DAR) structures which consolidate the signal generation, multiplication, filtering, and radiation in a single active electromagnetically coupled structure. As examples of distributed active radiators, we demonstrate 2x1 and 2x2 arrays of DAR structures radiating at 300GHz, which achieve three orders of magnitude higher total radiated power than previously reported.

Cancellation of the fundamental and radiation of the second harmonic can be done quasi-optically with low loss to perform several functions in one place. One key observation in devising a DAR structure capable of doing this is that two small electrical current elements in instantaneously opposite directions separated by a distance much less than the wavelength of interest cancel each other's EM radiations in the far-field, while two such elements propagating in phase will reinforce their radiated field as shown in Fig. 16.6.1. Therefore, to achieve quasi-optical filtration, the fundamental currents should be constrained to travel in opposite directions while the second harmonic should be in phase. This can be done effectively by forming a closed loop as shown in Fig. 16.6.1 where a traveling wave at the right frequency would produce these conditions. The radiation and ohmic energy losses of such system can be compensated by introduction of active elements that simultaneously sustain fundamental signal and generate harmonic content through their nonlinear transconductances and capacitances. This is a special kind of the traveling-wave oscillator [4,5] where the transmission lines and the ground plane are placed in such a way as to maximize the attenuation of the fundamental radiation and facilitate the radiation of the second harmonic.

The regenerative elements are realized as cross-coupled NMOS pairs that force the fundamental currents in the adjacent branches of the loop to be in opposite directions while they either source or sink second-harmonic currents in the same direction, as shown in Fig. 16.6.1. The return path of the second-harmonic currents goes through the source of the cross-coupled pair. Therefore in order for the second harmonic to radiate we 'separate' its return path removing the ground from underneath the loop and providing a local ground path for the second-harmonic current to propagate. This causes the same loop to behave as a coplanar stripline for the first harmonic but a distributed radiating structure for the second and therefore achieves *generation, radiation* and *filtering* simultaneously. The drains of all the transistors see the same second-harmonic radiative

impedance, thereby such a distributed arrangement overcomes the traditional narrowband-antenna impedance match for larger device sizes. In this implementation the loop is realized on a 2.1μm thick aluminum layer, where the inner loop diameter is 70μm while the ground plane diameter is 140μm. Four cross-coupled pairs are laid out equidistant along the loop circumference (Fig. 16.6.1). The fundamental frequency of oscillation is designed at 150GHz, radiating at 300GHz.

Such a traveling-wave radiating electromagnetic structure is particularly suitable for integrated implementation. The radiation is circularly polarized with very low fraction of the total power lost in substrate modes. The simulated radiation efficiency of 35% at 300 GHz from the backside of a 300μm thick silicon substrate of 10Ω-cm resistivity, without any lens, is much higher than the typical 5-10% efficiency of an on-chip half-wavelength dipole supported by a silicon lens [6].. Moreover such a distributed implementation eases the way for power combining where arrays of such radiating structures can be mutually locked, leading to lossless quasi-optical power combination and high E.I.R.P. Phase control in each individual element can also be potentially implemented for beam forming. Simulated radiation efficiency of the 2x1 and 2x2 arrays are 45% and 53% at 300GHz, respectively.

Figure 16.6.2 shows the mutual locking mechanism through a transmission line network, which also provides bias for the structure. The transmission line network provides an open circuit under phase-locked condition both at the fundamental (@150GHz) and second harmonic (@300GHz) so as not to load the structure. The multiple coupling networks at various points on the circumference impose the boundary conditions that ensure phase locking of corresponding points on all the radiators.

The measurement setup is show in Fig. 16.6.3. The radiation from the backside of the silicon die is captured by a WR-3 standard gain horn antenna and then down-converted in a harmonic mixer by the 16th harmonic of the LO. The IF is amplified by low noise amplifiers and analyzed using a spectrum analyzer. The heterodyne receiver setup is calibrated using a calibrated source from 290 to 300GHz with a calorimeter-based power meter giving absolute power measurements from 75 to 2000GHz. The total power radiated is calculated from the measurement of the far-field radiation pattern. EIRP is calculated directly from the far-field power-density captured at the receiver antenna with known aperture. The LO frequencies of 18.65GHz and 18.21GHz for down-conversion for the 2x1 and 2x2 arrays imply radiation at 299GHz and 292GHz respectively because of the 16th harmonic mixing. The absolute total power radiated from the backside for the 2x1 array at 299GHz is measured to be 12μW at an EIRP of -13 dBm. The measured output IF spectrum is shown in Fig. 16.6.4. Figure 16.6.5 shows its measured far-field radiation pattern. The measured boresight directivity of the array was found to be 6dB, which compares well against a simulated value of 5.6dB. This implies an EIRP of -13dBm. The 2x2 array radiated at 291GHz with a measured total output power of 80μW. The measured directivity of 10dB results in a net EIRP of -1dBm. The chips draw 22mA from 0.85V supply per DAR. The active area of the 2x1 array is 500x650μm² while that of the 2x2 array measures 800x800μm² as shown in Fig. 16.6.6.

Acknowledgement: The authors acknowledge J.Zmuidzinas, G.Blake, D.Miller, P.Siegel, S,Weinreb, Kaushik Dasgupta for their assistance and thank Ansoft, Zeland for software support.

References:
[1] D.Huang *et al.*, "324 GHz CMOS Frequency Generator Using Linear Superposition Technique," *ISSCC Dig. Tech. Papers*, pp. 476-477, Feb. 2008.
[2] E.Seok *et al.*, "A 410 GHz CMOS Push-Push Oscillator with an On-Chip Patch Antenna" *ISSCC Dig. Tech. Papers*, pp. 472-473, Feb. 2008.
[3] A.Hajimiri, "mm-Wave silicon ICs: An opportunity for holistic design," *RFIC Symp. Dig.*, pp. 357-360, June 2008.
[4] H. Wu and A. Hajimiri, "Silicon-based distributed voltage-controlled oscillators," *IEEE JSSC*, vol. 36, No.3 pp 493-502, March 2001.
[5] J. Wood *et al.*, "Rotary Traveling-Wave Oscillator Arrays: A New Clock Technology," *IEEE JSSC* vol. 36, No. 11, pp 1654-1665, Nov. 2001.
[6] A.Babakhani *et al.*, " A 77 GHz Phased-Array transceiver with on-chip antennas in silicon: receivers and antennas," *ISSCC Dig. Tech. Papers*, pp. 2795-2806, Feb. 2006.

978-1-61284-303-2/11 $26.00 © 2011 IEEE

ISSCC 2011 / February 22, 2011 / 3:15 PM

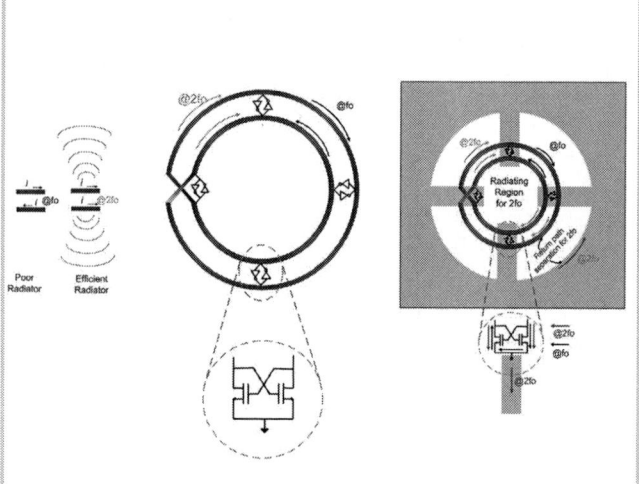

Figure 16.6.1: The conceptual idea depicting distributed active radiation.

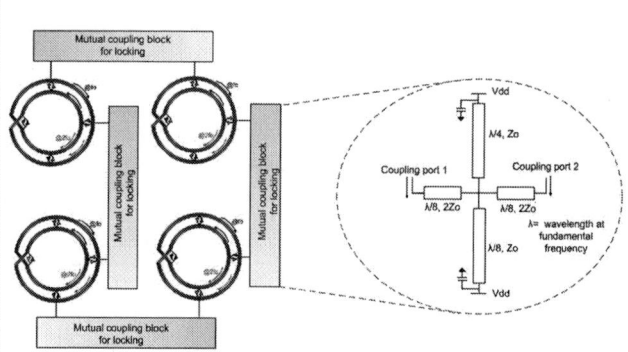

Figure 16.6.2: Mutual locking between different elements for coherent quasi-optical power combination generating high EIRP in array implementation.

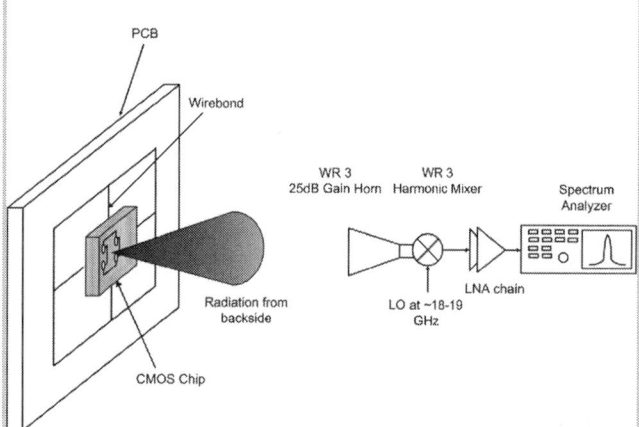

Figure 16.6.3: Chip, board and receiver setup configuration. Radiation is detected without lens from the backside of the 300μm thick silicon die.

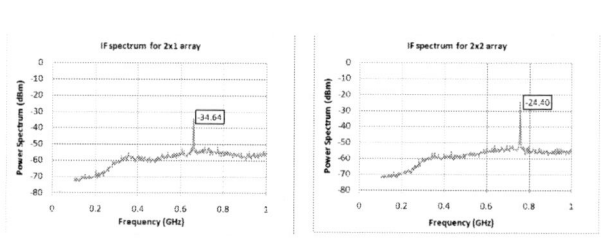

Figure 16.6.4: Detected output IF spectrum for the 2x1 and 2x2 array at the baseband respectively. LO frequencies at 18.65GHz and 18.21GHz imply radiation at 299GHz and 292GHz respectively because of the 16th harmonic mixing of the LO.

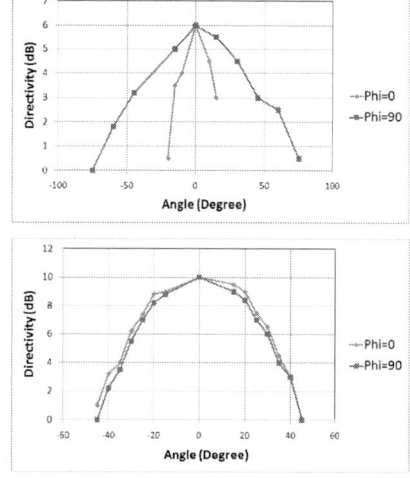

Figure 16.6.5: Measured far-field THz radiation patterns for the 2x1 and 2x2 arrays respectively.

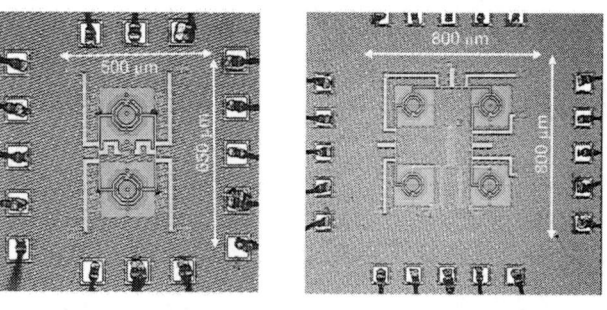

Figure 16.6.6: Die micrographs of the 2x1 and 2x2 arrays radiating at 300 GHz.

16

978-1-61284-303-2/11 $26.00 © 2011 IEEE

ISSCC 2011 / SESSION 16 / mm-WAVE DESIGN TECHNIQUES / 16.7

16.7 A 120GHz 10Gb/s Phase-Modulating Transmitter in 65nm LP CMOS

Noël Deferm, Patrick Reynaert

KU Leuven, Leuven, Belgium

This paper presents a 120GHz fully integrated 65nm low power (LP) CMOS transmitter that achieves data rates above 10Gb/s. At these high frequencies an extremely high bandwidth is available. This allows multi-gigabit-per-second communication which provides an answer to the ever-increasing demand for higher data rates in wireless systems. However, wideband modulation of a 120GHz signal in 65nm LP CMOS is a challenge.

To achieve a high data rate at these high frequencies, a straightforward and high-speed phase modulator architecture is implemented. Figure 16.7.1 shows the complete schematic of the transmitter. The LO input signal is first buffered by a 2-stage buffer amplifier. The phase generation is done by a novel differential branchline coupler that generates two differential quadrature signals. These 2 signals are fed into two 6-stage driver amplifiers. By means of differential splitters, each signal is split in an in-phase and an inverted signal. This results in 4 differential signals with relative phases of 0, 90, 180 and 270 degrees, allowing phase modulation. The actual modulation is carried out by a 4-to-1 analog multiplexer that selects the generated 120GHz phases depending on the modulation scheme. The high-speed MUX is directly driven by digital signals and takes over the function of a conventional DAC and upconverter. The output of the MUX is buffered and fed into a 9-stage differential PA.

The novel differential branchline coupler that generates the four phases is based on a single-ended branchline coupler. A 120GHz on-chip signal has a wavelength of about 1.2mm. Conventional branchline couplers have a width and length of about $\lambda/4$, which is thus about 300μm. The use of slow-wave transmission lines, which have a higher artificial relative permittivity, resulting in lower phase velocity and shorter wavelengths, leads to a very compact design. The differential branchline coupler consumes an area of 140μm by 180μm which is almost 4 times less than a conventional branchline coupler. The 50Ω differential lines of the coupler are designed in the top metal layers and only have a bottom shield. A branchline coupler also needs a transmission line with a value of 50Ω/$\sqrt{2}$. To implement this 35Ω differential slow-wave line, an extra top shield is added. The use of a differential structure results in a less complicated and more compact slow-wave line than the single-ended version. On top of that, the inverted signals are also available. The measured phases that are generated by this transmitter are shown in Fig. 16.7.3.

The analog multiplexer consists of 4 differential pairs that can be switched on or off by their tail-current source (See Fig. 16.7.2). The tail-current source is directly driven by a buffered differential digital signal. If two quadrature phase channels are switched on together, an additional 45° interpolated phase can be realized, resulting in an 8-point QAM constellation. The AM of this signal can be removed by pushing the PA into saturation, which also improves the overall power efficiency of the system. This will result in an 8PSK modulation scheme. Because of measurement setup limitations, the PA couldn't be driven into saturation and only QPSK and 8-QAM constellations were measured. The impedance-matching circuits between the switches and the drivers are realized with differential transmission lines. Series matching circuits, which are needed to connect active devices over long distance, are built with high-characteristic-impedance transmission lines. The parallel stubs are a combination of high Zc lines and slow-wave lines. This results in a length of 100μm, which is 3 times shorter than a conventional 300μm $\lambda/4$ stub. The complete area of the multiplexer is 700μm by 650μm. The two 6-stage driver chains between the branchline coupler and the MUX consume twice an area of 350μm by 150μm.

After the multiplexer, the signal is amplified by the differential power amplifier. Simulations point out that the Gmax at 120GHz of a pseudo-differential pair after gate-drain capacitive neutralization is about 6dB for small devices (W=8μm) and 5dB for larger devices (W=20μm). The loss in the differential parallel matching

stubs is about 2.5 to 3dB, depending on the size of the transistors to match to. When a gain of 20dB is desired this means that about 9 stages are needed. The 3dB bandwidth of the PA is 8GHz. The use of short parallel transmission line stubs with high characteristic impedance results in a very compact design. The 9-stage PA only consumes an area of about 550μm by 300μm.

Both input and output GSG probepads are single ended. The single-ended signals are converted to differential signals by creating a high impedance common-mode ground return path. In this way baluns, which are lossy and consume a lot of area, are not needed [5]. This technique is also used between the different stages in the amplifiers to overcome common-mode stability problems.

Different measurement setups were used to measure the complete behavior of the transmitter. The generation of the LO was done with an F-band source module. Signals were brought on chip via 50μm pitch GSG Picoprobes. Downconversion of the QPSK or 8QAM modulated 120GHz output signal was done with an external F-band mixer. The IF output signal of the mixer was fed into a high-speed oscilloscope to visualize the waveforms and to demodulate the output signal.

Figures 16.7.4 and 16.7.5 show the measured constellations, spectra and waveforms for different modulation schemes and data rates. For QPSK modulation, the results for a 4Gb/s and a 10Gb/s data rate are shown. The results for 8-QAM modulation are also shown and this for a data rate of 6Gb/s. Higher data rates are impeded by the limited bandwidth of the downconversion mixer and oscilloscope, resulting in a reduced demodulation performance.

The measured output power at 118GHz was -6.5dBm. The generator that was used to create the LO had a limited output power and as a result the saturated output power of the PA could not be reached. Simulations reveal a saturated output power higher than 0dBm over a frequency range of more than 8GHz, centered around 118GHz.

Figure 16.7.6 gives a summary of the results and a comparison with other mm-Wave transmitters. Compared to other mm-Wave transmitters it has the highest carrier frequency and the lowest supply voltage. By using the on-chip differential branchline coupler and high-speed multiplexer, the presented architecture enables multi-gigabit-per-second data rates at 120GHz. Furthermore, the architecture allows for both QPSK and 8-QAM. The transmitter consumes an area of about 1875μm by 940μm (See Fig. 16.7.7), operates from a 1V supply and consumes 200mW.

Acknowledgements:
The authors would like to thank NXP research Eindhoven for supporting this work. The IWT Flanders is acknowledged for their financial support. Rhode & Schwarz is also acknowledged for supporting the measurement setup.

References:
[1] K. Kawasaki, Y. Akiyama, K. Komori, M. Uno, H. Takeuchi, T. Itagaki, Y. Hino, Y. Kawasaki, K. Ito, A. Hajimiri, "A Millimeter-Wave Intra-Connect Solution," *ISSCC Dig. Tech. Papers*, pp. 414-415, Feb. 2010.
[2] D. Sandström, M. Varonen, M. Kärkkäinen, K. A. I. Halonen, "A W-Band 65nm CMOS Transmitter Front-End with 8GHz IF Bandwidth and 20dB IR-Ratio," *ISSCC Dig. Tech. Papers*, pp. 418-419, Feb. 2010.
[3] Y. Kawano, T. Suzuki, M. Sato, T. Hirose, K. Joshin, "A 77GHz Transceiver in 90nm CMOS," *ISSCC Dig. Tech. Papers*, pp. 310-311, Feb. 2009.
[4] C. Marcu, D. Chowdhury, C. Thakker, L. Kong, M. Tabesh, J. Park, Y. Wang, B. Afshar, A. Gupta, A. Arbabian, S. Gambini, R. Zamani, A. M. Niknejad, E. Alon, "A 90nm CMOS Low-Power 60GHz Transceiver with Integrated Baseband Circuitry," *ISSCC Dig. Tech. Papers*, pp. 314-315, Feb. 2009.
[5] T. LaRocca, M.-C.F. Chang, "60GHz CMOS differential and transformer-coupled power amplifier for compact design." *IEEE Radio Frequency Integrated Circuits Symposium,"* pp. 65-68, June 2008.

978-1-61284-303-2/11 $26.00 © 2011 IEEE

ISSCC 2011 / February 22, 2011 / 3:30 PM

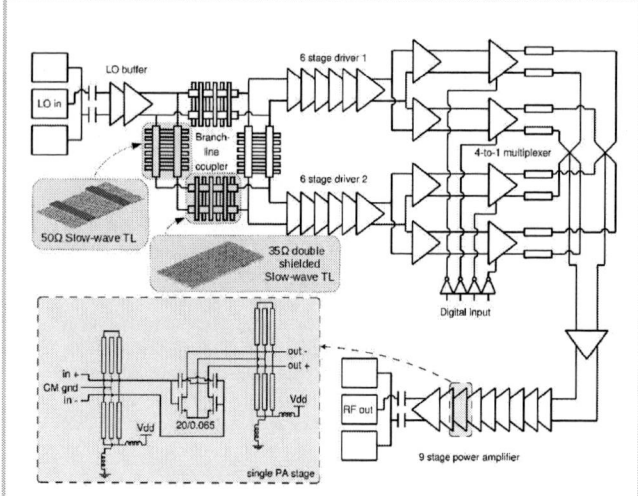

Figure 16.7.1: Circuit diagram of the 120GHz transmitter.

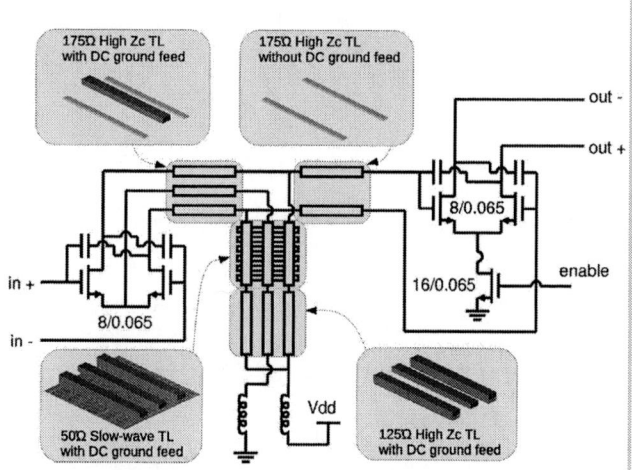

Figure 16.7.2: Circuit diagram of one multiplexer driver and switch.

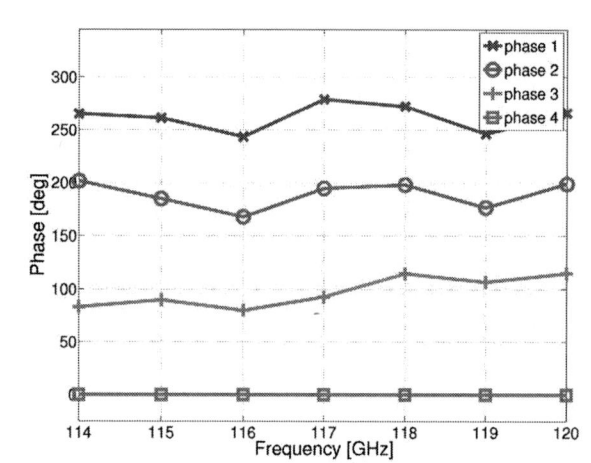

Figure 16.7.3: Measurement of the 4 generated phases vs. frequency.

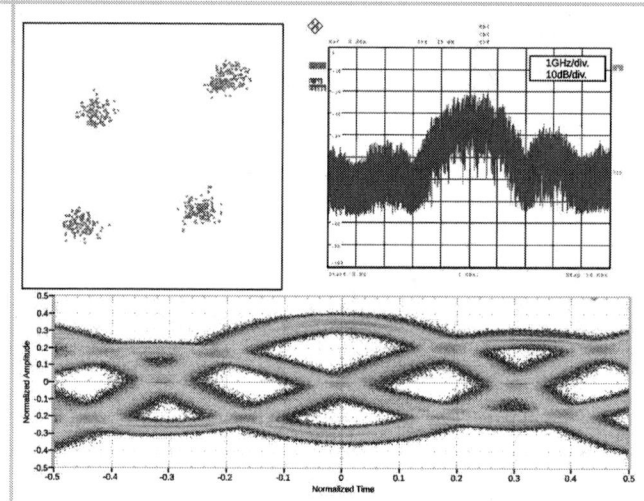

Figure 16.7.4: Measurement of 4Gb/s QPSK constellation, spectrum and modulated IF signal.

Figure 16.7.5: Constellation and spectrum of the modulated IF signal for 10Gb/s QPSK and 6Gb/s 8-QAM.

	[1]	[2]	[3]	[4]	This Work
Technology	40nm LP CMOS	65nm CMOS	90nm CMOS	90nm CMOS	65nm LP CMOS
Carrier Freq[GHz]	58	85	77	60	116
Data rate [Gbps]	>11	-	-	10	>10
Modulation	ASK	-	-	QPSK	BPSK/QPSK/ 8QAM
Bandwidth [GHz]	10	20	-	-	10*
Digital BB on chip	No	No	-	Yes	Yes
VCO on chip	Yes	No	Yes	Yes	No
Area [mm²]	0.06	1.2	2.88**	6.88**	1.55
Power consumption [mW]	29	120	390	170	200
Supply voltage [V]	1.1	1.2	-	1.2	1

* Estimation based on simulations. Exact numbers could not be measured because of measurement setup limitations.
** Area of complete transceivers (both transmitter and receiver are included).

Figure 16.7.6: Comparison with other work.

ISSCC 2011 PAPER CONTINUATIONS

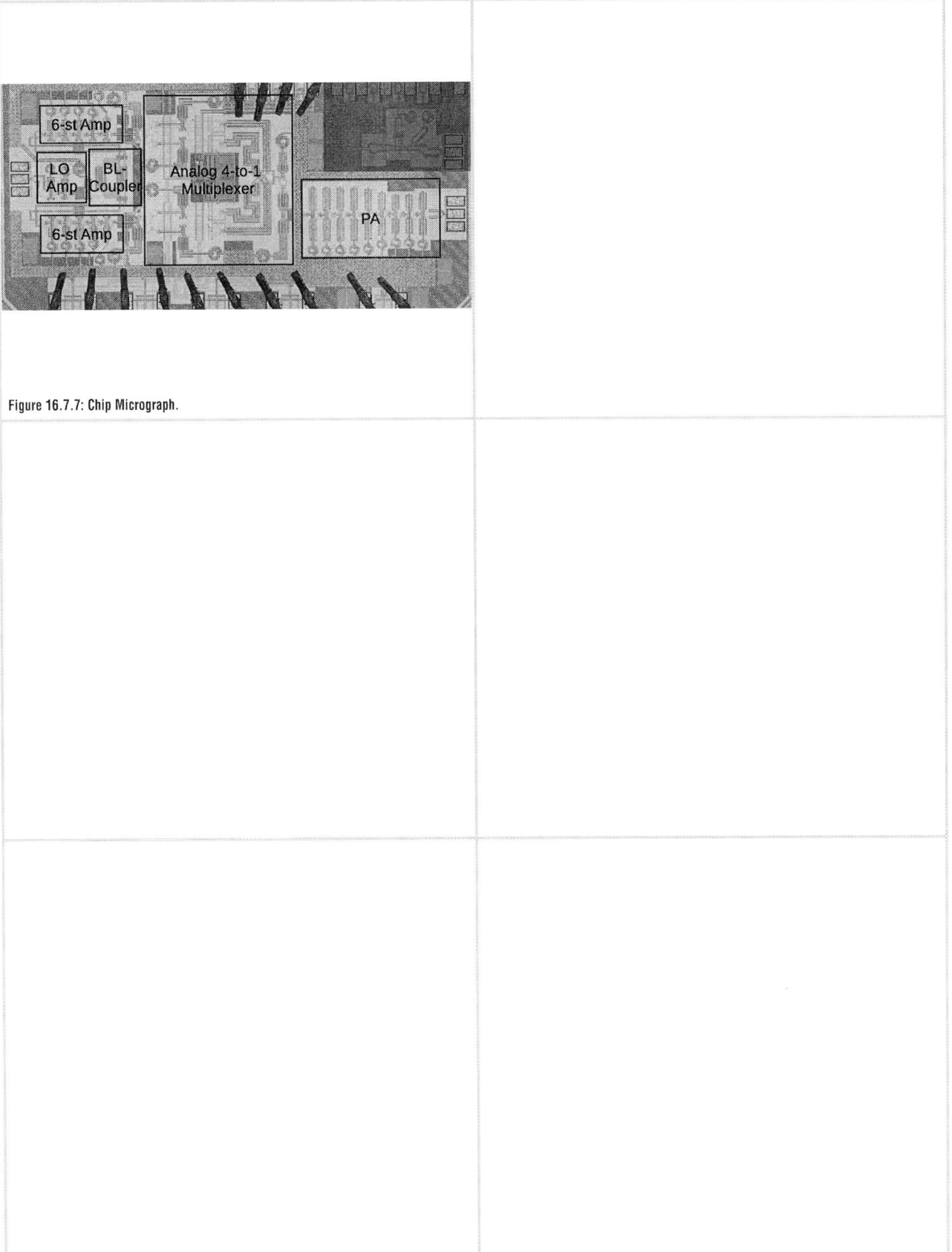

Figure 16.7.7: Chip Micrograph.

ISSCC 2011 / SESSION 16 / mm-WAVE DESIGN TECHNIQUES / 16.8

16.8 A 1.5GHz-Modulation-Range 10ms-Modulation-Period 180kHz$_{rms}$-Frequency-Error 26MHz-Reference Mixed-Mode FMCW Synthesizer for mm-Wave Radar Application

Hiroki Sakurai, Yuka Kobayashi, Toshiya Mitomo, Osamu Watanabe, Shoji Otaka

Toshiba, Kawasaki, Japan

A frequency modulated continuous-wave (FMCW) radar using triangular modulation is one of the promising candidates for realizing a CMOS radar IC [1-3]. Range and velocity resolutions of the FMCW radar are determined by the bandwidth and period of triangular modulation [4]. A short-range measurement requires wide (several GHz) bandwidth, while a long-range measurement with high-velocity resolution requires moderate (hundreds of MHz) bandwidth and long (several ms) period. Furthermore, since frequency error in the FMCW signal deteriorates range and velocity accuracy, a highly linear frequency chirp signal is required. However FMCW radars reported so far [2,3] exhibit a period up to 1.5ms because a long period degrades the chirp linearity in a conventional analog PLL. In this work, an analog/digital mixed-mode 82GHz FMCW synthesizer with 1.5GHz bandwidth, a period from 1ms to 10ms and less than 180kHz$_{rms}$ frequency error is described. The achieved performance corresponds to range and velocity resolutions of 10cm and 1.4km/h, respectively.

Figure 16.8.1 shows an implementation of the modulation in a conventional analog PLL and in the proposed FMCW synthesizer. In the conventional implementation [2,3], a modulated signal changes every step time, Δt, and the inverse of the step time, $1/\Delta t$, has to be higher than the loop bandwidth of the PLL to suppress the periodic frequency error. In order to shorten Δt, tens of kHz frequency resolution of the modulator is required. A high-resolution DDFS using large ROM table [2] or a high-speed delta-sigma modulator with more than 100MHz clock reference [3] is needed to realize such a high frequency resolution. Furthermore, this architecture exhibits inferior frequency error performance when the modulation period becomes long and the step time slow. In the proposed FMCW synthesizer, on the other hand, a digital modulation technique used in an all-digital PLL (ADPLL) architecture [5] is employed. A phase of a divided VCO output signal is captured and digitized by a digital phase detector (DPD) and the detected phase is converted to the frequency by a differentiator. The frequency is compared with a triangular frequency command word (FCW) and the difference between the two values is filtered by a digital low-pass filter. Because the FCW has enough bit length and operates in tens of MHz, a main source of a frequency error is a quantization noise caused by finite resolution of the DPD. Since the quantization noise added to the detected phase is differentiated, the frequency error is shifted to around a clock frequency by first-order noise shaping. A tens-of-MHz reference clock is sufficiently higher than 100kHz loop bandwidth, which has to be higher than the modulation frequency, $1/T_m$. Therefore low-frequency error is achieved with a standard XO reference. Moreover, a long modulation period can be achieved because the frequency error does not change according to the modulation period of the FMCW signal.

Figure 16.8.2 shows a signal flow from the FCW to the output frequency. In a conventional ADPLL, a digitally controlled oscillator (DCO) consists of a D/A converter (DAC) and a VCO because it is hard to implement capacitor banks in the millimeter-wave band. The DCO generates a stair-like triangular modulated frequency that has a periodic frequency error due to finite resolution of the DAC, which is the same as that shown in Fig. 16.8.1. Since the quantization noise caused by the DCO is not filtered by the loop filter, less than 10kHz frequency resolution is required. In addition, in order to achieve more than 1GHz modulation bandwidth, the DAC needs more than 18bit lengths. It consumes large power and chip area. Therefore an ADPLL architecture is not adopted.

In order to relax the DAC resolution and to achieve wide modulation bandwidth, we change the boundary between analog and digital. In the proposed FMCW synthesizer, the difference between the FCW and the feedback frequency is filtered by the digital low-pass filter, amplified by a gain stage and converted to analog signal by a DAC. Since the DAC output signal is integrated by an analog integrator, the input voltage of the VCO changes linearly with time. The DAC output signal decides the slope of the FMCW signal. We assume the maximum

slope is 1.5GHz/1ms and the minimum slope is 500MHz/10ms. The ratio of max and min is 30. Therefore only a 5-bit DAC can be applied. Moreover, since the low-pass filter and the gain stage are constructed in digital logic, the loop bandwidth of the synthesizer can be easily controlled in the same way as in the case of a conventional ADPLL.

Figure 16.8.3 shows the block diagram of the proposed analog/digital mixed-mode FMCW synthesizer and the circuit configuration of the DAC and the analog integrator. A 26MHz reference clock is used in the DPD, the digital logic and the DAC. The output signal of the VCO is divided by 32. The divide-by-32 circuit consists of 2 cascaded injection-locked dividers and 3 stages of flip-flop based divider. The divided signal is captured by the DPD, which consists of a counter and a time-to-digital converter (TDC). The TDC is configured by differential inverter chain [5] and its time resolution is about 20ps. The output phase is converted to a frequency and the frequency compared with 32bit FCW. The signal flow from the FCW to the DAC is the same as shown in Fig. 16.8.2. The DAC consists of a PMOS-based current steering 5-bit DAC for up chirp and an NMOS-based current steering 5-bit DAC for down chirp. An MSB of the input digital code, which indicates the chirp polarity, selects which DAC operates. The remaining 5 bits, which corresponds to the slope of the FMCW signal, are converted to thermometer code and scrambled by dynamic element matching (DEM). Since the DAC output current flows though an off-chip capacitor, the voltage of the capacitor changes linearly with time. An anti-aliasing filter is used to suppress the clock feedthrough of the DAC, and its cutoff frequency is set at 10MHz, which is located between the loop bandwidth of the PLL and the reference clock frequency.

Figure 16.8.4 shows the circuit configuration of the VCO and the first divider. The VCO consists of an NMOS cross-coupled pair with a resonator using transmission lines and an accumulation-mode varactor. The operating frequency of the VCO is 82.1GHz to 83.8GHz. It is higher than 77GHz and 79GHz band due to inaccurate transistor parameters. The VCO output is buffered by common-source amplifiers with an inductor load to drive the output buffer and the first divider. To extend the locking range of the first divider, a differential ac-coupled signal is used for injection.

Figure 16.8.5 shows the measured characteristics of the FMCW signal with different modulation bandwidth and period. The bandwidth and period of the triangular modulation are denoted as B and T_m. The rms frequency error without turn-around points (TAP) of the triangular chirp, which is not used for ranging, does not change according to the modulation bandwidth and period, and is less than 180kHz. The rms frequency error including TAP is 390kHz with B=500MHz, T_m=2ms. The performance of the proposed FMCW synthesizer is summarized in Fig. 16.8.6. The frequency error of this work is comparable to those of previous works. The phase noise is -84dBc/Hz at 1MHz offset. The shaded area of the figure denotes the operation range of the proposed circuit and those of previously reported circuits. The circuit achieves the widest operation range in the bandwidth and period. The maximum bandwidth of the FMCW modulation is 1.5GHz and the maximum modulation period with 500MHz bandwidth is 10ms. These performances correspond to range and velocity resolutions of 10cm and 1.4km/h, respectively [4]. Figure 16.8.7 shows the chip micrograph, which is implemented in 65nm CMOS technology with a core area of 1.7mm^2. The total power consumption is 152mW from a 1.2V supply. The VCO dissipates 24mW, the divider and the output buffer dissipate 120mW and the PLL block, which includes DPD, digital logic, and DAC dissipates 8mW.

References:

[1] Y. Kawano et al., "A 77 GHz transceiver in 90nm CMOS," *ISSCC Dig. Tech. Papers*, pp. 310-311, Feb. 2009.

[2] T. Mitomo et al., "A 77 GHz 90 nm CMOS Transceiver for FMCW Radar Applications," *J. Solid State Circuits*, vol. 45, no. 4, pp. 928-937, Apr. 2010.

[3] Yi-An Li et al., "A Fully Integrated 77GHz FMCW Radar System in 65nm CMOS," *ISSCC Dig. Tech. Papers*, pp. 216-217, Feb. 2010.

[4] K. Pourvoyeur et al., "Ramp Sequence Analysis to Resolve Multi Target Scenarios for a 77-GHz FMCW Radar Sensor," *Proc. International Conference on Information Fusion*, June, 2008.

[5] R. B. Staszewski, et al., "All-Digital PLL and Transmitter for Mobile Phones," *J. Solid State Circuits*, vol. 40, no. 12, pp. 2469-2482, Dec. 2005.

978-1-61284-303-2/11 $26.00 © 2011 IEEE

ISSCC 2011 / February 22, 2011 / 3:45 PM

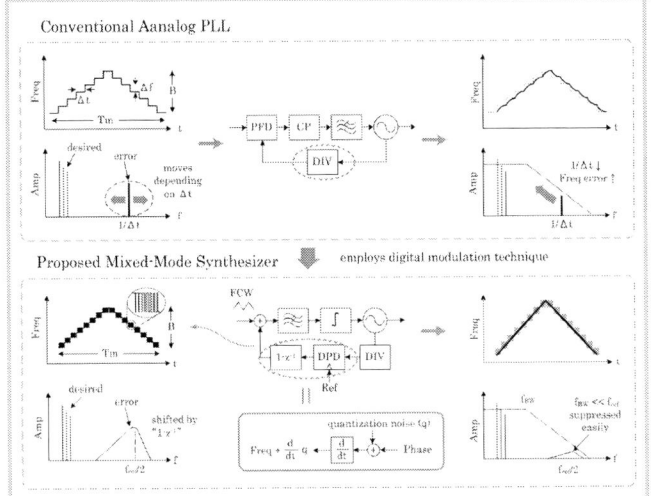

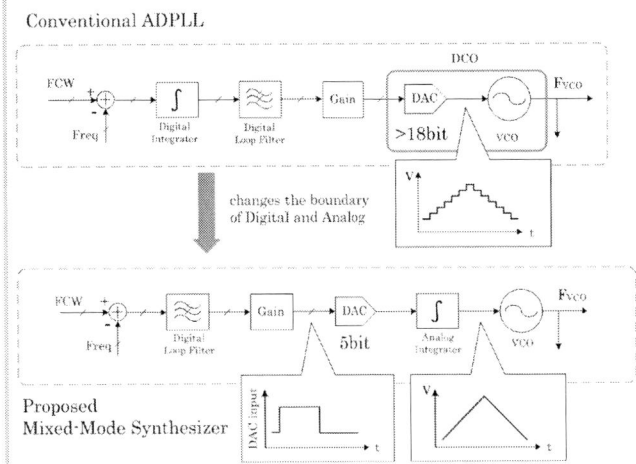

Figure 16.8.1: Modulation scheme of conventional PLL and proposed FMCW synthesizer.

Figure 16.8.2: Signal flow from the FCW to output frequency.

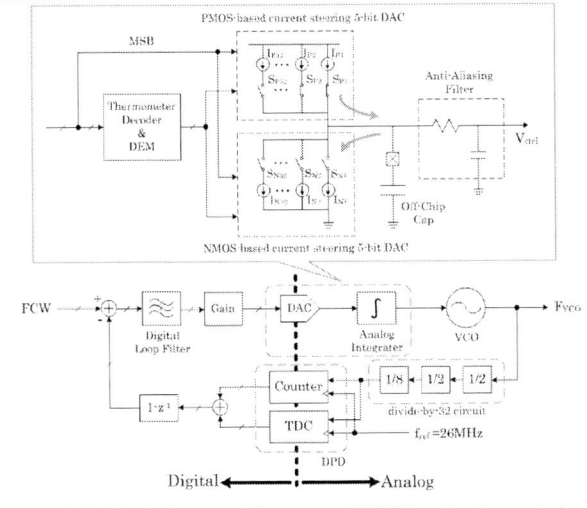

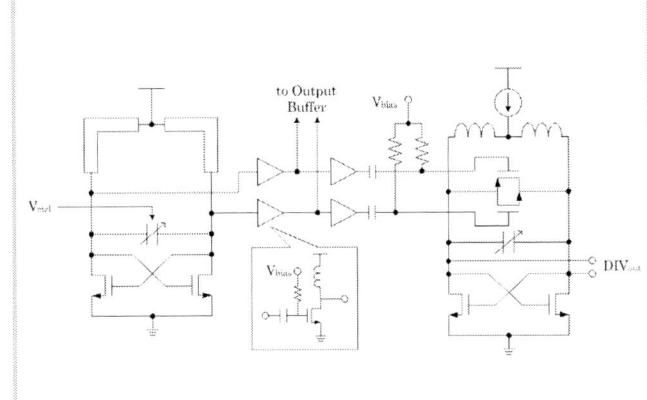

Figure 16.8.3: Block diagram of the proposed FMCW synthesizer and circuits configuration of the D/A converter and the analog integrator.

Figure 16.8.4: Circuit configuration of VCO and first divider.

16

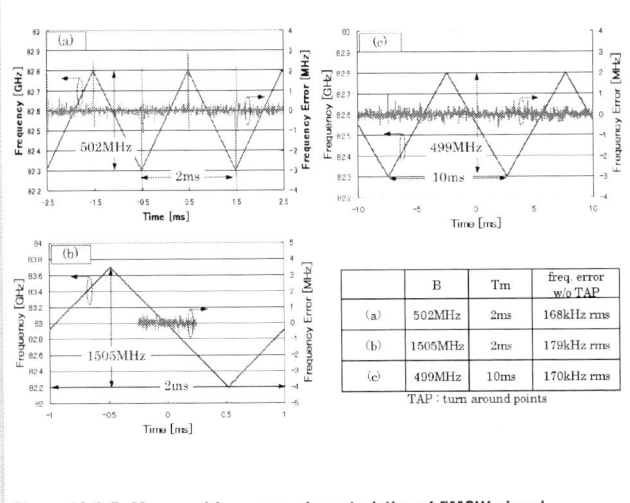

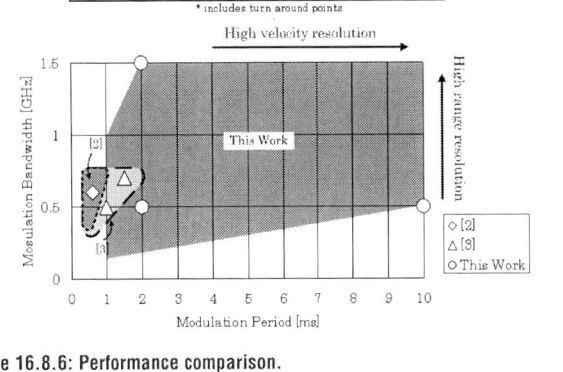

Figure 16.8.5: Measured frequency characteristics of FMCW signal.

Figure 16.8.6: Performance comparison.

ISSCC 2011 PAPER CONTINUATIONS

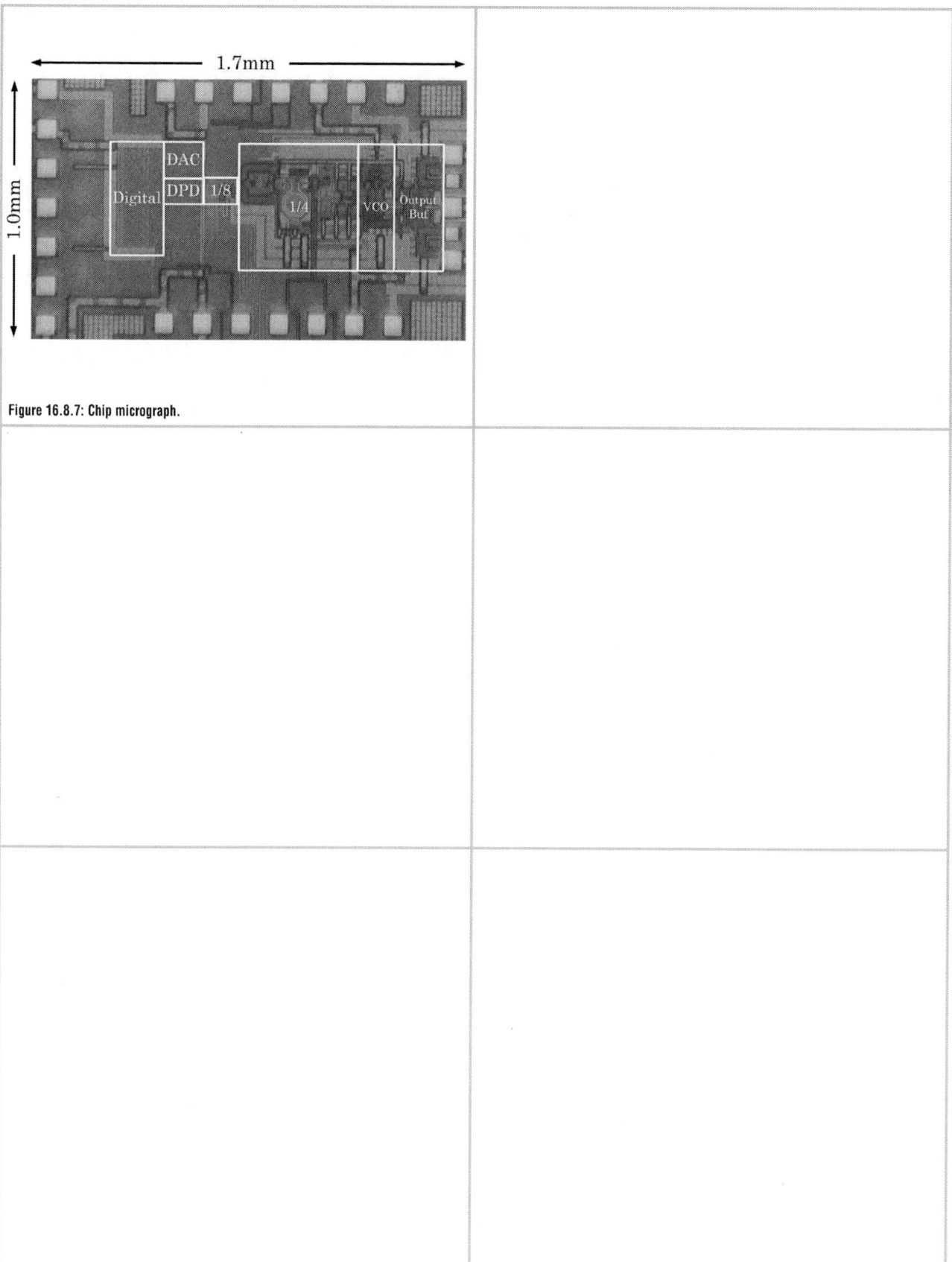

Figure 16.8.7: Chip micrograph.

ISSCC 2011 / SESSION 16 / mm-WAVE DESIGN TECHNIQUES / 16.9

16.9 A Short-Range UWB Impulse-Radio CMOS Sensor for Human Feature Detection

Ta-Shun Chu[1,2], Jonathan Roderick[1], SangHyun Chang[1],
Timothy Mercer[1], Chenliang Du[1], Hossein Hashemi[1]

[1]University of Southern California, Los Angeles, CA,
[2]National Tsing Hua University, Hsinchu, Taiwan

This paper presents a wireless non-contact sensor that enables detection, localization, and monitoring of people along with their specific features such as gait and cardiopulmonary activities. These types of sensors can be embedded in the environment and networked with the existing wireless infrastructure to create an intelligent and responsive ambient where the health of children, patients, and the elderly can be monitored without intrusion. None of the common sensing modalities including visible optical, infra-red, and ultra-sound can operate under different visibility conditions, without Line-Of-Sight (LOS), under environmental noise, and measure human-specific features simultaneously. Radio Frequency (RF) sensors have been used to localize humans, and monitor their gait and vital signs [1,2]. Most of these RF sensors transmit a Continuous Wave (CW) modulated or un-modulated waveform and detect the Doppler shift caused by the movement of hearts, lungs, and other body parts — the latter referred to as micro-Doppler technique. RF sensors do not require LOS and work under extreme weather and visibility conditions.

Ultra-WideBand (UWB) Impulse-Radio (IR) radar offers high-resolution ranging in dynamic environments without requiring LOS. The scattering pattern of UWB pulses from a time-varying environment (*e.g.*, moving human) is quite sensitive allowing for high-fidelity detection (Fig. 16.9.1) [3]. In order to acquire the maximum information from the environment and its objects, the scattered waveform should be completely reconstructed. Direct sampling of the received UWB waveform with sufficient dynamic range required in a typical sensing scenario requires unnecessarily large power consumption due to Analog-to-Digital Converters (ADC) and the subsequent Digital Signal Processing (DSP) [4,5]. This paper proposes a direct-sampling UWB IR radar that consumes less power than traditional wideband direct-sampling systems by employing techniques that take advantage of the static or slow-moving environments typically found in most health and safety sensing applications.

The proposed UWB IR radar system (Fig. 16.9.2) consists of a UWB waveform generator with tunable center frequency and bandwidth in the TX path and an SPDT switch, a Low-Noise Amplifier (LNA), and an Analog Signal Processing (ASP) Array in the RX path. The fast SPDT switch at the receiver input isolates the LNA from direct coupling with the transmitter or unwanted large reflections from close objects. The ASP front-end consists of a 16-way active power splitter, 16-channel time-interleaved samplers, and an integrator array. The ASP back-end consists of a sampler array and an analog multiplexer. The UWB IR radar utilizes a time-gated architecture for environmental sensing. The desired range of 15cm to 15m is divided into 128 coarse range bins, which corresponds to 11.72cm for each bin. Each coarse range bin is divided into 16 fine range bins with a resolution of 7.3mm. The ASP array utilizes equivalent-time sampling for the coarse range bin processing and the 16-channel time-interleaved samplers for the fine range bin processing. During the sensing of each coarse range bin, the environment in each fine range bin is sensed in the 16 different channels of the receiver. A pipelined architecture technique is implemented between the ASP front-end and back-end in an effort to reduce the time to cover all the range bins in a sequential operation. Two Delay-Locked Loops (DLL) generate the timing for the coarse and fine range bins. Each DLL consists of a Voltage-Controlled Delay Line (VCDL), Startup Circuitry (SC), Phase Detector (PD), and analog Charge Pump (CP). The first DLL is locked to an off-chip 10MHz reference clock. The VCDL in the coarse range bin DLL consists of 128 delay sections, each with 780ps nominal delay, and provides different delays covering a total of 100ns delay difference, which is required to encompass a 15m range. The MUX, at the output of the coarse range bin VCDL, is used for selecting the appropriate delay for triggering the waveform generator. A second MUX is used to control the SPDT switch at the input of the receiver to minimize direct coupling during transmission. The second DLL is used to set the timing for the fine range bin. There are 32 sections inside the fine-range-bin VCDL. These signals trigger the Track-And-Hold (T&H) amplifiers for the time-interleaved sampling. The second DLL is locked to the first DLL, and the time difference between adjacent channels of the

second DLL is 48.8ps. In an IR radar system that senses a stationary environment, the received Signal-to-Noise Ratio (SNR) is improved by $10\log N$ [dB] after averaging N received pulses. The integrator, in each time-interleaved channel, keeps adding sampled signals to improve the SNR. After a certain amount of time, the integrator is reset for the next sampling sequence. The output signal of each integrator has to be held in the next Sample-And-Hold (S&H) block and sent to the ASP output for channel-by-channel digitization through an analog multiplexer. In addition to SNR improvement from integration, the ASP array only can detect the signals which are synchronized with its clock. Since inferences are not synchronized with the clock of the ASP array, they will be rejected by the ASP array after integration.

The SPDT switch that is in the front of the LNA consists of a balun transformer and two parallel MOS switches. The LNA is a capacitively crossed-coupled common-gate amplifier. The 16-way active power splitter is a 3-stage design. The first stage, a shunt-peaked amplifier with resistive feedback, serves as an equalizer to flatten the gain response over a wide frequency range in an effort to minimize distortion. The following two T&H circuits together perform the S&H function. The first T&H amplifier is implemented as an active switched-source-follower configuration, and the second T&H block is implemented as a passive switched-series-transistor configuration with local gate-voltage boosting circuit. A transimpedance amplifier, a switched source follower, and a source follower connect the two T&H circuits (Fig. 16.9.3). Each path includes a VGA that is used for calibration of gain and DC offset mismatches between all 16 channels. The integrator is designed based on an Operational Transconductance Amplifier and Capacitor (OTA-C) architecture (Fig. 16.9.3). Identical unity gain buffers, following the S&H circuits, drive the analog multiplexer, which consists of 16 parallel bootstrapped switches. The analog output buffer is designed to drive an off-chip 12-bit slow ADC. The buffer gain can be varied to ensure the analog output signal magnitude can achieve the ADC full scale range.

The reported UWB IR radar chip has been implemented in a 0.13µm CMOS technology (Fig. 16.9.7), packaged, and integrated with a custom Data Acquisition (DAQ) system on one board. The board interfaces with a Personal Computer (PC) through the USB port. All signal-processing algorithms and sensor control functions (*e.g.*, range-bin select, power control) are implemented in MATLAB and run on the PC. The time-interleaved receiver is first calibrated for gain and offset mismatches between the channels. The measured received UWB waveform, after the calibration, in a loop-back configuration closely resembles the measured waveform using a real-time high-speed oscilloscope (Fig. 16.9.4). Various human-feature detection measurements have been performed. Figure 16.9.5 shows a representative wireless measurement for human breathing over 75cm distance captured over a period of 30s. Both discontinuous breathing (holding the breath for a while) and continuous breathing patterns are shown. Figure 16.9.6 summarizes the performance of the CMOS UWB IR radar.

Acknowledgements:
This work was partially supported by South Korea Agency for Advanced Development (ADD), LIG Nex1, and by an Alfred E. Mann Innovation in Engineering Doctoral Fellowship. Human feature specific signal processing algorithms are developed in collaboration with Prof. Burdick at the California Institute of Technology.

References:
[1] A.D. Droitcour, et al., "Range correlation and I/Q performance benefits in single-chip silicon Doppler radars for noncontact cardiopulmonary monitoring ," *IEEE Transactions on Microwave Theory and Techniques*, vol. 52, no.3, pp: 838-848, 2004.
[2] Y. Xiao, C, Li, J. Lin, "A Portable Non-Contact Heartbeat and Respiration Monitoring System Using 5-GHz Radar," *IEEE Sensors Journal*, Vol. 7, No. 7, pp. 1042-1043, July 2007.
[3] S. Chang, et al., "Human detection and tracking via ultra-wideband (UWB) radar," *Proc. IEEE International Conference on Robotics and Automation*, May 2010.
[4] H. Hashemi and H Krishnaswamy, "Challenges and opportunities in ultra wideband antenna array transceivers for imaging", *Proc. IEEE International Conference on Ultra Wideband*, pp. 586-591, September 2009.
[5] J. Cao, et al., "Digitally calibrated AFE in 65nm CMOS for 10Gb/s serial links over backplane and multimode fiber," *ISSCC Dig. Tech. Papers*, pp. 370-371, February 2009.

ISSCC 2011 / February 22, 2011 / 4:15 PM

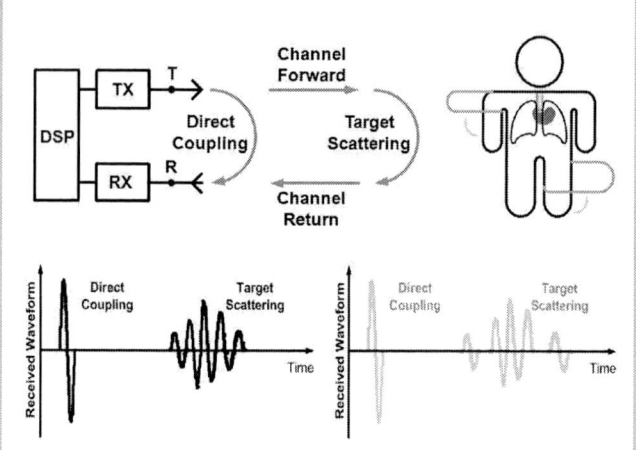

Figure 16.9.1: UWB impulse-radio sensor for human localization and activity monitoring.

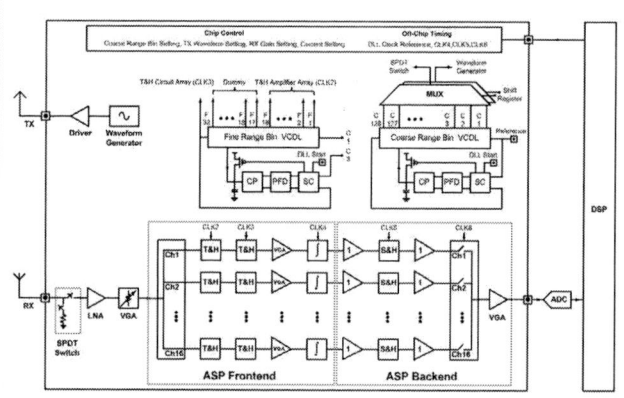

Figure 16.9.2: System-level block diagram of the proposed UWB impulse-radio radar transceiver.

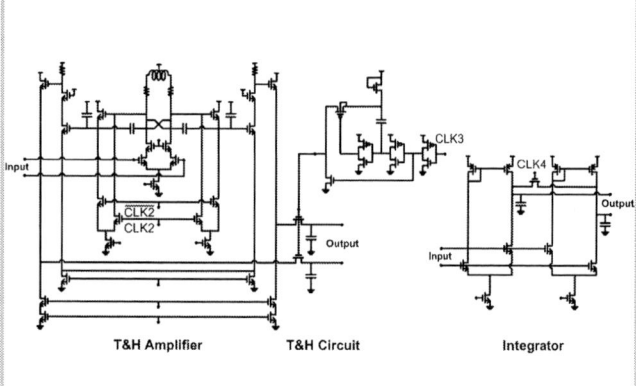

Figure 16.9.3: Circuit schematics of T&H amplifier, T&H circuit, and integrator.

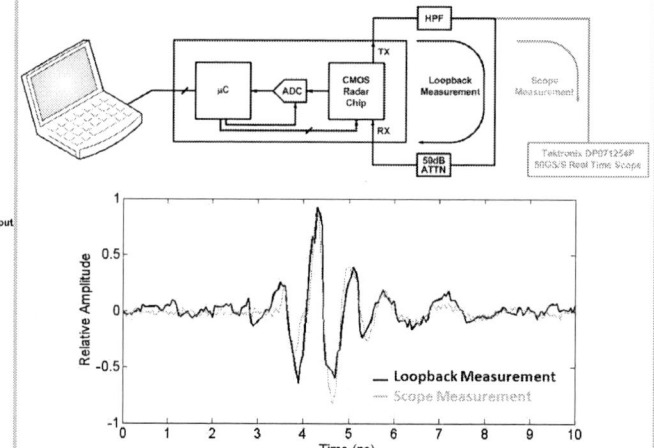

Figure 16.9.4: Measured waveforms of UWB impulse-radio radar from loopback measurement (black color) and scope measurement (gray color).

3D Human Breathing Pattern (a) / **3D Human Breathing Pattern** (b)

2D Human Breathing Pattern (a) No Breathing / Breathing

2D Human Breathing Pattern (b)

Figure 16.9.5: UWB impulse-radio radar measurement of (a) discontinuous and (b) continuous human breathing patterns.

Implementation	Technology	0.13μm CMOS
	Die Area	3.6mm X 3.3mm
	Supply Voltage	1.7V (Digital) & 1.9V(Analog)
Waveform Generator	Reference Clock Frequency	10MHz
	Pulse Repetition Frequency	10MHz
	Center Frequency	0.8GHz - 5GHz
	10 dB Bandwidth	1.6GHz - 2.5GHz
Receiver Frontend	Receiver Frontend -3dB bandwidth	2GHz - 5GHz
	Receiver Frontend Gain	12dB
	Noise Figure (Simulation)	4.5dB
	SPDT Switch Isolation	20dB
System Performance	Coarse Range Bin Resolution	11.72cm
	Fine Range Bin Resolution	0.73cm
	Radar Range	< 15m
	Max Integration Time (Analog Domain)	300us
	SNR Improvement (Theory)	34.7dB
	Max Integration Time (Digital Domain)	Unlimited
	RMS Jitter (Simulation)	< 20ps
	Receiver Gain for One Path (Simulation)	-18dB - 62dB
Power Consumption	Total Power Consumption	695 mW
	T&H Amplifier (X16)	58%
	RF 16-Way Power Splitter (X1)	18%
	Unity Gain Buffer (X32)	10%
	VGA (X16)	7%
	Timing Circuitry & Waveform Generation (X1)	4%
	RF Receiver Frontend (X1)	2%
	Output Analog Buffer (X16)	< 1%
	Integrator (X32)	< 1%

Figure 16.9.6: Performance summary of the short-range UWB impulse-radio CMOS sensor for human feature detection.

16

978-1-61284-303-2/11 $26.00 © 2011 IEEE

ISSCC 2011 PAPER CONTINUATIONS

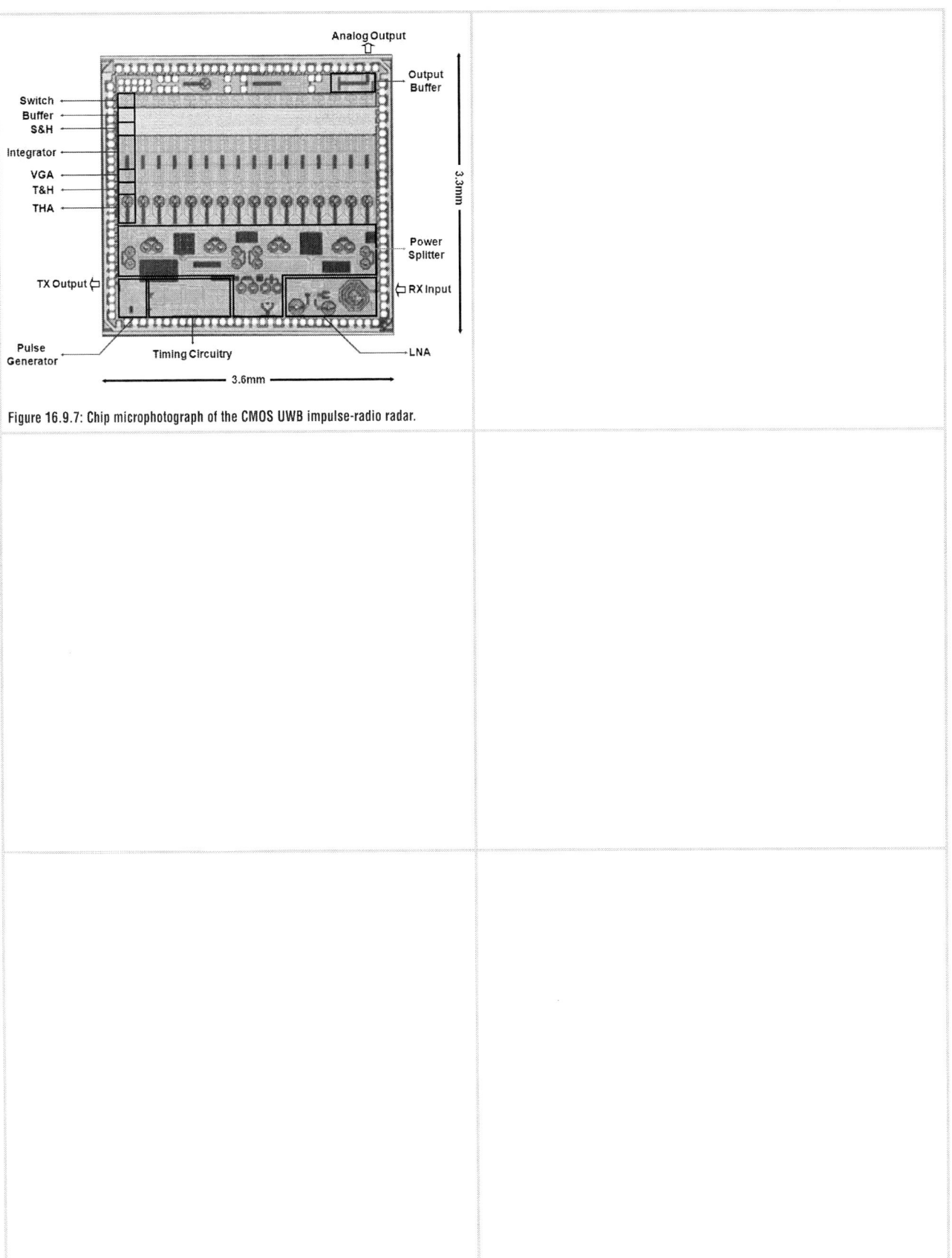

Figure 16.9.7: Chip microphotograph of the CMOS UWB impulse-radio radar.

ISSCC 2011 / SESSION 16 / mm-WAVE DESIGN TECHNIQUES / 16.10

16.10 183GHz 13.5mW/Pixel CMOS Regenerative Receiver for mm-Wave Imaging Applications

Adrian Tang, Mau-Chung Frank Chang

University of California, Los Angeles, CA

Terahertz- and mm-Wave-based imagers have recently gained interest for imaging in security screening and bio-imaging applications [1,2]. For these applications to become practical, the core pixel circuits employed in an imaging array must meet challenging constraints that originate from the system level design and the needs of constructing large array structures on-chip. The most critical of these constraints is that the pixel must consume very low power, as an array will inflate the total power by n^2, where n^2 is the total number of pixels in a square ($n \times n$) array. Pixel circuit area is the 2nd major constraint, as the single pixel area will be also inflated by n^2. This area constraint is critical because the cost-effective pixel array should ideally fit on a wafer to facilitate monolithic fabrication and avoid the need for complicated mechanical assembly of multiple array sections. A third system-level constraint similar to that experienced in CMOS image sensor arrays is the challenge of routing large numbers of analog signals between each pixel in the array and the sampling ADC.

For high-resolution systems (above 100x100 pixels) that operate in the mm-Wave spectrum, simultaneously meeting these three constraints becomes challenging, especially with traditional multiple-stage or heterodyne-based imaging receivers. This is because such receivers contain a large number of bias currents leading to increased power dissipation, and are constructed with a large number of passive devices, which consume a prohibitively large silicon area in the context of imaging array structures. A historically important alternative to the popular heterodyne and direct-conversion receiver architectures is the earlier developed super-regenerative receiver. Super-regenerative receivers are often used as non-coherent data receivers [3] in applications where power, cost, and area are extremely limited. Heterodyne and direct conversion architectures remain dominant in communications as they offer the advantages of coherent detection and recovery of phase information, enabling the quadrature signaling used by almost all modern wireless links. However, for mm-Wave imaging applications, there is no need or value in retaining phase information since imaging is based on power sensing only, making the super-regenerative receiver architecture a serious competitor in imaging applications.

In order for mm-Wave imaging to benefit from the advantages offered by the original super-regenerative architecture, we propose a new time-encoded regenerative receiver (TRR) architecture that is specifically adapted to meet the requirements of imaging array pixels. The proposed TRR uses digital CMOS circuitry to control a regenerative receiver and generate a time-encoded output signal. A block diagram of the proposed imaging receiver architecture with key waveforms is shown in Fig. 16.10.1. The proposed TRR receiver, when employed as the pixel element for an imaging array, directly addresses the three major constraints identified above:

1. TRR contains only two bias currents, greatly reducing power dissipation.
2. TRR is constructed with only two inductors, minimizing the pixel area.
3. TRR's time-encoded output can be routed by digital multiplexing rather than analog, greatly simplifying the pixel array interconnection.

Operation of the proposed TRR receiver is depicted in Fig. 16.10.1 and occurs as follows: When the digital clock edge arrives, the latch is set; engaging the oscillator. Once the oscillation envelope exceeds a designed threshold, the envelope detector is excited and resets the latch, terminating the oscillation. The super-regenerative principle (that an oscillator's startup time is inversely proportional to the log of injected power) suggests that the time between the latch set and reset is also inversely proportional to the input power.

Figure 16.10.2 shows the detailed schematic diagram of the TRR receiver implemented in 65nm CMOS technology. Transistor Q1 is biased by input transformer X1 and provides input injection into the oscillator stage from the antenna. Transistors Q2, Q3 and L1 form the oscillator tank and negative resistance element. The oscillator on/off control is provided at current source Q4, while Q5, Q6 and R1 form the envelope detector used to reset the digital latch. The latch itself

is implemented as a standard CMOS digital logic block. The receiver contains only two DC bias currents, one flowing through Q1 and the other through Q4. As R1 is large (200kΩ), the DC current consumption of the envelope detector is negligible.

Unlike conventional imaging square-law detectors or LNAs, the regenerative nature of the TRR allows it to provide high gain, even at frequencies approaching f_{max}. This enables superior suppression of the TRR's flicker and thermal noise, resulting in a lower NEP than traditional imaging detectors (listed in Fig. 16.10.6). The TRR also offers excellent spectrum selectivity of less than 1% of the RF bandwidth, improving imager sensitivity. For active imaging higher sensitivity relaxes source power requirements, increases maximum target range, and enables the possibility of "false color" imaging at multiple frequencies.

To demonstrate the operation of the TRR receiver, an oscilloscope in eye-diagram mode is used to capture the time-encoded output, as an applied 183GHz input tone is switched between two power levels (-30 dBm and -50 dBm). As a result, the corresponding time-encoded output signal varies in pulse width by 1.0ns, as shown in Fig. 16.10.3. The sensitivity (-72dBm) position is also indicated.

To determine the TRR's bandwidth, a -50dBm tone is applied to the input and frequency swept through the receiver's bandwidth while the output time-encoded difference is plotted at each frequency in Fig. 16.10.4. If we consider a 50% change in pulse width to be the 3dB bandwidth of the receiver, then a receive bandwidth of 1.4 GHz is demonstrated by the measurement data.

To demonstrate imaging operation of the TRR receiver, two easily identifiable items are concealed inside cardboard boxes and illuminated with a 183 GHz 0dBm source (VDI). We then scan the cardboard boxes by using the CMOS TRR mounted on a digitally controlled moving mechanical stage with a target distance of 10cm. The mechanical setup and captured images are shown in Fig 16.10.5. Clearly visible through the cardboard box are the metal and plastic components of a computer floppy disk and a metallic wrench.

The measured TRR receiver performance is summarized in Fig. 16.10.6 along with a comparison to other state-of-art mm-Wave imaging receivers. The TRR sensitivity is measured directly by reducing the source power until the time-encoded difference is zero, and then computing the effective path loss. The NF and NEP are determined directly from the sensitivity and bandwidth measurements. The total TRR power dissipation is measured as 13.5mW/pixel while occupying only 1.31×10^4 um^2/pixel of silicon area (not including antenna). Figure 16.10.7 shows the TRR die photo implemented in 65nm CMOS with an on-chip patch antenna. In the future, such an antenna can be constructed vertically above the TRR with spin-on polyimide as the interlayer dielectric.

Acknowledgements:
Authors are grateful to TSMC for their excellent foundry support. Also thanks to H. Jian, Q. Gu, Y. Wu, and Z. Xu for several excellent technical discussions.

References:
[1] K.B. Cooper, R. J. Dengler, G. Chattopadhyay, E. Schlect, J. Gill, A. Skalare, I. Mehdi, P.H. Siegel, "A High-Resolution Imaging Radar at 580 GHz", *IEEE Microwave and Wireless Component Letters*, Vol. 18, No. 1, pp. 64-66, Jan. 2008.
[2] Erik Ojefors, Ullrich R.Pfeiffer, Alvydas Lisauskas, Hartmut G. Roskos, "A 0.65 THz Focal-Plane Array in a Quarter-Micron CMOS Process Technology", *IEEE JSSC*, Vol. 44, No. 7, pp. 1968-1976, July 2009.
[3] Jia-Yi Chen, Michael P. Flynn, John P Hayes, "A Fully Integrated Auto-Calibrated Super-Regenerative Receiver in 0.13um CMOS", *IEEE JSSC*, Vol. 42, No 9, pp.1976-1985, Sept. 2007.
[4] A. Tomkins, P. Garcia and S. P. Voinigescu, "A passive W-band imager in 65nm bulk CMOS," *IEEE CSICS*, pp. 1-4, Oct. 2009.
[5] Gilreath L, Jain V, Hsin-Cheng Yao, Le Zheng, Heydari, P, "A 94-GHz passive imaging receiver using a balanced LNA with embedded Dicke switch", *IEEE RFIC*, pp. 79-82, May 2010.
[6] Tang, K.W, Khanpour M, Garcia P, Gamier, C, Voinigescu, S.P, "65-nm CMOS, W Band Receivers for Imaging Applications", *IEEE CICC*, pp. 749-752, Sept. 2007.

ISSCC 2011 / February 22, 2011 / 4:45 PM

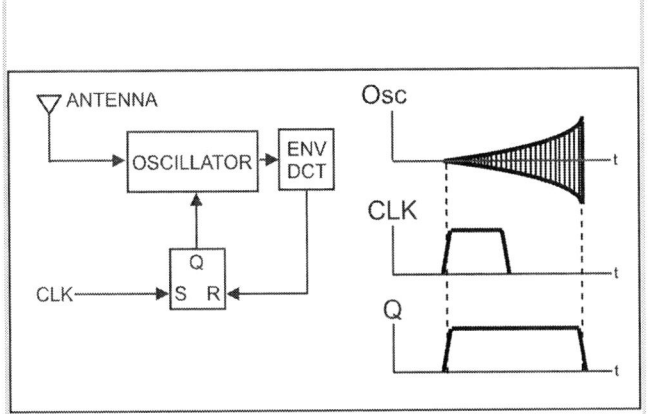

Figure 16.10.1: Time-encoded regenerative receiver (TRR) architecture and key waveforms.

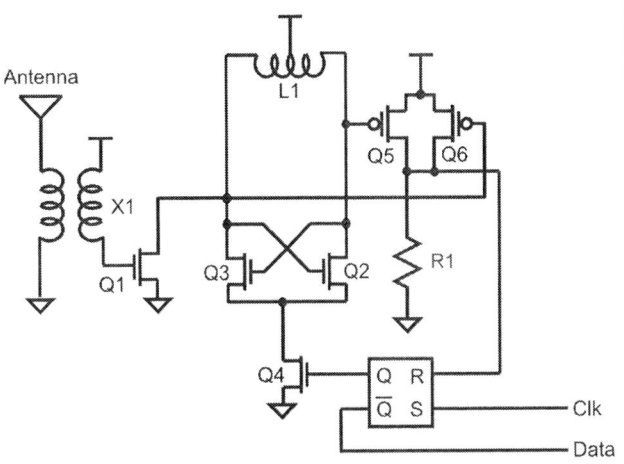

Figure 16.10.2: Time-encoded regenerative receiver (TRR) CMOS circuit implementation.

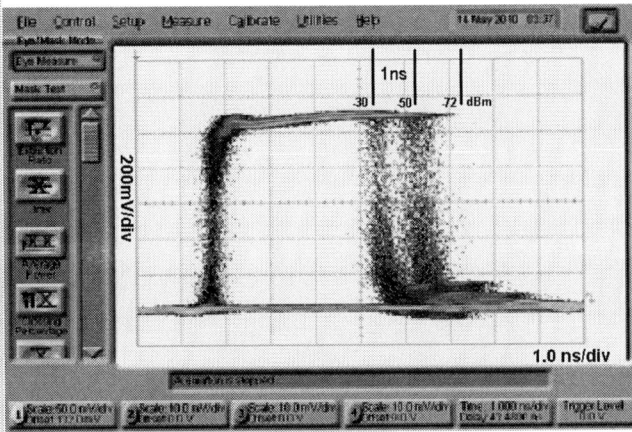

Figure 16.10.3: Eye-diagram waveform of TRR receiver for an applied 183GHz continuous-wave signal switching between −30 and −50dBm. Visible are two corresponding time-encoded pulse-widths (1.0ns difference).

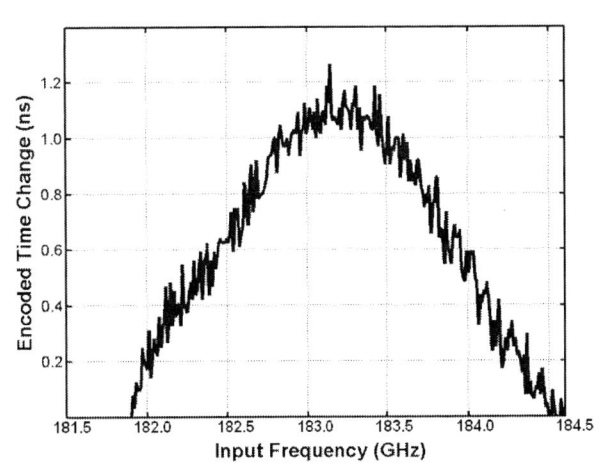

Figure 16.10.4: Frequency sweep of TRR showing receiver bandwidth. Power level of swept tone at -50dBm.

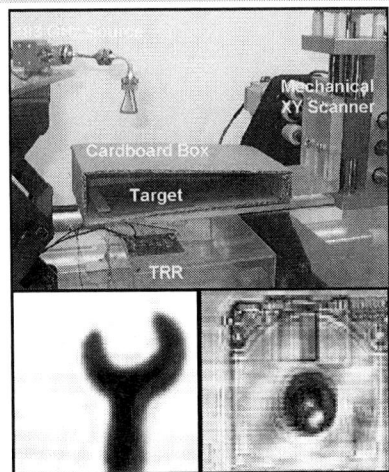

Figure 16.10.5: Mechanical setup and captured images of objects concealed in cardboard boxes when scanned by a TRR receiver pixel at 1mm/step resolution (left: adjustable wrench, right: 3.5" Floppy disk).

Receiver Characteristics	Value
Frequency	183 GHz
Power Dissipation	13.5mW / pixel
Peak Responsivity (time encoded)	1.3 ms/W
Pixel area	1.31×10^4 um^2 / pixel
Sensitivity	-72.5 dBm
Maximum Clock rate	150 MHz
3dB time-bandwidth	1.4 GHz
Noise Figure (NF)	9.9 dB (determined from sensitivity - KTB)
Noise Equivalent Power (NEP)	1.51 fW / Hz$^{0.5}$ (determined from KT[NF] B$^{0.5}$)

Receiver Characteristics	[1] MWCL 2008	[4] CSICS 2009	[5] RFIC 2010	[6] CICC 2007	This Work
Power Dissipation (mW/pixel)	250	39.6	200	93	13.5
Area (um^2/pixel)	Discrete	4.1×10^5	1.25×10^4	3.02×10^5	1.31×10^4
Output Format	Analog	Analog	Analog	Analog	Digital (Time-Encoded)
Frequency	580 GHz	94 GHz	94 GHz	94 GHz	183 GHz
NEP	N/A	200fW/Hz$^{0.5}$	10.3fW/Hz$^{0.5}$	N/A	1.51fW/Hz$^{0.5}$
Technology	GaAs	65nm CMOS	180nm SiGe	65nm CMOS	65nm CMOS

Figure 16.10.6: Performance summary and comparison with state-of-art imaging receivers.

16

978-1-61284-303-2/11 $26.00 © 2011 IEEE

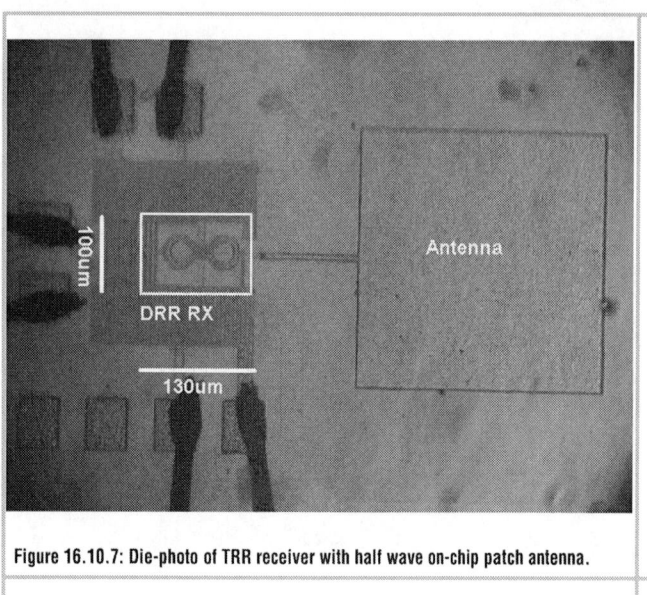

Figure 16.10.7: Die-photo of TRR receiver with half wave on-chip patch antenna.

ISSCC 2011 / SESSION 17 / BIOMEDICAL & DISPLAYS / OVERVIEW

Session 17 Overview / IMMD

Biomedical & Displays

Session Chair: *Sam Kavusi, Bosch Research and Technology Center, Palo Alto, CA*

Session Co-Chair: *Young-Sun Na, LG Electronics, Seoul, Korea*

This session features recent advances in biomedical and display circuits and systems. Biomedical highlights of this session include advances in dry electrode EEGs, neural signal acquisition and stimulation systems, and implantable pressure monitors. Another highlight involves fluorescence molecular imaging and lifetime-imaging ICs.

Paper 17.1 [IMEC; Delft U] presents an EEG system that employs gel-free active electrodes and achieves 82dB CMRR and 2GΩ input impedance. A chopper-stabilized amplifier is operated at 500Hz that leads to a total input-referred noise of 0.8µVrms (0.5 to 100Hz).

Paper 17.2 [UC Berkeley] discusses an effort to reduce the area of a DC-coupled neural-signal acquisition system by 3× compared to the state of the art. The chip is implemented in 65nm CMOS and consumes 5.04µW from a 0.5V supply and achieves 49nVrms/√Hz input noise. The system has been verified with live recordings from the motor cortex of an awake rat.

Paper 17.3 [IMMS; U Ulm] describes an AC-powered transcutaneous infrared data link for implantable systems that achieves 2Mb/s data transfer and sub-nJ/b energy consumption. Charge-balanced arbitrary stimulation waveform generation in neural stimulators intended for a 1024 electrode epiretinal stimulator is the subject of Paper 17.4 [U Ulm]. The stimulator achieves 50dB DR and is suitable for a 1024-electrode system.

Continuous epi-fluorescence-based molecular imaging is demonstrated in Paper 17.5 [Stanford U]. The system consists of a monolithically integrated GaAs detector, a vertical-cavity surface-emitting laser, and a 77dB DR readout circuit chip, implemented in 0.18µm CMOS, in a 0.7g 1cm^3 package.

Paper 17.6 [U Michigan] demonstrates a 1mm^3 intraocular pressure monitor allowing optimal implantation to track glaucoma. It achieves 0.5mmHg resolution with a MEMS sensor and $\Delta\Sigma$ converter. The system wirelessly transmits the data consuming 4.7nJ/b and harvests solar energy to recharge its battery.

The last medical paper in the session, Paper 17.7 [Delft U], presents a 160×128 pixel array, which can detect photons with 55ps resolution. Each pixel comprises a counter, a time-to-digital converter, and a 10b memory, while a frame is read out every 20µs. The sensor is well-suited for applications such as fast fluorescence lifetime imaging.

The second part of this session showcases 3 display papers. Active-matrix OLED (AMOLED) is a promising technology for small size applications due to its self-emitting and superior display characteristics. Accurate and fast current control is essential to fully reap the benefits of AMOLED.

Paper 17.8 [Fraunhofer Institute for Photonic Microsystems] presents a bidirectional microdisplay combining a near-to-eye AMOLED display with an eye-tracking image sensor, which enables gaze-based human-display interactions. Paper 17.9 [KAIST] proposes a 0.014mm^2 9b switched-current DAC structure with 1µs/b conversion rate for AMOLED displays.

Efforts to realize area-efficient DACs for higher resolution LCD source drivers are underway. Paper 17.10 [National Tsing Hua U] proposes a 10b Resistor-Resistor-String DAC (RRDAC) that eliminates loading effects with simple current sources and achieves 30% area savings with respect to an 8b RDAC.

978-1-61284-303-2/11 $26.00 © 2011 IEEE

ISSCC 2011 / February 22, 2011 / 1:30 PM

17.1 A 160µW 8-Channel Active Electrode System for EEG Monitoring   1:30 PM

J. Xu,

imec - Holst Centre, Eindhoven, The Netherlands; Delft University of Technology, Delft, The Netherlands

17.2 A 0.013mm² 5µW DC-Coupled Neural Signal Acquisition IC with 0.5V Supply   2:00 PM

R. Muller,

University of California, Berkeley, CA

17.3 An AC-Powered Optical Receiver Consuming 270µW for Transcutaneous 2Mb/s Data Transfer   2:15 PM

Maurits Ortmanns,

Ulm University, Ulm, Germany

17.4 A Neural Stimulator Front-End with Arbitrary Pulse Shape, HV Compliance and Adaptive   2:30 PM
  Supply Requiring 0.05mm² in 0.35µm HVCMOS

Maurits Ortmanns,

Ulm University, Ulm, Germany

17.5 A Low Noise Current Readout Architecture for Fluorescence Detection in Living Subjects   2:45 PM

R. T. Heitz,

Stanford University, Stanford, CA

17.6 A Cubic-Millimeter Energy-Autonomous Wireless Intraocular Pressure Monitor   3:15 PM

G. Chen,

University of Michigan, Ann Arbor, MI

17.7 A 160×128 Single-Photon Image Sensor with On-Pixel 55ps 10b Time-to-Digital Converter   3:45 PM

C. Veerappan,

Delft University of Technology, Delft, The Netherlands

17.8 Bidirectional OLED Microdisplay: Combining Display and Image   4:15 PM
  Sensor Functionality into a Monolithic CMOS Chip

U. Vogel,

Fraunhofer Institute for Photonic Microsystems, Dresden, Germany

17.9 A 0.014mm² 9b Switched-Current DAC for AMOLED Mobile Display Drivers   4:45 PM

H-S. Kim,

KAIST, Daejeon, Korea

17.10 A 10b Resistor-Resistor-String DAC with Current Compensation   5:00 PM
  for Compact LCD Driver ICs

C-W. Lu,

National Tsing Hua University, Hsinchu, Taiwan

17

ISSCC 2011 / SESSION 17 / BIOMEDICAL & DISPLAYS / 17.1

17.1 A 160µW 8-Channel Active Electrode System for EEG Monitoring

Jiawei Xu[1,2], Refet Firat Yazicioglu[3], Pieter Harpe[1], Kofi A.A. Makinwa[2], Chris Van Hoof[3]

[1]imec - Holst Centre, Eindhoven, The Netherlands,
[2]Delft University of Technology, Delft, The Netherlands, [3]imec, Leuven, Belgium

An important drawback of current biopotential monitoring systems is their dependence on gel electrodes, which can dry out, cause skin irritation, and necessitate skilled personnel. These associated drawbacks increase the running costs and significantly hamper their use in consumer healthcare and lifestyle applications. Unfortunately, the use of gel-free, or dry, electrodes increases the electrode-tissue contact impedance, thus exacerbating the effects of interference and cable motion artifacts. A solution is the use of active electrodes, i.e. electrodes in which an amplifier with high input impedance, low noise and good electrode offset rejection is co-integrated (Fig. 17.1.1). Previous active electrodes employed voltage buffers to facilitate the inter-channel gain matching necessary to achieve high CMRR [1]. However, low-noise buffers consume significant power and due to their lack of gain still require a low-noise and thus power-hungry back-end to keep the total integrated noise at acceptable levels. To reduce the total power dissipation, this paper proposes a biopotential monitoring system based on active electrodes with gain.

The proposed system consists of eight electrodes based on eight chopper-stabilized amplifiers with selectable gain (Fig. 17.1.2). To improve the CMRR between electrode pairs, which would otherwise be limited by gain mismatch, the back-end consists of a capacitive summing amplifier that feeds the average output CM voltage of all eight amplifiers back to their non-inverting inputs. This reduces the effective CM voltage applied to each amplifier and thus boosts the CMRR between electrode pairs by $20log(A_v)$, where A_v is the nominal voltage gain of each electrode. This is similar to the traditional Driven-Right-Leg (DRL) circuit [2], which can also be used to improve the CMRR of active electrodes. However, DRL circuits suffer from stability problems and increased power dissipation, since the CM signal must be fed back through an electrode with significant impedance.

State-of-the-art biopotential instrumentation amplifiers are not well suited for use in active dry electrodes. The input impedance of a chopper-stabilized capacitively coupled amplifier [4] is limited by the switched-capacitor impedance formed by the input choppers and the input capacitors. Moving the chopper to the amplifier's virtual ground solves this problem [6], at the expense of CMRR, which is now limited by capacitor mismatch [6]. A current-feedback amplifier has higher input impedance [5], but its DC-servo loop limits the maximum electrode offset (EO) that it can handle to tens of mV.

The proposed active electrode (Fig. 17.1.2) consists of a chopper-stabilized inverting amplifier, an input-impedance boosting loop, and two calibration loops to improve the input impedance and compensate the offset of the amplifier respectively. The amplifier's midband gain $A_v=C_2/C_1$ and different gains can be realized by switching between different values of C_2. The AC coupling capacitor C_1 effectively rejects any EO, while the pseudo-resistor R_2 and capacitors C_2 implement a high-pass characteristic. However, the switched-capacitor impedance formed by the input chopper and the amplifier's input capacitance accentuates its low-frequency noise [6]. To mitigate this C_1 must be large (300pF), but this reduces the input impedance of the amplifier. Therefore, a positive feedback loop (Fig. 17.1.3) is used to boost the amplifier's input impedance [3]. Via C_{fb}, the loop supplies part of the current in C_1 (I_{in}), thus reducing the current drawn from the recording electrode (I_{el}). C_{fb} is implemented as the combination of a coarse and fine capacitor array. At the various gain settings, the capacitors of the coarse array are switched in tandem with the value of C_2. The fine array can then be adjusted to compensate for the current that flows into the bond-pad and other parasitic capacitances.

Two design challenges associated with chopper-stabilized amplifiers are the reduction of their output ripple and residual offset. The output ripple is due to the amplifier's up-modulated offset, and resembles a low-pass filtered square wave (Fig. 17.1.4). Compared to the µV level biopotential signals, the ripple can be quite large and therefore limits the amplifier's headroom. The amplifier in [3]

employs a ripple-reduction loop to sense and compensate output ripple. However, it operates continuously and thus increases the amplifier's power consumption. A further challenge is the residual offset caused by the clock feedthrough or charge injection of the input chopper. This generates an offset current in pseudo-resistor R_2, which can lead to a large output DC voltage drift. In [6], this offset current is canceled by a G_m-C servo-loop. However, it uses a large off-chip capacitor (>10µF) to realize a sufficiently low cutoff frequency.

In the proposed active electrode, both the ripple and offset challenges are met by digitally assisted calibration loops: a ripple-reduction loop (RRL) and a DC servo loop (DSL) (Fig. 17.1.4). These exploit the fact that both the ripple and the offset are relatively static and so do not need to be continuously reduced. The calibration starts with the RRL: the peak ripple levels (V_a and V_b) are synchronously sampled, and its polarity is determined by a comparator. A logic circuit, implementing the SAR algorithm, then sets the input to a 7b current DAC in seven clock cycles. The DAC is implemented by two current mirror arrays CA_1 and CA_2. The DSL operates in a similar manner: V_{out} is sampled and compared to V_{ref} and the results are used to set the input of another pair of current DACs. Their outputs are chopped in order to generate a modulated compensation current. Once the calibration is finished, the inputs to the current DACs are frozen, the calibration loops are shut-down and normal signal amplification starts. Thereafter, the power dissipation of the RRL and DSL circuits is mainly determined by the DAC currents, and is less than 400nW.

The front and back-end of the eight-channel active electrode system were implemented in a standard 0.18µm CMOS process, and consumes a total 160µW from a 1.8V supply. Fig 17.1.7 shows the chip micrograph of one active electrode and back-end CMFB. Figure 17.1.5 shows the measurement results. The top-left plot shows the amplifier's noise. Chopping at 500Hz leads to a total input-referred noise of 0.8µVrms (0.5-100Hz). The top-right plot shows that at 50Hz the back-end CMFB improves the CMRR by 35dB, to 82dB. The bottom-left plot shows the 2GΩ input impedance achieved by the impedance-boosting loop. The bottom-right figure shows that the RRL reduced the output ripple from 40 to 2mV. Although not shown, the DSL reduces the output drift from 280 to 20mV. Figure 17.1.6 compares the performance of the active electrode system with the state-of-the-art. Occipital EEG measurement result using the presented active electrodes is shown below the table.

In conclusion, the proposed biopotential amplifier is well suited for use in gel-free active electrode applications. Its performance exceeds the state-of-the-art in terms of its CMRR, input impedance, offset rejection, and power-efficient offset compensation. This is realized through the use of an AC-coupled chopper-stabilized amplifier, in conjunction with back-end CM feedback, input impedance boosting and digital offset calibration.

Acknowledgment:
The authors would like to thank Q. Fan, C. van Liempd, M .Pertijs and J.H. Huijsing for valuable discussion and comments.

References:
[1] M. Fernandez and R. Pallas-Areny, "A simple active electrode for power line interference reduction in high resolution biopotential measurements," *IEEE EMBS*, vol.1, pp.97–98. 1997.
[2] Bruce B. Winter, John G. Webster, "Driven-Right-Leg Circuit Design", *IEEE Transaction on Biomedical Engineering*, Vol. 30, No. 1, pp.62-66, Jan. 1983
[3] Q. Fan et al., "A 1.8µW 1µV-Offset Capacitively-Coupled Chopper Instrumentation Amplifier in 65nm CMOS", to be published in *IEEE ESSCIRC*, 2010.
[4] T. Denison et al., "A 2.2µW 94nV/√Hz, Chopper-Stabilized Instrumentation Amplifier for EEG Detection in Chronic Implants," *ISSCC Dig. Tech. Papers*, pp. 162-594, Feb. 2007.
[5] R. F. Yazicioglu et al., "A 60 µW 60 nV/√Hz Readout Front-End for Portable Biopotential Acquisition Systems," *IEEE J. Solid-State Circuits*, pp. 1100-1110, May 2007.
[6] N. Verma et al., "A Micro-Power EEG Acquisition SoC With Integrated Feature Extraction Processor for a Chronic Seizure Detection System," *IEEE J. Solid-State Circuits*, pp. 804-816, April 2010.
[7] X. Zou et al., "A 1-V 450-nW Fully Integrated Programmable Biomedical Sensor Interface Chip," *IEEE J. Solid-State Circuits*, pp. 1067-1077, April 2009.

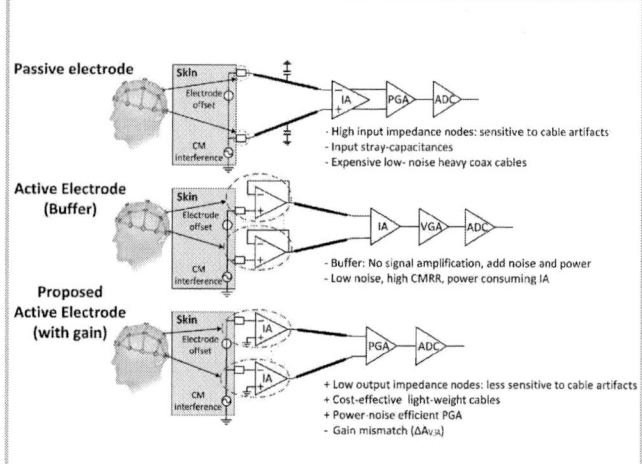

Figure 17.1.1: EEG monitoring systems based on passive electrodes, active electrodes (buffer) and the proposed active electrodes (with gain).

Figure 17.1.2: Block diagram of the proposed 8-channel active electrode system.

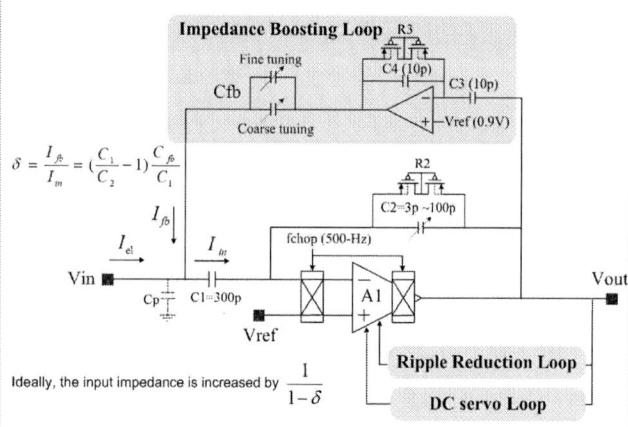

Figure 17.1.3: Input-impedance-boosting circuit.

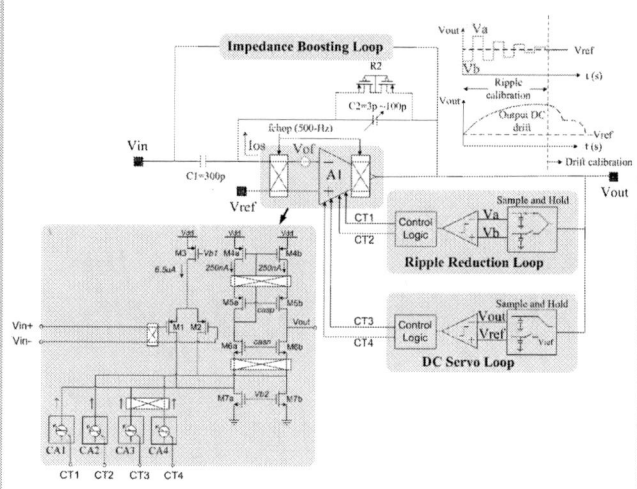

Figure 17.1.4: Current-DAC based offset-calibration circuit.

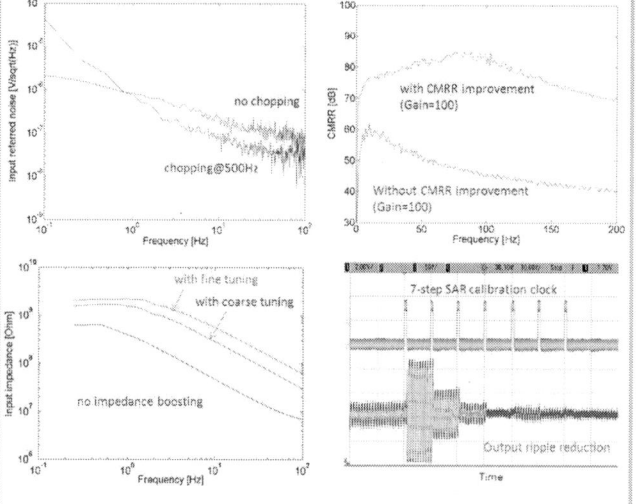

Figure 17.1.5: Measurement results.

Figure 17.1.6: Comparison table with published state-of-the-art and EEG signal measurement result.

Figure 17.1.7: Photograph of prototype ASIC (one active electrode and back-end CMFB).

ISSCC 2011 / SESSION 17 / BIOMEDICAL & DISPLAYS / 17.2

17.2 A 0.013mm² 5µW DC-Coupled Neural Signal Acquisition IC with 0.5V Supply

Rikky Muller[1], Simone Gambini[1,2], Jan M. Rabaey[1]

[1]University of California, Berkeley, CA, [2]Telegent Systems, Sunnyvale, CA

Recent success in brain-machine interfaces has provided hope for patients with spinal-cord injuries, Parkinson's disease, and other debilitating neurological conditions [1], and has boosted interest in electronic recording of cortical signals. State-of-the-art recording solutions [2-5] rely heavily on analog techniques at relatively high supply voltages to perform signal conditioning and filtering, leading to large silicon area and limited programmability. We present a neural interface in 65nm CMOS and operating at a 0.5V supply that obtains performance comparable or superior to state-of-the-art systems in a silicon area over 3× smaller. These results are achieved by using a scalable architecture that avoids on-chip passives and takes advantage of high-density logic. The use of 65nm CMOS eases integration with low-power digital systems, while the low supply voltage makes the design more compatible with wireless powering schemes [6].

The chip architecture is shown in Fig. 17.2.1. The input is DC-coupled, and the common mode is stabilized through a large (>1GΩ) on-chip resistor. This saves the area of large (often 10s of pF) AC-coupling capacitors. Since the resistor is not used to form a filter time constant, the system is agnostic to its variations. Mixed-signal feedback is used to cancel the electrode offset and separate the 1-to-300Hz Local Field Potential (LFP) band from the 300Hz-to-10kHz Action Potential (Spike) band. Two separate feedback loops are used: an outer low-noise DAC provides coarse offset suppression at 1Hz or less, while the inner loop performs LFP filtering and fine residual offset correction. The Spikes are available at the output of the ADC while the LFP signal becomes available at the input of the inner DAC. The mixed-signal feedback replaces traditionally analog filters and enables the system to operate with an 8b ADC while digitizing both the Spikes and LFPs. To minimize area and power consumption the feedback is split into a dual-loop architecture which reduces the very large (16b) DAC resolution requirement of filtering the offset and LFP together into two moderate resolution DACs. A low-pass digital IIR filter in the feedback path creates a high-pass filter response to separate the Spike and LFP bands. The closed-loop transfer function can be digitally configured, greatly increasing system flexibility.

Figure 17.2.2 shows the circuit schematics of the individual blocks. The first amplifier (LNA) of the signal chain needs to achieve a low input-referred noise and to provide high enough gain to suppress noise from the subsequent blocks, while rejecting the large electrode offset. At low frequency, both neural signals and transistors exhibit a 1/f power spectrum, therefore the low-frequency LFP band can absorb larger noise spectral density while maintaining SNR [7]. The key challenge is therefore to keep the Spike band noise low. Due to the low supply voltage, cascode topologies are impractical, and hence two differential pairs with PMOS triode loads are cascaded. Weak cross-coupled pairs are used to boost the LNA gain to 32dB. The offset is cancelled by changing the relative widths of the input differential pair transistors. For sub-threshold devices, the input-referred offset is given by Eq. 1:

$$\Delta V_{GS} = n \frac{kT}{q} \ln(k_w) \quad (1)$$

where k_w is the ratio of the widths of the input devices and n is the subthreshold slope factor. For sub-threshold devices g_m is independent of device width, thus the offset is canceled without compromising CMRR while maintaining a gain and input-referred noise voltage independent of DAC code. The measured characteristic of the offset DAC is shown in Fig. 17.2.3. The 7b, thermometer-coded offset DAC uses a nonlinear unit element coding to cancel ±50mV of electrode offset with an LSB of 0.8mV and a maximum DNL of 0.55LSB for a worst-case residual offset of 1.2mV.

Since thermal noise constraints of the second summing stage are relaxed by the gain of the LNA, a straightforward current summing DAC architecture is chosen (Fig. 17.2.2). To keep quantization noise from degrading the system SNR, a dynamic range of 9b is necessary to handle the LFP and residual offset. Oversampling and Sigma-Delta encoding are used to achieve this dynamic range in a small footprint by using a 4b thermometer-coded DAC with first-order noise shaping running at 2MHz.

To achieve minimal area, an 8b counter-based ADC is used (Fig. 17.2.2). The integrating transfer function of this ADC provides anti-aliasing and filters out-of-band noise from the sigma-delta DAC. A pseudo-differential, current driven architecture mitigates supply noise coupling and minimizes distortion [9]. A differential pair is used for V-I conversion, paired with an active load that provides a programmable, linear current-mode gain that tunes the full-scale range of the ADC from 870µV to 3.5mV when referred to the channel input. Measurements of a standalone ADC structure show 7.15 ENOB at 20kS/s with 240nW power consumption, leading to a FOM of 84fJ/step in 0.0018mm².

The prototype was fabricated in a 65nm 1P7M LP-CMOS technology. A microphotograph is shown in Fig. 17.2.7. The overall die area is pad-limited to 1.4mm², while the core channel area is 70×180µm². The entire channel consumes 5.04µW from a 0.5V supply. All signal processing is performed on chip, with the exception of the digital filter, which was implemented on an FPGA. Synthesized, the filter occupies 0.0017mm².

Figure 17.2.4 shows measurements of the closed-loop transfer function of both the LFP and Spike band outputs. The high-pass filter pole location is generated by integrator-based feedback and is digitally tunable. Measurements of the Spike and LFP band noise floors are also shown in Fig. 17.2.4. 4.9µV of input-referred noise is achieved in a 300Hz-to-10kHz bandwidth.

The system was further verified with the sensor in a realistic, 60Hz environment by performing live recordings from the motor cortex of an awake rat. The rat was implanted with two 2×8 Plexon Platinum-Iridium microwire arrays 2 months prior to recording. Figure 17.2.5 shows system outputs with separated Spike band (middle) and LFP band (below.) A high-pass filter pole frequency of 300Hz was chosen.

Figure 17.2.6 shows the performance summary and comparison with state-of-the-art neural signal acquisition systems. For this work, the specifications given in Fig. 17.2.6 are for the entire acquisition chain including the ADC, while the specifications for [2-5] are given for the LNA/Bandpass filter only. The total silicon area of this work is 3× smaller than that of the smallest amplifier previously reported [2]. High CMRR is also maintained despite the low V_{DD}. The power efficiency of this amplifier is compared through both the popular NEF metric [8] and the modified metric,

$$NEF^2 \cdot V_{DD} = V_{rms,in}^2 \left(\frac{2P_{tot}}{\pi \cdot kT/q \cdot 4kT \cdot BW} \right) \quad (2)$$

which is dependent on power rather than current. Given that power dissipation in neural amplifiers is typically limited by input-referred noise, and that this design achieves the lowest $NEF^2 \cdot V_{DD}$, utilizing the proposed techniques to reduce area and operating voltage does not come at the cost of system power efficiency.

Acknowledgments:
The authors thank the sponsors of BWRC, BDA, and STMicroelectronics for chip fabrication. Thanks to J. Carmena, A. Koralek and S. Venkatraman for rat cortical recordings, and to B. Boser and E. Alon for technical discussion.

References:
[1] M. Lebedev, et al., "Brain-machine interfaces: past, present and future," *Trends Neurosci*, vol. 29, no. 9, pp. 536-546, Sept. 2006.
[2] J. Aziz, et al., "256-Channel Neural Recording and Delta Compression Microsystem with 3-D Electrodes," *IEEE J. Solid-State Circuits*, vol. 44, no. 3, March 2009.
[3] W. Wattanapanitch, et al., "An Engergy-Efficient Micropower Neural Recording Amplifier," *IEEE Trans. BioCAS*, vol. 1, no. 2, June 2007.
[4] R. Harrison, et al., "A Low-Power Integrated Circuit for a Wireless 100-Electrode Neural Recording System," *IEEE J. Solid-State Circuits*, vol. 42, no. 1, Jan. 2007.
[5] Z. Xiao, et al.. "A 20µW Neural Recording Tage with Supply-Current-Modulated AFE in 0.13µm CMOS," *ISSCC Dig. Tech. Papers*, Feb. 2010.
[6] S.O'Driscoll, et al., "A mm-sized implantable power receiver with adaptive link compensation," *ISSCC Dig. Tech. Papers*, Feb. 2009.
[7] S. Venkatraman, et al., "Exploiting the 1/f Structure of Neural Signals for the Design of Integrated Neural Amplifiers," *IEEE EMBS*, Sep. 2009.
[8] M. Staeyaert, et al., "A Micropower Low-Noise Monolithic Instrumentation Amplifier for Medical Purposes," *IEEE J. Solid-State Circuits*, vol. SC-22, no. 6, Dec. 1987.
[9] U. Wismar, et al., 'Linearity of bulk-controlled inverter ring VCO in weak and strong inversion," *AICSP*, vol. 5, pp. 59-66, 2007.

ISSCC 2011 / February 22, 2011 / 2:00 PM

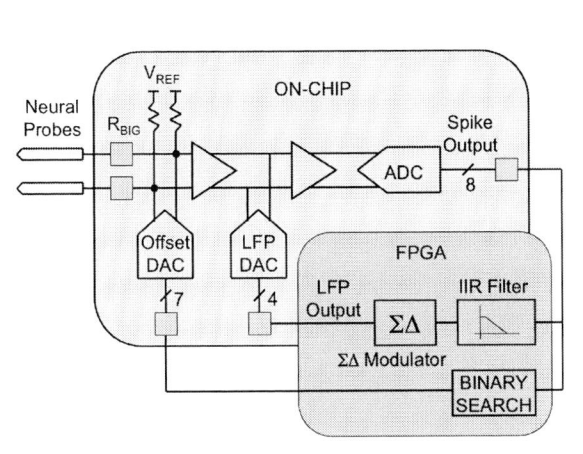

Figure 17.2.1: System block diagram.

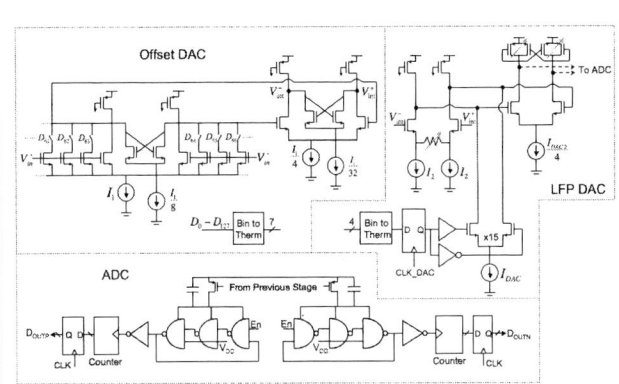

Figure 17.2.2: Circuit schematics.

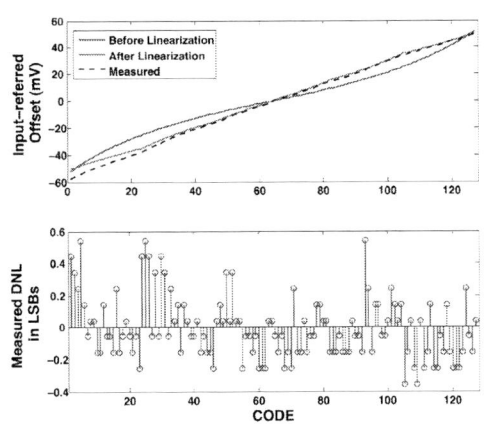

Figure 17.2.3: Simulated and measured offset DAC transfer function and measured DNL. Before linearization uses identically sized elements, while after linearization uses nonlinear coding.

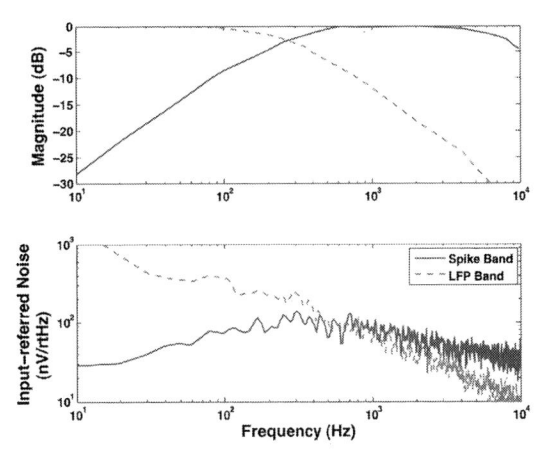

Figure 17.2.4: Measured neural signal acquisition system closed-loop magnitude (above), and Spike and LFP band input-referred noise (below).

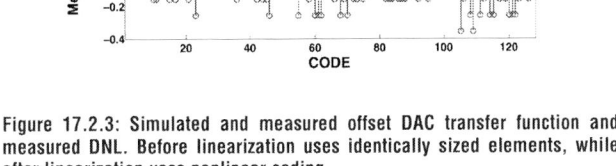

Figure 17.2.5: On-line recordings from the motor cortex of a live awake rat. Top shows complete input waveform before filtering. Middle and bottom show on-chip filtered LFP and Spike outputs.

	Table of Comparisons				
	[2]	[3]	[4]	[5]	This Work
Power (μW)	15	7.56	42.2	0.64	5.04
IRNoise (μV)	7.0	3.06	5.1	14	4.9
Bandwidth	5kHz	5.3kHz	5kHz	6.2kHz	10kHz
NEF	4.6	2.67	9.8	6.5	5.99
NEF^2-VDD	63.48	20	316.9	33.8	17.96
CMRR (dB)	-	66	-	59	75
PSRR (dB)	-	75	-	71	64
V$_{DD}$ (V)	3	2.8	3.3	0.8	0.5
Area (mm^3)	0.04	0.16	0.16	0.4*	0.013
Technology	0.35μm	0.5μm	0.5μm	0.13μm	65nm
Blocks included in comparison	LNA, BPF	LNA, BPF	LNA, BPF	LNA, BPF	LNA, BPF, ADC

*estimated

Figure 17.2.6: Performance summary and comparison with prior art.

ISSCC 2011 PAPER CONTINUATIONS

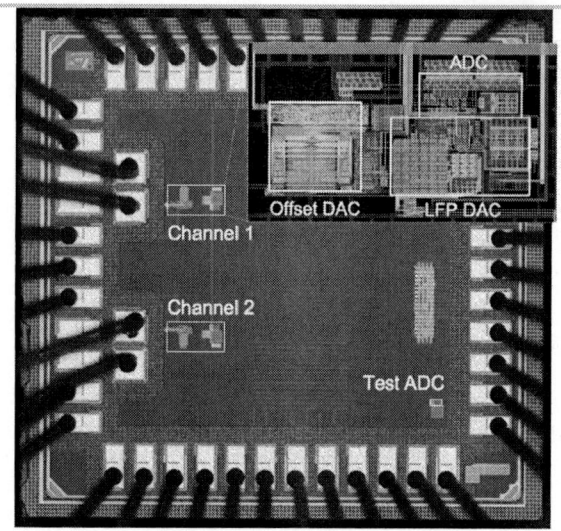

Figure 17.2.7: Die microphotograph; inset shows detailed layout.

ISSCC 2011 / SESSION 17 / BIOMEDICAL & DISPLAYS / 17.3

17.3 An AC-Powered Optical Receiver Consuming 270µW for Transcutaneous 2Mb/s Data Transfer

Steffen Lange[1], Hongcheng Xu[2], Christian Lang[1], Holger Pless[1],
Joachim Becker[2], Hans-Jürgen Tiedkte[3], Eckhard Hennig[1],
Maurits Ortmanns[2]

[1]Institute for Microelectronic and Mechatronics Systems, Ilmenau, Germany,
[2]Ulm University, Ulm, Germany, [3]Intelligent Medical Implants, Bonn, Germany

Improving communication with implantable systems remains an important topic of research due to the limitations in power dissipation and the simultaneous need for high data rates. Neural recorders generate well above 10Mb/s data [1], which needs to be transmitted out-of-body. Multichannel stimulators, such as epiretinal implants, need control data in the range of several Mb/s for the into-body link [2]. Up to now, RF has been the dominant form of transcutaneous communication. One major issue is the crosstalk between the RF power link and the data signal. Therefore, dual-band telemetry is common in order to spectrally separate the data and power transfer. The standards reach from UWB transmitters [1], MICS band [3], to customized [2] RF receivers, often using sophisticated digital encoding. Also, orthogonal alignment has been used for the data and power coils to suppress crosstalk. Such RF communication needs a 2nd pair of coils, and the state-of-the-art power consumption ranges from 1.5 to 3nJ/b at rates of 120kb/s to 2.5Mb/s [3].

Optical transcutaneous data links have rarely been researched, and mostly out-of-body communication with high data rates has been investigated: up to 40Mb/s has been achieved with high power consumption [4], and an obvious drawback was the implantation of a photoemitter as part of the system. In [5], optical communication was used to control an epiretinal stimulator, as illustrated in Fig. 17.3.1. The photoemitter sits in front of the lens, and a receiving discrete photodiode (PD) is placed in the line of sight. The PD is connected to a transimpedance amplifier (TIA), which is integrated on the stimulator ASIC.

This setup has major disadvantages: the discrete PD has a large junction capacitance of 5 to 10pF, which determines the power consumption of the TIA of several 100µW. Also, the PD is a bulky device with 1.4×1.4×0.8mm³, which resides at a very sensitive location, next to the attachment of the stimulation electrodes to the tissue. Most severely, the PD is DC-connected to the TIA, which works in the LV range between V_{SS} and 3.3V. Since the stimulator driver works between V_{SS} and $V_{DDP}\approx$20V, the body potential is connected to the center $V_{CM}\approx$10V. Thus, the 2 wire connections between the PD and the stimulator ASIC conduct V_{SS} (anode) and V_{SS}+1.3V (cathode) [5], both of which are ~10V$_{DC}$ under body potential, which is a major challenge for isolation.

This paper proposes a solution for a trancutaneous IR data link using an active device. The architecture is shown in Fig. 17.3.2. An integrated PD receives a Manchester-coded IR data stream up to 2Mb/s. The photocurrent is digitized and generates a bipolar output current $I_{out}=\pm$20µA which is then capacitively coupled to the stimulator ASIC. The DC potential of the output is controlled to V_{CM}. The system is AC-powered at 13.56MHz, with V_{AC} centered on body potential, and has an on-chip buffer capacitor. Thus, the optical receiver has no DC terminals, which is very favourable for wireline, inbody communication and the corresponding isolation [6].

Figure 17.3.3 shows the implementation of the high-efficiency rectifier. A voltage doubler is chosen – the forward diode is a PMOS-diode with bulk-regulation of the n-well, and the backward diode is an isolated NMOS-diode with 2 regulations for both isolated wells. In order to increase the voltage efficiency, a threshold-cancellation technique is applied: the reference diodes $M_{b1,2}$ are slightly biased, which sets V_{GS} of $M_{1,2}$ close to their threshold. Thus, the rectifying diodes operate at the border to turn on and a low forward drop is achieved. The efficiency of the rectifier allows the usage of 5V devices for $M_{1,2}$ to achieve a supply of $V_{DD}-V_{SS}<$3.5V. Figure 17.3.4 shows the output characteristic of the rectifier under a load current of 37µA, and a voltage efficiency of 75% over the whole range is achieved. The chip area is dominated by the on-chip buffer capacitor of $C_L\approx$5nF. This seems inefficient for an integrated circuit, but is very favourable compared to external solutions concerning weight and size as well as wiring of DC signals.

An integrated XFAB PIN PD is used as the sensor element for the IR signal. This device is prepared with a 14µm p-epi layer. The sensitivity of the PD at 850nm wavelength is about 0.3A/W. The PD is centered on the chip and the size is about 90,000µm². The photocurrent is fed into a mirror, amplified, and the output of the cascode is digitized using a comparator, which generates the data signal in Fig. 17.3.2. Two nonidealities need to be tolerated by the receiver: (1) a wide range of ambient light, which is quasistatic and adds a DC-signal onto the signal DC-value, and (2) the optical link can be disrupted e.g. by blinking. For DC-control, the recovered data is used as input to a current-integrating bang-bang controller. The integrator output V_{Ctrl} generates a positive current $I_{Ctrl\|DC}$ into the input node, which counteracts the negative DC photocurrent. The time constant of this control loop can be very large, typically τ_{DCtyp}=200µs, and thus the controller can be implemented with very low power.

In case of a link disruption, the output must not be supplied with a static current, but shorted to the body potential. Therefore, a no-signal detector is implemented as shown in Fig. 17.3.2 with its functionality illustrated in Fig. 17.3.5: a bias current of 50nA permanently charges a capacitor of 100fF, and the resulting voltage is digitized in a low-power inverter. Every edge of the recovered data resets the capacitor. In case of disruption, the capacitor fills up to V_{DD} within a few µs, and the inverter detects a no-signal condition. This causes the output to be shorted to V_{CM}. A major issue of link disruption is the detuning of the DC-control loop: since the bang-bang control is implemented with a large time constant, the continuing data needs to retune the DC-loop and only then is data received. This undesirably yields an even longer interruption of data reception. Therefore, the no-signal condition increases the biasing of the bang-bang-current, and its time constant is decreased to $\tau_{DCnosig}$=20µs until signal reception is achieved. The bipolar output current ideally achieves a DC-free terminal. To account for mismatch in time and amplitude, a DC-control loop is implemented, which steadily keeps the output at V_{CM}. A simple GmC integrator with class A/B output stage (Fig. 17.3.2) effectively counteracts a drift of the output.

The IR receiver works over a large supply range of $V_{DD}-V_{SS}$=2.2 to 3.5V, and thereby consumes 100 to 150µW from the internal supply, and 190 to 270µW from the AC supply. Figure 17.3.6 shows an eye diagram of the receiver with an optical input power of P_{IN}=5µW/mm², data at 2Mb/s and the AC supply at 3.2V$_{pp}$. Excellent reception with a timing jitter below 5ns is measured. The link is intended for 2Mb/s Manchester data transfer but can operate up to a rate of 3Mb/s. The AC-coupled IR receiver, including rectification, buffering, data reception and bipolar data generation consumes 1.4×1.4mm² in 0.35µm CMOS and is thinned to 300µm thickness. Including the output current, the energy per bit is as low as 0.135nJ/b at maximum supply and 2Mb/s, and can go as low as 66pJ/b for low supply at the maximum rate of 3Mb/s. This is an improvement of more than one order of magnitude compared to the state of the art of transcutaneous RF communication.

Acknowledgements:
This work has been partially funded by the German Federal Ministry of Education and Research (BMBF) under project *FutureRet.*

References:
[1] M. Chae, W. Liu, Z. Yang, T. Chen, J. Kim, M. Sivaprakasam, and M. Yuce, "A 128-Channel 6mW Wireless Neural Recording IC with On-the-Fly Spike Sorting and UWB Tansmitter", *ISSCC Dig. Tech. Papers*, Feb. 2008.
[2] K. Chen, Z. Yang, L. Hoang, J. Weiland, M. Humayun, and W. Liu, An Integrated 256-Channel Epiretinal Prosthesis, *IEEE J. Solid-State Circuits*, vol. 45, no. 9, pp. 1946-1956, Sept. 2010.
[3] J. Bae, N. Cho, and H.-J. Yoo, "A 490uW Fully MICS Compatible FSK Transceiver for Implantable Devices", *VLSI Symp.*, pp. 36-37, 2009.
[4] K.S. Guillory, A.K. Misener, and A. Pungor, "Hybrid RF/IR Transcutaneous Telemetry for Power and High-Bandwidth Data", *IEEE Eng. Med. Bio. Conf.*, Sept. 2004.
[5] M. Ortmanns, N. Unger, A. Rocke, M. Gehrke, and H.J. Tiedtke: "A 232-Channel Epiretinal Stimulator ASIC", *IEEE J. Solid-State Circuits*, vol. 42, no. 12, pp. 2946-2959, Dec. 2007.
[6] A. Rothermel, et al., "A 1600-pixel Subretinal Chip with DC-free Terminals and ±2V Supply Optimized for Long Lifetime and High Stimulation Efficiency", *ISSCC Dig. Tech. Papers*, Feb. 2006.

ISSCC 2011 / February 22, 2011 / 2:15 PM

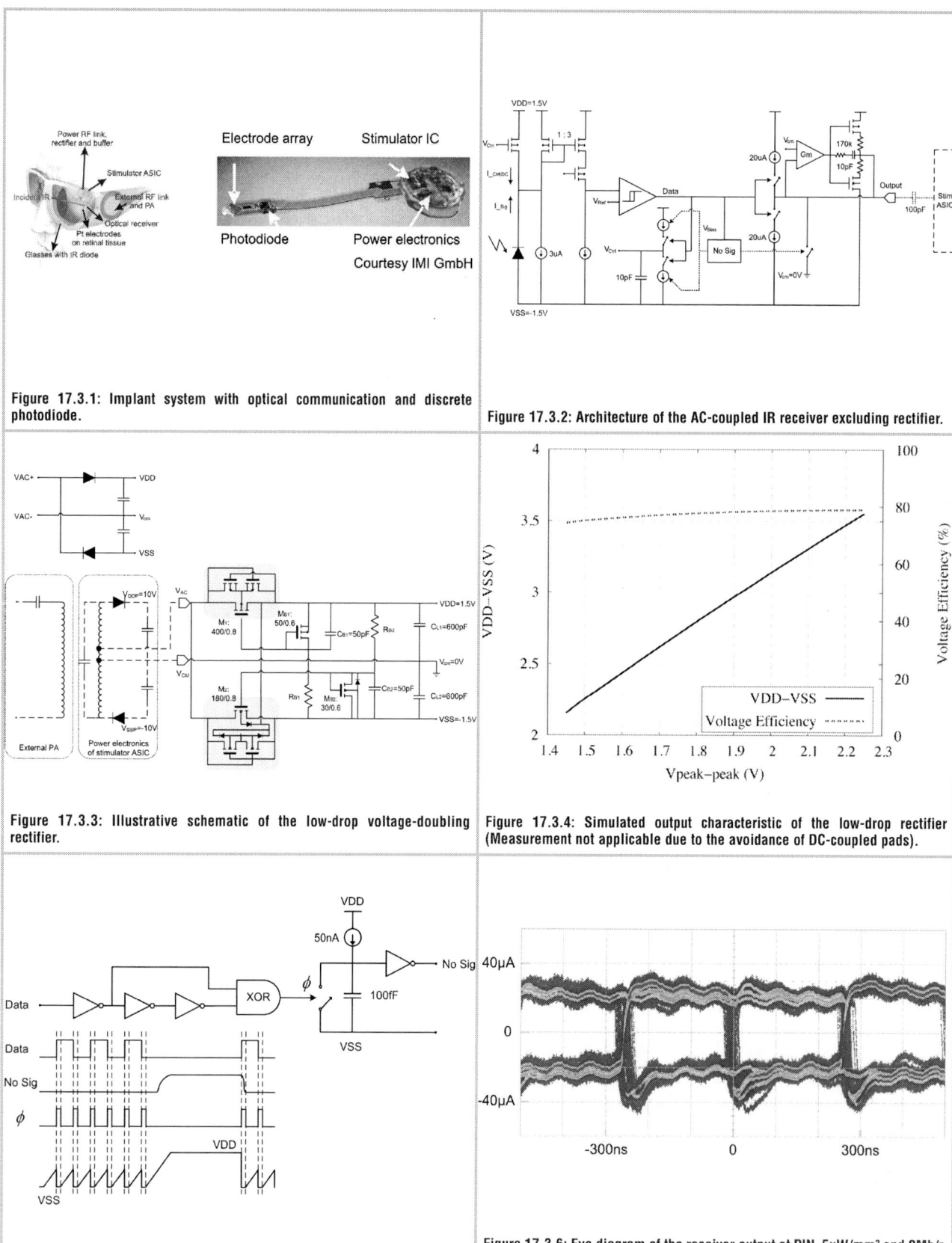

Figure 17.3.1: Implant system with optical communication and discrete photodiode.

Figure 17.3.2: Architecture of the AC-coupled IR receiver excluding rectifier.

Figure 17.3.3: Illustrative schematic of the low-drop voltage-doubling rectifier.

Figure 17.3.4: Simulated output characteristic of the low-drop rectifier (Measurement not applicable due to the avoidance of DC-coupled pads).

Figure 17.3.5: Illustrative schematic of the no-signal detector.

Figure 17.3.6: Eye diagram of the receiver output at PIN=5µW/mm² and 2Mb/s Manchester data.

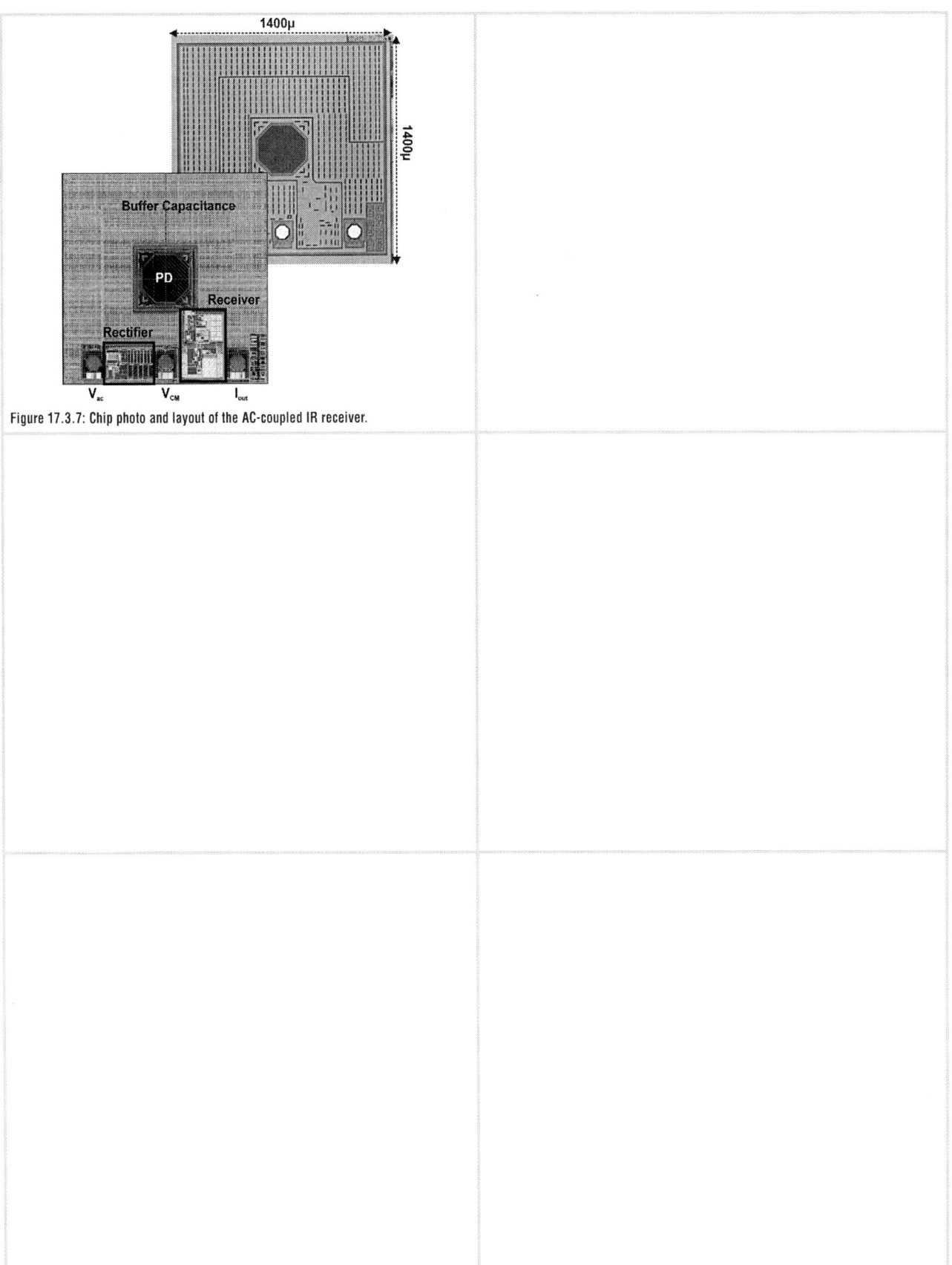

Figure 17.3.7: Chip photo and layout of the AC-coupled IR receiver.

ISSCC 2011 / SESSION 17 / BIOMEDICAL & DISPLAYS / 17.4

17.4 A Neural Stimulator Front-End with Arbitrary Pulse Shape, HV Compliance and Adaptive Supply Requiring 0.05mm² in 0.35µm HVCMOS

Kriangkrai Sooksood, Emilia Noorsal, Joachim Becker, Maurits Ortmanns

Ulm University, Ulm, Germany

The required number of channels for neural stimulation is increasing; while *in vitro* arrays reached more than 10000 sites [1], implantable systems, e.g. for retinal [2,3] or cortical stimulation [4], have not exceeded above 1600 [5] – which was achieved with a reduction to basic functionality and wireline control of rectangular stimulation pulses.

In contrast, experiments prove that a higher degree of flexibility of stimulators is needed: high-impedance applications require more than 10V output compliance [2], while low-voltage (LV) drivers satisfy low-threshold applications [5]. Also a higher flexibility in the waveshape has become of major interest: while most implantable systems feature rectangular waveforms, neural research reveals advantages for more sophisticated shapes, e.g. concerning charge injection or stimulation thresholds [6]. Flexibility is also beneficial to adapt the neural implant to new stimulation strategies or to an individual patient's needs.

The bottleneck for implantable multichannel systems with high flexibility is the size and power consumption of the stimulator frontend. Large memories, state machines or microprocessors in the stimulator sites are usually needed. In [4], the overhead was circumvented by controlling the stimulation wirelessly, which imposes a risk of harmful operation in case the link breaks. In most other designs, the large area consumption was caused by the local generation of the waveforms at each pixel.

This work presents a flexible frontend for a programmable implantable stimulator. It enables virtually arbitrary stimulation waveshapes, features a low-area and high-voltage (HV) output driver, and allows adaptation of the supply to the electrode impedance and stimulation current using a compliance monitor. The area efficiency is achieved with digital global waveform generation and its distribution to the local stimulators over a bus. This is based on the experience that once an optimal waveform is found, the shape remains unique for all sites [2], and the strength of the stimulation is selected over the amplitude at each electrode individually.

The system in Fig. 17.4.1 is an extended architecture of [2]. There are global functions such as power management, communication and a digital stimulation control (GSU), as well as an array of 10b digitally addressable stimulation pixels. Each local stimulation unit (LSU) features a 5b current DAC with 4 selectable output ranges [2], a high-compliance HV current driver with a compliance monitor, a S&H circuit for voltage measurement, charge balancers, a digital control (PCU), and 4 multiplexed electrodes. In the presented prototype, an integrated version of the frontend is realized by means of 2 LSUs, while the GSU is implemented in an FPGA for test.

The GSU receives a data stream to setup stimulation. The timing profile is sequentially received and stored within the GSU. 3 bits adjust the LSB of the stimulation timing from 4 to 512µs. Thereafter, a sequence of commands is received and stored as 11b packets: 5b representing an executable command and 6b for the associated time of execution. After the reception of the timing, the GSU accepts local stimulation data, which consists of 10b address and 9b amplitude values, and is immediately stored in the addressed PCU. One local data frame can contain up to 256 packets.

After successful reception of the local data, the stimulation starts. The stimulation state machine (Fig. 17.4.2) counts the time ΔT_i to zero in multiples of the programmed timing LSB. Then, the associated command is distributed over a bus, and executed by all previously programmed LSUs. Thus, the local drivers follow the global stimulation waveshape. Stimulation commands are used to start, stop and invert the stimulation current, as well as to modify the local DAC registers, which store the stimulation amplitude, by counting or shifting. Additionally, the stimulation commands can trigger single or combined versions of active and passive charge balancers as well as enable the S&H.

Voltage measurement via S&H is only intended at a single electrode at a time [2]. Thus, a 5th bit was added to the stimulation commands, which corresponds to a

dedicated selection bit stored in the local PCU data. Only PCUs with set selection bit listen to commands with a leading "1". This is intended for single S&H activation, but it can also be used to execute different waveshape commands at two LSU subsets.

A loop command can also be included into the timing profile, which enables an easy method to realize trains of pulses or other repetitive stimulation. Exemplary generated waveforms at 2 different electrodes are shown in Fig. 17.4.3: a simple biphasic pulse, next a double-exponentially shaped pulse and a simple rectangular pulse in parallel, and finally a train of pulses. For these waveforms, 3, 13 and 23 timing commands were stored, respectively.

Figure 17.4.4 shows the HV current driver, whose core and output stage work on the low 3V and variable HV supply V_{DDP}. The output stage consists of current sources $M_{1,3}$ biased in deep triode with regulated cascode [4]. In order to reduce the area, no digital-to-bias voltage converter is used, but the driver is based on a wide swing current mirror, where also the diode branch ($M_{1m,3m}$) is biased in deep triode. Figure 17.4.5 shows the measured output current versus output voltage. With $V_{DN}=200mV$, the driver only needs $|V_{out}|>250mV$ to achieve $r_{out}>10M\Omega$ over the current range of 4µA to 1mA. This enables a very large output swing and power efficient stimulation.

In order to adapt the supply V_{DDP} to the current load and drive strength, a compliance monitor is implemented. When a steadily reducing output voltage starts to push either $M_{2,4}$ into the triode region, the loop amplifiers increase their respective V_{GS}. This is monitored with a simple inverter and the result can be used to adapt V_{DDP}. Figure 17.4.6 shows an increasing biphasic stimulation current from 600 to 1000µA into a 1kΩ and 10kΩ load with $V_{DDP}=20V$. When the electrode voltage approaches the supply rails, the voltage-compliance monitor activates. The loop amps A_{1-4} consume 4.5µW, and the overall bipolar HV driver with 4 outputs, 5b DAC and compliance monitor occupies 0.06mm².

Safe stimulation requires charge-balanced pulses [2,4,5]. It was shown that for high-impedance microelectrodes passive discharge can become ineffective [7]. Active balancing is based on monitoring the electrode voltage with a window comparator after stimulation and applying balancing charge pulses [2]. In this prototype, offset-based charge balancing is introduced [7], where the electrode monitoring is used to adjust an offset in the driver output stage by means of $I_{Opull/push}$. Thus, the charge balance is not immediately achieved, but the drivers are matched over time. The LSU allows combinations of active spike [2], the new active offset, and simple passive balancing, the latter shortening the electrode to the body potential.

Figure 17.4.7 shows the prototype. One LSU consumes 0.2mm² with 20% needed by the 4 pads. It has the highest density of functions among published stimulators referenced herein. The frontend is intended for a 1024 epiretinal stimulator, which can be realized on less than 8×8mm² with all new features. This increases both dimensions by roughly 60% compared to [2], which is feasible for implantation.

Acknowledgements:
This work was partially funded by the German Ministry of Research and Education under BMBF project *FutureRet*.

References:
[1] P. Livi, F. Heer, U. Frey, D. Bakkum, and A. Hierlemann, "Compact Voltage and Current Stimulation Buffer for High-Density Microelectrode Arrays," *ISSCC Dig. Tech. Papers*, pp. 240-241, Feb. 2006.
[2] M. Ortmanns, N. Unger, A. Rocke, M. Gehrke and H.J. Tiedtke: "A 232-Channel Epiretinal Stimulator ASIC", *IEEE J. Solid-State Circuits*, vol. 42, no. 12, pp. 2946-2959, Dec. 2007.
[3] K. Chen, Z. Yang, L. Hoang, J. Weiland, M. Humayun, and W. Liu, "An Integrated 256- Channel Epiretinal Prosthesis", *IEEE J. Solid-State Circuits*, vol. 45, no. 9, pp. 1946-1956, Sept. 2010.
[4] M. Ghovanloo and K. Najafi, "A Modular 32-Site Wireless Neural Stimulation Microsystem", *ISSCC Dig. Tech. Papers*, pp. 226-227, Feb. 2004.
[5] Rothermel et al., "A 1600-pixel Subretinal Chip with DC-free Terminals and ±2V Supply Optimized for Long Lifetime and High Stimulation Efficiency", *ISSCC Dig. Tech. Papers*, pp. 144-145, Feb. 2008.
[6] M. Sahin and Y. Tie, "Non-rectangular waveforms for neural stimulation with practical electrodes", *J. Neural Eng.*, vol. 4, no. 3, 227-233, Sept. 2007.
[7] K. Sooksood, T. Stieglitz and M. Ortmanns, "An Active Approach for Charge Balancing in Functional Electrical Stimulation", *IEEE Trans. on Biomedical Circuits and Systems*, vol. 4, no. 3, pp. 162-170, June 2010.

978-1-61284-303-2/11 $26.00 © 2011 IEEE

ISSCC 2011 / February 22, 2011 / 2:30 PM

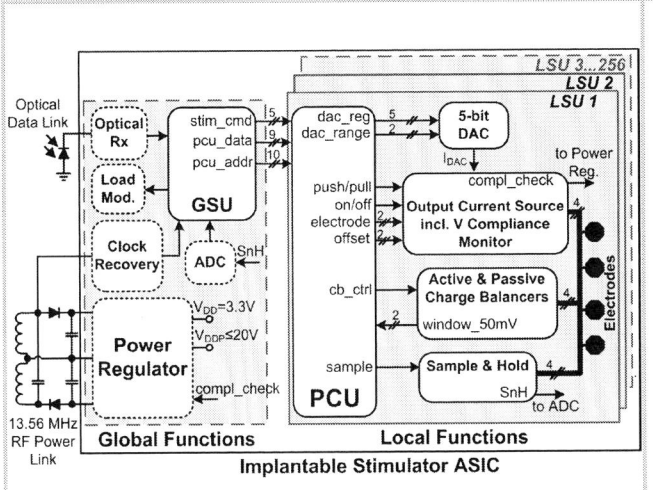

Figure 17.4.1: Global stimulator system and local frontend architecture (LSU).

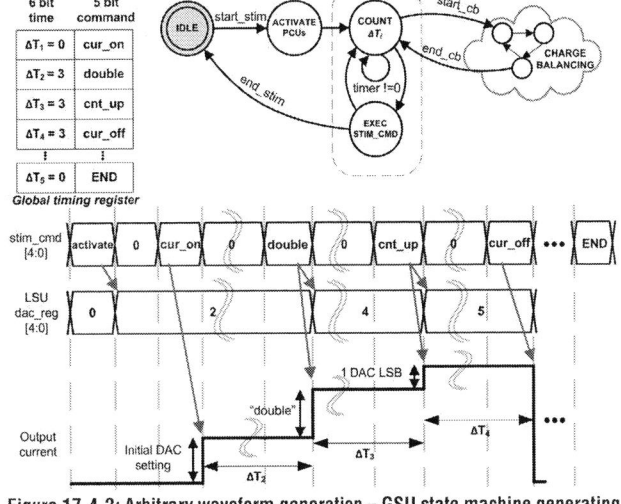

Figure 17.4.2: Arbitrary waveform generation – GSU state machine generating globally distributed stimulation commands (stim_cmd).

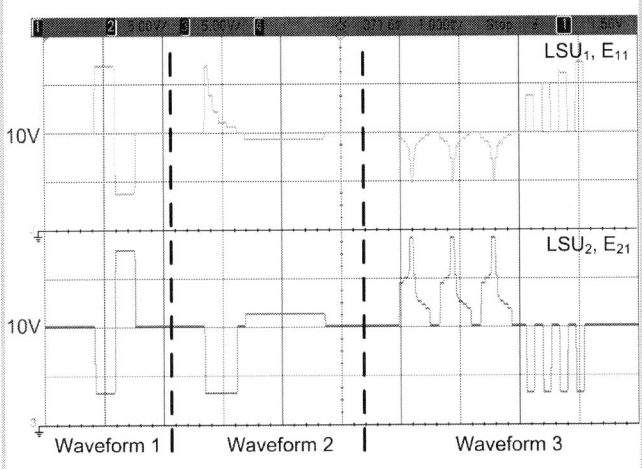

Figure 17.4.3: Measured arbitrary output waveforms.

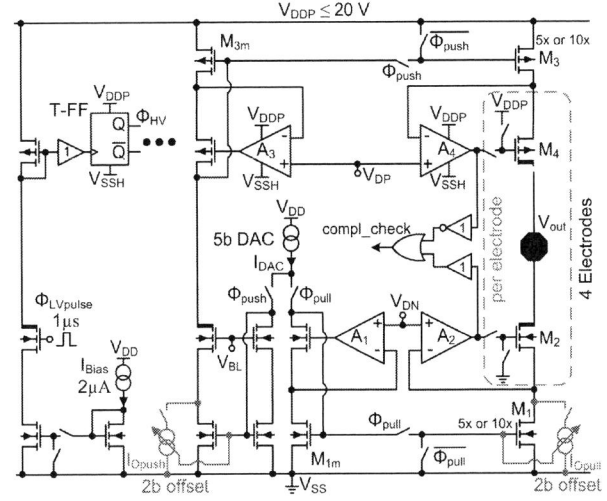

Figure 17.4.4: Illustrative schematic of the HV compliance output driver, including compliance monitor and offset balancer.

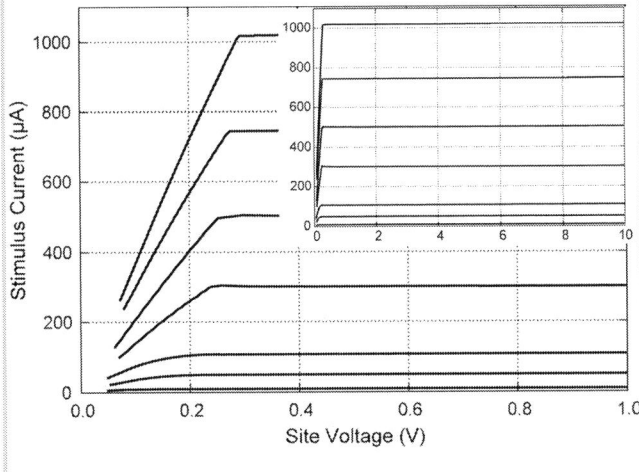

Figure 17.4.5: Output driver current vs. stimulation site voltage with V_{DN}=200mV, measured exemplarily for the cathodic driver.

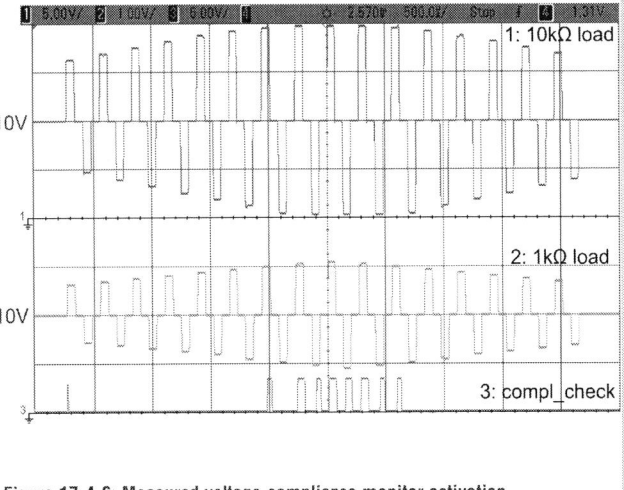

Figure 17.4.6: Measured voltage compliance monitor activation.

17

978-1-61284-303-2/11 $26.00 © 2011 IEEE

307

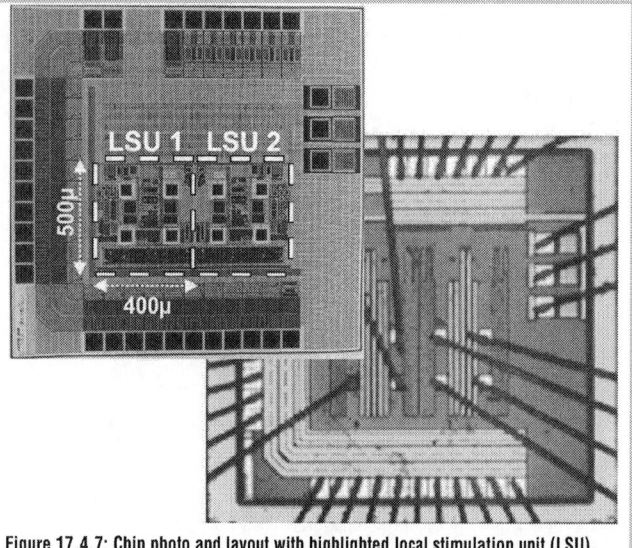

Figure 17.4.7: Chip photo and layout with highlighted local stimulation unit (LSU).

ISSCC 2011 / SESSION 17 / BIOMEDICAL & DISPLAYS / 17.5

17.5 A Low Noise Current Readout Architecture for Fluorescence Detection in Living Subjects

Roxana T. Heitz[1], David B. Barkin[2], Thomas D. O'Sullivan[1], Natesh Parashurama[1], Sanjiv S. Gambhir[1], Bruce A. Wooley[1]

[1]Stanford University, Stanford, CA,
[2]National Semiconductor, Santa Clara, CA

Optical molecular imaging is emerging as a powerful preclinical research tool for investigating and quantifying molecular events in living subjects, with applications including earlier detection of disease, therapeutic monitoring and understanding fundamental biology [1]. For example, imaging the fluorescent molecular probe RGD-Cy5.5, which specifically binds to molecules ($\alpha_v\beta_3$ integrin receptors) that regulate new blood vessel growth in tumors, can be used to quantify this growth [2]. Capturing the fluorescent signal in living subjects with an implanted biosensor would enable continuous monitoring of tumors in freely moving subjects. Continuous monitoring in the setting of cancer would give valuable information on tumor progression, both in assessing drug efficacy and detecting recurrent tumor growth after treatment. Presently, fluorescence imaging in living subjects is performed with bulky instrumentation that does not permit continuous monitoring of freely moving subjects over long time periods. In order to make a fluorescence-detection system implantable, and portable, a laser excitation source, a photodetector and a readout circuit for measuring and digitizing photocurrents are integrated in a single package, and continuous fluorescence detection is demonstrated in live animals.

This paper introduces a readout circuit designed to interface with the fluorescence sensor presented in [3], a monolithically integrated GaAs detector and vertical-cavity surface-emitting laser (VCSEL). The detector signals to be measured are low currents in the range of 5pA to 15nA, with bandwidths up to 100Hz. In order to capture binding dynamics of the fluorescent probe, as well as a wide range of possible tumor sizes and depths, the desired current resolution is 5pA. In applications with such low bandwidths and high sensitivity, a capacitive transimpedance amplifier (CTIA) can be used to provide high SNR through noise averaging [4]. The input current is integrated onto a capacitor C_{int} for a period of time T_{int}. The output of the CTIA is usually sampled and held at the end of T_{int}, before being digitized as a DC signal. The readout noise power introduced at the sampling instant, σ^2_{read}, limits the maximum achievable SNR of this architecture. If, instead, M samples are taken along the integration ramp and line-fitting is performed, the white noise component of σ^2_{read} is reduced by $M/12$, as shown in [5].

Figure 17.5.1 shows our readout system architecture, where $\Delta\Sigma$ modulation is used to acquire and digitize samples taken up the integration ramp. The output of a CTIA is sampled by an incremental $\Delta\Sigma$ modulator, without the need for a sample-and-hold stage at the CTIA output. Before each integration cycle, both the CTIA and the $\Delta\Sigma$ modulator are reset. After reset, the detector current accumulates on C_{int}, giving rise to a voltage ramp at the amplifier's output. This ramp is sampled by the $\Delta\Sigma$ modulator M times during the integration period, such that $M=T_{int}f_{\Sigma\Delta}$, where $f_{\Sigma\Delta}$ is the sampling frequency of the modulator. The output of the modulator is then processed by a digital filter $G(z)$ that implements a line-fitting operation similar to that described in [5].

The CTIA is designed with C_{int} of 3pF and T_{int} of 500µs. The output voltage range is 0.5 to 1.3V, and linearity of at least 10b must be maintained over this entire range. The detector must be biased close to 0V in order to avoid dark-current noise. These considerations result in the need for a high open-loop DC gain for the CTIA op-amp. A folded-cascode amplifier with gain boosting [6] is used to achieve a DC gain of 86dB. To achieve sufficiently low quantization noise, a second-order architecture is used for the $\Delta\Sigma$ modulator with $f_{\Sigma\Delta}$=2MHz, corresponding to an oversampling ratio of 1,000. The digital line-fitting is accomplished using an IIR filter with seven non-zero taps. In the experimental prototype, this filter is implemented in MATLAB.

To mitigate the large (>1V) voltage glitches that can occur due to charge sharing at the interface between the CTIA and the switched-capacitor (SC) integrator in the first stage of the $\Delta\Sigma$ modulator, an additional clock phase, ϕ_3, is intro-

duced between the charge-transfer phase, ϕ_2, and the sampling phase, ϕ_1. During ϕ_3, the sampling capacitor of the SC integrator is pre-charged to the CTIA output voltage, as illustrated in Fig. 17.5.2. A unity-gain buffer at the output of the CTIA provides the pre-charge voltage during ϕ_3 and is disconnected during the other clock phases. Since this buffer is outside the signal path, its performance is not critical, thus allowing for a simple, low-power implementation.

The dynamic range of this current readout system can be extended with a simple post-processing step in the digital domain, without any additional circuit elements. If large input currents cause voltage saturation at the CTIA output, the line-fitting filter can be dynamically adjusted to only use the samples before saturation. As illustrated in Fig. 17.5.3, fewer and fewer A/D samples are used (first 1/2 or first 1/4 of the samples) as long as the line-fitting operation produces increasing slope estimates. By this means, the maximum input current that can be sensed can be increased from 4.2 to 21nA.

A summary of the measured performance is shown in Fig. 17.5.4. These measurements were done using a Keithley current source at the input of the CTIA. The readout chip was ultimately interfaced with the sensor from [3] in an open cavity package, and in vitro experiments with Cy5.5 dye were performed. Figure 17.5.5 shows the measured results of the experiments. The measured current shows a linear relationship with dye concentrations between 10nM and 25µM, which is a broader range than the required 50nM to 25µM (dye saturation occurs beyond 25µM, and animal autofluorescence limits the useful range to above 50nM [3]).

In order to make the system implantable, the readout IC was packaged together with the sensor on a custom miniature PCB, shown in Fig. 17.5.6, using chip-on-board bonding. The package is covered with a lid and a lens for collimating the excitation laser, and coated in biocompatible epoxy. The assembly is approximately 1cm³ and weighs 0.7g. The miniature package was implanted in a live mouse, and fluorescence measurements were taken while a 50µL tail-vein injection of 50µM Cy5.5 dye was applied, as well as while the mouse was recovering from anesthesia and moving freely around its cage. The acquired signal is shown in Fig. 17.5.6. An increase of 500pA up to 1.2nA can be seen after injecting the dye. Repeated experiments have shown that the signal will increase by 2 to 3× in the presence of a tumor [2,3], yielding currents up to 10nA. With advances towards future use in humans, feasible dye concentrations may be as low as 100nM, corresponding to signal changes as low as 5pA. The expected signal range needed for preclinical research is thus within the performance achievable with the proposed readout circuit. This miniaturized system, constituting a fully implantable laser-based fluorescence detector with digital readout, is therefore a potentially enabling technology for preclinical research and the continuous monitoring of cancer progression. Other applications in small animal models and clinical translation are also being explored.

Acknowledgments:
The authors gratefully acknowledge National Semiconductor Corp. for fabrication of the chip, and partial funding support from NCI ICMIC P50 CA114747 (SSG).

References:
[1] T. F. Massoud and S. S. Gambhir, "Molecular imaging in living subjects: seeing fundamental biological processes in a new light," *Genes & Development*, vol. 17, pp. 545-80, Mar. 2003.
[2] Z. Cheng, et al., "Near-infrared fluorescent RGD peptides for optical imaging of integrin $\alpha_v\beta_3$ expression in living mice," *Bioconjugate Chemistry*, vol. 16, no. 6, pp. 1433-1441, 2005.
[3] T. O'Sullivan, et al., "Implantable semiconductor biosensor for continuous in vivo sensing of far-red fluorescent molecules," *Optics Express* 12513, vol. 18, no.12, June 2010.
[4] H. Ou and K. K. Chin, "Theory of gated multicycle integration (GMCI) for repetitive imaging of focal plane array," *IEEE Trans. Circuits and Systems II*, vol. 50, pp. 378-383, July 2003.
[5] A. M. Fowler and I. Gatley, "Noise reduction strategy for hybrid IR focal-plane arrays," *Proc. SPIE 1541*, pp. 127-133, 1991.
[6] K. Bult and G. J. G. M. Geelen, "The CMOS gain-boosting technique," *Analog Integrated Circuits and Signal Processing*, vol. 1, pp. 119-135, 1991.

ISSCC 2011 / February 22, 2011 / 2:45 PM

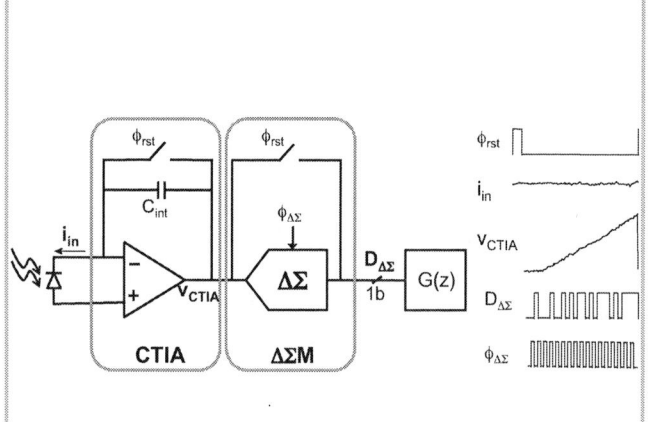

Figure 17.5.1: Block diagram of CTIA with ramp oversampling by a ΔΣ modulator.

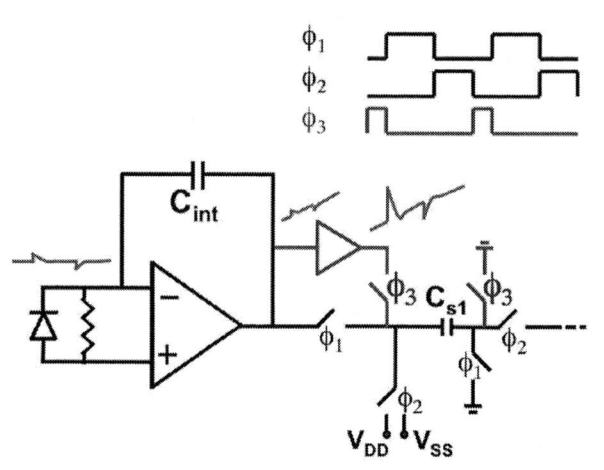

Figure 17.5.2: Schematic of charge-sharing mitigation scheme.

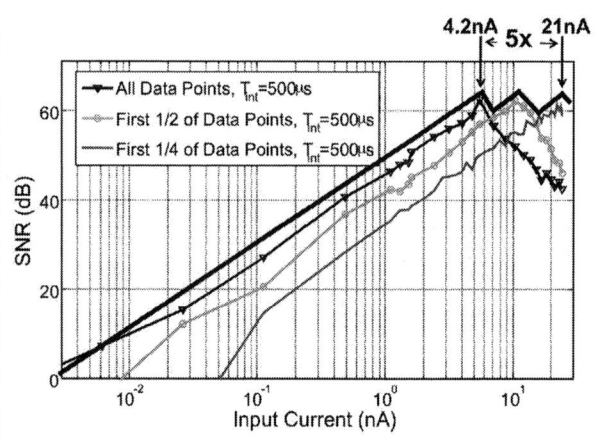

Figure 17.5.3: Measured dynamic range extension results.

Power Supply	1.8V
Technology	0.18μm CMOS
Die Size	0.9mm x 1.4mm
DNL	+/- 0.15LSB
INL	+/- 1LSB
Integration Time	500μs
Resolution	10bits (4pA)
I_{min}	4pA
I_{max}	21nA
DR	77dB
BW	800Hz
Power Consumption	2.4mW (VCSEL power consumption >15mW)

Figure 17.5.4: Summary of measured performance.

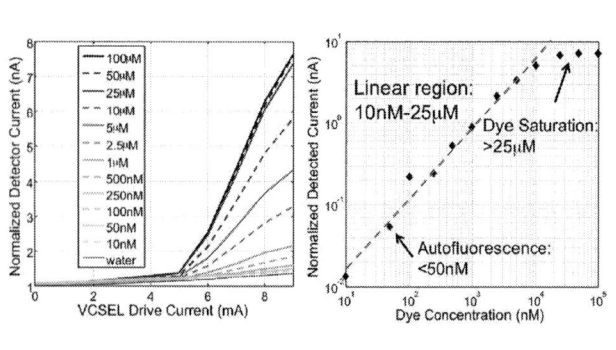

Figure 17.5.5: *in vitro* measurement results.

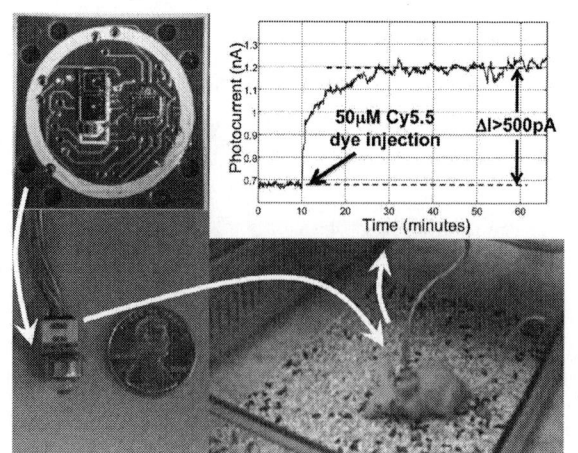

Figure 17.5.6: Implant packaging and *in vivo* measurements.

17

ISSCC 2011 PAPER CONTINUATIONS

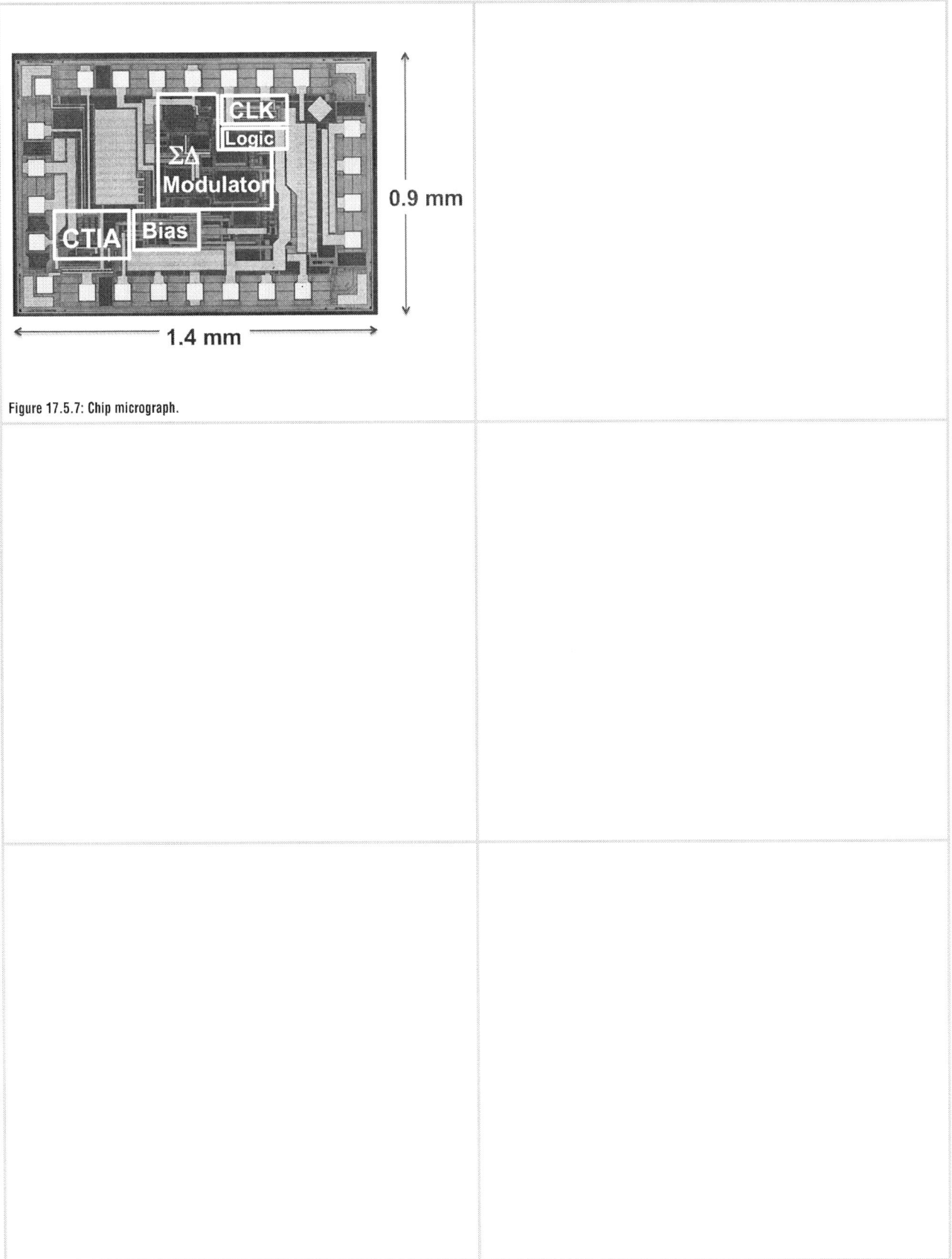

Figure 17.5.7: Chip micrograph.

17.6 A Cubic-Millimeter Energy-Autonomous Wireless Intraocular Pressure Monitor

Gregory Chen, Hassan Ghaed, Razi-ul Haque, Michael Wieckowski, Yejoong Kim, Gyouho Kim, David Fick, Daeyeon Kim, Mingoo Seok, Kensall Wise, David Blaauw, Dennis Sylvester

University of Michigan, Ann Arbor, MI

Glaucoma is the leading cause of blindness, affecting 67 million people worldwide [1]. The disease damages the optic nerve due to elevated intraocular pressure (IOP) and can cause complete vision loss if untreated. IOP is commonly assessed using a single tonometric measurement, which provides a limited view since IOP fluctuates with circadian rhythms and physical activity. Continuous measurement can be achieved with an implanted monitor to improve treatment regiments, assess patient compliance to medication schedules, and prevent unnecessary vision loss. The most suitable implantation location is the anterior chamber of the eye, which is surgically accessible and out of the field of vision. The desired IOP monitor (IOPM) volume is limited to 1.5mm³ (0.5×1.5×2mm³) by the size of a self-healing incision, curvature of the cornea, and dilation of the pupil. Previously, a 5.4mm³ (6×3×0.3mm³) sensor was demonstrated with a 27mm antenna [2]. The antenna size allows the sensor to be recharged wirelessly but may complicate implantation procedures [3].

The aggressive IOPM size constraint creates major challenges for achieving high-resolution capacitance measurements, wireless communication, and multi-year device lifetime. Little energy can be stored on the tiny microsystem, calling for ultra-low power operation and energy harvesting. The required millimeter antennas or inductors result in lower received power and higher transmission frequency, both increasing microsystem power. We present a cubic-millimeter IOPM with energy-autonomous operation and wireless communication. The IOPM targets implantation with a minimally invasive procedure through a tiny incision that is routinely used for outpatient cataract surgery. Glass haptics are designed to anchor the IOPM using the natural elasticity of the iris, preventing tissue damage and allowing for simple removal. The IOPM harvests solar energy that enters the eye through the transparent cornea to achieve energy-autonomy. The microsystem contains an integrated solar cell, thin-film Li battery, MEMS capacitive sensor, and integrated circuits vertically assembled in a biocompatible glass housing (Fig. 17.6.1). The circuits include a wireless transceiver, capacitance to digital converter (CDC), DC-DC switched capacitor network (SCN), microcontroller (μP), and memory fabricated in 0.18μm CMOS.

The IOPM measures IOP every 15 minutes using a MEMS capacitive pressure sensor connected to a 7μW 3.6V CDC with through-glass interconnects (Fig. 17.6.2) [4]. The measurement interval represents continuous monitoring, does not need to be exact for medical diagnosis [3], and is controlled by a slow timer in the wakeup controller (WUC) [5]. The CDC generates an IOP-dependent current by dropping $V_{DD}/2$-V_{TH} (V_{REF}) across an impedance generated by switching the MEMS pressure sensor (C_{MEMS}) at 50kHz. Simultaneously, a larger fixed current is generated in the same manner with the same clock and fixed capacitors (C_1, C_2). Two capacitors with out-of-phase clocks are used to generate a more constant current source. This fixed current is mirrored and compared to the IOP-dependent current using $\Delta\Sigma$ modulation to digitize IOP. The IOP-dependent current is integrated by discharging capacitor C_{INT}. The voltage on C_{INT} (V_{INT}) is compared to V_{REF} with a clocked comparator. When V_{INT} drops below V_{REF}, the fixed current is also integrated onto C_{INT}, increasing V_{INT}. The CDC achieves a pressure resolution of 0.5mmHg, which exceeds the 1mmHg resolution of typical tonometric measurements, using a decimation filter that counts the output bitstream over 10k cycles (Fig. 17.6.3). Since the CDC measures the ratio of two currents, it has low sensitivity to V_{DD}, clock, and temperature variations. After the CDC measurement, IOP data are logged into the 4kb SRAM using the 90nW 0.4V 8b μP. The microsystem can store 3 days of raw IOP data. The μP can also perform DSP or compression on the IOP data to extend storage capacity to over 1 week.

The user downloads IOP data using an external device (ED), placed near the eye. The microsystem is designed to respond to a wireless query by coupling RF energy from the ED onto an LC tank, rectifying the AC signal, and generating a

digital wakeup signal (U_0, U_1) with a variable offset comparator (Fig. 17.6.4). IOPM data are transmitted with an oscillator that acts as both a carrier generator and power amplifier (Fig. 17.6.4). The IOPM uses a dual-resonator tank to generate an FSK-modulated signal with two tones at 570MHz (f_0) and 690MHz (f_1). The large tone separation enables higher transmission distance by relaxing phase noise constraints. To transmit a *zero*, LC_1 is shorted by asserting D_1 and the oscillator runs at f_0 for 0.1μs using LC_0. A *one* is sent by oscillating at f_1 with LC_1. The signal is transmitted through the anterior chamber, 0.5mm cornea, and air [3]. The measured transmitter BER is 10^{-6} through 5mm of saline and 10cm of air (Fig. 17.6.5). This medium models the attenuation from aqueous humor in the eye and the distance from the eye to ED. The 4.7nJ/b 3.6V transmitter achieves a 4× improvement in energy efficiency over comparable work in highly-integrated biomedical implants [2][6]. The battery's peak current is 35 to 40μW, which cannot directly support wireless transmission. To prevent catastrophic V_{DD} droop, 1.6nF of integrated capacitance acts as a local power supply. The isolated V_{DD} drops by 0.5V when the radio transmits one bit every 131μs and is recharged between transmissions.

The desired IOPM lifetime is years to converge on a suitable glaucoma treatment. However, the anterior chamber volume limits lifetime by constraining the size and capacity of the microsystem's power sources [7]. The IOPM uses a custom 1μAh thin-film Li battery from Cymbet. The lifetime is 28 days with no energy harvesting. To extend lifetime, the IOPM harvests light energy entering the eye with an integrated 0.07mm² solar cell and recharges the battery. Given the ultra-small solar cell size, energy autonomy requires average power consumption of <10nW. Processor power is reduced using subthreshold operation and delivered using an SCN with 75% efficiency (Fig. 17.6.6). The SCN uses reduced swing clocks and level converters (LCs). While IOP measurements and wireless transmissions require μWs and mWs of power, these events are short and infrequent. When CDC and radio circuits are idle, their power consumption drops to 172.8pW and 3.3nW, respectively. For the majority of its lifetime the IOPM is in a 3.65nW standby mode where mixed-signal circuits are disabled, digital logic is power-gated, and 2.4fW/bitcell SRAM retains IOP instructions and data [5]. The average system power with pressure measurements every 15 minutes and daily wireless data transmissions, is 5.3nW. When sunny, the solar cells supply 80.6nW to the battery. The combination of energy harvesting and low-power operation allows the IOPM to achieve zero-net energy operation in low light. The IOPM requires 10 hours of indoor lighting or 1.5 hours of sunlight per day to achieve energy-autonomy.

Acknowledgments:
We gratefully acknowledge Cymbet Corporation for supplying a custom-size battery for this application.

References:
[1] K.C. Katuri, S. Asrani, and M.K. Ramasubramanian, "Intraocular pressure monitoring sensors," *IEEE Sensors Journal*, vol. 8, no. 1, pp. 12-19, Jan. 2008.
[2] E.Y. Chow, et al., "Mixed-signal integrated circuits for self-contained sub-cubic-millimeter biomedical implants," *ISSCC Dig. Tech. Papers*, pp.236-237, Feb. 2010.
[3] D.G. Vaughn, T. Asbury, P. Riordan-Eva, *General Ophthalmology*, 15th ed. Stamford, CT: Appleton & Lange, 1999.
[4] R. M. Haque and K. D. Wise, "An intraocular pressure sensor based on a glass reflow process," *Solid-State Sensors, Actuators, and Microsystems Workshop*, pp. 49–52, 2010.
[5] G. Chen, et al., "Millimeter-scale nearly perpetual sensor system with stacked battery and solar cells," *ISSCC Dig. Tech. Papers*, pp.288-289, Feb. 2010.
[6] R. Harrison, et al., "A low-power integrated circuit for a wireless 100-electrode neural recording system", *ISSCC Dig. Tech. Papers*, pp. 555-556, Feb. 2006.
[7] A.C.-W. Wong, et al., "A 1V, micropower system-on-chip for vital-sign monitoring in wireless body sensor networks", *ISSCC Dig. Tech. Papers*, pp. 138-139, Feb. 2008.

ISSCC 2011 / February 22, 2011 / 3:15 PM

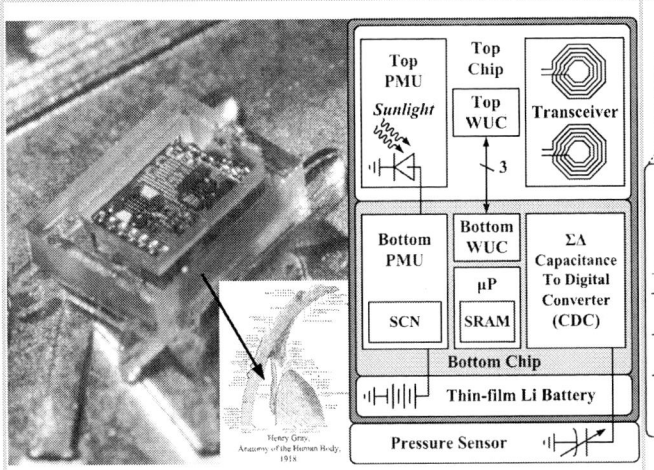

Figure 17.6.1: The IOPM contains a MEMS pressure sensor, integrated solar cell, and microbattery in a biocompatible enclosure. Its cubic-millimeter size enables implantation through a minimally invasive incision.

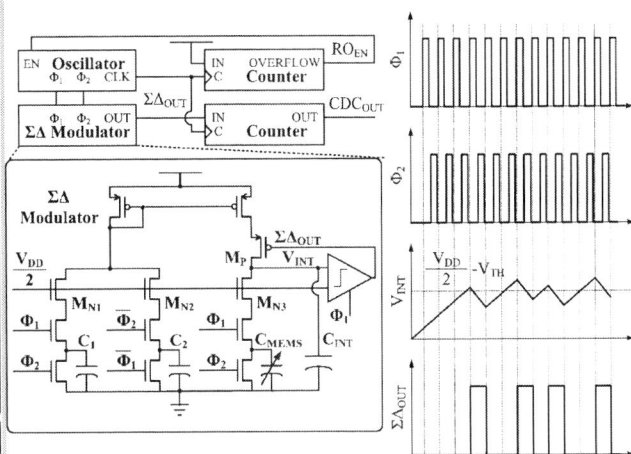

Figure 17.6.2: The capacitance to digital converter compares pressure-dependent and fixed currents using $\Delta\Sigma$ modulation. The design style provides independence to supply voltage and clock frequency uncertainty.

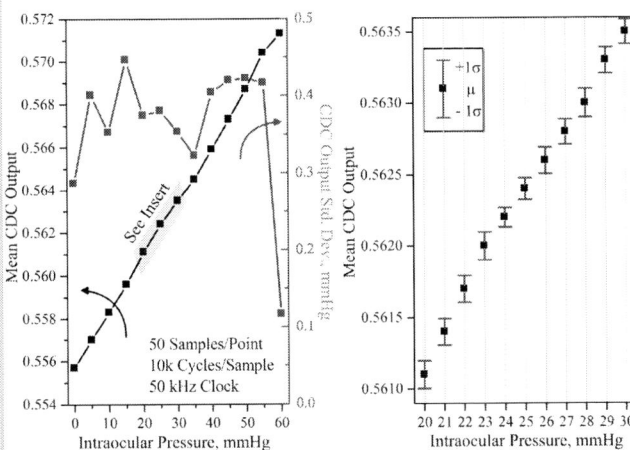

Figure 17.6.3: Measured results demonstrate CDC performance. The IOPM exceeds typical measurement techniques by achieving 0.5mmHg pressure resolution.

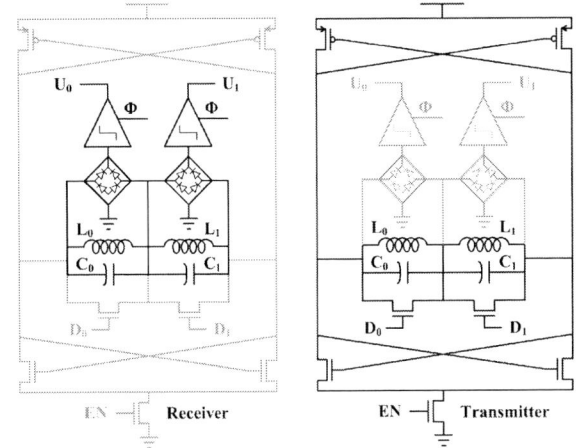

Figure 17.6.4: The series-connected LC tanks: (1) enable greater frequency separation than a single tank transmitter, relaxing phase noise requirements, and (2) reduce area compared to two separate LC tanks.

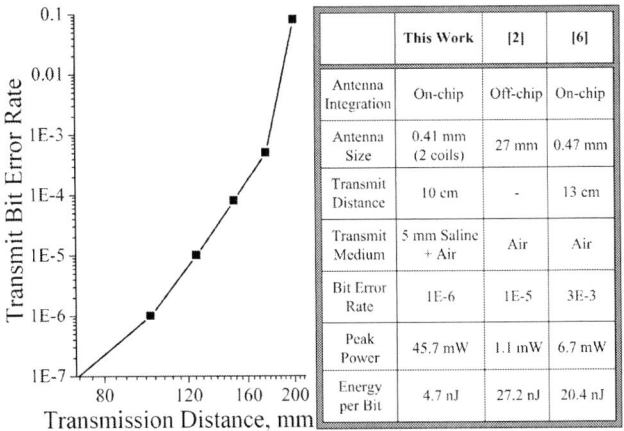

Figure 17.6.5: The IOPM is activated when it receives and rectifies the wireless wake up signal. The device then transmits pressure data with a BER of less than 10^{-6}.

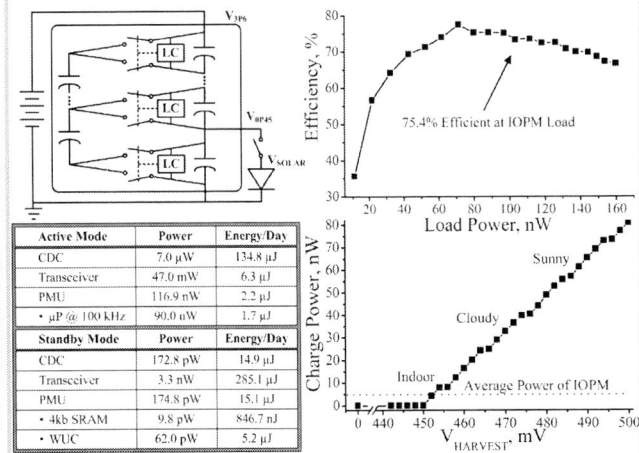

Figure 17.6.6: IOPM power consumption is 5.3nW with the expected usage model. Energy autonomy is achieved with a 0.07mm² solar cell that supplies 80.6nW to the battery. Battery life without recharge is 28 days.

978-1-61284-303-2/11 $26.00 © 2011 IEEE

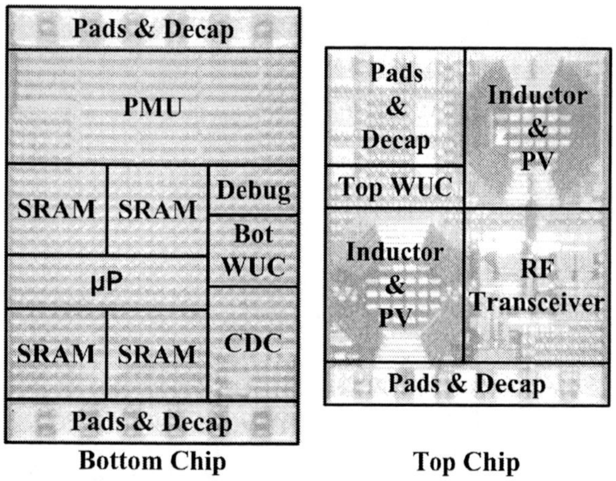

ISSCC 2011 PAPER CONTINUATIONS

Figure 17.6.7: Die photographs for the bottom and top chips as defined in Figure 17.6.1, both fabricated in 0.18μm CMOS.

ISSCC 2011 / SESSION 17 / BIOMEDICAL & DISPLAYS / 17.7

17.7 A 160×128 Single-Photon Image Sensor with On-Pixel 55ps 10b Time-to-Digital Converter

Chockalingam Veerappan[1], Justin Richardson[2], Richard Walker[2,3],
Day-Uey Li[3], Matthew W Fishburn[1], Yuki Maruyama[1], David Stoppa[4],
Fausto Borghetti[4], Marek Gersbach[5], Robert K Henderson[3],
Edoardo Charbon[1]

[1]Delft University of Technology, Delft, The Netherlands,
[2]STMicroelectronics, Edinburgh, United Kingdom,
[3]University of Edinburgh, Edinburgh, United Kingdom,
[4]Fondazione Bruno Kessler - IRST, Trento, Italy,
[5]EPFL, Lausanne, Switzerland

Image sensors capable of resolving the time-of-arrival (ToA) of individual photons with high resolution are needed in several applications, such as fluorescence lifetime imaging microscopy (FLIM), Förster resonance energy transfer (FRET), optical rangefinding, and positron emission tomography. In FRET, for example, typical fluorescence lifetime is of the order of 100 to 300ps, thus deep-subnanosecond resolutions are needed in the instrument response function (IRF). This in turn requires new time-resolved image sensors with better time resolution, increased throughput, and lower costs. Solid-state avalanche photodiodes operated in Geiger-mode, or single-photon avalanche diodes (SPADs), have existed for decades [1] but only recently have SPADs been integrated in CMOS. However, as array sizes have grown, the readout bottleneck has also become evident, leading to hybrid designs or more integration and more parallelism on-chip [2,3]. This trend has accelerated with the introduction of SPAD devices in deep-submicron CMOS, that have enabled the design of massively parallel arrays where the entire photon detection and ToA circuitry is integrated on-pixel [4,5].

In this paper, we present an array of 160×128 pixels capable of detecting the ToA of single photons, implemented in 0.13μm CMOS technology. The block diagram of the system is shown in Fig. 17.7.1. The sensor is partitioned into 4 identical quadrants that are served by a balanced clock tree so as to minimize skews and to ensure the fastest possible readout process. The pixels in the rows are read out in rolling shutter mode in two directions (top and bottom) simultaneously. The 10b content of each pixel is thus transferred to the exterior of the chip via 320 independent serializers, 2 for each column, working in parallel at a maximum rate of 160Mb/s, thus enabling in principle the readout of a complete frame in 4μs. In our implementation the frame rate is actually lower, due to current limitations in the readout firmware. The region-of-interest (RoI) is programmable, via vertical and horizontal registers that select which rows and columns are read out.

The pixel is shown in Fig. 17.7.2a; it comprises a SPAD, implemented as a p+/p-well/deep n-well junction, a TDC based on the architecture reported in [5], and a memory. The TDC is composed of a ring oscillator (RO) with fast start-up, shown in Fig. 17.7.2b, that is activated by the SPAD through falling signal R and a ripple-counter that is incremented upon each completion of a ring cycle via B_3. The ripple-counter, chosen for its compactness and simplicity, provides the 7 most significant bits of the measurement, for a coarse resolution of 440ps. The 3 least significant bits are provided by the eight intermediate RO outputs. These signals are a thermometer code representing the partial propagation of the circulating signal within the RO; they are transferred via control signal T, latched, and converted to a binary code (D_0: D_2), all within the pixel, as shown in Fig. 17.7.2c. The STOP is provided by S, freezing the ring oscillator, while a rising R resets the TDC (RO and counter). The overall resolution of the TDC is 55ps, its full scale is 55ns. The RO is a replica of a similar one implemented in a PLL outside the pixel array; the ROs are bound by global signal V_{PLL} that ensures that the propagation time of a cycle is constant and independent of PVT variations without the need to stop measurements for calibration as in [2]. Due to the negligible fixed-pattern noise measured on chip, pixel-to-pixel variability needs not be compensated.

The pixel supports both time-correlated single-photon counting (TCSPC) and time-uncorrelated photon counting (TUPC) modes. TUPC is achieved by direct counting; the dead time of the SPAD was chosen so as to keep afterpulsing at a reasonable value. In TCSPC mode, a ToA measurement is available in each pixel every 100ns. However, due to measurement sparsity, it is reasonable to retain only one of ten measurements in a minimum frame time of 1μs. Thanks to the use of the pixel-level memory, readout dead time is suppressed, while reverse START-STOP ensures that only those ROs are active that are in a measurement phase; this reduces power consumption, supply ripple, and substrate noise. An I²C module controls all of the chip's state.

The total number of transistors integrated in the chip is 60 million. A pixel pitch of 50μm was selected as a tradeoff between the functionality and optics feasibility. A microlens array was added as a post-processing step to improve the effective fill factor. The sensor was tested in 3 steps, at SPAD, TDC, and system levels. The SPAD performance is reported in the table of Fig. 17.7.6 for specific values of excess bias voltage V_e. The TDCs were characterized separately for jitter and nonlinearities. The performance is also reported in Fig. 17.7.6. At system level, the sensor was tested in TUPC mode to compute the DCR distribution over the chip and in TCSPC mode to measure the jitter distribution and its stability in temperature and supply voltage. The test was conducted in the intended photon-starved regime and with higher illumination levels, so as to determine the limits of operation within the specified accuracy. The chip's overall performance was characterized in a PCB-based system based on a dual FPGA Xilinx IV chip-set.

In Fig. 17.7.3a, the median DCR is plotted as a function of excess bias voltage and temperature, while the inset reports the DCR cumulative probability at room temperature for 0.73V excess bias. Figure 17.7.3b shows the measured DNL and INL over the whole range of 55ns. The results of the TCSPC experiment are shown in Fig. 17.7.4a and b, where the IRF is plotted for two laser source wavelengths. In the insets, the FWHM jitter is plotted for different values of excess bias. The laser sources (Advanced Laser Diode Systems, Germany) are pulsed at 40MHz, emitting blue (405nm) and red (637nm) beams with a pulse width (FWHM) of 40ps and 80ps, respectively. Figure 17.7.4c shows the FWHM jitter as a function of incoming light intensity. The plot shows that, when operating in medium exposure the jitter performance is negligibly degraded with respect to a photon-starved regime. Note that in all TCSPC experiments the power of the laser was adjusted so as to minimize pile-up distortion.

The sensor was mounted on a microscope (BX51IW, Olympus, Japan). In TCSPC mode, the blue laser source was used at an average optical power of 2mW. The sample, a Bisaccate Pine pollen grain (Carolina Biological Supply Company, NC, USA) was illuminated through a microscope objective (50×, 0.80 NA, MPlanFL N, Olympus, Japan) via a standard dichroic beam splitter and a 1× adapter lens. The reflected beam was redirected to the sensor via the beam splitter and filters. In TUPC mode, the illumination was a uniform collimated light source obtained from a standard broadband mercury lamp. The sample was stained using a 2-dye system, Harris hematoxylin and phloxinein, with different decay times. Figure 17.7.5a shows the structure of the pollen; Fig. 17.7.5b and c show the resulting intensity and FLIM images and scales, respectively. The FLIM color code corresponds to a decay time estimated using double exponential fit on 300,000 measurements per pixel; Fig. 17.7.5d shows the optical setup used in the experiments. Figure 17.7.6 summarizes the performance of the chip.

Acknowledgments:
This work has been supported by the EC in the FP6 IST FET Open Megaframe project (contract No. 029217-2, www.megaframe.eu). The authors are grateful to Claudio Favi, Xilinx Inc., and Enterpoint Ltd. C. Veerappan, J. Richardson, R. Walker, D.U. Li, M. Fishburn, D. Stoppa, and F. Borghetti have contributed equally to the design and test of the sensor chip.

References:
1] S. Cova et al., "Towards picosecond resolution with single-photon avalanche diodes", *Review of Scientific Instruments*, 52 (3), pp. 408-412, 1981.
[2] C. Niclass, et al., "A 128x128 Single-Photon Image Sensor with Column-Level 10-bit Time-to-Digital Converter Array", *IEEE J. Solid-State Circuits*, Vol. 43, N. 12, pp. 2977-2989, Dec. 2008.
[3] S. Tisa, et al., "100kframe/s 8 bit Monolithic Single-Photon Imagers", *IEEE European Solid-State Device Conference*, pp. 274-277, Sept. 2008.
[4] D. Stoppa, et al., "A 32x32-Pixel Array with In-Pixel Photon Counting and Arrival Time Measurement in the Analog Domain", *IEEE European Solid-State Device Conference*, Sept. 2009.
[5] J. Richardson, et al., "A 32x32 50ps Resolution 10 bit Time to Digital Converter Array in 130nm CMOS for time Correlated Imaging", *IEEE Custom Integrated Circuits Conference*, Sept. 2009.
[6] C.R. Bagnall, Jr., *Pollen morphology of Abies, Picea, and Pinus species of the U.S. Pacific Northwest using scanning electron microscopy*, Ph. D. dissertation, Washington State Univ., 1974.

978-1-61284-303-2/11 $26.00 © 2011 IEEE

ISSCC 2011 / February 22, 2011 / 3:45 PM

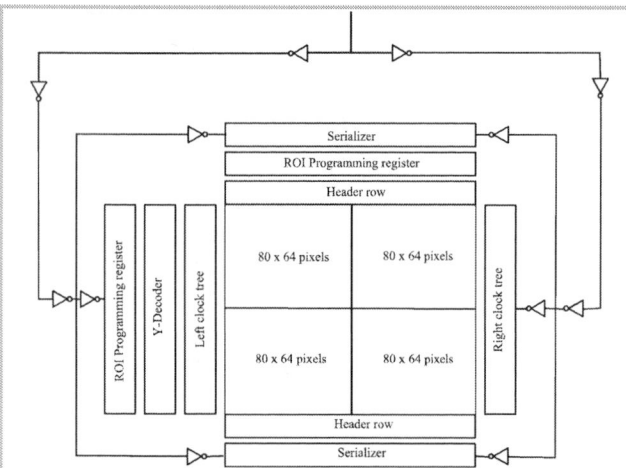

Figure 17.7.1: Block diagram of the sensor, split into 4 identical quadrants. Each pixel is read out in rolling shutter mode towards the top and the bottom simultaneously, with a balanced clock tree minimizing skews. The RoI-programming registers are used for selective row and column readout.

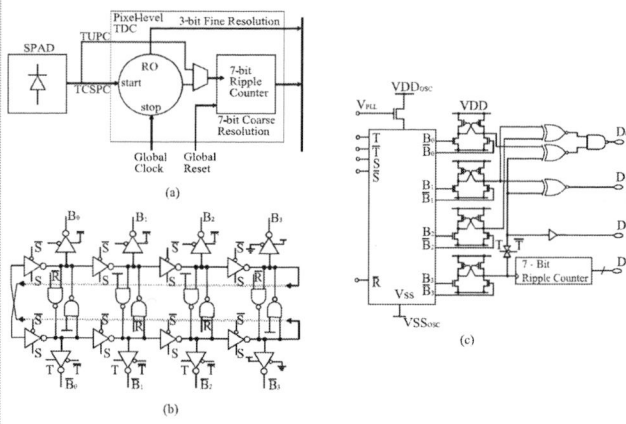

Figure 17.7.2: (a) Pixel schematic: the output of the SPAD acts as a start signal for a ring oscillator (RO) stopped by a global clock. (b) RO schematic. (c) TDC schematic: the frozen RO state is latched, converted from thermometer to binary code, and combined with the coarse measurement.

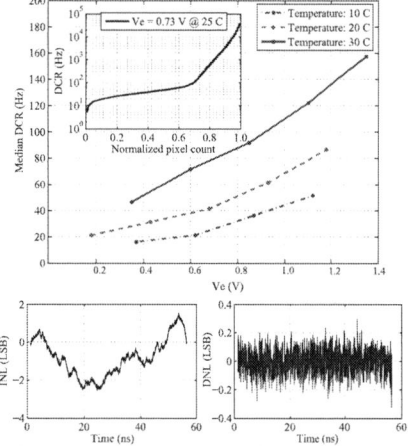

Figure 17.7.3: (a) Median dark count rate (DCR) as a function of excess bias voltage and temperature. In the inset the DCR cumulative probability is shown at 0.73V of excess bias. (b) Worst-case INL and DNL measured over the complete range of 55ns.

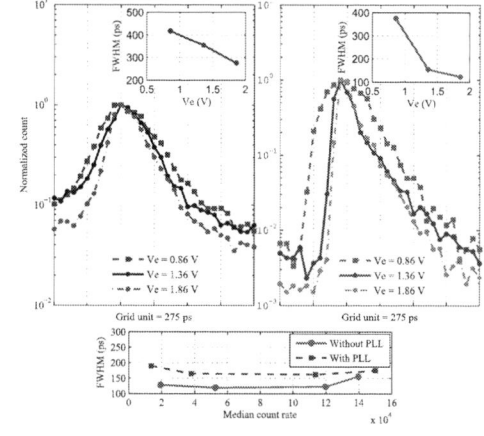

Figure 17.7.4: (a)-(b) Time-domain IRF at two wavelengths: 405nm (a), 637nm (b). In the insets, FWHM jitter as a function of excess bias voltage. (c) FWHM jitter as a function of incident photon flux with various levels of sensor activation, both with and without activating the PLL.

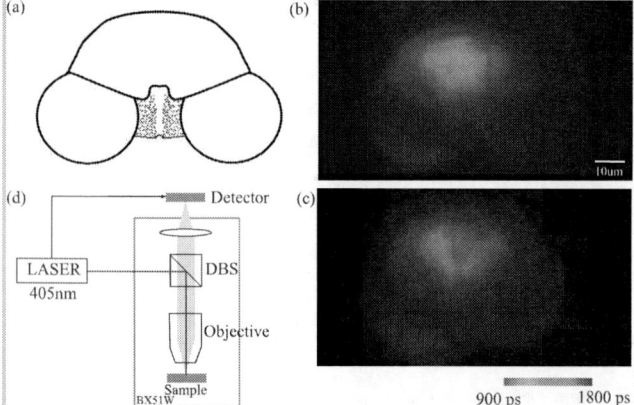

Figure 17.7.5: (a) Structure of a Pine Bisaccate pollen grain [6]. (b) Intensity image of the pollen grain obtained with a microscope with 50× magnification, 1x adapter lens and broadband illumination. (c) Reflection FLIM images of the same pollen obtained with 300,000 frames in 12s. The inset shows the time scale in ps. (d) Microscope setup comprising a 405nm laser source, a dichroic beam splitter, an objective, and the sensor mounted on a C-mount.

	Parameter	Condition	Min.	Typ.	Max.	Unit
Pixel	Photon detection probability	V_e=0.62V	3		19	%
	Photon detection probability	V_e=1.40V	3		27.5	%
	Sensitivity spectrum		350		900	nm
	Dead time			100		ns
	TDC measurement range			55		ns
	TDC resolution (1 LSB)			55		ps
	Measurement rate			1		MS/s
	TDC DNL / INL			±0.3/±2		LSB
	Nominal fill factor			1		%
System	Clock frequency			16	32	MHz
	Chips size			11.8x12.3		mm²
	Total I/O bandwidth			51.2		Gbps
	Power dissipation			550		mW
	CMOS Process			130um		-
	Median DCR	V_e=0.73V		50		Hz
	Mean FWHM jitter	λ=637nm		140		ps
	jitter non-uniformity	λ=637nm, PLL on, count rate = 50kHz		27		ps
FLIM Experiment	Number of measurements per pixel			3x10⁵		-
	Frame rate			25	50	kfps
	Laser source average power			2		mW
	Average count rate per pixel			15		kc/s
	Target area			125x80		µm²

Figure 17.7.6: Performance summary for the image sensor. All the parameters were measured at room temperature.

978-1-61284-303-2/11 $26.00 © 2011 IEEE

ISSCC 2011 PAPER CONTINUATIONS

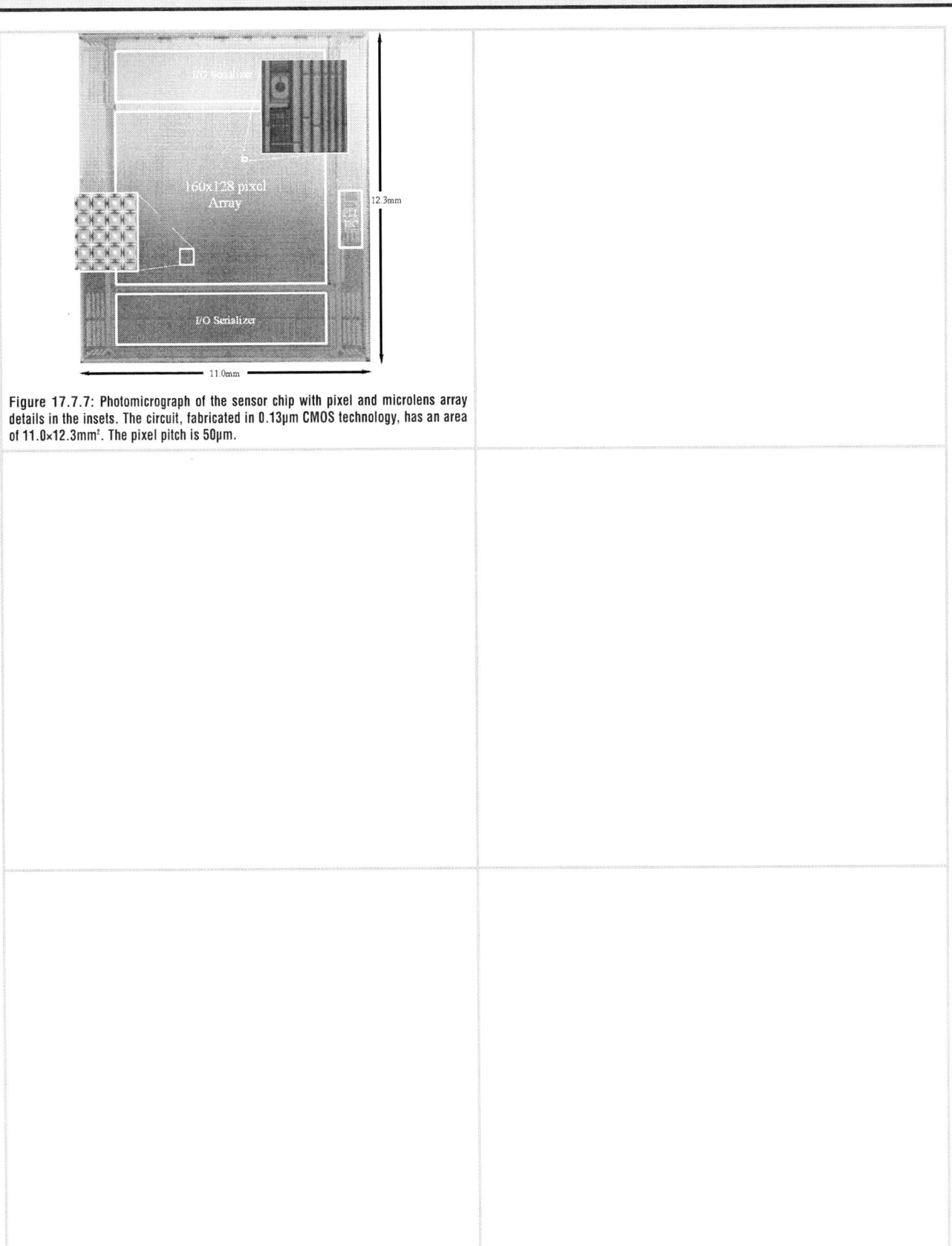

Figure 17.7.7: Photomicrograph of the sensor chip with pixel and microlens array details in the insets. The circuit, fabricated in 0.13µm CMOS technology, has an area of 11.0×12.3mm². The pixel pitch is 50µm.

ISSCC 2011 / SESSION 17 / BIOMEDICAL & DISPLAYS / 17.8

17.8 Bidirectional OLED Microdisplay: Combining Display and Image Sensor Functionality into a Monolithic CMOS Chip

Bernd Richter, Uwe Vogel, Rigo Herold, Karsten Fehse,
Stephan Brenner, Lars Kroker, Judith Baumgarten

Fraunhofer Institute for Photonic Microsystems, Dresden, Germany

Microdisplays based on organic light-emitting diodes (OLEDs) achieve high optical performance with excellent contrast ratio and large dynamic range at low power consumption. The direct light emission from the OLED enables small devices without additional backlight, making them suitable for mobile near-to-eye (NTE) applications such as viewfinders or head-mounted displays (HMD). In these applications the microdisplay acts typically as a purely unidirectional output device [1-3]. With the integration of an additional image sensor, the functionality of the microdisplay can be extended to a bidirectional optical input/output device. The major aim is the implementation of eye-tracking capabilities in see-through HMD applications to achieve gaze-based human-display-interaction.

This paper presents a bidirectional OLED microdisplay (BiMi) with a monochrome 320×240 (QVGA) display and a nested image sensor with 160×120 pixels in a commercially available 0.35μm 3.3V/12V CMOS process with customized top metal. The BiMi was designed as a universal prototype for evaluating the necessary display parameters in see-through HMDs and the eye-tracking capabilities in a nested setup with relatively low resolution. Because of the varying ambient light levels in see-through applications the main target was to achieve a wide range of possible display luminances from 100 to 10000Cd/m² instead of high resolution.

The monolithic integration of OLED-on-Silicon is realized by direct evaporation of the OLED stack on the silicon wafer through a shadow mask. Inside a BiMi, the OLED stack itself is unstructured and the pixels are defined by the shape of the top metal acting as the bottom electrode of the top-emitting OLED. Due to the sensitivity of the OLED materials against humidity and oxygen, a thin film encapsulation and a cover glass are applied at the end of the OLED post processing. Figure 17.8.2 shows a cross-section of a CMOS chip with an integrated OLED.

The active area of the BiMi consists of nested display and image sensor pixels surrounded by driving and control circuitry. The display and image sensor systems are electrically independent of one another, simply interacting via synchronization signals. The simplified system diagram is shown in Fig. 17.8.3. Since the luminance of an OLED is proportional to its current density, each display pixel comprises a current driver with analog modulation of the OLED current to realize 8b grayscale. The display pixel pitch is 36×36μm² and the pixel current ranges from pAs to μAs depending on the OLED efficiency and the desired luminance.

The pixel circuit of the display and the image sensor with a simplified driving scheme is shown in Fig. 17.8.4. The microdisplay is driven via a digital 8b video interface including synchronization signals. An integrated DAC converts the digital video data into a scaleable data and offset current to achieve a wide range of brightness and contrast. The DAC current is copied into the display data driver according to the principle of a dynamic current mirror. The display data driver is a dual set of 320 sample-and-hold (SH) stages, where the first set is being programmed by the DAC, while the second set is driving the previously stored current into the corresponding pixel cell of the selected line.

In this sample phase Md1 and Md2 are on, Md4 is off and thus Md3 is diode connected and the data current is driven through Md3. The established gate voltage is stored on Cd. In the following emit phase the switches Md1 and Md2 are turned off, Md4 is turned on and the data current flows from Md3 through the OLED.

This current copy principle compensates for CMOS process variations and achieves good matching between different pixels. A disadvantage of current programmed pixels is the different settling behavior over the entire signal range because of large parasitics on the data line. For low brightness levels and highly efficient OLEDs, the data current would be too small for sufficient settling in the sample phase. Since the current driver Md3 is working in the sub-threshold region, the programmed data current can be much larger in the sample phase and scaled down by shifting the reference voltage of the sampling capacitor Cd in the emit phase [1].

The complete display circuit operates on a 3.3V supply with an additional negative OLED cathode voltage. For evaluation purposes of different OLED stacks combined with high brightness, switch Md4 is implemented with a 12V transistor with rather large area consumption. The optical characterization of the display using a green OLED showed good linearity for peak luminances from 100 up to 24000 Cd/m² and a high contrast ratio of 3000:1@1000cd/m². The measured display transfer characteristic is presented in Fig. 17.8.5.

The integrated image sensor is composed of 160×120 pixels with 8×8μm² photodiodes in a pitch of 72×72μm². Thereby the photodiodes are placed between the OLED pixels and the active pixel circuitry is located underneath the OLED anode. The work function is global shutter with synchronous start and end of the exposure for all pixels to avoid moving artifacts during frame capture. During the exposure phase, an external IR-illumination of the eye-scene is activated to improve the signal-to-noise ratio.

Before the exposure phase the gate of Mc3 is connected to V_{dd} by turning Mc1 and Mc2 on. The exposure starts by turning the reset transistor Mc1 off, allowing the integration of the photocurrent from the photodiode, PD, on the gate of Mc3 and on Cc. The end of the exposure phase is determined by turning the transfer transistor Mc2 off and thus separating the PD from the integrating node. Subsequently, the stored picture information is read out line by line by turning Mc4 on and biasing Mc3 with a current via the data line. Now Mc3 works as a source follower and drives the stored gate voltage to the readout circuit.

To eliminate the influence of the threshold voltage of Mc3, the readout circuit performs correlated double sampling (CDS) by subtracting the reset value from the exposure value at each column. The offset corrected pixel values are latched on capacitors, get selected by the column addressing and are then transferred to an external ADC by an output driver.

With the exception of the external ADC for image readout, all control, configuration (I2C) and driving circuitry is integrated in the BiMi to allow the direct operation on a small mobile PDA sized host system. The setup of the bidirectional optic with a captured eye scene is shown in Fig. 17.8.6. Inside the optic the paths for display and camera are separated by filters for NIR and VIS spectrum to realize the different focal lengths between retina and eye scene. The realized see-through NTE display exhibits bright crisp images without motion artifacts and the captured images from the eye scene are suitable for basic eye tracking. This is the first documented work that presents an integrated OLED microdisplay with an image sensor in a standard CMOS process for see-through HMD applications.

Acknowledgements:
This work was partly funded by grants from the Federal Ministry for Education and Research (Bundesministerium für Bildung und Forschung, BMBF 01 BK 916-919, 16SV2283) of the German government and the Sächsische Aufbaubank (SAB) of the State of Saxony (11107/1733).

References:
[1] H. Lin, E. Naviasky, J. Ebner, et al., "An 852x600 Pixel OLED-on-Silicon Color Microdisplay Chip using CMOS Sub-Threshold-Voltage-Scaling Current Driver", *ISSCC Dig. Tech. Papers*, Feb. 2002.
[2] U. Vogel, I. Underwood, G. Notni, et al., "HYPOLED - VGA OLED Micro-display for HMD and Micro-projection", *IMID*, 2009.
[3] G. Kelly, R. Woodburn, I. Underwood, et al., "A Full-Color QVGA Microdisplay using Light-Emitting-Polymer on CMOS", *ICECS*, pp.760-763, 2006.

978-1-61284-303-2/11 $26.00 © 2011 IEEE

BiMi	
Technology	0.35μm CMOS + OLED post-processing
Supply	3.3V
Power consumption	~100mW @ 1000Cd/m² + CAM
Active area	11.52 x 8.64 mm²
Frame rate	10 ... 400 Hz
OLED microdisplay	
Resolution	320 x 240 (QVGA)
Pixel pitch	36 x 36 μm²
Peak luminance	100 ... 24000 Cd/m² @ green
Chromaticity	monochrome (green, orange, red, ...)
Gray levels	256
Contrast ratio	3,000:1 @ 1000cd/m²
OLED cathode voltage	0 ... -9 (-12)V
Data interface	8b digital data + digital synchronization
Image sensor	
Resolution	160x120 pixels (QQVGA)
Photoactive area per photodiode	8 x 8 μm²
Photodiode pitch	72 x 72 μm²
Data interface	analog data + digital synchronization

Figure 17.8.1: Performance summary of the bidirectional OLED microdisplay.

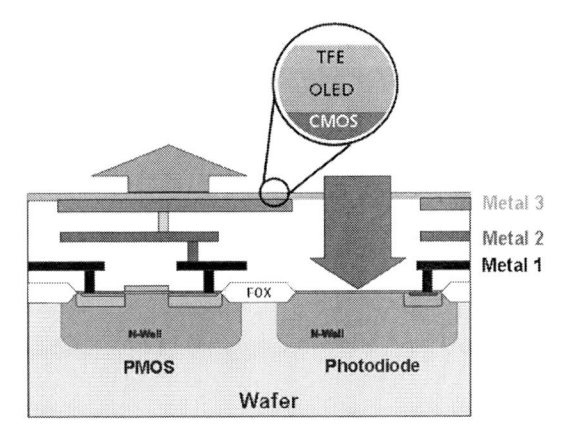

Figure 17.8.2: CMOS cross-section with OLED on top.

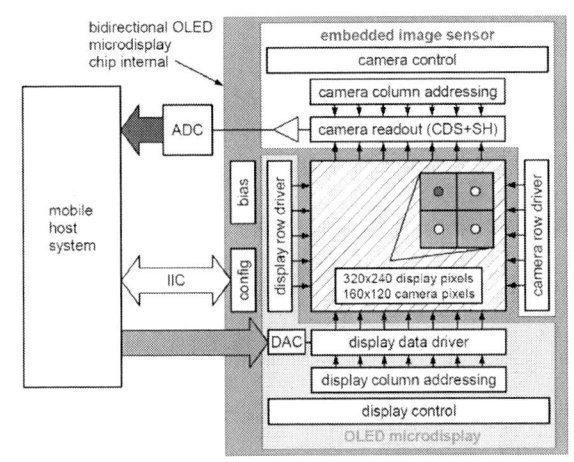

Figure 17.8.3: System diagram.

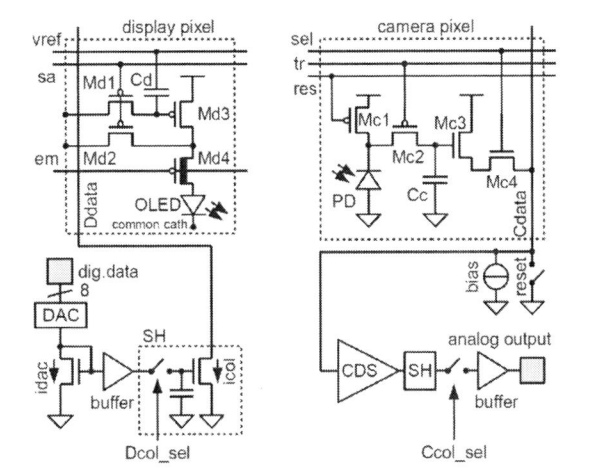

Figure 17.8.4: Display and camera pixel circuit with simplified driving/read-out scheme.

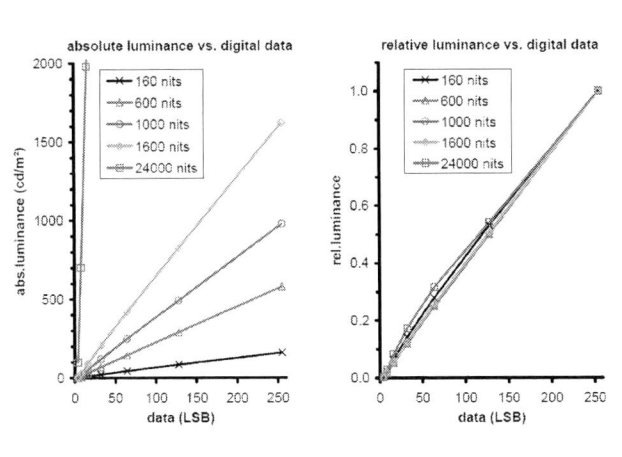

Figure 17.8.5: Measured display transfer characteristic for peak luminances from 160 to 24000 cd/m².

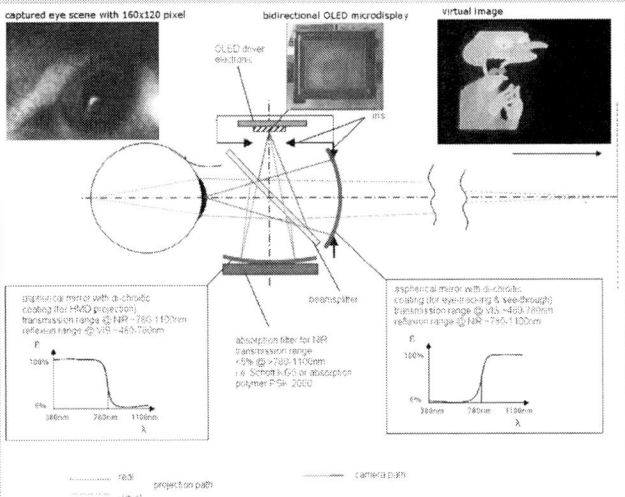

Figure 17.8.6: Setup of the realized bidirectional optic with separated optical paths for display (green) and camera (red).

ISSCC 2011 PAPER CONTINUATIONS

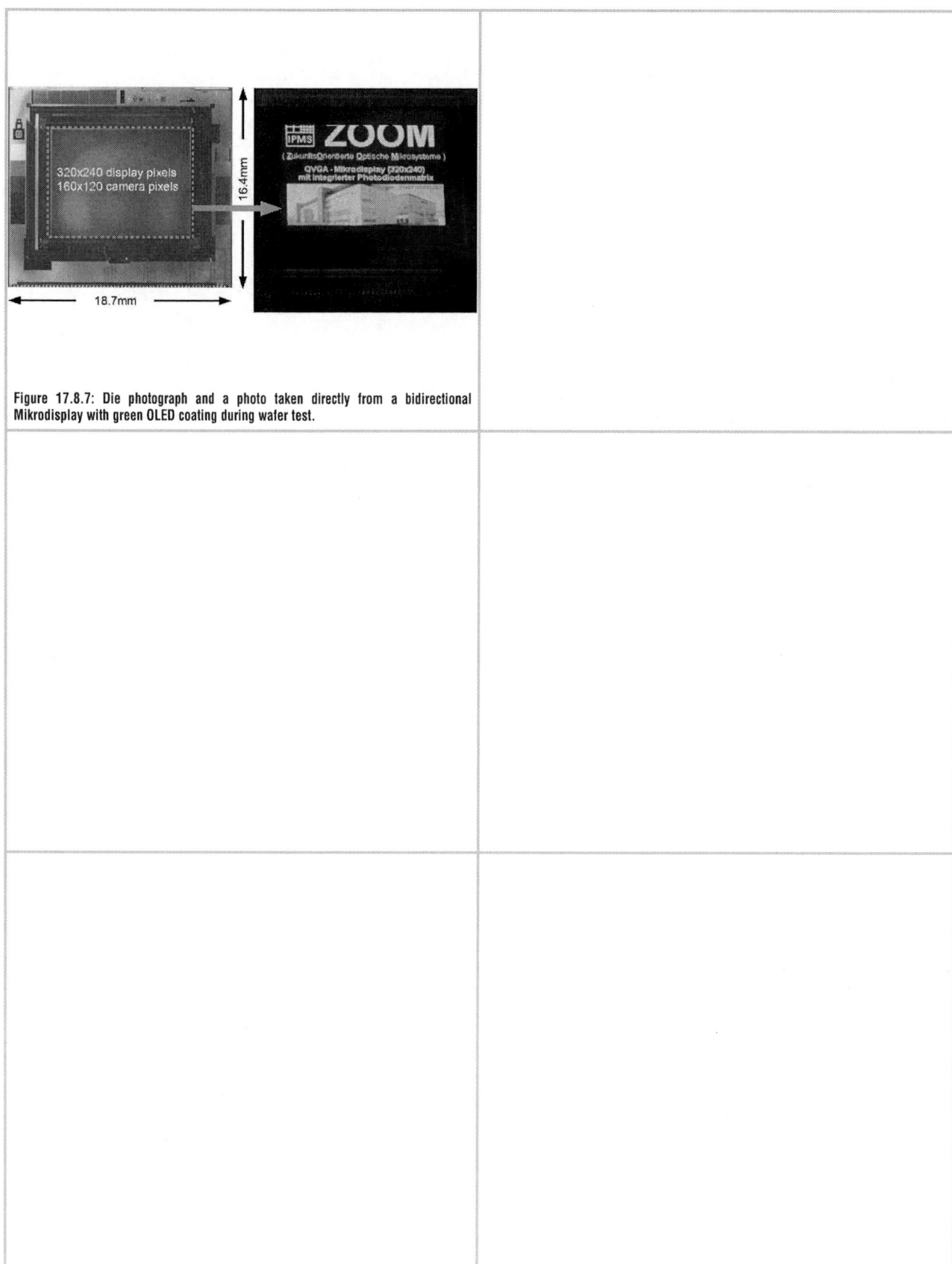

Figure 17.8.7: Die photograph and a photo taken directly from a bidirectional Mikrodisplay with green OLED coating during wafer test.

ISSCC 2011 / SESSION 17 / BIOMEDICAL & DISPLAYS / 17.9

17.9 A 0.014mm² 9b Switched-Current DAC for AMOLED Mobile Display Drivers

Hyun-Sik Kim, Jin-Yong Jeon, Sung-Woo Lee, Jun-Hyeok Yang, Seung-Tak Ryu, Gyu-Hyeong Cho

KAIST, Daejeon, Korea

Active-Matrix Organic LED (AMOLED) technology has attracted a considerable amount of attention as a very desirable display due to superior characteristics such as its wide viewing angle, fast response, thinness, and low power consumption. Among various driving techniques for AMOLED displays, the current-driving scheme deals well with the issues of spatial and temporal variation of the TFT as well as IR drops in the pixel circuits [1]. Several previous works have successfully demonstrated this [2-3]. Figure 17.9.1 shows a block diagram of an AMOLED display system adopting a current-driving scheme. The digital-to-analog converter (DAC) plays a key role in the column driver by reconstructing the analog signal for pixel-driving from digital data. Considering that the number of columns of pixels continues to increase, even in mobile-oriented displays, the chip area occupied by the DAC should be greatly reduced. Furthermore, as widescreen displays such as WXGA are coming into wider use, it is necessary for column drivers for these displays to have a narrow channel pitch.

In this paper, we present an area-efficient switched-current (SI) DAC structure for current-mode AMOLED display drivers. The designed 9b SI DAC reuses a single hardware unit up to N cycles to complete the N-bit conversion. Thus, the chip size of the DAC remains nearly constant regardless of the resolution, as long as the accuracy of the current division is guaranteed, resulting in excellent area-efficiency.

Figure 17.9.2 illustrates the operational principle of the DAC with its simplified schematic. Basic operation consists of two phases: a current-summing phase and a current-division phase. According to the operational phase, the NMOS transistor pair M_{1a}-M_{1b} functions as a current-summing or a current-dividing block by working in either the triode or saturation region. By repeating the above two phases, current DAC operation is achieved. In more detail, the digital input $b[k]$ is sequentially applied to the DAC from the LSB to the MSB. During the current-summing phase, the NMOS transistor pair M_{1a}-M_{1b} works in the triode region as current switches. The k-th bit current, $b[k] \cdot I_{ref}$, is transferred to S/H$_1$ via the switch M_{1a}. The current held in S/H$_2$ from the previous phase, $I_{SH2,k-1}$, is also supplied to S/H$_1$ through M_{1b} and is added to $b[k] \cdot I_{ref}$. Thus, the current sampled in S/H$_1$ can be described as Eq. (1). In the following current division phase, S/H$_1$ is in the hold mode and works as a current source, while S/H$_2$ is in the sampling mode. In order to divide the output current from S/H$_1$ in half, the identically designed transistor pair M_{1a}-M_{1b} now works in the saturation region as a current divider with the same gate bias, V_{bias}. Thus, half of the stored current in S/H$_1$ is sampled in S/H$_2$, while the other half of the current is simply wasted via M_{1a}. Through this operation, the sampled current in S/H$_2$ can be expressed as Eq. (2). This current division serves to implement the binary-weighted sum in the following summing phase. By iterating the above two operational phases for the entire N-bit input, S/H$_2$ will eventually have the final output current, expressed as Eq. (3). Hence, the N-bit D/A conversion is completed.

Figure 17.9.3 shows the detailed schematic of the 9b SI DAC. The DAC is composed of a reference current source, two current-mode S/Hs, analog switches (MUXs) for current path control, and a NMOS transistor pair M_{1a}-M_{1b} as in Fig. 17.9.2. In addition, because the matching accuracy of M_{1a} and M_{1b} is critical for DAC linearity, a path swapper is inserted above the M_{1a}-M_{1b} pair to exchange the current path at every current division phase. This shuffles the M_{1a}-M_{1b} mismatch error and enhances the overall linearity. V_{b1} connected to MUX$_3$ creates the drain voltage of M_{1a} similar to that of M_{1b} so as to ensure that the two current branches are under the same condition in the current-division phase. M_7 prevents the current source M_3 from entering triode region by bleeding its current to M_7 when the current through M_4 becomes 0 at $\Phi_1 = 0$ or $b[k] = 0$; this speeds up the transient response. A timing diagram of the control signals is shown in Fig. 17.9.3 as well. Before starting the conversion, S/H$_2$ samples zero current by the reset signal to MUX$_4$. After the conversion is done for a 9b input

data, the control signal Φ_2 activates and the D/A-converted current is provided to the output. While the output is active, Φ_1 remains high so that S/H$_2$ remains in the hold mode.

Through the above operations, the current-mode S/Hs serve key roles. In particular, their linearity and speed are very important. However, the performance of the conventional S/Hs is limited by linearity, speed, and stability problems [5]. To mitigate these problems, we introduce a high-performance S/H that employs a negative-feedback structure and a multi-functioning Miller capacitor.

Figure 17.9.4 shows a schematic of the current-mode S/H. During the sampling mode (SW = 1), the input current flows to node X and the charge corresponding to it is trapped on sampling capacitor C_H. The maximum sampling speed of the S/H is determined by the settling time constant of node X. In the S/H, a negative-feedback loop through M_3-M_8-M_2-M_1-M_{11} is applied during the sampling mode; as a result, node X is virtually grounded to V_{b1}. This allows the sampling speed of the S/H to be enhanced significantly. In addition, the Miller compensation capacitor C_H works as a sampling capacitor while guaranteeing a stable feedback loop. During the hold mode (SW = 0), the feedback loop is broken by turning off the switch, M_{SW}, leaving the S/H as a current source. The major limitation of the linearity of the S/H is due to charge injection when the M_{SW} turns off. In the S/H, mitigation of the charge injection is accomplished by two methods: a dummy switch and a boost of the sampling capacitance C_H by the Miller enhancement method. Additionally, the drain and source voltages of the M_{SW} always remain with the constant voltage V_{b1} regardless of the input current level due to the feedback loop. Therefore, it is possible to achieve signal-independent charge injection and improve the linearity of the current-mode S/H significantly. Lastly, the output impedance of the S/H during the hold mode is boosted by utilizing a regulated cascode circuit (M_9-M_{14}), which serves as an ideal current source. On the right hand side of Fig. 17.9.4, the linearity simulation result of the S/H at a sampling rate of 2.5MS/s is shown. The maximum error current is 0.1LSB (1nA) in 9b resolution, which is sufficient for the DAC.

A prototype of the 9b SI DAC was fabricated in a 0.35µm 3.3V CMOS process. Figure 17.9.5 shows the measured INL and DNL of the DAC. The maximum INL and DNL with path swapping (without path swapping) are 1.6LSB (3.8LSB) and 0.8LSB (2LSB), respectively. The deviation of current output (DCO) was also measured for 512 gray scales in 8 channels, and a maximum inter-channel DCO of 15nA was achieved. Figure 17.9.6 summarizes the measured performance of the prototype DAC and compares it with previous works. The chip size per channel DAC is 0.014mm² in this design, which is much smaller than those reported recently. The D/A conversion speed is 1µs/b with a static current of 10µA. A die micrograph is shown in Fig. 17.9.7.

Acknowledgement:
This work was supported by the ERC program of the Korea Science and Engineering Foundation (KOSEF) grant funded by the Korea Ministry of Science and Technology (MOST) (No. R11-2007-045-01004-0) and in part by the IC Design Education Center (IDEC).

References:
[1] A. Nathan, et al., "Driving Schemes for a-Si and LTPS AMOLED Displays," *IEEE J. Display Tech.*, vol. 1, no. 3, pp. 267-277, Dec. 2005.
[2] J. H. Baek et al., "A Current Driver IC using a S/H for QVGA Full-Color Active-Matrix Organic LED Mobile Displays," *ISSCC Dig. Tech. Papers*, pp. 609-618, Feb. 2006.
[3] J. -Y. Jeon, et al., "A Direct-Type Fast Current Driver for Medium- to Large-Size AMOLED Displays," *ISSCC Dig. Tech. Papers*, pp. 174-175, Feb. 2008.
[4] Y. -J. Jeon, et al., "Design Method for Area-efficient and Uniform Channel DACs in Current-mode AMOLED Display Data Drivers", *IEEE Trans. Consumer Electronics*, vol. 56, no. 2, pp. 271-279, May 2010.
[5] W. Guggenbuhl, et al., "Switched-Current Memory Circuits for High-Precision Applications", *IEEE J. Solid-State Circuits*, vol. 29, no. 9, pp. 1108-1116, Sep. 1994.
[6] I. Knausz, et al., "A 250µW 0.042mm² 2MS/s 9b DAC for Liquid Crystal Display Drivers", *ISSCC Dig. Tech. Papers*, pp. 599-608, Feb. 2006.

ISSCC 2011 / February 22, 2011 / 4:45 PM

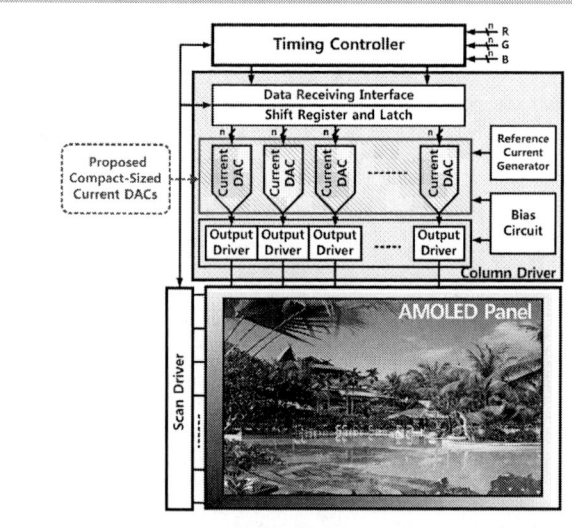

Figure 17.9.1: Block diagram of current-mode AMOLED display system.

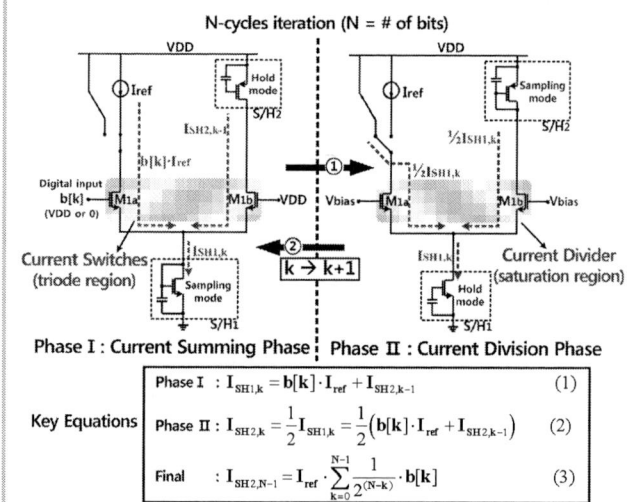

$$\text{Phase I} : I_{SH1,k} = b[k] \cdot I_{ref} + I_{SH2,k-1} \tag{1}$$

$$\text{Phase II} : I_{SH2,k} = \frac{1}{2} I_{SH1,k} = \frac{1}{2}\left(b[k] \cdot I_{ref} + I_{SH2,k-1}\right) \tag{2}$$

$$\text{Final} : I_{SH2,N-1} = I_{ref} \cdot \sum_{k=0}^{N-1} \frac{1}{2^{(N-k)}} \cdot b[k] \tag{3}$$

Figure 17.9.2: Conversion sequence of the SI DAC.

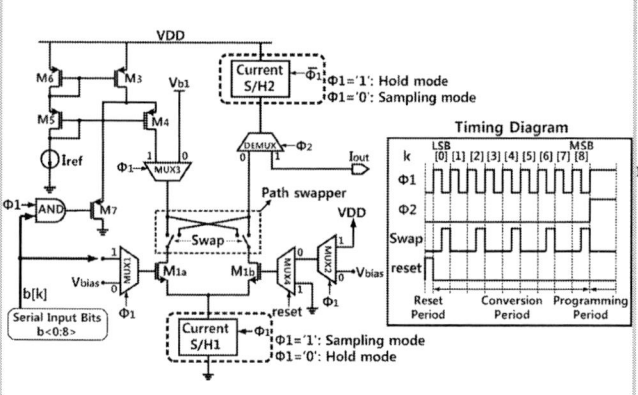

Figure 17.9.3: Schematic and timing diagram of the 9b SI DAC.

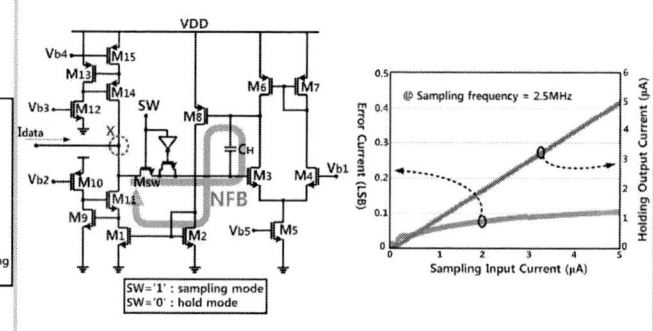

Figure 17.9.4: Schematic of the current-mode S/H and its linearity plot (simulation).

Figure 17.9.5: Measured results of DNL/INL and inter-channel DCO.

Ref.	This work	[3]	[4]	[6]
Application	AMOLED	AMOLED	AMOLED	AMLCD
Technology	0.35 μm (4M, 2P)	0.35 μm (4M, 2P)	0.35 μm (4M, 2P)	0.5 μm (4M, 2P)
Power Supply	3.3 V	3.3 V	3.3 V	5 V
Resolution	9 bits	8 bits	9 bits	9 bits
Static Power	33 μW	-	36.96 μW	250 μW*
INL	1.6 LSB	-	2.5 LSB	± 1 LSB
DNL	0.8 LSB	-	1.5 LSB	± 0.5 LSB
Conversion Speed	1 μs/bit	-	500 kS/s	2 MS/s
Chip Size (1 channel)	0.014 mm²	0.039 mm²	0.048 mm²	0.042 mm²

* including the power consumption of the output buffer.

Figure 17.9.6: Performance summary and comparison with previous works.

978-1-61284-303-2/11 $26.00 © 2011 IEEE

ISSCC 2011 PAPER CONTINUATIONS

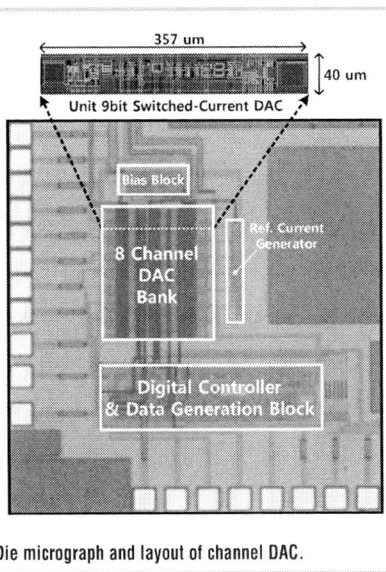

Figure 17.9.7: Die micrograph and layout of channel DAC.

ISSCC 2011 / SESSION 17 / BIOMEDICAL & DISPLAYS / 17.10

17.10 A 10b Resistor-Resistor-String DAC with Current Compensation for Compact LCD Driver ICs

Chih-Wen Lu[1], Ping-Yeh Yin[2], Ching-Min Hsiao[2],
Mau-Chung Frank Chang[3]

[1]National Tsing Hua University, Hsinchu, Taiwan,
[2]National Chi Nan University, Puli, Taiwan,
[3]University of California, Los Angeles, CA

Achieving a higher color depth for LCD drivers requires a higher DAC resolution and a larger circuit die area. Due to the stringent requirement on uniformity, a resistor-string DAC (RDAC) is predominantly used for LCD column drivers. However, the area of the RDAC and related routing lines are prohibitively large for a high-resolution data converter, making it impractical for column-driver ICs in high color depth displays [1].

To avoid the above-mentioned issue, the following DAC architectures were proposed in the past: a CDAC [2], an embedded DAC [3], DACs with current modulation/interpolation [4-5], and a resistor-resistor-string DAC (RRDAC) without unity-gain buffers [1]. However, the CDAC architecture suffers from a long D/A conversion time. The embedded DAC architecture requires a large number of input transistors with long widths and lengths for accurate matching. This results in a large area overhead for high bit interpolation. The DACs with current modulation/interpolation were implemented in 0.1μm CMOS technology, which requires many current switches for modulation/interpolation. If the DACs with current modulation/interpolation were implemented in a low-cost CMOS technology (wider channel length) or a high-voltage technology, they would occupy large die area.

A typical RRDAC, which contains two RDACs and two intermediate unity-gain buffers, may reduce the chip area. The unity-gain buffers can isolate these two RDACs. The buffers, however, have offset errors that can be further spread to the LCD driver output. Consequently, obtaining output uniformity for a high-color depth column driver is rather difficult. Furthermore, each output channel demands two additional buffers with increased power consumption. To reduce the area, researchers have used a RRDAC without unity-gain buffers [1]. Under such condition, parallel channel resistor strings have been connected directly to the global resistor string. This, in fact, affects the reference voltages of the global resistor string.

To overcome these issues, we introduce a type of 10b RRDAC with a current compensation scheme to provide good linearity and uniform channel performance, and simultaneously maintain the 10b DAC at a size smaller than that of a conventional 8b RDAC. A cascode class-AB control is devised to bias the output stage of the output buffer.

Figure 17.10.1 shows the 10b RRDAC by combining a 6b RDAC and a 4b RDAC without the need of unity-gain buffer to isolate parallel-connected resistor strings, yet with current compensation to offset the loading effect. In a column-driver chip, a 6b global resistor string ($64R_1$) is used. Each output channel has a 6b decoder, a 4b channel resistor string ($16R_2$), a 4b decoder, and an output buffer. Based on the higher 6b data signal, the 6b decoder selects two neighboring voltages (V_H and V_L) and connects them to the 4b channel resistor string. For the lower 4b data signal, the 4b channel resistor string divides the output voltage into 16 levels between V_H and V_L. The 4b decoder further selects a voltage from the channel resistor string and propagates it to the output buffer. The current flowing in the channel resistor string is $(V_H - V_L)/16R_2$. To minimize the loading effect, we can then inject currents of $(V_H - V_L)/16R_2$ into both the top and the bottom ends of the channel resistor at the same time. As a result, insignificant static current would flow between the global and channel resistor strings. The reference voltages for the global resistor string therefore remain intact.

All channel compensation currents for a column driver are mirrored from a global compensation current source to minimize the overhead area and power consumption. As shown in Fig. 17.10.2, the global compensation current source senses the voltage difference between two nodes at the middle of the global resistor string and generates the proper compensation current accordingly. Since the sensed voltage difference is $n \cdot (V_H - V_L)$ and the total value of the resistor string in the global compensation current source is n times that in the channel resistor string, the generated compensation current will be set at

$$I_{comp} = \frac{n \cdot (V_H - V_L)}{n \cdot 16R_2} = \frac{V_H - V_L}{16R_2}.$$

Figure 17.10.3 shows the schematic of the output buffer, where the cascode class-AB control, M11-M18, precisely controls the quiescent current of the output transistors. This makes the quiescent current insensitive to the supply voltage.

Using 0.35μm/0.5μm CMOS technology, an 18-channel prototype is fabricated for validating the 10b RRDAC's performance. The test pattern is applied simultaneously to all channel inputs. Figure 17.10.4 shows the measured results in terms of a linear 10b gray scale for RGB-separate gamma on five different chips. The maximum DNL and INL are measured as 0.14 LSB and 0.61 LSB, respectively, with 1LSB = 4.4mV. The DVO is also measured according to 1024 gray scales on five chips. Without applying any off-chip trimming, the maximum inter-chip DVO is 16mV. Figure 17.10.5 shows measured output waveform with a 30kΩ-resistance and 30pF-capacitance load, as the digital data change from "0000000000" to "1111111111". The time to settle within 0.2% of the final voltage is 4μs. Figure 17.10.6 summarizes the performance of the 10b RRDAC compared with the state-of-the-art. Figure 17.10.7 shows the area of the DAC compared with that of a conventional 8b RDAC. The 10b RRDAC occupies 70% of the conventional 8b RDAC area. Figure 17.10.7 also shows the die micrograph with sizes of 530×504μm² and 150×504μm² for 18 RRDACs and 18 buffers, respectively. The measured results show that the RRDAC with current compensation scheme is very suitable for both small- and large-size LCD applications.

Acknowledgements:
The authors would like to thank National Chip Implementation Center (CIC) for chip fabrication. This work is also supported by the National Science Council of Taiwan, R.O.C.

References:
[1] Y.-C. Sung, S.-M. So, J.-K. Kim, and O.-K. Kwon, "10bit Source Driver with Resistor-Resistor-String Digital to Analog Converter," *SID Symposium Digest 36*, pp. 1099-1101, 2005.
[2] Y.-K. Choi, et al, "A Compact Low-Power CDAC Architecture for Mobile TFT-LCD Driver ICs," *ISSCC Dig. Tech. Papers*, pp. 176-177, Feb. 2008.
[3] J.-S. Kang, et al, "A 10b Driver IC for a Spatial Optical Modulator for Full HDTV Applications," *ISSCC Dig. Tech. Papers*, pp. 138-139, Feb. 2007.
[4] Y.-J. Jeon, et al, "A Piecewise-Linear 10b DAC Architecture with Drain-Current Modulation for Compact AMLCD Driver ICs," *ISSCC Dig. Tech. Papers*, pp. 264-265, Feb. 2009.
[5] H.-M. Lee, et al, "A 10b Column Driver with Variable-Current-Control Interpolation for Mobile Active-Matrix LCDs," *ISSCC Dig. Tech. Papers*, pp. 266-267, Feb. 2009.

ISSCC 2011 / February 22, 2011 / 5:00 PM

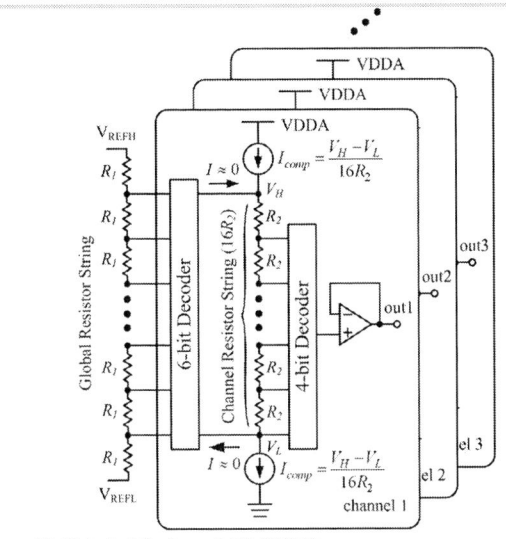

Figure 17.10.1: Architecture of 10b RRDAC.

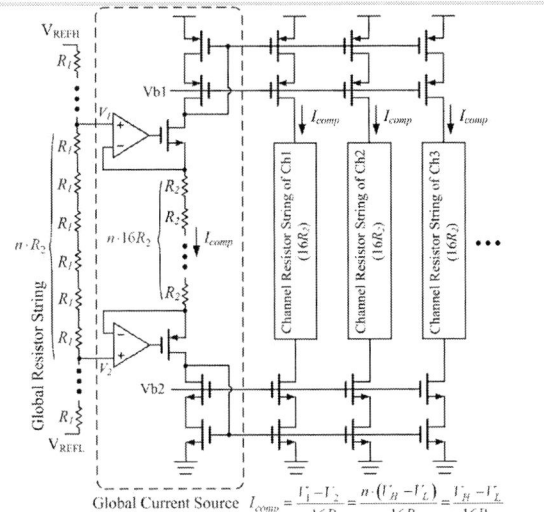

Figure 17.10.2: Schematic of the global compensation current source.

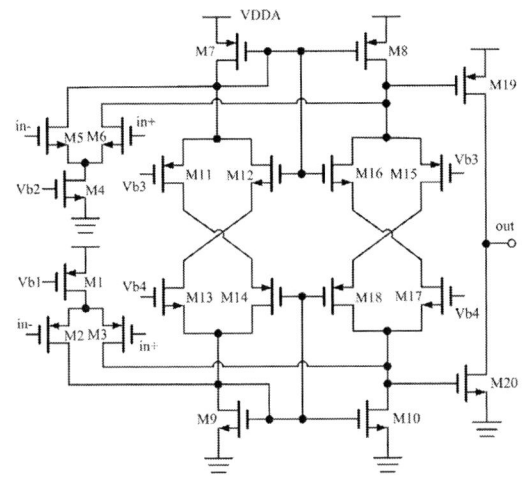

Figure 17.10.3: Schematic of output buffer.

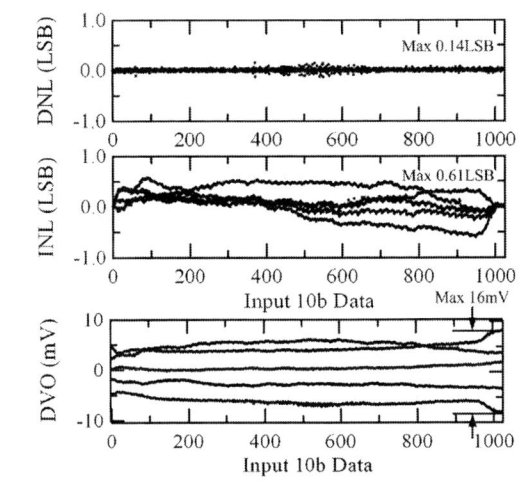

Figure 17.10.4: Measured DNL/INL and DVO from five different chips.

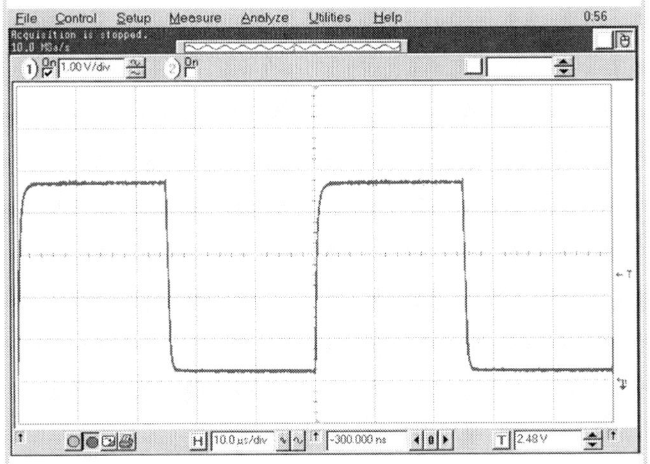

Figure 17.10.5: Measured output waveform.

	[4]	[5]	This work
Technology	0.1 μm CMOS (1P5M)	0.1 μm CMOS (1P5M)	0.35 μm/0.5 μm CMOS (2P4M)
VDD	1.5V (Logic) 5V (Analog)	1.5V (Logic) 5V (Analog)	5 V
Gray Scale	10b	10b	10b
Output Range	0V to 5V	0.25V to 4.75V	0.2V to 4.7V
DNL/INL	0.37LSB/1.71LSB	0.4LSB/0.7LSB	0.14LSB/0.61 LSB
Max. DVO	6.35 mV	20 mV	16 mV
Static Current Consumption	1.2 μA/channel	1 μA/Buffer	0.25 μA/DAC 0.93 μA/Buffer
DAC and Buffer Areas	206×14 μm²/DAC 127×14 μm²/Buffer	195×14 μm²/DAC 125×14 μm²/Buffer	530×28 μm²/DAC 150×28 μm²/Buffer
DAC Area Shrinkage (compared to 8b RDAC)	24 %	29 %	30 %

Figure 17.10.6: Performance summary.

978-1-61284-303-2/11 $26.00 © 2011 IEEE

ISSCC 2011 PAPER CONTINUATIONS

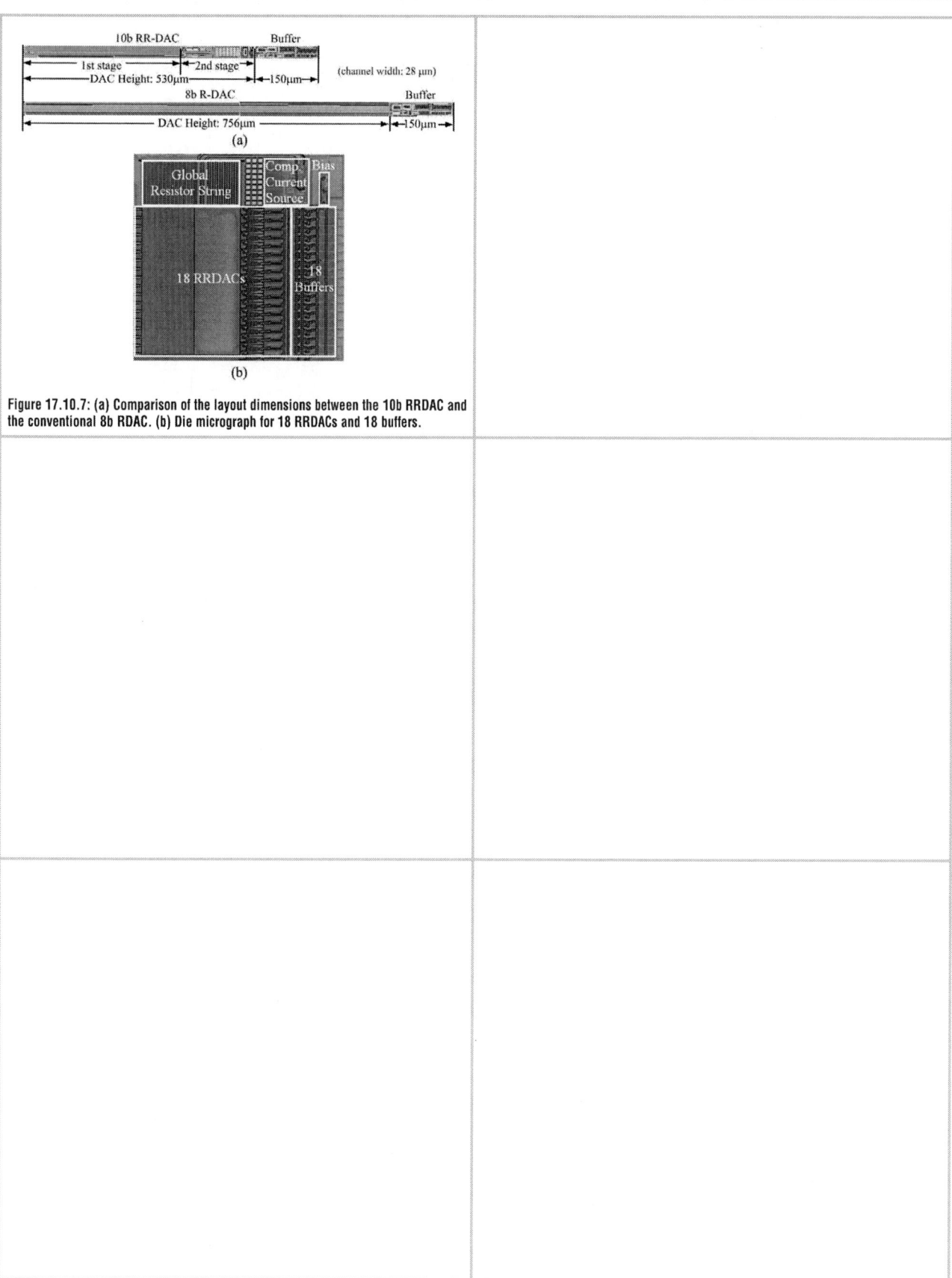

Figure 17.10.7: (a) Comparison of the layout dimensions between the 10b RRDAC and the conventional 8b RDAC. (b) Die micrograph for 18 RRDACs and 18 buffers.

ISSCC 2011 / SESSION 18 / ORGANIC INNOVATIONS / OVERVIEW

Session 18 Overview / *Technology Directions*

Organic Innovations

Session Chair: *Chris Van Hoof, imec, Leuven, Belgium*

Session Co-Chair: *Masaitsu Nakajima, Panasonic, Moriguchi, Japan*

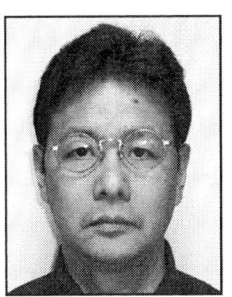

Implementing transistors, circuits and sensors on arbitrary substrates has great potential for a wide range of low-cost electronic products such as flexible displays, disposable biochemical sensors and large-area artificial skin. This vision is gradually becoming a reality thanks to continuing innovations in low-temperature-processed organic thin-film transistor (OTFT) technology. However, because of the limited palette of available devices and the large process variations in OTFT fabrication, the complexity achieved by organic circuits has so far been rather limited. Also, to achieve ultra-low-cost manufacturing, fully printed organic technology is needed, while most OTFT technologies are still using photolithography. The low-cost potential has therefore not been adequately demonstrated yet.

System-on-foil applications require distributed sensing and actuation over a large area. Such large-area sensors are typically organized in a matrix fashion, where data are collected and processed by crystalline silicon circuitry on the edges. To enable mechanical flexibility and low cost, data conversion and processing should be integrated on the foil. Last year's ISSCC already showcased the first organic ADCs. The first paper in this session demonstrates a microprocessor made with OTFTs that completes this vision. Similarly, large-area distributed actuators on foil can be substantially improved when digital-to-analog conversion is taking place at the pixel level. The second paper in this session addresses an organic DAC showing good speed and accuracy.

Paper 18.1 [IMEC; KU Leuven; Polymer Vision; TNO; KHLim], introduces the first 8b organic microprocessor in a dual-gate, pentacene (p-type only) TFT technology on 25μm thin plastic foils. The design consists of a microprocessor foil and an instruction foil, in which a program is hardcoded. The basic DSP functionality is demonstrated using a running average program at clock frequencies up to 6Hz for a 10V supply.

Paper 18.2 [IMS CHIPS; Max Planck Institute for Solid State Research; U Stuttgart] presents a 3.3V 6b binary-weighted current-steering D/A converter using organic p-type thin-film transistors (OTFTs). The circuit occupies 12mm^2 and is fabricated on a glass substrate using silicon stencil shadow masks. The converter consumes 260μW at an output swing of 2V and has a maximum update rate of 100kS/s. The measured DNL and INL at an update rate of 1kS/s are -0.69 LSB and 1.16 LSB, respectively.

Disposable sensors and circuits on foil will open up new applications such as low-cost RFID product labeling and sensors integrated in product wrappings. To achieve these functionalities, low-cost circuits and memories should be available: printing can be an effective manufacturing approach to achieve this goal. The third and fourth papers demonstrate innovations in this domain.

Paper 18.3 [U Minnesota; Optomec] presents the first DRAM cell realized in a printable, flexible and low-voltage ion-gel electrolyte organic TFT technology. At an operating voltage of 1.0V, the retention time of the implemented gain cell array can exceed one minute and the refresh power is less than 10nW per cell. Full read and write functionality is verified using an 8×8 printed organic DRAM test chip.

Paper 18.4 [CEA-LITEN; STMicroelectronics] presents a fully printed organic CMOS technology on flexible substrates that shows good device performance and enables both digital and analog circuit functionality. These are the natural stepping stones to more complex designs. A test chip includes single devices, inverters, ring oscillators and analog building blocks such as current mirrors and differential pairs. An additional significant contribution of this work is that both n- and p-type semiconductors are printed in a low-cost and scalable process on foil demonstrating the feasibility of fully-printed organic CMOS technology.

These four innovative achievements are very important steps towards a complete organic circuit technology platform integrating sensors, analog circuits, memory, and microprocessor functionality. A technology platform relying entirely on printing technology and roll-to-roll processing will ensure low-cost manufacturing. Such platform will enable new application domains such as large-area distributed sensing, actuation and organic smart labels.

978-1-61284-303-2/11 $26.00 © 2011 IEEE

ISSCC 2011 / February 23, 2011 / 8:30 AM

Presenters:

18.1 An 8b Organic Microprocessor on Plastic Foil   8:30 AM

K. Myny,

imec, Leuven, Belgium; KU Leuven, Leuven, Belgium

18.2 A 3.3V 6b 100kS/s Current-Steering D/A Converter Using   9:00 AM
Organic Thin-Film Transistors on Glass

T. Zaki,

Institute for Microelectronics Stuttgart (IMS CHIPS), Stuttgart, Germany; University of Stuttgart, Stuttgart, Germany

18.3 A 1V Printed Organic DRAM Cell Based on Ion-Gel Gated   9:30 AM
Transistors with a Sub-10nW-per-Cell Refresh Power

W. Zhang,

University of Minnesota, Minneapolis, MN

18.4 Fully Printed Organic CMOS Technology on Plastic Substrates   9:45 AM
for Digital and Analog Applications

A. Daami,

CEA-LITEN, Grenoble, France

18

978-1-61284-303-2/11 $26.00 © 2011 IEEE   321

ISSCC 2011 / SESSION 18 / ORGANIC INNOVATIONS / 18.1

18.1 An 8b Organic Microprocessor on Plastic Foil

Kris Myny[1,2,5], Erik van Veenendaal[3], Gerwin H. Gelinck[4], Jan Genoe[1,5], Wim Dehaene[1,2], Paul Heremans[1,2]

[1]imec, Leuven, Belgium, [2]KU Leuven, Leuven, Belgium,
[3]Polymer Vision, Eindhoven, The Netherlands,
[4]TNO Science and Industry, Eindhoven, The Netherlands,
[5]KHLim, Diepenbeek, Belgium

We introduce a microprocessor made by organic thin-film transistors processed directly onto flexible plastic foil. This is a direct realization of a microprocessor by thin-film technology, i.e., without transfer, on plastic. It paves the way to equip mundane supports and objects with low-cost computing power. We also demonstrate the correct execution of a digital signal-processing task, namely increasing the accuracy of a repetitive digital input by time-averaging.

Organic transistors are currently used for backplanes of displays, i.e., large arrays of identical pixel engines, each comprising a very limited number (typically 1 to 4) of transistors. The technology has been shown to be adequate for code generators for RFID tags [1-5], line drivers for rollable displays [6,7] and first analog circuits [8-10]. The digital circuits [1-7] all have a well-defined datapath and very limited control logic. Hence, the critical path was always predictable and it could be optimized by design for performance and yield.

Recently, an elegant way to achieve dual-VT logic with organic TFTs was shown. It makes use of a back-gate on the transistors to control VT of drive and load transistors of a basic inverter independently [11]. This organic transistor technology is developed by Polymer Vision for rollable active-matrix displays [12]. The organic gate insulator layer, the organic dielectric separating the channel from the back-gate and the p-type pentacene semiconductor all are processed from solution. The transistors have a typical channel length of 5μm and an average saturation mobility of 0.19cm^2/Vs. The top gate is intended to be the pixel electrode in the display backplane, but is used as back-gate to control VT in our digital circuit. Using this more robust dual-VT OTFT architecture, we explore in the present work the untrodden terrain of organic thin-film circuits with complex control logic and variable datapaths. In particular, we elaborate an 8b microprocessor comprising 4000 transistors. It is composed of two foils, a microprocessor foil and an instruction code foil, that can be connected 1-to-1. Our microprocessor is processed directly onto foil, following up on examples of microprocessors *transferred* to foil after processing at high temperatures on rigid substrates [13].

We modified the design strategy compared to the approach practiced so far in the field. Instead of simulating the schematic entry by an analog simulator (such as Spectre or Spice), we executed a functional simulator in Modelsim. First-order timing simulations have been done by using a fixed delay time for inverters and NAND gates, which led to a rough estimation of the expected clock frequency. Finally, the layout has been generated using an automatic P&R tool, having a library of 3 main cells: inverter, NAND and a buffer.

Figure 18.1.1 outlines the architecture of the microprocessor foil. The core of the processor is the 8b Arithmetic and Logical Unit (ALU), which implements classical logic (AND, OR, NOT), arithmetic (Add, subtract, increment, decrement) and shift (logic shift left, arithmetic shift right) operations. The ALU is controlled by the 3 least-significant bits among the 10 bits from the processor's opcode. The output of the ALU is stored in the accumulator register. Three additional working registers (C0, C1, and C2) and an output register are available to the microprocessor. The registers select bits [Regsel (1:0)] correspond to bits 7 and 8 of the opcode. Bit 10 of the opcode implements the jump instruction.

The microprocessor foil is designed using only NAND gates and invertors. The gate delay of the NAND gates and the invertors having a single load can be estimated to be about 238μs and 200μs respectively from measurements on ring oscillators. Modelsim simulations using the above-mentioned gate delays through the design, predict operation of the microprocessor foil at clock frequencies below 50Hz. To prevent the heavily loaded nodes from reducing clock frequency too much, buffers are added on several nodes in the design, but the effect of the load on logic gates cannot be fully reduced.

After processing, the correct operation of each of the instructions of the microprocessor foil is tested independently using a dedicated hardware testbench running on a PIC18F development board. The expected outcome (generated by the PIC18F testboard) and the measured output register value are plotted on a digital scope (Fig. 18.1.2), indicating that each instruction behaves as foreseen,

when appropriate power and back-gate voltages are applied. Figure 18.1.3 shows the Shmoo plots for both the back-gate voltage and the power voltage. The microprocessor foil is operational for supply voltages between 10 and 20V, and the backgate voltage can be varied between 45 and 65V at a supply voltage of 15V. A clock frequency up to 6Hz is obtained. A slight decrease in maximum frequency is obtained at higher backgate voltage, due to the lower pull-up current caused by higher backgate voltages. The discrepancy between the measured clock frequency and the estimated frequency by Modelsim is due to the fixed time delays used in the simulator for inverter and NAND gates, which can be improved in the future by including load-dependent delay times in the simulations.

Next, the instruction foil is tested. Figure 18.1.4 shows the architecture of the instruction foil comprising the program counter (PC) and the instruction matrix. Each instruction comprises 10b, from which 9b are passed further towards the microprocessor foil. When bit 10 of the instruction is 1, a jump is executed: the PC is loaded with the remaining bits of the instruction and a No-Operation (NOP) is passed to the microprocessor foil. A reset brings the PC back to 0.

We implemented a "running averager" algorithm out$_{new}$ = 0.5 round(in + out$_{old}$) on the instruction foil (Fig. 18.1.6b). This example is chosen because it is a typical digital signal-processing instruction, and the processing of sensor outputs is indeed a likely application for foil microprocessors. The running averager increases the accuracy of a digitized sensor output: when the processor input bits would be connected to a 6b analog-digital convertor on foil [9,10], the processor output would correspond to the averaged input signal, with one bit increase in resolution and a time constant equal to the sampling clock. Figure 18.1.5 shows the measured output when the microprocessor foil and the instruction foil are connected together. The input signal is manually set from 0 to 7 (using switches). The algorithm is executed twice each program loop and the input is also sampled twice each program loop, although the output pins are only updated after the second implementation. At the output, we then observe the sequences for 5.5, 6.5 and 7. Figure 18.1.7 shows the die picture of both the microprocessor foil (comprising 3381 transistors, 1.96×1.72cm^2) and the instruction foil (comprising 612 transistors 0.72×0.64cm^2). The power consumption of the microprocessor foil is typically 92μW at 10V V$_{DD}$.

Acknowledgment:
This work was performed in a collaboration between imec and TNO in the frame of the HOLST Centre. It was partially supported by the EU-Project COSMIC (IST-IP-247681).

References:
[1] E. Cantatore, et al., "A 13.56-MHz RFID System Based on Organic Transponders", *IEEE J. Solid-State Circuits* 42, pp. 8, 2007.
[2] K. Myny, et al., "Plastic circuits and tags for HF radio-frequency communication", *Solid-State Electronics* 53, pp. 1220, 2009.
[3] R. Blache, et al., "Organic CMOS Circuits for RFID Applications", *ISSCC Dig. Tech. Papers*, pp. 208-209, Feb. 2009.
[4] G.-J. Cho, "Roll-to-Roll Printed 13.56 MHz Operated 16-Bit RFID Tags and Smart RF Logos", *Printed Electronics & Photovoltaics Europe 2010*, Apr.13, 2010.
[5] M. Jung, "All-Printed and Roll-to-Roll-Printable 13.56-MHz-Operated 1-bit RF Tag on Plastic Foils", *IEEE Trans. Elec. Devs* 57, pp. 571, 2010.
[6] P. van Lieshout, et al., "A Flexible 240×320-Pixel Display with integrated Row Drivers Manufactured in Organic Electronics", *ISSCC Dig. Tech. Papers*, pp. 578-579, Feb. 2005.
[7] M. Noda, et al., "A Rollable AM-OLED Display Driven by OTFTs", *Proc. of SID2010*, paper 47.3, 2010.
[8] M.G. Kane, et al., "Analog and Digital Circuits Using Organic Thin-Film Transistors on Polyester Substrates", *IEEE Electron Dev.Lett.* 21, pp. 534, 2000.
[9] H. Marien, et al., "An Analog Organic First-Order CT ΔΣ ADC on a Flexible Plastic Substrate with 26.5dB Precision", *ISSCC Dig.Tech.Papers*, pp.136-137, Feb. 2010.
[10] W. Xiong, et al., "A 3V 6b Successive-Approximation ADC Using Complementary Organic Thin-Film Transistors on Glass", *ISSCC Dig. Tech. Papers*, pp. 134-135, Feb. 2010.
[11] K. Myny, et al., "Robust Digital Design in Organic Electronics by Dual-Gate Technology", *ISSCC Dig. Tech. Papers*, pp. 140-141, Feb. 2010.
[12] H.E.A. Huitema, et al., "Rollable Displays: The Start of a New Mobile Device Generation", *7th Annual USDC Flexible Electronics & Displays Conference*, 2008.
[13] N. Karaki, et al., "A Flexible 8b Asynchronous Microprocessor based on Low-Temperature Poly-Silicon TFT Technology", *ISSCC Dig. Tech. Papers*, pp. 272-273, Feb. 2005.

ISSCC 2011 / February 23, 2011 / 8:30 AM

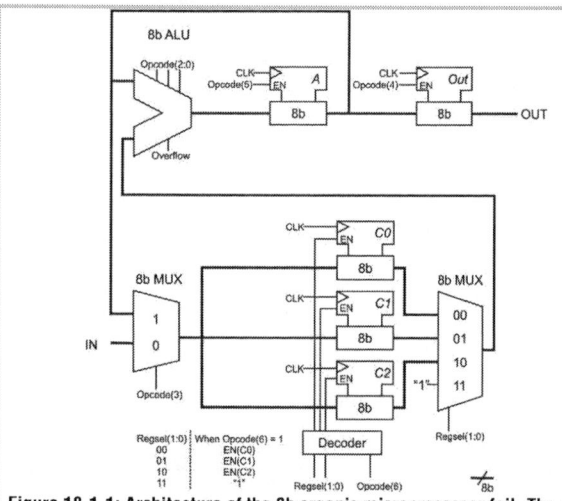

Figure 18.1.1: Architecture of the 8b organic microprocessor foil. The connector comprises 30 pins: 18 input pins [opcode(8,0); in(7,0); clk], 9 output pins [out(7,0), overflow] and power, ground and backgate voltage.

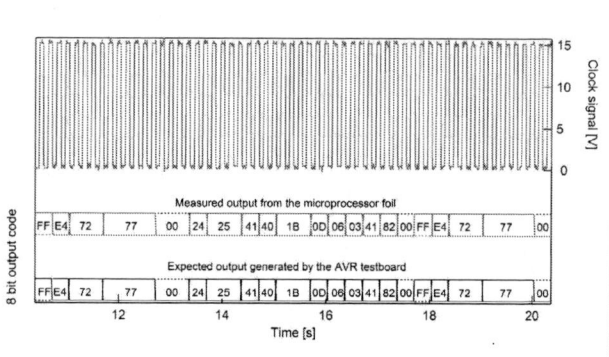

Figure 18.1.2: Hardware testbench measuring all individual instructions of the microprocessor foil at a clock frequency of 6Hz (V_{Back} = 50V and V_{DD} = 15V).

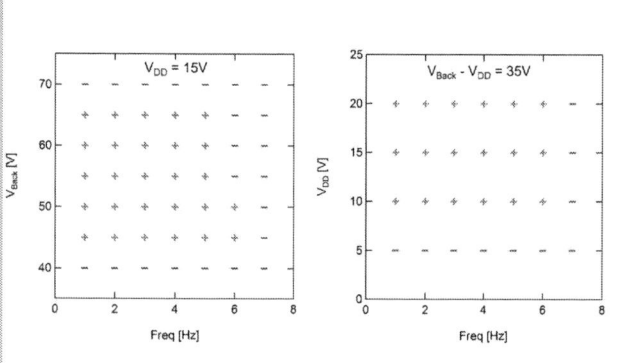

Figure 18.1.3: Shmoo plots of the microprocessor foil as a function of the clock frequency and backgate voltage (left) or power voltage (right).

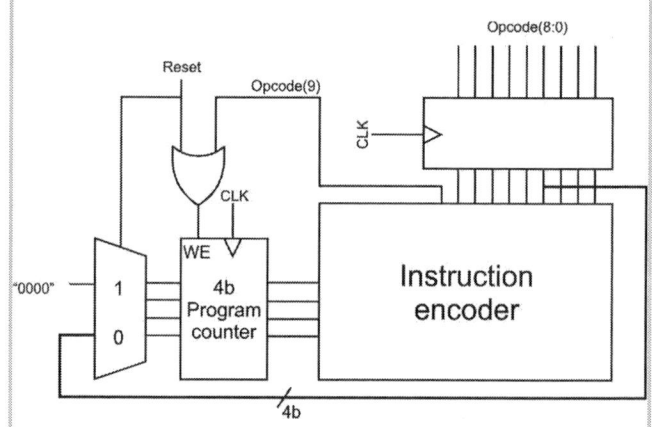

Figure 18.1.4: Architecture of instruction foil used in the implementation of the running averager. The connector uses 14 pins: 2 input pins [reset; clk], 9 output pins [opcode(8,0)] and power, ground and backgate voltage.

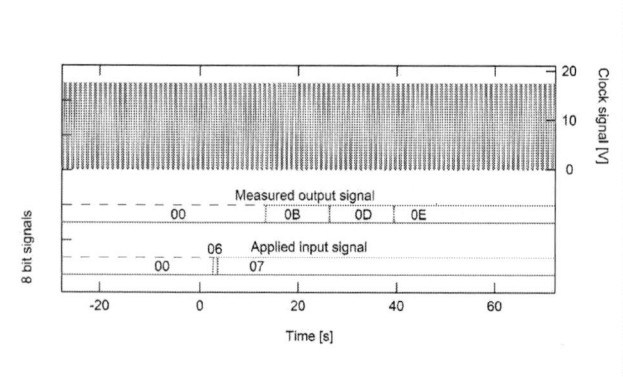

Figure 18.1.5: Measured output of the microprocessor foil connected to the running averager instruction foil. The switches at the input pins are manually changed from 0 to 7 (00000111). The output reaches 7.0 (00001110), with one additional bit of precision, after 3 program cycles.

Opcode(9:0)	Instruction
0RR0100000	AND A(0),C_{RR}
0RR0100001	OR A, C_{RR}
0XX0100010	NOT A
0RR0100011	LD A, C_{RR}
0RR0100100	ADD A, C_{RR}
0RR0100101	SUB A, C_{RR}
0XX0100110	LSR A
0XX0100111	LSL A
0XX000XXXX	NOOP
0RR1001XXX	LD C_{RR}, A
0RR1000XXX	LD C_{RR}, IN
0XX001XXXX	LD OUT,A
0110100100	INC A
0110100101	DEC A
10000AAAAA	JUMP to AAAAA

LD X,Y equals "Load X from Y"

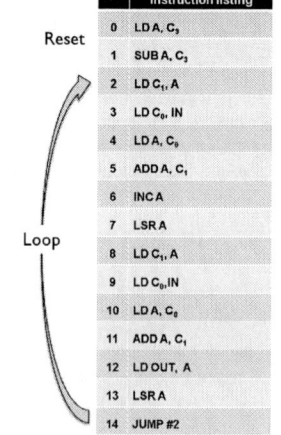

PC	Running averager instruction listing
0	LD A, C_3
1	SUB A, C_2
2	LD C_1, A
3	LD C_0, IN
4	LD A, C_0
5	ADD A, C_1
6	INC A
7	LSR A
8	LD C_1, A
9	LD C_0,IN
10	LD A, C_0
11	ADD A, C_1
12	LD OUT, A
13	LSR A
14	JUMP #2

Figure 18.1.6: (a) instruction table with corresponding opcodes. RR in the opcode represents the selected register RegSel (1,0). X is a don't care. (b) Program code for the running averager.

18

978-1-61284-303-2/11 $26.00 © 2011 IEEE

ISSCC 2011 PAPER CONTINUATIONS

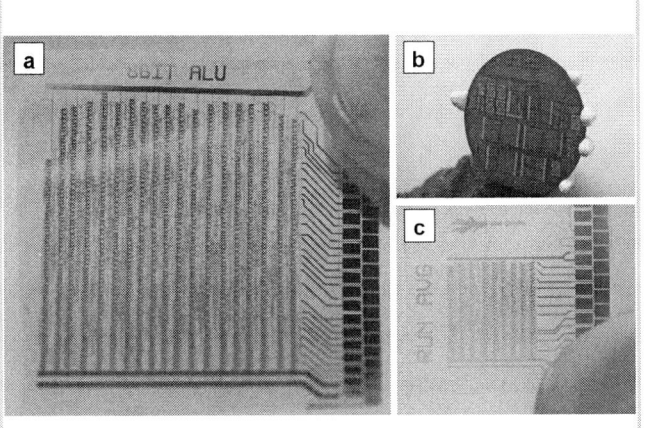

Figure 18.1.7: (a) die picture of the 8bit microprocessor foil; (b) foil comprising several microprocessor circuits laminated on a 6-inch wafer carrier during processing; (c) instruction generator foil for the running averager.

ISSCC 2011 / SESSION 18 / ORGANIC INNOVATIONS / 18.2

18.2 A 3.3V 6b 100kS/s Current-Steering D/A Converter Using Organic Thin-Film Transistors on Glass

Tarek Zaki[1,2], Frederik Ante[3], Ute Zschieschang[3], Joerg Butschke[1], Florian Letzkus[1], Harald Richter[1], Hagen Klauk[3], Joachim N. Burghartz[1,2]

[1]Institute for Microelectronics Stuttgart (IMS CHIPS), Stuttgart, Germany,
[2]University of Stuttgart, Stuttgart, Germany,
[3]Max Planck Institute for Solid State Research, Stuttgart, Germany

Organic thin-film transistors (OTFT) processed at low-temperatures offer prospects for a vast number of integrated circuit applications in mechanically flexible, inexpensive, large-area and biomedical electronics [1]. In addition, the low-voltage operation capability of recent OTFTs makes them well-suited for battery-powered or radio frequency-coupled portable devices [2]. In such applications, data conversion to interface the digital processors with the analog world is an essential necessity. Here, we demonstrate a compact 6b current-steering D/A converter (DAC) circuit, built in OTFT technology, which is 1000× faster and 30× smaller than the previously published data for a 6b DAC [3]. These considerable improvements result from an OTFT fabrication process based on silicon stencil masks that provide submicron channel length capability and excellent transistor matching [4], [5].

DACs can be classified into three main classes, namely resistive-based, capacitive-based and current-steering architectures. By comparing the three classes, the current-steering architecture is known to be superior in speed and compactness. However, excellent device matching is required for achieving a linear transfer characteristic of the DAC. Since OTFT fabrication processes so far have been prone to large device-to-device variations, the first approach to OTFT DACs used a switched-capacitor architecture (C-2C DAC), which is less sensitive to the poor transistor matching [3]. However, it has recently been reported in [4][5] that n-type and p-type top-contact OTFTs fabricated by using silicon stencil masks allow for submicron channel lengths and provide much improved transistor matching compared to devices fabricated with conventional shadow masks. This stencil mask-based fabrication process allows for designing current-steering DACs at considerably higher sample rate and smaller chip area consumption.

The schematic diagram of the 6b current-steering DAC is shown in Fig. 18.2.1. The converter consists of three main blocks: a current-source array (CSA) that contains binary-weighted current sources and an input current mirror, a switch array (SWA) that contains binary-weighted switches, and a linear current-to-voltage converter (CVC) at the output. The binary current-steering architecture is employed because of its lower complexity, smaller area and lower power dissipation compared to the unary current-steering architecture. This type of current-steering can be designed by exclusively using either n-type or p-type transistors, thus not requiring complementary OTFT technology. According to [3], the operational speed of the p-type OTFTs is approximately 10× higher than that of their n-type counterparts. This is mainly due to the higher carrier mobility in the p-type OTFT channel regions ($\mu_p \approx 10\mu_n$). For this reason, the 6b DAC is designed in p-type OTFT technology.

Referring to Fig. 18.2.1, strict requirements apply to the floorplan and the layout due to the large number of current sources that lead to complex routing. There are mainly two approaches for the floorplan, i.e. the SWA is integrated with the CSA or the SWA and the CSA are separated. The latter option offers a substantial advantage in the conversion performance. In this approach, the clear separation between the digital and the analog parts of the DAC minimizes the signal interference and eases power routing. Moreover, a matrix-like orientation of the transistors in the SWA and the CSA is utilized to arrange all identical transistors in closer proximity. This allows for exploiting the excellent feature size control by the stencil masks in spite of potential non-uniformity of material deposition through evaporation and so the circuit is more robust to gradient effects.

Transistor matching in the CSA is of high importance for obtaining a good linearity. Therefore, a sufficiently large total active area of 1500µm² is provided for the 63 current-source and the input current mirror transistors. The transistors in the SWA are designed to be always operating in the saturation mode during the full output voltage swing. The advantage of this approach is to avoid the use of

cascode current sources which requires larger chip area and limits the output voltage swing while maintaining the shielding effect of the cascode structure by isolating the drain nodes of the current-source transistors from the variant output node. Another advantage of this approach is that the nonlinearities of the output current-voltage characteristics of the OTFTs at low drain bias due to the non-ohmic contacts can be avoided [6]. In close agreement with simulation, the aspect ratio of the switch transistors are set to 40µm/4µm and the input ON voltage is set to be around 1V for proper circuit operation. Simulations were obtained using the OTFT compact DC model (Level 1) proposed in [6]. A simple linear CVC is realized at the output of the DAC. This CVC consists of only two p-type transistors (100µm/5µm) which minimizes the occupied chip area and ensures higher operational speed than with n-type transistors. One transistor is biased in its linear regime with a gate voltage of -2V, while the other transistor is diode-connected to serve as a variable current source to compensate for the nonlinearities of the output current-voltage characteristics of the OTFTs.

Using high-resolution silicon stencil shadow masks (Fig. 18.2.2), the converter has been fabricated on a glass substrate without the need of solvents and at a maximum process temperature of 90°C [4], [5]. The die photograph of the 6b DAC is shown in Fig. 18.2.3. Note that the fabrication process is also feasible for mechanically flexible plastic substrates. The converter employs 129 p-type OTFTs with a minimum feature length of 4µm. The total area of the DAC including the contact pads is only 2.6×4.6mm². The extracted hole mobility is 0.6cm²V⁻¹s⁻¹.

In order to achieve optimum linearity, the DAC is calibrated by adjusting the switch input ON voltage for the MSB to be 0V instead of 1V. This calibration method has a lower complexity compared to [3]. Figure 18.2.4(a) shows the measured DAC DC transfer function after calibration, where the maximum output swing is 1.94V, the power dissipation is 260µW, and the maximum DNL and INL are -0.29LSB and -1.21LSB, respectively. Before calibration the maximum DNL and INL were degraded to -0.97LSB and -1.64LSB, respectively. At the maximum output swing, the upper speed limit is about 20kS/s (see Fig. 18.2.4(b)). However, at a swing of 1V, the maximum update rate is found to be as high as 100kS/s. Figs. 18.2.4(c) and 18.2.4(d) show the measured DNL and INL at an update rate of 1kS/s and an output swing of 1.3V; the maximum DNL and INL are -0.69LSB and 1.16LSB, respectively. For the dynamic performance, Fig. 18.2.5 shows the spectrum of two output sinusoids at 31Hz (update rate of 1kS/s) and 3.1kHz (update rate of 100kS/s); the spurious-free dynamic range (SFDR) is 32dB for both outputs.

The table shown in Fig. 18.2.6 benchmarks the current-steering DAC in this work against the recently published switched-capacitor DAC [3]. The comparison clearly depicts that the 6b current-steering DAC fabricated by using silicon stencil masks consumes 30× smaller chip area and achieves a maximum update rate that is 1000× higher than that of the previously published 6b C-2C DAC at a power of only ~260µW.

References:
[1] K. Bock, "Polymer electronics systems – Polytronics," Proc. IEEE, vol. 93, pp. 1400-1406, Aug. 2005.
[2] H. Klauk, et al., "Ultralow-power organic complementary circuits," Nature, vol. 445, pp. 745-748, Feb. 2007.
[3] W. Xiong, et al., "A 3-V, 6-bit C-2C digital-to-analog converter using complementary organic thin-film transistors on glass," IEEE J. Solid-State Circuits, vol. 45, pp. 1380-1388, July 2010.
[4] F. Ante, et al., "Top-contact organic transistors and complementary circuits fabricated using high-resolution stencil masks," Device Research Conference, pp. 175-176, June 2010.
[5] F. Ante, et al., "Submicron low-voltage organic transistors and circuits enabled by high-resolution silicon stencil masks," IEDM Tech. Dig., Paper no. 21.6, Dec. 2010.
[6] O. Marinov, et al., "Organic thin-film transistors: Part I-compact DC modeling," IEEE Trans. Electron Devices, vol. 56, pp. 2952-2961, Dec. 2009.

978-1-61284-303-2/11 $26.00 © 2011 IEEE

ISSCC 2011 / February 23, 2011 / 9:00 AM

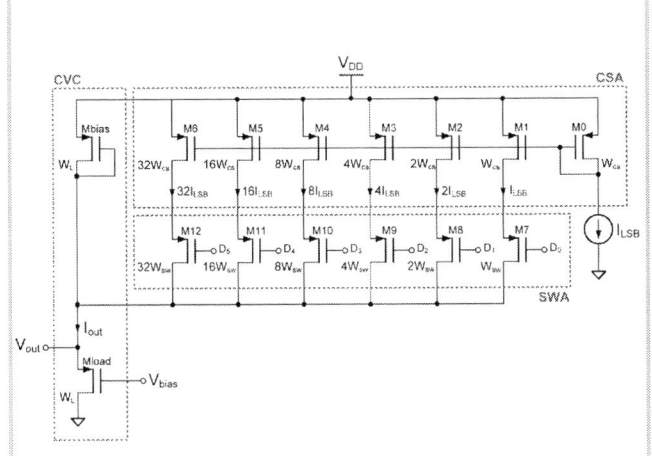

Figure 18.2.1: Schematic diagram of the 6b current-steering DAC.

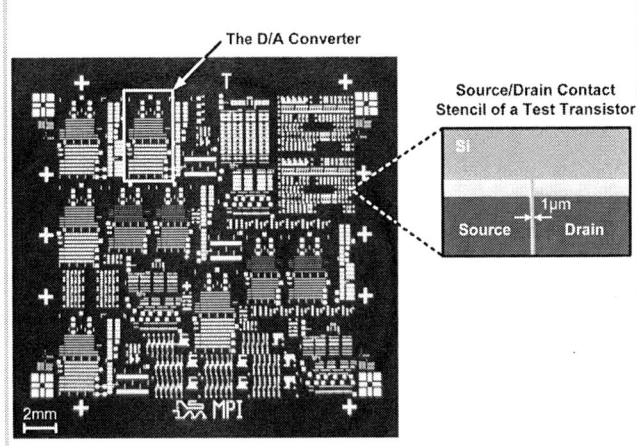

Figure 18.2.2: Photograph of the high-resolution silicon stencil mask for the top gold metal.

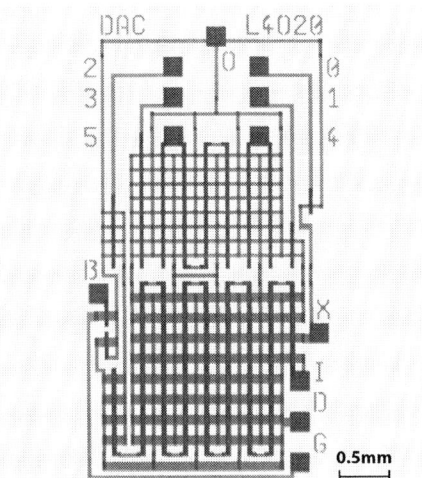

Figure 18.2.3: Photograph of the 6b current-steering DAC on a glass substrate.

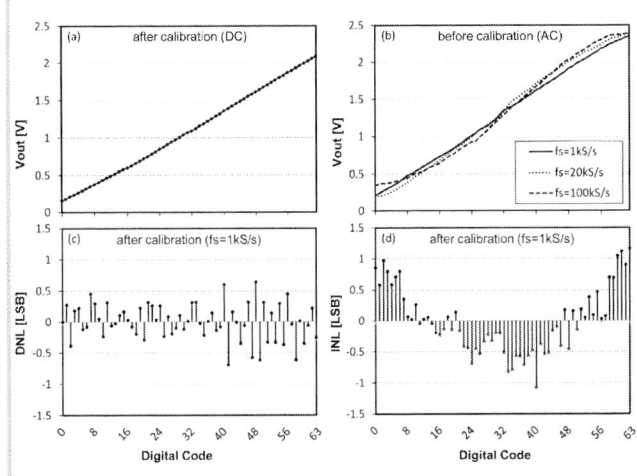

Figure 18.2.4: Measured DAC performance: (a) DC transfer function, (b) AC transfer function at various update rates, (c) DNL at 1kS/s, (d) INL at 1kS/s.

Figure 18.2.5: Measured DAC output spectrum for 31Hz and 3.1kHz sinusoids.

Design Parameter	[3]	This Work
Converter Architecture	Switched-Capacitor	Current-Steering
Technology	Complementary OTFTs	p-type OTFTs
Resolution	6-bit	6-bit
Min. Channel Length	20μm	4μm
Transistor/Capacitor Count	26 / 17	129 / 0
Chip Area	28mm × 14mm	2.6mm × 4.6mm
Supply Voltage	3V	3.3V
Max. Output Swing (V_{swing})	~1V	~2V
Power Dissipation	0.7μW (at V_{swing}=1V)	180μW (at V_{swing}=1V) 260μW (at V_{swing}=2V)
DNL_{max}	-0.6LSB (at 100S/s)	-0.69LSB (at 1kS/s)
INL_{max}	-0.8LSB (at 100S/s)	1.16LSB (at 1kS/s)
Min.-Max. Update Rate	10S/s-100S/s	DC-100kS/s
SFDR	24dB (at f_{sine}=10Hz) 29dB (at f_{sine}=45Hz)	32dB (at f_{sine}=31Hz and f_{sine}=3.1kHz)

Figure 18.2.6: Benchmarking of the previously published 6b C-2C DAC against the 6-bit current-steering DAC in this work.

ISSCC 2011 / SESSION 18 / ORGANIC INNOVATIONS / 18.3

18.3 A 1V Printed Organic DRAM Cell Based on Ion-Gel Gated Transistors with a Sub-10nW-per-Cell Refresh Power

Wei Zhang[1], Mingjing Ha[1], Daniele Braga[1], Michael J. Renn[2],
C. Daniel Frisbie[1], Chris H. Kim[1]

[1]University of Minnesota, Minneapolis, MN,
[2]Optomec, St. Paul, MN

Organic thin-film-transistors (OTFTs) are drawing much attention as they have attributes such as structural flexibility, low-temperature processing, large area coverage, and low cost, which make them attractive for large-area electronics. Various forms of OTFTs can enable applications that were not achievable using traditional inorganic transistors and/or surpass them in terms of performance and cost. OTFTs cannot match the performance of silicon-based transistors, but can complement them by enabling electronic flexible systems, which don't have to operate at high-speed. Recently, inkjet printing has become a popular method for low-cost manufacturing of OTFTs making product level implementations feasible. Despite these encouraging developments, the relatively high voltage needed to power up traditional OTFT devices and the lack of a good n-type device presents major circuit design challenges for OTFT-based systems.

Memory is a key component in many OTFT applications but only few attempts have been made on designing memory circuits suitable for OTFTs [1]. SRAM, which is the mainstream embedded memory in silicon chips, suffers from large power consumption in today's OTFT technologies, which only allow p-type devices to be fabricated. Although research on n-type OTFTs is currently underway [2], introduction of a viable n-type material has been delayed as existing materials have unstable characteristics, no good contact, and significantly lower mobility than their p-type counterpart. Since only p-type organic transistors are currently available for most designs, resistors or diode-connected transistors are required to achieve SRAM functionality, introducing large static power consumption. Moreover, unless the read operation can be sacrificed, the load resistor must be strong enough to avoid a destructive read, which prevents any further power-saving opportunities in SRAM. In this paper, we report a DRAM array in a printable, flexible and low-voltage ion-gel gated OTFT technology [3], which has no DC static power and achieves considerable performance improvement at operating voltages below 1V, overcoming the fundamental limitations of OTFT-based SRAMs.

DRAM cells can be implemented using three transistors, or even two transistors, in a standard logic process without a dedicated storage capacitor process [4]. Despite using a single type of device for the cell, this circuit (commonly referred to as "gain-cells") has no DC static current, making it an ideal memory candidate for OTFT technologies where only p-type devices are available. The main challenge for gain-cells is the short retention time, typically less than 100μs in silicon due to the small gate capacitance used for data storage [4]. Ion-gel gated OTFTs used in this work serve as an ideal device platform for DRAMs as their gate capacitances can exceed 100μF/cm^2, roughly 70× higher compared to that in 65nm CMOS [3]. This unusually high capacitance is derived from the ultra-thin electrical double layers formed upon electrical polarization as shown in Fig. 18.3.1 and can be used to enhance DRAM retention time and simultaneously achieve a much higher drive current at low supply voltages compared to other OTFT devices.

An 8×8 DRAM array was fabricated using an aerosol jet printer to demonstrate the feasibility of gain-cell DRAMs in a p-type-only OTFT technology. This printing method can accommodate functional inks with a variety of viscosities and has a patterning precision of 10μm. Our underlying design philosophy was to implement a general-purpose memory array with full read and write capability that can be utilized for a range of applications including electrochromic displays and/or sensor sheet arrays as shown in Fig. 18.3.2 (left). The basic read and write operations of the 3T OTFT memory cell are illustrated in Fig. 18.3.2 (right). Each transistor has a channel width of 500μm and a channel length of 25μm.

The leakage components during hold mode are illustrated in Fig. 18.3.2: namely the gate currents of the storage and write devices and the subthreshold leakage of the write device. A boosted WWL is preferred to suppress the subthreshold leakage as is the case in all practical DRAM designs. This boosted voltage can be efficiently generated using a single-stage charge pump consisting of a p-type gate capacitor and a diode-connected transistor. Note that the retention time of data '0' is more critical than data '1' for p-type cells as the storage node is surrounded by high voltages resulting in all leakage currents in the pull up direction. Moreover, the read current for driving the RBL is determined by the data '0' voltage, as it determines the gate overdrive of the p-type read device. Measurements were automated with LabVIEW™ software to collect reliable retention-time statistics. When read is enabled, a cell storing data '0' will have a large pull-up current and drive the RBL high. A pull-down resistor attached to the RBL resets the RBL signal after the read is completed. Measured waveforms in Fig. 18.3.3 verify full read and write functionality at 1V. WBL is kept high after the write access to test for the worst-case data '0' retention condition. The steady-state RBL voltage level is used to calculate the voltage stored in the cell at the time of access. We define retention time as the time it takes for the data '0' voltage inside the cell to increase to $0.5 \times V_{DD}$, and the resistor load is therefore selected so that a $0.5 \times V_{DD}$ voltage in the storage node translates into a RBL level of (V_{DD}-0.1V).

The retention-time statistics of the DRAM array at 1V are shown in Fig. 18.3.4. With a boosted WWL of 1.25V, the worst-case cell has a retention time of 30s while 72% of the cells have retention times longer than 1 minute. With a slightly higher boosted WWL of 1.30V, every cell in the array can have a retention time exceeding 1 minute. This translates into a retention time that is 5 orders of magnitude longer than recent gain-cell designs in 65nm CMOS [4], which is sufficiently long even for the slow OTFT circuits to perform a periodic refresh operation. Figure 18.3.5 (left) shows the impact of WWL pulse width on retention time indicating that a 20ms write pulse is needed to achieve a retention time greater than 1 minute. This can be further improved if a negative WWL voltage is available. The read delay, defined as the 50% rise time of the RBL signal when reading from a data '0' cell, was 12ms or less. The memory cells show robust read and write operation down to 0.8V.

Our printed DRAM array also demonstrates significant reduction in power consumption during active and standby modes. Due to the extremely long retention time, the refresh power of the proposed memory cell is in the nano-watt order. Figure 18.3.6 compares the measured power consumption of an SRAM and a gain-cell DRAM implemented using the same kind of ion-gel OTFT devices. Individual SRAM cell structures were fabricated and tested for this comparison. Results show that the overall power consumption of an OTFT DRAM (including active, static, and refresh power) is 12× lower than the static power consumption of an OTFT SRAM for a 50Hz operating frequency. In standby mode where all RBLs are left floating except for when the cells are refreshed, the DRAM power consumption per cell is reduced to below 10nW. Figure 18.3.7 shows the chip microphotograph and performance summary of the organic DRAM array chip.

Acknowledgements:
This work was funded by the National Science Foundation.

References:
[1] M. Takamiya, T. Sekitani, Y. Kato, et al., "An Organic FET SRAM With Back Gate to Increase Static Noise Margin and Its Application to Braille Sheet Display", *IEEE J. Solid-State Circuits*, vol. 42, no. 1, pp. 93-100, Jan. 2007.
[2] R. Blache, J. Krumm, and W. Fix, "Organic CMOS Circuits for RFID Applications", *ISSCC Dig. Tech. Papers*, pp. 208-109, Feb. 2009.
[3] Y. Xia, W. Zhang, M. Ha, et al., "Printed Sub-2 V Gel-Electrolyte-Gated Polymer Transistors and Circuits", *Advanced Functional Materials*, vol. 20, no. 4, pp. 587–594, Feb. 2010.
[4] K. Chun, P. Jain, T. Kim, et al., "A 1.1V, 667MHz Random Cycle, Asymmetric 2T Gain Cell Embedded DRAM with a 99.9 Percentile Retention Time of 110μsec", *Symposium on VLSI Circuits*, pp. 191-192, June 2010.

978-1-61284-303-2/11 $26.00 © 2011 IEEE

ISSCC 2011 / February 23, 2011 / 9:30 AM

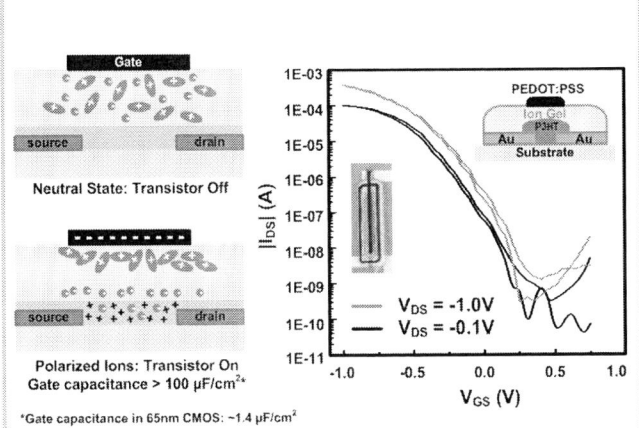

Figure 18.3.1: Basic operation of an ion-gel gated OTFT. I_{DS}-V_{GS} curve shows a 0.4mA drive current at a 1 V voltage (W=500μm, L=25μm).

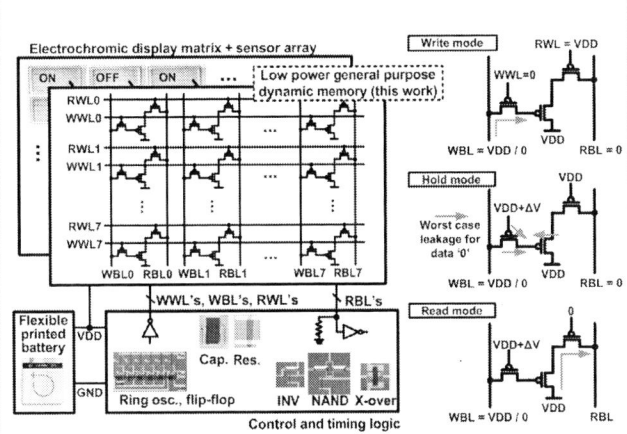

Figure 18.3.2: Conceptual diagram of a flexible display and sensor array system and detailed illustration of cell schematic and operations.

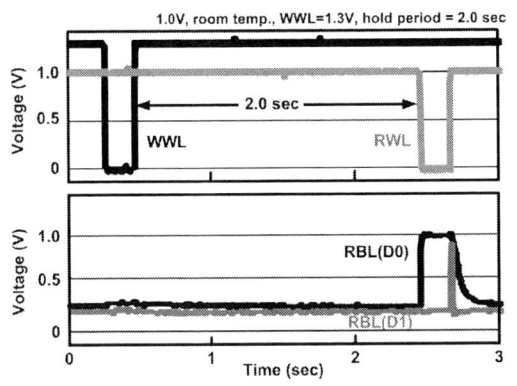

Figure 18.3.3: Measured waveforms of the 8×8 DRAM test chip for a 2.0s hold period.

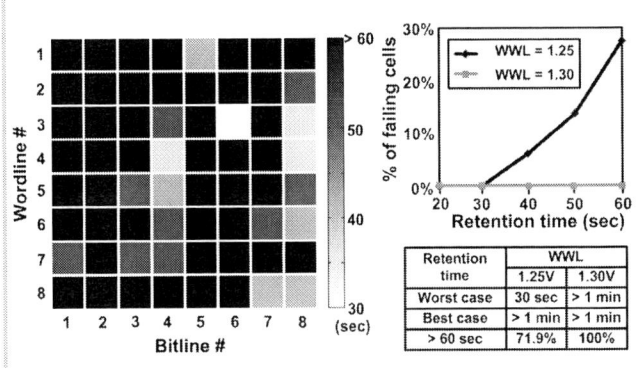

Figure 18.3.4: Measured retention time map for WWL=1.25V (left). Percentage of failures vs. retention time at different WWL voltages (right).

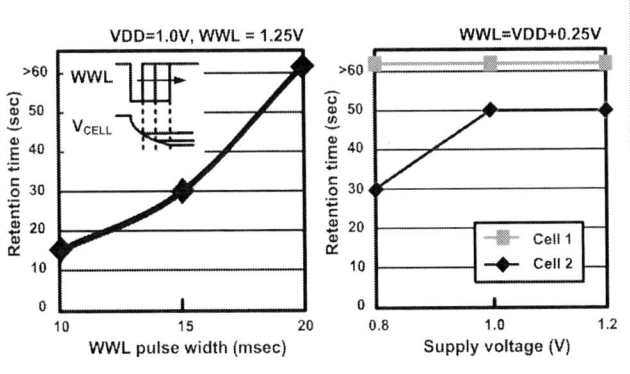

Figure 18.3.5: Retention time vs. WWL pulse width and retention time vs. supply voltage.

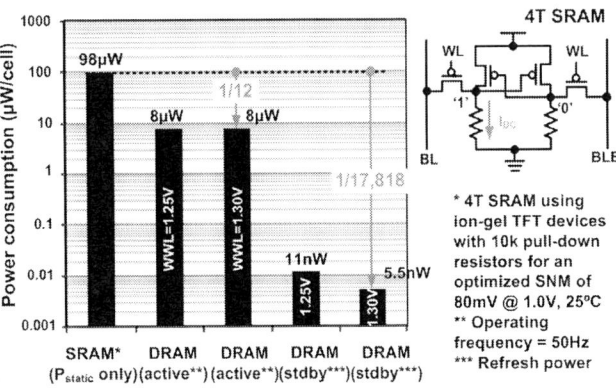

Figure 18.3.6: Measured power consumption of an OTFT SRAM (static power only) and DRAM for different WWL voltages and operation modes.

18

978-1-61284-303-2/11 $26.00 © 2011 IEEE   327

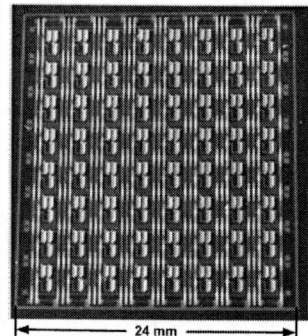

Process	Ion-gel organic TFT
Channel material	Poly(3-Hexylthiophene)
Array dimension	24 x 25 mm^2
Array size	64 bits (8 WLs, 8 BLs)
Read delay	< 12 msec
Write delay	< 20 msec
Supply	0.8V - 1.2V
*Retention time	30sec @ 1.25V WWL > 60sec @ 1.30V WWL
**Active power	8 µW/cell @ 1.25V WWL 8 µW/cell @ 1.30V WWL
**Standby power	11 nW/cell @ 1.25V WWL <5.5 nW/cell @ 1.30V WWL

* Measured @ 1.0V, 25 °C
** Measured @ 1.0V, 25 °C, refresh period of
30 sec (WWL=1.25V) and 60 sec (WWL=1.30V)

Figure 18.3.7: Chip photograph of the 8×8 printed organic DRAM and performance summary.

ISSCC 2011 / SESSION 18 / ORGANIC INNOVATIONS / 18.4

18.4 Fully Printed Organic CMOS Technology on Plastic Substrates for Digital and Analog Applications

Anis Daami[1], Cécile Bory[1], Mohamed Benwadih[1], Stéphanie Jacob[1], Romain Gwoziecki[1], Isabelle Chartier[1], Romain Coppard[1], Christophe Serbutoviez[1], Lidia Maddiona[2], Enzo Fontana[2], Antonino Scuderi[2]

[1]CEA-LITEN, Grenoble, France,
[2]STMicroelectronics, Catania, Italy

Drastic efforts have been realized these last years in order to develop complementary organic technology. This is the essential key to produce elementary low-cost circuits for digital and analog applications. Different techniques [1-3] are available nowadays to obtain both N- and/or P-type organic devices. Screen printing is one of the most highly awaited low-cost techniques that can be used to produce organic devices and circuits. It has been widely used in P-type organic technologies [4, 5]. Now that N-type semiconductors have become much more easily processed, developers are seeking a complete CMOS and lifetime robust technology. Many previous works have reported on a complete solution based on CMOS technology [6-8]. Large-area-compatible organic processes have also been demonstrated [9]. Nevertheless some of the technological steps in these latter reports are not fully printed and/or still present some lithography/vacuum deposition steps. We present here a complete fully printed CMOS technology on flexible substrates showing acceptable device performances and digital/analog circuit functionalities, which can lead to more complex designs.

Based on our design toolkit we have processed a testchip including single devices, inverters, ring oscillators and simple analog circuits such as current mirrors, differential pairs and cascodes (see Fig. 18.4.7) in order to show the feasibility of our organic CMOS technology.

Our organic CMOS top-gate design fabrication is carried out on a 10×10 cm^2 gold-plated 125µm-thick Polyethylene-naphtalate (PEN) substrate. The first step consists of a patterning by laser ablation of source/drain electrodes, which also serves as first level of interconnections, where we attain a 5µm line/space resolution. P-type and N-type semiconductors are then screen printed on the foil and then annealed at a temperature of 100°C in normal atmospheric conditions. The gate dielectric polymer is screen printed above both semiconductors leaving open vias for level interconnections and then annealed. Finally the gate and the second interconnection level are printed with same technique using a conductive silver ink. A final annealing step at 100°C is performed.

We have carried out all our measurements in ambient temperature and pressure. In Fig 18.4.1 we show the transfer and output characteristics of both types of devices for the geometry W/L = 2000/20µm. We observe that both types of devices show equivalent levels of ON and OFF state currents. Table 1 (see Fig. 18.4.2) summarizes the important electrical parameters for 2 different transistor geometries. Threshold voltage and saturation mobility for both types of devices have been monitored on the whole set of available transistors to evaluate their respective dispersions. Figure 18.4.3 shows the distributions of N- and P-type threshold voltage and mobility.

For digital applications, elementary inverters and ring oscillators have been tested. In Fig. 18.4.4, we present a fully functional inverter ($W_n = W_p = 1000$µm and L=20µm) and its corresponding 7-stage ring-oscillator time evolution. The inverter presents respectively a noise margin of 14V and 5.5V at a supply voltage V_{dd} of 40V and 20V, with gains around 13 and 6 at $V_{dd}/2$. In Table 2 (see Fig. 18.4.2), we summarize the important parameters of this inverter.

The 7-stage oscillator shows well-defined oscillation levels with oscillation frequencies varying from 70Hz @ 40V to 16Hz @ 20V, corresponding to delay/gate values of 1ms and 4.5ms, respectively.

Regarding analog applications, basic functions such as current mirrors and differential pairs are characterized. At a preliminary stage, we have carried out our measurements on equally sized devices of W/L=2000/20µm for both NMOS and PMOS transistors. Figure 18.4.5 presents the electrical results concerning a P-type current mirror tested at different supply-voltage values. Good matching properties are observed between measured and forced currents. In Fig. 18.4.6, we show a functional N-type differential pair loaded with a P-type current mirror and its corresponding characterization, showing acceptable maximum gains of 17 for V_+ = 10V and 9.6 for V_+ = 20V.

To summarize, we have developed a complete printed organic CMOS technology on a plastic foil using large-area-compatible printing processes, that shows good and stable electrical characteristics. In comparison to other large-area studies, we emphasize the fact that our technology uses a mass-printing-compatible process, which can be expandable to extra-large areas with potentially low-cost techniques. We also show elementary logic and analog circuits that are functional and present rather good performances. These results will allow the design and printing of higher-complexity organic CMOS circuits.

Acknowledgment:
Part of this work has been realized in the frame of the European funded project COSMIC.

References:
[1] I. Kymissis, A.I. Akinwande, V. Bulovic, "A Lithographic Process for Integrated Organic Field-Effect Transistors," *IEEE J. Disp. Tech.*, vol. 1, pp. 9-13, Jan. 2005.
[2] H. Yan, et al., "A high-mobility electron-transporting polymer for printed transistors", *Nature*, 457, 679-686, Feb. 5, 2009.
[3] R. Blache, et al., "Organic CMOS Circuits for RFID Applications", *ISSCC Dig. Tech. Papers*, pp. 208-210, Feb. 2009.
[4] M. Böhm, et al., "Printable Electronics for Polymer RFID Applications", *ISSCC Dig. Tech. Papers*, pp. 1034-1041, Feb. 2006.
[5] K. Myny, et al., "An Inductively-Coupled 64b Organic RFID Tag Operating at 13.56MHz with a Data Rate of 787b/s", *ISSCC Dig. Tech. Papers*, pp. 290-292, Feb. 2008.
[6] Wei Xiong, et al., "A 3V 6b successive-approximation ADC using complementary organic thin-film transistors on glass", *ISSCC Dig. Tech. Papers*, pp. 134-135, Feb. 2010.
[7] De Vusser, et al., " A 2V Organic Complementary Inverter", *ISSCC Dig. Tech. Papers*, pp. 1082-1091, Feb. 2006.
[8] Ishida, et al., "A Stretchable EMI Measurement Sheet with 8×8 Coil Array, 2V Organic CMOS Decoder, and -70dBm EMI Detection Circuits in 0.18µm CMOS", *ISSCC Dig. Tech. Papers*, pp. 472-474, Feb. 2010.
[9] Da He, et al., "An integrated Organic Circuit Array for Flexible Large-Area Temperature Sensing", *ISSCC Dig. Tech. Papers*, pp. 142-144, Feb. 2010.

978-1-61284-303-2/11 $26.00 © 2011 IEEE

ISSCC 2011 / February 23, 2011 / 9:45 AM

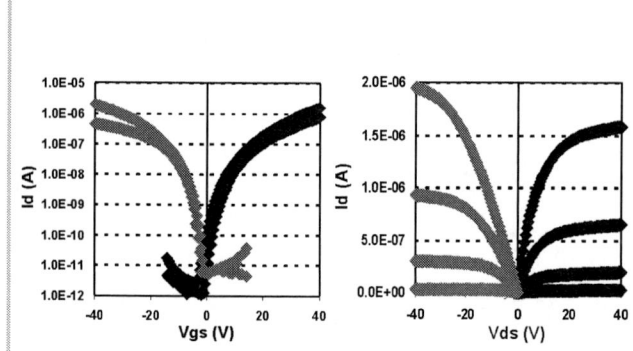

Type	W/L (μm/μm)	|VTlin| (V)	|VTsat| (V)	μsat (cm²V⁻¹s⁻¹)	Ioff (pA)	Ion (μA)
N	2000/100	14.9	13.6	0.065	2.7	0.66
N	2000/20	14.1	10.0	0.023	2.8	1.6
P	2000/100	12.4	10.3	0.039	5.1	0.51
P	2000/20	10.3	7.8	0.025	6.1	1.9

Table 1: NMOS and PMOS transistors electrical parameters

Vdd (V)	NMH (%)	NML (%)	Gain max	Voltage swing (V)
40	34.8	31.1	12.5	39.9
20	23.2	27.6	5.8	19.6

Table 2: Inverter Wn=Wp=1000 μm L=20μm important parameters

Figure 18.4.1: NMOS (black)/PMOS (red) transfer and output characteristics for W/L=2000/20μm at Vdslin = ± 5V and Vdsat = ± 40V.

Figure 18.4.2: Summary tables for NMOS, PMOS (W=2000μm, L=20/100μm) and inverter (W=1000μm, L=20μm) electrical parameters.

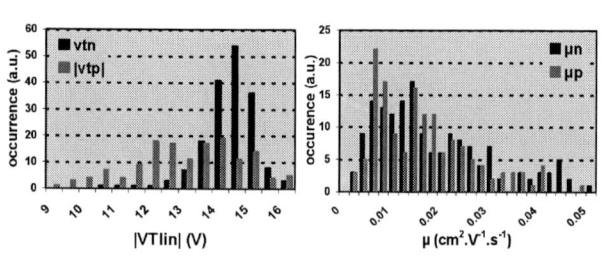

Figure 18.4.3: Threshold voltage and mobility distributions for N- and P-type transistors.

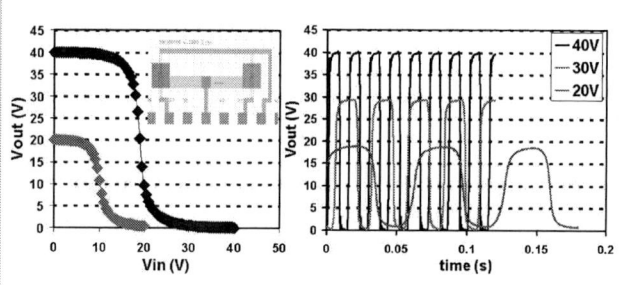

Figure 18.4.4: Inverter static characteristic and its corresponding ring oscillations at different power supplies. W_{N,P}=1000μm & L_{N,P}=20μm.

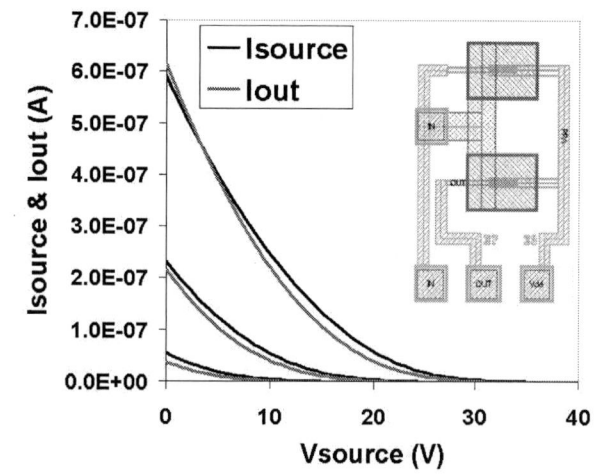

Figure 18.4.5: P-type current mirror characteristics. Transistors are equally sized with W=2000μm and L=20μm.

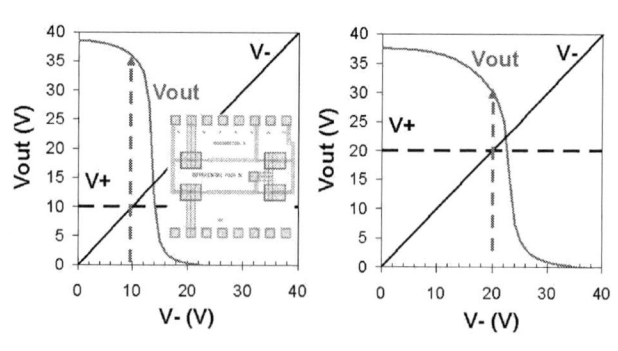

Figure 18.4.6: N-type differential pair with P-type current mirror as load. N and P-type transistors are all equally sized with W=2000μm & L=20μm.

ISSCC 2011 PAPER CONTINUATIONS

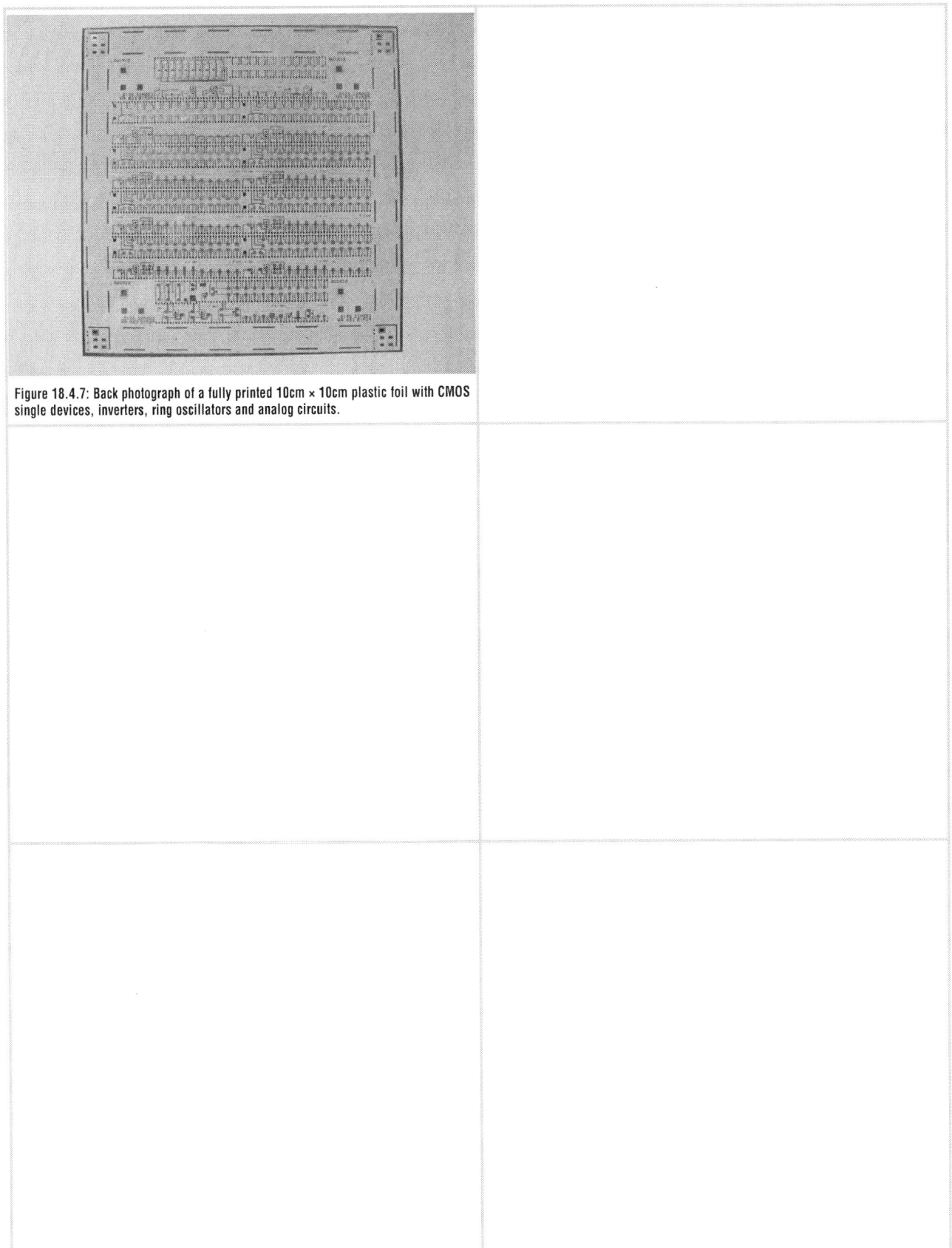

Figure 18.4.7: Back photograph of a fully printed 10cm × 10cm plastic foil with CMOS single devices, inverters, ring oscillators and analog circuits.

ISSCC 2011 / SESSION 19 / LOW-POWER DIGITAL TECHNIQUES / OVERVIEW

Session 19 Overview / *Energy-Efficient Digital*

Low-Power Digital Techniques

Session Chair: *Stephen Kosonocky, Advanced Micro Devices, Fort Collins, CO*

Session Co-Chair: *Ming-Yang Chao, Mediatek, Hsinchu, Taiwan*

Next-generation low-power applications such as biological signal processing, energy scavenging and wireless communications require new technologies and circuit techniques to achieve the lowest energy-per-operation. These applications are driving new techniques for efficient on-die clock generation, level conversion, non-volatile storage, low-power clocking, and reliable low-voltage operation. Energy-efficient systems must be optimized to maximize the utility of power shut down during idle and low-activity periods while also maintaining low-energy operation and high throughput during active modes to meet performance targets. The papers selected for this session highlight technology enhancements, novel circuit techniques, and system designs targeting these goals.

Biological signal-processing applications require very low energy and high performance to enable small form factors for non-intrusive human-body monitoring. Paper 19.1 [IMEC] describes a voltage-scalable 13pJ/cycle, 1MHz operation at 0.4V supply, event-driven wireless sensor node SoC for human-body signal processing, targeting applications such as electrocardiogram (ECG) and electroencephalogram (EEG) monitoring, allowing high-speed, energy-efficient data collection, while maintaining low-power idle modes using multiple scenario-based configurations. The SoC is constructed in 90nm LP CMOS and is voltage-scalable from 0.4 to 1.2V, running at 100MHz at the highest voltage.

Paper 19.2 [TI] introduces a low-power 16b microcontroller in an SoC designed for supporting applications that require energy harvesting. A dedicated 16KB FeRAM memory array is integrated on a 0.13µm CMOS die with fast fail detection ensuring uninterrupted refresh cycles, reducing active current consumption to 82µA/MHz for code executing from the NVM with less than 3µA standby current at 85°C.

Paper 19.3 [STMicorelectronics] compares a 65nm LP PD-SOI technology combined with an enhanced power gate device utilizing automatic adaptive body bias, to a standard LP CMOS bulk implementation, demonstrating an 802.11n LDPC codec. The authors show that a low resistivity power switch produced with forward body bias, combined with PD-SOI can reduce leakage current by 52.4% vs. bulk and increase the frequency by 20% at 1.2V, while decreasing power by 30% at 360MHz.

Paper 19.4 [Toshiba] describes a low clock capacitance single-phase D-flip/flop that saves 77% energy over a transmission-gate flop at zero-data activity, and 24% power reduction on a wireless LAN application. The new configuration uses a low transistor count and employs an adaptively weakening state-retention coupling scheme for reliable low-voltage operation under device variations in 40nm CMOS.

Paper 19.5 [U Freiburg] describes a standard-cell library utilizing Schmitt-trigger-enhanced logic cells to minimize leakage while demonstrating reliable circuit functionality at 62mV supply in standard 0.13µm CMOS. The digital circuits are tested using 8×8 multipliers demonstrating reduced minimum voltage up to a subthreshold-slope-limited point. This low single supply voltage is suitable for applications utilizing energy harvesting and low energy-per-operation.

Paper 19.6 [U Michigan] proposes circuit and architectural techniques to improve energy efficiency in aggressively voltage-scaled circuits. The approaches are validated in a high-throughput 1024-pt complex FFT core that utilizes super-pipelining in 65nm CMOS, to achieve 30MHz operation at 270mV supply, producing 234Ktransforms/s consuming only 17.7nJ per 1024-pt complex FFT operation.

978-1-61284-303-2/11 $26.00 © 2011 IEEE

ISSCC 2011 / February 23, 2011 / 10:15 AM

Presenters:

**19.1 A Voltage-Scalable Biomedical Signal Processor Running ECG
Using 13pJ/cycle at 1MHz and 0.4V**

M. Ashouei,
imec - Holst Centre, Eindhoven, The Netherlands

10:15 AM

**19.2 An 82µA/MHz Microcontroller with Embedded FeRAM for
Energy-Harvesting Applications**

M. Zwerg,
Texas Instruments, Freising, Germany

10:45 AM

**19.3 Comparison of 65nm LP Bulk and LP PD-SOI with Adaptive
Power Gate Body Bias for an LDPC Codec**

J. Le Coz,
STMicroelectronics, Crolles, France

11:00 AM

**19.4 A 77% Energy-Saving 22-Transistor Single-Phase-Clocking
D-Flip-Flop with Adaptive-Coupling Configuration in 40nm CMOS**

C. Teh,
Toshiba, Kawasaki, Japan

11:15 AM

**19.5 A 62mV 0.13µm CMOS Standard-Cell-Based Design Technique Using
Schmitt-Trigger Logic**

N. Lotze,
University of Freiburg - IMTEK, Freiburg, Germany and HSG-IMIT, Villingen-Schwenningen, Germany

11:30 AM

19

**19.6 A 0.27V 30MHz 17.7nJ/transform 1024-pt Complex
FFT Core with Super-Pipelining**

M. Seok,
University of Michigan, Ann Arbor, MI

11:45 AM

978-1-61284-303-2/11 $26.00 © 2011 IEEE

ISSCC 2011 / SESSION 19 / LOW-POWER DIGITAL TECHNIQUES / 19.1

19.1 A Voltage-Scalable Biomedical Signal Processor Running ECG Using 13pJ/cycle at 1MHz and 0.4V

Maryam Ashouei[1], Jos Hulzink[1], Mario Konijnenburg[1], Jun Zhou[1],
Filipa Duarte[1], Arjan Breeschoten[1], Jos Huisken[1], Jan Stuyt[1],
Harmke de Groot[1], Francisco Barat[2], Johan David[2],
Johan Van Ginderdeuren[2]

[1]imec - Holst Centre, Eindhoven, The Netherlands,
[2]NXP Semiconductors, Leuven, Belgium

Recent work on designing ultra-low-power systems has focused on the sub-threshold regime [1-3] and an energy efficiency of a few pJ/cycle was reported. While operating at the minimum energy point is attractive for energy-frugal devices like those used for wireless biomedical signal monitoring, the achieved clock frequency is usually in the kHz range. The low frequency combined with limited processing capacity, small on-chip memory, and low computation precision prevents the use of these systems for complex ambulatory monitoring beyond a simple ECG algorithm. Low-voltage systems with more computational power are demonstrated in [4] and [5].

In this paper, we present an event-driven system with resources to run applications with different degrees of complexity in an energy-aware way. The architecture uses effective system partitioning to enable duty cycling, SIMD instructions, power gating, voltage scaling, multi-clock domains, multi-voltage domains, and extensive clock gating. The system has sufficient computational power to run a complex ECG algorithm with feature extraction and motion artifact cancelation or multi-channel EEG processing. The system consumes an average of 13pJ/cycle running a CWT-based ECG application at 0.4V. The processor can run at voltage range of 0.4 to 1.2V and supports frequency range of 1 to 100MHz. The system has comparable energy/cycle, more computation capability, and larger available frequencies than the previously reported complex designs in [4] and [5].

The overall system architecture is shown in Fig. 19.1.1. The die photograph is shown in Fig. 19.1.7. The chip is designed in a standard 90nm LP process, has an area of 1.875×3.75 mm², and integrates 195K NAND2 equivalent gates and 2Mb memories. The system comprises a CoolFlux baseband processor, its memory sub-system with separated program and data memories, DMA controller (DMAC), interrupt controller, timers, JTAG, I²C, and 5 device interfaces Dev$_x$ (an 8-channel EEG sensor, ECG sensor, temperature sensor, 3D accelerometers for motion artifact cancelation, and SPI for radio transmission). The processor can self-boot from external flash. The program can also be loaded via the host processor or the JTAG interface.

Each device interface is associated to a DMA streaming shell (DSS) to collect streaming input/output data. The DSS communicates with the Dev$_x$ through point-to-point connections to transfer data to the memory. Such communication was chosen over a regular system bus to avoid excessive dynamic power dissipation on a long bus. The DMAC can access the memory subsystem even when the processor is shut off. This architecture allows for effective duty cycling where the processor and major parts of the memory can be shut off or put into retention mode to reduce power. Furthermore, the memory sub-system design allows both the DMAC and the processor to access the data memory simultaneously. The DMAC generates an interrupt to wake up the processor when a predefined number of samples is collected.

There are 2 potential clock sources: an external low-frequency clock and an on-chip generated clock. The on-chip clock generator (Fig. 19.1.2) can generate a 100MHz clock from a 32kHz source for all process corners. The on-chip clock generator is made completely with digital standard cells and consists of a "trombone oscillator" with a fixed delay line (for worst case) and a variable delay line (to adjust for process variability). A "walking-zero" shift register selects the portion of the variable delay line for the target frequency. The clock of each component can be derived from one of the two clock sources using a rate multiplier and can be programmed to be a fractional value between 0—f$_s$ where f$_s$ is the clock source frequency. To switch between two possible clock sources, a glitch-free clock switch is used (Fig.19.1.3). The switch uses a synchronizer-based clock

multiplexer to avoid output glitches when switching between two clocks. It first synchronizes the SELECT signal to the currently running clock to gate it and then the other clock is allowed to pass through. For most use cases, the device interfaces, the power manager (PWRM), and the clock manager (CLKM) are running at lower frequencies and the processor and memories are running at a higher frequency. When not used, a component is clock-gated and/or shut off. The clock manager controls the multi-clock domains, the frequency scaling, and the clock-domain crossings.

There are 15 power domains (Fig. 19.1.4). The data and program memory are divided into 13 different memory banks each in a separate power domain to be individually put in retention mode or turn on/off. The processor is also in a separate power domain. The rest of the system, including the periphery, is in one power domain (periphery domain), which is always on. The processor and periphery can have a voltage range of 0.4 to 1.2V, which is provided externally. Commercial memories used in the design cannot operate at voltages as low as 0.4V. Therefore, level shifters capable of up-converting from 0.36V to 1.32V for all process corners and temperature range of 0 to 125°C (Fig. 19.1.5) are placed between the logic domains and the memories. The level shifter, which has close to 50% duty cycle, uses its circuit architecture, transistor length sizing, and thick-oxide transistors to achieve the large operating voltage range with a performance of 1.5ns while up-converting from 0.36V to 1.32V at nominal condition [6]. All data memory banks share the same address, data, and control lines to communicate with the core logic to reduce the number of level shifters put between the memory and the logic. The same holds for the program memory banks.

There are 4 different power modes: (1) Data-collection mode: only the periphery domain and the required number of data memory banks are on running at 0.4V and 0.7V respectively, the program memory is in retention mode and the rest of the system is off. (2) Low-performance mode: the processor and periphery are on running at 0.4V, part of the program and data memory are on, running at 0.7V and the unused memory banks are off. (3) High-performance mode: All components are running at 1.2V and the unused memory banks are off. (4) Sleep mode: all power domains are off and no internal clock is running. The system can be woken up from the clock-less sleep mode by an external interrupt. The PWRM unit switches between different power modes when activated by the interrupt controller.

Fig. 19.1.6 shows the energy running a CWT-based ECG application. The energy/cycle and the clock frequency at 0.4V are 13pJ and 1MHz respectively. For this measurement one data memory bank and one program memory bank are on running at 0.7V. The energy/cycle when both logic and the memories are running at 1.2V is 95pJ executing the same application. It should be noted that the reported energy/cycle takes into account the time between samples where the peripheries collect the samples while the processor is in sleep mode.

Acknowledgement
The authors would like to thank the *invomec* division of imec for the back-end support.

References:
[1] J. Kwong, et al., "A 65nm Sub-Vt Microcontroller with Integrated SRAM and Switched-Capacitor DC-DC Converter", *ISSCC Dig. Tech. Papers*, Feb. 2008.
[2] M. Seok, et al., "Phoenix Processor: A 30pW Platform for Sensor Applications", *IEEE Symp. VLSI Circuits*, 2008.
[3] S. C. Jocke, et al., "A 2.6-µW Sub-threshold Mixed-signal ECG SoC", *IEEE Symp. VLSI Circuits*, 2009.
[4] S.R. Sridhara, et al., "Microwatt Embedded Processor Platform for Medical System-on-Chip Applications", *Symp. VLSI*, 2010.
[5] G. Chen, et al., "Millimeter-Scale Nearly Perpetual Sensor System with Stacked Battery and Solar Cells", *ISSCC Dig. Tech. Papers*, Feb. 2010.
[6] S. N. Wooters, et al., "An Energy-Efficient Sub threshold Level Converter in 130-nm CMOS", *IEEE Trans. on Circuits and Systems-II*, Vol. 57, No. 4, April 2010.

978-1-61284-303-2/11 $26.00 © 2011 IEEE

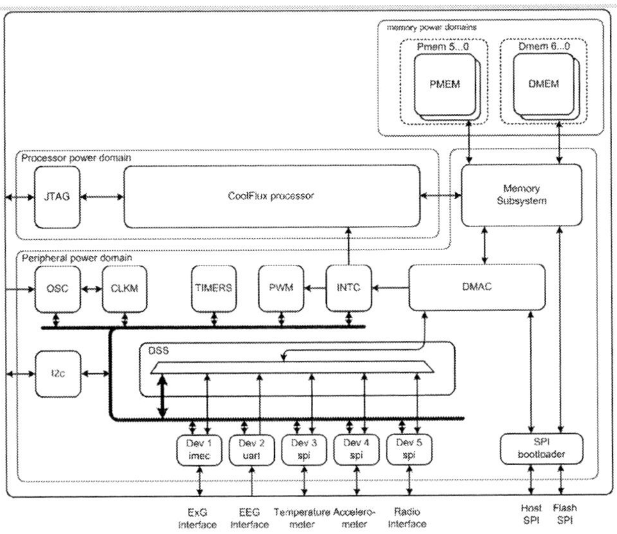

Figure 19.1.1: System block diagram.

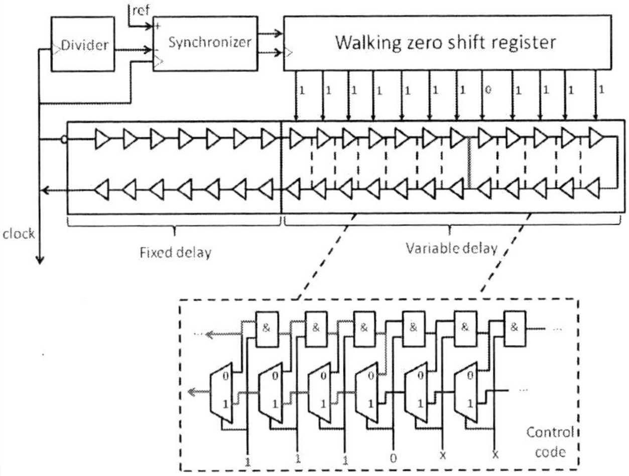

Figure 19.1.2: On-chip digital clock generator.

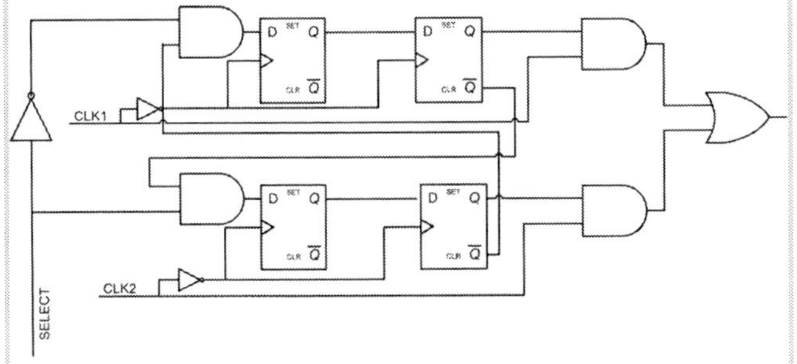

Figure 19.1.3: Two-input glitch-free clock switch.

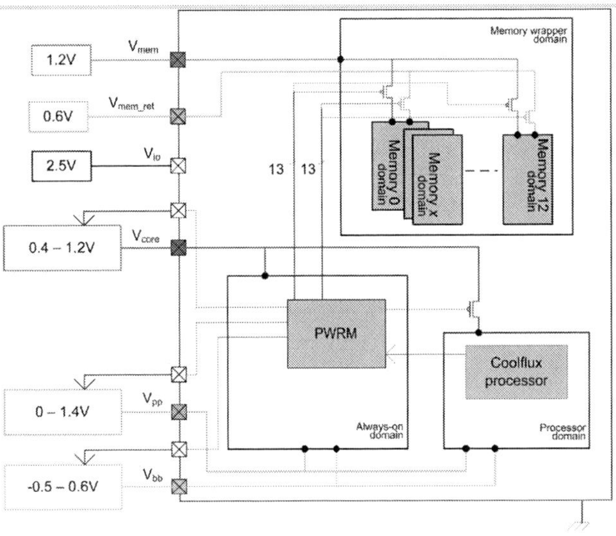

Figure 19.1.4: Voltage domains and supply and bias ranges.

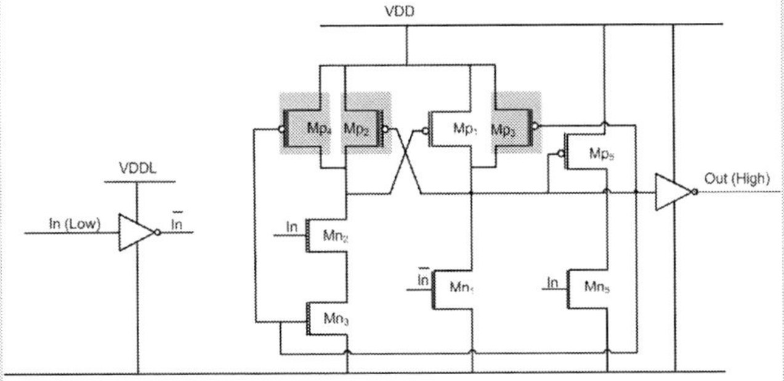

Figure 19.1.5: Near-threshold level shifter with non-conflicting rise and fall time paths.

Chip Summary			
Process Technology	CMOS 90nm LP		
Area	1.875 × 3.75 mm²		
Logic	195K NAND2 equivalent		
SRAM	2Mb		
Voltage	0.4V-1.2V		
Energy/Cycle (running ECG)	13pJ (0.4V logic, 0.7V SRAM) 95pJ (1.2V)		
Comparison with the state of the art			
	This Work	[Error! Reference source not found.]	[Error! Reference source not found.]
Technology	90nm	130nm	180nm
Performance	1MHz (0.4 logic, 0.7 SRAM) 100MHz (1.2V)	7kHz (0.5V) 5MHz (1.0V)	73kHz (0.4V) 1MHz (0.5V)
Energy/Cycle (pJ/Cycle)	12.8 (Data collection mode) 47 (Low performance mode) 145 (High performance mode)	34 (0.5V) 133 (1.0V)	28.9 (0.4V) 37.4 (0.5V)
Processor core	CoolFlux BSP	Cortex-M3	Cortex-M3

Figure 19.1.6: Chip features and measurement results.

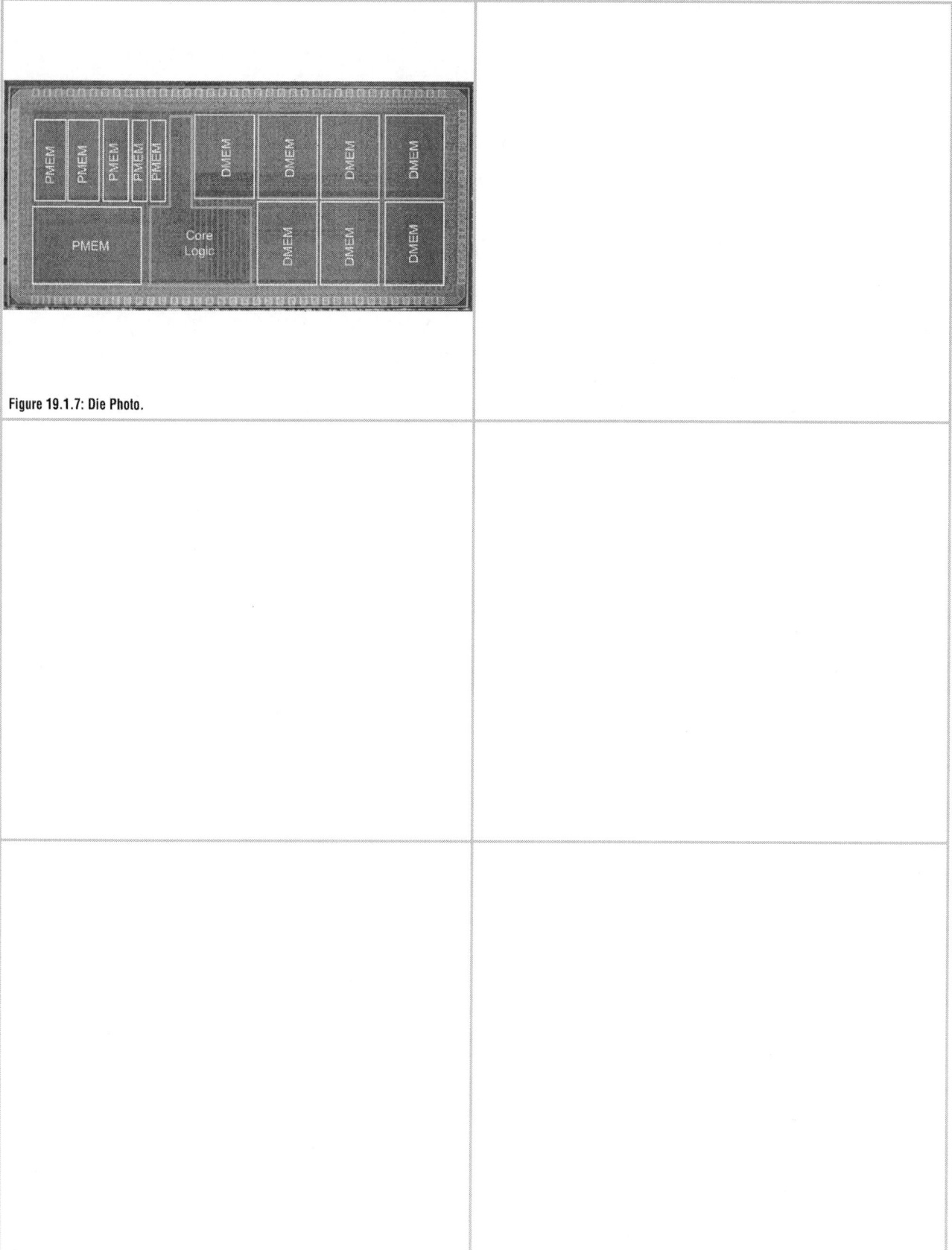

Figure 19.1.7: Die Photo.

ISSCC 2011 / SESSION 19 / LOW-POWER DIGITAL TECHNIQUES / 19.2

19.2 An 82µA/MHz Microcontroller with Embedded FeRAM for Energy-Harvesting Applications

Michael Zwerg[1], Adolf Baumann[1], Rüdiger Kuhn[1], Matthias Arnold[1], Ronald Nerlich[1], Marcus Herzog[1], Ralph Ledwa[1], Christian Sichert[1], Volker Rzehak[1], Priya Thanigai[2], Bjoern Oliver Eversmann[1]

[1]Texas Instruments, Freising, Germany,
[2]Texas Instruments, Dallas, TX

In recent years energy-harvesting technology has become much more mature. The energy source is free and widely available. However, there are some limitations that energy-harvesting devices need to cope with, to build a reliable system. First of all the peak current of typical energy-harvesting power sources is quite limited [1] and the system's active power must be as low as possible. A second limitation is the possibility of interruption or fluctuation of power source. For example, solar energy is only available during the day, and thermal or vibrational energy sources may change depending on environmental conditions. As a consequence, systems must retain state and information despite possible power outages. Flash or EEPROM as today's standard non-volatile memory is slow in write access and has high peak current demands due to the required charge pump for erase and write cycles. Flash technology has limited write endurance in the order of 10^5 write cycles requiring a Wear Leveling algorithm [2] for large cycle state storage. SRAM could be used for fast, low-power read/write [3], but it requires a separate battery for state retention.

This paper presents an ultra-low-power Micro Controller Unit (MCU) built as a System-on-Chip (SoC) that comprises of a 3-stage pipelined 16b MSP430 CPU, an integrated clock- and power-management unit, analog/digital peripherals and embedded FeRAM memory for fast write capability (Fig. 19.2.1). The SoC is fabricated in 0.13µm CMOS with 5 Cu metal layers. Two additional mask layers are required to include the FeRAM memory [4].

The CPU is capable of running up to speeds of 24MHz and supports 5 'low-power' modes using native CPU instructions. The clock system (CS) supports 2 crystal oscillators (32kHz and 4 to 24MHz), 2 internal oscillators (10kHz and 5MHz) and a configurable DCO (1 to 24MHz) allowing application-specific clock control. The Power Management Module (PMM) contains power supervision and control circuits. It supports 3 voltage domains with separate Low DropOut Regulators (LDO): (1) digital core logic, (2) real-time clock (RTC) domain, and (3) an independent LDO for the FeRAM. All clocks and LDOs are automatically controlled in conjunction with the selected low-power mode.

The following section outlines the differences between FeRAM-based and Flash-based memory systems. This is followed by an overview of the power-management concept used to ensure FeRAM typical refresh cycle under tough environmental conditions.

Figure 19.2.2 illustrates the datapath of the memory sub-system. The FeRAM array internal word size is 64b for data with an additional 8 parity bits for error detection and correction. The achieved memory access time is 55ns and cycle time is 110ns to enable up to 8MHz operation. Since the system operates up to 24MHz, a 2-way-2-line read cache from the internal 64b word to the 16b CPU increases the effective access rate to the FeRAM far beyond the 8MHz. A self-timed access mode directly interacts with the MCU clock system to automatically wait for data availability in case the actual CPU speed is higher than what the FeRAM memory supports. Compared to a typical Flash-based MCU system, the FeRAM-based system offers 1000× faster write capability at 100× lower power. To prevent software from accidental corruption of code and constant data space, a memory-protection unit (MPU) divides the unified memory structure into configurable segments. Each segment can be assigned with read, write and execute permissions. Data integrity is enhanced by Error Correction Circuit (ECC) logic and the memory array operating in 2T2C mode [5]. The integrated ECC for single-bit error correction and double-bit error detection is based on 8 redundant bits per 64b internal word. The ECC capability and the superior endurance of more than 10^{14} write cycles allows the FeRAM memory to offer the capability to support safety critical applications.

For energy-harvesting applications some data need retention even in case of a sudden power loss. To assure this, every memory read or write access must be completed under all circumstances, including hard failures of chip supply voltage. Therefore a dedicated power supply system for the FeRAM is built. The power domain for the FeRAM is kept internal to the chip, eliminating the possibility of external short-circuits (Fig. 19.2.3). The memory supply voltage is provided from an independent LDO and is buffered with a 2nF integrated capacitor. This capacitor is sized to hold a charge reserve to complete an ongoing memory access even under worst-case external voltage drops. A detector circuit at the LDO input voltage checks the overall chip supply. In case of an input voltage failure, the LDO is disconnected from the FeRAM supply, leaving the buffer capacitor as a source to complete an ongoing memory access. In parallel, the FeRAM's interface to the MCU is isolated, suppressing further access requests. The FeRAM memory performs self-timed memory accesses and completes an ongoing cycle. Note that for Flash-based systems, where the time taken to write data is in the range of milliseconds an extremely large on-chip capacitance (~100× larger) would be required to complete a pending write access under similar conditions. For immediate power-drop detection the circuit needs to react fast. Usually this requires support for high currents for the detector, which is in contradiction to a system-level low-power concept. To overcome this challenge a very fast detector response is assured only when needed, using an adaptively biased circuit scheme shown in Fig. 19.2.4. The current mirror provides a low quiescent bias current to the comparator for slow reaction time when supply voltage change slowly. In case of fast supply voltage drop the capacitor branch increases the comparator's tail current. This very-low-power scheme results in a comparator delay of typically 1ns at a quiescent current of 1µA.

In Fig. 19.2.5 the FeRAM power-supply system test is shown. The chip's supply voltage is periodically shorted with approximately 20ns glitches (500ns fall time on a board level, see trace DVDD). Before the actual test the memory is once filled with a known pattern. The short event is synchronized with a memory access (trace CE). After the short is released, the device starts up and performs a CRC, which is flagged (trace CRC_Pass/Fail). Sweeping the time point of shorting with respect to the CE signal produces supply failures at any time of memory access. In a loop, this test allows hundreds of tests per second for different addresses. No data corruption was found for any test condition.

Figure 19.2.6 shows the current consumption performing different types of memory access of the digital system versus clock speed. A typical current consumption of 82µA/MHz is achieved for an average cache hit/miss rate of 3/1, typical for linear code execution. To achieve such low numbers, extensive clock gating and power-aware synthesis are used. This number is significantly lower than other 16 or 32b MCU active power numbers reported so far [6].

The microcontroller with embedded FeRAM is an ideal platform for energy-harvesting systems. The low average execution current of 82µA/MHz and the unified memory structure with non-volatile data storage with a write endurance more than 10^{14} cycles is optimized to support harsh power environments and has excellent write speed and write power.

References:

[1] I. Doms, et al., "Integrated Capacitive Power-Management Circuit for Thermal Harvesters with Output Power 10 to 1000µW", *ISSCC Dig. Tech. Papers*, pp. 300-301, Feb. 2009.

[2] S. Kawai, et al., "An 8kB EEPROM-Emulation DataFLASH Module for Automotive MCU", *ISSCC Dig. Tech. Papers*, pp. 508-509, Feb. 2008.

[3] G. Chen, et al., "Millimeter-Scale Nearly Perpetual Sensor System with Stacked Battery and Solar Cells", *ISSCC Dig. Tech. Papers*, pp. 288-289, Feb. 2010.

[4] Hugh P. McAdams et al, "A 64-Mb Embedded FRAM Utilizing a 130-nm 5LM Cu/FSG Logic Process", in *IEEE J. Solid-State Circuits*, Vol. 39, No. 4, p.667, April 2004.

[5] John A. Rodriguez et. Al, "Reliability Properties of Low-Voltage Ferroelectric Capacitors and Memory Arrays", *IEEE Trans. Device and Material Reliability*, Vol. 4, No. 3, Sept. 2004.

[6] Energy Micro White Paper: EFM32 Introduction White Paper
http://downloads.energymicro.com/pdf/efm32_introduction_white_paper.pdf

978-1-61284-303-2/11 $26.00 © 2011 IEEE

ISSCC 2011 / February 23, 2011 / 10:45 AM

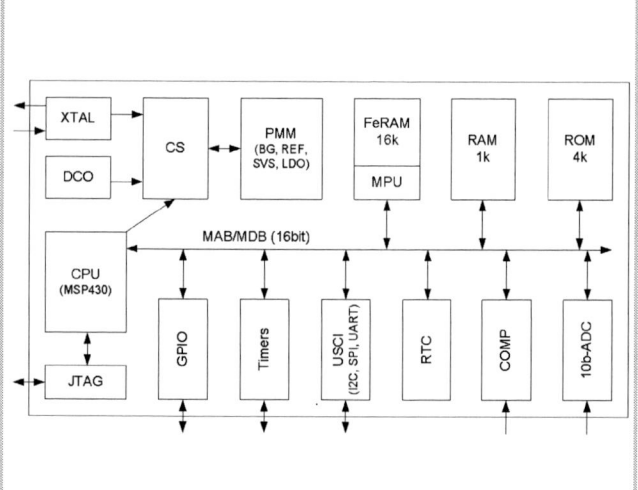

Figure 19.2.1: Microcontroller System Block Diagram.

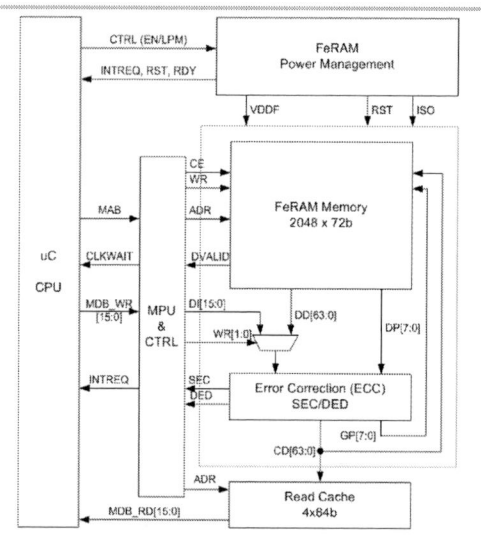

Figure 19.2.2: FeRAM Memory Sub-System.

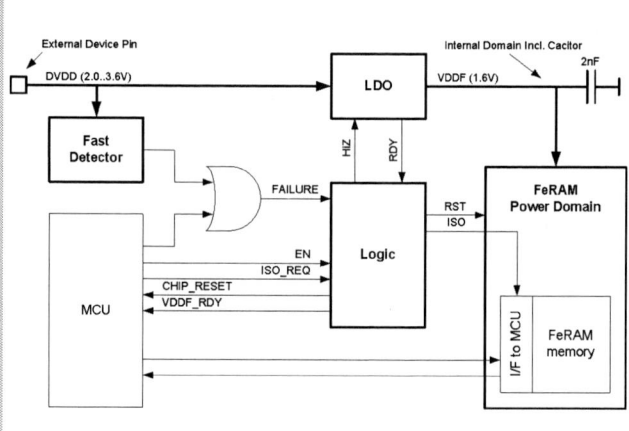

Figure 19.2.3: Block Diagram of the FeRAM Power Management System.

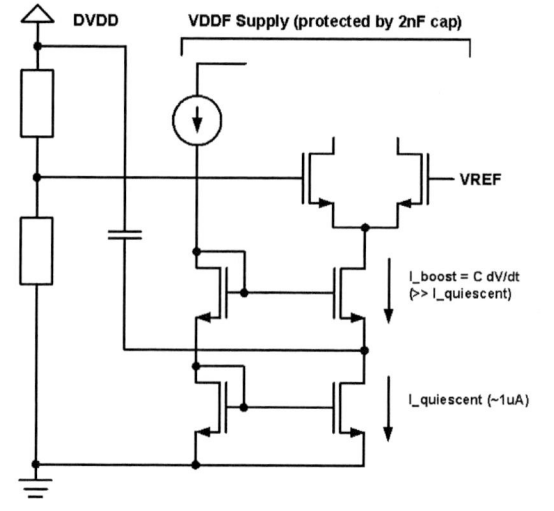

Figure 19.2.4: Circuit Diagram of a Input Voltage Detection Circuit.

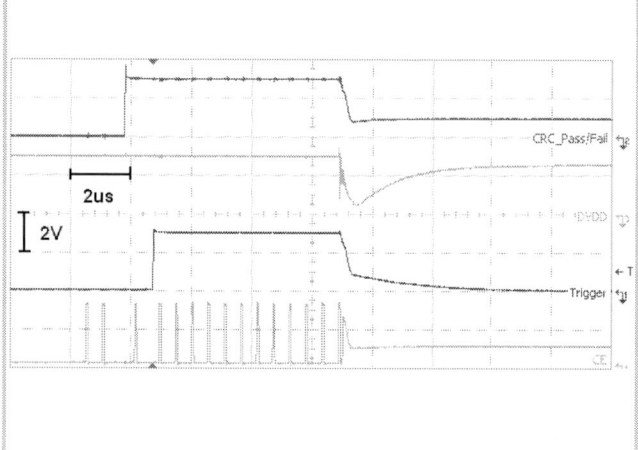

Figure 19.2.5: Oscilloscope Screen-Shot of FeRAM Supply System Testing.

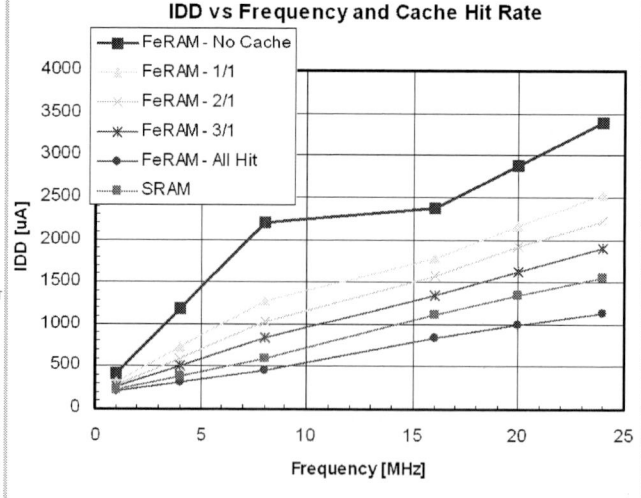

Figure 19.2.6: Active-Mode I_{DD} over Frequency and Cache Hit Rate.

19

978-1-61284-303-2/11 $26.00 © 2011 IEEE   335

ISSCC 2011 PAPER CONTINUATIONS

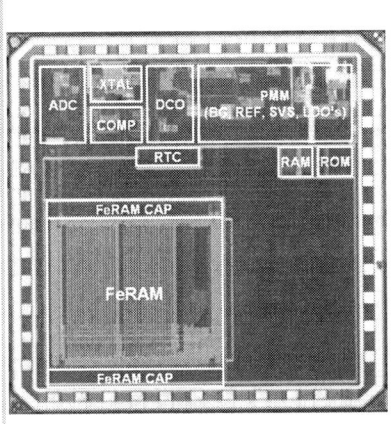

Technology	0.13μm CMOS, Embedded FeRAM, 5 Metal Cu
Chip Size	4.40mm²
External Supply	2.0V – 3.6V
Frequency	< 24Mhz (System) < 8MHz (FeRAM)
FeRAM	16kB with ECC
AM Current	183μA Base 82μA / MHz
Standby current (RTC mode)	<3μA @ 85°C
Logic	120k NAND2 Equi.

Figure 19.2.7: Chip micrograph and key parameters.

ISSCC 2011 / SESSION 19 / LOW-POWER DIGITAL TECHNIQUES / 19.3

19.3 Comparison of 65nm LP Bulk and LP PD-SOI with Adaptive Power Gate Body Bias for an LDPC Codec

Julien Le Coz[1], Philippe Flatresse[1], Sylvain Engels[1],
Alexandre Valentian[2], Marc Belleville[2], Christine Raynaud[1],
Damien Croain[1], Pascal Urard[1]

[1]STMicroelectronics, Crolles, France,
[2]CEA-LETI-MINATEC, Grenoble, France

A Low-Density Parity-Check (LDPC) codec circuit is implemented in a 65nm Low-Power Partially-Depleted SOI (LP PD-SOI) technology, as well as in a "conventional" Low-Power Bulk technology for a fair comparison. PD-SOI allows to increase maximum frequency versus bulk at a given voltage, and to decrease dynamic power versus bulk at a given frequency. Thanks to a digital power-switching technique specifically optimized for LP PD-SOI, we demonstrate a leakage current reduction versus bulk 65nm LP implementation, solving one of the most critical problems of PD-SOI technology, as identified in previous publications.

This LDPC codec supports the 12 modes required by the 802.11n standard (Wifi): 4 code rates (1/2, 2/3, 3/4, 5/6) over 3 possible block sizes (648, 1296 and 1944 bits). The maximum throughput achieved is 693Mb/s with a Packet Error Rate (PER) as low as 5×10^{-5} and a decoding latency lower than 6µs in all modes, which is fully compliant with the 802.11n standard. There are 2 instances of the LDPC decoder in LP PD-SOI design as described in Fig. 19.3.1, and one instance in 65LP Bulk. Each LDPC decoder has an area of 2mm² and implements a classical min-sum algorithm.

We introduce optimized power-gating techniques in PD-SOI to keep leakage current lower than bulk. We use the same power-gating strategy in Bulk and SOI, based on rings of high-V_t Multi-Threshold Header Power Switches (MTCMOS header power switches) [1].

We realize 2 different implementations of the LDPC decoder on LP PD-SOI. For the first decoder implementation, the design strategy is a block-level blinded porting from bulk technology to LP PD-SOI using the STMicroelectronics 65nm LP Bulk design flow, as described in Fig. 19.3.2. The block layout remains unchanged and standard bulk switches become Floating-Body (FB) switches in SOI. The sign-off is executed using LP PD-SOI timing values to check and fix timing constraints. The usual drawback of such an approach is the explosion of the leakage current on SOI: as an example 55× versus bulk is reported in [2]. This dramatic increase of the leakage can be explained by the fact that only the GP-PD-SOI technology was available for production up to now. For this implementation, we use the LP PD-SOI technology developed for battery-operated devices, which mitigates the leakage current increase versus 65nm LP Bulk to a mere factor of 5.6× (ten times less than previously published papers).

The second LDPC decoder implemented in PD-SOI is designed with a standard STMicroelectronics 65nm LP PD-SOI design flow, using LP PD-SOI libraries from RTL synthesis to sign-off. Additionally, some specific high-V_t Body-Contacted (BC) power switches are developed. These switches allow us to take advantage of our low-V_t Floating-Body CORE design when a switch is ON: the advantages are a lower dynamic power consumption and an higher speed compared to bulk. In the OFF mode however, those high-V_t switches achieve the same leakage level as bulk. According to the factor of merit we report in [3], the "BC" power switch is then demonstrated to be best trade-off, in terms of R_{on}/R_{off} value.

Additionally, we introduce a Forward-Body-Biasing circuit as described in Fig. 19.3.3 to improve significantly the R_{on} resistance of the power switches. We design an auto-adaptive biasing circuit to force FBB mode on the BC power switches ring only when switches are ON. When switches are in OFF mode, this biasing circuit becomes transparent by forcing Vdd onto the BC power switches'

bias. We call this specific solution "Auto-DTMOS" because it produces automatically a Dynamic Threshold MOS (DTMOS) effect [4]. Comparing to a previous publication on power-switch body-biasing results [5], our solution does not need a body bias supply (Vdds) in addition to a core supply (Vdd). This represents an improvement in terms of system complexity.

Figure 19.3.4 shows the silicon measurements of R_{on} and R_{off} implementing 3500 abutted power switches with a unitary width W=40µm and gate length L=200nm. These power-switch rings are implemented in a stand-alone test-chip. In the case of the auto-DTMOS, 4 body-biasing circuits are implemented (i.e.: one for every 1,000 power switches). 1 body-biasing circuit is equivalent to 5 power switches in terms of area. R_{on} is measured by the V_{drop}/I_{on} relation with I_{on} fixed at 100mA and V_{drop} measured thanks to probe pins at 3 temperatures (-40, 25, and 125°C) and 6 Vdd supplies (0.72, 0.8, 1.12, 1.2, 1.32, and 1.43V). R_{off} is measured under the same conditions. Thanks to the body-biasing circuit of auto-DTMOS, R_{on} is 20% lower than non-biased SOI BC and bulk solutions when we keep R_{off} unchanged. In our LDPC decoder implementation on LP PD-SOI, we decided to take advantage of the R_{on} decrease in auto-DTMOS solution to remove 20% of the unitary power switches, keeping the V_{drop} budget unchanged versus BC or bulk implementation. The total I_{off} value and power ring implementation area (which vary linearly with the number of switches) are reduced by 20%.

Figure 19.3.5 shows the dynamic power, static power, and maximum frequency comparison between LP Bulk and LP PD-SOI implementations, in the most representative cases. Concerning dynamic power, both LP PD-SOI switching strategies are equivalent and provide comparable results. For a given nominal Vdd = 1.2V and at room temperature (25°C), the SOI implementations reach 360MHz (693Mb/s throughput) when bulk maximum functional frequency is only 297MHz. This represents a gain of 20%. If we want to reach a given frequency (360MHz) we need to apply Vdd=1.4V in LP Bulk technology when only Vdd=1.2V is required in SOI implementation. This leads to a dynamic power decrease of 30% from 406mW in LP Bulk down to 288mW in LP PD-SOI. Silicon measurements show that auto-DTMOS has better results for leakage reduction than FB (floating body) by a factor of 3.6×. Comparing static power between bulk and auto-DTMOS, we finally achieve a leakage reduction of 20% for a given Vdd (1.2V) and 52.4% for a given maximum frequency (360MHz).

As described in Fig. 19.3.6, SOI performs better than bulk in all conditions, without adding any specific process steps to keep the lowest process cost: 20% maximum frequency increase (at a given voltage supply), 30% dynamic power decrease and up to 52.4% static power reduction for a given frequency target. A complete silicon validation demonstrates the validity of this approach for industrial PD-SOI circuit design because there is no need to develop a specific design platform. In this way, LP PD-SOI becomes very attractive and competitive versus LP Bulk for battery-operated devices.

Acknowledgements:
The authors would like to thank the overall design and test teams for their outstanding support, especially G. Thomas, R. Wilson from STMicroelectronics and L. Engels, G. Cogniard from Dolphin integration.

References:
[1] S. Mutoh, et al. "1-V power supply high-speed digital circuit technology with multithreshold-voltage CMOS," *IEEE J. Solid-State Circuits*, 1995.
[2] R. Pottier, et al. "ARM 1176 implementation in SOI 45nm technology and silicon measurement," SOICONF, 2010.
[3] J. Le-coz, et al. "Power Switch Optimization and Sizing in 65nm PD-SOI Considering Supply Voltage Noise," ICICDT, 2010.
[4] F. Assaderaghi, et al. "A dynamic threshold voltage MOSFET (DTMOS) for very low voltage operation," *IEEE Electron Device Letters*, 1994.
[5] S. V. Kosonocky, et al. "Low-power circuits and technology for wireless digital systems" *IBM Journal of Research and Development*, 2003.

978-1-61284-303-2/11 $26.00 © 2011 IEEE

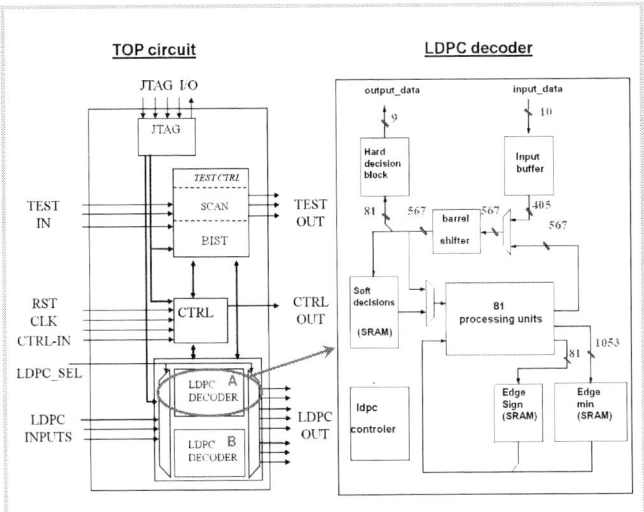

Figure 19.3.1: Top circuit and LDPC decoder microarchitecture.

LDPC implementation

CODE — RTL

Libraries — BULK / SOI — Same flow / Same tools

MASK — GDS: BULK version / GDS: SOI version

WAFER — BULK / SOI / SOI

BULK circuit | « Blind Porting » FB power switch | SOI design BC power switch

Figure 19.3.2: 65nm LP Bulk and PD-SOI implementation strategy.

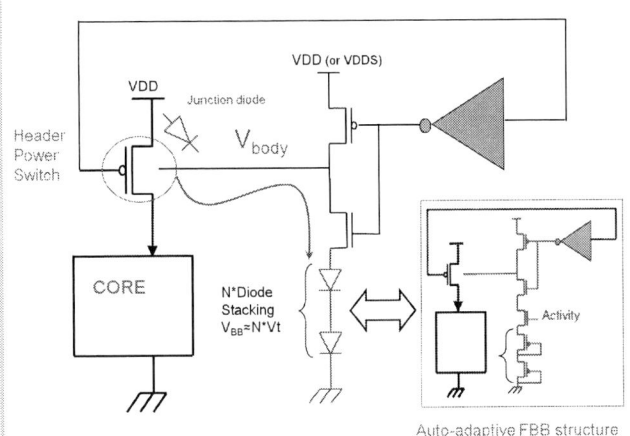

Figure 19.3.3: Auto-adaptive forward-body-biased power switching.

Ron at 25°C

- FB Ron
- BC Ron
- Auto-DTMOS Ron

Roff at 25°C

- FB Roff
- BC Roff
- Auto-DTMOS Roff

Figure 19.3.4: R_{on} and R_{off} values of the power switch ring.

Max Frequency and Dynamic Power Comparison @ 1.2V, 25°C

Silicon measurements	BULK 65LP	SOI 65LP (BC or auto-DTMOS)	SOI vs BULK
Max Frequency (MHz)	297	360	+ 20%
Dynamic Power (mW)	291	288	-1%

Dynamic Power Comparison @ 360MHz

Silicon measurments	BULK 65LP	SOI 65LP (BC or auto-DTMOS)	SOI vs BULK
VDD (V)	1.4	1.2	- 14%
Dynamic Power (mW)	406	288	- 30%

Static Power Comparison LP PD-SOI versus LP BULK

Silicon measurements	This work when working at1.2V	This work when targetting 360MHz	[1] GP-SOI
SOI (auto-DTMOS) vs LP BULK	-20%	-52.4%	x55

Figure 19.3.5: Frequency and power comparison LP PD-SOI versus LP Bulk.

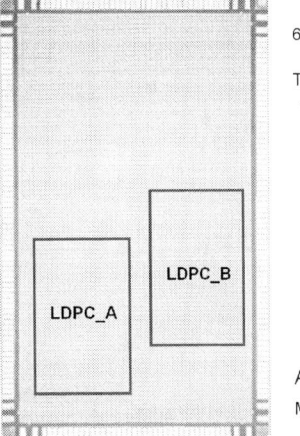

65 LP 7ML 3V₁ Bulk & SOI design

Testchip Content
* CORE Logic :
  * 2 LDPC blocks
    * Memories: SPREG, DPREG, ROM
    * LVT standard cells: CORE & CLOCK
  * Power Switch Solution
    * 1 variant in Bulk
    * 3 variants in PD-SOI:
      * FB (blind porting from Bulk)
      * BC
      * Auto-DTMOS
  * VDD@1.2 V
* I/O:
  * 224 pads / 2.5V 50A

Area 8mm² / LDPC decoder 2mm²

Minimum frequency of 200MHz

Figure 19.3.6: LP PD-SOI circuit microphotograph and content.

ISSCC 2011 / SESSION 19 / LOW-POWER DIGITAL TECHNIQUES / 19.4

19.4 A 77% Energy-Saving 22-Transistor Single-Phase-Clocking D-Flip-Flop with Adaptive-Coupling Configuration in 40nm CMOS

Chen Kong Teh, Tetsuya Fujita, Hiroyuki Hara, Mototsugu Hamada

Toshiba, Kawasaki, Japan

Flip-flops (FF) typically consume more than 50% of random-logic power in an SoC chip, due to their redundant transition of internal nodes, when the input and the output are in the same state. Several low-power techniques have been proposed [1-5], but all of them incur transistor-count penalties, leading to an increase in size, which is too costly since flip-flops typically account for 50% of random-logic area. In this work, we design and test a D-flip-flop, known as adaptive-coupling flip-flop (ACFF), which has a reduced transistor count compared to other low-power flip-flops, and 2 fewer transistors than the mainstream transmission-gate flip-flop (TGFF). ACFF features a single-phase clocking structure, with no local clock buffer and no precharging stage, enabling it to be more energy efficient than TGFF, where up to 77% energy saving is achieved at 0% data activity. ACFF also has an adaptive-coupling configuration, which weakens state-retention coupling during a transition, allowing it to be tolerant to process variations. Test chips are fabricated in a 40nm CMOS technology for 1.1V application, and 500k ACFFs are tested over all chips in 5 skew wafers. All tested ACFFs are fully functional down to 0.75V supply voltage, with spreads of timing parameters comparable to TGFF. We also demonstrate a P&R test by employing ACFF to a wireless LAN chip, and the results indicate chip power is reduced by as much as 24%.

Figure 19.4.1 explains the basic concept of ACFF. The conventional TGFF has 2 inverters of clock buffers, which persistently consume power in every clock cycle, even at low data activity. To remove these clock buffers, we consider a differential master-slave topology, shown in the lower left portion of the figure. However, the circuit is susceptible to process variations, because PMOS pass gates are too weak to pass through a substantially large drain current, in order to overcome the strong coupling in state-retention circuitry during a transition. We introduce a new method, the adaptive-coupling scheme, which configures ACFF such that the state-retention coupling is weakened if the input state is different to its internal state. This enables a transition to be easily performed, and allows ACFF to have a good tolerance to process variations. An adaptive-coupling element (ACE) is comprised of one PMOS and one NMOS, configured in parallel, and the gates are controlled by the same data signal. Consider the ACE circled in the right portion of the figure. If the gate level is high (BN node is high, B node is low), the PMOS is switched off, and the NMOS is switched on, weakening the charging ability of the G-F path. This enables the state of node F to be easily lowered to $V_{DD}-V_t$, during discharging through the F-B path. Since a PMOS pass gate is between the F-B path, node F cannot be completely discharged. When node G turns into a low state by charging of node FN, node F is completely discharged to 0V through the F-G path, since the ACE NMOS allows a strong discharge current.

Figure 19.4.2 shows a table of the simulated performance compared to other low-power FFs. ACFF has 22 transistors, which is the smallest transistor count. DMFF [1] also has the same transistor count, but for circuit stability it requires 5 more transistors in its NMOS stack and delay cell, dependent on process. As for energy efficiency, ACFF consumes the least energy among them, except around 100% activity. Note that the switching energy of FF inputs is considered in the energy calculation for a fair comparison. There are clock-buffer-free FFs, such as sense-amplifier FF and true-single-phase-clock FF [6], but they require a precharging phase in every clock cycle, leading to large energy consumption. DMFF, CCFF [4], and CPFF [5] have conditional circuitry embedded to disable the unnecessary precharging or discharging on the occasion of redundant transitions, but still consume comparatively large energy at low activity, due to their pulse-triggered structures that require clock buffers to generate a delayed local clock. CPSA [2] has no clock buffer and has conditional precharge circuitry, but consumes large switching energy on the clock pin, due to using of pass-gate inputs. CCKFF [3] is energy efficient even at low activity, but is bulky in size and has a poor D-Q delay time.

Figure 19.4.3 shows the measured process-variation tolerance of ACFF. The yields of ACFF are measured over 500k cells sampled from all chips in 5 skew wafers. The results demonstrate ACFF is operable down to 0.75V supply voltage. The right portion of the figure shows the simulated results verifying the effectiveness of ACE added to the master latch. Without ACE, the yield becomes dramatically poor, especially at the LH process corner (V_{tn}=low, V_{tp}=high) where the drain current of the PMOS pass gate is too weak to make a transition. The measured and simulated results indicate the critical process condition for ACFF is now the HL corner, led by failure of transition in the slave latch. An operating voltage below 0.4V, if required, can be achieved by employing ACE in the slave latch, the same way as for the master latch.

Figure 19.4.4 shows the measured average energy per clock cycle. ACFF consumes less energy compared to TGFF overall, and is more energy-efficient at a lower activity, up to 77% energy reduction at 0% activity, where the input and the output are persistently in the same state. Average data activity of flip-flops in an SoC chip is typically between 5 and 15%, and ACFF saves 68-to-54% power in this range. The right portion of the figure shows the schematics of the test chip configuration. A power/delay TEG and a setup/hold TEG are designed. The power/delay TEG features the ability to measure average power including switching power of FF inputs. The setup/hold TEG allows measuring setup time or hold time over an array of 198 DUTs at a time, which is more time-saving than the conventional TEG shown in [7]. This is realized by employing a differential voltage-controlled delay-line for each DUT, and a domino-toppling structure to connect multiple DUTs in an array. The TEG can be used to measure a large number of FFs for deciding the actual safety margin required to apply to a front-end library of a FF.

Figure 19.4.5 shows the measured clock-to-output (C-Q) delay, setup, and hold times, over process variations. ACFF has a smaller standard deviation (σ) of delay times, especially at 0.8V with 18ps of σ for ACFF compared to 34ps of σ for TGFF, contributed by its differential structure. The lower-right portion of the figure summarizes the measured mean-value comparison. In terms of performance, the total of setup time and delay time of ACFF is 6% higher than that of TGFF. In terms of race immunity, the hold time of ACFF is better than that of TGFF, indicating a better tolerance to races in pipeline stages, which offers a lesser requirement for hold buffers in a chip.

Figure 19.4.6 demonstrates the simulated power reduction for a wireless LAN chip [8] after employing ACFF. P&R is carried out, and power is calculated with normal-mode vectors as in [8]. After applying ACFF, the chip power is reduced by 24% from 70 to 53mW. About 84% of the FFs are replaced by ACFF, showing the majority of FFs meet the timing constraints despite ACFF having a larger setup time than TGFF. The random-logic area is reduced by 3% due to the ACFF'S smaller size. The average data activity of flip-flops is 6.7%.

Acknowledgments:
The authors thank Y. Unekawa, Y. Fujimura, and T. Shiozawa for their support.

References:
[1] S. Nomura, et al., "A 9.7mW AAC-Decoding, 620mW H.264 720p 60fps Decoding, 8-Core Media Processor with Embedded Forward-Body-Biasing and Power-Gating Circuit in 65nm CMOS Technology," *ISSCC Dig. Tech. Papers*, pp. 262-264, Feb. 2008.
[2] Y. Ueda, et al., "6.33mW MPEG Audio Decoding on a Multimedia Processor," *ISSCC Dig. Tech. Papers*, pp. 1636-1637, Feb. 2006.
[3] M. Hamada, et al., "A Conditional Clocking Flip-Flop for Low Power H. 264/MPEG-4 Audio/Visual Codec LSI," *IEEE Custom Integrated Circuits Conference*, pp. 527-530, Sept. 2005.
[4] B.-S. Kong, et al., "Conditional-capture flip-flop for statistical power reduction," *IEEE J. Solid-State Circuits*, vol. 36, pp. 1263–1271, Aug. 2001.
[5] N. Nedovic, et al., "Hybrid latch flip-flop with improved power efficiency," *Proc. Symp. Integrated Circuits and System Design*, pp. 211–215, Sep. 2000.
[6] J. Yuan, et al., "High-speed CMOS circuit technique", *IEEE J. Solid-State Circuits*, pp. 62-70, Feb. 1989.
[7] N. Nedovic, et al., "A test circuit for measurement of clocked storage element characteristics," *IEEE J. Solid-State Circuits*, vol. 39, pp. 1294–1304, Aug. 2004.
[8] T. Shiozawa, et al., "A New Timing Closure Methodology for an SoC with Multiple On-chip Regulators," *Design Automation Conference*, User Track 7U.4S, June 2010.

978-1-61284-303-2/11 $26.00 © 2011 IEEE

ISSCC 2011 / February 23, 2011 / 11:15 AM

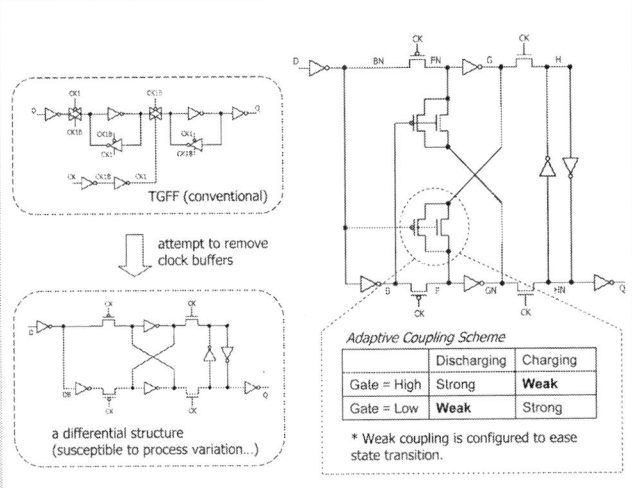

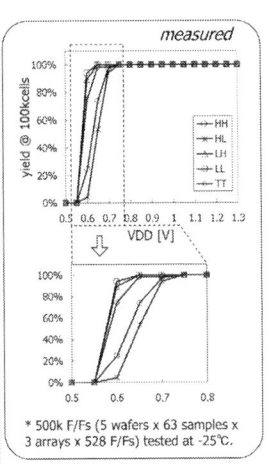

	transistor count	actual transistor count*1	minimum D-Q delay
this work	22	22	1.2
TGFF	24	24	1.0
DMFF [1]	22	27	0.8
CPSA [2]	28	30	1.1
CCKFF [3]	40	40	2.2
CCFF [4]	28	33	0.8
CPFF [5]	25	27	0.8

*1 Transistor count used in simulation. Scan circuitry not embedded.
* Flip-flops (FF) optimized for transistor count, power, and D-Q delay.
* Four fanouts of inverters loaded at the output.
* Switching energy of clock pin and data pin of FF included in energy calculation.
* DMFF : data mapping FF, CPSA : conditional-precharge sense-amplifier FF, CCKFF : conditional-clocking FF, CCFF : conditional-capture FF, CPFF : conditional-precharge FF

Figure 19.4.1: Concept of adaptive-coupling flip-flop (ACFF).

Figure 19.4.2: Simulated performance comparison.

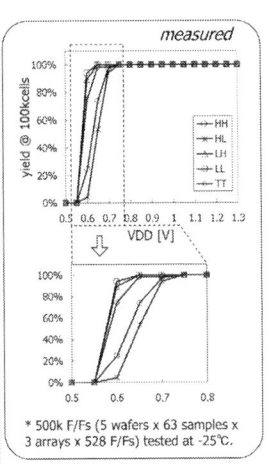

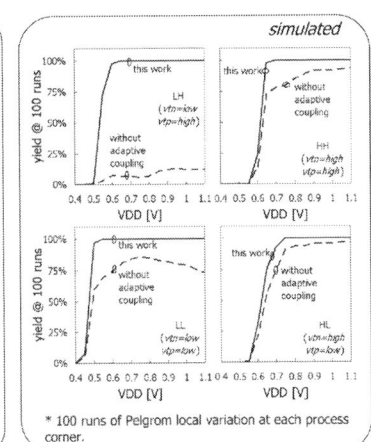

* 500k F/Fs (5 wafers x 63 samples x 3 arrays x 528 F/Fs) tested at -25°C.

* 100 runs of Pelgrom local variation at each process corner.

Figure 19.4.3: Measured and simulated process-variation tolerance.

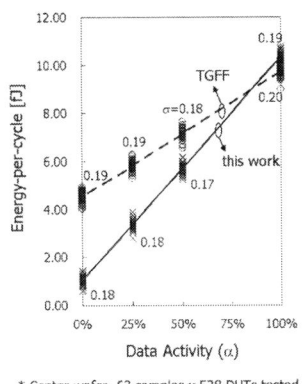

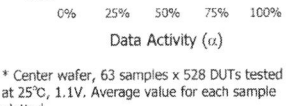

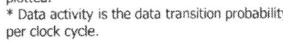

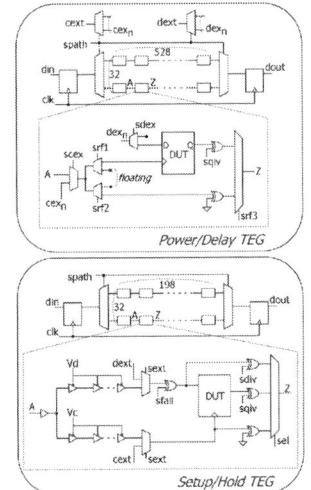

* Center wafer, 63 samples x 528 DUTs tested at 25°C, 1.1V. Average value for each sample plotted.
* Data activity is the data transition probability per clock cycle.

Figure 19.4.4: Measured average energy per clock cycle.

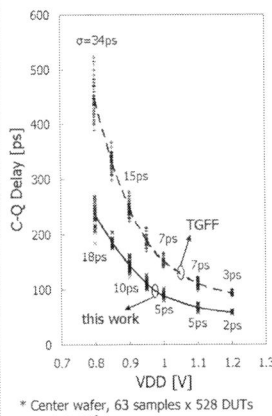

* Center wafer, 63 samples x 198 DUTs tested at -25°C, 1.1V.

Summary	TGFF	this work
Energy@10% activity	5.1 fJ	2.0 fJ
C-Q Delay *1	112 ps	67 ps
Setup *1	137 ps	197 ps
Hold *1	-14 ps	-73 ps

* Center wafer, 63 samples x 528 DUTs tested at -25°C.
* Average value for each sample plotted.
* Scan circuitry embedded in each flip-flop.

* 1.1V, mean value for center wafer.
*1 Each value is the average of values for D=1 and D=0.

Figure 19.4.5: Measured C-Q delay, setup, and hold times.

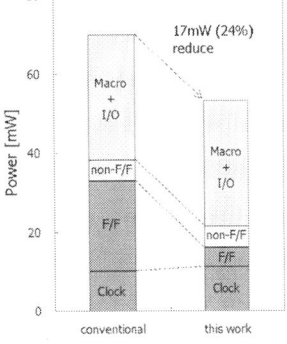

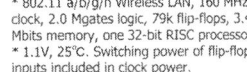

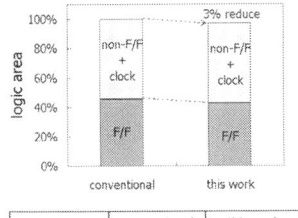

	conventional	this work
chip power	1.00	0.76
logic area	1.00	0.97
fmax	1.00	1.00
flip-flop ratio	100% TGFF	84% ACFF 16% TGFF
average data activity	6.7%	

* 802.11 a/b/g/n Wireless LAN, 160 MHz clock, 2.0 Mgates logic, 79k flip-flops, 3.4 Mbits memory, one 32-bit RISC processor
* 1.1V, 25°C. Switching power of flip-flop inputs included in clock power.

* Physical and timing constraints for both designs are identical. Gated clock applied for both designs.

Figure 19.4.6: Simulated power reduction for a wireless LAN chip.

Technology	40nm CMOS
Chip size	$0.87\ mm^2$ x 2 unit
Power/Delay Unit	32 arrays x 528 flip-flops
Setup/Hold Unit	32 arrays x 198 flip-flops
Wafer fabricated	4 skew-corner wafers + 1 center wafer
Chip per wafer	63 samples

Figure 19.4.7: Chip micrograph.

ISSCC 2011 / SESSION 19 / LOW-POWER DIGITAL TECHNIQUES / 19.5

19.5 A 62mV 0.13µm CMOS Standard-Cell-Based Design Technique Using Schmitt-Trigger Logic

Niklas Lotze[1,2], Yiannos Manoli[1,2]

[1]University of Freiburg - IMTEK, Freiburg, Germany,
[2]HSG-IMIT, Villingen-Schwenningen, Germany

Sub-threshold circuits have recently gained attention mainly due to the possibility of operating at the minimum energy per operation point [1]. There are applications where a supply voltage reduction below this point is advantageous though, even at the cost of increasing active energy per operation. Always-on circuits, e.g. wake-up circuitry for chips sleeping at ultra-low supply voltages, reduce power consumption with decreasing supply. Furthermore, energy-harvesting applications are often limited by the very low output voltages of the harvesting devices, thus the minimum V_{DD} of the electronic circuits dictates when active operation can start (e.g. thermoelectric harvesters [2]).

Numerous ultra-low-voltage circuits have been demonstrated: A 180mV FFT processor is shown in [1] and a SRAM cell applying feed-forward of inverter inputs to transistor bodies is reported to work at 135mV [5], but this circuit technique is not applicable to general logic. Many sub-threshold circuits rely on body biasing for post-silicon tuning, i.e. a microcontroller operational down to 160mV [3] (210mV without bias) and an 85mV FIR filter (160mV without bias) presented in [4]. This approach requires supply voltages higher than the digital V_{DD} to generate bias voltages though, limiting its usability in energy-harvesting applications. In this paper, we demonstrate a standard-cell-based circuit technique fully operational at supply voltages between 84mV and 62mV in standard 0.13µm bulk CMOS depending on the area overhead invested.

Supply voltage reduction is limited by the degradation of the on/off current-ratio of CMOS transistors with decreasing V_{DD}, causing the leakage currents through the off transistors to be on the same order of magnitude as the drive currents. The result is an output level degradation of logic gates due to a voltage-divider-like behavior, an effect emphasized by process variability. In this work, the output level degradation is mitigated by the use of Schmitt Trigger structures, which exhibit an effective leakage quenching in the off-path of a gate and have been proposed for low-voltage RAMs in [6]. The underlying mechanism is illustrated for a Schmitt Trigger inverter in Fig. 19.5.1 (w.l.o.g. for input low): The feedback transistor N2 in the off-path of the structure is on, tying node X to a high potential. This results in a negative V_{GS} for N1, giving an exponential reduction in the leakage current drawn from node Z and therefore an output level stabilization. This mechanism is applied to arbitrary CMOS logic gates as shown in Fig. 19.5.2 for a NAND gate: The pull-up/pull-down networks are duplicated, providing common nodes X/Y where the feedback transistors are connected.

Minimum achievable V_{DD} is optimized through transistor sizing, transistor area and choice of implemented gates. Regarding transistor sizing, the worst-case SNM of gate pairs is determined using an approach similar to [7] considering process variability via Monte Carlo simulations. Transistor dimensions are optimized to reach a maximum SNM. Even though the Schmitt Trigger gates show a very effective suppression of global NMOS vs. PMOS process variations in corner simulations, they still exhibit susceptibility to transistor-to-transistor variability. This results in trip point shifting and output level degradation caused by leakage through the feedback transistors. As random dopant fluctuation is the major contributor to this variability and directly correlates with transistor area, 3 different gate sizes are implemented, referred to as S1, S4 and S16. Transistor dimensions for S1 are given in Fig. 19.5.2, S4 and S16 each successively double both length and width, with the exact dimensions being optimized individually as to consider nonlinear scaling effects. Simulations show a reciprocal linear dependency between the root of transistor area and minimum V_{DD}, as shown in Fig. 19.5.5. The worst-case SNM of a gate is reduced with increasing number of inputs due to the increasing number of possible leakage current configurations. Therefore, only 2-input gates are used. As the clocked CMOS or transmission gate structures used in typical CMOS flip-flops also decrease SNM, a gate-based flip-flop is implemented.

A block diagram of the test chip is shown in Fig. 19.5.3. 8×8b multipliers are used as test structures for S1/S4/S16. A switchable feedback from the output MSB to the input LSB allows for a ring-oscillator mode for speed measurements. S16 furthermore incorporates an output register to test the flip-flop behavior at the lowest possible V_{DD}. To limit the I/O count, a digital serial-to-parallel interface operating at nominal V_{DD} is used, making input and output level converters necessary. To ensure reliable level-up conversion at the anticipated input V_{DD}, the leakage quenching principle is also applied in the first 2 stages of the level shifter, as shown in Fig. 19.5.2. Analog output of multiplier results via source followers is implemented to monitor level degradation. Signal routing from the multipliers to the primary outputs is realized using NMOS pass gates negatively biased in the off state to ensure reliable signal separation (AM in Fig. 19.5.3).

All low-voltage logic is implemented using a standard-cell approach with fully automatic place & route. Careful standard-cell layout is necessary to limit systematic V_t variations. As shown in Fig. 19.5.4 for the NAND gate, no transistor chaining is used to avoid design discrepancies due to length of diffusion (LOD) effects, and keep-out distances from well borders are maintained to limit well-proximity effects. The resulting area overhead of a NAND2 gate is 3/4/7.5× for S1/S4/S16 compared to a minimum-sized NAND2 of a commercial standard-cell library, which in SNM simulations shows functionality to approximately 140mV.

Out of 9 delivered chips, all are fully functional well below the limit V_{DD} predicted by the worst-case SNM simulations. Using a full test vector set, the minimum V_{DD} values shown in Fig. 19.5.5 are measured, giving minimum supply voltages as low as 84/68/62mV for the best chips (mean: 88/71/66mV) for multipliers S1/S4/S16. The minimum V_{DD} for S16 is located in the range predicted as lower bound in the simulations, indicating it is rather limited by sub-threshold slope than by variability, whereas the slope for the increase in minimum V_{DD} between S4 and S1 is in good accordance with the simulation results. Between room temperature and 80°C, minimum V_{DD} increases by approx. 3/9/11mV due to sub-threshold slope degradation (cf. Fig. 19.5.5). The flip-flops implemented in S16 are stable and fully functional in the same voltage range as the combinatorial logic and exhibit similar temperature dependencies as shown in Fig. 19.5.5. The level shifters are operated with intermediate supply voltages 160 and 460mV and convert signals with input V_{DD} down to 33mV in this configuration.

The expected exponential dependency on V_{DD} for circuit speed and quadratic dependency for active energy is identified in Fig. 19.5.6 (selected values in Fig. 19.5.7). At minimum V_{DD}, maximum frequencies are in the kHz range. The minimum active E/op of 47.7fJ occurs for S4. Leakage power (also shown in Fig. 19.5.6) is comparable for all implementations due to scaling both transistor length and width. The minimum value of 17.6nW is measured for S1. Compared to the value at the minimum E/operation point, leakage at minimum supply is reduced by a factor of 4.0/4.8/6.8×.

References:

[1] A. Wang, A. Chandrakasan, "A 180mV FFT Processor Using Subthreshold Circuit Techniques," *ISSCC Dig. Tech. Papers*, pp. 292-293, Feb. 2004.
[2] Y. Ramadass, A. Chandrakasan, "A Batteryless Thermoelectric Energy-Harvesting Interface Circuit with 35mV Startup Voltage," *ISSCC Dig. Tech. Papers*, pp. 486-487, Feb. 2010.
[3] S. Hanson, et al., "Exploring Variability and Performance in a Sub-200-mV Processor," *J. Solid-State Circuits*, vol. 43, pp. 881-891, Apr. 2008.
[4] M. Hwang, et al., "A 85mV 40nW Process-Tolerant Subthreshold 8x8 FIR Filter in 130nm Technology," *Dig. Symp. VLSI Circuits*, pp. 154-155, June 2007.
[5] M. Hwang, K. Roy, "A 135mV 0.13 µW Process Tolerant 6T Subthreshold DTMOS SRAM in 90nm Technology," *Proc. CICC*, pp. 419 -422, Sept. 2008.
[6] J.P. Kulkarni, K. Kim, K. Roy, "A 160 mV Robust Schmitt Trigger Based Subthreshold SRAM," *J. Solid-State Circuits*, vol. 42, pp. 2303-2313, Oct. 2007.
[7] J. Kwong, et al., "A 65 nm Sub-V_t Microcontroller With Integrated SRAM and Switched Capacitor DC-DC Converter," *J. Solid-State Circuits*, vol. 44, pp. 115-126, Jan. 2007.

ISSCC 2011 / February 23, 2011 / 11:30 AM

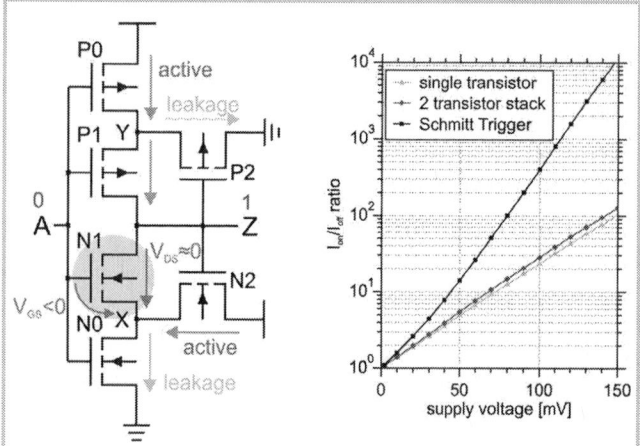

Figure 19.5.1: Illustration of leakage quenching in Schmitt Trigger: Increasing voltage at node X reduces leakage flowing from Z exponentially, thereby stabilizing the output voltage. Graph shows ratio between maximum on and off current in different topologies.

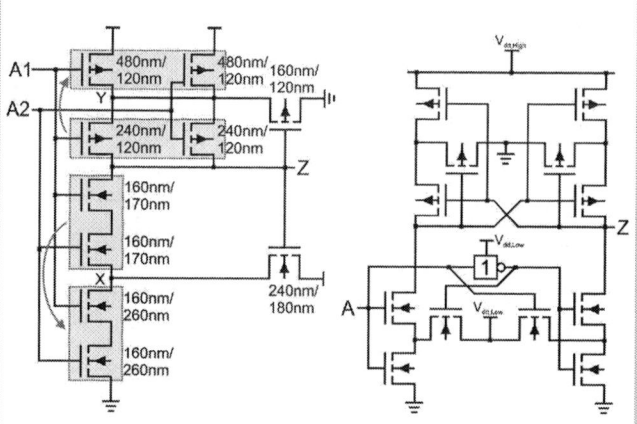

Figure 19.5.2: Application of leakage quenching in logic gates by duplication of pull-up/pull-down network and addition of feedback to middle nodes X/Y (NMOS with longer channel length have higher drive strength due to RSCE). Leakage quenching implementation of level shifter.

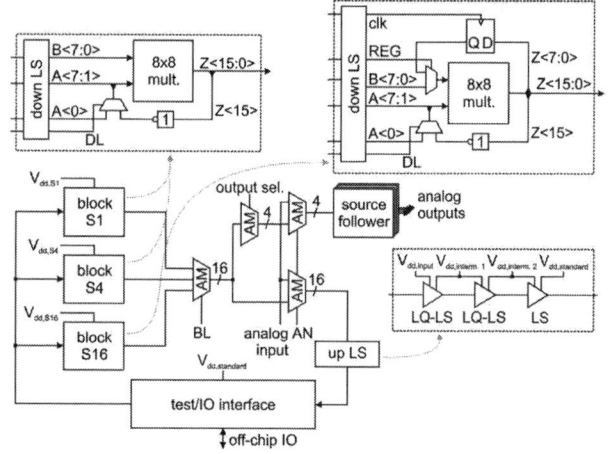

Figure 19.5.3: Top-level organization of test chip. Multipliers with optional feedback are implemented using three gate sizes (S1, S4, S16). S16 additionally incorporates a registered feedback. Analog input is used to characterize level shifters and source followers, and AM are analog multiplexers.

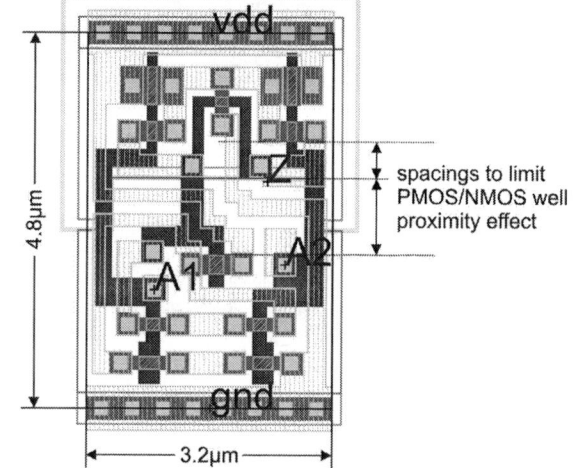

Figure 19.5.4: Layout of S1 NAND cell. To limit systematic V_t shifts, no transistor chaining is used (length of diffusion effect) and safety margins from well borders are kept (well proximity effect).

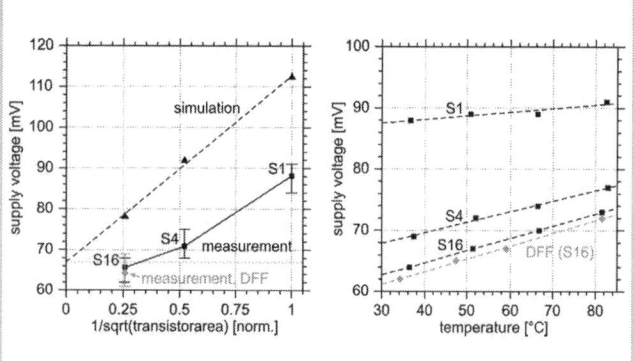

Figure 19.5.5: Minimum V_{DD} of multipliers and flip-flops versus gate transistor area (measurement and Monte Carlo SNM simulation results) as well as temperature dependence of V_{DD} (temperature dependence shown for single chip, similar results for other chips).

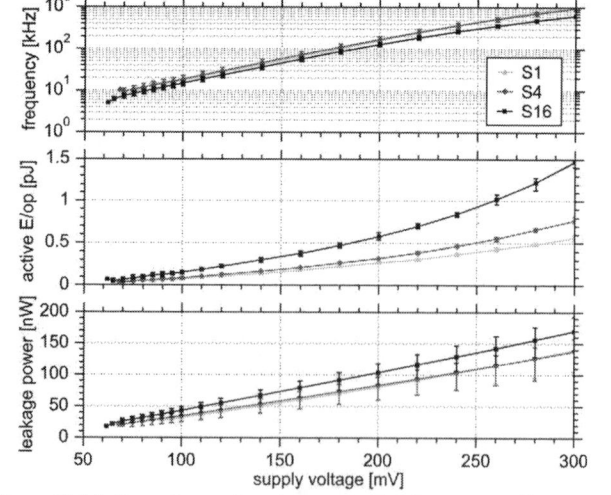

Figure 19.5.6: Dependence of maximum operation frequency, active energy per operation and leakage power on supply voltage.

19

978-1-61284-303-2/11 $26.00 © 2011 IEEE   341

ISSCC 2011 PAPER CONTINUATIONS

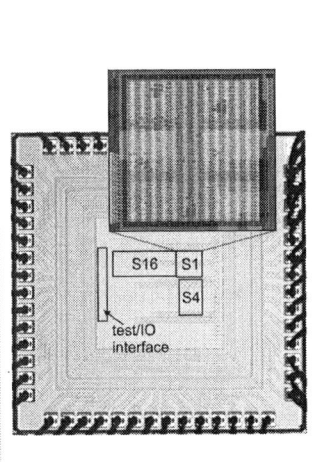

Process	0.13µm 8-metal CMOS	
Design	8x8 Multiplier	
Area (S1/S4/S16)	19176/23978/45002µm²	
Supply Voltage	min.	max.
(S1/S4/S16)	84/68/62mV	1.2V
Minimum Energy/Operation Point		
Frequency	S1	544.8kHz@260mV
	S4	328.3kHz@233mV
	S16	210.1kHz@226mV
Power	S1	346.6nW@260mV
	S4	243.8nW@233mV
	S16	275.6nW@226mV
Energy/	S1	0.63pJ@260mV
Operation	S4	0.74pJ@233mV
	S16	1.31pJ@226mV
Minimum Supply Voltage		
Frequency	S1	15.20kHz@84mV
	S4	10.35kHz@68mV
	S16	5.15kHz@62mV
Power	S1	29.9nW@84mV
	S4	21.1nW@68mV
	S16	17.9nW@62mV
Energy/	S1	1.97pJ@84mV
Operation	S4	2.04pJ@68mV
	S16	3.48pJ@62mV

Figure 19.5.7: Die mocrograph, layout of S1 and technical data.

ISSCC 2011 / SESSION 19 / LOW-POWER DIGITAL TECHNIQUES / 19.6

19.6 A 0.27V 30MHz 17.7nJ/transform 1024-pt Complex FFT Core with Super-Pipelining

Mingoo Seok[1], Dongsuk Jeon[1], Chaitali Chakrabarti[2], David Blaauw[1], Dennis Sylvester[1]

[1]University of Michigan, Ann Arbor, MI, [2]Arizona State University, Tempe, AZ

Recently, aggressive voltage scaling was shown as an important technique in achieving highly energy-efficient circuits. Specifically, scaling V_{dd} to near or sub-threshold regions was proposed for energy-constrained sensor systems to enable long lifetime and small system volume [1][2][4]. However, energy efficiency degrades below a certain voltage, V_{min}, due to rapidly increasing leakage energy consumption, setting a fundamental limit on the achievable energy efficiency. In addition, voltage scaling degrades performance and heightens delay variability due to large I_d sensitivity to PVT variations in the ultra-low voltage (ULV) regime. This paper uses circuit and architectural methods to further reduce the minimum energy point, or E_{min}, and establish a new lower limit on energy efficiency, while simultaneously improving performance and robustness. The approaches are demonstrated on an FFT core in 65nm CMOS.

Pipelining is a well-known method to improve performance or to enable limited energy savings by trading gained performance through lowering V_{dd}. However, in this paper, we make the counterintuitive observation that inserting additional pipeline latches improves both energy efficiency and performance in the ULV operating regime. Since pipelining shortens the clock period, it limits leakage energy consumed by idling gates, which reduces energy consumption and allows further voltage scaling. Simulations of inverter chains show that reducing stage depth from 65 to 11 fanout-of-four (FO4) delays yields 36% energy savings and a V_{min} reduction from 0.37 to 0.26V (Fig. 19.6.1 upper left). By applying this "super-pipelining" approach to the multipliers in an FFT core, we find that it consumes minimum energy when pipelined in 6 stages at a stage depth of 17 FO4 delays. This design approach differs radically from conventional ULV designs, which tend to use limited pipelining and typically have cycle times in the 50-to-200 FO4 range [1,2]. In this paper, we also show how clocking overhead can be reduced through circuit techniques to facilitate super-pipelining while process variation is addressed through the use of latch-based design. Additionally, architecture modifications are proposed to improve energy efficiency and throughput. Measurements show that the FFT core consumes 17.7nJ per 1024-pt complex FFT while operating at 30MHz at V_{dd}=0.27V, demonstrating an improvement over the FFT energy efficiency reported in [2-4].

An important principle driving our ULV design methodology is to suppress leakage energy, allowing for larger potential energy savings by enabling further voltage scaling. We first address this by architectural modifications through minimizing idling modules (Fig. 19.6.2). In a traditional memory-based FFT (Fig 19.6.2, bottom right), most memory cells idle while a single butterfly unit processes data word-by-word over many clock cycles. These idling cells increase leakage energy, harming energy efficiency and voltage scalability. On the other hand, conventional pipeline architectures such as MDC (Multi-path Delay Commutator) have high memory utilization but instead suffer from low butterfly unit activity [5]. We therefore propose a modified MDC that accepts 4 inputs concurrently using a commutator configuration, enabling full utilization of both butterflies and memory elements. Additionally, we use two of the modified MDC lanes to double throughput and halve memory counts per lane, reducing leakage energy consumption from memories. As shown in Fig. 19.6.1, these modifications improve energy efficiency and throughput by 2.8× and 6.2×, respectively, compared to a radix-4 memory-based FFT core.

Multipliers in the FFT are super-pipelined as shown in Fig 19.6.3. To successfully employ super-pipelining, sequential element overhead must be limited. Six latches share a local clock driver to reduce clock load. The drivers also use minimum-width fingers that enhance drivability at iso-input capacitance due to smaller V_{th} from inverse narrow width effects. Two latches are embedded in a mirror adder to save two transistors per latch. Latches are upsized from min-width for robustness such that they pass corners and 2 million Monte-Carlo mismatch simulations, providing an estimated 99% chip-level yield with 10k latch instances per chip at 0.2V. We implement the multiplier along with an unpipelined baseline multiplier, separately from the FFT core. Measured results (Fig. 19.6.1 upper right) show that the super-pipelined multiplier operates at 18MHz at 0.225V. It is 1.6× faster while consuming 30% less energy than an unpipelined multiplier. At iso-V_{dd}, it operates 3.6× faster and consumes 18% less energy.

The FIFOs in the commutators contribute as much as 29% of the total FFT energy consumption in this architecture. To address this, we replace the address decoder with a cyclic address generator for reduced energy and use logic-based readout paths for improved performance, as shown in Fig. 19.6.4. Simulation results show that the FIFO design consumes 12% lower energy while improving performance by 20% over a memory with MUX-based readout. Positive-edge read and negative-edge write operations for preventing hold time violation are used.

Although the techniques above improve energy efficiency and performance, we must pay attention to delay variability and overall design robustness given the ULV design point. To this end, we use 2-phase latches rather than flip-flops. Although the stage depth is drastically reduced in super-pipelined designs, time borrowing removes hard boundaries in the pipeline, re-establishing averaging of process variations along long paths that are present in unpipelined designs. Fig. 19.6.1 shows Monte Carlo simulations on latch and flip-flop pipelined multipliers indicating that a latch-pipelined multiplier can absorb delay variations, leading to higher performance yield. In addition, variability-induced hold time violations must also be avoided to ensure functionality. We identify short paths, aided by the regular structure of multipliers, and add delay elements that incur a marginal energy overhead of 2.4% per multiplier. Padded short paths were verified to satisfy hold times using 150k Monte-Carlo simulations under random process variations and corners.

The clock distribution network is designed to suppress process-variation-induced skew and its resulting hold time violations. Conventionally, many clock buffers are used to mitigate RC mismatch. However, at low V_{dd} the mismatch in these buffers is exacerbated and contributes significant skew, while RC delay is small compared to gate delay. Therefore, we design a 3-level clock network where a reduced number of large buffers and matched RC interconnect are used. The lowest and middle levels of the clock network are implemented with minimum width thin interconnect while the top level uses fish-bone shaped, thick metal interconnect for low RC delay and good slew. Fig. 19.6.4 shows that the simulated worst-case RC mismatch is less than 0.15ns (0.14×FO4 at V_{dd}=0.27V).

The FFT core is fabricated in 65nm CMOS using the above circuit and architectural techniques. Measurements in Fig. 19.6.6 show that it computes 234k 16b 1024-pt complex FFTs per second. The clock frequency is measured as 30MHz with V_{dd}=0.27V compared to frequencies of 10's of kHz for typical ULV designs at the same supply voltage. The FFT consumes 17.7nJ/transform, which is 4× smaller than prior work when scaled for word width, FFT size (2× from real to complex and 4× from 256 to 1024pt) and technology [3].

Acknowledgement:

The IC fabrication support of STMicroelectronics is gratefully acknowledged. Authors also acknowledge the Multiscale Systems Center and Army Research Laboratory for their support in this work.

References:

[1] G. K. Chen et al., "Millimeter-Scale Nearly-Perpetual Sensor System with Stacked Battery and Solar Cells," *ISSCC Dig. Tech. Papers*, Feb. 2010.
[2] A. Wang et al., "A 180mV FFT Processor using Subthreshold Circuit Techniques," *ISSCC Dig. Tech. Papers*, 2004.
[3] Y. Chen et al., "A 2.4-Gsample/s DVFS FFT Processor for MIMO OFDM Communication Systems," *IEEE J. Solid-State Circuits*, May 2008.
[4] S. Sridhara et al., "Microwatt Embedded Processor Platform for Medical System-on-Chip Applications," *Symp. on VLSI Circuits*, 2010.
[5] Y. Jung et al., "New Efficient FFT Algorithm and Pipeline Implementation Results for OFDM/DMT Applications," *Trans. Consumer Electronics*, Feb. 2003.

ISSCC 2011 / February 23, 2011 / 11:45 AM

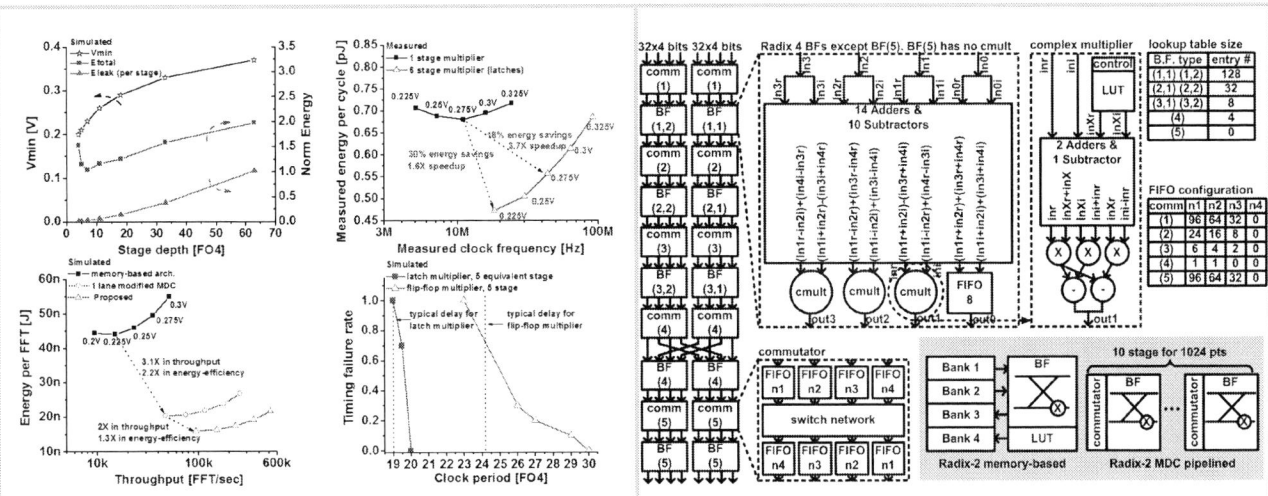

Figure 19.6.1: Inverter chain experiments, energy and performance of multipliers, architecture modification benefits, and multiplier variability.

Figure 19.6.2: Pipelined, 8×32b input, radix-4, 2-lane, 1024-pt, complex FFT along with conventional architectures.

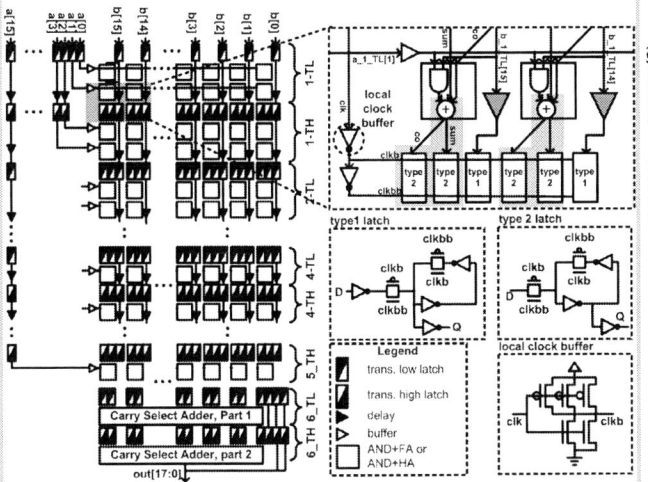

Figure 19.6.3: A 16b BW multiplier is pipelined with 12 banks of 2-phase latches. 5-4-3-2-2 length carry select adder is used for accumulation.

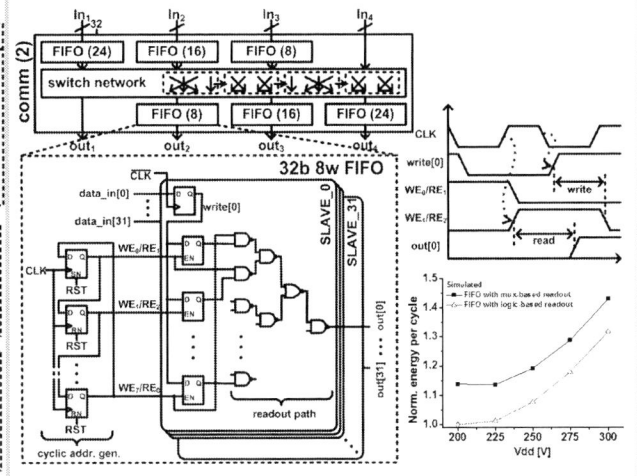

Figure 19.6.4: A commutator consists of a switch network and FIFOs. Positive-edge read and negative-edge write are described.

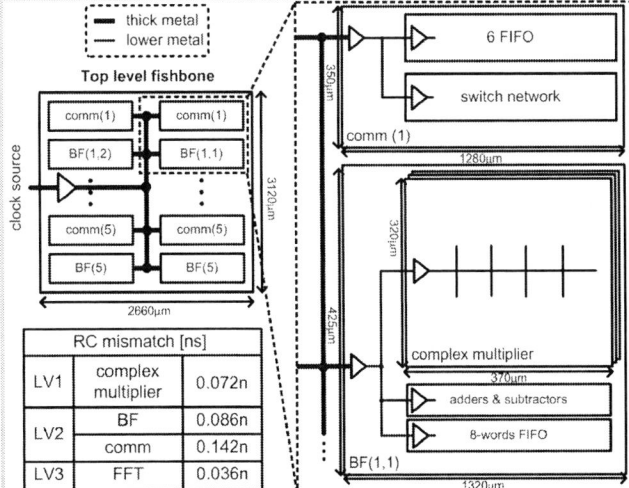

Figure 19.6.5: The clock network is designed with a limited number of buffers and matched interconnect to address key ULV skew sources.

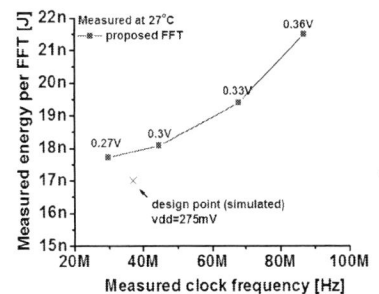

	Proposed	[2]	[3]	[4]
Technology	65nm	180nm	90nm	130nm
Low power config	1024-pt CV 0.27V, 30MHz	1024-pt RV 0.35V, 10KHz	256-pt CV 0.85V, 300MHz	256-pt RV 0.5V, 7KHz
Energy/FFT	17.7 nJ	155 nJ	12.8 nJ	100 nJ
Norm. energy/FFT	17.7 nJ	111.9 nJ	71.0 nJ	400 nJ

Figure 19.6.6: The measured energy efficiency and performance of the FFT and comparisons normalized to technology and FFT type.

19

978-1-61284-303-2/11 $26.00 © 2011 IEEE

ISSCC 2011 PAPER CONTINUATIONS

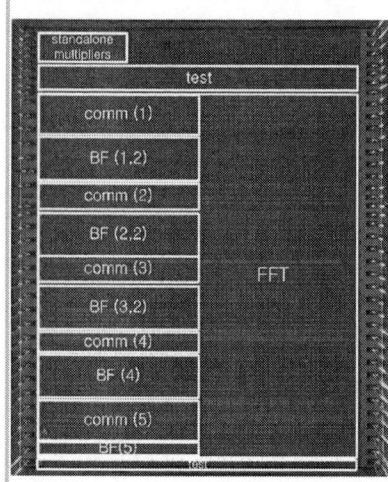

Type	1024pts complex
Vdd	270mV
Clock Freq.	30MHz
Energy / FFT	17.7nJ
Throughput	234k FFT/sec
Area	2.66x3.12mm^2
Technology	65nm CMOS

Figure 19.6.7: Die photo of the FFT core implemented in 65nm CMOS with summary table.

ISSCC 2011 / SESSION 20 / HIGH-SPEED TRANSCEIVERS & BUILDING BLOCKS / OVERVIEW

Session 20 Overview / Wireline

High-Speed Transceivers & Building Blocks

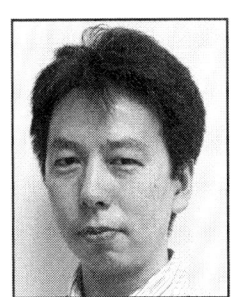

Session Chair: *Jae-Yoon Sim, Pohang University of Science and Technology, Pohang, Korea*

Session Co-Chair: *Masafumi Nogawa, NTT, Atsugi, Japan*

Transceivers designed for very high-speed wireline communication must contend with significant channel loss, crosstalk, and reflections by employing various equalization techniques in the transmitter and receiver. Confronting these challenges becomes even more difficult as data rates increase beyond 10Gb/s and designs become power-constrained. The first four papers in this session describe transceivers that address these concerns. The remaining papers describe key building blocks for next-generation transceivers with a focus on various receive equalizer adaptation techniques and spread-spectrum clock generation for EMI reduction.

Paper 20.1 from Fujitsu demonstrates a 10.3Gb/s transceiver in 90nm CMOS that achieves an adaptation for both of phase and amplitude distortions for the first time. It combines an analog linear equalizer, a decision-feedback equalizer (DFE), and a driver with pre-emphasis to adaptively compensate up to 41dB channel loss.

Paper 20.2 from LSI Corporation presents a multimedia transceiver implemented in 40nm CMOS. It proposes a new baseline wander correction scheme with a linear equalizer, a 10-tap DFE, and a 4-tap feed-forward equalizer (FFE) for operation over a wide range of data rates from 1.0625 to 14.025Gb/s.

Texas Instruments and Arda Technologies describe their latest transceiver in paper 20.3. This SerDes supports a data rate of 16Gb/s and uses a 14-tap DFE and analog equalizer along with an enhanced-swing voltage-mode driver to generate a 1.2V differential output to compensate 34dB of channel loss.

In Paper 20.4, SnowBush-Gennum demonstrates a 4-lane multistandard-compliant transceiver that supports data rates ranging from 1 to 12Gb/s to satisfy standards including PCIe, SATA, and 1-to-10Gb/s Ethernet without using on-chip inductors.

Paper 20.5 from University of Toronto and Fujitsu Laboratories describes an adaptive engine for a 6Gb/s DFE for a 2× blind ADC-based receiver. This engine digitally extracts the optimum coefficients for the DFE irrespective of the sampling phase of the blind clock.

Paper 20.6 from National Taiwan University presents a 6Gb/s receiver in 90nm CMOS with an adaptive IIR-based DFE compensating a channel loss of 32.7dB.

Yonsei University details a new receive equalizer adaptation technique using asynchronous-sampling histograms in Paper 20.7. The equalizer, implemented in 0.13μm CMOS, demonstrates successful adaptation at 5.4Gb/s over various lossy channels.

Finally, Paper 20.8 from Korea University demonstrates a 3.5GHz spread-spectrum clock generator in 0.13μm CMOS with a memoryless nonlinear Newton-Raphson modulation profile achieving an EMI reduction of 19.14dB.

978-1-61284-303-2/11 $26.00 © 2011 IEEE

ISSCC 2011 / February 23, 2011 / 8:30 AM

Presenters:

20.1 A 4-Channel 10.3Gb/s Transceiver with Adaptive Phase 8:30 AM
       Equalizer for 4-to-41dB Loss PCB Channel

Y. Hidaka,

Fujitsu Laboratories of America, Sunnyvale, CA

20.2 A 1.0625-to-14.025Gb/s Multimedia Transceiver with Full-rate Source-Series-Terminated Transmit 9:00 AM
       Driver and Floating-Tap Decision-Feedback Equalizer in 40nm CMOS

S. Quan,

LSI, Milpitas, CA

20.3 Analog-DFE-Based 16Gb/s SerDes in 40nm CMOS That Operates Across 34dB Loss Channels 9:30 AM
       at Nyquist with a Baud Rate CDR and 1.2V$_{pp}$ Voltage-Mode Driver

A. K. Joy,

Texas Instruments, Northampton, United Kingdom

20.4 An 8.4mW/Gb/s 4-Lane 48Gb/s Multi-Standard-Compliant 10:15 AM
       Transceiver in 40nm Digital CMOS Technology

M. Ramezani,

Snowbush-Gennum, Toronto, Canada

20.5 A Pattern-Guided Adaptive Equalizer in 65nm CMOS 10:45 AM

S. Shahramian,

University of Toronto, Toronto, Canada

20.6 A 6Gb/s Receiver with 32.7dB Adaptive DFE-IIR Equalization 11:15 AM

Y.-C. Huang,

National Taiwan University, Taipei, Taiwan

20.7 A 5.4Gb/s Adaptive Equalizer Using Asynchronous-Sampling Histograms 11:45 AM

W-S. Kim,

Yonsei University, Seoul, Korea

20.8 A 0.076mm^2 3.5GHz Spread-Spectrum Clock Generator with 12:00 PM
       Memoryless Newton-Raphson Modulation Profile in 0.13μm CMOS

S. Hwang,

Korea University, Seoul, Korea

20

ISSCC 2011 / SESSION 20 / HIGH-SPEED TRANSCEIVERS & BUILDING BLOCKS / 20.8

20.8 A 0.076mm² 3.5GHz Spread-Spectrum Clock Generator with Memoryless Newton-Raphson Modulation Profile in 0.13µm CMOS

Sewook Hwang, Minyoung Song, Young-Ho Kwak, Inhwa Jung, Chulwoo Kim

Korea University, Seoul, Korea

A spread-spectrum clock generator (SSCG) is a cost-effective solution to reduce EMI, which has become a serious problem in high-speed systems. In applications such as serial links, display drivers and consumer electronics, SSCG is essential or strongly recommended. Control options such as frequency deviation (δ) and modulation frequency (f_m) help to satisfy these demands. Furthermore, efficient modulation-profile generation is critical for achieving further EMI reduction and lowering fabrication cost. PLL-based SSCGs are reported in [1-3]. The ΔΣ modulator (ΔΣM) controls the division ratio [1,2] and the phase information of phase detector (PD) [3] to generate a spread-spectrum clock. The self-referenced clock generator uses a capacitor array to generate a spread-spectrum clock [7]. However, they do not have a way to control δ and f_m. Dual-loop direct VCO modulation [4] and digital period synthesizer with delay-line [5] are able to control δ and f_m. However, [4] requires an additional VCO, which increases the power consumption by 2×, and [5] suffers from a large deterministic jitter through the delay-line and logic circuits. A triangular profile is commonly used in many SSCGs [1,3,4]. Though the implementation of a triangular profile is very simple, its performance is poor. The chaotic PAM modulation in [2] requires complex analog circuits. Recently, a piecewise-linear profile with SRAM was presented in [5]. However, the additional memory consumes a large amount of power and occupies a large area. This paper presents a frequency-locked loop (FLL) based SSCG with frequency-to-voltage converter (FVC) [6], that saves area and provides multiple δ with low bandwidth variation. A memoryless Newton-Raphson modulation profile with multiple f_m is also described.

Figure. 20.8.1 shows the overall architecture of our SSCG. The SSCG is composed of an FLL and a digital spread-spectrum controller (DSSC). Two pulse generators, PG1 and PG2 make five control signals respectively. FVC1 and FVC2 convert the frequency information to voltage, V_{FVC1} and V_{FVC2}, respectively. The charge pump (CP) generates V_{CTRL} from the difference of these two voltages. The loop filter comprises a capacitor of tens of picofarad, sufficient to secure stability [6] and occupying a small area. DAC1 and DAC2 connect an interface between analog FLL and DSSC for frequency modulation. The Newton-Raphson profile generator produces a nonlinear profile with various modulation frequencies of f_m, $2f_m$ and $3f_m$. The 1-1-1 MASH ΔΣM transfers the modulation profile to the δ controller. The δ controller decodes the output of ΔΣM, DSM[3:0] and expands the control signal to control the spreading direction of up and down and spreading ratio (±0.5 to 3.5% with Δ=0.5%) of the output clock CLK_{OUT}.

Figure 20.8.2 shows the schematic of the proposed frequency modulation technique that uses a double binary-weighted DAC and FVC. The FVC detects the frequency with current source I_{M1} and the S/H circuit. The current source charges C_1, and during the charge sharing period, C_2 follows the sampling voltage level of C_1 and finally output voltage is settled to $I_{M1}/(C_1f)$, where f is the frequency of CLK_{FDIV}. If I_{M1} can be modulated with the DAC, the VCO output frequency can also be modulated. The double binary-weighted DAC has eight groups of current cells and 172 always-turned-on current cells. Each group has binary-weighted current cells but also seven groups, $1×G_0$, $2×G_1$ and $4×G_2$, except for a dotted line group that are built up as a binary-weight. Thus, it is possible to have seven current levels for the ΔΣ modulation and seven current step differences for the multiple δ. Each current cell has the same amount of current, I_{UNIT}, with bias voltage V_{BN}. When the SSCG operates with δ=-3.5%, SEL2[8:0] modulates the current of DAC2 from $172I_{UNIT}$ to $221I_{UNIT}$ with a step of $7I_{UNIT}$. Then, the SEL1[8:0] and SEL1_UP[2:0] fix the DAC1 current to $200I_{UNIT}$, hence 193/200-1=-3.5% of down spreading with 1-1-1 MASH ΔΣM is achieved. In the case of up spreading, δ=+3.5%, SEL2_UP[2:0] increases the DAC2 current by $7I_{UNIT}$ so that 207/200-1=+3.5% of up spreading can also be achieved. The op-amp A_1 is used in the CP for more precise matching of sink and source currents.

If f_m falls below the resolution bandwidth (RBW), a nonlinear profile, so-called Hershey-kiss, is an optimal modulation profile to achieve maximum EMI reduction. In addition, as f_m approaches RBW, peak amplitude can further increase [5]. Figure 20.8.3 shows the overall architecture of the DSSC and the measured modulation profiles. The profile generator applies the Newton-Raphson method for finding the square root to generate the nonlinear profile. The Newton-Raphson method generally requires multiple iterations to find a precise value. However, by selecting an appropriate starting point of the iteration, it is possible to get a square root value without iteration. The simple first-order function ((1/a)x+b) is added to the Newton-Raphson method to set up this starting point so that the formula can show the Hershey-kiss profile. The coefficient selector (CS) contains coefficients of the formula. In the implementation, numbers are rounded down to the nearest integer. A[4:0] determines the profile resolution, B[3:0] selects the profile slope and S[5:0] scales the profile maximum bit width. If A[4:0] and S[5:0] are multiple of 2 for simple implementation using shifters then we can eliminate the complex multiplications. The profile counter (PC) generates an input, X[9:0], of the square-root approximation (SRA) block. If we change the step size, STEP[1:0], of X[9:0] to 1, 2 and 3, then f_m will be changed to f_m, $2f_m$ and $3f_m$. The f_m is targeted to 31kHz with 0.5% down spreading. The SRA calculates a nonlinear shape Y[14:0]. Finally, the profile switch (PS) turns half of Y[14:0] upside down so that the Newton-Raphson nonlinear profile is generated without any memory. Figure 20.8.3 shows the measured modulation profiles of the SSCG. The top and bottom profiles are measured with δ=+2%, STEP[1:0]=2'b11 and δ=-3.5%, STEP[1:0]=2'b01, respectively. They show the exact frequency spreading of +2% and -3.5%. The top and bottom profile have f_m values of 95kHz and 31kHz, respectively. The DSSC consumes 400µW with maximum f_m.

Figure 20.8.4 shows the measured result of EMI reduction of various δ and f_m. The EMI is further reduced as |δ| and f_m increase, as expected in [5]. In the up-spreading case, EMI is reduced by an additional 0.2 to 1dB, because f_m is closer to RBW, 100kHz. The spectra of SSCG output clocks with δ=±2% are shown in Fig. 20.8.5. The spectra show harmonics of f_m, because f_m is much lower than RBW. Figure 20.8.6 is a comparison table with conventional SSCGs. This SSCG has a substantial EMI reduction and occupies small area of 0.076mm², considering the process technology with various δ and f_m. The power consumption of the SSCG is 23.72mW at 3.5GHz excluding I/O. Figure 20.8.7 shows the die micrograph with floor plan.

Acknowledgement :
This research was supported by the Ministry of Knowledge Economy, Korea, under the University ITRC support program supervised by the National IT Industry Promotion Agency (NIPA-2010-C1090-1001-0003). We would like to thank IDEC for supporting chip fabrication.

References :
[1] S.Y. Lin and S.I. Liu, "A 1.5GHz All-Digital Spread-Spectrum Clock Generator," *IEEE JSSC*, vol. 44, pp. 3111-3119, Nov. 2009.
[2] F. Pareschi *et al.*, "A 3 GHz Spread Spectrum Clock Generator for SATA Applications Using Chaotic PAM Modulation," *Proc. IEEE CICC*, pp. 451-454, Sep. 2008.
[3] W. Grollitsch *et al.*, "A 1.4ps$_{rms}$-Period-Jitter TDC-Less Fractional-N Digital PLL with Digitally Controlled Ring Oscillator in 65nm CMOS," *ISSCC Dig. Tech papers*, Feb. 2010.
[4] C.D. LeBlanc *et al.*, "Dual-Loop Direct VCO Modulation for Spread Spectrum Clock Generation," *Proc. IEEE CICC*, pp. 479-482, Sep. 2009.
[5] D.D. Caro *et al.*, "A 1.27GHz, All-Digital Spread Spectrum Clock Generator/Synthesizer in 65nm CMOS," *IEEE JSSC*, vol. 45, pp. 1048-1060, May 2010.
[6] H.T. Bui *et al.*, "Design of a High-Speed Differential Frequency-to-Voltage Converter and Its Application in a 5GHz Frequency-Locked Loop," *IEEE Trans. Circuits Syst. I*, vol. 55, pp. 766-774, Apr. 2008.
[7] M.S. McCorquodale *et al.*, "A 0.5-to-480MHz Self-Referenced CMOS Clock Generator with 90ppm Total Frequency Error and Spread-Spectrum Capability," *ISSCC Dig. Tech papers*, Feb. 2008.

978-1-61284-303-2/11 $26.00 © 2011 IEEE

ISSCC 2011 / February 23, 2011 / 12:00 PM

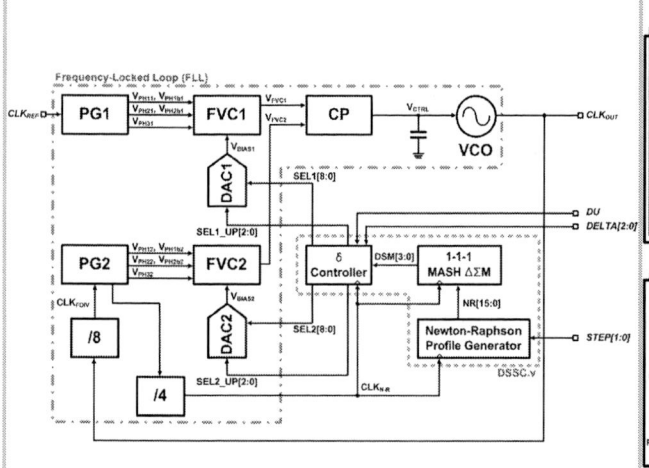

Figure 20.8.1: Block diagram of the SSCG.

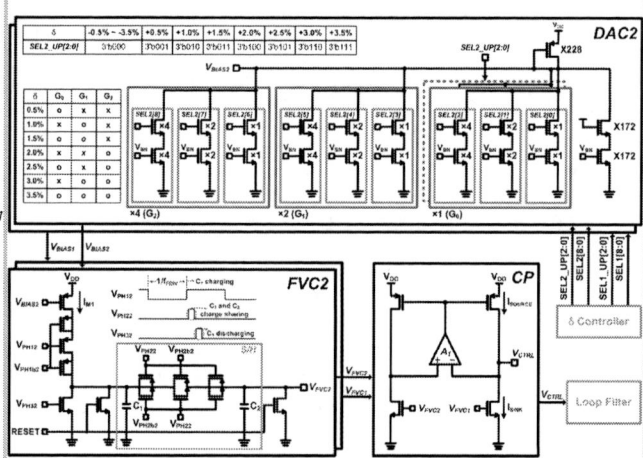

Figure 20.8.2: Schematics of the proposed double binary-weighted DAC for multiple δ, the frequency-to-voltage converter and charge pump.

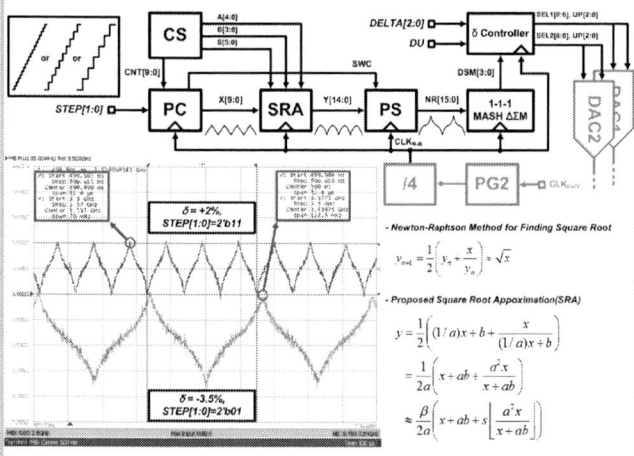

- Newton-Raphson Method for Finding Square Root

$$y_{n+1} = \frac{1}{2}\left(y_n + \frac{x}{y_n}\right) \approx \sqrt{x}$$

- Proposed Square Root Approximation (SRA)

$$y = \frac{1}{2}\left((1/a)x + b + \frac{x}{(1/a)x+b}\right)$$
$$= \frac{1}{2a}\left(x + ab + \frac{a^2x}{x+ab}\right)$$
$$\approx \frac{\beta}{2a}\left(x + ab + s\left\lfloor\frac{a^2x}{x+ab}\right\rfloor\right)$$

Figure 20.8.3: The digital spread-spectrum controller with multiple f_m, 1-1-1 MASH ΔΣM with δ controller and the measured modulation profiles of the SSCG.

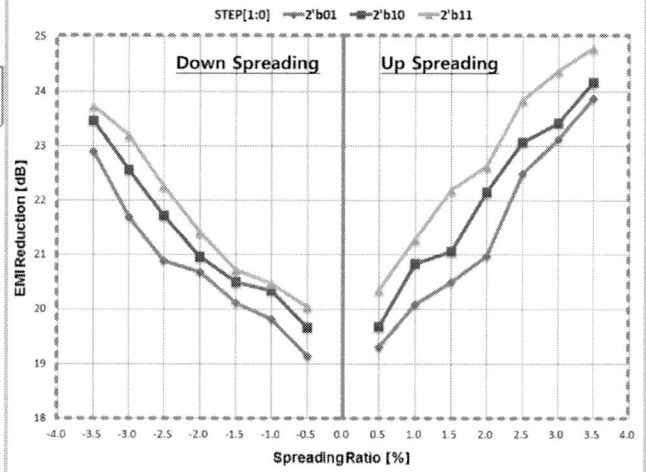

Figure 20.8.4: The measured EMI reduction of every frequency deviations and modulation frequencies.

- Measurement Condition

1. Center Frequency : 3.5GHz
2. Frequency Span : 350MHz
3. RBW : 100kHz
4. VBW : 100kHz
5. Attenuation : 20dB
6. Ref : 10dBm
7. Vertical Scale : 10 dB/div

Figure 20.8.5: The measured spectra with ±2% frequency deviations.

	Unit	This work	[1]	[2]	[3]	[4]	[5]
Modulation Method	-	ΔΣ-FLL, Newton-Raphson	ΔΣ-ADPLL, Triangular	ΔΣ-PLL, Chaotic	ADPLL, Triangular	Dual-loop Direct VCO, Triangular	AD-Delay Line, Piecewise
Operating Frequency	GHz	3.5	1.5	3	0.19~4.27	1	0.18~1.27
Modulation Frequency	kHz	31, 62, 92 with δ=-3.5% 32, 64, 95 with δ=+3.5%	31	33	-	33	3~703 @ 180MHz 19~4961 @ 1.27GHz
EMI Reduction	dB	19.14 ~ 23.73	10.48	14.5	18* **	8 ~ 12	5.4 ~ 20.5
Spreading Ratio	%	±0.5 ~ 3.5 (Δ = 0.5)	-0.5	-0.5	-0.5* **	-0.5 ~ 3 (Δ = 0.5)	0 ~ -50 (Δ = 0.2)
RMS Jitter	ps	2.44	4	5.9	1.4**	-	-
Peak-to-Peak Jitter	ps	16.15	28.4	-	15**	-	93***
Technology	nm	130	180	130	65	65	65
Area	mm²	0.076	0.20	0.210	0.046	0.129	0.044
Power	mW	23.72 @ 3.5GHz	15 @ 1.5GHz	14.7 @ 3GHz	15.94 @ 4.27GHz	14 @ 1GHz	44 @ 1.27GHz

* Estimated in the spectra figure
** Measured at 3.0GHz
*** Measured at 1.25GHz

Figure 20.8.6: Performance comparison with conventional works.

978-1-61284-303-2/11 $26.00 © 2011 IEEE

ISSCC 2011 PAPER CONTINUATIONS

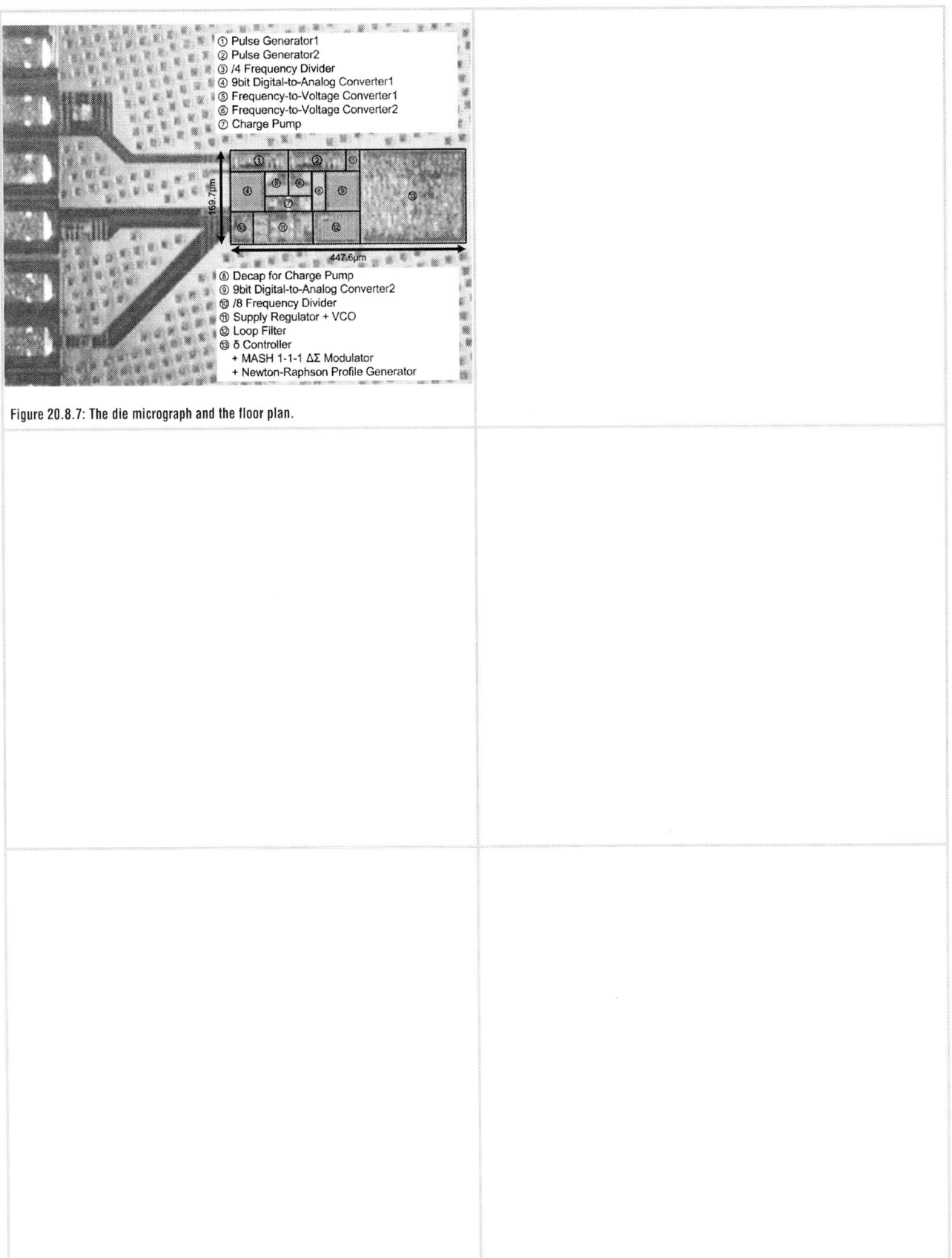

① Pulse Generator1
② Pulse Generator2
③ /4 Frequency Divider
④ 9bit Digital-to-Analog Converter1
⑤ Frequency-to-Voltage Converter1
⑥ Frequency-to-Voltage Converter2
⑦ Charge Pump

⑧ Decap for Charge Pump
⑨ 9bit Digital-to-Analog Converter2
⑩ /8 Frequency Divider
⑪ Supply Regulator + VCO
⑫ Loop Filter
⑬ δ Controller
   + MASH 1-1-1 ΔΣ Modulator
   + Newton-Raphson Profile Generator

Figure 20.8.7: The die micrograph and the floor plan.

ISSCC 2011 / SESSION 20 / HIGH-SPEED TRANSCEIVERS & BUILDING BLOCKS / 20.1

20.1 A 4-Channel 10.3Gb/s Transceiver with Adaptive Phase Equalizer for 4-to-41dB Loss PCB Channel

Yasuo Hidaka[1], Takeshi Horie[2], Yoichi Koyanagi[2], Takashi Miyoshi[2], Hideki Osone[1], Samir Parikh[1], Subodh Reddy[1], Toshiyuki Shibuya[1], Yasushi Umezawa[2], William W. Walker[1]

[1]Fujitsu Laboratories of America, Sunnyvale, CA,
[2]Fujitsu Laboratories, Kawasaki, Japan

In multi-Gb/s wireline communications, equalizers are used to compensate for channel-induced signal distortion in order to stretch the maximum distance of transmission. Both amplitude and phase can be distorted in a channel. Amplitude distortion is a frequency-dependent attenuation due to skin effect and dielectric loss, causing inter-symbol-interference (ISI). A transmitter (TX) discrete-time pre-emphasis (PE) filter, a receiver (RX) continuous-time Linear Equalizer (LE), and an RX Decision-Feedback Equalizer (DFE) are generally used to cancel ISI. At 10Gb/s or higher data rate, equalizers for up to 33 to 39dB Nyquist loss and up to 20 to 25dB adapted loss range were reported [1–3]. On the other hand, how to compensate phase distortion is not clearly understood in practical circuit design. Theoretically, if a channel has *minimum-phase-like* characteristics, phase distortion is automatically co-equalized with amplitude distortion by a *minimum-phase* equalizer [4]. While this is the case for high-speed cables [5], it is not for PCB traces, because a *non-minimum-phase* equalizer, e.g., a PE with a pre-cursor tap, produces lower BER over a high-loss PCB channel than a *minimum-phase* equalizer, e.g., a PE without a pre-cursor tap. Thus the IEEE 10Gb Ethernet standard for backplanes adopted 3-tap PE with a pre-cursor tap [6]. However, adaptive phase equalization in hardware has not been reported in the literature.

In this paper, a 4-channel 1.25 to 12.5Gb/s transceiver macro with both phase and amplitude adaptive equalization is presented. It will be shown that when loss at $f_S/2$ exceeds 20dB, phase distortion measurably closes eyes and increases BER. The proposed transceiver cancels 4.1 to 41.7dB loss for 10.3Gb/s or 4.9 to 34.9dB loss for 12.5Gb/s when both phase and amplitude adaptive equalization are enabled, extending both the length and the range of channels that can be adapted compared to when only amplitude adaptive equalization is used.

Figure 20.1.1 shows a block diagram of the transceiver. The TX uses a CML driver with 3-tap pre-emphasis (PE). The RX has an analog continuous-time LE whose output is sampled by a Data & Error detector with a 1-tap speculative DFE. The data values are sampled around the center of the eye, and the error values at transitions. After de-multiplexing, the data and error values are used for PE, LE, and DFE controls as well as clock recovery.

Figure 20.1.2 shows the RX front end circuit. The LE is a 2-stage differential buffer with capacitive and resistive source degeneration. Resistive degeneration in the first stage is controlled by the *LEGain* parameter which sets the DC gain and gain at $f_S/4$ relative to DC. Bandwidth of the LE is around $f_S/4$. The Data Detector is a 2-way interleaved decision circuit with a 1-tap speculative DFE. The Error Detector also uses speculation: if there is a data transition, *PHEn* (phase error) is selected, otherwise *LVEn* (level error) is selected. In contrast with [2], the comparator circuit (Fig. 20.1.2 inset) drives each input differential pair with one data signal and one reference signal to improve the sensitivity for high reference levels. The common-mode level of the reference signal is adjusted to match the LE output.

Figure 20.1.3 shows a block diagram of the sign-based Zero-Forcing (S-ZF) LE gain control [2] with extensions for PE and phase control. *ResISIs* at four time steps generated by filter pattern decoders (FPDs) for ISI [2] are multiplied with weight constants and integrated to give *QRGain* which represents the gain of LE and PE at $f_S/4$ relative to DC. Convergence forces the weighted sum of average *ResISIs* to zero. A single control loop is shared between LE and PE to equalize their strengths and avoid coupling between two similar control loops. *QRGain* is translated to *LEGain* and *PEGain* by table look-up, with anti-dithering embedded in the *PEGain* table. Block FPD for Phase Distortion (described below) generates *ResPD*, which is integrated to *PhaseEQ*, the required phase equalization. Convergence forces *ResPD* to zero. Here, we define *PEGain* and *PhaseEQ* as $C(0)/\{C(0)+C(-1)+C(1)\}$ and $-C(-1)/C(0)$, respectively. $C(0)$, $C(-1)$, and $C(1)$ are derived from *PEGain* and *PhaseEQ* using constraints on signs and a constant sum of magnitude.

Figure 20.1.4 shows detection of phase distortion by an FPD. Phase distortion is detected by comparing phase error at 000E111 and 100E110 patterns. Here, E denotes the location of a rising edge. (Inverted patterns with falling edges are used as well, but omitted here for simplicity.) If phase error is late/early at 000E111 and early/late at 100E110, phase *ResPD* is assigned +1/-1, indicating insufficient/excessive phase equalization. The FPD performs this comparison accurately by alternately checking each filter pattern for exactly the same number of times [2]. In addition, clock recovery uses phase error information only at a 00E11 pattern, because 1) the amplitude of 10E10 pattern may be too small, 2) the phase of 10E10 pattern may not be aligned with other low-frequency patterns, 3) locking clock phase on 00E11 pattern improves sensitivity to detect phase distortion with the above scheme.

The RX has built-in test circuitry allowing measurement of eyes at the DFE input. Figure 20.1.5 shows PRBS11 eyes and periodic waveforms measured at the DFE input for a 32.0dB loss PCB channel at 10.3Gb/s with best amplitude but various fixed phase equalizations. With minimum phase equalization (a and b), 4UI periodic pattern (00110011) comes later than 8UI (00001111) and 16UI (0*8/1*8) patterns. With optimal phase equalization (c and d), 4, 8, and 16UI patterns are aligned, and the eye has the thinnest trace width. With excessive phase equalization (e and f), patterns are misaligned again in the opposite order. Here, time lines are aligned between diagrams. As phase equalization increases, 4, 8, and 16UI patterns are delayed at different rates, but a 2UI periodic pattern (0101) is not delayed at all, because its pre-cursor tap is the same as its post-cursor tap. As a result, with optimal phase equalization, the 0101 pattern is out of phase with the other patterns. This is not a problem for data recovery using a DFE, because DFE recovers the Nyquist frequency component.

Figures 20.1.6(a) and (b) show TX output eye and jitter at 10.3Gb/s. Measured TJ was 10.2ps at BER 10^{-12}. To evaluate circuit performance, equalizer control was partially replaced with equivalent software and hardware which use real-time data measured with actual equalizer circuits and channel. Actual TX and RX circuits and PCB channels were used in the evaluation. Figure 20.1.6(c) and (d) show adaptation results and channel total insertion loss at $f_S/2$ for 10.3Gb/s and 12.5Gb/s. Test channels are up to 95" Nelco 4000-13SI stripline traces or up to 30" FR4 backplane with 2 Amphenol XCede connectors. Channel insertion loss includes the evaluation board, but not the chip package. Figure 20.1.6(e) shows measured (when > 10^{-14}) or estimated BER for PRBS31 after adaptation with three schemes: zero-forcing phase distortion (ZPD, the proposed scheme), zero-forcing pre-cursor ISI (ZISI), and minimum phase (MNP, no pre-cursor tap). With ZPD, target BER < 10^{-12} was achieved for 4.1 to 41.7dB or 4.9 to 34.9dB total channel loss at $f_S/2$ for 10.3Gb/s or 12.5Gb/s, respectively. ZISI was unstable, when channel-induced pre-cursor ISI as well as phase distortion is too low or too high. MNP had lower performance than ZPD, when channel has phase distortion. Figure 20.1.6(f) shows learning curves for 95" Nelco 4000-13SI stripline at 10.3Gb/s.

Figure 20.1.7 is a die micrograph. The macro integrates an LC-PLL and 4 channels of TX and RX, and occupies 3.03mm² in a 90nm CMOS technology. The TX and RX occupy 0.206mm² and 0.214mm², and consume 181mW and 135mW at 10.3Gb/s and 192mW and 156mW at 12.5Gb/s per channel from a 1.2V power supply, respectively.

References:
[1] Massimo Pozzoni, et al., "A 12Gb/s 39dB Loss-Recovery Unlocked-DFE Receiver with Bi-dimensional Equalization," *ISSCC Dig. Tech. Papers*, pp.164-165, Feb. 2010.
[2] Yasuo Hidaka, et al., "A 4-Channel 1.25–10.3Gb/s Backplane Transceiver Macro With 35dB Equalizer and Sign-Based Zero-Forcing Adaptive Control," *IEEE J. Solid-State Circuits*, vol. 44, no. 12, pp.3547-3559, Dec. 2009.
[3] John Bulzacchelli, et al., "A 10-Gb/s 5-Tap DFE/4-Tap FFE Transceiver in 90-nm CMOS Technology," *IEEE J. Solid-State Circuits*, vol. 41, no. 12, pp. 2885-2900, Dec. 2006.
[4] J. W. M. Bergmans, "Digital Baseband Transmission and Recording," Kluwer Academic Publishers, 1996.
[5] Yasuo Hidaka, et al., "Gain-Phase Co-Equalization for Widely-Used High-Speed Cables," *VLSI Circuits, Dig. Tech. Papers*, pp.194-197, June 2005.
[6] IEEE Standard 802.3-2008, Clause 72, "Physical Medium Dependent Sublayer and Baseband Medium, Type 10GBASE-KR," pp. 417-451, 2008.

978-1-61284-303-2/11 $26.00 © 2011 IEEE

ISSCC 2011 / February 23, 2011 / 8:30 AM

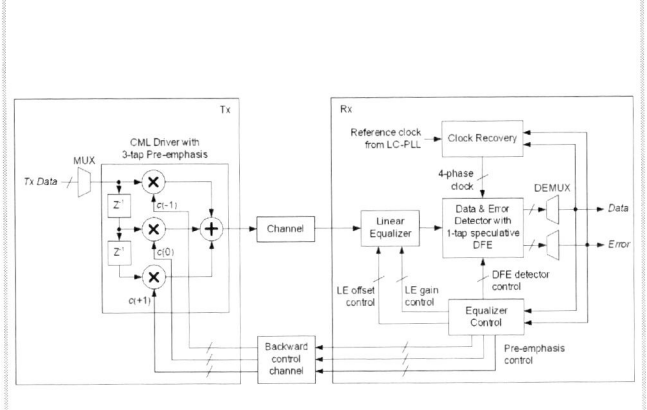

Figure 20.1.1: Transceiver block diagram.

Figure 20.1.2: Receiver front end circuit.

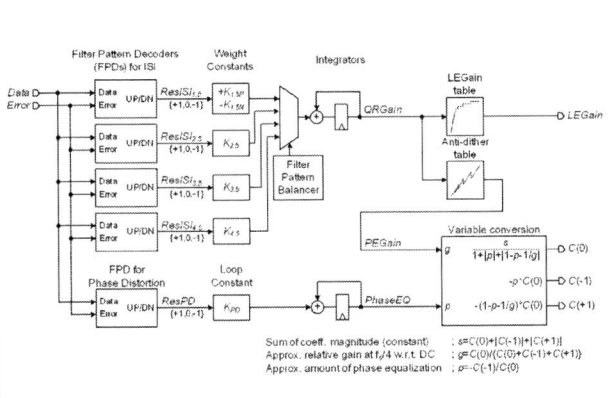

Figure 20.1.3: Block diagram of LE gain and pre-emphasis control.

Figure 20.1.4: Detection of phase distortion by filter pattern decoder for adaptive phase equalization.

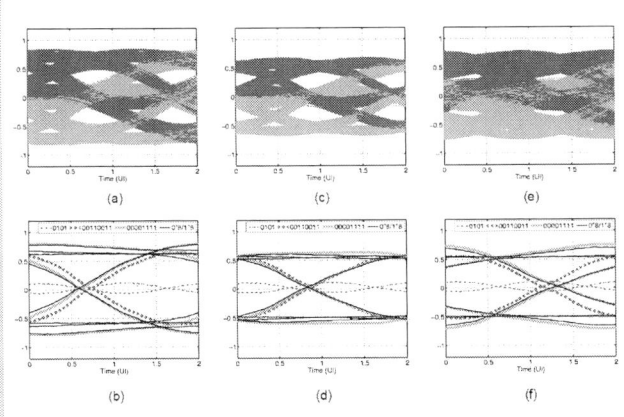

Figure 20.1.5: Measured wave forms at 1-tap DFE input with best amplitude and (a,b) minimum (c,d) optimal or (e,f) excessive phase equalization.

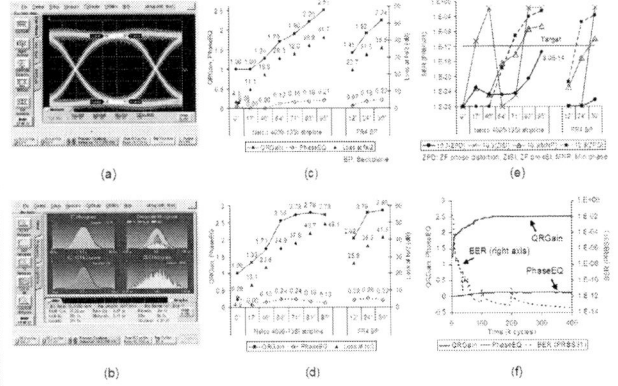

Figure 20.1.6: Tx output (a) eye and (b) jitter at 10.3Gb/s. Adaptation results at (c) 10.3 and (d) 12.5Gb/s. (e) BER after adaptation. (f) Learning curves.

978-1-61284-303-2/11 $26.00 © 2011 IEEE

ISSCC 2011 PAPER CONTINUATIONS

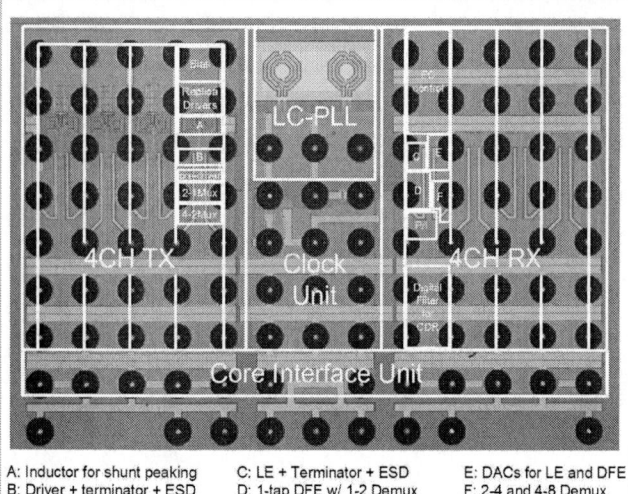

A: Inductor for shunt peaking   C: LE + Terminator + ESD   E: DACs for LE and DFE
B: Driver + terminator + ESD   D: 1-tap DFE w/ 1-2 Demux   F: 2-4 and 4-8 Demux

Figure 20.1.7: Micrograph of the transceiver macro.

ISSCC 2011 / SESSION 20 / HIGH-SPEED TRANSCEIVERS & BUILDING BLOCKS / 20.2

20.2 A 1.0625-to-14.025Gb/s Multimedia Transceiver with Full-rate Source-Series-Terminated Transmit Driver and Floating-Tap Decision-Feedback Equalizer in 40nm CMOS

Shaolei Quan, Freeman Zhong, Wing Liu, Pervez Aziz, Tai Jing, Jen Dong, Chintan Desai, Hairong Gao, Monica Garcia, Gary Hom, Tony Huynh, Hiroshi Kimura, Ruchi Kothari, Lijun Li, Cathy Liu, Scott Lowrie, Kathy Ling, Amaresh Malipatil, Ram Narayan, Tom Prokop, Chaitanya Palusa, Anil Rajashekara, Ashutosh Sinha, Charlie Zhong, Eric Zhang

LSI, Milpitas, CA

A robust transceiver designed for NRZ signaling beyond 10Gb/s over long-range physical media (including electrical backplanes, copper cables and optical modules) must contend with significant challenges from insertion loss, crosstalk, and reflection. For inter-symbol interference (ISI) cancellation, half-rate decision-feedback equalizer (DFE) with unrolled first tap [1-3] is widely used to avoid noise amplification and to relax timing for data sampling/feedback. However, tap-unrolling increases slicer count and entails half-rate multiplexers eating into timing margin. To remove reflection-induced ISI due to impedance discontinuities in the media, the DFE must cover tap positions higher than 30UI which is beyond tap range of the DFEs reported in previous work [1-4]. Fig. 1 shows an example of bumpy channel with reflection energy up to 40UI.

In order to work with a wide range of physical media, the transceiver presented in this paper employs a transmit (Tx) feed-forward equalizer (FFE), and a receive (Rx) linear equalizer (LE) and DFE to equalize up to 26 dB insertion loss at 14.025Gb/s. It supports baud rates from 1.0625Gb/s to 14.025Gb/s (8.5~14.025Gb/s in full rate).

Key contributions of this work are: 1) a 14.025Gb/s half-rate 10-tap DFE which employs direct feedback for the first tap, and four floating taps with positions independently adapted to any four out of the thirty-two tap positions ranging from 7 to 38, 2) a circuit that corrects for the baseline wonder (BLW) introduced by the on-chip Rx AC coupling, 3) a 14.025Gb/s source-series terminated (SST) Tx driver with 4-tap FFE, and 4) a calibrated dual-band LC PLL with tuning range of 9.5~15.5GHz over PVT.

A block diagram for the Rx is shown in Fig. 20.2.1. The use of on-chip AC coupling capacitors allows the input common-mode voltage of the variable gain amplifier (VGA) to be set independently of the common-mode at the package pins, thus optimizing the linearity and bandwidth of the VGA. The linear equalizer (LE) provides up to 10dB boost at 7GHz, and the DFE removes the remaining post-cursor ISI including that caused by reflection. The VGA gain, LE boost, DFE tap weights and floating tap positions are adapted simultaneously through the SS-LMS algorithm. For backplanes with severe crosstalk, the LE boost can be capped in adaptation to mitigate noise enhancement. The sample clocks are generated by a quadrature-output phase rotator, whose phase is controlled by a digital Bang-Bang (BB) phase detector (PD). Eye asymmetry resulting from residual precursor ISI is compensated by adjusting the phase difference between transition and data sampling clock outputs from the phase rotator.

The BLW effect is caused by the high-pass filter (HPF) consisting of the on-die AC coupling capacitors Cap/Can and common-mode feedback resistors Rap/Ran that attenuates Rx input signal at low frequency. For example, for a PRBS31 pattern at 14.025Gb/s, a HPF pole of 200KHz reduces the vertical eye opening by 18%. The proposed circuit (Fig. 20.2.2) corrects for the BLW effect by low-pass filtering the recovered symbols and adding the outcome to the signal at the HPF output. This is done specifically by first summing 32 consecutive recovered bits in a 5-bit accumulator after 32:1 decimation, and then using accumulator output to drive a resistively-loaded amplifier that produces a differential analog voltage (vcmop − vcmon) which is low-pass filtered by Cap/Can and Rap/Ran at the input to the VGA. In one measurement (Fig. 20.2.2) horizontal eye opening improves by 0.05UI at 14.025Gb/s with proposed scheme.

The proposed 10-tap half-rate floating-tap DFE (Fig. 20.2.3) employs direct feedback for the first tap [4], dynamic feedback for tap positions 2~6, and 4:2 multiplexed feedback for the remaining four taps which have adapted tap positions ranging from 7 to 38. To cancel first-tap (H1) ISI at 14.025Gb/s, the correction

signal originated from one data slicer must settle at input of the other data slicer within 65ps. This is achieved by optimizing the sense amplifier (SA) design for high speed, tapping H1 feedback directly at SR latch input in SA (in [5] SR latch input is tapped for 2nd-tap feedback in a quarter-rate DFE), and using a dedicated summation stage for H1 feedback (H1-SS) to reduce feedback latency; the H1-SS also isolates kickback noise from SA regeneration, and has 3dB gain to serve as SA preamplifier. The SA design (Fig. 20.2.3) has a "frequency doubler" input stage and only three devices cascoded from power to ground. Combined with the H1-SS, SA input sensitivity is better than 20mV for 7.0125GHz half rate sampling rate. DFE feedback for floating-tap positions is generated by delaying sliced data with a 32-bit shift register after the 2:4 DEMUX, and sending the post-cursor data bits to a summation stage via a 32:4 MUX and data sequencer. For floating tap positions less than 11, sliced data is sent to the data sequencer input directly with the 32-bit shift register and 32:4 MUX bypassed. As shown in Fig. 20.2.1, DFE with 6 fixed taps and 4 floating taps has vertical eye opening improved by 100% over a 10-fixed-tap DFE in system simulation.

Figure 20.2.4 shows the Tx block diagram and proposed SST driver. Source capacitive degeneration and AC-coupled input remove duty-cycle distortion (DCD) from the clock path. A DCD correction DAC with sub-ps resolution corrects residual DCD on 2:1 MUX clock using closed-loop calibration. A driver unit has a single transistor switch and a single resistor for pulling up/down; a pre-driver handles 2:1 multiplexing to avoid clock level-shifting, and to reduce drive size and the number of cascaded stages from power to ground compared with previous work [6]. A driver unit has three states: driving high, driving low, and high-impedance output; two units in the same half driver cancel each other in differential output by driving complementary data bits. Proposed 4-tap FFE divides driver units into five groups (besides a constant-on unit) dedicated to data bits of different phases d[k-2], d[k-1], d[k], inverted d[k] , and d[k+1], with all the units dynamically switching. Post-emphasis C(-2) of weight KM2 switches 2*KM2 among NM2 units to d[k-2] from d[k], and switches KM2 units from inverted d[k] to d[k], thereby keeping C(0) unchanged. Emphasis for the other taps are done in similar way to achieve independent tuning for FFE taps. The driver has step size of 14.6mV in differential output, and contains extra units for impedance trimming independent of equalization tuning.

Two PLLs are designed with a similar architecture (Fig. 20.2.5) but different frequency bands to cover the range of 8.5~14.025GHz. Each PLL features dual LC-VCOs and automatic VCO frequency band search to limit the range of the tuning voltage. With nominal processing, measured results indicate that PLL A alone covers the range of 9.3~15.8GHz (Fig. 20.2.6). Clock distribution employs active-inductor-based clock buffers, which reduce power by 30% over conventional CML buffers (Fig. 20.2.5).

The proposed transceiver has been fabricated in 40nm bulk CMOS technology. The SerDes hard macro consists of two PLLs and four Rx/Tx lanes with an area of 3.88mm^2 (Fig. 20.2.7). The silicon consumes 410mW under worst case condition and has passed 16G Fibre Channel (16GFC), 802.3ap-KR (10GKR) and 10G SFP+ compliance tests with performance margin. Fig. 20.2.6 shows a measured Tx output eye and the measured SJ tolerance from a 16GFC jitter tolerance test, both at 14.025Gb/s.

References:
[1] J. F. Bulzacchelli, et al., "A 10-Gb/s 5-tap DFE/4-Tap FFE Transceiver in 90-nm CMOS Technology," *IEEE J. Solid-State Circuits*, vol. 41, pp. 2885-2900, Dec., 2006

[2] Y. Hidaka, et al., "A 4-Channel 10.3Gb/s Backplane Transceiver Macro with 35dB Equalizer and Sign-Based Zero-Forcing Adaptive Control," *ISSCC Dig. Tech. Papers*, pp. 188-189, Feb., 2009

[3] B. S. Leibowitz, et al., "A 7.5Gb/s 10-Tap DFE Receiver with First Tap Partial Response, Spectrally Gated Adaptation, and 2nd-Order Data-Filtered CDR," *ISSCC Dig. Tech. Papers*, pp. 228-229, Feb., 2007

[4] R. Payne, et al., "A 6.25-Gb/s Binary Transceiver in 0.13um CMOS for Serial Data Transmission across High Loss Legacy Backplane Channels," *IEEE J. Solid-State Circuits*, vol. 40, pp. 2646-2657, Dec., 2005

[5] H. Sugita, et al., "A 16Gb/s 1st-Tap FFE and 3-Tap DFE in 90nm CMOS," *ISSCC Dig. Tech. Papers*, pp. 162-163, Feb., 2010

[6] M. Kossel, et al., "A T-Coil-Enhanced 8.5 Gb/s High-Swing SST Transmitter in 65 nm Bulk CMOS with <-16 dB Return Loss over 10 GHz Bandwidth," *IEEE J. Solid-State Circuits*, vol. 43, pp. 2905-2920, Dec., 2008.

978-1-61284-303-2/11 $26.00 © 2011 IEEE

ISSCC 2011 / February 23, 2011 / 9:00 AM

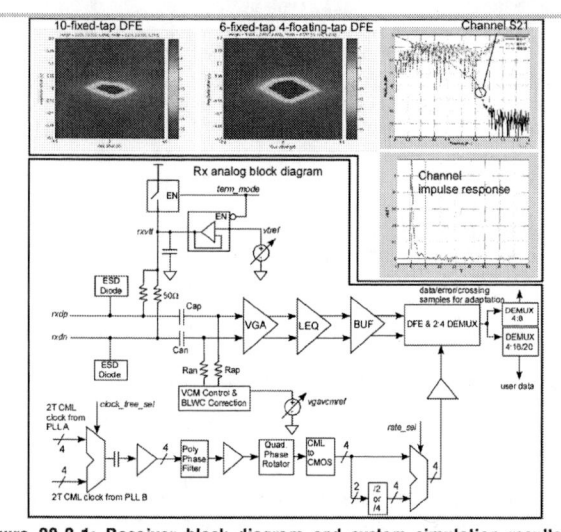

Figure 20.2.1: Receiver block diagram and system simulation results for floating-tap DFE.

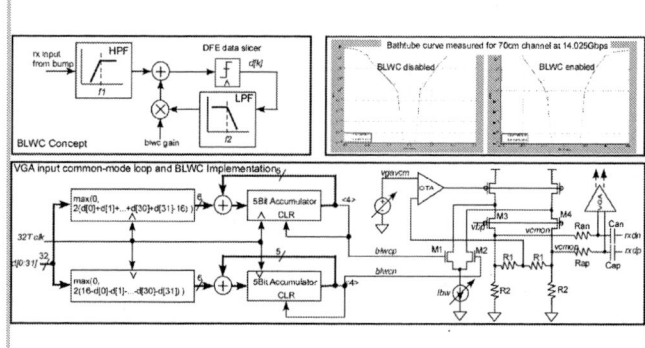

Figure 20.2.2: Concept, implementation, and measurement results for baseline wander correction.

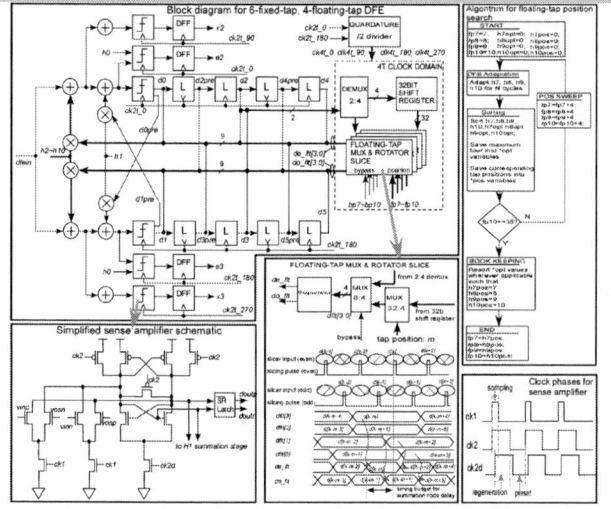

Figure 20.2.3: Block diagram for Floating-tap DFE, floating-tap MUX and rotator slice, sense amplifier circuit, and floating-tap search algorithm.

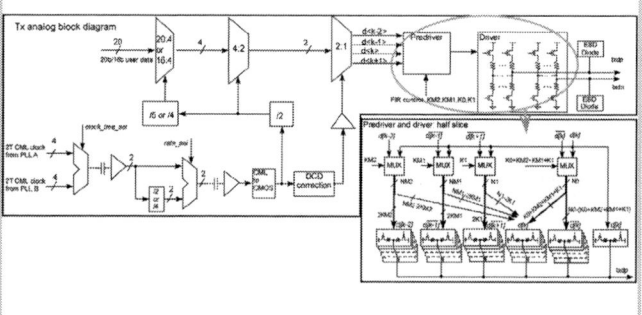

Figure 20.2.4: Transmitter block diagram and FFE scheme.

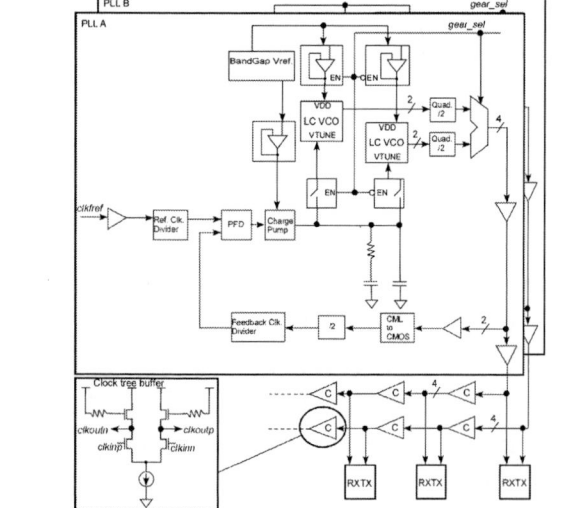

Figure 20.2.5: PLL and clock tree block diagram.

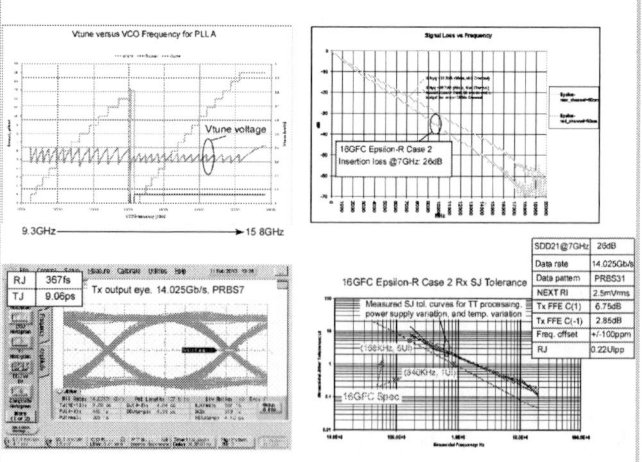

Figure 20.2.6: Measurement results for PLL frequency range, Tx output eye at 14.025Gb/s, and Rx SJ tolerance test for 16GFC.

20

978-1-61284-303-2/11 $26.00 © 2011 IEEE

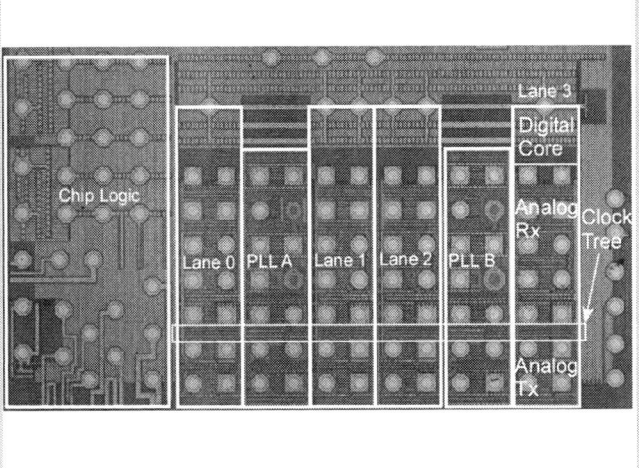

Figure 20.2.7: Micrograph for fabricated test chip in 40nm CMOS.

ISSCC 2011 / SESSION 20 / HIGH-SPEED TRANSCEIVERS & BUILDING BLOCKS / 20.3

20.3 Analog-DFE-Based 16Gb/s SerDes in 40nm CMOS That Operates Across 34dB Loss Channels at Nyquist with a Baud Rate CDR and 1.2V$_{pp}$ Voltage-Mode Driver

Andrew K. Joy[1], Hugh Mair[2], Hae-Chang Lee[3], Arnold Feldman[3],
Clemenz Portmann[3], Neil Bulman[1], Eugenia Cordero Crespo[1],
Peter Hearne[1], Patty Huang[3], Ben Kerr[1], Pulkit Khandelwal[1],
Franz Kuhlmann[2], Shaun Lytollis[1], Joaquim Machado[1],
Casey Morrison[2], Scott Morrison[2], Shahriar Rabii[3],
Dushmantha Rajapaksha[1], Vishnu Ravinuthula[2], Giuseppe Surace[1]

[1]Texas Instruments, Northampton, United Kingdom,
[2]Texas Instruments, Dallas, TX,
[3]Arda Technologies, Mountain View, CA

In networking systems today data rates are increasing beyond 15Gb/s and yet the installed backplanes are made of low cost materials with losses in excess of 30dB at 7.5GHz. Standards, such as IEEE802.3ap-10GBASE-KR and OIF-CEI25G, are specifying SerDes requirements for channels with 25dB loss at Nyquist and this has driven the development of SerDes with 4 or 5 tap DFEs [1]. Until now, solutions for 34dB or more channel loss have been limited to 10.3Gb/s or below [2,3], whereas this paper describes an adaptive 14-tap DFE that achieves a 10^{-17} BER across a 34dB loss channel at 16Gb/s for a power of 235mW/lane. A baud-rate CDR technique is specifically developed that gives excellent locking characteristics and alignment for use with a speculative DFE together with an enhanced swing TX voltage mode driver.

A block diagram of the complete transceiver is shown in Fig. 20.3.1. The transmitter utilizes an enhanced voltage mode driver as shown in Fig. 20.3.2. A standard voltage mode driver is commonly used in SerDes because of its higher power efficiency than a CML driver but has the problem of a limited output swing due to the series termination. This type of driver is normally limited to a maximum swing of ~800mV$_{pp}$ differential [4]. The addition of the CML type pull-up PMOS devices allows the common mode to be driven by the series driving stage but the swing to be increased to ~1200mV$_{pp}$ differential. This improves the output swing significantly at the cost of a small degradation of approximately 1dB in the TX output differential return loss.

The TX driver is made up of many slices of output stage that enables the design to have a 3-tap FIR filter to provide 1 pre-cursor and 1 post-cursor of de-emphasis. The pre-cursor can be up to ±17.5% and the post cursor ±37.5% in 2.5% steps. This has a high pass characteristic that compensates for some of the channel loss. The number of FIR taps is chosen to allow interoperability with other SerDes developed using the 10GBASE-KR and FC-16 standards. The increased swing capability makes the use of higher TX FIR pre- and post-cursor settings more feasible as these de-emphasize the overall output swing but allow the main cursor to remain higher.

The receiver utilizes an analog equalizer followed by a 2-stage VGA and a summer block to add in the DFE correction terms. The samplers are effectively multiple interleaved 3-bit ADCs with levels adapted to carry out 2-tap DFE speculation and error term slicing for DFE adaption and CDR adaption. This optimization of the slicing levels reduces the quantization noise that usually limits the performance of an ADC based solution.

The received input signal is terminated by an on-chip t-coil to give return loss that is more than 10dB down up to 10GHz. This is followed by an analog equalizer with variable DC attenuation, pole-zero spacing and bandwidth. This gives up to 14dB of peaking in steps of <0.5dB and is followed by two VGAs each with up to 7dB of gain to recover the amplitude of the data. The VGA gain is controlled by a 7b DAC and is adapted to match the output signal with the linear range of the following summers.

The samplers are run at one quarter of the incoming data rate and so the summers are interleaved and have to settle within 4UI. The CDR sampling point uses a Muller-Mueller baud rate technique but unlike other solutions [5] where h$_{-1}$ is forced to 0 this CDR uses h$_0$=h$_1$ for long reach applications. This has the advantage that the error slicing points fold over for a two-tap speculative DFE as shown in Fig. 20.3.3 and makes the slicing levels form an interleaved ADC. A further advantage is that the h$_{-1}$ term is suppressed by locking earlier than the peak in the symbol response. This does result in higher values for subsequent taps but this is not a problem as the DFE has a sufficient number of taps to cancel these. This locking criterion of the CDR is to the zero crossing points similar to Alexander phase detectors which have a high gain around the lock point and this results in low CDR jitter. However, with an analog summing and speculative DFE this also gives the optimum sampling point within the eye.

For shorter reach applications the locking point can be alternatively programmed to be related to the h$_0$, h$_1$ and h$_2$ taps in a similar manner that also exhibits the same folding of the error terms' slicing levels.

Two of the taps are speculative and the remaining 12 DFE taps are fed back to the summer with DAC current mirrors. The use of two taps of speculation allows the feedback loop to the slicer inputs 3UI to settle.

Hardware sequencing controls the front end auto-zeroing, slicer auto-zeroing and then performs a directed search for the region of operation where sufficient performance is achieved to give a sufficiently low BER for the CDR and DFE to lock. The RX is then released to adapt to the optimum CDR, DFE and AEQ settings.

The DFE adaptation uses a sign-sign LMS algorithm from the error slicer outputs. The AEQ is separately adapted to maximize the eye height at the cursor with additional samplers that can also be used for non-destructive eye height monitoring and plotting of eyes even in an asynchronous environment.

A B8P2 macro consisting of 4 transmit lanes, 4 receive lanes and 2 PLLs, is 1.796×1.195mm^2 in TSMC 40G 40nm CMOS technology. A B8P2 macro is shown in Fig. 20.3.4. It consumes 235mW per RX/TX pair including PLL overhead when running at 16Gb/s. The SerDes in silicon operates at 16Gb/s over 24" of Xcede FR-4 backplane plus evaluation card traces giving a total loss of -34dB. The transmitter eye is shown at 20Gb/s with a PRBS31 pattern in Fig. 20.3.5. The TX FIR map of performance of a complete link with a 10^{12} bit gate time, together with the RX eye plots and bathtub curves for the Xcede backplane trace above at 16Gb/s are shown in Fig. 20.3.6.

References:

[1] J. F. Bulzacchelli, T. O, Dickson, et al, "A 78mW 11.1Gb/s 5-Tap DFE Receiver with Digitally Calibrated Current-Integrating Summers in 65nm CMOS," *ISSCC Dig. Tech. Papers*, pp.368-369, Feb. 2009.
[2] K. Fukuda, H. Yamashita, et al, "An 8Gb/s Transceiver with 3×-Oversampling 2-Threshold Eye-Tracking CDR Circuit for -36.8dB-loss Backplane," *ISSCC Dig. Tech. Papers*, pp. 98-99, Feb. 2008.
[3] Y. Hidaka, W. Gai, et al, "A 4-Channel 10.3Gb/s Backplane Transceiver Macro with 35dB Equalizer and Sign-Based Zero-Forcing Adaptive Control," *ISSCC Dig. Tech. Papers*, pp.188-189, Feb. 2009.
[4] W. Dettloff, J. C. Eble, et al, "A 32mW 7.4Gb/s Protocol Agile Source-Series Terminated Transmitter in 45nm CMOS SOI," *ISSCC Dig. Tech. Papers*, pp. 370-371, Feb. 2010.
[5] M. Harwood, N.Warke, et al, "A 12.5Gb/s Serdes in 65nm CMOS Using a Baud-Rate ADC with Digital Receiver Equalization and Clock Recovery," *ISSCC Dig. Tech. Papers*, pp. 436-437, Feb. 2007.

ISSCC 2011 / February 23, 2011 / 9:30 AM

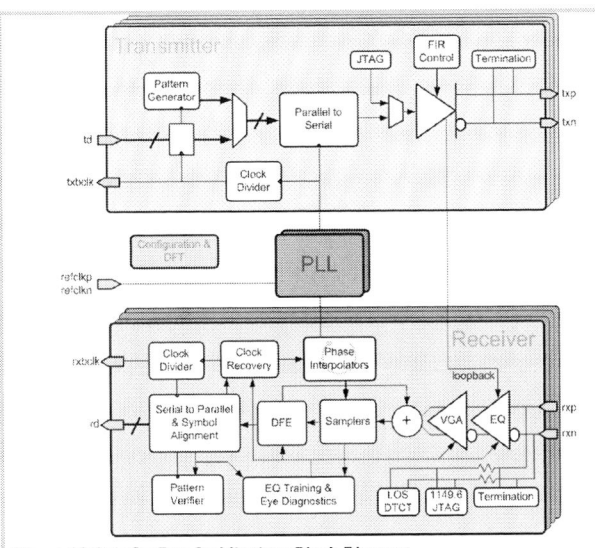

Figure 20.3.1: SerDes Architecture Block Diagram.

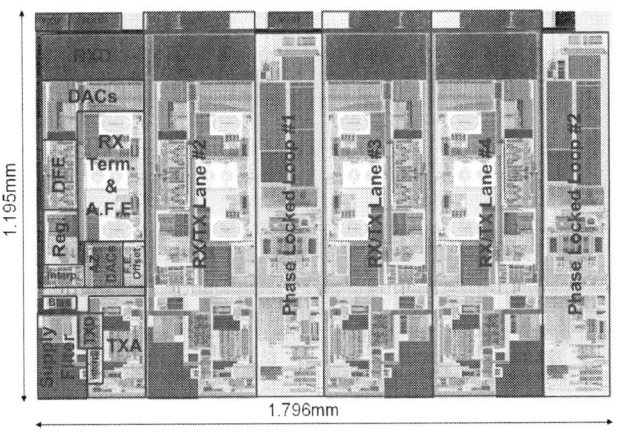

Figure 20.3.2: Extended Swing Transmitter Basic Schematic.

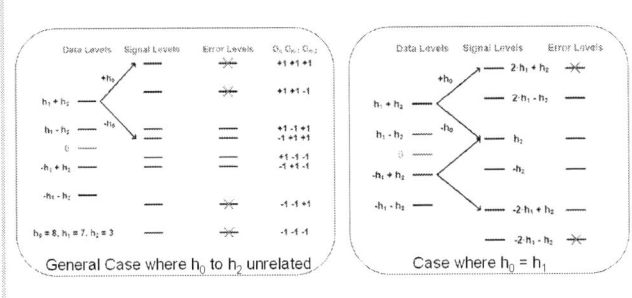

Figure 20.3.3: Two Tap Speculative DFE and Error Slicing Levels.

Figure 20.3.4: Macro Layout for 4 Lane, 2 PLL Version.

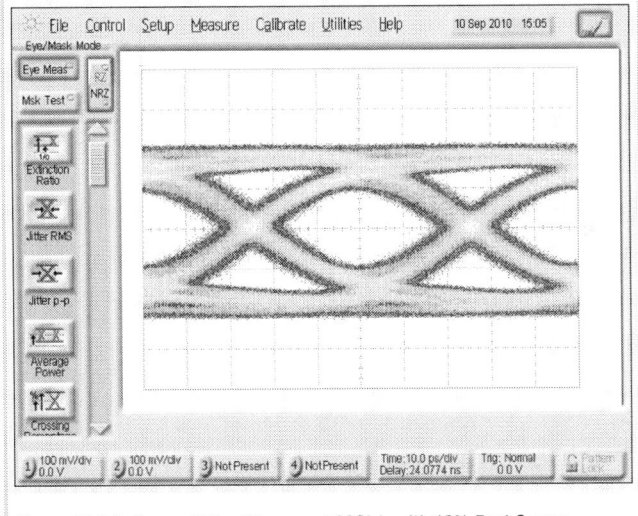

Figure 20.3.5: Transmit Eye Diagram at 20Gb/s with 10% Post Cursor

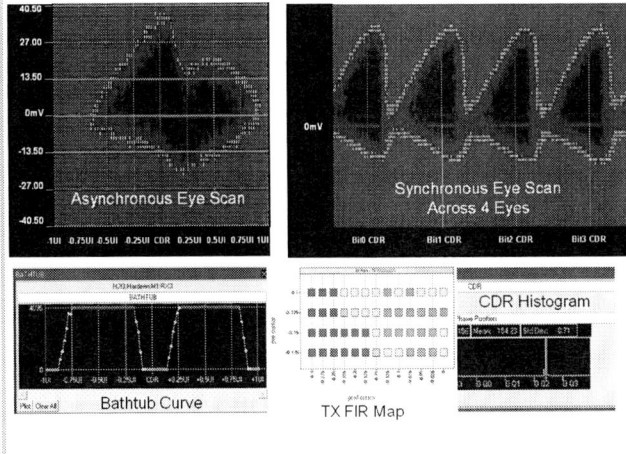

Figure 20.3.6: RX Results at 16.0Gb/s.

978-1-61284-303-2/11 $26.00 © 2011 IEEE   353

ISSCC 2011 / SESSION 20 / HIGH-SPEED TRANSCEIVERS & BUILDING BLOCKS / 20.4

20.4 An 8.4mW/Gb/s 4-Lane 48Gb/s Multi-Standard-Compliant Transceiver in 40nm Digital CMOS Technology

Mehrdad Ramezani[1], Mohamed Abdalla[1], Ayal Shoval[1],
Marcus Van Ierssel[1], Afshin Rezayee[2], Angus McLaren[1],
Chris Holdenried[1], Jennifer Pham[1], Eric So[1], David Cassan[1],
Saman Sadr[1]

[1]Snowbush-Gennum, Toronto, Canada,
[2]now with SecureKey, Toronto, Canada

The bandwidth limitation of existing backplanes has become an obstacle to meeting the increasing demand for high-data-rate wireline transmission. In order to compensate for this limitation, TX pre-emphasis, RX continuous-time linear equalizer (CTLE) and DFE are necessary [1,2]. This work presents a 4-lane transceiver implemented in 40nm CMOS technology that operates over a wide range of data rates from 1 to 12Gb/s (48Gb/s aggregated) using NRZ coding. The supply voltages are 0.9V and 1.8V. An algorithm is developed to adapt the CTLE and DFE to cancel the channel ISI. No inductors are used in the design and ring oscillators are used for both the TX and RX clock generation. This provides a wide frequency-tuning range, small layout area, and high design portability. With extensive use of digital programmability this transceiver is capable of meeting specifications of different standards, such as PCIe, SATA, and 1 to 10Gb/s Ethernet.

The RX front-end (Fig. 20.4.1) consists of a programmable attenuator (ATT), a CTLE and a 5-tap DFE that is followed by a CDR and a de-serializer. The ATT uses an AGC loop to set the desired signal level at the input of the DFE. A sign zero-forcing (S-ZF) adaptation algorithm with pattern filtering is used to adapt the CTLE and the DFE taps. The error signals for the S-ZF algorithm are generated from the same comparators used by the CDR, eliminating the need for additional comparators for adaptation. In addition, during test and debugging modes one of the edge comparators is re-used as an on-chip eye-monitor. The CDR is based on a half-rate bang-bang phase detector and a digital loop filter consisting of both a proportional and an integral path. The proportional signals from the phase detector directly control the CDR VCO. The integral path incorporates an up/down counter, which provides control over loop stability and programmability for optimum performance at different data rates. The VCO is based on a four-stage current-controlled ring oscillator with a 1 to 6GHz tuning range. The wide bandwidth of the bang-bang CDR loop significantly suppresses the ring oscillator inherent phase noise. The VCO coarse tuning is achieved by switching load capacitors at the delay cell outputs. The VCO fine control is achieved through a current DAC. At start up, digital calibration uses the same current DAC to pull the VCO frequency close to the locking condition. After that, the control is switched to real time CDR closed loop. The frequency error that the CDR loop can lock to is within ±5000ppm.

The ATT is realized using a passive capacitive divider (Fig. 20.4.2). A wide tuning range is achieved by using both programmable series and shunt capacitors. Switches are used to reconfigure series capacitors as shunt capacitors, which allows for a compact implementation. The CTLE is a 3 stage equalizer. Each stage consists of a differential-pair with an NMOS active inductor load (Fig. 20.4.2). Using the active inductors improves the portability of the design to other technologies/metal stacks. The DFE (Fig. 20.4.2) employs four low-power comparators: comprised of two data and two edge comparators. Each comparator contains a main branch, an identical offset branch, and 5 tap branches with tap 1 branch being ½ the main branch size and taps 2 to 4 being ¼ the main branch. This scaling helps reduce the DFE power consumption. The comparators use a common-source topology to allow for high-speed operation from a 0.9V supply. This requires the input common-mode to be accurately set, which is achieved through the CMFB in the CTLE. Furthermore, this common-mode reference voltage is used to set the mid-voltage of a resistive ladder which is used to generate the offset and tap voltages.

The DFE coefficients are adapted to minimize the ISI at the zero-crossings of the received data, which also improves the x-opening of the received eye, thus reducing the recovered clock jitter. The error signal for adaptation is the edge comparator output. However, the error signal is only used for specific patterns, which significantly simplifies the calibration procedure and allows calibrating the taps independently. Similar to the approach outlined in [3] for a 1-tap DFE, the approach used here selects the appropriate patterns to adapt each tap. However, as opposed to using 2 different patterns per tap and obtaining the error signal by combining the individual error signals from each pattern, the technique used here uses a single pattern per tap and computes the error signal from multiple edge decisions for a lone 1 or 0 following these patterns. This same technique is also used to adapt the CTLE by choosing a pattern that generates an error signal that is proportional to the sum of the first 5 edge-ISI components. Compared to a SS-LMS algorithm, this algorithm produces much less noise on the tap weights and does not diverge in the case of no transitions. The same technique was also extended to adapt the far-end Tx pre-emphasis taps by transmitting back these error signals.

The TX driver is based on an H-bridge architecture (Fig. 20.4.3), which uses a replica circuit and a buffer to set the driver common-mode output voltage. This method of common-mode control is essential for PCIe receiver-detect requirements. The output driver consists of sixteen identical driver/pre-driver cells placed in parallel. Each of these cells is controlled individually to facilitate four-tap TX FFE equalization (1-tap pre-cursor, 1 main tap, 2-tap post-cursor) and slew-rate control. Each cell includes a MUX that selects between the current and delayed data to achieve up to 3dB pre-emphasis and/or 12dB de-emphasis. The slew rate programmability is achieved by individually controlling the delay of the pre-driver cells. The differential output amplitude is programmable from 200 to 1000mV$_{pp}$, which is controlled by changing the bias of the H-bridge current sources. The VCO used for the TX CMU is also based on a four-stage ring oscillator architecture. To minimize the TX output jitter an on-chip regulated supply is used for the VCO and the TX clock path. Similar to the CDR VCO, the CMU VCO is calibrated at start up to achieve optimum tuning range for the desired data rate.

Figure 20.4.7 shows the die micrograph. The measured insertion and retun loss of a 52 inch FR4+ backplane is shown in Fig. 20.4.4. This channel is used to generate the RX bathtub curves for an 8Gb/s PRBS31 before and after the CTLE/DFE adaptation. It clearly shows the effectiveness of the adaptation in opening the eye at the input of the sampler. Furthermore, the RX performance is evaluated by measuring the BER for 2.5 to 11.3Gb/s PRBS31 inputs after 12 to 52 inch of the backplane. The test for 12Gb/s data is missing due to equipment limitation. The on-chip eye monitor is used to measure the internal eye after the CTLE as it is shown in Fig. 20.4.4. This is used to verify the results of the ATT and CTLE adaptation. The measured TX output eye diagram at 12Gb/s PRBS31 is shown in Fig. 20.4.5; the eye height is 660mV$_{pp}$. Figure. 20.4.6 provides the measured performance summary.

Acknowledgment:
Authors would like to thank the layout and testing groups, and also thank TSMC for fabricating this chip.

References:
[1] Ganesh Balamurugan, Frank O'Mahony, Mozhgan Mansuri, James E Jaussi, Joseph T Kennedy, Bryan Casper, "A 5-to-25Gb/s 1.6-to-3.8mW/(Gb/s) Reconfigurable Transceiver in 45nm CMOS", *IEEE Solid-State Circuits Conference*, pp. 372-373, Feb. 2010.
[2] Nagendra Krishnapura, Majid Barazande-Pour, Qasim Chaudhry,John Khoury, Kadaba Lakshmikumar, Akshay Aggarwal , "A 5Gb/s NRZ Transceiver with Adaptive Equalization for Backplane Transmission", *IEEE Solid-State Circuits Conference*, pp.60-61, Feb. 2005.
[3] Hidaka, Y. Weixin Gai Horie, T. Jian Hong Jiang Koyanagi, Y. Osone, H., "A 4-Channel 1.25–10.3 Gb/s Backplane Transceiver Macro With 35 dB Equalizer and Sign-Based Zero-Forcing Adaptive Control", *IEEE J. Solid-State Circuits*, vol. 44, pp. 3547-3559, Dec. 2009.

978-1-61284-303-2/11 $26.00 © 2011 IEEE

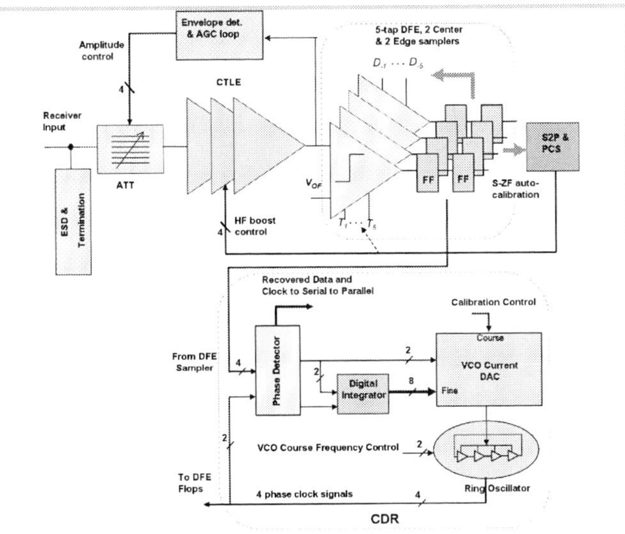

ISSCC 2011 / February 23, 2011 / 10:15 AM

Figure 20.4.1: Receiver block diagram.

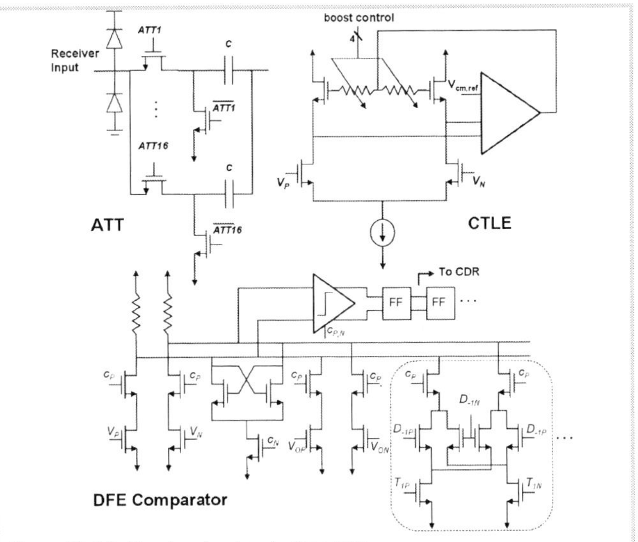

Figure 20.4.2: Receiver front-end schematics.

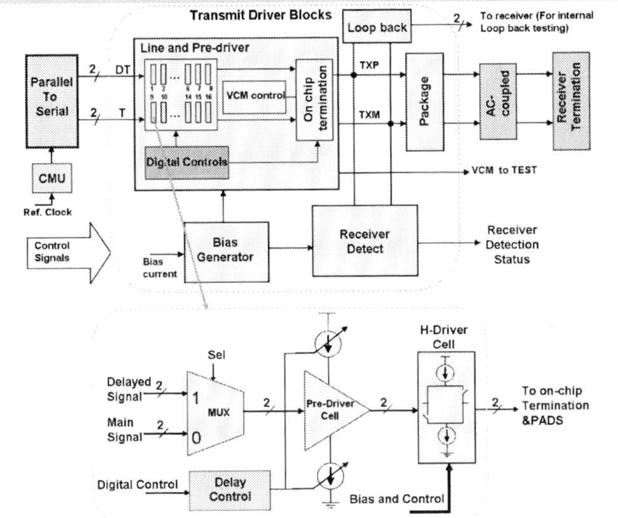

Figure 20.4.3: Transmitter block diagram.

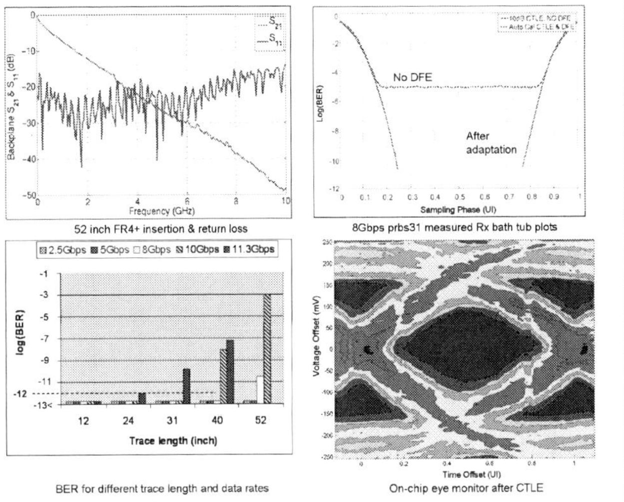

Figure 20.4.4: Receiver performance.

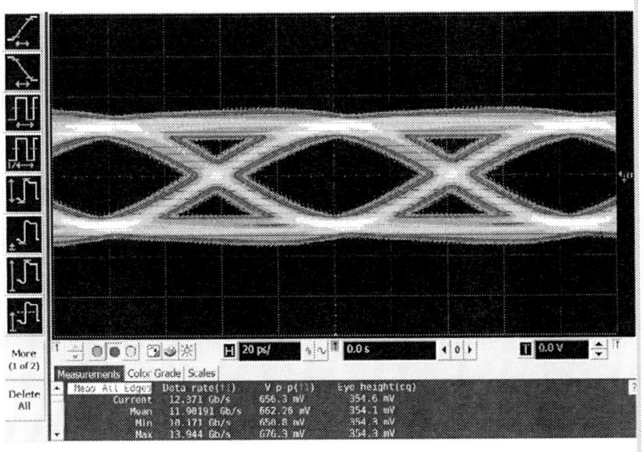

Figure 20.4.5: Transmitter performance at 12Gb/s.

Figure 20.4.6: Transceiver measured performance summary (per lane).

Technology		40nm CMOS
Package		Flip-chip BGA
Supply voltage	VDDA	0.9V
	VDDHA	1.8V
Speed		1-12Gbps
DFE taps		5
Tx FIR taps		4
Power per lane at 10.3125Gbps	800mV output differential swing	87mW
Power per lane at 8Gbps	800mV output differential swing	73mW
Open loop VCO phase noise @ 5GHz (10Gbps) (No LC tank)	1MHz offset	-80dBc/Hz
Tx driver jitter @10Gbps	800mV output swing	1.4ps (rms)
Area	Lane	360um x 750um
	CMU	180um x 880um

ISSCC 2011 PAPER CONTINUATIONS

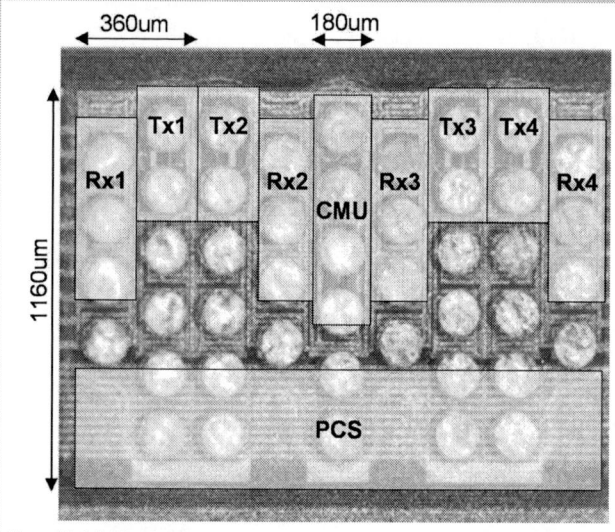

Figure 20.4.7: Die photo.

ISSCC 2011 / SESSION 20 / HIGH-SPEED TRANSCEIVERS & BUILDING BLOCKS / 20.5

20.5   A Pattern-Guided Adaptive Equalizer in 65nm CMOS

Shayan Shahramian[1], Clifford Ting[1], Ali Sheikholeslami[1],
Hirotaka Tamura[2], Masaya Kibune[2]

[1]University of Toronto, Toronto, Canada,
[2]Fujitsu Laboratories, Kawasaki, Japan

The use of adaptive equalizers at the front end of receivers is becoming a necessity as the data rates increase without channel improvements. Adaptive equalizers can be implemented using data-aided or non-data-aided schemes [1], with the latter requiring less area and power. Previous non-data-aided adaptive schemes [2-3] implement an asynchronous analog algorithm where the power spectrum of the received signal is checked for balance around a threshold frequency. Similarly, [4] proposes a digital adaptive algorithm which is based on the detection of specific 5-bit patterns. In all three works [2-4], however, adaptation is provided only for equalizers with a single coefficient, which are suitable for well-behaved channels. In contrast, this paper presents a digital adaptive engine for an equalizer with two coefficients: one adjusting the equalizer gain at the Nyquist frequency (f_N) and one at $f_N/2$. Furthermore, the proposed engine is asynchronous; it can function when driven by a blind clock at the receiver. This is useful as it allows the adaptation process to start even when the CDR has not yet achieved lock. This also avoids a deadlock situation where the CDR and equalizer require simultaneous access to the equalized data and the recovered clock. Our measured results of the proposed adaptive equalizer in 65nm CMOS confirm that the adaptation converges to within 2.6% of the optimal vertical eye opening in less than 400µs for two different channels at a data rate of 6Gb/s with a 25,000ppm frequency offset.

Figure 20.5.1 illustrates the concept behind the proposed adaptive equalizer. We assume an equalizer with controllable gains at f_N (C1) and $f_N/2$ (C2) to compensate for the channel loss at f_N and $f_N/2$, respectively. To control the equalizer gain at f_N, we count the number of 4-bit patterns whose signal power is concentrated around f_N (and have no signal power at $f_N/2$). It can be shown that the two 4-bit patterns that have this property are 0101 and 1010. We refer to these patterns as Type 1. Similarly, to control the gain at $f_N/2$, we count the number of 4-bit patterns whose signal power is concentrated around $f_N/2$ (and have no power at f_N). There are four 4-bit patterns among the possible 16 that have this property. We refer to these patterns as Type 2. It can be shown that the remaining ten 4-bit patterns have identical signal power at f_N and $f_N/2$. These patterns are not utilized in the proposed equalization scheme.

As shown in Fig. 20.5.2, the incoming data is attenuated by the channel and subsequently boosted by the equalizer with two independent, adjustable gains, C1 and C2. The equalized signal is sampled by two slicers (S1 and S2) whose thresholds differ by ΔV (mV). If the signal amplitude is above ΔV (mV), the outputs of S1 and S2 will be identical, signifying a vertical eye opening larger than ΔV (mV). The adaptive controller adjusts C1 and C2 to equalize the vertical eye opening to ΔV for both Type 1 and Type 2 patterns. This is achieved by counting each pattern at the output of the two slicers and forcing their respective differences to zero. This completes the equalization for a given ΔV. S1 and S2 are both driven by a clock signal, CK_{RX}. Ideally, CK_{RX} is the recovered clock from a CDR (CK_{REC}). However, our measured results confirm that the proposed architecture adapts to near-optimal coefficients even when there is a frequency offset between CK_{RX} and CK_{REC}. When the sampling phase deviates from the center of the eye, the slicers may underestimate the eye opening and cause the adaptation controllers to overestimate the required gains. To counteract this, the implemented adaptation algorithm has mechanisms (as explained later) to prevent coefficients from converging to over-equalizing gains.

For simplicity, Fig. 20.5.2 omits the two deserializers that convert serial data from the slicers into 16-bit words. The deserializers allow the counters to run at a frequency of $f_N/16$. To count patterns that span word boundaries, the design includes 4 sets of counters to cover all possible cases, but only selects the highest of the 4 counts. We have confirmed by measurement that the additional counters increase the algorithm's robustness and consistency as compared to when only one counter is used.

Figure 20.5.3 shows the C1, C2, and ΔV controllers. The C1 controller iteratively adjusts the gain at f_N until it converges to the lowest value such that the Type 1 count difference is less than or equal to an error tolerance that is programmable between 0 and 50. Eventually, C1 will reach steady state if it toggles between two adjacent values (e.g. {4, 5, 4, 5, 4, 5, 4, 5}), decreases to zero (i.e. {0, 0, 0, 0, 0, 0, 0, 0}), or increases to maximum (i.e. {7, 7, 7, 7, 7, 7, 7, 7}). This is identified via a 7-stage shift register as shown. The C2 controller is identical to the one for C1, except that it reads Type 2 counter differences.

When there is a frequency offset between CK_{RX} and the clock embedded in the data, CK_{RX} may sample the data at the edges (rather than at the center). This may cause the slicers to underestimate eye height, sometimes leading the controller to increase gain erroneously. To compensate for this, we employ two mechanisms in the C1 and C2 controllers. First, the programmable error tolerance allows a count difference up to 50 before the controller increases the gain. In measurement, the design is able to tolerate as much as 25,000ppm of offset with the error tolerance set at 20. Second, due to inherently larger signal slopes at the edge (compared to the center), the convergence checker does not flag convergence until CK_{RX} samples closer to the center of the eye.

The ΔV controller maximizes vertical eye opening by searching for the greatest ΔV that the C1 gain can compensate while avoiding bit errors on S2's output. It starts at the lowest setting (ΔV=1) and iteratively increments ΔV until the equalizer can no longer amplify the eye opening to ΔV (mV). The controller is based on C1 instead of C2 because the former compensates for higher signal attenuation. After all three coefficients converge, the controllers lock them to their final values.

Figure 20.5.4 shows the implementation of the analog equalizer, which consists of three different paths in order to create the desired frequency response [5]. Two bandpass filters are followed by variable gain amplifiers (VGA) that allow independent gain control at f_N and $f_N/2$. The buffers provide a unity-gain path for low frequency data. The bandpass filter is implemented by a differential pair with an RLC load, including a varactor that allows tuning of the center frequency. The VGA is implemented using a differential pair with resistive degeneration controlled by one-hot-encoded switches.

Figures 20.5.5 and 20.5.6 show the measured results of the packaged test chip. Figure 20.5.5 shows the eye diagrams for a 6 Gb/s PRBS 2^7-1 input before and after equalization. The eyes on the left and the right correspond to channel attenuations of 13dB and 17dB, respectively, at 3GHz. The adapted C1 and C2 for the 13dB channel are 7 and 1, respectively. The corresponding C1 and C2 for the 17dB channel are 6 and 2, respectively. Figure 20.5.6 (left) shows the vertical and horizontal eye openings for all possible C1 and C2 levels for a channel with 13dB of attenuation. The adapted eye opening is within 0.2% of the optimal vertical eye opening and within 5.4% of optimal horizontal eye opening. Figure 20.5.6 (right) shows the same results for a 17dB channel with the adapted eye opening being within 2.6% of the optimal vertical eye opening and within 7.0% of the optimal horizontal eye opening. We have also confirmed through measurements that the starting value of the coefficients, C1 and C2, do not affect the final converged state. The equalizer coefficients adapt to their final values in under 400µs. Figure 20.5.7 shows the die photo prior to packaging. The area of the equalizer (marked A) is 0.104 mm² with a power consumption of 60mW. The digital adaptation implementation (marked C) occupies an area of 0.101 mm² with a power consumption of 16.8mW.

References:

[1] J. W. M. Bergmans, "Non-data-aided adaptive equalization," in *Digital Baseband Transmission and Recording.* Kluwer Academic, 1996.

[2] J. Lee, "A 20Gb/s Adaptive Equalizer in 0.13µm CMOS Technology," *ISSCC Dig. Tech. Papers*, pp 273-282, Feb. 2006.

[3] H. Joo, K. Ha, L. Kim, "A data pattern-tolerant adaptive equalizer using spectrum balancing method," *Symp. on VLSI Circuits*, pp 220-221, 2009.

[4] Y. Hidaka, W. Gai, T. Horie, et al., "A 4-channel 10.3Gb/s backplane transceiver macro with 35dB equalizer and sign-based zero-forcing adaptive control" *ISSCC Dig. Tech. Papers*, pp 188-189, Feb. 2009.

[5] C. Liao, S. Liu., "A 40 Gb/s CMOS Serial-Link Receiver With Adaptive Equalization and Clock/Data Recovery," *IEEE J. Solid-State Circuits*, vol. 43, no. 11, pp 2492-2502, Nov. 2008.

ISSCC 2011 / February 23, 2011 / 10:45 AM

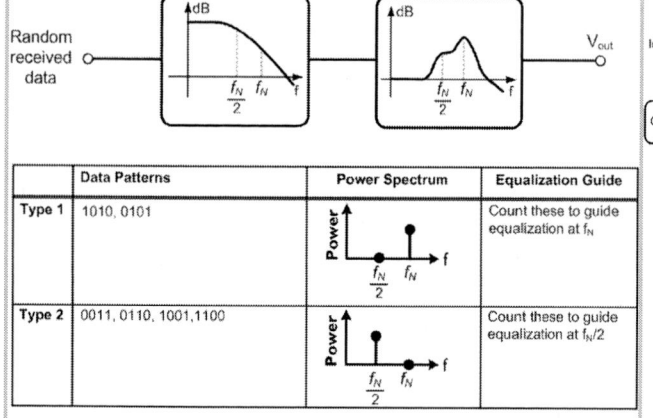

	Data Patterns	Power Spectrum	Equalization Guide
Type 1	1010, 0101		Count these to guide equalization at f_N
Type 2	0011, 0110, 1001, 1100		Count these to guide equalization at $f_N/2$

Figure 20.5.1: Six 4-bit data patterns are used to guide equalization. The remaining ten 4-bit patterns are not used because they have identical signal power at $f_N/2$ & f_N.

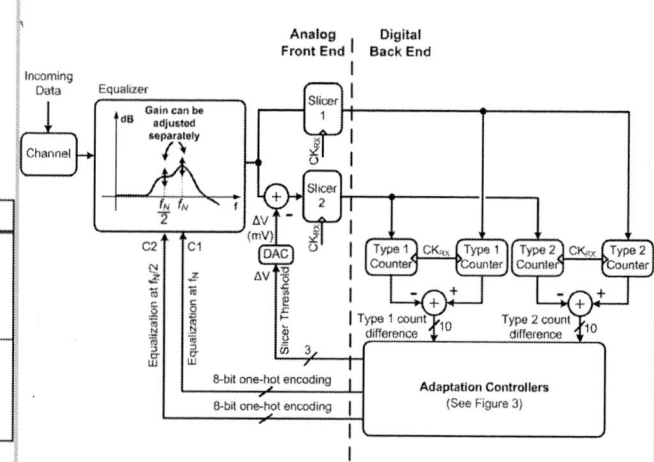

Figure 20.5.2: Block diagram of the proposed adaptive equalizer.

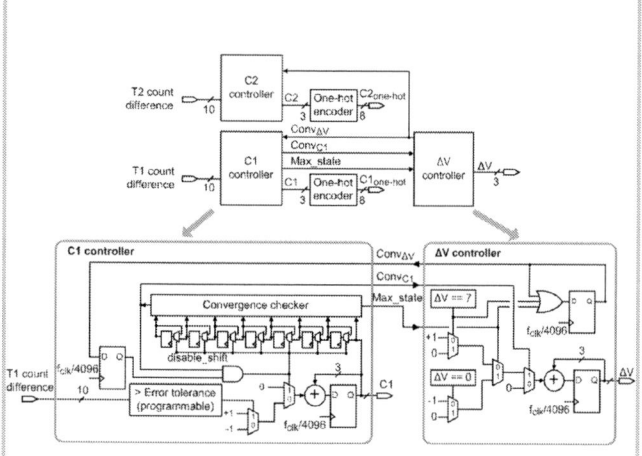

Figure 20.5.3: Implementation of digital adaptation controllers.

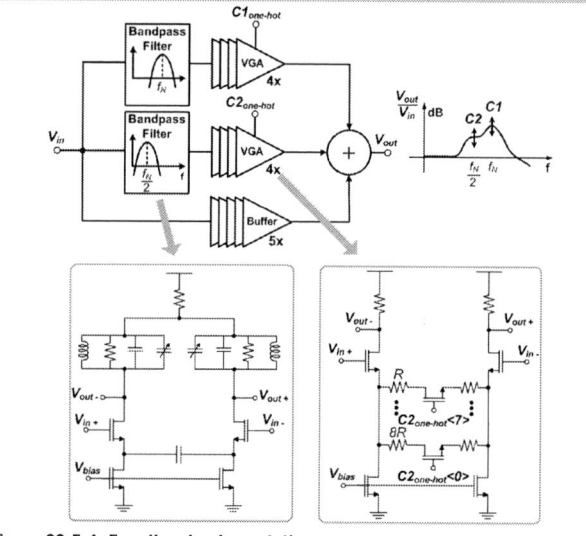

Figure 20.5.4: Equalizer implementation.

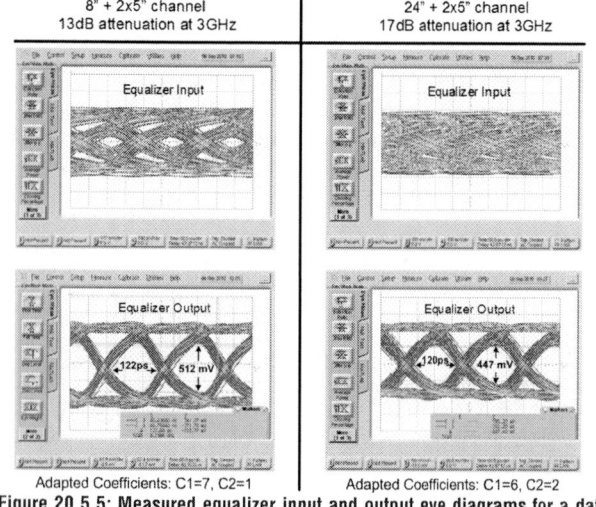

Figure 20.5.5: Measured equalizer input and output eye diagrams for a data rate of 6 Gb/s and a receiver clock with a frequency offset of 25,000ppm.

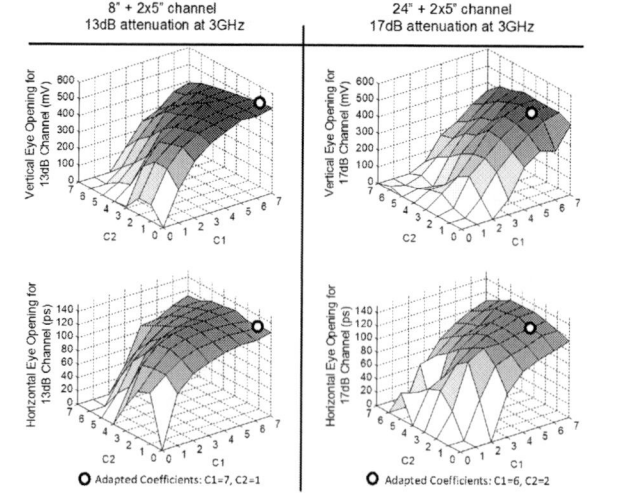

Figure 20.5.6: Measured vertical and horizontal eye openings for all possible gain coefficients for the 13dB and 17dB channels.

978-1-61284-303-2/11 $26.00 © 2011 IEEE

ISSCC 2011 PAPER CONTINUATIONS

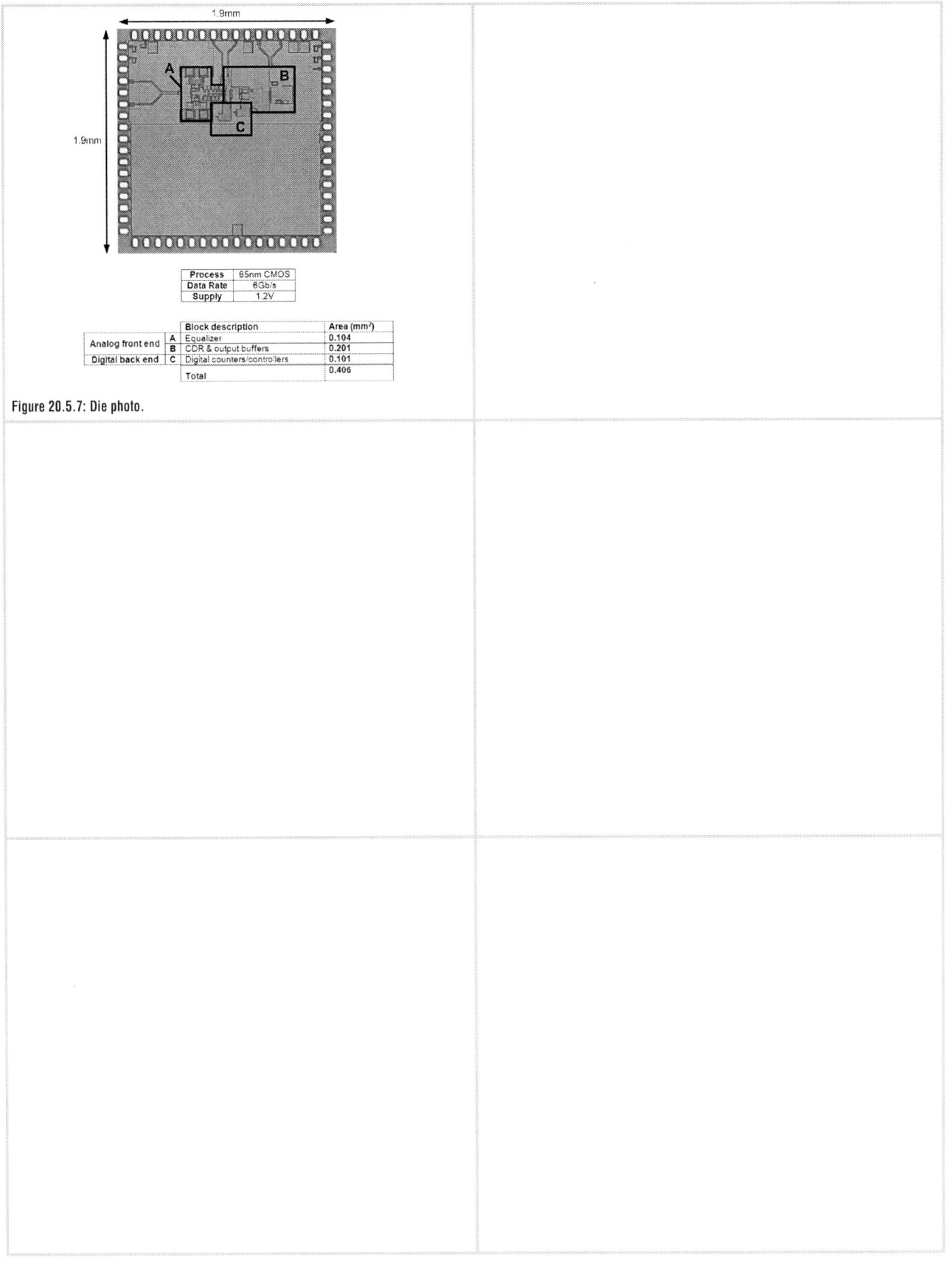

Process	65nm CMOS
Data Rate	6Gb/s
Supply	1.2V

		Block description	Area (mm²)
Analog front end	A	Equalizer	0.104
	B	CDR & output buffers	0.201
Digital back end	C	Digital counters/controllers	0.101
		Total	0.406

Figure 20.5.7: Die photo.

ISSCC 2011 / SESSION 20 / HIGH-SPEED TRANSCEIVERS & BUILDING BLOCKS / 20.6

20.6   A 6Gb/s Receiver with 32.7dB Adaptive DFE-IIR Equalization

Yi-Chieh Huang, Shen-Iuan Liu

National Taiwan University, Taipei, Taiwan

To ensure the signal integrity over a lossy channel, an analog equalizer and/or a decision-feedback equalizer (DFE) [2-6] are widely adopted in high-speed data transmission. An adaptive analog equalizer or adaptive DFE is also attractive to compensate the frequency-dependent loss due to the different channel lengths and environment variations. Conventionally, a multiple-tap DFE is adopted to compensate the inter-symbol interference (ISI), which is induced by postcursors due to the non-ideal channel impulse responses. To avoid the power and area penalty due to many postcursors, a DFE with infinite impulse response (IIR) filter feedback [2] has been presented. In [2], no adaptation scheme ensures that such IIR filter cancels the postcursors precisely, i.e., its RC time constant and amplitude need to be manually adjusted. In this work, a 6Gb/s receiver using a DFE with an adaptive continuous-time IIR filter and a clock/data recovery (CDR) circuit is presented. In a high loss environment, a conventional digital qaudricorrelator frequency detector (QFD) may fail due to the significant data dependent jitter. To integrate an adaptive DFE with a CDR circuit, a proposed frequency-sweeping frequency detector (FD) and a lock detector (LD) are presented in this work.

The receiver is shown in Fig. 20.6.1. It consists of a DFE with an IIR filter, a CDR circuit with the proposed frequency-sweeping FD and LD, an adaptive loop, and latches. When this CDR circuit does not lock, the LD disables the adaptive loop and the IIR filter. The proposed frequency-sweeping FD aids the CDR circuit to lock. When the output "Lock" of the LD goes high, it enables the adaptive loop and the IIR filter.

The DFE with a first-order continuous-time IIR filter is shown in Fig. 20.6.2. The summer is realized by a differential amplifier with resistive loads and RC source degeneration. The time constant and amplitude of this first-order continuous-time IIR filter are adjusted by a PMOS transistor and a NMOS one, respectively. The adaptive algorithm is realized by detecting data and edge information [1], as shown in Fig. 20.6.3. The normalized pulse responses before and after equalization are shown Fig. 20.6.3, where Tb is the bit time. To adjust the amplitude of this IIR filter, a special pattern of "110" is detected. For three sequent data (D2, D1, D0) is "110", the ISI at the edge E0 is given as

$$ISI_{E0}= h(1.5Tb)+h(0.5Tb)-h(-0.5Tb)   (1)$$

where the first two terms are postcursors and the last one is the precursor. If the sampled result ISI_{E0} is high, it implies the amplitude of this IIR filter should be increased and vice versa. To adjust the time constant of this IIR filter, a special pattern '1110' is detected. For four sequent data (D3, D2, D1, D0) is "1110", the ISI at the edge E0 is given as

$$ISI_{E0}=h(2.5Tb)+h(1.5Tb)+h(0.5Tb)-h(-0.5Tb)   (2)$$

If the sampled result ISI_{E0} is high, it implies the time constant of this IIR filter should be increased and vice versa. Once both the time constant and amplitude loops converge, Eq. (1) and Eq. (2) will be close to zero. Two equalities are obtained as

$$h(-0.5Tb)=h(0.5Tb)+h(1.5Tb) \& h(2.5Tb)=0   (3)$$

By h(-0.5Tb)=h(0.5Tb)+h(1.5Tb), to adjust the amplitude of this IIR filter is equivalent to tune the tap weighting to cancel the ISI. By h(2.5Tb)=0, a long pulse response tail due to a high-loss channel is compensated. To adjust the RC time constant of this IIR filter, it is helpful to decrease the ISI due to a long pulse response tail and it is equivalent to have a multiple-tap DFE.

The CDR circuit consists of a conventional Alexander phase detector, a voltage-to-current converter (VIC), a loop filter, a proposed frequency-sweeping FD, and the LD. In Fig. 20.6.1, the first 7 latches (L1~L7) and two XOR gates realize the

Alexander phase detector. The last 5 latches (L8~L12) with several logic gates, two voltage-to-current converters (VICs), two PMOS switches, and two capacitors realize the adaptive loop. When the signal "Lock" is low, the PMOS switches are on to disable the adaptive loop. As the channel loss exceeds 10dB, the conventional QFD may fail to work due to a large data dependent jitter. The proposed frequency-sweeping FD is presented to avoid this problem as shown in Fig. 20.6.4(a). When the signal "Lock" of the LD is low, this FD injects a constant current to charge or discharge the loop filter. It is equivalent to sweep the control voltage of a VCO within the high-level voltage of V_H and the low-level voltage of V_L. As the VCO's is sweeping to the vicinity of the target frequency, the LD disables the proposed FD and the Alexander PD takes over to complete the acquisition. This FD does not need a quadrature clock compared with a conventional QFD. The LD is shown in Fig. 20.6.4(b). The data are sampled by the clock CK and a delayed one CK_d with a time difference of 20ps. The results are compared and calculated by a 6-bits asynchronous counter. If the clock frequency is far from data rate, the clocks CK and CK_d will drift toward left or right in the eye diagram. As they drift to the edge of the eye diagram, their sampled results may be different within 64 bits. This asynchronous counter is reset, so the signal "Lock" still stays low. While their sampled results are the same within 64 bits, which means clock frequency is close to the data rate, the signal "Lock" goes high, and the LD disables the frequency-sweeping FD.

In this work, three different trace lengths of 60cm, 140cm, and 340cm are realized by the FR4 boards and their measured channel loss at 3GHz is 9.0dB, 15.0dB, and 32.7dB, respectively, as shown in Fig. 20.6.1. Figures 20.6.5(a) and Fig. 20.6.5(b) show the measured eye diagram before and after equalization under the above different channels. The measured rms jitters of the recovered data are 3.68ps, 3.97ps, and 4.16ps, respectively. The measured peak-to-peak jitters are 24.44ps, 26.67ps, and 27.78ps, respectively. The measured bit-error-rate is less than 10^{-12} for all the cases under a PRBS of 2^7-1.

The measured locking transient of the CDR circuit and the adaptive loop are shown in Fig. 20.6.6(a). This CDR circuit and the adaptive loop smoothly converge. The locking time of the CDR circuit, the time constant control, and the amplitude control of the adaptive loop are 9us, 160us, and 150us, respectively.

The die photo is shown in Fig. 20.6.7. It is fabricated in 90nm CMOS process. The total active area of this receiver is 0.089mm². The power consumption is 52mW, where the DFE and CDR dissipate 4mW and 48mW, respectively. Fig. 20.6.6(b) shows performance summary compared to that of the prior arts. The maximum loss is up to 32.7dB without an additional linear equalizer on a receiver and the pre-emphasis on a transmitter. This receiver has the outstanding equalization ability with the aid of the proposed adaptive algorithm.

Acknowledgements:
The authors thank TSMC, Hsinchu, Taiwan, for fabricating this chip through TSMC university shuttle program. This work is also supported by NTU-MediaTek Lab., and NSC, Taiwan.

References:
[1] Y. Hidaka, et al., "A 4-Channel 3.1/10.3Gb/s Transceiver Macro with a Pattern-Tolerant Adaptive Equalizer," *ISSCC Dig. Tech. Papers*, pp. 442-443, Feb. 2007.
[2] B. Kim, et al., "A 10-Gb/s Compact Low-Power Serial I/O With DFE-IIR Equalization in 65-nm CMOS ," *IEEE J. Solid-State Circuits*, vol. 44, pp. 3526-3538, Dec. 2009.
[3] J. F. Bulzacchelli, et al., "A 78mW 11.1Gb/s 5-Tap DFE Receiver with Digitally Calibrated Current-Integrating Summers in 65nm CMOS," *ISSCC Dig. Tech. Papers*, pp. 368-369, Feb. 2009.
[4] M. Park, et al., "A 7Gb/s 9.3mW 2-Tap Current-Integrating DFE Receiver," ISSCC Dig. Tech. Papers, pp. 230-231, Feb. 2007.
[5] T. Beukema, et al., "A 6.4-Gb/s CMOS SerDes Core With Feed-Forward and Decision-Feedback Equalization," *IEEE J. Solid-State Circuits*, vol. 40, pp. 2633-2645, Dec. 2005.
[6] R. Payne, et al., "A 6.25Gb/s Binary Adaptive DFE with First Post-Cursor Tap Cancellation for Serial Backplane Communications," *ISSCC Dig. Tech. Papers*, pp. 68–69, Feb. 2005.

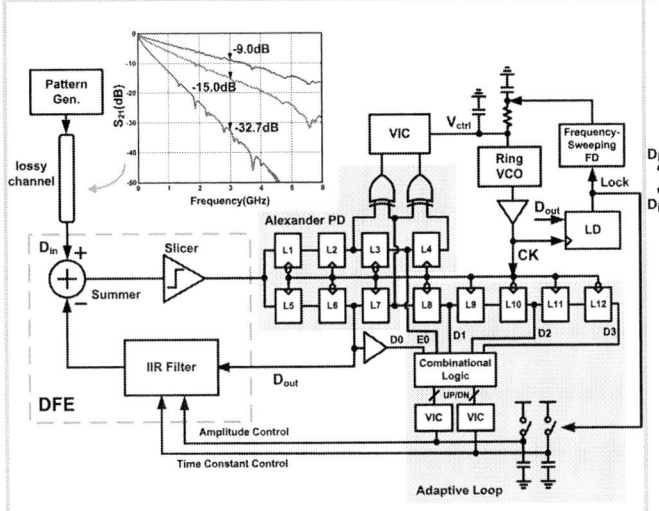

Figure 20.6.1: A receiver using a CDR circuit and an adaptive DFE-IIR.

Figure 20.6.2: A DFE with an adaptive IIR filter.

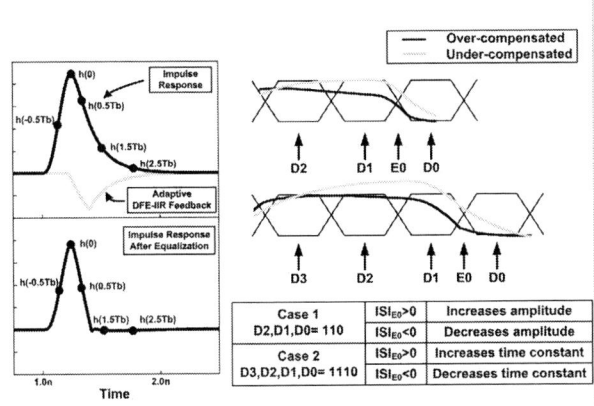

Figure 20.6.3: Adaptive algorithm.

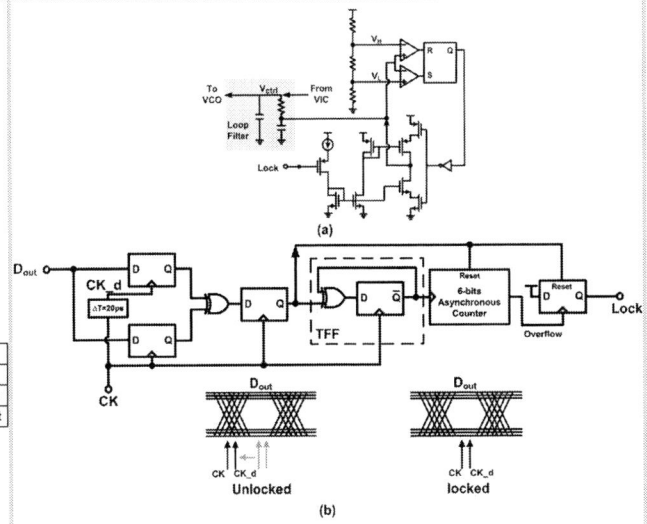

Figure 20.6.4: (a) Proposed frequency-sweeping FD, (b) proposed LD.

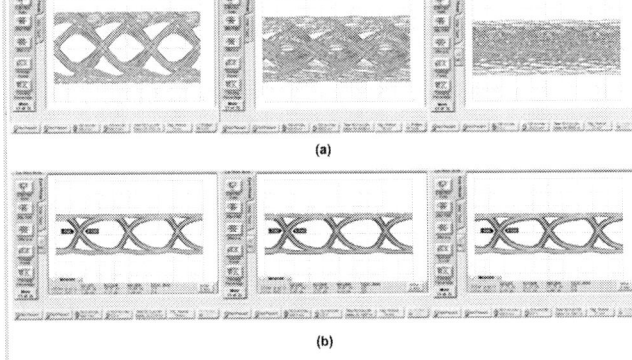

Figure 20.6.5: (a) Measured eye diagrams for 60cm, 140cm, and 340cm traces on FR4 board before equalization, (b) after equalization.

	[2]	[3]	[4]	[5]	[6]	This Work
Technology	65nm CMOS	65nm CMOS	90nm CMOS	130nm CMOS	130nm CMOS	90nm CMOS
DFE architecture	2 tap half-rate DFE-IIR	5 tap half-rate Current-Integrating	2 tap half-rate Current-Integrating	5 tap full-rate	4 tap full-rate	1 tap full-rate DFE-IIR
Adaptation	No	Yes	No	Yes	Yes	Yes
Data rate	10Gb/s	11.1Gb/s	7Gb/s	6.4Gb/s	6.25Gb/s	6Gb/s
Rx equalization	-27dB	-28dB	-15dB	-24.4dB	-20dB	-32.7dB
DFE power	7mW	N/A	9.3mW	N/A	N/A	4mW
Rx power	N/A	78.2mW	N/A	290mW*	180mW	52mW
Area	0.017mm^2	0.220mm^2	0.019mm^2	0.790mm^2*	0.240mm^2	0.089mm^2

*Include Tx

(b)

Figure 20.6.6: (a) Measured locking transient of the CDR circuit and the adaptive loop, (b) performance summary.

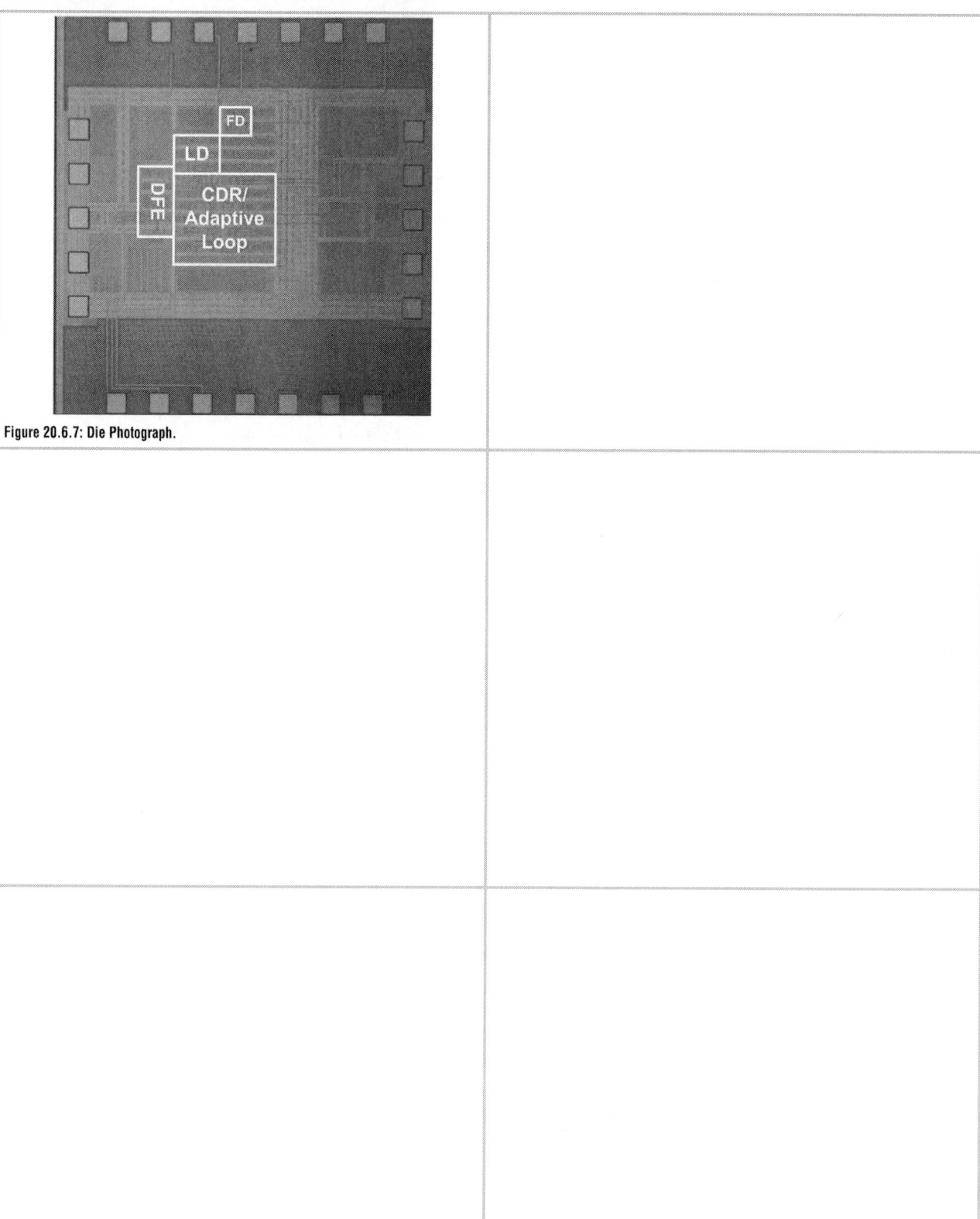

Figure 20.6.7: Die Photograph.

ISSCC 2011 / SESSION 20 / HIGH-SPEED TRANSCEIVERS & BUILDING BLOCKS / 20.7

20.7 A 5.4Gb/s Adaptive Equalizer Using Asynchronous-Sampling Histograms

Wang-Soo Kim, Chang-Kyung Seong, Woo-Young Choi

Yonsei University, Seoul, Korea

As the data rate requirements for many wireline applications increase, channel bandwidth limitation becomes a critical problem in serial interfaces. Equalizers are often used as a solution for this problem. In addition, many applications require the equalizer to be adaptive so that it can provide optimized equalization for different channel conditions. Various types of adaptive equalizers have been investigated for high-speed serial interface applications [1-6]. In the spectrum-balancing method, adaptive equalization is achieved by comparing high and low frequency components of signal power and generating feedback signals until the power spectrum is balanced [1]. Unfortunately, the precision of this scheme is easily affected by process variations, and capacitors both in filters and the feedback loop occupy a large Si area. Digital-signal processing based on maximum likelihood sequence detection can be used for adaptive equalization [2]. But, speed limitation and architecture complexity of ADC limits applicability of this scheme in high-speed applications. In the eye-opening monitoring (EOM) scheme, quality of the signal eye diagram is measured and used for equalizer adaptation [3-6]. For this method, a clock-recovery circuit is needed in order to generate clock signals synchronized to data for sampling. However, it can be difficult to recover clock signals from the initially closed eye diagram, limiting the applicability of this scheme. In this paper, we demonstrate an adaptive equalizer based on asynchronous-sampling histograms.

The adaptation scheme in our equalizer is based on the simple observation that the clearest eye diagram produces the largest peak value when data histograms are taken. This can be easily observed from Fig. 20.7.1, which shows Matlab simulated eye diagrams for three different cases of overequalization, optimal equalization, and underequalization, along with corresponding histograms representing the voltage distribution of equalizer output. For simulation, we use a 3-pole lowpass filter representing 3m DisplayPort cable and a high-frequency boosting filter for the equalizer. 5.4Gb/s 2^7-1 pseudo-random bit sequence (PRBS) data are transmitted through the channel and three different filters having different amounts of high-frequency boosting are applied. The resulting data are then asynchronously sampled and compared with 32-amplitude levels for obtaining a histogram for each case. As can be seen in the figure, the peak value for each histogram depends on equalization filter conditions, and the largest peak value is obtained from the clearest eye-diagram, which corresponds to the optimal equalization filter condition. If a circuit can produce histograms for various filter conditions, and select the filter condition that produces a histogram with the largest peak value, then it can determine the optimal filter condition for the given channel, thus, achieving adaptive equalization.

Figure 20.7.2 shows the block diagram of our adaptive equalizer. It has an equalizing filter with capacitive degeneration, sample-and-hold circuit, comparator, digital-to-analog converter (DAC), clock generator, and digital controller. A 2-stage capacitive degeneration equalizing filter is used for up to 18dB equalization gain. A 4b NMOS resistor array is used in the equalizer filter. This array provides 16 different levels of gain boosting with approximately 1.5dB increments controlled by digital codes. The comparator is a clocked sense amplifier, and the DAC generates 32-level reference voltages with which input signal is compared. The clock generator has a 5-stage inverter chain and generates an 114.166MHz clock that is asynchronous to the data rate. The digital controller determines the histogram for each code and selects the best condition having the largest peak value. The controller is digitally synthesized and integrated on the chip.

The adaptive equalizer operates as follows. An initial filter code is applied to the equalizing filter and the corresponding code is applied to DAC. Then, a reference voltage, V_{ref}, is applied to comparator from DAC. When signal level, V_{data}, is higher than V_{ref} at the sampling point, the comparator generates a high pulse. A counter counts up by one and the register stores the accumulated value. With a full scan of 32-level reference voltages, the cumulative distribution function (CDF) for the given filter code is obtained. By differentiating the CDF, the histogram is obtained and the peak value of the histogram is stored in a register.

The above operation is repeated for each value of filter codes. Finally, the filter code producing the largest peak value is chosen as the optimal equalizing filter code. According to our analysis, 3121 random samples are needed in order to achieve 2% variation in the largest histogram peak value with 99% confidence level using the approximation of normal distribution. Our prototype chip produces 4096 (12b) samples with the asynchronous clock for each filter code. The total time required for determining the optimal filter code is 18.37ms. It does take some scanning time to select the optimal equalizing filter for a given channel, but the fact that this scheme does not require any previous knowledge of the channel, and heavily relies on the digital block for its implementation makes it a promising solution for applications where the channel condition does not change once initial optimization is achieved.

A prototype chip is fabricated in 0.13μm CMOS technology. A pattern generator provides 5.4Gb/s 2^{31}-1 PRBS to DUT through four different channels: 40cm, 80cm, 120cm PCB trace, and 3m DisplayPort cable. The equalized output is observed by an oscilloscope. As a first verification, we measure the eye diagram and histogram of the equalize output for each of 16 filter codes when data are transmitted through 3m DisplayPort cable. The results are shown in Fig. 20.7.3. As can be seen, different filter codes with different amounts of high-frequency boosting produce different eye diagrams and histograms. However, filter code 0010 can be easily identified as the one producing the clearest eye diagram and this is the one having the largest peak value in its histogram. The digital controller selects this code to be the optimal filter code, successfully achieving adaptation to the given channel. Measurement with other 3 different channels shows the same result, verifying our scheme.

Figure 20.7.4 shows the channel response without any equalization for four different channels measured at 2.7GHz by a network analyzer. It also shows the optimal boosting gain provided by the equalizer measured by the same method. The total response with equalization can be determined by the sum of above two. As shown in the figure, the total response is close to zero indicating our equalizer provides the optimal amount of boosting gain that compensates channel loss for each of four different channels.

Figure 20.7.5 shows measured eye diagrams with the equalizer for 4 different channels with 5.4Gb/s 2^{31}-1 PRBS data. As can be seen, all the eyes are clearly open. The optimal filter code for each case is adaptively determined by the circuit and no external control is provided. The circuit consumes 35mW excluding output buffers from a 1.2V supply and its size is $340 \times 525 \mu m^2$ as shown in Fig. 20.7.6.

Acknowledgements:
This work was supported in part by the IT R&D program of MKE/KEIT (KI002145) and Samsung Electronics and IDEC for EDA software support and chip fabrication. The authors thank Dr. Pyung-Su Han for useful discussion.

References:
[1] J. Lee, "A 20-Gb/s adaptive equalizer in 0.13-μm CMOS technology," *IEEE J. Solid-State Circuits*, vol. 41, no. 9, pp. 2058–2066, Sep. 2006.
[2] O.E. Agazzi, *et al*, "A 90nm CMOS DSP MLSD transceiver with integrated AFE for electronic dispersion compensation of multimode optical fibers at 10 Gb/s," *IEEE J. Solid-State Circuits*, vol.43, no. 12, pp. 2939-2957, Dec. 2008.
[3] T. Ellermeyer, U. Langmann, B. Wedding and W. Pöhlmann, "A 10-Gb/s eye-opening monitor IC for decision-guided adaptation of the frequency response of an optical receiver," *IEEE J. Solid-State Circuits*, vol. 35, pp. 1958–1963, Dec. 2000.
[4] B. Analui, A. Rylyakov, S. Rylov, M. Meghelli and A. Hajimiri, "A 10-Gb/s two-dimensional eye-opening monitor in 0.13-μm standard CMOS," *IEEE J. Solid-State Circuits*, vol. 40, pp. 2689–2699, Dec. 2005.
[5] F. Gerfers, G.W. den Besten, P.V. Petkov, J.E. Conder, and A.J. Koellmann, " A 0.2–2 Gb/s 6x OSR receiver using a digitally self-adaptive equalizer," *IEEE J. Solid-State Circuits*, vol. 43, no. 6, pp. 1436–1448, June. 2008.
[6] H. Noguchi, N. Yoshida, H. Uchida, M. Ozaki, S. Kanemitsu, and S. Wada, " A 40-Gb/s CDR with adaptive decision-point control based on eye-opening monitor feedback," *IEEE J. Solid-State Circuits*, vol. 43, no. 12, pp. 2929–2938, Dec. 2008.

978-1-61284-303-2/11 $26.00 © 2011 IEEE

ISSCC 2011 / February 23, 2011 / 11:45 AM

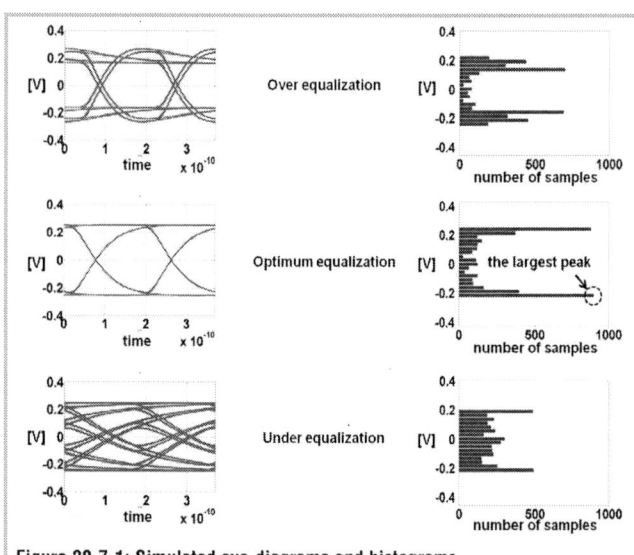

Figure 20.7.1: Simulated eye-diagrams and histograms.

Figure 20.7.2: Proposed adaptive equalizer architecture.

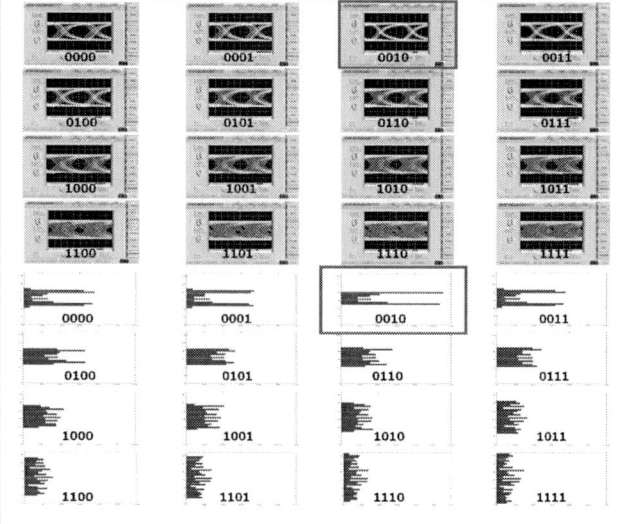

Figure 20.7.3: Measured eye-diagrams and histograms.

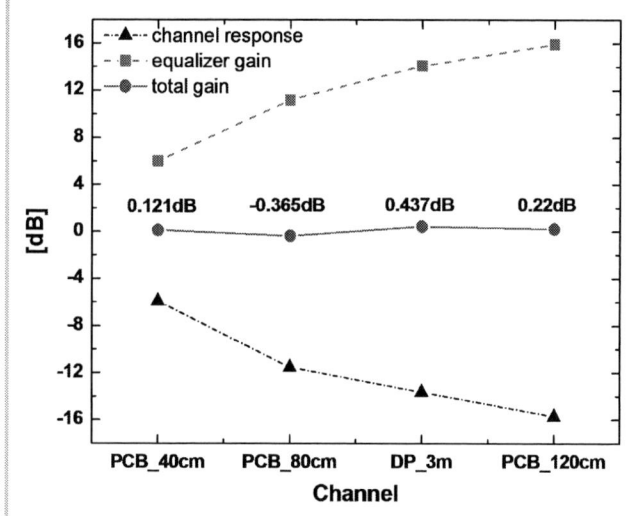

Figure 20.7.4: Channel response without equalizer and equalizer gain.

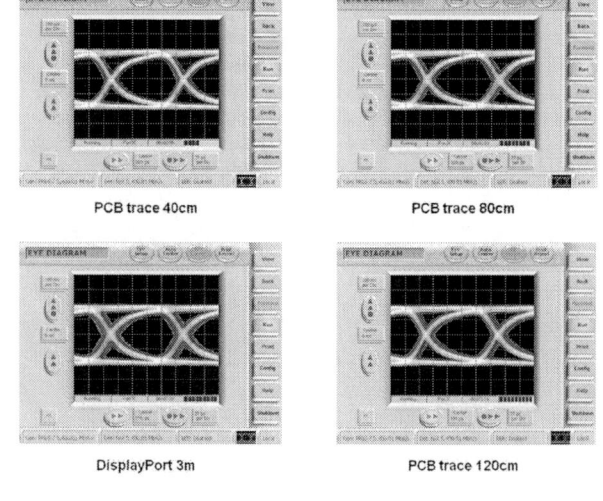

Figure 20.7.5: Eye-diagrams after adaptive equalization.

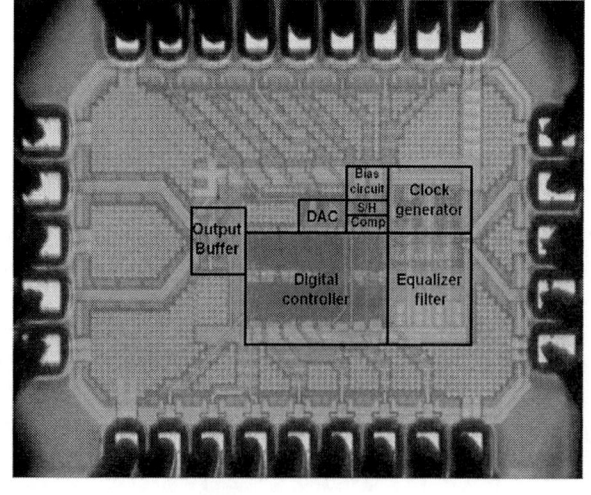

Figure 20.7.6: Chip micrograph.

20

ISSCC 2011 / SESSION 21 / CELLULAR / OVERVIEW

Session 21 Overview / Wireless

Cellular

Session Chair: *Myung-Woon Hwang, Future Communication IC(FCI), Sungnam, Korea*

Session Co-Chair: *Taizo Yamawaki, Renesas Electronics, Japan*

In most parts of the world the cellular mobile terminal market is starting to move from 2G and 3G to 4G systems in order to support higher date rates in the popular smart phones, netbooks and other mobile devices. These higher data rates are enabled by 3G standards such as WCDMA/HSPA and by 4G standards such as LTE . Another evolving direction is the removal of external components, aiming at lower-cost mobile devices and a higher integration level of functions, including a large number of bands and multistandard operation to guarantee global coverage.

The first three papers in this session will address multimode CMOS transceiver solutions.

Paper 21.1 [MediaTek] presents a 65nm CMOS SAW-less GSM/GPRS/EDGE receiver, embedded in SoC. By using a Class-AB low noise amplifier and passive mixer with current-mode LPF, the Rx achieves a high P_{1dB} of +1dBm without performance degradation of sensitivity or IIP2/IIP3, while consuming 58.9mA from 2.8V voltage supply.

A 9-band WCDMA/EDGE transceiver with full RX diversity will be presented in Paper 21.2 by ST-Ericsson. Realized in a 90nm RFCMOS process, the SAW-less 2G/3G RX has an NF of 2.3 to 2.5dB, with an IIP2 of +58dBm and an IIP3 of -6dBm. The transmitter achieves EVM below 1.5% for 2G and 4% for 3G, occupying a die area of 14.4 mm². The 2G RX/TX+LO consume 129mW/126mW, while the 3G TX+RX+LO consumes 269mW.

Paper 21.3 [Qualcomm] addresses the first SoC with embedded quad-band GSM/EDGE and triple-band 3G transceiver, mixed signal, audio, DSP and memory cores in a 65nm CMOS process. The SAW-less 3G Rx/Tx consumes battery current of 38.3mA/48.2mA in LB and 45.3mA/47.8mA in HB respectively with sensitivity of -111.4dBm. The RF circuitry runs at 1.3V, while analog baseband blocks use 2.2V.

The following 2 papers discuss an inductorless front-end topology with feedback techniques achieving small area.

Paper 21.4 [Media Tek] presents a 65nm CMOS 2G/3Greceiver with inductor-less front-end by shunt-shunt feedback topology, achieving an out-of-band IIP3 of -2dBm and an NF below 2.5dB up to 2.2GHz. The front-end area is 0.9mm². The implemented LNA consumes 21.7, 4.8, and 2mA in high-, mid-, and low-gain modes respectively, from a 1.5V supply.

Paper 21.5 [NXP Semiconductors] presents another technique in a 45nm CMOS inductor-less receiver using a translational loop for input matching without NF degradation. The receiver achieves 2.2dB NF in the 3G 900MHz band, 2.4dB NF in the GPS 1.5GHz band, and 2.7dB NF in the 3G 2.1GHz band, consuming 7.3 mA from a 1.3V supply.

The last 3 papers discuss a low-noise modulator targeted for 4G LTE and polar architecture for 3G application, aiming at removing external SAW filters.

Multiband LTE SAW-less modulator is presented in Paper 21.6 by IMEC in 40nm CMOS. The modulator achieves RX-band noise down to -162 dBc/Hz and OP_{1dB} up to +11dBm in all LTE FDD bands including the most challenging VII, XI, and XII bands, consuming 24.8 to 38.5 mW.

978-1-61284-303-2/11 $26.00 © 2011 IEEE   362

Paper 21.7 [Infineon Technologies] proposes a fully digital polar transmitter with 17bits (3G) and 19bits (2.5G) RFDAC supporting GSM /EDGE /EDGEEvo /WCDMA /HSPA+ using a 65nm CMOS process. The transmitter achieves -160 dBc/Hz far off noise at all 3GPP Rel.7 specified duplex distance and consumes 35mA DG09 weighted current in 3G mode (1.2V for digital and 2.5V for RFDAC).

Paper 21.8 [Broadcom] presents a 65nm polar transmitter based on a two-point PLL. The transmitter achieves -42dBc ACLR1, -159dBc/Hz noise at 45MHz, and 2.9% EVM at 0dBm with 40mA current from the battery by linearizing VCO gain with a wideband frequency-locked loop nested inside the PLL.

Highly integrated multiband multimode CMOS-based solutions presented in this session demonstrate advances in functionality, maturity, reduced area, power consumption, and performance required to realize current 2G and 3G as well as coming 4G systems.

Presenters:

8:30 AM

21.1 A SAW-less GSM/GPRS/EDGE Receiver Embedded in a 65nm CMOS SoC

I.S-C. Lu, MediaTek, San Jose, CA

9:00 AM

21.2 A 9-Band WCDMA/EDGE Transceiver Supporting HSPA Evolution

M. Nilsson, ST-Ericsson, Lund, Sweden

9:30 AM

21.3 A 65nm CMOS SoC with Embedded HSDPA/EDGE Transceiver, Digital Baseband and Multimedia Processor

A. Cicalini, Qualcomm, San Diego, CA

10:15 AM

21.4 A Receiver for WCDMA/EDGE Mobile Phones with Inductorless Front-End in 65nm CMOS

F. Beffa, MediaTek, West Malling, United Kingdom

10:45 AM

21.5 A Compact SAW-less Multiband WCDMA/GPS Receiver Front-End with Translational Loop for Input Matching

X. He, H. Kundur, NXP Semiconductors, Eindhoven, The Netherlands

11:00 AM

21.6 A Multiband LTE SAW-less Modulator with -160dBc/Hz RX-Band Noise in 40nm LP CMOS

V. Giannini, imec, Leuven, Belgium

11:15 AM

21.7 A Fully Digital Multimode Polar Transmitter Employing 17b RF DAC in 3G Mode

Z. Boos, Infineon Technologies, Neubiberg, Germany

11:45 AM

21.8 A Low-Power Wideband Polar Transmitter for 3G Applications

M. Youssef, Broadcom, Irvine, CA; University of California, Los Angeles, CA

978-1-61284-303-2/11 $26.00 © 2011 IEEE

ISSCC 2011 / SESSION 21 / CELLULAR / 21.1

21.1 A SAW-less GSM/GPRS/EDGE Receiver Embedded in a 65nm CMOS SoC

Ivan Siu-Chuang Lu[1], Chi-yao Yu[2], Yen-horng Chen[2], Lan-chou Cho[2], Chih-hao Eric Sun[2], Chih-Chun Tang[2], George Chien[1]

[1]MediaTek, San Jose, CA,
[2]MediaTek, Hsinchu, Taiwan

Over the last decade, significant progress has been made towards increasing integration and reducing bill of material (BOM) for GSM/GPRS/EDGE cellular systems. In modern cellular phones, transmit SAW filters have been largely eliminated with innovative TX architecture and circuits [1] while receive SAW filters are still present. This remains as one of the few bottlenecks in achieving a true low-cost single-chip solution which offers a genuine advantage for high-volume applications. The main challenges of eliminating SAW-based bandpass filters from multiband 2G/2.5G receivers lie in noise-figure (NF) degradation and gain desensitization induced by 0dBm blockers at merely 20MHz away from the -99dBm in-band signal at GSM850/900 band. Furthermore, these unfiltered interferers cause a strong reciprocal mixing effect and result in additional SNR degradation. Therefore, a SAW-less RX is required to have an extremely high dynamic range while maintaining best-in-class NF, LO wideband phase noise (PN) and spur performance.

In this paper, a receiver embedded in a single-chip quad-band GSM SoC implemented in 65nm CMOS technology is presented. Figure 21.1.1 shows the block diagram of the SoC. The chip includes an RF transceiver, modem, CPU, integrated memory, FM RX, mixed-signal audio front-end with an embedded class-D amplifier, power management unit, and all necessary peripheral blocks for multimedia and connectivity applications. The transceiver block diagram is shown in Fig. 21.1.2. The receiver consists of two RX front-ends that support the lowband (LB) GSM850/900 and high-band (HB) DCS/PCS. The two received signals are combined at mixer output stage and filtered by low-pass filter. Then, the signals are amplified by the transimpedance amplifier, and further filtered by a channel selection filter before being quantized and decoded by the digital signal processor.

Figure 21.1.3 shows a detailed diagram for one path of the SAW-less RX. Shunt and series inductors, L_p and L_s, provide a 50Ω in-band interface to the antenna. This matching topology provides better LO harmonic rejection than other topologies since the impedances of L_p and L_s become much larger than the LNTA input impedance at the LO harmonic frequency. The LNTA output current is directly applied to the mixer switching quad for frequency downconversion. Omitting the mixer Gm stage eliminates the dominant linearity bottleneck in conventional RXs. Without the SAW filter, the in-band signal v_{sig} and the out-of-band (OOB) blocker v_{blk}, located at f_{LO} and $f_{LO}+\Delta f$ respectively, are simultaneously received by the RX and then downconverted to baseband and Δf with virtually no selectivity. Under maximum OOB blocker conditions, $I_{blki/q}$ exceed the maximum tolerable input current of the TIA and must be filtered by the current-mode passive LPF. Apart from baseband filtering, C_{lpf} also suppresses OOB interferer's voltage swing on the RF side of the mixer switching quad and limits LNTA output compression. The LPF corner frequency is set to 2MHz to provide sufficient rejection for a 0dBm 20MHz-offset blocker while maintaining good in-band noise performance. The TIA, operating under a 2.5V supply, uses a two-stage op-amp with Class AB output stage to maximize dynamic range. Large input transistor sizes are chosen to achieve low flicker noise. A 3rd-order low-pass filter with a bandwidth of 700 kHz is used to filter out in-band interference. A similar approach of adding a frequency-translated BPF before the LNA output has been used to filter the blockers. In [2], the receiver performance with 0dBm 80MHz-offset blocker in the PCS band is demonstrated. However, having the frequency translated BPF directly at the LNA input degrades RX sensitivity and LO feedthrough performance.

Figure 21.1.3 also shows the differential L-degenerated LNTA. The 2.5V supply, along with L_{load} and mixer/C_{lpf} interface, maximizes the headroom available for active devices and suppresses output distortion. With the mixer Gm stage removed, the LNTA input V/I stage must provide sufficient Gm to suppress noise from later stages. In traditional Class-A amplifier designs, the required I_{DC} is >40mA for a specified Gm of 120mA/V and a blocker at 0dBm. To attain a low-power and high-dynamic-range input V/I stage, the input transistors M1 and M2

are biased in the weak inversion region to exploit the exponential drain current versus V_{gs} characteristic. The resulting Class-AB behavior provides gain expansion in the V-I transfer function and greatly improves the compression point of M1 and M2 [3]. Moreover, the Class-AB LNTA current consumption increases adaptively only with the strong RF input power, resulting in significant power savings under real-world operating conditions where strong blockers are not always present. To further reduce power consumption in the LO chain, Class-AB LNTA current is sensed to adaptively adjust the mixer switch size, divider supply voltage, and LO buffer size based on OOB blocker levels.

The fully integrated fractional-N frequency synthesizer, which has been optimized for low wideband phase noise and reference spur, is used to generate the LO signal. The generation of IQ LO signals consists of a frequency divider and a non-overlapping waveform generator operating under a 1.4V supply. The frequency divider consists of two cross-connected D flip-flops. It is driven by a differential VCO clock and generates non-overlapped quadrature outputs. The DFF is formed by clocked inverters in order to obtain large signal swing [4]. The non-overlapping waveform generator is a logic AND gate that takes the inputs and outputs of the frequency divider to generate non-overlapping signals. Divider outputs are re-timed by divider inputs. Therefore, noise from the frequency divider, which affects transition edges of divider outputs, does not contribute to the overall LO PN.

Figure 21.1.7 shows the die micrograph of the SAW-less RX portion of the 65nm CMOS SoC chip. The chip is packaged in a 179-ball FC CSP package and the total chip size is 8.6×8.1mm^2 where the transceiver occupies a die area of 2.6×1.9mm^2. Operating under a 3.8V external supply, the RX and SX draw 37.1 and 21.8mA respectively in the continuous RX mode. The RX current increases to ≈ 60mA when a 0dBm OOB blocker is present. To demonstrate true SAW-less operation, all measurements are performed on PCB with an L-C lattice-type balun and matching network only. When no blockers are present, the measured all-channel RX gain and sensitivity for both bands are shown in Fig. 21.1.4. The typical gain is ≈ 60dB; while the sensitivity is < -110dBm, which is comparable to the conventional RX with SAW filters. The RX's robustness to blocker interference is demonstrated in Fig. 21.1.5 where the BER remains < 2.4% for blocker power ≤ 4.3/4.7dBm at 20/80MHz offset for the GSM850/900 band. Similarly, the RX EVM remains < 11.5% for blocker power ≤ 2dBm when operating under MCS-9 in EDGE mode. The RX achieves OOB blocker P_{1dB} of +1dBm, in-band IIP2 and IIP3 of > +44dBm and > 0dBm, respectively.

Without the SAW filters, the measured RX $2f_{LO}$ and $3f_{LO}$ rejection is 68 and 44dB, respectively. A blocker around the f_{LO} harmonics can be downconverted by the harmonic LO signal to around the signal band, causing RX SNR degradation. Moreover, the close-in LO PN within the corresponding offset frequency range is also downconverted to the signal band and degrades the RX SNR further. When the BER is > 2.4%, exceptions must be taken as the desired signal cannot be decoded properly. The RX requires 4/4/6/8 exceptions in the GSM850/GSM900/DCS/PCS band, which satisfy the ETSI limit of 24 exceptions per band. Finally, the SX exhibits < -80dBc reference spur and results in negligible SNR degradation with blockers applied at frequencies equal to multiples of the 26MHz reference crystal. The RF transceiver performance results and required exception frequencies are summarized in Fig. 21.1.6.

This work demonstrates the possibility of replacing the discrete SAW filter function with inexpensive on-chip components to further reduce cost and form factor for today's cellular systems. The impact of strong OOB blockers specified in ETSI can be minimized with a careful design and budgeting of desired and blocking signal strengths through the RX signal path.

References:
[1] S. F. Chen, Y. B. Lee, and Bosen Tzeng et al., "A GSM/EDGE Transmitter in 0.13-μm CMOS Using Offset Phase Locked Loop and Direct Conversion Architecture" *IEEE RFIC Symp.*, pp. 581-584, June 2009.
[2] A. Mirzaie, A. Yazdi, and Z. Zhou et al. "A 65nm CMOS Quad-Band SAW-Less Receiver for GSM/GPRS/EDGE," *Symp. VLSI Circuits Dig. Tech. Papers*, pp. 179-180, June 2010.
[3] K. Fong, C. Hull and R. G. Meyer, "A Class AB Monolithic Mixer for 900-MHz Applications", *IEEE J. Solid State Circuits*, vol.32, pp.1166-1171, Aug. 1997.
[4] M. D. Tsai, "Low-noise Frequency Divider," U.S. patent no. 7719327.

978-1-61284-303-2/11 $26.00 © 2011 IEEE

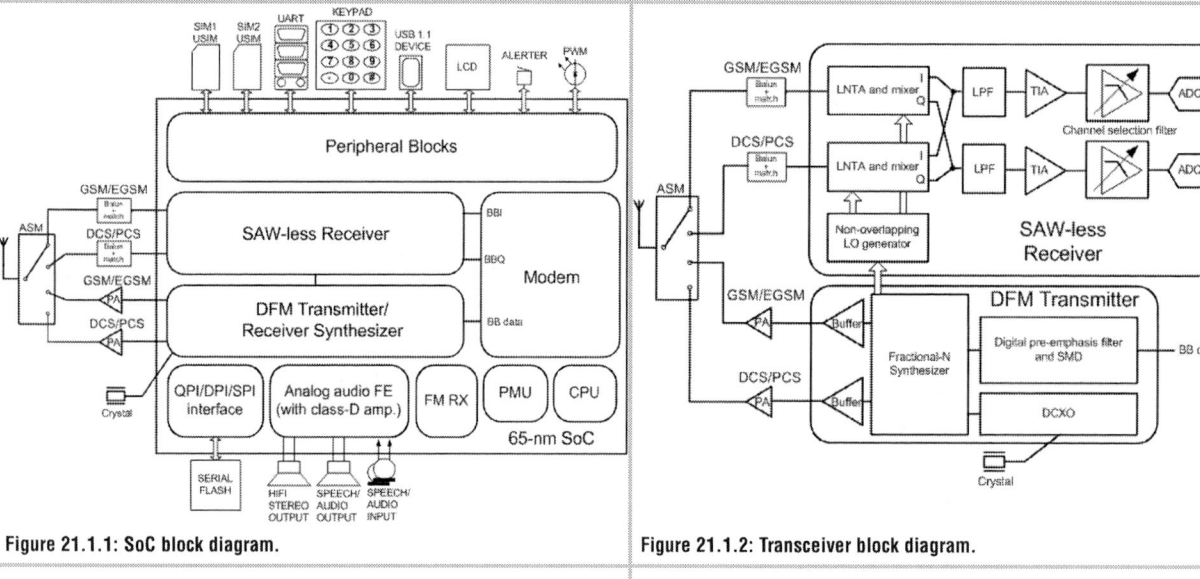

Figure 21.1.1: SoC block diagram.

Figure 21.1.2: Transceiver block diagram.

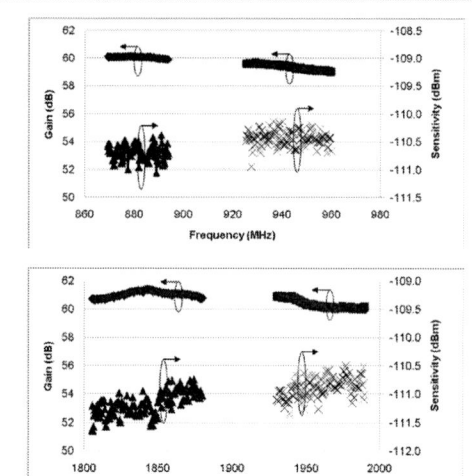

Figure 21.1.3: RX front-end circuit diagram.

Figure 21.1.4: RX gain and sensitivity.

Figure 21.1.5: RX GSM BER and EDGE EVM vs. blocker power. GSM signal power is -99dBm and EDGE signal power is -85dBm.

	This Work		[2]
Band Of Operation	GSM850 / GSM950	DCS / PCS	PCS
RX Gain (dB)	59.6	60.7	78
RX Sensitivity (dBm)	-110.1	-110.5	>-111
RX NF (dB)	2.7	2.9	3.1
RX NF $f_{blockoffset}=\pm20MHz$ (dB)	8.5	n/A	>17
RX NF $f_{blockoffset}=\pm80MHz$ (dB)	n/A	7	11.4
RX OOB Blocker P_{1db} (dBm)	+1	+1	+2+
RX In-Band IIP2 (dBm)	> 50	> 44.3	>45
RX In-Band IIP3 (dBm)	0	0	>-12.4
RX SNR $f_{blockoffset}=\pm3MHz$ (dB)	16.8	15.5	n/A
RX SNR $f_{blockoffset}=\pm20MHz$ (dB)	12.5	n/A	n/A
RX SNR $f_{blockoffset}=\pm80$ MHz (dB)	n/A	14	n/A
Exception Frequencies	$3f_{LO}$, $3f_{LO}\pm200k$, $5f_{LO}$ / $3f_{LO}$, $5f_{LO}$, $6f_{LO}\pm200k$	$2f_{LO}$, $3f_{LO}$, $3f_{LO}\pm200k$, $3f_{LO}\pm400k$ / $2f_{LO}$, $3f_{LO}$, $3f_{LO}\pm200k$, $3f_{LO}\pm400k$, $3f_{LO}\pm600k$	n/A
SX 26MHz Spur (dBc)	< -80		n/A
SX PN @ 20MHz (dBc/Hz), f_{LO}=3.5GHz	<-154		n/A
TX ±400-kHz Mask (dBc)	<-69	< -66.8	n/A
TX RMS Phase Error (deg)	<0.8	<0.95	n/A
RX Current (mA)	58.9*		55*
TX Current (mA)	39.6*		n/A
Supply Voltage (V)	2.8		n/A

+ $f_{blockoffset}=\pm80$ MHz

* includes SX

Figure 21.1.6: Performance summary and comparison with previously reported SAW-less RX.

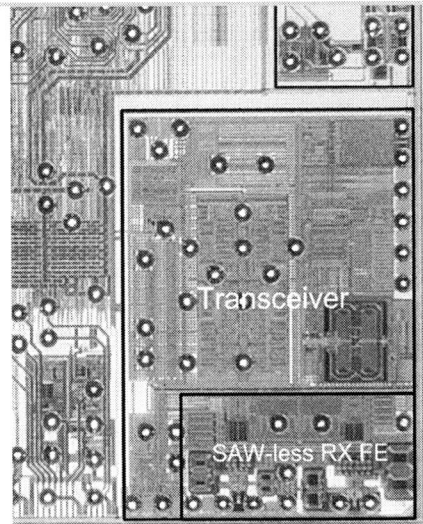

Figure 21.1.7: Die micrograph.

ISSCC 2011 / SESSION 21 / CELLULAR / 21.2

21.2 A 9-Band WCDMA/EDGE Transceiver Supporting HSPA Evolution

Magnus Nilsson[1], Sven Mattisson[2], Nikolaus Klemmer[3], Martin Anderson[2], Torkel Arnborg[1], Peter Caputa[1], Staffan Ek[2], Lin Fan[1], Henrik Fredriksson[1], Fabien Garrigues[3], Henrik Geis[1], Hans Hagberg[1], Joel Hedestig[1], Hu Huang[3], Yevgeniy Kagan[3], Niklas Karlsson[1], Henrik Kinzel[1], Thomas Mattsson[1], Thomas Mills[3], Fenghao Mu[2], Andreas Mårtensson[1], Lars Nicklasson[1], Filip Oredsson[1], Ufuk Ozdemir[1], Fitzgerald Park[3], Tony Pettersson[1], Tony Påhlsson[1], Markus Pålsson[1], Stephane Ramon[3], Magnus Sandgren[1], Per Sandrup[1], Anna-Karin Stenman[1], Roland Strandberg[2], Lars Sundström[2], Fredrik Tillman[2], Tobias Tired[1], Satish Uppathil[3], Joel Walukas[3], Eric Westesson[1], Xuhao Zhang[1], Pietro Andreani[1,4]

[1]ST-Ericsson, Lund, Sweden, [2]Ericsson, Lund, Sweden,
[3]Formerly with ST-Ericsson, Raleigh, NC, [4]Lund University, Lund, Sweden

The future of cellular radio ICs lies in the integration of an ever-increasing number of bands and channel bandwidths. Figure 21.2.1 shows the block diagram of our transceiver, together with the associated discrete front-end components. The transceiver supports 4 EDGE bands and 9 WCDMA bands (I-VI and VIII-X), while the radio can be configured to simultaneously support the 4 EDGE bands and up to 5 WCDMA bands: 3 high bands (HB) and 2 low bands (LB). The RX is a SAW-less homodyne composed of a main RX and a diversity RX. To reduce package complexity with so many bands, we chose to minimize the number of ports by using single-ended RF interfaces for both RX and TX. This saves several package pins, but requires careful attention to grounding. The main RX has 8 LNA ports and the diversity RX has 5, with some LNAs supporting multiple bands. On the TX side, 2 ports are used for all EDGE bands and 4 for the WCDMA bands.

To minimize the interference sensitivity of the LNAs we balance the currents flowing through the respective RF I/O. This is straightforward in a differential LNA, but in a single-ended LNA input and ground may carry different current levels, aggravating leakage. By introducing a dedicated on-chip ground node for the LNAs and returning the signal current from each LNA output and source degeneration to this ground, the LNA drain current is terminated on-chip and only the gate current, with its ground counterpart, loops through the RF I/O. With this grounding strategy, each LNA input and associated ground function as a balanced port, and common-mode interference will be canceled. All LB and HB LNAs, respectively, share the same on-chip degeneration inductor. Inactive LNA inputs are shorted to the LNA ground to further minimize noise pick up. Care has been taken to avoid that an LNA is located close to another LNA with an RX band in the vicinity of the former LNA's TX band (which is indeed possible, considering the allocation of the various TX/RX bands in e.g. the US, Europe, and Far East), which minimizes TX-to-RX leakage.

Finally, the LB (HB) LNA outputs are wired together via an LB (HB) cascode tree, which is connected to an on-chip passive LB (HB) balun (Fig. 21.2.2). Each balun is tuned with a capacitor bank. The LNA port of each balun is decoupled to the LNA ground rather than to the supply ground, thus closing the current loop on-chip at the LNA reference node. The balun IF port is balanced and feeds a passive voltage-mode mixer. A conversion gain of 30dB from LNA input to mixer output is obtained. The mixer has a large capacitive load at the IF port, which results in a relatively low first-order baseband (BB) pole. This load is also translated, by the reverse mixing process, into a narrow bandpass response at the RF port [1], which provides some 10dB of additional suppression for off-channel interferers. The dominant interferer is the TX signal leaking through the WCDMA duplexers, and at full TX power this leakage is close to the RX compression point, increasing the IIP2 requirements for the mixer. Unless the mixer IIP2 exceeds 50dBm, the RX sensitivity may be degraded when standard duplexers are used.

To enhance 2nd-order linearity, the balun AC couples the mixer to the LNA, thereby blocking low-frequency IM2 products that would otherwise leak to the BB I/Q outputs due to mixer imbalance. The mixer uses complementary MOS switches (Fig. 21.2.2), and the DC bias of the PMOS and NMOS switches can be set individually. By turning off the PMOS switches, power can be saved at 10dB IIP2 penalty. To increase the dynamic range, a 4-phase LO drive with approximately 25% duty cycle is used [2]. Supporting 64-QAM in RX, 16-QAM in TX, and 2x2 downlink MIMO for HSPA Evolution requires a very low EVM. Auto-calibration is used to minimize errors due to DC offsets, finite image rejection and BB filter rip-

ple. By controlling the RX/TX LO frequency error in the fractional-N PLLs, high-resolution digital AFC tuning is possible and the frequency error can be kept very small. The only major remaining source of EVM is the phase error from the LO, due to I/Q imbalance combined with phase noise.

The TX consists of separate EDGE and WCDMA paths. The WCDMA transmitter is a 4-phase IQ-modulator followed by a digitally-controlled amplifier. On-chip baluns are used to convert the output signal from differential to single-ended, and set the output impedance to 50Ω. To save power in the WCDMA TX I/Q modulator, a SAW filter is assumed.

The EDGE path adopts a polar modulation TX together with a polar PA. This provides sufficient phase-noise performance to allow SAW-less EDGE TX operation, and ensures an excellent EDGE TX EVM.

A key building block in the polar EDGE TX is the 2-point PLL, which allows for a phase modulation BW that is much higher than the PLL BW. This is necessary to meet the EDGE ACP requirements. As shown in Fig. 21.2.3, in a 2-point PLL the modulation is inserted both into the $\Delta\Sigma$ controlling the frequency divider (feedback path), and into the VCO itself (direct path).

A well-known key issue is gain matching, to ensure the correct modulation index in the direct path. Preferably, the modulation index should be estimated through a calibration algorithm just before the TX burst. Examples of this are proposed in [3,4], where a frequency jump in the PLL is used to estimate the VCO gain. This means that a small loop voltage change must be resolved on top of the much larger VCO tuning voltage, requiring a high-dynamic-range ADC for an accurate calibration. In our approach, a square wave is inserted at both modulation inputs, which mimics several step responses in the modulation paths. If the gain in the direct path is wrong, an error signal will develop in the loop filter, and the sign of this signal will indicate whether the gain is too low or too high. Moreover, by utilizing the built-in loop-filter zero, the VCO tuning voltage can be removed from the error signal. As a result, a simple and robust binary-search ADC can be implemented by using a sign detector. The estimated calibration error is as low as 1%.

Communication with the companion digital BB chip is implemented according to the DigRF v3.09 specifications. Only 6 pins are needed since both control signals and BB RX/TX data are multiplexed on the same wires. The clock frequency of 312MHz is sufficient to transfer both primary and diversity WCDMA data.

The transceiver has been implemented in a 90nm RF CMOS process and is currently in production.

Figure 21.2.4 shows the RX sensitivity of WCDMA band II with TX running at maximum output power. A small degradation is noticed when the TX is enabled, as well as when receiving close to multiples of the 26MHz XO frequency.

The RX EVM performance in WCDMA band II for 64-QAM is shown in Fig. 21.2.5. The EVM is below 3% across a very wide input signal power range.

The throughput gain using RX diversity has been measured in fading conditions (RX moving at 3, 30 and 120Km/h) at the system level. The throughput increases by 50 to 120% when diversity is enabled, clearly showing the merits of a diversity receiver.

A summary of the transceiver performance is given in Fig. 21.2.6 together with a comparison with relevant prior art. Worth noting is the excellent 2G TX EVM, limited only by the PLL noise floor in the polar transmitter. The die photo is shown in Fig. 21.2.7.

References:
[1] S. Vilhonen. "Transferred-impedance filtering in RF receivers", Patent US7187230(B2), 6 March 2007.
[2] R. S. Pullela et al. "Low Flicker-Noise Quadrature Mixer Topology", *IEEE ISSCC Dig. Tech. Papers*, pp. 1870-1871, Feb. 2006.
[3] R. B. Staszewski et al. "Just-In-Time Gain Estimation of an RF Digitally-Controlled Oscillator for Digital Direct Frequency Modulation", *IEEE TCAS-II*, vol. 50, no. 11, pp. 887–892, Nov. 2003.
[4] S.-A. Yu and P. Kinget. "A 0.65-V 2.5-GHz Fractional-N Synthesizer With Two-Point 2-Mb/s GFSK Data Modulation", *IEEE J. Solid-State Circuits*, vol. 44, no. 9, pp. 2411–2425, Sept. 2009.

978-1-61284-303-2/11 $26.00 © 2011 IEEE

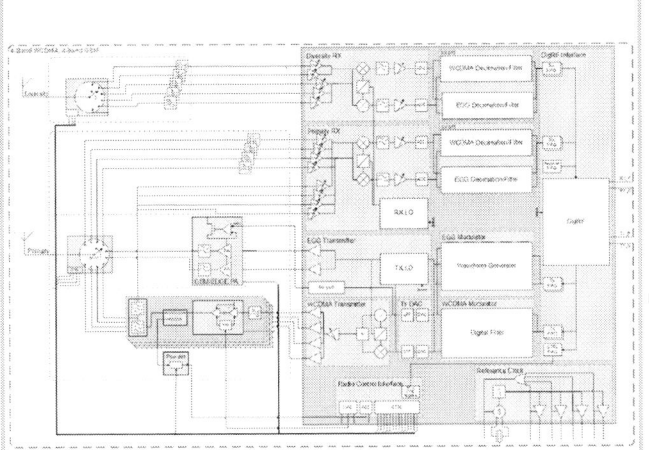

Figure 21.2.1: Block diagram of the transceiver (off-chip discrete components to the left and bottom).

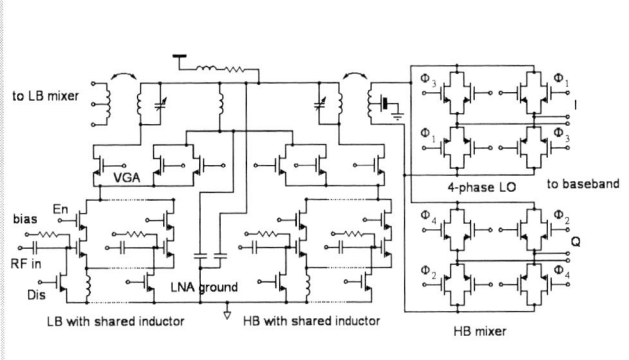

Figure 21.2.2: Simplified schematic view of the main RX front-end (LB mixer not shown for clarity). The diversity RX front-end is basically identical, and has its own dedicated LNA ground node.

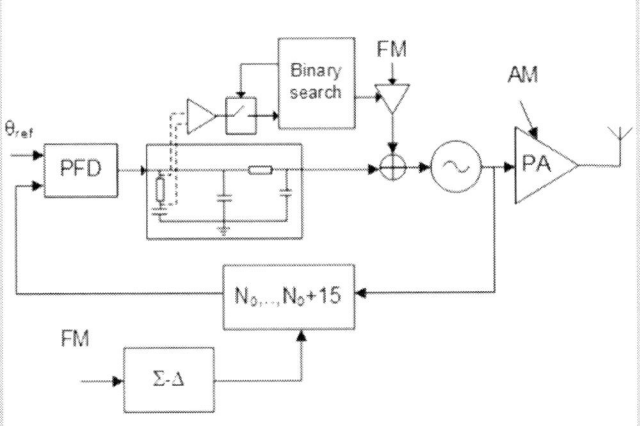

Figure 21.2.3: TX PLL with 2-point modulation and direct-path gain calibration.

Figure 21.2.4: Sensitivity for WCDMA RX band II.

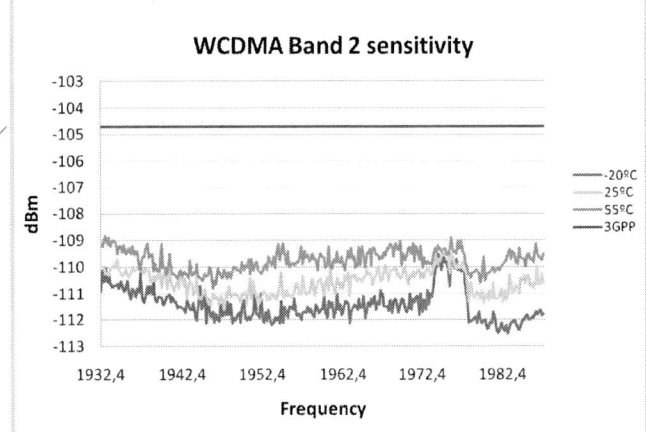

Figure 21.2.5: Left: 64-QAM RX constellation for the main RX in WCDMA band II; right: EVM vs. input power level.

Parameter	This work	Gaborieau et al, ISSCC '09	Sowlati et al, ISSCC '09	Hadjichristos et al, ISSCC '09	Yoshida et al, ESSCIRC '08
power consumption 2G RX mode	129 mW	90 mW (no LO)	---	---	317 mW
power consumption 2G TX mode	126 mW	---	---	---	407 mW
power consumption 3G RX+TX (0dBm)	269 mW	92 mW (RX only, no LO)	414 mW (@6 dBm)	255 mW	566 mW (@-4dBm)
RX NF (LB/HB)	2.3/2.5 dB	2.4/2.2 dB	2.0/2.5 dB	2.6/2.8 dB	1.8 dB (LNA)
3G RX IIP3 (LB/HB)	-6/-4 dBm	-5/-8 dBm	-2.7/-2.5 dBm	---	-6.1 dBm (LNA)
3G RX IIP2 (LB/HB)	62/58 dBm	62/59 dBm	>50 dBm	55 dBm	56.3 dBm
RX EVM (LB/HB)	2.5/3.0%	5.6%	---	4.0%	3.3%
TX 2G EVM (LB/HB)	1/1.5%	---	2.2/2.5%	---	1.8%
TX 3G EVM (LB/HB)	3.5/4%	---	2.1/2.5%	4.0%	3.1%
chip area	14.4 mm²	10 mm² (*)	25 mm²	31.4 mm²	42.2 mm²
technology	90nm CMOS	250nm BiCMOS	130nm CMOS	180nm CMOS	130nm CMOS
battery voltages	1.2 & 1.8V	1.8 & 2.7 V	1.8V	2.1 & 2.7 V	1.8 & 2.8 V
diversity	all bands	No	no	WCDMA	no
year	2011	2009	2009	2009	2008

(*) Some passive components mounted on a substrate inside the package.

Figure 21.2.6: Summary table of the transceiver performance and comparison with earlier EDGE/WCDMA transceivers.

ISSCC 2011 PAPER CONTINUATIONS

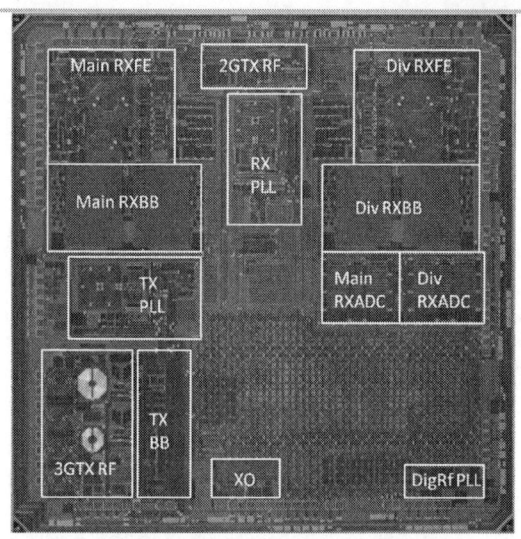

Figure 21.2.7: Die micrograph of the transceiver (3.8mm x 3.8mm).

ISSCC 2011 / SESSION 21 / CELLULAR / 21.3

21.3 A 65nm CMOS SoC with Embedded HSDPA/EDGE Transceiver, Digital Baseband and Multimedia Processor

Alberto Cicalini[1], Sankaran Aniruddhan[1], Rahul Apte[2], Frederic Bossu[1], Ojas Choksi[1], Dan Filipovic[1], Kunal Godbole[1], Tsai-Pi Hung[1], Christos Komninakis[1], David Maldonado[1], Chiewcharn Narathong[1], Babak Nejati[1], Deirdre O'Shea[1], Xiaohong Quan[1], Raj Rangarajan[1], Janakiram Sankaranarayanan[1], Andrew See[1], Ravi Sridhara[1], Bo Sun[1], Wenjun Su[1], Klaas van Zalinge[1], Gang Zhang[1], Kamal Sahota[1]

[1]Qualcomm, San Diego, CA, [2]Qualcomm, Santa Clara, CA

Cellular phones in emerging markets have continued to grow with multimedia features such as MP3 playback, video encode and decode, high-resolution cameras and web browsing. To efficiently support multimedia functionalities, high-performance modems are required. There is also a strong demand to reduce the cellular phone PCB footprint, and to enable integration of peripheral devices such as Bluetooth and Wireless LAN. Previously reported SoC or SiP solutions have integrated a GPRS/EDGE radio with modem and limited multimedia capabilities [1-4]. This paper presents a multimode UMTS/GSM RF transceiver integrated with a digital baseband having advanced multimedia functionalities. The SoC, designed in 65nm digital CMOS, supports quad-band GSM/GPRS/EDGE (3GPP R4, Class12) and tri-band UMTS/WCDMA (3GPP R99, R5 cat 5/6), including band 1-2-4-5-6-8-9-10.

Figure 21.3.1 shows the block diagram of the overall system. The SoC includes a transceiver, a mixed-signal (MS) section with house-keeping ADC (HKADC) and stereo audio CODEC. The digital baseband integrates GSM/GPRS/EDGE, WCDMA/HSDPA and GPS modem, as well as ARM9, application and modem DSP, internal ROM/RAM, external memory interface (EBI) and connectivity cores. The SoC die is packaged with a power management IC in a 12×12 BGA with 0.5mm pitch. The transceiver block diagram is shown in Fig. 21.3.2: it consists of 5 differential RX inputs, 5 single-ended TX outputs, a power detector for WCDMA TX power control, and a TX sense receiver to digitally calibrate receiver nonlinearity. The LO signals are generated by on-chip fractional PLLs that allow fast settling, required for EDGE multi-slot operation, and exhibit good integrated phase noise, required for HSDPA.

The transmitter supports 3 UMTS and 2 GSM outputs with 3 modulation schemes: UMTS uses direct IQ up-conversion, while GSM/EDGE employs either large-signal or small-signal Polar. Large-signal polar allows for EDGE amplitude modulation to be applied, through an integrated DAC, to an external PA with envelope modulation capability. In small-signal polar, instead, EDGE envelope restoration occurs at the IQ upconversion mixer: this scheme doesn't require PA amplitude modulation, enabling significant size, cost and current reduction as well as multimode operation through the use of shared GSM-UMTS PAs. Pre-distortion calibration is performed to further enhance the linearity and efficiency of the overall system. GSM/EDGE SAW-less quad-band operation is achieved in both Polar designs: phase modulation is generated using an on-chip PLL capable of two-point modulation. Good RMS phase error and EVM are achieved by performing on-the-fly K_{VCO} calibration at the beginning of each GSM/EDGE frame to compensate for PVT variations. In addition the loop response is digitally compensated to provide good EVM. Major portions of the TX front-end are shared between the three schemes. In UMTS mode direct I-Q upconversion is used: the baseband filter provides gain control and implements real poles to attenuate DAC aliases and noise at the duplex offset. The IQ upconverter employs an active Gilbert-cell topology based on 8 identical mixer cells that can be independently controlled: this architecture offers power control and current saving, since mixer current scales with output power. The output stage consists of two sections: a pre-driver amplifier (pDA) and driver amplifier (DA). An inductorless complementary push-pull topology is used for both pDA and DA. Good linearity performance is obtained in the output stage by employing gain expansion and class-AB biasing in the pDA, while achieving the desired output power at the DA output. Binary-weighted unit cells provide gain control in the pDA and DA respectively. Finally, a replica bias circuit sets pDA and DA operating points for optimum linearity and noise performance over PVT.

The receiver, based on a direct downconversion topology, has 5 differential inputs: 2 dedicated for GSM, and 3 shared between GSM and WCDMA. It simultaneously supports quad-band GSM and up to three UMTS bands, sharing duplexer filters between GSM and UMTS. This implementation, although degrading the GSM sensitivity due to higher duplexer insertion loss, further reduces BOM and PCB area, while satisfying most of the world's band configurations. All LNAs have inductive degeneration to improve linearity and input matching, and use PMOS active loads to minimize area. Each LNA is followed by a dedicated 25%-duty-cycle passive mixer, minimizing coupling between I and Q paths that is responsible for IIP2 and NF degradation [5]. The detailed IQ mixer block diagram is shown in Fig. 21.3.3. All mixers share a baseband filter with two real poles, implemented with a programmable gain transimpedance amplifier and a PGA, bypassed in WCDMA mode to save current. The BBF is followed by a passive alias-rejection filter, and a high-dynamic-range ADC (>92dB in GSM and >74dB in WCDMA) that allows for the majority of the selectivity to be done in the digital domain. The receiver includes an auxiliary TX sensing (TXS) path that taps a fraction of the LNA current to extract the TX leaked into the LNA through the duplexer. This TX signal is digitized, squared and correlated with the TX signal downconverted to the main RX baseband outputs through second-order nonlinearity (IM2). LO waveforms are adjusted on-the-fly to minimize this correlation, thereby minimizing in-band IM2 and increasing IIP2. In addition, an IM2 cancellation circuitry is implemented in the digital section.

The radio operates from two supply domains derived from the battery using SMPS, to improve power efficiency, and LDO, to boost supply isolation: RF runs at 1.3V, while analog baseband blocks use 2.2V. The transmitter achieves chip output power of 9dBm in GSM mode, 4dBm in WCDMA SAW-less mode and 7dBm in WCDMA SAW mode, while exceeding ORFS/ACLR requirements. In EGSM band, phone-level GSM phase error and 400kHz-ORFS are 0.8° and -67.7dBc respectively, and peak battery current consumption is 53mA (battery at 3.7V). All GSM TX output SAW filters are removed: RX band noise has more than 5dB margin, and only 2 out 5 allowed exceptions are used. GSM receiver NF is <2.4dB across all bands. A sample of GSM RX band noise, 400kHz-ORFS and RX sensitivity is shown in Fig. 21.3.4. In IMT band, phone-level WCDMA TX (QPSK) EVM and the ACLR1 are 4% and -41.7dBc respectively, while typical phone RX sensitivity is <-111dBm. At 0dBm PA output power, the TX and RX draw respectively 45.1mA and 45.3mA from the battery. Fig. 21.3.5 illustrates UMTS IMT ACLR1 and sensitivity performance. Receiver typical UMTS NF is <2.2dB, and it exceeds IIP2, IIP3 and RX LO phase noise specifications, while removing all inter-stage SAW.

A summary of radio stand-alone and system level performances measured at the antenna port of a reference phone are shown in Fig. 21.3.6: reported GSM sensitivity for EGSM and PCS is with shared UMTS duplexer filter, while TX HSDPA performance is according to 3GPP TS 34.121. A micrograph of the SoC is shown in Fig. 21.3.7: the total die area is 52.5mm², of which 13.4mm² is used by the transceiver. Currently in volume production, this SoC integrates, on the same die, a UMTS transceiver and digital baseband. The high level of integration reduces PCB area, power consumption, time-to-market, and BOM costs, providing an attractive solution for 2.5 and 3G feature phones.

References:
[1] P-H. Bonnaud, et al., "A Fully Integrated SoC for GSM/GPRS in 0.13µm CMOS," *ISSCC Dig. Tech. Papers*, pp. 482-483, Feb. 2006.
[2] R. B. Staszewski, et al., "All-Digital PLL and GSM/EDGE Transmitter in 90nm CMOS," *ISSCC Dig. Tech. Papers*, pp. 316-317, Feb., 2005.
[3] H. Darabi, et al., "A Fully Integrated Quad-Band GPRS/EDGE Radio in 0.13µm CMOS," *ISSCC Dig. Tech. Papers*, pp. 206–207, Feb. 2008.
[4] M. Hammes, et al., "A GSM Baseband Radio in 0.13µm CMOS with Fully Integrated Power-Management," *ISSCC Dig. Tech. Papers*, pp. 264-265, Feb. 2007.
[5] M. C. M. Soer, et al., "A 0.2-to-2.0GHz 65nm CMOS Receiver Without LNA Achieving >11dBm IIP3 and <6.5dB NF," *ISSCC Dig. Tech. Papers*, pp. 222-223, Feb. 2009.

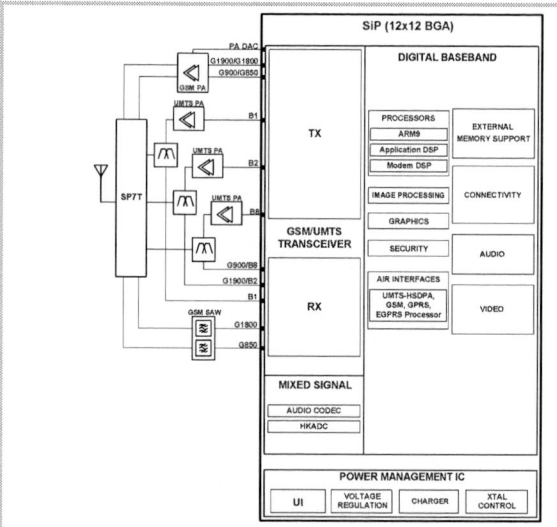

Figure 21.3.1: System Block Diagram.

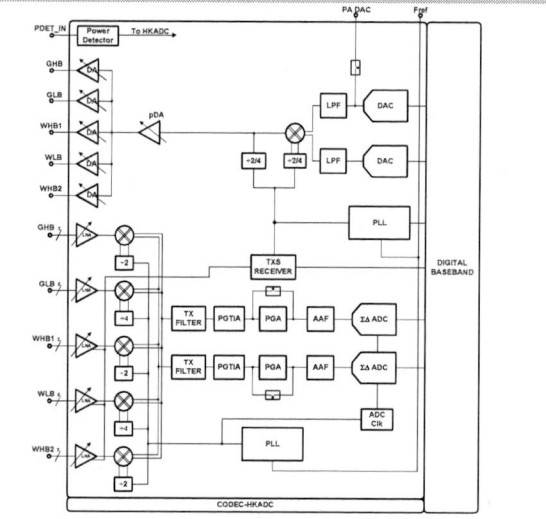

Figure 21.3.2: Block diagram of the RF transceiver.

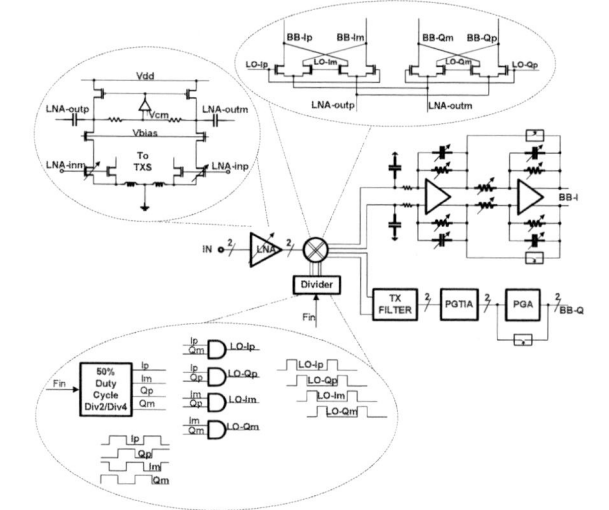

Figure 21.3.3: Receiver detailed block diagram.

Figure 21.3.4: GSM-EGSM phone level performance: RX band noise, 400kHz-ORFS and sensitivity (using B8 duplexer).

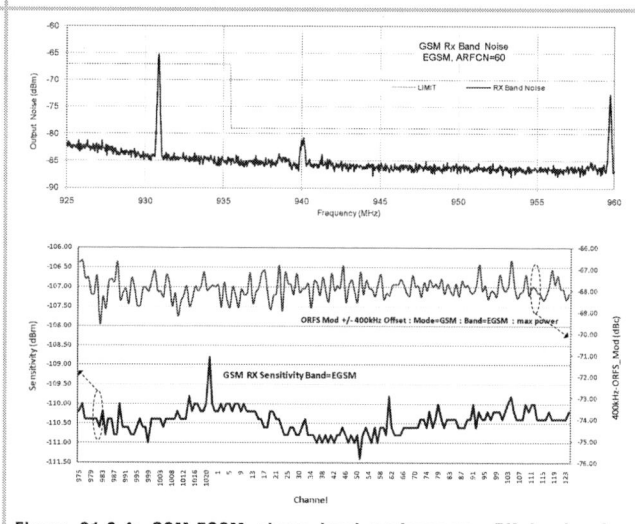

Figure 21.3.5: UMTS-IMT phone level performance: ACLR1 and sensitivity at max TX power.

BAND	SPEC	CELL	EGSM	DCS	PCS	IMT	Low Band 3GPP Requirements	High Band 3GPP Requirements
RX GSM	Chip NF [dB]	2.3	2.1	2.3	2.4	-	-	-
	Chip IIP2 [dBm]	>46	>46	>46	>46	-	-	-
	Chip IIP3 [dBm]	-3.8	-4.9	-4.1	-3.7	-	-	-
	Phone Typ. Sensitivity [dBm]	-111.3	-110.5	-110.6	-110.2	-	-102.0	-102.0
	Max Gain Peak Battery Current [mA]	50.7	46.1	56.4	53.4	-	-	-
TX GSM GMSK	TX output SAW	no	no	no	no	-	-	-
	Phone Typ. 400KHz mod [dBc]	-68.4	-67.7	-66.8	-65.7	-	-60.0	-60.0
	Phone Typ. Rms Phase Error [deg]	0.95	0.85	1.40	1.50	-	5.0	5.0
	Phone Typ. RX Band Noise [dBc/Hz]	-167.0	-168.0	-157.0	-159.0	-	-162.0	-151.0
	Max Pout Peak Battery Current [mA]	52.0	53.0	65.5	66.0	-	-	-
TX GSM EDGE	TX output SAW	no	no	no	no	-	-	-
	Phone Typ. 400KHz mod [dBc]	-65.0	-64.9	-63.7	-60.6	-	-54.0	-54.0
	Phone Typ. EVM rms [%]	2.0	2.0	2.3	2.9	-	9.0	9.0
	Max Pout Peak Battery Current [mA]	52.0	53.0	65.5	66.0	-	-	-
RX UMTS	Chip NF [dB]	1.9	2.0	-	1.9	2.2	-	-
	Chip IIP2 [dBm]	>50	>50	-	>50	>50	-	-
	Chip IIP3 [dBm]	-2.1	-2.4	-	-1.5	-0.5	-	-
	Phone Typ. Sensitivity [dBm]	-111.4	-112.0	-	-111.0	-111.4	-106.7	-106.7
	0dBm Peak Battery Current [mA]	37.5	38.3	-	43.8	45.3	-	-
TX UMTS	TX output SAW	no	yes	-	yes	no	-	-
	Phone Typ. R99 ACLR1 5MHz [dBc]	-44.7	-44.1	-	-43.0	-41.7	-33.0	-33.0
	Phone Typ. R99 ACLR2 10MHz [dBc]	-59.2	-59.3	-	-53.1	-55.9	-43.0	-43.0
	Phone Typ. R99 EVM rms [%]	4.9	4.3	-	4.1	4.0	17.5	17.5
	Phone Typ. HSDPA ACLR1 5MHz [dBc]	-44.6	-44.0	-	-40.8	-39.6	-33.0	-33.0
	Phone Typ. HSDPA ACLR1 10MHz [dBc]	-57.6	-58.0	-	-55.6	-54.1	-43.0	-43.0
	Phone Typ. HSDPA EVM rms [%]	5.3	5.3	-	3.9	4.5	17.5	17.5
	0dBm Peak Battery Current [mA]	48.2	44.8	-	47.8	45.1	-	-

Figure 21.3.6: Transceiver stand-alone and phone typical performance summary. Battery currents are measured at 3.7V.

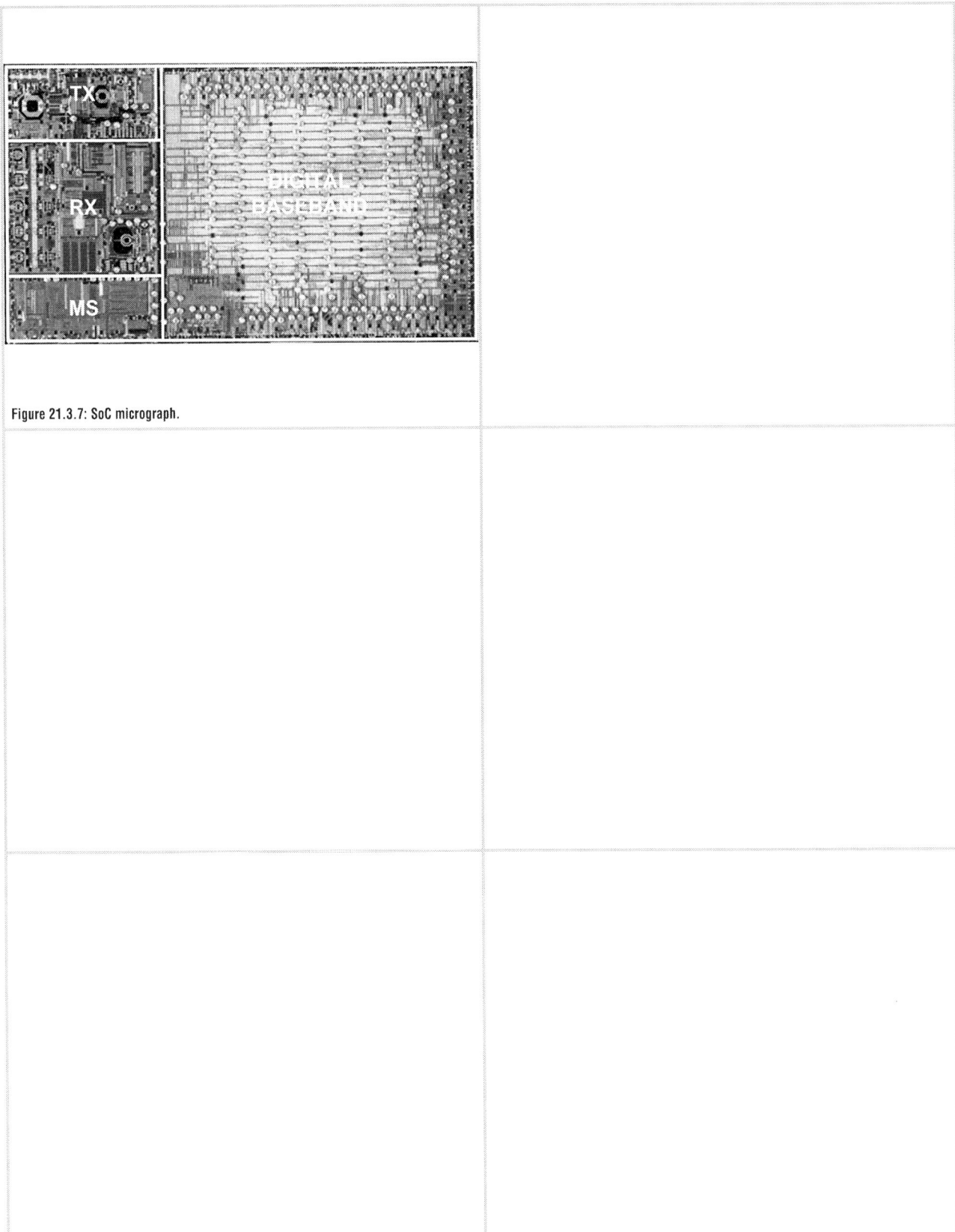

Figure 21.3.7: SoC micrograph.

ISSCC 2011 / SESSION 21 / CELLULAR / 21.4

21.4 A Receiver for WCDMA/EDGE Mobile Phones with Inductorless Front-End in 65nm CMOS

Federico Beffa[1], Tze Yee Sin[1], Alexander Tanzil[2], David Ivory[3], Bernard Tenbroek[1], Jon Strange[1], Walid Ali-Ahmad[2]

[1]MediaTek, West Malling, United Kingdom, [2]MediaTek, Singapore, Singapore, [3]MediaTek, Cambridge, United Kingdom

The spectrum allocated for the operation of cellular services is country dependent and fragmented in several frequency bands [1]. Mobile phones, to be usable globally, are therefore required to support many bands. The lack of suitable tuneable pre-selection filters and duplexers mandates the use of a large number of low-noise amplifiers (LNAs) in the receiver section of the transceiver. Minimization of the area occupied by the LNAs is therefore important. Virtually every transceiver for mobile phone applications published to date makes use of LNAs employing on-chip [3,5] or on-passivation [4] spiral inductors. Spiral inductors are large and do not scale with technology, leading to an increasing relative cost of the receiver front-end with newer technology nodes. Post-passivation and SiP technologies also add noticeably to the costs. In this paper we present the first receiver employing no on-chip or above-passivation spiral inductors on the receive signal path and with a linearity and noise performance suitable for WCDMA/EDGE applications. The eight channel direct-conversion receiver is part of a multimode transceiver requiring neither RX nor TX interstage SAW filters for FDD 3G operation. A block diagram of the receiver is shown in Fig. 21.4.1.

The core of the receiver is constituted by the front-end, a simplified schematic of which is shown in Fig. 21.4.2. The variable-gain LNA is based on the shunt-shunt feedback topology and provides a gain of 16, 7 or -5 dB. To break the otherwise fixed relation between input impedance and the transistor transconductance, a load impedance is used, consisting of two resistors connected in series with the core of a passive mixer. The impedance looking into the mixer RF port is made very low and the node is used as a virtual ground from DC up to GHz frequencies. The resistors R_{LIa}, R_{LIb}, R_{LQa} and R_{LQb} are not only used as load for the LNA, but they are also used (i) to transform the output voltage of the LNA into an input current for the passive mixer, (ii) to isolate the I channel from the Q channel and (iii) to improve the linearity of the mixer. The last point can be appreciated by noting that most of the mixer input voltage V_o^{LNA} is dropped across the load resistors and only a very small fraction of it is dropped across the switching transistors.

To more easily explain the working of the pseudo-differential variable gain LNA, Fig. 21.4.3 shows the simplified schematic of one-half of a two-gain-settings version of the amplifier. To increase power efficiency, the transconductance is implemented as a complementary stage and comprises transistors M1, M5, M4 and M10. In high-gain mode the signal current from the transconductance stage is steered by the cascodes to the output node A of the amplifier. In this mode the feedback resistor consists of the series connection of R_{Fa} and R_{Fb}. The load resistors R_{LIa}, R_{LQa} leading to the mixer core are shown for illustration purpose connected to AC ground. In low-gain mode the signal current from the transconductance stage is steered to the node B between R_{Fa} and R_{Fb}. In this mode the feedback resistor comprises just R_{Fa}. The load of the amplifier is now constituted by R_{Fb} in series with R_{LIa}, R_{LQa}. To maintain a constant input impedance, the effective transconductance of the forward stage is reduced. This is achieved by connecting the gate of M2 to ground and the gate of M3 to supply, thereby disabling a large part of the transconductor and allowing a substantial current reduction. The implemented LNA consumes 21.7, 4.8 and 2mA in high-, mid- and low-gain modes respectively from a 1.5V supply. No current is used to implement the RF portion of the mixer (apart from the LO driver). To maximize dynamic range, the output common mode of the amplifier is set to the mid-rail voltage by a servo loop built around OpAmp O1.

To reduce the size of the receiver, in the bands below 1GHz, two adjacent LNAs share a common I/Q mixer pair. At higher frequencies mixer sharing causes excessive loading with significant noise figure degradation. To avoid this, one mixer core pair per RF channel is used and the paths are merged at the baseband transimpedance amplifier virtual ground. The gates of unused mixer-core transistors are connected to ground thereby isolating the unused LNAs from the

active path. The mixer-core transistors effectively double in function as multiplexer switches.

The front-end is followed by a baseband strip extending the gain range of the receiver from 9 to 69dB programmable in 3dB steps and implementing three different 3rd-order low-pass filter characteristics suitable for the three supported modes of operation. An on-chip automatic cut-off frequency calibration engine ensures an accuracy of better than 5%. The output stage of the baseband strip provides a limiting function designed to prevent overloading of the $\Delta\Sigma$ A/D converter following that stage in the system.

A major challenge in the implementation of SAW-less receivers for operation in 3G FDD mode is the high TX power leaking to the input of the receiver. To limit desensitization to less than 0.2dB an IIP2 larger than 50dBm is required [2]. In our receiver this is achieved with no calibration or control loops by the described series-resistors-based interface between LNA and mixer. In addition, coupling between the RF and the LO paths was carefully minimized. The absence of on-chip inductors on the RF signal path was of great help in this respect. The typical TX leakage-induced desensitization performance of the receiver is shown in Fig. 21.4.4. The average measured IIP2 in the most challenging band 2 is 57dBm with a lowest value of greater than 50dBm.

The receiver achieves an IIP3 of better than -3dBm with 10/20MHz blockers in WCDMA mode and with 0.6/1.8MHz blockers in EDGE mode. At lower LNA gain settings the IIP3 increases up to 5.4dBm in WCDMA mode with 3.5/6.5MHz tones (see Fig. 21.4.5). This high level of linearity is the result of removing all nonlinear stages in front of the passive mixer core apart from the fundamental LNA input stage.

The noise figure of the receiver in WCDMA mode varies between a value of 2.2dB in band 2 to a value of 2.5dB in band 5. The flicker-noise corner frequency is 400Hz and has a negligible effect on the EDGE noise figure which is essentially identical to the one in WCDMA mode.

The performance of the receiver is summarised in Fig. 21.4.6. In spite of the fact that no inductors were used on the RX signal path, state-of-the-art performance was achieved as demonstrated by the comparison with other recently published SAW-less 3G transceivers. In addition, the inductorless strategy enabled us to achieve a very competitive die area, while efficient current-scaling techniques avoid significant power consumption penalty as compared with inductor-based designs. The area occupied by the 8 RF front-ends, including LNAs, mixers, frequency dividers, pads and ESD protection is smaller than 0.9mm². This is less than half the area occupied by the corresponding inductor based circuits used in [2] to implement just 3 RF channels. A micrograph of the receiver implemented in a digital 65nm CMOS technology is shown in Fig. 21.4.7. The IC is packaged in a dual-row QFN package.

Acknowledgement:
The authors would like to thank the many people who helped with the development and characterization of the presented IC.

References:
[1] 3GPP, TS 25.101, Release 8.6.0, www.3gpp.org.
[2] B. Tenbroek, et al., "Single-Chip Tri-Band WCDMA/HSDPA Transceiver without External SAW Filters and with Integrated TX Power Control" *ISSCC Dig. Tech. Papers*, pp. 202-607, Feb. 2008.
[3] Q. Huang. et al., "A Tri-Band SAW-Less WCDMA/HSPA RF CMOS Transceiver with On-Chip DC-DC Converter Connectable to Battery", *ISSCC Dig. Tech. Papers*, pp. 60-61, Feb. 2010.
[4] O. Gaborieau, et al., "A SAW-Less Multiband WEDGE Receiver", *ISSCC Dig. Tech. Papers*, pp. 114-105, Feb. 2009.
[5] T. Sowlati, et al., "Single-Chip Multiband WCDMA/HSDPA/HSUPA/EGPRS Transceiver with Diversity Receiver and 3G DigRF Interface Without SAW Filters in Transmitter / 3G Receiver Paths" *ISSCC Dig. Tech. Papers*, pp. 116-117, Feb. 2009.
[6] D. Kaczman, et al., "A Single-Chip 10-Band WCDMA/HSDPA 4-Band GSM/EDGE SAW-less CMOS Receiver With DigRF 3G Interface and +90 dBm IIP2", *IEEE J. Solid-State Circuits*, vol. 44, No. 3, Mar. 2009.

978-1-61284-303-2/11 $26.00 © 2011 IEEE

ISSCC 2011 / February 23, 2011 / 10:15 AM

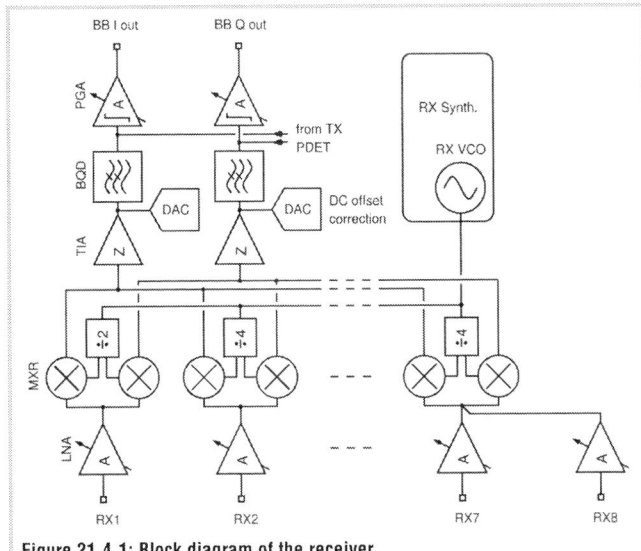

Figure 21.4.1: Block diagram of the receiver.

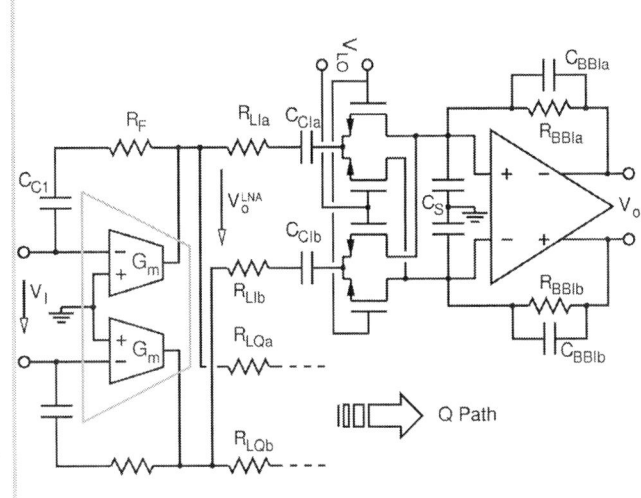

Figure 21.4.2: Overview of one RF channel.

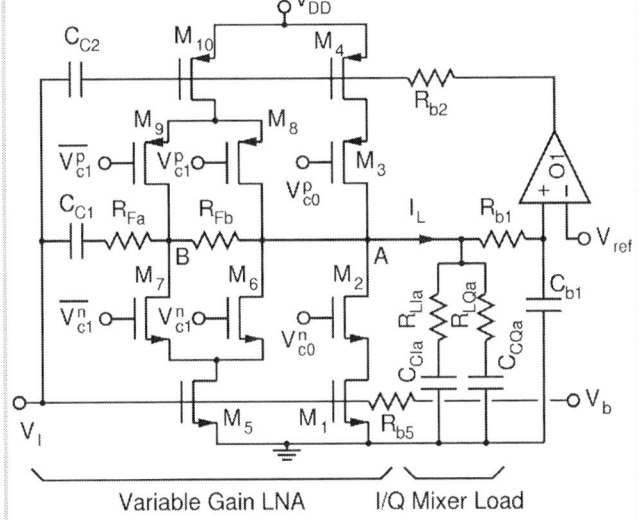

Figure 21.4.3: Half circuit schematic of a two-gain-setting version of the LNA.

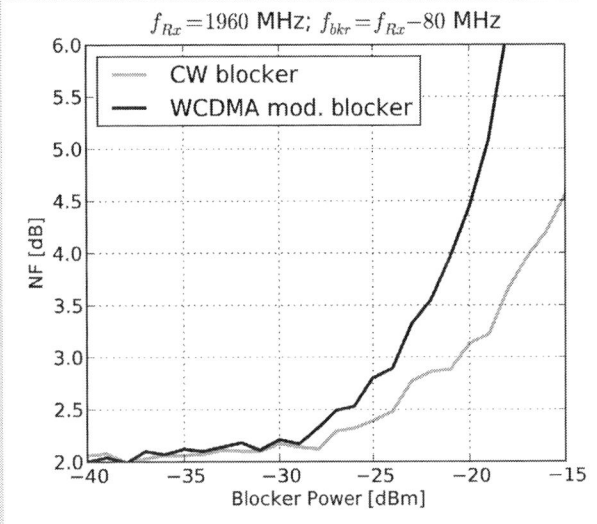

Figure 21.4.4: NF vs. blocker power.

Figure 21.4.5: Typical WCDMA mode IIP3.

Param.	This	[3]	[4]	[5]	[6]	Unit
RX batt. curr.	28.5 – 38.5[*2,3]	40[*2]	92mW@ 1.8/2.7V	135[*2] w. TX	15.1[*6]@ 1.5V	mA
Process	65nm	0.13µm	0.25µm[*1]	0.13µm	90nm	-
RX FE area[*5]/ch	0.11	0.28[*4]	0.12(Si)+4 (PICS)[*4]	0.28[*4]		mm²
RX WCDMA mode						
V. Gain	9 – 67	7 – 103	9 – 96	? – 63	? – 60	dB
DSB NF LB/HB	2.5/2.4	2.5/3.0	2.2/2.1	2.0/2.5	?/2.2	dB
IIP3 10 / 20MHz	-3		-8	-2.9	< -9	dBm
IIP2 DPX	>50	55	58 typ.	>50	90	dBm
RX GSM mode						
DSB NF LB/HB	2.5/2.4		2.4/2.2	2.1/2.6	?/3.2	dB
IIP3 0.8 / 1.6MHz	-2.5		-10	-6		dBm
IIP2 6MHz	>50		60	>48		dBm

[*1]: BiCMOS in SiP with high quality passives. [*2]: using a DC-DC converter.
[*3]: current scales with LNA gain. [*4]: estimated from micrograph.
[*5]: including pads and ESD protection. [*6]: core only

Figure 21.4.6: Summary and comparison with recent publications.

978-1-61284-303-2/11 $26.00 © 2011 IEEE

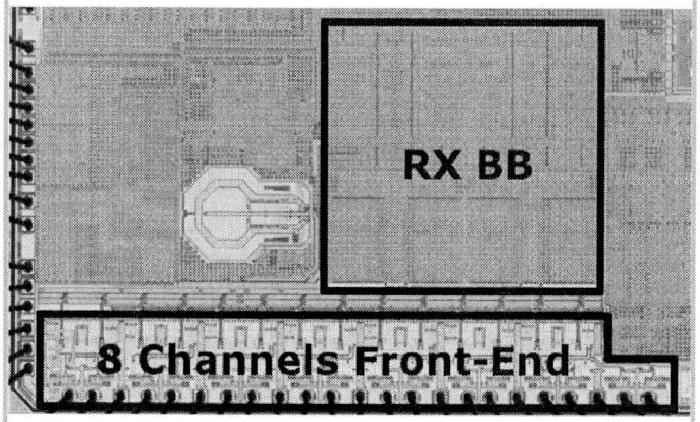

Figure 21.4.7: Micrograph of the receiver.

ISSCC 2011 / SESSION 21 / CELLULAR / 21.5

21.5 A Compact SAW-less Multiband WCDMA/GPS Receiver Front-End with Translational Loop for Input Matching

Xin He, Harish Kundur

NXP Semiconductors, Eindhoven, The Netherlands

In FDD systems such as WCDMA, strong TX leakage presented at RX imposes stringent RX out-of-band (OOB) IIP3 and IIP2 requirements in addition to low noise figure (NF) requirement, usually necessitating an inter-stage SAW filter in the RX chain. In a GPS receiver that co-exists with cellular applications in the same platform, such a filter is also often needed. Furthermore, to achieve low NF, conventional receivers adopt LNAs with inductive degeneration for input matching. This substantially increases the number of off-chip matching components as well as the chip size in multiband cellular applications. Another drawback is that the LNA current has to be fixed to provide a constant g_m over different gains for input matching, even when the wanted signal is strong.

SAW-less receiver design has been addressed in [1-6]. [1] and [2] rely on either an extra bonding wire or an additional Q-enhanced LC resonator to filter the blockers. The blocking filter technique presented in [3] sacrifices NF and introduces an LO leakage issue. A more attractive solution is to use a single stage transconductance amplifier (TCA) in combination with a passive current mixer switched by 25% (or effective) duty-cycle LO [4][5]. Also in [1-3], inductive degeneration is employed in the TCA for input matching.

This paper describes a compact SAW-less receiver front-end design targeting multiband multimode GSM/EDGE/WCDMA/GPS applications. Using a translational loop for input matching, the inductor-free receiver achieves high OOB linearity and low NF at low power consumption. The concept of the receiver is depicted in Fig. 21.5.1. First a TCA converts the RF input voltage to RF current. Subsequently a passive current mixer downconverts this RF current to IF current, which is then fed into a transimpedance amplifier (TIA) and converted to IF voltage with low-pass filtering. More gain control and filtering stages can be added after the TIA, with little impact on the OOB linearity and NF. Since the TCA is directly loaded by the low impedance virtual ground of the TIA, the voltage swing at the TCA output is low, leading to superior OOB IIP3 [5]. On top of the RX chain, a translational loop is formed by upconverting the inverted IF output voltage (naturally available from the differential TIA output), and then feeding it back to the RF input via the resistor R_f. Note that the antenna impedance and R_f form a voltage divider, which attenuates the LO leakage generated in the upconversion mixer.

Observing from the RF input, the receiver is equivalent to an amplifier with a shunt resistive feedback. The receiver input impedance can be approximated by $Z_{in} \approx R_f / G$, where G is the receiver voltage gain combined with the mixer upconversion gain. The characteristic of input matching with a translational loop is similar to that in the passive-mixer-first approach [7], where the S_{11} bandwidth is controlled by the IF bandwidth. When the RF input offset is smaller than the IF bandwidth, input matching is fulfilled by appropriate configuration of the voltage gain and R_f value. In the receiver, the noise contributed by R_f can be neglected under a large voltage gain. Thus this approach eliminates inductors without noise penalty. Moreover, it decouples the optimization of noise and linearity performance from input matching. Without the constraint of input matching, the TCA current can be adapted to the wanted input signal level.

Figure 21.5.2 details the receiver circuit implementation. First an inverter-type TCA converts the RF voltage to RF current. The inverter-type TCA takes advantage of current reuse to generate g_m, yielding high linearity and low noise at low power. To adapt to different wanted input signal levels, the TCA is split into multiple unit cells, which can be turned on/off individually to further save current. After the TCA, a passive current mixer driven by non-overlapping 25% duty-cycle LO is employed to downconvert the RF current to I/Q IF currents. In comparison with conventional mixers switched by a 50% duty-cycle LO, the passive current mixer switched by a 25% duty-cycle LO demonstrates significantly improved IIP2 [5]. To enhance the OOB linearity, capacitors are inserted at the virtual grounds of the I/Q TIA to absorb the OOB blockers. In the translational loop, a passive voltage mixer driven by a 25% duty-cycle LO is employed to

upconvert the I/Q IF voltage [8]. Owing to the high load impedance, there is little current flow in the voltage mixer. To minimize LO leakage at the low-gain mode, the translational loop can be switched off. By switching on $SW2$ and switching off $SW1$, the resistor R_L alternatively provides input matching, at the price of 3dB NF penalty. The degraded NF in the low-gain mode is not an issue because the wanted input signal level is higher.

In the receiver, the passive current mixer and the passive voltage mixer are driven by the same 25% duty-cycle LO. A re-clocking technique is adopted in the LO generation circuit to improve I/Q matching and phase noise, as presented in [8]. Since the size of the voltage mixer is much smaller than that of the current mixer, the current consumed by the buffer to drive the voltage mixer is negligible. Note that in implementation the LO duty cycle is chosen to be slightly smaller than 25% to avoid I/Q crosstalk.

Supplied with 1.3V, the receiver fabricated in a 45nm CMOS process consumes 4.3mA in the TCA, and total 3mA in the I/Q TIA. The receiver achieves 2.2dB NF at 900MHz (cellular low-band), 2.4dB NF at 1.575GHz (GPS band), and 2.7dB NF at 2.1GHz (high-band) with 37dB voltage gain. Using a fixed TIA current, the measured NF versus TCA current at low-band is shown in Fig. 21.5.3. Figure 21.5.4 plots the measured input S11 over two IF bandwidth settings at 960MHz LO frequency. The curve with wider matching bandwidth (<-10dB return loss over 8MHz) corresponds to the wider IF bandwidth. Due to parasitics, the S_{11} notch slightly shifts upwards by about 1MHz.

Figure 21.5.5 shows the measured IIP3 versus frequency offset with 37dB voltage gain at different bands. In particular, the receiver demonstrates -4.5/-3.5dBm half/full duplex IIP3 at low band with 45MHz duplex distance (measured with two tones at -45&-22.5MHz offsets for half duplex IIP3, and -45&-90MHz offsets for full duplex IIP3), and 0.5/1.5dBm half/full duplex IIP3 at high band with 190MHz duplex distance. Both low-band and high-band OOB IIP3 are sufficiently high to cope with TX leakage without needing a SAW filter. The high-band OOB IIP2 measured using two tones at -190MHz offset with 1MHz spacing is above 60dBm over different samples, which is high enough without calibration. The low-band OOB IIP2 measured with two tones at -45MHz offset is decreased to just above 40dBm, necessitating calibration to reach targeted 50dBm IIP2. The measured LO leakage at the RF input is lower than -80dBm over different bands and different samples, which complies with spectrum regulation for cellular systems. Figure 21.5.6 summarizes the performance comparison. Figure 21.5.7 shows the micrograph of the die (occupying 1mm²), with active circuit area less than 0.2mm².

Acknowledgements:
The authors would like to thank Jan van Sinderen and Frank Leong for technical discussion.

References:
[1] N. Yanduru et al., "A WCDMA, GSM/GPRS/EDGE receiver front end without interstage SAW filter", *RFIC Symp. Dig.*, pp.77–80, Jun. 2006.
[2] B. Tenbroek et al., "Single-Chip tri-band WCDMA/HSDPA transceiver without external SAW filters and with integrated TX power control", *ISSCC Dig. Tech. Papers*, pp. 202-203, Feb. 2008.
[3] H. Darabi, "A blocker filtering technique for SAW-less wireless receivers," *IEEE J. of Solid-State Circuits*, vol. 42, no. 12, pp. 2766–2773, Dec. 2007.
[4] T. Sowlati et al, "Single-Chip Multiband WCDMA/HSDPA/HSUPA/EGPRS Transceiver with Diversity Receiver and 3G DigRF Interface Without SAW Filters in Transmitter / 3G Receiver Paths", *ISSCC Digest Tech. Papers*, pp.116-117, Feb. 2009.
[5] D. Kaczman et al., "A Single–Chip 10-Band WCDMA/HSDPA 4-Band GSM/EDGE SAW-less CMOS Receiver With DigRF 3G Interface and +90dBm IIP2", *IEEE J. of Solid-State Circuits*, vol. 44, pp. 718-739, Mar. 2009.
[6] Q. Huang, et al., "A Tri-Band SAW-Less WCDMA/HSPA RF CMOS Transceiver with On-Chip DC-DC Converter Connectable to Battery", *ISSCC Dig. Tech. Papers*, pp. 60-61, Feb. 2010.
[7] C. Andrews, A. Molnar, "A Passive-Mixer-First Receiver with Baseband-Controlled RF Impedance Matching, < 6dB NF, and > 27dBm Wideband IIP3", *ISSCC Digest Tech. Papers*, pp. 46-47, Feb. 2010.
[8] X. He, J. van Sinderen, "A Low-Power, Low-EVM, SAW-Less WCDMA Transmitter Using Direct Quadrature Voltage Modulation", *IEEE J. of Solid-State Circuits*, vol. 44, pp. 3448-3458, Dec. 2009.

ISSCC 2011 / February 23, 2011 / 10:45 AM

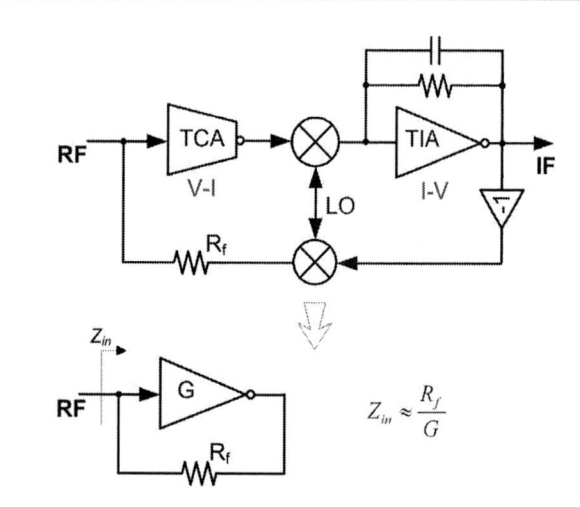

Figure 21.5.1: Concept of receiver input matching with a translational loop.

$$Z_{in} \approx \frac{R_f}{G}$$

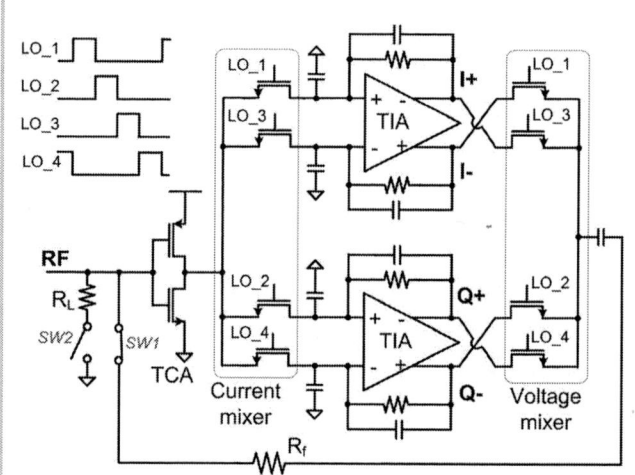

Figure 21.5.2: Receiver implementation.

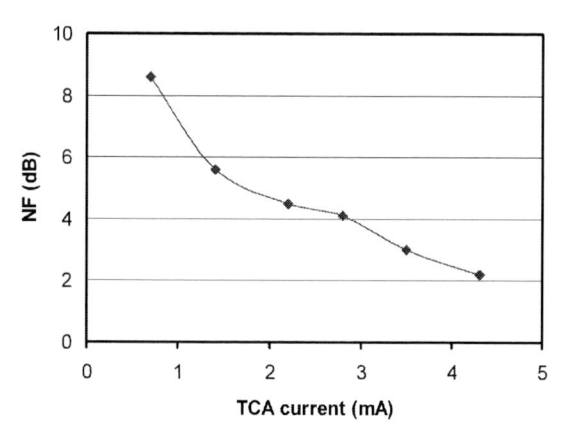

Figure 21.5.3: NF versus TCA current at low band.

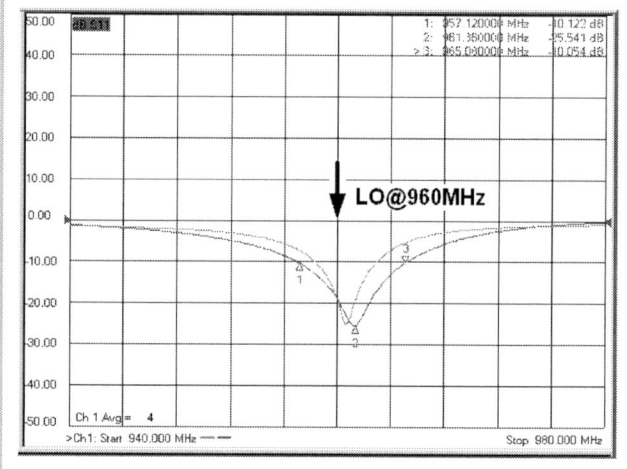

Figure 21.5.4: Low-band S_{11} measurement results over two IF bandwidths.

Figure 21.5.5: IIP3 versus frequency offset.

Ref	Process	Freq. band	NF (dB)	Duplex distance (MHz)	Duplex IIP3 (dBm)	2-tone IIP2 (dBm)	Current (mA)	Supply (V)	Inductors
[1]	0.18μm	High	2.9	80	-4	50	75mW	1.2/2.4	Yes
[2]	0.18μm	High	3.1	190	-2	>65	35*	2.9	Yes
[4]	0.13μm	High	2.4	NA	-0.5	>50	NA	1.8	Yes
[5]	90nm	High	2.2	190	5	>90**	15.1	1.5	Yes
This work	45nm	Low	2.2	45	-3.5	>40	7.3	1.3	No
		High	2.7	190	1.5	>60			

* current consumption including synthesizer.
** IIP2 after calibration.

Figure 21.5.6: Performance comparison for WCDMA applications.

ISSCC 2011 PAPER CONTINUATIONS

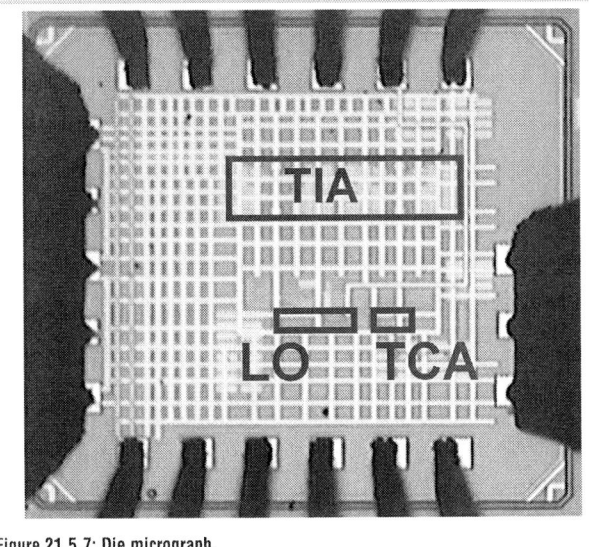

Figure 21.5.7: Die micrograph.

ISSCC 2011 / SESSION 21 / CELLULAR / 21.6

21.6 A Multiband LTE SAW-less Modulator with -160dBc/Hz RX-Band Noise in 40nm LP CMOS

Vito Giannini[1], Mark Ingels[1], Tomohiro Sano[2], Bjorn Debaillie[1], Jonathan Borremans[1], Jan Craninckx[1]

[1]imec, Leuven, Belgium, [2]Renesas Electronics, Itami, Japan

In FDD cellular standards, the transmitter's out of-band noise leaks into the receive band due to the finite duplexer TX to RX isolation. If this noise is not low enough, a SAW filter is needed before the Power Amplifier to preserve the RX sensitivity. Out-of-band noise is also an important concern for coexistence of cellular transmitters with standards like GPS, WLAN and/or WiMAX on the same smart phone, a very common scenario nowadays. The *SAW-less* RX-band noise challenge becomes even more acute in the Long-Term Evolution (LTE) standard [1] where transmitters will need to operate in multiple FDD bands, using wider channel bandwidths and higher Peak-to-Average Power Ratios (PAPR).

CMOS transmitters with RX-band Carrier-to-Noise-Ratio (CNR) down to -160dBc/Hz were presented for the most popular WCDMA bands (I, II, V) in [2-6]: high-voltage Gilbert mixers [2] are generally power hungry whereas feedback-based notching techniques [3] are hard to implement for wider bandwidths and low TX-RX offsets. A direct quadrature voltage modulator was proposed in [4]: this upconversion technique uses the combination of an RC pole and passive mixer to achieve good noise performance with low power consumption. However, losses of such mixers are proportional to the passive pole's resistor value and, especially at high carrier frequencies, shift the linearity requirements to the baseband driving stage which should provide large linear signal swings on a low impedance node.

In this paper, the potential of voltage sampling is conceptually re-examined in a complete TX chain to cover all LTE-FDD bands and beyond, resulting in a highly efficient and flexible architecture. The presented transmitter achieves CNR down to -162dBc/Hz for channel bandwidths up to 20MHz in most LTE-FDD bands, including band XI (1.4GHz carrier/ 48MHz TX-RX offset frequency), band XII (0.7GHz / 30MHz) and band VII (2.5GHz / 120MHz), without the aid of external inter-stage acoustic filters. Furthermore, the same transmitter provides legacy with all WCDMA FDD scenarios and can be used with other bands/modulations such as GMSK and OFDM-based standards up to 5.5GHz carrier frequency.

Figure 21.6.1 shows the block diagram of the presented transmitter: a flexible 3rd-order transimpedance low-pass filter (TILPF) removes the DAC aliases and out-of-band quantization noise. The TILPF is followed by a passive mixer which upconverts the baseband voltage on the pre-power-amplifier's (PPA) input capacitor. To ease the interfacing with different Power Amplifiers, 2 on-chip baluns centered around 1GHz and 2GHz are integrated, together with a wideband differential output.

Figure 21.6.2 shows the schematic of both baseband and upconversion blocks. The TILPF design is based on a flexible Tow-Thomas topology that offers independent programming of transimpedance gain, bandwidth and quality factor, whereas a quadrature voltage sampling mixer passively performs the upconversion using a low-noise 25%-duty-cycle LO driver. Filtering and mixing stages are designed to limit the impact on the transmitter CNR while keeping the power consumption minimal over the required RF range. At baseband, to achieve out-of-band noise lower than -180dBVrms/Hz with limited power consumption, a passive LPF is added after the TILPF. From a system perspective, good CNR and high output power over different bands can be achieved with a tuneable passive pole (RPx-CP in Fig. 21.6.2). Special care must be put in selecting the right RP value which finally determines the losses of the passive mixer [6] and therefore the transmitter maximal frequency range. Low cut-offs and big RP values are needed for the lower RF frequencies with small TX-RX separation; on the other hand, RP should be as small as possible at high RF frequency and/or wide channel bandwidths. As shown in Fig. 21.6.2, to avoid linearity degradation, the switches of the RP array are closed inside a multi-feedback loop generated from the previous filtering stage. When an SP switch is ON to activate a certain resistor RPx, a corresponding feedback loop is selected through the switch SF and feedback resistor RF. The nonlinear resistance of the active SP switch is thus divided by the open loop gain of the TILPF making its contribution to the distortion negligible. Overall the passive pole bandwidth can vary from about 7MHz up

to 50MHz without sacrificing linearity. In order to efficiently drive RP values as low as 40Ω, a Class-AB low-voltage op-amp topology is implemented that can trade power consumption for linearity where needed.

Figure 21.6.3 shows the schematic of the PPA: it consists of a cascoded differential Common-Source amplifier loaded with 2 on-chip baluns with programmable center frequency. A wideband differential output is provided as well. The amplifier transistor is split in binary scaled units that can be turned on or off by thick-oxide cascode transistors. At the lower gain end, additional gain steps are obtained by dumping the part of the signal current into the power supply rather than in the selected output. Overall, the PPA gain range is 60dB, achieved in 11 coarse steps of 6dB. Fine gain control can be provided in the digital domain by adding 1 extra bit (6dB) in the DAC.

The transmitter is fabricated in a 40nm LP CMOS process. The core area is $1.4 \times 0.7\text{mm}^2$ (Fig. 21.6.7), mainly dominated by the baseband capacitors and the on-chip baluns. Figure 21.6.4.a shows measurements results of the transmitter gain versus the RF frequency: as discussed previously, the right selection of the resistor RP is critical to achieve sufficient gain and CNR in all LTE FDD bands. The big dots represent the best possible configuration achieving flat gain up to 3.5GHz. Figure 21.6.4.b shows the measured EVM as a function of the output power. It is better than 2.2% in WCDMA bands VII, XI and XII. LO feedthrough better than -40dBc is shown in Fig. 21.6.4.c. The transmitter consumes 13 to 44mA from the 1.1V supply (TILPF + LO generation) depending on the selected bandwidth and LO frequency, whereas the PPA consumes less than 43mA from the 2.5V (PPA), proportionally to the required output power and linearity. Figure 21.6.4.d shows the current consumption versus the output power in LTE band XII mode. Thanks to the scalable PPA current, the DG09 weighted power consumption for most WCDMA scenarios can be lower than 30mW.

Figure 21.6.5 shows EVM/ACLR measurements for a 20MHz bandwidth SC-FDMA LTE uplink operating in band II. An extensive overview of the measured performance in various operating modes, including the toughest WCDMA/LTE bands as well as GSM, WLAN and WiMAX, is given in Fig. 21.6.6. Output P_{1dB} is better than 10dBm in all modes except GSM, where we show how lower PPA linearity can be traded for current consumption. An EVM better than 2.5% is measured in WCDMA, LTE GSM and WiMAX modes. Finally, a CNR better than -160dBc/Hz is obtained, which is sufficient for SAW-less operation in all WCDMA/LTE FDD bands.

Acknowledgment:
The authors would like to thank M. Libois, B. Verbruggen and H. Suys for their contribution to this work.

References:
[1] 3GPP TS 36.101, "Evolved Universal Terrestrial Radio Access (LTE): User Equipment (UE) Radio Transmission and Reception," v. 8.6.0, June 2009.
[2] M. Cassia et al., "A Low-Power CMOS SAW-Less Quad Band WCDMA/HSPA/HSPA+/ 1X/EGPRS Transmitter"', *IEEE Journal of Solid-State Circuits*, vol. 44, no. 7, pp. 1897-1906, July 2009.
[3] A. Mirzaei, H. Darabi; "A Low-Power WCDMA Transmitter With an Integrated Notch Filter", *IEEE Journal of Solid-State Circuits*, vol.43, no.12, pp.2868-2881, Dec. 2008.
[4] Xin He, J. van Sinderen, R. Rutten, "A 45nm WCDMA transmitter using direct quadrature voltage modulator with high oversampling digital front-end", *ISSCC Dig. Tech. Papers*, vol., no., pp.62-63, 7-11 Feb. 2010.
[5] K. Hausmann et al.; , "A SAW-less CMOS TX for EGPRS and WCDMA", *IEEE Radio Frequency Integrated Circuits Symposium (RFIC)*, pp.25-28, May 2010.
[6] M. Ingels, V. Giannini, J. Borremans, et al. , "A 5mm² 40nm LP CMOS 0.1-to-3GHz Multistandard Transceiver", *ISSCC Dig. Tech. Papers*, pp.458-459, Feb. 2010.

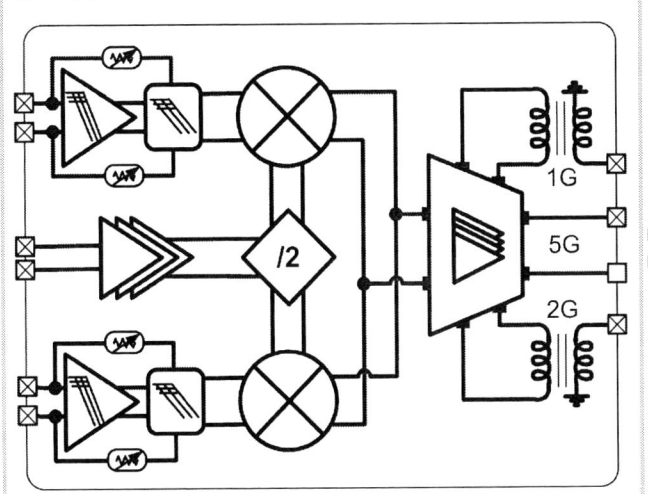

Figure 21.6.1: The multiband LTE Modulator block diagram.

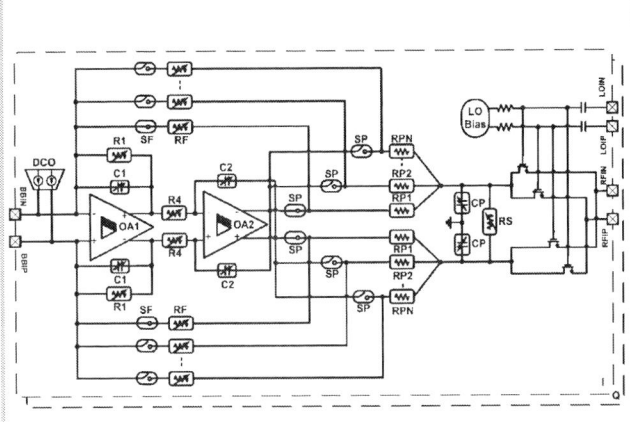

Figure 21.6.2: Flexible TIA and voltage-sampling mixer.

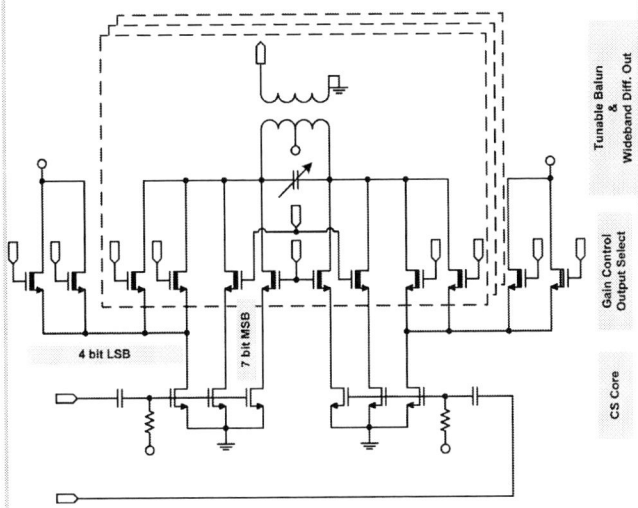

Figure 21.6.3: Multiband variable-gain Pre-Power Amplifier.

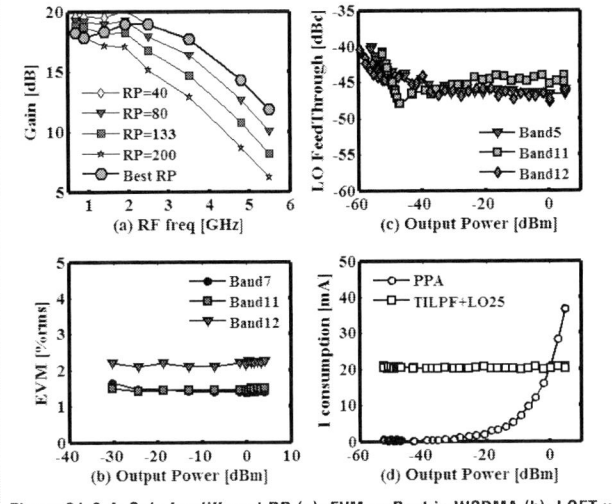

Figure 21.6.4: Gain for different RP (a); EVM vs Pout in WCDMA (b); LOFT vs Pout in LTE (c); I consumption vs Pout for LTE Band11 (d).

Figure 21.6.5: Spectral mask for LTE band II.

Band # Fcarrier ΔTX-RX	Mode	BW [MHz]	OP1dB [dBm]	Pout [dBm]	EVM [%]	ACLR1/2 [dBc]	CNR [dBc/Hz]	Imax [mA] 1.1V/2.5V	DG.09 [mW]
Band I 1.92GHz 190MHz	UMTS	4	10.4	3.79	2.0	-40.2/-67	-162	21/41	30.9
	LTE	20	10.4	2.1	2.5	-39/-58	-160	25/40	
Band II 1.85GHz 80MHz	UMTS	4	10.4	4.39	2.3	-40.2/-63	-164	20/40	30.2
	LTE	20	10.4	2.6	2.4	-38.4/-59	-162.5	24/40	
Band V 0.82GHz 45MHz	UMTS	4	10.9	4.45	1.7	-4U/-68	-161.7	14/37	24.8
	LTE	10	10.9	2.45	1.7	-41.4/-63	-160.5	17/37	
Band VII 2.5GHz 120MHz	UMTS	4	13.5	7.1	1.5	-46/-72	-159	24/40	38.5
	LTE	20	13.5	5.1	2	-42/-58	-158	28/40	
Band XI 1.42GHz 48MHz	UMTS	4	11.7	5.2	1.5	-41/-67	-162.6	18/37	27.8
	LTE	20	11.7	3.4	2.0	-42/-58	-160.4	21/36	
Band XII 0.7GHz 30MHz	UMTS	4	10.5	2.41	2.1	-39.5/-65	-160	13/34	24.8
	LTE	10	10.5	0.4	1.9	-41/-67	-159	17/33	
0.9GHz 20MHz	GSM	0.2	8.1	4.5	1.7	-54/-67	-160	18/25	
2.4GHz 100MHz	WLAN	40	13.6	4.6	3.3	-41/-54	-159	27/40	
3.5GHz 100MHz	WiMAX	20	12.6	3	1.8	-41/-54	-155	38/43	
4.8GHz 100MHz	WLAN	20	10.47	1	8.1	-42/-48	-156	44/39	

Figure 21.6.6: Performance summary.

Figure 21.6.7: Chip micrograph of the transmitter.

ISSCC 2011 / SESSION 21 / CELLULAR / 21.7

21.7 A Fully Digital Multimode Polar Transmitter Employing 17b RF DAC in 3G Mode

Zdravko Boos[1], Andreas Menkhoff[1], Franz Kuttner[2], Markus Schimper[1], Jose Moreira[1], Hans Geltinger[1], Timo Gossmann[1], Peter Pfann[1], Alexander Belitzer[1], Thomas Bauernfeind[3]

[1]Infineon Technologies, Neubiberg, Germany,
[2]Infineon Technologies, Villach, Austria, [3]DICE, Linz, Austria

The constantly growing quest for global roaming with a fast mobile internet access has further increased demand for multiband 3G cell phones supporting also 2G/2.5G standards, in order to provide full coverage any time, any place. In order to substantially reduce cost and size of the multimode, multiband handsets, it is necessary to develop a transmitter architecture that enables use of a multimode, multiband power amplifier, a key prerequisite to substantially reduce the number of external components, pins count, and PCB area, while at the same time significantly lowering power consumption.

A direct quadrature voltage modulator has been disclosed in [1] and current mode upconversion mixer with linearity boosting technique in [2] as preferred concepts for WCDMA transmitters (TX), enabling a solution for TX noise reduction at some duplex distances. Two problems remain unsolved: how to remove the TX SAW at the other duplexer distances, and how to support still highly desirable 2G and 2.5G modes, required for full coverage, without increasing power consumption and system complexity. A multimode solution was addressed in [3] but 2G and 2.5G far-off noise performance was not disclosed and in 3G high band an external local oscillator signal was still used. Further multimode work has been presented in [4] where required far off noise performance has been achieved but at the expense of the large chip area due to dedicated 2G/3G signal paths and still a high power consumption. The single-chain TX (all modes reuse the same signal path) has been focus of the work in [5], where other advantages of a digital concept have been demonstrated, but the TX SAW filter in 3G mode cannot be omitted due to higher-than-allowed far-off noise and remaining repetition spectrum. A further single-chain TX solution based on the analog polar modulator has been disclosed in [6], demonstrating reuse of the signal path for 2G/3G systems in an analog polar transmitter concept, but TX noise in the RX band has not been disclosed. Recently, a fully digital polar modulator for 2G and 2.5G modes has been published [7] with a low chip area and low power consumption, but still with unsolved far-off noise for the co-existence of cellular application with navigation and connectivity.

The digital polar transmitter concept disclosed in this work solves all above mentioned remaining obstacles. In order to employ the proposed digital polar TX two main challenges had to be solved: firstly, very stringent delay-matching requirements in WCDMA mode between amplitude (AM) and phase paths (PM) and secondly, far-off noise due to quantization, spurious and repetition spectrum. In order to achieve good EVM and ACLR requirements with margin for production, delay matching between AM and PM paths is required to be equal or lower than 1ns. The same delay matching is required between the split-phase signal paths in the two-point DPLL modulator.

As shown in the transmitter block diagram depicted in Fig. 21.7.1, digital data received from the baseband chip, via digital interface v3.09, are pre-processed in the I/Q domain (depending on system RRC filtered, mapped, etc.) and supplied to a Cartesian-to-polar converter (CORDIC). The phase signal after CORDIC is further filtered (FIL), differentiated, and fed to the digital PLL, which employs a two-point modulator due to a very wide 3G frequency spectrum. The composite signal after DPLL loop filter is interpolated (INT) to DCO/4 frequency, noise shaped (NS) and supplied to a DCO varactor field (TDC, loop filter and feedback divider are not shown). The modulated DCO output is divided by 2 or 4 and used as the LO input to the RFDAC. The magnitude signal after CORDIC has also been filtered, interpolated to DCO/4 frequency and supplied to the DAC input. As can be seen from Fig. 21.7.1, the transmitter uses two clock domains: a fixed clock of 104MHz for the synthesized part derived from the main system clock, where RF blocks are clocked with a modulated DCO clock divided by 4. As all RF blocks use the same clock and there is no analog filter in either amplitude or phase paths, the required delay matching of 1ns is inherently achieved. In order to solve

the second major challenge, far-off noise, spurious and repetition spectrum, it was decided to employ a high dynamic range RFDAC shown in Fig. 21.7.2. The DAC employs 10 thermometer- and 4 binary-coded bits (not shown in Fig. 21.7.2) with a high oversampling, using a GHz range clock and providing overall 17b DAC resolution in 3G mode (3.84MHz bandwidth) and 19b in EDGE mode (200kHz). This dynamic range is sufficient to fulfill -160dBc/Hz far-off noise specifications without using any analog filter and also without employing a noise shaper (NS) in the amplitude path at high signal levels, which is a preferred solution due to co-existence requirements of cellular applications with navigation and connectivity. In order to avoid an RF switch, a separate RFDAC was used in low- and high-frequency bands. The digital processing repetition spectrum is removed by using interpolators sharing the DCO/4 clock, where the interpolation ratio is derived from the RF channel programming. Clocking of the mixer/DAC with a modulated clock, with of course careful floor planning and layout design, further reduces spurious distortion to sufficiently low level at any 3GPP Rel. 7 specified duplex distance, thus enabling TX SAW removal, and co-existence in a navigation and connectivity application. The complete TX, including all building blocks from digital interface up to RF output, with biasing and clock distribution, consumes in low/high band 33mA/37mA DG09 weighted current from a 3.6V battery in 3G mode, using an external DC/DC converter with 93% efficiency. Due to shared DAC/mixer/load current in the output stage, the DAC's high dynamic range was achieved without consuming any extra current, except a few mA in the binary-to-thermometer decoder. The required gain-control range of 75dB in 3G mode was distributed with 50dB in the digital domain and 25dB in DAC bias, with an additional 10 to 15dB of gain control implemented in the external power amplifier. By using single-ended 1024 DAC cells connected to 32 mixers, beside increased efficiency, also low LO leakage has been achieved due to the absence of DC offset inherent to differential signal processing. The measured LO leakage was -90dBm and -96dBm in high and low band respectively. EDGE high-band far-off noise, measured with +6dBm output signal shown in Fig. 21.7.3 demonstrates low noise, low spurious performance by achieving better than -158dBc/Hz far-off noise at any distance from the carrier larger than 20MHz. For the highest 10dB of the power range far-off noise in the high band at 20MHz distance does not change relatively more than 0.75dB. Also in the 3G mode with a high AM signal dynamic and a high delay matching requirement, the proposed digital TX concept as shown in Fig. 21.7.4 achieves better than -160dBc/Hz far-off noise at all distances from the carrier higher than 45MHz, measured with +6dBm wanted signal level. Beside a low far-off noise design, a special attention has been taken to reduce spurious levels, and as shown in Fig. 21.7.5, when measured in GSM low band for all RF frequencies max hold spur TX + n*26MHz does not exceed -108dBc. The overall performance of all supported modes, summarized in Fig. 21.7.6 shows a large margin to any specification point. At +6dBm and -25dBm the RFDAC consumes 29mA and 2mA respectively from 2.5V, with a headroom for VSWR 2:1. The transmitter die micrograph shown in Fig. 21.7.7 includes bias, LDOs, ESDs, INT, NS, 2G/3G DCOs and LB/HB RFDACs, and measures 2mm^2, while the synthesized part not shown consumes 0.2mm^2 area .

Acknowledgment:
The authors would like to thank T. Maletz, T. Maierhofer and B. Kapfelsperger for detailed chip verification and to B. Adler for initiating and pursuing the multimode digital TX project.

References:
[1] X. He, J. van Sinderen et al., "A 45nm WCDMA Transmitter Using Direct Quadrature Voltage Modulator with High Oversampling Digital Front-End", *ISSCC Digest Tech. Papers*, pp. 62-63, Feb. 2010.
[2] Q. Huang et al., "A Tri-Band SAW-Less WCDMA/HSPA RF CMOS Transceiver with On-Chip DC-DC Converter Connectable to Battery", *ISSCC Digest Tech. Papers*, pp. 60-61, Feb. 2010.
[3] M. Ingels et al., "A 5mm^2 40nm LP CMOS 0.1-to-3GHz Multi-standard Transceiver" *ISSCC Digest Tech. Papers*, pp. 458-459, Feb. 2010.
[4] K. Hausmann at al., "A SAW-less CMOS TX for EGPRS and WCDMA", *IEEE RFIC Symposium*, pp. 25-28, 2010.
[5] P. Eloranta et al., "A Multimode Transmitter in 0.13 µm CMOS Using Direct-Digital RF Modulator", *IEEE JSSC*, vol. 42, no. 12, pp. 2774-2784, Dec. 2007.
[6] John Groe, "Polar Transmitters for Wireless Communications", *IEEE Communications Magazine*, vol.45, pp. 58-63, Sep. 2007.
[7] J. Mehta et al., "A 0.8mm^2 All-Digital SAW-Less Polar Transmitter in 65nm EDGE SoC", *ISSCC Digest Tech. Papers*, pp.58-59, Feb. 2010.

978-1-61284-303-2/11 $26.00 © 2011 IEEE

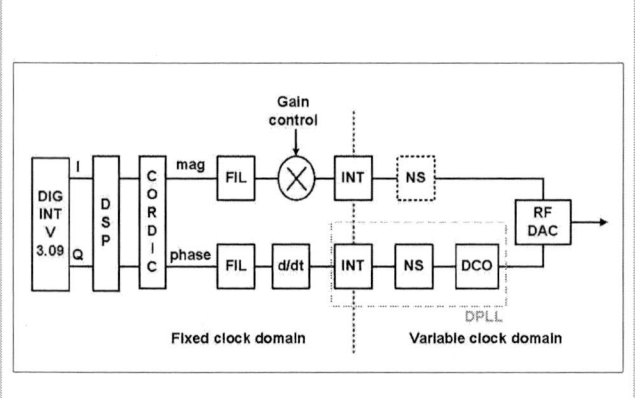

Figure 21.7.1: Diagram of the multimode digital transmitter.

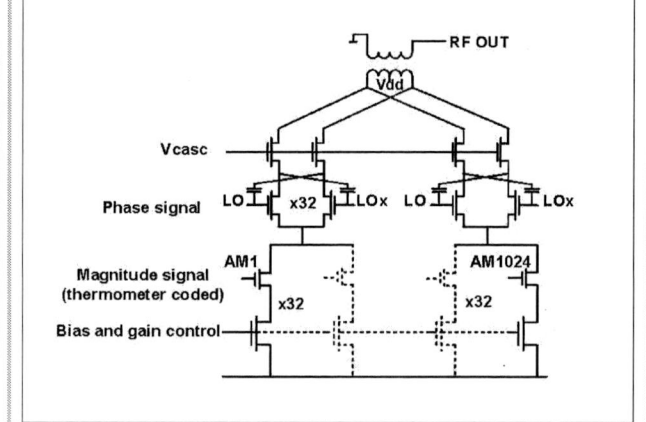

Figure 21.7.2: RFDAC diagram.

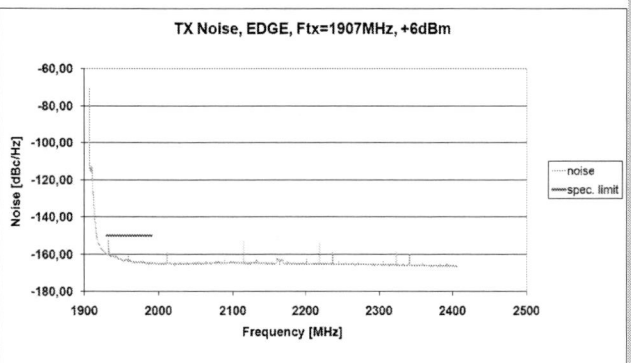

Figure 21.7.3: EDGE high-band far-off noise.

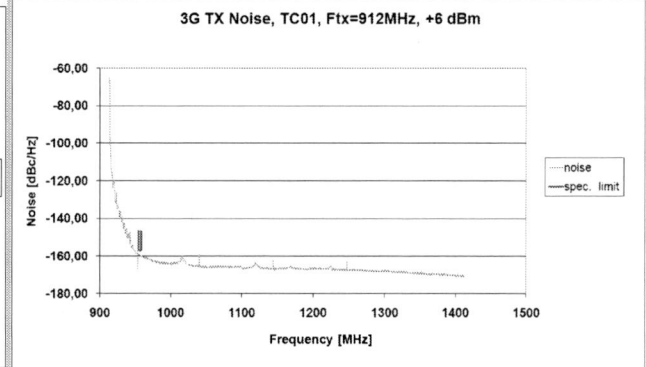

Figure 21.7.4: WCDMA low-band far-off noise.

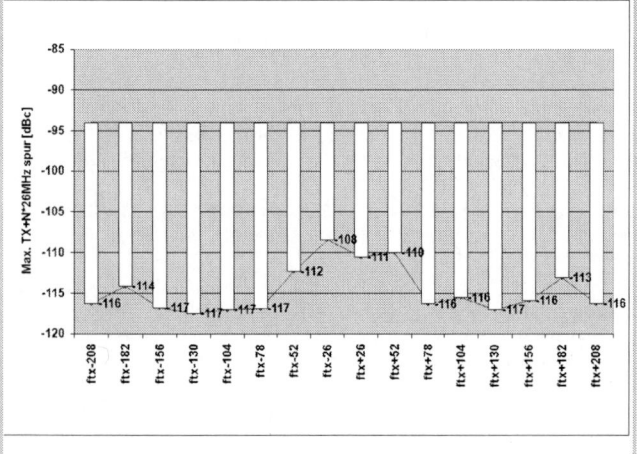

Figure 21.7.5: GSM low band TX+N*26MHz spurious.

TX Parameter	Measured		3GPP spec		Unit
	LB	HB	LB	HB	
GSM output power	12	12	N/A		dBm
GSM rms phase error	0,6	0,9	5		degree
GSM 400 kHz mask	-70	-68	-60	-58	dBc
GSM 20 MHz noise	-166	-160	-162	-151	dBc/Hz
EDGE output power	8	6	N/A		dBm
EDGE rms EVM	1,2	1,7	9		%
EDGE 400 kHz mask	-68	-65	-54		dBc
EDGE 20 MHz noise	-161	-158	-156	-146	dBc/Hz
EDGEEvo 16QAM NR out. pow.	6	5	N/A		dBm
EDGEEvo EVM rms	2	2,5	7		%
EDGEEvo 400 kHz mask	-63	-61	-54		dBc
EDGEEvo 20 MHz noise	-161	-158	-154	-144	dBc/Hz
WCDMA output power	6	6	N/A		dBm
WCDMA EVM rms	3,5	4	17,5		%
WCDMA ACLR1	-50	-49	-33		dBc
WCDMA 45/80 MHz noise	-160	-160	N/A		dBc/Hz
HSPA+ output power	4	4	N/A		dBm
HSPA+ EVM rms	4,4	4,7	14		%
HSPA+ ACLR1	-47	-45	-33		dBc
HSPA+ 45/80 MHz noise	-160	-159	N/A		dBc/Hz
WCDMA DG09 current	33	37	N/A		mA
Power supply (digital/RFDAC)	1,2/2,5		N/A		V

Figure 21.7.6: Transmitter performance summary.

ISSCC 2011 PAPER CONTINUATIONS

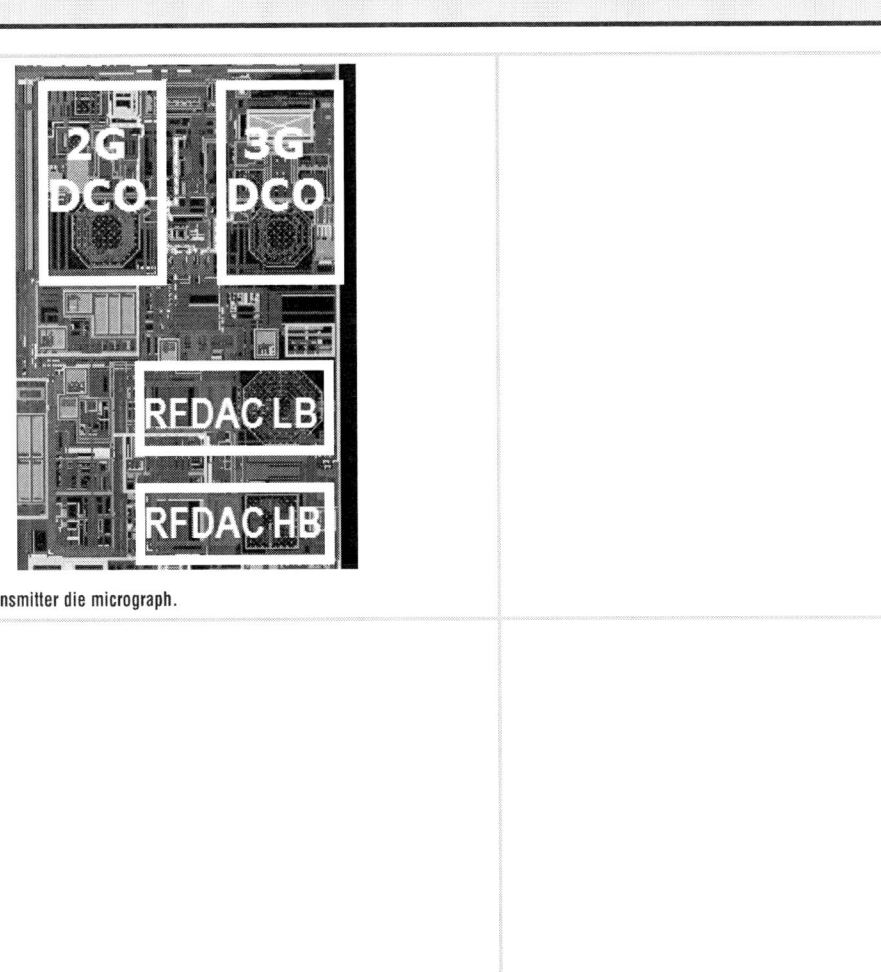

Figure 21.7.7: Transmitter die micrograph.

ISSCC 2011 / SESSION 21 / CELLULAR / 21.8

21.8 A Low-Power Wideband Polar Transmitter for 3G Applications

Michael Youssef[1,2], Alireza Zolfaghari[1], Hooman Darabi[1], Asad Abidi[2]

[1]Broadcom, Irvine, CA, [2]University of California, Los Angeles, CA

So far all mainstream transmitters for WCDMA are of the direct upconversion type. This architecture is versatile but requires calibration of the imbalance in its quadrature branches and DC offset at its inputs, and it is vulnerable to mixer noise. We believe it consumes more power and chip area than is warranted, and propose the polar transmitter as an alternative. Although it has been widely discussed as the ideal transmitter for EDGE, the polar architecture has yet to make significant inroads there, let alone into wideband CDMA. We will describe new circuits that enable a compact, largely self-calibrated polar transmitter, whose lower power and chip area put it ahead of state-of-the-art direct upconversion transmitters [1-3].

Polar's challenges lie in finding a reliable method of wideband phase modulation (PM), and in suppressing unequal delay in the two separate paths that the PM and AM waveforms will take before they unite in the pre-PA driver. Whereas EDGE requires a PM bandwidth of about 800kHz and can tolerate AM/PM delay mismatch on the order of tens of nanoseconds, WCDMA needs at least an 8MHz bandwidth with path delay mismatch less than 2ns. The solution we describe takes care of bandwidth and unequal delay with minimum intervention.

PM is transferred via a fractional-N phase-locked loop (PLL) on to a voltage-controlled oscillator (VCO). The PLL bandwidth is usually a small fraction of the reference frequency, otherwise spurs and fractional noise will leak into the output: in our circuit it is between 200 and 300kHz. Then to impose an 8MHz-wide modulation into this 300kHz-wide loop, one must resort to either pre-emphasis or two-point injection. Pre-emphasis is sensitive to variations in loop gain, and at large modulation bandwidths can push PLL components to the limits of their dynamic range. On the other hand two-point injection is very sensitive to the VCO gain (k_{VCO}), which changes across process, supply voltage, temperature, and operating frequency. Since in either method the oscillator is being modulated at rates well beyond the PLL bandwidth, feedback is unable to correct for VCO nonlinearity, and distortion will appear at high modulation frequencies.

But if a linear VCO with known and reproducible k_{VCO} were made available, two-point injection becomes feasible. Then by adding to it a fairly simple AM path and a power-efficient PA driver, a wideband polar modulator may be realized. What differential delay remains between the two paths must also be predictable, and more importantly, should be self-calibrated. All-digital PLL-based transmitters for GSM/EDGE are based on this method [4], but they are sensitive to the oscillator gain and their output is usually cluttered by spurious tones. Also, extensive calibration and signal processing make them power-hungry; it is not clear whether they can reach WCDMA bandwidths. We show that our approach which strikes a balance between analog and digital calibration is simpler, more robust, and consumes less power.

First we consider VCO linearization. A model-based pre-distortion [5] does not assure the well-controlled, stable, and linear VCO characteristic that we have shown at low frequencies with frequency-to-voltage (F/V) feedback [6]. With modern CMOS it becomes possible to extend this method to RF. Thus, as shown in Fig. 21.8.1, the VCO is surrounded by a local feedback loop, which we refer to as a frequency-locked loop (FLL). This measures the oscillation frequency continuously via a switched capacitor (C_S) that dispenses packets of charge from a bandgap voltage reference (V_{REF}) into a reference resistor R_L, and compares the resulting voltage with an applied voltage representing the desired PM. With frequency feedback the VCO gain ($k_{VCO—FB}$) is

$$k_{VCO—FB} = \frac{M}{(GV_{REF}C_S R_L)}$$

When the time constant $C_S R_L$ is stabilized against changes in process, voltage, temperature and frequency, $k_{VCO—FB}$ is also predictably constant. A number of methods have been developed for use for active filters that stabilize RC time constants against a crystal reference. They typically obtain accuracies better than 1%, limited by the discreteness of unit C and R. We employ this form of master-slave tuning.

A current-mode second-order passive lowpass filter suppresses spikes arising from the periodic charge dispensing, and an offset current centers the closed-loop VCO characteristic on zero. Figure 21.8.2 shows the F/V converter circuit. There is no need in the differential circuit to stabilize the quiescent point of the filter's input against PVT variations, and the size of the filter's capacitance is also halved. The sampling capacitance is chosen as a compromise between settling time and thermal noise. The output differential-to-single-ended stage boosts impedance relative to the (programmable) load resistor.

When embedded within the complete PLL, this FLL group delay sets the differential delay between PM and AM paths until they unite at the PA driver. By calibrating the F/V filter's poles with a master-slave scheme, the FLL bandwidth is held at 10MHz with better than 10% accuracy. The AM signal is time-retarded in digital baseband by this delay. Quantities that cannot be simply calibrated leave a small residual delay difference, but this is within the WCDMA 2ns delay margin.

A polar modulator using this delay compensation is built around the linearized VCO, as shown in Fig. 21.8.3. The PLL is a fractional-N, type II loop. The digital AM and PM signals are generated in an FPGA and applied to the transmitter using two sets of external DACs and reconstruction filters, which can be easily fabricated on-chip with superior matching. The reference clock of the PLL is supplied from an external generator, but in a complete transceiver implementation, it can be easily generated in the receiver PLL. Among the many candidates for a PA driver, we find the commutating differential driver amplifier to be convenient, simple and acceptably efficient. Its drain supply needs not be regulated. Gain is controlled with switchable unit cells in the PA driver and a two-step attenuation in the output transformer. The gain control range is 60dB and can be increased by adding more unit cells or using other techniques such as cascode current steering.

The measurement results of the FLL are shown in Fig. 21.8.4. The FLL suppresses deviations in the VCO characteristic from a straight line by 13× (measured as percentage deviation from a best-fit) across a sweep of ± 200mV in tuning voltage around mid-rail, and $k_{VCO—FB}$ remains within 1% of the designed value of 80MHz/V across temperature and supply voltage. These numbers are limited by spectrum analyzer accuracy. Spurs arising from sampling appear at 500MHz offset at a level of −85 dBc, well below the 3GPP specification of spurs less than −69dBc.

The transmitter's WCDMA constellation and spectrum are shown in Fig. 21.8.5, and compared in Fig. 21.8.6 with recently published linear transmitters. In WCDMA mode, the ACLR at 5/10MHz offsets is −42/−58dBc, noise at 45MHz offset is −159dBc/Hz, and EVM at 0dBm output power is 2.9%. The transmitter, including the VCO, LO chain, PLL, FLL, and PA driver, draws 40mA from the battery, of which the FLL takes only 1.5mA. The DG09 current is 25mA. The low power consumption stands out as the key advantage of the polar architecture: it arises from a simple, single-phase LO chain, a PA driver that always operates close to saturation, and far-away phase noise filtering by the loop rolloff. At 1dBm output power, the transmitter's EDGE modulation spectrum is down by −61dBc at 400kHz and the EVM is 2.4% at 32mA of battery current. Where direct upconversion transmitters would require extensive calibration, the measured as-is LO leakage of this device is −55dBc and scales favorably with gain control.

Figure 21.8.7 shows the die micrograph of the transmitter, which occupies 0.7mm² in 65nm CMOS. The low power consumption and chip area present a strong case in favor of the well-designed polar transmitter.

References:
[1] M. Cassia, A. Hadjichristos, et al., "A Low-Power CMOS SAW-Less Quad Band WCDMA/HSPA/HSPA+/1X/EGPRS Transmitter," IEEE J. Solid-State Circuits, July 2009.
[2] T. Sowlati, B. Agarwal, et al., "Single-Chip Multi-band WCDMA/HSDPA/HSUPA/EGPRS Transceiver with Diversity Receiver and 3G DigRF Interface without SAW filters in Transmitter / 3G Receiver," ISSCC Dig. Tech. Papers, Feb. 2009.
[3] C. Jones, B. Tenbroek, et al., "Direct-Conversion WCDMA Transmitter With -163dBc/Hz Noise at 190MHz Offset," ISSCC Dig. Tech. Papers, Feb. 2007.
[4] J. Mehta, R. Staszewski, et al., "A 0.8mm² All-Digital SAW-Less Polar Transmitter in 65nm EDGE SoC," ISSCC Dig. Tech. Papers, Feb. 2010.
[5] J. Oehm and D. Pham-Stabner, "Linear Controlled Temperature Independent Varactor Circuitry," Proc. ESSIRC, Sept. 2002.
[6] A. Abidi, "Linearization of Voltage-Controlled Oscillators Using Switched-Capacitor Feedback," IEEE J. Solid-State Circuits, June 1987.

978-1-61284-303-2/11 $26.00 © 2011 IEEE

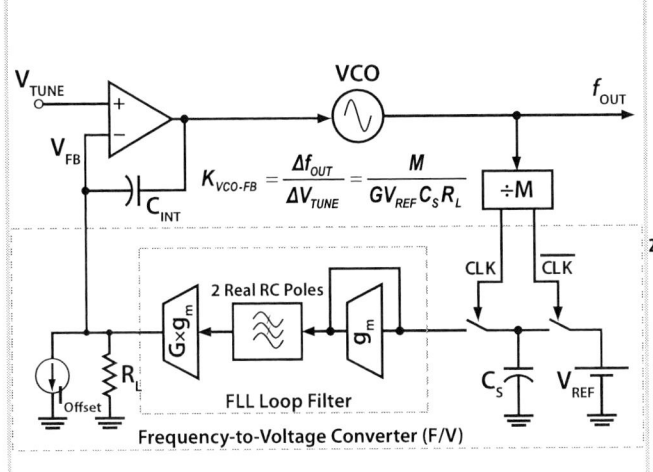

Figure 21.8.1: Linearizing the VCO by means of a negative feedback loop.

Figure 21.8.2: Frequency-to-Voltage converter circuit.

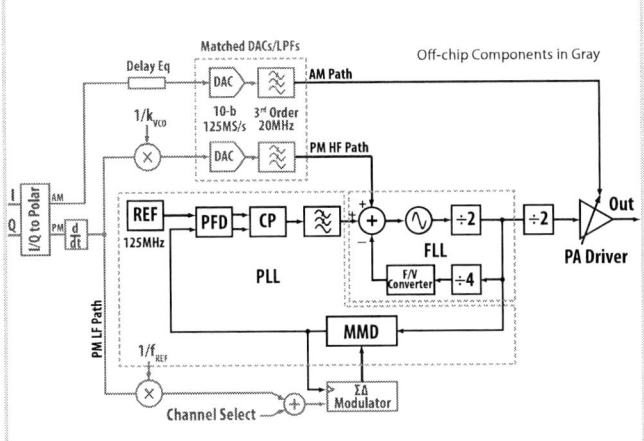

Figure 21.8.3: System block diagram.

Figure 21.8.4: Measured open- and closed-loop KVCO across temperature and supply corners.

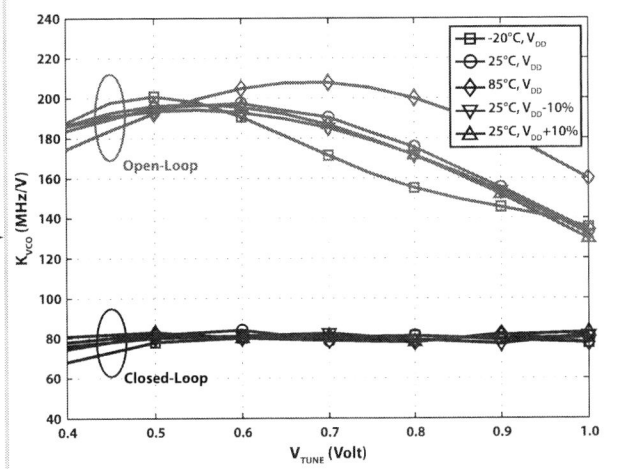

Figure 21.8.5: Measured WCDMA performance in GSM band.

Figure 21.8.6: WCDMA mode performance summary.

Parameter	[1]	[2]	[3]	This Work	3GPP Specifications
5-MHz ACLR (dBc)	−41[a]	−45	−46	−42	−33
10-MHz ACLR (dBc)	N/A	−70	−70	−58	−43
RMS EVM (%)	5.9[a]	2.1	3.7	2.9	17.5
Noise @ 45MHz (dBc/Hz)	N/A	−160	N/A	−159	--
DG09 Current (mA)[b]	N/A	N/A	N/A	25	--
Current @ 0-dBm (mA)[b]	48	67[c]	65	40	--
Area (mm²)	8.1	6.0[d]	5.6[e]	0.7	--
Process	180nm	130nm	180nm	65nm	--

[a] At 23-dBm output power

[b] Battery-referred current

[c] Estimated as half the reported full duplex current (135 mA)

[d] Estimated from the die micrograph

[e] Estimated from Fig. 10.2.7, ISSCC Dig. of Tech. Papers, 2008

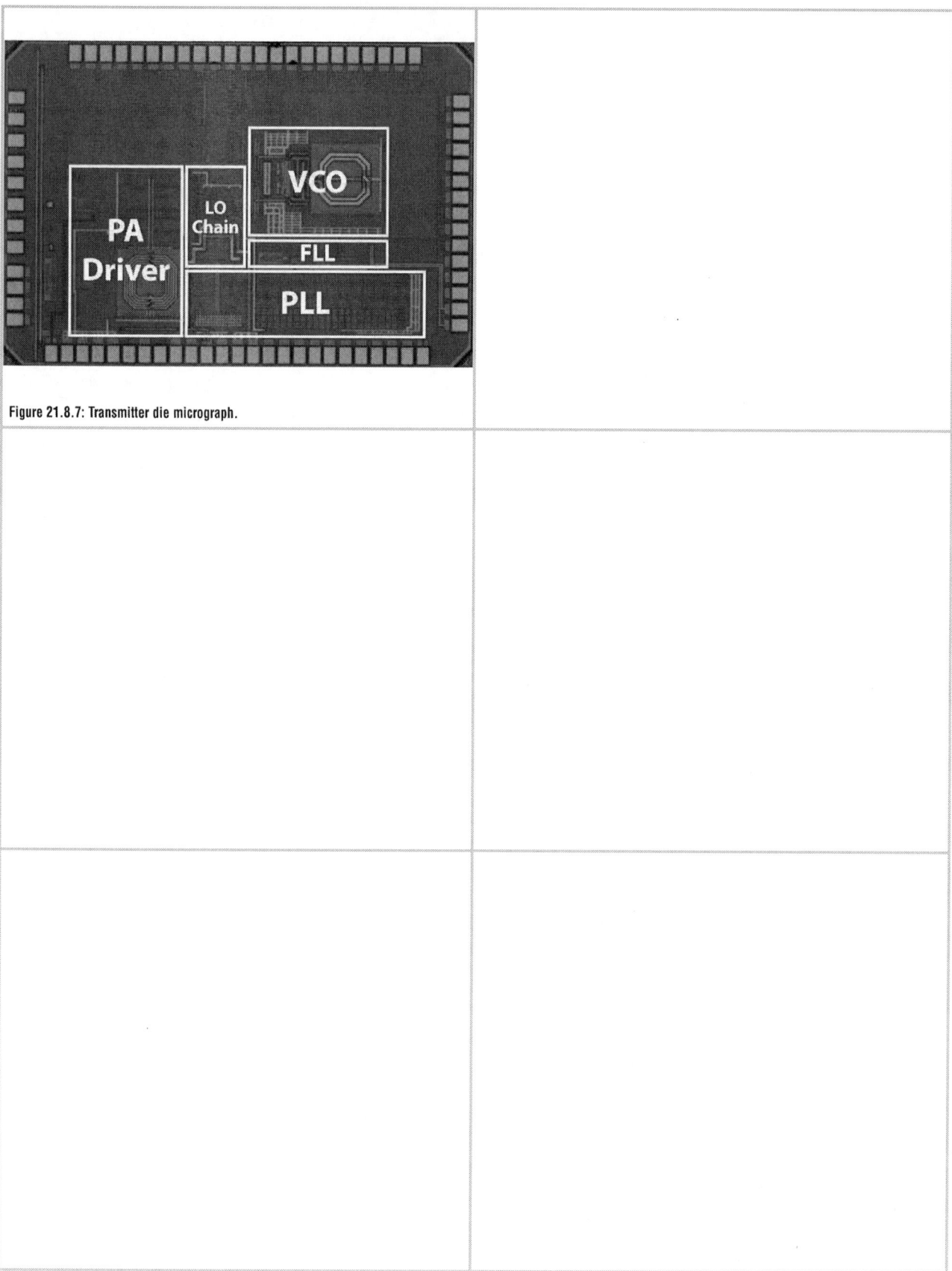

Figure 21.8.7: Transmitter die micrograph.

ISSCC 2011 / SESSION 22 / DC/DC CONVERTERS / OVERVIEW

Session 22 Overview / Analog

DC/DC Converters

Session Chair: *Francesco Rezzi, Marvell Semiconductors, Pavia, Italy*

Session Co-Chair: *Baher Haroun, Texas Instruments, Dallas, TX*

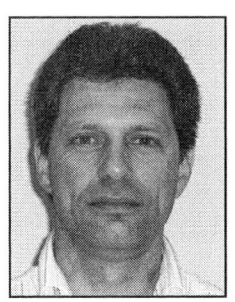

Market trends today are driving for highly efficient power conversion in high volume portable applications. This highlights the need for high-level single-chip integration of DC-DC converters in leading edge processes with other signal-path systems. This high-level of integration is driven by system cost and form factor. Moreover, system performance and cost and special design challenges at these advanced process nodes require taking into account the voltage and reliability limitations of nanometer technologies on switching converter designs. While power efficiency is of paramount importance, other system issues are driving new methods for design robustness and ease of test. The speakers in this session present the latest integration efforts and novel techniques to improve the performance and cost of DC-DC converters.

Paper 22.1 (ST Ericsson) demonstrates the tradeoffs involved in integrating a complete power management system enabling direct battery connection at 5.5V and charger control at 20V using 3V capable devices in a 65nm CMOS cellular handset chip.

Paper 22.2 (Infineon Technologies) demonstrates up to 90% efficient DC-DC converters in the most advanced CMOS process to date, a Hi-K 28nm metal gate CMOS, leading the way to enabling integration and direct battery connection up to 5.5V in future systems.

Paper 22.3 (MIT; Texas Instruments) demonstrates a digital pulse-width modulator and switched-capacitor bias techniques that provide high efficiency for a wide range of load conditions (20µA to 100mA) from 75% to 87.4% over that range. Both Papers 22.2 and 22.3 exploit the dense and power efficient digital capabilities of the advanced CMOS nodes to implement digital control algorithms.

However, digital techniques also enable near optimal implementation of the control loop and in Paper 22.4 (Arizona State University; JPL), a novel capability that uses an accurate inductor self-measurement technique is employed. The measured inductance and series resistance values are used to optimize the converter digital control loop parameters. That approach also provides a novel built-in self-test capability for the inductor used in the system. The paper demonstrates an accuracy of 2.1% for a range of 3.7µH to 22.3µH, 3.6% for a range of DCR 15mΩ to 75mΩ, and 2% for load current sensing for a range of 100mA to 750mA.

There are many other areas that require improvements in non-integrated systems to enhance power efficiency, to reduce system component count, to improve transient response and to reduce EMI. The rest of the papers in this session address these issues from different angles. In Paper 22.5 (KAIST), injection of a small saw-tooth waveform is used in a control loop of a boost converter to generate the timing signals needed to minimize the freewheeling time. This technique results in an 88% efficiency for a 3.7V input to 8V output and offers high efficiency over a wider load range.

Paper 22.6 (KAIST) improves efficiency by reducing, to near zero, the dead-time using adaptive delay gate drivers, dead-time sensing and control loop. The converter achieves 90% efficiency at 1MHz switching to generate a 1.5V output from a 3.3V input.

Paper 22.7 (Fudan University; Brandenburg University of Technology), on the other hand, presents a single inductor with dual output that presents a novel dual saw-tooth extended pulse-width modulator to allow multiple modes of buck/boost to cover a wider range of input and output. This approach offers high efficiency and automatic shift between modes with peak efficiency of 90% reached at a supply voltage of 3V and the two outputs set to 1.8V/214mA and 5V/150mA.

Another important angle is the coexistence of the switching power supplies with analog and RF subsystems. Spurious elimination becomes essential in such systems. Paper 22.8 (Iowa State University) addresses this issue with a novel frequency hopping scheme that eliminates spurs while minimally impacting the supply noise floor keeping it below -70dBm in any 100Hz bandwidth across a wideband measured to 60MHz.

978-1-61284-303-2/11 $26.00 © 2011 IEEE

ISSCC 2011 / February 23, 2011 / 8:30 AM

Presenters:

22.1 A Fully Integrated Power-Management Solution for a 65nm CMOS Cellular Handset Chip

A. J. D'Souza,
ST Ericsson, Bangalore, India

8:30 AM

22.2 A Digitally Controlled DC-DC Converter for SoC in 28nm CMOS

F. Kuttner,
Infineon Technologies, Villach, Austria

9:00 AM

22.3 20µA to 100mA DC-DC Converter with 2.8 to 4.2V Battery Supply for Portable Applications in 45nm CMOS

S. Bandyopadhyay,
Massachusetts Institute of Technology, Cambridge, MA

9:30 AM

22.4 A Digitally Controlled DC-DC Buck Converter with Lossless Load-Current Sensing and BIST Functionality

T. Liu,
Arizona State University, Tempe, AZ

10:15 AM

22.5 Zero-Order Control of Boost DC-DC Converter with Transient Enhancement Using Residual Current

T-H. Kong,
KAIST, Daejeon, Korea

10:45 AM

22.6 Robust and Efficient Synchronous Buck Converter with Near-Optimal Dead-Time Control

S. Lee,
KAIST, Daejeon, Korea

11:15 AM

22.7 A 90% Peak Efficiency Single-Inductor Dual-Output Buck-Boost Converter with Extended-PWM Control

W. Xu,
Fudan University, Shanghai, China

11:45 AM

22.8 Spurious-Noise-Free Buck Regulator for Direct Powering of Analog/RF Loads Using PWM Control with Random Frequency Hopping and Random Phase Chopping

C. Tao,
Iowa State University, Ames, IA

12:15 PM

978-1-61284-303-2/11 $26.00 © 2011 IEEE

ISSCC 2011 / SESSION 22 / DC/DC CONVERTERS / 22.1

22.1 A Fully Integrated Power-Management Solution for a 65nm CMOS Cellular Handset Chip

Arnold James D'Souza[1], Ravpreet Singh[1], Raja Prabhu J[1],
Gajendranath Chowdary[1], Ankit Seedher[1], Shyam Somayajula[1],
Nageswara Rao Nalam[1], Lionel Cimaz[2], Stephane Le Coq[2],
Praveen Kallam[3], Siddharth Sundar[3], Shanfeng Cheng[3],
Sanjay Tumati[3], Wenchang Huang[3]

[1]ST Ericsson, Bangalore, India,
[2]ST Ericsson, Rennes, France,
[3]ST Ericsson, Austin, TX

For cellular handset ICs, integrating the power management unit (PMU) in the same chip as the baseband and the radio is critical for footprint and cost reduction, however, it presents unique challenges in terms of device reliability, bill of materials (BOM) and PMU performance. In this paper, a PMU integrated as part of a GSM/EDGE chip in a 65nm deep-Nwell CMOS process (Fig. 22.1.1), comprising a battery charger, a regulated inverting charge pump, multiple regulators. and a White LED (WLED) driver is presented. The battery charger supports various charging profiles up to V_{CHG}=12V and can tolerate up to 20V. A regulated charge pump (CP) operating directly from the battery (V_{BAT}) provides negative supply for earphone class-AB audio power amplifiers (APA) for ground referred headset connectors. A buck converter and several linear regulators supply the various sub-systems on the chip as well as external modules (SIM Card/Camera/Interfaces etc.). An inductive-boost WLED driver drives up to 6 LEDs in series or up to 3 in parallel.

The battery charger supports charging of NiMH/NiCd batteries in two steps: with constant current (CC) first and constant voltage (CV) later when V_{BAT} is close to its final value. Charger subsystem is fully autonomous with capability to detect and handle deeply discharged batteries with CC trickle charging. Figure 22.1.2 shows the linear charging configuration with two external PFETs which route the 100mA to 1.4A charging current, limiting on-chip thermal dissipation. The transimpedance amplifier (TIA), V_{BAT} sensor, and CC-CV amplifier form a linear control loop. Transistors MN_{CV} and MN_{CC} control the loop dynamics in CV and CC modes respectively. The 20V tolerant circuits use device stacking for protection from V_{CHG}=20V. Transistors are biased such that the V_{GS} and V_{DS} of each stacked MOSFET is limited to less than 3.3V. Protection circuitry is aided by an all-pass gate biasing network to handle ramp rates for V_{CHG} of up to 20V in 10µs. A 2kV HBM ESD protection is achieved for the V_{CHG} pin with a RC trigger based clamp which is also stacked to handle the 20V supply and its ramp rate. For the I/O pins on the 20V domain, 3 series diodes ensure the reverse stress on each diode is within prescribed technology limits. For the circuitry exposed to 20V, junction diodes of stacked devices are protected using isolated wells. With no more than two devices in the same well the maximum voltage stress for these diodes is ensured to be below 6.6V. The only device that experiences full 20V across it is the Nwell P-substrate (NWPSUB) parasitic diode. Along the periphery of all such Nwells a 2µm wide lightly doped p-substrate region is created with no extra masks. This increases the reverse bias breakdown voltage of the NWPSUB diode beyond 20V. Including the 20V ESD circuitry, the charger system occupies 0.17mm².

The regulated CP generates a negative supply capable of sinking 150mA with low output ripple to ensure acceptable distortion performance for the APA at high power levels. The design (Fig. 22.1.3) employs a HV tolerant, peak current limited, charge pump power stage and pre-drivers for reliable operation using standard thick-oxide devices without gate oxide stress and hot-carrier induced degradation for V_{BAT} up to 5.5V. Pulse-skipping regulation schemes have large ripple and could produce load-dependent, audible spurious content. Constant frequency regulation [1] requires very low ripple since the output is observed and corrected in different phases. This requires $C_{out} \gg C_{fly}$ and a long power-up time. The present work avoids these issues by using a CT, constant frequency regulation scheme in the dump phase (clk$_{dump}$) with minimal power up time and output ripple. As the APA operates between V_{BAT} and $V_{NEG,EXT}$ and is designed to handle up to 7.5V, $V_{NEG,EXT}$ is chosen to be at –2V. Peak current limiting and slew rate control by sub-division of each of the non-overlapped phases minimizes undershoot and overshoot voltages due to bond wire inductance. Peak current

limiting also restricts excessive Joule heating in the metal interconnects. Bond wire ringing on the CP is isolated from the APA by using separate bond pads for CP and APA as shown. The CP occupies 0.12mm², operates with f_{clk}=300 kHz and C_{fly}=C_{out}=1µF, and regulates for loads up to 150mA. The output ripple is less than 300mV and power up time with I_L=150mA is less than 3T_{clk}.

The embedded WLED driver provides a generic solution for a wide range of display sizes and configurations. A hysteretic inductive boost WLED driver architecture (Fig. 22.1.4) with an external NMOS switch and an inductor, provides a switching current drive to the LEDs with the average current regulated to 2 to 20mA. The circuit operates in two phases: in the first phase (φ_1), switch MNEXT is turned ON for a fixed duration (T_{ON}) and during second phase (φ_2), MNEXT switches OFF releasing the inductor current into the LEDs. MNEXT is turned ON again when the inductor current falls below $I_{L,MIN}$ (Fig. 22.1.5). The primary loop comparator C1, senses the LED current and triggers a one-shot pulse of width T_{ON} whenever the inductor current falls below $I_{L,MIN}$. The secondary loop comparator C2 senses the average LED current and regulates the $I_{L,MIN}$, through the DAC, to achieve the desired LED average current. The secondary loop ensures regulation across a range of V_{BAT}, number of LEDs and inductor values. To avoid sub-harmonic oscillations, an off-time larger than the primary loop delay is enforced. Switching frequency of the converter can be dithered by modulating T_{ON}. The regulation loop can also be configured as a boost converter with an external Schottky diode [2-3] if needed to further minimize magnetic radiation. The driver supports up to 6 series LEDs producing average output voltages greater than 20V across the LED string, or up to 3 parallel LEDs where matching is obtained by degenerating the LEDs with series resistors as needed. The WLED driver occupies 0.05mm² with a quiescent current consumption of 850µA.

A 100dB PSRR (1kHz) coarse/fine cascaded BGR operating on the battery provides a production-trimmed reference voltage and bias currents for the PMU sub-system. To achieve 3µV A-weighted noise, the BGR voltage is externally filtered by a 200nF capacitor. A direct-to-battery modular LDO architecture is used where each module is capable of driving loads up to 50mA. For higher load currents, multiple LDO modules are connected in parallel to provide the desired drive. LDOs are externally compensated with 300nF per module. A hysteretic buck converter supplies the digital core of the SOC with 2.2µH and 4.7µF. The LDOs and DC/DC have low power modes that reduce the quiescent current in various modes of operation as defined by the system. The PMU consists of 10 LDOs and a DC/DC converter with a total load capability of 1.4A at an area of 0.64mm² (LDOs) and 0.25mm² (DC/DC). The PMU system is controlled autonomously by a Power Sequencer which runs on an always-on 32kHz clock and internal supply, handles all aspects of system power up and is intricately tied with the software to optimize system power consumption. The real-time clock generator (RTC) generates the 32kHz clock while consuming less than 1µA of quiescent current. The RTC supply is autonomously switched from the main battery to the backup battery when the former is not available. The measured results for the PMU components are summarized in Fig. 22.1.6.

References:

[1] B. Robert Gregoire, "A Compact Switched-Capacitor Regulated Charge Pump Power Supply," *IEEE J. Solid-State Circuits*, vol. 41, no. 8, pp. 1944-1953, Aug., 2006.

[2] Seok-in Hong, Jin-Wook Han, Dong-Hee Kim, and Oh-Kyong Kwon, "A Double-Loop Control LED Backlight Driver IC for Medium-Sized LCDs," *ISSCC Dig. Tech. Papers*, pp.116-117, Feb., 2010.

[3] W. Hollinger and M. Punzenberger, "An Asynchronous 1.8MHz DC/DC Boost Converter Implemented in the Current Domain for Cellular Phone Lighting Management," *ESSCIRC*, pp. 528-531, Sept., 2006.

ISSCC 2011 / February 23, 2011 / 8:30 AM

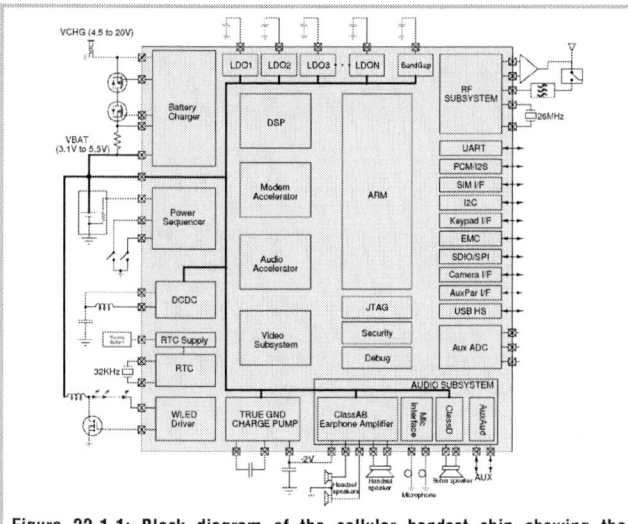

Figure 22.1.1: Block diagram of the cellular handset chip showing the integrated PMU.

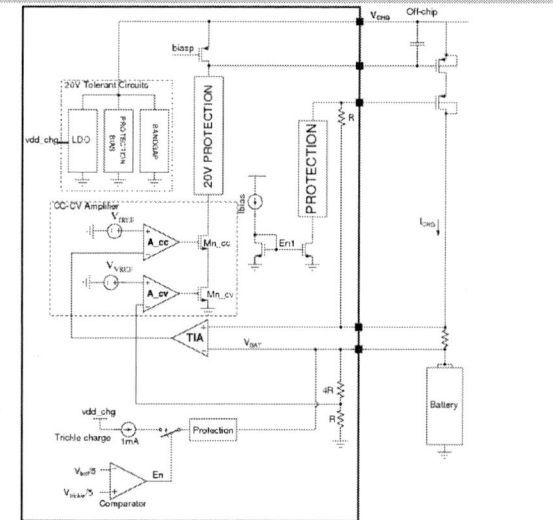

Figure 22.1.2: Block diagram of the 20V-compatible charger.

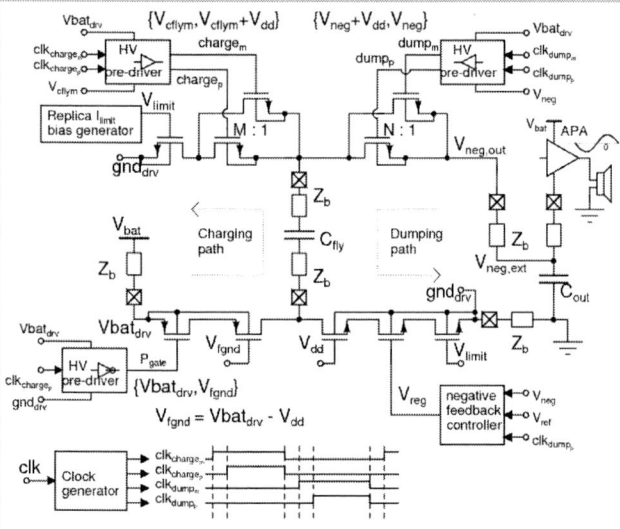

Figure 22.1.3: Regulated inverting charge pump.

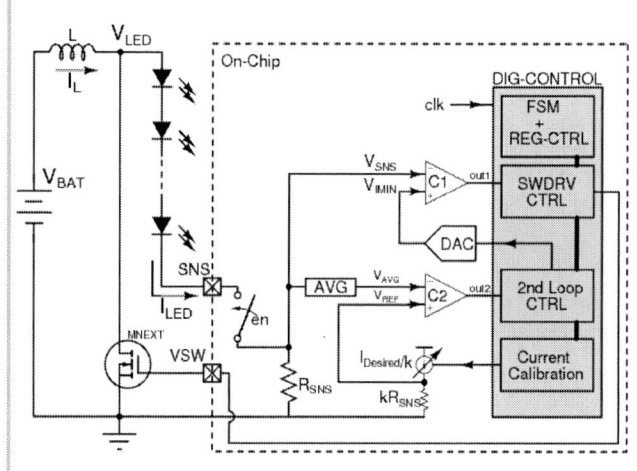

Figure 22.1.4: Inductive WLED driver block diagram.

$$I_{LED,avg} = \frac{V_{BAT}}{V_{LED}} \left(I_{L,min} + \frac{V_{BAT}\, T_{ON}}{2L} \right)$$

Figure 22.1.5: WLED driver waveforms.

SPECIFICATION	MEASUREMENT	COMMENTS
Switching DC-DC Regulator (Max I_LOAD=300mA) for Digital Core Supply		
Efficiency (%)	85	V_BAT=3.6V, V_OUT=1.2V, I_LOAD=150mA
Load Regulation (mV/A)	2.5	I_LOAD varies from 50mA to 300mA
Line Regulation (mV/V)	1.66	V_BAT varies from 3.1V to 5.5V, V_OUT=1.2V
Linear LDO (Results for 2X Module Supplying On-Chip Analog presented as a sample)		
Average Line Regulation (mV/V)	0.14	V_BAT varies from 3.1V to 5.5V
Load Regulation (mV/A)	290	I_LOAD from 1mA to 100mA
PSRR @ 1KHz (dB)	73.6	V_BAT= 3.6V, LDO Out=2.85V
Integrated Noise (μV_RMS)	54	25Hz – 20KHz
	118	10KHz-100KHz
Class-AB APAs Operating from ICP Negative Rail		
THD (%)	0.03	V_BAT=3.6V, Negative Rail =-2V, 2 APAs each with 40mW Power Delivery to 16Ω Load.
WLED Driver		
LED Current	2-20mA	16 Dimming Steps Provided by a PWM operation
Quiescent Current	850μA	

Figure 22.1.6: PMU measured performance summary.

978-1-61284-303-2/11 $26.00 © 2011 IEEE

ISSCC 2011 PAPER CONTINUATIONS

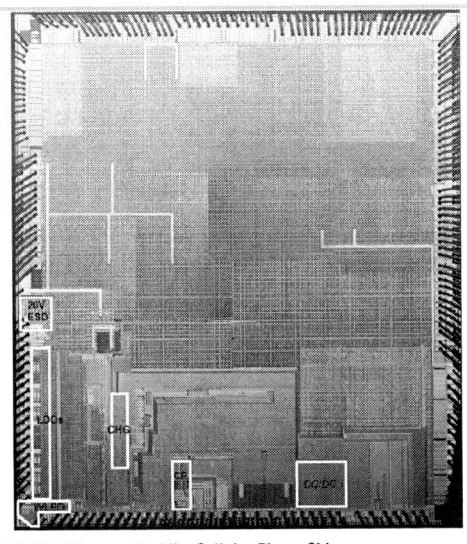

Figure 22.1.7: Die Micrograph of the Cellular Phone Chip.

ISSCC 2011 / SESSION 22 / DC/DC CONVERTERS / 22.2

22.2 A Digitally Controlled DC-DC Converter for SoC in 28nm CMOS

Franz Kuttner[1], Harun Habibovic[1], Thomas Hartig[1], Michael Fulde[1], Gernot Babin[1], Andreas Santner[1], Peter Bogner[1], Claus Kropf[1], Harald Riesslegger[1], Uwe Hodel[2]

[1]Infineon Technologies, Villach, Austria
[2]Infineon Technologies, Munich, Germany

Battery operation in mobile applications needs power efficient DC-DC converters which are able to handle battery voltages up to 5.5V. Normally, these DC-DC converters are built in special technologies. For decreased footprint and chip count of the overall system, a system-on-chip solution on modern CMOS technology with core supply voltages around 1V is preferred. The presented DC-DC buck converter generates 0.9 to 1.8V at output currents up to 500mA from a battery voltage of 2.4 to 5.5V with high efficiency in a high-k 28nm metal-gate CMOS technology.

Figure 22.2.1 shows the overall architecture of the DC-DC converter. Only two capacitors and one inductor are used as external components. The output voltage can be generated either in a pulse-frequency mode (PFM), which is used for startup and standby operation, or in pulse-width mode (PWM) for normal operation. PFM is operated with a current comparator and a voltage comparator only. If the output voltage falls below a certain level, current bursts from 50mA to 400mA are generated increasing the voltage at the output. This PFM operation is asynchronous, generating unwanted spurious in the frequency domain.

For normal operation PWM is used, where the location of the spurious is well known. In PWM, the output voltage is measured by a 4b A/D converter (ADC). The decision levels of this ADC are located ±60mV around the wanted output voltage. A power saving architecture with 40µA at 12MS/s is used for the ADC shown in Fig. 22.2.2. The averaged value of two samples taken at every PWM period is fed to a PID filter. The coefficients of this filter can be programmed, for adapting the PID behavior for different external components or clk frequencies. The calculated result of the PID filter is used for a digital PWM (DPWM) with 8b resolution. The master frequency of the DC-DC is 208MHz and DPWM output frequency is 1.6MHz.

For lower output currents this PWM mode is changed from continuous-current mode (CCM) to discontinuous-current mode (DCM) operation. At very small currents, a DCM with pulse skipping is used. Figure 22.2.3 shows the current waveforms for the different modes. To guarantee a stable operation in all modes the coefficients of the digital filter are changed on the fly, because the characteristic of the loop is changing with the mode.

The supply voltage of the DC-DC converter is 1V and all transistors used are single-gate-oxide transistors. In 28nm CMOS technologies with triple wells, special devices with extended drain (DeMOS) can be implemented that are able to handle higher voltages at their drain. To get a fast and low ohmic DeMOS without special process steps, these DeMOS are built with single gate oxide. This implies that the voltage from the gate to the channel has to be limited to the core voltage.

A traditional solution to inherently protect the gates of DeMOS devices is to supply the drivers by auxiliary voltage regulators which provide Vcore. Then the driver creates an output signal that is within a save operating range for the gates of the DeMOS devices [1]. This solution requires dedicated regulators to supply the driver of DeNMOS devices with Vcore and the driver of DePMOS devices with Vcore below the battery voltage VBAT. Driving large power switches, the regulators have to source huge dynamic current, which only can be provided only by huge internal or costly external capacitors.

In the realized DC-DC the limitation of DeMOS gate-source voltage is done by monitoring the voltages by comparators as shown in Fig. 22.2.4. The voltage at the gate of the P-channel DeMOS Transistor can be switched by a normal single gate oxide P-Channel transistor (P1) and an N channel DeMOS (N1). The high driver capability of the switched DeMOS transistor leads to high gate area, and

therefore the gate capacity of this transistor is large. Therefore the voltage swings at the gate of this transistor are relatively slow. A comparator, which is fast enough monitors this voltage and switches off transistor N1 when the comparator level is reached. Due to the high gate capacitance of the DeMOS, the voltage of node "Gate" remains at this comparator level, and therefore the DeMOS stays in conductive behavior until it is switched off by transistor P1.

This architecture moves the voltage limiting challenge to a fast comparator design. Preferably, this comparator is build with single-gate-oxide transistors, but the voltage, to be monitored is in the range of Vin-Vcore to Vin (typically 4.3 to 5.5V). A resistive divider can be used to bring the voltage in the right range but then the current through this resistor changes the voltage level after N1 is switched of. Another way is to do this shift in a capacitive way and not to monitor the absolute voltage level of the node "Gate", but to shift the voltage at the "Gate" node from Vin to Vin – Vcore. All switches in Fig. 22.2.4 can be built with single-gate-oxide transistors, because only voltages between 0 and Vcore have to be switched.

In the realized implementation also the N channel DeMOS transistor is driven by the new solution although the gate driving level for the N channel DeMOS is in the range from 0 to Vcore, which can be done with standard CMOS. But the gate capacitance of this switch is very large, which results in high driving current and a low ohmic supply voltage, stabilized by a large capacitance. Therefore, it is better to get the charge for driving the gate of the N channel DeMOS direct from the battery.

One issue of digitally controlled PWM converters is how to minimize the so-called steady-state limit cycles and increase the static accuracy. Publications such as [2] show the calculations on how to minimize the limit cycles. These limit cycles generate a pattern, which is repeated with a certain frequency. If the frequency of this pattern is in a sensitive range of the supplied electronic components, then it is necessary to avoid these cycles. The reason for limit cycles and worse static accuracy is a built-in hysteresis in the digital loop. The ADC is realized with limited resolution for area and power reason. It is obvious that if the output of the ADC is 0 the output of the filter does not change, because filtering 0 results in a 0 at the output of the filter, and integrating 0 results in a 0 at the output of the integrator. STherefore, in the range where the ADCs input delivers a 0 at its output the loop of the DC-DC is open, resulting in a static inaccuracy of the generated voltage of 1LSB of the ADC resolution. The "dead zone" of zero has to be avoided. Such dead zone can be avoided by adding ½ to the output of the ADC. Another possibility is to avoid the zero at the output of the ADC by realizing an ADC which changes its output value at rising analog input directly from –1 to +1. This however can change the stability considerations of the loop, because the transfer curve of the ADC is changed at 0, and therefore the loop gain is also influenced.

Figure 22.2.5 shows a chip micrograph. The DC-DC converter is fabricated in a high-k metal-gate 6M 28nm standard CMOS process and its core area occupies 0.25mm² [3].

Figure 22.2.6 shows measurement results for the efficiency of the DC-DC for the different operating modes (PFM, PWM&CCM, PWM&DCM). Efficiency is limited by small inductivity and high switching frequency, but exceeds 90% in the interesting range from 50 to 250mA.

References:
[1] B. Forejt, V. Rentala, J.D. Arteaga, and G. Burra, "A 700+-mW Class D Design with Direct Battery Hookup in a 90-nm Process," *IEEE J. Solid-State Circuit*s, vol. 40, no. 9, pp. 1880-1887, Sep., 2005.
[2] A.V. Peterchev and S.R. Sanders, "Quantization Resolution and Limit Cycling in Digitally Controlled PWM Converters," *IEEE Transaction on Power Electronics*, vol. 18, no. 1, pp. 301-308, Jan., 2003.
[3] F. Arnaud, A. Thean, M. Eller, et al., "Competitive and Cost Effective High-k Based 28nm CMOS Technology for Low Power Applications," *IEEE International Electron Devices Meeting* (IEDM), pp. 651-654, Mar., 2009.

978-1-61284-303-2/11 $26.00 © 2011 IEEE

ISSCC 2011 / February 23, 2011 / 9:00 AM

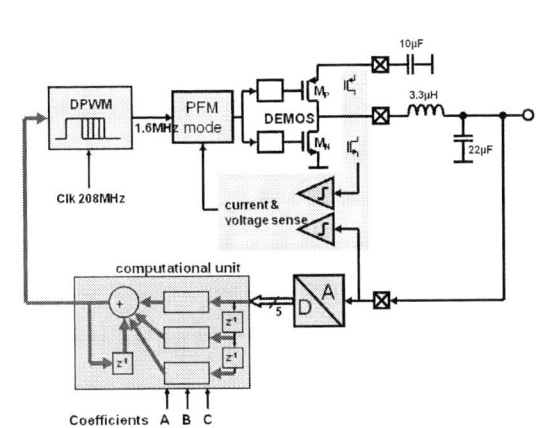

Figure 22.2.1: Overall architecture.

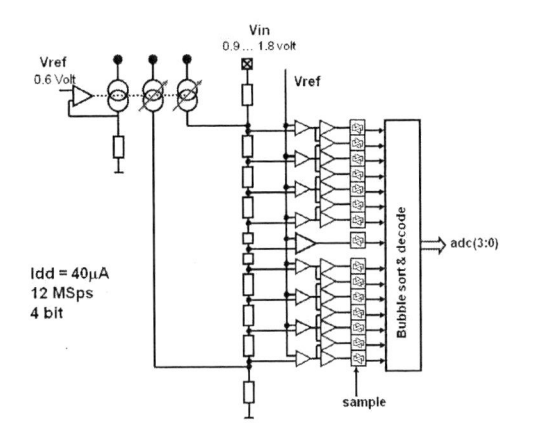

Figure 22.2.2: A/D converter of the DC-DC converter.

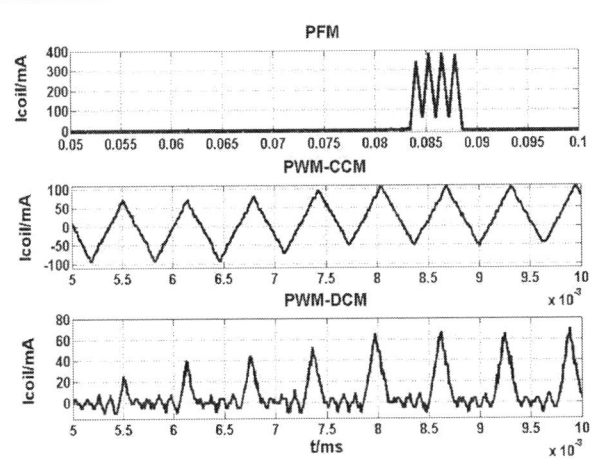

Figure 22.2.3: Measured current waveforms for different modes.

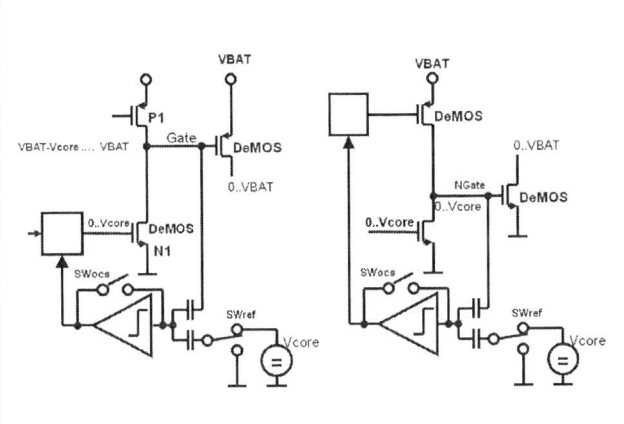

Figure 22.2.4: DEMOS driver.

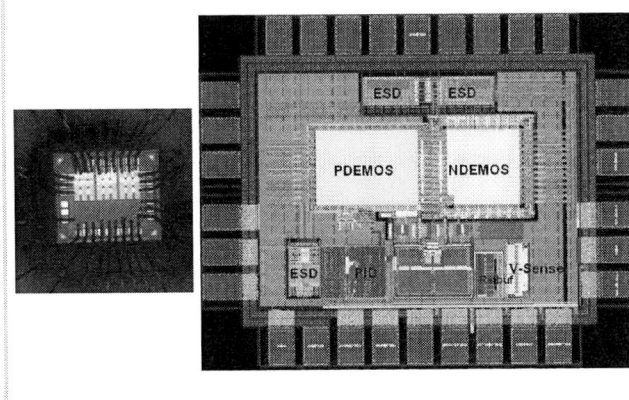

Figure 22.2.5: Chip micrograph.

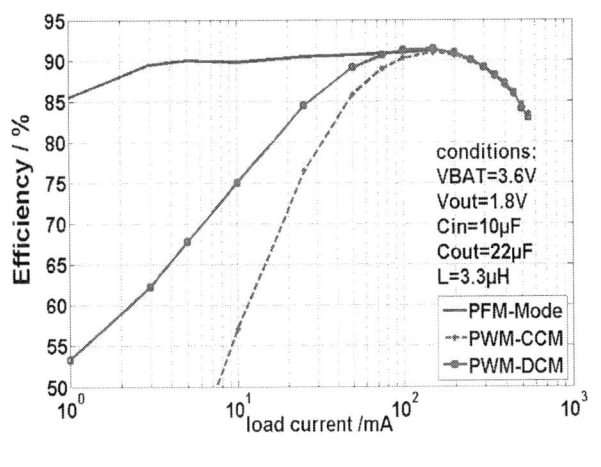

Figure 22.2.6: Measurement results.

978-1-61284-303-2/11 $26.00 © 2011 IEEE

ISSCC 2011 / SESSION 22 / DC/DC CONVERTERS / 22.3

22.3 20µA to 100mA DC-DC Converter with 2.8 to 4.2V Battery Supply for Portable Applications in 45nm CMOS

Saurav Bandyopadhyay[1], Yogesh K. Ramadass[1,2],
Anantha P. Chandrakasan[1]

[1]Massachusetts Institute of Technology, Cambridge, MA
[2]Texas Instruments, Dallas, TX

Digital baseband processors [1] in portable devices today are able to operate off voltages of 1V and less. Efficient DC-DC converters are required to power these ICs from Li-ion batteries with a voltage range of 2.8 to 4.2V. This is done by a discrete power management IC (PMIC) capable of handling the high battery voltage. However, there is significant push in integrating the PMIC module with the baseband processor implemented in scaled technologies thereby reducing the number of system-level components. This paper presents the circuit techniques used in a 45nm CMOS DC-DC converter with high battery-voltage handling capability.

The architecture of the DC-DC buck converter with the cascoded power train and control circuit is shown in Fig. 22.3.1. The process provides 1.8V devices. To support battery voltages more than 3.6V, three 1.8V PMOS devices have been stacked with their wells connected to their sources. This ensures each PMOS has 1.8V $V_{SD,(MAX)}$ when off and 1.8V V_{SG} when on. Since the process used does not allow isolated NMOS, stacking NMOS devices with their bulks tied to the substrate would cause the topmost NMOS in the stack to face V_{DB} of 4.2V. This is detrimental to device reliability. Instead, a high-voltage drain-extended NMOS capable of handling 4.2V V_{DS} and 1.8V V_{GS} is used. Although this hurts the efficiency due to its high resistance per unit width, the drain- extended NMOS ensures reliability. The corresponding 4.2V V_{SD} PMOS is not available in the process. Due to the aforesaid process constraints, the designed power train is unlike [2] where two isolated stacked devices suffice. The buck converter control circuit, level shifter, and gate drive circuits, shown in Fig. 22.3.1 are discussed later.

The gates of transistors P3 and N1 in Fig. 22.3.1 are driven by inverters. For P1, the gate voltage switches between V_{BAT} and $V_{BAT}-1.8V$. As the process does not allow any NMOS to have V_{GB} close to 4.2V, an inverter cannot be used in a gate drive circuit. Instead, a pass transistor implementation shown in left half of Fig. 22.3.2 is adopted for P1. This ensures no NMOS gets V_{BAT} across its gate and bulk terminals. Capacitive level shifters shown in the right half of Fig. 22.3.2 are used to shift the rails of the gate drive signals.

The buck converter requires different internal voltage supplies for the control circuit (1.2V), level shifters, power stage stack, and gate drive circuits (1.8V, $V_{BAT}-1.8V$, and $V_{BAT}-3.6V$). A switched-capacitor (SC) converter is used to convert the battery voltage to 1.8V. This SC, based on [3], has three reconfigurable gain settings, 1/2, 2/3, and 1 for different battery voltages. The converter in the 2/3 setting is shown on the left in Fig. 22.3.3. The capacitive level shifter is used for the switches that need different rails. A pulse-frequency modulation (PFM) loop regulates the output to 1.8V. The other internal supplies are generated from this SC output. A SC with a gain setting of 1 is used to convert 1.8V to 1.2V for the buck converter control circuit. This ensures a battery-independent supply for the control circuit. As shown in the right half of Fig. 22.3.3, for the $V_{BAT}-1.8V$ and $V_{BAT}-3.6V$ converters, average current flows into the converters. When the comparator (C1/C2) senses that the voltage is above the reference (ref1/ref2), it sends a pulse to the switch (N1/N2) which brings the voltage down below the reference thereby regulating the output close to the reference. To prevent overstress in the PMOS stack during startup, additional sequencing logic in SC and stacking converters can be implemented as in [7]. These have not been included in the present design.

The control circuit of the buck converter uses both pulse-width modulation (PWM) and PFM schemes to support the wide load range. The PWM loop consists of a 4b ADC, a PID controller, and a digital pulse-width modulator (DPWM). The PFM loop is a one bit control scheme consisting of a comparator and the DPWM. For a µW to mW DC-DC converter, the counter-based DPWM [4] cannot be used as it needs a high-frequency clock (~100MHz). A DLL-based DPWM [2] is low in power but is area consuming for low/medium (~1MHz) switching fre-

quencies. Here, a 5b I-C DAC-based DPWM with sleep mode is used which has significantly smaller area and comparable power with the DLL based DPWM. To create pulses of desired widths, the analog technique of charging a capacitor with a constant current and comparing the voltage to a reference is used. This normally requires a continuous time analog comparator with a quiescent current thus, increasing the power consumption. Here, a *sleep control* mechanism switches off the comparator when the comparison is over. The DPWM circuit is shown in Fig. 22.3.4. The capacitor to be charged is digitally controlled by switches D0 to D4 and their compliments thereby controlling the pulse widths. The analog comparator operation can be split into a *comparison phase* and a *sleep phase*. The rising edge of the system clock (at switching frequency) is used to wake up the analog comparator, thus starting the *comparison phase*. The reset signal to N1 forces the top plate of the effective capacitor to ground. To ensure the node is pulled to ground, N1 is designed to be wider than P2. The current source P1 then starts to charge the capacitor. When the voltage exceeds the reference, the comparator output starts to flip. P2 further increases the differential input to the comparator further helping the comparator to flip. This is the end of the *comparison phase*. Since, the comparator's second stage is essentially a latch that stores the comparison result, using Sleep Control Logic, sleep_p and sleep_n are generated that power off the comparator thus preventing any further power consumption. This is the beginning of the *sleep phase*. Thus, by leveraging the fact that the time instant of comparison is known, we can reduce the overall power consumption. The DPWM comparators and capacitor bank occupy 0.01mm^2 as compared to the estimated 0.042mm^2 with delays and multiplexers of the DLL-based DPWM (without the routing overhead). The simulated power of the proposed DPWM is 15µW in PWM and 159nW in PFM for 20µW output power. The estimated power of the DLL-based DPWM is 24µW in PWM and 156nW in PFM for 20µW. A well-controlled current reference and matched capacitors also ensure DPWM linearity.

Figure 22.3.5 shows the measured efficiency plots for both PFM and PWM modes. The peak efficiency for 20µW output in PFM mode is 75%. This is higher than previously published direct-battery µW-to-mW sub-micron converters [2]. This efficiency is limited by the leakage in the power stage estimated to be 1.35µA at 3V input and 1V output resulting in 10% loss in efficiency. The efficiency is 87.4% for 12mW in PFM mode and 87.2% for 53mW in PWM mode. For an integrated 45nm CMOS power stage where cascoding three PMOS devices is necessary, switching losses are high as two stacked devices have to switch. Conduction losses also go up by 3× as compared to one PMOS power stage [4,5,6]. The conduction and switching losses in the power stage cause 10 to 11% degradation in peak efficiency. Left side of Fig. 22.3.6 shows the load transient when the load changes from 100µA to 5mA at 0.8V output in PFM and 10mA to 50mA in PWM for 1V output. The summary of the design is shown in the right half of Fig. 22.3.6.

Acknowledgements:
This work was funded by DARPA. We thank Baher Haroun, Ayman Fayed and Alice Wang of Texas Instruments for discussions and support.

References:
[1] G. Gammie, A. Wang, M. Chau, et al., "A 45nm 3.5G Baseband-and-Multimedia Application Processor using Adaptive Body-Bias and Ultra-Low-Power Techniques," *ISSCC Dig. Tech. Papers*, pp. 258-259, Feb., 2008.
[2] J. Xiao, A. Peterchev, J. Zhang, and S. Sanders, "A 4-µA Quiescent-Current Dual-Mode Digitally Controlled Buck Converter IC for Cellular Phone Applications," *IEEE J. Solid-State Circuits*, vol.39, no.12, pp. 2342-2348, Dec., 2004.
[3] J. Kwong, Y. Ramadass, N. Verma, and A. Chandrakasan, "A 65nm Sub-V$_t$ Microcontroller with Integrated SRAM and Switched Capacitor DC-DC Converter," *IEEE J. Solid-State Circuits*, vol.44, no.1, pp. 115-126, Jan., 2009.
[4] E. Soenen, A. Roth, J. Shi, et al., "A Robust Digital DC-DC Converter with Rail-to-Rail Output Range in 40nm CMOS," *ISSCC Dig. Tech. Papers*, pp. 198-199, Feb., 2010.
[5] Texas Instruments, "TPS62200 Datasheet," Accessed August, 2010, *http://focus.ti.com/lit/ds/symlink/tps62200.pdf.*
[6] Linear Technology, "LTC3388 Datasheet," Accessed August, 2010, *http://cds.linear.com/docs/Datasheet/338813f.pdf.*
[7] Texas Instruments, "TPS54521 Datasheet," Accessed September, 2010, *http://focus.ti.com/lit/ds/symlink/tps54521.pdf.*

978-1-61284-303-2/11 $26.00 © 2011 IEEE

ISSCC 2011 / February 23, 2011 / 9:30 AM

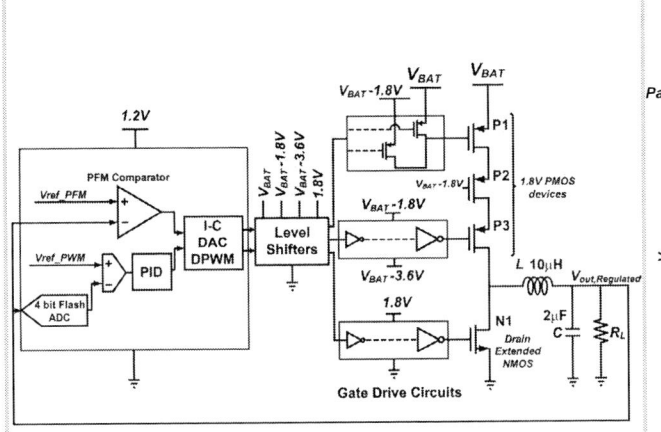

Figure 22.3.1: Architecture of the DC-DC converter (internal SCs for 1.2V, 1.8V, VBAT−1.8V and VBAT−3.6V not shown in this figure).

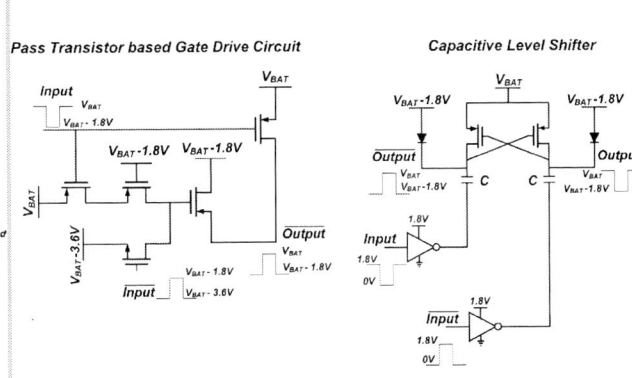

Figure 22.3.2: PMOS gate drive with pass transistors and level shifters used in power train.

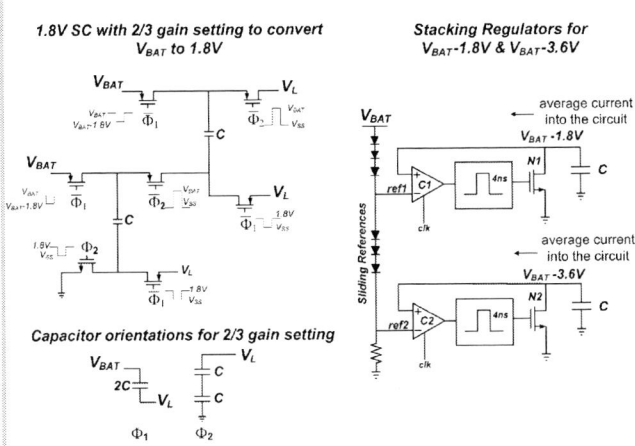

Figure 22.3.3: 1.8V switched-capacitor converter and stacking regulators.

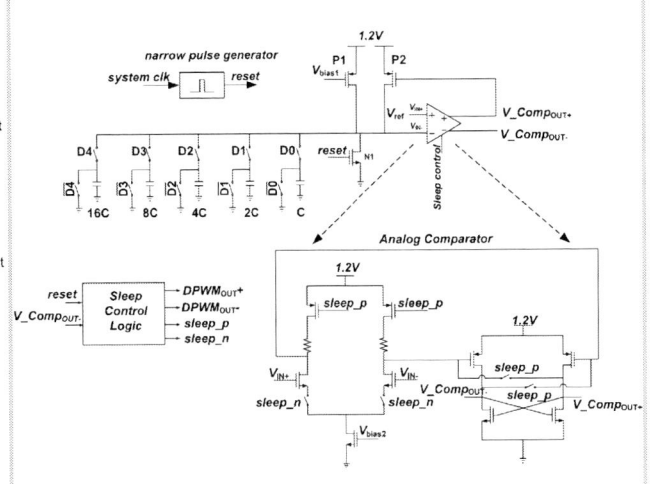

Figure 22.3.4: I-C DAC DPWM with analog comparator and sleep control.

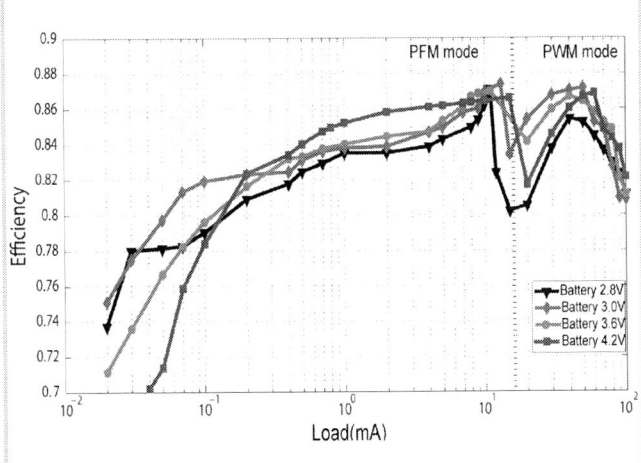

Figure 22.3.5: Efficiency curves for PFM and PWM modes including buck converter power stage, SCs, stacking regulators and control.

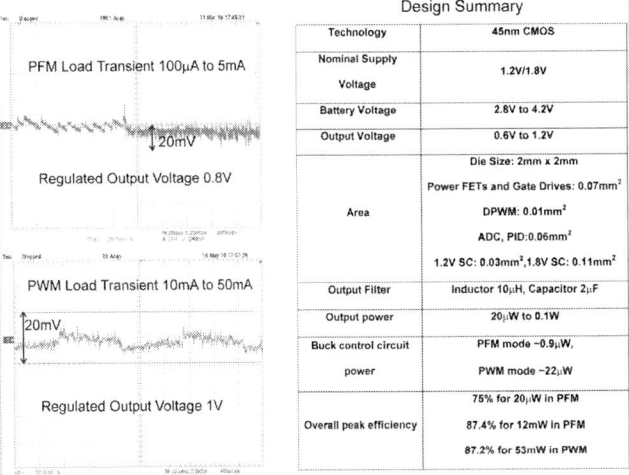

Figure 22.3.6: Transient response and design summary.

ISSCC 2011 PAPER CONTINUATIONS

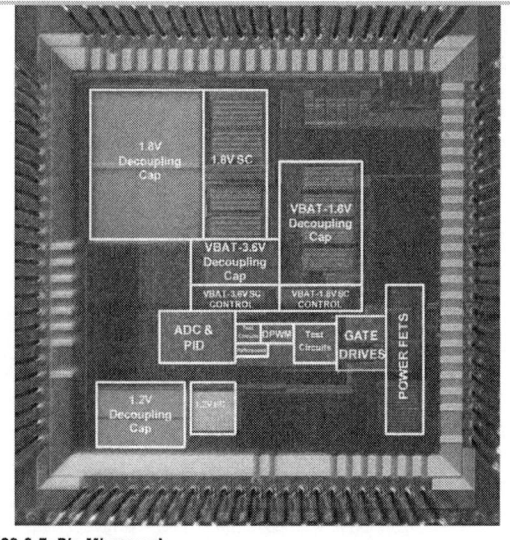

Figure 22.3.7: Die Micrograph.

ISSCC 2011 / SESSION 22 / DC/DC CONVERTERS / 22.4

22.4 A Digitally Controlled DC-DC Buck Converter with Lossless Load-Current Sensing and BIST Functionality

Tao Liu[1], Hyunsoo Yeom[1], Bert Vermeire[1], Philippe Adell[2], Bertan Bakkaloglu[1]

[1]Arizona State University, Tempe, AZ
[2]Jet Propulsion Laboratory, Pasadena, CA

Lossless load current sensing ability is one of the most desirable features of contemporary current- or voltage-mode-controlled DC-DC converters. Current sensing can be used for short circuit detection, multi-stage converter load balancing, thermal control, and load-independent control of DC-DC converters [1]. Recently, current sensing techniques using the existing inductor series DC resistance (DCR) are gaining attention due to their reduced complexity and minimized loss [2][4]. These techniques increase the need for inductor build-in self-test (BIST) ability to measure DCR. Inductor BIST is also critical for high-reliability applications such as automotive and aerospace systems where component variation and drift need to be closely monitored. An analog inductor characterization and current sensing technique utilizing G_m-C filtering is proposed in [2], where the gain errors and sensing offset are cancelled in analog domain. The DC-DC buck converter presented here uses an offset- independent inductor characterization enabling a digital continuous lossless load-current-sensing scheme. The proposed inductor BIST and current-sensing techniques can be extended to current-mode-controlled converters and multi-stage parallel converters as well.

Figure 22.4.1 shows the proposed digitally controlled DC-DC buck converter with the load current sensing circuitry. The converter is configured in voltage-controlled mode where a PID compensator calculates the required pulse duty cycle based on the difference of the digitized DC-DC feedback voltage V_{FB} and the reference voltage V_{REF}. A 9b DLL-based digital PWM (DPWM) generator described in [3] generates the desired duty cycle. The digitization is carried out by digitally intensive frequency-domain $\Delta\Sigma$ ADCs. The proposed ADC is composed of a VCO followed by a first-order $\Delta\Sigma$ frequency discriminator ($\Delta\Sigma$FD) and a CIC decimator [3]. The digital feedback loop is designed with a cross-over frequency of 50kHz with a switching frequency of 500kHz. The sensed load current can be used to select prestored PID coefficient sets and update compensator, thereby minimizes the closed-loop response variation due to the load-current deviation.

One of the key contributions in this DC-DC converter is the accurate lossless digital load-current-sensing capability using DCR of the inductor. It is obtained by an average-current-sensing technique [4]. An RC network set by R_S and C_S is added in parallel to the power-stage output. R_S and C_S can be integrated on chip because their absolute values do not affect the measurement. The load current can be obtained by $I_{LOAD}=[V_C(Avg)-V_{OUT}]/R_{DCR}$, here V_C is the voltage at the RC filter output and $\Delta V_C \propto 1/(R_S \cdot C_S)$ can be controlled by R_S and C_S. R_{DCR} is the inductor DCR. This approach does not require a current transformer or sensing resistor. At start-up, inductor BIST mode characterizes the R_{DCR} as described later. A non-inverting differential sensing amplifier (SA) amplifies the difference of V_C and V_{OUT}. The gain of SA is set by resistor ratio R_F/R_G. A $\Delta\Sigma$ ADC is used to digitize the SA output V_{DIFF}. After the digitization, the I_{LOAD} computation is accomplished in the digital domain. This sensing scheme can update I_{LOAD} value in every switching clock. To eliminate the effects of SA offset and DC level, the SA input is shorted by turning on switch *Offset_Cal* before load current measurement and readout chain offset are digitally restored. Since digitized analog voltages are based on frequency domain digitization, the ADC gain is proportional to f_{VCO} and VCO gain. VCO gain drift will result in digitized code drift. By comparing the digitized codes $V_{DIFF,D}$ and $V_{REF,D}$ from V_{DIFF} and V_{REF}, the digital code $V_{DIFF,real}=V_{DIFF,D}/V_{REF,D}\times V_{REF}$ is normalized across temperature and process variations with a matched pair of VCOs. Due to the digital offset cancellation, only one sensing unit is needed.

The required digital measurement of the inductor DCR is performed by the proposed offset independent BIST circuit as shown in Fig. 22.4.2. The BIST module shares the same SA and $\Delta\Sigma$ ADC as in current sensing. During BIST mode, the power FETs are turned off and the load capacitor is shorted to ground at the

feedback point. Due to differential measurement, the R_{dson} of the FET shorting the filter capacitor does not need to be low. Upon startup, a symmetric triangular current I_{TRI} is applied to the inductor. The voltage across the inductor V_{IND} is amplified and sensed by the digitization chain. Let *slope* be the slope of current I_{TRI}, and thus V_{IND} is defined as $V_{IND}=L\times slope+I_{TRI}\times R_{DCR}$. Consider the rising part of I_{TRI} from its minimum, $I_{TRI,min}$, to its maximum, $I_{TRI,max}$, as *slope* maintains the same, the change of V_{IND} is only contributed by R_{DCR}, i.e., $(I_{TRI,max}-I_{TRI,min})\times R_{DCR}$. Similarly, at the peak I_{TRI} instance, I_{TRI} maintains $I_{TRI,max}$ but slope changes from rising part *slope*$_+$ to falling part $-slope_-$, and V_{IND} has a change of $(slope_++slope_-)\times L$. Correspondingly, L can be obtained by measuring the V_{DIFF} change ($V_{DIFF,B}-V_{DIFF,A}$) resulting from I_{TRI} slope change, and DCR by measuring V_{DIFF} change ($V_{DIFF,C}-V_{DIFF,B}$) resulting from I_{TRI} value change:

$$L=\frac{V_{DIFF,B}-V_{DIFF,A}}{slope_++slope_-}\cdot\frac{R_G}{R_F}, \quad R_{DCR}=\frac{V_{DIFF,C}-V_{DIFF,B}}{I_{TRI,max}-I_{TRI,min}}\cdot\frac{R_G}{R_F}.$$

Therefore, the measured L and DCR do not depend on the common-mode level or input-referred offset of the SA. After digitization, V_{DIFF} is converted to 13b data updated at 500kHz rate. L and R_{DCR} computation are then accomplished in the digital domain. To obtain steep edges of V_{DIFF} required by L and DCR measurement, I_{TRI} frequency is limited to 10kHz to avoid the potential resonance resulting from large inductance and parasitic capacitance of power FETs. The measured L can be used to update PID compensator to minimize inductor variation effect and DCR is used in I_{LOAD} computation.

This BIST technique needs a controllable triangular current generator and an edge-detection circuit as shown in Fig. 22.4.3. Triangular current I_{TRI} generator consists of a triangular voltage V_{TRI} generator followed by a linear V-I converter. V_{TRI} is generated by applying a matched charge and discharge bias current pair I_B to a known capacitor C_{TRI}. The charge and discharge time is controlled by the comparators. V_{TRI} maximum voltage, V_H, and minimum voltage, V_L, are precisely set by matched series resistors. I_{TRI} slope is controlled by bias current, I_B, and V-I converter current mirror ratio. The comparator output, V_{SEL}, and its one clock delayed signal $V_{SEL,Delay}$ are used as clock signals to detect and differentiate the maximum and minimum values of the output V_{DIFF} at peak triangular current instance.

A maximum 5.5V input, maximum 1A load current buck converter with the proposed BIST and current-sensing scheme is fabricated in a 0.7µm power CMOS process. As shown in Fig. 22.4.4, the BIST module achieves an average measurement error of 2.1% for inductance range of 3.7µH to 22.3µH, and 3.6% for DCR range of 15mΩ to 75mΩ The load-current-sensing scheme obtains average 1.5% error for current range of 100 to 750mA. Figure 22.4.5 shows the regulator response when the load changes from 10µA to 100mA at 1kHz where overshoot voltage is 95mV and settling time is around 85µS. Figure 22.4.6 summarizes the chip performance. The die micrograph is provided in Fig. 22.4.7. The extra hardware cost for BIST and current sensing including related pads is 5.2% of the total chip area.

Acknowledgements:
This research is supported by Jet Propulsion Laboratory, NASA Electronic Parts and Packaging (NEPP) and Director's Research and Development Fund (DRDF) programs.

References:
[1] Y. Woo, et al., "Load-Independent Control of Switching DC-DC Converters with Freewheeling Current Feedback," *ISSCC Dig. Tech. Papers*, pp. 446 - 626, Feb., 2008.
[2] H. Forghani-zadeh and G. Rincón-Mora, "An Accurate, Continuous, and Lossless Self-Learning CMOS Current-Sensing Scheme for Inductor-Based DC-DC Converters," *IEEE J. Solid-State Circuits*, vol. 42, no. 3, pp. 665-679, Mar., 2007.
[3] H. Ahmad and B. Bakkaloglu, "A 300mA 14mV-Ripple Digitally Controlled Buck Converter Using Frequency Domain $\Delta\Sigma$ ADC and Hybrid PWM Generator," *ISSCC Dig. Tech. Papers*, pp. 202-203, Feb., 2010.
[4] X. Zhou, P. Xu, and F. Lee, "A Novel Current-Sharing Control Technique for Low-Voltage High-Current Voltage Regulator Module Applications," *IEEE Tran. Power Electronics*, vol. 15, no. 6, pp.1153-1162, Nov., 2000.

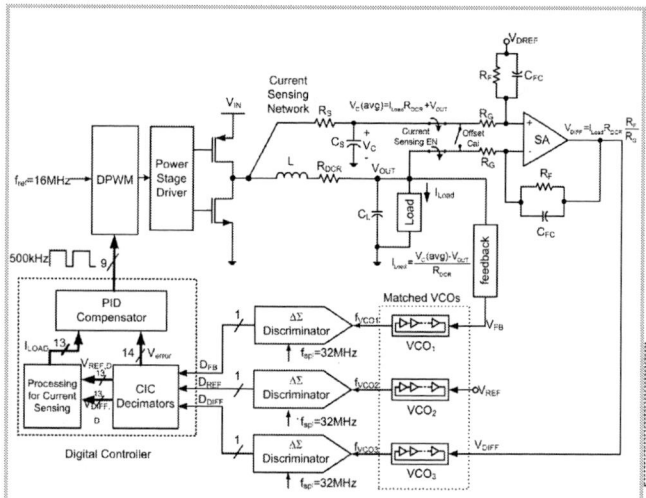

Figure 22.4.1: Proposed DC-DC buck converter architecture with lossless load current sensing circuitry.

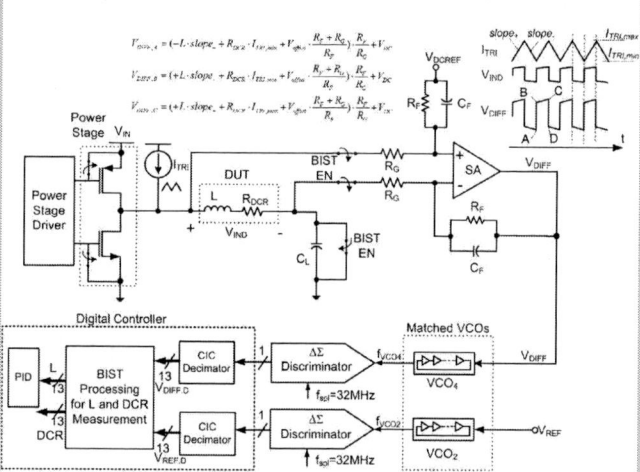

Figure 22.4.2: BIST-mode signal chain for inductor inductance and DCR characterization.

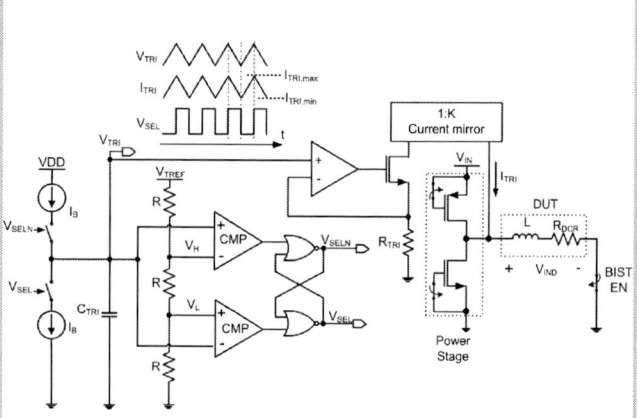

Figure 22.4.3: Circuitry to generate triangular current in inductor BIST mode.

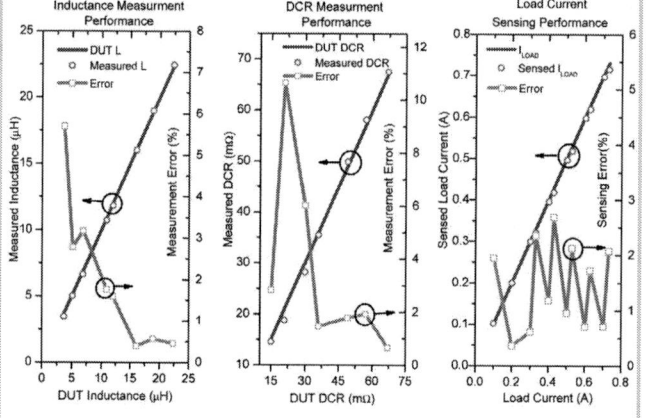

Figure 22.4.4: Inductor measurement and load current sensing performance, achieving 2.1% and 3.6% average error for inductance and DCR measurement, respectively, and 1.5% average error for load current sensing. Here current sensing is based on a DCR=63mΩ, L=18.8µH inductor.

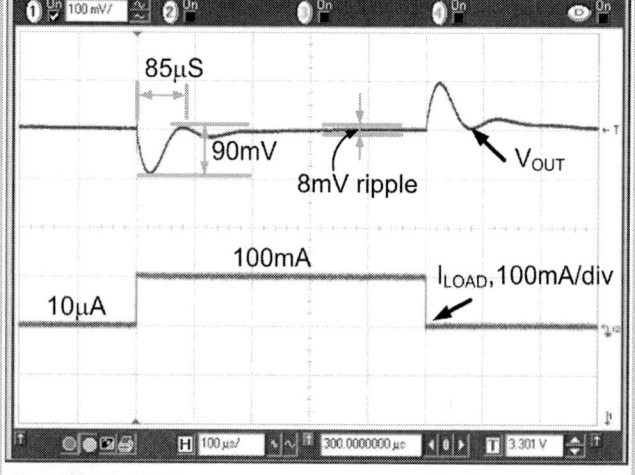

Figure 22.4.5: DC-DC buck converter transient response for 100mA step, 1kHz switching load current. V_{IN}=5V, V_{OUT}=3.3V, C_L=22µF, ESR=70mΩ, L=18.8µH, and DCR=63mΩ.

DC-DC Converter Parameters	
Technology	AMI i2t100 0.7µm CMOS
V_{IN}	1V-5.5V (5V typical)
V_{OUT}	1V-5.5V
Max load current(I_{LOAD})	1A
Output voltage ripple	≤10mV
Switching frequency	500 KHz
Cross-over frequency	50 KHz
Off-chip C_L	22 µF
ESR (R in series with C_L)	70 mΩ
Efficiency	
I_{LOAD} [0.1A – 1A] @V_{IN}=5V, V_{OUT}=4V	89.7% ≤ η ≤ 94.5%
I_{LOAD} [0.1A – 1A] @V_{IN}=5V, V_{OUT}=3.3V	81.7% ≤ η ≤ 91.8%
BIST Performance	
Inductance [3.7µH – 22.3µH]	Average Error 2.1%
DCR [15mΩ – 80mΩ]	Average Error 3.6%
Load Current Sensing Performance	
I_{LOAD} [100mA – 750mA] (@DCR=63 mΩ, V_{IN}=5V V_{OUT}=3.3V)	Average Error 1.5%
Chip Area	
Die	3.5mm × 3.5mm
Inductor BIST & Current Sensing	5.2% of Die Area
Quiescent Current Consumption	
BIST mode	40.85mA for max 200µs
Current sensing	390 µA
Normal regulation	610 µA

Figure 22.4.6: Performance summary.

978-1-61284-303-2/11 $26.00 © 2011 IEEE

ISSCC 2011 PAPER CONTINUATIONS

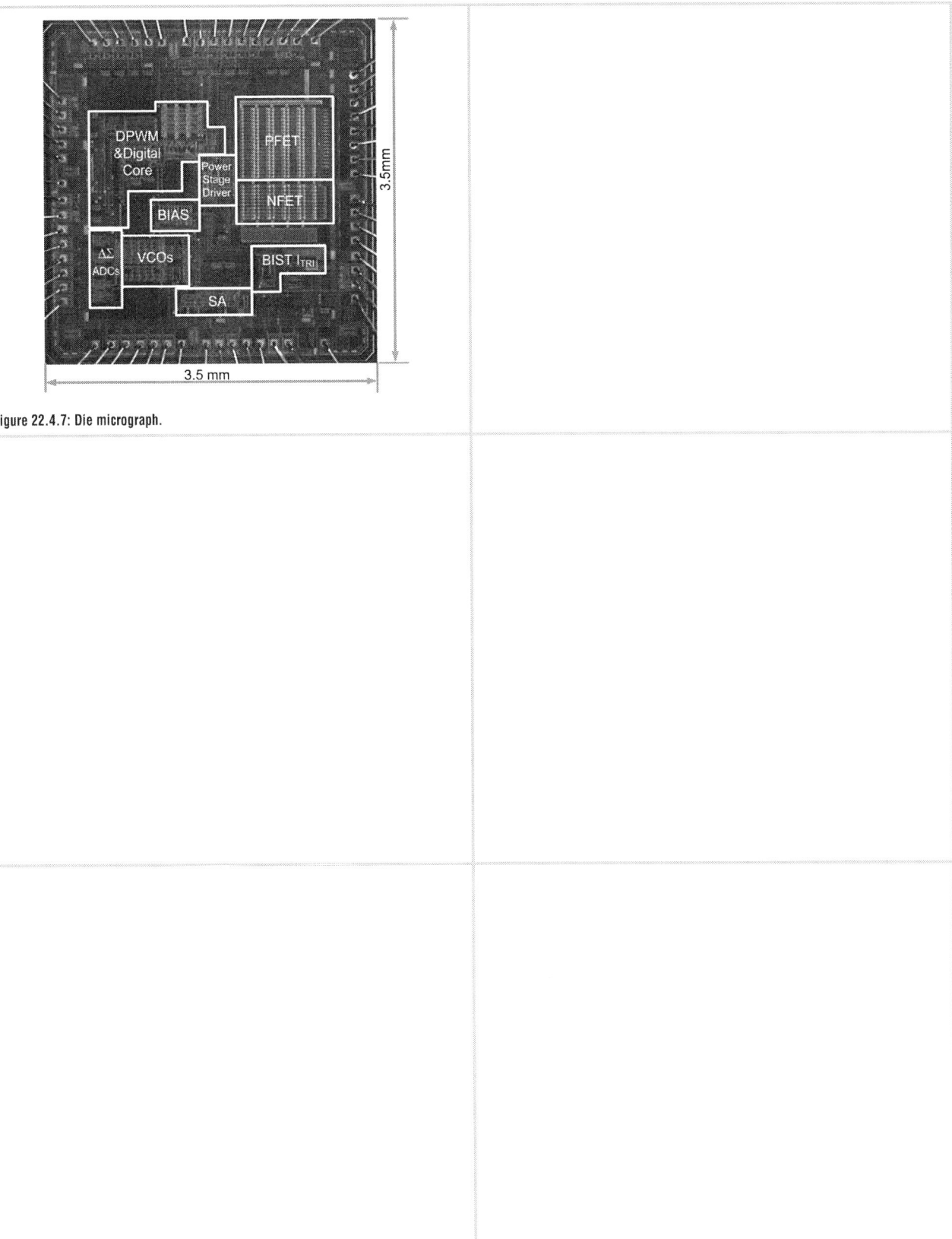

Figure 22.4.7: Die micrograph.

ISSCC 2011 / SESSION 22 / DC/DC CONVERTERS / 22.5

22.5 Zero-Order Control of Boost DC-DC Converter with Transient Enhancement Using Residual Current

Tae-Hwang Kong, Young-Jin Woo, Se-Won Wang, Sung-Wan Hong, Gyu-Hyeong Cho

KAIST, Daejeon, Korea

A variety of controllers are used in DC-DC converters. Among them, voltage-mode control and current-programmed-mode control are widely used in industrial applications. In such controllers, however, values of inductor, output capacitor, and/or load condition usually affect loop stability and limit the performance of switching converter. Recently, load-independent-control (LIC) method is reported, where freewheeling current is fed back to overcome such a limitation [1]. While this is a viable solution in principle, it still has a vulnerable aspect that must be addressed: the feedback control is affected by the level of freewheeling current and an extra power switch is needed for freewheeling current flow which lowers power efficiency. Another LIC method using vestigial current control is reported in [2]. The weak points of vestigial control are that it needs an auxiliary output and power is consumed in steady state to regulate vestigial current. In this paper, we present a zero-order-controlled (ZOC) boost DC-DC converter that has a robust control loop and does not consume any extra power in the steady state.

In the voltage mode, converter dynamics are second order due to the inductor and output capacitor. In contrast, the dynamics are reduced to first order in the current mode, since inductor dynamics are removed, leaving only capacitor dynamics. The proposed converter reduces converter dynamics further by removing both inductor and capacitor dynamics, and is thus dubbed zero-order control. The ZOC concept is previously proposed for LIC, however, we present a new control algorithm that overcomes the weak points of LIC converters.

Figure 22.5.1 shows the block diagram of the ZOC boost converter. The operation principle can be explained using timing diagram of Fig. 22.5.2. The key idea of this controller is to find the optimum condition for balancing inductor energy at a minimum value during each switching cycle. Depending on whether inductor energy is excessive or not, duty cycle is adjusted to make freewheeling current (I_W) zero when a balanced condition is reached. The energy status of inductor is detected by adding a small sawtooth waveform (V_{ST}) to the output voltage (V_0). Owing to this V_{ST}, inductor energy status can be reflected to on-time of Φ_P, which is generated from the comparator (CMP_0) by comparing the composite signal (V_{0X}) of V_0 and V_{ST} with V_{REF}. The V_0 is regulated by comparator (CMP_0) operation, and main current loop is controlled by rising edge difference between Φ_P and driving signal (Φ_N) of NMOS switch, S_N. When inductor energy is excessive, Φ_P goes high during energy transfer period (D_T) before Φ_N does. Therefore, the difference of rising edge between Φ_P and Φ_N indicates how much energy level has exceeded necessary energy, when Φ_P leads Φ_N. On the other hand, when inductor energy is lower than the necessary level, Φ_P cannot go high during both D_T and energy build-up period (D_B) without the help of V_{ST}. In other words, in this case, Φ_P goes high after Φ_N only with the help of V_{ST}, thus accounting for the need for a V_{ST} in the proposed control. Therefore, the difference of rising edge between Φ_N and Φ_P shows the amount of energy deficiency, when Φ_P lags Φ_N. Φ_P and Φ_N are used to generate UP/DOWN pulses which adjust V_{IC} to limit the upper magnitude of the inductor current in comparison with V_{ISN}. During D_B, the inductor energy is increased until V_{ISN} reaches V_{IC}, which is maintained at a constant level when inductor energy is balanced. In this procedure, however, a small DC offset error occurs at output voltage due to the effect of adding V_{ST}, but it can be corrected by using modified comparator concept given in [1]. The above operation is also valid in the discontinuous conduction mode (DCM) except for the extended freewheeling period.

As described above, inductor dynamics are removed, because inductor current is controlled within the feedback loop, similar to the conventional current mode. Capacitor dynamics are removed as well, since output voltage is regulated by comparator operation. In some transients, excessive energy can be freewheeled through PMOS switch, S_W. However, this freewheeling current approaches zero in the steady state where on-time duty of UP/DOWN pulses remains the same,

while V_{IC} is settled without freewheeling period. Thus, the ZOC control requires small switch S_W with negligible power loss.

Figure 22.5.3 shows the schematic diagram of the UP/DOWN pulse-generation block. When the load changes rapidly, Φ_P stays in a high or low state for several periods and causes lost of control due to inaccurate DOWN pulse. Hence, by using Φ_{ADD}, on-time for DOWN pulse is limited to MAX (or MIN) under load transient conditions. Figure 22.5.3 shows the schematic diagram of the V_{IC} control block. UP/DOWN pulses operate switches S_1 to S_4 making I_C flow into or from integrating capacitor C_C to generate V_{IC}. In this case, a first-order control loop can be simply designed by using C_C since the converter is zero order. However, additional R_Z is needed to insert one zero to compensate the pole that intervenes during the voltage-to-phase conversion process as in PLLs. The role of C_P is to remove high-frequency noise.

Figure 22.5.4 shows the transient enhancement block of the proposed scheme. For heavy to light load transient, output voltage remains almost constant owing to freewheeling operation. On the other hand, for light to heavy load transient, output voltage drops and recovering the original voltage level takes some time. To improve this transient response, the transient enhancement block is added. If a heavy load is abruptly applied to output, I_{TRAN} increases rapidly as shown in Fig. 22.5.4, and flows into node V_A of Fig. 22.5.3 to rapidly increase the control voltage V_{IC} [1]. If a load changes from heavy to light, freewheeling switch S_W turns on (as shown in Fig. 22.5.4). This residual current is sensed using circuit of Fig. 22.5.4 and subtracted from node V_A, causing V_{IC} and inductor current level to decrease [1]. Unlike the conventional transient improvement method, the presented scheme achieves fast transient response by using Φ_P and residual current, and without adding any other output voltage feedback loop.

The proposed boost converter with the ZOC scheme is implemented in a 0.35μm BCD process and occupies 3.15mm^2. The input voltage varies from 2.7V to 4.5V and the output voltage is regulated to 8V. Figure 22.5.5 shows measured waveforms when I_{LOAD} is 60mA, 300mA, 10mA, and also when it changes abruptly between 60 and 300mA. When the converter operates in continuous-conduction mode (CCM), the switching node voltages of Fig. 22.5.5 show that in the steady state Vx does not have freewheeling period. When the converter operates in DCM, small gap between exact zero point of inductor current and Φ_P causes ringing as shown in Fig. 22.5.5. When load current changes abruptly between 60 and 300mA, output voltage dip is within 50mV and settling time is measured to be 80μs for rising and 180μs for falling. The converter performance is summarized in Fig. 22.5.6. Maximum efficiency of 88% is achieved at the total output power of 480mW with 8V (from a 3.7V supply). Figure 22.5.6 shows the efficiency is good over a wide load range. This enhancement of the performance can be obtained because unlike [1,2], the proposed converter does not use extra energy to control main feedback loop and unlike [3,4] does not require freewheeling period. Figure 22.5.7 shows the chip micrograph.

Acknowledgement :

This work was supported by the ERC program of the Korea Science and Engineering Foundation (KOSEF) grant funded by the Korea Ministry of Science and Technology (MOST) (No. R11-2007-045-01004-0) and in part by the IC Design Education Center (IDEC).

References :

[1] Y.-J. Woo, H.-P. Le, G.-H. cho, G.-H. Cho, S.-I. Kim, "Load-Independent Control of Switching DC-DC Converters with Freewheeling Current Feedback," *IEEE J. Solid-state Circuits*, vol. 43, no. 12, pp. 2798–2808, Dec., 2008.
[2] K.-S. Seol, Y.-J. Woo, G.-H. Cho, G.-H. Cho, J.-W. Lee, S.-I. Kim, "Multiple-Output Step-Up/Down Switching DC-DC Converter with Vestigial Current Control," *ISSCC Dig. Tech. Papers*, pp. 442–443, Feb., 2009.
[3] D. Ma, W.-H. Ki, and C.-Y. Tsui, "A Pseudo-CCM/DCM SIMO Switching Converter With Freewheeling Switching," *IEEE J. Solid-State Circuits*, vol. 38, no. 6, pp. 1007-1014, Jun., 2003.
[4] S. Kapat, A. Patra, and S. Banerjee, "A current-Controlled Tristate Boost Converter With Improved Performance Through RHP Zero Elimination," *IEEE Transactions on Power Electronics*, vol. 24, no. 3, pp. 776–786, Mar., 2009.

ISSCC 2011 / February 23, 2011 / 10:45 AM

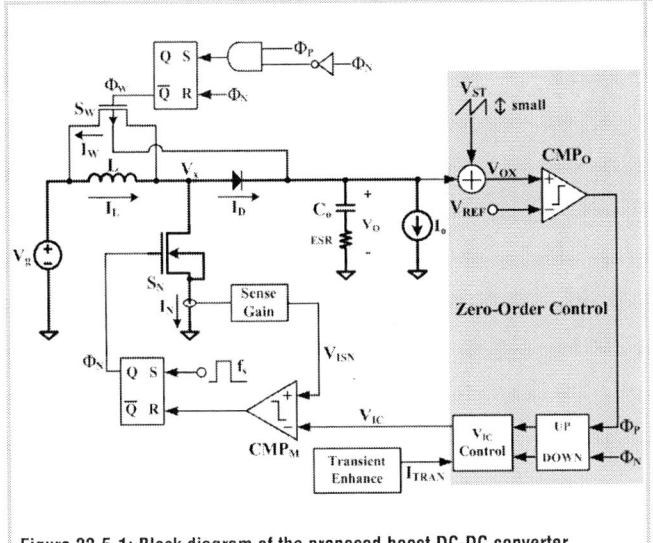

Figure 22.5.1: Block diagram of the proposed boost DC-DC converter.

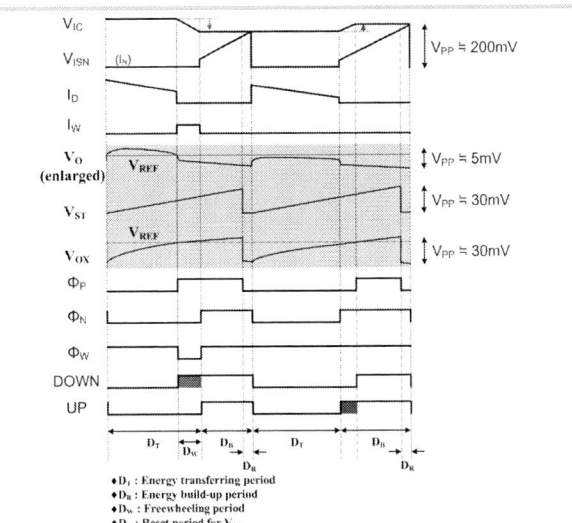

Figure 22.5.2: Timing diagram of the proposed Boost DC-DC converter.

- D_T : Energy transferring period
- D_B : Energy build-up period
- D_W : Freewheeling period
- D_R : Reset period for V_{ST}

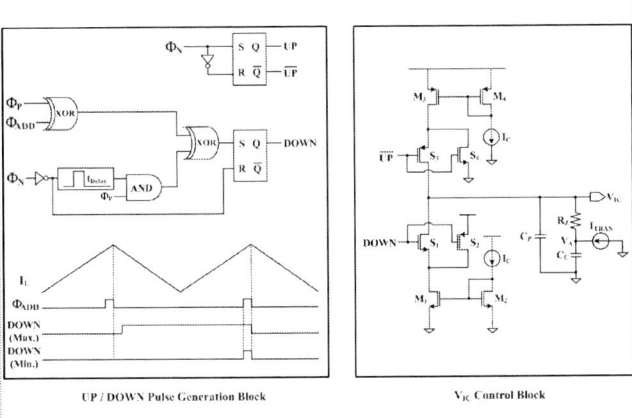

Figure 22.5.3: Generating UP/DOWN pulse and control voltage V_{IC}.

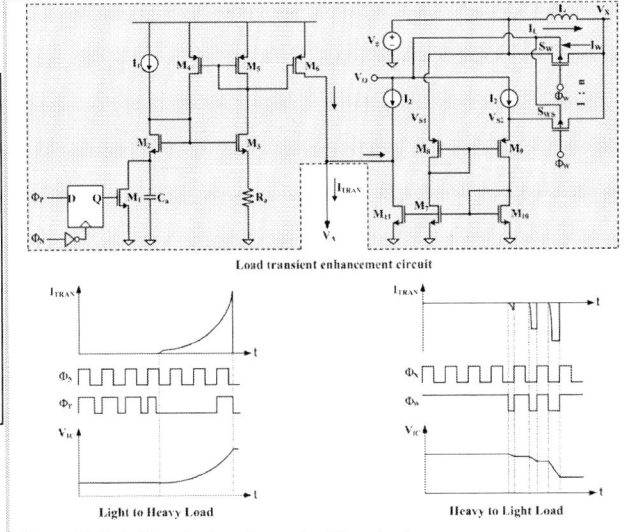

Figure 22.5.4: Circuitry to enhance load transient response.

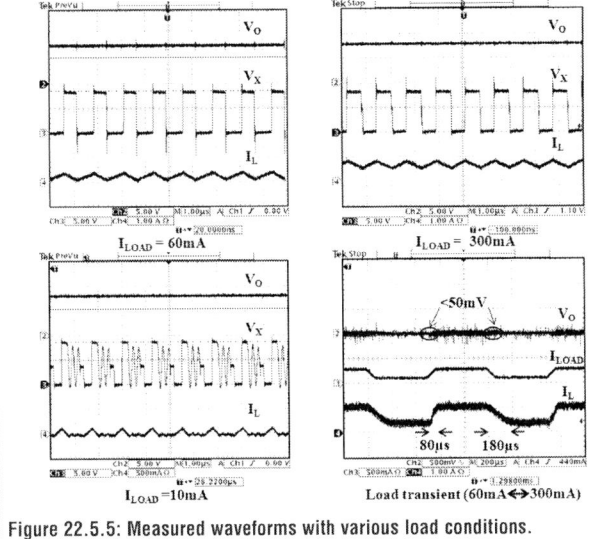

Figure 22.5.5: Measured waveforms with various load conditions.

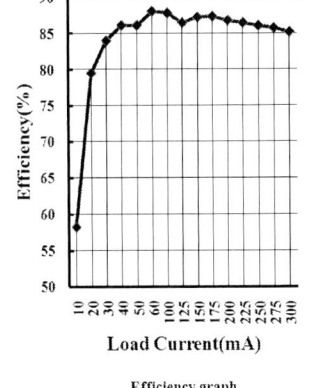

Process	0.35μm Dongbu BCD
Type	Boost
Supply Voltage	3.7V nominal (2.7 ~ 4.5V)
Output Voltage	8V
Inductor /DCR	10μH / 53mΩ
Switching Frequency	833KHz
Filtering Capacitor / ESR	10μF / 20mΩ
Load Current(Max)	60mA nominal(300mA)
Efficiency	88% *
Line regulation	0.625% / V ** [0.781% / V]
Load regulation	0.01% / mA *** [0.018% / mA]

* Load Current = 60mA ** Vg = 2.7 to 4.5 V
*** 60mA to various load current @ Vg = 3.7 V

Performance summary

Efficiency graph

Figure 22.5.6: Performance summary table and efficiency graph.

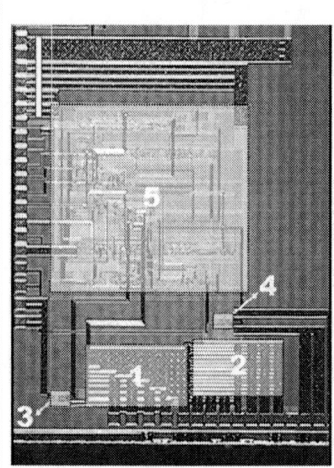

1. Main NMOS switch(S_N)

2. PMOS switch(S_W)

3. Gate driver for S_N

4. Gate driver for S_W

5. BIAS & Control Blocks

Figure 22.5.7: Chip micrograph.

ISSCC 2011 / SESSION 22 / DC/DC CONVERTERS / 22.6

22.6 Robust and Efficient Synchronous Buck Converter with Near-Optimal Dead-Time Control

Sungwoo Lee, Seungchul Jung, Jin Huh, Changbyung Park, Chun-Taek Rim, Gyu-Hyeong Cho

KAIST, Daejeon, Korea

In switching power converters, the turn on/off process of switches is crucial for the reliability and efficiency of the converter. In general, optimum switching is challenging, and it is particularly difficult for hard switching. The MOSFET synchronous buck converter having wide applications due to high switching speed and low loss, however, operates in hard switching and needs a good timing control for its switching. On/off commutation dead-times should be adjusted carefully in accordance with load current change. Previous work on the dead-time control includes predictive gate drive technique [1], load current sensing [2][7], sensor-less optimization technique [3], and delay-locked loops [4-6]. These techniques have problems of sensing noisy switching node [1, 4-6] or load current [2, 7], and requiring high quantizing resolution [3]. In this paper, a near optimum dead-time control method is proposed thate resolves such problems.

Figure 22.6.1 shows the proposed synchronous buck converter with near optimum switching conditions for powering mode. The key idea is to block the reverse recovery current of the body-diode (I_{rr}) by preventing it from turning on during switching. To do this, providing optimized dead-times for different load conditions is an important issue. The gate signals for near-optimum on/off dead-times (on/off: with respect to the PMOS power switch S_P) are also illustrated in Fig. 22.6.1. It shows the condition for preventing the body-diode from turning on, which necessiates precise control of time delays for given load conditions.

The adaptive-delay gate driver as shown in Fig. 22.6.1 controls the on/off dead-times. It is composed of two adaptive delay units (ADUs), t_{pdA} & t_{pdB}, one offset delay unit (t_{od}), and logic gates. It controls the propagation delays of input signals (v_A and v_B) according to the delay control signals (v_{rP} and v_{rN}). The on dead-time $t_{d,on}$ is determined by the difference between the upper side adaptive delay (t_{pdB}) and the offset delay (t_{od}). The off dead-time $t_{d,off}$ is directly determined by the lower side adaptive delay (t_{pdA}). The equivalent capacitance (C_{eq}), crucial for explaining zero voltage switching, represents internal capacitances of the MOS switches. In this way, the on dead-time can be controlled near zero as required.

The proposed integral feedback control of on/off dead-time error (DTE) is depicted in Fig. 22.6.2. The control of the synchronous buck converter includes the outer voltage regulation loop and the inner dead-time control loop. A dead-time control loop comprises four units: an on/off DTE sensing, a DTE integration and adaptive-delay control signal generation, an adaptive-delay gate driver, and a dv/dt-induced turn-on prevention switch (S_{aux}). The on/off DTE is sensed from the drain- source voltage of S_N, which is the same as the switching node voltage v_M. The DTE integration unit integrates the sensed DTE and generates a precise delay-control signal for the optimum delay between two gate pulses (v_{gP} and v_{gN}). The small switch S_{aux} is turned on to prevent S_N from turning on when the switching node voltage (v_M) rises sharply.

On/off DTE circuit, DTE integrator and delay-control circuit along with their signals are shown in Fig. 22.6.3. On/off DTE is sensed from the drain-source voltage of S_N. The on DTE sampling starts when S_N is turned off at the falling edge of Φ_1, and ends when V_{DS} rises to $0.5v_{IN}$ at the falling edge of Φ_2. The duration of on DTE sampling is expressed as t_{sp} in the upper three waveforms. If the on dead-time is excessive, body-diode turns on and I_2 becomes large during t_{sp} due to the negative V_{DS}. In this case, the node voltage $v_S(t)$ at the capacitor C decreases significantly. The charge across the C is integrated during Φ_3 by the error integrator with 20C. The node voltage v_{INTP} becomes the same as the integrator output by an appropriate feedback, where a corresponding delay-control signal v_{rP} (or v_{rN}) is generated. On the other hand, if the on dead-time is too short, the operation is vice versa since negative voltage never appears on V_{DS}. Using this integral feedback operation, the on dead-time can be set to a near optimum value by adjusting the magnitude of I_1. Under this condition, V_{DS} drops slightly negative during t_{sp} but less than the body-diode turn-on voltage, and the average value of $v_S(t)$ equals to V_R (=$0.5v_{IN}$) with charge/discharge balancing at the capacitor C. Therefore, on dead-time becomes unchanged in the steady-

state. On the other hand, the off DTE is sampled by the same principle with the on DTE sampling as discussed.

In addition to the dead-time control, the dv/dt-induced turn-on prevention scheme is considered to maximize the converter efficiency as shown in Fig. 22.6.4 [8]. Due to the gate-drain capacitance (C_{gdN}) coupling effect, the gate-source voltage (V_{gsN}) of S_N increases sharply when its drain-source voltage rises steeply. It causes simultaneous turn on of S_N and S_P, and additional power loss occurs due to shoot-through current. Since the gate-source voltage of S_N is just below its turn-on threshold voltage, in case of the near-optimum dead-time control, the additional power consumption caused by the dv/dt-induced turn on is significantly large when the drain-source voltage of S_N starts to increase sharply (lower left waveform of Fig. 22.6.4). To prevent this situation, an auxiliary switch (S_{aux}) is inserted between the gate and source nodes of S_N. The switch S_{aux} is on for 20ns during the transition of S_N off and S_P on. In this way, the near-optimum on-dead-time control is possible without any additional power loss as illustrated in the lower right of Fig. 22.6.4. During the turning off case, the soft switching occurs in the powering mode, however, no dv/dt-induced turn on phenomenon occurs.

Figure 22.6.5 shows the test results of drain-source voltages of the S_N for several load currents, which are compared with the adaptive dead-time control (ADTC) scheme in [1]. The drain-source voltages of the proposed scheme do not go below 0.7V with small negative peaks (about one half of ADTC scheme), and the body-diode of S_N is never turned-on (for all load current conditions).

The effect of S_{aux} on switching waveform is shown in the left side of Fig. 22.6.6. V_{dsN} rises sharper with S_{aux} since there is no dv/dt-induced turn on of the NMOS. The on/off DTE integrated voltages (v_{INTP} in Fig. 22.6.3) are also shown in Fig. 22.6.6 for load current change between 50 and 300mA. It shows that the on/off dead-times are automatically optimized for the load change.

The measured efficiency of the proposed on/off dead-time optimization scheme shows wider operation range with higher efficiency up to 4.6% as compared to the ADTC scheme. Both the proposed and ADTC schemes are implemented in a 0.35μm 3.3V 4M CMOS technology in a die area of 1.73mm². The proposed near-optimum dead-time switching synchronous buck converter shows 90% efficiency at 1MHz switching frequency and fast responses for various loads.

Acknowledgment:
This work was supported by the ERC program of the Korea Science and Engineering Foundation (KOSEF) grant funded by the Korea Ministry of Science and Technology (MOST) (No. R11-2007-045-01004-0) and in part by the IC Design Education Center (IDEC).

References:
[1] S. Mappus, "Predictive Gate Drive Boosts Synchronous DC/DC Power Converter Efficiency," TI Application Reports, SLUA281, April, 2003.
[2] H.-W. Huang, K.-H. Chen, and S.-Y. Kuo, "Dithering Skip Modulation, Width and Dead Time Controller in Highly Efficient DC-DC Converters for System-On-Chip Applications," *IEEE J. Solid-State Circuits*, vol. 42, no 11, pp. 2451-2465, Nov., 2007.
[3] V. Yousefzadeh and D. Maksimovic, "Sensorless Optimization of Dead Times in DC-DC Converters With Synchronous Rectifiers," *IEEE Trans. Power Electronics*, vol. 21, no 4, pp. 994-1002, July, 2006.
[4] B. Acker, C.R. Sullivan, and S.R. Sanders, "Synchronous Rectification with Adaptive Timing Control," *26th Annual Power Electronics Specialists conf.*, pp. 88-95, June, 1995.
[5] W. Lau and S.R. Sanders, "An Integrated Controller for a High Frequency Buck Converter," *28th Annual Power Electronics Specialists Conf.*, pp. 246-254, June, 1997.
[6] O. Trescases, W.T. Ng, and S. Chen, "Precision Gate Drive Timing in a Zero-Voltage-Switching DC-DC Converter," *16th International Symposium on Power Semiconductor Devices and ICs*, pp. 55-58, May, 2004.
[7] Y.U. Hong, B.K. Choi, Y.J. Woo, et al., "Optimum Efficiency-Tracking Gate Driver Using Adaptive Dead Time Control for Single Chip DC-DC Converter," *37th IEEE Power Electronics Specialists Conf.*, pp. 1-5, June, 2006.
[8] Q. Zhao and G. Stojcic, "Characterization of Cdv/dt Induced Power Loss in Synchronous Buck DC-DC Converters," *IEEE Trans. Power Electronics*, vol. 22, no 4, pp. 1508-1513, July, 2007.

978-1-61284-303-2/11 $26.00 © 2011 IEEE

ISSCC 2011 / February 23, 2011 / 11:15 AM

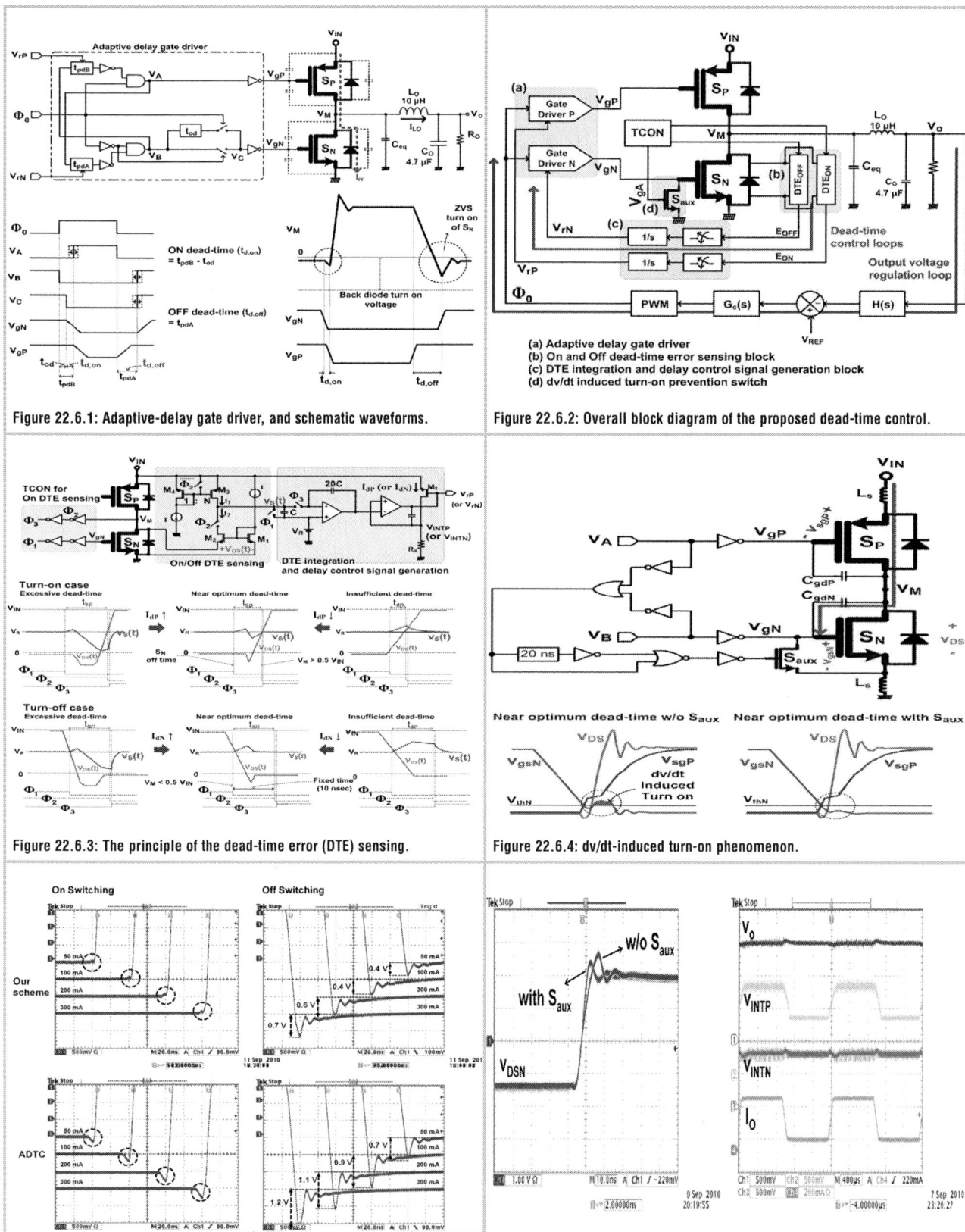

Figure 22.6.1: Adaptive-delay gate driver, and schematic waveforms.

Figure 22.6.2: Overall block diagram of the proposed dead-time control.

(a) Adaptive delay gate driver
(b) On and Off dead-time error sensing block
(c) DTE integration and delay control signal generation block
(d) dv/dt induced turn-on prevention switch

Figure 22.6.3: The principle of the dead-time error (DTE) sensing.

Figure 22.6.4: dv/dt-induced turn-on phenomenon.

Figure 22.6.5: The V_{DS} waveforms, S_N for various load currents.

Figure 22.6.6: The effect of dv/dt-induced turn-on prevention switch (left) and the load change response from 50mA to 300mA (right).

978-1-61284-303-2/11 $26.00 © 2011 IEEE

393

22

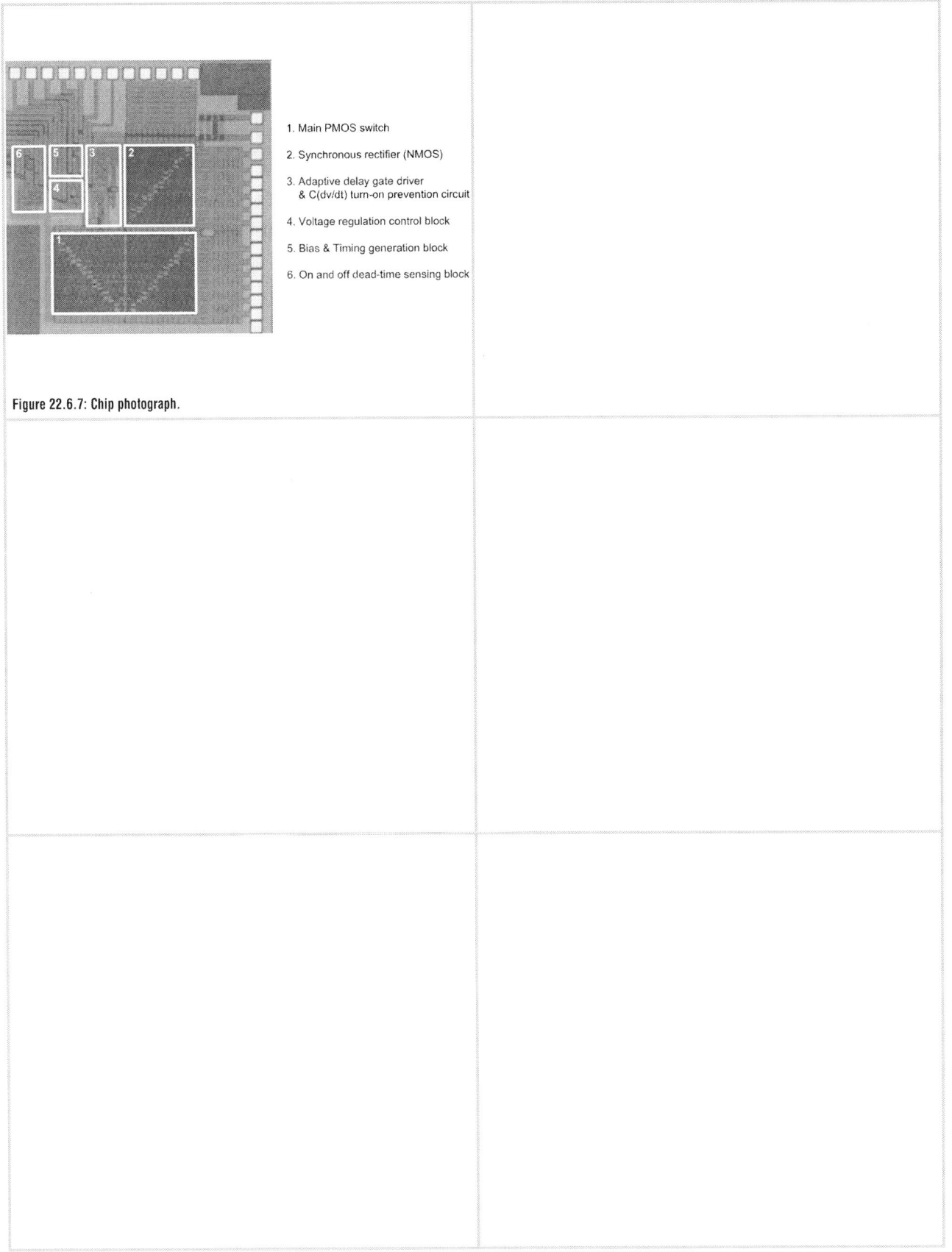

1. Main PMOS switch

2. Synchronous rectifier (NMOS)

3. Adaptive delay gate driver
   & C(dv/dt) turn-on prevention circuit

4. Voltage regulation control block

5. Bias & Timing generation block

6. On and off dead-time sensing block

Figure 22.6.7: Chip photograph.

ISSCC 2011 / SESSION 22 / DC/DC CONVERTERS / 22.7

22.7 A 90% Peak Efficiency Single-Inductor Dual-Output Buck-Boost Converter with Extended-PWM Control

Weiwei Xu[1], Ye Li[1], Zhiliang Hong[1], Dirk Killat[2]

[1]Fudan University, Shanghai, China
[2]Brandenburg University of Technology, Cottbus, Germany

Power management in portable devices demands small size, low cost as well as long battery lifetime, which in turn drive the development of single-inductor multiple-output (SIMO) converters [1-5]. Due to the battery voltage variation during usage and the wide-range dynamic voltage scaling (DVS) applied for power reduction, high-efficiency buck-boost conversion is required to extend the battery lifetime. The buck-boost converter in [6] selects the operation mode by comparing the output with the supply voltage, which is not suitable for multi-output converters. The reported single-inductor dual-output (SIDO) buck-boost converter in [4] uses a state machine with sophisticated current sense for mode selection and requires a freewheeling state that dissipates energy. The converter in [3] uses one additional auxiliary inductor for step-up/down mode adjustment. This paper proposes an extended-PWM (EPWM) control which automatically selects buck or boost mode and facilitates smooth mode transition. It is suitable for flexible outputs and maintains a high efficiency in buck and in boost converters.

Figure 22.7.1 shows the overall architecture of the SIDO buck-boost converter. It consists of five switches in the power stage: S1, S2, and S3 regulate the total transferred energy by buck/boost conversion; S4 and S5 distribute the energy stored in the inductor to both outputs. All switches are controlled by PWM signals which are generated in two control loops. The converter is specified for two outputs from 1 to 5V with the maximum total power of 2.5W and supply voltage ranging from 2.5 to 5V. The switching frequency is 2MHz. The external inductor is 2.2µH and the capacitor is 20µF each.

Controlling the five power switches in the topology of Fig. 22.7.1 requires three independent duty-ratio signals. A switching converter with PWM control usually generates a duty-ratio signal by comparing the error amplifier output V_{ea1} with a sawtooth waveform $Saw1$. In a buck converter the PWM signal D_{1a} also represents the relation between output and supply voltage, thus $D_{1a}=V_{ea1}/V_{saw1}=V_{out}/V_g$. However, this equation is only valid when $V_{ea1}<V_{saw1}$. If $V_{ea1}>V_{saw1}$, the transferred energy is not enough to supply the outputs in buck mode. The converter has to enter into its boost mode, where $D_{1a}=1$ and D_{1b} is modulated. The proposed EPWM control achieves automatic buck/boost mode selection by considering the relation of V_{ea1} and V_{saw1}. As shown in Fig. 22.7.2, there are two sawtooth waveforms used for D_{1a} and D_{1b} generation. When $V_{ea1}<V_{saw1}$, $Saw2$ is just the same as $Saw1$. When $V_{ea1}>V_{saw1}$, $Saw2$ ramps up to the amplitude of V_{ea1} while $Saw1$ turns to ground. The buck signal D_{1a} is always generated by comparing V_{ea1} with $Saw1$, while the boost signal D_{1b} is only high when V_{ea1} is higher than V_{saw1}. An equivalent PWM signal D_{eq} can be expressed by the two signals: $D_{eq}=D_{1a}+D_{1b}=V_{ea1}/V_{saw1}=V_{out}/V_g$, which extends the conversion ratio concept from a buck to a general switching converter.

Due to parasitic effects of the PWM signal generation, there would be a pulse-skipping problem when D_{eq} is close to 1. A small voltage is added on $Saw1$ as level-shift to overcome this problem. The level-shift leads to an overlap of buck and boost mode, which results in a buck-boost mode for smooth transition. Figure 22.7.2 shows the duty-ratio signal generation of the SIDO buck-boost converter with EPWM control. The buck switches S1 and S2 are controlled by D_{1a}, while the boost switch S3 is controlled by D_{1b}. The output switches S4 and S5 are controlled by D_2 and are only valid when S3 is off. The distribution PWM signal D_2 is generated by comparing V_{ea2} with $Saw1$. As shown in Fig. 22.7.2, the total duration of charging outputs is kept constant (T_C) whether in buck or in boost mode. The mode transition has little influence on the V_{ea2} loop, which helps to attenuate the interaction between the two control loops of the SIDO system. Hence, the proposed EPWM control is also suitable for multiple-output converters.

Figure 22.7.3 shows the schematic of the sawtooth waveform generator of the EPWM control. At the rising edge of the $Pulse$ signal, V_{ea1} and $Saw2$ are compared and the result is used for mode selection. When in buck mode, the sawtooth pull-down signal $dn2$ is the same as $dn1$. In boost mode, $dn2$ is only triggered when $Saw2$ reaches V_{ea1}. As described in Fig. 22.7.2, the EPWM control combines constant-frequency buck mode and constant-off-time boost mode. By adding a phase and frequency detector (PFD) and charge-pump block, a PLL can be built to keep the switching frequency synchronized with an external clock. The bandwidth of the PLL is designed to be lower than that of the SIDO control loop, which provides a pseudo constant-off-time boost mode to make the system compensation easier.

The SIDO converter is a multi-loop feedback system. The power-stage small-signal modeling and system decoupling analysis have been described in [5]. As illustrated in Fig. 22.7.1, a VLX filter is used for the V_{ea1} loop compensation. $VLX1$ and $VLX2$ have a direct response to the PWM signals without the delay of inductor current integration, which makes it suitable for system compensation in switching converters. Low-pass filters are added at both nodes and an active feedback amplifier is used for the differential signal extraction. The VLX-filter-based compensation method in Fig. 22.7.4 is suitable for both buck and boost situations.

The converter is fabricated in a 0.25µm 2P4M CMOS process. The die area is 2.6×2.9mm². Each output can be used as step down or up conversion. Figure 22.7.5 shows the measured waveforms of output ripples and $VLX1$ and $VLX2$ node voltages in different modes. The output ripple including its spike is lower than 80mV for all situations. $VLX1$ is switched to ground in buck mode, while $VLX2$ is switched to ground in boost mode. In the buck-boost mode, both $VLX1$ and $VLX2$ are switched, which results in more power loss (about 4% efficiency deterioration). The operation mode of the SIDO converter is also determined by the load currents when the converter has one buck and one boost outputs. A load response measurement (I2=42mA → 140mA) in Fig. 22.7.6 shows the transient mode transition between buck and boost mode with load current change. The dynamic measurement also reveals that the EPWM-controlled buck-boost converter may achieve faster response than the conventional buck and boost converters due to the wide modulation range of D_{eq}. The peak power efficiency of 90% is reached at a supply voltage of 3V and when the two outputs are set to 1.8V/214mA and 5V/150mA. By using the proposed EPWM control, the SIDO converter keeps a high efficiency over 80% over a wide output range and includes an automatic mode transition.

Acknowledgement:
This work was supported by Dialog Semiconductor. The authors would like to thank Horst Schleifer for the help and support.

Reference:
[1] M. Belloni, E. Bonizzoni, E. Kiseliovas, et al., "A 4-Output Single-Inductor DC-DC Buck Converter with Self-Boosted Switch Drivers and 1.2A Total Output Current," *ISSCC Dig. Tech. Papers*, pp. 444-445, Feb., 2008.
[2] K-C. Lee, C-S. Chae, G-H. Cho, et al., "A PLL-Based High-Stability Single-Inductor 6-channel Output DC-DC Buck Converter," *ISSCC Dig. Tech. Papers*, pp. 200-201, Feb., 2010.
[3] K-S. Seol, Y-J. Woo, G-H. Cho, et al., "Multiple-Output Step-Up/Down Switching DC-DC Converter with Vestigial Current Control," *ISSCC Dig. Tech. Papers*, pp. 442-443, Feb., 2009.
[4] M-H. Huang and K-H. Chen, "Single-Inductor Dual Buck-Boost Output (SIDB-BO) Converter with Adaptive Current Control Mode (ACCM) and Adaptive Body Switch (ABS) for Compact Size and Long Battery Life in Portable Devices," *Dig. Symp. VLSI Circuits*, pp. 164-165, Jun., 2009.
[5] W. Xu, Y. Li, X. Gong, et al., "A Dual-Mode Single-Inductor Dual-Output Switching Converter with Small Ripple," *IEEE Trans. Power Electronics*, vol. 25, no. 3, pp. 614-623, Mar., 2010.
[6] C. Zheng and D. Ma, "A 10-MHz 92.1%-Efficiency Green-Mode Automatic Reconfigurable Switching Converter with Adaptively Compensated Single-Bound Hysteresis Control", *ISSCC Dig. Tech. Papers*, pp. 204-205, Feb., 2010.

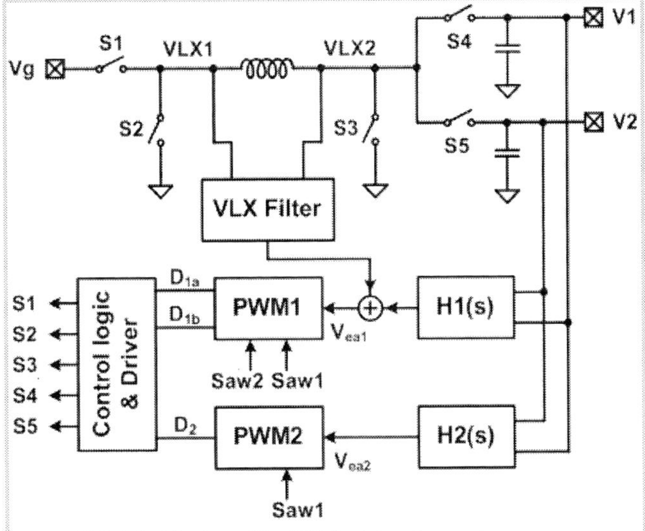

Vg
S1
VLX1
VLX2
S2
S3
S4
S5
V1
V2
VLX Filter
D1a
D1b
PWM1
H1(s)
Veal
Saw2 Saw1
Control logic & Driver
S1
S2
S3
S4
S5
D2
PWM2
H2(s)
Vea2
Saw1

Figure 22.7.1: Architecture of the SIDO buck-boost converter.

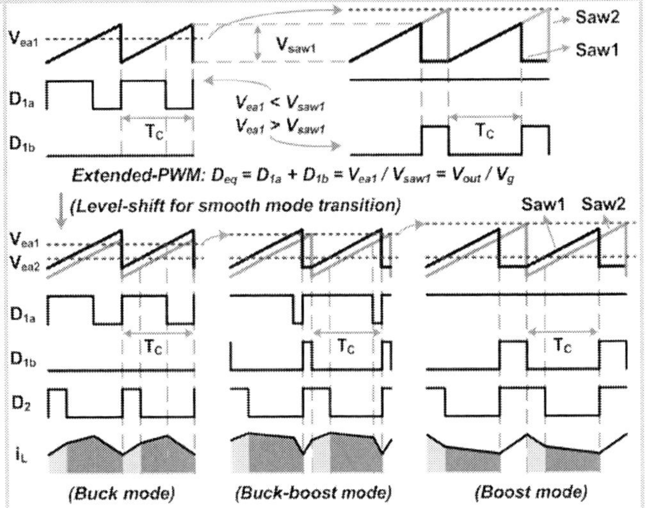

Saw2
Saw1
Vea1
Vsaw1
D1a
D1b
Vea1 < Vsaw1
Vea1 > Vsaw1
TC
Extended-PWM: Deq = D1a + D1b = Vea1 / Vsaw1 = Vout / Vg
(Level-shift for smooth mode transition)
Saw1 Saw2
Vea1
Vea2
D1a
D1b
D2
iL
(Buck mode)
(Buck-boost mode)
(Boost mode)

Figure 22.7.2: Scheme of extended-PWM (EPWM) control for buck-boost conversion.

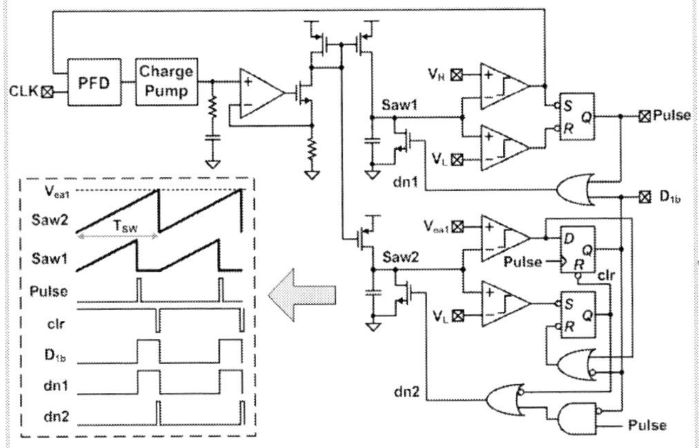

CLK
PFD
Charge Pump
VH
Saw1
VL
Pulse
dn1
D1b
Vea1
Saw2
Pulse
clr
dn2
Pulse
Vea1
Saw2
TSW
Saw1
Pulse
clr
D1b
dn1
dn2

Figure 22.7.3: Schematic of sawtooth waveform generator for EPWM control.

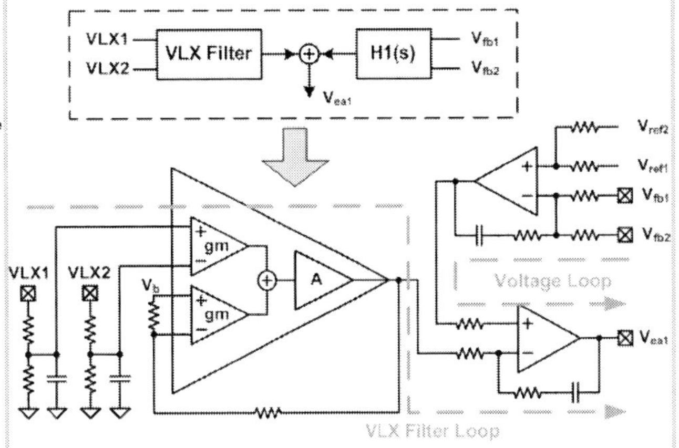

VLX1
VLX2
VLX Filter
H1(s)
Vfb1
Vfb2
Vea1
Vref2
Vref1
Vfb1
Vfb2
Voltage Loop
VLX1
VLX2
Vb
gm
gm
A
Vea1
VLX Filter Loop

Figure 22.7.4: Schematic of VLX-filter-based compensation.

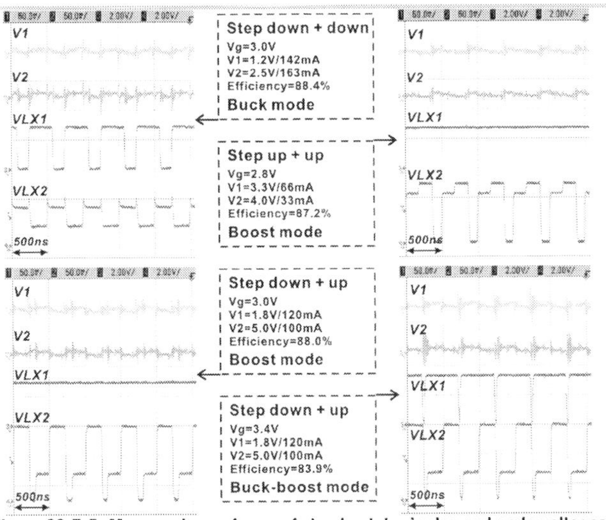

Step down + down
Vg=3.0V
V1=1.2V/142mA
V2=2.5V/163mA
Efficiency=88.4%
Buck mode
Step up + up
Vg=2.8V
V1=3.3V/66mA
V2=4.0V/33mA
Efficiency=87.2%
Boost mode
Step down + up
Vg=3.0V
V1=1.8V/120mA
V2=5.0V/100mA
Efficiency=88.0%
Boost mode
Step down + up
Vg=3.4V
V1=1.8V/120mA
V2=5.0V/100mA
Efficiency=83.9%
Buck-boost mode
V1
V2
VLX1
VLX2
500ns

Figure 22.7.5: Measured waveforms of steady-state ripples and node voltages in different modes.

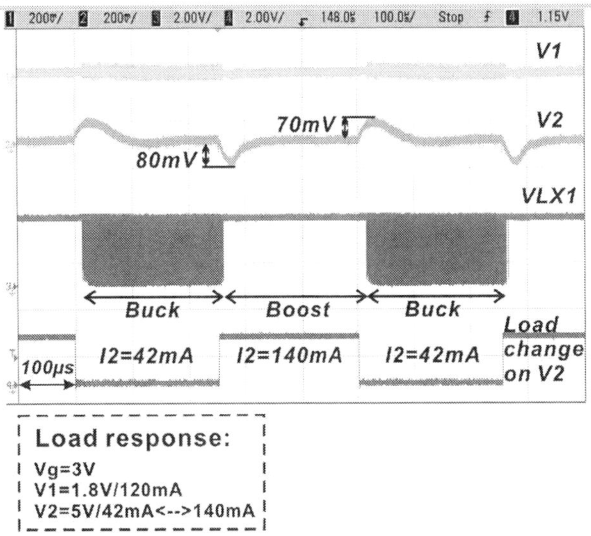

V1
V2
70mV
80mV
VLX1
Buck
Boost
Buck
I2=42mA
I2=140mA
I2=42mA
Load change on V2
100μs
Load response:
Vg=3V
V1=1.8V/120mA
V2=5V/42mA<-->140mA

Figure 22.7.6: Measured waveform of dynamic response and mode transition.

ISSCC 2011 PAPER CONTINUATIONS

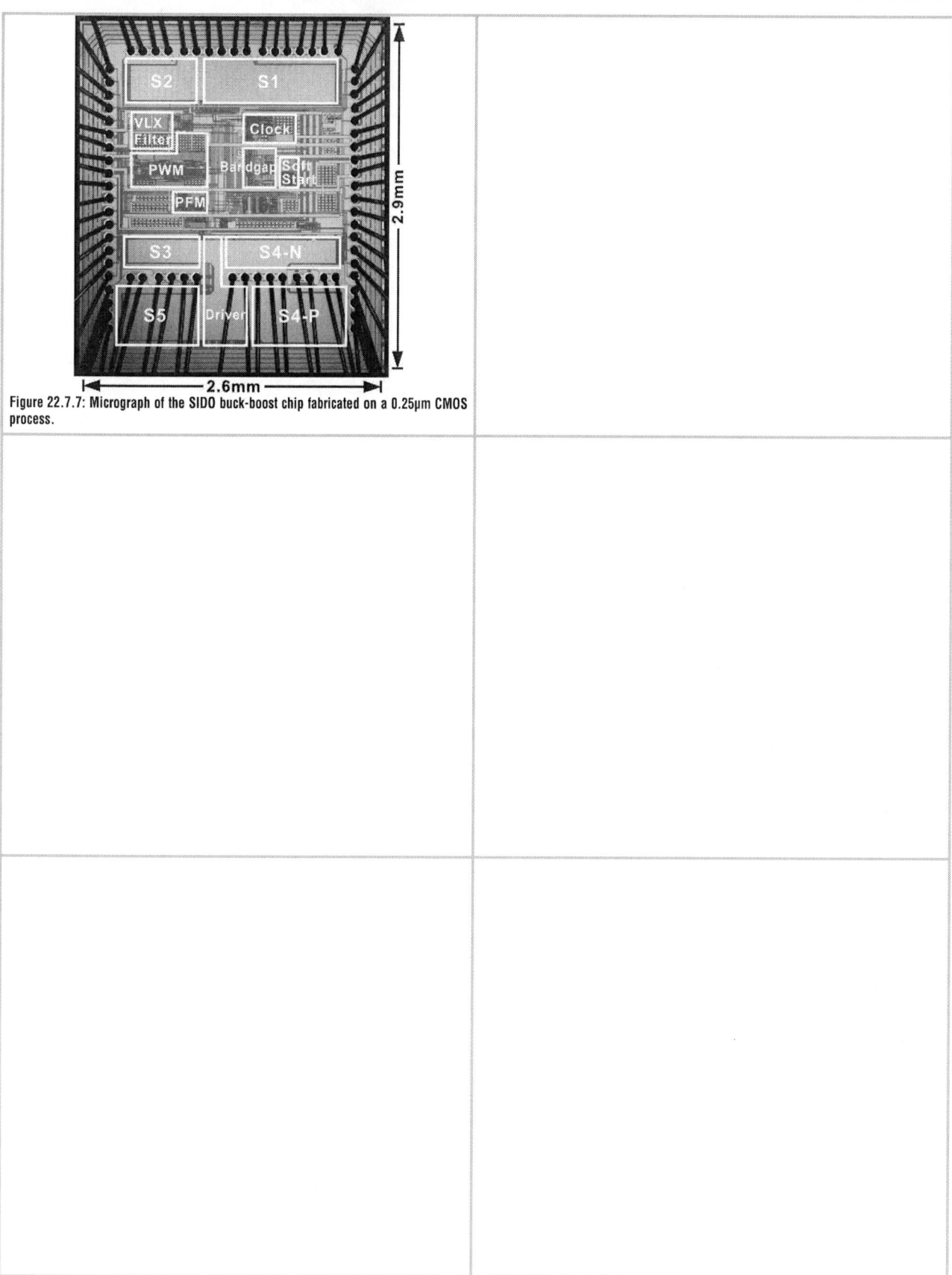

Figure 22.7.7: Micrograph of the SIDO buck-boost chip fabricated on a 0.25µm CMOS process.

ISSCC 2011 / SESSION 22 / DC/DC CONVERTERS / 22.8

22.8 Spurious-Noise-Free Buck Regulator for Direct Powering of Analog/RF Loads Using PWM Control with Random Frequency Hopping and Random Phase Chopping

Chengwu Tao, Ayman A. Fayed

Iowa State University, Ames, IA

Buck regulators are widely employed in portable devices due to their high power-conversion efficiency. However, due to their spurious output noise, they are not directly used to power sensitive analog/RF modules, and subsequent linear low-dropout regulators (LDOs) are needed to generate secondary low-noise supply rails for these modules. This results in lower efficiency, and increased size and cost. Moreover, as switching frequencies increase to reduce passive components, LDOs become less effective in filtering the switching noise due to their poor power-supply rejection (PSR) beyond 1MHz [1]. Several techniques for reducing the spurious noise of buck regulators by manipulating their switching behavior have been studied. This includes using $\Delta\Sigma$ or Δ modulators in the control loop [2, 3], which although reduce the spurs, result in large increases in the noise floor that mandates subsequent LDOs [1]. Other techniques redistribute the power of each spur into multiple smaller ones using random frequency hopping [4], or periodic monotonic frequency stepping [5]. However, the resulting spectrum continues to be spurious and the reduction reported in the largest spur is limited to 10 to 12dB. This limited reduction, coupled with the fact that many extra spurs are generated, leaves the load circuitry vulnerable to performance degradation. This paper proposes a pulse- width modulation (PWM) control scheme for buck converters based on combining random frequency hopping with phase chopping. The technique results in full elimination of spurs, elimination of hopping transients, very low noise floor, and minimalist implementation with little overhead on area and power.

Figure 22.8.1 shows the proposed converter with PWM control and type-III compensation. The ramp generator uses two reference voltages to generate the ramp signal by integrating one of the ramp currents I_1 to I_M, resulting in corresponding ramp frequencies f_1 to f_M, respectively. The End_of_Cycle signal produced at the end of each ramp cycle triggers a count-to-N counter, which triggers an h-bit pseudo-random number generator every N ramp cycles. This determines the ramp frequency for the next N ramp cycles, hence implementing random frequency hopping. Moreover, whenever a given frequency f_i is selected, it starts a new full ramp cycle from scratch, and thus it experiences a phase shift from when it was last selected. This phase shift is also random due to the random selection of frequencies, hence implementing random phase chopping. Furthermore, the proposed scheme ensures that whenever a new frequency is selected, it always starts at the end of a full ramp cycle of the previous frequency. This, along with the fact that the ramp retains the same V_H and V_L levels, will eliminate switching duty-cycle disturbance due to the hopping activity. Note that some inductor current disturbance still persists due to the mismatch between the rising and falling slopes of the ramp. This normally results in transients that appear as low frequency noise [5], yet, due to randomization, this noise is non-spurious and spreads across a wide frequency band leading to minimal increase in the low-frequency noise floor.

The basis of the proposed scheme can be explained using Fig. 22.8.2. For an intuitive analysis, we will consider the case with two frequencies. It can be shown that the results hold for M frequencies. The spectrum of the switching node signal V_{sw} (t) can be analyzed by adding the spectra of V_{sw1} (t) $=P_1$(t) $\times V_{sw}$ (t) and V_{sw2} (t) = P_2(t) $\times V_{sw}$ (t) (similar to binary FSK), where P_1(t) and P_2(t) are non-overlapping unipolar random digital sequences. Starting with V_{sw1} (t), a virtual signal V_{sw1}'' (t) that matches the phase of V_{sw1} (t) at time zero is constructed. Using V_{sw1}'' (t), one can write V_{sw1} (t) $=P_1$(t) $\times V_{sw1}''$ (t $- k \cdot N\Delta T_h$), where $\Delta T_h = (1/f_2 - 1/f_1)$, and k is a running integer that counts the multiples of (N/f_2) that are elapsed since time zero (at which f_2 is applied and not f_1). Note that k can increase indefinitely. The time delay applied to V_{sw1}'' (t) is to account for the fact that each time f_1 is applied beyond the first occurrence, V_{sw1}'' (t) resets its phase, and thus its phase no longer matches the phase of V_{sw1}'' (t) (phase chopping). Since is a periodic function, we can use its Fourier

series to write:

$$V_{sw1}(t) = \left[\left(\frac{f_2}{f_1+f_2} \right) + P_1''(t) \right] \times \left[\left(V_{in} \cdot dc \right) + \sum_{j=1}^{\infty} \gamma_j \sin\left(2\pi j \cdot f_1 \left(t - k \cdot N\Delta T_h \right) + \phi_j \right) \right] \quad (1)$$

where γ_j and ϕ_j are the coefficient and phase of the jth harmonic of V_{sw1}'' (t), respectively, dc is the switching duty-cycle, and P_1'' (t) is the non-spurious part of P_1(t). Note that P_1(t) does not contain any impulses (spurs) except at DC due to randomization. All the terms in Eq. 1 are non-spurious except for

$$\left(f_2/\left(f_1 + f_2 \right) \right) \times \sum_{j=1}^{\infty} \gamma_j \sin\left(2\pi j \cdot f_1 \left(t - k \cdot N\Delta T_h \right) + \phi_j \right),$$

which can be rewritten as:

$$\left(\frac{f_2}{f_1+f_2} \right) \times \sum_{j=1}^{\infty} \gamma_j \times \begin{bmatrix} \cos\left(2\pi j \cdot k \cdot f_1 N\Delta T_h \right) \times \sin\left(2\pi j f_1 t + \phi_j \right) - \\ \sin\left(2\pi j \cdot k \cdot f_1 N\Delta T_h \right) \times \cos\left(2\pi j f_1 t + \phi_j \right) \end{bmatrix} \quad (2)$$

However, ΔT_h (phase chopping) can be used to eliminate spurs at $(j \times f_1)$ by setting $f_1 N \Delta T_h = (m/n) \neq Integer$, where m and n are arbitrary integers and ($n \neq 1$). First, this ensures that the running integer k assumes a value from a finite set of integers [1,2,...,n] with an equal possibility to be any value in the set. Second, using the Dirichlet kernel [6], it can be shown that the average of the spur at $(j \times f_1)$, resulting from Eq. 2, will be eliminated for all $j \neq (Integer \times Nn)$. Thus, the first spur appears at $(Nn \times f_1)$ when $(j = Nn)$. However, if n is chosen large enough, this spur drops below the noise floor since γ_j will be very small, and therefore, all spurs are practically eliminated. Using the same analysis for V_{sw2} (t) , and observing that setting $f_1 N \Delta T_h = m/n$ results in $f_2 N \Delta T_h = (N/m)/(N/n + m)$, spurs at $(j \times f_2)$ are also eliminated for all $j \neq Integer \times (Nn + m)$, provided that $(Nm)/(Nn+m) \neq Integer$. Since the proposed design eliminates duty cycle disturbance, one can afford very rapid hopping by making $N = 1$. In this case, $(Nm)/(Nn+m)$ becomes $m/(n+m)$, which is guaranteed to be a non-integer number, thus satisfying the condition of spur elimination for V_{sw2} (t) as well. The rapid hopping resulting from $N = 1$ also spreads P_1'' (t) and P_2'' (t) across the maximum possible frequency range, which minimizes the peaking in the noise floor. The condition for spur elimination then becomes $(f_1 - f_2)/f_2 = m/n$, which corresponds to $I_1/I_2 = (n+m)/n$ in the ramp implementation, and is implemented using segmented current mirrors. In the case of M frequencies, spur elimination conditions become $f_1/f_i = (n + m_i)/n$ for $i = 2$ to M, where m_i are arbitrary positive integers.

The proposed converter is implemented in a 0.35µm standard CMOS technology. The converter's output spectrum is measured using a real-time spectrum analyzer. Figures 22.8.3 and 22.8.4 show the output spectrum with 8 switching frequencies up to 16MHz and 60MHz. Full elimination of spurs is achieved with a peak noise floor of −70dBm/100Hz (≈−90dBm/Hz). The peaking in the noise floor is confined to narrow pockets and is much lower than in $\Delta\Sigma$ and Δ techniques [1-3]. Figure 22.8.5 shows the results using 2 and 4 frequencies. As expected, noise floor peaking is higher, but full elimination of spurs is accomplished even with only 2 frequencies. Figure 22.8.6 compares the proposed technique with other published work, and Fig. 22.8.7 shows the chip micrograph and key performance parameters of the regulator. With spurs eliminated and a very low noise floor, this design can enable powering many analog/RF loads directly from the converter resulting in significant improvements in efficiency, size, and cost.

References:
[1] J.N. Kitchen, C. Chu, S. Kiaei, and B. Bakkaloglu, "Combined Linear and Δ-Modulated Switch-Mode PA Supply Modulator for Polar Transmitters," *IEEE J. Solid-State Circuits*, vol. 44, no. 2, pp. 404-413, Feb., 2009.
[2] S. Dunlap and T. Fiez, "A Noise-Shaped Switching Power Supply Using a Delta–Sigma Modulator," *TCAS-I*, vol. 51, no. 6, pp. 1051-1061, Jun., 2004.
[3] J.N. Kitchen, I. Deligoz, S. Kiaei, and B. Bakkaloglu, "Polar SiGe Class E and F Amplifiers Using Switch-Mode Supply Modulation," *IEEE TMTT*, vol. 55, no. 5, pp. 845-856, May, 2007.
[4] J.-H. Chen, P.-J. Liu, and Y.-J.E. Chen, "A Spurious Emission Reduction Technique for Power Amplifiers Using Frequency Hopping DC-DC Converters," *IEEE RFIC Symposium*, pp. 145-148, Jun., 2009.
[5] E.J. Kim, C.-H. Cho, W. Kim, et al., "Spurious Noise Reduction by Modulating Switching Frequency in DC-to-DC Converter for RF Power Amplifier," *IEEE RFIC Symposium*, pp. 43-46, May, 2010.
[6] A. M. Bruckner, J.B. Brukner, and B.S. Thomson, *Real Analysis*, Prentice Hall, NJ 1997.

ISSCC 2011 / February 23, 2011 / 12:15 AM

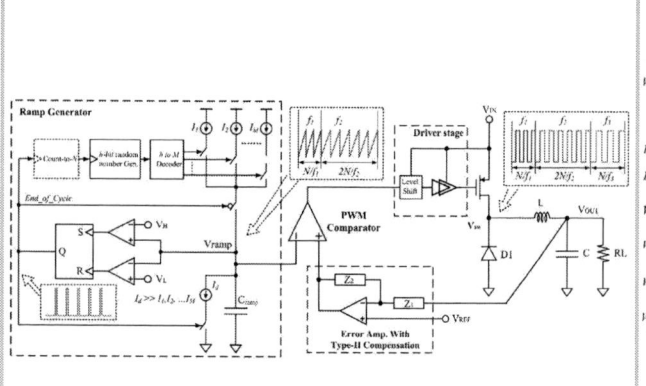

Figure 22.8.1: Proposed buck converter with random frequency hopping combined with random phase Chopping.

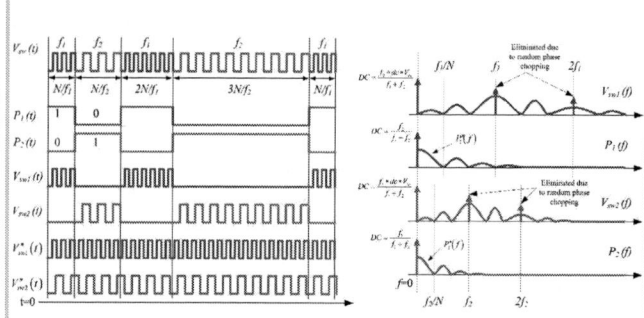

Figure 22.8.2: Time and frequency domain representations used to derive the spectrum at the output of the proposed regulator. Note the phase shift (chopping) in both $V_{sw1}(t)$ and $V_{sw2}(t)$ relative to the periodic versions at the bottom.

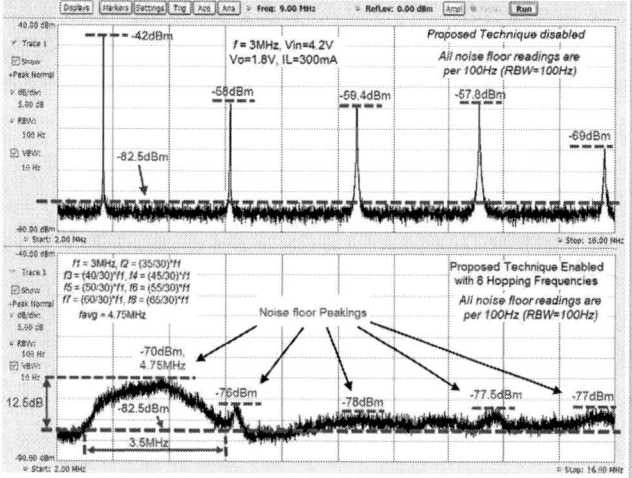

Figure 22.8.3: Spectrum comparison up to 16MHz. All spurs are eliminated with −70dBm/100Hz (≈ −90dBm/Hz) noise floor peaking at the average of all the switching frequencies.

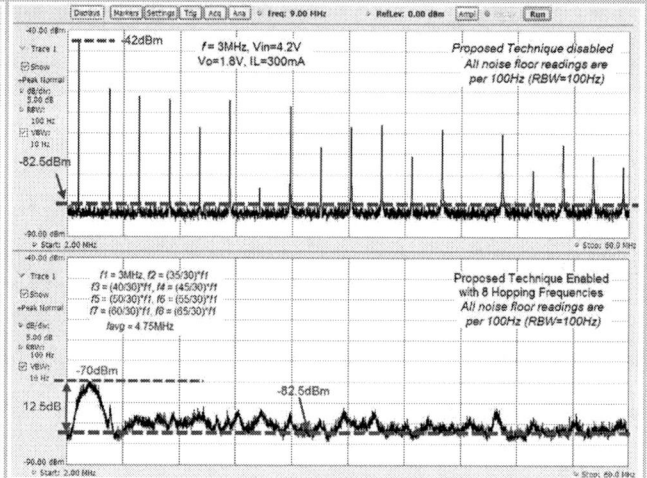

Figure 22.8.4: Spectrum comparison up to 60MHz. Consistent elimination of spurs up to very high frequencies with noise floor peaking always less than −70dBm/100Hz (≈ −90dBm/Hz).

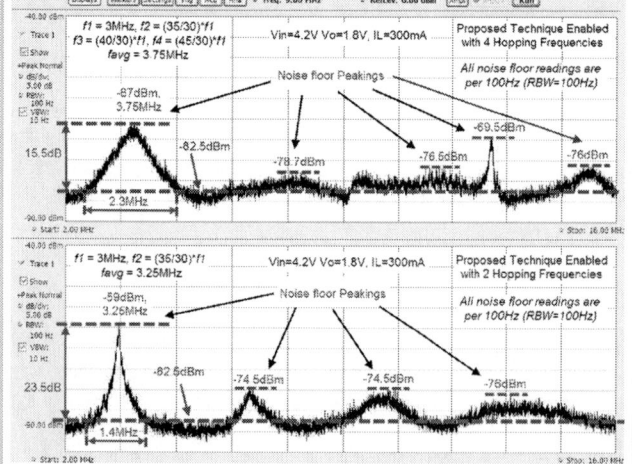

Figure 22.8.5: Spectrum comparison up to 16MHz, with the proposed technique using 4 and 2 frequencies. Higher noise floor peaking is observed as expected.

	Proposed Work			[2]	[3]		[4]	[5]
Technology	0.35um			Discrete	0.18um		Discrete	0.18um
Control Method	PWM + Combined frequency hopping & phase chopping			Δ-Σ modulation	Δ-Σ modulation	Δ Modulation	PWM + frequency hopping	PWM + Monotonic Frequency stepping
Number of Frequencies	2	4	8	NA	NA	NA	8	64
Switching Frequencies (MHz)	3, 3.5	3, 3.5, 4, 4.5	3, 3.5, 4, 4.5, 5, 5.5, 6, 6.5	0.2−0.4	5	5	Between 1.6−2	Between 2.2−4.4
Maximum Spur reduction (db)	Eliminated	Eliminated	Eliminated	15−33	26	21	11	12
Peak Noise Floor	−59 dBm/100Hz (≈ −79 dBm/Hz)	−67 dBm/100Hz (≈ −87 dBm/Hz)	−70 dBm/100Hz (≈ −90 dBm/Hz)	−53.2 dBm/Hz	−26 dBm/Hz	−28 dBm/Hz	Not Reported	−56 dBm/Hz
Area overhead (compared to PWM case)	less than 8% of total active area			Not Reported	Not Reported	Not Reported	Not Reported	10% of total active area

Figure 22.8.6: Comparison between the proposed regulator and other published work.

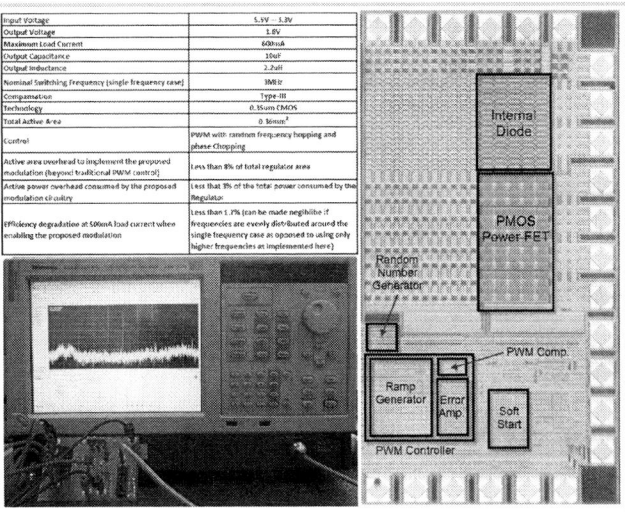

Input Voltage	5.5V – 3.3V
Output Voltage	1.8V
Maximum Load Current	600mA
Output Capacitance	10uF
Output Inductance	2.2uH
Nominal Switching Frequency (single frequency case)	3MHz
Compensation	Type-III
Technology	0.35um CMOS
Total Active Area	0.36mm²
Control	PWM with random frequency hopping and phase Chopping
Active area overhead to implement the proposed modulation (beyond traditional PWM control)	Less than 8% of total regulator area
Active power overhead consumed by the proposed modulation circuitry	Less than 3% of the total power consumed by the Regulator
Efficiency degradation at 500mA load current when enabling the proposed modulation	Less than 1.3% (can be made negligible if frequencies are evenly distributed around the single frequency case as opposed to using only higher frequencies as implemented here)

Figure 22.8.7: Die Micrograph, and test setup of the proposed regulator and its key performance parameters.

ISSCC 2011 / SESSION 23 / IMAGE SENSORS / OVERVIEW

Session 23 Overview / IMMD

Image Sensors

Session Chair: *Tetsuo Nomoto, Sony, Kanagawa, Japan*

Session Co-Chair: *Jan Bosiers, DALSA Professional Imaging, Eindhoven, The Netherlands*

Higher speeds, increased dynamic range and improved performance for small pixels are clearly driving the imaging industry. This session introduces several interesting new approaches, based on combinations of imager design, technology and architecture, to achieve better performance on these competitive imaging aspects.

The first two papers present approaches to achieve low readout noise and high dynamic range for CMOS imagers.

Paper 23.1 [Shizuoka U] presents a column-parallel folding-integration and cyclic ADC to achieve a variable gray-scale resolution of 13b through 19b by changing the number of samples. $1.2e^-_{rms}$ temporal noise is achieved when using 64-fold sampling, resulting in 82dB DR in a $7.5 \times 7.5 \mu m^2$ 4T pixel.

Paper 23.2 [CSEM] introduces a CMOS imager with 256x256 11µm pixels that achieves $0.86e^-_{rms}$ readout noise. The pixels have PMOS transistors in an open-loop voltage amplification architecture. This results in a dynamic range of 90dB, with a conversion factor of $300\mu V/e^-$.

Paper 23.3 [CEA-LETI — MINATEC] introduces a readout IC with a 15b pixel-level ADC for cooled hybrid infrared image sensors. When bonded with indium bumps to a long-wave infrared HgCdTe detector array, an SNR of 90dB is measured on the detector that has 320×256 pixels with a 25µm pixel pitch.

Then, Paper 23.4 [Kinki U; DALSA PI; U Arizona; NHK] introduces a backside-illuminated CCD imager capable of storing 117 consecutive images recorded at up to 16Mfps. The CCD has 362×456 pixels of $43.2 \times 43.2 \mu m^2$. The high fill factor resulting from backside imaging combined with an electron-multiplication CCD readout significantly increases the sensitivity, as low as 7 photons/pixel.

Paper 23.5 [Canon] introduces a 12-inch wafer-scale 1.6Mpixel stitched CMOS imager with $160 \times 160 \mu m^2$ pixels with in-pixel 0-to-24dB variable-gain voltage amplifier, fabricated in 0.35µm CMOS. By simultaneously reading out the reset and integrated signals through a pair of column lines a random noise of $13e^-_{rms}$ is achieved at 100fps in global shutter mode.

Paper 23.6 [U Edinburgh] presents a 128×96 pixel, 44.65µm pitch, digital 3D camera SoC. Each pixel comprises an SPAD (single photon avalanche diode) and phase-domain $\Delta\Sigma$ loop for on-chip computation of distance. 3D images are obtained at 20fps with sub-16cm repeatability error and ±0.5cm linearity over a range of 0.4 to 2.4m.

Paper 23.7 [Cornell U] presents a new approach to 3D imaging by providing local diffraction gratings over each pixel. The imager with 400×384 pixels of $7.5 \times 7.5 \mu m^2$ fabricated in a 0.18µm CMOS process enables post-capture refocus and range finding with ±1.3cm accuracy at 50cm by using only one lens and ambient light.

The last four papers present several innovations to improve the performance of small-pixel CMOS imagers for mobile imaging, and high resolution and high speed for digital still and video-camera applications.

Paper 23.8 [Aptina] describes a 1/13-inch VGA SoC CMOS image sensor, with a 1.75µm pixel pitch capable of outputting 30fps at full resolution. To limit the size of the imaging core, a very small separate reference pixel array is used to generate the dark reference. The sampling capacitors for signal and reset are laid out with double pitch and stacked on top of each other. There is only

978-1-61284-303-2/11 $26.00 © 2011 IEEE

ISSCC 2011 / February 23, 2011 / 8:30 AM

one horizontally routed metal wire per pixel alternating in time as RST and TX. The result is an imaging core of only 1.77mm². The pixel has a dynamic range of 63.8dB with 3400e⁻ full well and a conversion factor of 272µV/e⁻.

Paper 23.9 [Samsung Electronics] presents a backside-illuminated 1/2.33-inch, 1.4µm pixel pitch, 14.6Mpixel CMOS image sensor. A floating diffusion boosting scheme is implemented with an additional row-wise metal line, with no penalty on QE because of the backside illumination. The QE is 71%, which is 30% higher compared to front-side illumination, and a SNR of 10 is achieved at 87lux.

Paper 23.10 [Aptina] describes a 16Mpixel CMOS image sensor with 28.26mm diagonal optical format with 14b SAR-ADC and 8-lane LVDS output. A dynamic response pixel with 4.78µm pitch is realized by adding an additional transistor switch to the standard 4T pixel. By closing or opening the connection of a physical capacitor to the FD node, the pixel operates in low- or high-sensitivity mode. In low-sensitivity mode, the full well capacity is 50ke⁻ at 16e⁻$_{rms}$ readout noise; in high-gain mode the full well is 16ke⁻, with 2.2e⁻$_{rms}$ readout noise. The pixel with ring gate transistors and no STI achieves 62% QE in green.

The last paper, Paper 23.11 [Sony] presents a 17.7Mpixel CMOS image sensor with a 27.5mm diagonal optical format, using 90nm CMOS. By employing 16 channels of scalable low-voltage signaling interface with embedded clock operating at 2.376Gb/s combined with a single-slope ADC at 2.376GHz, a data rate of 34.8Gb/s is realized. This allows 120fps imaging at 12b full resolution.

Presenters:

8:30 AM

23.1 An 80µV$_{rms}$-Temporal-Noise 82dB-Dynamic-Range CMOS Image Sensor with a 13-to-19b Variable-Resolution Column-Parallel Folding-Integration/Cyclic ADC

M-W. Seo, Shizuoka University, Hamamatsu, Japan

9:00 AM

23.2 A Sub-Electron Readout Noise CMOS Image Sensor with Pixel-Level Open-Loop Voltage Amplification

C. Lotto, Heliotis, Root Längenbold, Switzerland; CSEM, Zurich, Switzerland

9:15 AM

23.3 A 320×256 90dB SNR and 25µm-Pixel-Pitch Infrared Image Sensor

A. Peizerat, CEA-LETI-MINATEC, Grenoble, France

9:30 AM

23.4 A 16 Mfps 165kpixel Backside-Illuminated CCD

T. G. Etoh, Kinki University, Higashi-Osaka, Japan

9:45 AM

23.5 A 300mm Wafer-Size CMOS Image Sensor with In-Pixel Voltage-Gain Amplifier and Column-Level Differential Readout Circuitry

Y. Yamashita, Canon, Kawasaki, Japan

10:15 AM

23.6 A 128×96 Pixel Event-Driven Phase-Domain ΔΣ-Based Fully Digital 3D Camera in 0.13µm CMOS Imaging Technology

R. J. Walker, University of Edinburgh, Edinburgh, United Kingdom; STMicroelectronics, Edinburgh, United Kingdom

10:45 AM

23.7 An Angle-Sensitive CMOS Imager for Single-Sensor 3D Photography

A. Wang, Cornell University, Ithaca, NY

11:15 AM

23.8 A 1/13-inch 30fps VGA SoC CMOS Image Sensor with Shared Reset and Transfer-Gate Pixel Control

R. Johansson, Aptina, Oslo, Norway

11:30 AM

23.9 A 1/2.33-inch 14.6M 1.4µm-Pixel Backside-Illuminated CMOS Image Sensor with Floating Diffusion Boosting

S. Lee, Samsung Electronics, Yong-In, Korea

11:45 AM

23.10 An APS-C Format 14b Digital CMOS Image Sensor with a Dynamic Response Pixel

D. Pates, Aptina, San Jose, CA

12:00 PM

23.11 A 17.7Mpixel 120fps CMOS Image Sensor with 34.8Gb/s Readout

T. Toyama, Sony, Kanagawa, Japan

23

ISSCC 2011 / SESSION 23 / IMAGE SENSORS / 23.1

23.1 An 80µV$_{rms}$-Temporal-Noise 82dB-Dynamic-Range CMOS Image Sensor with a 13-to-19b Variable-Resolution Column-Parallel Folding-Integration/Cyclic ADC

Min-Woong Seo[1], Sungho Suh[1], Tetsuya Iida[2], Hiroshi Watanabe[3],
Taishi Takasawa[1], Tomoyuki Akahori[2], Keigo Isobe[2], Takashi Watanabe[2],
Shinya Itoh[1], Shoji Kawahito[1,2]

[1]Shizuoka University, Hamamatsu, Japan,
[2]Brookman Technology, Hamamatsu, Japan,
[3]Sanei Hytechs, Hamamatsu, Japan

Low-noise CMOS image sensors (CIS) employing column-parallel amplifiers that significantly reduce temporal noise, as well as electron-multiplication CCD (EM-CCD) image sensors are becoming popular for very-low-light-level imaging. These low-noise imagers with high-gain amplification in either the charge or voltage domains sacrifice the intra-scene dynamic range. Scientific applications of solid-state imagers strongly require very low temporal noise and wide intra-scene dynamic range as well as very high gray-scale resolution. A column-parallel analog-to-digital converter (ADC) and column-level signal processing in CISs are key techniques to meet these requirements. Single-slope [1,2], successive-approximation [3] and cyclic ADCs [4] are widely used for the column-parallel ADC in CMOS imagers. However, these ADCs require additional gain enhancements to achieve very low temporal noise. A recently reported [5] delta-sigma ($\Delta\Sigma$) ADC has an attractive feature that low temporal noise and high resolution can be simultaneously attained by an oversampling technique. However, for very high resolution, a high number of samplings per pixel output, e.g., more than 360 samplings for 16b, is required.

This paper presents a column-parallel ADC for CMOS imagers using a successive operation of folding-integration ADC (FI-ADC) and cyclic ADC for attaining very low noise (1e$^-$), high gray-scale resolution (>16b) and resulting wide dynamic range (>80dB).

Figure 23.1.1 shows a block diagram of our column-parallel ADC. It consists of an analog core for the ADC, a digital counter for the folding-integration ADC, and a register for the cyclic ADC. The analog core used for both the folding-integration and cyclic ADCs consists of an amplifier, 2 capacitors (C_1 and C_2), 2 comparators for a 1.5b sub-ADC, a 1.5b digital-to-analog converter (DAC) for reference subtraction. The folding-integration technique is also called an extended counting [6]. In our design, the folding-integration is embedded into the cyclic ADC without any increase of analog components. The operation of the ADC is shown in Fig. 23.1.2. In the FI-ADC mode, the pixel outputs are sampled multiple times and the samples are integrated over in a switched-capacitor (SC) integrator. The operation of M samplings in the FI-ADC has a gain of M and it leads to a great reduction of the input-referred thermal noise values due to the pixel source follower and SC integrator itself. To maintain the input dynamic range while reducing the noise with the integration, a negative feedback technique using a sub-ADC and DAC is used. The integrator output is digitized into 3 zones using 2 references of $+V_{REF}/2$ and $-V_{REF}/2$ in the 1.5b sub-ADC, where V_{REF} is a common reference signal of the A/D conversion. The output codes of the 1.5b sub-ADC which takes 0, 1 or 2 are used for a counter and also the 1.5b DAC in the next phase. While a new input sample is added in the integrator, the DAC output is subtracted from the integrator output such that the integrator output swing is limited by folding the curve for multiple times. A digital code that corresponds to the total number of reference subtractions is stored in a digital counter. Figure 23.1.3 shows the characteristics of the FI-ADC. In this example with M=16, the output is folded 14 times and the counter output which shows the number of folds linearly increase as the input signal increases. In Fig. 23.1.2, an operation for the reset level samplings in the FI-ADC mode is shown. For the signal-level samplings, the connections of input signal and DAC to the capacitor C_1 are switched to perform an analog CDS operation by changing the polarity of charge transfer between the reset and signal samplings.

After the folding-integration ADC, the operation of the analog core is switched to the cyclic ADC mode to perform the A/D conversion of the final integrator output of the folding-integration ADC mode. The cyclic ADC is performed with the same analog core as of the folding-integration ADC without any additional analog circuits, but with a digital counter to count the number of folds. The operation of the cyclic ADC is the same as for the previous design [4]. In order to obtain 13b resolution from the cyclic ADC mode, the feedback sampling and amplification are repeated for 12 times. In the cyclic ADC, 2 references of $+V_{REF}/4$ and $-V_{REF}/4$ are used for the 1.5b sub-ADC.

The folding-integration ADC has m-bit resolution by sampling the pixel output M times, where m=$\log_2$ M. The successive operation of the folding-integration with M samplings, and 13b cyclic ADCs has a total equivalent resolution of 13+m-1 bits. Though the actual resolution and resulting dynamic range are determined by the temporal noise level, this extremely high gray-scale resolution allows us to realize an all-digital gain control without any programmable-gain amplifiers and quantization-noise-free imaging at extremely low light level as well as a wide dynamic range imaging.

A 1Mpixel (1024×1032) CMOS image sensor chip (Fig. 23.1.7) is implemented with standard pinned photodiode 0.18µm CMOS technology. The measured input-referred noise as a function of the number of samplings in the FI-ADC is shown in Fig. 23.1.4. The noise and dynamic range have a dependency to V_{REF} of the A/D conversion. The input-referred noise values of 80µV$_{rms}$ (1.2e$^-_{rms}$) with M=64 and V_{REF}=1.0V, and 64µV$_{rms}$ (0.95e$^-_{rms}$) with M=128 and V_{REF}=0.5V are obtained with a pixel conversion gain of 67µV/e$^-$. The measured dynamic range for the condition of M=64 and V_{REF}=1V is 82dB. The sensor performance and characteristics are summarized in Fig. 23.1.5. The image sensor has a pixel size of 7.5×7.5µm^2 with a fill factor of 52% and a sensitivity of 10V/lx-s.

Figure 23.1.6 shows a sample image at very low light level of 0.005lx with M=64 (m=6b) for the FI-ADC. In this setting, the total ADC resolution is 18b using the 13b cyclic ADC. A very high digital gain of 2048× (11b) is applied to amplify the very small signal level. Each pixel has a maximal signal level 7.3e$^-$ (=14,900/2048), but a sufficient gray-scale resolution of 7b (=18b-11b) in each pixel is attained without any additional analog gain.

Acknowledgment:
This work was partly supported by the Knowledge Cluster Initiative of Ministry of Education, Culture, Sports, Science and Technology.

References:
[1] Y. Nitta, Y. Muramatsu, A. Amano, et al., "High-speed digital double sampling with analog CDS on column parallel ADC architecture for low-noise active pixel sensor," *ISSCC Dig. Tech. Papers*, pp. 500-501, Feb. 2006.
[2] Y. Lim, K. M. Koh, K.M. Kim, et al., "A 1.1e- temporal noise 1/3.2-inch 8Mpixel CMOS image sensor using pseudo-multiple sampling," *ISSCC Dig. Tech. Papers*, pp.396-397, Feb. 2010.
[3] S. Matsuo, T. Bales, M. Shoda, et al., "A very low column FPN and row temporal noise 8.9M-pixel, 60 fps CMOS image sensor with 14bit column parallel SA-ADC," *Symposium on VLSI Circuits*, pp.138-139, June 2008.
[4] J.H. Park, S. Aoyama, T. Watanabe, et al., "A 0.1e- vertical FPN 4.7e- read noise 71dB DR CMOS image sensor with 13b column-parallel single-ended cyclic ADCs," *ISSCC Dig. Tech. Papers*, pp.268-269, Feb. 2009.
[5] Y.C. Chae, J.M. Cheon, S.H. Lim, et al., "A 2.1Mpixel 120frame/s CMOS image sensor with column-parallel $\Delta\Sigma$ ADC architecture," *ISSCC Dig. Tech. Papers*, pp. 394-395, Feb. 2010.
[6] C. Jansson, "A High-Resolution, Compact, and Low-Power ADC Suitable for Array Implementation in Standard CMOS," *IEEE Trans. Circuits and Systems*, vol. 42, no. 11, pp. 904-912, Nov. 1995.

978-1-61284-303-2/11 $26.00 © 2011 IEEE

ISSCC 2011 / February 23, 2011 / 8:30 AM

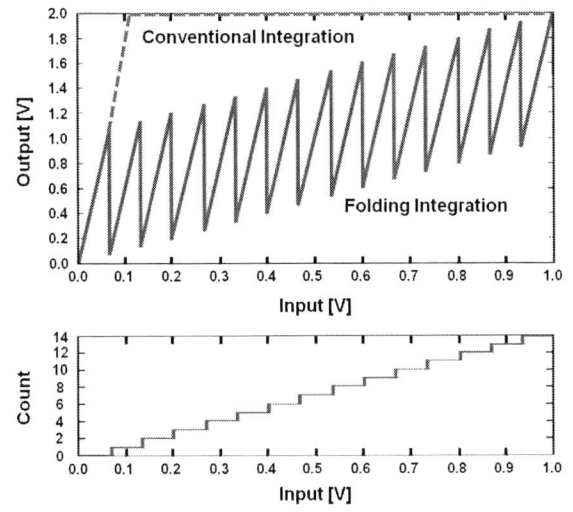

Figure 23.1.1: Block diagram of the folding integration/cyclic cascaded ADC.

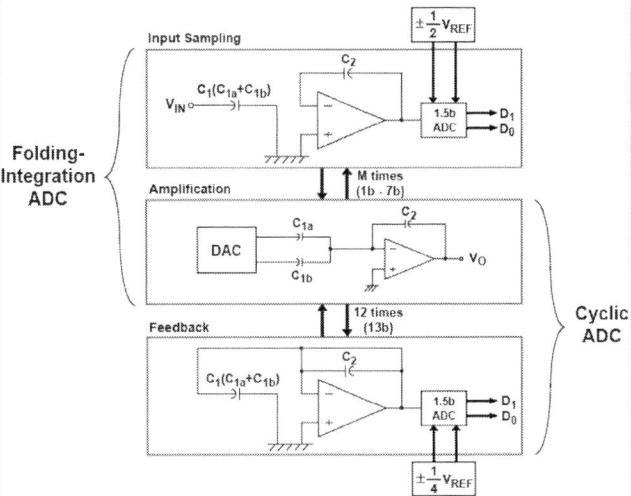

Figure 23.1.2: Circuit connections for each operating mode.

Figure 23.1.3: Transfer curve of the folding-integration ADC (M=16).

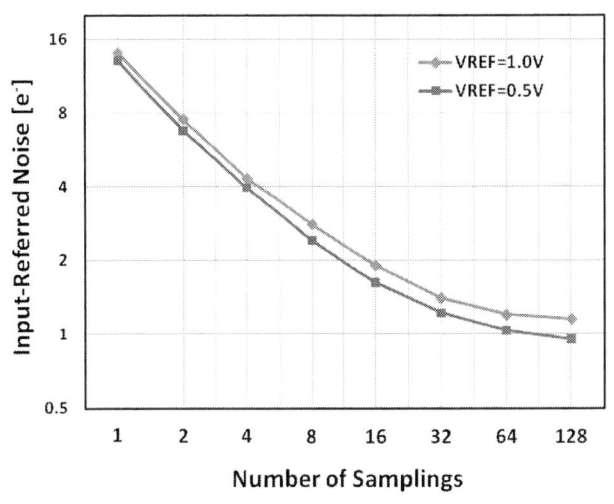

Figure 23.1.4: Measured noise as a function of the number of samplings.

Parameter	Value
Technology	0.18 μm 1P4M CIS process
Total area	10.0 (H) mm X 13.5 (V) mm
Power supplies	1.8V (Digtal), 3.3V (Analog, Digital)
Number of pixels	1024 (H) X 1032 (V)
Pixel type	4-TR (Pinned Photodiode)
Pixel size	7.5 μm X 7.5 μm
Fill factor	52 %
Sensitivity	10 V/lx·s (@ 3746K light sourse with IR cut filter)
ADC resolution	13b - 19b
Input referred noise	1.2 e- (# of samplings= 64)
Conversion gain	67 μV/e-
Full Well Capacity	14,900 e-
Dynamic range	82 dB (# of samplings= 64)
Frame rate	4 fps (# of samplings= 64)

Figure 23.1.5: Sensor performance and characteristics.

Figure 23.1.6: Captured image with the image sensor at 0.005lx, F=1.4, Gain=2048×.

978-1-61284-303-2/11 $26.00 © 2011 IEEE

ISSCC 2011 PAPER CONTINUATIONS

Figure 23.1.7: Chip micrograph.

ISSCC 2011 / SESSION 23 / IMAGE SENSORS / 23.2

23.2 A Sub-Electron Readout Noise CMOS Image Sensor with Pixel-Level Open-Loop Voltage Amplification

Christian Lotto[1,2], Peter Seitz[3,4], Thomas Baechler[2]

[1]Heliotis, Root Längenbold, Switzerland,
[2]CSEM, Zurich, Switzerland,
[3]CSEM, Landquart, Switzerland,
[4]EPFL, Neuchâtel, Switzerland

State-of-the-art low-noise CMOS image sensors use source-followers for pixel-level buffering of the sense node voltage. Due to the near-unity gain of such pixel-level circuits and the typical values of the column conductor's load capacitance the thermal noise on the column conductor generated by the pixel-level source-follower reaches amplitudes that are far from single-electron resolution. Conventional image sensors reduce the bandwidth of the source-follower's noise by employing column-level low-pass filters [1], commonly implemented using switched-capacitor amplifiers [2,3]. Given the near-unity gain of the pixel, the criteria applying to the input-referred noise voltage of the column-level amplifier are as demanding as those applying to the pixel-level circuit. As a consequence, the conventionally used column-level amplifiers contribute a significant part of the overall noise budget in conventional image sensors [4].

An ultra-low-noise CMOS image sensor based on an alternative pixel circuit featuring pixel-level voltage amplification is reported. Besides a significant reduction in the contribution of electronic noise generated in column-level circuits, pixel-level voltage amplification achieves sub-electron noise of the pixel-level circuit even without any column-level low-pass filter.

Figure 23.2.1 shows the schematic diagram of a sensor for scientific imaging applications, fabricated in UMC's 0.18µm CIS technology. Our 4T-pixel structure features pixel-level common-source amplifiers with column-wise load resistor sharing. This pixel-level amplifier uses one single transistor per pixel and offers moderate voltage gain that provides an attractive compromise between low amplifier noise as well as high pixel conversion factor on one hand and adequate linear swing on the other hand. The image sensor uses buried photodiodes and employs CDS. During the reset phase of each row operating sequence, negative feedback is applied to the pixel-level common-source amplifier by activating the reset transistor M_{res}, as illustrated on the left in Fig. 23.2.2. Switching off the reset transistor sets the common-source amplifier to an open-loop configuration for the subsequent sampling of the reset level by the column-level sample-and-hold (S/H) circuit, as well as for the sampling of the signal level after charge transfer from the buried photodiode to the sense node.

In the open-loop configuration, the voltage amplification, which has a nominal value of 10, is defined by the product of the common-source transistor's transconductance g_m times the load resistance R_l. In the case of the chosen gain value, the thermal noise of the amplifier is dominated by the contribution of the common-source transistor, and the equivalent thermal noise charge is accordingly expressed as $C_{sn} \sqrt{\gamma kT / (g_m R_l C_{col})}$, where C_{sn} is the sense node capacitance, γ is the noise excess factor and C_{col} is the load capacitance of the common-source amplifier. This result illustrates that, thanks to the bandwidth reduction resulting from voltage amplification, the load capacitance required to reach a desired noise level is divided by the voltage gain in comparison to the capacitance required by a unity-gain pixel circuit. In contrast to conventional architectures using pixel-level source-followers, the proposed architecture achieves equivalent thermal noise charge significantly below one electron using the parasitic load capacitance of the column net and the input capacitance of the S/H circuit without any additional column-level low-pass filter. Furthermore, the noise components of the column-level S/H circuit, the output drivers, and off-chip ADCs are negligible thanks to the pixel-level voltage amplification. The predicted excellent theoretical noise performance is confirmed by a measured readout noise of 0.86e- at room temperature. This result was obtained by operating the image sensor in the dark with a reduced exposure time of approximately 50µs.

The conversion factor used to calculate the equivalent noise charge is determined by a photon transfer curve measurement. Investigations of the pixel histogram of the readout noise as well as the measured amplitude distributions of selected pixels with noise values above average indicate that the observed noise performance is limited by low-frequency noise, particularly RTS noise [5].

The Miller effect can lead to a significant increase of sense node capacitance of a pixel making use of an inverting amplifier. The presented pixel circuit avoids the Miller effect concerning the gate-drain overlap capacitance of the common-source transistor by operating the select transistor M_{sel} as a cascode transistor, i.e. by achieving a common-source transistor's drain voltage virtually independent of the column voltage. For this purpose, M_{sel} is operated in the saturated region rather than in the triode region by choice of an adequate gate voltage during the select state of the pixel.

The pixel-to-pixel spread of the amplifier's offset due to threshold voltage mismatch between the common-source transistors of different pixels is eliminated by CDS. The combination of voltage amplification and threshold variations due to mismatch, temperature change, or process parameter spread would, however, result in unacceptable variations in the amplifier's linear output swing, assuming the sense nodes of all pixels were reset to single common reset voltage. This problem is solved though by the application of negative feedback during the sense node reset phase. Thanks to the feedback path established by activating M_{res}, each pixel's amplifier is driving the pixel's sense node to its self-defined operating point. As a result, the impact of threshold variations on the output reset voltage, and hence the swing of the pixel is attenuated by a factor equal to the voltage gain.

The compensation of transistor offset by the described self-biased reset method allows the use of open-loop amplification while competing with conventional pixel circuits in terms of linearity and uniformity. Accordingly, a peak linearity error as low as 1.7% over a linear output range of approximately 1V and a PRNU as low as 2.5% have been measured, as illustrated in Figs. 23.2.3 and 23.2.4, respectively.

A linear dynamic range of 73dB is achieved. In addition to the linear swing, the pixel provides a compressed region thanks to the transition of the common-source transistor's operating condition from the saturated to the triode region for high pixel output voltages. This compressed region of the common-source amplifier's characteristics extends the intra-scene dynamic range to 90dB, obtained with a single readout step using a low-complexity readout circuit.

The achieved results illustrate that pixel-level open-loop voltage amplification provides flicker noise limited sub-electron noise performance using simple and elegant circuits, by (1) effective reduction of noise generated in column-level circuits, and by (2) reduction of the pixel-level electronic circuit's noise bandwidth without any additional area-consuming load capacitance or column-level low-pass filtering circuit.

References:
[1] Y. Chae, et al., "A 2.1Mpixel 120frame/s CMOS Image Sensor with Column-Parallel ΔΣ ADC Architecture," *ISSCC Dig. Tech. Papers*, pp. 394-395, Feb. 2010.
[2] B. Fowler, et al., "Wide Dynamic Range Low Light Level CMOS Image Sensor," *in Proc. of 2009 International Image Sensor Workshop*, June 2009.
[3] A. Krymski, et al., "A 2e- Noise 1.3Megapixel CMOS Sensor," *in Proc. of 2003 IEEE Workshop on Charge-Coupled Devices and Advanced Image Sensors*, May 2003.
[4] N. Kawai and S. Kawahito, "Measurement of low-noise column readout circuits for CMOS image sensors," *IEEE Trans. Electron Devices*, vol. 53 no 3, pp. 1737-1739, July 2006.
[5] X. Wang, et al., "A CMOS Image Sensor with a Buried-Channel Source Follower," *ISSCC Dig. Tech. Papers*, pp. 62-63, Feb. 2008.

ISSCC 2011 / February 23, 2011 / 9:00 AM

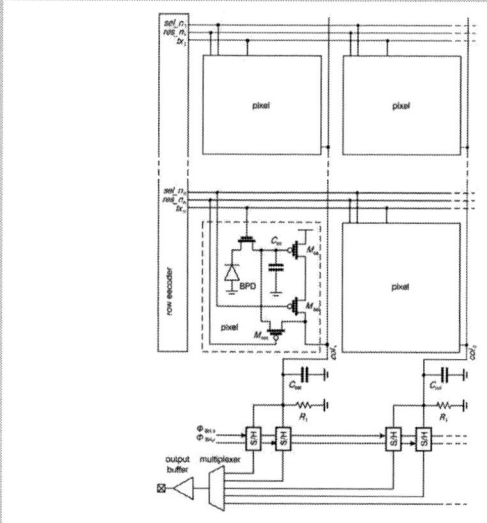

Figure 23.2.1: Sensor with pixel-level open-loop amplifiers.

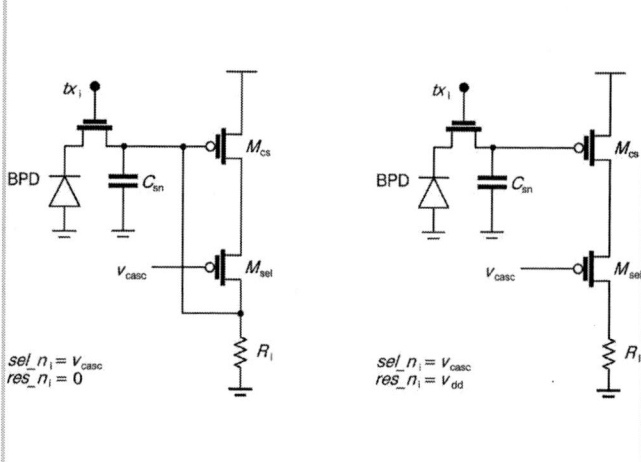

$$sel\_n_i = v_{casc}$$
$$res\_n_i = 0$$

$$sel\_n_i = v_{casc}$$
$$res\_n_i = v_{dd}$$

Figure 23.2.2: Reset configuration (left) and amplifying configuration (right).

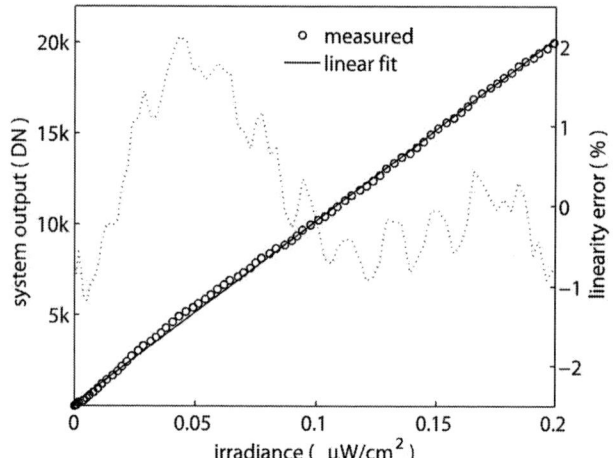

Figure 23.2.3: Linear part of the photoresponse and linearity error.

Figure 23.2.4: Histogram of the responsivity of 65536 pixels.

Parameter	Value	Conditions / Remarks
readout noise	0.86 e-	room temperature, 60 frames/sec
pixel conversion factor	300uV/e-	
pixel size	11um x 11um	
fill factor	50%	
number of pixels	256 x 256	prototype sensor
Row readout time	15 us	
Maximum frame rate	60 frames / sec	sensor scaled up to 1000 rows
Peak linearity error	1.7%	
PRNU	2.5%	

Figure 23.2.5: Most relevant parameters.

Figure 23.2.6: Image with a maximum of 24 and an average of 6 photoelectrons per pixel.

23

Figure 23.2.7: Packaged prototype sensor with a die size of 5mm by 5mm.

Due to formatting issues there is a gap in pagination.

Pages 404 - 405

ISSCC 2011 / SESSION 23 / IMAGE SENSORS / 23.4

23.4 A 16 Mfps 165kpixel Backside-Illuminated CCD

Takeharu G. Etoh[1], Dung H. Nguyen[1], Son V. T. Dao[1], Cuong L. Vo[1], Masatoshi Tanaka[1], Kohsei Takehara[1], Tomoo Okinaka[1], Harry van Kuijk[2], Wilco Klaassens[2], Jan Bosiers[2], Michael Lesser[3], David Ouellette[3], Hirotaka Maruyama[4], Tetsuya Hayashida[4], Toshiki Arai[4]

[1]Kinki University, Higashi-Osaka, Japan,
[2]DALSA PI, Eindhoven, The Netherlands,
[3]University of Arizona, Tucson, AZ,
[4]NHK Science and Technical Laboratories, Tokyo, Japan

In 2002, we reported a CCD image sensor with 260×312 pixels capable of capturing 103 consecutive images at 1,000,000 frames per second (1Mfps) [1]. We named the sensor "ISIS-V2", for In-situ Storage Image Sensor Version 2. 103 memory elements are attached to every pixel; generated image signals were instantly and continuously stored in the in-situ storage without being read out of the sensor. The ultimate high-speed recording was enabled by this parallel recording at all pixels. In 2006, the color version, ISIS-V4, was reported [2]. In 2009, we developed ISIS-V12, a backside-illuminated image sensor mounting the ISIS structure and the CCM, charge-carrier multiplication, on the front side [3]. The CCM is a CCD-specific efficient signal-amplification device. CCM, combined with the BSI structure and cooling, achieved very high sensitivity. The ISIS-V12 was a test sensor intended to prove the technical feasibility of the structure. The maximum frame rate was 250kfps for a charge-handling capacity of Q_{max}=10,000e⁻ and 1Mfps for a reduced Q_{max}. The pixel count was 489×400 pixels. For backside-illuminated (BSI) image sensors, metal wires can be placed on the front surface to increase the frame rate without reducing fill factor or violating uniformity of the pixel configuration. It has been proved by simulations that 100Mfps is achievable by introducing innovative technologies including a special wiring method [4]. We now report on ISIS-V16, developed by incorporating technologies to increase the frame rate with those to achieve very high sensitivity, which was confirmed by evaluation of ISIS-V12. The performance specification of ISIS-V16 is summarized in Fig. 23.4.1.

Figure 23.4.2 shows the global planar structure of ISIS-V16. The imaging area is divided into 4 rectangular subareas. A set of driving voltages used in the image-capturing operation, which requires very high frequency, is transferred from the left and right, toward the vertical center-line through metal inner bus lines. The very wide inner bus lines, significantly reduce the resistance. The inner bus lines are connected to the outer metal bus lines with a special shape, named "Thunderbolt bus lines," which also serve to reduce the resistance in transferring the driving voltages.

Figs. 23.4.3 and 23.4.4 depict the plane structure, installed on the front side, and a cross-section taken along the A-A' line in Fig. 23.4.3. In Fig. 23.4.4, incident photons generate electron-hole pairs in the thick p⁻ generation layer. The generated electrons travel to the collection gate on the front side to form a signal charge packet. The charge packet is then transferred along an n⁻ CCD channel, which is a memory device extending linearly in a slightly slanted direction to the orthogonal direction to the sheet, as shown in Fig. 23.4.3.

In Fig. 23.4.3, a signal charge packet is transferred from the collection gate to the memory CCD channel, carried downward and drained from the drain at the end of the CCD channel. Therefore, a sequence of the latest image signals is always stored in the in-situ storage of each pixel.

The chip has a thickness of 33μm, consisting of a 10μm-thick n-epi layer and a 23μm-thick p-epi layer. In the n-epi layer on the front side, n-type CCD memories are installed, which are protected by the p-wells. A special 3-mask injection is applied to the p-well design to increase the frame rate [4]. This configuration provides the following advantageous functions [5]:

(1) To prevent incident photons from directly reaching the memory CCD to generate additional electrons,
(2) To prevent signal electrons generated in the generation layer from migrating to the memory CCD, and
(3) To deplete the layers all the way along the path of the signal electron.

The special cross-section configuration serves to create functional BSI image sensors by replacing the multi-memory elements of each pixel in the p-well by functional circuitry. The simplest example is a global-shutter BSI imager with a single memory in the p-well. Another possibility is installation of an ADC in each pixel with a stacked digital memory chip.

The frame rate was measured by images of a rotating laser beam chopper shown in the left image of Fig. 23.4.6. The middle image shows an example image taken at 16Mfps. Next, the chopper is replaced by an LED to measure the charge-handling capacity Q_{max}. The pulse duration of the LED was adjusted so as to fit in the 51st frame. If the image of the LED appears only in one frame, the whole charge packet of the LED image is perfectly transferred. The charge-handling capacity is defined as the highest number of electrons satisfying this condition. Figure 23.4.5 shows the dependency of the charge handling capacity Q_{max} on the frame rate. The result is summarized as follows:

(1) Q_{max} keeps a constant value 22,000 e⁻ at frame rates up to 4Mfps, and
(2) Q_{max} decreases to 16,000e⁻ and 8,000e⁻, respectively, at 8Mfps and 16Mfps.

The right image of Fig. 23.4.6 is captured in a very dark condition at 16fps by activating the CCM. The noise floor of the hastily assembled evaluation camera is about 30e⁻/pixel, which is much larger than the original signal level for the imaging condition with the CCM inactivated. To evaluate the original signal level for this condition, we averaged 117 noise images in which the chopper image at rest should be submerged; then, the chopper image showed up. The average signal level was measured at the gaps of the blades, i.e., at the brighter parts of the chopper image, which was 7e⁻/pixel. Assuming that the quantum efficiency of thick BSI imagers is larger than 80%, ISIS-V16 captures images for incident light with the intensity less than 9 photons/pixel (=7e⁻/0.8).

Large differences in brightness of the 4 blocks in the right image of Fig. 23.4.6 are caused by slight instability or unbalance of the voltage set to drive CCMs. In the current design of ISIS-V16, a single voltage set drives the CCMs on the 4 blocks. Further modification of the sensor is necessary, including introduction of separate voltage-delivery to the CCMs.

An ultra-high-speed and ultra-high-sensitivity image sensor is developed. The frame rate reaches 16Mfps for Q_{max}=8,000e⁻. For frame rates up to 4Mfps, Q_{max} remains at the full-well capacity of 22,000e⁻. Images can be captured for incident light of 9 photons/pixel. The pixel count is 456×362 (=165,072) pixels. The number of consecutive frames is 117. Consequently, the frame rate and Q_{max} of ISIS-V16 are respectively 16× and 2.2× higher than those of the ISIS-V12, keeping the very high sensitivity and a fixed pixel count independent of the frame rates.

Acknowledgments:
The ISIS-V16 was developed in the "Ultra-high-speed bio-nano scope" project of JST-SENTAN, and evaluated in an "Academic Frontier" project for private universities of MEXT, Japan.

References:
[1] T.G. Etoh, D. Poggemann, A. Ruckelshausen, et al, "A CCD Image Sensor of 1,000,000 fps for Continuous Image Capturing of More Than 100 Consecutive Frames", *ISSCC Dig. Tech. Papers*, pp. 46-47, Feb. 2002.
[2] H. Ohtake, T. Hayashida, K. Kitamura, et al, "300,000-pixel ultrahigh-speed high-sensitivity CCD and a single-chip color camera mounting this CCD", *Broadcast Technology*, NHK STRL, vol. 28, pp. 1-8, 2006.
[3] C. Vo Le, T.G. Etoh, H.D. Nguyen, et al, "A Backside-Illuminated Image Sensor with 200,000 Pixels Operating at 250,000 Frames per Second", *IEEE Trans. Electron Devices*, vol. 56, no. 11, pp. 2556-2562, 2009.
[4] D.V.T. Son, T.G. Etoh, M. Tanaka, et al, "Toward 100 Mega-frames per second: Design of an Ultimate Ultra-High-Speed Image Sensor", *Sensors*, no. 10, pp. 16-35, 2010.
[5] T.G. Etoh, T. Hayashida, H. Maruyama, et al, "Backside Illuminated Image Sensors manufactured with Gradated Double Epitaxial Layers: an Application to a High-speed High-sensitivity Image Sensor", *Proc. of IISW*, 2009.

ISSCC 2011 / February 23, 2011 / 9:30 AM

	Performance
Highest frame rate	16,000,000fps
Pixel count	362x456(=165,072)pixels
Full well capacity	22,000 e-
Pixel size	43.2x43.2 micron²
Size of CCD Element	3.0x3.6 micron²
Fill Factor	100 % (1,866 micron²)
Average QE	about 80%
Number of frames	117 frames
Transfer scheme	Four-phase transfer
Operation temperature	-50 degree C
Sensitive wavelength	350 ~ 650 nm
Overwriting operation	Installed
CCM	Installed
Interlaced imaging	Installed

Figure 23.4.1: Specification of ISIS-V16.

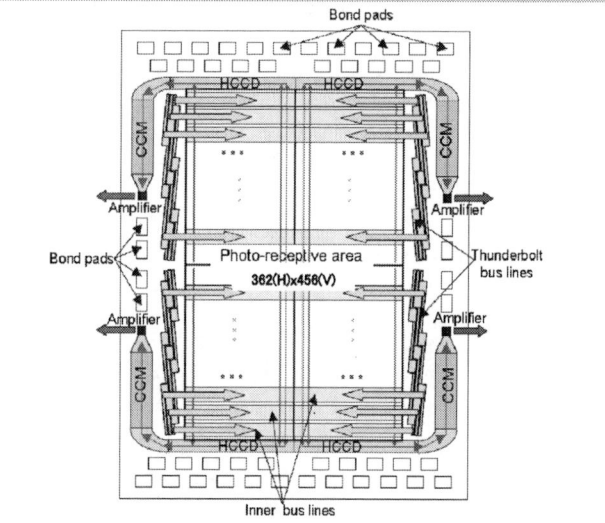

Figure 23.4.2: Planar structure of ISIS-V16.

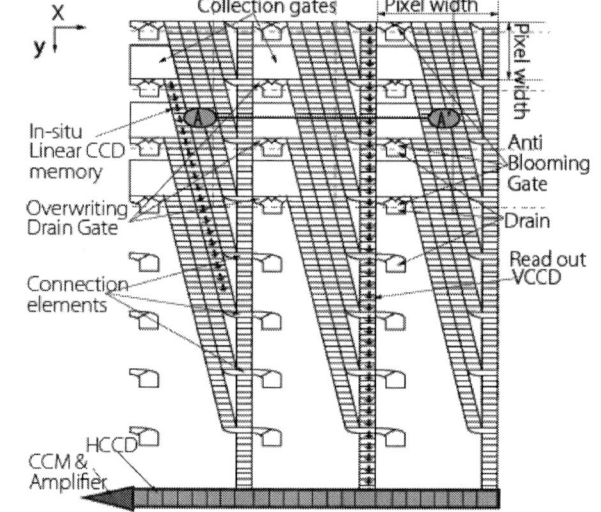

Figure 23.4.3: Schematic diagram of the front-side circuitry.

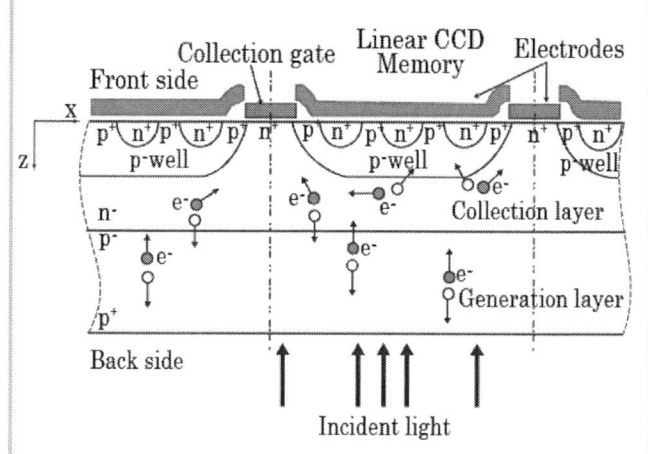

Figure 23.4.4: Cross Section (A-A' cross section in Figure 23.4.3).

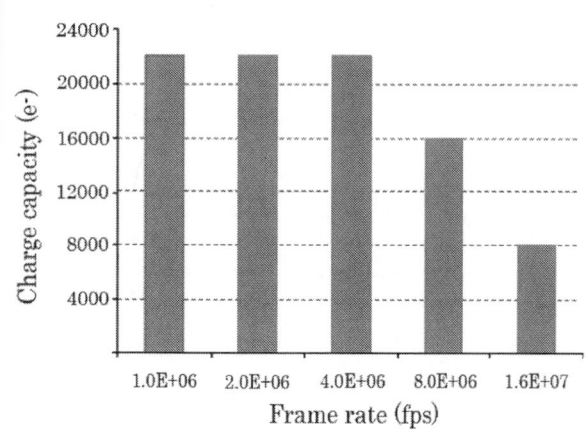

Figure 23.4.5: Qmax vs. Frame rate.

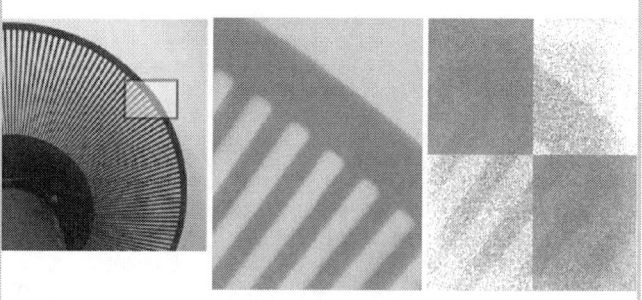

Figure 23.4.6: Images of a 200-blade laser beam chopper Left: Appearance, Middle: A magnified image taken at 16Mfps, Right: An image captured under a dark condition and amplified by CCM; frame rate 16fps; signal level: 7e-/pixel.

978-1-61284-303-2/11 $26.00 © 2011 IEEE

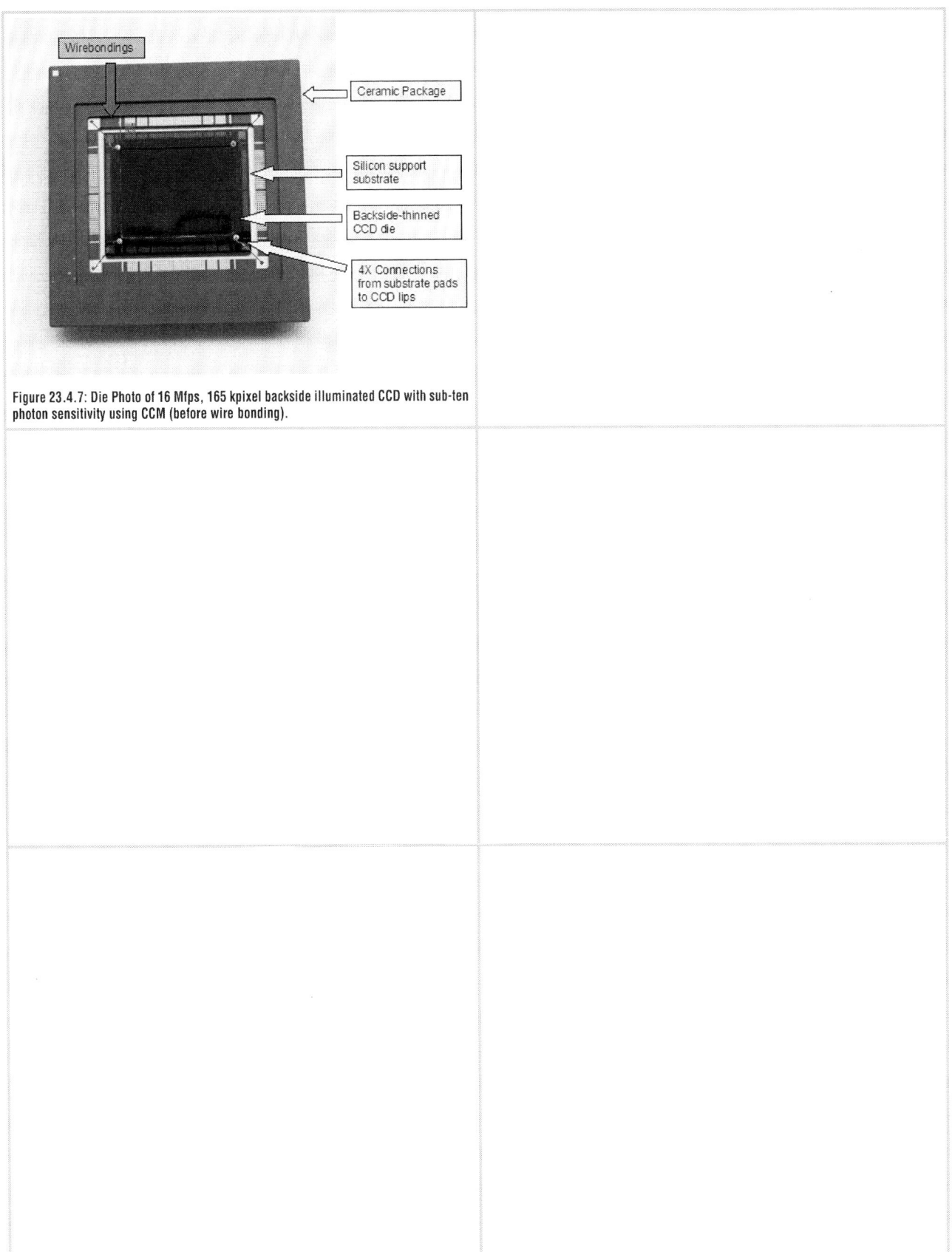

Figure 23.4.7: Die Photo of 16 Mfps, 165 kpixel backside illuminated CCD with sub-ten photon sensitivity using CCM (before wire bonding).

ISSCC 2011 / SESSION 23 / IMAGE SENSORS / 23.5

23.5 A 300mm Wafer-Size CMOS Image Sensor with In-Pixel Voltage-Gain Amplifier and Column-Level Differential Readout Circuitry

Yuichiro Yamashita, Hidekazu Takahashi, Shin Kikuchi, Keisuke Ota, Masato Fujita, Satoshi Hirayama, Taikan Kanou, Sakae Hashimoto, Genzo Momma, Shunsuke Inoue

Canon, Kawasaki, Japan

Large-format image sensors provide us with a new form of vision in several areas such as astronomy and industry. The sensors commonly comprise thin-film transistors (TFTs) and photodiodes (PDs) on amorphous silicon. The capabilities of the amorphous silicon sensors, however, are insufficient due to the low carrier mobility of the TFTs. Recently several large-format CCDs [1] and CMOS image sensors [2,3] have been developed on crystal silicon wafers for faster readout speed, reduced image lag, high sensitivity and reduced noise.

In this paper, we describe the architecture of a wafer-size CMOS image sensor enabling to enlarge the size of the large-format sensor while maintaining good signal quality. The good signal quality is the key for a low-noise and high-frame-rate image sensor. For this purpose, each pixel of our sensor has a programmable voltage amplifier. In addition, the differential readout circuitry on the column signal path ensures tolerance to common-mode noise and the drift of power/ground voltage.

A 202×205mm², 1.6Mpixel, 100 frame-per-second (fps) CMOS image sensor is fabricated on a 300mm wafer in 0.25μm 1P3M CMOS. The diagonal of the sensor is 1.5× larger than that of the conventional wafer-size imager on a 200mm wafer.

Figure 23.5.1 shows a conceptual view of the sensor. The chip is formed by stitching 3 fragments named block A, B and C because the area of the chip is larger than the maximum field of view of existing steppers. Stitching one block of A, nineteen blocks of B and one block of C yields one subarray. The subarray includes pixels of 128 columns × 1248 rows, a pair of output buffers for reset and integrated signals, a horizontal scanning circuit (HSC), and a vertical scanning circuit (VSC). Ten subarrays compose the wafer-size image sensor. The VSC is embedded in the array of pixels, therefore the pixel pitch between adjacent pixels at the border of subarrays is the same as that in the subarray. In addition, the sensor can be arranged to 2-by-N tiles to form larger image area because there is no margin both on the edge of block A and on the long sides of the subarray.

Figure 23.5.2 shows the layout concept of a pixel array. In the array, a row is flipped horizontally onto its adjacent rows, which yields a space between every 2 rows. Unit cells of both the VSC and the HSC are placed at the spaces between rows. Repeaters are inserted between unit cells of the VSC to regenerate a vertical shift pulse. The body of the PD is fully depleted, and connected with a conductor at its center.

Figure 23.5.3 is a schematic view of the signal path from a PD to output terminals. A unit pixel consists of a PD, a source follower (SF), a programmable in-pixel voltage amplifier (PGA), and 2 S/H amplifiers for reset and integrated signals. Figure 23.5.4 describes its operation as follows: (1) signal integration, (2) holding an integrated signal at the S/H A, (3) resetting the PGA, (4) resetting the PD while resetting the PGA, and (5) holding the reset signal. All pixels are stimulated at the same time, achieving synchronous shutter operation. The reset signal and the integrated signal held at the S/Hs do not contain kT/C noise of the PD because it is filtered out by the PGA. The kT/C noise generated when resetting the PGA, however, cannot be canceled. The PGA is an inverting voltage amplifier with capacitive feedback. Its gain is configured to 0, 18, or 24dB, which enables a maximum conversion gain of 318μV/e-. The PGA consists of an nMOS-input single-staged operational amplifier with 40dB DC open-loop gain.

The reset signal and the integrated signal from a pixel are read out to a pair of column lines of 200mm long each, and transmitted simultaneously to a column buffer within a period of 2.0μs. Unlike a conventional column signal line that transmits reset and integrated signals sequentially, the differential signal line cancels the drift of ground/power line due to insufficient settling time while transferring high-amplitude integrated signals. It enables almost the same column read time as that of a small-format CMOS image sensor.

A synchronous shutter operation is important to remove the effect of V_{ref} drift at the PGA. The change of V_{ref} multiplied by the gain appears on the output of the PGA. If the sensor is in rolling-shutter operation, signals of the pixel obtained from a different row i.e. at a different time have a large difference due to the V_{ref} drift. It results in line noise flickering in the horizontal direction. Human vision is very sensitive to the line flicker. It is, however, very difficult to remove the line flicker because averaging an insufficient number of horizontal optical black pixels in each row does not give the true value of V_{ref}. Under synchronous operation, the change of V_{ref} affects the entire pixel identically, which can be removed by frame-offset compensation using all the vertical optical black pixels of more than one row.

Figure 23.5.5 summarizes the specifications and performance of the sensor. Due to high conversion gain, measured full-well capacity is limited to 77ke- at 1× pixel gain. Minimum random noise of 13e-$_{rms}$ is measured at 16× pixel gain. A conversion gain of 318μV/e-, and a voltage sensitivity of 7.8kV/lx-s are obtained with a 2856K light source and IR cut filter. Electron sensitivity is 25Me-/lx-s. Maximum power consumption is 320mW per subarray for a total of 3.2W. The maximum power occurs during a V-blanking period when the SFs and the PGAs of all the pixels are activated. The average power is very small because the two amplifiers above are in power-down mode when the data are read out from S/H amplifiers.

Figure 23.5.6 is a 1.6Mpixel image taken by the sensor under 0.3lx ambient light with Sinaron SE 5.6/300 lens. F-number and exposure time are set to 6.8 and (1/60)s, respectively. Neither low-pass filtering nor other filtering to reduce random noise is employed. Line defect is not found. Pixel defects are not compensated. Note that details of the owl are preserved under very low illumination. This image indicates that the sensor has higher sensitivity than an FEA-HARP image sensor [4], because a clear image can be taken using the lens with a fifth-magnitude smaller aperture compared to the previous work. The wafer-size CMOS sensor enables to identify eighth-magnitude stars of the starry heavens in a 60fps movie. In order to remove the offset variation of column buffers, the image was compensated by the column offset data generated by projecting the vertical optical black pixel data of the image itself. To remove the offset voltage variation of S/H amplifiers, the averaged dark image, which was prepared beforehand, was subtracted from the raw image.

Figure 23.5.7 shows photographs of the sensor on a 300mm wafer and a package. The lines between subarrays are visible, implying irregularity on the chip. The image taken by the sensor, however, is not affected by the irregularity because the size of PD and the pixel pitch, i.e. the sensitivity and the spatial frequency are kept constant among all the pixels.

Acknowledgement:
The authors would like to thank T. Aoki and K. Shigeta at Canon, for their contributions to this work.

References:
[1] E.-J.P. Manoury, et al., "A 36×48mm² 48M-pixel CCD imager for professional DSC applications," *IEDM Dig. Tech. Papers*, pp. 263-266, Dec. 2008.
[2] L. Korthout, et al., "A wafer-scale CMOS APS imager for medical X-ray applications," *International Image Sensor Workshop*, June 2009.
[3] R. Reshef, et al., "Large-Format Medical X-Ray CMOS Image Sensor for High Resolution High Frame Rate Applications," *International Image Sensor Workshop*, June 2009.
[4] N. Egami, et al., "Highly Sensitive VGA FEA-HARP Image Sensor," *International Image Sensor Workshop*, June 2009.

ISSCC 2011 / February 23, 2011 / 9:45 AM

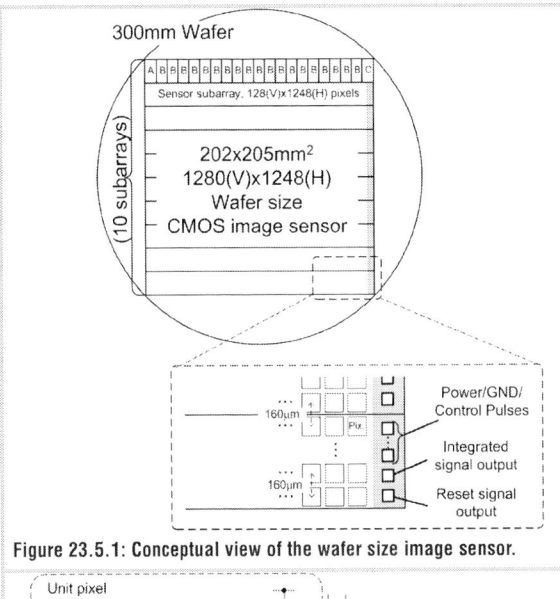

Figure 23.5.1: Conceptual view of the wafer size image sensor.

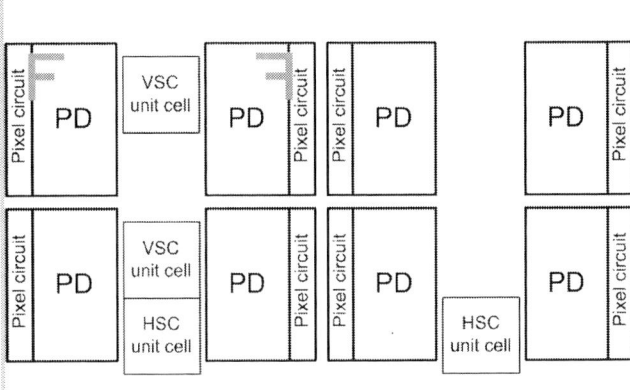

Figure 23.5.2: Layout concept of pixel array to embed VSC and HSC among PDs.

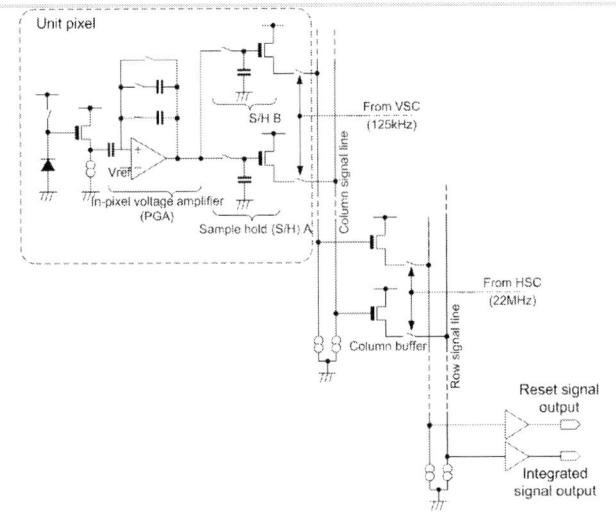

Figure 23.5.3: Schematic view of the signal path from a PD to reset/integrated signal output terminals.

Figure 23.5.4: Timing diagram.

Process	0.25μm 1P3M CMOS
Chip size	202(H)mm x205(V)mm
Pixel size	160μm x 160μm
Number of total pixels	1280(H) x 1248(V)
Number of effective pixels	1280(H) x 1128(V)
Maximum Frame rate	100 frames/s
Pixel rate	220MHz (22MHz x 10ch)
Pixel gain setting	0[dB](x1), 18[dB](x8), 24[dB](x16)
Conversion gain @ x16	318 μV/electron
Sensitivity @ x16	7800 V/lx-sec
	25M electrons/lx/sec
Full well capacity @ x1	77000 electrons
Random noise	13 e⁻$_{rms}$
Max. power consumption	3.2 W

Figure 23.5.5: Summarized specifications and performance of the wafer size image sensor.

Figure Figure 23.5.6: Sample image taken under 0.3lx ambient light.

23

978-1-61284-303-2/11 $26.00 © 2011 IEEE   409

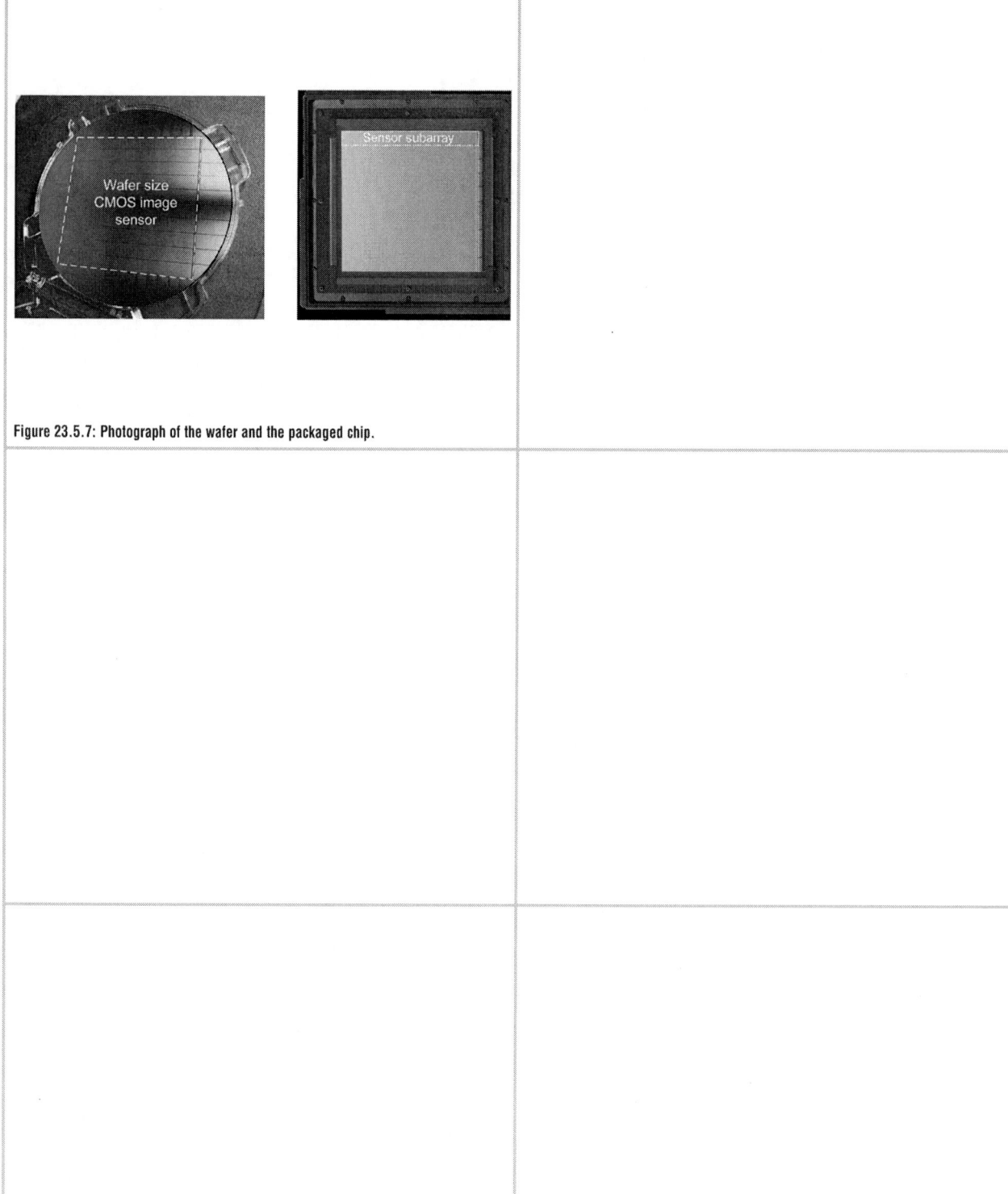

Figure 23.5.7: Photograph of the wafer and the packaged chip.

ISSCC 2011 / SESSION 23 / IMAGE SENSORS / 23.6

23.6 A 128×96 Pixel Event-Driven Phase-Domain ΔΣ-Based Fully Digital 3D Camera in 0.13μm CMOS Imaging Technology

Richard J. Walker[1,2], Justin A. Richardson[2], Robert K. Henderson[1]

[1]University of Edinburgh, Edinburgh, United Kingdom,
[2]STMicroelectronics, Edinburgh, United Kingdom

Low-cost 3D image capture devices are enabling new applications in the gaming, robotics, automotive and surveillance industries. A number of approaches are competing for a share of these markets. Stereoscopic cameras employ intensive image processing to interpret distance from the correlation of two separate image streams [1], structured light systems analyse the deformation of patterned light projected over the scene, while time-of-flight (TOF) cameras require custom frequency-modulated image sensors and optical sources to measure the phase or return time of reflected light pulses. As rapid progress is made on compact, high-frequency NIR LEDs, much research is being devoted to improvements in the size, sensitivity and resolution of TOF image sensors. Analog pixel approaches provide compact pixel implementations but accuracy is limited by noise sources and nonlinearities of the analog electronics [2]. Single Photon Avalanche Diodes (SPADs) circumvent these issues and enable fully digital distance computation down to millimetric accuracy [3] by either direct or indirect demodulation schemes.

However, SPAD-based 3D imagers have several drawbacks; large pixels produce high volumes of raw data, which must be transmitted off chip and processed externally in order to generate the 3D image, consuming I/O power and external system resources [3,4]. In this paper, we introduce a compact pixel structure (44.65μm pitch) that is able to directly compute distance (phase) based on a fully digital, event-driven Phase-Domain Delta-Sigma (PDΔΣ) approach. Together with 10b decimation and serialization circuits, this 128×96 pixel sensor generates 3D images requiring no further external processing (beyond standard defect correction algorithms and inter-frame averaging). The ΔΣ loop further serves to increase the dynamic range and linearity of the sensor over other phase demodulation approaches, which employ trigonometric computation in software.

A PDΔΣ loop measures the mean phase of an input signal, Φ_{sig} with respect to a pair of reference signals, Φ_1 and Φ_2. Analog PDΔΣ loops have been demonstrated in temperature-to-digital converters, where the phase shift induced by a sensing element is digitized [5]. This work demonstrates an all-digital PDΔΣ loop embedded in a ranging pixel. Within each pixel, the incoming optical waveform is detected by a SPAD, providing a train of digital pulses. The PDΔΣ loop converges to measure the phase difference between the envelope of these pulses and the outgoing light. This is accompanied with decimation filters, split between in-pixel and in-column logic, providing a 128×96 pixel 3D image sensor capable of generating depth maps directly.

Figure 23.6.1 illustrates the architecture of the sensor with a block diagram of the PDΔΣ loop, while Fig. 23.6.2 presents a schematic of the pixel. SPAD pulses are fed into the ΔΣ loop integrator, implemented as a 6b 2's compliment up/down counter, providing ample headroom to handle multiple SPAD pulses per modulation cycle (which may be the case with a long illumination period.) As the timing diagram in Fig. 23.6.3 shows, this integrator is driven to count up in phase with the outgoing light while the bitstream is in the high state, and driven to count down out of phase with the illumination light while the bitstream is in the low state. The counter's operation is controlled by a separate pair of up/down signals for each bitstream state, providing flexibility to implement various timing modes. A single flip-flop sampling the counter MSB (sign bit) acts as the quantizer, and is updated only at the end of cycles in which a photon was detected, as recorded by a 'fired' latch in the pixel, which is reset at the end of every cycle. This activity dependent operation saves power, and ensures the output code is not weighted by the incident light level.

The bitstream is then partially decimated to a 6b value within the pixel using a simple accumulate and dump regime: a bit counter flags when 64 valid bitstream states have been sampled into a decimation counter. The pixel data may be read

out directly at up to 70kHz frame rate (10.24Gb/s), or further decimated on-chip to reduce the required readout bandwidth. The upper 4b forming a 10b decimation filter for each pixel are implemented in column logic to avoid compromising pixel fill factor excessively. The array is periodically scanned and the in-pixel MSBs are cleared and added to the column counters. To complete a 10b code, the pixel is instructed to halt acquisition when its internal bit counter has been completely filled, ensuring 2^{10} bitstream values are decimated to produce each range estimate. This eliminates the division operation typically required for this ranging technique. Also in the column structures, a frame buffer may be used to store the last complete 10b range value for each pixel. These are updated as soon as new values become available. Readout may be performed from the frame buffer at any desired frame rate, asynchronously to the internal operation of the sensor if necessary, providing the most up to date range estimate for each pixel without requiring external computation.

Four Osram SFH4236 LEDs are used as an illumination source, driven with a 3.33MHz square wave in order to improve the linearity of our system, avoiding the harmonic distortion typically inherent in sinewave-modulated LED light sources. The 20ns rise and fall slopes of the modulation light only introduce distortion if these non-ideal portions are in the vicinity of the boundary between the up and down counting periods. This can easily be avoided by aligning the pulse such that the distance range of interest lies within the linear portion of the illumination waveform, since this low repetition rate provides far greater unambiguous range than is required for an indoor setting. The sensor can also operate with sinusoidal modulation, but the pixel will then have a trigonometric instead of linear response.

Figure 23.6.4 shows a plot of measured range-finding performance. The sensor operates with a lower illumination power and repetition rate than previous work [2,4] at the expense of repeatability error. However, an improved INL of ±5mm is provided over a 0.4-to-2.4m range along with the significant advantages that come with generating the depth maps internally instead of in software. While the device in [4] has a readout bandwidth of over 2Gb/s, 20fps images can be obtained from our system with a typical IO bandwidth of 60Mb/s and simple firmware frame averaging to produce 16b depth maps. This is achieved with a comparable power consumption of 40mW, but relates to over 4× as many pixels.

Figure 23.6.7 shows a micrograph of the SoC fabricated in a 0.13μm CMOS imaging process. The pixel benefits from a highly optimized SPAD design [6], providing a median DCR of below 100Hz and compact guard ring, resulting in a pixel pitch of 44.65μm, with 3.17% fill factor. The 128×96 pixel sensor die measures 7.2×7.5mm². Example 3D images are shown in Fig. 23.6.5 illustrating that the sensor is able to image both near and distant objects in a single frame, while Fig. 23.6.6 presents a comparison with the prior art.

Acknowledgements:
This work was supported by the UK EPSRC. The authors would like to thank ST Microelectronics for chip fabrication. The assistance of David Renshaw, Bruce Rae and David Tyndall is gratefully acknowledged.

References:
[1] R.M. Philipp and R. Etienne-Cummings, "A 128×128 33mW 30frames/s Single-Chip Stereo Imager", *ISSCC Dig. Tech. Papers*, pp. 506-507, Feb. 2006.
[2] D. Stoppa, et. al., "An 80×60 Range Image Sensor based on 10μm 50MHz Lock-In Pixels in 0.18μm CMOS", *ISSCC Dig. Tech. Papers*, pp 406-407, Feb. 2010.
[3] C. Niclass, et. al., "A 128×128 Single-Photon Imager with On-Chip Column-Level 10b Time-to-Digital Converter Array Capable of 97ps Resolution", *ISSCC Dig. Tech. Papers*, pp. 44-45, Feb. 2008.
[4] C. Niclass, et al., "Single-Photon Synchronous Detection", *IEEE J. Solid-State Circuits*, vol 44, pp. 1977-1989, July 2009.
[5] C. P. L. van Vroonhoven and K. A. A. Makinwa, "A CMOS Temperature-to-Digital Converter with an Inaccuracy of ±0.5°C (3σ) from -55 to 125°C", *ISSCC Dig. Tech. Papers*, pp. 576-577, Feb 2008.
[6] J. A. Richardson, et. al., "Low Dark Count Single-Photon Avalanche Diode Structure Compatible With Standard Nanometer Scale CMOS Technology," *IEEE Photonics Technology Letters*, vol.21, no.14, pp.1020-1022, July 2009.

978-1-61284-303-2/11 $26.00 © 2011 IEEE

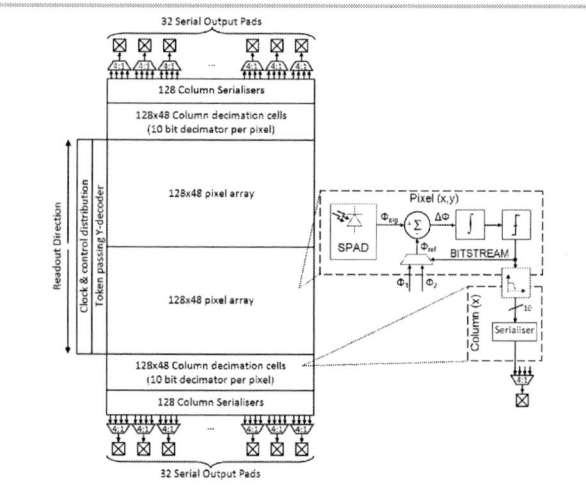

Figure 23.6.1: System overview showing block diagram of in-pixel Phase-Domain Delta-Sigma (PDΔΣ) loop with bitstream decimation split between the pixel and column.

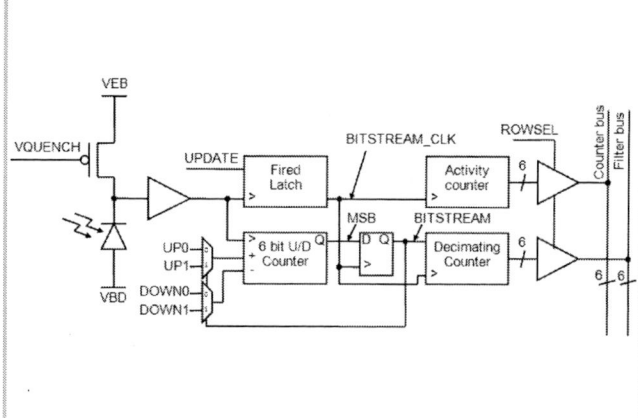

Figure 23.6.2: Pixel diagram showing all-digital PDΔΣ implementation with 6 bit counters for partial in-pixel decimation.

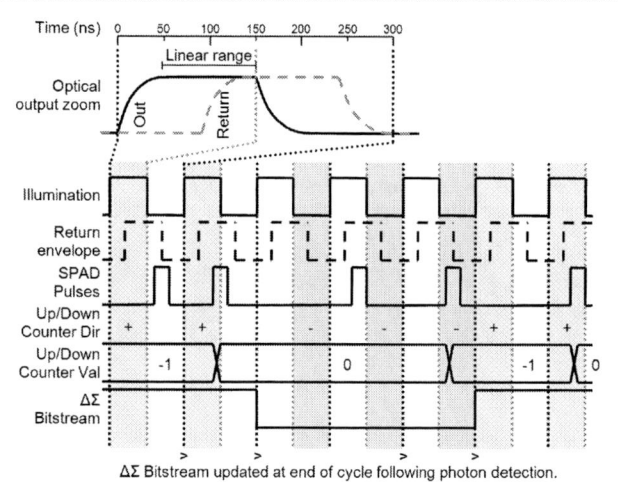

Figure 23.6.3: Pixel timing diagram, with inset showing a typical LED optical output waveform as measured using a fast photodiode.

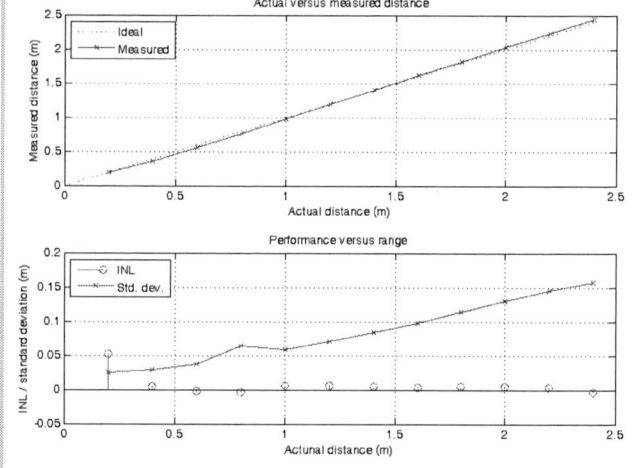

Figure 23.6.4: Plots showing measured versus actual distance, non-linearity (INL) and standard deviation over 128,000 measurements under conditions described in Fig. 23.6.6. Note: photon pile up occurs at 0.2m.

Figure 23.6.5: Example 3D images.

Parameter	[4]	[2]	This work	Unit
Sensor resolution	60 x 48	80 x 60	128 x 96	Pixels
Pixel pitch	85	10	44.65	μm
Process node	0.35	0.18 (imaging)	0.13 (imaging)	μm
Median DCR	245	-	100 [a]	Hz
SPAD dead time	40	-	50	ns
Pixel fill factor	0.5	24	3.17	%
Imaging lens f-number	1.4	1.4	1.4	-
Illumination wavelength	850	850	850	nm
Narrowband optical filter width	40	Not stated	40	nm
Illumination modulation frequency	30	20	3.33	MHz
Illumination field of view	50	Not stated	40	°
Illumination optical power on scene	800	80	50	mW
PDP at illumination wavelength	3	-	5	%
Unambiguous range	5	7.5	45	m
Integration time	45	50	50	ms
Target reflectivity	Not stated	White	White	-
Ambient illumination	150	Not stated	110	lux
Maximum distance INL up to 2.4m	110	40	5	mm
Maximum 1σ distance resolution at 2.4m	38	140	160	mm
Maximum readout bandwidth	2,133	Not stated	Typical: 61.4 Test: 10,240	Mbps
Supply voltage		1.8 / 3.3	1.2 core / 2.8 IO	V
Chip power dissipation	35	18	40	mW

Figure 23.6.6: Performance comparison table. a) Measured at SPAD excess bias voltage of 1.2V.

ISSCC 2011 PAPER CONTINUATIONS

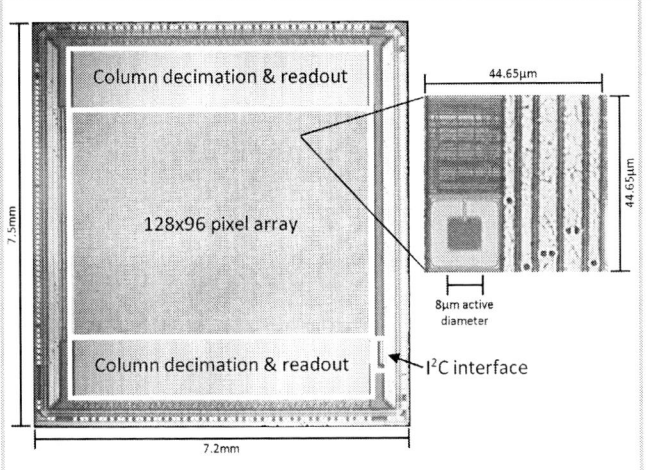

Figure 23.6.7: Chip photomicrograph with pixel inset. The sensor die is fabricated in a 0.13μm CMOS imaging process and measures 7.2×7.5mm, with a pixel pitch of 44.65μm.

ISSCC 2011 / SESSION 23 / IMAGE SENSORS / 23.7

23.7 An Angle-Sensitive CMOS Imager for Single-Sensor 3D Photography

Albert Wang, Patrick R. Gill, Alyosha Molnar

Cornell University, Ithaca, NY

Conventional cameras capture 2D photographs at a single plane of focus. Acquisition of a 3D photograph with multiple planes of focus typically requires scanning the focus of a single camera [1], or using arrays of cameras or lenses [2]. This paper demonstrates a 150kpixel image sensor, which captures 3D scenes. The chip is manufactured in commodity 0.18µm CMOS and requires only a single convex lens at a fixed focal length in ambient light.

As shown in Fig. 23.7.1a, a full description of the light striking a surface requires not only a map of light intensity at each location, but also a map of the angular distribution of light at each location [3]. Figure 23.7.1b demonstrates how angular information permits extraction of 3D depth information about imaged objects. Also, while in-focus images have high spatial resolution but contain little angular information, out-of-focus images have degraded spatial information, but richer angular information. As traditional image sensors only capture the intensity of the light at each location, they lose significant information about out-of-focus objects. Recovering the angular distribution of incident light permits recovery of the blurred, out-of-focus objects [1]. This paper presents a CMOS imager that locally captures both the intensity and incident angle of the light striking it.

The key components in this design are "angle sensitive pixels" (ASPs) [4], which respond to both the intensity and angular distribution of light striking them. The physical principle behind an ASP is the Talbot effect [5]: An incident light wave striking a uniform, periodic diffraction grating generates intensity patterns with the same period as the grating at specific distances beyond that grating (see Fig. 23.7.2b). When the incident angle of this light shifts, the intensity pattern shifts laterally. To measure these shifts, we insert a second grating with identical periodicity, called an analyzer grating, behind the diffraction grating. When the intensity patterns align with the gaps of the analyzer grating, the total light flux through the stack of two gratings is high. When the intensity patterns align with the bars of the second grating, the light flux is low. A photodiode placed behind the two gratings thus measures a periodic response to incident angle (Fig. 23.7.2d). This response, as a function of incident angle, θ, and intensity, I_0, can be approximated as $I = I_0(1+m \cos(\beta\theta+\alpha))$ where m $(0<m<1)$, β, and α, are coefficients that depend on the geometry of the two gratings. β defines the "angular gain" of the ASP: high β provides better resolution, but generates more peaks in the periodic response of the ASP. β is set by the pitch and vertical separation of the gratings [4]. α defines which angles give a peak response, and depends on the lateral offset between diffraction and analyzer gratings.

To fully distinguish changes in incident angle from changes in intensity, 4 ASPs are required, with α values of 0, $\pi/2$, π, and $3\pi/2$. These responses form two differential signals whose common mode encodes light intensity, but whose differences encode angle information along one orientation. For more complete information, more grating orientations are needed. Choosing β involves a trade-off between angular resolution and ambiguity due to the periodic nature of an ASP's response. Our image sensor uses tiled blocks of 64 sensors which extract 32 different angle measures (see Fig 23.7.3a). The 64 ASPs have 4 periodicities ($\beta=8, 12, 16, 24$) with 4 orientations (0, 90, ±45 degrees) and 4 α's, encoded as differential pairs. Arraying large blocks of diverse ASPs takes advantage of the trade-off between spatial and angular information: for in-focus images, light strikes single ASPs from a wide range of angles, blurring angular information but maintaining spatial intensity resolution. For out-of-focus objects, intensity is correlated across space but distributed heterogeneously across incident angle, such that each ASP provides independent information about incident angle, and so about the object. The array comprises a 50×48 grid of these tiles for a total of 153,600 ASPs.

ASPs are easily manufactured in standard CMOS processes. In our device, fabricated in an 0.18µm 1P6M mixed-signal process, gratings were implemented in the interconnect metallization layers. Grating pitches of 1µm and 0.7µm and interlayer separations of 1 to 4µm provided ASPs with four values of β: see Fig. 23.7.3b. ASPs incorporating local gratings and 3T active CMOS pixels with n-well/p-substrate photodiodes were arrayed at a pitch of 7.5µm. For row/column addressing, ASPs were grouped into 4×4 blocks of equal β, and the ASPs' gratings routed 8 simultaneous signals from each block. Column-parallel sample and hold circuits, programmable gain amplifiers (PGAs) and ADCs were integrated on the manufactured die (see Fig. 23.7.4).

The PGAs provide both differential and common-mode outputs from each pair of complementary ASPs, and allow for independent gain control of these two signals. Since the sum of the signals tends to be significantly larger than their difference, this gain control permits optimum use of the ADC dynamic range.

The column-parallel ADCs use a 2-stage algorithmic architecture (see Fig. 23.7.5). A shared-comparator, non-binary radix design (gain=1.8) was chosen for maximum compactness. Functionally, each cycle of the conversion begins with a 1b digitization of the previous cycle's output (or the input for the first cycle). While the comparator settles, the result is sampled differentially onto a pair of capacitors. Amplification and bit-subtraction are performed at the end of the cycle in the charge domain by switching the capacitors from a parallel to series configuration through two reference voltages set by the comparator. This signal is buffered into the other stage by source followers [6]. All timing and control circuitry are shared across columns, providing an extremely compact 160×15µm² column-level design. Measured ENOB was 8.2b at 60kS/s, supporting video frame rates of 30fps with correlated double sampling.

Figure 23.7.6 demonstrates the 3D capture capabilities of this imager. We placed a single chip behind a convex lens, and arranged a scene containing two blocks with recognizable numerals ("1" and "3") separated by 5cm at a distance of 50cm (see Fig. 23.7.6a). We then captured an image with the ASP array. Using only intensity information yields the image in Fig. 23.7.6b: the "3" is out-of-focus and unrecognizable. By correlating gradients in intensity with the angular inhomogeneity detected by ASPs, we estimated the distance from each block to the focal plane (Fig. 23.7.6c) for various focal depths. These computed range estimates had a mean-squared error of 1.3cm (2.7%). Once depth was known, information from the ASPs was used to computationally refocus the image (Fig. 23.7.6d). Thus, the ASP array acquired sufficient angular and spatial information to create a 3D reconstruction of the scene shown in Fig. 23.7.6e.

The chip consumes 50mA from a 1.8V supply, measures 20mm² (see Fig. 23.7.7) and demonstrates a 3-D imager built entirely in standard CMOS without any specialized off-chip optics.

Acknowledgement:
This work was supported under NIH grant 5R21EB009841 and the DARPA ITMARS program.

References:
[1] M. Levoy and P. Hanrahan "Light Field Rendering," *Proc. SIGGRAPH*, pp. 31-42, 1996.
[2] K. Fife, et al., "A 3MPixel Multi-Aperture Image Sensor with 0.7µm Pixels in 0.11µm CMOS," *ISSCC Dig. Tech. Papers*, pp. 47-48, Feb. 2008.
[3] A. Gershun, "The light field," *J. Math. Phys.*, vol. 18, pp. 51-151, 1939.
[4] A. Wang, P. Gill, and A. Molnar, "Light field image sensors based on the Talbot effect," *Appl. Opt.*, vol. 48, pp. 5897-5905, 2009.
[5] H. F. Talbot, "Facts relating to optical science. No. IV," *Philos. Mag.*, vol. 9, pp. 401-407, 1836.
[6] I. Ahmed, et al., "A 50MS/s 9.9mW pipelined ADC with 58dB SNDR in 0.18µm CMOS using capacitive charge-pumps," *ISSCC Dig. Tech. Papers*, pp. 164-165, Feb. 2009.

ISSCC 2011 / February 23, 2011 / 10:45 AM

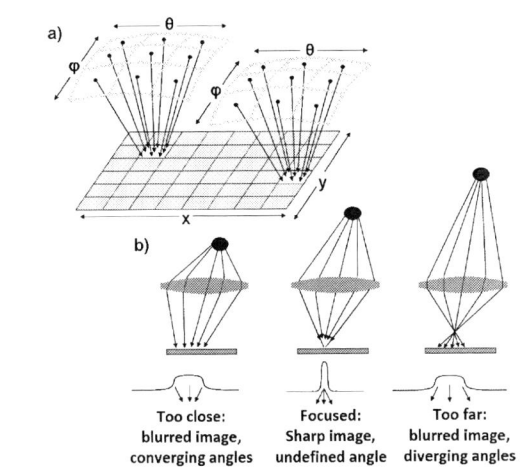

Figure 23.7.1: Conceptual drawing of a) the light field in four dimensions x,y,θ,φ, and b) how incident angle can be used to extract depth.

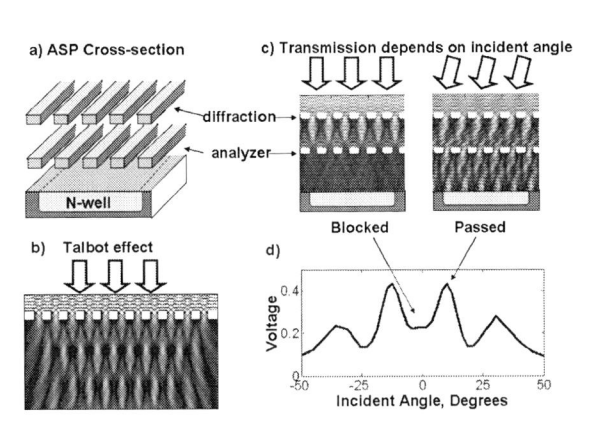

Figure 23.7.2: Structure a) and physics of ASPs: b,c) FDTD simulations of wave propagation and d) measured response to swept incident angle.

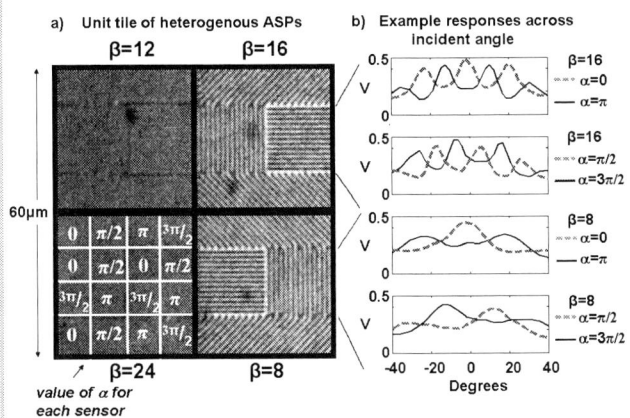

Figure 23.7.3: Implemented ASPs: a) Microphotograph of 64-pixel tile and b) measured angle-response curves for ASPs with various α and β values.

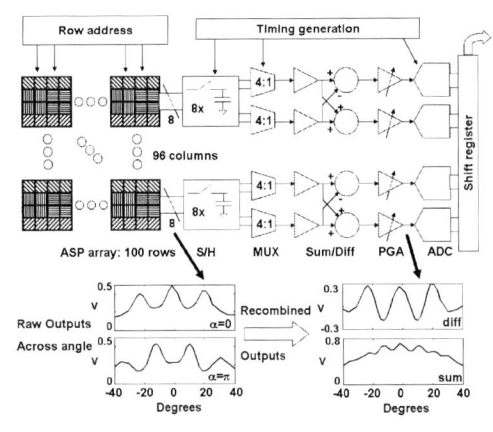

Figure 23.7.4: Top-level block diagram, curves show complementary measured ASP outputs before and after recombination.

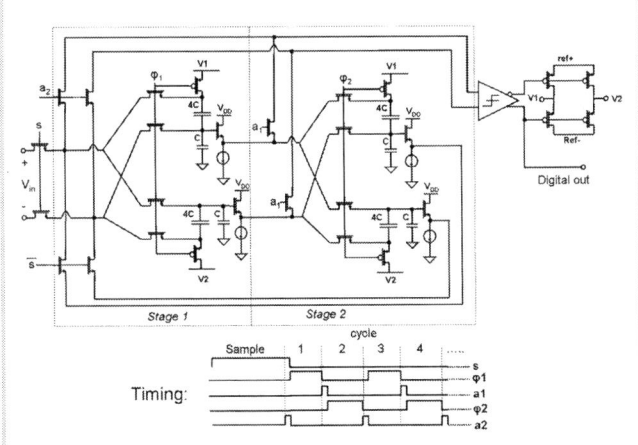

Figure 23.7.5: ADC schematic and timing diagram.

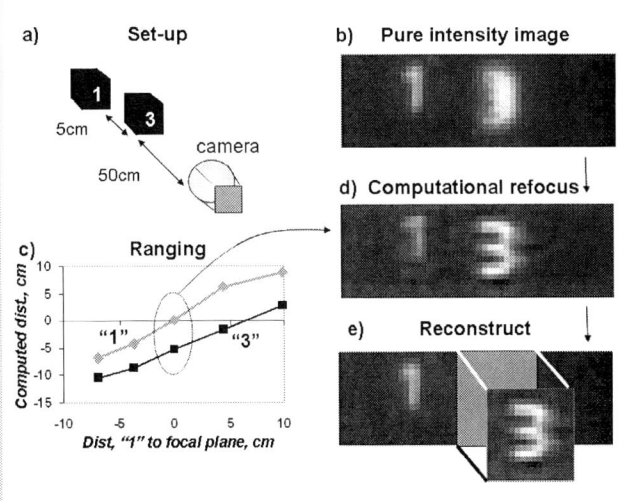

Figure 23.7.6: Demonstration of 3D imaging function.

23

978-1-61284-303-2/11 $26.00 © 2011 IEEE

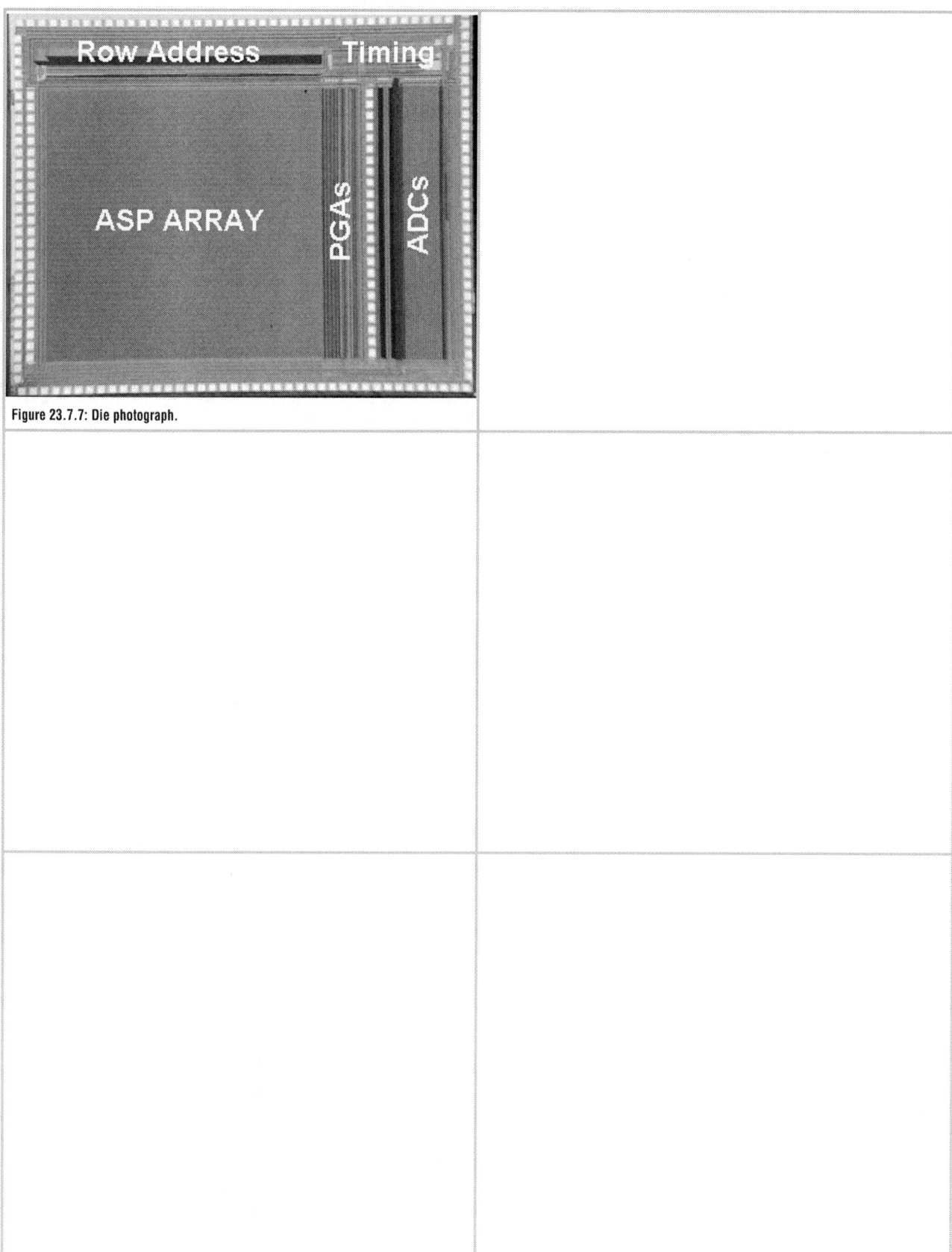

Figure 23.7.7: Die photograph.

ISSCC 2011 / SESSION 23 / IMAGE SENSORS / 23.8

23.8 A 1/13-inch 30fps VGA SoC CMOS Image Sensor with Shared Reset and Transfer-Gate Pixel Control

Robert Johansson[1], A. Storm[1], C. Stephansen[1], S. Eikedal[1],
T. Willassen[1], S. Skaug[1], T. Martinussen[1], D. Whittlesea[2], G. Ali[3],
J. Ladd[3], X. Li[3], S. Johnson[3], V. Rajasekaran[3], Y. Lee[4], J. Bai[3],
M. Flores[3], G. Davies[2], H. Samiy[2], A. Hanvey[2], D. Perks[2]

[1]Aptina, Oslo, Norway,
[2]Aptina, Bracknell, United Kingdom,
[3]Aptina, San Jose, CA,
[4]Aptina, Corvallis, OR

This paper presents a VGA 1/13-inch system-on-chip (SoC), primarily targeted for the consumer camera mobile phone market. In this market low cost, ease of product integration, low module height, and low light image quality are important features. The SoC simplifies integration of the sensor into the final product by providing camera functions such as: automatic exposure control, automatic white balance, flicker detection and avoidance, etc. This reduces cost and saves power since no companion chip is required, and the internal image-processing data rate of the SoC is higher than the output data rate [1], which keeps the inter-chip communication data rate as low as possible. The small optical format enables a low camera module height. However, the choice of the small optical format also restricts the pixel pitch to 1.75µm, which influences both image quality and die size. The low light image quality, in this sensor, is improved by incorporating a 1.75µm pixel architecture, and the size of the sensor core is greatly reduced by removing the dark pixel reference rows and columns, using a single side dual bank of double-pitch sampling capacitors, a class-AB programmable-gain amplifier (PGA), followed by an inter-stage shared pipeline ADC.

Figure 23.8.1 shows a block diagram of the analog core. The pixel array consists of 488 rows and 648 columns. A very small reference pixel array, separated from the main array, provides inputs for the black level algorithm and the analog row noise correction loop. A row of pixels are sampled onto two capacitors per column, one for the pixel reset voltage and one for the signal voltage. The sampling capacitors are laid out with double pitch and stacked on top of each other with a shared column decoder in between the two capacitor banks. Each capacitor bank has a common output bus, connected via a MUX to a single PGA. The PGA consists of two gain stages with a shared class-AB operational transconductance amplifier (OTA) that, together with the currently selected pair of sampling capacitors, implements voltage gain from 0.0625 to 31.5V/V in 255 steps. The ADC is an 11b pipeline ADC with inter-stage shared OTAs as well as inter-stage shared 1.5b AD/DA converters. Half of the ADC output range is allocated for the output signal, and the rest for analog signal chain offset. The maximum output data rate of the ADC is 11MS/s, which is sufficient to output the full field-of-view at 30fps.

Figure 23.8.2 shows a transistor schematic of the shared pixel architecture, which is a further development of the architecture described in [2]. However in this case the floating diffusions (FD$_X$) are not shared [3], instead the pixel reset control signal (RST) of row N is shared with the pixel charge transfer control signal (TX) of row N-1, for all rows. There is only one horizontally routed metal wire per pixel (rstx<N>), which alternates in time as RST and TX, and two vertically routed metal wires: one for the power supply (VAAPIX), and one for pixel output and voltage control (pixel output line). The rolling shutter readout operation of row N and N+M is carried out as follows: the reset pointer for row N+M+1 asserts both rstx<N+M> and rstx<N+M+1> while clamping the pixel output to VAAPIX. This resets both FD$_{N+M}$ and PD$_{N+M}$. Then rstx<N+M> is pulled low followed by the pixel output line being clamped to a low voltage, to discharge FD$_{N+M+1}$ to a voltage low enough to turn off the pixel source follower of row N+M+1. The reset of row N+M is followed by the readout of row N according to the following sequence: rstx<N> is asserted while the pixel output line is clamped to VAAPIX, which resets FD. The pixel output reset voltage is sampled after pulling rstx<N> low. The integrated charges are transferred from PD$_N$ to FD$_N$ by asserting rstx<N+1> while the pixel output line is clamped to a low volt-

age, in order to turn off the pixel source follower of row N+1. After pulling rstx<N+1> low followed by releasing the clamp from the pixel output, the pixel signal of row N is sampled at FD$_N$. Finally the voltage at FD$_N$ is restored to a low voltage by asserting rstx<N> while clamping the pixel output.

The improvement in low light performance, of this pixel architecture, originates mainly from the large metal opening and the high conversion gain. The non-shared FD makes it possible to further increase the conversion gain, if so required, but it also makes it possible to optimize the size and shape of the TX gate. As a result, no image lag is detected in this design while operating the TX control signal at the voltage VAAPIX. A drawback of this pixel architecture is that reading out the rows in opposite order will not produce a vertically mirrored image. This is a consequence of the shared TX and RST control signal.

The miniature dark pixel reference array consists of 17 rows and 48 columns of regular pixels covered by metal. One row is used as input for the analog row noise correction loop and is addressed whenever the main array is addressed. This configuration is shown in Fig. 28.3.3. The other 16 rows are dedicated to the integration of dark current. Half of the columns of these dark current rows have the TX pulse suppressed. The average within a row of the difference between pixels with or without a TX pulse is an estimate of the dark current, virtually free from row noise. The dark current estimate is improved by calculating the average over a number of rows, typically 2 to 8 rows. Figure 28.3.4 shows how the reference array and signal sampling circuitry are organized during the readout of the dark current rows. To get a full-sized row (648 columns) the rest of the pixel values are filled with the data obtained from the main array with SHR = SHS. This data is used to extract the analog signal chain offset. The use of this miniature reference dark pixel array significantly reduces the pixel array size overhead for a VGA sensor, and the analog row noise correction reduces the power consumption. The latter because the reference columns are not read out, resulting in shorter horizontal blanking period and, thus, lower clock frequency for a given frame rate. This is especially beneficial for the SoC core, where the power consumption is almost proportional to the clock frequency.

The lower left part of Fig. 23.8.5 shows an image taken with the previous generation VGA SoC, comprising a 2.2µm pixel, and the lower-right part the same image taken with this sensor. Although the area and power consumption of this sensor core is much lower than the previous generation sensor core, (1.77mm^2 and 17mW versus 3.52mm^2 and 30mW), the image quality is equally good or better. This is further emphasized in the table in the upper part of Fig. 23.8.5. A significant improvement in pixel performance exists when compared to the previous 1.75µm pixel architecture [4].

Figure 23.8.6 shows a die photograph of this sensor, manufactured in a 0.11µm 4-metal-layer image sensor process, where the outline of the 1.77mm^2 large sensor core has been marked with a white line. The total area of this VGA SoC is 7.0mm^2 and the total power consumption is less than 55mW.

Acknowledgements:
The authors would like to thank all of the staff and management at Aptina, for their contribution to this design.

References:
[1] S. Lim and A. El Gamal, "Integration of image capture and processing: beyond single chip digital camera," in *Sensors and Camera Systems for Scientific, Industrial, and Digital Photography Applications II*, vol. 4306 of *Proceedings of SPIE*, pp. 219-226, Jan. 2001.
[2] J. Moholt, et al., "A 2-Megapixel, 1/4-inch CMOS Image Sensor with Enhanced Pixel Architecture for Camera Phones and PC Cameras", *ISSCC Dig. Tech. Papers*, Feb. 2008.
[3] N. Tanaka et. al. "A 1/2.5-inch 8Mpixel CMOS image sensor with a staggered shared-pixel architechture and an FD-boost operation", *ISSCC Dig. Tech. Papers*, Feb. 2009.
[4] K. B. Cho, et. al., "A 1/2.5 inch 8.1 Mpixel CMOS Image Sensor for Digital Cameras", *ISSCC Dig. Tech. Papers*, Feb. 2007.

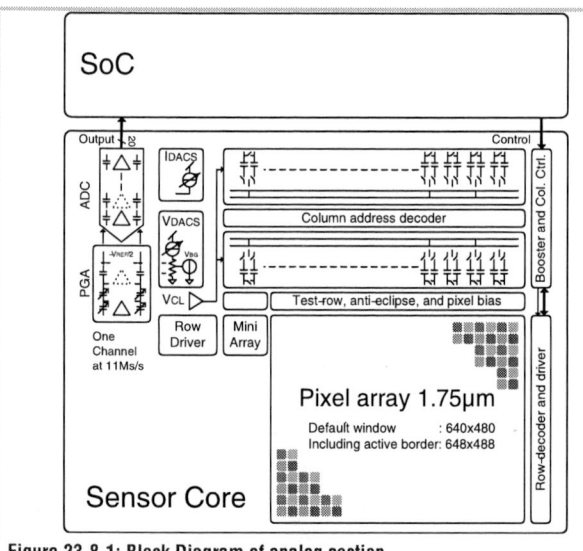

Figure 23.8.1: Block Diagram of analog section.

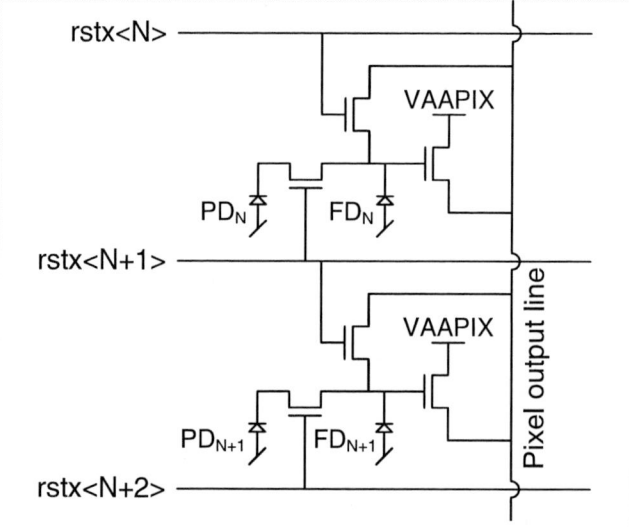

Figure 23.8.2: Pixel schematic.

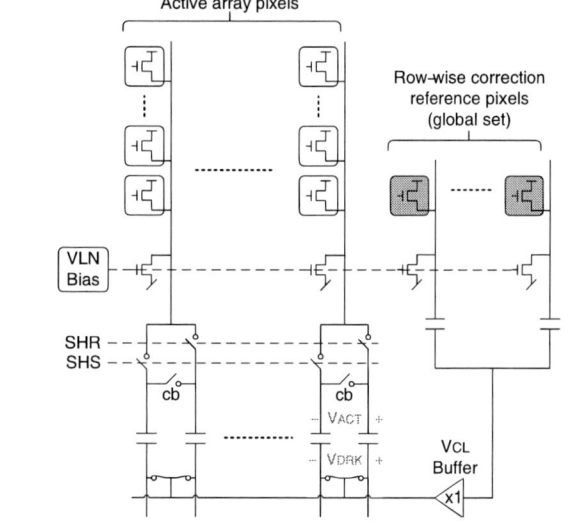

Figure 23.8.3: Miniature dark pixel reference array readout of row noise.

Figure 23.8.4: Miniature dark pixel reference array readout of dark current.

Parameter	A	B
Responsivity @ FD	0.99 V/lux*s	1.86 V/lux*s
Conversion gain @ FD	137 µV/e-	272 µV/e-
Pixel dynamic range	65.5 dB	65.1 dB
SNR max	38.1 dB	35.3 dB
Read noise at max gain	3.21 e-	1.88 e-
Column-wise and row-wise PRNU	0.40, 0.30 %	0.36, 0.25 %
Quantum efficiency (@ max, green)	45.3 %	53.3 %
Pixel Capacity (@ max linear range)	6500 e-	3400 e-

A) VGA SoC (2.2µm pixel) B) This work (1.75µm pixel)

Luminance SNR = 20.5dB   Luminance SNR = 22.2dB

Figure 23.8.5: Performance summary.

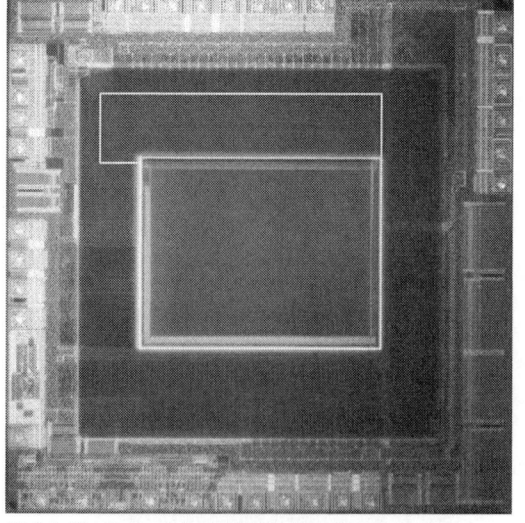

Figure 23.8.6: Die photograph with highlighted sensor core.

ISSCC 2011 / SESSION 23 / IMAGE SENSORS / 23.9

23.9 A 1/2.33-inch 14.6M 1.4μm-Pixel Backside-Illuminated CMOS Image Sensor with Floating Diffusion Boosting

Sangjoo Lee[1], Kyungho Lee[1], Jongeun Park[1], Hyungjun Han[1],
Younghwan Park[1], Taesub Jung[1], Youngheup Jang[1], Bumsuk Kim[1],
Yitae Kim[1], Shay Hamami[2], Uzi Hizi[2], Mickey Bahar[2], Changrok Moon[1],
JungChak Ahn[1], Duckhyung Lee[1], Hiroshige Goto[1], Yun-Tae Lee[1]

[1]Samsung Electronics, Yong-In, Korea,
[2]Samsung Semiconductor, Ramat-Gan, Israel

As pixel sizes continue to scale down, backside-illuminated (BSI) technology has been recently adopted as a solution to improve pixel SNR performance [1,2]. In addition, as the application of image sensors widens from digital still cameras to digital camcorders, high-resolution and high-speed operation are required. This paper presents 1/2.33-inch 14.6Mpixel CMOS image sensor employing a 1.4μm BSI pixel architecture with a floating-diffusion (FD) boosting scheme that enables high SNR and high speed read-out.

In a high-resolution and high-speed CMOS image sensor with small size pixel, one horizontal readout time period is too short to get enough margin for the duration time of charge transfer from photodiode to FD region. Unless excellent charge-transfer capability is achieved, it could cause severe image degradation such as a noisy image due to the pixel-to-pixel nonuniform charge transfer in the low-light region and color distortion at the edge of a moving object in video mode. On the other hand, while for a small pixel the capacitance of a FD is designed small enough to get higher conversion gain and thus reduces all kinds of analog circuit noise as well as source-follower flicker noise, the full-well capacity of a pixel is limited by the reduced FD capacitance. In order to solve the trade-off between dark random noise and the FD signal capacity, and also to achieve excellent charge-transfer capability, several FD-boosting schemes are reported [3,4]. The FD-boosting schemes in conventional image sensors use the capacitive coupling between the source follower transistor's gate and the signal line. However these schemes might require relatively long settling time to charge the high capacitance of an output line because the signal line of each column output consists of junction and metallic parasitic capacitances of output nodes of thousands of pixels. In this sensor, the FD-boosting (FDB) is achieved through the fast capacitive coupling between the FD jumper meal-line and the additionally incorporated row-wise FDB metal-line by making the best use of metal wiring flexibility of the BSI sensor architecture.

The pixel schematic and the FD-boosting timing are illustrated in Fig. 23.9.1. The pixel is a 1×2 shared architecture. In this 2-shared pixel, 2 photodiodes share a floating diffusion (FD), a reset transistor (RG), a source follower transistor (SF), a row-select transistor (SEL) and an FD-booster (FDB). The FDB line is in parallel with the FD jumper line using the same metal layer to maximize the capacitive coupling. The amount of FD-boosting potential can be simply controlled by 2 alternatives: (1) one is to vary the capacitance between the FD jumper and the FDB line through adjusting the overlap length of these 2 lines in the point of view of pixel layout; (2) the other way is to vary the voltage difference between high and low values of the FDB pulse. For the FD-boosting operation, the FDB is pulsed high in accordance with TG pulse both in the shutter and read operation. In order to boost FD potential effectively in the shutter operation, the RG is pulsed low before FDB is pulsed high. By applying FD-boosting both in the shutter and the read operation, there exist no lag electrons in the shutter operation, and complete charge transfer in the read operation are achieved, respectively.

Figure 23.9.2 shows measured image-lag characteristics with an analog gain of 8×. At the typical VPIX of 2.8V, the image lag is 0.06% when the FDB is enabled and 0.31% when the FDB is disabled with a default TG on-time of 0.9μs. The image lag when the FDB is disabled increases if the TG on-time is shorter than 0.7μs and degrades to 1.44% at 0.2μs. The image lag degradation is negligible even when the TG on-time is decreased to 0.2μs if the FDB is enabled. Furthermore, if the VPIX is deceased to 2.6V, the image lag is degraded abruptly as the shutter TG on-time is shortened when the FDB is disabled. However

image lag degradation is suppressed under when the FDB is enabled. The FD-boosting scheme in this sensor allows high-speed read-out without image-lag degradation.

The cross-sectional view of the fabricated 1.4μm BSI pixel is shown in Fig. 23.9.3. The sensor is fabricated in 90nm CMOS with BSI technologies. While the previous BSI technology uses a metal shield to reduce optical crosstalk [2], BSI technology in this work employs no obstacle in the backside optical track from micro-lens to the silicon back-surface as well as the lowest optical stack so that both high quantum efficiency (QE) and low optical crosstalk can be achieved. The anti-reflection coating with an optimized thickness is adopted on top of the back silicon surface to reduce the incident photon loss due to optical reflection. Furthermore, the thick silicon with deep photodiode and narrow isolation are employed for lower electrical crosstalk as well as higher QE. To minimize the recombination loss while maintaining low dark current and low dark defect, the back-surface is passivated with shallow P+ implantation and laser annealing.

Figure 23.9.4 shows the normalized QE spectrum of the fabricated 1.4μm BSI pixel compared to that of a 1.4μm frontside-illuminated (FSI) pixel. A green sensitivity of 5320e-/lx-s is measured, which is 30% higher than that of the FSI sensor while maintaining lower crosstalk as well, with the help of low optical stack height. It is also noted that high green and red QE are due to thick silicon and low crosstalk in the red wavelength regime is due inherently to the absence of substrate in the BSI structure. Low crosstalk in the blue wavelength regime reveals that a deep photodiode and narrow isolation are successfully made.

The sensor performance is summarized in Fig. 23.9.5. The sensor supports 30fps and 10fps at full-resolution for 10b and 12b ADC resolution, respectively for still mode, and 60fps at 6.72Mpixel for HD video mode. The sensor supports 2 high-speed serial interface 4-lane MIPI with 1Gb/s per lane and 12-channel SLVDS with 650b/s/channel. The power consumption of 512mW is measured in full-resolution mode @ 30fps. This power consumption is owing to supporting 12-channel SLVDS with high data bandwidth. The Y-SNR is measured by applying white balance and color-correction matrix to captured raw images from an 18% gray patch under 3200K light source and an F#2.8 lens [6]. The Y-SNR=10 is achieved at 87lux, which is 36% better than FSI of 135lux. In addition, the pixel has a saturation signal of 5700e- with no image lag, dark current of 7e-/s@55°C and a dark random noise of 1.4e- with conversion gain of 110μV/e-. Figure 23.9.6 is a reproduced image at 14.6Mpixel taken by the sensor. Figure 23.9.7 shows a chip micrograph.

In summary, a 1.4×1.4μm^2 BSI pixel with an FD-boosting scheme, which is implemented in a 1/2.33-inch 14.6Mpixel backside-illuminated CMOS image sensor, is presented. The FD-boosting is achieved by the fast capacitive coupling between the FD jumper line and the row-wise FDB line using the same layer. By applying FD boosting, complete charge transfer with negligible signal lag is achieved with fast read-out. The pixel performance shows 30% higher sensitivity than that of a FSI sensor while maintaining low crosstalk, resulting that SNR=10 is achieved at 87lux.

References:
[1] J.-C. Ahn, et al., "Advanced Image Sensor Technology for Pixel Scaling Down Toward 1.0um(Invited)," *IEDM Dig. Tech. Papers*, pp.1-4, 2008.
[2] H. Wakabayashi, et al., "A 1/2.3-inch 10.3Mpixel 50 frame/s Back-Illuminated CMOS Image Sensor," *ISSCC Dig. Tech. Papers*, pp. 410-412, Feb. 2010.
[3] K. Mabuchi, et al., "CMOS Image Sensor Using a Floating Diffusion Driving Buried Photodiode," *ISSCC Dig. Tech. Papers*, pp.112-113, Feb. 2004.
[4] N. Tanaka, et al., "A 1/2.5-inch 8Mpixel CMOS Image Sensor with a Staggered Shared-Pixel Architecture and an FD-Boost Operation," *ISSCC Dig. Tech. Papers*, pp.44-46, Feb. 2009.
[5] S. Iwabuchi, et al., "A Back-Illuminated High-Sensitivity Small-Pixel Color CMOS Image Sensor with Flexible Layout of Metal Wiring," *ISSCC Dig. Tech. Papers*, pp.302-303, Feb. 2006.
[6] J. Alakarhu, "Image Sensors and Image Quality in Mobile Phones," *Int. Image Sensor Workshop*, p.1-4, 2007.

ISSCC 2011 / February 23, 2011 / 11:30 AM

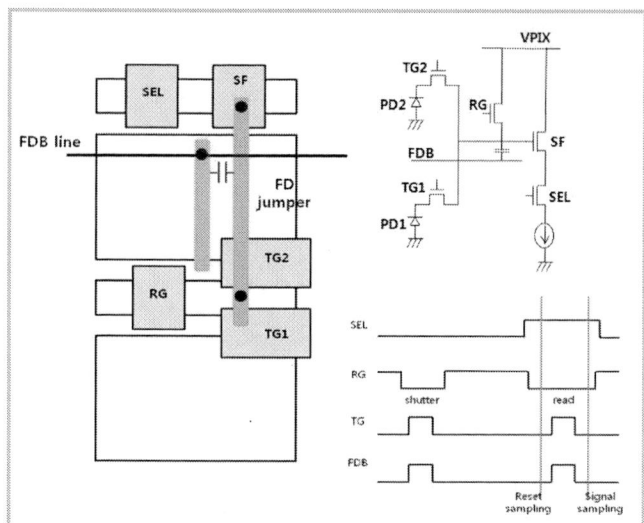

Figure 23.9.1: Pixel schematics and FDB timing.

Figure 23.9.2: Image lag characteristics.

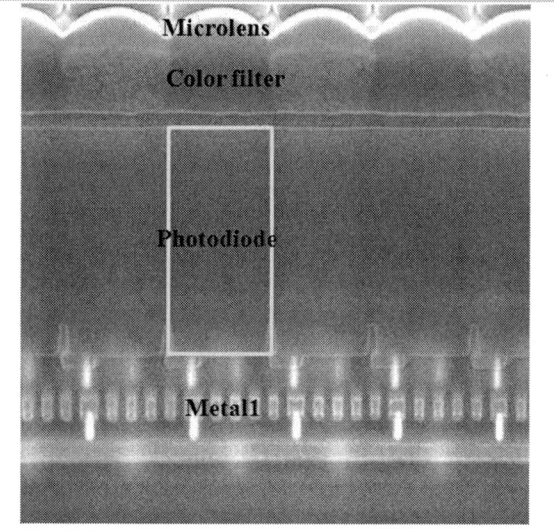

Figure 23.9.3: Cross-sectional view.

Figure 23.9.4: Measured QE spectra.

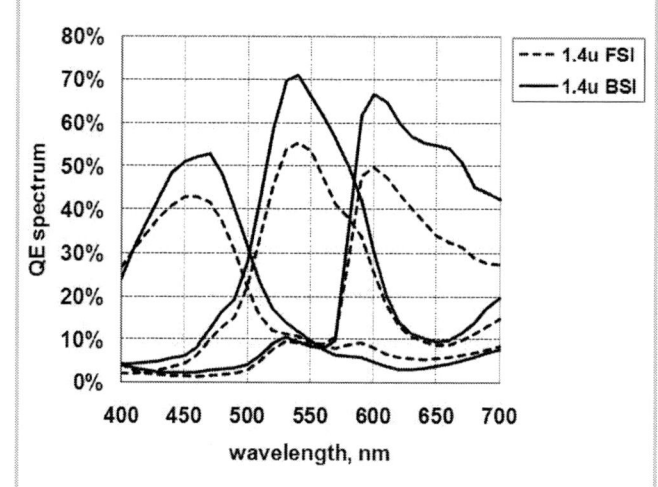

Item	Feature
Resolution	14.6MP
Pixel size	1.4um X1.4um, Backside Illumination
Optical Format	1/2.33-inch
Shutter type	Electronics rolling shutter /Global reset
Supported Frame Rate	14.6MP 30fps , 6.72MP 60fps
Full HD movie	1080p & 720p 60fps
Output format	10/12bit , SLVDS 12ch and MIPI 4ch
Supply Voltage	Analog : 2.8V
	Digital : 1.2V
	I/O voltage : 1.8V~2.8V
Saturation signal	5700e-
Conversion gain	110uV/e-
Sensitivity	5320e-·lux*sec @3200K light source
Illumination for SNR=10	8 lux @3200K light source, F=2.8
Dark current	7e-/sec (55℃)
Image lag	No image lag
Dark random noise	1.4e- rms

Figure 23.9.5: Sensor performance.

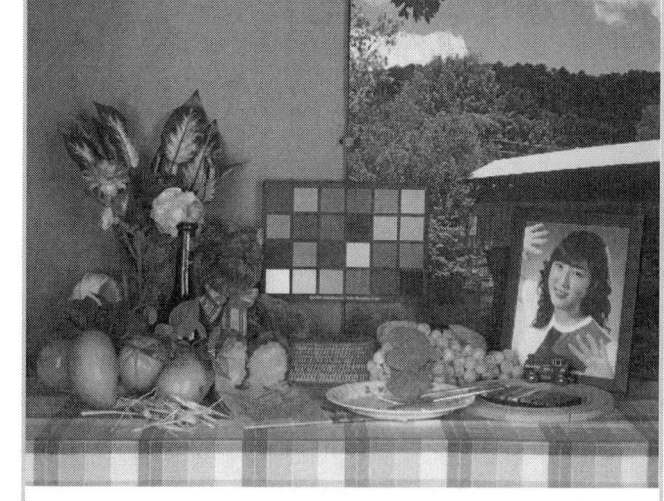

Figure 23.9.6: Reproduced image (F2.8, D65, 33ms, Gain is 1×).

23

ISSCC 2011 PAPER CONTINUATIONS

Figure 23.9.7: The chip micrograph.

ISSCC 2011 / SESSION 23 / IMAGE SENSORS / 23.10

23.10 An APS-C Format 14b Digital CMOS Image Sensor with a Dynamic Response Pixel

Dan Pates[1], Jeong-Ho Lyu[1], Shinji Osawa[2], Isao Takayanagi[2], Toshiaki Sato[2], Tim Bales[3], Katsuyuki Kawamura[2], Eduard Pages[2], Shinichiro Matsuo[2], Tetsuji Kawaguchi[2], Tadashi Sugiki[2], Norio Yoshimura[2], Junichi Nakamura[2], John Ladd[1], Zhiping Yin[1], Russell Iimura[1], Xiaofeng Fan[1], Scott Johnson[1], Aditya Rayankula[1], Rick Mauritzson[1], Gennadiy Agranov[1]

[1]Aptina, San Jose, CA,
[2]Aptina, Tokyo, Japan,
[3]Aptina, Bracknell, United Kingdom

CMOS image sensors are now used in a variety of camera applications, such as mobile phones, camcorders, and digital still cameras (DSCs). In particular, most digital single-lens-reflex (DSLR) cameras and so called "mirror-less" cameras that require a large-format sensor also employ CMOS image sensors due to their lower power consumption and higher readout speed compared to their CCD counterparts.

This paper presents an APS-C optical format (image array diagonal is 28.3 mm), 14b digital output CMOS image sensor for DSLR applications. Although its basic architecture and operation are the same as those of a 1.25-inch 8.9M-pixel CMOS image sensor for an ultra-high definition television application [1], several improvements and modifications have been made to optimize it for DSLR applications. These include: higher pixel count of 16.2Mpixels in the APS-C format, a use of a dynamic response pixel (DR-Pix), built on a pseudo-n-substrate, several operating modes that are required or desirable to the DSLR application, and improvements to the analog circuit performance to reduce noise.

The sensor architecture and operating sequence is shown in Fig. 23.10.1. The pixel array consists of 4928 (H) × 3280 (V) dynamic response pixels with a two-way common-element pixel architecture (CEPA) [2]. The signals from each pixel are processed by a column signal chain that is laid out in double the pixel pitch, resulting in a top and bottom readout scheme. The column signal chain includes a column amplifier, a 14b successive approximation register (SAR) ADC, a digital signal processor for digital CDS (correlated double sampling), and a line buffer memory [1]. This top and bottom readout architecture offers a symmetric layout that allows the center of the pixel array to be aligned with the center of the die, and reduces the potential for localized heat generation. The digitized pixel signals are fed to 1 of 8 data paths, where a digital pedestal control is carried out. Finally the pixel signals are read out through an 8-lane low-voltage differential signaling transmitter (4 lanes on top for G_B/G_R channels and 4 lanes on bottom for B/R channels).

In a full-resolution still-image capture mode, the sensor receives an external "TRIGGER" signal that controls the start of the exposure (charge integration) and the signal readout. The rising edge of the TRIGGER sets the sensor to start the global reset for the entire array. After a pre-determined period when the global reset is completed, a mechanical shutter is opened, and charge integration begins. After the falling edge of TRIGGER is detected, exposure time is completed, the mechanical shutter is closed, and the sensor starts the readout sequence. The pixel signal readout is performed on a row-by-row basis as shown in the Fig. 23.10.1. There are several other operation modes, which include a 1080/30p HD video mode with 2×2 binning, a pre-view mode at >60fps, and a pre-flash mode at >120fps. Both pre-view and pre-flash modes cover full field of view.

The pixel structure is shown on Fig. 23.10.2. The starting substrate is p-type silicon and the sensor is fabricated at the top of a p-epi region, with a n-epi region embedded between the p-sub and p-epi to reduce the electrical cross-talk. The n-epi region prevents the diffusion of electrons from the adjacent pixels and blocks any thermally or defect generated electrons from the p-type substrate, thus achieving low dark current and cross-talk. The sensor utilizes 2-way row-shared CEPA [2] to increase the fill factor and PD area. All pixel transistors have ring shaped gates, which allow elimination of the STI (shallow trench isolation)

within the pixel array, thus eliminating a major source of hot pixels and dark current. Also, 1/f and RTS noise of the source follower is reduced due to the larger gate width and fewer defect sites associated with the STI elimination. The TX gate also has a ring shape with the floating diffusion (FD) surrounded by this TX poly gate. MOS capacitor structure is used to electrically isolate PD and transistor regions.

Figure 23.10.3 presents the quantum efficiency (QE) spectral characteristic. The pixel achieves high QE (62%) and low crosstalk. It also provides very good angular performance when working with interchangeable lenses with a wide range of F-numbers.

One very important feature of the pixel is its ability to significantly boost low-light performance and achieve high ISO Speed via its Dynamic Response Pixel (DR-Pix) architecture. DR-Pix effectively switches FD conversion gain to higher value at low-light conditions thus increasing the sensitivity of the FD node, and reducing the input-referred read noise. Differently from [3] and [4] the DR-Pix is realized by adding a transistor to the standard 4T pixel, which is used as a switch to toggle the connection of a physical capacitor to the FD node. When the switch is closed, the capacitor is connected to the FD, which results in a low conversion gain (LCG) value, thus maintaining the large full-well capacity and high max SNR. When the switch is open, then the capacitor is disconnected from the FD, which results in high conversion gain (HCG) and smaller full-well capacity. Figure 23.10.4 illustrates two modes of pixel operation showing signal at the FD as a function of exposure for both low and high conversion gains.

DR-Pix allows for much improved sensitivity when shooting at low-light / high-ISO speed conditions. The remarkable benefit is that the high-sensitivity mode is added to the pixel, while the high-light performance is un-compromised, allowing pixel capacity of more than 50ke. With the 3-to-1 conversion gain ratio, and the maximum 8× column gain, the effective gain factor becomes 24×, which realizes the input-referred noise floor of 2.2 e. In combination with high QE, small read noise enables high ISO speed with high SNR.

Figure 23.10.5 presents images from a fabricated sensor at low ISO-100 and high ISO-12,800 speed. Major sensor characteristics are summarized on Fig. 23.10.6. The chip micrograph is shown on Fig. 23.10.7.

In summary, a 16M-pixel image sensor having 4.78×4.78um² pixel size with APS-C optical format is successfully fabricated and characterized. Pixel uses original design with ring-shaped gate transistors, no STI, and dual conversion gain. Sensor features high QE of 62%, pixel capacity of 50ke, and readout noise of 2.2e. Column FPN is measured to 0.11e$_{rms}$ at HCG and column gain of 8×, which is 20× lower than the noise floor. The measured dark current level is 17e-/s at 60°C die temperature.

Acknowledgements:
The authors would like to thank many people and groups at Aptina and Micron Technology Inc. for their support and contribution to this investigation.

References:
[1] S. Matsuo, et al., "8.9-megapixel video image sensor with 14-b column parallel SA-ADC", *IEEE T-ED*, vol. 56, no. 11, pp. 2380-2389, Nov. 2009.
[2] G. Agranov, et al., "Optical-electrical characteristics of small, sub-4μm and sub-3μm pixels for modern CMOS image sensors," IEEE CCD and AIS Workshop Proceeding, pp. 206-209, 2005.
[3] S. Adachi, et al., "A 200-μV/e CMOS image sensor with 100-ke full well capacity," *IEEE J. Solid-State Circuits*, vol. 43, no. 4, pp. 823–830, Apr. 2008.
[4] S. Sugawa, N. Akahane, S. Adachi, and K. Mori, "A 100dB Dynamic Range CMOS Image Sensor Using a Lateral Overflow Integration Capacitor," *ISSCC Dig. Tech. Papers*, pp.352-353, Feb. 2005.

ISSCC 2011 / February 23, 2011 / 11:45 AM

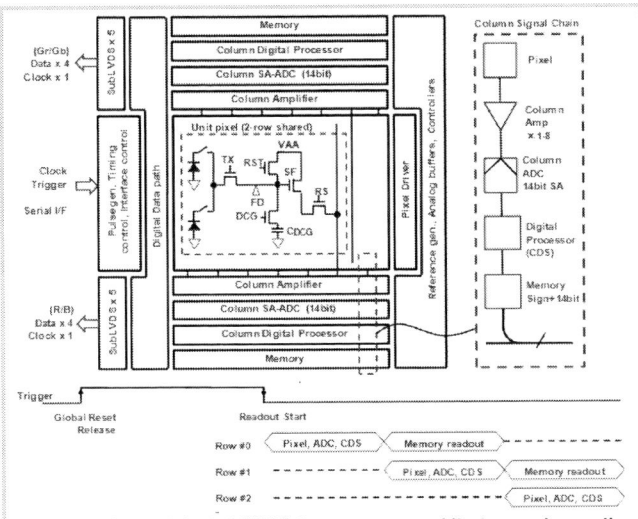

Figure 23.10.1: Fabricated CMOS image sensor architecture and operating sequence.

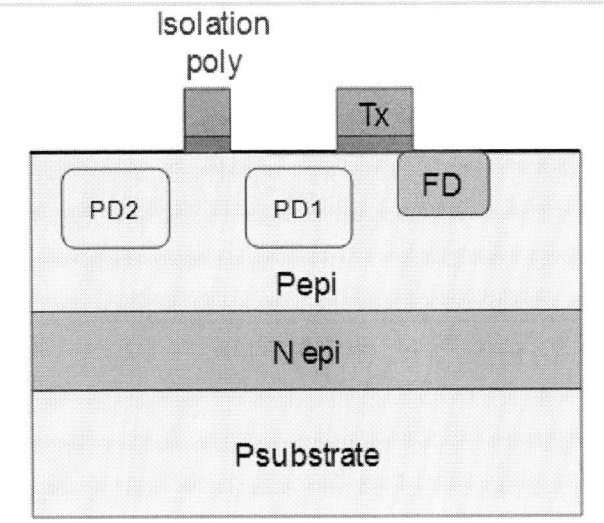

Figure 23.10.2: Pixel cross-section structure.

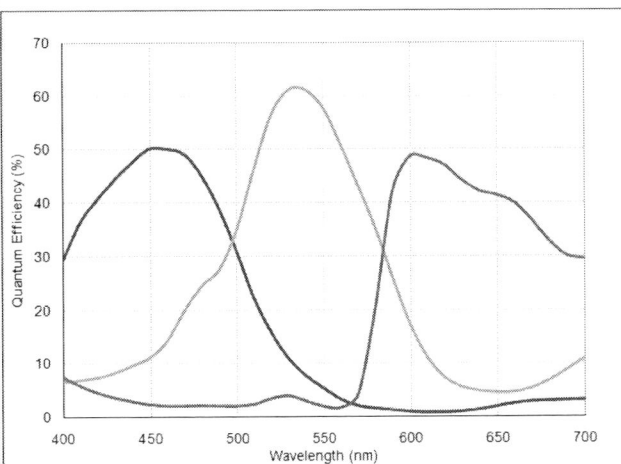

Figure 23.10.3: Quantum efficiency spectrum characteristic.

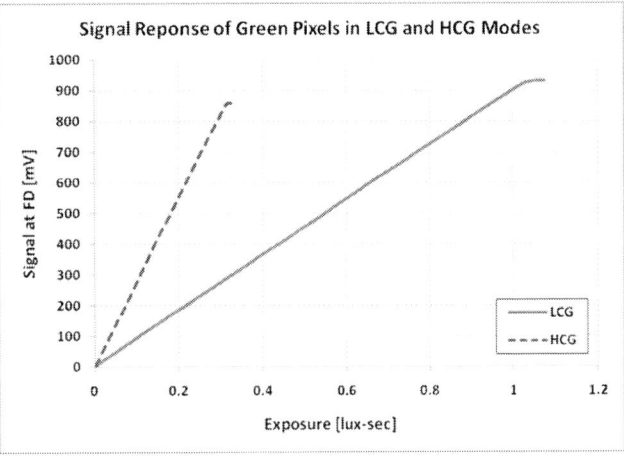

Figure 23.10.4: Photo-conversion characteristic of green pixel for low and high conversion gain modes.

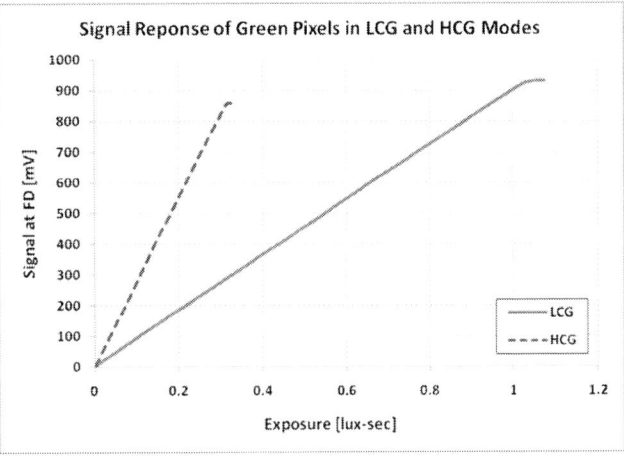

Figure 23.10.5: Images taken with manufactured sensor at ISO-100 (top and bottom-left) and ISO-12,800 (bottom-right).

Item	Specification	Item	Characteristics
Process technology	130nm, 2P3M CMOS	Conversion gain (H, L)	LCG: 18 uV/e HCG: 54 uV/e
Number of effective pixels	4928 (H) × 3280 (V) (16.2 M-pixels)	Full well capacity (ke⁻)	LCG: 50 HCG: 16
Pixel size	4.78 um	Responsivity (ke⁻/lux-s)	49.5
Optical format (Aspect ratio)	APS-C (28.26mm diagonal) 3:2	ADC resolution	14b
Supply voltages	Analog : 3.3V Digital : 1.8V I/F : 2.8V	Column analog gain	1x, 2x, 4x, 8x
Input clock frequency	46.406 MHz	Noise floor (e⁻)	16.0 (LCG, 1x) 2.2 (HCG, 8x)
Frame rate	10.48 fps (Full resolution)	Column FPN (e⁻rms)	0.11 (HCG, 8x)
Data rate	2.592 Gbps	Column PRNU (%)	± 0.2
Operation modes	Full resolution still HD video PreView Pre-flash	Dark current (e⁻/s)	17 at 60°C
Output interface	8 lane Sub-LVDS	Power consumption (full resolution @ 10.48FPS)	840 mV
Package	116-pin CLCC		

Figure 23.10.6: Summary table of sensor characteristics.

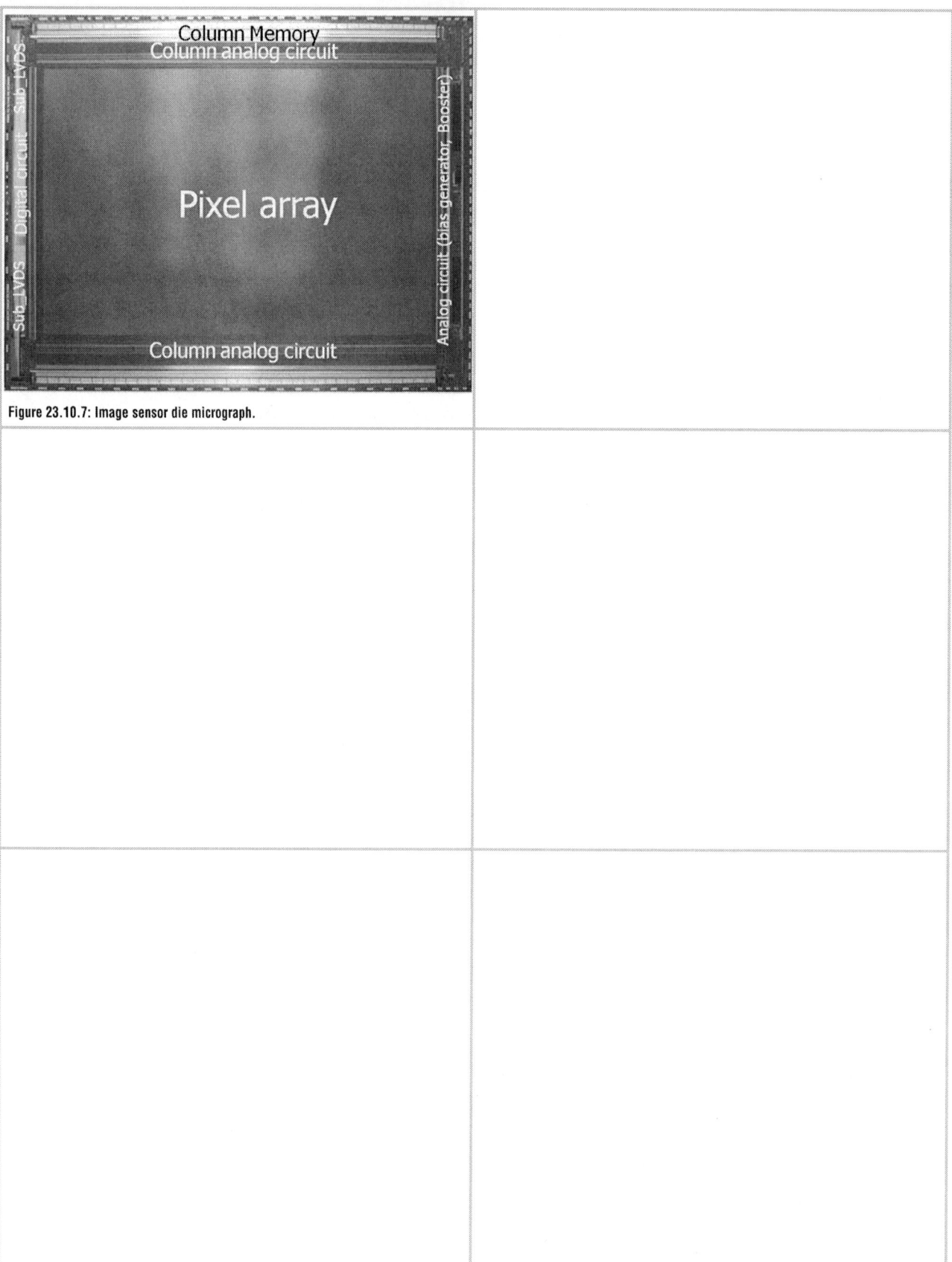

Figure 23.10.7: Image sensor die micrograph.

ISSCC 2011 / SESSION 23 / IMAGE SENSORS / 23.11

23.11 A 17.7Mpixel 120fps CMOS Image Sensor with 34.8Gb/s Readout

Takayuki Toyama[1], Koji Mishina[1], Hiroyuki Tsuchiya[1], Tatsuya Ichikawa[1], Hiroyuki Iwaki[1], Yuji Gendai[1], Hirotaka Murakami[1], Kenichi Takamiya[2], Hiroshi Shiroshita[1], Yoshinori Muramatsu[1], Toshihiro Furusawa[1]

[1]Sony, Kanagawa, Japan,
[2]Sony LSI Design, Kanagawa, Japan

Recently, the demands to achieve both high-speed and high-quality imaging – including high A/D resolution – have increased. The new target specification is 60fps Ultra-High-Definition with 12b resolution. This imaging requires 24Gb/s, while reported CMOS image sensors have reached up to 6.5Gb/s [1-4]. This paper presents a 34.8Gb/s CMOS image sensor with high image quality, which realizes 17.7M pixels at 120fps with 12b resolution and a dynamic range of over 75dB (one of the highest image qualities, compared with data from [2]). Generally there exists a trade-off between high-speed imaging and high image quality. This is because high-speed imaging brings high power with high thermal noise, which also degrades the signal quality, and because fast data transfer may require more output pins, bringing more interference.

We address the issue using a single-slope ADC (SS-ADC), which is a very popular architecture for modern CMOS image sensors because the analog circuit area per column is smaller and scaling of fabrication processes contributes to diminish the digital circuit area. The first solution uses hybrid column counters for the SS-ADC that enable very fast A/D conversion with the least power consumption. The second solution is a Scalable Low Voltage Signaling interface with Embedded Clock (SLVS-EC) that achieves 2.376Gb/s/channel.

Figure 23.11.1 shows the overall architecture of the CMOS image sensor. This sensor consists of a pixel array, column-parallel ADCs with upper and lower side allocation, control logic circuits with left and right side allocation, and SLVS-EC composed of 4 channels at each corner. These 4 SLVS-EC circuits receive the sensor input clock individually. The effective pixel array consists of 8192(H) × 2160(V) 4T pixels placed in a zigzag manner; the pixel unit cell size is 4.2μm(H) × 4.2μm(V), the pixel pitch is 5.94μm(H) × 5.94μm(V), the column circuit pitch is 2.97μm. The size of the imaging region is 24.3mm(H) × 12.8mm(V) and is compatible with a 27.5mm optical format. In the sensor, vertical pixel readout lines are divided into 2 sides – up and down image area – to realize high-speed analog readout so that a single row-signal of both sides is read out and converted into digital data simultaneously. Then, ADC outputs are transferred to the left or right SLVS-EC circuits at the same time through sense-amplifier circuits.

Since sensor readout data are output from the SLVS-EC circuits at 4 corners simultaneously, they are forwarded into frame memory out of the sensor, if raster scanning is needed for data processing.

Figure 23.11.2 shows the block diagram of this SS-ADC. The SS-ADC is composed of comparators, hybrid column counters including data memories, and single-slope generators with regulation circuits for an imaging offset between up and down image area. The imaging offset regulation has a 16b full-scale resolution, which is sufficient to make the border invisible between the divided imaging area. In SS-ADCs, it is generally thought to be difficult to get both high speed and high A/D resolution [2,4,5], because the A/D resolution is restricted by the number of clock cycles. Concretely speaking, SS-ADCs with high-speed column-counters have 2 big problems: (1) power consumption of column-counters and (2) count-clock signal quality. First, power of column-counters is consumed more in proportion to the number of columns and/or ADC count-clock frequency increase. This degrades image quality due to a chip thermal rise and/or counting errors by the IR-drop of the count-clock line. This results from the fact that lower-bit column counters that consume more power exist in every single column. Secondly, quality of fast count-clock signals is difficult to maintain especially at the end of column-counters, because the count-clock signal from the PLL is delivered to every single column-counter and degrades gradually. In the

other way to achieve high speed, a multiple-ramp SS-ADC has been reported [5], but this architecture is difficult to implement and it consumes more area per column and more power, which reduces advantage of the SS-ADC. Here, we realize a SS-ADC operating up to 2.376GHz by re-structuring column-counters (hybrid column counters), which consist of 2 blocks as shown in Fig. 23.11.2: (1) one is a lower 5b counter in every 248 columns, and (2) the other is an upper 9b counter in every single column, connected with lower bit counter outputs. This re-structured counter provides a solution to the 2 big problems as mentioned before, by decreasing both power consumption of column-counters and degradation of clock signal quality. The pixel signal can be converted to the aggregate 14b digital code by the column-parallel slope ADCs.

Figure 23.11.3 shows its timing diagram of 1 horizontal scanning period. A dual row-signal readout expands one horizontal scanning period to double of the conventional readout, and A/D conversion and horizontal data transfer are done simultaneously. It occurs at most once in one scanning vertical period that the offset regulation for upper and lower image areas is applied to single-slope generators through a 3-wire serial interface from the digital processing unit (DSP) out of the sensor. The horizontal scanning time is set to 7.4μs, allowing 120fps for 17.7Mpixel with a 12b ADC, where hybrid column counters count at 2.376GHz and the SLVS-EC circuit operates up to 2.376Gb/s/channel at an input clock of 148.5MHz. Chip specifications are summarized in Fig. 23.11.4. This sensor is fabricated in 90nm 1P4M CMOS sensor. The supply voltages are 2.9V for pixel, 2.7V for the analog circuits, 1.2V for the digital circuits, and 0.4V for the SLVS-EC circuits. The sensor performance is effective 17.7Mpixel 120fps at 12b, and 60fps at 14b resolution. Random noise of $2.75e^-_{rms}$ and a saturation signal of $21,000e^-$ are achieved at 120fps at 12b resolution, so the dynamic range is 77.6dB. A sensitivity of 40ke$^-$/lx-s is achieved. The measured power consumption is typically 3W.

Figure 23.11.5 shows the signal quality of SLVS-EC. The differential voltage and the common-mode voltage of SLVS-EC signals are 400mV and 200mV, respectively. The other advantage of this SLVS-EC is that it is unnecessary for channel-skew regulation of transmission lines and it can translate data at half the power of a conventional sub-LVDS, totalling around 200mW at 34.8Gb/s. The total jitter 42.4ps (around 0.1UI) at 2.376Gb/s through a wire length of 300mm is sufficient for 10^{-15} bit error rate, which is necessary for highly reliable image data transmission.

Figure 23.11.6 shows a sample image taken at 34.8Gb/s operation, which is 8192(H) × 4320(V) after pixel interpolation in the vertical direction using software treatment. The other software treatments than the above pixel interpolation have not adopted for this image. In addition, image distortion due to the rolling shutter is negligible at 120fps compared to that of 48fps.

A 17.7Mpixel CMOS image sensor realizes 120fps at 12b ADC, and maximum data rate 34.8Gb/s. Dual row-signal readout, a SLVS with Embedded Clock interfaces (2.376Gb/s × 16ch) and fast column counters (2.376GHz counts) are reported. The dynamic range and the power consumption are 77.6dB and 3W, respectively at 34.8Gb/s. The chip micrograph is shown in Fig. 23.11.7.

References:
[1] Y. Nitta, et al., "High-Speed Digital Double Sampling with Analog CDS on Column Parallel ADC Architecture for Low-Noise Active Pixel Sensor," *ISSCC Dig. Tech. Papers*, pp. 500-501, Feb. 2006.
[2] S. Kawahito, "Column readout circuit design for high-speed low-noise imaging" *ISSCC Forum: High-Speed Image-Sensor Technologies*, Feb. 2010.
[3] I. Takayanagi, et al., "A 1.1/4-inch 8.3M pixel digital-output CMOS APS for UDTV application, " *ISSCC Dig. Tech. Papers*, pp. 216-217, Feb. 2003.
[4] Y. Chae, et al., "A 2.1Mpixel 120frame/s CMOS Image Sensor with Column-Parallel ΔΣADC Architecture," *ISSCC Dig. Tech. Papers*, pp. 394-395, Feb. 2010.
[5] M.F. Snoeij, et al., "A CMOS Image Sensor with a Column-Level Multiple-Ramp Single-Slope ADC," *ISSCC Dig. Tech. Papers*, pp. 506-507, Feb. 2007.

978-1-61284-303-2/11 $26.00 © 2011 IEEE

ISSCC 2011 / February 23, 2011 / 12:00 AM

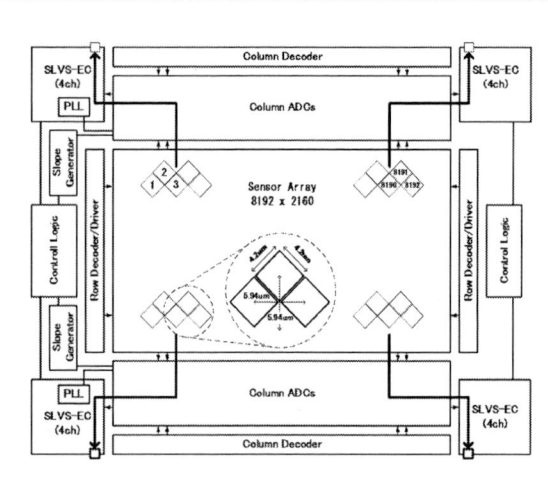

Figure 23.11.1: Block diagram.

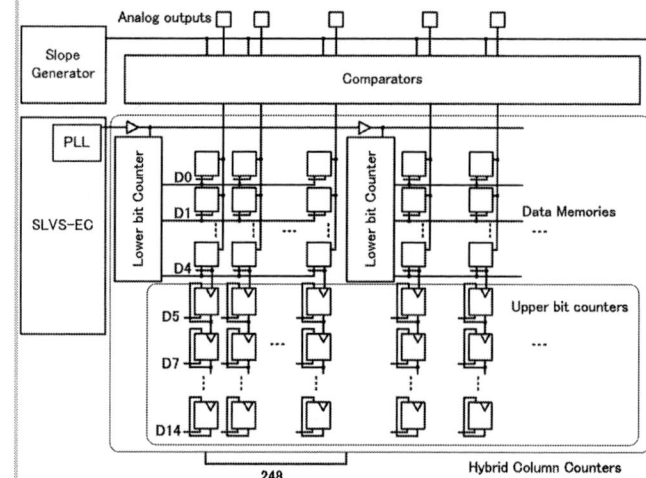

Figure 23.11.2: Block diagram of this SS-ADC.

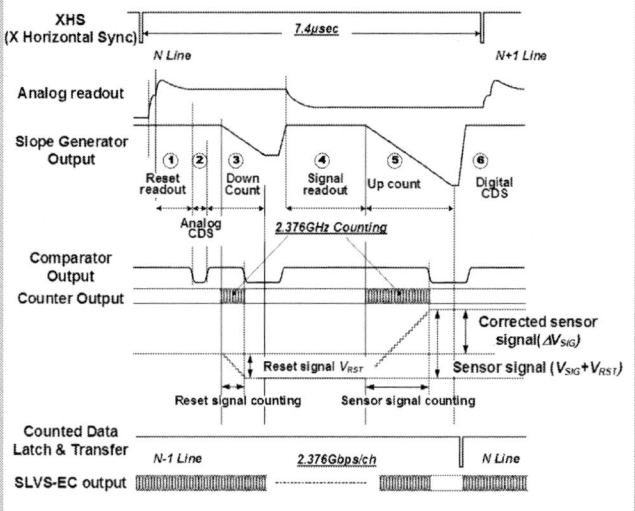

Figure 23.11.3: Timing diagram for one horizontal scanning period.

Fabrication Process	90nm 1P 4M
Supply Voltage	2.9V / 2.7V / 1.2V / 0.4V
Number of effective pixels	8192 (H) x 2160 (V)
Pixel size	4.2 μm (H) x 4.2 μm (V)
Aperture ratio	51% w/o on-chip microlens
Max. data rate	34.8Gbits/s w/ data clocks
Max. frame rate	120 frames/s(fps) at 12b ADC
Power consumption	3.0W at 120fps
Saturation signal	21,000e⁻ at 60 C
Sensitivity	40,000e⁻/lx-s At 3200K light source with IR cut filter of 650nm cut-off
Image lag	Below measurement threshold
RMS random noise	2.75e⁻rms at 120fps (AnalogGain:21dB)
RMS vertical FPN	0.033e⁻rms at 120fps w/o additional correction circuit
Dynamic range	77.6dB at 120fps

Figure 23.11.4: Chip characteristics.

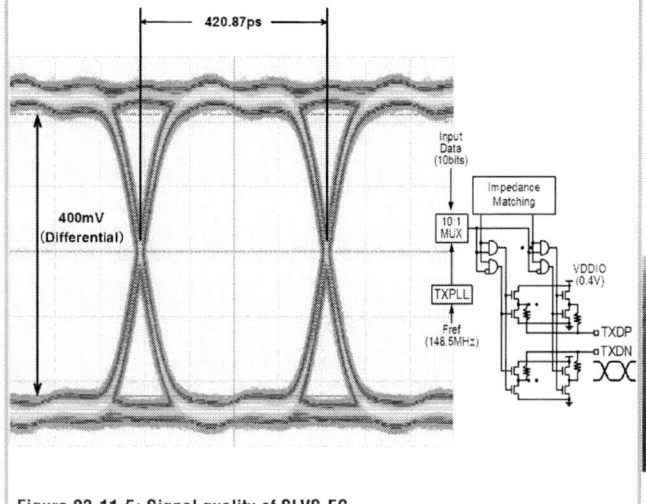

Figure 23.11.5: Signal quality of SLVS-EC.

Figure 23.11.6: Sample images.

ISSCC 2011 PAPER CONTINUATIONS

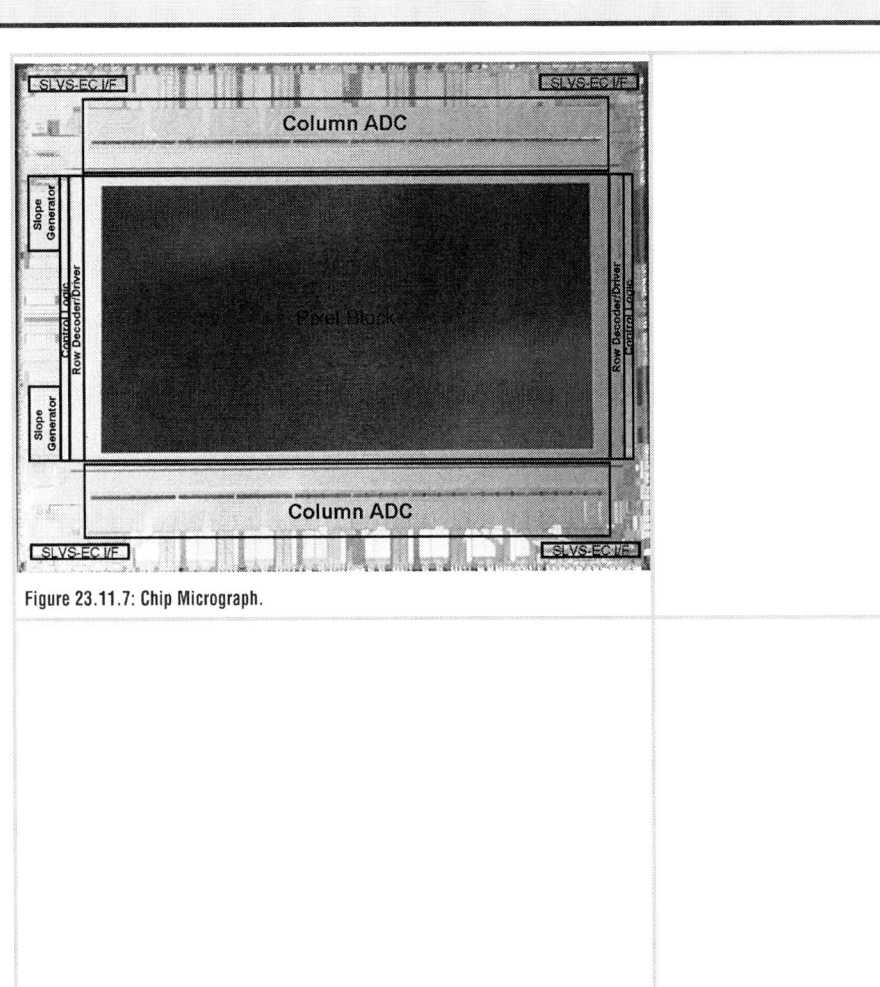

Figure 23.11.7: Chip Micrograph.

ISSCC 2011 / SESSION 24 / TRANSMITTER BLOCKS / OVERVIEW

Session 24 Overview / RF

Transmitter Blocks

Session Chair: *Francesco Svelto, Università degli Studi di Pavia, Italy*

Session Co-Chair: *Shoji Otaka, Toshiba, Kawasaki, Japan*

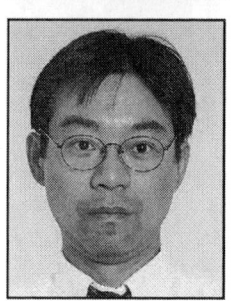

Over the recent past, we have witnessed an impressive progress towards the integration of analog/RF circuit blocks together with digital circuits for wireless communication SoCs. Digitally assisted calibration of RF parameters is nowadays extensively applied to ensure performance over process, voltage and temperature variations. Among transceiver circuit blocks, power amplifiers have been difficult to integrate and CMOS has been limited to applications with constant envelope modulation such as GSM. The successful implementation of commercially available switching PAs has further motivated the investigation of PA solutions with a robust linearity in the presence of amplitude-modulated signals.

This session presents advances in state-of-the-art design. It opens with a wideband transmitter with built-in auto-calibration, then presents three linear RF PAs and concludes with a highly efficient 60GHz CMOS realization.

Paper 24.1 (Broadcom) introduces a TX architecture for calibrating OFDM transmitter non-idealities leveraging RX resident hardware. Transmitter image, local oscillator feed-through (LOFT) and output power are calibrated by sub-sampling the transmitter spectrum. The TX attains ACPR>55dBc and IM3<-64dBc up to 200MHz baseband frequency. The digitally assisted system assures an IR of 55dBc, a 40dBc LOFT and a ±0.6dB gain accuracy over PVT up to 1.6GHz LO frequency.

A Class-AB PA with shielded concentric transformer in 32nm CMOS is presented in Paper 24.2 (Intel). The device is flip-chip packaged and achieves 28dBm P_{sat} with peak PAE of 31.9%, and P_{1dB} of 26.5dBm. Average power is 21dBm with 16% PAE meeting -25dB EVM for OFDM 64-QAM without digital pre-distortion linearization.

Paper 24.3 (University of Washington) presents a 90nm PA achieving EER/Polar operation via switched-capacitor techniques. The PA delivers a peak output power of 25.2dBm with 55.2% PAE. Average power and PAE are 17.7dBm and 32.1%, respectively. An EVM of 2.9% with a 64-QAM OFDM-modulated signal is measured in the 2.4GHz band.

Paper 24.4 (Samsung Electro-Mechanics) reports the last RF PA in the session, which is a dual mode quad-band CMOS PA with integrated passive device, assembled in a 5x5mm² QFN package. The linear PA for EDGE mode achieves an average output power of 28.5dBm, a PAE of 22%, an ACPR of -57dBc and an EVM_{rms} of 1.6% for GSM/EGSM bands. The presented PA satisfies requirements for power Class-E2 operation. As a switching amplifier, the PA shows an output power of 34.5dBm with PAE of 55% for GSM application.

An effort toward compact, low area mm-Wave PAs still able to deliver output powers in the tens to hundreds of mW is underway. Paper 24.5 (University of Berkeley) presents a 65nm digital CMOS PA with 1V supply, using an efficient 4-input transformer power combiner. The PA achieves 18.6dBm P_{sat}, 15dBm P_{1dB} and 15.1% PAE while using only 0.28mm² silicon area. The presented solution is 2x smaller in size and achieves highest PAE at peak power of 18.6dBm over prior art. The PA maintains 17.8dBm P_{sat} and 12.6% PAE over the IEEE band from 58GHz to 64GHz.

The techniques and implementation approaches presented in this session pave the way to advances in wireless transmitters, ranging from DSP-assisted calibration, to robust and linear RF PAs, to high-output-power mm-Wave PAs.

978-1-61284-303-2/11 $26.00 © 2011 IEEE

ISSCC 2011 / February 23, 2011 / 1:30 PM

Presenters:

24.1 A 40nm Wideband Direct-Conversion Transmitter with Sub-Sampling-Based Output Power, LO Feedthrough and I/Q Imbalance Calibration

E. Lopelli,
Broadcom, Bunnik, The Netherlands

1:30 PM

24.2 A Flip-Chip-Packaged 1.8V 28dBm Class-AB Power Amplifier with Shielded Concentric Transformers in 32nm SoC CMOS

Y. Tan,
Intel, Hillsboro, OR

2:00 PM

24.3 A Switched-Capacitor Power Amplifier for EER/Polar Transmitters

S-M. Yoo,
University of Washington, Seattle, WA

2:30 PM

24.4 An EDGE/GSM Quad-Band CMOS Power Amplifier

W. Kim,
Samsung Electro-Mechanics, Atlanta, GA

3:15 PM

24.5 A Compact 1V 18.6dBm 60GHz Power Amplifier in 65nm CMOS

J. Chen,
University of California, Berkeley, CA

3:45 PM

24

978-1-61284-303-2/11 $26.00 © 2011 IEEE

ISSCC 2011 / SESSION 24 / TRANSMITTER BLOCKS / 24.1

24.1 A 40nm Wideband Direct-Conversion Transmitter with Sub-Sampling-Based Output Power, LO Feedthrough and I/Q Imbalance Calibration

Emanuele Lopelli, Silvian Spiridon, Johan van der Tang

Broadcom, Bunnik, The Netherlands

In the last decade the amount of digital data generated in connection with digital devices such as cameras, media players and high-definition TVs has seen a significant growth. This requires tuners for home networking such as MoCA with increasingly large bandwidth. Though advanced CMOS technology allows for the design of high-speed circuits and systems that can meet the need for more bandwidth, 40nm feature sizes and beyond introduce new challenges in analog circuit design [1]. Moreover, dependence on environmental conditions of device spread and matching performance with parasitic coupling can drastically reduce the overall system performance.

The integration of analog/RF circuits with the digital part in an SoC enables the use of inexpensive DSP power to calibrate non-idealities. Digitally-assisted RF allows for more power- and area-efficient systems that achieve good performance over process, temperature and supply (PVT) variations. This paper presents a calibration scheme that uses the available DSP power to perform the transmitter image, local oscillator feed-through (LOFT) and output power calibration by sub-sampling the transmitter spectrum. The system achieves an IR of 55dBc, a 40dBc LOFT and a ±0.6dB gain accuracy up to a 1.6GHz LO frequency.

OFDM signals require an accurate control of LOFT, image level and delivered output power to meet the required performance. Though LOFT does not harm signal reception, a high LOFT level results in transmission of redundant energy and it reduces the receiver's dynamic range. A high image level due to I/Q imbalances affects the reception of the signal, degrading the receiver SNR. Finally, an accurate control of the transmitter output power removes the need for sophisticated power estimation procedures, reducing the system complexity.

Transmitter LOFT or I/Q calibration can be performed at factory level [2] at the expense of a higher production cost or by dedicated mixed-signal circuitry [3,4] at the expense of extra chip area. The proposed calibration scheme reuses part of the receiver (the ADC) to perform all the calibration steps, minimizing chip area and production cost.

Figure 24.1.1 shows the transmitter block diagram including the calibration path. The transmitter signal path consists of an I/Q baseband filter with a 1dB bandwidth tunable between 22MHz and 200MHz, a highly linear single-sideband mixer and a PA driver (PAD). The local oscillator (LO), used to up-convert the baseband signal, covers the frequencies between 50MHz and 1.6GHz.

During calibration, either the upconversion mixer output (I/Q and LOFT calibration) or the PAD output (TX gain calibration) are looped-back to the RX ADC. A variable bandwidth buffer (VBB) is used to drive the RX ADC input and it is the only extra block required in the proposed calibration. The 12b ADC is clocked by an integer-N PLL driven by a 50MHz crystal-based reference signal. In the current prototype, CML buffers are used to drive the ADC outputs off-chip for data analysis. During calibration, a test tone (f_{BB}) is generated in baseband and upconverted around the LO frequency by the mixer. The spectrum at the mixer output consists of the wanted tone, an image tone and a tone at the LO frequency due to LOFT (see Fig. 24.1.1).

The three tones are detected by performing a 2048-points FFT. A simple algorithm [2] minimizes the image tone by varying the phase and the gain difference between I and Q baseband signals or the LOFT by adjusting the output currents of the IDACs in Fig. 24.1.1. Finally, the TX gain calibration is performed by varying the PAD gain.

The mixer is based on a current-switching topology. The wanted and the image tones are replicated around each odd harmonic of the LO. The tones above the Nyquist frequency fold within the Nyquist band after sub-sampling. This can cause a destructive folding if, for example, a copy of the wanted tone around one

of the LO harmonics folds on top of the image tone. Indeed, these tones can be up to 45dB above the image tone making the detection impossible.

To avoid this situation the ADC sampling frequency (f_s) must be properly chosen. The choice of f_{BB} is also crucial in order to relax the jitter specification of the ADC PLL. Indeed, the phase noise around large folding tones can cause a severe SNR reduction. This in turn lowers the maximum achievable rejection of the image or the LOFT. On the other hand, it might be very difficult to find a suitable f_s and f_{BB} combination that guarantees no destructive folding and reasonable PLL jitter specifications over the whole LO range.

To alleviate this problem, it is necessary to reduce the number of harmonics that can cause a destructive folding or a large SNR degradation. Therefore, a variable bandwidth buffer (VBB) is used before the signal is looped-back to the RX ADC. Considering that f_s must be a multiple of the PLL comparison frequency, the problem reduces to finding the optimal f_s and f_{BB} combination within a finite space of possibilities. This problem can be solved using a computer routine to map all the aliasing tones.

The schematic block diagram of the VBB is shown in Fig. 24.1.2. It consists of binary-weighted source follower slices that can be connected or disconnected from the signal path. The bandwidth is, therefore, changed by varying the g_m of the source follower. To guarantee that the VBB gain spread over PVT does not harm the gain-calibration accuracy, a self-biasing structure is used. Reducing the VBB bandwidth by reducing its g_m harms both the linearity and the SNR. While linearity is not a concern for the calibration, SNR is important to minimize the averaging time for an accurate estimation. Therefore, the system has been designed to assure a >10dB SNR for the lowest bandwidth setting and for a 55dBc image rejection. Process variation can change the 3dB bandwidth of the VBB. The system has been designed to avoid any destructive folding up to the 31st LO harmonic. In this way, the impact of bandwidth variation due to PVT variations on the system performance is minimized.

The chip is fabricated in a 40nm CMOS process. Figure 24.1.3(a) shows the transmitter output spectrum at 1.6GHz LO frequency. The measured adjacent-channel power rejection (ACPR) is better than 55dBc. Figure 24.1.3(b) shows the two-tone test result versus the baseband frequency. The two input tones have a -20dBm power, and they are separated by 1MHz. The generated third-order intermodulation products are more than 64dBc down from the signal tones up to a 200MHz baseband frequency.

Figure 24.1.4 shows the spectrum of the transmitter and ADC output before and after calibration for f_{sample}=85MHz, f_{BB}=6.25MHz, an LO frequency of 1.6GHz and a 2048-points FFT. The post calibration IR is better than 55dBc while the LOFT is better than 40dBc, limited by the resolution of the IDACs.

In Fig. 24.1.5(a) an example calibration curve for IR is shown. The x axis is the phase unbalance applied at baseband while the y axis shows the image level in dBc. The curve C is derived by choosing a 55MHz sampling frequency that causes a destructive folding of the main tone around the 7th harmonic of the LO and consequently the failure of the calibration. The curves A and B refer to the transmitter and ADC output when the proper sampling frequency is chosen.

Figure 24.1.6 shows the transmitter output power after calibration over PVT. The accuracy is within ± 0.6dB over PVT for f_{LO}=1.6GHz. Figure 24.1.7 shows the chip micrograph.

References:

[1] A.-J. Annema, B. Nauta, R. van Langevelde and H. Tuinhout, "Analog Circuits in Ultra-deep-submicron CMOS", *IEEE Journal of Solid-State Circuits*, vol. 40, no.1, pp. 132-143, January 2005.

[2] P. Zhang et al., "A 5-GHz Direct Conversion CMOS Transceiver", *IEEE Journal of Solid-State Circuits*, vol.38, no.12, pp. 2232-2238, January 2003.

[3] C. Paul Lee, A. Behzad, D. Ojo, M. Kappers, S. Au, M.-A. Pan, K. Carter and S. Tian, "A Highly-linear Direct-conversion Transmit Mixer Transconductance Stage with Local Oscillator Feedthrough and I/Q imbalance Cancellation Scheme", *ISSCC Dig. Tech. Papers*, pp. 1450-1459, 2006.

[4] Y.-H.Hsieh, W.-Y. Hu, S.-M. Lin, C.-L. Chen, W.-K. Li, S.-J. Chen and D.-J. Chen, "An Auto-I/Q Calibrated CMOS Transceiver for 802.11g", *ISSCC Dig. Tech. Papers*, pp. 92-93, 2005.

ISSCC 2011 / February 23, 2011 / 1:30 PM

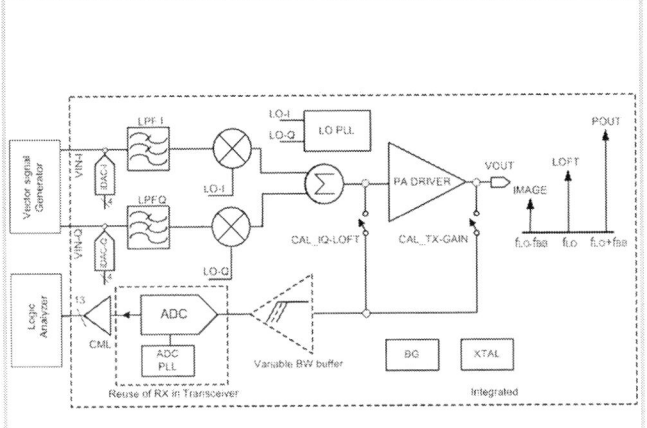

Figure 24.1.1: Transmitter block diagram with sub-sampling-based calibration path.

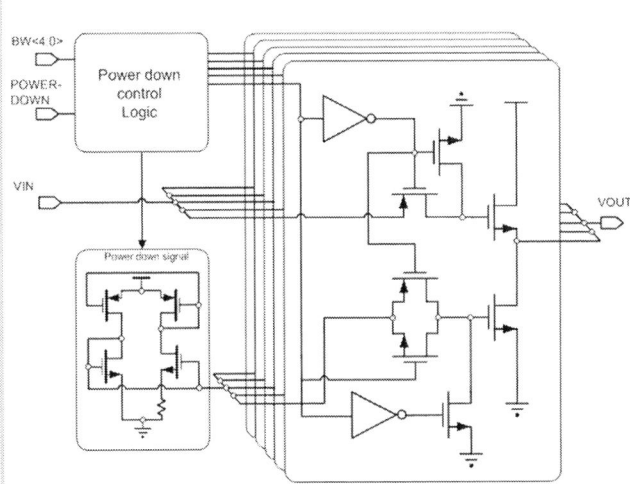

Figure 24.1.2: Circuit-level implementation of the VBB including biasing.

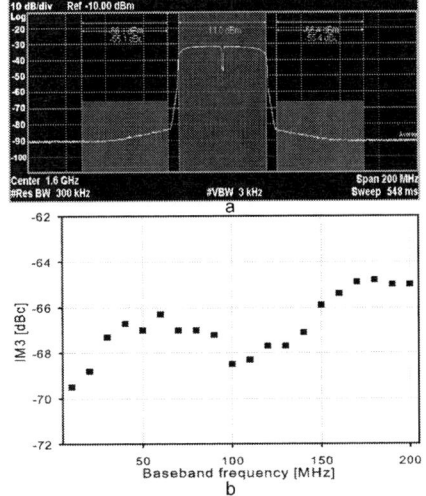

Figure 24.1.3: (a) Measured transmitter ACPR. (b) Transmitter IM3 versus baseband frequency for a 1.6GHz LO frequency.

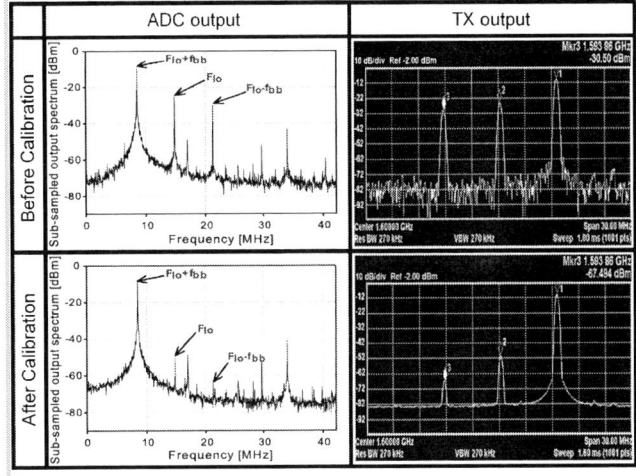

Figure 24.1.4: Measured transmitter output spectrum and ADC output spectrum before and after image and LOFT calibration (ADC output spectrum averaged 5 times).

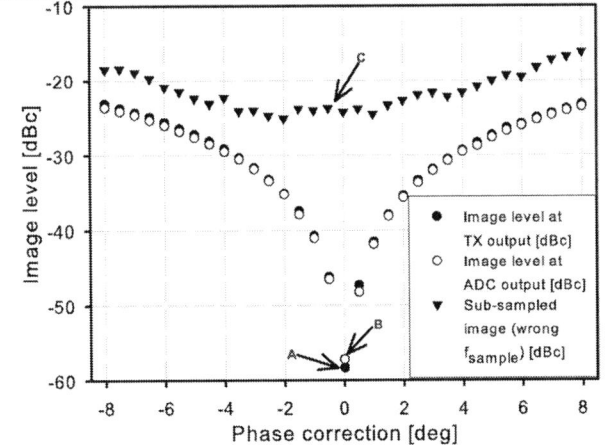

Figure 24.1.5: Image calibration curves for a 400 MHz LO frequency. Curve A is the TX image level. Curve B is the sub-sampled image level for f_{sample}=85MHz and f_{BB}=9MHz. Curve C is the sub-sampled image level for f_{sample}=55MHz and f_{BB}=5MHz.

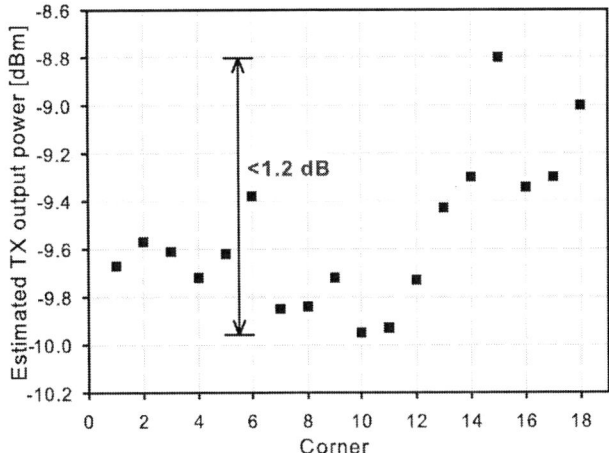

Figure 24.1.6: Gain calibration accuracy over PVT for f_{LO}=1.6GHz: testing temperature is hot, room, cold, supply voltage is 2.75V and 2.25V, process is typical, slow and fast.

24

978-1-61284-303-2/11 $26.00 © 2011 IEEE

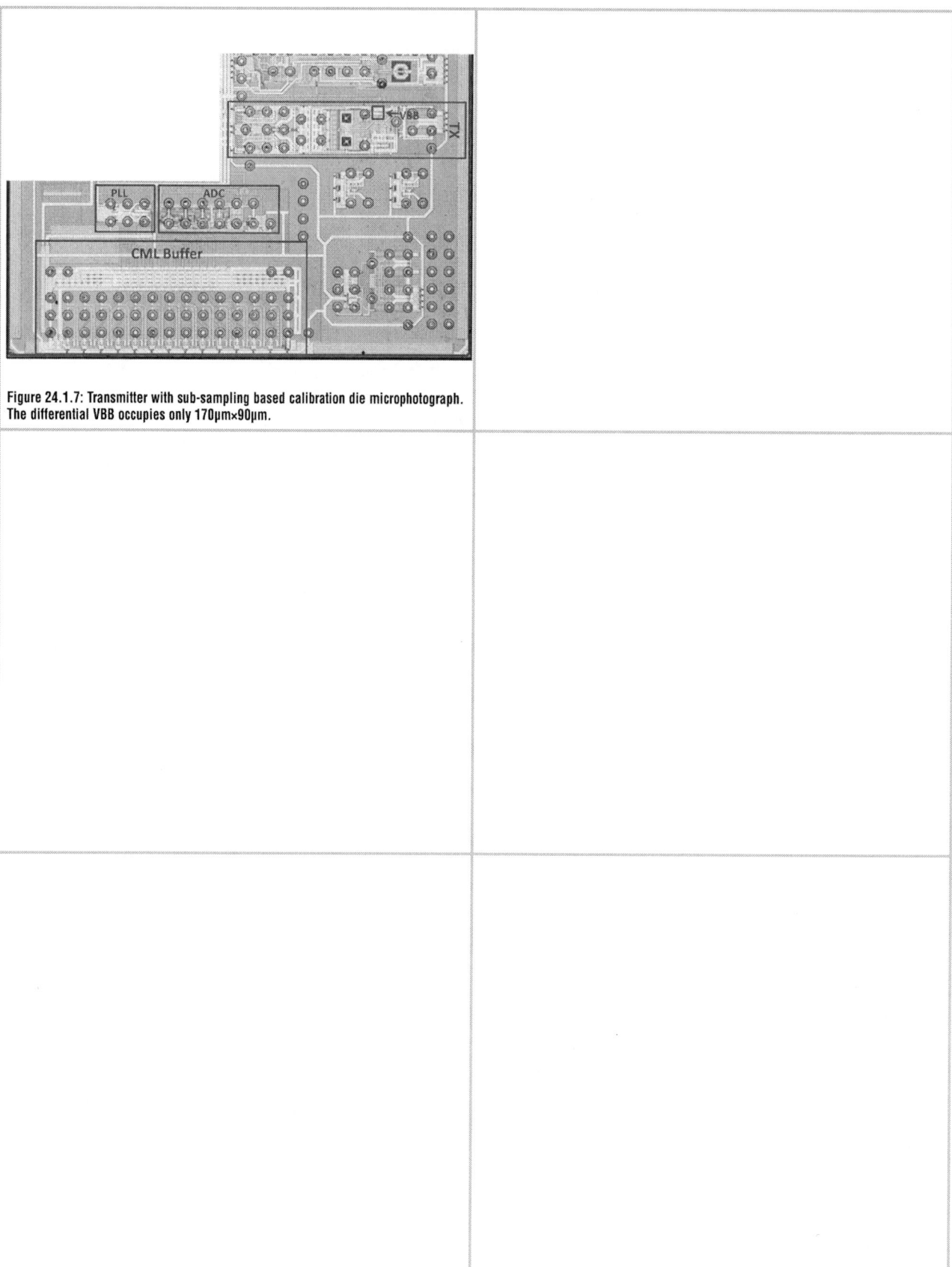

Figure 24.1.7: Transmitter with sub-sampling based calibration die microphotograph. The differential VBB occupies only 170μm×90μm.

ISSCC 2011 / SESSION 24 / TRANSMITTER BLOCKS / 24.2

24.2 A Flip-Chip-Packaged 1.8V 28dBm Class-AB Power Amplifier with Shielded Concentric Transformers in 32nm SoC CMOS

Yulin Tan, Hongtao Xu, Mohammed A El-tanani, Stewart Taylor, Hasnain Lakdawala

Intel, Hillsboro, OR

As CMOS technology continues to scale for SoC applications, significant challenges to implement a monolithic linear high-power amplifier have emerged. This results from the low breakdown voltage of transistors, high on-chip passive loss on a conductive substrate, stringent bump-pattern constraints and thermal dissipation requirements of flip-chip packages. Most fully-integrated linear CMOS PAs reported to date were either fabricated on a relatively mature process (65nm or above) in a wire-bonded die with relatively high supply voltage (3.3V) [1-3], or exhibited low power efficiency when backed-off. In many cases, digital pre-distortion was implemented to improve linearity and efficiency [2,3]. We present a 1.8V single-chip CMOS PA with 21dBm average output power and 16% PAE while meeting -25dB EVM for 64-QAM OFDM signal without digital pre-distortion linearization in 32nm SoC CMOS on a flip-chip package. This performance is enabled by: 1) On-chip shielded concentric transformers with power-splitting/combining structures to fit the design within the bump pattern constraint and thermal dissipation requirement of an SoC flip-chip package and to enhance power handling capability; 2) Minimum phase distortion by optimized inter-stage power matching allowed by the high f_{max} of the minimum-channel-length device.

Figure 24.2.1 is a simplified schematic diagram showing the two-stage PA with an input transformer acting as matching network to the source and splitting the power into two pairs of differential driver devices. Each pair drives an inter-stage transformer that interfaces to corresponding differential pairs of output devices. The output transformer converts the output impedance of the output devices and combines their power to drive a 50Ω load. The power-splitting and combining structures allow each power device to operate at a 3dB lower level, enhancing operation at a relatively low 1.8V supply. This combined with a shielded concentric transformer structure mitigates heat localization that is usually inherent in a conventional PA.

To facilitate designing over the bump-patterned die for the flip-chip SoC package that does not allow bump de-population and for the uniform thermal dissipation, critical physical elements such as transformers, power devices and their associated decoupling networks, etc., are distributed evenly over four symmetric quadrants with respect to a common origin as shown in the PA layout diagram. Figure 24.2.2 illustrates the two-fold symmetrical layout of the PA. Each quadrant of each transformer corresponds to one positive or negative arm of that transformer for each differential pair signal in Fig. 24.2.1. Three concentric transformers are placed between bumps to fully utilize the limited space for top metal layer. To leverage the ground distribution in the SoC flip-chip package, all bumps are connected or AC-coupled (for supply bumps) to package ground plane through bump via. Solid ground shield between transformers is created from package ground, bumps, top-layer metal, then through each lower via/metal layer down to the conductive substrate (SoC chip ground). It helps to reduce electromagnetic coupling between transformers and stabilize the circuit.

For the best quality factor, each transformer uses transmission-line-based structure to reduce loss due to metal resistance and substrate coupling. The transistor biases are fed in through the center tap of corresponding transformers. The output transformer is designed with two 20um primary coils in parallel with a 10um secondary strip sitting in between to handle high current density of the output stage [4]. To leverage the higher cutoff frequency of the 32nm device, the inductance of the inter-stage transformers is optimized to increase the tuning capacitance (C4, in Fig. 24.2.1) at the input of the common-source device of the output stage. This reduces the impact of the nonlinearity due to device capacitance variation.

The structure of power transistors is customized to achieve minimum parasitic capacitance and moderately high gate resistance, rather than using a typical RF transistor structure. The common-source output devices are thin-gate 32nm transistors cascoded with thick-gate 180nm transistors [5]. The driver device is sized as 1/12 of the output device size while biased at about 2x current density. A resistive bias with an RC decoupling network (Cg in Fig. 24.2.1) at the common gate is employed to achieve minimum phase distortion and to enhance stability. One pair of staggered-RC decoupling networks is deployed at each center tap of the primary of the inter-stage transformer and the output transformer to ensure common-mode stability [1].

The common-source devices are biased to Class-AB and are optimized for the max average output power and PAE at the targeted EVM for 64-QAM OFDM signals, rather than for the max-flatness of power gain for single-tone signals. At 1.8V supply, the driver bias current is optimized to be 20mA (5mA each device). Figure 24.2.3 shows the measured average output power and PAE at -25dB EVM together with small-signal gain over the decreasing bias current of output devices (from 200mA to 100mA), for a 64-QAM OFDM signal at 2.75GHz. The average output power and PAE are saturated at 20.8dBm and 15.5% respectively, while the small-signal gain ranges from 19.5dB to 14.5dB.

Figure 24.2.4 shows the measured average output power and PAE over carrier frequency at -25dB EVM for a 64-QAM OFDM signal, at 120mA/20mA bias for output/driver devices. The max output power is 21dBm with 16% PAE at 2.8GHz. The 3dB bandwidth is 450MHz. A speculative model of the RF transistor and packaging together with poor extraction accuracy at the time of design contributed to the gain being centered at 2.75GHz rather than the targeted 2.45GHz. At output/driver bias=120mA/20mA, P_{sat} is 28dBm with a peak PAE of 31.9% and P_{1dB} of 26.5dBm for a 2.75GHz CW input signal as shown in Fig. 24.2.5. The maximum AM-AM gain expansion for a CW signal is about 1dB while AM-PM is within 3 degrees for output power up to 21dBm. Figure 24.2.6 shows the EVM and PAE over average output power for a 64-QAM OFDM signal with the output spectrum at -25dB EVM. Continuous testing for 500 hours at max average output power for -25dB EVM with elevated supply voltage shows no sign of power degradation or reliability concern.

Figure 24.2.7 shows the 1.60x1.25mm^2 die with 8x6 bumps. The PA employs a low cost flip-chip molded-matrix BGA package, and was mounted and tested on an FR4 circuit board.

Acknowledgments:
The authors would like to thank C.T. Fu, J. Duster, B. Carlton, R. Bishop, H. Alavi, K. Soumyanath, of Intel Labs, and T. Nguyen, Q. Fan, W. Kwong, L. Haggis, J. Rizk, T. Kamgaing, V. Rao, J. Lin, P. Vandervoorn, C.H. Jan, G. Pandya, R. Kevin, of Intel TMG group, and other members at Intel, for their supportive work or constructive discussions.

References:
[1] D. Chowdhury, et al., "A Single-Chip Highly Linear 2.4GHz 30dBm Power Amplifier in 90nm CMOS", *IEEE ISSCC Dig. Tech.*, 2009, pp. 378-379, Feb. 2009.
[2] A. Afsahi, et al., " Linearized Dual-Band Power Amplifiers With Integrated Baluns in 65 nm CMOS for a 2x2 802.11n MIMO WLAN SoC", *IEEE JSSC vol.45, no.5*, pp. 955-965, May 2010.
[3] M. Terrovitis, et al., "A 1x1 802.11n WLAN SoC with fully integrated RF front-end utilizing PA linearization", *in Proc. of ESSCIRC*, pp. 224-227, Sept. 2009.
[4] H. Xu, et al.,"A Highly Linear 25dBm Outphasing Power Amplifier in 32nm CMOS for WLAN application", *Proc. of ESSCIRC*, pp.306-309, Sept. 2010.
[5] C.-H. Jan, et al., "A 32nm SoC platform technology with 2nd generation high-k/metal gate transistors optimized for ultra low power, high performance, and high density product applications ", *IEEE IEDM Tech. Dig.*, pp. 1-4, Dec. 2009.

ISSCC 2011 / February 23, 2011 / 2:00 PM

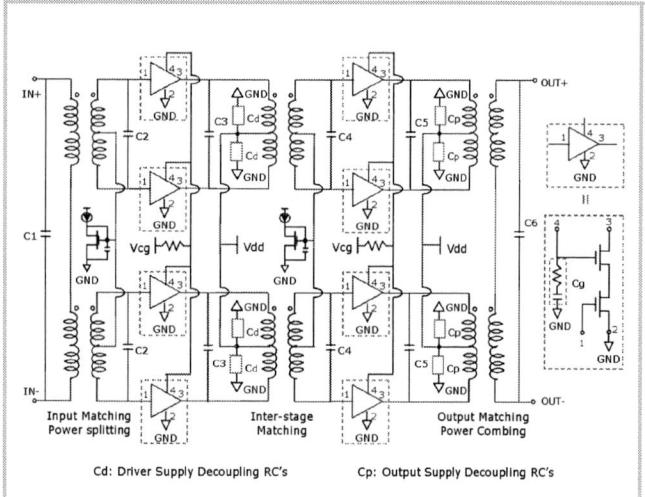

Figure 24.2.1: Simplified PA schematic diagram.

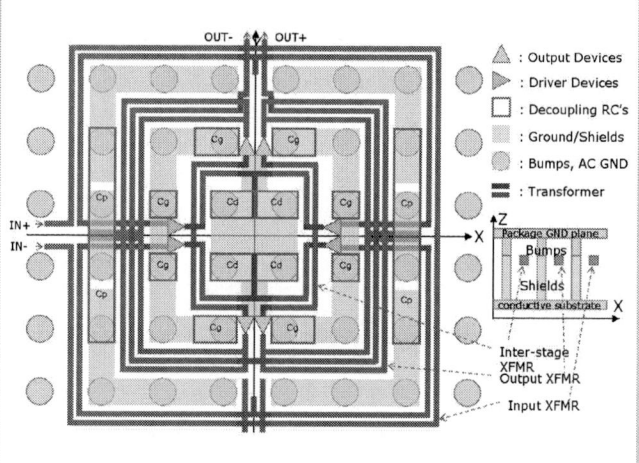

Figure 24.2.2: Simplified PA layout implementation diagram.

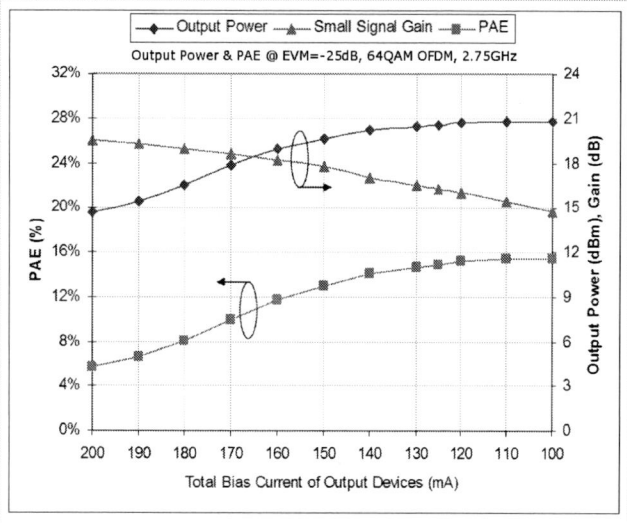

Figure 24.2.3: Measured gain, output power & PAE, over bias current.

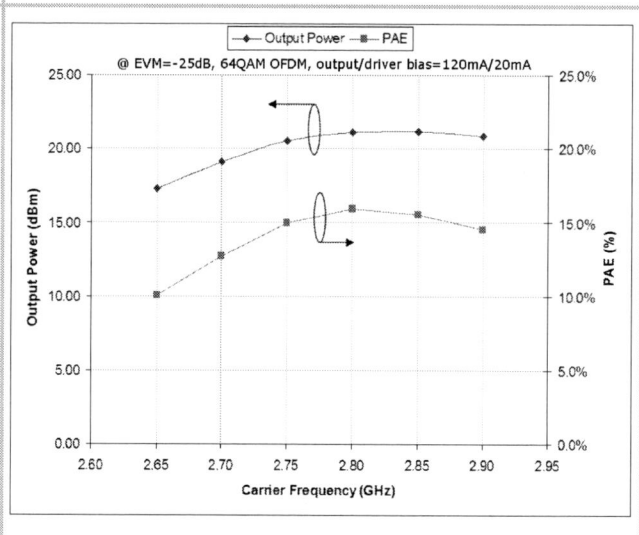

Figure 24.2.4: Measured output power & PAE, over carrier frequency.

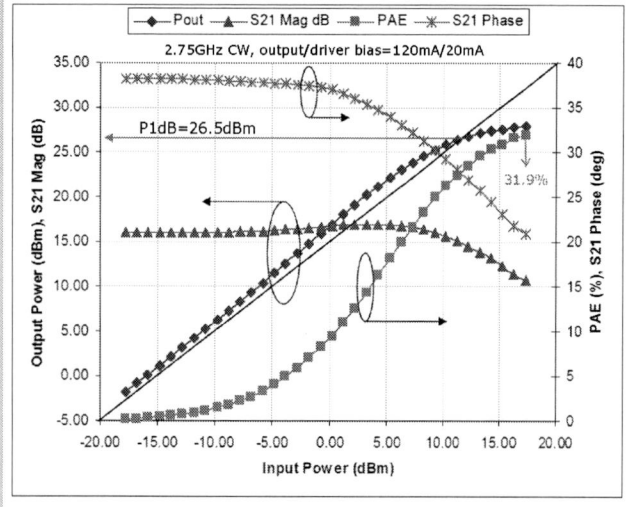

Figure 24.2.5: Measured transfer characteristic & PAE of CW signal.

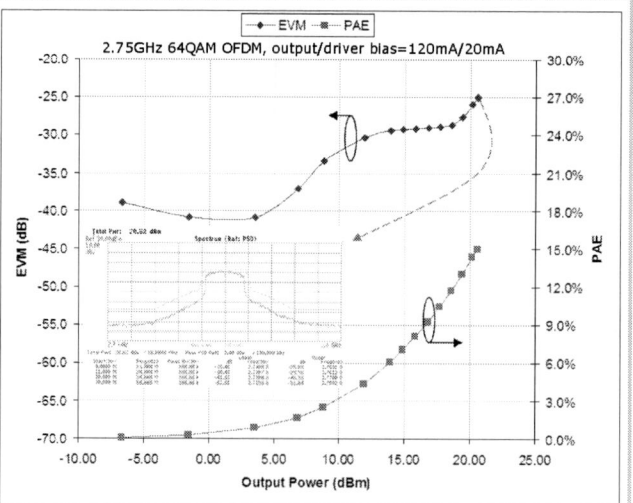

Figure 24.2.6: Measured EVM & PAE over output power.

24

978-1-61284-303-2/11 $26.00 © 2011 IEEE

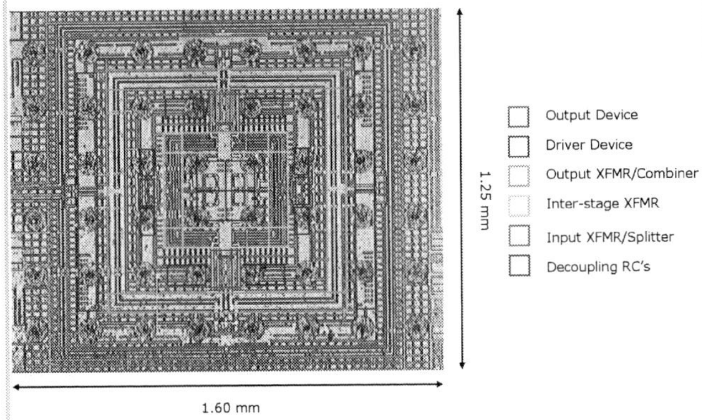

1.25 mm

1.60 mm

Output Device
Driver Device
Output XFMR/Combiner
Inter-stage XFMR
Input XFMR/Splitter
Decoupling RC's

Figure 24.2.7: Die Micrograph.

ISSCC 2011 / SESSION 24 / TRANSMITTER BLOCKS / 24.3

24.3 A Switched-Capacitor Power Amplifier for EER/Polar Transmitters

Sang-Min Yoo, Jeffrey S. Walling, Eum Chan Woo, David J. Allstot

University of Washington, Seattle, WA

Wireless high-speed communication standards such as *WiFi*, *WiMax* and *LTE* use spectrally-efficient OFDM modulation that encodes signal information in both amplitude and phase. Use of this non-constant envelope modulation requires a linear PA, operating at a less-than-peak signal level to realize higher linearity and inherently reduced efficiency. Because the PA is the dominant power consumer in most RF transceivers, operation with reduced efficiency leads to short battery lifetime and reduced mobility. Consequently, many efforts to utilize more efficient switching amplifiers with linearization circuitry have been made, notably through pulse-width modulation [1], outphasing [2] and envelope elimination and restoration (EER) [3,4]. Of the three, EER offers the best performance tradeoff between linearity, output power and efficiency; however, most previous implementations have come at the cost of large, power-hungry analog supply modulators. Additionally, conventional EER techniques are subject to nonlinearity induced by delay mismatch between the amplitude- and phase-modulated signal components. An alternative solution modulated the output power by selecting multiple PA unit cells [5], but this exhibits low efficiency at low output power levels because the power control is achieved by changing the total PA transconductance through switching of inefficient cells.

This paper introduces an EER 90nm CMOS experimental prototype switched-capacitor power amplifier (SCPA) that achieves high output power, efficiency and linear output-power control using a switched-capacitor-based switching PA without the use of a supply modulator. While amplifying 64-QAM OFDM modulation with a 20MHz signal bandwidth it achieves an average output power of 17.7dBm, an average PAE of 32.1%, and an EVM of 2.9%.

Switched-capacitor circuit techniques are widely used in analog/mixed-signal design because capacitors are area-efficient native devices and CMOS transistors are excellent switches [6]. High-accuracy capacitor ratios coupled with digital signal processing techniques are easily applied to switched-capacitor circuits. These techniques can now be adopted directly at RF frequencies because of the higher operating speeds with scaled CMOS. In a switched-capacitor circuit, any voltage can be generated based on the ratio of the capacitors switched to V_{DD} or ground (V_{GND}). It is important to note that there is no loss of energy ideally in the charge redistribution among capacitors (Fig. 24.3.1). To efficiently generate a desired output voltage, capacitors are selectively connected to either V_{GND} or switched between V_{GND} and V_{DD}. Hence, the ratio of capacitors switching (ΣC_{on}) between V_{GND} and V_{DD} compared to the total capacitance ($\Sigma C_{on} + \Sigma C_{off}$) defines the output voltage. The capacitors are switched at the desired RF carrier frequency; a bandpass filter (BPF) (e.g., matching network) is created by connecting an inductive reactance in series with the capacitor array to select the RF signal to be broadcast by the SCPA. Because V_{DD} and V_{GND} are AC grounds, the total capacitance seen by the matching network remains constant regardless of the number of capacitors being switched; thus, the frequency response of the SCPA is constant and independent of the output voltage amplitude.

An SCPA operating as part of an EER transmitter is depicted in Fig. 24.3.2. A polar-modulated signal is generated using baseband signal processing and the digital envelope signal (A) input to a thermometer decoder controls the number of unit capacitors to be switched, while the digital phase signal (ϕ), upconverted to an RF carrier frequency, synchronizes all switches selected by A; i.e., delay mismatch is easily accounted for in DSP. The SCPA can be understood as capacitive power-combining of multiple individual switching PAs; i.e., the output power is combined via charge redistribution on the capacitors and then filtered by the bandpass matching network.

A single-ended implementation is shown in Fig. 24.3.3 for clarity, though it is implemented differentially. The envelope code $B_{in}(A)$ is processed in a binary-to-thermometer decoder that selects the capacitors to be switched while a dedicated buffer drives the switches connected to each capacitor. Digital logic gates

subsequently combine the envelope and phase information. To achieve higher output power, a supply voltage of $2V_{DD}$ is adopted by cascoding the output switches. Inverter chains operating between ground and V_{DD} drive NMOS switches, and inverters operating between V_{DD} and $2V_{DD}$ drive PMOS switches. A level shifter converts the PMOS drive signal from the nominal logic level to the higher supply voltage. To achieve high efficiency, a non-overlapping clock is used to prevent crowbar currents in the NMOS and PMOS switches, thus avoiding unnecessary power dissipation. The digital power consumption in the inverter buffer chains is proportional to the output voltage, due to fewer logic stages switching; hence, the roll-off in efficiency versus power backoff is less dramatic than in other PAs. Finally, the voltage generated by the switched capacitors is delivered to the BPF output matching network. The impedance presented to the capacitor array by the matching network is inductive; i.e., its reactance is used to negate the reactance of the capacitor array. Although the SCPA is designed for 6b resolution, more resolution can be achieved with more unary/binary bits or using other switched-capacitor (e.g., C-2C ladder) or signal processing techniques (e.g., $\Delta\Sigma$ or pulse-width modulation).

The measured output power versus input code and PAE versus output power are shown in Fig. 24.3.4. The peak output power and PAE are 25.2dBm and 55.2%, respectively, with a fully-integrated output matching network. As mentioned, the PAE has slower roll-off at power backoff than typical CMOS PAs; the PAE is 35.1% at power backoff of -6dB from peak power. For comparison the efficiency characteristic of an ideal Class-B PA scaled to have similar losses as the SCPA is also plotted. The linearity of the SCPA is characterized in Fig. 24.3.5; the measured output voltage and AM-PM distortion versus input code show weak second-order nonlinearities that are correctible by digital predistortion. The nonlinearity owes to the finite switch performance and power line impedance. Efficiency, output power and linearity are coupled in the design of the SCPA; optimum efficiency and output power are achieved with a reasonable tradeoff in linearity as demonstrated from dynamic measurements. A measured constellation for a 64-QAM OFDM modulated signal with a 20MHz signal bandwidth is shown along with the measured output power spectral density (Fig. 24.3.5). A performance summary is given in Fig. 24.3.6.

A chip micrograph of the 90nm prototype is shown in Fig. 24.3.7. It is noted that both linearity and efficiency can be improved using a more advanced CMOS technology due to faster switching performance and lower digital power consumption in the driving buffers. Because the operation of the SCPA is based only on digital circuits such as logic gates, switches and capacitors, it is ideally suited for future scaled CMOS technologies.

Acknowledgement:
The authors gratefully acknowledge the constant encouragement and support of the late Dr. Krishnamurthy Soumyanath of Intel Corp.

References:
[1] J. S. Walling, H. Lakdawala, Y. Palaskas, A. Ravi, O. Degani, K. Soumyanath, and D. J. Allstot, "A 28.6dBm 65nm class-E PA with envelope restoration by pulse-width and pulse-position modulation," *ISSCC Dig. Tech. Papers*, pp. 566-567,636, 2008.
[2] H. Xu, Y. Palaskas, A. Ravi, M. Sajadieh, M. Elmala, and K. Soumyanath, "A 28.1dBm class-D outphasing power amplifier in 45nm LP digital CMOS," *VLSI Circ. Dig. Tech. Papers*, pp. 206-207, 2009.
[3] P. Reynaert and M. S. J. Steyaert, "A 1.75-GHz polar modulated CMOS RF power amplifier for GSM-EDGE," *IEEE J. Solid-State Circuits*, vol. 40, pp. 2598-2608, Dec. 2005.
[4] P. Cruise, C.-M. Hung, R. B. Staszewski, O. Eliezer, S. Rezeq, K. Maggio, and D. Leipold, "A digital-to-RF-amplitude converter for GSM/GPRS/EDGE in 90-nm digital CMOS," *IEEE Radio Frequency Integrated Circuits (RFIC) Symp.*, pp. 21-24, 2005.
[5] A. Kavousian, D. K. Su, and B. A. Wooley, "A digitally modulated polar CMOS PA with 20MHz signal BW," *ISSCC Dig. Tech. Papers*, pp. 78-79, 588, 2007.
[6] S.-M. Yoo, J.-B. Park, H.-S. Yang, H.-H. Bae, K.-H. Moon, H.-J. Park, S.-H. Lee, and J.-H. Kim, "A 10 b 150 MS/s 123 mW 0.18 um CMOS pipelined ADC," *ISSCC Dig. Tech. Papers*, pp. 326-327, 2003.

978-1-61284-303-2/11 $26.00 © 2011 IEEE

ISSCC 2011 / February 23, 2011 / 2:30 PM

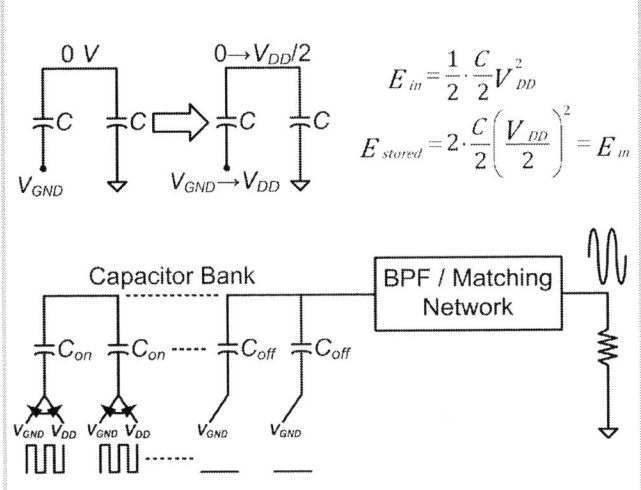

Figure 24.3.1: Switched-capacitor circuit for voltage modulation.

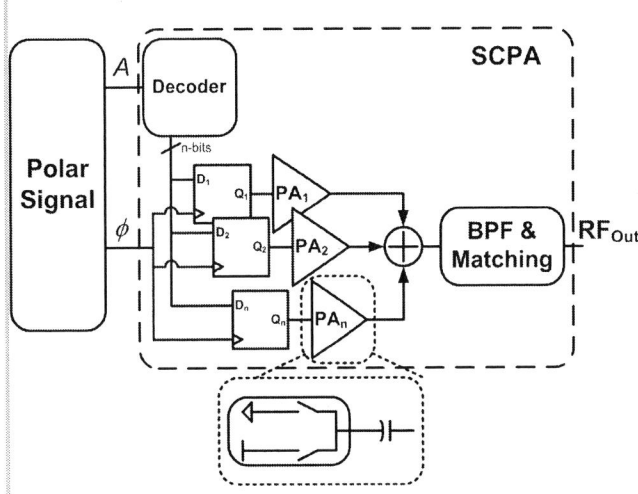

Figure 24.3.2: Top-level implementation of an SCPA in a polar transmitter.

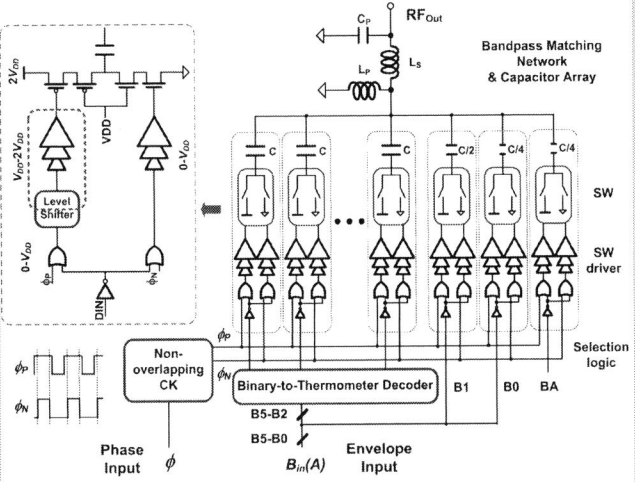

Figure 24.3.3: Single-ended 6b SCPA; actual implementation is differential.

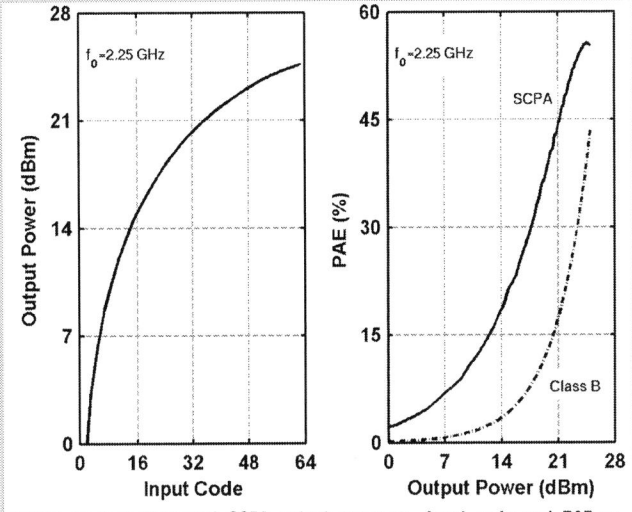

Figure 24.3.4: Measured SCPA output power vs. input code and PAE vs. output power.

Figure 24.3.5: (Clockwise from top left) Measured distortion vs. input code, frequency response, power spectral density and constellation for a 20MHz 64-QAM OFDM signal.

Technology	90nm CMOS
Resolution	6 bits
Supply Voltage	1.25V / 2.5V
Output Power (Peak / 64 QAM OFDM)	25.2dBm / 17.7dBm
PAE (Peak / 64 QAM OFDM)	55.2% / 32.1%
EVM (64 QAM OFDM)	2.9%
Frequency	1.8GHz – 2.8GHz
Die Area	1.04mm^2 (0.73mm × 1.43mm)

Figure 24.3.6: Performance Summary.

978-1-61284-303-2/11 $26.00 © 2011 IEEE

ISSCC 2011 PAPER CONTINUATIONS

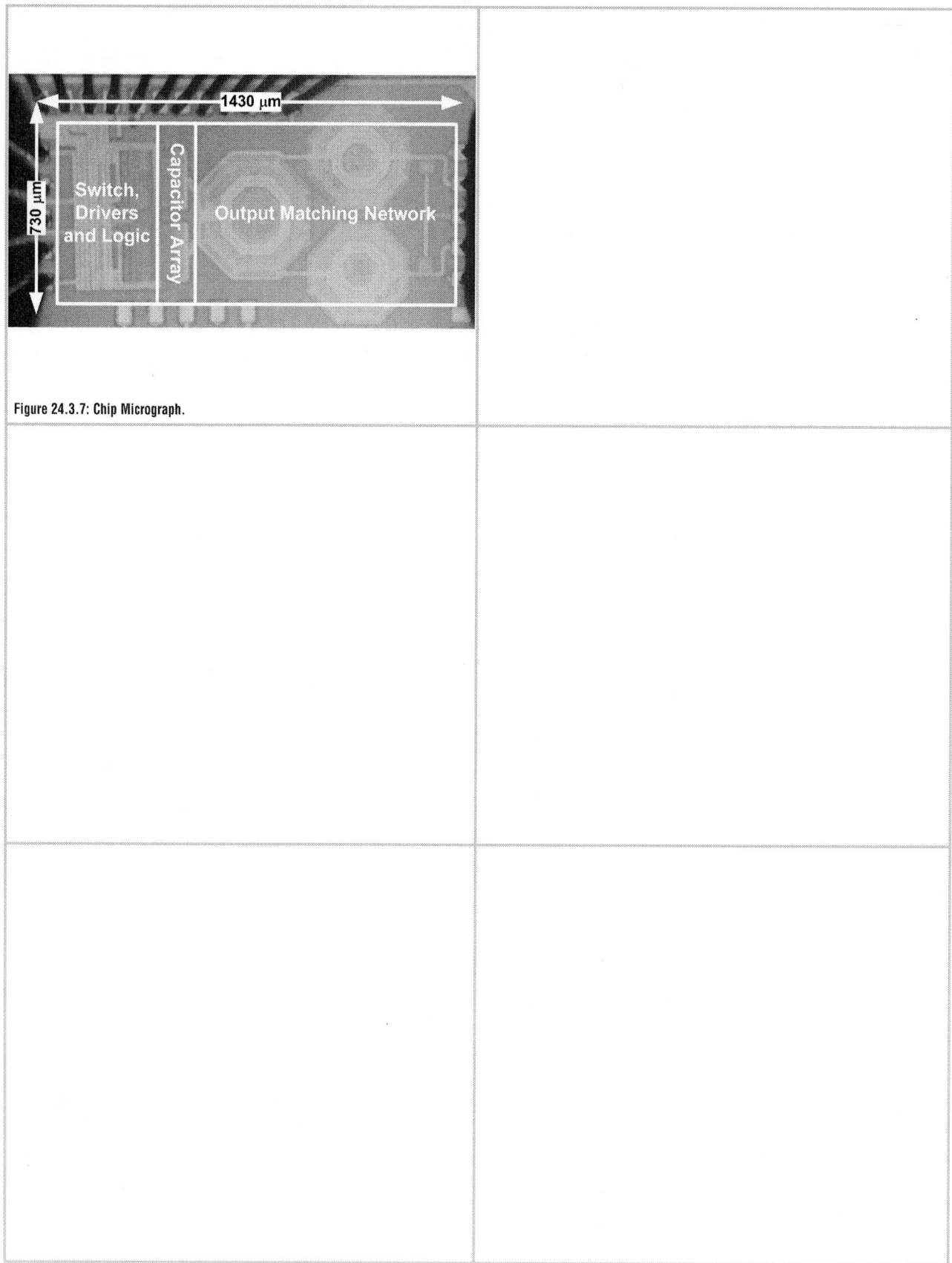

Figure 24.3.7: Chip Micrograph.

ISSCC 2011 / SESSION 24 / TRANSMITTER BLOCKS / 24.4

24.4 An EDGE/GSM Quad-Band CMOS Power Amplifier

Woonyun Kim, Ki Seok Yang, Jeonghu Han, Jaejoon Chang,
Chang-Ho Lee

Samsung Electro-Mechanics, Atlanta, GA

Although Si CMOS PAs for mobile applications have demonstrated specification-compliant performance over the last several years, Si CMOS has not been widely employed in cellular PA applications due to certain inferior properties of its power capability, PAE, and breakdown to its counterparts such as GaAs and SiGe BJTs [1]. However, research conducted in the past decade has enabled commercially available cellular switching CMOS PAs for constant envelope modulation such as GSM applications [2,3]. This successful implementation of a GSM PA called forth a demand for a combination of linear EDGE operation with GSM operation in order to increase data rate up to 384kb/s while using legacy infrastructures [1,4]. The challenges of implementation of a dual-mode CMOS PA arise from required linearity for high peak-to-average-power ratio, which forces the PA to operate at power back-off from the P_{1dB} and results in inevitable low efficiency. Although a CMOS EDGE/GSM PA was reported in the past, the application was intended for Class-E3 operation [5]. The presented PA is an EDGE/GSM dual-mode, quad-band CMOS cellular application PA satisfying the requirement for power Class-E2 operation [4].

A CMOS quad-band RF PA module (PAM) for wireless GSM/EDGE applications has been designed with two integrated passive devices (IPDs) on a high-resistivity Si substrate. Figure 24.4.1 shows the schematic diagram of the two-stage PAs with IPD-based transformers. The PA IC comprises two RF PAs for the low-band (GSM850/EGSM) and the high-band (DCS/PCS), a programmable power LDO shared by both paths, two LDOs for the driver stages, an RF power controller, four reference voltage (Vref) generators for adaptive bias circuitry (ABC) of the driver and power stages for both bands, and two input baluns in a substrate. Three and two parallel units of power stages for both bands, respectively, are combined by parallel combining transformers (PCTs) for efficient power transfer and impedance transformation using Samsung's proprietary IPD technology, which is well described in [3].

The differential driver and power stages are designed in a cascode topology to prevent excessive voltage stress on the devices for a supply voltage up to 4.5V. Each power stage depicted in Fig. 24.4.2 is a stack of 0.4μm thick-oxide transistors and 0.18μm thin-oxide transistors with parallel-capacitive feedback. The driver stage with a differential inductor load has the same structure as the power stage, and is connected to the power stage through an inter-stage matching network.

In a CMOS linear PA, AM-PM distortion is caused by the gate-drain capacitance (C_{gd}) as well as the gate-source capacitance (C_{gs}) when the input signal is large enough to constantly turn devices on and off. This work is focused on reducing the nonlinearity of C_{gd} in the CG stage since the signal is largest at the output node, which periodically puts the CG stage into the triode and saturation regions. For a normal differential cascode amplifier, the gate of its common-gate (CG) stage is virtually grounded due to its generic differential nature. Therefore, C_{gd} seen from the output node is C_{gd} itself and is highly nonlinear in a high power range. To prevent the gate node of the CG stage from a virtual ground, the transistor M5 is turned off for linear mode, and MIM capacitances C1 and C2 (less than 1pF) are added to the gate as shown in Fig. 24.4.2. The effective capacitance seen from the output is a series capacitance of C_{gd} and C1, which is much more linear than C_{gd}. As a result, the AM-PM distortion in a high output power range can remarkably decrease.

To reduce AM-AM distortion and power back-off from the P_{1dB} for EDGE application, an ABC is proposed. The ABC boosts up the dc bias voltage to the gate of the CS stage as the input signal increases, thus increasing the PAE for a given linearity requirement. As shown in Fig. 24.4.3, the presented ABC consists of two

diode-connected NMOS transistors in series connected between a reference voltage (Vref) and ground, respectively. The dc bias of Vbias is determined by the sizes of diodes and Vref when the RF input signal is small enough or not applied. When the RF signal is large enough to turn off the Mu and Md, it begins operation in the same way as a charge pump in a PLL. For the lower half cycle of the RF signal, the upper transistor, Mu, charges C_{gs} of the CS stage in the amplifier, while Md is turned off. For the upper half cycle, Md discharges C_{gs}, while Mu is turned off. If the size of Mu is larger than that of Md, Vbias will increase as the RF signal level increases, since the charging current is larger than the discharging one. Carefully optimizing the size of Mu and Md, and Vref controls the dc level of Vbias in the low power range and the increasing slope in high power operation. It should be noted that the sizes of Mu and Md are carefully chosen so that the ABC consumes under 0.5mA at an output power of 28.5dBm for each branch.

In order to minimize the performance variation from environmental change such as PVT, the Vref voltage that is used to set the dc operating point of the ABC is produced from the circuit depicted in Fig. 24.4.3. The circuit is comprised of a replica of the ABC, a replica of the CS stage in the amplifier, a current mirror, and an op-amp. A PTAT current, I_{ptat}, is supplied as the reference current to reduce the gain variation depending on temperature. The width of the CS stage in the main amplifier is M times that of the replica, while the replica of the ABC is the same size as that of the original. Therefore, Vref can be determined and supplied to each ABC such that the dc current in the power stage, I_{PA}, is calculated as $M \times N \times I_{ptat}$. The values of R1 and R2 are carefully chosen because of the channel-length modulation effect of the PMOS current mirror and the replica of the CS stage. Another noteworthy point of the design of the block is bandwidth. The bandwidth has to be optimized between the factors of RX band noise and AM bandwidth.

In the GSM mode of Class-E PA operation, it is required to control the output power from the PA with a dynamic range of more than 30dB and a tolerance of ±2dB. In this work, the power control scheme is implemented using gate voltage control of the driver stage and closed-loop current control of the power stage. As shown in Fig. 24.4.4, two different Vramp shaping circuits with a temperature compensation feature are used for the driver and power stages to have a gentle slope of power output versus Vramp. In the power stage control, the current through the power PMOS is sensed by a replica of the PMOS and a resistive component to be compared with the output of the power cell control shaped Vramp signal. The result of the comparison, which is the error amplifier output, goes to the gate of the power amplifier core cell and adjusts the current from the power PMOS.

The active IC is implemented in a 0.18μm CMOS technology. Measured RF performance for low-band is shown in Fig. 24.4.5. The right figure is the output power and PAE for GSM mode, and the other is for EDGE mode. Figure 24.4.6 gives a summary of PAM performance. The measured PAM efficiency is 23% in EDGE mode and 48 to 55% in GSM mode depending on the band. The receiver Band noise with a 100kHz RBW at 20MHz offset is -82 and -86dBm for GSM and EDGE modes, respectively. Figure 24.4.7 shows the micrograph of the PAM with the transformers assembled in a 32-pin QFN package.

References:
[1] H. S. Bennett, et al., "Device and Technology Evolution for Si-Based RF Integrated Circuits," *IEEE Trans. Electron Devices*, vol. 52, no. 7, pp.1235-1258, July 2005.
[2] I. Aoki, et al., "A Fully-Integrated Quad-Band GSM/GPRS CMOS Power Amplifier," *IEEE J. Solid-State Circuits*, vol. 43, no. 12, pp 2747-2758, Dec. 2008.
[3] C.-H. Lee, et al., "A highly efficient GSM/GPRS quad-band CMOS PA Module," *IEEE Radio Frequency Integrated Circuits Symposium*, pp. 229-232, June 2009.
[4] 3GPP TS 45.005, Radio transmission and reception, V9.3.0, July 2010.
[5] P. Reynaert, M. Steyaert, "A 1.75GHz GSM/EDGE polar modulated CMOS RF power amplifier," *ISSCC Dig. Tech.Papers*, pp. 312-313, Feb. 2005.

978-1-61284-303-2/11 $26.00 © 2011 IEEE

ISSCC 2011 / February 23, 2011 / 3:15 PM

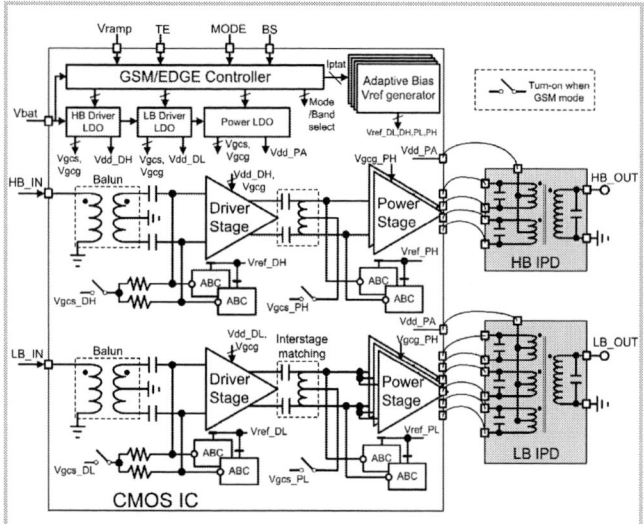

Figure 24.4.1: Block diagram of GSM/EDGE Quad-band CMOS PA Module.

Figure 24.4.2: Simplified circuit schematic of PA stage.

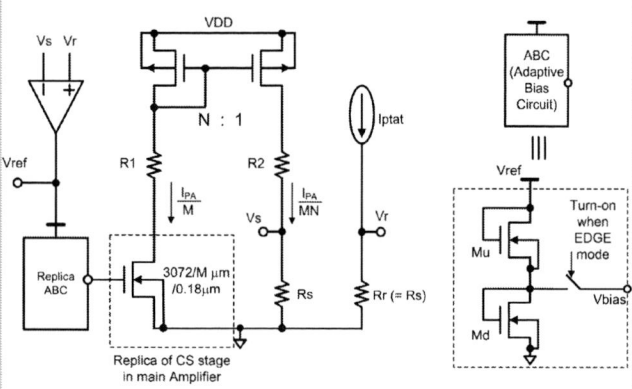

Figure 24.4.3: Simplified circuit schematic of reference voltage generator for adaptive bias circuitry.

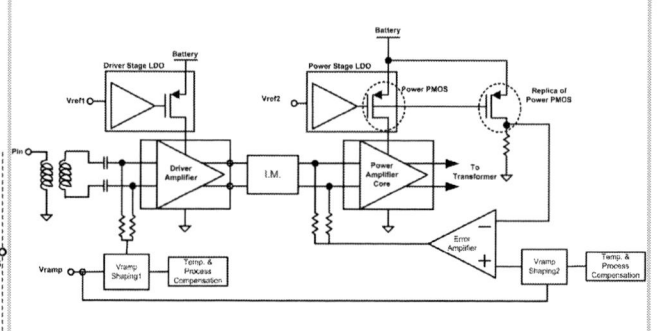

Figure 24.4.4: Power control scheme of driver stage and closed loop current control of power stage for GSM mode.

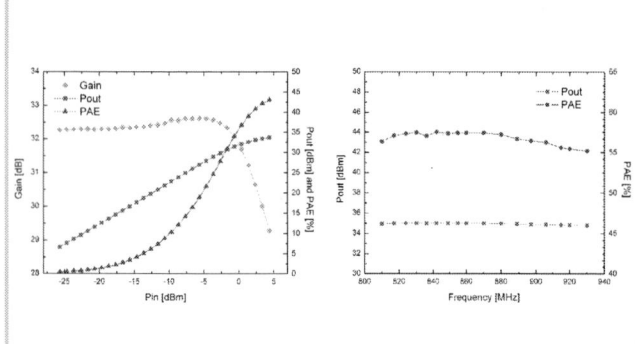

Figure 24.4.5: Measured RF performance for low-band. The right figure is the output power and PAE for GSM mode, and the other is for EDGE mode.

			This work	[5]
	Technology		0.18μm CMOS & IPDs	0.18μm CMOS
	Silicon Area [mm²]		1.1×2.8, 1.1×1.2, 1.5×1.4	1.8×3.6
	Package size [mm²]		5×5	-
	Supply Voltage [V]		3.0-4.2	-
EDGE	GSM/EGSM	average linear Pout [dBm]	28.5	-
		average PAE [%]	22	-
		ACPR [dBc] @ 400kHz	-57	-
		ACPR [dBc] @ 600kHz	-75	-
		EVM-rms [%]	1.6	-
		Rx band noise @ 20MHz offset	-86 dBm/100kHz	-
	DCS/PCS	average linear Pout [dBm]	27.5	23.8
		average PAE [%]	23	22
		ACPR [dBc] @ 400kHz	-56	-56.7
		ACPR [dBc] @ 600kHz	-74	-64
		EVM-rms [%]	2.5	1.67
		Rx band noise @ 20MHz offset	-86 dBm/100kHz	-
GSM	GSM/EGSM	Input power [dBm]	0~+6	-
		Output power [dBm]	34.5	-
		PAE [%]	55	-
		Rx band noise @ 20MHz offset	-82 dBm/100kHz	-
	DCS/PCS	Input power [dBm]	0~+6	-3
		Output power [dBm]	32.5	27
		PAE [%]	48	34
		Rx band noise @ 20MHz offset	-82 dBm/100kHz	-

-Unless otherwise specified: Vbat=3.5V, Pin=+3dBm, Duty Cycle=25%, Zin=Zout=50Ω, Tc=25℃, Vramp=1.8V

Figure 24.4.6: PAM measurement summary for low bands and high bands.

ISSCC 2011 PAPER CONTINUATIONS

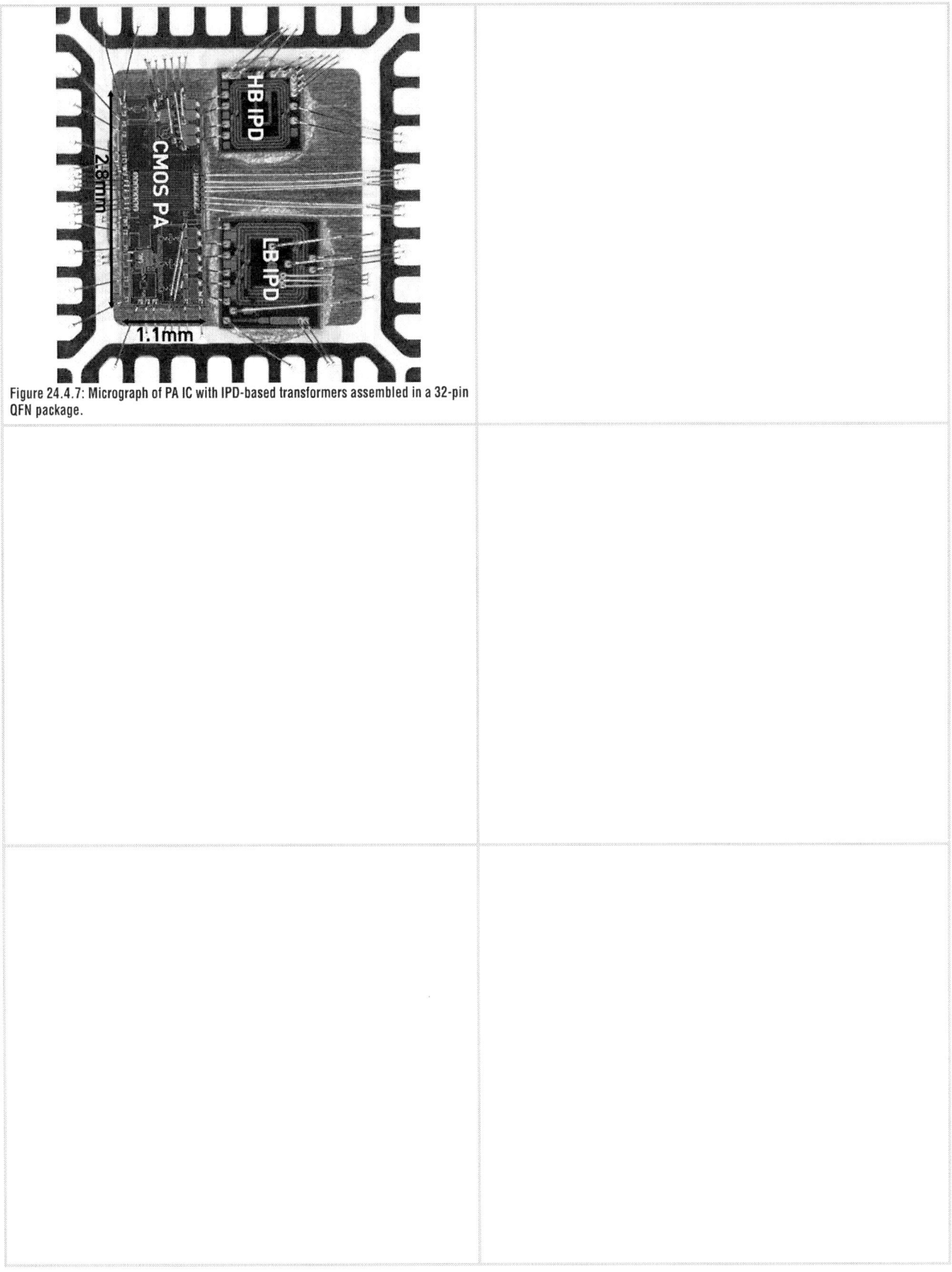

Figure 24.4.7: Micrograph of PA IC with IPD-based transformers assembled in a 32-pin QFN package.

ISSCC 2011 / SESSION 24 / TRANSMITTER BLOCKS / 24.5

24.5 A Compact 1V 18.6dBm 60GHz Power Amplifier in 65nm CMOS

Jiashu Chen, Ali M Niknejad

University of California, Berkeley, CA

One of the remaining challenges in implementing CMOS 60GHz radios is to cover longer communication distance as the high path loss at mm-Wave frequencies demands higher EIRP, which in turn requires considerable design effort on the transmitter. In addition, to comply with the OFDM transmitting mode of the IEEE 802.15.3c standards, the power amplifier (PA) must be capable to handle a peak power level 6~9dB higher than the average without sacrificing reliability. With the low supply voltage limitation of deeply scaled CMOS technologies, efficient power-combining techniques are essential. Spatial power combining is an attractive solution to meet the EIRP goal, but at the cost of spending extra power on the complex signal distribution and in order to compensate for the phase-shifter loss. Spatial power-combining solutions also occupy larger area due to the requirement of multiple antennas, minimum antenna spacing, and the transmission line feed network. On the other hand, CMOS PAs with on-chip power-combining structures have achieved 18dBm of output power [1-3]. However, all the combiners comprise multiple stages to achieve both impedance transformation and power combining which not only increase the insertion loss and directly degrade the efficiency, but also consume a significant amount of silicon area, rendering them far less appealing for system integration. This paper presents a fully integrated 18.6dBm CMOS PA based on an efficient and compact on-chip power combiner.

Transformers have the advantage of being very compact [4]. However, a standard 1:1 transformer can only perform differential to single-ended conversion and therefore has a limited voltage enhancement ratio of 2 and power enhancement ratio of 4. To achieve greater voltage and power enhancement ratio, a 4-input transformer combiner is proposed as shown in Fig. 24.5.1. The primary side of the transformer combiner has two pairs of differential excitation ports located on the opposite sides of the metal winding while the secondary side is the same as the standard transformer. The operation resembles that of a Distributed Active Transformer [5]: the two pairs of differential ports are excited in such a way that the ac current circulates in the primary winding and induces the magnetic flux for the secondary winding while the center points of the two half-octagon traces provide convenient dc access. Since the voltage swings are effectively stacked, the impedance seen at each input port is one quarter of the load. Consequently, this structure has a voltage enhancement ratio of 4 and a power enhancement ratio of 16, a significant improvement compared to the standard 1:1 transformer.

Two of the most critical design aspects of the transformer combiner are the insertion loss and the input impedance. The insertion loss is mainly determined by the magnetic coupling factor and the inductor Q factor. A vertical broadside-coupled structure with primary winding on the top and secondary winding on the bottom is used and a coupling factor of 0.87 is achieved at 60GHz. The thin metals in digital CMOS processes usually limit the achievable Q factor of on-chip inductors. In this design, the aluminum capping layer, which is usually used to cover the copper bondpads, is strapped together with the top copper layer to form the primary winding. The aluminum capping layer has a thickness of 1.2µm while the top two copper layers have the same thickness of 0.9µm. By forming a much thicker primary metal winding, the Q factor is improved by 50%, reaching 15.5 at 60GHz. The insertion loss of the combiner is further evaluated as a function of transformer radius and trace width using EM simulators. On the other hand, the input impedance of the transformer must be taken into account when deciding the physical dimensions. Since the input susceptance is a strong function of the transformer radius, an iterative output stage optimization process involving load-pull simulation with different transistor sizes is performed to maximize the combined output power. With 28µm radius and 12µm trace width, the final transformer combiner has an insertion loss of 0.63dB.

Characterization of power transistor behaviors at mm-Wave frequencies is important to ensure the PA performance. The iterative output stage design

results in an optimal unit power transistor size of 140µm with 2µm finger width. A test chip with a stand-alone 140µm power transistor was fabricated and two-port S-parameters were measured up to 110GHz for different bias points. A simple wrap-around model with lumped inductors and resistors is used to achieve better device π-model fitting at mm-Wave frequencies.

Figure 24.5.2 shows the complete schematic of the three-stage PA. A power splitter is needed preceding the output stage in order to feed two differential pairs and the same transformer combiner structure is once more utilized with the inputs and outputs swapped to perform the task. Due to the large mismatch between the input conductance of the output stage and the output conductance of the driver stage, an extra matching network is required to compensate the limited conductance transformation range of 1:1 transformers. To accomplish this goal, two pairs of differential strip-lines are inserted between the transformer splitter and the output differential pairs. These strip-line pairs also conveniently cover the geometric distance of the transistors separated by the transformers. A similar conductance mismatch at the input of the PA is solved by adding a series differential inductor to the transformer. To ensure the PA stability, a parallel RC combination is added to introduce resistive loss at lower frequencies [4].

The prototype PA is fabricated in ST 1V 65nm GP 1P7M digital CMOS process. The measured S-parameters are shown in Fig. 24.5.3. With 1V supply, the PA achieves a power gain of 20.2dB and a 3dB bandwidth of 9GHz (56 to 65GHz). The amplifier is unconditionally stable over the entire measured frequency range with stability factor greater than unity. The measured S-parameters are compared to the simulation results based on pre-measured transistor S-parameters. It can be observed that the measurement results are very close to the simulation.

The 60GHz power measurement results are shown in Fig. 24.5.4. With 1V supply voltage, the measured 1dB gain compressed output power (P_{1dB}) is 15dBm and the saturated output power (P_{SAT}) is 18.6dBm. The measured peak PAE is 15.1% and the peak drain efficiency is 16.4%. At the saturated output power level, the amplifier still has 11dB of power gain, which significantly relieves the design of the preceding block. The power performance is also measured over the IEEE 802.15.3c band. As shown in Fig. 24.5.5, the PA maintains over 17.8dBm P_{SAT}, 13.8dBm P_{1dB} and 12.6% PAE from 58GHz to 64GHz.

A chip micrograph of the PA is shown in Fig. 24.5.7. Due to the use of compact power combiner and splitter, the PA only occupies an area of 0.28mm² including the GSG RF pads. Compared to the reported PAs with comparable power levels, the size is significantly smaller and hence it achieves the highest output power per area. Figure 24.5.6 summarizes a comparison with the state-of-the-art 60GHz CMOS PAs.

Acknowledgements:
The authors acknowledge BWRC sponsors, NSF Infrastructure Grant No. 0403427, chip fabrication donation by STMicroelectronics and Fulbright S&T Fellowship.

References:
[1] Lai, J.-W.; Valdes-Garcia, A. , "A 1V 17.9dBm 60GHz power amplifier in standard 65nm CMOS," *ISSCC Dig. Tech. Papers*, pp.424-425, 7-11 Feb. 2010.
[2] Martineau, B.; Knopik, V.; Siligaris, A.; Gianesello, F.; Belot, D.; , "A 53-to-68GHz 18dBm power amplifier with an 8-way combiner in standard 65nm CMOS," *ISSCC Dig. Tech. Papers*, pp.428-429, 7-11 Feb. 2010.
[3] Law, C.Y.; Pham, A.-V. , "A high-gain 60GHz power amplifier with 20dBm output power in 90nm CMOS," *ISSCC Dig. Tech. Papers*, pp.426-427, 7-11 Feb. 2010.
[4] Chowdhury, D.; Reynaert, P.; Niknejad, A.M.; "A 60GHz 1V + 12.3dBm Transformer-Coupled Wideband PA in 90nm CMOS," *ISSCC Dig. Tech. Papers*, pp.560-635, 3-7 Feb. 2008.
[5] Aoki, I.; Kee, S.D.; Rutledge, D.B.; Hajimiri, A., "Fully integrated CMOS power amplifier design using the distributed active-transformer architecture," *J. Solid-State Circuits*, vol.37, no.3, pp.371-383, Mar 2002.

978-1-61284-303-2/11 $26.00 © 2011 IEEE

ISSCC 2011 / February 23, 2011 / 3:45 PM

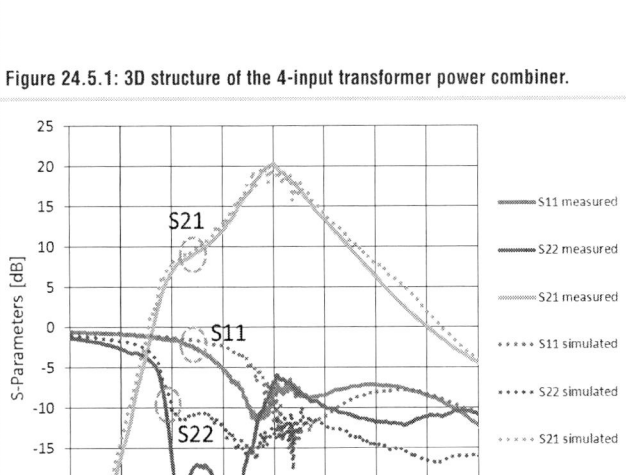

Figure 24.5.1: 3D structure of the 4-input transformer power combiner.

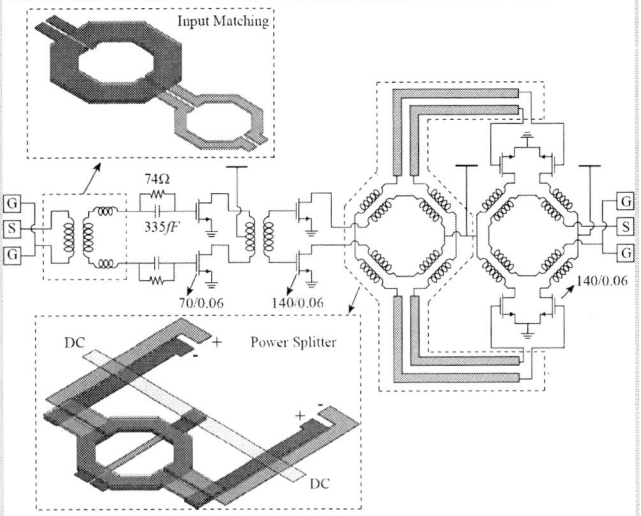

Figure 24.5.2: Schematic of the 3-stage 60GHz PA.

Figure 24.5.3: S-parameters of the PA (solid line: measured; dotted: simulated based on pre-measured transistor S-parameters).

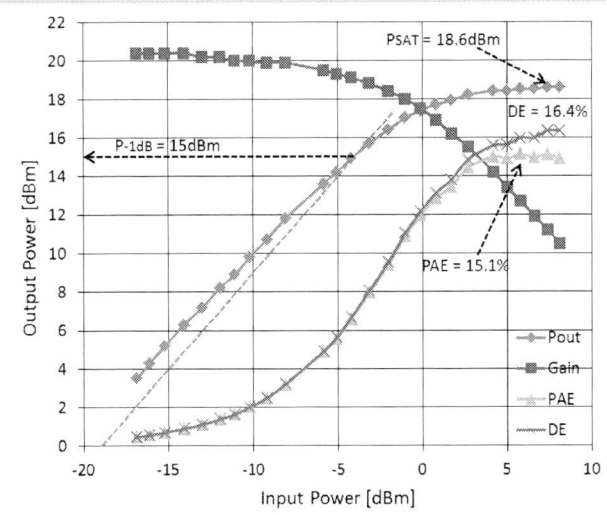

Figure 24.5.4: Measured PA output power, gain, drain efficiency and PAE.

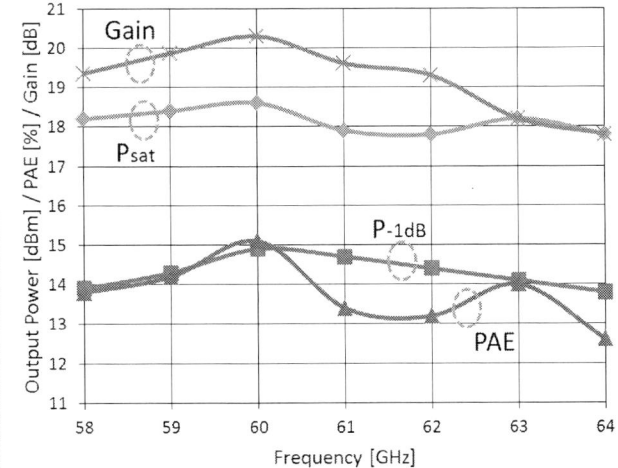

Figure 24.5.5: Measured PA output power, gain and PAE over the IEEE 802.15.3c band.

REF	This Work	[1]	[2]	[3]	[4]
CMOS Technology	1V 65nm	1V 65nm	1.2V 65nm	1V 90nm	1V 90nm
Supply Voltage [V]	1	0.9/1	1.2/1.8	1.2	1
Combiner Loss [dB]	0.63	1	1	1.1	0.9
S21 [dB]	20.3	18.9/19.2	14.3/15.5	20.6	5.6
P$_{-1dB}$ [dBm]	15	13/15.1	11/11.5	18.2	9
P$_{SAT}$ [dBm]	18.6	16.8/17.7	16.6/18.1	19.9	12.3
Peak PAE [%]	15.1	10.8/11.1	4.9/3.6	14.2	8.8
Area [mm^2]	0.28	0.83	0.46	1.67	0.25
P$_{SAT}$/Area [mW/ mm^2]	256	59/72.3	99/140	58.7	68

Figure 24.5.6: Table of comparison to state-of-the-art 60GHz CMOS PAs.

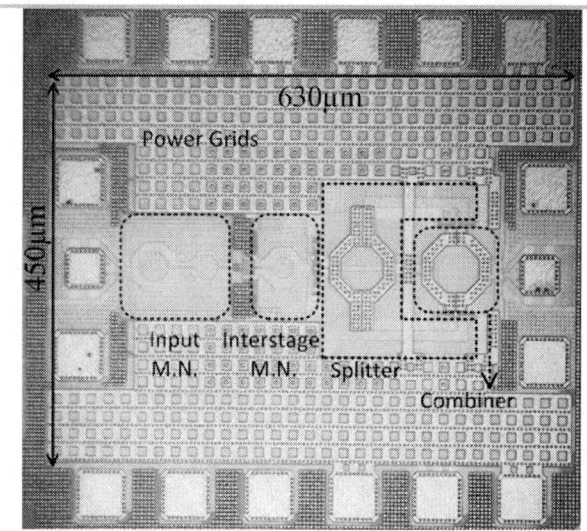

Figure 24.5.7: Chip micrograph of the PA.

ISSCC 2011 / SESSION 25 / CDRs & EQUALIZATION TECHNIQUES / OVERVIEW

Session 25 Overview / Wireline

CDRs & Equalization Techniques

Session Chair: *SeongHwan Cho, KAIST, Daejon, Korea*

Session Co-Chair: *Tatsuya Saito, Hitachi, Tokyo, Japan*

The explosive demand for high-bandwidth low-power chip-to-chip communication in severe channel conditions calls for innovations in transceiver architectures and circuits. Clock-and-data recovery (CDR) and equalization are both essential techniques in wireline communications to produce low-jitter clock and low-BER data. The first four papers in this session address challenges in CDRs by employing digital-friendly techniques for nanometer CMOS processes. The next three papers present innovative equalization techniques to address the adverse effects of lossy channels such as inter-symbol interference, crosstalk, and dispersion. The last paper addresses issues in emerging video interface using an embedded clock technique.

Paper 25.1 (University of Toronto; Fujitsu Laboratories) presents an adaptive DFE for a 2× blind ADC-based receiver. The work employs a triangular waveform to adjust the equalization coefficients in order to shape the equalizer output with blind sampling. The adaptive engine restores a 5Gb/s received eye over a channel with 13dB of attenuation at Nyquist frequency to an equivalent of 320mV of vertical opening. The digital CDR is implemented in 65nm CMOS and consumes 78mW from a 1.2V supply.

Paper 25.2 (Oregon State University) shows a reference-less half-rate digital CDR with very wide frequency acquisition range, while improving the tolerance to input duty cycle errors. The recovered clock is generated by passing the random input data sequence through a series of dividers which results in a sub-harmonic spectral peak. Fabricated in 0.13μm CMOS, the CDR consumes 6.1mW at 2Gb/s from a 1.2V supply and operates from 0.5 to 2.5Gb/s.

Paper 25.3 (Oregon State University) describes a digital CDR that combines linear and bang-bang phase detectors to decouple the design trade-off between jitter generation and jitter transfer. Fabricated in 0.13μm CMOS, it achieves an operating range of 0.5 to 3.2Gb/s. The chip consumes 7mW from a 1.2V supply when operating at 2.5Gb/s with the recovered rms jitter of 6ps.

Paper 25.4 (Broadcom) presents a digital CDR using a double-edge triggered shift register to speed up the phase rotation and a latency-minimized loop filter to widen the bandwidth. It achieves ±1.56% tracking range and up to 10MHz tracking bandwidth at 8Gb/s in 40nm CMOS. At 8Gb/s, the CDR consumes 12mW from a 0.81V supply.

Paper 25.5 (National Taiwan University) presents an adaptive equalizer using blind sampling, which consists of an analog equalizer and a DFE. Fabricated in 65nm CMOS, it achieves a data rate of 20Gb/s with BER<10^{-12} over a 35cm FR4 board while consuming 52mW from a 1.2V supply.

Paper 25.6 (Caltech) demonstrates a 2-tap DFE receiver by using a switched-capacitor S/H/summer front-end, which enables FEXT cancellation with 33μW/Gb/s/lane power overhead. Implemented in a 45nm SOI technology, it equalizes 15Gb/s data over a link with >14dB loss and dissipates 7.5mW from a 1.2V supply.

Paper 25.7 (NetLogic Microsystems) presents a transmit equalizer designed for 10Gb/s serial communication which features half-UI IIR and FIR taps. Both SFI transmitter waveform dispersion penalty and data-dependent jitter specifications are met with a single transmitter configuration. The circuit, fabricated in 40nm CMOS, dissipates 125mW and occupies 0.22 mm².

Paper 25.8 (Seoul National University) presents a 3.0Gb/s video-data interface that minimizes the complexity in the receiver. The DLL generates multiphase clocks that sample 8b data from a data stream with 3b-wide delimiter of zero-to-one transition. Fabricated in 0.13μm CMOS, the receiver operates from 1.36 to 3.0Gb/s with a BER of <10^{-12} while consuming 13.8mW from a 1.2V supply at 3.0Gb/s.

The architectures and circuit innovations presented in this session provide important techniques to satisfy the demand for higher aggregate bandwidth and lower power consumption.

978-1-61284-303-2/11 $26.00 © 2011 IEEE

Presenters:

25.1 A 5Gb/s Adaptive DFE for 2× Blind ADC-Based CDR in 65nm CMOS
                      1:30 PM

B. Abiri,
University of Toronto, Toronto, Canada

25.2 A 0.5-to-2.5Gb/s Reference-less Half-Rate Digital CDR with Unlimited Frequency Acquisition Range and Improved Input Duty-Cycle Error Tolerance
    2:00 PM

R. Inti,
Oregon State University, Corvallis, OR

25.3 A TDC-less 7mW 2.5Gb/s Digital CDR with Linear Loop Dynamics and Offset-Free Data Recovery
    2:30 PM

W. Yin,
Oregon State University, Corvallis, OR

25.4 A Digital Wideband CDR with ±15.6Kppm Frequency Tracking at 8Gb/s in 40nm CMOS
    3:15 PM

H. Pan,
Broadcom, Irvine, CA

25.5 A 20Gb/s Digitally Adaptive Equalizer/DFE with Blind Sampling
    3:45 PM

Y-M. Ying,
National Taiwan University, Taipei, Taiwan

25.6 A 15Gb/s 0.5mW/Gb/s 2-Tap DFE Receiver with Far-End Crosstalk Cancellation
    4:15 PM

M. Honarvar Nazari,
California Institute of Technology, Pasadena, CA

25.7 A 10Gb/s Half-UI IIR-Tap Transmitter in 40nm CMOS
    4:45 PM

H. Cirit,
Netlogic Microsystems, Santa Clara, CA

25.8 A 13.8mW 3.0Gb/s Clock-Embedded Video Interface with DLL-Based Data-Recovery Circuit
    5:00 PM

S. Jang,
Seoul National University, Seoul, Korea

ISSCC 2011 / SESSION 25 / CDRs & EQUALIZATION TECHNIQUES / 25.1

25.1 A 5Gb/s Adaptive DFE for 2× Blind ADC-Based CDR in 65nm CMOS

Behrooz Abiri[1], Ali Sheikholeslami[1], Hirotaka Tamura[2], Masaya Kibune[2]

[1]University of Toronto, Toronto, Canada
[2]Fujitsu Laboratories, Kawasaki, Japan

ADC-based receivers allow for extensive equalization in the digital domain and therefore can easily compensate for channel loss at higher data rates [1]. Digital equalization can be implemented as an FFE [2] or DFE [3]. An adaptive FFE is straight forward to implement [2], as it relies on magnitudes only (not phases) of the blind samples, however, it enhances the quantization noise of the ADC. A DFE has better noise immunity [3], but challenges remain in designing an adaptive DFE for receivers with blind clocks. A typical adaptive engine for a DFE compares the equalized center samples against a desired level to calculate error in each sample. The power of this error is then minimized to guide equalization. In a blind receiver, however, the center samples are not known; the blind samples may deviate from the eye's center and move closer to the zero crossings. As a result, one cannot use a single desired level independent of the blind sampling phase; the desired level must change according to the sampling phase. In this paper, we use a *desired waveform*, instead of a *desired level*, to perform the adaptation. Our measurement results confirm that this approach adapts the coefficients within 80µs and opens an otherwise closed eye for a channel attenuation of 13.3dB at 2.5GHz.

A simplified block diagram of a 2× blind ADC-based receiver with an adaptive DFE is demonstrated in Fig. 25.1.1. The received 5Gb/s signal is blindly sampled at 10GS/s using four interleaved 5b flash ADCs. The ADC uses a 4-phase 2.5GHz clock that is generated from an external 5GHz reference clock. The 5-bit samples are DMUXed further to produce 16 samples of 5-bit each (corresponding to a total of 8 UIs). All other blocks in this diagram operate at the reduced clock rate of 625MHz. The 5-bit samples are then passed to a 1-tap speculative DFE, where the contribution of the previous bit is subtracted (or added) from each sample. This contribution, which defines the DFE coefficient (α), depends on the sampling phase of the blind clock (which is not fixed). Accordingly, we define eight such coefficients (α_1 to α_8) corresponding to eight phase bins in one UI [3] and choose one for each blind sample based on the sampling phase extracted in the digital CDR. The task of the adaptive engine is to determine these eight coefficients automatically while the link is in full operation (i.e., no training sequence). The equalized samples (S'_{EQ} and S_{EQ}) are then passed to the digital CDR [4] where the average phase of the zero crossings (Φ_{AVE}) is extracted and the data is recovered.

The detailed implementation of the adaptive engine is presented in Fig. 25.1.2. The speculative DFE produces two sets of 16 equalized samples corresponding to 8 UIs. For adaptation, 2 samples from each set ($S'_{EQ-8,9}$ and $S_{EQ-8,9}$) corresponding to 1 UI out of 8 UIs are used. A 2:1 MUX selects either $S'_{EQ-8,9}$ or $S_{EQ-8,9}$ based on the previous recovered bit and denotes these by $S_{EQ-8,9}$, which are the equalized signals corresponding to two blind samples in 1 UI. These equalized signals should be compared against their corresponding reference levels ($d_{ref-1,2}$) which are provided by the desired waveform generator block (discussed later). The result of subtracting $d_{ref-1,2}$ from $S_{EQ-8,9}$ is 2 equalization error signals whose values depend on the remaining inter-symbol interference (ISI) in the equalized samples. The goal of adaptation engine is to find coefficients α_1 to α_8 such that these errors are minimized regardless of the sampling phase of blind clock. To this end, we multiply the error by the previous bit to obtain the residual 1st post-cursor ISI. We then accumulate these residual ISI to obtain the ISI, or equivalently the DFE coefficient.

The desired waveform generator produces valid desired references whenever there is a data transition. If no transition occurs, the error values are incorrect and are discarded. This transition filtering, implemented by 2:1 MUX after the multiplication in Fig. 25.1.2, reduces the adaptation speed but makes the design of desired waveform generator simpler. The 1:8 DMUX selects the accumulators that correspond to the current sampling phase of the blind clock. The outputs of accumulators (α_1 to α_8) are the DFE coefficients corresponding to 8-phase bins of 1 UI. Depending on the sampling phase of the clock, two of these coefficients

(α_i, α_{i+4}) are provided to the DFE to be subtracted from the two adjacent samples of ADC, which are ½ UI apart.

The PD in the digital CDR uses linear interpolation to recover zero-crossings [4]. For this PD, a triangular waveform is the ideal waveform as its interpolated zero-crossings coincide with the actual zero-crossings. Figure 25.1.3 shows how the amplitude of the desired waveform is generated. Two adjacent ADC samples (S_8 and S_9) are fed to a 2:1 MUX which selects one of the two based on proximity to eye's center. Ideally the value at the center of the eye should be interpolated to provide an estimate of the eye height, however, this increases hardware complexity. The absolute value of the selected sample is averaged using a first-order digital lowpass filter to produce S_{AVE}. Desired references are then generated from S_{AVE} and Φ_{AVE} in a dynamic lookup table with the stored shape of the triangular waveform. We use two different dynamic look up tables so that $d_{ref-1,2}$ correspond to 2 blind samples of ½ UI apart, corresponding to two different reference levels.

We use two Tyco FR4 channels with 9.9 and 13.3dB attenuation at 2.5GHz (26-inch and 34-inch channels) for measurement. Figure 25.1.4 shows the measured learning curves of adaptation engine for these channels. The adaptation converges in less than 80µs. As expected, the DFE coefficients are higher for the channel with higher attenuation.

Figure 25.1.5 shows the measured eye diagrams at the ADC input (analog), ADC output (prior to equalization), and the equalized output for two distinct channels described earlier. The first row shows the eye diagram at the output of the channels. The second row shows the eye diagram at the output of the ADC and prior to the DFE. The third row shows the eye diagram after equalization with adaptive 1-tap DFE (S_{EQ-8} in Fig. 25.1.2). For the 13.3dB channel, the adaptation engine is capable of creating a 320mV opening from a completely closed eye. To generate these eye diagrams, a frequency offset is added to the blind clock. This offset causes the sampling clock to scan the entire input eye once every period of the offset. ADC samples are then overlapped accordingly to reconstruct the eye diagrams.

Figure 25.1.6 shows the jitter tolerance of receiver for PRBS7 with BER<10^{-12} and adapted coefficients for two mentioned channels. Sinusoidal jitter frequencies below 80kHz and above 8MHz could not be applied due to equipment limitations.

The chip micrograph as well as the power and area breakdowns is shown in Fig. 25.1.7. The chip is fabricated in a 65nm CMOS and occupies 0.4mm^2. The receiver operates from 1.2V supply and consumes 192mW, of which 114mW is consumed by the flash ADC and 78mW by the digital CDR and equalization/adaptation.

Acknowledgement:
The authors would like to thank Ravi Shivnaraine for his help with the measurements.

References:
[1] M. Harwood, N. Warke, R. Simpson, et al., "A 12.5Gb/s SerDes in 65nm CMOS Using a Baud-Rate ADC with Digital Receiver Equalization and Clock Recovery," *ISSCC Digest of Tech. Papers*, pp. 436-591, Feb., 2007.
[2] H. Yamaguchi, H. Tamura, Y. Doi, et al., "A 5Gb/s transceiver with an ADC-based feedforward CDR and CMA adaptive equalizer in 65nm CMOS," *ISSCC Digest of Tech. Papers*, pp. 168-169, Feb., 2010.
[3] S. Sarvari, T. Tahmoureszadeh, A. Sheikholeslami, et al., "A 5Gb/s Speculative DFE for 2x Blind ADC-based Receivers in 65-nm CMOS," *IEEE Symposium on VLSI Circuits*, pp. 69-70, June, 2010.
[4] O. Tyshchenko, A. Sheikholeslami, H. Tamura, et al., "A 5Gb/s ADC-Based Feed-Forward CDR in 65nm CMOS," *IEEE J. Solid-State Circuits*, vol. 45, no. 6, pp. 1091-1098, June, 2010.

ISSCC 2011 / February 23, 2011 / 1:30 PM

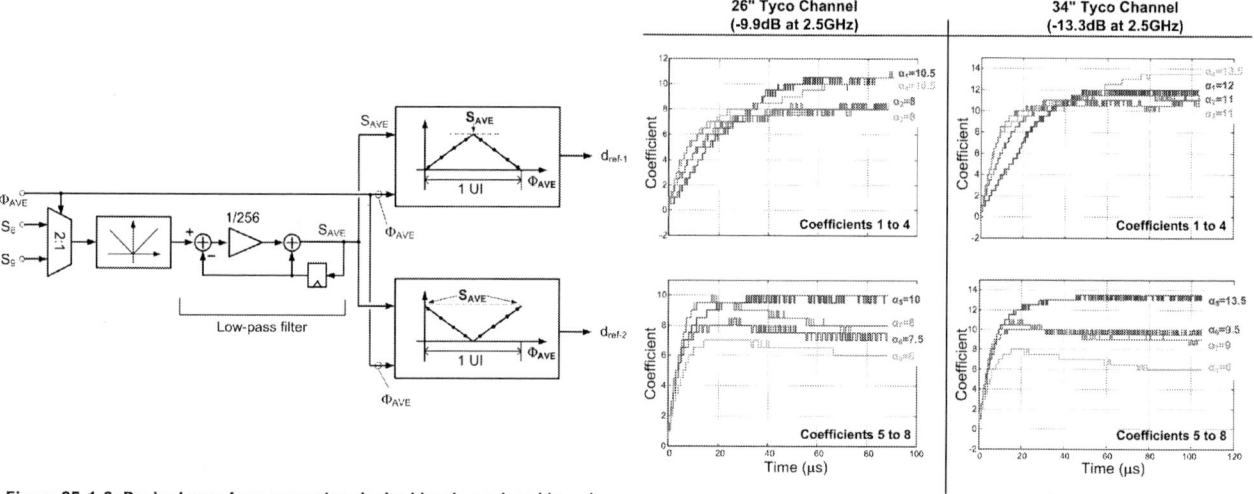

Figure 25.1.1: Adaptive DFE for 2× blind ADC-based receiver.

Figure 25.1.2: Adaptation engine implementation.

Figure 25.1.3: Desired waveform generator: desired levels produced based on average phase.

Figure 25.1.4: Measured learning curves for 26" and 34" Tyco channels.

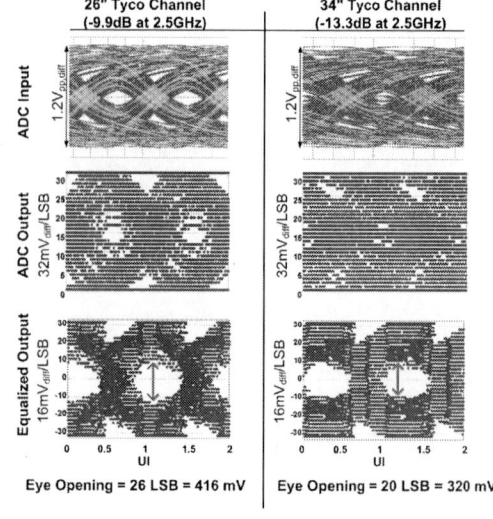

Figure 25.1.5: Measured eye diagrams for a 26" and 34" Tyco channels.

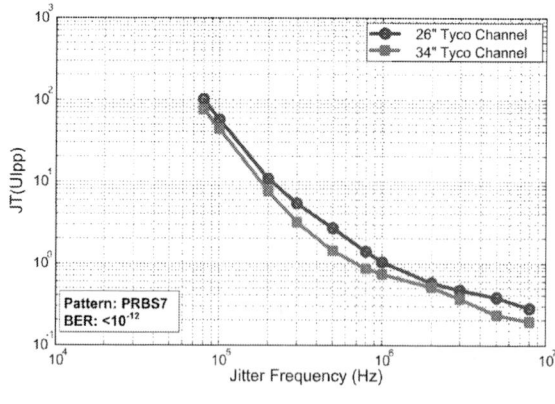

Figure 25.1.6: Measured jitter tolerance for 26" and 34" Tyco channels.

978-1-61284-303-2/11 $26.00 © 2011 IEEE

ISSCC 2011 PAPER CONTINUATIONS

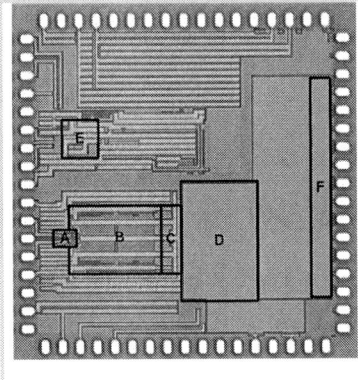

Process	65-nm CMOS
Data Rate	5 Gb/s
Supply	1.2V
ADC Power	114mW
Digital Power	78mW
Total Power	192mW

A	Input Buffers	$50\times60\mu m^2$
B	4 × 2.5GSa/s ADC	$400\times490\mu m^2$
C	4:16 DeMUX	$60\times490\mu m^2$
D	Digital CDR/DFE + Test Structures	$420\times640\mu m^2$
E	BGR and Bias Gen.	$170\times140\mu m^2$
F	Pad Drivers	$60\times1200\mu m^2$

Figure 25.1.7: Die micrograph.

ISSCC 2011 / SESSION 25 / CDRs & EQUALIZATION TECHNIQUES / 25.2

25.2 A 0.5-to-2.5Gb/s Reference-less Half-Rate Digital CDR with Unlimited Frequency Acquisition Range and Improved Input Duty-Cycle Error Tolerance

Rajesh Inti, Wenjing Yin, Amr Elshazly, Naga Sasidhar, Pavan Kumar Hanumolu

Oregon State University, Corvallis, OR

Clock and data recovery (CDR) circuits with wide frequency acquisition range offer flexibility in optical communication networks, help reduce link power through activity-based rate adaptation, and minimize cost with a single-chip multi-standard solution. Extracting the bit rate from the incoming random data stream is the main challenge in implementing reference-less CDRs. A conventional rotational frequency detector has a limited acquisition range of about ±50% of the VCO frequency, consumes large power, and is susceptible to harmonic locking. Extending its range requires additional high-speed circuitry and a complex state machine [1]. The DLL-based architecture in [2] requires passing high-speed data through a long string of power-hungry buffers, imposes stringent matching requirements, and works only with ring oscillators. Other approaches require detailed statistical [3] or timing analysis [4]. Further, all the above techniques are only suitable for full-rate CDRs. In this paper, we present a reference-less half-rate CDR that uses a sub-harmonic extraction method to achieve unlimited frequency acquisition range. This technique is capable of locking the CDR to within 40ppm of any sub-rate of the data (making it applicable for any sub-rate CDR architecture), while being immune to undesirable harmonic locking. This CDR also integrates a calibration loop to improve robustness to input duty cycle error.

Figure 25.2.1 shows the digital CDR architecture. It is composed of: a frequency-locking loop (FLL), a phase-locking loop (PLL), and data duty-cycle-correction loop (DCCL). The FLL consisting of a frequency discriminator (FD), an accumulator, and a $\Delta\Sigma$ DAC drives the VCO towards frequency lock. When the frequency error is within the PLL's pull-in range, it acquires phase lock. Because both the FLL and the PLL are driven by the incoming data, they can operate simultaneously. However, to prevent any interaction between the two loops, the FLL bandwidth is made much lower than that of the PLL.

The digital PLL consists of a half-rate bang-bang phase detector (BBPD), a digital loop filter, and a $\Delta\Sigma$ DAC to drive the VCO. The BBPD compares input data edges and the VCO clock phases and generates two pairs of early/late (E/L) signals that directly drive the VCO through a 4-level current-mode DAC. The E/L data are decimated by a factor of 16 before feeding to a low frequency accumulator. A $\Delta\Sigma$ DAC consisting of a digital modulator and a lowpass filter drives the VCO. The digital DCCL estimates the input data duty-cycle error and calibrates the VCO clock phases for optimal sampling of both the EVEN and ODD data bits.

Figure 25.2.2 illustrates the block diagram of the FD circuit. The power spectral density (PSD) of a random binary NRZ data, D_{IN}, is a squared *sinc* function with nulls at the data rate (F_B in Fig. 25.2.2). When D_{IN} is passed through a divider its PSD is modified resulting in a sub-harmonic spectral peak at $F_B/8$. As illustrated in Fig. 25.2.2, the PSD after the first divide-by-2 stage exhibits a peak at $F_B/8$ and its strength at $F_B/16$ increases by 6dB at the output of the subsequent stage. It can be shown that each divide-by-2 stage increases the sub-harmonic peak strength, eventually resulting in a clock-like output (REF) tone after the 10th stage at $F_B/4096$. Intuitively, a 2^{10} divider acts as an asynchronous modulo-2^{10} counter with its output toggling whenever the number of low-to-high data transitions reaches 2^{10}. Therefore, for binary random data with equal low-to-high and high-to-low transitions, the average frequency of the divider output is equal to $0.5F_B/2^{10+1}$.

Frequency difference between the sub-harmonic tone at $F_B/4096$ and the divided VCO output is determined using a counting-type frequency detector. The 14-bit counter output is sampled on the positive edge of REF, and the difference between two consecutive samples gives the frequency error (F_{ERR}). By virtue of subtracting 128 from the output, F_{ERR} equals $F_B/4096 - F_{VCO}/2048$, thereby forcing F_{VCO} to $F_B/2$ in steady state, as desired in a half-rate CDR. This architecture allows the VCO frequency to be locked to any sub-harmonic of the data rate by appropriately choosing the feedback/feed-forward divider values.

Figure 25.2.3 shows the block diagram of the duty cycle estimator and illustrates the calibration principle of the DCCL. At start-up, E_{EVEN}/L_{EVEN} signals corresponding to EVEN data bits are masked and the I-phase of the VCO is aligned with the edge of the ODD data bit using only E_{ODD}/L_{ODD} information. Ideally, under this condition, the Ib'-phase aligns with the edge of the EVEN data bit. However, due to the input duty cycle error, if the ODD data bit is larger (or smaller) than the EVEN data bit, Ib'-phase is always early (or late) (Fig. 25.2.3). The sign of the duty-cycle error determined by the slope detector is used to either advance or delay Ib so as to align it to the edge of the EVEN data bit. Considering the eye diagrams shown in Fig. 25.2.3, when the ODD data eye is larger by αUI, for optimal sampling, Ib and Q/Qb have to be delayed by αUI and 0.5αUI, respectively. This is achieved using a linear digital-to-delay converter (DDC).

Figure 25.2.4 shows the schematic of a linear DDC. Since the delay of a current-starved inverter is inversely proportional to the control current, a DDC implemented by digitally scaling the current exhibits 1/x nonlinearity. This makes it impossible to realize αUI and 0.5αUI delays by simply scaling the input. Pre-warping the control current by 1/x function eliminates this non-linearity. To this end, a digitally controlled resistor is used to make the output current vary inversely with the input digital control word ($I_{OUT} \propto 1/D_C$), thus linearizing the DDC transfer curve. A delay of 0.5αUI is obtained by bit shifting D_C (Fig. 25.2.4). The delay range can be adjusted by using an appropriate reference voltage, V_{REF}, and is chosen to cover up to ±20% input data duty cycle error. The simulated delay characteristics of the DDC shown in Fig. 25.2.4 illustrate the linear behavior and range variation with V_{REF}.

The digital CDR is implemented in 0.13µm CMOS and operates from 0.5Gb/s to 2.5Gb/s with a 1.2V supply. The operating range is limited only by the VCO tuning range and can be extended using programmable dividers as in [1]. Figure 25.2.5 illustrates the measured performance of the FLL. Because the low-to-high and high-to-low transitions of a PRBS sequence are not equally likely, the FLL locks with a frequency offset equal to the difference in their probabilities. For instance, PRBS10 and PRBS15 sequences lock with a frequency offset of about 1000ppm ($1/2^{10}$) and 20ppm ($1/2^{15}$), respectively. Scrambling with a polynomial $p(x)=1+x^{18}+x^{23}$ randomizes the input data and improves the accuracy to better than 40ppm, irrespective of the input sequence periodicity (Fig. 25.2.5). The recovered clock spectrums for PRBS10/PRBS15 input patterns are also shown in Fig. 25.2.5. With a 2Gb/s PRBS10 input data, the CDR achieves BER < 10^{-12} and recovered clock jitter is 5.4ps_{rms} (44ps_{pp}). It dissipates 6.1mW (from a 1.2V supply) of which the FLL consumes 1.2mW.

The proposed CDR is compared with state-of-the-art reference-less/digital CDRs in Fig. 25.2.6. The highly digital nature of the CDR allows it to operate with a lower supply voltage of 0.8V and 0.2-to-1Gb/s data rate. Power efficiency is 10 times better than all reported reference-less CDRs. Die micrograph is shown in Fig. 25.2.7.

Acknowledgments:
NSF under CAREER EECS-0954969 and Intel for financial support, Dongbu HiTek for IC fabrication and Ganesh Balamurugan of Intel for help with testing.

References:
[1] D. Dalton, K. Chai, E. Evans, M. Ferriss, D. Hitchcox, P. Murray, S. Selvanayagam, P. Shepherd, L. DeVito, "A 12.5Mb/s to 2.7Gb/s Continuous-Rate CDR with Automatic Frequency Acquisition and Data-Rate Read Back," *ISSCC Dig. Tech. Papers*, pp. 230-231, Feb., 2005.
[2] S. K. Lee, Y. S. Kim, H. Ha, Y. Seo, H. J. Park, J. Y. Sim, "A 650Mb/s-to-8Gb/s Referenceless CDR Circuit with Automatic Acquisition of Data Rate," *ISSCC Dig. Tech. Papers*, pp. 184-185, Feb., 2009.
[3] M. H. Perrott, Y. Huang, R. T. Baird, B. W. Garlepp, L. Zhang, J. P. Hein, "A 2.5Gb/s Multi-Rate 0.25µm CMOS CDR Utilizing a Hybrid Analog/Digital Loop Filter," *ISSCC Dig. Tech. Papers*, pp. 328-329, Feb., 2006.
[4] R. Yang, K. Chao, S. Hwu, C. Liang, and S. Liu, "A 155.52 Mbps-3.125 Gb/s Continuous-Rate Clock and Data Recovery Circuit," *IEEE J. Solid-State Circuits*, vol. 41, no. 6, pp. 1380-1390, June. 2006.
[5] D. H. Oh, D. S. Kim, S. Kim, D. K. Jeong, W. Kim, "A 2.8 Gb/s All-Digital CDR with a 10b Monotonic DCO," *ISSCC Dig. Tech. Papers*, pp. 222-223, Feb., 2007.

ISSCC 2011 / February 23, 2011 / 2:00 PM

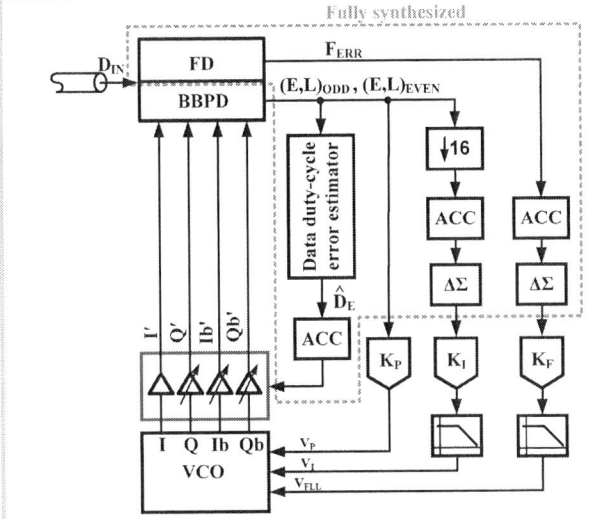

Figure 25.2.1: Proposed digital CDR architecture.

Figure 25.2.2: Frequency discriminator block diagram and representative intermediate random data PSDs.

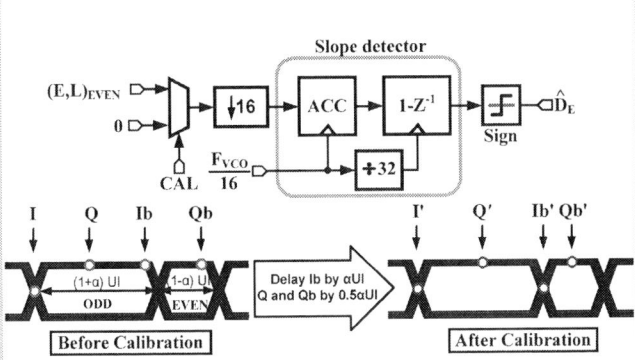

Figure 25.2.3: Input data duty-cycle-error estimator and calibration principle.

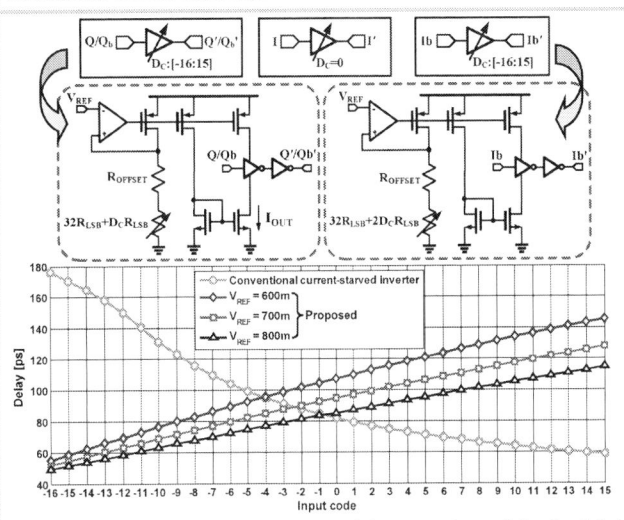

Figure 25.2.4: Schematic of the digital-to-delay converter and its simulated transfer characteristics.

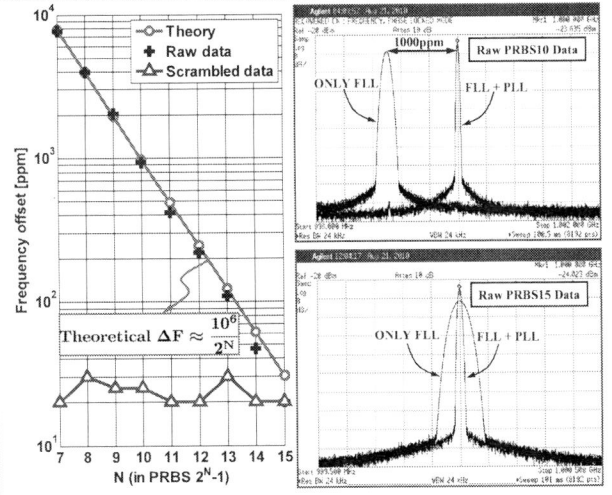

Figure 25.2.5: Measured steady-state FLL frequency offset (ΔF) and recovered clock spectrum (FLL/PLL modes) for raw PRBS10 and PRBS15 data patterns.

Figure 25.2.6: Performance summary and comparison to the prior art.

	ISSCC'05 [1]	ISSCC'09 [2]	ISSCC'06 [3]	JSSC'06 [4]	ISSCC'07 [5]	This work
Technology	0.13μm	65nm	0.25μm	0.25μm	0.13μm	0.13μm
Supply voltage	3.3V	1.2	1.8V	1.2V		0.8V/1.2V
Architecture	Full-rate	Full-rate	Full-rate	Full-rate	Full-rate	Half-rate
Data rate	0.0125-2.7Gbps	0.65-8Gbps	0.15/0.6/1.2/ 2.5Gbps	0.15-3.1Gbps	2.8Gbps	0.2-1.0Gbps/ 0.5-2.5Gbps
Filter	Analog	Analog	Digital	Analog	Digital	Digital
Acquisition	Reference-less	Reference-less	Reference-less	Reference-less	Reference	Reference-less
Jitter [rms/pp]	N/A	94.7ps/578ps @0.65Gbps	1.2ps/N/A @2.5Gbps	6.4ps/48.9ps @3.125Gbps	7.2ps/47.2ps @2.5Gbps	5.4ps/44ps @2Gbps
Acquisition time	1ms	N/A	25ms	N/A	N/A	0.25ms
Power	775.5mW @2.5Gbps	20.6mW @0.65Gbps	425mW @2.5Gbps	95mW @3.125Gbps	13.2mW @2.5Gbps	6.1mW @2Gbps
Power FOM [mW/Gbps]	310.2	31.7	170	30.4	4.72	3.05
Area	9mm²	0.11mm²	20mm²	0.88mm²	0.13mm²	0.36mm²

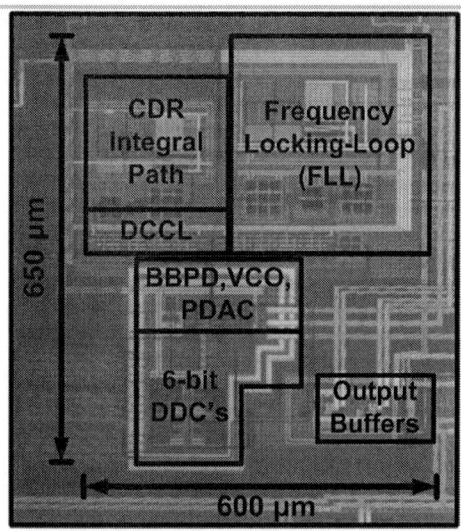

Figure 25.2.7: Die micrograph.

ISSCC 2011 / SESSION 25 / CDRs & EQUALIZATION TECHNIQUES / 25.3

25.3 A TDC-less 7mW 2.5Gb/s Digital CDR with Linear Loop Dynamics and Offset-Free Data Recovery

Wenjing Yin, Rajesh Inti, Amr Elshazly, Pavan Kumar Hanumolu

Oregon State University, Corvallis, OR

A clock and data recovery (CDR) circuit is the key building block in all serial communication systems. A classical CDR is implemented using a Type-2 phase-locked loop (PLL) wherein a passive lead-lag analog loop filter is used to set the loop response. Large capacitors needed to achieve low jitter transfer bandwidth and a highly over-damped response to reduce jitter peaking prohibit monolithic integration of the analog loop filter [1, 2]. Digital loop filters (DLFs) that are robust to process and temperature variations have recently emerged as an alternate solution to implementing fully integrated CDRs [3-5].

A bang-bang phase detector provides a simple interface to the DLF without any inherent static phase offset (SPO) [1, 3]. However, it introduces a large phase quantization error that limits the jitter transfer bandwidth, and its nonlinear behavior makes loop dynamics difficult to control. A time-to-digital converter (TDC) that is realized using a linear phase detector followed by an ADC could provide fixed gain that enables well-controlled loop dynamics and achieve the desired jitter transfer bandwidth [4, 5]. However, it is susceptible to ADC non-idealities that manifest as large SPO, as well as additional power for the ADC. In this paper, the advantages of bang-bang and linear phase detectors are combined in a digital CDR to achieve error-free operation and fixed jitter transfer bandwidth, independent of input jitter amplitude. The CDR decouples the tradeoff between jitter generation and jitter transfer by eliminating the phase quantization error in the proportional path, which allows wide loop bandwidth to suppress DCO phase noise and frequency quantization error.

Figure 25.3.1 illustrates the proposed CDR architecture. It consists of a frequency-locking loop (FLL) and a Type-2 digital PLL. At start-up, the FLL drives the DCO frequency to within 500ppm of any incoming data rate in a 0.5 to 3.2Gb/s range. The PLL consisting of a linear proportional path and a bang-bang digital integral path achieves frequency and phase lock. A bang-bang Alexander detector recovers the data without any systematic SPO and generates the early/late information to drive the accumulator in the digital integral path. A decimator allows lower clock frequency operation of the digital circuitry.

The PLL proportional path is implemented by directly controlling the DCO with a linear Hogge detector, thus eliminating the non-linearity and quantization error of a bang-bang detector [3] and the need for a high resolution TDC [4, 5]. Because, only a 3-level DAC is needed to interface Hogge detector outputs to the oscillator, the circuit realization of the proportional path incurs minimal area and power penalty. The infinite DC gain of the digital accumulator forces the bang-bang PD to its metastability point and locks the recovered clock (RCK) to the incoming data without any static phase offset. As a result, any offset in the Hogge detector causes periodic modulation at the data rate and introduces a fixed frequency offset and ripple in the proportional control voltage. The digital integral path automatically compensates the frequency offset, and the deterministic jitter due to the ripple is minimized by introducing a higher order pole at the virtual supply of the DCO.

The CDR is designed for a heavily over-damped response by making the digital integral gain very small compared to the proportional gain. Consequently, jitter peaking is almost completely eliminated and the well-controlled proportional path alone sets the jitter transfer bandwidth. Unlike in a conventional DPLL, because of the fixed gain of the linear PD, a constant jitter transfer bandwidth is achieved. Since bang-bang detector is employed only in the digital integral path, transfer bandwidth can be increased without increasing dither jitter generation.

Besides the dither jitter, jitter generation in conventional digital CDRs arises from the frequency quantization error and the phase noise of the DCO. The finite DCO resolution manifests as deterministic jitter due to jitter accumulation and its magnitude is determined both by the number of consecutive identical digits

(CIDs) and the amount of frequency quantization error. The finite slew-rate of a bang-bang CDR compounds this effect further. In the proposed architecture, sensitivity to the DCO quantization error and CIDs is greatly reduced by the linear proportional path.

The DCO shown in Fig. 25.3.2 is a current-controlled ring oscillator whose frequency is separately controlled by an FLL, digital integral path, and the proportional path through two 14b DACs and a 3-level DAC. Using separate DACs for the frequency control and the integral control relaxes the stringent quantization error requirements otherwise present in a shared DAC architecture [3]. To further ease the precision requirements, high resolution DACs are implemented using a delta-sigma modulator that truncates the 14b input to 15 levels, a thermometer-coded current-mode DAC and a lowpass filter. The proportional control is realized by a 3-level current-mode DAC driven by the pulse width modulated PD outputs D_E and D_R. The DAC current is set to achieve a jitter transfer bandwidth of about 4.5MHz.

The prototype digital CDR is fabricated in a 0.13μm CMOS technology and achieves error-free operation (BER < 10^{-12}) for PRBS data sequences ranging from 2^7-1 to 2^{31}-1 sequence lengths over 0.5Gb/s to 3.2Gb/s data rates. At 2.5Gb/s, the CDR consumes 7mW power from a single 1.2V supply. Figure 25.3.3 shows the measured transfer bandwidth plots for input jitter amplitudes varying from 0.01UI to 0.31UI. The proposed CDR achieves a near-constant bandwidth of 4.5MHz. When configured in conventional bang-bang mode, its bandwidth varies from 3.8MHz to 9.5MHz.

Figure 25.3.4 depicts the measured phase noise of the 2.5GHz recovered clock. The rms jitter, obtained by integrating the phase noise from 10kHz to 1GHz, is 6.1ps and 8.3ps for the proposed and conventional CDRs, respectively. In the conventional CDR, increasing the integral path gain, K_I, increases the DCO frequency quantization error and degrades the jitter from 8.3ps to 20.5ps (Fig. 25.3.4). In contrast, the jitter remained unchanged for the proposed digital CDR, thus minimizing the sensitivity to PVT-induced DCO resolution variation. Figure 25.3.5 depicts the measured recovered clock jitter as a function of input PRBS sequence length. When the length is increased from 2^7-1 to 2^{15}-1, jitter degrades from 6ps to 10.7ps and 9ps to 23ps for the proposed and conventional cases, respectively, thus validating the proposed CDR's improved immunity to CIDs. In the conventional CDR increasing jitter transfer bandwidth degraded recovered clock jitter due to increased dither jitter in the proportional path (Fig. 25.3.5). However, in the proposed CDR the bandwidth can be increased without introducing dither jitter, thereby reducing DCO phase noise-induced recovered clock jitter. The measured jitter tolerance is greater than 0.41UI. Figure 25.3.6 summarizes the performance and comparison to the prior art. Figure 25.3.7 shows the micrograph of the chip. The active die area is 0.2mm^2.

Acknowledgments:
Semiconductor Research Corporation (SRC) provided partial financial support under contract 2007-HJ-1597. Dongbu HiTek provided IC fabrication. Kawasaki Microelectronics America provided partial testing support.

References:
[1] D. Dalton, K. Chai, E. Evans, M. Ferriss, D. Hitchcox, P. Murray, S. Selvanayagam, P. Shepherd, and L. DeVito, "A 12.5-Mb/s to 2.7-Gb/s Continuous-Rate CDR with Automatic Frequency Acquisition and Data-Rate Read Back", *ISSCC Dig. Tech. Papers*, pp. 230-231, Feb., 2005.
[2] J. Kenney, D. Dalton, M. Eskiyerli, et al., "A 9.95 to 11.1Gb/s XFP Transceiver in 0.13μm CMOS," *ISSCC Dig. Tech. Papers*, pp. 864-873, Feb., 2006.
[3] D. H. Oh, D. S. Kim, S. Kim, D. K. Jeong, and W. Kim, "A 2.8 Gb/s All-Digital CDR with a 10b Monotonic DCO," *ISSCC Dig. Tech. Papers*, pp. 222-223, Feb., 2007.
[4] M. H. Perrott, Y. Huang, R. T. Baird, B. W. Garlepp, L. Zhang, and J. P. Hein, "A 2.5 Gb/s Multi-Rate 0.25μm CMOS CDR Utilizing a Hybrid Analog/Digital Loop Filter," *ISSCC Dig. Tech. Papers*, pp. 328-329, Feb., 2006.
[5] K. Fukuda, H. Yamashita, F. Yuki, et al., "10Gb/s Receiver with Track-and-Hold-Type Linear Phase Detector and Charge-Redistribution 1st-Order ΔΣ Modulator," *ISSCC Dig. Tech. Papers*, pp. 186-187, Feb., 2009.

ISSCC 2011 / February 23, 2011 / 2:30 PM

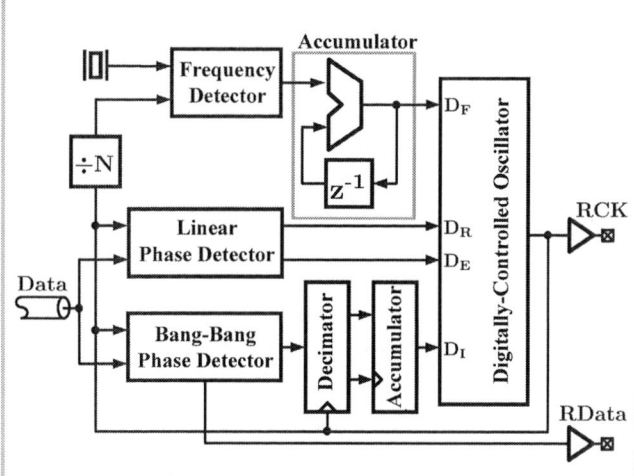

Figure 25.3.1: Block diagram of the proposed digital CDR.

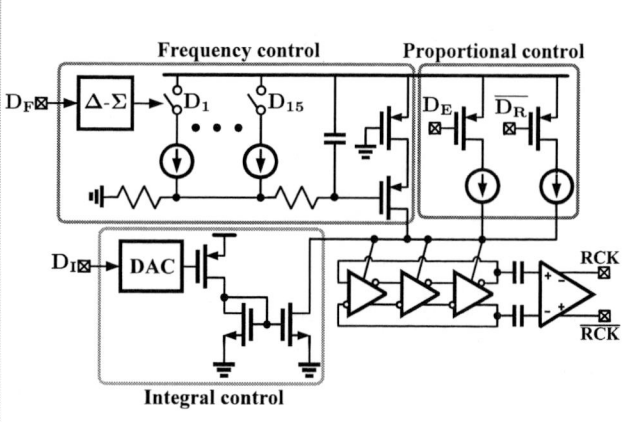

Figure 25.3.2: Schematic of the DCO.

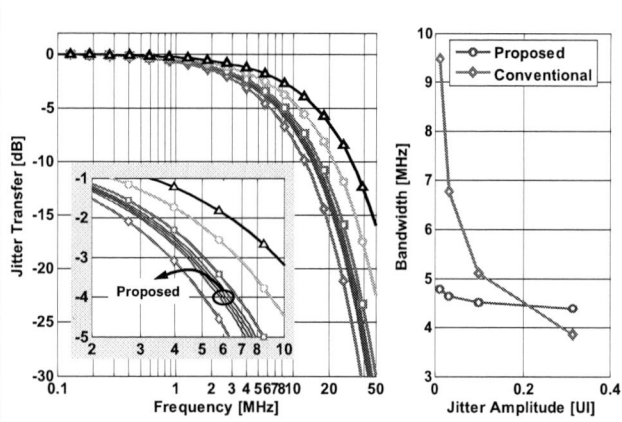

Figure 25.3.3: Measured jitter transfer bandwidth of the proposed and conventional CDRs for different input jitter amplitudes.

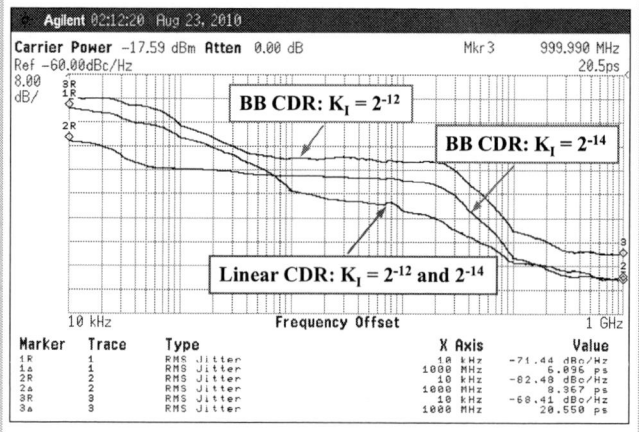

Figure 25.3.4: Measured phase noise of the proposed and conventional CDRs for two integral path gains.

Figure 25.3.5: Measured recovered clock jitter as a function of jitter transfer bandwidth and input PRBS sequence length.

Figure 25.3.6: Performance summary and comparison with prior art.

	ISSCC 05 [1]	ISSCC 07 [3]	ISSCC 06 [4]	ISSCC 09 [5]	This work
Technology	0.35μm	0.13μm	0.25μm	65nm	0.13μm
Supply voltage	3.3V	1.2V	2.5V/3.3V	1.2V	1.2V
Data rate	0.0125-2.7Gbps	2.8Gbps	0.15/0.6/1.2/2.5Gbps	10Gbps	0.5-3.2Gbps
Filter	Analog	Digital	Hybrid	Digital	Digital
Phase Detector	N/A	Binary	Linear	Linear	Linear
JTRAN Bandwidth	0.5MHz	N/A	1.4MHz	20MHz	4.5MHz
Jitter [rms/pp]	N/A	7.2ps/47.2ps @2.5Gbps	1.2ps/N/A @2.5Gbps	N/A/14ps @10Gbps	5.7ps/68ps @2.5Gbps
Power	775.5mW @2.5Gbps	13.2mW @2.5Gbps	425mW @2.5Gbps	65mW @10Gbps	7mW @2.5Gbps
Area	9mm²	0.13mm²	25mm²	0.11mm²	0.2mm²

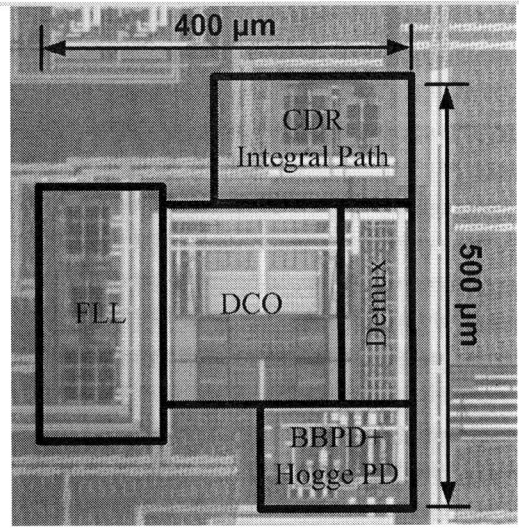

Figure 25.3.7: Die micrograph.

ISSCC 2011 / SESSION 25 / CDRs & EQUALIZATION TECHNIQUES / 25.4

25.4 A Digital Wideband CDR with ±15.6kppm Frequency Tracking at 8Gb/s in 40nm CMOS

Hui Pan[1], Magesh Valliappan[2], Wei Zhang[1], Kambiz Vakilian[1],
Seong-Ho Lee[1], Hamid Hatamkhani[1], Mario Caresosa[1],
Karo Khanoyan[1], Haitao Tong[1], Duke Tran[1], Anthony Brewster[1],
Ichiro Fujimori[1]

[1]Broadcom, Irvine, CA
[2]Broadcom, Austin, TX

It has been well understood that the digital clock and data recovery (CDR) architecture has many system merits over the analog counterpart for multi-Gb/s transceivers [1]. However, the applications have been limited in systems where the clock is forwarded or has small frequency offset [2, 3], due to the finite frequency and jitter tracking capability of the digitally controlled phase rotation. Recently, tracking range up to ±7800ppm has been reported [4] to extend the applications to the SATA/SAS interfaces that require 5000ppm spread spectrum clocking (SSC) to suppress electromagnetic emissions. To enable broad acceptance in high-speed applications, the digital CDRs must have much wider tracking range.

First, it is highly desirable to use an on-chip (CMOS) clock available at any frequency F_{ref} as the reference for the fixed frequency PLL. This requires the CDR to track any fractional frequency offset within $\pm F_{ref}$ assuming the integer feedback divider ratio N programmable at a step size of 2. The ratio $\pm F_{ref} / F_{baud}$ amounts to ±5kppm, given F_{ref} in the low tens MHz range (e.g., 27MHz) and the data rate F_{baud} in the multi-Gb/s range (e.g., 6Gb/s). Second, the CDR needs to be wideband to track out the low frequency jitter from the clock reference, the VCO, the power supply and other noise sources in the system. The jitter can be tens of unit intervals (UIs) as is specified by many sinusoidal jitter (SJ) tolerance masks that roll-off at 20dB/dec and settle to $SJ_{min} \sim 0.5UI$ at frequencies $F_{SJ} > F_{baud}/1000$. The SJ translates to a maximum frequency variation of $\pm(2\pi F_{baud}/1000) \times SJ_{min} \sim \pm3000ppm$. Finally, there is an increasing demand for ±5kppm and more SSC. The overall frequency tracking range spec is therefore ±5kppm (offset) ±3kppm (SJ) ±5kppm (SSC) = ±13kppm. This paper presents an 8Gb/s digital CDR that achieves a tracking range of ±15.6kppm and a tracking bandwidth up to 10MHz.

Figure 25.4.1 shows the block diagram of a 1.25 to 8Gb/s SerDes transceiver in 40nm CMOS that incorporates the digital wideband CDR. Along with the adaptive peaking equalization, TX pre-emphasis, and SSC generation in the transceiver, the digital CDR supports applications such as SATA/SAS, USB3.0, PCIe, DisplayPort, EPON, etc. Special features such as the host tracking for HDD SATA and the loop timing for passive optical network (PON) applications are realized by looping the digital phase rotation signal from the CDR to the TX side phase interpolator (PI) through a programmable IIR filter without the need for a costly analog clean-up filter.

The CDR performance is achieved through architecture optimization and circuit innovations. The first two architecture decisions are on the demux ratio and the resolution of phase rotation, which together determine the tracking range. Demultiplexing by 1-to-4 maximizes the speed without causing substantial latency overhead from pipelining the digital operations; a resolution of 32 phase steps is chosen as a result of trade-off between the quantization noise and the rotation speed. As is shown in Fig. 25.4.2, the 32 steps are derived by interpolating on the 4 phases of the full-speed clock generated by a pseudo differential CMOS ring VCO. The CDR loop is closed with the RX PI output clock driving the "data" and "phase" slicers.

A conventional implementation of the digital CDR would involve mapping (decoding) the 5b MSBs of a high resolution (>10b) phase accumulator output to a 32b cyclic code for PI phase selection [1]. The high resolution additions would cause latency and speed bottlenecks. Figure 25.4.3 proposes a CDR loop filter that avoids the bottlenecks using a 32b cyclic shift register as the decoder. The register is loaded with a "bubble" of 8 active (zero) bits surrounded by 24 inactive (one) bits; each 1b shift of the bubble causes one LSB = (1/32) UI phase rotation in the PI output (see Figure 25.4.2). Since the register takes only the

phase rotation signal instead of the phase position as its input, the required resolution is reduced. The speed is further enhanced by distributing the phase accumulator in the proportional and integral paths, resulting in an accumulator of only 4b and a pipeline of only 5 stages running at $F_{baud}/4 = 2GHz$ in the critical path from the bang-bang phase detector through the proportional path to the decoder output. The non-critical integral path runs at half the frequency. The overflow of the 4b accumulator is linearly combined with the integral output over two clock cycles such that the loop filter output LF does not exceed the maximum shift that the register can execute per clock cycle. The linear combination in time domain provides another mechanism to adjust the proportional and integral gains.

The maximum shift per clock cycle must be 2b to meet the ±15.6kppm target as the tracking range is given by ±(demux ratio) × (1/total number of phase steps) × (max phase steps per cycle) = ±(1/4) × (1/32) × 2 = ±15.6kppm. A simple solution is double the step size [4]. However, this doubles the quantization noise and worsens the limit cycle amplitude. Figure 25.4.4 shows a double-edge-triggered shift register that effectively doubles the shift frequency to $F_{baud}/2 = 4GHz$ while being synthesizable at $F_{baud}/4 = 2GHz$. This is how it works.

When the cyclic code in the master register M shifts by 2b at the rising clock edge, the 2 consecutive active bits at one end of the bubble and the 2 inactive bits concatenating the other end of the bubble flip simultaneously while the rest remain unchanged. The simultaneous transitions of the 2 consecutive bits are spread out in time as one in each half clock cycle by transferring the shifted code to a slave register S at the falling edge and taking the first transitioning bit from M and the second from S as the decoder output. A mux makes the bit selection on the basis of the transitioning bit index parity, which is determined by the next shift direction D and the current parity P of the bubble location. The multiplexing is glitch-less because it happens after the register transfer and before the next shift in register M.

Figure 25.4.5 shows the jitter tolerance measured in a real SATA application environment with the PLL bandwidth set low (<1MHz) to filter the reference noise. At 6Gb/s with the adaptive equalization (AEQ) turned on, the high frequency SJ tolerance increases by about the same amount of the ISI jitter injected by the J-BERT. When the data rate is offset from the PLL output clock frequency by 8kppm with SSC = ±2.5kppm applied to the PRBS7 test pattern, the slope changes from 40dB/dec to 20dB/dec for $F_{SJ} < 1MHz$ due to saturation of the CDR integral register. The measured maximum frequency deviation at saturation verifies the ±15.6kppm tracking range. At 8Gb/s, the 3-dB tracking bandwidth scales up to 8~10MHz.

The digital CDR occupies 72×72μm² and dissipates 12mW at 8Gb/s from a 0.81V power supply. The transceiver characteristics are summarized in Fig. 25.4.6. The die micrograph of the dual port transceiver is shown in Fig. 25.4.7 with the digital CDR, the VCO, and each lane highlighted.

Acknowledgements:
The authors thank Fang Lu, David Gee and Rajesh Divecha for digital synthesis and layout; Charlie Rocio and Justin Nguyen for test.

References:
[1] J. L. Sonntag and J. Stonick, "A digital clock and data recovery architecture for multi-gigabit/s binary links," *IEEE J. Solid-State Circuits*, vol. 41, no. 8, pp. 1867–1875, Aug., 2006.
[2] Y.-B. Luo, P. Chen, Q.-T. Chen, et al., "A 250Mb/s-to-3.4Gb/s HDMI Receiver with Adaptive Loop Updating Frequencies and an Adaptive Equalizer," *ISSCC Dig. Tech. Papers*, pp. 190-191, Feb., 2009.
[3] F. Spagna, L. Chen, M. Deshpande, et al., "A 78mW 11.8Gb/s Serial Link Transceiver with Adaptive RX Equalization and Baud-Rate CDR in 32nm CMOS," *ISSCC Dig. Tech. Papers*, pp. 366-367, Feb., 2010.
[4] M. Pozzoni, S. Erba, P. Viola, et al., "A Multi-Standard 1.5 to 10Gb/s Latch-Based 3-Tap DFE Receiver With a SSC Tolerant CDR for Serial Backplane Communication," *IEEE J. Solid-State Circuits*, vol. 44, no. 4, pp. 1306–1314, April, 2009.

978-1-61284-303-2/11 $26.00 © 2011 IEEE

ISSCC 2011 / February 23, 2011 / 3:15 PM

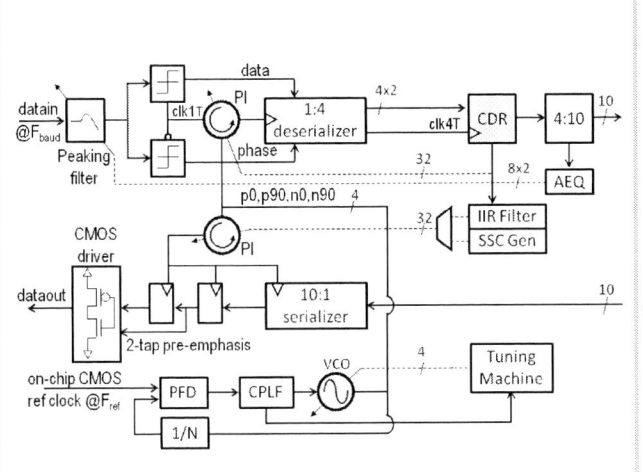

Figure 25.4.1: Block diagram of the transceiver using digital CDR.

Figure 25.4.2: Circuit schematic of the 32-step phase interpolator.

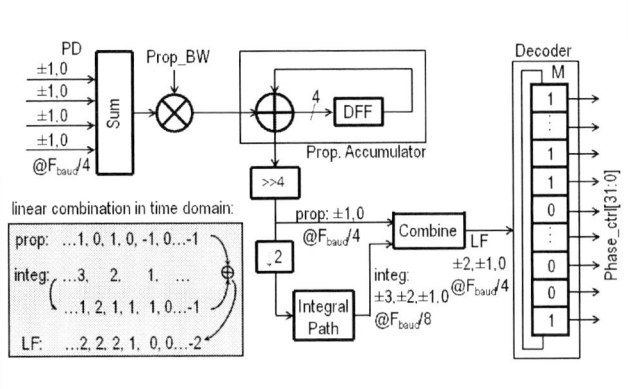

Figure 25.4.3: High-speed and low-latency loop filter and decoder.

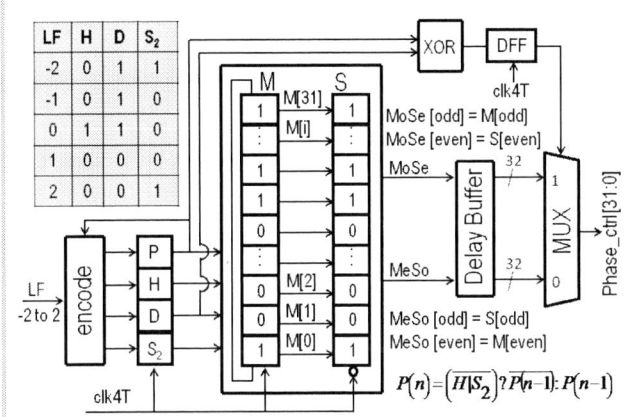

Figure 25.4.4: Double-edge-triggered shift register as decoder.

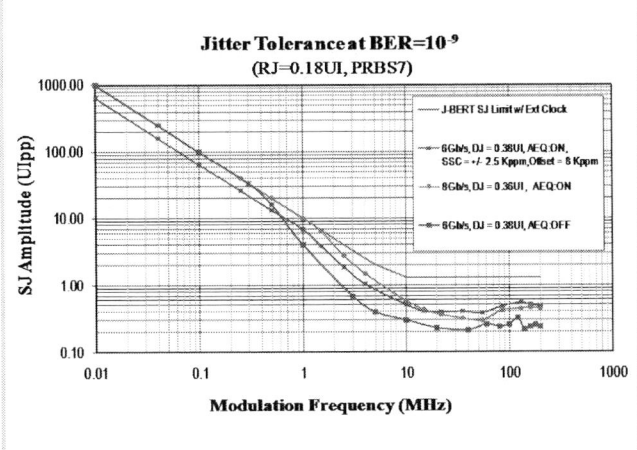

Figure 25.4.5: Measured jitter tolerance at 6Gb/s and 8Gb/s.

Technology	40nm CMOS	
Supply voltage	0.81V	
Data rates	1.25 ~ 8Gbps	
Tracking range	±15.6Kppm	
Tracking BW	8 ~ 10MHz @8Gbps	
RX CTLE	0 ~ 10dB (adaptive)	
Offset Calibration	±24mV @3.4mV/step	
TX pre-emphasis	2 taps	
SSC generation	±15.6Kppm with host tracking	
Power @8Gbps / area	CDR	12mW / 72x72um²
	VCO	16mW / 54x36um²
	RX	24mW / 150x630um²
	TX	19mW / 150x630um²

Figure 25.4.6: Transceiver performance summary.

25

Figure 25.4.7: Die micrograph of the dual port transceiver.

ISSCC 2011 / SESSION 25 / CDRs & EQUALIZATION TECHNIQUES / 25.5

25.5 A 20Gb/s Digitally Adaptive Equalizer/DFE with Blind Sampling

Yu-Ming Ying, Shen-Iuan Liu

National Taiwan University, Taipei, Taiwan

As data rates increase, the backplane communication systems suffer from serious inter-symbol interference (ISI). Due to different channel lengths, loss, and environment variations, an adaptive equalizer is an attractive and robust circuit to equalize the received data in high-speed data communications. Several techniaques are presented for adaptive equalizers. In [1], a spectrum balancing method is presented for an analog equalizer. However, this method is valid only for the random data with fixed data rate. In [5], an eye-opening-monitor (EOM) method adopts a two-dimensional map to detect the signal quality. This EOM method needs a synchronous sampling clock and high-speed comparators. It results in high power consumption and furthermore it requires accurate analog circuits.

In this paper, a 20Gb/s digitally adaptive equalizer with blind sampling is presented. Shown in Fig. 25.5.1, this adaptive equalizer consists of an analog equalizer and a decision-feedback equalizer (DFE), a detection circuit, and an adaptive controller. This adaptive equalizer works as follows. First, the boosting gain of an analog equalizer is set to the lowest level. The analog equalizer is composed of three gain stages and a buffer,and its boosting gain is controlled by EQ[3:0]. A 4b digital-to-analog converter (DAC) realizes a differential threshold voltage, V_{REF}. The comparator compares the output voltage of the analog equalizer with the reference voltage, V_{REF}. The result, which is blindly sampled by an asynchronous clock, is accumulated in the adaptive controller. If the output swing of the analog equalizer is larger than the reference voltage within a specified time interval, the adaptive controller increases the digital code, REF[3:0], to raise the reference voltage until the reference voltage is just over the output voltage of this analog equalizer. Finally, the digital code, REF[3:0], of this DAC is recorded. This action which works like a peak detector to search the low-frequency magnitude of the received data is shown in Fig. 25.5.2(a). Second, once the digital code, REF[3:0], of this DAC is fixed, the boosting gain of analog equalizer is altered from high to low by the digital code, EQ[3:0]. For the received data, due to the limited bandwidth, the high-frequency components experience more attenuation than their low-frequency counterparts. If the output of the analog equalizer is larger than the expected reference voltage, the boosting gain is reduced until the equalized voltage is just lower than the reference voltage. This is equivalent to equalizing the magnitude of high-frequency and low-frequency components. Based on the above procedures, the final digital code, EQ[3:0], is fixed and the adaptation for this analog equalizer is finished. This action is shown in Fig. 25.5.2(b). However, it is difficult to have precise equalization, especially by digital control. Third, an adaptive DFE is included. The digital code, DFE[3:0], controls the tap weighting α of this DFE by comparing the average of the rectified clock output with the rectified received data at the data center as Fig. 25.5.2(c). The asynchronous clock must be retimed by the synchronous clock to ensure the comparison at the data center.

Figure 25.5.3 shows the block diagram of detection circuit and the adaptive controller. Initially, all digital codes, REF[3:0], EQ[3:0], and DFE[3:0], are set to zero. So, V_{REF} is minimum and there is no boosting gain for the analog equalizer. In the first step, we circuit searches for the maximal voltage of the analog equalizer by increasing the digital code, REF[3:0], bit by bit. If $V_{EQ} > V_{REF}$, V_{out1} goes high which in turn increases the digital code, REF[3:0], and resets the 12b counter for next round. When $V_{REF} > V_{EQ}$, V_{out1} stays low until the overflow signal of the 12b counter rises. The overflow signal will stop the REF counter to latch the digital code, REF[3:0]. Simultaneously, it enables the EQ counter and changes the digital code, EQ[3:0], from 0000 to 1111 by a multiplexer. In the second step, the 12b counter is restarted and searches for the optimum boosting gain by decreasing the digital code, EQ[3:0], bit by bit. If V_{out1} goes high, it means the analog equalizer is over-boosting. The digital code, EQ[3:0], is decreased until V_{out1} goes low. When the overflow signal of the 12b counter goes high again, it stops the EQ counter and enables the DFE counter. In the above steps, the comparator is blindly sampled by an asynchronous clock. Since the asynchronous clock is much slower than the data rate, the required time for

the measurement should be long enough to meet the accuracy. In this work, a 12b counter and an asynchronous clock of 620MHz are used so that the convergence time is 13.2µs for the analog equalizer adaptation. Note that the synchronous clock should not be divisible by this asynchronous clock. In the third step, the summer output of the DFE is rectified and is compared with the average of the rectified 20GHz clock. The asynchronous clock is retimed by the 20GHz clock to ensure the comparison at the data center. While the voltage V_{out2} goes high the digital code, DFE[3:0], is increased. When DFE[3:0] reaches "1111", the DFE counter will stop.

The gain stage of this analog equalizer using inductor peaking and negative capacitor (NC) [6] is shown in Fig. 25.5.4(a). Its boosting gain is controlled by 4b digital codes, EQ[3:0], in a step of 1.25dB per 1LSB. In the analog equalizer, two PMOS varactor arrays are used to adjust the boosting gain. Figure 25.5.4(b) shows the DFE and its tap weighting α is digitally controlled by DFE[3:0].

As mentioned, this work uses a clock of 20GHz for the DFE and another asynchronous clock of 620MHz for the adaptive controller. The analog equalizer and the DFE consume 37mW from a 1.2V supply. The detection circuit and the adaptive controller consume 15mW from a 1V supply. The measured eye diagrams are shown in Fig. 25.5.5. For a 15cm FR4 board, the measured peak-to-peak jitter of the recovered data is 8.44ps. For a 35cm FR4 board, the measured peak-to-peak jitter of the recovered data is 14.22ps. For the above measurements, the measured BER is less than 10^{-12} with 20Gb/s 2^7-1 PRBS input data.

Figure 25.5.6 shows the performance summary and comparison with prior works. There are few adaptive equalizers and DFEs which operate at 20Gb/s. This work proposes two digital techniques for implementation of high-speed low-power adaptive equalizers.

The proposed adaptive equalizer successfully operates for different channel lengths (from 3cm to 35cm) on FR4 boards. Figure 25.5.7(a) shows the measured S_{21} for channels of length 15cm and 35cm on FR4 boards. At 10GHz, the measured channel loss is −11.9dB and −22.3dB, respectively. The die micrograph is shown in Fig. 25.5.7(b). The chip is fabricated in a 65nm CMOS process. The area of the adaptive equalizer is 0.1mm².

Acknowledgements:
The authors thank TSMC, Hsinchu, Taiwan, for fabricating this chip through TSMC university shuttle program. This work is also supported by NTU-MediaTek Lab., and NSC, Taiwan.

References:
[1] J. Lee, "A 20Gb/s Adaptive Equalizer in 0.13-µm CMOS Technology", *IEEE J. Solid-State Circuits*, vol. 41, no. 9, pp. 2058-2066, Sep., 2006.
[2] H. Wang, C.-C. Lee, A.-M. Lee, and J. Lee, "A 21-Gb/s 87-mW Transceiver with FFE/DFE/Linear Equalizer in 65-nm CMOS Technology," *IEEE Symp. VLSI Circuits*, pp. 50-51, June, 2009.
[3] D.Z. Turker, A. Rylyakov, D. Friedman, et al., "A 19Gb/s 38mW 1-Tap Speculative DFE Receiver in 90nm CMOS," *IEEE Symp. VLSI Circuits*, pp. 216-217, June 2009.
[4] S. A. Ibrahim and B. Razavi, "A 20Gb/s 40mW Equalizer in 90nm CMOS Technology," *ISSCC Dig. Tech. Papers*, pp. 170-171, Feb., 2010.
[5] H. Noguchi, N. Yoshida, H. Uchida, et al., "A 40-Gb/s CDR with Adaptive Decision-Point Control Using Eye-Opening-Monitor Feedback," *ISSCC Dig. Tech. Papers*, pp. 228-229, Feb., 2008.
[6] D. Lee, J. Han, G. Han, S.M. Park, "10 Gbit/s 0.0065 mm² 6mW Analogue Adaptive Equaliser Utilising Negative Capacitance," *Electronics letters*, vol. 45, no. 17, pp. 863-865, Aug., 2009.

978-1-61284-303-2/11 $26.00 © 2011 IEEE

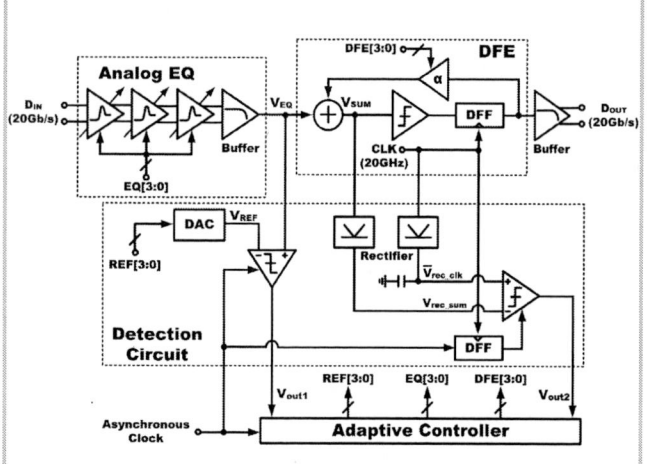

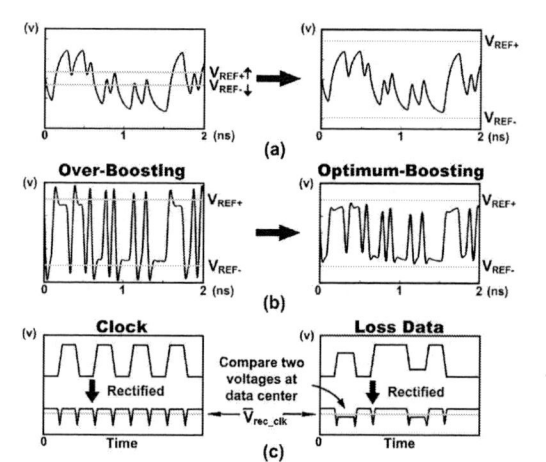

Figure 25.5.1: Block diagram of the proposed digitally adaptive equalizer with blind sampling.

Figure 25.5.2: Adaptation procedure for the proposed adaptive equalizers.

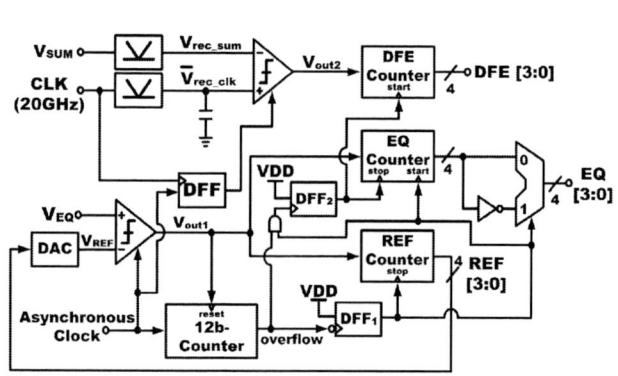

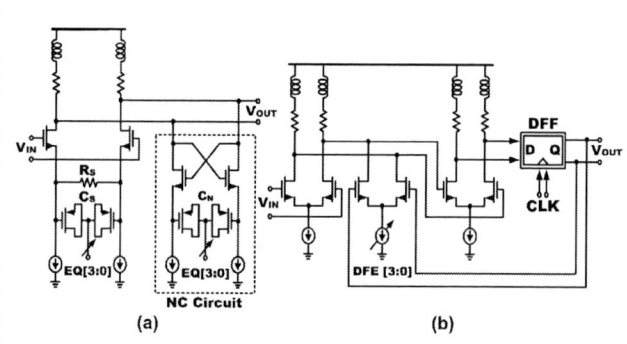

Figure 25.5.3: Block diagram of the detection circuit and the adaptive controller.

Figure 25.5.4: Schematic of: (a) a gain stage of an equalizer filter, and (b) decision feedback equalizer.

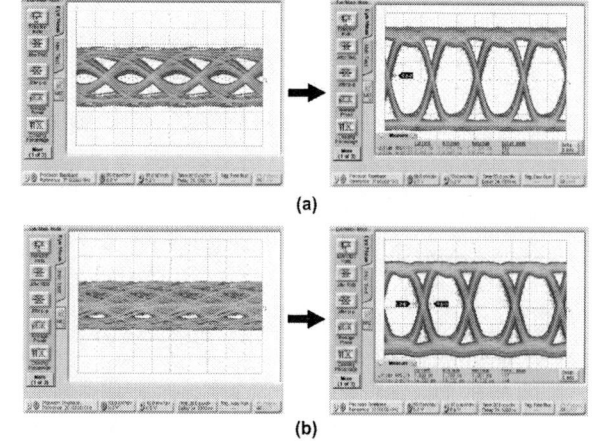

Figure 25.5.5: The measured eye diagrams with 2⁷-1 PRBS before and after equalization for channels of length (a) 15cm and (b) 35cm on FR4 board.

	[1]	[2]	[3]	[4]	This Work
Data Rate	20 Gb/s	21 Gb/s	19 Gb/s	20 Gb/s	20 Gb/s
Architecture	Analog EQ	Analog EQ + 1-Tap DFE	1-Tap Speculative DFE	Analog EQ + 1-Tap Speculative DFE	Analog EQ + 1-Tap DFE
Supply	1.5 V	1.2 V	1 V	1 V	1.2 V
DFE Clock	N/A	Full-rate	Half-rate	Half-rate	Full-rate
Max Loss	-15 dB	-11.7 dB *	-11 dB	-24 dB	-22 dB
Boost Gain	Adaptive	Manual	Manual	Manual	Adaptive
EQ+DFE power	60 mW	42 mW	38 mW	40 mW	37 mW
Adaptive power		N/A	N/A	N/A	15 mW
Area	0.2 mm²	0.04 mm²	0.02 mm²	0.09 mm²	0.1 mm²
Technology	0.13um-CMOS	65nm-CMOS	90nm-CMOS	90nm-CMOS	65nm-CMOS

* Since the transmitter in [2] has the equalization of 9.5 dB, the channel loss 21.2 dB is discounted by this amount for fair comparison.

Figure 25.5.6: Performance summary and comparison.

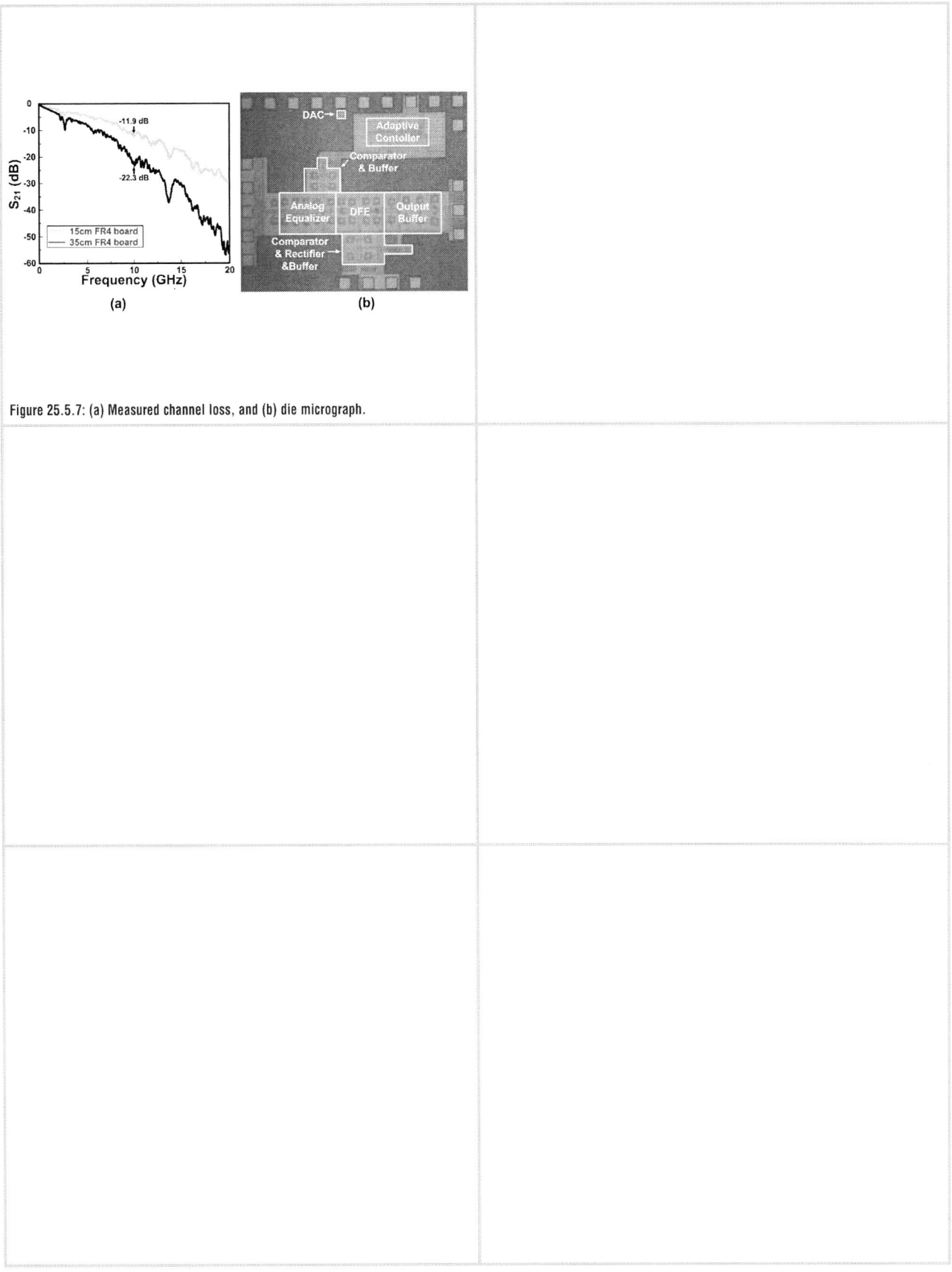

Figure 25.5.7: (a) Measured channel loss, and (b) die micrograph.

ISSCC 2011 / SESSION 25 / CDRs & EQUALIZATION TECHNIQUES / 25.6

25.6 A 15Gb/s 0.5mW/Gb/s 2-Tap DFE Receiver with Far-End Crosstalk Cancellation

Meisam Honarvar Nazari, Azita Emami-Neyestanak

California Institute of Technology, Pasadena, CA

The increasing demand for high-bandwidth interconnection between integrated circuits requires large numbers of I/Os per chip as well as high data rates per I/O. Key limitations in meeting these requirements include channel characteristics and I/O power consumption. Even in short interconnects, the channel attenuation at very high data rates is significant, and using receiver equalization can greatly improve the link performance [1-5]. However, compensating a high level of loss requires many taps of equalization, which can significantly reduce the power efficiency of the link. Parallel data transmission increases the aggregate data rate, but compact traces placed in close proximity suffer from a high level of crosstalk interference. This problem is exacerbated when transmit pre-emphasis techniques are exploited to boost the high frequency gain. While the use of differential signaling can mitigate the effect of crosstalk, it requires twisting pairs leading to area and bandwidth penalties.

In this paper, we present a low-power receiver that supports high data rates over bandwidth-limited and coupled links. The receiver employs a half-rate 2-tap speculative DFE architecture with a far-end cross-talk (FEXT) cancellation technique. Figure 25.6.1 shows the top-level architecture of the DFE receiver. Conventionally, analog taps of the equalizer are implemented using current-mode summers, thus the power consumption of the DFE increases proportionally with the number of taps. In the proposed architecture, a switched-capacitor S/H is employed to sample the input signal and combine it with the feedback coefficients at the front-end of the receiver [1], as shown in Fig. 25.6.2 (S/H/summer). In this design, the switched-capacitor network is modified to support two taps of DFE without any signal loss. This technique can be further extended to realize more number of taps. The extra power due to sampling capacitors, switches and voltage-mode DACs is very small. The S/H/summer operates in two phases as shown in Fig. 25.6.2. In the first, sample/sum phase, the input is sampled into capacitor C_1 and the first tap coefficient (αV_{REF1}) is added to (or subtracted from) this sample. During this phase, as will be discussed later, the crosstalk canceling signal is stored into capacitor C_2. In the second, sum/hold phase, the result of the first phase is added to the second tap coefficient (βV_{REF2}) and applied to the slicer. A delayed version of the clock (CK_d) is used to sample the input in the first phase to minimize the input dependent charge injection. Considering the trade-off between the kT/C noise and the required RC time-constant of the S/H, the sampling capacitor is optimized to be 20fF. Two 4b current-steering DACs generate the equalization coefficients (αV_{REF1}, βV_{REF2}) while drawing <900μA (Fig. 25.6.2). A 1pF capacitor at the output of the DAC reduces the high frequency switching noise. A combined slicer/MUX, shown in Fig. 25.6.1, is used to implement the loop unrolling and resolve the current bit based on the previous bit. In order to cancel the kickback from latch output to the sensitive sampling nodes, small metal capacitors cross-couple the output and the input. These capacitors also reduce the loss of the S/H/summer due to the charge sharing between the sampling capacitors (C_1, C_2) and the slicer/MUX input parasitic capacitor.

The interference from the adjacent aggressor line (FEXT) appears at the front-end of the receiver and is proportional to the derivative of the transmitted signal. The FEXT signal can have the same or opposite polarity as the aggressor signal if the link is capacitive or inductive, respectively. The incoming aggressor signal is sent through an adjustable high-pass filter, shown in Fig. 25.6.3, to approximate FEXT for different levels of coupling. This method does not involve resolving the aggressor signal to compensate for its effect on the victim signal. The effect of FEXT can be removed by addition of the mimicked FEXT signal (V_{X-TALK}) to the sampled input signal during the sum/hold phase. As addition and subtraction have minimal power overhead in this architecture, the extra hardware results in only 5% (33μW/Gb/s) extra power dissipation, primarily due to the additional clock buffers. As the effect of FEXT signal and the mimicked FEXT signal are sampled at the same time, this scheme is not sensitive to the phase offset between the aggressor and the victim. The measurements were performed for varying lengths of the victim and the aggressor traces to prove this fact. The

design in [6] cancels the crosstalk-induced jitter but it does not compensate for the crosstalk-induced amplitude ISI and consumes 80mW and occupies 0.014μm².

The prototype is fabricated in 45nm SOI technology. Figure 25.6.6 shows the performance of this design and compares it to a DFE-IIR receiver for Si carrier channel compensation [2], an FFE-DFE receiver [3] and 5- and 2-tap current-integrating DFE receivers [4, 5]. To evaluate the performance of the receiver at high data rates, first an input data with low level of ISI was used. The receiver operates error-free (BER<10^{-12}) up to 20Gb/s with an input sensitivity of ±100mV$_{ppd}$ which reduces to ±50mV$_{ppd}$ at 15Gb/s. The input-referred offset was measured to be 20mV at 15Gb/s. The equalization capability of the receiver was tested by transmitting data over 5, 10 and 20inch FR-4 PCB traces. The channels characteristics including the connecting SMA cables are shown in Fig. 25.6.3. With 15Gb/s PRBS7 data, the received eye is closed for all these channels. The closed eyes are shown as insets in Fig. 25.6.4. The 5inch channel exhibits a loss of 14.5dB at 7.5GHz. Employing the 2-tap DFE, while consuming 7.5mW from a 1.2V supply, 34% horizontal eye opening (BER=10^{-8}) with BER<10^{-12} in the center is achieved. The DFE receiver, consisting of clock buffers, S/H/summer, slicer/MUX and DACs, occupies an area of 220×65μm². The DFE was also tested with 10 and 20inch channels. 13Gb/s data was transmitted over the 10inch channel with 17dB of loss at 6.5GHz. Under these conditions the DFE achieved 43% horizontal eye opening while dissipating 6.1mW. Over a 20inch link with 21dB roll-off at 5.5GHz the DFE receiver operates at 11Gb/s with a 37.5% eye opening while consuming 5.5mW.

The crosstalk cancellation scheme is evaluated by transmitting uncorrelated victim and aggressor data over a 5inch long, 32mil-wide coupled trace with 40mil separation on an FR-4 board. The amount of coupling (FEXT) at 6.25GHz is −15dB (Fig. 25.6.3). The channel also has 12.5dB of loss at 6.25GHz, which results in a closed input eye. With no aggressor, the DFE generates a 47.5% open eye at BER<10^{-8}. Applying the aggressor closes the eye completely. The crosstalk canceller restores the horizontal eye opening to 24%. The table in Fig. 25.6.5 summarizes the performance of the crosstalk cancellation technique. The 2-tap DFE receiver with the FEXT cancellation capability consumes less than 0.5mW/Gb/s of power. The proposed architecture is well-suited for implementation in highly scaled technologies. Experimental results validate the feasibility of the DFE receiver for ultra-low-power high-data-rate highly parallel I/O links.

Acknowledgements:
The authors acknowledge the support of NSF, Intel, and the C2S2 Focus Center, funded under the Focus Center Research Program.

References:
[1] A. Emami-Neyestanak, A. Varzaghani, J.F. Bulzacchelli, R. Rylyakov, C.K. Yang, and D.J. Friedman "A 6.0 mW, 10.0 Gb/s Receiver with Switched-Capacitor Summation DFE," *IEEE J. Solid-State Circuits*, vol. 42, no. 4, pp. 889-896, April, 2007.
[2] Y. Liu, B. Kim, T.O. Dickson, J.F. Bulzacchelli, and D.J. Friedman, "A 10Gb/s Compact Low-Power Serial I/O with DFE-IIR Equalization in 65nm CMOS," *ISSCC Dig. Tech. Papers*, pp. 182-183, Feb., 2009.
[3] H. Sugita, K. Sunaga, K. Yamaguchi, and M. Mizuno, "A 16Gb/s 1st-Tap FFE in 90nm CMOS," *ISSCC Dig. Tech. Papers*, pp. 162-163, Feb., 2010.

[4] T.O. Dickson, J.F. Bulzacchelli, D.J. Friedman, "A 12-Gb/s 11-mW Half-Rate Sampled 5-Tap Decision Feedback Equalizer With Current-Integrating Summers in 45-nm SOI CMOS Technology," *IEEE J. Solid-State Circuits*, vol. 44, no. 4, pp. 1298-1305, April, 2009.
[5] M. Park, J.F. Bulzacchelli, M. Beakes, and D.J. Friedman, "A 7Gb/s 9.3mW 2-Tap Current-Integrating DFE Receiver," *ISSCC Dig. Tech. Papers*, pp. 230-231, Feb., 2007.
[6] J. F. Buckwalter and A. Hajimiri, "Cancellation of Crosstalk-Induced Jitter," *IEEE J. Solid-State Circuits*, vol. 41, no. 3, pp. 621-632, Mar., 2006.

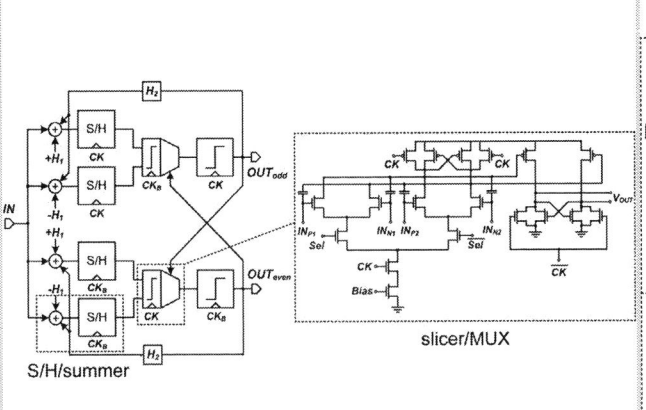

Figure 25.6.1: Top-level architecture of the 2-tap DFE, and the slicer/MUX circuit schematic.

Figure 25.6.2: Front-end S/H/summer block operation (single-ended version is illustrated for simplicity), and the coefficient generating DAC.

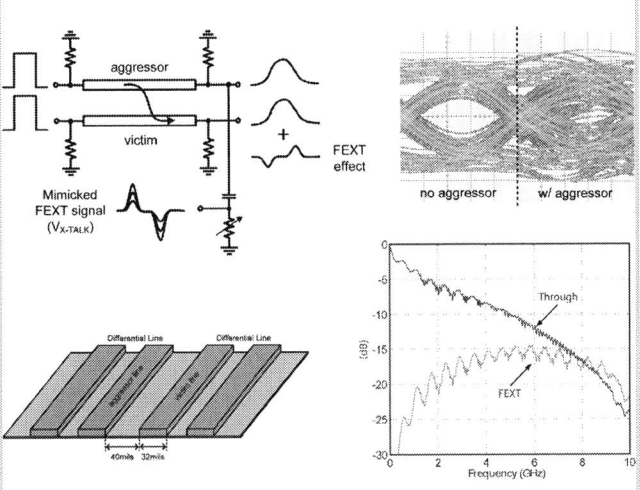

Figure 25.6.3: FEXT cancellation technique.

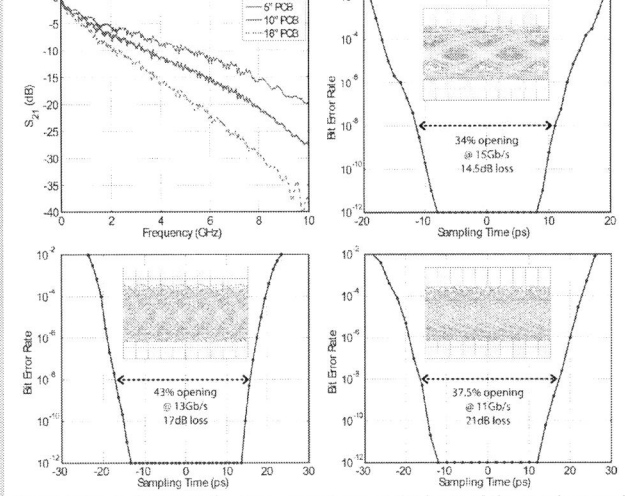

Figure 25.6.4: Channels S_{21} plot, closed eye at the input of the receiver, and equalized bathtub curve at different data rates.

Crosstalk cancellation performance table

Data Rate	12.5Gb/s
Channel loss @ Nyquist	12.5dB
Channel coupling @ Nyquist	-15dB
Horizontal eye opening (BER <10^{-8}) no aggressor, DFE ON	47.5%
Horizontal eye opening (BER <10^{-8}) w/o cancellation, DFE ON	0%
Horizontal eye opening (BER <10^{-8}) w/ cancellation, DFE ON	24%

Figure 25.6.5: Crosstalk cancellation performance.

Reference	This work*			[2]	[3]	[4]		[5]
Process	45nm SOI			65nm bulk CMOS	90nm bulk CMOS	45nm SOI		90nm bulk CMOS
Architecture	2-tap speculative DFE			DFE-IIR	1st-tap FFE 3-tap DFE	5-tap DFE		2-tap DFE
Supply (V)	1.2			1.0	1.4	1.0		1.0
Power (mW)	7.5	6.1	5.5	6.8	69	11	10.1	9.3
Data Rate (Gb/s)	15	13	11	10	16	12	9.0	7.0
Channel loss @ Nyquist (dB)	14.5	17	21	23.2	22	15	25	15
Horizontal eye opening (BER<)	34% <10^{-8}	43% <10^{-8}	37.5% <10^{-8}	45% <10^{-9}	30% <10^{-12}	32% <10^{-8}	44% <10^{-8}	45% <10^{-8}
FOM (pJ/bit/dB)	0.034	0.028	0.024	0.029	0.196	0.061	0.045	0.088
Sensitivity	±50mV$_{ppd}$ (20mV offset)			±32mV$_{ppd}$	N/A	±93mV$_{ppd}$		±31mV$_{ppd}$
Area	0.014 mm^2			0.017mm^2	0.063mm^2	0.004mm^{2**}		0.019mm^2

* Excluding FEXT cancellation

** Excluding DACs for the five taps

Figure 25.6.6: Performance summary and comparison with prior art.

Figure 25.6.7: Die micrograph.

ISSCC 2011 / SESSION 25 / CDRs & EQUALIZATION TECHNIQUES / 25.7

25.7 A 10Gb/s Half-UI IIR-Tap Transmitter in 40nm CMOS

Halil Cirit, Marc J Loinaz

Netlogic Microsystems, Santa Clara, CA

Two commercially important standards for 10Gb/s serial data transfer include the SFI specification, associated with SFP+ optical modules and copper twinaxial cable [1], and the 10GBASE-KR specification for backplane channels in computer servers and networking equipment [2]. The SFI specification requires the transmitter to operate with both low data-dependent jitter (DDJ) and low transmitter waveform dispersion penalty (TWDPc where the "c" denotes direct-attach copper cable). The SFI TWDPc specification is associated with the copper twinaxial cable channel in [1] and is defined as the ratio (in dB) between the signal-to-noise ratio (SNR) of a matched-filter receiver and the SNR at the slicer input of an ideal, adapted, FFE+DFE receiver [3]. In data center applications it is necessary to use the same serial transmitter IC (with no change in transmitter programming) with both hot-pluggable twinaxial cable and optical modules as shown in Fig. 25.7.1. Because of the use of limiting laser drivers in optical modules, DDJ produced by the transmitter cannot be equalized by the receiver at the far-end of an optical fiber link. Therefore it is necessary for the transmitter to satisfy *both* the SFI TWDPc and DDJ specifications at the same time.

To illustrate how to improve TWDPc without increasing DDJ, Fig. 25.7.2 illustrates waveforms through a low pass filter (LPF) with a cut off frequency of 3.5GHz. This LPF is a crude approximation of the FR-4 host board trace in Fig. 25.7.1. Fig. 25.7.2(a) illustrates a unit-step waveform and the resulting step response and eye diagram with a data rate of 10Gb/s. In Fig. 25.7.2(b) the data pattern is conditioned by using a one-UI post-cursor FIR filter. If the post-cursor tap weight is chosen correctly the DDJ is eliminated. Figure 25.7.2(c) illustrates a third waveform conditioned by a half-UI post-cursor FIR filter tap [5]. With a properly chosen tap weight, at one-half UI after the data transition the output signal is settled and all the ISI and DDJ are cancelled. Using half-UI, post-cursor FIR filtering maximizes the ratio Vac/Vdc and consequently reduces the TWDPc (9.64dB in Fig. 25.7.2(c)) relative to the one-UI post-cursor filter (10.13dB in Fig. 25.7.2(b)).

In order to fully minimize the DDJ seen at the compliance point in Fig. 25.7.1 it is important to study in more detail the effects of the FR-4 trace. The FR-4 host board trace exhibits frequency-dependent loss due to "skin effect" and dielectric absorption. These effects slow down the initial edge of the step response and produce a long tail component that is much like a slow, RC settling transient. To replicate observed 6-inch FR-4 trace behavior, simulations were performed with a model that consists of the sum of two low-pass filters, the first having a 3dB cut-off frequency at 3.5GHz and the second having a 3dB cut-off at 0.35GHz. Figure 25.7.3(a) shows the step response and 10Gb/s eye diagram produced by the trace model when driven with an ideal driver with no transmit equalization. Figure 25.7.3(b) illustrates that a conventional single-tap, post-cursor transmit equalizer is not capable of completely eliminating the ISI introduced by the 0.35GHz pole. Eliminating that ISI can be achieved by implementing an IIR tap employing a continuous time, low-pass filter described in the next paragraph. As illustrated in Fig. 25.7.2(c), combining one-UI, post-cursor FIR and IIR taps almost completely eliminates the DDJ. In this work the combination of half-UI post-cursor FIR tap filtering (shown in Fig. 25.7.2(c)) and post-cursor IIR tap filtering (shown in Fig. 25.7.3(c)) is proposed as a solution for achieving simultaneous TWDPc and DDJ optimization for SFI.

The architecture of the transmitter is shown in Fig. 25.7.1 [6]. In order to cancel internal ISI due to power supply noise sensitivity, all of the circuitry is implemented with differential current-mode logic (CML). The 3-tap transmitter architecture specified for 10GBASE-KR, **mode_1**, is enabled when the *delay_select* input is set high, the half-UI shift register block is powered down and the programmable LPF is bypassed. 10Gb/s data is shifted through the three-stage shift register and creates pre-cursor, cursor and post-cursor signals. In the output driver the pre-cursor signal and post-cursor signal are subtracted from the cursor signal to produce the output signal. In **mode_2**, the pre-cursor FIR tap in **mode_1** is converted into a pre-cursor infinite impulse response ("IIR") tap by enabling the LPF on the pre-cursor path. As will be shown below, **mode_2** is useful for long, lossy channels where strict SFI compliance is not required. In **mode_3**, half-UI post-cursor FIR filtering and half-UI post-cursor IIR filtering are employed. This mode is meant for SFI applications where SFI TWDPc and DDJ requirements must be satisfied simultaneously. **mode_3** is enabled when the

delay_select input is set low, the one-UI shift register block is powered down, the LPF is enabled, and the half-UI shift register block is powered up and generates two signals: the half-UI delayed signal is fed to Mux0 and Mux2 and the cursor signal is fed to Mux1.

A schematic of the pad driver is shown in Fig. 25.7.4. During normal operation, the outputs are AC-coupled to an off-chip 100Ω differential termination. Thick-oxide NMOS cascode transistors M_{b1}-M_{b6} are used to protect the differential pair core transistors from ESD damage. The voltage V_{bias} is set to keep the cascode devices in saturation so as to isolate the driver load capacitance from the input device capacitances and results in no speed degradation. Taps 0 and 2 are designed to produce up to 25% and 50%, of the full-scale current of tap1 respectively. To achieve faster rise and fall times at the transmitter outputs, active-peaked buffers are used in the pre-driver buffer chain. A schematic of a buffer with a tunable LPF is shown in Fig. 25.7.4. Transistors M_4 and M_5 in combination with resistors R_1 and R_2 provide an active shunt-peaked load [4]. Capacitors M_{14}-M_{19}, in conjunction with the active peaked loads, produce a tunable LPF for the IIR tap in Fig. 25.7.1. With M_8-M_{13} turned off the LPF is effectively bypassed.

The transmitter is integrated as part of a quad-channel, 10Gb/s Ethernet transceiver fabricated in 40nm CMOS. The transmitter circuit operates from a nominal core supply of 0.9V and an I/O supply of 1.8V. Transmitter performance measurements were performed on a PBGA-packaged device with 5% lowered power supplies at high temperature so as to exhibit worst-case performance. To evaluate performance for SFI, the transmitter is connected to a host board trace that exhibits the worst-case loss according to the SFI specification [1]. This trace consists of a 6-inch differential FR-4 microstrip (with a characteristic impedance deliberately mismatched at 110Ω) and is followed by an SFP+ connector and a test board to model the short PCB traces inside the SFP+ module. Data is taken at 85°C device case temperature. The resulting output eyes in **mode_1** and **mode_3** are shown in Fig. 25.7.5. Pre-emphasis tap settings are optimized for minimum DDJ in both these modes. In **mode_3** the LPF bandwidth is set to 0.5GHz. The total jitter (Tj), DDJ and random jitter (RJ) are measured using an Agilent 86100C oscilloscope with an 86108A acquisition module. In addition to TWDPc, pulse-width shrinkage (PWS) and voltage modulation amplitude (VMA) values are calculated using Matlab as described in [1] from a waveform captured by the oscilloscope. Use of **mode_1** violates both SFI DDJ and TWDPc requirements. However, the use of **mode_3** allows all of the SFI specifications to be met.

In 10GBASE-KR backplane links [2] there is no specification for TWDPc or DDJ at the transmitter. In such applications transmit equalization can be utilized more freely than in SFI. Because backplane links tend to exhibit long tails associated with frequency-dependent loss, the use of an IIR tap is an efficient means of equalizing the channel with small power and area overhead. To show the benefit of the filtering in **mode_2**, Fig. 25.7.6 shows data eyes for a 40-inch test backplane [7]. Use of the IIR tap with the backplane in **mode_2** opened up an eye that is closed when **mode_1** is used.

The transmitter, which includes a 16:1 multiplexer plus associated clock dividers (in addition to the circuits shown in Fig. 25.7.1), dissipates 125mW. Die area including ESD diodes, clamps and bypass capacitors is 0.22mm². A die micrograph is shown in Fig. 25.7.7.

Acknowledgements:
The authors thank S. Sidiropoulos and S. Verma for circuit discussions, and H. Liaw, M. Luschas and A. Orphanou for their signal integrity expertise.

References:
[1] *SFF-8431*, Rev4.1, July, 2009.
[2] *IEEE Standard 802.3ap*, May, 2007.
[3] N. Swenson, P. Voois, T. Lindsay, and S. Zeng, "Explanation of IEEE 802.3, Clause 68 TWDP," Jan., 2006.
[4] A. Lin and M. Loinaz, "A Serial Data Transmitter for Multiple 10Gb/s Communication Standards in 0.13 µm CMOS," *ISSCC Dig. Tech. Papers*, pp. 108-109, Feb., 2008.
[5] M. Bichan and A. Carusone, "A 6.5Gb/s Backplane Transmitter with 6-tap FIR Equalizer and Variable Tap Spacing," *Proc. IEEE CICC*, pp. 611-5614, Sep., 2008.
[6] H. Cirit, USPTO patent application pending.
[7] GbX I-Trac Standard Evaluation Backplane, Molex, 2008.

ISSCC 2011 / February 23, 2011 / 4:45 PM

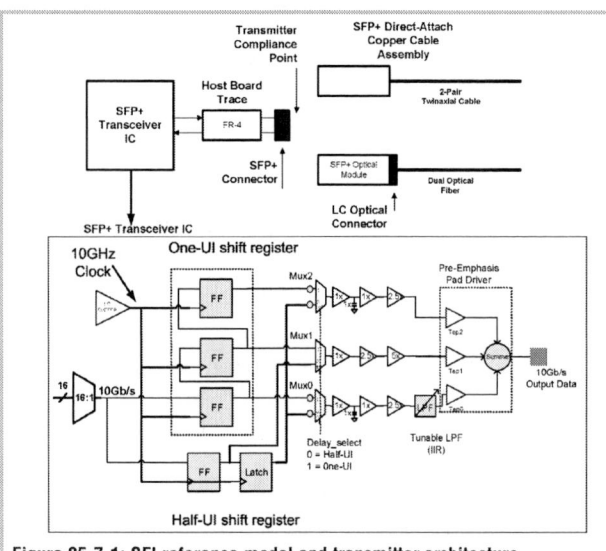

Figure 25.7.1: SFI reference model and transmitter architecture.

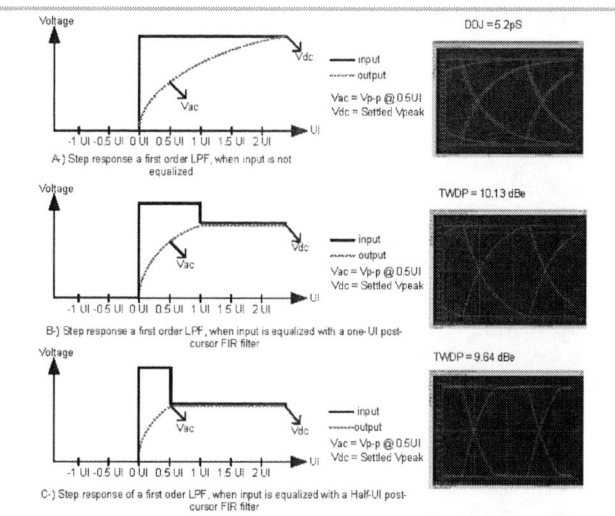

Figure 25.7.2: Simulation results for an ideal transmitter driving an LPF with a cut-off frequency of 3.5GHz.

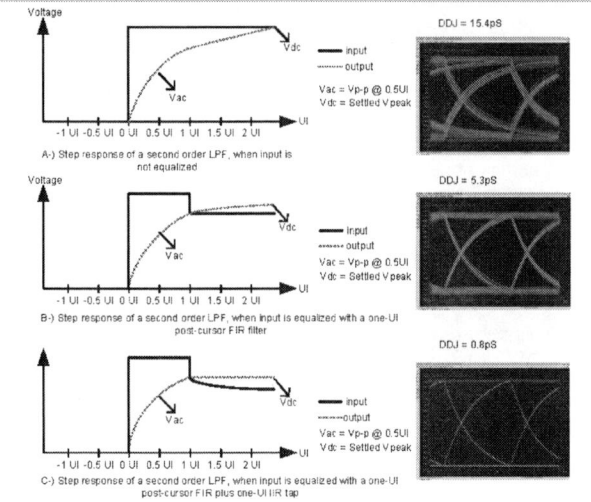

Figure 25.7.3: DDJ simulation results for an ideal transmitter driving a second order LPF with 3.5GHz and 0.35GHz.

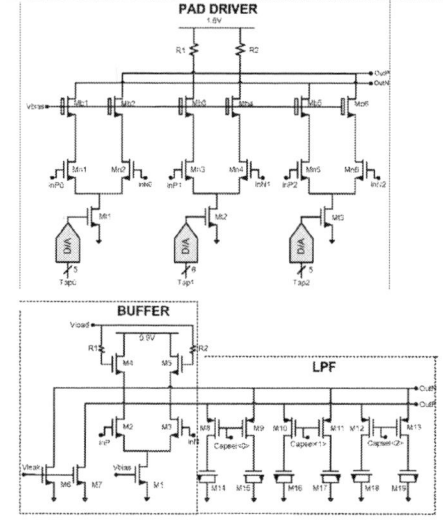

Figure 25.7.4: Pad driver and active-peaked buffer with LPF.

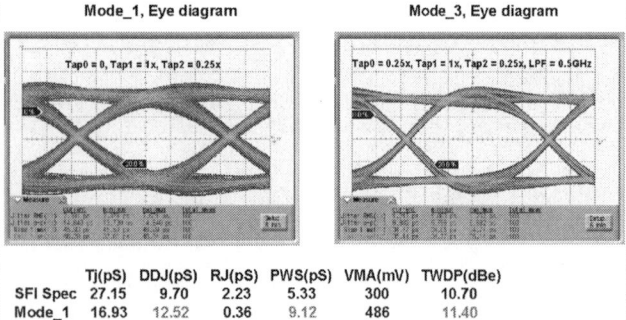

	Tj(pS)	DDJ(pS)	RJ(pS)	PWS(pS)	VMA(mV)	TWDP(dBe)
SFI Spec	27.15	9.70	2.23	5.33	300	10.70
Mode_1	16.93	12.52	0.36	9.12	486	11.40
Mode_3	10.50	5.93	0.38	4.05	322	10.30

Figure 25.7.5: Performance comparison between mode_1 and mode_3 when driving 6-inch 110Ω trace and SFP+ connector.

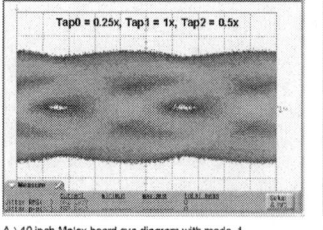
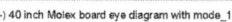
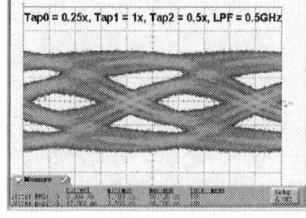

Figure 25.7.6: Eye measurements for a 40-inch Molex backplane channel with mode_1 versus mode_2.

978-1-61284-303-2/11 $26.00 © 2011 IEEE

ISSCC 2011 PAPER CONTINUATIONS

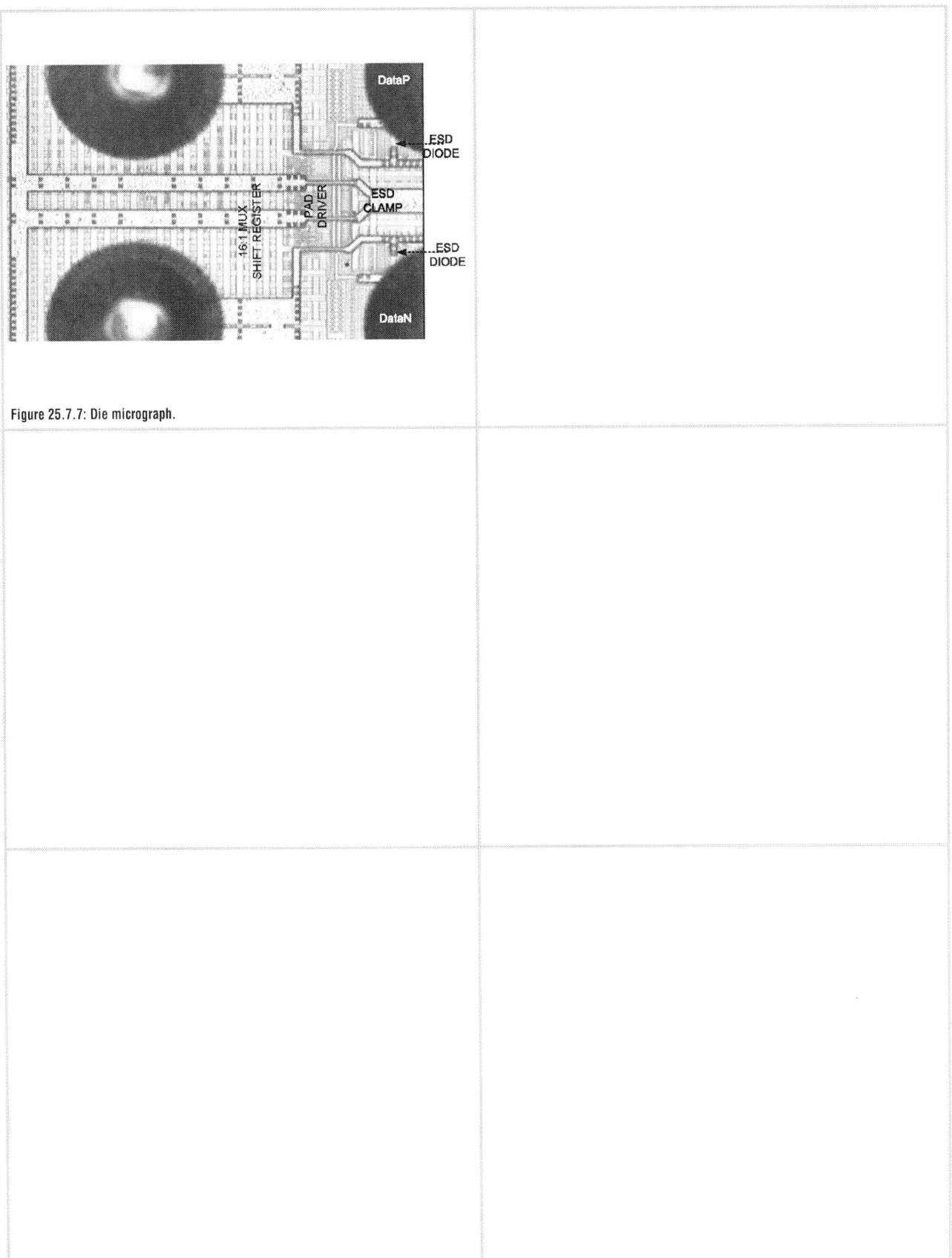

Figure 25.7.7: Die micrograph.

ISSCC 2011 / SESSION 25 / CDRs & EQUALIZATION TECHNIQUES / 25.8

25.8 A 13.8mW 3.0Gb/s Clock-Embedded Video Interface with DLL-Based Data-Recovery Circuit

Sungchun Jang, Heesoo Song, Seokmin Ye, Deog-Kyoon Jeong

Seoul National University, Seoul, Korea

As the panel technology continues to offer displays with higher resolution, greater color depth, and increased frame rate, the amount of video data to display driver ICs (DDIs) inside the panel keeps on expanding. Since the conventional intra-panel interfaces with multi-drop configurations, such as RSDS and mini-LVDS, increase the cost of overall systems at high bandwidth, new intra-panel interfaces have been proposed to meet the bandwidth requirement with point-to-point configurations [1-5]. This paper presents a new high-speed video interface that offers significant complexity reduction in the receiver. It is because receivers are integrated in a DDI with relatively slow high-voltage processes, while transmitters in host controllers are implemented with the more advanced deep-submicron processes. Compared to the PLL-based clock recovery circuits in [1-2], the DLL-based data recovery circuit occupies a smaller area with lower power consumption and offers unconditionally stable characteristics along with higher jitter tolerance.

In video interfaces, the pixel data and blank data are transmitted alternately. DDIs receive valid pixel data of a row sequentially during an active period and they receive blank data while driving the panel during a blank period. In the proposed scheme, 8-bit valid data followed by zero and one are transmitted during active periods whereas clocks are transmitted during blank periods (Fig. 25.8.1(a)). Since the input makes a transition from zero to one at the same position, the DLL can generate multiphase clocks from the gated input. The detailed operation of the DLL will be described in the following paragraphs. Compared with previous serial interfaces in which transmitters send an evenly spaced data stream so that receivers require twice finer timing resolution for sampling data at the center of the eye, the proposed interface sends an unevenly spaced data stream at the transmitter. The width of zero and one near the embedded-clock transition is not 1UI but 1.5UI, which reduces the required phases of the receiver from 20 to 11.

Figure 25.8.1(b) shows the overall block diagram of the proposed receiver. The receiver recovers data by sampling the input with equally spaced 11-phase clocks from the DLL. Since the paths to the DLL and the samplers are not symmetric and the setup and hold times of the samplers are varied by PVT variations, the samplers unavoidably sample the data at a non-optimum point. The skew compensation unit adjusts the delay of the path to the samplers so that the samplers sample the input with a maximum margin. For this purpose, the delay is adjusted by a 3-bit code from a digital logic, which collects the polarity of the skew detected by sampling the delayed input for samplers with the VCDL input, CLK0. For example, when the sampled output is high, the input is delayed more by increasing the capacitance of node INP_D.

Figure 25.8.2 illustrates the block diagram of the proposed DLL which is similar to [6]. During the blank period, the DLL is locked to one clock period of the input clock. Figure 25.8.3(a) shows the timing diagram of the locked DLL. Only the input around the rising edge is transferred to the VCDL by a WINDOW signal, which is generated by AND operation of CLK6 and CLK10. After the rising edge passes through the first delay cell, the WINDOW signal goes low to prevent the transfer of the falling edge of the input. The falling edge of the VCDL output is made by feeding back the output of the last delay cell to the first delay cell of the VCDL. When the CLK10 goes high, the WINDOW signal goes high to transfer the next rising edge of input. However, when the delay of the VCDL is smaller than one clock period shown in Fig. 25.8.3(b), the WINDOW signal is activated earlier than a locked situation. If WINDOW signal goes high before the input falls, the input makes the VCDL output transit twice. To prevent this wrong transition, the dummy delay cell before the VCDL, which generates CLK0, is also gated by the SEL signal which is activated when CLK10 rises with a high input. On the other hand, when the VCDL has a much longer delay than one clock period, all the transition of the input is not transferred to the VCDL since the WINDOW signal is activated after the one clock period from rising of CLK0, as shown in Fig. 25.8.3(c). In this case, the VCDL delay is lowered by the generation of UP signal from the PFD since the frequency of the VCDL output is lower than that of the

input. Therefore, the proposed DLL can be properly locked regardless of the initial delay of the VCDL. In addition, the duty-cycle of the VCDL output is always 50% regardless of the duty-cycle of the input since the falling edge is generated using the middle-phase clock. Before the DLL is coarsely locked, the PFD compares the VCDL output and the input, and it locked during a blank period. Once the DLL is coarsely locked, the DLL tracks the input phase even during active periods since the PFD compares the VCDL output and the gated input by the WINDOW signal.

Figure 25.8.4(a) shows the schematic of the PFD and the coarse-lock detector (CLD). Conventional CLDs check the sequence of multiphase clocks as well as counting the number of multiphase clocks in one clock period in order to avoid the harmonic lock. When the DLL is in 2nd-harmonic lock state as shown in Fig. 25.8.4(b), the phase of CLK0 and CLK11 is aligned but the multiphase clocks are not in order since the rising edge of the input transfers to the VCDL before the previous rising edge reaches to the last delay cell. However, in the proposed DLL, the rising edges of the multiphase clocks are always arranged in order since the WINDOW signal prevents the progress of the input before the previous rising edge arrives at the last delay cell. Therefore, it is sufficient for the proposed CLD to avoid the harmonic lock that it only checks the location of CLK1, CLK10, CLK12. The DLL is coarsely locked when the 10 or 11 phases of total 12 phases are in one clock periods. When less than 10-phase clocks or more than 11-phase clocks are in one clock period, the CLD generates COARSE_DN or COARSE_UP, respectively. The proposed combination of the PFD and CLD also avoids the stuck problem by comparing the outputs of the PFD and the CLD. In the stuck state, the state of the PFD become reset by a UP output from the PFD and a COARSE_DN generation from the CLD, and then the PFD will generate DN, as shown in Fig. 25.8.4(c).

Figure 25.8.7 shows a die micrograph of the prototype chip which is fabricated in a 0.13μm CMOS process. The receiver occupies 0.064mm²; the DLL and samplers occupy 0.053mm² and, the input buffer and skew compensation logic occupy 0.011mm². The receiver operates from 1.36 to 3.0Gb/s with a BER of less than 10^{-12}. Figure 25.8.5 shows the measured jitter tolerance curve and recovered clock with 10MHz, 220ps$_{pp}$ sinusoidal jitter at a 3.0Gb/s 2^{10}-1 PRBS. The receiver tolerates the maximum jitter generated by the test equipment, Agilent N4903A, below 10MHz. The proposed receiver recovers the clock with a 50% duty cycle even when the input is not a clock, as shown in Fig 25.8.6(a). The measured jitter is 5.85ps$_{rms}$ and 48.4ps$_{pp}$ in response to a 3.0Gb/s 2^{10}-1 PRBS. At 3.0Gb/s, the receiver dissipates 13.8mW from a 1.2V supply, which includes the power consumption of the input buffer (8mW). Figure 25.8.6(b) summarizes the performance of this work and compares it with other works for DDI interfaces.

Acknowledgements:
The chip fabrication was supported by IC Design Education Center (IDEC).

References:
[1] K. Yamguchi, Y. Hori, K. Nakajima, et al., "A 2.0Gb/s Clock-Embedded Interface for Full-HD 10b 120Hz LCD Drivers with 1/5-Rate Noise-Tolerant Phase and Frequency Recovery," *ISSCC Dig. Tech. Papers*, pp. 192-193, Feb., 2009.
[2] I. Jung, D. Shin, T. Kim, and C. Kim, "A 140-Mb/s to 1.82-Gb/s Continuous-Rate Embedded Clock Receiver for Flat-Panel Displays," *IEEE Trans. on Circuits and Systems II*, vol. 56, no. 10, pp. 773-777, Oct., 2009.
[3] H. Nam, K.Y. Oh, S.K. Kim, et al., "Cost Effective 60Hz FHD LCD with 800Mbps AiPi Technology," *SID Symp. Dig.*, vol. 39, no. 1, pp. 677-680, May, 2008.
[4] H.-K. Jeon, Y.-W. Moon, J.-I. Seo, et al., "A Clock Embedded Differential Signaling (CEDS™) for the Next Generation TFT-LCD Applications," *SID Symp. Dig.*, vol. 40, no. 1, pp. 975-978, June, 2009.
[5] W.-J. Choe, B.-J. Lee, J. Kim, et al., "A Single-Pair Serial Link for Mobile Displays With Clock Edge Modulation Scheme," *IEEE J. Solid-State Circuits*, vol. 42, no. 9, pp. 2012-2020, Sep., 2007.
[6] R. Farjad-Rad, W. Dally, H.-T. Ng, et al., "A 0.2-2GHz 12mW Multiplying DLL for Low-Jitter Clock Synthesis in Highly-Integrated Data Communication Chips," *ISSCC Dig. Tech. Papers*, pp. 76-77, Feb., 2002.

ISSCC 2011 / February 23, 2011 / 5:00 PM

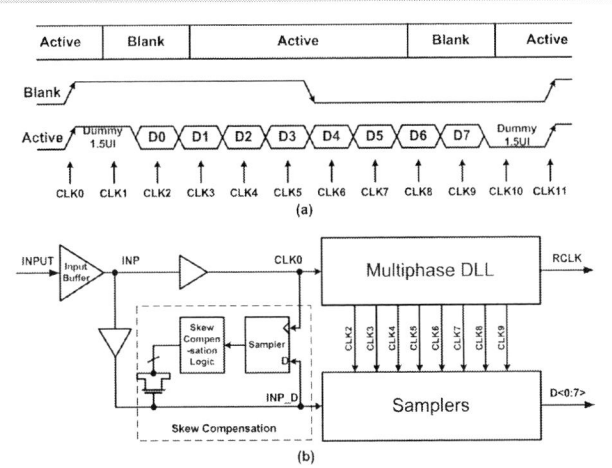

Figure 25.8.1: (a) Clock is transmitted during blank periods and clock-embedded data is transmitted during active periods. (b) The DLL-based receiver architecture.

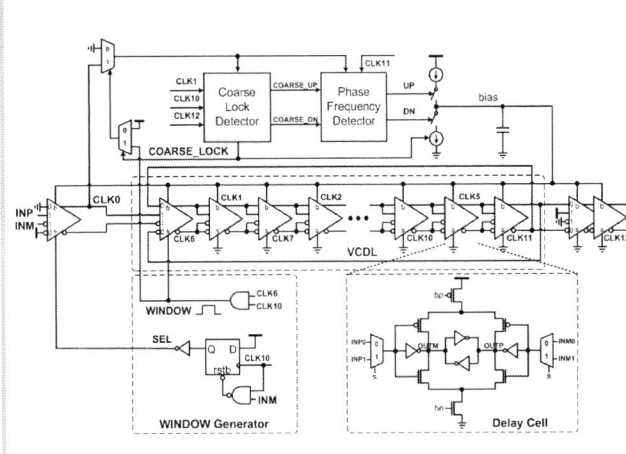

Figure 25.8.2: Proposed DLL architecture.

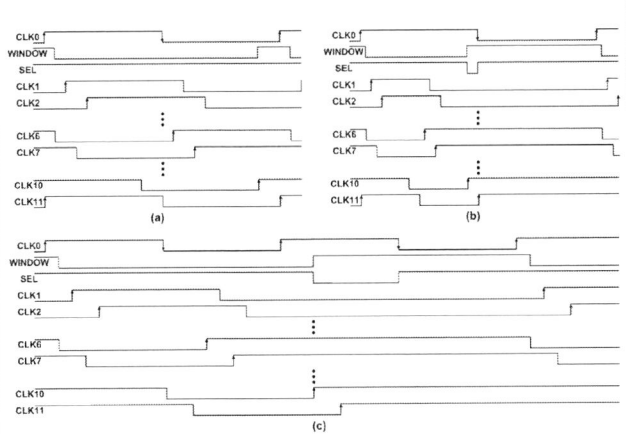

Figure 25.8.3: Timing diagram when the DLL (a) is locked, (b) has shorter delay than one clock period, and (c) has longer delay than one clock period.

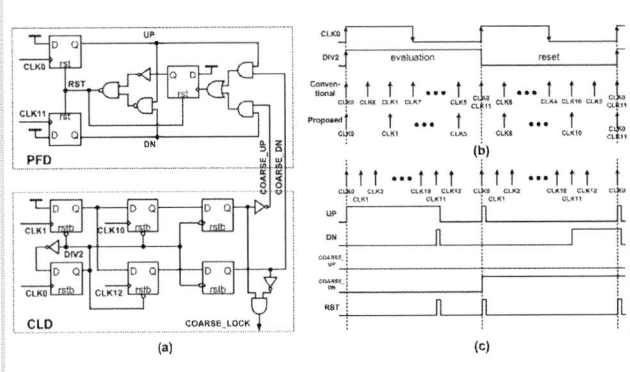

Figure 25.8.4: (a) The schematic of PFD and CLD, and the timing diagram when the DLL is in (b) 2nd-harmonic lock and (c) stuck state.

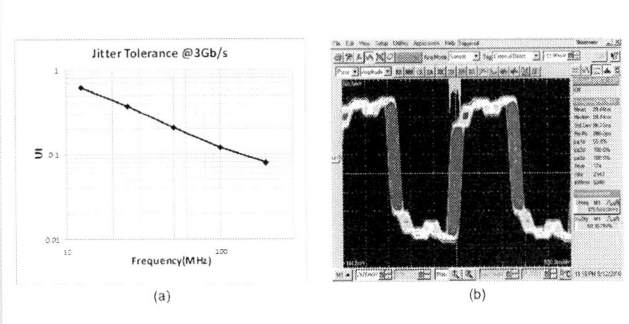

Figure 25.8.5: (a) Measured jitter tolerance curve and (b) recovered clock with 10MHz 220ps_pp sinusoidal jitter at a 3.0Gb/s 2^10-1 PRBS.

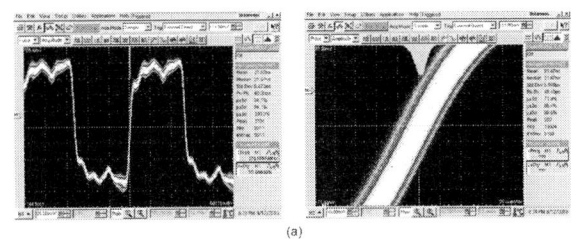

	[1]	[2]	This work
Technology	0.25μm CMOS	0.25μm CMOS	0.13μm CMOS
Supply Voltage (V)	3.0	2.5	1.2
Data Rate (Gb/s)	1.25 ~ 3.0	0.14 ~ 1.82	1.36 ~ 3.0
RMS-jitter (ps)	11@2.0Gb/s	14.96@1.82Gb/s	5.85@3.0Gb/s
Power Consumption (mW)	93@2.0Gb/s (CDR only)	342.5@1.82Gb/s	13.8@3.0Gb/s
Area (mm²)	0.45	2	0.064

(b)

Figure 25.8.6: (a) Measured recovered clock and jitter histogram. (b) Performance summary.

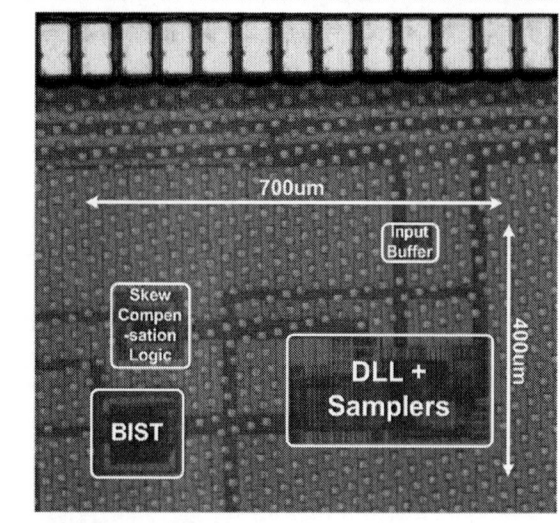

Figure 25.8.7: Die micrograph.

ISSCC 2011 / SESSION 26 / LOW-POWER WIRELESS / OVERVIEW

Session 26 Overview / Wireless

Low-Power Wireless

Session Chair: *Jan Crols, AnSem, Leuven, Belgium*

Session Co-Chair: *Stefan Heinen, RWTH Aachen University, Aachen, Germany*

RFID and remotely powered sensors will be ubiquitous in the future in applications as diverse as health-care and inventory management. These applications represent a new paradigm for data transfer in the information age in that this technology combines remote powering and wireless data transmission resulting in the need for low-power solutions. A key component is the sensor node, which monitors and transfers the data from a remote site. A reader might provide the radio frequency energy to remotely power up the sensor, as well as detect its information. This session features major breakthroughs related to key aspects of the implementation.

Paper 26.1 [Graz University of Technology and Infineon Technologies] presents the first sensor remotely powered by the electro-magnetic field in the frequency range from 13MHz to 2.45GHz. The analog front-end uses a new concept for lowering the forward voltage drop in the AC/DC rectifier. Also, a new concept is used to control the combined shunt and modulator device. The DC current consumption of the system is 70% of previous published designs. The multipurpose sensor block enables on-chip voltage monitoring, temperature sensing, RSSI measurement and has an interface to supply off-chip sensors with a power consumption up to the mW range. The ADC of the sensor block achieves an FOM of 419fJ/conversion step. A fully passive multi-frequency EPC compatible RFID sensor node is fabricated in a 0.13µm CMOS process. The chip operates from 13.56MHz to 2.45GHz, the RF sensitivity is -10.3dBm, and the overall power consumption is 7.9µW.

Paper 26.2 [KAIST and PHYCHIPS] presents an isolator-less, low-power RF front-end architecture for mobile UHF RFID readers. The work presented is the first to replace the commonly used bulky isolator by a small 45° phase shifter. A self-correlation receiver aligns the LO phase and the TX leakage to eliminate the additive noise from a strong TX leakage. A highly efficient system is realized by implementing a polar ASK TX with envelope feedback linearization. A high dynamic range exceeding 30dB is achieved, while a good PA efficiency up to 33% is maintained. This front-end is the first mobile RFID reader solution that achieves a high system efficiency and integration level simultaneously.

Paper 26.3 [Holst Centre / imec] reports on the first super-regenerative RX that integrates RF front-end, quench generator, analog baseband, digital baseband and PLL. The presented ultra-low-power OOK single-chip transceiver for WBAN applications in 90nm CMOS is intended for operation in the 2.4GHz medical BAN and ISM bands. The transmitter uses pulse-shaped OOK with 0dBm peak power, and consumes 2.59mW with 50% OOK. Including the digital part, the Rx consumes 715µW at 1Mb/s data rate, which is oversampled at 3MHz. The work presented achieves the highest published data rate (5Mb/s) of a super-regenerative RX frontend at a sensitivity of -75dBm.

Paper 26.4 [University of Washington] reports a 120µW MICS/ISM band receiver for wireless sensing using a 130nm CMOS technology. The low-IF FSK receiver achieves a -90dBm sensitivity at 200kp/s data rate. A low-power mode reduces the power to 44µW while maintaining -70dBm sensitivity. 9x frequency multiplication is used, allowing a 44.5MHz LO as a foundation for the low power consumption. The injection-locked LO has a settling time below 100ns. The proposed receiver represents a 3x improvement over previously published state-of-the art MICS band receivers both in terms of energy/bit and total power consumption

Paper 26.5 [MediaTek] presents a host-based GPS/Galileo SoC, which achieves 2.0dB NF resulting in a -165dBm chip-in tracking sensitivity. The device can sustain out-of band blockers as high as +16dBm at its antenna port without the need of external LNA and inter-stage SAW filter. By using the proposed scheme of in-band blocker cancellation, this SOC can withstand 12 inband CW blockers simultaneously. The chip occupies 6.6mm² in a 65nm CMOS process. The receiver exhibits a -5dBm out-of-band IIP3 while consuming a total power of 18mW, where 8.6mW are due to the RF portion which is the lowest power consumption ever reported.

ISSCC 2011 / February 23, 2011 / 1:30 PM

Paper 26.6 [HKUST] presents a frequency synthesizer for software-defined radios covering all wireless standards from 47MHz to 10GHz (including 14-band MB-OFDM UWB), as well as the 802.15.3c standard from 57 to 66GHz. The synthesizer is implemented in a 0.13µm CMOS process and occupies an active area of 3mm². Power consumption ranges from 33mW to 83mW, for a phase noise of -139.6dBc/Hz at 3MHz offset from a 1.7GHz carrier. Employing novel circuits, including a dual-band quadrature VCO, a x3/x5/x7 injection-locked frequency multiplier, single-sideband mixers for UWB, and x2/x3 sub-harmonic injection-locked oscillators at mm-Wave with automatic peak calibration, the synthesizer offers a superior performance compared to existing solutions in terms of frequency coverage, phase noise, spur level, and power consumption.

In Paper 26.7 [University of California] a low-spur, multiplying, injection-locked delay locked loop using a 90nm CMOS process is presented. Spurs and jitter due to reference clock injection mismatch are minimized by tuning the aperture timing, using a phase interpolator in the feedback path and an adaptive tuning algorithm based on duty-cycle estimation. With these adaptation techniques, the MDLL achieves a -46dBc of reference spur, 1ps of rms and 1.1ps of peak-to-peak jitter, while operating at a reference clock frequency of 570MHz and output clock frequency of 4.6GHz.

The papers in this session show the continuing achievements in reducing the power consumption of integrated wireless sensor nodes as well of wireless building block as part of SoC implementations. This session contributes to the progress of low-power wireless systems that drive low-cost consumer applications.

Presenters:

1:30 PM

26.1 A 7.9µW Remotely Powered Addressed Sensor Node Using EPC HF and UHF RFID Technology with -10.3dBm Sensitivity

H. Reinisch, Graz University of Technology, Graz, Austria

2:00 PM

26.2 An Isolator-less CMOS RF Front-End for UHF Mobile RFID Reader

E-H. Kim, KAIST, Daejeon, Korea

2:30 PM

26.3 A 2.4GHz ULP OOK Single-Chip Transceiver for Healthcare Applications

M. Vidojkovic, Holst Centre / imec, Eindhoven, The Netherlands

3:15 PM

26.4 A 120µW MICS/ISM-Band FSK Receiver with a 44µW Low-Power Mode Based on Injection-Locking and 9x Frequency Multiplication

J. Pandey, University of Washington, Seattle, WA

3:30 PM

26.5 A GPS/Galileo SoC with Adaptive In-Band Blocker Cancellation in 65nm CMOS

C-H. Wu, MediaTek, Hsinchu, Taiwan

3:45 PM

26.6 A 0.05-to-10GHz 19-to-22GHz and 38-to-44GHz SDR Frequency Synthesizer in 0.13µm CMOS

S. Rong, HKUST, Hong Kong, China

4:15 PM

26.7 A 4.6GHz MDLL with -46dBc Reference Spur and Aperture Position Tuning

T. A. Ali, University of California, Los Angeles, CA; Oracle, Menlo Park, CA

26

978-1-61284-303-2/11 $26.00 © 2011 IEEE

ISSCC 2011 / SESSION 26 / LOW-POWER WIRELESS / 26.1

26.1 A 7.9µW Remotely Powered Addressed Sensor Node Using EPC HF and UHF RFID Technology with -10.3dBm Sensitivity

Hannes Reinisch[1], Martin Wiessflecker[2], Stefan Gruber[2],
Hartwig Unterassinger[1], Günter Hofer[2], Michael Klamminger[1],
Wolfgang Pribyl[1], Gerald Holweg[2]

[1]Graz University of Technology, Graz, Austria,
[2]Infineon Technologies, Graz, Austria

The combination of remote powering and wireless data transmission enables the monitoring of data in harsh and difficultly accessible environments, where batteries cannot be replaced or where it is not possible to supply the sensor system by wire or by thermal or photovoltaic energy harvesting. Equipping sensor nodes with RFID functionality not only enables identification and logistic applications but also an easy integration of the sensing tag into existing RFID systems. Exploiting the different characteristics of HF and UHF RFID systems, namely the large operating range in UHF and the high available power in HF, increases the flexibility of wireless sensor nodes.

This paper presents the first fully passive multifrequency RFID transponder according to the EPC HF and EPC Class 1 Gen 2 UHF standard, containing an on-chip temperature sensor and an interface for off-chip sensors. The chip is operating in the frequency range from 13MHz to 2.45GHz, has only two interface pins to a broadband antenna and can power off-chip sensors (e.g. pressure sensors) with a power consumption up to the mW range. The input sensitivity is -10.3dBm and is comparable to [1-3] despite the additional functionalities.

Figure 26.1.1 shows the block diagram of the wireless node. The AC/DC converter rectifies the incoming RF signal from 13MHz to 2.45GHz and supplies the chip with power. Shunt and modulator are combined in one device. A frequency detection unit distinguishes between HF (13.56MHz) and UHF (860MHz to 2.45GHz) modes. The clock for the digital unit and the sensor block is either extracted by the clock recovery unit from the RF input signal in HF mode or provided by a relaxation oscillator in UHF mode as published in [3].

Figure 26.1.2 shows the single-stage full-wave multifrequency rectifier. L_A and L_B are the differential inputs with the common-mode potential V_{SS} which is generated by the cross-coupled transistors M_{N1} and M_{N2}. For further considerations V_{SS} is the reference. To achieve full operation from input power levels of about -12dBm in UHF mode a special circuitry is needed to lower the forward voltage drop of M_{P1}/M_{P2}. M_{P5} provides a static bias voltage at the node gate_control that is one V_{GS} below V_{DD}. M_{N3} and M_{N4} are in the linear region. M_{P1} is conductive when the RF input signal L_A is in its positive phase and its amplitude is one V_t higher than its gate potential. The gate potential of M_{P1} (V_{gate_P1}) is lowered due to coupling of L_B (L_B is in negative phase) by C_1. Simultaneously the gate potential of M_{P2} is increased, which reduces the losses caused by the direct path between L_A and L_B. Vice-versa M_{P2} conducts when L_B is in its positive phase and its amplitude is one V_t higher than its gate potential and the other part of the circuit is working according to the same principle. So M_{P1} and M_{P2} are working as switches instead of diodes as in conventional rectifiers, which increases the efficiency. The current flow is depicted in Fig. 26.1.2.

The combined TX and shunt block is presented in Fig. 26.1.3. M_{N8} and M_{N9} are diode connected and generate the sense voltage V_{sense1}. V_{DD} is limited to a maximum of 1.2V in UHF and 1.6V in HF mode. M_{P6} is sensitive to changes in V_{GS} (W/L>), so the gate potential of MN7 (V_{gate}) changes very quickly and overshoots are prevented. At high field strength, M_{N7} dissipates not only the excessive power, but also transforms the chip impedance, which results in a mismatch between antenna and chip. For load modulation in HF or backscattering in UHF the control of the shunt/modulator is different. In UHF mode the incoming RF signal is backscattered by shorting the chip's input pins. In HF mode a V_{in} of at least 600mV is needed to keep the clock recovery working during modulation. So V_{in} cannot be shorted as in UHF mode. M_{N10} and M_{N11} provide V_{sense2} that is one diode voltage below V_{in_peak}. During the modulation pulse (mod_data=1) M_{N13} charges the node V_{mod} and so V_{gate} increases. If mod_data=0, M_{N16} restores the former V_{gate}. The current source M_{N14} is switched off by M_{N15} to freeze V_{gate} during receiving and backscattering.

Figure 26.1.4 shows the multipurpose sensor block. Embedded on-chip voltage monitoring, temperature sensing, RSSI measurement and the controlling of a variety of off-chip sensors are implemented. The on-chip temperature sensor is based on the principle of comparing a temperature-dependent voltage to a temperature-independent voltage. The measurable temperature range is from -40°C to 130°C. An SC BGR based on the principle described in [4] provides these two voltages, which are buffered and fed to a 10b SAR ADC with a split capacitive DAC and a time-domain comparator according to the principle in [5]. The split-capacitor array is formed by two equally-sized 5b arrays. The conversion speed of the ADC is 36kS/s at a clock frequency of 1MHz and it achieves a FOM of 419fJ/conversion step. To reduce the power consumption, power gating is implemented. Different sensors are activated by writing specific data via an EPC write command into the distinct memory bank. The typical operation sequence is: The control unit starts the BGR and the buffer provides V_{Temp} and V_{ref}. After settling of the BGR the ADC starts sampling the analog data and then the signal conversion_complete goes to high and signalizes correct data in the memory.

Driving a pressure sensor or other off-chip sensors with a power consumption in the mW range is only possible for a short distance between the reader and the tag. To increase the operating range up to a few meters, an interface for off-chip sensors is developed. It is also depicted in Fig. 26.1.4. The interface is based on the principle shown in [6] and includes an RF input power detector, a VCO, a DC/DC charge pump based on the Dickson topology and a level detection unit (LDU). If the RF input power is sufficient a VCO is switched on and the DC output voltage of the RFID rectifier is pumped up to 3V. The energy is stored in an on-chip or off-chip buffer capacitor. The LDU activates the off-chip sensor if the voltage level in the buffer capacitor is high enough. The current consumption of the LDU is 310nA. For this load the DC/DC charge pump has an efficiency of 32%. The analog sensor data is fed into the ADC via a multiplexer. After conversion, the sensor data can be read out by an EPC read command.

Figure 26.1.5 shows the contact-based measurement setup for the input sensitivity, the curve of the DC supply voltage of the chip versus the RF input power and the current consumption at different output voltages of the PSU. Neglecting all measurement losses an input sensitivity of -10.3dBm for the operation mode as UHF RFID Sensor tag was measured. Compared to [1] the input sensitivity is lower due to the fact that the system is designed to work at both HF and UHF. If the sensor is disabled the input sensitivity is -12.5dBm. An input power of at least -7.9dBm is needed to activate the off-chip sensor interface. A field strength of 36mA/m is sufficient for RFID functionality in HF mode. The overall current consumption with activated temperature sensor in continuous mode is 7.9µA at 1.0V whereas the RFID analog front-end and the digital core consume 4.1µA. The remotely powered sensor tag is fabricated in an Infineon 0.13µm CMOS process. All key facts are given in Fig. 26.1.6 and the micrograph is shown in Fig. 26.1.7.

Acknowledgement:
This work has been partly funded by the Austrian Research Promotion Agency (FFG) within the project iTire.

References:
[1] D. Yeager, F. Zhang, A. Zarrasvand, B. P Otis, "A 9.2µA Gen 2 compatible UHF RFID sensing tag with -12dBm Sensitivity and 1.25µVrms input-referred noise floor," *ISSCC Dig. Tech. Papers*, pp.52-53, 7-11 Feb. 2010.
[2] Jun Yin et al., "A system-on-chip EPC Gen-2 passive UHF RFID tag with embedded temperature sensor," *ISSCC Dig. Tech. Papers*, pp.308-309, 7-11 Feb. 2010.
[3] A. Missoni, C. Klapf, W. Pribyl, G. Hofer, G. Holweg, "A Triple-Band Passive RFID Tag," *ISSCC Dig. Tech. Papers*, pp.288-614, 3-7 Feb. 2008.
[4] S. Chen and B. J. Blalock, "Switched capacitor bandgap voltage reference for sub-1-V operation," *IEICE Electron. Express*, vol. 3, no. 24, pp.529-533, 2006.
[5] A. Agnes et al., "A 9.4-ENOB 1V 3.8µW 100kS/s SAR ADC with Time-Domain Comparator," *ISSCC Dig. Tech. Papers*, pp.246-610, 3-7 Feb. 2008.
[6] H. Reinisch et al., "An Electro-Magnetic Energy Harvester with 190nW Idle Mode Power Consumption for Wireless Sensor Nodes," *Proc. ESSCIRC*, pp.234-237, 14-16 Sep. 2010.

978-1-61284-303-2/11 $26.00 © 2011 IEEE

ISSCC 2011 / February 23, 2011 / 1:30 PM

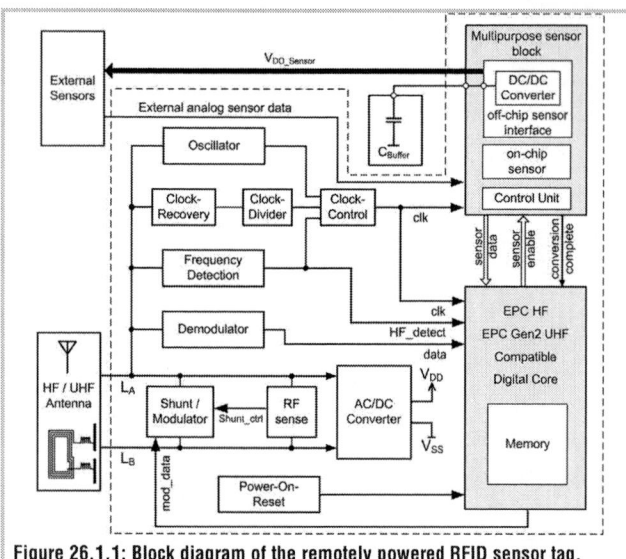

Figure 26.1.1: Block diagram of the remotely powered RFID sensor tag.

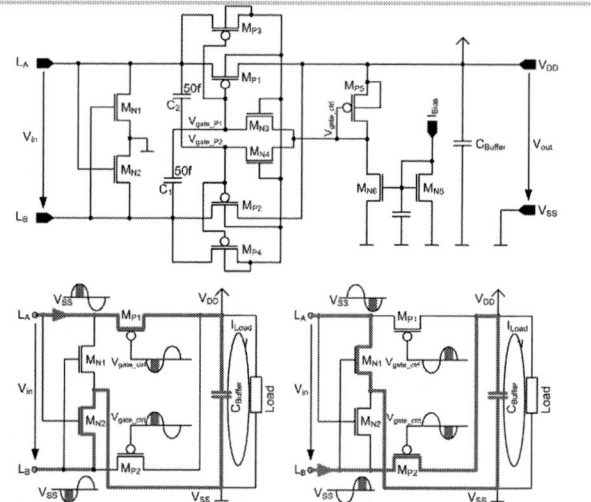

Figure 26.1.2: Simplified schematic of the full-wave rectifier including the current flow for the positive and the negative phase of V_{in}.

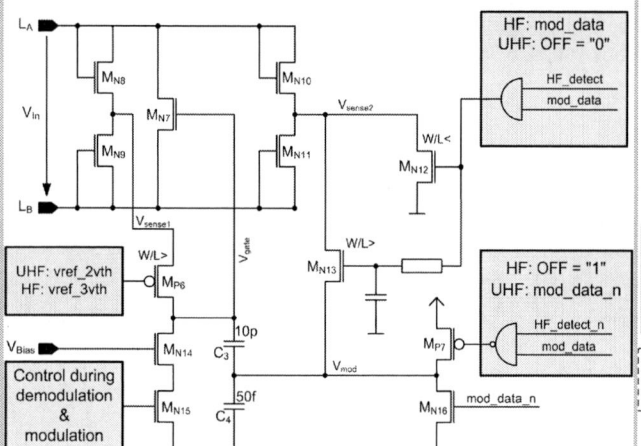

Figure 26.1.3: Combined TX and shunt module for HF and UHF.

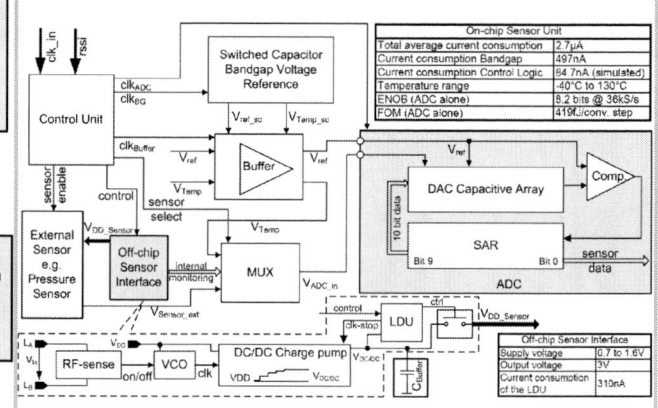

Figure 26.1.4: Multipurpose sensor block showing the principle of the on-chip temperature sensor and the off-chip sensor interface.

Figure 26.1.5: Measurement setup, VP-characteristic (f=900MHz) and IV-characteristic for different operating modes.

	This work	[1]	[2]
Technology	Infineon 0.13µm CMOS	0.13µm CMOS	0.18µm CMOS
Active Area	0.95mm²	2.0mm²	1.1mm²
Frequency Range	HF (13.56MHz), UHF (860MHz to 2.45GHz)	UHF (900MHz)	UHF (860 to 960MHz)
Protocol Standard	EPC HF, EPC Gen2 UHF	EPC Gen2 UHF	EPC Gen2 UHF
Sensitivity HF EPC alone	36mA/m¹	not applicable	not applicable
Sensitivity UHF EPC alone	-12.5dBm¹	not available	not available
Sensitivity UHF EPC & Sensor	-10.3dBm¹	-12dBm	-6dBm
Sensitivity UHF EPC & off-chip Sensor	-7.9dBm¹	not applicable	not applicable
DC Loading Power EPC only	3.1µW	not available	not available
DC Loading Power EPC & Sensor	7.9µW	11µW (9.2µA/1.2V)	12µW
DC Loading Power EPC & off-chip Sensor	10.2µW	not applicable	not applicable
Sensor Type on-chip	Temperature, RSSI, on-chip monitoring	Temperature	Temperature
Supply voltage for off-chip Sensors	3V	not applicable	not applicable

¹ measured at 900MHz

Figure 26.1.6: Performance summary and comparison with prior art.

978-1-61284-303-2/11 $26.00 © 2011 IEEE

ISSCC 2011 PAPER CONTINUATIONS

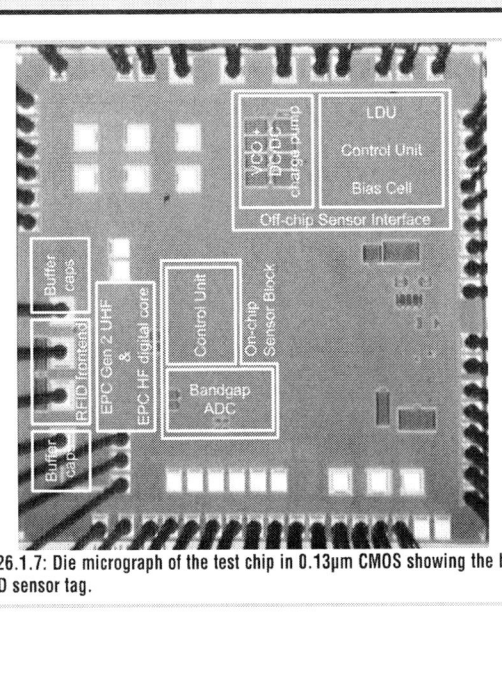

Figure 26.1.7: Die micrograph of the test chip in 0.13μm CMOS showing the blocks of the RFID sensor tag.

ISSCC 2011 / SESSION 26 / LOW-POWER WIRELESS / 26.2

26.2 An Isolator-less CMOS RF Front-End for UHF Mobile RFID Reader

Eun-Hee Kim[1], Kwyro Lee[1], Jinho Ko[2]

[1]KAIST, Daejeon, Korea,
[2]PHYCHIPS, Daejeon, Korea

RFID systems are based on backscattering communications, and this encourages research on how to guarantee a reliable RX performance under simultaneous TX/RX operation. To isolate RX from the TX self-jammer, the RFID transceivers are generally accompanied by an off-chip circulator or isolator. This has advantages in selectivity between tag signal and blocker, but it results in a low level of integration and bulky system [1,2,4]. In addition, to meet the stringent linearity requirement, the RFID reader needs to adopt a linear PA with sufficient power back-off, which degrades a global efficiency. Considering that battery life time is the critical issue particularly for mobile applications, existing UHF RFID readers are unsuitable for the mobile RFID technology due to their low system efficiency and low integration level.

Figure 26.2.1 describes the mechanisms of DC offset and PN-induced noise, which degrades the RX sensitivity in the presence of a TX leakage. The DC-offset problem is caused by a downconverted TX leakage, which can saturate the following stages of RX. Moreover, although the TX leakage and the LO signal are originated from an identical local oscillator, their cross-correlation exhibits a non-zero and finite value due to a time delay as well as a phase difference between the two paths. The random phase noise of TX leakage is transformed into extra additive noise after the downconversion process, which is called a PN-induced noise. Especially, when a strong TX leakage exists, it dominates the RX noise floor. As in the equation of V_N, shown in Fig. 26.2.1, representing an additive noise from the two ways, there is no ultimate solution to simultaneously remove noise sources of the DC offset and PN-induced noise because of θ. Alternatively, a TX leakage cancellation method, providing the coefficient α close to zero, has been reported as a possible solution, which uses a replica TX leakage [3]. However, there still remains a drawback in that it requires extra hardware and a complex algorithm.

In this work, we present a new approach, called a self-correlation RX by replacing an LO with a clamped version of the RX input, which is a summation of the TX leakage and wanted tag signal. It results in the zero phase difference θ between the LO and TX leakage, instead of the coefficient α. It implies that the PN-induced noise can be completely eliminated by aligning the phase of LO and TX leakage, rendering an external component, bulky H/W, and a complex algorithm unnecessary. Then, the remaining DC offset problem can be easily handled by DC blocking technique.

Figure 26.2.2 shows the proposed isolator-less, low-power RF front-end architecture for mobile UHF RFID readers, exploiting a quadrature self-correlation RX and polar ASK TX. Recall that the incoming RF signal is a superposition of the PA output and backscattered tag signal, the conventional quadrature LO generation is inapplicable to the proposed RX. In this work, a new quadrature demodulation scheme is proposed, based on an external RF phase shifter. Compared to a conventional RX, downconverting an RF signal with (0 and 90°) shifted LO signals, the proposed RX applies two signal pairs of (V_{TAG}, $V_{TX}\angle45°$) and ($V_{TAG}\angle45°$, V_{TX}) for I and Q output, respectively. As shown in the vector diagram, when the phase difference between an input tag and TX signal is exactly 90°, it is unable to detect the tag signal with one signal pair of (V_{TAG}, $V_{TX}\angle45°$), however large an input signal comes from the tag. On the contrary, for the other pair of ($V_{TAG}\angle45°$, V_{TX}), the tag signal is aligned with the TX leakage, and therefore it becomes the most probably detectable.

Figure 26.2.3 shows the circuit implementation of the self-correlation RX, and its operational principle is similar to a voltage-mode switching mixer. The PMOS transistors, M1 and M2, form a positive full-wave rectifier, while the NMOS transistors, M3 and M4, form a negative one. Its complementary configuration doubles the conversion gain and output swing, allowing differential signaling in the following stages. To further increase the input P_{-1dB}, the DC bias for the positive and negative output is set to be GND and V_{DD}, separately. Under this bias condition, it extends the maximum allowable input voltage up to V_{DD} and provides a conversion gain of $(2/\pi)$.

To address the low-power requirement for mobile RFID readers, the polar ASK TX is presented, shown in Fig. 26.2.4. The PA uses a saturated Class-A topology, and a third cascode stage is added to support the phase-reversal (PR) ASK-modulated signal. When delivering amplitude-modulated signals, the resultant average efficiency is much higher than that of the linear PA because it dissipates a DC current proportional to its output amplitude.

To suppress a non-linear behavior of the PA, an envelope-feedback linearization is used. The feedback loop is achieved by a linear envelope detector and differential-difference integrator to form a type-I feedback system. Ideally, as no error signal occurs in the type-I feedback system, the linear dynamic range of an overall TX is mainly determined by that of the envelope detector. In this paper, to enlarge the linear range of the envelope detector, an accurate passive-type envelope detector is proposed. By using transistors (M1~M2) as switches, an RF input signal is rectified to an envelope output with a constant conversion gain of $(1/\pi)$. To maintain the linear relationship between (input-output) signals even for a small RF input, the gate-bias voltage of transistors is set to be the same as a threshold voltage of V_{TH}. As a result, the proposed TX enlarges the linear dynamic range over 30dB while offering a good PA efficiency.

The RFID reader prototype is fabricated in a standard 0.13μm CMOS process. Figure 26.2.5 shows the measured output SNR of the proposed RX, when a modulated tag signal of -60dBm is applied without any isolator or additional leakage cancellation circuit. Considering that the required SNR is 15dB [3], the RX obtains a sensitivity of -65dBm when suffering from a TX leakage of up to 20dBm. Especially in the TX leakage range from 2 to 10dBm, it realizes the sensitivity of -75dBm, which is compatible with that of the conventional I-Q RX adopting an off-chip isolator [4].

As shown in Fig. 26.2.6, under a supply voltage of 3.3V, the implemented TX delivers 23dBm CW output power with current consumption of 198mA. The linear control range of the TX is measured to be over 25dB. In the case of DSB-ASK modulation with 85% modulation index, the ACPR performance meets the requirement of UHF RFID reader applications, which provides margin more than 10dB.

Acknowledgement:
This work was partially supported by the National Research Foundation of Korea (NRF) grant funded by the Korea government (MEST).

References:
[1] I. Kwon et al., "A single-chip CMOS transceiver for UHF mobile RFID reader," *ISSCC Dig. Tech. Papers*, pp. 216-218, Feb. 2007.
[2] W. Wang et al., "A single-chip UHF RFID reader in 0.18 um CMOS process," *IEEE J. Solid-State Circuits*, vol. 4, pp. 1741-1754, Aug. 2008.
[3] A. Safarian et al., "RF identification (RFID) reader front ends with active blocker rejection," *IEEE Trans. Microwave Theory and Techniques*, vol. 57, no. 5, pp. 1320-1329, May 2009.
[4] L. Ye et al., "A single-chip CMOS UHF RFID reader transceiver for Chinese mobile applications," *IEEE J. Solid-State Circuits*, vol. 45, no. 7, pp. 1316-1329, Jul. 2010.

ISSCC 2011 / February 23, 2011 / 2:00 PM

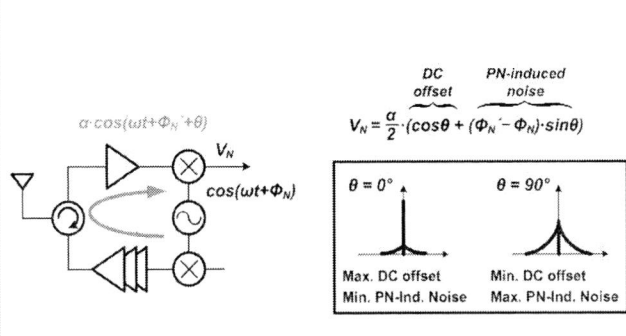

Figure 26.2.1: RX noise mechanisms from TX leakage.

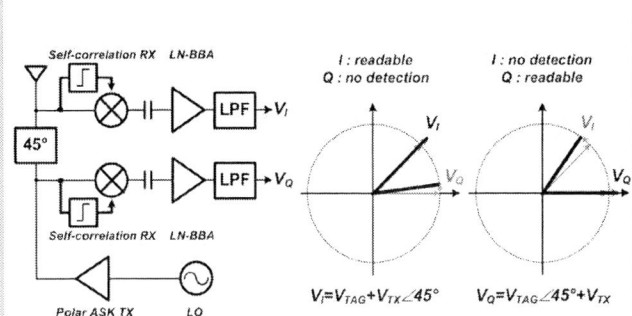

Figure 26.2.2: The proposed RF front-end architecture.

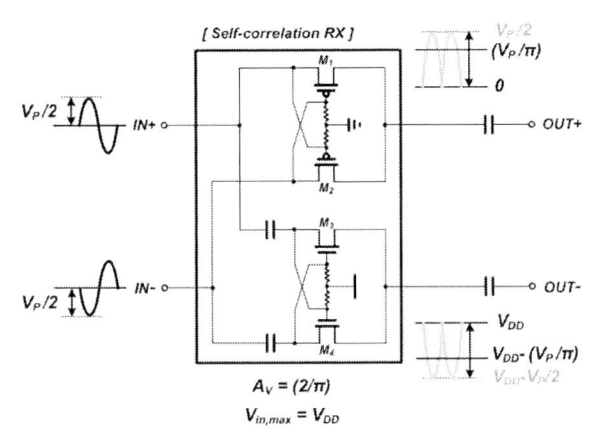

Figure 26.2.3: Circuit implementation of the self-correlation RX.

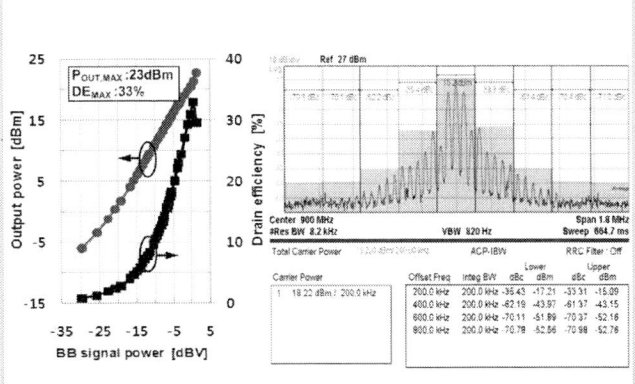

Figure 26.2.4: TX block diagram and circuit schematic.

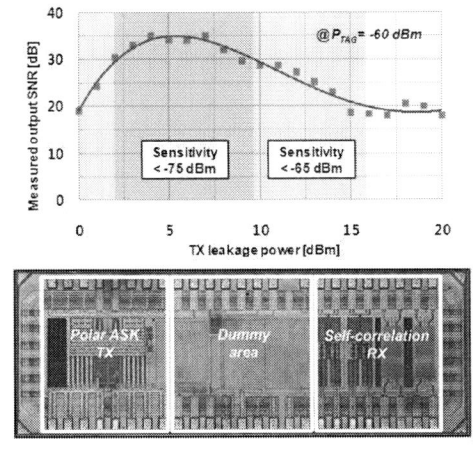

Figure 26.2.5: Measured output SNR of the proposed RX and die photo.

Figure 26.2.6: Output power, drain efficiency, and ACPR of the TX.

978-1-61284-303-2/11 $26.00 © 2011 IEEE

References			[1] ISSCC 2007	[2] JSSC 2008	[3] T-MTT 2009	[4] JSSC 2010	This work
Technology			0.18 um CMOS	0.18 um CMOS	0.18 um CMOS	0.18 um CMOS	0.13 um CMOS
Integration			RF Analog Digital baseband	RF Analog Data converts	RF (only RX)	RF Analog Data converts	RF (TRX)
Supply voltage			1.8V	1.8V	3.3V	1.8V 3.3V only for PA	3.3V
RX	TX leakage handling		Directional coupler	Two antennas IM2 injection Active trap	Using a replica TX leakage	Directional coupler AC coupling	Self-correlation RX
RX	Sensitivity		-70 dBm	-70 dBm	-82 dBm	-79 dBm	-75 dBm
RX	Test condition for sensitivity		0 dBm self-jammer	-5 dBm self-jammer	10 dBm self-jammer	22 dBm PA output (be attenuated by directional coupler)	10 dBm self-jammer
TX	Output power		4 dBm	10.4 dBm	–	22 dBm	23 dBm
TX	ACPR	±1 ch	-23.2 dBc	-30 dBc	–	-43.7 dBc	-33.3 dBc
TX	ACPR	±2 ch	-56.6 dBc	–	–	-59.51 dBc	-61.4 dBc
TX	ACPR	±3 ch	-62.7 dBc	–	–	-66.1 dBc	-70.1 dBc
TX	ACPR	±4 ch	-66.1 dBc	–	–	–	-70.8 dBc
TX	ACPR	BW_{CH}	200 KHz	–	–	250 KHz	200 KHz
Current consumption			4 mA (RX) 31 mA (TX)	300 mA (TRX)	40 mA (RX)	> 200 mA (TRX)	10 mA (RX) 198 mA (TX)

Figure 26.2.7: Performance summary and comparison.

ISSCC 2011 / SESSION 26 / LOW-POWER WIRELESS / 26.3

26.3 A 2.4GHz ULP OOK Single-Chip Transceiver for Healthcare Applications

Maja Vidojkovic[1], Xiongchuan Huang[1], Pieter Harpe[1],
Simonetta Rampu[1], Cui Zhou[1], Li Huang[1], Koji Imamura[2],
Ben Busze[1], Frank Bouwens[1], Mario Konijnenburg[1],
Juan Santana[1], Arjan Breeschoten[1], Jos Huisken[1],
Guido Dolmans[1], Harmke de Groot[1]

[1]Holst Centre / imec, Eindhoven, The Netherlands,
[2]Panasonic, Osaka, Japan

Wireless body-area networks (WBAN) are used for communication among sensor nodes operating on, in or around the human body, e.g. for healthcare purposes. In view of energy autonomy, the total energy consumption of the sensor nodes should be minimized. Because of their low complexity, a combination of the super-regenerative (SR) principle [1-3] and OOK modulation enables ultra-low power (ULP) consumption. This work presents a 2.4GHz ULP OOK single-chip transceiver for WBAN applications. A block diagram of the implemented transceiver is shown in Fig. 26.3.1. Next to the direct modulation TX [4] and SR RF [5] front-ends, this work integrates analog and digital baseband, PLL functionality and additional programmability for flexible data rates, and achieves ultra-low power consumption for the overall system.

Based on the direct modulation transmitter front-end from [4], this transceiver also implements digital pulse-shaping for improved spectral efficiency. The digital baseband oversamples the raw bits by 6X, and shapes the OOK pulse with a raised-cosine FIR filter. To ensure optimal power efficiency despite PVT variations, a duty-cycled swing detector is designed to monitor and adjust the driving level of the PA.

In the proposed SR RX front-end [5], the time-variant bias current of the VCO (Iquench) is generated in a combined analog and digital fashion to achieve both lower power consumption and higher data rate. In this work the added programmability of the analog part (AQG) and digital part (DAC), as well as their control clocks, provides high flexibility for variable data rates up to 5Mb/s. To complete the system, the analog baseband (VGA and ADC) is also integrated. The VGA is partially open-loop to reduce the power consumption. The 8b SAR ADC [6] uses asynchronous dynamic logic and custom-build 0.5fF capacitors to achieve low power at high speed.

Both the TX and RX VCO are controlled by 4b digital codes for coarse tuning and one analog tuning voltage for fine tuning as a trade-off between PVT robustness and accuracy. The analog tuning voltage is generated by a charge pump and a loop filter in the PLL (Fig. 26.3.2). Each VCO has its own buffer and 30/32 prescaler, which consists of a divider-by-2 and a divider-by-15/16. The signals are further divided by a factor M and applied to a phase/frequency detector (PFD). With multiplexers MUX1 and MUX2, the PLL can be activated for either the TX or the RX. Also, external analog tuning can be applied instead of using the PLL. Because of the quenched operation, the PLL and the SR RX cannot work simultaneously. Therefore, during the RX frequency-locking mode, the RX VCO is biased with a constant instead of a quenched current [1]. After that, the PLL must be disabled, and the RX starts the receiving mode. In the transmission mode, the PLL can also be disabled to further decrease power consumption. An algorithm to optimize timing of the PLL during the TX/RX frequency-locking modes and the transmitting/receiving modes is being developed.

On the RX side (Fig. 26.3.3), the received signal VIN is 3X oversampled by the quench signal to achieve better peak synchronization. By omitting oversampling, the data rate can be increased to the quench rate, but additional synchronization is needed. Furthermore, the timing is important and it is controlled by 3 clocks from the digital baseband: an_clk_qch, dig_clk_qch and ADC_clk. The shape of Iquench is controlled by dig_clk_qch and an_clk_qch. In this way the VCO on-time is controlled. The ADC samples the output of the VGA on the rising edge of the ADC_clk. Each of the control clocks is derived from the 6MHz baseband clock (BBCLK) by a dedicated clock generator. For each clock, two edge selectors from both delay lines (T1 and T2) are combined with standard-cell-based delay elements in order to generate two phase-shifted 3MHz signals. An edge-triggered set-reset latch is used to combine both 3MHz signals. In order to maximize the performance, the phase and duty cycle of each clock is programmable with a resolution of 750ps by controlling the delay in the delay lines. The digital

baseband also provides data spreading/de-spreading for scalable data rates, reliable timing synchronization, data detection algorithms, CRC-16-CCITT encoding and decoding for packet validation. In addition, it enables the automatic control loop for the VGA, and has the enhanced timing tracking capability to deal with the clock offset between transmitter and receiver.

Figure 26.3.7 shows a die photo of the transceiver implemented in 90nm CMOS. It occupies 2.4x1.85mm². Figure 26.3.4 shows the measured results of the TX front-end. The upper graph shows the spectrum efficiency. When transmitting a 1Mb/s OOK modulated signal, most of the signal power (-20dBc) is within ±1MHz around the carrier. Due to 6X oversampling of the data rate in the digital pulse-shaping filter, there are sidebands at multiples of 6MHz away from the carrier, all below –30dBc. The lower graph of Fig. 26.3.4 shows the power efficiency of the TX. In this measurement, the digital PA control word is used to adjust the output power. At nominal 0dBm output power, the overall efficiency of the TX is 24% when transmitting continuous 1's, while the efficiency for both 0's and 1's reaches 40%. The upper part of Fig. 26.3.5 shows the measured BER and selectivity of the RX front-end. BER is plotted for various data rates without oversampling. For a BER of 10⁻³, the achieved sensitivity is –75dBm for 5Mb/s. For selectivity measurements, a –75dBm 500Kb/s OOK input signal with different carrier frequencies is applied to the RX. The carrier frequency is swept around the receiver operating frequency. From the measurements, it can be concluded that outside the 4MHz bandwidth the interferer will be rejected taking BER=10⁻³ as a detection level. Both the TX and RX measurements use external tuning of the VCOs, while the PLL is tested separately. With a 500kHz reference frequency (fref), the PLL locks the VCO frequency from 2.36 to 2.485GHz with increments of 15 or 16MHz, dependent on the setting of the divider-by-15/16. In the lower part of Fig. 26.3.5 the measured outputs of the prescaler (79MHz) and the divider-by-M (500kHz) are shown. For the total 30x158 division ratio the VCO output frequency is 2.37GHz.

With 50% OOK 0dBm transmitting power, the TX front-end consumes 2.53mW from 1V. At 1Mb/s, oversampled at 3MHz, the RF part of the RX consumes 390µA from 1.2V, while the analog baseband consumes 48µA from 1V. The digital TX and RX basebands consume 51µA and 166µA at 1.2V, respectively, running at 6MHz clock. The peak current of the PLL is 915µA from 1.2V. Since the duration of the TX/RX frequency locking modes is negligible compared to the duration of the transmitting/receiving modes, the contribution of the PLL to the total transceiver power can be neglected. Figure 26.3.6 compares the presented TX and RX front-ends (without RX-oversampling) to state-of-the-art. The presented work achieves a better sensitivity for higher data rates with lower or comparable power consumption.

Acknowledgement:
The authors would like to thank Ronald van Langevelde from Philips Research for useful discussions on the PLL design.

References
[1] J. Y. Chen, et al., "A fully integrated auto-calibrated super-regenerative receiver in 0.13µm CMOS," *IEEE J. Solid-State Circuits*, vol. 42, pp. 1976 – 1985, Sept. 2007.
[2] D. Shi et al., "A 5GHz fully integrated super-regenerative receiver with on-chip slot antenna in 0.13µm CMOS," *IEEE Symposium on VLSI Circuits*, pp. 34–35, June 2008.
[3] J. Ayers et al., "A 0.4nJ/b 900MHz CMOS BFSK Super-Regenerative Receiver," *IEEE CICC*, pp. 591 – 594, Sep. 2008.
[4] X. Huang, et al., "A 0dBm 10Mbps 2.5GHz Ultra-Low Power ASK/OOK Transmitter with Digital Pulse-Shaping," *IEEE RFIC Symposium*, pp. 263-266, May 2010.
[5] M. Vidojkovic, et al., "A 500µW 5Mbps ULP Super-regenerative RF Front-End," *IEEE ESSCIRC*, Sep. 2010.
[6] P. Harpe, et al., "A 12fJ/Conversion-Step 8bit 10MS/s Asynchronous SAR ADC for Low Energy Radios," *IEEE ESSCIRC*, Sep. 2010.
[7] M.K. Raja, et al., "A 52 pj/bit OOK transmitter with adaptable data rate," *ASSCC 2008*, pp.341-344, Nov 2008.
[8] Y.H. Chee et al., "An Ultra-Low-Power Injection Locked Transmitter for Wireless Sensor Networks," *IEEE J. Solid-State Circuits*, vol. 41, no.8, pp. 1740–1748, Aug. 2006.
[9] D.C. Daly et al., "An Energy-Efficient OOK Transceiver for Wireless Sensor Networks" *IEEE J. Solid-State Circuits*, vol. 42, no.5, pp. 1003–1011, May 2007.

978-1-61284-303-2/11 $26.00 © 2011 IEEE

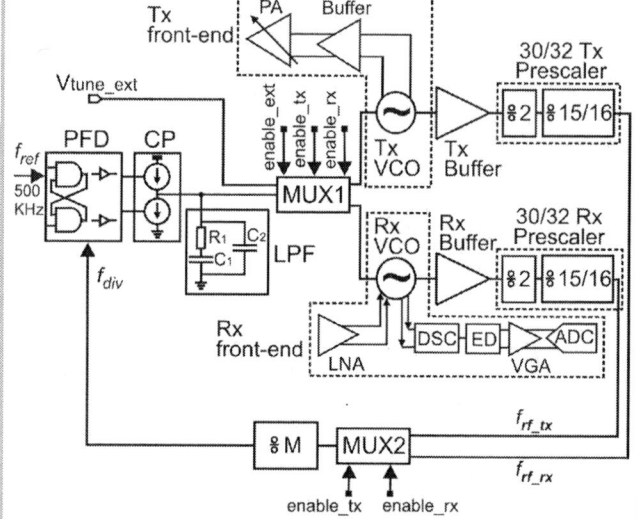

Figure 26.3.1: Block diagram of the ultra-low-power single-chip transceiver.

Figure 26.3.2: Block diagram of the transceiver PLL.

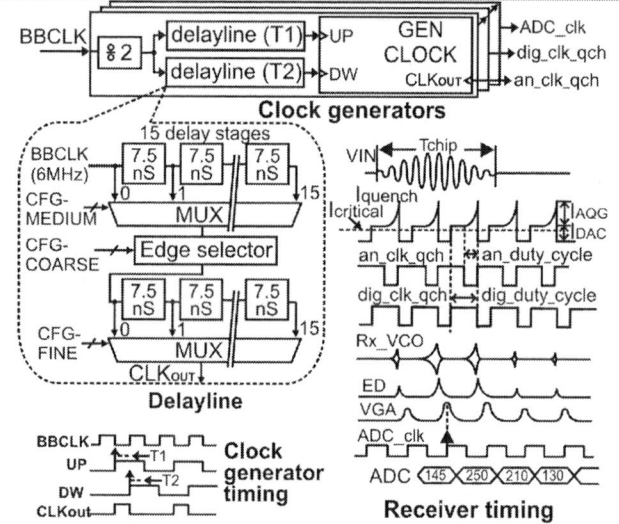

Figure 26.3.3: RX timing illustration.

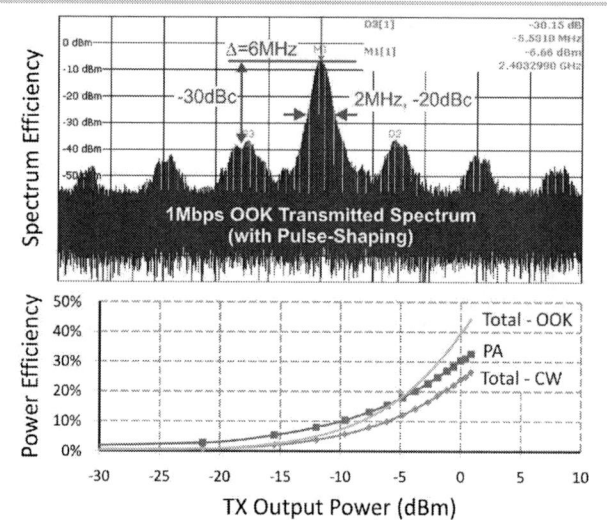

Figure 26.3.4: Measured TX spectrum and power efficiency.

Figure 26.3.5: Measured RX front-end BER / selectivity and the prescaler/divider-by-M output.

Tx front-end	Freq. (GHz)	P_out (mW)	η_{TX} (%)	Data-rate (Mbps)	FOM (nJ/bit/mW)
This work	2.4	1	24	1/10	2.53/0.253
[7]	0.433	0.054	4.5	10	0.97
[8]	1.9	1	28	0.156	11.54
[9]	0.9	0.6	6.9	1	14.4

RX front-end	Freq. (GHz)	Sensitivity (dBm)	Selectivity	Data-rate (Mbps)	Power (mW)
This work	2.4	-75	4MHz	5	0.534 @1/1.2V
[1]	2.4	-60	900kHz @-3dB	1	2.8 @1.2V
[2]	5	-60	10MHz @-30dB	1.2	6.6 @1.2V
[3]	0.9	-70	/	1	0.4 @1.3V

Figure 26.3.6: Performance summary and comparison with the state-of-the-art TX and RX front-ends.

ISSCC 2011 PAPER CONTINUATIONS

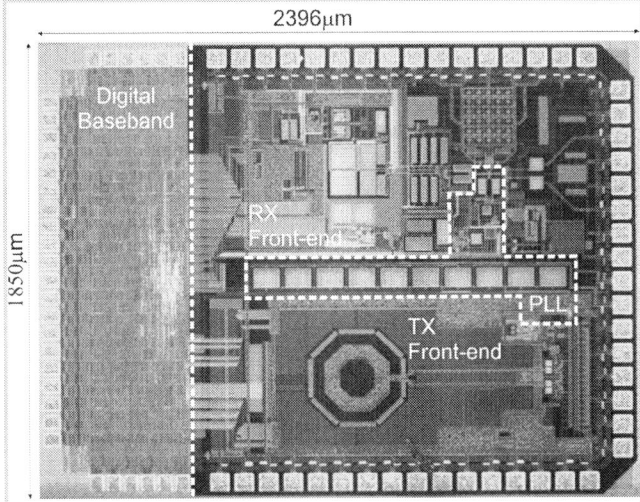

Figure 26.3.7: Micrograph of the single-chip transceiver. The chip package is plastic QFN80 with wirebonding.

ISSCC 2011 / SESSION 26 / LOW-POWER WIRELESS / 26.4

26.4 A 120µW MICS/ISM-Band FSK Receiver with a 44µW Low-Power Mode Based on Injection-Locking and 9x Frequency Multiplication

Jagdish Pandey, Jianlei Shi, Brian Otis

University of Washington, Seattle, WA

For true low-power peer-to-peer wireless links for sensing, link symmetry must be maintained unlike in [1,2] where the burden is shifted to the receiver. In addition, the transceiver power dissipation and performance should be adaptive to save power when the link distance is short. We present a 120µW MICS/ISM band receiver with -90dBm sensitivity at a data-rate of 200kb/s with BER<0.1%. The receiver incorporates a 44µW low-power (LP) mode that achieves -70dBm sensitivity at 200kb/s and 0.1% BER. At a lower data-rate of 20kb/s, the LP mode can be used as a wake-up receiver with increased sensitivity of -75dBm with 1% BER and 38µW power consumption.

In traditional receiver architectures, the LO generation circuitry and the LNA typically consume the bulk of the power consumption [3]. As a result, the super-regenerative architecture has been a popular choice to achieve high sensitivity at lower power dissipation [2,4,5]. The free-running LO employed in typical super-regenerative receivers needs frequent calibration for frequency stability, introducing asymmetry in the radio link. The proposed architecture overcomes this limitation by creating a "virtual LO" from a stable quartz reference.

Figure 26.4.1 shows the block diagram of the proposed ultra-low-power FSK receiver. We utilize a low-IF (1.5MHz) architecture to reduce the influence of 1/f noise on the noise figure of the receiver. We perform 9x frequency multiplication in the mixer to allow LO operation at 44.5MHz. Carrier generation at a fraction of the RF frequency allows extremely low-power operation. Using multistage, multiphase injection locking, we generate nine equally spaced phases from the 44.5MHz ring oscillator locked to a quartz oscillator at the same frequency with a settling time of <100ns permitting aggressive duty-cycling of the receiver [6]. The 9x subharmonic mixer utilizes these nine phases to effectively switch the RF current at 401.5MHz, creating a virtual LO. The proposed receiver system is fully integrated except for a crystal, matching network, and input balun. The chip represents a 3x improvement over previously-published MICS band receivers both in terms of energy/bit and total power consumption [2-5].

Figure 26.4.2 shows the receiver front-end schematic. We use a g_m-boosted common-gate LNA with stacked subharmonic mixer to save power. The high-Q input matching network provides voltage gain and helps reduce the system noise figure while matching the high impedance LNA input (due to very low bias current) to the 50Ω source. The LNA input transistors operate in the subthreshold region. We use a differential RF path in the mixer to reject LO feedthrough in the IF signal present due to random mismatches in the nine-phase LO. The 4-stage IF amplifiers provide 80dB of voltage gain and are AC-coupled to reject DC offsets.

The receiver front-end conversion gain, noise figure (NF) and linearity are plotted vs. the IF frequency in Fig. 26.4.3. At a 1.5MHz IF, the front-end exhibits >30dB of voltage conversion gain, a 13dB NF and a P_{IIP3} of -23dBm. In LP mode, the DC current though the mixer and LNA is disabled, effectively converting the front-end to a passive mixer. The low conversion gain (5dB) and higher NF (18dB) in the front-end lead to approximately 20dB higher system NF and 20dB reduced sensitivity in this mode.

Figure 26.4.4 presents the schematic and measured results of the 44.5MHz injection-locked ring oscillator. The use of frequency multiplication by a large factor (9x) reduces the LO power and permits direct locking to a crystal reference, completely avoiding the need for a PLL/DLL. However, direct single-phase injection into a nine-stage oscillator disturbs its phase symmetry, leading to increased 44.5MHz LO feedthrough to the IF. We solve this problem by first lock-

ing a three-stage ring oscillator (ILRO1). These mismatched edges from ILRO1 are symmetrically injected into a nine-stage oscillator (ILRO2) that reduces the phase mismatch. The injection-locked oscillator can be modeled as a first order PLL with the one-sided lock range determining the equivalent bandwidth [7]. The proposed LO architecture eliminates the phase/delay-locked-loops saving associated loop filter area and power while avoiding stability issues and a long settling time. The measured lock-range of the cascaded oscillator stages is 32 to 52MHz with a settling time <100ns. The phase noise and spectra of the free-running and locked oscillators are shown in Fig. 26.4.4. The phase noise of the locked oscillator is limited by the noise floor of the Agilent E4446A used to measure it. The total power dissipation of the crystal and both ring oscillators is 22µW.

Figure 26.4.5 explains the operation of our all-digital FSK demodulator. The output of the crystal oscillator (44.5MHz) is fed into the "measurement" counter (counter2) in the FSK demodulator as the reference clock. The IF output is fed into the "window" counter (counter1), which operates for N_1 cycles and gates the measurement counter. This counter measures the number of periods of the reference clock in each measurement window. The time of measurement, T_{meas}, is a multiple of the unknown IF frequency and can be controlled by the predefined value N_1. $T_{meas}=N_1/f_{if}$. Also, $T_{meas}=N_2/f_{ref}$. It follows that the measurement counter output N_2 contains the unknown IF frequency. $N_2=N_1*f_{ref}/f_{if}$. A change in the IF frequency will change T_{meas} and N_2. The FSK-modulated IF signal can be demodulated according to the changes in N_2. The resolution of this frequency-detection scheme is determined by f_{ref}, T_{meas}, and the FSK deviation. The asynchronous nature of sampling reduces the error margin for the comparator and limits the maximum data-rate to 100kb/s when the input SNR drops to 10dB. The measured output of the demodulated FSK signal with the input pseudo-random data is also shown in Fig. 26.4.5.

Figure 26.4.6 presents the performance comparison with the existing work and the power breakdown of the system. Our main mode receiver achieves a data-rate of 200kb/s at 600pJ/bit with BER<0.1%. At the expense of 20dB loss in sensitivity, the LP-mode receiver achieves 220pJ/bit. For MICS-band transmitters with a -16dBm output, the LP mode can permit wireless operation over 2m of distance. The main mode can be turned on in case of additional losses such as body tissue attenuation or fading losses in a hospital scenario. The front-end and IF limiting amplifiers consume 75µW and 11µW from a 1V supply, respectively. The total die area of the receiver is 0.5mm² in a 0.13µm CMOS process (Fig. 26.4.7). With our previously-reported sub-100µW FSK transmitter [6], this receiver will enable a fully autonomous symmetric wireless link in a peer-to-peer network with an energy per transceived bit of 1nJ/bit.

Acknowledgment:
We would like to thank Fan Zhang, Manohar Nagaraju, and Intel Research Labs Seattle for assistance in testing.

References:
[1] D. Yeager et al., "NeuralWISP: A wirelessly powered neural interface with 1m range", *IEEE TBioCAS*, vol.3, no.6, pp. 379-387, June 2009.
[2] Bohorquez et al., "A 350µW CMOS MSK transmitter and 400µW OOK super-regen. receiver," *IEEE JSSC*, vol.44, no.4, pp.248-259, April 2009.
[3] P. Bradley, "An ultra low power, high performance MICS transceiver," *IEEE BioCAS*, Nov. 2006.
[4] J. Bae et al., "A 490µW fully MICS compatible FSK transceiver", *IEEE Symp. on VLSI Circuits*, June 2009.
[5] Y.-H. Liu et al., "A super-regen. ASK receiver with ΔΣ pulse-width digitizer and SAR-based fast frequency calibration," *IEEE Symp. on VLSI Circuits*, June 2009.
[6] J. Pandey et al., "A 90µW MICS/ISM band transmitter with 22% global efficiency", *IEEE RFIC*, May 2010.
[7] R. Adler, "A study of locking phenomenon in oscillators", *Proc. of the IEEE*, vol.61, no.10, Oct.1973.

ISSCC 2011 / February 23, 2011 / 3:15 PM

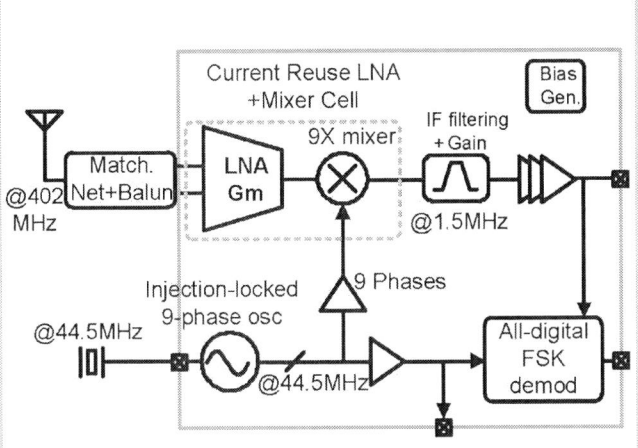

Figure 26.4.1: Block diagram of the proposed ultra-low-power receiver.

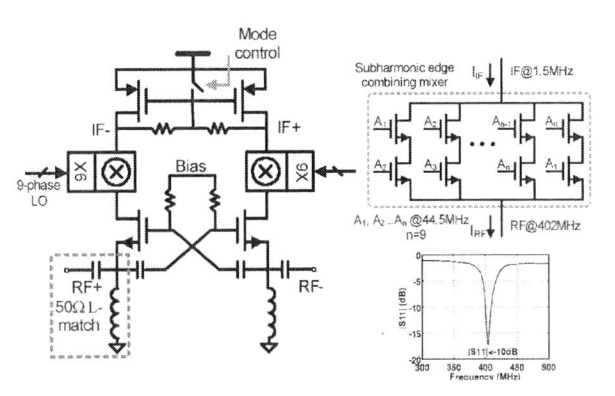

Figure 26.4.2: The LNA and the sub-harmonic mixer. The mode-control switch, when ON, switches off the LNA current, turning it into a sub-harmonic passive mixer. The LNA input return loss is greater than 10dB.

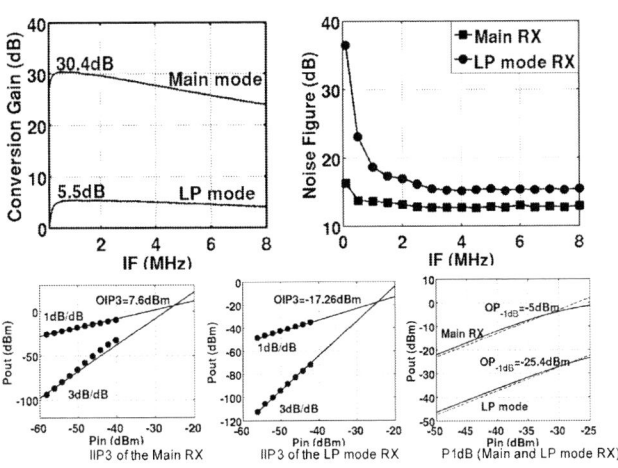

Figure 26.4.3: Voltage conversion gain, IIP3 and Noise Figure of the main and the LP-mode receivers.

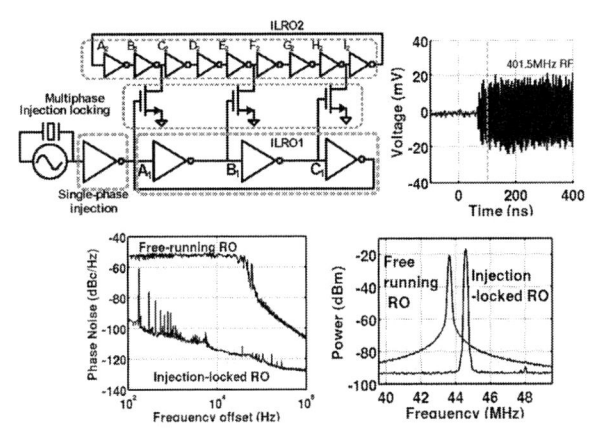

Figure 26.4.4: Multiphase injection-locked ring oscillator. The close-in phase noise is improved by 45dB over the free-running state. The lock-bandwidth is >20MHz which leads to a settling time <100ns.

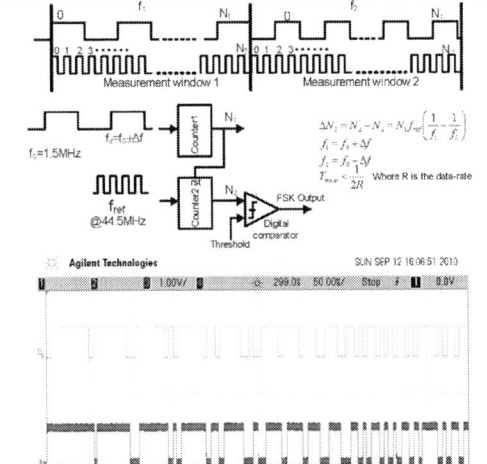

Figure 26.4.5: Operation of the FSK demodulator based on a frequency-counter algorithm. Input pseudo-random data and the resulting demodulator output are shown for a 200kb/s data-rate and -80dBm input.

Performance comparison table

	Power	Date-rate	Sensitivity	Energy/bit	Process	Arch.	Mod. type
[2]	400μW	120kbps	-93dBm	10nJ/bit	90nm	Super-reg.	OOK
[3]	5mW	800kbps	-110dBm	6.3nJ/bit	180nm	Low-IF	FSK
[4]	490μW	250kbps	-98dBm	2nJ/bit	180nm	Low-IF	FSK
[5]	910μW	156kbps	-80dBm	5.8nJ/bit	180nm	Super-reg.	OOK
This Work (Main mode)	120μW	200kbps	-90dBm*	0.6nJ/bit	130nm	Low-IF	FSK
This Work (LP mode)	44μW	200kbps	-70dBm*	0.22nJ/bit	130nm	Low-IF	FSK

*Using our on-chip demodulator, the maximum data-rate is limited to 100kbps for both main and LP mode receivers.

Power consumption breakdown

Circuit block	Power dissipation (Main mode)	Power dissipation (LP mode)
Front-end(LNA+Mixer)	75μW	0
LO+Crystal	22μW	22μW
IF Amplifiers, Limiters	11μW	10μW
Demodulator	12μW	12μW
Total	120μW	44μW

Figure 26.4.6: Performance comparison and power breakdown of the receiver modes. Our receiver improves the state-of-the-art by >3X both in terms of energy/bit and total power consumption.

978-1-61284-303-2/11 $26.00 © 2011 IEEE

ISSCC 2011 PAPER CONTINUATIONS

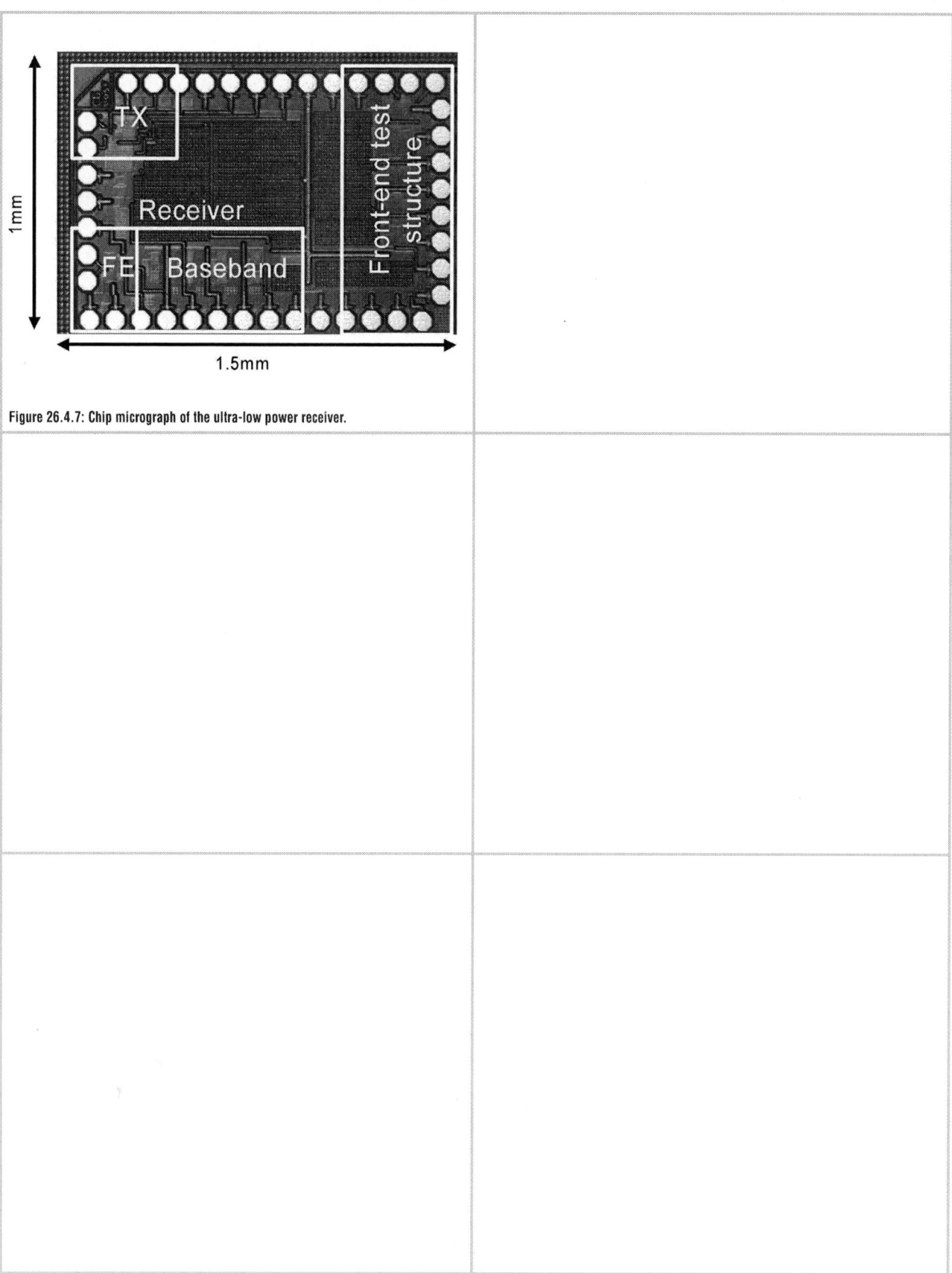

Figure 26.4.7: Chip micrograph of the ultra-low power receiver.

ISSCC 2011 / SESSION 26 / LOW-POWER WIRELESS / 26.5

26.5 A GPS/Galileo SoC with Adaptive In-Band Blocker Cancellation in 65nm CMOS

Chia-Hsin Wu[1], Wen-Chieh Tsai[1], Chun-Geik Tan[2], Chun-Nan Chen[1], Kuan-I Li[1], Jui-Lin Hsu[1], Chi-Lun Lo[1], Hsin-Hua Chen[1], Sheng-Yuan Su[1], Kun-Tso Chen[1], Min Chen[1], Osama Shana'a[2], Shu-Hung Chou[1], George Chien[3]

[1]MediaTek, Hsinchu, Taiwan, [2]MediaTek, Singapore, Singapore,
[3]MediaTek, San Jose, CA

The proliferation of location-based applications inside various handheld electronic devices, such as mobile phones and internet tablets, demands the GPS system to have low power consumption, small form-factor and be co-located on the same device with other radio systems, such as cellular, BT, and WLAN. The conventional GPS solution often uses two SAW filters, before and after an external LNA, to meet the requirements of low noise and multi-radio coexistence. Nevertheless, it is highly desirable to remove the external LNA and interstage SAW filter due to size and cost, which presents a great design challenge to achieve high out-of-band linearity with very low power consumption. To fulfill these stringent requirements, a more comprehensive approach is needed to target a radio architecture with a proper RX system budgeting and optimal circuit design. In addition, a GPS system can be desensitized by unexpected in-band blockers generated from other subsystems on the same platform, such as LCD display, PMU, CPU system clocks, etc. The GPS digital baseband processor must possess the capability to withstand in-band blockers without significant performance degradation. This paper presents a GPS/Galileo SoC with an adaptive in-band blocker cancellation scheme, which is implemented in a 65nm CMOS process.

GPS and Galileo are two positioning systems that are developed by the USA and EU. Both systems utilize direct-sequence spread-spectrum CDMA modulation technique at 1575.42MHz with bandwidths of 2.5M/4M, respectively. Figure 26.5.1 shows the block diagram of the host-based GPS/Galileo SoC, which consists of a receiver using an IF frequency of 4.092MHz, a fractional-N frequency synthesizer with digitally-assisted calibrations, and a dual-mode GPS/Galileo digital baseband processor with in-band blocker cancellation. All blocks operate under a 1.2V supply voltage.

To remove an external LNA mandates that the on-chip LNA must have a sufficiently low noise figure (NF). To obtain a low NF of the receiver chain, the on-chip LNA is designed to provide high gain and low-noise matching simultaneously. Figure 26.5.2 shows the single-ended LNA with on-chip stacked inductor of 30nH as its LC tank to provide a high load impedance. Two package bondwires are used as the LNA degeneration inductor to achieve low-noise input matching. These LC resonators also provide additional out-of-band filtering. The on-chip LNA has a NF of 1.2dB, while consuming only 1.6mA. To remove the inter-stage SAW filter, the overall receiver linearity needs to be improved. Using the PCS cellular TX signal of +30dBm as a blocker, considering the antenna isolation of 15dB and the front-end SAW filter rejection of 50dB, the residual blocker signal level is around -35dBm at the LNA input. The blocker is first amplified by the on-chip LNA, and the linearity of the mixer and baseband channel-select filter (CSF) must be large enough to handle this blocker level. For example, a mixer-CSF input P_{1dB} of >-12dBm at the PCS band is required to accommodate blockers from this band. Figure 26.5.2 illustrates the mixer-CSF topology. A highly-linear single-balanced passive mixer with push-pull g_m input stage downconverts the RF input signal. The signal into the CSF is in current mode, and the current-to-voltage conversion is performed by the CSF itself without the need of additional transimpedance amplifiers. The CSF first stage opamp ensures that the input terminals remain as a virtual ground with no AC swing. Traditionally, the opamp loop bandwidth is designed as large as possible to improve out-of-band linearity, however it increases current consumption. To obtain high out-of-band linearity with less current consumption, an 80MHz RC filter is implemented at the mixer output before the first stage of CSF. The out-of-band blocker signal current can be attenuated to enhance the linearity over a wider range of frequencies. As depicted in Fig. 26.5.2, the CSF first stage is an active second-order complex bandpass filter responsible for current-to-voltage conversion and provides moderate image rejection. This is followed by a second-order passive RC polyphase filter. By using the proposed hybrid filter topology, the total image rejection ratio can reach around 40dB. The final CSF stage is a single-opamp real bandpass filter that provides more out-of-band rejection and performs I/Q com-

bining. The proposed hybrid CSF provides high stop-band attenuation while achieving excellent linearity.

The LO signals are generated by a third-order type-II fractional-N frequency synthesizer with VCO oscillating at twice the LO frequency. The VCO uses an NMOS cross-coupled pair with constant-g_m biasing. The fully-integrated frequency synthesizer with no external loop filter exhibits rms integrated phase error of 1.3°/2.4° at 3142.656MHz for reference clock frequencies of 16.368MHz/26MHz, respectively, while consuming 1.2mA.

Figure 26.5.1 also shows the block diagram of the digital baseband. The main algorithm and circuit topologies are disclosed in [3], except for the in-band blocker-cancellation scheme. The unexpected in-band blockers are generally produced by other subsystems in the same platform and seriously degrade GPS system performance, such as sensitivity and the Time-To-First-Fix (TTFF). Conventional digital approaches to deal with the unexpected in-band blockers include digital notch filter and de-weighting schemes with FFT engines. However, these approaches require additional hardware and provide limited performance improvement. In this paper, a low-cost self-learning scheme of adaptive in-band blocker cancellation is proposed. First, the received GPS IF signal can be described by

$$x(t) = s(t) + n(t) = j(t) \qquad (1)$$

where s(t), n(t), and j(t) represent the signal, noise, and in-band blocker respectively. In general, the in-band blocker j(t) consists of multiple narrow-band CW tones. The thermal noise n(t) is AWGN with zero mean. The GPS signal s(t) is coded and its power spectral density is smaller than the noise n(t). Since s(t) and n(t) are with zero mean, Eq(1) can be rewritten as

$$y(t) = x(t) - j(t) = s(t) + n(t)$$
$$E\{y(t)\} = E\{x(t) - j(t)\} = E\{s(t) + n(t)\} = 0 \qquad (2)$$

The blocker learning unit is as shown in Fig. 26.5.3(a). According to Eq(2), the cascaded blocker learning units adjust the frequency, phase and amplitude of local CW replicas to learn and predict the blocker j(t). After removing the predicted j(t) from the received IF signal x(t), the output signal y(t) will consist of only GPS signal and AWGN. The proposed design of in-band blocker cancellation can improve the in-band blocking performance by 18dB, as shown in Fig. 26.5.3(b), where the blocker level is shown at which the received GPS CNR is degraded by 3dB.

The die micrograph of the GPS SOC, implemented in 65nm CMOS process, is shown in Fig. 26.5.7. The die area is 6.6mm², of which 1.4mm² is occupied by RF radios. The measured out-of-band IIP3 is -5dBm referred to the LNA input. The two out-of-band blockers are located at 2GHz WCDMA and 2.5GHz WLAN bands. Figure 26.5.4 shows the measured results of the received blocker level at which the GPS signal CNR is degraded by 3dB. These blocking signals include typical radio transmitters found in a mobile device with an antenna isolation of 15dB. In this blocking test, a GPS front-end SAW filter is used between the antenna and the SoC RF input. The receiver shows good immunity to out-of-band blockers with levels as high as +16dBm at its antenna port. Figure 26.5.5 shows the comparison of in-band anti-blocking performance implemented in this work and other GPS solutions with signal power of -130dBm and 12 CW blocking signals. The curve shows that the device with the proposed anti-blocking scheme can achieve an immunity of in-band blockers at least 12dB better than other solutions under cold-start TTFF. Figure 26.5.6 summarizes the measured performance of this work compared to previously reported papers. The entire SoC achieves a chip-in -165dBm tracking sensitivity while consuming the lowest reported power consumption of 18mW.

References:
[1] Hyunwon Moon, et al. "A 23mW fully integrated GPS receiver with robust interferer rejection in 65nm CMOS," *ISSCC Dig. Tech. Papers*, pp. 68-69, Feb., 2010.
[2] Kuang-Wei Cheng, et al. "A 7.2mW quadrature GPS receiver in 0.13um CMOS," *ISSCC Dig. Tech. Papers*, pp. 422-423, Feb., 2009.
[3] J-M Wei, et al. "A 110nm RFCMOS GPS SoC with 34mW -165dBm tracking sensitivity," *ISSCC Dig. Tech. Papers*, pp. 254-255, Feb., 2009.
[4] V. D. Torre, et al, "A 20mW 3.24mm² Fully Integrated GPS Radio for Location Based Service," *IEEE J. Solid-State Circuits*, vol. 42, pp. 602-612, March 2007.

978-1-61284-303-2/11 $26.00 © 2011 IEEE

ISSCC 2011 / February 23, 2011 / 3:30 PM

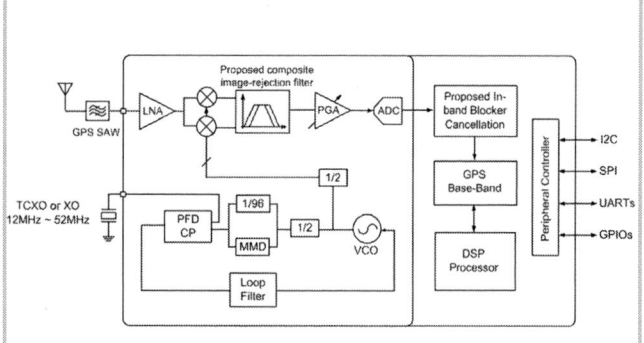

Figure 26.5.1: The host-based GPS/Galileo SoC architecture and block diagram.

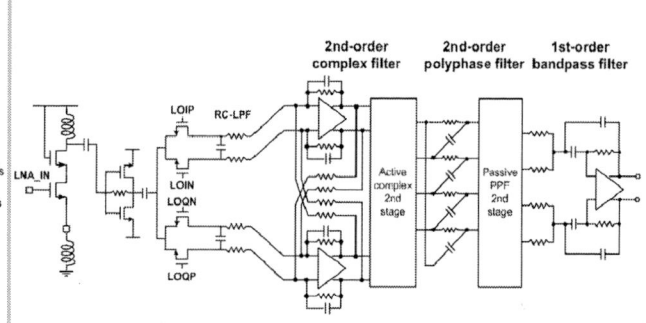

Figure 26.5.2: The proposed hybrid topology of mixer-CSF of the receiver.

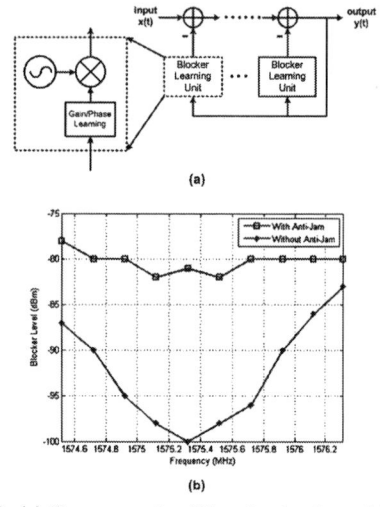

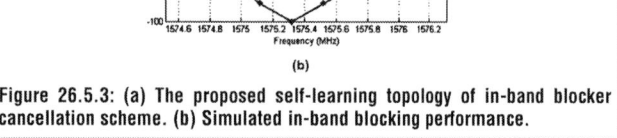

Figure 26.5.3: (a) The proposed self-learning topology of in-band blocker cancellation scheme. (b) Simulated in-band blocking performance.

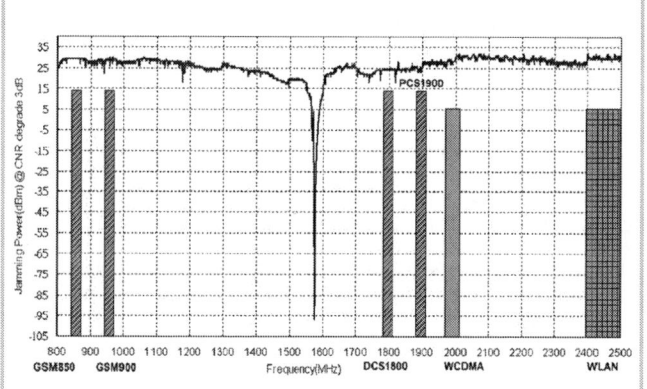

Figure 26.5.4: Measured out-of-band blocking performance (with front-end GPS SAW) for possible aggressors of other wireless radios.

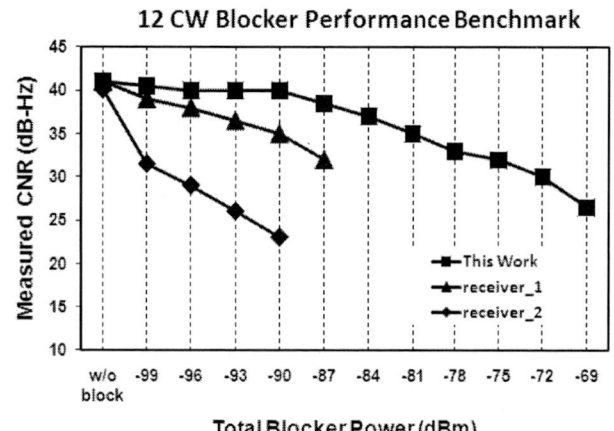

Figure 26.5.5: Measured in-band blocking performance benchmark with 12 CW in-band blocker signals among this work and the competitors.

	This Work	ISSCC 10 [1]	ISSCC 09 [2]	ISSCC 09 [3]	JSSC 07 [4]	JSSC 06 [5]
RF Architecture	GPS/Galileo SoC	GPS SoC	GPS RFRX Only	GPS SoC	GPS RF Only	GPS SoC
Technology	65nm CMOS	65nm CMOS	0.13um CMOS	0.11um CMOS	0.18um SiGe	0.18um CMOS
RF Area (mm²)	1.4	2.5	3.1	2.4	3.2	4.1
RX NF (dB)	2.0	2.3	6.5	3.2	5.0	4.8
Image Rejection (dB)	40	30	N/A	40	23	30
LO Phase Noise @ 1MHz offset (dBc/Hz)	-118	-114	-110	N/A	-110	-112
RF-only Power Consumption (mW)	8.6	23	7.2	19.5	19.8	27.2
SoC Power Consumption (mW) @tracking	18	N/A	N/A	38	N/A	56

Figure 26.5.6: Measured SoC performance summary and comparison with other previously reported papers.

Figure 26.5.7: Die micrograph.

ISSCC 2011 / SESSION 26 / LOW-POWER WIRELESS / 26.6

26.6 A 0.05-to-10GHz 19-to-22GHz and 38-to-44GHz SDR Frequency Synthesizer in 0.13μm CMOS

Sujiang Rong, Howard C. Luong

HKUST, Hong Kong, China

Extensive research has been focused on generating LO signals for software-defined radios (SDRs) with ultra-wide tuning range over several frequency decades and sufficiently high spectrum purity to support diverse wireless standards [1-5]. This work presents an integrated frequency synthesizer (FS) that is able to cover not only all the wireless standards from 47MHz to 10GHz including 14-band MB-OFDM UWB but also the 802.15.3c standard from 57 to 66GHz.

Figure 26.6.1 shows the architecture of the FS. A dual-band (DB) Q-VCO generates 3-to-4.2GHz and 8.4-to-12GHz IQ signals and enables IQ signals continuously from 47MHz to 6GHz to be obtained simply with only divide-by-2 operations. The 14-band carriers for MB-OFDM UWB are generated by single-side-band (SSB) mixing the 8448/4224MHz outputs from the Q-VCO or its following divider-A, with the 264/792/1320/1848MHz signals generated by a tri-mode injection-locked frequency multiplier (ILFM). A distributed multiplexer with shunt-peaking bandwidth enhancement selects the desired IQ signals from 47MHz to 10GHz as the LO signals for direct-conversion transceivers, whose IQ accuracy is measured by an on-chip SSB mixer. In addition, employing the synthesized IQ signals around 2.3GHz, x3 and x2 sub-harmonic injection-locked oscillators (SH-ILOs) with automatic peak calibration (APC) are used to generate the 20GHz IQ and 40GHz differential LO signals for dual-conversion transceivers operating from 57 to 66GHz. The Q-VCO's frequency is controlled and stabilized by a PLL with adjustable dead-zone period, CP current, loop-filter bandwidth, V2I converter's transconductance and thus K_{VCO}.

Transformers (TFs) are utilized to implement the DB Q-VCO with little area penalty compared to the conventional LC Q-VCO. Figure 26.6.2(a) shows the schematic. In the low-band mode, with only I_{core1} and I_{tune1} on, the Q-VCO operates as a 2-port oscillator mainly for good phase noise. A large L_1/L_2 ratio ensures the voltage swing at the drain of $M_{1/2}$ to be much smaller than the gate voltage swing to prevent $M_{1/2}$ from operating into the triode region even with rail-to-rail gate swing. Such a design minimizes the noise contributions from M_{1-2} and enables the Q-VCO to meet the most stringent phase noise specifications for all the cellular standards around 900MHz and 1.8GHz. On the other hand, in the high-band mode, because the phase noise requirement is more relaxed, the circuit operates as a 1-port oscillator with only I_{core2} and I_{tune2} on to save the power consumption. In both modes, the fine tuning is realized by varying the bias current $I_{tune1-2}$ instead of using varactors to minimize AM-PM noise and avoid the tank Q degradation. M_{5-6} are added in the coupling paths to eliminate any bimodal oscillation [6]. To reduce the IQ mismatch, the current sources are connected as shown in the dotted lines.

Figure 26.6.3 shows the ILFM for UWB. Sinusoidal input signals are converted into narrow pulses with rich harmonics to injection-lock a ring oscillator (RO). Differential circuitry is used to cancel all even harmonics, and the spectrum purity is further improved by utilizing the RO's property that those injected harmonics having inconsistent phase sequence with the RO's natural phase sequence are automatically rejected. For example, when the ILFM operates at x5 mode, only the top injection cells are enabled. As illustrated in the phase-sequence table in Fig. 26.6.3, the unwanted 3rd and 7th harmonics of the injected current i_{inj_A-B}, are rejected by the RO because their phase sequences are contradicted to that of the RO's internal current i_{int_A-B}. The 1st and 9th harmonics are also selected but significantly suppressed because they are far away. The ILFM's multiplication ratio 3/5/7 is selected by controlling I_{sense} and I_{latch} of the delay cells, and by enabling the appropriate injection cells. Experimentally, spur rejections of larger than 30dB are measured at the ILFM's output. Figure 26.6.2(b) shows the UWB SSB mixer. A single-coil TF-based LC-tank is used to filter out the spurs introduced by the mixing in an area-efficient way. The TF is designed to be tightly coupled and with a small turns ratio so that both of the lower and higher frequency peaks can become dominant and be utilized for the filtering by controlling the ratio C_1/C_2. A tunable capacitor C_3 is added between L_1 and L_2 to further extend the tuning range of the higher frequency peak to fully cover the 3-to-10GHz band.

Figure 26.6.4(a) shows the x3 ILO with IQ outputs. The IQ input signals at ω hard-switch M_{1-4} to generate strong 3rd harmonics that lock a Q-VCO self-oscillating around 3ω. The x2 ILO is shown in Fig. 26.6.4(b). IQ signals provided by the preceding ILO are applied to two push-push pairs. By doing so, the differential oscillator can be naturally locked by the differentially injected currents, which avoids the mismatch problem as in [7]. To prevent the ILO operating outside or at the boundary of the locking range (LR), which would result in limited output swing or even malfunction of the ILO chain, it is desirable to align the LC-tank's peak frequency to the desired multiple of the input frequency for each ILO. Another assisted PLL could help, but its LR would be limited by the dividers operating at such high frequencies. Alternatively, making use of the fact that the output swing is maximized when the peak frequency alignment is achieved, the peak calibration can be done simply by detecting and maximizing the output swing. To guarantee the peak calibration, I_b is much reduced during the calibration to greatly enlarge the injection ratio and thus the LR of the ILO. Figure 26.6.4 shows the APC circuit. At the rising edge of clock-1, V_{in}, provided by a squaring circuit, is sampled and stored in C_S. At the rising edge of clock-2, the counter counts up or down once to change the LC-tank's peak frequency by tuning capacitors, and V_{in} is sampled again when clock-2 is high. At the falling edge of clock-2, the comparator delivers the comparison result between the two V_{in} samples to the shift register (SR). If the result is positive, the counter keeps counting towards the right direction. Otherwise, the toggling register changes the counting direction. The states stored in the SR control the logic to hold the counter when $V_{in,max}$ is found. In the measurement, the APC improves the minimum single-ended swings from 400mV to 600mV and from 50mV to 250mV for the 20GHz and 40GHz band signals, respectively.

The FS is fabricated in a 0.13μm CMOS. Figure 26.6.5 plots the measured close-loop phase noise, RMS and IQ phase errors, and the associated power consumption versus the output frequency. At the cellular modes, phase noises of -141.7dBc/Hz and -139.6dBc/Hz are measured at 3MHz offset from 900MHz and 1.7GHz carriers, respectively. The RMS phase error is below or around 1° for frequencies below 2.5GHz and gradually increases to around 20° at 40GHz. With coarse IQ calibrations, the measured IQ phase error is <3.4° from 47MHz to 10GHz. With a 1.2V supply, the FS consumes around 33mW for ZigBee, 50mW for cellular, 60mW for WiFi, and 80mW for MB-OFDM and 802.15.3c UWB. The measured spur is <-57dBc and <-31dBc for non-UWB and MB-OFDM UWB modes, respectively. The channel-hopping time is measured to be <3.7ns for MB-OFDM UWB. Figure 26.6.6 summarizes the measured performance and compares with that of the recently reported SDR FS. Figure 26.6.7 shows the die micrograph.

Acknowledgement:
This project was jointly funded by Hong Kong Innovation and Technology Fund ITS/169/09 and General Research Fund 617609.

References:
[1] A. Koukab, Y. Lei, and M. Declercq, "A GSM-GPRS/UMTS FDD-TDD/WLAN 802.11a-b-g Multi-Standard Carrier Generation System," *IEEE J. Solid-State Circuits*, vol. 41, pp. 1513-1521, Jul. 2006.
[2] J. Borremans, et al., "A Compact Wideband Front-End Using a Single-Inductor Dual-Band VCO in 90 nm Digital CMOS," *IEEE J. Solid-State Circuits*, vol. 43, pp. 2693-2705, Dec. 2008.
[3] S. Yu, et al., "A Single-Chip 0.126-26GHz Signal Source in 0.18μm SiGe BiCMOS," *IEEE RFIC Symp.*, pp.427-430, Jun. 2009.
[4] B. Razavi, "Cognitive Radio Design Challenges and Techniques," *IEEE J. Solid-State Circuits*, vol. 45, pp. 1542-1553, Aug. 2010.
[5] S. Osmany, F. Herzel, and J. Scheytt, "An Integrated 0.6-4.6 GHz, 5-7 GHz, 10-14 GHz, and 20-28 GHz Frequency Synthesizer for Software-Defined Radio Applications," *IEEE J. Solid-State Circuits*, vol. 45, pp. 1657-1668, Sept. 2010.
[6] S. Li, I. Kipnis, and M. Ismail, "A 10-GHz CMOS Quadrature LC-VCO for Multirate Optical Applications," *IEEE J. Solid-State Circuits*, vol. 38, pp. 1626-1634, Oct. 2003.
[7] E. Monaco, et al, "Injection-Locked CMOS Frequency Doublers for μ-Wave and mm-Wave Applications," *IEEE J. Solid-State Circuits*, vol. 45, pp. 1565-1574, Aug. 2010.

ISSCC 2011 / February 23, 2011 / 3:45 PM

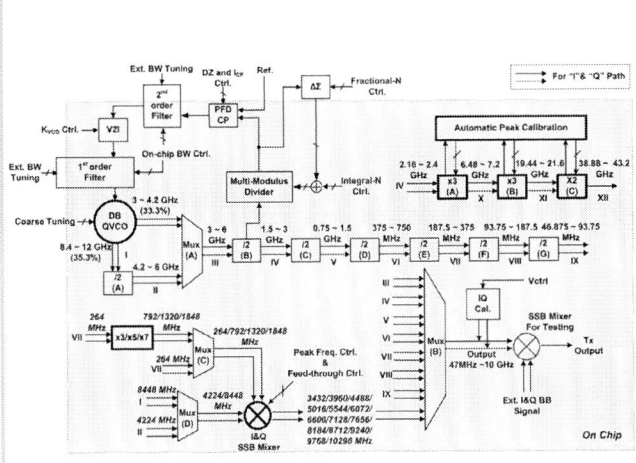

Figure 26.6.1: SDR frequency synthesizer architecture.

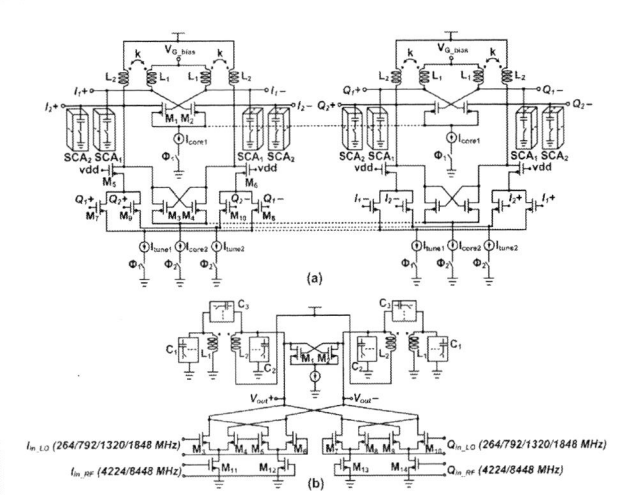

Figure 26.6.2: Schematics of TF-based (a) DB Q-VCO, and (b) SSB mixer.

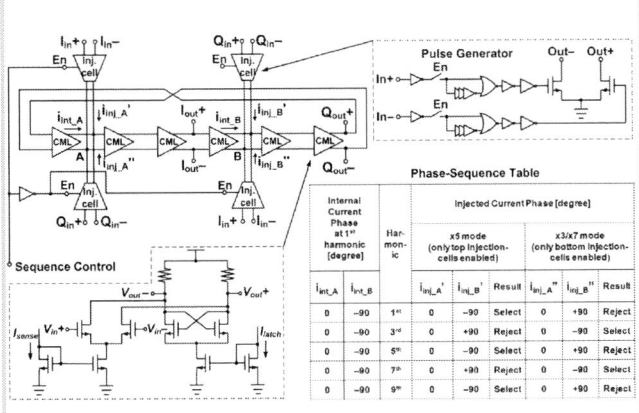

Figure 26.6.3: Schematic and phase sequence table of x3/x5/x7 ILFM.

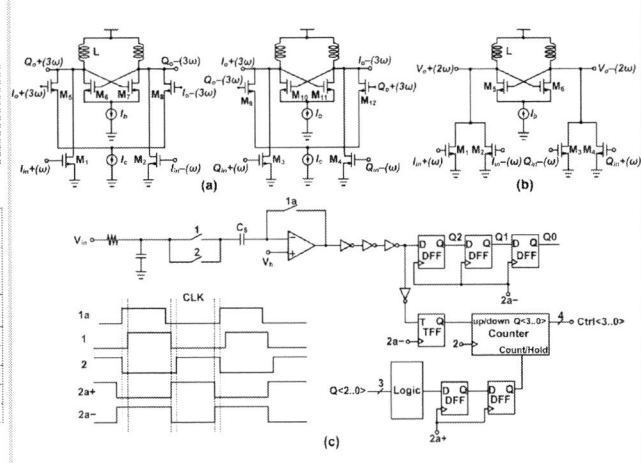

Figure 26.6.4: Schematics of (a) x3 ILO, (b) x2 ILO, and (c) APC circuit.

Figure 26.6.5: Measured phase noise, phase errors, and power.

Figure 26.6.6: Table of performance summary and comparison.

Ref.	[1]	[2]	[3]	[4]	[5]	This work
Generated carrier frequencies [GHz]	0.8~1.1 1.5~2.1 2.3~3.1 4.7~6.2	0.8~1 1.6~2 2.2~2.8 4.4~5.6	0.125~26	1.4/1.75/2/2. 19/2.3/2.9 /3.5/4.38/4.7 /5.83/7/8.75	0.6~4.6 5~7 10~14 20~28	0.047~10 19~22 38~44
In-band phase noise @ 10KHz offset (f_c=1.7GHz) [dBc/Hz]	-79.8[4]	N/A	-91.6[4]	N/A	-109.9[4]	-91~-98[4]
Out-band phase noise @ 3MHz offset (f_c=1.7GHz) [dBc/Hz]	-138.5	-129.6[3]	-137.2[3,4]	-129.8[3]	-136.5[3,4]	-139.6
Power [mW]	6.2 (VCO only)	60[1]	1283	31	680	33~83
Area [mm²]	2.55	0.06[1]	4.4[2]	0.29	4.8[2]	3.0
Technology	0.25µm BiCMOS	90nm CMOS	0.18µm BiCMOS	90nm CMOS	0.25µm BiCMOS	0.13µm CMOS

1. Power/area of the receiver front-end
2. Area including pads
3. Normalized to 3MHz from values reported at other frequency offsets
4. Normalized to 1.7GHz from values reported at other carrier frequencies

26

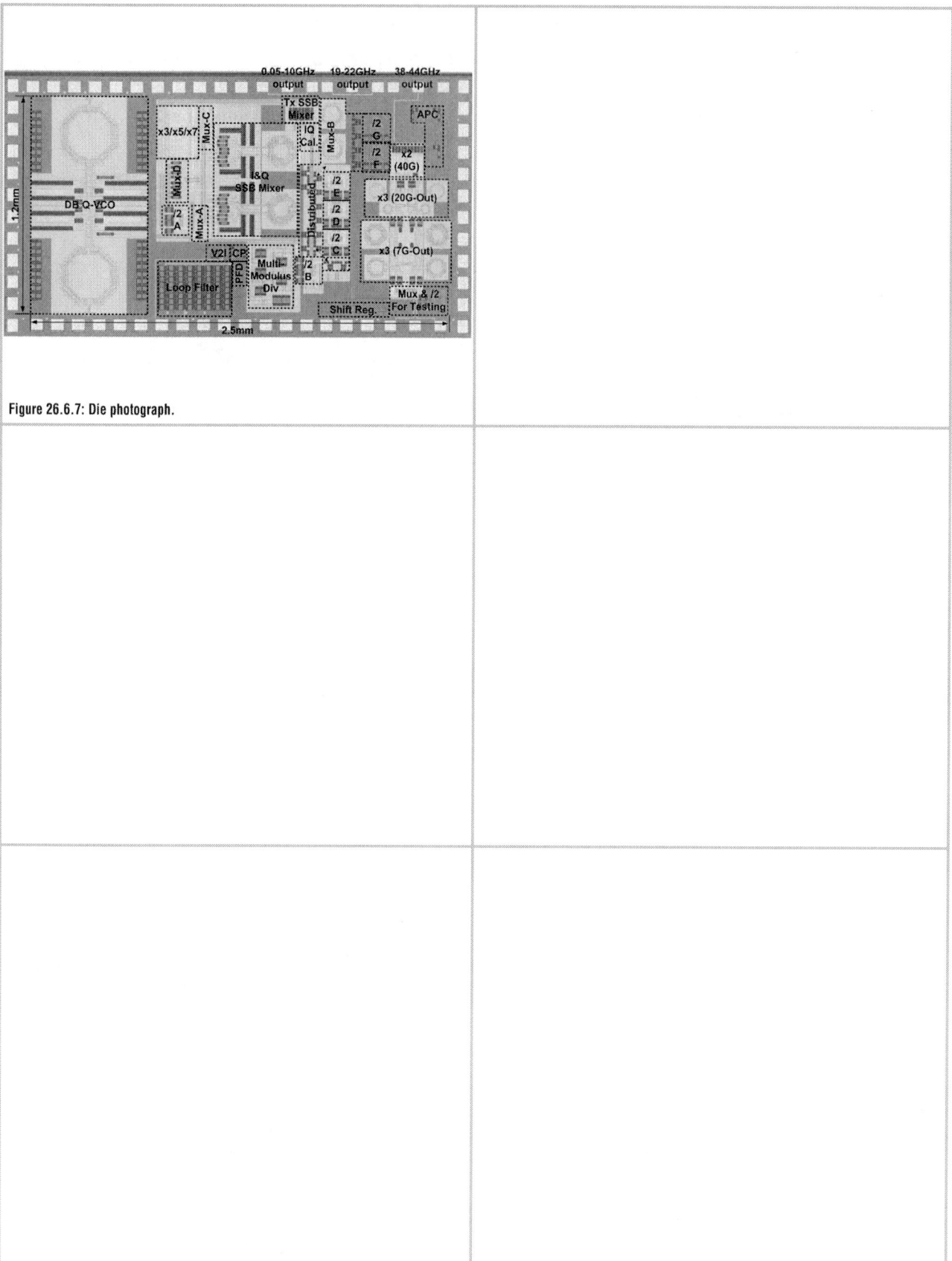

Figure 26.6.7: Die photograph.

ISSCC 2011 / SESSION 26 / LOW-POWER WIRELESS / 26.7

26.7 A 4.6GHz MDLL with -46dBc Reference Spur and Aperture Position Tuning

Tamer A. Ali[1,2], Amr A. Hafez[1], Robert Drost, Ronald Ho[2], Chih-Kong Ken Yang[1]

[1]University of California, Los Angeles, CA,
[2]Oracle, Menlo Park, CA

Multiplying delay-locked loops (MDLLs) [1-5] have been shown to have improved jitter accumulation and tracking over VCO-based PLLs. By injecting the reference clock edge into the VCO at each reference cycle, an MDLL removes the accumulated jitter of the VCO. The principal challenge in MDLL design is to align the injected reference edge with the loop feedback signal. Timing mismatch between the reference edge and the VCO feedback edge, or offsets in the charge pump, would introduce a phase error in the injected edge. The error manifests as a period jitter or reference spur in the frequency domain. This effect limits the minimum jitter attained by the MDLL.

In previously published techniques [1-5], a select logic block generates an SEL pulse to briefly open an aperture for reference injection. A necessary attribute of the SEL pulse is that it is sufficiently sharp and well-positioned to select the next reference edge. However, at high frequencies, the position of the SEL pulse impacts the delay of the MUX and hence introduces pattern jitter and spurs. This paper minimizes the spur by introducing a calibrated phase delay to properly position the SEL pulse with respect to the reference edge, and by minimizing the inherent error of the charge pump.

Figure 26.7.1 shows the block diagram of an MDLL with the proposed enhancements shown in bold. The SEL pulse is generated every Mth VCO cycle [1], where M is the division ratio. The pulse-generation delay includes the latency of a level- restoration clock buffer plus the select logic delay. This delay, under worst-case PVT conditions, must be shorter than a VCO's clock cycle. A simulated sweep of the SEL signal with respect to input reference and VCO signal (Fig. 26.7.3) shows substantial variation of the MUX delay similar to the setup/hold response of a latch. The window of minimal delay variation reduces substantially with higher frequency outputs.

The proposed MDLL design adds a 360° phase interpolator to vary the delay of the SEL pulse by up to a VCO's clock cycle. This adjustment range compensates for any latency in the select logic generation and enables precise positioning of the SEL pulse. The VCO comprises of four delay stages. A MUX stage configured as a delay cell serves as the VCO's third stage in order to match the first stage, thus maintaining a clean quadrature relationship and ensuring more linear phase steps of the SEL pulse tuning. A calibration loop uses the duty-cycle error of the VCO output (similar to [4]) as a measure of the period jitter and controls a simple descent algorithm. The algorithm advances the phase interpolator setting toward minimal duty cycle error, which corresponds to the optimal aperture position.

The phase interpolator design is shown in the inset of Fig. 26.7.1. The design opportunistically alters the low-to-high swing converter to also digitally interpolate. The interpolation takes place at the common terminal of a bank of weighted capacitors, a low-bandwidth node. The capacitor weight determines the degree of interpolation. 4fF unit capacitors are used to achieve 4b matching while maintaining low power dissipation.

The charge pump is a critical component of this design. Any mismatches between the up and down currents in the charge pump appear as a static phase error. While the interpolator can compensate for phase errors, a charge-pump mismatch would result in undesirable control voltage ripple. The charge pump shown in Fig. 26.7.2 minimizes static and dynamic mismatches between I_{UP} and I_{DN} to less than 0.5%. Static mismatch is corrected by transistors M1-M4, which form a singled-ended replica of the charge pump. Amplifier 1 forces the output node of the replica to be equal to the control voltage by controlling I_{DN}. The feedback matches I_{UP} and I_{DN} to within the limits of the 50dB loop gain over a wide range of control voltages.

A voltage-follower amplifier connected between the differential outputs of the charge pump (VCTRL and VCTRL1 in Fig. 26.7.2) has been shown to reduce dynamic mismatch between I_{UP} and I_{DN} by minimizing the charge sharing between the current-source capacitance and the loop filter capacitance during switching. Instead of using a follower amplifier that can consume considerable power, this design corrects dynamic mismatch by another replica charge pump M5-M10. This replica provides a current path for the charge pump branches formed by M12 and M14. Since the path is identical to the single-ended replica, the feedback loop controls VCTRL1 to be equal to VCTRL. This replica consumes only tens of microamperes. To minimize area, the MDLL loop filter capacitor is implemented using core thin-oxide devices. Due to the high gate leakage of these devices, an additional compensation technique similar to [6] is used.

Overlaid onto Fig. 26.7.3 are the measurement results for the period jitter versus SEL aperture position at 4.6GHz. Period jitter drops to 1ps at the optimal setting. By sliding the aperture to the right, period jitter increases slightly until the setup time of the MUX with respect to the VCO edge, after which the period jitter increases sharply and the loop eventually loses lock. Similarly, sliding the aperture to the left approaches the hold time of the MUX. Note that the period jitter sensitivity is shallow near the optimum and can absorb residual mismatch in the loop. The graph indicates the aperture position with and without the aperture calibration. After enabling the calibration loop by measuring the duty cycle, the period jitter reduces from 25ps to 1ps at 4.6GHz.

Figure 26.7.4 shows the reference spur for the MDLL with the control loop active. The spur level is -46dBc at 4.6GHz, or the equivalent to 1ps jitter. Figure 26.7.5 shows a jitter histogram of the MDLL output. The readout shows 2ps RMS of random jitter. Because the reference source has 1.5ps RMS jitter and the scope has a 0.86ps RMS jitter floor, a 1ps RMS random jitter is inferred in the MDLL. Figure 26.7.6 shows a comparison of the MDLL with published results. Our MDLL achieves more than twice the output frequency of published MDLLs with comparable spur levels.

Fabricated in a 90nm CMOS technology, the MDLL consumes a total of 6.8mW from a 1.2V analog supply and a 1V digital supply. Clock buffering consumes 1.5mW from this total current. The capacitive phase interpolator and the charge pump (including all feedback loops) consume 1.25mW and 1.2mW, respectively. Figure 26.6.7 shows the chip micrograph, the MDLL occupies an area of 0.025mm².

Acknowledgments:
The authors thank UMC and Prof. M.F. Chang for their support, Wonho Park and Henry Park for their help in synthesizing digital blocks.

References:
[1] Farjad-Rad, R.; Dally, W.; Hiok-Tiaq Ng; et al., "A low-power multiplying DLL for low-jitter multigigahertz clock generation in highly integrated digital chips," *IEEE Journal of Solid-State Circuits*, vol.37, no.12, pp. 1804-1812, Dec 2002.
[2] Lee, M.-J.E.; Dally, W.J.; Poulton, J.; et al., "A second-order semi-digital clock recovery circuit based on injection locking," *ISSCC Dig. Tech. Papers*, pp. 74-75, vol.1, 2003.
[3] Qingjin Du; Jingcheng Zhuang; Kwasniewski, T., "A Low-Phase Noise, Anti-Harmonic Programmable DLL Frequency Multiplier With Period Error Compensation for Spur Reduction," *IEEE Transactions on Circuits and Systems II: Express Briefs*, vol.53, no.11, pp.1205-1209, Nov. 2006.
[4] Helal, B.M.; Straayer, M.Z.; Gu-Yeon Wei; Perrott, M.H., "A Highly Digital MDLL-Based Clock Multiplier That Leverages a Self-Scrambling Time-to-Digital Converter to Achieve Subpicosecond Jitter Performance," *IEEE Journal of Solid-State Circuits*, vol.43, no.4, pp.855-863, April 2008.
[5] Maulik, P.C.; Mercer, D.A., "A DLL-Based Programmable Clock Multiplier in 0.18-μm CMOS With -70 dBc Reference Spur," *IEEE Journal of Solid-State Circuits*, vol.42, no.8, pp.1642-1648, Aug. 2007.
[6] Yohan Frans; Nhat M. Nguyen; "Compensator for Leakage Through Loop Filter Capacitors in Phase-Locked Loops"; U.S. Patent 6,963,232 B2; November 8 2005.

ISSCC 2011 / February 23, 2011 / 4:15 PM

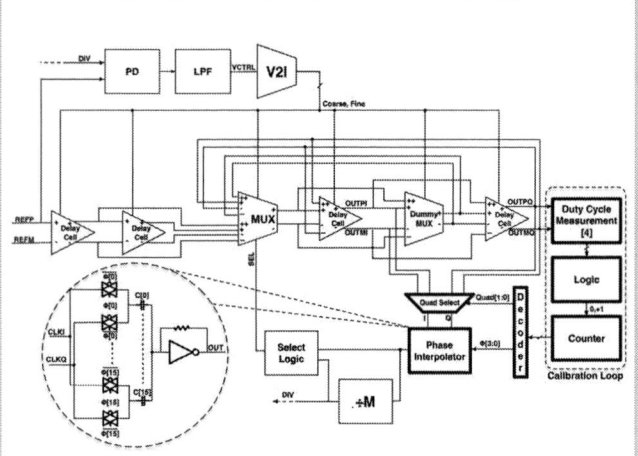

Figure 26.7.1: MDLL with proposed Aperture Tuning Mechanism. Phase interpolator is shown in the inset to the left.

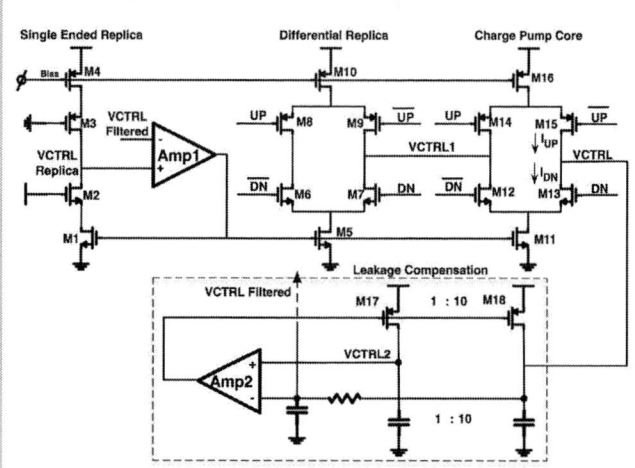

Figure 26.7.2: Charge Pump design with static/dynamic offset compensation and leakage compensation circuits.

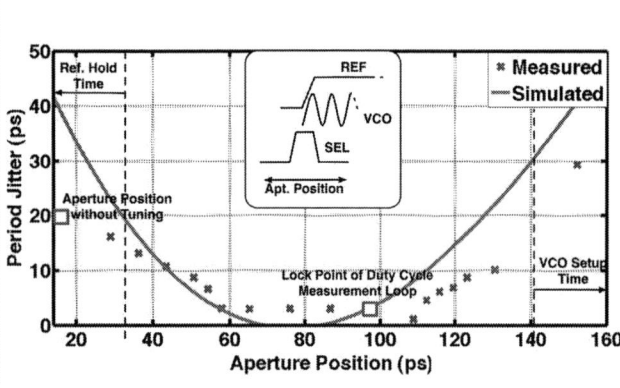

Figure 26.7.3: Measured tuning characteristic of aperture circuit with simulation results overlaid. Aperture position without tuning is shown.

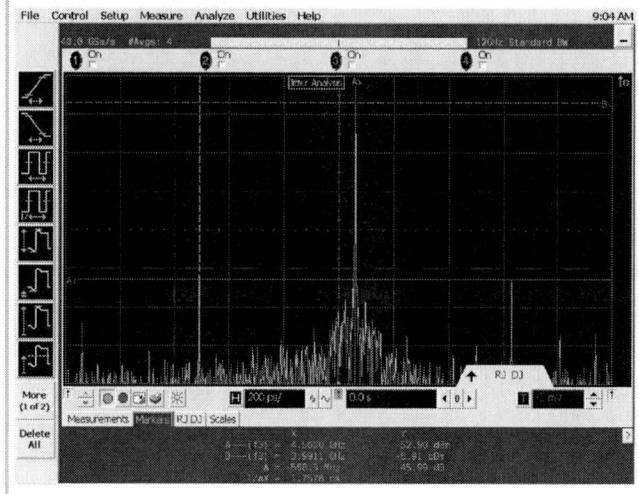

Figure 26.7.4: Measured frequency spectrum of the MDLL output at optimal tuning position.

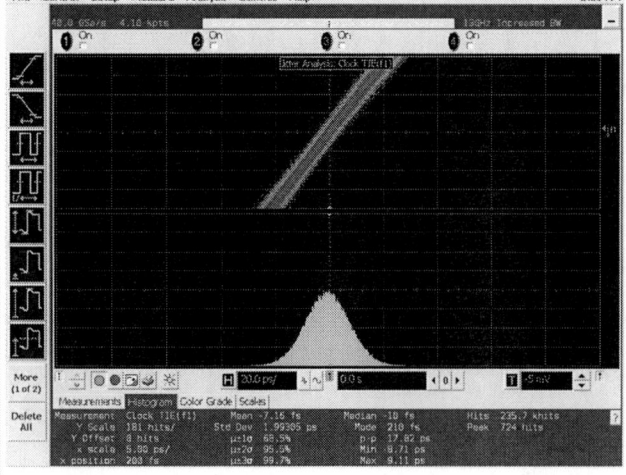

Figure 26.7.5: Measured jitter histogram.

	[1]	[3]	[4]	This Work
Output Frequency (GHz)	2	1.2	1.6	4.6
Reference Frequency (MHz)	250	64	50	570
Reference Spur (dBc)	-37	-46.5	-58.3	-46
Deterministic Jitter (ps pk) (Estimated from Spur)	3.53	1.95	0.38	1.1
Random Jitter (ps rms)	N/A	N/A	0.68	1
Technology	0.18μm	0.18μm	0.13μm	90nm
Power (mW)	12	19.8	8.2	6.8

Figure 26.7.6: Comparison table with prior work.

26

978-1-61284-303-2/11 $26.00 © 2011 IEEE

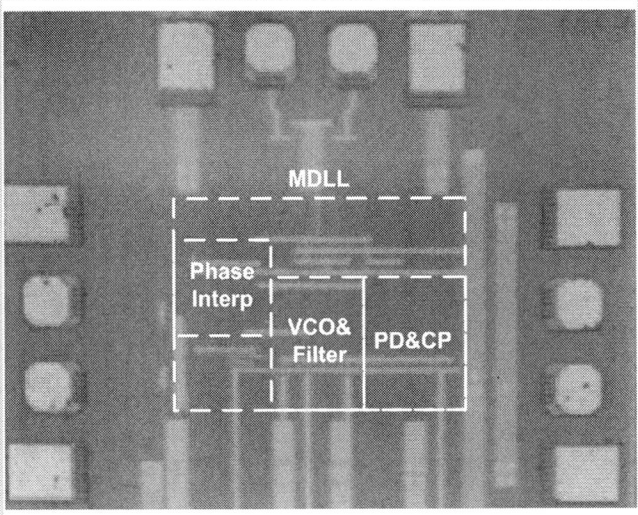

Figure 26.7.7: Chip Micrograph.

ISSCC 2011 / SESSION 27 / OVERSAMPLING CONVERTERS / OVERVIEW

Session 27 Overview / Data Converters

Oversampling Converters

Session Chair: *Kong-Pang Pun, Chinese University of Hong Kong, Hong Kong, China*

Session Co-Chair: *Lucien Breems, NXP Semiconductors, Eindhoven, The Netherlands*

Oversampling delta-sigma ($\Delta\Sigma$) data converters are widely used for high-resolution A/D and D/A conversions of low-to-medium bandwidth signals with applications ranging from sensor networks and audio interfaces to wireless communications. The performance of these converters is advancing in several fronts including resolution, bandwidth and power efficiency. This session presents papers that highlight such advancements to unprecedented levels, including continuous-time (CT) $\Delta\Sigma$ A/D converters with signal bandwidths up to 125MHz, discrete-time (DT) $\Delta\Sigma$ A/D converters with power efficient architecture and functional blocks, and audio $\Delta\Sigma$ D/A converters with improved accuracy-enhancement methods. Three unconventional $\Delta\Sigma$ converters, one demonstrating a new technique to realize the noise shaping, one operating under a 0.25V supply, and one presenting a new architecture for time-to-digital conversion, expand the boundaries of oversampling converters in other dimensions.

The session starts with Paper 27.1 from NXP Semiconductors and Delft University of Technology, which describes a high-speed multi-bit CT $\Delta\Sigma$ A/D converter that is sampled at 4GHz. Stable operation is accomplished by using a high-speed filter topology that absorbs the pole originating from the input capacitance of the 4b quantizer in combination with a direct feedback loop around the quantizer. The ADC realizes 70dB DR and -74dBFS THD over a 125MHz bandwidth while consuming 256mW from a dual supply of 1V and 1.8V. The 45nm CMOS chip occupies 0.9mm^2.

In Paper 27.2 from Ulm University, a 25MHz bandwidth 63.5dB SNDR CT $\Delta\Sigma$ modulator is presented that instead of using dynamic element matching uses a digital background calibration to compensate for the feedback DAC errors and enhance the SFDR to -81dB. A pseudo-random binary test signal is injected in the DAC and the error estimation is done by means of correlation. All feedback amplifiers are compensated for finite gain-bandwidth effects. The 3rd-order single-loop modulator has a FOM of 125fJ/conversion-step and is sampled at 500MHz while consuming 8mW from a 1.2V core supply and occupying 0.15mm^2 in 90nm CMOS.

A third-order DT $\Delta\Sigma$ modulator with a single-slope quantizer using modified bi-directional discharging is demonstrated next in Paper 27.3 by Oregon State University. The single-slope quantizer provides one additional order of noise-shaping of the quantization noise and produces a multi-bit output while using only a single comparator. Therefore, it does not require offset calibration. The multi-bit feedback DAC employs DWA for linearization. The 0.18μm CMOS $\Delta\Sigma$ ADC realizes 78.2dB SNDR in 1MHz while consuming 2.9mW from a 1.5V supply and achieving a FOM of 210fJ/conversion-step.

Paper 27.4 from KU Leuven presents an ultra-low-voltage switched-capacitor $\Delta\Sigma$ modulator in 0.13μm CMOS which is accomplished using a near-threshold-voltage biased CMOS inverter technique. This approach guarantees reliable operation of the inverter-based integrators over temperature while running at a supply voltage of only 0.25V. The ADC achieves 61dB SNDR over a 10kHz bandwidth at a power consumption of 7.5μW.

Paper 27.5 from the University of Pavia shows an energy-efficient 84dB SNDR 100kHz bandwidth $\Delta\Sigma$ modulator. The noise-shaping characteristic is third-order due to a second-order loop filter in combination with an additional first-order error feedback loop. A low power consumption of 140μW is achieved by a time-interleaved two integrators scheme with a single slew-rate boosted amplifier. The 0.18μm CMOS ADC occupies 0.5mm^2, consumes 140μW from a 1.5V supply, and has a FOM of only 54fJ/conversion-step.

978-1-61284-303-2/11 $26.00 © 2011 IEEE

ISSCC 2011 / February 23, 2011 / 1:30 PM

The time-to-digital converter described in Paper 27.6, from K.U. Leuven, SCK-CEN, and KH Kempen, is based on a 1-1-1 MASH $\Delta\Sigma$ architecture. The $\Delta\Sigma$ TDC has a time resolution of 5.6ps while consuming 1.7mW. At an oversampling ratio of 250, the ENOB is 11b over a 100kHz bandwidth. Radiation hardness of the 0.13μm CMOS TDC is verified for future nuclear and space applications.

The session concludes with two $\Delta\Sigma$ audio DAC papers that utilize different mismatch shaping techniques. Paper 27.7 from Analog Devices demonstrates a CT 8b oversampling DAC architecture which measures 120dB SNR and 100dB THD+N. The audio DAC employs a 3-level rotational data shuffling scheme. The DAC consumes 21.5mW from a dual supply of 5V and 2.25V, and occupies 1.35mm^2 area per channel and is fabricated in a 0.35μm DPQM CMOS process.

Finally, Paper 27.8 from Texas Instruments presents an algorithm that concurrently shapes static mismatch errors and dynamic inter-symbol interference (ISI) errors. The audio DAC achieves a THD better than -120dBFS with spurious tone free operation at low signal levels. The 45nm CMOS DAC occupies 0.16mm^2 and consumes 0.875mW from a dual supply of 1.45V and 1.1V with a 0.5V$_{rms}$ output.

The oversampling converter session demonstrates the state-of-the-art advances in $\Delta\Sigma$ architectures, circuit designs and digital enhancement techniques. The boundaries of $\Delta\Sigma$ A/D and D/A converters are pushed to much higher bandwidths and resolutions, extremely low supply voltages and high power efficiencies. The plurality of benefits of $\Delta\Sigma$ modulation is also finding its way in other application domains such as time-to-digital converters used in radiation hostile environments.

Presenters:

1:30 PM

27.1 A 4GHz CT $\Delta\Sigma$ ADC with 70dB DR and -74dBFS THD in 125MHz BW

M. Bolatkale,
NXP Semiconductors, Eindhoven, The Netherlands

2:00 PM

27.2 An 8mW 50MS/s CT $\Delta\Sigma$ Modulator with 81dB SFDR and Digital Background DAC Linearization

J. G. Kauffman, Ulm University, Ulm, Germany

2:30 PM

27.3 A Third-Order DT $\Delta\Sigma$ Modulator Using Noise-Shaped Bidirectional Single-Slope Quantizer

N. Maghari, Oregon State University, Corvallis, OR

3:15 PM

27.4 A 250mV 7.5μW 61dB SNDR CMOS SC $\Delta\Sigma$ Modulator Using a Near-Threshold-Voltage-Biased CMOS Inverter Technique

F. Michel, KU Leuven, Leuven, Belgium

3:45 PM

27.5 A 84dB SNDR 100kHz Bandwidth Low-Power Single Op-Amp Third-Order $\Delta\Sigma$ Modulator Consuming 140μW

A. Pena Perez, University of Pavia, Pavia, Italy

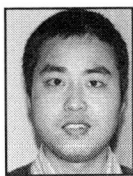

4:15 PM

27.6 A 1.7mW 11b 1-1-1 MASH $\Delta\Sigma$ Time-to-Digital Converter

Y. Cao, KU Leuven, Leuven, Belgium;
SCK-CEN, Mol, Belgium

4:45 PM

27.7 A 120dB-SNR 100dB-THD+N 21.5mW/Channel Multibit CT $\Delta\Sigma$ DAC

A. Bandyopadhyay, Analog Devices, Wilmington, MA

5:15 PM

27.8 A 108dB-DR 120dB-THD and 0.5V$_{rms}$ Output Audio DAC with Inter-Symbol-Interference-Shaping Algorithm in 45nm CMOS

L. Risbo, Texas Instruments, Copenhagen, Denmark

27

978-1-61284-303-2/11 $26.00 © 2011 IEEE

ISSCC 2011 / SESSION 27 / OVERSAMPLING CONVERTERS / 27.1

27.1 A 4GHz CT $\Delta\Sigma$ ADC with 70dB DR and −74dBFS THD in 125MHz BW

Muhammed Bolatkale[1], Lucien J. Breems[1], Robert Rutten[1],
Kofi A.A. Makinwa[2]

[1]NXP Semiconductors, Eindhoven, The Netherlands
[2]Delft University of Technology, Delft, The Netherlands

In this paper, a high-speed continuous-time (CT) $\Delta\Sigma$ ADC topology is proposed that absorbs the pole normally caused by the quantizer's input capacitance, while a local feedback loop compensates for the quantizer's excess delay. These measures allow a high-resolution multi-bit $\Delta\Sigma$ ADC to operate at GHz sampling rates. The bandwidth of this CMOS $\Delta\Sigma$ ADC is 6× wider than the state-of-the-art [1, 2]. Compared to a state-of-the-art pipeline BiCMOS ADC [3], it achieves similar power efficiency and bandwidth, but it only occupies 0.9mm² in 45nm CMOS, which is essential for low-cost integration. The 4b 3rd-order CT $\Delta\Sigma$ ADC is sampled at 4GHz and achieves 70dB DR and −74dBFS THD in a 125MHz BW while consuming 256mW. This prototype enables the use of $\Delta\Sigma$ ADCs in applications such as GSM base-stations and HD video systems.

Achieving bandwidths in excess of 100MHz requires GHz sampling frequencies. In a given technology, the maximum sampling frequency will usually be limited by the quantizer's latency and the parasitic pole caused by its input capacitance. A 1b quantizer is most suitable for high-speed operation as it has relaxed offset requirements and thus it can be very small. For example, a 35GHz 1b 2nd-order $\Delta\Sigma$ modulator has been demonstrated in SiGe BiCMOS [4] with 55dB DR in 100MHz BW. However, increasing the sampling rate, to achieve higher DR, is currently impractical in CMOS. Therefore, a higher-order loop filter and multi-bit quantization must be used, even though the use of the latter reduces the maximum achievable sampling rate, due to the quantizer's increased latency and input capacitance. In this paper, a high-speed filter topology that overcomes these limitations and enables much higher sampling rates is proposed.

Figure 27.1.1a illustrates the block diagram of the 3rd-order single-loop $\Delta\Sigma$ modulator. It consists of three integrators with feed-forward coefficients a_1 and a_2 for high-frequency stabilization, a feedback coefficient b_1 to minimize the in-band quantization noise, and a feed-forward signal path (a_0) to relax the loop filter's linearity requirements [5]. To compensate for the quantizer's latency, a second feedback path comprising a multi-bit D/A converter (DAC2) is employed (Fig. 27.1.1a) [6]. The loop filter is preferably implemented with RC integrators, since they can operate at low supply voltages. The last integrator's virtual ground node can then be used as the loop filter's output summation node (Fig. 27.1.1b). However, the quantizer's input capacitance (C_Q) loads the last integrator (Fig. 27.1.1b), introducing an extra pole that limits the maximum sampling rate. To overcome this, the last integrator is implemented as a G_m-C integrator which absorbs C_Q, as well as the DAC2's output capacitance (C_{D2}), into its integration capacitance (Fig. 27.1.1c). However, the last integrator's input is then no longer available for use as a summing node. Alternatively, the output of the G_m-C stage can be used as a wideband passive summation node for *differentiated* input signals. The feed-forward coefficients can then be implemented as capacitor ratios, i.e., $a_0=C_{a0}/C_{total}$, $a_1=C_{a1}/C_{total}$ and $a_2=C_{a2}/C_{total}$ where $C_{total}=C_{a0}+C_{a1}+C_{a2}+C_Q+C_{D2}$. The passive summation does not attenuate the quantizer's input signal if $(a_0+a_1+a_2)<1-(C_Q+C_{D2})/C_{total}$, which can be guaranteed by design.

Figure 27.1.2 shows the architecture of the ADC in more detail. The first two integrators are implemented as RC integrators. To cancel the right-half plane zero introduced by the first integrator, a resistor (R_z) is employed. The OTA's are implemented as two-stage amplifiers with feed-forward frequency compensation [7]. They achieve 35dB DC gain and 8GHz UGBW, while drawing 23mA from a 1.1V supply. To compensate for RC spread, C_1, C_2 and R_3 can be individually calibrated via 5-bit networks. The bias current of the G_m-C integrator can also be programmed to calibrate its unity-gain frequency (ω_3). The input of DAC2 is differentiated (in the digital domain), because its output current will be integrated by the last integrator [1]. Since it is in the high-speed-path (Fig. 27.1.2), the latency of the 4b quantizer must be less than half a clock period (125ps) to ensure loop stability. Moreover, the combination of the 4b DAC1 and its driver (Fig. 27.1.2) must achieve similar latency while still meeting the linearity and noise requirements. The ADC also includes a binary-to-thermometer decoder, decimation filter, and LVDS buffers.

The quantizer is a 4b flash converter that consists of 15 unit elements (Fig. 27.1.3a), whose reference voltages are generated from a 15-tap resistive ladder. Each quantizer consists of a preamplifier, a latch and a D-FF. The preamplifier is a resistively loaded NMOS pair with a reset switch that enables fast overdrive recovery. The latch is realized as a differential pair that drives a cross-coupled latch. The D-FF consists of two stages: a double-tail sense amplifier [8] and a symmetrical slave latch (SL) [9]. The use of a symmetrical SL means that each of the D-FF's outputs has equal delay, which makes it possible to drive DAC2 directly, and thus minimizes the excess delay associated with re-clocking the data. DAC1 driver uses the same D-FF architecture. The ADC's timing is shown in Fig. 27.1.3b. The comparator outputs (Dq) are valid after half a clock period, while the output of the DAC1 driver (D1), which drives the unit current sources, is valid in less than 1 clock period.

A unit element of the DAC driver is shown in Fig. 27.1.3a. It consists of a D-FF, a switch driver, and a data buffer. The thermometer output of each quantizer is directly connected to each unit element, where it is re-clocked on the rising edge of CLK_{DAC1} (Fig. 27.1.3b). The additional clocking of the data minimizes the jitter introduced by the D-FF's data-dependent delay and metastability. DAC1 is a 4b current-steering DAC designed for 11b intrinsic matching. One unit element of the DAC is shown in Fig. 27.1.3c. It consists of a resistively degenerated PMOS current source, which has better matching and lower noise than a MOS-only current source. By using a higher supply voltage for DAC1 (1.8V), R_1 can be made larger, effectively reducing the noise contribution of DAC1 and reducing the ADC's overall power consumption. Since the voltage drop on R_1 is about 0.7V, M_{1-8} can still be implemented using thin-oxide transistors.

The 45nm CMOS ADC (Fig. 21.1.7) occupies 0.9mm² including the modulator, clock buffers, and decimation filter. All high-speed digital signals use differential routing to minimize substrate noise. Its total power consumption is 256mW from a 1.1V supply. The decimated output for a 41MHz input signal at −0.5dBFS is captured in real-time, and its FFT is shown in Fig. 27.1.4. The achieved THD is −74dBFS. As shown in Fig. 27.1.5, the ADC achieves 70dB DR in a 125MHz BW. The peak SNR and SNDR are 65.5 and 65dB at −0.5dBFS input, respectively. For large signals, the residual nonlinearity of DAC1 causes harmonic components and quantization errors to fold into the signal band, thus increasing the in-band noise. Figure 27.1.6 summarizes the performance of the ADC. 16 samples are measured with similar results. The high-speed loop filter topology, the low-latency 4b quantizer, and DAC enable the operation of the CT $\Delta\Sigma$ ADC at 4GHz, thus achieving 70dB DR and −74dB THD in 125MHz BW.

Acknowledgments:
The authors would like to thank Harish Kundur and Jingjing Hu for their careful layout

References:
[1] G. Mitteregger, C. Ebner, S. Mechnig, et al., "A 14b 20mW 640MHz CMOS CT $\Sigma\Delta$ ADC with 20MHz Signal Bandwidth and 12b ENOB," *ISSCC Dig. Tech. Papers*, pp. 62-63, Feb., 2006.
[2] M. Park and M. Perrott, "A 0.13μm CMOS 78dB SNDR 87mW 20MHz BW CT $\Delta\Sigma$ ADC with VCO-based Integrator and Quantizer," *ISSCC Dig. Tech. Papers*, pp. 170-171, Feb., 2009.
[3] A.M.A. Ali, A. Morgan, C. Dillon, et al., "A 16b 250MS/s IF-sampling Pipelined A/D Converter with Background Calibration," *ISSCC Dig. Tech Papers*, pp. 292-293, Feb., 2010.
[4] A. Hart and S.P. Voinigescu , "A 1 GHz Bandwidth Low-Pass $\Sigma\Delta$ ADC With 20-50 GHz Adjustable Sampling Rate," *IEEE J. Solid-State Circuits*, vol. 44, no. 5, pp. 1401-1414, May, 2009.
[5] J. Silva, U.-K. Moon, J. Steensgaard, and G.C. Temes, "Wideband Low-Distortion $\Delta\Sigma$ ADC Topology," *Electronic Letters*, vol. 37, no. 12, pp. 737-738, Jun., 2001.
[6] P. Benabes, M. Keramat, and R. Kielbasa, "A Methodology for Designing Continuous-Time Sigma-Delta Modulators," *Proc. IEEE ED & TC.*, pp. 46–50, Mar., 1997.
[7] L.J. Breems, R. Rutten, R.H.M. van Veldhoven, and G. van der Weide, "A 56 mW Continuous-Time Quadrature Cascaded $\Sigma\Delta$ Modulator With 77 dB DR in a Near Zero-IF 20 MHz Band," *IEEE J. Solid-State Circuits*, vol. 42, no. 12, pp. 2696-2705, Dec., 2007.
[8] D. Schinkel, E. Mensink, E. Kiumperink, et al., "A Double-Tail Latch-Type Voltage Sense Amplifier with 18ps Setup+Hold Time," *ISSCC Dig. Tech, Papers*, pp. 314-315, Feb., 2007.
[9] B. Nikolic, V.G. Oklobdzija, V. Stojanovic, et al., "Improved Sense-Amplifier-Based Flip-Flop: Design and Measurements," *IEEE J. Solid-State Circuits*, vol. 35, no. 6, pp. 876-884, Jun., 2000.

978-1-61284-303-2/11 $26.00 © 2011 IEEE

ISSCC 2011 / February 23, 2011 / 2:00 PM

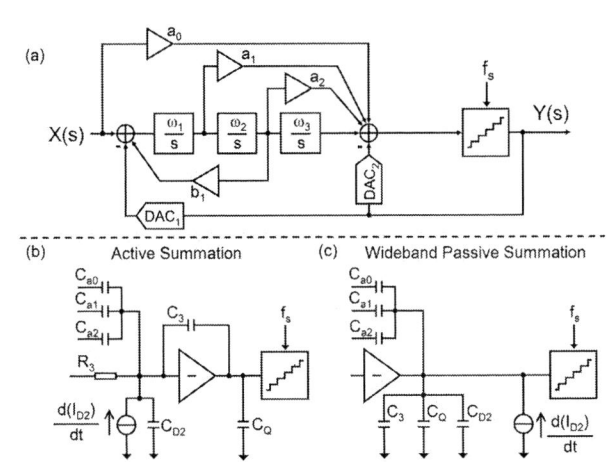

Figure 27.1.1: A 3rd-order ΔΣ modulator (a), conventional virtual-ground summation (b), and the proposed wideband passive summation (c).

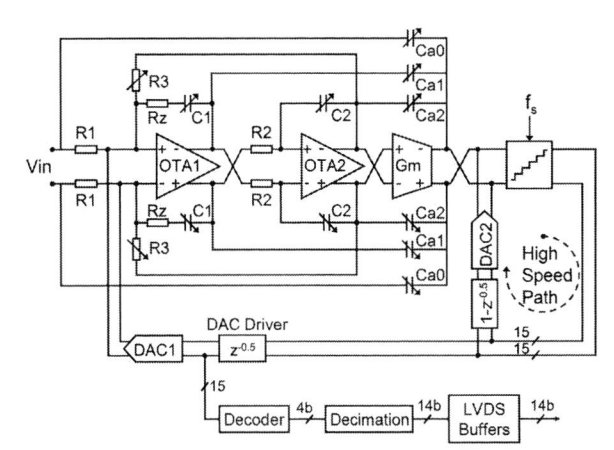

Figure 27.1.2: Architecture of the ADC.

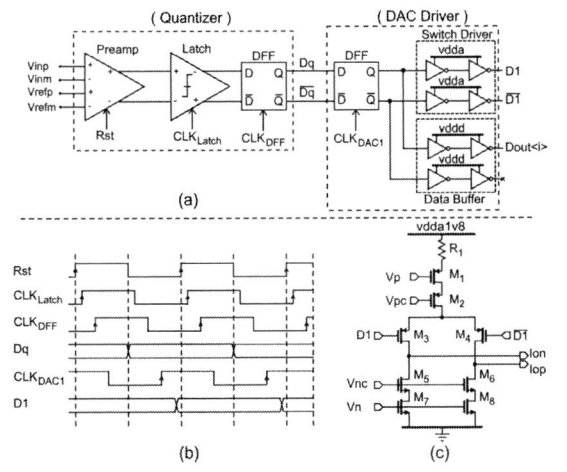

Figure 27.1.3: A unit element of the (a) 4b quantizer and DAC1 Driver, (b) the timing diagram of the ADC, and (c) DAC1.

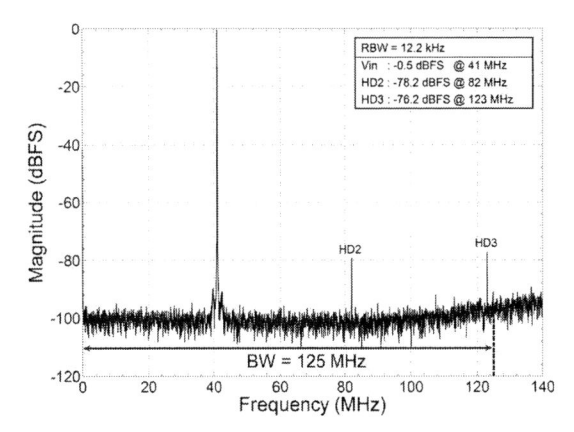

Figure 27.1.4: An FFT of measured decimated output for an input signal of −0.5dBFS at 41MHz.

Figure 27.1.5: Measured SNR and SNDR versus input power with a 41MHz input.

	This Work	[1]	[2]	[4]	[3]
Architecture	ΣΔ	ΣΔ	ΣΔ	ΣΔ	Pipeline
Fs (GHz)	4	0.64	0.9	35	0.25
BW (MHz)	125	20	20	100	125
DR (dB)	70	80	81.2	59	77.5
Peak SNR (dB)	65.5	76	81.2	59	77.5
Peak SNDR (dB)	65	74	78	53	77.5
THD (dBc)	-74	-78	-	-	-
Power (mW)	256	20	87	650	1000
VDD (V)	1.1/1.8	1.2	1.5	2.5	1.8/3.3
Area (mm²)	0.9	1.2	0.45	4	50
Technology	45nm CMOS	130nm CMOS	130nm CMOS	130nm SiGe	180nm BiCMOS
FOM (pJ/s)¹	0.70	0.12	0.33	8.9	0.55
FOM (pJ/s)²	0.40	0.06	0.23	4.4	0.55

Figure 27.1.6: Performance table and comparison to prior work. FOM¹=P/(2*BW*2^((SNDR-1.76)/6.02)). FOM²=P/(2*BW*2^((DR-1.76)/6.02)).

ISSCC 2011 PAPER CONTINUATIONS

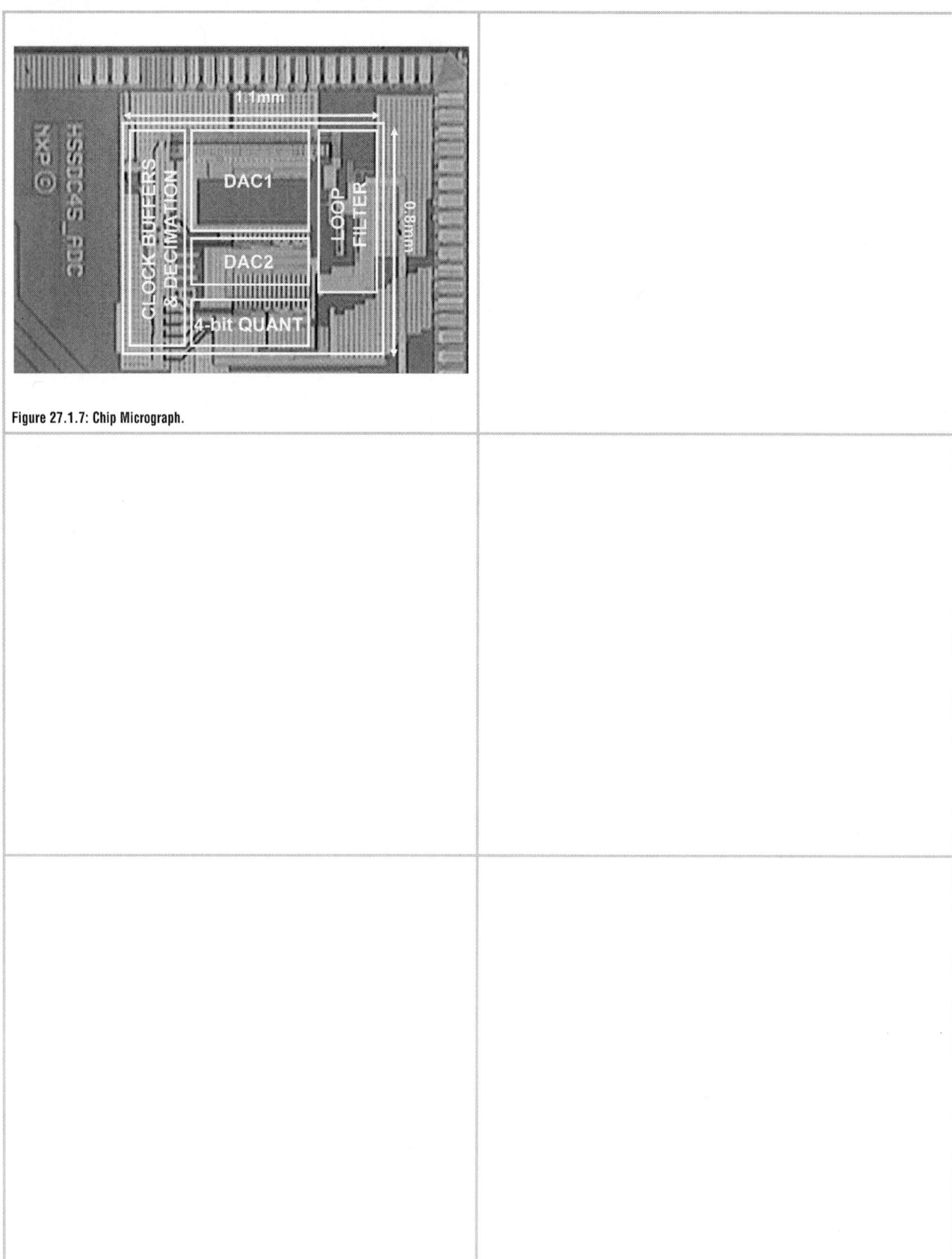

Figure 27.1.7: Chip Micrograph.

27.2 An 8mW 50MS/s CT $\Delta\Sigma$ Modulator with 81dB SFDR and Digital Background DAC Linearization

John G. Kauffman, Pascal Witte, Joachim Becker, Maurits Ortmanns

Ulm University, Ulm, Germany

There is ongoing effort to realize low-power $\Delta\Sigma$ ADCs with more than 10MHz bandwidth (BW) – especially for wireless transceivers. Besides the trend to make these ADCs more reconfigurable [1], recent advances in the design of CT $\Delta\Sigma$ modulators focused on design improvements [2], on replacing analog by digital circuits [3, 4], or on avoiding costly multi-bit DACs with time-domain or single-bit feedback [1, 4]. Apart from that, the power versus performance advances in $\Delta\Sigma$ ADCs have slowed down, and the best FOM for above 10MHz $\Delta\Sigma$ modulators ($\Delta\Sigma$Ms) is still 120fJ/conv [5]. Among other reasons, the further advance of $\Delta\Sigma$Ms to higher conversion BWs with compatible FOM faces two hurdles: first, high BW requires a low OSR to achieve a feasible sampling frequency f_S. Since the filter order rarely exceeds three, multi-bit operation is commonly employed. This requires linearization of the multilevel feedback DAC, e.g., by using DEM, which is however ineffective at low OSR [6]. Therefore, a large, linear multilevel DAC is needed, which introduces a significant parasitic load for the amplifiers and increases their power consumption. Proposed time domain multi-bit DACs still worsen the jitter sensitivity [4]. Second, the operation at high f_S makes the excess loop-delay (ELD) a large fraction of the sampling period. This ELD is caused by the quantizer, the phase shift due to finite amplifier BW [1][2], and the DEM. While many different techniques for ELD compensation have been proposed, the basic idea remains to realize a fast feedback loop around the quantizer. Alternative implementations with a summer, integrators with differentiating DAC [5], or compensation within the quantizer [1] only change the required location of the fast settling speed and high power. When implemented with the last integrator, ELD compensation makes its speed requirements the highest within the $\Delta\Sigma$ loop filter [1].

The intention of this work is to realize a power- and area-efficient 50MS/s $\Delta\Sigma$ modulator with an OSR of only 10, an effective linearization without DEM, and a very low power ELD compensation. This is achieved by using a digital DAC error estimation and correction, and a compensation for finite gain-bandwidth (FGBW) in all feedback amplifiers. The chosen architecture is illustrated in Fig. 27.2.1. A multi-bit 3rd-order, mixed feedforward–feedback compensation is used, which avoids the second DAC, limiting the swing of the first opamp, while reducing the STF peaking compared to an all feedforward topology [5]. The direct feedback path around the quantizer for ELD compensation is realized by extending the last integrator with an additional proportional forward path – called PI [8]. The modulator works around V_{CM}=600mV with ±400mV full scale and a 4b NRZ feedback, relaxing clock jitter requirements to $\sigma_{rms}\approx$10ps. A simplified schematic of the loop filter is shown in Fig. 27.2.2 – the PI transfer function is implemented with R_{PI}, and the quantizer is a 4b flash with auto-zeroed preamplifiers. The influence of FGBW in all feedback amplifiers is compensated. As shown in Fig. 27.2.1, the step response of an integrator with FGBW can be modelled as a delayed ramp with reduced integrator gain [2]. For low-power ELD compensation, the step response of the PI is modelled by a delayed step, reduced step size, and reduced slope gain. This results in reduced gain and step values which can be compensated by properly readjusting the loop scaling factors k_i. The delays $\tau_{1..3}$ correspond to the different phase shifts due to FGBW. They are summed with the quantizer delay τ_D=50%·T_S to combined ELDs (Fig. 27.2.1). Thus, the non-ideal 3rd-order loop filter, including FGBW in all amplifiers and PI-based ELD compensation, is matched to an ideal NTF, while the amplifiers are designed with only GBW$_{1,2,3}$=500, 300, 600MHz. Particularly for the 3rd amplifier, which realizes the ELD compensation, the speed and power requirements are drastically relaxed compared to similar implementations [1, 5]. The 1st and 3rd integrators are realized using three-stage positive feedforward compensated amplifiers with class AB output, while the 2nd integrator uses a two-stage symmetrical amplifier with class AB output.

Figure 27.2.3 shows the schematic of a DAC$_1$ unit cell. The cascodes operate in subthreshold, and the current-source transistors are intentionally made small since their intrinsic mismatches are compensated by the background DAC correction. The simulated DAC$_1$ linearity is designed to only 9 bit. DAC$_3$ is implemented even smaller than DAC$_1$, and needs no correction since its nonlinearity is suppressed by the preceding loop filter. For the error estimation, a 16th unit

element is inserted in addition to 15 DAC$_1$ cells. This test source is switched by a slow pseudo-random binary noise signal (prn) generated by an LFSR. This prn is converted by the modulator into the digital output y_d in Figs. 27.2.1 and 2. A cross-correlation (CCF) of y_d and the digital input prn yields a value, which is unique for the DAC unit element under test and proportional to its current. Swapping the spare element successively with all DAC unit elements – i.e., using the spare element in the DAC and multiplexing prn into the respective DAC element – yields a unique CCF for each unit cell. This reveals the relative mismatch among the unit elements [7], and can be used to estimate digital correction factors c_i [6] as shown in Fig. 27.2.1. The method works fully in background, since the inserted prn is digitally subtracted from the modulator output data stream. In the presented design, the estimation and correction is implemented in an FPGA; no DEM has been used. The mismatch estimation of 4 samples, as displayed in Fig. 27.2.3, shows the random but unique variation of each DAC element. Unexpectedly, the DACs of all samples tested are more linear than what expected from simulations. The measured SNR and SNDR in Fig. 27.2.4 shows the successful operation of the $\Delta\Sigma$M as well as the intended extension of the linear input range and the increase of the maximum SNDR from 59.5dB to 63.5dB with background correction. Since the DAC matching is better than expected, the demonstrable nonlinearity improvement by digital correction is limited by the noise of the designed circuit. To emphasize the potential of the background correction, Fig. 27.2.5 compares the SFDR at P_{in}=−3dBFS for both the digitally estimated and corrected DAC as well as the uncorrected DAC. By using the background estimation and correction, the harmonics are well suppressed by up to 15dB, and the digitally enhanced circuit achieves an SFDR=81dB. Such suppression of nonlinearity is not achievable with a DEM linearization at an OSR=10 [6, 7].

The chip micrograph is shown in Fig. 27.2.7 with the modulator occupying 0.2mm² (0.15mm² active) in a 90nm CMOS process. The modulator including loop filter (amp$_1$=1mW, amp$_2$=516μW, amp$_3$=1.8mW), 4b flash quantizer (3mW), DAC$_1$ (737μW), DAC$_3$ (320μW), and all biasing circuits (620μW) consumes an overall power of 8mW from a 1.2V supply. Figure 27.2.6 reveals that the presented design – with FOM=125fJ/conv – compares favorably with the best published work. The background estimation and correction displays major advantages over DEM by achieving an SFDR=81dB from an only 9b linear DAC. The presented analog and digital compensation techniques open a feasible way to further increase the conversion rates of CT $\Delta\Sigma$Ms, while pushing their FOM below 100fJ/conv.

Acknowledgements:
This work was supported by German Research Foundation under DFG grant OR 245/1-2.

References:
[1] P. Crombez, G. van der Plas, M. Steyaert, M. Craninckx, "Single-Bit 500 kHz-10 MHz Multimode Power-Performance Scalable 83-to-67 dB DR CT $\Delta\Sigma$ for SDR in 90 nm Digital CMOS," *IEEE J. Solid-State Circuits*, vol. 45, no. 6, pp. 1159 – 1171, June, 2010.
[2] K.Reddy and S. Pavan, "A 20.7 mW Continuous-Time Delta-Sigma Modulator with 15 MHz Bandwidth and 70dB Dynamic Range," *ESSCIRC*, pp. 210-213, Sept., 2009.
[3] M. Park and M. Perrott, "A 0.13μm CMOS 78dB SNDR 87mW 20MHz BW CT $\Delta\Sigma$ ADC with VCO-Based Integrator and Quantizer," *ISSCC Dig. Tech. Papers*, pp. 170-171, Feb., 2009.
[4] V. Dhanasekaran, M. Gambhir, M.M. Elsayed, et al., "A 20MHz BW 68dB DR CT $\Delta\Sigma$ ADC Based on a Multi-Bit Time-Domain Quantizer and Feedback Element," *ISSCC Dig. Tech. Papers*, pp. 174-175, Feb., 2009.
[5] G. Mitteregger, C. Ebner, S. Mechnig, et al., "A 14b 20mW 640MHz CMOS CT DS ADC with 20MHz Signal Bandwidth and 12b ENOB," *ISSCC Dig. Tech. Papers*, pp. 62-63, Feb., 2006.
[6] J. Silva, X. Wang, P. Kiss, U.-K. Moon, and G.C. Temes, "Digital Techniques for Improved $\Delta\Sigma$ Data Conversion," *IEEE CICC*, pp. 183-190, May, 2002.
[7] P. Witte and M. Ortmanns, "Background DAC Error Estimation Using a Pseudo Random Noise Based Correlation Technique for Sigma-Delta Analog-to-Digital Converters," *IEEE TCAS-I*, vol. 57, no. 7, pp. 1500-1512, July, 2010.
[8] M. Vadipour, C. Chen, C. Yazdi, et al., "A 2.1 mW/3.2 mW Delay-Compensated GSM/WCDMA $\Sigma\Delta$ Analog–Digital Converter," *IEEE Symp. VLSI Circuits*, pp. 180–181, June, 2008.

ISSCC 2011 / February 23, 2011 / 2:00 PM

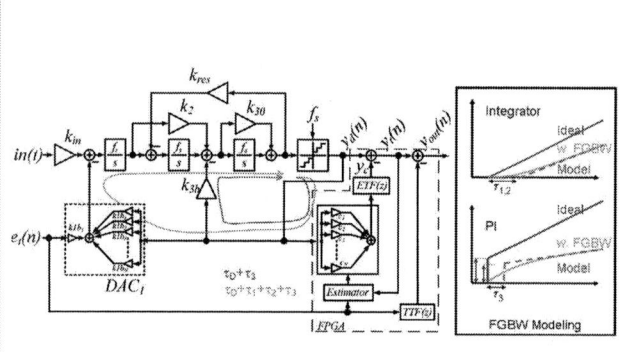

Figure 27.2.1: 3rd-order $\Delta\Sigma$ architecture including digital DAC error correction, bandwidth compensation for integrator and PI (k_{30}). Feedback delay due to quantizer ELD and amplifier FGBW exemplarily shown for 2 of the 6 paths.

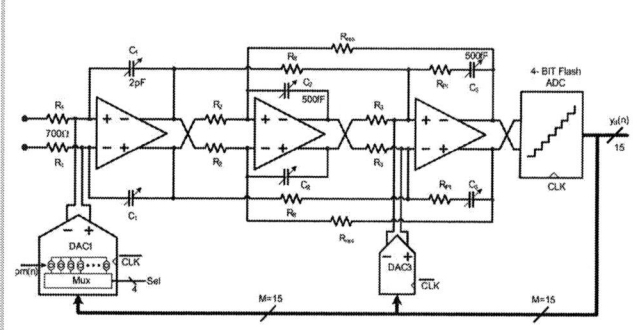

Figure 27.2.2: Simplified schematic of the 3rd-order loop filter illustrating the DAC$_1$ unit cell swapping and PI based ELD compensation.

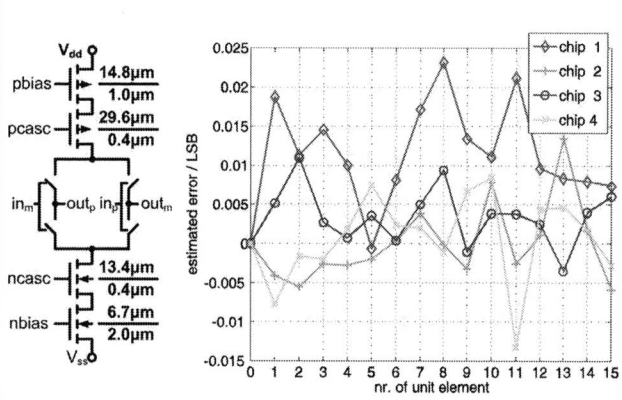

Figure 27.2.3: Simplified schematic of a unit cell of DAC$_1$ and estimated percentage DAC unit cell variation with reference to one DAC LSB measured for 4 chip samples.

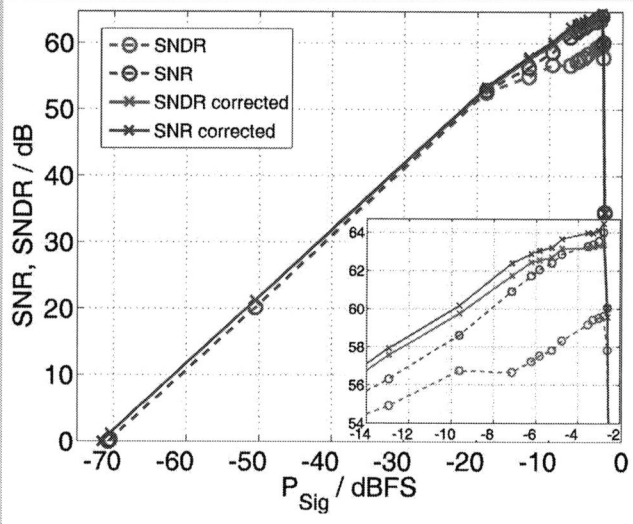

Figure 27.2.4: SNDR measured at f_{in}=2MHz, with and without correction.

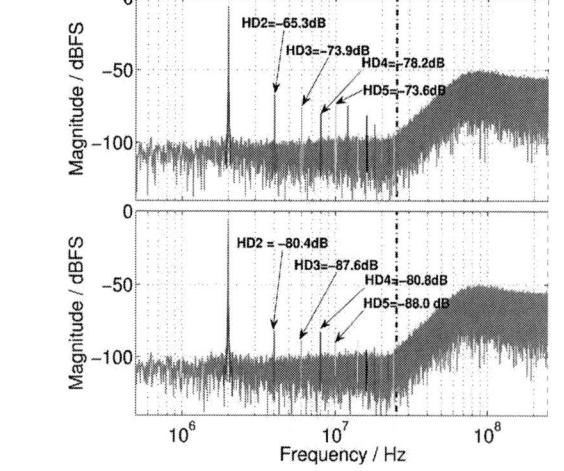

Figure 27.2.5: Harmonic distortion and spectra without and with digital DAC correction measured at P_{in}=−3dBFS @ f_{in}=2MHz with 65536-point Blackman window.

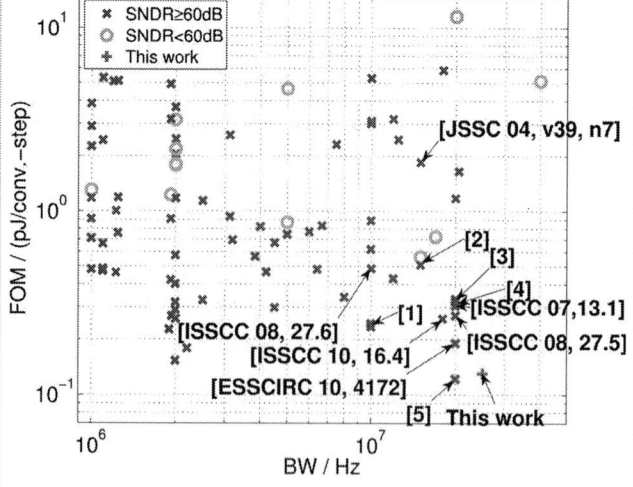

Figure 27.2.6: FOM comparison with the state-of-the-art.

978-1-61284-303-2/11 $26.00 © 2011 IEEE

ISSCC 2011 PAPER CONTINUATIONS

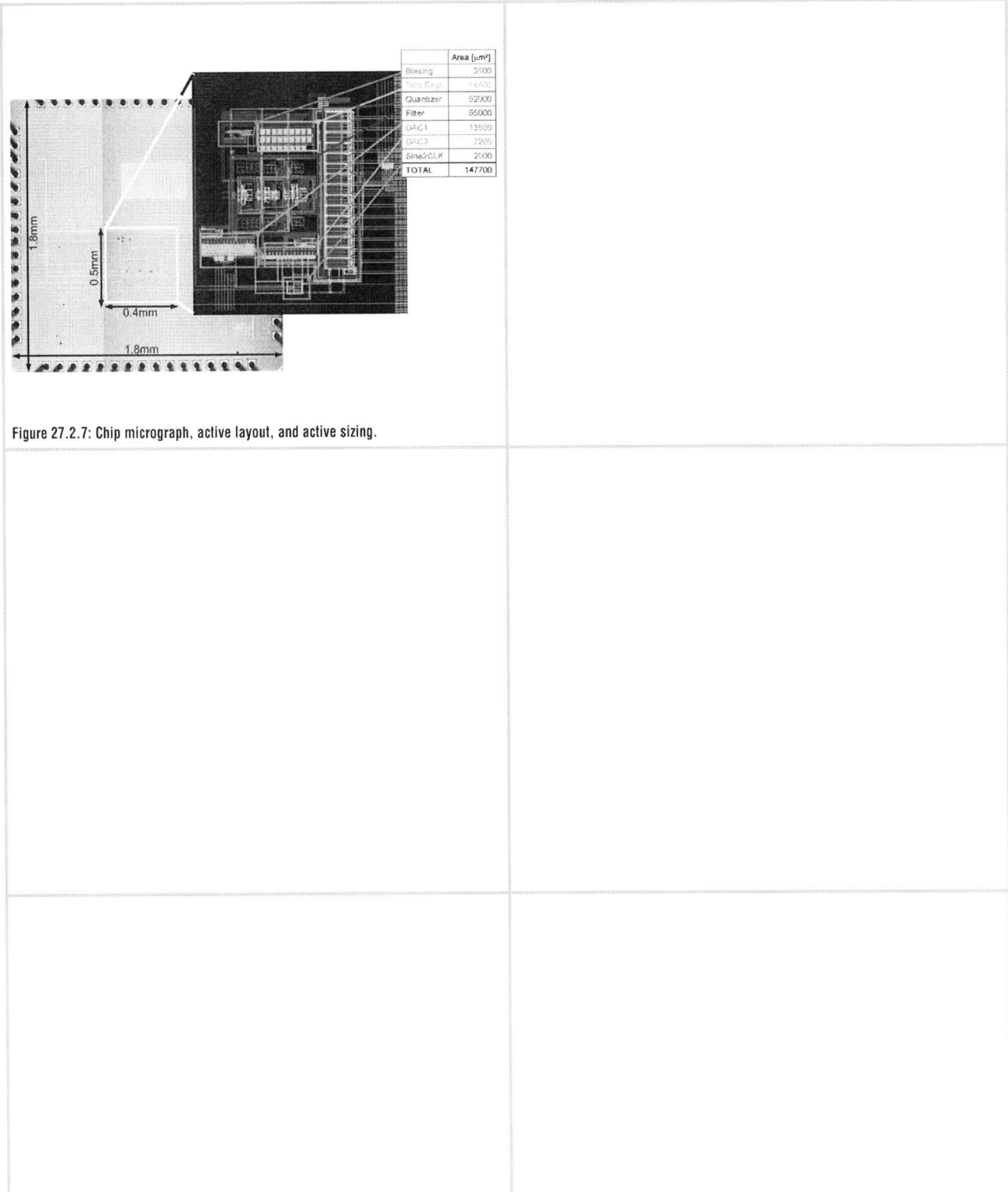

Figure 27.2.7: Chip micrograph, active layout, and active sizing.

ISSCC 2011 / SESSION 27 / OVERSAMPLING CONVERTERS / 27.3

27.3 A Third-Order DT $\Delta\Sigma$ Modulator Using Noise-Shaped Bidirectional Single-Slope Quantizer

Nima Maghari, Un-Ku Moon

Oregon State University, Corvallis, OR

The aspirations for power efficient ADCs have led to many improvements in this area. In delta-sigma modulators, techniques such as VCO-based quantizer [1, 2] and time-domain quantization [3] have been proposed to enhance the overall performance. The former provides additional noise shaping and the latter eliminates the need for the flash ADC. However, in addition to process and temperature variations, the inherent nonlinear behavior of the VCO requires careful attention. The time-domain quantization technique is an efficient way to utilize the quantizer but it does not improve the noise-shaping property. In this paper, we describe modified dual-slope and single-slope ADCs using bidirectional discharging. With a small modification in the discharging phase of these types of ADCs, a first-order quantization noise shaping is attained. These structures may be used as standalone ADCs or as the quantizer in a delta-sigma loop. Moreover, because the modified discharging extracts the quantization error in the analog domain, possibilities exist for further performance improvements [4].

The basic operations of the traditional unidirectional dual-slope ADC and the proposed noise-shaped one are shown in Fig. 27.3.1. The input signal is sampled via a resistor in Φ_2 and is discharged in Φ_1. During the discharging period, a high-speed clock is fed to a digital counter and is incremented. A continuous-time comparator is used to detect the zero crossing of the output and terminates the discharging period. The number of clock pulses fed to the digital counter during the discharging represents the quantized input signal. In the traditional discharging, the termination of the discharge pulse is right at the zero-crossing instance. The proposed discharging technique will continue the discharging until the next edge of the counting clock after the zero crossing. In this fashion, the value stored on the integrator is half-LSB minus the quantization error of the current sample. The next sample of the signal is added to this shifted quantization error before the next discharging phase. Hence, the final digital output code will be

$$D_{out}(n) = X(n) + q(n) - q(n-1) + \frac{LSB}{2} \qquad (1)$$

showing a first-order noise shaping of the quantization error. The half-LSB offset is a fixed value and can be removed either in the digital or the analog domain. The input sampling via a resistor can provide first-order anti-aliasing for a standalone ADC. However, if this quantizer is to be used in a delta-sigma modulator, it requires a fixed sampled data during its sampling period so that the loop characteristics are not affected. If the discharging is always done in one direction as illustrated, careful common-mode biasing and input sampling are also needed to make sure the output of the opamp always stays above the comparator reference.

To overcome these issues, single-slope sampling with bidirectional discharging is proposed. In this implementation, the input is sampled via a switched-capacitor branch to avoid unnecessary windowing of the input. In the discharging phase, the differential output polarity of the discharging opamp is first assessed. Depending on this polarity, the discharging will be either in the positive or the negative direction. This discharging operation is illustrated in Fig. 27.3.2. The positive and negative discharging can be seen in the first and second discharging phases. This technique will also halve the speed requirements of the counting clock since the MSB (polarity) is first resolved. Moreover, input common-mode sensitivity is avoided. The bidirectional discharging results in a half-LSB offset that is signal polarity dependent, which is easily fixed in the next sampling phase by subtracting the half-LSB based on the polarity of the previous sample. This cancelation need not be very accurate because the relatively significant delta-sigma quantization noise at the input of this bidirectional quantizer will randomize this error.

The low-distortion implementation of the second-order loop filter uses an active adder before the quantizer. As in traditional structures, input signal and outputs of the loop filter are sampled on to the active adder in one phase. However, in the next phase, instead of being reset, the output of the active adder is discharged and quantized via the proposed bidirectional discharging. This technique not only eliminates the need for the flash ADC but also provides an extra order of noise shaping for free. In addition, the active adder does not need to drive a large capacitive load that is typical of a multi-bit flash ADC.

Figure 27.3.3 shows the merged active adder and the quantizer where the second-order loop filter is omitted for simplicity. By the end of the sampling phase (Φ_2), the continuous-time comparator, which is connected at all times, will read the direction before discharging starts in Φ_1. During the non-overlapping time between these two clock phases, the switches are set in the proper direction for discharging to take place. Instead of a high-speed counting clock, a coarse DLL is used to provide five edges during the discharging phase. This will form a 9-level quantizer considering the mid-level is shared between the positive and negative polarities. By the end of the discharging phase, the digital output is ready to be fed to the DWA which drives the front-end DAC of the modulator. The quantization error is kept on the active adder (in the feedback capacitor) and it is naturally subtracted from the next sample. During the next sampling phase, the polarity dependent half-LSB offset is removed via an additional resistive discharging path. The RC mismatch in the sampling and discharging of the quantizer only results in a fixed gain error and is mostly suppressed by the loop filter. Any imperfection that occurs in the correction of the polarity dependent half-LSB offset will only slightly increase the quantization error and is shaped by the modulator.

As a proof of concept, a prototype delta-sigma ADC is implemented in a 0.18μm 2P5M CMOS process. The input signal sampling capacitors are shared with the front-end DAC capacitors. Figure 27.3.4 shows two measured output spectrums. The sampling frequency is 50MHz and oversampling ratio is 24. The out-of-band peaking is deliberately set to help the stability and to allow larger input signals to be processed by the loop. This modulator achieves 78.2dB peak SNDR and 79.3dB peak SNR while consuming 1.35mW analog and 1.55mW digital power from 1.5V supplies. The major portion of the digital power (1.3mW) is consumed by an overdesigned generic clock generator to provide flexibility in testing with various sampling frequencies. The rest of the digital power (0.25mW) includes the DLL, digital counter, DWA and the comparator. The achieved minimal analog power is the direct result of the extra order of noise shaping, and the elimination of the flash ADC and the typical large capacitive loading that comes with it. The FoM is 210fJ/conversion-step and it can easily be reduced further with redesign (i.e., eliminating the wasted clock generator power).

Acknowledgment:
The authors would like to thank National Semiconductor for providing fabrication. This work was funded by the Semiconductor Research Corporation.

References:
[1] M.Z. Straayer and M.H. Perrott, "A 12-bit 10-MHz Bandwidth, Continuous-Time Sigma-Delta ADC With a 5-bit, 950-MS/s VCO-Based Quantizer," *IEEE J. Solid-State Circuits*, vol. 43, no. 4, pp. 805-814, Apr., 2008.
[2] M. Park and M. Perrott, "A 0.13μm CMOS 78dB SNDR 87mW 20MHz BW CT $\Delta\Sigma$ ADC with VCO-Based Integrator and Quantizer," *ISSCC Dig. Tech Papers*, pp. 170-171, Feb., 2009.
[3] V. Dhanasekaran, M. Gambhir, M.M. Elsayed, et al., "A 20MHz BW 68dB DR CT $\Delta\Sigma$ ADC Based on a Multi-Bit Time-Domain Quantizer and Feedback Element," *ISSCC Dig. Tech. Papers*, pp.174-175, Feb., 2009.
[4] K. Lee, J. Chae, M. Aniya, et al., "A Noise-Coupled Time-Interleaved $\Delta\Sigma$ ADC with 4.2MHz BW, -98dB THD, and 79dB SNDR," *ISSCC Dig. Tech Papers*, pp. 494-495, Feb., 2008.

ISSCC 2011 / February 23, 2011 / 2:30 PM

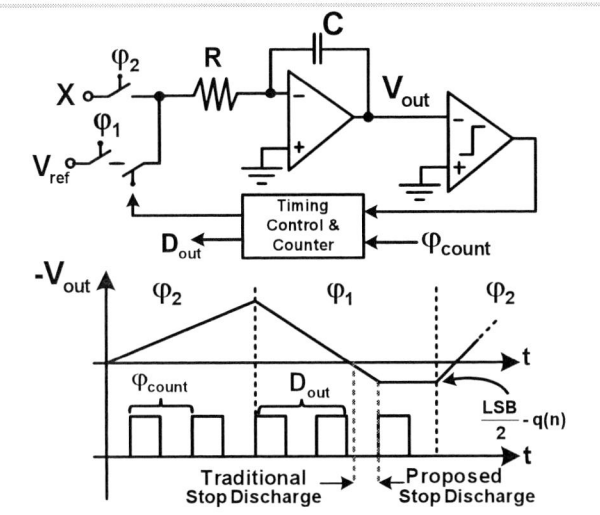

Figure 27.3.1: Basic operation of the traditional and the proposed noise-shaped unidirectional dual-slope ADC.

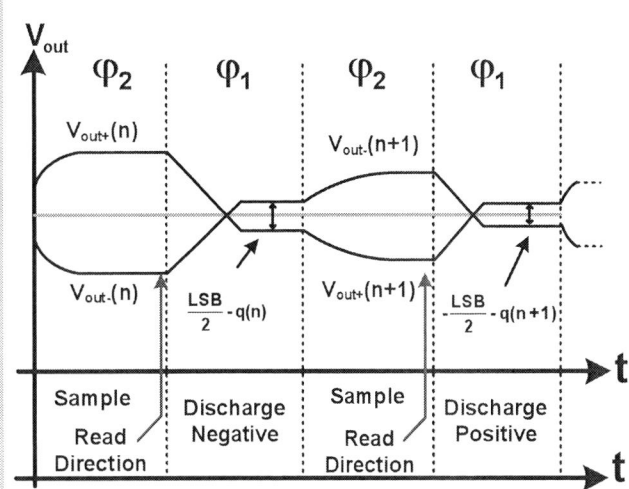

Figure 27.3.2: Bidirectional noise-shaped discharging concept.

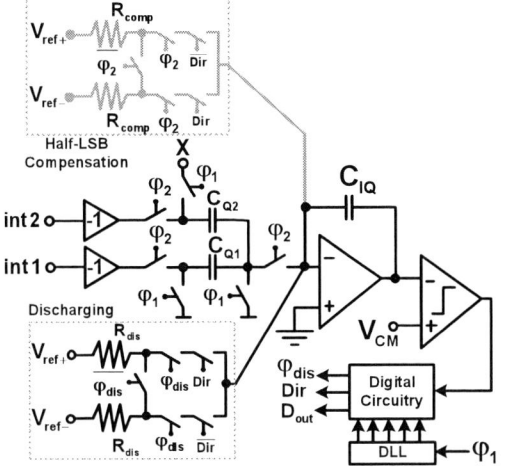

Figure 27.3.3: The proposed structure merged with the active-adder of a low-distortion second-order delta-sigma loop (single-ended structure shown for simplicity).

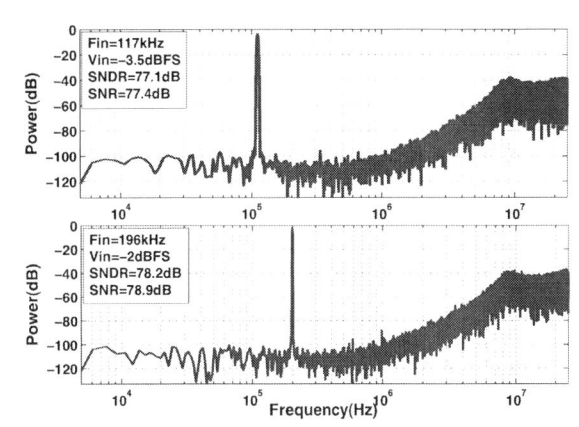

Figure 27.3.4: Measured 32K-samples FFT output spectrums.

Figure 27.3.5: Measured output SNDR and SNR versus input signal amplitude for a 196kHz sine wave.

Technology		0.18 μm CMOS
Supply Voltage		1.5 V
Sampling Frequency		50 MHz
Oversampling Ratio		24
Peak SNDR		78.2 dB
Peak SNR		79.3 dB
Peak SFDR		86.3 dB
Power	Analog	1.35 mW
	Clocking	1.3 mW
	DWA & Logic	0.25 mW
Active Area		0.44 mm^2
FOM		210 fJ/Conv.Step

Figure 27.3.6: Table of measured results.

27

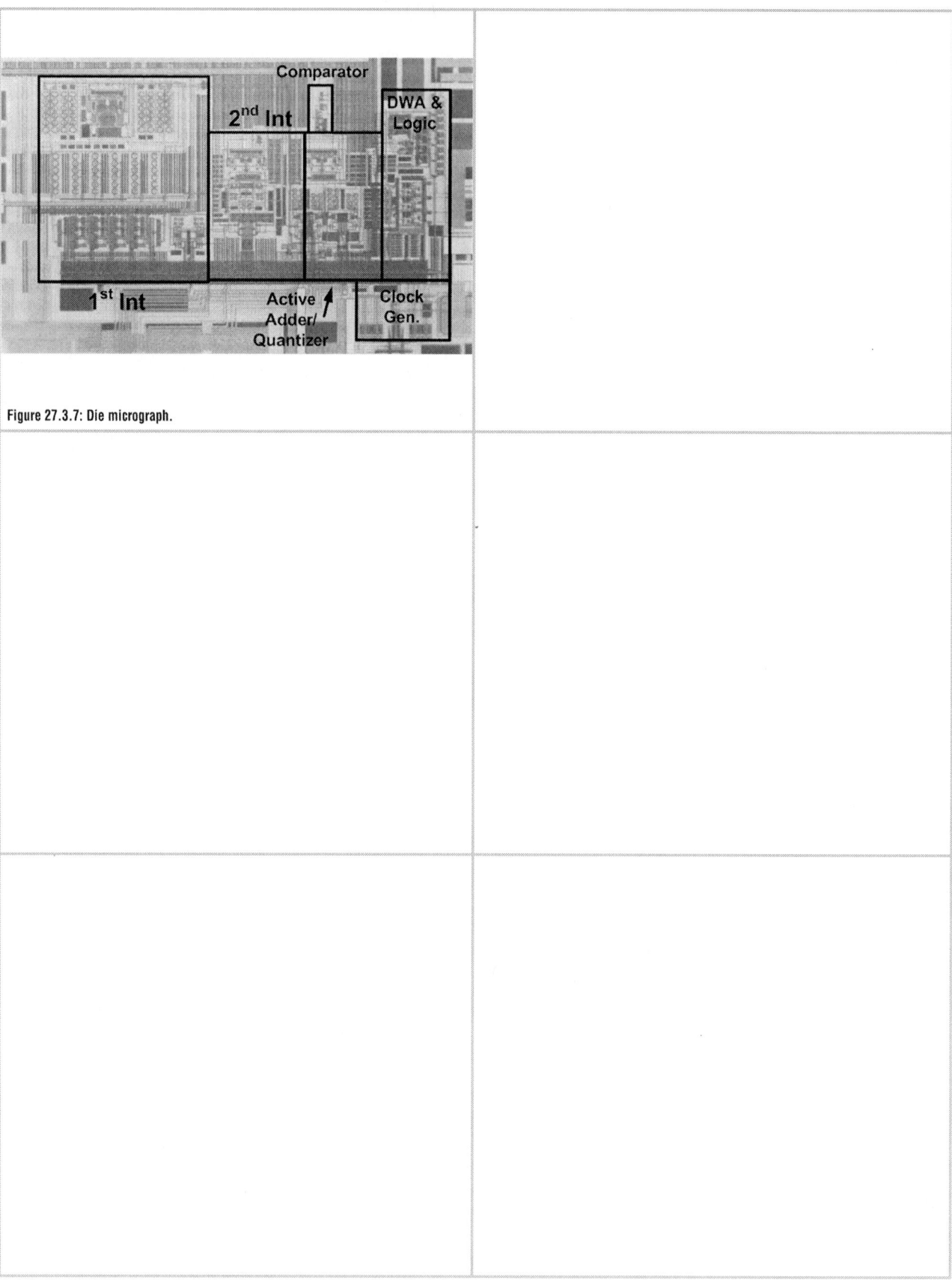

Figure 27.3.7: Die micrograph.

ISSCC 2011 / SESSION 27 / OVERSAMPLING CONVERTERS / 27.4

27.4 A 250mV 7.5µW 61dB SNDR CMOS SC ΔΣ Modulator Using a Near-Threshold-Voltage-Biased CMOS Inverter Technique

Fridolin Michel, Michiel Steyaert

KU Leuven, Leuven, Belgium

One of the most continuous trends in solid-state circuits is the decrease in power supply as a direct consequence of technology scaling. The fact that V_t does not scale linearly with supply voltage has encouraged several low-voltage design techniques recently [1-6]. The received voltage level in a wireless power transfer system decays rapidly with distance and also medical portable systems are calling for low-voltage circuitries.

Major milestones in low-voltage ΔΣ converters include a 0.5V CT ΔΣ converter [1] as well as a 0.7V SC converter [2]. Whereas achieving excellent figures of merit in the bound of ultra low supply voltage, these designs trade performance with robustness to process and temperature variation, because a clear bias current definition is omitted. The current is either defined by the input common-mode (CM) [1] or inherently undefined by driving the gates of PMOS and NMOS of a CMOS inverter both with the supply midlevel [2]. Besides, for a 250mV design biasing the inverter gates at the supply midlevel would result in an extremely low overdrive voltage of only 125mV. Apart from low overdrive voltage, which is the limiting factor above 0.5V, the saturation voltage V_{sat} becomes the key challenge below 0.5V. For a CMOS inverter the minimum possible power supply for a required output swing V_{swing} is $2V_{sat} + V_{swing}$. Therefore, this work describes design techniques that can cope with internal voltage swings of ±35mV, so that a ΔΣ converter operating with 200mV$_{pp}$ input signals at a record low voltage power supply of only 250mV can be presented. Moreover, a near V_t biasing technique is used that results in small transistor sizes and accurate bias current definition, so that a peak SNDR of 61dB is achieved for a signal bandwidth of 10kHz while offering tolerance to temperature variation (Fig. 27.4.6).

In CT circuits wide input CM range can only be realized using bulk input [3], which comes with shortcomings in noise, gain, and offset performance. This is different in SC solutions, where a charged capacitor can be connected in series to the gate in order to set the bias point and compensate offset [4]. The most attractive technique for ultra low voltage is the inverter-based technique [2], for it reduces the circuitry to the absolute minimum. As a result it presents the optimum in terms of available voltage headroom. Therefore, a pseudo- differential inverter-based architecture [2] has been adopted for this design.

The ΔΣ converter design (Fig. 27.4.1) includes an on-chip clock controller to boost the 250mV$_{pp}$ input clock as well as on-chip biasing. Reserving headroom for a peak V_t variation of $|\Delta V_t|$, the maximum nominal biasing voltage for an NMOS and PMOS transistor is $V_{DD} - |\Delta V_t|$ and $V_{SS} + |\Delta V_t|$, respectively. However, this requires moving to the power rails for slow corners, which is obviously not possible for a conventional constant-g_m biasing circuit due to V_{sat}. Therefore, a level-shifting circuit is proposed (Fig. 27.4.2), that shifts the constant-g_m bias by 50mV via pushing transistor M_x into triode region in a feedback configuration.

To reduce voltage swings, the ΔΣ loop is based on a feed-forward structure. This is necessary, since V_{sat} is about 90mV which reduces the maximum integrator swing to ±35mV. Using a 3rd-order loop makes low OSR possible, thereby relaxing the clock controller design. The converter is a two-phase system that consists of a biasing and offset compensation phase (φ_2) as well as an amplification phase (φ_1). One of the main innovations in this design takes effect during the biasing phase (φ_1), where the CMOS inverter is biased near V_t and offset compensated simultaneously using a new SC biasing approach (Fig. 27.4.3). As V_t variations for PMOS and NMOS are uncorrelated, both gates are biased independently. Offset compensation is achieved via feedback of the inverter output. However, since the PMOS gate is biased 100mV below the ground reference, the output is level shifted down to V_{bp} via C_{ls}, which is charged during the amplification phase. In this way, high overdrive voltage can be accomplished and the residual offset due to gain error is reduced because the output remains near ground level during biasing. This is one of the major key points why robust operation near V_t at 250mV power supply (V_{DD}–V_{SS}) is achieved.

A disadvantage of gate series capacitance is a penalty in gain accuracy. The Miller capacitances C_{p1} and C_{p2} are charged to V_{out} during the amplification phase (Fig. 27.4.3). The charge flowing onto these capacitors will also flow through C_{c1} and C_{c2}, thereby introducing signal dependent offset. An obvious solution is guaranteeing $C_c \gg C_p$, which yields a trade-off between area and performance. However, owing to near V_t biasing the integrators can be designed small, resulting in less parasitic capacitance.

Due to the limited available voltage headroom, CM swings must be controlled accurately. As CM signals are not fed back in the pseudo-differential ΔΣ loop, even small CM signals will accumulate and push the inverter out of the saturation region. Consequently, CM unity feedback has been implemented for each integrator (Fig. 27.4.1), reducing the integrator CM gain to 1. Whereas this technique prevents accumulation over time, it does not give sufficient protection to large CM inputs, so that a CM cancellation circuit (Fig. 27.4.2) has been added to the first integrator stage.

The voltage at the summing node is fed to a latched comparator that has to meet several design challenges. As the integrator swing must be highly scaled down due to the limited headroom, the comparator input is small, so that a preamplifier is mandatory to insure proper switching of the latch. Moreover, due to the feed-forward path the comparator will see CM signals despite the cancellation in the first integrator. To meet these requirements at 250mV power supply, an improved comparator structure is designed (Fig. 27.4.4). The CM input signals are removed by CM cancellation (Fig. 27.4.2), so that gate input is possible for high preamplifier gain. Because residual CM signals might saturate the high gain preamplifier, some inherent CM rejection is realized by a bulk cross-coupled load that reduces the CM gain to $g_m/g_{mb} \approx 5$. The latch stage requires the load transistors to be diode-connected during the tracking phase and cross-coupled during the latch phase. Using PMOS gate input with bulk diode connection does not add enough positive feedback for successful latching because of $g_{mb} < g_m$. Therefore, the NMOS gate is diode connected, leaving only the PMOS bulk as an input terminal, because the PMOS gate must be used for bias current definition. This results in an attenuation of g_{mb}/g_m during the tracking phase, another reason to use a preamplifier.

The converter is implemented in 0.13µm standard CMOS (Fig. 27.4.7) and is fed by a 250mV power supply and a 250mV$_{pp}$ input clock at 1.4MHz. With an OSR of 70, a BW of 10kHz with 61dB SNDR for a differential-mode (DM) input signal of 200mV$_{pp}$ is achieved. A performance summary is provided in Fig. 27.4.6. As is indicated in the table, raising the power supply to 300mV allows increasing the temperature range and bandwidth. Furthermore, the circuit achieves the performance up to 600mV power supply. As a result the complete converter can operate under process and temperature variation at 250mV (the lowest reported value for ΔΣ in bulk CMOS) by coping with limited voltage headroom and the developed near-threshold-voltage biasing technique.

References:

[1] K.-P.Pun, S. Chatterjee, and P. Kinget, "A 0.5V 74dB SNDR 25kHz CT Delta-Sigma Modulator with Return-to-Open DAC," *ISSCC Dig. Tech. Papers*, pp. 72-73, Feb., 2006.

[2] Y. Chae, I. Lee, and G. Han, "A 0.7V 36µW 85dB-DR Audio ΔΣ Modulator using class-C inverter," *ISSCC Dig. Tech. Papers*, pp. 490-491, Feb., 2008.

[3] S. Chatterjee, Y. Tsividis, and P. Kinget, "0.5-V analog circuit techniques and their application in OTA and filter design," *IEEE J. Solid-State Circuits*, vol. 40, no. 12, pp. 2373–2387, Dec., 2005.

[4] Y. Tang and R. L. Geiger, "A 0.6 V Ultra Low Voltage Operational Amplifier," *ISCAS*, vol. 3, pp. 611-614 , May, 2002.

[5] K. Lee, Q. Meng, T. Sugimoto, K. Hamashita, K. Takasuka, S. Takeuchi, U.-K. Moon, and G.C. Temes, "A 0.8 V, 2.6 mW, 88 dB Dual-Channel Audio Delta-Sigma D/A Converter With Headphone Driver," *IEEE J. Solid-State Circuits*, vol. 44, no. 3, pp. 916–927, March, 2009.

[6] G.-C. Ahn, D.-Y. Chang, M.E. Brown, et al., "A 0.6V 82dB Audio ADC Using Switched-RC Integrators," *ISSCC Dig. Tech. Papers*, pp. 2398–2407, Feb., 2005.

978-1-61284-303-2/11 $26.00 © 2011 IEEE

ISSCC 2011 / February 23, 2011 / 3:15 PM

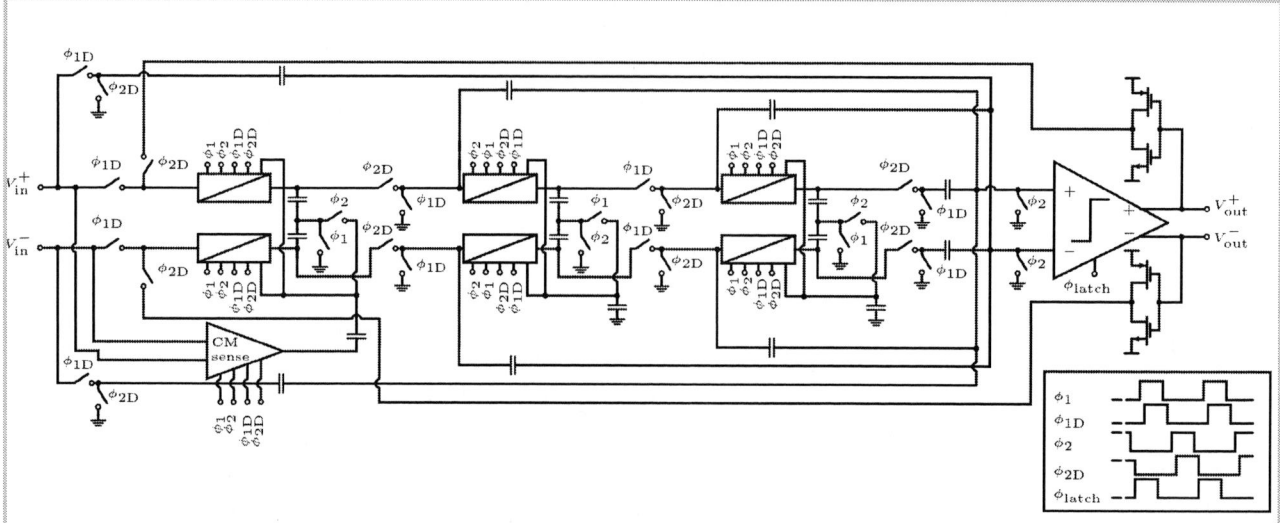

Figure 27.4.1: Pseudo-differential SC ΔΣ modulator with clock timing.

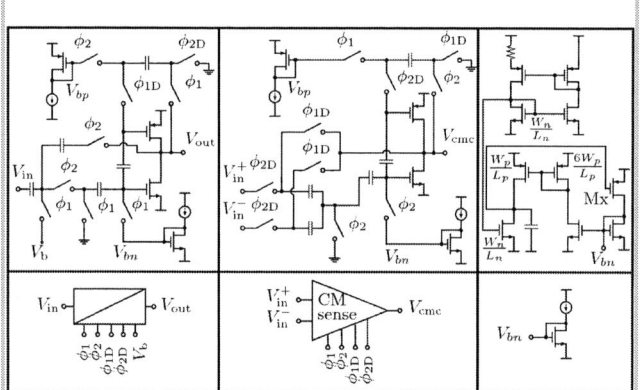

Figure 27.4.2: Integrator, CM sense, and level-shifted biasing circuits.

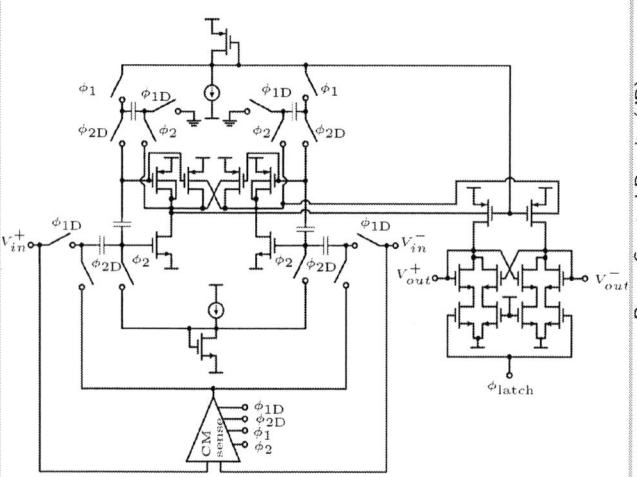

Figure 27.4.4: Latched comparator with CM cancellation.

Figure 27.4.3: Illustration of biasing + offset compensation (φ1) and amplification phase (φ2).

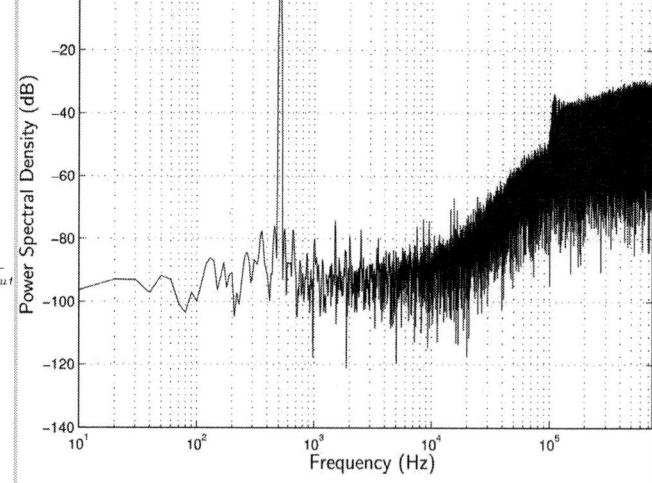

Figure 27.4.5: Measured power spectral density for a 500Hz sinusoidal input with 200 mV$_{pp}$.

27

978-1-61284-303-2/11 $26.00 © 2011 IEEE   477

ISSCC 2011 PAPER CONTINUATIONS

Supply Voltage ($V_{DD} - V_{SS}$)	250 mV	300 mV
Technology	standard CMOS 0.13 μm twin well	
Threshold voltage	$V_{t_n} = 270$ mV, $V_{t_p} = -280$ mV	
DM Input range	200 mV	240 mV
CM Input range	rail to rail	
Total Power consumption	7.5 μW	18.3 μW
Power consumption analog	4.825 μW	9.36 μW
Power consumption digital	2.675 μW	8.94 μW
Sampling Frequency	1.4 MHz	2.8 MHz
Bandwidth	10 kHz	20 kHz
OSR	70	70
Peak SNDR	61 dB	61.4 dB
Peak SNR	64 dB	70 dB
Die Area	0.3375 mm^2	
Temperature Range	20 - 100 °C	0 - 100 °C

Figure 27.4.6: Performance Summary.

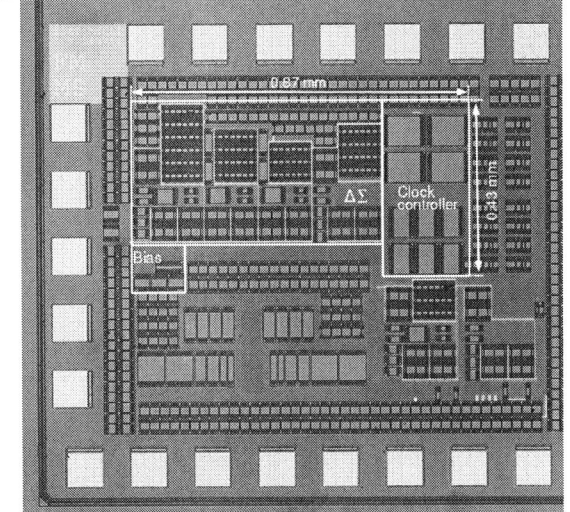

Figure 27.4.7: Chip micrograph.

ISSCC 2011 / SESSION 27 / OVERSAMPLING CONVERTERS / 27.5

27.5 A 84dB SNDR 100kHz Bandwidth Low-Power Single Op-Amp Third-Order ΔΣ Modulator Consuming 140µW

Aldo Pena Perez, Edoardo Bonizzoni, Franco Maloberti

University of Pavia, Pavia, Italy

This third-order ΔΣ modulator [1, 2], suitable for high-resolution low-power sensor systems, consumes 140µW to obtain 84dB SNDR with OSR=16 and 100kHz signal bandwidth. The achieved FoM is 54fJ/conversion-step

The DACs use a single resistive divider to generate 32 differential 5b reference voltages. The proposed scheme totally cancels the error caused by gradient in the resistance values. Resistor sizes and layout of the resistive DAC limit the high-order distortion terms and obtain an SFDR of 96dB at −4dB$_{FS}$ without the need for digital calibration or dynamic element matching (DEM).

The scheme uses a second-order architecture with delays in its integrators to obtain a third-order noise shaping. This is thanks to an auxiliary block that injects $(1-z^{-1})\varepsilon_Q$ (ε_Q is the quantization error) at the input of the second integrator of the second-order delta-sigma modulator, as shown in Fig. 27.5.1. The gain of two of the second integrator, transferred to the quantizer, reduces the swing of the second op-amp at the cost of a higher flash sensitivity. The three zeros of the noise transfer function (NTF) depend on the additional path accuracy. For the target resolution, a precision of 0.2% is enough.

The use of analog feed-forward reduces the swing of the first integrator. A 5b quantizer results in a relatively low quantization error at the output of the second integrator. This provides a good correlation between successive output samples [3]. This scheme exploits the feature to obtain a further reduction of the second integrator output that goes down to $V_2(nT)-V_{DAC3}(nT-T)/2$, approximately ±0.56V$_{ref}$ around the zero reference. To obtain the result, the circuit injects the D/A conversion of $Y'(1/2z^{-1}-1)$ at input of the second integrator, as shown in Fig. 27.5.2. Y' is the output of the flash. A simple processing in the digital domain after the flash provides the quantized output as the sum of the current output and the previous digital output. The relatively small amplitude of the op-amp results in a moderate/low power consumption and limits the number of comparators needed in the flash to 18. Two extra comparators detect possible overload and activate the reset of the two integrators. This technique also avoids possible ringing at the start-up. The multiplexing of a single operational amplifier [4] realizes both integrators.

The schematic of Fig. 27.5.3 illustrates the circuit diagram of the sampled-data modulator. The op-amp serves as the first integrator during phase 1. During the complementary phase, the op-amp and the connected networks operate as the second integrator. The first integrator uses 10C$_U$ and 20C$_U$ for complying the kT/C noise (C$_U$=80fF) requirement. The second integrator uses a unity capacitance for the ½z^{-1} path and 2 unity capacitances for other branches. Note that the path feeding back the analog part of the quantization error needs two analog delays because of the z^{-1} term and the time multiplexing operation. Figure 27.5.3 also shows the clock phases that are used to control switches.

A single resistive divider generates 32 reference voltages as Kelvin division of the V$_{ref}$ to −V$_{ref}$ range. As shown in Fig. 27.5.3, three digitally controlled selection networks provide the differential output of the three DACs shown in Fig. 27.5.2.

The use of calibration or DEM in delta-simga modulators depends on the required accuracy of unity elements which as we know improves with the area. For high-resolution targets, the area of capacitors should become much more than what dictated by the kT/C noise. However, large capacitances require high power. On the contrary, with resistive-based DACs, one can find an optimal unity value of resistance and increase the matching accuracy by just increasing the area without affecting power consumption. In this case, while matching improves, an error caused by gradient occurs. This design exploits a simple property of linear strings of resistors that cancels the gradient error. It can be verified that the differential signal ($V_{N-i}-V_i$) derived from taps of the same

resistive divider made by N resistances with a linear gradient error $R_k=R_u+k\delta_R$

$$V_{diff} = V_{ref}\frac{(N-i)R_u+\delta_R(N-i)(N-i+1)-iR_u+\delta_R i(i+1)}{NR_u+\delta_R N(N+1)}$$

is independent on δ_R. Therefore, it is possible to use a linear string of resistors with large area without caring about the gradient limit and without affecting the power performance.

In this design, the three DACs obtain their differential outputs from a single resistive string. The unity resistors have area equal to 250µm^2, a large but reasonable value. The solution also allows the use of 5b flash converter, an impractical value for digital calibration or DEM techniques. The expected matching is 0.1%, leading to harmonic distortion of more than 86dB without any calibration. Three intermediate taps on the resistive string connected to output pins allow us to measure the gradient in the poly resistor values.

Figure 27.5.4 shows the schematic of the operational amplifier. It is a conventional two-stage amplifier with 65dB of gain and 60µW of total power consumption. It has a simulated bandwidth of 25MHz with 67° of phase margin. The 3.2MHz clock frequency and the multiplexed operation require a slew-rate much higher than what such op-amp can provide. The use of the slew-rate booster shown in the top section of the schematic solves this problem at the cost of an extra current of 10µA (25% of the op-amp bias). The bias current of the differential pair established by V$_{BB}$ is 10µA. The bias of the PMOS loads in the slew-rate booster circuit gives rise to 7µA in saturation. Thus, with balanced inputs, the current in the diode-connected transistors M$_{M5}$ and M$_{M6}$ is zero. Slewing directs all the 10µA in one branch and one of the diode-connected transistors drains 3µA. This current and its mirrored version multiplied by suitable factors augments the slewing currents of the main amplifier. The slew rate goes from 4 to 15.5V/µsec.

The single op-amp third-order modulator is fabricated in a conventional 0.18µm CMOS technology with dual poly and 6 metal layers. Figure 27.5.5 shows the measured 65536-point FFT output spectrum with 1.6kHz input signal at −4dB$_{FS}$. An OSR of 16 gives SNDR=84dB corresponding to an effective number of bits equal to 13.6. The sampling frequency is 3.2MHz and the signal band 100kHz. The measured power consumption is 140µW from a 1.5V supply. The third and the fifth harmonic tones are at −100dB$_{FS}$ and −112dB$_{FS}$, respectively. The achieved SFDR is 96dB. The FoM is 54fJ/conversion-step.

Figure 27.5.6 depicts the measured SNDR versus input amplitude with OSR equal to 16 and 32. The higher oversampling ratio results in an increase in the SNDR peak (88.9dB with −4dB$_{FS}$ input). The dynamic ranges for the OSR of 16 and 32 are 88dB and 94dB, respectively. The higher resolution achieved with OSR=32 does not benefit the noise shaping because the noise floor is slightly higher than expected, likely because of a slight shift of the NTF zeros. The FoM with 50kHz bandwidth is 61fJ/conversion-step.

Figure 27.5.7 shows the chip micrograph with main circuit blocks highlighted. The active area is 1200×410µm^2 (the chip area including pads is 1880×1880µm^2).

Acknowledgment:
The authors thank National Semiconductor for chip fabrication and FIRB, Italian National Program #RBAP06L4S5, for funding support.

References:
[1] L. Bos, G. Vandersteen, J. Ryckaert, P. Rombouts, Y. Rolain, and G. Van der Plas, "A Multirate 3.4-to-6.8mW 85-to-66dB DR GSM/Bluetooth/UMTS Cascade DT ΔΣM in 90nm digital CMOS," *ISSCC Dig. Tech. Papers*, pp. 176-177, Feb., 2009.
[2] Hyunsik Park, Ki Young Nam, D.K. Su, K. Vleugels, and B.A. Wooley, "A 0.7-V 100-dB 870-µW Digital Audio ΣΔ Modulator," *Proc. IEEE Symp. VLSI Circuits*, pp. 178-179, June, 2008.
[3] E. Bonizzoni, A. Perez, F. Maloberti, and M. Garcia-Andrade, "Third-Order ΣΔ Modulator with 61-dB SNR and 6-MHz Bandwidth Consuming 6 mW," *ESSCIRC*, pp. 218-221, Sept., 2008.
[4] J. Koh, Y. Choi, and G. Gomez, "A 66dB DR 1.2V 1.2mW Single-Amplifier Double-Sampling 2nd-Order ΔΣ ADC for WCDMA in 90nm CMOS," *ISSCC Dig. Tech. Papers*, pp. 170-171, Feb., 2005.

978-1-61284-303-2/11 $26.00 © 2011 IEEE

ISSCC 2011 / February 23, 2011 / 3:45 PM

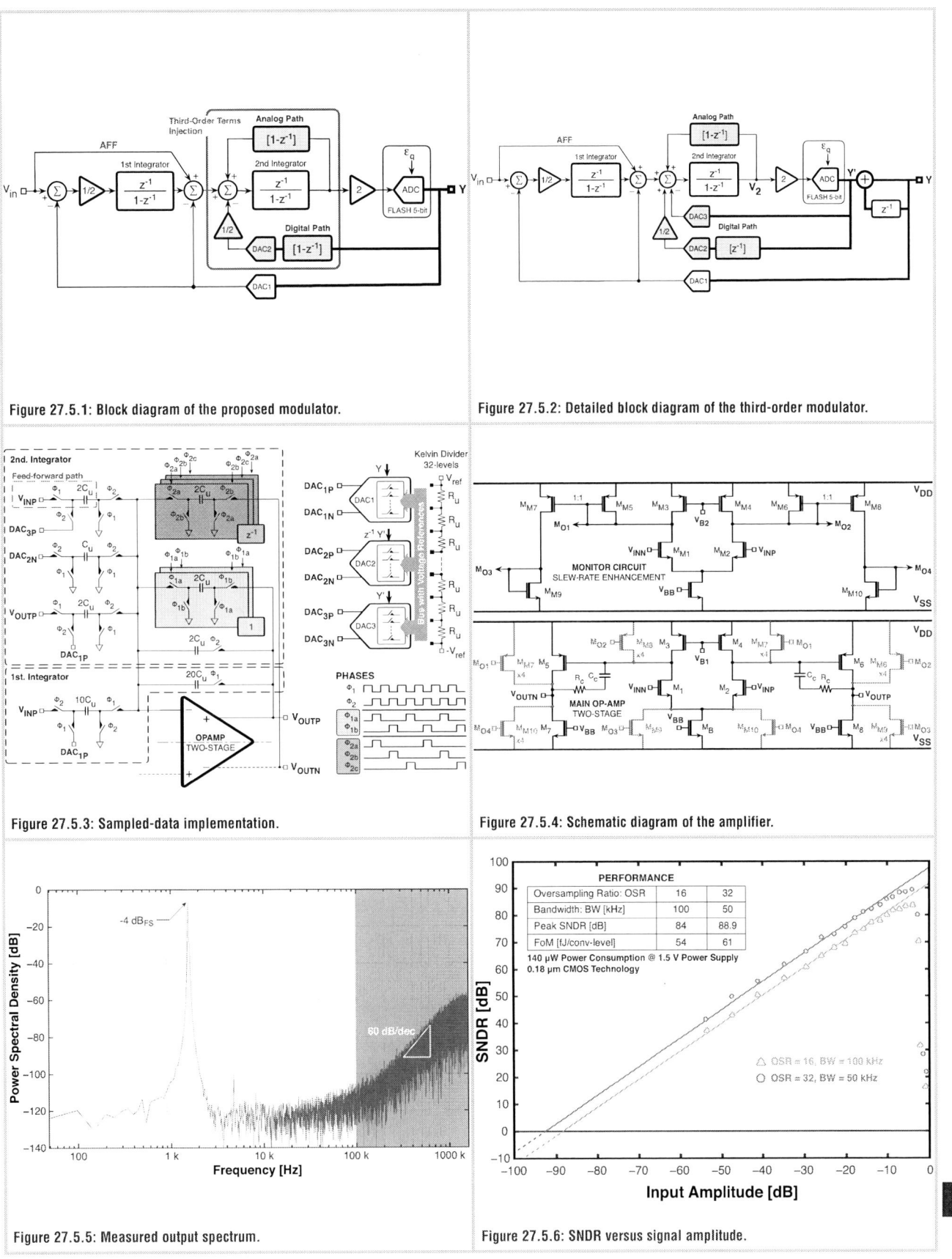

Figure 27.5.1: Block diagram of the proposed modulator.

Figure 27.5.2: Detailed block diagram of the third-order modulator.

Figure 27.5.3: Sampled-data implementation.

Figure 27.5.4: Schematic diagram of the amplifier.

Figure 27.5.5: Measured output spectrum.

Figure 27.5.6: SNDR versus signal amplitude.

PERFORMANCE		
Oversampling Ratio: OSR	16	32
Bandwidth: BW [kHz]	100	50
Peak SNDR [dB]	84	88.9
FoM [fJ/conv-level]	54	61

140 µW Power Consumption @ 1.5 V Power Supply
0.18 µm CMOS Technology

27

978-1-61284-303-2/11 $26.00 © 2011 IEEE

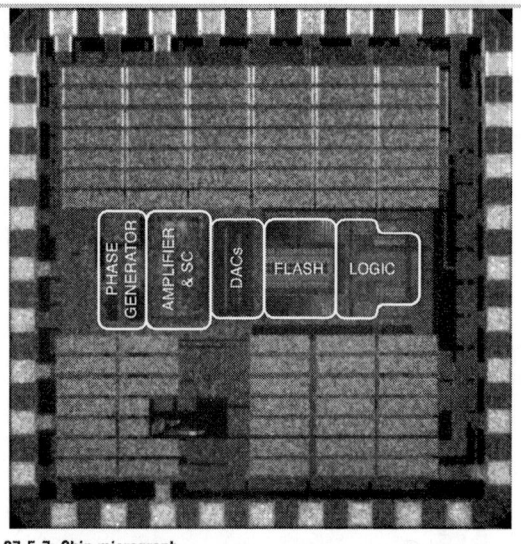

Figure 27.5.7: Chip micrograph.

ISSCC 2011 / SESSION 27 / OVERSAMPLING CONVERTERS / 27.6

27.6 A 1.7mW 11b 1-1-1 MASH ΔΣ Time-to-Digital Converter

Ying Cao[1,2], Paul Leroux[1,3], Wouter De Cock[2], Michiel Steyaert[1]

[1]KU Leuven, Leuven, Belgium
[2]SCK-CEN, Mol, Belgium
[3]KH Kempen, Geel, Belgium

Recently, high-resolution TDCs have gained more and more popularity due to their increasing implementation in digital PLLs, ADCs, jitter measurement and time-of-flight measurement units. Similar to ADCs, existing architectures of TDCs can be divided into several categories: flash TDCs [1, 3], pipeline TDCs [2], and SAR TDCs [4]. The highest achievable time resolution of a TDC is mainly limited by the CMOS gate delay. In order to achieve sub-gate-delay resolution, the Vernier method is commonly used. However, the mismatch problem caused by process variation limits its effectiveness, and the same holds for the time amplification method. The gated-ring-oscillator (GRO) method [5] is introduced to achieve sub-ps time resolution, but it still requires an equivalent CMOS gate delay as low as 6ps. Upcoming applications in 4th-generation nuclear reactors, space, and high-energy physics such as the large Hadron collider (LHC), require the TDC to achieve a high time resolution in harsh environments with high temperature and radiation, where the threshold voltage, transconductance, and delay of a transistor undergo dramatic changes. In these cases, the high accuracy and robustness of the TDC need to be inherent to the design rather than by employing a fast CMOS technology.

The noise-shaping technique is widely used in design of ADCs, which can improve the effective resolution of a coarse quantizer and does not require precisely matched analog components. The same concept can principally be ported to the design of TDCs. The commonly used ΔΣ structure which consists of integrators can not be directly adopted by a TDC, due to the difficulty of realizing a time integrator. In this work, however, the first MASH ΔΣ TDC is reported. A first-order noise-shaping TDC has been built in an error-feedback manner, as shown in Fig. 27.6.1. The error-feedback structure is impractical for a ΔΣ ADC since its performance is limited by the inaccuracy of the analog subtractors. However, time subtraction can be easily realized by nand/nor operation, and moreover, the error-feedback structure does not require an explicit integrator in the loop.

Directly preserving the quantization error in the time regime is still impossible with current technologies: the time information has to be converted into another intermediate physical quantity such as voltage or charge. A relaxation oscillator can generate a clock by alternatively charging and discharging two capacitors. The phase of the clock corresponds to the voltage on each capacitor. The time can be measured by enabling the oscillator during the measurement interval and counting the number of periods of the generated clock. When the oscillation stops, the phase of the clock, which refers to the quantization error, can be stored on the capacitor as a residue voltage. Principally, it is similar with the GRO [5], but the skew error caused by charge redistribution during the start and stop of the oscillator is negligible here due to the large capacitance. In a GRO-based TDC, it can significantly deteriorate the performance.

By cascading several error-feedback stages, a higher-order noise shaping can be obtained. A third-order MASH TDC is shown in Fig. 27.6.1. All stages have the same architecture and are followed by digital processing blocks. It is algebraically equivalent to a conventional 1-1-1 MASH ΔΣM. The MASH TDC works as follows: In each stage, the time signal *tin* controls a current to charge two capacitors. When it is active high, one of the two capacitors starts charging. For instance, *vinp1* starts rising when *vinn1* stays at *vlow*, as illustrated in Fig. 27.6.2. When *vinp1* reaches *vhigh*, the comparator output becomes '1'. This reverses the state of the SR-latch, and triggers the oscillation. After the stop signal arrives, the phase of the clock is preserved; the counter is first read out and then reset to 0. Due to the fact that the counter is only driven by the rising edge of the clock, a time which equals the quantization error *q[1]* will be subtracted from the next input. The overall quantization error can then be described as *q[2]-q[1]*. The time signal which feeds into the following stage is generated by taking the first rising edge of the counting clock as a start signal, and keeping the same falling edge as the TDC's input.

The frequency of the relaxation oscillator can be expressed as *IREF/(VREF·2C)*. By correlating *VREF* and *IREF* as *VREF = IREF·R*, its frequency becomes only dependant on passive components, that is, *1/(2·RC)*. Thus, it exhibits inherent PVT-variation tolerance and the matching between stages is better than for its MASH ADC counterparts.

Figure 27.6.3 shows the schematics in the 1-1-1 MASH TDC. The main origin of noise in the TDC is the comparator delay. When the oscillator is disabled, the comparator state may not be perfectly preserved due to the hysteresis. This will introduce extra noise into the preserved quantization error, and it can only be suppressed by oversampling. A four-stage threshold detection comparator is adopted in this design. Each of the first three stages has a gain of 10dB and draws 40µA from a 1.2V supply. The last stage provides a higher gain of 20dB with a current draw of 80µA. This comparator has a delay of 0.8ns. Two comparators in each stage are turned off alternately to save power, when its connected capacitor is not being charged. Charge injection when turning on/off input switches and kT/C noise also contribute to the overall noise level. In the conversion-to-time noise the local noise voltage is divided by the charging slope of the capacitor. Thus, a larger slope is mostly desirable. In this design, *IREF*=50µA, *VREF*=650mV, and *C*=0.64pF, which gives a charging slope of 80µV/ps.

A PWM signal, modulated by a sine wave, is employed to evaluate the performance of the TDC, and the bandwidth is set to 100kHz. An 18kHz −3dBFS signal is used for large signal measurements. The full-scale input range is 100ns, when the OSR is 25, and the carrier frequency is 5MHz. The output spectrum and waveform are shown in Fig. 27.6.4a. It shows an SNDR of 60.3dB. Figure 27.6.4b shows the results of a small-signal measurement, when the full scale input range is 10ns, and a 22kHz −40dBFS signal is applied. The carrier frequency is 50MHz and the OSR is 250. An SNDR of 28dB and a resolution of 5.6ps are achieved. The system is also compatible with other OSR values, such as 50 and 100. Figure 27.6.5 shows the DR of the TDC, which is 68dB. Note that, there is no apparent drop on SNDR when the input level is close to full scale, since in a TDC system, the input dynamic range is only limited by the depth of the counter, which can be easily extended to avoid any overloading of the system.

A radiation assessment with a dose rate of 30kGy/h has been performed, as shown in Fig. 27.6.6, proving the TDC's robustness. The current consumption is almost not affected and even after an extremely high radiation dose of 3.4MGy, the ENOB drops only by 1 bit and, for an OSR of 250, a 10.5ps time resolution is still achieved.

This first reported 1-1-1 MASH ΔΣ TDC is implemented in 0.13µm CMOS. It consumes only 1.7mW from a 1.2V supply and achieves an ENOB of 11b. The SNDR is mainly limited by the comparator delay. It can principally be reduced by adding more power to the comparator or applying calibration. The developed MASH TDC architecture gives full flexibility to design high-resolution low-power TDCs even for extreme environments.

References:
[1] R. B. Staszewski, S. Vemulapalli, P. Vallur, et al., "1.3 V 20 ps Time-to-Digital Converter for Frequency Synthesis in 90-nm CMOS," *IEEE Trans. on Circuits and Systems II*, vol. 53, no. 3, pp. 220-224, Mar., 2006.
[2] M. Lee, and A. Abidi, "A 9 b, 1.25 ps Resolution Coarse-Fine Time-to-Digital Converter in 90 nm CMOS that Amplifies a Time Residue," *IEEE J. Solid-State Circuits*, vol. 43, no. 4, pp. 769-777, Apr., 2008.
[3] S. Henzler, S. Koeppe, W. Kamp, et al., "90nm 4.7ps-Resolution 0.7-LSB Single-Shot Precision and 19pJ-per-Shot Local Passive Interpolation Time-to-Digital Converter with On-Chip Characterization," *ISSCC Dig. Tech. Papers*, pp. 548-549, Feb., 2008.
[4] A. Mäntyniemi, T. Rahkonen, and J. Kostamovaara, "A CMOS Time-to-Digital Converter (TDC) Based On a Cyclic Time Domain Successive Approximation Interpolation Method," *IEEE J. Solid-State Circuits*, vol. 44, no. 11, pp. 3067-3078, Nov., 2009.
[5] M. Z. Straayer and M. H. Perrott, "A Multi-Path Gated Ring Oscillator TDC With First-Order Noise Shaping," *IEEE J. Solid-State Circuits*, vol. 44, no. 4, pp. 1089-1098, Apr., 2009.

978-1-61284-303-2/11 $26.00 © 2011 IEEE

ISSCC 2011 / February 23, 2011 / 4:15 PM

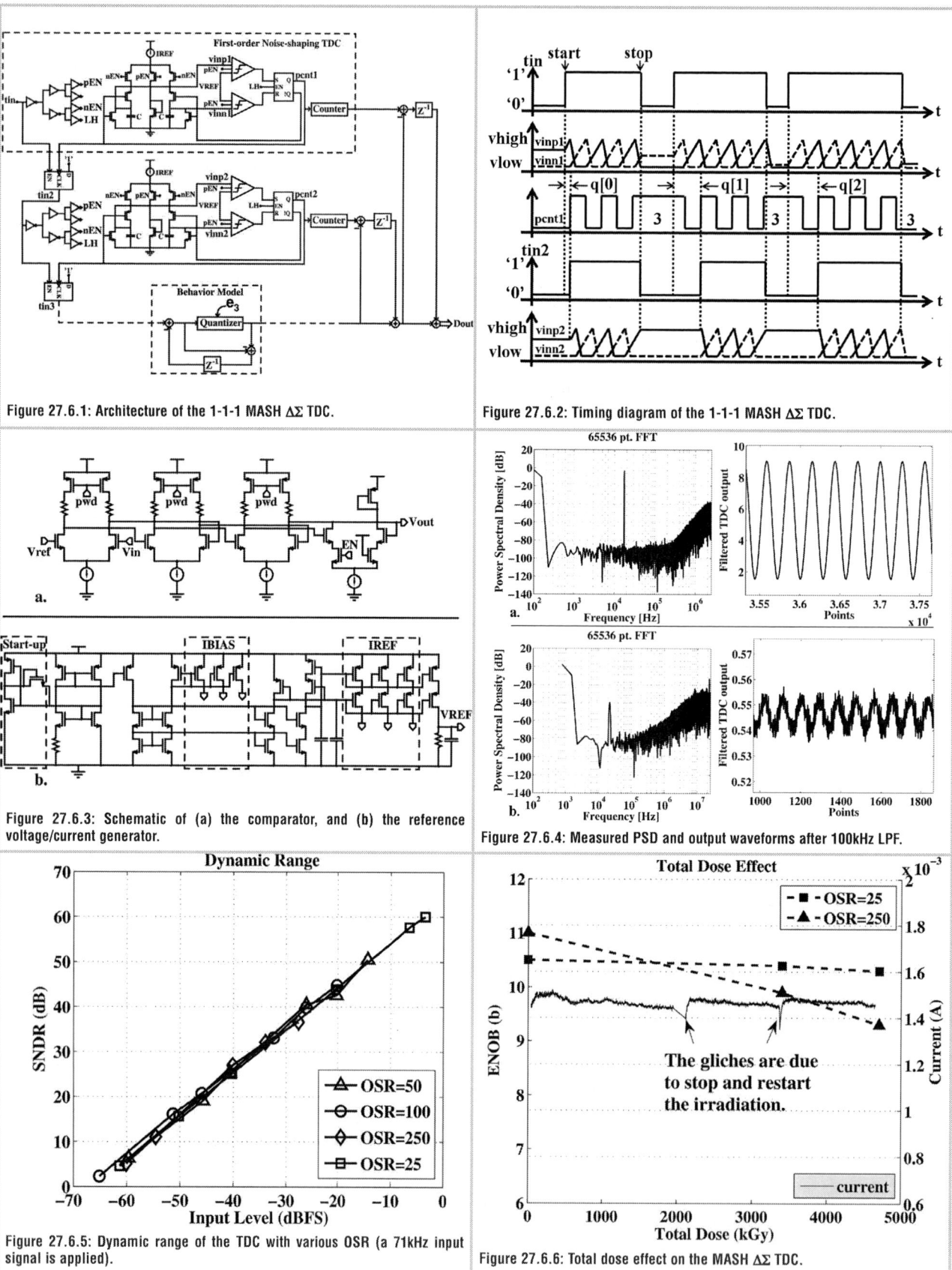

Figure 27.6.1: Architecture of the 1-1-1 MASH ΔΣ TDC.

Figure 27.6.2: Timing diagram of the 1-1-1 MASH ΔΣ TDC.

Figure 27.6.3: Schematic of (a) the comparator, and (b) the reference voltage/current generator.

Figure 27.6.4: Measured PSD and output waveforms after 100kHz LPF.

Figure 27.6.5: Dynamic range of the TDC with various OSR (a 71kHz input signal is applied).

Figure 27.6.6: Total dose effect on the MASH ΔΣ TDC.

27

Figure 27.6.7: Die micrograph. The core area is 0.41x0.27mm².

ISSCC 2011 / SESSION 27 / OVERSAMPLING CONVERTERS / 27.7

27.7 A 120dB-SNR 100dB-THD+N 21.5mW/Channel Multibit CT ΔΣ DAC

Abhishek Bandyopadhyay, Michael Determan, Sejun Kim, Khiem Nguyen

Analog Devices, Wilmington, MA

Automotive and consumer multi-channel 24b audio systems have demanded low-cost digital-to-analog converters (DACs) which offer wide dynamic range, high linearity, small die size, and low power consumption such that the system can be housed in a small low-cost plastic package. Several 120dB SNR audio ΔΣ DACs have been reported using either switched-capacitor or continuous-time techniques [1-3]. This paper presents a continuous time (CT) area- optimized multibit DAC which achieves 120dB SNR and 100dB THD+N at 21.5mW/channel. This performance is achieved by using a new 3-level rotational data shuffling scheme which achieves small area and low digital activity at low signal level, and by applying low-power low-noise analog techniques.

Assuming a thermal resistance of 42.3°C/W (typical low-cost LQFP) and a maximum junction temperature of 150°C, a 16 channel DAC would need to have a power consumption of less than 36.9mW/channel for an operational temperature of 125°C. This limit is more stringent than that of any of the previously reported high-end DACs [1-3]. High performance and low power are enabled in the present design by using both architectural- and circuit-level considerations. The first architecture-level consideration is choosing a CT approach so that the analog section of the DAC is less prone to pick up on-chip digital noise. Second, a 2nd-order 8b modulator is used to reduce clock jitter requirements and to achieve low out-of-band noise. Third, a 3-level rotational dynamic element matching (DEM) scheme is introduced which achieves better SNR performance and lower power consumption than that of previously presented work [4]. Fourth, a signed-magnitude approach is taken at the boundary between the digital and the analog domains such that combined with the 3-level rotational DEM technique would ensure the digital activities at the boundary are proportional to the input signal amplitude which reduces digital coupling effects at low input levels. This approach would also reduce the operational power. At the circuit level the key considerations are the power, noise, and area tradeoffs of the DAC output stage, design of the 3-level current sources (I-DACs) for reduced element-to-element mismatch, and of low-noise voltage reference generators.

Figure 27.7.1 shows the block diagram of one channel of the DAC. Digital audio is received at the input by a serial audio interface capable of receiving all standard audio formats. The signal is then passed to a 4× interpolation filter that uses canonical signed-digit arithmetic for low power consumption. The filter also performs volume control and de-emphasis functions. The filter engine output is then further interpolated to 128× via a linear interpolation, and is then presented to the 2nd-order 8b modulator. For area and power considerations a noise-shaped segmentation technique is used, to split the 8b modulator data into a 4b word with a weight of 16×, and a pair of 3b words with weights of 4× and 1×, respectively. The gain errors between the weighted segments are 1st-order noise shaped [4]. After a sign-magnitude thermometer code conversion individual rotational shuffling is applied to each of the segmented data. This data is then presented to the drivers for the respective I-DACs. The output current is converted to voltage using an output stage consisting of an inter-symbol interference (ISI) free switching scheme and a low-quiescent-current low-offset class AB operational amplifier. The currents are generated from low-noise reference generators. A clock generator generates the required signals for the ISI-free operation of the output stage [4].

Figure 27.7.2 shows the operation of the 3-level rotational data shuffler. For simplicity, the signed-magnitude data of just one of the 3b shufflers is shown here. The two pointers (positive and negative) indicate the first element to be used for each set of data and they are advanced separately by their associated data only.

It can be seen that since the pointers advance in the same direction and wrap around, over time, all elements will be used equally thereby ensuring that the long term average errors contributed by each element approach zero. For the first cycle, let's assume that a+1 comes and this would advance just the positive pointer by 1. Similarly, a negative data (−2, cycle 3) would advance the negative pointer by 2. This would go on until the two pointers are pointing to the same location (cycle 7). At this point, all the elements have been used equally and the cumulative contributed mismatch error is zero. To eliminate the low-frequency pseudo-periodic behavior of the outputs, usually associated with rotational shufflers, both of these pointers can either be set to start from a random position or be advanced by a fixed value. Once the positions have been changed, the pointers are advanced by their respective data values and normal operation ensues. Both approaches are implemented on silicon and they show the same 120dB performance, since by design, the analog operational amplifier is made the dominant noise source in the present implementation. Figure 27.7.3 shows a simple hardware implementation of the new 3-level rotational DEM logic shuffler using a single barrel shifter. The input of the DEM block is the signed binary data. The DEM logic block outputs the positive or negative pointers, which indicate the current location of the sequence of elements to be used.

The challenges on the circuit level are designing a low-offset low-power low-noise class AB amplifier, and the design of low-noise current sources and bias generators. For power consumption and performance tradeoff a two-stage folded-cascode input stage with a feed-forward class AB output stage is used, as shown in Fig. 27.7.4. Typically, in a modern process node the low-voltage- tolerant devices usually have a better 1/f noise performance while the large- voltage-tolerant devices have a better thermal noise performance. Over-voltage protected low-voltage-tolerant devices are used as the input pair of the opamp to reduce overall 1/f noise without any area impact. Layout design of the input pair ensures low offset. The same technique is used for lowering the noise of the bias generators and the voltage references. The current elements are cascoded and are laid out in a common-centroid fashion, and this along with low offset of the opamp ensures that the mismatch of the elements are within 0.5%.

The 1.35mm²/channel DAC is fabricated in 0.35μm DPQM CMOS due to 3Vrms differential output swing requirements. The DAC achieves 120dB SNR, and 100dB THD+N with a load of 3kΩ, as shown in Fig. 27.7.5. Due to the nature of the 3-level unit element architecture the noise floor is slightly modulated by the signal amplitude. The third harmonic is dominated by self-heating effects of the feedback resistors of the output stage. The measured performance summary table is shown in Fig. 27.7.6a, and Fig. 27.7.6b compares this architecture with recently published chips reporting comparable SNR. The chip has a power consumption of 21.5mW/channel and consumes 3.5 times less power as compared to previously reported high-end DACs [1-3]. The chip micrograph is shown in Fig 27.7.7.

Acknowledgements:
The authors would like to thank K. Sweetland, R. DeRosa, N. Caporale, G. Shull, and K. Tofalo.

References:
[1] I. Fujimori, A. Nogi, and T. Sugimoto, "A Multi-Bit Audio DAC with 120dB Dynamic Range," *ISSCC Dig. Tech. Papers*, pp. 152-153, Feb. 1999.
[2] X.M. Gong, E. Gaalaas, M. Alexander, et al., "A 120dB Multi-bit SC Audio DAC with second order noise shaping," *ISSCC Dig. Tech. Papers*, pp. 344-345, Feb. 2000.
[3] E. van Tuijl, J. van den Homberg, D. Reefman, et al., "A 128fs Multi-bit ΣΔ CMOS Audio DAC with Real-Time DEM and 115dB SFDR," *ISSCC Dig. Tech. Papers*, pp. 368-369, Feb. 2004.
[4] K. Nguyen, A. Bandyopandhyay, B. Adams, et al., "A 108dB SNR 1.1mW oversampling audio DAC with a three-level DEM technique," *ISSCC Dig. Tech. Papers*, pp. 488-489, Feb. 2008.

ISSCC 2011 / February 23, 2011 / 4:45 PM

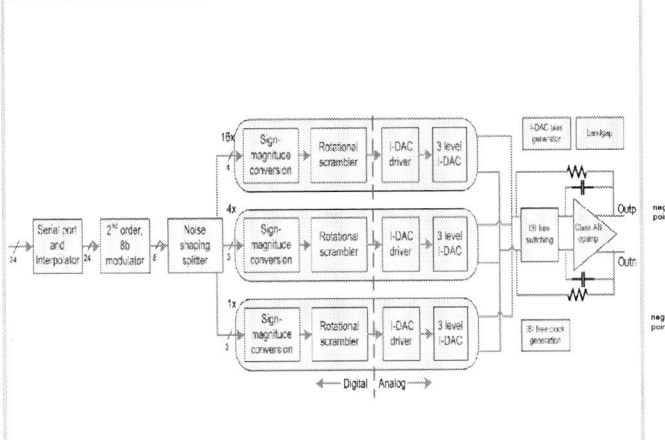

Figure 27.7.1: System-level architecture.

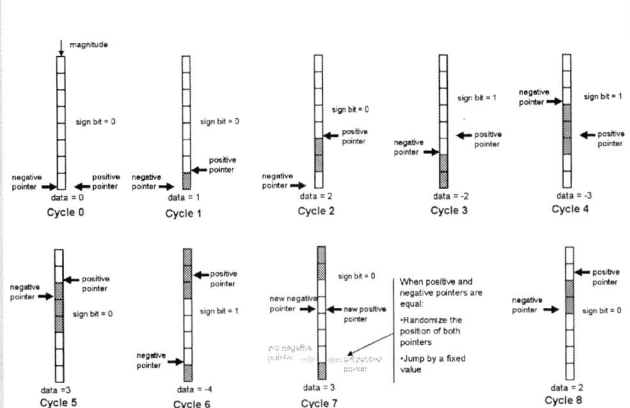

Figure 27.7.2: Three-level rotational data scrambler scheme.

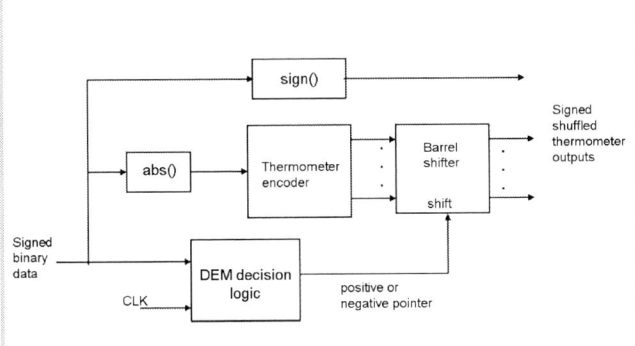

Figure 27.7.3: A simple hardware implementation of the new three-level rotational DEM.

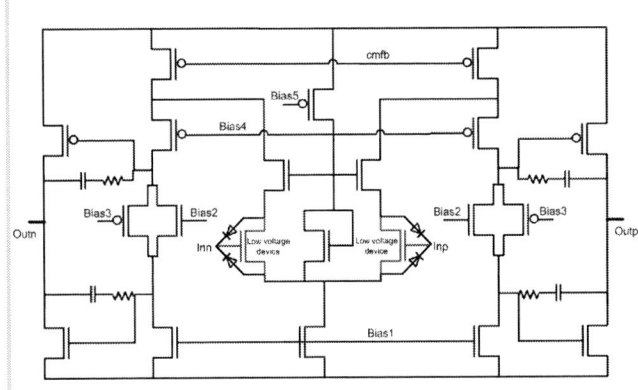

Figure 27.7.4: A simplified schematic of a low-noise two-stage class-AB amplifier.

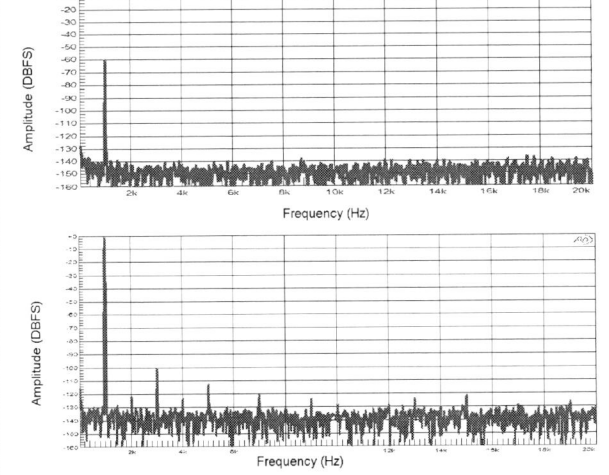

Figure 27.7.5: Measured 8K point FFT plots for 1kHz tone with −60dBFS (top) and −1dBFS (bottom) input signals.

Process	0.35um, DPQM
Supply	AVDD: 3.13V – 5.5V; DVDD: 2.25V-3.63V
Area/channel	1.35 mm²
OSR	128
Power consumption	21.5mW = 17.5 mW(analog)+4 mW(digital)
SNR	120dB (A- Weighted), 117.5dB (unweighted)
THD+N	-100dB
Clock frequency	6.144MHz
Output Swing (Differential)	3Vrms (AVDD=5V), 2Vrms (AVDD=3.3V)

(a)

Design	Fujimori[1]	Gong[2]	Tuijl[3]	This work
Year	ISSCC,1999	ISSCC,2000	ISSCC,2004	ISSCC,2011
Technology	0.5um, DPTM	0.35um, DPTM	0.18um	0.35um, DPQM
Technique	SC	SC	CT	CT
Supply(Analog, digital)	5V, 3.3V	5V, 3.3V	3.3V, 1.8V	5V, 2.25V
SNR(A-wt)(dB)	120	120	119(un-weighted)	120
Power/channel (mW)	145	100	75	21.5

(b)

Figure 27.7.6: (a) DAC performance Summary; and (b) comparison between recently reported audio DACs (BW=20kHz and SNR>117dB).

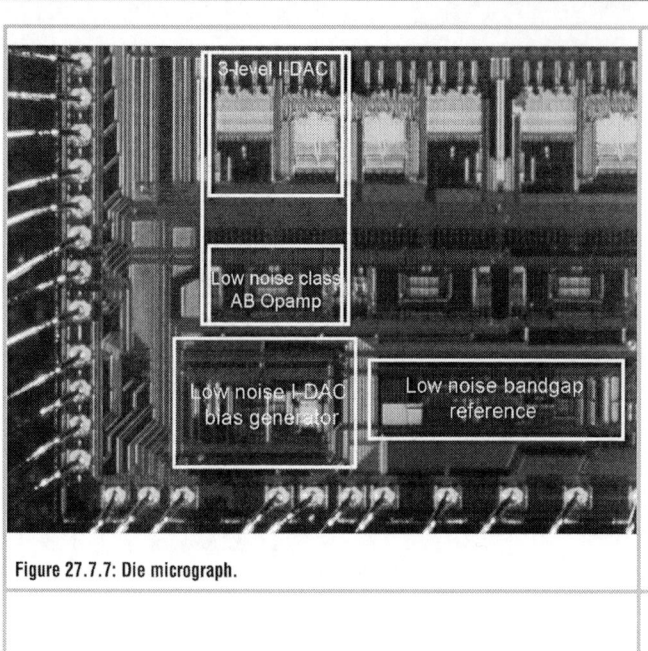

Figure 27.7.7: Die micrograph.

ISSCC 2011 / SESSION 27 / OVERSAMPLING CONVERTERS / 27.8

27.8 A 108dB-DR 120dB-THD and 0.5V$_{rms}$ Output Audio DAC with Inter-Symbol-Interference-Shaping Algorithm in 45nm CMOS

Lars Risbo[1], Rahmi Hezar[2], Burak Kelleci[2], Halil Kiper[2], Mounir Fares[2]

[1]Texas Instruments, Copenhagen, Denmark
[2]Texas Instruments, Dallas, TX

The current trend in high-performance audio DACs is to use fine-resolution quantization to reduce the out-of-band noise (OBN), reduce jitter sensitivity, and simplify analog filtering. Recent techniques achieve this goal by using a mix of DAC elements with different weights, e.g., segmenting [1] or cascading [2]. Unlike 1b modulation, the multi-level DACs need mismatch shaping algorithms to compensate for the typical 0.1 to 1% on-die mismatch. In addition to the element mismatch, dynamic error sources such as asymmetrical switching, clock skew, and parasitic memory are major hindrances to achieve distortion and dynamic range targets. The resulting dependence of present symbol error to the past symbol is referred to as inter-symbol-interference (ISI), and is a function of the switching activity of all the individual DAC elements. Unfortunately, the popular mismatch-shaping algorithms (e.g., rotation-DWA) are addressing only the static mismatch problem. They typically increase the switching activity thus amplify ISI errors. Furthermore, the error comes often in the form of spurious tones with signal-dependent frequency (FM modulation [3]) that ruins the low-amplitude performance (harmonics for a −60dB signal) which is critical for the perceived sound quality. A common remedy is to add a digital DC offset to shift the tones out of band. However, this merely moves the problematic amplitude region, and does not solve the problem. Moreover, ISI errors often limit the large signal THD as a result of a strong signal-dependent modulation of the element transition rate.

One popular analog solution to reduce ISI errors is to use return-to-zero (RTZ) which eliminates the dependence on previous symbol. However, it results in increased current consumption, more high-frequency components, and higher sensitivity to circuit timing/jitter. Another analog solution [1] is to use a track-and-hold circuit to deglitch the waveform. However, such sampling circuits are sensitive to aliasing of high-frequency noise. Other proposed techniques [4] use modified digital algorithms with various trade-offs between mismatch shaping and the element transition rate. The pulse-width modulation (PWM) DAC method in [2] forces a fixed transition rate of the DAC elements, resulting in a significantly reduced sensitivity to ISI. However, this comes at the expense of a high clock rate which may not be available in the system.

We propose a new digital algorithm that can shape element mismatch and ISI errors simultaneously. This method fully shapes ISI and the mismatch errors outside audio band and eliminates the need for layout critical and non-automated analog design methods that often require multiple design iterations and are hard to migrate over processes. This is enabled by using digital processing circuits that are simple and predictable with modern EDA tools and come at low cost in power and area using deep-submicron CMOS. In addition, the proposed solution works at regular DSM clock rate and does not require another PLL and clock management circuit.

The new ISI-shaping algorithm [5] is shown in Fig. 27.8.1 and consists of a traditional mismatch-shaping loop (top box) using a vectorized mismatch loop filter $H_{MLF}(z)$ and a vector quantizer (VQ) that is sorting and selecting the elements with the highest input values. This circuit is equivalent to the rotation scheme when the loop filter is a first-order integrator and there is no dither. However, the simple rotation is a strong source of tones. Vectorized dithering or other randomization is needed to break up these tones. In the proposed algorithm, concurrent mismatch and ISI shaping is achieved by adding another loop (bottom box). The independent ISI loop filter $H_{ILF}(z)$ accumulates the deviation of actual transition density signal $St(n)$ from a fixed target rate R_{tran}. The ISI loop adds a vectorized contribution, Sd, to the VQ input. When making the element selection, the VQ takes both the time accumulated relative static usage history as well as the dynamic transition density history into account. The Sd signal is formed as either plus or minus the loop filter output, Se, depending on the previous selec-

tion state, Ss. For example, a positive Se indicates that more transitions are needed. This can be achieved by increasing the likelihood of selection (positive Sd) for elements that were not selected in the prior cycle and reduce the likelihood of the negative Sd. The ISI-shaping loop will try to keep the transition rate (i.e., the mean of St_n) equal to R_{tran} while highpass shaping any AC deviations due to a high loop filter gain at low frequencies. Since the ISI error is closely correlated with the St_n, this combined loop suppresses both noise and distortion from ISI errors while shaping static-mismatch errors with the traditional loop. Different loop-filter characteristics and weighting of the mismatch and ISI loops can be used depending on which error source dominates in the analog domain. Several variations of the ISI shaper are possible [5] allowing to trade off between gate count and performance. The achieved constant transition rate eliminates the large signal distortion due to ISI. The AC highpass-shaping addresses the noise and low-level tones. The choice of R_{tran} is a balance between a good high amplitude performance which requires a low value that can be kept constant to a high modulation depth and a good AC shaping at low amplitude that requires a high rate which secures a wider loop bandwidth. An R_{tran} value of 1/8 together with 2nd-order mismatch and 1st-order ISI shaping are chosen for the current design.

The signal chain, shown in Fig. 27.8.2, is implemented using the cascaded-modulator [2] with a primary 33-level and a secondary 33-level bank of current-steering elements in a 10:1 current ratio which results in 330 quantization levels. Both the primary and secondary paths deploy the new ISI shaper of Fig. 27.8.1. A differential amplifier (I2V) sums the currents of all the 64 sources and converts them to a voltage output. The I2V speed requirements are relaxed thanks to the reduced OBN achieved by cascading. The schematic for the I2V is shown in Fig. 27.8.3. It is a classical class-AB design with floating battery configuration, optimized for low noise with source degeneration and without chopping. No feedback capacitor is needed again thanks to the low OBN. The feedback resistors are Nwell type without trimming. The supply voltage is 1.45V.

Figures 27.8.4 (a) and (b) compare the large signal response of rotation and ISI-shaping algorithms. The inner plots in (b) is the current-mode output of the DAC. It shows the perfect linearity achieved by a constant switching rate. The voltage-mode output linearity is limited by the I2V output-stage and the matching of the resistors for large signals. This can further be improved by trimming and increased headroom if necessary. The small-signal tone test (voltage mode) in (c) displays the highly nonlinear behavior of DWA. In (d), this problem is completely resolved by the ISI-shaper's highpass-shaping of AC transitions.

Figure 27.8.5 shows amplitude sweeps versus THD. We include 2nd-order DWA mode by disabling the ISI-loop in Fig. 27.8.1 and use a 2nd-order $H_{MLF}(z)$. ISI algorithm provides excellent linearity. THD is limited by I2V amplifier only close to the full scale. Rotation causes dips at low swings due to tones and large-signal THD hits a ceiling. 2nd-order DWA is only slightly better. Fig. 27.8.6 summarizes the performance summary.

Acknowledgements:
The authors would like to thank Chen Kung, Viral Parikh, Anker Josefsen and Srinadh Madhavapeddi for their valuable help and assistance.

References:
[1] K. Nguyen, A. Bandyopadhyay, R. Adams, K. Sweetland, and P. Baginski, "A 108dB SNR, 1.1mW Oversampling Audio DAC with A Three-level DEM Technique," *IEEE J. Solid-State Circuits*, vol. 42, no.12, pp. 2592-2600, Dec., 2008.
[2] R. Hezar, L. Risbo, H. Kiper, M. Fares, B. Haroun, G. Burra, and G. Gomez "A 110dB SNR and 0.5mW Current-Steering Audio DAC Implemented in 45nm CMOS," *ISSCC Dig. Tech. Papers*, pp 304-305, Feb., 2010.
[3] L. Risbo, "Low-level distortion, tones, tone-extraction & FM modulation in D/A converters", Workshop W9 - The Hows and Whys of Sigma Delta Converters, *Audio Engineering Society 122nd Convention*, Vienna, May, 2007.
[4] T. Shui, R. Schreier, and F. Hudson, "Mismatch Shaping for a Current-Mode Multi-bit Delta-Sigma DAC, " *IEEE J. Solid-State Circuits*, vol. 34, no. 3, March, 1999.
[5] L. Risbo, R. Hezar, B. Kelleci, and A.B. Josefsen, "Shaping Inter-symbol Interference in Sigma Delta Converter," filed to U.S. Patent Office April 28, 2010, application number 12/769,629.

978-1-61284-303-2/11 $26.00 © 2011 IEEE

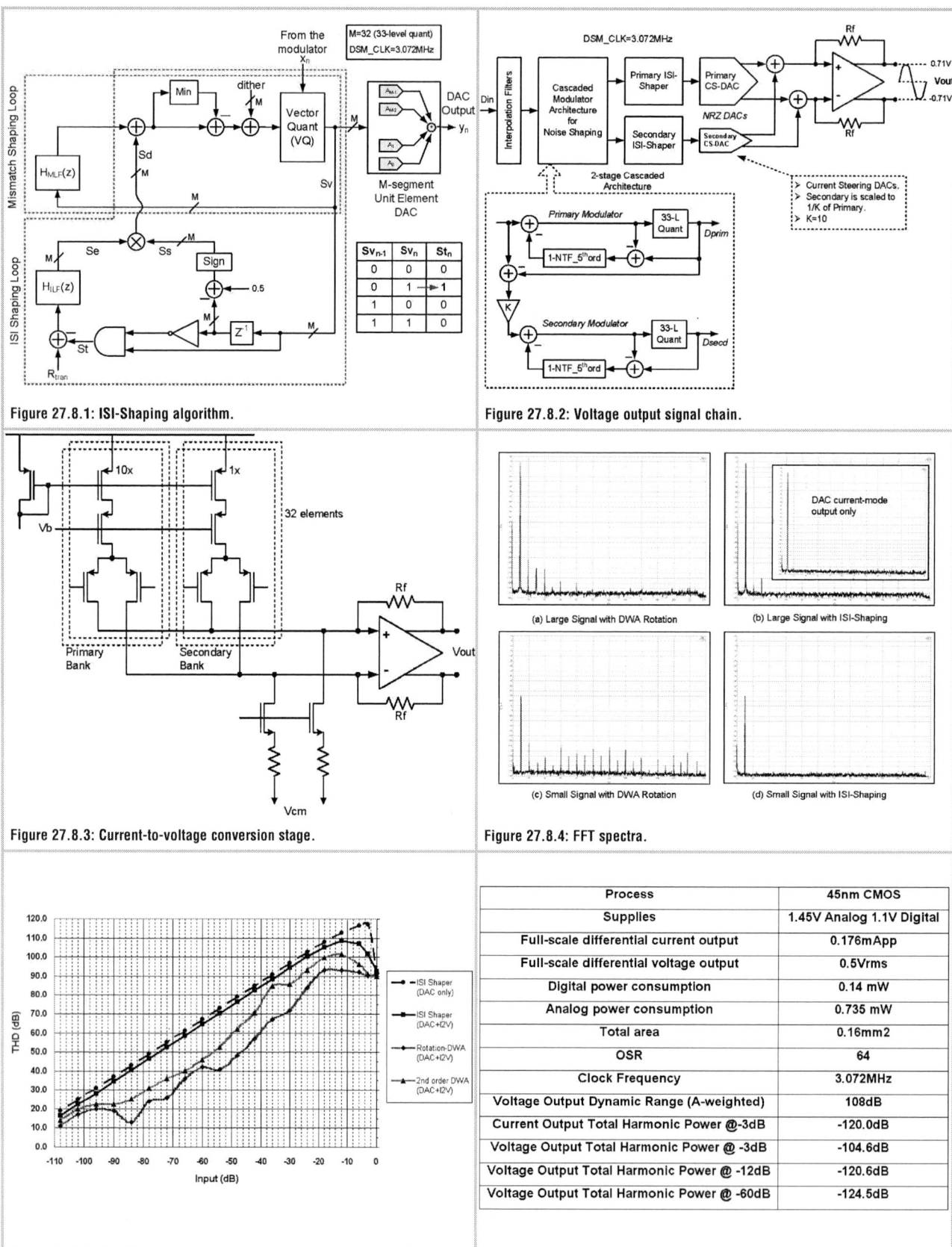

Figure 27.8.1: ISI-Shaping algorithm.

Figure 27.8.2: Voltage output signal chain.

Figure 27.8.3: Current-to-voltage conversion stage.

Figure 27.8.4: FFT spectra.

(a) Large Signal with DWA Rotation

(b) Large Signal with ISI-Shaping

(c) Small Signal with DWA Rotation

(d) Small Signal with ISI-Shaping

Figure 27.8.5: Amplitude sweeps.

Process	45nm CMOS
Supplies	1.45V Analog 1.1V Digital
Full-scale differential current output	0.176mApp
Full-scale differential voltage output	0.5Vrms
Digital power consumption	0.14 mW
Analog power consumption	0.735 mW
Total area	0.16mm2
OSR	64
Clock Frequency	3.072MHz
Voltage Output Dynamic Range (A-weighted)	108dB
Current Output Total Harmonic Power @-3dB	-120.0dB
Voltage Output Total Harmonic Power @ -3dB	-104.6dB
Voltage Output Total Harmonic Power @ -12dB	-120.6dB
Voltage Output Total Harmonic Power @ -60dB	-124.5dB

Figure 27.8.6: Performance summary per channel.

ISSCC 2011 / SESSION 28 / DRAM & HIGH-SPEED I/O / OVERVIEW

Session 28 Overview / Memory

DRAM & High-Speed I/O

Session Chair: *Yasuhiro Takai, Elpida,Sagamihara, Japan*

Session Co-Chair: *Heinz Hoenigschmid, Elpida Europe, Munich, Germany*

In this session advancements in memory and high-speed I/O interfaces are presented that significantly increase system performance and reduce power. Recently, demand for high-speed I/O interface has increased in various memory application areas such as mobile applications, solid-state disc (SSD), digital appliances and server or cloud computing. This year, 4 papers will be discussed reaching data rates up to 12Gb/s by introducing dual-band interconnect, inductive-coupling techniques and impedance matching.

On the memory side, bit density and shrinking process technology are the key market demands for further industry growth. A new record in bit density and smallest process technology is shown this year in the GDDR5 graphics DRAM area. Furthermore, a LPDDR2-N interface has been implemented in a 1Gb phase change memory bridging DRAM and Flash worlds.

Paper 28.1 [UCLA] reveals 8.4Gb/s 2.5pJ/b dual-band signalling for mobile memory I/O interface using simultaneous bi-directional data communication. The chip is designed in a 1V 65nm CMOS technology and uses not only conventional baseband but also RF band on transmission line.

Paper 28.2 [Keio U] discloses 2.7Gb/s/mm^2 0.9pJ/b/chip inductive-coupling interface for NAND flash memory in 0.18µm. The design utilizes 0.1× area-efficient 1 coil/channel interface and 0.5× energy-efficient coupled resonant based CDR compared with the latest work.

Paper 28.3 [Keio U] reveals 12Gb/s non-contact interface for memory card using transmission lines as coils implemented in 90nm. Well-designed termination of transmission lines, transmitter and receiver improves data rate 5× faster than the previous study.

Paper 28.4 [Seoul National U] discloses impedance-matched bi-directional multi-drop DQ bus for large density DRAM server system using 0.13µm process technology. This eliminates reflections at stubs of a 4-slot 8-drop channel, and shows the possibility of 4.8Gb/s bus

In Paper 28.5 [Samsung] a 1.2V 1Gb mobile SDRAM with 4×128 I/Os is discussed for usage in portable electronic devices. This wide-I/O DRAM shows 4× higher data bandwidth than existing LPDDR2, exhibits nearly same standby power and shows 90% reduction of I/O power compared to previous designs. A stack of 2 dies is fabricated using a 7.5µm via diameter TSV technology.

Paper 28.6 [Samsung] demonstrates the first 1.5V 2Gb GDDR5 graphics DRAM in 40nm technology for high speed graphic cards and game consoles. Using a crosstalk equalizing scheme for the transmitter which has a programmable signal ordering capability, a data rate of 7Gb/s/pin with 10% jitter reduction is achieved.

Paper 28.7 [Samsung] exhibits a 1.8V 1Gb PRAM in 58nm technology using a LPDDR2-N interface for low power mobile applications. A data comparison write concept with inversion flag to enhance the core write performance enables 6.4MB/s programming bandwidth with core-write performance 1.5× faster than previous designs. For a faster read, a mid-array pre-charge scheme is presented.

978-1-61284-303-2/11 $26.00 © 2011 IEEE

ISSCC 2011 / February 23, 2011 / 1:30 PM

Paper 28.8 [Hynix] demonstrates a self dynamic voltage scaling technique for consumer DRAMs. The design is fabricated in 44nm CMOS technology and uses 1.8V power supply. Using this concept an operating frequency of consumer DRAM at 1.4Gb/s/pin across process skew and operating frequency is maintained, while minimizing the standby power by 12%.

In Paper 28.9 [Toshiba] an embedded 32KB DRAM macro for high-performance NAND Flash memories is discussed. For the first time, an embedded DRAM macro is using a 32nm standard NAND flash memory process. 100mV cell signal and 90ns cycle time are achieved.

Paper 28.10 [University of Minnesota] presents the first gain cell eDRAM macro implemented in 65nm CMOS technology without boosted supplies. A truly logic compatible implementation is enabled achieving 700MHz random-access frequency.

Presenters:

1:30 PM

28.1 An 8.4Gb/s 2.5pJ/b Mobile Memory I/O Interface Using Simultaneous Bidirectional Dual (Base+RF)-Band Signaling

G-S. Byun, University of California, Los Angeles, CA

2:00 PM

28.2 A 2.7Gb/s/mm² 0.9pJ/b/Chip 1Coil/Channel ThruChip Interface with Coupled-Resonator-Based CDR for NAND Flash Memory Stacking

N. Miura, Keio University, Yokohama, Japan

2:30 PM

28.3 A 12Gb/s Non-Contact Interface with Coupled Transmission Lines

T. Takeya, Keio University, Yokohama, Japan

2:45 PM

28.4 A 4.8Gb/s Impedance-Matched Bidirectional Multi-Drop Transceiver for High-Capacity Memory Interface

W-Y. Shin, Seoul National University, Seoul, Korea

3:15 PM

28.5 A 1.2V 12.8GB/s 2Gb Mobile Wide-I/O DRAM with 4×128 I/Os Using TSV-Based Stacking

C. Oh, Samsung Electronics, Hwasung, Korea

3:45 PM

28.6 A 40nm 2Gb 7Gb/s/pin GDDR5 SDRAM with a Programmable DQ Ordering Crosstalk Equalizer and Adjustable Clock-Tracking BW

S-J. Bae, Samsung Electronics, Hwasung, Korea

4:15 PM

28.7 A 58nm 1.8V 1Gb PRAM with 6.4MB/s Program BW

H. Chung, Samsung Electronics, Hwasung, Korea

4:45 PM

28.8 A 1.6V 1.4Gb/s/pin Consumer DRAM with Self-Dynamic Voltage-Scaling Technique in 44nm CMOS Technology

H-W. Lee, Hynix Semiconductor, Icheon, Korea; Korea University, Seoul, Korea

5:15 PM

28.9 An Embedded DRAM Technology for High-Performance NAND Flash Memories

D. Takashima, Toshiba, Yokohama, Japan

5:30 PM

28.10 A 700MHz 2T1C Embedded DRAM Macro in a Generic Logic Process with No Boosted Supplies

K. Chun, University of Minnesota, Minneapolis, MN

ISSCC 2011 / SESSION 28 / DRAM & HIGH-SPEED I/O / 28.1

28.1 An 8.4Gb/s 2.5pJ/b Mobile Memory I/O Interface Using Simultaneous Bidirectional Dual (Base+RF) Band Signaling

Gyung-Su Byun[1], Yanghyo Kim[1], Jongsun Kim[2], Sai-Wang Tam[1], H-H Hsieh[3], P-Y Wu[3], C Jou[3], Jason Cong[1], Glenn Reinman[1], Mau-Chung Frank Chang[1]

[1]University of California, Los Angeles, CA,
[2]Hongik University, Seoul, Korea,
[3]TSMC, Hsinchu, Taiwan

Power and bandwidth requirements have become more stringent for DRAMs in recent years. This is largely because mobile devices (such as smart phones) are more intensively relying on the use of graphics. Current DDR memory I/Os operate at 5Gb/s with a power efficiency of 17.4mW/Gb/s (i.e., 17.4pJ/b)[1], and graphic DRAM I/Os operate at 7Gb/s/pin [3] with a power efficiency worse than that of DDR. High-speed serial links [5], with a better power efficiency of ~1mW/Gb/s, would be favored for mobile memory I/O interface. However, serial links typically require long initialization time (~1000 clock cycles), and do not meet mobile DRAM I/O requirements for fast switching between active, stand-by, self-refresh and power-down operation modes [4]. Also, traditional base-band-only (or BB-only) signaling tends to consume power super-linearly [4] for extended bandwidth due to the need of power hungry pre-emphasis, and equalization circuits.

To overcome aforementioned technical limitations, we propose to implement a Dual (Base+RF) Band Interconnect (DBI) to enable high throughput data rate and low power operation in a mobile DRAM I/O interface. Unlike the conventional BB-only signaling, the proposed DBI signaling, as shown in Fig. 28.1.1, uses both BB and RF bands for simultaneous dual data stream communications, but shares the common transmission line (T-Line). Instead of limiting the baseband operation within its linear-power-consumption region versus the bandwidth, we can now double the interface bandwidth by using DBI and still maintain the linear-power-consumption versus the bandwidth in each of the dual bands. Additionally with forwarded clocking that adds a small overhead to DRAM I/O (Fig. 28.1.1), the DBI enables simultaneous bidirectional data links [6] as well. By applying such links to DRAM I/O data (DQ) and command/address (C/A), we can greatly reduce the DRAM access time by requesting DRAM read/write-operations simultaneously. Consequently, we can implement bidirectional DRAM I/Os with a much higher aggregate data rate (up to 10Gb/s) and lower power operation (2.5mW/Gb/s).

Figure 28.1.2 shows the DBI transceiver schematic of the memory controller side with an RF-band transmitter (RFTX) and a baseband receiver (BBRX). The RFTX contains an LC tank VCO, an amplitude-shift keying (ASK) modulator and a frequency-selective transformer. In RFTX, the VCO first generates RF carrier at 23GHz and continuously modulates M1 and M2 for ASK communication. The data stream $D_{1(RF)}$ modulates the 23GHz carrier by switching on/off the current flow through M3 and M4 to complete the ASK modulation. The modulated output is then inductively coupled into an off-chip T-Line by way of an on-chip differential transformer. The BBRX amplifies the incoming data stream $D_{2(BB)}$ using buffers with On-Die Termination (ODT) to set the common mode voltage at the transformer center tap and remove the impedance mismatch. As a result, we transmit and receive $D_{1(RF)}$ and $D_{2(BB)}$ data streams concurrently under both differential (RF-band) and common (BB) modes. While DBI's dual band streams are simultaneously transmitted and received, the inter-band interference can be suppressed by spectra separation and the orthogonal property between the differential and common mode signaling.

Figure 28.1.3 shows the DBI transceiver of the DRAM side with a RF-band receiver (RFRX) and a baseband transmitter (BBTX). The RFRX first splits data streams into the BB and RF-band by using an on-chip frequency-selective transformer. The band-pass filtered RF-band data stream is then injected to the receiver differential mutual-mixer with differential input signals and down-converted to the baseband data $D_{1(RF)}$. The termination voltage and the tail source current determine the operating point of the mixer. We utilize a pair of resistor

loaded switching devices and a class-AB amplifier with resistive feedback to further filter out the residue of the RF carrier with a 12.7dB gain. By taking both active device and passive component parasitic into account, we can design the mixer with high signal integrity and high immunity to supply noise without extensive phase/frequency synchronization circuit. The BBTX utilizes a low common-mode push-pull output driver with Off-Chip Driver (OCD) based on digital impedance control logic to evade impedance mismatch and reduce sensitivity to PVT variations. In the meantime, the BB output driver couples the data stream $D_{2(BB)}$ via the common mode (i.e. the center tap of the differential transformer) to the off-chip T-Line.

Since the RF-band in our proposed DBI can easily use a high microwave frequency carrier to minimize the inter-band interference (in this case 23GHz), its signal bandwidth to carrier ratio becomes relatively small so that equalization is generally unnecessary. This dual (BB+RF) band concept can further extended to Base+Multiple-RF bands in the future, such that multiple data streams can be simultaneously transmitted through a shared memory I/O interface T-Line, as long as a multi-band coupling scheme can be devised. Furthermore, since the receiver mixer with differential input signals only senses the incoming signal's amplitude, the frequency and phase synchronizations between RF TX and RX are not required. This greatly simplifies the overall memory I/O interface design. For the same reason, the BER is also expected to be better than that of phase sensitive modulation schemes.

The measured waveforms for DBI input and recovered data streams, and the frequency spectrum of the RF carrier at 23GHz are depicted in Fig. 28.1.4. Figure 28.1.5 shows measured eye diagrams of aggregate 8.4Gb/s (4.6Gb/s BB + 3.8Gb/s RF-band) data throughput over a 10cm T-Line on a FR4 board and 10Gb/s (5Gb/s BB + 5Gb/s RF-band) over the same distance T-Line on a Roger 4003C board, respectively. Both are with <14ps jitter performance.

In summary, we have designed and fabricated a DBI for mobile DRAM I/O interface in 65nm CMOS to obtain an aggregate data throughput of 8.4Gb/s and 10Gb/s on FR4 and Roger test boards, respectively, with power consumptions of 21mW and 25mW. The BERs for both test boards are measured as $<1\times10^{-15}$ by using $2^{23}-1$ PRBS from the Agilent-70843C. Figure 28.1.6 compares the DBI performance to that of prior memory I/O interfaces. Among all, the DBI exhibits the highest aggregate data throughput, best energy efficiency (~2.5pJ/b) and smallest active die area (0.14mm^2 with die photo shown in Fig. 28.1.7).

References:

[1] Kwang-Il Oh, et al., "A 5-Gb/s/pin Transceiver for DDR Memory Interface with a Crosstalk Suppression Scheme," *IEEE J. Solid-State Circuits*, vol. 44, pp. 2222-2232, Aug. 2009.

[2] Kyung-Soo Ha, et al., "A 6Gb/s/pin Pseudo-Differential Signaling Using Common-Mode Noise Rejection Techniques Without Reference Signal for DRAM Interfaces," *ISSCC Dig. Tech. Papers*, pp.138-139, Feb. 2009.

[3] Tae-Young Oh, et al., "A 7Gb/s/pin GDDR5 SDRAM with 2.5ns Bank-to-Bank Active Time and No Bank-Group Restriction," *ISSCC Dig. Tech. Papers*, pp. 138-139, Feb. 2010.

[4] Brian Leibowitz, et al., "A 4.3 GB/s Mobile Memory Interface With Power-Efficient Bandwidth Scaling," *IEEE J. Solid-State Circuits*, vol. 45, no. 4, pp. 889-898, Apr. 2010.

[5] Koji Fukuda, et al., "A 12.3mW 12.5Gb/s Complete Transceiver in 65nm CMOS," *ISSCC Dig. Tech. Papers*, pp. 368-369, Feb. 2010.

[6] Jae-Kwan Kim, et al., "A 3.6 Gb/s/pin simultaneous bidirectional (SBD) I/O interface for high-speed DRAM," *ISSCC Dig. Tech. Papers*, pp.414-415, Feb. 2004.

ISSCC 2011 / February 23, 2011 / 1:30 PM

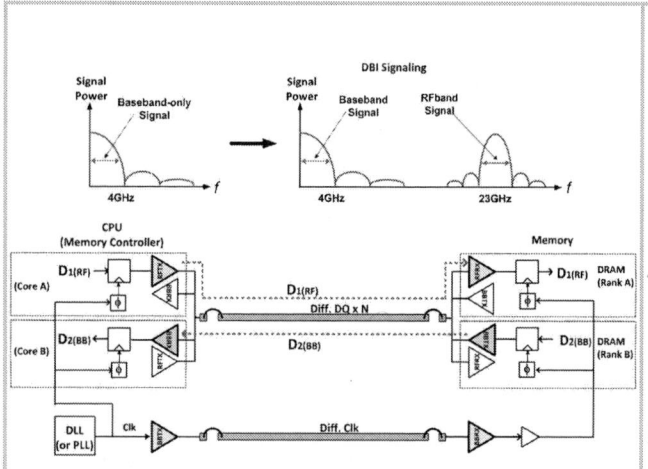

Figure 28.1.1: DBI-based mobile memory I/O interface with forwarded-clock for simultaneous bidirectional signaling.

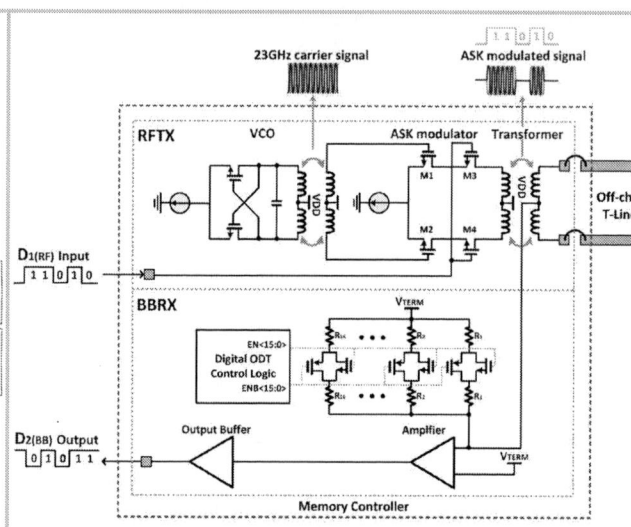

Figure 28.1.2: DBI transceiver schematic of the memory controller side.

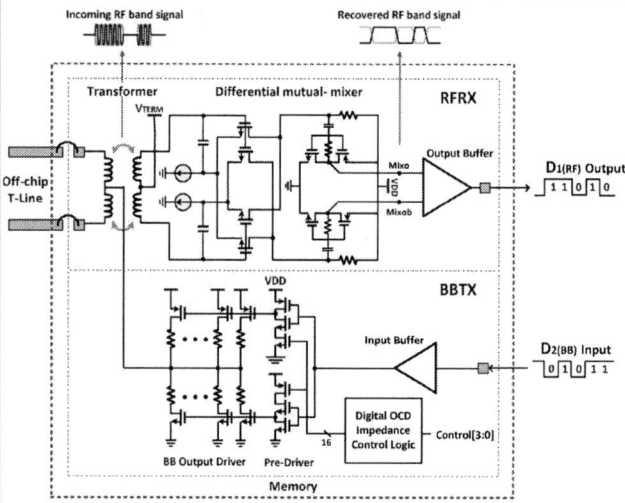

Figure 28.1.3: DBI transceiver schematic of the memory side.

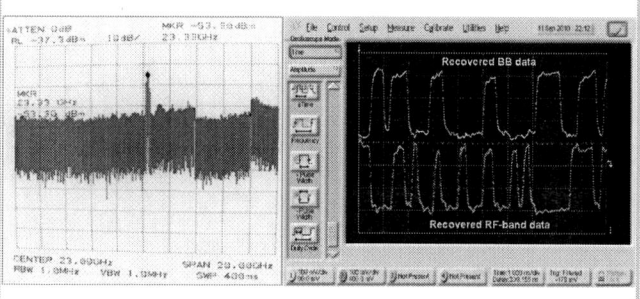

Figure 28.1.4: Measured 23.3GHz RF-band carrier and simultaneous bidirectional 8.4Gb/s Dual (Base+RF)-band waveforms.

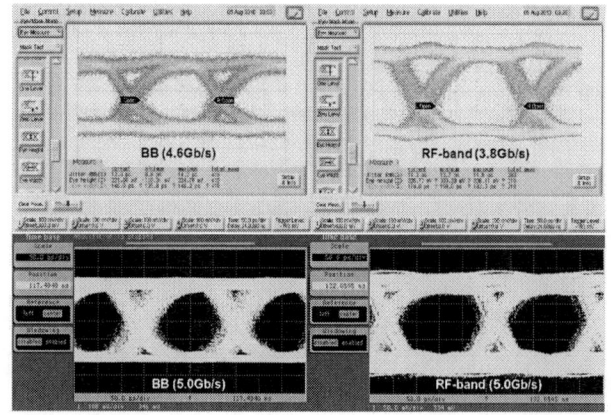

Figure 28.1.5: Measured eye diagrams of aggregate 8.4Gb/s (4.6Gb/s BB + 3.8Gb/s RF-band) and 10Gb/s (5Gb/s BB + 5Gb/s RF-band) data rate, respectively, on FR4 and Roger 4003C test boards.

	[1] JSSC 2009	[2] ISSCC 2009	[4] JSSC 2010	This Work	
				FR4 DBI	Roger DBI
Technology	0.18μm CMOS	0.13μm CMOS	40nm CMOS	65nm CMOS	65nm CMOS
Bands	BB	BB	BB	BB+RF(23GHz)	BB+RF(23GHz)
Supply	1.8V	1.2V	1.1V	1.0V	1.0V
T-Line Length	10cm (FR4)	5cm (N/A)	7cm (FR4)	10cm (FR4)	10cm (Roger)
Aggregate data rate	(BB-only) 5Gb/s	(BB-only) 6.0Gbps	(BB-only) 4.3Gbps	(RF+BB) 8.4Gb/s	(RF+BB) 10Gb/s
Communication	Bidirectional	Bidirectional	Bidirectional	Simultaneous bidirectional	Simultaneous bidirectional
Energy per bit	17.4pJ/bit	15.8pJ/bit	3.3pJ/bit	2.5pJ/bit	2.5pJ/bit
Total power	87mW	95mW	14.4mW	11mW (BB) 10mW (RF)	13mW (BB) 12mW (RF)
Chip Area	0.52mm²	0.30mm²	N/A	0.14mm²	0.14mm²
Measured BER	10^{-12} (PRBS2^{15}-1)	10^{-12} (PRBS2^{15}-1)	N/A	$<10^{-15}$ (PRBS2^{23}-1)	$<10^{-15}$ (PRBS2^{23}-1)

Figure 28.1.6: DBI performance summary and comparison with prior arts.

28

978-1-61284-303-2/11 $26.00 © 2011 IEEE

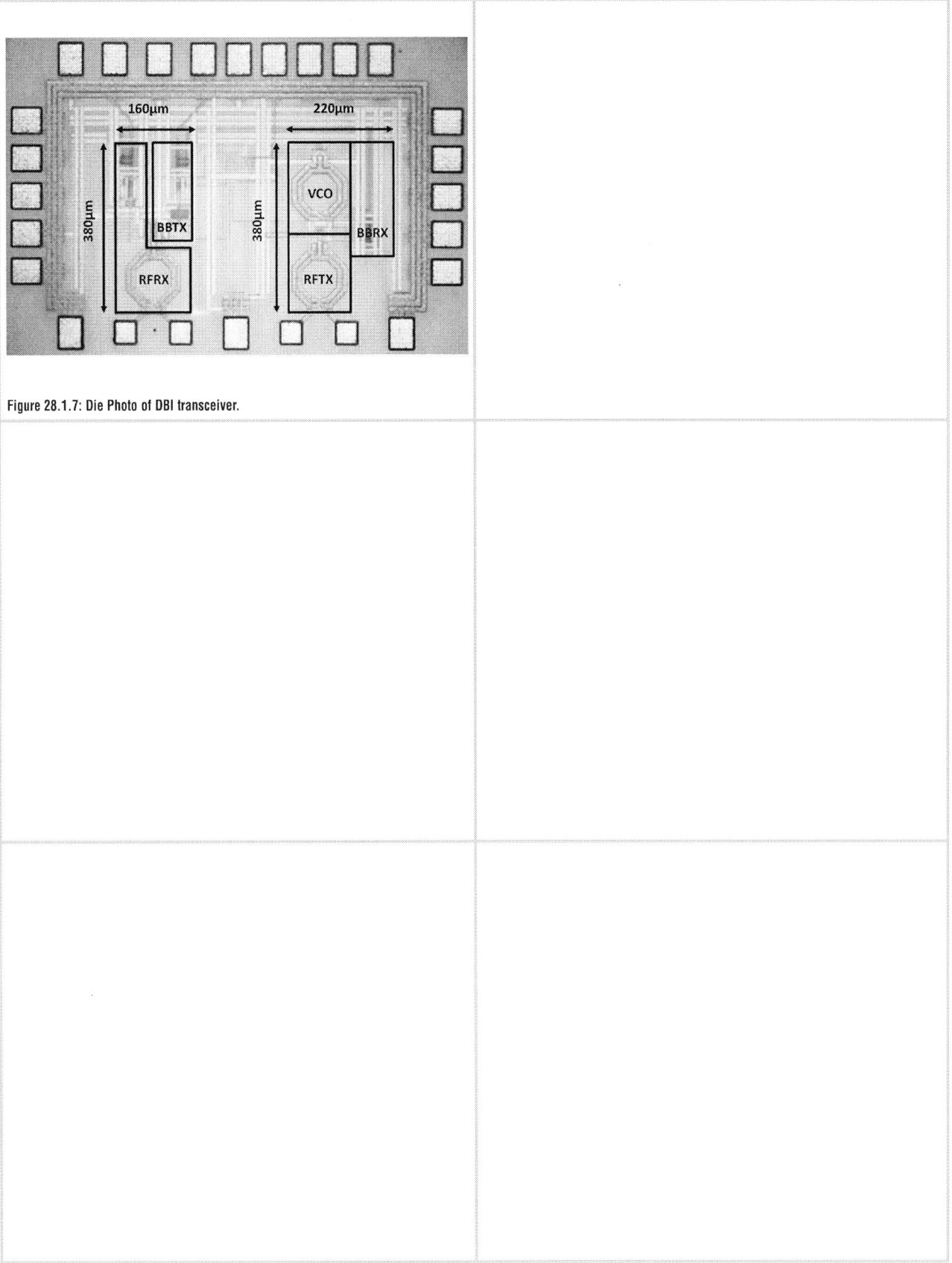

Figure 28.1.7: Die Photo of DBI transceiver.

ISSCC 2011 / SESSION 28 / DRAM & HIGH-SPEED I/O / 28.2

28.2 A 2.7Gb/s/mm² 0.9pJ/b/Chip 1Coil/Channel ThruChip Interface with Coupled-Resonator-Based CDR for NAND Flash Memory Stacking

Noriyuki Miura, Yasuhiro Take, Mitsuko Saito, Yoichi Yoshida, Tadahiro Kuroda

Keio University, Yokohama, Japan

This paper presents an inductive-coupling interface for NAND Flash memory stacking whose bandwidth per unit area is 2.7Gb/s/mm² and energy consumption per chip is 0.9pJ/b/chip. The bandwidth is increased by 10× (in other words, layout area is reduced to 1/10 for the same data rate), and the energy consumption is reduced by half, both compared to the latest research results [1]. A relayed transmission scheme using one coil is proposed to reduce the number of coils in a data link. Coupled resonation is utilized for clock and data recovery (CDR) for the first time in the world, resulting in elimination of a source synchronous clock link. As a result, total number of coils needed to form a channel is reduced from 6 to 1, yielding the significant improvement in data rate, layout area and energy consumption.

Figure 28.2.1 depicts circuit diagram of the data link based on 1coil/channel ThruChip Interface (TCI). The interface consists of asynchronous inductive-coupling transceivers and a CML buffer. They have switchable current sources that program function of the interface: a transmitter, a receiver, or a repeater. Figure 28.2.1 illustrates a case when a controller is reading data from Memory1 by assistance of the other memory chips as repeaters. Grey lines indicate circuits that are disabled at the moment by the program. When Memory1 transmits signal S_{12}, small pulsed V_{R2} is induced and detected by Rx2 in Memory2. The detected signal is amplified by the CML buffer and transmitted to Memory3 by Tx2 as signal S_{23}. The S_{23} induces undesired 2nd pulse in V_{R2}. It is caused by self-crosstalk (C_{22}) and larger than the 1st signal pulse due to closer distance between Rx2 and Tx2, but it doesn't cause malfunction. The Rx is implemented by a hysteresis comparator whose input threshold has already changed to receive next signal with the opposite polarity. The large crosstalk with the same polarity shifts the operating voltages in the comparator and degrades its sensitivity. To recover the sensitivity quickly, internal time constant of the circuits is designed to be small enough. The Rx2 also receives the 3rd pulse crosstalk (C_{32}) when Memory3 repeats the data to Memory4. After this the Rx2 is ready to receive the next signal. The shorter period of the pulses (T_P), the higher data rate, and the shorter latency. Circuits are designed to operate with low-swing signal and achieve 0.4ns of T_P which corresponds to data rate of 2.5Gb/s. In this way transceivers can be placed concentrically in the chip stacking and the number of coils to form a channel is reduced to 2 (one for a data link and the other for a clock link) from 6 that were needed to cope with the crosstalk problem in the conventional interface [1,2].

Burst data transfer by SerDes (serializer in a transmitter and de-serializer in a receiver) requires retiming in the receiver. In the conventional source synchronous scheme [1,2], timing signal was relayed from the transmitter to the receiver by a clock link that was placed in parallel with a data link. In this work, clock and data recovery (CDR) [3] is employed to eliminate the source synchronous clock link. This results in further reduction in the number of required coils from 2 to 1. However, the CDR scheme requires global reference clock distribution among all the stacked chips. In addition, since this distribution network must be always active for random memory access, the power consumption should be reduced. In this investigation, coupled resonation [4] is utilized for the reference clock link. Figure 28.2.2 illustrates circuit diagram of the clock link based on coupled resonation. LC oscillators generate clocks in each chip. Their frequency and phase are all synchronized by inductive coupling between the oscillating coils with high Q factor. Since the coupling efficiency is very high due to the high Q factor, the power dissipation can be significantly reduced. The clock outputs of the coupled resonators are used as transmission timing (Txc) and a reference frequency of a replica PLL in a receiver. The reference frequency is copied to a VCO by the PLL and its phase is locked by injection at edges of received data [3]. The graph in upper right of Fig. 28.2.2 shows simulated impedance (Z_{CR}) of the coupled resonator when four LC oscillators are stacked. Four peaks appear in the impedance characteristic. In the first peak the LC oscillators are coupled in the same phase. Since the oscillation condition of crossed-coupled LC oscillator is

given by $g_m Z_{CR} > 1$, Z_{CR} must exceed $1/g_m$ in order to oscillate the LC oscillators. The g_m can be designed such that the first peak only meets the oscillation condition even under large process variations. The coupled resonator oscillates just at the fundamental frequency and doesn't move to the other frequencies. The lower right graph of Fig. 28.2.2 demonstrates that the proposed resonator has sufficient design margin for coupling coefficient and tail current. In addition, the coupled resonator can reduce its phase noise, due to higher Q by a factor of $(1+k)^{0.5}$ compared to the single LC oscillator. The coupled resonator and the replica PLL are shared among multiple parallel data links to save the power consumption and the layout area.

A test chip was designed and fabricated in 0.18μm CMOS (Fig. 28.2.3). 8-bit parallel data channels are implemented with a coupled resonator and a replica PLL. Sixteen chips are stacked by terraced stacking with an offset of 80μm. Silver (Ag) paste is used to supply power and ground. Cure temperature is 150°C which is as low as that for conventional Die-Attachment Films (DAF). Processing time is only 1 minute. Sidewall of the chips is oxidized in order to prevent short circuit between the Ag paste and the chip substrate. Previously wire bonding was used where the chip offset should be larger than 150μm due to large pad size and wire loop control. A large coil, typically diameter of 1.1mm in this case, was therefore required to compensate for degradation in coupling caused by coil misalignment. The Ag paste is flexible enough to sweep in a small power pad of 20μm width. The chip offset is reduced by half, and hence, the coil diameter to be reduced to 0.9mm in the data link and 0.6mm in the resonator. Measured impedance in Fig. 28.2.3 demonstrates that the Ag paste exhibits comparable power supply integrity with the wire bonding.

Figure 28.2.4 shows measured oscillation frequency (f_{CR}) dependence on tail current (I_{TAIL}) in the coupled resonator. The coupled resonator successfully locks these four clocks though the largest discrepancy among clocks (in this case, between $Clk2$ and $Clk4$) is initially more than 10%. For the range of I_{TAIL} from 0.21mA to 0.39mA, four LC oscillators oscillate in the same frequency (f_{CR0}) and phase with very small discrepancy of less than 0.01%. The measurement also demonstrates that the coupled resonator has a tolerance against V_{DD} change by up to ±25%. Measured RMS timing jitter is less than 2.5ps (<2.4% U.I.), which is low enough to ensure BER of lower than 10^{-12}. Power dissipation for clock generation and distribution is reduced to 1/9 compared with the conventional scheme [1,2] where clock was generated at one chip and distributed to the other chips by relay transmission.

In order to evaluate oscillation frequency dependence on LC variation, many capacitors with different capacitance were implemented in the test chip. As shown in Fig. 28.2.5, measured lock range on capacitance mismatch is ±17.5% which is smaller than standard variation of Metal-Insulator-Metal (MIM) capacitors (±10%) in mass production. Frequency shift is less than 0.03% within the range of the ±10% capacitance variation.

Figure 28.2.6 shows measured BER dependence on data rate. All 8 channels in the test chip were tested by using 2^7-1 PRBS data. BER<10^{-12} operation is confirmed at data rate of higher than 2.4Gb/s for all the channels. The chip performance is summarized in Fig. 28.2.7.

Acknowledgement:
This work is supported by CREST/JST. The Ag paste power supply process was provided by Sony Chemical & Information Device Corporation. The Si chip sidewall oxidization was processed at the Nano-Processing Facility, supported by IBEC Innovation Platform, AIST.

References:
[1] M. Saito, et al., "A 2Gb/s 1.8pJ/b/chip Inductive-Coupling Through-Chip Bus for 128-Die NAND-Flash Memory Stacking," *ISSCC Dig. Tech. Papers*, pp.440-441, Feb. 2010.
[2] Y. Sugimori, et al., "A 2Gb/s 15pJ/b/chip Inductive-Coupling Programmable Bus for NAND Flash Memory Stacking," *ISSCC Dig. Tech. Papers*, pp.244-246, Feb. 2009.
[3] N. Miura, et al., "An 8Tb/s 1pJ/b 0.8mm²/Tb/s QDR Inductive-Coupling Interface Between 65nm CMOS and 0.1um DRAM," *ISSCC Dig. Tech. Papers*, pp.436-437, Feb. 2010.
[4] R. Adler, "A study of locking phenomena in oscillators," *Proc. IEEE*, vol.60, pp.1380-1385, Oct. 1973.

978-1-61284-303-2/11 $26.00 © 2011 IEEE

ISSCC 2011 / February 23, 2011 / 2:00 PM

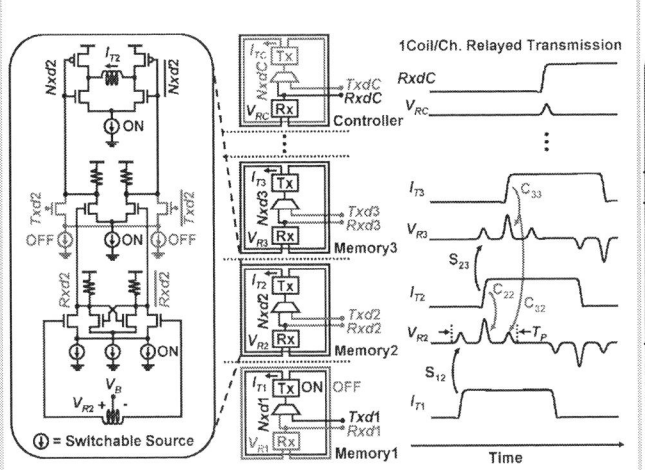

Figure 28.2.1: Data link circuit based on 1coil/channel ThruChip Interface (TCI).

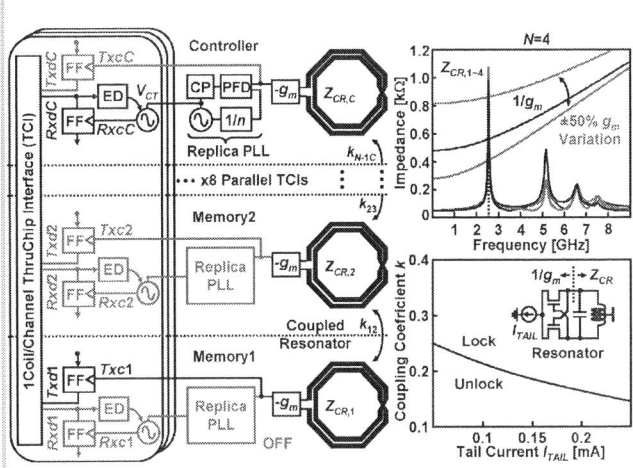

Figure 28.2.2: Clock link circuit based on coupled resonation.

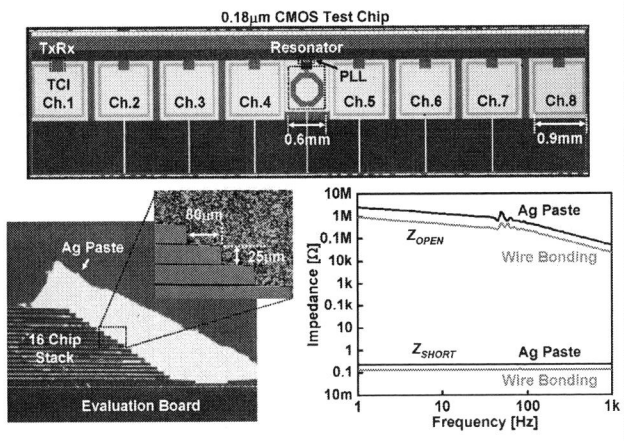

Figure 28.2.3: Stacked test chips with Ag paste for power supply and measured supply impedance.

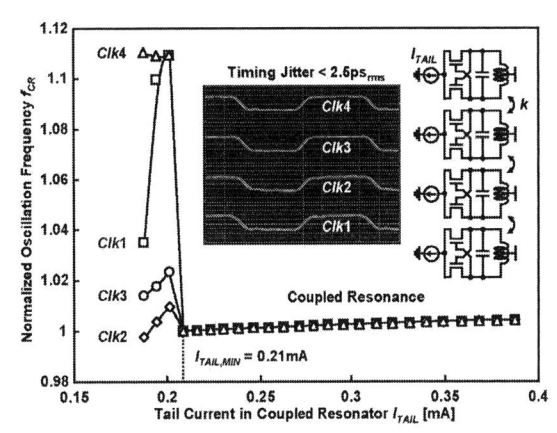

Figure 28.2.4: Measured oscillation frequency dependence on tail current in coupled resonator.

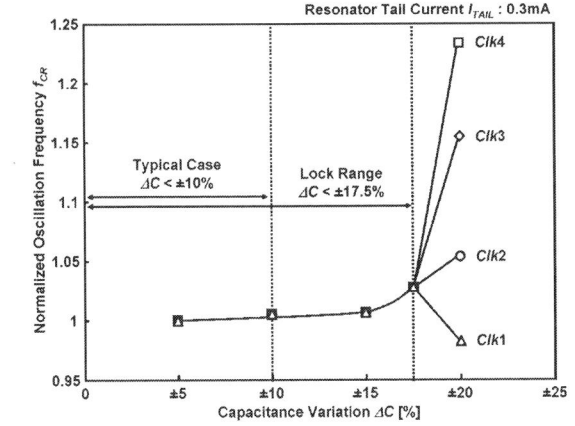

Figure 28.2.5: Measured oscillation frequency dependence on capacitance variation.

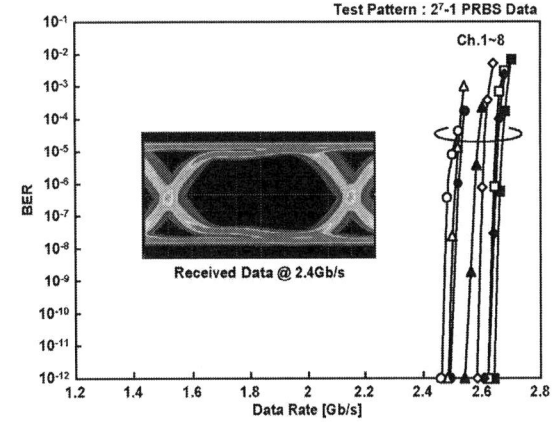

Figure 28.2.6: Measured BER dependence on data rate.

28

ISSCC 2011 PAPER CONTINUATIONS

	This Work	[1]Previous Work
Aggregated Bandwidth	19.2Gb/s (9.6)	2.0Gb/s (1)
Data Rate	2.4Gb/s/ch	2.0Gb/s/ch
Number of Coils	1Coil/ch	6Coils/ch
Coil Diameter	0.9mm (Data) 0.6mm (Clk)	1.1mm (Data, Clk)
Total Layout Area	7.0mm²	7.3mm²
Bandwidth / Area	2.7Gb/s/mm² (10)	0.27Gb/s/mm² (1)
Energy Dissipation / Chip — Data	0.8pJ/b	0.9pJ/b
Energy Dissipation / Chip — Clock	0.1pJ/b	0.9pJ/b
Energy Dissipation / Chip — Total	0.9pJ/b (1/2)	1.8pJ/b (1)
Clock Recovery	CDR w/ Coupled Resonator	Source Synchronous

[1]M. Saito (ISSCC'10)

Figure 28.2.7: Performance summary and comparison.

ISSCC 2011 / SESSION 28 / DRAM & HIGH-SPEED I/O / 28.3

28.3 A 12Gb/s Non-Contact Interface with Coupled Transmission Lines

Tsutomu Takeya, Lan Nan, Shinya Nakano, Noriyuki Miura, Hiroki Ishikuro, Tadahiro Kuroda

Keio University, Yokohama, Japan

The expanding capacity of today's memory cards and increasing speed of processors have created demands for high data rate interfaces between memory cards and processors. Compared to conventional contact pins, wireless interfaces have received tremendous interest for reasons of more convenience, higher reliability, and higher data rate. Two major types of near-field wireless communication techniques using capacitive coupling and inductive coupling channels have been investigated. Tens of Gb/s/ch is achieved using both methods with a communication distance of tens of microns in TCI (thru-chip-interface) applications [1-2]. However, when the communication distance d enters the mm range in such applications as non-contact memory cards, the sizes of capacitors or inductors must be up scaled to detect enough electrical flux or magnetic flux, whose magnitude decay as $1/d^n$ ($n>1$). As a result, the self-resonance frequency f_{SR} is reduced significantly, which becomes the dominant limiting factor for the achievable data rate, since the maximum data rate is usually chosen as 1/2 or 1/3 of f_{SR} in order to avoid signal peaking [3]. Although multi-channel solutions are viable to increase the total data rate, complex systems are required to address the skew issues in synchronization, and low area efficiency is resulted to reduce crosstalk interferences as shown in Fig. 28.3.1.

This work aims to push the state-of-the-art performance of the wireless data links with mm range using a single channel. The proposed channel exploits broadside coupled transmission lines with proper terminations which provide wideband coupling characteristics. The comparison of the single channel characteristics between the proposed channel and inductive coupling channel in Fig. 28.3.1 indicates the considerable expansion of the available bandwidth for high data rate communication. In the case of inductive coupling channels, for example, the inductor size should be larger than 1.0mm if the communication distance longer than 1mm is required. In this case, the self resonant frequency becomes lower than 3GHz and the maximum data rate is limited around 1Gb/s/ch. If more than tens of Gb/s data rate is required, more than 10 parallel inductor channels are required, which makes it difficult to deal with the channel skew problem. Furthermore, the long wiring between each inductors and transceiver chip degrades the signal quality because of the multiple signal reflection between the inductors and transceiver chip. In the case of the proposed coupled transmission lines, a single channel can achieve a data rate faster than 10Gbps which solves the channel skew problem. Signal degradation by multiple reflection can be avoided by proper termination.

The proposed channel is illustrated in Fig. 28.3.2(a). It consists of four coupled transmission lines on two FPC boards with a separation of 1mm. The input differential signals are coupled to output port on the same side of the link, while the other two ports are terminated with 100Ω resistors for differential-mode signaling. The design parameters of the directional coupling link include line length L, width W and spacing. Given the relative permittivity ε_r of the medium in between the memory card and processor, typically 3-5 for the plastics packages, the length of the transmission lines L can be determined as a quarter wavelength at the desired center frequency f_c, i.e., $L=c/4f_c\sqrt{\varepsilon_r}$, where c is the speed of light in vacuum. As shown in Fig. 28.3.2(b), as the center frequency is inversely proportional to L. Although wider bandwidth can be easily achieved by up scaling f_c, the lower 3dB frequency f_L gets higher, which increases difficulty in recovering low frequency components of the signals in the receiver. In the chosen of W and S, trade-offs between the coupling level, bandwidth and area consumption need to be made based on system requirements. Consider a pair of vertically coupled lines, the line width W determines the single-end coupling coefficient. In order to maintain the same coupling coefficient for the differential-mode signals, a sufficient spacing S is required to reduce the cross-coupling of the lines on the same FPC board. The geometry parameters are finally designed as: L=6mm, W=0.5mm, and S=1.5mm.

The performance tolerance to the horizontal offset and variations in the communication distance d is examined as shown in Fig. 28.3.2(c) and (d) respectively. As the offset is increased to 100% of the line width, the coupling level only decreases by 1.7dB. The coupling level can be changed by 5 dB when and communication distance varies by 0.5mm. However, the 3dB bandwidth does not change with the horizontal offset or variation in the communication distance.

The schematics of the transmitter and the receiver are shown in Fig. 28.3.3. Txdata is input to the driver in the transmitter through the repeaters because of the long distance between the pads and the driver. The driver outputs and the receiver inputs are terminated by termination resisters (50Ω for impedance matching). The first stage amplifiers in the receiver have wideband characteristics by using inductive peaking. Due to the band pass filter characteristics of the channel, the low frequency components of the baseband NRZ are degraded. As a result, the received signal level droops if the run length of the transmitted data becomes long. Data recovery circuits are required for data communication. In this study, a hysteresis buffer is applied to prevent the undesired signal droop by holding the signal level.

Figure 28.3.4 shows the chip micrograph and the evaluation board for measurement. The test chip was fabricated in a 90nm CMOS process. The core areas of the transmitter and the receiver are 90x240µm^2 and 300x700µm^2 respectively. For probing purpose, one chip containing a transmitter and a receiver is used and bonded to the designed channel. The coupling effect of the bonding wires is simulated and confirmed to be negligible.

Figure 28.3.5 shows the simulated and measured differential-mode coupling coefficient S_{21} of the designed link at 1mm. It achieves a coupling level of 16dB at 6GHz and a 3dB bandwidth from 2.6GHz to 9GHz. The measured result coincides well with the simulated result.

The system BER is evaluated at various data rates for d=0.5mm and d=1mm respectively and plotted in Fig. 28.3.6. The transceiver successfully achieves BER<10^{-13} at 12.5Gb/s with d=0.5mm and 12Gb/s with d=1mm. The eye diagram for 12Gb/s with d=1mm is also shown.

To evaluate the interference due to the wireless power delivery, a 20mm×20mm coil is designed outside the data link on the same FPC boards for power delivery at 13.56MHz. The measured bath tub curves with and without power delivery are presented in Fig. 28.3.7. With a current of 75mA$_{rms}$ at 13.56MHz surrounding the data channel, the timing margin with power delivery is slightly smaller but still sufficient. It confirms the directional coupling link can be implemented together with power delivery coils.

Acknowledgement:
This work is supported by CREST/JST.

References:
[1] N. Miura, et al., "An 11Gb/s Inductive-Coupling Link with Burst Transmission," *ISSCC Dig. Tech. Papers*, pp.298-299, Feb 2008.
[2] Q. Gu, et al., "Two 10Gb/s/pin Low-Power Interconnect Methods for 3D ICs," *ISSCC Dig. Tech. Papers*, pp.448-449, Feb 2007.
[3] S. Kawai, et al., "A 2.5Gb/s/ch Inductive-Coupling Transceiver for Non-Contact Memory Card," *ISSCC Dig. Tech. Papers*, pp.264-265, Feb 2010.
[4] K. Kasuga, et al., "Electromagnetic Interference and Susceptibility in Inductive-Coupling Link," *SSDM Dig. Tech. Papers*, Oct. 2009.

ISSCC 2011 / February 23, 2011 / 2:30 PM

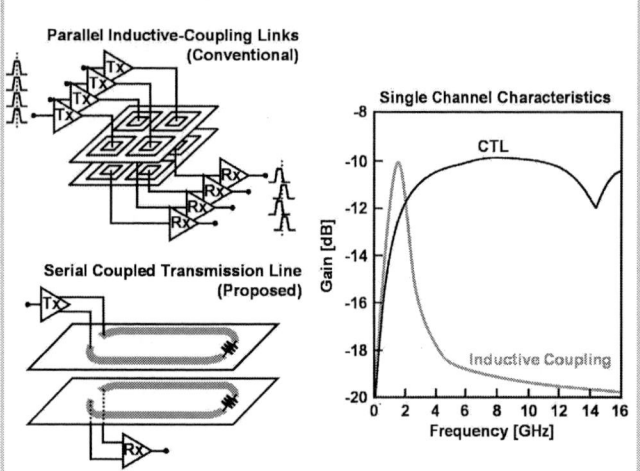

Figure 28.3.1: Conventional parallel inductive coupling link and proposed interface with coupled transmission line.

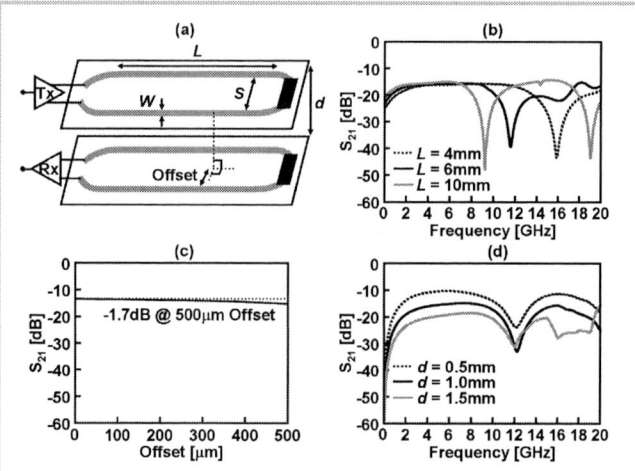

Figure 28.3.2: (a) Design parameters of CTL and channel characteristics depending on (b) line length, (c) offset, and (d) communication distance.

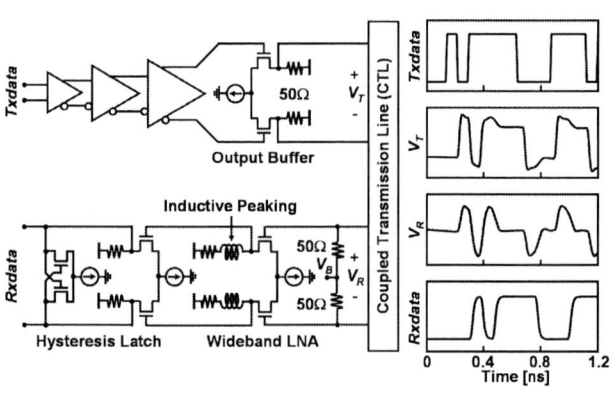

Figure 28.3.3: Transceiver circuit and its simulated waveforms.

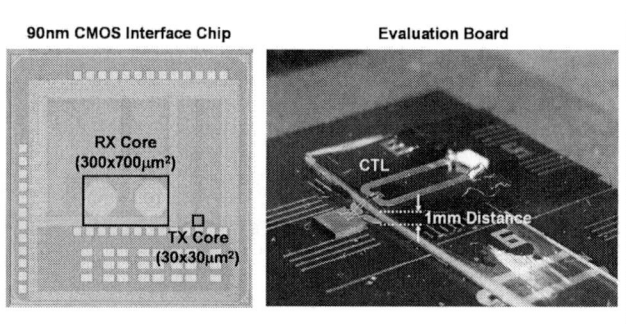

Figure 28.3.4: Die photo and measurement setup.

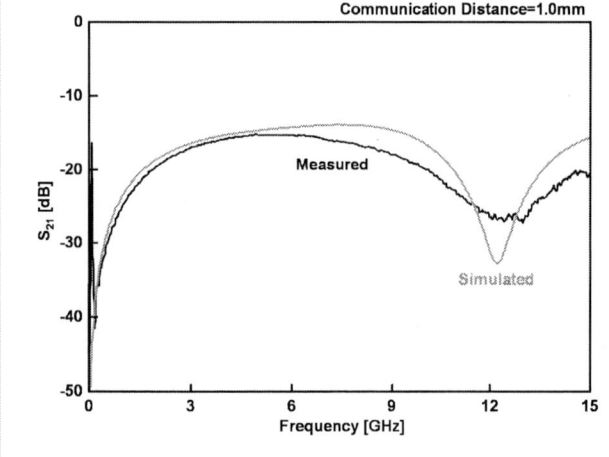

Figure 28.3.5: Measured and simulated S_{21} of coupled transmission line.

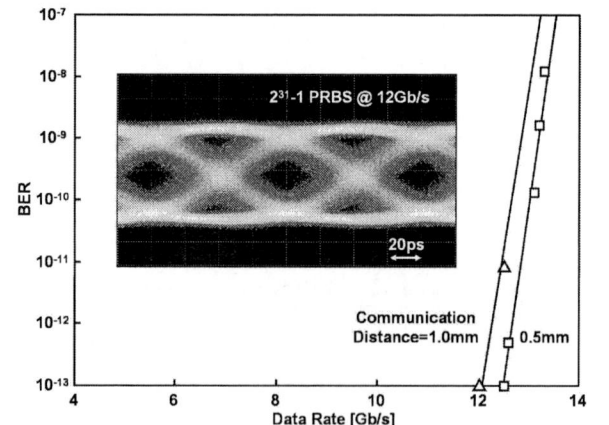

Figure 28.3.6: Measured BER dependence on data rate.

28

978-1-61284-303-2/11 $26.00 © 2011 IEEE

ISSCC 2011 PAPER CONTINUATIONS

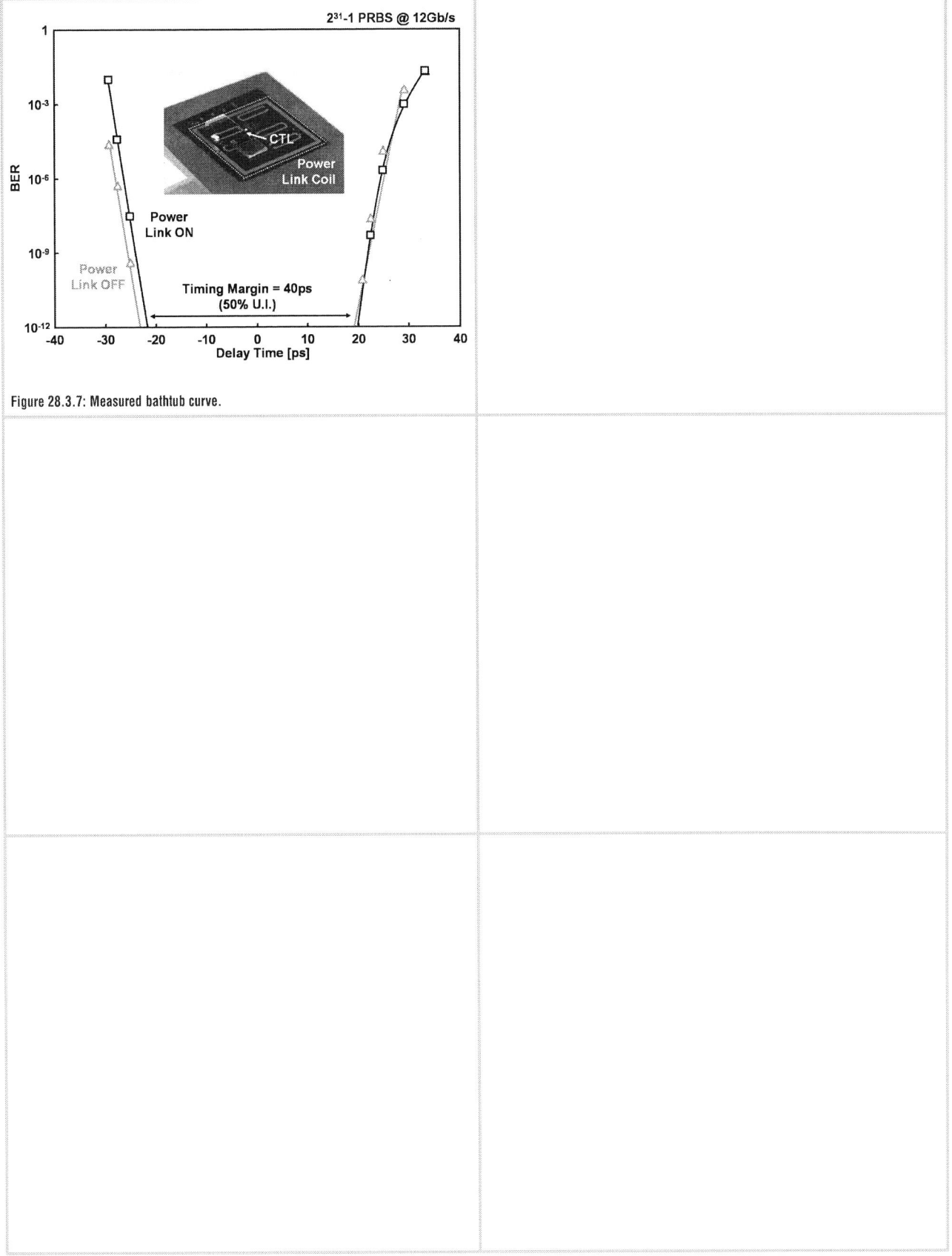

Figure 28.3.7: Measured bathtub curve.

ISSCC 2011 / SESSION 28 / DRAM & HIGH-SPEED I/O / 28.4

28.4 A 4.8Gb/s Impedance-Matched Bidirectional Multi-Drop Transceiver for High-Capacity Memory Interface

Woo-Yeol Shin[1], Gi-Moon Hong[1], Hyongmin Lee[1], Jae-Duk Han[1],
Sunkwon Kim[1], Kyu-Sang Park[1], Dong-Hyuk Lim[1], Jung-Hoon Chun[2],
Deog-Kyoon Jeong[1], Suhwan Kim[1]

[1]Seoul National University, Seoul, Korea,
[2]Sungkyunkwan University, Suwon, Korea

With the scaling of CMOS transistors and advance in I/O circuitry, the data rate of memory interfaces has recently reached 16Gb/s per channel [1], in which a point-to-point channel is required rather than a multi-drop channel for the high data rate. While point-to-point channels are advantageous in achieving higher data rates because of the absence of undesired reflections that occur at each stub of multi-drop channels, they are not suitable for high-capacity, high-throughput memory systems such as transaction servers or cloud computing nodes due to their prohibitively large PCB routing area connecting the memory chips. FBDIMM [2] and the cascading memory architecture [3] aim to reduce the routing area by the use of daisy-chained configurations, but they suffer from increased latency problems. This is why the recent DDR2/3 memory interface still uses the multi-drop bus architecture called stub series terminated logic (SSTL), and a number of proposals have been made to mitigate the problem of stub reflections in SSTL. For instance, a decision feedback equalizer has been used [4] to cancel the inter-symbol interference (ISI) due to stub reflections; but this requires a large number of filter taps, resulting in a limited speed under 3Gb/s. Another approach to eliminate impedance discontinuity is to use a $2Z_0$ ohm transmission line [5], but this scheme is only applicable to 2-slot configurations.

In this paper, first, we describe an impedance-matched bi-directional multi-drop (IMBM) DQ bus that can handle up to 4 slots, 8 drops at a data-rate of up to 4.8Gb/s. In the case of the SSTL DQ bus, as shown in Fig. 28.4.1, the series resistor of $Z_0/2$ can suppress ringing and attenuate reflections within the channel. But, the SSTL DQ bus is still not entirely free from reflections among the slots because the reflection coefficient of the SSTL DQ bus at the stub junctions is -1/4. However, the IMBM DQ bus that we propose makes the reflection coefficient 0 as can be seen from the equations presented in Fig. 28.4.1. Therefore, the IMBM DQ bus generates no reflection signal at each stub. Moreover, the IMBM DQ bus can send write signals of the same current level to every module according to its current division relation of I_k and I_{k+1}. The transceiver also receives the read signal from every module with the identical level, according to the reciprocity theorem [4]. This characteristic produces identical transfer responses for all the modules regardless of their positions.

Figure 28.4.2 shows the 4-slot, 8-drop IMBM DQ bus that we have implemented. Instead of matching the impedance in every direction, the IMBM DQ bus matches impedance only in the left-to-right direction at the upper transmission lines (TL 1,2,3,4 in Fig. 28.4.2). Since there is no reflection from the right-end of the upper TLs on the motherboard during the write operation, the write data signal from the memory controller to the memory modules at positions 0 through 7 can be transmitted without any reflection. During a read operation, however, reflection signals may be generated at each stub; but these reflections proceed from the stubs to right-end of the upper TLs, and are finally absorbed by the ODT resistors. Thus they do not reach the memory controller and hence do not degrade the integrity of the desired signal. Therefore, the IMBM DQ bus is able to transmit and receive both read and write signals without reflective ISI.

Second, we design a memory transceiver architecture that is suitable for the IMBM DQ bus, which attenuates the transmitted signals by the reciprocal of the number of modules. Direct data sampling using the received strobe is not recommended since the limiting amplifier chains necessary to recover the original signal swings would draw too much power in this architecture. Instead, we use the memory transceiver and RX clocking architecture, as shown in Fig. 28.4.3. This transceiver consists of 4 DQ channels, a DQS channel, a PLL and clock trees. A current-mode driver with a 4-tap FIR filter in the TX of each DQ enables de-emphasis for equalization. The PLL and a clock tree provide a TX clock for the serializer and multiphase clocks for the strobe recovery unit (SRU) of the DQS

block. The received strobe signal is used for timing recovery in RX mode. A dual-loop architecture is used in the SRU for timing recovery. To generate the sampling clock for every DQ with the proper phase, the SRU generates the RX sampling clock with a phase interpolator, a half-rate bang-bang phase detector and control logics. A duty cycle corrector (DCC) compensates for the duty cycle distortions that may arise in the clock tree and the phase interpolator. To ensure that all DQs receive the recovered sampling clock signal with the equal phase, their clocks are shortened together to reduce possible on-chip clock skews. To reduce the skew between each DQ and SRU, the sampling clock for the PD is also delivered along the same clock tree.

The IMBM memory transceiver is fabricated in 0.13μm CMOS process and occupies 1400×1200μm². The die microphotograph and the performance summary are shown in Fig. 28.4.7. The chip board and channel board were implemented separately to enable measurements with different channel configurations. Both boards are fabricated in NELCO material instead of FR4 to reduce the PCB insertion loss. MicroTCA [6] connectors are used on the channel board. The transceiver consumes 14.24mW/Gb/s/DQ at 4.8Gb/s in TX mode and 13.69mW/Gb/s/DQ at 4.8Gb/s in RX mode.

Figure 28.4.4 illustrates the 4.8Gb/s single-bit responses (SBRs) and the eye diagrams of a conventional SSTL DQ bus and the new IMBM DQ bus. Due to the reflections between connectors, the SBRs of the SSTL DQ bus vary vastly depending on the module positions. However, all the SBRs of the IMBM DQ bus are almost identical regardless of the module positions. Even when compared to the scaled SBR of the transmitted signal measured only with the chip board, the SBR of the IMBM DQ bus generates little reflection. The measured eye diagrams in Fig. 28.4.4 also substantiate the difference between the SSTL DQ bus and the IMBM DQ bus. Figure 28.4.5 shows the measured TX BER results of both the SSTL and IMBM DQ buses. Without TX equalization, SSTL modules #5 and #7 have no timing margin, whereas modules #1 and #3 have large margins. However, the BER of the IMBM DQ bus is nearly 10^{-7} at the optimum sampling point. When TX de-emphasis is enabled (by setting the precursor tap to 0, the main cursor tap to 1, the first post-cursor tap to -0.33, and the second post-cursor tap to -0.08), the IMBM DQ bus achieves 10^{-9} BER with a timing margin of 0.39UI, as shown in Fig. 28.4.5. In comparison, we can see that some of the SSTL DQ modules fail to reach a BER of 10^{-9} at any sampling positions. In the BER measurement of the RX without equalization, the recovered data exhibits many errors beyond reliable operation at SSTL module #5 and #7 as was the case with the TX BER measurement, as shown in Fig. 28.4.6. Un-equalized IMBM DQ bus, however, has a 0.61UI timing margin with a BER threshold of 10^{-9}. Timing margin of the RX BER test is larger than that of the TX BER test because of the clean data transmitted from the PBERT. When the memory transceiver turns on the linear equalizer to boost its high-frequency gain, the IMBM DQ bus has a 0.73UI timing margin, which is 0.21UI higher than that of the SSTL DQ bus.

References:
[1] H. Lee, et al., "A 16 Gb/s/Link, 64 GB/s Bidirectional Asymmetric Memory Interface," *IEEE J. Solid-State Circuits*, vol. 44, no. 4, pp. 1235-1247, April 2009.
[2] E. Prete, D. Scheideler, and A. Sanders, "A 100mW 9.6Gb/s Transceiver in 90nm CMOS for Next-Generation Memory Interfaces," *ISSCC Dig. Tech. Papers*, pp. 253–262, Feb. 2006.
[3] Z. Gu, et al., "Cascading Techniques for a High-Speed Memory Interface," *ISSCC Dig. Tech. Papers*, pp. 234–235, Feb. 2007.
[4] H. Fredriksson and C. Svensson, "Improvement Potential and Equalization Example for Multidrop DRAM Memory Buses," *IEEE Trans. Adv. Packag.*, vol. 32, no. 3, pp. 675-682, Aug. 2009.
[5] S.-J. Bae, H.-J. Chi, H.-R. Kim, and H.-J. Park, "A 3Gb/s 8b Single-Ended Transceiver for 4-Drop DRAM Interface with Digital Calibration of Equalization Skew and Offset Coefficients," *ISSCC Dig. Tech. Papers*, pp. 520–521, Feb. 2005.
[6] MicroTCA Standard, http://www.picmg.org/v2internal/microtca.htm

ISSCC 2011 / February 23, 2011 / 2:45 PM

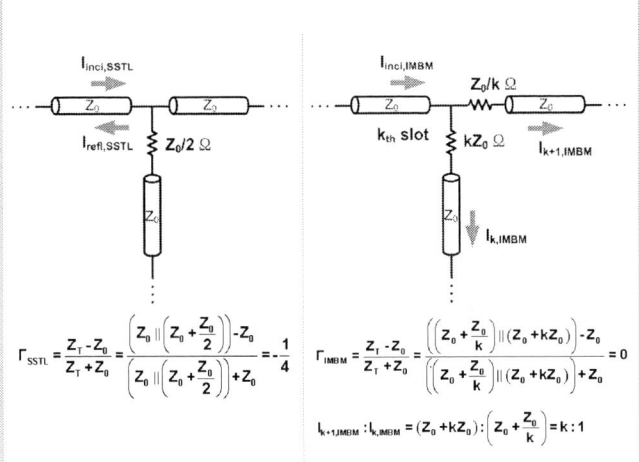

Figure 28.4.1: Conventional SSTL DQ bus and proposed IMBM DQ bus.

Figure 28.4.2: Implemented 4-slot, 8-drop IMBM DQ bus.

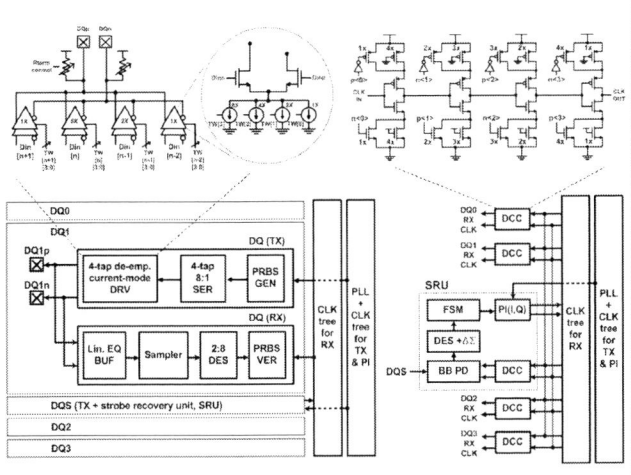

Figure 28.4.3: Overall memory transceiver and RX clocking architecture.

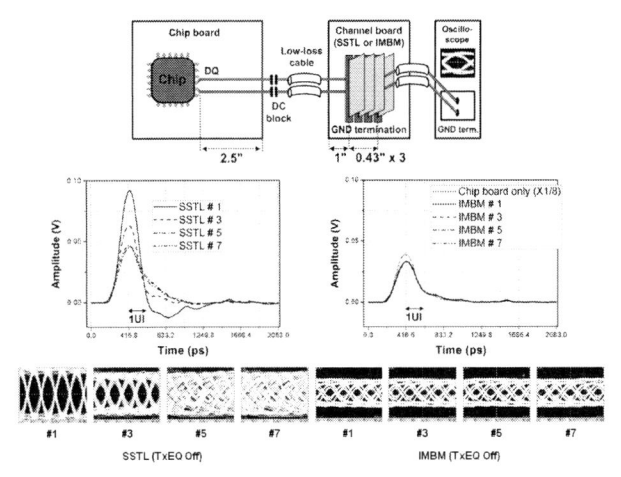

Figure 28.4.4: Measured single bit responses and eye diagrams.

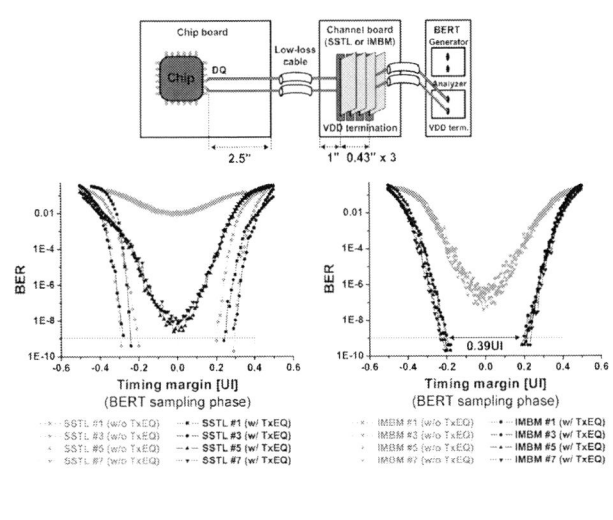

Figure 28.4.5: Measured TX BER.

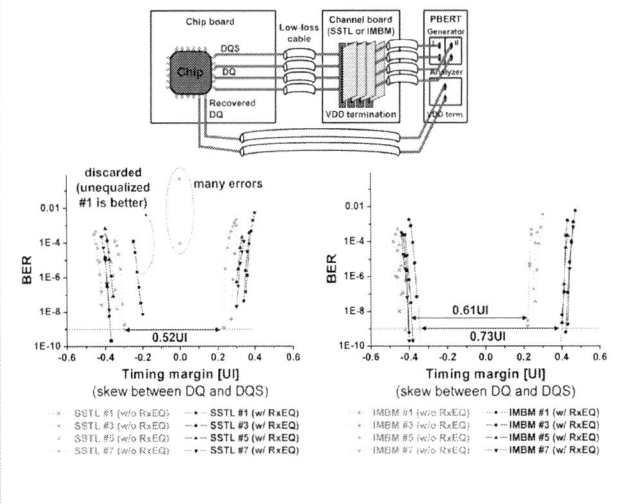

Figure 28.4.6: Measured RX BER.

28

978-1-61284-303-2/11 $26.00 © 2011 IEEE

ISSCC 2011 PAPER CONTINUATIONS

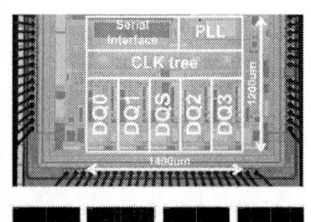

Data rate	4.8Gb/s
Technology	0.13μm 1P8M CMOS
Channel	IMBM, 4 slots (8 drops)
Connector	MicroTCA
Package	TQFP 100p
Timing margin (@4.8Gb/s, TX)	0.39UI @ 10^{-9}
Timing margin (@4.8Gb/s, RX)	0.73UI @ 10^{-9}
Energy efficiency (@4.8Gb/s, per DQ)	14.24mW/Gb/s (TX mode) 13.69mW/Gb/s (RX mode)

Figure 28.4.7: Chip microphotograph, eye diagrams, and performance summary.

ISSCC 2011 / SESSION 28 / DRAM & HIGH-SPEED I/O / 28.5

28.5 A 1.2V 12.8GB/s 2Gb Mobile Wide-I/O DRAM with 4×128 I/Os Using TSV-Based Stacking

Jung-Sik Kim, Chi Sung Oh, Hocheol Lee, Donghyuk Lee,
Hyong-Ryol Hwang, Sooman Hwang, Byongwook Na,
Joungwook Moon, Jin-Guk Kim, Hanna Park, Jang-Woo Ryu,
Kiwon Park, Sang-Kyu Kang, So-Young Kim, Hoyoung Kim,
Jong-Min Bang, Hyunyoon Cho, Minsoo Jang, Cheolmin Han,
Jung-Bae Lee, Kyehyun Kyung, Joo-Sun Choi, Young-Hyun Jun

Samsung Electronics, Hwasung, Korea

Mobile DRAM is widely employed in portable electronic devices due to its feature of low power consumption. Recently, as the market trend renders integration of various features in one chip, mobile DRAM is required to have not only low power consumption but also high capacity and high speed. To attain these goals in mobile DRAM, we designed a 1Gb single data rate (SDR) Wide-I/O mobile SDRAM with 4 channels and 512 DQ pins, featuring 12.8GB/s data bandwidth.

Figure 28.5.1 shows the chip architecture with 4-channels and 16 segmented 64Mb arrays. The whole chip is made up of 4 partitions which are symmetric with respect to the chip center, and each partition consists of 4×64Mb arrays, peripheral circuits and microbumps. Each channel has its own input pins while external power pins and internal voltage generators are shared with the other channels. In a single channel, 128 data lines are controlled to feed 128 DQs. Four 64Mb arrays in 1-channel can be configured in 4 banks with bank addresses BA [0:1] and row addresses RA [0:11], or in 2 banks with BA [0] and RA [0:12]. In a 4-bank structure, each bank has 4k row depth and 8k page depth.

To reduce power consumption in 512b I/O operations and to support high data bandwidth, I/O driver loading is reduced by adoption of 44×6 microbump pads per channel, which are located in the middle of the chip. Figure 28.5.1 also shows a SEM image of the fabricated microbumps with $20×17\mu m^2$ size and max 50μm pitch. To detect bump connection failure with other devices, a simple boundary scan test mode is implemented. It is a subset of IEEE Standard 1149.1 [1] in pin configurations and in operational modes to reduce chip size burden. Each channel has its own scan chain and scan clock input. To reduce the number of ballouts, this mode scans parallel data I/O and the scanned data propagates through the dedicated pins. Normal device operation is performed after the boundary scan test is finished. Figure 28.5.2 shows boundary scan block and AC timing diagram. This boundary scan chain is enabled when /SEN, scan enable pin, is low. The data input from the pad is captured when /SSH, scan shift pin, is low, and it is shifted along the chain when /SSH is high. SDI, SDO and SCK pins are for scan input, output and clock, respectively.

We also adopt typical metal pads for test purposes since it is difficult to probe small microbumps directly. These pads are aligned in vertical direction at the chip center to allow precise correlation with microbumps and to reduce skews between channels. With test pads, this DRAM is handled as SDR×16, but internally as SDR×128 per channel. Selections of 16 out of 512 data can be done with column addresses, which are for the read-out of result data through test pads. Figure 28.5.3 shows correlation scheme between microbumps and test pads in data write/read operations. Through test pads, 9.3ns delayed outputs of microbumps, 2ns for clock propagation from pad and 7.3ns for data transfer from microbump to pad, are measured. Because we redirect outputs from microbumps and inputs from test pads to each other, there is no timing margin point for data write/read operations in the path between microbumps and test pads. With this scheme, validity of all the internal functions are checked through test pads. Direct access (DA) mode, for the test of DRAM after package assembly with chipset, is also possible with additional pins that are directly connected to memory. In this mode, each channel can be operated as SDR×8 with separate chip select (/CS) pin, and by merging 4 /CS pins in all channels, this device can be operated as SDR×32 with 1 /CS pin.

To reduce self refresh current (IDD6), a dual-period based refresh scheme is adopted. Various methods with differentiated self refresh periods have been tried with additional registers [2, 3, 4], but the additional registers and row-based mapping result in chip size increase. Figure 28.5.4 shows allocation of memory banks in 1-channel of Wide-I/O DRAM for dual data retention scheme. Each bank is subdivided into 32 sub-blocks according to row addresses, and with designated 32 bits of refresh information data in 1 channel. Block-based mapping of cell array and minimization of the number of blocks made it possible to minimize additional chip size. As a result, the area of relevant circuits, including metal fuses, registers and logics, occupies 0.11% of total chip area. All the blocks can be classified into 2 types: ×1 and ×2 refresh period. These blocks are discriminated by their refresh characteristics, which are recorded by fuse cutting at the test stage in electrical die sorting or stored at internal registers during the BIST period after power-up sequence, as shown in Fig. 28.5.4. Refresh data at fuses and internal registers are combined to determine the refresh period of the corresponding memory block. Refresh operation in blocks with ×2 refresh period is skipped at alternate turns of 8k refresh, i.e. during Period 1 in Fig. 28.5.4.

To resolve frequency limitations from speed delay and variation due to low operation voltage at 1.2V, read strobe function (QS) is added, playing a similar role as DQS of DDR. Figure 28.5.5 shows a timing diagram of QS function. As shown in Fig. 28.5.5, adoption of QS increases timing margin to controllers, because controllers can fix and optimize input setup/hold time, increasing the timing window to fetch output data from memory. Because QS is an additive function in SDR, the existing DM pins accommodate the functionalities of QS during read operation. Operational mode of QS is chosen as SDR mode, which means QS is aligned with the rising edge of the clock and gives 50% power reduction compared to DDR mode. The number of QS can be controlled by extended mode register set (EMRS) as 1ea per 8DQs, 16DQs or 32DQs to reduce power consumption as shown in Fig. 28.5.5.

Figure 28.5.6 shows measured voltage vs. frequency shmoo from test pads in SDR×16 4-channel operation mode. It shows stable operation at 200MHz clock frequency. Measurement results show t_{CK}, t_{RCD} and t_{RP} of 4.6ns, 12.8ns and 14.0ns, respectively, at V_{DD2}=1.14V.

Figure 28.5.6 also shows comparison of features with 1.8V 1.6GB/s 1Gb mobile DDR (MDDR) SDRAM and 1.2V 3.2GB/s 1Gb LPDDR2, fabricated with the same 50nm technology. The measured total read power including DQ power amounts to 330.6mW, almost equal to MDDR and LPDDR2, and the measured I/O power per 1 bit data transfer is 0.78mW/Gbps, corresponding to 4.5% of MDDR's. Reduction of I/O power comes from voltage and I/O loading reduction together with data bandwidth increase. The die area is $64.34mm^2$, about a 25% increase when compared with 1Gb LPDDR2. This comes mostly from the increase in number of circuits to support 4-channel and 512-DQ feature. However, the Wide-I/O mobile DRAM shows almost the same standby power as in MDDR and LPDDR2 due to the optimization in number of transistors and circuits.

Figure 28.5.7 shows the micrograph of the fabricated chip with microbumps and a vertical section of 2-stacked Wide-I/O DRAMs. With microbumps and TSVs (Through-Silicon Via), it is possible to stack 2 Wide-I/O DRAM dies. A fabricated TSV has 7.5μm diameter, 0.22 to 0.24Ω resistance and 47.4fF capacitance. The connectivity of TSVs can be checked also by boundary scan method. The overall yield of these TSV processes is confirmed to be about 67 to 76% from daisy chain packages and function die packages.

References:

[1] IEEE Standard Test Access Port and Boundary-Scan Architecture, IEEE Std. 1149.1-2001, 2001.

[2] Y. Idei, et al., "Dual-Period Self-Refresh Scheme for Low-Power DRAM's with On-Chip PROM Mode Register," *IEEE J. Solid-State Circuits*, vol. 33, no. 2, pp. 253-259, Feb. 1998.

[3] J. Kim and M. C. Papaefthymiou, "Block-Based Multiperiod Dynamic Memory Design for Low Data-Retention Power," *IEEE Trans. VLSI Systems*, vol. 11, no. 6, pp. 1006-1018, Dec. 2003.

[4] J.-H. Ahn, et al., "Adaptive Self Refresh Scheme for Battery Operated High-Density Mobile DRAM Applications," *Proc. ASSCC*, pp. 319-322, Nov. 2006.

ISSCC 2011 / February 23, 2011 / 3:15 PM

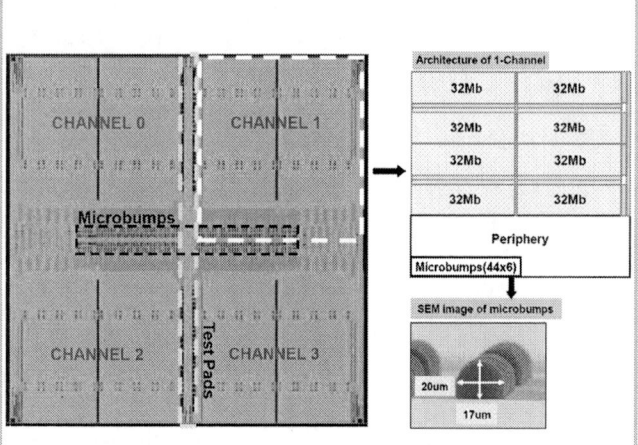

Figure 28.5.1: Chip architecture of 1Gb Wide-I/O DRAM and SEM image of microbumps.

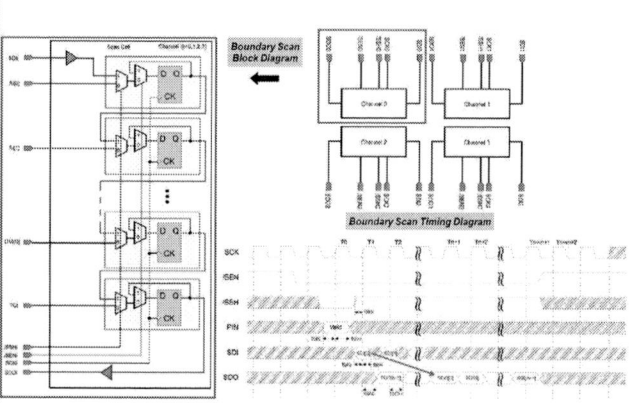

Figure 28.5.2: Block and timing diagram of boundary scan test.

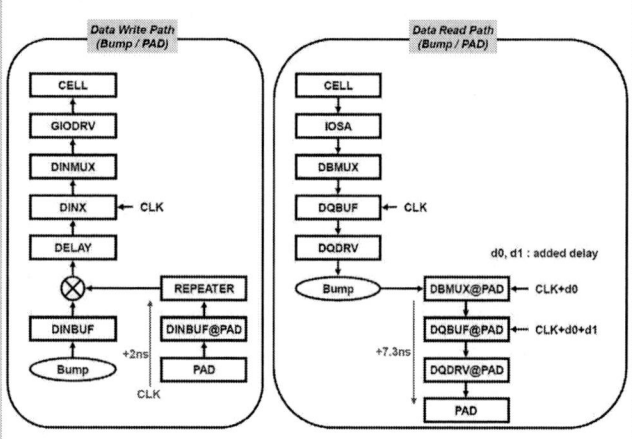

Figure 28.5.3: Correlation scheme for data write / read operations through microbumps and test pads.

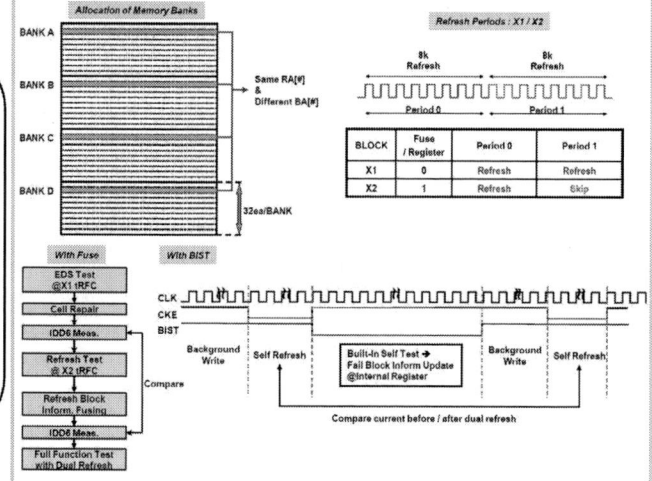

Figure 28.5.4: Scheme for dual data retention: Memory blocks allocation, dual refresh periods and block indexing procedure with fuse and BIST.

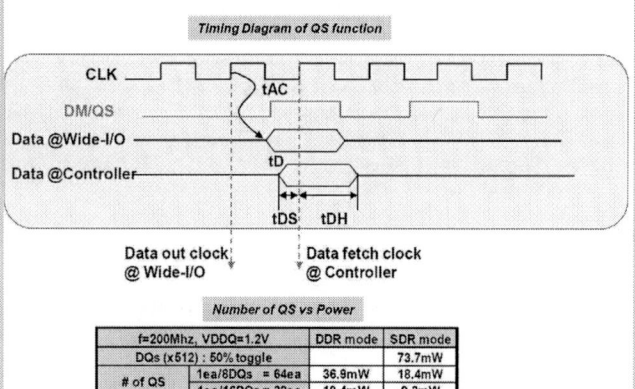

Figure 28.5.5: Timing diagram of QS function and power reduction according to the number of QS pins.

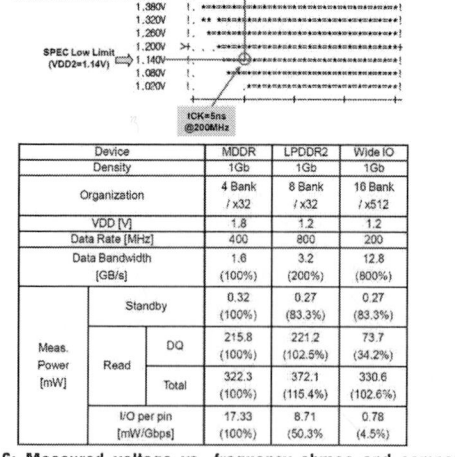

Figure 28.5.6: Measured voltage vs. frequency shmoo and comparison of features in Wide-I/O DRAM with MDDR and LPDDR2.

28

978-1-61284-303-2/11 $26.00 © 2011 IEEE

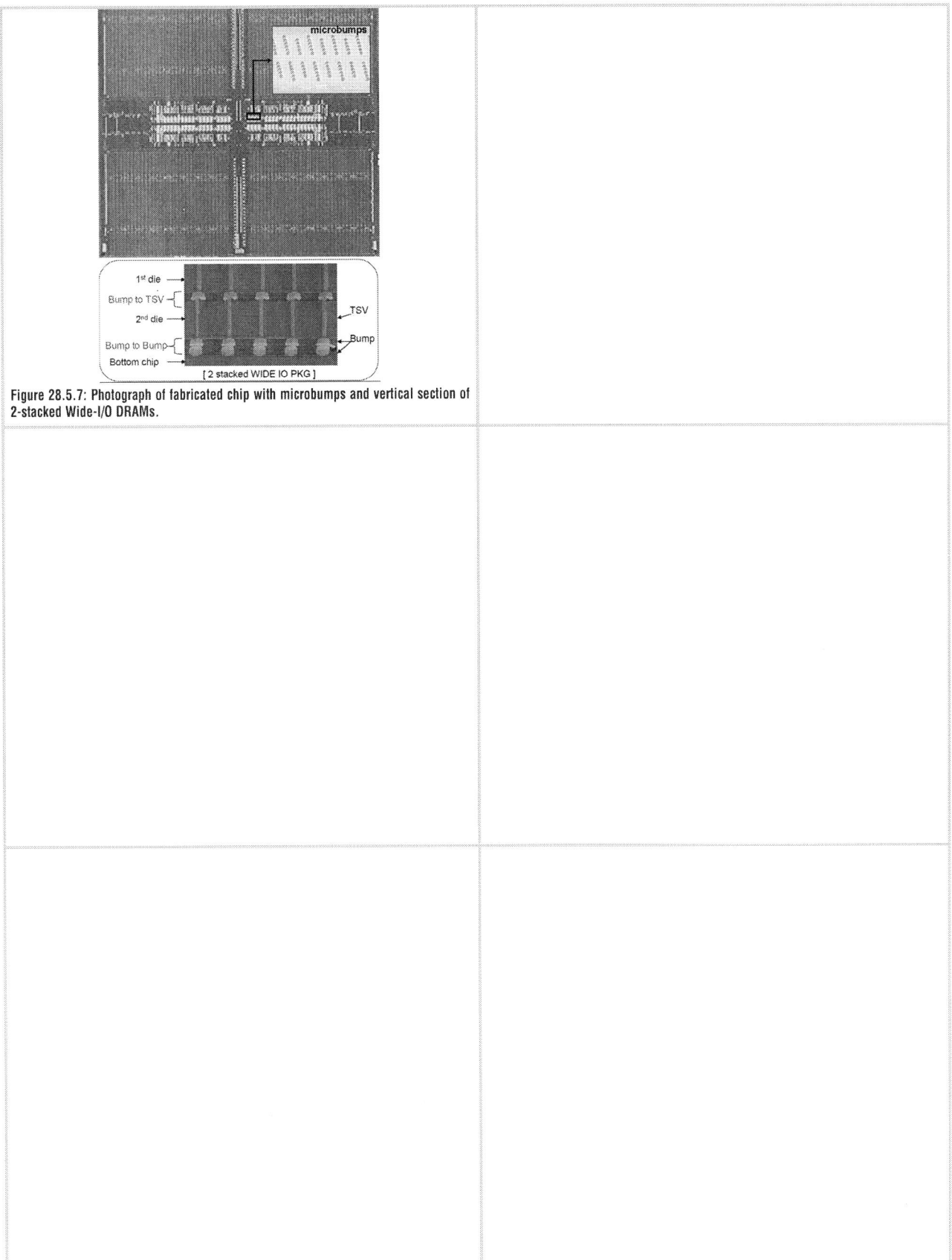

Figure 28.5.7: Photograph of fabricated chip with microbumps and vertical section of 2-stacked Wide-I/O DRAMs.

28.6 A 40nm 2Gb 7Gb/s/pin GDDR5 SDRAM with a Programmable DQ Ordering Crosstalk Equalizer and Adjustable Clock-Tracking BW

Seung-Jun Bae, Young-Soo Sohn, Tae-Young Oh, Si-Hong Kim, Yun-Seok Yang, Dae-Hyun Kim, Sang-Hyup Kwak, Ho-Seok Seol, Chang-Ho Shin, Min-Sang Park, Gong-Heom Han, Byeong-Cheol Kim, Yong-Ki Cho, Hye-Ran Kim, Su-Yeon Doo, Young-Sik Kim, Dong-Seok Kang, Young-Ryeol Choi, Sam-Young Bang, Sun-Young Park, Yong-Jae Shin, Gil-Shin Moon, Cheol-Goo Park, Woo-Seop Kim, Hyang-Ja Yang, Jeong-Don Lim, Kwang-Il Park, Joo Sun Choi, Young-Hyun Jun

Samsung Electronics, Hwasung, Korea

Most DRAM interfaces such as GDDR5 and DDR3 use parallel single-ended signaling due to pin-count restriction and backward compatibility. Notwithstanding poor signal and power integrity issues, GDDR5 speed reached beyond 5Gb/s in recent years by utilizing data bus inversion, error-detection coding, data training and channel equalization [1-3]. However, channel crosstalk is becoming a major barrier to further speed improvement. A common solution for channel crosstalk reduction at the system level is to use a shielding line or wide spacing between signal lines, but increasing the number of layers in a chip package and PCB increase system cost. To remove the shielding lines and increase speed, this paper presents a channel crosstalk equalizer with programmable signal ordering capability for the DRAM transmitter. In addition, this paper addresses tri-mode clocking to reduce the system jitter for better timing margin: PLL off, LC-PLL and injection-locked oscillator.

Figure 28.6.1 shows how the crosstalk equalizer removes crosstalk in victim DQ with a single bit response and the implemented 2-tap equalizing filter. Since equalizers for removing ISI are usually designed based on channel pulse response, this work also uses pulse response for analyzing crosstalk and designing crosstalk equalizing filter. The single bit pulse is applied on the aggressor DQ and the crosstalk response is captured at receiver side of the victim DQ as shown in Fig. 28.6.1. Since inductive coupling is larger than capacitive coupling in a DRAM channel, the crosstalk pulse on the victim occurs in the opposite direction of the aggressor. Since the main cursor crosstalk occurs at T_D and the post cursor crosstalk occurs at $T+T_D$, crosstalk equalizer requires 2-tap filter as shown in right of Fig. 28.6.1 where T_D is determined by channel characteristic. Since the output of the crosstalk equalizer should be the inversion of the victim signal, tap coefficients of main and post cursor are $+\alpha 0$ and $-\alpha 1$, respectively.

Although package ball-out and pad location of DQ are fixed at the DRAM side, DQ signal ordering of the whole system can be switched across board and controller package designs. For example, while the DQ ordering of one system is DQ0-DQ1-DQ2-DQ3, that of other system can be DQ0-DQ2-DQ3-DQ1. To deal with this issue, this work proposes a 4-DQ group transmitter with programmable crosstalk equalizer as shown in Fig. 28.6.2. Output driver of DQ0 consists of a main driver for DQ0 and three crosstalk equalizers for the three aggressors: DQ1, DQ2, and DQ3. Controller programs the crosstalk equalizers through the 6 bit DQ0_EN<0:5> code, where each consecutive two bits are assigned to each aggressors to either fully turn off the equalizer or control the equalizing coefficient. The other three DQs are also implemented similar to DQ0. To minimize the extra signal lines and circuit area for the crosstalk equalizer, the outputs, instead of the inputs, of the 4 to 1 MUX (DO0~DO3) are transmitted to every DQs as inputs to the crosstalk equalizers. The long routing (~1000μm) of DO0~DO3 nodes causes ISI at 7Gb/s, which is effectively removed by applying on-chip de-emphasis technique [3] as shown in Fig. 28.6.2.

Figure 28.6.3 shows the detailed circuit schematic of the output driver with crosstalk equalizer. In a current mode driver, equalizing filter is easily implemented by summing current at the output pin while maintaining output impedance [4-5]. In a voltage mode driver, equalizer filter is implemented by dynamically changing the pull-up and pull-down impedance and the combination of the main driver and the equalizer driver ratio determines the output level. In this work, this

is achieved by changing the shunted equalizer driver impedance, while maintaining the asymmetric main driver impedance as 60Ω pull-up and 40Ω pull-down. The size ×1 in Fig. 28.6.3 is calculated to match the target equalization output levels. Due to the asymmetric impedance of the main driver, the size ×1 yields different values for the pull-up and pull-down case. As a result, minor difference exists in the equalizing output levels (~10mV) between high and low data of the victim. In addition, the output impedance is reduced (60Ω → ~48Ω) due to the additional shunt impedance. However, in short channel systems such as graphics, this does not seriously hurt signal integrity. For a 2-tap equalizing filter, the second tap input (DO1B) is generated through three inverters for T_D. For coefficient control, two binary legs are used in both pull-up and pull-down equalizer driver. The equalizer driver adds an additional 100fF pin capacitance.

This work adopts 3 modes for flexible clocking to minimize system jitter: PLL off, LC-PLL on, and injection-locked oscillator (ILO) as shown in Fig. 28.6.4. PLL off mode is effective when the correlated low frequency jitter components of WCK and DQ in the controller are large. But when high frequency jitter is large, PLL off mode amplifies it by 2 due to the delay mismatch between the clock tree path and DQ path. PLL on mode is ideal to mitigate the high frequency jitter amplification; high loop BW PLL is better to match the loop BW to the minimum jitter frequency where amplification occurs. But when uncorrelated jitter is large, low loop BW PLL is better. In this work, LC-PLL [3] is implemented as a low BW (~5MHz) PLL to minimize random jitter. Furthermore, injection-locked oscillator mode [6] is added in LC-PLL to increase the loop BW up to 100MHz for maximizing the correlated jitter reduction. In injection-locked oscillator mode, outputs of WCK buffer are directly applied to the LC-VCO through input differential pairs and interpolated with the self-oscillated VCO output. The VCO bias current is adjusted proportional to coarse lock code to satisfy the start-up constraints [3]. In this work, the bias current of injection path is also adjusted proportional to coarse lock code to obtain constant injection ratio independent of lock frequency. Figure 28.6.5 shows measured output jitter results where 40ps input clock jitter of 100MHz is injected. Input jitter is bypassed in PLL off mode while the LC-PLL mode filters most of the input jitter. In ILO mode, the output jitter is in between PLL off and PLL on modes.

Figure 28.6.6 shows the measured eye patterns of read operation at a data rate of 7Gb/s with and without crosstalk equalizer. Data sequence of adjacent DQ is different from that of the measured DQ to inject both even and odd mode crosstalk. When crosstalk equalizer is fully turned on, the output jitter is reduced by 6ps at 7Gb/s while 10ps is reduced at 5Gb/s. Parasitic capacitance and process variation cause undesired increase of T_D value in Fig. 28.6.3, by which the equalizer performance is relatively impaired more at 7Gb/s compared to 5Gb/s. Figure 28.6.7 shows the chip micrograph and summary of the 2Gb GDDR5 SRAM fabricated in a 40nm CMOS DRAM process with 3-layer metal technology.

References:
[1] Seung-Jun Bae, Young-Soo Sohn, Kwang-Il Park, et al, "A 60nm 6Gb/s/pin GDDR5 Graphics DRAM with Multifaceted Clocking and ISI/SSN-Reduction Techniques," *ISSCC Dig. Tech. Papers*, pp.278-613, Feb. 2008.
[2] Kho, R., Boursin, D., Brox, M, et al, "75nm 7Gb/s/pin 1Gb GDDR5 graphics memory device with bandwidth-improvement techniques," *ISSCC Dig. Tech. Papers*, pp.134-135, Feb. 2009.
[3] Seung-Jun Bae, Young-Soo Sohn, Tae-young Oh, et al, "A 40nm 7Gb/s/pin Single-ended Transceiver with Jitter and ISI Reduction Techniques for High-Speed DRAM Interface," *Symp. VLSI Circuits, Dig. Tech. Papers*, pp.193-194, Jun. 2010.
[4] Zerbe, J.L., Chau, P.S., et al, "A 2 Gb/s/pin 4-PAM parallel bus interface with transmit crosstalk cancellation, equalization, and integrating receivers," *ISSCC Dig. Tech. Papers*, pp.66-67, Feb. 2001
[5] Sham, K., Ahmadi, M., et al, "FEXT Crosstalk Cancellation for High-Speed Serial Link Design," *IEEE CICC*, pp. 405-406, Sept. 2006.
[6] F. O'Mahony, S. Shekhar, M. Mansuri, et al., "A 27 Gb/s forwarded-clock I/O receiver using an injection-locked LC-DCO in 45nm CMOS," *ISSCC Dig. Tech. Papers*, pp. 452–453, Feb. 2008.

ISSCC 2011 / February 23, 2011 / 3:45 PM

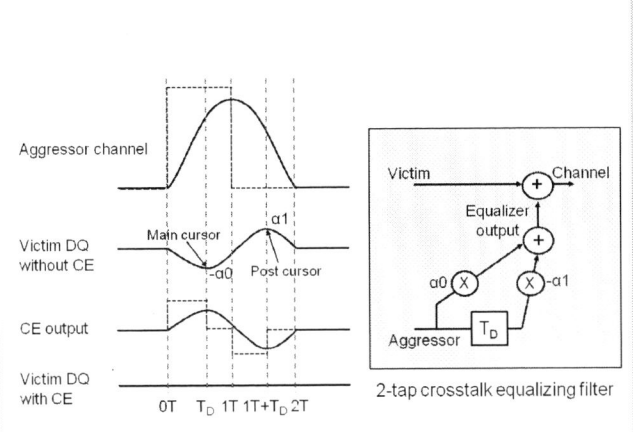

Figure 28.6.1: Crosstalk equalization (CE) in single bit pulse.

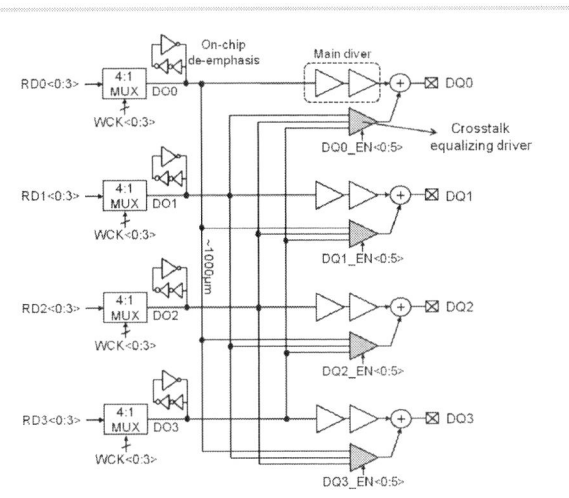

Figure 28.6.2: 4-DQ group transmitter with programmable crosstalk equalizer.

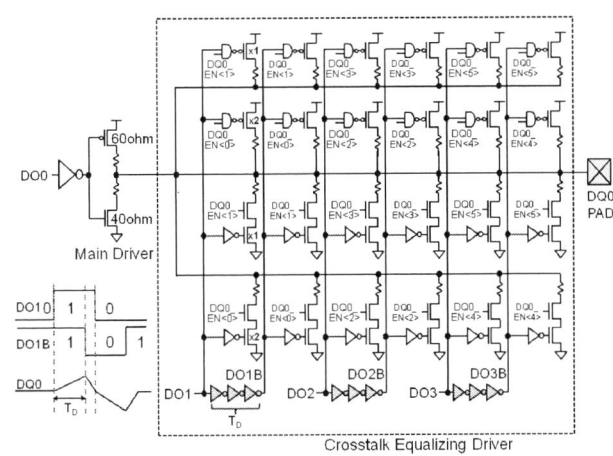

Figure 28.6.3: Voltage mode output driver with crosstalk equalizer.

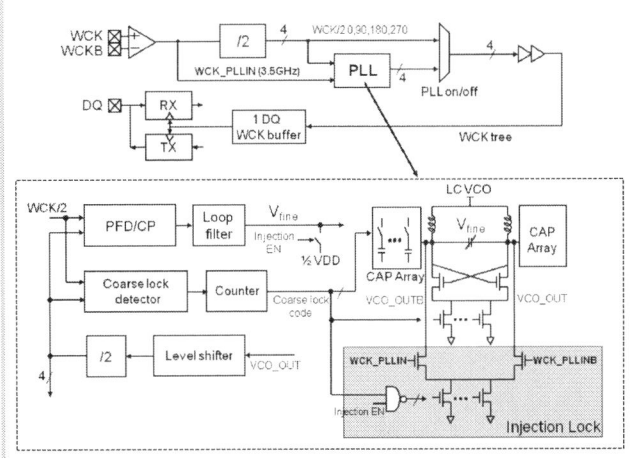

Figure 28.6.4: BW controllable clocking architecture with injection-locked oscillator.

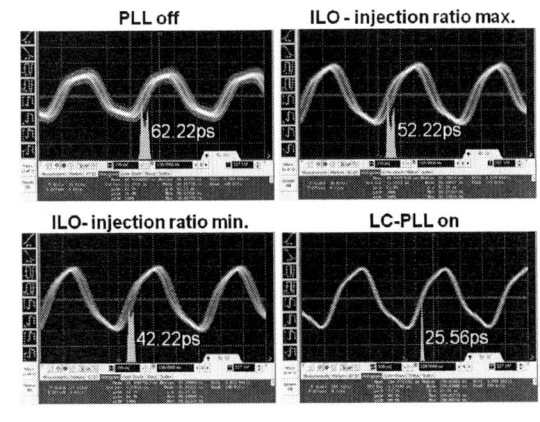

Figure 28.6.5: Output jitter of each clocking mode for 100MHz input jitter of 40ps.

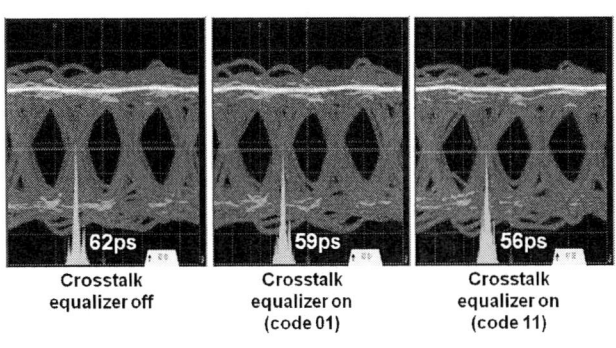

Figure 28.6.6: Measured output eye diagram with and without crosstalk equalizer at 7Gb/s.

28

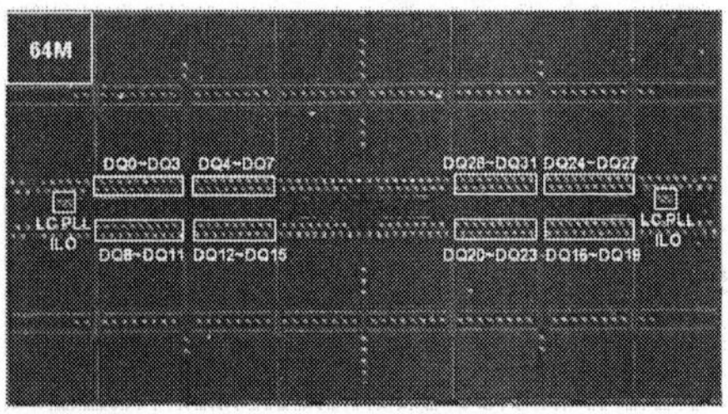

40nm CMOS 3 metal
170ball Flip-chip
7Gb/s/pin
16bank
64Mx32b
X32 IO
1.5V
80.6 mm²

Figure 28.6.7: Chip photograph and summary.

ISSCC 2011 / SESSION 28 / DRAM & HIGH-SPEED I/O / 28.7

28.7 A 58nm 1.8V 1Gb PRAM with 6.4MB/s Program BW

Hoeju Chung, Byung Hoon Jeong, ByungJun Min, Youngdon Choi, Beak-Hyung Cho, Junho Shin, Jinyoung Kim, Jung Sunwoo, Joon-min Park, Qi Wang, Yong-jun Lee, Sooho Cha, Dukmin Kwon, Sangtae Kim, Sunghoon Kim, Yoohwan Rho, Mu-Hui Park, Jaewhan Kim, Ickhyun Song, Sunghyun Jun, Jaewook Lee, KiSeung Kim, Ki-won Lim, Won-ryul Chung, ChangHan Choi, HoGeun Cho, Inchul Shin, Woochul Jun, Seokwon Hwang, Ki-Whan Song, KwangJin Lee, Sang-whan Chang, Woo-Yeong Cho, Jei-Hwan Yoo, Young-Hyun Jun

Samsung Electronics, Hwasung, Korea

In mobile systems, the demand for the energy saving continues to require a low power memory sub-system. During the last decade, the floating-gate flash memory has been an indispensable low power memory solution. However, NOR flash memory has begun to show difficulties in scaling due to the device's reliability and yield issues. Over the past few years, phase-change random access memory (PRAM) has emerged as an alternative non-volatile memory (NVM) owing to its promising scalability and low cost process [1,2]. In this paper, a PRAM, implemented in a 58nm PRAM process with a low power double-data-rate non-volatile memory (LPDDR2-N) interface, is presented [3].

Figure 28.7.1 shows the functional block diagram of the designed LPDDR2-N PRAM with 1Gb diode-switch cell array. In LPDDR2-N interface, to reduce the system routing complexity and pin count, the DRAM and PRAM share the command-address (CA) and data (DQ) channel. The PRAM that has significantly low programming bandwidth compared to the DRAM-write has a SRAM-based 1KB program buffer with 800Mb/s write throughput and internally controls the programming operation to cope with such a huge bandwidth difference. The PRAM gets an embedded command set through the DQ channel when a designated addressing mode enabled, that is, an overlay window mode. If the overlay window mode is enabled and the address are matched with the stored address, the overlay window match (OWMTCH) signal is asserted and the data from DQs are used as an embedded command set. For the programming operation, as an example, the memory stores the address, data count, data for programming, and the mask information through DQs and DMs. After the execution is issued, the PRAM enters into the embedded programming mode and the internal FSM performs the programming sequence. The memory controller could check the PRAM's status by reading the status register. For the data read out, the PRAM gets a row address at pre-active and active phase. Then the core data sensed and stored at the row data buffer (RDB) after sensing time. The data stored in RDB can be read out by read command with column address.

In the LPDDR2-N interface, the performance requirements for the core-write bandwidth is tight compared to the conventional PRAM with NOR-flash interface. To meet the required core-write performance, the number of the simultaneous program bits needs to be increased and the program current injected to core, accordingly. Since such an amount of program current can cause a significant voltage bouncing and can be a burden for the pumping capacity during the operation, the number of simultaneous activating core-write driver (WRDRV) is limited. This work proposes a method to manipulate the ground bouncing. All WRDRVs are grouped into two, the main and the skew, and for each group, pulses of separate timing by tSKEW are applied as depicted in Fig. 28.7.2. By doing this, superpositioning of the program peak current is prevented without significant performance degradation.

In addition, we design a data comparison write with inversion flag (DCWI) scheme. A 256b sized row (32 bytes) is divided into four groups, and each group of 64b has a flag bit that indicates whether the group of the data is inverted or not. By the way, the difference between 0→1 (RESET to SET) and 1→0 (SET to RESET) time is so large that we have considered the effective number of change in making a decision for the better core-write performance. Figure 28.7.3 shows the circuitry for making a decision of inversion flag. First, when the core-write execution is issued, the embedded controller starts to read the stored data from PRAM core and compares them with the data to be programmed. After the data comparison, the counters calculate the number of required changes 0→1 and 1→0 for the case of both non-inversion and inversion. Then, by multiplying the required changes and the ratio of SET period and RESET period (α), the number of effective changes is obtained. Based on these results, the embedded controller performs the programming operation with the smaller effective changes. In ideal case, the DCWI scheme enhances the worst case write bandwidth by two if α is 1. In this work, the ratio α can be chosen by register setting. In the case of erase operation, all the changes from 0 to 1 are enabled and subsequent sequence is the same as overwrite (erase-less core-write) case. Notice that the core-read operation prior to DCWI decision is carried out via core-write data bus.

Figure 28.7.4 shows the comparison of the core-write bandwidth between DCWI enabled and disabled. The worst case performance is enhanced by 1.5× in the case of DCWI enabled for both programming and overwriting. Note that the performance isn't enhanced by 2 because the DCWI scheme requires the additional programming time for the inversion flag and control sequence. The program bandwidth is measured through the device ready bit (DRB) under 50% data transition.

The stacked partition architecture gives us high cell efficiency in the PRAM. Figure 28.7.5 shows the proposed sensing scheme designed for high read speed operation. Special care has been taken to finding appropriate solution to cope with the big RC delay of global bit line (GBL) in 1Gb 16 stacked partition architecture. Compared with the conventional architecture in the left figure, the proposed architecture enhances the efficiency of pre-charge operation through the additional pre-charge path in the middle of the PRAM cell array, named as mid-array pre-charge scheme, presented in the gray blocks. As a result, the RC delay of GBL line is decreased to 1/16 level and the fabricated chip achieves the successful RAS to CAS delay time (tRCD) <80ns operation.

Figure 28.7.6 shows the test results of this work. The output data valid window is measured to be 3.4ns at VDD1 is 1.8V, 400Mb/s. The tRCD value is measured to be 76ns at 85°C, VDD1 is 1.6V. Figure 28.7.7 shows a chip micrograph of the 1Gb diode switch LPDDR2-N PRAM and the summary, fabricated in a 58nm PRAM process with stacked partition architecture. If the proposed DCWI enables, the program and overwrite bandwidth are measured by 6.4MB/s and 2.3MB/s respectively.

Acknowledgement:
We appreciate the Samsung's PRAM process group for the chip fabrication.

References:
[1] K.Lee, et al., "A 90nm 1.8V 512Mb Diode-Switch PRAM with 266MB/s Read Throughput," *ISSCC Dig. Tech. Papers*, pp.474-473 , Feb., 2007.
[2] Corrado Villa, et al., "14.8 A 45nm 1Gb 1.8V Phase-Change Memory," *ISSCC Dig. Tech. Papers*, pp.270-271 , Feb., 2007.
[3] "Low Power Double Data Rate 2 (LPDDR2)", *JEDEC STANDARD, JESD 209-2D*, Apr., 2010.

ISSCC 2011 / February 23, 2011 / 4:15 PM

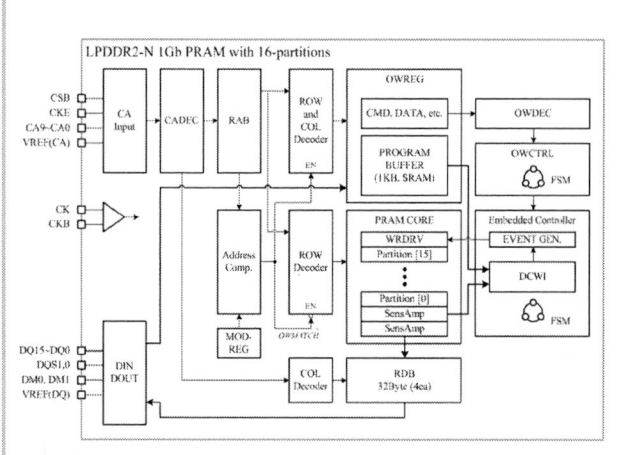

Figure 28.7.1: The functional block diagram of LPDDR2-N PRAM including embedded controller.

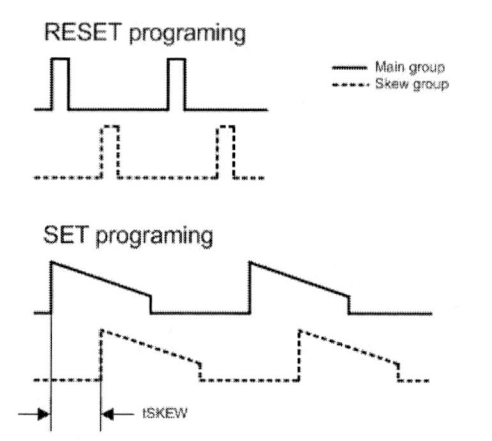

Figure 28.7.2: Core-write driver skewing concept.

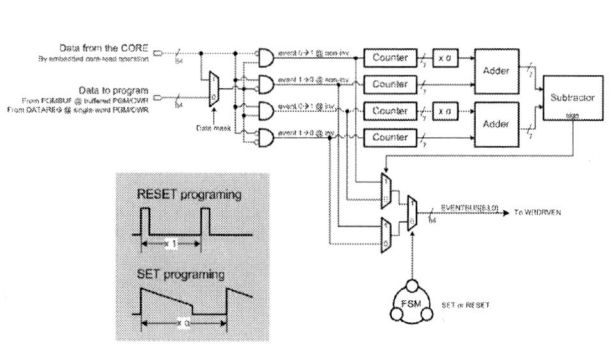

Figure 28.7.3: Decision circuits for the data comparison write with inversion flag.

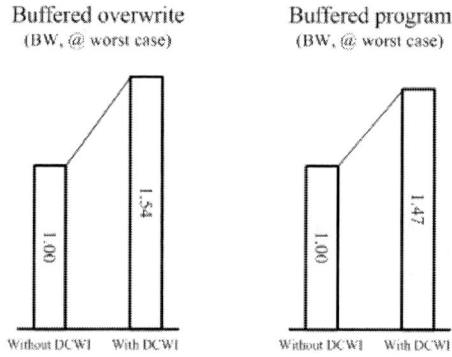

Figure 28.7.4: Write performance comparison between with and without DCWI.

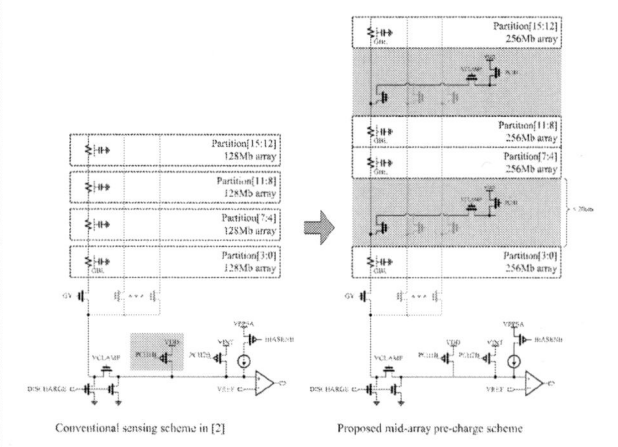

Figure 28.7.5: Mid-array pre-charge sensing scheme.

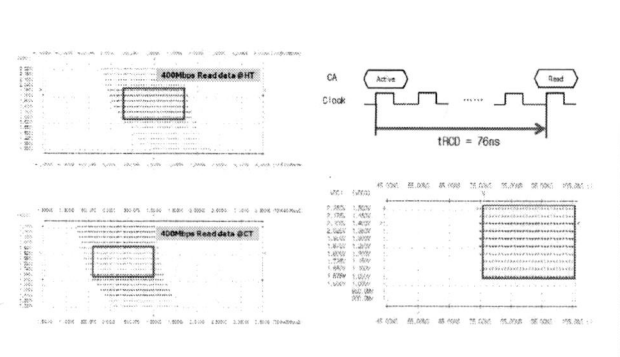

Figure 28.7.6: Test results for the output valid data window and tRCD.

28

978-1-61284-303-2/11 $26.00 © 2011 IEEE   501

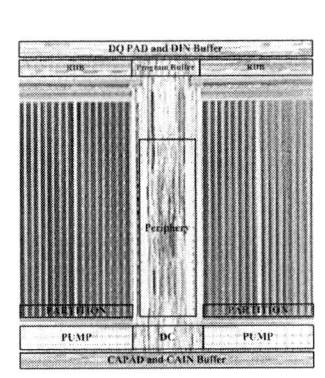

Process technology	58nm PRAM Process
Chip size	63.4 (mm²), 4F²
Cell switch	Diode-switch
Power supply	VDD 1.8V (typical) VDDQ, VDDCA 1.2V (typical)
Temperature range	-25 ~ 85 °C
Architecture	16-stacked partition architecture
Core-write Performance (with DCWL enabled)	Single word program: 20us Single word overwrite: 20us Buffered program: 6.4MByte/s Buffered overwrite: 2.3MByte/s Erase: 22.2KByte/s
Core-read Performance	(RCD = 76nsec (at 85°C, VDD = 1.8V)
IO speed	800 Mbps/pin

Figure 28.7.7: Chip micrograph and summary.

ISSCC 2011 / SESSION 28 / DRAM & HIGH-SPEED I/O / 28.8

28.8 A 1.6V 1.4Gb/s/pin Consumer DRAM with Self-Dynamic Voltage-Scaling Technique in 44nm CMOS Technology

Hyun-Woo Lee[1,2], Ki-Han Kim[1], Young-Kyoung Choi[1], Ju-Hwan Shon[1], Nak-Kyu Park[1], Kwan-Weon Kim[1], Chulwoo Kim[2], Young-Jung Choi[1], Byong-Tae Chung[1]

[1]Hynix Semiconductor, Icheon, Korea,
[2]Korea University, Seoul, Korea

DRAM's process technology has been scaled down rapidly and the size of the wafer has reached 300mm. Despite being fabricated on the same wafer, two chips may have very different characteristics if the one is from the center and the other is from the edge of wafer. Therefore, the process skew reduction is becoming more important as the process is scaled down under 100nm. The dynamic voltage scaling scheme (DVS) has already won huge popularity in mobile applications with limited battery life. Various dynamic voltage scaling techniques for μ-processors have also been developed during the last decade [1]. However, selection of the power supply voltage for DRAM is dictated by the application or the worst process skew that guarantees the performance of DRAM. This paper proposes a self-dynamic voltage scaling (SDVS) technique for DRAM to overcome the process variation and reduce the power consumption according to the operating frequency.

DRAM has several internal voltage levels which are VPP for driving word line, VBLP for bit line pre-charge, VCP for cell capacitor plate, VCORE for data store and VBBW for off-state bias of word line to reduce the off-leakage. These internal voltages have to be used to satisfy the data store operating and keep the charges during retention time. Therefore, it is hard to change these internal voltages for DRAM core. On the other hand, performance of peripheral circuits is closely connected with the FO3 (Fan-Out 3) delay. Therefore, the internal voltage level for peripheral circuits can be easily adapted according to the operating frequency and the process skew.

Recently, application systems with DVS are being widely used to reduce power consumption. However, they did not take the process condition of DRAM into account. The proposed SDVS technique uses two parameters; one is the process skew of DRAM component and the other is the frequency of the CLK fed to DRAM. The proposed SDVS block diagram is shown in Figure 28.8.1. The SDVS generates the signals UP and DN for the voltage regulator after measuring the tRDLY in terms of the number of CLK cycles that comes from the controller. The tRDLY copies the timing delay between the CLK input and the DQ output. In case of slow skew, the signal UP is enabled. So the internal voltage for peripheral circuits (VPERI) is increased to improve the performance. Otherwise, if the operating frequency is slow enough to ignore the reduced timing margin, the SDVS maintains the previous VPERI or lowers the VPERI by the signal DN.

By the help of SVDS, each DRAM can have an optimal VPERI that lowers the leakage and active standby current. The SDVS determines the VPERI during power up sequence or self-refresh exit condition.

Figure 28.8.2(a) shows the relative FO3 delay distribution within a wafer. Randomly selected 450 samples are used for measurement. The slowest FO3 delay is improved by 20% when VPERI is raised from 1.1V to 1.2V. At VPERI of 1.1V, the FO3 delay variation is almost 30% even if the measured samples are from the same wafer. The more wafers fabricated, the more variation occurred. To satisfy the given specification of DRAM and to improve the yield, it is necessary to improve the performance of slow transistors. However, this may result in increased leakage and operating current because fast transistors become faster. The proposed SDVS improves the performance of slow transistors without affecting the very fast skew and reduces the leakage current of fast transistors by selecting the optimal VPERI. The normalized gate leakage is shown in Figure 28.8.2(b). The reliability of DRAM is also improved due to the SDVS by changing VPERI according to process skew during the lifetime of DRAM. Figure 28.8.2(c) shows the lifetime measurement results. The lifetime of transistor with 1.5V is 23.7 times longer than that with 1.8V. The tAA can be changed by the SDVS as shown in Figure 28.8.2(d). The SDVS can increase the tAA by about 9% of nominal value by lowering the VPERI and decrease the tAA by about 8% by raising the VPERI.

Generally, DLL performance depends highly on the supply voltage fluctuation [2]. The architectures of the proposed DLL and output enable (OE) control logic are shown in Figure 28.8.3. Because the VPERI is changed during initializing sequence by SDVS, the DLL needs a wide bandwidth to obtain fast lock. However, the DLL bandwidth is limited by the length of the clock path from the input node (iCLK) of the variable delay line to the output node (fbCLK) of the phase detector (K×tCK-tD1+tD1=K×tCK, K is an integer) as shown in Figure 28.8.4(a). The maximum bandwidth that the DLL can have is 1/(tCK×K). To increase the bandwidth of DLL, the parameter "K" needs to be decreased. To guarantee performance in all process corners and various operating frequencies, "K" has to be a large integer number [3]. The proposed SDVS calculates the "K" value on the fly according to the process skew and the operating frequency. The proposed DLL uses CODE<0:N> that is calculated from tRDLY. CODE<0:N> is binary code for "K". By using CODE<0:N>, the DLL can be operated with optimal bandwidth.

The fine delay makes a slow update to compensate the tDQSCK. For low frequency operation, low resolution is better for fast tracking. Fine unit delay is changed by changing the VPERI. The measured fine unit delay is shown in Figure 28.8.4(b). The power consumption of DLL with SDVS by the calculation from "IDD3PF – IDD3PS" is 2.5mW [6.25pJ/Hz] at 400MHz and 1.8V with 90°C.

To adjust the Cas Latency of DRAM [4], the conventional OE control logic needs a DLL clock for "domain crossing". Though the DLL clock is only needed for read operation, the DLL clock should always be running when the ACT (active) command is issued because of lack of timing margin between the delay from DLL output to input of DFF in OE control logic (tD2) and the RD (read) command. It is shown in Figure 28.8.4(c). It indicates that the IDD3 condition needs more current. To assure the timing margin of OE, the DLL clock must arrives (K-1)×tCK time ahead of RD command to align DLL clock with RD command for "domain crossing". The proposed DLL compares (K-1)×tCK and tD2, if tD2 is smaller than (K-1)×tCK, the DLL controls the DLL clock driver with read command. If not, the DLL controls it with active command. tD2 is almost the same amount of delay as tRDLY.

This work introduces the first ever DVS technique for DRAM considering both the process skew and the operating frequency which is adopted for the Consumer DDR2 SDRAM. The SDVS itself is a very powerful technique to stretch the battery life and increase the reliability of DRAM. Figure 28.8.5(a) shows the shmoo plot of tCK vs. tAC at 25°C. Figure 28.8.5(b) is a signal scope under one of the blue-ray players. By the SDVS, the variation of normalized IDD3P is reduced from 0.106 to 0.047 as shown in Figure 28.8.6(a). The SDVS decrease the VPERI if DRAM is under the fast skew. It seems that two out of 8 samples are the faster skew. Figure 28.8.6(b) shows normalized IDD3N comparison graph according to OE control scheme on and off. When tCK is 2ns, the DLL clock driver is rarely controlled by RD command to improve the speed performance. But when tCK is 5ns, the IDD3N is reduced by 12%. The SDVS selects the optimal VPERI for each operating frequency. The micrograph of the chip, fabricated in 44nm CMOS technology, is shown in Figure 28.8.7 and the SDVS, DLL and OE occupy the active area of 0.147mm².

Acknowledgements:
We appreciate Debashis Dhar, Yun-Hyung Hoe and test team for their helpful discussion and support.

References:
[1] Dighe. S, et al., "Within-die variation-aware dynamic-voltage-frequency scaling core mapping and thread hopping for an 80-core processor" *ISSCC Dig. Tech. Papers*, pp. 174-175, Feb. 2010.
[2] H. W. Lee, et al., "An 1.6V 3.3Gbps Dual-Mode Phase and Delay Locked Loop using power noise management technique with unregulated power supply for pseudo-rank DRAM in 54nm CMOS technology" *ISSCC Dig. Tech. Papers*, pp. 140-141, Feb. 2009.
[3] W. J. Yun, et al., "A 0.1-to-1.5GHz 4.2mW ALL-Digital DLL with Dual Duty-Cycle Correction Circuit and Update Gear Circuit for DRAM in 66nm CMOS Technology," *ISSCC Dig. Tech. Papers*, pp. 282-283, Feb. 2008.
[4] D. U. Lee, et al., "A 2.5Gb/s/pin 256Mb GDDR3 SDRAM with Series Pipelined CAS Latency Control and Dual-Loop Digital DLL," *ISSCC Dig. Tech. Papers*, pp. 547-556, Feb. 2006.

ISSCC 2011 / February 23, 2011 / 4:45 PM

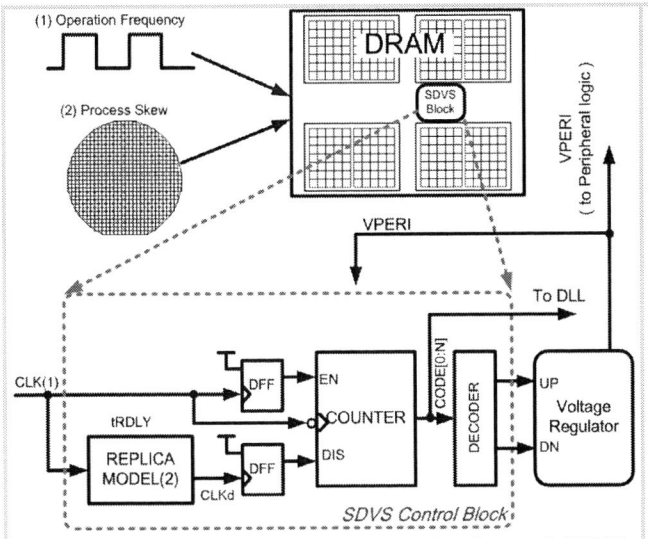

Figure 28.8.1: Block diagram of the proposed dynamic voltage scaled DRAM.

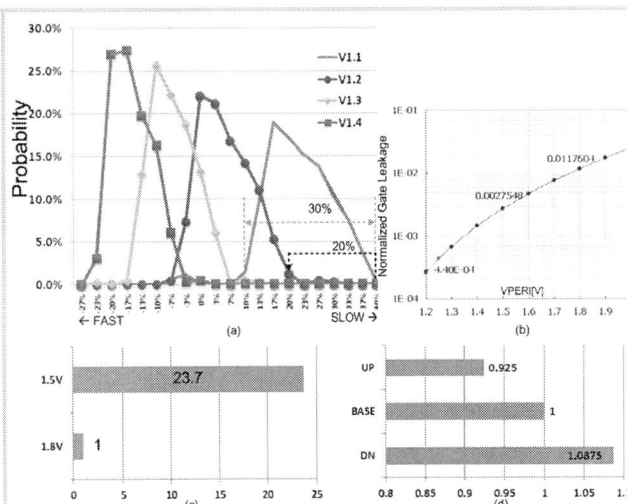

Figure 28.8.2: (a) Relative FO3 delay distribution (b) Normalized Gate Leakage current (c)Normalized AC life time (d)Normalized tAA trend.

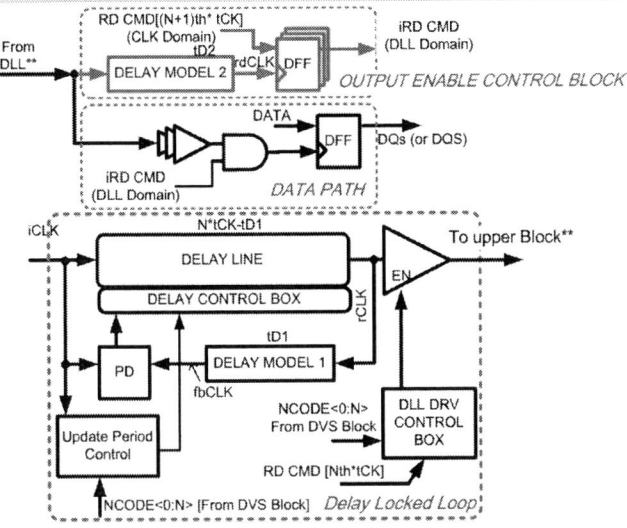

Figure 28.8.3: The architecture of DLL and Output Enable control logic.

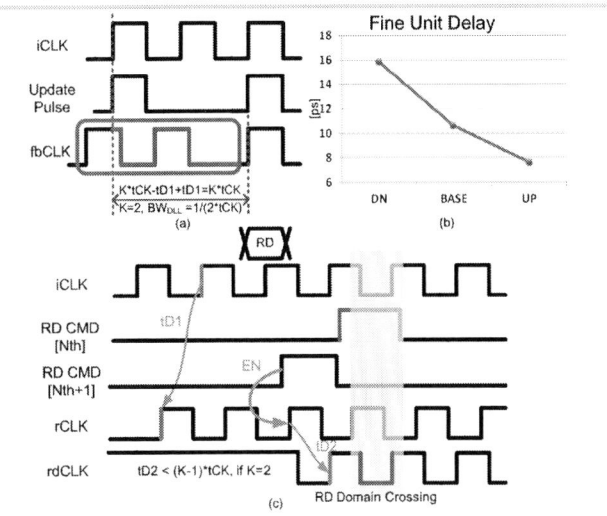

Figure 28.8.4: (a) Timing diagram of bandwidth of DLL(b) Fine unit delay measurement results (c) Timing diagram of domain crossing.

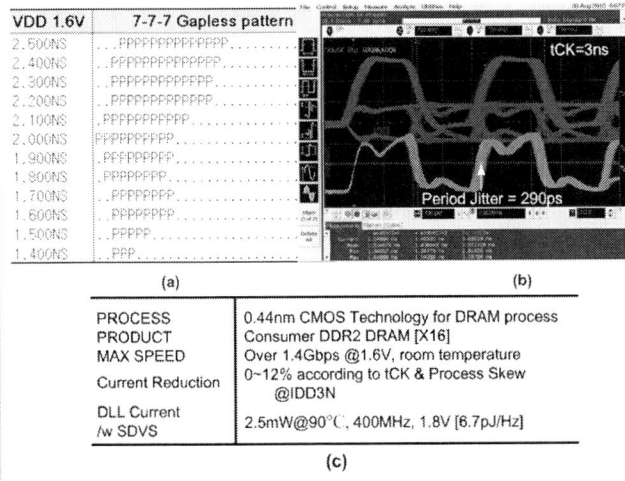

PROCESS	0.44nm CMOS Technology for DRAM process
PRODUCT	Consumer DDR2 DRAM [X16]
MAX SPEED	Over 1.4Gbps @1.6V, room temperature
Current Reduction	0~12% according to tCK & Process Skew @IDD3N
DLL Current /w SDVS	2.5mW@90℃, 400MHz, 1.8V [6.7pJ/Hz]

(c)

Figure 28.8.5: (a)tCK vs tAC Shmoo Plot at 25°C by ATE (x-axis: tAC, y-axis: tCK, 50ps step) (b)DQS Jitter @Blue-ray player (c)Performance Summary.

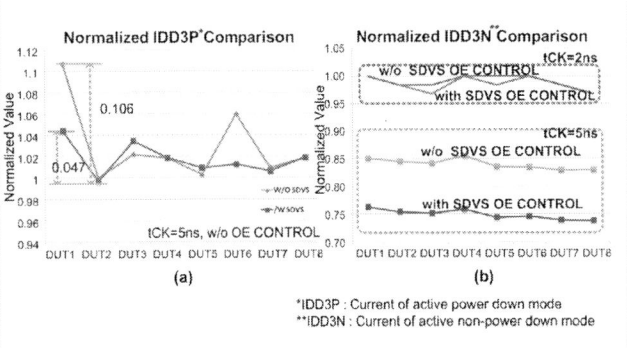

*IDD3P : Current of active power down mode
**IDD3N : Current of active non-power down mode

Figure 28.8.6: (a) Normalized IDD3P comparison (b) Normalized IDD3N comparison.

28

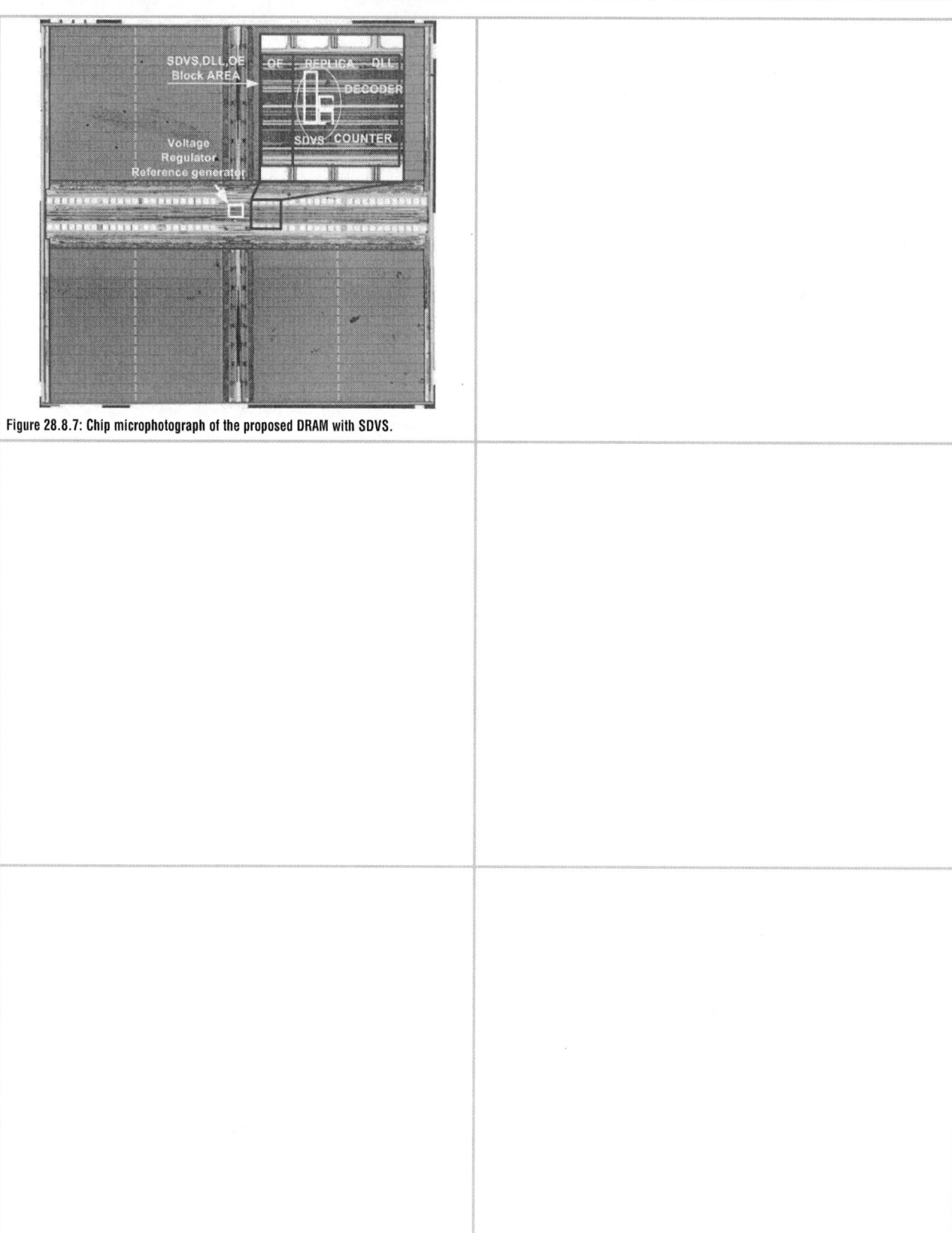

Figure 28.8.7: Chip microphotograph of the proposed DRAM with SDVS.

ISSCC 2011 / SESSION 28 / DRAM & HIGH-SPEED I/O / 28.9

28.9 An Embedded DRAM Technology for High-Performance NAND Flash Memories

Daisaburo Takashima, Mitsuhiro Noguchi, Noboru Shibata,
Kazushige Kanda, Hiroshi Sukegawa, Shuso Fujii

Toshiba, Yokohama, Japan

The increase of NAND flash memory capacity has strongly contributed to market growth of a variety of mobile equipment such as visual/audio electronics, smart phones, and SSD/USB devices. This memory capacity increase up to 64Gb [1] is attributed to multi-level cell (MLC) technology up to 4b/cell as well as shrinking process. However, the realization of MLC requires tight cell V_t distribution by the precise and gradual programming [2] and pre-programming to the neighboring wordlines to compensate the floating-gate coupling [1,3]. This gives rise to poor program throughput due to long program time when increasing from 1b/cell to 4b/cell, as shown in Fig. 28.9.1(a). Therefore, a lot of page buffers beside a cell array have been introduced to increase the number of simultaneously programmed cells to maintain the program throughput. Moreover, multiple data latches in a page buffer are required to latch all bits from/to each MLC [3], as shown in Fig. 28.9.1(b). Consequently, the total page buffer size per chip has increased remarkably and has reached 64KB (512Kb) in a 4b/cell 64Gb chip as shown in Fig. 28.9.1(c). These results imply that numerous page buffers based on SRAM flip-flops are mandatory and their area overhead is a serious problem in scaled NAND flash memories.

This paper demonstrates a prototype of embedded DRAM using a standard NAND flash memory process. This embedded DRAM can replace on-chip SRAM page buffers and other caches with minimum cost to enhance NAND flash memory performance. This opens up a new technical direction and wide applications for NAND flash memories. Figure 28.9.2(a) shows the concept of the embedded DRAM. An obstacle to embed DRAM in a NAND flash memory chip is a lack of cell charge, because the cell capacitance C_s using a planar MOS capacitor is limited to 3fF at the highest, whereas that of an ordinary embedded DRAM is around 20fF by using extra trenched or stacked capacitor process steps. However, even with a small 3fF cell, a cell node bias up to 4V, which is 4 times as high as the ordinary DRAM bias by taking advantage of the high-voltage NAND flash process, and a low parasitic capacitance bitline of 60fF, together enable sufficient ±100mV cell signals. A memory cell is composed of a cell transistor and a depletion-type MOS capacitor, and the cell size is 1.5μm², as shown in Figs. 28.9.2 (b). The large 4V bias to the cell node is realized by adopting a self-boost technique [2,4], which is widely used for the program inhibit of NAND flash memory, as shown in Fig. 28.9.2(c). After writing the bitline BL voltage of 2V V_{int} for "1" data and 0V for "0" data into a cell node and pulling down the wordline WL from 3.4V V_{pp} to 2V V_{int}, the plateline PL is boosted to 3.4V V_{pp}. Therefore, the cell node voltage of "1" data is self-boosted to around 4V due to cell transistor cut-off by negative V_{gs}-V_t, while that of "0" data is kept at ground due to cell transistor being turned-on.

A small bitline BL capacitance is realized by the zigzag column-select line (CSL) of Fig. 28.9.3(a) in addition to the initial sensing using only pMOS sense amplifier SA [5]. A CSL is shared with 8 SAs and 8 BL pairs, and is arranged in a zigzag fashion to have 1/8 of parasitic capacitance equally with each BL pair in order to minimize parasitic coupling between BL and CSL. This technique reduces total BL capacitance C_b from 75fF to 60fF at 128WL/BL, and achieves ±100mV cell signals, as shown in Fig. 28.9.3(b).

A concern of DRAM embedded into a NAND flash memory is that the sensing operation when adopting the usual ½ V_{int} BL precharge scheme is marginal at worst PVT due to the relatively high-V_t required for low-standby. Therefore, a

grounded-BL precharge scheme is preferable. However, this causes an undershoot problem of "0" data cell nodes, as shown in Fig. 28.9.4(a). When cell data read out to bitlines are mostly "1", their boosted cell nodes CNs are discharged, and the plateline PL, which should be set to 3.4V V_{pp}, are pulled down for an instant, due to cell capacitor coupling. This unwanted PL pull-down causes undershoot of minority "0" data cell nodes to negative bias. The 2-step-rise/fall wordline scheme shown in Fig. 28.9.4(b) overcomes this problem. The wordlines WLs are raised to 2V V_{int} until cell charges of "1" data are read out. After that, WLs are fully pulled up to 3.4V V_{pp}. This suppresses the unwanted PL pull-down and undershoot of "0" data cell nodes.

A key design issue for embedding DRAM in a flash process is how to set the reference BL voltage to an intermediate level between "1" and "0" data, because the readout signal of "1" data varies with variations of V_t of cell MOS capacitor, BL swing voltage V_{int}, and boosted WL and PL voltage V_{pp}. A perfect solution is to adopt a dummy cell to obtain an average of "1" and "0" data. But the dummy cell requires 7 transistors. A-half-of-"1"-data dummy cell scheme, shown in Fig. 28.9.5(a), minimizes layout area and compensates for the variations. The dummy-cell node DC, which is precharged to V_{int} by activating DWW signal during stand-by, is boosted by pulling up dummy plateline DPL to V_{pp}. After that, the cell charge of DC is delivered to two reference bitlines /BL0 and /BL1 by enabling DWL signal. There is a signal mismatch between (1) a half of "1" data and (2) an average of "1" and "0" data. This mismatch is expressed approximately as $\frac{1}{4}(C_s/C_b)^2(V_{int}+V_{pp}) + \frac{1}{2}(C_s/C_b)V_{pre}$, where V_{pre} is BL precharge voltage. However, this mismatch is minimized to 3mV when the grounded BL precharge is applied (V_{pre}=0V) as shown in Fig. 28.9.5(b), because the first term: $\frac{1}{4}(C_s/C_b)^2(V_{int}+V_{pp})$ is negligible, and the second term: $\frac{1}{2}(C_s/C_b)V_{pre}$ is 0 when V_{pre} is 0V. This means that the mismatch of inflowing charge into the dummy cell with different capacitance from BL precharged to V_{pre} is solved. Figure 28.9.5(c) shows the simulated dependency of "1" data, "0" data, and reference BL signals on V_t of the depletion-type MOS capacitor, and the dependency of those on V_{pp} and V_{int}. The reference BL voltage maintains an intermediate value between "1" and "0" data at any fluctuated conditions.

The embedded DRAM macro is fabricated using a 32nm standard NAND flash memory process. Figure 28.9.6(a) shows the micrograph of a 32KB (256Kb) DRAM macro with die size of 2×0.6mm². This realizes 2.4mm²/Mb macro density. The entire simulation result of DRAM macro at the worst PVT is also shown in Fig. 28.9.6(b). A random read/write cycle is 90ns and a burst cycle is 15ns (66Mb/s/pin) at 8-I/Os. The maximum burst page length is 256B. These results suggest that 32b-I/Os design of 266MB/s by four 32KB macro instantiations in a chip is enough for page buffer and cache applications. The active current of 32KB macro is 6.3mA at 90ns random read/write cycle, and 1.6mA with practical use of 15ns burst with 256B page access.

References:

[1] C. Trinh et al., "A 5.6MB/s 64Gb 4b/cell NAND flash memory in 43nm CMOS," in ISSCC Dig. Tech. Papers, Feb. 2009, pp. 246-247.
[2] K. D. Shu et al.,"A 3.3 V 32 Mb NAND flash memory with incremental step pulse programming scheme, "IEEE J. Solid-State Circuits, Vol. 30, No. 11, pp.1149-1156, Nov. 1995.
[3] N. Shibata et al., "A 70nm 16Gb 16-level-cell NAND flash memory," in symposium on VLSI Circuits Dig. Tech. Papers, June 2007, pp. 190-191.
[4] M. Aoki et al., "A 1.5-V DRAM for battery-based applications," IEEE J. Solid-State Circuits, Vol. 24, No. 5, pp. 1206-1212, Oct. 1989.
[5] H. Shiga et al., "A 1.6GB/s DDR2 128Mb chain FeRAM with scalable octal bitline and sensing schemes," in ISSCC Dig. Tech. Papers, Feb. 2009, pp.464-465.

ISSCC 2011 / February 23, 2011 / 5:15 PM

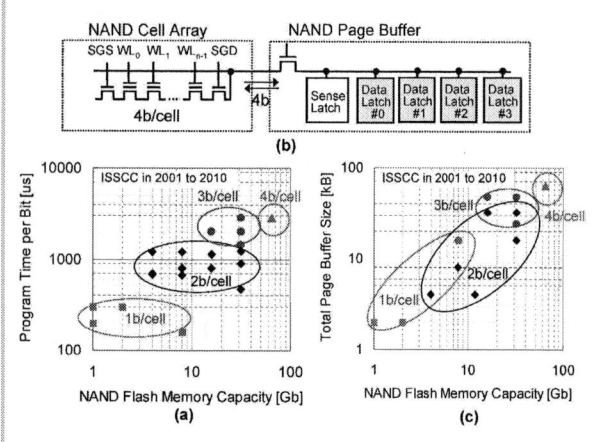

Figure 28.9.1: Trends of (a) program time per bit and (c) total page buffer size in a NAND flash memory, and (b) page buffer.

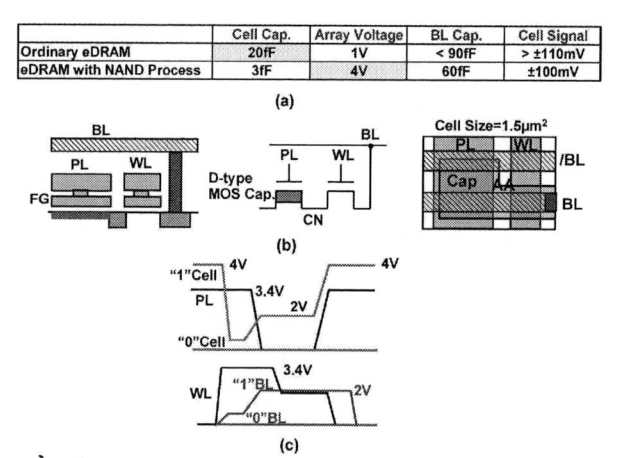

	Cell Cap.	Array Voltage	BL Cap.	Cell Signal
Ordinary eDRAM	20fF	1V	< 90fF	> ±110mV
eDRAM with NAND Process	3fF	4V	60fF	±100mV

(a)

Figure 28.9.2: (a) Embedded DRAM using standard NAND flash process, (b) cell configuration, and (c) operating principle.

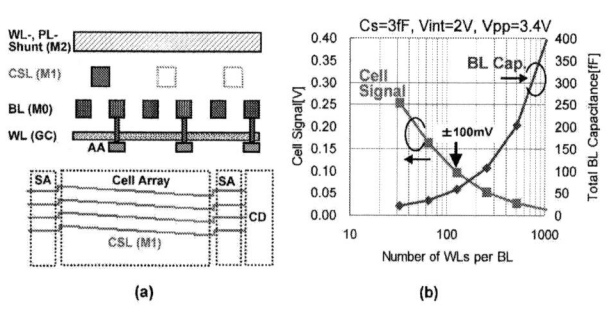

Figure 28.9.3: (a) Zigzag column selecting line for small bitline capacitance, and (b) cell signal vs. number of WLs per BL.

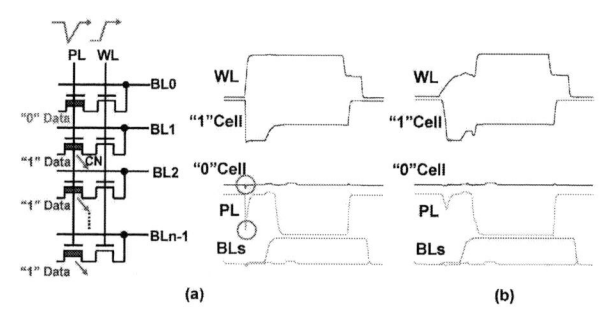

Figure 28.9.4: (a) Undershoot problem of cell node, and (b) 2-step-rise/fall wordline scheme.

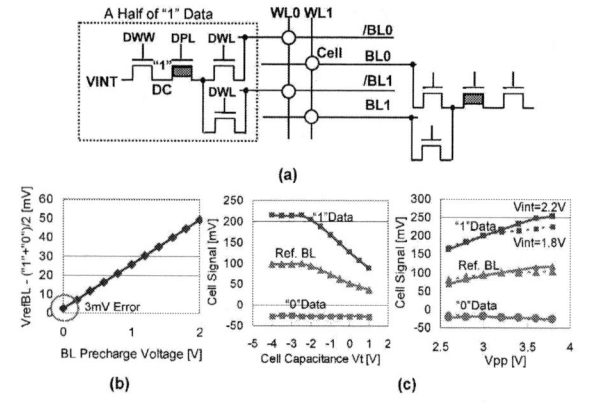

Figure 28.9.5: (a) A-half-of-1-data dummy cell, and its dependency on (b) BL precharge voltage, (c) cell cap. Vt, Vint and Vpp.

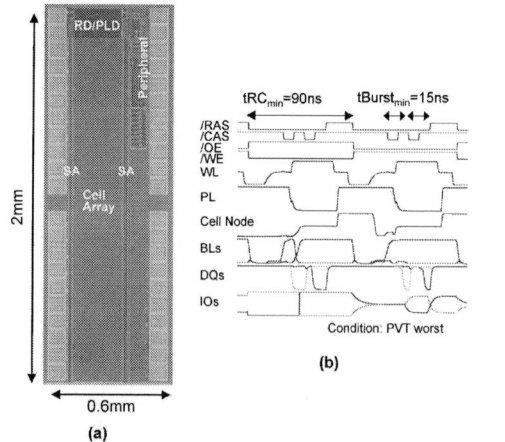

Figure 28.9.6: (a) Chip microphotograph of 32KB embedded DRAM, and (b) chip simulation result.

978-1-61284-303-2/11 $26.00 © 2011 IEEE

ISSCC 2011 / SESSION 28 / DRAM & HIGH-SPEED I/O / 28.10

28.10 A 700MHz 2T1C Embedded DRAM Macro in a Generic Logic Process with No Boosted Supplies

Ki Chul Chun, Wei Zhang, Pulkit Jain, Chris H. Kim

University of Minnesota, Minneapolis, MN

6T SRAMs have been the embedded memory of choice for modern microprocessors due to their logic compatibility, high speed, and refresh-free operation. The relatively large cell size and conflicting requirements for read and write at low operating voltages make aggressive scaling of 6T SRAMs challenging in sub-22nm. Recently, 1T1C embedded DRAMs (eDRAMs) have replaced SRAMs in several server applications reducing the cache area and improving performance [1]. Difficulties in scaling the trench capacitor and the additional process steps involved in manufacturing the thick oxide access devices are currently limiting the wide spread adoption of 1T1C technology. Gain cells have features such as decoupled read and write paths, a nondestructive read, and a 2X higher bit-cell density than a 6T SRAM, making them a strong contender for future embedded memories [2-4]. However, the boosted supplies needed for robust operation necessitates thick oxide devices to prevent gate reliability issues in gain cells. Although this would lead to a larger bit-cell size and decreased macro performance, these limitations have been overlooked in the past. In this paper, we present the following circuit techniques for realizing a truly logic compatible (i.e. thin oxide only implementation) gain cell eDRAM with no boosted supplies; (i) a 2T1C gain cell featuring a beneficial couple-up read and a preferential couple-down write, (ii) a single-ended 7T SRAM to repair weak gain cells, and (iii) a storage voltage monitor capable of tracking PVT and cell retention time for adaptive refresh control. The 64kb test macro in Fig. 28.10.1 achieves a random cycle frequency of 700MHz and a retention time of 500μsec.

DRAMs typically require two boosted supplies: a boosted high voltage to suppress the subthreshold leakage (consider a PMOS write device) and a boosted low voltage to prevent V_{TH} drop during write. The proposed 2T1C gain cell shown in Fig. 28.10.2 can operate reliably with no boosted supplies allowing the gain cell to be implemented using regular thin oxide devices. The new cell structure consists of an asymmetric 2T cell [4] and a coupling MOS capacitor controlled by the PCOU signal. PCOU is pre-discharged to 0V during hold mode introducing only a small amount of gate-overlap leakage through the coupling device (PC). At the beginning of the read access when the Read Word-Line (RWL) is activated, PCOU is also switched to VDD. This couples up both data '1' and '0' storage voltages. The higher voltage levels increase the drive current for the NMOS read access device (PS) enhancing the read performance. After the Sense Amplifier (S/A) samples the Read Bit-Line (RBL) data, a write-back operation follows which drives the Write Word-Line (WWL) to 0V instead of the usual negative boosted supply. Using a data '1' Write Bit-Line (WBL) voltage that is slightly lower than VDD (i.e. VDD-α in Fig. 28.10.2), the sub-threshold leakage in the unselected cell can be effectively cut off without using a boosted high supply for WWL [3]. Data '1' can be easily written back to the cell with a PMOS write device (PW). However, without a boosted low supply, data '0' will not be fully restored due to the V_{TH} drop in PW. To resolve this issue, PCOU is switched to 0V immediately after write back. This couples down the data '0' voltage but not the data '1' voltage since PW remains on when WBL is high. Finally, WWL is switched back to its precharge level of VDD and this couples up both data '1' and '0' voltages using the gate-overlap capacitance, fully restoring the cell storage levels.

Outlier cells with poor retention times can be replaced using 7T SRAM cells with decoupled read and write paths implemented as part of the array (Fig. 28.10.3). The single-ended read and differential write of the 7T SRAM share the same control signals with the main 2T1C array minimizing the circuit overhead. Gain cells have a very small storage capacitance and therefore the probability of having a failing cell in a redundant row or column is high compared to 1T1C DRAMs. In fact, measured results in Fig. 28.10.3 show that this probability is 3.13% when a target retention time is longer than 200μsec and repair cells are identical to normal cells. In order to guarantee that all repair cells work with negligible array

overhead (1.23% in case of a single-WL repair per 128 WL's), a redundancy scheme utilizing the 7T SRAM repair cells is proposed. The redundancy scheme was implemented for replacing weak word-lines to evaluate the stability of 7T SRAMs under dummy cell and BL-S/A variations. Measured retention bitmap of a 1kb 2T1C sub-array shows weak bit-lines as well as randomly located weak cells. The 128b 7T SRAM shows very stable operation under variation and mismatches. The proposed redundancy scheme can be easily adopted to replace weak bit-lines which improves the retention time from 200μsec to 500μsec.

Retention time of commodity DRAMs varies exponentially with temperature since it is highly sensitive to junction leakage and subthreshold leakage. Therefore, DRAM products have on-chip temperature sensors to adaptively control the refresh period according to the chip operating temperature [5]. Similarly, retention time of gain cells is also highly dependent on operating temperature as shown in Fig. 28.10.4 where there is a 5X difference in the measured retention time between 25°C and 85°C. This means that the refresh power can be reduced at low operating temperatures without sacrificing read performance. However, retention time of an asymmetric 2T gain cell is highly dependent on the gate leakage that has a weaker dependency on temperature [4]. Moreover, the various coupling effects illustrated in Fig. 28.10.2 makes a simple temperature sensor based refresh control ineffective for our gain cell. To overcome this problem, we propose a gain cell based temperature sensor that directly measures the storage node voltage at different temperatures using a cell access pattern generator. The VCO-1 output indicates the current retention time setting and the VCO-2 output provides the average storage node voltage for the 256 cells. To remove any systematic error that may have been introduced while merging the 256 cells, a calibration step is needed to obtain the relationship between the measured storage voltage and the actual retention characteristic. Measured storage voltages including the coupling effects at different temperatures and retention times are shown in Fig. 28.10.4.

A 128kb test macro implemented in a 1.2V, 65nm low-leakage logic CMOS process contains a conventional 3T array and the proposed 2T1C array for performance comparison. Figure 28.10.5 shows the cumulative retention time distribution while varying the boosted low supply of the 3T array. Our design achieves a 1.4nsec (714MHz) random cycle and a 500μsec retention time (after a single-BL repair scheme) without using a boosted supply. Figure 28.10.6 shows the measured VDD shmoo of retention time and random cycle time. Static current consumption of the proposed gain cell including the refresh operation is 147.1μA/Mb at 1.1V while a 6T SRAM consumes 307.5μA/Mb at a power-down data retention voltage of 0.6V. The chip microphotograph and key features of the 65nm eDRAM test chip are shown in Fig. 28.10.7.

Acknowledgements:
This work was supported in part by Broadcom Corporation, an IBM faculty partnership award, and a scholarship from Samsung Electronics.

References:
[1] J. Barth, D. Plass, E. Nelson, C. Hwang, et al., "A 45nm SOI Embedded DRAM Macro for POWER7™ 32MB On-Chip L3 Cache," *International Solid-State Circuits Conference*, pp. 342-343, Feb. 2010.
[2] D. Somasekhar, Y. Ye, P. Aseron, S. L. Lu, et al., "2GHz 2Mb 2T Gain-Cell Memory Macro with 128GB/s Bandwidth in a 65nm Logic Process," *International Solid-State Circuits Conference*, pp. 274-275, Feb. 2008.
[3] K. Chun, P. Jain, J. Lee, and C. H. Kim, "A Sub-0.9V Logic-compatible Embedded DRAM with Boosted 3T Gain Cell, Regulated Bit-line Write Scheme and PVT-tracking Read Reference Bias," *Symposium on VLSI Circuits*, pp. 134-135, June 2009.
[4] K. Chun, P. Jain, T. Kim, and C. H. Kim, "A 1.1V, 667MHz Random Cycle, Asymmetric 2T Gain Cell Embedded DRAM with a 99.9 Percentile Retention Time of 110μsec," *Symposium on VLSI Circuits*, pp. 191-192, June 2010.
[5] J. Sim, H. Yoon, K. Chun, H. Lee, et al., "Double Boosting Pump, Hybrid Current Sense Amplifier, and Binary Weighted Temperature Sensor Adjustment Schemes for 1.8V 128Mb Mobile DRAMs", *Symposium on VLSI Circuits*, pp. 294-297, June 2002.

978-1-61284-303-2/11 $26.00 © 2011 IEEE

ISSCC 2011 / February 23, 2011 / 5:30 PM

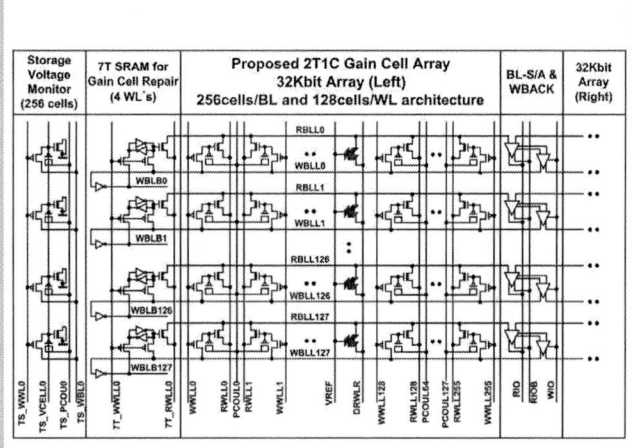

Figure 28.10.1: A 64kb gain cell eDRAM macro with no boosted supplies.

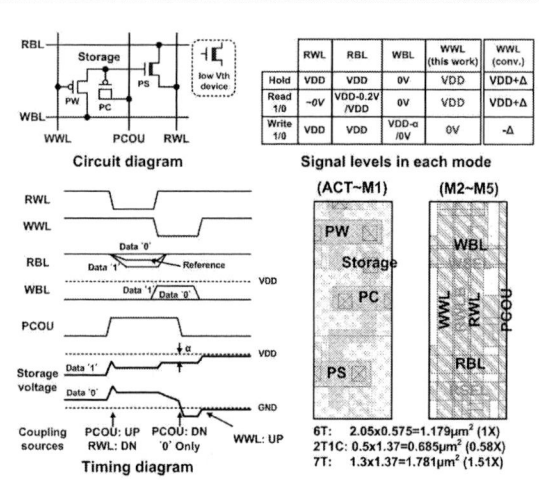

Figure 28.10.2: Proposed 2T1C cell featuring a beneficial couple-up read and a preferential couple-down write.

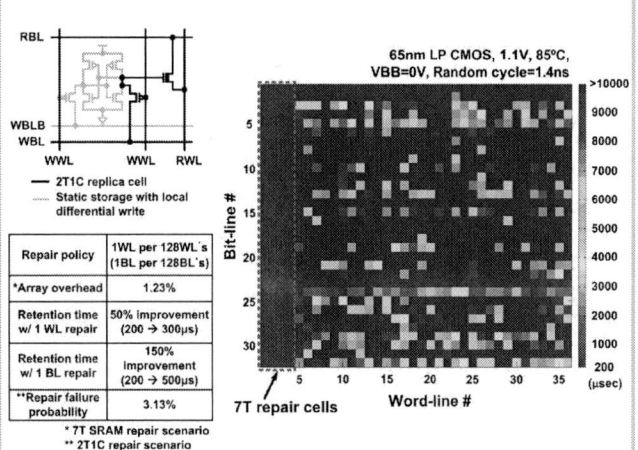

Figure 28.10.3: Proposed 7T SRAM repair cell and redundancy scheme (left). Measured retention statistics map (right).

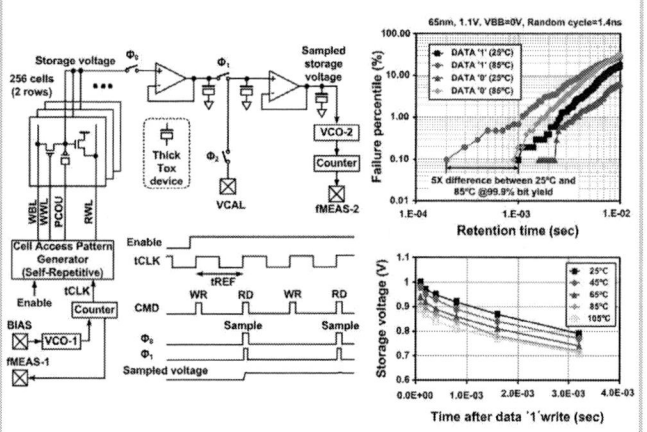

Figure 28.10.4: Proposed storage voltage monitor (left). Measured retention statistics of 2T1C array and storage voltage (right).

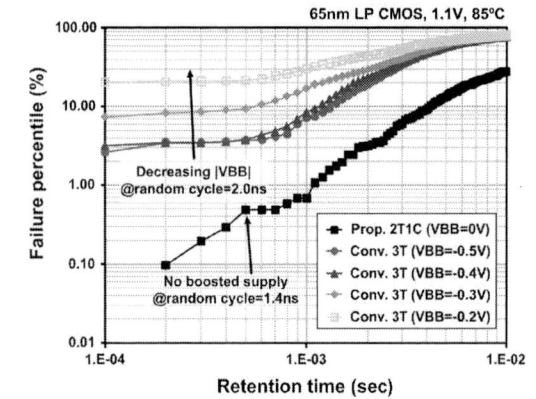

Figure 28.10.5: Retention time distribution of the proposed 2T1C and conventional 3T.

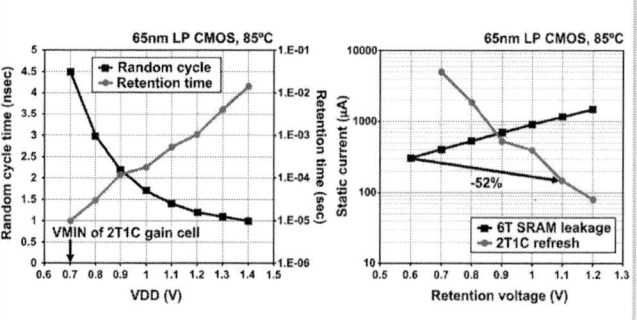

Figure 28.10.6: Measured VDD shmoo and static power comparison between 6T SRAM and 2T1C eDRAM.

28

ISSCC 2011 PAPER CONTINUATIONS

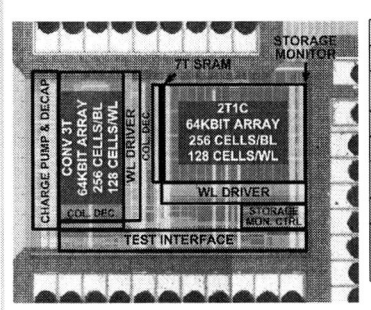

Process	65nm LP CMOS
Ckt dimension	$555.5 \times 345 \mu m^2$
Array size	2x64kbits (Conv. 3T & Prop. 2T1C)
Cell size	58% of 6T SRAM
Retention time	500µs @ 1.1V, 85°C
Random cycle time	1.40ns (714MHz) @ 1.1V
VMIN	0.7V @ 10µs retention
*Refresh power	161.8µW per Mb (**0.48X of 6T SRAM)

*@ 1.1V, 85°C, 500µsec refresh rate
**@ Retention voltage of 0.6V

Figure 28.10.7: Test chip microphotograph and chip feature summary.

Due to formatting issues there is a gap in pagination.

Pages 508-509

ISSCC 2011 / TUTORIALS

T1: Integrated LC Oscillators

This tutorial will go through the fundamentals of LC oscillator and LC VCO design, such as basic phase-noise theory, design for low power, low phase noise and, large tuning range (including varactor choice). It will also include other key issues such as supply pushing, LDO-VCO co-design, routing/buffering the oscillator signals in large SoCs, and PLL-VCO co-design for fully integratable frequency synthesis. The goal it to give a thorough overview that will be easy to follow yet comprehensive and in touch with the latest significant research results (e.g. DCOs with extremely fine tuning steps for use in DPLLs) of one of the truly key blocks in today's and tomorrow's radios.

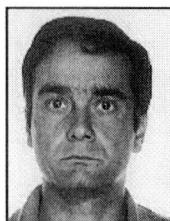

Instructor: Pietro Andreani received the M.S.E.E. degree from the University of Pisa, Italy, in 1988, and the Ph.D. degree from the Dept. of Electrical and Information Technology (EIT), Lund University, Sweden, in 1999. Between 2001 and 2007, he was Chair Professor at the Center for Physical Electronics, Technical University of Denmark. Since 2007, he has been Associate Professor at EIT, Lund University, working in analog/mixed-mode/RF IC design. He is also a part-time IC designer at ST-Ericsson in Lund. He is a TPC member of both ISSCC and ESSCIRC. He has published several papers on oscillator phase noise and VCOs.

T2: Embedded Memories for SoC: Overview of Design, Test, and Applications and Challenges in the Nano-Scale CMOS

This tutorial provides a detailed description of the workhorse static memory, the 6T SRAM. Multi-port and content-addressable memories are introduced next with special design considerations beyond the 6T cell. Embedded DRAM is described, as well as its advantages in the SoC ecosystem. Special considerations for SOI memory design are described. State-of-the-art industry techniques that improve power and voltage scaling in the nanoscale regime are also reviewed. Finally, an overview of Built-In-Self-Test is provided.

Instructor: Harold Pilo is a Senior Technical Staff member at IBM Systems and Technology Group. He joined IBM in 1993 to develop OEM SRAM products for the IT industry. He currently leads the circuit IP development for ASIC SRAM Technology Development. Prior to joining IBM, he worked at Motorola from 1989 to 1993. Harold has presented many papers at the ISSCC, VLSI and ITC. He holds over 50 US Patents and is currently a member of the ISSCC Memory Sub-committee. He graduated with a BSEE from the University of Florida in 1989.

T3: Ultra Low-Power and Low-Voltage Digital-Circuit Design Techniques

Until some 15 years ago there was no attention to low-power CMOS design, due to the 'C'omplementary nature of logic gates. Despite this beautiful property power dissipation has become the limiting factor in many fields of CMOS design, from high-performance computing to wireless autonomous transducer systems (wireless sensor nodes).

In this tutorial an overview will be given of low-power and low-voltage digital circuit design techniques with a focus on truly energy-limited systems, such as wireless sensor nodes. Circuit techniques for super-threshold toward sub-threshold, impact on speed, area, and power, and consequently on architectures will be highlighted.

Instructor: Jos Huisken joined Philips Research after graduation from the University of Twente in digital signal processor design in 1984. Since then he has been involved in architectural synthesis for digital signal processors and has applied these techniques to the first Digital Audio Broadcast (ETSI-DAB) ICs in the 1990s. Since then he has been driving low-power design from an architectural point of view. After investigating turbo and LDPC decoders, and being involved in creating a spinoff company from Philips, he joined Holst Centre / imec Netherlands in 2007 to work on ultra-low-power DSP for wireless sensor nodes, specifically for body-area networks, with a strong focus on low-voltage and low-power circuit design.

T4 : Layout – The Other Half of Nanometer Analog Design

The layout of analog blocks has for a long time been a critical aspect to achieving the theoretical performance of a circuit. Perhaps more importantly, when the layout fails to match the design, the project can fail in schedule, cost, or performance. In nanometer CMOS there are many layout-induced effects that alter the transistor characteristics, and this means that matching the layout to the design is becoming more difficult and critical. This tutorial provides an overview of those effects, and provides some strategies and approaches to combat them. Even if you don't do the layout yourself, every designer should be able to guide the process through to success.

Instructor : Jed Hurwitz received his Electronics BEng from Nottingham University, United Kingdom, in 1987. He joined Plessey Semiconductors, working on mixed-signal CMOS and design-related process issues. From 1990, he worked on videotelephony circuits at MatraMHS. In 1995, he joined a start-up, Vision (later acquired by STMicroelectronics), which became one of the pioneers of CMOS imaging, were he led all aspects of the architecture, design, and development of CMOS image-sensor systems (from photons to applications) and their optimization for the mobile cellular market. In 2005 he co-founded Gigle Networks, which has since successfully introduced powerline and anywire solutions on nanometer technologies.

T5: DPLL-Based Clock and Data Recovery

The purpose of this tutorial is to introduce attendees to Digital Phase-Locked Loop (DPLL) based Clock-and-Data Recovery (CDR). The talk will start with an overview of different types of CDRs to frame where DPLL-based CDRs fit into the overall landscape. Next, the basic theory behind DPLL-based CDRs will be presented with an eye towards practical application. Following this, the performance of DPLL-based CDRs in the face of various practical impairments such as ISI, random jitter, deterministic jitter, slicer offset, etc will be explored. This part of the talk will also tie into jitter budgets and electrical tables found in standards. Finally, the talk will conclude with an example to show how all the concepts come together into a design.

Instructor: John T. Stonick received his Ph.D. in ECE from North Carolina State University in 1992. From 1993 to 1997 he held a postdoctoral research position in the ECE department at Carnegie Mellon University. From 1997 to 2000, he was an Assistant Professor with the ECE department at Oregon State University and a co-director of the NSF Center for the Design of Analog-Digital Integrated Circuits (CDADIC). Starting in 2000 he was a Principal Design Engineer with Accelerant Networks until they were acquired by Synopsys in 2004.

978-1-61284-303-2/11 $26.00 © 2011 IEEE

Since 2004 he has remained with Synopsys where he holds the title of Synopsys Scientist. His interests include system architecture and simulation, clock-and-data recovery, and using adaptive digital techniques to compensate for analog circuit imperfections in transceivers.

T6: Practical Power-Delay Design Trade-offs

Design of high-speed, power-optimized circuits is an increasingly large part of the digital circuit designer's responsibilities. Circuit and block design trade-offs between delay, active, and leakage power are vital to meeting power and delay goals. This tutorial begins with basics of digital circuit delay and power consumption. Discussion turns to methods for reducing power and delay, active/leakage power trade-offs, and a review of circuit styles and their power-delay characteristics. Design and synthesis tools and methods to meet timing and power goals will also be discussed. Finally, system-level power management problems, solutions, and trade-offs are discussed.

Instructor: Tim Fischer has worked in high-speed digital circuit design for 22 years. He earned the Masters Degree in Computer Engineering from the University of Cincinnati in 1989. Tim then worked for Digital Equipment in Hudson, Massachusetts until 1998 designing VAX and Alpha CPUs. From 1998 to 2006, he worked on Itanium CPU design with Hewlett-Packard and Intel in Fort Collins, Colorado. Since 2006 Tim has been an AMD Fellow working on CPU circuit design and methodologies. His interests include high-speed CMOS circuits, latching/clocking structures, and power-efficient design.

T7: Distortion in Cellular Receivers

In this tutorial we discuss how noise and distortion limits the dynamic range of cellular receivers. The connection between the narrow-band cubic nonlinearity and the properties of typical radio receivers, both for narrow- and wide-band signals, is explained. First the properties of the cubic nonlinearity and its relation to several important narrow-band distortion types are explained. The concept of intercept points and their use to characterize weak MOS and BJT device nonlinearities as well as amplifiers entering compression, with and without feedback, is highlighted. Some linearization techniques are introduced, and the impact of third-order intermodulation for general receivers and second-order intermodulation for low- and zero-IF receivers is explained with circuit examples. Finally, the effects of more wide-band signals are coupled to the narrow-band approximations.

Instructor: Sven Mattisson received his PhD in Applied Micro Electronics from Lund University in 1986. From 1987 through 1994 he was an Associate Professor in Applied Micro Electronics in Lund where his research was focused on circuit simulation and analog ASIC design. 1995 he joined Ericsson in Lund to work on cellular hand-set development. Presently he is with Ericsson in Lund, where he holds a position as senior expert in analog system design. Since 1996 he is also an Adjunct Professor at Lund University. Dr. Mattisson is one of the principal developers of the Bluetooth concept.

T8: Noise Analysis in Switched-Capacitor Circuits

Switched-capacitor (SC) circuits are ubiquitous in CMOS mixed-signal ICs. The most fundamental performance limitation in these circuits stems from the thermal noise introduced by MOSFET switches and active amplifier circuitry. This tutorial reviews hand analysis techniques that allow the designer to predict the noise performance of switched-capacitor circuits at various levels of complexity. The material will focus on practical examples ranging from basic passive and active track-and-hold stages, integrators and examples of SC filters. Simulation examples using periodic noise analysis tools are included throughout the tutorial to complement and verify the theory.

Instructor: Boris Murmann is an Associate Professor in the Department of Electrical Engineering, Stanford, CA. He received the Ph.D. degree in electrical engineering from the University of California at Berkeley in 2003. From 1994 to 1997, he was with Neutron Mikrolektronik GmbH, Hanau, Germany, where he developed low-power and smart-power ASICs in automotive CMOS technology. His research interests are in the area of mixed-signal integrated circuit design, with special emphasis on data converters and sensor interfaces. In 2008, Dr. Murmann was a co-recipient of the Best Student Paper Award at the VLSI Circuit Symposium and the recipient of the Best Invited Paper Award at the Custom Integrated Circuits Conference (CICC). In 2009, he received the Agilent Early Career Professor Award.

T9: Interfacing Silicon with the Human Body: A Primer on Applications, Interface Circuits and Technologies for the Medical Market

This tutorial will provide a holistic overview of modern circuit techniques for interfacing to the human body: both sensing of key biomarkers as well as delivery of stimulation. A brief summary of constraints for safety and the biological environment will be provided to motivate general considerations for medical circuit techniques. Special emphasis will be put on power and noise issues as well for portable, ambulatory devices. A review of exemplary circuit blocks for sensing and design techniques such as dynamic offset compensation will then be discussed, with representative applications from cardiac sensing, brain-machine interfacing, electrochemical monitoring and cochlear prosthesis to drive home key points. A short review of future trends will close out the tutorial.

Instructor: Tim Denison is the Director of Neuroengineering for Medtronic Neuromodulation, and a Technical Fellow for Medtronic, the world's leading manufacturer of medical devices. In this role Tim helps guide the creation of sensor, actuator and algorithmic building blocks and architectural frameworks for future devices intended to treat nervous system disorders. Tim received his A.B. in Physics with honors from the University of Chicago, and an S.M. and Ph.D. in Electrical Engineering from MIT. He continues to actively support academic students through the MIT 6-A co-op program, as well as serving as a frequent guest instructor for courses such as TU Delft Smart Sensor Systems and the NYAS Brain Forum. His extramural roles include serving on the IMMD section of the ISSCC technical program committee and the organization committee for the National Academy of Engineering's "Frontiers in Engineering."

978-1-61284-303-2/11 $26.00 © 2011 IEEE

ISSCC 2011 / FORUM / F1

F1: Advanced Transmitters for Wireless Infrastructure

Organizers: **Gabriele Manganaro,** *Analog Devices, Wilmington, MA*

Domine Leenaerts, *NXP Semiconductors, Eindhoven, The Netherlands*

Chair: **Francesco Dantoni,** *Texas Instruments, Dallas, TX*

Committee: **Andrea Baschirotto,** *University of Milan-Bicocca, Milan, Italy*
Bogdan Staszewski, *TU Delft, Delft, The Netherlands*
Nikolaus Klemmer, *Texas Instruments, Dallas, Texas*
Seongchol Hong, *KAIST, Daejeon, Korea*

Digital cellular standards have emerged over the past 20 years. As a consequence modern cellular handheld radios make use of highly digitized transceiver architectures. This is in contrast to wireless infrastructure systems where nowadays still traditional analog-intensive radio architecture concepts are used, partly owing to more demanding performance requirements. However, future wireless infrastructures will move to more flexible and digitally-intensive architectures to address the increasing number of standards, modes and bands in cellular communication.

This forum will cover present and future transmitters in wireless infrastructure applications, like base stations for cellular mobile communication. Future digitized transmitter architectures will most likely be based upon the use of advanced RF DACs or ultra-high speed DACs in combination with up conversion techniques. These digital transmitter architectures need flexible (re-configurable) power amplifiers which make use of advanced techniques like multi-way Doherty or envelope tracking. Enhanced linearization techniques need to be applied in the transmitter to cope with the stringent linearity demands, while advanced calibration techniques and other design techniques will be needed to overcome circuit block impairments.

The target participants are analogue/RF circuit designers and architecture engineers working on wireless (infrastructure) systems, who want to learn about current and future transmitters. This all-day forum encourages open information exchange.

Begin at the Beginning for Transmitter Design
Earl McCune, *RF Communications Consulting, Santa Clara, CA*
Earl has nearly 40 years of wireless system and circuit design experience. His bachelor, master, and doctorate degrees were earned at UC Berkeley, Stanford University, and UC Davis respectively. In addition to being a named inventor on more than 50 US patents, he is also a Silicon Valley serial entrepreneur, having founded two successful wireless technology companies beginning in 1986. In 2009 he retired from Panasonic as a Technology Fellow. He now operates his own consulting business, is an author of and contributor to wireless reference books, and is a visiting instructor in both industrial and academic settings.

Flexible Digital-Centric RF-DAC Based Transmitter
Renato Negra, *RWTH Aachen University, Aachen, Germany*
Renato Negra received the M.Sc. degree in telematics from Graz University of Technology, Austria, in 1999 and the Ph.D. degree in electrical engineering from ETH Zurich, Switzerland in 2006. From 1998 to 2000 he was with Alcatel Space Norway AS, Horten, Norway where he was involved in the design and characterisation of space-qualified RF equipment. In 2000 he joined the Laboratory for Electromagnetic Fields and Microwave Electronics at ETH Zurich. His Ph.D. research was focused on power-efficient linear amplification of wireless communication signals. From 2006 to May 2008 he was a Post-Doctoral Fellow at the iRadio Lab, University of Calgary, Canada. Since June 2008, he is Assistant Professor at the RWTH Aachen University where he heads the Mixed-Signal CMOS Circuits group. His research interests are linearisation techniques, highly efficient power amplifiers, advance transmitter architectures, software defined and reconfigurable radios.

978-1-61284-303-2/11 $26.00 © 2011 IEEE

Current-Steering DACs for Direct RF Transmission
Klaas Bult, *Broadcom, Bunnik, The Netherlands*

Klaas Bult (MSc. 1984, PhD. 1988) joined Broadcom in 1996, after 6 years at Philips Research Labs and 2 years at UCLA. At Broadcom he started the Analog Design Group in Irvine, CA and later Broadcom's Design Center in Bunnik, The Netherlands. He is now Vice President and CTO of Operations and Central Engineering within Broadcom. He is also a Broadcom Fellow. Klaas Bult is an author of many international publications and holds more than 50 patents. He was awarded the Lewis Winner Award for Outstanding Conference Paper on ISSCC, 1990, 1992 and 1997 and the Jan van Vessem Award for Best European Conference Paper on ISSCC 2004.

Dynamic Element Matching and Calibration for Nyquist-Rate DACs
Ian Galton, *University of California, San Diego, CA*

Ian Galton is a professor of electrical engineering at the University of California, San Diego, (UCSD) where he conducts research on mixed-signal integrated circuits. He also consults at several semiconductor companies and teaches industry-oriented short courses on mixed-signal integrated circuits. He has served on a corporate board of directors, on several corporate technical advisory boards, as the Editor-in-Chief of the IEEE Transactions on Circuits and Systems II, and as a member of several IEEE boards and committees including the ISSCC Technical Program Committee. He is an IEEE Solid-State Circuits Society "Distinguished Lecturer," he and his graduate students received the 2008 IEEE Transactions on Circuits and Systems Darlington Best Paper Award and the 2008 ISSCC Jack Kilby Award for Outstanding Student Paper, and he received the 2005-2006 UCSD Electrical and Computer Engineering Department "Best Teacher Award."

Low Power, RF domain PA Linearization for Infrastructure Transmitters
Olivier Charlon, *Scintera Networks, Sunnyvale, CA*

Olivier Charlon attended "Ecole Supérieure d'Ingénieurs en Electrotechnique et Electronique", Paris, France, and received his engineering diploma in 1996. After joining Philips Semiconductors, UK, he developed high speed Analog-to-Digital Converters for TV and CD/DVD application IC's. In 2000, he joined Philips Semiconductors, CA, USA where he worked on or led several RF BiCMOS/CMOS SoC projects in the field of: CDMA, Bluetooth, and low power WLAN transceivers for mobile applications. Over the years, he developed expertise in high-speed ADC's, various RF building blocks, deep submicron RF CMOS monolithic SoC integration and PA linearization. He holds 15 patents granted or pending in the field of semiconductor and wireless communications. He is author and co-author of 11 international papers/conference presentations. Since December 2007, he develops infrastructure Power Amplifier Linearizers at Scintera Networks where he manages the engineering programs as VP of IC Design.

Flexible Doherty Power Amplifiers for Base Station Application – Multi-way and ET Operations
Bumman Kim, *Pohang University of Science and Technology, Pohang, Korea*

Bumman Kim is a professor of electrical engineering at Pohang University of Science and Technology (POSTECH) in Korea, where he conducts research on microwave power amplifiers for mobile base-station and handset. Before he joined POSTECH, he was with Central Research Labs. of Texas Instrument Inc. as a senior member of technical staff, working on MMIC. He developed the first MMIC in millimeter-wave frequency and the first power operation of GaAs HBT at microwave frequency. He also consults at several semiconductor companies and served on a corporate board of directors (Hynix). He is a fellow of IEEE and IET, was an associate editor of IEEE Transactions on microwave theory and technique, AdCom member of IEEE MTT, Distinguished Lecturer of the same society. He is TPC member of IMS, RFIC symposium, EuMC, et al. He and his graduate students received two times for the first prize, one times for the second and the third places in PA contest at IMS. He was a dean of academic affairs, dean of graduate school, Namko chaired professor and is fellow of POSTECH.

Case Study: System and Design Aspects of State of the Art of Communication DACs
Martin Clara, *Lantiq, Villach, Austria*

Martin Clara was born in Innsbruck, Austria, in 1970. He received the M.Sc. degree in electrical engineering from Vienna University of Technology, Austria, in 1996 and the Ph.D. degree in electronics from Graz University of Technology, Austria, in 2009. In 1997 he joined Siemens Microelectronics' Design Center in Villach, Austria, as an Analog Design Engineer, mainly working on BiCMOS and CMOS linear circuits. From 2000 to 2009 he was with Infineon Technologies' Design Center in Villach, Austria, where he designed data converters, linear circuits and RF building blocks in deep-submicron and nanometer CMOS technologies. Since 2009 he holds the position of principal engineer for analog/mixed-signal and RF-design at LANTIQ's design center, based in Villach, Austria. His main interests include the implementation of low-voltage and high dynamic range analog front-ends in advanced CMOS technologies, the concept and design of high-performance data converters, as well as RF-CMOS design.

ISSCC 2011 / FORUM / F2

F2: Ultra-Low Voltage VLSIs for Energy Efficient Systems

Organizer / Chair: **Ken Takeuchi**, *University of Tokyo, Tokyo, Japan*

Co-chair: **Ken Chang**, *Xilinx, San Jose, CA*

Committee: **Ken Takeuchi**, *University of Tokyo, Tokyo, Japan*
Ken Chang, *Xilinx, San Jose, CA*
Kevin Zhang, *Intel, Hillsboro, OR*
Tadashi Yamauchi, *Renesas Electronics, Itami, Japan*
Roberto Gastaldi, *Numonyx, Brianza, Italy*

Energy efficient VLSIs with an ultra-low voltage power supply down to 0.5V are in growing demand in various applications, e.g., secure card, sensor node, and medical systems where power is supplied by RF wave, solar cells and small batteries. Various technological challenges, including PVT variations, low operating voltage margin, increased stand-by power consumption and low driving current, must be addressed. This forum provides an overview of the technical challenges as well as most recent circuit advances in key building blocks for digital/analog VLSI applications. The forum starts with the overview on microprocessor design for smarphones. The microprocessor used in personal computers is redesigned and optimized, delivering over 2-4x performance in the smartphone power envelope. High performance and variability resistant device technologies are presented for low power and multimedia products operating at very low V_{DD}. The forum also has three presentations to discuss the design challenges for logic and memory in scaled technologies with increased parameter variations. For the digital design, pitfalls in deep-low-voltage circuits are summarized with emphasis on V_{DDmin} issues with experimental results. SRAM design trade-offs with respect to area, read/write stability, and access time are discussed for different applications. Embedded non-volatile memory with zero power is also presented. Design tradeoffs involved in building blocks for low-power signal path along with on-die voltage regulators are discussed. Key analog circuits such as low-voltage voltage generator, VCO, PLL, ADC and DAC are examined for voltage scaling. Conventional bangap reference is no longer feasibile down to 0.5V regime. Different low-voltage voltage generator circuits and their pros and cons will be discussed. The impact of V_{DD} scaling on the various performance metrics for timing related circuits, such as voltage-controlled oscillators and phase-locked loops is presented with an emphasis on their application in input/output interfaces. Low-voltage data converters are reviewed, covering the current status, enhancement solutions and future direction for higher energy efficiency.

Squeezing a IA Computer in a Smartphone
Ticky Thakkar, *Intel, Hillsboro, OR*

Shreekant (Ticky) Thakkar is an Intel Fellow and director of UMG Platform Architecture for the Intel Architecture Group at Intel Corporation. In this role, Thakkar serves as the lead architect for the Mobile Internet Device Platform, a computing and communications device representing a strategic growth area for Intel. Thakkar joined Intel in 1993 as the P6 Multiprocessor architect. He has held several architecture and engineering management positions at the company in the areas of multiprocessor, CPU, mobile computing and low power architecture. Previously, Thakkar held positions with Sequent Computer Systems, the Oregon Graduate Institute and Metheus Corp. Thakkar holds 64 patents with another 15 patents pending, and has published 29 technical papers. He received his bachelor's degree in statistics and computer science from the University of London in 1977. Thakkar earned masters and doctoral degrees from the University of Manchester in computer science in 1978 and 1982, respectively.

FDSOI: An Innovative Technology for Low-Vdd / High-Performance Logic
Frederic Boeuf, *STMicroelectronics, Crolles, Rhone Alpes, France*

Frédéric Boeuf obtained his M.Eng. and M.Sc. degree from Institut National Polytechnique de Grenoble in 1996 and Ph.D. from the University Joseph Fourier of Grenoble (France) in 2000. He joined STMicroelectronics in 2000, where he worked on the pre-development phase of 65nm and 45nm CMOS technologies. He actively participated to the development of the MASTAR model, used for the definition of the 2005 and 2007 editions of the "International Technology Roadmap for Semiconductors" roadmap to which he collaborated. He is currently managing the Advanced Devices Research Group working towards the 20/16nm CMOS thin-films technology and is industrial advisor of several Ph.D. theses in the field of device integration and modelling. His fields of expertise are semiconductor physics and CMOS device physics. He is member of the "Device Physics" subcommittee of the "Solid State Devices and Materials" (SSDM) conference since 2006, and participated to the Technical Program Committee of the European Solid State Device Research Conference (ESS-DERC) in 2004-2005. He authored and co-authored over 135 publications.

Pitfalls in Deep-Volt Logic Design
Takayasu Sakurai, *University of Tokyo, Tokyo, Japan*

Takayasu Sakurai received Ph.D in EE from University of Tokyo in 1981 and joined Toshiba Corporation, where he designed CMOS DRAM, SRAM, RISC processors, DSPs, and SoC solutions. He has worked on interconnect and MOSFET models such as alpha power-law MOS model. From 1988 through 1990, he was a visiting researcher at the University of California Berkeley. From 1996, he has been a professor at University of Tokyo, working on low-power high-speed VLSI designs and large-area electronics. He is the executive committee chair for VLSI Symposia and a steering committee chair for the IEEE A-SSCC and a committee member of numerous IEEE conferences. He is a recipient of 2010 IEEE Donald Pederson Award, 2010 IEEE Paul Rappaport award, 2010 IEICE Electronics Society award, 2009 IEICE achievement award, 2005 IEEE ICICDT award, 2004 IEEE Takuo Sugano award and four product awards. He is IEICE Fellow and IEEE Fellow.

978-1-61284-303-2/11 $26.00 © 2011 IEEE

ISSCC 2011 / February 20, 2011 / 8:00 AM

Ultra Low Voltage Logic and Embedded Memories
Kaushik Roy, *Purdue University, West Lafayette, IN*

Kaushik Roy received B.Tech. degree in electronics and electrical communications engineering from the Indian Institute of Technology, Kharagpur, India, and Ph.D. degree from the electrical and computer engineering department of the University of Illinois at Urbana-Champaign in 1990. He was with the Semiconductor Process and Design Center of Texas Instruments, Dallas, where he worked on FPGA architecture development and low-power circuit design. He joined the electrical and computer engineering faculty at Purdue University, West Lafayette, IN, in 1993, where he is currently a Professor and holds the Roscoe H. George Chair of Electrical & Computer Engineering. His research interests include Spintronics, VLSI design/CAD for nano-scale Silicon and non-Silicon technologies, low-power electronics for portable computing and wireless communications, VLSI testing and verification, and reconfigurable computing.

Low Energy Consumption as a Qualifying Factor for Embedded Non Volatile Memories
Guido De Sandre, *STMicroelectronics, Agrate Brianza, Italy*

Guido De Sandre received the Engineering Degree (Laude) in electronic engineering from Politecnico di Milano, Milano, Italy, in 1995. In the same year he joined STMicroelectronics, Agrate Brianza, Italy, where he worked in the test program development for hard disk controllers. Since 1998 he is in the Central R&D organization, where he has been involved in the design and test of flash memory test vehicles aimed at technology development. Since 2006 he also leads the project of silicon demostrators for embedded PCM technology. He holds 19 patents and is author of several papers in the area of NVM memories. His research interests are in the field of analog and mixed-mode design and non-volatile memories design and test in the context of embedded applications.

System Tradeoffs for Efficient Embedded Power Management
Baher Haroun, *Texas Instruments, Dallas, TX*

Baher Haroun joined TI in 1995 to work in the Mixed Signal audio, video and speech R&D design organization. He later built and managed a worldwide team to integrate analog, mixed signal and RF integration into digital CMOS in TI's Wireless business unit where his team shipped billions of analog IP into cell phones. Today he is a chief technologist in the Wireless BU, and member of the technical board of TI Kilby Labs. Prior to TI, Baher was an assistant, then associate professor of electrical and computer engineering at Concordia University in Montreal, Canada. An IEEE and a TI Fellow, Baher has tens of internal and external published presentations, papers and patents to his credit, and is interested in research areas such as THz signal paths, ultra low power and low voltage mixed signal wireless ICs, GHz serial interfaces and DSP custom architectures. He received his bachelor's and master's degrees in electrical engineering from Ain Shams University in Egypt, and a Ph.D. from University of Waterloo in Ontario, Canada.

Voltage Generator with Supply below the Bandgap Voltage
Philip K.T. Mok, *The Hong Kong University of Science and Technology, Kowloon, Hong Kong, China*

Philip K. T. Mok (S'86-M'95-SM'02) received his B.A.Sc., M.A.Sc., and Ph.D. degrees in electrical and computer engineering from the University of Toronto, Toronto, ON, Canada, in 1986, 1989 and 1995, respectively. In January 1995, he joined the Department of Electronic and Computer Engineering at the Hong Kong University of Science and Technology, Hong Kong, China, where he is currently a Professor. His current research interests include power management integrated circuits and low-voltage analogue integrated circuits design. He received the Henry G. Acres Medal, the W.S. Wilson Medal and a Teaching Assistant Award from the University of Toronto, and the Teaching Excellence Appreciation Award three times from The Hong Kong University of Science and Technology. He is also a co-recipient of the Best Student Paper Award in the 2002 and 2009 IEEE Custom Integrated Circuits Conference. In addition, he has been a member of the International Technical Program Committees of the ISSCC from 2005 to 2010 and he has served as an associate editor for IEEE Transactions on Circuits and Systems – II from 2005 to 2007 and IEEE Transactions on Circuits and Systems – I from 2007 to 2009 and IEEE Journal of Solid-State Circuits since 2006.

Voltage Scaling Trade-Offs in VCOs, PLLs and I/Os
Peter Kinget, *Columbia University, New York, NY*

Peter R. Kinget received an engineering degree in electrical and mechanical engineering and the Ph.D. in electrical engineering from the Katholieke Universiteit Leuven, Belgium. He has worked in industrial research and development at Bell Laboratories, Broadcom, Celight and Multilink before joining the faculty of the Department of Electrical Engineering, Columbia University, NY in 2002. His research interests are in analog and RF integrated circuits and signal processing using nanoscale CMOS technologies. He has been an Associate Editor of the IEEE Journal of Solid State Circuits (2003-2007) and the IEEE Transactions on Circuits and Systems II (2008-2009). He is also serving on the Technical Program Committees of the International Solid-State Circuits Conference and the European Solid-State Circuits Conference. He is a "Distinguished Lecturer" for the IEEE Solid-State Circuits Society.

Make the Low Voltage ADC/DAC Energy Efficient
Akira Matsuzawa, *Tokyo Institute of Technology, Tokyo, Japan*

Akira Matsuzawa received B.S., M.S., and Ph. D. degrees in electronics engineering from Tohoku University, Sendai, Japan, in 1976, 1978, and 1997 respectively. In 1978, he joined Panasonic. Since then, he has been working on research and development of analog and Mixed Signal LSI technologies, such as ultra-high speed ADCs. On April 2003, he joined Tokyo Institute of Technology and he is professor on physical electronics. Currently he is researching in mixed signal technologies. He served a guest editor in chief for special issue on analog LSI technology of IEICE transactions on electronics in 1992, 1997, and 2003, and committee member for analog technology in ISSCC. Recently he served IEEE SSCS elected Adcom and IEEE SSCS Distinguished lecturer, educational program chair for A-SSCC and now he serves chapter chair of IEEE SSCS Japan Chapter. He received the IR100 award in 1983, the R&D100 award and the remarkable invention award in 1994, and the ISSCC evening panel award in 2003 and 2005. He is an IEEE Fellow since 2002 and IEICE Fellow since 2010.

978-1-61284-303-2/11 $26.00 © 2011 IEEE

ISSCC 2011 / FORUM / F3

F3: Towards Personalized Medicine and Monitoring for Healthy Living

Organizer/Co-Chair: **Christian Enz,** *CSEM, Neuchâtel, Switzerland*

Co-Chair: **Andreia Cathelin,** *STMicroelectonics, Crolles, France*

Committee: **Maysam Ghovanloo,** *Georgia Institute of Technology, Atlanta, GA*
Stefan Heinen, *RWTH Aachen University, Aachen, Germany*
Minkyu Je, *Institute of Microelectronics, A*STAR, Singapore*
David Scott, *TSMC, Plano, TX*

The ever-increasing cost of healthcare and the aging of population are driving the development of diagnostic tests and drug delivery systems that determine an individual's therapeutic responsiveness and enable a more efficient therapy, saving healthcare dollars and giving patients better choices. The targeted personalized medicine also represents a huge market (estimated at about $232 billion for the US only today) with an annual growth rate of 11% in the coming years [1]. Besides the regulatory and new business models to be developed, one of the many challenges ahead for this to happen, lies in the various new technologies to bring together for the successful realization and production of the devices required for a truly personalized medicine.

This forum will give the opportunity to engineers to better understand the technical issues and challenges faced when designing devices for personalized medicine. It is organized in 3 parts: the first part is focused on the basic components required for sensing and drug delivery. In particular, we will be looking at the basic sensing principles of bio-molecules, the related electronics interface and the miniaturized drug delivery systems using MEMS. The second part is dedicated to health monitoring systems with a particular emphasis on the aspects of wireless communications. Finally, the third part is more focused on cell-inspired electronics and will give some perspectives on the future of personalized medicine.

[1] PriceWaterhouseCoopers (PWC) report titled "The New Science of Personalized Medicine: Translating the Promise into Practice."

Bio-Sensing on Chip
Carlotta Guiducci, *EPFL, Lausanne, Switzerland*
Carlotta Guiducci holds her PhD in Electrical Engineering from the University of Bologna (I). She was a postdoc at the Nanobiophysics Lab at Ecole Supérieure de Physique et Chimie Industrielles Paris (F) between 2005 and 2007. Later at University of Bologna she led a joint research group of electrical engineers, physicists and biologists funded by an Integrated Project of The EU (DiNamICS) and by national projects. She recently joined the Institute of Bioengineering at the Swiss Federal Institute of Technology in Lausanne (CH) where she holds a position as Tenure-Track Assistant Professor. Her research activity spans from the characterization of MOS in quantic regime to the development of novel techniques for sensing biological affinity reactions on surfaces by means of semiconductor sensors and electronic transducers. She developed in collaboration with Infineon technologies two test chips for DNA detection by capacitance measurements. She has been working on electrical, electrochemical and optical techniques. She demonstrated and patented the measurement of DNA by UV absorption on non volatile memory cells. Her laboratory team is focused on the design and application of electronic biosensors and is at the forefront of electronic engineering and bioengineering. The sensors address a wide range of applications, from nucleic acid, protein and drug detection to the measurements of bacterial metabolism and they are based on detection principles supporting electronic transduction, in order to couple directly and integrate the sensors themselves with electronic circuitry for data acquisition.

Electronic Sensing of DNA: CMOS-Based Active Microarrays and Single-Molecule Levels of Detection
Ken Shepard, *Columbia University, New York, NY*
Ken Shepard received the B.S.E. degree from Princeton University, Princeton, NJ, in 1987 and the M.S. and Ph.D. degrees in electrical engineering from Stanford University, Stanford, CA, in 1988 and 1992, respectively. From 1992 to 1997, he was a Research Staff Member and Manager with the VLSI Design Department, IBM T. J. Watson Research Center, Yorktown Heights, NY, where he was responsible for the design methodology for IBM's G4 S/390 microprocessors. Since 1997, he has been with Columbia University, New York, where he is now Professor. He also was Chief Technology Officer of CadMOS Design Technology, San Jose, CA, until its acquisition by Cadence Design Systems in 2001. His current research interests focus primarily on "more than Moore" CMOS IC design.

Engineering Pain Relief: Microsystems for Intrathecal Drug Delivery
Yogesh Gianchandani, *University of Michigan, Ann Arbor, MI*
Professor Yogesh B. Gianchandani serves as the deputy director for the Center for Wireless Integrated Microsystems (WIMS) at the University of Michigan. He has published more than 200 papers in microsystems-related conferences and journals and has more than 30 patents issued or pending. From 2007 to 2009, he served at the National Science Foundation, as the program director for Micro and Nano Systems within the Electrical, Communication, and Cyber Systems (ECCS) Division. Dr. Gianchandani was a Chief Co-Editor of Comprehensive Microsystems: Fundamentals, Technology, and Applications, published in 2008, and served as a General Co-Chair for the IEEE/ASME International Conference on Micro Electro Mechanical Systems (MEMS) in 2002. Dr. Gianchandani is a Fellow of IEEE.

978-1-61284-303-2/11 $26.00 © 2011 IEEE

ISSCC 2011 / February 24, 2011 / 8:00 AM

Wearable Body Area Network: Towards Preemptive and Proactive Healthcare Applications
Jerald Yoo, *Masdar Institute of Science and Technology, Abu Dhabi, United Arab Emirates*

Jerald Yoo (S'05-M'10) received the B.S., M.S., and Ph.D. degrees in Department of Electrical Engineering from the Korea Advanced Institute of Science and Technology (KAIST), Daejeon, Korea, in 2002, 2007, and 2010, respectively. In May 2010, he joined the faculty of Microsystems Engineering, Masdar Institute, Abu Dhabi, United Arab Emirates, where he is an assistant professor. He is currently also with Technology and Development Program, Massachusetts Institute of Technology (MIT), Cambridge, MA, USA, as a visiting scholar. He developed low-energy Body Area Network (BAN) transceivers and wearable body sensor network using Planar-Fashionable Circuit Board (P-FCB) for continuous health monitoring system. His research focuses on low energy circuit technology for wearable bio signal sensors, wireless power transmission, SoC design to system realization for wearable healthcare applications, and energy-efficient biomedical circuit techniques. He is an author of a book chapter in Biomedical CMOS ICs (Springer, 2010). Dr. Yoo is a co-recipient of the Asian Solid-State Circuits Conference (A-SSCC) Outstanding Design Awards in 2005.

Miniaturized Wireless CMOS Drug Delivery SoC
Shey-shi Lu, *National Taiwan University, Taipei, Taiwan*

Shey-Shi Lu received his B.S. degree, M.S. Degree, and Ph.D. Degree from National Taiwan University, Cornell University, and University of Minnesota, all in electrical engineering, in 1985, 1988, and 1991, respectively. He joined the Department of Electrical Engineering, National Taiwan University in August of 1991 as associated professor and was promoted to full professor in 1995. From 1991 to 2000, he endeavored in the development of InGaP/GaAs HBT devices and processing techniques for cell phones. Since 2000 he has switched his research directions into areas of CMOS RFICs, Bio-MEMS and SoCs for medical electronics. He received the outstanding research award from National Science Council of Taiwan in 2009.

Power and Data Telemetry for Implantable Biomimetic Systems
Wentai Liu, *University of California, Santa Cruz, CA*

Wentai Liu received his PhD from the University of Michigan. In 1983, he joined North Carolina State University, where he held the Alcoa Chair Professorship in ECE and was the founder of the Analog/Mixed-Mode Design Consortium. Since 2003, he has been a professor at the University of California, Santa Cruz, where he is also the Campus Director and Thrust Leader of the NSF Engineering Research Center on Biomimetic Microelectronic Systems. His research interests include neuro-engineering, neural prosthesis, brain-machine interface, bioelectronics, transceiver, sensors and actuators, timing/clock optimization, vision/image processing. He has been working on the neural implants dealing with nerves and muscles for retina, epilepsy, muscle, eyelids, spinal cord, and bladder. Since its inception, he has been leading the engineering efforts of the retinal prosthesis to restore vision, finally leading to successful implant trials in blind patients. He has published more than 300 technical papers and two books. He received 2010 Breakthrough Award from Popular Mechanics, RD-100 Award, two IEEE Outstanding Paper Awards, Alcoa Foundation's Distinguished Engineering Research Award, NASA Group Achievement Award, and Outstanding Alumni Award from National Chiao-Tung University, where he also holds a Chair Professorship as the Founder and Honorary Director of Biomimetic Systems Research Center. He has served as Associate Editor for IEEE Trans. on Biomedical Engineering, Guest Editor for both IEEE Proceedings Special Issue of Biomimetic Systems and IEEE-MTT Special Issue of Wireless IC for Biomedical Applications. He is the founder of the International Conference on Neuroprosthetic Devices, which has been held twice in 2009 and 2010.

Bionic Ear: Hearing Health Care Rehabilitation in the 21st Century
Stefan Launer, *Phonak/Sonova, Staefa, Switzerland*

Stefan Launer is VP of Advanced Concepts and Technologies leading the overall Research & Technology development activities at Phonak/Sonova. In his role as classically called "CTO" he is in charge of driving the group wide technology roadmap of Sonova, the holding company of several companies active in hearing health care rehabilitation. Sonova is today's world market leader in hearing instruments, it includes companies developing and producing classical hearing instruments, implantable hearing instruments, extended wear hearing instruments as well as wireless communication systems. During his 15 years with Phonak/Sonova, Stefan Launer has lead several departments including the DSP and Microelectronics teams. His early work included the development of signal processing strategies for hearing instruments applying "bionic" concepts based on applying auditory models for various problems.

Ultra Low Power Biomedical and Bio-inspired Systems
Rahul Sarpeshkar, *MIT, Cambridge, MA*

Rahul Sarpeshkar obtained Bachelor's degrees in Electrical Engineering and Physics at MIT. After completing his PhD at CalTech, he joined Bell Labs as a member of the technical staff. Since 1999, he has been on the faculty of MIT's Electrical Engineering and Computer Science Department, where he heads a research group on Analog VLSI and Biological Systems (www.rle.mit.edu/avbs). He holds over 25 patents and has authored more than 100 publications, including one that was featured on the cover of Nature. His book, Ultra Low Power Bioelectronics: Fundamentals, Biomedical Applications, and Bio-inspired Systems was released in February 2010 at the ISSCC conference and contains a broad and deep treatment of the fields of bioelectronics and ultra low power electronics. He has won several awards for his interdisciplinary bioengineering research including the Packard Fellow award given to outstanding faculty.

Towards Disposable Healthcare – a Paradigm Shift
Chris Toumazou, *Imperial College, London, United Kingdom*

Professor Christofer Toumazou (FRS, FREng, FIEEE, FIET, DEng, Ceng) is currently the Founding Director and Chief Scientist at The Institute of Biomedical Engineering Imperial College. He has published over 500 research papers and holds 30 patents in the field many of which are now fully granted PCT. Chris founded the IEEE BIOCAS society in 2000. Chris is the founder of three technology based companies with applications spanning ultra low-power mobile technology and wireless glucose monitors (Toumaz Technology Ltd, UK) Digital Audio Broadcasting (Future-Waves Pte Taiwan) and DNA Sequencing (DNA Electronics Ltd, UK). He has received many awards including: The Royal Society Clifford Patterson prize Lecture, entitled "The Bionic Man", for which he received The Royal Society Clifford Patterson bronze medal in 2003. He is the recipient of the 2005 IEEE CAS Education Award for pioneering contributions to circuits and systems for biomedical applications. He received the Royal Academy of Engineering Silver Medal in 2007 for pioneering contributions to British industry. The IET Premium best paper award and the IEEE CAS outstanding young author award. Elected in 2006 to Academia Europea. Chris is also the recipient of 2007 Royal Academy of Engineering Silver Medal. In 2008 he was appointed the Fellowship of the Royal Academy of Engineering and the Fellowship of the Royal Society, which is the highest honour in UK science. Chris received the 2009 World Technology Award sponsored by Time Magazine for the Health & Medicine category.

978-1-61284-303-2/11 $26.00 © 2011 IEEE

ISSCC 2011 / FORUM / F4

F4: Design of "Green" High-Performance Processor Circuits

Organizer: **Tobias Noll,** *RWTH Aachen University, Aachen, Germany*

Co-Organizer: **Raney Southerland,** *ARM, Austin, TX*

Chair: **Vladimir Stojanovic,** *MIT, Cambridge, MA*

Committee: **Sonia Leon,** *Oracle, Santa Clara, CA*
                     Vladimir Stojanovic, *MIT, Cambridge, MA*
                     Lew Chua-Eoan, *Qualcomm, San Diego, CA*
                     Alice Wang, *Texas Instruments, Dallas, TX*
                     Byeong-Gyu Nam, *Samsung Electronics, Gyeonggi-Do, Korea*
                     Masaya Sumita, *Panasonic, Osaka, Japan*

Energy efficiency today is one of the most important design criteria of high-performance processors for practically every application: whether the final chip dissipates more than 100 Watts for general purpose processing or a few Watts for application processing on a SoC and whether a single or multi core architecture is applied best practice design techniques to achieve high performance at low power are essentially the same. Design targets become more and more challenging with every new technology generation which shows with increasing variability how to keep energy efficiency high in every mode of operation, i.e. sleep, standby, regular, and peak performance. General techniques being applied here are gating, adaptivity, and calibration. Finally, in the future really successful improvements of energy efficiency will bring the dilemma that the classical "flexibility vs. energy conflict" is replaced by a new "energy vs. reliability conflict" due to an increasing rate of transient faults.

The objective of this Forum is to present a comprehensive overview of energy efficient optimization methodologies for the different kinds of processors (general purpose / application / embedded SoC, single / multi core processors) and to give exemplary examples. While power optimization generally has to be performed at every design level and architectural components including memories this Forum focuses on micro-architecture, logic and circuit, down to the physical implementation level as well as state-of-the-art clocking and supply techniques. The Forum concludes with an outlook on future options, developments, challenges and issues; possible way-outs will be discussed.

Microarchitectural Features for a Low-Power Multiprocessor System
Thomas B. Berg, *MIPS, Sunnyvale, CA*

Thomas Berg is a staff engineer at MIPS Technologies, where he is a lead hardware developer of multicore technology. He has 25 years of industry experience developing computer systems with a focus on coherent multiprocessing, cache hierarchy, and SoC architecture. Thomas received a BS in computer and electrical engineering from the University of Michigan and an MS in electrical engineering from Purdue University.

Energy-Efficient Computing
William Dally, *NVIDIA, Santa Clara, CA*

Bill is Chief Scientist and Senior VP of Research at NVIDIA and the Willard R. and Inez Kerr Bell Professor of Engineering at Stanford University. Bill and his group have developed system architecture, network architecture, signalling, routing, and synchronization technology that can be found in most large parallel computers today. At Bell Labs he contributed to the BELLMAC32 uP and the MARS accelerator; at Caltech he designed the MOSSIM Simulation Engine and the Torus Routing Chip. While a Professor of EECS at MIT his group built the J- and the M-Machine and at Stanford University the Imagine processor. Bill has worked with Cray Research and Intel to incorporate many of his innovations in commercial parallel computers, with Avici Systems to incorporate it into Internet routers, co-founded Velio Communications and Stream Processors, Inc. He is a member of the National Academy of Engineering, a Fellow of IEEE, of ACM, and of the American Academy of Arts and Sciences. He has received numerous honours including the ACM Eckert-Mauchly Award, the IEEE Seymour Cray Award, and the ACM Maurice Wilkes award. He currently leads projects on computer architecture, network architecture, and programming systems. He has published over 200 papers in these areas, holds over 75 issued patents, and is an author of two textbooks.

978-1-61284-303-2/11 $26.00 © 2011 IEEE

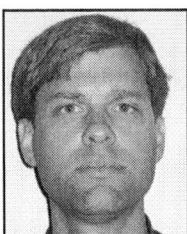

Power Management in the Wireless Communications Ecosystem
Anthony Hill, *Texas Instruments, Dallas, TX*

Anthony Hill is a distinguished member of technical staff for Texas Instruments. He currently leads the design implementation and signoff team for communications infrastructure products and is responsible for new technology deployment into SoC designs. He joined TI in 1996 after taking his PhD from the University of Illinois Urbana Champaign.

State Retention and Advanced Power Gating Techniques Applied to SoC Subsystems — Design Approaches and Silicon Evaluation
David Flynn, *ARM, Cambridge, United Kingdom*

Dr. David Flynn, a Fellow in R&D at ARM Ltd, has been with the company since 1991, specializing in System-on-Chip IP deployment and methodology. He is the original architect behind ARM's synthesizable CPU family and the AMBA on-chip interconnect standard. His current research focus is low-power system-level design. He holds a BSc in Computer Science from Hatfield Polytechnic, UK and a Doctorate in Electronic Engineering from Loughborough University, UK. He is currently part-time Visiting Professor with the Electronics and Computer Science Department at Southampton University, UK. David is a primary author of the Low Power Methodology Manual co-developed with Synopsys and launched in 2007 and a contributing author to the VMM-LP launched 2009.

An H.264 Full-HD Application Processor with 10GB/s x512b Stacked DRAM and a Power Management System with Multiple Power Domains
Masafumi Takahashi, *Toshiba, Kawasaki, Japan*

Masafumi Takahashi received the B.S. and M.S. degrees in information engineering from the University of Tsukuba, Japan, in 1985 and 1987, respectively. In 1987 he joined the Research and Development Center, Toshiba Corporation, where he was engaged in research on microprocessors and a single-chip multiprocessor architecture. From 1996 to 2005 he was engaged in the research of multimedia processor architecture in the Center of Semiconductor Research and Development, Semiconductor Company, Toshiba Corporation. Since 2005 he has been involved in the development of multimedia SoCs for audiovisual and 3D graphics applications and currently he is a chief specialist of System LSI Division, Semiconductor Company, Toshiba Corporation, Kawasaki, Japan. He was a member of ISSCC Signal Processing Subcommittee from ISSCC2005 to 2007. He served ISSCC as an associated chair for ISSCC2005 and a tutorial speaker for ISSCC2006.

Design and Process Optimization for Complex SoCs
Ronald Preston, *Intel, Austin, TX*

Ron Preston is a Senior Principal Engineer with the Atom and SOC Development Group at Intel in Austin Texas. His primary responsibilities are early path finding, process technology selection, and circuit technology optimization for Atom family products. Prior to joining Intel, Ron was part of the Alpha Development Group at Compaq and Digital Equipment. Ron holds Masters and Bachelor's Degrees in Electrical Engineering from Rensselaer Polytechnic Institute. He's a Senior Member of the IEEE, has coauthored over 25 technical publications, and is a past member of the ISSCC Technical Program Committee.

Near-Threshold Computing: Trade-Offs in Energy Efficiency, Reliability and Adaptivity
David Blaauw, *Michigan University, Ann Arbor, MI*

David Blaauw received his B.S. in Physics and Computer Science from Duke University in 1986, and his Ph.D. in Computer Science from the University of Illinois, Urbana, in 1991. Until August 2001, he worked for Motorola, Inc. in Austin, TX, were he was the manager of the High Performance Design Technology group. Since August 2001, he has been on the faculty at the University of Michigan where he is a Professor. He has published over 300 papers and hold 30 patents. His work has focussed on VLSI design with particular emphasis on ultra low power and high performance design. He was the Technical Program Chair and General Chair for the International Symposium on Low Power Electronic and Design. He was also the Technical Program Co-Chair of the ACM/IEEE Design Automation Conference and a member of the ISSCC Technical Program Committee.

A 1-TFLOPS 100-mW Processor
Sung Bae Park, *Samsung Electronics, Yougin, Korea*

Sung Bae Park was born in Seoul, Korea, on August 12, 1958. He received the B.S. and M.S. degrees in electronics engineering from Korea University, Seoul, Korea, in 1981 and 1989, respectively. He joined ETRI (Electronics and Telecommunication Research Institute), Daejeon, Korea, in 1982, where he worked on the design and development of NMOS/CMOS MCU and CPU. In 1991, he joined Samsung Electronics Company, Giheung, Korea, where he worked on the design of CPU and was leading the SBDC (Samsung Boston Design Center) for joint development with DEC for 21164 and 21264 CPUs, where his contributions included Cu-SOI optimization. He led the ARM CPU design team by 2009 in Samsung and now a Vice President of SAIT (Samsung Advanced Institute of Technology) leading the next generation processor R&D. He was awarded the Haedong Prize from IEEK(Institute of Electronics Engineers of Korea) for the acknowledgement of his contribution to the CPU technology development in 1999. He served as a Member of Program Committee for ISSCC from 2003 to 2006, and as an Executive Member for ISSCC in 2006. Also he served the A-SSCC 2005 as a Co-Chair of Technical Program Committee and A-SSCC 2006 as a Chair of Technical Program Committee.

ISSCC 2011 / FORUM / F5

F5: Image Sensors for 3D Capture

Organizer: Johannes Solhusvik, *Aptina, Oslo, Norway*

Chair: Albert Theuwissen, *Harvest Imaging, Bree, Belgium*
and Delft University of Technology, Delft, The Netherlands

Committee: Sam Kavusi, *Bosch, Palo Alto, CA*
Tetsuo Nomoto, *Sony, Kanagawa, Japan*
Iliana Chen, *Analog Devices, Somerset, NJ*

Low-cost 3D image capture is a popular R&D topic these days driven by a rapid growth in 3D viewing applications such as gaming and 3D movies. This forum commences with an overview of 3D display technologies and applications. It is followed by a 3D capture technology overview, how each method compares in performance, and its key design challenges to be successful in cost driven consumer and industrial applications. World experts on image sensor design will present pixel architectures, devices, circuits and technologies used to build 3D imager chips. Time-of-flight, stereo vision, structured light and multi-aperture imaging will be covered. The forum concludes with a panel discussion providing the opportunity for participants to give feedback and ask questions. The forum is aimed at circuit designers and engineers working in the imaging industry.

3D Display Technologies and Applications
Ian Underwood, *University of Edinburgh, Edinburgh, Scotland*
Ian Underwood is Professor of Electronic Displays at The University of Edinburgh where he is currently Head of the Institute for Micro and Nano Systems (IMNS). He was a co-founder of MicroEmissive Displays (MED) and co-inventor of its P-OLED microdisplay technology. Major honors include - Ernst & Young Emerging Entrepreneur of the Year, Scotland (2003), Fellow of the Royal Society of Edinburgh (2004), Gannochy Medal for Innovation winner (2004), Fellow of the Royal Academy of Engineering (2008) and Fellow of the Institute of Physics (2008).

He is recognized worldwide as an authority on microdisplay technology, systems and applications. Until recently he sat on the Council of the Scottish Optoelectronics Association and the Steering Committee of ADRIA (Europe's Network in Advanced Displays). He is co-author of a recently released book entitled Introduction to Microdisplays (Wiley, 2006). In late 2008 he was appointed to the Scottish Science Advisory Council.

Review of Optical 3D Ranging Techniques and Key Design Challenges for Required Image Sensors
Peter Seitz, *CSEM, Landquart, Switzerland*
Peter Seitz received his M.Sc. and Ph.D. degree in physics from the Swiss Federal Institute of Technology ETH in Zurich, Switzerland, in 1980 and 1984 respectively. He then worked for the RCA research laboratory in Zurich and the David Sarnoff Research Center in Princeton, NJ, in the domains of optical metrology and digital image processing. In 1987 he joined the Paul Scherrer Institute in Zurich, where he created and headed the research group for solid-state image sensing. In 1997 he transferred to CSEM, the Swiss Center for Electronics and Microtechnology, in Neuchatel and Zurich, Switzerland, where he became Vice President and head of the Photonics Division in 2000. In 2007 he created CSEM's new Nanomedicine Division in Landquart, Switzerland, which he is heading still today. Since 1998 he has also been professor in optoelectronics at the University of Neuchâtel and the Swiss Federal Institute of Technology EPFL, and he continues research and teaching in this capacity. He is a life member of the Swiss Academy of Technical Sciences SATW, and he is a Fellow of the European Optical Society EOS. He has authored and co-authored 180 scientific publications, and he is holding 35 patents in semiconductor imaging and optical measurement techniques.

3D Time-Of-Flight Image Capture with Pulsed Illumination
Pierre Magnan, *ISAE, Toulouse, France*
Pierre Magnan was born in Nevers, France. He received the Agregation Degree in Electrical Engineering from the ENS Cachan and the MS and D.E.A. degrees in integrated circuit design from the University of Paris, France in 1982. From 1984 to 1993, he was a research scientist at LAAS, Toulouse, France, involved in CMOS analog and semi-custom design. In 1995, he joined the CMOS Imagers Research Group at SUPAERO, also in Toulouse, where he was involved in active-pixels sensors research and development activities. In 2002, he became full Professor at SUPAERO, now called ISAE, where he is currently Head of CMOS Imagers Group working on both PhD subjects in cooperation with companies and development of custom image sensor dedicated to space instruments. He has supervised 12 Ph.D candidates in the field of image sensors and has been serving in the TPC of the International Image Sensor Workshop in 2007 and 2009, also being Associate Guest Editor of the IEEE Trans. Electron Devices for the Special Issue on Solid-State Image Sensors in Nov. 2009. His research interests include solid state image sensors design, modeling and technology and circuit design for imaging applications.

978-1-61284-303-2/11 $26.00 © 2011 IEEE

3D TOF Image Capture with Drift Field Pixel Structures
Bernhard Buettgen, *MESA Imaging, Zurich, Switzerland*

Bernhard Büttgen received the Diploma Degree in Electrical Engineering from the University of Siegen, Germany, in 2002, and the PhD degree from the University of Neuchatel, Switzerland, in 2006. He pursued his PhD studies on the topic of 3D time-of-flight imaging at the Swiss Centre for Electronics and Microtechnology (CSEM) in Zurich, Switzerland. Since 2006, he has been working as a research and development engineer on pixel and system level optimization of 3D time-of-flight imagers at MESA Imaging in Zurich, Switzerland. From 2009 till 2010, Bernhard has been a Post-Doc researcher within the image sensor group of Prof. Dr. Albert Theuwissen at the Delft University of Technology, the Netherlands. Bernhard holds several patents on 3D time-of-flight pixel architectures.

3D Time-Of-Flight Image Capture Based on SPADs
Edoardo Charbon, *TU Delft, Delft, The Netherlands*

Edoardo Charbon received the Diploma from ETH Zurich in 1988, the M.S. degree from UCSD in 1991, and the Ph.D. degree from UC-Berkeley in 1995, all in Electrical Engineering and EECS. From 1995 to 2000 he was with Cadence Design Systems and from 2000 to 2002 he was the chief architect of Canesta Inc., where he developed high-speed image sensors. In 2002 he joined the Faculty of EPFL, where he founded the AQUA Group, a laboratory devoted to the study of CMOS quantum sensors for biophotonics and 3D imaging. In 2008 he was appointed professor at TU Delft, where he holds the Chair of VLSI Design. He has authored over 150 technical papers, 13 issued patents, and two books in VLSI design, noise, and high-speed SPAD image sensors. His current research interests include 3D sensing, biomedical imaging, and SPAD fundamentals.

High-Speed 3D Image Capture Using 1D Structured Light
Shingo Mandai, *University of Tokyo, Tokyo, Japan*

Shingo Mandai received the B.S. degree in department of electronic engineering and the M.S. degree in department of electrical engineering and information systems from the University of Tokyo in 2008 and 2010, respectively. His current research interests include architecture and design of smart image sensors and time-to-digital converter.

Single-Chip Stereo Vision Cameras
Ralph Etienne-Cummings, *Johns Hopkins University, Baltimore, MD*

Ralph Etienne-Cummings received his B. Sc. in physics, 1988, from Lincoln University, Pennsylvania. He completed his M.S.E.E. and Ph.D. in electrical engineering at the University of Pennsylvania in December 1991 and 1994, respectively. Currently, Dr. Etienne-Cummings is a professor of electrical and computer engineering, and computer science at Johns Hopkins University (JHU). His research interest includes mixed signal VLSI systems, computational sensors, computer vision, neuromorphic engineering, smart structures, mobile robotics, legged locomotion and neuroprosthetic devices. He has published ~180 technical articles, 1 book, 7 book chapters and holds 3 patents plus 4 pending on his work in these subjects.

Multi-Aperture Image Sensor for 3D Capture
Keith Fife, *Ubixum, Palo Alto, CA*

Keith Fife received his B.S. and M.Eng. degrees in Electrical Engineering and Computer Science from Massachusetts Institute of Technology in 1999 and his Ph.D. in Electrical Engineering from Stanford University in 2009. He co-founded SMaL Camera Technologies in 1999 to develop consumer and automotive cameras with extended dynamic range and low power operation. One product was recognized as "Best of CES" in 2001 and as "World's Thinnest Camera" by Guinness World Records in 2002. In 2003, he returned to graduate school as a Hertz Foundation Fellow to work on devices and architectures to enable new imaging systems and finished by implementing sub-micron CCDs in a CMOS process for use in multi-aperture imaging. In 2007, he co-founded Ubixum to develop image sensors for high-sensitivity, long-read applications. His current interest is in design automation of complex sensors.

Arrays of Angle-Sensitive Pixels in Standard CMOS for 3D Light-Field Capture
Alyosha Molnar, *Cornell University, Ithaca, New York*

Alyosha Molnar received a BS in engineering with highest honors from Swarthmore College in 1997. After working as a deckhand on a fishing boat, he joined Conexant Systems Inc in Newport Beach CA in 1998. At Conexant, he worked as a RFIC design engineer and co-led the design of their first generation direct conversion GSM transceiver, which has sold more than 20 million parts to date. He entered graduate school at UC Berkeley in 2001 and received his MSEE in 2003 for his design of an ultra-low power RF transceiver for "Smart Dust". He then joined a retinal neurobiology lab where he completed his doctoral work (still in electrical engineering), focusing on dissecting the neuronal circuitry and signal processing of the mammalian retina using a combination of electrophysiology, and computational modelling. After receiving his PhD in May 2007, Alyosha joined the ECE department at Cornell as an assistant professor where continues his interdisciplinary research, focusing on RF integrated circuits, neural interfaces, and CMOS optics for optical image processing. He has published more than 25 technical papers, holds more than 20 patents, and has won multiple teaching awards.

ISSCC 2011 / FORUM / F6

F6: High-Speed Transceivers:
Standards, Challenges, and Future

Organizer/Chair: Ali Sheikholeslami, *University of Toronto, Toronto, Canada*

Committee: **Franz Dielacher,** *Infineon Technologies, Austria*
Miki Moyal, *Intel, Haifa, Israel*
Jafar Savoj, *Xilinx, San Jose, CA*
John Stonick, *Synopsys, Hillsboro, OR*
Takuji Yamamoto, *Fujitsu Laboratories, Kawasaki, Japan*

Multi-Gbps transceivers have evolved from what were once exotic special-purpose cells to ubiquitous building blocks that are expected to function as the glue between critical compute, communications, and storage elements. As such they are expected to function flawlessly, interoperate cleanly, and consume little power, area, and system design mindshare. As part of the 'commoditizing' of transceivers, the pressure to achieve multiple standards compliance from individual designs has increased dramatically - no longer are ASIC or system designers content with a single PHY for a single standard. The consequential impact on the design space means link designers must often 'right size' a swiss-army-knife design including power and area optimization; you don't want too many or too few 'blades'. Sometimes conflicting standards requirements can make customization and optimization a real challenge in this space. **Jared Zerbe** of Rambus will review some of the fundamental transceiver challenges and techniques used to address them, including a discussion of the impact of multi-standard compliance on the transceiver design effort and ultimate performance. Following this, **Thomas Toifl** of IBM will discuss the design challenges and solutions to data rates above 20Gb/s. Main challenges are achieving the required bandwidth with full ESD protection, the DFE loop timing, and overall design complexity.

Marcus van Ierssel of Snowbush IP will discuss implementation challenges and the benefits of a multi-standard PHY supporting Ethernet/PCIe/SATA/USB. **Takeshi Horie** from Fujitsu will present Ethernet standards and the technologies required to realize 10GB serial backplane transfer for 10GB Ethernet.

When it comes to buidling standards compliant PHYs, IP providers face challenges that arise from the need to proliferate the design across many process geometries/foundries and also from the large variability in customer's expertise in package design and signal integrity analysis. **Dan Weinlader** of Synopsys will discuss how standards influence these challenges.

The main focus of standards over the last few years has understandably been on addressing the bandwidth curve. The need to address the rising power struggle of SoCs is usually recognized, but limited to simple power targets. The final two presentations in this forum will focus on energy efficiency by **Fulvio Spagna** of Intel and on low-power solutions by **Anthony Sanders** of Lantiq.

Overview of High-Speed Transceiver Techniques and Their Challenges
Jared Zerbe, *Rambus, Los Altos, CA*
Jared Zerbe graduated from Stanford University, Stanford, CA, in 1987. From 1987 to 1992 he worked at VLSI Technology and MIPS Computer Systems. In 1992 he joined Rambus Inc. where he has since specialized in the design of high-speed I/O, PLL/DLL clock-recovery, and equalization and data-synchronization circuits. He has authored multiple papers and patents in the area of high-speed data transmission and clocking and has taught courses at Berkeley and Stanford in high-speed link design. He is currently a Technical Director at Rambus where he is focused on development of future low-power signaling technologies.

Design Challenges, Latest Achievements and Future Directions of High-Speed I/Os
Thomas Toifl, *IBM Research, Zurich, Switzerland*
Thomas Toifl received the Dipl.-Ing. (M.S.) degree and the Ph.D. degree (with highest honors) from Vienna University of Technology, Austria, in 1995 and 1999, respectively. In 1996, he joined the Microelectronics Group of the European Research Center for Particle Physics (CERN), Geneva, Switzerland, where he developed radiation-hard circuits for detector synchronization and data transmission, which were integrated in the four particle detector systems of the new Large Hadron Collider (LHC). In 2001, he joined the IBM Zurich Research Laboratory in Ruschlikon, Switzerland, where since then he has been working on multi-gigabit per second, low-power communication circuits in advanced CMOS technologies. In that area he authored or co-authored fourteen patents and numerous technical publications. Since July 2008 he manages the I/O Link technology group at the IBM Zurich Research Laboratory. Dr. Toifl received the Beatrice Winner Award for Editorial Excellence at the 2005 IEEE International Solid-State Circuits Conference (ISSCC).

978-1-61284-303-2/11 $26.00 © 2011 IEEE

Ethernet/PCIe/SATA/USB PHY Implementation Challenges and Tradeoffs
Marcus van Ierssel, *Snowbush-Gennum, Toronto, Canada*

Marcus van Ierssel received his Ph.D. in electrical engineering from the University of Toronto in 2007. He currently works for Snowbush IP, a division of Gennum where he is a senior engineer working on multi-standard SerDes. One of his recent roles was as technical lead in a 1G - 12.5G Multi-standard SerDes development in 40nm CMOS. Dr. van Ierssel has authored several publications in major conferences and journals including the ISSCC, and the JSSC.

Ethernet Standards and Their Implementation Challenges
Takeshi Horie, *Fujitsu Laboratories, Kawasaki, Japan*

Takeshi Horie was born in Tokyo, Japan, in 1962. He received B.S. in Electrical Engineering, M.S. in Electronics Engineering, and PhD in Engineering from the University of Tokyo, Tokyo, Japan in 1984, 1986, and 2003 respectively. In 1986, he joined Fujitsu Laboratories Ltd., Kawasaki, Japan, where he worked on the development of highly parallel computers. In 1994, he was a Visiting Scholar at Stanford University. In 1995, he worked on the development of scalar parallel servers at Fujitsu Limited, Kawasaki, Japan. In 1997, he joined HAL computer systems Inc., Campbell, CA, where he worked on the development of high-speed interconnects. In 2001, he joined Fujitsu Laboratories of America, Sunnyvale, CA, where he worked on 10G Ethernet technologies. Since 2007, he has been with Fujitsu Laboratories Ltd., Kawasaki, Japan, where he is currently the Director of Server Technologies Laboratory. His current research interest is computer systems and high-speed interconnects. He is a member of Information Processing Society of Japan (IPSJ). He received the IEICE outstanding paper award in 1993 from the Information and Communication Engineers (IEICE) of Japan, Sakai Memorial Award and Industrial Achievement Award from IPSJ in 1995 and 2005 respectively, and OHM Technology Award from the Promotion Foundation for Electrical Science and Engineering in 2004.

Standards and Their IP Perspectives
Daniel Weinlader, *Synopsys, Allentown, PA*

Daniel Weinlader receiver the M.S. and Ph.D. degrees in Electrical Engineering from Stanford University in 1993 and 1999 respectively. From 2000-2004 he was a design engineer and project lead at Accelerant Networks focusing on high-speed backplane I/O links. In 2004, as part of an acquisition, he became a Principal Design Engineer at Synopsys Inc. focusing on PHY IP. His main role in this capacity is technical project lead and he is involved with multiple Serial I/O standards and was part of the USB 3.0 PHY Working Group. His research interests include PHY product test and characterization, equalization techniques, low-power design and design portability. He has authored or co-authored numerous papers on high-speed PHY design and hold multiple patents on PHY related technologies.

Energy-Efficiency Considerations for High-Speed Transceivers
Fulvio Spagna, *Intel, Santa Clara, CA*

Fulvio Spagna is a Senior Principal Engineer at INTEL, Santa Clara. He is responsible for architecture of Serial IO interfaces for high density applications. Prior to this, he worked on Read Channel circuit development for Hard Disk Drive applications at Silicon Systems and, later, Texas Instruments. Fulvio graduated summa cum laude at the University of Naples, Italy, with a thesis on "Phase Noise in Microwave Oscillators". Fulvio has participated in the development of several industry and proprietary standard for Serial IO interfaces.

How Can Standards Provide Low Power Solutions for SoC?
Anthony Fraser Sanders, *Lantiq, Neubiberg, Germany*

Anthony Fraser Sanders is Senior Principal at Lantiq, former Infineon Wireline in Munchen, Germany. Anthony is responsible for providing Lantiq's products with high speed electrical and optical interface solutions, and has been responsible for the development of Infineon's first 10Gbps products in CMOS. Anthony earned a Bachelor Degree with Honours in Electrical and Electronic Engineering from the University of Nottingham in 1991 and joined Infineon Technologies, formerly Siemens HL, in 1996, after previously working for Phoenix Technologies, and GPT in Great Britain. During this time he has focused on high speed electrical and optical interfaces and now owns twelve patents. He has written a number of papers and contributed to two books on, the statistical analysis of electrical channels, statistical static timing analysis and equalization techniques for band limited systems, and is a founder of the Stateye open source statistical analysis tool. His current research activities focus on the modeling of jitter and statistical phenomena based on laboratory observation, and understanding the limits of CMOS technology for the use of implementing high speed interfaces in high volume manufacturing. Anthony was chair of the JEDEC FBDIMM2 Standard Link Signalling Task Group, the IEEE adhoc on XAUI Jitter for 10Gigabit Ethernet, editor of the OIFs Common Electrical I/O (CEI) standard, and has contributed to the PCIe Gen III, SATA Gen III & MIPI PHY standards.

ISSCC 2011 / SHORT COURSE

ISSCC 2011 / February 24, 2011 / 8:00 AM

Cellular and Wireless LAN Transceivers: From Systems to Circuit Design

Organizer: John R. Long, *Delft University of Technology, Delft, The Netherlands*

Instructors: **Hooman Darabi,** *Broadcom, Irvine, CA*
Frank Op't Eynde, *Audax Technologies, Leuven, Belgium*
Behzad Razavi, *University of California, Los Angeles, CA*
Bogdan Staszewski, *Technical University of Delft, Delft, The Netherlands*

OVERVIEW

RF design has matured to the point where almost any wireless standard (e.g., GSM, WiFi, Bluetooth, etc.) and/or multiple standards operating in different bands can be integrated onto a single chip in deep submicron CMOS. This course is an in-depth tutorial that discusses many of the challenges facing the designer of a single-chip radio, and their solutions.

The topics addressed in the course range from the system right down to the circuit level, including: system requirements and the demands they place on transceiver circuit specs, the design of RF front-ends with >100dB dynamic range at GHz frequencies, direct conversion architectures, autocalibration methods, sub-systems robust to substrate/supply coupling and crosstalk, fast responding PLL synthesizers with low spurious content, and the latest 'more digital than analog' developments.

This course is aimed at circuit designers who are familiar with the basic terminology of RF and mixed-signal/analog circuits and wish to acquire further depth in RF IC design.

OUTLINE

System Requirements of RF Transceivers

This lecture offers a system-level analysis of advanced RF transceivers for wireless and mobile applications. An overview of RF standards is then presented, followed by a detailed discussion of various radio transceiver architectures. Key radio requirements are derived and translated to their corresponding circuit specifications, giving an overview of a practical top-down radio design. Several non-idealities and limitations of RF-CMOS are discussed, and architectural and calibration techniques that overcome these limitations are described. Finally, advanced topics such as handset calibration, the evolution to broadband, RF diversity, and next generation mobile standards and their requirements are presented.

Instructor: Hooman Darabi was born in Tehran, Iran in 1972. He received the BS and MS degrees in electrical engineering from Sharif University of Technology, Tehran in 1994 and 1996, respectively. He received the Ph.D. degree in electrical engineering from the University of California, Los Angeles, in 1999. He is currently a senior technical director with Broadcom, Irvine, CA. His interests include analog and RF IC design for wireless communications. Dr. Darabi holds over 60 issued, and 80 pending patents with Broadcom, and has published over 40 IEEE peer-reviewed and conference papers.

Front-end Circuit Design for RF Transceivers

This lecture deals with the transistor-level design of RF front-end circuits, including low-noise amplifiers, downconversion and upconversion mixers, VCOs, and PA drivers. The role of each block in the transceiver is described, and design procedures are offered and carried out in a 65-nm CMOS technology. State-of-the-art examples are also presented.

Instructor: Behzad Razavi received the B.S.E.E. degree from Sharif University of Technology, Tehran, Iran, in 1985 and the M.S.E.E. and Ph.D.E.E. degrees from Stanford University, Stanford, CA, in 1988 and 1992, respectively. He was with AT&T Bell Laboratories and Hewlett-Packard Laboratories until 1996. Since 1996, he has been Associate Professor and subsequently Professor of Electrical Engineering at the University of California, Los Angeles. His current research includes wireless transceivers, frequency synthesizers, phase-locking and clock recovery for high-speed data communications, and data converters. Professor Razavi is a Distinguished Lecturer for the Solid-State Circuits Society and an IEEE Fellow.

Implementation of Direct-Conversion Transceivers

The direct-conversion transceiver principle offers the advantages of a high degree of integration, low overall system cost, and low power consumption. However, the principle is difficult to implement due to practical design limitations such as calibration of static and dynamic offset, and I/Q imbalance. In this lecture, the principle and practical implementation of direct conversion receivers and transmitters are described. Different direct-conversion architectures are compared, and strategies for static and dynamic offset removal and autocalibration methods are presented. Special attention is paid to secondary effects such as crosstalk. Design problems are highlighted and explained by means of real design examples.

Instructor: Frank Op't Eynde received the E.E. and the Ph.D. degrees from the Catholic University of Leuven in 1986 and 1990, respectively. From 1990 to 1994 he was design project leader with Alcatel Mietec in Brussels, Belgium. From 1994 until 1997, he was CTO of Mixed Silicon Structures (Roubaix, France) and a member of the board of Misil Design (Rungis, France). From 1997 to 2001, he was Development Manager for xDSL front-ends and for wireless circuits at Alcatel Microelectronics in Brussels, Belgium and was promoted to Corporate R&D Director. In 2002, he co-founded AsicAhead SRL, a Wireless Product Development company based in Bucharest, Romania and in Genk, Belgium. Since 2007, Dr. Op't Eynde has been self-employed. He has approximately 40 publications and thirteen patents in the field of analog and RFIC design.

Recent Advances in RF Synthesis

Frequency synthesizers are an integral part of digital, mixed-signal, and RF systems-on-chip. As CMOS processes continue to scale, raw transistor performance and power consumption improve dramatically, but difficulties arise when implementing traditional phase-locked loop architectures. This lecture first reviews design challenges such as operation from a 1V supply, gate and off-channel leakage, flicker noise, effect of non-linear device characteristics, and poor isolation from digital logic. Next, well-known workarounds to these problems are presented. Finally, new solutions for RF synthesis that are amenable to nanometer-scale technologies are described.

Instructor: R. Bogdan Staszewski received his B.S.E.E. (summa cum laude), M.S.E.E. and Ph.D. degrees from the University of Texas at Dallas in 1991, 1992 and 2002, respectively. He joined Texas Instruments in Dallas in 1995. In 1999, he co-started the Digital RF Processor (DRPTM) group at TI with a mission to invent digitally intensive approaches to traditional RF functions. Since July 2009, he has been an Associate Professor at Delft University of Technology in the Netherlands. He has co-authored one book, two book chapters, 110 journal and conference publications, and holds 60 issued and 40 pending US patents. Professor Staszewski is an IEEE Fellow.

978-1-61284-303-2/11 $26.00 © 2011 IEEE

ISSCC 2011 / EVENING DISCUSSION SESSION /EP1   ISSCC 2011 / February 21, 2011 / 8:00 PM

EP1: Good, Bad, Ugly - 20 Years of Broadband Evolution: What's Next?

Organizer: **Jerry Lin,** *Ralink Technology, Hsinchu, Taiwan*

Co-Organizers: **Franz Dielacher,** *Infineon, Villach, Austria*
                    Jing-Hong Conan Zhan, *MediaTek,*
                                       Hsinchu, Taiwan

Moderator: **Robert Payne,** *Texas Instruments, Dallas, TX*

Internet users today have extremely high demands for bandwidth consuming content wherever they are and on multiple platforms, ranging from tiny screened phones to high-definition 3D home theaters. This has driven the rapid deployment of multiple broadband access solutions including legacy infrastructure driven DSL and cable, the almost limitless bandwidth provided by optical fiber to the home (FTTH), and the on-the-go appeal of wireless solutions including 3G/4G cellular or WiMAX.

Over the past two decades, each of these technologies has successfully competed for the consumer's dollar, partly due to the differentiated ways data is accessed – televisions were naturally connected to the CATV infrastructure and cell phones were linked to a wireless basestation. These traditional boundaries are blurring as cell phone calls are placed over WiFi networks and video content is delivered via an ISP. The common denominator is data access and the delivery method is irrelevant to the end user.

As DSL continues to compete with cable and FTTH in the wireline space, more and more users are choosing wireless broadband solutions such as 3G, 4G, or WiMAX even for their home access. In this panel, we will review the good, the bad, and the ugly aspects of the recent history of broadband evolution and provide a vision of the future. What are the pros and cons of each technology? Do we need so many competing solutions? What will be the successful business models in the future – will everyone (or anyone) make money? Which ones will be the winners ten years from now? Experts from academia, chip suppliers, and broadband systems vendors will share their visions on future broadband markets and technologies. Take your seat and enjoy the debate.

Panelist's Statements

Michiel Steyaert, *K.U.Leuven, Heverlee, Belgium*
What's next? Only good things. Broadband evolution will go on, and the discussion or competition between wireless and wireline will go on: turn after turn they overtake each other, like a pendulum goes from left to right and back. And with every swing the speed is improved by an order of magnitude. We have seen that in telecom, radio and TV-broadcasting, and we will see it in internet connections. Eg. television broadcasting: first wireless, then cable (but more channels and thus more bandwidth), then wireless again (satellite with hundreds of channels), and now FTTH (fiber to the home and let's consider that again as wireline) with even more bandwidth for the many HD channels. And after that? Again wireless for the 3D HD movies... but at 120GHz or higher.

Larry DeVito, *Analog Devices, Wilmington, MA*
Tutto bene! It's all good! Except DSL.
The bewildering variety of Broadband options evolved as response to specific market needs. DSL shares a key attribute with the (truly ugly) plain old dial-up that it replaced: use the installed infrastructure of twisted pair - get it quick, get it cheap; that's just about all it is good for. Even back in those dark old days everyone knew fiber optics was the only answer and everything else was an interim technology. Today PON fiber optics is the only growth area and the only option to slake the unlimited bandwidth thirst.

While PON is good for dense population, rural access needs wireless for lower cost of infrastructure. But - surprise! - wireless took on a life of its own with teenagers using cellphones for everything but voice calls. Whose idea was it to put a camera in a cellphone? Despite the mobility appeal, wireless will never have enough capacity to meet video demands of the consumer.
0Cable is instructive not for broadband access, but for deployment of the original infrastructure. The invisible hand brought cable tv to wealthy neighborhoods and only government intervention altered this trajectory. Similar predictions for "fiber to the rich" delayed optical fiber until equivalent services could be delivered to all neighborhoods.

Sven Mattisson, *Ericsson, Lund, Sweden*
The introduction of 3G was not perceived by all users as a step forward initially because of the inferior talk and standby times compared to 2G handsets. In fact, many users were quite disappointed. If scalable performance, both in speed and power consumption is not properly handled user acceptance will be at stake, delaying introduction of more efficient standards enabling better spectrum usage. With 50 billion wireless connections we will have to support both voice and infrequent machine generated as well as high-rate media streams in the same scattered frequency bands. The scattering of bands also aggravates the hand-set front-end complexity and we must find ways of keeping this complexity reasonable.

David Borison, *Ralink Technology, Cupertino, CA*
DSL plus WiFi is the solution! With the erosion of traditional voice services and the ever-increasing competition amongst service providers, major shifts are occurring in the telecommunications industry business model. Carriers are not only working to expand service coverage and increase subscriber penetration rates, but they are also exploring how to significantly improve interoperability and network performance, increase bandwidth, and lower latencies for demanding applications such as IPTV, interactive gaming, video chat, etc. In light of these changing dynamics, carriers must accelerate the introduction of next-generation services and products. Integrated DSL plus 802.11n Wi-Fi combo solutions enable service providers to introduce cost-effective Integrated Access Device (IAD) products with higher bandwidth, better service-level agreements (SLAs), and enhanced quality of service (QoS) needed to deliver high-performance residential video and data connectivity solutions.

Eric Yeh, *MediaTek, Hsinchu, Taiwan*
WiMAX as 4G technology, and its future in developed and emerging markets is promising.

Wireless broadband promises to deliver Internet and multimedia content anytime and anywhere. Consumers demand data services on-the-go and all wireless service providers look forward to the explosive revenue growth coming from data applications. WiMAX is the first OFDMA-based 4th generation wireless broadband technology to become commercially available for deployment, and it is the first to materialize multi-megabit-per-second throughput on handsets and other portable/mobile devices. Including trial networks, today there are a few hundred deployments worldwide. From a chipset vendor standpoint, there are still challenges facing current WiMAX deployments. However, chipset vendors, system vendors, and service providers will work together to resolve those hurdles in order to take the technology to the next level of mass adoption. We expect to see the promising results on the future of WiMAX technology, especially in both developed and emerging markets.

Stephen Palm, *Broadcom, Irvine, CA*
Although there have been strong advances in wireless broadband, it appears wireline broadband may continue to dominate for delivering entertainment throughout the home. Every successful specification or standard has enjoyed a clear market changing benefactor to fund development and then deployment; entertainment demands reliable capacity. Hollywood looks to amplify its requirements over yesterday's data and voice applications. And if the trend continues, the industry's nimble specification groups will supply the needed advancements by supplanting the traditional standards development organizations. Low latency and high throughput are also compulsory to the groups that create them.

978-1-61284-303-2/11 $26.00 © 2011 IEEE   525

ISSCC 2011 / EVENING DISCUSSION SESSION / EP2   ISSCC 2011 / February 22, 2011 / 8:00 PM

EP2: 20-22nm Technology Options and Design Implications

Organizer / Moderator: Don Draper, *True Circuits, Los Altos, CA*

The move to 22/20nm comes at the cost of an increased level of leakage and variability that both technology and product developers need to address. Lithography, still at 193nm, uses liquid immersion, double patterning, mask and source optimization, and increasingly-restrictive design rules. High-k Metal Gate methods, targeted to reducing gate leakage and Gate-Induced Drain Leakage (GIDL), have proponents of Gate-First (IBM, GLOBALFOUNDRIES) and Gate-Last (TSMC, Intel), also using different metals for p-channel and n-channel for Vt adjustment. Partially-depleted SOI continues to be promoted by IBM and GLOBALFOUNDRIES. Under consideration for possible later introduction at 22/20nm are EUV, multi-beam E-beam and finfets. R&D dollars are increasing, once at 12% of revenue in 1998, now at 18% in 2010. As product design costs rise, the number of projects is declining, needing more synchronization of process and design development. Restrictive design rules (unidirectional poly, no jogs, dummy insertion, structured layout), adaptive architectures and expanded design for manufacturability are needed to address variability. Accurate layout-aware modeling for local variability, well proximity effects, STI stress (LOD) and gate implementation are needed. Design enablement and technology-design collaboration are increasingly emphasized.

Panelist's Statements

Bill Liu, *Global Foundries, Sunnyvale, CA*
With risk production set to begin in 2H 2012, GLOBALFOUNDRIES is well on its way to delivering 22/20nm technology to customers for product introduction in 2013. The 20nm technology offerings will come in two varieties: a High Performance (HP) technology with low leakage capability (or low power with high performance) for either wired applications such as servers and media processors or mobile/SoC applications such as smart phones and wireless portable devices, and for high performance a 20nm Super Low Power (SLP) technology designed for power-sensitive cost-optimized mobile applications. GLOBALFOUNDRIES also will have access to a 22nm Super High Performance (SHP) technology designed for devices requiring the utmost in performance. The 22/20nm technologies are planned to be a full node shrink from 32/28nm, and will utilize next-generation HKMG technology, immersion lithography and strain engineering to enable the area and die cost scaling the industry has come to expect with each technology generation. With V0.01 models/PDK released, test chip shuttles for customers will begin running in Fab 1 in 2H2011.

Mark Bohr, *Intel, Hillsboro, OR*
Once again we find ourselves preparing to introduce a new generation of process technology and recognizing that innovative approaches are needed to meet product requirements. The challenge of patterning 22nm dimensions using 193nm light is profound, but double patterning will be used on more layers to provide both dimensional scaling and improved dimensional control. Transistor material improvements, such as used for strained silicon, high-k and metal gate, will continue to play a major role in scaling, but the focus will shift from increasing transistor speed to improving power efficiency and variability. 22nm logic products will incorporate an increasingly diverse set of circuits requiring a diverse set of process features, making SoC technologies more mainstream than in the past. The increased use of non-traditional scaling techniques on recent process generations mandates a close collaboration between process and design teams to successfully bring leading edge products into high volume manufacturing.

Ghavam Shahidi, *IBM, Fishkill, NY*
20/22 nm node development is well underway (i.e. device, interconnect, features, and ground-rules are mostly locked down), toward a GA of 2011-12 (depending on the chip size) To the first order, it will be a straight shrink of 32/28 nm node (i.e. planar device, bulk or non-fully depleted SOI, about the same performance enhancing elements as 32 nm). High K inversion thickness will be shrunk (~ 1 nm) for better short channel and performance (most probably both gate first and last offerings). For the low power, the minimum device leakage will creep-up (to ~100 pA/um). Variability will be only slightly worse than 32 nm, thus SRAM scaling will continue. The cell area will be well below 0.1µm². 22/20 nm will be using 193 immersion lithography. One key questions has been the trade-off between double patterning and shrink factor (i.e. cost vs. productivity). Some offerings may be less than the historical 70% shrink.

MIn Cao, *TSMC, Hsinchu, Taiwan*
Variation and leakage control will become more challenging as we scale toward the 22/20nm node, driven by voltage scaling, dimension scaling, pattern loading effect, and layout dependent effect. New process techniques, design-process co-optimization, and new design techniques are all necessary for us to combat variation and leakage. More than ever, manufacturers and designers need to work closer with each other to meet the challenges. On the process technology front, new HKMG, patterning, junction and CMP techniques are being developed to combat variation and leakage. FinFET, under consideration for 22/20nm, are both opportunities and challenges from variation control standpoint. There are also plenty of opportunities for designers and manufacturers to collaborate in design-process co-optimization, restrictive design rules (RDR), dummy insertion techniques to reduce pattern density variation, layout-aware models, more structured layout, EDA tool enablement, early Si validations, just to name a few.

Koichiro Ishibashi, *Renesas, Electronics, Tokyo, Japan*
It is predicted that 20/22nm node technologies have a lot of difficulties about GIDL and variability of threshold voltages. Adaptive techniques address these issues. An effective adaptive technique is to apply suitable supply voltage depending on device performance. GIDL and gate leakage are strongly depend on applied voltage, so that adaptive voltage can also reduce leakage. Variation of LSI performance due to variation of device characteristics can also be compensated by adaptive techniques. As for UTB FD SOI technology, adaptive body bias technique is also effective because large forward body bias can be applied and the forward bias can reduce the local variation of threshold voltage, thereby lowering the supply voltage. Leakage can also be reduced thanks to the low supply voltages.

978-1-61284-303-2/11 $26.00 © 2011 IEEE

ISSCC 2011 / EVENING SESSION / ESO
February 20, 2011 / 7:30 PM

ESO: Student Research Preview

Chair: Jan Van der Spiegel, *University of Pennsylvania, PA*

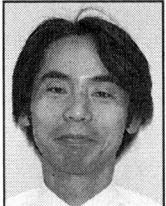

Session 1: Analog & Mixed-Signal Circuits
Session Chair: **Takayuki Kawahara**, *Hitachi, Tokyo, Japan*
Associate Chair: **Kofi Makinwa**, *Technical University of Delft, Delft, The Netherlands*

1.1:   **A New Class of Highly Efficient Integrated Rectifiers**
        Hongcheng Xu, Ulm University, Germany

1.2:   **Design of a DLL with Embedded Phase Interpolation**
        Steven Callender, UC Berkeley, CA

1.3:   **High PSRR LDO with Embedded Ripple Feed-Forward Path**
        Jianping Guo, The Chinese University, Hong Kong

1.4:   **A High Performance OTA for Oversampled Data Converters**
        Uğur Sönmez, Middle East Technical University, Turkey

1.5:   **Synchronized Multi Level Self-Oscillating Class D Amplifier**
        Cellier Rémy, CPE Lyon, France

Session 2: RF & Wireless Circuits
Session Chair: **Marian Verhelst**, *Intel, USA*
Associate Chair: **SeongHwan Cho**, *KAIST, Daejon, Korea*

2.1:   **Reconfigurable Multilevel Outphasing Transmitter**
        Ahmed Aref, RWTH Aachen University, Germany

2.2:   **A Triple-mode Balanced Linear CMOS Power Amplifiers Using Switched Quadrature Coupler**
        Hamhee Jeon, Georgia Institute of Technology, GA

2.3:   **A Pulse Shaping Technique for Spur Suppression in Injection Locked Frequency Synthesizers**
        Mehran Izad, National University of Singapore, Singapore

2.4:   **A 4-Element RF Beamforming Array in 65nm CMOS with Off-Chip Antennas**
        Saihua Lin, Stanford University, CA

2.5:   **A 55-GHz, 17% FTR, Low Phase Noise VCO Using an Artificial Grounded Metal Guard Ring**
        Pen-Li You, National Cheng Kung University, Taiwan

2.6:   **A Sub-mW All-Digital Signal Component Separator with Branch Mismacth Compensation for OFDM LINC Transmitters**
        Ping Yuan Tsai, National Chiao-Tung University, Taiwan

Session 3: Technology Directions and Biomedical Applications
Session Chair: **Vincent Gaudet**, *University of Waterloo, Waterloo, Canada*
Associate Chair: **Eugenio Cantatore**, *Eindhoven University of Technology, Eindhoven, The Netherlands*

3.1:   **Independent L and Q Adjustment of On-Chip Inductors by Above-CMOS Processing for Rapid Prototyping of RF SoCs**
        Yuki Sasaki, Tohoku University, Japan

3.2:   **An Integrated Omnidirectional Wireless Power Receiver and Its Helix Transmitter for Wireless Endoscopy**
        Tianjia Sun, Tsinghua University, China

3.3:   **On-Chip Autonomous Axonal Elongation**
        Faisal Abu-Nimeh, Michigan State University, MI

3.4:   **An Autonomous micro-Digital Sun Sensor Achieving 0.004° Resolution @ 21mW**
        Ning Xie, Delft University, The Netherlands

3.5:   **Adaptive Analog Front End for Neural Recording Application**
        Vikram Chaturvedi, IISc Bangalore, India

3.6:   **A 65nm 2.97GHz Self Synchronous FPGA with 42% Power Bounce Tolerance**
        Benjamin Devlin, University of Tokyo, Japan

3.7:   **A 92mW 76.8GOPS Vector Matching Processor with Parallel Huffman Decoder and Query Re-ordering Buffer for Real-time**
        Seungjin Lee, KAIST, Korea

978-1-61284-303-2/11 $26.00 © 2011 IEEE

ISSCC 2011 / EVENING DISCUSSION SESSION / ES1　　　**ISSCC 2011 / February 20, 2011 / 8:00 PM**

ES1: Data Converter Breakthroughs in Retrospect

Organizer:　　**Boris Murmann,** *Stanford University, Stanford, CA*

Chair:　　**Venu Gopinathan,** *Texas Instruments, Bangalore, India*

The performance of data converters has been pushed relentlessly over the years, leveraging advancements in scaling and design techniques that exploit the high density and speed of modern process technology. However, most of the underlying architectures in use today were conceived decades ago, and are nowadays regarded as fundamental in their nature. Were these architectures viewed as fundamental, potentially long lasting breakthroughs when they were first demonstrated? In this session, we bring together four pioneers of data converter design to review the invention and progression of the basic data converter architectures.

Oversampling analog-to-digital and digital-to-analog conversion has become the dominant approach in high-resolution, moderate-bandwidth interfaces. In the first talk, Bruce Wooley will review the beginnings of oversampling converters and discuss the various architectural refinements that have emerged over time. Our second talk will focus on two equally important developments of the 1970s and 1980s: charge-based A/D conversion and digital self-calibration. Hae-Seung Lee will examine these inventions in retrospect and explain why these techniques remain equally valuable today.

In the third talk of this session, Stephen Lewis will lead us back to the early days of high-speed pipelined conversion. While today's pipelines can easily sample at several hundred megahertz, the challenge back then was to cope with the bandwidth of video signals. Pipelining the digitization enabled the designer to extract very high speeds from CMOS transistors that were relatively slow and still getting ready for prime time. The session concludes with a presentation from Doug Mercer on high-speed digital-to-analog converters, which will review design philosophy changes that occurred in the 1990s. With the emergence of broadband data communication, new application requirements came into play, forcing many designers of high-speed DACs back to the drawing board.

Abstract

The Evolution of Oversampling ADCs

Bruce A. Wooley, *Stanford University, Stanford, CA*

For the digital encoding of analog signals, so-called oversampling modulators combine sampling at well above the Nyquist rate with feedback to exchange resolution in time for that in amplitude. This concept first emerged in the mid-twentieth century as delta modulation, with potential applications in digital communications. However, because delta modulators encode the rate of change of a signal, rather than the signal itself, their implementation presented significant practical challenges. The subsequent evolution of noise shaping modulators in various forms led to robust high-resolution ADC architectures that are ideally suited to realization in modern VLSI technologies. Commonly referred to as delta-sigma, or sigma-delta, modulators, these architectures have come to dominate precision interfaces between analog and digital signals at frequencies below a few megahertz.

MOS A/D Converters: Development of Capacitor Array ADCs and Digital Self-Calibration

Hae-Seung Lee, *Massachusetts Institute of Technology, Cambridge, MA*

In the mid 1970s, capacitor array successive approximation ADCs revolutionized data converters. Unlike resistor or current-source based ADCs in bipolar technology, these converters were a perfect match for MOS technologies. Capacitors offered much better matching while MOS switches were capable of switching capacitors at high speed and without the IR drop. A clever input sampling operation removed the need for sample-and-hold amplifier and eliminated the effect of parasitic capacitance.

Until the mid 1980s, high accuracy ADCs were predominantly individually laser-trimmed hybrids that were bulky and expensive. Therefore, the digital self-calibration technique which measured and calibrated the ADC's own accuracy without a 'golden standard' was a revolutionary concept. The ADC measured the ratio errors of the capacitors and recorded them in digital memory during calibration for use to cancel the error during the regular conversion. Ensuing CMOS scaling rendered the calibration logic circuitry trivial and popularized digital calibration in many data converters.

Early Monolithic Pipelined ADCs

Stephen H. Lewis, *University of California, Davis, CA*

The idea of pipelining has been around for a long time, at least since when assembly lines were first used in manufacturing. In pipelined analog-to-digital converters (ADCs), the conversion is broken up into steps separated by analog sample-and-hold amplifiers (SHAs). In the 1980s, CMOS technologies advanced to the point where errors in SHAs became small enough to build monolithic pipelined ADCs. This talk will review the history of some early pipelined ADCs. The effects of various error sources will be considered. The talk will review the use of redundancy to reduce the impact of comparator offsets. Also, redundancy reduces the significance of SHA offsets, and these characteristics turned out to be important in practice because they allow pipelines to be built without offset cancellation. Also, the talk will review the 1.5-bit/stage architecture, which was able to handle analog video standards using 0.9-micron CMOS in the 1990s.

High Sample Rate Signal-Reconstruction DACs - a Retrospective

Doug Mercer, *Analog Devices, Wilmington, MA*

In the late 1980s and early 1990s the principal application for high clock rate DACs was in graphic and computer CRT displays, both raster and vector scan. These DACs were optimized for time domain specifications such as settling time and glitch impulse. When communications system engineers attempted to use these DACs in their designs they found that the frequency domain performance was rather poor. With more of a frequency domain optimization mindset, DAC designers started to fundamentally rethink their approach to designing DACs for digital communications applications. This paper looks back over the evolution of high speed signal reconstruction DACs which enabled much of the broad band wireline and wireless communications explosion since the early 1990's. In addition to new circuit techniques, this new class of high sample rate DACs benefited from process evolution starting with ECL-based bipolar to BiCMOS to sub-micron and even deep sub-micron CMOS.

978-1-61284-303-2/11 $26.00 © 2011 IEEE

ISSCC 2011 / EVENING DISCUSSION SESSION / ES2　　　　**ISSCC 2011 / February 20, 2011 / 8:00 PM**

ES2: Wireless Sensor Systems: Solution & Technology

Chair / Co-organizer:　　**Pascal Urard,** *STMicroelectronics, Crolles, France*
Co-organizer:　　　　　　**Jun Ohta,** *NAIST, Nara, Japan*

Recently, we start to see Wireless Sensor Network (WSN) solutions on the market, which are mostly based on 802.15.4 standard or Zigbee equivalent. Most of these solutions integrate an 8-bit to 32-bit low power microcontroller, one or multiple sensors and run on batteries. However, business is not yet there massively. Does it mean we need additional innovation to boost market adoption? The aim of this session is to provide the ISSCC audience comprehensive and innovative information. It will allow discussion to understand what new techniques are required for the next generation of wireless sensor systems.

This Special Evening Topic session will present current wireless sensor system solutions, and explore what could be the next innovations to come. We will review what changes could be operated at designer level, and which low power design techniques will allow the system to run the extra mile in terms of energy efficiency. MEMS are at the heart of the wireless sensor system, and its integration must be realized at the lowest cost. For this purpose, some heterogeneous techniques for industrial sensors integration on CMOS substrate will be presented.

Realization of autonomous Wireless Sensor Networks seems to be a trend with first announcements at industry level. Next generation energy storage techniques will be discussed to enable longer autonomy and easier integration into Wireless Sensor Systems. Finally, we will consider innovative solutions for energy scavenging from light, temperature, vibration or even radiofrequency and their power level compatibility with wireless systems.

Abstracts

Commercial WSN: Low Power Hardware, Efficient Stacks
Kris Pister, *Dust Networks, Hayward, CA*
The Wireless Sensor Network industry has crossed the $100M/year barrier and is well on it's way toward $1B/year. Until recently, this growth had been gated by the lack of low power hardware, efficient and reliable communication stacks, and international standards. Now all of the technical problems have been solved, and the IEEE and IETF have developed open, global standards for WSN communication. Transmit and receive currents for some 802.15.4 radios are in the low single-digit mA range, and several medium access protocols allow even routing nodes to duty cycle their radios below 0.1%. Low-power 32-bit cores are now common even in low-end devices. Combining all of these means that standards-compliant IP routers can burn less than 10uA from 3V while hosting a variety of internet application protocols. The result is that reliable, secure WSN nodes can run on batteries for over a decade, and on scavenged power indefinitely.

Ultra-Low Power Design for WSN
Dennis Sylvester, *University of Michigan, Ann Arbor, MI*
The proliferation of miniature wireless sensors is challenging due to the need to simultaneously achieve unprecedented small form factor and very long lifetime. While better power sources are certainly needed, including energy scavenging techniques, smarter low-power circuit design approaches are best prepared to meet these challenges. Recognizing that this is a new class of computing in the spirit of Bell's Law, circuit designers should turn their attention to the "nW challenge", which replaces the previous GHz race focusing on high performance. By doing so, order of magnitude improvements at both block and system levels are achievable, enabling truly ubiquitous wireless sensing. Key techniques include near-threshold supply voltages, judicious technology and device selection/tuning, and a systematic approach to total energy reduction that focuses on minimizing wireless communication needs (e.g., via enhanced on-chip signal processing capabilities, possibly using low-voltage accelerators) and aggressively reducing sleep mode power given most applications are heavily duty-cycled.

Heterogeneous Integration for Sensing and Wireless Systems
Masayoshi Esashi, *Tohoku University & Memsas, Sendai, Japan*
The integration of heterogeneous components on a Si chip (heterogeneous integration) is required for value added devices. MEMS (Micro Electro Mechanical Systems) should be encapsulated in cavities to ensure their motion. Wafer level packaging is required to enable encapsulation with low cost. Micromechanical components such as switches and variable capacitors or micromechanical resonators for oscillators and filters should be integrated on small feature size LSI for RF applications in order to be suited for wireless systems. The materials for these micromechanical components should have excellent mechanical properties as high Q and creep free or should be piezoelectric materials as LiNbO3, AIN or PZT. Following three processes are discussed as surface micromachining by deposition and etching on the LSI wafer, bonding and thinning of the material wafer on the LSI wafer followed by etching process, and wafer level transfer of micromechanical components on LSI.

Advanced Energy Storage Techniques
Raphaël Salot, *CEA-LITEN, Grenoble, France*
Today's miniaturised systems go beyond monolithically integrated or hybrid systems which combine measurement, data processing and storage functions. The future will consist of integrated smart systems which are able to sense and diagnose a situation, which are predictive, and therefore are able to make decisions. According to the strategic research agenda of the main technology platform in these domains (ENIAC, EPoSS, and ARTEMIS), the main functionalities that are often requested for these smart systems are: increased miniaturization, high reliability, networking and communication capability, integration and energy-autonomous performance. Recent developments in energy storage techniques designed for this kind of application will be presented. Among those techniques, thin film lithium microbatteries are all solid state technology with high integration capabilities (volume, temperature resistance…) and good electrical characteristics (no self discharge, high number of cycles without capacity fading…). Both, state-of-the-art performance and technological road map will be discussed.

Advanced Energy Harvesting Techniques
Ruud Vullers, *IMEC/Holst Centre, Eindhoven, The Netherlands*
Today, the batteries that are needed to power wireless autonomous transducer systems limit the possibilities of this emerging technology. Aim is to generate and store power at the micro-scale to improve autonomy and reduce size. Energy harvesters fabricated by micro-system technology can realize this goal. The choice of harvesting principle depends on the application and vibration, thermal, photovoltaic and radiofrequency power conversions are investigated. An overview of latest results and remaining challenges will be given with the focus on thermal and vibrational energy harvesters fabricated with micro-system technology. Also the importance of a complete module will be discussed, including an efficient power management circuit and energy storage system. Energy harvesting modules used in demonstrators are highlighted showing the feasibility to be applicable in wireless autonomous sensor systems. Application fields for these sensor systems are machine monitoring, predictive maintenance and body area networks.

978-1-61284-303-2/11 $26.00 © 2011 IEEE　　　　529

ES3: Future System and Memory Architectures: Transformations by Technology and Applications

Chair/Co-organizer:	**Nicky Lu,** *Etron Technology, Hsinchu, Taiwan*
Co-organizer:	**Leland Chang,** *IBM, Yorktown Heights, NY*
Co-organizer:	**Daisaburo Takashima,** *Toshiba, Yokohama, Japan*

The emergence of new enabling technologies and applications paradigms will likely drive radical changes in the memory architecture of future systems. With multi-core CPU dies sporting embedded DRAM caches, ever-improving NAND flash storage densities for SSD and SCM, and 3D-integration technologies to bring everything together into a single package, possibilities abound for system enhancements throughout the memory hierarchy. At the same time, applications needs are rapidly evolving as the world shifts from a product-centric economy to a service- and experience-oriented economy focused on hardware such as smartphones, set-top boxes, and 3D digital TV. This evening session will discuss future system and memory architectures from perspectives spanning the 3 C's: computing, consumer electronics, and communications – considering both what new technology might offer and what new applications might need.

Abstracts

Using New Technologies in Post Scaling Era
Jim Kahle, *IBM, Austin, TX*
New approaches to Innovation will be required as standard CMOS scaling slows. A number of promising technologies such as storage class memory and 3D stacked memory are appearing on the horizon. These new technologies have promising attributes for storage and memory usage and how we approach hybrid configurations with accelerators. First we must look at the markets and usage scenarios that will align with these new technologies. Then we must look at the system architecture implications of these technologies, and then or course look at the down stream effects into the Software programming models and other implications. With a continued demand for more function from Consumer electronics to cloud based servers, we must look into numerous domains that a new technology will be driven from.

Digital TV System Design and Future TV Direction
Tomofumi Shimada, *Toshiba, Tokyo, Japan*
A key design issue of digital TV is how to process concurrently a variety of multimedia image streaming including broadcast and data via network by not only optimized system design using system LSIs and memories, but also software design treating multi-core processor and managing memories. The latest high-definition (HD) TV including Cell TV realizes real-time high-quality multimedia image processing by super-resolution and an HD distortion removal technology, fine 3D picture indication, conversion of 2D to natural and beautiful 3D image, simultaneous multi-channel broadcast indications, simultaneous multi-channel recording / multi-window reproduction to/from TB HDD, and high dynamic contrast by controlling precisely a lot of high luminance LEDs. These superior functions have been realized by co-design of high-performance processor, high-bandwidth memories to minimize bottleneck of multi-GB/s streaming and intelligent software. Furthermore, futuristic TV features and direction such as up-conversion to cinema level pixels of 4K x 2K, network services on demand and interactive human interface, and system architecture direction to realize these new functions are addressed.

Smartphone Memory Architecture Challenges and Opportunities
Raj Talluri, *Qualcomm, San Diego, CA*
Smartphones are one of the fastest growing segments of the cell phone market. The main reasons for the popularity of the smartphones is the variety of applications that they can perform. High resolution camera, extremely low power MP3 playback, GPS-based navigation, HD video playback/recording, e-mail, calendar and other productivity applications are standard in these gadgets. However, in order to deliver a compelling user experience, there are significant challenges that the memory architectures used in these devices need to address including extremely low power, high bandwidth, small form factors. This talk will focus on the various use cases in the emerging and future smartphones and the stringent requirements that these impose on the memory architectures. It will also address the ways the device makers are solving these problems today and the challenges moving forward.

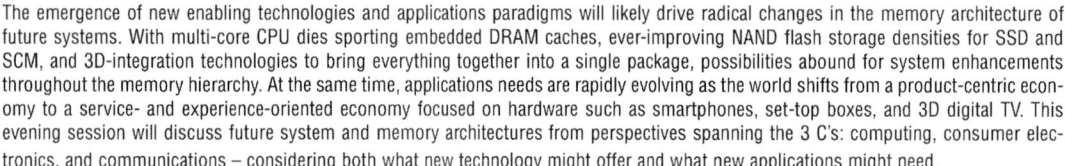

Memory Architectures in the Petascale Era and Beyond: Challenges and Opportunities
Stephen Pawlowski, *Intel, Hillsboro, OR*
Moore's law has enabled an unprecedented pace of increased compute density allowing the computing industry to potentially deliver terascale performance on every laptop in the future. However, the scaling of memory bandwidth has been much more modest, creating an imminent memory bottleneck in future CPUs. In addition, current DRAM and flash based NVM technologies are pushing their scaling limits. These three aspects put together have introduced several key challenges namely, power, reliability, diminished bytes/flop, security, and cost - that will require significant innovation in technology, architecture and design. In this talk, Steve will discuss these trends and challenges in memory system design and provide insight into approaches to address them.

ISSCC 2011 / EVENING DISCUSSION SESSION / ES4   ISSCC 2011 / February 21, 2011 / 8:00 PM

ES4: Body Area Network: Technology, Solutions, and Standardization

Organizer   Hoi-Jun Yoo, *KAIST, Daejeon, Korea*

Chair   Alison Burdett, *Toumaz, Abingdon, United Kingdom*

Recently, wireless protocols related to BAN (Body Area Networks) are under standardization by the IEEE 802.15 (Personal Area Networks, PAN) committee, to enable interoperability of a wide range of applications in the areas of medical support, healthcare monitoring and consumer wellness electronics. BAN requirements are closely related to PAN or WSN (wireless sensor network) technologies; however a major difference in BAN applications is that human body should be carefully considered not only as a possible communication medium or an obstacle to the signal transport, but also taking into account the possibility of physiological effects resulting from the chosen EM wave frequency.

This Special Evening Topic session will present the current status of Body Area Network standard development, and explore proposed solutions and applications with a strong focus on integrated circuit implementations. As an introduction, the IEEE BAN standard process will be reviewed, including its historical background and current status. Following this, further technical details including MAC and PHY layers and security protocols will be addressed. Integrated circuit solutions for BAN applications will be disclosed, and examples of BAN application requirements, especially for ambulatory patient monitoring, will be discussed.

Abstracts

Standardiztion of BAN: History, Major Issues & Current Status of TG6
Arthur Astrin, *Astrin Radio, Palo Alto, CA*
Body Area Networks (BAN) devices operate in close vicinity to, on, or inside body and can enable a wide range of applications, including medical support, healthcare monitoring and consumer electronics with increased convenience or comfort. Due to strong demands of medical, healthcare and information technology industries, IEEE was requested to standardize the Body

Area Network. IEEE 802.15 task group 6 (TG6) was set up to develop an IEEE international standard for BAN in January 2008. This talk reviews major issues, history and current status of TG6. Early on, the TG6 invited representatives from industry to present applications which require body area networks. We then developed an application matrix, and summarized it into a single document, which was issued to proposers. The proposers were asked to propose a communication protocol that would accommodate this application summary document. The other issue facing TG6 was to have a detail understanding of available spectrum for BANs. And to complete the design an accurate model of the channel; in this case the human body was needed. This channel is much more difficult than free space/air to measure and to correctly model. Thanks to several laboratories around the world the measurements taken and a detail model of the body channel was developed.

MAC & Security Network Solutions for BAN
Okundu Omeni, *Toumaz, Abingdon, United Kingdom*
A key requirement for Body area networks is low power. The challenge has been to develop a MAC and security specification which would serve the requirements of the wide range of BAN applications, while also being amenable to a low power implementation. The IEEE 802.15 task group 6 has worked for the last few years developing such a specification and this task is ongoing. This has considered the MAC and security aspects together to ensure the resulting specification provides the required MAC functionality as well as addressing security concerns of many of the medical applications robustly. The resulting specification features a compact core feature set with optional extensions to cater for specific application requirements. The details of this MAC and Security specification are presented together with the key considerations for low power implementation.

PHY Realization for BAN Using Ultra-Wideband Technology
Huang-Bang Li, *NICT, Yokosuka, Japan*
The first draft of the IEEE802.15.6 BAN standard was completed in May 2010. To meet various technical requirements, a narrow band PHY, an ultra-wideband (UWB) PHY, and a PHY using human body communication (HBC) are specified in the draft. Characteristics of BAN and the related PHY and MAC technologies will be addressed with emphasis being put on UWB technology. The updated regulation status on UWB will be overviewed. Although UWB is a good candidate to build BAN in the sense of providing high data rate with less emission power and low consumption power, some regulations present strong restriction on UWB. To show the suitability of UWB for BAN, two prototype BAN systems developed with UWB will be illustrated.

PHY Layers Issues and Narrowband Solutions for BAN
Anuj Batra, *Texas Instruments, Dallas, TX*
A body area network (BAN) is a collection of devices that are located on, in or around a body that can be used to collect or transfer data. The wireless link in a BAN is referred to as the "last meter" connection, for which the IEEE802.15.6 task group has proposed three vastly different physical (PHY) layers (narrowband, ultra-wideband and human body communications). An overview of these PHYs is presented. A popular application for BANs is wireless medical monitoring, in which medical data is collected from sensor nodes and relayed to a hub device. Such sensors have stringent limits on both their peak and average power consumption. Only the narrowband PHY is well suited to this application, since it supports the required range, robustness and scalable data rates while enabling very low power solutions. Details of this PHY are presented, with discussion on selection of the modulation, coding, packet structure, etc.

BAN Realization with Body-Channel Communication
Seong-Jun Song, *Samsung, Suwon, Korea*
As various functions such as a camera and a global positioning system (GPS) are included to mobile devices, request of information interchange between mobile devices is increasing. Electric-field communication (EFC) is one of the PHY candidates for BAN technologies that utilize the dielectric material itself, such as air or human body, and detects feeble changes of electric field induced on the dielectric material. Power efficiency, data scalability, coverage, and convenience of use are significantly considered for the frequent usage of data connectivity in mobile environments. The EFC technology has the potential to serve energy-efficient and intuitive connectivity solutions for a wide range of BAN applications. This presentation will introduce concept, architecture, and protocol of the EFC. In addition, this will cover low-power design strategies in architecture-level implementation and feasible applications between BAN devices using the EFC.

Wireless Propagation and Coexistence for Medical BAN
David M. Davenport, *GE Global Research, Niskayuna, NY*
Medical Body Sensor Networks represent a new tool for health care providers to utilize in order to address the rising level of hospital acuity resulting from an aging United States population. Technical requirements and system issues for wireless Medical Body Sensor Networks (BSNs) will be presented. Design guidelines are driven by the need to improve ambulatory patient monitoring and care while reducing logistic constraints for patients as well as healthcare professionals. We present a study on three key components of Medical BSN: On-body wireless link (to characterize the RF channel for body worn wireless devices), Coupling between bodies (to characterize the RF interaction between bodies) and Coexistence of Medical BSNs in the RF spectrum. Results and conclusions are presented through simulation and measurement studies. Discussion will also address strategic enablement of Medical Body Sensor Networks including IEEE standardization and spectrum allocation via Federal Communications Commission.

978-1-61284-303-2/11 $26.00 © 2011 IEEE

ISSCC 2011 / EVENING DISCUSSION SESSION / ES5   ISSCC 2011 / February 22, 2011 / 8:00 PM

ES5: Gb/s+ Portable Wireless Communications

Organizer: **Didier Belot,** *STMicroelectronics, Crolles, France*

Chair: **George Chien,** *MediaTek, San Jose, CA*

In the last few years, several gigabit-class wireless communication standards have reached commercial status, these are mainly driven by the higher demand of fast data transfer in multimedia applications (e.g. wireless HDMI, video streaming, etc.) At the same time, the number of portable devices has been increasing drastically; in the coming decade, the adoption of gigabit-class wireless communication in these portable devices will become a standard. In order to meet this demand, the electronic industry has to address the limitations of power consumption and form factor in the present solutions.

Let us define gigabit class as physical layer data rate to be greater than 1Gb/s; while portability as the battery capacity limited to 25W-hour lithium polymer battery at 3.75V (similar to the Apple iPad) or about 1000mA-hour at 3.75V for a typical cellular phone. These devices offer different classes of multi-media experience, however, are both very popular among consumers. The key question is: how to be energy efficient in data-rate/power consumption for the Gb/s+ device? For practical wireless application, the range needs to be at least a size of a room.

In this special evening session, we have assembled a group of industrial experts in gigabit wireless communication to speak about their respective technology; and their views on how these standards can achieve lowest power AND highest data rate simultaneously. These standards include WirelessHD, Wireless Gigabit Alliance (802.11ad), 802.11ac and a proprietary short range radio. In addition, after 4 individual presentations, a panel discussion will be held to provide everyone in the session an opportunity to ask questions to these experts.

Abstracts

Opportunities for 60GHz Short-Range Link
Jri Lee, *National Taiwan University, Taipei, Taiwan*
Recent research on low-power low-cost 60GHz RF makes high-speed (>1Gb/s) short-range (<2m) wireless link between mobile devices become possible. Providing data rate up to tens or hundreds times higher than Bluetooth EDR 2.0, this line-of-sight communication connects portable devices such as cell phones, laptop, and digital camera at tremendously high speed. With power consumption less than 200mW, large volume of data such as video can be easily shared. To realize miniature transceivers without baseband and digitizing circuitry, it is fully possible to adopt non-coherent demodulation for OOK and FSK, and Costas loop recovery for BPSK and QPSK. Modern technique also suggests that printed-circuit antenna on board can be co-designed and assembled with the TRx chip in an economic way.

WirelessHD 60 GHz Technology for Multi-Gbps Portable Wireless Applications
Jeff Gilbert, *SiBEAM, Sunnyvale, CA*
60 GHz is the best technology for high speed portable wireless applications based on its scalability up to multiple Gbps for full in-room non-line-of-sight (NLOS) applications, and down to hundreds of mW power consumption at Gbps rates. Line-of-sight limitations have been overcome by adaptive array antennas. Highly integrated cost-effective 60 GHz solutions have been developed leveraging standard digital CMOS radios packaged with embedded antenna arrays occupying less area than a single 2.4 or 5 GHz antenna.

The WirelessHD Consortium (including Broadcom, Intel, LG, Panasonic, Philips, NEC, Samsung, SiBEAM, Sony, and Toshiba) created the industry's first specification for multi-Gbps wireless transmission of lossless high-definition A/V and data for consumer electronics, PC, and portable products. Products available since 2009 have demonstrated the viability of NLOS transmissions at 4 Gbps. WirelessHD v1.1 includes enhancements ranging from optimizations for mobile devices consuming only a few hundred mW to maximum data rates exceeding 10 Gbps

Defining the Future of Multi-Gigabit Wireless Communications
Ali S. Sadri, *Intel, Santa Clara, CA*
The widespread availability and use of digital multimedia content has created a need for faster wireless connectivity that current commercial standards cannot support. This has driven demand for a single standard that can support advanced applications such as wireless display and docking, as well as more established usages such as network access. The Wireless Gigabit (WiGig) Alliance was formed to meet this need by establishing a unified specification for wireless communication at multi-gigabit speeds; this specification is designed to drive a global ecosystem of interoperable products.

The WiGig MAC and PHY Specification enables data rates up to 7 Gbps, more than 10 times the speed of the fastest Wi-Fi networks based on IEEE 802.11n. It operates in the unlicensed 60 GHz frequency band, which has much more spectrum available than the 2.4 GHz and 5 GHz bands used by existing Wi-Fi products. This allows wider channels that support faster transmission speeds.

The WiGig specification is based on the existing IEEE 802.11 standard, which is at the core of hundreds of millions of Wi-Fi products deployed worldwide. The specification includes native support for Wi-Fi over 60 GHz; new devices with tri-band radios will be able to seamlessly integrate into existing 2.4 GHz and 5 GHz Wi-Fi networks.

The specification enables a broad range of advanced uses, including wireless docking and connection to displays, as well as virtually instantaneous wireless backups, synchronization and file transfers between computers and handheld devices. For the first time, consumers will be able to create a complete computing and consumer electronics experience without wires.

802.11ac Standard Toward Portable Applications
Rolf De Vegt, *Qualcomm, Santa Clara, CA*
This presentation will focus on the high speed mobile potential for products based on the IEEE802.11ac standard under development. The 802.11ac standard is targeting system level throughputs of over 1 Gbps in the 5 GHz band. The presentation will describe the use cases analyzed prior to starting the .11ac project, the functional requirements and the technologies that are being included in the 802.11ac standard. For example, the use of wider bandwidths, multi-channel operation and Multi User MIMO (MU-MIMO) will be discussed. In addition, the talk will address potential future extensions and high level implications for IC vendor's roadmaps; and it will also include feature and design implications for battery operated devices. Lastly, the presentation will cover a possible market place roll out scenario for 802.11ac based products

978-1-61284-303-2/11 $26.00 © 2011 IEEE

ISSCC 2011 / EVENING DISCUSSION SESSION / ES6   **ISSCC 2011 / February 22, 2011 / 8:00 PM**

ES6: Technologies for Smart Grid and Smart Meter

Co-organizer/Chair: **Jed Hurwitz,** *Gigle Networks, Edinburgh, United Kingdom*

Co-organizer: **Wing-Hung Ki,** *HKUST, Hong Kong, China*

A Smart Grid is a required infrastructure in the near future to meet the world's energy usage and generation demands. The US government has just injected 4.5B$ of stimulus into creating Smart Grid technologies, and this has created a huge momentum to implement the infrastructure today.

With a variety of diverse renewable power sources (wind, wave, solar) and new challenging loads (e.g. electric vehicle) and usage models (e.g. peak shaving), the role of Smart Meters and communications among meters and appliances of your home is becoming increasingly important.

As consumers are becoming more aware of their own economic and environmental responsibilities regarding green power consumption and generation, connecting their appliances in the home to the WWW will also open up new opportunities in Energy 2.0 applications independent of the traditional utility companies.

This special topic session aims at introducing to the audience the market requirements and three of the key candidate technologies for the Smart Grid and Energy 2.0 application: accurate power metering, powerline communications, and low-power wireless communications for power management.

Abstracts

Smart Gird – An Overview
Martin Manniche, *GreenWave Reality, Irvine, CA*
Today, across both Asia and Europe, utility companies face the challenge of the deregulation of energy retailing, giving the consumers a choice of energy providers. The concept of Smart Gird emerges as a compelling energy management solution. The success of the Smart Grid hinges on providing an affordable, easy-to-use energy management solution that is engaging and has minimal impact on the user's lifestyle. Several specific points can be noted. (1) Solution providers should build the Smart Grid on an open standards-based network architecture that can work seamlessly on different physical network layers. This will ensure scalability and compatibility into the future, a key objective of most governments and utilities. (2) The Smart Grid should provide a high level of data security for consumers and utilities while maintaining great respect for data privacy requirements. (3) The platform should enable a meaningful, two-way communication between utilities and their customers, creating value for consumers thereby keeping them engaged and loyal. (4) The platform must be affordable, easy-to-use, and highly localized, serving consumers and spanning the global market. (5) Last but not least, think simplicity and usability! It is easy to get excited about new technology but we should also remember to ask ourselves "How do we create a fun and engaging experience that truly helps consumers save energy?"

Metrology - Smart Meter Reading Devices
Mick Mueck, *Analog Devices, Wilmington, MA*
Energy meters have come a long way in the last 15 years when they were virtually all electro-mechanically based and only provided an accuracy of around 1-2% of the real power. This presentation will address the state-of-the-art and beyond in electronics-based energy metrology. The IC at the heart of a modern energy meter will directly connect to a wide variety of transducers and provide superior measurements of the magnitudes and phases of voltages, currents and power, including the quality of that power (reactive and apparent power, and harmonics up to several kHz) and, of course, the energy consumed (and delivered back to the grid if applicable). These measurements require precision analog circuitry and sophisticated digital computation engines which will be discussed. Other challenges faced by these deep submicron ICs include high ESD/EMC and surge protection, line isolation up to several kV, and tamper protection and detection.

HomePlug Green PHY
Keith Findlater, *Gigle Networks, Edinburgh, United Kingdom*
The demand for improved energy efficiency worldwide is driving the development of the Smart Grid. Both utilities and consumers require intelligent energy management solutions to monitor costs, improve grid balancing, and reduce consumption. HomePlug Green PHY (GP) offers a robust networking technology for Home Area Network (HAN) Smart Grid communications. It is interoperable with the broadband capable HomePlug AV that is the basis of the IEEE 1901 standard, enabling network deployments that are for mixed broadband and smart energy use. The requirements of low cost, low power, reliability, and compatibility with existing standards are well addressed by this technology. This presentation will give an overview of the HomePlug GP approach to the Smart Grid HAN along with some of the key challenges associated with its implementation. Some examples from the development of a HomePlug GP single-chip SoC in nanometer CMOS technology will be presented.

Wireless Communications - Zigbee for Energy Management
Sang-Gug Lee, *KAIST, Daejeon, Korea*
ZigBee is a low-power wireless communication technology designed for monitoring and control of devices, which are the functions needed for smart energy management. ZigBee technology, based on the 802.15.4 standard, can provide robust and reliable data communication in noisy radio frequency (RF) environments, featuring energy detection, clear channel assessment, channel agility and multiple levels of security. ZigBee is also able to cover large areas with routers through mesh networking, which differentiate it from other technologies. ZigBee is a key technology for Smart Grid considering interoperability, automated controllability of appliances, ability to query/control devices, and lower installation and upgrade cost. ZigBee can offer scheduling/canceling of messages, mechanism for time-of-use pricing and price tiers, meter-to-meter communication ability for sub metering, remote monitoring ability of whole home conditions, etc. In this talk, the latest research and development of ZigBee in the application of energy management will be discused.

978-1-61284-303-2/11 $26.00 © 2011 IEEE

AUTHOR INDEX

Abdalla, Mohamed..................................354
Abdollahi-Alibeik, Shahram 170
Abidi, Asad 378
Abiri, Behrooz 154, 436
Acar, Mustafa.......................................58
Adell, Philippe 388
Afshari, Ehsan 286
Agarwal, Anil 74
Agarwal, Shilpa 138
Agranov, Gennadiy 418
Aguirre, Jorge 194
Ahmed, Nayera...................................114
Ahn, Jungchak 416
Akahori, Tomoyuki..............................400
Akiyama, Yoshiyuki.............................150
Aktakka, Ethem Erkan 120
Ali, Ershad...164
Ali, G...414
Ali, Tamer A.......................................466
Ali-Ahmad, Walid 370
Allstot, David J....................................428
Alon, Elad 166
Amadi, Christophe...............................240
Anderson, Martin.................................366
Andreani, Pietro..................................366
Aniruddhan, Sankaran 368
Ante, Frederik 324
Antonello, Riccardo 104
Apostolidou, Melina...............................58
Apte, Rahul..368
Arai, Toshiki.......................................406
Arekapudi, Srikanth 78, 80
Arimoto, Kazutami 266
Arnborg, Torkel..................................366
Arnold, Matthias.................................334
Arsovski, Igor.....................................254
Asada, Hiroki 160
Asahina, Katsushi...............................266
Ashouei, Maryam................................332
Assefa, Solomon.................................222
Averill, R..70
Ay, Suat U..116
Ayranci, Emre 184
Aziz, Pervez.......................................350
Babin, Gernot.....................................384
Bae, Joonsung......................................34
Bae, Seung-Jun..................................498
Baechler, Thomas...............................402
Baek, Kwang-Ho 202
Bahar, Mickey.....................................416
Bai, J...414
Bakkaloglu, Bertan 388
Balachandran, Ganesh K. 246
Balakrishnan, Bargav 72
Baldwin, Greg.....................................132

Bales, Tim 418
Bandyopadhyay, Abhishek 482
Bandyopadhyay, Saurav 386
Bang, Jong-Min...................................496
Bang, Sam-Young................................498
Bang, Sarvesh 238
Banno, Naoki 228
Barat, Francisco 332
Barkin, David B. 308
Barlow, Allen 186
Bartling, Steven 208
Batson, Kevin 254
Bauernfeind, Thomas............................376
Baumann, Adolf...................................334
Baumgarten, Judith 314
Baytekin, Burcin 170
Becker, Joachim 304, 306, 472
Beffa, Federico 370
Belitzer, Alexander 376
Belleville, Marc 336
Belot, Didier 162
Ben-Hamida, Naim 194
Benwadih, Mohamed 328
Bergervoet, Jos 54
Besson, Marinette 194
Bhatia, Rohit 84
Biganzoli, Fabio 104
Biro, Larry 84
Blaauw, David 46, 310, 342
Blanken, Peter G. 118
Bogner, Peter 384
Boku, Taisuke 266
Bolatkale, Muhammed 470
Bonizzoni, Edoardo 478
Boos, Zdravko 376
Borghetti, Fausto 312
Borremans, Jonathan 62, 374
Bory, Cécile 328
Bosiers, Jan 406
Bossu, Frederic...................................368
Bouwens, Frank 458
Bowhill, Bill 84
Braceras, Geordie 254
Braendli, Matthias 156
Braga, Daniele 326
Bray, Jeff 186
Breems, Lucien J. 470
Breeschoten, Arjan 332, 458
Brenner, Stephan 314
Brewster, Anthony 442
Bright, William 182
Brooks, David M. 268
Brown, Phil 186
Brunsilius, Janet..................................186
Buchmann, Peter 156

AUTHOR INDEX

Bulman, Neil 352
Bult, Klaas 184
Bunsen, Keigo 160
Burghartz, Joachim N. 324
Buss, Dennis 132
Busson, Pierre 162
Busta, Eric 78
Busze, Ben 458
Butschke, Joerg 324
Byun, Gyung-Su 488
Calvillo-Cortes, David A. 58
Caminada, Carlo 104
Cao, Jun 142
Cao, Ying 480
Caputa, Peter 366
Caresosa, Mario 442
Cassan, David 354
Cathelin, Andreia 162
Cazzaniga, Gabriele 104
Cha, Sooho 500
Chae, Youngcheol 106
Chaivipas, Win 160
Chaix, Fabrice 162
Chakrabarti, Chaitali 342
Chan, Chi-Hang 188
Chan, Yuen H. 70, 72
Chandra, Gaurav 144
Chandrakasan, Anantha P. 126, 132, 190, 208, 260
Chang, Andrew 170
Chang, Jaejoon 430
Chang, Leland 256
Chang, Mau-Chung Frank 296, 318, 488
Chang, Meng-Fan 200, 206
Chang, Richard 170
Chang, Sanghyun 294
Chang, Sang-Whan 500
Chao, Ming-Yang 136
Charbon, Edoardo 312
Chartier, Isabelle 328
Chen, Chun-Nan 462
Chen, Frederick T. 200
Chen, Gregory 310
Chen, Hsin-Hua 462
Chen, Jiashu 166, 432
Chen, Jin-Ru 136
Chen, Kuan-Yu 124
Chen, Kun-Tso 462
Chen, Liang-Gee 124
Chen, Liang-Yu 254
Chen, Mike Shuo-Wei 170
Chen, Min 136, 462
Chen, Ming-Shuan 146
Chen, Pang-Ning 168
Chen, Phoebe 170
Chen, Po-Hung 136, 216

Chen, Tung-Chien 124
Chen, Wei-Yin 124
Chen, Yen-Horng 364
Chen, Yunji 76
Chen, Yu-Sheng 200
Chen, Y-Y. 44
Cheng, Chih-Chi 124
Cheng, Kuo-Hsing 200
Cheng, Shanfeng 382
Cherepacha, Don 270
Chiang, Pei-Chia 200
Chien, George 364, 462
Chien, Shao-Yi 124
Chih, Yu-Der 206
Chio, U-Fat 188
Chiu, Pi-Feng 200
Chiueh, Tzi-Dar 134
Cho, Beak-Hyung 500
Cho, Gyehyun 236
Cho, Gyu-Hyeong 316, 390, 392
Cho, Hogeun 500
Cho, Ho-Youb 202
Cho, Hyunwoo 34
Cho, Hyunyoon 496
Cho, Kunhee 236
Cho, Lan-Chou 364
Cho, Seonghwan 176
Cho, Woo-Yeong 500
Cho, Yong-Ki 498
Choi, Changhan 500
Choi, Joo Sun 496, 498
Choi, Myung-Hoon 212
Choi, Woo-Young 360
Choi, Yoon-Hee 212
Choi, Youngdon 500
Choi, Young-Jung 502
Choi, Young-Kyoung 502
Choi, Young-Ryeol 498
Choksi, Ojas 368
Chou, Chia-Hua 136
Chou, Shu-Hung 462
Chowdary, Gajendranath 382
Chu, Ta-Shun 294
Chuang, Gene C. H. 134
Chuang, Tzu-Der 124
Chun, Jung-Hoon 494
Chun, Ki Chul 506
Chung, Byong-Tae 502
Chung, Hayun 230
Chung, Hoeju 500
Chung, Hyun 202
Chung, Won-Ryul 500
Cicalini, Alberto 368
Cimaz, Lionel 382
Ciraula, Michael 258

AUTHOR INDEX

Cirit, Halil..................................448
Clinton, Michael.............................208
Condemine, Cyril.............................112
Cong, Jason.................................488
Cong, Peng....................................9
Contaldo, Matteo.............................48
Coppard, Romain.............................328
Coquillat, Dominique.........................42
Coronato, Luca..............................104
Corsi, Marco................................182
Craninckx, Jan...........................62, 374
Crespo, Eugenia Cordero......................352
Croain, Damien..............................336
Cui, Delong.................................142
Curran, B....................................70
D'Souza, Arnold James........................382
Daami, Anis.................................328
Dakshinamurthy, Sampath......................74
Dance, Sherman M............................256
Dao, Son V. T...............................406
Darabi, Hooman...........................60, 378
Datla, Satyendra............................132
David, Johan................................332
Davies, G...................................414
De Boeck, Jo.................................15
De Cock, Wouter.............................480
De Graauw, Anton............................278
De Groot, Harmke.........................332, 458
De Mercey, Gregoire Le Grand.................100
De Vreede, Leo C. N..........................58
Debaillie, Bjorn............................374
Debarnot, Miguel............................112
Decanis, Ugo.............................280, 282
Deferm, Noël................................290
Defernez, Arnaud............................114
Dehaene, Wim................................322
Dehos, Cedric...............................162
Desai, Chintan..............................350
Desai, Utpal.................................74
Determan, Michael...........................482
Dierickx, Bart..............................114
Dietz, Carl..................................78
Ding, Hong..................................198
Ding, Li-Fu.................................124
Ditlow, Gary S..............................256
Doan, Chinh H...............................164
Doany, Fuad.................................222
Dogan, Hakan................................170
Dolmans, Guido..............................458
Dong, Jen...................................350
Doo, Su-Yeon................................498
Doris, Kostas...............................180
Doulcier-Verdier, Marion.....................274
Dreesen, Michael............................258
Drost, Robert...............................466

Du, Chenliang...............................294
Duarte, Filipa..............................332
Dulger, Fikret...............................52
Dupont, Benoit..............................114
Dupret, Antoine..............................42
Dussopt, Laurent............................162
Dutertre, Jean-Max..........................274
Edahiro, Toshiaki...........................198
Ehrenreich, Sebastian.......................256
Eikedal, S..................................414
Eisen, L.....................................70
Ek, Staffan.................................366
El-Chammas, Manar...........................182
Eliezer, Oren................................52
Elshazly, Amr.................92, 152, 438, 440
El-Tanani, Mohammed A.......................426
Emami, Sohrab...............................164
Emami-Neyestanak, Azita.....................446
Engels, Sylvain.............................336
Esumi, Atsushi..............................204
Etoh, Takeharu G............................406
Eun, Chang-Gyu..............................212
Eversmann, Bjoern Oliver.....................334
Ezekwe, Chinwuba D..........................246
Fallahi, Siavash............................142
Falt, Chris.................................194
Fan, Baoxia..................................76
Fan, Chi-Wei................................192
Fan, Lin....................................366
Fan, Xiaofeng...............................418
Fang, Q......................................44
Fares, Mounir...............................484
Fayed, Ayman A..............................396
Fehse, Karsten..............................314
Feldman, Arnold.............................352
Ferragut, Romain............................162
Fetzer, Eric.................................84
Fick, David.................................310
Filipovic, Dan..............................368
Fischer, Tim.................................78
Fishburn, Matthew W.........................312
Flatresse, Philippe.........................336
Fleischer, Bruce M..........................256
Flemeke, Philip.............................194
Flores, M...................................414
Foley, Denis................................270
Fontana, Enzo...............................328
Foo, Zhiyoong................................46
Forbes, Mark G..............................164
Fournier, Jacques...........................274
Fox, Thomas W...............................256
Fredriksson, Henrik.........................366
Freese, Ryan................................258
Frisbie, C. Daniel..........................326
Fu, Chang-Tsung..............................56

AUTHOR INDEX

Fujii, Shuso 504
Fujimori, Ichiro 442
Fujita, Masato 408
Fujita, Tetsuya 338
Fukaishi, Muneo 220
Fukuda, Koichi 198
Fukuda, Koji 148
Fukuda, Satoshi 150
Fulde, Michael 384
Furusawa, Toshihiro 420
Furuta, Yuka 198
Gabric, John 254
Gambhir, Sanjiv S. 308
Gambini, Simone 302
Gammie, Gordon 132
Gan, Haitao 170
Gao, Hairong 350
Gao, Xiang 76
Garcia, Monica 350
Garg, Adesh 142
Garrigues, Fabien 366
Geis, Henrik 366
Gelinck, Gerwin H. 322
Geltinger, Hans 376
Gendai, Yuji 420
Genoe, Jan 322
Gerrish, Paul 9
Gersbach, Marek 312
Ghaed, Hassan 310
Ghilioni, Andrea 280, 282
Giannini, Vito 62, 374
Gibbins, Robert 194
Gieschke, Pascal 108
Giffard, Benoît 42
Gilbert, Jeffery M. 164
Gill, Patrick R. 412
Giridhar, Bharan 46
Godbole, Kunal 368
Golden, Michael 78, 80
Gordon, Michael 164
Gossmann, Timo 376
Goto, Hiroshige 416
Green, William 222
Greshishchev, Yuriy M. 194
Gribben, Tony 240
Gronowski, Paul 84
Gruber, Stefan 454
Grutkowski, Tom 84
Grzyb, Janus 224
Gu, J. 132
Gu, Pei-Yi 200
Guan, Xiang 164
Gutta, Srinivasa Rao 270
Gwoziecki, Romain 328
Ha, Mingjing 326

Habibovic, Harun 384
Hada, Hiromitsu 228
Haesler, Jacques 48
Hafez, Amr A. 466
Hagberg, Hans 366
Hajimiri, Ali 150, 288
Hamada, Mototsugu 174, 338
Hamami, Shay 416
Han, Bong-Seok 202
Han, Cheolmin 496
Han, Gong-Heom 498
Han, Hyungjun 416
Han, Jae-Duk 494
Han, Jeonghu 430
Han, Jin-Man 212
Hanawa, Toshihiro 266
Hanumolu, Pavan Kumar 92, 152, 238, 438, 440
Hanvey, A. 414
Haque, Razi-Ul 310
Hara, Hiroyuki 338
Hara, Takahiko 198
Harpe, Pieter 300, 458
Hartig, Thomas 384
Hashemi, Hossein 294
Hashimoto, Sakae 408
Hassibi, Mahnaz 100
Hatamkhani, Hamid 442
Hayakawa, Yasushi 266
Hayashida, Tetsuya 406
He, Xin 372
Hearne, Peter 352
Hedestig, Joel 366
Heijden, Edwin V. D. 278
Heinemann, Bernd 224
Heitz, Roxana T. 308
Helder, Edward 82
Helt, Chris 258
Hemink, Gertjan 198
Henderson, Robert K. 312, 410
Hennig, Eckhard 304
Henrion, Carson 258
Heremans, Paul 322
Herold, Rigo 314
Herzog, Marcus 334
Hezar, Rahmi 484
Hidaka, Yasuo 348
Higashitani, Masaaki 198
Hijikata, Keijiro 226
Hilker, Scott 78
Hilzensauer, Sascha Alexander 108
Hino, Yasufumi 150
Hirayama, Satoshi 408
Hizi, Uzi 416
Ho, Lam 132
Ho, Ronald 466

AUTHOR INDEX

Hode, Uwe..............................384
Hofer, Günter..........................454
Holdenried, Chris......................354
Holmes, Kyle M.........................256
Holweg, Gerald.........................454
Hom, Gary..............................350
Honda, Kentaro.....................216, 218
Hong, Gi-Moon..........................494
Hong, J-H...............................44
Hong, Sung-Wan.........................390
Hong, Sunjoo............................36
Hong, Zhiliang.........................394
Honnavara-Prasad, Sushma...............132
Horie, Takeshi.........................348
Horiuchi, Aaron.........................78
Hosomura, Yoshikazu....................198
Houle, Robert..........................254
Hsiao, Ching-Min.......................318
Hsiao, Keng-Jan.........................90
Hsiao, Ying-Chuan......................134
Hsieh, C-H..............................44
Hsieh, H-H.............................488
Hsu, Jen-Yuan..........................134
Hsu, Jui-Lin...........................462
Hsueh, Ching-Wen.......................136
Hu, Weiwu...............................76
Huang, Chia-En.........................206
Huang, Hu..............................366
Huang, Kuo-Ken.........................284
Huang, Li..............................458
Huang, Patty...........................352
Huang, Shih-Jou........................168
Huang, Tsung-Ching.....................218
Huang, Wenchang........................382
Huang, Wen-Tso.........................136
Huang, Xiongchuan......................458
Huang, Yi-Chieh........................358
Huff, Bill.............................278
Huh, Jin...............................392
Huijsing, Johan H..................110, 244
Huisken, Jos.......................332, 458
Hulzink, Jos...........................332
Hung, Chih-Ming.........................52
Hung, Chinglin.........................204
Hung, Hao-Wei..........................146
Hung, Tsai-Pi..........................368
Hurd, Kevin A...........................78
Huynh, Tony............................350
Hwang, Hyong-Ryol......................496
Hwang, Jong Tae........................236
Hwang, Kyu-Dong........................128
Hwang, Seokwon.........................500
Hwang, Sewook..........................346
Hwang, Sooman..........................496
Hyvonen, Sami...........................96

Ichikawa, Tatsuya......................420
Ickes, Nathan..........................132
Ikeuchi, Katsuyuki.....................216
Ikeya, Atsuyuki........................266
Imamura, Koji..........................458
Ingels, Mark.......................62, 374
Inoue, Shunsuke........................408
Inti, Rajesh...............92, 152, 438, 440
Ishibe, Manabu.........................226
Ishida, Koichi.....................216, 218
Ishida, Shinji.........................228
Ishihara, Hiroaki......................226
Ishikuro, Hiroki...................230, 492
Isobe, Keigo...........................400
Itakura, Tetsuro...................98, 226
Ito, Kimihiko..........................228
Ito, Mitsuyoshi........................204
Ito, Shogo.............................160
Itoh, Shinya...........................400
Ivanov, Mikhail V......................248
Ivory, David...........................370
Iwai, Makoto...........................198
Iwaki, Hiroyuki........................420
Iwata, Shigeyasu.......................226
Jacob, Stéphanie.......................328
Jahnes, Christopher....................222
Jain, Pulkit...........................506
Jang, Eun-Seong........................202
Jang, Minsoo...........................496
Jang, Sungchun.........................450
Jang, Youngheup........................416
Janssen, Erwin.........................180
Jee, Dong-Woo...........................94
Jeon, Dongsuk..........................342
Jeon, Jin-Yong.........................316
Jeong, Byung Hoon......................500
Jeong, Deog-Kyoon..................450, 494
Jia, Vivian.............................82
Jiao, Dong.............................272
Jing, Tai..............................350
Johansson, Robert......................414
Johnson, Dave...........................78
Johnson, Scott.....................414, 418
Jou, C.................................488
Joy, Andrew K..........................352
Juang, Den-Kai.........................136
Jun, Sunghyun..........................500
Jun, Woochul...........................500
Jun, Young-Hyun...............496, 498, 500
Jung, Inhwa............................346
Jung, Minho............................236
Jung, Seungchul........................392
Jung, Taesub...........................416
Jung, Tae-Sung.........................212
Kaczynski, Brian.......................170

AUTHOR INDEX

Kagan, Yevgeniy 366
Kallam, Praveen 382
Kambe, Akihiro 148
Kanagawa, Naoaki 198
Kanda, Kazushige 504
Kaneko, Satoshi 266
Kang, Dong-Seok 498
Kang, Sang-Kyu 496
Kang, Shinwon 166
Kanou, Taikan 408
Kao, Kaipon 136
Kao, Ming-Jer 200
Karegar, Anita 258
Karlsson, Niklas 366
Kash, Jeffrey 222
Kato, Yosuke 198
Kauffman, John G. 472
Kawaguchi, Tetsuji 418
Kawahito, Shoji 400
Kawakami, Koichi 198
Kawamura, Katsuyuki 418
Kawasaki, Kenichi 150
Kelleci, Burak 484
Kern, Alexandra M. 96
Kerr, Ben 352
Khan, Qadeer 238
Khandelwal, Pulkit 352
Khanoyan, Karo 442
Kibune, Masaya 154, 356, 436
Kikuchi, Shin 408
Killat, Dirk 394
Kim, Bo-Geun 212
Kim, Bumsuk 416
Kim, Byeong-Cheol 498
Kim, Chris H. 272, 326, 506
Kim, Chulwoo 346, 502
Kim, Dae-Hyun 498
Kim, Daeyeon 310
Kim, Donghwan 236
Kim, Eun-Hee 456
Kim, Gyeonghoon 130
Kim, Gyouho 310
Kim, Hoyoung 496
Kim, Hyeong-Jun 212
Kim, Hye-Ran 498
Kim, Hyo-Eun 128
Kim, Hyun-Sik 316
Kim, In-Mo 212
Kim, Jaewhan 500
Kim, Jaewook 176
Kim, Jin-Guk 496
Kim, Jinyoung 500
Kim, Jongsun 488
Kim, Jong-Woo 202
Kim, Jong-Young 212

Kim, Jung-Sik 496
Kim, Keon-Soo 212
Kim, Ki-Han 502
Kim, Kiseung 500
Kim, Kwan-Weon 502
Kim, Lee-Sup 128
Kim, Min-Seok 212
Kim, Min-Su 202
Kim, Sangtae 500
Kim, Sejun 482
Kim, Si-Hong 498
Kim, So-Young 496
Kim, Suhwan 494
Kim, Sunghoon 500
Kim, Sunkwon 494
Kim, Taehyun 212
Kim, Tae-Yun 202
Kim, Wang-Soo 360
Kim, Wonyoung 268
Kim, Woonyun 430
Kim, Woo-Seop 498
Kim, Yanghyo 488
Kim, Yejoong 310
Kim, Yitae 416
Kim, Young-Jun 128
Kim, Young-Sik 498
Kimura, Hiroshi 350
King, Ya-Chin 206
Kinzel, Henrik 366
Kiper, Halil 484
Kitagawa, Makoto 210
Klaassens, Wilco 406
Klamminger, Michael 454
Klauk, Hagen 324
Klemmer, Nikolaus 366
Klumperink, Eric A. M. 64
Knap, Wojciech 42
Knoll, Ernest 264
Ko, Jinho 456
Ko, Shang-Bin 254
Ko, Uming 132
Kobayashi, Hiroyuki 174
Kobayashi, Yuka 292
Kocaman, Namik 142
Koh, Yo-Hwan 202
Komai, Hiromitsu 198
Komijani, Abbas 170
Komninakis, Christos 368
Komori, Kenji 150
Kondo, Hiroyuki 266
Kong, Lingkai 166
Kong, Tae-Hwang 390
Konijnenburg, Mario 332, 458
Kosic, Steve 186
Kossel, Marcel 152, 156

AUTHOR INDEX

Kothari, Ruchi	350	Lee, Jae-Ho	202
Kousai, Shouhei	174	Lee, Jaewook	500
Koyanagi, Yoichi	348	Lee, Jri	146, 168
Krishnakumar, Ashwin	82	Lee, Jung-Bae	496
Kroker, Lars	314	Lee, Jun-Hee	212
Kropf, Claus	384	Lee, Kwangjin	500
Ku, Tzu-Kun	200	Lee, Kwyro	456
Kuhl, Matthias	108	Lee, Kyungho	416
Kuhlmann, Franz	352	Lee, Meelan	170
Kuhn, Rüdiger	334	Lee, Sang-Don	202
Kundur, Harish	372	Lee, Sangjoo	416
Kuo, Chia-Chen	200	Lee, Sang-Min	170
Kuo, Tzu-Chieh	138	Lee, Seong-Ho	442
Kurchuk, Mariya	232	Lee, Seulki	36
Kuroda, Tadahiro	230, 490, 492	Lee, Seungjin	130
Kurts, Tsvika	264	Lee, Shuenn-Yuh	44
Kusuda, Yoshinori	242	Lee, Sungsoo	212
Kuttner, Franz	376, 384	Lee, Sung-Woo	316, 392
Kwak, Pansuk	212	Lee, Tsai-Pao	136
Kwak, Sang-Hyup	498	Lee, Y.	414
Kwak, Young-Ho	346	Lee, Yen-Po	136
Kweon, Tae-Min	212	Lee, Yong-Jun	500
Kwon, Dukmin	500	Lee, Yoonmyung	46
Kwon, Oh-Hyun	22	Lee, Yun-Tae	416
Kwon, Ohsuk	212	Leenaerts, Domine	58
Kyung, Kyehyun	496	Lei, Bo	198
Lacaita, Andrea L.	88	Leroux, Paul	480
Ladd, John	414, 418	Lesser, Michael	406
Lai, Han-Chao	206	Letzkus, Florian	324
Lai, H-Y.	44	Levantino, Salvatore	88
Lai, Jiun-You	134	Lhermet, Helene	112
Lai, Thriven	138	Li, Day-Uey	312
Lakdawala, Hasnain	56, 426	Li, Kai	204
Lamphier, Steve	254	Li, Kuan-I	462
Lang, Christian	304	Li, Lijun	350
Lange, Steffen	304	Li, Ning	160
Lanteri, Jerome	162	Li, Shenggao	82
Le Coq, Stephane	382	Li, X.	414
Le Coz, Julien	336	Li, Ye	394
Lecomte, Steve	48	Liang, Che-Fu	90
Ledwa, Ralph	334	Liang, M-C.	44
Lee, Benjamin	222	Liang, Paul C. P.	100
Lee, Chang-Ho	430	Liao, Chieh-Wei	136
Lee, Dana	198	Liao, Hung-Jen	206
Lee, Donghyuk	496	Liao, Shih-Hsien	136
Lee, Duckhyung	416	Liao, Shyuan	136
Lee, Hae-Chang	352	Liao, Yu-Te	38
Lee, Heng-Yuan	200	Lien, Chen-Hsin	200
Lee, Hocheol	496	Lim, Dong-Hyuk	494
Lee, Ho-Chul	212	Lim, Jeong-Don	498
Lee, Hyongmin	494	Lim, Ki-Won	500
Lee, Hyung-Jin	96	Lim, Sang-O	202
Lee, Hyungwoo	34	Lim, Young-Ho	212
Lee, Hyun-Woo	502	Limotyrakis, Sotirios	170
Lee, Jae-Eun	212	Lin, Che-He	200

AUTHOR INDEX

Lin, Chen-Lun 146
Lin, Chorng-Jung 206
Lin, Eric Chien-Chih 170
Lin, Jian-Bang 136
Lin, J-W. 44
Lin, Ku-Feng 200
Lin, Ping-Chih 124
Lin, Wen-Pin 200
Lin, Yu-Fan 206
Ling, Kathy 350
Liu, Cathy 350
Liu, Chia-Chi 206
Liu, Chin-Tai 136
Liu, Shen-Iuan 358, 444
Liu, Tao 388
Liu, Wing 350
Lo, Chi-Lun 462
Lo, Shun-Chang 134
Lo, Steve 164
Loinaz, Marc J. 448
Lopelli, Emanuele 424
Lotto, Christian 402
Lotze, Niklas 340
Lowrie, Scott 350
Lu, Chih-Wen 318
Lu, Ivan Siu-Chuang 364
Luff, Gwilym F. 250
Luong, Howard C. 464
Luu, Binh 186
Lytollis, Shaun 352
Lyu, Jeong-Ho 418
Machado, Joaquim 352
Maddiona, Lidia 328
Maeda, Hiroki 218
Maghari, Nima 474
Mair, Hugh 132, 260, 352
Mak, Alex 198
Mak, P.-K. 70
Mak, Pui-In 172
Maki, Yasunori 226
Makino, Eiichi 198
Makinwa, Kofi A. A. 106, 110, 244, 300, 470
Maldonado, David 368
Malgioglio, F. 70
Malipatil, Amaresh 350
Malkin, Moshe 144
Maloberti, Franco 188, 478
Malone, D. 70
Mandal, Gunjan 62
Manoli, Yiannos 108, 340
Marcu, Cristian 166
Mårtensson, Andreas 366
Martineau, Baudouin 162
Martins, Rui 172, 188
Martinussen, T. 414

Maruhashi, Kenichi 220
Maruyama, Hirotaka 406
Maruyama, Yuki 312
Marzin, Giovanni 88
Mas, Patrick 112
Masuda, Noboru 148
Matsunaga, Yasuhiko 198
Matsuo, Shinichiro 418
Matsushita, Kota 160
Matsuzawa, Akira 160
Mattisson, Sven 366
Mattsson, Thomas 366
Mauritzson, Rick 418
Mayer, G. 70
Mayo, M. 70
Mazzanti, Andrea 280, 282
McElwee, Patrick T. 164
McIntyre, Hugh 78
McIntyre, William 238
McLaren, Angus 354
McPherson, T. 70
Meaney, P. 70
Mehalel, Moty 264
Mehta, Srenik 170
Mendis, Sunetra 170
Menkhoff, Andreas 376
Menolfi, Christian 156
Mercer, Timothy 294
Michel, Fridolin 476
Mihara, Junichi 256
Miles, Tommy 258
Mills, Thomas 366
Min, Byungjun 500
Minami, Ryo 160
Minamoto, Takatoshi 198
Mincica, Martina 40
Mirzaei, Ahmad 60
Mishina, Koji 420
Mitomo, Toshiya 292
Miura, Noriyuki 230, 490, 492
Miura, Shin'Ichi 266
Miwa, Toru 198
Miyamura, Makoto 228
Miyata, Koji 210
Miyoshi, Takashi 348
Molnar, Alyosha 412
Momeni, Omeed 286
Momma, Genzo 408
Momtaz, Afshin 142
Monaco, Enrico 280, 282
Montoye, Robert K. 256
Moon, Changrok 416
Moon, Gil-Shin 498
Moon, Joungwook 496
Moon, Un-Ku 474

978-1-61284-303-2/11 $26.00 © 2011 IEEE   541

AUTHOR INDEX

Morche, Dominique 232
Moreira, Jose 376
Morf, Thomas 156
Moritsuka, Fumi 226
Morrison, Casey 352
Morrison, Scott 352
Motomura, Masato 228
Mounet, Christopher 162
Mu, Fenghao 366
Mulder, Jan 184
Muller, Rikky 302
Murakami, Hirotaka 420
Murakami, Rui 160
Muramatsu, Yoshinori 420
Muramoto, Mai 198
Murden, Frank 186
Murphy, Brian 198
Murphy, David 60
Musa, Ahmed 160
Musha, Junji 198
Muto, Takashi 148
Myny, Kris 322
Na, Byongwook 496
Naffziger, Samuel 78
Nagao, Osamu 198
Naini, Ajay 270
Najafi, Khalil 120
Nakajima, Hiroyoshi 218
Nakamichi, Masaru 198
Nakamura, Dai 198
Nakamura, Junichi 418
Nakamura, Yutaka 256
Nakano, Shinya 492
Nakaya, Shogo 228
Nalam, Nageswara Rao 382
Namise, Tomohiro 210
Nan, Lan 492
Nani, Claudio 180
Narathong, Chiewcharn 368
Narayan, Ram 350
Narayanan, Rajagopal 74
Naruke, Kiyomi 198
Nathawad, Lalitkumar 170
Nauta, Bram 64
Nazari, Meisam Honarvar 446
Nejati, Babak 368
Nemoto, Ryo 148
Nerlich, Ronald 334
Nguyen, Dung H. 406
Nguyen, Khiem 482
Nicholson, Roan 82
Nicklasson, Lars 366
Niknejad, Ali M. 166, 432
Nilsson, Magnus 366
Nishimura, Shinji 148

Noguchi, Mitsuhiro 504
Nonomura, Itaru 266
Noorsal, Emilia 306
O'Shea, Deirdre 368
O'Sullivan, Thomas D. 308
Oboe, Roberto 104
Oesterle, Stephen 9
Ogata, Kentaro 210
Ogawa, Takeshi 198
Oh, Chi Sung 496
Oh, Jinwook 130
Oh, Tae-Young 498
Oh, Yangjin 170
Ohashi, Sho 150
Ohno, Katsuya 226
Öjefors, Erik 224
Ojha, Mukund 138
Okada, Kenichi 160
Okamoto, Koichiro 228
Okinaka, Tomoo 406
Okuma, Yasuyuki 216
Onishi, Shohji 256
Ono, Goichi 148
Onodera, Keith 170
Oredsson, Filip 366
Ortmanns, Maurits 304, 306, 472
Osawa, Shinji 418
Oshita, Takeshi 266
Osone, Hideki 348
Osorio, Juan F. 278
Ota, Keisuke 408
Otaka, Shoji 292
Otani, Sugako 266
Otis, Brian 38, 460
Otsuka, Wataru 210
Ouellette, David 406
Ozdemir, Ufuk 366
Pages, Eduard 418
Påhlsson, Tony 366
Pålsson, Markus 366
Palusa, Chaitanya 350
Pan, Hui 442
Pan, Jyh-Shin 136
Pandey, Jagdish 460
Parashurama, Natesh 308
Parikh, Samir 348
Park, Changbyung 392
Park, Cheol-Goo 498
Park, Fitzgerald 366
Park, Hanna 496
Park, Hong-June 94
Park, Hyunsik 170
Park, Jin-Su 202
Park, Jongeun 416
Park, Joon-Min 500

AUTHOR INDEX

Park, Joo-Yong.................................212
Park, Jun-Seok................................128
Park, Junyoung................................130
Park, Ki-Tae..................................212
Park, Kiwon...................................496
Park, Kwang-Il................................498
Park, Kyu-Sang................................494
Park, Min-Sang................................498
Park, Mu-Hui..................................500
Park, Nak-Kyu.................................502
Park, Paul....................................170
Park, Sun-Young...............................498
Park, Younghwan...............................416
Parker, James.................................164
Parviz, Babak..................................38
Patel, Pradip..................................72
Pates, Dan....................................418
Paul, Oliver..................................108
Pavlik, Frank.................................254
Pavlovic, Nenad................................54
Payne, Robert.................................182
Pelella, Antonio R.............................72
Pepe, Domenico.................................40
Perez, Aldo Pena..............................478
Perks, D......................................414
Peterson, Rebecca L...........................120
Pettersson, Tony..............................366
Pfann, Peter..................................376
Pfeiffer, Ullrich R...........................224
Pham, Jennifer................................354
Pi, Deyi......................................142
Pilard, Romain................................162
Pilo, Harold..................................254
Piovaccari, Alessandro.........................66
Plass, D.......................................70
Pless, Holger.................................304
Pollex, Daniel................................194
Portmann, Clemenz.............................352
Prabhu, Raja J................................382
Prandi, Luciano...............................104
Pribyl, Wolfgang..............................454
Prokop, Tom...................................350
Qazi, Masood..................................208
Qi, Zichu......................................76
Qiu, Yifeng...................................118
Quan, Shaolei.................................350
Quan, Xiaohong................................368
Rabaey, Jan M..............................31, 302
Rabii, Shahriar...............................352
Radecki, Andrzej..............................230
Radens, Carl..................................254
Rafi, Aslam A..................................66
Raghavan, Bharath.............................142
Rajapaksha, Dushmantha........................352
Rajasekaran, V................................414

Rajashekara, Anil.............................350
Rajavi, Yashar................................170
Ramadass, Yogesh K............................386
Ramezani, Mehrdad.............................354
Ramon, Stephane...............................366
Rampu, Simonetta..............................458
Ranade, Mandar.................................74
Rangarajan, Raj...............................368
Rao, Arun.....................................238
Rao, Sachin...................................238
Ravinuthula, Vishnu...........................352
Rayankula, Aditya.............................418
Raynaud, Christine............................336
Reddy, Subodh.................................348
Reinisch, Hannes..............................454
Reinman, Glenn................................488
Renn, Michael J...............................326
Reynaert, Patrick.............................290
Rezayee, Afshin...............................354
Rho, Yoohwan..................................500
Richard, Olivier..............................162
Richardson, Justin........................312, 410
Richter, Bernd................................314
Richter, Harald...............................324
Riedlinger, Reid J.............................84
Riesslegger, Harald...........................384
Rim, Chun-Taek................................392
Risbo, Lars...................................484
Rithe, Rahul..................................132
Roderick, Jonathan............................294
Rodko, Daniel..................................72
Roh, Taehwan...................................36
Rong, Bing....................................132
Rong, Sujiang.................................464
Rossbach, Daniel..............................108
Rostaing, Jean-Pierre..........................42
Rothan, Frederic..............................112
Rude, D..70
Rueegg, Michael...............................156
Ruffieux, David................................48
Russell IImura................................418
Ruther, Patrick...............................108
Rutten, Robert................................470
Rylyakov, Alexander...........................222
Ryu, Jang-Woo.................................496
Ryu, Seung-Tak................................316
Ryu, Yoshikatsu...............................216
Rzehak, Volker................................334
Saccamango, M. J...............................70
Sadr, Saman...................................354
Sahota, Kamal.................................368
Sai, Akihide...................................98
Saito, Mitsuko................................490
Saito, Tatsuya................................148
Sakai, Manabu.................................198

978-1-61284-303-2/11 $26.00 © 2011 IEEE   543

AUTHOR INDEX

Sakamoto, Toshitsugu 228
Sakimura, Noboru 228
Sakowicz, Maciej 42
Sakurai, Hiroki 292
Sakurai, Kiyofumi 198
Sakurai, Takayasu 216, 218
Samavati, Hirad 170
Samiy, H. 414
Samori, Carlo 88
Sandgren, Magnus 366
Sandrup, Per 366
Sankaranarayanan, Janakiram 368
Sano, Tomohiro 62, 374
Santana, Juan 458
Santner, Andreas 384
Sasidhar, Naga 438
Sato, Junpei 198
Sato, Mitsuhisa 266
Sato, Takahiro 160
Sato, Toshiaki 418
Sawant, Shankar 74
Schimper, Markus 376
Schneller, Bryan 258
Schow, Clint 222
Schreiber, Russell 258
Schuster, Franz 42
Schvan, Peter 194
Schwarz, E. 70
Scuderi, Antonino 328
See, Andrew 368
Seedher, Ankit 382
Seferagic, Adnan 254
Seitz, Peter 402
Sekitani, Tsuyoshi 218
Sel, Jongsun 212
Sengupta, Kaushik 288
Seo, Min-Woong 400
Seok, Mingoo 310, 342
Seol, Ho-Seok 498
Seong, Chang-Kyung 360
Serbutoviez, Christophe 328
Serton, Richard E. 72
Sestok, Charles 182
Shahramian, Shayan 356
Shamanna, Gururaj 74
Shana'A, Osama 462
Sharma, Lokesh 74
Shearer, Robert 256
Sheikholeslami, Ali 154, 356, 436
Shen, Shin-Jang 206
Sheu, Shyh-Shyuan 200
Shi, Jianlei 460
Shi, Kai 170
Shibata, Noboru 504
Shibuya, Toshiyuki 348

Shidei, Tsunaaki 230
Shiga, Hitoshi 198
Shih, Chih-Heng 136
Shih, Yu-Nan 146
Shikata, Go 198
Shim, Dong-Kyo 212
Shin, Chang-Ho 498
Shin, Inchul 500
Shin, Ji-Yeon 212
Shin, Junho 500
Shin, Seung-Hwan 212
Shin, Woo-Yeol 494
Shin, Yong-Jae 498
Shindo, Yoshihiko 198
Shinke, Satoru 150
Shiroshita, Hiroshi 420
Shivnaraine, Ravi 154
Shon, Ju-Hwan 502
Shor, Joseph 264
Shoval, Ayal 354
Si, William W. 170
Sibuet, Henri 112
Sichert, Christian 334
Sigal, L. 70
Siligaris, Alexandre 162
Sim, Jae-Yoon 94
Sin, Sai-Weng 188
Sin, Tze Yee 370
Sinangil, Mahmut E. 132, 260
Singh, Ravpreet 382
Singh, Ullas 142
Sinha, Ashutosh 350
Siragusa, Eric 186
Skaug, S. 414
Smith, David 182
Smith, H. 70
Snoeij, Martijn F. 248
So, Eric 354
Soer, Michiel C. M. 64
Sohn, Young-Soo 498
Somayajula, Shyam 382
Someya, Takao 218
Song, Heesoo 450
Song, Ickhyun 500
Song, Kiseok 34
Song, Ki-Whan 500
Song, Minyoung 346
Song, Youngson 212
Sonsky, Jan 58
Sooksood, Kriangkrai 306
Soumyanath, Krishnamurthy 56
Spiridon, Silvian 424
Sridhara, Ravi 368
Staszewski, Robert Bogdan 52
Stenman, Anna-Karin 366

AUTHOR INDEX

Stephansen, C. 414
Steyaert, Michiel 476, 480
Stoppa, David 312
Storino, Salvatore N. 256
Storm, A. 414
Strach, T. 70
Strandberg, Richard 100
Strandberg, Roland 366
Strange, Jon 370
Stuyt, Jan 332
Su, David 170
Su, Keng-Li 200
Su, Sheng-Yuan 462
Su, Wenjun 368
Su, Y-C. 44
Sugibayashi, Tadahiko 228
Sugiki, Tadashi 418
Suh, Sungho 400
Suh, Yunjae 94
Sukegawa, Hiroshi 504
Sun, Bo 368
Sun, Chih-Hao Eric 364
Sundar, Siddharth 382
Sundström, Lars 366
Sunwoo, Jung 500
Surace, Giuseppe 352
Suzuki, Eiichi 148
Suzuki, Yuya 198
Svelto, Francesco 280, 282
Swank, Damian 238
Sylvester, Dennis 46, 310, 342
Sze, Vivienne 126
Szilagyi, Stefan 194
Tabesh, Maryam 166
Tada, Munehiro 228
Takagiwa, Teruo 198
Takahashi, Hidekazu 408
Takahashi, Shingo 220
Takamiya, Kenichi 420
Takamiya, Makoto 216, 218
Takasawa, Taishi 400
Takashima, Daisaburo 504
Takayama, Naoki 160
Takayanagi, Isao 418
Take, Yasuhiro 490
Takeda, Takahiro 150
Takehara, Kohsei 406
Takemoto, Takashi 148
Takeuchi, Ken 204
Takeya, Tsutomu 492
Takinami, Koji 100
Tal, Noam 182
Tam, Sai-Wang 488
Tamura, Hirotaka 154, 356, 436
Tan, Chun-Geik 462

Tan, Yulin 426
Tanaka, Masatoshi 406
Tanaka, Rieko 198
Tanakamaru, Shuhei 204
Tang, Adrian 296
Tang, Chih-Chun 364
Tang, Keith 142
Tani, Jon R. 164
Tanzil, Alexander 370
Tao, Chengwu 396
Tasca, Davide 88
Taylor, Stewart 56, 426
Tchagaspanian, Michaël 42
Teh, Chen Kong 338
Tenbroek, Bernard 370
Teplechuk, Mykhaylo 240
Terao, Yasuhiro 210
Terrovitis, Manolis 170
Tetsuya, IIda 400
Thanigai, Priya 334
Tiedkte, Hans-Jürgen 304
Tillack, Bernd 224
Tillman, Fredrik 366
Ting, Clifford 356
Ting, Pang-An 134
Tired, Tobias 366
Toifl, Thomas 152, 156
Tokiwa, Naoya 198
Tong, Haitao 142, 442
Toyama, Takayuki 420
Toyoda, Hidehiro 148
Tran, Duke 442
Tsai, Ming-Jinn 200
Tsai, Wen-Chieh 462
Tseng, Wei-Hsin 192
Tsividis, Yannis 232
Tsuchiya, Hiroyuki 420
Tsung, Pei-Kuei 124
Tsushima, Tomohito 210
Tsutsui, Keiichi 210
Tumati, Sanjay 382
Tung, Chien-Hsun 136
Tuttle, Tyson 66
Uemura, Minoru 266
Umeda, Toshiyuki 226
Umezawa, Yasushi 348
Uno, Masahiro 150
Unterassinger, Hartwig 454
Uppathil, Satish 366
Urard, Pascal 336
Vakili-Amini, Babak 170
Vakilian, Kambiz 442
Valentian, Alexandre 336
Valliappan, Magesh 442
Van Campenhout, Joris 222

AUTHOR INDEX

Van Der Goes, Frank M. L.........184
Van Der Heijden, Mark P. 58
Van Der Tang, Johan.........424
Van Der Weide, Gerard.........180
Van Ginderdeuren, Johan.........332
Van Hoof, Chris.........118, 300
Van Ierssel, Marcus.........354
Van Kuijk, Harry.........406
Van Liempd, Chris.........118
Van Veenendaal, Erik.........322
Van Vliet, Frank E.........64
Van Vroonhoven, Caspar P. L.........106
Van Zalinge, Klaas.........368
Vancorenland, Peter.........66
Vanderhaegen, Johan P.........246
Vasani, Anand.........142
Vaucher, Cicero S.........278
Vecchi, Davide.........184
Veerappan, Chockalingam.........312
Veld, Bert Op Het.........118
Vemulapalli, Sudheer.........52
Verbruggen, Bob.........62
Vermeire, Bert.........388
Videlier, Hadley.........42
Vidojkovic, Maja.........458
Vincent, Pierre.........162
Vinh, James.........78, 80
Vlasov, Yurii.........222
Vo, Cuong L.........406
Vogel, Uwe.........314
Waheed, Khurram.........52
Walker, Richard.........312, 410
Walker, William W.........348
Wallberg, John.........52
Walling, Jeffrey S.........428
Walukas, Joel.........366
Wan, Jiansong.........184
Wang, Albert.........412
Wang, Alice.........132
Wang, Huaide.........168
Wang, Qi.........500
Wang, Ru.........76
Wang, Se-Won.........390
Wang, Shing-Chi.........194
Wang, Sum-Min.........200
Ward, Christopher M.........184
Wasmuth, Robert.........270
Watanabe, Hiroshi.........400
Watanabe, Keiki.........148
Watanabe, Mitsuyuki.........198
Watanabe, Osamu.........292
Watanabe, Takashi.........400
Watanabe, Yoshihisa.........198
Webb, C.........70
Weber, David.........170

Wei, Gu-Yeon.........268
Wei, Hegong.........188
Weiss, Don.........258
Weitzel, S.........70
Weltin-Wu, Colin.........232
Wendel, Dieter.........256
Wentzloff, David D.........284
Westesson, Eric.........366
Westra, Jan R.........184
White, Jonathan.........78
Whittlesea, D.........414
Wieckowski, Michael.........310
Wiessflecker, Martin.........454
Wilcox, Kathryn.........78
Willassen, T.........414
Willson, Alan.........138
Wise, Kensall.........310
Wiser, Robert F.........164
Witte, Pascal.........472
Wong, Tony.........100
Woo, Eum Chan.........428
Woo, Young-Jin.........390
Wood, M.........70
Wooley, Bruce A.........308
Wu, Che-Wei.........200, 206
Wu, Chia-Hsin.........136, 462
Wu, Hsin-Tun.........200
Wu, Jianfeng.........106
Wu, Jieh-Tsorng.........192
Wu, P-Y.........488
Wu, Rong.........110, 244
Wu, Shang-Chi.........206
Wuu, John.........258
Xing, Xinyu.........246
Xu, Hongcheng.........304
Xu, Hongtao.........426
Xu, Jiawei.........300
Xu, Weiwei.........394
Yagyu, Masayoshi.........148
Yamaji, Takafumi.........98
Yamamoto, Silas D.........162
Yamashita, Hiroki.........148
Yamashita, Yuichiro.........408
Yamauchi, Hiroyuki.........206
Yan, Long.........34, 36
Yanagidaira, Kosuke.........198
Yang, Chang-Won.........202
Yang, Chih-Kong Ken.........466
Yang, C-M.........44
Yang, Hsin-Jung.........124
Yang, Hyang-Ja.........498
Yang, Jerry Jian-Ming.........170
Yang, Jun-Hyeok.........316
Yang, Ki Seok.........430
Yang, Min.........222

AUTHOR INDEX

Yang, Seunguk 236
Yang, Shun-An 136
Yang, Wen-Wei 136
Yang, Xu 76
Yang, Yih-Shan 200
Yang, Yun-Seok 498
Yao, Huanfen 38
Yazicioglu, Refet Firat 300
Ye, Seokmin 450
Yeh, Eric 136
Yeh, Jen-Hao 136
Yeh, Yu-Ching 168
Yeom, Hyunsoo 388
Yetis, Ege 186
Yin, Ping-Yeh 318
Yin, Wenjing 92, 152, 438, 440
Yin, Zhiping 418
Ying, Yu-Ming 444
Yip, Marcus 190
Yoo, Hoi-Jun 34, 36, 130
Yoo, Jei-Hwan 500
Yoo, Sang-Min 428
Yoon, Hyun-Jun 212
Yoon, Jae-Sung 128
Yoon, Sangyong 212
Yoshida, Nobuhide 220
Yoshida, Yoichi 230, 490
Yoshihara, Hiroshi 210
Yoshihara, Yoshiaki 174
Yoshimura, Norio 418
You, Byoung-Sung 202
Young, Brian 92, 152
Young, Ian A. 96
Youssef, Michael 378
Yu, Chi-Yao 364
Yu, Hyun-Kyu 176
Yu, Wonsik 176
Yuffe, Marcelo 264
Yuki, Fumio 148
Yuminaka, Ayako 198
Yun, Misun 202
Zaki, Tarek 324
Zanikopoulos, Athon 180
Zanuso, Marco 88
Zargari, Masoud 170
Zhang, Bo 142
Zhang, Eric 350
Zhang, Gang 368
Zhang, Wei 142, 326, 442, 506
Zhang, Xin 216
Zhang, Xuhao 366
Zhao, Yan 224
Zhong, Charlie 350
Zhong, Freeman 350
Zhong, Shiqiang 76

Zhou, Cui 458
Zhou, Jun 332
Zito, Domenico 40
Zito, Fabio 40
Zolfaghari, Alireza 378
Zongo, Brice 112
Zschieschang, Ute 324
Zwerg, Michael 334

ISSCC GLOSSARY

1
1P1M	1-Polysilicon layer 1-Metal layer
1T1C	1-Transistor 1-Capacitor

3
3D	3-Dimensional
3G	Third-Generation (Wireless)
3T	3-Transistor

6
6T	6-Transistor

8
8T	8-Transistor

ΔΣ
ΔΣ	Delta-Sigma
ΔΣM	Delta-Sigma Modulator

ΣΔ
ΣΔ	ΔΣ is preferred.
ΣΔM	ΔΣM is preferred.

A
a-Si	Amorphous Silicon
AC	Alternating Current
A/D	Analog-to-Digital Converter
AAC	Advanced Audio Coding
ACI	Adjacent-Channel Interface
ACLR	Adjacent-Channel Leakage Power Ratio
ACPR	Adjacent-Channel Power Ratio
ADC	Analog-to-Digital Converter
ADDLL	All-Digital DLL
ADPLL	All-Digital PLL
ADSL	Asynchronous Digital Subscriber line
ADU	Analog-to-Digital Unit
AES	Advanced Encryption Standard
AFC	Automatic Frequency Control
AFE	Analog Front End
AGC	Automatic Gain Control
AGU	Address-Generation Unit
AIP	Artificial-Intelligent Partner
ALU	Arithmetic Logic Unit
AM	Amplitude Modulation
AMLCD	Active-Matrix LCD
AMOLED	Active-Matrix OLED
AMP	Asymmetric Multi-Processing
AMPS	Advanced Mobile-Phone Service
AMS	Analog Mixed-Signal (System)
APD	Avalanche Photo-Diode
APG	Algorithmic Pattern Generator
APSK	Amplitude Phase-Shift Keying
ARM	Advanced RISC Machine
ASIC	Application-Specific Integrated Circuit
ASK	Amplitude Shift Keying
ASP	Average Selling Price
ASP	Advanced Simple Profile (MPEG-4 Video)
ATA	Advanced Technology Attachment
ATD	Address-Transition Detection
ATE	Automatic Test Equipment
ATM	Asynchronous Transfer Mode
ATSC	Advanced Television Systems Committee
AVC	Audio-Visual CODEC
AVC	Automatic Volume Control (use AGC)
AWG	Arrayed-Waveguide Grating

B
BAW	Bulk Acoustic Wave
BB	Baseband
BBT	Band-to-Band Tunneling
BCD	Bipolar-CMOS-DMOS Process
BCD	Binary-Coded Decimal
BCH	Bose-Chaudhuri-Hocquenghem (a type of error-correcting code)
BD	Blu-ray disc
BER	Bit-Error Rate
BGA	Ball-Grid Array
BGR	Band-Gap Reference
BiCMOS	Bipolar Complementary-MOS
BIOS	Basic Input/Output System
BIST	Built-in Self-Test
BJT	Bipolar Junction Transistor
BL	Bitline
BOM	Bill of Materials
BPF	Bandpass Filter
BPSK	Binary Phase-Shift Keying
B-VOP	Bidirectional-Video Object Planes
BW	Bandwidth

C
C4	Controlled-Collapse Chip Connection
CAD	Computer-Aided Design
CAM	Content-Addressable Memory
CAN	Controller Area Network
CAS	Column-Address Strobe
CCCS	Current-Controlled Current Source
CCD	Charge-coupled Device
CCK	Complementary Code Keying
CCO	Current-Controlled Oscillator
CCVS	Current-Controlled Voltage Source
CDAC	Capacitor DAC
CDMA	Code-Division Multiple Access
CDR	Clock and Data Recovery
CDS	Correlated Double Sampling
CF	Compact Flash
CFA	Color Filter Array
CHE	Channel Hot-Electron (Injection)
CIS	Complex-Instruction-Set Computer
CIS	CMOS Image Sensor
CML	Current-Mode Logic
CMOS	Complementary MOS
CMRR	Common-Mode Rejection Ratio
CMU	Clock Multiplier Unit
CNFET	Carbon Nanotube NFET
CO	Central-Office (hardware)
CODEC	Coder-Decoder
COFDM	Coded FDM
CPE	Customer Premises Equipment
CPU	Central Processing Unit
CPW	CoPlanar Waveguide
CRC	Cyclic Redundancy Check
CSMA	Carrier-Sense Multiple Access
CT	Continuous Time (system)
CUI	Command User Interface
CVD	Chemical Vapor Deposition

D
D/A	Digital-to-Analog Converter
DAB	Digital-Audio Broadcasting
DAC	Digital-to-Analog Converter
dBFS	dB relative to Full Scale
DBS	Direct-Broadcast Satellite
DCC	Duty-Cycle Corrector

ISSCC GLOSSARY

DCO	Digitally Controlled Oscillator
DCT	Discrete Cosine Transform
DCVS	Differential Cascode Voltage Switch
DCVS	Digitally-Controlled Voltage Source
DCXO	Digitally Controlled Crystal Oscillator
DDFS	Direct-Digital Frequency Synthesis (or synthesizer)
DDR	Dual Data Rate
DDS	Direct Digital Synthesis
DECT	Digitally Enhanced Cordless Communication
DEM	DEModulator
DEM	Dynamic Element Matching
DEMOS	Depletion MOS
DEMUX	Demultiplexer
DES	Data-Encryption Standard
DEVM	Differential Error Vector Magnitude
DFE	Decision-Feedback Equalizer
DFF	D-type Flip Flop
DFT	Design for Testability
DFT	Discrete Fourier Transform
DfY	Design-for-Yield
DIMM	Dual In-Line Memory Module
DIP	Dual In-line Package
DLL	Delay-Locked Loop
DMA	Direct Memory Access
DMB	Digital Multimedia Broadcasting
DMIPS	Dhrystone Million Instructions Per Second
DMOS	(Double-)Diffused MOS
DNA	Deoxyribonucleic Acid
DNL	Differential Non-Linearity
DNR	Dynamic Range (DR preferred)
DOM	Digital Optical Module
DR	Dynamic Range (see also DNR)
DRAM	Dynamic Random-Access Memory
DRC	Design-Rule Check
DSB	Double Side Band
DSC	Digital Still Camera
DSL	Digital Subscriber Line
DS-OFDM	Direct-Sequence Orthogonal Frequency Division Multiplexing
DSP	Digital Signal Processing
DSSS	Direct-Sequence Spread-Spectrum
DT	Discrete Time
DTL	Diode-Transistor Logic
DTV	Digital Television
DUT	Device Under Test
DVB	Digital-Video Broadcasting
DVB-C	Digital-Video Broadcasting - Cable
DVB-H	Digital-Video Broadcasting - Handhelds
DVB-S	Digital-Video Broadcasting - Satellite
DVB-T	Digital-Video Broadcasting - Terrestrial
DVD	Digital Video Disc
DVS	Dynamic Voltage Scaling
DWA	Data Weighted Averaging
DWDM	Dense-Wavelength-Division Multiplexing
DWMT	Discrete Wavelet Multi-Tone

E

ECC	Error-Correcting Code
ECG	Electro-CardioGram
ECL	Emitter-Coupled Logic
ECP	Emitter-Coupled Pair
EDGE	Enhanced Data rates for Global Evolution
EDR	Enhanced Data Rate
EEG	Electro-EncephaloGram
EEPROM	Electrically Erasable Programmable Read-Only Memory
EFR	Enhanced Full-Rate {GSM}
EIRP	Effective Isotropic Radiated Power

EKG	Electro-CardioGram (see ECG)
EMG	ElectroMyoGram
EMI	Electro Magnetic Interference
ENOB	Effective Number o f Bits
EOT	Electrical Oxide Thickness
EPON	Ethernet-based Passive Optical Network
EPROM	Erasable Programmable Read-Only Memory
ERBW	Effective-Resolution Bandwidth
ESD	ElectroStatic Discharge
EVDO	EVolution Data Optimized (in the context of CDMA)
EVM	Error-Vector Magnitude
EWC	Enhanced Wireless Consortium

F

f_{max}	Unity power gain frequency.
f_s	Sampling frequency
f_t	Transit frequency
FAMOS	Floating-gate Avalanche-injection MOS Transistor
FBAR	Film Bulk Acoustic Resonator
FBDIMM	Fully Buffered DIMM
FCC	Federal Communications Commission (U.S.)
FDM	Frequency-Division Multiplexing
FDMA	Frequency-Division Multiple-Access
FDNR	Frequency-Dependent Negative Resistor
FEC	Forward Error Checking
FeRAM	Ferro-electric Random Access Memory
FET	Field-Effect Transistor
FF	Flip-Flop
FFE	Feed-Forward Equalizer
FFT	Fast Fourier Transform
FIB	Focused Ion Beam
FIFO	First In-First Out
FinFET	A MOSFET with the gate on two sides {acronym describes the physical shape}
FIR	Finite Impulse Response {filter}
FLOPS	Floating-Point Operations Per Second
FLOTOX	Floating-gate Tunnel OXide
FM	Frequency Modulation
FMCW	Frequency Modulated Continuous Wave
FN	Fowler-Nordheim
FO4	Fan-Out of 4
FOM	Figure Of Merit
FPGA	Field-Programmable Gate Array
FPN	Fixed Pattern Noise
FPU	Floating Point Unit
FSG	FluoroSilicate glass {dielectric}
FSG	Fluorine-doped Silicate Glass
FSK	Frequency-Shift Keying
FSM	Finite-State Machine

G

GBW	Gain-BandWidth {product}
GCA	Gain-Controlled Amplifier
GDDR	Graphics Double-Data-Rate
GDfM	Generalized Design-for-Manufacturability
GFSK	Gaussian Frequency-Shift Keying
GFLOPS	Giga Floating-Point Operations Per Second
GIDL	Gate-Induced Drain Leakage
GMSK	Gaussian Minimum-Shift Keying
GOPS	Giga-Operations Per Second
GPRS	General Packet-Radio Service
GPS	Global Positioning System
GSM	Global Standard for Mobile Communication
GUI	Graphical User Interface
GVCO	Gated VCO

ISSCC GLOSSARY

H

HBT	Hetero-junction Bipolar Transistor
HBM	Human Body Model
HCI	Host-Controller Interface
HCI	Human-to-Computer Interface
HD	High-Density
HDD	Hard-Disk-Drive
HDL	Hardware-Description Language
HDTV	High-Definition TeleVision
HD2	2nd order Harmonic Distortion
HD3	3rd order Harmonic Distortion
HiFi	High Fidelity
HPF	High-Pass Filter
HSPDA	High-Speed Downlink Packet Access
HT3	Hypertransport 3 (I/O standard)
HTM	Hierarchical Temporal Memory
HV	High Voltage
HVCMOS	High-Voltage Complementary MOS
HVMOS	High-Voltage MOS

I

I/O	Input-Output
I/Q	In Phase and Quadrature
IBOC	In-Band Out-of-Channel
IC	Integrated Circuit
ID	Identification
IF	Intermediate Frequency
IIP2	Input-referred Input 2nd-order Intercept Point
IIP3	Input-referred Input 3rd-order Intercept Point
IIR	Infinite Impulse Response Filter
IMD	Inter-Modulation Distortion
IM2	2nd-order InterModulation distortion
IM3	3rd-order InterModulation distortion
INL	Integral Non-Linearity
InP	Indium Phosphide
IP	Intellectual Property
IPSEC	Internet (Network) Protocol for Security
IR	Image Rejection
IR	InfraRed
ISDB	Integrated Services Digital Broadcasting
ISDB-C	Integrated Services Digital Broadcasting - Cable
ISDB-S	Integrated Services Digital Broadcasting - Satellite
ISDB-T	Integrated Services Digital Broadcasting - Terrestrial
ISFET	Ion-Sensitive Field-Effect Transistor
ISI	Inter-Symbol Interference
ISM	Industrial, Scientific and Medicine Band
ITU-T	International Telecommunications Union

J

JPEG	Joint Photographic Expert Group
JTAG	Joint Test-Automation Group

L

LAGS	Locally-Asynchronous Globally-Synchronous
LAN	Local-Area Network
LBS	Location-Based Services
LCD	Liquid-Crystal Display
LCOS	Liquid-Crystal On Silicon
LDI	LVDS Display Interface
LDCMOS	Laterally Diffused Complementary Metal-Oxide Silicon
LDMOS	Laterally Diffused Metal-Oxide Silicon
LDO	Low Drop-Out (regulator)
LDPC	Low-Density Parity Check
LED	Light-Emitting Diode
LEP	Light-Emitting Polymer

LFCSP	LeadFrame Chip-Scale Package
LFSR	Linear-Feedback Shift Register
LIN	Local Interconnect Network
LMS	Least Mean Square
LNA	Low-Noise Amplifier
LNB	Low-Noise Blockers
LO	Local Oscillator
LPCVD	Low-Pressure Chemical-Vapor Deposition
LPF	Low-Pass Filter
LSB	Least-Significant Bit
LSI	Large-Scale Integration
LTPS	Low-Temperature-Poly Silicon
LV	Low Voltage
LVDS	Low-Voltage Differential Signaling
LVS	Layout Verification to Schematic

M

MAC	Media-Access Controller
MAC	Mutiply-Accumulate
MASH	Multi-stAge noise SHaping
MBOA	Multi-Band OFDM Alliance
MB-OFDM	Multi-Band OFDM
MCM	Multi-Chip Module
MCU	MicroController Unit
MDAC	Multiplying DAC
MEMS	Micro-Electro-Mechanical System
MER	Modulation Error Ratio
MfD	Manufacturing-for-Design
microSD	micro Secure Digital
MICS	Medical Implant Communication Band
MIM	Metal-Insulator-Metal
MIMO	Multipl- Input, Multiple-Output
MIPI	Mobile Industry Processor Interface
MIPS	Million Instructions Per Second
ML	Matchline
MLC	Multi-Level Cell
MLSD	Maximum-Likelihood-Sequence Detection
MLSE	Maximum-Likelihood-Sequence Estimation
MMAC	Mega Mutliply-Accumulate
MMIC	Monolithic Microwave Integrated Circuit
MODEM	Modulator-Demodulator
MOS	Metal-Oxide-Semiconductor (Silicon)
MOSFET	Metal-Oxide-Semiconductor FET
MOST	MOS Transistor
MOST	Media-Oriented Systems Transport
MP	Multi-Processor
MP3	MPeg-1 audio layer 3 (lossy compression algorithm)
MPEG	Motion-Picture Expert Group
Mpps	Million packets per second {VoIP}
MSB	Most Significant Bit
MRAM	Magnetic Random-Access Memory
MRAM	Magnetoresistive Random-Access Memory
MRC	Maximum-Ratio Combining
MSB	Most-Significant Bit
MTJ	Magnetic Tunnel Junction
MTPR	Multi-Tone Power Ratio
MUX	Multiplexer

N

NBTI	Negative-Bias Temperature Instability
NEF	Noise Efficiency Factor
NEM	Nano-Electro-Mechanical system
NF	Noise Figure
NMOS	N-channel MOS
NMOST	NMOS Transistor
NoC	Network on (a) Chip
NPN	N-type-P-type-N-type bipolar (transistor)
NRTZ	Non Return-To-Zero (see also NRZ)

ISSCC GLOSSARY

NRZ	Non-Retun- to-Zero (see also NRTZ)
NTF	Noise Transfer Function
NVM	Non-Volatile Memory
NVRAM	Non-Volatile Random-Access Memory

O

ODT	On-die Termination
OEM	Original-Equipment Manufacturer
OFDM	Orthogonal Frequency-Division Multiplexing
OIF	Optical-Internetworking Forum
OIP2	Output-referred intercept point for 2nd-order distortion
OIP3	Output-referred intercept point for 3rd-order distortion
OLED	Organic LED
ONO	Oxide-Nitride-Oxide
OOK	On-Off Keying
OSR	Over-Sampling Ratio
OTA	Operational Transconductance Amplifier
OTP	One Time Programmable

P

P_{1dB}	1dB gain-compression Point
PA	Power Amplifier
PA	Public-Address (system)
PAE	Power-Added Efficiency
PAM	Pulse-Amplitude Modulation
PAN	Personal-Area Network
PCB	Printed-Circuit Board
PCI	Peripheral-Component Interconnect
PCI-X	PCI Express
PCM	Pulse-Code Modulation
PCS	Personal Communication Services
PD	Phase Detector
PD-SOI	Partially-Depleted SOI
PDA	Personal Data Assistant
PDP	Power-Delay Product
PFD	Phase and Frequency Detector
PGA	Programmable-Gain Amplifier
PGA	Programmable Gate Array
PHEMT	Pseudomorphic High-Electron-Mobility Transistor
PHY	PHYsical layer (of a communications protocol)
PID	Proportional, Integral, Derivative (a type of control loop)
PLA	Programmable Logic Array
PLC	Power-Line Communication
PLD	Programmable Logic Device
PLL	Phase-Locked Loop
PMOS	P-channel MOS
PMOST	PMOS transistor
PMU	Power Management Unit
PNP	P-type-N-type-P-type bipolar (transistor)
PON	Passive Optical Network
POTS	Plain-Old Telephone Service
ppm	parts per million
PPM	Pulse-Position Modulation
PR	Pseudo Random
PR	Partial Response
PRAM	Phase-Change RAM
PRBS	Psuedo-Random Binary Sequence
PRML	Partial-Response, Maximum-Likelihood
PROM	Programmable Read-Only Memory
PSD	Power Spectral Density
PSK	Phase-Shift Keying
PSNR	Peak SNR
PSRR	Power-Supply Rejection Ratio
PTAT	Proportional To Absolute Temperature
PVD	Physical Vapor Deposition

PVT	Process, Voltage, Temperature
PWM	Pulse-Width Modulation

Q

QAM	Quadrature Amplitude Modulation
QDR	Quad Data Rate
QoS	Quality of Service
QPSK	Quadrature Phase-Shift Keying
QVCO	Quadrature Voltage-Controlled Oscillator
QVGA	Quarter Video Graphics Array

R

R/W	ReadWrite
RAM	Random-Access Memory
RBW	Resolution BandWidth
RDAC	Resistor DAC
RF	Radio Frequency
RFID	RF ID (tag)
RISC	Reduced-Instruction-Set Computer
ROM	Read-Only Memory
rms	root mean square
ROM	Read-Only Memory
RSA	A public-key cryptographic system, named after: Ron Rivest, Adi Shamir, and Leonard Adleman
RSSI	Received-Signal-Strength Indicator
RTL	Resistor-Transistor Logic
RTS	Random Telegraph Signal
RTZ	Return-To-Zero
RX	Receiver
RZ	Return-to-Zero (see also RTZ)

S

SAR	Successive-Approximation-Register
SAW	Surface Acoustic Wave
SATA	Serial Advanced-Technology Attachment
SC	Switched-Capacitor
SCL	Source-Coupled Logic
SCP	Source-Coupled Pair
SCR	Silicon Controlled Rectifier
S-DMB	Satellite Digital-Multimedia Broadcasting
SD	Secure Digital (package type)
SDR	Software Defined Radio
SDRAM	Synchronous Dynamic Random-Access Memory
SEM	Scanning Electron Microscope
SER	Soft-Error Rate
SER	Symbol-Error Rate
SerDes	Serializer/Deserializer
SFDR	Spurious-Free Dynamic Range
SFI	Serdes Framer Interface
SFP	Small Form-factor Pluggable
S/H	Sample-and-Hold
SHA	Sample-and-Hold Amplifier
SiGe	Silicon Germanium
SiGe:C	Silicon, Germanium, Carbon
SIL	Safety Integrity Level
SILC	Stress-Induced Leakage Current
SIMD	Single-Instruction Multiple-Data
SINAD	SIgnal-to-Noise And Distortion (ratio)
SIO	Synchronous I/O
SIP	Single-Inline Package
SiP	System in (a) Package
SL	Searchline
SMP	Symmetric Multi-Processing
SMS	Short-Messaging Service
SNDR	Signal-to-Noise and Distortion Ratio
SNM	Static Noise Margin
SNR	Signal-to-Noise Ratio

ISSCC GLOSSARY

SNS	Social-Network Service
SoC	System on (a) Chip
SOI	Semiconductor on Insulator
SONET	Synchronous Optical NETwork
SONOS	Silicon-Oxide-Nitride-Oxide-Silicon
SOS	Silicon On Saphire
SP	Simple Profile
SPAD	Single-Photon Avalanche Diode
SPDT	Single Pole Double Throw
SPI	System Packet Interface
SRAM	Static Random-Access Memory
SSB	Single Side-Band
SSC	Spread Spectrum Clocking
SSCG	Spread Spectrum Clock Generator
SSD	Sold State Disk
SSN	Simultaneous Switching Noise
SSO	Simultaneous Switching Output
SSTL	Stub Series Terminated Logic
SXGA	Super Extended Graphics Array

T

TC	Temperature Coefficient
TCAM	Ternary Content-Addressable Memory
TCP	Transmission Control Protocol
TDC	Time-to-Digital Converter
TDDB	Time-Dependent Dielectric Breakdown
TDM	Time-Division Multiplexing
TDMA	Time-Division Multiple-Access
TD-SCDMA	Time-Division Synchronous Code-Division Multiple Access
TEM	Tunneling-Electron Microscope
TFLOPS	Tera Floating-Point Operations Per Second
TFT	Thin-Film Transistor
TIA	TransImpedance Amplifier
T/H	Track and Hold
THA	Track-and-Hold Amplifier
THD	Total Harmonic Distortion
THD+N	THD plus Noise
TLB	Translation Lookaside Buffer
TOPS	Tera-Operations Per Second
TTL	Transistor-Transistor Logic
TV	Television
TX	Transmitter

U

UD	Ultra-high Definition
UDTV	Ultra-high-Definition TeleVision
UGC	User-Generated Contents
UHDV	Ultra-High-Definition Video
UHF	Ultra-High Frequency
UI	Unit Interval
UI	User Interface
UIPP	UI_{pp} (peak-to-peak)
U-NII	Unlicensed National Information Infrastructure
UMPC	Ultra-Mobile-PC
UMTS	Universal Mobile-Telecommunication System
UPROM	Unerasable Programmable Read-Only Memory
USB	Universal Serial Bus
UTP	Unshielded Twisted Pair
UWB	Ultra-WideBand
UXGA	Ultra-eXtended Graphics Array
UPF	Unified Power Format

V

V_t	MOS transistor threshold voltage
V_T	thermal voltage
VCCS	Voltage-Controlled Current Source

VCDL	Voltage-Controlled Delay Line
VCO	Voltage-Controlled Oscillator
VCVS	Voltage-Controlled Voltage-Source
VCXO	Voltage-Controlled Crystal Oscillator
VCSEL	Vertical-Cavity Surface-Emitting Laser
VDMOS	Vertically Diffused MOS
VDSL	Very high bit-rate Digital Subscriber Line
VGA	Variable-Gain Amplifier
VGA	Video Graphics Array
VLIW	Very Long Instruction Word
VLF	Very Low Frequency
VLSI	Very Large-Scale Integration
VoD	Vision on Demand
VoIP	Voice over IP
VSB	Vestigial Side Band
VSWR	Voltage Standing-Wave Ratio

W

WAN	Wide-Area Network
WCDMA	Wideband Code-Division Multiple- Access
WDM	Wavelength-Division Multiplexing
WEP	Wired-Equivalent Privacy
WiFi	Wireless Fidelity; (an interoperability certification for IEEE 802.11 WLAN products)
WiMax	Worldwide Interoperability for Microwave Access (IEEE802.16)
WL	Wordline
WLAN	Wireless Local-Area Network

X

XAUI	(10 Gigabit) eXtended Attachment Unit Interface
XDR	Extreme Data Rate
XENPAK	10 Gb Ethernet-compatible fiber-optic standard
XFMR	TransForMeR
XFP	Small form-factor pluggable
XGA	Extended Graphics Array
XO	Crystal Oscillator

ISSCC 2012 Call for Papers

IEEE INTERNATIONAL SOLID-STATE CIRCUITS CONFERENCE

Sunday–Thursday, February 19–23, 2012 • San Francisco Marriott Marquis Hotel, San Francisco, CA

ISSCC 2012 CONFERENCE THEME:
"SILICON SYSTEMS FOR SUSTAINABILITY"

What can we do for earth's sustainability? Sustainabiity must be the paramount theme for the future of human society! Electronics will play a primary role: more robust and sustainable silicon systems will emphaszie re-usability, re-organization, re-configurability, self-repair, and self-organization. Currently, environmentally supportive sustainable systems are exemplified by silicon technology used in electric-automotive and smart-power-grid systems. Potentially, silicon technology and its applications will provide solutions for the realization of smarter recycling sustainable global systems. To fulfill this goal, system-level approaches, as well as technology and circuit advancements will play important roles! Contributions to ISSCC 2012 are encouraged in support of the theme "Silicon Systems for Sustainability".

Innovative and original papers are solicited in subject areas including (but not limited to) the following:

ANALOG — Op-amps and instrumentation amps, baseband amplifiers, comparators, multipliers, voltage references, power-control circuits, regulators & DC-DC converters; continuous-time & discrete-time filters; consumer electronics, non-linear analog circuits, switched-capacitor circuits; synthesizers, PLLs.

DATA CONVERTERS — Nyquist-rate and oversampling A/D and D/A converters; sample-and-hold circuits; TDC's.

ENERGY-EFFICIENT DIGITAL — Smart phone IC and application processor, digital baseband (eg LTE, WSN, body-area network); innovative multimedia ICs (eg 3D TV, quad HD); vision/graphics processing IC; > 1GHz embedded processors, energy-efficient multi-core processors, energy-efficient wide-operating-range processors; digital IC for personal e-health devices, digital IC for automotive applications, energy-efficient sensor system; intelligent power-management methods for digital VLSI; adaptive voltage and frequency scaling, adaptive body bias circuits, on-chip monitoring/sensing circuits for energy-efficient applications; variation-tolerant circuit and architecture, techniques for timing margin reduction; energy-efficient circuit techniques.

HIGH-PERFORMANCE DIGITAL — Microprocessors; graphics processors; many-core and thread-rich processors; network processors; high-speed digital circuits; intra-chip communication circuits; soft error, variation, and fault-tolerant circuits; reconfigurable logic arrays; security circuits; high-speed CAMs and register files; clock generation and distribution circuits and architectures; high-performance-logic microarchitectures and circuit techniques; implementation methodologies for high-performance digital VLSI; power and leakage management techniques for high performance processors and graphics; adaptive digital circuits; thermal and wear-out sensors.

IMAGERS, MEMS, MEDICAL & DISPLAY — Image sensors and companion chips; image sensor SoC; smart sensors; MEMS for analog, RF, and sensor applications, integrated sensors and transducers; organic sensors; sensor-interface circuits; neural interfaces; biosensors, microarrays and lab-on-a-chip; environmental and wearable biomedical electronics; display drivers, controllers, and companion chips; organic LED and liquid-crystal-display interface circuits; organic sensors; flat-panel and projection displays; circuits for print-heads.

MEMORY — Static, dynamic, non-volatile, read-only memory, and content-addressable-memory; memory subsystem and array architecture along with related circuits; memory I/O interface design and circuit techniques, including 3D memory integration; phase-change, magnetic, spin-torque-transfer, ferroelectric, resistive, and other emerging memory design and architecture; embedded memory architecture and design, including single- and multi-port cache memory or register files, for computing, consumer electronics, and emerging applications such as biomedical devices; advanced circuit techniques to enable high-performance and low-voltage memory design; advanced architecture and design to improve memory reliability and fault-tolerance, e.g., novel error-correction (ECC) and redundancy schemes; memory controllers and solid-state-disk controllers.

RF — mm-wave/RF/IF/baseband circuits and sub-systems, both narrowband and wideband, including receiver and transmitter front-ends, modulators/demodulators, power amplifiers/detectors, RF switches and integrated antennas/MIMO/phased arrays, frequency generators; circuits for communications, networking, sensing, RADAR, RF imaging, SiP integration and emerging RF applications. Also, circuits that achieve increased frequency range, tunability, selectivity, dynamic range, power consumption, power efficiency, configurability, silicon scaling, environmental robustness, mixed-signal and digital-assist functionality.

TECHNOLOGY DIRECTIONS —- Advanced circuit technologies and techniques; ultra-low-voltage and sub-threshold logic design; molecular-, organic-, and nano-electronics; flexible substrates and printable electronics; 3D-integration and novel packaging technologies; compound-semiconductor, superconductive, and micro-photonic technologies and circuits; energy sources and energy harvesting; emerging applications such as biomedical and ambient-intelligence; emerging wireless applications and circuits; advanced signal-processing and microprocessor architectures; design for manufacturability; analog and optical processors, non-transistor-based analog and digital circuits and their system architectures; advanced memory technologies; spintronics; quantum storage; emerging sensor-network concepts such as body-area and body-sensor networks.

WIRELESS — Receivers, transmitters, transceivers, and SoCs, for connectivity, cellular, broadcast, and radar applications including multi-standard and multi-band solutions. Examples include (but are not limited to) WLAN, WiMax, cellular base stations and handsets, GPS, DVB/DMB, UWB, ISM, and mm-wave-band systems. Also, highly-integrated solutions for emerging wireless applications targeting sensing, imaging, etc. are encouraged.

WIRELINE — Receivers/transmitters/transceivers for wireline systems; backplane transceivers and chip-to-chip communications. Examples include (but not limited to) LAN, WAN, FDDI, Ethernet, token ring, fiber channel, SONET, SDH, PON, ATM, ISDN, xDSL, cable-modem; optical/electrical data links, power-line/phone-line home networks, subscriber-line circuits and modems. Also wireline transceiver building blocks like AGC, oscillators, PLLs, line-drivers and hybrids, etc. are encouraged.

Submission Deadline is Monday, September 12, 2011 • 3:00PM Eastern Daylight Time (19:00 GMT)

STUDENT INITIATIVES

Graduate students are invited to participate in events that provide an opportunity to showcase ongoing work and exchange experiences with other students and researchers from academia and industry. Events include the Student Research Preview, DAC/ISSCC Student Design Contest, and the Silkroad Award presented to first-time student presenting authors at ISSCC from an emerging region in the Far East.

Further information including submission procedures, formats, student initiatives, and deadlines can be found at http://www.isscc.org